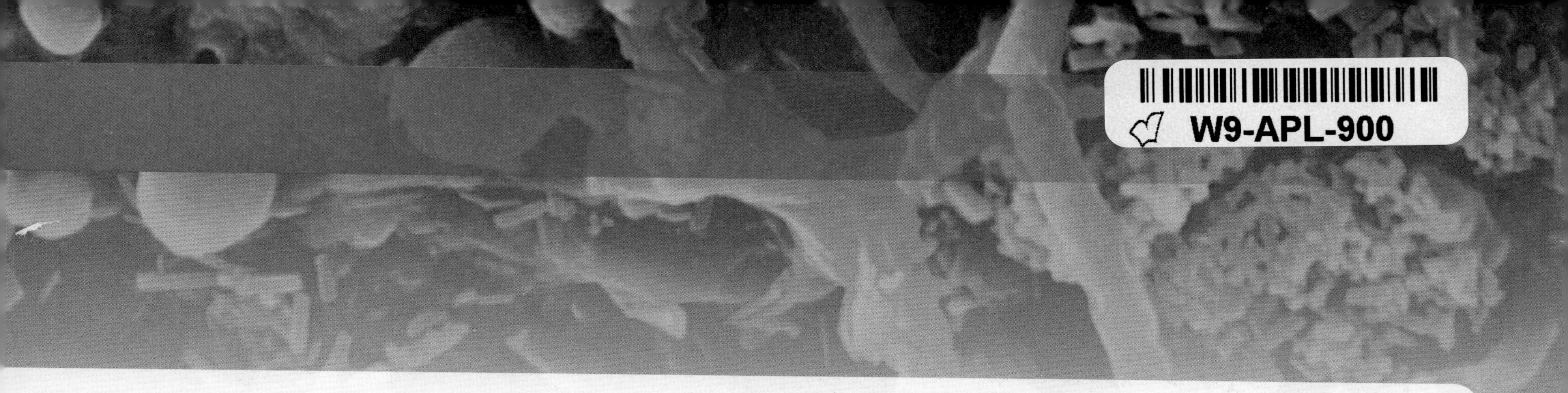

Animations

Access animations of key microbiological processes to bring these concepts to life! 3-D animations help you visualize and comprehend difficult and important topics. You can download animations to your computer or portable media player, or view them online with attached quizzing to help retain the concepts.

Look for the icon in the text. It will highlight animations you can find on the website.

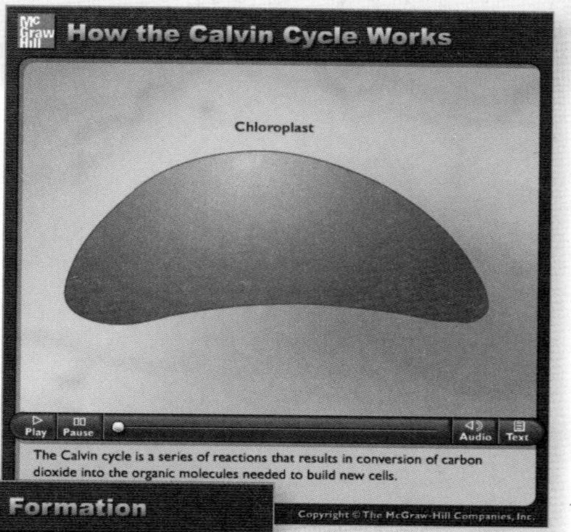
How the Calvin Cycle Works
Chloroplast
Play
Pause
Audio
Text
The Calvin cycle is a series of reactions that results in conversion of carbon dioxide into the organic molecules needed to build new cells.
Copyright © The McGraw-Hill Companies, Inc.

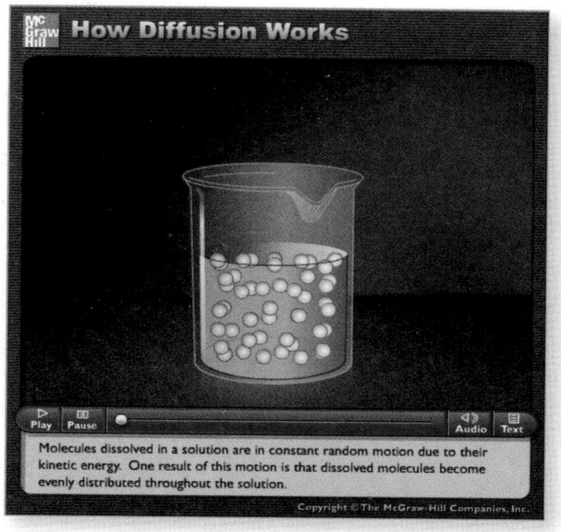
How Diffusion Works
Play
Pause
Audio
Text
Molecules dissolved in a solution are in constant random motion due to their kinetic energy. One result of this motion is that dissolved molecules become evenly distributed throughout the solution.
Copyright © The McGraw-Hill Companies, Inc.

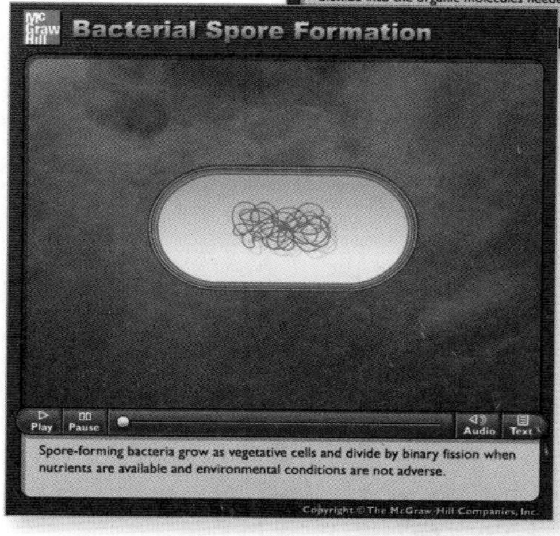
Bacterial Spore Formation
Play
Pause
Audio
Text
Spore-forming bacteria grow as vegetative cells and divide by binary fission when nutrients are available and environmental conditions are not adverse.
Copyright © The McGraw-Hill Companies, Inc.

Electronic Books

GO GREEN!

Go Green.
Save Green.
McGraw-Hill eBooks.

Green. . . it's on everybody's minds these days. It's not only about saving trees; it's also about saving money. At 55% of the bookstore price, **McGraw-Hill eBooks** are an eco-friendly and cost-saving alternative to the traditional print textbook. So, you do some good for the environment. . . and you do some good for your wallet.

Visit **www.mhhe.com/ebooks** for details.

PRESCOTT'S

MICROBIOLOGY

EIGHTH EDITION

JOANNE M. WILLEY
HOFSTRA UNIVERSITY

LINDA M. SHERWOOD
MONTANA STATE UNIVERSITY

CHRISTOPHER J. WOOLVERTON
KENT STATE UNIVERSITY

McGraw Hill
Connect Learn Succeed™

The McGraw·Hill Companies

PRESCOTT'S MICROBIOLOGY, EIGHTH EDITION

Published by McGraw-Hill, a business unit of The McGraw-Hill Companies, Inc., 1221 Avenue of the Americas, New York, NY 10020.

Some ancillaries, including electronic and print components, may not be available to customers outside the United States.

This book is printed on acid-free paper.

1 2 3 4 5 6 7 8 9 0 DOW/DOW 1 0 9 8 7 6 5 4 3 2 1 0

ISBN 978-0-07-337526-7
MHID 0-07-337526-8

Vice President & Editor-in-Chief: *Marty Lange*
Vice President, EDP: *Kimberly Meriwether David*
Director of Development: *Kristine Tibbetts*
Sponsoring Editor: *Lynn M. Breithaupt*
Developmental Editor: *Fran Schreiber*
Marketing Manager: *Amy L. Reed*
Senior Project Manager: *Sandy Wille*
Lead Production Supervisor: *Sandy Ludovissy*
Senior Media Project Manager: *Jodi K. Banowetz*
Senior Designer: *David W. Hash*
Cover/Interior Designer: *Greg Nettles/Squarecrow Design*
Cover Image: *Kitchen sponge microbes, ©Dennis Kunkel Microscopy, Inc.*
Lead Photo Research Coordinator: *Carrie K. Burger*
Photo Research: *Mary Reeg*
Compositor: *Aptara, Inc.*
Typeface: *10/12 Minion Pro*
Printer: *R. R. Donnelley*

Library of Congress Cataloging-in-Publication Data

Willey, Joanne M.
Prescott's microbiology. -- 8th ed. / Joanne M. Willey, Linda M. Sherwood, Christopher J. Woolverton.
p. cm.
Rev. ed. of: Prescott, Harley, and Klein's microbiology. 7th ed.© 2008.
Includes index.
ISBN 978-0-07-337526-7 — ISBN 0-07-337526-8 (hard copy : alk. paper) 1. Microbiology--Textbooks. I. Sherwood, Linda. II. Woolverton, Christopher J. III. Prescott, Lansing M. Microbiology. IV. Willey, Joanne M., Prescott, Harley, and Klein's microbiology. V. Title. VI. Title: Microbiology.
QR41.2.P74 2011
616.9'041--dc22

2009033823

www.mhhe.com

ABOUT THE AUTHORS

JOANNE M. WILLEY has been a professor at Hofstra University on Long Island, New York, since 1993, where she is Professor of Microbiology; she holds a joint appointment with the Hostra University School of Medicine. Dr. Willey received her B.A. in Biology from the University of Pennsylvania, where her interest in microbiology began with work on cyanobacterial growth in eutrophic streams. She earned her Ph.D. in biological oceanography (specializing in marine microbiology) from the Massachusetts Institute of Technology-Woods Hole Oceanographic Institution Joint Program in 1987. She then went to Harvard University where she spent her postdoctoral fellowship studying the filamentous soil bacterium *Streptomyces coelicolor*. Dr. Willey continues to investigate this fascinating microbe and has co-authored a number of publications that focus on its complex developmental cycle. She is an active member of the American Society for Microbiology, served on the editorial board of the journal *Applied and Environmental Microbiology* for nine years, and served as Chair of the Division of General Microbiology. Dr. Willey regularly teaches microbiology to biology majors as well as medical students. She also teaches courses in cell biology, marine microbiology, and laboratory techniques in molecular genetics. Dr. Willey lives on the north shore of Long Island with her husband and two sons. She is an avid runner and enjoys skiing, hiking, sailing, and reading. She can be reached at joanne.m.willey@hofstra.edu.

LINDA M. SHERWOOD is a member of the Department of Microbiology at Montana State University. Her interest in microbiology was sparked by the last course she took to complete a B.S. degree in Psychology at Western Illinois University. She went on to complete an M.S. degree in Microbiology at the University of Alabama, where she studied histidine utilization by *Pseudomonas acidovorans*. She subsequently earned a Ph.D. in Genetics at Michigan State University where she studied sporulation in *Saccharomyces cerevisiae*. She briefly left the microbial world to study the molecular biology of *dunce* fruit flies at Michigan State University before her move to Montana State University. Dr. Sherwood has always had a keen interest in teaching, and her psychology training has helped her to understand current models of cognition and learning and their implications for teaching. Over the years, she has taught courses in general microbiology, genetics, biology, microbial genetics, and microbial physiology. She has served as the editor for ASM's *Focus on Microbiology Education*, and has participated in and contributed to numerous ASM Conferences for Undergraduate Educators (ASMCUE). She also has worked with K–12 teachers to develop a kit-based unit to introduce microbiology into the elementary school curriculum, and has coauthored with Barbara Hudson a general microbiology laboratory manual, *Explorations in Microbiology: A Discovery Approach*, published by Prentice-Hall. Her association with McGraw-Hill began when she prepared the study guides for the fifth and sixth editions of *Microbiology*. Her non-academic interests focus primarily on her family. She also enjoys reading, hiking, gardening, and traveling. She can be reached at lsherwood@montana.edu.

CHRISTOPHER J. WOOLVERTON is professor of Environmental Health Science and a founding member of the College of Public Health at Kent State University (Kent, OH). Dr. Woolverton served in the Department of Biological Sciences at Kent State for 14 years and continues to serve as the Director of the Kent State University Center for Public Health Preparedness, overseeing its BSL-3 Training Facility. He holds a joint appointment at Akron Children's Hospital (Akron, OH). He earned his B.S. in Biology from Wilkes College, Wilkes-Barre, PA, and his M.S. and a Ph.D. in Medical Microbiology from West Virginia University, College of Medicine. He spent two years as a postdoctoral fellow at UNC-Chapel Hill elucidating the role of bacteria on the regulation of immunity in germ-free rodents. Dr. Woolverton's research interests today focus on real time detection and identification of pathogens using a liquid crystal biosensor that he patented in 2001. Dr. Woolverton has published and lectured widely on the mechanisms by which liquid crystals are used as biosensors. Professor Woolverton has taught Microbiology to undergraduates, along with Immunology and Microbial Physiology, to graduate students. He also teaches laboratory safety, risk assessment, and bioterrorism readiness to laboratory professionals. Woolverton is an active member of ASM, recently serving as the editor-in-chief of ASM's *Journal of Microbiology and Biology Education*. Woolverton resides in Kent, Ohio, with his wife Nancy and daughters Samantha and Abbey. Daughter Lyssa married Ken and currently resides in Arizona. Dr. Woolverton enjoys hiking and camping, and is an avid cyclist and Spinning instructor. His email address is cwoolver@kent.edu.

CONNECTING STUDENTS TO MICROBIOLOGY THROUGH...

INNOVATIVE TABLE OF CONTENTS:

- An **introduction to the entire microbial world** is now covered in Chapters 3, 4, and 5. Structure and function of Bacteria, Archaea, and Eukaryotes precede an introduction to the viruses that is extremely accessible to students.
- A more detailed look at specific viruses, replication, and classification is found later in the book (Chapter 25).

CONTENTS

- **Microbial diversity** is now integrated throughout the text. Chapters 17–25 have been thoroughly updated, with many new figures and physiological and genetic information added where appropriate.

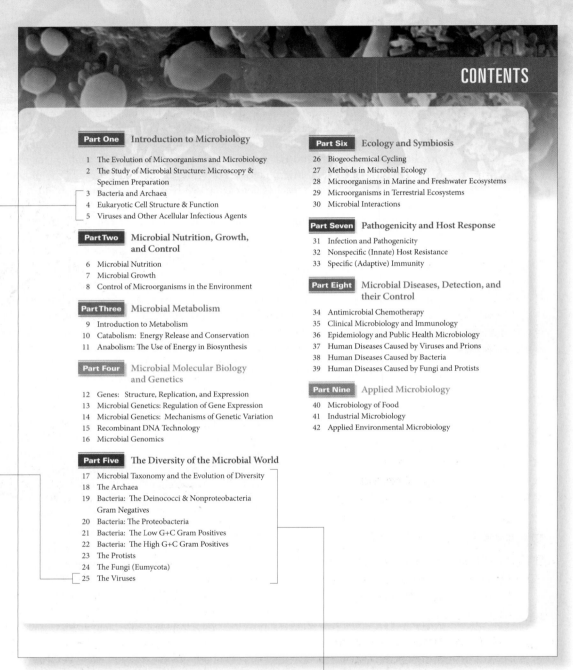

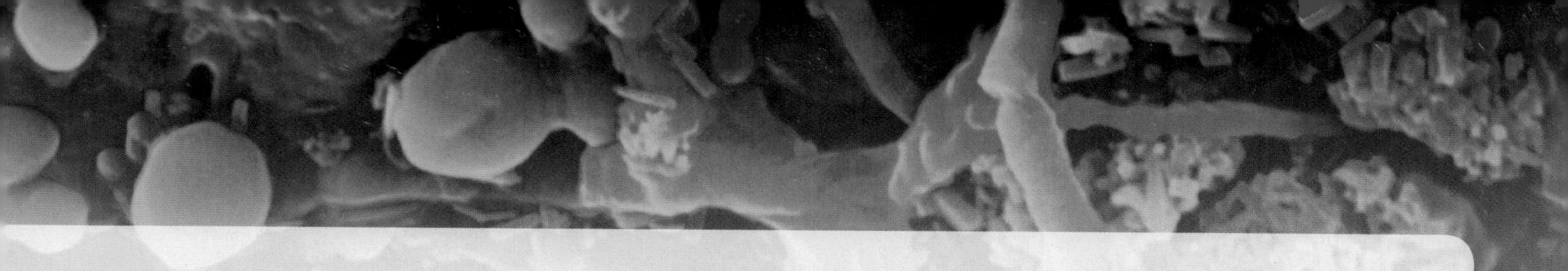

THEMATIC INTEGRATION

- **Evolution**—Introduced immediately in Chapter 1 and used as an overarching theme throughout, evolution helps tie the pieces together and provide a framework upon which students can build their knowledge of microbiology.

- **Microbial diversity**—Now integrated through the entire text, students are constantly reminded of the diversity of the microbial world.

HOT TOPICS

- **The GREEN text!** Global climate change, public health, alternative energy, and biofuels have all been added, making *Prescott's Microbiology* a very green text that will spark students' interest. **For more, see Chapters 26 and 41.**

- **H1N1**—This virus is ubiquitous in the news and sure to be a topic of interest to students. **For more, see Chapters 36 and 37.**

- **Updated Genomics**—21st-century genomic sequencing techniques are now covered in **Chapter 16.**

1
The Evolution of Microorganisms and Microbiology

Modern stromatolites from Western Australia. Each stromatolite is a rocklike structure, typically 1 m in diameter, containing layers of cyanobacteria. The existence of fossilized stromatolites is evidence of the existence of photosynthetic organisms early in the history of life on Earth.

CHAPTER GLOSSARY

Archaea The domain of life containing anucleate cells that have unique lipids in their membranes, distinctive rRNA sequences, and cell walls that lack peptidoglycan.

sequencing the genome, identifying genes, and assigning functions to the genes.

Koch's postulates A set of rules for proving that a specific microorganism

protists Mostly unicellular eukaryotic organisms that lack cellular differentiation into tissues. Cell differentiation is limited to cells involved in sexual reproduction, alternate vegetative morphology, or cysts; includes ...rred to as algae and

...eration An early ...ed, that living ...elop from nonliving

...ents of plants ...gle-stranded RNA. ...ents having a simple ... with a protein coat ...nome, lacking inde- ... and reproducing ...st cells.

...s agents composed ...ed RNA. They are ...thout the aid of ...oinfect the host cell.

...wly formed star in the ... bodies reached planet ...robes. Since that time, ...nal vents in the ocean ...being indispensable for ... webs. Most important, ...forming organic matter, 1

16
Microbial Genomics

Each dot in the microarray pictured here consists of an oligonucleotide fragment of a single gene bound to a glass slide. Gene expression of two types of cells (e.g., one wild type and one mutant) can be compared by labeling the cDNA from each cell with either a red or green fluorescent label and allowing the cDNA to bind to homologous sequences attached to the microarray. The color of each spot reveals the relative level of expression of each gene.

CHAPTER GLOSSARY

bioinformatics The interdisciplinary field that manages and analyzes large biological data sets, including genome and protein sequences.

coding sequence (CDS) A nucleotide sequence that encodes or is thought to encode a protein.

comparative genomics The analysis of genomes from different ... for significant differences ...

core genome The com... found in all genomes in a ...

DNA microarrays Soli... have DNA attached in an ... pattern; used to evaluate ... or microbial population ...

functional genomics ... genome transcripts and t... encode.

genome annotation ... determining the location ... function of genes and ge... in a genome sequence.

genomics The study of ... organization of genomes, ...

tion content, and the gene products they encode.

in silico analysis The study of physiology and genetics through the examination of nucleic acid and amino acid sequences.

metagenomics The study of genomes recovered from natural samples without first isolating members of the microbial

through duplication of a common ancestral gene.

phylotype A taxon that is characterized only by its nucleic acid sequence; generally discovered during metagenomic analysis.

proteome The complete collection of proteins that an organism produces.

single-cell genomic sequencing A

G enomics is an exc...
...ics, ecology, and ...
identification of a smal...
study the cell in a holis...

36
Epidemiology and Public Health Microbiology

This laboratory worker at the Centers for Disease Control and Prevention is in the highest level of isolation (level 4), to avoid contact with microorganisms and to prevent their escape into the environment. Extensive training is required to work in these labs.

CHAPTER GLOSSARY

antigenic drift A small change in the immunogenic character of an organism that allows it to evade immune system attack.

antigenic shift A major change in the immunogenic character of an organism; it is unrecognized by immune mechanisms.

attack rate The proportional number of cases that develop in a population that was exposed to an infectious agent.

common-source epidemic An epidemic characterized by a sharp rise to a peak and then less rapid decline in the number of individuals infected; it usually involves a single infectious source.

communicable disease A disease associated with a pathogen that can be transmitted from one host to another.

endemic disease A disease that is constantly present in a population, usually at a steady low frequency.

epidemic A disease that suddenly increases in occurrence above the normal level in a given population.

epidemiology The science that evaluates the occurrence, determinants, distribution, and control of health and disease in a defined human population.

herd immunity The resistance of a population to infection and spread of an infectious agent due to the immunity of a high percentage of the population.

incidence The number of new cases of a disease in a population at risk during a specified time period.

incubation period The period after pathogen entry into a host and before signs and symptoms appear.

morbidity rate The number of individuals who become ill as a result of a particular disease within a susceptible population during a specific time period.

mortality rate The ratio of the number of deaths from a given disease to the total number of cases of the disease.

nosocomial infection An infection that is acquired while a patient is in a hospital or other clinical care facility.

pandemic An increase in the occurrence of a disease within a large and geographically widespread population (often refers to a worldwide epidemic).

public health surveillance The systematic collection, analysis, and interpretation of outcome-based health data so as to control or prevent disease or injury.

toxoid A bacterial exotoxin that has been modified so that it is no longer toxic but will still stimulate antitoxin formation when injected into a person or animal.

vaccine A preparation given to induce an immune response and protect the individual against a pathogen or a toxin.

vector A living organism, usually an arthropod or other animal, that transfers an infective agent between hosts.

vehicle An inanimate substance or medium that transmits a pathogen.

zoonosis A disease of animals that can be transmitted to humans.

I n this chapter, we describe the practical goal of epidemiology: to establish effective disease recognition, control, prevention, and eradication measures within a given population. Because emerging and reemerging diseases and pathogens, hospital-acquired (nosocomial) infections, and bioterrorism are major concerns for public health, these topics are also covered here.

CONNECTING STUDENTS TO MICROBIOLOGY WITH...

INNOVATIVE CHAPTER FEATURES

NEW! **Chapter Glossary**—Each chapter begins with a glossary—a list of key terms discussed in the chapter.

Review and Reflection—Questions within the narrative in each chapter assist students in mastering section concepts before moving on to other topics.

Concept Maps—Key chapters include a concept map that outlines critical themes to help students understand important relationships.

Cross-Referenced Notes—In-text references refer students to other parts of the book to review.

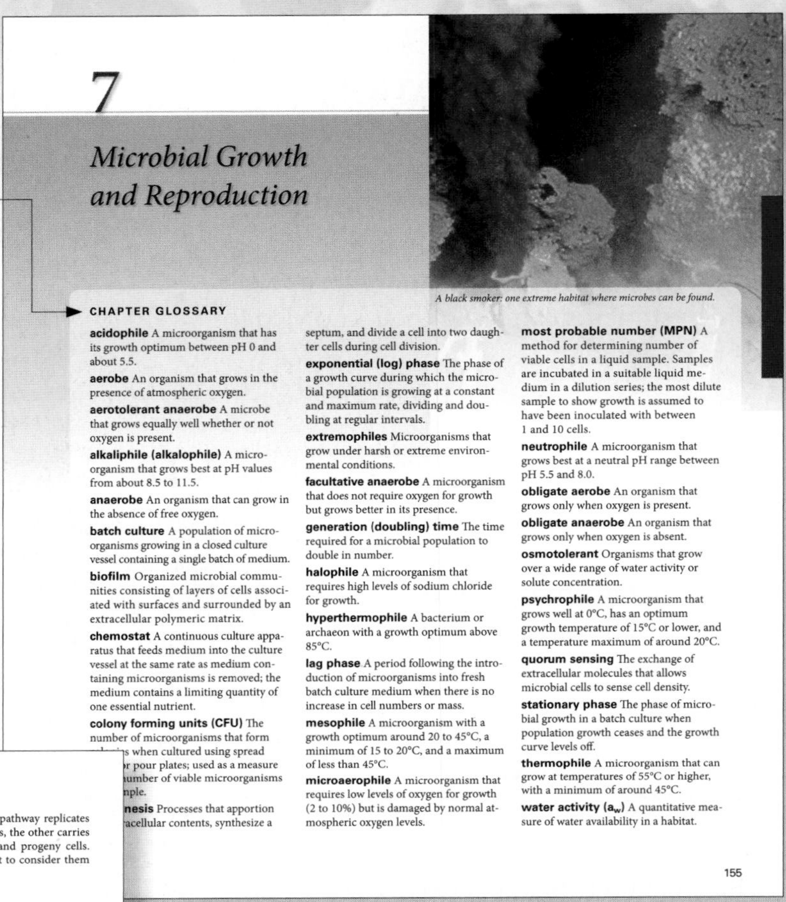

7

Microbial Growth and Reproduction

A black smoker: one extreme habitat where microbes can be found.

CHAPTER GLOSSARY

acidophile A microorganism that has its growth optimum between pH 0 and about 5.5.
aerobe An organism that grows in the presence of atmospheric oxygen.
aerotolerant anaerobe A microbe that grows equally well whether or not oxygen is present.
alkaliphile (alkalophile) A microorganism that grows best at pH values from about 8.5 to 11.5.
anaerobe An organism that can grow in the absence of free oxygen.
batch culture A population of microorganisms growing in a closed culture vessel containing a single batch of medium.
biofilm Organized microbial communities consisting of layers of cells associated with surfaces and surrounded by an extracellular polymeric matrix.
chemostat A continuous culture apparatus that feeds medium into the culture vessel at the same rate as medium containing microorganisms is removed; the medium contains a limiting quantity of one essential nutrient.
colony forming units (CFU) The number of microorganisms that form ...s when cultured using spread ...r pour plates; used as a measure ...umber of viable microorganisms ...ple.
...**nesis** Processes that apportion ...acellular contents, synthesize a septum, and divide a cell into two daughter cells during cell division.
exponential (log) phase The phase of a growth curve during which the microbial population is growing at a constant and maximum rate, dividing and doubling at regular intervals.
extremophiles Microorganisms that grow under harsh or extreme environmental conditions.
facultative anaerobe A microorganism that does not require oxygen for growth but grows better in its presence.
generation (doubling) time The time required for a microbial population to double in number.
halophile A microorganism that requires high levels of sodium chloride for growth.
hyperthermophile A bacterium or archaeon with a growth optimum above 85°C.
lag phase A period following the introduction of microorganisms into fresh batch culture medium when there is no increase in cell numbers or mass.
mesophile A microorganism with a growth optimum around 20 to 45°C, a minimum of 15 to 20°C, and a maximum of less than 45°C.
microaerophile A microorganism that requires low levels of oxygen for growth (2 to 10%) but is damaged by normal atmospheric oxygen levels.
most probable number (MPN) A method for determining number of viable cells in a liquid sample. Samples are incubated in a suitable liquid medium in a dilution series; the most dilute sample to show growth is assumed to have been inoculated with between 1 and 10 cells.
neutrophile A microorganism that grows best at a neutral pH range between pH 5.5 and 8.0.
obligate aerobe An organism that grows only when oxygen is present.
obligate anaerobe An organism that grows only when oxygen is absent.
osmotolerant Organisms that grow over a wide range of water activity or solute concentration.
psychrophile A microorganism that grows well at 0°C, has an optimum growth temperature of 15°C or lower, and a temperature maximum of around 20°C.
quorum sensing The exchange of extracellular molecules that allows microbial cells to sense cell density.
stationary phase The phase of microbial growth in a batch culture when population growth ceases and the growth curve levels off.
thermophile A microorganism that can grow at temperatures of 55°C or higher, with a minimum of around 45°C.
water activity (a_w) A quantitative measure of water availability in a habitat.

155

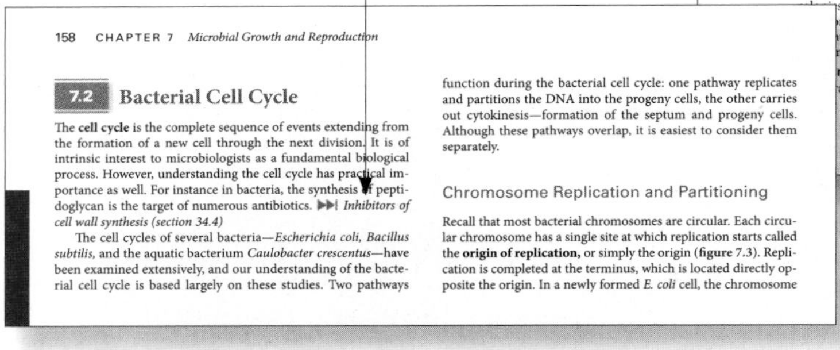

158 CHAPTER 7 *Microbial Growth and Reproduction*

7.2 Bacterial Cell Cycle

The **cell cycle** is the complete sequence of events extending from the formation of a new cell through the next division. It is of intrinsic interest to microbiologists as a fundamental biological process. However, understanding the cell cycle has practical importance as well. For instance in bacteria, the synthesis of peptidoglycan is the target of numerous antibiotics. ▶▶| *Inhibitors of cell wall synthesis (section 34.4)*

The cell cycles of several bacteria—*Escherichia coli, Bacillus subtilis*, and the aquatic bacterium *Caulobacter crescentus*—have been examined extensively, and our understanding of the bacterial cell cycle is based largely on these studies. Two pathways function during the bacterial cell cycle: one pathway replicates and partitions the DNA into the progeny cells, the other carries out cytokinesis—formation of the septum and progeny cells. Although these pathways overlap, it is easiest to consider them separately.

Chromosome Replication and Partitioning

Recall that most bacterial chromosomes are circular. Each circular chromosome has a single site at which replication starts called the **origin of replication**, or simply the origin (figure 7.3). Replication is completed at the terminus, which is located directly opposite the origin. In a newly formed *E. coli* cell, the chromosome

END OF CHAPTER MATERIAL

Chapter Summaries are organized by numbered headings and provide a snapshot of important chapter concepts.

Critical Thinking Questions taken from current literature supplement the questions for review and reflection found throughout each chapter; they are designed to stimulate analytical problem-solving skills.

NEW! **Concept Mapping Challenge** encourages students to build their own concept maps based on what they have learned in that chapter.

NEW!
Micro Inquiry
Select figures throughout every chapter contain thought questions adding another assessment opportunity for the student.

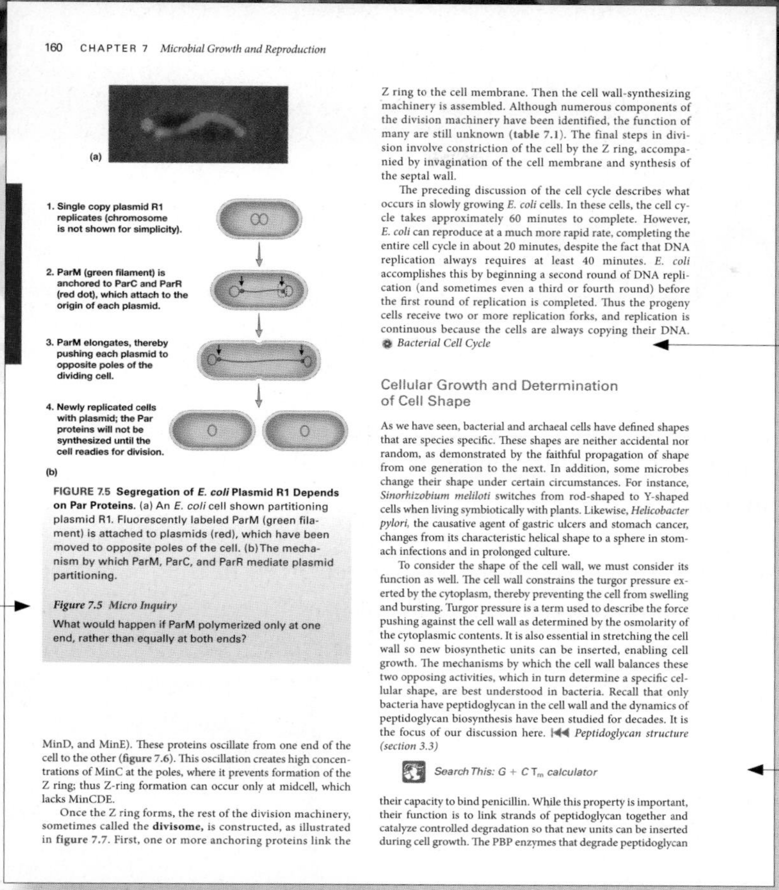
160 CHAPTER 7 *Microbial Growth and Reproduction*

(a)

1. Single copy plasmid R1 replicates (chromosome is not shown for simplicity).
2. ParM (green filament) is anchored to ParC and ParR (red dot), which attach to the origin of each plasmid.
3. ParM elongates, thereby pushing each plasmid to opposite poles of the dividing cell.
4. Newly replicated cells with plasmid; the Par proteins will not be synthesized until the cell readies for division.

(b)

FIGURE 7.5 **Segregation of *E. coli* Plasmid R1 Depends on Par Proteins.** (a) An *E. coli* cell shown partitioning plasmid R1. Fluorescently labeled ParM (green filament) is attached to plasmids (red), which have been moved to opposite poles of the cell. (b) The mechanism by which ParM, ParC, and ParR mediate plasmid partitioning.

Figure 7.5 Micro Inquiry

What would happen if ParM polymerized only at one end, rather than equally at both ends?

MinD, and MinE). These proteins oscillate from one end of the cell to the other (**figure 7.6**). This oscillation creates high concentrations of MinC at the poles, where it prevents formation of the Z ring; thus Z-ring formation can occur only at midcell, which lacks MinCDE.

Once the Z ring forms, the rest of the division machinery, sometimes called the **divisome,** is constructed, as illustrated in **figure 7.7**. First, one or more anchoring proteins link the Z ring to the cell membrane. Then the cell wall-synthesizing machinery is assembled. Although numerous components of the division machinery have been identified, the function of many are still unknown (**table 7.1**). The final steps in division involve constriction of the cell by the Z ring, accompanied by invagination of the cell membrane and synthesis of the septal wall.

The preceding discussion of the cell cycle describes what occurs in slowly growing *E. coli* cells. In these cells, the cell cycle takes approximately 60 minutes to complete. However, *E. coli* can reproduce at a much more rapid rate, completing the entire cell cycle in about 20 minutes, despite the fact that DNA replication always requires at least 40 minutes. *E. coli* accomplishes this by beginning a second round of DNA replication (and sometimes even a third or fourth round) before the first round of replication is completed. Thus the progeny cells receive two or more replication forks, and replication is continuous because the cells are always copying their DNA. *Bacterial Cell Cycle*

Cellular Growth and Determination of Cell Shape

As we have seen, bacterial and archaeal cells have defined shapes that are species specific. These shapes are neither accidental nor random, as demonstrated by the faithful propagation of shape from one generation to the next. In addition, some microbes change their shape under certain circumstances. For instance, *Sinorhizobium meliloti* switches from rod-shaped to Y-shaped cells when living symbiotically with plants. Likewise, *Helicobacter pylori,* the causative agent of gastric ulcers and stomach cancer, changes from its characteristic helical shape to a sphere in stomach infections and in prolonged culture.

To consider the shape of the cell wall, we must consider its function as well. The cell wall constrains the turgor pressure exerted by the cytoplasm, thereby preventing the cell from swelling and bursting. Turgor pressure is a term used to describe the force pushing against the cell wall as determined by the osmolarity of the cytoplasmic contents. It is also essential in stretching the cell wall so new biosynthetic units can be inserted, enabling cell growth. The mechanisms by which the cell wall balances these two opposing activities, which in turn determine a specific cellular shape, are best understood in bacteria. Recall that only bacteria have peptidoglycan in the cell wall and the dynamics of peptidoglycan biosynthesis have been studied for decades. It is the focus of our discussion here. ◀◀ *Peptidoglycan structure (section 3.3)*

Search This: G + C T_m calculator

their capacity to bind penicillin. While this property is important, their function is to link strands of peptidoglycan together and catalyze controlled degradation so that new units can be inserted during cell growth. The PBP enzymes that degrade peptidoglycan

NEW! Animation Icon
This symbol indicates that material presented in the text is also accompanied by an animation on the text website at www.mhhe.com/willey8.

NEW! Search This
Icons throughout the chapter highlight topics students can search on their own.

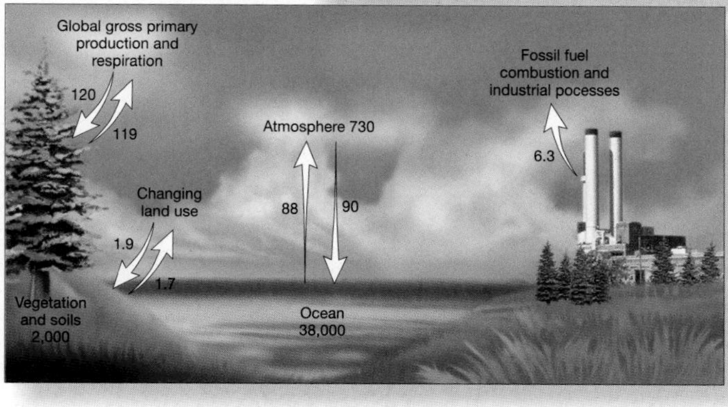

VIVID INSTRUCTIONAL ART PROGRAM

- Three-dimensional renditions and bright, attractive colors.
- Annotation of key pathways.
- Concept Maps.

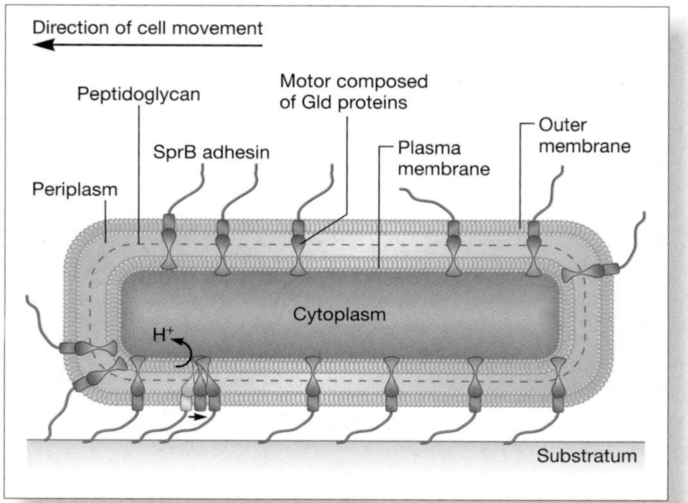

INCORPORATE ENGAGING MATERIALS FOR LECTURE AND LAB

connect™ | MICROBIOLOGY

Create customized lectures, visually enhanced tests and quizzes, compelling course website, or attractive printed support materials using **McGraw-Hill's Presentation Tools.**

3-D animations that bring difficult topics to life are cross-referenced in the text and now available completely embedded in PowerPoint for total ease of use!

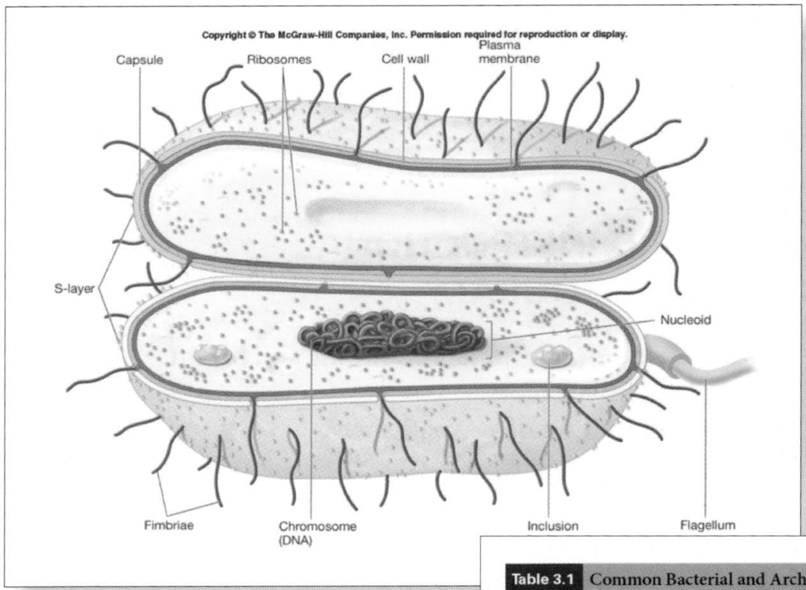

Table 3.1 Common Bacterial and Archaeal Structures and Their Functions

Plasma membrane	Selectively permeable barrier, mechanical boundary of cell, nutrient and waste transport, location of many metabolic processes (respiration, photosynthesis), detection of environmental cues for chemotaxis
Gas vacuole	Buoyancy for floating in aquatic environments
Ribosomes	Protein synthesis
Inclusions	Storage of carbon, phosphate, and other substances
Nucleoid	Localization of genetic material (DNA)
Periplasmic space	In gram-negative bacteria, contains hydrolytic enzymes and binding proteins for nutrient processing and uptake; in gram-positive bacteria and archaeal cells, may be smaller or absent
Cell wall	Provides shape and protection from osmotic stress
Capsules and slime layers	Resistance to phagocytosis, adherence to surfaces; rare in the *Archaea*
Fimbriae and pili	Attachment to surfaces, bacterial conjugation and transformation, twitching and gliding motility
Flagella	Swimming motility
Endospore	Survival under harsh environmental conditions; only observed in *Bacteria*

Enhanced Image PowerPoint files include line art, photos, and tables. Make it easy to customize your classroom presentations!

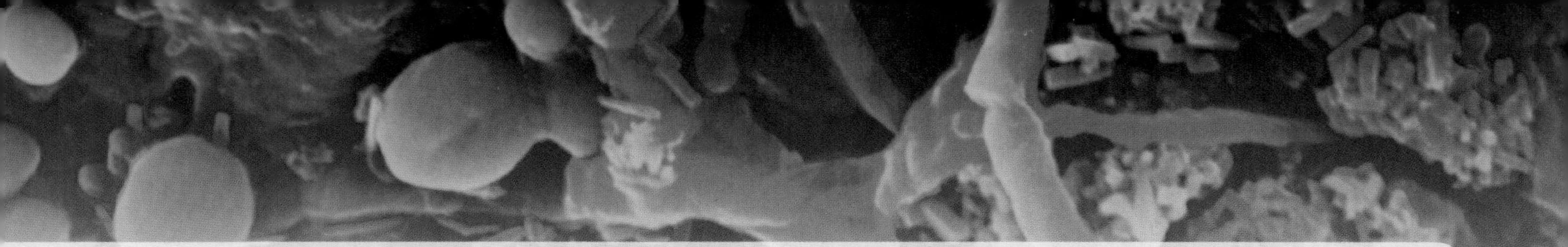

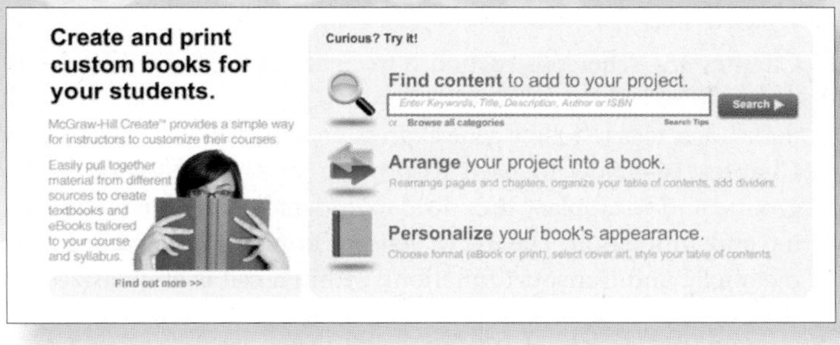

Craft your teaching resources to match the way you teach! With McGraw-Hill Create, www.mcgrawhillcreate.com, you can easily rearrange chapters, combine material from other content sources, and quickly upload content you have written like your course syllabus or teaching notes. Find the content you need in Create by searching through thousands of leading McGraw-Hill textbooks. Arrange your book to fit your teaching style. Create even allows you to personalize your book's appearance by selecting the cover and adding your name, school, and course information. Order a Create book and you'll receive a complimentary print review copy in 3–5 business days or a complimentary electronic review copy (eComp) via email in about one hour. Go to www.mcgrawhillcreate.com today and register. Experience how McGraw-Hill Create empowers you to teach your students your way.

NEW! **Microbiology Prep**, also available on the text website at www.mhhe.com/willey8, helps students to prepare for their upcoming coursework in microbiology. This website enables students to perform self-assessments, conduct self-study sessions with tutorials, and perform a post-assessment of their knowledge in the following areas:

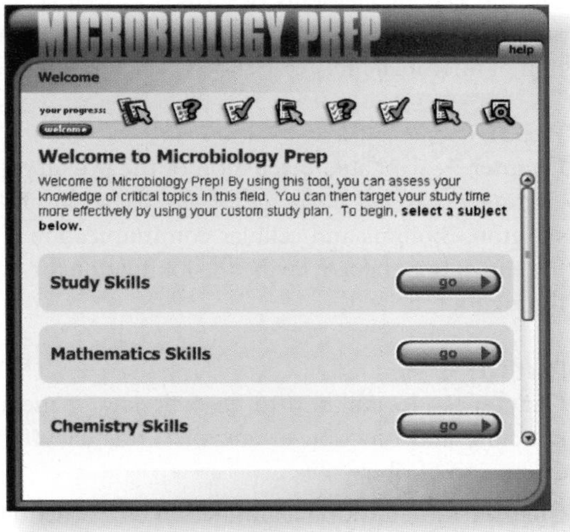

- Study Skills
- Mathematics Skills
- Chemistry Skills
- Biology Skills
- Metric System Skills
- Lab Reports and Referencing

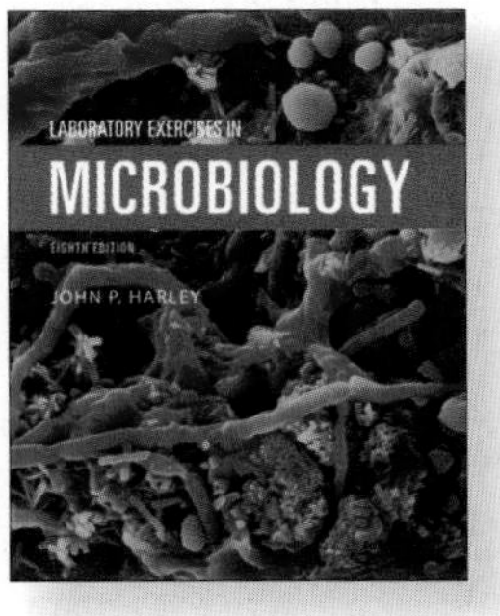

Other Resources Available!

Laboratory Exercises in Microbiology Eighth Edition by John P. Harley has been prepared to accompany the text. Like the text, the laboratory manual provides a balanced introduction in each area of microbiology. The class-tested exercises are modular and short so that an instructor can easily choose those exercises that fit his or her course.

The **Student Study Guide** is a valuable resource that provides learning objectives, study outlines, learning activities, and self-testing material to help students master course content. It is available on the text website at www.mhhe.com/willey8.

Visit the text website at www.mhhe.com/willey8

LIST OF CHANGES

CONTENT CHANGES BY PART

Each chapter has been thoroughly reviewed and almost all have undergone significant revision. The highlights include:

Part I

Chapter 1—We open the text with the essential concepts of microbial evolution to underscore the role of evolution as the driving force of all biological systems.

Chapter 3—Discussion of archaeal and bacterial cellular structure is integrated throughout.

Chapter 4—The essentials of protist and fungal morphology are presented here, so that chapters 23 (The Protists) and 24 (The Fungi) now focus specifically on the diversity of these microbes.

Chapter 5—This new chapter, entitled *Viruses and Other Acellular Infectious Agents*, surveys the essential morphological, physiological, and genetic elements of viruses as well as viroids, virusoids, and prions. This chapter completes our three-chapter introduction of microbial life.

Part II

Chapter 7—Updated discussion of the prokaryotic cell cycle, including expanded coverage of the bacterial and archaeal cytoskeleton. Biofilms and cellular communication are introduced at the end of this chapter, laying the foundation for further discussion throughout the text.

Part III

Chapter 9—In addition to an overview of metabolism and the structure and function of enzymes, this chapter now includes a discussion of ribozymes.

Chapter 10—Additional annotation of biochemical pathways has been included. Rhodopsin-based phototrophy is introduced as well as oxygenic and anoxygenic photosynthesis.

Part IV

Chapter 12—Protein folding and secretion now completes this chapter on genome structure and replication, gene structure, and gene expression.

Chapter 13—Focus continues to be exclusively on the regulation of gene expression, presented according to the level at which regulation occurs; updated and expanded discussion of riboswitches and regulation by small RNA molecules. Quorum sensing and its role in microbial physiology and biofilm formation are given in-depth coverage.

Chapter 14—Covers mutation, repair, and recombination in the context of processes that introduce genetic variation into populations. The consequences of mutating protein coding genes, regulatory sequences, and RNA genes are given individual consideration.

Chapter 15—The construction of recombinant DNA molecules is now followed with a discussion of protein purification and the use of fluorescent tags to follow gene expression and protein localization.

Chapter 16—Now focused to provide students with the background and vocabulary they need to understand genome sequencing and annotation. The use of genome annotation to reconstruct metabolic and transport functions within a cell is emphasized as such figures are now featured in the diversity chapters. The growing importance of metagenomics is introduced here.

Part V

Chapter 17—Rewritten and re-titled *Microbial Taxonomy and the Evolution of Diversity,* this chapter features expanded coverage of the construction and utility of phylogenetic trees. Microbial evolution, introduced in chapter 1, is expanded with a complete discussion of the concept and definition of microbial species, the core genome and the importance of horizontal gene transfer and the development of the pan-genome.

Chapter 18—Expanded coverage of archaeal genetics and thermostabilty. Archaeal metabolism has been expanded with new figures presenting alternate, archaeal-specific modifications of the Embden-Meyerhoff and Entner-Douderoff pathways. Genome-based metabolic reconstructions based on genomic sequencing of a crenarchaeote (*Sulfolobus*) and a euryarchaeote (*Halobacterium*) are now included.

Chapter 19—Updated and expanded coverage of photosynthetic pigments, the anammox reaction, and exciting new discoveries within the phylum *Verrucomicrobia* are now included.

Chapter 20—Increased coverage of proteobacterial physiology includes discussion of topics such as photosynthetic pigment apparatus in purple bacteria, nitrification, methylotrophy, dissimilatory sulfate reduction, and gliding motility. Also featured are genome-based, metabolic reconstructions of several *Proteobacteria.*

Chapter 21—New discussions include the proposed mechanism of phototrophy in *Heliobacterium* and biofilm formation by *Bacillus subtilis.*

Part VI

Chapter 26—Re-titled *Biogeochemical Cycling,* this chapter now focuses only on biogeochemical cycling and the role of microbes in global climate change.

Chapter 27—This new chapter, entitled *Methods in Microbial Ecology*, surveys the most commonly used techniques in the field. The importance and difficulty of culturing microbes in the laboratory is stressed with coverage of novel approaches that have been developed for isolating and growing bacteria and archaea. Molecular and biogeochemical methods for the analysis of microbial diversity and community activity are also reviewed.

Chapter 28—While the focus remains on microbial communities found in the major biomes within marine and freshwater environments, the importance of microbial primary production and its potential role in modulating global climate change is stressed.

Chapter 30—Microbial interactions are presented along with human-microbe relationships, helping to convey the notion that the human body is an ecosystem. The concept and study of the human microbiome is introduced.

Part VII

Chapter 31—This chapter has been re-titled *Infection and Pathogenicity*. It also now follows the chapter on microbial interactions to stress that pathogens represent one half of a parasite-host relationship. The essential elements required for a pathogen to establish infection are introduced followed by common virulence mechanisms.

Chapter 32—Reorganized and updated, this chapter on nonspecific (innate) host resistance provides in-depth coverage of physical and chemical components of the nonspecific host response followed by an overview of cells, tissues, and organs of the immune system. This includes a step-by-step discussion of phagocytosis and inflammation. The groundwork is laid for a full appreciation of the connections between the specific and nonspecific arms of the immune system.

Chapter 33—Reorganized and updated to enhance linkages between innate and acquired immune activities; integrated medical immunology concepts, including insights into regulatory mechanisms.

Part VIII

Chapter 34—Content focuses on the mechanism of action of each antimicrobial agent, with an additional emphasis on the development of antibiotic resistance.

Chapter 36—Expanded focus on the important role of epidemiology in preventative medicine and the role of the public health system.

Chapter 37—Updated and expanded coverage of viral pathogenesis; select (potential bioterrorism) agents highlighted.

Chapter 38—Expanded coverage of bacterial pathogenesis; select (potential bioterrorism) agents highlighted.

Chapter 39—Updated to reflect disease transmission routes (similar to chapters 37 and 38.

Part IX

Chapter 40—Expanded discussion of the control of food spoilage and probiotics.

Chapter 41—Some material in this new chapter, called *Biotechnology and Industrial Microbiology*, was covered in chapter 42 of the 7th edition. Now expanded and updated, the development of industrial strains of microbes is discussed, followed by some of the more important industrial microbial applications. This includes the use of microbes in producing biofuels and the development of microbial fuel cells.

Chapter 42—Updated and expanded discussion of water purification, wastewater treatment, and bioremediation.

ACKNOWLEDGEMENTS

We would like to thank the Board of Advisors and Reviewers, who provided constructive reviews of every chapter. Their specialized knowledge helped assimilate more reliable sources of information, and find more effective ways of expressing an idea for the student reader.

Board of Advisors
David A. Battigell, *University of North Carolina–Greensboro*
Mary B. Farone, *Middle Tennessee State University*
Sandra Gibbons, *University of Illinois at Chicago*
Michael C. Hudson, *The University of North Carolina* at Charlotte
Dr. Tamara L. McNealy, *Clemson University*
John Steiert, *Missouri State University*
Lori Zeringue Crow, *Louisiana State University*

Reviewers
Elizabeth Wheeler Alm, *Central Michigan University*
Shivanthi Anandan, *Drexel University*
Penny P. Antley, *University of Louisiana at Lafayette*
Larry L. Barton, *University of New Mexico–Albuquerque*
Linda D. Bruslind, *Oregon State University*
Mary Burke, *Oregon State University*
Joseph P. Caruso, *Florida Atlantic University*
Carlton Rodney Cooper, *University of Delaware*
Ellen C. Cover, *Lamar University*
Sidney A. Crow Jr., *George State University*
Richard Crowther, *University of Wisconsin–Stevens Point*
James S. Dickson, *Iowa State University*
Lehman L. Ellis, *Our Lady of Holy Cross College*
Mark Farinha, *Richland College*
Amy E. Fleishman Littlejohn, *Northern Arizona University*
Ken Flint, *Department of Biological Sciences,* University of Warwick
Bernard Lee Frye, *The University of Texas at Arlington*
Phillip E. Funk, *DePaul University*
Joseph J. Gauthier, *University of Alabama at Birmingham*
Sandra Gibbons, *University of Illinois, Chicago*
Darryl V. Grennell, *Alcorn State University*
Helmut Hirt, *Kansas State University*
Gilbert H. John, *Oklahoma State University*
Karen E. Kesterson, *LSTF–Training Branch*
Jeff Kingsbury, *Mohave Community College*
Duncan C. Krause, *University of Georgia*
Jennifer Kraft Leavey, *Georgia Institute of Technology*
Jean Lu, *Kennesaw State University*
John Makemson, *Florida International University*
D. C. Ghislaine Mayer, *Virginia Commonwealth University*
Vance J. McCracken, *Southern Illinois University Edwardsville*
Robert J. C. McLean, *Texas State University*
Richard L. Myers, *Missouri State University*
Nick Nagle, *Metropolitan State College of Denver*
Karen G. Nakaoka, *Weber State University*
Carolyn Peters, *Spoon River College*
Marcia M. Pierce, *Eastern Kentucky University*
Don V. Plantz, *Jr., Mohave Community College*
Todd P. Primm, *Sam Houston State University*
S. N. Rajagopal, *University of Wisconsin–La Crosse*
Jackie Reynolds, *Richland College*
Timberley Roane, *University of Colorado at Denver* and Health Sciences Center
Ben Rowley, *University of Central Arkansas*
Pratibha Saxena, *University of Texas at Austin*
Peter P. Sheridan, *Idaho State University*
Garriet W. Smith, *University of South Carolina, Aiken*
Geoffrey Battle Smith, *New Mexico State University*
Lisa Y. Stein, *University of California, Riverside*
Michael A. Sulzinski, *University of Scranton (Pennsylvania)*
Stephen Wagner, *Stephen F. Austin State University*
Kathryn G. Zeiler, *Red Rocks Community College*

The authors wish to extend their gratitude to our editors, Jim Connoly, Fran Schreiber, Sandy Wille, and Lynn Breithaupt. We would also like to thank our photo editor Mary Reeg and the tremendous talent and patience displayed by the artists. We are also very grateful to Professors Mary Ann Moran and Celine Brochier for helpful discussions, and the many reviewers who provided helpful criticism and analysis. Finally, we thank our spouses and children who provided support and tolerated our absences (mental if not physical) while we completed this demanding project.

BRIEF CONTENTS

CONTENTS

CONTENTS

CONTENTS

CONTENTS

CONTENTS

1

The Evolution of Microorganisms and Microbiology

Modern stromatolites from Western Australia. Each stromatolite is a rocklike structure, typically 1 m in diameter, containing layers of cyanobacteria. The existence of fossilized stromatolites is evidence of the existence of photosynthetic organisms early in the history of life on Earth.

CHAPTER GLOSSARY

Archaea The domain of life containing anucleate cells that have unique lipids in their membranes, distinctive rRNA sequences, and cell walls that lack peptidoglycan.

Bacteria The domain of life that contains anucleate cells having distinctive rRNA sequences and cell walls that contain the structural molecule peptidoglycan.

Eukarya The domain of life that features organisms made of cells having a membrane-delimited nucleus and differing in many other ways from *Archaea* and *Bacteria*; includes protists, fungi, plants, and animals.

fungi A diverse group of eukaryotic microorganisms that range from unicellular forms (yeasts) to multicellular molds and mushrooms.

genome The entire genetic makeup of an organism.

genomic analysis An approach to studying organisms that involves sequencing the genome, identifying genes, and assigning functions to the genes.

Koch's postulates A set of rules for proving that a specific microorganism causes a particular disease.

microbiology The study of organisms that are usually too small to be seen with the naked eye; special techniques are required to isolate and grow them.

microorganism An organism that is too small to be seen clearly with the naked eye and lacks highly differentiated cells and distinct tissues.

prions Infectious agents, composed only of protein, that cause spongiform encephalopathies such as scrapie in sheep.

prokaryotic cells Cells having a type of structure characterized by the lack of a true, membrane-enclosed nucleus. All known members of *Archaea* and most members of *Bacteria* exhibit this type of cell structure.

protists Mostly unicellular eukaryotic organisms that lack cellular differentiation into tissues. Cell differentiation is limited to cells involved in sexual reproduction, alternate vegetative morphology, or resting states such as cysts; includes organisms often referred to as algae and protozoa.

spontaneous generation An early belief, now discredited, that living organisms could develop from nonliving matter.

viroids Infectious agents of plants composed only of single-stranded RNA.

viruses Infectious agents having a simple acellular organization with a protein coat and a nucleic acid genome, lacking independent metabolism, and reproducing only within living host cells.

virusoids Infectious agents composed only of single-stranded RNA. They are unable to replicate without the aid of specific viruses that coinfect the host cell.

Beginning billions of years ago, by a process that was chaotic and violent, the dust circulating a newly formed star in the Milky Way galaxy began to aggregate into larger bodies. About 4.5 billion years ago, one of these bodies reached planet size: Earth had formed. Within the next 1 billion years, the first cellular life forms appeared. They were microbes. Since that time, microorganisms have evolved and diversified to occupy virtually every habitat on Earth: from geothermal vents in the ocean depths to the coldest Arctic ice. Today they are major contributors to the functioning of the biosphere, being indispensable for cycling the elements essential for life. They also are a source of nutrients at the base of all ecological food webs. Most important, certain microorganisms carry out photosynthesis, rivaling plants in their role of capturing carbon dioxide, forming organic matter, and releasing oxygen into the atmosphere. Microbes currently are estimated to contain 50% of the biological carbon and 90% of the biological nitrogen on Earth, and they greatly exceed in number every other group of organisms on the planet.

Microorganisms have also evolved to use other organisms as habitats, including humans. Microbes can be found in various environments provided by the human body, such as skin, intestines, and mouth. Indeed, more microbial cells are found in and on the human body than are human cells. These microbes begin to colonize humans shortly after birth. As the microbes establish themselves, they contribute to the development of the body's immune system. Those microbes that inhabit the large intestine help the body digest food and produce vitamins B and K. In these and other ways, microbes help maintain the health and well-being of their human hosts.

The diversity of cellular microorganisms is best exemplified by their metabolic capabilities, and humans have learned to harness the metabolism of microbes to achieve a variety of goals. Microbes are used to produce bread, cheese, beer, antibiotics, vaccines, vitamins, enzymes, and many other products. Their ability to produce biofuels such as ethanol is being intensively explored. These alternative fuels are renewable and may help decrease pollution associated with burning fossil fuels.

Although most microbes play beneficial or benign roles, some harm the large organisms they live with, causing disease in both animals and plants. Human pathogens have disrupted society over the millennia. Microbial diseases undoubtedly played a major role in the decline of the Roman Empire and the conquest of the New World. In 1347, plague (Black Death), an arthropod-borne disease, struck Europe with brutal force, killing one-third of the population (about 25 million people) within four years. Over the next 80 years, the disease struck repeatedly, eventually wiping out 75% of the European population. The plague's effect was so great that some historians believe it changed European culture and prepared the way for the Renaissance. Today the struggle against killers such as AIDS and malaria continues.

Our focus in this chapter is on the evolution of microorganisms and the nature and development of microbiology—the scientific discipline that studies microbes. We begin by introducing modern microbes. Then we describe how microbiologists, geologists, and evolutionary biologists are piecing together a history of how these simple, yet amazing, organisms arose. Finally, we turn our attention to microbiology—the tools it uses, its past, and its present.

1.1 Members of the Microbial World

Microorganisms are defined as those organisms and acellular biological entities too small to be seen clearly by the unaided eye (**figure 1.1**). They are generally 1 millimeter or less in diameter. Although small size is an important characteristic of microbes, it alone is not sufficient to define them. Some cellular microbes, such as bread molds and filamentous photosynthetic microbes, are actually visible without microscopes. These macroscopic microbes are often colonial, consisting of small aggregations of cells. Some macroscopic microorganisms are multicellular. They are distinguished from other multicellular life forms such as plants and animals by their lack of highly differentiated tissues. Although most unicellular microbes are microscopic, there are interesting exceptions, as we describe in chapter 3. In summary, cellular microbes are usually smaller than 1 millimeter in diameter, often unicellular, and if multicellular, lack differentiated tissues. ▶▶| *Microbial Diversity & Ecology 3.1: Microbes That Have Babies*

The diversity of microorganisms has always presented a challenge to microbial taxonomists. The early descriptions of cellular microbes as either plants or animals were too simple. For instance, some microbes are motile like animals but also have cell walls and are photosynthetic like plants. Such microbes cannot be placed easily into one kingdom or another. An important breakthrough in microbial taxonomy arose from studies of their cellular architecture, when it was discovered that cells exhibited one of two possible "floor plans." Cells that came to be called **prokaryotic cells** (Greek *pro,* before, and *karyon,* nut or kernel; organisms with a primordial nucleus) have an open floor plan. That is, their contents are not divided into compartments ("rooms") by membranes ("walls"). The most obvious characteristic of these cells is that they lack the membrane-delimited nucleus observed in **eukaryotic cells** (Greek *eu,* true, and *karyon,* nut or kernel). Eukaryotic cells not only have a nucleus but also many other membrane-bound organelles that separate some cellular materials and processes from others.

These observations eventually led to the development of a classification scheme that divided organisms into five kingdoms: the *Monera, Protista, Fungi, Animalia,* and *Plantae.* Microorganisms (except for viruses and other acellular infectious agents, which have their own classification system) were placed in the first three kingdoms. In this scheme, all organisms with prokaryotic cell structure were placed in the *Monera.* Although the five-kingdom system was an important development in microbial taxonomy, it is no longer accepted by microbiologists. This is because it was eventually determined that not all "prokaryotes" are the same and therefore should not be grouped together in a single kingdom. Furthermore, it is

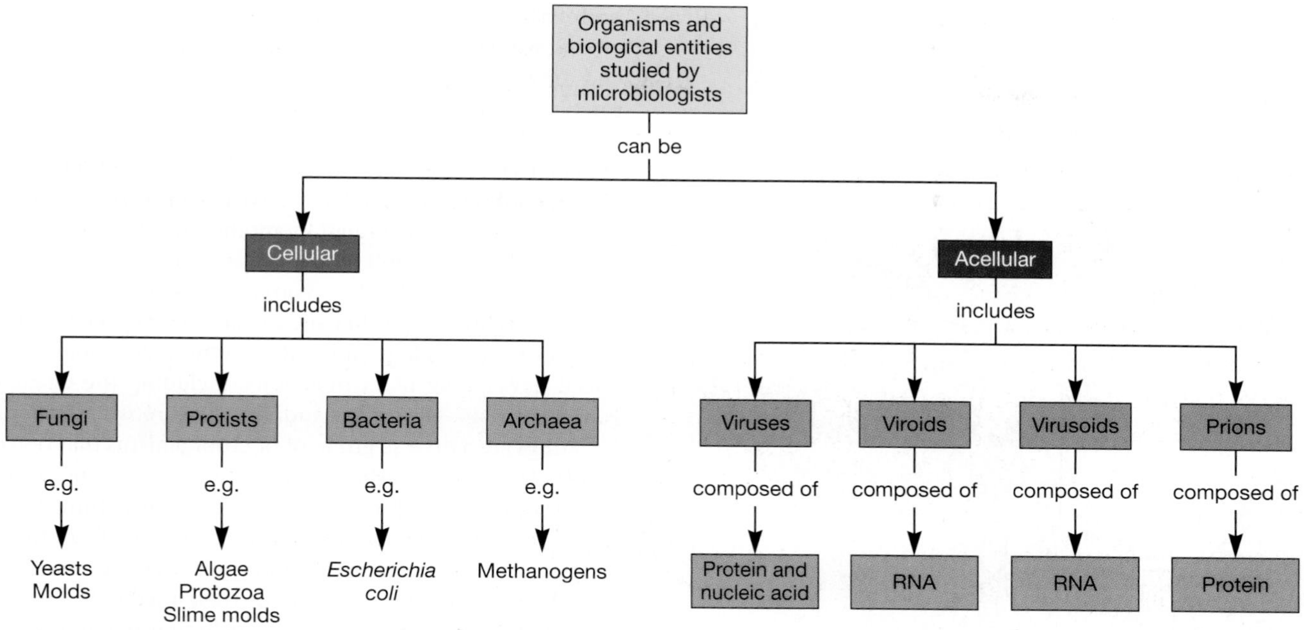

FIGURE 1.1 Concept Map Showing the Types of Biological Entities Studied by Microbiologists.

Figure 1.1 ***Micro Inquiry***

How would you alter this concept map so that it also distinguishes the cellular organisms from each other?

currently argued that the term prokaryote is not meaningful and should be abandoned. As we describe next, this discovery required several advances in the tools used to study microbes.

In the last few decades, great progress has been made in three areas that profoundly affect microbial classification. First, much has been learned about the detailed structure of microbial cells from the use of electron microscopy. Second, microbiologists have determined the biochemical and physiological characteristics of many different microorganisms. Third, the sequences of nucleic acids and proteins from a wide variety of organisms have been compared. The comparison of ribosomal RNA (rRNA), begun by Carl Woese in the 1970s, was instrumental in demonstrating that there are two very different groups of organisms with prokaryotic cell architecture: *Bacteria* and *Archaea*. Later studies based on rRNA comparisons showed that *Protista* is not a cohesive taxonomic unit (i.e., taxon) and that it should be divided into three or more kingdoms. These studies and others have led many taxonomists to reject the five-kingdom system in favor of one that divides cellular organisms into three domains: *Bacteria* (sometimes referred to as the true bacteria or eubacteria), *Archaea* (sometimes called archaeobacteria or archaebacteria) and *Eukarya* (all eukaryotic organisms) (**figure 1.2**). We use this system throughout the text. However, a brief description of the three domains and of the microorganisms placed in them follows. ▶▶| *Nucleic acids (appendix I); Proteins (appendix I)*

Members of the domain ***Bacteria*** are usually single-celled organisms.[1] Most have cell walls that contain the structural molecule peptidoglycan. Although most bacteria exhibit typical prokaryotic cell structure (i.e., they lack a membrane-bound nucleus), a few members of the unusual phylum *Planctomycetes* have their genetic material surrounded by a membrane. This inconsistency is another reason why it is argued that use of the term prokaryote should be abandoned. Bacteria are abundant in soil, water, and air, and are major inhabitants of our skin, mouth, and intestines. Some bacteria live in environments that have extreme temperatures, pH, or salinity. Although some bacteria cause disease, many more play beneficial roles such as cycling elements in the biosphere, breaking down dead plant and animal material, and producing vitamins. Cyanobacteria (once called blue-green algae) produce significant amounts of oxygen through the process of oxygenic photosynthesis. ▶▶| *Phylum Planctomycetes (section 19.4)*

Members of ***Archaea*** are distinguished from the *Bacteria* by many features, most notably their distinctive rRNA sequences, lack of peptidoglycan in their cell walls, and unique membrane lipids. Some have unusual metabolic characteristics, such as the methanogens, which generate methane gas. Many archaea are

[1] In this text, the term bacteria (s., bacterium) is used to refer to those microbes belonging to domain *Bacteria,* and the term archaea (s., archaeon) is used to refer to those that belong to domain *Archaea.* In some publications, the term bacteria is used to refer to all cells having prokaryotic cell structure. That is not the case in this text.

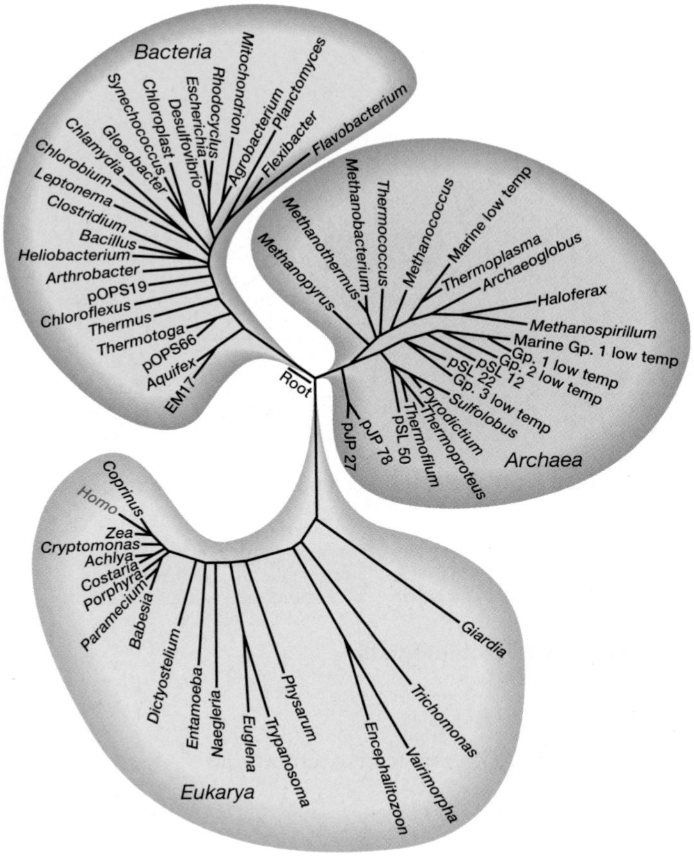

FIGURE 1.2 Universal Phylogenetic Tree. These evolutionary relationships are based on rRNA sequence comparisons. The human genus (*Homo*) is highlighted in red.

Figure 1.2 Micro Inquiry

How many of the taxa listed in the figure include microbes?

found in extreme environments, including those with high temperatures (thermophiles) and high concentrations of salt (extreme halophiles). Although some archaea are members of a community of microbes involved in gum disease in humans, their role in causing disease has not been clearly established.

Domain ***Eukarya*** includes microorganisms classified as protists or fungi. Animals and plants are also placed in this domain. **Protists** are generally unicellular but larger than most bacteria and archaea. They have traditionally been divided into protozoa and algae. However, none of these terms has taxonomic value as protists, algae, and protozoa do not form cohesive taxa. However, for convenience, we use these terms here.

The major types of protists are algae, protozoa, slime molds, and water molds. **Algae** are photosynthetic, and together with the cyanobacteria, they produce about 75% of the planet's oxygen and are the foundation of aquatic food chains. **Protozoa** are unicellular, animal-like protists that are usually motile. Many free-living protozoa function as the principal hunters and grazers of the microbial world. They obtain nutrients by ingesting organic matter and other microbes. They can be found in many different environments, and some are normal inhabitants of the intestinal tracts of animals, where they aid in digestion of complex materials such as cellulose. A few cause disease in humans and other animals. **Slime molds** are protists that behave like protozoa in one stage of their life cycle but like fungi in another. In the protozoan phase, they hunt for and engulf food particles, consuming decaying vegetation and other microbes. **Water molds** are protists that grow on the surface of freshwater and moist soil. They feed on decaying vegetation such as logs and mulch. Some water molds have produced devastating plant infections, including the Great Potato Famine of 1846–1847 in Ireland. ▶▶| *The protists (chapter 23)*

Fungi are a diverse group of microorganisms that range from unicellular forms (yeasts) to molds and mushrooms. Molds and mushrooms are multicellular fungi that form thin, threadlike structures called hyphae. They absorb nutrients from their environment, including the organic molecules that they use as sources of carbon and energy. Because of their metabolic capabilities, many fungi play beneficial roles, including making bread rise, producing antibiotics, and decomposing dead organisms. Some fungi associate with plant roots to form mycorrhizae. Mycorrhizal fungi transfer nutrients to the roots, improving growth of the plants, especially in poor soils. Other fungi cause plant diseases (e.g., rusts, powdery mildews, and smuts) and diseases in humans and other animals. ▶▶| *Fungi (chapter 24)*

The microbial world also includes numerous acellular infectious agents. **Viruses** are acellular entities that must invade a host cell to replicate. The simplest viruses are composed only of proteins and a nucleic acid, and can be extremely small (the smallest is 10,000 times smaller than a typical bacterium). However, their small size belies their power—they cause many animal and plant diseases and have caused epidemics that have shaped human history. Viral diseases include smallpox, rabies, influenza, AIDS, the common cold, and some cancers. Viruses also play important roles in aquatic environments, and their role in shaping aquatic microbial communities is currently being explored. **Viroids** and **virusoids** are infectious agents composed only of ribonucleic acid (RNA). Viroids cause numerous plant diseases, whereas virusoids cause some important animal diseases such as hepatitis. Finally, **prions**, infectious agents composed only of protein, are responsible for causing a variety of spongiform encephalopathies such as scrapie and "mad cow disease." ▶▶| *Viruses and other acellular infectious agents (chapter 5)*

1. How did the methods used to classify microbes change, particularly in the last half of the twentieth century? What was the result of these technological advances?
2. Microscopic organisms such as rotifers are not studied by microbiologists. Why do you think this is so?
3. Compare and contrast *Bacteria*, *Archaea*, protists, fungi, viruses, viroids, virusoids, and prions. Why do you think viruses, viroids, virusoids, and prions are not included in the three domain system?

TECHNIQUES & APPLICATIONS

1.1 The Scientific Method

Microbiologists and other scientists employ the scientific method to understand natural phenomena. They first gather observations of the process to be studied and then develop a tentative hypothesis—an educated guess—to explain the observations (**box figure**). This step often is inductive and creative because there is no detailed, automatic technique for generating hypotheses. Next they decide what information is required to test the hypothesis and collect this information through observation or carefully designed experiments. After they have collected the information, they decide whether the hypothesis has been supported or falsified. If it has failed to pass the test, the hypothesis is rejected, and a new explanation or hypothesis is constructed. If the hypothesis passes the test, it is subjected to more severe testing. The procedure often is made more efficient by constructing and testing alternative hypotheses and then refining the hypothesis that survives the test. This general approach is often called the hypothetico-deductive method. One deduces predictions from the currently accepted hypothesis and tests them. In deduction the conclusion about specific cases follows logically from a general premise ("if . . . , then . . ." reasoning). Induction is the opposite. A general conclusion is reached after considering many specific examples. Both types of reasoning are used by scientists.

When carrying out an experiment, it is essential to use a control group as well as an experimental group. The control group is treated precisely the same as the experimental group except that the experimental manipulation is not performed on it. In this way, one can be sure that any changes in the experimental group are due to the experimental manipulation rather than to some other factor not taken into account.

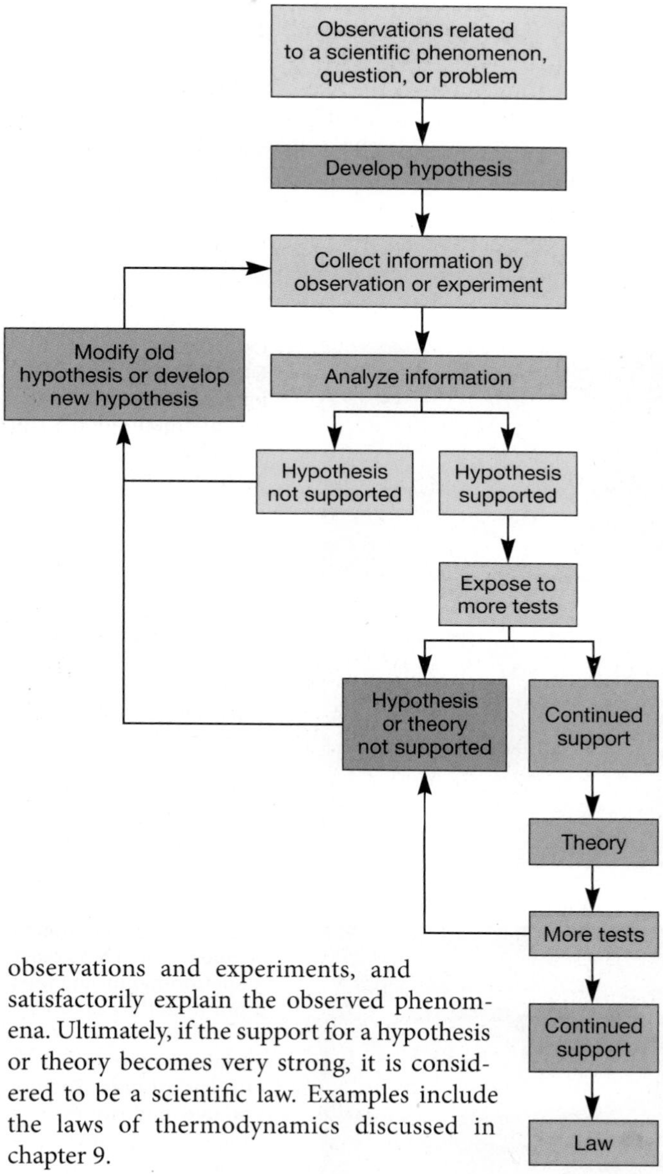

If a hypothesis continues to survive testing, it may be accepted as a valid theory. A theory is a set of propositions and concepts that provides a reliable, systematic, and rigorous account of an aspect of nature. It is important to note that hypotheses and theories are never absolutely proven. Scientists simply gain more and more confidence in their accuracy as they continue to survive testing, fit with new observations and experiments, and satisfactorily explain the observed phenomena. Ultimately, if the support for a hypothesis or theory becomes very strong, it is considered to be a scientific law. Examples include the laws of thermodynamics discussed in chapter 9.

1.2 Microbial Evolution

A review of figure 1.2 reminds us that in terms of the number of taxa, microbes are the dominant organisms on Earth. How has microbial life been able to radiate to such an astonishing level of diversity? To answer this question, we must consider microbial evolution. The field of microbial evolution, like any other scientific endeavor, is based on the formulation of hypotheses, the gathering and analysis of data, and the reformation of hypotheses based on newly acquired evidence. That is to say, the study of microbial evolution is based on the scientific method (**Techniques & Applications 1.1**). To be sure, it is sometimes more difficult to amass evidence when considering events that occurred millions, and often billions, of years ago, but the advent of molecular methods in biology has offered scientists a living record of life's ancient history. This section describes the outcome of this scientific research.

Evidence for the Origin of Life

Dating meteorites through the use of radioisotopes places our planet at an estimated 4.5 to 4.6 billion years old. However, conditions on Earth for the first 100 million years or so were far too harsh to sustain any type of life. Eventually bombardment by meteorites decreased, water appeared on the planet in liquid form, and gases were released by geological activity to form Earth's atmosphere. These conditions were amenable to the origin of the first life forms. But how did this occur, and what did these life forms look like?

Clearly, in order to find evidence of life and to develop hypotheses about its origin and subsequent evolution, scientists must be able to define life. Although even very young children can examine an object and correctly determine whether it is living or not, defining life in a succinct statement has proven elusive for scientists. Thus most definitions of life consist of a set of attributes (**figure 1.3**). The attributes of particular importance to paleobiologists are an orderly structure, the ability to obtain and use energy (i.e., metabolism), and the ability to reproduce. Scientists often start to understand the origin of life by examining **extant organisms**, those organisms present today. Extant organisms can have structures and molecules that represent "relics" of ancient life forms. Furthermore, they can provide scientists with ideas about the type of evidence to seek when shaping hypotheses.

The first direct evidence of primitive cellular life was the 1977 discovery of microbial fossils in the Swartkoppie chert, a granular type of silica. These microbial fossils as well as those from the Archaean Apex chert of Australia have been dated at about

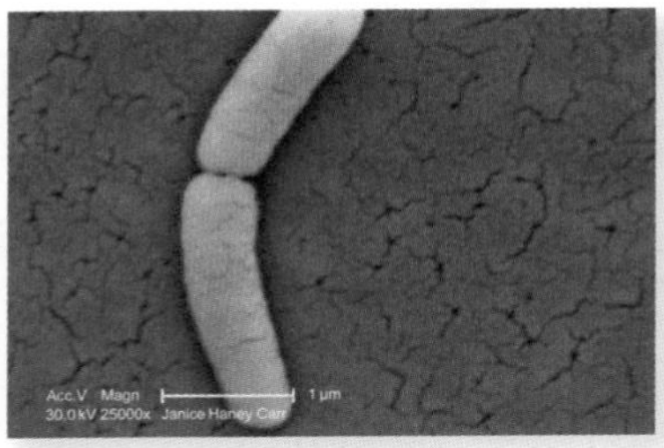

(a) Cells and organization: Organisms maintain an internal order. The simplest unit of organization is the cell. Shown are two bacterial cells.

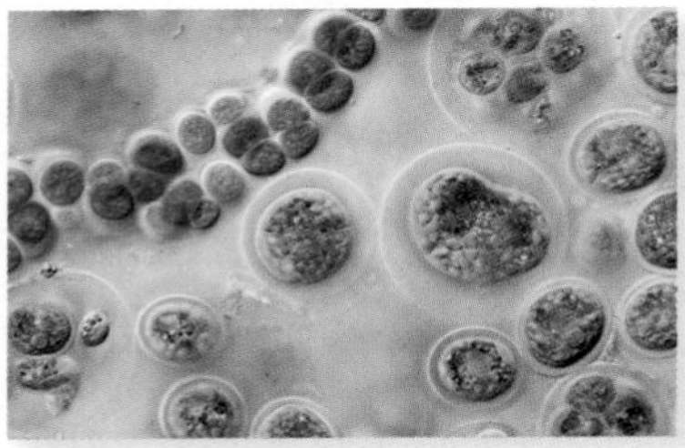

(b) Energy use and metabolism: To maintain their internal order, energy is needed by organisms. Energy is utilized in chemical reactions collectively known as metabolism. Shown are two photosynthetic organisms: a cyanobacterium and a green alga.

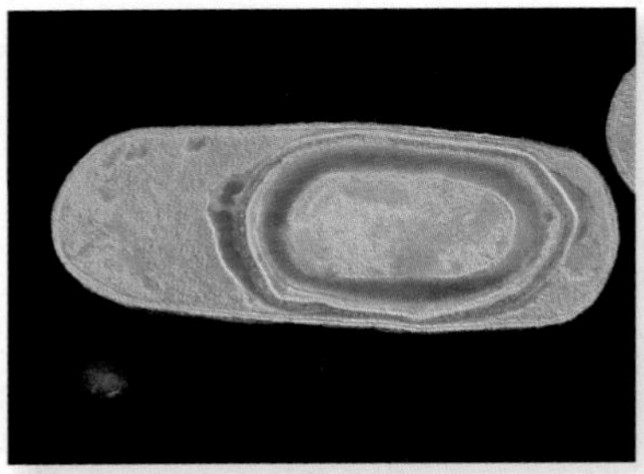

(c) Response to environmental changes: Organisms react to environmental changes to promote their survival. In response to diminishing nutrients, some bacteria form endospores. Shown is an oval endospore within the vegetative mother cell.

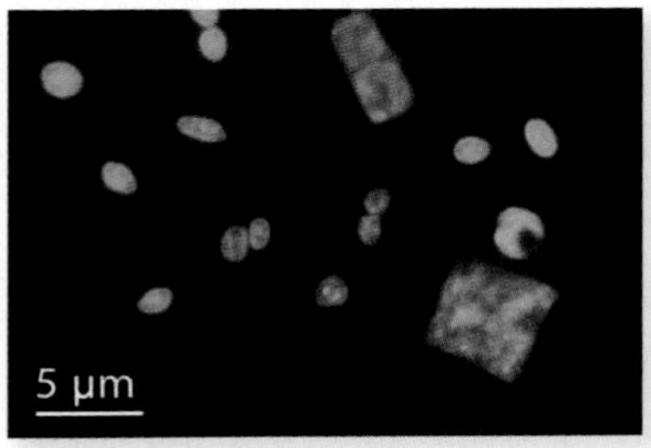

(d) Regulation and homeostasis: Organisms regulate their cells to maintain relatively stable internal conditions, a process called homeostasis. Shown are halophiles that regulate their metabolism in response to oxygen levels.

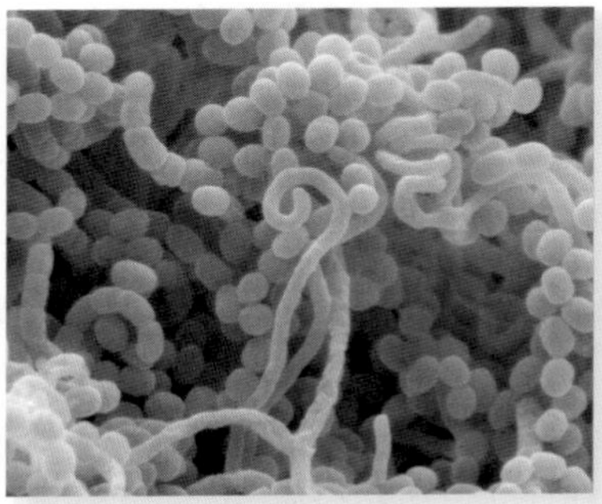

(e) Growth and development: Growth produces more or larger cells, while development produces organisms with a defined set of characteristics. Shown is an actinomycete that forms substrate and aerial hyphae. Cells in the aerial hyphae differentiate into spores.

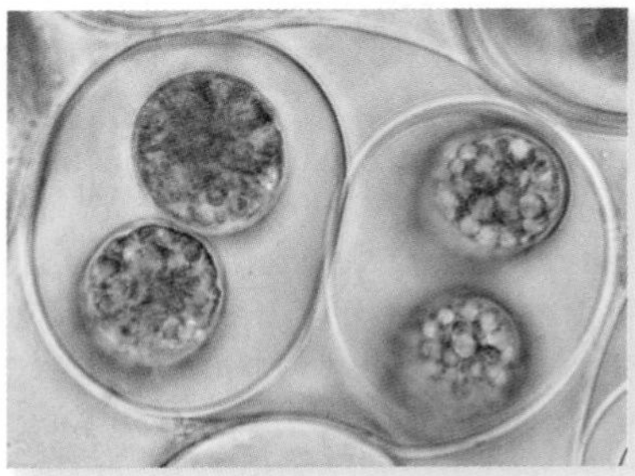

(f) Reproduction: To sustain life over many generations, organisms must reproduce.

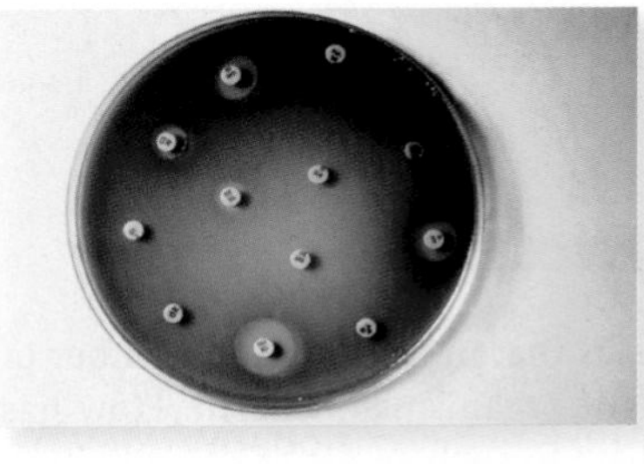

(g) Biological evolution: Populations of organisms change over the course of many generations. Evolution results in traits that promote survival and reproductive success. Bacteria often evolve to become resistant to antibiotics.

FIGURE 1.3 Important Attributes of Life.

3.5 billion years old (**figures 1.4** and **1.5**). Despite these findings, the microbial fossil record is understandably sparse. Thus to piece together the very early events that led to the origin of life, biologists must rely primarily on indirect evidence. Each piece of evidence must fit together as in a jigsaw puzzle for a coherent picture to emerge.

RNA World

The origin of life rests on a single question: How did early cells arise? Modern cells consist at a minimum of a plasma membrane enclosing water in which numerous chemicals are dissolved and subcellular structures float. It seems likely that the first self-replicating entity was much simpler than even the most primitive modern living cells. Before there was life, Earth was a very different place: hot and anoxic, with an atmosphere rich in gases such as hydrogen, methane, carbon dioxide, nitrogen, and ammonia. Earth's surface was a prebiotic soup in which chemicals reacted with one another, randomly "testing" the usefulness of the reaction and the stability of the resulting molecules. Some reactions released energy and would eventually become the basis of modern cellular metabolism. Other reactions generated molecules that could function as catalysts, some aggregated with other molecules to form the predecessors of modern cell structures, and others were able to replicate and act as units of hereditary information.

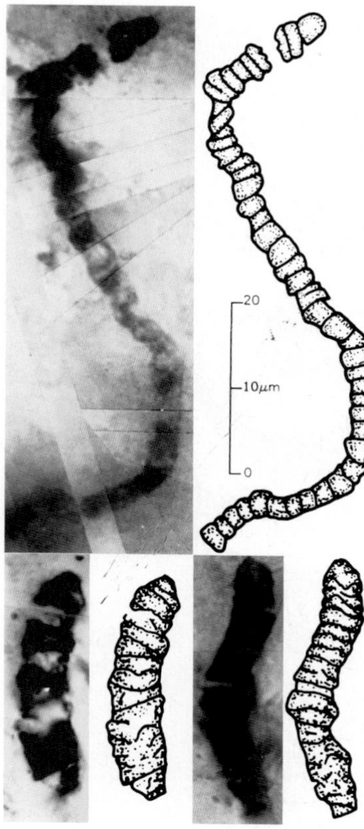

FIGURE 1.4 Microfossils of the Archaeon Apex Chert of Australia. These microfossils are similar to modern filamentous cyanobacteria.

In modern cells, three different molecules fulfill these roles (**figure 1.6**). Proteins have two major roles in modern cells: structural and catalytic. Catalytic proteins are called **enzymes**, and they speed up the myriad of chemical reactions that occur in cells. Thus enzymes are the workhorses of the cell. DNA stores hereditary information and can be replicated to pass the information on to the next generation. RNA is involved in converting the information stored in DNA into protein. Any hypothesis about the origin of life must account for the evolution of these molecules, but the very nature of their relationships to each other in modern cells complicates all attempts to imagine how they evolved. As demonstrated in figure 1.6, proteins can do cellular work, but their synthesis involves other proteins and RNA, and uses information stored in DNA. DNA can't do cellular work. It stores genetic information and serves as the template for its own replication, a process that requires proteins. RNA is synthesized using DNA as the template and proteins as the catalysts for the reaction.

Based on these considerations, it seemed to evolutionary biologists that at some time in the evolution of life there must have been a single molecule that could do both cellular work and replicate itself. A possible solution to the nature of this molecule was suggested in 1981 when Thomas Cech discovered an RNA molecule in the protist *Tetrahymena* that could cut out an internal section of itself and splice the remaining sections back together. Since then, other catalytic RNA molecules have been discovered, including an RNA found in ribosomes that is responsible for forming peptide bonds—the bonds that hold together amino acids, the building blocks of proteins. Catalytic RNA molecules are now called **ribozymes**.

The discovery of ribozymes suggested the possibility that RNA at some time had the ability to catalyze its own replication, using itself as the template. In 1986 Walter Gilbert coined the term **RNA world** to describe a precellular stage in the evolution of life in which RNA was capable of storing, copying, and expressing genetic information, as well as catalyzing other chemical reactions. However, for this precellular stage to proceed to the evolution of cellular life forms, a lipid membrane must have formed around the RNA (**figure 1.7**). This important evolutionary step is easier to imagine than other events in the origin of cellular life forms because it is well known that lipids, major structural components of the membranes of modern organisms, spontaneously form liposomes—vesicles bounded by a lipid bilayer. A fascinating experiment performed by Marin Hanczyc, Shelly Fujikawa, and Jack Szostak in 2003 showed that clay triggers the formation of liposomes that actually grow and divide. Considering this together with the data on ribozymes, it seems plausible that early cells may have been liposomes containing RNA molecules (figure 1.7). ▶▶| *Lipids (appendix I)*

Apart from its ability to perform catalytic activities, the function of RNA suggests its ancient origin. Consider that much of

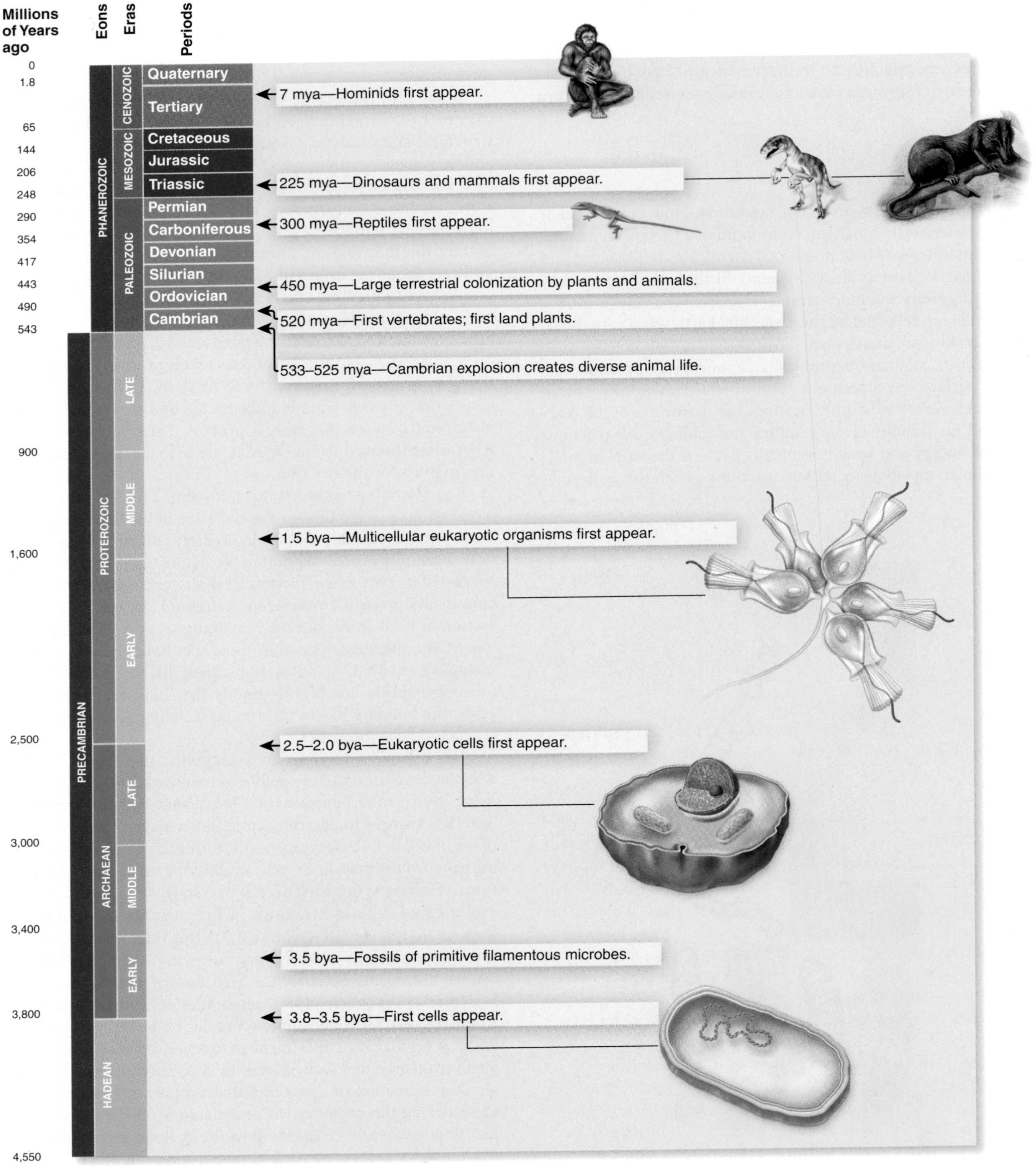

FIGURE 1.5 An Overview of the History of Life on Earth. mya = million years ago; bya = billion years ago.

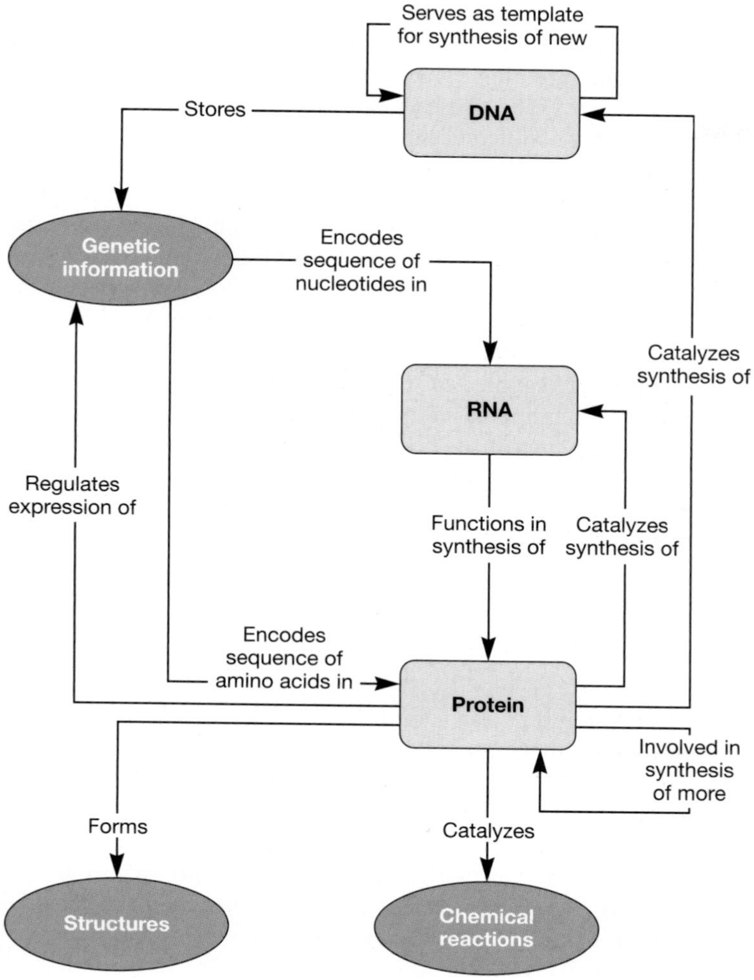

FIGURE 1.6 Functions of DNA, RNA, and Protein, and Their Relationships to Each Other in Modern Cells.

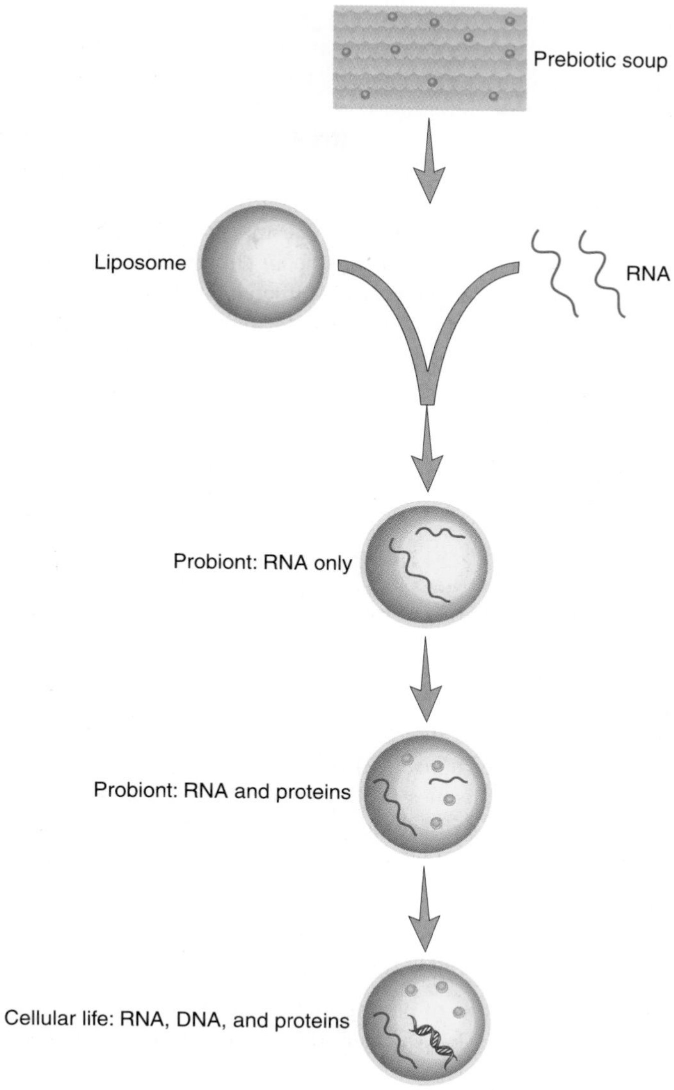

FIGURE 1.7 The RNA World Hypothesis for the Origin of Life.

Figure 1.7 Micro Inquiry

Why are the probionts pictured above not considered cellular life?

the cellular pool of RNA in modern cells exists in the ribosome, a structure that consists largely of rRNA and uses messenger RNA (mRNA) and transfer RNA (tRNA) to construct proteins. Also recall that rRNA itself catalyzes peptide bond formation during protein synthesis. Thus RNA seems to be well poised for its importance in the development of proteins. Because RNA and DNA are structurally similar, RNA could have given rise to double-stranded DNA. It is suggested that once DNA evolved, it became the storage facility for genetic information because it provided a more chemically stable structure. Two other pieces of evidence support the RNA world hypothesis: the fact that the energy currency of the cell, ATP, is a ribonucleotide and the more recent discovery that RNA can regulate gene expression. So it would seem that proteins, DNA, and cellular energy can be traced back to RNA. ▶▶| *ATP (section 9.4); Riboswitches (sections 13.3 and 13.4)*

Despite the evidence supporting the hypothesis of an RNA world, it is not without problems, and many argue against it. Another area of research is also fraught with considerable debate: the evolution of metabolism, in particular, the evolution of energy-conserving metabolic processes. Recall that early Earth was a hot environment that lacked oxygen. Thus the cells that arose there must have been able to use the available energy sources under these harsh conditions. Today there are heat-loving archaeal species capable of using inorganic molecules such as FeS as a source of energy. Some suggest that this interesting metabolic capability is a remnant of the first form of energy metabolism.

Another metabolic strategy, oxygen-releasing photosynthesis, appears to have evolved perhaps as early as 2.5 billion years ago. Fossils of cyanobacteria-like cells found in rocks dating to that time support this hypothesis, as does the discovery of ancient stromatolites (**figure 1.8**). Stromatolites are layered rocks, often domed, that are formed by the incorporation of mineral sediments into layers of microorganisms growing as thick mats on surfaces (**chapter opener figure**). The appearance of cyanobacteria-like cells was an important step in the evolution of life on Earth. The oxygen they released ultimately altered Earth's atmosphere to its current oxygen-rich state, allowing the evolution of additional energy-capturing strategies such as aerobic respiration, the oxygen-consuming metabolic process that is used by many microbes and animals.

Search This: RNA world/EvoWiki

Evolution of the Three Domains of Life

As noted in section 1.1, an important breakthrough in the classification of microbes also provides insights into the evolutionary history of all life. What began with the examination of rRNA from relatively few organisms has been expanded by the work of many others, including Norman Pace. The **universal phylogenetic tree** developed by Pace (figure 1.2) is based on comparisons of small subunit rRNA molecules (SSU rRNA), the rRNA found in the small subunit of the ribosome. Here we examine how these comparisons are made and what the universal phylogenetic tree tells us. ▶▶| *Ribosomes (section 3.5)*

Comparing SSU rRNA Molecules

Although the details of phylogenetic tree construction are discussed in chapter 17, the general concept is not difficult to understand. In one commonly used approach, the sequences of nucleotides in the genes that encode SSU rRNAs from diverse organisms are aligned and pair-wise comparisons of the sequences are made. For each pair of SSU rRNA gene sequences, the number of differences in the nucleotide sequences is counted (**figure 1.9**). This value serves as a measure of the evolutionary distance between the organisms. The evolutionary distances from many comparisons are used by sophisticated computer programs to construct the tree. Each tip of a branch in the tree represents one of the organisms used in the comparison. The distance from the tip of one branch to the tip of another is the evolutionary distance calculated when the differences in sequence were counted for the two organisms represented by the branches. It is important to note that this distance is a measure of relatedness, not of time. If the distance along the lines is very long, then the two organisms are more evolutionarily diverged (i.e., less related). However, we do not know when they diverged from each other. This concept is analogous to a map that accurately shows the distance between two cities but because of many factors (traffic, road conditions, etc.) cannot show the time needed to travel that distance.

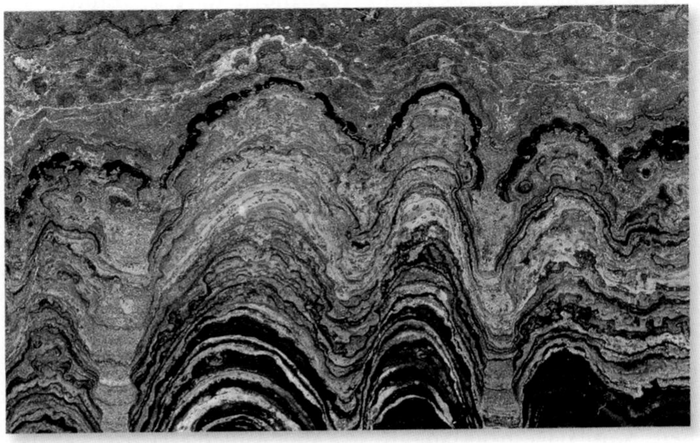

FIGURE 1.8 Section of a Fossilized Stromatolite. Evolutionary biologists think the layers of material were formed when mats of cyanobacteria, layered one on top of each other, became mineralized.

LUCA

What does the universal phylogenetic tree tell us about the origin of life? Close to the center of the tree is a line labeled "Root" (figure 1.2). This is where the data indicate the *l*ast *u*niversal *c*ommon *a*ncestor (LUCA) to all three domains should be placed (there are no branches here because there is no such extant organism). The root, or origin of modern life, is on the bacterial branch, so it appears that *Archaea* and *Eukarya* evolved independently, separate from *Bacteria*. Following the lines of descent away from the root, toward *Archaea* and *Eukarya*, it is evident that they shared common ancestry but diverged and became separate domains. The common evolution of these two forms of life is still evident in the manner in which they process genetic information. For instance, archaeal and eukaryal RNA polymerases, the enzymes that catalyze RNA synthesis, resemble each other to the exclusion of those of *Bacteria*. Thus the universal phylogenetic tree presents a picture in which all life, regardless of eventual domain, arose from a single common ancestor. One can envision the universal tree of life as a real tree that grows from a single seed.

Unfortunately, the nature of LUCA is still not known. Some argue that it most closely resembled modern bacteria; others argue that it was more like modern eukaryotes, though probably without a nucleus. The observation that the *Archaea* have some features that are most similar to their counterparts in the *Bacteria* (e.g., mechanisms for conserving energy) and others more like their counterparts in *Eukarya* (e.g., processing of genetic information) has further complicated and fueled the debate. The evolution of the nucleus is also at the center of many controversies and remains unresolved. However, hypotheses regarding the evolution of other membrane-bound organelles are more widely accepted and are considered next.

FIGURE 1.9 The Construction of Phylogenetic Trees Using a Distance Method.

Figure 1.9 Micro Inquiry

Why does the branch length indicate amount of evolutionary change but not the time it took for that change to occur?

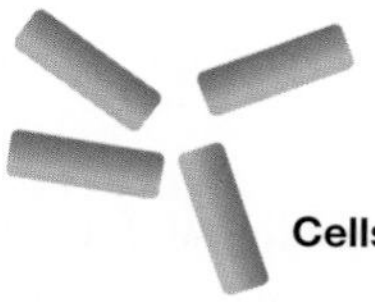

Cells from organism 1

Lyse cells to release contents and isolate DNA.

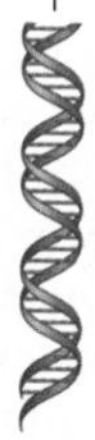

DNA

Use polymerase chain reaction to amplify and purify SSU rRNA genes.

SSU rRNA genes

Sequence genes.

ATGCTCAAGTCA

Repeat process for other organisms.

Align sequences to be compared.

Organism	SSU rRNA sequence
1	ATGCTCAAGTCA
2	TAGCTCGTGTAA
3	AAGCTCTAGTTA
4	AACCTCATGTTA

Count the number of nucleotide differences between each pair of sequences and calculate evolutionary distance (E_D).

Pair compared	E_D	Corrected E_D
1 → 2	0.42	0.61
1 → 3	0.25	0.30
1 → 4	0.33	0.44
2 → 3	0.33	0.44
2 → 4	0.33	0.44
3 → 4	0.25	0.30

For organisms 1 and 2, 5 of the 12 nucleotides are different: $E_D = 5/12 = 0.42$.

The initial ED calculated is corrected using a statistical method that considers for each site the probability of a mutation back to the original nucleotide or of additional forward mutations.

Feed data into computer and use appropriate software to construct phylogenetic tree.

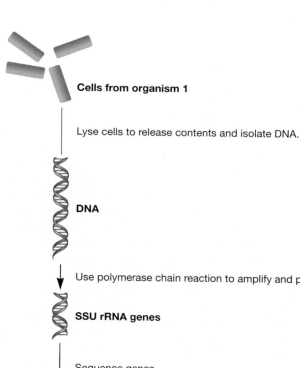

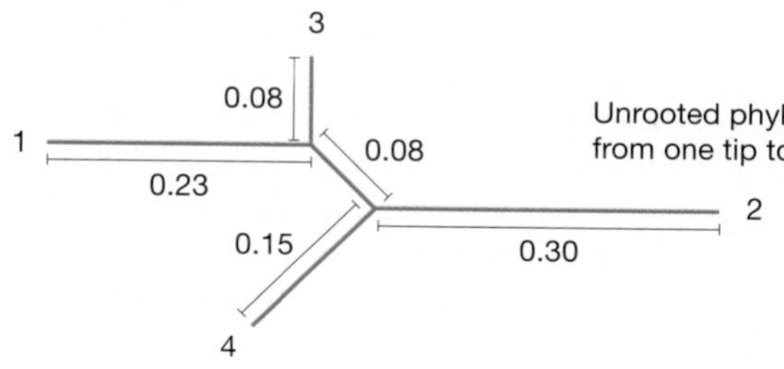

Unrooted phylogenetic tree. Note that distance from one tip to another is proportional to the E_D.

Endosymbiotic Origin of Mitochondria, Chloroplasts, and Hydrogenosomes

Although the origin of the nucleus remains unresolved, the **endosymbiotic hypothesis** is generally accepted as the origin of three eukaryotic organelles: mitochondria, chloroplasts, and hydrogenosomes. Endosymbiosis is an interaction between two organisms in which one organism lives inside the other. The initial statement of the endosymbiotic hypothesis proposed that over time a bacterial endosymbiont of an ancestral eukaryote lost its ability to live independently, becoming either a mitochondrion, if the intracellular bacterium used aerobic respiration, or a chloroplast, if the endosymbiont was a photosynthetic bacterium.

Although the mechanism by which the endosymbiotic relationship was established is unknown, there is considerable evidence to support the hypothesis. Mitochondria and chloroplasts contain DNA and ribosomes; both are similar to bacterial DNA and ribosomes. Indeed inspection of figure 1.2 shows that both organelles belong to the bacterial lineage based on SSU rRNA analysis. Further evidence for the origin of mitochondria comes from the genome sequence of the bacterium *Rickettsia prowazekii*, an obligate intracellular parasite and the cause of epidemic (lice-borne) typhus. Its genome is more closely related to that of modern mitochondrial genomes than to any other bacterium. The chloroplasts of plants and green algae are thought to have descended from an ancestor of the cyanobacterial genus *Prochloron*, which contains species that live within marine invertebrates.

Recently the endosymbiotic hypothesis for mitochondria has been modified by the **hydrogen hypothesis.** This asserts that the endosymbiont was an anaerobic bacterium that produced H_2 and CO_2 as end products of its metabolism. Over time, the host became dependent on the H_2 produced by the endosymbiont. Ultimately the endosymbiont evolved into one of two organelles. If the endosymbiont developed the capacity to perform aerobic respiration, it evolved into a mitochondrion. However, if the endosymbiont did not develop this capacity, it evolved into a hydrogenosome—an organelle found in some extant protists that produces ATP by a process called fermentation (*see figure 4.17*).

Evolution of Cellular Microbes

Although the history of early cellular life forms may never be known, we know that once they arose, they were subjected to the same evolutionary processes as modern organisms. The ancestral bacteria, archaea, and eukaryotes possessed genetic information that could be duplicated, lost, or in some other way mutated. These mutations could have many outcomes. Some led to the death of the mutant microbe, but others allowed new functions and characteristics to evolve. Those mutations that allowed the organism to increase its reproductive ability were selected for and passed on to subsequent generations. Over time new collections of genes (i.e., genotypes) arose and many species evolved.

The evolutionary process just described is similar for all organisms: mutation creates the fodder on which selection can act. In addition to mutation, organisms have other mechanisms for continually reconfiguring the genotypes of the members of a species. Most eukaryotic species increase their genetic diversity by reproducing sexually. Thus each offspring of the two parents has a mixture of parental genes and a unique genotype. The *Bacteria* and *Archaea* do not reproduce sexually. Bacterial and archaeal species increase their genetic diversity by the process of horizontal (lateral) gene transfer (HGT). During HGT, genetic information from a donor organism is transferred to a recipient, creating a new genotype. Thus genetic information need not be passed from one generation to the next but between individuals of the same generation and even between different microbial species. Genome sequencing has revealed that HGT has played an important role in the evolution of bacterial and archaeal species. Furthermore, HGT still occurs and continues to shape their genomes, leading to the evolution of species with antibiotic resistance, new virulence properties, and novel metabolic capabilities.

Microbial Species

All students of biology are introduced early in their careers to the concept of a species. But the term has different meanings depending on whether the organism is sexual or not. Taxonomists working with plants and animals define a **species** as a group of interbreeding or potentially interbreeding natural populations that is reproductively isolated from other groups. This definition also is appropriate for the many eukaryotic microbes that reproduce sexually. However, bacterial and archaeal species cannot be defined by this criterion, since they do not reproduce sexually. An appropriate definition is currently the topic of considerable discussion. A common definition is that bacterial and archaeal species are a collection of strains that share many stable properties and differ significantly from other groups of strains. A **strain** consists of the descendants of a single, pure microbial culture. Strains within a species may be described in a number of different ways. Biovars are variant strains characterized by biochemical or physiological differences, morphovars differ morphologically, serovars have distinctive properties that can be detected by antibodies (p. 19), and pathovars are pathogenic strains distinguished by the plants in which they cause disease.
▶▶| *Evolutionary processes and the concept of a microbial species (section 17.5)*

Microbiologists name microbes using the binomial system of the eighteenth-century biologist and physician Carl Linnaeus. The Latinized, italicized name consists of two parts. The first part, which is capitalized, is the generic name (i.e., the name of the genus to which the microbe belongs), and the second is the uncapitalized species epithet. For example, the bacterium that causes plague is called *Yersinia pestis*. Often the name of an organism will be shortened by abbreviating the genus name with a single capital letter (e.g., *Y. pestis*).

1. Why is RNA thought to be the first self-replicating biomolecule?
2. Explain the endosymbiotic hypothesis of the origin of mitochondria, hydrogenosomes, and chloroplasts. List two pieces of evidence that support this hypothesis.
3. What is the difference between a microbial species and a strain?
4. What is the correct way to write this microbe's name: *bacillus subtilis*, Bacillus subtilis, *Bacillus Subtilis*, or *Bacillus subtilis*? Identify the genus name and the species epithet.

1.3 Microbiology and Its Origins

Even before microorganisms were seen, some investigators suspected their existence and responsibility for disease. Among others, the Roman philosopher Lucretius (about 98–55 BCE) and the physician Girolamo Fracastoro (1478–1553) suggested that disease was caused by invisible living creatures. However, until microbes could actually be seen or studied in some other way, their existence remained a matter of conjecture. Therefore **microbiology** is defined not only by the organisms it studies but also by the tools used to study them. Because microbes are usually microscopic, the development of microscopes was the critical first step in the evolution of the discipline. However, microscopy alone is unable to answer the many questions microbiologists ask about microbes. A distinct feature of microbiology is that microbiologists often remove microorganisms from their normal habitats and culture them isolated from other microbes. Although the development of techniques for isolating microbes in pure culture was another critical step in microbiology's history, it is now recognized as having limitations. Microbes in pure culture are in some ways like animals in a zoo; just as a zoologist cannot fully understand the ecology of animals by studying them in zoos, microbiologists cannot fully understand the ecology of microbes by studying them in pure culture. Today molecular genetic techniques and genomic analyses are providing new insights into the lives of microbes.

Here we describe how the tools used by microbiologists have influenced the development of the field. As microbiology evolved as a science, it contributed greatly to the well-being of humans. The historical context of some of the important discoveries in microbiology is shown in **figure 1.10**.

Microscopy and the Discovery of Microorganisms

The earliest microscopic observations of organisms appear to have been made between 1625 and 1630 on bees and weevils by the Italian Francesco Stelluti (1577–1652), using a microscope probably supplied by Galileo (1564–1642). Robert Hooke (1635–1703) is credited with publishing the first drawings of microorganisms in the scientific literature. In 1665 he published a highly detailed drawing of the fungus *Mucor* in his book *Micrographia*. *Micrographia* is important not only for its exquisite drawings but also for the information it provided on building microscopes. One design discussed in *Micrographia* was probably a prototype for the microscopes built and used by the amateur microscopist Antony van Leeuwenhoek (1632–1723) of Delft, the Netherlands (**figure 1.11*a***). Leeuwenhoek earned his living as a draper and haberdasher (a dealer in men's clothing and accessories) but spent much of his spare time constructing simple microscopes composed of double convex glass lenses held between two silver plates (figure 1.11*b*). His microscopes could magnify around 50 to 300 times, and he may have illuminated his liquid specimens by placing them between two pieces of glass and shining light on them at a 45° angle to the specimen plane. This would have provided a form of dark-field illumination in which the organisms appeared as bright objects against a dark background and made bacteria clearly visible (figure 1.11*c*). Beginning in 1673, Leeuwenhoek sent detailed letters describing his discoveries to the Royal Society of London. It is clear from his descriptions that he saw both bacteria and protists.

1. Give some examples of the kind of information you think can be provided by microscopic observations of microorganisms.
2. Give some examples of the kind of information you think can be provided by isolating microorganisms from their natural environment and culturing them in the laboratory.

Culture-Based Methods for Studying Microorganisms

As important as Leeuwenhoek's observations were, the development of microbiology essentially languished for the next 200 years until techniques for isolating and culturing microbes in the laboratory were formulated. Many of these techniques began to be developed as scientists grappled with the conflict over the theory of spontaneous generation. This conflict and the subsequent studies on the role played by microorganisms in causing disease ultimately led to what is now called the golden age of microbiology.

Spontaneous Generation

From earliest times, people had believed in **spontaneous generation**—that living organisms could develop from nonliving matter. This view finally was challenged by the Italian physician Francesco Redi (1626–1697), who carried out a series of experiments on decaying meat and its ability to produce maggots spontaneously. Redi placed meat in three containers. One was uncovered, a second was covered with paper, and the third was covered with fine gauze that would exclude flies. Flies laid their

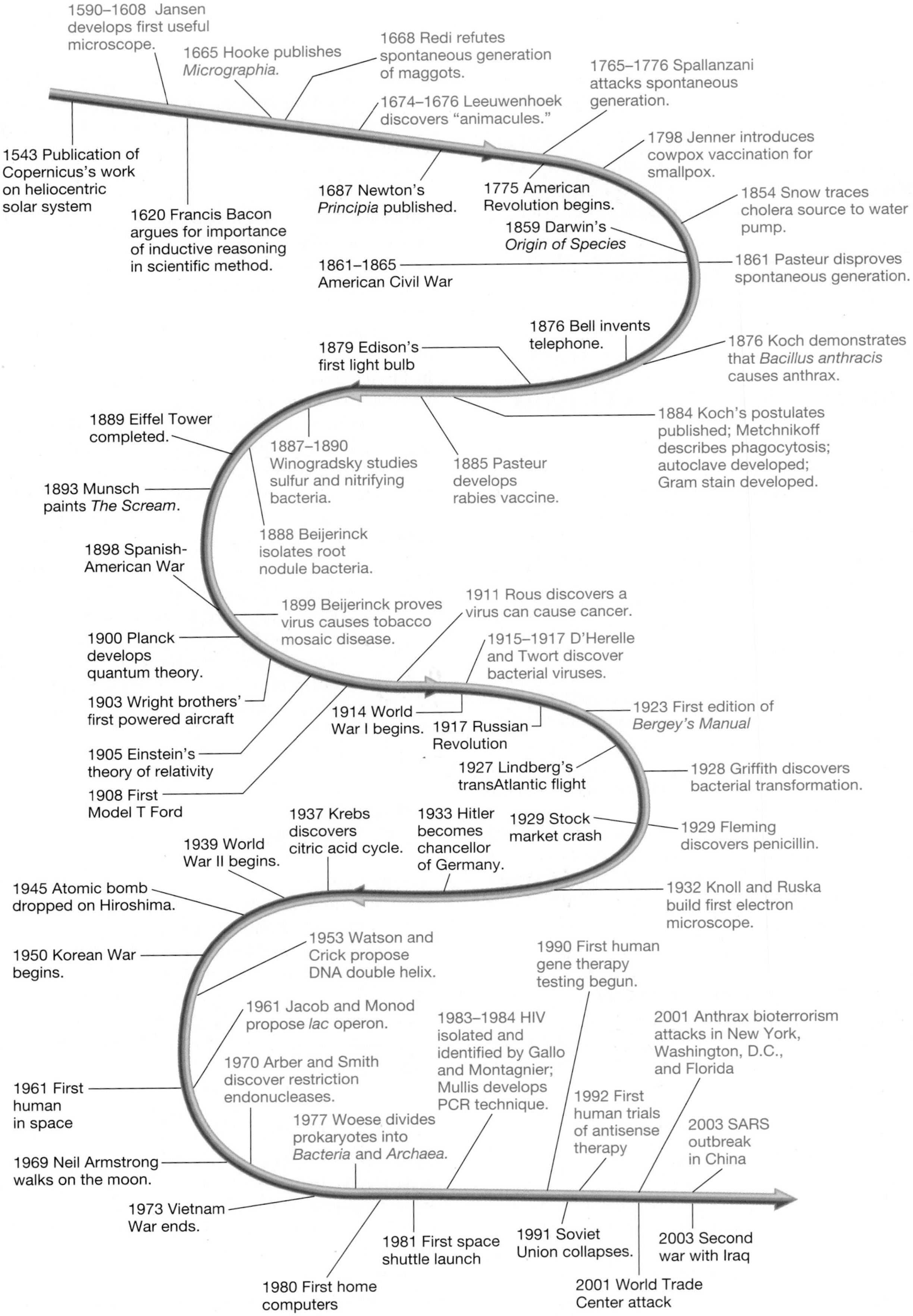

FIGURE 1.10 Some Important Events in the Development of Microbiology. Milestones in microbiology are marked in red; other historical events are in black.

eggs on the uncovered meat and maggots developed. The other two pieces of meat did not produce maggots spontaneously. However, flies were attracted to the gauze-covered container and laid their eggs on the gauze; these eggs produced maggots. Thus the generation of maggots by decaying meat resulted from the presence of fly eggs, and meat did not spontaneously generate maggots, as previously believed. Similar experiments by others helped discredit the theory for larger organisms.

Leeuwenhoek's communications on microorganisms renewed the controversy. Some proposed that microorganisms arose by spontaneous generation even though larger organisms did not. They pointed out that boiled extracts of hay or meat gave rise to microorganisms after sitting for a while. Indeed, such extracts were the forerunners of the culture media still used today in many microbiology laboratories.

In 1748 the English priest John Needham (1713–1781) reported the results of his experiments on spontaneous generation. Needham boiled mutton broth in flasks that he then tightly stoppered. Eventually many of the flasks became cloudy and contained microorganisms. He thought organic matter contained a vital force that could confer the properties of life on nonliving matter. A few years later, the Italian priest and naturalist Lazzaro Spallanzani (1729–1799) improved on Needham's experimental design by first sealing glass flasks that contained water and seeds. If the sealed flasks were placed in boiling water for three-quarters of an hour, no growth took place as long as the flasks remained sealed. He proposed that air carried germs to the culture medium but also commented that the external air might be required for growth of animals already in the medium. The supporters of spontaneous generation maintained that heating the air in sealed flasks destroyed its ability to support life.

Several investigators attempted to counter such arguments. Theodore Schwann (1810–1882) allowed air to enter a flask containing a sterile nutrient solution after the air had passed through a red-hot tube. The flask remained sterile. Subsequently Georg Friedrich Schroder (1810–1885) and Theodor von Dusch (1824–1890) allowed air to enter a flask of heat-sterilized medium after it had passed through sterile cotton wool. No growth occurred in the medium even though the air had not been heated. Despite these experiments, the French naturalist Felix Pouchet (1800–1872) claimed in 1859 to have carried out experiments conclusively proving that microbial growth could occur without air contamination.

(a)

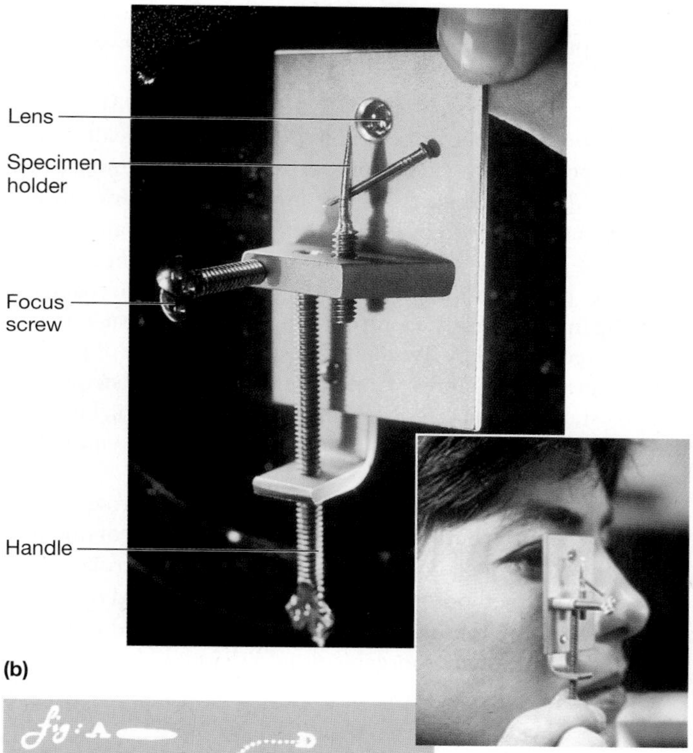

(b)

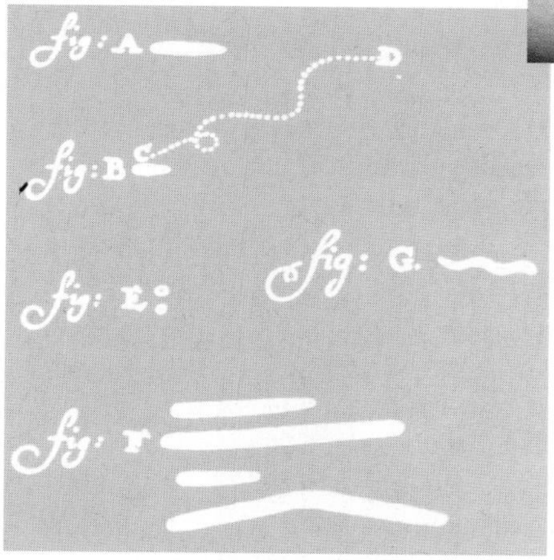

(c)

FIGURE 1.11 Antony van Leeuwenhoek. (a) An oil painting of Leeuwenhoek. (b) A brass replica of the Leeuwenhoek microscope. Inset photo shows how it is held. (c) Leeuwenhoek's drawings of bacteria from the human mouth.

Pouchet's claim provoked Louis Pasteur (1822–1895) to settle the matter of spontaneous generation. Pasteur (**figure 1.12**) first filtered air through cotton and found that objects resembling plant spores had been trapped. If a piece of the cotton was placed in sterile medium after air had been filtered through it, microbial growth occurred. Next he placed nutrient solutions in flasks, heated their necks in a flame, and drew them out into a variety of curves. The swan-neck flasks that he produced in this way had necks open to the atmosphere. Pasteur then boiled the solutions for a few minutes and allowed them to cool. No growth took place even though the contents of the flasks were exposed to the air (**figure 1.13**). Pasteur pointed out that no growth occurred because dust and germs had been trapped on the walls of the curved necks. If the necks were broken, growth commenced immediately. Pasteur had not only resolved the controversy by 1861 but also had shown how to keep solutions sterile.

The English physicist John Tyndall (1820–1893) and the German botanist Ferdinand Cohn (1828–1898) dealt a final blow to spontaneous generation. In 1877 Tyndall demonstrated that dust did indeed carry germs and that if dust was absent, broth remained sterile even if directly exposed to air. During the course of his studies, Tyndall provided evidence for the existence of exceptionally heat-resistant forms of bacteria. Working independently, Cohn discovered that the heat-resistant bacteria recognized by Tyndall were species capable of producing bacterial endospores. Cohn later played an instrumental role in establishing a classification system for bacteria based on their morphology and physiology. ▶▶| *Bacterial endospores (section 3.8)*

Clearly, these early microbiologists not only disproved spontaneous generation but also contributed to the rebirth of microbiology. They developed liquid media for culturing microbes. They also developed methods for sterilizing these media and maintaining their sterility. These techniques were next applied to understanding the role of microorganisms in disease.

FIGURE 1.12 Louis Pasteur.

1. How did Pasteur, Tyndall, and Cohn finally settle the spontaneous generation controversy?
2. Why was the belief in spontaneous generation an obstacle to the development of microbiology as a scientific discipline?
3. What did Pasteur prove when he showed that a cotton plug that had filtered air would trigger the microbial growth when transferred to the medium? What argument made previously was he addressing?

Microorganisms and Disease

Although Fracastoro and a few others had suggested that invisible organisms produced disease, most people believed that disease was due to causes such as supernatural forces, poisonous vapors called miasmas, and imbalances among the four humors thought to be present in the body. The role of the four humors (blood, phlegm, yellow bile [choler], and black bile [melancholy]) in disease had been widely accepted since the time of the Greek physician Galen (129–199). Support for the idea that microorganisms cause disease—that is, the germ theory of disease—began to accumulate in the early nineteenth century from diverse fields. Agostino Bassi (1773–1856) first showed a microorganism could cause disease when he demonstrated in 1835 that a silkworm disease was due to a fungal infection. He also suggested that many diseases were due to microbial infections. In 1845 M. J. Berkeley (1803–1889) proved that the great potato blight of Ireland was caused by a water mold (then thought to be a fungus), and in 1853 Heinrich de Bary (1831–1888) showed that smut and rust fungi caused cereal crop diseases.

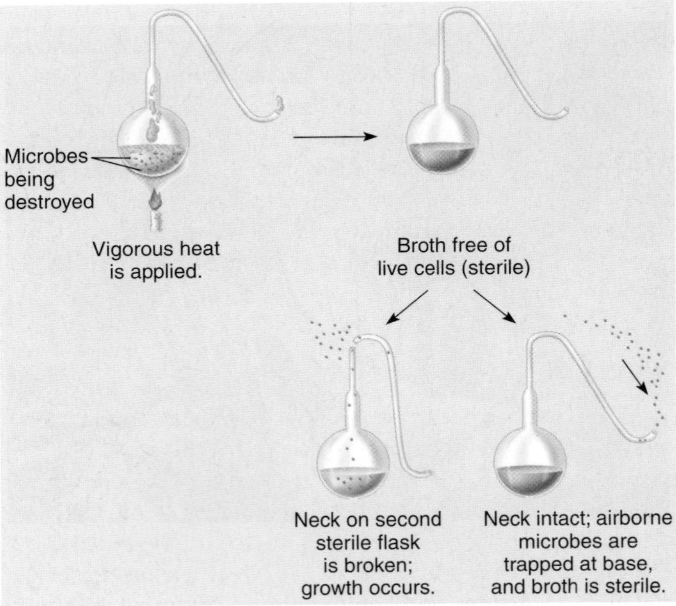

Microbes being destroyed
Vigorous heat is applied.
Broth free of live cells (sterile)
Neck on second sterile flask is broken; growth occurs.
Neck intact; airborne microbes are trapped at base, and broth is sterile.

FIGURE 1.13 Pasteur's Experiments with Swan-Neck Flasks.

Pasteur also contributed to this area of research in several ways. His contributions began in what may seem an unlikely way. Pasteur was trained as a chemist and spent many years studying the alcoholic fermentations that yield ethanol and are used in the production of wine and other alcoholic beverages. When he began his work, the leading chemists believed microorganisms were not involved in alcoholic fermentations. They were convinced that fermentation was due to a chemical instability that degraded the sugars in grape juice and other substances to alcohol. Pasteur did not agree; he believed that fermentations were carried out by living organisms.

In 1856 M. Bigo, an industrialist in Lille, France, where Pasteur worked, requested Pasteur's assistance. His business produced ethanol from the fermentation of beet sugars, and the alcohol yields had recently declined and the product had become sour. Pasteur discovered that the fermentation was failing because the yeast normally responsible for alcohol formation had been replaced by bacteria that produced acid rather than ethanol. In solving this practical problem, Pasteur demonstrated that all fermentations were due to the activities of specific yeasts and bacteria, and he published several papers on fermentation between 1857 and 1860.

Pasteur was also called upon by the wine and silk industries in France for help. The wine industry was having problems resulting in the production of poor quality wines. Pasteur referred to the wines as diseased and demonstrated that particular wine diseases were linked to particular microbes contaminating the wine. He eventually suggested a method for heating the wines to destroy the undesirable microbes. The process is now called pasteurization. The silk industry asked Pasteur to investigate the *pèbrine* disease of silkworms. After several years of work, he showed that the disease was due to a protozoan parasite.

Indirect evidence for the germ theory of disease came from the work of the English surgeon Joseph Lister (1827–1912) on the prevention of wound infections. Lister, impressed with Pasteur's studies on the involvement of microorganisms in fermentation and putrefaction, developed a system of antiseptic surgery designed to prevent microorganisms from entering wounds. Instruments were heat sterilized, and phenol was used on surgical dressings and at times sprayed over the surgical area. The approach was remarkably successful and transformed surgery. It also provided strong indirect evidence for the role of microorganisms in disease because phenol, which kills bacteria, also prevented wound infections.

Koch's Postulates

The first direct demonstration of the role of bacteria in causing disease came from the study of anthrax by the German physician Robert Koch (1843–1910). Koch (**figure 1.14**) used the criteria proposed by his former teacher Jacob Henle (1809–1885) and others to establish the relationship between *Bacillus anthracis* and anthrax; he published his findings in 1876. Koch injected healthy mice with material from diseased animals, and the mice became ill. After transferring anthrax by inoculation through a series of 20 mice, he incubated a piece of spleen containing the anthrax bacillus in beef serum. The bacteria grew, reproduced, and produced endospores. When isolated bacteria or their spores were injected into healthy mice, anthrax developed. His criteria for proving the causal relationship between a microorganism and a specific disease are known as **Koch's postulates.** Koch's proof that *B. anthracis* caused anthrax was independently confirmed by Pasteur and his coworkers. They discovered that after burial of dead animals, anthrax spores survived and were brought to the surface by earthworms. Healthy animals then ingested the spores and became ill.

After completing his anthrax studies, Koch fully outlined his postulates in his work on the cause of tuberculosis (**figure 1.15**). In 1884 he reported that this disease was caused by the rod-shaped bacterium *Mycobacterium tuberculosis,* and he was awarded the Nobel Prize in Physiology or Medicine in 1905 for his work. Koch's postulates were quickly adopted by others and used to connect many diseases to their causative agent.

While Koch's postulates are still widely used, their application is at times not feasible. For instance, organisms such as *Mycobacterium leprae,* the causative agent of leprosy, cannot be isolated in pure culture. Some human diseases are so deadly (e.g., Ebola hemorrhagic fever) that it would be unethical to use humans as the experimental organism; if an appropriate animal model does not exist, the postulates cannot be fully met. To avoid some of these difficulties, microbiologists sometimes use molecular and genetic evidence. For instance, molecular methods might be used to detect the nucleic acid of a virus in body tissues, rather than isolating the virus in pure culture. Or the genes thought to be associated with the virulence of a pathogen might be mutated.

FIGURE 1.14 Robert Koch. Koch examining a specimen in his laboratory.

Postulate	Experimentation
1. The microorganism must be present in every case of the disease but absent from healthy organisms.	Koch developed a staining technique to examine human tissue. *Mycobacterium tuberculosis* could be identified in diseased tissue.
2. The suspected microorganisms must be isolated and grown in a pure culture.	Koch grew *M. tuberculosis* in pure culture on coagulated blood serum.
3. The same disease must result when the isolated microorganism is inoculated into a healthy host.	Koch injected cells from the pure culture of *M. tuberculosis* into guinea pigs. The guinea pigs subsequently died of tuberculosis.
4. The same microorganisms must be isolated again from the diseased host.	Koch isolated *M. tuberculosis* in pure culture on coagulated blood serum from the dead guinea pigs.

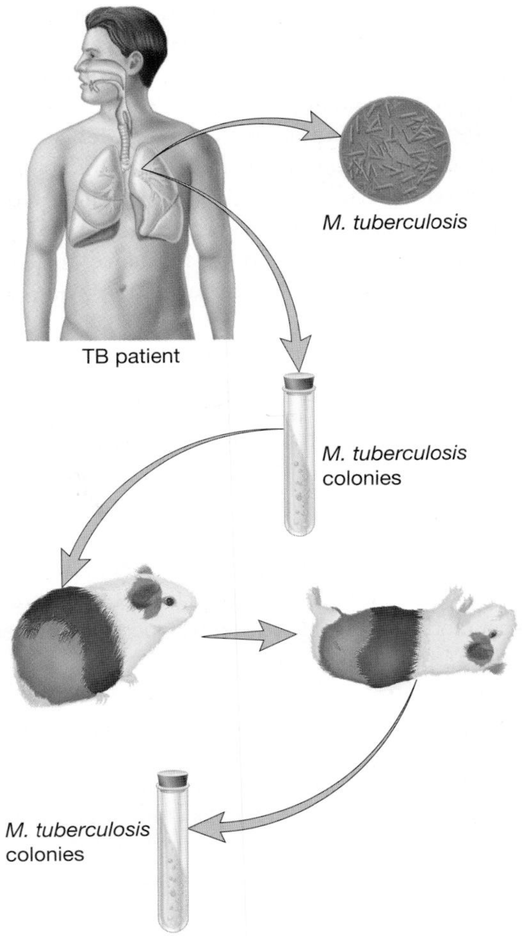

FIGURE 1.15 Koch's Postulates Applied to Tuberculosis.

In this case, the mutant organism should have decreased ability to cause disease. Introduction of the normal gene back into the mutant should restore the pathogen's virulence.

Pure Culture Methods

During Koch's studies on bacterial diseases, it became necessary to isolate suspected bacterial pathogens in pure culture—a culture containing only one type of microorganism. At first Koch cultured bacteria on the sterile surfaces of cut, boiled potatoes, but the bacteria would not always grow well. Eventually he developed culture media using meat extracts and protein digests, reasoning these were similar to body fluids. Initially he tried to solidify the media by adding gelatin. Separate bacterial colonies developed after the surface of the solidified medium had been streaked with a bacterial sample. The sample could also be mixed with liquefied gelatin medium. When the medium hardened, individual bacteria produced separate colonies.

Despite its advantages, gelatin was not an ideal solidifying agent because it can be digested by many microbes and melts at temperatures above 28°C. A better alternative was provided by Fanny Eilshemius Hesse (1850–1934), the wife of Walther Hesse (1846–1911), one of Koch's assistants. She suggested the use of agar, which she used to make jellies, as a solidifying agent. Agar was not attacked by most bacteria. Furthermore, it did not melt until reaching a temperature of 100°C and, once melted, did not solidify until reaching a temperature of 50°C; this eliminated the need to handle boiling liquid. Some of the media developed by Koch and his associates, such as nutrient broth and nutrient agar, are still widely used. Another important tool developed in Koch's laboratory was a container for holding solidified media—the Petri dish (plate), named after Richard Petri (1852–1921), who devised it. These developments directly stimulated progress in all areas of microbiology. ▶▶| *Culture media (section 6.7); Isolation of pure cultures (section 6.8)*

Our focus thus far has been on the development of methods for culturing bacteria. But viral pathogens were also being studied during this time, and methods for culturing them were also being developed. The discovery of viruses and their role in disease was made possible when Charles Chamberland (1851–1908), one of Pasteur's associates, constructed a porcelain bacterial filter in 1884. Dimitri Ivanowski (1864–1920) and Martinus Beijerinck (pronounced "by-a-rink"; 1851–1931) used the filter to study tobacco mosaic disease. They found that plant extracts and sap from diseased plants were infectious, even after being filtered with Chamberland's filter. Because the infectious agent passed through a filter that was designed to trap bacterial cells, they reasoned that the agent must be something smaller than a bacterium. Beijerinck proposed that the agent was a "filterable virus." Eventually viruses were shown to be tiny, acellular infectious agents.

1. Discuss the contributions of Lister, Pasteur, and Koch to the germ theory of disease and the treatment or prevention of diseases. What other contributions did Koch make to microbiology?
2. Describe Koch's postulates. What is a pure culture? Why are pure cultures important to Koch's postulates?
3. Would microbiology have developed more slowly if Fanny Hesse had not suggested the use of agar? Give your reasoning.
4. Some individuals can be infected by a pathogen yet not develop disease. In fact, some become chronic carriers of the pathogen. How does this observation affect Koch's postulates? How might the postulates be modified to account for the existence of chronic carriers?

Immunology

The ability to culture microbes also played an important role in early immunological studies. During studies on chicken cholera, Pasteur and Pierre Roux (1853–1933) discovered that incubating the cultures for long intervals between transfers would attenuate the bacteria, which meant they had lost their ability to cause the disease. If the chickens were injected with these attenuated cultures, they remained healthy and developed the ability to resist the disease when exposed to virulent cultures. Pasteur called the attenuated culture a vaccine (Latin *vacca,* cow) in honor of Edward Jenner (1749–1823) because, many years earlier, Jenner had used material from cowpox lesions to protect people against smallpox. Shortly after this, Pasteur and Chamberland developed an attenuated anthrax vaccine. ▶▶| *Vaccines and immunizations (section 36.7)*

Pasteur also prepared a rabies vaccine using an attenuated strain of *Rabies virus.* During the course of these studies, Joseph Meister, a nine-year-old boy who had been bitten by a rabid dog, was brought to Pasteur. Since the boy's death was certain in the absence of treatment, Pasteur agreed to try vaccination. Joseph was injected 13 times over the next 10 days with increasingly virulent preparations of the attenuated virus. He survived. In gratitude for Pasteur's development of vaccines, people from around the world contributed to the construction of the Pasteur Institute in Paris, France. One of the initial tasks of the institute was vaccine production.

Search This: Institut Pasteur EN

These early advances in immunology were made without any concrete knowledge about how the immune system works. The immune system uses chemicals produced by the body and several types of blood cells to provide protection. Among the chemicals are soluble proteins called antibodies, which can be found in blood, lymph, and other body fluids. The role of soluble substances in preventing disease was recognized by Emil von Behring (1854–1917) and Shibasaburo Kitasato (1852–1931). After the discovery that diphtheria was caused by a toxin released by a bacterium, they injected inactivated diphtheria toxin into rabbits. The inactivated toxin induced rabbits to produce an antitoxin, which protected against the disease. Antitoxins are now known to be antibodies that specifically bind toxins, neutralizing them. The first immune system cells were discovered when Elie Metchnikoff (1845–1916) found that some white blood cells could engulf disease-causing bacteria. He called these cells phagocytes and the process phagocytosis (Greek *phagein,* eating).

Microbial Ecology

Culture-based techniques were also applied to the study of microbes in soil and aquatic habitats. Early microbial ecologists studied microbial involvement in the carbon, nitrogen, and sulfur cycles. The Russian microbiologist Sergei Winogradsky (1856–1953) made many contributions to soil microbiology. He discovered that soil bacteria could oxidize iron, sulfur, and ammonia to obtain energy and that many of these bacteria could incorporate CO_2 into organic matter much as photosynthetic organisms do. Winogradsky also isolated anaerobic nitrogen-fixing soil bacteria and studied the decomposition of cellulose. Martinus Beijerinck was one of the great general microbiologists who made fundamental contributions not only to virology but to microbial ecology as well. He isolated the aerobic nitrogen-fixing bacterium *Azotobacter,* a root nodule bacterium also capable of fixing nitrogen (later named *Rhizobium*), and sulfate-reducing bacteria. Beijerinck and Winogradsky also developed the enrichment-culture technique and the use of selective media, which have been of great importance in microbiology. ▶▶| *Biogeochemical cycling (section 26.1); Culture media (section 6.7)*

1. How did Jenner, Pasteur, von Behring, Kitasato, and Metchnikoff contribute to the development of immunology? How was the ability to culture microbes important to their studies?
2. How did Winogradsky and Beijerinck contribute to the study of microbial ecology? What new culturing techniques did they develop in their studies?

1.4 Microbiology Today

Microbiology today is as diverse as the organisms it studies. It has both basic and applied aspects. The basic aspects are concerned with the biology of microorganisms themselves. The applied aspects are concerned with practical problems such as disease, water and wastewater treatment, food spoilage and food production, and industrial uses of microbes. The basic and applied aspects of microbiology are intertwined. Basic research is often conducted in applied fields, and applications often arise out of basic research.

An important recent development in microbiology is the increasing use of molecular and genomic methods to study microbes and their interactions with other organisms. These methods have led to a time of rapid advancement that rivals the golden age of microbiology. Indeed, many feel that microbiology is in its second golden age. Here we describe some of the important advances that have enabled microbiologists to use molecular and genomic techniques. We then discuss some of the important research being done in the numerous subdisciplines of microbiology.

Molecular and Genomic Methods of Studying Microbes

Molecular and genomic methods of studying microbes rely on the ability of scientists to manipulate the genes and the genomes of the organisms being studied. An organism's **genome** is all the genetic information that organism contains. To study single genes or the entire genome, microbiologists must be able to isolate DNA and RNA, cut DNA into smaller pieces, insert one piece of DNA into another, and determine the sequence of nucleotides in DNA (and sometimes RNA).

Cutting DNA into smaller pieces was made possible by the discovery in the 1960s by Werner Arber and Hamilton Smith that certain bacterial enzymes cut double-stranded DNA. These enzymes are now known as restriction endonucleases, or simply restriction enzymes. This discovery was followed relatively quickly by the report in 1972 that David Jackson, Robert Symons, and Paul Berg had successfully generated recombinant DNA molecules—molecules made by combining two or more different DNA molecules together. They did this by cutting DNA from two different organisms with the same restriction enzyme, mixing the two DNA molecules together, and linking them together with an enzyme called DNA ligase. ▶▶| *Key developments in recombinant DNA technology (section 15.1)*

The next major breakthrough was the development of methods to determine the sequence of nucleotides in RNA and DNA. RNA sequencing techniques were important in the work of Carl Woese, as described in section 1.1. However, in most studies, DNA rather than RNA is sequenced. In the late 1970s, Frederick Sanger introduced a method that is the most commonly used DNA sequencing method. Over the last 30 years, the method has been modified and adapted for use in automated systems, so that entire genomes of organisms can be sequenced in a matter of days. ▶▶| *Genome sequencing (section 16.2)*

Genome sequencing is only the first step in **genomic analysis.** Once the genome sequence is in hand, microbiologists must decipher the information found in the genome. This involves identifying potential protein-coding genes, determining what they code for, and identifying other regions of the genome that may have important functions other than coding for proteins (e.g., genes encoding tRNA and rRNA or sequences playing a role in regulating the function of genes). This work requires the use of computers, and a new scientific discipline has arisen as a result. Bioinformatics is the discipline that manages the ever-increasing amount of genetic information available for analysis. It also is involved in determining the function of genes and can be used to generate hypotheses that can be tested either in silico (i.e., in the computer) or in the laboratory. ▶▶| *Bioinformatics (section 16.3)*

Major Fields in Microbiology

Although pathogenic microbes are the minority, they garner considerable interest. Thus one of the most active and important fields in microbiology is medical microbiology, which deals with diseases of humans and animals. Medical microbiologists identify the agents causing infectious diseases and plan measures for their control and elimination. Frequently they are involved in tracking down new, unidentified pathogens such as those causing variant Creutzfeldt-Jakob disease (the human version of "mad cow disease"), hantavirus pulmonary syndrome, and West Nile encephalitis. These microbiologists also study the ways microorganisms cause disease. As described in section 1.3, our understanding of the role of microbes in disease began to crystallize when we were able to isolate them in pure culture. Today clinical laboratory scientists, the microbiologists who work in hospital and other clinical laboratories, use microscopy and a variety of culture-based techniques to provide information needed by physicians to diagnose infectious disease. Increasingly, molecular genetic techniques are also being used.

As noted in the chapter opening, major epidemics have regularly affected human history. The 1918 influenza pandemic is of particular note; it killed more than 50 million people in about one year. Public health microbiology is concerned with the control and spread of such communicable diseases. Public health microbiologists and epidemiologists monitor the amount of disease in populations. Based on their observations, they can detect outbreaks and developing epidemics, and implement appropriate control measures. They also conduct surveillance for new diseases as well as bioterrorism events. Public health microbiologists working for local governments monitor community food establishments and water supplies in an attempt to keep them safe and free from infectious disease agents.

To understand, treat, and control infectious disease, it is important to understand how the immune system protects the body from pathogens; this question is the concern of immunology.

Immunology is one of the fastest growing areas in science. Much of the growth began with the discovery of human immunodeficiency virus (HIV), which specifically targets cells of the immune system. Immunology also deals with health problems such as the nature and treatment of allergies and autoimmune diseases such as rheumatoid arthritis. ▶▶| *Techniques & Applications 29.1: Monoclonal Antibody Technology*

Microbial ecology is another important field in microbiology. Microbial ecology developed when early microbiologists such as Winogradsky and Beijerinck chose to investigate the ecological role of microorganisms rather than their role in disease. Today, a variety of approaches are being used to describe the vast diversity of microbes in terms of their morphology, physiology, and relationships with organisms and the components of their habitats. Although the importance of microbes in global and local cycling of carbon, nitrogen, and sulfur cycles is well documented, many questions are still unanswered. Of particular interest is the role of microbes in both the production and removal of greenhouse gases such as carbon dioxide and methane. The study of pollution effects on microorganisms also is important because of the impact these organisms have on the environment. Microbial ecologists are employing microorganisms in bioremediation to reduce pollution. The study of the microbes normally associated with the human body has become a new frontier in microbial ecology. Scientists are currently trying to identify all the members of the communities found on the skin and in the large intestine using molecular techniques that grew out of Woese's pioneering work to establish the phylogeny of microbes. ▶▶| *Global climate change (section 26.2); Biodegradation and bioremediation in natural communities (section 42.3)*

Agricultural microbiology is a field related to both medicine and microbial ecology. Agricultural microbiology is concerned with the impact of microorganisms on agriculture. Microbes such as nitrogen-fixing bacteria play critical roles in the nitrogen cycle and affect soil fertility. Other microbes live in the digestive tracts of ruminants such as cattle and break down the plant materials these animals ingest. There are also plant and animal pathogens that can have significant economic impacts if not controlled. Agricultural microbiologists work on methods to increase soil fertility and crop yields, study rumen microorganisms in order to increase meat and milk production, and try to combat plant and animal diseases. Currently many agricultural microbiologists are studying the use of bacterial and viral insect pathogens as substitutes for chemical pesticides. ▶▶| *Agricultural biotechnology (section 41.4)*

Agricultural microbiology has contributed to the ready supply of high-quality foods, as has the discipline of food and dairy microbiology. Numerous foods are made using microorganisms. On the other hand, some microbes cause food spoilage or are pathogens spread through food. An excellent example of the latter is *Escherichia coli* O157:H7, which in 2006 caused a widespread outbreak of disease when it contaminated a major source of spinach in the United States. Food and dairy microbiologists explore the use of microbes in food production. They also work to prevent microbial spoilage of food and the transmission of food-borne diseases. This involves monitoring the food industry for the presence of pathogens. Increasingly, molecular methods are being used to detect pathogens in meat and other foods. Food and dairy microbiologists also conduct research on the use of microorganisms as nutrient sources for livestock and humans. ▶▶| *Microbiology of food (chapter 40)*

Humans unknowingly exploited microbes for thousands of years. However, the systematic and conscious use of microbes in industrial microbiology did not begin until the 1800s. Industrial microbiology developed in large part from Pasteur's work on alcoholic fermentations, as described in section 1.3. His success led to the development of pasteurization to preserve wine during storage. Pasteur's studies on fermentation continued for almost 20 years. One of his most important discoveries was that some fermentative microorganisms were anaerobic and could live only in the absence of oxygen, whereas others were able to live either aerobically or anaerobically. ▶▶| *Controlling food spoilage (section 40.2)*

Another important advance in industrial microbiology occurred in 1929 when Alexander Fleming discovered that the fungus *Penicillium* produced what he called penicillin, the first antibiotic that could successfully control bacterial infections. Although it took World War II for scientists to learn how to mass-produce penicillin, scientists soon found other microorganisms capable of producing additional antibiotics as well as compounds such as citric acid, vitamin B_{12}, and monosodium glutamate (MSG). Today industrial microbiologists use microorganisms to make products such as antibiotics, vaccines, steroids, alcohols and other solvents, vitamins, amino acids, and enzymes. Industrial microbiologists identify microbes of use to industry. They also utilize techniques to improve production by microbes and devise systems for culturing them and isolating the products they make.

The advances in medical microbiology, agricultural microbiology, food and dairy microbiology, and industrial microbiology are in many ways outgrowths of the labor of many microbiologists doing basic research in areas such as microbial physiology, microbial genetics, molecular biology, and bioinformatics. Microbes are metabolically diverse and can employ a wide variety of energy sources, including organic matter, inorganic molecules (e.g., H_2 and NH_3), and sunlight. Microbial physiologists study many aspects of the biology of microorganisms, including their metabolic capabilities. They may also study the synthesis of antibiotics and toxins, the ways in which microorganisms survive harsh environmental conditions, and the effects of chemical and physical agents on microbial growth and survival. Microbial geneticists, molecular biologists, and bioinformaticists study the nature of genetic information and how it regulates the development and function of cells and organisms. The bacteria *E. coli* and *Bacillus subtilis,* the yeast *Saccharomyces cerevisiae* (baker's yeast), and bacterial viruses such as T4 and lambda continue to be important model organisms used to understand biological phenomena.

Because of the practical importance of microbes, their use as model organisms, and the advent of genomic analysis, the future of microbiology is bright. Genomics in particular is revolutionizing microbiology, as we are now beginning to understand organisms in toto, rather than in a reductionist, piecemeal manner. How the genomes of microbes evolve, the nature of host-pathogen interactions, the minimum set of genes required for an organism to survive, and many more topics are aggressively being examined by molecular and genomic analyses. This is an exciting time to be a microbiologist. Enjoy the journey.

Search This: Careers in microbiology/ASM

1. Since the 1970s, microbiologists have been able to study individual genes and whole genomes at the molecular level. What advances made this possible?
2. Briefly describe the major subdisciplines in microbiology. Which do you consider to be applied fields? Which basic?
3. Why do you think microorganisms are useful to biologists as experimental models?
4. List all the activities or businesses you can think of in your community that directly depend on microbiology.

Summary

1.1 Members of the Microbial World

a. Microbiology studies microscopic cellular organisms that are often unicellular or, if multicellular, do not have highly differentiated tissues. Microbiology also focuses on biological entities that are acellular. That is, they are not composed of cells (**figure 1.1**).

b. Prokaryotic cells differ from eukaryotic cells in their lack of a membrane-delimited nucleus and in other ways as well.

c. Microbiologists divide organisms into three domains: *Bacteria, Archaea,* and *Eukarya* (**figure 1.2**).

d. Domains *Bacteria* and *Archaea* consist of prokaryotic microorganisms. The eukaryotic microbes (protists and fungi) are placed in *Eukarya.* Viruses, viroids, and virusoids are acellular entities that are not placed in any of the domains but are classified by a separate system.

1.2 Microbial Evolution

a. The fossil record of microbial evolution is very sparse. Therefore evolutionary biologists and others interested in the origin of life must rely on many types of evidence.

b. Earth is approximately 4.5 billion years old. Within the first 1 billion years of its existence, life arose (**figure 1.5**).

c. The RNA world hypothesis posits that the earliest self-replicating entity on the planet used RNA both to store genetic information and to conduct cellular processes. The discovery of ribozymes and regulatory RNA molecules support this hypothesis (**figure 1.7**).

d. Examination of the universal phylogenetic tree of life provides information about the evolution of life after it arose. The tree is based on comparisons of small subunit (SSU) rRNA genes.

e. The last universal common ancestor (LUCA) is placed on the bacterial branch of the universal phylogenetic tree. Thus, *Bacteria* diverged first, and *Archaea* and *Eukarya* shared a common line of descent until they diverged from each other, becoming separate domains.

f. Considerable evidence supports the endosymbiotic hypothesis that mitochondria, chloroplasts, and hydrogenosomes evolved from bacterial endosymbionts of ancestral eukaryotic cells. This hypothesis has recently been modified by the hydrogen hypothesis, which proposes that the original bacterial endosymbiont was an anaerobe. In some ancestral eukaryotes, the endosymbiont developed the ability to carry out aerobic respiration, becoming a mitochondrion. In other eukaryotes, the endosymbiont remained an anaerobe and evolved into a hydrogenosome.

g. The concept of a bacterial or archaeal species is difficult to define and is the source of considerable debate because these microbes do not reproduce sexually. However, species are named using the binomial system of Linnaeus.

1.3 Microbiology and Its Origins

a. Microbiology is defined not only by the organisms it studies but also by the tools it uses. Microscopy and culture-based techniques have played and continue to play important roles in the evolution of the discipline.

b. Antony van Leeuwenhoek was the first person to extensively describe microorganisms. He used simple microscopes and described both bacteria and protists (**figure 1.11**).

c. Culture-based techniques for studying microbes began to develop as scientists debated the theory of spontaneous generation. Experiments by Redi and others disproved the theory of spontaneous generation of larger organisms. The spontaneous generation of microorganisms was disproved by Spallanzani, Pasteur, Tyndall, Cohn, and others.

d. The availability of culture-based techniques played an important role in the study of the microbes as the causative agents of disease. Support for the germ theory of disease came from the work of Bassi, Pasteur, Koch, and others.

Lister provided indirect evidence with his development of antiseptic surgery.

e. Koch's postulates are used to prove a direct relationship between a suspected pathogen and a disease. As Koch and his coworkers applied Koch's postulates to the study of anthrax and tuberculosis, they developed the techniques required to grow bacteria on solid media and to isolate pure cultures of pathogens (**figure 1.15**).

f. Viruses were discovered and methods for culturing them were also developed. Dimitri Ivanowski and Martinus Beijerinck were important contributors to the field of virology.

g. The ability to culture bacteria and viruses contributed to the development of immunology. Pasteur generated attenuated cultures of *Bacillus anthracis* and *Rabies virus*, and used them to create vaccines against anthrax and rabies. These advances were made without prior knowledge of how the immune system worked. Antibodies, soluble products of the immune system, were discovered by von Behring and Kitasato in their work on diphtheria. Metchnikof discovered white blood cells that could phagocytose and destroy bacterial pathogens.

h. Microbial ecology grew out of the work of Winogradsky and Beijerinck. They studied the role of microorganisms in carbon, nitrogen, and sulfur cycles and developed enrichment culture techniques and selective media.

1.4 Microbiology Today

a. Today molecular and genomic analyses have paved the way for understanding microbes as biological systems. These types of analyses were made possible by scientists who developed techniques for isolating DNA, cutting DNA molecules, splicing DNA molecules together, and determining the sequence of nucleotides in DNA.

b. There are many fields in microbiology. These include the more applied disciplines such as medical, public health, industrial, and food and dairy microbiology. Microbial ecology, physiology, and genetics are examples of basic fields. Increasingly, molecular and genomic methods for studying microbes are being used in microbiology.

Critical Thinking Questions

1. Consider the impact of microbes on the course of world history. History is full of examples of instances or circumstances under which one group of people lost a struggle against another. In fact, when examined more closely, the "losers" often had the misfortune of being exposed to, more susceptible to, or unable to cope with an infectious agent. Thus weakened in physical strength or demoralized by the course of a devastating disease, they were easily overcome by human "conquerors."

   a. Choose an example of a battle or other human activity such as exploration of new territory and determine the impact of microorganisms, either indigenous or transported to the region, on that activity.

   b. Discuss the effect that the microbe(s) had on the outcome in your example.

   c. Suggest whether the advent of antibiotics, food storage and preparation technology, or sterilization technology would have made a difference in the outcome.

2. Leeuwenhoek is often referred to as the father of microbiology. However, many historians feel that Louis Pasteur, Robert Koch, or perhaps both, deserve that honor. Who do you think is the father of microbiology? Why?

3. Consider the discoveries described in sections 1.3 and 1.4. Which do you think were the most important to the development of microbiology? Why?

4. Vaccinations against various childhood diseases have contributed to the entry of women, particularly mothers, into the full-time workplace.

   a. Is this statement supported by data comparing availability and extent of vaccination with employment statistics in different places or at different times?

   b. What is the incubation time and duration for measles, mumps, and chickenpox? Before vaccinations became available, what impact would these childhood diseases have had on mothers who had several elementary school-age children if the mothers had full-time jobs and lacked substantial child care support?

   c. What would be the consequence if an entire generation of children (or a group of children in one country) were not vaccinated against any diseases? What do you predict would happen if these children went to college and lived in a dormitory in close proximity with others who had received all of the recommended childhood vaccines?

5. Scientists are very interested in understanding when cyanobacteria first emerged because, as the first organisms capable of oxygenic photosynthesis, they triggered a sharp rise in atmospheric oxygen. For many years, the oldest fossil evidence for cyanobacteria was dated at 2.15 billion years ago (bya). In the late twentieth century, lipid biomarkers were

reported from geological samples from Western Australia that indicated that cyanobacteria were present 2.7 bya. In 2008, however, it was determined that these lipids were contaminants and the accepted value for the rise of cyanobacteria reverted to 2.15 bya. Discuss the specific challenges encountered in the study of microbial evolution. Contrast these with the advent of the use of molecular markers such as small subunit rRNA nucleotide sequences to assess microbial evolution.

Read the original paper: Rasmussen, B., et al. 2008. Reassessing the first appearance of eukaryotes and cyanobacteria. Nature 455:1101.

Learn More

Learn more by visiting the text website at www.mhhe.com/willey8, where you will find a complete list of references.

2

The Study of Microbial Structure: Microscopy and Specimen Preparation

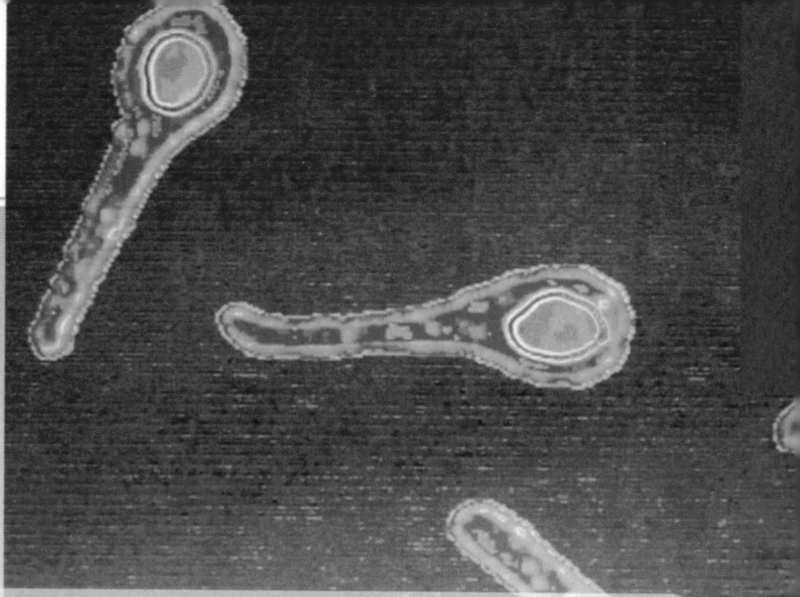

Clostridium tetani *is a rod-shaped bacterium that forms endospores and releases tetanus toxin, the cause of tetanus. In this colorized phase-contrast micrograph, the endospores are the bright, oval objects located at the ends of the rods.*

CHAPTER GLOSSARY

atomic force microscope A type of scanning probe microscope. It can be used to study surfaces of objects at high magnification by moving a probe over the specimen's surface.

bright-field microscope A microscope that illuminates the specimen directly with bright light and forms a dark image on a brighter background.

confocal scanning laser microscope (CSLM) A light microscope in which laser-derived light scans across the specimen at a single plane of focus; stray light from other parts of the specimen is blocked out to give an image with excellent contrast and resolution.

dark-field microscope A microscope that brightly illuminates the specimen while leaving the background dark.

differential interference contrast (DIC) microscopy Microscopy that combines two beams of plane-polarized light after passing through a specimen; their interference is used to create the image.

differential staining Procedures that divide microbes into separate groups based on staining properties.

fixation The process in which the internal and external structures of cells and organisms are preserved and fixed in position.

fluorescence microscopy A type of microscopy that exposes a specimen to light of a specific wavelength and then forms an image from the fluorescent light produced; usually the specimen is stained with a fluorescent dye (**fluorochrome**).

Gram stain A differential staining procedure that divides bacteria into gram-positive and gram-negative groups based on their ability to retain crystal violet when decolorized with an organic solvent such as ethanol.

negative staining A staining procedure in which a dye is used to make the background dark while the specimen is unstained.

parfocal Term used to describe a microscope that retains proper focus when the objectives are changed.

phase-contrast microscope A microscope that converts slight differences in refractive index and cell density into easily observed differences in light intensity.

refractive index A measure of how much a substance deflects a light ray from a straight path as it passes from one medium (e.g., glass) to another (e.g., air).

resolution The ability of a microscope to separate or distinguish between small objects that are close together.

scanning electron microscope (SEM) An electron microscope that scans a beam of electrons over the surface of a specimen and forms an image of the surface from the electrons that are emitted by it.

simple staining Staining a specimen with a single dye.

transmission electron microscope (TEM) A microscope in which an image is formed by passing an electron beam through a specimen and focusing the scattered electrons with magnetic lenses.

Microbiology usually is concerned with organisms so small they cannot be seen distinctly with the unaided eye. The organisms and entities studied by microbiologists typically range in size from viruses, which are measured in nanometers (nm), to protists, the largest of which are about 200 micrometers (μm) in diameter (**table 2.1**). Thus the microscope is of crucial importance for seeing microorganisms and understanding their structure. Furthermore, some types of microscopy also provide valuable information about the functions of structures. Therefore it is important to understand how microscopes work and how specimens are prepared for examination.

Today's microscopes look quite different than those built by van Leeuwenhoek in the 1600s (*see figure 1.11*). His microscopes were light microscopes, and light microscopes are still the most widely used. In this chapter, we begin with a detailed treatment of the standard bright-field light microscope and then describe other common types of light microscopes, including confocal microscopy. Next we discuss preparation and staining of specimens for examination with the light microscope. This is followed by a description of transmission and scanning electron microscopes, both of which are used extensively in current microbiological research. We close the chapter with a brief introduction to scanning probe microscopy.

Table 2.1 Common Units of Measurement

Unit	*Abbreviation*	*Value*
1 centimeter	cm	10^{-2} meter or 0.394 inches
1 millimeter	mm	10^{-3} meter
1 micrometer	μm	10^{-6} meter
1 nanometer	nm	10^{-9} meter
1 Angstrom	Å	10^{-10} meter

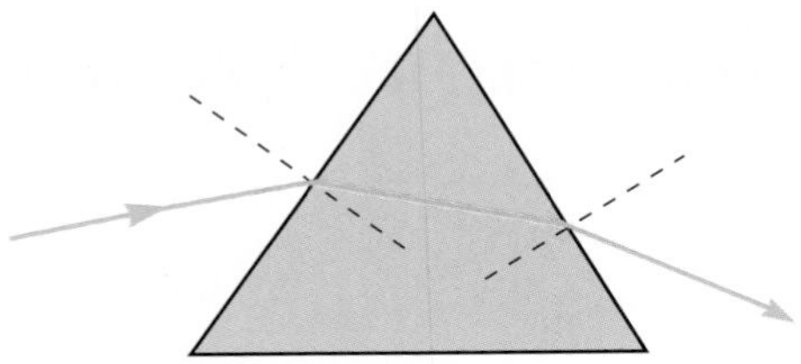

FIGURE 2.1 The Bending of Light by a Prism. Normals (lines perpendicular to the surface of the prism) are indicated by dashed lines. As light enters the glass, it is bent toward the first normal. When light leaves the glass and returns to air, it is bent away from the second normal. As a result, the prism bends light passing through it.

2.1 Lenses and the Bending of Light

To understand how a light microscope operates, we must consider the way lenses bend and focus light to form images. When a ray of light passes from one medium to another, refraction occurs—that is, the ray is bent at the interface. The **refractive index** is a measure of how greatly a substance slows the velocity of light; the direction and magnitude of bending are determined by the refractive indices of the two media forming the interface. For example, when light passes from air into glass, a medium with a greater refractive index, it is slowed and bent toward the normal, a line perpendicular to the surface (**figure 2.1**). As light leaves glass and returns to air, a medium with a lower refractive index, it accelerates and is bent away from the normal. Thus a prism bends light because glass has a different refractive index from air and the light strikes its surface at an angle.

Lenses act like a collection of prisms operating as a unit. When the light source is distant so that parallel rays of light strike the lens, a convex lens focuses these rays at a specific point, the focal point (F in **figure 2.2**). The distance between the center of the lens and the focal point is called the focal length (f in figure 2.2).

Our eyes cannot focus on objects nearer than about 25 cm (i.e., about 10 inches). This limitation may be overcome by using a convex lens as a simple magnifier (or microscope) and holding it close to an object. A magnifying glass provides a clear image at much closer range, and the object appears larger. Lens strength is related to focal length; a lens with a short focal length magnifies an object more than a lens having a longer focal length.

1. Define refraction, refractive index, focal point, and focal length.
2. Describe the path of a light ray through a prism or lens.
3. How is lens strength related to focal length? How do you think this principle is applied to corrective eyeglasses?

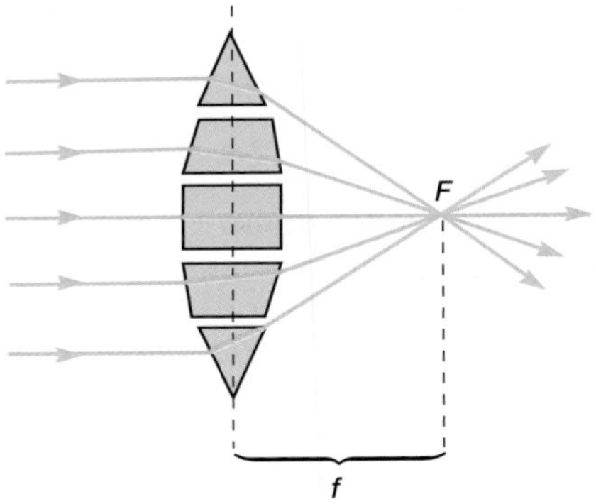

FIGURE 2.2 Lens Function. A lens functions somewhat like a collection of prisms. Light rays from a distant source are focused at the focal point F. The focal point lies a distance f, the focal length, from the lens center.

Figure 2.2 Micro Inquiry

How would the focal length change if the lens shown here were thicker?

2.2 Light Microscopes

Microbiologists currently employ a variety of light microscopes in their work; bright-field, dark-field, phase-contrast, fluorescence, and confocal microscopes are most commonly used. Each is useful for certain applications. Modern microscopes are all compound microscopes—microscopes having two sets of lenses. The **objective lens** is the lens closest to the specimen. It forms a magnified image that is further enlarged by one or more additional lenses.

Bright-Field Microscope

The **bright-field microscope** is routinely used in microbiology labs, where it can be employed to examine both stained and unstained specimens. It is called a bright-field microscope because it forms a dark image against a brighter background. It consists of a sturdy metal stand composed of a base and an arm to which the remaining parts are attached (**figure 2.3**). A light source, either a mirror or an electric illuminator, is located in the base. Two focusing knobs, the fine and coarse adjustment knobs, are located on the arm and can move either the stage or the nosepiece to focus the image.

The stage is positioned about halfway up the arm. It holds microscope slides either by simple slide clips or by a mechanical stage clip. A mechanical stage allows the operator to move a slide smoothly during viewing by use of stage control knobs. The **substage condenser** (or simply, condenser) is mounted within or beneath the stage and focuses a cone of light on the slide. Its position often is fixed in simpler microscopes but can be adjusted vertically in more advanced models.

The curved upper part of the arm holds the body assembly, to which a nosepiece and one or more **ocular lenses** (also called **eyepieces**) are attached. More advanced microscopes have eyepieces for both eyes and are called binocular microscopes. The body assembly contains a series of mirrors and prisms so that the barrel holding the eyepiece may be tilted for ease in viewing. The nosepiece holds three to five objective lenses of differing magnifying power and can be rotated to position any objective lens beneath the body assembly. Ideally a microscope should be **parfocal**—that is, the image should remain in focus when objective lenses are changed.

The image one sees when viewing a specimen with a compound microscope is created by the objective and ocular lenses working together. Light from the illuminated specimen is focused by the objective lens, creating an enlarged image within the microscope. The ocular lens further magnifies this primary image. The total magnification is calculated by multiplying the objective and eyepiece magnifications together. For example, if a 45× objective lens is used with a 10× eyepiece, the overall magnification of the specimen is 450×.

Microscope Resolution

The most important part of the microscope is the objective lens, which must produce a clear image, not just a magnified one. Thus resolution is extremely important. **Resolution** is the ability

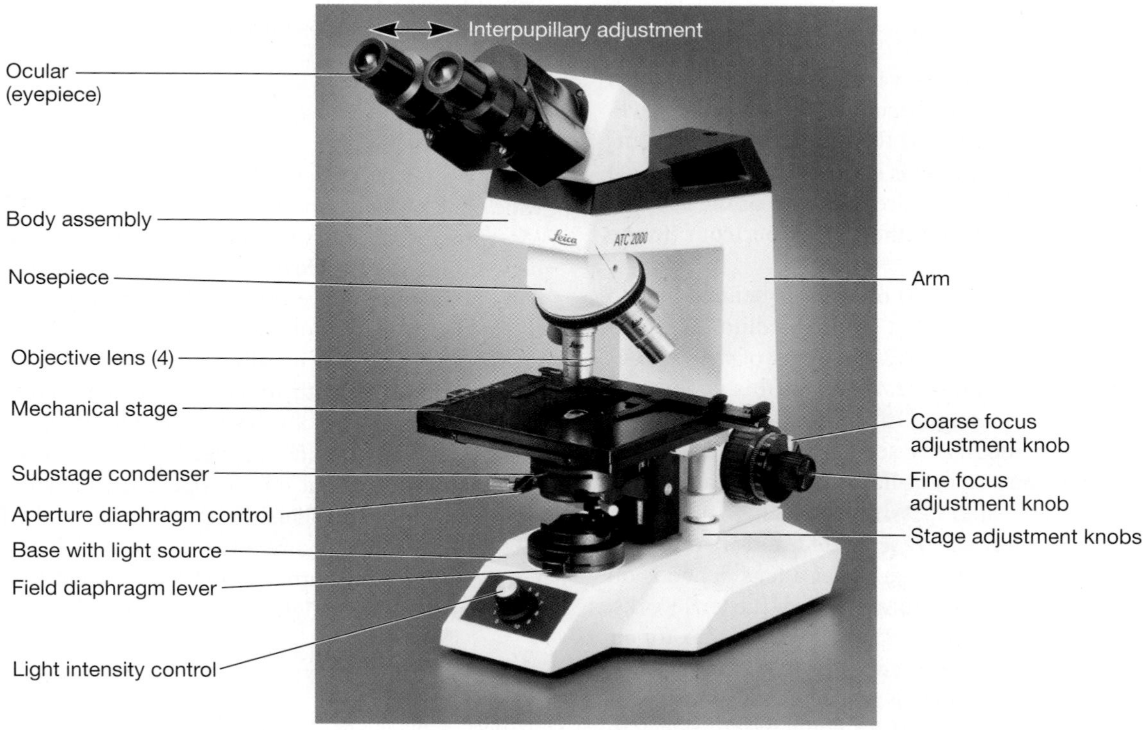

FIGURE 2.3 A Bright-Field Microscope. The parts of a modern bright-field microscope. The microscope pictured is somewhat more sophisticated than those found in many student laboratories. For example, it is binocular (has two eyepieces) and has a mechanical stage, an adjustable substage condenser, and a built-in illuminator.

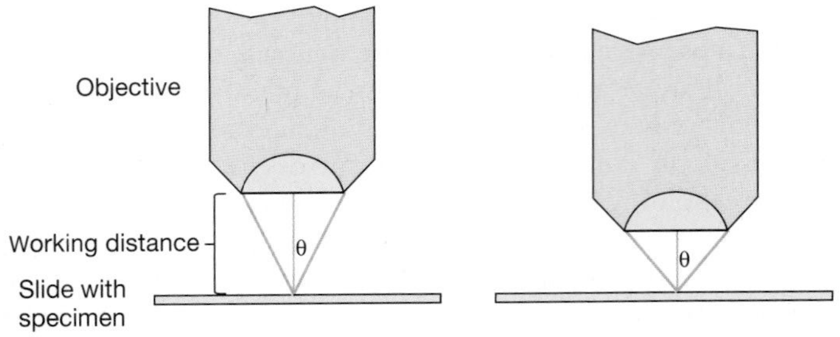

FIGURE 2.4 Numerical Aperture in Microscopy. The angular aperture θ is 1/2 the angle of the cone of light that enters a lens from a specimen, and the numerical aperture is *n* sin θ. In the right-hand illustration, the lens has larger angular and numerical apertures; its resolution is greater and its working distance smaller.

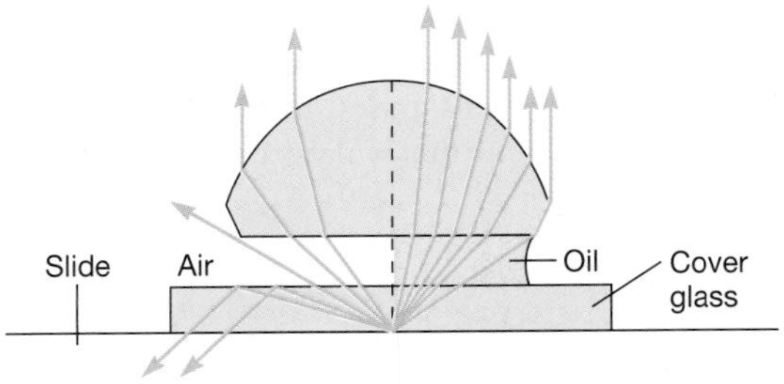

FIGURE 2.5 The Oil Immersion Objective. An oil immersion objective operating in air and with immersion oil.

of a lens to separate or distinguish between small objects that are close together.

Resolution is described mathematically by an equation developed in the 1870s by Ernst Abbé, a German physicist responsible for much of the optical theory underlying microscope design. The Abbé equation states that the minimal distance (d) between two objects that reveals them as separate entities depends on the wavelength of light (λ) used to illuminate the specimen and on the **numerical aperture** of the lens ($n \sin \theta$), which is the ability of the lens to gather light.

$$d = \frac{0.5\lambda}{n \sin \theta}$$

As d becomes smaller, the resolution increases, and finer detail can be discerned in a specimen; d becomes smaller as the wavelength of light used decreases and as the numerical aperture increases. Thus the greatest resolution is obtained using a lens with the largest possible numerical aperture and light of the shortest wavelength, light at the blue end of the visible spectrum (in the range of 450 to 500 nm).

The numerical aperture ($n \sin \theta$) of a lens is defined by two components: n is the refractive index of the medium in which the lens works (e.g., air) and θ is 1/2 the angle of the cone of light entering an objective (**figure 2.4**). When this cone has a narrow angle, it does not spread out much after leaving the slide and therefore does not adequately separate images of closely packed objects. If the cone of light has a very wide angle, the light spreads out rapidly after passing through a specimen, closely packed objects appear widely separated, and are resolved. The angle of the cone of light that can enter a lens depends on the refractive index (n) of the medium in which the lens works, as well as on the objective lens itself. The refractive index for air is 1.00 and sin θ cannot be greater than 1 (the maximum θ is 90° and sin 90° is 1.00). Therefore no lens working in air can have a numerical aperture greater than 1.00. The only practical way to raise the numerical aperture above 1.00, and therefore achieve higher resolution, is to increase the refractive index with immersion oil, a colorless liquid with the same refractive index as glass (**table 2.2**). If air is replaced with immersion oil, many light rays that did not enter the objective due to reflection and refraction at the surfaces of the objective lens and slide will now do so (**figure 2.5**). This results in an increase in numerical aperture and resolution.

Numerical aperture is related to another characteristic of an objective lens, the working distance. The working distance of an objective is the distance between the front surface of the lens and the surface of the cover glass (if one is used) or the specimen when it is in sharp focus (figure 2.4). Objectives with large numerical apertures and great resolving power have short working distances (table 2.2).

The preceding discussion has focused on the resolving power of the objective lens. The resolution of an entire microscope must take into account the numerical aperture of its condenser, as is evident from the following equation, where NA is the numerical aperture.

$$d_{\text{microscope}} = \frac{\lambda}{(\text{NA}_{\text{objective}} + \text{NA}_{\text{condenser}})}$$

The condenser is a large, light-gathering lens used to project a wide cone of light through the slide and into the objective lens. Most microscopes have a condenser with a numerical aperture between 1.2 and 1.4. However, the condenser's numerical aperture will not be much above about 0.9 unless the top of the condenser is oiled to the bottom of the slide. During routine microscope operation, the condenser usually is not oiled, and this limits the overall resolution, even with an oil immersion objective.

Although the resolution of the microscope must consider both the condenser and the objective lens, in most cases the limit of resolution of a light microscope is calculated using the Abbé equation, which considers the objective lens only. The maximum theoretical resolving power of a microscope when viewing a specimen using an oil immersion objective (numerical aperture of 1.25) and blue-green light is approximately 0.2 μm.

$$d = \frac{(0.5)(530 \text{ nm})}{1.25} = 212 \text{ nm or } 0.2\ \mu\text{m}$$

At best, a bright-field microscope can distinguish between two dots about 0.2 μm apart (the same size as a very small bacterium). Thus the vast majority of viruses cannot be examined with a light microscope.

Table 2.2 Properties of Objective Lenses

	Objective			
Property	*Scanning*	*Low Power*	*High Power*	*Oil Immersion*
Magnification	4×	10×	40–45×	90–100×
Numerical aperture	0.10	0.25	0.55–0.65	1.25–1.4
Approximate focal length (f)	40 mm	16 mm	4 mm	1.8–2.0 mm
Working distance	17–20 mm	4–8 mm	0.5–0.7 mm	0.1 mm
Approximate resolving power with light of 450 μm (blue light)	2.3 μm	0.9 μm	0.35 μm	0.18 μm

Given the limit of resolution of a light microscope, the largest useful magnification—the level of magnification needed to increase the size of the smallest resolvable object to be visible with the light microscope—can be determined. Our eye can just detect a speck 0.2 mm in diameter. When the acuity of the eye and the resolution of the microscope are considered together, it is calculated that the useful limit of magnification is about 1,000 times the numerical aperture of the objective lens. Most standard microscopes come with 10× eyepieces and have an upper limit of about 1,000× with oil immersion. A 15× eyepiece may be used with good objective lenses to achieve a useful magnification of 1,500×. Any further magnification does not enable a person to see more detail. Indeed, a light microscope can be built to yield a final magnification of 10,000×, but it would simply be magnifying a blur. Only the electron microscope provides sufficient resolution to make higher magnifications useful (p. 38).

Search This: MicroscopyU

Dark-Field Microscope

Although unpigmented living cells can be viewed in the bright-field microscope, they are not clearly visible because there is little difference in contrast between the cells and water. As discussed in section 2.3, one solution to this problem is to stain cells before observation to increase contrast and create variations in color between cell structures. But what if an investigator must view living cells to observe a dynamic process such as movement or phagocytosis? Three types of light microscopes create detailed, clear images of living specimens: the dark-field microscope, the phase-contrast microscope, and the differential interference contrast microscope.

The **dark-field microscope** produces detailed images of living, unstained cells and organisms by simply changing the way in which they are illuminated. A hollow cone of light is focused on the specimen in such a way that unreflected and unrefracted rays do not enter the objective. Only light that has been reflected or refracted by the specimen forms an image (**figure 2.6**). The field surrounding a specimen appears black, while the object itself is brightly illuminated (**figure 2.7**). The dark-field microscope can reveal considerable internal structure in larger eukaryotic microorganisms (figure 2.7*b*). It also is used to identify certain bacteria such as the thin and distinctively shaped *Treponema pallidum*, the causative agent of syphilis (figure 2.7*a*).

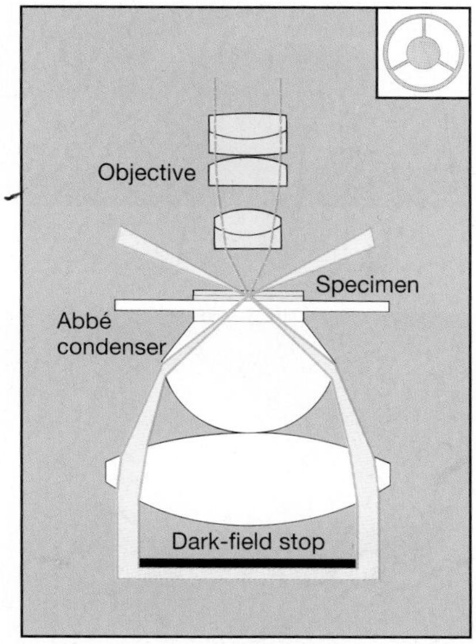

FIGURE 2.6 Dark-Field Microscopy. The simplest way to convert a microscope to dark-field microscopy is to place a dark-field stop (inset) underneath the condenser lens system. The condenser then produces a hollow cone of light so that the only light entering the objective comes from the specimen.

Phase-Contrast Microscope

A **phase-contrast microscope** converts slight differences in refractive index and cell density into easily detected variations in light intensity (**figure 2.8**). The condenser of a phase-contrast microscope has an annular stop, an opaque disk with a thin transparent

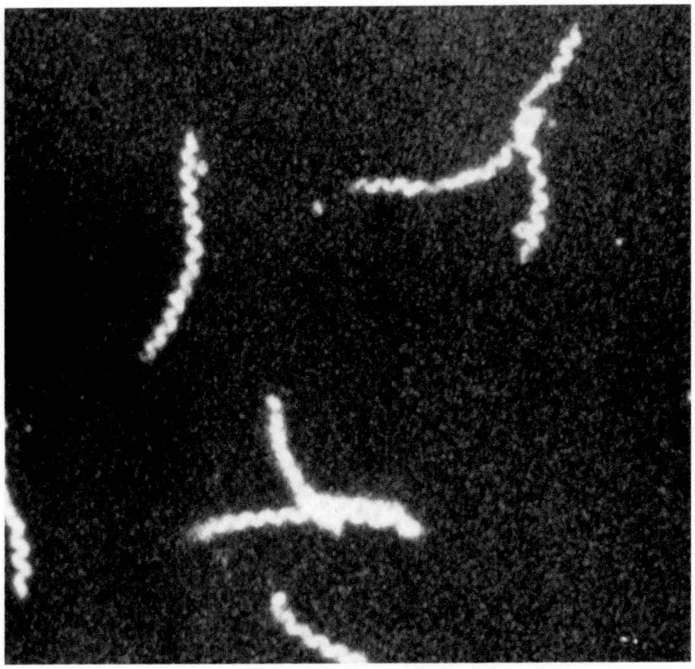

(a) *T. pallidum*

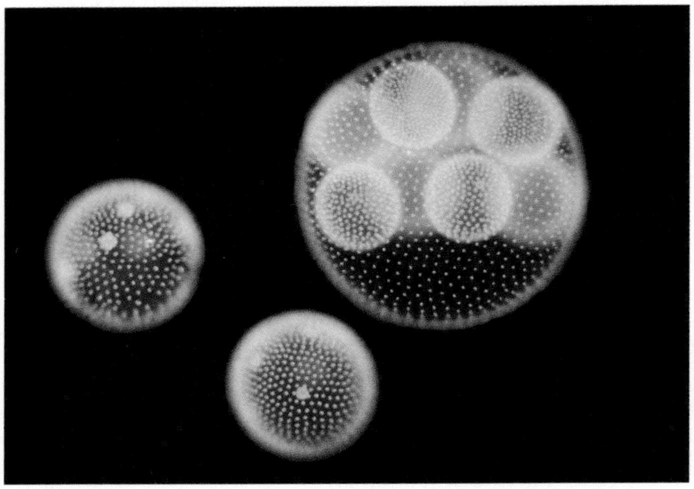

(b) *Volvox*

FIGURE 2.7 Examples of Dark-Field Microscopy. (a) *Treponema pallidum,* the spirochete that causes syphilis. (b) *Volvox*. Note daughter colonies within the mature *Volvox* colony.

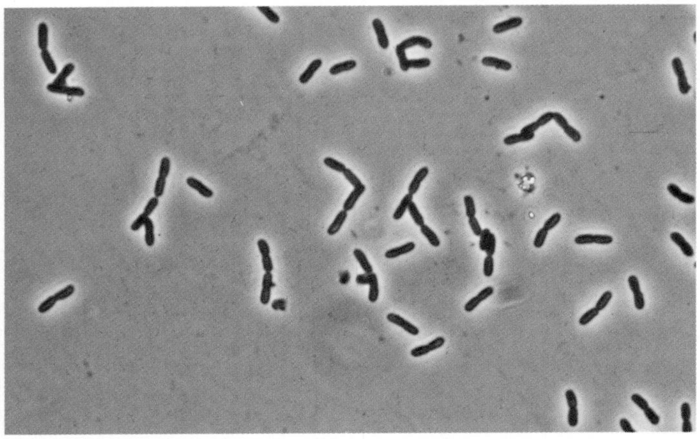

(a) *Pseudomonas*

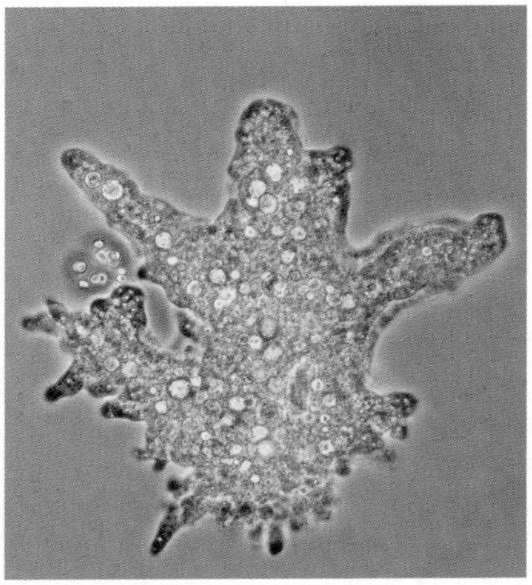

(b) An amoeba

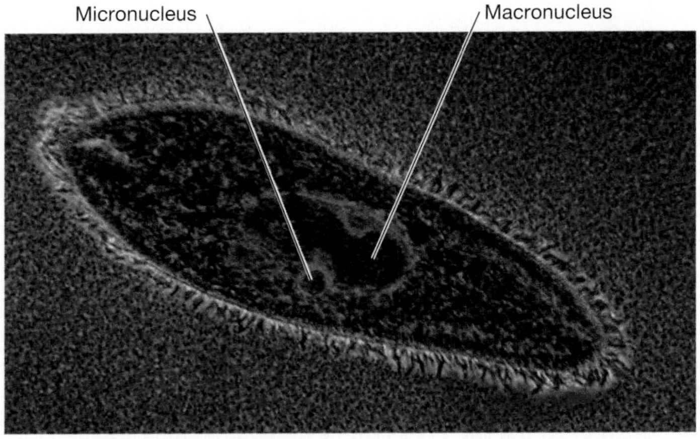

(c) *Paramecium*

FIGURE 2.8 Examples of Phase-Contrast Microscopy. (a) *Pseudomonas* cells, which range from 1–3 μm in length. (b) An amoeba, a eukaryotic microbe that moves by means of pseudopodia, which extend out from the main part of the cell body. (c) *Paramecium* stained to show a large central macronucleus with a small spherical micronucleus at its side (×100).

ring, that produces a hollow cone of light (**figure 2.9**). As this cone passes through a cell, some light rays are bent due to variations in density and refractive index within the specimen, and are retarded by about 1/4 wavelength. The deviated light is focused to form an image of the object. Undeviated light rays strike a phase ring in the phase plate, an optical disk located in the objective, while the deviated rays miss the ring and pass through the rest of the plate. If the

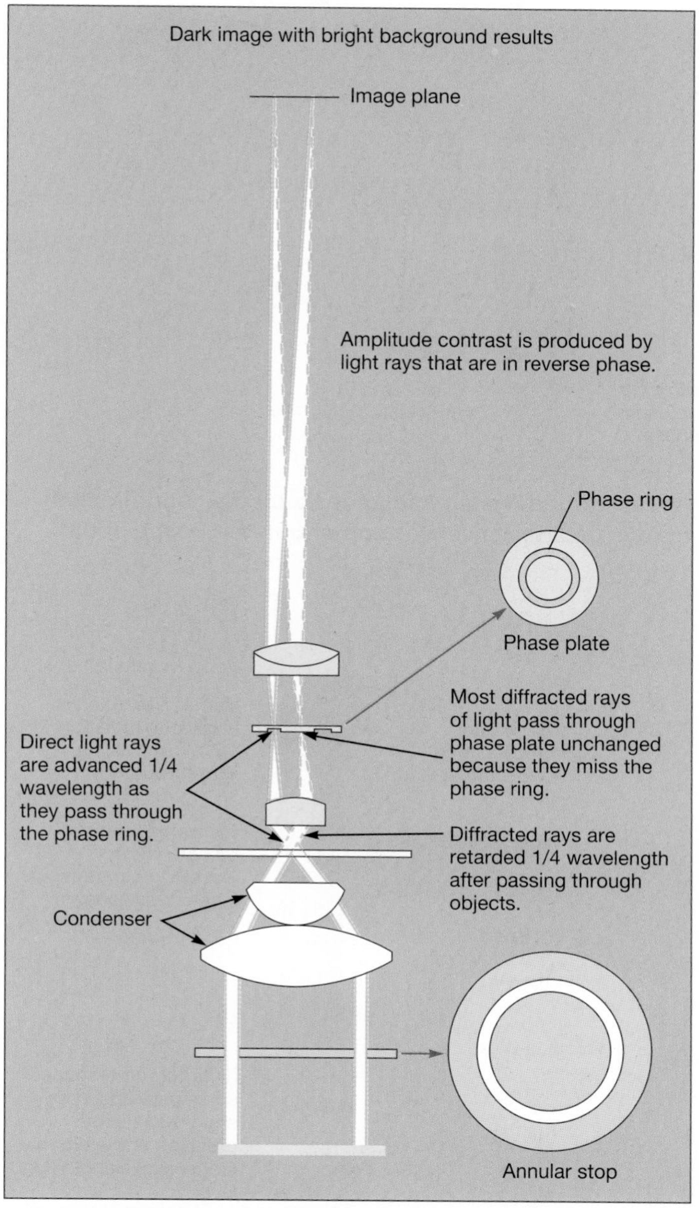

Dark image with bright background results
Image plane
Amplitude contrast is produced by light rays that are in reverse phase.
Phase ring
Phase plate
Most diffracted rays of light pass through phase plate unchanged because they miss the phase ring.
Direct light rays are advanced 1/4 wavelength as they pass through the phase ring.
Diffracted rays are retarded 1/4 wavelength after passing through objects.
Condenser
Annular stop

FIGURE 2.9 Phase-Contrast Microscopy. The optics of a dark-phase-contrast microscope.

Figure 2.9 Micro Inquiry

What is the purpose of the annular stop in a phase-contrast microscope? Is it found in any other kinds of light microscopes?

phase ring is constructed in such a way that the undeviated light passing through it is advanced by 1/4 wavelength, the deviated and undeviated waves will be about 1/2 wavelength out of phase and will cancel each other when they come together to form an image (**figure 2.10**). The background, formed by undeviated light, is bright, while the unstained object appears dark and well defined. This type of microscopy is called dark-phase-contrast microscopy. Color filters often are used to improve the image.

Phase-contrast microscopy is especially useful for studying microbial motility, determining the shape of living cells, and detecting bacterial structures such as endospores and inclusion bodies. These are clearly visible because they have refractive indices markedly different from that of water. Phase-contrast microscopes also are widely used to study eukaryotic cells. ▶▶| *Bacterial endospores (section 3.8); Inclusions (section 3.5)*

Differential Interference Contrast Microscope

The **differential interference contrast (DIC) microscope** is similar to the phase-contrast microscope in that it creates an image by detecting differences in refractive indices and thickness. Two beams of plane-polarized light at right angles to each other are generated by prisms. In one design, the object beam passes through the specimen, while the reference beam passes through a clear area of the slide. After passing through the specimen, the two beams are combined and interfere with each other to form an image. A live, unstained specimen appears brightly colored and three-dimensional (**figure 2.11**). Structures such as cell walls, endospores, granules, vacuoles, and nuclei are clearly visible.

Search This: Nomarski (DIC) microscopy/JIC

1. List the parts of a light microscope and describe their functions.
2. If a specimen is viewed using a 5× objective in a microscope with a 15× eyepiece, how many times has the image been magnified?
3. How does resolution depend on the wavelength of light, refractive index, and numerical aperture? How are resolution and magnification related?
4. What is the function of immersion oil?
5. Why don't most light microscopes use 30× ocular lenses for greater magnification?

Fluorescence Microscope

The microscopes thus far considered produce an image from light that passes through a specimen. An object also can be seen because it emits light: this is the basis of fluorescence microscopy. When some molecules absorb radiant energy, they become excited and release much of their trapped energy as light. Any light emitted by an excited molecule has a longer wavelength (i.e., has lower energy) than the radiation originally absorbed. **Fluorescent light** is emitted very quickly by the excited molecule as it gives up its trapped energy and returns to a more stable state.

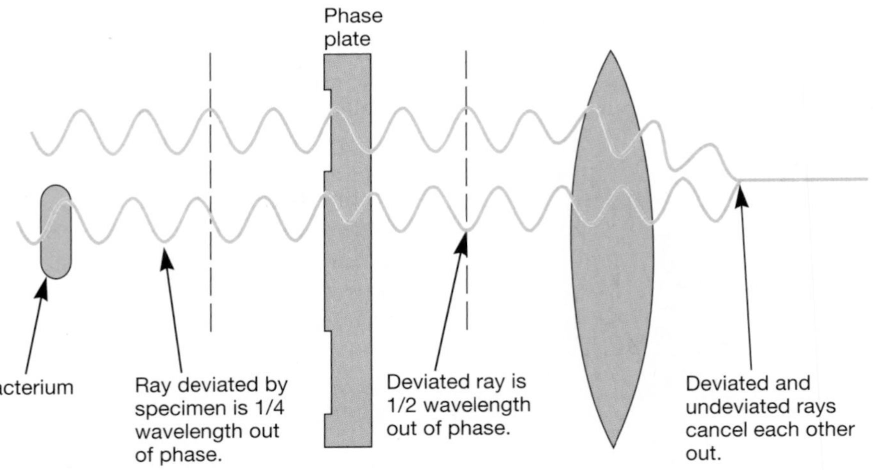

FIGURE 2.10 The Production of Contrast in Phase Microscopy. The behavior of deviated and undeviated (i.e., undiffracted) light rays in the dark-phase-contrast microscope. Because the light rays tend to cancel each other out, the image of the specimen will be dark against a brighter background.

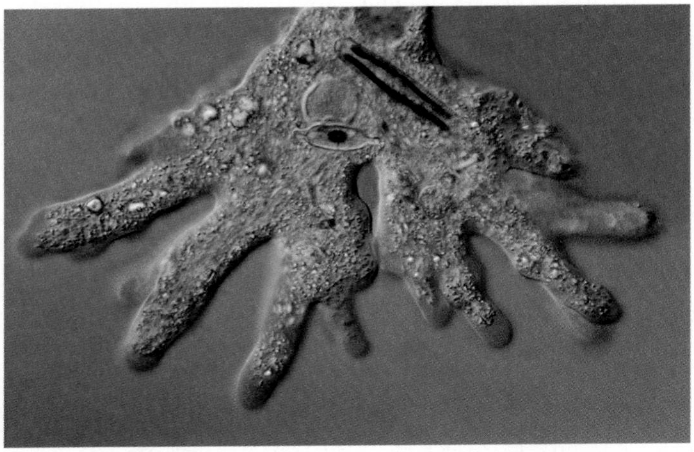

FIGURE 2.11 Differential Interference Contrast Microscopy. A micrograph of the protozoan *Amoeba proteus.* The three-dimensional image contains considerable detail.

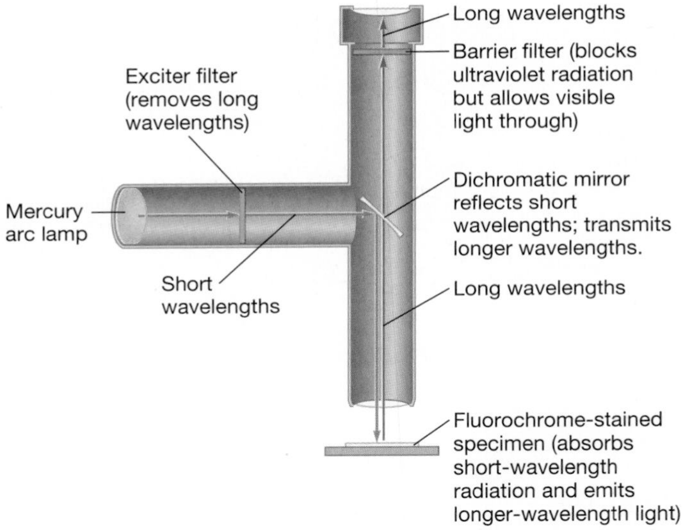

FIGURE 2.12 Epifluorescence Microscopy. The principles of operation of an epifluorescence microscope.

The **fluorescence microscope** exposes a specimen to ultraviolet, violet, or blue light and forms an image of the object with the resulting fluorescent light. The most commonly used fluorescence microscopy is epifluorescence microscopy, also called incident light or reflected light fluorescence microscopy. Epifluorescence microscopes employ an objective lens that also acts as a condenser (**figure 2.12**). A mercury vapor arc lamp or other source produces an intense beam of light that passes through an exciter filter. The exciter filter transmits only the desired wavelength of excitation light. The excitation light is directed down the microscope by the dichromatic mirror. This mirror reflects light of shorter wavelengths (i.e., the excitation light) but allows light of longer wavelengths to pass through. The excitation light continues down, passing through the objective lens to the specimen, which is usually stained with molecules called **fluorochromes** (**table 2.3**). The fluorochrome absorbs light energy from the excitation light and fluoresces brightly. The emitted fluorescent light travels up through the objective lens into the microscope. Because the emitted fluorescent light has a longer wavelength, it passes through the dichromatic mirror to a barrier filter, which blocks out any residual excitation light. Finally, the emitted light passes through the barrier filter to the eyepieces.

The fluorescence microscope has become an essential tool in microbiology. Bacterial pathogens can be identified after staining with fluorochromes or specifically tagging them with fluorescently labeled antibodies using immunofluorescence

Table 2.3 Commonly Used Fluorochromes

Fluorochrome	*Uses*
Acridine orange	Stains DNA
Diamidino-2-phenyl indole (DAPI)	Stains DNA
Fluorescein isothiocyanate (FITC)	Often attached to antibodies that bind specific cellular components or to DNA probes
Tetramethyl rhodamine isothiocyanate (TRITC or rhodamine)	Often attached to antibodies that bind specific cellular components

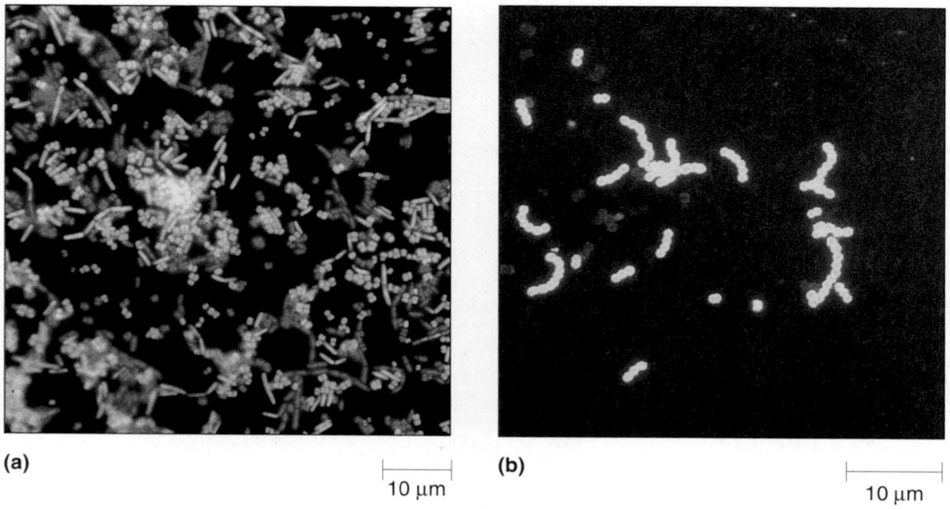

FIGURE 2.13 Fluorescent Dyes and Tags. (a) Dyes that cause live cells to fluoresce green and dead ones red. (b) Fluorescent antibodies tag specific molecules. In this case, the antibody binds to a molecule that is unique to *Streptococcus pyogenes,* the bacterium that causes strep throat.

procedures. In ecological studies, the fluorescence microscope is used to observe microorganisms stained with fluorochrome-labeled probes or fluorochromes that bind specific cell constituents (table 2.3). In addition, microbial ecologists use epifluorescence microscopy to visualize photosynthetic microbes, as their pigments naturally fluoresce when excited by light of specific wavelengths. It is even possible to distinguish live bacteria from dead bacteria by the color they fluoresce after treatment with a specific mixture of stains (**figure 2.13*a***). Thus the microorganisms can be viewed and directly counted in a relatively undisturbed ecological niche. ▶▶| *Microscopy (section 35.2)*

Another important use of fluorescence microscopy is the localization of specific proteins within cells. One approach is to use genetic engineering techniques that fuse the gene for the protein of interest to a gene isolated from jellyfish belonging to the genus *Aequorea.* The jellyfish gene encodes a protein that naturally fluoresces green when exposed to light of a particular wavelength and is called green fluorescent protein (GFP). Since its isolation, the gene has been altered to create proteins that fluoresce different colors. GFP and its variants have been used extensively in studies on bacterial cell division and related phenomena (**figure 2.14**). In fact, the 2008 Nobel Prize in chemistry was awarded to Osamu Shimomura of Japan and Americans Martin Chalfie and Roger Tsien for their development of this important tool. ▶▶| *Fluorescent labeling (section 15.7)*

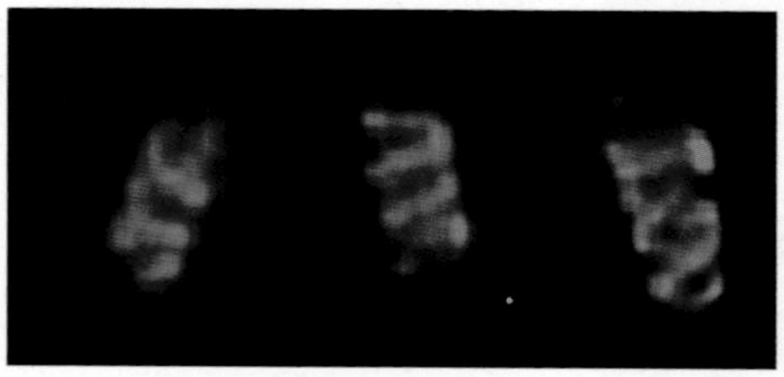

FIGURE 2.14 Green Fluorescent Protein. Visualization of Mbl, a cytoskeletal protein of *Bacillus subtilis.* The helical Mbl protein has been fused with green fluorescent protein and therefore fluoresces green.

Confocal Microscopy

Like the large and small beads illustrated in **figure 2.15*a***, biological specimens are three-dimensional. When three-dimensional objects are viewed with traditional light microscopes, light from all areas of the object, not just the plane of focus, enters the microscope and is used to create an image. The resulting image is murky and fuzzy (figure 2.15*b*). This problem has been solved by the **confocal scanning laser microscope (CSLM),** or simply, confocal microscope. The confocal microscope uses a laser beam to illuminate a specimen, usually one that has been fluorescently stained. A major component of the confocal microscope is an aperture placed above the objective lens. The aperture eliminates stray light from parts of the specimen that lie above and below the plane of focus (**figure 2.16**). Thus the only light used to create the image is from the plane of focus, and a much sharper image is formed.

Computers are integral to the process of creating confocal images. A computer interfaced with the confocal microscope receives digitized information from each plane in the specimen that is examined. This information can be used to create a composite image that is very clear and detailed (figure 2.15*c*) or to create a three-dimensional reconstruction of the specimen (figure 2.15*d*). Images of x-z plane cross sections of the specimen can also be generated, giving the observer views of the specimen from three perspectives (figure 2.15*e*). Confocal microscopy has numerous applications, including the study of biofilms, which can form on many different types of surfaces, including indwelling medical devices such as hip joint replacements. As shown in figure 2.15*f*,

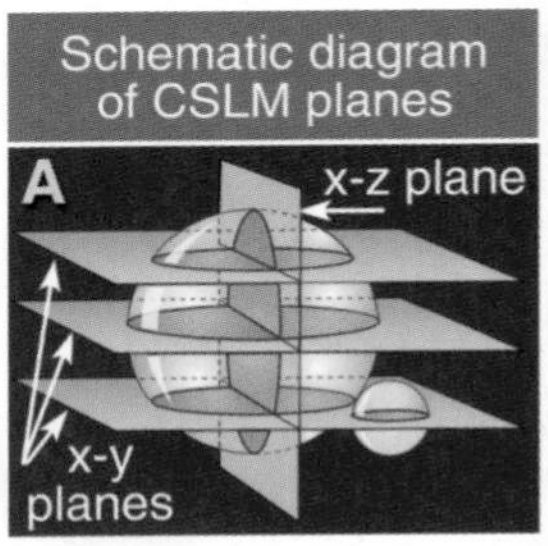

(a) All three-dimensional objects are defined by three axes: x, y, and z. The confocal scanning light microscope (CSLM) is able to create images of planes formed by the x and y axes (x-y planes) and planes formed by the x and z axes (x-z planes).

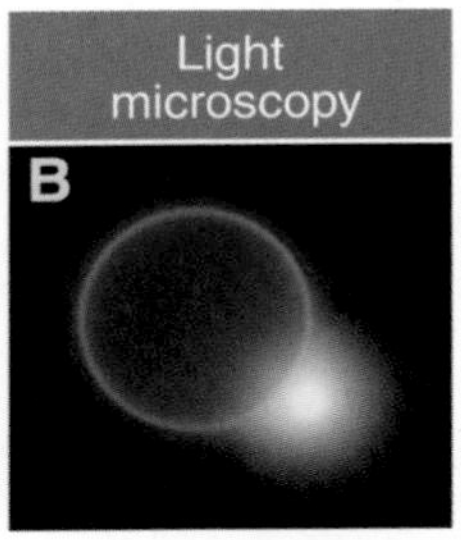

(b) The light microscope image of the two beads shown in **(a)**. Note that neither bead is clear and that the smaller bead is difficult to recognize as a bead. This is because the image is generated from light emanating from multiple planes of focus.

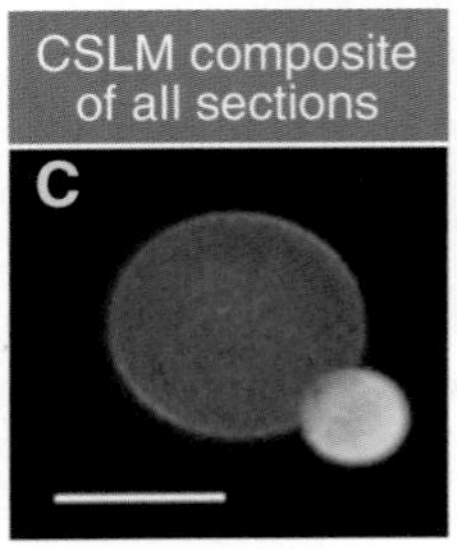

(c) A CSLM uses light from a single plane of focus to generate an image. A computer connected to a CSLM can make a composite image of the two beads using digitized information collected from multiple planes within the beads. The result is a much clearer and more detailed image.

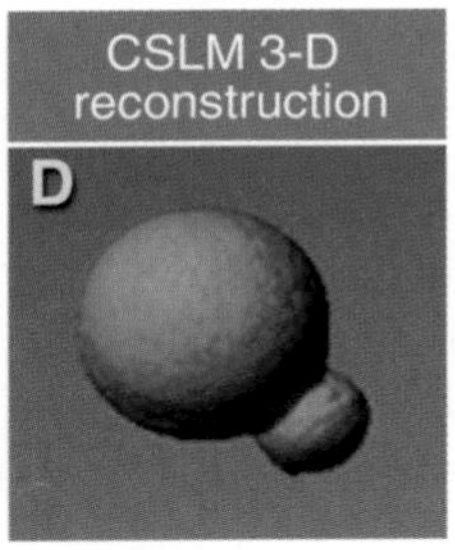

(d) The computer can also use digitized information collected from multiple planes within the beads to generate a three-dimensional reconstruction of the beads.

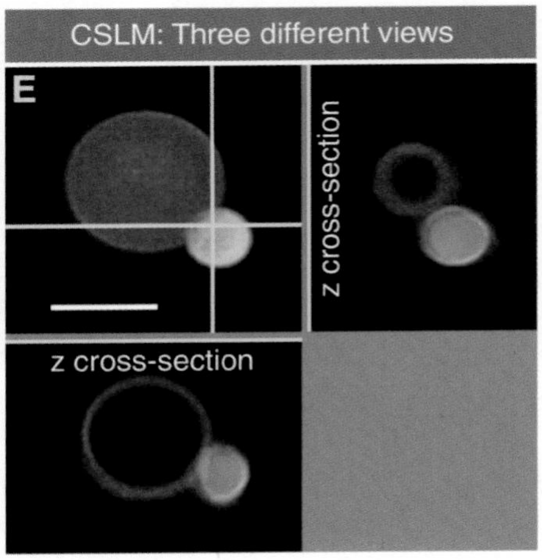

(e) The computer can also generate views of the specimen using different planes. The top left panel is the image of a single x-y plane (i.e., looking down from the top of the beads). The two lines represent the two x-z planes imaged in the other two panels. The vertical line indicates the x-z plane shown in the top right panel (i.e., a view from the right side of the beads) and the horizontal line indicates the x-z plane shown in the bottom panel (i.e., a view from the front face of the beads).

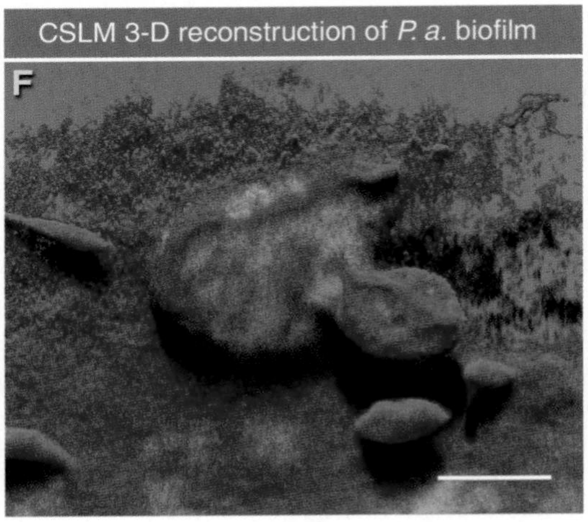

(f) A three-dimensional reconstruction of a *Pseudomonas aeruginosa* biofilm. The biofilm was exposed to an antibacterial agent and then stained with dyes that distinguish living (green) from dead (red) cells. The cells on the surface of the biofilm have been killed, but those in the lower layers are still alive. This image clearly demonstrates the difficulty of killing all the cells in a biofilm.

FIGURE 2.15 Light and Confocal Microscopy. (a–e) Two beads examined by light and confocal microscopy. (f) Three-dimensional reconstruction of a biofilm.

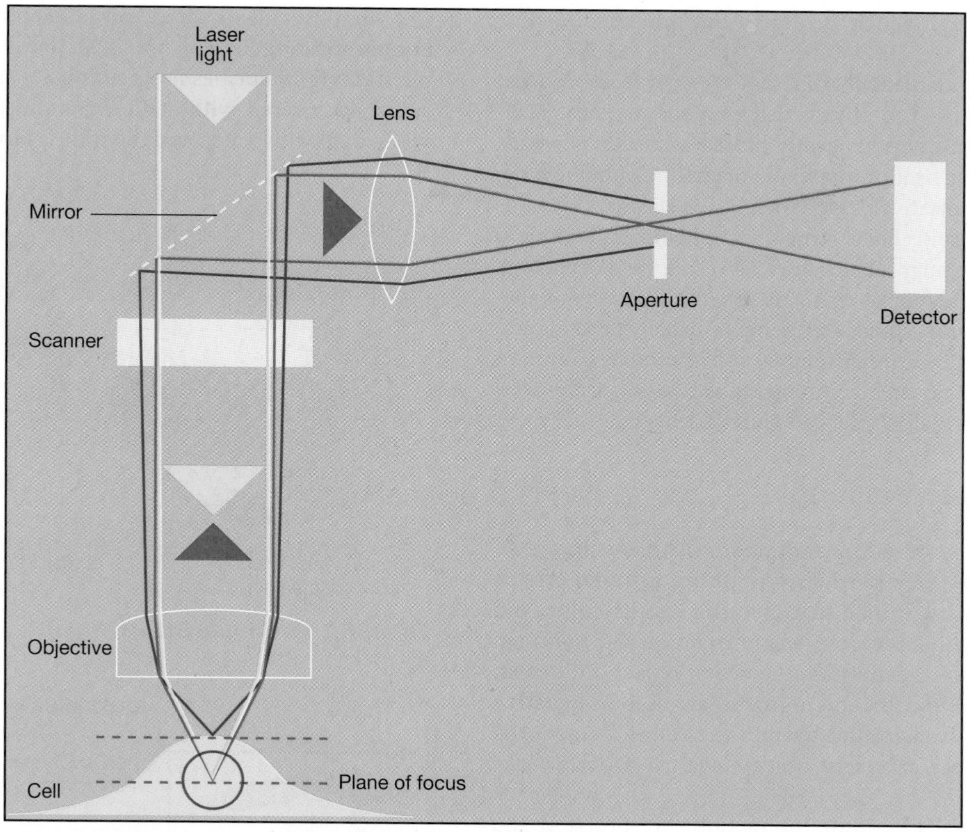

FIGURE 2.16 A Ray Diagram of a Confocal Laser Scanning Microscope. The yellow lines represent laser light used for illumination. Red lines symbolize the light arising from the plane of focus, and the blue lines stand for light from parts of the specimen above and below the focal plane.

Figure 2.16 Micro Inquiry

How does the light source differ between a confocal light microscope and other light microscopes?

it is difficult to kill all cells in a biofilm. This makes biofilms a particular concern to the medical field because their formation on medical devices can result in infections that are difficult to treat. ▶▶| *Biofilms (section 7.7)*

1. Do all specimens examined by fluorescence microscopy need to be stained with a fluorochrome? Explain.
2. How is GFP similar to a fluorochrome? How is it different? What advantage does GFP have over a traditional fluorochrome?
3. How does a confocal microscope operate? Why does it provide better images of thick specimens than does the standard compound light microscope?
4. Briefly describe how dark-field, phase-contrast, differential interference contrast, epifluorescence, and confocal microscopes work and the kind of image or images provided by each. Give a specific use for each type.

2.3 Preparation and Staining of Specimens

Although living microorganisms can be directly examined with the light microscope, they often are fixed and stained. Such preparation serves to increase the visibility of the microorganisms, accentuate specific morphological features, and preserve them for future study.

Fixation

The stained cells seen in a microscope should resemble living cells as closely as possible. **Fixation** is the process by which the internal and external structures of specimens are preserved and fixed in position. It inactivates enzymes that might disrupt cell morphology and toughens cell structures so that they do not change during staining and observation. A microorganism usually is killed and attached firmly to the microscope slide during fixation.

There are two fundamentally different types of fixation. **Heat fixation** is routinely used to observe bacteria and archaea. Typically, a film of cells (a smear) is gently heated as a slide is passed through a flame. Heat fixation preserves overall morphology. Although heat fixation inactivates enzymes, it also destroys proteins that may be part of subcellular structures. **Chemical fixation** is used to protect fine cellular substructure as well as the morphology of larger, more delicate microorganisms. Chemical fixatives penetrate cells and react with cellular components, usually proteins and lipids, to render them inactive, insoluble, and immobile. Common fixative mixtures contain such components as ethanol, acetic acid, mercuric chloride, formaldehyde, and glutaraldehyde.

Dyes and Simple Staining

The many types of dyes used to stain microorganisms have two features in common: they have **chromophore groups,** groups with conjugated double bonds that give the dye its color, and they can bind cells with ionic, covalent, or hydrophobic bonding. Most dyes are used to directly stain the cell or object of interest, but some dyes (e.g., India ink and nigrosin) are used in **negative staining,** where the background but not the cell is stained; the unstained cells appear as bright objects against a dark background.

Dyes that bind cells by ionic interactions are probably the most commonly used dyes. These ionizable dyes may be divided into two general classes based on the nature of their charged group.

1. **Basic dyes**—methylene blue, basic fuchsin, crystal violet, safranin, malachite green—have positively charged groups (usually some form of pentavalent nitrogen) and are generally sold as chloride salts. Basic dyes bind to negatively charged molecules such as nucleic acids, many proteins, and the surfaces of bacterial and archaeal cells.
2. **Acidic dyes**—eosin, rose bengal, and acid fuchsin—possess groups such as carboxyls (—COOH) and phenolic hydroxyls (—OH). Acidic dyes, in their ionized form, have a negative charge and bind to positively charged cell structures.

The staining effectiveness of ionizable dyes may be altered by pH, since the nature and number of the charged moieties on cell components change with pH. Thus acidic dyes stain best under acidic conditions when proteins and many other molecules carry a positive charge; basic dyes are most effective at higher pH values.

Dyes that bind through covalent bonds or that have certain solubility characteristics are also useful. For instance, DNA can be stained by the Feulgen procedure in which the staining compound (Schiff's reagent) is covalently attached to its deoxyribose sugars. Sudan III (Sudan Black) selectively stains lipids because it is lipid soluble but does not dissolve in aqueous portions of the cell.

Microorganisms often can be stained very satisfactorily by **simple staining,** in which a single dye is used (**figure 2.17**). Simple staining's value lies in its simplicity and ease of use. The fixed smear is covered with stain for a short period, excess stain is washed off with water, and the slide is blotted dry. Basic dyes such

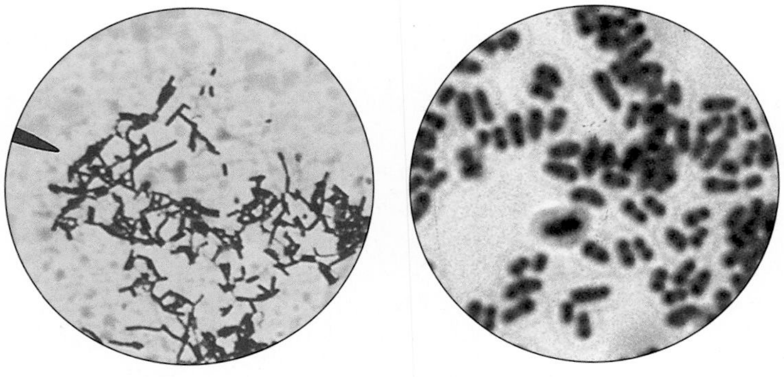

(a) Crystal violet stain of *Escherichia coli*

(b) Methylene blue stain of *Corynebacterium*

FIGURE 2.17 Simple Stains.

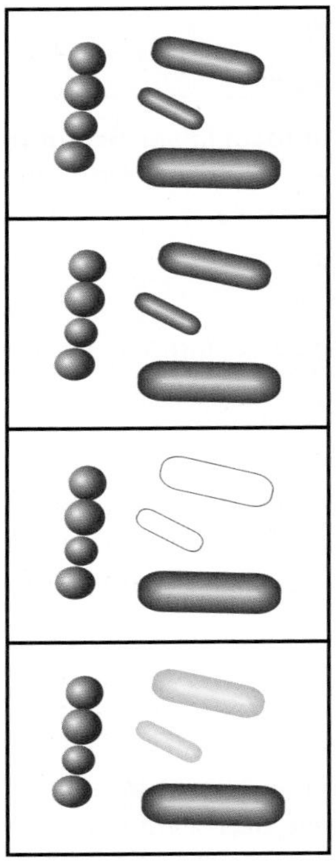

Steps in Staining
State of Bacteria
Step 1: Crystal violet (primary stain) for 1 minute. Water rinse.
Cells stain purple.
Step 2: Iodine (mordant) for 1 minute. Water rinse.
Cells remain purple.
Step 3: Alcohol (decolorizer) for 10–30 seconds. Water rinse.
Gram-positive cells remain purple. Gram-negative cells become colorless.
Step 4: Safranin (counterstain) for 30–60 seconds Water rinse. Blot dry.
Gram-positive cells remain purple. Gram-negative cells appear red.

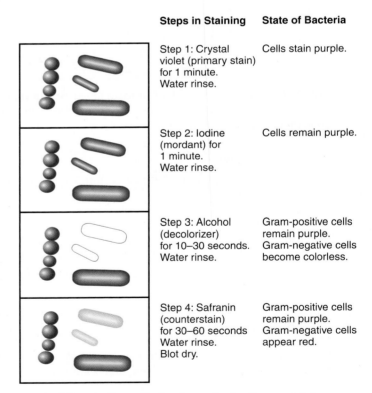

FIGURE 2.18 Gram Stain. Steps in the Gram-staining procedure.

Figure 2.18 Micro Inquiry

Why is the decolorization step considered the most critical in the Gram-staining procedure?

as crystal violet, methylene blue, and carbolfuchsin are frequently used in simple staining to determine the size, shape, and arrangement of bacterial and archaeal cells.

Differential Staining

The **Gram stain,** developed in 1884 by the Danish physician Christian Gram, is the most widely employed staining method in bacteriology. It is an example of **differential staining**—procedures that are used to distinguish organisms based on their staining properties. Use of the Gram stain divides most bacteria (but not archaea) into two groups—those that stain gram negative and those that stain gram positive.

The Gram-staining procedure is illustrated in **figure 2.18.** In the first step, the smear is stained with the basic dye crystal violet, the primary stain. This is followed by treatment with an iodine solution functioning as a **mordant,** a substance that helps bind the dye to a target molecule. Iodine increases the interaction between the cell and the dye so that the cell is stained more strongly. The smear is next decolorized by washing with ethanol or acetone. This step generates the differential aspect of the Gram stain; gram-positive bacteria retain the crystal violet, whereas gram-negative bacteria lose the crystal violet and become colorless. Finally, the smear is counterstained with a simple, basic dye different in color from crystal violet. Safranin, the most common counterstain, colors gram-negative bacteria pink to red and leaves gram-positive bacteria dark purple (**figure 2.19*a***). ▶▶| *Bacterial cell walls (section 3.3)*

Search This: Gram stain/microbiologybytes

Acid-fast staining is another important differential staining procedure. It is most commonly used to identify *Mycobacterium tuberculosis* and *M. leprae* (figure 2.19*b*), the pathogens responsible for tuberculosis and leprosy, respectively. These bacteria have cell walls containing lipids constructed from mycolic acids, a group of branched-chain hydroxy fatty acids, which prevent dyes from readily binding to the cells (*see figure 22.12*). However, *M. tuberculosis* and *M. leprae* can be stained by harsh procedures such as the Ziehl-Neelsen method, which uses heat and phenol to drive basic fuchsin into cells. Once basic fuchsin has penetrated, *M. tuberculosis* and *M. leprae* are not easily decolorized by acidified alcohol (acid-alcohol) and thus are said to be acid-fast. Non-acid-fast bacteria are decolorized by acid-alcohol and thus are stained blue by methylene blue counterstain.

Many differential staining procedures are based on detecting the presence of specific structures. **Endospore staining,** like acid-fast staining, requires heat to drive dye into a target, in this case an

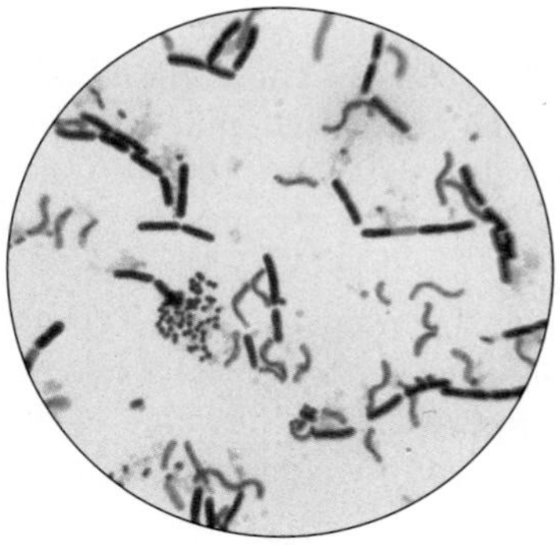

(a) Gram stain
Purple cells are gram positive.
Red cells are gram negative.

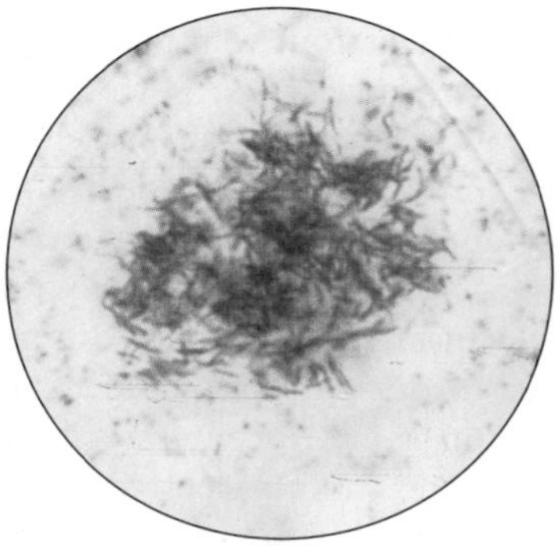

(b) Acid-fast stain
Red cells are acid-fast.
Blue cells are non-acid-fast.

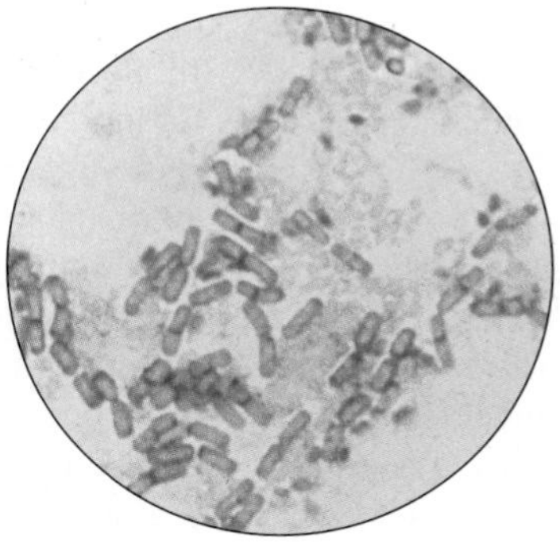

(c) Endospore stain of *Bacillus subtilis,* showing endospores (red) and vegetative cells (blue)

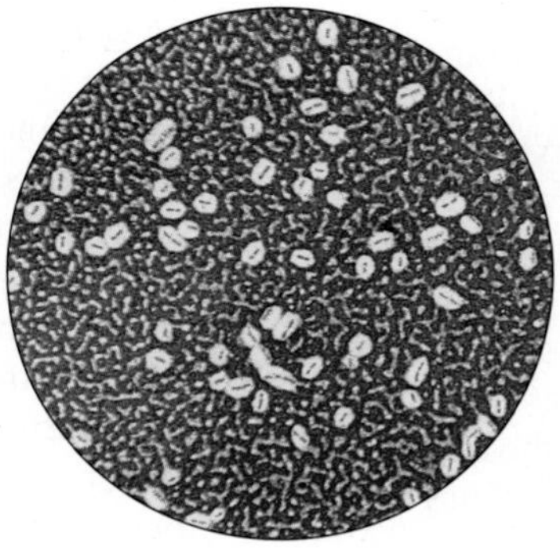

(d) Capsule stain of bacteria

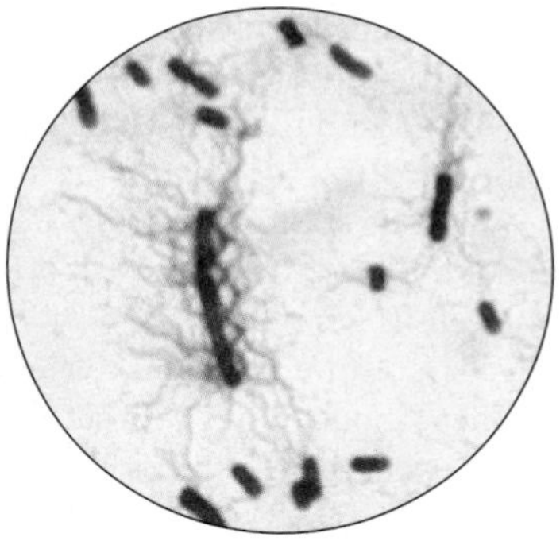

(e) Flagellar stain of *Proteus vulgaris.* A basic stain was used to coat the flagella.

FIGURE 2.19 Differential Stains.

endospore. An endospore is an exceptionally resistant structure produced by some bacterial genera (e.g., *Bacillus* and *Clostridium*). It can survive for long periods in an unfavorable environment, and it is called an endospore because it develops within the parent bacterial cell. Endospore morphology and location vary with species and often are valuable in identification; endospores may be spherical to elliptical and either smaller or larger than the diameter of the parent bacterium. Endospores are not stained well by most dyes, but once stained, they strongly resist decolorization. This property is the basis of most endospore staining methods (figure 2.19*c*). In the Schaeffer-Fulton procedure (not shown), endospores are first stained with malachite green in the presence of heat. After malachite green treatment, the rest of the cell is washed free of dye with water and is counterstained with safranin. This technique stains endospores bluish-green and the vegetative cells pink to red. ▶▶| *Bacterial endospores (section 3.8); Class* Clostridia *(section 21.2); and Class* Bacilli *(section 21.3)*

One of the simplest staining procedures is **capsule staining** (figure 2.19*d*), a technique that reveals the presence of capsules, a network usually made of polysaccharides that surrounds many bacteria and some fungi. Cells are mixed with India ink or nigrosin dye and spread out in a thin film on a slide. After air-drying, the cells appear as lighter bodies in the midst of a blue-black background because ink and dye particles cannot penetrate either the cell or its capsule. Thus capsule staining is an example of negative staining. The extent of the light region is determined by the size of the capsule and of the cell itself. There is little distortion of cell shape, and the cell can be counterstained for even greater visibility. ▶▶| *Capsules and slime layers (section 3.3)*

Flagella staining provides taxonomically valuable information about the presence and distribution pattern of flagella on prokaryotic cells (figure 2.19*e*; *see also figure 3.41*). Bacterial and archaeal flagella are fine, threadlike organelles of locomotion that are so slender (about 10 to 30 nm in diameter) they can only be seen directly using the electron microscope (although bundles of flagella can be visualized by dark-field microscopy). To observe bacterial flagella with the light microscope, their thickness is increased by coating them with mordants such as tannic acid and potassium alum, and then staining with pararosaniline (Leifson method) or basic fuchsin (Gray method). ▶▶| *Flagella (section 3.6)*

1. Describe the two general types of fixation. Which would you use when Gram staining a bacterium? Which would you use before observing the organelles of a protist?
2. Why would basic dyes be more effective under alkaline conditions?
3. Describe the Gram-staining procedure and explain what happens to a bacterial cell at each step. What step in the procedure could be omitted without losing the ability to distinguish between gram-positive and gram-negative bacteria? Why?
4. What other differential staining procedures can be useful in identifying a bacterium? What information do they provide?

2.4 Electron Microscopy

For centuries the light microscope has been the most important instrument for studying microorganisms. However, even the best light microscopes have a resolution limit of about 0.2 μm, which greatly compromises their usefulness for detailed studies of many microorganisms. Viruses, for example, are too small to be seen with light microscopes. Bacteria and archaea can be observed, but because they are usually only 1 to 2 μm in diameter, only their general shape and major morphological features are visible. The detailed internal structure of larger microorganisms also cannot be effectively studied by light microscopy. These limitations arise from the nature of visible light waves, not from any inadequacy of the light microscope itself. Electron microscopes have much greater resolution. Their use has transformed microbiology and added immeasurably to our knowledge. Here, we briefly review the nature of the electron microscope and the ways in which specimens are prepared for observation.

Transmission Electron Microscope

Recall that the resolution of a light microscope increases with a decrease in the wavelength of the light it uses for illumination. In electron microscopes, electrons replace light as the illuminating beam. The electron beam can be focused, much as light is in a light microscope, but its wavelength is about 100,000 times shorter than that of visible light. Therefore electron microscopes have a practical resolution roughly 1,000 times better than the light microscope; with many electron microscopes, points closer than 0.5 nm can be distinguished, and the useful magnification is well over 100,000× (**figure 2.20**). The value of the electron microscope is evident on comparison of the photographs in **figure 2.21**—microbial morphology can now be studied in great detail.

A modern **transmission electron microscope (TEM)** is complex, but the basic principles behind its operation can be readily understood (**figure 2.22**). A heated tungsten filament in the electron gun generates a beam of electrons that is focused on the specimen by the condenser (**figure 2.23**). Since electrons cannot pass through a glass lens, doughnut-shaped electromagnets called magnetic lenses are used to focus the beam. The column containing the lenses and specimen must be under high vacuum to obtain a clear image because electrons are deflected by collisions with air molecules. The specimen scatters some electrons, but those that pass through are used to form an enlarged image of the specimen on a fluorescent screen. A denser region in the specimen scatters more electrons and therefore appears darker in the image since fewer electrons strike that area of the screen; these regions are said to be "electron dense." In contrast, electron-transparent regions are brighter. The image can be captured digitally or on photographic film as a permanent record.

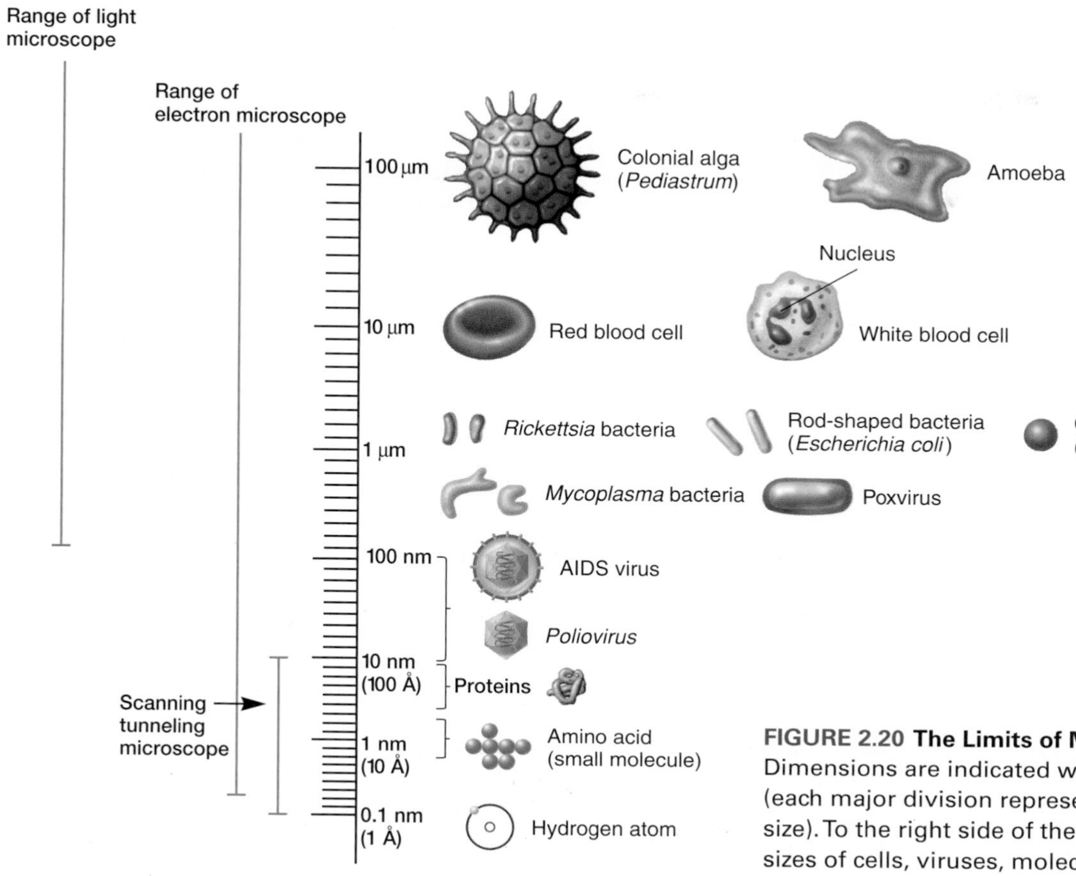

FIGURE 2.20 The Limits of Microscopic Resolution. Dimensions are indicated with a logarithmic scale (each major division represents a tenfold change in size). To the right side of the scale are the approximate sizes of cells, viruses, molecules, and atoms.

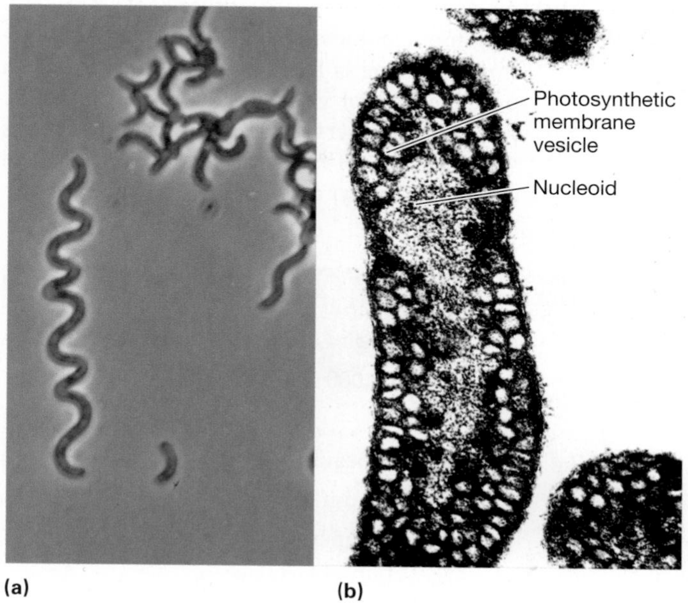

FIGURE 2.21 Light and Electron Microscopy. A comparison of light and electron microscopic resolution. (a) *Rhodospirillum rubrum* in phase-contrast light microscope (×600). (b) A thin section of *R. rubrum* in transmission electron microscope (×100,000).

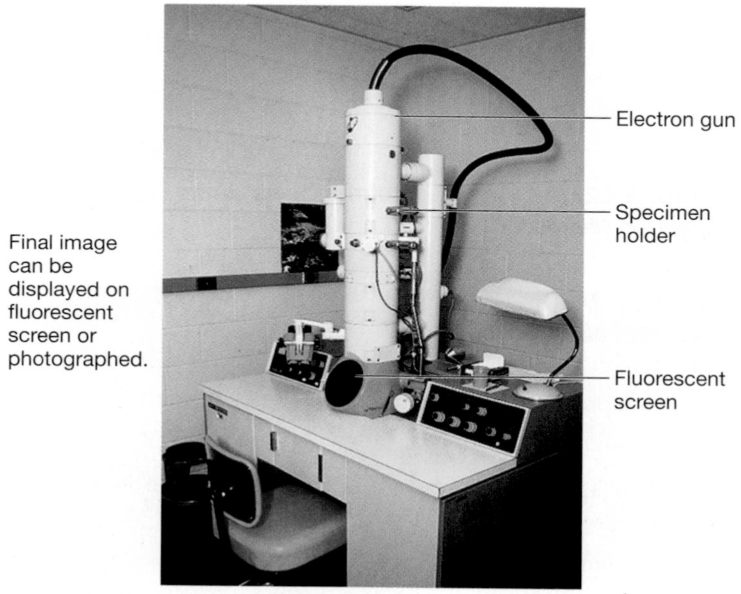

FIGURE 2.22 A Transmission Electron Microscope. The electron gun is at the top of the central column, and the magnetic lenses are within the column. The image on the fluorescent screen may be viewed through a magnifier positioned over the viewing window. The camera is in a compartment below the screen.

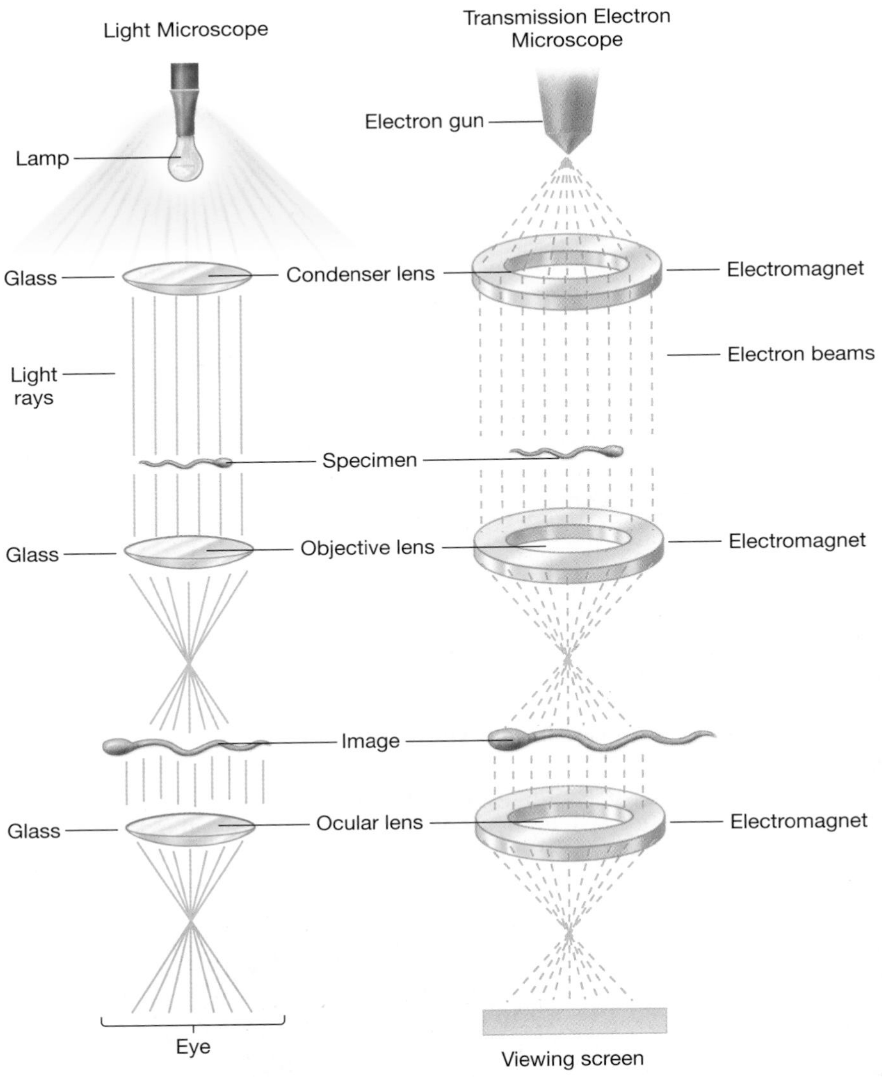

FIGURE 2.23 A Comparison of the Operation of Light and Transmission Electron Microscopes.

Table 2.4 compares some of the important features of light and transmission electron microscopes. The TEM has distinctive features that place harsh restrictions on the nature of samples that can be viewed and the means by which those samples must be prepared. As already mentioned, electrons are deflected by air molecules, and specimens must be viewed in a vacuum. The electron beams are also easily absorbed and scattered by solid matter. Therefore only extremely thin slices (20 to 100 nm) of a specimen can be viewed in the average TEM. Such a thin slice cannot be cut unless the specimen has support of some kind; the necessary support is provided by plastic. After fixation with chemicals such as glutaraldehyde and osmium tetroxide to stabilize cell structure, the specimen is dehydrated with organic solvents (e.g., acetone or ethanol). Complete dehydration is essential because most plastics used for embedding are not water soluble. Next the specimen is soaked in unpolymerized, liquid epoxy plastic until it is completely permeated, and then the plastic is hardened to form a solid block. Thin sections are cut from this block with a glass or diamond knife using a device called an ultramicrotome.

As with bright-field light microscopy, cells usually must be stained before they can be seen clearly with a TEM. The probability of electron scattering is determined by the density (atomic number) of atoms in the specimen. Biological molecules are composed primarily of atoms with low atomic numbers (H, C, N, and O), and electron scattering is fairly constant throughout an unstained cell or virus. Therefore specimens are prepared for observation by soaking thin sections with solutions of heavy metal salts such as lead citrate and uranyl acetate. The lead and uranium

Table 2.4 Characteristics of Light and Transmission Electron Microscopes

Feature	*Light Microscope*	*Transmission Electron Microscope*
Highest practical magnification	About 1,000–1,500	Over 100,000
Best resolution[a]	0.2 μm	0.5 nm
Radiation source	Visible light	Electron beam
Medium of travel	Air	High vacuum
Type of lens	Glass	Electromagnet
Source of contrast	Differential light absorption	Scattering of electrons
Focusing mechanism	Adjust lens position mechanically	Adjust current to the magnetic lens
Method of changing magnification	Switch the objective lens or eyepiece	Adjust current to the magnetic lens
Specimen mount	Glass slide	Metal grid (usually copper)

[a] The resolution limit of a human eye is about 0.2 mm.

ions bind to structures in the specimen and make them more electron opaque, thus increasing contrast in the material. Heavy osmium atoms from the osmium tetroxide fixative also stain specimens and increase their contrast. The stained thin sections are then mounted on tiny copper grids and viewed.

Two other important techniques for preparing specimens are negative staining and shadowing. In negative staining, the specimen is spread out in a thin film with either phosphotungstic acid or uranyl acetate. Just as in negative staining for light microscopy, heavy metals do not penetrate the specimen but render the background dark, whereas the specimen appears bright in photographs. Negative staining is an excellent way to study the structure of viruses, bacterial gas vacuoles, and other similar objects (**figure 2.24*a***). In shadowing, a specimen is coated with a thin film of platinum or other heavy metal by evaporation at an angle of about 45° from horizontal so that the metal strikes the microorganism on only one side. In one commonly used imaging method, the area coated with metal appears dark in photographs, whereas the uncoated side and the shadow region created by the object are light (figure 2.24*b*). This technique is particularly useful in studying virus morphology, bacterial and archaeal flagella, and DNA.

The shapes of organelles within cells can be observed by TEM if specimens are prepared by the freeze-etching procedure. When cells are rapidly frozen in liquid nitrogen, they become very brittle and can be broken along lines of greatest weakness, usually down the middle of internal membranes (**figure 2.25**). The exposed surfaces are then shadowed and coated with layers of platinum and carbon to form a replica of the surface. After the specimen has been removed chemically, this replica is studied in the TEM, providing a detailed, three-dimensional view of intracellular structure. An advantage of freeze-etching is that it minimizes the danger of artifacts because the cells are frozen quickly rather than being subjected to chemical fixation, dehydration, and embedding in plastic.

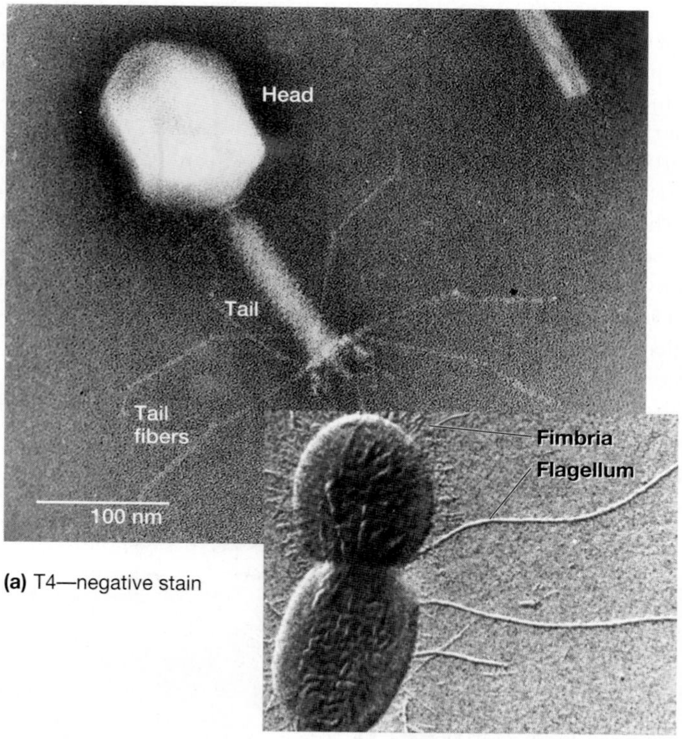

(a) T4—negative stain

(b) *P. mirabilis*—after shadowing

FIGURE 2.24 Staining Microorganisms for the TEM. (a) T4, a virus that infects *E. coli*. (b) *Proteus mirabilis* (×42,750); note flagella and fimbriae.

Figure 2.24 Micro Inquiry

Why are all electron micrographs black and white (although they are sometimes artificially colorized after printing)?

Scanning Electron Microscope

Transmission electron microscopes form an image from radiation that has passed through a specimen. The **scanning electron microscope (SEM)** works in a different manner. It produces an image from electrons released from atoms on an object's surface. The SEM has been used to examine the surfaces of microorganisms in great detail; many SEMs have a resolution of 7 nm or less.

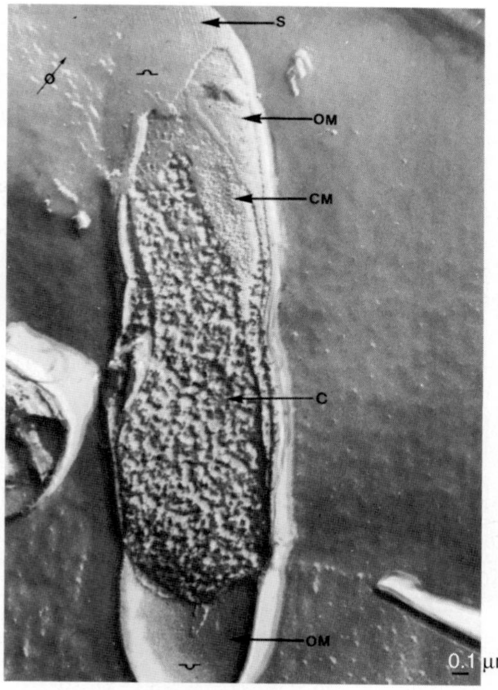

FIGURE 2.25 Example of Freeze-Etching. A freeze-etched preparation of the bacterium *Thiobacillus kabobis*. Note the differences in structure between the outer surface, S; the outer membrane of the cell wall, OM; the cytoplasmic membrane, CM; and the cytoplasm, C.

Specimen preparation for SEM is relatively easy, and in some cases, air-dried material can be examined directly. However, microorganisms usually must first be fixed, dehydrated, and dried to preserve surface structure and prevent collapse of the cells when they are exposed to the SEM's high vacuum. Before viewing, dried samples are mounted and coated with a thin layer of metal to prevent the buildup of an electrical charge on the surface and to give a better image.

To create an image, the SEM scans a narrow, tapered electron beam back and forth over the specimen (**figure 2.26**). When the beam strikes a particular area, surface atoms discharge a tiny shower of electrons called secondary electrons, and these are trapped by a detector. Secondary electrons entering the detector strike a scintillator, causing it to emit light flashes that a photomultiplier converts to an electrical current and amplifies. The signal is sent to a cathode-ray tube and produces an image like a television picture, which can be viewed or photographed.

The number of secondary electrons reaching the detector depends on the nature of the specimen's surface. When the electron beam strikes a raised area, a large number of secondary electrons enter the detector; in contrast, fewer electrons escape a depression in the surface and reach the detector. Thus raised areas appear lighter on the screen and depressions are darker. A realistic three-dimensional image of the microorganism's surface results (**figure 2.27**). The actual in situ location of microorganisms in ecological niches such as the human skin and the lining of the gut also can be examined.

Search This: Scanning electron microscope/MOS

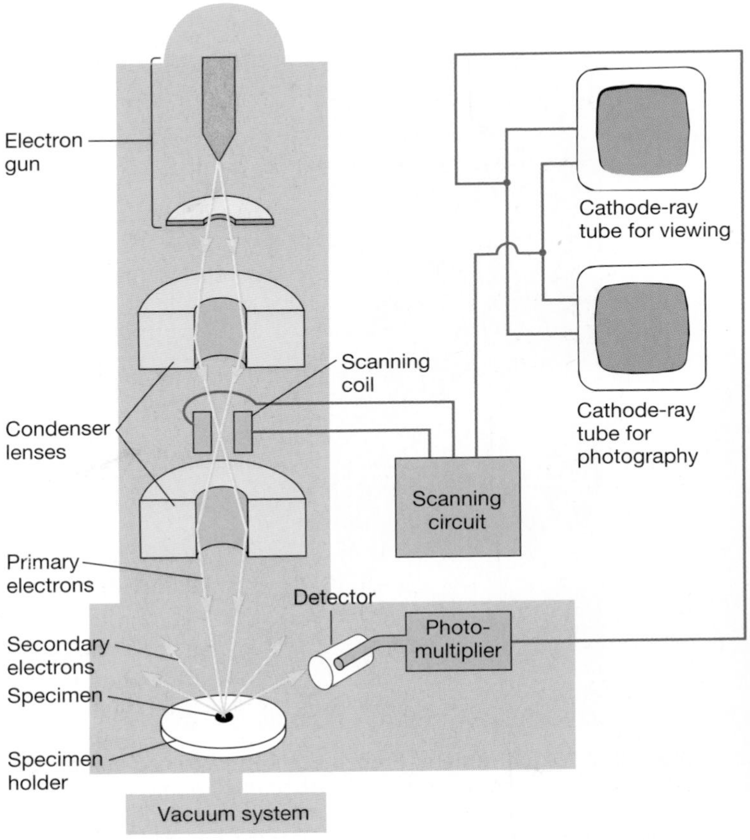

FIGURE 2.26 The Scanning Electron Microscope.

Electron Cryotomography

Beginning approximately 40 years ago, a series of advances in electron microscopy paved the way for the development of **electron cryotomography,** a technique that since the 1990s has been providing exciting insights into the structure and function of cells and viruses. Cryo- refers to the methods used to prepare and view specimens. Samples are rapidly plunged into an extremely cold liquid (e.g., ethane) and then kept frozen while being examined. The rapid freezing of the sample results in the formation of vitreous ice rather than ice crystals. Vitreous ice is a glasslike solid that preserves the native state of structures and immobilizes the specimen so that it can be viewed in the high vacuum of the electron microscope. Tomography refers to the method used to create images. Images of an object are recorded from many directions, referred to as a tilt series. These individual images are processed by computer programs and merged to form a three-dimensional reconstruction of the object. Three-dimensional views, slices, and other types of images can be derived from the reconstruction (**figure 2.28**). The ultrastructure of bacterial and archaeal cells has been the focus of several recent studies using electron cryotomography. Some of these studies revealed new cytoskeletal elements, such as those associated with magnetosomes, the inclusions used by some bacteria to orient themselves in magnetic fields (*see figure 3.37*); others have provided beautifully detailed images of flagellar motors. ▶▶I *Inclusions (section 3.5); Flagella (section 3.6)*

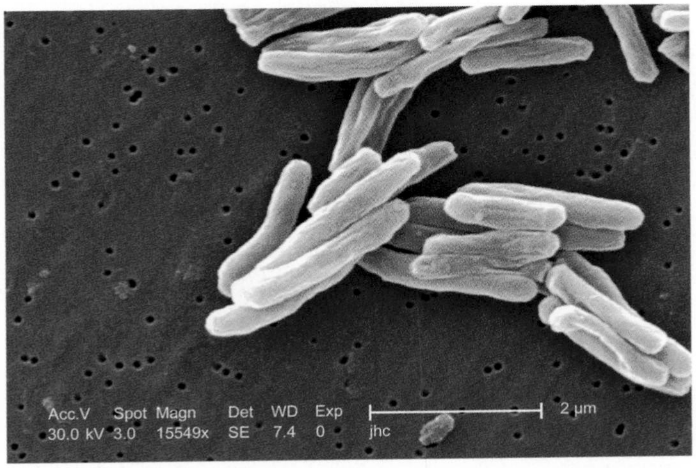

FIGURE 2.27 Scanning Electron Micrograph of *Mycobacterium tuberculosis*. Colorized image (×15,549).

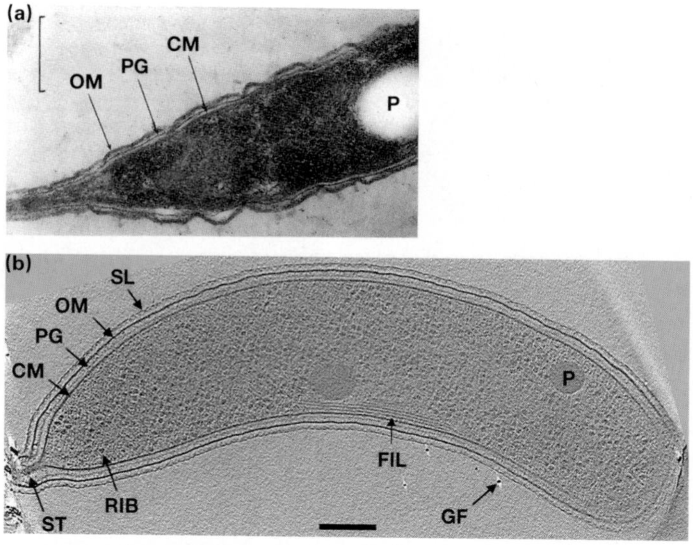

FIGURE 2.28 TEM and Electron Cryotomography. A comparison of (a) a thin section of a *Caulobacter crescentus* cell that has been prepared using conventional TEM procedures and (b) a central slice from a three-dimensional reconstruction of an intact *C. crescentus* cell. CM, cytoplasmic membrane; FIL, a bundle of filaments; OM, outer membrane; P, a presumed inclusion; PG, peptidoglycan; RIB, ribosome; SL, S-layer; ST, stalk.

1. Why does the transmission electron microscope have much greater resolution than the light microscope?
2. Describe in general terms how the TEM functions. Why must the TEM use high vacuum and very thin sections?
3. Material is often embedded in paraffin before sectioning for light microscopy. Why can't this approach be used when preparing a specimen for the TEM?
4. Under what circumstances would it be desirable to prepare specimens for the TEM by use of negative staining? Shadowing? Freeze-etching?
5. How does the scanning electron microscope operate, and in what way does its function differ from that of the TEM? The SEM is used to study which aspects of morphology?
6. How are freeze etching and electron cryotomography similar? How do they differ? What is the advantage of electron cryotomography?

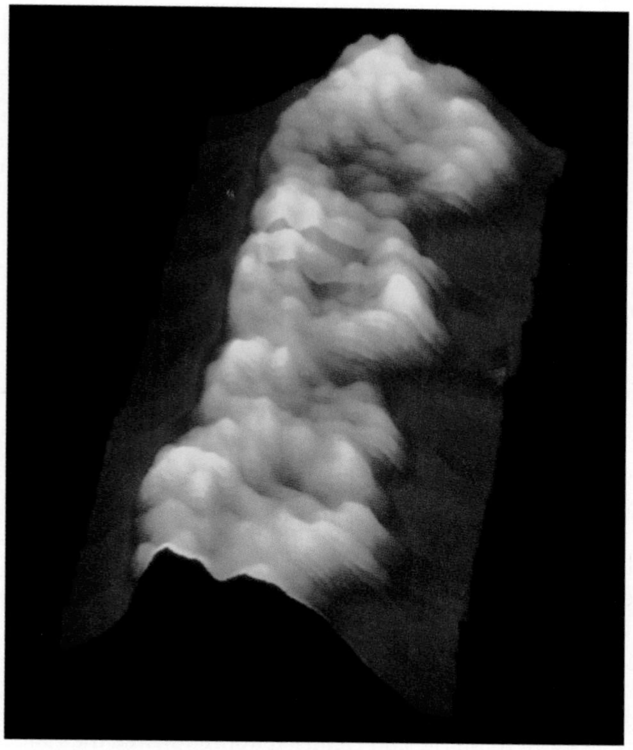

FIGURE 2.29 Scanning Tunneling Microscopy of DNA. The DNA double helix with approximately three turns shown (false color; ×2,000,000).

Figure 2.29 Micro Inquiry

Compare the magnification of this image with that shown in figures 2.24 (TEM) and 2.27 (SEM). Which has the highest magnification and which has the lowest?

2.5 Scanning Probe Microscopy

Although the first microscopes were developed about 400 years ago, they continue to be improved and new types of microscopy continue to be developed. Among the most powerful new microscopes are **scanning probe microscopes.** These microscopes measure surface features of an object by moving a sharp probe over the object's surface. The **scanning tunneling microscope** was invented in 1980. It can achieve magnifications of 100 million times, and it allows scientists to view atoms on the surface of a solid. The scanning tunneling microscope has a needlelike probe with a point so sharp that often there is only one atom at its tip. The probe is lowered toward the specimen surface until its electron cloud just touches that of the surface atoms. If a small voltage is applied between the tip and specimen, electrons flow through a narrow channel in the electron clouds. This tunneling current, as it is called, is extraordinarily sensitive to distance and will decrease about a thousandfold if the probe is moved away from the surface by a distance equivalent to the diameter of an atom.

The arrangement of atoms on the specimen surface is determined by moving the probe tip back and forth over the surface while keeping the probe at a constant height above the specimen. As the tip moves up and down while following the surface contours, its motion is recorded and analyzed by a computer to create an accurate three-dimensional image of the surface atoms. The surface map can be displayed on a computer screen or plotted on paper. The resolution is so great that individual atoms are observed easily. Also because the microscope can examine objects when they are immersed in water, it can be used to study biological molecules such as DNA (**figure 2.29**). The microscope's

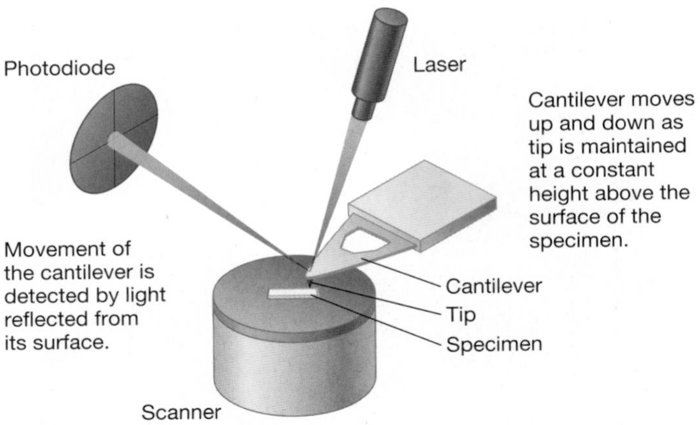

FIGURE 2.30 The Basic Elements of an Atomic Force Microscope. The tip used to probe the specimen is attached to a cantilever. As the probe passes over the "hills and valleys" of the specimen's surface, the cantilever is deflected vertically. A laser beam directed at the cantilever is used to monitor these vertical movements. Light reflected from the cantilever is detected by the photodiode and used to generate an image of the specimen.

inventors, Gerd Binnig and Heinrich Rohrer, shared the 1986 Nobel Prize in Physics for their work, together with Ernst Ruska, the designer of the first transmission electron microscope.

More recently, a second type of scanning probe microscope has been developed. The **atomic force microscope** moves a sharp probe over the specimen surface while keeping the distance between the probe tip and the surface constant. It does this by exerting a very small amount of force on the tip, just enough to maintain a constant distance but not enough force to damage the surface. The vertical motion of the tip usually is followed by measuring the deflection of a laser beam that strikes the lever holding the probe (**figure 2.30**). Unlike the scanning tunneling microscope, the atomic force microscope can be used to study surfaces that do not conduct electricity well. The atomic force microscope has been used to study the interactions of proteins, to follow the behavior of living bacteria and other cells, and to visualize membrane proteins such as aquaporins (**figure 2.31**).

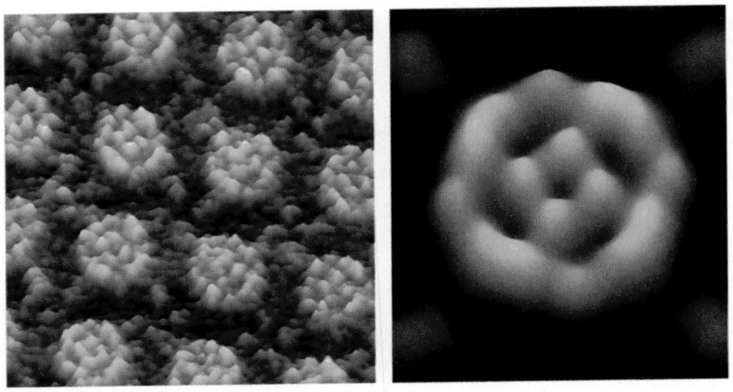

(a) (b)

FIGURE 2.31 The Membrane Protein Aquaporin Visualized by Atomic Force Microscopy. Aquaporin is a membrane-spanning protein that allows water to move across the membrane. (a) Each circular structure represents the surface view of a single aquaporin protein. (b) A single aquaporin molecule observed in more detail and at higher magnification.

Search This: Atomic force microscopy/nanoscience

1. Briefly describe the scanning probe microscope and compare and contrast its most popular versions—the scanning tunneling microscope and the atomic force microscope. What are these microscopes used for?
2. Scanning probe microscopy can be applied to samples that have not been dehydrated, while electron microscopy cannot. Why do you think this is an important advantage?

Summary

2.1 Lenses and the Bending of Light

a. A light ray moving from air to glass or vice versa is bent in a process known as refraction.
b. Lenses focus light rays at a focal point and magnify images (**figure 2.2**).

2.2 Light Microscopes

a. In a compound microscope such as the bright-field microscope, the primary image is an enlarged image formed by the objective lens. The primary image is further enlarged by the ocular lens to yield the final image. A substage condenser focuses a cone of light on the specimen (**figure 2.3**).
b. Microscope resolution increases as the wavelength of radiation used to illuminate the specimen decreases. The maximum resolution of a light microscope is about 0.2 μm.
c. The dark-field microscope uses only refracted light to form an image, and objects glow against a black background (**figure 2.6**).
d. The phase-contrast microscope converts variations in the refractive index and density of cells into changes in light intensity and thus makes colorless, unstained cells visible (**figure 2.9**).
e. The differential interference contrast microscope uses two beams of light to create high-contrast, three-dimensional images of live specimens.

f. The fluorescence microscope illuminates a fluorochrome-labeled specimen and forms an image from its fluorescence (**figure 2.12**).

g. The confocal scanning laser microscope is used to study thick, complex specimens. It creates an image by using only the light emanating from the plane of focus, while blocking out light from above and below the plane of focus (**figures 2.15** and **2.16**).

2.3 Preparation and Staining of Specimens

a. Specimens usually must be fixed and stained before viewing them in the bright-field microscope.

b. Most dyes are either positively charged basic dyes or negatively charged acidic dyes that bind to ionized parts of cells.

c. In simple staining, a single dye is used to stain microorganisms (**figure 2.17**).

d. Differential staining procedures such as the Gram stain and acid-fast stain distinguish between microbial groups by staining them differently (**figures 2.18** and **2.19*a,b***). Other differential staining techniques are specific for particular structures such as endospores, bacterial capsules, and flagella (**figure 2.19*c–e***).

2.4 Electron Microscopy

a. The transmission electron microscope (TEM) uses magnetic lenses to form an image from electrons that have passed through a very thin section of a specimen (**figure 2.23**). Resolution is high because the wavelength of a beam of electrons is very short.

b. Specimens for TEM are usually prepared by methods that increase contrast. Specimens can be stained by treatment with solutions of heavy metals such as osmium tetroxide, uranium, and lead. They can also be prepared for TEM by negative staining, shadowing with metal, or freeze-etching (**figures 2.24** and **2.25**).

c. The scanning electron microscope is used to study external surface features of microorganisms (**figure 2.26**).

d. Electron cryotomography is a relatively new technique that involves freezing specimens rapidly, keeping them frozen while being examined, and creating images from a series of directions that are combined and processed to form a three-dimensional reconstruction of the object (**figure 2.28**).

2.5 Scanning Probe Microscopy

a. Scanning probe microscopes reach very high magnifications that allow scientists to observe biological molecules (**figures 2.29** and **2.31**).

b. Scanning tunneling microscopy enables the visualization of molecular surfaces using electron interaction between the probe and the specimen, whereas atomic force microscopy can scan the surface of molecules that do not conduct electricity well (**figure 2.30**).

Critical Thinking Questions

1. If you prepared a sample of a specimen for light microscopy, stained it with the Gram stain, and failed to see anything when you looked through your light microscope, list the things that you may have done incorrectly.

2. In a journal article, find an example of a light micrograph, a scanning or transmission electron micrograph, or a confocal image. Discuss why the figure was included in the article and why that particular type of microscopy was the method of choice for the research. What other figures would you like to see used in this study? Outline the steps that the investigators would take to obtain such photographs or figures.

3. Electron cryomicroscopy (cryo-EM) employs a process called single particle reconstruction to generate images of macromolecular assemblies and their components (e.g., flagella, ribosomes, and protein assemblages). A major challenge in the application of cryo-EM is the level of background "noise" relative to the actual signal emitted by the structure in question. In 2008, a method of image processing was reported that reduces this signal-to-noise ratio problem such that the clarity of a viral protein was comparable to a 3.9 angstrom resolution map obtained from X-ray crystallography. This resolution is sufficient so that most amino acid side chains produced a detectable signal. How do you think the development of a cryomicroscopy technique capable of such high structural resolution will impact biology? Suggest two specific areas in microbiology where this could be especially helpful.

   Read the original paper: Zhang, X., et al. 2008. Near-atomic resolution using electron cryomicroscopy and single-particle reconstruction. *Proc. Nat. Acad. Sci., USA*. 105:1867.

Learn More

Learn more by visiting the text website at www.mhhe.com/willey8, where you will find a complete list of references.

3

Bacteria and *Archaea*

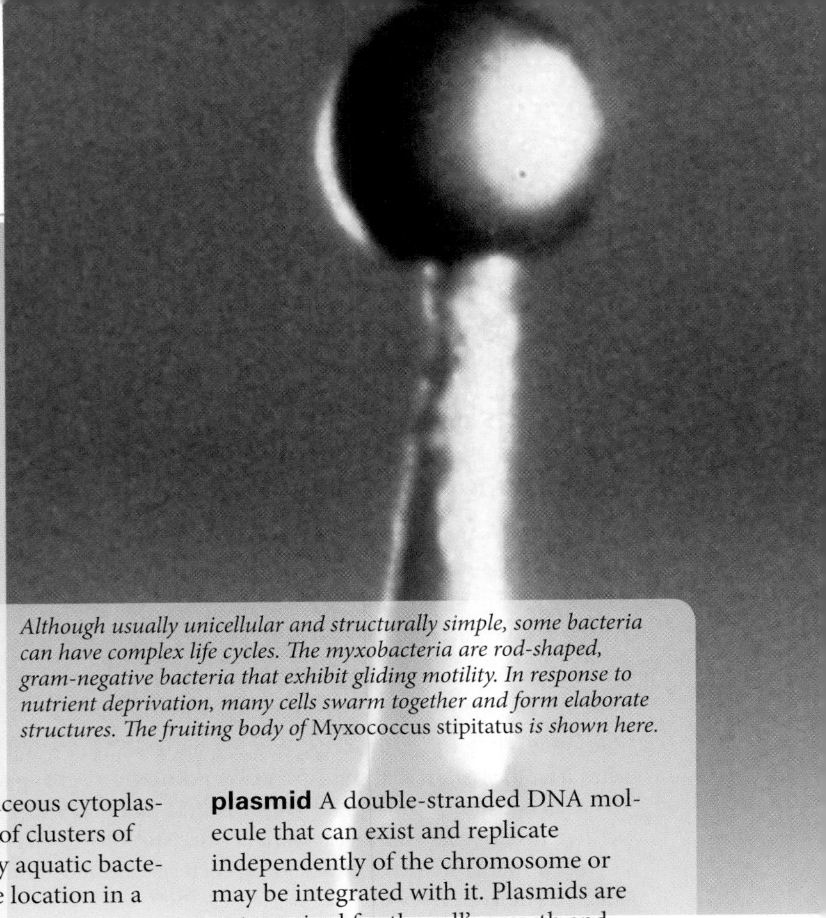

Although usually unicellular and structurally simple, some bacteria can have complex life cycles. The myxobacteria are rod-shaped, gram-negative bacteria that exhibit gliding motility. In response to nutrient deprivation, many cells swarm together and form elaborate structures. The fruiting body of Myxococcus stipitatus *is shown here.*

CHAPTER GLOSSARY

bacillus A rod-shaped bacterial or archaeal cell.

capsule A layer of well-organized material, not easily washed off, lying outside the cell wall of some microbes.

cell envelope The plasma membrane and all surrounding layers external to it.

chemotaxis The movement of a microorganism toward chemical attractants and away from chemical repellents.

coccus A spherical bacterial or archaeal cell.

endospore An extremely heat- and chemical-resistant, dormant, thick-walled spore that develops within some gram-positive bacteria.

fimbriae Fine, hairlike protein appendages on some bacterial and archaeal cells; some help attach cells to surfaces, and others are involved in a type of twitching motility; also called pili.

fluid mosaic model The currently accepted model of cell membranes in which the membrane is a lipid bilayer with integral proteins buried in the lipid and peripheral proteins more loosely attached to the membrane surface.

gas vacuole A proteinaceous cytoplasmic organelle composed of clusters of gas-filled vesicles; used by aquatic bacteria and archaea to change location in a water column.

glycocalyx A network of polysaccharides extending from the surface of some cells.

inclusions Granules of organic or inorganic material in the cytoplasm of many bacteria and archaea.

lipopolysaccharides (LPSs) Molecules in the outer membrane of the cell wall of gram-negative bacteria; they contain both lipid and polysaccharide.

nucleoid An irregularly shaped region where the genetic material of bacteria and archaea is found.

peptidoglycan A component of bacterial cell walls. It is composed of long chains of alternating *N*-acetylglucosamine and *N*-acetylmuramic acid residues, with the chains linked to each other by short chains of amino acids.

periplasmic space The space between the plasma membrane and outermost layers of the bacterial cell wall.

plasmid A double-stranded DNA molecule that can exist and replicate independently of the chromosome or may be integrated with it. Plasmids are not required for the cell's growth and reproduction.

porin proteins Proteins that form channels for the transport of small molecules across the outer membrane of gram-negative bacterial cell walls.

sex pili Thin protein appendages required for bacterial conjugation; cells with sex pili donate DNA to recipient cells.

S-layer A regularly structured layer composed of protein or glycoprotein that lies on the surface of many bacteria and archaea.

slime layer A layer of diffuse, unorganized, easily removed material lying outside the cell wall.

spirillum A rigid, spiral-shaped bacterial or archaeal cell.

spirochete A flexible, spiral-shaped bacterium with periplasmic flagella.

Most bacteria and archaea are smaller than eukaryotic organisms and therefore may seem insignificant. But these organisms, with their simple cell structure, must do all the things eukaryotes do: obtain energy from their environment and convert that energy into a useful form so that they can do work such as movement and biosynthesis. To accomplish this, they must interact with their environment and respond to the stimuli they encounter there. As we discuss in chapters 9 through 14, many bacteria and archaea carry out metabolic and related processes that are similar to those in eukaryotes. Because of this, several species of bacteria have served as model organisms for understanding complex processes that would be difficult, if not impossible, to study in eukaryotes. Furthermore, members of *Bacteria* and *Archaea* have metabolic capabilities not seen in the eukaryotic world. Their unique biochemical characteristics are critical to the cycling of nutrients, and the *Bacteria* and *Archaea* play major roles in maintaining the health of our planet.

To understand these amazing microbes, we must first examine their cell structure and begin to relate it to the functions they carry out. At a superficial level, most bacteria and archaea share a common cellular architecture and therefore are referred to as prokaryotes, with the implication that they are evolutionarily closely related. This view is being challenged. Thus we begin the chapter with a brief discussion of the "prokaryote controversy," followed by a brief consideration of the structural features common to both domains of life. We then consider specific cell structures, pointing out the major differences between the two groups. The *Bacteria* and *Archaea* also differ in terms of their physiology. These differences are discussed in other chapters throughout the book and are considered more completely in chapter 18. ◀◀ *Members of the microbial world (section 1.1)*

As we consider the *Bacteria* and *Archaea*, it is important to keep in mind that probably only about 1% of bacterial and archaeal species have been cultured. Of the cultivated species, only a few have been studied in great detail. From this small sample, many generalizations about the two domains are made, and it is presumed that others in a domain should be like the well-studied model organisms. However, part of the wonder and fun of science is that nature is full of surprises. As the genomes of more and more microbes are analyzed, our understanding of these two domains may change in interesting and exciting ways. ▶▶ *Microbial genomics (chapter 16)*

3.1 Prokaryotes

Bacteria and *Archaea* have long been lumped together and referred to as prokaryotes. Although the term was first introduced early in the twentieth century, the concept of a prokaryote was not fully outlined until 1962, when R. Stanier and C. B. vanNiel described prokaryotes in terms of what they lacked in comparison to eukaryotic cells. For instance, Stanier and vanNiel pointed out that prokaryotes lack a membrane-bound nucleus, a cytoskeleton, membrane-bound organelles, and internal membranous structures such as the endoplasmic reticulum and Golgi apparatus.

Since the 1960s, biochemical, genetic, and genomic analyses, coupled with improved methods for imaging bacterial and archaeal cells, have shown that the *Bacteria* and *Archaea* are distinct taxa. Because of this and other discoveries, some microbiologists question our traditional ways of thinking about prokaryotes. For instance, some members of the bacterial phylum *Planctomycetes* have their genetic material enclosed in a membrane; other members of this interesting taxon have a membrane-bound organelle called the anammoxosome. This organelle is the site of anoxic ammonia oxidation, an unusual metabolic process that is important in the nitrogen cycle. Other discoveries include the finding that cytoskeletal elements are widespread in both domains and that some bacteria have extensive intracytoplasmic membranes. Because of these discoveries, some microbiologists think the term prokaryote should be abandoned. ▶▶ *Phylum* Planctomycetes *(section 19.4); Nitrogen cycle (section 26.1)*

The current controversy creates a dilemma for both the textbook author and the student of microbiology—to use or not to use the term prokaryote. However, it also illustrates that microbiology is an exciting, dynamic, and rapidly changing field of study. All of us are waiting to see the outcome of this debate. Until then, we will try to limit our use of the term prokaryote and to be as explicit as possible about which characteristics are associated with *Bacteria*, which with *Archaea*, and which with both taxa.

3.2 Common Features of Bacterial and Archaeal Cell Structure

Much of this chapter is devoted to a discussion of individual cell components. Therefore a preliminary overview of the features common to both the *Bacteria* and *Archaea* is in order. We begin by considering overall cell morphology and then move to the cell structures seen in members of both domains. Keep in mind that even though a structure may be found in both taxa, at the molecular level these structures may be quite distinct.

Shape, Arrangement, and Size

It might be expected that small, relatively simple organisms such as the *Bacteria* and *Archaea* would be uniform in shape and size. This is not the case, as the microbial world offers almost endless variety in terms of morphology. However, the two most common shapes in both domains are cocci and rods (**figure 3.1**). **Cocci** (s., **coccus**) are

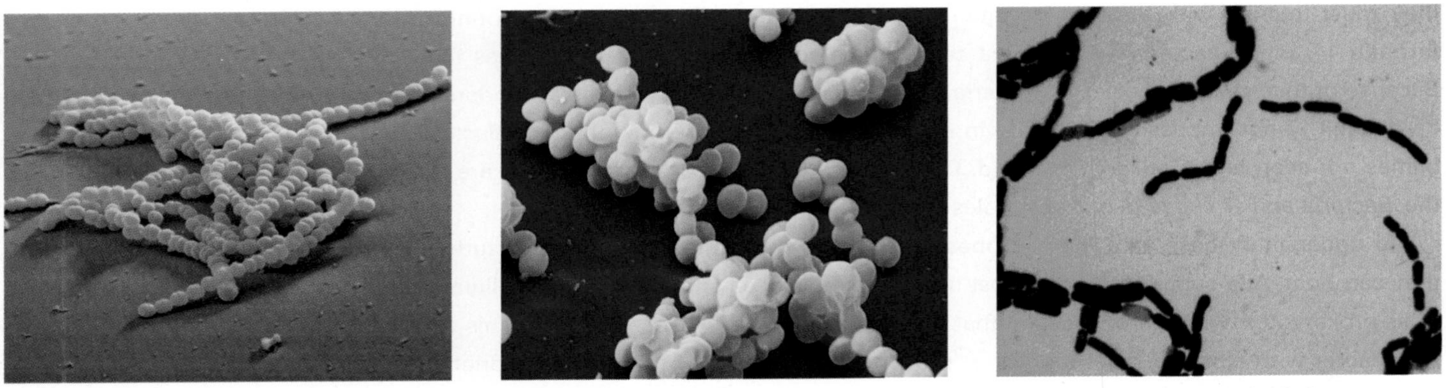

(a) *S. agalactiae*—cocci in chains **(b)** *S. aureus*—cocci in clusters **(c)** *B. megaterium*—rods in chains

FIGURE 3.1 Cocci and Rods. (a) *Streptococcus agalactiae,* the cause of Group B streptococcal infections; color-enhanced scanning electron micrograph (×4,800). (b) *Staphylococcus aureus*; color-enhanced scanning electron micrograph; average cell diameter is about 1 μm. (c) *Bacillus megaterium,* Gram stain (×600).

roughly spherical cells. They can exist singly or can be associated in characteristic arrangements that are frequently useful in their identification. **Diplococci** (s., **diplococcus**) arise when cocci divide and remain together to form pairs. Long chains of cocci result when cells adhere after repeated divisions in one plane; this pattern is seen in the bacterial genera *Streptococcus, Enterococcus,* and *Lactococcus* (figure 3.1*a*). Members of the bacterial genus *Staphylococcus* divide in random planes to generate irregular, grapelike clumps (figure 3.1*b*). Divisions in two or three planes can produce symmetrical clusters of cocci. Those in the bacterial genus *Micrococcus* often divide in two planes to form square groups of four cells called tetrads. In the bacterial genus *Sarcina,* cocci divide in three planes, producing cubical packets of eight cells.

Bacillus megaterium is an example of a bacterium with a **rod** shape (figure 3.1*c*). Rods, sometimes called **bacilli** (s., **bacillus**), differ considerably in their length-to-width ratio, the coccobacilli being so short and wide that they resemble cocci. The shape of the rod's end often varies between species and may be flat, rounded, cigar-shaped, or bifurcated. Although many rods occur singly, some remain together after division to form pairs or chains (e.g., *Bacillus megaterium* is found in long chains).

Other bacterial cell shapes and arrangements are also observed. **Vibrios** most closely resemble rods, as they are comma-shaped (**figure 3.2*a***). **Spirilla** are rigid, spiral-shaped cells that usually have tufts of flagella at one or both ends (figure 3.2*b*). **Spirochetes** are flexible, spiral-shaped bacteria that have a unique, internal flagellar arrangement (figure 3.2*c*). These bacteria are distinctive in other ways, and all belong to a single phylum, *Spirochaetes.* Actinomycetes typically form long filaments called hyphae. The hyphae may branch to produce a network called a **mycelium** (figure 3.2*d*), and in this sense, they are similar to filamentous fungi, a group of eukaryotic microbes. The oval- to pear-shaped members of the genus *Hyphomicrobium* (figure 3.2*e*) produce a bud at the end of a long hypha. The myxobacteria are of particular note. These bacteria sometimes aggregate to form complex structures called fruiting bodies (**chapter opener figure**). Finally, some bacteria are **pleomorphic,** being variable in shape and lacking a single, characteristic form. ▶▶| *Phylum* Spirochaetes *(section 19.6);* Caulobacteraceae *(section 20.1);* Hyphomicrobiaceae *(section 20.1); Order* Mxyococcales *(section 20.4)*

The additional forms observed in *Archaea* include curved rods and spiral shapes. Some archaea are pleomorphic (many shaped) and others exhibit unique shapes such as the branched form of *Thermoproteus tenax* (figure 3.2*f*). A few archaea actually are flat. For example, *Haloquadratum walsbyi*, an archaeon that lives in salt ponds, is shaped like a square-to-rectangular box about 2 μm by 2 to 4 μm and only 0.25 μm thick (figure 3.2*g*).

Bacterial and archaeal cells vary in size as much as in shape (**figure 3.3**). *Escherichia coli* is a rod-shaped bacterium of about average size, 1.1 to 1.5 μm wide by 2.0 to 6.0 μm long. Near the small end of the size continuum are members of the bacterial genus *Mycoplasma* (0.3 μm in diameter), an interesting group of bacteria that lack cell walls, and the parasitic archaeon *Nanoarchaeum equitans* (0.4 μm in diameter). At the other end of the continuum are bacteria such as the spirochetes, which can reach 500 μm in length, and the photosynthetic bacterium *Oscillatoria,* which is about 7 μm in diameter (the same diameter as a red blood cell). The huge bacterium *Epulopiscium fishelsoni* lives in the intestine of the brown surgeonfish (*Acanthurus nigrofuscus*). *E. fishelsoni* grows as large as 600 by 80 μm, a little smaller than a printed hyphen and clearly larger than the well-known eukaryote *Paramecium* (**figure 3.4**). An even larger bacterium, *Thiomargarita namibiensis,* has been discovered in ocean sediment. Thus a few bacteria are much larger than the average eukaryotic cell (typical plant and animal cells are around 10 to 50 μm in diameter).

The variety of sizes and shapes observed in *Bacteria* and *Archaea* raises a fundamental question. What causes a species to have a particular size and shape? Although far from being answered, recent discoveries have fueled a renewed interest in this question, and it is clear that size and shape determination are related and have selective value. For many years it was thought that microbes were small (and indeed, had to be small) because

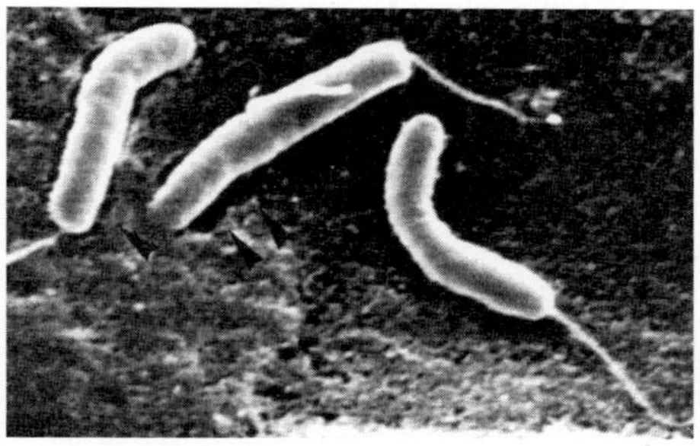

(a) *V. cholerae*—comma-shaped vibrios

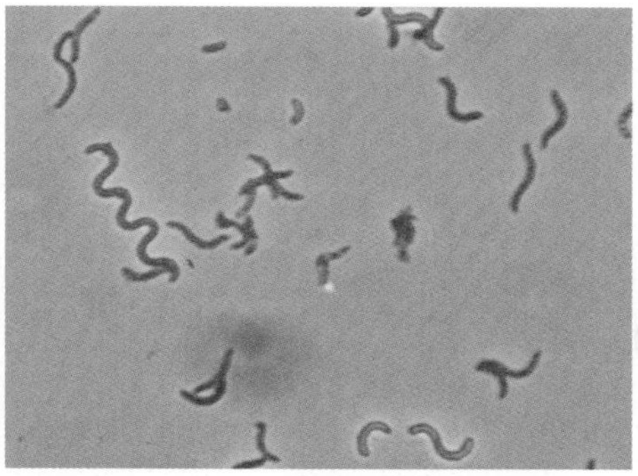

(b) *R. rubrum*—spiral-shaped spirilla

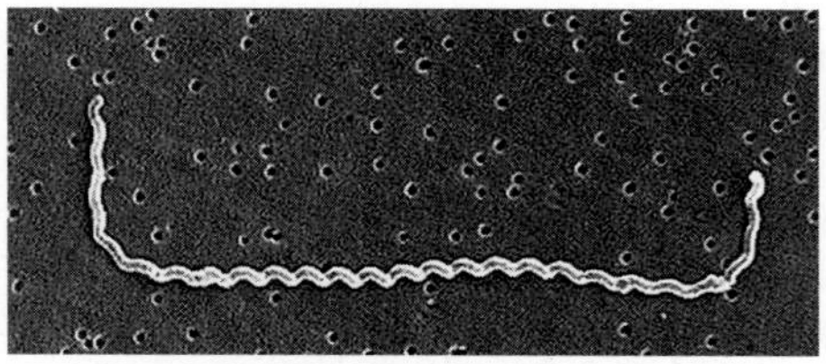

(c) *Leptospira interrogans*—a spirochete

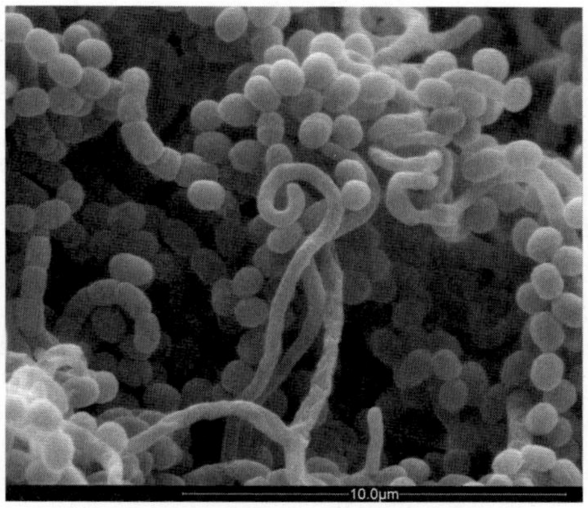

(d) *Streptomyces*—a filamentous bacterium

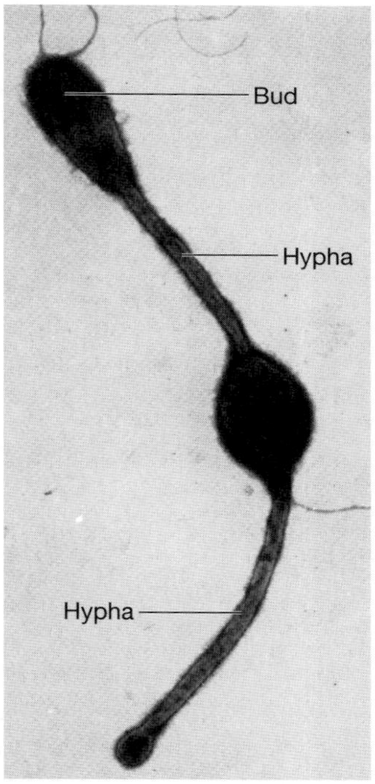

(e) *Hyphomicrobium*

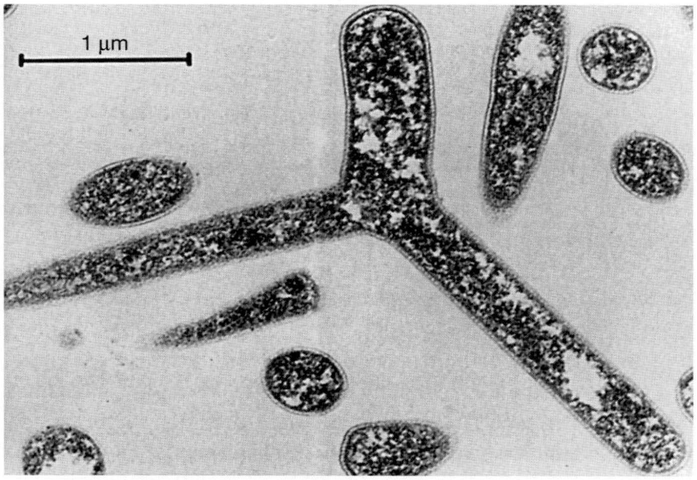

(f) *Thermoproteus tenax*—a branched archaeal cell

(g) *Haloquadratum walsbyi*—a square archaeon

FIGURE 3.2 Other Cell Shapes. (a) *Vibrio cholerae,* note polar flagella; scanning electron micrograph (SEM). (b) *Rhodospirillum rubrum,* phase contrast (×500). (c) *Leptospira interrogans,* the spirochete that causes the waterborne disease leptospirosis. (d) *Streptomyces,* SEM. (e) *Hyphomicrobium,* electron micrograph with negative staining. (f) *Thermoproteus tenax,* electron micrograph. (g) *Haloquadratum walsbyi,* SEM (×12,000).

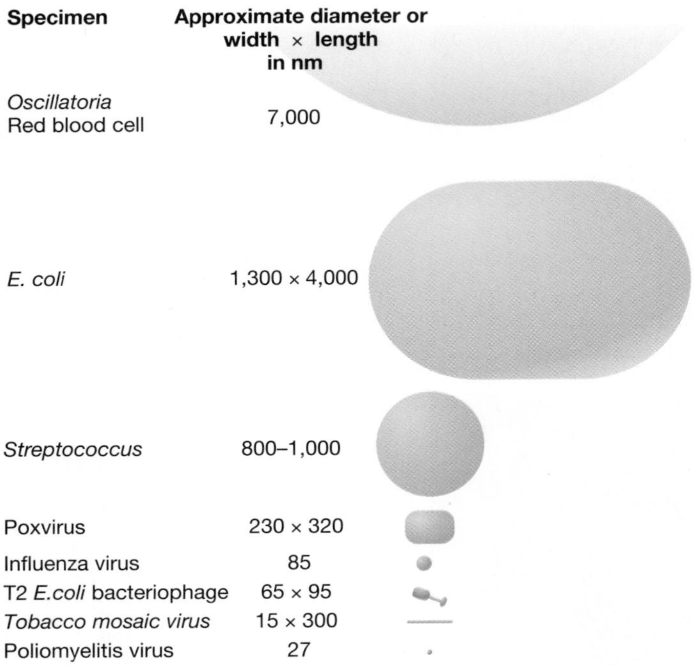

Specimen
Approximate diameter or width × length in nm
Oscillatoria
Red blood cell
7,000
E. coli
1,300 × 4,000
Streptococcus
800–1,000
Poxvirus
230 × 320
Influenza virus
85
T2 E.coli bacteriophage
65 × 95
Tobacco mosaic virus
15 × 300
Poliomyelitis virus
27

FIGURE 3.3 Sizes of Bacteria Relative to a Red Blood Cell and Viruses. The size range of archaeal cells is similar to that of *Bacteria*. Recall that 1,000 nm = 1 μm. Thus *E. coli* is 1.3 × 4 μm.

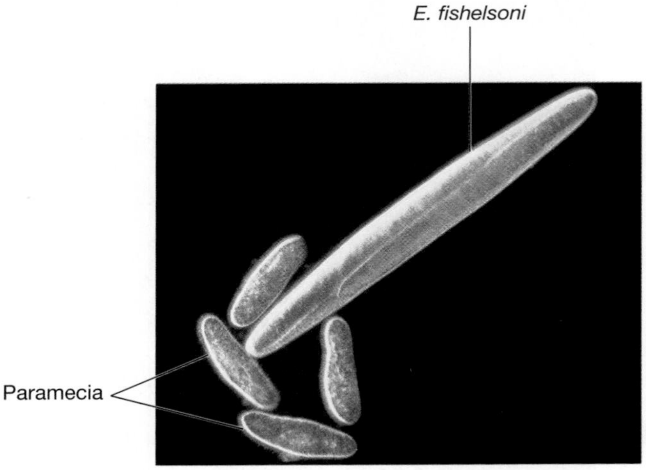

FIGURE 3.4 A Giant Bacterium. This photograph, taken with pseudo dark-field illumination, shows *Epulopiscium fishelsoni* at the top of the figure dwarfing the paramecia at the bottom (×200).

being small increases the surface area-to-volume ratio (S/V ratio; **figure 3.5**). As this ratio increases, nutrient uptake and diffusion of molecules within the cell become more efficient, which in turn facilitates a rapid growth rate. Shape affects the S/V ratio. A rod with the same volume as a coccus has a higher S/V ratio than does the coccus. This means that a rod can have

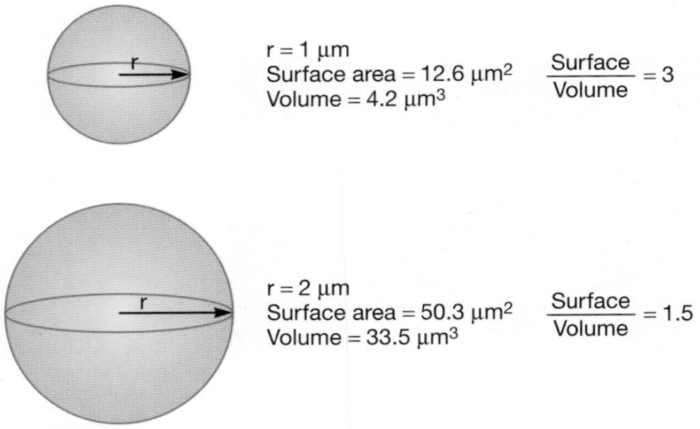

FIGURE 3.5 The Surface-to-Volume Ratio. Surface area is calculated by the formula $4\pi r^2$. Volume is calculated by the formula $4/3\pi r^3$. Shape also affects the S/V ratio; rods with the same diameter as a coccus have a greater S/V ratio.

greater nutrient flux across its plasma membrane. Furthermore, the discovery of *E. fishelsoni* clearly demonstrates that bacteria can be very large. For bacteria to be large, they must have other characteristics that maximize their S/V ratio, or their size must be of some selective value. For instance, *E. fishelsoni* has a highly convoluted plasma membrane, which increases its S/V ratio. In addition, large cells are less likely to be eaten by predatory protists. Cells that are filamentous, have prostheca (e.g., the hypha and bud of *Hyphomicrobium*), or are oddly shaped also are less susceptible to predation. Finally, the renewed interest in the molecular mechanisms by which shape is determined has led to the discovery of numerous cytoskeletal elements. These are discussed in more detail in section 3.5. Prokaryotic Cell Shapes

Cell Organization

Bacteria and *Archaea* share a common cell organization. Structures observed in members of both domains are summarized and illustrated in **table 3.1** and **figure 3.6**. Note that no single organism possesses all of these structures at all times. Some are found only in certain cells in certain conditions or in certain phases of the life cycle. It is also important to remember that bacterial and archaeal versions of the same structure can differ dramatically at the molecular level.

Bacterial and archaeal cells are surrounded by several layers, which are collectively called the cell envelope. The innermost layer of the cell envelope is the plasma membrane, which surrounds the cytoplasm and its contents. Most bacteria and archaea have a chemically complex cell wall, which covers the plasma membrane. Many bacteria are surrounded by a capsule or slime layer external to the cell wall. Capsules are not widespread among those archaea examined thus far. Because most bacteria and archaea do not contain internal, membrane-bound organelles, their interior appears morphologically simple. The genetic material is localized in

Table 3.1 Common Bacterial and Archaeal Structures and Their Functions

Plasma membrane	Selectively permeable barrier, mechanical boundary of cell, nutrient and waste transport, location of many metabolic processes (respiration, photosynthesis), detection of environmental cues for chemotaxis
Gas vacuole	Buoyancy for floating in aquatic environments
Ribosomes	Protein synthesis
Inclusions	Storage of carbon, phosphate, and other substances
Nucleoid	Localization of genetic material (DNA)
Periplasmic space	In gram-negative bacteria, contains hydrolytic enzymes and binding proteins for nutrient processing and uptake; in gram-positive bacteria and archaeal cells, may be smaller or absent
Cell wall	Provides shape and protection from osmotic stress
Capsules and slime layers	Resistance to phagocytosis, adherence to surfaces; rare in the *Archaea*
Fimbriae and pili	Attachment to surfaces, bacterial conjugation and transformation, twitching and gliding motility
Flagella	Swimming motility
Endospore	Survival under harsh environmental conditions; only observed in *Bacteria*

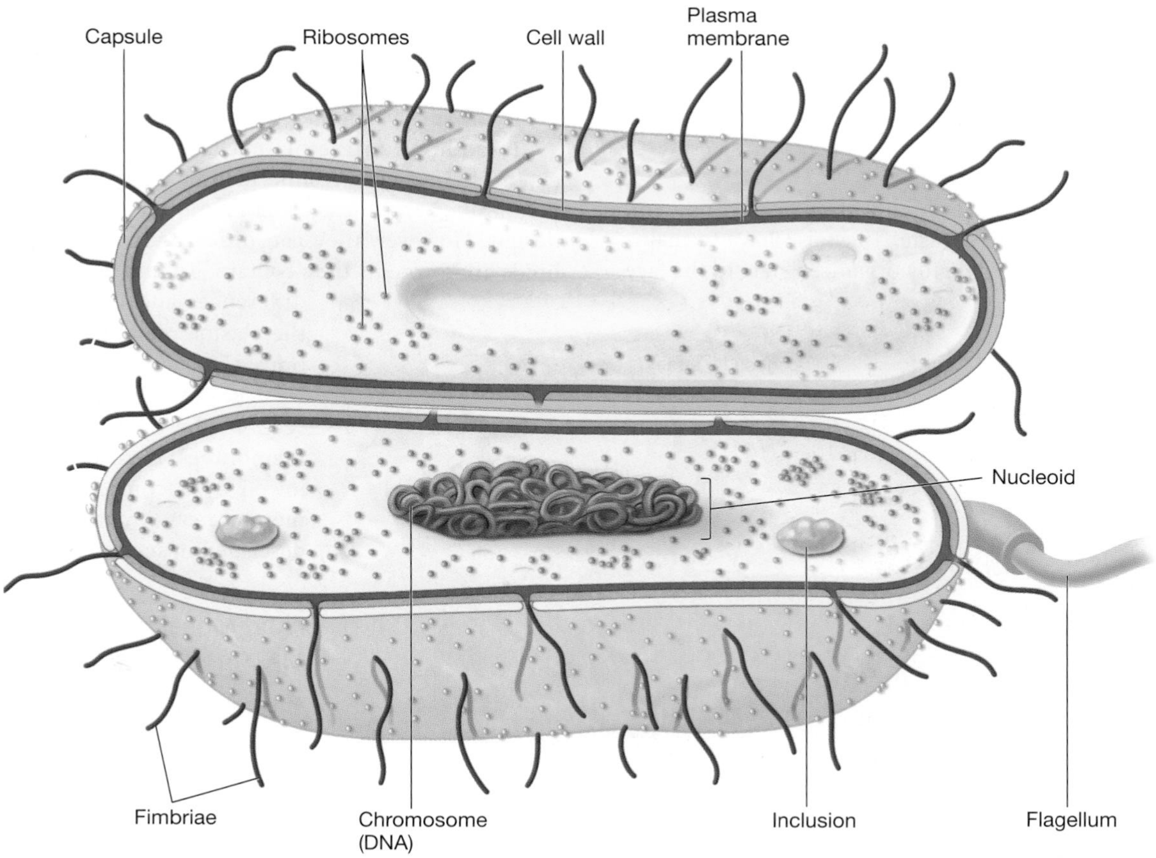

FIGURE 3.6 Morphology of a Bacterial Cell. Archaeal cells are similar, except that capsules are rare.

a discrete region, the nucleoid, and usually is not separated from the surrounding cytoplasm by membranes. Ribosomes and larger masses called inclusions are scattered about the cytoplasm. Many bacteria and archaea use flagella for locomotion.

In the remaining sections of this chapter, we describe the major structures observed in *Bacteria* and *Archaea* in more detail. We begin with the cell envelope. We then proceed inward to consider structures located within the cytoplasm. Next the discussion moves outward, to a variety of appendages that are involved in attachment to surfaces, motility, or both. Finally, we consider a structure unique to bacteria, the bacterial endospore.

1. Why is the term prokaryote considered an inadequate descriptor by some microbiologists?
2. What characteristic shapes can bacteria and archaea assume? Describe the ways in which bacterial cells cluster together.
3. What advantages might a microbial species that forms multicellular arrangements (e.g., clusters or chains) have that are not afforded unicellular microbes?
4. What is the relevance of the surface area-to-volume ratio?

3.3 Bacterial Cell Envelopes

The **cell envelope** is defined as the plasma membrane and all the surrounding layers external to it. The cell envelopes of many bacteria consist of the plasma membrane, cell wall, and at least one additional layer. In this section, we describe these layers, beginning with the plasma membrane and moving outward.

Plasma Membrane

Cells must interact in a selective fashion with their environment, acquire nutrients, and eliminate waste. They also have to maintain their interior in a constant, highly organized state in the face of external changes. Plasma membranes are an absolute requirement for all living organisms because they are involved in carrying out these cellular tasks.

The **plasma membrane** encompasses the cytoplasm of all cells. It is the chief point of contact with the cell's environment and thus is responsible for much of its relationship with the outside world. Receptor molecules that help cells detect and respond to chemicals in their surroundings are involved in this interaction. The plasma membrane also serves as a selectively permeable barrier: it allows particular ions and molecules to pass either into or out of the cell, while preventing the movement of others. Thus the membrane prevents the loss of essential components through leakage while allowing the movement of other molecules.

The plasma membranes of bacterial and archaeal cells are particularly important because they must fill an incredible variety of roles. Because the *Bacteria* and *Archaea* do not carry out endocytosis, the nutrients they need and the wastes they dispose of must cross the plasma membrane. However, many of these substances cannot cross the plasma membrane without assistance. Thus the plasma membrane contains transport systems used for nutrient uptake, waste excretion, and protein secretion. The plasma membrane also is the location of several crucial metabolic processes: respiration, photosynthesis, and the synthesis of lipids and cell wall constituents. Clearly the plasma membrane is essential to the survival of these microorganisms.

In addition to the plasma membrane, some bacteria have extensive intracytoplasmic membrane systems. These internal membranes and the plasma membrane have a common, basic design. However, they can differ significantly in terms of the lipids and proteins they contain. To understand these chemical differences and the many functions of the plasma membrane and other membranes, it is necessary to become familiar with membrane structure. In this section, the fundamental design of all membranes is discussed. This is followed by a consideration of bacterial membranes. Archaeal membranes are described in section 3.4.

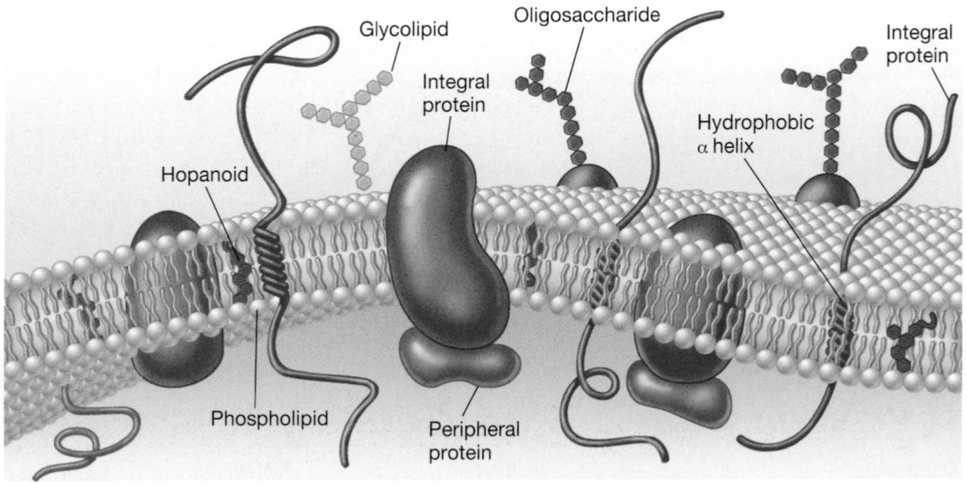

FIGURE 3.7 The Fluid Mosaic Model of Bacterial Membrane Structure. This diagram shows the integral proteins (blue) floating in a lipid bilayer. Peripheral proteins (purple) are associated loosely with the inner membrane surface. Small spheres represent the hydrophilic ends of membrane phospholipids, and wiggly tails are the hydrophobic fatty acid chains. Other membrane lipids such as hopanoids (red) may be present. For the sake of clarity, phospholipids are shown in proportionately much larger size than in real membranes.

Fluid Mosaic Model of Membrane Structure

The most widely accepted model for membrane structure is the **fluid mosaic model** of Singer and Nicholson, which proposes that membranes are lipid bilayers within which proteins float (**figure 3.7**). The model is based on studies of eukaryotic and bacterial membranes, and was established using a variety of experimental approaches. Transmission electron microscopy (TEM) studies were particularly important. When cell membranes are stained and examined by TEM, they are revealed as very thin structures, about 5 to 10 nm thick, that appear as two dark lines on either side of a light interior. This characteristic appearance has been interpreted to mean that the membrane is composed of two sheets of molecules arranged end-to-end (figure 3.7). When membranes are cleaved by the freeze-etching technique, they can be split down the center of the lipid bilayer, exposing the complex internal structure. Within the lipid bilayer, small globular particles are visible; these have been suggested to be membrane proteins lying within the membrane lipid bilayer. The use of atomic force microscopy has provided powerful images to support this interpretation. ◀◀ *Electron microscopy (section 2.4); Scanning probe microscopy (section 2.5)*

The chemical nature of membrane lipids is critical to their ability to form bilayers. Most membrane-associated lipids (e.g., the phospholipids shown in figure 3.7) are **amphipathic**—they are structurally asymmetric, with polar and nonpolar ends (**figure 3.8**). The polar ends interact with water and are **hydrophilic;**

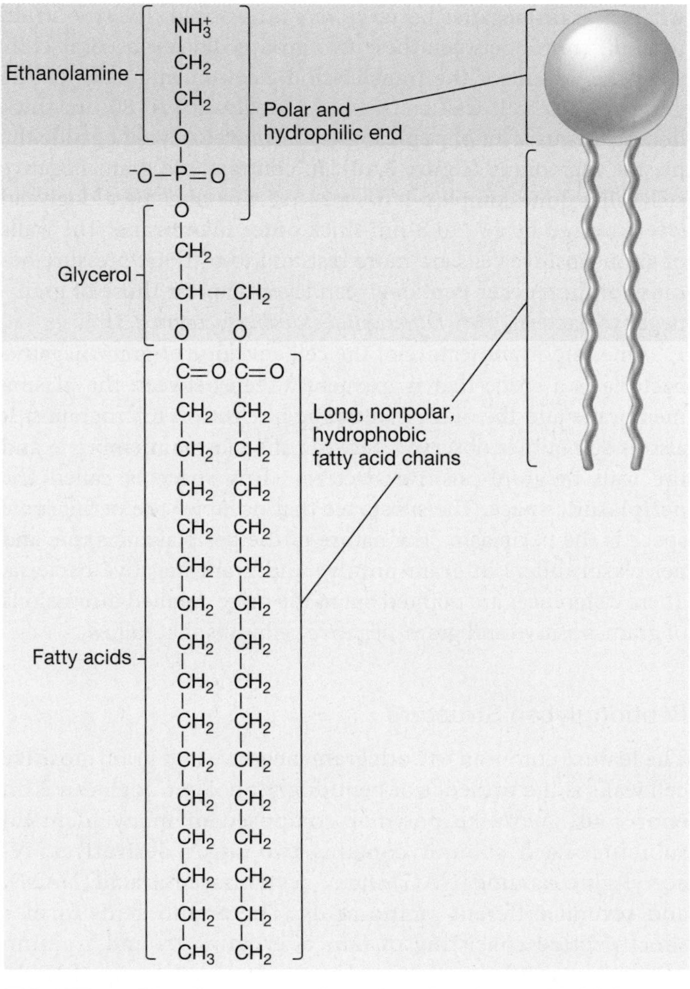

FIGURE 3.8 The Structure of a Polar Membrane Lipid. Phosphatidylethanolamine, an amphipathic phospholipid often found in bacterial membranes.

the nonpolar **hydrophobic** ends are insoluble in water and tend to associate with one another. In aqueous environments, amphipathic lipids can interact to form a bilayer. The outer surfaces of the bilayer are hydrophilic, whereas hydrophobic ends are buried in the interior away from the surrounding water (figure 3.7). ▶▶ *Lipids (appendix I)*

Two types of membrane proteins have been identified based on their ability to be separated from the membrane. **Peripheral proteins** are loosely connected to the membrane and can be easily removed (figure 3.7). They are soluble in aqueous solutions and make up about 20 to 30% of total membrane protein. The remaining proteins are **integral proteins.** These are not easily extracted from membranes and are insoluble in aqueous solutions when freed of lipids. Integral proteins, like membrane lipids, are amphipathic; their hydrophobic regions are buried in the lipid while the hydrophilic portions project from the membrane surface (figure 3.7). Carbohydrates often are attached to the outer surface of plasma membrane proteins. The integral proteins carry out some of the most important functions of the membrane. Many are transport proteins used to move materials either into or out of the cell. Some are part of protein secretion systems. Others are involved in energy-conserving processes, such as the proteins found in electron transport chains. Those integral proteins with regions exposed to the outside of the cell enable the cell to interact with its environment. The fluidity of the membrane enables these proteins to function properly. It also allows many integral proteins to move laterally in the membrane. ▶▶ *Proteins (appendix I); Uptake of nutrients (section 6.6); Protein maturation and secretion (section 12.9)*

Although most aspects of the fluid mosaic model are well supported by experimentation, some are being questioned. The fluid mosaic model suggests that membrane lipids are homogeneously distributed and that integral membrane proteins are free to move laterally within the membrane. However, the presence of microdomains enriched for certain lipids and the observation that some integral proteins are present at only certain sites do not support this view. Although the term mosaic initially referred to the clusters of proteins embedded in a homogeneous lipid bilayer, it may be more accurate to use the term mosaic to refer to the patchwork of lipid microdomains found in membranes. Research is ongoing to determine the physiological role of these microdomains.

Bacterial Lipids

Bacterial membranes are similar to eukaryotic membranes in that they are lipid bilayers and many of their amphipathic lipids are phospholipids (figure 3.8). The plasma membrane is very dynamic: the lipid composition varies with environmental temperature in such a way that the membrane remains fluid during growth. For example, bacteria growing at lower temperatures have more unsaturated fatty acids in their membrane phospholipids—that is, there are one or more double covalent bonds in the long hydrocarbon chain. At higher temperatures,

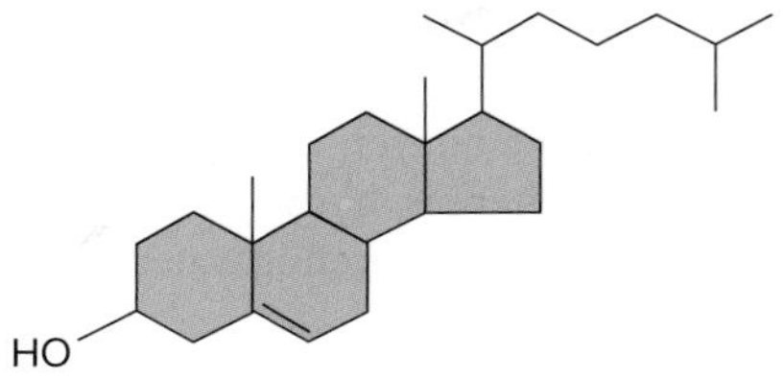

(a) Cholesterol (a steroid) is found in eucaryotes

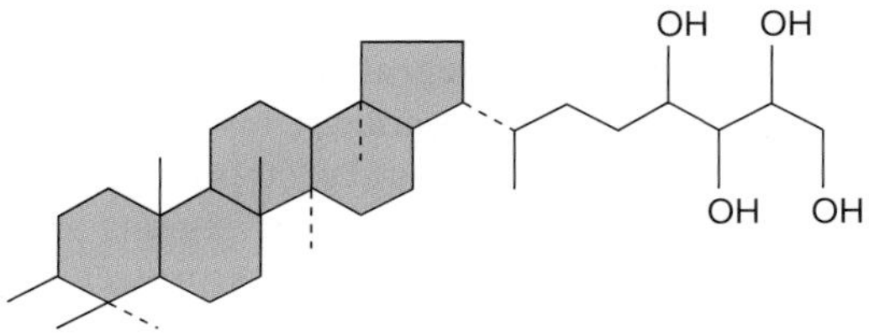

(b) A bacteriohopanetetrol (a hopanoid) is found in bacteria

FIGURE 3.9 Membrane Steroids and Hopanoids. Common examples.

their phospholipids have more saturated fatty acids—those in which the carbon atoms are connected only with single covalent bonds. Other factors also affect the lipid composition of membranes. For instance, some pathogens change the lipids in their plasma membranes in order to protect themselves from antimicrobial peptides produced by the immune system. ▶▶| *Influences of environmental factors on growth (section 7.6); Antimicrobial peptides (section 32.3)*

Bacterial membranes usually differ from eukaryotic membranes in lacking sterols (steroid-containing lipids) such as cholesterol (**figure 3.9*a***). However, many bacterial membranes contain sterol-like molecules called hopanoids (figure 3.9*b*). **Hopanoids** are synthesized from the same precursors as steroids, and like the sterols in eukaryotic membranes, they probably stabilize the membrane. Hopanoids are also of interest to ecologists and geologists: the total mass of hopanoids in sediments is estimated to be around 10^{11-12} tons—about as much as the total mass of organic carbon in all living organisms (10^{12} tons)—and evidence exists that hopanoids have contributed significantly to the formation of petroleum.

1. List the functions of bacterial and archaeal plasma membranes. Why must their plasma membranes carry out more functions than the plasma membranes of eukaryotic cells?
2. Describe in words and with a labeled diagram the fluid mosaic model for cell membranes. What aspects of this model are currently being challenged. Why?

Bacterial Cell Walls

The **cell wall** is the layer, usually fairly rigid, that lies just outside the plasma membrane. It is one of the most important structures for several reasons: it helps determine the shape of the cell; it helps protect the cell from osmotic lysis; it can protect the cell from toxic substances; and in pathogens, it can contribute to pathogenicity. Cell walls are so important that most bacteria have them. Those that do not have other features that fulfill cell wall function. The bacterial cell wall also is the site of action of several antibiotics. Therefore it is important to understand its structure. ▶▶| *Antibacterial drugs (section 34.4)*

The cell walls of *Bacteria* and *Archaea* are distinctive and serve as another example of the fundamental difference between these organisms. An overview of bacterial cell wall structure is provided first. This is followed by more detailed discussions. Archaeal cell walls are discussed in section 3.4.

Overview of Bacterial Cell Wall Structure

After Christian Gram developed the Gram stain in 1884, it soon became evident that most bacteria could be divided into two major groups based on their response to the Gram-staining procedure (*see table 17.6*). Gram-positive bacteria stained purple, whereas gram-negative bacteria were pink or red. The true structural difference between these two groups did not become clear until the advent of the transmission electron microscope. The gram-positive cell wall consists of a single, 20- to 80-nm-thick homogeneous layer of **peptidoglycan (murein)** lying outside the plasma membrane (**figure 3.10**). In contrast, the gram-negative cell wall is quite complex. It has a 2- to 7-nm-thick peptidoglycan layer covered by a 7- to 8-nm-thick **outer membrane.** The walls of gram-positive cells are more resistant to osmotic pressure because of the thicker peptidoglycan layer than are those of gram-negative bacteria. |◀◀ *Differential staining (section 2.3)*

One important feature of the cell envelope of gram-negative bacteria is a space that is frequently seen between the plasma membrane and the outer membrane in electron micrographs. It also is sometimes observed between the plasma membrane and the wall in gram-positive bacteria. This space is called the **periplasmic space.** The substance that occupies the periplasmic space is the periplasm. The nature of the periplasmic space and periplasm differs in gram-positive and gram-negative bacteria. These differences are pointed out in the more detailed discussions of gram-positive and gram-negative cell walls that follow.

Peptidoglycan Structure

The feature common to both gram-negative and gram-positive cell walls is the presence of peptidoglycan. Peptidoglycan is an enormous, meshlike polymer composed of many identical subunits. Each subunit contains two sugar derivatives, *N*-acetyl-glucosamine (NAG) and *N*-acetylmuramic acid (NAM), and several different amino acids. The amino acids form a short peptide consisting of four alternating D- and L-amino acids; the peptide is connected to the carboxyl group of NAM (**figure 3.11**). Three of the amino acids are not found in proteins: D-glutamic acid, D-alanine, and *meso*-diaminopimelic acid. The presence of D-amino acids protects against degradation by most

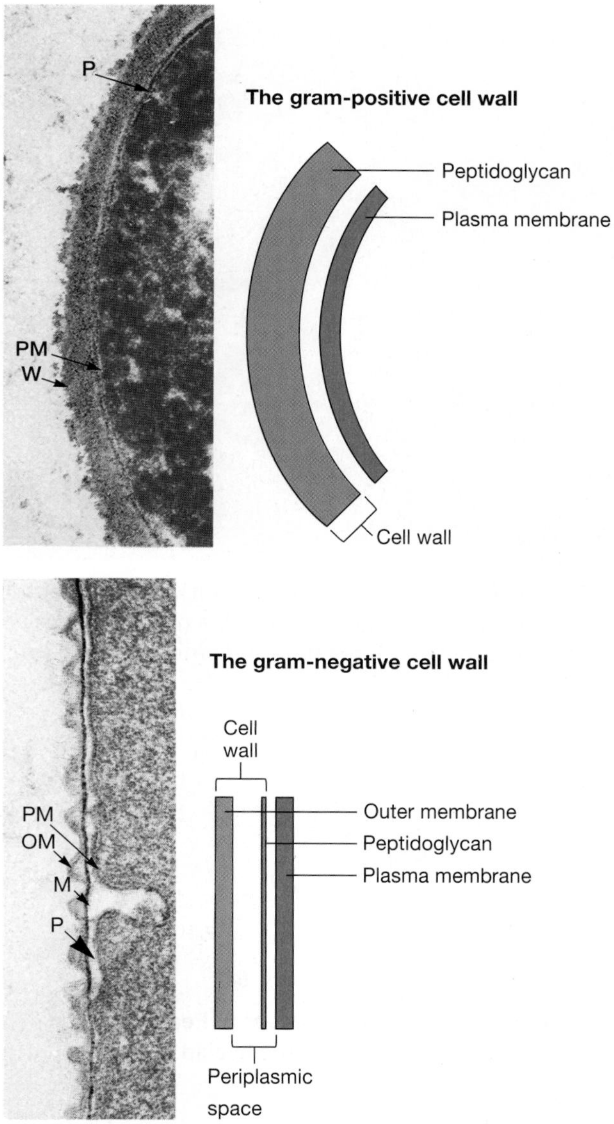

FIGURE 3.10 Gram-Positive and Gram-Negative Cell Walls. The gram-positive envelope is from *Bacillus licheniformis* (top), and the gram-negative micrograph is of *Aquaspirillum serpens* (bottom). M, peptidoglycan or murein layer; OM, outer membrane; PM, plasma membrane; P, periplasmic space; W, gram-positive peptidoglycan wall.

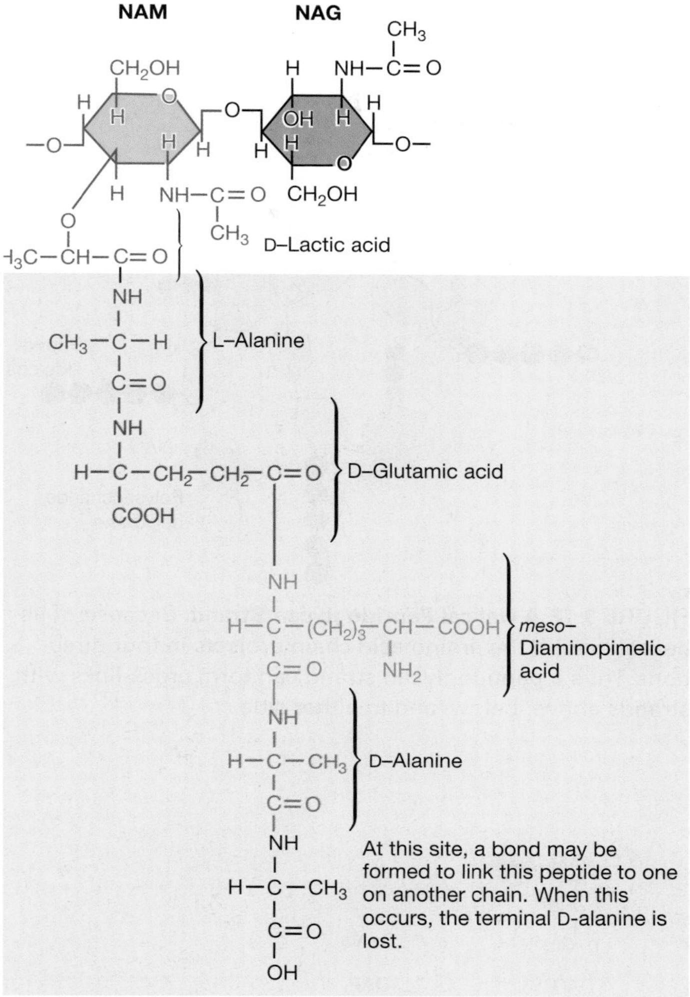

FIGURE 3.11 Peptidoglycan Subunit Composition. The peptidoglycan subunit of *E. coli,* most other gram-negative bacteria, and many gram-positive bacteria. NAG is *N*-acetylglucosamine. NAM is *N*-acetylmuramic acid (NAG with lactic acid attached by an ether linkage). The tetrapeptide side chain is composed of alternating D- and L-amino acids since *meso*-diaminopimelic acid is connected through its L-carbon. NAM and the tetrapeptide chain attached to it are shown in different shades of color for clarity.

peptidases, which recognize only the L-isomers of amino acid residues. Many bacteria replace *meso*-diaminopimelic acid with another diamino acid. The peptidoglycan subunit present in most gram-negative and many gram-positive bacteria is shown in figure 3.11. ▶▶| *Carbohydrates (appendix I); Proteins (appendix I)*; Protein structure *(section 12.3)*

The meshlike peptidoglycan polymer is formed by linking the peptidoglycan subunits together to form a peptidoglycan strand. These strands are then cross-linked to each other. The backbone of a peptidoglycan strand is thus composed of alternating NAG and NAM residues. The peptidoglycan strands are joined by cross-links between the peptides of adjacent strands. Each peptidoglycan strand is helical, and the peptides extend out from the backbone at right angles to each other (**figure 3.12**). Thus each strand can become crosslinked to strands at each side, above, and below. Many bacteria cross-link peptidoglycan by connecting the carboxyl group of the terminal D-alanine (position 4) directly to the amino group of diaminopimelic acid (position 3). Some other bacteria use a **peptide interbridge** instead (**figure 3.13**). With or without a peptide interbridge, cross-linking results in an enormous peptidoglycan sac that is actually one dense,

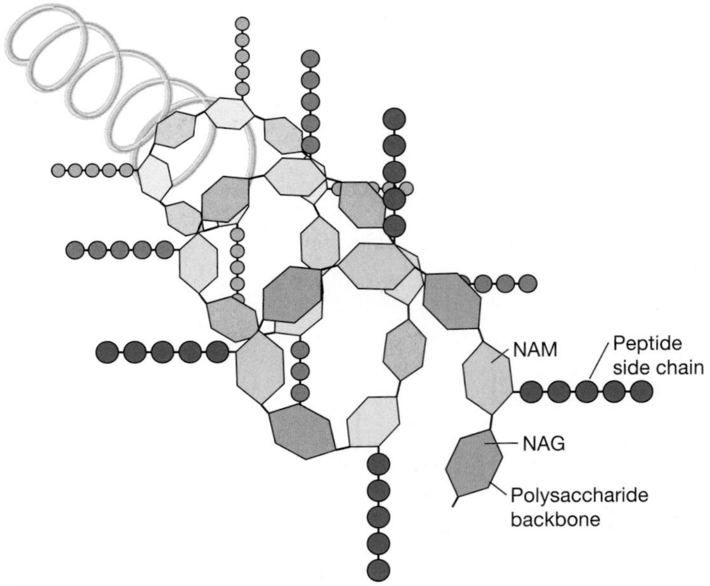

FIGURE 3.12 A Helical Peptidoglycan Strand. Because of its helical nature, the amino acid chain projects in four directions. Thus a peptidoglycan strand can form cross-links with strands above, below, and to either side.

FIGURE 3.14 Isolated Gram-Positive Cell Wall. The peptidoglycan wall from *Bacillus megaterium,* a gram-positive bacterium. The latex spheres have a diameter of 0.25 μm.

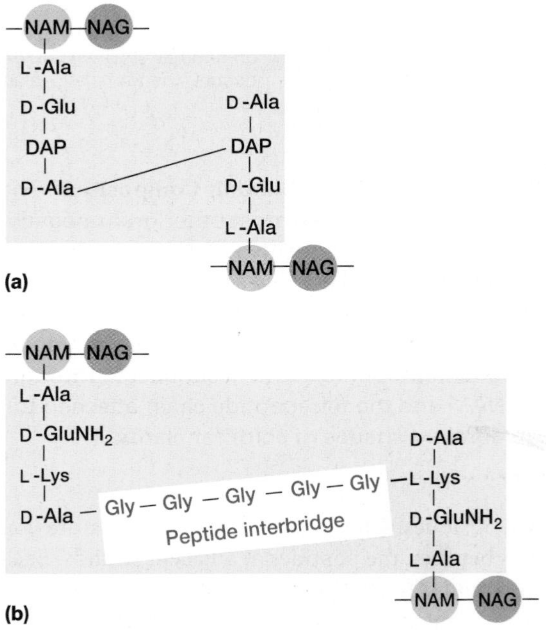

FIGURE 3.13 Peptidoglycan Cross-Links. (a) *E. coli* peptidoglycan with direct cross-linking, typical of many gram-negative bacteria. (b) *Staphylococcus aureus* peptidoglycan. *S. aureus* is a gram-positive bacterium. NAM is *N*-acetylmuramic acid. NAG is *N*-acetylglucosamine. Gly is glycine. D-GluNH$_2$ is D-glutamic acid amidated on the α carbon, as found in some gram-positive bacteria.

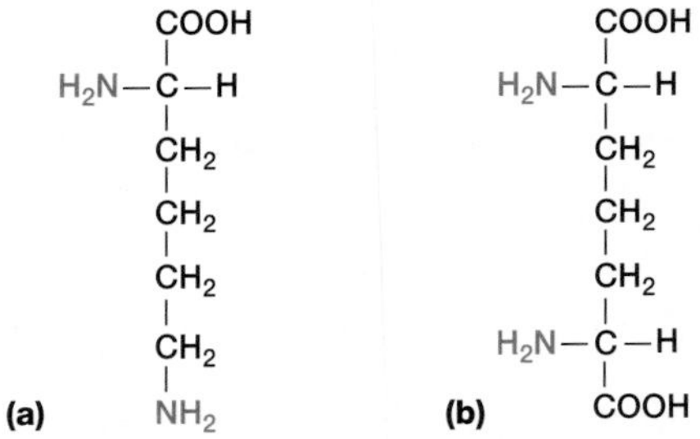

FIGURE 3.15 Diamino Acids Present in Peptidoglycan. (a) L-lysine. (b) *meso*-diaminopimelic acid.

interconnected network. These sacs have been isolated from gram-positive bacteria and are strong enough to retain their shape and integrity, yet they are relatively porous and elastic (**figure 3.14**).

Variants of peptidoglycan are often characteristic of particular groups and are therefore of some taxonomic value. Most gram-negative bacteria have the amino acid composition and cross-linking shown in figure 3.13*a*. This variant is also observed in many gram-positive genera such as *Bacillus, Clostridium, Corynebacterium, Mycobacterium,* and *Nocardia.* Other gram-positive bacteria substitute the diamino acid lysine for *meso*-diaminopimelic acid (**figure 3.15**) and cross-link chains via interpeptide bridges. These interpeptide bridges can vary considerably (**figure 3.16*a***). A very distinctive form of peptidoglycan is observed in some of the plant pathogenic corynebacteria. In these bacteria, an interpeptide bridge consisting of a single diamino acid (often ornithine) connects the carboxyl group of glutamic acid in position 2 with the carboxyl group of the D-alanine in position 4 of the other peptide (figure 3.16*b*).

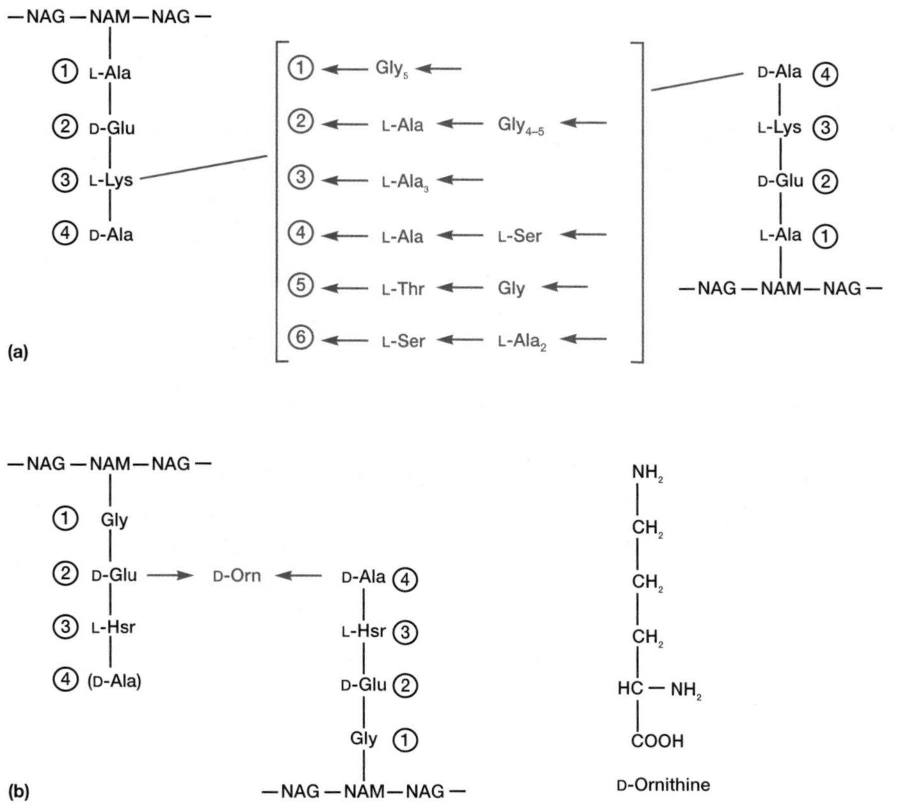

FIGURE 3.16 Different Forms of Peptidoglycan Structure. Most variation in peptidoglycan structure occurs in the composition of the peptide chain and the method by which peptidoglycan strands are cross-linked. (a) Examples of interpeptide bridges. All bridges are used to link D-alanine in position 4 with L-lysine in position 3. The bracket contains six typical bridges: (1) *Staphylococcus aureus,* (2) *S. epidermidis,* (3) *Micrococcus roseus* and *Streptococcus thermophilus,* (4) *Lactobacillus viridescens,* (5) *Streptococcus salvarius,* and (6) *Leuconostoc cremoris.* The arrows indicate the polarity of peptide bonds running in the C to N direction. (b) An interpeptide bridge observed in *Corynebacterium poinsettiae.* The bridge extends between positions 2 and 4 and consists of the D-diamino acid ornithine. Note that L-homoserine (L-Hsr) is in position 3 rather than *meso*-diaminopimelic acid or L-lysine.

Peptidoglycan can also vary in terms of the amount of cross-linking. Members of the genus *Bacillus* and most gram-negative bacteria have fewer cross-links than do gram-positive bacteria such as *Staphylococcus aureus.*

Gram-Positive Cell Walls

Gram-positive bacteria belong to the phyla *Firmicutes* and *Actinobacteria.* Most of these bacteria have cell walls that are thick and composed primarily of peptidoglycan (**figure 3.17**). In addition, their cell walls usually contain large amounts of secondary cell wall polymers, including teichoic acids. **Teichoic acids** are polymers of glycerol or ribitol joined by phosphate groups (**figure 3.18**). Amino acids such as D-alanine or sugars such as glucose are attached to the glycerol and ribitol groups. The teichoic acids are covalently connected to peptidoglycan or to plasma membrane lipids; in the latter case, they are called lipoteichoic acids. Teichoic acids appear to extend to the surface of the peptidoglycan. Because they are negatively charged, they help give the gram-positive cell wall its negative charge. Teichoic acids are not present in gram-negative bacteria.

The functions of teichoic acids are still unclear, but there is increasing evidence that they play several important roles. One is to help create and maintain the structure of the cell envelope and to protect the cell from harmful substances in the environment (e.g., antibiotics and host defense molecules). Some teichoic acids may be involved in binding pathogenic species to host tissues, thus initiating the infectious disease process.

The periplasmic space of gram-positive bacteria lies between the plasma membrane and the cell wall, and is smaller than that of gram-negative bacteria. The periplasm has relatively few proteins; this is probably because the peptidoglycan sac is porous and any proteins secreted by the cell usually pass through it. Enzymes secreted by gram-positive bacteria are called **exoenzymes.** They often serve to degrade polymeric nutrients that would otherwise be too large for transport across the plasma membrane. Those proteins that remain in the periplasmic space are usually attached to the plasma membrane.

Staphylococci and most other gram-positive bacteria have a layer of proteins on the surface of the peptidoglycan. These proteins are involved in interactions of the cell with its environment. Some are noncovalently attached by binding to teichoic acids or other cell wall polymers. For example, S-layer proteins (p. 62) bind noncovalently to polymers scattered throughout the cell wall. Enzymes involved in peptidoglycan synthesis and turnover also seem to interact noncovalently with the cell wall. Other surface proteins are covalently attached to the peptidoglycan. Many covalently attached proteins have roles in virulence. For example, the M protein of pathogenic streptococci aids in adhesion to host tissues and interferes with host defenses. In staphylococci, these surface proteins are covalently joined to the pentaglycine interbridge of the peptidoglycan (figure 3.13*b*). An enzyme called sortase catalyzes the attachment of these surface proteins to the peptidoglycan. Sortases are attached to the plasma membrane of the cell.

The acid-fast bacteria, a taxon within the phylum *Actinobacteria,* have somewhat different cell wall structure than the typical gram-positive wall just described. These bacteria include members

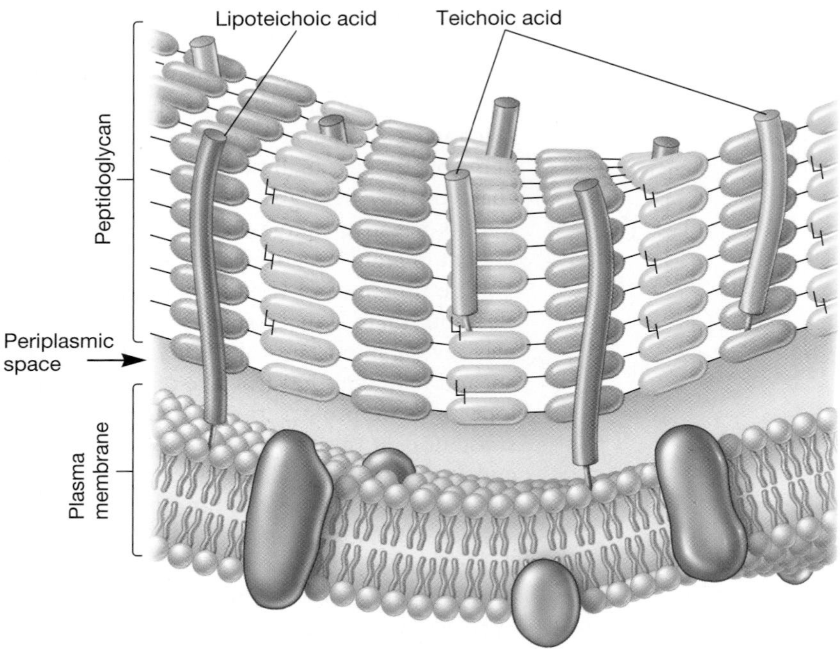

FIGURE 3.17 The Gram-Positive Envelope.

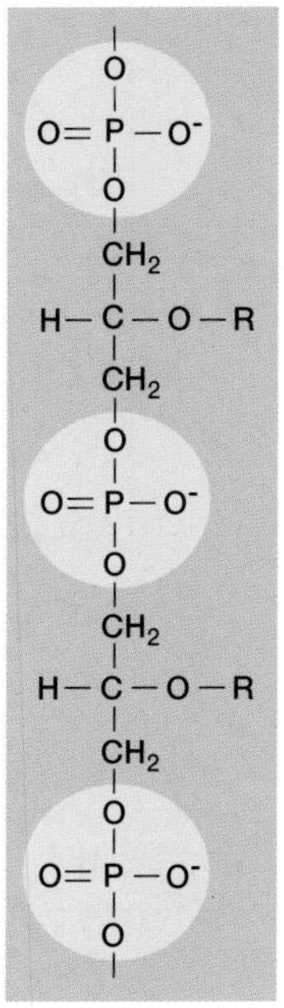

FIGURE 3.18 Teichoic Acid Structure. The segment of a teichoic acid made of phosphate, glycerol, and a side chain, R. R may represent D-alanine, glucose, or other molecules.

of the genus *Mycobacterium* (e.g., *M. tuberculosis*). Their cell walls are unique among the gram-positive bacteria because they consist of other layers in addition to peptidoglycan. An important component of their walls is a group of fatty acids called mycolic acids (*see figure 22.12*). Some have proposed that the mycolic acids form a bilayer structure that is analogous to the outer membrane of gram-negative bacteria. ▶▶| *Suborder* Corynebacterineae *(section 22.4)*

Gram-Negative Cell Walls

Even a brief inspection of figure 3.10 shows that gram-negative cell walls are much more complex than gram-positive walls. The thin peptidoglycan layer next to the plasma membrane and bounded on either side by the periplasmic space usually constitutes only 5 to 10% of the wall weight. In *E. coli,* it is about 2 nm thick and contains only one or two sheets of peptidoglycan.

The periplasmic space of gram-negative bacteria is also strikingly different from that of gram-positive bacteria. It ranges in width from 1 nm to as great as 71 nm. Some studies indicate that it may constitute about 20 to 40% of the total cell volume, and it is usually 30 to 70 nm wide. When cell walls are disrupted carefully or removed without disturbing the underlying plasma membrane, periplasmic enzymes and other proteins are released and may be easily studied. Some periplasmic proteins participate in nutrient acquisition—for example, hydrolytic enzymes and transport proteins. Some periplasmic proteins are involved in energy conservation. For example, the denitrifying bacteria, which convert nitrate to nitrogen gas, and bacteria that use inorganic molecules as energy sources (chemolithotrophs) have electron transport proteins in their periplasm. Other periplasmic proteins are involved in peptidoglycan synthesis and the modification of toxic compounds that could harm the cell. ▶▶| *Chemolithotrophy (section 10.11); Nitrogen cycle (section 26.1)*

The outer membrane lies outside the thin peptidoglycan layer and is thought to be linked to the cell in two ways (**figure 3.19**). The first is by Braun's lipoprotein, the most abundant protein in the outer membrane. This small lipoprotein is covalently joined to the underlying peptidoglycan and is embedded in the outer membrane by its hydrophobic end. The second linking mechanism is not as widely accepted. It involves contact sites that appear to join the outer membrane and the plasma membrane. In *E. coli,* 20 to 100 nm areas of contact between the two membranes can be seen in transmission electron micrographs of cells fixed by chemical methods. These contact sites are not observed when cells are fixed by rapid freezing (i.e., methods similar to those used in electron cryotomography), and it is argued that they may be artifacts of chemical fixation. ▶▶| *Electron cryotomography (section 2.4)*

Possibly the most unusual constituents of the outer membrane are its **lipopolysaccharides (LPSs).** These large, complex molecules contain both lipid and carbohydrate, and consist of three parts: (1) lipid A, (2) the core polysaccharide, and (3) the O side chain. The LPS from *Salmonella* has been studied most, and its general structure is described here (**figure 3.20**). The **lipid A** region contains two glucosamine sugar derivatives, each with three fatty acids and phosphate or pyrophosphate attached. The

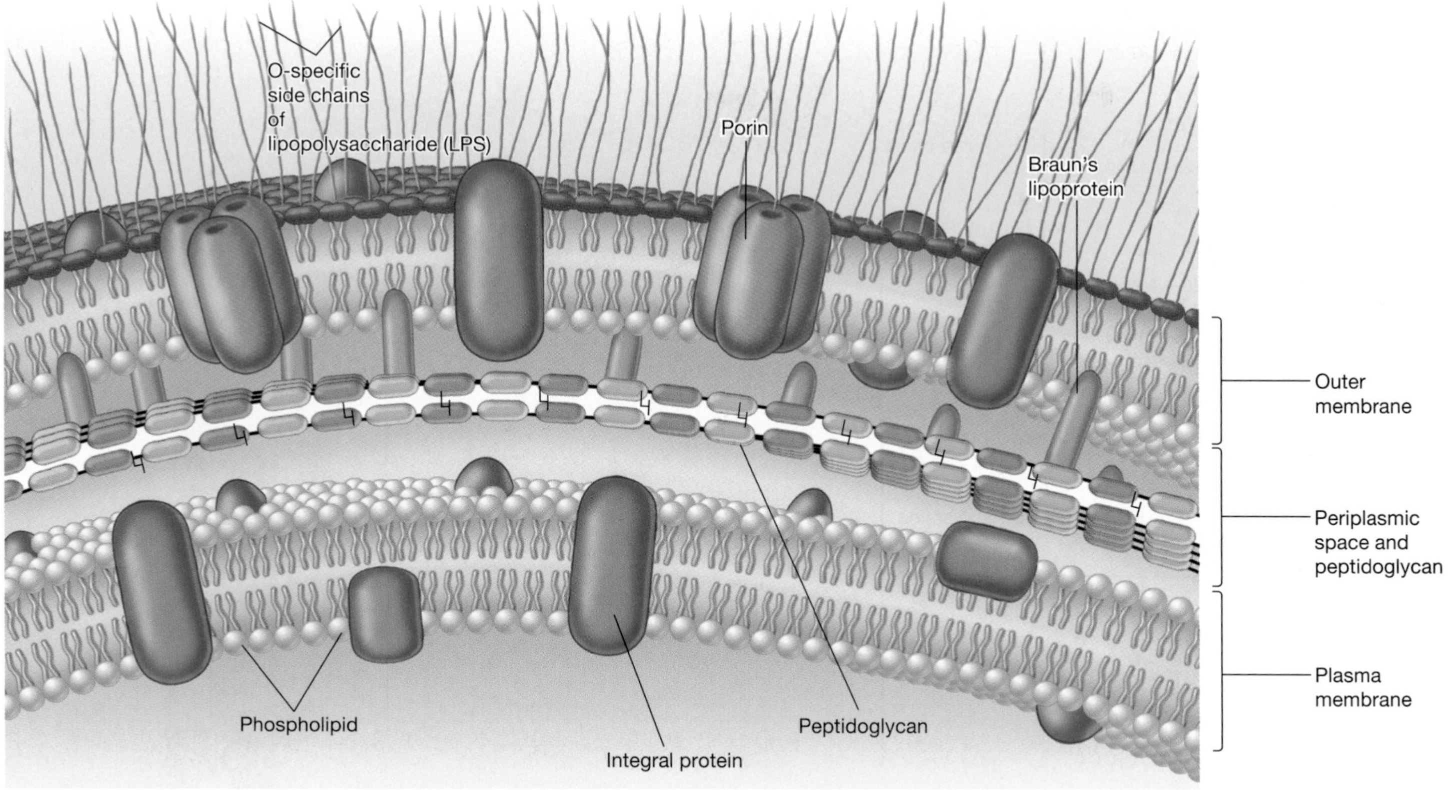

FIGURE 3.19 The Gram-Negative Envelope.

fatty acids of lipid A are embedded in the outer membrane, while the remainder of the LPS molecule projects from the surface. The **core polysaccharide** is joined to lipid A. In *Salmonella,* it is constructed of 10 sugars, many of them unusual in structure. The O side chain or **O antigen** is a polysaccharide chain extending outward from the core. It has several peculiar sugars and varies in composition between bacterial strains.

LPS has many important functions. (1) LPS contributes to the negative charge on the bacterial surface because the core polysaccharide usually contains charged sugars and phosphate (figure 3.20). (2) LPS helps stabilize outer membrane structure because lipid A is a major constituent of the exterior leaflet of the outer membrane. (3) LPS may contribute to bacterial attachment to surfaces and biofilm formation. (4) A major function of LPS is that it helps create a permeability barrier. The geometry of LPS (figure 3.20*b*) and interactions between neighboring LPS molecules are thought to restrict the entry of bile salts, antibiotics, and other toxic substances that might kill or injure the bacterium. (5) LPS also plays a role in protecting pathogenic gram-negative bacteria from host defenses. The O side chain of LPS is also called the O antigen because it elicits an immune response by an infected host. This response involves the production of antibodies that bind the strain-specific form of LPS that elicited the response. However, many gram-negative bacteria can rapidly change the antigenic nature of their O side chains, thus thwarting host defenses. (6) Importantly, the lipid A portion of LPS is toxic; as a result, LPS can act as an endotoxin and cause some of the symptoms that arise in gram-negative bacterial infections. If LPS or lipid A enters the bloodstream, a form of septic shock develops, for which there is no direct treatment. ▶▶| *Toxins (section 31.3); Antibodies (section 33.7)*

Despite the role of LPS in creating a permeability barrier, the outer membrane is more permeable than the plasma membrane and permits the passage of small molecules such as glucose and other monosaccharides. This is due to the presence of **porin proteins.** Most porin proteins cluster together to form a trimer in the outer membrane (figure 3.19 and **figure 3.21**). Each porin protein spans the outer membrane and is more or less tube-shaped; its narrow, water-filled channel allows passage of molecules smaller than about 600 daltons. However, larger molecules such as vitamin B_{12} also cross the outer membrane. Such large molecules do not pass through porins; instead, specific carriers transport them across the outer membrane. ▶▶| *Uptake of nutrients (section 6.6)*

Mechanism of Gram Staining

The difference between those bacteria that stain gram positive and those that stain gram negative is due to the physical nature

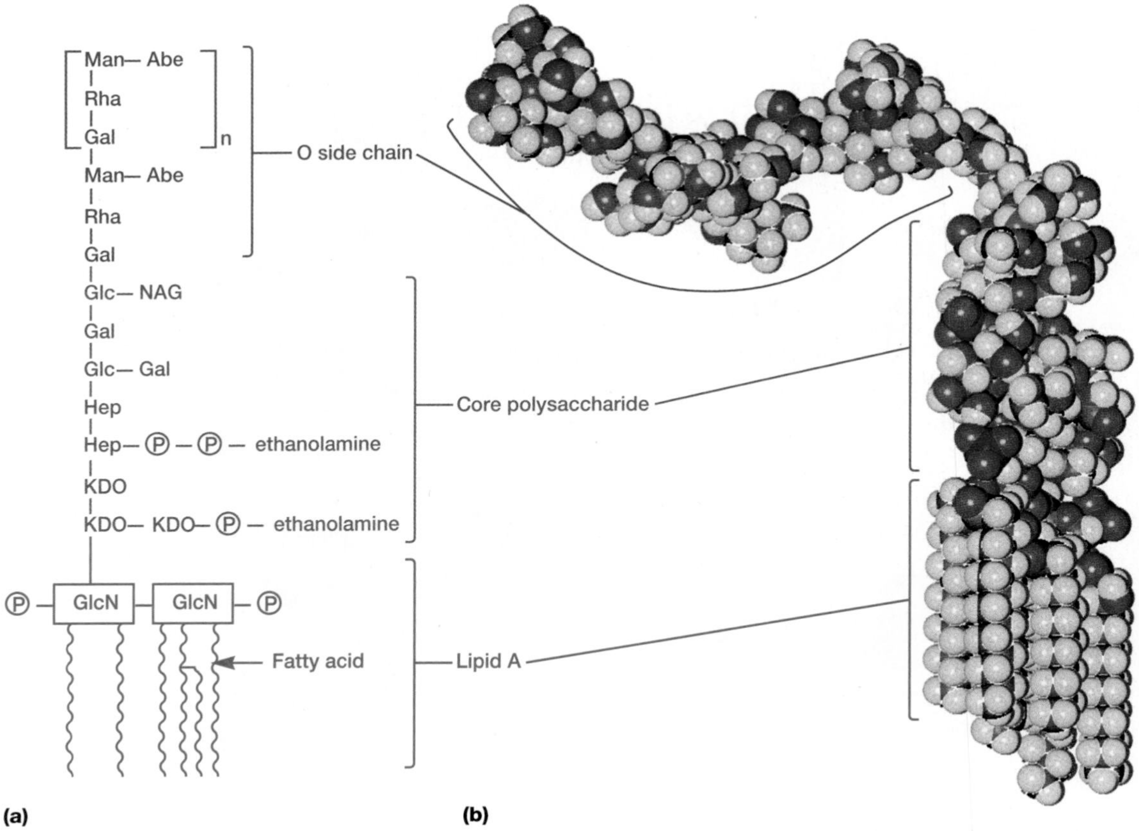

FIGURE 3.20 Lipopolysaccharide Structure. (a) The lipopolysaccharide from *Salmonella*. This slightly simplified diagram illustrates one form of LPS. Abbreviations: Abe, abequose; Gal, galactose; Glc, glucose; GlcN, glucosamine; Hep, heptulose; KDO, 2-keto-3-deoxyoctonate; Man, mannose; NAG, *N*-acetylglucosamine; P, phosphate; Rha, L-rhamnose. Lipid A is buried in the outer membrane. (b) Molecular model of an *Escherichia coli* lipopolysaccharide. The lipid A and core polysaccharide are straight; the O side chain is bent at an angle in this model.

of their cell walls. If the cell wall is removed from gram-positive bacteria, they stain gram negative. Furthermore, bacteria that never make cell walls, such as the mycoplasmas, also stain gram negative. During the procedure, bacteria are first stained with crystal violet and next treated with iodine to promote dye retention. When bacteria are treated with ethanol in the decolorization step, the alcohol is thought to shrink the pores of the thick peptidoglycan found in the cell walls of most gram-positive bacteria, causing the peptidoglycan to act as a permeability barrier that prevents loss of crystal violet. Thus the dye-iodine complex is retained during the decolorization step and the bacteria remain purple. In contrast, the peptidoglycan in gram-negative cell walls is very thin, not as highly cross-linked, and has larger pores. Alcohol treatment also may extract enough lipid from the outer membrane to increase the cell wall's porosity further. For these reasons, alcohol more readily removes the crystal violet-iodine complex from gram-negative bacteria. Thus gram-negative bacteria are easily stained red or pink by the counterstain safranin.

Cell Walls and Osmotic Protection

Microbes have several mechanisms for responding to changes in osmotic pressure. This pressure arises when the concentration of solutes inside the cell differs from that outside, and the responses work to equalize the solute concentrations. However, in certain situations, osmotic pressure can exceed the cell's ability to acclimate. In these cases, additional protection is provided by the cell wall. When cells are in hypotonic solutions—ones in which the solute concentration is less than that in the cytoplasm—water diffuses into the cell, causing it to swell. Without the cell wall, the pressure on the plasma membrane would become so great that the membrane would be disrupted and the cell would burst—a process called **lysis.** Conversely, in hypertonic solutions, water flows out and the cytoplasm shrivels up—a process called **plasmolysis.**

The protective nature of the cell wall is most clearly demonstrated when bacterial cells are treated with lysozyme or penicillin. The enzyme **lysozyme** attacks peptidoglycan by hydrolyzing the glycosidic bond that connects *N*-acetylmuramic acid with

(a) Porin trimer

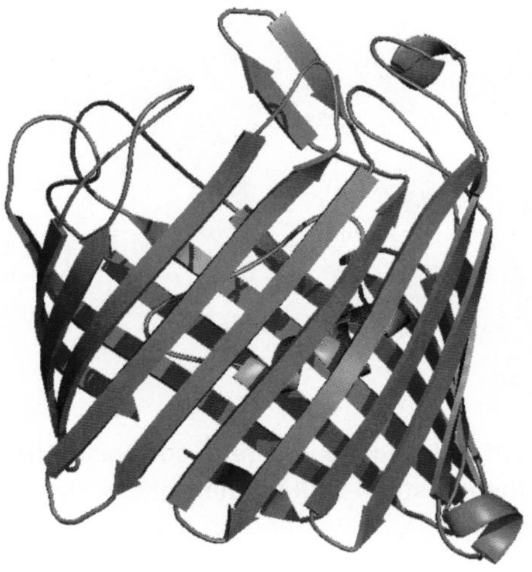

(b) OmpF side view

FIGURE 3.21 Porin Proteins. Two views of the OmpF porin of *E. coli.* (a) Porin structure observed when looking down at the outer surface of the outer membrane (i.e., top view). The three porin proteins forming the trimer each form a water-filled channel. Each OmpF porin can be divided into three loops: the green loop forms the channel, the blue loop interacts with other porin proteins to help form the trimer, and the orange loop narrows the channel. The arrow indicates the area of a porin molecule viewed from the side in panel (b). Side view of a porin monomer showing the β-barrel structure characteristic of porin proteins.

N-acetylglucosamine (figure 3.11). Penicillin works by a different mechanism. It inhibits the enzyme transpeptidase, which is responsible for making the cross-links between peptidoglycan chains. If bacteria are treated with either of these substances while in a hypotonic solution, they lyse. However, if they are in an isotonic solution, they can survive and grow normally. If they are gram positive, treatment with lysozyme or penicillin results in the complete loss of the cell wall, and the cell becomes a protoplast. When gram-negative bacteria are exposed to lysozyme or penicillin, the peptidoglycan layer is lost, but the outer membrane remains. These cells are called **spheroplasts.** Because they lack a complete cell wall, both protoplasts and spheroplasts are osmotically sensitive. If they are transferred to a hypotonic solution, they lyse due to uncontrolled water influx (**figure 3.22**). ▶▶| *Antibacterial drugs (section 34.4)*

Although most bacteria require an intact cell wall for survival, some have none at all. For example, the mycoplasmas lack a cell wall and are osmotically sensitive yet often can grow in dilute media or terrestrial environments because their plasma membranes are more resistant to osmotic pressure than those of bacteria having walls. The precise reason for this is not clear, although the presence of sterols in the membranes of many species may provide added strength. Without a rigid cell wall, mycoplasmas tend to be pleomorphic (*see figure 21.3*).

1. List the functions of the cell wall.
2. Describe in detail the composition and structure of peptidoglycan. Why does peptidoglycan contain the unusual D-isomers of alanine and glutamic acid rather than the L-isomers observed in proteins?
3. Compare and contrast the cell walls of gram-positive bacteria and gram-negative bacteria. Include labeled drawings in your discussion.
4. When protoplasts and spheroplasts are made, the shape of the cell becomes spherical regardless of the original cell shape. Why does this occur?
5. Design an experiment that illustrates the cell wall's role in protecting against lysis.
6. With a few exceptions, the cell walls of gram-positive bacteria lack porins. Why is this the case?

Layers Outside the Cell Wall

Many bacteria have another layer in their cell envelopes that lies outside the cell wall. This layer is given different names depending on its makeup and how it is organized.

Capsules and Slime Layers

Capsules are layers that are well organized and not easily washed off (**figure 3.23*a***). Capsules are most often composed of polysaccharides, but some are constructed of other materials. For example, *Bacillus anthracis* has a proteinaceous capsule composed of poly-D-glutamic acid. Capsules are clearly visible in the light microscope when negative stains or specific capsule stains are employed (figure 3.23*a*); they also can be studied with the electron microscope.

Although capsules are not required for growth and reproduction in laboratory cultures, they confer several advantages when bacteria grow in their normal habitats. They help pathogenic

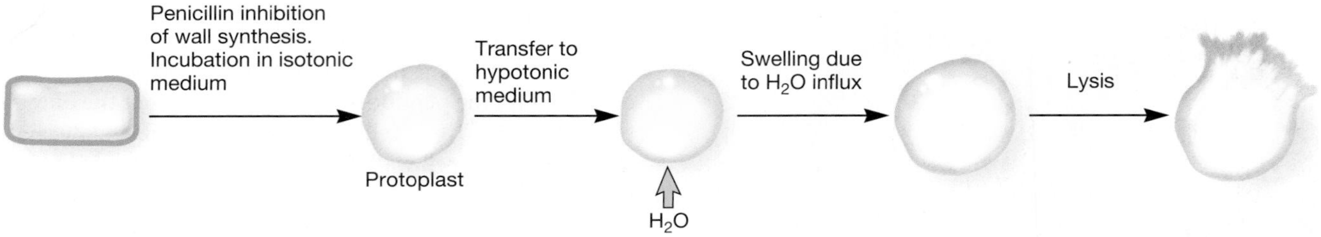

FIGURE 3.22 Protoplast Formation and Lysis. Protoplast formation induced by incubation with penicillin in an isotonic medium. Transfer to hypotonic medium will result in lysis.

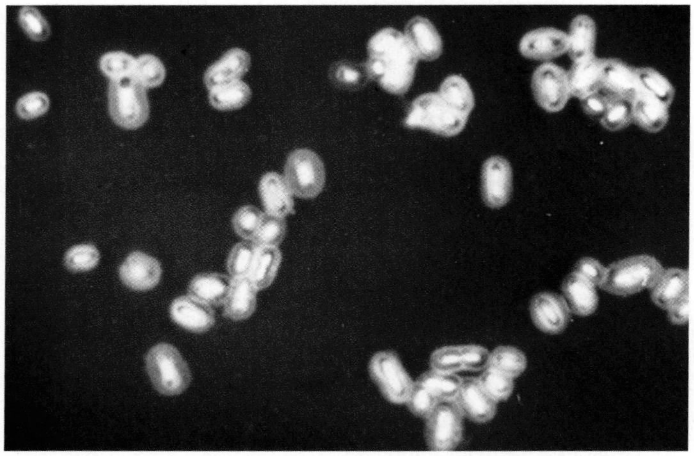

(a) *K. pneumoniae*

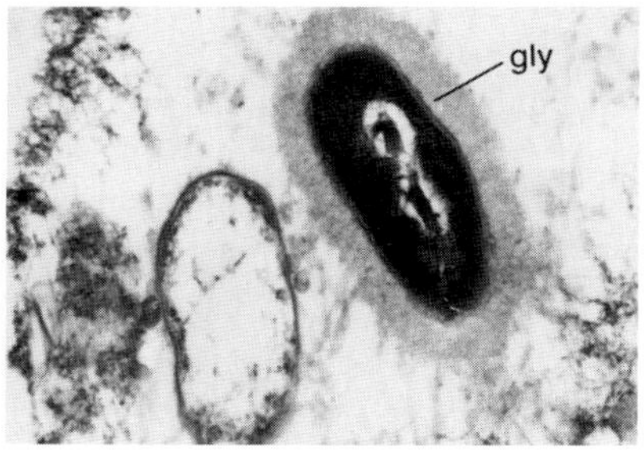

(b) *Bacteroides*

FIGURE 3.23 Bacterial Capsules. (a) *Klebsiella pneumoniae* with its capsule stained for observation in the light microscope (×1,500). (b) *Bacteroides* glycocalyx (gly), TEM (×71,250).

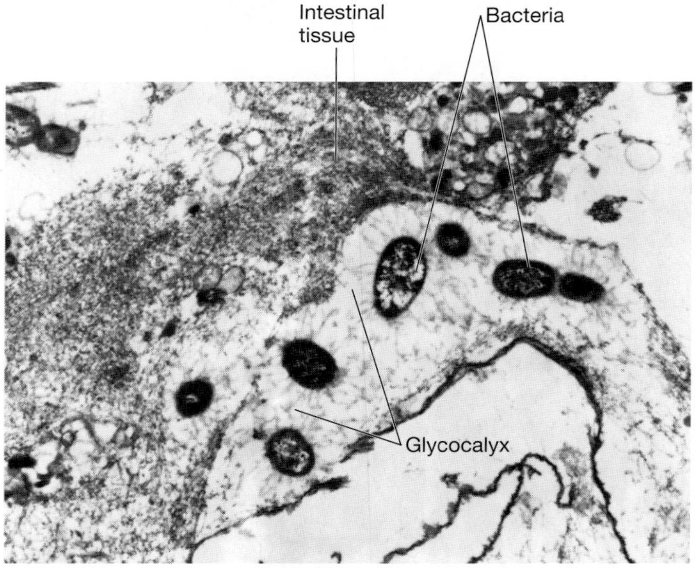

FIGURE 3.24 Bacterial Glycocalyx. Bacteria connected to each other and to the intestinal wall by their glycocalyxes, the extensive networks of fibers extending from the cells (×17,500).

bacteria resist phagocytosis by host phagocytes. *Streptococcus pneumoniae* provides a dramatic example. When it lacks a capsule, it is destroyed easily and does not cause disease. On the other hand, the capsulated variant quickly kills mice. Capsules contain a great deal of water and can protect against desiccation. They exclude viruses and most hydrophobic toxic materials such as detergents.

A **slime layer** is a zone of diffuse, unorganized material that is removed easily. It is usually composed of polysaccharides, but is not as easily observed by light microscopy. Gliding bacteria often produce slime, which in some cases has been shown to facilitate motility (section 3.7).

The term **glycocalyx** refers to a layer consisting of a network of polysaccharides extending from the surface of the cell (figure 3.23*b*). The term can encompass both capsules and slime layers because they usually are composed of polysaccharides. The glycocalyx aids in attachment to solid surfaces, including tissue surfaces in plant and animal hosts (**figure 3.24**). ▶▶| *Virulence factors (section 31.3)*

S-Layers

Many bacteria have a regularly structured layer called an **S-layer** on their surface. The S-layer has a pattern something like floor tiles and is composed of protein or glycoprotein (**figure 3.25**). In gram-negative bacteria, the S-layer adheres directly to the outer membrane; it is associated with the peptidoglycan surface in gram-positive bacteria.

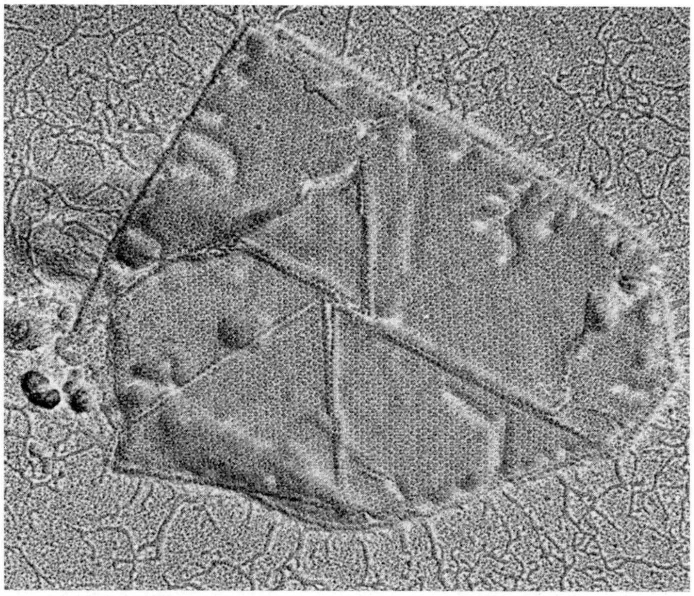

FIGURE 3.25 The S-Layer. An electron micrograph of the S-layer of the bacterium *Deinococcus radiodurans* after shadowing.

Currently S-layers are of considerable interest not only for their biological roles but also in the growing field of nanotechnology. Their biological roles include protecting the cell against ion and pH fluctuations, osmotic stress, enzymes, or predacious bacteria. The S-layer also helps maintain the shape and envelope rigidity of some cells, and it can promote cell adhesion to surfaces. Finally, the S-layer seems to protect some bacterial pathogens against host defenses, thus contributing to their virulence. The potential use of S-layers in nanotechnology is due to the ability of S-layer proteins to self-assemble. That is, the S-layer proteins contain the information required to spontaneously associate and form the S-layer without the aid of any additional enzymes or other factors. Thus S-layer proteins could be used as building blocks for the creation of technologies such as drug-delivery systems and novel detection systems for toxic chemicals or bioterrorism agents.

3.4 Archaeal Cell Envelopes

One of the most distinctive features of the *Archaea* is the nature of their cell envelopes. They differ not only with respect to their molecular makeup but also in terms of organization. For many archaea, an S-layer is the major, and sometimes only, component of the cell wall. However, unlike bacterial S-layers, archaeal S-layers are usually the layer just outside the plasma membrane. Some archaea, like some bacteria, lack cell walls but have a glycocalyx lying outside the cell membrane. Capsules and slime layers are also rare among archaea. Because of the relative rarity of these external layers, they are not discussed here.

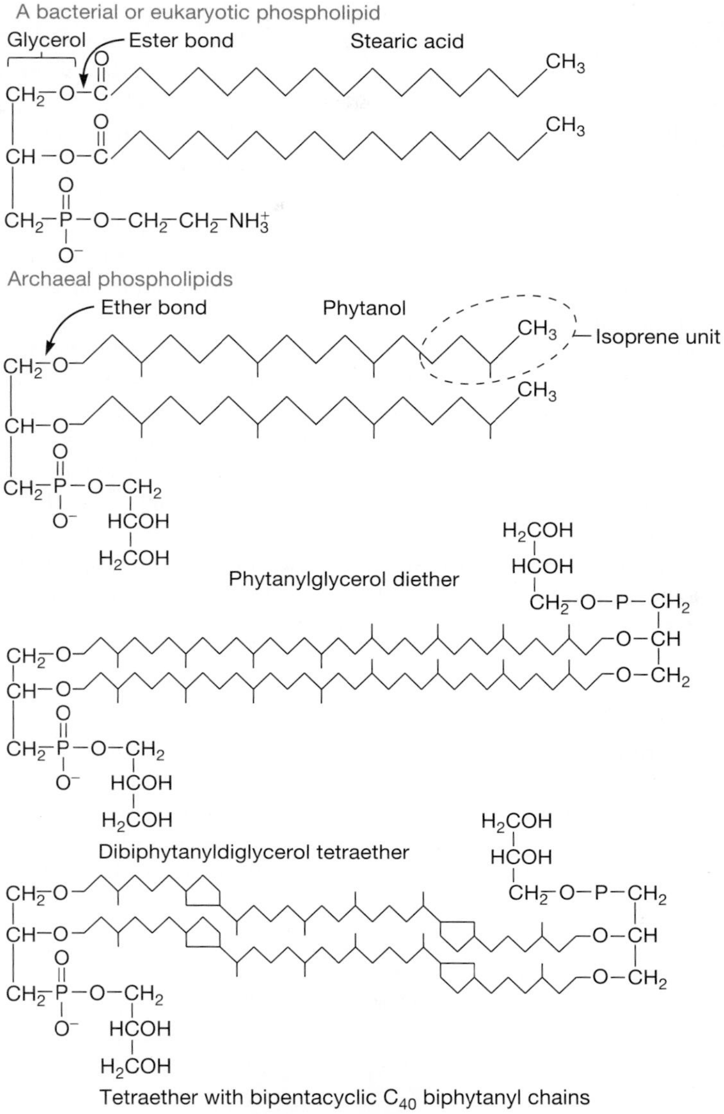

FIGURE 3.26 Archaeal Membrane Lipids. An illustration of the difference between archaeal lipids and those of *Bacteria.* In *Archaea,* the hydrocarbon side chains are derived from isoprene units (indicated by dashed line). Thus archaeal lipids are isoprenyl glycerol ethers rather than glycerol fatty acid esters, as in *Bacteria.*

Archaeal Plasma Membranes

Archaeal membranes are composed primarily of lipids that differ from bacterial and eukaryotic lipids in two ways. First, they contain hydrocarbons derived from isoprene units—five-carbon, branched molecules. Thus the hydrocarbons are branched as shown in **figure 3.26**. Second, the hydrocarbons are attached to glycerol by ether links rather than ester links. When two hydrocarbons are attached to glycerol, the lipids are called diether lipids. Usually the diether hydrocarbon chains are 20 carbons in length. Sometimes tetraether lipids are formed when two glycerol residues are linked by two long hydrocarbons that are 40 carbons in length. Cells can adjust the overall length of the tetraethers by cyclizing the chains to form pentacyclic rings (figure 3.26). Phosphate-, sulfur-, and sugar-containing groups can be attached to the third carbons of the glycerol moieties in the diethers and tetraethers, making them polar lipids.

Despite these significant differences in membrane lipids, the basic design of archaeal membranes is similar to that of bacterial and eukaryotic membranes—there are two hydrophilic surfaces and a hydrophobic core. When C_{20} diethers are used, a typical bilayer membrane is formed (**figure 3.27*a***). When the membrane is constructed of C_{40} tetraethers, a monolayer membrane with much more rigidity is formed (figure 3.27*b*). As might be expected from their need for stability, the membranes of extreme thermophiles such as *Thermoplasma* and *Sulfolobus,* which grow best at temperatures over 85°C, are almost completely tetraether monolayers. Archaea that live in moderately hot environments have membranes containing some regions with monolayers and some with bilayers. ▶▶| *Archaeal cell surfaces and membranes (section 18.1); Phylum* Crenarchaeota *(section 18.2)*

Archaeal Cell Walls

Before they were distinguished as a unique domain of life, the *Archaea* were characterized as being either gram positive or gram negative. However, their staining reaction does not correlate reliably with a particular cell wall structure. Archaeal wall structure and chemistry differ from those of the *Bacteria.* Archaeal cell walls lack peptidoglycan and exhibit considerable variety in terms of their chemical makeup.

The most common type of archaeal cell wall is an S-layer composed of either glycoprotein or protein (**figure 3.28*a***). The layer may be as thick as 20 to 40 nm. Some methanogens (*Methanolobus* and *Methanococcus*), salt-loving archaea (*Halobacterium*), and extreme thermophiles (*Sulfolobus, Thermoproteus,* and *Pyrodictium*) have S-layer cell walls.

Other archaea have additional layers of material outside the S-layer. For instance, *Methanospirillum* has a protein sheath external to the S-layer (figure 3.28*b*). Another methanogen, *Methanosarcina,* has a layer of chondroitin-like material covering the S-layer (figure 3.28*c*). This material, called methanochondroitin, is similar to the chondroitin sulfate of animal connective tissue.

In some archaea, the S-layer is the outermost layer and is separated from the plasma membrane by an interesting molecule called pseudomurein (figure 3.28*d*). Pseudomurein is a peptidoglycan-like molecule. It differs from peptidoglycan in that it has L-amino acids instead of D-amino acids in its cross-links, *N*-acetyltalosaminuronic acid instead of *N*-acetylmuramic acid, and $\beta(1 \rightarrow 3)$ glycosidic bonds instead of $\beta(1 \rightarrow 4)$ glycosidic bonds (**figure 3.29**).

The last type of archaeal cell wall does not include an S-layer. Instead these archaea have a wall with a single, thick homogeneous layer resembling that in gram-positive bacteria (figure 3.28*e*). These archaea often stain gram positive. Their wall chemistry varies from species to species but usually consists of complex polysaccharides such as pseudomurein.

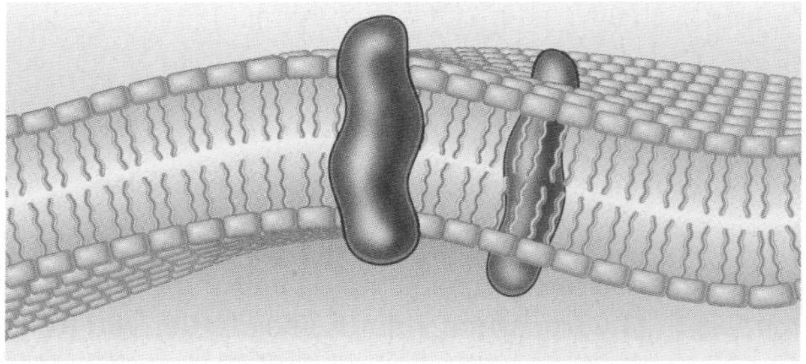

(a) Bilayer of C_{20} diethers

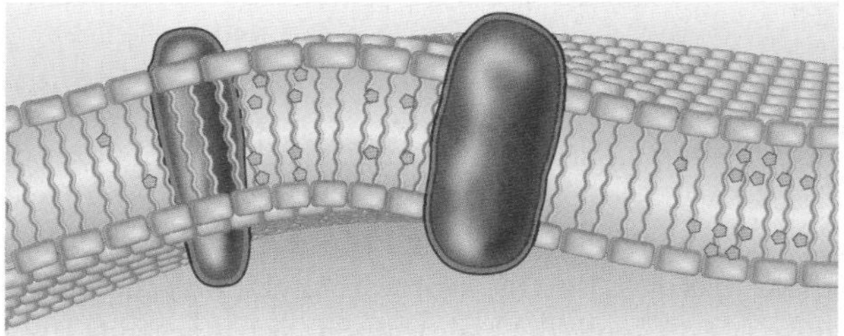

(b) Monolayer of C_{40} tetraethers

FIGURE 3.27 Examples of Archaeal Membranes.

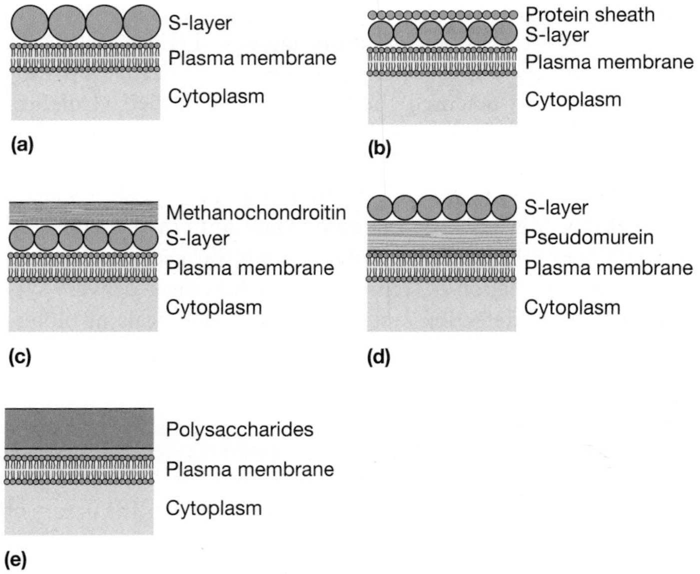

FIGURE 3.28 Cell Envelopes of *Archaea*. (a) *Methanococcus, Halobacterium, Pyrodictium, Sulfolobus,* and *Thermoproteus* cell envelope. (b) *Methanospirillum* cell envelope. (c) *Methanosarcina* cell envelope. (d) *Methanothermus* and *Methanopyrus* cell envelope. (e) *Methanobacterium, Methanospaera, Methanobrevibacter, Halococcus,* and *Natronococcus* cell envelope.

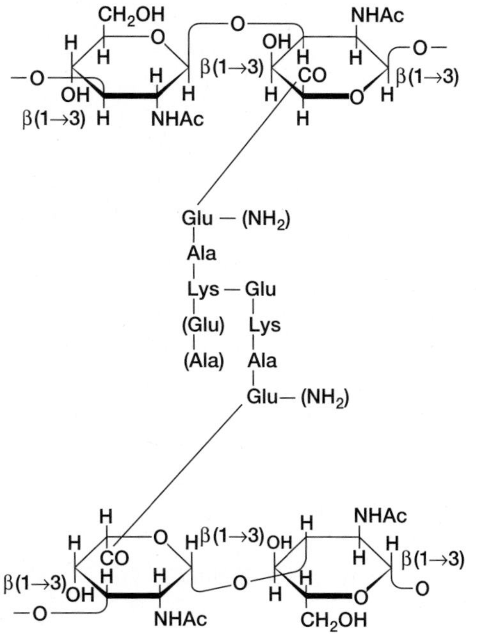

FIGURE 3.29 The Structure of Pseudomurein. The amino acids and amino groups in parentheses are not always present. Ac represents the acetyl group.

1. How do the cell envelopes of *Archaea* differ from those of *Bacteria*?
2. Compare and contrast bacterial and archaeal membranes.
3. Discuss the ways bacteria and archaea adjust the lipid content of their membranes in response to environmental conditions.
4. What is pseudomurein? How is it similar to peptidoglycan? How is it different?
5. Archaea with cell walls consisting of a thick, homogeneous layer of complex polysaccharides often retain the crystal violet dye when stained using the Gram-staining procedure. Why do you think this is so?
6. The structure of archaeal membrane lipids is considered additional evidence that these microbes belong to their own domain. Why is this so?

3.5 Cytoplasm of *Bacteria* and *Archaea*

The plasma membrane and everything within is called the **protoplast.** The **cytoplasm** is the material bounded by the plasma membrane; thus the cytoplasm is a major part of the protoplast. For many years bacterial and archaeal cells were thought of as bags of water. Randomly distributed in the water were structures such as inclusions, ribosomes, the nucleoid, and plasmids. But the exciting discovery of cytoskeletal proteins in both bacterial and archaeal cells has forever changed that view. It is now clear that the cytoplasm of bacterial cells is highly organized, just as it is in eukaryotic cells. The bacterial cytoskeleton and its many functions are the focus of many studies, and we begin with a discussion of it. We then consider a variety of structures found in the cytoplasm of both bacterial and archaeal cells.

Bacterial Cytoskeleton

The eukaryotic cytoskeletal elements are called microfilaments, microtubules, and intermediate filaments. Microfilaments and microtubules are formed when specific proteins come together to form filamentous structures. Microfilaments are made from actin, and microtubules are made from tubulin. Intermediate filaments are composed of a mixture of one or more members of different classes of proteins. Homologues of all three types of eukaryotic proteins have been identified in bacteria, and two have been identified in archaea (**table 3.2**). The bacterial cytoskeletal proteins are structurally similar to their eukaryotic counterparts and carry out similar functions: they participate in cell division, localize proteins to certain sites in the cell, and determine cell shape (table 3.2). In addition, some cytoskeletal proteins appear to be unique to the *Bacteria*. Thus it is likely that the evolution of the cytoskeleton was an early event in the history of life on Earth. ▶▶| *Cytoplasm of eukaryotes (section 4.3); Bacterial cell cycle (section 7.2)*

The cytoskeletons of *Escherichia coli, Bacillus subtilis,* and *Caulobacter crescentus* are the best studied and are the focus of this discussion. These three organisms are important bacterial model systems for several reasons. *E. coli* is a gram-negative rod that has been extensively studied and can be easily manipulated. *B. subtilis* is a gram-positive rod found in soil. It is an endospore-forming bacterium making it a good model for cellular differentiation. *C. crescentus* is a curved rod found in aquatic habitats. It is of interest in part because it exhibits an interesting life cycle that includes two different stages: a motile swarmer cell and a sessile, stalked cell that attaches to surfaces by a holdfast (*see figure 20.10*). ▶▶| Caulobacteraceae (*section 20.1*)

The best studied bacterial cytoskeletal proteins are FtsZ, MreB, and CreS (also known as crescentin). FtsZ was one of the first cytoskeletal proteins identified in bacteria and has since been found in most bacteria and many archaea. FtsZ is a homologue of the eukaryotic protein tubulin. It forms a ring at the center of a dividing cell and is required for the formation of the septum that will separate the daughter cells (**figure 3.30*a*).** MreB and its relative Mbl are actin homologues. Their major function is to maintain cell shape in rod-shaped cells; indeed, MreB is not found in cocci. The shape of a bacterial cell is the same as that of the peptidoglycan sac lying outside the plasma membrane. MreB and Mbl maintain cell shape by properly positioning the machinery needed for peptidoglycan synthesis (figure 3.30*b, c*). CreS (crescentin) was discovered in *C. crescentus* and is responsible for its curved shape (figure 3.30*d*). CreS is a homologue of lamin and keratin, two intermediate filament proteins. ▶▶| *Cellular growth and determination of cell shape* (*section 7.2*)

Table 3.2 Bacterial Cytoskeletal Proteins

Type	*Function*	*Comments*
Tubulin homologs		
FtsZ	Cell division	Widely observed in *Bacteria* and *Archaea*
BtubA/BtubB	Unknown	Observed only in *Prosthecobacter*; thought to be encoded by eukaryotic tubulin genes obtained by horizontal gene transfer
Actin homologs		
FtsA	Cell division	Observed in many bacterial species
MamK	Positioning magnetosomes	Observed in magnetotactic species
MreB/Mbl	Maintains cell shape, segregates chromosomes, localizes proteins	Most rod-shaped bacteria
Intermediate filament homologs		
CreS (crescentin)	Induces curvature in curved rods	*Caulobacter crescentus*
Unique bacterial cytoskeletal proteins		
MinD	Prevents polymerization of FtsZ at cell poless	Many rod-shaped bacteria
ParA (chromosome-encoded form)	Segregates chromosomes	Observed in many species including *Vibrio cholerae* and *C. crescentus*

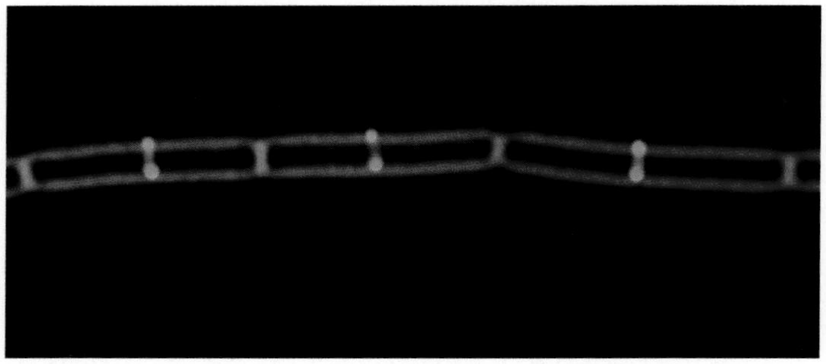

(a) FtsZ

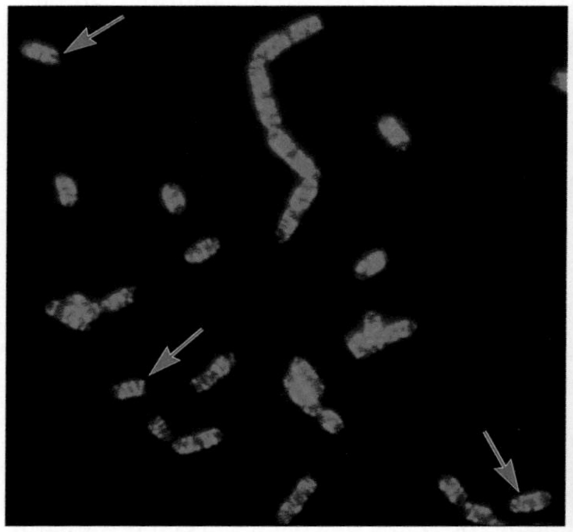

(b) Mbl

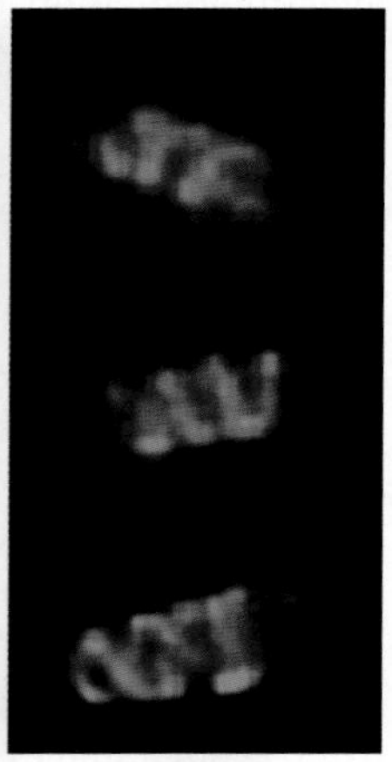

(c) Mbl

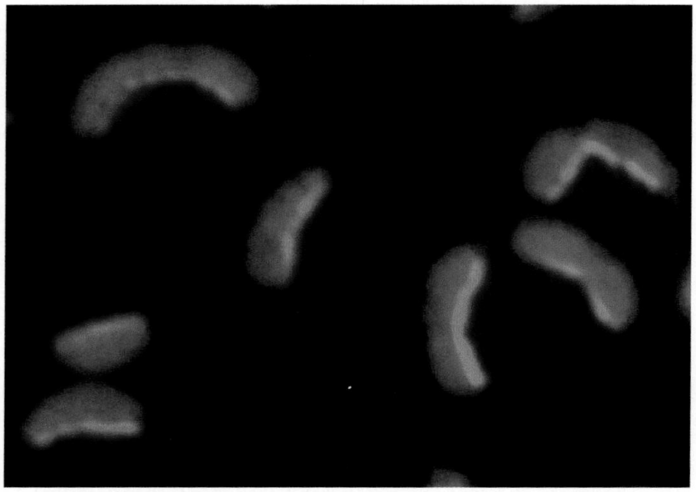

(d) Crescentin

FIGURE 3.30 The Bacterial Cytoskeleton. (a) FtsZ protein of *Bacillus subtilis*; FtsZ-green fluorescent (GFP) fusion protein viewed by fluorescence microscopy. (b) Visualization of the MreB-like cytoskeletal protein (Mbl) of *Bacillus subtilis;* Mbl-GFP in live cells was examined by fluorescence microscopy. Arrows point to the helical cytoskeletal cables that extend the length of the cells. (c) Three of the cells from (b) are shown at a higher magnification. (d) CreS (crescentin), in red, of *Caulobacter crescentus.* The DNA in the cells was stained blue with DAPI.

Intracytoplasmic Membranes

Although *Bacteria* and *Archaea* do not contain complex membranous organelles like mitochondria or chloroplasts, internal membranous structures are observed in some bacteria (**figure 3.31**). These can be extensive and complex in photosynthetic bacteria and in bacteria with very high respiratory activity, such as the nitrifying bacteria. The internal membranes of the photosynthetic cyanobacteria are called thylakoids and are analogous to the thylakoids of chloroplasts. ▶▶| *Photosynthetic bacteria (section 19.3);*

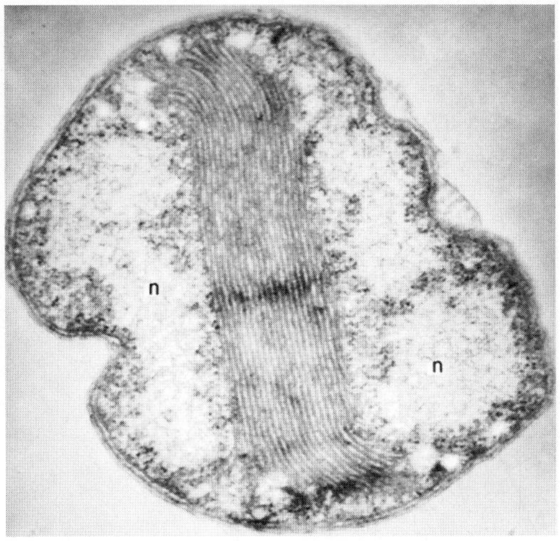

(a)

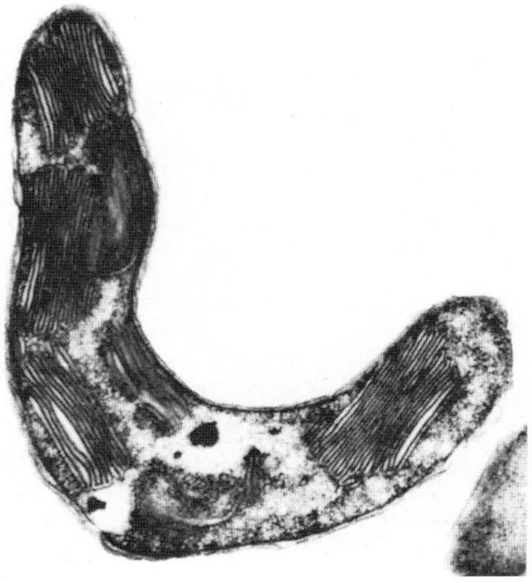

(b)

FIGURE 3.31 Internal Bacterial Membranes. Membranes of nitrifying and photosynthetic bacteria. (a) *Nitrocystis oceanus* with parallel membranes traversing the whole cell. Note nucleoid (n) with fibrillar structure. (b) *Ectothiorhodospira mobilis* with an extensive intracytoplasmic membrane system (×60,000).

Nitrifying bacteria (section 20.1); Organelles involved in energy conservation (section 4.6)

The internal membranous structures observed in bacteria may be aggregates of spherical vesicles, flattened vesicles, or tubular membranes. They are often connected to the plasma membrane and are thought to arise from it by invagination. However, these internal membranes differ from the plasma membrane by being enriched for proteins and other molecules involved in energy conservation. For instance, the thylakoids of cyanobacteria contain the chlorophyll and photosynthetic reaction centers responsible for converting light energy into ATP, the energy currency used by cells. Thus the function of internal membranes may be to provide a larger membrane surface for greater metabolic activity.

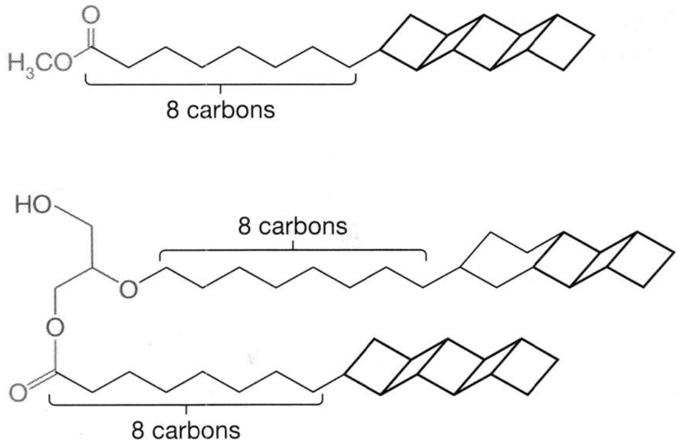

FIGURE 3.32 Ladderane Lipids.

One internal membrane deserves additional comment. Recall that in some members of the phylum *Planctomycetes,* a membrane-bound organelle called the anammoxosome is observed (*see figure 19.10*). This organelle is the site of anaerobic ammonia oxidation and is unique to these bacteria. The membrane of this organelle contains an unusual group of lipids called ladderane lipids (**figure 3.32**). Ladderane lipids are defined by the presence of two or more fused cyclobutane rings. They pack together very tightly, and this characteristic is thought to be important to anammoxosome function.

Inclusions

Inclusions are common in all cells. They are formed by the aggregation of substances that may be either organic or inorganic. The first bacterial inclusions were discovered in the late 1800s. Since then much has been learned about their structure and function. Inclusions can take the form of granules, crystals, or globules; some are amorphous. Some inclusions lie free in the cytoplasm. Other inclusions are enclosed by a shell or membrane that is single-layered and may consist of proteins or of both proteins and phospholipids. Some inclusions are surrounded by invaginations of the plasma membrane. Many inclusions are used for storage (e.g., of carbon compounds, inorganic substances, and energy) or to reduce osmotic pressure by tying up molecules in particulate form. The quantity of inclusions used for storage varies with the nutritional status of the cell. Some inclusions are so distinctive that they are increasingly being referred to as microcompartments. A brief description of several important inclusions follows.

Storage Inclusions

Cells have a wide variety of storage inclusions. Many are formed when one nutrient is in ready supply but another nutrient is not. Some store end products of metabolic processes. In some cases these end products are used by the microbe when it is in different environmental conditions. The most common storage inclusions are glycogen inclusions, polyhydroxyalkonate granules, sulfur globules, and polyphosphate granules. Some storage inclusions, such as the cyanophycin granules in cyanobacteria, are observed only in certain organisms.

Glycogen is a long branched chain of glucose units. Glycogen inclusions are found in both bacterial and archaeal cells but have been best studied in bacteria. Glycogen inclusions usually form when bacteria are growing in an environment that is limited for an important nutrient (e.g., phosphate) but contains excess carbon. Thus glycogen inclusions serve to store carbon until the missing nutrient becomes available. Most glycogen inclusions are not bound by a membrane, but some are surrounded by a single-layered membrane. Other glycogen inclusions that are not membrane bound are dispersed evenly throughout the cytoplasm as small granules (about 20 to 100 nm in diameter) and often can be seen only with the electron microscope. If cells contain a large amount of glycogen, staining with an iodine solution will turn the cells reddish-brown. ▶▶| *Carbohydrates (appendix I)*

Carbon is also stored as polyhydroxyalkonate (PHA) granules. Several types of PHA granules have been identified, but the most common granules contain **poly-β-hydroxybutyrate (PHB).** PHB contains β-hydroxybutyrate molecules joined by ester bonds between the carboxyl and hydroxyl groups of adjacent molecules. PHB accumulates in distinct bodies, around 0.2 to 0.7 μm in diameter, that are readily stained with Sudan black for light microscopy and are seen as empty "holes" in the electron microscope (**figure 3.33*a***). This is because PHB is hydrophobic, so it is dissolved by the solvents used to prepare specimens for electron microscopy. The structure of PHB inclusions has been well studied, and PHB granules are now known to be surrounded by a single-layered membrane composed of proteins and a small amount of phospholipids (figure 3.33*b*). Much of the interest in PHB and other PHA granules is due to their industrial use in making biodegradeable plastics. ▶▶| *Biopolymers (section 41.3)*

Polyphosphate granules and sulfur globules are inorganic inclusions observed in many organisms. Many bacteria store phosphate as **polyphosphate granules,** also called volutin granules or metachromatic granules. Polyphosphate is a linear polymer of orthophosphates joined by ester bonds. Thus polyphosphate granules store the phosphate needed for synthesis of important cell constituents such as nucleic acids. In some cells they act as an energy reserve, and polyphosphate also can serve as an energy source in some reactions. Polyphosphate granules are called metachromatic granules because they show the metachromatic effect; that is, they appear red or blue when stained with the blue dyes methylene blue or toluidine blue. Sulfur globules are formed by bacteria that use reduced sulfur-containing compounds as a source of electrons during their energy-conserving metabolic processes **(figure 3.34).** For example, some photosynthetic bacteria can use hydrogen sulfide (rather than water) as an electron donor and accumulate the resulting sulfur either externally or internally. ▶▶| *Light reactions in anoxygenic photosynthesis (section 10.12);* Gammaproteobacteria *(section 20.3)*

Cyanophycin granules are observed in cyanobacteria, a group of photosynthetic bacteria. These inclusions are composed of large polypeptides containing approximately equal amounts of the amino acids arginine and aspartic acid. The formation of these granules is of particular interest because the cyanophycin granule polypeptide is not encoded by mRNA and is not synthesized by ribosomes. The granules often are large

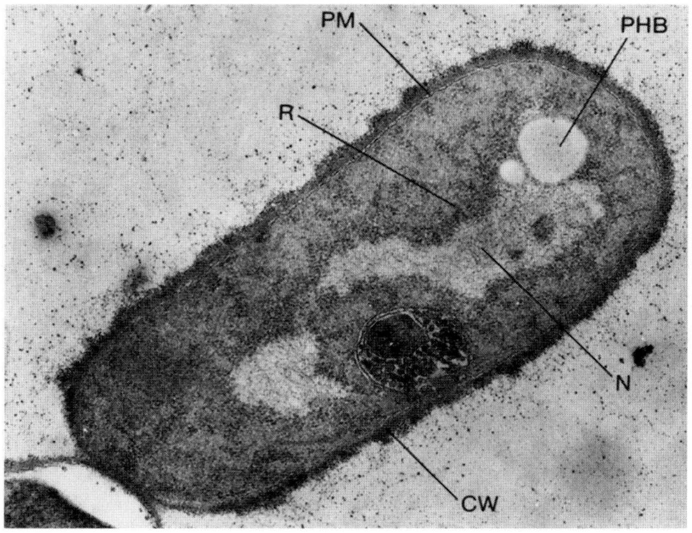

(a)

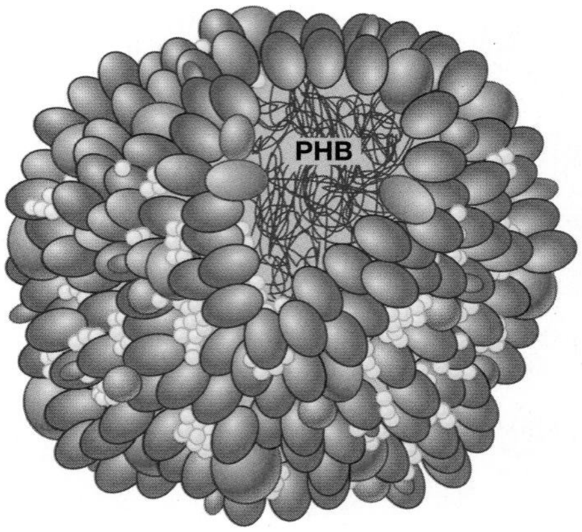

(b)

FIGURE 3.33 PHB Inclusions in Bacteria. (a) Electron micrograph of *Bacillus megaterium* (×30,500). PHB, poly-β-hydroxybutyrate inclusion; CW, cell wall; N, nucleoid; PM, plasma membrane; and R, ribosomes. (b) Structure of a PHB granule. PHB is enclosed by a membrane composed of several different proteins, including the PHB-synthesizing enzyme (red sphere) and the PHB-degrading enzyme (green sphere). Yellow spheres represent the phospholipids that are also found in the membrane. Note that the membrane is not a phospholipid bilayer.

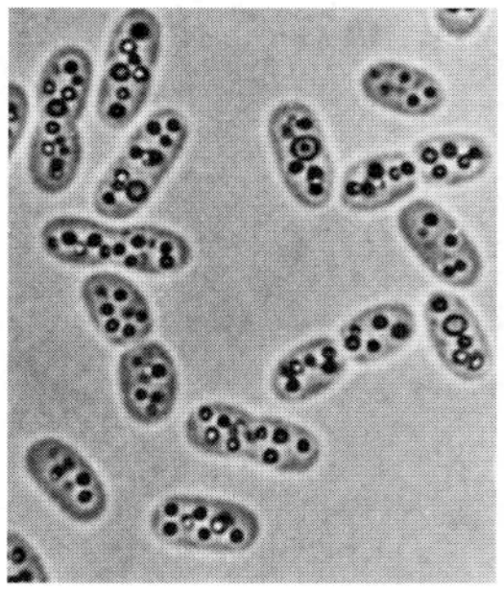

FIGURE 3.34 Sulfur Globules. *Chromatium vinosum*, a purple sulfur bacterium, with intracellular sulfur globules, bright-field microscopy (×2,000).

enough to be visible in the light microscope and store extra nitrogen for the bacteria.

Microcompartments

Some bacterial inclusions are very unique and serve functions other than simply storing substances for later use by the cell. Many researchers now refer to these inclusions as microcompartments. Although microcompartments are not bound by a lipid bilayer, some scientists feel they are analogous to membrane-bound organelles such as mitochondria. The best studied microcompartment is the carboxysome.

Carboxysomes are present in many cyanobacteria and other CO_2-fixing bacteria (**figure 3.35**). They, like other microcompartments, consist of a protein coat that is polyhedral, a shape similar to that of certain viruses. The polyhedral coat is composed of 6 to 10 different proteins and is about 100 nm in diameter. One of the proteins in the shell is the enzyme carbonic anhydrase, which converts carbonic acid into CO_2 and releases it into the lumen of the carboxysome. The nature of the carboxysome shell prevents the CO_2 from escaping; thus the carboxysome concentrates CO_2. Also enclosed within the polyhedron is the enzyme ribulose-1, 5-bisphosphate carboxylase (Rubisco). Rubisco is the critical enzyme for CO_2 fixation, the process of converting CO_2 into sugar. Thus the carboxysome also serves as a site for CO_2 fixation. ▶▶| *CO_2 fixation (section 11.3)*

Other Inclusions

Inclusions can be used for functions other than storage or as microcompartments. Two of the most remarkable inclusions are gas vacuoles and magnetosomes. Both are involved in the movement of microbes.

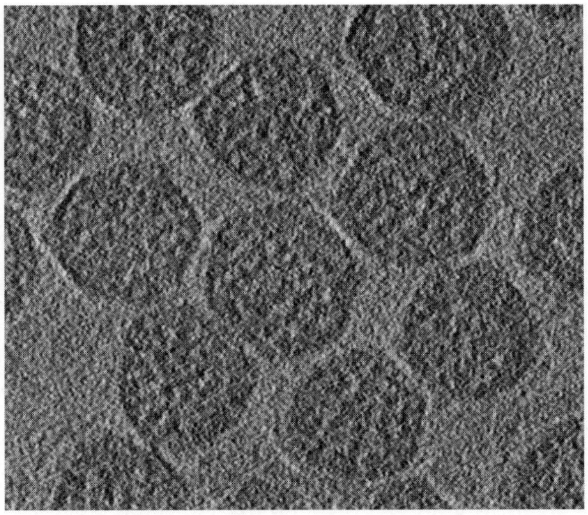

FIGURE 3.35 Carboxysomes. Carboxysomes in the bacterium *Halothiobacillus neapolitanus.* This is one image of a tilt series taken for electron cryotomography. Each carboxysome is approximately 100 nm in diameter.

The **gas vacuole** provides buoyancy to some aquatic bacteria and archaea. Gas vacuoles are present in many photosynthetic bacteria, aquatic archaea such as *Halobacterium* (a salt-loving archaeon), and some aquatic bacteria that are not photosynthetic (e.g., *Thiothrix,* a filamentous bacterium). Gas vacuoles are aggregates of enormous numbers of small, hollow, cylindrical structures called **gas vesicles (figure 3.36).** Gas vesicle walls are composed entirely of a single small protein. These protein subunits assemble to form a rigid cylinder that is impermeable to water but freely permeable to atmospheric gases. Cells with gas vacuoles can regulate their buoyancy to float at the depth necessary for proper light intensity, oxygen concentration, and nutrient levels. They descend by simply collapsing vesicles and float upward when new ones are constructed.

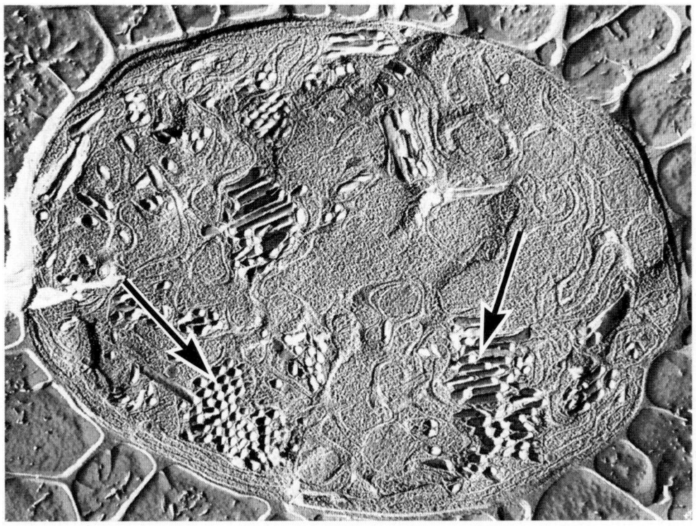

(a)

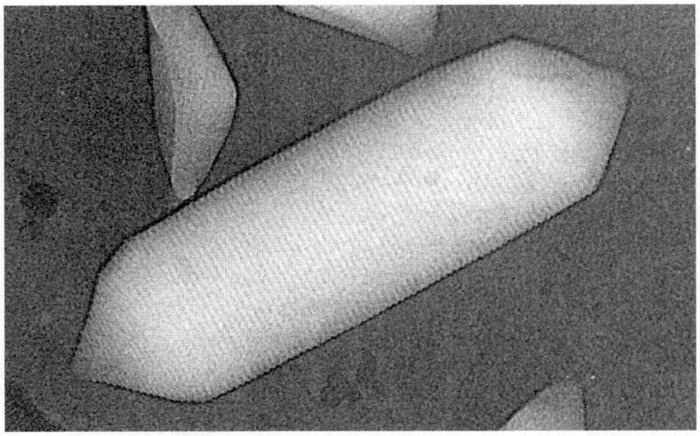

(b)

FIGURE 3.36 Gas Vacuoles and Gas Vesicles. (a) A freeze-fracture preparation of *Anabaena flosaquae* (×89,000) showing gas vesicles and gas vacuoles. Clusters of the cylindrical vesicles form gas vacuoles. Both longitudinal and cross-sectional views of gas vesicles are indicated by arrows. (b) Gas vesicles of *Halobacterium salinarum* (~ ×150,000).

Aquatic magnetotactic bacteria use **magnetosomes** to orient themselves in Earth's magnetic field. Magnetosomes are intracellular chains of magnetite (Fe_3O_4) particles (**figure 3.37**). They are around 35 to 125 nm in diameter and enclosed within invaginations of the plasma membrane. Since each iron particle is a tiny magnet, the Northern Hemisphere bacteria use their magnetosome chain to determine northward and downward directions, and swim down to nutrient-rich sediments or locate the optimum depth in freshwater and marine habitats. Magnetotactic bacteria in the Southern Hemisphere generally orient southward and downward, with the same result. For the cell to move properly within a magnetic field, the magnetosomes must be arranged in a chain. A cytoskeletal protein called MamK is currently thought to be responsible for establishing a framework upon which the chain can form (figure 3.37*b*).

Ribosomes

Ribosomes are the site of protein synthesis, and large numbers of them are found in nearly all cells. The cytoplasm of bacterial and archaeal cells is often packed with ribosomes, and other ribosomes may be loosely attached to the plasma membrane. The cytoplasmic ribosomes synthesize proteins destined to remain within the cell, whereas plasma membrane-associated ribosomes make proteins that will reside in the cell envelope or are transported to the outside.

Translation, the process of protein synthesis, is amazingly complex and is discussed in detail in chapter 12. This complexity is evidenced in part by the structure of ribosomes, which are made of numerous proteins and several ribonucleic acid (RNA) molecules. Although the overall morphology and makeup of ribosomes is similar across all domains, there are some important differences. The similarities and some of the differences between bacterial and archaeal ribosomes are discussed here.

Bacterial and archaeal ribosomes are called 70S ribosomes (as opposed to 80S in eukaryotes) and are constructed of a 50S and a 30S subunit (**figure 3.38**). The S in 70S and similar values stands for **Svedberg unit.** This is the unit of the sedimentation coefficient, a measure of the sedimentation velocity in a centrifuge; the faster a particle travels when centrifuged, the greater its Svedberg value or sedimentation coefficient. The sedimentation coefficient is a function of a particle's molecular weight, volume, and shape. Heavier and more compact particles normally have larger Svedberg numbers and sediment faster. Thus bacterial and archaeal ribosomes are smaller than the ribosomes of eukaryotic cells.

Despite their similar size, the makeup of bacterial ribosomes and their shape are somewhat different than archaeal ribosomes. Both have ribosomal RNA (rRNA) molecules of similar size: 16S in the small subunit, and 23S and 5S in the large subunit. However, at least one archaeon has an additional rRNA, a 5.8S rRNA, in the large subunit. This is of interest because the large subunit of eukaryotic ribosomes contains both 5S and 5.8S rRNA molecules. The protein composition of bacterial and archaeal ribosomes also differs. Bacterial ribosomes have about 55 proteins, archaeal ribosomes about 68, and eukaryotic ribosomes about 78. Some of the ribosomal proteins are similar across the three domains, but others are observed only in archaeal and eukaryotic ribosomes. All the ribosomal proteins present in both *Archaea* and *Bacteria* are also seen in the *Eukarya*. Thus

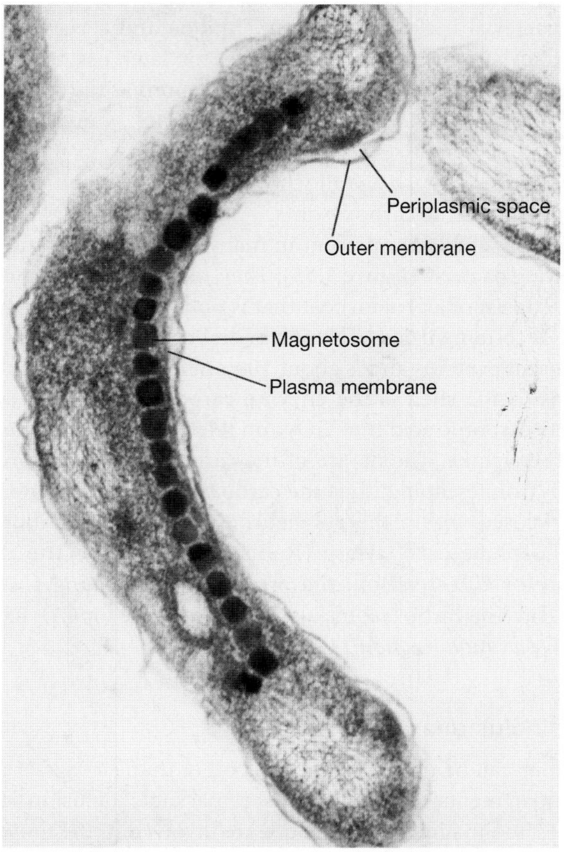

(a)

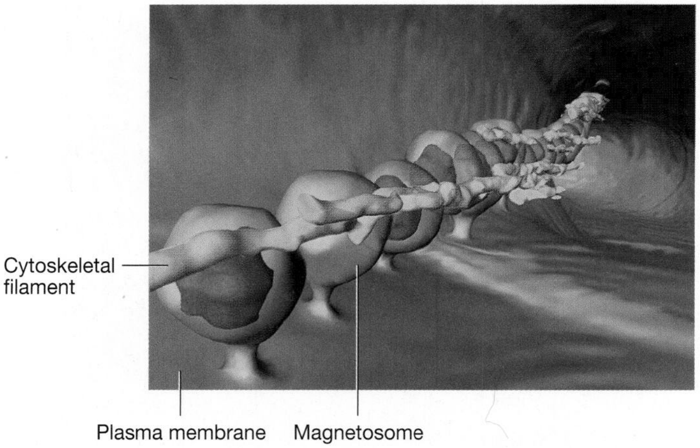

(b)

FIGURE 3.37 Magnetosomes. (a) Transmission electron micrograph of the magnetotactic bacterium *Aquaspirillum magnetotacticum* (×123,000). (b) An electron cryotomography three-dimensional reconstruction of the magnetosomes of *Magnetospirillum magneticum.*

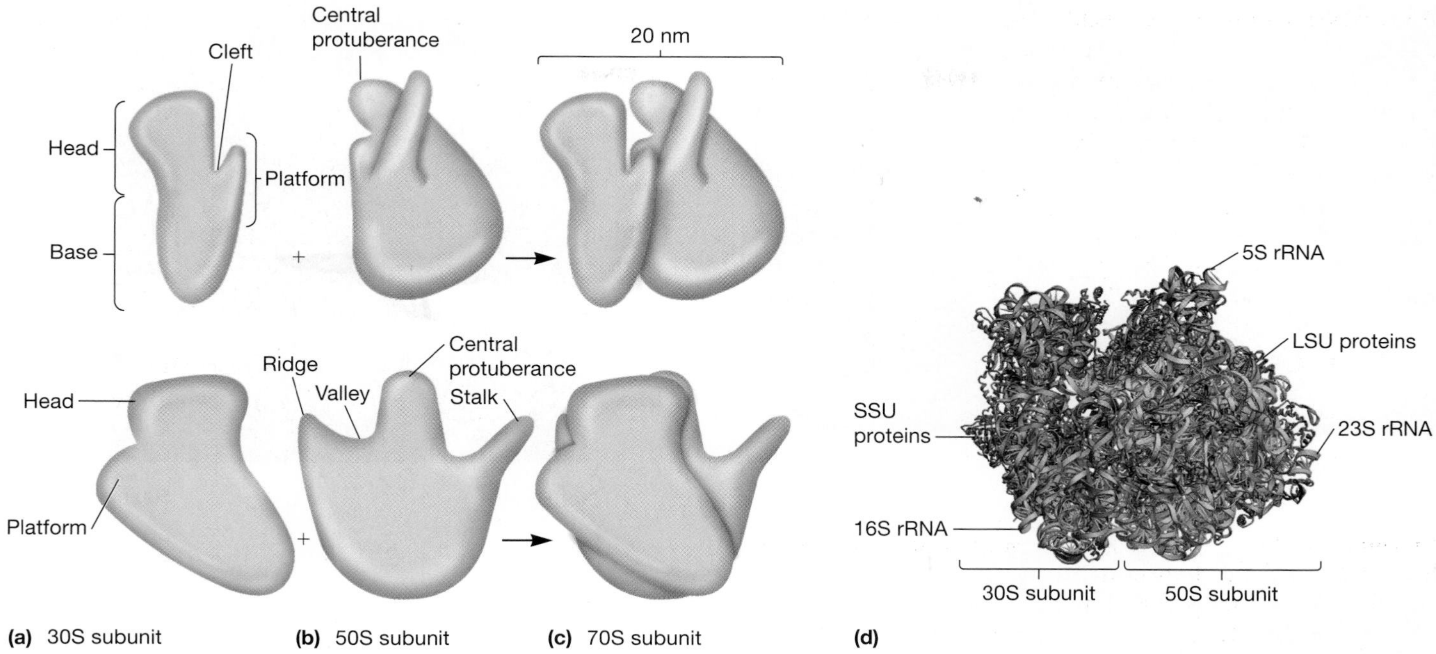

FIGURE 3.38 Bacterial Ribosomes. (a–c) Schematic representation of the two subunits and the complete 70S ribosome of *Escherichia coli.* (d) The molecular structure of the 70S ribosome of *Thermus thermophilus.* The 50S subunit (LSU) includes 23S rRNA (gray) and 5S rRNA (lavender), while 16S rRNA (turquoise) is found in the 30S subunit (SSU). A molecule of tRNA (gold) is shown in the A site. To generate this ribbon diagram, crystals of purified bacterial ribosomes were grown, exposed to X rays, and the resulting diffraction pattern analyzed.

there are no ribosomal proteins that might be referred to as prokaryotic (i.e., present only in the ribosomes of *Archaea* and *Bacteria*). The similarities in protein makeup between archaeal and eukaryotic ribosomes are consistent with the observation that the overall shape of archaeal ribosomes is more similar to that of eukaryotic ribosomes.

Nucleoid

The **nucleoid** (other names are also used: the nuclear body, chromatin body, and nuclear region) is an irregularly shaped region that contains the cell's chromosome and numerous proteins (**figure 3.39**). The chromosomes of most bacteria and all known archaea are a single circle of double-stranded **deoxyribonucleic acid (DNA),** but some bacteria have a linear chromosome, and some bacteria, such as *Vibrio cholerae* and *Borrelia burgdorferi* (the causative agents of cholera and Lyme disease, respectively), have more than one chromosome.

Bacterial and archaeal chromosomes are longer than the length of the cell. Thus an important and still unanswered question is how these microbes manage to fit their chromosomes into the relatively small space occupied by the nucleoid. For instance, *Escherichia coli*'s circular chromosome measures approximately 1,400 μm, or about 230–700 times longer than the cell (figure 3.39*b*). Thus the chromosome must be compacted in some way. It is thought that much of the compaction is the result of supercoiling, which produces a dense, central core of DNA with loops extending out from the core. There is evidence that some of the proteins found in the nucleoid also contribute to packing the DNA into a smaller space. In bacteria, the protein HU is thought to be important. An HU homologue is also found in some archaea. Most other archaea have histones associated with their chromosomes. These histones form nucleosomes that are similar to the nucleosomes observed in eukaryotes (*see figure 4.11*). During cell division, bacterial and archaeal chromosomes are further condensed by proteins called condensins. This extra level of packing is important for proper segregation of daughter chromosomes during cell division. ▶▶| *Nucleus (section 4.5)*

For most bacteria and all known archaea, the nucleoid is simply a region in the cytoplasm; it is not separated from other components of the cytoplasm by a membrane. However, there are a few exceptions. Membrane-bound DNA-containing regions are present in at least two genera of the unusual bacterial phylum *Planctomycetes (see figure 19.10). Pirellula* has a single membrane that surrounds a region called the pirellulosome, which contains a fibrillar nucleoid and ribosome-like particles. The nuclear body of *Gemmata obscuriglobus* is bounded by two membranes. More work is required to determine the functions of these membranes and how widespread this phenomenon is. ▶▶| *Phylum* Planctomycetes *(section 19.4)*

Plasmids

In addition to the genetic material present in the nucleoid, many bacteria, archaea, and some yeasts and other fungi contain extrachromosomal DNA molecules called plasmids. Indeed, most of

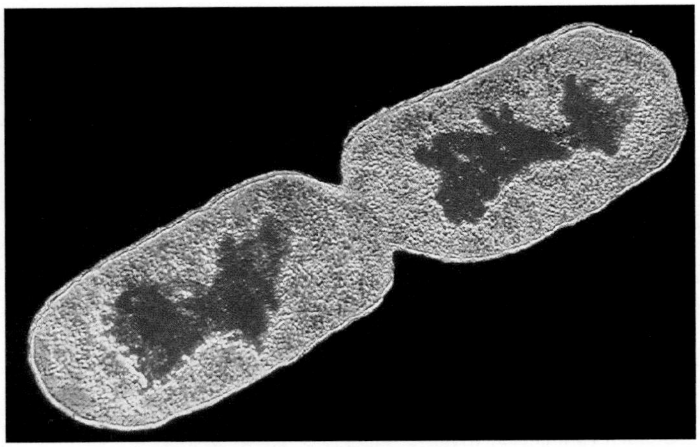

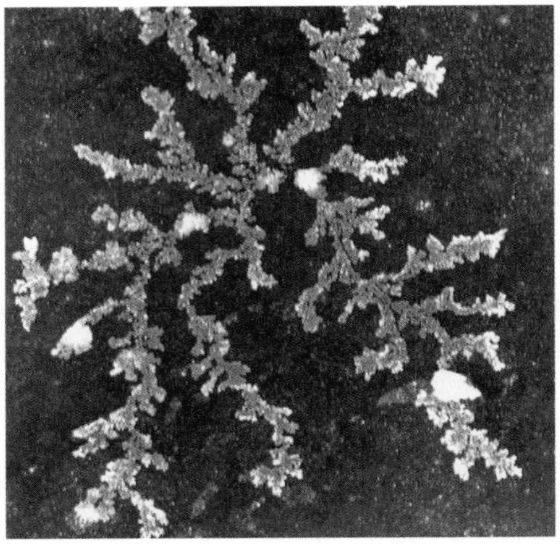

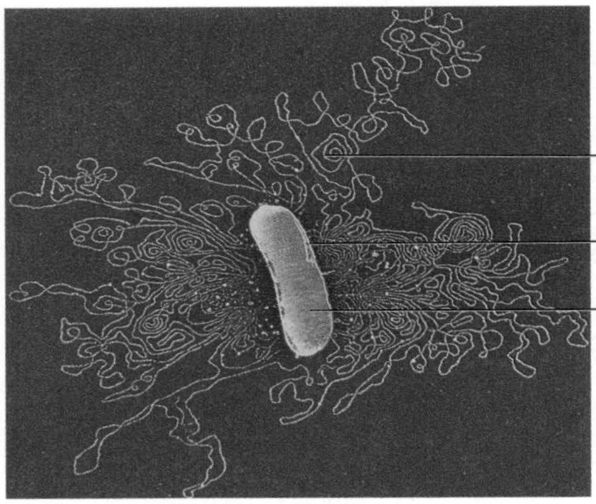

FIGURE 3.39 ***E. coli* Nucleoids and Chromosomes.** Bacterial and archaeal chromosomes are located in the nucleoid, an area in the cytoplasm. (a) A color-enhanced transmission electron micrograph of a thin section of a dividing *E. coli* cell. The red areas are the nucleoids present in the two daughter cells. (b) Chromosome released from a gently lysed *E. coli* cell. Note how tightly packaged the DNA must be inside the cell. (c) Atomic force micrograph of an *E. coli* nucleoid. The image was prepared from a culture in the stationary phase of the growth cycle. In that phase, the rate of cell division is the same as the rate of cell death; thus the culture does not increase in size.

the bacterial and archaeal genomes sequenced thus far include plasmids. In some cases, numerous different plasmids within a single species have been identified. For instance, *Borrelia burgdorferi,* which causes Lyme disease, carries 12 linear and nine circular plasmids. Plasmids play many important roles in the lives of the organisms that have them. They also have proved invaluable to microbiologists and molecular geneticists in constructing and transferring new genetic combinations and in cloning genes, as described in chapter 15.

Plasmids are small, double-stranded DNA molecules that can exist independently of the chromosome. Both circular and linear plasmids have been documented, but most known plasmids are circular. Plasmids have relatively few genes, generally less than 30. Their genetic information is not essential to the host, and cells that lack them usually function normally. However, many plasmids carry genes that confer a selective advantage to their hosts in certain environments.

Plasmids are able to replicate autonomously. That is, plasmid and chromosomal replication are independent. Single-copy plasmids produce only one copy per host cell. Multicopy plasmids may be present at concentrations of 40 or more per cell. Some plasmids are able to integrate into the chromosome and thus are replicated with the chromosome. Such plasmids are called **episomes.** Plasmids are inherited stably during cell division, but they are not always equally apportioned into daughter cells and sometimes are lost. The loss of a plasmid is called **curing.** It can occur spontaneously or be induced by treatments that inhibit plasmid replication but not host cell reproduction. Some commonly used curing treatments are acridine mutagens, ultraviolet and ionizing radiation, thymine starvation, antibiotics, and growth above optimal temperatures.

Plasmids may be classified in terms of their mode of existence, spread, and function. A brief summary of the types of bacterial plasmids and their properties is given in **table 3.3.** **Conjugative plasmids** are of particular note because they can transfer copies of themselves to other bacteria during conjugation. Perhaps the best-studied conjugative plasmid is the **F factor** (fertility factor or F plasmid) of *E. coli,* which is discussed in detail in chapter 14. Some conjugative plasmids are also **R plasmids (resistance factors, R factors).** R plasmids confer antibiotic resistance to the cells that contain them. Conjugative R factors are therefore important in the spread of antibiotic resistance among bacteria. ▶▶| *Bacterial conjugation (section 14.7)*

Table 3.3 Major Types of Bacterial Plasmids

Type	*Representatives*	*Approximate Size (kbp)*	*Copy Number (Copies/ Chromosome)*	*Hosts*	*Phenotypic Features*[a]
Conjugative Plasmids[b]	F factor	95–100	1–3	*E. coli, Salmonella, Citrobacter*	Sex pilus, conjugation
R Plasmids	RP4	54	1–3	*Pseudomonas* and many other gram-negative bacteria	Sex pilus, conjugation, resistance to Amp, Km, Nm, Tet
	pSH6	21		*Staphylococcus aureus*	Resistance to Gm, Tet, Km
Col Plasmids	ColE1	9	10–30	*E. coli*	Colicin E1 production
	CloDF13	10	50–70	*E. coli*	Cloacin DF13
Virulence Plasmids	Ent (P307)	83		*E. coli*	Enterotoxin production
	Ti	200		*Agrobacterium tumefaciens*	Tumor induction in plants
Metabolic Plasmids	CAM	230		*Pseudomonas*	Camphor degradation
	TOL	75		*Pseudomonas putida*	Toluene degradation

[a]Abbreviations used for resistance to antibiotics: Amp, ampicillin; Gm, gentamycin; Km, kanamycin; Nm, neomycin; Tet, tetracycline.
[b]Many R plasmids, metabolic plasmids, and others are also conjugative.

Several other important types of plasmids have been discovered. These include bacteriocin-encoding plasmids, virulence plasmids, and metabolic plasmids. Bacteriocin-encoding plasmids may give the bacteria that harbor them a competitive advantage in the microbial world. Bacteriocins are proteins that destroy other, closely related bacteria. Col plasmids contain genes for the synthesis of bacteriocins known as colicins, which are produced by and directed against strains of *E. coli.* Plasmids in other bacteria carry genes for bacteriocins against other species. **Virulence plasmids** encode factors that make their hosts more pathogenic. For example, enterotoxigenic strains of *E. coli* cause traveler's diarrhea because they contain a plasmid that codes for an enterotoxin. **Metabolic plasmids** carry genes for enzymes that degrade substances such as aromatic compounds (toluene), pesticides (2,4-dichlorophenoxyacetic acid), and sugars (lactose). Metabolic plasmids even carry the genes required for some strains of *Rhizobium* to induce legume nodulation and carry out nitrogen fixation. ▶▶| *Order Rhizobiales (section 20.1)*

1. Briefly describe the nature and function of the cytoplasm, and the regions and structures within it.
2. What is the importance of bacterial cytoskeletal proteins? What do you think would be the outcome if you were able to "transplant" CreS into a rod-shaped bacterium such as *Bacillus subtilis*?
3. List the most common kinds of inclusions. How are they similar to eukaryotic organelles such as mitochondria and chloroplasts? How do they differ?
4. Relate the structure of a gas vacuole to its function. Why do you think gas vacuoles are bounded by proteins rather than a lipid bilayer membrane?
5. List three genera that are exceptional in terms of their chromosome or nucleoid structure. Suggest how the differences observed in these genera might impact how they function.
6. Give the major features of plasmids. How do they differ from chromosomes? What is an episome?
7. Describe each of the following plasmids and explain their importance: conjugative plasmid, F factor, R factor, Col plasmid, virulence plasmid, and metabolic plasmid.

3.6 External Structures

Many bacteria and archaea have structures that extend beyond the cell envelope. These external structures can function in protection, attachment to surfaces, horizontal gene transfer, and cell movement. Several are discussed in this section.

Pili and Fimbriae

Many bacteria and archaea have short, fine, hairlike appendages that are thinner than flagella. These are usually called **fimbriae** (s., **fimbria**) or **pili** (s., **pilus**). The terms are synonymous, although certain structures are always called pilus (e.g., sex pilus), while others are always called fimbriae. We will use the terms interchangeably, except in those instances. A cell may be covered with up to 1,000 fimbriae, but they are only visible in an electron microscope due to their small size (**figure 3.40**). They are slender tubes composed of helically arranged protein subunits and are about 3 to 10 nm in diameter and up to several micrometers long. Several different types of fimbriae have been identified in gram-negative bacteria. Most function to attach bacteria to solid surfaces such as rocks in streams and host tissues. One type, called type IV pili, are involved in motility (section 3.7) and the uptake of DNA during the process of bacterial transformation. Gram-positive bacteria have at least two types of pili; both are involved in attaching the bacteria to surfaces. ▶▶| *Transformation (section 14.8)*

Many bacteria have up to 10 **sex pili** (s., **sex pilus**) per cell. These hairlike structures differ from other pili in the following ways. Sex pili often are larger than other pili (around 9 to 10 nm in diameter). They are genetically determined by conjugative plasmids and are required for conjugation. Some bacterial viruses attach specifically to receptors on sex pili at the start of their reproductive cycle. ▶▶| *Bacterial conjugation (section 14.7)*

Flagella

Many motile bacteria and archaea move by use of **flagella** (s., **flagellum**), threadlike locomotor appendages extending outward from the plasma membrane and cell wall. Although the main function of flagella is motility, they can have other roles. Flagella are important for certain types of swarming behavior. They can be involved in attachment to surfaces, and in some bacteria they are virulence factors. Bacterial flagella are the best studied and are considered first. ▶▶| *Microbial growth on solid surfaces (section 6.8)*

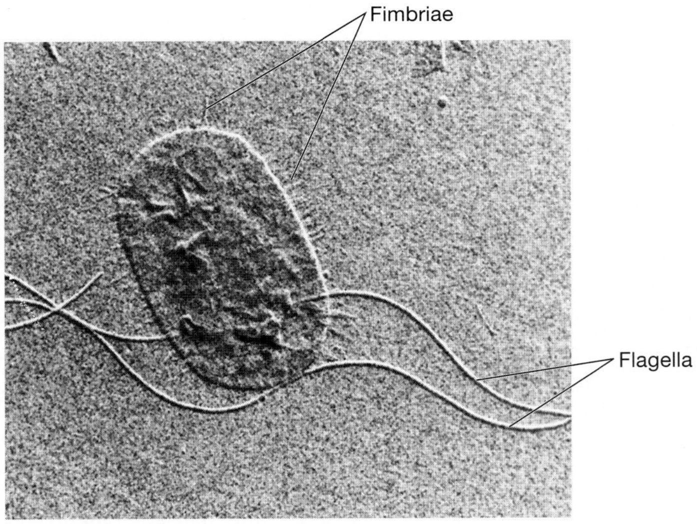

FIGURE 3.40 Flagella and Fimbriae. The long flagella and numerous shorter fimbriae are evident in this electron micrograph of the bacterium *Proteus vulgaris* (×39,000).

Bacterial Flagella

Bacterial flagella are slender, rigid structures about 20 nm across and up to 20 µm long. Flagella are so thin they cannot be observed directly with a bright-field microscope but must be stained with techniques designed to increase their thickness. The detailed structure of a flagellum can only be seen in the electron microscope.

Bacterial species often differ distinctively in their patterns of flagella distribution, and these patterns are useful in identifying bacteria. **Monotrichous** bacteria (*trichous* means hair) have one flagellum; if it is located at an end, it is said to be a **polar flagellum** (**figure 3.41*a***). **Amphitrichous** bacteria (*amphi* means on both sides) have a single flagellum at each pole. In contrast, **lophotrichous** bacteria (*lopho* means tuft) have a cluster of flagella

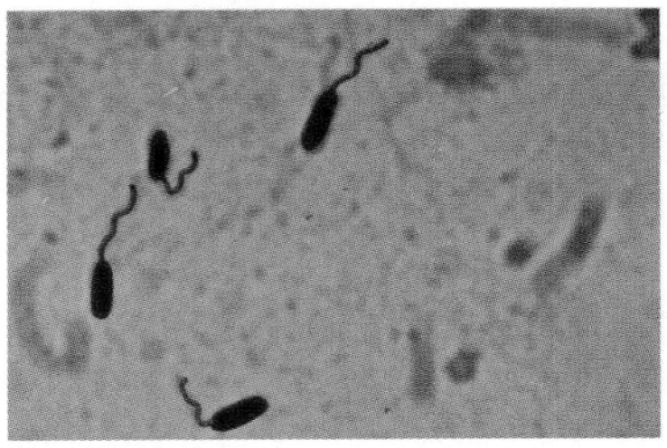

(a) *Pseudomonas*—monotrichous polar flagellation

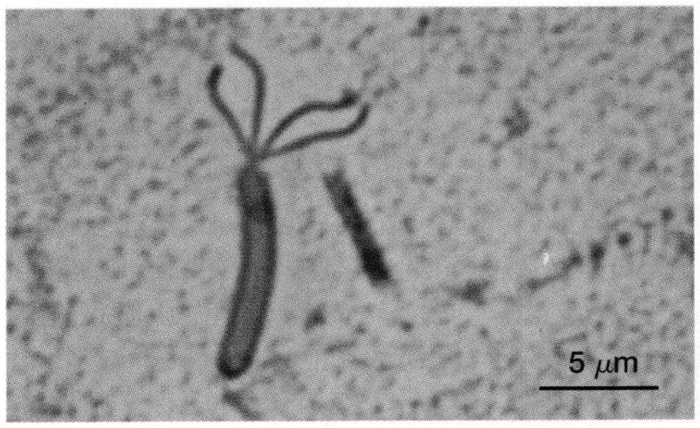

(b) *Spirillum*—lophotrichous flagellation

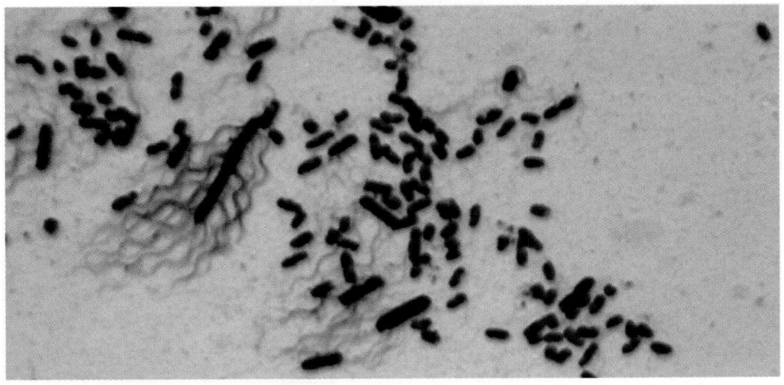

(c) *P. vulgaris*—peritrichous flagellation

FIGURE 3.41 Flagellar Distribution. Examples of various patterns of flagellation as seen in the light microscope. (a) Monotrichous polar (*Pseudomonas*). (b) Lophotrichous (*Spirillum*). (c) Peritrichous (*Proteus vulgaris,* ×600).

at one or both ends (figure 3.41*b*). Flagella are spread evenly over the whole surface of **peritrichous** (*peri* means around) bacteria (figure 3.41*c*).

Transmission electron microscope studies have shown that the bacterial flagellum is composed of three parts (**figure 3.42**). (1) The longest and most obvious portion is the **filament,** which extends from the cell surface to the tip. (2) The **basal body** is embedded in the cell; and (3) a short, curved segment, the **hook,** links the filament to its basal body and acts as a flexible coupling. The filament is a hollow, rigid cylinder constructed of subunits of the protein **flagellin,** which ranges in molecular weight from 30,000 to 60,000 daltons, depending on the bacterial species. The filament ends with a capping protein. Some bacteria have sheaths surrounding their flagella. For example, *Vibrio cholerae* has a lipopolysaccharide sheath.

The hook and basal body are quite different from the filament (figure 3.42). Slightly wider than the filament, the hook is made of different protein subunits. The basal body is the most complex part of a flagellum. In the earliest-made transmission electron micrographs of the basal bodies of *E. coli* and most other gram-negative bacteria, the basal body appeared to have four rings—L ring, P ring, S ring, and M ring—connected to a central rod (figure 3.42*a*). It is now known that the S ring and M ring are different portions of the same protein, and they are now referred to as the MS ring. A later discovery was the C ring, which is on the cytoplasmic side of the MS ring. Gram-positive bacteria have only two rings—an inner ring connected to the plasma membrane and an outer one probably attached to the peptidoglycan (figure 3.42*b*).

The synthesis of bacterial flagella is a complex process involving at least 20 to 30 genes. Besides the gene for flagellin, 10 or more genes code for hook and basal body proteins; other genes are concerned with the control of flagellar construction or function. How the cell regulates or determines the exact location of flagella is not known.

Because many components of the flagellum lie outside the cell wall, they must be transported across the plasma membrane and cell wall. Interestingly, evidence suggests that components of the basal body are evolutionarily related to a type of protein secretion system observed in gram-negative bacteria. This system, called a type III secretion system, has a needlelike structure through which proteins are secreted. The needle is thought to be analogous to the filament of the flagellum. Thus the flagellin subunits are transported by way of a type III-like secretion process through the filament's hollow internal core. When the subunits reach the tip, they spontaneously aggregate under the direction of a protein called the filament cap; thus the filament grows at its tip rather than at the base (**figure 3.43**). Filament synthesis, like S-layer formation, is an example of **self-assembly.** ▶▶| *Protein maturation and secretion (section 12.9)*

Archaeal Flagella

Archaeal flagella have not been as thoroughly studied as bacterial flagella. They are superficially similar to their bacterial counterparts, but important differences have been identified. Archaeal flagella are thinner than bacterial flagella (10 to 13 nm rather than 20 nm) and are composed of more than one type of flagellin subunit (**figure 3.44**). The flagellum is not hollow. Archaeal

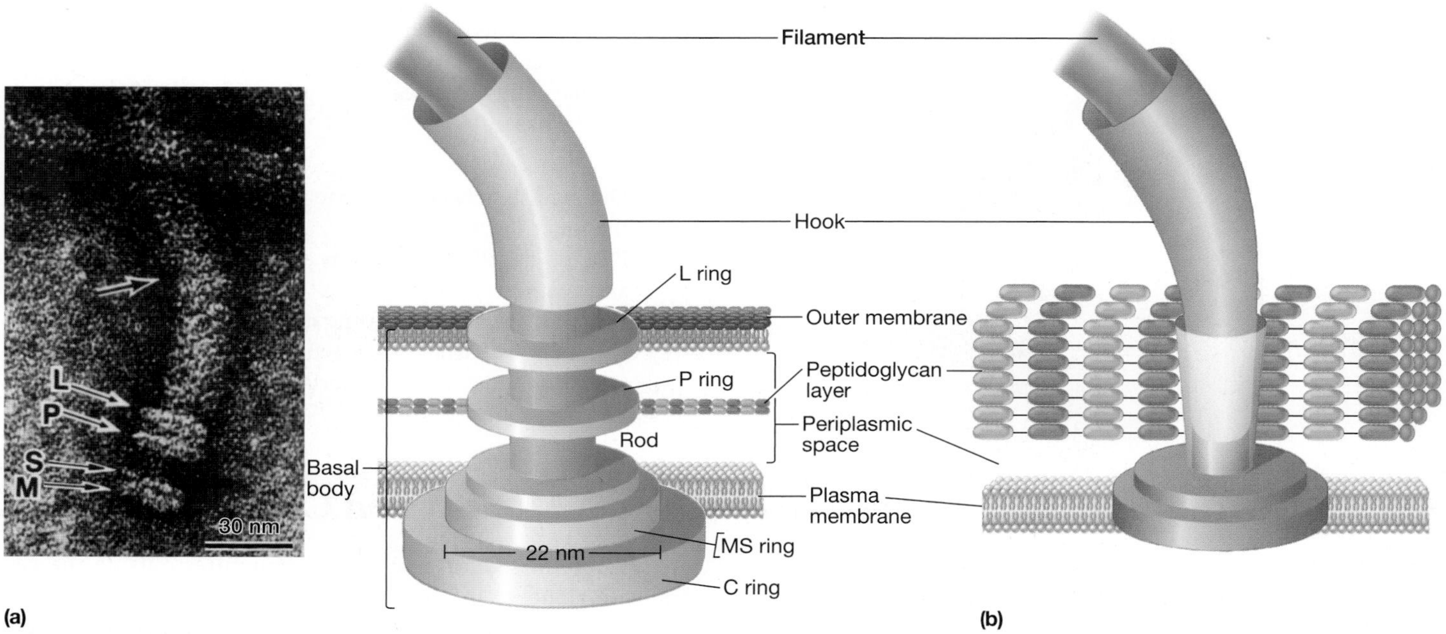

FIGURE 3.42 The Ultrastructure of Bacterial Flagella. Flagellar basal bodies and hooks in (a) gram-negative and (b) gram-positive bacteria. The photo shows an enlarged view of the basal body of an *E. coli* flagellum. All three rings (L, P, and MS) can be clearly seen. The uppermost arrow is at the junction of the hook and filament.

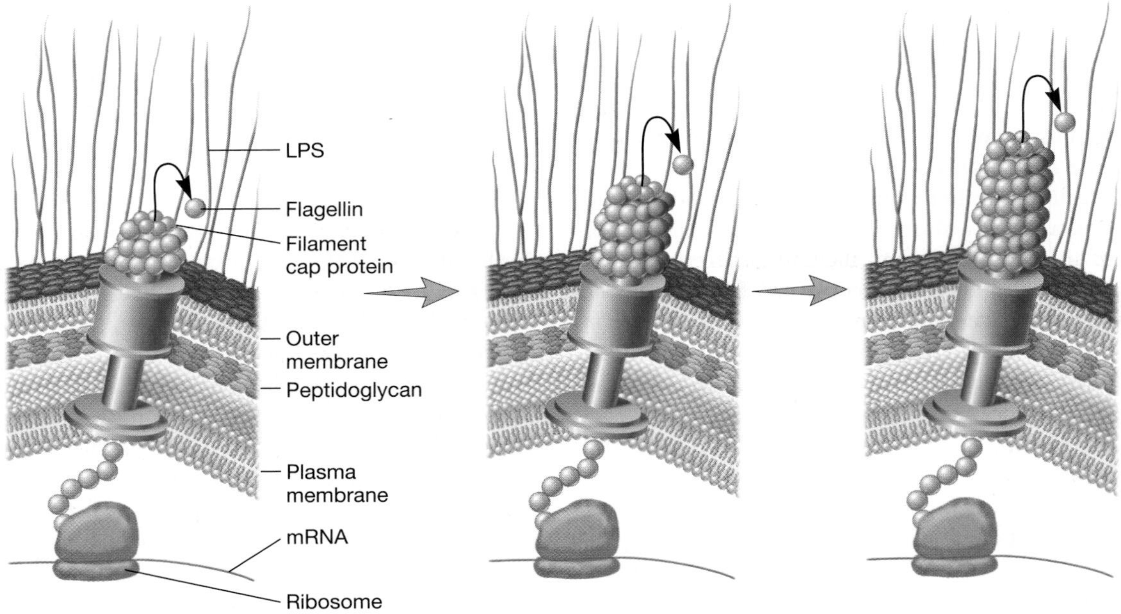

FIGURE 3.43 Growth of Flagellar Filaments. Flagellin subunits travel through the flagellar core and attach to the growing tip. Their attachment is directed by the filament cap protein.

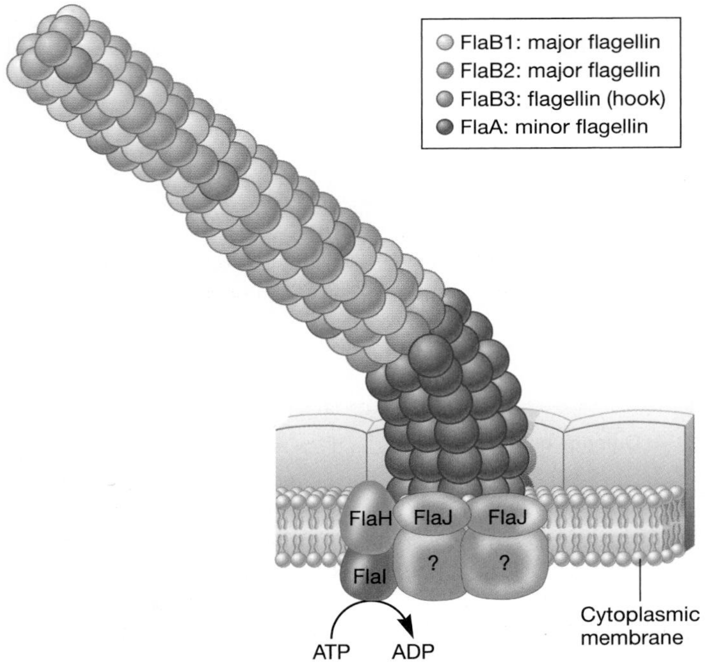

FIGURE 3.44 Archaeal Flagellum. The different shades of blue in the filament illustrate that the filament is composed of more than one type of flagellin. Clear images of the basal body have not been obtained, although some electron micrographs show a knob at the cell end of the flagellum, as illustrated here.

hooks are difficult to distinguish from the filament; they also tend to be longer than bacterial hooks. A basal body has not been identified, but some preparations of archaeal flagella have a knoblike structure at the end embedded in the cell. One interesting feature of the flagellar proteins thus far characterized is that they are more related to type IV bacterial pili than they are to the proteins in bacterial flagella. Like type IV pili, the filament of the archaeal flagellin increases in length as flagellin subunits are added at the base.

1. Distinguish between fimbriae (pili) and sex pili, and give the function of each.
2. Discuss flagella distribution patterns and the structure and synthesis of flagella.
3. What is self-assembly? Why does it make sense that the filament of a flagellum is assembled in this way?
4. Compare and contrast bacterial and archaeal flagella.

3.7 Motility and Chemotaxis

As we note in section 3.6, several structures outside the cell wall contribute to motility. Four major methods of movement have been observed in *Bacteria:* the swimming movement conferred by flagella; the corkscrew movement of spirochetes; the twitching motility associated with type IV pili; and gliding motility.

Motile bacteria and archaea do not move aimlessly. Rather, motility is used to move toward nutrients such as sugars and amino acids and away from many harmful substances and bacterial waste products. Motile bacteria also can respond to environmental cues such as temperature (thermotaxis), light (phototaxis), oxygen (aerotaxis), osmotic pressure (osmotaxis), and gravity. Movement toward chemical attractants and away from repellents is known as chemotaxis. Motile archaea also exhibit a variety of different types of directed movements, including chemotaxis.

We begin this section with a discussion of flagellar movement. We consider the *Bacteria* and *Archaea* separately, pointing out the similarities and differences. We then focus on three other types of motility observed in *Bacteria*.

Flagellar Movement

Flagellar movement is common in both the *Bacteria* and the *Archaea*. Some aspects of flagellar motility in the two domains of life are superficially similar. We begin with an examination the mechanism of flagellar movement in *Bacteria*.

Bacterial Flagellar Movement

Bacterial flagella operate differently from eukaryotic flagella. Eukaryotic flagella flex and bend, resulting in a whiplash that moves the cell. The filament of a bacterial flagellum is in the shape of a rigid helix, and the cell moves when this helix rotates like a propeller on a boat. The flagellar motor can rotate very rapidly. The *E. coli* motor rotates 270 revolutions per second (rps); *Vibrio alginolyticus* averages 1,100 rps. ▶▶| *External structures (section 4.7)*

The direction of flagellar rotation determines the nature of bacterial movement. Monotrichous, polar flagella rotate counterclockwise (when viewed from outside the cell) during normal forward movement, whereas the cell itself rotates slowly clockwise. The rotating helical flagellar filament thrusts the cell forward with the flagellum trailing behind (**figure 3.45**). This smooth swimming movement is often called a **run**. Monotrichous bacteria stop and **tumble** randomly by reversing the direction of flagellar rotation. Peritrichously flagellated bacteria operate in a somewhat similar way. To move forward in a run, the flagella rotate counterclockwise. As they do so, they bend at their hooks to form a rotating bundle that propels the cell forward. Clockwise rotation of the flagella disrupts the bundle and the cell tumbles.

The motor that drives flagellar rotation is located at the base of the flagellum, where it is associated with the basal body. Torque generated by the motor is transmitted by the basal body to the hook and filament. The motor is composed of two components: the rotor and the stator. It is thought to function like an electrical motor, where the rotor turns in the center of a ring of electromagnets, the stator. In gram-negative bacteria, the rotor is composed of the MS ring and the C ring (**figure 3.46**). The C ring protein FliG is a particularly important component of the rotor as it is thought to interact with the stator. The stator is composed of the proteins MotA and MotB, which form a channel through the plasma membrane. MotB also anchors MotA to cell wall peptidoglycan.

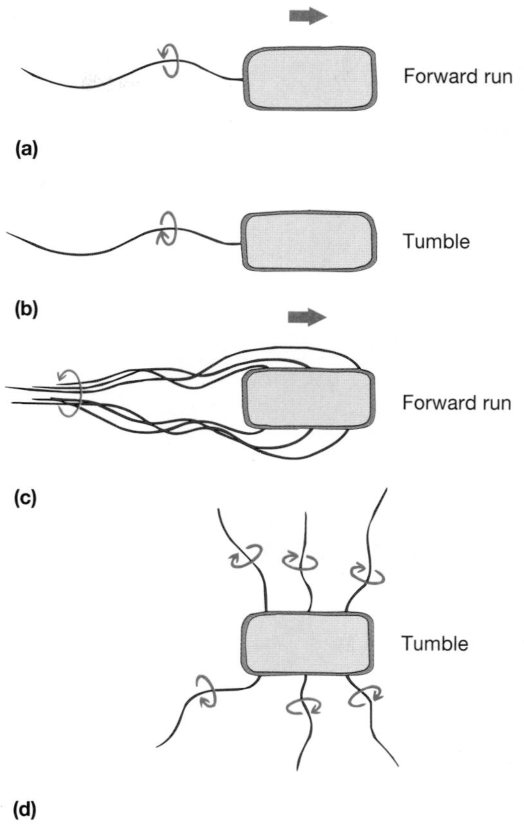

FIGURE 3.45 Flagellar Motility. The relationship of flagellar rotation to bacterial movement. Parts (a) and (b) describe the motion of monotrichous, polar bacteria. Parts (c) and (d) illustrate the movements of peritrichous organisms.

As with all motors, the flagellar motor must have a power source that allows it to generate torque and cause flagellar rotation. The power used by most flagellar motors is a difference in charge and pH across the plasma membrane. This difference is called the proton motive force (PMF). PMF is largely created by the metabolic activities of organisms, as described in chapter 10. One important metabolic process is the transfer of electrons from an electron donor to a terminal electron acceptor via a chain of electron carriers called the electron transport chain (ETC). In *Bacteria* and *Archaea*, the ETC is located in the plasma membrane. As electrons are transported down the ETC, protons are transported from the cytoplasm to the outside of the cell. Because there are more protons outside the cell than inside, the outside has more positively charged ions (the protons) and a lower pH. PMF is a type of potential energy that can be used to do work: mechanical work, as in the case of flagellar rotation; transport work, the movement of materials into or out of the cell; or chemical work such as the synthesis of ATP, the cell's energy currency.

So how can PMF be used to power the flagellar motor? The channels created by the MotA and MotB proteins allow protons to move across the plasma membrane from the outside to the inside (figure 3.46). Thus the protons move down the charge and pH

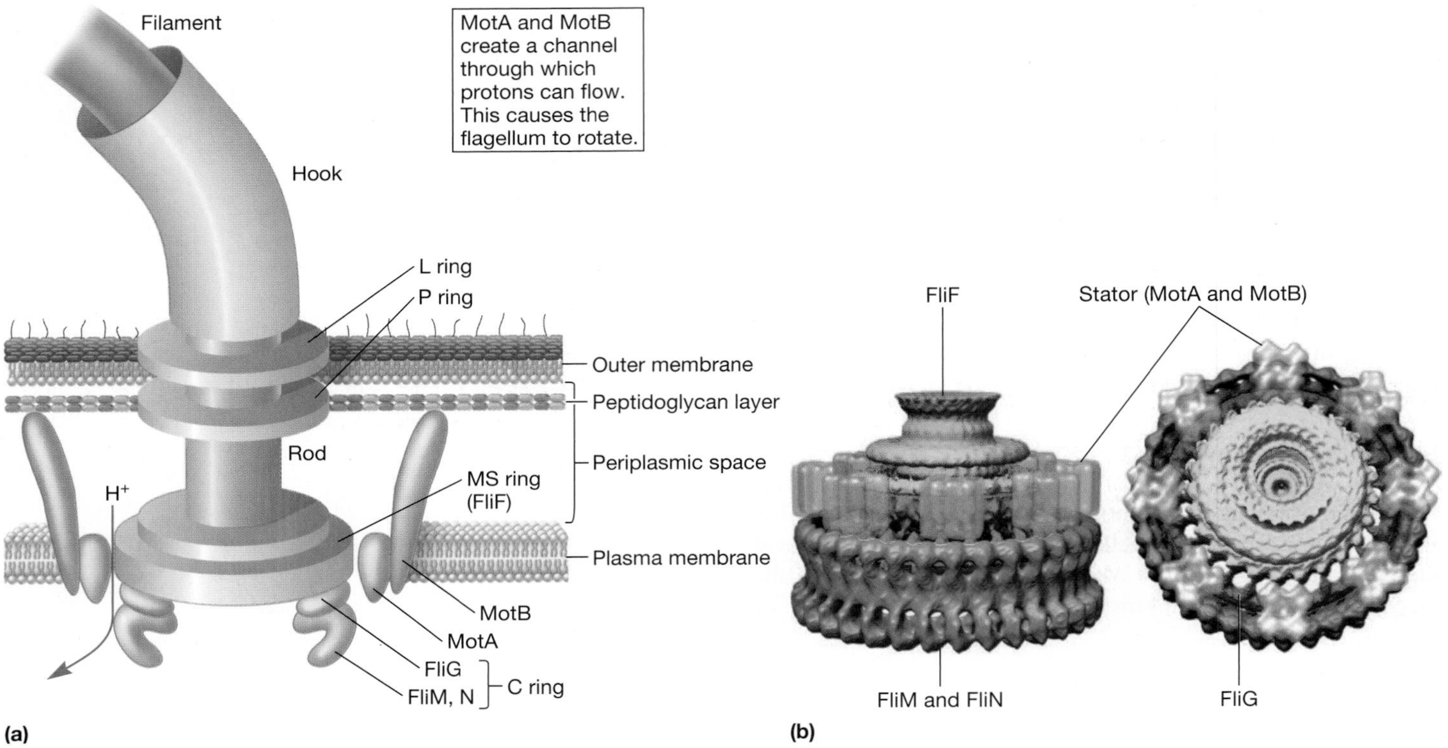

FIGURE 3.46 Mechanism of Flagellar Movement. (a) This diagram of a gram-negative flagellum shows some of the more important components and the flow of protons that drives rotation. Five of the many flagellar proteins are labeled (MotA, MotB, FliF, FliG, FliM, FliN). (b) A three-dimensional electron cryotomographic reconstruction of the flagellar motor.

gradient. This movement releases energy that is used to rotate the flagellum. In essence, the entry of a proton into the channel is like the entry of a person into a revolving door. The "power" of the proton generates torque, rather like a person pushing the revolving door. Indeed, the speed of flagellar rotation is proportional to the magnitude of the PMF. ▶▶| *Electron transport and oxidative phosphorylation (section 10.5)*

The flagellum is a very effective swimming device. From the bacterium's point of view, swimming is quite a difficult task because the surrounding water seems as viscous as molasses. The cell must bore through the water with its corkscrew-shaped flagella, and if flagellar activity ceases, it stops almost instantly. Despite such environmental resistance to movement, bacteria can swim from 20 to almost 90 μm/second. This is equivalent to traveling from 2 to over 100 cell lengths per second. In contrast, an exceptionally fast human might be able to run around 5 to 6 body lengths per second.

Archaeal Flagellar Movement

Recall that archaeal flagella are more similar to type IV bacterial pili than they are to bacterial flagella. Despite this, archaeal and bacterial flagella work in a somewhat similar manner: rotation propels the cell. *Halobacterium salinarum* flagellar movement is the best studied. In this archaeon, clockwise rotation of the flagella pushes the cell forward and counterclockwise rotation pulls the cell (i.e., the flagella are in front of the cell as it moves). Thus an alternation between forward movement and tumbles is not observed.

Spirochete Motility

Although spirochetes have flagella, they work in a different manner. In many spirochetes, multiple flagella arise from each end of the cell and associate to form an axial fibril, which winds around the cell (**figure 3.47**). The flagella do not extend outside the cell wall but rather remain in the periplasmic space and are covered by an outer sheath. The way in which axial fibrils propel the cell has not been fully established. They are thought to rotate like the external flagella of other bacteria, causing the corkscrew-shaped outer sheath to rotate and move the cell through the surrounding liquid, even very viscous liquids. Flagellar rotation may also flex or bend the cell and account for the creeping or crawling movement observed when spirochetes are in contact with a solid surface. ▶▶| *Phylum* Spirochaetes *(section 19.6)*

Twitching and Gliding Motility

Twitching and gliding motility occur when cells are on a solid surface. Both types of motility can involve type IV pili, the production of slime, or both. Thus they are considered together. The

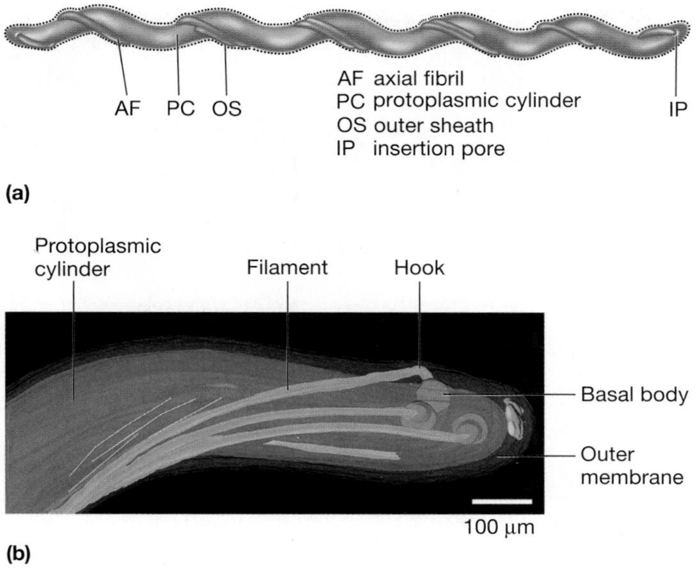

FIGURE 3.47 Spirochete Flagella. (a) Numerous flagella arise from each end of the spirochete. These intertwine to form an axial fibril. The axial fibril winds around the cell, usually overlapping in the middle. (b) Electron cryotomographic surface view of the spirochaete *Treponema denticola* showing three flagella arising from the tip of the cell.

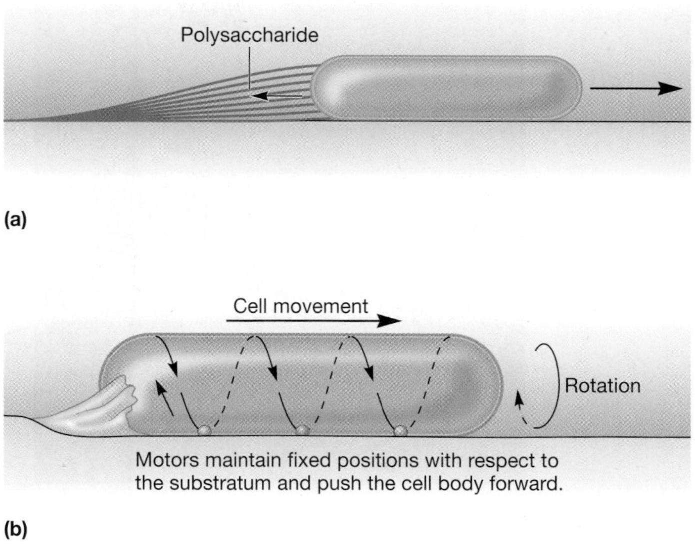

FIGURE 3.48 Possible Mechanisms for the Adventurous Motility of *Myxococcus xanthus*. (a) Polysaccharide secretion model. (b) Focal adhesion model.

gliding motility of *Flavobacterium* spp. and *Mycoplasma* spp. are distinctive and are considered in detail in chapters 19 and 21, respectively. ▶▶| *Phylum* Bacteroidetes *(section 19.7); Class* Mollicutes *(The Mycoplasmas) (section 21.1)*

Type IV pili are present at one or both poles of some bacteria and are involved in twitching motility and in the gliding motility of some bacteria. **Twitching motility** is characterized by short, intermittent, jerky motions of up to several micrometers in length and is normally seen on very moist surfaces. It occurs only when cells are in contact with each other; isolated cells rarely move by this mechanism. Considerable evidence exists that the pili alternately extend and retract to move bacteria during twitching motility. The extended pilus contacts the surface at a point some distance from the cell body. When the pilus retracts, the cell is pulled forward. Hydrolysis of ATP is thought to power the extension/retraction process.

Gliding motility is smooth and varies greatly in rate (from 2 to over 600 μm per minute) and in the nature of the motion. Although first observed over 100 years ago, the mechanism by which many bacteria glide remains a mystery. Some glide along in a direction parallel to the longitudinal axis of their cells. Others travel with a screwlike motion or even move in a direction perpendicular to the long axis of the cells. Still others rotate around their longitudinal axis while gliding. Such diversity in gliding movement correlates with the observation that more than one mechanism for gliding motility exists. Some types involve type IV pili, some involve slime, and some involve mechanisms that have not yet been elucidated.

The motility of the bacterium *Myxococcus xanthus* is well studied because it exhibits two types of motility. One type of motility is called social (S) motility because it occurs when large groups of cells move together in a coordinated fashion. It is mediated by type IV pili, as we have just described. The other type of motility is by gliding and is called adventurous (A) motility; it is observed when single cells move independently. Gliding motility is not as well understood in this microbe, and two quite different hypotheses have been made to explain it. One hypothesis is that the cells contain pores through which slime is secreted and that this propels the cell forward (**figure 3.48*a***). A more recent hypothesis is that adhesion complexes are located along the length of the cell and that these attach the cell to the surface (figure 3.48*b*, *also see figure 20.37*). The adhesion complexes are thought to span all the layers of the cell envelope, such that some portions are external and in contact with the surface and other portions are in the cytoplasm. The adhesion complexes remain stationary relative to the surface on which the cell is gliding but move along a "track" within the cell. This track is presumably provided by a cytoskeletal element, which has not yet been identified. ▶▶| *Order* Myxococcales *(section 20.4)*

Chemotaxis

As noted earlier, bacteria and archaea exhibit taxes to a variety of stimuli, including light and oxygen. However, the movement of cells toward chemical attractants or away from chemical repellents (**chemotaxis**) is the best studied type of taxis, and bacterial chemotaxis is best understood. Chemotaxis is readily observed in Petri dish cultures. If bacteria are placed in the center of a dish of

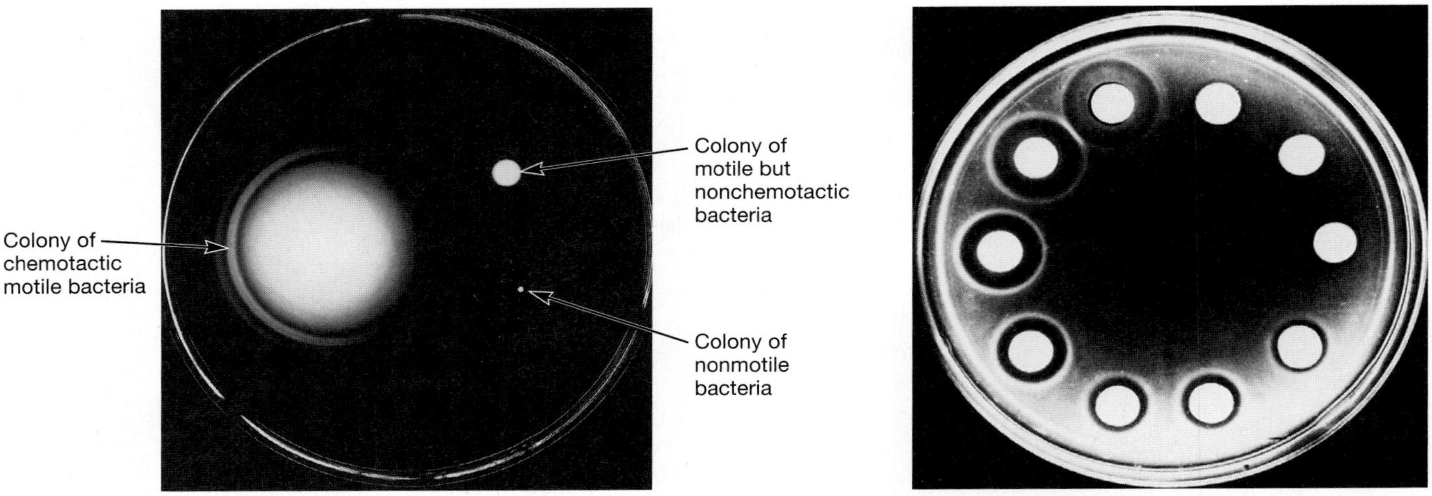

(a) Positive chemotaxis

(b) Negative chemotaxis

FIGURE 3.49 Positive and Negative Bacterial Chemotaxis. (a) Positive chemotaxis can be demonstrated on an agar plate that contains various nutrients; positive chemotaxis by *E. coli* is shown on the left. The outer ring is composed of bacteria consuming serine. The second ring was formed by *E. coli* consuming aspartate, a less powerful attractant. (b) Negative chemotaxis by *E. coli* in response to the repellent acetate. The bright disks are plugs of concentrated agar containing acetate that have been placed in dilute agar inoculated with *E. coli*. Acetate concentration increases from zero at the top right to 3 M at top left. Note the increasing size of bacteria-free zones with increasing acetate. The bacteria have migrated for 30 minutes.

semisolid agar containing an attractant, the bacteria will exhaust the local supply of the nutrient and swim outward following the attractant gradient they have created. The result is an expanding ring of bacteria (**figure 3.49*a***). When a disk of repellent is placed in a Petri dish of semisolid agar and bacteria, the bacteria will swim away from the repellent, creating a clear zone around the disk (figure 3.49*b*).

Attractants and repellents are detected by **chemoreceptors,** proteins that bind chemicals and transmit signals to other components of the chemosensing system. The chemosensing systems are very sensitive and allow the cell to respond to very low levels of attractants (about 10^{-8} M for some sugars). In gram-negative bacteria, the chemoreceptor proteins are located in the periplasmic space or in the plasma membrane. Some receptors also participate in the initial stages of sugar transport into the cell.

The chemotactic behavior of bacteria has been studied using the tracking microscope, a microscope with a moving stage that automatically keeps an individual bacterium in view. In the absence of a chemical gradient, bacteria move randomly, switching back and forth between a run and a tumble. During a run, the bacterium travels in a straight or slightly curved line. After a few seconds, the flagella "fly apart" and the bacterium stops and tumbles. The tumble randomly reorients the bacterium so that it often is facing in a different direction. Therefore when it begins the next run, it usually goes in a different direction (**figure 3.50*a***). In contrast, when the bacterium is exposed to an attractant, it tumbles less frequently (or has longer runs) when traveling toward the attractant. Although the tumbles can still orient the bacterium away from the attractant, over time, the bacterium gets closer and closer to the attractant (figure 3.50*b*). The opposite response occurs with a repellent. Tumbling frequency decreases (the run time lengthens) when the bacterium moves away from the repellent.

Clearly, the bacterium must have some mechanism for sensing that it is getting closer to the attractant (or moving away from the repellent). The behavior of the bacterium is shaped by temporal changes in chemical concentration. The bacterium moves toward the attractant because it senses that the concentration of the attractant is increasing. Likewise, it moves away from a repellent because it senses that the concentration of the repellent is decreasing. The bacterium's chemoreceptors play a critical role in this process. The molecular events that enable bacterial cells to sense a chemical gradient and respond appropriately are presented in chapter 13.

Though the story of chemotaxis is less complete in the *Archaea*, some interesting discoveries have been made. In the archaeal phylum *Euryarchaeota,* the components of the chemosensing system are similar to those found in bacterial systems. Furthermore the regulation of the activity of the components is similar. Thus far, no member of the archaeal phylum *Crenarchaeota* has been found to possess bacteria-like chemosensing components. However, these archaea do carry out chemotaxis and therefore must have a unique system for doing so.

1. Describe the way flagella operate to move a bacterium. How does flagellar movement differ in *H. salinarum*?
2. Explain in a general way how bacteria move toward substances such as nutrients and away from toxic materials.
3. Why do you think chemotaxis is sometimes called a "biased random walk"?

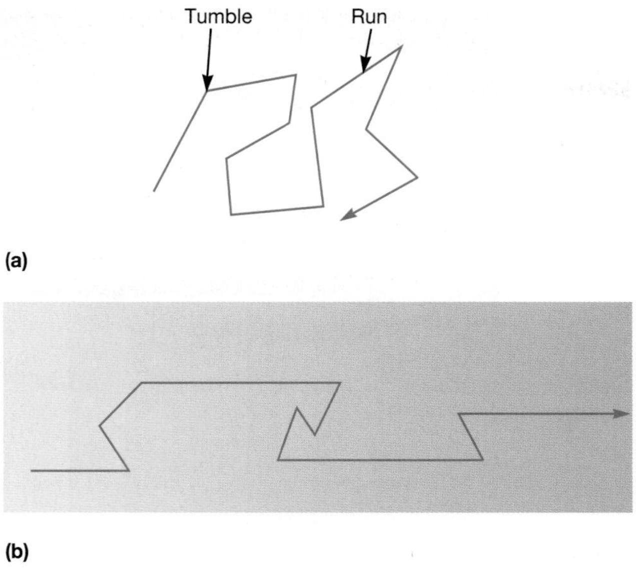

FIGURE 3.50 Directed Movement in Bacteria. (a) Random movement of a bacterium in the absence of a concentration gradient. Tumbling frequency is fairly constant. (b) Movement in an attractant gradient. Tumbling frequency is reduced when the bacterium is moving up the gradient. Therefore runs in the direction of increasing attractant are longer.

3.8 Bacterial Endospores

Several genera of gram-positive bacteria, including *Bacillus* and *Clostridium* (rods), and *Sporosarcina* (cocci), can form a resistant, dormant structure called an **endospore.** Endospore-forming bacteria are common in soil, where they must be able to withstand fluctuating levels of nutrients. Endospore formation (sporulation) normally commences when growth ceases due to lack of nutrients. Thus it is a survival mechanism that allows the bacterium to produce a dormant cell that can survive until nutrients are again available and vegetative growth can resume. Interestingly, some bacteria have modified the sporulation process and use it to produce live offspring within themselves (**Microbial Diversity & Ecology 3.1**).

Endospores are extraordinarily resistant to environmental stresses such as heat, ultraviolet radiation, gamma radiation, chemical disinfectants, and desiccation. In fact, some endospores have remained viable for around 100,000 years. They are of both practical and theoretical interest. Because of their resistance and the fact that several species of endospore-forming bacteria are dangerous pathogens, endospores are of great practical importance in food, industrial, and medical microbiology. In these areas, it is essential to be able to sterilize solutions and solid objects. Endospores often survive boiling for an hour or more; therefore autoclaves must be used to sterilize many materials. Endospores are of considerable theoretical interest to scientists studying the construction of complex biological structures. Bacteria

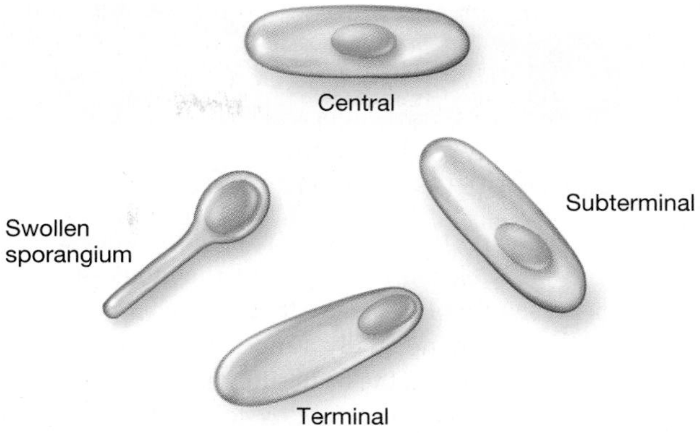

FIGURE 3.51 Examples of Endospore Location and Size.

manufacture these intricate structures in a very organized fashion over a period of a few hours. Thus spore formation is well suited for this type of research, and the endospore-forming *Bacillus subtilis* has become an important model organism. ▶▶| *Heat (section 8.4); Sporulation in* Bacillus subtilis *(section 13.6)*

Endospores can be examined with both light and electron microscopes. Because endospores are impermeable to most stains, they often are seen as colorless areas in bacteria treated with methylene blue and other simple stains; staining procedures specific for endospores are used to make them clearly visible. Endospore position in the mother cell (**sporangium**) frequently differs among species, making it of value in identification. Endospores may be centrally located, close to one end (subterminal), or terminal (**figure 3.51**). Sometimes an endospore is so large that it swells the sporangium. |◀◀ *Preparation and staining of specimens (section 2.3)*

Electron micrographs show that endospore structure is complex (**figure 3.52**). The spore often is surrounded by a thin, delicate covering called the exosporium. A coat lies beneath the exosporium. It is composed of several protein layers and may be fairly thick. The cortex, which may occupy as much as half the spore volume, rests beneath the coat. It is made of a peptidoglycan that is less cross-linked than that in vegetative cells. The core wall is inside the cortex and surrounds the core. The core has normal cell structures such as ribosomes and a nucleoid but has a very low water content and is metabolically inactive.

The various layers of the spore are thought to contribute to its resistance to heat and other lethal agents. The exosporium and spore coat are both thought to protect the spore from chemicals, although the mechanisms by which they do so are not completely understood. It is known that the spore coat is impermeable to many toxic molecules. The inner membrane, which separates the cortex from the core, is also impermeable to various chemicals, including those that cause DNA damage. The core plays a major role in resistance. Several factors may play a part in resistance (e.g., very low water content, high amounts of dipicolinic acid complexed with calcium ions, and a slightly lower pH). However, the major core-related factor is the protection of the spore's DNA by small, acid-soluble DNA-binding proteins (SASPs), which

MICROBIAL DIVERSITY & ECOLOGY

3.1 Bacteria That Have Babies

Epulopscium fishelsoni (Latin, *epulum,* a feast or banquet, and *piscium,* fish) first gained the attention of biologists because of its large size. It can reach a size of 80 μm by 600 μm and normally ranges from 200 to 500 μm in length. However, the bacterium has other interesting characteristics. One is its ability to produce living offspring within itself (viviparity). Usually two offspring are formed.

E. fishelsoni is related to the gram-positive, endospore-forming bacterial genus *Clostridium.* It appears that its mechanism of reproduction is a modified version of sporulation. The process begins with division of the cytoplasm into three compartments: two small compartments at each end and a larger, centrally located compartment. The central compartment is the mother cell, and the two smaller compartments will become the offspring (**box figure**). The mother cell engulfs the offspring, just as the mother cell of an endospore former engulfs a developing endospore. Initially both the offspring and mother cell grow, but later the mother cell's growth begins to slow. Eventually the mother cell dies and the offspring are released.

A central question being addressed by Esther Angert and others studying this bacterium is which came first: viparity or endospore formation. Currently it is thought that endospore formation evolved first. One line of evidence is that *E. fishelsoni* is closely related to another interesting bacterium, *Metabacterium polyspora.* As its name implies, this bacterium produces multiple endospores rather than just one, as is the case for members of the genera *Bacillus* and *Clostridium.*

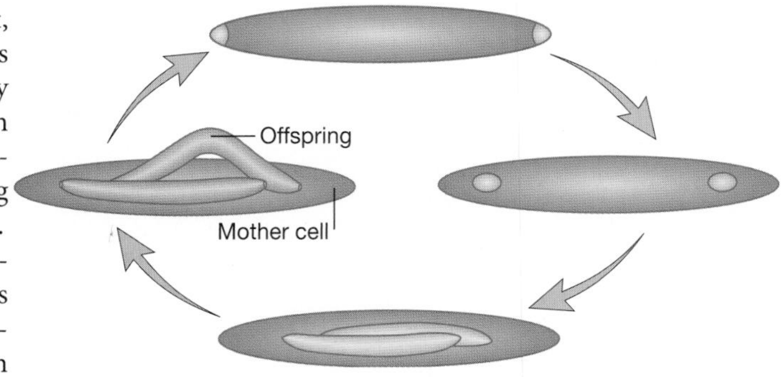

Production of Intracellular Living Offspring by *E. fishelsoni.* The life cycle begins with the formation of two small compartments within the mother cell and ends with the death of the mother cell and release of the offspring.

Source: Angert, E. R. 2005. Alternatives to binary fission in bacteria. *Nature Rev. Microbiol.* 3:214–24.

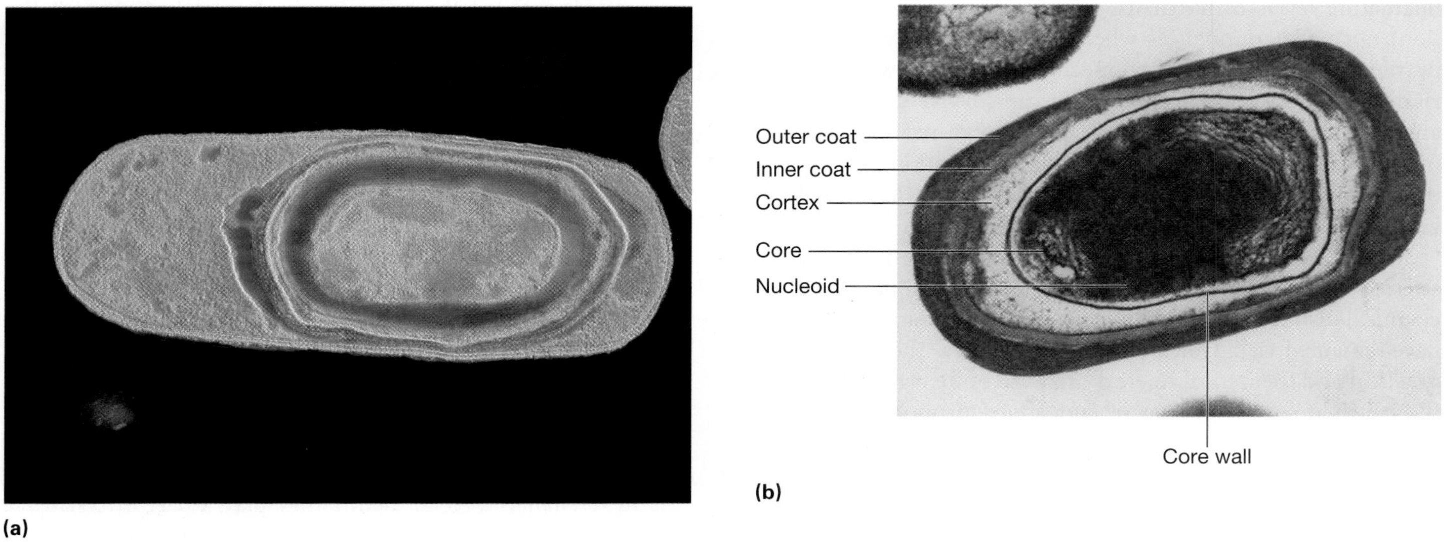

FIGURE 3.52 Bacterial Endospores. (a) A colorized cross section of a *Bacillus subtilis* cell undergoing sporulation. The oval in the center is an endospore that is almost mature; when it reaches maturity, the mother cell will lyse to release it. (b) A cross section of a mature *B. subtilis* spore showing the cortex and spore coat layers that surround the core. The endospore in (a) is 1.3 μm; the spore in (b) is 1.2 μm.

saturate spore DNA. There are several types of SASPs. The α/β type plays a major role in resistance. Cells that have been mutated and do not make α/β SASPs are considerably more sensitive to heat, UV radiation, dessication, and a variety of chemicals but are still resistant to other types of DNA damage. Thus other mechanisms for protecting the DNA must exist.

Sporulation is a complex process and may be divided into seven stages (**figure 3.53**). An axial filament of nuclear material

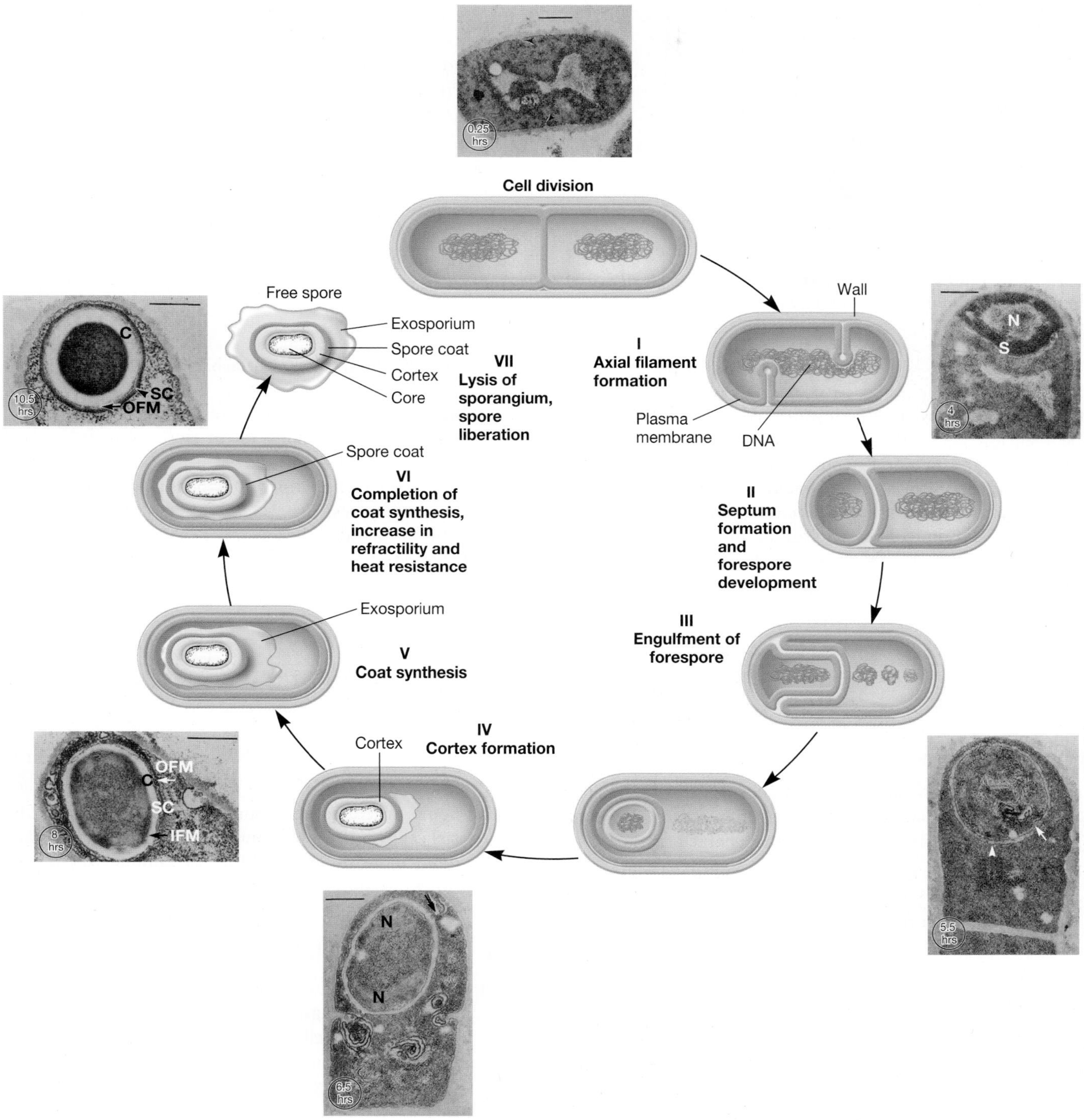

FIGURE 3.53 Endospore Formation: Life Cycle of *Bacillus megaterium*. The stages are indicated by Roman numerals. The circled numbers in the photographs refer to the hours from the end of the logarithmic phase of growth: 0.25 h—a typical vegetative cell; 4 h–stage II cell, septation; 5.5 h–stage III cell, engulfment; 6.5 h–stage IV cell, cortex formation; 8 h–stage V cell, coat formation; 10.5 h–stage VI cell, mature spore in sporangium. Abbreviations used: C, cortex; IFM and OFM, inner and outer forespore membranes; N, nucleoid; S, septum; SC, spore coats. Bars = 0.5 µm.

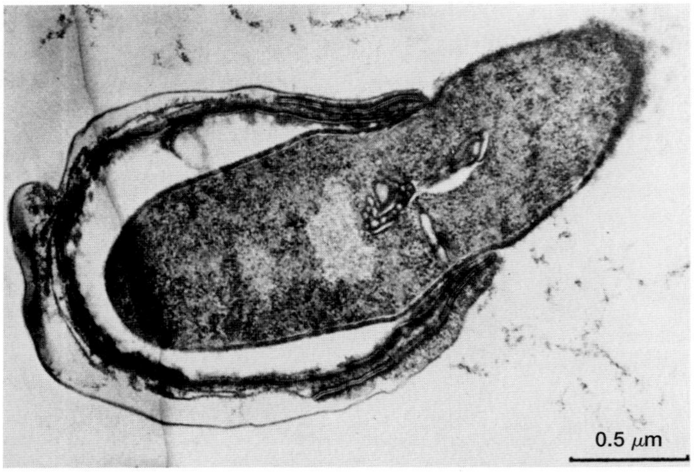

FIGURE 3.54 Endospore Germination. *Clostridium pectinovorum* emerging from the spore during germination.

forms (stage I), followed by an inward folding of the cell membrane to enclose part of the DNA and produce the forespore septum (stage II). The membrane continues to grow and engulfs the immature endospore in a second membrane (stage III). Next, cortex is laid down in the space between the two membranes, and both calcium and dipicolinic acid are accumulated (stage IV). Protein coats are formed around the cortex (stage V), and maturation of the endospore occurs (stage VI). Finally, lytic enzymes destroy the sporangium, releasing the spore (stage VII). Sporulation requires about 10 hours in *Bacillus megaterium.* Bacterial Spore Formation

The transformation of dormant spores into active vegetative cells seems almost as complex a process as sporulation. It occurs in three stages: (1) activation, (2) germination, and (3) outgrowth (**figure 3.54**). Activation is a process that prepares spores for germination and can result from treatments such as heating. This is followed by **germination,** the breaking of the spore's dormant state. Proteins in the exosporium, spore coat, and inner membrane are all thought to play a role. Some are involved in detecting the presence of certain compounds such as sugars and amino acids. Germination is characterized by spore swelling, rupture, or absorption of the spore coat, loss of resistance to heat and other stresses, loss of refractility, release of spore components, and increase in metabolic activity. Germination is followed by the third stage, outgrowth. The spore protoplast makes new components, emerges from the remains of the spore coat, and develops again into an active bacterium.

1. Describe the structure of the bacterial endospore using a labeled diagram.
2. Briefly describe endospore formation and germination. What is the importance of the endospore? What might account for its heat resistance?
3. How might one go about showing that a bacterium forms true endospores?
4. Why do you think the low water content of the spore contributes to its dormancy and resistance? Why are SASPs important?

Summary

3.1 Prokaryotes

a. *Bacteria* and *Archaea* are often considered to be related because they share a common cell architecture. However, they are now known to have distinct characteristics.

b. Recent discoveries have challenged traditional notions of what a prokaryote is. These include the discovery of membrane-bound organelles in some bacteria and cytoskeletal elements in both *Bacteria* and *Archaea.*

3.2 Common Features of Bacterial and Archaeal Cell Structure

a. Rods (bacilli) and cocci (spheres) are the most common bacterial and archaeal shapes. In addition, bacteria may be comma-shaped (vibrio), spiral (spirillum and spirochaetes), or filamentous; they may form buds and stalks; or they may have no characteristic shape (pleomorphic). Some archaea are curved rods, spirals, branched, square, and pleomorphic (**figures 3.1** and **3.2**).

b. Some cells remain together after division to form pairs, chains, and clusters of various sizes and shapes.

c. Some structures are common to both archaeal and bacterial cells (**figure 3.6**). However, their molecular makeup is often quite different. **Table 3.1** summarizes the major functions of these cell structures.

3.3 Bacterial Cell Envelopes

a. The cell envelope consists of the plasma membrane and all external coverings, including cell walls and other layers. Most bacterial cell envelopes consist of the plasma membrane, cell wall, and one additional layer. Archaeal cell envelopes usually consist of only the plasma membrane and the cell wall.

b. The plasma membrane fulfills many roles, including acting as a semipermeable barrier, carrying out respiration and photosynthesis, and detecting and responding to chemicals in the environment.

c. The fluid mosaic model proposes that cell membranes are lipid bilayers in which integral proteins are buried. Peripheral proteins are loosely associated with the membrane (**figure 3.7**).

d. Bacterial membranes are bilayers composed of phospholipids constructed of fatty acids connected to glycerol by ester linkages (**figure 3.8**). Bacterial membranes usually lack sterols but often contain hopanoids (**figure 3.9**).

e. The vast majority of bacteria and archaea have a cell wall outside the plasma membrane to give them shape and protect them from osmotic stress.

f. Bacterial walls are chemically complex and usually contain peptidoglycan (**figures 3.10–3.16**). Bacteria often are classified as either gram positive or gram negative based on differences in cell wall structure and their response to Gram staining. Gram-positive walls have thick, homogeneous layers of peptidoglycan and teichoic acids (**figure 3.17**). Gram-negative bacteria have a thin peptidoglycan layer surrounded by a complex outer membrane containing lipopolysaccharides (LPSs) and other components (**figure 3.19**).

g. The mechanism of the Gram stain is thought to depend on the thickness of the peptidoglycan, which binds crystal violet tightly, preventing the loss of crystal violet during the ethanol wash. Thus gram-positive bacteria retain crystal violet during alcohol decolorization, while gram-negative bacteria lose the stain. When counterstained, only the gram-negatives take up the second dye.

h. Capsules, slime layers, and glycocalyxes are layers of material lying outside the cell wall. They are common in *Bacteria* but rare in *Archaea*. They can protect cells from certain environmental conditions, allow cells to attach to surfaces, and protect pathogenic bacteria from host defenses (**figures 3.23** and **3.24**).

i. S-layers are the external-most layer in some bacteria. They are composed of proteins or glycoprotein and have a characteristic geometric shape (**figure 3.25**).

3.4 Archaeal Cell Envelopes

a. Archaeal membranes are composed of glycerol diether and diglycerol tetraether lipids (**figure 3.26**). Membranes composed of glycerol diether are lipid bilayers. Membranes composed of diglycerol tetraethers are lipid monolayers (**figure 3.27**). The overall structure of a monolayer membrane is similar to that of the bilayer membrane in that the membrane has a hydrophobic core and its surfaces are hydrophilic.

b. Archaeal cell walls do not contain peptidoglycan, and they exhibit great diversity in their cell wall makeup. The most common type of cell wall is one consisting of an S-layer only (**figure 3.28**).

3.5 Cytoplasm of *Bacteria* and *Archaea*

a. The cytoplasm of both bacterial and archaeal cells contains proteins that are similar in structure and function to the cytoskeletal proteins observed in eukaryotes (**figure 3.30 and Table 3.2**).

b. Some bacteria have simple internal membrane systems containing photosynthetic and respiratory machinery (**figure 3.31**).

c. Inclusions are observed in all cells. Most are used for storage (glycogen inclusions, PHB inclusions, cyanophycin granules, and polyphosphate granules) (**figures 3.33** and **3.34**), but some are used for other purposes (e.g., magnetosomes and gas vacuoles) (**figures 3.36** and **3.37**). Microcompartments such as carboxysomes contain enzymes that catalyze important reactions (e.g., CO_2 fixation; **figure 3.35**).

d. Bacterial and archaeal ribosomes are 70S in size but differ slightly in their morphology. They also differ in terms of their protein content, with many archaeal ribosomal proteins being more similar to those in eukaryotic ribosomes than to those in bacterial ribosomes (**figure 3.38**).

e. The genetic material of bacterial and archaeal cells is located in an area within the cytoplasm called the nucleoid. The nucleoid is not usually enclosed by a membrane (**figure 3.39**). In most bacteria and all known archaea, the nucleoid contains a single chromosome. The chromosome usually consists of a double-stranded, covalently closed, circular DNA molecule.

f. Plasmids are extrachromosomal DNA molecules found in many bacteria and archaea. Some are episomes—plasmids that are able to exist freely in the cytoplasm or can be integrated into the chromosome. Although plasmids are not required for survival in most conditions, they can encode traits that confer selective advantage in some environments. Many types of plasmids have been identified. Conjugative plasmids encode genes that promote their transfer from one cell to another. Resistance factors have genes conferring resistance to antibiotics. Col plasmids contain genes for the synthesis of colicins, proteins that kill *E. coli*. Other plasmids encode virulence factors or metabolic capabilities (**table 3.3**).

3.6 External Structures

a. Many bacteria and archaea have short, hairlike appendages called fimbriae. They are also known as pili. Fimbriae function primarily in attachment to surfaces, but type IV pili are involved in twitching motility. Sex pili participate in the transfer of DNA from one bacterium to another (**figure 3.40**).

b. Many bacteria and archaea are motile, often by means of threadlike, locomotory organelles called flagella.

c. Bacterial species differ in the number and distribution of their flagella (**figure 3.41**). Each bacterial flagellum is composed of a filament, hook, and basal body (**figure 3.42**).

d. Archaeal flagella appear to be similar to bacterial flagella, but they are more related to type IV bacterial pili than they are to bacterial flagella (**figure 3.44**).

3.7 Motility and Chemotaxis

a. Several types of bacterial motility have been observed. These include movement by flagella, spirochete motility, twitching motility, and gliding motility.

b. The bacterial flagellar filament is a rigid helix that rotates like a propeller to push the bacterium through water (**figures 3.45** and **3.46**). Bacteria usually alternate between two types of movement: a smooth swimming motion called a run and tumbling.

c. Archaeal flagella are also rigid helices that rotate. However, these cells do not alternate between runs and tumbles.

d. Spirochete motility is brought about by flagella that are wound around the cell and remain within the periplasmic space. When they rotate, the outer sheath of the spirochete is thought to rotate, thus moving the cell (**figure 3.47**).

e. Twitching and gliding motility are similar in that both occur on moist surfaces and can involve type IV pili and the secretion of slime. Twitching motility is a jerky movement, whereas gliding motility is smooth.

f. Motile cells can respond to gradients of attractants and repellents, a phenomenon known as chemotaxis. A peritrichously flagellated bacterium accomplishes movement toward an attractant by increasing the length of time it spends moving toward the attractant and shortening the time it spends tumbling (**figure 3.50**). Conversely, a bacterium increases its run time when it moves away from a repellent.

3.8 Bacterial Endospores

a. Some bacteria survive adverse environmental conditions by forming endospores, dormant structures resistant to heat, desiccation, and many chemicals (**figure 3.52**).

b. Both endospore formation and germination are complex processes that begin in response to certain environmental signals and involve numerous stages (**figures 3.53** and **3.54**).

Critical Thinking Questions

1. Propose a model for the assembly of a flagellum in a gram-positive cell envelope. How would that model need to be modified for the assembly of a flagellum in a gram-negative cell envelope?

2. The peptidoglycan of bacteria has been compared with the chain mail worn beneath a medieval knight's suit of armor. It provides both protection and flexibility. Can you describe other structures in biology that have an analogous function? How are they replaced or modified to accommodate the growth of the inhabitant?

3. Enterohemorrhagic *E. coli* (EHEC) O157:H7 has caused severe food poisoning outbreaks in the United States and elsewhere. In order for EHEC to colonize the host intestinal epithelium, it must have a mechanism for attachment. When EHEC are grown in vitro with cultured human epithelial cells, the bacteria form fibrils that attach them to the epithelial cells. After 3 hours of incubation, changes in the morphology of the human cells consistent with the changes seen during infection are noted. Based solely on their function, would you hypothesize that these fibrils are flagella, fimbriae, or pili? Explain your answer. How would you determine the identity of the fibrils? How would you determine if they were necessary for infection?

   Read the original paper: Rendon, M. A., et al. 2008. Commensal and pathogenic *Escherichia coli* use a common pilus adherence factor for epithelial cell colonization. *Proc. Nat. Acad. Sci., USA*. 105:10637.

4. The gram-positive bacterium *Bacillus thuringiensis* subsp. *israelensis* is one of the many bacteria that accumulates PHB. To harvest the energy stored in this inclusion body, the cell must dismantle the PHB. This is accomplished by depolymerizing PHB into smaller 3-hydroxybutyrate molecules. The enzyme that catalyzes PHB depolymerization in *B. thuringiensis* subsp. *israelensis* was originally thought to degrade a different lipid that is not associated with PHB granules. How do you think it was shown that the enzyme in question was specific for PHB? Do you think the cell produces this enzyme all the time? Why or why not?

   Read the original paper: Tseng, C. L., et al. 2006. Identification and characterization of the *Bacillus thuringiensis phaZ* gene, encoding new intracellular poly-3-hydroxybutyrate depolymerase. *J. Bacteriol.* 188:7592.

5. Although the assembly of lipopolysaccharide (LPS) in gram-negative bacteria is well understood, its transport across and its insertion into the outer membrane remains unclear. It is known that lipid A attached to the core moiety is moved

from the inner leaflet of the plasma membrane to the outer leaflet by a "flippase that resides in the membrane." Once in the periplasm, polymerized O-antigen units are attached to the lipid A core. Five candidate proteins, LptA, B, C, D, and E, were identified as essential team members responsible for ferrying the newly assembled LPS from the periplasm to the outer membrane. Where do you think these proteins are located within the cell or cell envelope to do this job? How do you think these proteins ensure that the orientation of the LPS in the outer membrane is correct? (Hint: It appears these proteins may work sequentially in time and space.)

Read the original paper: Sperandeo, P., et al. 2008. Functional analysis of the protein machinery required or transport of lipopolysaccharide to the outer membrane of *Escherichia coli. J. Bacteriol.* 190:4460.

Concept Mapping Challenge

Construct a concept map that allows you to compare and contrast *Bacteria* and *Archaea.* Use the concepts that follow, any other concepts or terms you need, and your own linking words between each pair of concepts in your map. For guidance in builidng a concept map, see appendix 3.

anucleate	cell wall	membrane	lipids
ribosomes	inclusions	S-layer	nucleoid
capsule	flagella	cell shapes	endospore

Learn More

Learn more by visiting the text website at www.mhhe.com/willey8, where you will find a complete list of references.

4

Eukaryotic Cell Structure and Function

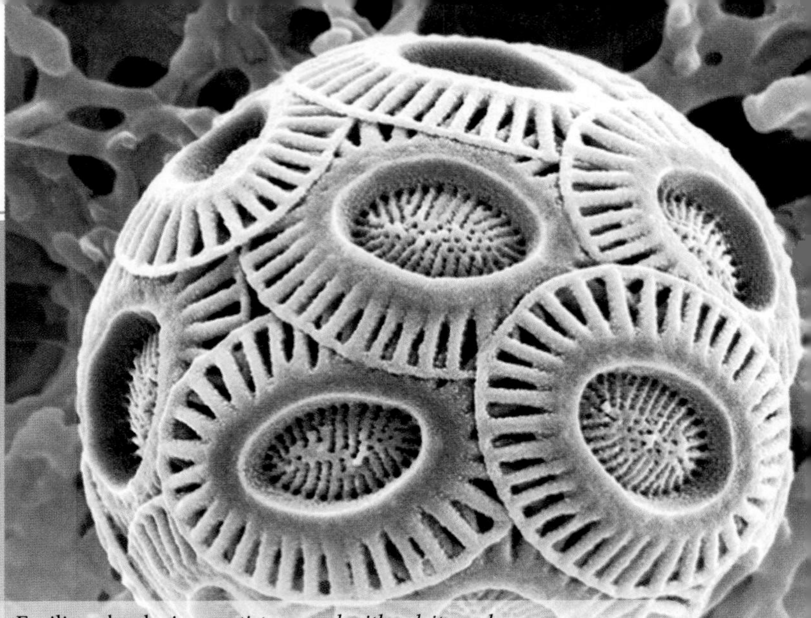

Emiliana huxleyi: *a protist covered with calcite scales.*

CHAPTER GLOSSARY

cilia Threadlike appendages extending from the surface of some protists that beat rhythmically to propel them. They are membrane-bound cylinders with a complex internal array of microtubules, usually in a 9 + 2 pattern.

cristae Infoldings of the inner mitochondrial membrane.

cytoskeleton A network of microfilaments, intermediate filaments, microtubules, and other components in the cytoplasm of eukaryotic cells that helps give them shape, functions during cell division, and helps move materials in the cytoplasm.

endocytosis The process in which a cell takes up solutes or particles by enclosing them in vesicles pinched off from its plasma membrane.

endoplasmic reticulum (ER) A system of membranous tubules and flattened sacs (cisternae) in the cytoplasm of eukaryotic cells; **rough endoplasmic reticulum (RER)** bears ribosomes on its surface; **smooth endoplasmic reticulum (SER)** lacks ribosomes.

flagellum A thin, threadlike appendage on many cells that is responsible for their motility.

Golgi apparatus A membranous eukaryotic organelle composed of stacks of flattened sacs (cisternae) that is involved in many processes including packaging and modifying materials for secretion.

lysosome A spherical membranous eukaryotic organelle that contains hydrolytic enzymes and is responsible for the intracellular digestion of substances.

microfilaments (actin filaments) Protein filaments, about 4 to 7 nm in diameter, that are present in the cytoplasm of eukaryotic cells and play a role in cell structure and motion.

microtubules Small cylinders, about 25 nm in diameter, made of tubulin proteins and present in the cytoplasm, cilia, and flagella of eukaryotic cells. They are involved in cell structure and movement.

mitochondrion The eukaryotic organelle that is the site of cellular respiration. It provides most of a nonphotosynthetic cell's energy under oxic conditions.

nuclear envelope The complex double-membrane structure forming the outer boundary of the nucleus.

nucleolus An organelle located within the nucleus and not bounded by a membrane; it is the location of ribosomal RNA synthesis and the assembly of ribosomal subunits.

nucleus The eukaryotic organelle enclosed by a double-membrane that contains the cell's chromosomes.

organelle A structure within or on a cell that performs specific functions and is related to the cell in a way similar to that of an organ to the body of a multicellular organism.

phagocytosis The endocytotic process in which a cell encloses large particles in a membrane-delimited phagocytic vacuole (phagosome) and engulfs them.

proteasome A cylindrical complex found in the cytoplasm of eukaryotic cells that is responsible for the degradation of proteins.

secretory pathway The process used by eukaryotic cells to synthesize proteins and lipids, followed by secretion or delivery to organelles or the plasma membrane; involves the endoplasmic reticulum, Golgi apparatus, and secretory vesicles.

septa Cross walls that divide microbial cells.

thylakoid A flattened sac in the chloroplast stroma that contains photosynthetic pigments and the proteins and other molecules that convert light energy into ATP.

Although the *Bacteria* and *Archaea* are immensely important in microbiology and have occupied a large portion of microbiologists' attention in the past, eukaryotic microbes also have been extensively studied (**figure 4.1**). Like *Bacteria* and *Archaea*, they are prominent members of ecosystems. Many are important model organisms, as well as being exceptionally

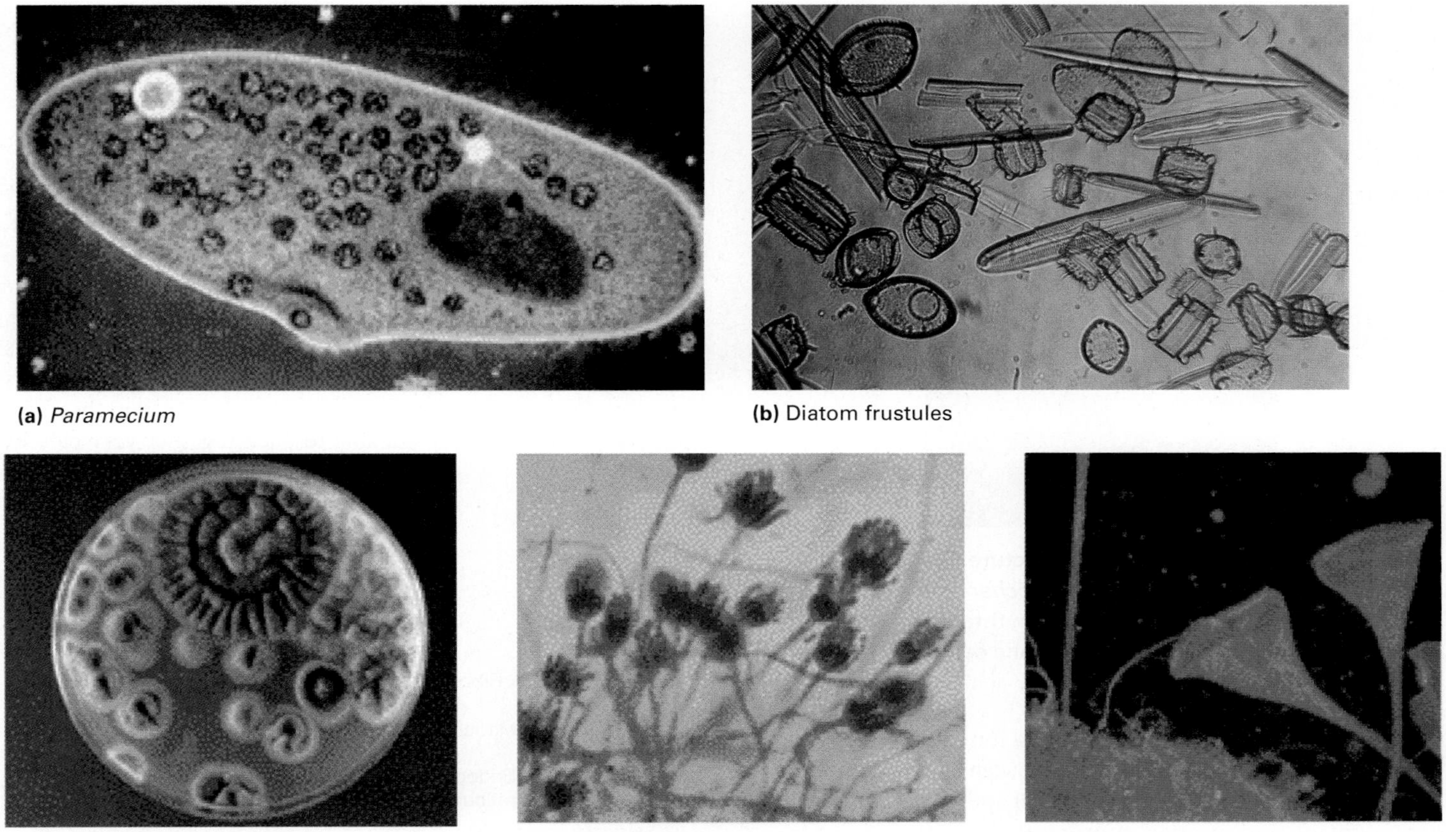

(a) *Paramecium* (b) Diatom frustules (c) *Penicillium* (d) *Penicillium* (e) *Stentor*

FIGURE 4.1 Some Eukaryotic Microorganisms. (a) *Paramecium* as seen with interference-contrast microscopy (×115). (b) Mixed diatom frustules. (c) *Penicillium* colonies, and (d) a microscopic view of the mold's hyphae and conidia (×220). (e) *Stentor.* The ciliated protozoa are extended and actively feeding, dark-field microscopy (×100).

useful in industrial microbiology. A number are also major human pathogens; one only need think of candidiasis, malaria, or African sleeping sickness to appreciate the significance of eukaryotic microbes in medical microbiology.

Eukaryotic microbes can be divided into two major groups: protists and fungi. Protists are very diverse. Some use organic molecules as energy sources just as animals do; others are photosynthetic like plants. Each protist lineage has a distinct common ancestor. Thus protists are said to be paraphyletic. Because of this, the term protist is a common name for these microbes rather than a valid taxon. All fungi share a common evolutionary history and are monophyletic.

Chapter 4 introduces eukaryotic microbes with an initial emphasis on common cell structures and their functions. In this discussion, comparisons are made to eukaryotic organisms other than microorganisms; that is, plants and animals. Following this discussion, the three domains of life are compared. We end the chapter with a more detailed look at structures specific to protists and fungi. Chapters 23 and 24 review the diversity of these organisms, and chapter 39 covers some of the important diseases they cause.

4.1 Common Features of Eukaryotic Cells

Eukaryotic cells are distinctive because of their use of membranes. They have membrane-delimited nuclei, and membranes play a prominent part in the structure of many other organelles (**figures 4.2** and **4.3**). **Organelles** are intracellular structures that perform specific functions in cells analogous to the functions of organs in the body of a multicellular organism. A comparison of figures 4.2 and 4.3 with figures 3.6 and 3.39*a* shows how structurally complex the eukaryotic cell is. This complexity is due chiefly to the use of internal membranes for several purposes. The partitioning of the eukaryotic cell interior by membranes makes possible the placement of

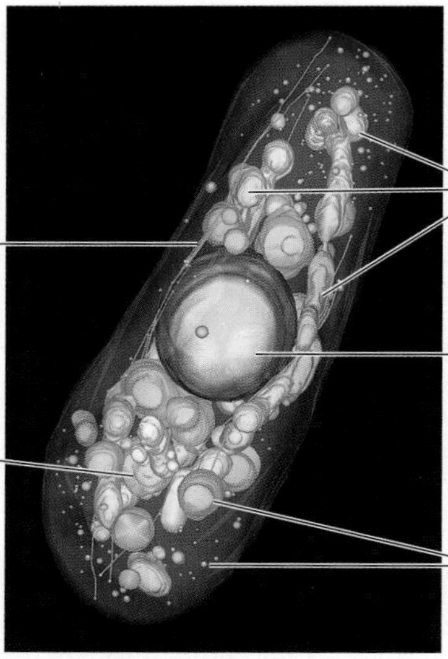

FIGURE 4.2 Eukaryotic Cell Ultrastructure. An electron cryotomogram of the yeast *Schizosaccharomyces pombe*. This image is the first high-resolution, three-dimensional reconstruction of a complete eukaryotic cell.

different biochemical and physiological functions in separate compartments so that they can more easily take place simultaneously under independent control and proper coordination. Large membrane surfaces make possible greater respiratory and photosynthetic activity because these processes are located exclusively in membranes. The intracytoplasmic membrane complex also serves as a transport system to move materials between different cell locations. Thus abundant membrane systems probably are necessary in eukaryotic cells because of their large volume and the need for adequate regulation, metabolic activity, and transport.

Figures 4.2 and 4.3 illustrate most of the organelles to be discussed here. **Table 4.1** briefly summarizes the functions of the major organelles observed in most eukaryotes, including plants and animals. Our detailed discussion of eukaryotic cell structure begins with eukaryotic cell envelopes. We then proceed to the cytoplasm, organelles within the cytoplasm, and finally to structures used for motility.

1. What is an organelle? How are organelles analogous to organs of a multicellular organism?
2. Why is the compartmentation of the cell interior advantageous to eukaryotic cells?

4.2 Eukaryotic Cell Envelopes

As we discuss in chapter 3, the cell envelope consists of the plasma membrane and all coverings external to it. Eukaryotic microorganisms differ greatly from *Bacteria* and *Archaea* in the structures they have external to the plasma membrane. Many eukaryotic microbes lack a cell wall. When cell walls are present, they are chemically distinctive.

The plasma membrane of eukaryotes is a lipid bilayer composed of a high proportion of sphingolipids and sterols (e.g., cholesterol and ergosterol) in addition to the phosphoglycerides observed in bacterial membranes (**figure 4.4**). The large amounts of sphingolipids and sterols contribute to the strength of the plasma membrane; they are able to pack together very closely,

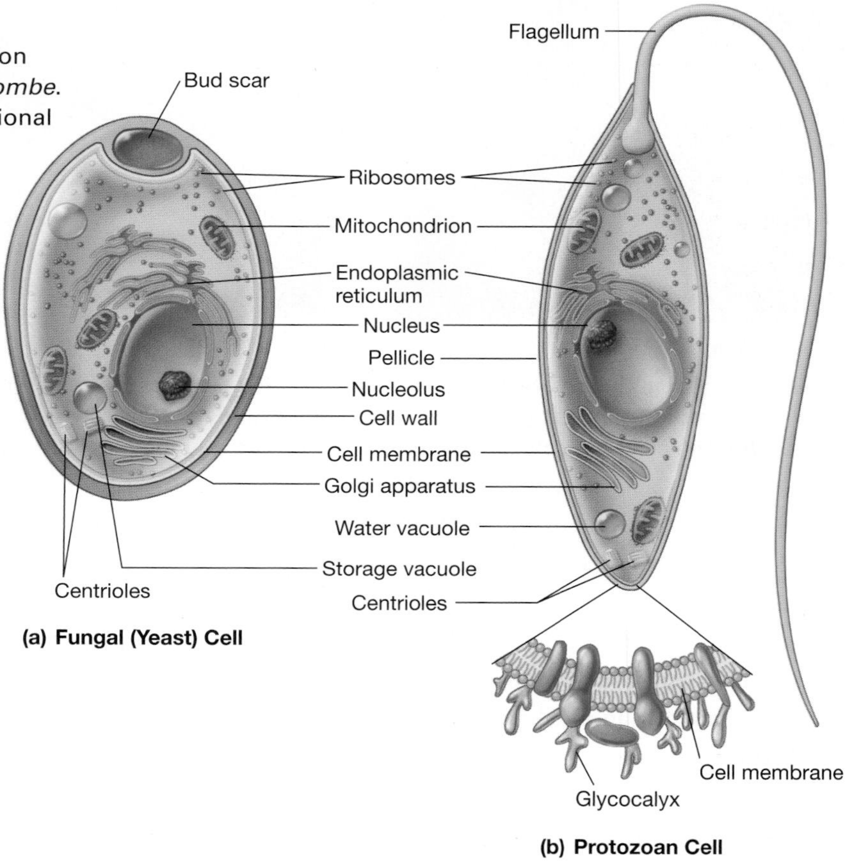

FIGURE 4.3 The Structure of Two Representative Eukaryotic Microbes. Illustrations of a yeast cell (fungus) (a) and the flagellated protist *Peranema* (b).

Figure 4.3 Micro Inquiry

In addition to separating organelles from the cytoplasm, what other functions do internal membranes serve?

Table 4.1 Functions of Eukaryotic Organelles

Plasma membrane	Mechanical cell boundary; selectively permeable barrier with transport systems; mediates cell-cell interactions and adhesion to surfaces; secretion; signal transduction
Cytoplasm	Environment for other organelles; location of many metabolic processes
Microfilaments, intermediate filaments, and microtubules	Cell structure and movements; form the cytoskeleton
Endoplasmic reticulum	Transport of materials; lipid synthesis
Ribosomes	Protein synthesis
Golgi apparatus	Packaging and secretion of materials for various purposes; lysosome formation
Lysosomes	Intracellular digestion
Mitochondria	Energy production through use of the tricarboxylic acid cycle, electron transport, oxidative phosphorylation, and other pathways
Chloroplasts	Photosynthesis—trapping light energy and formation of carbohydrate from CO_2 and water
Nucleus	Repository for genetic information
Nucleolus	Ribosomal RNA synthesis; ribosome construction
Cell wall and pellicle	Strengthen and give shape to the cell
Cilia and flagella	Cell movement
Vacuole	Temporary storage and transport; digestion (food vacuoles); water balance (contractile vacuole)

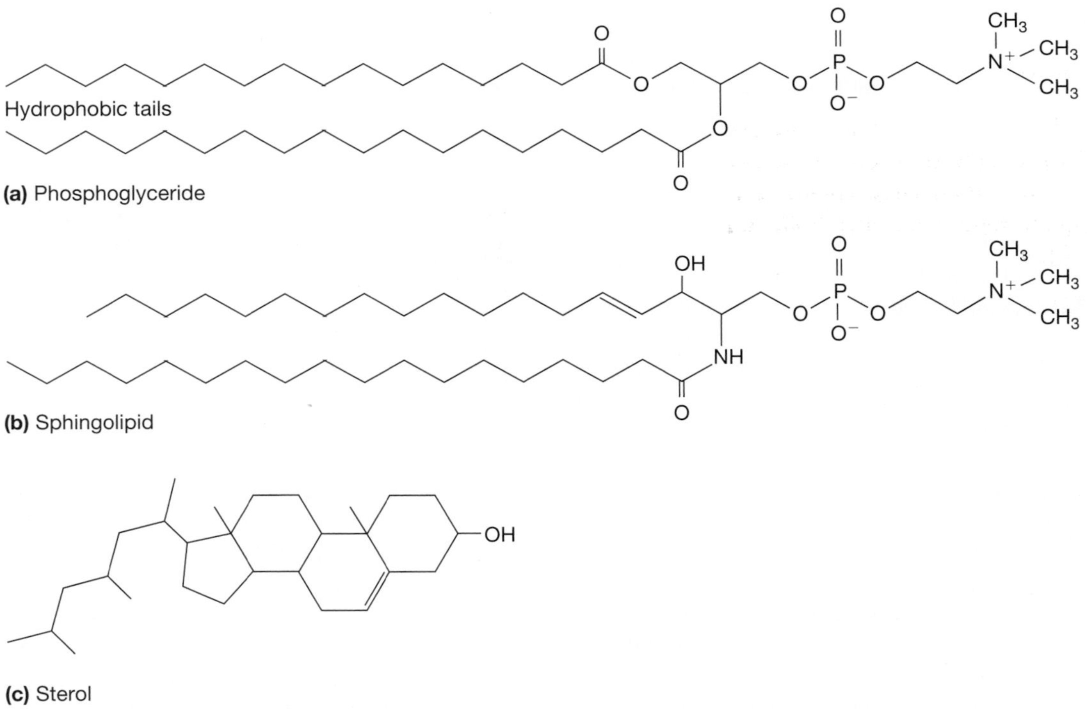

FIGURE 4.4 Examples of Eukaryotic Membrane Lipids. (a) Phosphatidylcholine, a phosphoglyceride. (b) Sphingomyelin, a sphingolipid. (c) Cholesterol, a sterol.

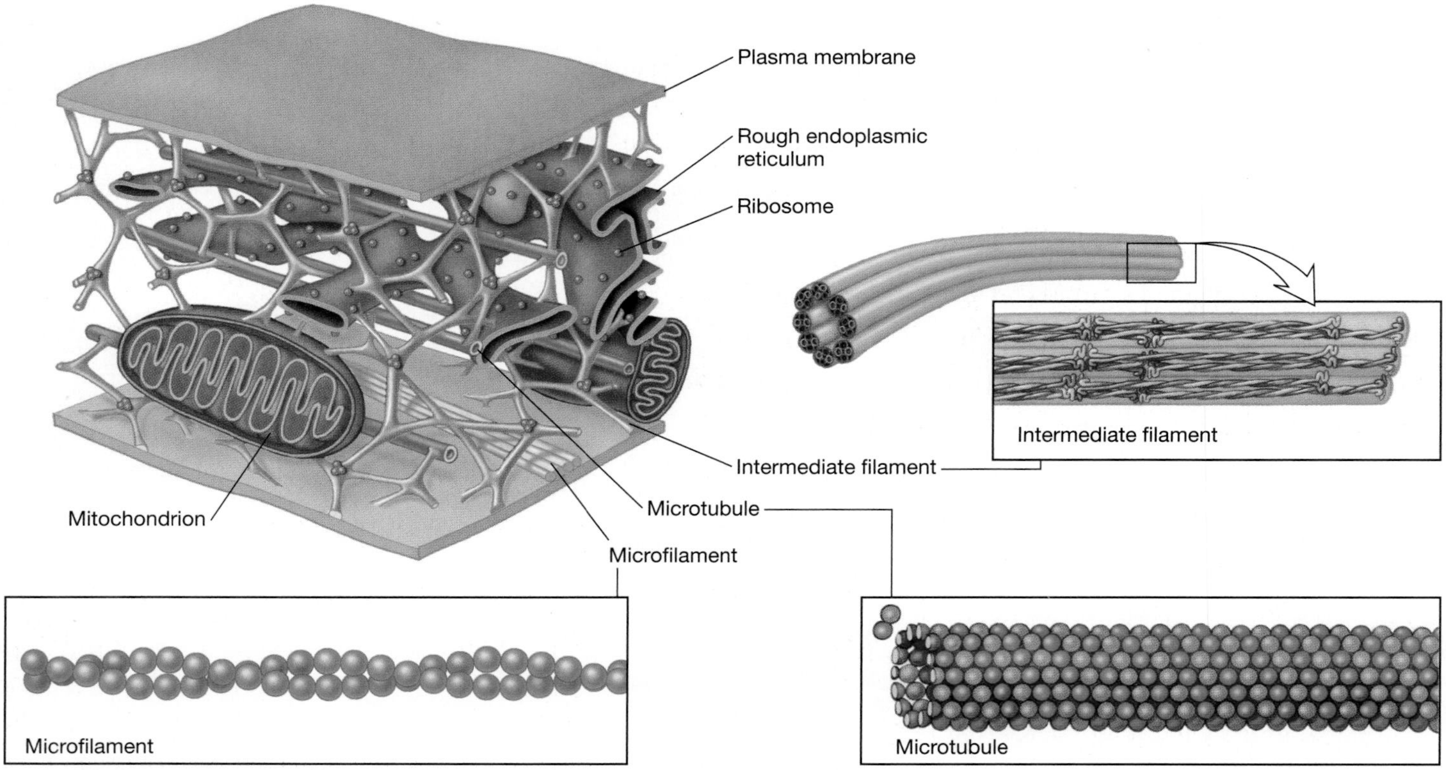

FIGURE 4.5 The Eukaryotic Cytoplasm and Cytoskeleton. The cytoplasm of eukaryotic cells contains many important organelles. The cytoskeleton helps form a framework within which the organelles lie. The cytoskeleton is composed of three elements: microfilaments, microtubules, and intermediate filaments.

Figure 4.5 Micro Inquiry

Which cytoskeletal filament is made of the protein actin and which is composed of α- and β-tubulin?

more so than phosphoglycerides can. This explains why so many eukaryotic microbes are able to resist osmotic pressure and other mechanical stresses even though they lack cell walls.

The distribution of lipids in the plasma membrane is asymmetric. Lipids in the outer monolayer differ from those of the inner monolayer. Although most lipids in individual monolayers mix freely with each other, there are microdomains that differ in lipid and protein composition. These microdomains appear to participate in a variety of cellular processes (e.g., cell movement, cell division, and signal transduction). They also may be involved in the entrance of some viruses into their host cells and assembly of progeny viruses before they are released.

The chemical composition of the cell walls of eukaryotic microbes varies considerably. The cell walls of photosynthetic protists (commonly called algae) usually have a layered appearance and contain large quantities of polysaccharides such as cellulose and pectin. In addition, inorganic substances such as silica (in diatoms) or calcium carbonate may be present. Fungal cell walls normally are rigid. Their exact composition varies with the organism; usually cellulose, chitin, or glucan (a glucose polymer different from cellulose) are present. Despite their nature, the rigid materials in eukaryotic cell walls are chemically simpler than bacterial peptidoglycan. ◀◀ *Bacterial cell walls (section 3.3)*

4.3 Cytoplasm of Eukaryotes

The **cytoplasm** is one of the most important and complex parts of a cell. It consists of a liquid component, the cytosol, in which many organelles are located. It is the location of many important biochemical processes, and several physical changes seen in cells (e.g., viscosity changes and cytoplasmic streaming) also are due to cytoplasmic activity. Because so many different kinds of organelles are observed in the eukaryotic cytoplasm, we address them separately in sections 4.4–4.6. Here we focus on the cytoskeleton, which helps organize the contents of the cytoplasm.

The **cytoskeleton** is a vast network of interconnected filaments. Three types of filaments form the eukaryotic cytoskeleton: microfilaments (also called actin filaments), intermediate filaments, and microtubules (**figure 4.5**). Motor proteins (myosin,

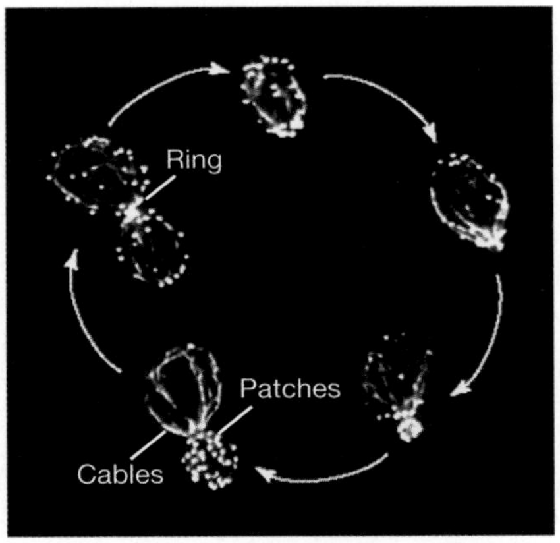

FIGURE 4.6 The Actin Cytoskeleton of the Yeast *Saccharomyces cerevisiae*. Cells from different stages in the yeast cell cycle were stained with the fluorescent dye rhodamine phalloidin. Three kinds of structures made from actin flilaments can be observed: actin patches, actin rings, and actin cables.

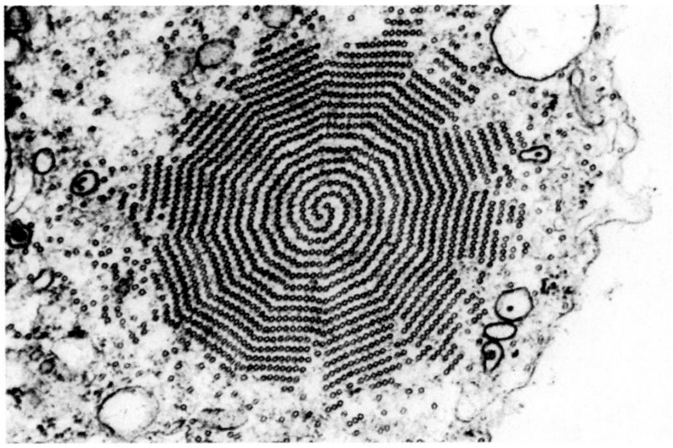

FIGURE 4.7 Microtubules. Electron micrograph of a transverse section through the axopodium of a protist known as a heliozoan (×48,000). Note the parallel array of microtubules organized in a spiral pattern.

kinesin, and dynein) are often associated with microfilaments and microtubules. These cellular motors track along these cytoskeletal filaments, helping to move cell structures from one location to another.

Microfilaments are minute protein filaments, 4 to 7 nm in diameter, that are organized to form a variety of structures depending on the cell type (**figure 4.6**). Microfilaments are composed of an actin protein that is similar to the actin contractile protein of muscle tissue. Microfilaments are involved in amoeboid movement, endocytosis, cytokinesis, and the movement of some structures within the cell. Interestingly, some bacterial pathogens use the actin proteins of their eukaryotic hosts to move rapidly through the host cell and to propel themselves into new host cells (*see figure 31.9*).

Intermediate filaments are heterogeneous elements of the cytoskeleton that play structural roles. They are about 10 nm in diameter and are assembled from a group of proteins that can be divided into several classes (e.g., keratin and vimentin). Intermediate filaments having different functions are assembled from one or more of these classes of proteins. The role of intermediate filaments in eukaryotic microorganisms is unclear. Thus far, they have been identified and studied only in animals: some intermediate filaments form the nuclear lamina, a structure that provides support for the nuclear envelope (p. 97); some help position organelles within the cell; and other intermediate filaments help link cells together to form tissues.

Microtubules are shaped like thin cylinders about 25 nm in diameter. They are complex structures constructed of two spherical protein subunits—α-tubulin and β-tubulin. The two proteins are the same molecular weight and differ only slightly in terms of their amino acid sequence and tertiary structure. Each tubulin is approximately 4 to 5 nm in diameter. These subunits are assembled in a helical arrangement to form a cylinder with an average of 13 subunits in one turn (figure 4.5).

Microtubules have several important functions. They form the spindle apparatus that separates chromosomes during mitosis and meiosis. They form tracks along which numerous types of organelles and vesicles are moved about the cell, and they are found in cilia and flagella, two organelles that confer motility. Microtubules also are found in long, thin cell structures requiring support such as the axopodia (long, slender, rigid pseudopodia) of protists (**figure 4.7**).

Search This: Cytoskeleton movie

1. Compare the membranes of *Eukarya, Bacteria,* and *Archaea.* How are they similar? How do they differ?
2. How do eukaryotic microorganisms differ from *Bacteria* and *Archaea* with respect to supporting or protective structures external to the plasma membrane?
3. Compare bacterial cytoskeletal elements to the eukaryotic cytoskeleton. Which elements perform similar functions?

4.4 Organelles of the Secretory and Endocytic Pathways

The cytoplasm of eukaryotic cells is permeated with an intricate complex of membranous organelles and vesicles that move materials into the cell from the outside (endocytic pathway) and from the inside of the cell out, as well as from location to location within the cell (secretory pathway). In this section, some of these organelles are described. This is followed by a summary of how the organelles function in the secretory and endocytic pathways.

Endoplasmic Reticulum

The **endoplasmic reticulum (ER)** (figure 4.3) is an irregular network of branching and fusing membranous tubules, around 40 to 70 nm in diameter, and many flattened sacs called cisternae

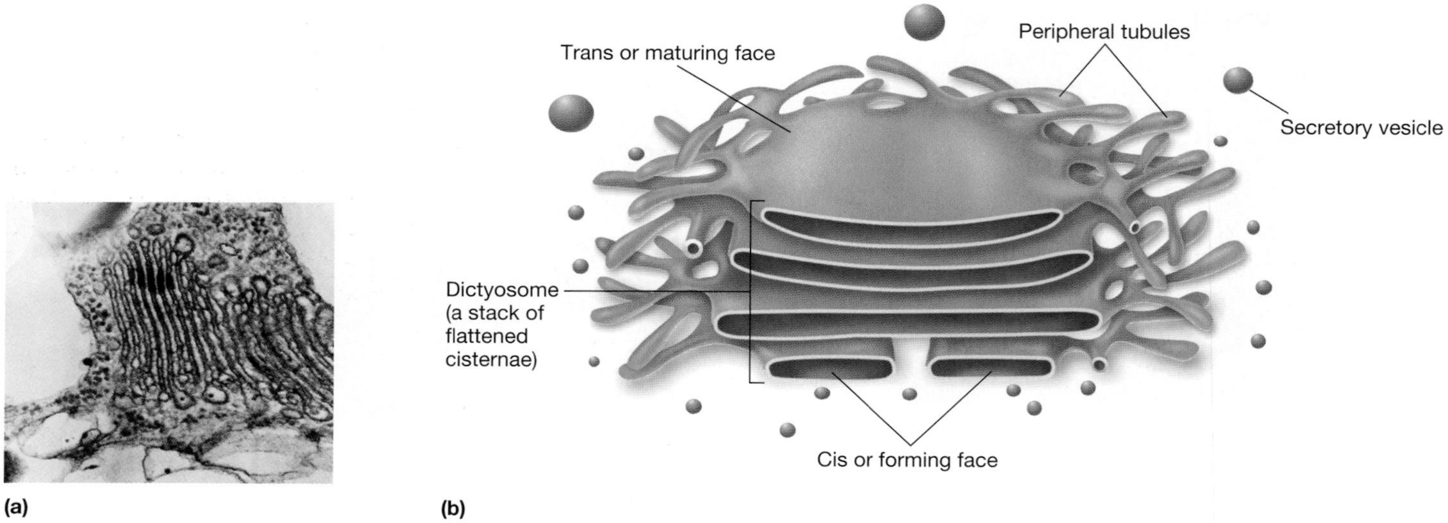

FIGURE 4.8 Golgi Apparatus Structure. Golgi apparatus of *Euglena gracilis.* Cisternal stacks are shown in the electron micrograph (×165,000) in (a) and diagrammatically in (b).

(s., cisterna). The nature of the ER varies with the functional and physiological status of the cell. In cells synthesizing a great deal of protein that will be secreted, a large part of the ER is studded on its outer surface with ribosomes and is called **rough endoplasmic reticulum (RER).** Other cells, such as those producing large quantities of lipids, have ER that lacks ribosomes. This is **smooth endoplasmic reticulum (SER).**

The endoplasmic reticulum has many important functions. Not only does it transport proteins, lipids, and other materials through the cell, it is also involved in the synthesis of many of the materials it transports. Lipids and proteins are synthesized by ER-associated enzymes and ribosomes. Polypeptide chains synthesized on RER-bound ribosomes may be inserted either into the ER membrane or into its lumen for transport elsewhere. The ER is also a major site of cell membrane synthesis.

Golgi Apparatus

The **Golgi apparatus** is composed of flattened, saclike cisternae stacked on each other (**figure 4.8**). These membranes, like the smooth ER, lack bound ribosomes. Usually around four to eight cisternae are in a stack, although there may be many more. Each is 15 to 20 nm thick and separated from other cisternae by 20 to 30 nm. A complex network of tubules and vesicles (20 to 100 nm in diameter) is located at the edges of the cisternae. The stack of cisternae has two faces that are quite different from one another. The sacs on the cis or forming face often are associated with the ER and differ from the sacs on the trans or maturing face in thickness, enzyme content, and degree of vesicle formation.

The Golgi apparatus is present in most eukaryotic cells, but many fungi and ciliate protists lack a well-formed structure. Sometimes the Golgi consists of a single stack of cisternae; however, many cells may contain 20 or more separate stacks. These stacks of cisternae, often called dictyosomes, can be clustered in one region or scattered about the cell.

The Golgi apparatus packages materials and prepares them for secretion, the exact nature of its role varying with the organism. For instance, the surface scales of some flagellated photosynthetic and radiolarian protists appear to be constructed within the Golgi apparatus and then transported to the surface in vesicles. The Golgi often participates in the development of cell membranes and the packaging of cell products. The growth of some fungal hyphae occurs when Golgi vesicles contribute their contents to the wall at the hyphal tip. ▶▶| *The Protists (chapter 23)*

Lysosomes

Lysosomes or lysosome-like organelles are found in most eukaryotic organisms, including protists, fungi, plants, and animals. Lysosomes are roughly spherical and enclosed in a single membrane; they average about 500 nm in diameter but range from 50 nm to several μm in size. They are involved in intracellular digestion and contain the enzymes needed to digest all types of macromolecules. These enzymes, called hydrolases, catalyze the hydrolysis of molecules and function best under slightly acidic conditions (usually around pH 3.5 to 5.0). Lysosomes maintain an acidic environment by pumping protons into their interior. *Lysosomes*

Secretory Pathway

The **secretory pathway** is used to move materials to various sites within the cell, as well as to either the plasma membrane or cell exterior. The process is complex and not fully understood. The movement of proteins is of particular importance and is the focus of this discussion.

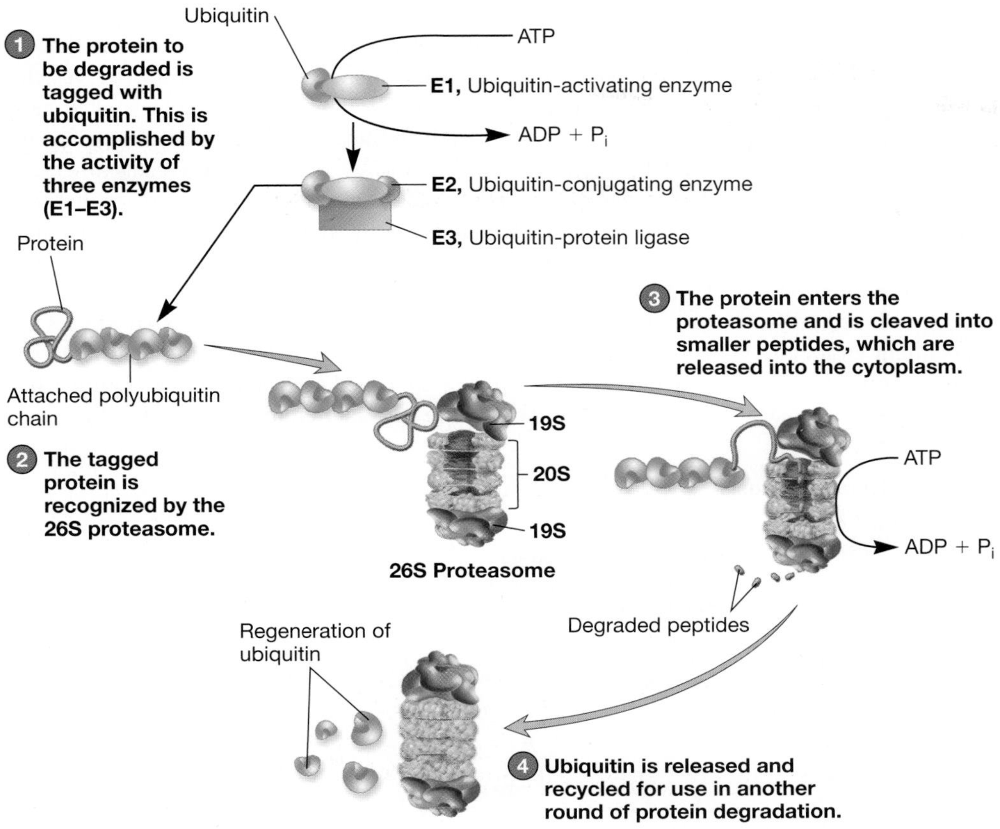

FIGURE 4.9 Proteasome Degradation of Proteins. The protein degradation pathway.

***Figure 4.9** Micro Inquiry*

Why must proteins be unfolded when they enter the proteasome?

Proteins destined for the cell membrane, endosomes and lysosomes, or secretion are synthesized by ribosomes attached to the rough endoplasmic reticulum (RER). These proteins have sequences of amino acids that target them to the lumen of the RER through which they move until released in small vesicles that bud from the ER. As the proteins pass through the ER, they are often modified by the addition of sugars—a process known as glycosylation.

The vesicles released from the ER travel to the cis face of the Golgi apparatus (figure 4.8). One popular model of the secretory pathway posits that these vesicles fuse to form the cis face of the Golgi. How this occurs is still unresolved. As the proteins proceed from the cis to the trans side, they are further modified. Some of these modifications target the proteins for their final location.

Transport vesicles are released from the trans face of the Golgi. Some deliver their contents to endosomes and lysosomes. Others deliver proteins and other materials to the cell membrane. Two types of vesicles transport materials to the cell membrane. One type constitutively delivers proteins in an unregulated manner, releasing them to the outside of the cell as the transport vesicle fuses with the plasma membrane. Other vesicles, called secretory vesicles, are found only in multicellular eukaryotes, where they are observed in secretory cells such as mast cells and other cells of the immune system. Secretory vesicles store the proteins to be released until the cell receives an appropriate signal. Once received, the secretory vesicles move to the plasma membrane, fuse with it, and release their contents to the cell exterior. ▶▶| *Cells, tissues, and organs of the immune system (section 32.4)*

One interesting and important feature of the secretory pathway is its quality-assurance mechanism. Proteins that fail to fold or have misfolded are not transported to their intended destination. Instead they are secreted into the cytosol, where they are targeted for destruction by the attachment of several small ubiquitin polypeptides, as detailed in **figure 4.9**. Ubiquitin marks the protein for degradation, which is accomplished by a huge, cylindrical complex called a 26S **proteasome.** The protein is broken down to smaller peptides using energy supplied by adenosine triphosphate (ATP) when the terminal phosphate is removed. As the protein is broken down, ubiquitins are released. In animal cells, the proteasome is involved in producing peptides for antigen presentation during many immunological responses described in chapter 33. ▶▶| *ATP (section 9.4)*

Endocytic Pathway

Endocytosis is observed in all eukaryotic cells. It is used to bring materials into the cell from the outside. During endocytosis, a cell takes up solutes (pinocytosis) or particles (phagocytosis) by enclosing them in vesicles pinched off from the plasma membrane. In most cases, these materials are delivered to a lysosome where they are digested. Endocytosis occurs regularly in all cells as a mechanism for recycling molecules in the membrane. In addition, some cells have specialized **endocytic pathways** that allow them to concentrate materials outside the cell before bringing them in. Others use endocytic pathways as a feeding mechanism. Many viruses and other intracellular pathogens use endocytic pathways to enter host cells.

Mammalian cells have several types of endocytic pathways (**figure 4.10**). **Phagocytosis** involves the use of protrusions from the cell surface to surround and engulf particulates. The endocytic vesicles formed by phagocytosis are called **phagosomes.** **Clathrin-dependent endocytosis** begins with coated pits, which are specialized membrane regions coated with the protein clathrin on the cytoplasmic side. The endocytic vesicles formed when these regions invaginate are called coated vesicles. Coated pits have receptors on their extracellular side that specifically bind macromolecules, concentrating them before they are endocytosed. Therefore this endocytic mechanism is referred to as **receptor-mediated endocytosis.** Clathrin-dependent endocytosis is used to internalize hormones, growth factors, iron, and cholesterol. **Caveolin-dependent endocytosis** involves caveolae ("little caves"), tiny, flask-shaped invaginations of the plasma membrane (about 50 to 80 nm in diameter) that are enriched in cholesterol and the membrane protein caveolin. The vesicles formed when caveolae pinch off are called caveolin-coated vesicles. Caveolin-dependent endocytosis has been implicated in signal transduction, transport of small molecules such as folic acid, as well as transport of macromolecules. Evidence exists that toxins such as cholera toxin enter their target cells via caveolae. Caveolae also appear to be used by many viruses, bacteria, and protozoa to enter host cells.

Clathrin-coated vesicles and some caveolin-coated vesicles deliver their contents to small organelles containing hydrolytic enzymes. These organelles are called early endosomes (figure 4.10). Early endosomes mature into late endosomes, which fuse with

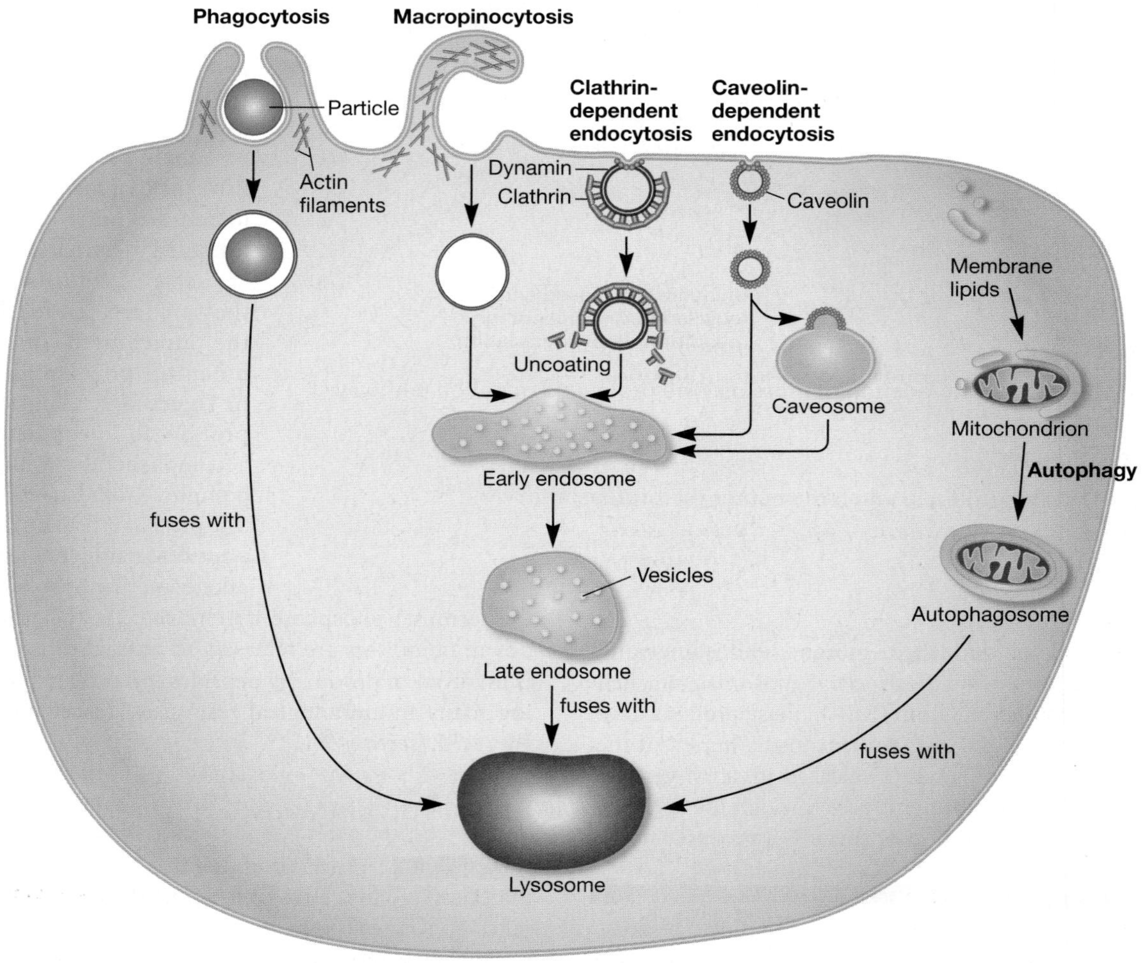

FIGURE 4.10 Some Endocytic Pathways Observed in Mammalian Cells. Materials ingested by endocytic processes are delivered to lysosomes. The pathway to lysosomes differs, depending on the type of endocytosis. In addition, cell components are recycled when autophagosomes deliver them to lysosomes for digestion. This process is called autophagy.

Figure 4.10 Micro Inquiry

Which endocytic process is considered receptor mediated? Why?

lysosomes. The development of early endosomes into late endosomes is not well understood. It appears that maturation involves the selective retrieval of membrane proteins. Phagosomes and some caveolin-coated vesicles take a different route to lysosomes. Phagosomes fuse directly with lysosomes (figure 4.10) to form a phagolysome. The caveolin-coated vesicles deliver their contents to an organelle termed a caveosome. The caveosome then fuses with an early endosome.

Endocytosis in eukaryotic microbes has not been studied as extensively as it has in mammals. It is known that some eukaryotic microbes carry out phagocytosis and probably macropinocytosis (figure 4.10). Pathways similar to clathrin-dependent endocytosis have been observed in fungi. These pathways have been best characterized in the yeast *Saccharomyces cerevisiae*. Although some aspects of the endocytic process differ between this yeast and mammals, considerable similarity has been discovered, especially in terms of the mechanisms used to sort materials among early endosomes, late endosomes, lysosomes, and the plasma membrane.

Lysosomes are also involved in another process that is increasingly attracting the attention of scientists: autophagy. Several types of autophagy have been identified. Perhaps the best studied is **macroautophagy.** Macroautophagy is used by cells to selectively digest and recycle cytoplasmic components (including organelles such as mitochondria). This is a normal process that helps maintain cellular homeostasis. The cell components to be digested are surrounded by a double membrane, the source of which is unknown. The resulting **autophagosome** fuses with a lysosome, delivering materials to be digested to it.

No matter the route taken, digestion occurs once the lysosome is formed. Amazingly, the lysosome accomplishes this without releasing its digestive enzymes into the cytoplasm. As the contents of the lysosome are digested, small products of digestion leave the lysosome and are used as nutrients or for other purposes. The resulting lysosome containing undigested material is sometimes called a **residual body.** In some cases, the residual body can release its contents to the cell exterior by a process called lysosome secretion.

Search This: Endocytosis movie

1. How do RER and SER differ from one another in terms of structure and function? List the processes in which the ER is involved.
2. What is a proteasome? Why is it important to the proper functioning of the ER?
3. Describe the structure of a Golgi apparatus. How do the cis and trans faces of the Golgi apparatus differ? List the major Golgi apparatus functions.
4. What are lysosomes? How do they participate in intracellular digestion? What might happen if lysosomes released their enzymatic contents into the cytoplasm?
5. Describe the secretory pathway. To what destinations does this pathway deliver proteins and other materials?
6. Define endocytosis. Describe the endocytic pathway and the routes that deliver materials to lysosomes for digestion.
7. Why do you think eukaryotic cells use different endocytic pathways for different substrates?

4.5 Organelles Involved in Genetic Control of the Cell

DNA is the molecule that houses the genetic blueprint of the cell. Eukaryotic cells differ dramatically from bacterial and archaeal cells in the way DNA is stored and used. In this section, the organelles involved with these important cellular functions are introduced.

Nucleus

The nucleus is by far the most visually prominent organelle in eukaryotic cells. It was discovered early in the study of cell structure and was shown by Robert Brown in 1831 to be a constant feature of eukaryotic cells. The **nucleus** is the repository for the cell's genetic information.

Nuclei are membrane-delimited spherical bodies about 5 to 7 μm in diameter (figures 4.2 and 4.3). Dense fibrous material called **chromatin** can be seen within the nucleus of a stained cell. Chromatin is a complex of DNA and proteins, with histones being the most important. Histones are small basic proteins rich in the amino acids lysine, arginine, or both. There are five types of histones in most eukaryotic cells: H1, H2A, H2B, H3, and H4. Eight histone molecules (two each of H2A, H2B, H3, and H4) form an ellipsoid about 11 nm long and 6.5 to 7 nm in diameter **(figure 4.11).** In nondividing cells, chromatin is dispersed, but it condenses during cell division to become visible as **chromosomes.** Some chromatin, the euchromatin, is loosely organized, and it contains those genes that are actively expressed. In contrast, heterochromatin is coiled more tightly, appears darker in the electron microscope, and is genetically inactive most of the time.

The nucleus is bounded by the **nuclear envelope** (figure 4.3), a complex structure consisting of inner and outer membranes separated by a 15 to 75 nm perinuclear space. The envelope is continuous with the ER at several points, and its outer membrane is covered with ribosomes. A network of intermediate filaments, called the nuclear lamina, is observed in animal cells. It lies against the inner surface of the nuclear envelope and supports it. Chromatin usually is associated with the inner membrane. Many nuclear pores penetrate the envelope, and each pore is formed by about 30 proteins; each pore plus the associated proteins is called a **nuclear pore complex**

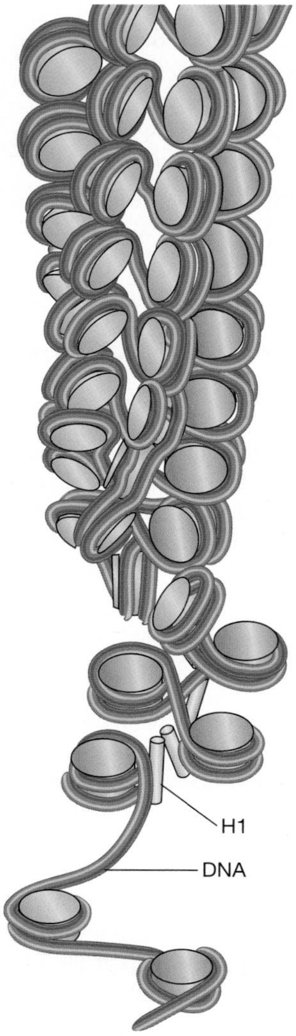

FIGURE 4.11 Nucleosomes. An illustration of how a string of nucleosomes, each associated with a histone H1, might be organized to form a highly supercoiled chromatin fiber. The nucleosomes are drawn as cylinders.

(**figure 4.12**). Pores are about 70 nm in diameter and collectively occupy about 10 to 25% of the nuclear surface. The nuclear pore complexes serve as transport routes between the nucleus and surrounding cytoplasm. Small molecules move through the nuclear pore complex unaided. However, large molecules are transported through the nuclear pore complex. Some nuclear pore complex proteins are involved in these transport processes.

Often the most noticeable structure within the nucleus is the **nucleolus** (**figure 4.13**). A nucleus may contain from one to many nucleoli. Although the nucleolus is not membrane-enclosed, it is a complex organelle with separate granular and fibrillar regions. It is present in nondividing cells but frequently disappears during mitosis. After mitosis, the nucleolus reforms around the nucleolar organizer, a particular part of a specific chromosome.

The nucleolus plays a major role in ribosome synthesis. The DNA of the nucleolar organizer directs the production of ribosomal RNA (rRNA). This RNA is synthesized in a single long piece that is cut to form the final rRNA molecules. The processed rRNAs combine with ribosomal proteins (which have been synthesized in the cytoplasm) to form partially completed ribosomal subunits. The granules seen in the nucleolus are probably these subunits. Immature ribosomal subunits then leave the nucleus, presumably by way of the nuclear pore complexes, and mature in the cytoplasm.

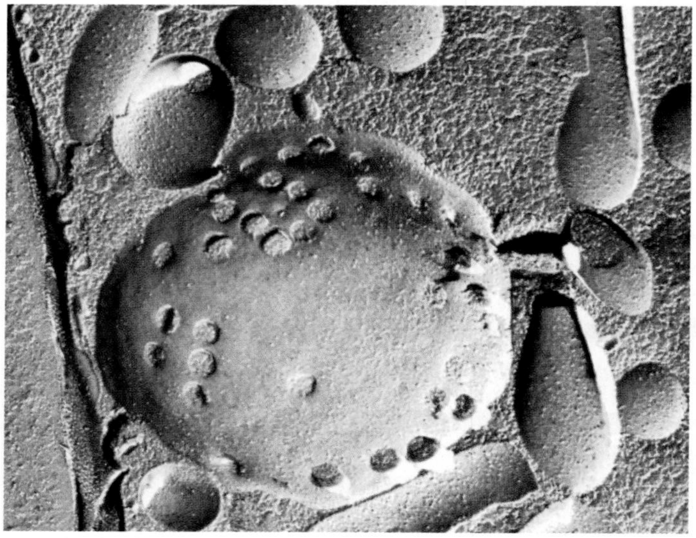

(a)

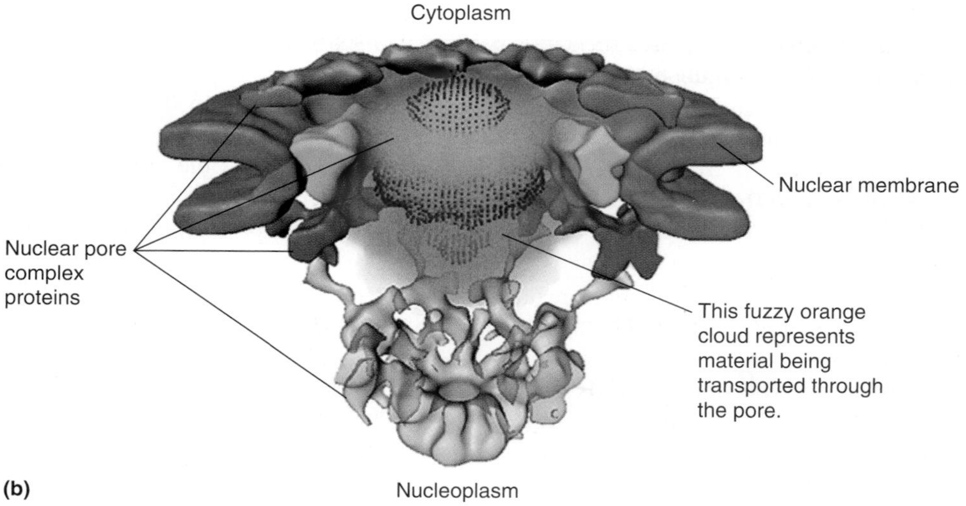

(b)

FIGURE 4.12 The Nucleus and Nuclear Pore Complex. (a) A freeze-etch preparation of a conidiospore (p. 109) of the fungus *Geotrichum candidum* (×44,600). Note the large, convex nuclear surface with pores scattered over it. (b) Electron cryotomogram of the nuclear pore complex.

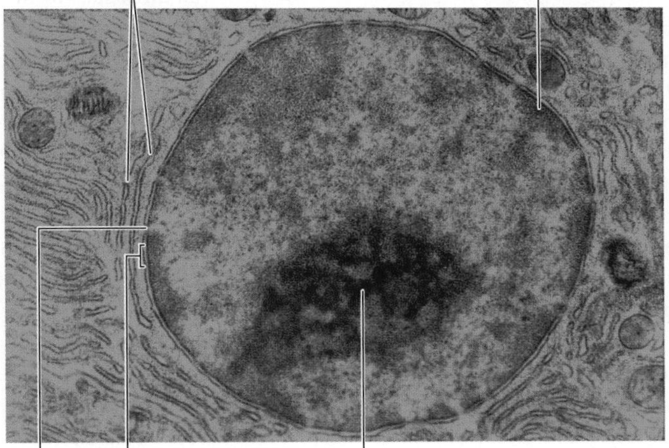

FIGURE 4.13 The Nucleolus. The nucleolus is a prominent feature of the nucleus. It functions in rRNA synthesis and the assembly of ribosomal subunits. Chromatin, nuclear pores, and the nuclear envelope are also visible in this electron micrograph of an interphase nucleus.

Figure 4.13 Micro Inquiry

What are the granules within the nucleolus thought to be?

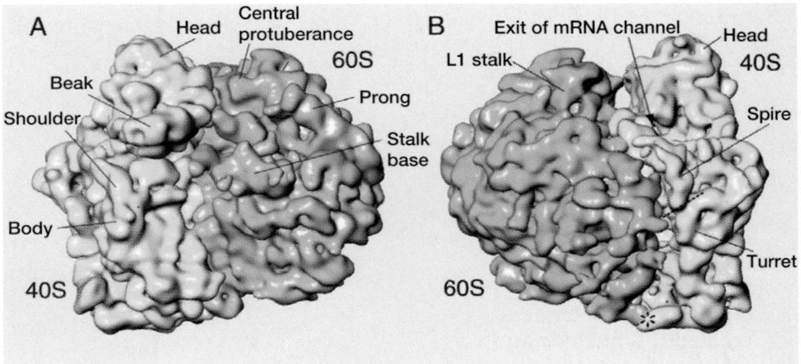

FIGURE 4.14 Eukaryotic Ribosome. This image of the 80S ribosome of *Trypanosoma cruzi* was generated using electron cryomicroscopy. The image is a density map with a resolution to 12 Å. Landmarks of the ribosome are shown; compare these to bacterial 70S ribosomes shown in figure 3.38.

Figure 4.14 Micro Inquiry

Which subunit attaches to the rough ER?

Eukaryotic Ribosomes

Eukaryotic ribosomes (i.e., those not found in mitochondria and chloroplasts) are larger than bacterial and archaeal 70S ribosomes. Each ribosome is a dimer of a 60S and a 40S subunit, is about 22 nm in diameter, and has a sedimentation coefficient of 80S and a molecular weight of 4 million (**figure 4.14**). Eukaryotic ribosomes are either associated with the endoplasmic reticulum or free in the cytoplasm. When bound to the endoplasmic reticulum to form rough ER, they are attached through their 60S subunits.

Both free and ER-bound ribosomes synthesize proteins. Proteins made on the ribosomes of the RER are often secreted or are inserted into the ER membrane as integral membrane proteins. Free ribosomes are the sites of synthesis for nonsecretory and nonmembrane proteins. Some proteins synthesized by free ribosomes are inserted into organelles such as the nucleus, mitochondrion, and chloroplast. As we discuss in chapter 12, proteins called molecular chaperones aid the proper folding of proteins after synthesis. They also assist the transport of proteins into eukaryotic organelles such as mitochondria.

1. What is the difference between chromatin and chromosomes?
2. Why do you think the nucleolus disappears during mitosis?
3. Describe the structure of the eukaryotic 80S ribosome and contrast it with bacterial and archaeal ribosomes.
4. How do free ribosomes and those bound to the ER differ in function?

4.6 Organelles Involved in Energy Conservation

Three important energy-conserving organelles have been identified in eukaryotes: mitochondria, hydrogenosomes, and chloroplasts. They are of scientific interest not only for this role but also because of their evolutionary history. All are thought to be derived from bacterial cells that invaded or were ingested by early ancestors of eukaryotic cells (**figure 4.15**). This hypothesis, called the **endosymbiotic hypothesis,** is supported by several lines of evidence, including the fact that mitochondria and chloroplasts contain their own chromosomes and ribosomes. Mitochondrial and chloroplast chromosomes bear striking similarity that of certain extant (living) bacteria and cyanobacteria, respectively. Their ribosomes are the same size as bacterial ribosomes, and the 16S ribosomal RNA sequences are most similar to those of *Bacteria (see figure 1.2)*. Other bacteria-like features are noted in the discussions that follow. |◀◀ *Ribosomes (section 3.5)*

Mitochondria

Found in most eukaryotic cells, **mitochondria** (s., **mitochondrion**) frequently are called the "powerhouses" of the cell (**figure 4.16**). Metabolic processes such as the tricarboxylic

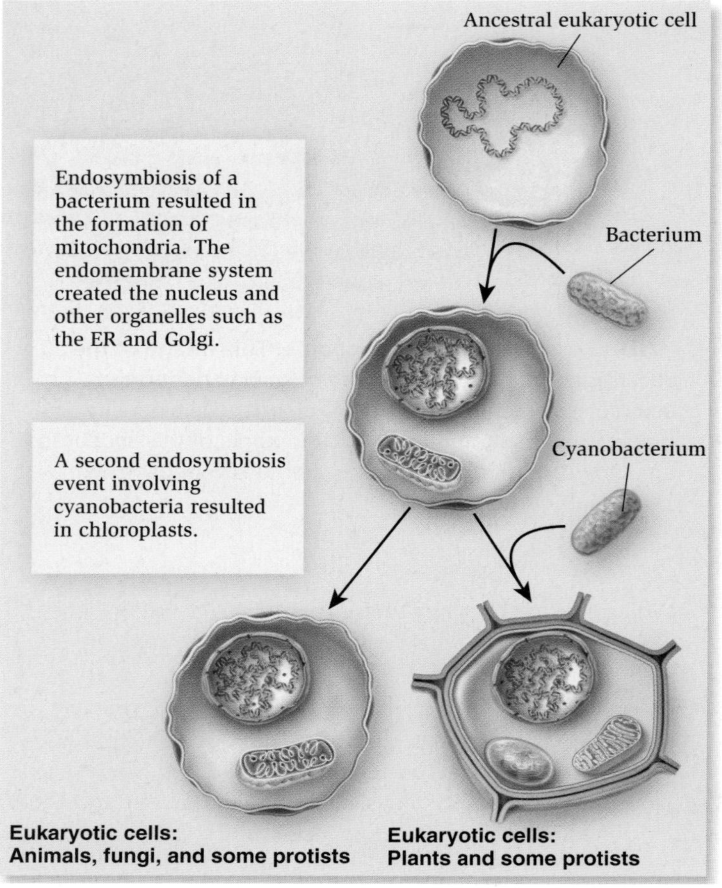
Ancestral eukaryotic cell
Endosymbiosis of a bacterium resulted in the formation of mitochondria. The endomembrane system created the nucleus and other organelles such as the ER and Golgi.
Bacterium
A second endosymbiosis event involving cyanobacteria resulted in chloroplasts.
Cyanobacterium
Eukaryotic cells: Animals, fungi, and some protists
Eukaryotic cells: Plants and some protists

FIGURE 4.15 The Endosymbiotic Hypothesis for Evolution of Mitochondria and Chloroplasts. Hydrogenosomes are thought to have also evolved by this mechanism. It is thought that the original endosymbiotic bacterium was an anaerobe. It if remained an anaerobe, it evolved into a hydrogenosome. If it developed the ability to carry out aerobic respiration, it evolved into a mitochondrion.

acid cycle and the generation of ATP, the energy currency of all life forms, take place here. When viewed with a transmission electron microscope, many mitochondria are cylindrical structures and measure approximately 0.3 to 1.0 μm by 5 to 10 μm. (In other words, they are about the same size as bacterial cells.) Some cells possess 1,000 or more mitochondria; others (some yeasts, unicellular algae, and trypanosome protists) have a single, giant, tubular mitochondrion twisted into a continuous network permeating the cytoplasm. ▶▶| *Tricarboxylic acid cycle (section 10.4); Electron transport and oxidative phosphorylation (section 10.5)*

The mitochondrion is bounded by two membranes, an outer mitochondrial membrane separated from an inner mitochondrial membrane by a 6 to 8 nm intermembrane space (figure 4.16). The outer mitochondrial membrane contains porins and thus is similar to the outer membrane of gram-negative bacteria. The inner membrane has infoldings called **cristae** (s., **crista**), which greatly increase its surface area. The shape of cristae differs in mitochondria from various species. Fungi have platelike (laminar) cristae, whereas euglenoid flagellates may have cristae shaped like disks. Tubular cristae are found in a variety of eukaryotes. Some amoebae possess mitochondria with cristae in the shape of vesicles. The inner membrane encloses the mitochondrial matrix, a dense material containing ribosomes, DNA, and often large calcium phosphate granules. In many organisms, mitochondrial DNA is a closed circle, like most bacterial DNA. However, in some protists, mitochondrial DNA is linear.

Each mitochondrial compartment is different from the others in chemical and enzymatic composition. For example, the outer and inner mitochondrial membranes possess different lipids. Enzymes and electron carriers involved in electron transport and oxidative phosphorylation (the formation of ATP during respiration) are located only in the inner membrane. Enzymes of the tricarboxylic acid cycle and those involved with the catabolism (breaking down) of fatty acids are located in the matrix. ▶▶| *Lipid catabolism (section 10.9)*

The mitochondrion uses its DNA and ribosomes to synthesize some of its own proteins. In fact, mutations in mitochondrial DNA often lead to serious diseases in humans. However, most mitochondrial proteins are manufactured under the direction of the nucleus. Mitochondria reproduce by binary fission, a reproductive process used by many bacteria. ▶▶| *Bacterial cell cycle (section 7.2)*

Hydrogenosomes

Hydrogenosomes are small organelles involved in energy-conservation processes in some anaerobic protists (**figure 4.17**). Like mitochondria, hydrogenosomes are bound by a double membrane. However, they often lack cristae and usually lack DNA. They also differ from mitochondria in terms of the method used to generate ATP. Within hydrogenosomes, pyruvate is catabolized by a fermentative process rather than respiration, and CO_2, H_2, and acetate are formed. In some hydrogenosome-bearing protists, these metabolic products are consumed by symbiotic bacteria and archaea living within the protist. The symbiotic archaea include methanogens that consume the CO_2 and H_2 and generate methane (CH_4). ▶▶| *Mutualism (section 30.1)*

Despite the differences between mitochondria and hydrogenosomes, evidence suggests that mitochondria and hydrogenosomes are descended from a common ancestral organelle. Supporting this hypothesis is the discovery of an organelle in the protist *Nyctotherus ovalis*, an inhabitant of the hindgut of cockroaches, that has both mitochondrial and hydrogenosome features. The organelles contain DNA, like mitochondria, but produce hydrogen like hydrogensomes. Furthermore, their DNA is similar to mitochondrial DNA in that it encodes ribosomal proteins and components of a typical mitochondrial electron transport chain.

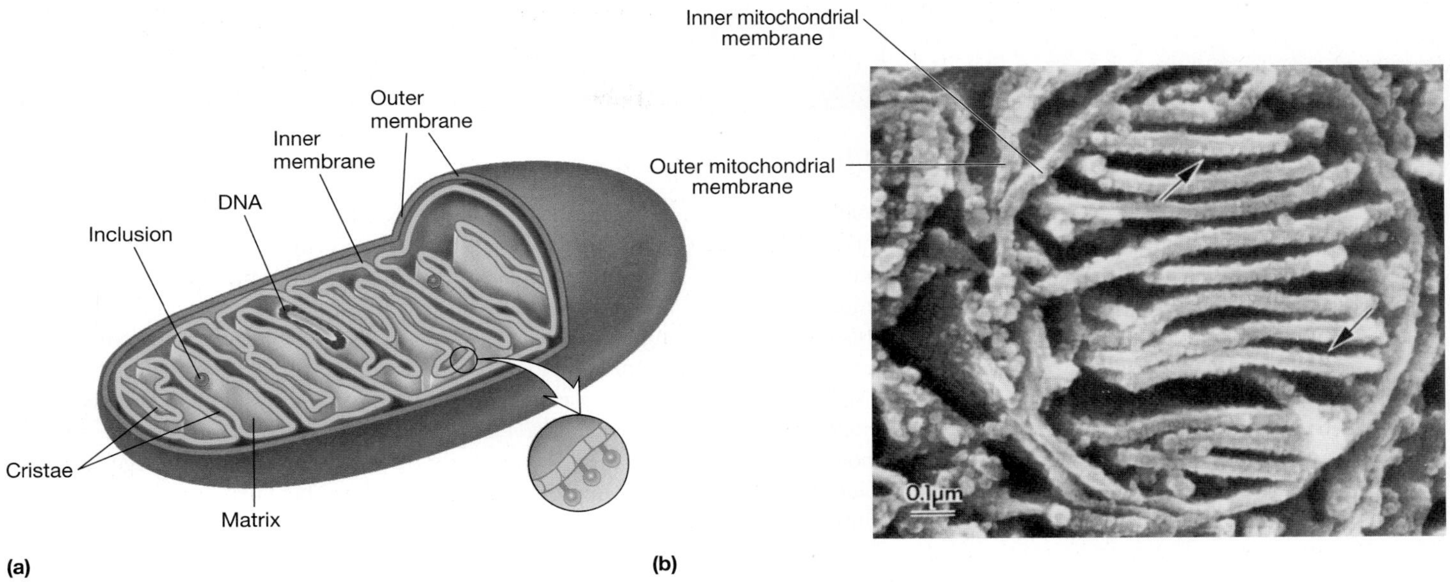

FIGURE 4.16 Mitochondrial Structure. (a) A diagram of mitochondrial structure. The insert shows the ATP-synthesizing enzyme ATP synthase lining the inner surface of the cristae. (b) Scanning electron micrograph (×70,000) of a freeze-fractured mitochondrion showing the cristae (arrows). The outer and inner mitochondrial membranes also are evident.

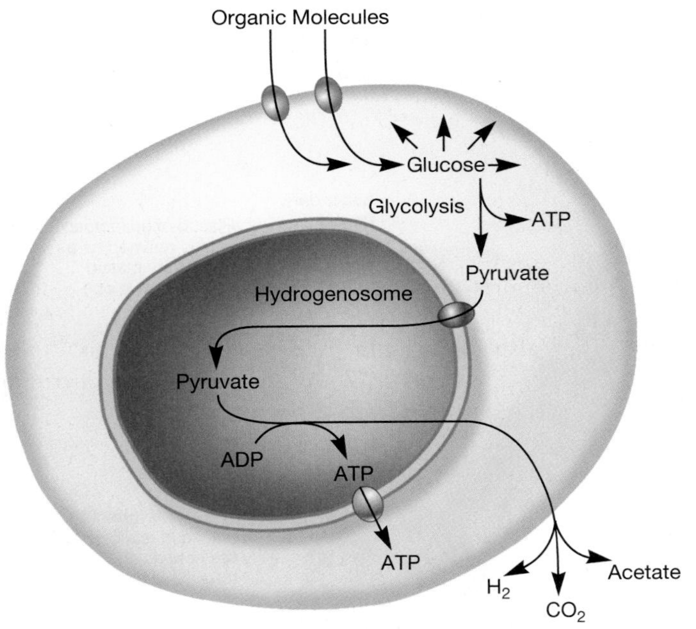

FIGURE 4.17 A Hydrogenosome and Associated Metabolic Activities.

Chloroplasts

Plastids are cytoplasmic organelles of photosynthetic protists and plants. They often possess pigments such as chlorophylls and carotenoids, and are the sites of synthesis and storage of food reserves. The most important type of plastid is the chloroplast. **Chloroplasts** contain chlorophyll and use light energy to convert CO_2 and water to carbohydrates and O_2. That is, they are the site of photosynthesis. Two major types of chloroplasts have been identified: those that evolved from a primary endosymbiotic event and those that evolved from a secondary (or tertiary) event (**Microbial Ecology & Diversity 4.1**). The chloroplasts of plants and some photosynthetic protists are primary plastids and are the focus of this discussion.

Although chloroplasts are quite variable in size and shape, they share many structural features. Most are oval with dimensions of 2 to 4 μm by 5 to 10 μm, but some photosynthetic protists possess one huge chloroplast that fills much of the cell. Like mitochondria, chloroplasts are encompassed by two membranes (**figure 4.18**). A matrix called the stroma is enclosed by the inner membrane. It contains DNA, ribosomes, lipid droplets, starch granules, and a complex internal membrane system whose most prominent components are flattened, membrane delimited sacs called **thylakoids.** Clusters of two or more thylakoids are dispersed within the stroma of most algal chloroplasts (figure 4.18*b*). In some photosynthetic protists, several disklike thylakoids are stacked on each other like coins to form grana (s., granum).

Photosynthetic reactions are separated structurally in the chloroplast just as electron transport and the tricarboxylic acid cycle are in the mitochondrion. The trapping of light energy to generate ATP, NADPH, and O_2 is referred to as the light reactions. These reactions are located in the thylakoid membranes, where chlorophyll and electron transport components are also found. The ATP and NADPH formed by the light reactions are used to form carbohydrates from CO_2 and water in the dark reactions. The dark reactions take place in the stroma.

▶▶| *Phototrophy (section 10.12)*

MICROBIAL DIVERSITY & ECOLOGY

4.1 There Was an Old Woman Who Swallowed a Fly

The children's song "There Was an Old Woman Who Swallowed a Fly" describes the unusual eating habits of an old woman. Her story began when she ate a fly but continued as she ate a series of other animals, all in an attempt to get rid of the fly. The evolutionary history of chloroplasts also appears to involve a series of meals. Molecular studies of the chloroplasts of a wide variety of photosynthetic protists has revealed that some chloroplasts arose when a phagocytic ancestor of a eukaryotic cell engulfed the ancestor of a cyanobacterium, as we discuss in chapter 1 (**box figure**). This evolutionary event eventually gave rise to plants and the photosynthetic protists commonly called green algae and red algae. These organisms are now classified together in the supergroup *Archaeplastida (see chapter 23)*. The chloroplasts that evolved from this endosymbiosis are the typical chloroplasts that biology students are familiar with. These primary plastids have two membranes. Subsequently, a phagocytic plastid-free eukaryote ingested an archaeplastid, in particular, a red alga. The ingested archaeplastid evolved into a chloroplast surrounded by three membranes. These chloroplasts are known as secondary plastids and are found in photosynthetic protists in the *Straminopila*. They include diatoms and brown algae. The final meal in the series was eaten by some dinoflagellates. These interesting protists can be either chemoorganotrophic or phototrophic. It appears that some photosynthetic dinoflagellates ingested a protist containing secondary plastids. The ingested protist evolved into a tertiary plastid surrounded by four membranes.

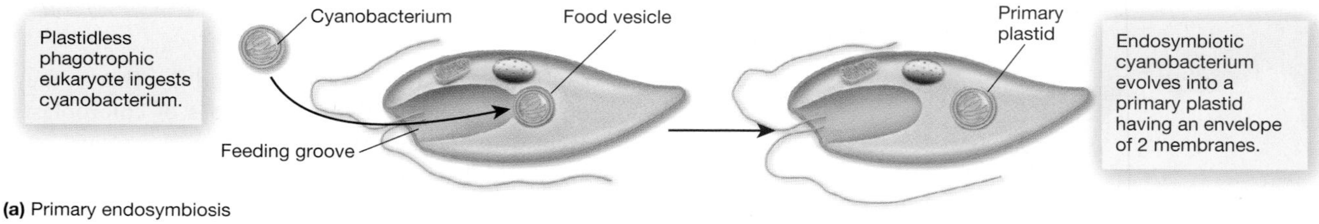

Plastidless phagotrophic eukaryote ingests cyanobacterium.
Cyanobacterium
Food vesicle
Feeding groove
Primary plastid
Endosymbiotic cyanobacterium evolves into a primary plastid having an envelope of 2 membranes.

(a) Primary endosymbiosis

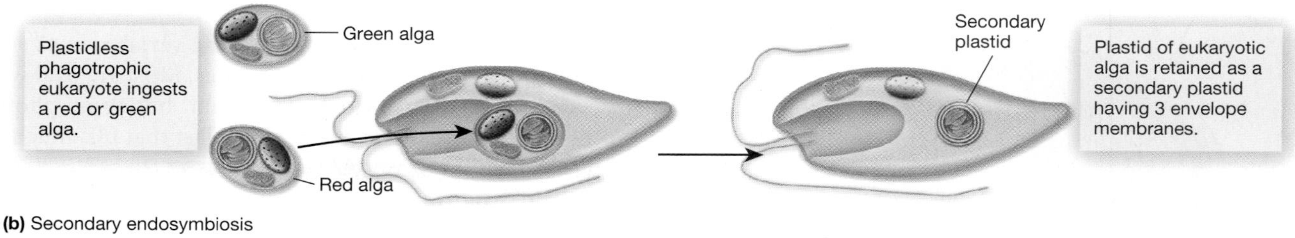

Plastidless phagotrophic eukaryote ingests a red or green alga.
Green alga
Red alga
Secondary plastid
Plastid of eukaryotic alga is retained as a secondary plastid having 3 envelope membranes.

(b) Secondary endosymbiosis

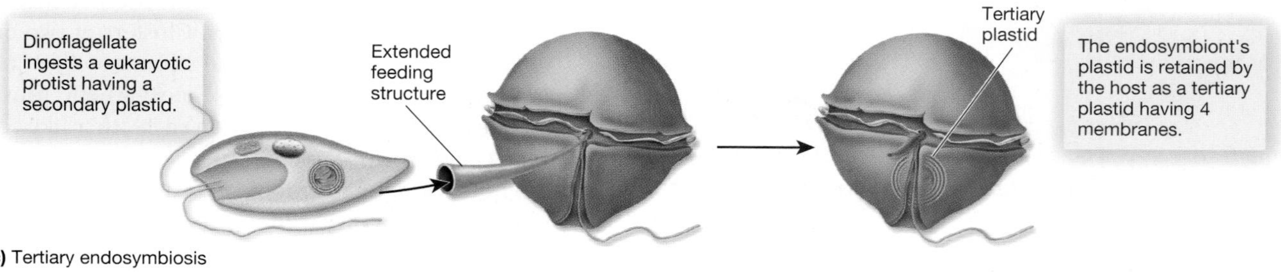

Dinoflagellate ingests a eukaryotic protist having a secondary plastid.
Extended feeding structure
Tertiary plastid
The endosymbiont's plastid is retained by the host as a tertiary plastid having 4 membranes.

(c) Tertiary endosymbiosis

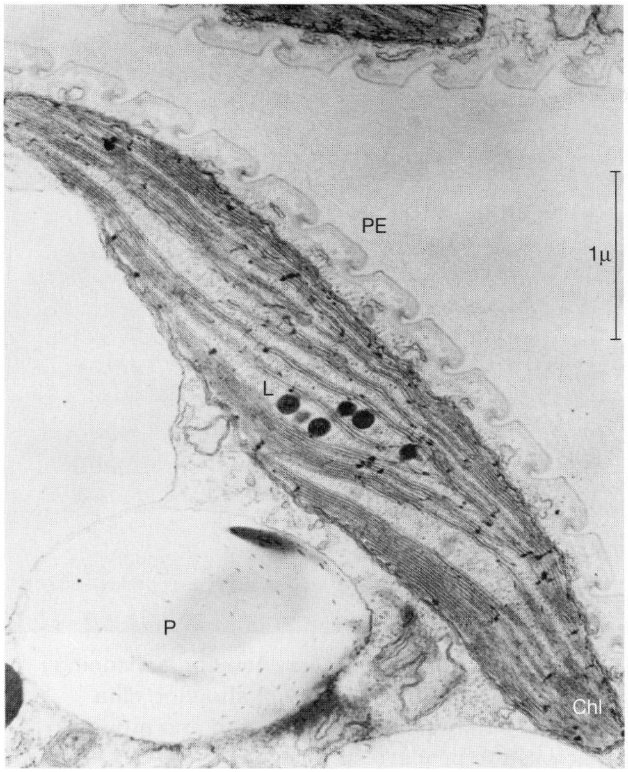

(a)

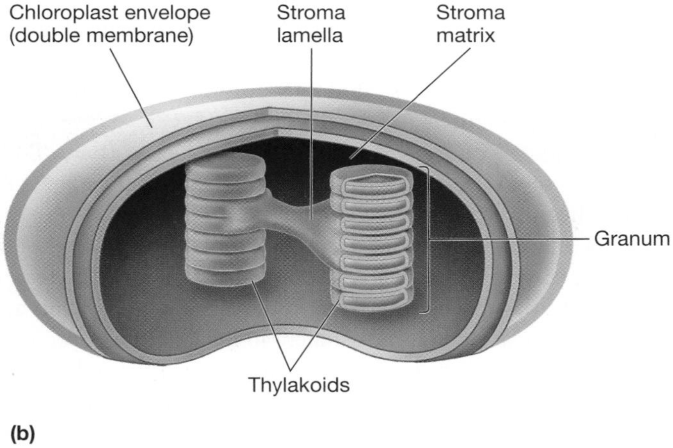

(b)

FIGURE 4.18 Chloroplast Structure. (a) The chloroplast (Chl) of the euglenoid flagellate *Colacium cyclopicolum*. The chloroplast is bounded by a double membrane and has its thylakoids in groups of three or more. A paramylon granule (P), lipid droplets (L), and the pellicular strips (PE) can be seen (×40,000). (b) A diagram of chloroplast structure.

Figure 4.18 Micro Inquiry

In what regions of the chloroplast does the light reaction occur? The dark reaction?

1. Describe in detail the structure of mitochondria, hydrogenosomes, and chloroplasts. Where are the different components of mitochondria and chloroplast energy-trapping systems located?
2. Define plastid, dark reactions, and light reactions.
3. What is the role of mitochondrial DNA?
4. What features of mitochondria, hydrogenosomes, and chloroplasts support the endosymbiotic hypothesis of their evolution?

4.7 External Structures

Cilia (s., **cilium**) and **flagella** (s., **flagellum**) are the most prominent external structures; they are associated with motility in eukaryotes. Although both are whiplike and beat to move the microorganism along, they differ from one another in two ways. First, cilia are typically only 5 to 20 μm in length, whereas flagella are 100 to 200 μm long. Second, their patterns of movement are usually distinctive (**figure 4.19**). Flagella move in an undulating fashion and generate planar or helical waves originating at either the base or the tip. If the

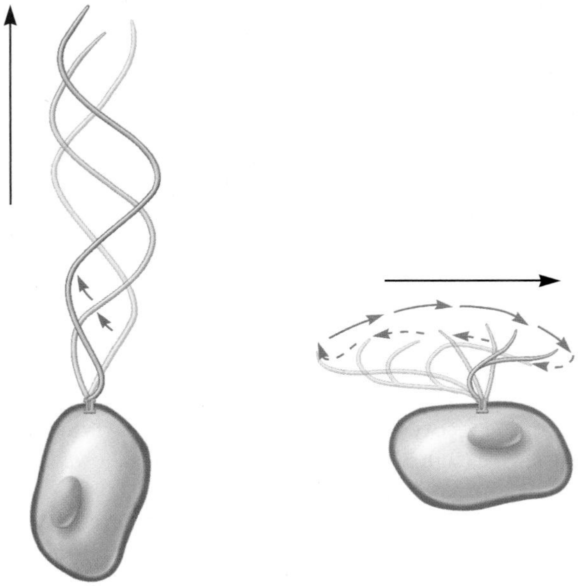

FIGURE 4.19 Patterns of Flagellar and Ciliary Movement. Flagellar and ciliary movement often takes the form of waves. Flagella (left illustration) move either from the base of the flagellum to its tip or in the opposite direction. The motion of these waves propels the organism along. The beat of a cilium (right illustration) may be divided into two phases. In the effective stroke, the cilium remains fairly stiff as it swings through the water. This is followed by a recovery stroke in which the cilium bends and returns to its initial position. The black arrows indicate the direction of water movement in these examples.

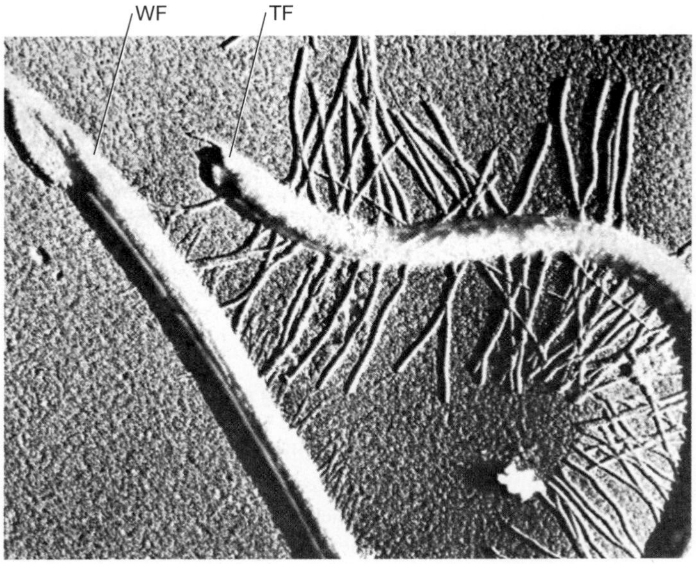

FIGURE 4.20 Whiplash and Tinsel Flagella. Transmission electron micrograph of a shadowed whiplash flagellum, WF, and a tinsel flagellum, TF, with mastigonemes.

wave moves from base to tip, the cell is pushed along; a beat traveling from the tip toward the base pulls the cell through the water. Sometimes the flagellum has lateral hairs called flimmer filaments (thicker, stiffer hairs are called mastigonemes). These filaments change flagellar action so that a wave moving down the filament toward the tip pulls the cell along instead of pushing it. Such a flagellum often is called a tinsel flagellum, whereas the naked flagellum is referred to as a whiplash flagellum (**figure 4.20**). Cilia, on the other hand, normally have a beat with two distinctive phases. In the effective stroke, the cilium strokes through the surrounding fluid like an oar, thereby propelling the organism. The cilium next bends along its length while it is pulled forward during the recovery stroke in preparation for another effective stroke (figure 4.19). A ciliated microorganism actually coordinates the beats so that some of its cilia are in the recovery phase while others are carrying out their effective stroke (**figure 4.21**). This coordination allows the organism to move smoothly through the water.

Despite their differences, cilia and flagella are very similar in ultrastructure. They are membrane-bound cylinders about 0.2 μm in diameter. Located in the matrix of the organelle is the axoneme, which consists of nine pairs of microtubule doublets arranged in a circle around two central tubules (**figure 4.22**). This is called the 9 + 2 pattern of microtubules. Each doublet has pairs of arms projecting from subtubule A (the complete microtubule) toward a neighboring doublet. A radial spoke extends from subtubule A toward the internal pair of microtubules with their central sheath. These microtubules are similar to those found in the cytoplasm. ◀◀ *Flagella (section 3.6)*

A basal body lies in the cytoplasm at the base of each cilium or flagellum. It is a short cylinder with nine microtubule triplets around its periphery (a 9 + 0 pattern) and is separated from the rest of the organelle by a basal plate. The basal body directs the construction of these organelles. Cilia and flagella appear to grow through the addition of preformed microtubule subunits at their tips.

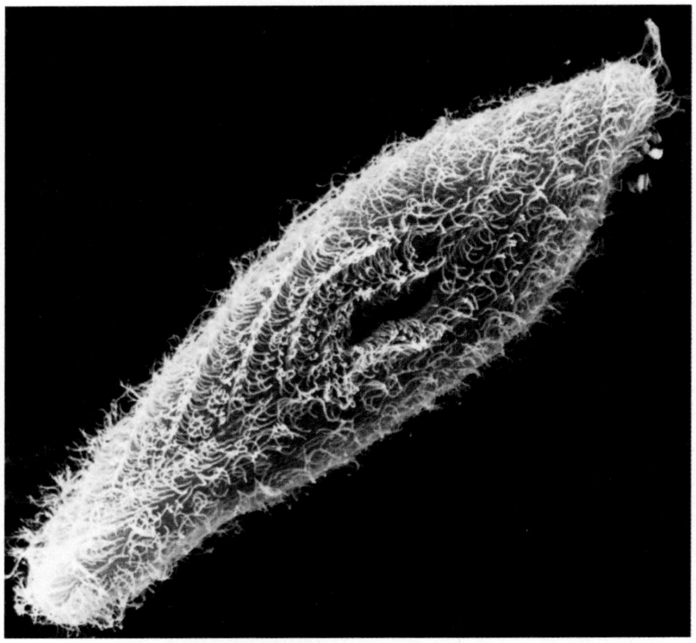

FIGURE 4.21 Coordination of Ciliary Activity. A scanning electron micrograph of *Paramecium* showing cilia (×1,500). The ciliary beat is coordinated and moves in waves across the protist's surface.

Cilia and flagella bend because adjacent microtubule doublets slide along one another while maintaining their individual lengths. The doublet arms, about 15 nm long, are made of the protein dynein (figure 4.22). It appears that dynein arms interact with the B subtubules of adjacent doublets to cause the sliding. The radial spokes also participate in this sliding motion. ATP powers the movement of cilia and flagella.

Cilia and flagella beat at a rate of about 10 to 40 strokes or waves per second and propel microorganisms rapidly. The record holder is the flagellate protist *Monas stigmatica,* which swims at a rate of 260 μm/second (approximately 40 cell lengths per second); the euglenoid flagellate *Euglena gracilis* travels at around 170 μm or 3 cell lengths per second. The ciliate protist *Paramecium caudatum* swims about 2,700 μm/second (12 lengths per second). Such speeds are equivalent to or much faster than those seen in higher animals but not as fast as those in bacteria.

Search This: Ciliate movie

1. Prepare and label a diagram showing the detailed structure of a cilium or flagellum.
2. How do cilia and flagella move, and what is dynein's role in the process? Contrast the ways in which flagella and cilia propel eukaryotic microorganisms through water.
3. Compare the structure and mechanism of action of bacterial and eukaryotic flagella.

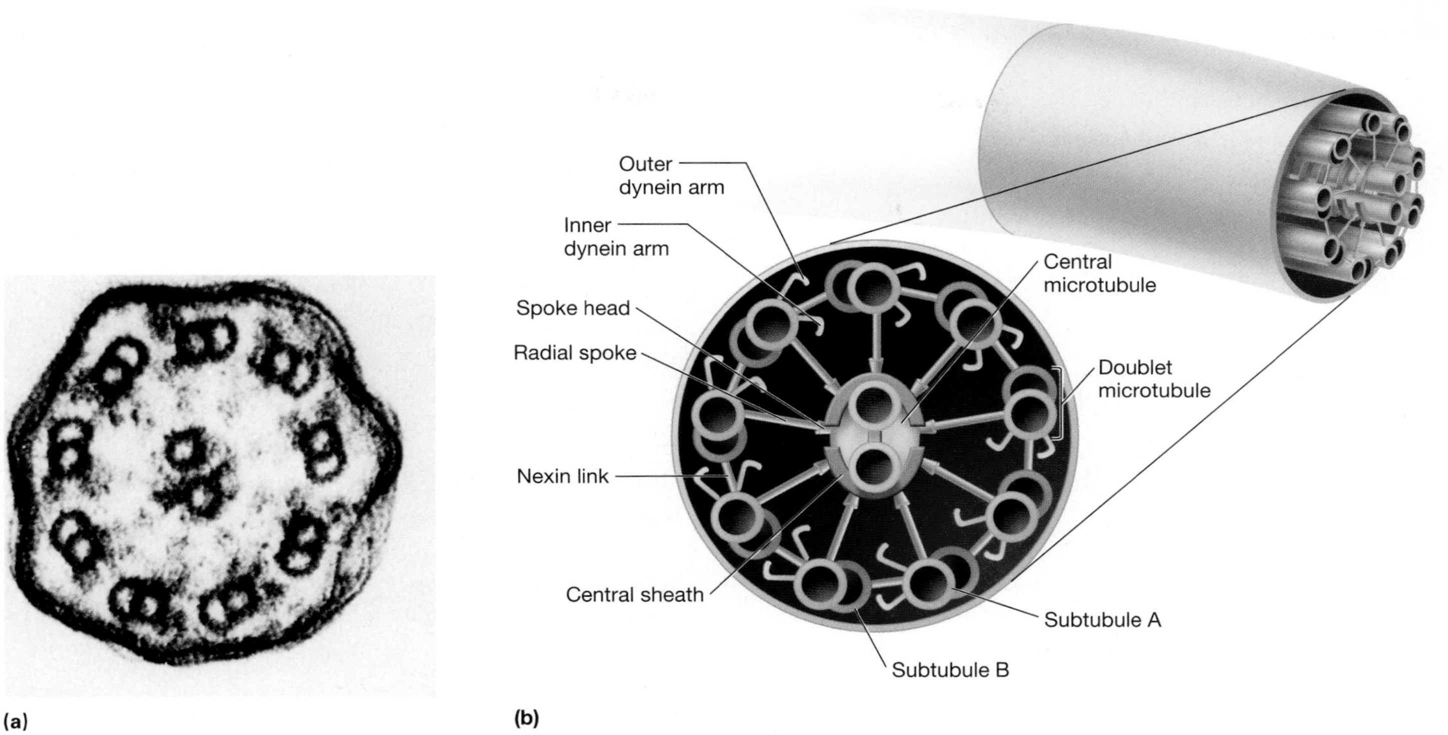

FIGURE 4.22 Cilia and Flagella Structure. (a) An electron micrograph of a cilium cross section. Note the two central microtubules surrounded by nine microtubule doublets (×160,000). (b) A diagram of cilia and flagella structure.

4.8 Comparison of Bacterial, Archaeal, and Eukaryotic Cells

A comparison of the bacterial and eukaryotic cells in **figure 4.23** demonstrates that there are many fundamental differences between these cells. These differences are also observed between archaeal and eukaryotic cells, as archaea are similar to bacteria at the gross structural level. Eukaryotic cells have a membrane-enclosed nucleus. In contrast, bacterial and archaeal cells lack a true, membrane-delimited nucleus. Most bacteria and archaea are smaller than eukaryotic cells, often about the size of eukaryotic mitochondria and chloroplasts.

The presence of the eukaryotic nucleus is the most obvious difference between these cell types, but many other major distinctions exist. It is clear from **table 4.2** that bacterial and archaeal cells are much simpler structurally. In particular, an extensive and diverse collection of membrane-delimited organelles is missing. Furthermore, bacterial and archaeal cells are functionally simpler in several ways. They lack mitosis and meiosis, and have a simpler genetic organization. Many complex eukaryotic processes are absent in the *Bacteria* and *Archaea*: endocytosis, intracellular digestion, directed cytoplasmic streaming, and ameboid movement are just a few.

Despite the many significant differences between these cell forms, they are remarkably similar on the biochemical level, as we discuss in succeeding chapters. With a few exceptions, the genetic code is the same in all, as is the way in which the genetic information in DNA is expressed. The principles underlying metabolic processes and many important metabolic pathways are identical. Thus beneath the profound structural and functional differences between bacterial, archaeal, and eukaryotic cells, there is an even more fundamental unity: a molecular unity that is basic to all known life processes.

1. Outline the major differences between bacterial, archaeal, and eukaryotic cells. How are they similar?
2. What characteristics make members of the *Archaea* more like eukaryotes? What features make them more like the *Bacteria?*

4.9 Overview of Protist Structure and Function

As noted earlier, protists are not a cohesive taxon. Rather, the term protist is used to informally group together many eukaryotic lineages. As seen in **figure 4.24**, some of these lineages contain unicellular microbes that are chemoorganotrophs—that is, they use organic molecules as energy sources. These animal-like microbes are also called protozoa. Other lineages, the slime molds and the water molds, also are chemoorganotrophs but are usually distinguished from other protists because of their unique life cycles

Table 4.2 Comparison of Bacterial, Archaeal, and Eukaryotic Cells

Property	*Bacteria*	*Archaea*	*Eukarya*
Organization of Genetic Material			
True membrane-bound nucleus	No	No	Yes
DNA complexed with histones	No	Some	Yes
Chromosomes	Usually one circular chromosome	One circular chromosome	More than one; chromosomes are linear
Plasmids	Very common	Very common	Rare
Introns in genes	Rare	Rare	Yes
Nucleolus	No	No	Yes
Mitochondria	No	No	Yes
Chloroplasts	No	No	Yes
Plasma Membrane Lipids	Ester-linked phospholipids and hopanoids; some have sterols	Glycerol diethers and diglycerol tetraethers	Ester-linked phospholipids and sterols
Flagella	Submicroscopic in size; composed of one protein fiber	Submicroscopic in size; composed of a fiber made from multiple different flagellin proteins	Microscopic in size; membrane bound; usually 20 microtubules in 9 + 2 pattern
Endoplasmic Reticulum	No	No	Yes
Golgi Apparatus	No	No	Yes
Peptidoglycan in Cell Walls	Yes	No	No
Ribosome Size	70S	70S	80S
Lysosomes	No	No	Yes
Cytoskeleton	Rudimentary	Rudimentary	Yes
Gas Vesicles	Yes	Yes	No

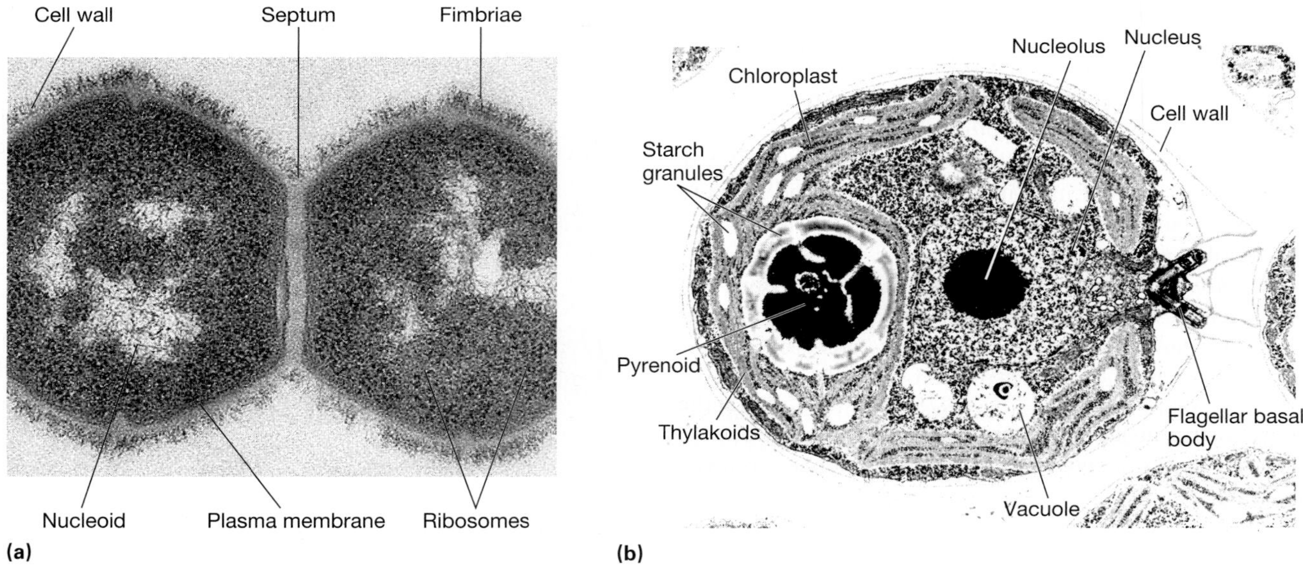

FIGURE 4.23 Comparison of Bacterial and Eukaryotic Cell Structure. (a) The bacterium *Streptococcus pyogenes* undergoing cell division. The septum separates the two daughter cells (×20,000). (b) The eukaryotic alga *Chlamydomonas reinhardtii.* Note the large chloroplast with its pyrenoid body (×30,000).

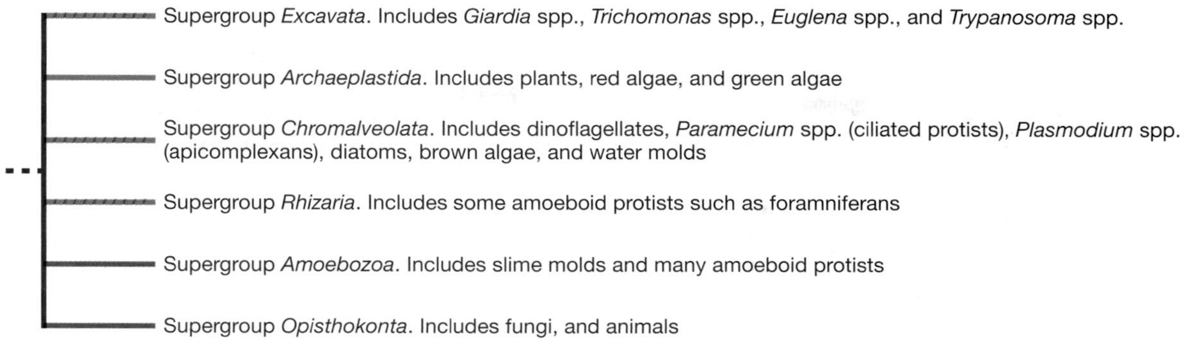
Supergroup *Excavata*. Includes *Giardia* spp., *Trichomonas* spp., *Euglena* spp., and *Trypanosoma* spp.
Supergroup *Archaeplastida*. Includes plants, red algae, and green algae
Supergroup *Chromalveolata*. Includes dinoflagellates, *Paramecium* spp. (ciliated protists), *Plasmodium* spp. (apicomplexans), diatoms, brown algae, and water molds
Supergroup *Rhizaria*. Includes some amoeboid protists such as foramniferans
Supergroup *Amoebozoa*. Includes slime molds and many amoeboid protists
Supergroup *Opisthokonta*. Includes fungi, and animals

FIGURE 4.24 A Recently Proposed Taxonomy of Eukaryotes. Eukaryote phylogeny and taxonomy are in a state of flux in part because of the limited molecular data available for many eukaryotic organisms. Therefore it is likely that this phylogenetic tree will change in the future. However, all current schemes group eukaryotes into supergroups based on molecular and other characteristics. Microbes are found in all supergroups. Brown lines represent lineages in which the members are chemoorganotrophs. Green lines are lineages that contain photosynthetic organisms. Striped green and brown lines are lineages that contain some photosynthetic organisms and some chemoorganotrophic organisms. The dashed line to the left of the tree is the root. It is arbitrarily placed in the drawing because at this time the root of Domain Eukarya is unknown. Notice that protists are found in many of the lineages; therefore they are not monophyletic.

or other characteristics. Some lineages contain photosynthetic microbes often referred to as algae. Finally, some lineages contain both chemoorganotrophs and photosynthetic organisms.

Despite these differences, protists of all types share many common features. In many respects, their morphology and physiology are the same as the cells of multicellular plants and animals. However, because many protists are unicellular, all of life's various functions must be performed within a single cell. Those that are multicellular lack highly differentiated tissues. Therefore, the structural complexity observed in protists arises at the level of specialized organelles. We now discuss some of the fundamental features of protist morphology.

Protist Morphology

The protist cell membrane is called the **plasmalemma** and is identical to that of multicellular organisms. In some protists, the cytoplasm immediately under the plasmalemma is divided into an outer gelatinous region called the **ectoplasm** and an inner fluid region, the **endoplasm.** The ectoplasm imparts rigidity to the cell body.

Many protists have a supportive mechanism called the **pellicle** (figure 4.18*a*). The pellicle consists of the plasmalemma and a relatively rigid layer of components just beneath it. The pellicle may be simple in structure. For example, *Euglena*, a photosynthetic protist, has a series of overlapping strips with a ridge at the edge of each strip fitting into a groove on the adjacent one. In contrast, the pellicles of ciliate protists are exceptionally complex with two membranes and a variety of associated structures. Although pellicles are not as strong and rigid as cell walls, they give their possessors a characteristic shape.

One or more vacuoles are usually present in the cytoplasm of protozoa. These are differentiated into contractile, secretory, and food vacuoles. **Contractile vacuoles** function as osmoregulatory organelles in those protists that live in hypotonic environments, such as freshwater lakes. Osmotic balance is maintained by continuous water expulsion. **Phagocytic vacuoles** are conspicuous in protists that ingest food by phagocytosis (holozoic protists) and in parasitic species. Phagocytic vacuoles are the sites of food digestion. In some organisms, they may occur anywhere on the cell surface, while others have a specialized structure for phagocytosis called the **cytostome** (cell mouth). When digestion commences, the phagocytic vacuole is acidic, but as it proceeds, the vacuolar pH increases and the membrane forms small blebs. These pinch off and carry nutrients throughout the cytoplasm. The undigested contents of the original phagocytic vacuole are expelled from the cell either at a random site on the cell membrane or at a designated position called the **cytoproct.**

Several energy-conserving organelles are observed in protists. Most aerobic chemoorganotrophic protists have mitochondria. The majority of anaerobic chemoorganotrophic protists (such as *Trichonympha,* which lives in the gut of termites) lack mitochondria; some of these organisms have hydrogenosomes**.** Photosynthetic protists have chloroplasts. A dense proteinaceous area, the **pyrenoid,** which is associated with the synthesis and storage of starch, may be present in the chloroplasts (figure 4.23*b*).

Many protists feature cilia or flagella at some point in their life cycle. Their formation is associated with a basal body-like organelle called the kinetosome. In addition to aiding in motility, these organelles may be used to generate water currents for feeding and respiration.

Encystment and Excystment

Many protists are capable of **encystment.** During encystment, the organism becomes simpler in morphology and develops into a resting stage called a cyst. The cyst is a dormant form marked by the presence of a cell wall and very low metabolic activity. Cyst formation is particularly common among aquatic, free-living protists and parasitic forms. Cysts serve three major functions: (1) they protect against adverse changes in the environment, such as nutrient deficiency, desiccation, adverse pH, and low levels of O_2; (2) they are sites for nuclear reorganization and cell division (reproductive cysts); and (3) they serve as a means of transfer between hosts in parasitic species. Protists escape from cysts by a process called **excystment.** Although the exact stimulus for excystment is unknown, it is generally triggered by a return to favorable environmental conditions. For example, cysts of parasitic species excyst after ingestion by the host.

Protist Reproductive Cells and Structures

Most protists have both asexual and sexual reproductive phases in their life cycles. The most common method of asexual reproduction is **binary fission.** During this process, the nucleus first undergoes mitosis, and then the cytoplasm divides by cytokinesis to form two identical individuals (**figure 4.25**). Multiple fission is also common, as is budding. Some filamentous, photosynthetic protists undergo fragmentation so that each piece of the broken filament grows independently.

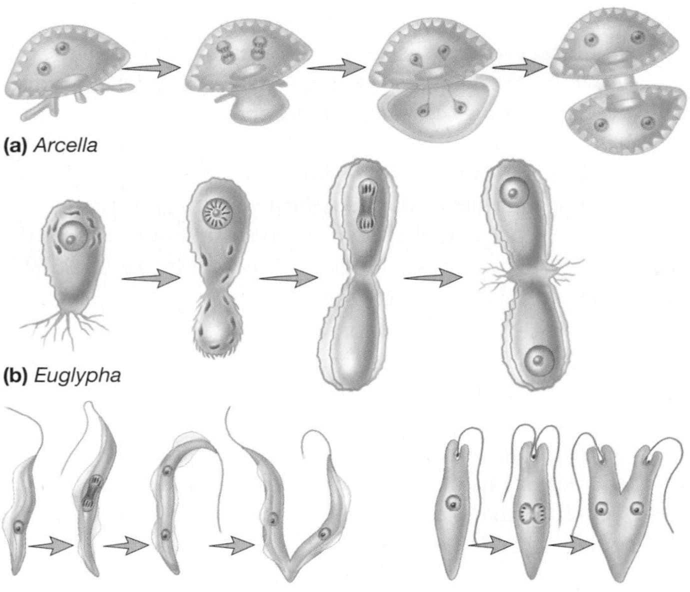

FIGURE 4.25 Binary Fission in Protists. (a) The two nuclei of the testate (shelled) amoeba *Arcella* divide as some of its cytoplasm is extruded and a new test for the daughter cell is secreted. (b) In another testate amoeba, *Euglypha,* secretion of new platelets is begun before cytoplasm begins to move out of the aperture. The nucleus divides while the platelets are used to construct the test for the daughter cell. For many protists, all organelles must be replicated before the cell divides. This is also the case with (c) *Trypanosoma* and (d) *Euglena.*

Sexual reproduction involves the formation of gametes. Protist cells that produce gametes are termed **gamonts.** The fusion of haploid gametes is called **syngamy.** Among protists, syngamy can involve the fusion of two morphologically similar gametes (isogamy) or two morphologically different types (anisogamy). Meiosis may occur before the formation and union of gametes, as in most animals, or just after fertilization, as is the case with lower plants. Furthermore, the exchange of nuclear material may occur in the familiar fashion—between two different individuals **(conjugation)**—or by the development of a genetically distinct nucleus within a single individual (autogamy).

With this level of reproductive complexity, perhaps it is not surprising that the nuclei among protists show considerable diversity. Most commonly, a vesicular nucleus is present. This is 1 to 10 μm in diameter, spherical, and has a distinct nucleolus and uncondensed chromosomes. Ovular nuclei are up to 10 times this size and possess many peripheral nucleoli. Still others have chromosomal nuclei, in which the chromosomes remain condensed throughout the cell cycle. Finally, the *Ciliophora* have two types of nuclei: a large macronucleus with distinct nucleoli and condensed chromatin, and a smaller, diploid micronucleus with dispersed chromatin but lacking nucleoli. Macronuclei are engaged in trophic activities and regeneration processes, whereas micronuclei are involved only in genetic recombination during sexual reproduction and the regeneration of the macronucleus. ▶▶| Chloroplastida *(section 23.6)*

1. Describe the pellicle. What is its function?
2. Trace the path of a food item from the phagocytic vacuole to the cytoproct.
3. What functions do cysts serve for a typical protist? What causes excystment to occur?
4. What is a gamont? What is the difference between anisogamy, isogamy, and syngamy?
5. How does conjugation differ from autogamy?
6. Describe vesicular, ovular, and chromosomal nuclei. Which is most like the nuclei found in higher eukaryotes?

4.10 Overview of Fungal Structure and Function

The term fungus (pl., fungi; Latin *fungus,* mushroom) describes eukaryotic organisms that are spore-bearing, have absorptive nutrition, lack chlorophyll, and reproduce sexually and asexually.

Fungal Structure

The body or vegetative structure of a fungus is called a **thallus** (pl., **thalli**). It varies in complexity and size, ranging from the single-cell microscopic yeasts to multicellular molds, macroscopic puffballs, and mushrooms. The fungal cell usually is encased in a cell wall of **chitin.** Chitin is a strong but flexible nitrogen-containing polysaccharide consisting of *N*-acetylglucosamine residues.

A **yeast** is a unicellular fungus that has a single nucleus and reproduces either asexually by budding and transverse division or sexually through spore formation. Each bud that separates can grow into a new cell, and some group together to form colonies. Generally yeast cells are larger than bacteria, vary considerably in size, and are commonly spherical to egg shaped. They lack flagella and cilia but possess most other eukaryotic organelles (figure 4.3*a*).

The thallus of a **mold** consists of long, branched, threadlike filaments of cells called **hyphae** (s., hypha; Greek *hyphe,* web) that form a **mycelium** (pl., **mycelia**), a tangled mass or tissuelike aggregation of hyphae. In some fungi, protoplasm streams through hyphae, uninterrupted by cross walls. These hyphae are called **coenocytic** or **aseptate hyphae** (**figure 4.26*a***). The hyphae of other fungi (figure 4.26*b*) have cross walls called **septa** (s., **septum**) with either a single pore (figure 4.26*c*) or multiple pores (figure 4.26*d*) that enable cytoplasmic streaming. These hyphae are termed **septate hyphae.**

Hyphae are composed of an outer cell wall and an inner lumen, which contains the cytosol and organelles. A plasma membrane surrounds the cytoplasm and lies next to the cell wall. The filamentous nature of hyphae results in a large surface area relative to the volume of cytoplasm. This makes adequate nutrient absorption possible.

Fungal Reproductive Cells and Structures

Reproduction in fungi can be either asexual or sexual. Asexual reproduction is accomplished in several ways: (1) a parent cell can undergo mitosis and divide into two daughter cells by a central constriction and formation of a new cell wall (**figure 4.27*a***), and (2) mitosis in vegetative cells may be concurrent with budding to produce a daughter cell. This is very common in the yeasts. Often accompanying asexual reproduction is the formation of asexual spores. Many asexual spores are formed as a means of dispersal. These spores are generally small and easily released from the fungus by air currents. There are many types of asexual spores, each with its own name. **Arthroconidia (arthrospores)** are formed when hyphae fragment through splitting of the cell wall or septum (figure 4.27*b*). **Sporangiospores** develop within a sac (sporangium; pl., sporangia) at a hyphal tip (figure 4.27*c*). **Conidiospores** are spores that are not enclosed in a sac but produced at the tips or sides of the hypha (figure 4.26*d*). **Blastospores** are produced from a vegetative mother cell by budding (figure 4.27*e*).

Sexual reproduction in fungi involves the fusion of compatible nuclei. Homothallic fungal species are self-fertilizing and produce sexually compatible gametes on the same mycelium. Heterothallic species require outcrossing between different but sexually compatible mycelia. It has long been held that sexual reproduction must occur between mycelia of opposite **mating types (MAT).** However, one instance of same-sex mating was discovered following an outbreak of the pathogenic yeast *Cryptococcus gatti* in Canada. Depending on the species, sexual fusion may occur between haploid gametes, gamete-producing bodies called **gametangia,** or hyphae. Sometimes both the cytoplasm and haploid nuclei fuse immediately to produce the diploid zygote. Usually, however, there is a delay between cytoplasmic and nuclear fusion. This produces a **dikaryotic stage** in which cells contain two separate haploid nuclei (N + N), one from each parent (**figure 4.28**). After a period of dikaryotic

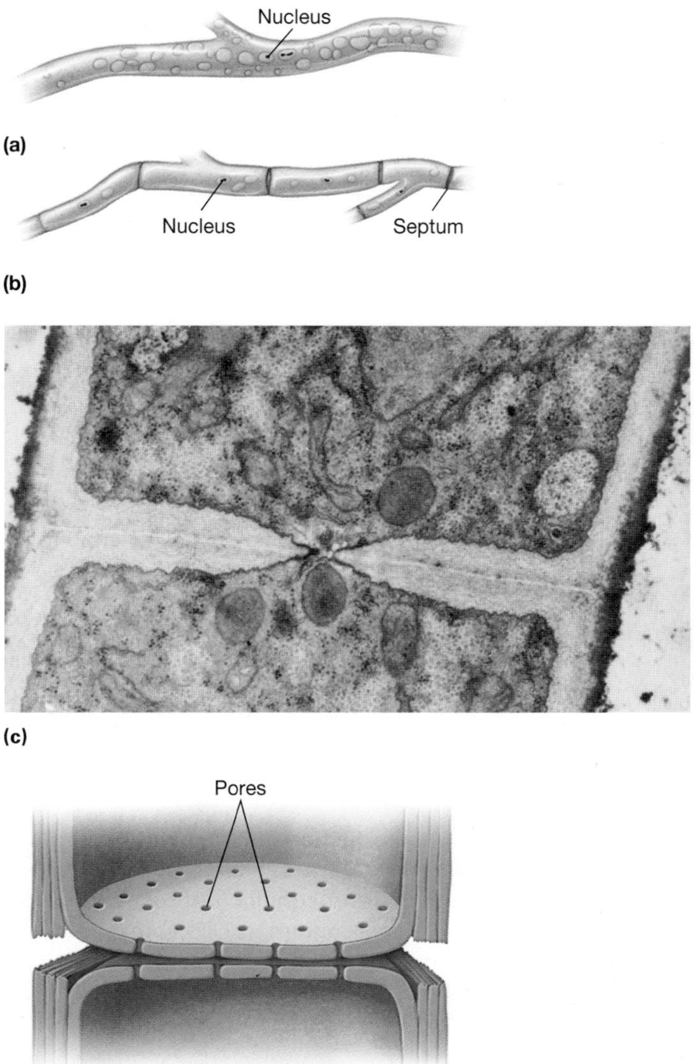

(a)
(b)
(c)
(d)

FIGURE 4.26 Hyphae. Drawings of (a) coenocytic hyphae (aseptate) and (b) hyphae divided into cells by septa. (c) Electron micrograph (×40,000) of a section of *Drechslera sorokiniana* showing wall differentiation and a single pore. (d) Drawing of a multiperforate septal structure.

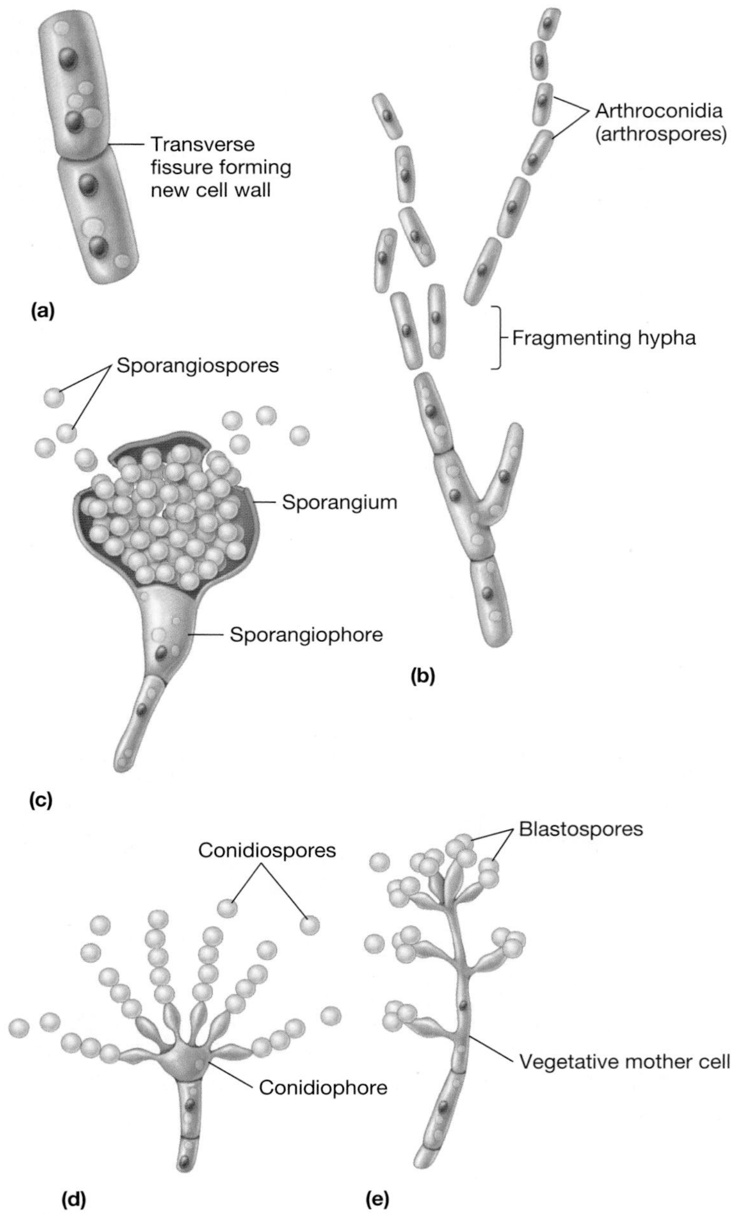

FIGURE 4.27 Asexual Reproduction in Fungi and Some Representative Asexual Spores. (a) Transverse fission. (b) Hyphal fragmentation resulting in arthroconidia (arthrospores). (c) Sporangiospores in a sporangium. (d) Conidiospores arranged in chains at the end of a conidiophore. (e) Blastospores are formed from buds off of the parent cell.

existence, the two nuclei fuse and undergo meiosis to yield haploid spores.

Fungal spores, both asexual and sexual, are important for several reasons. They enable fungi to survive environmental stresses such as desiccation, nutrient limitation, and extreme temperatures, although they are not as stress resistant as bacterial endospores. They aid in fungal dissemination, which helps explain their wide distribution. Because spores are often small and light, they can remain suspended in air for long periods. Thus fungal spores often spread by adhering to the bodies of insects and other animals. The bright colors and fluffy textures of many molds often are due to their aerial hyphae and spores. Finally, the size, shape, color, and number of spores are useful in the identification of fungal species.

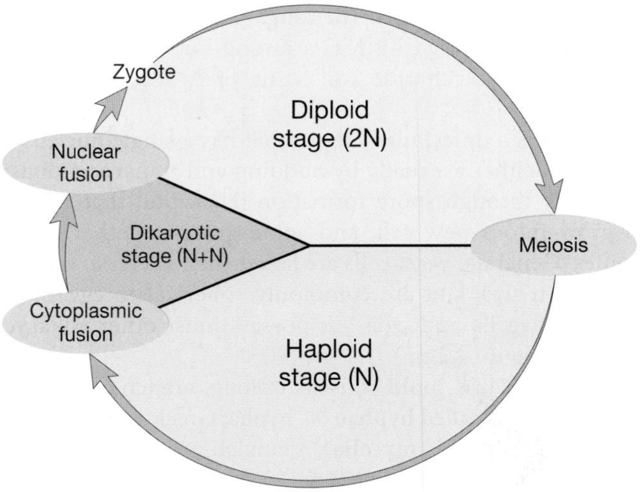

FIGURE 4.28 Reproduction in Fungi. A drawing of the generalized life cycle for fungi showing the alternation of haploid and diploid stages. Some fungal species do not pass through the dikaryotic stage indicated in this drawing. The asexual (haploid) stage is used to produce spores that aid in the dissemination of the species. The sexual (diploid) stage involves the formation of spores that survive adverse environmental conditions (e.g., cold, dryness, heat).

Figure 4.28 Micro Inquiry

What do you think happens to the nuclear envelopes when an organism leaves the dikaryotic stage and enters the diploid stage?

Search This: Fungus of the month

1. How can a fungus be defined? What is the structural difference between a yeast and a mold?
2. What organelles would you expect to find in the cytoplasm of a typical fungus?
3. Describe each of the following types of asexual fungal spores: sporangiospore, arthrospore, conidiospore, and blastospore.
4. How are fungi dispersed in the environment?

Summary

4.1 Common Features of Eukaryotic Cells

a. The eukaryotic cell has a true, membrane-delimited nucleus and many membranous organelles (**table 4.1; figures 4.2** and **4.3**).

b. The membranous organelles compartmentalize the cytoplasm of the cell. This allows the cell to carry out a variety of biochemical reactions simultaneously. It also provides more surface area for membrane-associated activities such as respiration.

4.2 Eukaryotic Cell Envelopes

a. Eukaryotic membranes are similar in structure and function to bacterial membranes. The two differ in terms of their lipid composition.

b. Eukaryotic membranes contain microdomains that are enriched for certain lipids and proteins, and participate in a variety of cellular processes.

c. When a cell wall is present, it is constructed from polysaccharides (e.g., cellulose) that are chemically simpler than peptidoglycan, the molecule found in bacterial cell walls.

4.3 Cytoplasm of Eukaryotes

a. The cytoplasm consists of a cytosol in which are found the organelles and cytoskeleton.

b. The cytoskeleton is organized from microfilaments, intermediate filaments, and microtubules. These cytoskeletal elements are partly responsible for cell structure, endocytosis, movement of cell structures, and motility (**figure 4.5**).

c. Microfilaments and microtubules have been observed in eukaryotic microbes. Microfilaments are composed of actin proteins; microtubules are composed of α-tubulin and β-tubulin (**figures 4.6** and **4.7**).

d. Intermediate filaments are assembled from a heterogeneous family of proteins. They are found in animal cells and some bacteria but have not been identified or studied in eukaryotic microbes.

4.4 Organelles of the Secretory and Endocytic Pathways

a. The cytoplasm is permeated with a complex of membranous organelles and vesicles. Some are involved in the synthesis and secretion of materials (secretory pathway). Some are involved in the uptake of materials from the extracellular milieu (endocytic pathway).

b. The endoplasmic reticulum (ER) is an irregular network of tubules and flattened sacs (cisternae). The ER may have attached ribosomes and may be active in protein synthesis (rough ER), or it may lack ribosomes (smooth ER).

c. The ER can donate materials to the Golgi apparatus, an organelle composed of one or more stacks of cisternae (**figure 4.8**). This organelle prepares and packages cell products for secretion.

d. The Golgi apparatus also buds off vesicles that deliver hydrolytic enzymes and other proteins to lysosomes. Lysomes are organelles that contain digestive enzymes and aid in intracellular digestion of extracellular materials delivered to them by endocytosis.

e. Mammalian cells use several kinds of endocytosis (**figure 4.10**). These include phagocytosis, clathrin-dependent endocytosis, and caveolin-dependent endocytosis. Some macromolecules are bound to receptors prior to endocytosis in a process called receptor-mediated endocytosis.

f. Endocytic pathways in eukaryotic microbes are not as well studied as are mammalian pathways. However, similar mechanisms are employed.

4.5 Organelles Involved in Genetic Control of the Cell

a. The nucleus is a large organelle containing the cell's chromosomes. It is bounded by a complex, double-membrane envelope perforated by pores through which materials can move (**figure 4.12**).

b. The nucleolus lies within the nucleus and participates in the synthesis of ribosomal RNA and ribosomal subunits (**figure 4.13**).

c. Eukaryotic ribosomes are found either free in the cytoplasm or bound to the ER. They are 80S in size (**figure 4.14**).

4.6 Organelles Involved in Energy Conservation

a. Mitochondria are organelles bounded by two membranes, with the inner membrane folded into cristae (**figure 4.16**). They are responsible for energy conservation by cell respiration, which involves the tricarboxylic acid cycle, electron transport, and oxidative phosphorylation.

b. Hydrogenosomes are thought to be related to mitochondria. They have a single membrane and carry out metabolic processes that generate hydrogen gas (**figure 4.17**).

c. Chloroplasts are pigment-containing organelles that serve as the site of photosynthesis (**figure 4.18**). The trapping of light energy takes place in the thylakoid membranes of the chloroplast, whereas CO_2 fixation occurs in the stroma.

4.7 External Structures

a. Many eukaryotic cells are motile because of cilia and flagella, membrane-enclosed organelles with nine microtubule doublets surrounding two central microtubules (**figure 4.22**).

b. Cilia and flagella exhibit a whiplike motion. This is brought about when the microtubule doublets slide along each other, causing the cilium or flagellum to bend.

4.8 Comparison of Bacterial, Archaeal, and Eukaryotic Cells

a. Bacterial and archaeal cells are similar in appearance but differ at the molecular levels. A major difference between these cells and eukaryotic cells is that eukaryotic cells contain a nucleus and numerous membranous structures.

b. Despite the fact that eukaryotic, bacterial, and archaeal cells differ structurally in many ways (**table 4.2**), they are quite similar biochemically.

4.9 Overview of Protist Structure and Function

a. Protists include protozoa, algae, slime molds, and water molds. They are found in most eukaryotic supergroups (**figure 4.24**).

b. Because protists are eukaryotic cells, in many respects their morphology and physiology resemble those of multicellular plants and animals. However, because all their functions must be performed within the individual protist, many morphological and physiological features are unique.

c. The protistan cell membrane is called the plasmalemma, and the cytoplasm can be divided into the ectoplasm and the endoplasm. A pellicle lies beneath the plasma membrane of some protists. The cytoplasm contains contractile, secretory, and food (phagocytic) vacuoles.

d. Energy metabolism occurs within mitochondria, hydrogenosomes, or chloroplasts.

e. Some protists can secrete a resistant covering and go into a resting stage (encystment) called a cyst. Cysts protect the organism against adverse environments, function as a site for nuclear reorganization, and serve as a means of transmission in parasitic species.

f. Most protists reproduce asexually by binary or multiple fission or budding (**figure 4.25**).

g. Some protists also use a variety of sexual reproductive strategies, including syngamy and autogamy.

4.10 Overview of Fungal Structure and Function

a. A fungus is a eukaryotic, spore-bearing organism that has absorptive nutrition and lacks chlorophyll; reproduces asexually, sexually, or by both methods; and normally has a cell wall containing chitin.

b. The body or vegetative structure of a fungus is called a thallus.

c. Yeasts are unicellular fungi that have a single nucleus and reproduce either asexually by budding and transverse division or sexually through spore formation (**figure 4.3a**).

d. A mold consists of long, branched, threadlike filaments of cells, the hyphae, that form a tangled mass called a mycelium. Hyphae may be either septate or coenocytic (nonseptate). The mycelium can produce reproductive structures (**figure 4.26**).

e. Asexual reproduction in the fungi often leads to the production of specific types of spores that are easily dispersed (**figure 4.27**).

f. Sexual reproduction is initiated in fungi by the fusion of hyphae (or cells) of different mating strains. In some fungi, the nuclei in the fused hyphae immediately combine to form a zygote. In others, the two genetically distinct nuclei remain separate, forming pairs that divide synchronously. Eventually some nuclei fuse (**figure 4.28**).

Critical Thinking Questions

1. Discuss the statement: "The most obvious difference between eukaryotic, bacterial, and archaeal cells is in their use of membranes." Specifically, compare the roles of the bacterial and archaeal plasma membrane with those of the organelle membranes in eukaryotes.

2. Protist encystment is usually triggered by changes in the environment. How might this be similar to or different from endospore formation in gram-positive bacteria?

3. Yeasts and some protists tend to reproduce asexually when nutrients are plentiful and conditions are favorable for growth but reproduce sexually when environmental or nutrient conditions are not favorable. Why is this an evolutionarily important and successful strategy?

4. In recent years there have been advances in understanding longevity through research on three model organisms: mice, nematode worms, and the yeast *Saccharomyces cerevisiae*. It is clear that caloric restriction is linked to increased life span in multicellular organisms (mice and worms). This phenomenon can be studied at the molecular level in yeast, where loss of the protein Sch9p increases the yeast life span by increasing

the rate of mitochondrial activity. In this way, caloric restriction is mimicked. List the pros and cons of the use of eukaryotic microorganisms as models for complex physiological and genetic questions that are relevant to higher organisms, including humans. What do you think is the most significant advantage and the most unfortunate disadvantage?

Read the original paper: Lovoie, L., and Whiteway, M. 2008. Increased respiration in the *sch9Δ* mutant is required for increasing chronological life span but not replicative life span. *Euk. Cell* 7:1127.

5. The parasitic protozoan *Trypanosoma brucei* causes African sleeping sickness, which is transmitted to humans and other animals by the tsetse fly. It has been discovered that the flagella of this protist are required for cell division, and thus viability, when in the bloodstream of the animal. Interestingly, the loss of just one flagellar protein results in binary fission that gives rise to misshapen "monster" cells. This is despite the fact that the parasites still replicate their DNA, divide their nuclei, and even form new flagella and other organelles. It appears that the monster cells form because the protists use their flagella to mark the polarity of the cell during division. Which proteins do you think were targeted in this study? How do you think the researchers figured out that flagella serve as a tag for cell polarity? Finally, what are the potential public health implications of this research?

Read the original paper: Broadhead, R., et al. 2006. Flagellar motility is required for the viability of the bloodstream trypansosome. *Nature* 440:224.

Concept Mapping Challenge

Construct a concept map that allows you to describe the molecular makeup and function of eukaryotic cell structures. Use the concepts that follow, any other concepts or terms you need, and your own linking words between each pair of concepts in your map. ▶▶| *Concept Mapping (appendix III)*

cell wall	plasma membrane	cytoplasm
chloroplast	mitochondrion	endoplasmic reticulum
Golgi	lysosome	nucleus
ribosomes		

Learn More

Learn more by visiting the text website at www.mhhe.com/willey8, where you will find a complete list of references.

5

Viruses and Other Acellular Infectious Agents

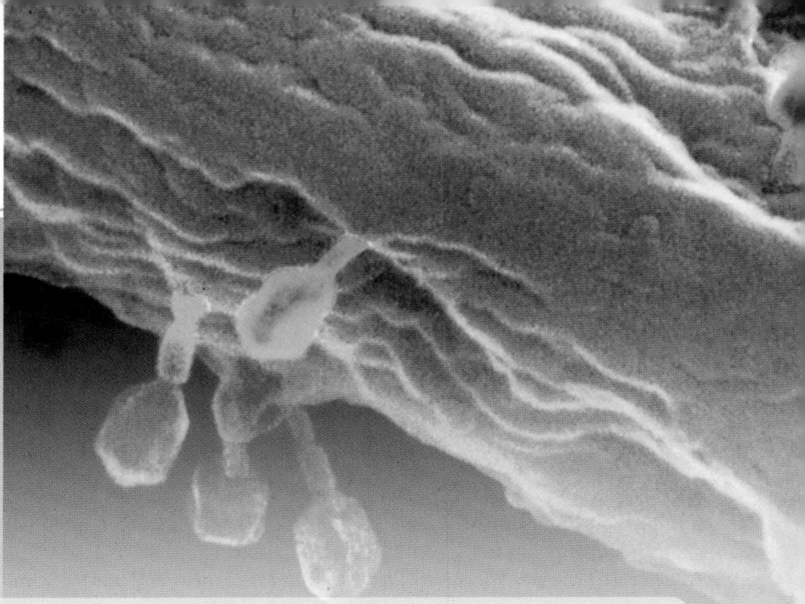

A scanning electron micrograph of T-even bacteriophages infecting E. coli. *The phages are colored blue.*

CHAPTER GLOSSARY

bacteriophage A virus that uses bacteria as its host; often called a phage.

capsid The protein coat or shell that surrounds a virion's nucleic acid.

enveloped virus A virus with an outer membranous layer that surrounds the nucleocapsid.

induction An event in the life cycle of some viruses (e.g., temperate bacteriophages) that results in the provirus initiating synthesis of mature virions and entering the lytic cycle.

lysogen A bacterium or archaeal cell that is infected with a temperate virus.

lysogeny The state in which a viral genome remains within a bacterial or archaeal cell after infection and reproduces along with it, rather than taking control of the host cell and producing new virions.

lytic cycle The infection of a host cell by a virus that induces cell death through lysis and release of progeny viruses.

nonenveloped virus A virus that lacks an envelope; also called a **naked virus.**

nucleocapsid That portion of a virus that consists of viral nucleic acid surrounded by a protein coat called the capsid.

peplomers or **spikes** Proteins that project from a viral envelope.

plaque A clear area in a lawn or layer of host cells that results from the lysis or killing of the host cells by viruses.

prion An infectious agent composed only of protein that is responsible for causing a variety of spongiform encephalopathies (e.g., scrapie).

provirus (prophage) The latent form of a virus that remains within a host cell, usually integrated into the host chromosome; in archaeal and bacterial hosts, this occurs during the lysogenic cycle; when the virus is a bacteriophage, the provirus is called a prophage.

retrovirus A member of a group of viruses with RNA genomes that carry the enzyme reverse transcriptase and form a double-stranded DNA copy of their genome during their multiplication cycle.

segmented genome A viral genome that is divided into several fragments, each usually coding for a single polypeptide.

temperate phage A bacteriophage that infects bacteria and establishes a lysogenic relationship with its host rather than causing immediate lysis.

virion A complete virus particle.

viroid An infectious agent of plants that consists only of RNA.

virulent viruses Viruses that lyse their host cells during the multiplication cycle.

virus An infectious agent having a simple acellular organization, often just a protein coat and a nucleic acid genome; lacking independent metabolism; and reproducing only within living host cells.

virusoid An infectious agent composed only of RNA; its RNA encodes one or more proteins, but it can only replicate in cells also infected by a virus.

The microbial world consists not only of cellular organisms but also of acellular infectious agents. In this chapter, we turn our attention to these infectious agents: viruses, viroids, virusoids, and prions. These entities are composed simply of protein and nucleic acid (viruses), RNA only (viroids and virusoids), or protein only (prions). Yet they are major causes of disease. For instance, many human diseases are caused by viruses, and more are discovered every year, as demonstrated by the appearance of SARS in 2003, new avian influenza viruses over the last 5 to 6 years, and the HIN1 (swine) influenza virus in 2009.

Although viruses are most often discussed in terms of their ability to cause disease, it is important to remember that viruses are significant for other reasons. Recent ecological studies have shown that viruses are vital members of aquatic ecosystems. There they interact with cellular microbes and contribute to the movement of organic matter from particulate forms to dissolved forms. They also affect population sizes of cellular microbes in these habitats. Finally, bacterial viruses transfer genes from bacterium to bacterium at a high rate, thus contributing to the evolution of bacteria. Bacterial viruses are also receiving renewed interest as therapies for infections caused by the increasing number of drug-resistant bacterial pathogens, including methicillin-resistant *Staphylococcus aureus* (MRSA). Viruses also serve as models for understanding processes such as DNA replication, RNA synthesis, and protein synthesis. Therefore the study of viruses has contributed significantly to the discipline of molecular biology. In fact, the field of genetic engineering is based in large part on the use of viruses and viral enzymes such as the retroviral enzyme reverse transcriptase, discovered by David Baltimore, Howard Temin, and Satoshi Mitzutani. ▶▶| *Aquatic viruses (section 28.2); Drug resistance (section 34.9); Recombinant DNA technology (chapter 15)*

We begin this chapter by reviewing the general properties and structure of viruses and the ways in which viruses are cultured and studied. The chapter ends with a brief introduction to viroids, virusoids, and prions.

5.1 Viruses

The discipline of **virology** studies **viruses,** a unique group of infectious agents whose distinctiveness resides in their simple, acellular organization and pattern of multiplication. A complete virus particle, called a **virion,** consists of one or more molecules of DNA or RNA enclosed in a coat of protein. Some viruses have additional layers that can be very complex and contain carbohydrates, lipids, and additional proteins (**figure 5.1**). Viruses can exist either extracellularly or intracellularly. When extracellular, they are inactive (with one interesting acception) because they possess few, if any, enzymes and cannot reproduce outside of living cells (**Microbial Diversity & Ecology 5.1**). When intracellular, viruses exist primarily as replicating nucleic acids that induce the host to synthesize viral components from which virions are assembled and eventually released.

Viruses can infect all cell types. Numerous viruses infect bacteria. They are called **bacteriophages,** or **phages** for short. Fewer archaeal viruses have been identified. Most known viruses infect eukaryotic organisms, including plants, animals, protists, and fungi.

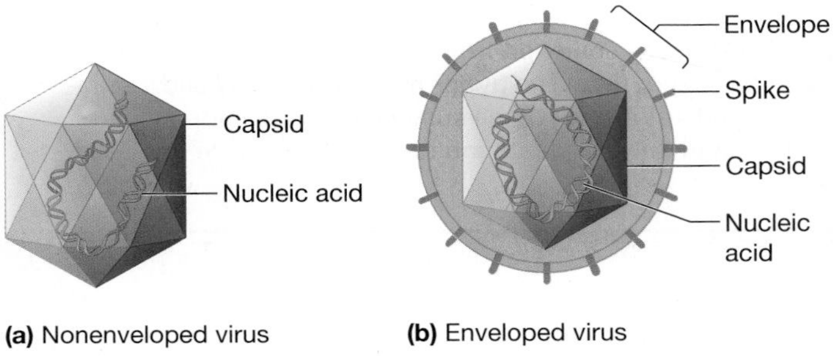

FIGURE 5.1 Generalized Structure of Viruses. (a) The simplest virus is a nonenveloped virus (nucleocapsid) consisting of a capsid assembled around its nucleic acid. (b) An enveloped virus is composed of a nucleocapsid surrounded by a flexible membrane called an envelope. The envelope usually has viral proteins called spikes inserted into it.

Viruses have been classified into numerous families based primarily on genome structure, life cycle, morphology, and genetic relatedness. These families have been designated by the International Committee for the Taxonomy of Viruses (ICTV), the agency responsible for standardizing the classification of viruses. ▶▶| *Virus taxonomy and phylogeny (section 25.1)*

In this chapter, we introduce the structure and multiplication strategies of viruses. Their taxonomic diversity is presented in chapter 25, and their ecological relevance is discussed in chapter 26.

5.2 Structure of Viruses

Viral morphology has been intensely studied over the past decades because of the importance of viruses and the realization that their structure was simple enough to be understood in detail. Progress has come from the use of several different techniques: electron microscopy, X-ray diffraction, biochemical analysis, and immunology. Although our knowledge is incomplete due to the large number of different viruses, we can discuss the general nature of viral structure.

Virion Size

Virions range in size from about 10 to 400 nm in diameter (**figure 5.2**). The smallest viruses are a little larger than ribosomes, whereas poxviruses (e.g., *Variola virus,* the causative agent of small pox) are about the same size as the smallest bacteria and can be seen in the light microscope. Most viruses, however, are too small to be visible in the light microscope and must be viewed with electron microscopes. |◀◀ *Electron microscopy (section 2.4)*

General Structural Properties

The simplest viruses are constructed of a **nucleocapsid.** The nucleocapsid is composed of a nucleic acid, either DNA or RNA, held within a protein coat called the **capsid.** The viral nucleic acid

MICROBIAL DIVERSITY & ECOLOGY

5.1 Host-Independent Growth of an Archaeal Virus

The fact that viruses cannot multiply without first infecting a host cell has resulted in their classification as "acellular entities" or "forms"—they are not cells. So it was quite a surprise when an archaeal virus that develops long tails only when outside its host was discovered (**box figure**). This archaeal virus was found in acidic hot springs (pH 1.5, 85–93°C) in Italy, where it infects the hyperthermophilic archaeon *Acidianus convivator.* When the virus infects its host, lemon-shaped virions are assembled. Following host lysis, two "tails" begin to form at either end of the virion. These projections continue to assemble until they reach a length at least that of the viral capsid. Curiously, tails are only produced if virions are incubated at high temperature. The virus is called ATV for *Acidianus* two-tailed virus.

To find out more about the structure of the tails, the ATV genome was sequenced. ATV is a double-stranded DNA virus that encodes only nine structural proteins. The tail protein is an 800 amino acid protein that bears homology to eukaryotic intermediate filament proteins. Both intermediate filaments and purified ATV tail proteins assemble into filamentous structures without additional energy or cofactors. ⏮ *Cytoplasm of eukaryotes (section 4.3)*

It is suspected that the development of tails only at high temperatures may be a survival strategy for the virus when host cell density is low. So far, ATV is the only virus that induces lysis rather than lysogeny in bacteria or archaea living in acidic hot springs. It would thus seem that all other such viruses have evolved lysogeny as a means to survive these harsh conditions. Why this virus has evolved lysis and tail development is unknown, but it suggests that viruses may be more complicated than simple "entities."

Source: Häring, M.; Vestergaard, G.; Rachel, R.; Chen, L.; Garret, R. A.; and Prangishvili, D. 2005. Independent virus development outside a host. Nature *436:1101–02.*

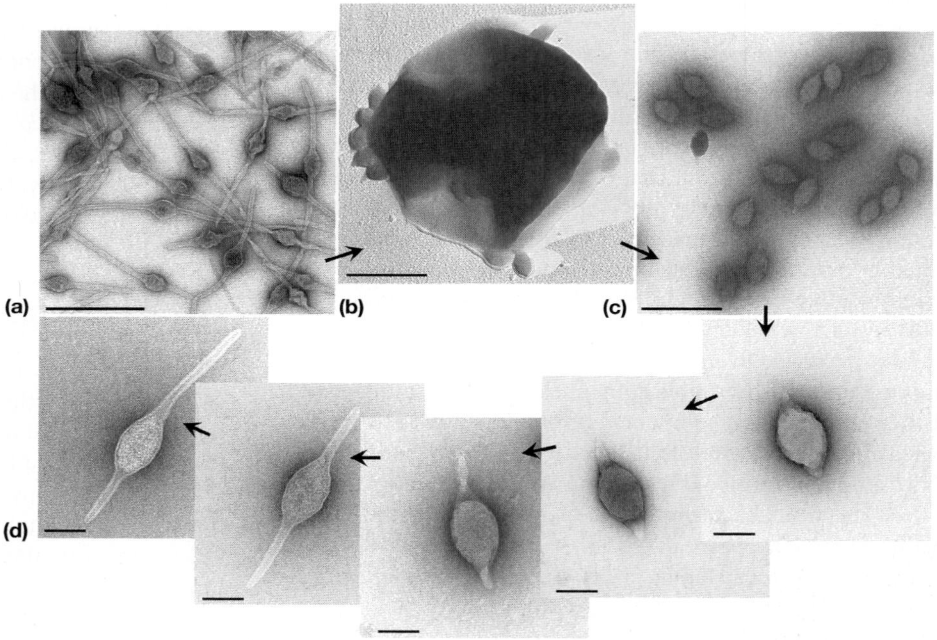

***Acidianus* Two-Tailed Virus, ATV.** (a) Virions collected from an acidic hot spring bear two long projections, or tails, at each end. (b, c) The hyperthermophilic archaeon *A. convivator* extrudes lemon-shaped virions. (d) These subsequently develop tail-like structures independent of the host and only when at high temperature. Scale bars: *a–c,* 0.5 μm; *d,* 0.1 μm.

encodes viral proteins, some of which are used to form the capsid. The capsid protects the viral genome and aids in its transfer between host cells. The proteins that form the capsid are called **protomers.** Capsids self-assemble by a process that is not fully understood. Some viruses use noncapsid proteins as scaffolding upon which the capsids are assembled.

Probably the most important advantage of this design strategy is that the viral genome is used with maximum efficiency. For example, the *Tobacco mosaic virus* (TMV) capsid is constructed using a single type of protomer (**figure 5.3**). Recall that the building blocks of proteins are amino acids and that each amino acid is encoded by three nucleotides, the building blocks of nucleic acids. The TMV

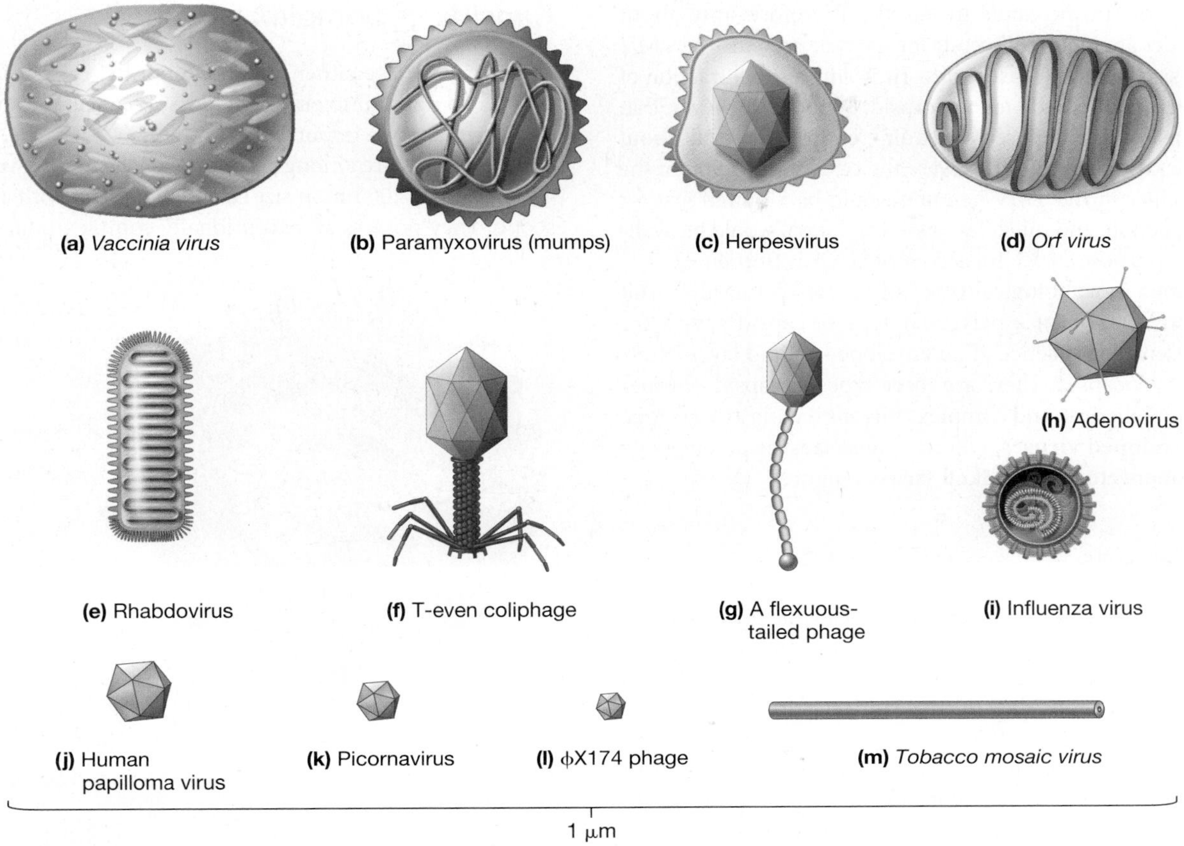

FIGURE 5.2 The Size and Morphology of Selected Viruses. The viruses are drawn to scale. A 1 μm line is provided at the bottom of the figure.

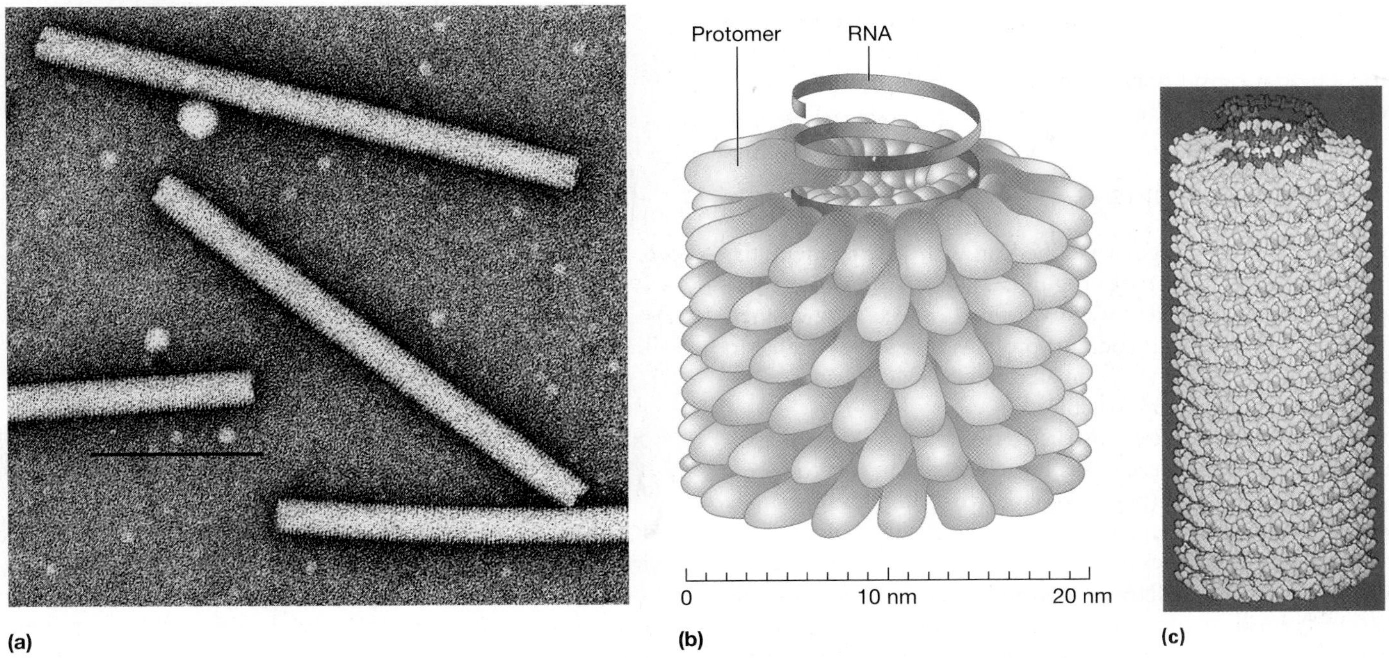

FIGURE 5.3 *Tobacco Mosaic Virus* Structure. (a) An electron micrograph of the negatively stained helical capsid (×400,000). (b) Illustration of TMV structure. Note that the nucleocapsid is composed of a helical array of protomers with the RNA spiraling on the inside. (c) A model of TMV.

protomer is 158 amino acids in length. Therefore only about 474 nucleotides are required to code for the coat protein. The TMV genome consists of 6,400 nucleotides. Thus only a small fraction of the genome is used to code for the capsid. Suppose, however, that the TMV capsid was composed of six different protomers all about 150 amino acids in length. If this were the case, about 2,900 of the 6,400 nucleotides in the TMV genome would be required just for capsid construction, and much less genetic material would be available for other purposes. ▶▶| Tobacco mosaic virus *(section 25.5)*

The various morphological types of viruses primarily result from the combination of a particular type of capsid symmetry with the presence or absence of an envelope—a lipid layer external to the nucleocapsid. There are three types of capsid symmetry: helical, icosahedral, and complex. Virions having an envelope are called **enveloped viruses,** whereas those lacking an envelope are called **nonenveloped** or **naked viruses** (figure 5.1).

Helical Capsids

Helical capsids are shaped like hollow tubes with protein walls. The *Tobacco mosaic virus* is a well-studied example of helical capsid structure (figure 5.3). In this virus, the self-assembly of protomers in a helical arrangement produces a long, rigid tube, 15 to 18 nm in diameter by 300 nm long. The capsid encloses an RNA genome, which is wound in a spiral and lies within a groove formed by the protein subunits. Not all helical capsids are as rigid as the TMV capsid. The influenza virus genome is enclosed in thin, flexible helical capsids that are folded within an envelope (**figure 5.4**). ▶▶| Tobacco mosaic virus *(section 25.5)*

The size of a helical capsid is influenced by both its protomers and the nucleic acid enclosed within the capsid. The diameter of the capsid is a function of the size, shape, and interactions of the protomers. The nucleic acid appears to determine its length because a helical capsid does not extend much beyond the end of the viral genome.

Icosahedral Capsids

The icosahedron is a regular polyhedron with 20 equilateral triangular faces and 12 vertices (figure 5.2*h, j–l*). The **icosahedral capsid** is the most efficient way to enclose a space. A few genes, sometimes only one, can code for proteins that self-assemble to form the capsid. In this way, a small number of genes can specify a large three-dimensional structure.

Icosahedral capsids are constructed from ring- or knob-shaped units called **capsomers,** each usually made of five or six protomers (**figure 5.5**). Pentamers (pentons) have five protomers; hexamers (hexons) possess six. Pentamers are usually at the vertices of the icosahedron, whereas hexamers generally form its edges and triangular faces. In some RNA viruses, both the pentamers and hexamers of a capsid are constructed with only one type of subunit. In other viruses, pentamers are composed of different proteins than are the hexamers. Although many icosahedral capsids contain both pentamers and hexamers, some have only pentamers.

Capsids of Complex Symmetry

Most viruses have either icosahedral or helical capsids, but some viruses do not fit into either category. Poxviruses and large bacteriophages are two important examples.

Poxviruses are among the largest of the animal viruses (about 400 by 240 by 200 nm in size) and can be seen with a light microscope. They possess an exceptionally complex internal structure

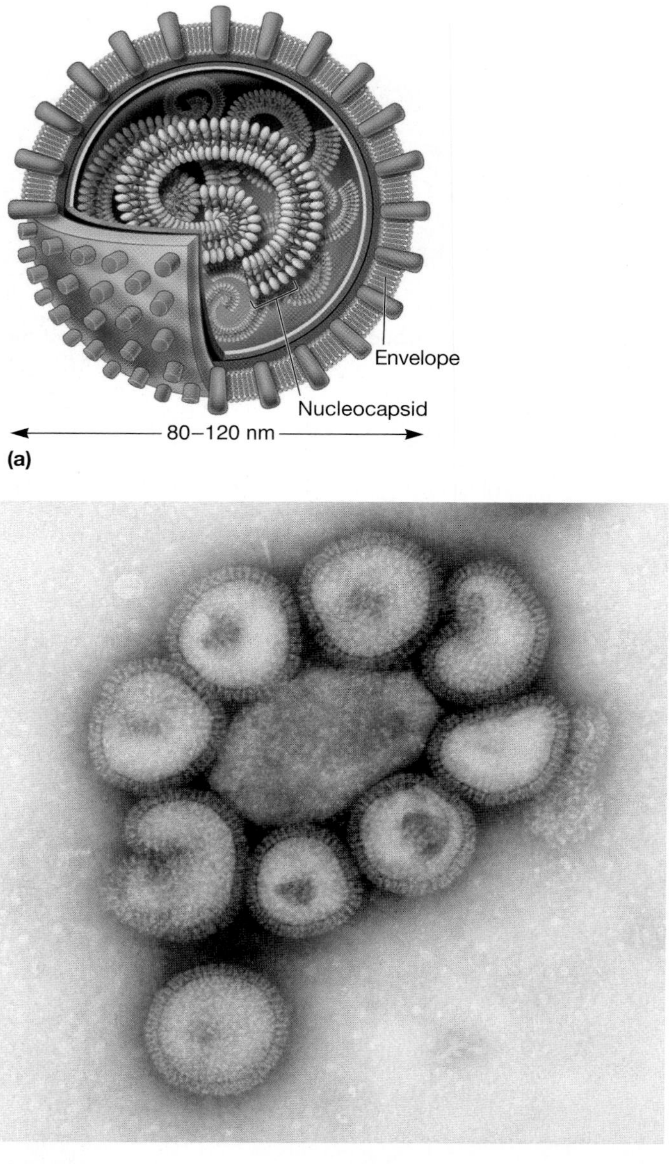

FIGURE 5.4 Influenza Virus. Influenza virus is an enveloped virus with a helical nucleocapsid. (a) Schematic view. Influenza viruses have segmented genomes consisting of seven to eight different RNA molecules. Each is coated by capsid proteins. (b) Because there are seven to eight flexible nucleocapsids enclosed by an envelope, the virions are pleomorphic. Electron micrograph (×350,000).

with an ovoid- to brick-shaped exterior. **Figure 5.6** shows the morphology of *Vaccinia virus,* a poxvirus. Its double-stranded DNA genome is associated with proteins and contained in the core, a central structure shaped like a biconcave disk and surrounded by a membrane. Two lateral bodies lie between the core and the virus's outer envelope.

Search This: Vaccinia virion 3D tour

Some large bacteriophages are even more elaborate than the poxviruses. The T2, T4, and T6 phages (T-even phages) that infect *Escherichia coli* are said to have **binal symmetry** because they have a head that resembles an icosahedron and a tail that is helical. The icosahedral head is elongated by one or two rows of hexamers in the middle and contains the DNA genome (**figure 5.7**). The tail is composed of a collar joining it to the head, a central hollow tube, a sheath surrounding the tube, and a complex baseplate. In T-even phages, the baseplate is hexagonal and has a pin and a jointed tail fiber at each corner. ▶▶| *Bacteriophage T4: A virulent bacteriophage (section 25.2)*

Considerable variation in structure exists among the large bacteriophages, even those infecting a single host. In contrast

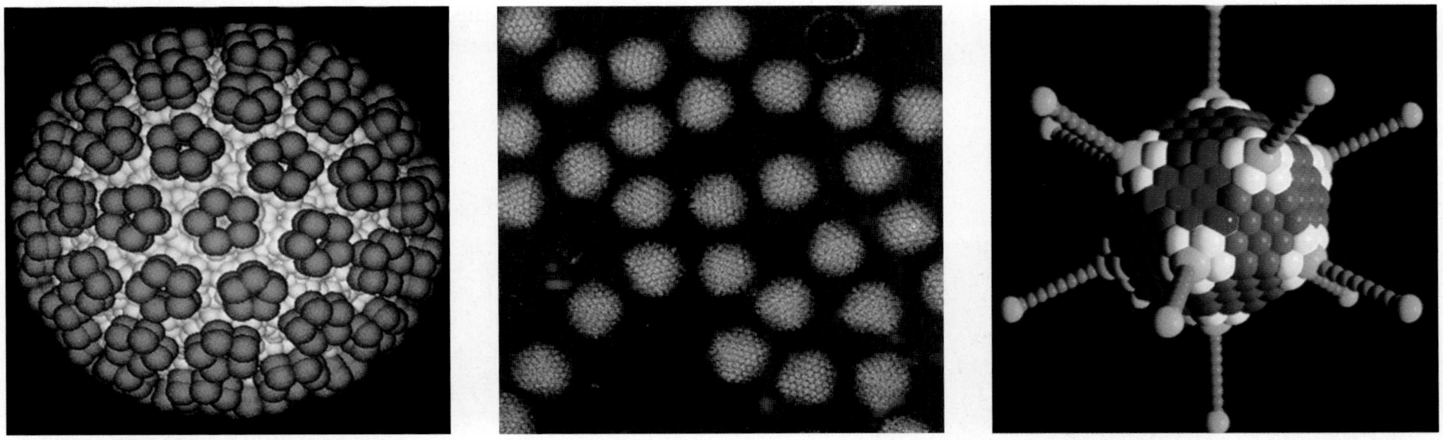

(a) *Polyomavirus* **(b)** Adenovirus **(c)** Adenovirus

FIGURE 5.5 Examples of Icosahedral Capsids. (a) Computer-simulated image of *Polyomavirus* (72 capsomers) that causes a rare demyelinating disease of the central nervous system. (b) Adenovirus, 252 capsomers (×171,000). (c) Computer-simulated model of an adenovirus.

Figure 5.5* *Micro Inquiry

What is the difference between a pentamer and hexamer?

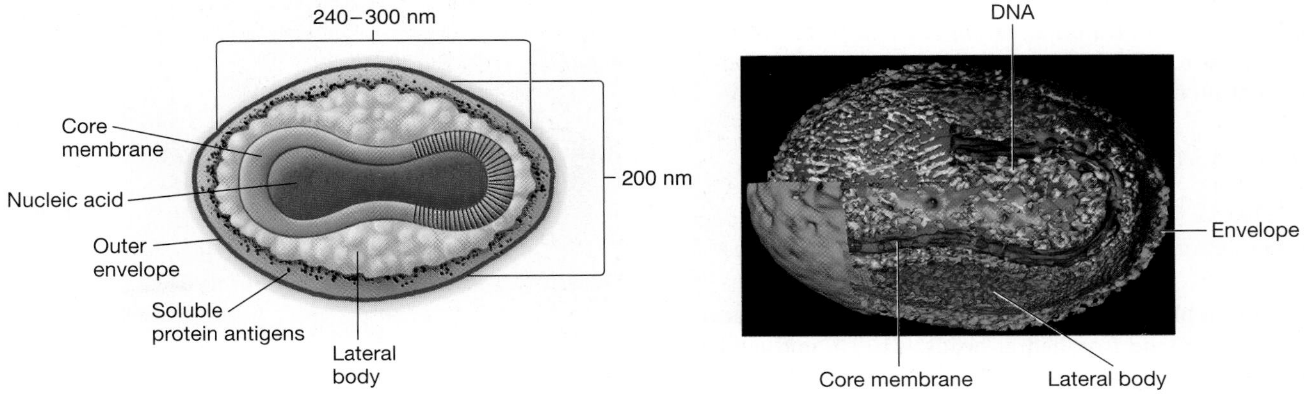

(a) **(b)**

FIGURE 5.6 *Vaccinia Virus* Morphology. (a) Diagram of vaccinia structure. (b) Electron cryotomogram of the virus.

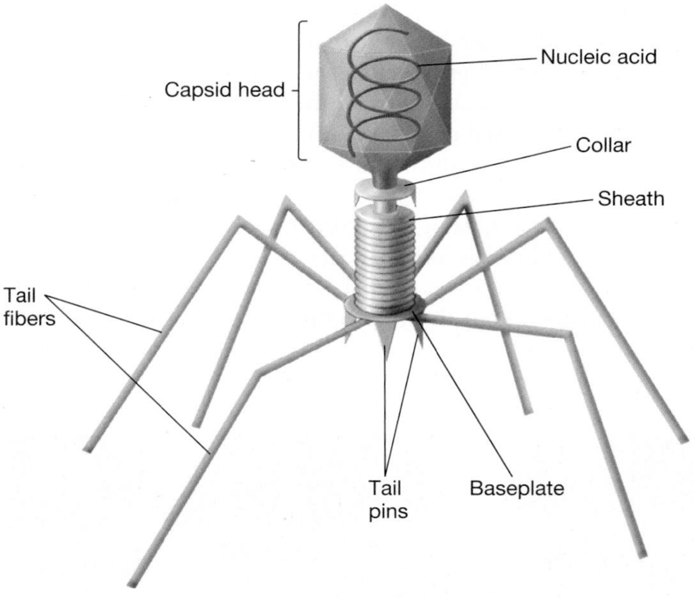

(a)

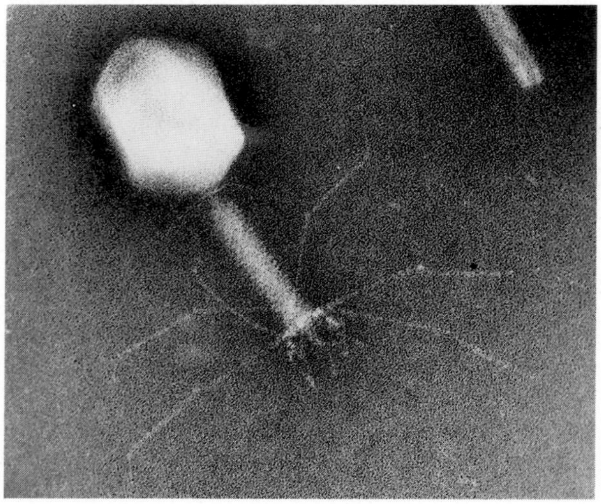

(b)

FIGURE 5.7 T-Even Coliphages. (a) The structure of T4 bacteriophage. (b) Micrograph showing the phage before injection of its DNA.

Figure 5.7 Micro Inquiry

Why are T-even phages said to have binal symmetry?

with the T-even phages, many other coliphages (phages that infect *E. coli*) have true icosahedral heads. T1, T5, and lambda phages have sheathless tails that lack a baseplate and terminate in rudimentary tail fibers. Coliphages T3 and T7 have short, noncontractile tails without tail fibers. ▶▶| *Bacteriophage lambda: A temperate bacteriophage (section 25.2)*

Viral Envelopes and Enzymes

Many animal viruses, some plant viruses, and at least one bacterial virus are bounded by an outer membranous layer called an **envelope** (**figure 5.8**). Animal virus envelopes usually arise from host cell plasma or nuclear membranes. Envelope lipids and carbohydrates are therefore acquired from the host. In contrast, envelope proteins are coded for by viral genes and may even project from the envelope surface as **spikes,** which are also called **peplomers** (figure 5.8). In many cases, these spikes are involved in virus attachment to the host cell surface. Because they differ among viruses, they also can be used to identify some viruses. Many enveloped viruses have a somewhat variable shape and are called pleomorphic. However, the envelopes of viruses such as the bullet-shaped *Rabies virus* are firmly attached to the underlying nucleocapsid and endow the virion with a constant, characteristic shape (figure 5.8*a*).

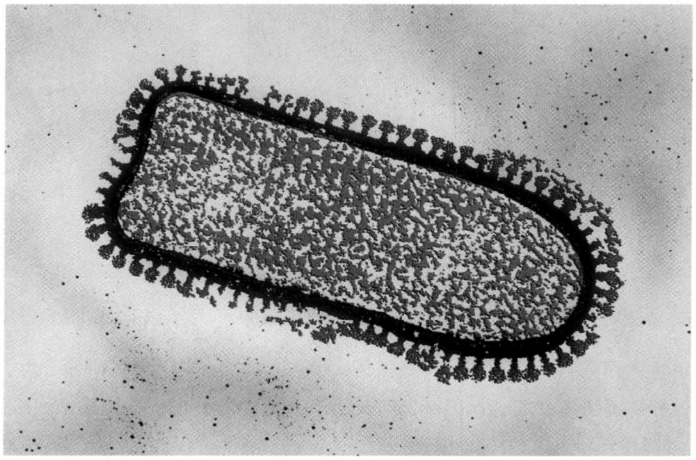

(a) *Rabies virus*

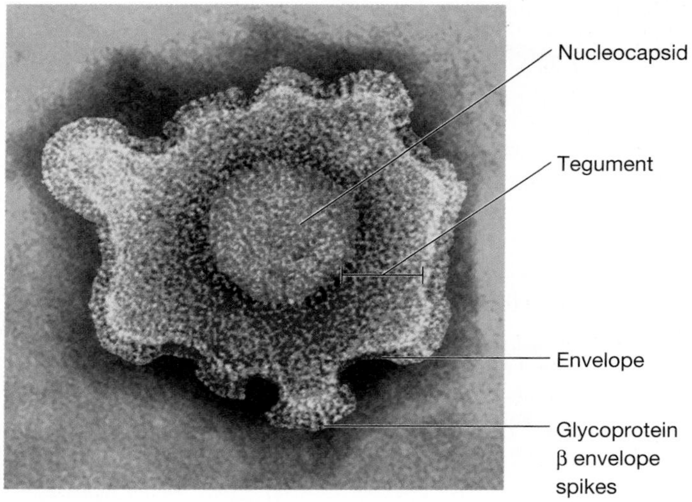

(b) Herpesvirus

FIGURE 5.8 Examples of Enveloped Viruses. (a) Negatively stained *Rabies virus*. (b) Herpesvirus. Images are artificially colorized.

Influenza virus (figure 5.4) is a well-studied enveloped virus. Spikes project about 10 nm from the surface at 7 to 8 nm intervals. Some spikes consist of the enzyme **neuraminidase,** which functions in the release of mature virions from the host cell. Other spikes are **hemagglutinin** proteins, so named because they bind virions to red blood cells and cause the cells to clump together—a process called hemagglutination. Hemagglutinins participate in virion attachment to host cells. Most of the influenza virus's envelope proteins are glycoproteins—proteins that have carbohydrate attached to them. A nonglycosylated protein, the M (matrix) protein, is found on the inner surface of the envelope and helps stabilize it.

It was originally thought that virions lacked enzymes. However, this is not the case. In some instances, enzymes are associated with the envelope or capsid (e.g., influenza neuraminidase). Enzymes within the capsid are usually involved in nucleic acid replication. For example, the influenza virus has an RNA genome and carries an enzyme that synthesizes RNA using an RNA template. Thus although viruses lack true metabolism and cannot reproduce independently of living cells, they may carry one or more enzymes essential to the completion of their life cycles.

Viral Genomes

Viruses are exceptionally diverse with respect to the nature of their genomes. They employ all four possible nucleic acid types: single-stranded (ss) DNA, double-stranded (ds) DNA, ssRNA, and dsRNA. All four types are found in animal viruses. Most plant viruses have ssRNA genomes, and most bacterial viruses contain dsDNA. The size of viral genomes also varies greatly. Very small genomes (e.g., those of the MS2 and Qβ phages) are around 4,000 nucleotides. These genomes are just large enough to code for three or four proteins. MS2, Qβ, and some other viruses save additional space by using overlapping genes. At the other extreme, T-even bacteriophages, herpesviruses, and *Vaccinia virus* have genomes of 1.0 to 2.0 $\times 10^5$ nucleotides and may be able to direct the synthesis of over 100 proteins. Among the largest known viral genomes is the genome of *Mimivirus*, a parasite of a protist. It is about 1.2 $\times 10^6$ nucleotides, rivaling some bacteria and archaea in coding capacity. ▶▶| *Gene structure (section 12.5); Bacteriophages MS2 and Qβ (section 25.5); Herpesviruses (section 25.2); A virus with a virus (Microbial Ecology & Diversity 25.1)*

Most DNA viruses use dsDNA as their genetic material. However, some have ssDNA genomes (e.g., ϕX174 and M13). In both cases, the genomes may be either linear or circular. Some DNA genomes can switch from one form to the other. For instance, the *E. coli* phage lambda has a genome that is linear in the capsid but is converted into a circular form once the genome enters the host cell. Another important characteristic of DNA bacteriophages is that their genomes often contain unusual nitrogenous bases. ▶▶| *ϕX174 and fd (section 25.3)*

RNA viruses also can be either dsRNA or ssRNA. Although relatively few RNA viruses have dsRNA genomes, dsRNA viruses are known to infect animals, plants, fungi, and at least one bacterial species (e.g., ϕ6 and rotaviruses). More common are viruses with ssRNA genomes. Polio, tobacco mosaic, SARS, rabies, mumps, measles, influenza, human immunodeficiency (HIV), and brome mosaic viruses are all RNA viruses.

Many RNA viruses have **segmented genomes**—genomes that consist of more than one piece (segment) of RNA. In many cases, each segment codes for one protein and there may be as many as 10 to 12 segments. Usually all segments are enclosed in the same capsid; however, this is not always the case. For example, the genome of *Brome mosaic virus,* a virus that infects certain grasses, is composed of three segments distributed among three different virions. Despite this complex and seemingly inefficient arrangement, the different brome mosaic virions successfully manage to infect the same host.

1. How are viruses similar to cellular organisms? How do they differ?
2. What is the difference between a nucleocapsid and a capsid?
3. Compare the structure of an icosahedral capsid with that of a helical capsid. How do pentamers and hexamers associate to form a complete icosahedron? What determines helical capsid length and diameter?
4. What is an envelope? What are spikes (peplomers)? Why are some enveloped viruses pleomorphic? Give two functions spikes might serve in the viral life cycle and the proteins that the influenza virus uses in these processes.
5. All four nucleic acid forms can serve as viral genomes. Describe each. What is a segmented RNA genome?
6. The RNA genomes of some RNA viruses resemble the messenger RNA (mRNA) of their eukaryotic hosts. What advantage would an RNA virus gain by having this type of genome?

5.3 Viral Multiplication

The differences in viral structure and genomes have important implications for the mechanism a virus uses to multiply within its host cell. Indeed, even among viruses with similar structures and genomes, each can exhibit unique life cycles. Despite these differences, a general pattern of viral life cycles can be discerned. Because viruses need a host cell in which to multiply, the first step in the life cycle of a virus is attachment (often called adsorption) to a host (**figure 5.9**). This is followed by entry of either the nucleocapsid or the viral nucleic acid into the host. If the nucleocapsid enters, uncoating of the genome usually occurs before the life cycle continues. Once inside the host cell, the synthesis stage begins. During this stage, genes encoded by the viral genome are expressed. That is, the viral genes are transcribed and translated. This allows the virus to take control of the host cell, forcing it to manufacture the viral genome and

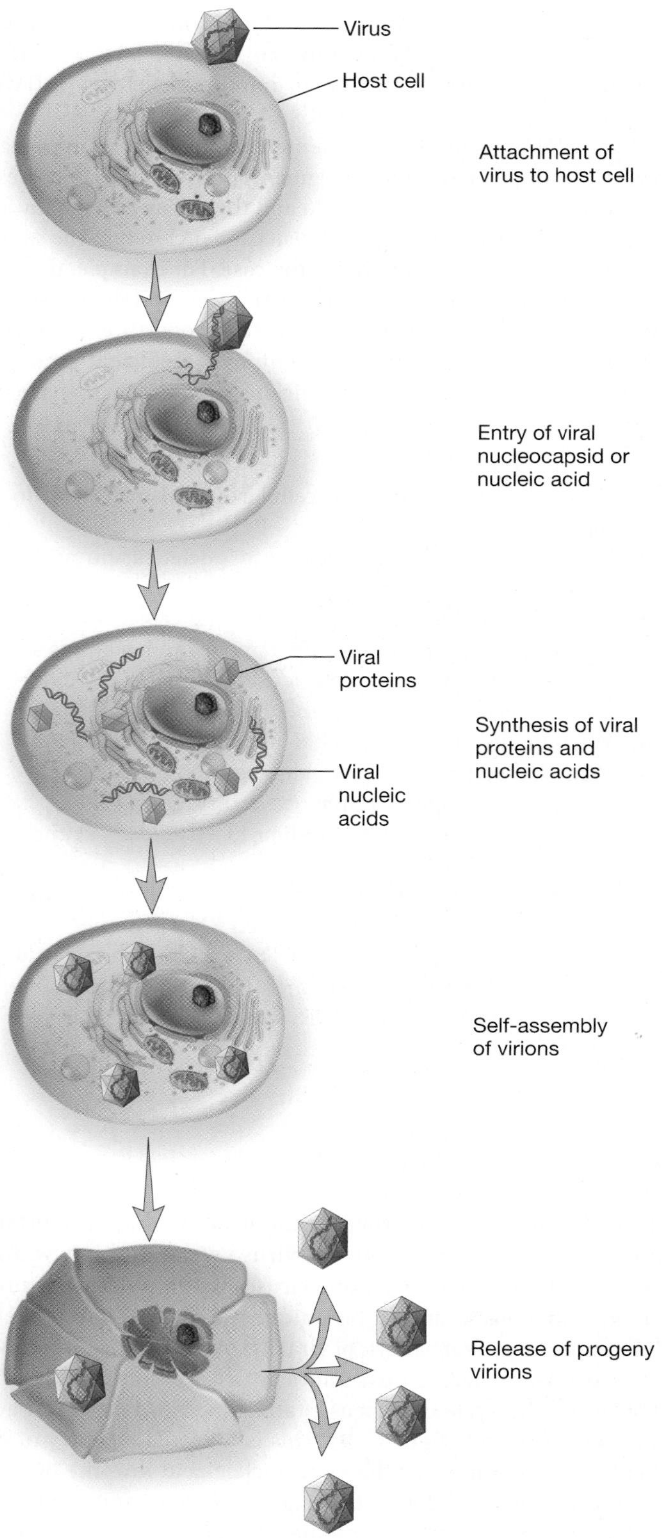

FIGURE 5.9 Generalized Illustration of Virus Multiplication. There is great variation in the details of virus multiplication for individual virus species. Viruses are not drawn to scale.

viral proteins. This is followed by the assembly stage, during which new nucleocapsids are constructed by self-assembly of coat proteins with the nucleic acids. Finally, during the release step, mature viruses escape the host.

Attachment (Adsorption)

Encounters of the virus and the host cell surface are thought to occur through a random collision of the virion with a potential host. However, viruses do not randomly attach to the surface of a host cell; rather, adsorption to the host is mediated by an interaction between **receptors** on the surface of the host cell and molecules (ligands) on the surface of the virion. The nature of cellular receptors varies. For instance, some bacteriophages use cell wall lipopolysaccharides and proteins as receptors, while others use teichoic acids, flagella, or pili.

Variation in receptors used by viruses for attachment is at least partly responsible for host preferences. Bacteriophages not only infect a particular species but often infect only certain strains within a given species. Receptor specificity also accounts for the observation that viruses of eukaryotes (i.e., eukaryotic viruses) infect specific organisms and, in some cases, only particular tissues within that host. However, if the receptor recognized by a virus is present in numerous organisms, then the virus will infect more than one host organism. Such is the case with *Rabies virus*. It is important to remember that the receptors on the host cell have specific cellular functions. For instance, they may normally bind hormones or other molecules essential to the cell's function. In some cases, it appears that two or more host cell receptors are involved in attachment. For instance, *h*uman *i*mmunodeficiency *v*irus (HIV) binds to two different proteins on human cells (e.g., CD4 and CCR5). Both of these host molecules normally bind chemokines—signaling molecules used by the immune system. ▶▶| *Cytokines (section 32.3)*

The distribution of the host cell receptors to which animal viruses attach also varies at the cellular level. Eukaryotic cell membranes have microdomains often called lipid rafts that seem to be involved in both virion entrance and assembly. For example, the receptors for enveloped viruses such as HIV and Ebola are concentrated in lipid rafts. Distribution at the tissue level plays a crucial role in determining the tropism of the virus and the outcome of infection. For example, *Poliovirus* receptors are found only in the human nasopharynx, gut, and anterior horn cells of the spinal cord. Therefore *Poliovirus* infects these tissues, causing gastrointestinal disease in its milder forms and paralytic disease in its more serious forms. In contrast, *Measles virus* receptors are present in most tissues and disease is disseminated throughout the body, resulting in the widespread rash characteristic of measles. |◀◀ *Eukaryotic cell envelopes (section 4.2); Human diseases caused by viruses and prions (chapter 37)*

Entry into the Host

After attachment to the host cell, the virus's genome or the entire nucleocapsid enters the cytoplasm. Many bacteriophages inject their nucleic acid into the cytoplasm of their host, leaving the capsid outside and attached to the cell wall. In contrast to phages, the nucleocapsid of most eukaryotic viruses penetrates the plasma membrane and enters the cytoplasm with the genome still enclosed. Once inside the cytoplasm, some animal viruses shed some or all of their capsid proteins, in a process called uncoating, whereas other viruses remain encapsidated. Because penetration and uncoating are often coupled, we consider them together.

The mechanisms of penetration and uncoating vary with the type of virus, and for many eukaryotic viruses, detailed mechanisms of penetration are unclear; however, it appears that one of three different modes of entry is usually employed (**figure 5.10**).

1. Fusion of the viral envelope with the host cell membrane—The envelopes of some viruses fuse directly with the host cell plasma membrane (figure 5.10*a*). Fusion may involve envelope glycoproteins that bind to plasma membrane proteins. For example, after attachment of paramyxoviruses (negative-strand RNA viruses), membrane lipids rearrange, the adjacent halves of the contacting membranes merge, and a proteinaceous fusion pore forms. The nucleocapsid then enters the host cell cytoplasm, where a viral enzyme carried within the nucleocapsid begins synthesizing viral mRNA while it is still within the capsid.
2. Entry by endocytosis—Nonenveloped viruses and some enveloped viruses enter cells by endocytosis. They may be engulfed by receptor-mediated endocytosis (figure 5.10*b*). The resulting endocytic vesicles are filled with viruses and fuse with endosomes; depending on the virus, escape of the nucleocapsid or its genome may occur either before or after fusion. Endosomal enzymes can aid in virus uncoating, and low pH often triggers the uncoating process. In at least some instances, the viral envelope fuses with the endosomal membrane, and the nucleocapsid is released into the cytoplasm (the capsid proteins may have been partially removed by endosomal enzymes). Once in the cytoplasm, viral nucleic acid may be released from the capsid upon completion of uncoating or may function while still attached to capsid components. Nonenveloped eukaryotic viruses cannot employ the membrane fusion mechanism for release from the endosome (figure 5.10*c*). In this case, it appears that the low pH of the endosome causes a conformational change in the capsid. The altered capsid contacts the vesicle membrane and either releases the viral nucleic acid into the cytoplasm or ruptures the membrane to release the intact virion. ⏮ *Organelles of the secretory and endocytic pathways (section 4.4)*
3. Injection of nucleic acid—Evidence exists that some nonenveloped viruses (e.g., *Poliovirus*) may inject their RNA genome into the cytoplasm of the host cell, leaving the capsid outside. *Entry of Animal Viruses into Host Cells*

Synthesis Stage

This stage of the viral life cycle differs dramatically among viruses because the genome of each virus dictates the events that occur. For dsDNA viruses, the synthesis stage can be very similar to the typical flow of information in cells. That is, the genetic information is stored in DNA and replicated by enzymes called DNA polymerases, recoded as mRNA (transcription), and decoded during protein synthesis (translation). Because of this similarity, some dsDNA viruses have the luxury of depending solely on their host cells' biosynthetic machinery to replicate their genomes and synthesize their proteins.

The same is not true for RNA viruses. Cellular organisms (except for plants) lack the enzymes needed to replicate RNA or to synthesize mRNA from an RNA genome. Therefore RNA viruses must carry in their nucleocapsids the enzymes needed to complete the synthesis stage or they must be synthesized during the infection process.

No matter what the genome, synthesis of viral proteins is tightly regulated. Some proteins, often called early proteins, are synthesized early in the infection, whereas other proteins are synthesized later. Early proteins are most often involved in taking over the host cell. Late proteins usually include capsid proteins and other proteins involved in self-assembly and release.

Assembly

Several kinds of late proteins are involved in the assembly of mature viruses. Some are nucleocapsid proteins, some are not incorporated into the nucleocapsid but participate in its assembly, and still other late proteins are involved in virus release. In addition, proteins and other factors synthesized by the host may be involved in assembling mature viruses, as is the case in bacteriophage T4.

The assembly process can be quite complex with multiple subassembly lines functioning independently and converging in later steps to complete nucleocapsid construction. As shown in **figure 5.11**, bacteriophage T4 assembles the baseplate, tail fibers, and head components separately. Once the baseplate is finished, the tail tube is built on it and the sheath is assembled around the tube. The phage prohead (procapsid) is assembled with the aid of scaffolding proteins that are degraded or removed after construction is completed. DNA is incorporated into the prohead by a complex of proteins sometimes called the "packasome." The packasome consists of a protein called the portal protein, which is located at the base of the prohead, and a set of proteins called the terminase complex, which moves DNA into the prohead. The movement of DNA consumes energy in the form of ATP, which is supplied by the metabolic activity of the host bacterium. After the head is completed, it spontaneously combines with the tail assembly.

While bacteriophages are assembled in the host cytoplasm, the site of morphogenesis varies with plant and animal viruses. Some are assembled in the nucleus; others such as poxviruses are

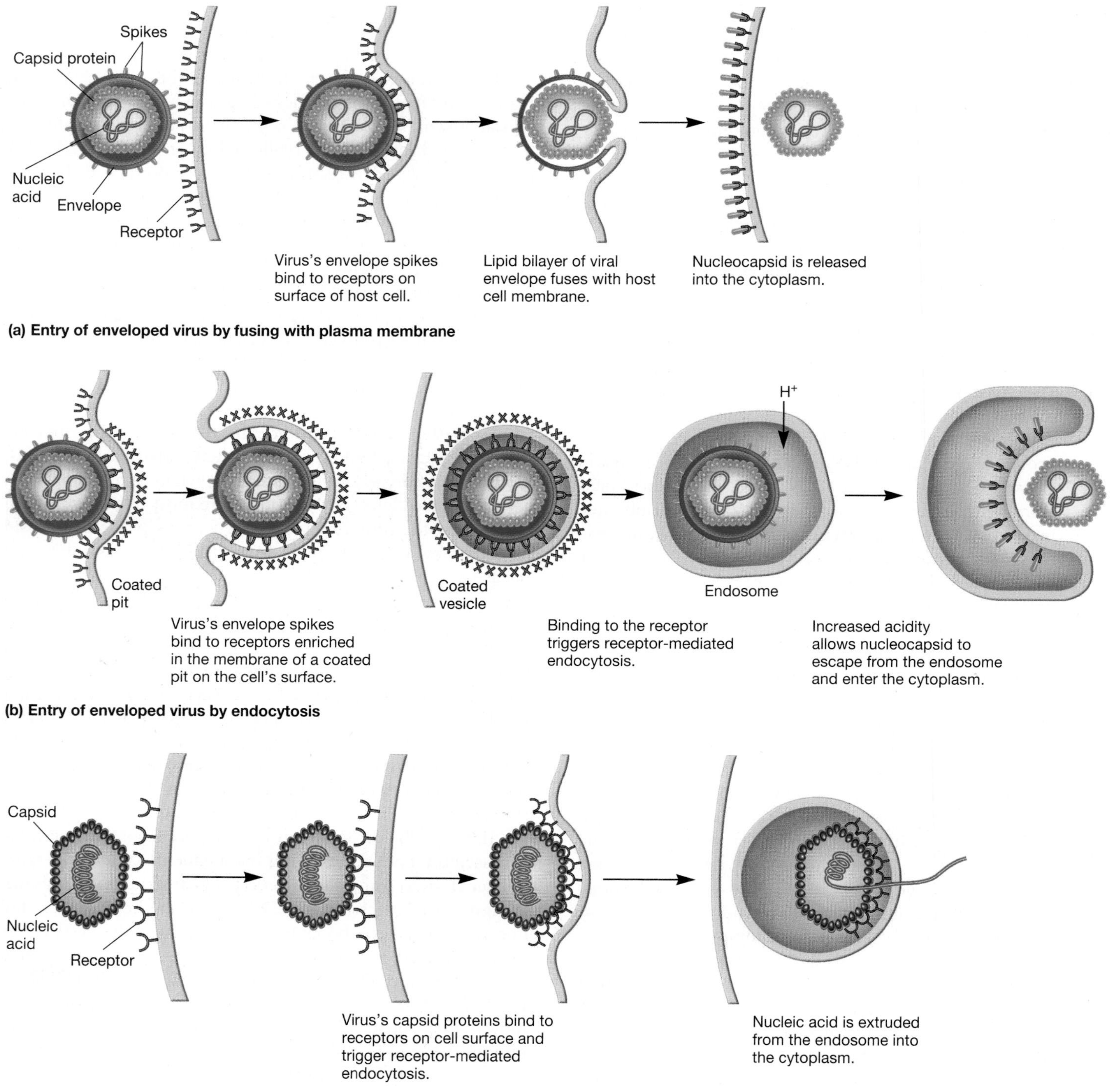
Spikes
Capsid protein
Nucleic acid
Envelope
Receptor
Virus's envelope spikes bind to receptors on surface of host cell.
Lipid bilayer of viral envelope fuses with host cell membrane.
Nucleocapsid is released into the cytoplasm.
(a) Entry of enveloped virus by fusing with plasma membrane
H+
Coated pit
Coated vesicle
Endosome
Virus's envelope spikes bind to receptors enriched in the membrane of a coated pit on the cell's surface.
Binding to the receptor triggers receptor-mediated endocytosis.
Increased acidity allows nucleocapsid to escape from the endosome and enter the cytoplasm.
(b) Entry of enveloped virus by endocytosis
Capsid
Nucleic acid
Receptor
Virus's capsid proteins bind to receptors on cell surface and trigger receptor-mediated endocytosis.
Nucleic acid is extruded from the endosome into the cytoplasm.
(c) Entry of nonenveloped virus by endocytosis

FIGURE 5.10 Animal Virus Entry. Examples of animal virus attachment and entry into host cells. Enveloped viruses can (a) enter after fusion of the envelope with the plasma membrane or (b) escape from the vesicle after endocytosis. (c) Nonenveloped viruses such as *Poliovirus*, a picornavirus, may be taken up by endocytosis and then insert their nucleic acid into the cytoplasm through the vesicle membrane. It also is possible that they insert the nucleic acid directly through the plasma membrane.

Figure 5.10 Micro Inquiry

Which of these mechanisms involves the production of a vesicle within the host cell? Which involves the interaction between viral proteins and host cell membrane proteins?

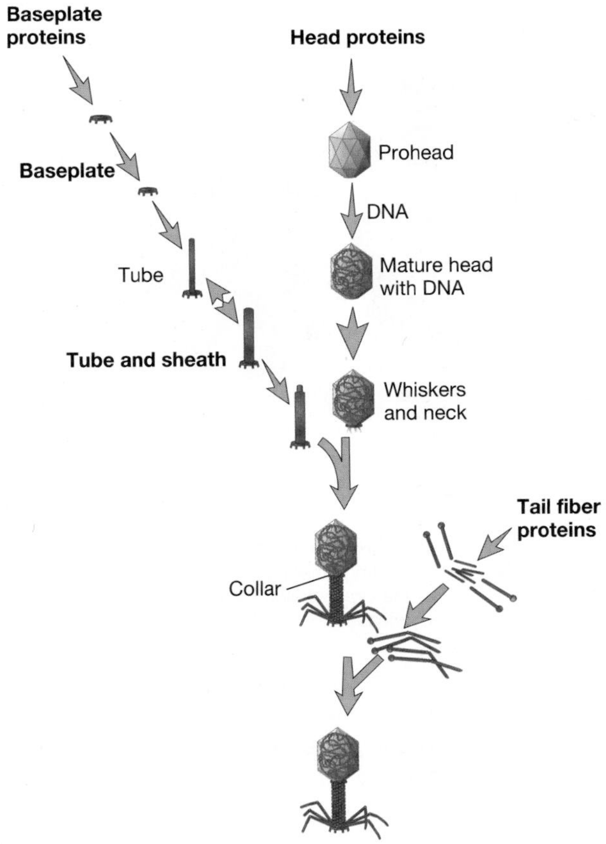

FIGURE 5.11 The Assembly of T4 Bacteriophage. Note the subassembly lines for the baseplate, tail tube and sheath, tail fibers, and head.

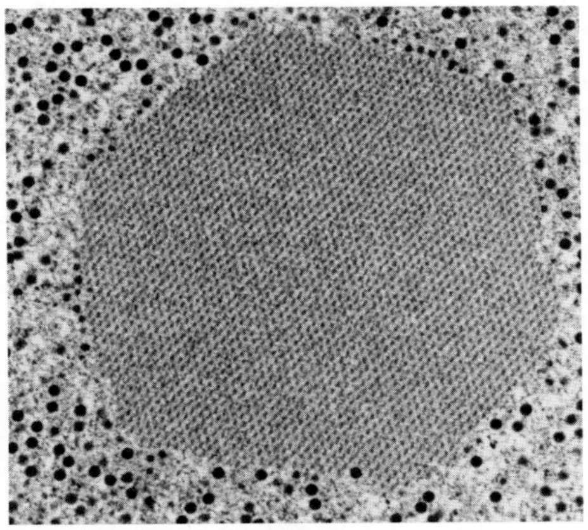

FIGURE 5.12 Paracrystalline Clusters. A crystalline array of adenoviruses within the cell nucleus (×35,000).

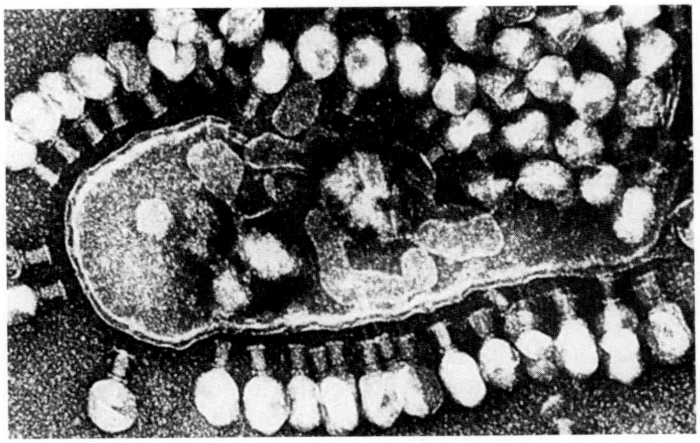

FIGURE 5.13 Release of T4 Bacteriophages by Lysis of the Host Cell. The host cell has been lysed (upper right portion of the cell) and virions have been released into the surroundings. Progeny virions also can be seen in the cytoplasm. In addition, empty capsids of the infecting phages coat the outside of the cell (×36,500).

Figure 5.13 Micro Inquiry

Why do you think the empty capsids remain attached to the cell after the viral genome enters the host cell?

assembled in the cytoplasm. Large paracrystalline clusters of either complete virions or procapsids are often seen at the site of virus maturation (**figure 5.12**).

Virion Release

Many viruses, especially nonenveloped viruses, lyse their host cells at the end of the intracellular phase. This process involves the activity of additional viral proteins. For instance, the lysis of *E. coli* by T4 requires two specific proteins (**figure 5.13**). One is lysozyme, an enzyme that attacks peptidoglycan in the host's cell wall. Another T4 protein creates holes in the *E. coli* plasma membrane, enabling T4 lysozyme to move from the cytoplasm to the peptidoglycan.

Another common release mechanism is budding. This is frequently observed in enveloped viruses—in fact, envelope formation and virus release are usually concurrent processes. When viruses are released by budding, the host cell may survive and continue releasing virions for some time. All envelopes of eukaryotic viruses are derived from host cell membranes by a multistep process. First, virus-encoded proteins are incorporated into the membrane. Then the nucleocapsid is simultaneously released and the envelope formed by membrane budding (**figure 5.14**). In several virus families, a matrix (M) protein attaches to the plasma membrane and aids in budding. Most envelopes arise from the plasma membrane. The endoplasmic reticulum, Golgi apparatus, and other internal membranes also can be used to form envelopes. *Mechanism for Releasing Enveloped Virions*

Interestingly, it has been discovered that actin filaments can aid in the release of many eukaryotic viruses. These viruses alter the actin microfilaments of the host cell cytoskeleton. For example, *Vaccinia virus* appears to form long actin tails and uses

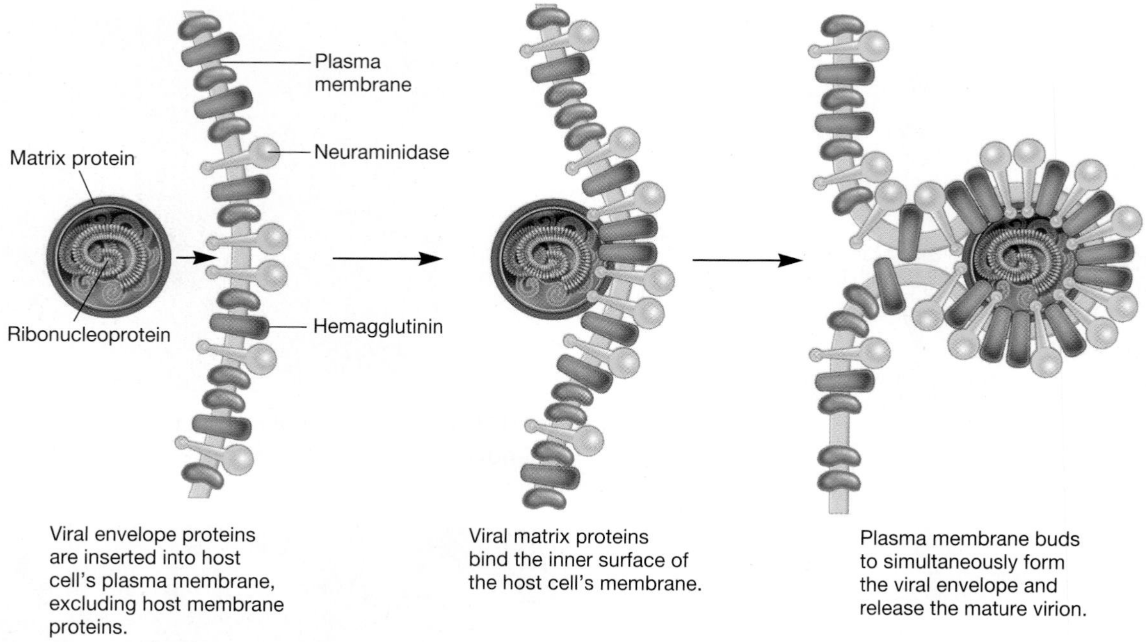
Plasma membrane
Matrix protein
Neuraminidase
Ribonucleoprotein
Hemagglutinin
Viral envelope proteins are inserted into host cell's plasma membrane, excluding host membrane proteins.
Viral matrix proteins bind the inner surface of the host cell's membrane.
Plasma membrane buds to simultaneously form the viral envelope and release the mature virion.

FIGURE 5.14 Release of Influenza Virus by Plasma Membrane Budding. First, viral envelope proteins (hemagglutinin and neuraminidase) are inserted into the host plasma membrane. Then the nucleocapsid approaches the inner surface of the membrane and binds to it. At the same time, viral proteins collect at the site and host membrane proteins are excluded. Finally, the plasma membrane buds to simultaneously form the viral envelope and release the mature virion.

them to move intracellularly at speeds up to 2.8 μm per minute. The actin filaments also propel vaccinia through the plasma membrane. In this way, the virion escapes without destroying the host cell and infects adjacent cells. This is similar to the mechanism of pathogenesis used by some intracellular pathogenic bacteria such as *Listeria monocytogenes.* ◀◀ *Cytoplasm of eukaryotes (section 4.3)*

1. Why do you think viruses have evolved to use host surface proteins that otherwise serve very important, and sometimes essential, functions for the host cell?
2. What probably plays the most important role in determining the tissue and host specificity of viruses? Give some specific examples.
3. How do you think the complexity of the viral assembly process correlates with viral genome size?
4. In general, DNA viruses can be much more dependent on their host cells than can RNA viruses. Why is this so?
5. Consider the origin of viral envelopes. Why do you think enveloped viruses that infect plants and bacteria are rare?

5.4 Types of Viral Infections

So far, our discussion has focused on viral structures and modes of multiplication with little mention of the cost to host cells. The dependence of viruses on their host cells has many consequences—as anyone who has ever had a cold or the flu well understands. Next, we discuss the interaction between virus and host cell.

Infections of Bacterial and Archaeal Cells: Lysis and Lysogeny

Figure 5.9 illustrates the life cycle of a **virulent phage**—one that has only one option: to begin multiplying immediately upon entering its bacterial host, followed by release from the host by lysis. T4 is an example of a virulent phage. Other phages are **temperate phages** that have two options: Upon entry into the host, they can multiply like the virulent phages and lyse the host cell, or they can remain within the host without destroying it (**figure 5.15**). Many temperate phages accomplish this by integrating their genome into the host cell's chromosome. Bacteriophage lambda is an example of this type of phage. The life cycles of both T4 and lambda are described in more detail in chapter 25. *Steps in the Replication of T4 Phage in* E. coli; *Lambda Phage Replication Cycle*

The relationship between a temperate phage and its host is called **lysogeny.** The form of the virus that remains within its host is called a **prophage,** and the infected bacteria are called **lysogens** or **lysogenic bacteria.** Lysogenic bacteria reproduce and in most other ways appear to be perfectly normal. However, they have two distinct characteristics. The first is that they cannot be reinfected by the same virus—that is, they have immunity to superinfection. The second is that they can switch from the lysogenic cycle to the **lytic cycle.**

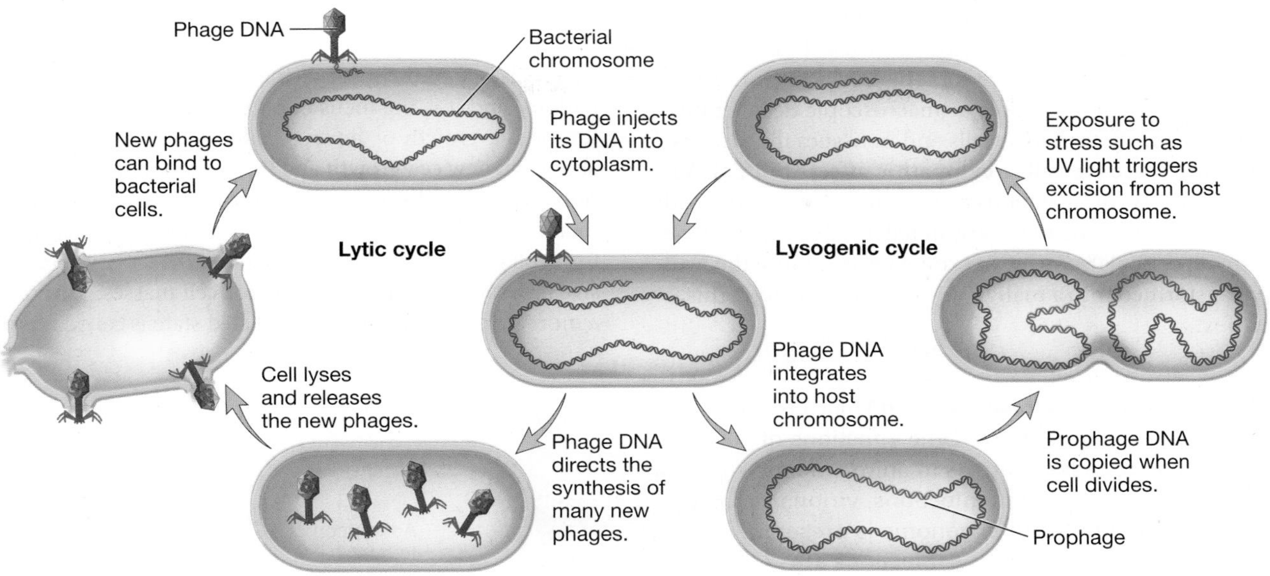

FIGURE 5.15 Lytic and Lysogenic Cycles of Temperate Phages. Temperate phages have two phases to their life cycles. The lysogenic cycle allows the genome of the virus to be replicated passively as the host cell's genome is replicated. Certain environmental factors such as UV light can cause a switch from the lysogenic cycle to the lytic cycle. In the lytic cycle, new virus particles are made and released when the host cell lyses. Virulent phages are limited to just the lytic cycle.

***Figure 5.15** Micro Inquiry*

Why is a lysogen considered a new or different strain of a given bacterial species?

This results in host cell lysis and release of phage particles. This occurs when conditions within the cell cause the prophage to initiate synthesis of phage proteins and to assemble new virions, a process called **induction.** Induction is commonly caused by changes in growth conditions or ultraviolet irradiation of the host cell.

Another important outcome of lysogeny is **lysogenic conversion.** This occurs when a temperate phage changes the phenotype of its host. Lysogenic conversion often involves alteration in surface characteristics of the host. For example, when *Salmonella* is infected by epsilon phage, the phage changes the activities of several enzymes involved in construction of the carbohydrate component of the bacterium's lipopolysaccharide. This eliminates the receptor for epsilon phage, so the bacterium becomes immune to infection by another epsilon phage. Many other lysogenic conversions give the host pathogenic properties. This is the case when *Corynebacterium diphtheriae,* the cause of diphtheria, is infected with phage β. The phage genome encodes diphtheria toxin, which is responsible for the disease. Thus only those strains of *C. diphtheriae* that are infected by the phage (i.e., lysogens) cause disease.

Clearly, the infection of a bacterium by a temperate phage has significant impact on the host, but why would viruses evolve this alternate life cycle? Two advantages of lysogeny have been recognized. The first is that lysogeny allows a virus to remain viable within a dormant host. Bacteria often become dormant due to nutrient deprivation, and while in this state, they do not synthesize nucleic acids or proteins. In such situations, a prophage would survive but most virulent bacteriophages would not be replicated, as they require active cellular biosynthetic machinery. The second advantage arises when there are many more phages in an environment than there are host cells, a situation virologists refer to as a high multiplicity of infection (MOI). In these conditions, lysogeny enables the survival of host cells so that the virus can continue to reproduce.

Archaeal viruses can also be virulent or temperate. Most archaeal viruses discovered thus far are temperate. Unfortunately, little is known about the mechanisms they use to establish lysogeny.

1. Define the terms lysogeny, temperate phage, lysogen, prophage, immunity, and induction.
2. What advantages might a phage gain by being capable of lysogeny?
3. Describe lysogenic conversion and its significance.

Infection of Eukaryotic Cells

Viruses can harm their eukaryotic host cells in many ways. An infection that results in cell death is a cytocidal infection. As with bacterial and archaeal viruses, this can occur by lysis of the host (**figure 5.16*a***). Viral growth does not always result in the lysis of host cells. Some viruses (e.g., herpesviruses) can establish persistent infections lasting many years (figure 5.16*b,c*). Eukaryotic viruses can cause microscopic or macroscopic degenerative changes or abnormalities in host cells and in tissues that are distinct from lysis. These are called **cytopathic effects (CPEs).** Viruses use a variety of mechanism to cause cytopathic and cytocidal effects. Many of these are noted in chapter 37. One mechanism of particular note is that some viruses cause the host cell to be transformed into a malignant cell (figure 5.16*d*). This is discussed next.

Viruses and Cancer

Cancer is one of the most serious medical problems in developed nations, and it is the focus of an immense amount of research. A tumor is a growth or lump of tissue resulting from **neoplasia**—abnormal new cell growth and reproduction due to loss of regulation. Tumor cells have aberrant shapes and altered plasma membranes that may contain distinctive tumor antigens. Their unregulated proliferation and loss of differentiation result in invasive growth that forms unorganized cell masses. This reversion to a more primitive or less differentiated state is called **anaplasia.**

FIGURE 5.16 Types of Viral Infections and Their Effects on Eukaryotic Host Cells. (a) Lytic infections of host cells can lead to disease states in animal hosts called acute infections. In latent infections, virions can not be detected, but the viral genome is present. (b) Latent infections and (c) chronic infections are two different types of persistent infections. (d) Some infections of animal cells cause the cell to be transformed into a malignant cell, which can cause cancer in the animal host.

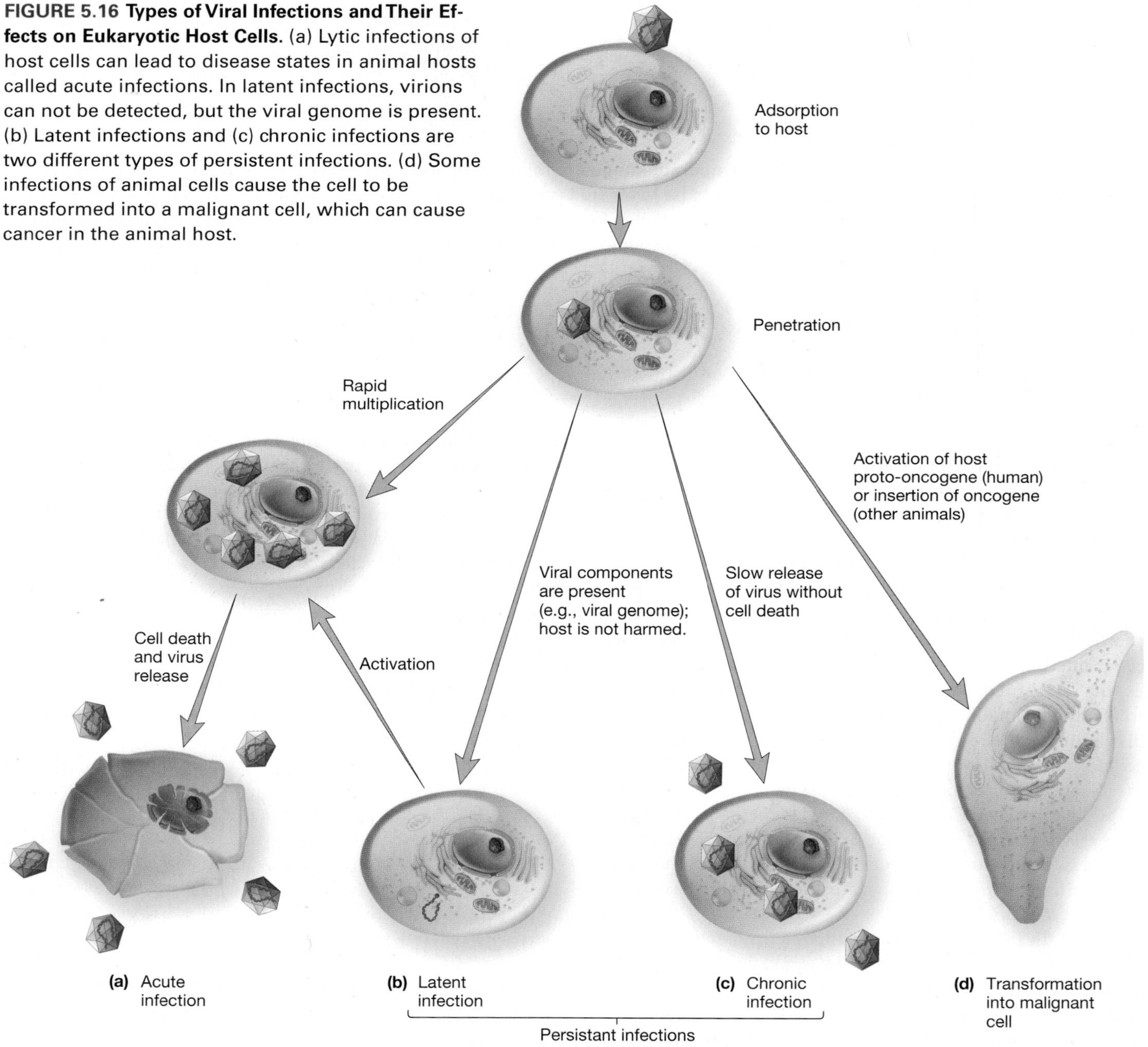

Two major types of tumor growth patterns exist. If the tumor cells remain in place to form a compact mass, the tumor is benign. In contrast, cells from malignant or cancerous tumors actively spread throughout the body in a process known as metastasis. Some cancers are not solid but cell suspensions. For example, leukemias are composed of undifferentiated malignant white blood cells that circulate throughout the body. Indeed, dozens of kinds of cancers arise from a variety of cell types and afflict all kinds of organisms.

As one might expect from the wide diversity of cancers, cancer has many causes, only a few of which are directly related to viruses. Carcinogenesis is a complex, multistep process that involves the mutation of multiple genes. Genes involved in carcinogenesis are called **oncogenes.** Some oncogenes are contributed to a cell by viruses; others arise from normal genes within the cell called **proto-oncogenes.** Proto-oncogenes are cellular genes required for normal growth, but when mutated or overexpressed, they become oncogenes. That is, their products contribute to the malignant transformation of the cell. Many oncogenes are involved in the regulation of cell growth and signal transduction; for example, some code for growth factors that regulate cell reproduction. Proto-oncogenes can be transformed into oncogenes by spontaneous mutation or through the activity of a mutation-causing agent, called a mutagen. ▶▶| *Mutations and their chemical basis (section 14.1)*

Viruses are associated with cancers in animals, including humans. In some human cancers, the virus has been established as the cause of the cancer. In others, causality has not been established. For these cancers, the virus often functions as a cofactor. Most human viruses associated with cancer have dsDNA genomes. Some of the viruses involved in human cancers are discussed here.

1. Two herpesviruses, *Human herpesvirus 8* (HHV8) and Epstein-Barr virus (EBV), are linked to cancer. HHV8 causes Kaposi's sarcoma, which has a high incidence in AIDS patients, and is associated with other cancers as well. EBV is linked to several cancers, including Burkitt's lymphoma and nasopharyngeal carcinoma. Burkitt's lymphoma is a malignant tumor of the jaw and abdomen found in children of central and western Africa. Interestingly, evidence exists that a person also must have had malaria to develop Burkitt's lymphoma. This is supported by the low rates of EBV-associated cancer in the United States where EBV is prevalent but there is a low incidence of malaria. ▶▶| *Herpesviruses (section 25.2)*
2. Two viruses that cause hepatitis are associated with human cancers. *Hepatitis B virus* is linked with one form of liver cancer (hepatocellular carcinoma). *Hepatitis C virus* causes cirrhosis of the liver, which can lead to liver cancer.
3. Some strains of human papillomaviruses (HPV) cause cervical cancer.
4. The retrovirus human T-cell lymphotropic virus I (HTLV-1) is associated with adult T-cell leukemia.

Viruses known to cause cancer are called **oncoviruses.** Most known human dsDNA oncoviruses trigger cancerous transformation of cells by a similar mechanism. They encode proteins that bind to and thereby inactivate host proteins known as **tumor suppressor proteins.** Tumor-suppressor proteins regulate cell cycling, or monitor or repair DNA damage. Two tumor suppressors known to be targets of human oncovirus proteins are called Rb and p53. Rb has multiple functions in the nucleus, all of which are critical to normal cell cycling. When Rb molecules are rendered inactive by an oncoviral protein, cells undergo uncontrolled reproduction and are said to be hyperproliferative. The protein p53 is often referred to as "the guardian of the genome." This is because p53 normally initiates either cell cycle arrest or programmed cell death in response to DNA damage. However, when p53 is inactivated by the binding of an oncoviral protein, it cannot do so and genetic damage persists. From the point of view of the virus, hyperproliferation and the lack of programmed cell death are beneficial. For the cell, however, it can be catastrophic. Cells can rapidly accumulate the additional mutations needed for oncogenic transformation. *Tumor Suppressor Genes*

Viruses related to HIV (the retroviruses) exert their oncogenic powers in a different manner. Some carry oncogenes captured from host cells many, many generations ago. Thus they transform the host cell by bringing the oncogene into the cell. For example, the retrovirus HTLV-1 transforms a group of immune system cells called T cells by producing a regulatory protein that sometimes activates genes involved in cell division and stimulates viral multiplication. The second transformation mechanism used by retroviruses such as *Avian leukosis virus* involves the integration of a viral genome into the host chromosome such that strong, viral regulatory elements are near a cellular proto-oncogene. These elements cause the nearby proto-oncogene to be transcribed at a high level, causing the gene to be considered an oncogene.

1. What is a cytocidal infection? What is a cytopathic effect?
2. Define the following terms: tumor, neoplasia, anaplasia, metastasis, and oncogene.
3. Distinguish the mechanism by which DNA viruses cause cancer from that of retroviruses. Why do you think there is an environmental component to many kinds of cancer?

5.5 Cultivation and Enumeration of Viruses

Because they are unable to reproduce outside of living cells, viruses cannot be cultured in the same way as cellular microorganisms. It is relatively simple to grow lytic bacterial and archaeal viruses as long as the host cell is easily grown in culture. The cells and viruses are simply mixed together in a broth culture. Over time, more and more cells are infected and lysed, releasing the viruses into the broth. Culturing temperate viruses requires an additional step of inducing the lytic phase of their life cycles.

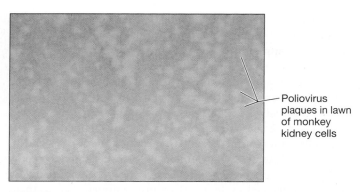

FIGURE 5.17 Viral Plaques. *Poliovirus* plaques in a monkey kidney cell culture.

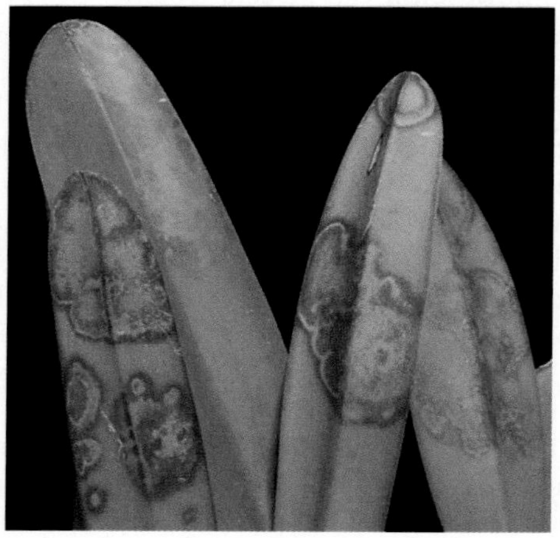

FIGURE 5.18 Necrotic Lesions on Plant Leaves. *Tobacco mosaic virus* infection of an orchid showing leaf color changes.

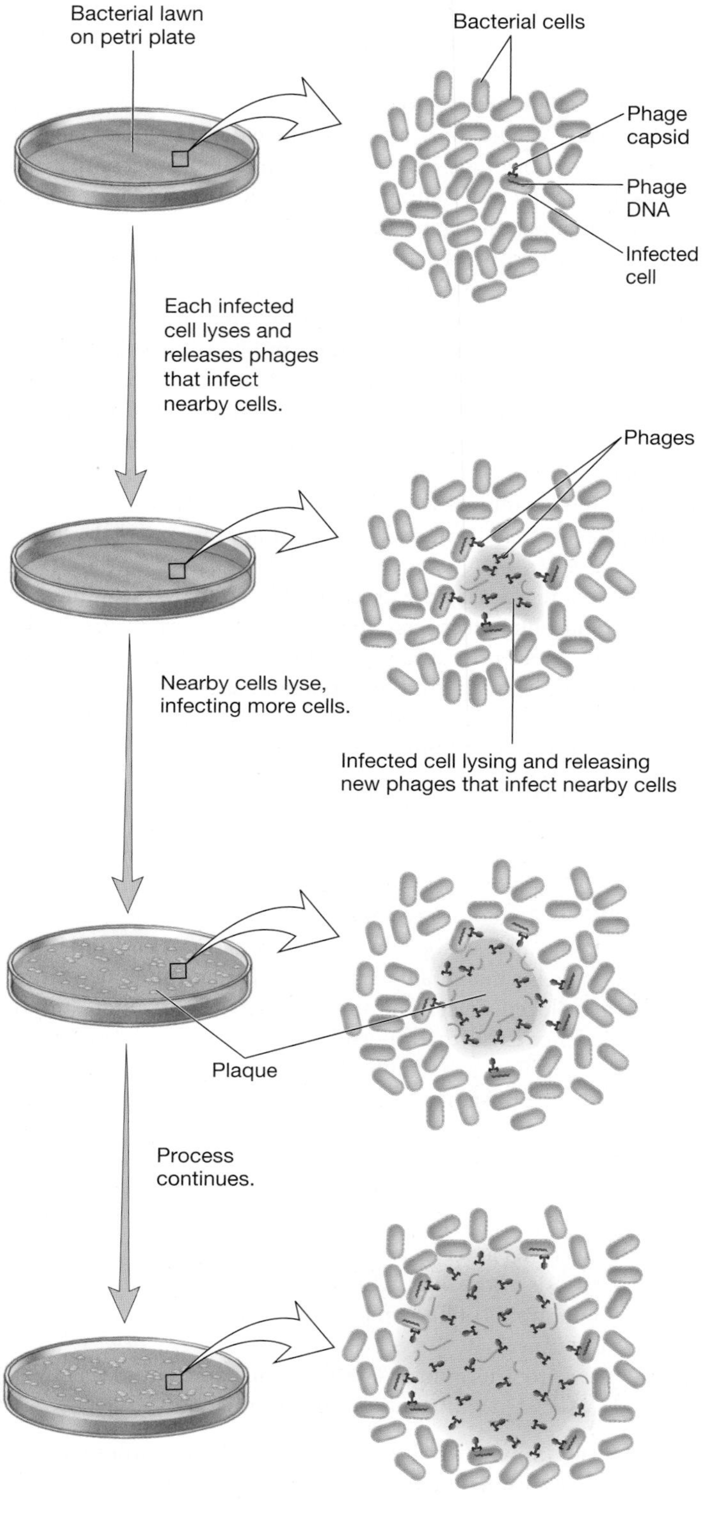

FIGURE 5.19 Formation of Phage Plaques. When phages and host bacterial cells are mixed at an appropriate ratio, only a portion of the cells are initially infected. When this mixture is plated, the infected cells will be separated from each other. The infected cells eventually lyse, releasing progeny phages. They infect nearby cells, which eventually lyse, releasing more phages. This continues and ultimately gives rise to a clear area within a lawn of bacteria. The clear area is a plaque.

Figure 5.19 Micro Inquiry

What would happen to a plate with individual plaques if it were returned to an incubator so that the viral infection could continue?

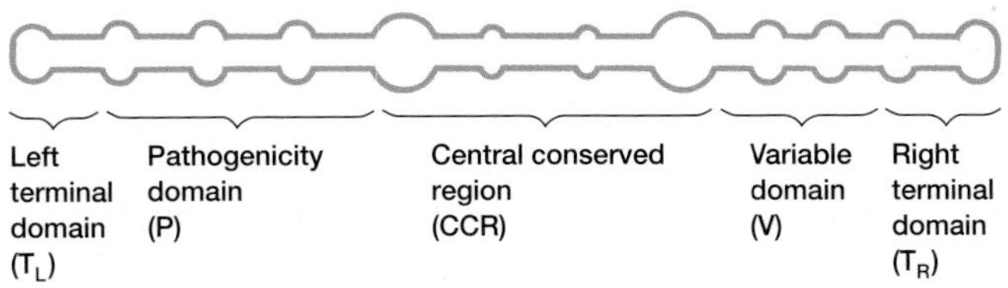

FIGURE 5.23 Viroid Structure. This schematic diagram shows the general organization of a viroid. The closed single-stranded RNA circle has extensive intrastrand base pairing and interspersed unpaired loops. Viroids have five domains. Most changes in viroid pathogenicity seem to arise from variations in the P and T_L domains.

that particular regions of the RNA are required; studies have shown that removing these regions blocks the development of disease (figure 5.23). Some data suggest that viroids cause disease by triggering a eukaryotic response called **RNA silencing,** which normally functions to protect against infection by dsRNA viruses. During RNA silencing, the cell detects the presence of dsRNA and selectively degrades it. Viroids may usurp this response by hybridizing to specific host mRNA molecules to which they have a complementary nucleotide sequence. Formation of the viroid-host hybrid dsRNA molecule is thought to elicit RNA silencing. This results in destruction of the host mRNA and therefore silencing of the host gene. Failure to express a required host gene leads to disease in the host plant.

Search This: Viroids/USDA

Virusoids, also called satellite RNAs, are similar to viroids in that they also consist only of covalently closed, circular ssRNA molecules with regions capable of intrastrand base pairing. In contrast to viroids, they encode one or more gene products, and they typically need a helper virus to infect host cells. The helper virus supplies gene products and other materials needed by the virusoid for completion of its replication cycle. The best-studied virusoid is the human hepatitis D virusoid, which is 1,700 nucleotides long. It uses *Hepatitis B virus*, a virus with an interesting gapped dsDNA genome, as its helper virus. If a host cell contains both *Hepatitis B virus* and the hepatitis D virusoid, the virusoid RNA and its gene product, called delta antigen, can be packaged within the envelope of the virus. These enveloped virusoids and delta antigens are capable of entering other host cells and initiating infection. ▶▶| *Viruses with gapped DNA genomes (Group VII) (section 25.8)*

5.7 Prions

Prions (for *pro*teinaceous *in*fectious particle) cause a variety of neurodegenerative diseases in humans and animals. The best-studied prion is the scrapie prion, which causes the disease scrapie in sheep. Afflicted animals lose coordination of their movements, tend to scrape or rub their skin, and eventually cannot walk.

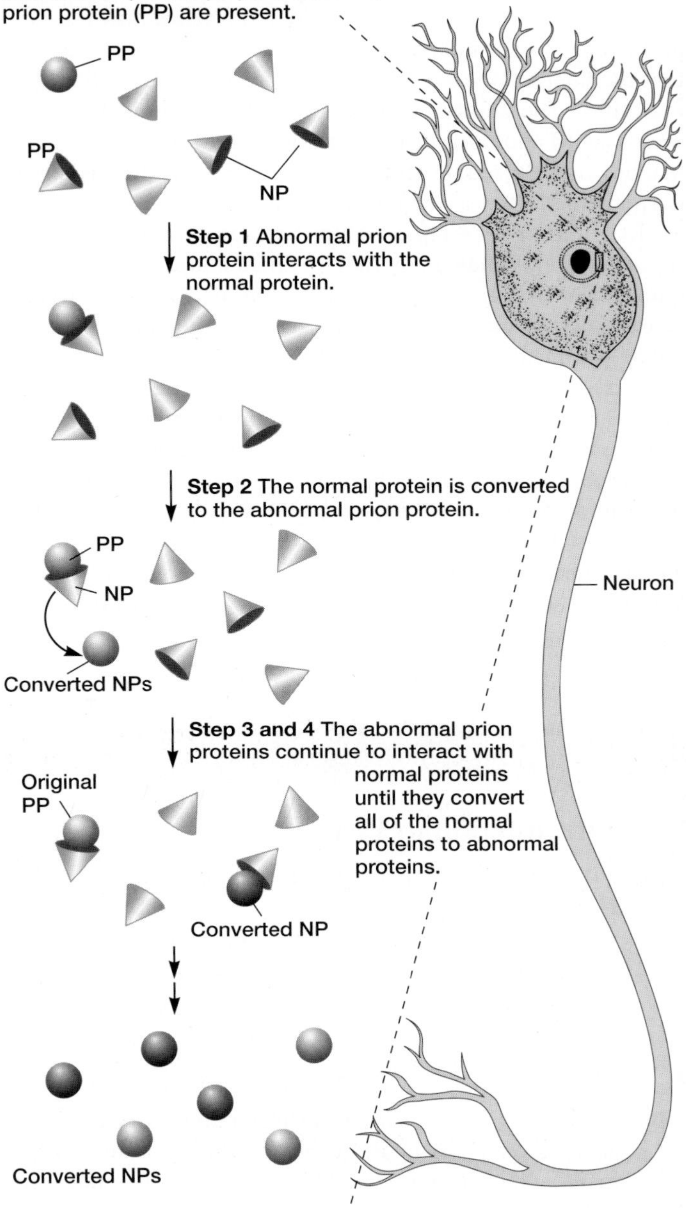

FIGURE 5.24 Proposed Mechanism by Which Prions Replicate. The normal and prion proteins differ in their tertiary structures.

Researchers have shown that scrapie is caused by an abnormal form of a cellular protein. The abnormal form is called PrP^{Sc} (for *s*crapie-associated *pr*ion *p*rotein), and the normal cellular form is called PrP^{C}. Evidence supports a model in which entry of PrP^{Sc} into the brain of an animal causes the PrP^{C} protein to change from its normal conformation to the abnormal form (**figure 5.24**). The newly produced PrP^{Sc} molecules then convert more PrP^{C} molecules into the abnormal PrP^{Sc} form. How the PrP^{Sc} causes this conformational change is unclear. However, the best-supported model is that the PrP^{Sc} directly interacts with PrP^{C}, causing the change. It is noteworthy that mice lacking the PrP gene cannot be infected with PrP^{Sc}. Although the evidence is strong that PrP^{Sc} causes PrP^{C} to fold abnormally, how this triggers neuron loss is poorly understood. Recent evidence suggests that the interaction of PrP^{Sc} with PrP^{C} serves to cross-link PrP^{C} molecules. The cross-linked PrP^{C} molecules trigger a series of events called apoptosis or programmed cell death. Thus the normal but cross-linked protein causes neuron loss, whereas the abnormal protein acts as the infectious agent.

In addition to scrapie, prions are responsible for bovine spongiform encephalopathy (BSE or "mad cow disease") and the human diseases kuru, fatal familial insomnia, Creutzfeldt-Jakob disease (CJD), and Gerstmann-Strässler-Scheinker syndrome (GSS). All result in progressive degeneration of the brain and eventual death. At present, no effective treatment exists. Mad cow disease reached epidemic proportions in Great Britain in the 1990s and initially spread because cattle were fed meal made from all parts of cattle, including brain tissue. It has now been shown that eating meat from cattle with BSE can cause a variant of Creutzfeldt-Jakob disease in humans (vCJD). More than 90 people have died in the United Kingdom and France from this source. Variant CJD differs from CJD in origin only: people acquire vCJD by eating contaminated meat, while CJD is an extremely rare condition caused by spontaneous mutation of the gene that encodes the prion protein. CJD and GSS are rare and cosmopolitan in distribution among middle-aged people, while kuru has been found only in the Fore, an eastern New Guinea tribe. This tribe had a custom of consuming dead kinsmen. Women and children were given the less desirable body parts to eat, including the brain, and were thus infected. Cannibalism was stopped many years ago, and kuru has been eliminated.

Search This: BSE/CDC

1. What are viroids and why are they of great interest?
2. How does a viroid differ from a virus?
3. What is a prion? In what way does a prion appear to differ fundamentally from viruses and viroids?
4. Prions are difficult to detect in host tissues. Why do you think this is so? Why do you think we have not been able to develop effective treatments for prion-caused diseases?

Summary

5.1 Introduction to Viruses

a. Virology is the study of viruses and other acellular infectious agents.

b. Viruses are composed of either DNA or RNA enclosed in a coat of protein (and sometimes other substances as well). They cannot reproduce independently of living cells.

5.2 Structure of Viruses

a. All virions have a nucleocapsid composed of a nucleic acid, either DNA or RNA, held within a protein capsid made of one or more types of protein subunits called protomers (**figure 5.1**).

b. There are four types of viral morphology: nonenveloped icosahedral, nonenveloped helical, enveloped icosahedral and helical, and complex.

c. Helical capsids resemble long, hollow protein tubes and may be either rigid or quite flexible. The nucleic acid is coiled in a spiral on the inside of the cylinder (**figure 5.3**).

d. Icosahedral capsids are usually constructed from two types of capsomers: pentamers (pentons) at the vertices and hexamers (hexons) on the edges and faces of the icosahedron (**figure 5.5**).

e. Complex viruses (e.g., poxviruses and large phages) have complicated morphology not characterized by icosahedral and helical symmetry (**figure 5.6**). Large phages often have binal symmetry: their heads are icosahedral and their tails are helical (**figure 5.7**).

f. Viruses can have a membranous envelope surrounding their nucleocapsid. The envelope lipids usually come from the host cell; in contrast, many envelope proteins are viral and may project from the envelope surface as spikes or peplomers (**figure 5.8**).

g. Viral nucleic acids can be either single stranded (ss) or double stranded (ds), DNA or RNA.

h. Most DNA viruses have dsDNA genomes that may be linear or closed circles.

i. RNA viruses usually have ssRNA. Some RNA genomes are segmented.

j. Although viruses lack true metabolism, some contain a few enzymes necessary for their multiplication.

5.3 Viral Multiplication

a. Viral life cycles can be divided into five steps: (1) attachment to host; (2) entry into host; (3) synthesis of viral nucleic acids and proteins; (4) self-assembly of virions; and (5) release from host (**figures 5.9** to **5.11** and **5.13** and **5.14**).

b. The synthesis stage of a virus's life cycle depends on the nature of its genome. DNA viruses use enzymes that are similar to host enzymes. In some cases, they can rely solely on their hosts for synthesis of their nucleic acids and proteins.

c. RNA viruses must either encode the enzymes they need to make mRNA and replicate their genomes or carry these enzymes in their capsids.

5.4 Types of Viral Infections

a. Virulent phages and archaeal viruses lyse their host. In addition to host cell lysis, temperate bacterial and archaeal viruses can enter the lysogenic cycle in which they remain dormant in the host cell. This is often accomplished by integrating the viral genome into that of the host (**figure 5.15**).

b. Viruses that infect eukaryotic cells can cause either host cell lysis or a more insidious cellular death; such viruses cause cytocidal infection. One relatively rare outcome of animal virus infection is the transformation of normal host cells into malignant or cancerous cells (**figure 5.16**).

5.5 Cultivation and Enumeration of Viruses

a. Viruses are cultivated using tissue cultures, embryonated eggs, bacterial cultures, and other living hosts.

b. Phages produce plaques in bacterial lawns. Sites of animal viral infection may be characterized by cytopathic effects such as pocks and plaques. Plant viruses can cause localized necrotic lesions in plant tissues (**figures 5.17** to **5.20**).

c. Virions can be counted directly with the transmission electron microscope or indirectly by hemagglutination and plaque assays.

d. Infectivity assays can be used to estimate virus numbers in terms of plaque-forming units, lethal dose (LD_{50}), or infectious dose (ID_{50}) (**figure 5.21**).

5.6 Viroids and Virusoids

a. Infectious agents that are simpler than viruses exist. For example, several plant diseases are caused by small, circular ssRNA molecules called viroids (**figures 5.22** and **5.23**).

b. Virusoids are infectious RNAs that encode one or more gene products. They require a helper virus for replication.

5.7 Prions

a. Prions are small proteinaceous agents associated with at least six degenerative nervous system disorders: scrapie, bovine spongiform encephalopathy, kuru, fatal familial insomnia, Gerstmann-Strässler-Scheinker syndrome, and Creutzfeldt-Jakob disease.

b. Most evidence supports the hypothesis that prion proteins exist in two forms: the infectious, abnormally folded form and a normal cellular form. The interaction between the abnormal form and the cellular form converts the cellular form into the abnormal form (**figure 5.24**).

Critical Thinking Questions

1. Many classification schemes are used to identify bacteria. These start with Gram staining, progress to morphology and arrangement characteristics, and include a battery of metabolic tests. Build an analogous scheme that could be used to identify viruses. You might start by considering the host, or you might start with viruses found in a particular environment, such as a marine filtrate.

2. The origin and evolution of viruses is controversial. Discuss whether you think viruses evolved before the first cell or whether they have coevolved and are perhaps still coevolving with their hosts.

3. Consider the separate stages of an animal virus life cycle. Assemble a short list of structures and processes that are unique to the virus and would make good drug targets for an antiviral agent. Explain your rationale for each choice.

4. Large dsDNA viruses such as most bacteriophages and the human herpesviruses package their DNA into preassembled procapsids. The DNA exists as a very long molecule that contains multiple, adjacent genomes. The DNA is pumped into the capsid via a "headful" mechanism. That is, when a specified number of nucleotides has been pumped in, the procapsid is "full" and DNA packaging is stopped by cleaving the DNA. The pressure within the capsid during this process is estimated to exceed that of bottled champagne. In fact, it has been hypothesized that a pressure-sensing mechanism determines when packaging should stop. For this to occur, the DNA must interact with a protein complex at the entrance to the capsid called the portal. The portal must then "tell" the DNA-cleaving enzyme to cut the genome. How do you think the portal protein might detect a proper "headful" of DNA and communicate this message to the DNA cleaving enzyme? (Hint: Think about how increasing pressure might affect the portal protein.)

   Read the original paper (and see stunning images of the P22 bacteriophage): Lander, G. C., et al. 2006. The structure of an infectious P22 virion shows the signal of headful DNA packaging. *Science* 312:1791.

5. *Mimivirus* is the largest known virus, infecting the protist *Acanthamoeba polyphaga*. Mature virions are 400 nm (0.4 μm) in diameter and carry dsDNA in a 1.2 million base pair genome, which is more than twice the size of some parasitic

bacteria. It also encodes four aminoacyl-tRNA synthetases—the enzymes that attach amino acids to tRNA—as well as other proteins involved in protein synthesis such as peptide release factor, translation elongation factor EF-TU, and translation initiation factor 1. The presence of these genes has sparked a debate about viruses and the evolution of the nucleus. Why do you think this is the case? Recall that endosymbiotic bacteria and cyanobacteria are thought to be the origins of mitochondria and chloroplasts respectively. Propose a series of events that accounts for the development of these energy-conserving organelles and the nucleus. Be sure to indicate which event you think happened first (mitochondria/chloroplast development versus nucleus evolution) and explain your reasoning.

Read the original paper: Raoult, D., et al. 2004. The 1.2 megabase genome sequence of *Mimivirus*. *Science* 306:1344.

Concept Mapping Challenge

Construct a concept map that describes the five stages of viral life cycles and illustrates some specific examples of how that stage is accomplished. Use the stages of the life cycle, any other concepts or terms you need, and your own linking words between each pair of concepts in your map. ▶▶I Concept mapping (Appendix III)

Learn More

Learn more by visiting the text website at www.mhhe.com/willey8, where you will find a complete list of references.

6

Microbial Nutrition

Beautiful snowflakelike colonies produced by Bacillus subtilis *when grown on nutrient-poor agar.*

CHAPTER GLOSSARY

active transport The transport of solute molecules across a membrane against a gradient; it requires a carrier protein and the input of energy.

agar A complex sulfated polysaccharide that is used as a solidifying agent in the preparation of culture media.

autotroph An organism that uses CO_2 as its sole or principal source of carbon.

chemolithoautotroph A microorganism that oxidizes reduced inorganic compounds to derive both energy and electrons; CO_2 is the carbon source.

chemolithoheterotroph A microorganism that uses reduced inorganic compounds to derive both energy and electrons; organic molecules are used as the carbon source.

chemoorganoheterotroph An organism that uses organic compounds as sources of energy, electrons, and carbon.

chemotroph An organism that uses chemicals, either organic or inorganic, as its source of energy.

colony An assemblage of microorganisms growing on a solid surface.

facilitated diffusion Diffusion across a membrane through channels created by transmembrane proteins. The solutes move down their concentration gradient, and no cellular energy is required for their movement.

group translocation A transport process in which a molecule is moved across a membrane by carrier proteins while being chemically altered at the same time (e.g., **phosphoenolpyruvate: sugar phosphotransferase system**).

heterotroph An organism that uses reduced, preformed organic molecules as its principal carbon source.

lithotroph An organism that uses reduced inorganic compounds as its electron source.

macroelements Nutrients such as carbon, hydrogen, oxygen, and nitrogen that are required in relatively large amounts; also called macronutrients.

micronutrients Nutrients such as manganese, zinc, and copper that are required in very small amounts; also called **trace elements**.

organotroph An organism that uses reduced organic compounds as its electron source.

permease A membrane-bound carrier protein or a system of two or more proteins that transports a substance across the plasma membrane.

photoautotroph An organism that uses light for energy and CO_2 as its carbon source.

photolithoautotroph An organism that uses light for energy, an inorganic electron source (e.g., H_2O, H_2, H_2S), and CO_2 as its carbon source.

photoorganoheterotroph A microorganism that uses light energy, organic electron sources, and organic molecules as a carbon source.

phototroph An organism that uses light as its source of energy.

pour plate A method by which microbes are inoculated into a molten agar medium so that colonies form within the agar matrix. Commonly used for isolation and enumeration of microbes from natural environments.

pure culture A population of cells that are identical because they arose from a single cell.

siderophore A small molecule that complexes with ferric iron and supplies it to a cell by aiding in its transport across the plasma membrane.

streak plate A technique by which microbes are diluted on an agar plate, using an inoculating loop, so that single colonies will form.

Just as a newborn infant must eat soon after birth, so too must a newborn microbe. If the microbe does not obtain nutrients from its environment, it will quickly exhaust its supply of amino acids, nucleotides, and other molecules needed to survive. In addition, if the microbe is to thrive and reproduce, it must have a source of energy. The energy source is used to generate the cell's energy currency—the high-energy molecule ATP. ATP is used to fuel the various types of work done by the cell, including biosynthesis. The nutrients needed by the microbe provide the elements that are used to construct the molecules of life. Clearly, obtaining and utilizing energy and nutrient sources is one of the most important jobs an organism has.

In this chapter, we describe the nutritional requirements of microorganisms and how nutrients are acquired. This information is critical to the microbiologist trying to cultivate a microbe. Thus we also discuss the types of media used to cultivate microbes in the laboratory. How microbes make ATP from their energy sources is the focus of chapter 10. The use of nutrients and ATP in biosynthesis is discussed in chapter 11.

6.1 Elements of Life

Organisms are composed of a variety of elements called **macroelements** or macronutrients because they are required in relatively large amounts. The macroelements include carbon, oxygen, hydrogen, nitrogen, sulfur, and phosphorus, which are found in organic molecules such as proteins, lipids, nucleic acids, and carbohydrates (**table 6.1**). Other macroelements are potassium, calcium, magnesium, and iron. They exist as cations and play a variety of roles. For example, potassium (K^+) is required for activity by a number of enzymes, including some involved in protein synthesis. Calcium (Ca^{2+}), among other functions, contributes to the heat resistance of bacterial endospores. Magnesium (Mg^{2+}) serves as a cofactor for many enzymes, complexes with ATP, and stabilizes ribosomes and cell membranes. Iron (Fe^{2+} and Fe^{3+}) is a part of some molecules involved in the synthesis of ATP by electron transport-related processes. ▶▶| *Enzymes (section 9.7); Electron transport chains (section 9.6)*

In addition to macroelements, all microorganisms require several nutrients in small amounts—amounts so small that in the lab they are often obtained as contaminants in water, glassware, and growth media. Likewise in nature, they are ubiquitous and usually present in adequate amounts to support the growth of microbes. These nutrients are called **micronutrients** or **trace elements.** The micronutrients—manganese, zinc, cobalt, molybdenum, nickel, and copper—are needed by most cells. Micronutrients are a part of certain enzymes and cofactors, and they aid in the catalysis of reactions and maintenance of protein structure. For example, zinc (Zn^{2+}) is present at the active site of some enzymes but can also be involved in the association of different subunits of a multimeric protein. Manganese (Mn^{2+}) aids many enzymes that catalyze the transfer of phosphate. Molybdenum (Mo^{2+}) is required for nitrogen fixation, and cobalt (Co^{2+}) is a component of vitamin B_{12}. ▶▶| *Proteins (appendix I); Nitrogen fixing bacteria (section 29.3)*

Table 6.1 Elemental Makeup of Important Cell Molecules

Molecule	*Elements*
Proteins	C, H, O, N, S
Lipids	C, H, O, P
Carbohydrates	C, H, O
Nucleic acids	C, H, O, N, P

Besides the common macroelements and trace elements, some microorganisms have particular requirements that reflect their specific morphology or metabolic capabilities. For instance, diatoms, members of the eukaryotic taxon *Stramenopila,* need silicic acid (H_4SiO_4) to construct their beautiful cell walls of silica [$(SiO_2)_n$] (*see figure 23.16*). But no matter what their nutritional requirements, microbes require a balanced mixture of nutrients. If an essential nutrient is in short supply, microbial growth will be limited regardless of the concentrations of other nutrients. The importance of the most critical macroelements is discussed next. ▶▶| Stramenopila *(section 23.5)*

6.2 Carbon, Hydrogen, Oxygen, and Electrons

All organisms need carbon, hydrogen, oxygen, and a source of electrons. Carbon is needed to synthesize the organic molecules from which organisms are built. Hydrogen and oxygen are also important elements found in many organic molecules. Electrons are needed for two reasons. As we describe more completely in chapter 10, the movement of electrons through electron transport chains and during other oxidation-reduction reactions can provide energy for use in cellular work. Electrons also are needed to reduce molecules during biosynthesis (e.g., the reduction of CO_2 to form organic molecules). ▶▶| *Oxidation-reduction reactions (section 9.5); CO_2 fixation (section 11.3)*

The requirements for carbon, hydrogen, and oxygen usually are satisfied together because molecules serving as carbon sources often contribute hydrogen and oxygen as well. For instance, many **heterotrophs**—organisms that use reduced, preformed organic molecules as their carbon source—can also obtain hydrogen, oxygen, and electrons from the same molecules (**table 6.2**). Because the electrons provided by these organic carbon sources can be used in electron transport as well as in other oxidation-reduction reactions, many heterotrophs also use their carbon source as an energy source. Indeed, the more reduced the organic carbon source (i.e., the more electrons it carries), the higher its energy

Table 6.2 Sources of Carbon, Energy, and Electrons

Carbon Sources	
Autotrophs	CO_2 sole or principal biosynthetic carbon source (*section 11.3*)
Heterotrophs	Reduced, preformed, organic molecules from other organisms (*chapters 10 and 11*)
Energy Sources	
Phototrophs	Light (*section 10.12*)
Chemotrophs	Oxidation of organic or inorganic compounds (*chapter 10*)
Electron Sources	
Lithotrophs	Reduced inorganic molecules (*section 10.11*)
Organotrophs	Organic molecules (*chapters 9 and 10*)

content. Thus lipids have a higher energy content than carbohydrates. ▶▶| *Carbohydrates (appendix I); Lipids (appendix I)*

A most remarkable characteristic of heterotrophic microorganisms is their extraordinary flexibility with respect to carbon sources. Laboratory experiments indicate that all naturally occurring organic molecules can be used as a source of carbon or energy or both by at least some microorganisms. Actinomycetes, common soil bacteria, degrade amyl alcohol, paraffin, and even rubber. The bacterium *Burkholderia cepacia* can use over 100 different carbon compounds. Microbes can degrade even relatively indigestible human-made substances such as pesticides. This is usually accomplished in complex microbial communities. These molecules sometimes are degraded in the presence of a growth-promoting nutrient that is metabolized at the same time—a process called cometabolism. Other microorganisms can use the products of this breakdown process as nutrients. In contrast to these bacterial omnivores, some microbes are exceedingly fastidious and catabolize only a few carbon compounds. Cultures of methylotrophic bacteria metabolize methane, methanol, carbon monoxide, formic acid, and related one-carbon molecules. Parasitic members of the genus *Leptospira* use only long-chain fatty acids as their major source of carbon and energy. ▶▶| *Biodegradation and bioremediation by natural communities (section 42.3)*

Other microbes are **autotrophs**—organisms that use carbon dioxide (CO_2) as their sole or principal source of carbon (table 6.2). Although CO_2 is plentiful, its use as a carbon source presents a problem to autotrophs. CO_2 is the most oxidized form of carbon, lacks hydrogen, and is unable to donate electrons during oxidation-reduction reactions. Therefore CO_2 cannot be used as a source of hydrogen, electrons, or energy. Because CO_2 cannot supply their energy needs, autotrophs must obtain energy from other sources, such as light or reduced inorganic molecules.

1. What are nutrients? On what basis are they divided into macroelements and trace elements?
2. What are the six most important macroelements? How do cells use them?
3. List two trace elements. How do cells use them?
4. Define heterotroph and autotroph. Which one are you? How do you know?

6.3 Nutritional Types of Microorganisms

Because the need for carbon, energy, and electrons is so important, biologists use specific terms to define how these requirements are fulfilled. We have already seen that microorganisms can be classified as either heterotrophs or autotrophs with respect to their preferred source of carbon (table 6.2). Only two sources of energy are available to organisms: (1) light energy, and (2) energy derived from oxidizing organic or inorganic molecules.

Phototrophs use light as their energy source; **chemotrophs** obtain energy from the oxidation of chemical compounds (either organic or inorganic). Microorganisms also have only two sources for electrons. **Lithotrophs** (i.e., "rock-eaters") use reduced inorganic substances as their electron source, whereas **organotrophs** extract electrons from reduced organic compounds.

Despite the great metabolic diversity seen in microorganisms, most may be placed in one of five nutritional classes based on their primary sources of carbon, energy, and electrons (**table 6.3**). The majority of microorganisms thus far studied are either photolithoautotrophic or chemoorganoheterotrophic.

Photolithoautotrophs (often called simply **photoautotrophs**) use light energy and have CO_2 as their carbon source. Photosynthetic protists and cyanobacteria employ water as the electron donor and release oxygen (**figure 6.1*a***). Other photolithoautotrophs, such as the purple sulfur bacteria and the green sulfur bacteria (figure 6.1*b*), cannot oxidize water but extract electrons from inorganic donors such as hydrogen, hydrogen sulfide, and elemental sulfur. Photoautotrophs are important primary producers in ecosystems. That is, they convert light energy into chemical energy that can sustain the chemoorganoheterotrophs that share their habitats.

Chemoorganoheterotrophs (sometimes called **chemoheterotrophs** or chemoorganotrophs) use organic compounds as sources of energy, hydrogen, electrons, and carbon. Frequently the same organic nutrient will satisfy all these requirements. Chemoorganotrophs contribute to biogeochemical cycles such as the carbon cycle and nitrogen cycle, in which elements are converted into different forms. In addition, they are of considerable practical importance. Many chemoorganotrophs are used industrially to make foods (e.g., yogurt, pickles, cheese), medical products (e.g., antibiotics), and beverages (e.g., beer and wine). Nearly all pathogenic microorganisms are chemoorganoheterotrophs. ▶▶| *Biogeochemical cycling (section 26.1)*

Table 6.3 Major Nutritional Types of Microorganisms

Nutritional Type	*Carbon Source*	*Energy Source*	*Electron Source*	*Representative Microorganisms*
Photolithoautotroph	CO_2	Light	Inorganic e^- donor	Purple and green sulfur bacteria, cyanobacteria, diatoms
Photoorganoheterotroph	Organic carbon	Light	Organic e^- donor	Purple nonsulfur bacteria, green nonsulfur bacteria
Chemolithoautotroph	CO_2	Inorganic chemicals	Inorganic e^- donor	Sulfur-oxidizing bacteria, hydrogen-oxidizing bacteria, methanogens, nitrifying bacteria, iron-oxidizing bacteria
Chemolithoheterotroph	Organic carbon	Inorganic chemicals	Inorganic e^- donor	Some sulfur-oxidizing bacteria (e.g., *Beggiatoa*)
Chemoorganoheterotroph	Organic carbon	Organic chemicals, often same as C source	Organic e^- donor, often same as C source	Most nonphotosynthetic microbes, including most pathogens, fungi, and many protists and archaea

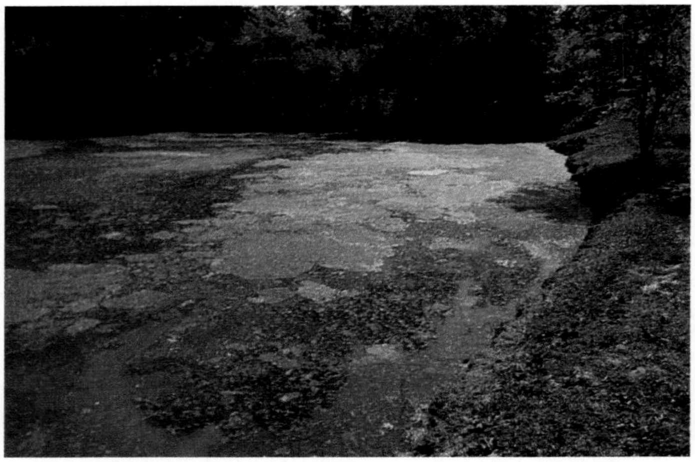

(a) Bloom of cyanobacteria (photolithoautotrophic bacteria)

(b) Purple sulfur bacteria (photolithoautotrophs)

FIGURE 6.1 Phototrophic Bacteria. Phototrophic microbes play important roles in aquatic ecosystems, where they can cause blooms. (a) A cyanobacterial bloom in a eutrophic pond. (b) Purple sulfur bacteria growing in a bog.

The other nutritional types have fewer known microorganisms but are very important ecologically. Some photosynthetic bacteria (purple nonsulfur and green bacteria) use organic matter as their electron donor and carbon source. These **photoorganoheterotrophs** are common inhabitants of polluted lakes and streams. **Chemolithoautotrophs** oxidize reduced inorganic compounds such as iron, nitrogen, or sulfur molecules to derive both energy and electrons for biosynthesis (**figure 6.2*a***). Carbon dioxide is the carbon source. **Chemolithoheterotrophs** use reduced inorganic molecules as their energy and electron source but derive their carbon from organic sources (figure 6.2*b*). Chemolithotrophs contribute greatly to the chemical transformations of elements (e.g., the conversion of ammonia to nitrate or sulfur to sulfate) that continually occur in ecosystems. ▶▶| *Photosynthetic bacteria (section 19.3); Nitrifying bacteria (section 20.1)*

Although we have sorted microbes into a particular nutritional type, microbes are not always so easily categorized. Some show great metabolic flexibility and alter their metabolism in response to environmental changes. For example, many purple nonsulfur bacteria act as photoorganoheterotrophs in the absence of oxygen but oxidize organic molecules and function chemoorganotrophically at normal oxygen levels. When oxygen is low, phototrophic and chemoorganotrophic metabolism may function simultaneously. This allows the organism to gain energy from both light and organic molecules, and still supply the carbon it needs for biosynthesis. Some of these bacteria also can grow as photolithoautotrophs with molecular hydrogen as an electron

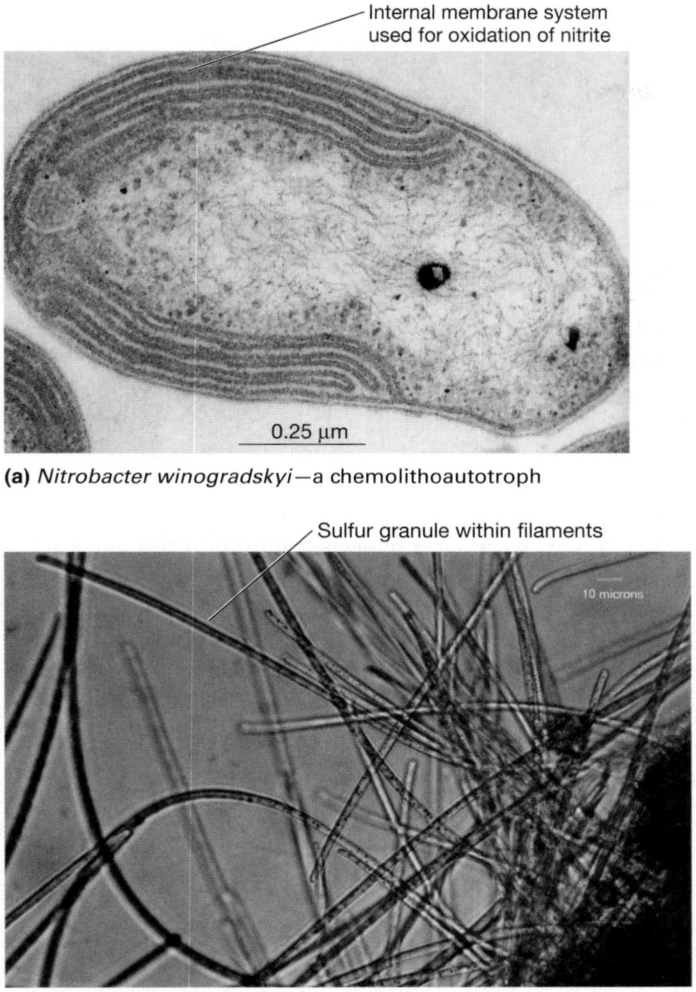

(a) *Nitrobacter winogradskyi*—a chemolithoautotroph

(b) *Beggiatoa alba*—a chemolithoheterotroph

FIGURE 6.2 Chemolithotrophic Bacteria. (a) Transmission electron micrograph of *Nitrobacter winogradskyi,* an organism that uses nitrite as its source of energy (×213,000). (b) Light micrograph of *Beggiatoa alba,* an organism that uses hydrogen sulfide as its energy source and organic molecules as carbon sources.

donor. This sort of flexibility gives these microbes a distinct advantage if environmental conditions frequently change.

1. Discuss the ways in which organisms are classified based on their requirements for energy, carbon, and electrons.
2. Describe the nutritional requirements of the major nutritional groups and give some microbial examples of each.
3. Compare photolithoautotrophy with chemolithoautotrophy. Do you think it is possible for an ecosystem to exist solely on the organic carbon generated by chemolithotrophy (i.e., without any contribution by photosynthetic organisms)? Explain your reasoning.

6.4 Nitrogen, Phosphorus, and Sulfur

To grow and reproduce, a microorganism must be able to incorporate large quantities of nitrogen, phosphorus, and sulfur. Although these elements may be acquired from the same nutrients that supply carbon, microorganisms usually employ inorganic sources as well.

Nitrogen is needed for the synthesis of amino acids, purines, pyrimidines, some carbohydrates and lipids, enzyme cofactors, and other substances. Many microorganisms can use the nitrogen in amino acids as their source of N in other macromolecules. Others can incorporate ammonia directly through the action of enzymes such as glutamate dehydrogenase or glutamine synthetase and glutamate synthase (*see figures 11.16 to 11.18*). Most phototrophs and many chemotrophic microorganisms reduce nitrate to ammonia and incorporate the ammonia in a process known as assimilatory nitrate reduction. A variety of bacteria (e.g., many cyanobacteria and the symbiotic bacterium *Rhizobium*) can assimilate atmospheric nitrogen (N_2) by reducing it to ammonia (NH_3). This is called nitrogen fixation. ▶▶| *Synthesis of amino acids (section 11.5); Nitrogen cycle (section 26.1)*

Phosphorus is present in nucleic acids, phospholipids, nucleotides such as ATP, several cofactors, some proteins, and other cell components. Almost all microorganisms use inorganic phosphate as their phosphorus source and incorporate it directly. Low phosphate levels can limit microbial growth in aquatic environments. Some microbes, such as *Escherichia coli,* use both organic and inorganic phosphate. Some organophosphates such as hexose 6-phosphates are taken up directly by the cell. Other organophosphates are hydrolyzed in the periplasm to produce inorganic phosphate, which then is transported across the plasma membrane. ▶▶| *Synthesis of purines, pyrimidines, and nucleotides (section 11.6)*

Sulfur is needed for the synthesis of substances such as the amino acids cysteine and methionine, some carbohydrates, biotin, and thiamine. Most microorganisms use sulfate as a source of sulfur after reducing it; a few microorganisms require an organic form of sulfur such as the amino acid cysteine.

1. Briefly describe how microorganisms use the various forms of nitrogen, phosphorus, and sulfur.
2. Why do you think ammonia (NH_3) can be directly incorporated into amino acids, whereas other forms of combined nitrogen (e.g., NO_2^- and NO_3^-) are not?

6.5 Growth Factors

Some microbes are able to synthesize all organic molecules from a single carbon source and inorganic salts. Others are more metabolically limited and require specific **growth factors** to support growth. Thus growth factors are organic compounds

Table 6.4 Functions of Some Common Vitamins in Microorganisms

Vitamin	Functions
Biotin	Carboxylation (CO_2 fixation) One-carbon metabolism
Cyanocobalamin (B_{12})	Molecular rearrangements One-carbon metabolism—carries methyl groups
Folic acid	One-carbon metabolism
Lipoic acid	Transfer of acyl groups
Pantothenic acid	Precursor of coenzyme A—carries acyl groups (pyruvate oxidation, fatty acid metabolism)
Pyridoxine (B_6)	Amino acid metabolism (e.g., transamination)
Niacin (nicotinic acid)	Precursor of NAD^+ and $NADP^+$—carry electrons and hydrogen atoms
Riboflavin (B_2)	Precursor of FAD and FMN—carry electrons or hydrogen atoms
Thiamine (B_1)	Aldehyde group transfer (pyruvate decarboxylation, α-keto acid oxidation)

that cannot be synthesized by an organism but are essential for its growth. The organism must obtain the needed growth factors from the environment. There are three major classes of growth factors: (1) amino acids, (2) purines and pyrimidines, and (3) vitamins. Amino acids are used for protein synthesis, and purines and pyrimidines for nucleic acid synthesis. **Vitamins** are small organic molecules that usually make up all or part of enzyme cofactors. They are needed in only very small amounts to sustain growth. The functions of selected vitamins are given in **table 6.4.** Some microorganisms require many vitamins; for example, *Enterococcus faecalis* needs eight different vitamins for growth. Other growth factors include heme (for the synthesis of cytochromes), which is required by *Haemophilus influenzae,* and cholesterol, which is needed by some mycoplasmas. ▶▶| *Enzymes (section 9.7)*

Some microorganisms are able to synthesize large quantities of the molecules that are the vitamins acquired by humans in their diets. We can use these microbes to manufacture these vitamins for dietary supplements. Several water-soluble and fat-soluble vitamins are industrially produced partly or completely using microbes. Examples are riboflavin (*Clostridium, Candida*), coenzyme A (*Brevibacterium*), vitamin B_{12} (*Streptomyces, Propionibacterium, Pseudomonas*), vitamin C (*Gluconobacter, Erwinia, Corynebacterium*), β-carotene (*Dunaliella*), and vitamin D (*Saccharomyces*). Current research focuses on improving yields and finding microorganisms that can produce large quantities of other vitamins. ▶▶| *Major products of industrial microbiology (section 41.3)*

1. What are growth factors? What are vitamins?
2. List the growth factors that microorganisms produce industrially.
3. Why do you think amino acids, purines, and pyrimidines are often growth factors, whereas glucose is not?

6.6 Uptake of Nutrients

The first step in nutrient use is their uptake by the microbial cell. Microbes can only take in dissolved molecules. Uptake mechanisms must be specific—that is, the necessary substances, and not others, must be acquired. It does a cell no good to take in a substance that it cannot use. Because microorganisms often live in nutrient-poor habitats, they must be able to transport nutrients from dilute solutions into the cell against a concentration gradient. Finally, nutrient molecules must pass through a selectively permeable plasma membrane that prevents the free passage of most substances. In view of the enormous variety of nutrients and the complexity of the task, it is not surprising that microorganisms make use of several different transport mechanisms. Members of *Bacteria* and *Archaea* use facilitated diffusion, active transport, and group translocation for nutrient uptake. Eukaryotic microorganisms use facilitated diffusion, active transport, and endocytosis. They do not employ group translocation. |◀◀ *Organelles of the secretory and endocytic pathways (section 4.4)*

Passive Diffusion

Passive diffusion, often called diffusion or simple diffusion, is the process by which molecules move from a region of higher concentration to one of lower concentration. The rate of passive diffusion depends on the size of the concentration gradient between a cell's exterior and its interior (**figure 6.3**). A large concentration gradient is required for adequate nutrient uptake by passive diffusion (i.e., the external nutrient concentration must be high while the internal concentration is low). Unless the nutrient is used immediately upon entry, the rate of diffusion decreases as more nutrient accumulates in the cell.

Cells must be able to take up nutrients from their environment, but it is also imperative that the plasma membrane be impermeable to most substances. If the membrane were freely permeable to all substances, then the cell would not be able to concentrate needed nutrients in the cytoplasm. It also would not be able to maintain other important concentration differences across the membrane. However, some metabolic processes require that certain gases easily diffuse across the plasma membrane, including O_2 and CO_2, which are able to enter the cell by passive diffusion. H_2O also moves across membranes by passive diffusion. This is important because it allows the cell to adjust to differences in solute concentrations. Thus the only molecules that enter the cell by passive diffusion are certain gases and small molecules such as H_2O. Larger molecules, ions, and polar substances must enter the cell by other mechanisms. *How Diffusion Works*

Facilitated Diffusion

The rate of diffusion across selectively permeable membranes is greatly increased by using carrier proteins, sometimes called **permeases,** which are embedded in the plasma membrane and create channels through which the substance passes. Diffusion involving carrier proteins is called **facilitated diffusion.** The rate of facilitated diffusion increases with the concentration gradient much more rapidly and at lower concentrations of the diffusing molecule than that of passive diffusion (figure 6.3). Note that the diffusion rate reaches a plateau above a specific gradient value because the carrier protein is saturated—that is, it is transporting as many solute molecules as possible. The resulting curve resembles an enzyme-substrate curve (*see figure 9.18*) and is different from the linear response seen with passive diffusion. Permeases also resemble enzymes in their specificity for the substance to be transported; each permease is selective and transports only closely related solutes. Some permeases are related to the major intrinsic protein (MIP) family of proteins. MIPs facilitate diffusion of small polar molecules. They are observed in virtually all organisms. The two most widespread MIP channels in bacteria are aquaporins (*see figure 2.31*), which transport water, and glycerol facilitators, which aid glycerol diffusion.

Although a carrier protein is involved, facilitated diffusion is truly diffusion. A concentration gradient spanning the membrane drives the movement of molecules, and no metabolic energy input is required. If the concentration gradient disappears, net inward movement ceases. The gradient can be maintained by transforming the transported nutrient to another compound. Eukaryotic cells can maintain a gradient by moving the nutrient to another membranous compartment.

Although much work has been done on the mechanism of facilitated diffusion, the process is not understood completely. After the solute molecule binds to the outside, the carrier is thought to change conformation and release the molecule on the cell interior (**figure 6.4**). The carrier subsequently changes back to its original shape and is ready to pick up another molecule. The net effect is that a hydrophilic molecule can enter the cell in response to its concentration gradient. Remember that the mechanism is driven by concentration gradients and therefore is reversible. If the solute's concentration is greater inside the cell, it will move outward. However, because the cell metabolizes nutrients upon entry, influx is favored.

Facilitated diffusion processes have been documented in some bacteria and archaea. However, facilitated diffusion does not seem to be the major uptake mechanism for these microbes.

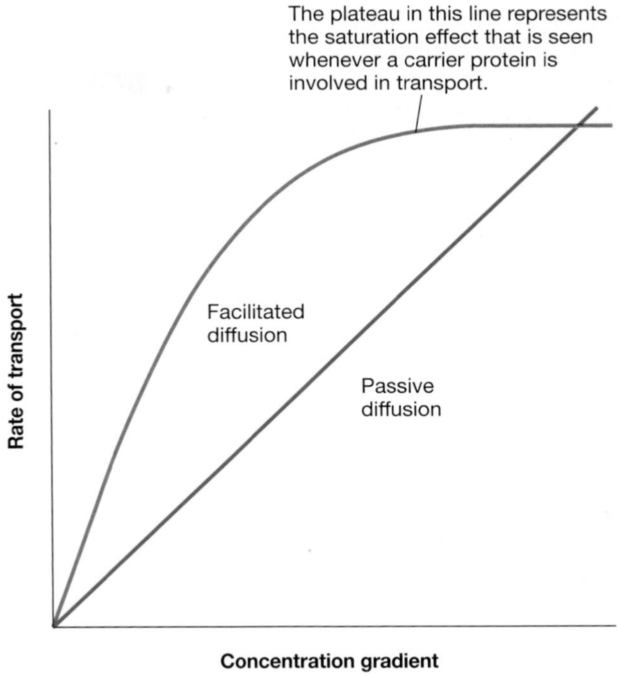

FIGURE 6.3 Passive and Facilitated Diffusion. The dependence of diffusion rate on the size of the solute's concentration gradient (the ratio of the extracellular concentration to the intracellular concentration).

Figure 6.3 Micro Inquiry

What happens to the slope of the passive diffusion curve when the concentration of solute is equal on both sides of the microbe's plasma membrane?

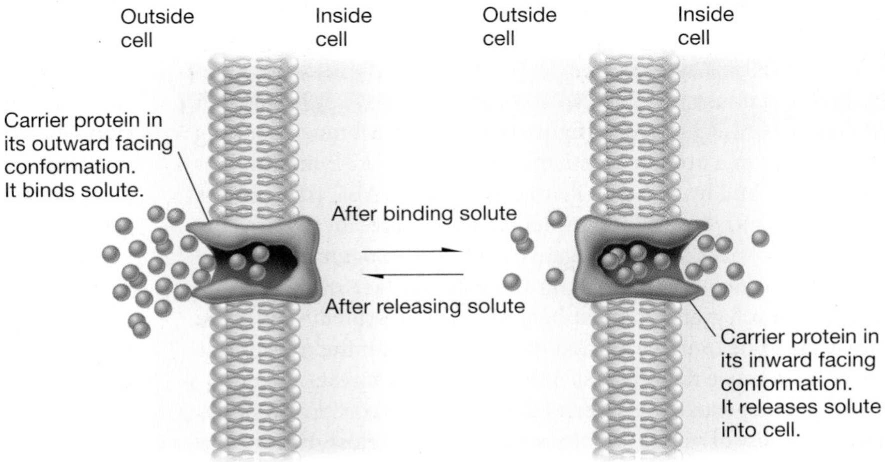

FIGURE 6.4 A Model of Facilitated Diffusion. Because there is no energy input, molecules continue to enter only as long as their concentration is greater on the outside.

This is because many bacteria and archaea live in environments where nutrient concentrations are low, and facilitated diffusion cannot concentrate the nutrients inside the cells. Facilitated diffusion is much more prominent in eukaryotic cells, where it is used to transport a variety of sugars and amino acids.

Active Transport

Because facilitated diffusion can efficiently move molecules to the interior only when the solute concentration is higher on the outside of the cell, microbes must have transport mechanisms that can move solutes against a concentration gradient. Microbes use two important transport processes in such situations: active transport and group translocation. Both are energy-dependent processes.

Active transport is the transport of solute molecules to higher concentrations (i.e., against a concentration gradient) with the input of metabolic energy. Because active transport involves transport proteins, it resembles facilitated diffusion in some ways. The transport proteins bind particular solutes with great specificity. Similar solute molecules can compete for the same carrier protein in both facilitated diffusion and active transport. Active transport is also characterized by the carrier saturation effect at high solute concentrations (figure 6.3). Nevertheless, active transport differs from facilitated diffusion in its use of metabolic energy and its ability to concentrate substances. Metabolic inhibitors that block energy production inhibit active transport but do not immediately affect facilitated diffusion.

Active transport proteins are divided into two types: primary transporters and secondary transporters. Primary active transporters use the energy provided by ATP hydrolysis to move substances against a concentration gradient. Secondary active transporters couple the potential energy of ion gradients to transport of substances.

ATP-*b*inding *c*assette transporters (ABC transporters) are important primary active transporters. They are observed in *Bacteria, Archaea,* and eukaryotes. Some are used for import of substances (observed in *Bacteria* and *Archaea*), and others are used for export of substances (observed in all three domains of life). Usually ABC transporters consist of two hydrophobic membrane-spanning domains associated on their cytoplasmic surfaces with two ATP-binding domains (**figure 6.5**). The membrane-spanning domains form a pore in the membrane, and the ATP-binding domains bind and hydrolyze ATP to drive uptake. ABC transporters employ substrate-binding proteins, which are located in the periplasmic space of gram-negative bacteria (*see figure 3.19*) or are attached to membrane lipids on the external face of the plasma membrane of gram-positive bacteria. These proteins bind the molecule to be transported and then interact with the transporter proteins to move the molecule into the cell. Because a single molecule is transported, this is termed uniport transport. *E. coli* transports a variety of sugars (arabinose, maltose, galactose, ribose) and amino acids (glutamate, histidine, leucine) by this mechanism.

Recall that gram-negative bacteria have an outer membrane in addition to the plasma membrane. Thus substances entering gram-negative bacteria must pass through the outer membrane

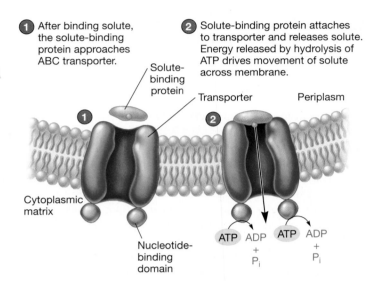

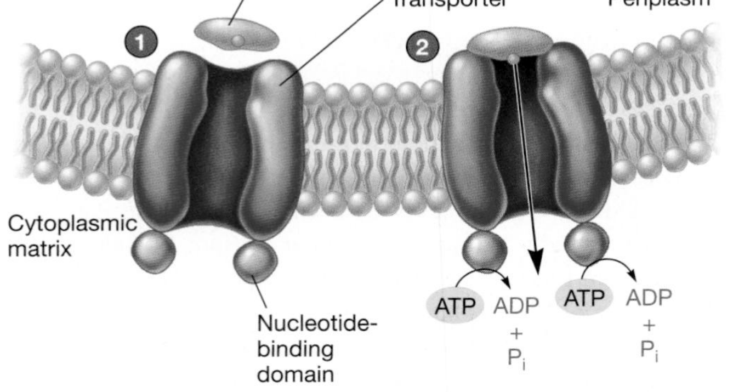
1 After binding solute, the solute-binding protein approaches ABC transporter.
2 Solute-binding protein attaches to transporter and releases solute. Energy released by hydrolysis of ATP drives movement of solute across membrane.
Solute-binding protein
Transporter
Periplasm
1
2
Cytoplasmic matrix
Nucleotide-binding domain
ATP
ADP + Pi
ATP
ADP + Pi

FIGURE 6.5 ABC Transporter Function.

before ABC transporters and other active transport systems can take action. This is accomplished in several ways. When the substance is small, a porin protein such as OmpF (outer membrane protein) can be used (*see figure 3.21*). The transport of larger molecules, such as vitamin B_{12}, requires the use of specialized, high-affinity outer-membrane receptors that function in association with specific transporters in the plasma membrane.

Secondary active transporters include the major facilitator superfamily (MFS) proteins. MFS transporters use ion gradients, some of which are created by microbes during their metabolic processes. For instance, electron transport during energy-conserving processes generates a proton gradient (in members of the *Bacteria* and *Archaea*, the protons are at a higher concentration outside the cell than inside). The proton gradient can be used to do cellular work, including secondary active transport.

The uptake of lactose by the lactose permease of *E. coli* is a well-studied example of a secondary active transporter. The lactose permease is a single protein that transports a lactose molecule inward as a proton simultaneously enters the cell. The proton is moving down a proton gradient, and the energy released drives solute transport. Such linked transport of two substances in the same direction is called **symport** (**figure 6.6**). Although the mechanism of lactose symport is not completely understood, X-ray diffraction studies show that the transport protein exists in outward- and inward-facing conformations. When lactose and a proton bind to separate sites on the outward-facing conformation, the protein changes to its inward-facing conformation. Then the sugar and proton are released into the cytoplasm. *E. coli* also uses proton symport to take up amino acids and organic acids such as succinate and malate.
▶▶| *Electron transport and oxidative phosphorylation (section 10.5)*

A proton gradient also can power secondary active transport indirectly, often through the formation of a sodium ion gradient. For example, an *E. coli* sodium transport system pumps sodium outward in response to the inward movement of protons. Such linked transport in which the transported substances move in

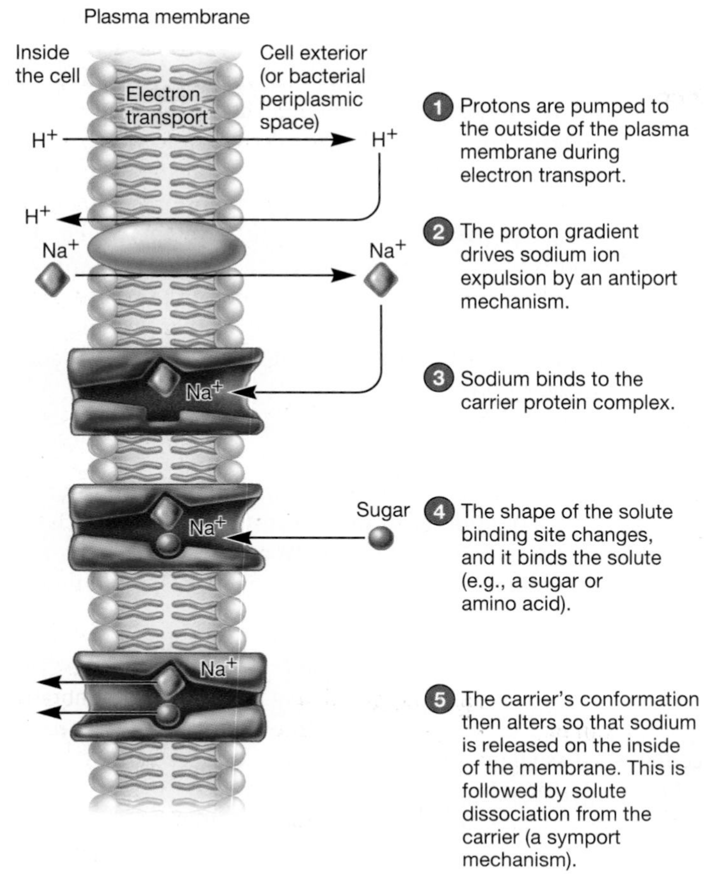

Plasma membrane
Inside the cell
Cell exterior (or bacterial periplasmic space)
Electron transport
H+
H+
H+
Na+
Na+
Na+
Na+
Sugar
Na+
1 Protons are pumped to the outside of the plasma membrane during electron transport.
2 The proton gradient drives sodium ion expulsion by an antiport mechanism.
3 Sodium binds to the carrier protein complex.
4 The shape of the solute binding site changes, and it binds the solute (e.g., a sugar or amino acid).
5 The carrier's conformation then alters so that sodium is released on the inside of the membrane. This is followed by solute dissociation from the carrier (a symport mechanism).

FIGURE 6.6 Active Transport Using Proton and Sodium Gradients.

Figure 6.6 Micro Inquiry

Could a proton also be used to cotransport an uncharged substrate in a manner analogous to that shown for Na^+?

opposite directions is termed **antiport** (figure 6.6). The sodium gradient generated by this proton antiport system drives the uptake of sugars and amino acids. It is thought that a sodium ion attaches to a carrier protein, causing it to change shape. The carrier then binds the sugar or amino acid tightly and orients its binding sites toward the cell interior. Because of the low intracellular sodium concentration, the sodium ion dissociates from the carrier, and the other molecule follows. *E. coli* transport proteins carry the sugar melibiose and the amino acid glutamate when sodium simultaneously moves inward. Sodium symport also is an important process in eukaryotic cells, where it is used in sugar and amino acid uptake. However, ATP, rather than proton motive force, usually drives sodium transport in eukaryotic cells. *Cotransport (Symport and Antiport)*

Microorganisms often have more than one transport system for a nutrient, as can be seen with *E. coli.* This bacterium has at least five transport systems for the sugar galactose, three systems each for the amino acids glutamate and leucine, and two potassium transport complexes. When several transport systems exist for the same substance, the systems differ in such properties as their energy source, their affinity for the solute transported, and the nature of their regulation. This diversity gives the microbe an added competitive advantage in a variable environment.

Group Translocation

In active transport, solute molecules move across a membrane without modification. Another type of energy-dependent transport, called **group translocation,** chemically modifies the molecule as it is brought into the cell. The best-known group translocation system is the **phosphoenolpyruvate: sugar *phos*pho*trans*ferase *s*ystem (PTS),** which is observed in many bacteria. The PTS transports a variety of sugars while phosphorylating them, using phosphoenolpyruvate (PEP) as the phosphate donor.

$$\text{PEP} + \text{sugar (outside)} \rightarrow \text{pyruvate} + \text{sugar-phosphate (inside)}$$

PEP is an important intermediate of a biochemical pathway used by many chemoorganoheterotrophs to extract energy from organic energy sources. PEP is a high-energy molecule that can be used to synthesize ATP, the cell's energy currency. However, when it is used in PTS reactions, the energy present in PEP is used to energize uptake rather than ATP synthesis. ▶▶| *ATP (section 9.4); Glycolytic pathways (section 10.3)*

The transfer of phosphate from PEP to the incoming molecule involves several proteins and is an example of a **phosphorelay system.** In *E. coli* and *Salmonella,* the PTS consists of two enzymes and a low molecular weight heat-stable protein (HPr). A phosphate is transferred from PEP to enzyme II with the aid of enzyme I and HPr (**figure 6.7**). Then a sugar molecule is phosphorylated as it is carried across the membrane by enzyme II. Enzyme II transports only specific sugars and varies with the PTS, whereas enzyme I and HPr are common to all PTSs. ▶▶| *Enzymes (section 9.7)*

PTSs are widely distributed in bacteria. Most members of the genera *Escherichia, Salmonella,* and *Staphylococcus,* as well as many other facultatively anaerobic bacteria (bacteria that grow in either the presence or absence of O_2), have PTSs; some obligately anaerobic bacteria (e.g., *Clostridium*) also have PTSs. However, most aerobic bacteria lack PTSs. Many carbohydrates are transported by PTSs. *E. coli* takes up glucose, fructose, mannitol, sucrose, *N*-acetylglucosamine, cellobiose, and other carbohydrates by group translocation. Besides their role in transport, PTS proteins can bind chemical attractants, toward which bacteria move by the process of chemotaxis. ▶▶| *Oxygen concentration (section 7.6); Motility and chemotaxis (section 3.7)* *Active Transport by Group Translocation*

Iron Uptake

Almost all microorganisms require iron for use in cytochromes and many enzymes. Iron uptake is made difficult by the extreme insolubility of ferric iron (Fe^{3+}) and its derivatives, which leaves

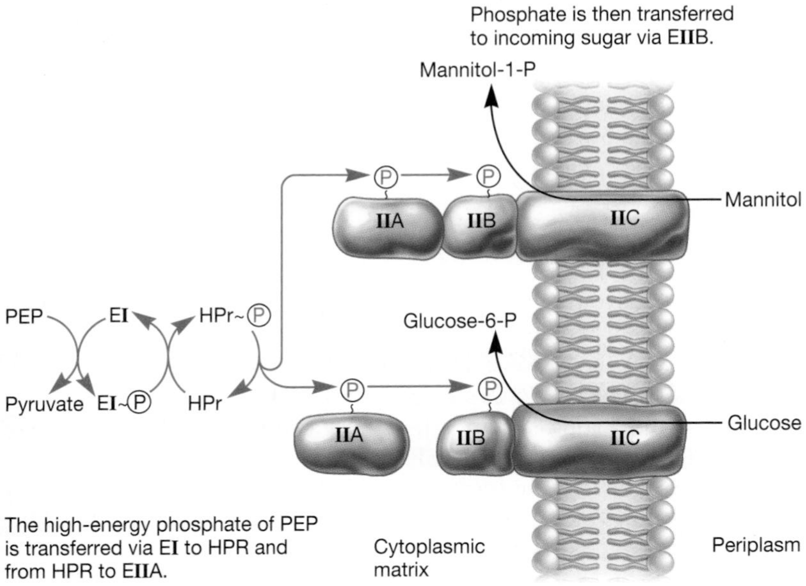

FIGURE 6.7 Group Translocation: Bacterial PTS Transport. Two examples of the phosphoenolpyruvate: sugar phosphotransferase system (PTS) are illustrated. The following components are involved in the system: phosphoenolpyruvate (PEP), enzyme I (EI), the low molecular weight heat-stable protein (HPr), and enzyme II (EII). EIIA is attached to EIIB in the mannitol transport system and is separate from EIIB in the glucose system.

***Figure 6.7* Micro Inquiry**

Why is this system considered an example of a phosphorelay?

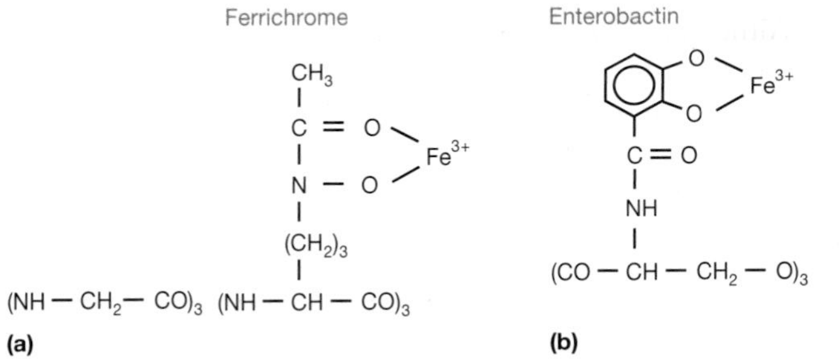

FIGURE 6.8 Siderophore Ferric Iron Complexes. (a) Ferrichrome is a cyclic hydroxamate [—CO—N(O⁻)—] molecule formed by many fungi. (b) *E. coli* produces the cyclic catecholate derivative, enterobactin.

little free iron available for transport. Many bacteria and fungi have overcome this difficulty by secreting **siderophores** (Greek for iron bearers). Siderophores are low molecular weight organic molecules that bind ferric iron and supply it to the cell. Two examples of siderophores are ferrichrome, which is produced by many fungi, and enterobactin, which is formed by *E. coli* (**figure 6.8**).

Microorganisms secrete siderophores when iron is scarce in the medium. Once the iron-siderophore complex has reached the cell surface, it binds to a siderophore-receptor protein. Then either the iron is released to enter the cell directly or the whole iron-siderophore complex is transported inside by an ABC transporter. In *E. coli,* the siderophore receptor is in the outer membrane of the cell envelope; when the iron reaches the periplasmic space, it moves through the plasma membrane with the aid of the transporter. Iron is so crucial to microorganisms that they may use more than one route of iron uptake to ensure an adequate supply.

1. Describe facilitated diffusion, primary and secondary active transport, and group translocation in terms of their distinctive characteristics and mechanisms. What advantage does a microbe gain by using active transport rather than facilitated diffusion?
2. What are uniport, symport, and antiport?
3. What two mechanisms allow the passage of nutrients across the outer membrane of gram-negative bacteria before they are actively transported across the plasma membrane?
4. Why might a microbe have more than one uptake system for certain substrates?
5. What is the difference between an ABC transporter and a porin protein in terms of function and cellular location?
6. What are siderophores? Why are they important?

6.7 Culture Media

Microbiology research depends largely on the ability to grow and maintain microorganisms in the laboratory, and this is possible only if suitable culture media are available. A culture medium is a solid or liquid preparation used to grow, transport, and store microorganisms. To be effective, the medium must contain all the nutrients the microorganism requires for growth. Specialized media are essential in the isolation and identification of microorganisms, the testing of antibiotic sensitivities, water and food analysis, industrial microbiology, and other activities. Although all microorganisms need sources of energy, carbon, nitrogen, phosphorus, sulfur, and various minerals, the precise composition of a satisfactory medium depends on the species being cultivated due to the great variety of nutritional requirements. Knowledge of a microorganism's normal habitat often is useful in selecting an appropriate culture medium because its nutrient requirements reflect its natural surroundings. Frequently a medium is used to select and grow specific microorganisms or to help identify a particular species. Media can also be specifically designed to facilitate the growth of one type of microbe present in a

Table 6.5 Types of Media

Basis for Classification	*Types*
Chemical composition	Defined (synthetic), complex
Physical nature	Liquid, semisolid, solid
Function	Supportive (general purpose), enriched, selective, differential

sample from nature. The resulting culture is called an **enrichment culture,** and it can be used to isolate a species of interest for study in the lab. ▶▶| *Culturing techniques (section 27.1)*

Culture media can be classified based on several parameters: the chemical constituents from which they are made, their physical nature, and their function (**table 6.5**). The types of media defined by these parameters are described here.

Chemical and Physical Types of Culture Media

A medium in which all chemical components are known is a **defined** or **synthetic medium.** It can be in a liquid form (broth) or solidified by an agent such as agar. Defined media are often used to culture photolithoautotrophs such as cyanobacteria and photosynthetic protists. They can be grown on media containing CO_2 as a carbon source (often added as sodium carbonate or bicarbonate), nitrate or ammonia as a nitrogen source, sulfate, phosphate, and other minerals (**table 6.6**). Many chemoorganoheterotrophs also can be grown in defined media with glucose as a carbon source and an ammonium salt as a nitrogen source. Not all defined media are as simple as the examples in table 6.6 but may be constructed from dozens of components. Defined media are used widely in research, as it is often desirable to know exactly what the microorganism is metabolizing.

Media that contain some ingredients of unknown chemical composition are **complex media.** They are very useful, as a single complex medium may be sufficiently rich to meet all the nutritional requirements of many different microorganisms. In addition, complex media often are needed because the nutritional requirements of a particular microorganism are unknown, and thus a defined medium cannot be constructed. Complex media are also used to culture fastidious microbes, microbes with complex nutritional or cultural requirements.

Most complex media contain undefined components such as peptones, meat extract, and yeast extract. Peptones are protein hydrolysates prepared by partial proteolytic digestion of meat, casein, soya meal, gelatin, and other protein sources. They serve as sources of carbon, energy, and nitrogen. Beef extract and yeast extract are aqueous extracts of lean beef and brewer's yeast, respectively. Beef extract contains amino acids, peptides, nucleotides, organic acids, vitamins, and minerals. Yeast extract is an excellent source of B vitamins as well as nitrogen and carbon compounds. Three commonly used complex media are nutrient broth, tryptic soy broth, and MacConkey agar (**table 6.7**).

Table 6.6 Examples of Defined Media

BG–11 Medium for Cyanobacteria	*Amount (g/liter)*
$NaNO_3$	1.5
$K_2HPO_4 \cdot 3H_2O$	0.04
$MgSO_4 \cdot 7H_2O$	0.075
$CaCl_2 \cdot 2H_2O$	0.036
Citric acid	0.006
Ferric ammonium citrate	0.006
EDTA (Na_2Mg salt)	0.001
Na_2CO_3	0.02
Trace metal solution[a]	1.0 ml/liter
Final pH 7.4	
***Medium for* Escherichia coli**	***Amount (g/liter)***
Glucose	1.0
Na_2HPO_4	16.4
KH_2PO_4	1.5
$(NH_4)_2SO_4$	2.0
$MgSO_4 \cdot 7H_2O$	200.0 mg
$CaCl_2$	10.0 mg
$FeSO_4 \cdot 7H_2O$	0.5 mg
Final pH 6.8–7.0	

Sources: Data from Rippka, R., et al. 1979. Journal of General Microbiology, *111:1–61; and Cohen, S. S., and Arbogast, R. 1950.* Journal of Experimental Medicine, *91:619.*

[a] The trace metal solution contains H_3BO_3, $MnCl_2 \cdot 4H_2O$, $ZnSO_4 \cdot 7H_2O$, $Na_2Mo_4 \cdot 2H_2O$, $CuSO_4 \cdot 5H_2O$, and $Co(NO_3)_2 \cdot 6H_2O$.

Although both liquid and solidified media are routinely used, solidified media are particularly important because they can be used to isolate different microbes from each other to establish pure cultures. As we discuss in chapter 1, this is a critical step in demonstrating the relationship between a microbe and a disease using Koch's postulates. **Agar** is the most commonly used solidifying agent. It is a sulfated polymer composed mainly of D-galactose, 3,6-anhydro-L-galactose, and D-glucuronic acid. It usually is extracted from red algae. Agar is well suited as a solidifying agent for several reasons. One is that it melts at about 90°C but, once melted, does not harden until it reaches about 45°C. Thus after being melted in boiling water, it can be cooled to a temperature that is tolerated by human hands as well as microbes. Furthermore, microbes growing on agar medium can be incubated at a wide range of temperatures. Finally, agar is an excellent hardening agent because most microorganisms cannot degrade it.

Functional Types of Media

Media such as tryptic soy broth and tryptic soy agar are called general purpose or **supportive media** because they sustain the

Table 6.7 Some Common Complex Media

Nutrient Broth	*Amount (g/liter)*
Peptone (gelatin hydrolysate)	5
Beef extract	3
Tryptic Soy Broth	
Tryptone (pancreatic digest of casein)	17
Peptone (soybean digest)	3
Glucose	2.5
Sodium chloride	5
Dipotassium phosphate	2.5
MacConkey Agar	
Pancreatic digest of gelatin	17.0
Pancreatic digest of casein	1.5
Peptic digest of animal tissue	1.5
Lactose	10.0
Bile salts	1.5
Sodium chloride	5.0
Neutral red	0.03
Crystal violet	0.001
Agar	13.5

growth of many microorganisms. Blood and other nutrients may be added to supportive media to encourage the growth of fastidious microbes. These fortified media (e.g., blood agar) are called **enriched media** (**figure 6.9**).

Selective media favor the growth of particular microorganisms (**table 6.8**). Bile salts or dyes such as basic fuchsin and crystal violet favor the growth of gram-negative bacteria by inhibiting the growth of gram-positive bacteria; the dyes have no effect on gram-negative organisms. Endo agar, eosin methylene blue agar, and MacConkey agar (tables 6.7 and 6.8) are three media widely used for the detection of *E. coli* and related bacteria in water supplies and elsewhere. These media contain dyes that suppress the growth of gram-positive bacteria. MacConkey agar also contains bile salts. Bacteria also may be selected by incubation with nutrients that they specifically can use. A medium containing only cellulose as a carbon and energy source is quite effective in the isolation of cellulose-digesting bacteria from samples such as soil. Thus selective media are useful as media for enrichment cultures. The possibilities for selection are endless, and dozens of selective media are in use (**Techniques & Applications 6.1**).

Differential media are media that distinguish among different groups of microbes and even permit tentative identification of microorganisms based on their biological characteristics. Blood agar is both a differential medium and an enriched one. It distinguishes between hemolytic and nonhemolytic bacteria. Some hemolytic bacteria (e.g., many streptococci and staphylococci isolated from

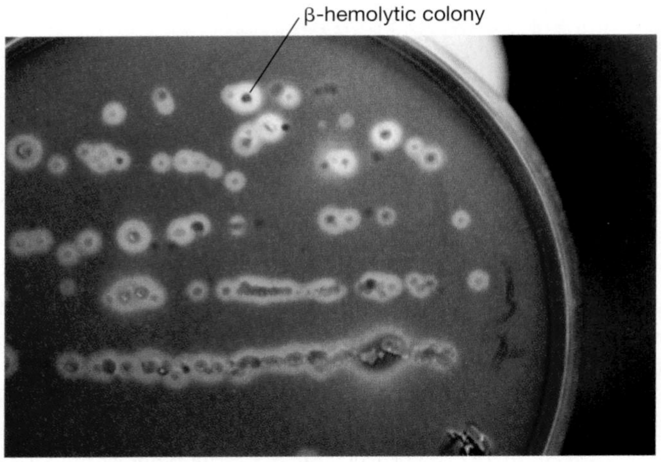

(a) Blood agar

(b) Chocolate agar

FIGURE 6.9 Enriched Media. (a) Blood agar culture of bacteria from the human throat. (b) Chocolate agar is used to grow fastidious organisms such as *Neisseria gonorrhoeae*. The brown color is the result of heating red blood cells and lysing them before adding them to the medium. It is called chocolate agar because of its chocolate brown color.

Figure 6.9 Micro Inquiry

What growth factor(s) do you think the red blood cells provide?

TECHNIQUES AND APPLICATIONS

6.1 Enrichment Cultures

A major practical problem for microbiologists is the preparation of pure cultures when microorganisms are present in very low numbers in a sample. Plating methods can be combined with the use of selective or differential media to enrich and isolate rare microorganisms. A good example is the isolation of bacteria that degrade the herbicide 2,4-dichlorophenoxyacetic acid (2,4-D). Bacteria able to metabolize 2,4-D can be obtained with a liquid medium containing 2,4-D as its sole carbon source and the required nitrogen, phosphorus, sulfur, and mineral components. When this medium is inoculated with soil, only bacteria able to use 2,4-D will grow. After incubation, a sample of the culture is transferred to a fresh flask of selective medium for further enrichment of 2,4-D metabolizing bacteria. A mixed population of 2,4-D degrading bacteria will arise after several such transfers. Pure cultures can be obtained by plating this mixture on agar containing 2,4-D as the sole carbon source. Only bacteria able to grow on 2,4-D form visible colonies, and these can be subcultured. This same general approach is used to isolate and purify a variety of bacteria by selecting for specific physiological characteristics.

The preceding techniques require the use of special culture dishes named Petri dishes after their inventor Julius Richard Petri, a member of Robert Koch's laboratory; Petri developed these dishes around 1887, and they immediately replaced agar-coated glass plates. They consist of two round halves, the top half overlapping the bottom. Petri dishes are very easy to use, may be stacked on each other to save space, and are one of the most common items in microbiology laboratories.

Table 6.8 Mechanisms of Action of Selective and Differential Media

Medium	*Functional Type*	*Mechanism of Action*
Blood agar	Enriched and differential	Blood agar supports the growth of many fastidious bacteria. These can be differentiated based on their ability to produce hemolysins—proteins that lyse red blood cells. Hemolysis appears as a clear zone (β-hemolysis) or greenish halo around the colony (α-hemolysis) (e.g., *Streptococcus pyogenes,* a β-hemolytic streptococcus).
Eosin methylene blue (EMB) agar	Selective and differential	Two dyes, eosin Y and methylene blue, inhibit the growth of gram-positive bacteria. They also react with acidic products released by certain gram-negative bacteria when they use lactose or sucrose as carbon and energy sources. Colonies of gram-negative bacteria that produce large amounts of acidic products have a green, metallic sheen (e.g., fecal bacteria such as *E. coli*).
MacConkey (MAC) agar	Selective and differential	The selective components in MAC are bile salts and crystal violet, which inhibit the growth of gram-positive bacteria. The presence of lactose and neutral red, a pH indicator, allows the differentiation of gram-negative bacteria based on the products released when they use lactose as a carbon and energy source. The colonies of those that release acidic products are red (e.g., *E. coli*).
Mannitol salt agar	Selective and differential	A concentration of 7.5% NaCl selects for the growth of staphylococci. Pathogenic staphylococci can be differentiated based on the release of acidic products when they use mannitol as a carbon and energy source. The acidic products cause a pH indicator (phenol red) in the medium to turn yellow (e.g., *Staphylococcus aureus*).

throats) produce clear zones around their colonies because of red blood cell destruction (figure 6.9*a*). Blood agar is an enriched growth medium in that blood (usually sheep blood) provides protein, carbohydrate, lipid, iron, and a number of growth factors and vitamins necessary for the cultivation of fastidious organisms.

MacConkey agar is both differential and selective. Because it contains lactose and neutral red dye, bacteria that catabolize lactose by fermenting it release acidic waste products that make colonies appear pink to red in color. These are easily distinguished from colonies of bacteria that do not ferment lactose.

In addition to providing nutrients, some media must exclude substances that are harmful to the organism. An excellent example is O_2. Numerous microbes are strict anaerobes that require anoxic media to survive. This can be accomplished by including reducing agents such as thioglycollate and cysteine. The medium is boiled during preparation to drive off oxygen. The reducing agents then eliminate any residual dissolved O_2 in the medium.

1. Describe the following kinds of media and their uses: defined media, complex media, supportive media, enriched media, selective media, and differential media. Give an example of each kind.
2. What are peptones, yeast extract, beef extract, thioglycollate, and agar? Why are they used in media?

6.8 Isolation of Pure Cultures

In natural habitats, microorganisms usually grow in complex, mixed populations with many species. This presents a problem for microbiologists because a single type of microorganism cannot be studied adequately in a mixed culture. One needs a **pure culture,** a population of cells arising from a single cell, to characterize an individual species. Pure cultures are so important that the development of pure culture techniques by the German bacteriologist Robert Koch transformed microbiology. Within about 20 years after the development of pure culture techniques, most pathogens responsible for the major human bacterial diseases had been isolated. Pure cultures can be prepared in several ways; a few of the more common approaches are reviewed here.

Streak Plate

If cells from a mixture of microbes can be spatially isolated from each other, each cell will give rise to a completely separate **colony**—a macroscopically visible cluster of microorganisms in or on a solid medium. Because each colony arises from a single cell, each colony represents a pure culture. One method for separating cells is the **streak plate.** In this technique, cells are transferred to the edge of an agar plate with an inoculating loop or swab and then streaked out over the surface in one of several patterns (**figure 6.10**). After the first sector is streaked, the inoculating loop is sterilized and an inoculum for the second sector is obtained from the first sector. A similar process is followed for streaking the third sector, except that the inoculum is from the second sector. Thus this is essentially a dilution process. Eventually very few cells will be on the loop, and single cells will drop from it as it is rubbed along the agar surface. These develop into separate colonies.

Spread Plate and Pour Plate

Spread-plate and pour-plate techniques are similar in that they both dilute a sample of cells before separating them spatially. They differ in that the spread plate spreads the cells on the surface of the agar, whereas the pour plate embeds the cells within the agar.

For the **spread plate,** a small volume of a diluted mixture containing around 30 to 300 cells (25 to 250 cells if examining microbes in food or water samples) is transferred to the center of an agar plate and spread evenly over the surface with a sterile bent rod (**figure 6.11**). The dispersed cells develop into isolated colonies.

The **pour plate** is extensively used with bacteria, archaea, and fungi. The original sample is diluted several times to reduce the microbial population sufficiently to obtain separate colonies when plating (**figure 6.12**). Then small volumes of several diluted samples are mixed with liquid agar that has been cooled to about 45°C, and the mixtures are poured immediately into sterile culture dishes. Most microbes survive a brief exposure to the warm agar. Each cell becomes fixed in place to form an individual colony after the agar hardens. Like the spread plate, the pour plate can be used to determine the number of cells in a population. For both methods, the total number of colonies equals the number of viable microorganisms in the sample that are capable of growing in the medium used. Colonies

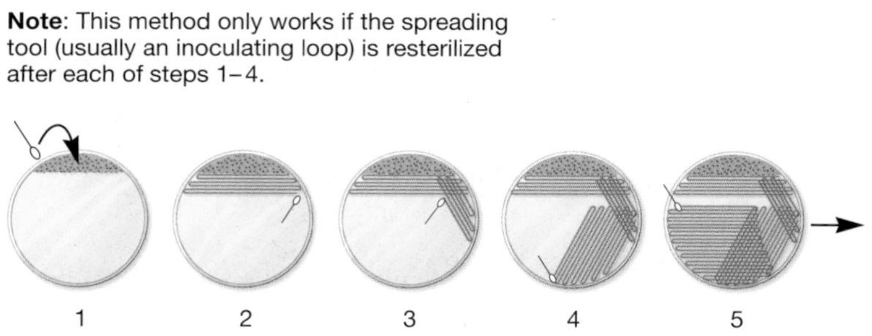

Note: This method only works if the spreading tool (usually an inoculating loop) is resterilized after each of steps 1–4.
1
2
3
4
5

(a) Steps in a streak plate

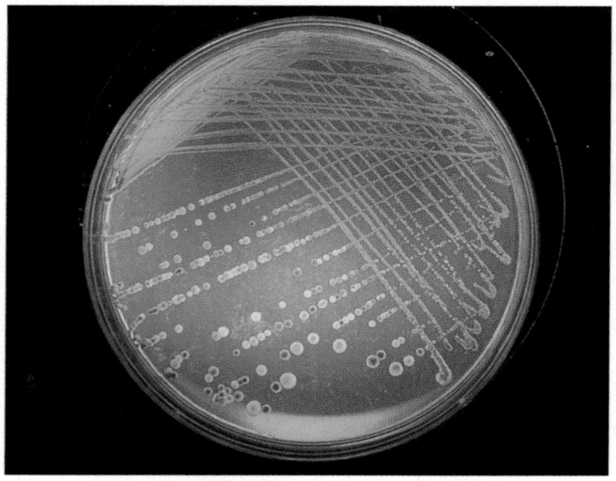

(b)

FIGURE 6.10 Streak-Plate Technique. A typical streaking pattern is shown (a), as well as an example of a streak plate (b).

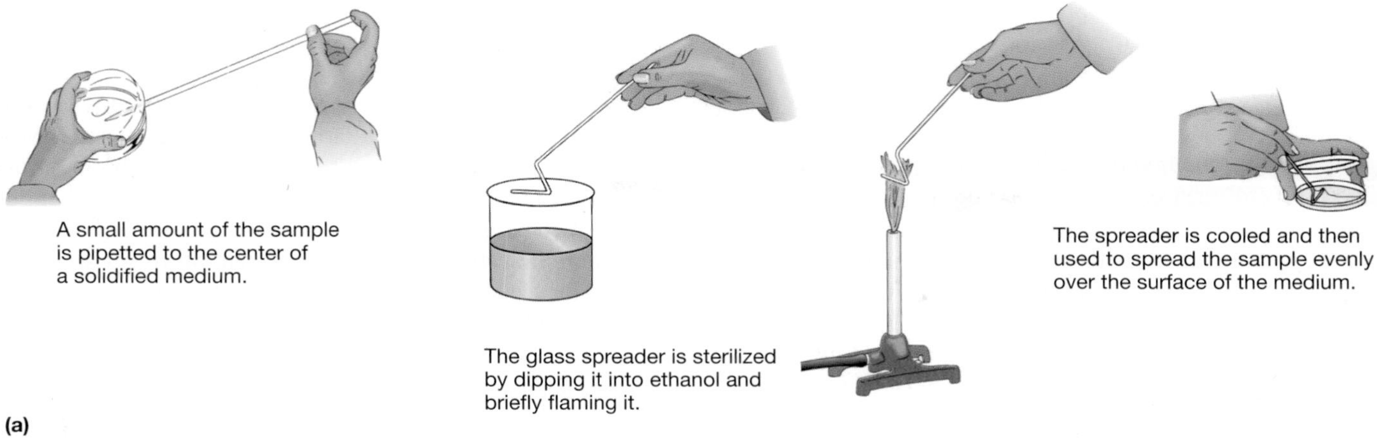

A small amount of the sample is pipetted to the center of a solidified medium.
The glass spreader is sterilized by dipping it into ethanol and briefly flaming it.
The spreader is cooled and then used to spread the sample evenly over the surface of the medium.
(a)

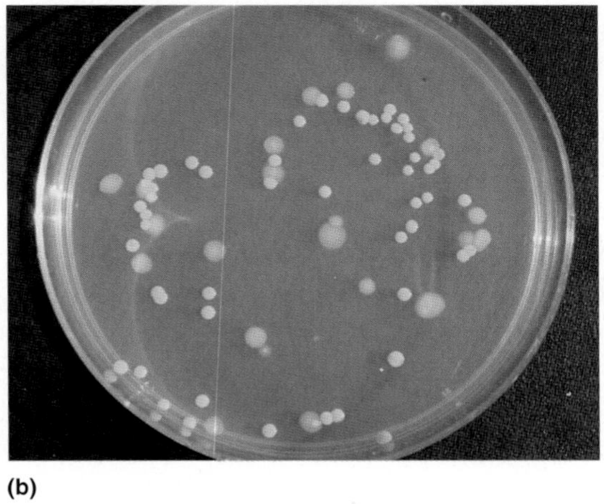

FIGURE 6.11 Spread-Plate Technique. (a) The preparation of a spread plate. (b) Typical result of spread-plate technique.

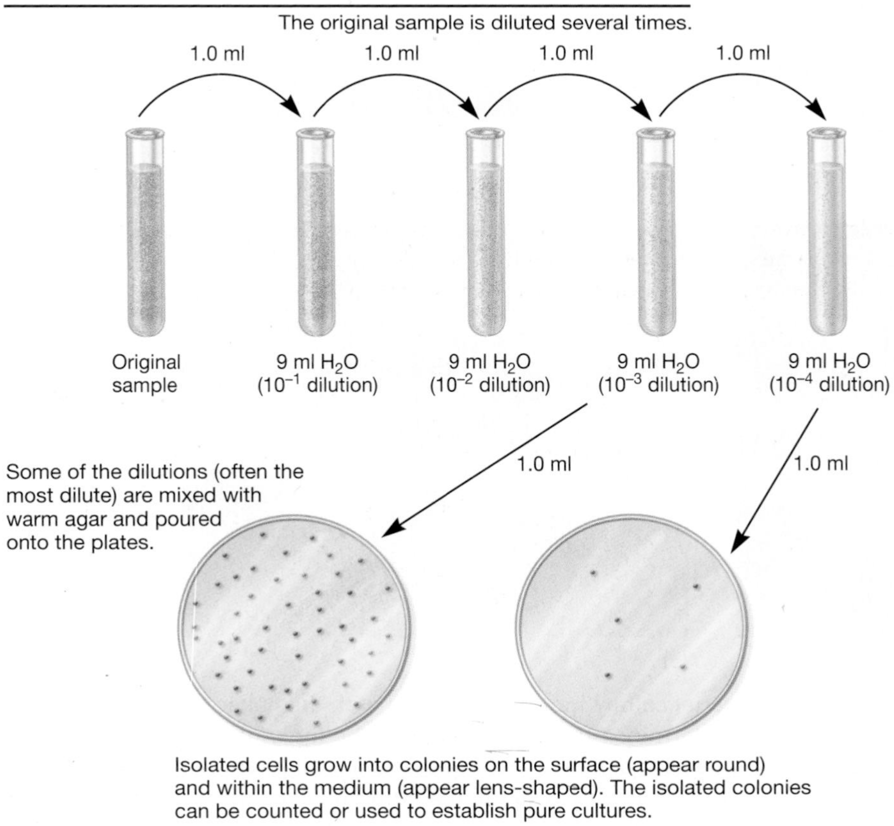

9 ml H_2O (10^{-1} dilution)
9 ml H_2O (10^{-2} dilution)
9 ml H_2O (10^{-3} dilution)
9 ml H_2O (10^{-4} dilution)

Some of the dilutions (often the most dilute) are mixed with warm agar and poured onto the plates.
Isolated cells grow into colonies on the surface (appear round) and within the medium (appear lens-shaped). The isolated colonies can be counted or used to establish pure cultures.

FIGURE 6.12 The Pour-Plate Technique.

growing on the surface also can be used to inoculate fresh medium and prepare pure cultures (Techniques & Applications 6.1).

Although the preparation of serial dilutions is the bane of many microbiology students, it has many applications other than the spread-plate and pour-plate methods. The numbers of cells in a solution can be diluted to the point where no cells are present in the dilution tube. If replicate dilutions are prepared and the number of tubes yielding growth determined, then the number of viable cells in the sample can be estimated by the most probable number method (MPN; *see figure 27.3*). Similarly, the number of viruses in a solution or the number of antibodies in a blood sample can be determined by first diluting the sample and then testing for the presence of viruses or antibodies. ◀◀ *Cultivation and enumeration of viruses (section 5.5)* ▶▶ *Measurement of microbial growth (section 7.4); Culturing techniques (section 27.1); Antibodies (section 33.7)*

Microbial Growth on Solid Surfaces

Colony development on agar surfaces aids microbiologists in identifying microorganisms because individual species often form colonies of characteristic size and appearance (**figure 6.13**). When a mixed population has been plated properly, it sometimes is possible to identify the desired colony based on its overall appearance and use it to obtain a pure culture.

In nature, microorganisms often grow on surfaces in biofilms—slime-encased aggregations of microbes. Growth of microorganisms in biofilms is often similar to growth of microbes in colonies. Generally in colonies, the most rapid cell growth occurs at the colony edge. Growth is much slower in the center, and cell autolysis takes place in the older central portions of some colonies. These differences in growth are due to gradients of oxygen, nutrients, and toxic products within the colony. At the colony edge, oxygen and nutrients are plentiful. The colony center is much thicker than the edge. Consequently oxygen and nutrients do not diffuse readily into the center, toxic metabolic products cannot be quickly eliminated, and growth in the colony center is slowed or stopped. Because of these environmental variations within a colony, cells on the periphery can be growing at maximum rates, while cells in the center are dying. Likewise in biofilms, gradients of nutrients and oxygen exist. Thus cells in one area of the biofilm may be dead or dormant, while cells in another area are actively reproducing. ▶▶| *Biofilms (section 7.7)*

It is obvious from the colonies pictured in figure 6.13 that bacteria growing on solid surfaces such as agar can form quite complex and intricate colony shapes. These patterns vary with nutrient availability and the hardness of the agar surface. The size and shape of a colony depend on many factors. Nutrient diffusion and availability, bacterial chemotaxis, and the presence of liquid on the surface all appear to play a role in pattern formation. Cell-cell communication is important as well. Much research is currently focused on understanding the formation of bacterial colonies and biofilms. ▶▶| *Cell-cell communication (section 7.7)*

1. What are pure cultures, and why are they important? How are spread plates, streak plates, and pour plates prepared?
2. In what ways does microbial growth vary within a colony? What factors might cause these variations in growth?
3. How might an enrichment culture be used to isolate bacteria capable of growing photoautotrophically from a mixed microbial assemblage?

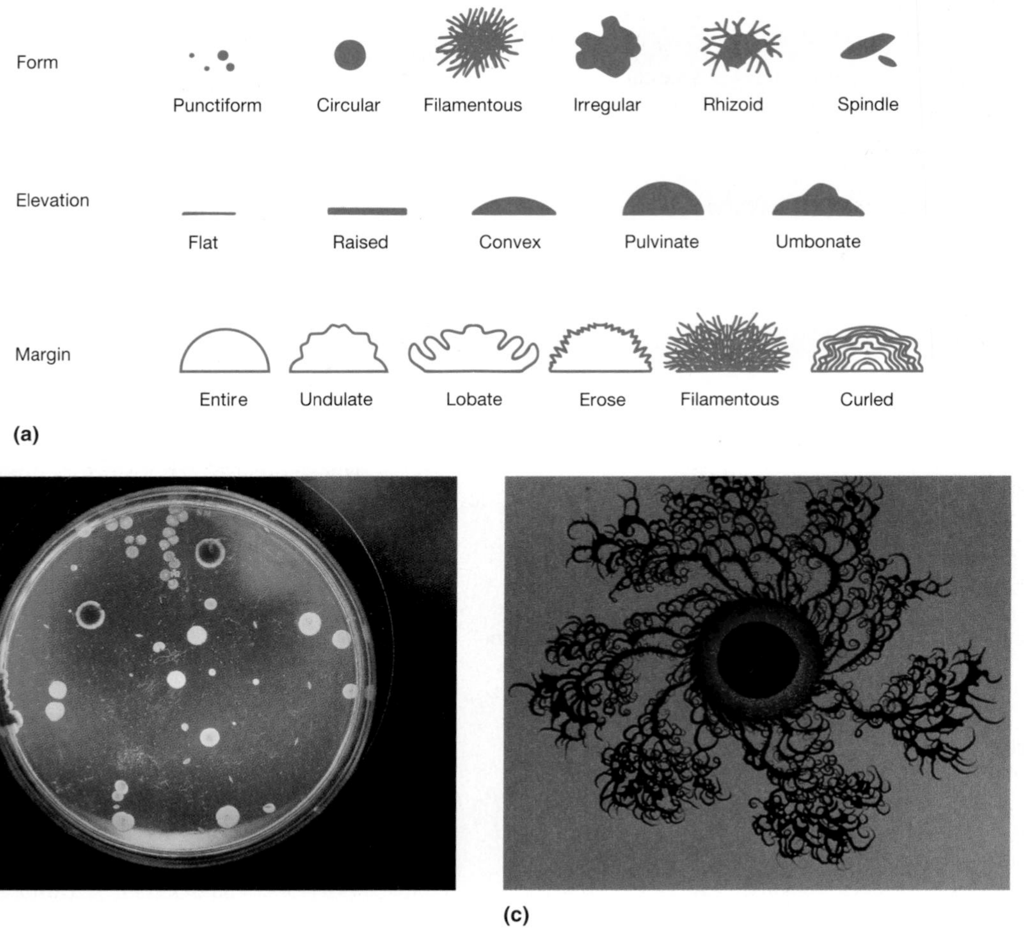

FIGURE 6.13 Bacterial Colony Morphology. (a) Variations in bacterial colony morphology seen with the naked eye. The general form of the colony and the shape of the edge or margin can be determined by looking down at the top of the colony. The nature of colony elevation is apparent when viewed from the side as the plate is held at eye level. (b) Examples of commonly observed colony morphologies. (c) Colony morphology can vary dramatically with the medium on which the bacteria are growing. These beautiful snowflakelike colonies were formed by *Bacillus subtilis* growing on nutrient-poor agar. The bacteria behave cooperatively when confronted with poor growth conditions.

Summary

6.1 Elements of Life

a. Microorganisms require nutrients, materials that are used in energy conservation and biosynthesis (**table 6.1**).

b. Macronutrients (macroelements: C, O, H, N, S, P, K, Ca, Mg, and Fe) are needed in relatively large quantities.

c. Micronutrients (trace elements: e.g., Mn, Zn, Co, Mo, Ni, and Cu) are used in very small amounts.

6.2 Carbon, Hydrogen, Oxygen, and Electrons

a. All organisms require a source of carbon, hydrogen, oxygen, and electrons (**table 6.2**).

b. Heterotrophs use reduced organic molecules as their source of carbon. These molecules often supply hydrogen, oxygen, and electrons as well. Some heterotrophs also derive energy from their organic carbon source.

c. Autotrophs use CO_2 as their primary or sole carbon source; they must obtain hydrogen, energy, and electrons from other sources.

6.3 Nutritional Types of Microorganisms

a. Microorganisms can be classified based on their energy and electron sources (**table 6.2**). Phototrophs use light energy, and chemotrophs obtain energy from the oxidation of chemical compounds.

b. Electrons are extracted from reduced inorganic substances by lithotrophs and from organic compounds by organotrophs (**table 6.2**).

c. Most organisms can be placed into one of five nutritional classes: photolithoautotroph, photoorganoheterotroph, chemolithoautotroph, chemolithoheterotroph, and chemoorganoheterotroph (**table 6.3**).

6.4 Nitrogen, Phosphorus, and Sulfur

a. Nitrogen, phosphorus, and sulfur may be obtained from the same organic molecules that supply carbon.

b. Nitrogen, phosphorus, and sulfur also may be obtained from the direct incorporation of ammonia and phosphate, and by the reduction and assimilation of oxidized inorganic molecules.

6.5 Growth Factors

a. Many microorganisms need growth factors (**table 6.4**).

b. The three major classes of growth factors are amino acids, purines and pyrimidines, and vitamins. Vitamins are small organic molecules that usually are components of enzyme cofactors.

6.6 Uptake of Nutrients

a. Although some nutrients can enter cells by passive diffusion, a membrane carrier protein is usually required.

b. In facilitated diffusion, the transport protein simply carries a molecule across the membrane in the direction of decreasing concentration, and no metabolic energy is required (**figure 6.4**).

c. Active transport systems use metabolic energy and membrane carrier proteins to concentrate substances by transporting them against a gradient. ATP is used as an energy source by ABC transporters, a type of primary active transporter (**figure 6.5**). Gradients of protons and sodium ions also drive solute uptake across membranes in secondary active transport systems (**figure 6.6**).

d. *Bacteria* also transport organic molecules while modifying them, a process known as group translocation. For example, many sugars are transported and phosphorylated simultaneously (**figure 6.7**).

e. Iron is accumulated by the secretion of siderophores, small molecules able to complex with ferric iron. When the iron-siderophore complex reaches the cell surface, it is taken inside and the iron is reduced to the ferrous form (**figure 6.8**).

6.7 Culture Media

a. Culture media can be constructed completely from chemically defined components (defined media or synthetic media) or constituents, such as peptones and yeast extract, whose precise composition is unknown (complex media).

b. Culture media can be solidified by the addition of agar, a complex polysaccharide from red algae.

c. Culture media are classified based on function as supportive media, enriched media, selective media, and differential media. Supportive media are used to culture a wide variety of microbes. Enriched media are supportive media that contain additional nutrients needed by fastidious microbes. Selective media contain components that select for the growth of some microbes. Differential media contain components that allow microbes to be differentiated from each other, usually based on some metabolic capability.

6.8 Isolation of Pure Cultures

a. Pure cultures usually are obtained by isolating individual cells with any of three plating techniques: the streak-plate, spread-plate, and pour-plate methods.

b. The streak-plate technique uses an inoculating loop to spread cells across an agar surface (**figure 6.10**).

c. The spread-plate (**figure 6.11**) and pour-plate (**figure 6.12**) methods usually involve diluting a culture or sample and then plating the dilutions. In the spread-plate technique, a specially shaped rod is used to spread the cells on the agar surface; in the pour-plate technique, the cells are first mixed with cooled agar-containing media before being poured into a Petri dish.

d. Microorganisms growing on solid surfaces tend to form colonies with distinctive morphology (**figure 6.13**). Colonies usually grow most rapidly at the edge, where larger amounts of required resources are available.

Critical Thinking Questions

1. Several differences between eukaryotes, *Bacteria*, and *Archaea* have been noted in this chapter. For instance, group translocation is not observed in eukaryotes, nor are ABC import systems. However, major facilitator superfamily proteins, facilitated diffusion, and endocytosis are observed in eukaryotes. Why do you think such differences exist between eukaryotes and *Bacteria* and *Archaea*?

2. If you wished to obtain a pure culture of bacteria that could degrade benzene and use it as a carbon and energy source, how would you proceed?

3. *Vibrio cholerae* is the causative agent of cholera. When not attached to human gut epithelial cells, it lives in aquatic environments, where it attaches to plants, insects, and crustaceans. It is known to form multicellular communities called biofilms on all these surfaces. The nutrients available to *V. cholera* differ depending on its environment, and like many bacteria, *V. cholerae* relies on its phosphoenolpyruvate phosphotransferase system (PTS) for sugar uptake. It has been shown that Enzyme I (EI) of *V. cholerae* is involved in at least two processes. In addition to its role in accepting a phosphate from PEP (figure 6.7), EI also helps trigger biofilm formation when glucose is present. Why do you think *V. cholerae* evolved so that EI also regulates bacterial behavior in response to glucose availability? In which *V. cholerae* environment do you think glucose is most abundant, and why would it be advantageous for the microbe to form a biofilm there?

   Read the original paper: Houot, L., and Watnick, P. I. 2008. A novel role for Enzyme I of the *Vibrio cholerae* phosphoenolpyruvate phosphotransferase system in regulation of growth in a biofilm. *J. Bacteriol.* 190:311.

4. Melioidosis, also known as glanders, is a life-threatening infection caused by *Burkholderia pseudomallei*. Over 70 years ago, it was noted that *B. pseudomallei* changes its colony morphology when grown on agar plates. It was later observed that *B. pseudomallei* isolated from patients also showed different colony morphologies when grown on a single agar medium. Some of these colony morphologies are clinically relevant because they correlate with bacterial persistence and levels of virulence (i.e., how ill the patient becomes). Furthermore, some colony morphologies are descendents of others. For instance, rough dark and rough light purple colonies both arise from rough pink colonies. How would you go about cataloguing all the colony morphologies? How might the relationship between colony morphology and virulence be used to improve patient outcome?

   Read the original paper (and see color images of all the colony morphologies): Chantratita, N., et al. 2007. Biological relevance of colony morphology and phenotypic switching by *Burkholderia pseudomallei*. *J. Bacteriol.* 189:807.

Concept Mapping Challenge

Construct a concept map that differentiates the mechanisms by which molecules in a microbe's environment enter the cell. Use the concepts that follow, any other concepts you need, and your own linking words between each pair of concepts in your map. ▶▶| *Appendix III*

Passive diffusion	Concentration gradient	Permeases
Facilitated diffusion	Active transport	Symport
Siderophore	Group translocation	Antiport
ATP transporters	Energy	

Learn More

Learn more by visiting the text website at www.mhhe.com/willey8, where you will find a complete list of references.

7

Microbial Growth

A black smoker: one extreme habitat where microbes can be found.

CHAPTER GLOSSARY

acidophile A microorganism that has its growth optimum between pH 0 and about 5.5.

aerobe An organism that grows in the presence of atmospheric oxygen.

aerotolerant anaerobe A microbe that grows equally well whether or not oxygen is present.

alkaliphile (alkalophile) A microorganism that grows best at pH values from about 8.5 to 11.5.

anaerobe An organism that can grow in the absence of free oxygen.

batch culture A population of microorganisms growing in a closed culture vessel containing a single batch of medium.

biofilm Organized microbial communities consisting of layers of cells associated with surfaces and surrounded by an extracellular polymeric matrix.

chemostat A continuous culture apparatus that feeds medium into the culture vessel at the same rate as medium containing microorganisms is removed; the medium contains a limiting quantity of one essential nutrient.

colony forming units (CFU) The number of microorganisms that form colonies when cultured using spread plates or pour plates; used as a measure of the number of viable microorganisms in a sample.

cytokinesis Processes that apportion the intracellular contents, synthesize a septum, and divide a cell into two daughter cells during cell division.

exponential (log) phase The phase of a growth curve during which the microbial population is growing at a constant and maximum rate, dividing and doubling at regular intervals.

extremophiles Microorganisms that grow under harsh or extreme environmental conditions.

facultative anaerobe A microorganism that does not require oxygen for growth but grows better in its presence.

generation (doubling) time The time required for a microbial population to double in number.

halophile A microorganism that requires high levels of sodium chloride for growth.

hyperthermophile A bacterium or archaeon with a growth optimum above 85°C.

lag phase A period following the introduction of microorganisms into fresh batch culture medium when there is no increase in cell numbers or mass.

mesophile A microorganism with a growth optimum around 20 to 45°C, a minimum of 15 to 20°C, and a maximum of less than 45°C.

microaerophile A microorganism that requires low levels of oxygen for growth (2 to 10%) but is damaged by normal atmospheric oxygen levels.

most probable number (MPN) A method for determining number of viable cells in a liquid sample. Samples are incubated in a suitable liquid medium in a dilution series; the most dilute sample to show growth is assumed to have been inoculated with between 1 and 10 cells.

neutrophile A microorganism that grows best at a neutral pH range between pH 5.5 and 8.0.

obligate aerobe An organism that grows only when oxygen is present.

obligate anaerobe An organism that grows only when oxygen is absent.

osmotolerant Organisms that grow over a wide range of water activity or solute concentration.

psychrophile A microorganism that grows well at 0°C, has an optimum growth temperature of 15°C or lower, and a temperature maximum of around 20°C.

quorum sensing The exchange of extracellular molecules that allows microbial cells to sense cell density.

stationary phase The phase of microbial growth in a batch culture when population growth ceases and the growth curve levels off.

thermophile A microorganism that can grow at temperatures of 55°C or higher, with a minimum of around 45°C.

water activity (a_w) A quantitative measure of water availability in a habitat.

Earth provides a multitude of habitats, some populated by both macroorganisms and microorganisms, and others the domain only of microbes. The former habitats are relatively benign, providing adequate nutrients and moderate conditions. The latter often are extreme in one or more ways, such as temperature and pH. How is it that microbes can survive, grow, and reproduce in so may different habitats? That question is the focus of this chapter.

We begin our discussion by examining the variety of reproductive strategies used by the *Bacteria* and *Archaea*. We then consider binary fission, the type of cell division most frequently observed among the *Bacteria* and *Archaea*. Cell reproduction leads to an increase in population size, so next we consider growth and the ways in which it can be measured. Measurements of population growth can provide important information about how microbes are responding to different environmental conditions. Microbiologists also use continuous culture techniques to understand microbial responses to nutrient deprivation or other conditions. Thus our discussion of microbial growth will also include a brief examination of these methods. These initial sections in the chapter will provide the foundation needed for comprehending the influence of environmental factors on microbial growth. This topic and the topic of microbial growth in natural environments complete the chapter.

7.1 Reproductive Strategies

Eukaryotic microbes differ dramatically from the *Bacteria* and *Archaea* in their reproductive strategies. Many eukaryotic microbes exhibit both asexual reproduction, involving mitosis, and sexual reproduction, involving meiosis to produce gametes or gametelike cells. Thus a single eukaryotic microbe can be both haploid and diploid, depending on its life cycle stage. Some eukaryotic strategies are described in chapters 4, 23, and 24. Here our focus is on the types of cell division observed in bacterial and archaeal cells.

Unlike eukaryotes, all bacterial and archaeal cells are haploid. Most reproduce by **binary fission** (**figure 7.1**). Binary fission is a relatively simple type of cell division: the cell elongates, replicates its chromosome, and separates the newly formed DNA molecules so there is one chromosome in each half of the cell. Finally, a septum (cross wall) is formed at midcell, dividing the parent cell into

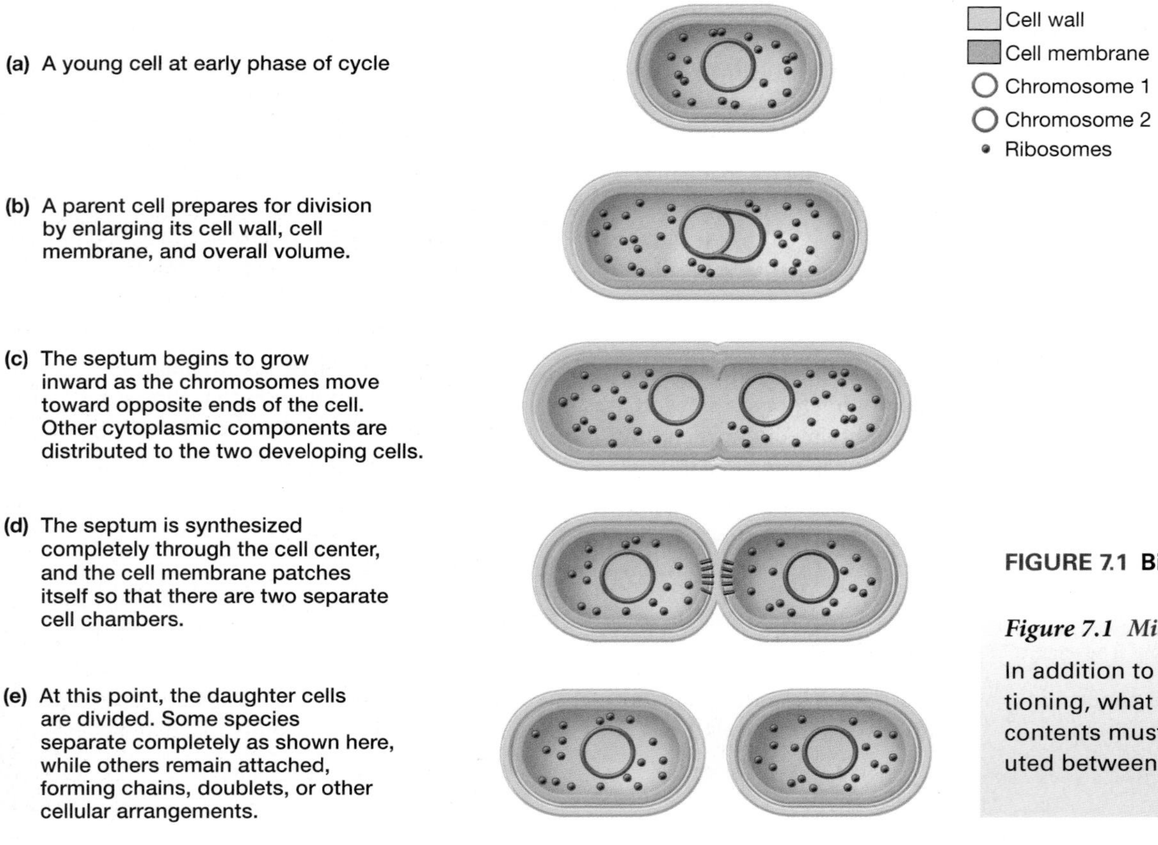

(a) A young cell at early phase of cycle
(b) A parent cell prepares for division by enlarging its cell wall, cell membrane, and overall volume.
(c) The septum begins to grow inward as the chromosomes move toward opposite ends of the cell. Other cytoplasmic components are distributed to the two developing cells.
(d) The septum is synthesized completely through the cell center, and the cell membrane patches itself so that there are two separate cell chambers.
(e) At this point, the daughter cells are divided. Some species separate completely as shown here, while others remain attached, forming chains, doublets, or other cellular arrangements.
Cell wall
Cell membrane
Chromosome 1
Chromosome 2
Ribosomes

FIGURE 7.1 Binary Fission.

Figure 7.1 Micro Inquiry

In addition to chromosome partitioning, what other cytoplasmic contents must be equally distributed between daughter cells?

two progeny cells, each having its own chromosome and a complement of other cellular constituents.

Several other reproductive strategies have been identified in *Bacteria* (**figure 7.2**). Some prosthecate bacteria such as *Hyphomonas* reproduce by forming a bud at the end of their prostheca. Certain cyanobacterial cells undergo multiple fission. The progeny cells, called baeocytes, are held within the cell wall of the parent cell until they are released. Other bacteria, such as members of the genus *Streptomyces*, form multinucleoid filaments that eventually divide to form uninucleoid spores. These spores are readily dispersed, much like the dispersal spores formed by filamentous fungi.

Despite the diversity of bacterial and archaeal reproductive strategies, they share certain features. In all cases, the genome of the cell must be replicated and segregated to form distinct nucleoids. At some point during reproduction, each nucleoid becomes enclosed within its own plasma membrane and eventually its own cell wall. These processes are the major steps of the cell cycle. In section 7.2, we examine the bacterial cell cycle in more detail.

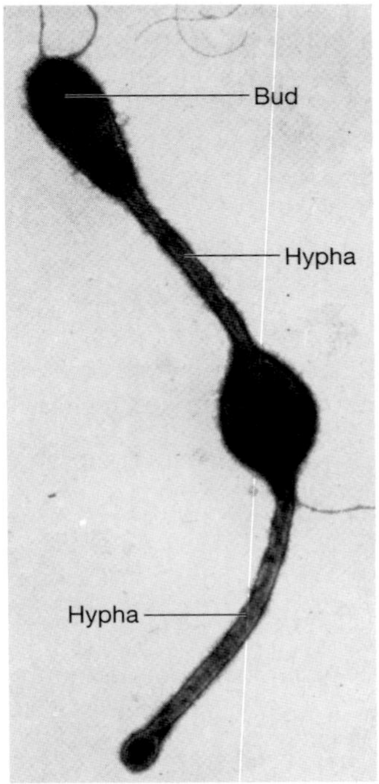

(a) *Hyphomonas* mother cell and bud

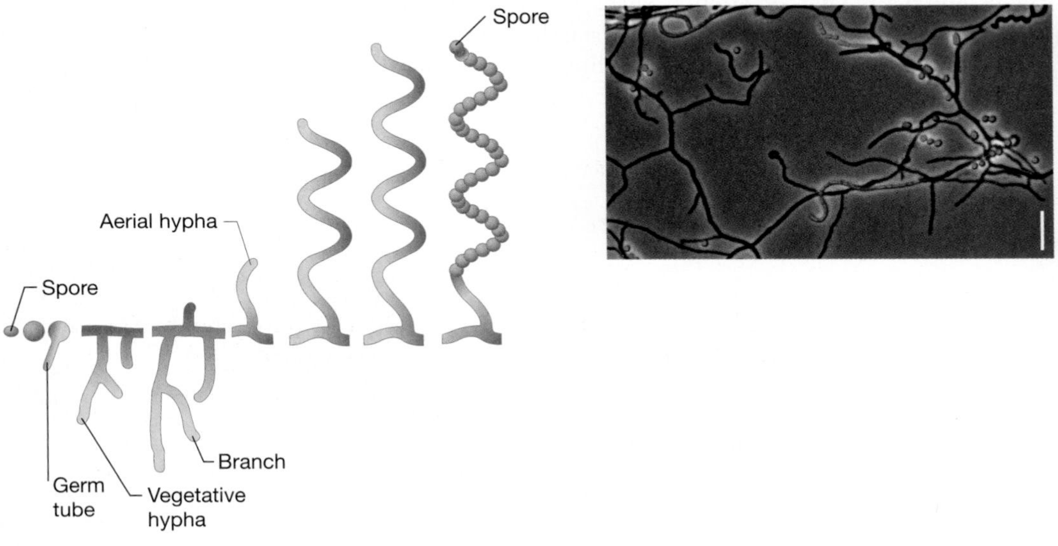

(c) *Streptomyces coelicolor* spore formation

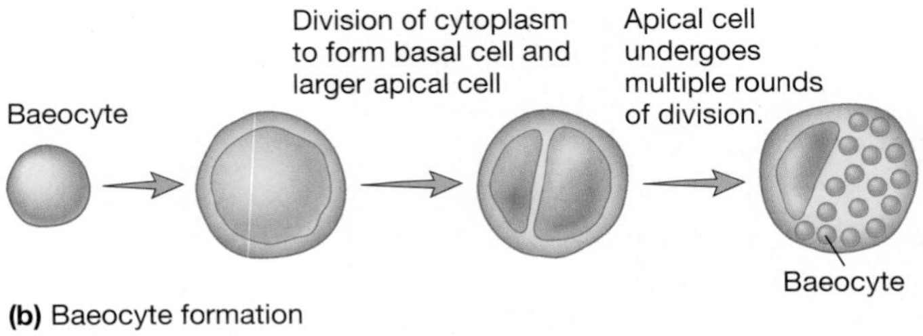

Apical cell undergoes multiple rounds of division.

(b) Baeocyte formation

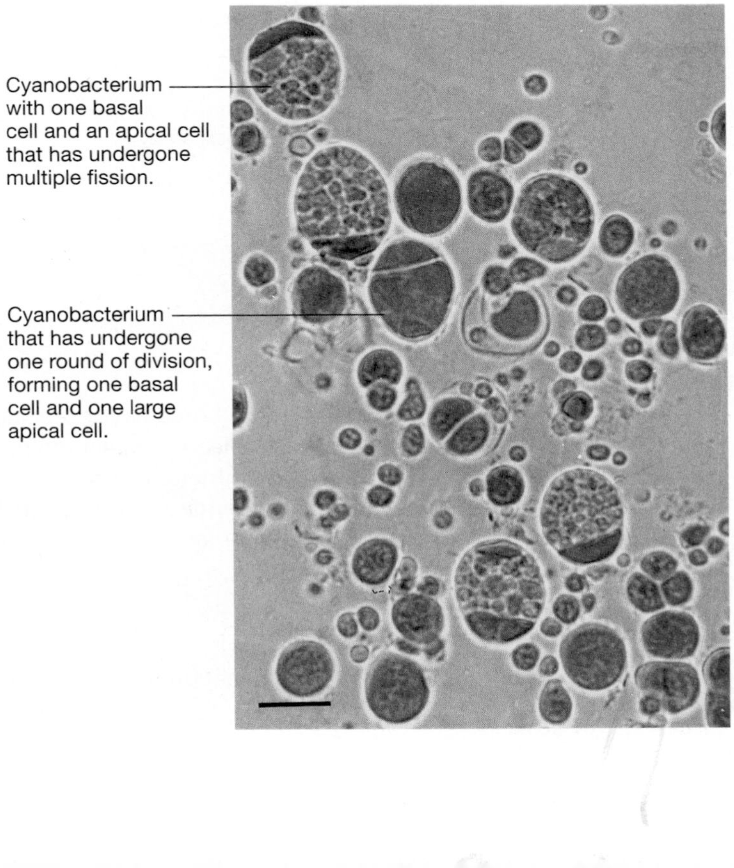

Cyanobacterium with one basal cell and an apical cell that has undergone multiple fission.
Cyanobacterium that has undergone one round of division, forming one basal cell and one large apical cell.

FIGURE 7.2 Other Bacterial Reproductive Strategies. (a) A budding *Hyphomonas* cell. (b) On the left is a drawing illustrating baeocyte formation. On the right is a photo of baeocytes produced by the cyanobacterium *Dermocarpella* (×1,000). (c) Spore formation by *Streptomyces coelicolor*. The micrograph on the right is a false-colored image.

7.2 Bacterial Cell Cycle

The **cell cycle** is the complete sequence of events extending from the formation of a new cell through the next division. It is of intrinsic interest to microbiologists as a fundamental biological process. However, understanding the cell cycle has practical importance as well. For instance in bacteria, the synthesis of peptidoglycan is the target of numerous antibiotics. ▶▶| *Inhibitors of cell wall synthesis (section 34.4)*

The cell cycles of several bacteria—*Escherichia coli, Bacillus subtilis,* and the aquatic bacterium *Caulobacter crescentus*—have been examined extensively, and our understanding of the bacterial cell cycle is based largely on these studies. Two pathways function during the bacterial cell cycle: one pathway replicates and partitions the DNA into the progeny cells, the other carries out cytokinesis—formation of the septum and progeny cells. Although these pathways overlap, it is easiest to consider them separately.

Chromosome Replication and Partitioning

Recall that most bacterial chromosomes are circular. Each circular chromosome has a single site at which replication starts called the **origin of replication,** or simply the origin (**figure 7.3**). Replication is completed at the terminus, which is located directly opposite the origin. In a newly formed *E. coli* cell, the chromosome

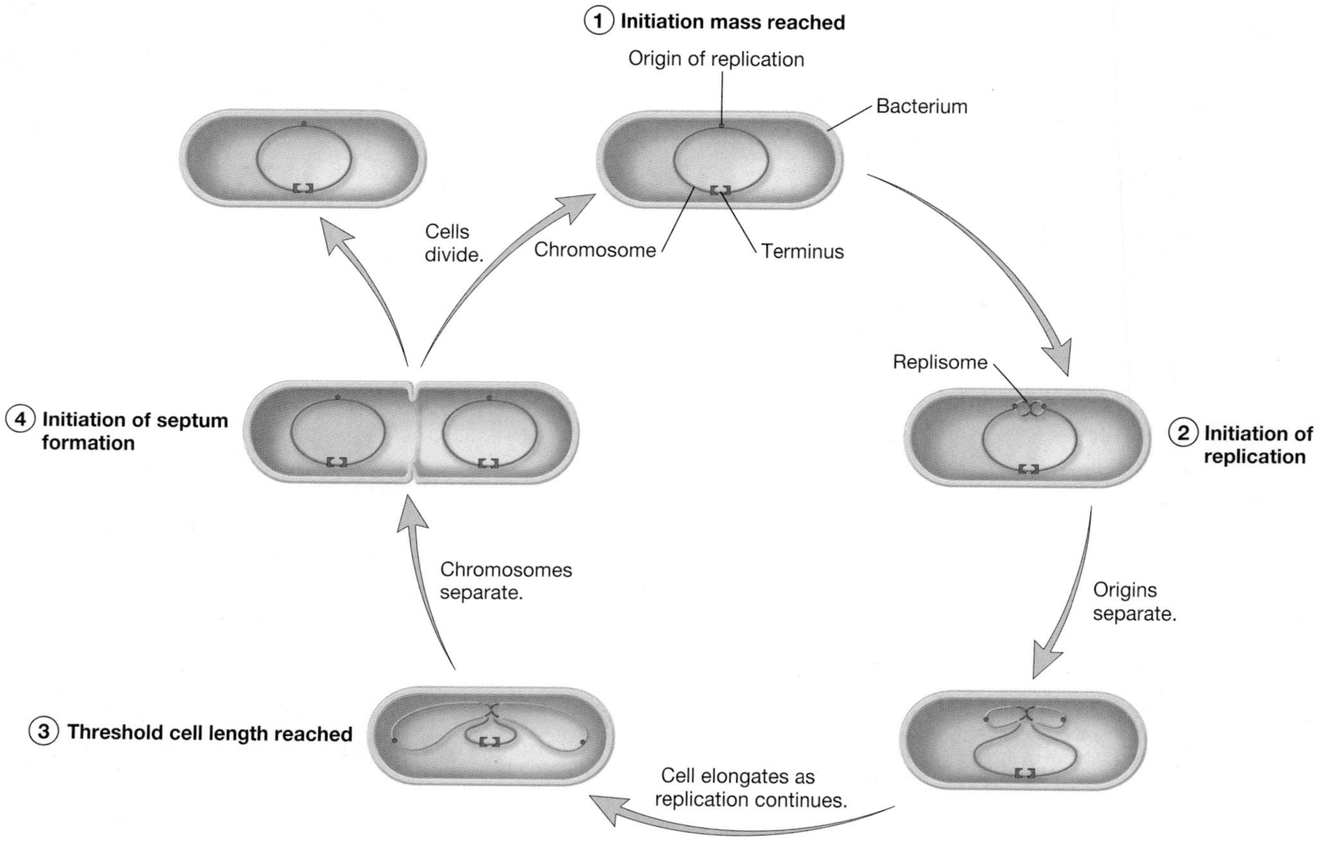

FIGURE 7.3 Cell Cycle of *E. coli*. As the cell readies for replication, the origin migrates to the center of the cell and proteins that make up the replisome assemble. As replication proceeds, newly synthesized chromosomes move toward poles so that upon cytokinesis, each daughter cell inherits only one chromosome. In this illustration, one round of DNA replication is completed before the cell divides. In rapidly growing cultures (generation times of about 20 minutes), second and third rounds of replication are initiated before division of the original cell is completed. Thus the daughter cells inherit partially replicated DNA.

Figure 7.3 Micro Inquiry

Why is it important that the origin of replication migrate to the center of the cell prior to replication?

is compacted and organized so that the origin and terminus are in opposite halves of the cell. Early in the cell cycle, the origin and terminus move to midcell, and a group of proteins needed for DNA synthesis assemble at the origin to form the **replisome.** DNA replication proceeds in both directions from the origin. As progeny chromosomes are synthesized, the two newly formed origins move toward opposite ends of the cell, and the rest of the chromosome follows in an orderly fashion.

Although the process of DNA synthesis and movement seems rather straightforward, the mechanism by which chromosomes are partitioned to each daughter cell has not been fully elucidated. Several models have been proposed. Some involve the daughter chromosomes being pushed to opposite sides of the cell by either the replisome or the enzyme RNA polymerase. Others propose that condensation of the daughter chromosomes pulls the DNA to each end. Evidence also exists for involvement of the cytoskeletal protein **MreB** (*mur*ein cluster *B*) in rod-shaped microorganisms. MreB is similar to eukaryotic actin and polymerizes to form a spiral around the inside periphery of the cell (**figure 7.4**). This model suggests that the origin of each newly replicated chromosome associates with MreB. Each chromosome then moves to opposite poles of the cell perhaps by following the helical MreB tracks. The model is supported by the fact that if MreB is mutated so that it can no longer utilize ATP, its source of energy, chromosomes fail to segregate properly. ◀◀ *Bacterial cytoskeleton (section 3.5)* *Binary Fission; Bidirectional DNA Replication*

It has been much easier to dissect the events needed for plasmid partitioning in bacterial cells, and such studies serve as models for building hypotheses regarding chromosome segregation. Recall that plasmids are DNA elements that are separate from the chromosome; that is, they are extrachromosomal. Plasmid replication is independent of the chromosome, and the number of copies of plasmid per cell, the copy number, is determined by the plasmid's origin of replication. A single copy plasmid has only one copy per cell. Because plasmids independently replicate, it is not surprising that they bear genes that encode proteins responsible for their segregation to each daughter cell. The single copy *E. coli* plasmid called RI produces three proteins that are essential for its inheritance. The first is ParM (Par for *par*tition, M for *m*otor). ParM, like MreB, is an actin homologue, and it polymerizes in an ATP-dependent way to form long filaments within the cell. ParR (R for repressor) and ParC (C for centromere-like) bind to the origin of replication sequences of each R1 plasmid and link ParM to the plasmid (**figure 7.5**). Once the three proteins are attached to the plasmids, both ends of the ParM filament elongate, thereby moving each plasmid to opposite ends of the cell. ◀◀ *Plasmids (section 3.5)*

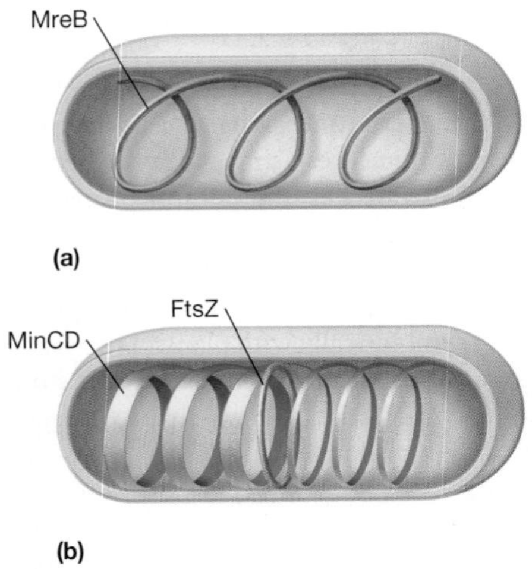

FIGURE 7.4 Cytoskeletal Proteins Involved in Cytokinesis in Rod-Shaped Bacteria. (a) The actin homologue MreB forms spiral filaments around the inside of the cell that help determine cell shape and may serve to move chromosomes to opposite cell poles. (b) The tubulin-like protein FtsZ assembles in the center of the cell to form a Z ring, which is essential for septation. MinCD, together with other Min proteins, oscillates from pole to pole, thereby preventing the formation of an off-center Z ring (see figure 7.6).

Figure 7.4 Micro Inquiry

What would be the outcome if FtsZ formed a Z ring that was not in the center of the cell? Consider both the morphology of the daughter cells and the partitioning of chromosomes.

Cytokinesis

Septation is the process of forming a cross wall between two daughter cells. **Cytokinesis,** a term that has traditionally been used to describe the formation of two eukaryotic daughter cells, is now used to describe this process in all cells. Septation is divided into several steps: (1) selection of the site where the septum will be formed; (2) assembly of the Z ring; (3) linkage of the Z ring to the plasma membrane and perhaps components of the cell wall; (4) assembly of the cell wall–synthesizing machinery (i.e., for synthesis of peptidoglycan and other cell wall constituents); and (5) constriction of the cell and septum formation. ▶▶ *Synthesis of peptidoglycan (section 11.4)*

The assembly of the Z ring is a critical step in septation, as it must be formed if subsequent steps are to occur. The FtsZ protein, a tubulin homologue found in most bacteria and many archaea, forms the Z ring. FtsZ, like tubulin, polymerizes to form filaments, which are thought to create the meshwork that constitutes the Z ring. Numerous studies show that the Z ring is very dynamic, with portions being exchanged constantly with newly formed, short FtsZ polymers from the cytosol.

In *E. coli,* the MinCDE system limits Z-ring formation to the center of the cell. Three proteins compose the system (MinC,

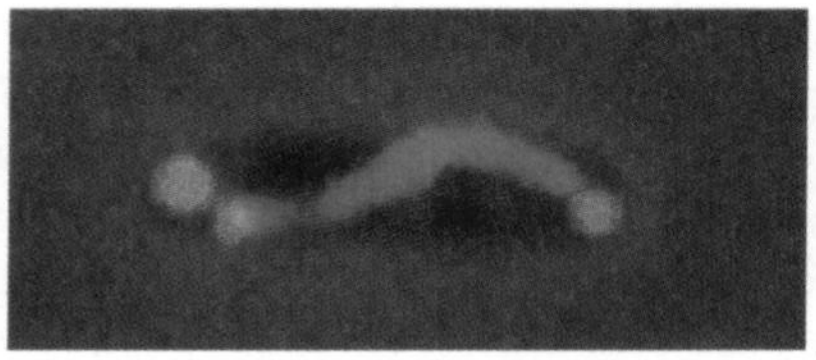

(a)

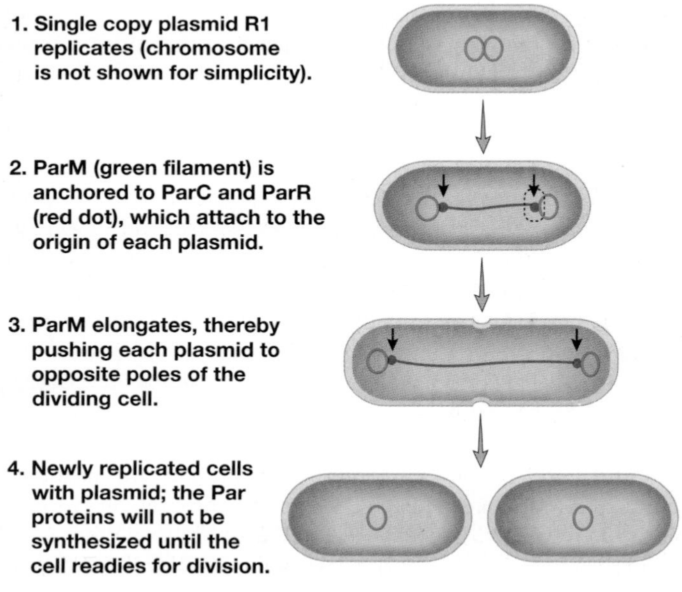
1. Single copy plasmid R1 replicates (chromosome is not shown for simplicity).
2. ParM (green filament) is anchored to ParC and ParR (red dot), which attach to the origin of each plasmid.
3. ParM elongates, thereby pushing each plasmid to opposite poles of the dividing cell.
4. Newly replicated cells with plasmid; the Par proteins will not be synthesized until the cell readies for division.

(b)

FIGURE 7.5 Segregation of *E. coli* Plasmid R1 Depends on Par Proteins. (a) An *E. coli* cell shown partitioning plasmid R1. Fluorescently labeled ParM (green filament) is attached to plasmids (red), which have been moved to opposite poles of the cell. (b) The mechanism by which ParM, ParC, and ParR mediate plasmid partitioning.

Figure 7.5 Micro Inquiry

What would happen if ParM polymerized only at one end, rather than equally at both ends?

MinD, and MinE). These proteins oscillate from one end of the cell to the other (**figure 7.6**). This oscillation creates high concentrations of MinC at the poles, where it prevents formation of the Z ring; thus Z-ring formation can occur only at midcell, which lacks MinCDE.

Once the Z ring forms, the rest of the division machinery, sometimes called the **divisome,** is constructed, as illustrated in **figure 7.7**. First, one or more anchoring proteins link the Z ring to the cell membrane. Then the cell wall-synthesizing machinery is assembled. Although numerous components of the division machinery have been identified, the function of many are still unknown (**table 7.1**). The final steps in division involve constriction of the cell by the Z ring, accompanied by invagination of the cell membrane and synthesis of the septal wall.

The preceding discussion of the cell cycle describes what occurs in slowly growing *E. coli* cells. In these cells, the cell cycle takes approximately 60 minutes to complete. However, *E. coli* can reproduce at a much more rapid rate, completing the entire cell cycle in about 20 minutes, despite the fact that DNA replication always requires at least 40 minutes. *E. coli* accomplishes this by beginning a second round of DNA replication (and sometimes even a third or fourth round) before the first round of replication is completed. Thus the progeny cells receive two or more replication forks, and replication is continuous because the cells are always copying their DNA. *Bacterial Cell Cycle*

Cellular Growth and Determination of Cell Shape

As we have seen, bacterial and archaeal cells have defined shapes that are species specific. These shapes are neither accidental nor random, as demonstrated by the faithful propagation of shape from one generation to the next. In addition, some microbes change their shape under certain circumstances. For instance, *Sinorhizobium meliloti* switches from rod-shaped to Y-shaped cells when living symbiotically with plants. Likewise, *Helicobacter pylori,* the causative agent of gastric ulcers and stomach cancer, changes from its characteristic helical shape to a sphere in stomach infections and in prolonged culture.

To consider the shape of the cell wall, we must consider its function as well. The cell wall constrains the turgor pressure exerted by the cytoplasm, thereby preventing the cell from swelling and bursting. Turgor pressure is a term used to describe the force pushing against the cell wall as determined by the osmolarity of the cytoplasmic contents. It is also essential in stretching the cell wall so new biosynthetic units can be inserted, enabling cell growth. The mechanisms by which the cell wall balances these two opposing activities, which in turn determine a specific cellular shape, are best understood in bacteria. Recall that only bacteria have peptidoglycan in the cell wall and the dynamics of peptidoglycan biosynthesis have been studied for decades. It is the focus of our discussion here. ◀◀ *Peptidoglycan structure (section 3.3)*

Peptidoglycan synthesis involves a number of proteins, including a group of enzymes called **penicillin binding proteins (PBPs)**. They bear this name because they were first noted for their capacity to bind penicillin. While this property is important, their function is to link strands of peptidoglycan together and catalyze controlled degradation so that new units can be inserted during cell growth. The PBP enzymes that degrade peptidoglycan

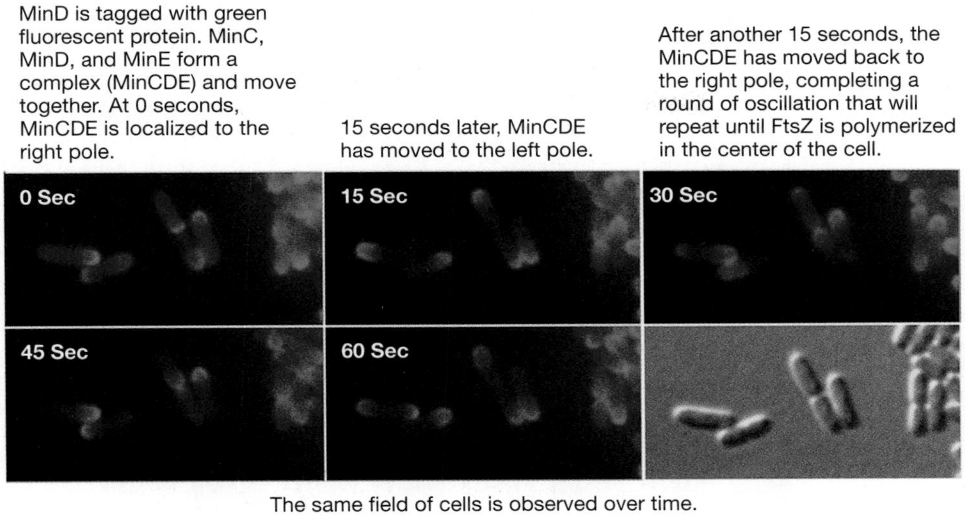

MinD is tagged with green fluorescent protein. MinC, MinD, and MinE form a complex (MinCDE) and move together. At 0 seconds, MinCDE is localized to the right pole.
15 seconds later, MinCDE has moved to the left pole.
After another 15 seconds, the MinCDE has moved back to the right pole, completing a round of oscillation that will repeat until FtsZ is polymerized in the center of the cell.
0 Sec
15 Sec
30 Sec
45 Sec
60 Sec
The same field of cells is observed over time.

FIGURE 7.6 MinCDE Proteins and Establishment of the Site of Septum Formation. A GFP-MinD fusion protein is shown oscillating from one end of an *E. coli* cell to the other. MinC blocks septum formation. It oscillates with MinD, and since concentrations of MinC are highest at the poles, septum formation is forced to occur at the center of the cell.

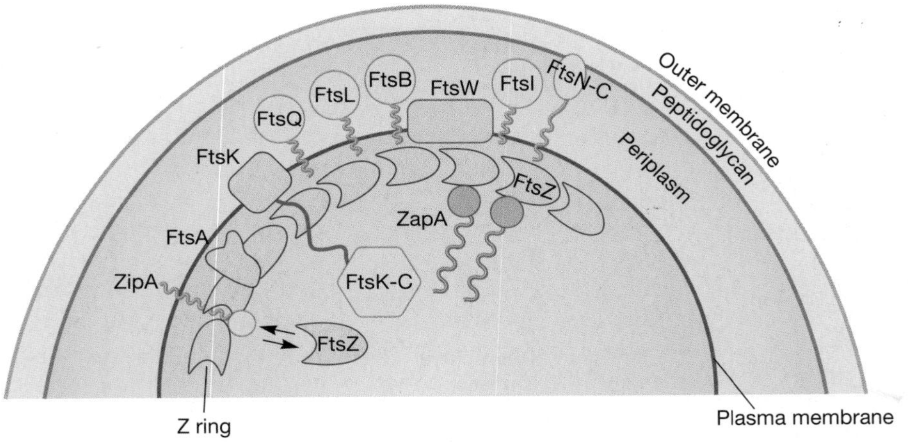

The *E. coli* divisome

FIGURE 7.7 Formation of the Cell Division Apparatus in *E. coli*. The cell division apparatus is composed of numerous proteins. The first step in divisome formation is the polymerization of FtsZ to form the Z ring. Then FtsA and ZipA (possibly ZapA in *Bacillus subtilis*) proteins anchor the Z ring to the plasma membrane. Although numerous proteins are known to be part of the cell division apparatus, the functions of relatively few are known.

Table 7.1 Divisome Proteins and Their Functions

Divisome Protein[a]	*Function*
FtsA, ZipA (ZapA)	Anchor Z ring to plasma membrane
FtsE, FtsX	Unknown
FtsK	Chromosome segregation and separation of chromosome dimers
FtsQLB	Unknown
FtsI, FtsW	Peptidoglycan synthesis
AmiC, EnvC	Amidase and hydrolase activity, respectively

[a] Proteins are listed according to the order they join the forming divisome.

are called **autolysins.** **Figure 7.8** shows a general scheme of peptidoglycan synthesis (*see figure 11.13* for a more detailed diagram). Note that individual NAM and NAG units are linked in the cytoplasm and are then ferried across the plasma membrane by a specific lipid-soluble carrier called bactoprenol. Once in the periplasmic space, the units can be added where autolysins have degraded the bonds between existing NAM and NAG molecules. The cellular location of autolysin activity and peptidoglycan export is not random and plays an important role in determining cell shape.

We begin our discussion with the simplest shape, the sphere or coccus. While it has been stated that this is the "default" cellular shape, the growth of a spherical cell is more complicated than once thought. New peptidoglycan forms only at the central septum during growth of the cocci *Enterococcus faecalis* and *Staphylococcus aureus* (**figure 7.9*a***). This region of new cell wall growth

1. Peptidoglycan synthesis starts in the cytoplasm with the attachment of uridine diphosphate (UDP) to the sugar molecules *N*-acetylmuramic acid (NAM) and *N*-acetylglucosamine (NAG). Amino acid addition to NAM is not shown for simplicity.

2. NAM-UDP is linked to NAG. Bactoprenol carries this unit to the periplasm.

3. Autolysins (cyan balls labeled "A") located at the divisome degrade the glycosidic linkages between specific NAG-NAM molecules and amino acid cross bridges. This permits the insertion of new peptidoglycan subunits.

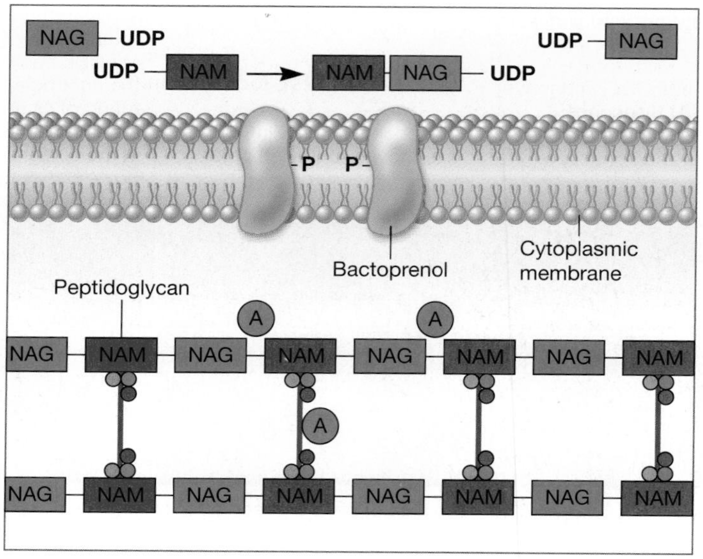

FIGURE 7.8 **Overview of Peptidoglycan Growth.** Only the essential steps needed for cell wall growth are shown. See figure 11.13 for a more detailed diagram.

AI. Spherical cells build new peptidoglycan only at the septum during division. This leads to the daughter cells that consist of one old and one new cell wall hemisphere.

BI. During growth, prior to division, new cell wall is made along the side of the cell but not at the poles. This placement is thought to be determined by the position of MreB homologues, which are present in rod-shaped bacteria and archaea.

BII. As division begins, FtsZ polymerization forms a Z ring and new cell wall growth is confined to the midcell.

BIII. As division occurs, rod-shaped daughter cells are formed with one new pole and one old pole.

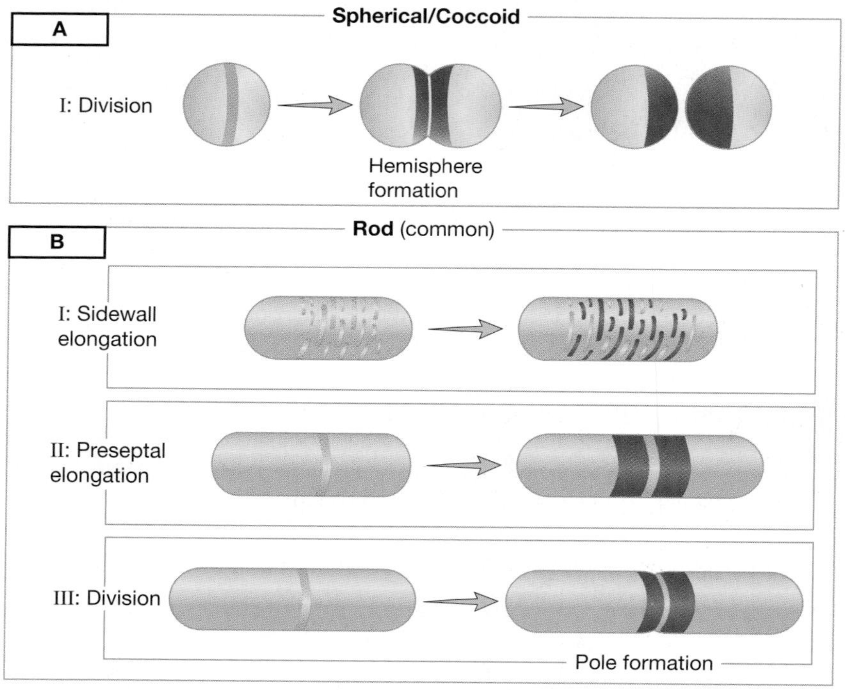

FIGURE 7.9 **Cell Wall Biosynthesis and Determination of Cell Shape in Spherical and Rod-Shaped Cells.**

***Figure 7.9** Micro Inquiry*

Which step in the development of rod-shaped cells is essential in determining cell morphology?

is sometimes called the divisome. When daughter cells separate, each has one new and one old hemisphere. As in most microbes, proper placement of the septum depends on FtsZ localization. In mutant cells without this tubulin homologue, peptidoglycan synthesis occurs in a random pattern around the cell, leading to bloated cells that lyse. It is thus thought that FtsZ placement determines the site of cell wall growth, perhaps by recruiting PBPs and other enzymes needed for peptidoglycan synthesis to the septum.

Most rod-shaped cells undergo a similar process; however, prior to cell division, they elongate (figure 7.9*b*). In these cells, proteins in the actin homologue MreB family play an essential role in determining cell diameter during elongation. MreB proteins polymerize to form helical ribbons that rotate inside the cell, along the cytoplasmic face of the cell wall (figure 7.4). Although it is not clear how MreB controls cell diameter and elongation, much information has been gained by studying *Bacillus subtilis,* which has three MreB proteins—MreB, Mbl, and MreBH. Prior to division, cell wall growth is patchy (or sometimes helical) along the length of the cell but does not occur at the poles. During this time, it has been suggested that MreB functions much like a scaffold inside the cytoplasm upon which cell wall remodeling components assemble. As the FtsZ ring forms at the midcell, it specifies peptidoglycan synthesis in this region. In some cells, MreB proteins have been shown to redeploy to the midcell, or divisome, so it may also contribute to cell wall synthesis during cytokinesis. In any event, cell wall growth switches from the side wall to the septum at this time. Despite the uncertain mechanisms by which MreB proteins function, their importance in determining cell shape is demonstrated by two observations. First, rod-shaped cells in which MreB has been chemically depleted assume a spherical shape. In addition, while almost all rod-shaped bacteria and archaea synthesize at least one MreB homologue, coccoid-shaped cells lack proteins in the MreB family.

The last cell shape we consider is that of comma-shaped cells as studied in the aquatic bacterium *Caulobacter crescentus.* In addition to the actin homologue MreB and the tubulin-like protein FtsZ, these cells (and other vibrioid-shaped cells) produce a cytoskeletal protein called crescentin, a homologue of eukaryotic intermediate filaments. This protein localizes specifically to the short, curved side of the cells (**figure 7.10**). The way in which crescentin confers a vibrioid shape is not known, but one idea is similar to that discussed for MreB: that crescentin also functions like a scaffold, but only negative regulators of cell wall growth are localized along its length. The resulting asymmetric cell wall growth would give rise to the inner curvature that characterizes this cell shape. ▶▶| Caulobacteraceae *and* Hyphomicrobiaceae *(section 20.1)*

It is clear from these examples that common microbial cell shapes require cytoskeletal proteins that bear startling homology with those of eukaryotes. However, some bacterial shapes appear to be determined by other mechanisms. For example, the spirochete *Borrelia burgdorferi,* which causes Lyme disease, establishes and retains its spiral shape by the placement of its periplasmic flagellar filaments, called axial fibrils (*see figure 19.14*). Perhaps one of the most interesting cases is that of *Spiroplasma,* a citrus pathogen, which has no cell wall but maintains its spiral shape by the positioning of contractile cytoplasmic fibrils. ▶▶| *Phylum* Spirochaetes *(section 19.6); Class* Mollicutes *(section 21.1)*

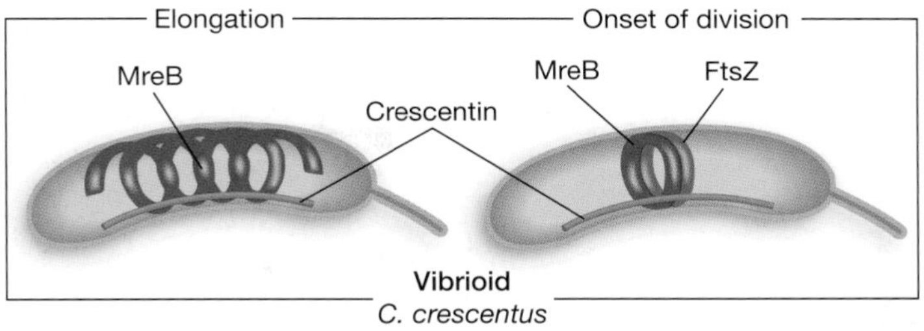

FIGURE 7.10 Vibrioid Shape Is Determined by Crescentin in *Caulobacter crescentus*. The intermediate filament homologue crescentin polymerizes along the length of the inner curvature of the cell. Cells depleted of crescentin are rod-shaped. Note that *C. crescentus* forms all three types of cytoskeleton protein homologues.

While the study of bacterial cytoskeletal elements and cell shape is inherently interesting, it also has practical aspects. Bacterial cytoskeletal homologues provide a tractable model for the study of more complicated eukaryotic cytoskeletal proteins and their assembly. In addition, these proteins may prove to be valuable targets for drug development. Indeed, a small molecule that blocks FtsZ polymerization and shows strong inhibition of *Staphylococcus aureus* growth was reported in 2008. ▶▶| *Antimicrobial chemotherapy (chapter 34)*

1. What two pathways function during the bacterial cell cycle?
2. How does the bacterial cell cycle compare with the eukaryotic cell cycle? List two ways they are similar and two ways they differ.
3. Do you think MinCDE functions in coccoid-shaped cells? Explain your answer.
4. In what ways are FtsZ and MreB similar? In what ways are they different?
5. Do you think *Spiroplasma* produces FtsZ? What about MreB? Explain your reasoning.

7.3 Growth Curve

The term growth has different meanings to a microbiologist. Growth sometimes refers to an increase in cellular constituents. This often leads to the cells growing longer and larger, and is usually accompanied by some type of cell division. Binary fission and other cell division processes bring about an increase in the number of cells in a population. Therefore the term growth is also used to refer to the growth in size of a population.

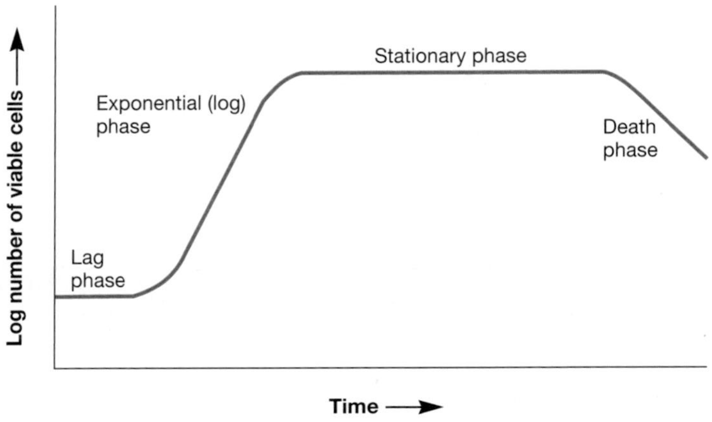

FIGURE 7.11 Microbial Growth Curve in a Closed System. The four phases of the growth curve are identified on the curve.

Figure 7.11 Micro Inquiry

Identify the regions of the growth curve in which (1) nutrients are rapidly declining and (2) wastes accumulate.

Population growth is often studied by analyzing the growth curve of a microbial culture. When microorganisms are cultivated in liquid medium, they usually are grown in a **batch culture**—that is, they are incubated in a closed culture vessel with a single batch of medium. Because no fresh medium is provided during incubation, nutrient concentrations decline and concentrations of wastes increase. The growth of a population of microbes reproducing by binary fission in a batch culture can be plotted as the logarithm of the number of viable cells versus the incubation time. The resulting curve has four distinct phases (**figure 7.11**).

Lag Phase

When microorganisms are introduced into fresh culture medium, usually no immediate increase in cell number occurs. However, cells are synthesizing new components. This period is called the **lag phase.** A lag phase can be necessary for a variety of reasons. The cells may be old and depleted of ATP, essential cofactors, and ribosomes; these must be synthesized before growth can begin. The medium may be different from the one the microorganism was growing in previously. In this case, new enzymes would be needed to use different nutrients. Possibly the microorganisms have been injured and require time to recover. Whatever the causes, eventually the cells begin to replicate their DNA, increase in mass, and finally divide.

Exponential Phase

During the **exponential (log) phase,** microorganisms are growing and dividing at the maximal rate possible given their genetic potential, the nature of the medium, and the environmental conditions. Their rate of growth is constant during the exponential phase; that is, they are completing the cell cycle and doubling in number at regular intervals (figure 7.11). The population is most uniform in terms of chemical and physiological properties during this phase; therefore exponential phase cultures are usually used in biochemical and physiological studies.

Exponential (logarithmic) growth is **balanced growth.** That is, all cellular constituents are manufactured at constant rates relative to each other. If nutrient levels or other environmental conditions change, **unbalanced growth** results. During unbalanced growth, the rates of synthesis of cell components vary relative to one another until a new balanced state is reached. Unbalanced growth is readily observed in two types of experiments: shift-up, where a culture is transferred from a nutritionally poor medium to a richer one; and shift-down, where a culture is transferred from a rich medium to a poor one. In a shift-up experiment, there is a lag while the cells first construct new ribosomes to enhance their capacity for protein synthesis. In a shift-down experiment, there is a lag in growth because cells need time to make the enzymes required for the biosynthesis of unavailable nutrients. Once the cells are able to grow again, balanced growth is resumed and the culture enters the exponential phase. These shift-up and shift-down experiments demonstrate that microbial growth is under precise, coordinated control and responds quickly to changes in environmental conditions.

When microbial growth is limited by the low concentration of a required nutrient, the final net growth or yield of cells increases with the initial amount of the limiting nutrient present (**figure 7.12*a***). The rate of growth also increases with nutrient concentration (figure 7.12*b*) but in a hyperbolic manner much like that seen with many enzymes *(see figure 9.18)*. The shape of the curve seems to reflect the rate of nutrient uptake by microbial transport proteins. At sufficiently high nutrient levels, the transport systems are saturated, and the growth rate does not rise further with increasing nutrient concentration. ◀◀ *Uptake of nutrients (section 6.6)*

1. Define growth.
2. Why would cells that are vigorously growing when inoculated into fresh culture medium have a shorter lag phase than those that have been stored in a refrigerator?
3. Define balanced growth and unbalanced growth. Why do shift-up and shift-down experiments cause cells to enter unbalanced growth?
4. What effect does increasing a limiting nutrient have on the yield of cells and the growth rate?

Stationary Phase

In a closed system such as a batch culture, population growth eventually ceases and the growth curve becomes horizontal (figure 7.11). This **stationary phase** is attained by most bacteria at a population

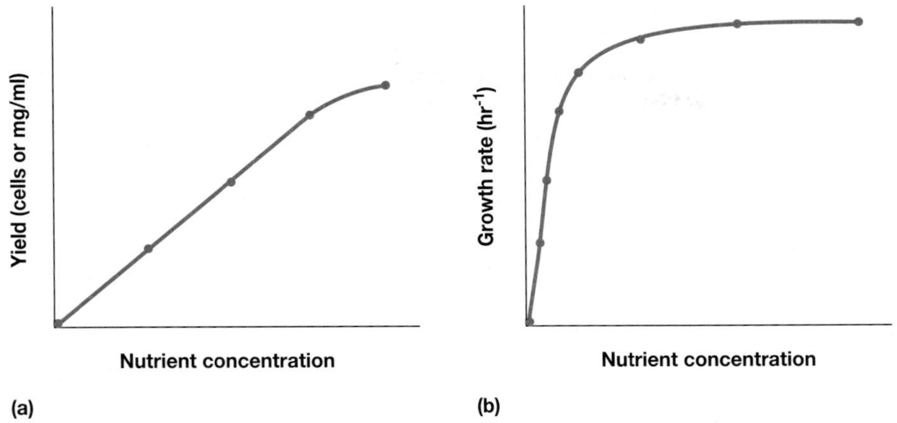

FIGURE 7.12 Nutrient Concentration and Growth. (a) The effect of changes in limiting nutrient concentration on total microbial yield. At sufficiently high concentrations, total growth will plateau. (b) The effect on growth rate.

level of around 10^9 cells per milliliter. However, some microorganisms do not reach such high population densities. For instance, protist cultures often have maximum concentrations of about 10^6 cells per milliliter. Final population size depends on nutrient availability and other factors, as well as the type of microorganism being cultured. In the stationary phase, the total number of viable microorganisms remains constant. This may result from a balance between cell division and cell death, or the population may simply cease to divide but remain metabolically active.

Microbial populations enter the stationary phase for several reasons. One obvious factor is nutrient limitation; if an essential nutrient is severely depleted, population growth will slow. Aerobic organisms often are limited by O_2 availability. Oxygen is not very soluble and may be depleted so quickly that only the surface of a culture will have an O_2 concentration adequate for growth. The cells beneath the surface will not be able to grow unless the culture is aerated in some way. Population growth also may cease due to the accumulation of toxic waste products. This factor seems to limit the growth of many cultures growing in the absence of O_2. For example, streptococci can produce so much lactic acid and other organic acids from sugar fermentation that their medium becomes acidic and growth is inhibited. Finally, some evidence exists that growth may cease when a critical population level is reached. Thus entrance into the stationary phase may result from several factors operating in concert.

Entry into stationary phase in response to starvation probably occurs often in nature. Many environments have low nutrient levels, and microbes have evolved a number of strategies to survive starvation. Some bacteria respond with obvious morphological changes such as endospore formation, but many only decrease somewhat in overall size. This is often accompanied by protoplast shrinkage and nucleoid condensation.

In many bacteria, the action of a protein called RpoS is central to starvation survival strategies. RpoS is a component of RNA polymerase holoenzyme, the enzyme responsible for binding DNA and initiating RNA synthesis. RpoS directs the other subunits of the enzyme (collectively called RNA polymerase core enzyme) to the appropriate locations along the chromosome, so that transcription can begin. Specifically, RpoS directs the core enzyme to genes encoding proteins that will help the bacterium survive starvation. ▸▸| *Transcription (section 12.6)*

The proteins made in response to starvation are called **starvation proteins.** They make the cell much more resistant to damage by starvation and other stressful conditions. Some increase peptidoglycan cross-linking and cell wall strength. The protein Dps (*DNA*-binding *p*rotein from *s*tarved cells) protects DNA. Proteins called chaperone proteins prevent protein denaturation and renature damaged proteins. Because of these and many other mechanisms, starved cells become harder to kill and more resistant to starvation, damaging temperature changes, oxidative and osmotic damage, and toxic chemicals such as chlorine. These changes are so effective that some members of the bacterial population survive starvation for years. These cells are called persister cells. There is even evidence that *Salmonella enterica* serovar Typhimurium (*S.* Typhimurium) and some other bacterial pathogens become more virulent when starved. Clearly these considerations are of great practical importance in medical and industrial microbiology.

Senescence and Death

Cells growing in batch culture cannot remain in stationary phase indefinitely. Eventually they enter a phase that for many years was described simply as the "death phase" (figure 7.11). During this phase, the number of viable cells often declines at an exponential rate. It was assumed that detrimental environmental changes such as nutrient deprivation and the buildup of toxic wastes caused irreparable harm to the cells. That is, even when bacterial cells were transferred to fresh medium, no cellular growth was observed. Because loss of viability was often not accompanied by a loss in total cell number, it was assumed that cells died but did not lyse.

This view is currently being debated. There are two alternative hypotheses (**figure 7.13**). Some microbiologists think the cells are temporarily unable to grow, at least under the laboratory conditions used. This phenomenon, in which the cells are called **viable but nonculturable (VBNC),** is thought to be the result of a genetic response triggered in starving, stationary phase cells. Just as some bacteria form endospores as a survival mechanism, it is argued that others are able to become dormant without changes in morphology (figure 7.13*b*). Once the appropriate conditions are available (e.g., a change in temperature or passage through an animal), VBNC microbes resume growth. VBNC microorganisms could pose a public health threat, as many assays that test for food and drinking water safety are culture-based.

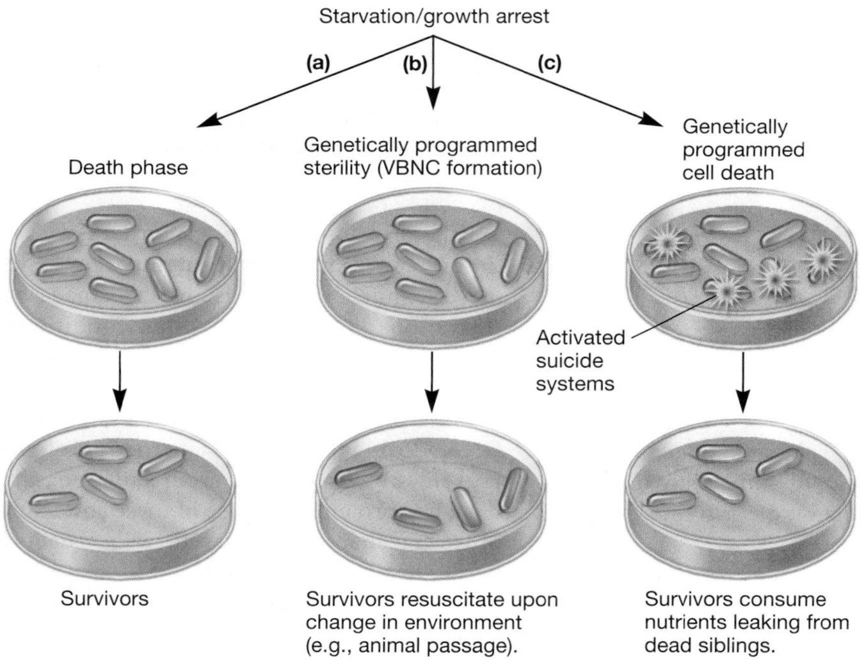

FIGURE 7.13 Loss of Viability. (a) It has long been assumed that as cells leave stationary phase due to starvation or toxic waste accumulation, the exponential decline in culturability is due to cellular death. (b) The viable but nonculturable (VBNC) hypothesis posits that when cells are starved, they become temporarily nonculturable under laboratory conditions. When exposed to appropriate conditions, some cells will regain the capacity to reproduce. (c) Some believe that a fraction of a microbial population dies due to activation of programmed cell death genes. The nutrients that are released by dying cells support the growth of other cells.

Figure 7.13 Micro Inquiry

Would each of these possibilities result in a sharp decline in growth as shown in a growth curve (figure 7.11)? That is, is each hypothesis consistent with what is observed when microbes are grown in batch culture?

The second alternative to a simple death phase is **programmed cell death** (figure 7.13*c*). In contrast to the VBNC hypothesis whereby cells are genetically programmed to survive, programmed cell death predicts that a fraction of the microbial population is genetically programmed to die after growth ceases. In this case, some cells die and the nutrients they leak enable the eventual growth of those cells in the population that did not initiate cell death. The dying cells are thus "altruistic"—they sacrifice themselves for the benefit of the larger population.

The view that the death phase is best described as an exponential decline in viability is also being questioned. Long-term growth experiments reveal that some microbes have a very gradual decline in the number of culturable cells. This decline can last months to years (**figure 7.14**). During this time, the bacterial population continually evolves so that actively reproducing cells are those best able to use the nutrients released by their dying brethren and best able to tolerate the accumulated toxins. This dynamic process is marked by successive waves of genetically distinct variants. Thus natural selection can be witnessed within a single culture vessel.

1. Describe the four phases of the growth curve and discuss the causes of each.
2. List two physiological changes that are observed in stationary cells. How do these changes impact the organism's ability to survive?
3. Contrast and compare the viable but nonculturable status of microbes with that of programmed cell death as a means of responding to starvation.

Mathematics of Growth

Knowledge of microbial growth rates during the exponential phase is indispensable to microbiologists. Growth rate studies contribute to basic physiological and ecological research, and are applied in industry. The quantitative aspects of exponential phase growth discussed here apply to microorganisms that divide by binary fission.

During the exponential phase, each microorganism is dividing at constant intervals. Thus the population doubles in number during a specific length of time called the **generation (doubling) time.** This can be illustrated with a simple example. Suppose that a culture tube is inoculated with one cell that divides every 20 minutes (**table 7.2**). The population will be 2 cells after 20 minutes,

Table 7.2 An Example of Exponential Growth

Time[a]	*Division Number*	2^n	*Population*[b] $(N_0 \times 2^n)$	$\log_{10}N_t$
0	0	$2^0 = 1$	1	0.000
20	1	$2^1 = 2$	2	0.301
40	2	$2^2 = 4$	4	0.602
60	3	$2^3 = 8$	8	0.903
80	4	$2^4 = 16$	16	1.204

[a] The hypothetical culture begins with one cell having a 20-minute generation time.
[b] Number of cells in the culture.

4 cells after 40 minutes, and so forth. Because the population is doubling every generation, the increase in population is always 2^n where n is the number of generations. The resulting population increase is exponential—that is, logarithmic (**figure 7.15**).

The mathematics of growth during the exponential phase are illustrated in **figure 7.16**, which shows the calculation of two important values. The **mean growth rate (μ)** is the number of generations per unit time and is often expressed as generations per hour. It can be used to calculate the **mean generation (doubling) time (g).** As can be seen in figure 7.16, the mean generation time is simply the reciprocal of the mean growth rate. The mean generation time can also be determined directly from a semilogarithmic plot of growth curve data (**figure 7.17**). Once this is done, it can be used to calculate the mean growth rate, again due to the reciprocal relationship between these two values.

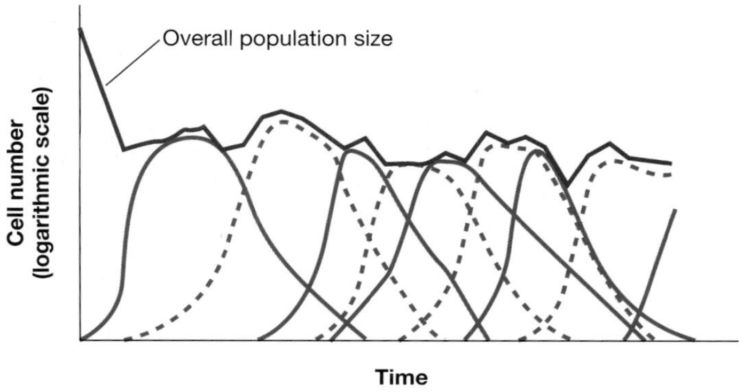

FIGURE 7.14 Prolonged Decline in Cell Numbers. Instead of a distinct death phase, successive waves of genetically distinct subpopulations of microbes better able to use the released nutrients and accumulated toxins survive. Each successive solid or dashed blue curve represents the growth of a new subpopulation.

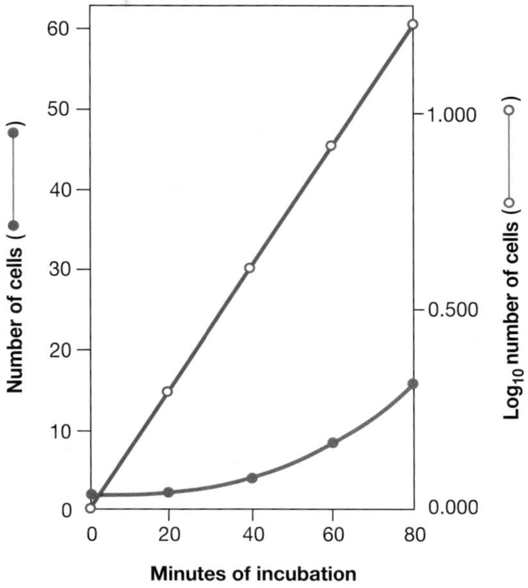

FIGURE 7.15 Exponential Microbial Growth. Four generations of growth are plotted directly (●—●) and in the logarithmic form (○—○). The growth curve is exponential, as shown by the linearity of the log plot.

Calculation of the mean growth rate

Let N_0 = the initial population number

N_t = the population at time t

n = the number of generations in time t

For populations reproducing by binary fission

$$N_t = N_0 \times 2^n$$

Solving for n, the number of generations, where all logarithms are to the base 10,

$$\log N_t = \log N_0 + n \cdot \log 2, \text{ and}$$

$$n = \frac{\log N_t - \log N_0}{\log 2} = \frac{\log N_t - \log N_0}{0.301}$$

The mean growth rate (μ) is the number of generations per unit time $\left(\frac{n}{t}\right)$. Thus

$$\mu = \frac{n}{t} = \frac{\log N_t - \log N_0}{0.301t}$$

Calculation of the mean generation (doubling) time

If a population doubles, then

$$N_t = 2N_0$$

Substitute $2N_0$ into the mean growth rate equation and solve for

$$\mu = \frac{\log (2N_0) - \log N_0}{0.301g} = \frac{\log 2 + \log N_0 - \log N_0}{0.301g}$$

$$\mu = \frac{1}{g}$$

The mean generation time is the reciprocal of the mean growth rate.

$$g = \frac{1}{\mu}$$

FIGURE 7.16 Calculation of Mean Growth Rate and Mean Generation Time. The calculations are only valid for the exponential phase of growth, when the growth rate is constant.

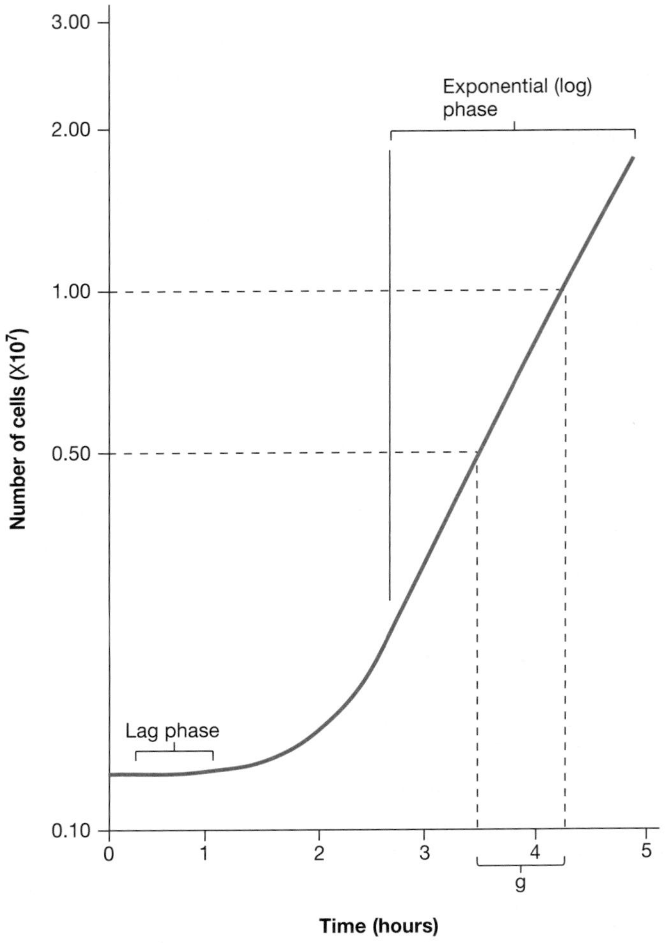

FIGURE 7.17 Generation Time Determination. The generation time can be determined from a microbial growth curve. The population data are plotted with the logarithmic axis used for the number of cells. The time to double the population number is then read directly from the plot. The log of the population number can also be plotted against time on regular axes.

Generation times vary markedly with the species of microorganism and environmental conditions. They range from less than 10 minutes (0.17 hours) to several days (**table 7.3**). Generation times in nature are usually much longer than in culture.

1. Define generation (doubling) time and mean growth rate. Calculate the mean growth rate and generation time of a culture that increases in the exponential phase from 5×10^2 to 1×10^8 in 12 hours.
2. Suppose the generation time of a bacterium is 90 minutes and the number of cells in a culture is 10^3 cells at the start of the log phase. How many bacteria will there be after 8 hours of exponential growth?

Table 7.3 Examples of Generation Times[a]

Microorganism	*Incubation Temperature (°C)*	*Generation Time (Hours)*
Bacteria		
Escherichia coli	40	0.35
Bacillus subtilis	40	0.43
Staphylococcus aureus	37	0.47
Pseudomonas aeruginosa	37	0.58
Clostridium botulinum	37	0.58
Mycobacterium tuberculosis	37	≈12
Treponema pallidum	37	33
Protists		
Tetrahymena geleii	24	2.2–4.2
Chlorella pyrenoidosa	25	7.75
Paramecium caudatum	26	10.4
Euglena gracilis	25	10.9
Giardia lamblia	37	18
Ceratium tripos	20	82.8
Fungi		
Saccharomyces cerevisiae	30	2
Monilinia fructicola	25	30

[a] Generation times differ depending on the growth medium and environmental conditions used.

7.4 Measurement of Microbial Growth

There are many ways to measure microbial growth to determine growth rates and generation times. Either population number or mass may be followed because growth leads to increases in both. Here the most commonly employed techniques for determining population size are examined and the advantages and disadvantages of each noted. No single technique is always best; the most appropriate approach depends on the experimental situation.

Direct Measurement of Cell Numbers

The most obvious way to determine microbial numbers is by **direct counts** using a counting chamber. This approach is easy, inexpensive, and relatively quick. It also gives information about the size and morphology of microorganisms. Petroff-Hausser counting chambers can be used for counting bacterial and archaeal cells; hemocytometers can be used for all cell types. Counting chambers consist of specially designed slides and coverslips;

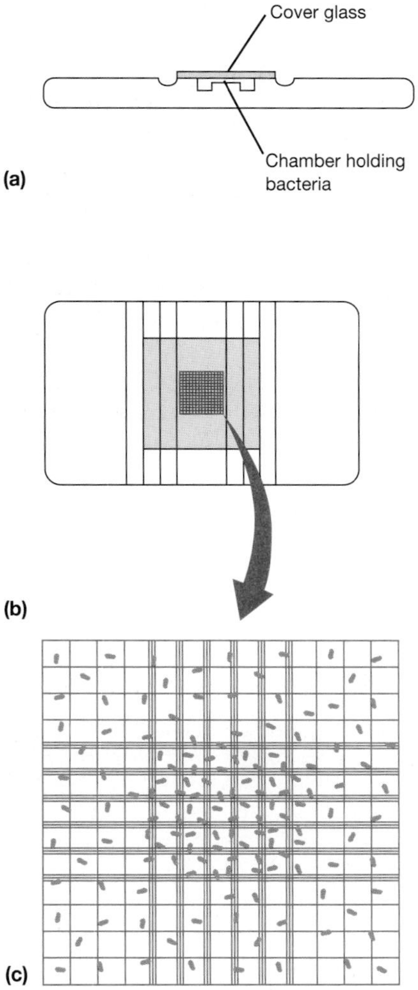

FIGURE 7.18 The Petroff-Hausser Counting Chamber. (a) Side view of the chamber showing the cover glass and the space beneath it that holds a bacterial suspension. (b) A top view of the chamber. The grid is located in the center of the slide. (c) An enlarged view of the grid. The bacteria in several of the central squares are counted, usually at ×400 to ×500 magnification. The average number of bacteria in these squares is used to calculate the concentration of cells in the original sample. Since there are 25 squares covering an area of 1 mm^2, the total number of bacteria in 1 mm^2 of the chamber is (number/square)(25 squares). The chamber is 0.02 mm deep and therefore,

$$\text{bacteria/mm}^3 = (\text{bacteria/square})(25 \text{ squares})(50).$$

The number of bacteria per cm^3 is 10^3 times this value. For example, suppose the average count per square is 28 bacteria:

$$\text{bacteria/cm}^3 = (28 \text{ bacteria})(25 \text{ squares})(50)(10^3) = 3.5 \times 10^7.$$

the space between the slide and coverslip creates a chamber of known depth. On the bottom of the chamber is an etched grid that facilitates counting the cells (**figure 7.18**). The number of microorganisms in a sample can be calculated by taking into account the chamber's volume and any dilutions made of the sample before counting. One disadvantage of using counting chambers is that to determine population size accurately, the microbial population must be relatively large and evenly dispersed because only a small volume of the population is sampled.

The number of bacteria in aquatic samples is frequently determined from direct counts after the bacteria have been trapped on membrane filters. In the membrane filter technique, the sample is first filtered through a black polycarbonate membrane filter. Then the bacteria are stained with nucleic acid fluorescent stains such as acridine orange or DAPI and observed microscopically *(see figure 27.6)*. Alternatively, fluorescently labeled dyes that are specific for a given taxon may be used. The stained cells are easily observed against the black background of the membrane filter and can be counted when viewed with an epifluorescence microscope. Direct cell counts of environmental samples almost invariably result in higher cell densities than do methods that rely on culturing. This is because only a small percentage (about 1%) of cells growing in nature can be cultivated in the laboratory. ◀◀ *Fluorescence microscope (section 2.2)*

Flow cytometry is increasingly being used to directly count microbes and to gain detailed information about them. A flow cytometer creates a stream of cells so narrow that one cell at a time passes through a beam of laser light. As each cell passes through the beam, the light is scattered. Scattered light is detected by the flow cytometer. Because cells are separated in space, each light scattering event is detected independently. Thus the number of light-scattering events represents the number of cells in the sample. Cells of differing size, internal complexity, and other characteristics within a population can also be counted. This usually involves the use of fluorescent dyes or fluorescently labeled antibodies. These more sophisticated uses of flow cytometry can provide valuable information about characteristics of the population of cells. ▶▶ *Flow cytometry (section 35.3)*

Search This: Flow cytometry tutorial

Microorganisms also can be directly counted with electronic counters such as the Coulter counter. In the Coulter counter, a microbial suspension is forced through a small hole. Electrical current flows through the hole, and electrodes placed on both sides of the hole measure electrical resistance. Every time a microbial cell passes through the hole, electrical resistance increases (i.e., the conductivity drops), and the cell is counted. ▶▶ *Identification of microorganisms from specimens (section 35.2)*

Traditional methods for directly counting microbes in a sample usually yield cell densities that are much higher than the plating methods described next because direct counting procedures do not distinguish dead cells from culturable cells. Newer methods for direct counts avoid this problem. Commercial kits that use fluorescent reagents to stain live and dead cells differently are now available, making it possible to count directly the number of live and dead microorganisms in a sample *(see figures 2.13a and 27.2)*.

Viable Counting Methods

Several plating methods can be used to determine the number of viable microbes in a sample. These are referred to as either **viable counting methods** or **plate counts** because they count only those cells that are able to reproduce when cultured. Two commonly used procedures are the spread-plate and the pour-plate techniques. In both of these methods, a diluted sample of microorganisms is dispersed over or within agar. If each cell is far enough away from other cells, then each cell will reproduce, generating a distinct colony. The samples should yield between 30 and 300 colonies for most accurate counting (counts of 25 to 250 are used in some applications), and the count is made more accurate by use of a colony counter. Once the number of colonies is known, the original number of viable microorganisms in the sample can be calculated from that number and the sample dilution. For example, if 1.0 milliliter of a solution diluted by a factor of 1×10^6 yielded 150 colonies (i.e., 1.5×10^2 colonies), then the original sample contained around 1.5×10^8 cells per milliliter. However, because it is not possible to be certain that each colony arose from an individual cell, the results are often expressed in terms of **colony forming units (CFU),** rather than the number of microorganisms. ⏮ *Spread plate and pour plate (section 6.8)*

Another commonly used plating method first traps bacteria in aquatic samples on a membrane filter. The filter is then placed on an agar medium or on a pad soaked with liquid media (**figure 7.19**) and incubated until each cell forms a separate colony. A colony count gives the number of microorganisms in the filtered sample, and selective media can be used to select for specific microorganisms (**figure 7.20**). This technique is especially useful in analyzing water

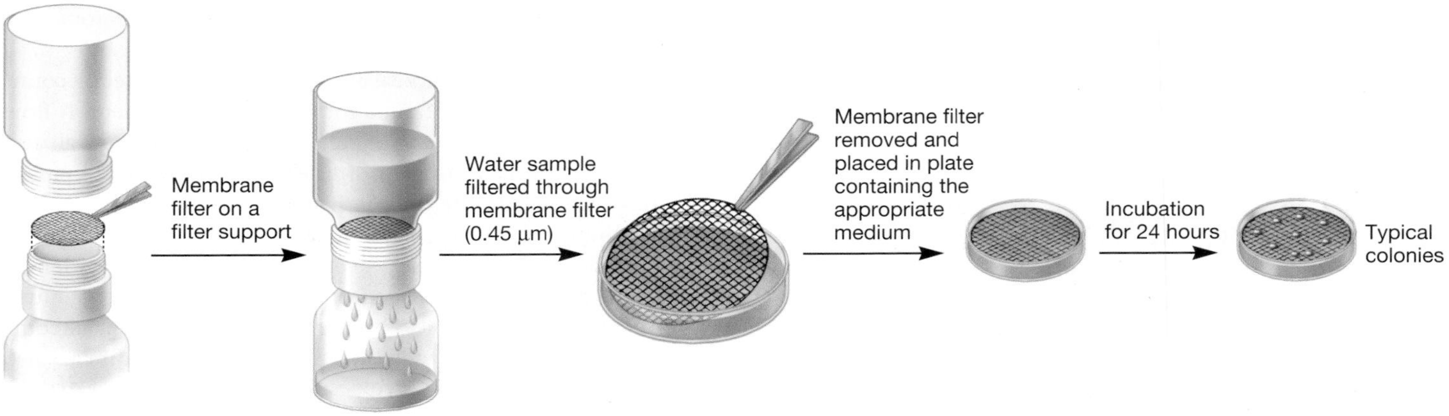

FIGURE 7.19 The Membrane Filtration Procedure. Membranes with different pore sizes are used to trap different microorganisms. Incubation times for membranes also vary with the medium and microorganism.

Figure 7.19 Micro Inquiry

Why is it important to have no more than about 250 colonies on the plate when counting colonies to determine population size?

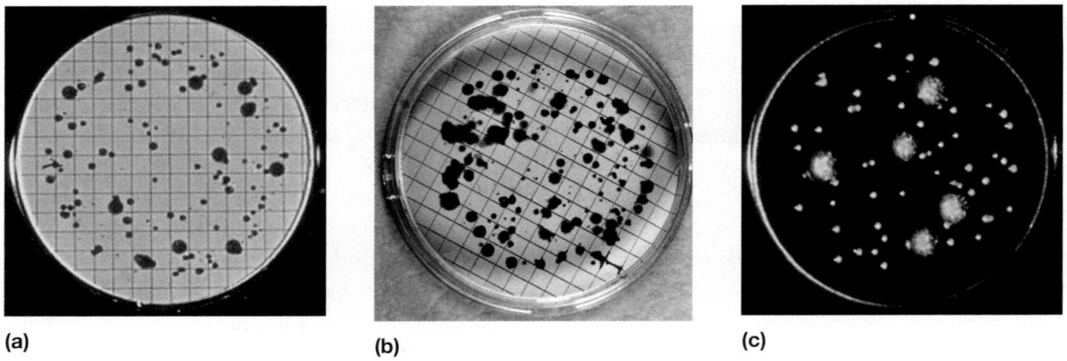

FIGURE 7.20 Colonies on Membrane Filters. Membrane-filtered samples grown on a variety of media. (a) Standard nutrient media for a total bacterial count. An indicator colors colonies dark red for easy counting. (b) Fecal coliform medium for detecting fecal coliforms that form blue colonies. (c) Wort agar for the culture of yeasts and molds.

purity. ⏮ *Culture media (section 6.7);* ⏭ *Water purification and sanitary analysis (section 42.1).*

Plating techniques are simple, sensitive, and widely used for viable counts of bacteria and other microorganisms in samples of food, water, and soil. Several problems, however, can lead to inaccurate counts. Low counts will result if clumps of cells are not broken up and the microorganisms well dispersed. The hot agar used in the pour-plate technique may injure or kill sensitive cells; thus spread plates sometimes give higher counts than pour plates. A major problem is that plate counts will be low if the medium employed cannot support growth of all the viable microorganisms present. This is a common problem encountered by microbial ecologists trying to understand the makeup of microbial communities. In such studies, the number of cells observed by microscopy is often much higher than the population size determined by plate counts. This discrepancy is sometimes referred to as "the great plate count anomaly" and is discussed in detail in chapter 27. ⏭ *Assessing microbial diversity (section 27.2)*

Sometimes plate counts cannot be used to measure population size. For instance, plate counts are not helpful if the microbe cannot be cultured on solid media or if its colonies spread across the plate, making it impossible to get an accurate count. In these cases, another approach is used: **most probable number (MPN)** determination. In this method, numerous replicates of several dilutions of a culture are made and added to tubes containing a suitable liquid growth medium. The volume of inoculum to each tube and a tenfold dilution series is usually used. After incubation, each tube is examined to determine if growth occurred. It is assumed that the last tube in the dilution series that demonstrates growth was inoculated with between one and 10 cells, while the next tube had between 11 and 100 cells, and so on. If no growth is observed, then the tube is assumed not to have received any cells. In this way, the number of cells in the original sample is estimated (*for example, see figure 27.3*). MPN values are most commonly used when a selective medium can be employed that supports the growth of a specific type of microbe. It is often used in assessing microbial density in water samples. Because results can vary widely due to sampling errors, it is usually necessary to set up many replicates of each dilution. ⏭ *Culturing techniqes (section 27.1)*

Measurement of Cell Mass

Techniques for measuring changes in cell mass also can be used to follow growth. One approach is the determination of microbial dry weight. Cells growing in liquid medium are collected by centrifugation, washed, dried in an oven, and weighed. This is an especially useful technique for measuring the growth of filamentous fungi. It is time-consuming, however, and not very sensitive. Because bacteria weigh so little, it may be necessary to centrifuge several hundred milliliters of culture to collect a sufficient quantity.

A more rapid and sensitive method for measuring cell mass is spectrophotometry (**figure 7.21**). Spectrophotometry depends on the fact that microbial cells scatter light that strikes them. Because microbial cells in a population are of roughly constant size, the amount of scattering is directly proportional to the biomass of cells present and indirectly related to cell number. When the concentration of bacteria reaches about a million (10^6) cells per milliliter, the medium appears slightly cloudy or turbid. Further increases in concentration result in greater turbidity, and less light is transmitted through the medium. The extent of light scattering (i.e., decrease in transmitted light) can be measured by a spectrophotometer and is called the absorbance (optical density) of the medium. Absorbance is almost linearly related to cell concentration at absorbance levels less than about 0.5. If the sample exceeds this value, it must first be diluted and then absorbance measured. Thus population growth can be easily measured as long as the population is high enough to give detectable turbidity.

Cell mass can also be estimated by measuring the concentration of some cellular substance, as long as its concentration is constant in each cell. For example, a sample of cells can be analyzed for total protein or nitrogen. An increase in the microbial population will be reflected in higher total protein levels. Similarly, chlorophyll determinations can be used to measure phototrophic protist and cyanobacterial populations, and the quantity of ATP can be used to estimate the amount of living microbial mass.

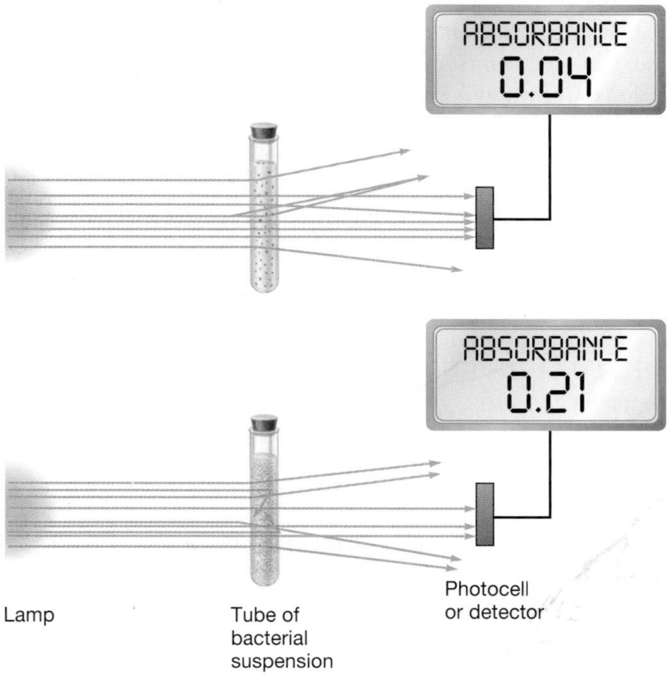

FIGURE 7.21 Turbidity and Microbial Mass Measurement. Determination of microbial mass by measurement of light absorption. As the population and turbidity increase, more light is scattered and the absorbance reading given by the spectrophotometer increases.

1. Briefly describe each technique by which microbial population numbers may be determined and give its advantages and disadvantages.
2. When using direct cell counts to follow the growth of a culture, it may be difficult to tell when the culture enters the phase of senescence and death. Why?
3. Why are plate count results expressed as colony forming units?
4. For each of the following, which enumeration technique would you use? Explain your choice. (a) A pure culture of *Staphylococcus aureus;* (b) a water sample that needs to be checked for *E. coli* contamination; (c) a sample of yogurt.

7.5 Continuous Culture of Microorganisms

Thus far, our focus has been on closed systems called batch cultures in which nutrients are not renewed nor wastes removed. Exponential (logarithmic) growth lasts for only a few generations and soon stationary phase is reached. However, it is possible to grow microorganisms in a system with constant environmental conditions maintained through continual provision of nutrients and removal of wastes. Such a system is called a **continuous culture system.** These systems can maintain a microbial population in exponential growth, growing at a known rate and at a constant biomass concentration for extended periods. Continuous culture systems make possible the study of microbial growth at very low nutrient levels, concentrations close to those present in natural environments. These systems are essential for research in many areas, including ecology. For example, interactions between microbial species in environmental conditions resembling those in a freshwater lake or pond can be modeled. Continuous culture systems also are used in food and industrial microbiology. Two major types of continuous culture systems commonly are used: chemostats and turbidostats. ▶▶| *Microbiology of food (chapter 40); Industrial microbiology (chapter 41)*

Chemostats

A **chemostat** is constructed so that the rate at which sterile medium is fed into the culture vessel is the same as the rate at which the media containing microorganisms is removed (**figure 7.22**). The culture medium for a chemostat possesses an essential nutrient (e.g., a vitamin) in limiting quantities. Because one nutrient is limiting, growth rate is determined by the rate at which new medium is fed into the growth chamber; the final cell density depends on the concentration of the limiting nutrient. The rate of nutrient exchange is expressed as the dilution rate (D), the rate at which medium flows through the culture vessel relative to the vessel volume, where f is the flow rate (milliliter/hr) and V is the vessel volume (milliliter).

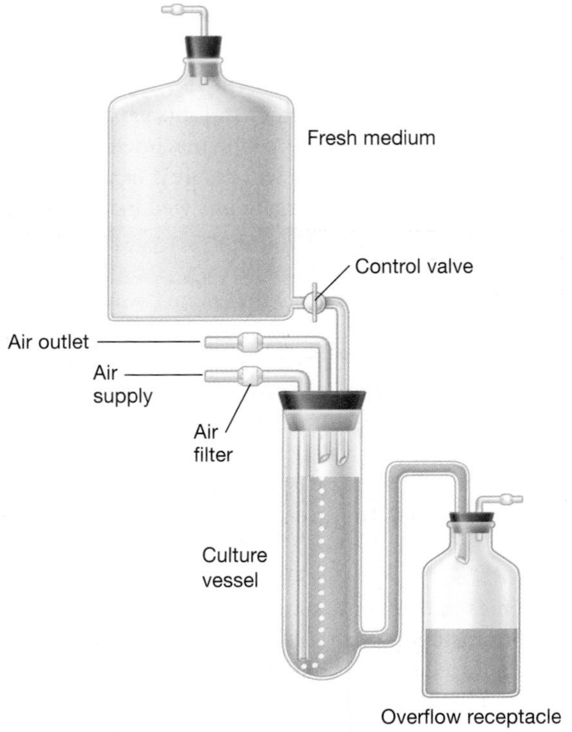

FIGURE 7.22 The Chemostat. Schematic diagram of the system. The fresh medium contains a limiting amount of an essential nutrient. Although not shown, usually the incoming medium is mechanically mixed with that already in the culture vessel. Growth rate is determined by the rate of flow of medium through the culture vessel.

$$D = f/V$$

For example, if f is 30 milliliter/hr and V is 100 milliliter, the dilution rate is 0.30 hr^{-1}.

Both population size and generation time are related to the dilution rate (**figure 7.23**). When dilution rates are very low, only a limited supply of nutrient is available and the microbes can conserve only a limited amount of energy. Much of that energy must be used for cell maintenance, not for growth and reproduction. Slightly higher dilution rates make more nutrients available to the microbes. When the dilution rate provides enough nutrients for both maintenance and reproduction, the cell density will begin to rise. In other words, the growth rate can increase (and the generation time can decrease) when the total available energy provided by the nutrient supply exceeds the **maintenance energy.** Notice in figure 7.23 that for a wide range of dilution rates, the cell density remains stable. This is because the limiting nutrient is almost completely depleted by the rapid reproductive rate of the cells (as seen by the decreasing generation times). However, if the dilution rate is too high,

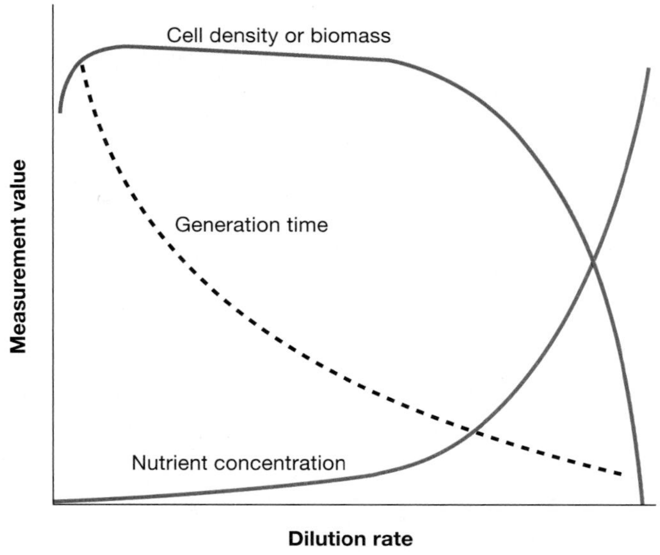

FIGURE 7.23 Chemostat Dilution Rate and Microbial Growth. The effects of changing the dilution rate in a chemostat.

Figure 7.23 Micro Inquiry

Why does the cell density stay relatively constant over such a large range of dilution rates?

microorganisms can actually be washed out of the culture vessel before reproducing because the dilution rate is greater than the maximum growth rate.

Turbidostats

The second type of continuous culture system, the **turbidostat,** has a photocell that measures the turbidity (defined as the amount of light scattered) of the culture in the growth vessel. The flow rate of media through the vessel is automatically regulated to maintain a predetermined turbidity. Because turbidity is related to cell density, the turbidostat maintains a desired cell density. The turbidostat differs from the chemostat in several ways. The dilution rate in a turbidostat varies, rather than remaining constant, and a turbidostat's culture medium contains all nutrients in excess. That is, none of the nutrients is limiting. The turbidostat operates best at high dilution rates; the chemostat is most stable and effective at lower dilution rates.

1. How does a continuous culture system differ from a closed culture system (i.e., a batch culture)?
2. Describe how chemostats and turbidostats operate. How do they differ?
3. What is the dilution rate? What is maintenance energy? How are they related?

7.6 Influences of Environmental Factors on Growth

As we have seen, microorganisms are able to respond to variations in nutrient levels. Microorganisms also are greatly affected by the chemical and physical nature of their surroundings. An understanding of environmental influences aids in the control of microbial growth and the study of the ecological distribution of microorganisms.

The adaptations of some microorganisms to extreme and inhospitable environments are truly remarkable. Microbes are present virtually everywhere on Earth. Many habitats in which microbes thrive would kill most other organisms. Bacteria such as *Bacillus infernus* are able to live over 2.4 km below Earth's surface, without oxygen and at temperatures above 60°C. Other microbes live in acidic hot springs, at great ocean depths, or in lakes such as the Great Salt Lake in Utah (USA) that have high sodium chloride concentrations. Microorganisms that grow in such harsh conditions are called **extremophiles.** ▶▶| *Microorganisms in benthic marine environments (section 28.2); The subsurface biosphere (section 29.4)*

In this section, we briefly review the effects of the most important environmental factors on microbial growth. Major emphasis is given to solutes and water activity, pH, temperature, oxygen level, pressure, and radiation. **Table 7.4** summarizes how microorganisms are categorized in terms of their response to these factors. It is important to note that for most environmental factors, a range of levels supports growth of a microbe. For example, a microbe might exhibit optimum growth at pH 7 but grows, though not optimally, at pH values down to pH 6 (its pH minimum) and up to pH 8 (its pH maximum). Furthermore, outside this range, the microbe might cease reproducing but remain viable for some time. Clearly each microbe must have evolved adaptations that allow it to adjust its physiology within its ecological range, and it may also have adaptations that protect it in environments outside this range. These adaptations also are discussed in this section.

Solutes and Water Activity

Because a selectively permeable plasma membrane separates microorganisms from their environment, they can be affected by changes in the osmotic concentration of their surroundings. If a microorganism is placed in a hypotonic solution (one with a lower osmotic concentration), water will enter the cell and cause it to burst unless something is done to prevent the influx or inhibit plasma membrane expansion. Conversely, if it is placed in a hypertonic solution (one with a higher osmotic concentration), water will flow out of the cell. In microbes that have cell walls, the membrane shrinks away from the cell wall—a process called plasmolysis. Dehydration of the cell in hypertonic environments may damage the cell membrane and cause the cell to become metabolically inactive.

Table 7.4 Microbial Responses to Environmental Factors

Descriptive Term	*Definition*	*Representative Microorganisms*
Solute and Water Activity		
Osmotolerant	Able to grow over wide ranges of water activity or osmotic concentration	*Staphylococcus aureus, Saccharomyces rouxii*
Halophile	Requires high levels of sodium chloride, usually above about 0.2 M, to grow	*Halobacterium, Dunaliella, Ectothiorhodospira*
pH		
Acidophile	Growth optimum between pH 0 and 5.5	*Sulfolobus, Picrophilus, Ferroplasma, Acontium*
Neutrophile	Growth optimum between pH 5.5 and 8.0	*Escherichia, Euglena, Paramecium*
Alkalophile	Growth optimum between pH 8.0 and 11.5	*Bacillus alcalophilus, Natronobacterium*
Temperature		
Psychrophile	Grows at 0°C and has an optimum growth temperature of 15°C or lower	*Bacillus psychrophilus, Chlamydomonas nivalis*
Psychrotroph	Can grow at 0–7°C; has an optimum between 20 and 30°C and a maximum around 35°C	*Listeria monocytogenes, Pseudomonas fluorescens*
Mesophile	Has growth optimum between 20–45°C	*Escherichia coli, Trichomonas vaginalis*
Thermophile	Can grow at 55°C or higher; optimum often between 55 and 65°C	*Geobacillus stearothermophilus, Thermus aquaticus, Cyanidium caldarium, Chaetomium thermophile*
Hyperthermophile	Has an optimum between 85 and about 113°C	*Sulfolobus, Pyrococcus, Pyrodictium*
Oxygen Concentration		
Obligate aerobe	Completely dependent on atmospheric O_2 for growth	*Micrococcus luteus,* most protists and fungi
Facultative anaerobe	Does not require O_2 for growth but grows better in its presence	*Escherichia, Enterococcus, Saccharomyces cerevisiae*
Aerotolerant anaerobe	Grows equally well in presence or absence of O_2	*Streptococcus pyogenes*
Obligate anaerobe	Does not tolerate O_2 and dies in its presence	*Clostridium, Bacteroides, Methanobacterium*
Microaerophile	Requires O_2 levels between 2–10% for growth and is damaged by atmospheric O_2 levels (20%)	*Campylobacter, Spirillum volutans, Treponema pallidum*
Pressure		
Barophile	Growth more rapid at high hydrostatic pressures	*Photobacterium profundum, Shewanella benthica*

Clearly it is important that microbes be able to respond to changes in the osmotic concentrations of their environment. Microbes in hypotonic environments can reduce the osmotic concentration of their cytoplasm. Microbes use several mechanisms to achieve this. For example, some bacteria have mechanosensitive (MS) channels in their plasma membrane. In a hypotonic environment, the membrane stretches due to an increase in hydrostatic pressure and cellular swelling. MS channels then open and allow solutes to leave. Thus MS channels act as escape valves to protect cells from bursting. Because many protists do not have a cell wall, they use contractile vacuoles to expel excess water.

Many microorganisms, whether in hypotonic or hypertonic environments, keep the osmotic concentration of their cytoplasm somewhat above that of the habitat so that the plasma membrane is always pressed firmly against their cell wall. They do so by the use of **compatible solutes,** which are chemicals that can be kept at high intracellular concentrations without interfering with metabolism and growth. Most bacteria and archaea increase their internal osmotic concentration in a hypertonic environment through the synthesis or uptake of the compatible solutes choline and betaine. They also use amino acids such as proline and glutamic acid, and elevated levels of potassium and chloride ions. Photosynthetic protists and fungi employ sucrose and polyols (e.g., arabitol, glycerol, and mannitol) as compatible solutes. Polyols and amino acids are ideal compatible solutes because they normally do not disrupt enzyme structure and function. ◀◀ *Inclusions (section 3.5)*

Some microbes are adapted to extreme hypertonic environments. **Halophiles** require the presence of NaCl or other salts at a concentration above about 0.2 M (**figure 7.24**). Extreme

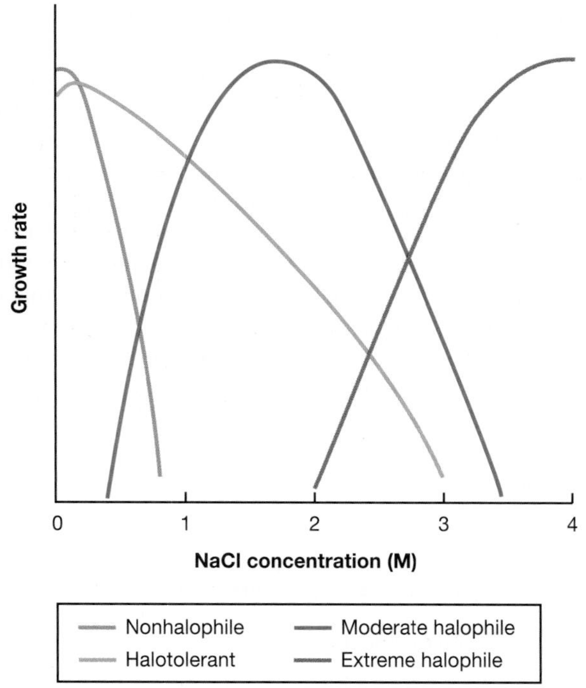

FIGURE 7.24 The Effects of Sodium Chloride on Microbial Growth. Four different patterns of microbial dependence on NaCl concentration are depicted. The curves are only illustrative and are not meant to provide precise shapes or salt concentrations required for growth.

Figure 7.24 Micro Inquiry

What is the difference between halophilic and halotolerant?

halophiles have adapted so completely to hypertonic, saline conditions that they require high levels of sodium chloride to grow—concentrations between about 2 M and saturation (about 6.2 M). The archaeon *Halobacterium* can be isolated from the Dead Sea (a salt lake between Israel and Jordan), the Great Salt Lake in Utah, and other aquatic habitats with salt concentrations approaching saturation. *Halobacterium* and other extremely halophilic archaea accumulate enormous quantities of potassium and chloride ions in order to remain hypertonic to their environment; the internal potassium concentration may reach 4 to 7 M. Furthermore, their enzymes, ribosomes, and transport proteins require high potassium levels for stability and activity. In addition, the plasma membrane and cell wall of *Halobacterium* are stabilized by high concentrations of sodium ion. If the sodium concentration decreases too much, the wall and plasma membrane disintegrate. Extreme halophiles have successfully adapted to environmental conditions that would destroy most organisms. In the process, they have become so specialized that they have lost ecological flexibility and can prosper only in a few extreme habitats. ▶▶| *Phylum Euryarchaeota: Halobacteria (section 18.3)*

Because the osmotic concentration of a habitat has such profound effects on microorganisms, it is useful to express quantitatively the degree of water availability. Microbiologists generally use **water activity (a_w)** for this purpose (water availability also may be expressed as water potential, which is related to a_w). The water activity of a solution is 1/100 the relative humidity of the solution (when expressed as a percent). It is also equivalent to the ratio of the solution's vapor pressure (P_{soln}) to that of pure water (P_{water}).

$$a_w = \frac{P_{soln}}{P_{water}}$$

Thus while distilled water has an a_w of 1, milk has an a_w of 0.97, a saturated salt solution has an a_w of 0.75, and the a_w of dried fruits is only about 0.5.

Microorganisms differ greatly in their ability to acclimate to habitats with low water activity. A low water activity indicates that most of the water is chemically or structurally bound to other components in the medium. A microorganism must expend extra effort to grow in a habitat with a low a_w value because it must maintain a high internal solute concentration to retain water. Some microorganisms can do this and are **osmotolerant;** they grow over wide ranges of water activity. For example, *Staphylococcus aureus* is halotolerant, can be cultured in media containing sodium chloride concentration up to about 3 M, and is well adapted for growth on the skin. The yeast *Saccharomyces rouxii* grows in sugar solutions with a_w values as low as 0.6. The photosynthetic protist *Dunaliella viridis* tolerates sodium chloride concentrations from 1.7 M to a saturated solution. Some microbes (e.g., *Halobacterium*) are true xerophiles. That is, they grow best at low a_w. However, most microorganisms only grow well at water activities around 0.98 (the approximate a_w for seawater) or higher. This is why drying food or adding large quantities of salt and sugar effectively prevents food spoilage. ▶▶| *Controlling food spoilage (section 40.2)*

1. How do microorganisms adapt to hypotonic and hypertonic environments? What is plasmolysis?
2. Define water activity and briefly describe how it can be determined. Why is it difficult for microorganisms to grow at low a_w values?
3. What are halophiles and why does *Halobacterium* require sodium and potassium ions?

pH

pH is a measure of the relative acidity of a solution and is defined as the negative logarithm of the hydrogen ion concentration (expressed in terms of molarity).

$$pH = -\log[H^+] = \log(1/[H^+])$$

The pH scale extends from pH 0.0 (1.0 M H^+) to pH 14.0 (1.0×10^{-14} M H^+), and each pH unit represents a tenfold change in hydrogen ion concentration. **Figure 7.25** shows that microbial habitats vary widely in pH—from pH 0 to 2 at the acidic end to alkaline lakes and soil with pH values between 9 and 10.

Each species has a definite pH growth range and pH growth optimum. **Acidophiles** have their growth optimum between pH 0 and 5.5; **neutrophiles,** between pH 5.5 and 8.0; and **alkaliphiles (alkalophiles),** between pH 8.0 and 11.5. Extreme alkaliphiles have growth optima at pH 10 or higher. In general, different microbial groups have characteristic pH preferences. Most known bacteria and protists are neutrophiles. Most fungi prefer more acidic surroundings, about pH 4 to 6; photosynthetic protists also seem to favor slight acidity. Many archaea are acidophiles. For example, the archaeon *Sulfolobus acidocaldarius* is a common inhabitant of acidic hot springs; it grows well from pH 1 to 3 and at high temperatures. The archaea *Ferroplasma acidarmanus* and *Picrophilus oshimae* can actually grow very close to pH 0. Alkaliphiles are distributed among all three domains of life. They include bacteria belonging to the genera *Bacillus, Micrococcus, Pseudomonas,* and *Streptomyces;* yeasts and filamentous fungi; and numerous archaea. Because seawater has a pH of about 8.3, marine microorganisms are alkaliphilic.

Although microorganisms often grow over wide ranges of pH and far from their optima, there are limits to their tolerance. When the external pH is low, the concentration of H^+ is much greater outside than inside, and H^+ will move into the cytoplasm and lower the cytoplasmic pH. Drastic variations in cytoplasmic pH can harm microorganisms by disrupting the plasma membrane or inhibiting the activity of enzymes and membrane transport proteins. Most microbes die if the internal pH drops much below 5.0 to 5.5. Changes in the external pH also might alter the ionization of nutrient molecules and thus reduce their availability to the organism.

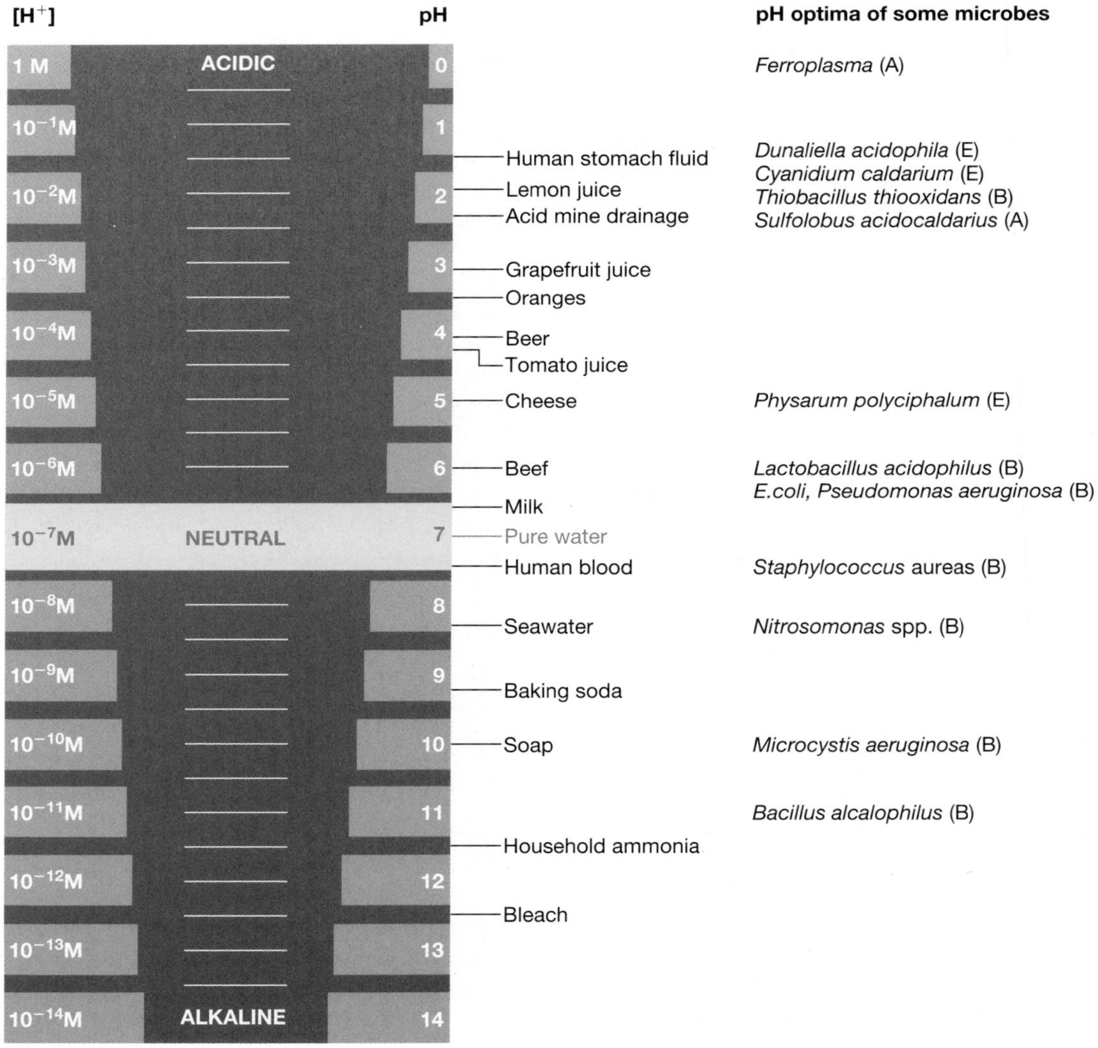

FIGURE 7.25 The pH Scale. The pH scale and examples of substances with different pH values. The microorganisms are placed at their growth optima.

Microorganisms respond to external pH changes using mechanisms that maintain a neutral cytoplasmic pH. Several mechanisms for adjusting to small changes in external pH have been proposed. Neutrophiles appear to exchange potassium for protons using an antiport transport system. Internal buffering also may contribute to pH homeostasis. However, if the external pH becomes too acidic, other mechanisms come into play. When the pH drops below about 5.5 to 6.0, *Salmonella enterica* serovar Typhimurium and *E. coli* synthesize an array of new proteins as part of what has been called their acidic tolerance response. A proton-translocating ATPase enzyme contributes to this protective response by pumping protons out of the cell. If the external pH decreases to 4.5 or lower, acid shock proteins and heat shock proteins are synthesized. These prevent the denaturation of proteins and aid in the refolding of denatured proteins in acidic conditions. ◀◀ *Uptake of nutrients (section 6.6);* ▶▶ *Protein maturation and secretion (section 12.9)*

What about microbes that live at pH extremes? Extreme alkaliphiles such as *Bacillus alcalophilus* maintain their internal pH close to neutrality by exchanging internal sodium ions for external protons. Acidophiles use a variety of measures to maintain a neutral internal pH. These include the transport of cations (e.g., potassium ions) into the cell, thus decreasing the movement of H^+ into the cell; proton transporters that pump H^+ out if they get in; and highly impermeable cell membranes (*see figure 18.4*).

1. Define pH, acidophile, neutrophile, and alkaliphile.
2. Classify each of the following organisms as an alkaliphile, a neutrophile, or an acidophile: *Staphylococcus aureus, Microcystis aeruginosa, Sulfolobus acidocaldarius,* and *Pseudomonas aeruginosa.* Which might be pathogens? Explain your choices.
3. Describe the mechanisms microbes use to maintain an internal neutral pH. Explain how extreme pH values might harm microbes.

Temperature

Microorganisms are particularly susceptible to external temperatures because they cannot regulate their internal temperature. An important factor influencing the effect of temperature on growth is the temperature sensitivity of enzyme-catalyzed reactions. Each enzyme has a temperature at which it functions optimally. At some temperature below the optimum, it ceases to be catalytic. As the temperature rises from this low point, the rate of catalysis increases to that observed for the optimal temperature. The velocity of the reaction roughly doubles for every 10°C rise in temperature. When all enzymes in a microbe are considered together, as the rate of each reaction increases, metabolism as a whole becomes more active and the microorganism grows faster. However, beyond a certain point, further increases actually slow growth, and sufficiently high temperatures are lethal. High temperatures denature enzymes, transport carriers, and other proteins. Temperature also has a significant effect on microbial membranes. At very low temperatures, membranes solidify. At high temperatures, the lipid bilayer simply melts and disintegrates. Thus when organisms are above or below their optimum temperature, both function and cell structure are affected.

Because of these opposing temperature influences, microbial growth has a characteristic temperature dependence with distinct **cardinal temperatures**—minimum, optimum, and maximum growth temperatures (**figure 7.26**). Although the shape of temperature dependence curves varies, the temperature optimum is always closer to the maximum than to the minimum. The cardinal temperatures are not rigidly fixed. Instead they depend to some extent on other environmental factors such as pH and available nutrients. For example, *Crithidia fasciculate,* a flagellated protist living in the gut of mosquitoes, grows in a simple medium at 22 to 27°C. However, its medium must be supplemented with extra metals, amino acids, vitamins, and lipids for it to grow at 33 to 34°C.

The cardinal temperatures vary greatly among microorganisms (**table 7.5**). Optima usually range from 0°C to 75°C, whereas microbial growth occurs at temperatures extending from less than −20°C to over 120°C. Some archaea even grow at 121°C (250°F), the temperature normally used in autoclaves (**Microbial Diversity & Ecology 7.1**). A major factor determining growth range seems to be water. Even at the most extreme temperatures, microorganisms need liquid water to grow. The growth temperature range for a particular microorganism usually spans about 30 degrees. Some species (e.g., *Neisseria gonorrhoeae*) have a small range; others, such as *Enterococcus faecalis,* grow over a wide range of temperatures. The major microbial groups differ from one another regarding their maximum growth temperatures. The upper limit for protists is around 50°C. Some fungi grow at temperatures as high as 55 to 60°C. *Bacteria* and *Archaea* can grow at much higher temperatures than eukaryotes. It has been suggested that eukaryotes are not able to manufacture stable and functional

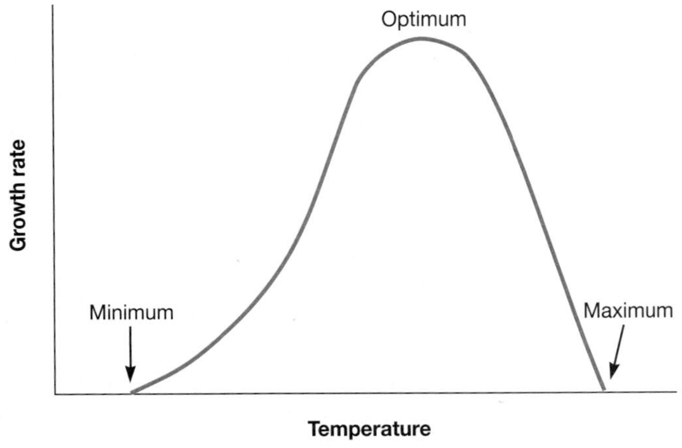

FIGURE 7.26 Temperature and Growth. The effect of temperature on growth rate.

Table 7.5 Temperature Ranges for Microbial Growth

Microorganism	Cardinal Temperatures (°C) Minimum	Optimum	Maximum
Nonphotosynthetic Bacteria and Archaea			
Bacillus psychrophilus	−10	23–24	28–30
Pseudomonas fluorescens	4	25–30	40
Enterococcus faecalis	0	37	44
Escherichia coli	10	37	45
Neisseria gonorrhoeae	30	35–36	38
Thermoplasma acidophilum	45	59	62
Thermus aquaticus	40	70–72	79
Pyrococcus abyssi	67	96	102
Pyrodictium occultum	82	105	110
Pyrolobus fumarii	90	106	113
Photosynthetic Bacteria			
Anabaena variabilis	ND[a]	35	ND
Synechococcus eximius	70	79	84
Protists			
Chlamydomonas nivalis	−36	0	4
Amoeba proteus	4–6	22	35
Skeletonema costatum	6	16–26	>28
Trichomonas vaginalis	25	32–39	42
Tetrahymena pyriformis	6–7	20–25	33
Cyclidium citrullus	18	43	47
Fungi			
Candida scotti	0	4–15	15
Saccharomyces cerevisiae	1–3	28	40
Mucor pusillus	21–23	45–50	50–58

[a] ND, not determined.

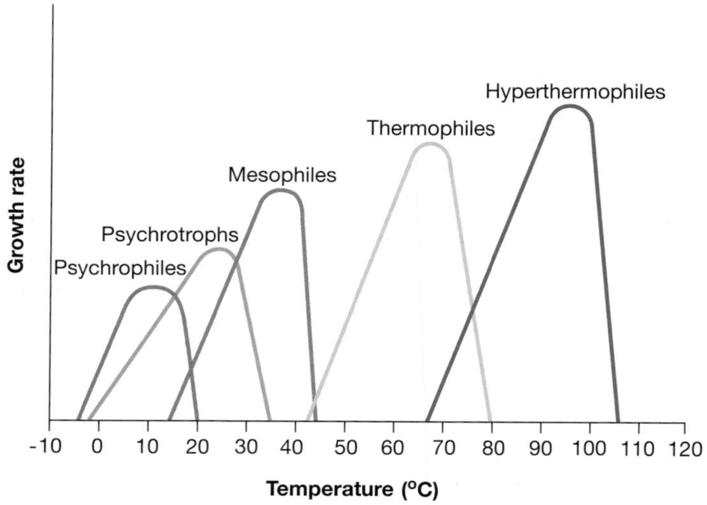

FIGURE 7.27 Temperature Ranges for Microbial Growth. Microorganisms are placed in different classes based on their temperature ranges for growth. They are ranked in order of increasing growth temperature range as psychrophiles, psychrotrophs, mesophiles, thermophiles, and hyperthermophiles. Representative ranges and optima for these five types are illustrated.

organellar membranes at temperatures above 60°C. The photosynthetic apparatus also appears to be relatively unstable because photosynthetic organisms are not found growing at very high temperatures.

Several terms are used to describe microbes based on their temperature ranges for growth (**figure 7.27**). Microbes that can grow in cold environments are either psychrophiles or psychrotrophs. **Psychrophiles** grow well at 0°C and have an optimum growth temperature of 15°C; the maximum is around 20°C. They are readily isolated from Arctic and Antarctic habitats. Oceans constitute an enormous habitat for psychrophiles because 90% of ocean water is 5°C or colder. The psychrophilic protist *Chlamydomonas nivalis* can actually turn a snowfield or glacier pink with its bright red spores. Psychrophiles are widespread among bacterial taxa and are found in such genera as *Pseudomonas, Vibrio, Alcaligenes, Bacillus, Photobacterium,* and *Shewanella.* A variety of psychrophilic archaea have been isolated from Antarctica, the Arctic, and other cold habitats. Psychrophilic microorganisms have adapted to their environment in several ways. Their enzymes, transport systems, and protein synthetic machinery function well at low temperatures. The cell membranes of psychrophilic microorganisms have high levels of unsaturated fatty acids and remain semifluid when cold. Indeed, many psychrophiles begin to leak cellular constituents at temperatures higher than 20°C because of cell membrane disruption. **Psychrotrophs (psychrotolerants)** grow at 0 or higher and typically have maxima at about 35°C. Psychrophilic bacteria and fungi are major causes of refrigerated food spoilage.

Mesophiles are microorganisms that grow in moderate temperatures. They have growth optima around 20 to 45°C and often have a temperature minimum of 15 to 20°C and a maximum of about 45°C. Most microorganisms probably fall within this category. Almost all human pathogens are mesophiles, as might be expected because the human body is a fairly constant 37°C.

Microbes that grow best at high temperatures are the thermophiles and hyperthermophiles. **Thermophiles** grow at temperatures between 55 and 85°C. Their growth minimum is usually around 45°C, and they often have optima between 55 and 65°C. The vast majority are members of *Bacteria* or *Archaea,* although a few photosynthetic protists and fungi are thermophilic (table 7.4). Thermophiles flourish in many habitats including composts, self-heating hay stacks, hot water lines, and hot springs. **Hyperthermophiles** have growth optima between 85°C and about 113°C. They usually do not grow below 55°C. *Pyrococcus abyssi* and *Pyrodictium occultum* are examples of marine hyperthermophiles found in hot areas of the seafloor.

MICROBIAL DIVERSITY & ECOLOGY

7.1 Life Above 100°C

Until recently, the highest reported temperature for bacterial or archaeal growth was 105°C. It seemed that the upper temperature limit for life was about 100°C, the boiling point of water. Now thermophilic bacteria and archaea have been reported growing in sulfide chimneys or "black smokers," located along rifts and ridges on the ocean floor, that spew sulfide-rich, superheated vent water with temperatures above 350°C (**chapter opener;** *see figure 30.4*). Evidence suggests that these microbes can grow and reproduce at 121°C and can survive temperatures to 130°C for up to 2 hours. The pressure present in their habitat is sufficient to keep water liquid (at 265 atm, seawater doesn't boil until 460°C).

The implications of this discovery are many. The proteins, membranes, and nucleic acids of these microbes are remarkably temperature stable and provide ideal subjects for studying the ways in which macromolecules and membranes are stabilized. In the future, it may be possible to design enzymes that operate at very high temperatures. Some thermostable enzymes from these organisms have important industrial and scientific uses. For example, the Taq polymerase from the thermophile *Thermus aquaticus* is used extensively in the polymerase chain reaction. ▶▶| *Polymerase chain reaction (section 15.2)*

Thermophiles and hyperthermophiles differ from mesophiles in many ways. They have heat-stable enzymes and protein synthesis systems that function properly at high temperatures. These proteins are stable for a variety of reasons. Heat-stable proteins have highly organized hydrophobic interiors and more hydrogen and other noncovalent bonds. Larger quantities of amino acids such as proline also make polypeptide chains less flexible and more heat stable. In addition, the proteins are stabilized and aided in folding by proteins called chaperone proteins. Evidence exists that nucleoid-associated proteins stabilize the DNA of thermophilic bacteria. The membrane lipids of thermophiles and hyperthermophiles are also quite temperature stable. They tend to be more saturated, more branched, and of higher molecular weight. This increases the melting points of membrane lipids. Archaeal thermophiles have membrane lipids with ether linkages, which protect the lipids from hydrolysis at high temperatures. Sometimes archaeal lipids actually span the membrane to form a rigid, stable monolayer. |◀◀ *Archaeal cell envelopes (section 3.4)* ▶▶| *Thermostability (section 18.1); Proteins (appendix I)*

1. What are cardinal temperatures?
2. Why does the growth rate rise with increasing temperature and then fall again at higher temperatures?
3. Define psychrophile, psychrotroph, mesophile, thermophile, and hyperthermophile.
4. What metabolic and structural adaptations for extreme temperatures do psychrophiles and thermophiles have? How might these protect the organisms from their environment?

Oxygen Concentration

The importance of oxygen to the growth of an organism correlates with its metabolism—in particular, with the processes it uses to conserve the energy supplied by its energy source. Almost all energy-conserving metabolic processes involve the movement of electrons through a series of membrane-bound electron carriers called the electron transport chain (ETC). For chemotrophs, an externally supplied terminal electron acceptor is critical to the functioning of the ETC. The nature of the terminal electron acceptor is related to an organism's oxygen requirement. ▶▶| *Electron transport chains (section 9.6)*

An organism able to grow in the presence of atmospheric O_2 is an **aerobe,** whereas one that can grow in its absence is an **anaerobe.** Almost all multicellular organisms are completely dependent on atmospheric O_2 for growth—that is, they are **obligate aerobes** (table 7.4). Oxygen serves as the terminal electron acceptor for the ETC in the metabolic process called aerobic respiration. In addition, aerobic eukaryotes employ O_2 in the synthesis of sterols and unsaturated fatty acids. **Microaerophiles** such as *Campylobacter* are damaged by the normal atmospheric level of O_2 (20%) and require O_2 levels in the range of 2 to 10% for growth. **Facultative anaerobes** do not require O_2 for growth but grow better in its presence. In the presence of oxygen, they use O_2 as the terminal electron acceptor during aerobic respiration. **Aerotolerant anaerobes** such as *Enterococcus faecalis* grow equally well whether O_2 is present or not; chemotrophic aerotolerant anaerobes are often described as having strictly fermentative metabolism. In contrast, strict or **obligate anaerobes** (e.g., *Bacteroides, Clostridium pasteurianum, Methanococcus*) are usually killed in the presence of O_2.

Strict anaerobes cannot generate energy through aerobic respiration and employ other metabolic strategies such as fermentation or anaerobic respiration, neither of which requires O_2. The nature of bacterial O_2 responses can be readily determined by growing the bacteria in a solid culture medium or a medium such as thioglycollate broth, which contains a reducing agent to lower O_2 levels (**figure 7.28**). ▶▶| *Aerobic respiration (section 10.2); Anaerobic respiration (section 10.6); Fermentation (section 10.7)*

A microbial group may show more than one type of relationship to O_2. All five types are found among the *Bacteria, Archaea,* and protists. Fungi are normally aerobic, but a number of species—particularly among the yeasts—are facultative anaerobes. Photosynthetic protists are usually obligate aerobes. Although obligate anaerobes are killed by O_2, they may be recovered from habitats that appear to be oxic. In such cases, they associate with facultative anaerobes that use up the available O_2 and thus make the growth of strict anaerobes possible. For example, the strict anaerobe *Bacteroides gingivalis* lives in the mouth, where it grows in the anoxic crevices around the teeth. Clearly the ability to grow in both oxic and anoxic environments provides considerable flexibility and is an ecological advantage.

The different relationships with O_2 are due to several factors, including the inactivation of proteins and the effect of toxic O_2 derivatives. Enzymes can be inactivated when sensitive groups such as sulfhydryls are oxidized. A notable example is the nitrogen-fixation enzyme nitrogenase, which is very oxygen sensitive. However, even molecules that have evolved to function aerobically can be damaged by O_2. This is because the unpaired electrons in the outer shell of oxygen makes it inherently unstable (*see figure AI.2*). Thus, toxic O_2 derivatives are formed when proteins such as flavoproteins promote oxygen reduction. These toxic O_2 derivatives are called *r*eactive *o*xygen *s*pecies (ROS), and they can damage proteins, lipids, and nucleic acids. ROS include the superoxide radical, hydrogen peroxide, and the most dangerous hydroxyl radical.

$$O_2 + e^- \rightarrow O_2^{\bar{\bullet}} \text{ (superoxide radical)}$$

$$O_2^{\bar{\bullet}} + e^- + 2H^+ \rightarrow H_2O_2 \text{ (hydrogen peroxide)}$$

$$H_2O_2 + e^- + H^+ \rightarrow H_2O + OH\bullet \text{ (hydroxyl radical)}$$

A microorganism must be able to protect itself against ROS or it will be killed. Indeed, neutrophils and macrophages, two important immune system cells, use ROS to destroy invading pathogens. ▶▶| *Nitrogen assimilation (section 11.5); Phagocytosis (section 32.5)*

Many microorganisms possess enzymes that protect against toxic O_2 products (figure 7.28). Obligate aerobes and facultative anaerobes usually contain the enzymes **superoxide dismutase (SOD)** and **catalase,** which catalyze the destruction of superoxide radical and hydrogen peroxide, respectively. Peroxidase also can be used to destroy hydrogen peroxide.

$$2O_2^{\bar{\bullet}} + 2H^+ \xrightarrow[\text{dismutase}]{\text{superoxide}} O_2 + H_2O_2$$

$$2H_2O_2 \xrightarrow{\text{catalase}} 2H_2O + O_2$$

$$H_2O_2 + NADH + H^+ \xrightarrow{\text{peroxidase}} 2H_2O + NAD^+$$

Aerotolerant microorganisms may lack catalase but usually have superoxide dismutase. All strict anaerobes lack both enzymes

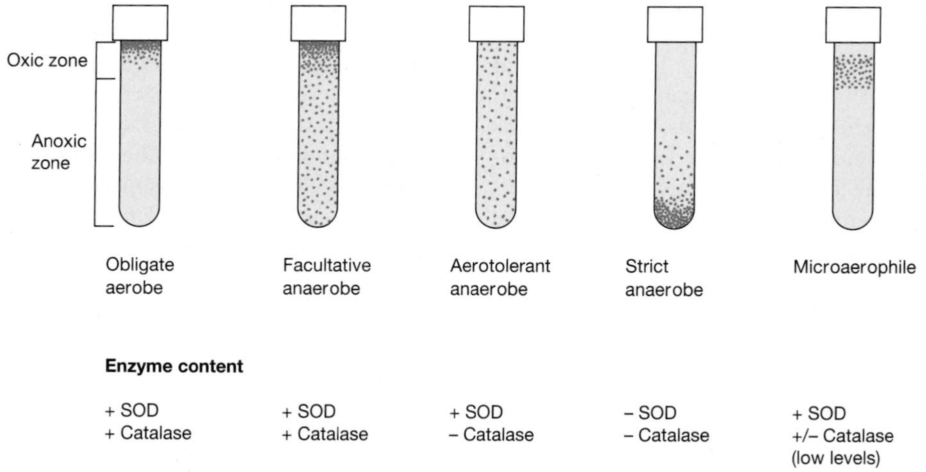

FIGURE 7.28 Oxygen and Bacterial Growth. Each dot represents an individual bacterial colony within the agar or on its surface. The surface, which is directly exposed to atmospheric oxygen, is oxic. The oxygen content of the medium decreases with depth until the medium becomes anoxic toward the bottom of the tube. The presence and absence of the enzymes superoxide dismutase (SOD) and catalase for each type are shown.

Figure 7.28 Micro Inquiry

Why do facultative anaerobes grow best at the surface of the tube while aerotolerant microbes demonstrate a uniform growth pattern throughout the tube?

or have them in very low concentrations and therefore cannot tolerate O_2. However, some microaerophilic bacteria and anaerobic archaea protect themselves from the toxic effects of O_2 with the enzymes superoxide reductase and peroxidase. Superoxide reductase reduces superoxide to H_2O_2 without producing O_2. The H_2O_2 is then converted to water by peroxidase. ▶▶| *Oxidation-reduction reactions (section 9.5)*

Because aerobes need O_2 and anaerobes are killed by it, radically different approaches must be used when they are cultured. When large volumes of aerobic microorganisms are cultured, either they must be shaken to aerate the culture medium or sterile air must be pumped through the culture vessel. Without aeration, the low solubility of O_2 in liquid would prevent them from obtaining an adequate supply. Precisely the opposite problem arises with anaerobes—all O_2 must be excluded. This is accomplished in several ways. (1) Anaerobic media containing reducing agents such as thioglycollate or cysteine may be used. The medium is boiled during preparation to dissolve its components and drive off oxygen. The reducing agents eliminate any residual dissolved O_2 in the medium so that anaerobes can grow beneath its surface. (2) Oxygen also may be eliminated from an enclosed work area, often called an anaerobic chamber or anaerobic workstation. Most of the air is removed with a vacuum pump followed by purges with nitrogen gas (**figure 7.29**). A gas mix containing hydrogen is then introduced into the workstation. In the presence of a palladium catalyst, the hydrogen and last remaining molecules of O_2 react to form water, creating an anoxic environment. Often CO_2 is added to the chamber because many anaerobes require a small amount of CO_2 for best growth. (3) One of the most popular ways of culturing small numbers of anaerobes is by use of a GasPak jar, which also uses hydrogen and a palladium catalyst to remove O_2 (**figure 7.30**). (4) A similar approach uses plastic bags or pouches containing calcium carbonate and a catalyst, which produce an anoxic, carbon dioxide–rich atmosphere.

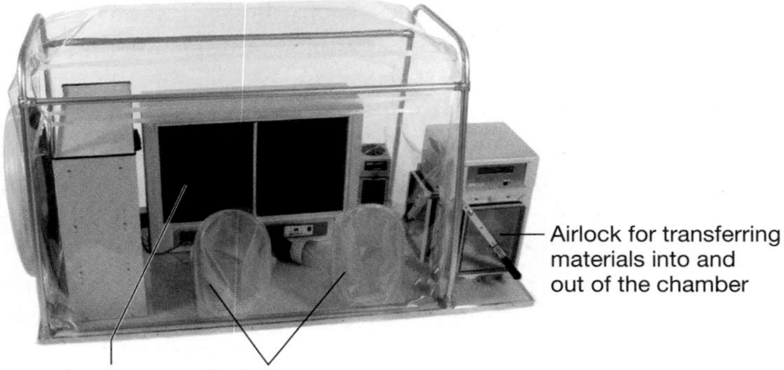
Airlock for transferring materials into and out of the chamber
Incubator
Work gloves. The user reaches into the anoxic chamber using these gloves.

FIGURE 7.29 An Anaerobic Workstation and Incubator. This system contains an oxygen-free work area and an incubator. The interchange compartment on the right of the work area allows materials to be transferred inside without exposing the interior to oxygen. The anoxic atmosphere is maintained largely with a vacuum pump and nitrogen purges. The remaining oxygen is removed by a palladium catalyst and hydrogen. The oxygen reacts with hydrogen to form water, which is absorbed by a desiccant.

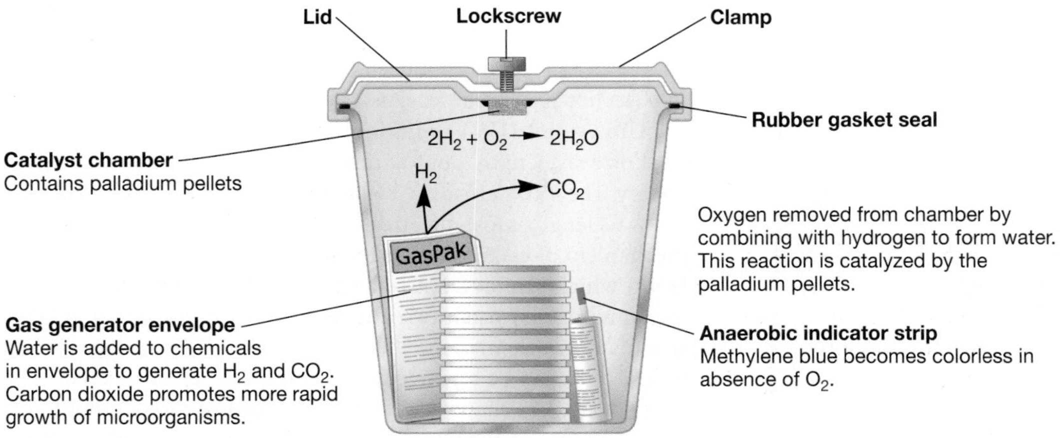
Lid
Lockscrew
Clamp
Rubber gasket seal
Catalyst chamber
Contains palladium pellets
2H2 + O2 → 2H2O
H2
CO2
GasPak
Oxygen removed from chamber by combining with hydrogen to form water. This reaction is catalyzed by the palladium pellets.
Gas generator envelope
Water is added to chemicals in envelope to generate H2 and CO2. Carbon dioxide promotes more rapid growth of microorganisms.
Anaerobic indicator strip
Methylene blue becomes colorless in absence of O2.

FIGURE 7.30 The GasPak Anaerobic System. Hydrogen and carbon dioxide are generated by a GasPak envelope. The palladium catalyst in the chamber lid catalyzes the formation of water from hydrogen and oxygen, thereby removing oxygen from the sealed chamber.

1. Describe the five types of O_2 relationships seen in microorganisms.
2. What are the toxic effects of O_2? How do aerobes and other oxygen-tolerant microbes protect themselves from these effects?
3. Describe four ways in which anaerobes may be cultured.

Pressure

Organisms that spend their lives on land or the surface of water are always subjected to a pressure of 1 atmosphere (atm; ~0.1 MPa) and are never affected significantly by pressure. Other organisms, including many bacteria and archaea, live in the deep sea (ocean depths of 1,000 m or more), where the hydrostatic pressure can reach 600 to 1,100 atm and the temperature is about 2 to 3°C. These high hydrostatic pressures affect membrane fluidity and membrane-associated function. How do microbes survive under such conditions?

Many microbes found at great ocean depths are **barotolerant:** increased pressure adversely affects them but not as much as it does nontolerant microbes. Some are truly **piezophilic (barophilic)**—they grow more rapidly at high pressures. A piezophile is defined as an organism that has a maximal growth rate at pressures greater than 1 atm but less than about 590 atm (60 MPa). For instance, a piezophile recovered from the Mariana trench near the Philippines (depth about 10,500 m) grows only at pressures between about 400 to 500 atm when incubated at 2°C. Some microbes are hyperpiezophiles, having growth rate maxima greater than 590 atm.

An important adaptation observed in piezophiles is that they change their membrane lipids in response to increasing pressure. For instance, bacterial piezophiles increase the amount of unsaturated fatty acids in their membrane lipids as pressure increases. They may also shorten the length of their fatty acids. Piezophiles are thought to play important roles in nutrient recycling in the deep sea. Thus far, they have been found among several bacterial genera (e.g., *Photobacterium, Shewanella, Colwellia*). Some archaea are thermopiezophiles (e.g., *Pyrococcus* spp., *Methanocaldococcus jannaschii*). ▶▶| *Microorganisms in benthic marine environments (section 28.2)*

Radiation

Our world is bombarded with electromagnetic radiation of various types (**figure 7.31**). Radiation behaves as if it were composed of waves moving through space like waves traveling on the surface of water. The distance between two wave crests or troughs is the wavelength. As the wavelength of electromagnetic radiation decreases, the energy of the radiation increases; gamma rays and X rays are much more energetic than visible light or infrared waves. Electromagnetic radiation also acts like a stream of energy packets called photons, each photon having a quantum of energy whose value depends on the wavelength of the radiation.

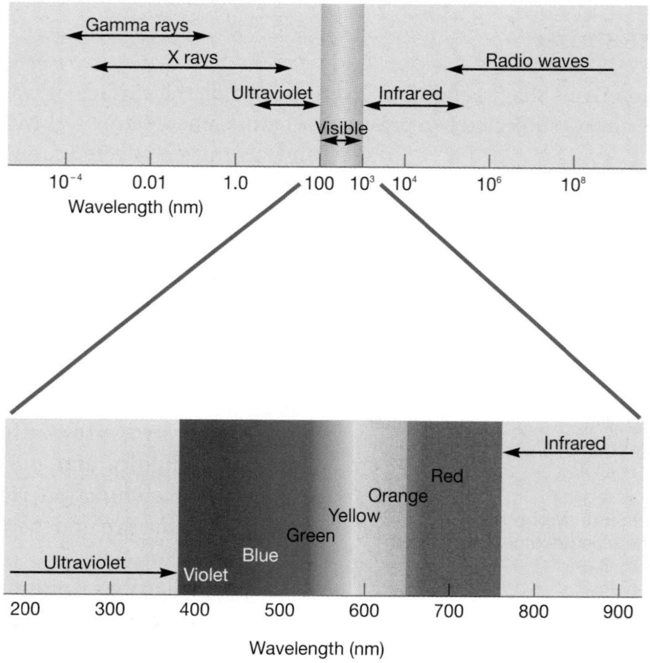

FIGURE 7.31 The Electromagnetic Spectrum. A portion of the spectrum is expanded at the bottom of the figure.

Sunlight is the major source of radiation on Earth. It includes visible light, ultraviolet (UV) radiation, infrared rays, and radio waves. Visible light is a most conspicuous and important aspect of our environment: most life depends on the ability of photosynthetic organisms to trap the light energy of the sun. Almost 60% of the sun's radiation is in the infrared region rather than the visible portion of the spectrum. Infrared is the major source of Earth's heat. At sea level, one finds very little ultraviolet radiation below about 290 to 300 nm. UV radiation of wavelengths shorter than 287 nm is absorbed by O_2 in Earth's atmosphere; this process forms a layer of ozone between 40 and 48 km above Earth's surface. The ozone layer absorbs somewhat longer UV rays and reforms O_2. The even distribution of sunlight throughout the visible spectrum accounts for the fact that sunlight is generally "white." ▶▶| *Phototrophy (section 10.12)*

Many forms of electromagnetic radiation are very harmful to microorganisms. This is particularly true of **ionizing radiation,** radiation of very short wavelength and high energy, which can cause atoms to lose electrons (ionize). Two major forms of ionizing radiation are (1) X rays, which are artificially produced, and (2) gamma rays, which are emitted during radioisotope decay. Low levels of ionizing radiation may produce mutations and may indirectly result in death, whereas higher levels are directly lethal. Ionizing radiation causes a variety of changes in cells. It breaks hydrogen bonds, oxidizes double bonds, destroys ring structures, and polymerizes some molecules. Oxygen enhances these destructive effects, probably through the generation of hydroxyl radicals (OH•). Although many types of constituents can be affected, the destruction of DNA is probably the most important cause of death.

Although microorganisms are more resistant to ionizing radiation than larger organisms, they are still destroyed by a sufficiently large dose. Indeed, ionizing radiation can be used to sterilize items. Bacterial endospores and bacteria such as *Deinococcus radiodurans* are extremely resistant to large doses of ionizing radiation. *D. radiodurans* is of particular note. This amazing microbe is able to piece together its genome after it is blasted apart by massive doses of radiation. How it does this is a matter of intense interest to microbiologists and still has not been completely elucidated. ▶▶| *Radiation (section 8.4);* Deinococcus-Thermus *(section 19.2)*

Ultraviolet (UV) radiation can kill microorganisms due to its short wavelength (approximately from 10 to 400 nm) and high energy. The most lethal UV radiation has a wavelength of 260 nm, the wavelength most effectively absorbed by DNA. The primary mechanism of UV damage is the formation of thymine dimers in DNA, which inhibit DNA replication and function. Thymine dimers are formed when two adjacent thymines in the same DNA strand are covalently joined (*see figure 14.4*). The damage caused by UV light can be repaired by several DNA repair mechanisms, as we discuss in chapter 14. Excessive exposure to UV light outstrips the organism's ability to repair the damage and death results. Longer wavelengths of UV light (near-UV radiation; 325 to 400 nm) can also harm microorganisms because they induce the breakdown of the amino acid tryptophan to toxic photoproducts.

It appears that these toxic photoproducts plus the near-UV radiation itself produce breaks in DNA strands. The precise mechanism is not known, although it is different from that seen with 260 nm UV. ▶▶| *Mutations and their chemical basis (section 14.1)*

Even visible light, when present in sufficient intensity, can damage or kill microbial cells. Usually pigments called photosensitizers and O_2 are involved. Photosensitizers include pigments such as chlorophyll, bacteriochlorophyll, cytochromes, and flavins, which can absorb light energy and become excited or activated. The excited photosensitizer (P) transfers its energy to O_2, generating singlet oxygen ($^1O_2$).

$$P \xrightarrow{\text{light}} P\ (\text{activated})$$

$$P\ (\text{activated}) + O_2 \rightarrow P + {}^1O_2$$

Singlet oxygen is a very reactive, powerful oxidizing agent that quickly destroys a cell.

Many microorganisms that are airborne or live on exposed surfaces use carotenoid pigments for protection against photooxidation. Carotenoids effectively quench singlet oxygen—that is, they absorb energy from singlet oxygen and convert it back into the unexcited ground state. Both photosynthetic and nonphotosynthetic microorganisms employ pigments in this way.

1. What are barotolerant and piezophilic bacteria? Where would you expect to find them?
2. List the types of electromagnetic radiation in the order of decreasing energy or increasing wavelength.
3. What is the importance of ozone formation?
4. How do ionizing radiation, ultraviolet radiation, and visible light harm microorganisms? How do microorganisms protect themselves against damage from UV and visible light?

7.7 Microbial Growth in Natural Environments

The microbial environment is complex and constantly changing. It often contains low nutrient concentrations (**oligotrophic environment**) and exposes microbes to many overlapping gradients of nutrients and other environmental factors. The growth of microorganisms depends on both the nutrient supply and their tolerance of the environmental conditions present in their habitat at any particular time. Inhibitory substances in the environment can also limit microbial growth. For instance, rapid, unlimited growth ensues if a microorganism is exposed to excess nutrients. Such growth quickly depletes nutrients and often results in the release of toxic products. Both nutrient depletion and the toxic products limit further growth.

Biofilms

Although ecologists observed as early as the 1940s that more microbes in aquatic environments were found attached to surfaces (sessile) than were free-floating (planktonic), only relatively recently has this fact gained the attention of microbiologists. These attached microbes are members of complex, slime-encased communities called **biofilms.** Biofilms are ubiquitous in nature, where they are most often seen as layers of slime on rocks or other objects in water or at water-air interfaces (**figure 7.32*a***). When they form on the hulls of boats and ships, they cause corrosion, which limits the life of the ships and results in economic losses. Of major concern is the formation of biofilms on medical devices such as hip and knee implants (figure 7.32*b*). These biofilms often cause serious illness and failure of the medical device. Biofilm formation is apparently an ancient ability among the

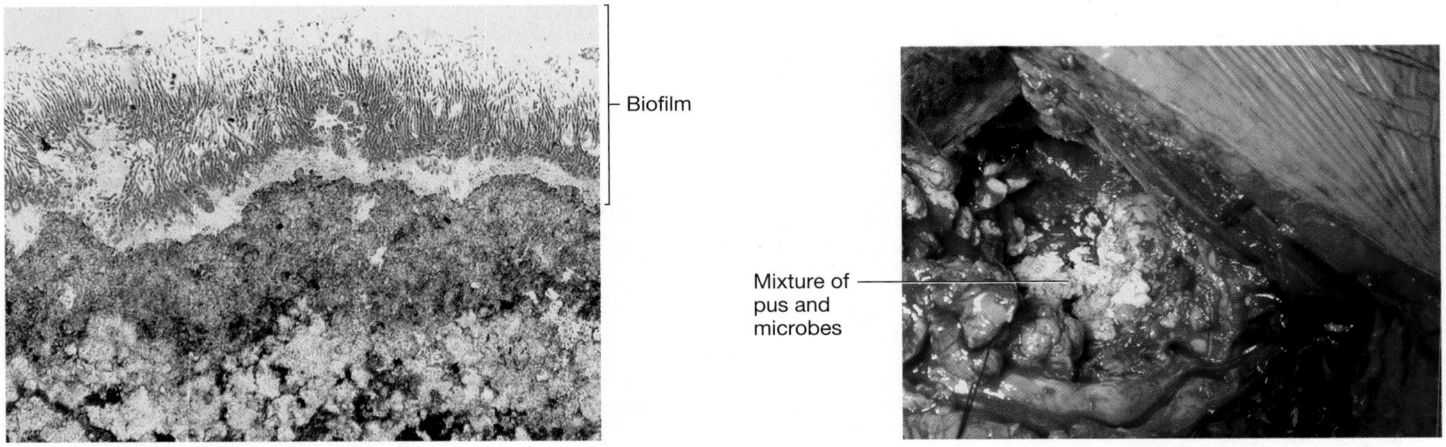

(a) Biofilm on surface of a stromatolite

(b) Infected tissue after hip replacement

FIGURE 7.32 Examples of Biofilms. Biofilms form on almost any surface exposed to microorganisms. (a) Biofilm on the surface of a stromatolite in Walker Lake (Nevada, USA), an alkaline lake. The biofilm consists primarily of the cyanobacterium *Calothrix.* (b) Photograph taken during surgery to remove a biofilm-coated artificial joint. The white material is composed of pus, bacterial and fungal cells, and the patient's white blood cells.

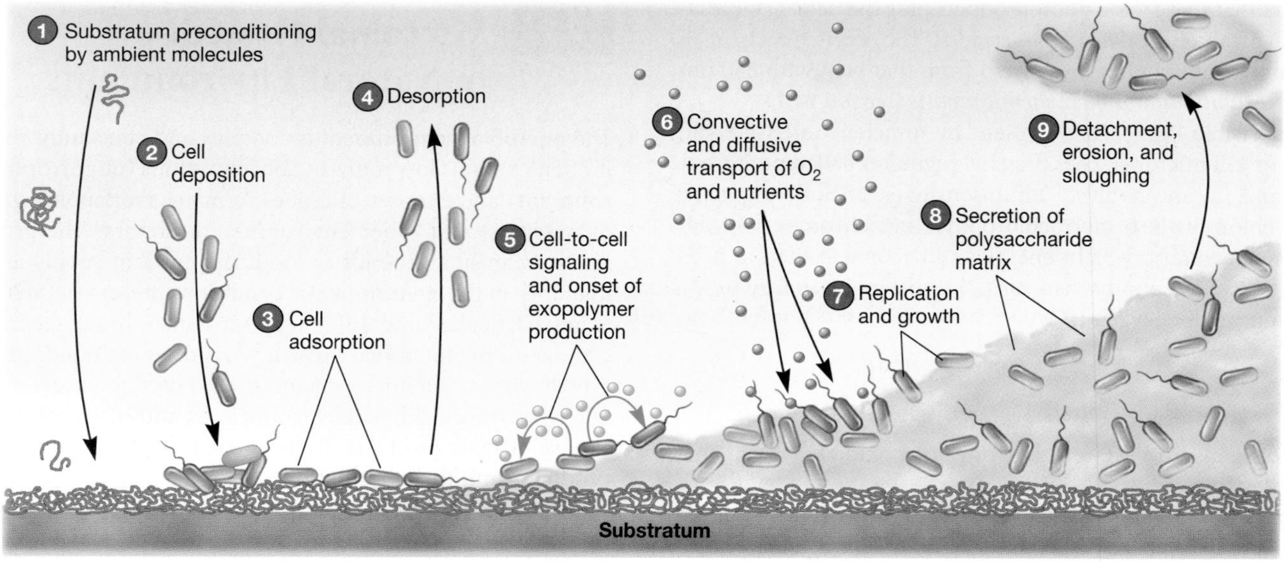

FIGURE 7.33 Biofilm Formation.

Figure 7.33 Micro Inquiry
What biomolecules make up the extracellular polymeric matrix, and what functions does the matrix serve?

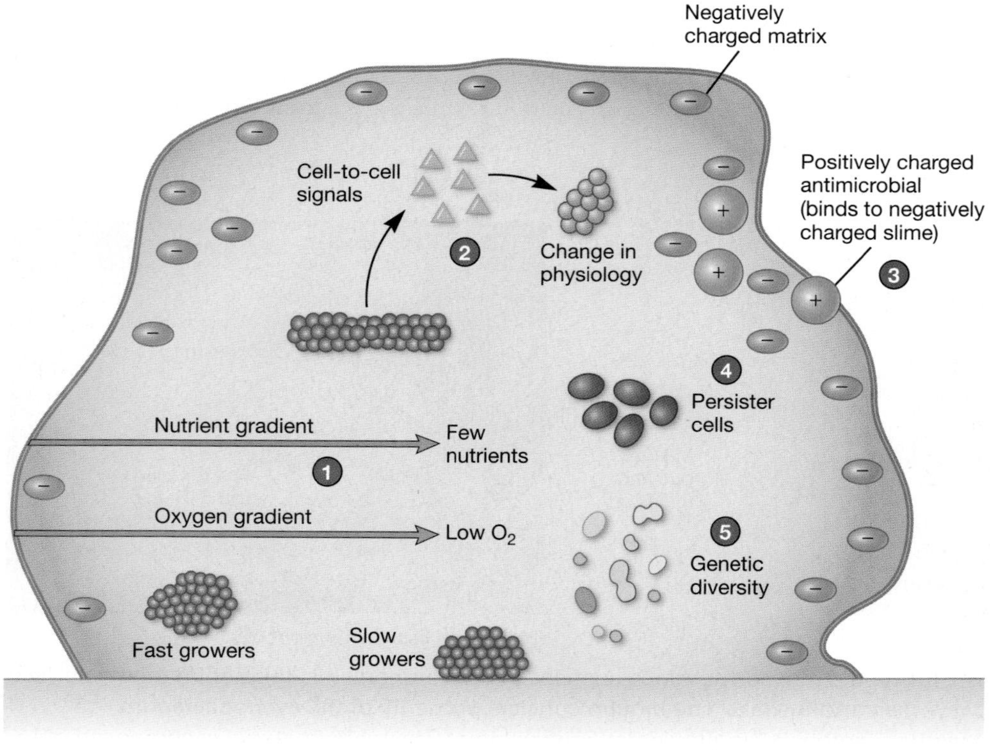

FIGURE 7.34 Biofilm Heterogeneity.

microbes, as evidence for biofilms can be found in the fossil record from about 3.4 billion years ago.

Biofilms can form on virtually any surface, once it has been conditioned by proteins and other molecules present in the environment (**figure 7.33**). Initially microbes attach to the conditioned surface but can readily detach. Eventually they form a matrix made up of polysaccharides, proteins, glycoproteins, glycolipids, and DNA. These are collectively called extracellular polymeric substances (EPS). The EPS matrix allows the microbes to stick more stably to the surface. As the biofilm thickens and matures, the microbes reproduce and secrete additional polymers.

A mature biofilm is a complex, dynamic community of microorganisms. It exhibits considerable heterogeneity due to differences in the metabolic activity of microbes at various locations within the biofilm (**figure 7.34**). The microbes interact in a variety of ways. For instance, the waste products of one microbe may be the energy source for another microbe. The cells also communicate with each other, as we describe next. Finally, DNA present in the EPS can be taken up by members of the biofilm community. Thus genes can be transferred from one cell (or species) to another.

While in the biofilm, microbes are protected from numerous harmful agents such as UV light, antibiotics, and other antimicrobial agents. This is due in part to the EPS in which they are embedded (figure 7.34), but it also is due to physiological changes. Indeed, numerous proteins synthesized or activated in biofilm cells are not observed when these cells are free-living, planktonic cells, and vice versa. The resistance of biofilm cells to antimicrobial agents has serious consequences. When biofilms form on a medical device such as a hip implant (figure 7.32*b*), they are difficult to kill and can cause serious illness. Often the only way to treat patients in this situation is by removing the implant. Another problem with biofilms is that cells are regularly sloughed off (figure 7.33). This can have many consequences. For instance, biofilms in a city's water distribution pipes can serve as a source of contamination after the water leaves a water treatment facility. *Biofilms*

Search This: Biofilm movies

Cell-Cell Communication Within Microbial Populations

For decades, microbiologists tended to think of bacterial populations as collections of individual cells growing and behaving independently. But about 40 years ago, two examples of bacterial cells using molecular signals to communicate with each other in a density-dependent manner were discovered. This is now referred to as **quorum sensing;** a quorum usually refers to the minimum number of members in an organization, such as a legislative body, needed to conduct business.

The first example of quorum sensing was observed in the gram-positive, pathogenic bacterium *Streptococcus pneumoniae*. These bacteria produce a small protein (i.e., a peptide) that increases in concentration as population size increases. Eventually the concentration of the peptide is high enough to convert some cells in the population from a noncompetent state to a competent state. In the competent state, cells are able to take up DNA, a process called transformation. In addition, the competent cells release a chemical called bacteriocin that causes lysis of the cells in the population that did not become competent. When the noncompetent cells lyse, they release DNA and virulence factors that allow the competent cells to invade tissues in a host organism, causing serious diseases such as pneumonia and meningitis. ▶▶| *Bacterial transformation (section 14.8)*

Although the quorum-sensing system of *S. pneumoniae* was the first discovered, the term was not coined until a few years later when the quorum-sensing system of a very different bacterium was discovered. The marine luminescent bacterium *Vibrio fischeri* lives within the light organ of certain fish and squid. It controls its ability to glow by producing a small, diffusible substance called autoinducer. The autoinducer molecule was later identified as an ***N*-acylhomoserine lactone (AHL).** It is now known that many gram-negative bacteria make AHL molecular signals that vary in length and substitution at the third position of the acyl side chain (**figure 7.35**). In many of these species, AHL is freely diffusible across the plasma membrane. Thus at a low cell density, it diffuses out of the cell. However, when the cell population increases and AHL accumulates outside the cell, the diffusion gradient is reversed so that the AHL enters the cell. Because the influx of AHL is cell density dependent, it enables individual cells to assess population density. When AHL reaches a threshold level inside the cell, it induces the expression of target genes that regulate a

Signal and Structure	Representative Organism	Function Regulated
N-acylhomoserine lactone (AHL) 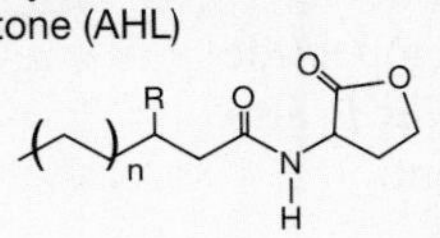	*Vibrio fischeri* *Agrobacterium tumefaciens* *Erwinia carotovora* *Pseudomonas aeruginosa* *Burkholderia cepacia*	Bioluminescence Plasmid transfer Virulence and antibiotic production Virulence and biofilm formation Virulence
Furanosylborate (AI-2) 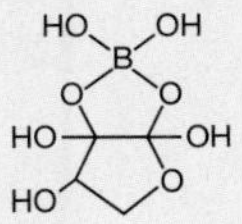	*Vibrio harveyi* [a]	Bioluminescence
Cyclic thiolactone (AIP-II) 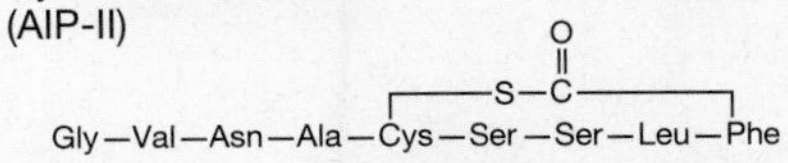	*Staphylococcus aureus*	Virulence
Hydroxy-palmitic acid methyl ester (PAME) 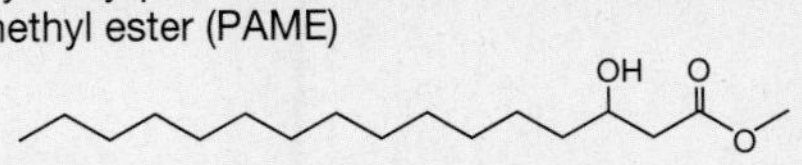	*Ralstonia solanacearum*	Virulence
Methyl dodecenoic acid 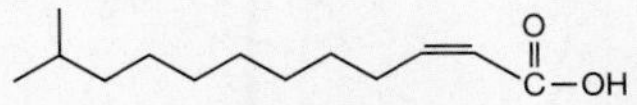	*Xanthomonas campestris*	Virulence
Farnesoic acid 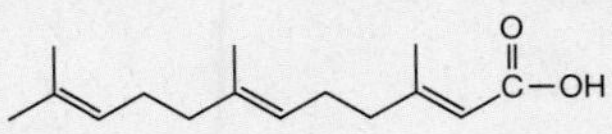	*Candida albicans*	Dimorphic transition and virulence
3-hydroxytridecan-4-one 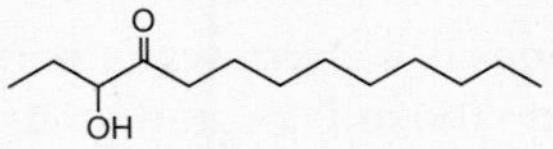	*Vibrio cholerae*	Virulence

[a] Other bacteria make a form of AI-2 that lacks boron.

FIGURE 7.35 Representative Cell-Cell Communication Molecules.

number of functions, depending on the microbe. These functions are most effective only if a large number of microbes are present. For instance, the light produced by one *V. fischeri* cell is not visible, but cell densities within the light organ of marine fish and squid reach 10^{10} cells per milliliter. This provides the animal with a flashlight effect while the microbes have a safe and nutrient-enriched habitat (**figure 7.36**). ▶▶| *Quorum sensing (section 13.6)*

Since these discoveries, scientists have learned that many of the processes regulated by quorum sensing involve host-microbe interactions such as symbioses and pathogenicity. For instance, the gram-negative, opportunistic bacterial pathogens *Burkholderia cepacia* and *Pseudomonas aeruginosa* use AHLs to regulate biofilm formation and the expression of virulence factors (figure 7.35). These bacteria cause debilitating pneumonia in people who are immunocompromised and are important pathogens in cystic fibrosis patients. The plant pathogens *Agrobacterium tumefaciens* will not infect a host plant and *Erwinia carotovora* will not produce antibiotics without AHL signaling. Finally, *B. cepacia* and *P. aeruginosa* use AHL intercellular communication to control biofilm formation, an important strategy to evade the host's immune system.

Like *S. pneumoniae*, other gram-positive bacteria communicate using short peptides called oligopeptides. Examples include *Enterococcus faecalis,* whose oligopeptide signal is used to determine the best time to conjugate (transfer genes). Oligopeptide communication by *Staphylococcus aureus* and *B. subtilis* is also used to trigger the uptake of DNA from the environment.

The discovery of additional molecular signals made by a variety of microbes underscores the importance of cell-cell communication in regulating cellular processes. For instance, while only gram-negative bacteria are known to make AHLs, both gram-negative and gram-positive bacteria make autoinducer-2 (AI-2). The soil microbe *Streptomyces griseus* produces a γ-butyrolactone known as A-factor. This small molecule regulates both morphological differentiation and the production of the antibiotic streptomycin. Eukaryotic microbes also rely on cell-cell communication to coordinate key activities within a population. For example, the pathogenic fungus *Candida albicans* secretes farnesoic acid to govern morphology and virulence (figure 7.35).

These examples of cell-cell communication demonstrate what might be called multicellular behavior in that many individual cells communicate and coordinate their activities to act as a unit. Other examples of such complex behavior are pattern formation in colonies and fruiting body formation in the myxobacteria. |◀◀ *Microbial growth on solid surfaces (section 6.8);* ▶▶| *Order* Myxococcales *(section 20.4)*

1. What is a biofilm? Why might life in a biofilm be advantageous for microbes?
2. What medical challenges do biofilms present?
3. What is quorum sensing? Describe how it occurs and briefly discuss its importance to microorganisms.

(a) *E. scolopes*, the bobtail squid

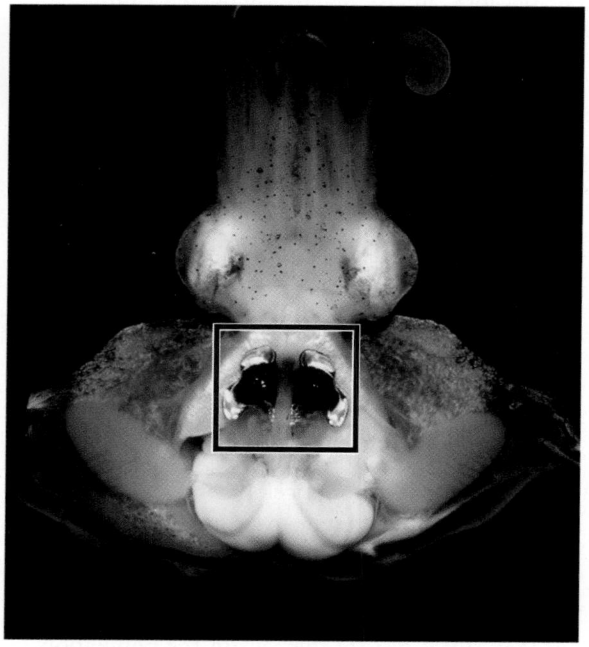

(b) Light organ

FIGURE 7.36 ***Euprymna scolopes.*** (a) *E. scolopes* is a warm-water squid that remains buried in sand during the day and feeds at night. (b) When feeding it uses its light organ (boxed, located on its ventral surface) to provide camouflage by projecting light downward. Thus the outline of the squid appears as bright as the water's surface to potential predators looking up through the water column. The light organ is colonized by a large number of *Vibrio fischeri* so autoinducer accumulates to a threshold concentration, triggering light production.

Summary

7.1 Reproductive Strategies

a. Many eukaryotic microbes are capable of carrying out both sexual and asexual reproduction. The former involves meiosis and the latter, mitosis.

b. The most common type of cell division observed among *Bacteria* and *Archaea* is binary fission. During binary fission, the cell elongates, the chromosome is replicated, and then it segregates to opposite poles of the cell prior to the formation of a septum, which divides the cell into two progeny cells (**figure 7.1**).

c. Other bacterial reproductive strategies include budding, and baeocyte and spore formation (**figure 7.2**).

7.2 Bacterial Cell Cycle

a. Two overlapping pathways function during the bacterial cell cycle: the pathway for chromosome replication and segregation, and the pathway for septum formation (**figure 7.3**).

b. In bacterial cells, chromosome replication begins at a single site called the origin of replication. Many models for partitioning of the progeny chromosomes have been developed. Some suggest that homologues of eukaryotic cytoskeletal proteins are involved (**figure 7.4**).

c. Septum formation in most bacteria is accomplished by Z ring formation and the subsequent assembly of cell wall-synthesizing machinery (**figure 7.7**).

d. In rapidly dividing cells, initiation of DNA synthesis may occur before the previous round of synthesis is completed. This allows the cells to shorten the time needed for completing the cell cycle.

e. The shape of bacterial and archaeal cells is determined by the cell wall. This has been best studied in bacteria, where the cytoskeletal proteins FtsZ and MreB (in rod-shaped bacteria) are thought to form scaffoldlike structures upon which peptidoglycan remodeling components assemble. Thus new cell wall is confined to the midcell in cocci, while in rods it occurs along the length of cell as well as midcell during division (**figures 7.8** and **7.9**).

7.3 Growth Curve

a. Growth is an increase in cellular constituents and results in an increase in cell size, cell number, or both.

b. When microorganisms are grown in a batch culture, the resulting growth curve usually has four phases: lag, exponential (log), stationary, and death (**figure 7.11**).

c. Exponential growth is balanced growth; that is, cell components are synthesized at constant rates relative to one another. Changes in culture conditions (e.g., in shift-up and shift-down experiments) lead to unbalanced growth. A portion of the available nutrients is used to supply maintenance energy.

d. In the exponential phase, the population number of cells undergoing binary fission doubles at a constant interval called the generation or doubling time (**figure 7.15**). The mean growth rate (μ) is the reciprocal of the generation time (**figure 7.16**).

7.4 Measurement of Microbial Growth

a. Microbial populations can be counted directly with counting chambers, flow cytometers, electronic counters, or fluorescence microscopy (**figure 7.18**).

b. Viable counting techniques (plate counts) such as the spread plate, the pour plate, or the membrane filter method can be employed (**figure 7.19**).

c. Population changes also can be followed by determining variations in microbial mass through the measurement of dry weight, turbidity, or the amount of a cell component (**figure 7.21**).

7.5 Continuous Culture of Microorganisms

a. Microorganisms can be grown in an open system in which nutrients are constantly provided and wastes removed.

b. A continuous culture system can maintain a microbial population in log phase. There are two types of these systems: chemostats and turbidostats (**figure 7.22**).

7.6 Influences of Environmental Factors on Growth

a. Most bacteria, photosynthetic protists, and fungi have rigid cell walls and are hypertonic relative to the habitat because of solutes such as amino acids, polyols, and potassium ions. The amount of water actually available to microorganisms is expressed in terms of the water activity (a_w).

b. Although most microorganisms do not grow well at water activities below 0.98 due to plasmolysis and associated effects, osmotolerant organisms survive and even flourish at low a_w values. Halophiles actually require high sodium chloride concentrations for growth (**figure 7.24** and **table 7.4**).

c. Each microbial species has an optimum pH for growth and can be classified as an acidophile, neutrophile, or alkalophile (**figure 7.25**).

d. Microorganisms have distinct temperature ranges for growth with minima, maxima, and optima—the cardinal temperatures. These ranges are determined by the effects of temperature on the rates of catalysis, protein denaturation, and membrane disruption (**figure 7.26**).

e. There are five major classes of microorganisms with respect to temperature preferences: (1) psychrophiles, (2) psychrotrophs (psychrotolerants), (3) mesophiles, (4) thermophiles, and (5) hyperthermophiles (**figure 7.27** and **table 7.4**).

f. Microorganisms can be placed into at least five different categories based on their response to the presence of O_2: obligate aerobes, microaerophiles, facultative anaerobes, aerotolerant anaerobes, and strict or obligate anaerobes (**figure 7.28** and **table 7.4**).

g. Oxygen can become toxic because of the production of hydrogen peroxide, superoxide radical, and hydroxyl radical. These are destroyed by the enzymes superoxide dismutase, catalase, and peroxidase. In some organisms, superoxide reductase and peroxidase are used instead.

h. Most deep-sea microorganisms are barotolerant, but some are piezophilic and require high pressure for optimal growth.

i. High-energy or short-wavelength radiation harms organisms in several ways. Ionizing radiation—X rays and gamma rays—ionizes molecules and destroys DNA and other cell components. Ultraviolet (UV) radiation induces the formation of thymine dimers and strand breaks in DNA.

j. Visible light can provide energy for the formation of reactive singlet oxygen, which will destroy cells.

7.7 Microbial Growth in Natural Environments

a. Microbial growth in natural environments is profoundly affected by nutrient limitations and other adverse factors.

b. Many microbes form biofilms, aggregations of microbes growing on surfaces and held together by extracellular polymeric substances (EPS) (**figures 7.33** and **7.34**). Life in a biofilm has several advantages, including protection from harmful agents.

c. Bacteria often communicate with one another in a density-dependent way and carry out a particular activity only when a certain population density is reached. This phenomenon is called quorum sensing (**figure 7.35**).

Critical Thinking Questions

1. As an alternative to diffusible signals, suggest another mechanism by which bacteria can quorum sense.

2. Design an enrichment culture medium and a protocol for the isolation and purification of a soil bacterium (e.g., *Bacillus subtilis*) from a sample of soil. Note possible contaminants and competitors. How will you adjust conditions of growth, and what conditions will be adjusted to enhance preferentially the growth of the *Bacillus*?

3. Design an experiment to determine if a slow-growing microbial culture is exiting lag phase or is in exponential phase.

4. Why do you think the cardinal temperatures of some microbes change depending on other environmental conditions (e.g., pH)? Suggest one specific mechanism underlying such change.

5. Consider cell-cell communication: bacteria that "subvert" and "cheat" have been described. Describe a situation in which it would be advantageous for one species to subvert another, that is, degrade an intercellular signal made by another species. Also, describe a scenario whereby bacterial cheaters—defined as bacteria that do not make a molecular signal but profit by the uptake and processing of signal made by another microbe—might have a growth advantage.

6. *Vibrio parahaemolyticus* is a gram-negative marine bacterium with the typical comma or vibrioid shape. It contains all three cytoskeletal proteins: MreB, FtsZ, and CreB (crescentin). When *V. parahaemolyticus* is starved or enters the viable but nonculturable (VBNC) state, it switches from a vibrioid shape to a nearly spherical shape. Why do you think the cells change shape when stressed? What do you think happens to each of the cytoskeletal proteins? Consider their rates of polymerization or depolymerization and whether or not they will be needed at all.

   Read the original paper: Chiu, S.-W., et al. 2008. Localization and expression of MreB in *Vibrio parahaemolyticus* under different stresses. *Appl. Environ. Microbiol.* 74:7016.

7. Because many persistent bacterial infections involve biofilms, there is great interest in developing strategies to prevent their formation and eliminate them once they have formed. The red alga *Delisea pulchra* produces a class of compounds known as halogenated furanones that inhibit biofilm formation by *Pseudomonas aeruginossa, E. coli, B. subtilis, Staphylococcus epidermidis, and Streptococcus* sp. In these microbes, the algal furanones inhibit cell-cell signaling. While furanones also inhibit biofilm formation in the food-borne pathogen *Salmonella enterica* serovar Typhimurium, they do not alter cell-cell signaling. How does the inhibition of intercellular communication inhibit biofilm formation in *P. aeruginosa* and others? Do you think the fact that the furanones fail to interrupt cell-cell communication in *Salmonella* means that these bacteria do not use intercellular signaling when constructing

biofilms? What alternative explanations might explain the loss of biofilm formation without influencing cell-cell communication?

Read the original paper: Janssens, J. C., et al. 2008. Brominated furanones inhibit biofilm formation by *Salmonella enterica* serovar Typhimurium. *Appl. Environ. Microbiol.* 74:6639.

Concept Mapping Challenge

Construct a concept map that summarizes the environmental factors that affect growth and the types of responses made to these factors by microbes (see table 7.4). Use the concepts that follow, any other concepts or terms you need, and your own linking words between each pair of concepts in your map.

Solutes	Water activity	pH	Temperature
Oxygen	Pressure	Radiation	Compatible solutes

Learn More

Learn more by visiting the text website at www.mhhe.com/willey8, where you will find a complete list of references.

8

Control of Microorganisms in the Environment

Showering in the decontamination booth with two different quaternary ammonium chloride solutions is required prior to exiting the BSL-4 lab.

CHAPTER GLOSSARY

antiseptic Agents that prevent infection or sepsis.

autoclave An apparatus for sterilizing objects by the use of steam under pressure.

bactericide An agent that kills bacteria.

bacteriostatic Inhibiting the growth and reproduction of bacteria.

biocide A chemical or physical agent, usually broad spectrum, that inactivates microorganisms.

biological safety cabinets Specialized containers that use HEPA filters to project a curtain of sterile air across their opening, preventing microbes from entering or exiting the cabinet.

chemotherapy The use of chemical agents to kill or inhibit microbial growth in tissues.

decimal reduction time (D value) The time required to kill 90% of microorganisms or spores at a specific temperature.

depth filter Fibrous or granular materials bound as a thick layer and used to retain microbial cells when contaminated liquids are passed through it.

detergent An organic molecule, other than a soap, that serves as a wetting agent and emulsifier; it is normally used as a cleanser, but some may be used as antimicrobial agents.

disinfectant An agent that kills, inhibits, or removes microorganisms that may cause disease. It usually refers to the use of chemicals for the treatment of inanimate objects.

fungicide An agent that kills fungi.

fungistatic Inhibiting the growth and reproduction of fungi.

high-efficiency particulate air (HEPA) filter Depth filter constructed to remove 99.97% of particles that are 0.3 μm or larger.

iodophor Antiseptic formed as a complex of iodine and an organic carrier.

ionizing radiation Radiation of very short wavelength and high energy that causes atoms to lose electrons or ionize.

pasteurization The process of heating milk and other liquids to destroy microorganisms that can cause spoilage or disease.

sanitizer An agent that reduces the microbial population on an inanimate object to levels judged safe by public health standards.

sterilant An agent that destroys or removes all living cells, viable spores, viruses, viroids, virusoids, and prions from an object or habitat.

tyndallization The process of repeated heating and incubation of liquids to destroy bacterial spores.

use dilution test Test method used to determine the useful dilution of an antimicrobial agent by exposing specific bacteria to various dilutions of the agent.

viricide An agent that inactivates viruses so that they cannot reproduce within host cells.

Z value The temperature change at a given D value required to reduce the microorganism population by 90% (one log unit).

In this chapter, we address the subject of the control and destruction of microorganisms, a topic of immense practical importance. Although most microorganisms are beneficial, some microbial activities have undesirable consequences, such as food spoilage and disease. Therefore it is essential to be able to kill a wide variety of microorganisms or inhibit their growth to minimize their destructive effects. The goal is twofold: (1) to destroy pathogens and prevent their transmission, and (2) to reduce or eliminate microorganisms responsible for the contamination of water, food, and other substances.

TECHNIQUES & APPLICATIONS

8.1 Standard Microbiological Practices

The identification of potentially fatal, blood-borne infectious microbes (HIV, *Hepatitis B virus*, and others) spurred the codification of standard microbiological practices to limit exposure to such agents. These standard microbiological practices are *minimum* guidelines that should be supplemented with other precautions based on the potential exposure risks and biosafety level regulations for the lab. Briefly:

1. Eating, drinking, manipulation of contact lenses, and the use of cosmetics, gum, and tobacco products are strictly prohibited in the lab.
2. Hair longer than shoulder length should be tied back. Hands should be kept away from face at all times. Items (e.g., pencils) should not be placed in the mouth while in the lab. Protective clothing (lab coat, smock, etc.) is recommended while in the lab. Exposed wounds should be covered and protected.
3. Lab personnel should know how to use the emergency eyewash and shower stations.
4. Work space should be disinfected at the beginning and completion of lab time. Hands should be washed thoroughly after any exposure and before leaving the lab.
5. Precautions should be taken to prevent injuries caused by sharp objects (needles, scalpels, etc.). Sharp instruments should be discarded for disposal in specially marked containers.

Recommended guidelines for additional precautions should reflect the laboratory's biosafety level (BSL). The following table defines the BSL for the four categories of biological agents and suggested practices.

BSL	*Agents*	*Practices*
1	Not known to consistently cause disease in healthy adults (e.g., *Lactobacillus casei, Vibrio fischeri*)	Standard Microbiological Practices
2	Associated with human disease, potential hazard if percutaneous injury, ingestion, mucous membrane exposure occurs (e.g., *Salmonella enterica* serovar Typhi, *E. coli* O157:H7, *Staphylococcus aureus*)	BSL-1 practice plus: • Limited access • Biohazard warning signs • "Sharps" precautions • Biosafety manual defining any needed waste decontamination or medical surveillance policies
3	Indigenous or exotic agents with potential for aerosol transmission; disease may have serious or lethal consequences (e.g., *Coxiella burnetti, Yersinia pestis,* herpesviruses)	BSL-2 practice plus: • Controlled access • Decontamination of all waste • Decontamination of lab clothing before laundering • Baseline serum values determined in workers using BSL-3 agents
4	Dangerous/exotic agents that pose high risk of life-threatening disease; aerosol-transmitted lab infections; or related agents with unknown risk of transmission (e.g., Variola major [smallpox virus], hemorrhagic fever viruses)	BSL-3 practices plus: • Clothing change before entering • Shower on exit • All material decontaminated on exit from facility

Microorganism control continues to be a hot topic as microorganisms evolve to resist current strategies. Control efforts have a substantial role in public health to prevent disease as well as in therapeutic use to treat disease. The techniques described in this chapter are also essential for the personal safety of the laboratorians who work with microorganisms (**Techniques & Applications 8.1**). Thus this chapter focuses on the control of microorganisms by physical, chemical, and biological agents. Chemotherapeutic agents are discussed in chapter 34.

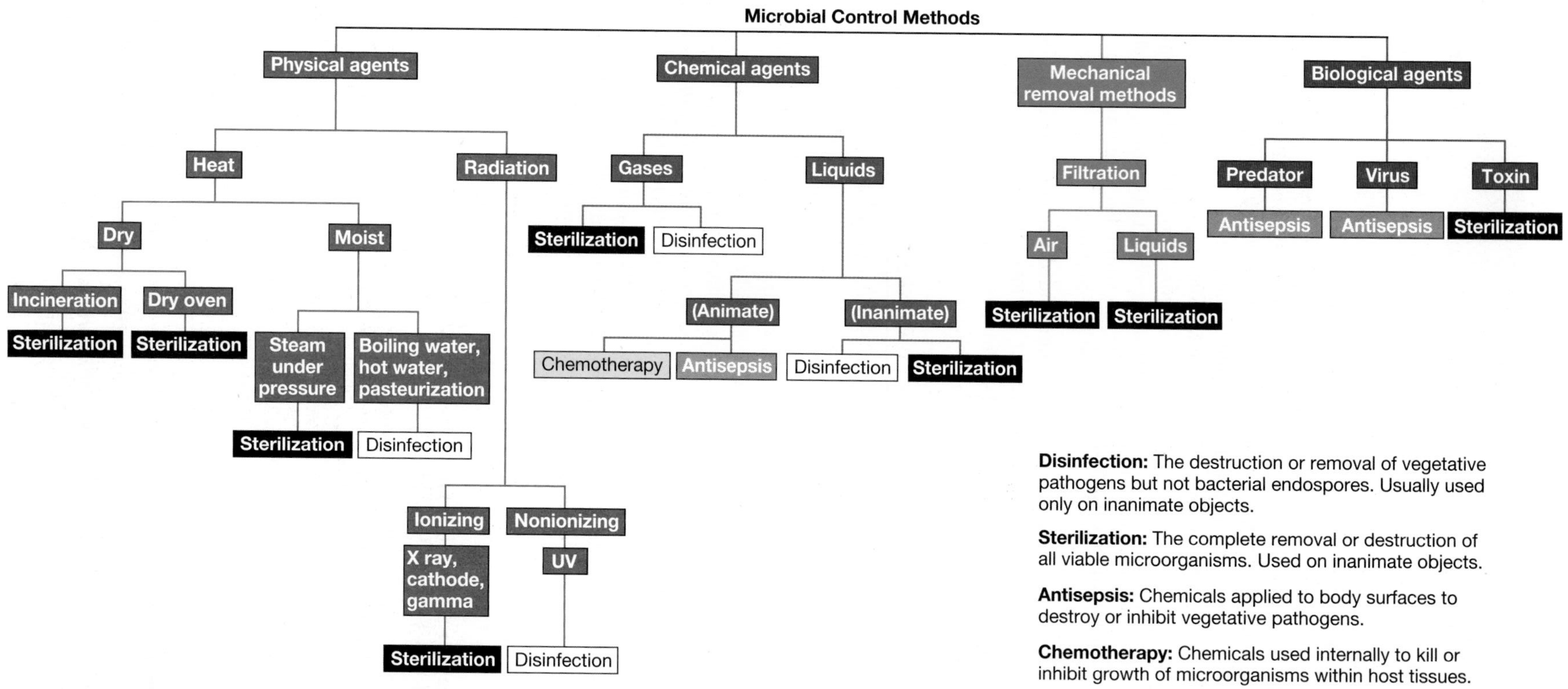

Disinfection: The destruction or removal of vegetative pathogens but not bacterial endospores. Usually used only on inanimate objects.

Sterilization: The complete removal or destruction of all viable microorganisms. Used on inanimate objects.

Antisepsis: Chemicals applied to body surfaces to destroy or inhibit vegetative pathogens.

Chemotherapy: Chemicals used internally to kill or inhibit growth of microorganisms within host tissues.

FIGURE 8.1 Microbial Control Methods.

Figure 8.1 Micro Inquiry

Which types of agents can be used for sterilization? Which can be used for antisepsis? What is the difference?

8.1 Definitions of Frequently Used Terms

Terminology is especially important when the control of microorganisms is discussed because words such as disinfectant and antiseptic often are used loosely. The situation is even more confusing because a particular treatment can either inhibit growth or kill, depending on the conditions. The types of control agents and their uses are outlined in **figure 8.1**. In general, to control microorganisms a **biocide** must be evaluated so as to determine the specific parameters under which it is to be effective.

Sterilization (Latin *sterilis,* unable to produce offspring or barren) is the process by which all living cells, spores, and acellular entities (e.g., viruses, viroids, virusoids, and prions) are either destroyed or removed from an object or habitat. A sterile object is totally free of viable microorganisms, spores, and other infectious agents. When sterilization is achieved by a chemical agent, the chemical is called a sterilant. In contrast, **disinfection** is the killing, inhibition, or removal of microorganisms that may cause disease; disinfection is the substantial reduction of the total microbial population and the destruction of potential pathogens. **Disinfectants** are agents, usually chemical, used to carry out disinfection and normally used only on inanimate objects. A disinfectant does not necessarily sterilize an object because viable spores and a few microorganisms may remain. **Sanitization** is closely related to disinfection. In sanitization, the microbial population is reduced to levels that are considered safe by public health standards. The inanimate object is usually cleaned as well as partially disinfected. For example, sanitizers are used to clean eating utensils in restaurants. ◀◀ *Viroids and virusoids (section 5.6); Prions (section 5.7)*

It also is frequently necessary to control microorganisms on or in living tissue with chemical agents. **Antisepsis** (Greek *anti,* against, and *sepsis,* putrefaction) is the destruction or inhibition of microorganisms on a living tissue; it is the prevention of infection or sepsis. **Antiseptics** are chemical agents applied to tissue to prevent infection by killing or inhibiting pathogen growth; they also reduce the total microbial population. Because they must not destroy too much host tissue, antiseptics are generally not as toxic as disinfectants. The exposure of microorganisms to increasing biocide concentrations decreases the number of viable organisms. **Figure 8.2** shows three possible population reduction curves resulting from three different biocides. The shape of the curve reflects various conditions that influence biocide effectiveness. Note that in each case, the eventual decline in viable microorganisms can occur as a staged decline of viability from antisepsis to sterilization. **Chemotherapy** is the use of chemical agents to kill or inhibit the growth of microorganisms within host tissue. ▶▶ *Antimicrobial chemotherapy (chapter 34)*

A suffix can be employed to denote the type of antimicrobial agent. Substances that kill organisms often have the suffix *–cide* (Latin *cida,* to kill); a cidal agent kills pathogens (and many nonpathogens) but not necessarily endospores. A disinfectant or antiseptic can be particularly effective against a specific group, in which case it may be called a **bactericide, fungicide,** or **viricide.** Other chemicals do not kill but rather prevent growth. If these agents are removed, growth will resume. Their names end in *–static* (Greek *statikos,* causing to stand or stopping)—for example, **bacteriostatic** and **fungistatic.**

1. Define the following terms: sterilization, sterilant, disinfection, disinfectant, sanitization, antisepsis, antiseptic, chemotherapy, biocide.
2. What is the difference between bactericidal and bacteriostatic? To which category do you think most household cleaners belong? Why?

8.2 The Pattern of Microbial Death

A microbial population is not killed instantly when exposed to a lethal agent. Population death is generally exponential (logarithmic)—that is, the population will be reduced by the same fraction at constant intervals (**table 8.1**). If the logarithm of the population number remaining is plotted against the time of exposure of the microorganism to the agent, a straight-line plot will result (**figure 8.3**). When the population has been greatly reduced, the rate of killing may slow due to the survival of a more resistant strain of the microorganism.

It is essential to have a precise measure of an agent's killing efficiency. One such measure is the **decimal reduction time** (D) or **D value.** The decimal reduction time is the time required to kill 90% of the microorganisms or spores in a sample under specified conditions. For example, in a semilogarithmic plot of the population remaining versus the time of heating, the D value is the time required for the line to drop by one log cycle or tenfold (figure 8.3*a*). It is also possible to determine the temperature change at a given D value that decreases the microbial population by one log cycle (90%). This temperature change is referred to as the Z value and is predicted from a semilogarithmic plot of D values versus temperature (figure 8.3*b*).

To study the effectiveness of a lethal agent, one must be able to decide when microorganisms are dead, which may present some challenges. A microbial cell is often defined as dead if it does not grow when inoculated into culture medium that would normally support its growth. In like manner, an inactive virus cannot infect a suitable host. This definition has flaws, however. It has been demonstrated that when bacteria are exposed to certain conditions, they can remain alive but are temporarily unable to reproduce. When in this state, these persister cells are often referred to as viable but nonculturable (VBNC) *(see figure 7.13)*. In conventional tests to demonstrate killing by an antimicrobial agent, VBNC bacteria would be thought to be dead. This is a serious problem because the bacteria may regain their ability to reproduce and cause infection after a period of recovery. ◀◀ *Senescence and death (section 7.3)*

Table 8.1 A Theoretical Microbial Heat-Killing Experiment

Minute	Microbial Number at Start of Minute[a]	Microorganisms Killed in 1 Minute (90% of Total)[a]	Microorganisms at End of 1 Minute	Log_{10} of Survivors
1	10^6	9×10^5	10^5	5
2	10^5	9×10^4	10^4	4
3	10^4	9×10^3	10^3	3
4	10^3	9×10^2	10^2	2
5	10^2	9×10^1	10	1
6	10^1	9	1	0
7	1	0.9	0.1	−1

[a] Assume that the initial sample contains 10^6 vegetative microorganisms per milliliter and that 90% of the organisms are killed during each minute of exposure. The temperature is 121°C.

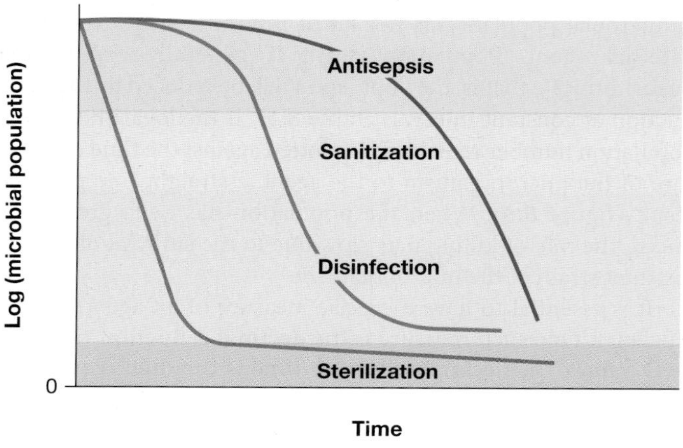

FIGURE 8.2 Impact of Biocide Exposure. Three exponential plots of survivors versus time of biocide exposure, indicating the potential kinetics of biocide action. Note how the general terms referencing microbial control are reflected by decreasing numbers of microbes. For example, sterilization refers to the absence of viable organisms regardless of biocide kinetics.

8.3 Conditions Influencing the Effectiveness of Antimicrobial Agents

Destruction of microorganisms and inhibition of microbial growth are not simple matters because the efficiency of an **antimicrobial agent** (an agent that kills microorganisms or inhibits their growth) is affected by at least six factors.

1. **Population size.** Because an equal fraction of a microbial population is killed during each interval, a larger population requires a longer time to die than a smaller one (table 8.1 and figure 8.3).
2. **Population composition.** The effectiveness of an agent varies greatly with the nature of the organisms being treated because microorganisms differ markedly in susceptibility. Bacterial spores are much more resistant to most antimicrobial agents than are vegetative forms, and younger cells are usually more readily destroyed than mature organisms. Some species are able to withstand adverse conditions better than others. For instance, *Mycobacterium tuberculosis*, which causes tuberculosis, is much more resistant to antimicrobial agents than most other bacteria.
3. **Concentration or intensity of an antimicrobial agent.** Often, but not always, the more concentrated a chemical agent or intense a physical agent, the more rapidly microorganisms are destroyed. However, agent effectiveness usually is not directly related to concentration or intensity. Over a short range, a small increase in concentration leads to an exponential rise in effectiveness; beyond a certain point, increases may not raise the killing rate much at all. Sometimes an agent is more effective at lower concentrations. For example, 70% ethanol is more bacteriocidal than 95% ethanol because the activity of ethanol is enhanced by the presence of water.
4. **Contact time.** The longer a population is exposed to a microbicidal agent, the more organisms are killed (figures 8.2 and 8.3). To achieve sterilization, contact time should be long enough to reduce the probability of survival by at least 6 logs.
5. **Temperature.** An increase in the temperature at which a chemical acts often enhances its activity. Frequently a lower concentration of disinfectant or sterilizing agent can be used at a higher temperature.
6. **Local environment.** The population to be controlled is not isolated but surrounded by environmental factors that may either offer protection or aid in its destruction. For example, because heat kills more readily at an acidic pH, acidic foods and beverages such as fruits and tomatoes are easier to pasteurize than more alkaline foods such as milk. A second important environmental factor is organic

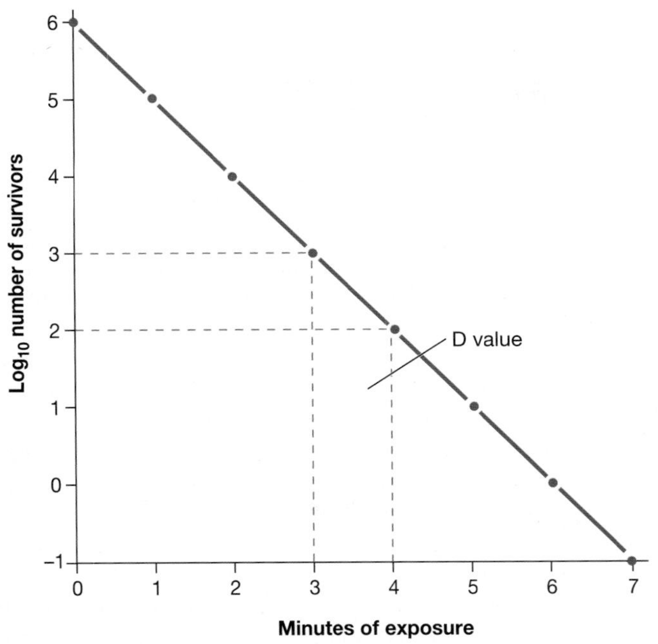

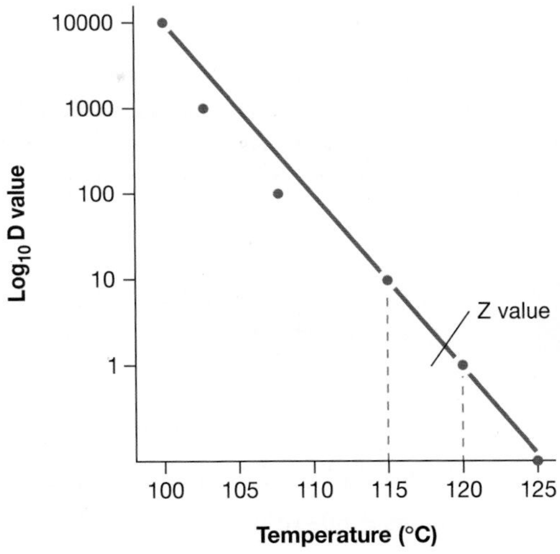

FIGURE 8.3 The Pattern of Microbial Death. (a) An exponential plot of the survivors versus the minutes of exposure to heating at 121°C. In this example, the *D* value is 1 minute. The data are from table 8.1. (b) An exponential plot of D values versus temperature. The temperature change at a given D value that reduces the population by one log unit is known as the Z value.

Figure 8.3 Micro Inquiry

Examine graph (a). How long would it take to kill one-half of the original population of microorganisms? Keep in mind that the Y axis is exponential.

matter, which can protect microorganisms against physical and chemical disinfecting agents. Biofilms are a good example. The organic matter in a biofilm protects the biofilm's microorganisms. Furthermore, it has been clearly documented that bacteria in biofilms are altered physiologically, and this makes them less susceptible to many antimicrobial agents. Because of the impact of organic matter, it may be necessary to clean objects, especially medical and dental equipment, before they are disinfected or sterilized. ◀◀ *Biofilms (section 7.7)*

1. Briefly explain how the effectiveness of antimicrobial agents varies with population size, population composition, concentration or intensity of the agent, contact time, temperature, and local environmental conditions.
2. How does being in a biofilm affect an organism's susceptibility to antimicrobial agents?
3. Suppose hospital custodians have been assigned the task of cleaning all showerheads in patient rooms to prevent the spread of infectious disease. What two factors would have the greatest impact on the effectiveness of the disinfectant the custodians use? Explain what that impact would be.

8.4 Physical Control Methods

Heat and other physical agents are normally used to control microbial growth and sterilize objects, as can be seen from the operation of the autoclave. The most frequently employed physical agents are heat, filtration, and radiation.

Heat

Moist heat readily destroys viruses, bacteria, and fungi (**table 8.2**). Moist heat kills by degrading nucleic acids and denaturing enzymes and other essential proteins. It also disrupts cell membranes. Exposure to boiling water for 10 minutes is sufficient to destroy vegetative cells and eukaryotic spores. Unfortunately, the temperature of boiling water (100°C or 212°F at sea level) is not sufficient to destroy bacterial spores, which may survive hours of boiling. Therefore boiling can be used for disinfection of drinking water and objects not harmed by water, but boiling does not sterilize.

To destroy bacterial endospores, moist heat sterilization must be carried out at temperatures above 100°C, and this requires the use of saturated steam under pressure. Steam sterilization is carried out with an **autoclave** (**figure 8.4**), a device somewhat like a fancy pressure cooker. The development of the autoclave by Chamberland in 1884 tremendously stimulated the growth of microbiology as a science. Water is boiled to produce steam, which is released into the autoclave's chamber (figure 8.4*b*). The air initially present in the chamber is forced out until the chamber is filled with saturated steam and the outlets are closed. Hot, saturated steam continues to enter until the chamber reaches the desired temperature and pressure, usually 121°C and 15 pounds of pressure. At this temperature, saturated steam destroys all vegetative cells and spores in a small volume of liquid within 10 to 12 minutes. Treatment is continued for at least 15 minutes to provide a margin of safety. Of course larger containers of liquid such as flasks and carboys require much longer treatment times.

Table 8.2 Approximate Conditions for Moist Heat Killing

Organism	*Vegetative Cells*	*Spores*
Yeasts	5 minutes at 50–60°C	5 minutes at 70–80°C
Molds	30 minutes at 62°C	30 minutes at 80°C
Bacteria[a]	10 minutes at 60–70°C	2 to over 800 minutes at 100°C 0.5–12 minutes at 121°C
Viruses	30 minutes at 60°C	

[a] Conditions for mesophilic bacteria.

Autoclaving must be carried out properly or the processed materials will not be sterile. If all air has not been flushed out of the chamber, it will not reach 121°C, even though it may reach a pressure of 15 pounds. The chamber should not be packed too tightly because the steam needs to circulate freely and contact everything in the autoclave. Bacterial spores will be killed only if they are kept at 121°C for 10 to 12 minutes. When a large volume of liquid must be sterilized, an extended sterilization time is needed because it takes longer for the center of the liquid to reach 121°C; 5 liters of liquid may require about 70 minutes. In view of these potential difficulties, a biological indicator is often autoclaved along with other material. This indicator commonly consists of a culture tube containing a sterile ampule of medium and a paper strip covered with spores of *Geobacillus stearothermophilus*. After autoclaving, the ampule is aseptically broken and the culture incubated for several days. If the test bacterium does not grow in the medium, the sterilization run has been successful. Sometimes either indicator tape or paper that changes color upon sufficient heating is autoclaved with a load of material. These approaches are convenient and save time but are not as reliable as the killing of bacterial spores.

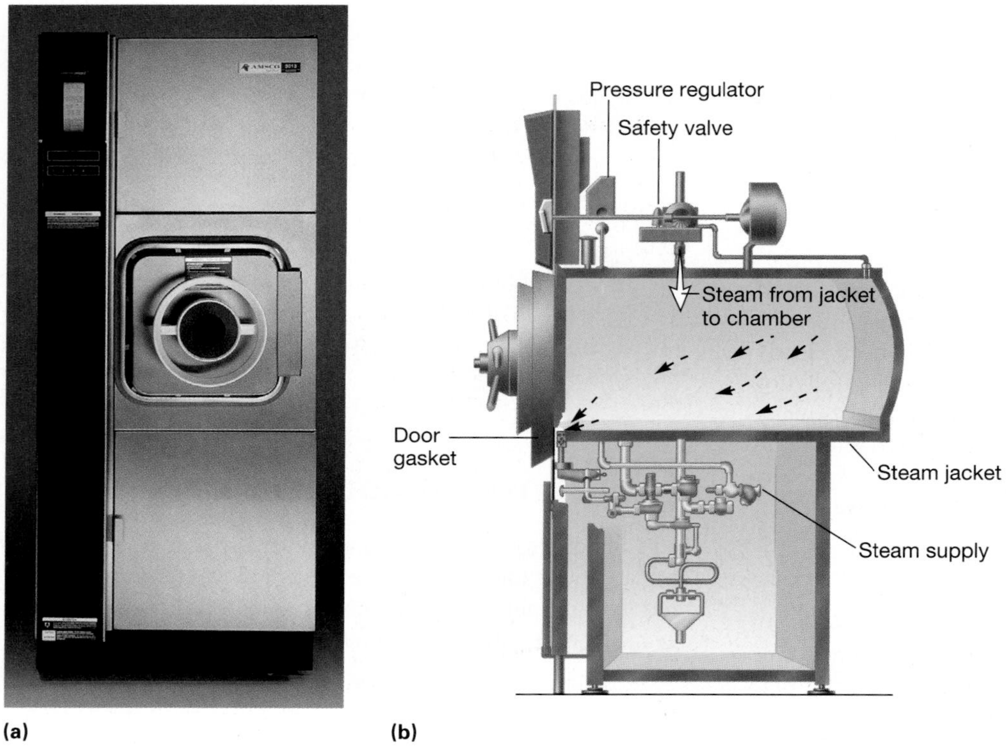

FIGURE 8.4 The Autoclave. (a) A modern, automatically controlled autoclave or sterilizer. (b) Longitudinal cross section of a typical autoclave showing some of its parts and the pathway of steam. *From John J. Perkins,* Principles and Methods of Sterilization in Health Science, *2d ed., 1969. Courtesy of Charles C. Thomas, Publisher, Springfield, Illinois.*

Many heat-sensitive substances, such as milk, are treated with controlled heating at temperatures well below boiling, a process known as **pasteurization** in honor of its developer, Louis Pasteur. In the 1860s the French wine industry was plagued by the problem of wine spoilage, which made wine storage and shipping difficult. Pasteur examined spoiled wine under the microscope and detected microorganisms that looked like the bacteria responsible for lactic acid and acetic acid fermentations (which do not form spores). He then discovered that a brief heating at 55 to 60°C would destroy these microorganisms and preserve wine for long periods. In 1886 the German chemists V. H. and F. Soxhlet adapted the technique for preserving milk and reducing milk-transmissible diseases. Milk pasteurization was introduced in the United States in 1889. Milk, beer, and many other beverages are now pasteurized. Pasteurization does not sterilize a beverage, but it does kill any pathogens present and drastically slows spoilage by reducing the level of nonpathogenic spoilage microorganisms.

Some materials cannot withstand the high temperature of the autoclave, and spore contamination precludes the use of other methods to sterilize them. For these materials, a process of intermittent sterilization, also known as **tyndallization** (for John Tyndall, the British physicist who used the technique to destroy heat-resistant microorganisms in dust) is used. The process also uses steam (30–60 minutes) to destroy vegetative bacteria. However, steam exposure is repeated for a total of three times with 23- to 24-hour incubations between steam exposures. The incubations permit remaining spores to germinate into heat-sensitive vegetative cells that are then destroyed upon subsequent steam exposures.

Many objects are best sterilized in the absence of water by dry heat sterilization. Some items are sterilized by incineration. For instance, inoculating loops, which are used routinely in the laboratory, can be sterilized in a small, bench-top incinerator (**figure 8.5**). Other items are sterilized in an oven at 160 to 170°C for 2 to 3 hours. Microbial death results from the oxidation of cell constituents and denaturation of proteins. Dry air heat is less effective than moist heat. The spores of *Clostridium botulinum,* the cause of botulism, are killed in 5 minutes at 121°C by moist heat but only after 2 hours at 160°C with dry heat. However, dry heat has some definite advantages. It does not corrode glassware and metal instruments as moist heat does, and it can be used to sterilize powders, oils, and similar items. Despite these advantages, dry heat sterilization is slow and not suitable for heat-sensitive materials such as many plastic and rubber items.

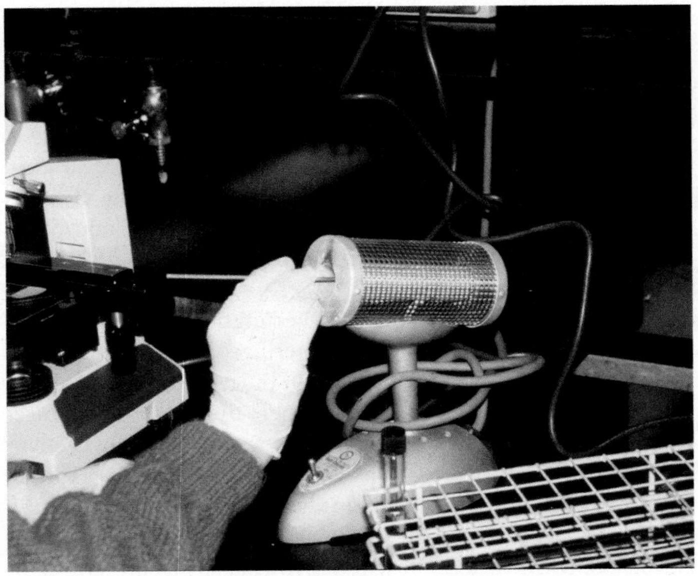

FIGURE 8.5 Dry Heat Incineration. Bench-top incinerators are routinely used to sterilize inoculating loops used in microbiology laboratories.

1. Describe how an autoclave works. What conditions are required for sterilization by moist heat? What three things must one do when operating an autoclave to help ensure success?
2. In the past, spoiled milk was responsible for a significant proportion of infant deaths. Why is untreated milk easily spoiled?

Filtration

Filtration is an excellent way to reduce the microbial population in solutions of heat-sensitive material and can be used to sterilize various liquids and gases (including air). Rather than directly destroying contaminating microorganisms, the filter simply removes them. There are two types of filters. **Depth filters** consist of fibrous or granular materials that have been bonded into a thick layer filled with twisting channels of small diameter. The solution containing microorganisms is sucked through this layer under vacuum, and microbial cells are removed by physical screening or entrapment and by adsorption to the surface of the filter material. Depth filters are made of diatomaceous earth (Berkefield filters), unglazed porcelain (Chamberland filters), asbestos, or other similar materials.

Membrane filters have replaced depth filters for many purposes. These filters are porous membranes, a little over 0.1 mm thick, made of cellulose acetate, cellulose nitrate, polycarbonate, polyvinylidene fluoride, or other synthetic materials. Although a wide variety of pore sizes are available, membranes with pores about 0.2 μm in diameter are used to remove most vegetative cells, but not viruses, from liquids ranging in volume from less than 1 milliliter to many liters. The membranes can be held in special holders (**figure 8.6**), and their use is often preceded by the use of depth filters made of glass fibers to remove larger particles that might clog the membrane filter. The liquid is pulled or forced through the filter with a vacuum or with pressure from a syringe, peristaltic pump, or nitrogen gas, and collected in previously sterilized containers. Membrane filters remove microorganisms by screening them out much as a sieve separates large sand particles from small ones (**figure 8.7**). These filters are used to sterilize pharmaceuticals, ophthalmic solutions, culture media, oils, antibiotics, and other heat-sensitive solutions.

Air also can be filtered to remove microorganisms. Two common examples are N-95 disposable masks used in hospitals and labs, and cotton plugs on culture vessels that let air in but keep microorganisms out. N-95 masks exclude 95% of particles that are larger than 0.3 μm. Other important examples are **biological safety cabinets,** which employ **high-efficiency particulate air (HEPA) filters** (a type of depth filter made from fiberglass) to

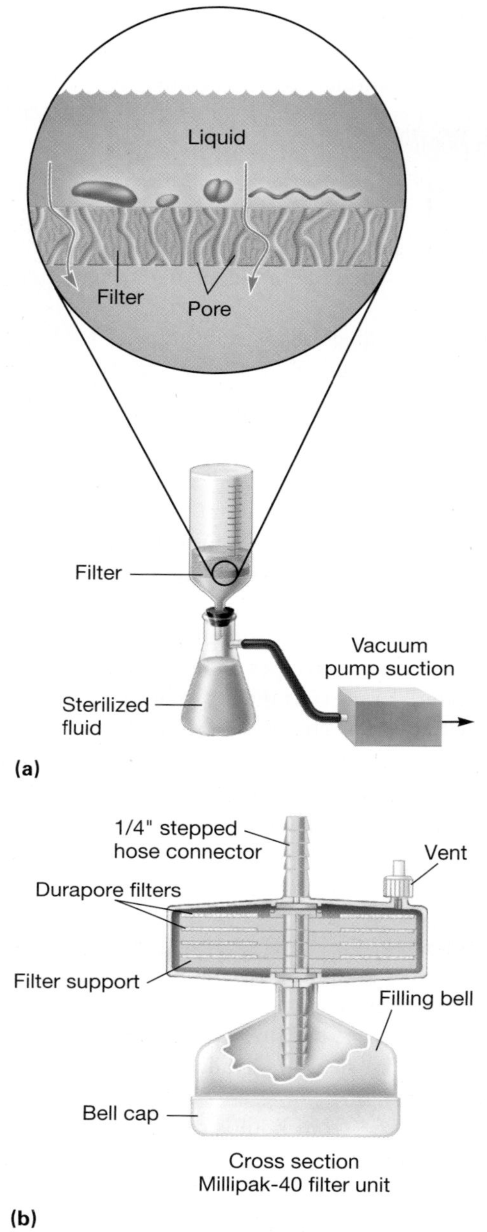

FIGURE 8.6 Membrane Filter Sterilization. The liquid to be sterilized is pumped through a membrane filter and into a sterile container. (a) Schematic representation of a membrane filtration setup that uses a vacuum pump to force liquid through the filter. The inset shows a cross section of the filter and its pores, which are too small for microbes to pass through. (b) Cross section of a membrane filtration unit. Several membranes are used to increase its capacity.

Figure 8.6 Micro Inquiry

How might one verify that filtration removed all microorganisms?

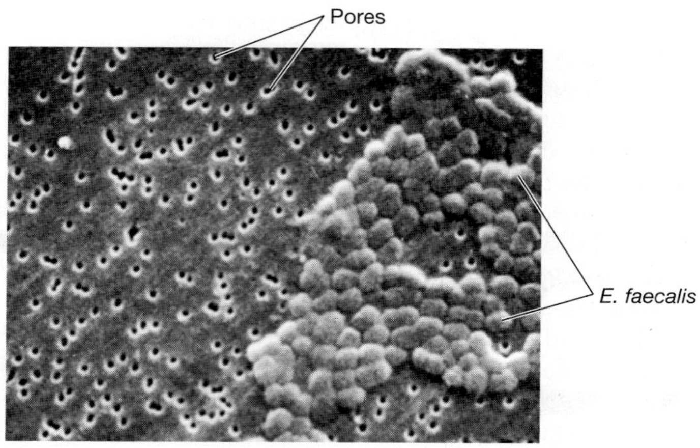

FIGURE 8.7 Membrane Filter. *Enterococcus faecalis* resting on a polycarbonate membrane filter with 0.4 μm pores (×5,900).

remove 99.97% of particles 0.3 μm or larger by both physical retention and electrostatic interactions. HEPA filters can also be made to adsorb viruses. HEPA filters are available that remove viruses that are 0.1 μm and smaller; they are used to sterilize air. Laminar flow biological safety cabinets or hoods force air through HEPA filters, then project a vertical curtain of sterile air across the cabinet opening. This protects a worker from microorganisms being handled within the cabinet and prevents contamination of the room (**figure 8.8**). A person uses these cabinets when working with dangerous agents such as *M. tuberculosis,* pathogenic fungi, or tumor viruses. They are also employed in research labs and industries, such as the pharmaceutical industry, when a sterile working environment is needed.

Radiation

Ultraviolet (UV) radiation around 260 nm *(see figure 7.31)* is quite lethal. UV radiation causes thymine-thymine dimerization of DNA, preventing replication and transcription. However, UV does not penetrate glass, dirt films, water, and other substances very effectively. Because of this disadvantage, UV radiation is used as a sterilizing agent only in a few specific situations. UV lamps are sometimes placed on the ceilings of rooms or in biological safety cabinets to sterilize the air and any exposed surfaces. Because UV radiation burns the skin and damages eyes, the UV lamps are off when the areas are in use. Commercial UV units are available for water treatment. Pathogens and other microorganisms are destroyed when a thin layer of water is passed under the lamps.

Ionizing radiation is an excellent sterilizing agent and penetrates deep into objects. It will destroy bacterial spores and all microbial cells; however, ionizing radiation is not always effective against viruses. Gamma radiation from a cobalt 60 source and accelerated electrons from high-voltage electricity are used

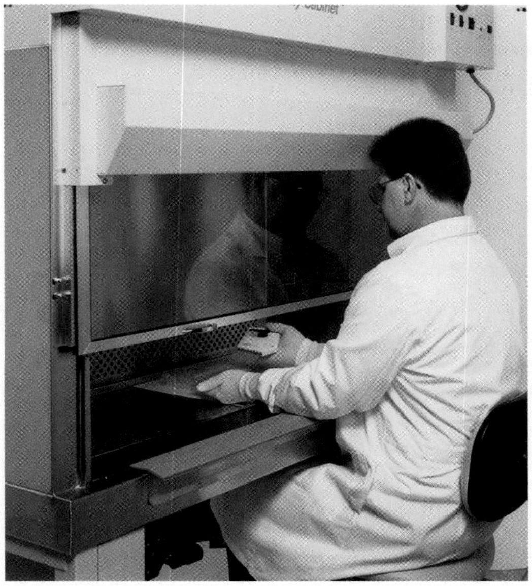

(a)

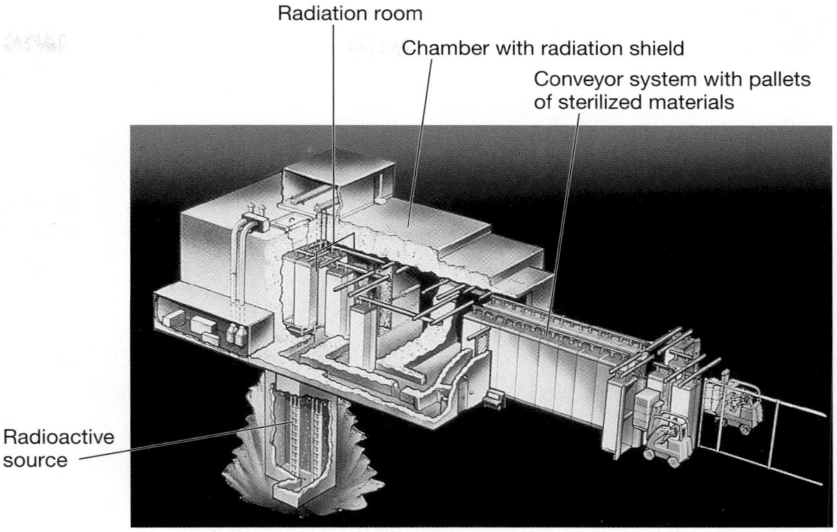

FIGURE 8.9 Sterilization with Ionizing Radiation. An irradiation machine that uses radioactive cobalt 60 as a gamma radiation source to sterilize fruits, vegetables, meats, fish, and spices.

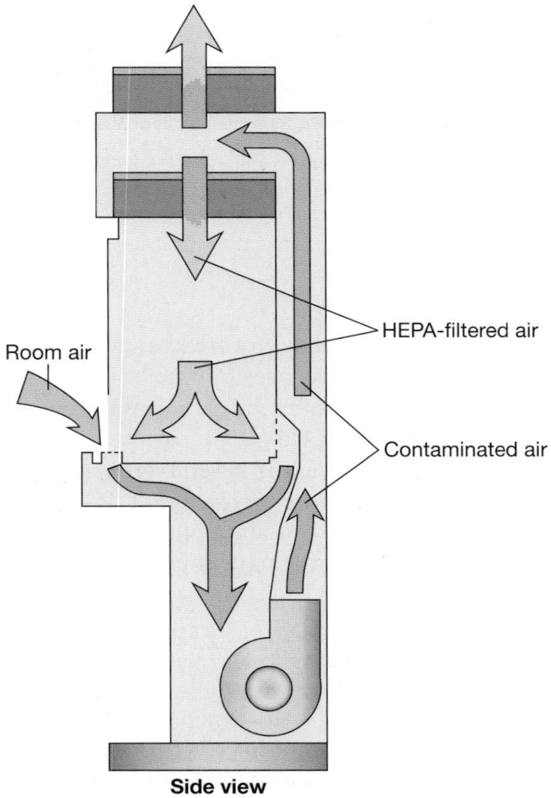

(b)

FIGURE 8.8 A Biological Safety Cabinet. (a) A technician pipetting potentially hazardous material in a safety cabinet. (b) A schematic diagram showing the airflow pattern within a class II safety cabinet.

in the cold sterilization of antibiotics, hormones, sutures, and plastic disposable supplies such as syringes. Gamma radiation and electron beams have also been used to sterilize and "pasteurize" meat and other foods (**figure 8.9**). Irradiation can eliminate the threat of such pathogens as *Escherichia coli* O157:H7, *Staphylococcus aureus,* and *Campylobacter jejuni.* Based on the results of numerous studies, both the U.S. Food and Drug Administration and the World Health Organization have approved irradiated food and declared it safe for human consumption. Currently irradiation is being used to treat poultry, beef, pork, veal, lamb, fruits, vegetables, and spices. ▶▶| *Controlling food spoilage (section 40.2)*

1. What are depth filters and membrane filters, and how are they used to sterilize liquids? Describe the operation of a biological safety cabinet.
2. Give the advantages and disadvantages of ultraviolet light and ionizing radiation as sterilizing agents. Provide a few examples of how each is used for this purpose.

8.5 Chemical Control Agents

Physical agents are generally used to sterilize objects. Chemicals, on the other hand, are more often employed in disinfection and antisepsis. The proper use of chemical agents is essential to laboratory and hospital safety (Techniques & Applications 8.1). Chemicals also are employed to prevent microbial growth in food, and certain chemicals are used to treat infectious disease. The use of chemical agents for chemotherapy in humans is covered in chapter 34. Next we discuss chemicals used outside the body.

Table 8.3 Activity Levels of Selected Biocides

Class	*Use Concentration of Active Ingredient*	*Activity Level*[a]
Gas		
Ethylene oxide	450–500 mg/l[b]	High
Liquid		
Glutaraldehyde, aqueous	2%	High to intermediate
Formaldehyde + alcohol	8 + 70%	High
Stabilized hydrogen peroxide	6–30%	High to intermediate
Formaldehyde, aqueous	6–8%	High to intermediate
Iodophors, high concentration	750–5,000 mg/l[c]	High to intermediate
Iodophors, low concentration	75–150 mg/l[c]	Intermediate to low
Iodine + alcohol	0.5 + 70%	Intermediate
Chlorine compounds	0.1–0.5%[d]	Intermediate
Phenolic compounds, aqueous	0.5–3%	Intermediate to low
Iodine, aqueous	1%	Intermediate
Alcohols (ethyl, isopropyl)	70%	Intermediate
Quaternary ammonium compounds	0.1–0.2% aqueous	Low
Chlorhexidine	0.75–4%	Low
Hexachlorophene	1–3%	Low
Mercurial compounds	0.1–0.2%	Low

Source: From Seymour S. Block, Disinfection, Sterilization and Preservation. *Copyright © 1983 Lea & Febiger, Malvern, Pa. Reprinted by permission.*
[a] High-level disinfectants destroy vegetative bacterial cells including *M. tuberculosis,* bacterial endospores, fungi, and viruses. Intermediate-level disinfectants destroy all of the above except spores. Low-level agents kill bacterial vegetative cells except for *M. tuberculosis,* fungi, and medium-sized lipid-containing viruses (but not bacterial endospores or small, nonlipid viruses).
[b] In autoclave-type equipment at 55 to 60°C.
[c] Available iodine.
[d] Free chlorine.

Many different chemicals are available for use as disinfectants, and each has its own advantages and disadvantages. Ideally the disinfectant must be effective against a wide variety of infectious agents (gram-positive and gram-negative bacteria, acid-fast bacteria, bacterial spores, fungi, and viruses) at low concentrations and in the presence of organic matter. Although the chemical must be toxic for infectious agents, it should not be toxic to people or corrosive for common materials. In practice, this balance between effectiveness and low toxicity for animals is hard to achieve. Some chemicals are used despite their low effectiveness because they are relatively nontoxic. The ideal disinfectant should be stable upon storage, odorless or with a pleasant odor, and soluble in water and lipids for penetration into microorganisms; have a low surface tension so that it can enter cracks in surfaces; and be relatively inexpensive.

One potentially serious problem is the overuse of antiseptics. For instance, the antibacterial agent triclosan is found in products such as deodorants, mouthwashes, soaps, cutting boards, and baby toys. Unfortunately, the emergence of triclosan-resistant bacteria has become a problem. For example, *Pseudomonas aeruginosa* actively pumps the antiseptic out of the cell. There is now evidence that extensive use of triclosan also increases the frequency of bacterial resistance to antibiotics. Thus overuse of antiseptics can have unintended harmful consequences. ▶▶| *Drug resistance (section 34.9)*

The properties and uses of several groups of common disinfectants and antiseptics are surveyed next. Many of the characteristics of disinfectants and antiseptics are summarized in **tables 8.3** and **8.4**. Structures of some common agents are shown in **figure 8.10**.

Phenolics

Phenol was the first widely used antiseptic and disinfectant. In 1867 Joseph Lister employed it to reduce the risk of infection during surgery. Today phenol and phenolics (phenol derivatives) such as cresols, xylenols, and orthophenylphenol are used as disinfectants in laboratories and hospitals. The commercial disinfectant Lysol® is made of a mixture of phenolics. Phenolics act by denaturing proteins and disrupting cell membranes. They have some real advantages as disinfectants: phenolics are tuberculocidal, effective in the presence of organic

Table 8.4 Relative Efficacy of Commonly Used Disinfectants and Antiseptics

Class	*Disinfectant*	*Antiseptic*	*Comment*
Gas			
Ethylene oxide	3–4[a]	0[a]	Sporicidal; toxic; good penetration; requires relative humidity of 30% or more; microbicidal activity varies with apparatus used; absorbed by porous material; dry spores highly resistant; presoaking is most desirable
Liquid			
Glutaraldehyde, aqueous	3	0	Sporicidal; active solution unstable; toxic
Stabilized hydrogen peroxide	3	0	Sporicidal; solution stable up to 6 weeks; toxic orally and to eyes; mildly skin toxic; little inactivation by organic matter
Formaldehyde + alcohol	3	0	Sporicidal; noxious fumes; toxic; volatile
Formaldehyde, aqueous	1–2	0	Sporicidal; noxious fumes; toxic
Phenolic compounds	3	0	Stable; corrosive; little inactivation by organic matter; irritates skin
Chlorine compounds	1–2	0	Fast action; inactivation by organic matter; corrosive; irritates skin
Alcohol	1	3	Rapidly microbicidal except for bacterial spores and some viruses; volatile; flammable; dries and irritates skin
Iodine + alcohol	0	4	Corrosive; very rapidly microbicidal; causes staining; irritates skin; flammable
Iodophors	1–2	3	Somewhat unstable; relatively bland; staining temporary; corrosive
Iodine, aqueous	0	2	Rapidly microbicidal; corrosive; stains fabrics; stains and irritates skin
Quaternary ammonium compounds	1	0	Inactivated by soap and anionics; compounds absorbed by fabrics; old or dilute solution can support growth of gram-negative bacteria
Hexachlorophene	0	2	Insoluble in water, soluble in alcohol; not inactivated by soap; weakly bactericidal
Chlorhexidine	0	3	Soluble in water and alcohol; weakly bactericidal
Mercurial compounds	0	±	Greatly inactivated by organic matter; weakly bactericidal

Source: From Seymour S. Block, Disinfection, Sterilization and Preservation. *Copyright © 1983 Lea & Febiger, Malvern, Pa. Reprinted by permission.*

[a] Subjective ratings of practical usefulness in a hospital environment—4 is maximal usefulness; 0 is little or no usefulness; ± signifies that the substance is sometimes useful but not always.

material, and remain active on surfaces long after application. However, they have a disagreeable odor and can cause skin irritation. The newer phenolic, tricolsan (figure 8.10), is often used in hand sanitizers due to its effective blockage of bacterial fatty acid synthesis.

Alcohols

Alcohols are among the most widely used disinfectants and antiseptics. They are bactericidal and fungicidal but not sporicidal; some enveloped viruses are also destroyed. The two most popular alcohol germicides are ethanol and isopropanol, usually used in about 60 to 80% concentration. They act by denaturing proteins and possibly by dissolving membrane lipids. A 10- to 15-minute soaking is sufficient to disinfect small instruments.

Halogens

A halogen is any of the five elements (fluorine, chlorine, bromine, iodine, and astatine) in group VIIA of the periodic table. They exist as diatomic molecules in the free state and form saltlike compounds with sodium and most other metals. The halogens iodine and chlorine are important antimicrobial agents. Iodine is used as a skin antiseptic and kills by oxidizing cell constituents and iodinating cell proteins. At higher concentrations, it may even kill some spores. Iodine often has been applied as tincture of iodine, 2% or more iodine in a water-ethanol solution of potassium iodide. Although it is an effective antiseptic, the skin may be damaged, a stain is left, and iodine allergies can result. Iodine has been complexed with an organic carrier to form an **iodophor.** Iodophors are water soluble, stable, and nonstaining, and release iodine slowly to minimize skin burns and irritation. They are used in

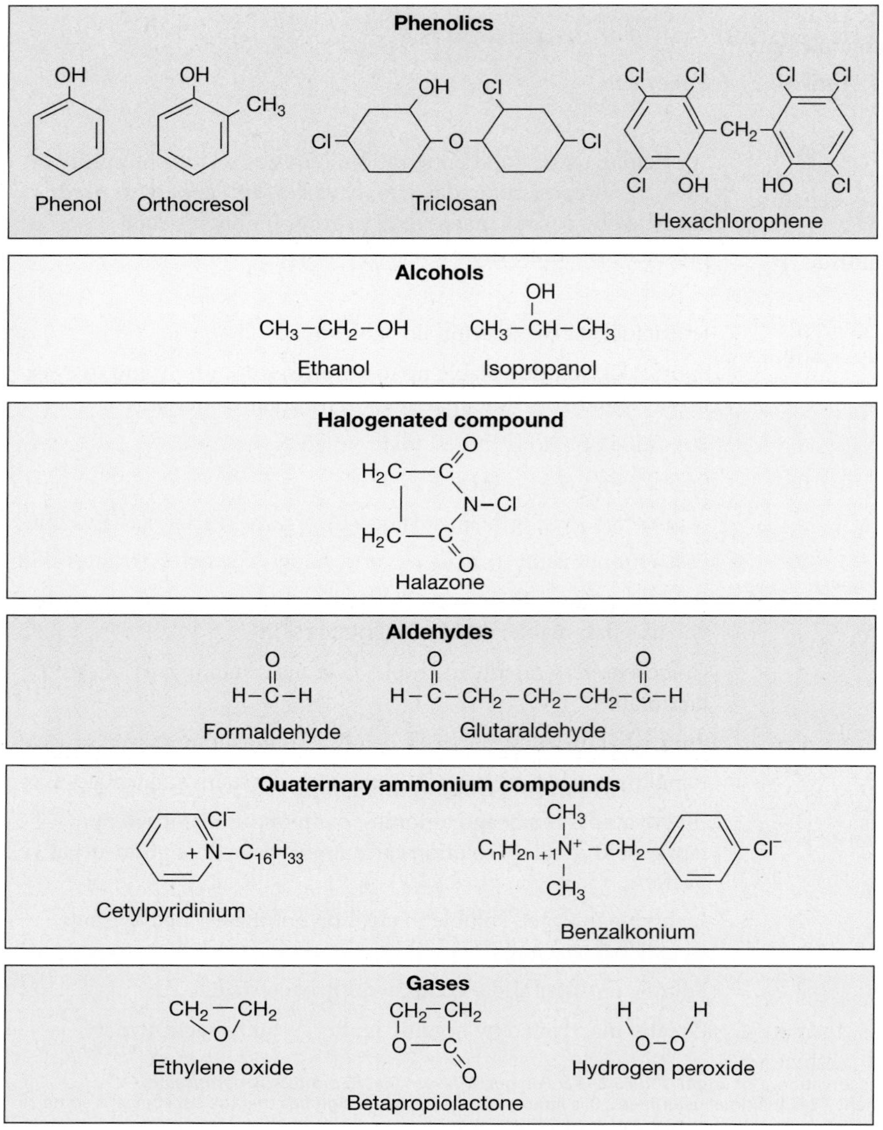

FIGURE 8.10 Disinfectants and Antiseptics. The structures of some frequently used disinfectants and antiseptics.

Figure 8.10 Micro Inquiry

Why do you think it is important that all of these compounds are relatively hydrophobic?

hospitals for cleansing preoperative skin and in hospitals and laboratories for disinfecting. Some popular brands are Wescodyne for skin and laboratory disinfection, and Betadine for wounds.

Chlorine is the usual disinfectant for municipal water supplies and swimming pools, and is employed in the dairy and food industries. It may be applied as chlorine gas (Cl_2), sodium hypochlorite (bleach, NaOCl), or calcium hypochlorite [$Ca(OCl)_2$], all of which yield hypochlorous acid (HOCl) (see chemical reactions that follow). The result is oxidation of cellular materials and destruction of vegetative bacteria and fungi.

$$Cl_2 + H_2O \rightarrow HCl + HOCl$$

$$NaOCl + H_2O \rightarrow NaOH + HOCl$$

$$Ca(OCl)_2 + 2H_2O \rightarrow Ca(OH)_2 + 2HOCl$$

Death of almost all microorganisms usually occurs within 30 minutes. One potential problem is that chlorine reacts with organic compounds to form carcinogenic trihalomethanes, which must be monitored in drinking water. Some wastewater treatment facilities recover the chlorine from the water prior to discharge to prevent the formation of trihalomethanes.

Search This: CDC safe drinking water podcast

Chlorine is also an excellent disinfectant for individual use because it is effective, inexpensive, and easy to employ. Small quantities of drinking water can be disinfected with halazone tablets. Halazone (parasulfone dichloramidobenzoic acid) slowly releases chloride when added to water and disinfects it in about a half hour. It is frequently used by campers lacking access to uncontaminated drinking water. Of note is the fact that household bleach (diluted to 10% in water, 10 minute contact time) can be used to disinfect surfaces contaminated by human body fluids and that it is made more effective by the addition of household vinegar.

Heavy Metals

For many years the ions of heavy metals such as mercury, silver, arsenic, zinc, and copper were used as germicides. These have now been superseded by other less toxic and more effective germicides (many heavy metals are more bacteriostatic than bactericidal). There are a few exceptions. In some hospitals, a 1% solution of silver nitrate is added to the eyes of infants to prevent ophthalmic gonorrhea. Silver sulfadiazine is used on burns. Copper sulfate is an effective algicide in lakes and swimming pools. Heavy metals combine with proteins, often with their sulfhydryl groups, and inactivate them. They may also precipitate cell proteins.

Quaternary Ammonium Compounds

Quaternary ammonium compounds are detergents that have broad spectrum antimicrobial activity and are effective disinfectants that are used for decontamination purposes (**opening figure**).

Detergents (Latin *detergere,* to wipe away) are organic cleansing agents that are amphipathic, having both polar hydrophilic and nonpolar hydrophobic components. The hydrophilic portion of a quaternary ammonium compound is a positively charged quaternary nitrogen; thus quaternary ammonium compounds are cationic detergents. Their antimicrobial activity is the result of their ability to disrupt microbial membranes; they may also denature proteins.

Cationic detergents such as benzalkonium chloride and cetylpyridinium chloride kill most bacteria but not *M. tuberculosis* or spores. They have the advantages of being stable and nontoxic, but they are inactivated by hard water and soap. Cationic detergents are often used as disinfectants for food utensils and small instruments, and as skin antiseptics.

Aldehydes

Both of the commonly used aldehydes, formaldehyde and glutaraldehyde (figure 8.10), are highly reactive molecules that combine with nucleic acids and proteins, and inactivate them, probably by cross-linking and alkylating molecules (**figure 8.11**). They are sporicidal and can be used as chemical sterilants. Formaldehyde is usually dissolved in water or alcohol before use. A 2% buffered solution of glutaraldehyde is an effective disinfectant. It is less irritating than formaldehyde and is used to disinfect hospital and laboratory equipment. Glutaraldehyde usually disinfects objects within about 10 minutes but may require as long as 12 hours to destroy all spores.

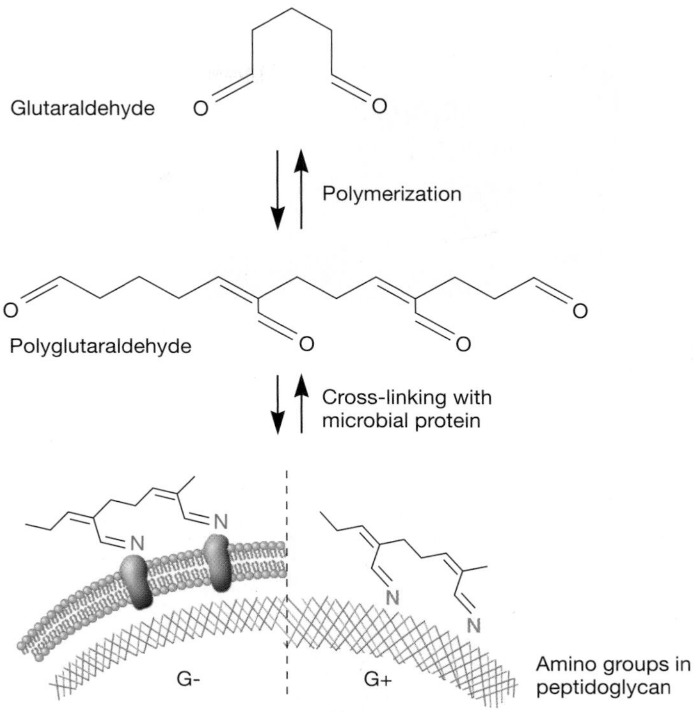

FIGURE 8.11 Effects of Glutaraldehyde. Glutaraldehyde polymerizes and then interacts with amino acids in proteins (left) or in peptidoglycan (right). As a result, the proteins are alkylated and cross-linked to other proteins, which inactivates them. The amino groups in peptidoglycan are also alkylated and cross-linked, which prevents them from participating in other chemical reactions such as those involved in peptidoglycan synthesis.

Figure 8.11 ***Micro Inquiry***

Why do you think that cross-linking agents such as glutaraldehyde are often called "fixatives" or are said to "fix the cells"?

Sterilizing Gases

Many heat-sensitive items such as disposable plastic Petri dishes and syringes, heart-lung machine components, sutures, and catheters are sterilized with ethylene oxide gas (figure 8.10). Ethylene oxide (EtO) is both microbicidal and sporicidal. It is a very strong alkylating agent that kills by reacting with functional groups of DNA and proteins to block replication and enzymatic activity. It is a particularly effective sterilizing agent because it rapidly penetrates packing materials, even plastic wraps.

Sterilization is carried out in an ethylene oxide sterilizer, which resembles an autoclave in appearance. It controls the EtO concentration, temperature, and humidity (**figure 8.12**). Because pure EtO is explosive, it is usually supplied in a 10 to 20% concentration mixed with either CO_2 or dichlorodifluoromethane. The EtO concentration, humidity, and temperature influence the rate of sterilization. A clean object can be sterilized if treated for 5 to 8 hours at 38°C or 3 to 4 hours at 54°C when the relative humidity is maintained at 40 to 50% and the EtO concentration at 700 mg/l. Because it is so toxic to humans, extensive aeration of the sterilized materials is necessary to remove residual EtO.

Betapropiolactone (BPL) is occasionally employed as a sterilizing gas. In the liquid form, it has been used to sterilize vaccines and blood products. BPL decomposes to an inactive form after several hours and is therefore not as difficult to eliminate as EtO. It also destroys microorganisms more readily than ethylene oxide but does not penetrate materials well and may be carcinogenic. For these reasons, BPL has not been used as extensively as EtO.

Vaporized hydrogen peroxide (VHP) can also be used to decontaminate biological safety cabinets, operating rooms, and other large facilities. VPH is produced from a solution of hydrogen peroxide in water that is passed over a vaporizer to achieve a vapor concentration between 140 and 1400 parts per million (ppm), depending on the agent to be destroyed. VPH is then introduced as a sterilizing vapor into the enclosure for some time, depending on the size of the enclosure and the materials within. Hydrogen peroxide and its oxy-radical by-products are toxic (75 ppm are dangerous to human health) and kill a wide variety of microorganisms. During the course of the decontamination

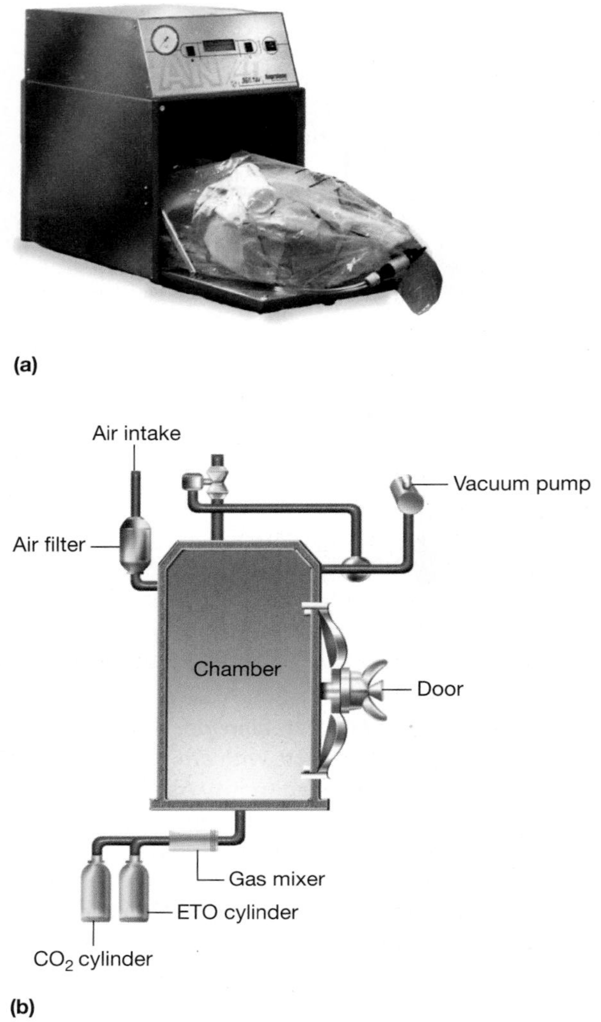

FIGURE 8.12 An Ethylene Oxide Sterilizer. (a) An automatic ethylene oxide (EtO) sterilizer. (b) Schematic of an EtO sterilizer. Items to be sterilized are placed in the chamber, and EtO and carbon dioxide are introduced. After the sterilization procedure is completed, the EtO and carbon dioxide are pumped out of the chamber and air enters.

process, VPH breaks down to water and oxygen, both of which are harmless. Other advantages of VPH are that it can be used at a wide range of temperatures (4 to 80°C) and does not damage most nonliving materials.

1. Why are most antimicrobial chemical agents disinfectants rather than sterilants? What general characteristics should one look for in a disinfectant?
2. Describe each of the following agents in terms of its chemical nature, mechanism of action, mode of application, common uses and effectiveness, and advantages and disadvantages: phenolics, alcohols, halogens, heavy metals, quaternary ammonium compounds, aldehydes, and ethylene oxide.
3. Which disinfectants or antiseptics would be used to treat the following: laboratory bench top, drinking water, patch of skin before surgery, small medical instruments (probes, forceps, etc.)? Explain your choices.
4. How do phenolic agents differ from the other chemical control agents described in this chapter?
5. Which physical or chemical agent would be the best choice for sterilizing the following items: glass pipettes, tryptic soy broth tubes, nutrient agar, antibiotic solution, interior of a biological safety cabinet, wrapped package of plastic Petri plates? Explain your choices.

8.6 Evaluation of Antimicrobial Agent Effectiveness

Testing of antimicrobial agents is a complex process regulated by two different federal agencies. The Environmental Protection Agency regulates disinfectants, whereas agents used on humans and animals are under the control of the Food and Drug Administration. Testing of antimicrobial agents often begins with an initial screening to see if they are effective and at what concentrations. This may be followed by more realistic in-use testing.

The best-known disinfectant screening test is the **phenol coefficient test** in which the potency of a disinfectant is compared with that of phenol. A series of dilutions of phenol and the disinfectant being tested are prepared. Standard amounts of *Salmonella enterica* serovar Typhi and *Staphylococcus aureus* are added to each dilution; the dilutions are then placed in a 20 or 37°C water bath. At 5-minute intervals, samples are withdrawn from each dilution and used to inoculate a growth medium, which is incubated and examined for growth. Growth in the medium indicates that the dilution at that particular time of sampling did not kill the bacteria. The highest dilution (i.e., the lowest concentration) that kills the bacteria after a 10-minute exposure but not after 5 minutes is used to calculate the phenol coefficient. The higher the phenol coefficient value, the more effective the disinfectant under these test conditions. A value greater than 1 means that the disinfectant is more effective than phenol.

The phenol coefficient test is a useful initial screening procedure, but the phenol coefficient can be misleading if taken as a direct indication of disinfectant potency during normal use. This is because the phenol coefficient is determined under carefully controlled conditions with pure bacterial cultures, whereas disinfectants are normally used on complex populations in the presence of organic matter and with significant variations in environmental factors such as pH, temperature, and presence of salts.

To more realistically estimate disinfectant effectiveness, other tests are often used. The rates at which selected bacteria are destroyed with various chemical agents may be experimentally determined and compared. A **use dilution test** can also be carried out. Stainless steel carriers are contaminated with one of three specific bacterial species under carefully controlled conditions. The carriers are dried briefly, immersed in the test disinfectants for 10 minutes, transferred to culture media, and incubated for two days. The disinfectant concentration that kills the bacteria on at least 59 out of 60 carriers (a 95% level of confidence) is determined. Disinfectants also can be tested under conditions designed to simulate normal in-use situations. In-use testing techniques allow a more accurate determination of the proper disinfectant concentration for a particular situation.

1. Briefly describe the phenol coefficient test.
2. Why might it be necessary to employ procedures such as the use dilution and in-use tests?

8.7 Biological Control of Microorganisms

The emerging field of biological control of microorganisms demonstrates great promise. Scientists are learning to exploit natural control processes such as predation of one microorganism on another, viral-mediated lysis, and toxin-mediated killing. While these control mechanisms occur in nature, their approval and use by humans is relatively new. Studies evaluating control of *Salmonella, Shigella,* and *E. coli* by gram-negative predators such as *Bdellovibrio* suggest that poultry farms may be sprayed with the predator to reduce potential contamination. Another biological control method had its start in the early 1900s at the Pasteur Institute in France. Felix d'Herelle isolated bacteriophage from patients recovering from bacillary dysentery. After numerous tests in vitro, d'Herelle concluded that the bacteriophage participated in the destruction of the bacteria causing dysentery. Bacteriophage therapies were well on the way of development when penicillin ushered in the age of antibiotics. The control of human pathogens using bacteriophage is regaining wide support and appears to be effective in the eradication of a number of bacterial species by lysing the pathogenic host. It is well known today how the phage produce lytic enzymes (lysins) that attack bacterial cell walls, facilitating their release from their specific bacterial host. A large research effort is underway to evaluate phage and phage lysins as bacterial control methods. This seems intuitive, knowing that the virus lyses its specific bacterial host, yet unnerving when one thinks about maybe swallowing, injecting, or applying a virus (albeit a bacteriophage) to the human body. The use of microbial toxins (such as bacteriocins) to control susceptible populations suggests yet another method for potential control of other microorganisms. ⏮ *Viruses and other acellular infectious agents (chapter 5)*

Summary

8.1 Definitions of Frequently Used Terms

a. Sterilization is the process by which all living cells, viable spores, viruses, virusoids, prions, and viroids are either destroyed or removed from an object or habitat. Disinfection is the killing, inhibition, or removal of microorganisms (but not necessarily endospores) that can cause disease.

b. The main goal of disinfection and antisepsis is the removal, inhibition, or killing of pathogenic microbes. Both processes also reduce the total number of microbes. Disinfectants are chemicals used to disinfect inanimate objects; antiseptics are used on living tissue.

c. Antimicrobial agents that kill organisms often have the suffix *-cide,* whereas agents that prevent growth and reproduction have the suffix *-static.*

8.2 The Pattern of Microbial Death

a. Microbial death is usually exponential or logarithmic (**figure 8.3**).

b. The decimal reduction time measures an agent's killing efficiency. It represents the time needed to kill 90% of the microbes under specified conditions.

8.3 Conditions Influencing the Effectiveness of Antimicrobial Agents

a. The effectiveness of a disinfectant or sterilizing agent is influenced by population size, population composition, concentration or intensity of the agent, exposure duration, temperature, and nature of the local environment.

b. The presence of a biofilm can dramatically alter the effectiveness of an antimicrobial agent.

8.4 Physical Control Methods

a. Moist heat kills by degrading nucleic acids, denaturing enzymes and other proteins, and disrupting cell membranes.

b. Although treatment with boiling water for 10 minutes kills vegetative forms, an autoclave must be used to destroy endospores by heating at 121°C and 15 pounds of pressure (**figure 8.4**).

c. Glassware and other heat-stable items may be sterilized by dry heat at 160 to 170°C for 2 to 3 hours.

d. Microorganisms can be efficiently removed by filtration with either depth filters or membrane filters (**figure 8.6**).

e. Biological safety cabinets with high-efficiency particulate filters sterilize air by filtration (**figure 8.8**).

f. Radiation of short wavelength or high-energy ultraviolet and ionizing radiation can be used to sterilize objects (**figure 8.9**).

8.5 Chemical Control Agents

a. Chemical agents usually act as disinfectants because they cannot readily destroy bacterial spores. Disinfectant effectiveness depends on concentration, treatment duration, temperature, and presence of organic material (**tables 8.3** and **8.4**).

b. Phenolics and alcohols are popular disinfectants that act by denaturing proteins and disrupting cell membranes (**figure 8.10**).

c. Halogens (iodine and chlorine) kill by oxidizing cellular constituents; cell proteins may also be iodinated. Iodine is applied as a tincture or iodophor. Chlorine may be added to water as a gas, hypochlorite, or an organic chlorine derivative.

d. Heavy metals tend to be bacteriostatic agents. They are employed in specialized situations such as the use of silver nitrate in the eyes of newborn infants and copper sulfate in lakes and pools.

e. Cationic detergents are often used as disinfectants and antiseptics; they disrupt membranes and denature proteins.

f. Aldehydes such as formaldehyde and glutaraldehyde can sterilize as well as disinfect because they kill spores.

g. Ethylene oxide gas penetrates plastic wrapping material and destroys all life forms by reacting with proteins. It is used to sterilize packaged, heat-sensitive materials.

h. Vaporized hydrogen peroxide is used to decontaminate enclosed spaces (e.g., safety cabinets and small rooms). The vaporized hydrogen peroxide is a vapor that can be circulated throughout the space. The peroxide and its oxyradical by-products are toxic to most microorganisms.

8.6 Evaluation of Antimicrobial Agent Effectiveness

a. The phenol coefficient test is frequently used to evaluate the effectiveness of antimicrobial agents. However, it does so using conditions that do not replicate real-life use.

b. Other procedures used to determine the effectiveness of disinfectants include measurement of killing rates with germicides, use dilution testing, and in-use testing.

8.7 Biological Control of Microorganisms

a. Control of microorganisms by natural means such as through predation, viral lysis, and toxins is emerging as a promising field.

Critical Thinking Questions

1. Throughout history, spices have been used as preservatives and to cover up the smell or taste of food that is slightly spoiled. The success of some spices led to a magical, ritualized use of many of them, and possession of spices was often limited to priests or other powerful members of the community.

   a. Choose a spice and trace its use geographically and historically. What is its common use today?

   b. Spices grow and tend to be used predominantly in warmer climates. Explain.

2. Design an experiment to determine whether an antimicrobial agent is acting as a cidal or static agent. How would you determine whether an agent is suitable for use as an antiseptic rather than as a disinfectant?

3. Suppose that you are testing the effectiveness of disinfectants with the phenol coefficient test and obtained the following results. What disinfectant can you safely say is the most effective? Can you determine its phenol coefficient from these results?

Bacterial Growth after Treatment

Dilution	Disinfectant A	Disinfectant B	Disinfectant C
1/20	–	–	–
1/40	+	–	–
1/80	+	–	+
1/160	+	+	+
1/320	+	–	+

4. The death of *Bacillus* endospores is routinely used to assess the effectiveness of sterilization procedures. In this procedure, endospores are exposed to the sterilization process (e.g., autoclaving) and then plated on germination medium. If sterilization has been successful, no growth occurs. This

requires about two days. Yung and Ponce have developed a new approach that yields results within 15 minutes after sterilization. Their technique is based on the release of dipicolinic acid (DPA) from the spore coat by germinating cells. DPA release is monitored microscopically after endospores are placed in media containing alanine and terbium ion ($Tb3+$). Alanine triggers germination and $Tb3+$ binds DPA, which then fluoresces green when illuminated with UV light. Design an experiment in which this new method could be validated, that is, shown it is as reliable as the culture-based approach. What other sterilization procedures besides autoclaving could this technique be used to monitor?

Read the original paper: Yung, P. T., and Ponce, A. 2008. Fast sterility by germinable-endospore biodensimetry. *Appl. Environ. Microbiol.* 74: 7669.

Concept Mapping Challenge

Using Figure 8.1 as reference, design a concept map that focuses on the mechanism of action of the various microbial control methods.

Learn More

Learn more by visiting the text website at www.mhhe.com/willey8, where you will find a complete list of references.

9

Introduction to Metabolism

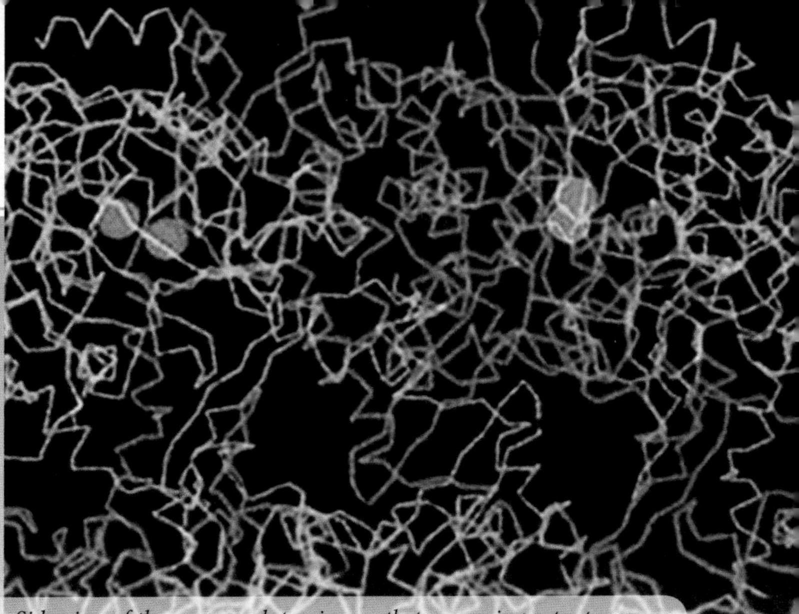

Side view of the enzyme glutamine synthetase, an important enzyme in nitrogen assimilation.

CHAPTER GLOSSARY

activation energy The energy required to bring reacting molecules together to reach the transition state in a chemical reaction.

active site The part of an enzyme that binds the substrate to form an enzyme-substrate complex and catalyze the reaction; also called the **catalytic site.**

allosteric enzyme An enzyme whose activity is altered by noncovalent binding of a small effector molecule at a regulatory site separate from the catalytic site.

anabolism The synthesis of complex molecules from simpler molecules with the input of energy.

apoenzyme The protein part of an enzyme that also has a nonprotein component (i.e., a cofactor such as a coenzyme or prosthetic group).

catabolism The breakdown of larger, more complex molecules into smaller, simpler molecules with the release of energy.

catalyst A substance that accelerates a reaction without being permanently changed itself.

coenzyme A loosely bound cofactor that often dissociates from the enzyme active site after product has been formed.

denaturation A change in protein shape that destroys the protein's activity; also can refer to changes in nucleic-acid structure.

electron transport chain (ETC) A series of electron carriers that operate together to transfer electrons from donors such as the energy source of a chemotroph to acceptors such as oxygen; also called an electron transport system (ETS).

endergonic reaction A reaction that does not spontaneously go to completion as written; the standard free energy change is positive, and the equilibrium constant is less than one.

entropy A measure of the randomness or disorder of a system; a measure of that part of the total energy in a system that is unavailable for useful work.

enzyme A protein catalyst with specificity for both the reaction catalyzed and its substrates.

equilibrium The state of a system in which no net change is occurring and free energy is at a minimum; in a chemical reaction at equilibrium, the rates in the forward and reverse directions are equal.

exergonic reaction A reaction that spontaneously goes to completion as written; the standard free energy change is negative, and the equilibrium constant is greater than one.

feedback inhibition A regulatory mechanism in which the end product of a biochemical pathway inhibits the activity of one of the pathway's enzymes.

free energy change The total energy change in a system that is available to do useful work as the system goes from its initial state to its final state at constant temperature and pressure.

holoenzyme A complete enzyme, including any nonprotein components.

metabolism The total of all chemical reactions carried out by a cell.

Michaelis constant (K_m) A kinetic constant for an enzyme reaction that equals the substrate concentration required for the enzyme to operate at half-maximal velocity.

phosphorelay system A mechanism for altering enzyme (or other protein) activity that involves the transfer of phosphate from one molecule to another.

prosthetic group A cofactor that is tightly attached to an apoenzyme.

reducing power Molecules such as NADH and NADPH that store electrons until they are used in anabolic reactions.

reversible covalent modification A mechanism of enzyme regulation in which the enzyme's activity is altered by the reversible covalent addition of a chemical group such as phosphate.

standard reduction potential A measure of the tendency of a chemical to lose electrons in an oxidation-reduction (redox) reaction.

In the early chapters of this text, we focus on a series of "what" questions about microorganisms: What are they? What do they look like? What are they made of? We now begin to consider a number of "how" questions: How do microbes extract energy from their energy source? How do they use the nutrients obtained from their environment? How do they build themselves? To begin to answer these "how" questions, we must turn our attention more fully to the chemistry of cells; that is, their metabolism. Chapters 9 through 11 consider metabolism, focusing on those processes that conserve the energy supplied by an organism's energy source and how that energy is used to synthesize the building blocks from which an organism is constructed.

Microbial metabolism is extremely important to the well-being of humans, so we begin the chapter by reviewing how microbial metabolism affects and is exploited by humans. To understand metabolism, the nature of energy and the laws of thermodynamics must be considered, so we also examine these topics. Microorganisms display an amazing array of metabolic diversity, especially in terms of the energy sources and energy-conserving processes they employ. Despite this diversity, several basic principles and processes are common to the metabolism of all microbes. These are the focus of most of the chapter. The chapter ends with a discussion of metabolic regulation.

9.1 Microbial Metabolism and Its Importance

Metabolism is the total of all chemical reactions occurring in the cell. These chemical reactions are summarized in **figure 9.1.** Some metabolic reactions are **fueling reactions.** The fueling reactions are part of **catabolism.** They conserve (capture) energy from the organisms' energy source, generate a ready supply of electrons **(reducing power),** and generate precursors for biosynthesis. The products of the fueling reactions are used in another set of metabolic reactions that build new organic molecules from smaller inorganic and organic compounds. These biosynthetic reactions are called **anabolism.**

The metabolic prowess of microbes is at the heart of their success on Earth. They have representatives among all five major nutritional types (*see table 6.3*) and therefore contribute to the cycling of elements in ecosystems. The nitrogen cycle is of particular note. Four of the eight transformations of nitrogen illustrated in **figure 9.2** are done only by microbes, and microbes contribute to the remaining four. ◀◀ *Nutritional types of microorganisms (section 6.3);* ▶▶ *Biogeochemical cycling (chapter 26)*

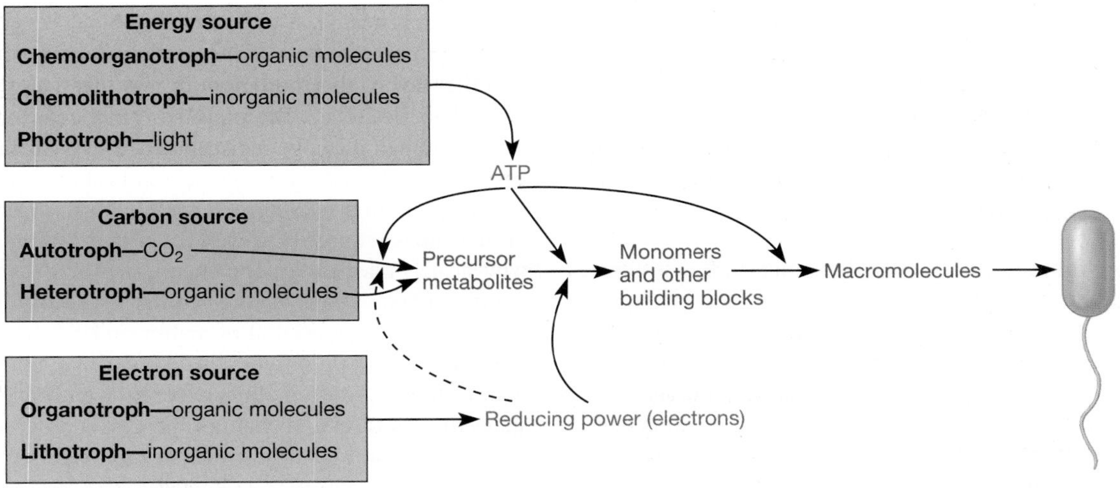

FIGURE 9.1 Overview of Metabolism. The cell structures of organisms are assembled from various macromolecules (e.g., nucleic acids and proteins). Macromolecules are synthesized from monomers and other building blocks (e.g., nucleotides and amino acids), which are the products of biochemical pathways that begin with precursor metabolites (e.g., pyruvate and α-ketoglutarate). In autotrophs, the precursor metabolites arise from CO_2-fixation and related pathways; in heterotrophs, they arise from reactions of the central metabolic pathways (see figure 11.3). Reducing power and ATP are consumed in many metabolic pathways. As indicated in the colored boxes, all organisms can be defined metabolically in terms of their energy source, carbon source, and electron source.

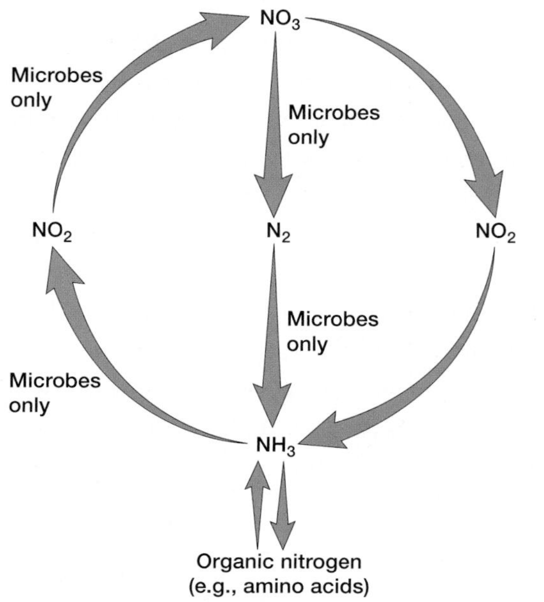

FIGURE 9.2 Contributions of Microbes to the Nitrogen Cycle. This is a simplified representation of the nitrogen cycle. A more complete cycle is presented in chapter 26.

As microbes carry out their metabolic processes, they generate numerous metabolites that humans have found useful. Ethanol is important in the production of beer, ale, and wine. Propionic acid flavors Swiss cheese. Carbon dioxide leavens bread, and antibiotics save the lives of humans every day. But the impact of microbial metabolites goes beyond their use in industry and food production. It is now increasingly recognized that the microbes living in and on the human body contribute metabolic products that are needed for human metabolic processes. Indeed, it is thought that the metabolic capabilities of the human normal microbial flora and the metabolic capabilities of their human hosts are highly integrated and are the result of coevolution. ▶▶| *Microbiology of food (chapter 40); Industrial microbiology (chapter 41); Human-microbe interactions (section 30.2)*

However, microbes don't conduct their metabolic activities for the benefit of humans. Rather, they use their vast repertoire of chemical reactions to survive and reproduce. To do this, cells must do work. Cells carry out three major types of work. **Chemical work** involves the synthesis of complex biological molecules from much simpler precursors (i.e., anabolism); energy is needed to increase the molecular complexity of a cell. **Transport work** requires energy to take up nutrients, eliminate wastes, and maintain ion balances. Energy input is needed because molecules and ions often must be transported across cell membranes against an electrochemical gradient. The third type of work is **mechanical work,** perhaps the most familiar of the three. Energy is required for cell motility and the movement of structures within cells, such as partitioning chromosomes during cell division.

1. Define metabolism, catabolism, and anabolism.
2. What kinds of work are carried out in a cell? Suppose a bacterium was doing the following: synthesizing peptidoglycan, rotating its flagellum and swimming, and secreting siderophores. What type of work is the bacterium doing in each case?

9.2 Thermodynamics

For cells to do work, they must have energy. Indeed, **energy** may be defined most simply as the capacity to do work. This is because all physical and chemical processes are the result of the application or movement of energy. Organisms obtain the energy they need from an energy source present in their environment. They convert the energy provided by the energy source into a useful form. The most commonly used form of cellular energy is the nucleoside triphosphate, ATP. In addition, other nucleoside triphosphates and other high-energy molecules are required for specific processes.

To understand how energy is conserved in ATP and how ATP is used to do cellular work, some knowledge of the basic principles of thermodynamics is required. The science of **thermodynamics** analyzes energy changes in a collection of matter (e.g., a cell or a plant) called a system. All other matter in the universe is called the surroundings. Thermodynamics focuses on the energy differences between the initial state and the final state of a system. It is not concerned with the rate of the process. For instance, if a pan of water is heated to boiling, only the condition of the water at the start and at boiling is important in thermodynamics, not how fast it is heated.

Two important laws of thermodynamics must be understood. The **first law of thermodynamics** says that energy can be neither created nor destroyed. The total energy in the universe remains constant, although it can be redistributed, as it is during the many energy exchanges that occur during chemical reactions. For example, heat is given off by exothermic reactions and absorbed during endothermic reactions. However, the first law alone cannot explain why heat is released by one chemical reaction and absorbed by another. Explanations for this require the **second law of thermodynamics** and a condition of matter called entropy. **Entropy** is a measure of the randomness or disorder of a system. The greater the disorder of a system, the greater is its entropy. The second law states that physical and chemical processes proceed in such a way that the randomness or disorder of the universe (the system and its surroundings) increases. However, even though the entropy of the universe increases, the entropy of any given system within the universe can increase, decrease, or remain unchanged.

It is necessary to specify the amount of energy used in or evolving from a particular process, and two types of energy units are employed. A **calorie** (cal) is the amount of heat energy needed to raise 1 gram of water from 14.5 to 15.5°C. The amount of energy also may be expressed in terms of **joules** (J), the units of work capable of being done. One cal of heat is equivalent to 4.1840 J of work. One thousand calories, or a kilocalorie (kcal), is enough energy to boil 1.9 milliliters of water. A kilojoule is enough energy to boil about 0.44 milliliters of water or enable a person weighing 70 kilograms to climb 35 steps.

9.3 Free Energy and Reactions

The first and second laws of thermodynamics can be combined in a useful equation, relating the changes in energy that can occur in chemical reactions and other processes.

$$\Delta G = \Delta H - T\,\Delta S$$

ΔG is the change in free energy, ΔH is the change in enthalpy, T is the temperature in Kelvin (°C + 273), and ΔS is the change in entropy occurring during the reaction. The change in **enthalpy** is the change in heat content. Cellular reactions occur under conditions of constant pressure and volume. Thus the change in enthalpy is about the same as the change in total energy during the reaction. The **free energy change** is the amount of energy in a system (or cell) available to do useful work at constant temperature and pressure. Therefore the change in entropy (ΔS) is a measure of the proportion of the total energy change that the system cannot use in performing work. Free energy and entropy changes do not depend on how the system gets from start to finish. A reaction will occur spontaneously—that is, without any external cause—if the free energy of the system decreases during the reaction or, in other words, if ΔG is negative. It follows from the equation that a reaction with a large positive change in entropy will normally tend to have a negative ΔG value and therefore occur spontaneously. A decrease in entropy will tend to make ΔG more positive and the reaction less favorable.

It is helpful to think of the relationship between entropy (ΔS) and change in free energy (ΔG) in terms that are more concrete. Consider the Greek myth of Sisyphus, king of Corinth. For his assorted crimes against the gods, he was condemned to roll a large boulder to the top of a steep hill for all eternity. This represents a very negative change in entropy—a boulder poised at the top of a hill is neither random nor disordered—and this activity (reaction) has a very positive ΔG. That is to say, Sisyphus had to put a lot of energy into the system. Unfortunately for Sisyphus, as soon as the boulder was at the top of the hill, it spontaneously rolled back down the hill. This represents a positive change in entropy and a negative ΔG. Sisyphus did not need to put energy into the system. He probably just stood at the top of the hill and watched the reaction proceed.

The change in free energy also has a definite, concrete relationship to the direction of chemical reactions. Consider this simple reaction.

$$A + B \rightleftharpoons C + D$$

If molecules A and B are mixed, they will combine to form the products C and D. Eventually C and D will become concentrated enough to combine and produce A and B at the same rate as C and D are formed from A and B. The reaction is now at **equilibrium:** the rates in both directions are equal and no further net change occurs in the concentrations of reactants and products. This situation is described by the **equilibrium constant (K_{eq})**, relating the equilibrium concentrations of products and substrates to one another.

$$K_{eq} = \frac{[C]\,[D]}{[A]\,[B]}$$

If the equilibrium constant is greater than one, the products are in greater concentration than the reactants at equilibrium—that is, the reaction tends to go to completion as written.

The equilibrium constant of a reaction is directly related to its change in free energy. When the free energy change for a process is determined at carefully defined standard conditions of concentration, pressure, pH, and temperature, it is called the **standard free energy change** (ΔG°). If the pH is set at 7.0 (which is close to the pH of living cells), the standard free energy change is indicated by the symbol $\Delta G^{\circ\prime}$. The change in standard free energy may be thought of as the maximum amount of energy available from the system for useful work under standard conditions. Using $\Delta G^{\circ\prime}$ values allows one to compare reactions without worrying about variations in ΔG due to differences in environmental conditions. The relationship between $\Delta G^{\circ\prime}$ and K_{eq} is given by this equation.

$$\Delta G^{\circ\prime} = -2.303RT \cdot \log K_{eq}$$

R is the gas constant (1.9872 cal/mole-degree or 8.3145 J/mole-degree), and T is the absolute temperature. Inspection of this equation shows that when $\Delta G^{\circ\prime}$ is negative, the equilibrium constant is greater than one and the reaction goes to completion as written. It is said to be an **exergonic reaction** (**figure 9.3**). In an **endergonic reaction,** $\Delta G^{\circ\prime}$ is positive and the equilibrium constant is less than one. That is, the reaction is not favorable, and little product will be formed at equilibrium under standard conditions. Keep in mind that the $\Delta G^{\circ\prime}$ value shows only where the reaction lies at equilibrium, not how fast the reaction reaches equilibrium.

Exergonic reactions

$$A + B \rightleftharpoons C + D$$

$$K_{eq} = \frac{[C]\,[D]}{[A]\,[B]} > 1.0$$

$\Delta G^{\circ\prime}$ **is negative.**

Endergonic reactions

$$A + B \rightleftharpoons C + D$$

$$K_{eq} = \frac{[C]\,[D]}{[A]\,[B]} < 1.0$$

$\Delta G^{\circ\prime}$ **is positive.**

FIGURE 9.3 $\Delta G^{\circ\prime}$ and Equilibrium. The relationship of $\Delta G^{\circ\prime}$ to the equilibrium of reactions. Note the differences between exergonic and endergonic reactions.

Figure 9.3 Micro Inquiry

Which reaction would release heat? Explain your answer.

9.4 ATP

Considerable metabolic diversity exists in the microbial world. However, several biochemical principles are common to all types of metabolism. These are (1) the use of ATP to store (conserve) energy released during most exergonic reactions, so it can be used to drive endergonic reactions; (2) the organization of metabolic reactions into pathways and cycles; (3) the catalysis of metabolic reactions by enzymes or ribozymes; and (4) the importance of oxidation-reduction reactions in energy conservation. This section considers the role of ATP in metabolism.

Energy is released from a cell's energy source in exergonic reactions (i.e., those reactions with a negative ΔG). Rather than wasting this energy, much of it is trapped in a practical form that allows its transfer to the cellular systems doing work. These systems carry out endergonic reactions (e.g., anabolism), and the energy captured by the cell is used to drive these reactions to completion. In living organisms, the most commonly used practical form of energy is **adenosine 5′-triphosphate (ATP; figure 9.4).** In a sense, cells carry out certain processes so that they can "earn" ATP and carry out other processes in which they "spend" their ATP. Thus ATP is often referred to as the cell's energy currency. In the cell's economy, ATP serves as the link between exergonic reactions and endergonic reactions (**figure 9.5**).

What makes ATP suited for this role as energy currency? ATP is a high-energy molecule. That is, it hydrolyzes almost completely to the products **adenosine diphosphate (ADP)** and orthophosphate (P_i), and is strongly exergonic, having a $\Delta G^{\circ\prime}$ of −7.3 kcal/mole (−30.5 kJ/mole).

$$ATP + H_2O \rightleftharpoons ADP + P_i + H^+$$

Because ATP readily transfers its phosphate to water, it is said to have a high phosphate transfer potential, defined as the negative of $\Delta G^{\circ\prime}$ for the hydrolytic removal of phosphate. A molecule with a higher transfer potential donates phosphate to one with a lower potential. Thus ATP readily donates a phosphate to molecules such as glucose and glucose 6-phosphate in reactions such as those found in some catabolic pathways. ▶▶| *Glycolytic pathways (section 10.3)*

Although the free energy change for hydrolysis of ATP is quite large, metabolic reactions exist that release even greater amounts of free energy. This energy is used to resynthesize ATP from ADP and P_i during fueling reactions. Likewise, catabolism can generate molecules with a phosphate transfer potential that is even higher than that of ATP. Phosphoenolpyruvate (PEP) is an important example. Cells use these other high-energy molecules to regenerate ATP from ADP by a mechanism called substrate-level phosphorylation. Thus ATP, ADP, and P_i form an energy cycle (**figure 9.6**). The fueling reactions conserve energy released from an energy source by using it to synthesize ATP from ADP and P_i. When ATP is hydrolyzed, the energy released drives endergonic processes such as anabolism, transport, and mechanical work. The mechanisms for synthesizing ATP are described in more detail in chapter 10.

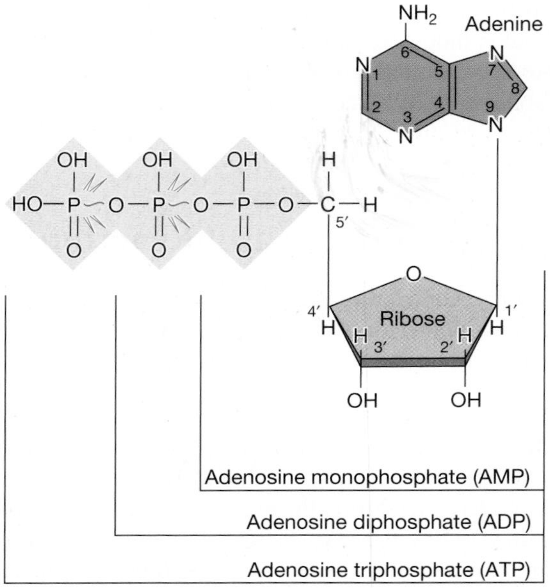

(a) ∼ Bond that releases energy when broken

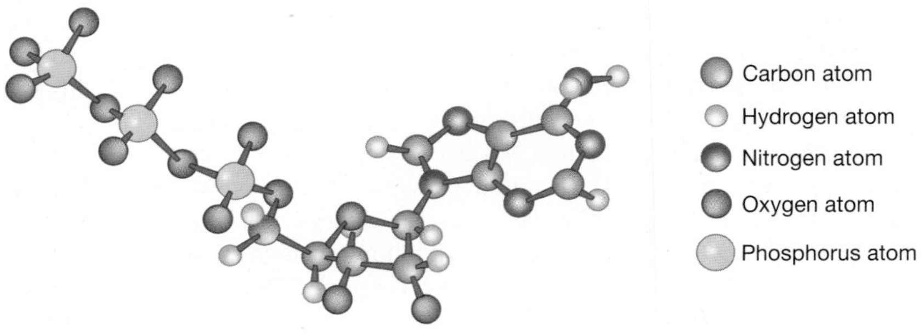

(b)

FIGURE 9.4 Adenosine Triphosphate. (a) Structure of ATP, ADP, and AMP. The two red bonds (∼) are more easily broken and have a high phosphate transfer potential. The pyrimidine ring atoms have been numbered as have the carbon atoms in ribose. (b) A model of ATP.

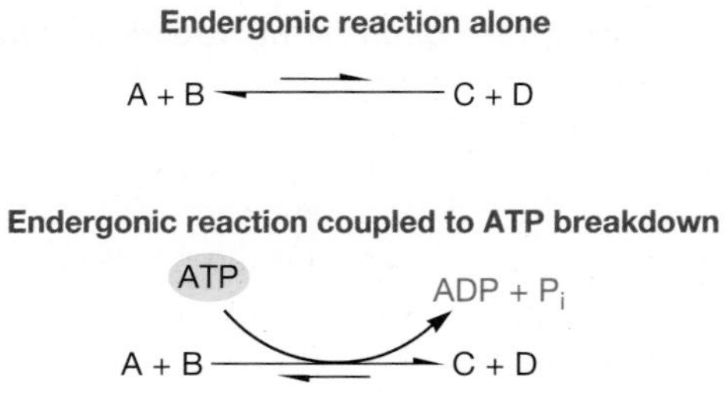

FIGURE 9.5 ATP as a Coupling Agent. The use of ATP to make endergonic reactions more favorable. It is formed by exergonic reactions and then used to drive endergonic reactions.

Finally, it should be noted that other nucleoside triphosphates (NTPs) have major roles in metabolic processes. Guanosine 5′-triphosphate (GTP) supplies some of the energy used during protein synthesis. Cytidine 5′-triphosphate (CTP) is used during lipid

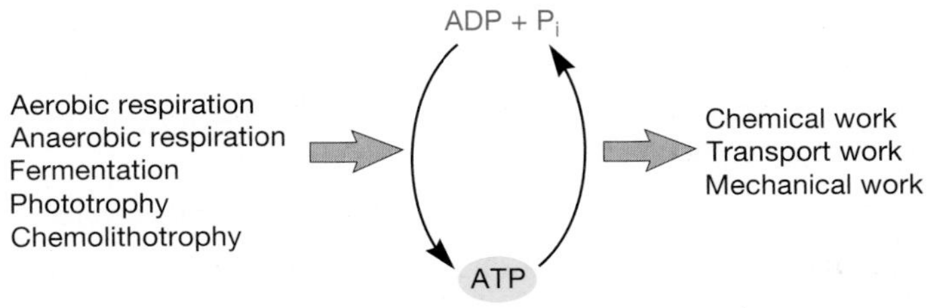

FIGURE 9.6 The Cell's Energy Cycle. ATP is formed from energy made available during aerobic respiration, anaerobic respiration, fermentation, chemolithotrophy, and phototrophy. Its breakdown to ADP and phosphate (P_i) makes chemical, transport, and mechanical work possible.

synthesis, and uridine 5′-triphosphate (UTP) is used for the synthesis of peptidoglycan and other polysaccharides. ▶▶| *Translation (section 12.8); Lipid synthesis (section 11.7); Synthesis of sugars and polysaccharides (section 11.4)*

1. What is thermodynamics? Summarize the first and second laws of thermodynamics.
2. Define entropy and enthalpy. Do living cells increase entropy within themselves? Do they increase entropy in the environment?
3. Define free energy. What are exergonic and endergonic reactions?
4. Suppose that a chemical reaction had a large negative $\Delta G^{\circ\prime}$ value. Is the reaction endergonic or exergonic? What would this indicate about its equilibrium constant?
5. Describe the energy cycle and ATP's role in it. What characteristics of ATP make it suitable for this role? Why is ATP called a high-energy molecule?

Table 9.1 Selected Biologically Important Half Reactions

Half Reaction	*E'_0 (Volts)*[a]
$2H^+ + 2e^- \rightarrow H_2$	−0.42
Ferredoxin (Fe^{3+}) + e^- → ferredoxin (Fe^{2+})	−0.42
$NAD(P)^+ + H^+ + 2e^- \rightarrow NAD(P)H$	−0.32
$S + 2H^+ + 2e^- \rightarrow H_2S$	−0.27
Acetaldehyde + $2H^+ + 2e^-$ → ethanol	−0.20
Pyruvate$^-$ + $2H^+ + 2e^-$ → lactate^{2-}	−0.19
$FAD + 2H^+ + 2e^- \rightarrow FADH_2$	−0.18[b]
Oxaloacetate^{2-} + $2H^+ + 2e^-$ → malate^{2-}	−0.17
Fumarate^{2-} + $2H^+ + 2e^-$ → succinate^{2-}	0.03
Cytochrome *b* (Fe^{3+}) + e^- → cytochrome *b* (Fe^{2+})	0.08
Ubiquinone + $2H^+ + 2e^-$ → ubiquinone H_2	0.10
Cytochrome *c* (Fe^{3+}) + e^- → cytochrome *c* (Fe^{2+})	0.25
Cytochrome *a* (Fe^{3+}) + e^- → cytochrome *a* (Fe^{2+})	0.29
Cytochrome a_3 (Fe^{3+}) + e^- → cytochrome a_3 (Fe^{2+})	0.35
$NO_3^- + 2H^+ + 2e^- \rightarrow NO_2^- + H_2O$	0.42
$NO_2^- + 8H^+ + 6e^- \rightarrow NH_4^- + 2H_2O$	0.44
$Fe^{3+} + e^- \rightarrow Fe^{2+}$	0.77[c]
$1/2O_2 + 2H^+ + 2e^- \rightarrow H_2O$	0.82

[a] E'_0 is the standard reduction potential at pH 7.0.
[b] The value for $FAD/FADH_2$ applies to the free cofactor because it can vary considerably when bound to an apoenzyme.
[c] The value for free Fe, not Fe complexed with proteins (e.g., cytochromes).

9.5 Oxidation-Reduction Reactions

Free energy changes are related to the equilibria of all chemical reactions, including the equilibria of oxidation-reduction reactions. The release of energy from an energy source normally involves oxidation-reduction reactions. **Oxidation-reduction (redox) reactions** are those in which electrons move from an **electron donor** to an **electron acceptor.**[1] As electrons move from the donor to acceptor, the donor becomes less energy rich and the acceptor becomes more energy rich. Thus electrons can be thought of as packets of energy. The more electrons a molecule has and is able to donate in a redox reaction, the more energy rich the molecule is. This explains why molecules such as glucose, which can donate up to 24 electrons in redox reactions, are such excellent sources of energy for chemoorganotrophs.

Each redox reaction consists of two half reactions. One half reaction functions as the electron-donating half reaction (i.e., an oxidation reaction), and the other functions as the electron-accepting half reaction (i.e., the reduction). By convention, half reactions are written as reductions. Thus each half reaction consists of a molecule that can accept electrons (on the left side of the chemical equation), the number (n) of electrons (e^-) it accepts, and the molecule it becomes after accepting the electrons. The latter is placed on the right side of the chemical equation and is referred to as a donor, because it has electrons it can give up. The acceptor and donor of a half reaction are referred to as a **conjugate redox pair.**

$$\text{Acceptor} + ne^- \rightleftharpoons \text{donor}$$

The equilibrium constant for a redox reaction is called the **standard reduction potential** (E_0) and is a measure of the tendency of the donor of a half reaction to lose electrons. By convention, the standard reduction potentials for half reactions, such as those in **table 9.1,** are determined at pH 7 and are represented by

[1] In redox reactions, the electron donor is often called the reducing agent or reductant because it is donating electrons to the acceptor and thus reducing it. The electron acceptor is called the oxidizing agent or oxidant because it is removing electrons from the donor and oxidizing it.

E'_0. Standard reduction potentials are measured in volts, a unit of electrical potential or electromotive force. Therefore conjugate redox pairs are a potential source of energy.

The reduction potential has a concrete meaning. Conjugate redox pairs with more negative reduction potentials will spontaneously donate electrons to pairs with more positive potentials and greater affinity for electrons. Thus electrons tend to move from donors at the top of the list in table 9.1 to acceptors at the bottom because the latter have more positive potentials. This may be expressed visually in the form of an electron tower in which the most negative reduction potentials are at the top (**figure 9.7**). Electrons move from donors to acceptors down the potential gradient or fall down the tower to more positive potentials.

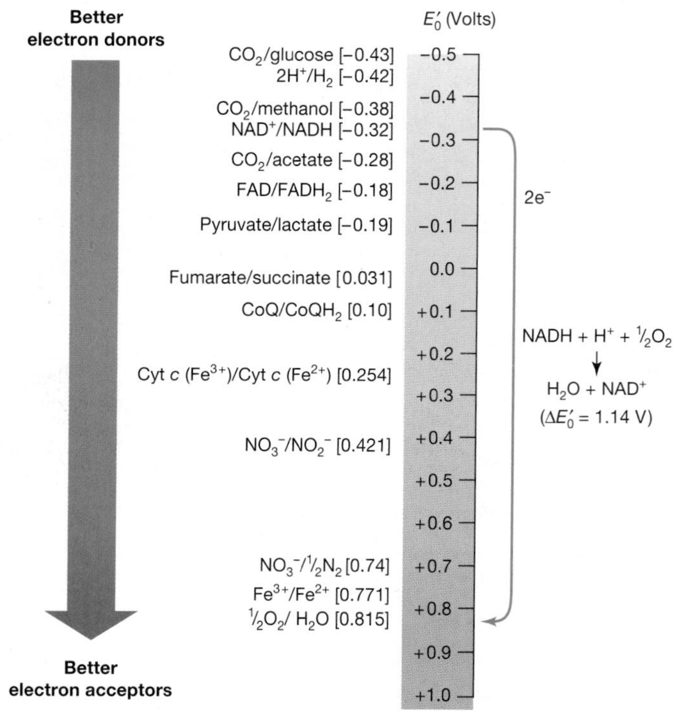

FIGURE 9.7 Electron Movement and Reduction Potentials. Electrons spontaneously move from donors higher on the tower (more negative potentials) to acceptors lower on the tower (more positive potentials). That is, the donor is always higher on the tower than the acceptor. For example, NADH will donate electrons to oxygen and form water in the process. Some typical donors and acceptors are shown on the left, and their reduction potentials are given in brackets.

***Figure 9.7* Micro Inquiry**

Why would energy be required to move electrons "up" the tower, from water to nitrate, for example?

Consider the case of the electron acceptor **nicotinamide adenine dinucleotide (NAD^+)**. The NAD^+/NADH conjugate redox pair has a very negative E'_0 and can therefore give electrons to many acceptors, including O_2.

$$NAD^+ + 2H^+ + 2e^- \rightleftharpoons NADH + H^+ \quad E'_0 = -0.32 \text{ volts}$$

$$1/2O_2 + 2H^+ + 2e^- \rightleftharpoons H_2O \quad E'_0 = +0.82 \text{ volts}$$

Because the reduction potential of the NAD^+/NADH conjugate redox pair is more negative than that of $1/2O_2/H_2O$, electrons flow from NADH (the donor) to O_2 (the acceptor), as shown in figure 9.7.

$$NADH + H^+ + 1/2O_2 \rightarrow H_2O + NAD^+$$

The relatively negative E'_0 of the NAD^+/NADH pair also means that the pair stores more potential energy than redox pairs with less negative (or more positive) E'_0 values. It follows that when electrons move from a donor to an acceptor with a more positive redox potential, free energy is released. The $\Delta G^{\circ\prime}$ of the reaction is directly related to the magnitude of the difference between the reduction potentials of the two couples ($\Delta E'_0$). The larger the $\Delta E'_0$, the greater the amount of free energy made available, as is evident from the equation

$$\Delta G^{\circ\prime} = -nF \cdot \Delta E'_0$$

in which n is the number of electrons transferred, F is the Faraday constant (23,062 cal/mole-volt; 96,480 J/mole-volt), and $\Delta E'_0$ is the E'_0 of the acceptor minus the E'_0 of the donor. For every 0.1 volt change in $\Delta E'_0$, there is a corresponding 4.6 kcal (19.3 kJ) change in $\Delta G^{\circ\prime}$ when a two-electron transfer takes place. This is similar to the relationship of $\Delta G^{\circ\prime}$ and K_{eq} in other chemical reactions—the larger the equilibrium constant, the greater the $\Delta G^{\circ\prime}$. The difference in reduction potentials between NAD^+/NADH and $1/2O_2/H_2O$ is 1.14 volts, a large $\Delta E'_0$ value. When electrons move from NADH to O_2, a large amount of free energy is made available and can be used to synthesize ATP and do other work. *How NAD^+ Works*

9.6 Electron Transport Chains

We have focused our attention on the reduction of O_2 by NADH because NADH plays a central role in the metabolism of many organisms. For instance, many chemoorganotrophs use glucose as a source of energy. As glucose is catabolized, it is oxidized. Many of the electrons released from glucose are accepted by NAD^+, reducing it to NADH. NADH next transfers the electrons to O_2. However, it does not do so directly. Instead, the electrons are transferred to O_2 via a series of electron carriers that are organized into a system called an **electron transport chain (ETC)**. An ETC is similar to a bucket brigade. Each carrier represents a person receiving a pail of water (electrons) that are destined for the fire (terminal electron acceptor). The pail of water is passed down the line, just as electrons are passed from carrier to carrier. As soon as one person hands off a pail of water to the person after him in the line, he can receive a new bucket from the person before him in the line. Likewise, as electrons flow through the ETC, each carrier is sequentially reduced (given the pail full of water) and then reoxidized (passes the pail on to the next person in the brigade), and is ready to accept more electrons as catabolism continues.

The first electron carrier in an ETC has the most negative E'_0, and each successive carrier is slightly less negative (**figure 9.8**). Thus electrons are transferred spontaneously from one carrier to the next. The carriers direct the electrons to the terminal electron acceptor (in this case, O_2). This protects the cells from random, nonproductive reductions of other molecules in the cell. The use of several carriers in a chain also releases the energy from the redox reaction in a controlled manner. In this way, the potential energy stored in the conjugate redox pair whose electrons initiate electron flow is released and used to form ATP.

The electron transport chains of chemoorganotrophs are located in the plasma membrane of bacterial and archaeal cells and in the internal mitochondrial membranes in eukaryotes. ETCs also play a pivotal role in the metabolism of chemolithotrophs and most phototrophs, where they are used to conserve energy from inorganic chemicals and light, respectively. These ETCs are located in the plasma membrane or internal membrane systems of chemolithotrophs, which are all members of either *Bacteria* or *Archaea*. They are located in the plasma membrane and internal membrane systems of some bacterial phototrophs and in the thylakoid membranes of cyanobacteria and chloroplasts in eukaryotic phototrophs (figure 9.8). ◀◀ *Nutritional types of microorganisms (section 6.3)*

The carriers that make up ETCs differ in terms of their chemical nature and the way they carry electrons. NAD^+ and its

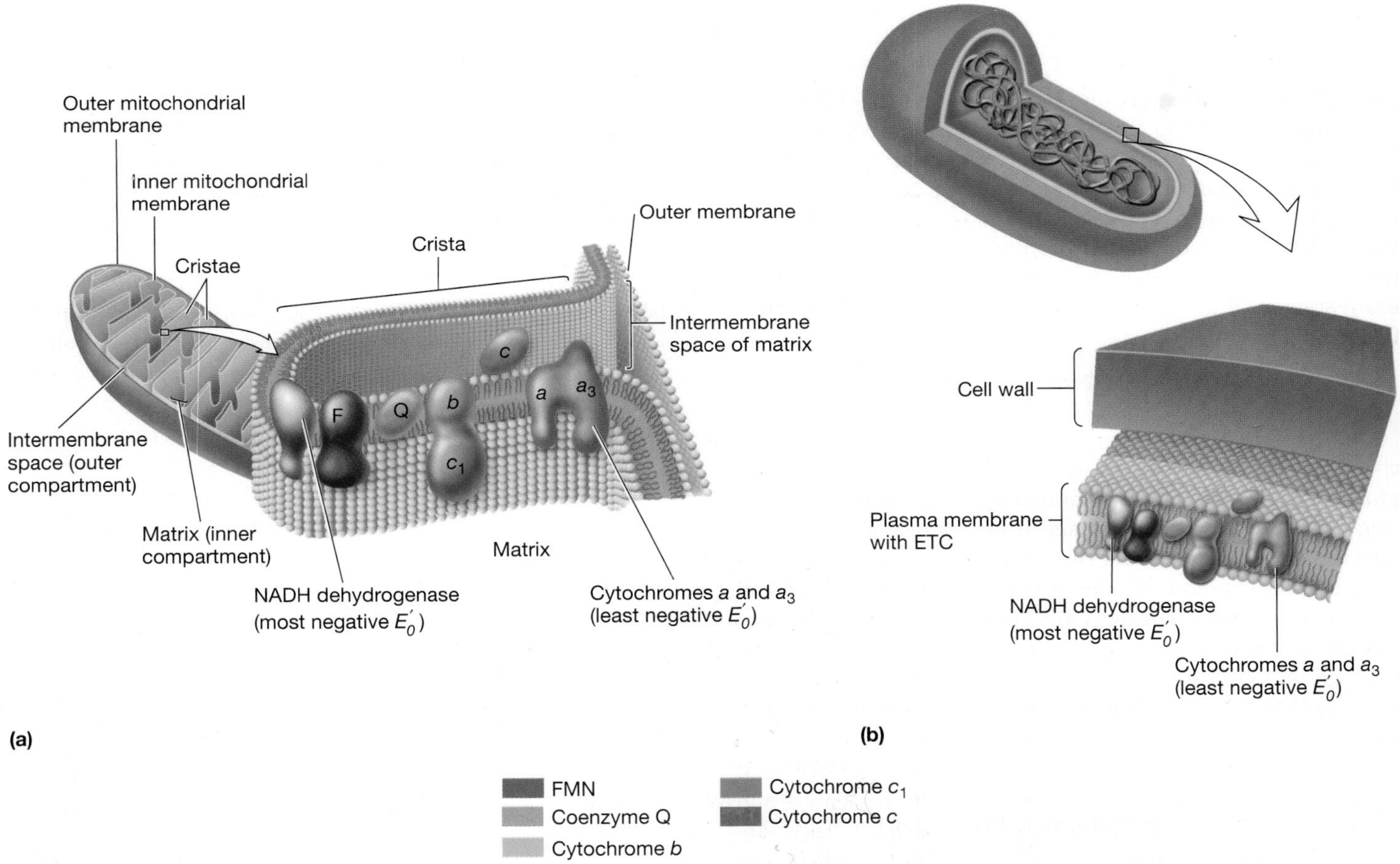

FIGURE 9.8 Electron Transport Chains. Electron transport chains (ETCs) are located in membranes. Electrons flow from the electron carrier having the most negative reduction potential to the carrier having the most positive reduction potential. During respiratory processes (aerobic respiration, anaerobic respiration, and chemolithotrophy), an exogenous molecule such as oxygen serves as the terminal electron acceptor. (a) The mitochondrial ETC. (b) A typical bacterial ETC.

Figure 9.8 Micro Inquiry

Refer to figure 9.7 and determine the E°′ for NAD^+/NADH and coenzyme Q/$CoQH_2$. Suggest a plausible E°′ value for FMN.

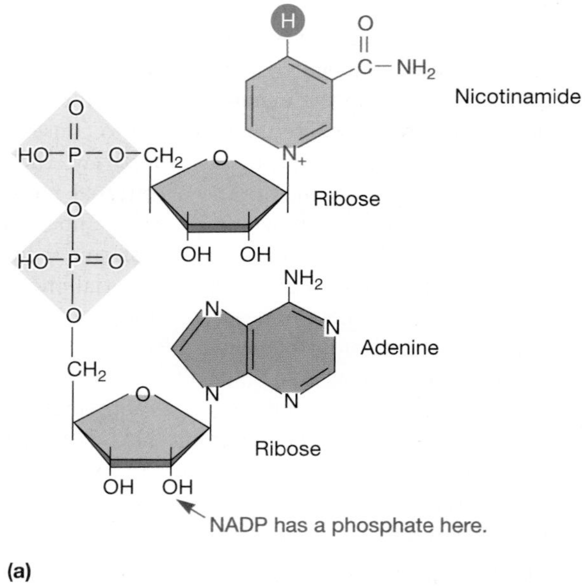

(a)

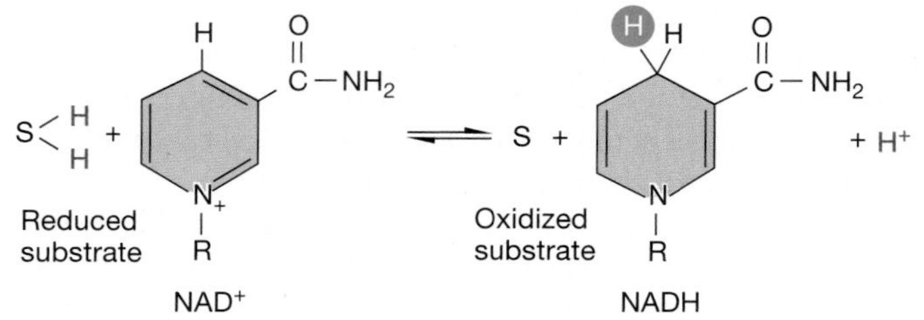

(b)

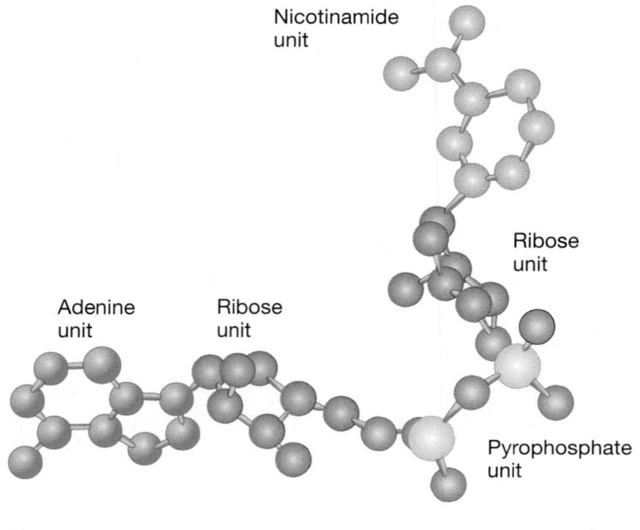

(c)

FIGURE 9.9 The Structure and Function of NAD. (a) The structure of NAD^+ and $NADP^+$. $NADP^+$ differs from NAD^+ in having an extra phosphate on one of its ribose sugar units. (b) NAD^+ can accept two electrons and one proton from a reduced substrate (e.g., SH_2 of a cysteine residue). In this example, the electrons and proton are supplied by the hydrogen atoms of the reduced substrate; recall that each hydrogen atom consists of one proton and one electron. (c) Stick-and-ball model of NAD^+.

chemical relative **nicotinamide adenine dinucleotide phosphate ($NADP^+$)** contain a nicotinamide ring (**figure 9.9**). This ring accepts two electrons and one proton from a donor (e.g., an intermediate formed during the catabolism of glucose), and a second proton is released. **Flavin adenine dinucleotide (FAD)** and **flavin mononucleotide (FMN)** bear two electrons and two protons on the complex ring system shown in **figure 9.10**. Proteins bearing FAD and FMN are often called flavoproteins. **Coenzyme Q (CoQ)** or **ubiquinone** is a quinone that transports two electrons and two protons (**figure 9.11**). **Cytochromes** and several other carriers use iron atoms to transport one electron at a time. In cytochromes, the iron atoms are part of a heme group or other similar iron-porphyrin rings (**figure 9.12**). There are several different cytochromes, each consisting of a protein and an iron-porphyrin ring. Some iron-containing electron-carrying proteins lack a heme group and are called **nonheme iron proteins.** They are also commonly called **iron-sulfur (Fe-S) proteins** because the iron is associated with sulfur atoms (**figure 9.13**). The sulfur atoms are often present in the cysteine residues of the protein. **Ferredoxin** is an Fe-S protein active in photosynthetic electron transport and several other electron transport processes. Like cytochromes, Fe-S proteins carry only one electron at a time. These differences in the number of electrons and

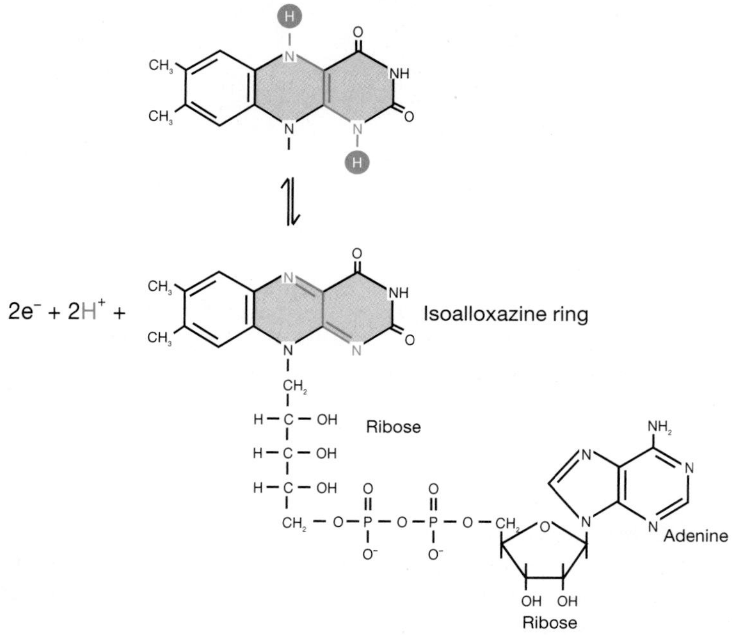

FIGURE 9.10 The Structure and Function of FAD. The vitamin riboflavin is composed of the isoalloxazine ring and its attached ribose sugar. FMN is riboflavin phosphate. The portion of the ring directly involved in oxidation-reduction reactions is in color.

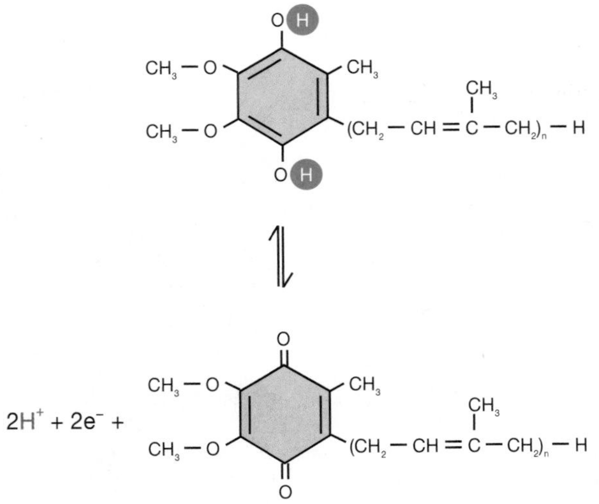

FIGURE 9.11 The Structure and Function of Coenzyme Q (Ubiquinone). The length of the side chain varies among organisms from n = 6 to n = 10.

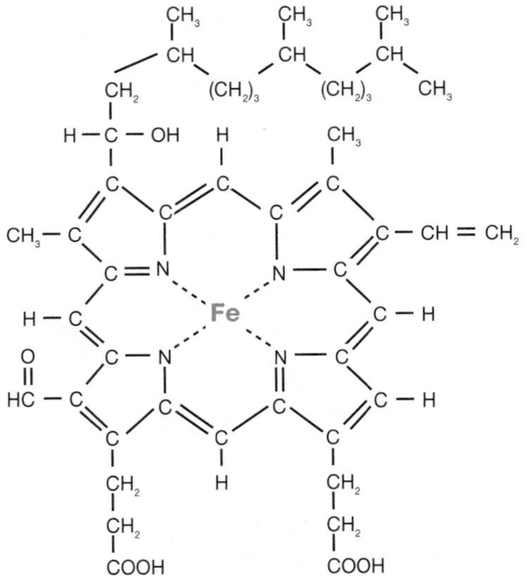

FIGURE 9.12 The Structure of Heme. Heme is composed of a porphyrin ring and an attached iron atom. It is the nonprotein component of many cytochromes. The iron atom alternatively accepts and releases an electron.

protons transported by carriers in ETCs are of great importance in their operation, and we discuss them further in chapter 10. *Proton Pump*

Search This: Electron transport chain

1. How is the direction of electron flow between conjugate redox pairs related to the standard reduction potential and the release of free energy?

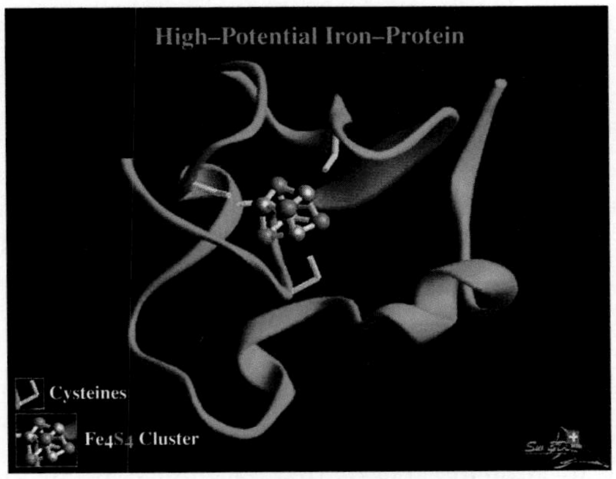

FIGURE 9.13 An Iron-Sulfur Protein. The protein portion of the molecule is represented by the green ribbon. The sulfur atoms (orange) are contributed by cysteine residues (yellow) in the protein. The iron atoms (gray) are associated with the sulfur to form an iron-sulfur cluster. The iron atoms are responsible for electron transfer, transferring one electron at a time.

2. When electrons flow from the $NAD^+/NADH$ conjugate redox pair to the $1/2O_2/H_2O$ redox pair, does the reaction begin with NAD^+ or with NADH? What is produced—O_2 or H_2O?
3. Which among the following would be the best electron donor? Which would be the worst? Ubiquinone/ubiquinone H_2, $NAD^+/NADH$, $FAD/FADH_2$, NO_3^-/NO_2^-. Explain your answers.
4. In general terms, how is $\Delta G^{\circ\prime}$ related to $\Delta E'_0$? What is the $\Delta E'_0$ when electrons flow from the $NAD^+/NADH$ redox pair to the Fe^{3+}/Fe^{2+} redox pair? How does this compare to the $\Delta E'_0$ when electrons flow from the Fe^{3+}/Fe^{2+} conjugate redox pair to the $1/2O_2/H_2O$ pair? Which will yield the largest amount of free energy to the cell?
5. Name and briefly describe the major electron carriers found in cells. Why is NADH a good electron donor? Why is ferredoxin an even better electron donor?

9.7 Enzymes

Recall that an exergonic reaction is one with a negative $\Delta G^{\circ\prime}$ and an equilibrium constant greater than one. An exergonic reaction proceeds to completion in the direction written (i.e., toward the right of the equation). Nevertheless, reactants for an exergonic reaction often can be combined with no obvious result. For instance, if a polysaccharide such as starch is mixed in water, the hydrolysis of the starch into its component monosaccharides (glucose) is exergonic and will occur spontaneously—that is, it will occur on its own, given enough time. However, the time needed is very long. Even if an organic chemist carried out this reaction in 6 moles/liter (M) HCl and at 100°C, it would still take several hours to go to completion. A cell, on the other hand, can accomplish the same reaction at neutral pH, at a much lower temperature, and in just fractions of a second. Cells can do this because they manufacture proteins called enzymes and RNA molecules called ribozymes that speed up chemical reactions. Enzymes and ribozymes are critically important to cells, since most biological reactions occur very slowly without them. Indeed, enzymes and ribozymes make life possible. Here we focus on enzymes. We consider ribozymes in section 9.8.

Structure and Classification of Enzymes

Enzymes are protein catalysts that have great specificity for the reaction catalyzed, the molecules acted on, and the products they yield. A **catalyst** is a substance that increases the rate of a chemical reaction without being permanently altered itself. Thus enzymes speed up cellular reactions. The reacting molecules are called **substrates,** and the substances formed are the **products.** ▶▶| *Proteins (appendix I)*

Many enzymes are composed only of proteins. However, some enzymes are composed of two parts: a protein component called the **apoenzyme** and a nonprotein component called a **cofactor.** The complete enzyme consisting of the apoenzyme and its cofactor is called the **holoenzyme.** If the cofactor is firmly attached to the apoenzyme, it is a **prosthetic group.** If the cofactor is loosely attached and can dissociate from the apoenzyme after products have been formed, it is called a **coenzyme.** Many coenzymes carry one of the products to another enzyme or transfer chemical groups from one substrate to another (**figure 9.14**). For example, NAD^+ is a coenzyme that carries electrons within the cell. Many vitamins required by humans serve as coenzymes or as their precursors. For example, niacin is incorporated into NAD^+ and riboflavin into FAD. Metal ions may also be bound to apoenzymes and act as cofactors.

Enzymes may be placed in one of six general classes and usually are named in terms of the substrates they act on and the type of reaction catalyzed (**table 9.2**). For example, lactate dehydrogenase (LDH) removes hydrogens from lactate. Lactate dehydrogenase can also be given a more complete and detailed name, L-lactate: NAD^+ oxidoreductase. This name describes the substrates and reaction type with even more precision.

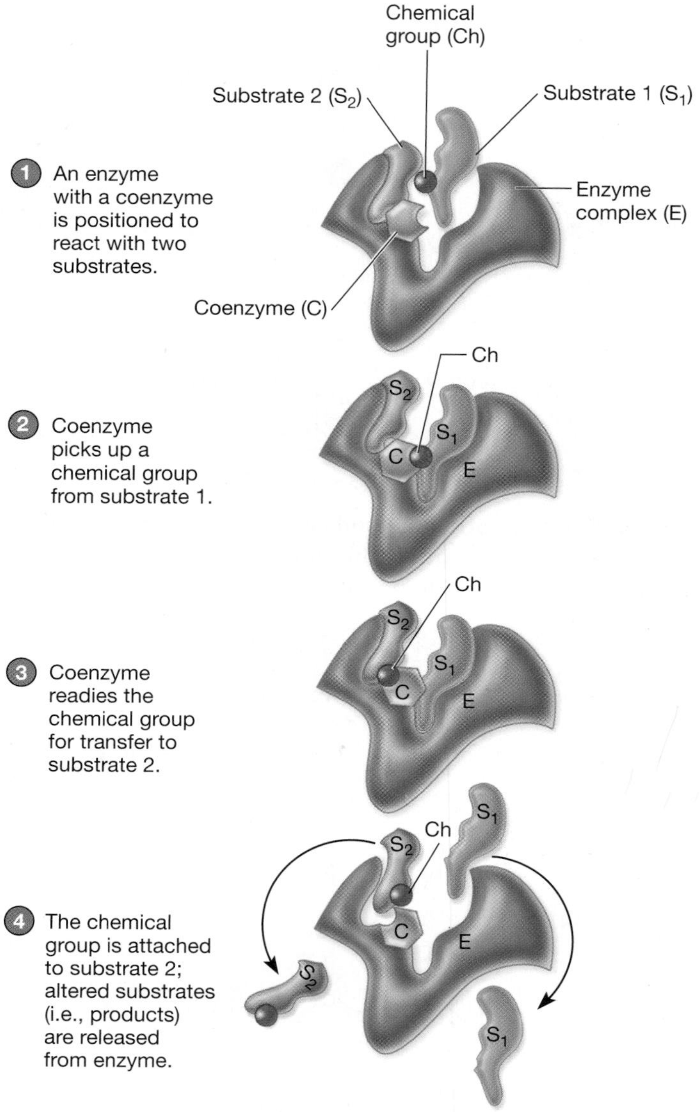

Chemical group (Ch)
Substrate 2 (S_2)
Substrate 1 (S_1)
Enzyme complex (E)
1 An enzyme with a coenzyme is positioned to react with two substrates.
Coenzyme (C)
Ch
S_2
S_1
C
E
2 Coenzyme picks up a chemical group from substrate 1.
3 Coenzyme readies the chemical group for transfer to substrate 2.
4 The chemical group is attached to substrate 2; altered substrates (i.e., products) are released from enzyme.

FIGURE 9.14 Coenzymes as Carriers.

Mechanism of Enzyme Reactions

It is important to keep in mind that enzymes increase the rates of reactions but do not alter their equilibrium constants. If a reaction is endergonic, the presence of an enzyme will not shift its equilibrium so that more products are formed. Enzymes simply speed up the rate at which a reaction proceeds toward its final equilibrium. *How Enzymes Work*

How do enzymes catalyze reactions? Some understanding of the mechanism can be gained by considering the course of a simple exergonic chemical reaction.

$$A + B \rightleftharpoons C + D$$

When molecules A and B approach each other to react, they form a transition-state complex, which resembles both the substrates and the products (**figure 9.15**). **Activation energy** is required to bring the reacting molecules together in the correct way to reach the transition state. The transition-state complex can then resolve to yield the products C and D. The difference in free energy level between reactants and products is $\Delta G^{\circ\prime}$. Thus the equilibrium in our example lies toward the products because $\Delta G^{\circ\prime}$ is negative (i.e., the products are at a lower energy level than the substrates).

As seen in figure 9.15, A and B will not be converted to C and D if they are not supplied with an amount of energy equivalent to the activation energy. Enzymes accelerate reactions by lowering the activation energy; therefore more substrate molecules will have sufficient energy to come together and form products. Even

Table 9.2 Enzyme Classification

Type of Enzyme	*Reaction Catalyzed by Enzyme*	*Example of Reaction*
Oxidoreductase	Oxidation-reduction reactions	Lactate dehydrogenase: Pyruvate + NADH + H^+ ⇌ lactate + NAD^+
Transferase	Reactions involving the transfer of chemical groups between molecules	Aspartate carbamoyltransferase: Aspartate + carbamoylphosphate ⇌ carbamoylaspartate + phosphate
Hydrolase	Hydrolysis of molecules	Glucose-6-phosphatase: Glucose 6-phosphate + H_2O → glucose + P_i[a]
Lyase	Breaking of C-C, C-O, C-N and other bonds by a means other than hydrolysis	Fumarase: L-malate ⇌ fumarate + H_2O
Isomerase	Reactions involving isomerizations	Alanine racemase: L-alanine ⇌ D-alanine
Ligase	Joining of two molecules using ATP (or the energy of other nucleoside triphosphates)	Glutamine synthetase: Glutamate + NH_3 + ATP → glutamine + ADP + P_i

[a]P_i is inorganic phosphate.

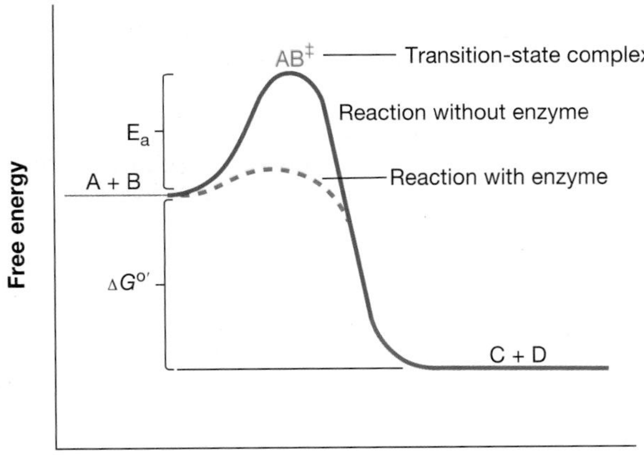

FIGURE 9.15 Enzymes Lower the Energy of Activation. This figure traces the course of a chemical reaction in which A and B are converted to C and D. The transition-state complex is represented by AB‡, and the activation energy required to reach it, by E_a. The red line represents the course of the reaction in the presence of an enzyme. Note that the activation energy is much lower in the enzyme-catalyzed reaction.

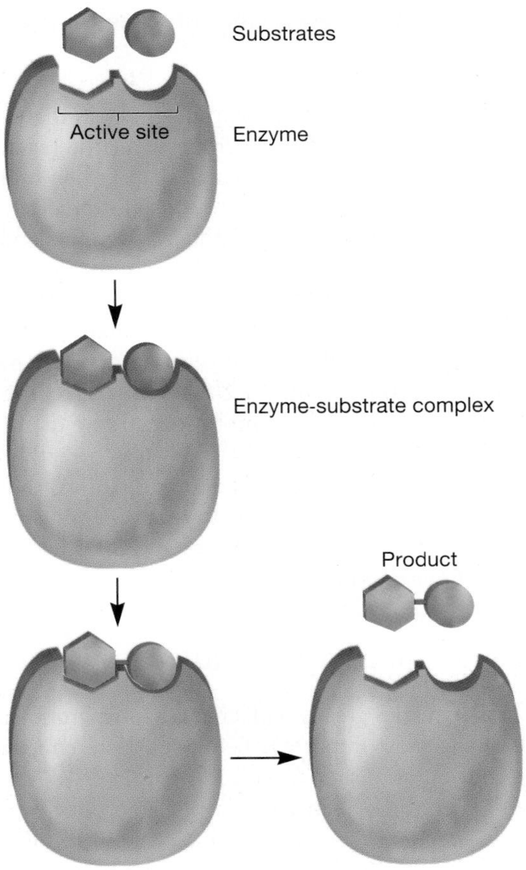

FIGURE 9.16 Lock-and-Key Model of Enzyme Function. In this model, the active site is a relatively rigid structure that accommodates only those molecules with the correct corresponding shape.

though the equilibrium constant (or $\Delta G^{\circ\prime}$) is unchanged, equilibrium is reached more rapidly in the presence of an enzyme because of this decrease in activation energy.

Researchers have worked hard to discover how enzymes lower the activation energy of reactions, and the process is becoming clearer. Enzymes bring substrates together at a specific location in the enzyme called the **active** or **catalytic site** to form an enzyme-substrate complex (**figures 9.16** and **9.17**; *see also appendix figure AI.16*). An enzyme can interact with its substrate in two general ways. In the lock-and-key model, the active site is

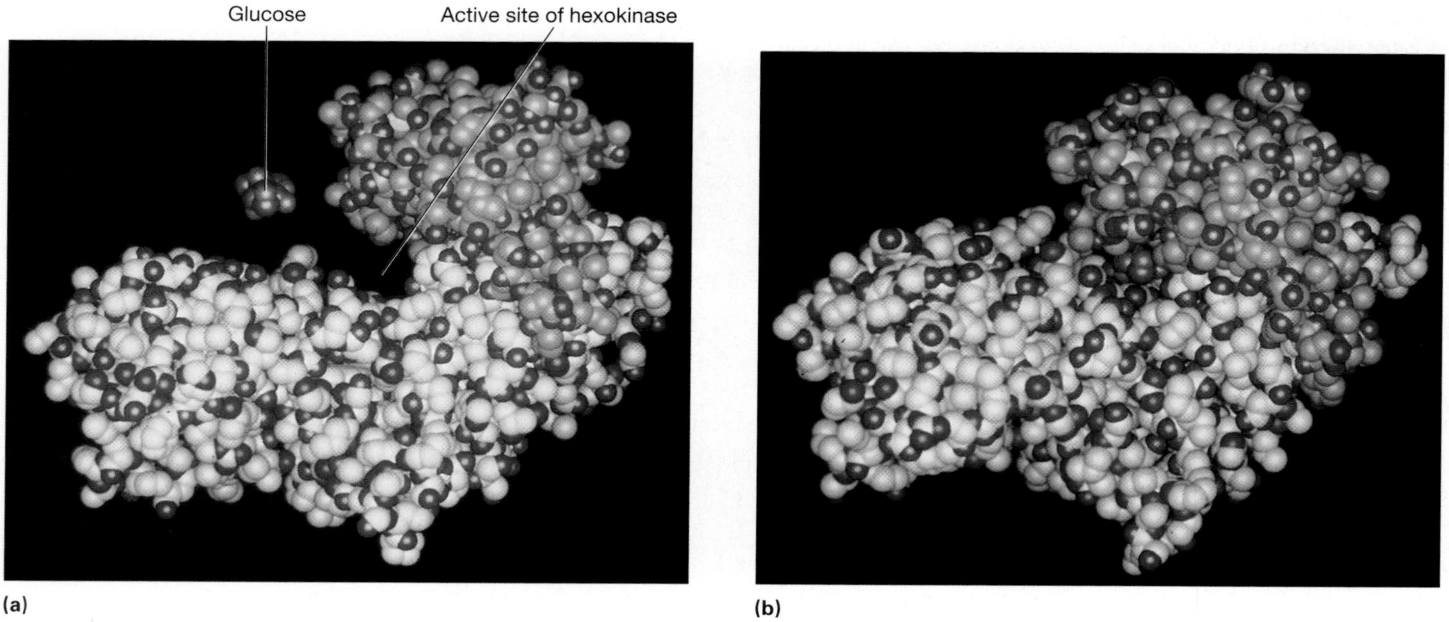
Glucose
Active site of hexokinase
(a)
(b)

FIGURE 9.17 The Induced Fit Model of Enzyme Function. (a) A space-filling model of yeast hexokinase and its substrate glucose (purple). The active site is in the cleft formed by the enzyme's small lobe (green) and large lobe (blue). (b) When glucose binds to form the enzyme-substrate complex, hexokinase changes shape and surrounds the substrate.

rigid and precisely shaped to fit the substrate, so that a specific substrate binds and is positioned properly for the reaction (figure 9.16). An enzyme also may change shape when it binds the substrate so that the active site surrounds and precisely fits the substrate. This has been called the induced fit model and is used by hexokinase, the enzyme illustrated in figure 9.17, as well as many other enzymes. The formation of an enzyme-substrate complex can lower the activation energy in many ways. For example, by bringing the substrates together at the active site, the enzyme is, in effect, concentrating them and speeding up the reaction. An enzyme does not simply concentrate its substrates, however. It also binds them so that they are correctly oriented with respect to each other. Such an orientation lowers the amount of energy that the substrates require to reach the transition state. These and other catalytic site activities speed up a reaction by hundreds of thousands of times.

Environmental Effects on Enzyme Activity

Enzyme activity varies greatly with changes in environmental factors, one of the most important being substrate concentration. Substrate concentrations are usually low within cells. At very low substrate concentrations, an enzyme makes product slowly because it seldom contacts a substrate molecule. If more substrate molecules are present, an enzyme binds substrate more often, and the velocity of the reaction (usually expressed in terms of the rate of product formation) is greater than at a lower substrate concentration. Thus the rate of an enzyme-catalyzed reaction increases with substrate concentration (**figure 9.18**). Eventually

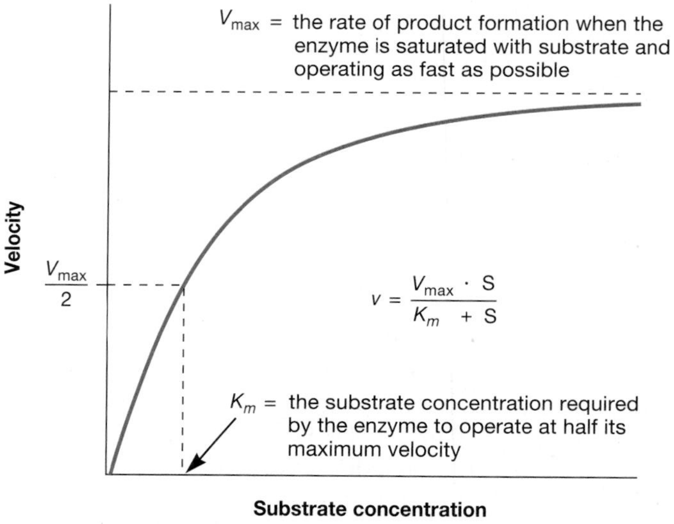
V_{max} = the rate of product formation when the enzyme is saturated with substrate and operating as fast as possible
Velocity
$\frac{V_{max}}{2}$
$v = \frac{V_{max} \cdot S}{K_m + S}$
K_m = the substrate concentration required by the enzyme to operate at half its maximum velocity
Substrate concentration

FIGURE 9.18 Michaelis-Menten Kinetics. The dependence of enzyme activity upon substrate concentration. This substrate curve fits the Michaelis-Menten equation given in the figure. The Michaelis-Menten equation was derived by biochemists Leonor Michaelis (1875–1949) and Maud Menten (1879–1960), and it relates reaction velocity (v) to the substrate concentration (S) using the maximum velocity and the Michaelis constant (K_m).

Figure 9.18 Micro Inquiry

Will an enzyme with a relatively high K_m have a high or low affinity for its substrate? Explain.

further increases in substrate concentration do not result in a greater reaction velocity because the available enzyme molecules are binding substrate and converting it to product as rapidly as possible. That is, the enzyme is saturated with substrate and operating at maximal velocity (V_{max}). The resulting substrate concentration curve is a hyperbola (figure 9.18).

It is useful to know the substrate concentration an enzyme needs to function adequately. Usually the **Michaelis constant (K_m)**, the substrate concentration required for the enzyme to achieve half-maximal velocity, is used as a measure of the apparent affinity of an enzyme for its substrate. The lower the K_m value, the lower the substrate concentration at which an enzyme catalyzes its reaction. Enzymes with a low K_m value are said to have a high affinity for their substrates. Since the concentrations of substrates in cells are often low, enzymes with lower K_m values are able to function better.

Enzyme activity is changed not only by substrate concentration but also by alterations in pH and temperature. Each enzyme functions most rapidly at a specific pH optimum. When the pH deviates too greatly from an enzyme's optimum, activity slows and the enzyme may be damaged. Enzymes likewise have temperature optima for maximum activity. If the temperature rises too much above the optimum, an enzyme's structure will be disrupted and its activity lost. This phenomenon is known as **denaturation.** The pH and temperature optima of a microorganism's enzymes often reflect the pH and temperature of its habitat. Not surprisingly, bacteria and archaea that grow best at high temperatures often have enzymes with high temperature optima and great heat stability. ◀◀ *Influences of environmental factors on growth (section 7.6)*; ▶▶ *Thermostability (section 18.1)*

Enzyme Inhibition

Microorganisms can be poisoned by a variety of chemicals, and many of the most potent poisons are enzyme inhibitors. A **competitive inhibitor** directly competes with the substrate at an enzyme's catalytic site and prevents the enzyme from forming product (**figure 9.19**). Competitive inhibitors usually resemble normal substrates, but they cannot be converted to products.

Competitive inhibitors are important in the treatment of many microbial diseases. Sulfa drugs such as sulfanilamide (figure 9.19*b*) resemble *p*-aminobenzoate (PABA), a molecule used in the formation of the coenzyme folic acid. Sulfa drugs compete with PABA for the catalytic site of an enzyme involved in folic acid synthesis. This blocks the production of folic acid and inhibits the growth of organisms that require its synthesis. Humans are not harmed because they do not synthesize folic acid but rather obtain it in their diet. ▶▶ *Metabolic antagonists (section 34.4)*

Noncompetitive inhibitors affect enzyme activity by binding to the enzyme at some location other than the active site. This

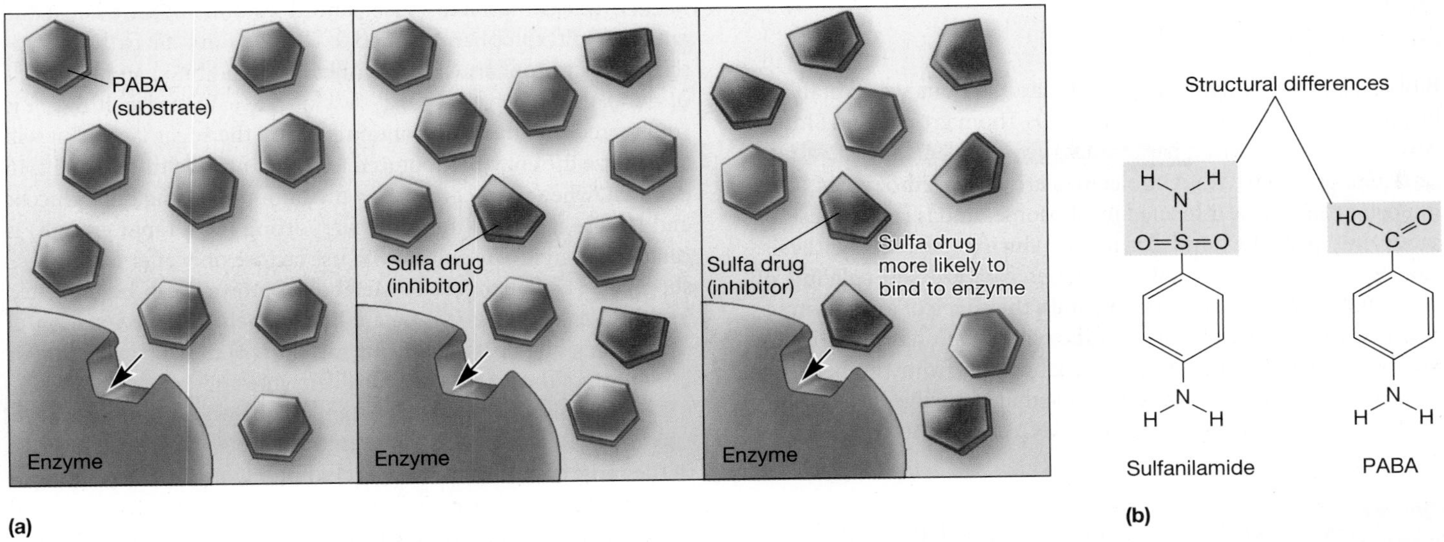

FIGURE 9.19 Competitive Inhibition of Enzyme Activity. (a) A competitive inhibitor is usually similar in shape to the normal substrate of the enzyme and therefore can bind the active site of the enzyme. This prevents the substrate from binding, and the reaction is blocked. (b) Structure of sulfanilamide, a structural analog of PABA. PABA is the substrate of an enzyme involved in folic acid biosynthesis. When sulfanilamide binds the enzyme, activity of the enzyme is inhibited and synthesis of folic acid is stopped.

Figure 9.19 Micro Inquiry

Based on this figure, explain the notion of a minimum inhibitory concentration of sulfanilamide.

alters the enzyme's shape, rendering it inactive or less active. These inhibitors are called noncompetitive because they do not directly compete with the substrate. Heavy metal poisons such as mercury frequently are noncompetitive inhibitors of enzymes.

1. What is an enzyme? How does it speed up reactions? How are enzymes named? Define apoenzyme, holoenzyme, cofactor, coenzyme, prosthetic group, active or catalytic site, and activation energy.
2. Draw a diagram showing how enzymes catalyze reactions by altering the activation energy. What is a transition state complex? Use the diagram to explain why enzymes do not change the equilibria of the reactions they catalyze.
3. What is the difference between the lock-and-key and the induced-fit models of enzyme-substrate complex formation?
4. Define the terms Michaelis constant and maximum velocity. How does enzyme activity change with substrate concentration, pH, and temperature?
5. What special properties might an enzyme isolated from a psychrophilic bacterium have? Will enzymes need to lower the activation energy more or less in thermophiles than in psychrophiles?
6. What are competitive and noncompetitive inhibitors, and how do they inhibit enzymes?

9.8 Ribozymes

Biologists once thought that all cellular reactions were catalyzed by proteins. However, in the early 1980s Thomas Cech and Sidney Altman discovered that some RNA molecules also can catalyze reactions. Catalytic RNA molecules are called **ribozymes.** One important ribozyme is located in ribosomes and is responsible for catalyzing peptide bond formation during protein synthesis. However, the best-studied ribozymes catalyze self-splicing of RNA (**table 9.3**). Just as with enzymes, the shape of a ribozyme is essential to catalytic efficiency. Ribozymes even have Michaelis-Menten kinetics (figure 9.18). The ribozyme from the hepatitis delta virusoid catalyzes RNA cleavage that is involved in its replication. It is unusual in that the same RNA can fold into two shapes with quite different catalytic activities: the regular RNA cleavage activity and an RNA ligation reaction. ▶▶| *Translation (section 12.8)*

Search This: Ribozymes/HHMI

Table 9.3 Examples of Self-Splicing Ribozymes

Function	*Where observed*
Splicing of pre-rRNA	Tetrahymena (protist)
Splicing of mitochondrial rRNA and mRNA	Numerous fungi
Splicing of chloroplast tRNA, rRNA, and mRNA	
Splicing of viral mRNA	Bacterial viruses (e.g., T4)
Splicing of virusoid mRNA	Hepatitis delta virusoid

9.9 Regulation of Metabolism

Microorganisms must regulate their metabolism to conserve raw materials and energy and to maintain a balance among various cell components. Because they live in environments where the nutrients, energy sources, and physical conditions often change rapidly, they must continuously monitor internal and external conditions and respond accordingly. This involves activating or inactivating metabolic pathways as needed. For instance, if a particular energy source is unavailable, the enzymes required for its use are not needed and their further synthesis is a waste of carbon, nitrogen, and energy. Similarly it would be extremely wasteful for a microorganism to synthesize the enzymes required to manufacture a certain end product, such as an amino acid, if that end product were already present in adequate amounts.

The drive to maintain balance and conserve energy is evident in the regulatory responses of a bacterium such as *Escherichia coli.* If the bacterium is grown in a very simple medium containing only glucose as a carbon and energy source, it will synthesize all needed cell components in balanced amounts. However, if the amino acid tryptophan is added to the medium, the pathway synthesizing tryptophan will be immediately inhibited and synthesis of the pathway's enzymes will slow or cease. Likewise, if *E. coli* is transferred to a medium containing only the sugar lactose, it will synthesize the enzymes required for catabolism of this nutrient. In contrast, when *E. coli* grows in a medium possessing both glucose and lactose, glucose (the sugar supporting most rapid growth) is catabolized first. The culture will use lactose only after the glucose supply has been exhausted. Metabolic pathways can be regulated in three major ways: (1) metabolic channeling, (2) regulation of the synthesis of a particular enzyme, and (3) direct stimulation or inhibition of the activity of critical enzymes.

Metabolic channeling influences pathway activity by localizing metabolites and enzymes into different parts of a cell. One of the most common metabolic channeling mechanisms is **compartmentation,** the differential distribution of enzymes and metabolites among separate cell structures or organelles. Compartmentation is particularly important in eukaryotic microorganisms with their many membrane-bound organelles. For example, fatty acid catabolism is located within the mitochondrion, whereas fatty acid synthesis occurs in the cytoplasm. The periplasm in gram-negative bacteria can also be considered an example of compartmentation. Compartmentation makes possible the simultaneous but separate operation and regulation of similar pathways. Furthermore, pathway activities can be coordinated through regulation of the transport of metabolites and coenzymes between cell compartments.

Metabolic channeling can generate marked variations in metabolite concentrations and therefore directly affect enzyme activity. Substrate levels are generally around 10^{-3} M to 10^{-6} M or even lower. Thus they may be in the same range as enzyme concentrations and equal to or less than the Michaelis constants (K_m) of many enzymes (figure 9.18). Under these conditions, the concentration of an enzyme's substrate may control its activity because the substrate concentration is in the rising portion of the hyperbolic substrate saturation curve. As the substrate level increases, it is converted to product more rapidly; a decline in substrate concentration automatically leads to lower enzyme activity. If two enzymes in different pathways use the same metabolite, they may directly compete for it. The pathway winning this competition—the one with the enzyme having the lowest K_m value for the metabolite—will operate closer to full capacity. Thus channeling within a cell compartment can regulate and coordinate metabolism through variations in metabolite and coenzyme levels.

In the second regulatory mechanism—regulation of the synthesis of a particular enzyme—transcription and translation can be regulated to control the amount of an enzyme present in the cell. Regulation at this level is relatively slow, but it saves the cell considerable energy and raw materials. In contrast, the direct stimulation or inhibition of the activity of critical enzymes rapidly alters pathway activity. It is often called **posttranslational regulation** because it occurs after the enzyme has been synthesized. This type of regulation is discussed next. ▶▶| *Transcription (section 12.6); Translation (section 12.8); Regulation of gene expression (chapter 13)*

1. Briefly describe the three ways a metabolic pathway may be regulated.
2. Define the terms metabolic channeling and compartmentation. How are they involved in the regulation of metabolism?

9.10 Posttranslational Regulation of Enzyme Activity

Adjustment of the activity of regulatory enzymes and other proteins controls the functioning of many metabolic pathways and cellular processes. A number of posttranslational regulatory mechanisms are known. Some are irreversible—for instance, cleavage of a protein can either activate or inhibit its activity. Others, such as allosteric regulation and covalent modification, are reversible.

Allosteric Regulation

Most regulatory enzymes are **allosteric enzymes.** The activity of an allosteric enzyme is altered by a small molecule known as an **allosteric effector.** The effector binds reversibly by noncovalent forces to a **regulatory site** separate from the catalytic site and causes a change in the shape (conformation) of the enzyme (**figure 9.20**). The activity of the catalytic site is altered as a result. A positive

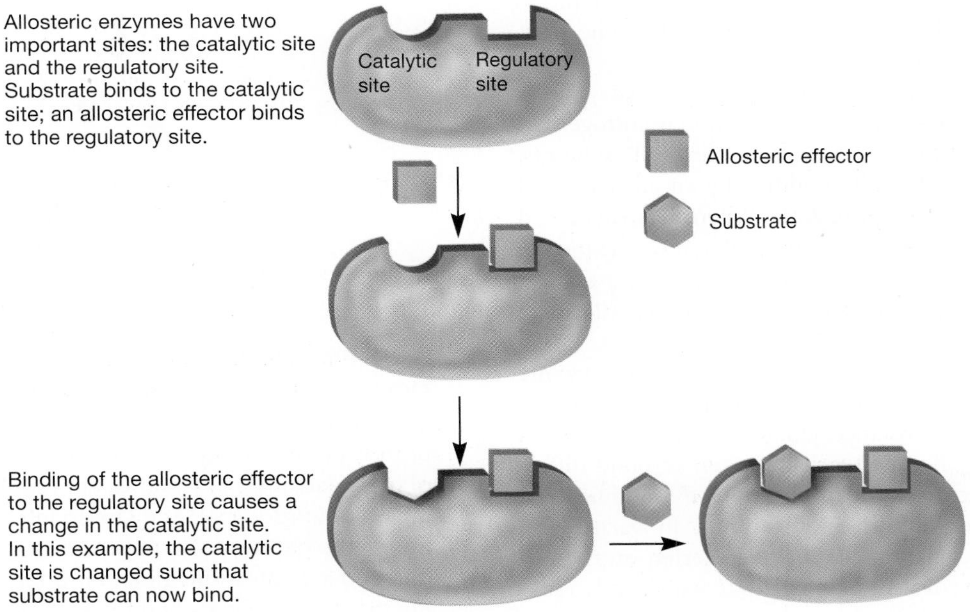

Allosteric enzymes have two important sites: the catalytic site and the regulatory site. Substrate binds to the catalytic site; an allosteric effector binds to the regulatory site.
Catalytic site
Regulatory site
Allosteric effector
Substrate
Binding of the allosteric effector to the regulatory site causes a change in the catalytic site. In this example, the catalytic site is changed such that substrate can now bind.

FIGURE 9.20 Allosteric Regulation. The structure and function of an allosteric enzyme.

***Figure 9.20** Micro Inquiry*

Is this allosteric effector a positive or negative effector?

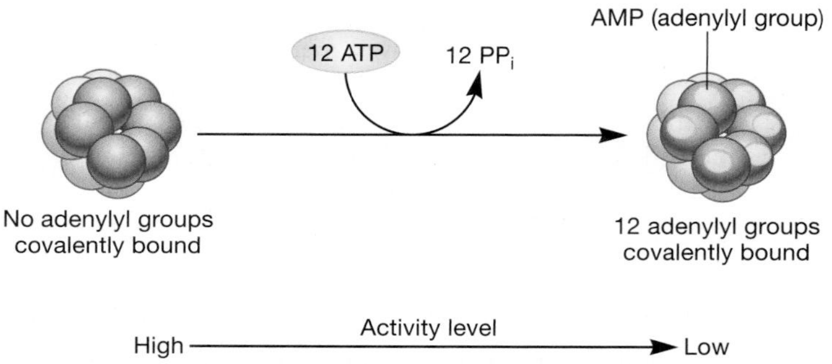

FIGURE 9.21 Regulation of Glutamine Synthetase Activity by Covalent Modification. Glutamine synthetase consists of 12 subunits. Six subunits form a ring, and one ring (darker color) is above the other (lighter color). Not all of the lower ring subunits are shown. Each subunit can be adenylylated. Only the adenylyl groups on the upper ring are shown. As the number of adenylyl groups increases, the activity of the enzyme decreases.

effector increases enzyme activity, whereas a negative effector decreases activity (i.e., inhibits the enzyme). These changes in activity often result from alterations in the apparent affinity of the enzyme for its substrate, but changes in maximum velocity also can occur.

Covalent Modification of Enzymes

Regulatory enzymes also can be switched on and off by **reversible covalent modification.** Usually this occurs through the addition and removal of a particular chemical group, typically a phosphoryl, methyl, or adenylyl group.

One of the most intensively studied regulatory enzymes is *E. coli* glutamine synthetase, an enzyme involved in nitrogen assimilation. It is a large, complex enzyme consisting of 12 subunits, each of which can be covalently modified by an adenylic acid residue (**chapter opener** and **figure 9.21**). When an adenylic acid residue is attached to all of its 12 subunits, glutamine synthetase is not very active. Removal of AMP groups produces more active deadenylylated glutamine synthetase, and glutamine is formed. ▶▶| *Nitrogen assimilation (section 11.5)*

Using covalent modification for the regulation of enzyme activity has some advantages. These interconvertible enzymes often are also allosteric. For instance, glutamine synthetase also is regulated allosterically. Because each form can respond differently to allosteric effectors, systems of covalently modified enzymes are able to respond to more stimuli in varied and sophisticated ways. Regulation can also be exerted on the enzymes that catalyze the covalent modifications, which adds a second level of regulation to the system.

Feedback Inhibition

The rate of many metabolic pathways is adjusted through control of the activity of the regulatory enzymes described in the sections on allosteric regulation and covalent modification. Every pathway has at least one pacemaker enzyme that catalyzes the slowest or rate-limiting reaction in the pathway. Because other reactions proceed more rapidly than the pacemaker reaction, changes in the activity of this enzyme directly alter the speed with which a pathway operates. Usually the first step in a pathway is a reaction catalyzed by a pacemaker enzyme. The end product of the pathway often inhibits this regulatory enzyme, a process known as **feedback inhibition** or **end product inhibition.** Feedback inhibition ensures balanced production of a pathway end product. If the end product becomes too concentrated, it inhibits the regulatory enzyme and slows its own synthesis. As the concentration of the end product decreases, pathway activity again increases and more product is formed. In this way, feedback inhibition automatically matches end product supply with the demand. *A Biochemical Pathway, Feedback Inhibition of Biochemical Pathways*

Frequently a biosynthetic pathway branches to form more than one end product. In such a situation, the synthesis of all end products must be coordinated precisely. It would not do to have one end product present in excess while another is lacking. Branching biosynthetic pathways usually achieve a balance among end products by using regulatory enzymes at branch points (**figure 9.22**). If an end product is present in excess, it

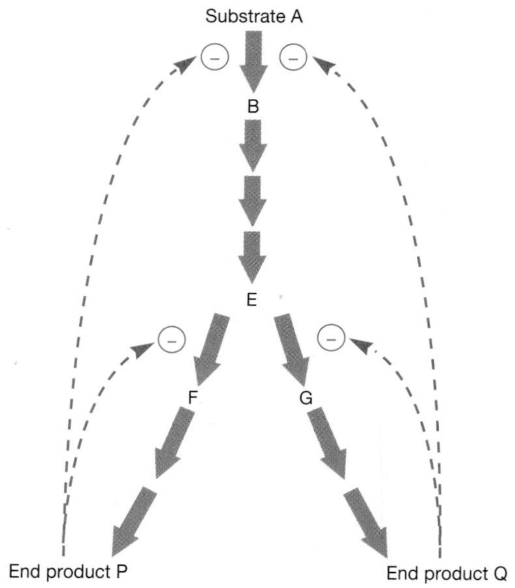

FIGURE 9.22 Feedback Inhibition. Feedback inhibition in a branching pathway with two end products. The branch-point enzymes, those catalyzing the conversion of intermediate E to F and G, are regulated by feedback inhibition. Products P and Q also inhibit the initial reaction in the pathway. A colored line with a minus sign at one end indicates that an end product, P or Q, is inhibiting the enzyme catalyzing the step next to the minus.

Figure 9.22 Micro Inquiry

What would happen to product formation if P and Q inhibited the formation of intermediate E independently of one another?

often inhibits the branch-point enzyme on the sequence of reactions leading to its formation, in this way regulating its own formation without affecting the synthesis of other products. In figure 9.22, notice that both products also inhibit the initial enzyme in the pathway. An excess of one product slows the flow of carbon into the whole pathway while inhibiting the appropriate branch-point enzyme. Because less carbon is required when only one branch is functioning, feedback inhibition of the initial pacemaker enzyme helps match the supply with the demand in branching pathways. The regulation of multiple branched pathways is often made even more sophisticated by the presence of **isoenzymes,** different forms of an enzyme that catalyze the same reaction. The initial pacemaker step may be catalyzed by several isoenzymes, each under separate control. In this case, an excess of a single end product reduces pathway activity but does not completely block pathway function because some isoenzymes are still active.

1. Define allosteric enzyme and allosteric effector.
2. How can regulatory enzymes be influenced by reversible covalent modification? What group is used for this purpose with glutamine synthetase, and which form of this enzyme is active?
3. What is a pacemaker enzyme? Feedback inhibition? How does feedback inhibition automatically adjust the concentration of a pathway end product?
4. What is the significance of the fact that regulatory enzymes often are located at pathway branch points? What are isoenzymes, and why are they important in pathway regulation?

Summary

9.1 Microbial Metabolism and Its Importance

a. Metabolism is the total of all chemical reactions that occur in cells. It can be divided into two parts: the energy-conserving fueling reactions (called catabolism) and anabolism (**figure 9.1**).

b. Microbial metabolism contributes significantly to biogeochemical cycles (**figure 9.2**). It also generates metabolites that humans value (e.g., ethanol and antibiotics).

c. Living cells carry out three major kinds of work: chemical work of biosynthesis, transport work, and mechanical work. Energy is required to do cellular work.

9.2 Thermodynamics

a. Energy can be defined as the capacity to do work.

b. Thermodynamics is the field that analyzes energy changes in systems such as cells.

c. The first law of thermodynamics states that energy is neither created nor destroyed.

d. The second law of thermodynamics states that changes occur in such a way that the randomness or disorder of the universe increases to the maximum possible. That is, entropy always increases during spontaneous processes.

9.3 Free Energy and Reactions

a. The first and second laws of thermodynamics can be combined to determine the amount of energy made available for useful work.

$$\Delta G = \Delta H - T \cdot \Delta S$$

In this equation the change in free energy (ΔG) is the energy made available for useful work, the change in enthalpy (ΔH) is the change in heat content, and the change in entropy is ΔS.

b. The standard free energy change ($\Delta G^{\circ\prime}$) for a chemical reaction is directly related to the equilibrium constant.

c. In exergonic reactions, $\Delta G^{\circ\prime}$ is negative and the equilibrium constant is greater than one; the reaction goes to completion as written. Endergonic reactions have a positive $\Delta G^{\circ\prime}$ and an equilibrium constant less than one (**figure 9.3**).

9.4 ATP

a. ATP is a high-energy molecule that transports energy in a useful form from one reaction or location in a cell to another (**figure 9.4**).

b. ATP is readily synthesized from ADP and P_i using energy released from exergonic reactions; when hydrolyzed back to ADP and P_i, it releases the energy, which is used to drive endergonic reactions. This cycling of ATP with ADP and P_i is called the cell's energy cycle (**figure 9.6**).

9.5 Oxidation-Reduction Reactions

a. In oxidation-reduction (redox) reactions, electrons move from an electron donor to an electron acceptor. Each redox reaction consists of two half reactions: one is the electron-donating half reaction; the other is the electron-accepting half reaction.

b. Each half reaction of a redox reaction consists of a molecule that can accept electrons and one that can donate electrons. These two molecules are called a conjugate redox pair.

c. The standard reduction potential measures the tendency of the donor of a conjugate redox pair to give up electrons.

d. Conjugate redox pairs with more negative reduction potentials donate electrons to those with more positive potentials, and energy is made available during the transfer (**figure 9.7** and **table 9.1**).

9.6 Electron Transport Chains

a. Many metabolic processes use a series of electron carriers to transport electrons from the primary electron donor (e.g., a chemotroph's energy source) to a final electron acceptor (e.g., oxygen). A series of electron carriers is called an electron transport chain (ETC). ETCs are located in membranes (**figure 9.8**).

b. Some of the most important electron carriers in cells are NAD^+, $NADP^+$, FAD, FMN, coenzyme Q, cytochromes, and the nonheme iron (Fe-S) proteins (**figures 9.9–9.13**). They differ in several ways, including how many electrons and protons they transfer.

9.7 Enzymes

a. Enzymes are protein catalysts that catalyze specific reactions.

b. Many enzymes consist of a protein component, the apoenzyme, and a nonprotein cofactor that may be a prosthetic group, a coenzyme, or a metal activator.

c. Enzymes speed reactions by binding substrates at their active sites and lowering the activation energy (**figure 9.15**).

d. The rate of an enzyme-catalyzed reaction increases with substrate concentration at low substrate levels and reaches a plateau (the maximum velocity) at saturating substrate concentrations. The Michaelis constant is the substrate concentration that the enzyme requires to achieve half maximal velocity (**figure 9.18**).

e. Enzymes have pH and temperature optima for activity.

f. Enzyme activity can be slowed by competitive and noncompetitive inhibitors (**figure 9.19**).

9.8 Ribozymes

a. Some RNA molecules have catalytic activity.

b. Some ribozymes alter either their own structure or that of other RNAs. An important ribozyme is located in the ribosome, where it links amino acids together during protein synthesis.

9.9 Regulation of Metabolism

a. The regulation of metabolism keeps cell components in proper balance and conserves metabolic energy and material.

b. Metabolic channeling localizes metabolites and enzymes in different parts of the cell and influences pathway activity. A common channeling mechanism is compartmentation.

9.10 Posttranslational Regulation of Enzyme Activity

a. Many regulatory enzymes are allosteric enzymes, enzymes in which an allosteric effector binds noncovalently and reversibly to a regulatory site separate from the catalytic site and causes a conformational change in the enzyme to alter its activity (**figure 9.20**).

b. Enzyme activity also can be regulated by reversible covalent modification. Usually a phosphoryl, methyl, or adenylyl group is attached to the enzyme.

c. The first enzyme in a pathway and enzymes at branch points often are subject to feedback inhibition by one or more end products. Excess end product slows its own synthesis (**figure 9.22**).

Critical Thinking Questions

1. Suppose that a chemical reaction had a large negative ΔG° value. What would this indicate about its equilibrium constant? If displaced from equilibrium, would it proceed rapidly to completion? Would much or little free energy be made available?

2. Examine the structures of macromolecules in appendix I. Which type has the most electrons to donate? Why are carbohydrates usually the primary source of electrons for chemoorganotrophic bacteria?

3. Most enzymes do not operate at their biochemical optima inside cells. Why not?

4. Examine the branched pathway shown here for the synthesis of the amino acids aspartate, methionine, lysine, threonine, and isoleucine.

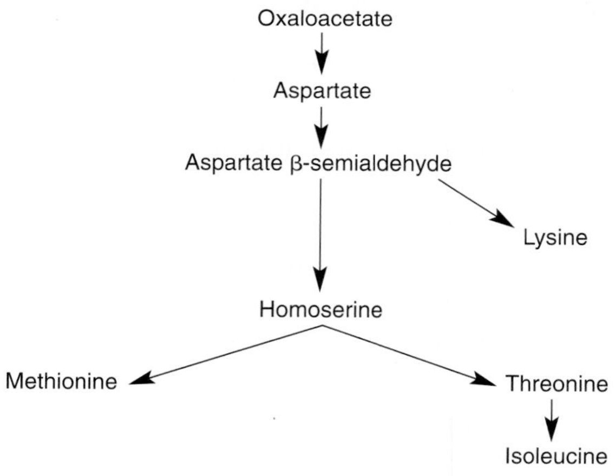

Scenario 1: The microbe is cultured in a medium containing aspartate and lysine but lacking methionine, threonine, and isoleucine.

Scenario 2: The microbe is cultured in a medium containing a rich supply of all five amino acids.

For each of these two scenarios, answer the following questions:

a. Which portion(s) of the pathway would need to be shut down in this situation?

b. How might allosteric control be used to accomplish this?

5. *Chlamydiales* are pathogenic bacteria that must be grown within a eukaryotic host cell. They rely on the host cell for many nutrients, including nucleotides, lipids, and amino acids. It is thus no surprise that when grown in coculture with a host depleted of any of these metabolites, chlamydial growth declines. However, investigators were surprised to find that high levels of certain amino acids also inhibit chlamydial growth. Further investigation found that the amino acids leucine, isoleucine, methionine, and phenylalanine could retard growth, but this effect could be reversed if valine was added at the same time. The authors discovered that the amino acid transporter BrnQ was blocked by the four growth inhibitory amino acids.

   Develop a model whereby BrnQ transports valine and the observed growth inhibition is due to competitive inhibition by high levels of these four amino acids. Your model should take into account the similarity between valine, leucine, and isoleucine (see appendix I) and the fact that BrnQ transports methionine under normal conditions.

   Read the original paper: Braun, P. R., et al. 2008. Competitive inhibition of amino acid uptake suppresses chlamydial growth: Involvement of the chlamydial amino acid transporter BrnQ. *J. Bacteriol.* 190:1822.

Concept Mapping Challenge

Construct a concept map that summarizes metabolic processes. Use the concepts that follow, any other concepts or terms you need, and your own linking words between each pair of concepts in your map.

anabolism	catabolism	ATP
endergonic reaction	exergonic reaction	entropy
free energy change	redox reactions	ETC
enzyme		

Learn More

Learn more by visiting the text website at www.mhhe.com/willey8, where you will find a complete list of references.

10

Catabolism: Energy Release and Conservation

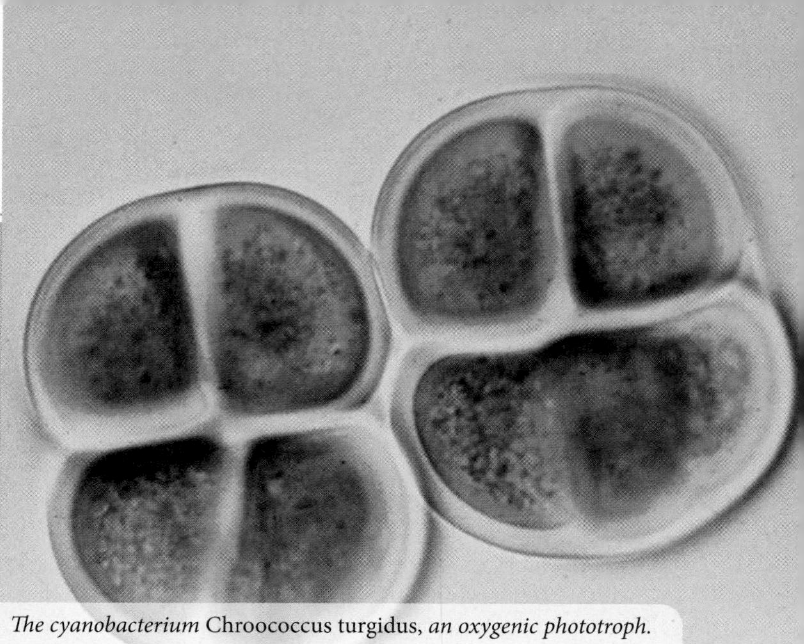

The cyanobacterium Chroococcus turgidus, *an oxygenic phototroph.*

CHAPTER GLOSSARY

aerobic respiration An energy-yielding process in which molecules, often organic, are oxidized with oxygen as the final electron acceptor.

anaerobic respiration An energy-yielding process in which the terminal acceptor for an electron transport chain is a molecule other than oxygen.

anoxygenic photosynthesis Photosynthesis that does not use water as an electron donor, thus no oxygen is produced.

ATP synthase A membrane-bound enzyme that synthesizes ATP from ADP + P_i using the energy derived from the proton motive force.

bacteriochlorophyll A modified chlorophyll that is the primary light-trapping pigment in anoxygenic photosynthetic bacteria.

bacteriorhodopsin A transmembranous protein to which retinal is bound; it functions as a light-driven proton pump.

chemiosmotic hypothesis The hypothesis that a proton and electrochemical gradient are generated by electron transport and then used to perform work (e.g., drive ATP synthesis).

chemolithotroph A microorganism that oxidizes reduced inorganic compounds to derive both energy and electrons.

chlorophyll A green photosynthetic pigment that consists of a tetrapyrrole ring with a central magnesium atom.

cyclic photophosphorylation The formation of ATP when light energy is used to move electrons cyclically through an electron transport chain.

dark reactions Pathways in which photosynthetically derived energy is used to fuel CO_2 fixation.

Embden-Meyerhof pathway A glycolytic pathway that degrades glucose to pyruvate and also generates several precursor metabolites.

Entner-Doudoroff pathway A glycolytic pathway that converts glucose to pyruvate and glyceraldehyde 3-phosphate.

fermentation An energy-yielding process in which an organic molecule is oxidized without an exogenous electron acceptor or participation of an electron transport chain.

glycolysis The conversion of glucose to pyruvic acid by use of the Embden-Meyerhof pathway, pentose phosphate pathway, or Entner-Doudoroff pathway.

light reactions Photochemical events that lead to the synthesis of ATP and, in some cases, the reduction of $NAD(P)^+$ to NAD(P)H.

noncyclic photophosphorylation The process in which light energy is used to make ATP and reducing power when electrons are moved from water to $NADP^+$ during oxygenic photosynthesis.

oxidative phosphorylation The synthesis of ATP from ADP + P_i using energy made available during electron transport initiated by the oxidation of a chemical energy source.

oxygenic photosynthesis Photosynthesis that oxidizes water to form oxygen; the form of photosynthesis characteristic of plants, protists, and cyanobacteria.

pentose phosphate pathway A glycolytic pathway that forms reducing power (NADPH) for biosynthesis and several precursor metabolites.

photophosphorylation The synthesis of ATP from ADP + P_i using energy made available during electron transport initiated by the absorption of light energy.

photosynthesis The trapping of light energy and its conversion to chemical energy, which is then used to reduce CO_2 and incorporate it into organic molecules.

proton motive force (PMF) The potential energy arising from a chemical and charge gradient of protons across a membrane.

substrate-level phosphorylation The synthesis of ATP from ADP by phosphorylation coupled with the exergonic breakdown of a high-energy organic molecule.

tricarboxylic acid (TCA) cycle The cycle that oxidizes acetyl coenzyme A to CO_2 and generates NADH and $FADH_2$ for oxidation in the electron transport chain; the cycle also supplies precursor metabolites; also called citric acid cycle and Kreb's cycle.

Microbes are the most successful organisms on Earth as witnessed by their growth under almost every conceivable condition. In large part, their success results from the diversity of their fueling reactions—those reactions that convert energy from an organism's energy source into ATP, provide reducing power, and generate precursor metabolites. The energy-conserving reactions of the major nutritional types of organisms and the mechanisms they use to provide reducing power are the focus of this chapter. We also briefly introduce the provision of precursor metabolites by chemoorganotrophic processes. However, the generation of precursor metabolites is considered in more detail in chapter 11.

Recall from chapter 6 that there are five major nutritional types. Animals and many microbes are chemoorganoheterotrophs. These organisms use organic molecules as their source of energy, carbon, and electrons. In other words, the same molecule that supplies them with energy also supplies them with carbon and electrons. Chemoorganoheterotrophs (often simply referred to as chemoorganotrophs or chemoheterotrophs) can use one or more of the following catabolic processes: aerobic respiration, anaerobic respiration, or fermentation. Chemolithoautotrophs use CO_2 as a carbon source and reduced inorganic molecules as sources of both energy and electrons. Their energy-conserving processes are sometimes referred to as respiration because they are similar to the respiratory processes carried out by chemoorganoheterotrophs. Photolithotrophic microbes use light as their source of energy and inorganic molecules as a source of electrons. Some use water as their electron source, as do plants, and release oxygen into the atmosphere in a process called oxygenic photosynthesis. Some photosynthetic bacteria use molecules such as hydrogen sulfide (H_2S) rather than water as an electron source, so do not release oxygen into the atmosphere; they are called anoxygenic phototrophs. Photolithotrophs are often autotrophic, using CO_2 as a carbon source. However, some phototrophic microbes are heterotrophic.

We begin our consideration of fueling reactions with an overview of the metabolism of chemoorganotrophs. This is followed by an introduction to the oxidation of carbohydrates, especially glucose, and a discussion of ATP synthesis during aerobic and anaerobic respiration. Fermentation is then described, followed by a survey of the breakdown of other carbohydrates and organic substances. We end the chapter with sections on chemolithotrophy and phototrophy.

10.1 Chemoorganotrophic Fueling Processes

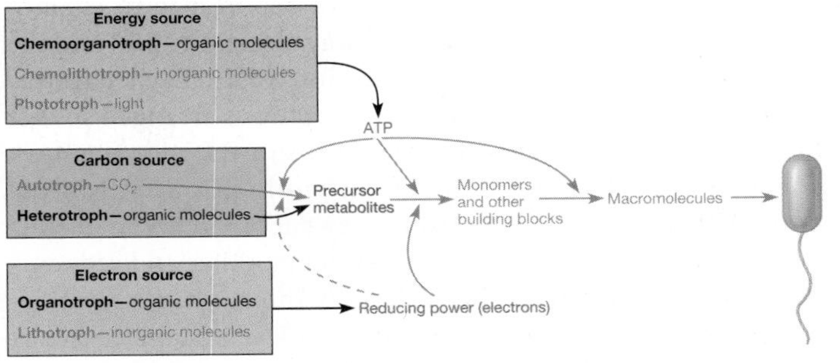

When chemoorganotrophs oxidize an organic energy source, the electrons released are accepted by electron carriers such as NAD^+ and FAD. When these reduced electron carriers (e.g., NADH, $FADH_2$) in turn donate the electrons to an electron transport chain, the metabolic process is called **respiration** and may be divided into two different types (**figure 10.1**). In aerobic respiration, the final electron acceptor is oxygen, whereas the terminal acceptor in anaerobic respiration is a different molecule such as NO_3^-, SO_4^{2-}, CO_2, Fe^{3+}, and SeO_4^{2-}. Organic acceptors such as fumarate and humic acids also may be used. As just noted, respiration involves the activity of an electron transport chain. As electrons pass through the chain to the final electron acceptor, a type of potential energy called the proton motive force (PMF) is generated and used to synthesize ATP from ADP and P_i. In contrast, **fermentation** (Latin *fermentare,* to cause to rise) uses an electron acceptor that is endogenous (from within the cell) and does not involve an electron transport chain. The endogenous electron acceptor is usually an intermediate (e.g., pyruvate) of the catabolic pathway used to degrade and oxidize the organic energy source. During fermentation, ATP is synthesized almost exclusively by substrate-level phosphorylation, a process in which a phosphate is transferred to ADP from a high-energy molecule (e.g., phosphoenolpyruvate) generated by catabolism of the energy source.

By convention, aerobic respiration, anaerobic respiration, and fermentation are usually described with glucose as the energy

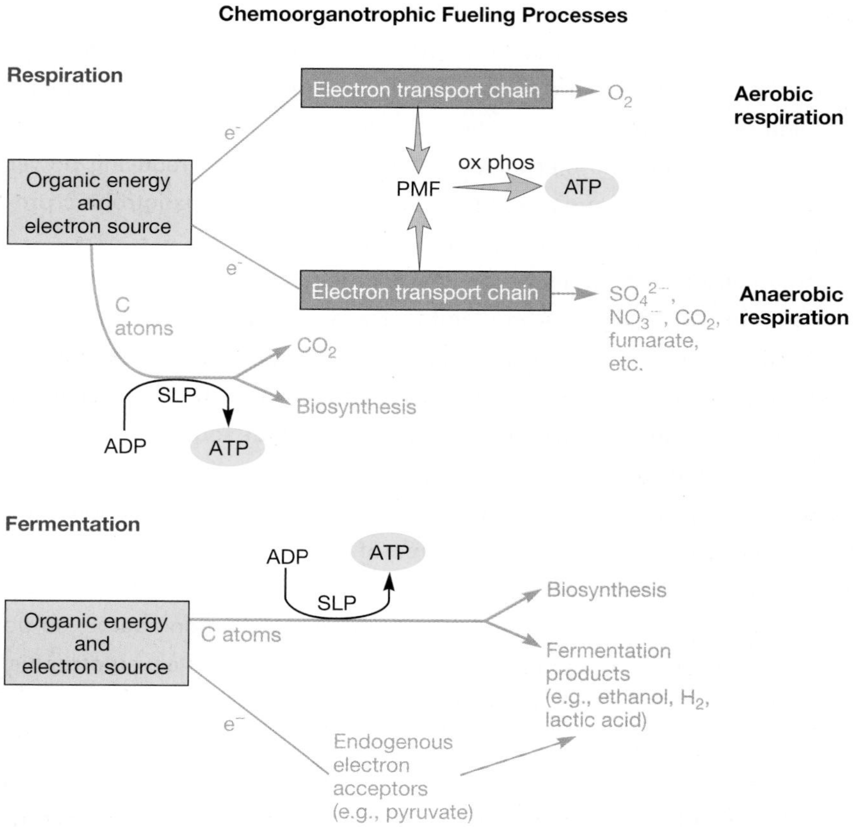

FIGURE 10.1 Chemoorganotrophic Fueling Processes. Organic molecules serve as energy and electron sources for all three fueling processes used by chemoorganotrophs. In aerobic respiration and anaerobic respiration, the electrons pass through an electron transport chain. This generates a proton motive force (PMF), which is used to synthesize most of the cellular ATP by a mechanism called oxidative phosphorylation (ox phos); a small amount of ATP is made by a process called substrate-level phosphorylation (SLP). In aerobic respiration, O_2 is the terminal electron acceptor, whereas in anaerobic respiration, exogenous molecules other than O_2 serve as electron acceptors. During fermentation, endogenous organic molecules act as electron acceptors, electron transport chains do not function, and most organisms synthesize ATP only by substrate-level phosphorylation.

source. This is done for several reasons. One is that glucose is used by many chemoorganotrophs as an energy source. But perhaps more important is the way catabolic pathways are organized. Most chemoorganotrophs can use a wide variety of organic molecules as energy sources (**figure 10.2**). They are degraded by pathways that either generate glucose or intermediates of the pathways used in glucose catabolism. Thus nutrient molecules are funneled into ever fewer metabolic intermediates. Indeed a common pathway often degrades many similar molecules (e.g., several different sugars). The existence of a few metabolic pathways, each degrading many nutrients, greatly increases metabolic efficiency by avoiding the need for a large number of less metabolically flexible pathways.

The diversity of organic molecules used as energy sources by chemoorganotrophic microbes contributes to their ecology. They are found in habitats rich in organic molecules, particularly those molecules they can use as energy sources. For instance, lactose catabolizers are common in the intestinal tracts of mammals, where they may have a ready supply of this sugar. Clostridia and other soil bacteria that use amino acids as energy sources are important decomposers of proteins in protein-rich habitats. One-carbon molecules such as methane can also be catabolized by microbes; some do so aerobically, others anaerobically. Furthermore, some bacteria such as the soil-dwelling pseudomonads are able to degrade unusual and often complex organic molecules, including pesticides, polystyrene, polychlorinated biphenyls (PCBs), and even the phenolic resins used as glue in the production of plywood and fiberboard. These microbes are useful in bioremediation. ▶▶| *Biodegradation and bioremediation processes (section 42.3)*

The catabolic pathways of greatest importance to chemoorganotrophs are the glycolytic pathways (section 10.3) and the tricarboxylic acid (TCA) cycle (section 10.4). Each pathway consists of a set of enzyme-catalyzed reactions arranged so that the product of one reaction serves as a substrate for the next. They are important not only for their role in catabolism but also for their roles in anabolism. They supply material needed for biosynthesis, including precursor metabolites and reducing power. Precursor metabolites serve as the starting molecules for biosynthetic pathways. Reducing power is used in redox reactions that reduce the precursor metabolites as they are transformed into amino acids, nucleotides, and other small molecules needed for synthesis of macromolecules. Furthermore, some enzymes of these pathways are reversible and can function either catabolically or anabolically. ▶▶| *Precursor metabolites (section 11.2)*

Pathways with enzymes that function both catabolically and anabolically are often called **amphibolic pathways** (Greek

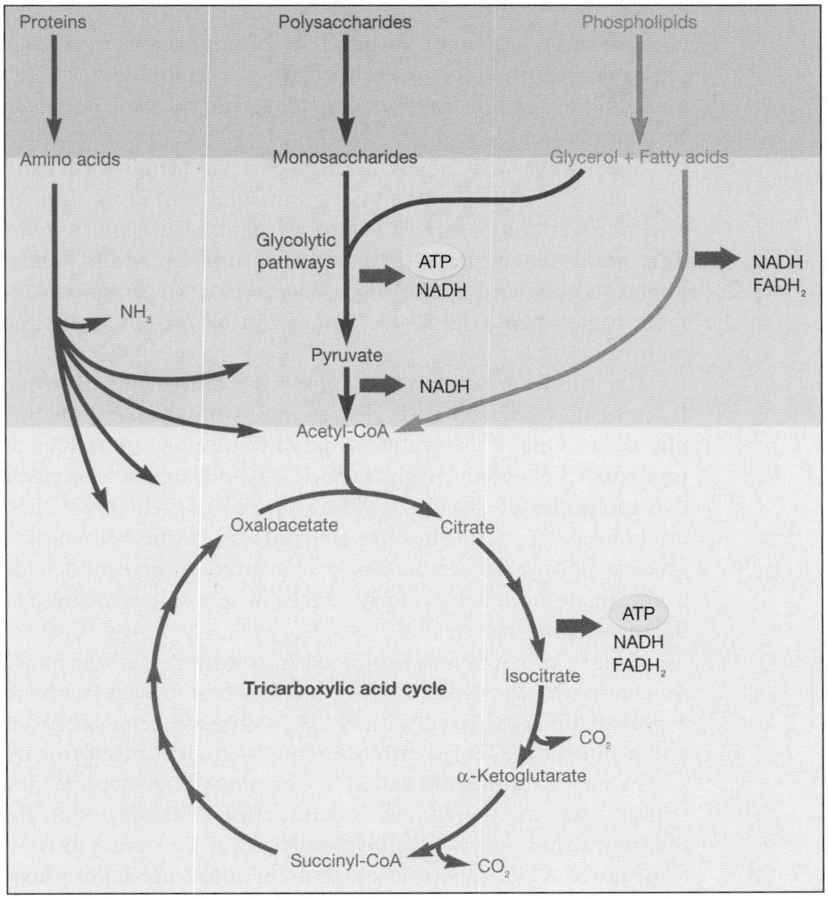

FIGURE 10.2 Pathways Used by Chemoorganotrophs to Catabolize Organic Energy Sources. Notice that these pathways funnel metabolites into the glycolytic pathways and the tricarboxylic acid cycle, thus increasing metabolic efficiency and flexibility.

amphi, on both sides). For instance, many of the enzymes of the Embden-Meyerhof pathway (a glycolytic pathway) are freely reversible. They function catabolically during glycolysis but anabolically during gluconeogenesis, a pathway that generates glucose from pyruvate and other small molecules (**figure 10.3**). Other reactions in glycolysis are not reversible. Thus, during glycolysis, an enzyme catalyzes the reaction in a catabolic direction, and during gluconeogenesis, a different enzyme catalyzes the reverse, anabolic reaction. Thus, although glycolysis and gluconeogenesis share many enzymes, they are two distinct pathways.

In the following sections, we explore the metabolism of chemoorganotrophs in more detail. The discussion begins with aerobic respiration and introduces the glycolytic pathways, TCA cycle, and other important processes. Many of these also occur during anaerobic respiration. Thus in many cases, anaerobic respiration differs from aerobic respiration mainly in terms of the terminal electron acceptor. Fermentation is quite distinct from respiration. It involves only a subset of the reactions that function during respiration and only partially catabolizes the energy source. However, it is widely used by microbes and has important practical applications, including the production of many foods and the identification of pathogenic microbes. ▶▶| *Microbiology of food (chapter 40); Clinical microbiology and immunology (chapter 35)*

1. Compare and contrast fermentation and respiration. Give examples of the types of electron acceptors used by each process. What is the difference between aerobic respiration and anaerobic respiration?
2. Why is it to the cell's advantage to catabolize diverse organic energy sources by funneling them into a few common pathways?
3. What are amphibolic pathways? Why are they important?

10.2 Aerobic Respiration

Aerobic respiration is a process that can completely catabolize an organic energy source to CO_2 using the glycolytic pathways and TCA cycle with O_2 as the terminal electron acceptor for an electron transport chain (figure 10.1). We focus in this section and the following three sections on the aerobic respiration of glucose. The catabolism of glucose begins with one or more of the glycolytic pathways that yield pyruvate (section 10.3). These pathways produce some ATP as well as NADH, $FADH_2$, or both. Next, the partially oxidized carbon is fed into the TCA cycle and oxidized completely to CO_2 with the production of some GTP or ATP, NADH, and $FADH_2$ (section 10.4). The NADH and $FADH_2$ that have been formed by glycolysis and the TCA cycle are oxidized by an electron transport chain, using O_2 as the terminal electron acceptor (section 10.5). It is the functioning of the electron transport chain that yields most of the ATP during aerobic respiration.

10.3 Glycolytic Pathways

Microorganisms employ several metabolic pathways to catabolize glucose to pyruvate, including (1) the Embden-Meyerhof pathway, (2) the Entner-Doudoroff pathway, and (3) the pentose phosphate pathway. We refer to these pathways collectively as **glycolytic pathways** or as **glycolysis** (Greek *glyco,* sweet, and *lysis,* a loosening). However, in some texts, the term glycolysis refers only to the Embden-Meyerhof pathway. For the sake of simplicity, the detailed structures of some metabolic intermediates are not used in pathway diagrams. However, these can be found in appendix II.

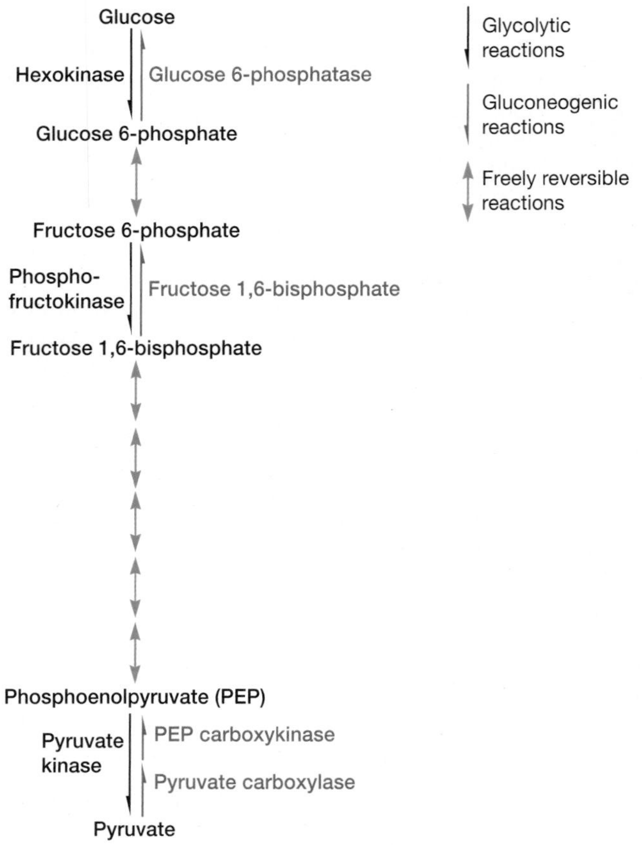

FIGURE 10.3 An Amphibolic Pathway. The Embden-Meyerhof pathway is an example of an amphibolic pathway. Many reactions of this pathway are catalyzed by enzymes that function in glycolysis and in an anabolic pathway called gluconeogenesis. Gluconeogenesis reverses the glycolytic process and allows cells to synthesize glucose from smaller molecules such as pyruvate. Note that some glycolytic reactions are catalyzed by enzymes unique to that pathway. Likewise, some gluconeogenic reactions are catalyzed by enzymes unique to gluconeogenesis. Therefore, although they share several enzymes, the two pathways are distinct.

Figure 10.3 Micro Inquiry

Is NAD^+ reduced to NADH in the catabolic or anabolic direction of this pathway?

Embden-Meyerhof Pathway

The **Embden-Meyerhof pathway** is observed in almost all organisms (including multicellular eukaryotes) and is undoubtedly the most common pathway for glucose degradation to pyruvate. It is found in all major groups of microorganisms, as well as plants and animals, and functions in the presence or absence of O_2. As noted earlier, it is an important amphibolic pathway and provides several precursor metabolites, NADH, and ATP for the cell. The Embden-Meyerhof pathway occurs in the cytoplasm.

The pathway may be divided into two parts (**figure 10.4** and *appendix II*). In the initial six-carbon phase, ATP is used to phosphorylate glucose twice, yielding fructose 1,6-bisphosphate. This preliminary phase "primes the pump" by adding phosphates to each end of the sugar. In essence, the organism invests some of its ATP so that more can be made later in the pathway.

The three-carbon, energy-conserving phase begins when fructose 1,6-bisphosphate is cleaved into two halves, each with a phosphate. One of the products, dihydroxyacetone phosphate, is immediately converted to glyceraldehyde 3-phosphate. This yields two molecules of glyceraldehyde 3-phosphate, which are then catabolized to pyruvate in a five-step process. Because dihydroxyacetone phosphate can be easily converted to glyceraldehyde 3-phosphate, both halves of fructose 1,6-bisphosphate are used in the three-carbon phase.

NADH and ATP are produced in the three-carbon phase of the pathway. NADH is formed when glyceraldehyde 3-phosphate is oxidized with NAD^+ as the electron acceptor, and a phosphate (P_i) is simultaneously incorporated to give a high energy molecule called 1,3-bisphosphoglycerate. This reaction sets the stage for ATP production. The phosphate on the first carbon of 1,3-bisphosphoglycerate is donated to ADP to produce ATP. This is an example of **substrate-level phosphorylation** because ADP phosphorylation is coupled with the exergonic hydrolysis of a high energy molecule having a higher phosphate transfer potential than ATP. A second ATP is made by substrate-level phosphorylation when the phosphate on phosphoenolpyruvate (the last intermediate of the pathway) is donated to ADP. This reaction also yields pyruvate, the final product of the pathway. ◀◀ *ATP (section 9.4)*; *How NAD^+ Works*

The yields of ATP and NADH by the Embden-Meyerhof pathway may be calculated. In the six-carbon phase, two ATPs are used to form fructose 1,6-bisphosphate. For each glyceraldehyde 3-phosphate transformed into pyruvate, one NADH and two ATPs are formed. Because two glyceraldehyde 3-phosphates arise from a single glucose (one by way of dihydroxyacetone phosphate), the three-carbon phase generates four ATPs and two NADHs per glucose. Subtraction of the ATP used in the six-carbon phase from that produced by substrate-level phosphorylation in the three-carbon phase gives a net yield of two ATPs per glucose. Thus the catabolism of glucose to pyruvate can be represented by this simple equation.

$$\text{Glucose} + 2\text{ADP} + 2\text{P}_i + 2\text{NAD}^+ \rightarrow 2\text{ pyruvate} + 2\text{ATP} + 2\text{NADH} + 2\text{H}^+$$

The NADH is used during aerobic respiration to transport electrons to an electron transport chain. *How Glycolysis Works*

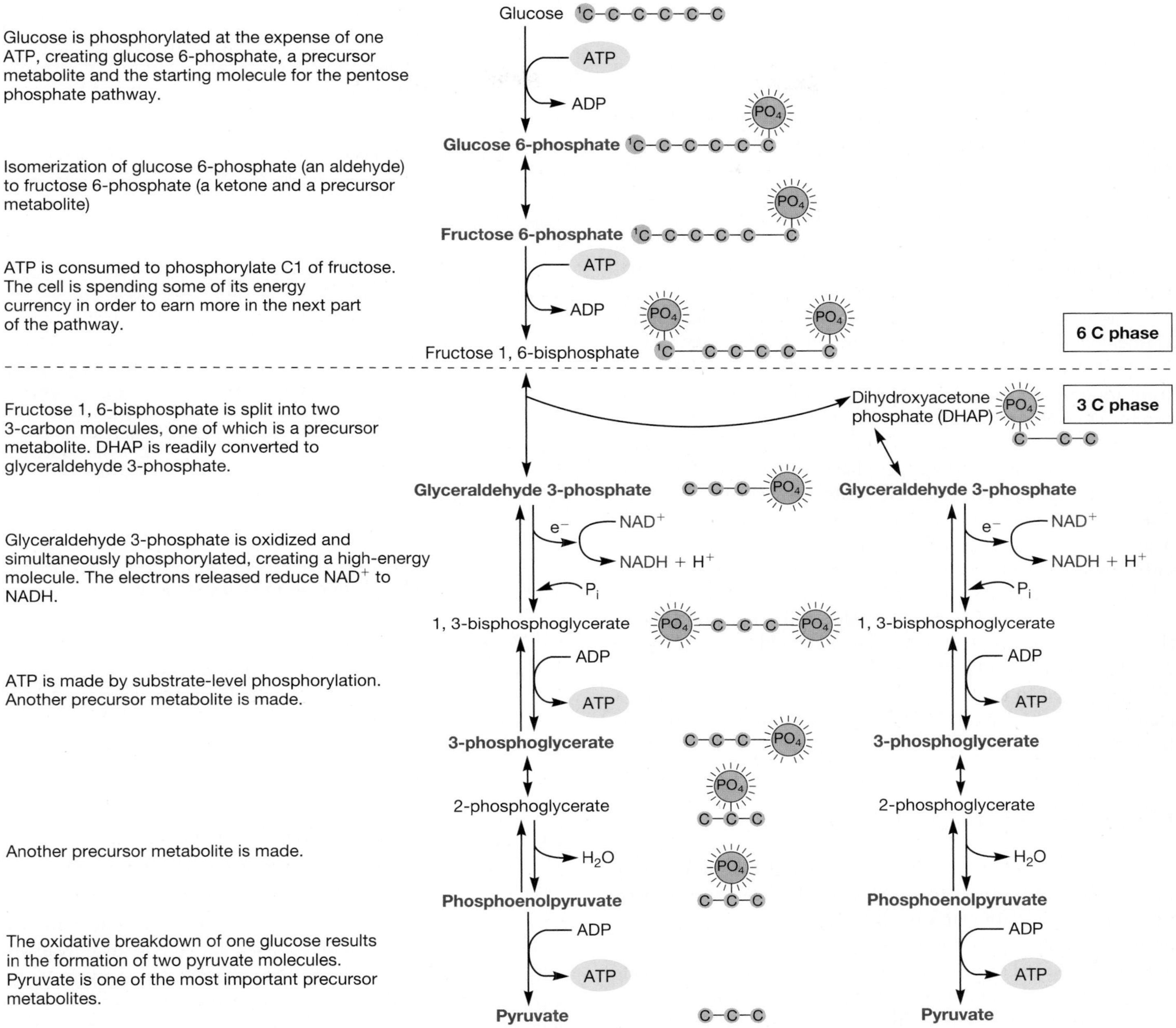

Glucose is phosphorylated at the expense of one ATP, creating glucose 6-phosphate, a precursor metabolite and the starting molecule for the pentose phosphate pathway.
Isomerization of glucose 6-phosphate (an aldehyde) to fructose 6-phosphate (a ketone and a precursor metabolite)
ATP is consumed to phosphorylate C1 of fructose. The cell is spending some of its energy currency in order to earn more in the next part of the pathway.
Fructose 1, 6-bisphosphate is split into two 3-carbon molecules, one of which is a precursor metabolite. DHAP is readily converted to glyceraldehyde 3-phosphate.
Glyceraldehyde 3-phosphate is oxidized and simultaneously phosphorylated, creating a high-energy molecule. The electrons released reduce NAD+ to NADH.
ATP is made by substrate-level phosphorylation. Another precursor metabolite is made.
Another precursor metabolite is made.
The oxidative breakdown of one glucose results in the formation of two pyruvate molecules. Pyruvate is one of the most important precursor metabolites.
Glucose
ATP
ADP
Glucose 6-phosphate
Fructose 6-phosphate
Fructose 1, 6-bisphosphate
6 C phase
3 C phase
Dihydroxyacetone phosphate (DHAP)
Glyceraldehyde 3-phosphate
e−
NAD+
NADH + H+
Pi
1, 3-bisphosphoglycerate
3-phosphoglycerate
2-phosphoglycerate
H2O
Phosphoenolpyruvate
Pyruvate
PO4
C

FIGURE 10.4 Embden-Meyerhof Pathway. This is one of three glycolytic pathways used to catabolize glucose to pyruvate, and it can function during aerobic respiration, anaerobic respiration, and fermentation. When used during respiration, the electrons accepted by NAD^+ are transferred to an electron transport chain and are ultimately accepted by an exogenous electron acceptor. When used during fermentation, the electrons accepted by NAD^+ are donated to an endogenous electron acceptor (e.g., pyruvate). The Embden-Meyerhof pathway also generates several precursor metabolites (shown in blue).

Figure 10.4 Micro Inquiry

Which reactions are examples of substrate-level phosphorylation?

Entner-Doudoroff Pathway

Although the Embden-Meyerhof pathway is the most common route for the conversion of six-carbon sugars (hexoses) to pyruvate, the **Entner-Doudoroff pathway** is used by some soil bacteria (e.g., *Pseudomonas, Rhizobium, Azotobacter,* and *Agrobacterium*) and a few other gram-negative bacteria. To date, very few gram-positive bacteria have been found to use this pathway, with the intestinal bacterium *Enterococcus faecalis* being a rare exception. It is not used by eukaryotes.

The Entner-Doudoroff pathway essentially replaces the first phase of the Embden-Meyerhof pathway, yielding pyruvate and glyceraldehyde 3-phosphate, rather than two molecules of glyceraldehyde 3-phosphate. A key intermediate of the pathway is 2-keto-3-deoxy-6-phosphogluconate (KDPG), which is formed from glucose by three reactions that consume one ATP and produce one NADPH (**figure 10.5** and *appendix II*). The KDPG is then cleaved to pyruvate and glyceraldehyde 3-phosphate. Bacteria that use this pathway also have the enzymes that function in the second phase of the Embden-Meyerhof pathway. These enzymes may be used to catabolize the glyceraldehyde 3-phosphate to form a second pyruvate molecule. If this occurs, then two ATP and one NADH are formed. Thus the catabolism of one glucose molecule to two pyruvates by way of the Entner-Doudoroff pathway coupled with the second half of the Embden-Meyerhof pathway has a net yield of one ATP, one NADH, and one NADPH. During aerobic respiration, the NADH is used to transport electrons to an electron transport chain (ETC); the NADPH is used as reducing power for anabolic reactions.

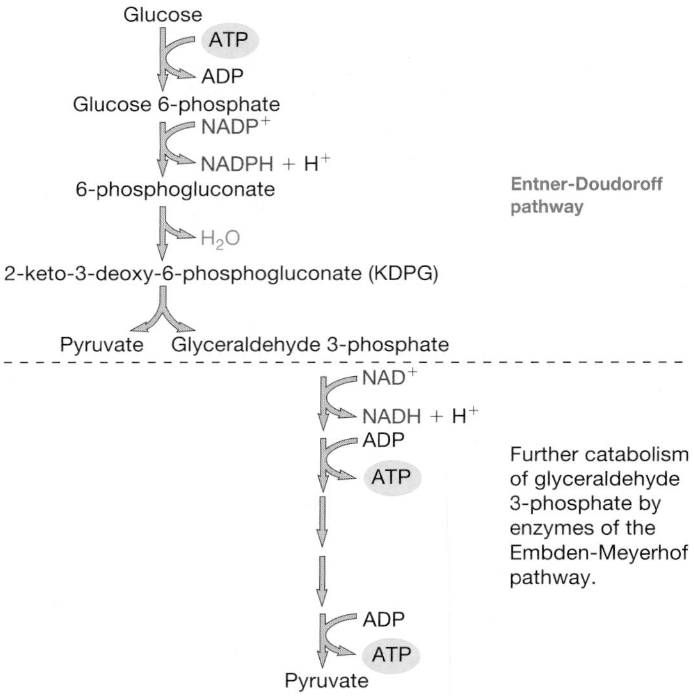

FIGURE 10.5 The Entner-Doudoroff Pathway.

Figure 10.5 *Micro Inquiry*

For what kinds of reactions is NADPH used?

Pentose Phosphate Pathway

The **pentose phosphate,** or **hexose monophosphate, pathway** may be used at the same time as either the Embden-Meyerhof or the Entner-Doudoroff pathways. It can operate either aerobically or anaerobically and is important in both biosynthesis and catabolism. It is used by all organisms because of its role in providing reducing power and important precursor metabolites.

The pentose phosphate pathway begins with the oxidation of glucose 6-phosphate to 6-phosphogluconate followed by the oxidation of 6-phosphogluconate to the five-carbon sugar (i.e., pentose sugar) ribulose 5-phosphate and CO_2 (**figure 10.6** and *appendix II*). NADPH is produced during these oxidations. Ribulose 5-phosphate is then converted to a mixture of three- through seven-carbon sugar phosphates. Two enzymes play a central role in these transformations: (1) transketolase catalyzes the transfer of two-carbon groups, and (2) transaldolase transfers a three-carbon group from sedoheptulose 7-phosphate (a seven-carbon molecule) to glyceraldehyde 3-phosphate (a three-carbon molecule). The overall result is that three glucose 6-phosphates are converted to two fructose 6-phosphates, glyceraldehyde 3-phosphate, and three CO_2 molecules, as shown in this equation.

$$\text{3 glucose 6-phosphate} + 6\text{NADP}^+ + 3\text{H}_2\text{O} \rightarrow \text{2 fructose 6-phosphate} + \text{glyceraldehyde 3-phosphate} + 3\text{CO}_2 + 6\text{NADPH} + 6\text{H}^+$$

These intermediates are used in two ways. The fructose 6-phosphate can be changed back to glucose 6-phosphate while glyceraldehyde 3-phosphate is converted to pyruvate by enzymes of the Embden-Meyerhof pathway. Alternatively two glyceraldehyde 3-phosphates may combine to form fructose 1,6-bisphosphate, which is eventually converted back into glucose 6-phosphate. This results in the complete degradation of glucose 6-phosphate to CO_2 and the production of a great deal of NADPH.

$$\text{Glucose 6-phosphate} + 12\text{NADP}^+ + 7\text{H}_2\text{O} \rightarrow 6\text{CO}_2 + 12\text{NADPH} + 12\text{H}^+ + \text{P}_\text{i}$$

The pentose phosphate pathway is an important amphibolic pathway because: (1) NADPH produced by the pathway serves as a source of electrons for the reduction of molecules during biosynthesis. Indeed, the pentose phosphate pathway is the major source of reducing power for cells, yielding 2 NADPH molecules for each glucose metabolized to pyruvate in this way. (2) The pathway produces two important precursor metabolites:

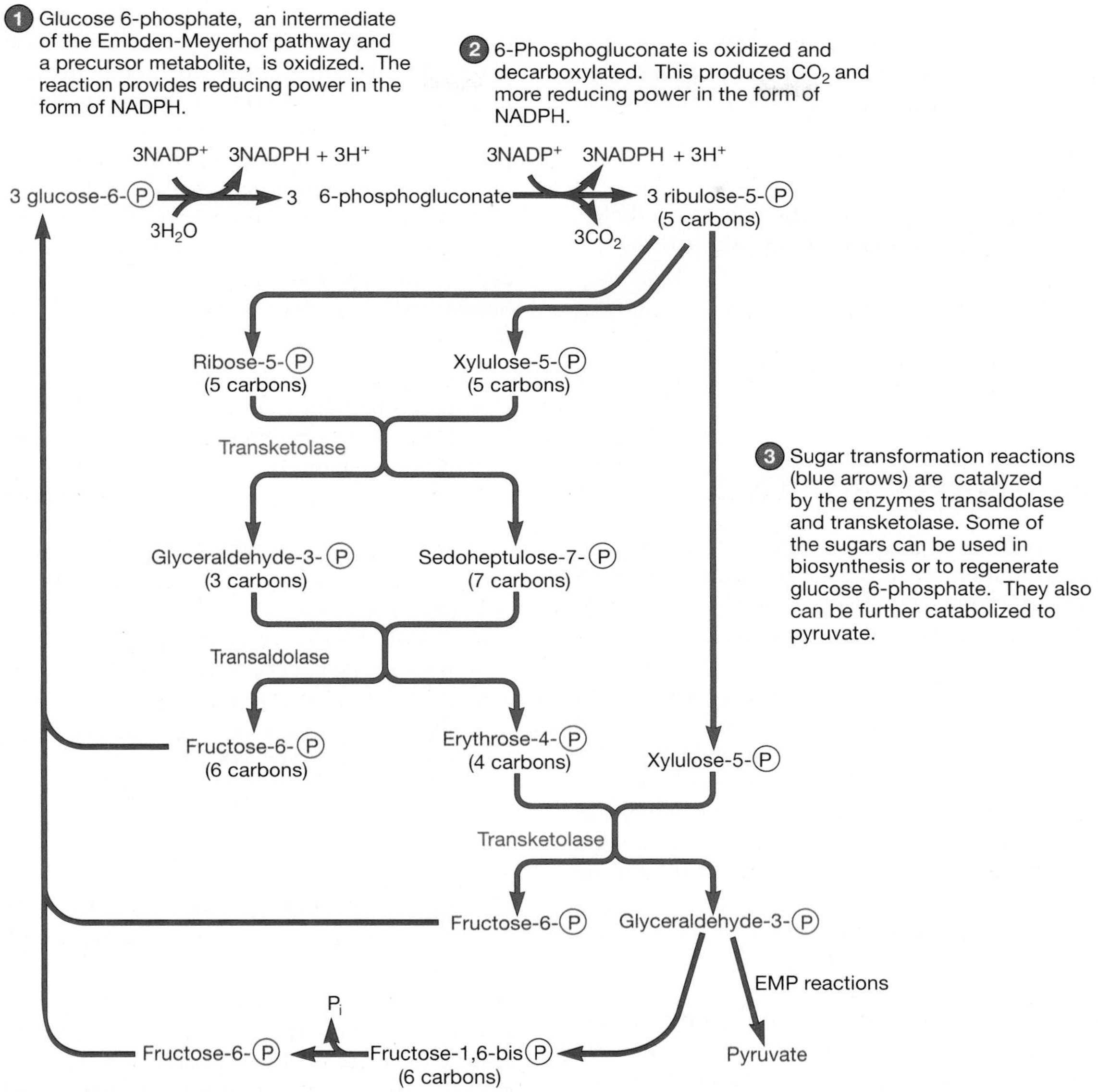

FIGURE 10.6 The Pentose Phosphate Pathway. The catabolism of three glucose 6-phosphate molecules to two fructose 6-phosphates, a glyceraldehyde 3-phosphate, and three CO_2 molecules is traced. Note that the pentose phosphate pathway generates several intermediates that are also intermediates of the Embden-Meyerhof pathway (EMP). These intermediates can be fed into the EMP with two results: (1) continued degradation to pyruvate or (2) regeneration of glucose 6-phosphate by gluconeogenesis. The pentose phosphate pathway also plays a major role in producing reducing power (NADPH) and several precursor metabolites (shown in blue). The sugar transformations are indicated with blue arrows.

Figure 10.6 Micro Inquiry

For what macromolecule is ribose 5-phosphate a precursor?

erythrose 4-phosphate and ribose 5-phosphate. Erythrose 4-phosphate is used to synthesize aromatic amino acids and vitamin B_6 (pyridoxal); ribose 5-phosphate is a major component of nucleic acids. Furthermore, when a microorganism is growing on a five-carbon sugar, the pathway can function biosynthetically to supply hexose sugars (e.g., glucose needed for peptidoglycan synthesis). (3) Intermediates in the pathway may be used to produce ATP. For instance, glyceraldehyde 3-phosphate can enter the three-carbon phase of the Embden-Meyerhof pathway. As it is degraded to pyruvate, two ATPs are formed by substrate-level phosphorylation. ▶▶| *Synthesis of amino acids (section 11.5); Synthesis of purines, pyrimidines, and nucleotides (section 11.6)*

1. Summarize the major features of the Embden-Meyerhof, Entner-Doudoroff, and pentose phosphate pathways. Include the starting points, the products of the pathways, the ATP yields, and the metabolic roles each pathway has.
2. What is substrate-level phosphorylation?

10.4 Tricarboxylic Acid Cycle

In the glycolytic pathways, glucose is oxidized to pyruvate. During aerobic respiration, the catabolic process continues by oxidizing pyruvate to three CO_2. The first step of this process employs a multienzyme system called the pyruvate dehydrogenase complex. It oxidizes and cleaves pyruvate to form one CO_2 and the two-carbon molecule **acetyl-coenzyme A (acetyl-CoA)** (**figure 10.7**). Acetyl-CoA is energy rich because hydrolysis of the bond that links acetic acid to coenzyme A (a thioester bond) has a large negative change in free energy, just as hydrolysis of the bond in many phosphate-containing molecules does. As shown in figure 10.2, carbohydrates as well as fatty acids (section 10.9) and amino acids (section 10.10) can be converted to acetyl-CoA.

Acetyl-CoA then enters the **tricarboxylic acid (TCA) cycle,** which is also called the **citric acid cycle** or the **Krebs cycle** (figure 10.7 and *appendix II*). In the first reaction, acetyl-CoA is condensed with (i.e., added to) the four-carbon intermediate oxaloacetate to form citrate, a molecule with six carbons. Citrate is rearranged to give isocitrate, a more readily oxidized alcohol. Isocitrate is subsequently oxidized and decarboxylated twice to yield α-ketoglutarate (five carbons) and then succinyl-CoA (four carbons), another high-energy molecule containing a thioester bond. At this point, two NADH molecules have been formed and two carbons lost from the cycle as CO_2. The cycle continues when succinyl-CoA is converted to succinate. This involves hydrolysis of the thioester bond in succinyl-CoA and using the large amount of energy released to form one GTP by substrate-level phosphorylation. GTP is also a high-energy molecule, and it is functionally equivalent to ATP. It is used in protein synthesis and to make other nucleoside triphosphates, including ATP. Two oxidation steps follow, yielding one $FADH_2$ and one NADH. The last oxidation step regenerates oxaloacetate, and as long as there is a supply of acetyl-CoA, the cycle can repeat itself. Inspection of figure 10.7 shows that the TCA cycle generates two CO_2 molecules, three NADH molecules, one $FADH_2$, and one GTP for each acetyl-CoA molecule oxidized.
How the Kreb's Cycle Works

TCA cycle enzymes are widely distributed among microorganisms. In bacteria and archaea, they are located in the cytoplasm. In eukaryotes, they are found in the mitochondrial matrix. The complete cycle appears to be functional in many aerobic bacteria, free-living protists, and fungi. This is not surprising because the cycle plays an important role in energy conservation by producing numerous NADH and $FADH_2$ (section 10.5). Even those microorganisms that lack the complete TCA cycle usually have most of the cycle enzymes, because the TCA cycle is also a key source of precursor metabolites for use in biosynthesis. ▶▶| *Anaplerotic reactions and amino acid biosynthesis (section 11.5)*

1. Give the substrate and products of the TCA cycle. Describe its organization in general terms. What are its major functions?
2. What chemical intermediate links pyruvate to the TCA cycle?
3. How many times must the TCA cycle be performed to oxidize one molecule of glucose completely to six molecules of CO_2? Why?
4. In what eukaryotic organelle is the TCA cycle found? Where is the cycle located in bacterial and archaeal cells?
5. Why is it desirable for a microbe with the Embden-Meyerhof pathway and the TCA cycle also to have the pentose phosphate pathway?
6. Why is GTP functionally equivalent to ATP?

10.5 Electron Transport and Oxidative Phosphorylation

During the oxidation of glucose to six CO_2 molecules by glycolysis and the TCA cycle, as many as four ATP molecules are generated by substrate-level phosphorylation. Thus at this point, the work done by the cell has yielded relatively little ATP. However, in oxidizing glucose, the cell has also generated numerous molecules of NADH and $FADH_2$. Both of these molecules have a relatively negative E'_0 and can be used to conserve energy. In fact, most of the ATP generated during respiration comes from the oxidation of these electron carriers in the electron transport chain. We examine the mitochondrial electron transport chain first because it has been well studied and because it functions in fungi and many protists. Recall, however, that some protists do not have mitochondria, as we discuss in chapter 4. After completing the discussion of mitochondrial electron transport, we turn to bacterial chains and finish with a discussion of ATP synthesis.

Electron Transport Chains

The mitochondrial **electron transport chain (ETC)** is composed of a series of electron carriers that operate together to transfer electrons from donors, such as NADH and $FADH_2$, to O_2

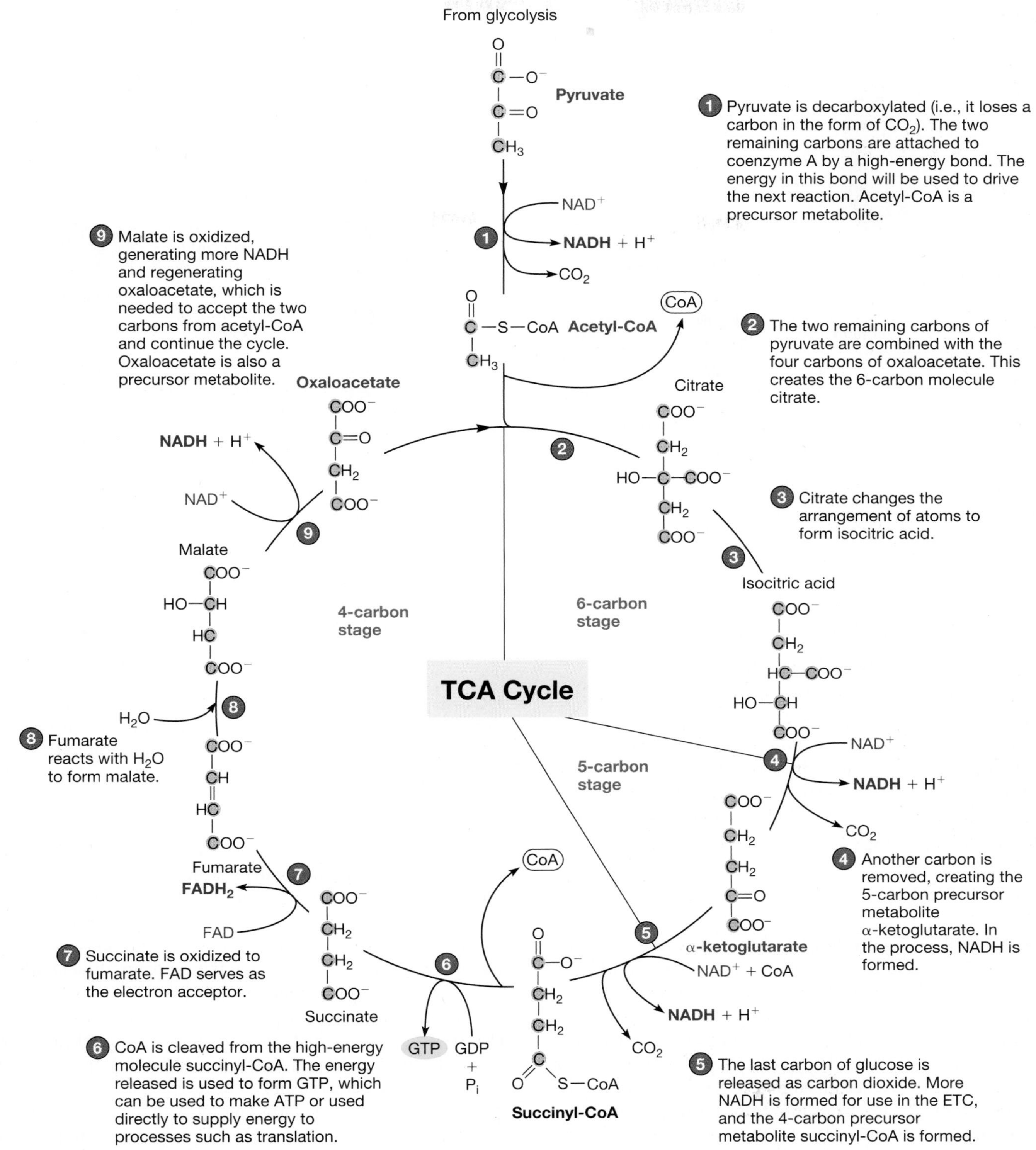

From glycolysis
Pyruvate
1 Pyruvate is decarboxylated (i.e., it loses a carbon in the form of CO2). The two remaining carbons are attached to coenzyme A by a high-energy bond. The energy in this bond will be used to drive the next reaction. Acetyl-CoA is a precursor metabolite.
NAD+
NADH + H+
CO2
CoA
Acetyl-CoA
2 The two remaining carbons of pyruvate are combined with the four carbons of oxaloacetate. This creates the 6-carbon molecule citrate.
9 Malate is oxidized, generating more NADH and regenerating oxaloacetate, which is needed to accept the two carbons from acetyl-CoA and continue the cycle. Oxaloacetate is also a precursor metabolite.
Oxaloacetate
Citrate
3 Citrate changes the arrangement of atoms to form isocitric acid.
Malate
Isocitric acid
4-carbon stage
6-carbon stage
TCA Cycle
5-carbon stage
8 Fumarate reacts with H2O to form malate.
4 Another carbon is removed, creating the 5-carbon precursor metabolite α-ketoglutarate. In the process, NADH is formed.
Fumarate
FADH2
FAD
α-ketoglutarate
NAD+ + CoA
7 Succinate is oxidized to fumarate. FAD serves as the electron acceptor.
Succinate
GTP
GDP + Pi
Succinyl-CoA
6 CoA is cleaved from the high-energy molecule succinyl-CoA. The energy released is used to form GTP, which can be used to make ATP or used directly to supply energy to processes such as translation.
5 The last carbon of glucose is released as carbon dioxide. More NADH is formed for use in the ETC, and the 4-carbon precursor metabolite succinyl-CoA is formed.

FIGURE 10.7 The Tricarboxylic Acid Cycle. The TCA cycle is linked to glycolysis by a connecting reaction catalyzed by the pyruvate dehydrogenase complex. The reaction decarboxylates pyruvate (removes a carboxyl group as CO_2) and generates acetyl-CoA. The cycle may be divided into three stages based on the size of its intermediates. The three stages are separated from one another by two decarboxylation reactions. Precursor metabolites, carbon skeletons used in biosynthesis, are shown in blue. NADH and $FADH_2$ are shown in purple; they can transfer electrons to the electron transport chain (ETC).

Figure 10.7 Micro Inquiry

What reaction provides the energy to fuel the condensation (joining) of the 4-carbon oxaloacetate with the 2-carbon acetyl-CoA to form citrate?

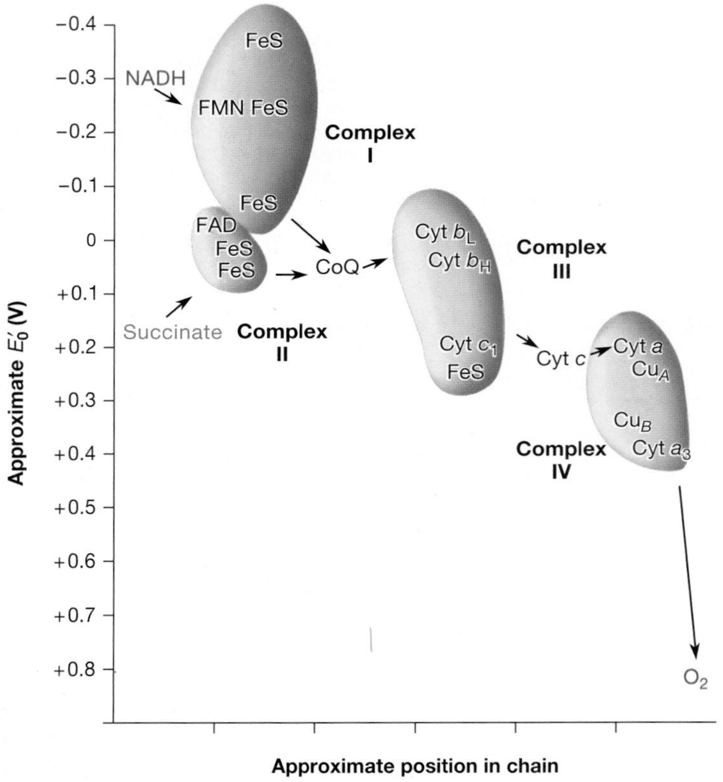

FIGURE 10.8 The Mitochondrial Electron Transport Chain. Many of the more important carriers are arranged at approximately the correct reduction potential and sequence. In mitochondria, they are organized into four complexes that are linked by coenzyme Q (CoQ) and cytochrome *c* (Cyt *c*). Electrons flow from NADH and succinate down the reduction potential gradient to oxygen.

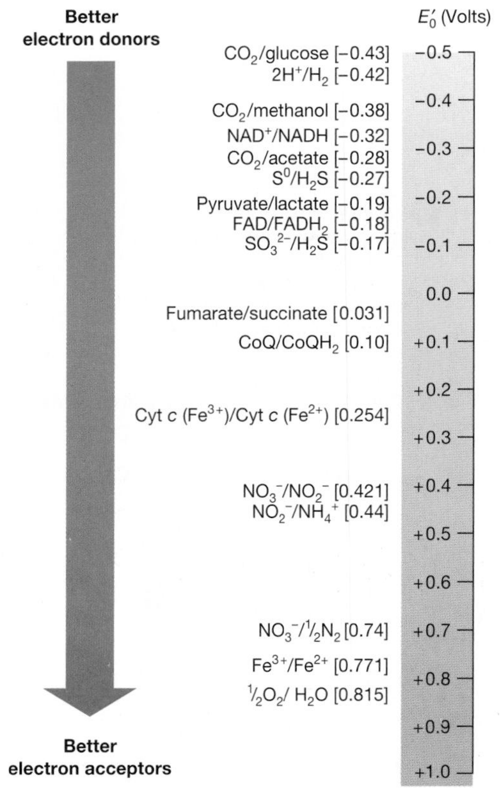

FIGURE 10.9 The Electron Tower. Electrons move spontaneously from donors with more negative reduction potentials to acceptors with more positive reduction potentials.

(**figure 10.8**). The electrons flow from carriers with more negative reduction potentials to those with more positive potentials and eventually combine with O_2 and H^+ to form water. Each carrier is in turn reduced and then reoxidized. Thus the carriers are constantly recycled as electron transport through the chain occurs. The pattern of electron flow is exactly the same as seen in the electron tower that is described in chapter 9 and reproduced here (**figure 10.9**). The electrons move down this potential gradient much like water flowing down a series of rapids. The difference in reduction potentials between O_2 and NADH is large, about 1.14 volts, which makes possible the release of a great deal of energy. The differences in reduction potential at several points in the chain are large enough to provide sufficient energy for ATP production, much as the energy from waterfalls can be harnessed by waterwheels and used to generate electricity. Thus the ETC breaks up the large overall energy release into small steps. As will be seen shortly, electron transport generates proton and electrical gradients. These gradients can drive ATP synthesis and perform other work.

In eukaryotes, the ETC carriers reside within the inner membrane of the mitochondrion where they are arranged into four complexes, each capable of transporting electrons part of the way to O_2 (**figure 10.10**). Coenzyme Q and cytochrome *c* connect the complexes with each other. Although some bacterial and archaeal ETCs resemble the mitochondrial chain, they are frequently very different. First, their ETCs are located within the plasma membrane. They also can be composed of different electron carriers and may be branched. That is, electrons may enter the chain at several points and leave through several terminal oxidases. Bacterial and archaeal ETCs also may be shorter, resulting in the release of less energy. Although microbial ETCs differ in details of construction, they operate using the same fundamental principles. *Electron Transport System and ATP Synthesis*

The ETC of *Paracoccus denitrificans* and *Escherichia coli* will serve as examples of bacterial chains. *P. denitrificans* is a gram-negative, soil bacterium that is a facultative anaerobe and is extremely versatile metabolically. Under oxic conditions, it carries out aerobic respiration using the ETC shown in **figure 10.11**. This chain has four complexes and is very similar to the mitochondrial chain. NADH generated by the oxidation of organic substrates (during glycolysis and the TCA cycle) is oxidized to

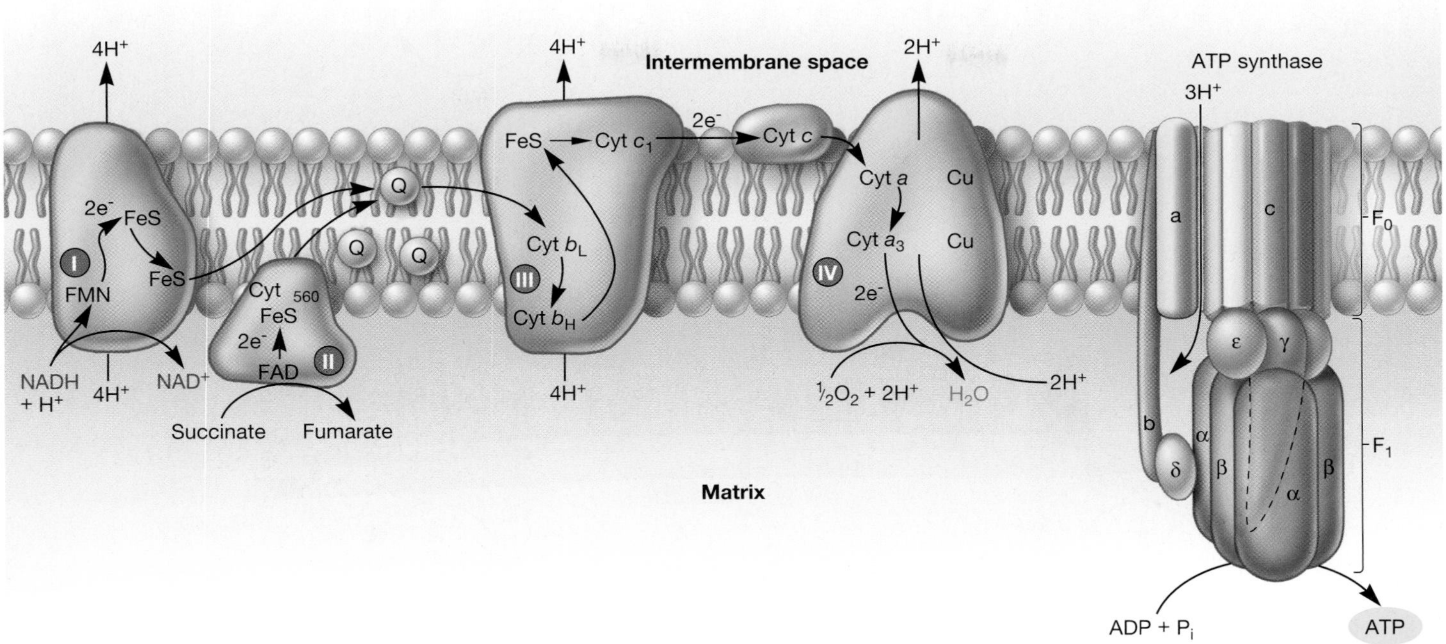

FIGURE 10.10 The Chemiosmotic Hypothesis Applied to Mitochondria. In this scheme, the carriers are organized asymmetrically within the inner membrane so that protons are transported across as electrons move along the chain. Proton release into the intermembrane space occurs when electrons are transferred from carriers, such as FMN and coenzyme Q (Q), that carry both electrons and protons to components such as nonheme iron proteins (FeS proteins) and cytochromes (Cyt) that transport only electrons. Complex IV pumps protons across the membrane as electrons pass from cytochrome *a* to oxygen. Coenzyme Q transports electrons from complexes I and II to complex III. Cytochrome *c* moves electrons between complexes III and IV. The number of protons moved across the membrane at each site per pair of electrons transported is still somewhat uncertain; the current consensus is that at least 10 protons must move outward during NADH oxidation. One molecule of ATP is synthesized and released from the enzyme ATP synthase for every three protons that cross the membrane by passing through it.

NAD^+ by the first component in the ETC, membrane-bound NADH dehydrogenase. The electrons from NADH are transferred to carriers with progressively more positive reduction potentials. As electrons move through the carriers, protons are moved across the plasma membrane to the periplasmic space (i.e., outside the cell), rather than to an intermembrane space, as seen in the mitochondrion (compare figures 10.10 and 10.11). The metabolic diversity of the bacterium is also demonstrated in figure 10.11. When the microbe is growing on an energy source such as glucose, electrons are transferred to the ETC by NADH, as just described. However, *P. denitrificans* is also able to use one-carbon molecules such as methanol as a source of energy. In this case, NADH is not involved, and electrons are donated directly to the chain from this one-carbon compound at the level of cytochrome *c*.

A simplified view of the *E. coli* chain is shown in **figure 10.12**. This chain differs from the mitochondrial chain in part because it contains a different array of cytochromes. Furthermore, *E. coli* has evolved two branches of the ETC that operate under different oxygen levels. When oxygen is readily available, the cytochrome *bo* branch is used (lower half of figure 10.12). When oxygen is less plentiful, the cytochrome *bd* branch is used because it has a higher affinity for oxygen (upper half of figure 10.12). However, it is less efficient than the *bo* branch because the *bd* branch moves fewer protons into the periplasmic space. *Electron Transport Systems and Formation of ATP*

Oxidative Phosphorylation

Oxidative phosphorylation, sometimes called respiratory phosphorylation, is the process by which ATP is synthesized as the result of electron transport driven by the oxidation of a chemical energy source. The mechanism by which oxidative phosphorylation takes place has been studied intensively for years. The most widely accepted hypothesis is the **chemiosmotic hypothesis,** which was formulated by British biochemist Peter Mitchell. According to the chemiosmotic hypothesis, the ETC is organized so that protons move outward from the mitochondrial matrix as electrons are transported down the chain (figure 10.10). In bacteria and archaea, the protons are moved across the plasma membrane from the cytoplasm to the periplasmic space.

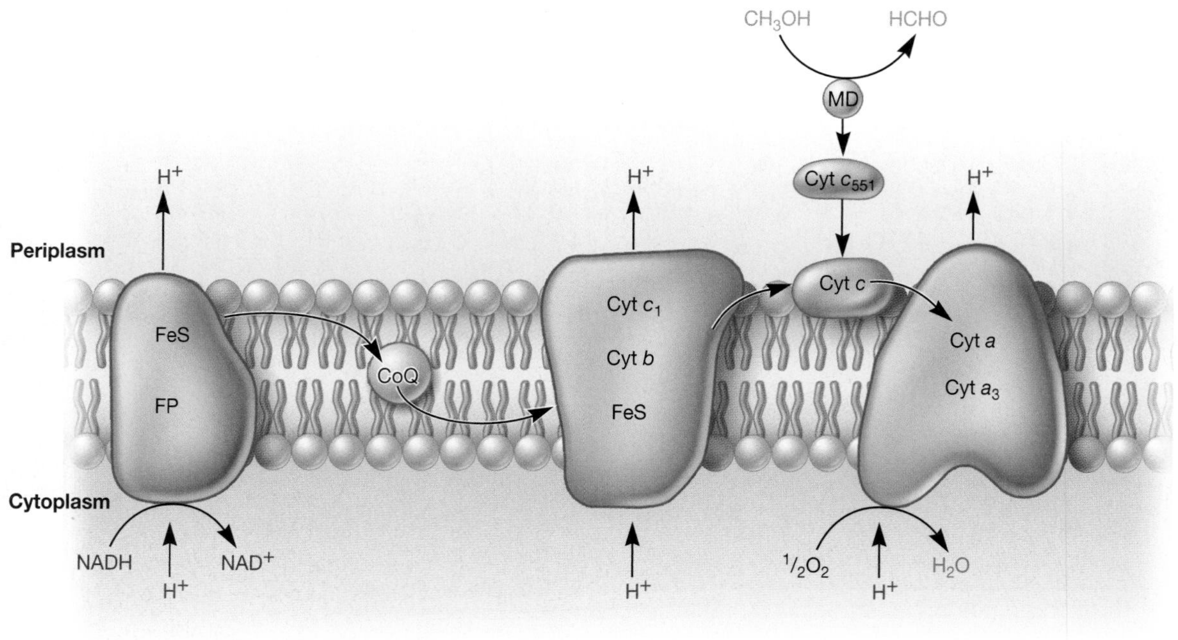

FIGURE 10.11 *Paracoccus denitrificans* Electron Transport Chain Used During Aerobic Respiration. *P. denitrificans* can respire both aerobically and anaerobically (see figure 10.17). The aerobic ETC resembles a mitochondrial ETC and uses oxygen as the terminal electron acceptor. This metabolically versatile bacterium can also use methanol and methylamine as electron donors. They donate electrons to the chain at the level of cytochrome *c*. Flavoprotein (FP), methanol dehydrogenase (MD).

Figure 10.11 Micro Inquiry

How does the number of H^+ moved across the membrane as a result of NADH oxidation compare with the number of H^+ moved following methanol oxidation?

The movement of protons across the membrane is not completely understood. However, in some cases, the protons are actively pumped across the membrane (e.g., by complex IV of the mitochondrial chain; figure 10.10). In other cases, translocation of protons results from the juxtaposition of carriers that accept both electrons and protons with carriers that accept only electrons. For instance, coenzyme Q accepts two electrons and two protons but delivers only two electrons to complex III of the mitochondrial chain. This transfer of electrons is complicated by the fact that cytochrome *b* and the Fe-S protein of complex III accept only one electron at a time and do not accept protons. This difference in protons and electrons carried sets the stage for a phenomenon called the **Q cycle** that ultimately moves four protons across the membrane (**figure 10.13**). The Q cycle involves the sequential oxidation of two reduced coenzyme Q molecules (QH_2), the transfer of two of the four electrons released by these oxidations to one molecule of the oxidized form of coenzyme Q (located nearby in the membrane) via the cytochrome *b* located in complex III, and the transfer of the remaining two electrons to the Fe-S protein of complex III.

The result of proton expulsion during electron transport is the formation of a concentration gradient of protons (Δ pH; chemical potential energy) and a charge gradient ($\Delta\psi$; electrical potential energy). Thus the mitochondrial matrix is more alkaline and more negative than the intermembrane space. Likewise with bacterial and archaeal cells, the cytoplasm is more alkaline and more negative than the periplasmic space. The combined chemical and electrical potential differences make up the **proton motive force (PMF).** The PMF is used to perform work when protons flow back across the membrane, down the concentration and charge gradients, and into the mitochondrial matrix (or the cytoplasm of bacterial and archaeal cells). This flow is exergonic and is often used to phosphorylate ADP to ATP. The PMF is also used to transport molecules into the cell directly (i.e., without the hydrolysis of ATP) and to rotate the

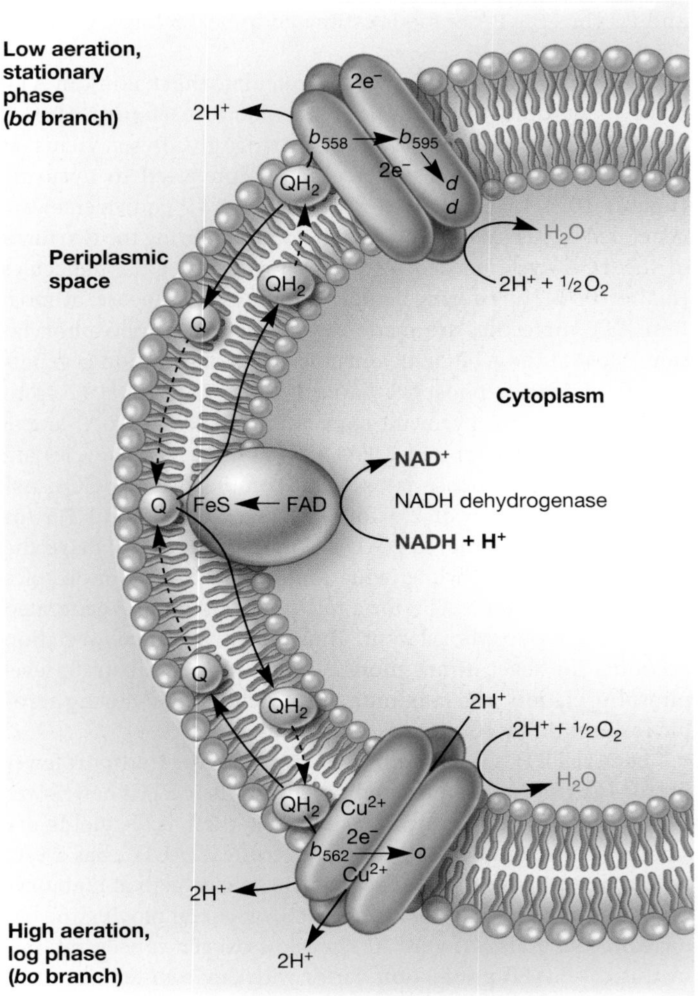

FIGURE 10.12 The Electron Transport Chain of *E. coli*. NADH transfers electrons from the organic energy and electron source to the electron transport chain. Ubiquinone-8 (Q) connects the NADH dehydrogenase with two terminal oxidase systems. The upper branch operates when the bacterium is in stationary phase and there is little oxygen. At least five cytochromes are involved: b_{558}, b_{595}, b_{562}, *d*, and *o*. The lower branch functions when *E. coli* is growing rapidly with good aeration.

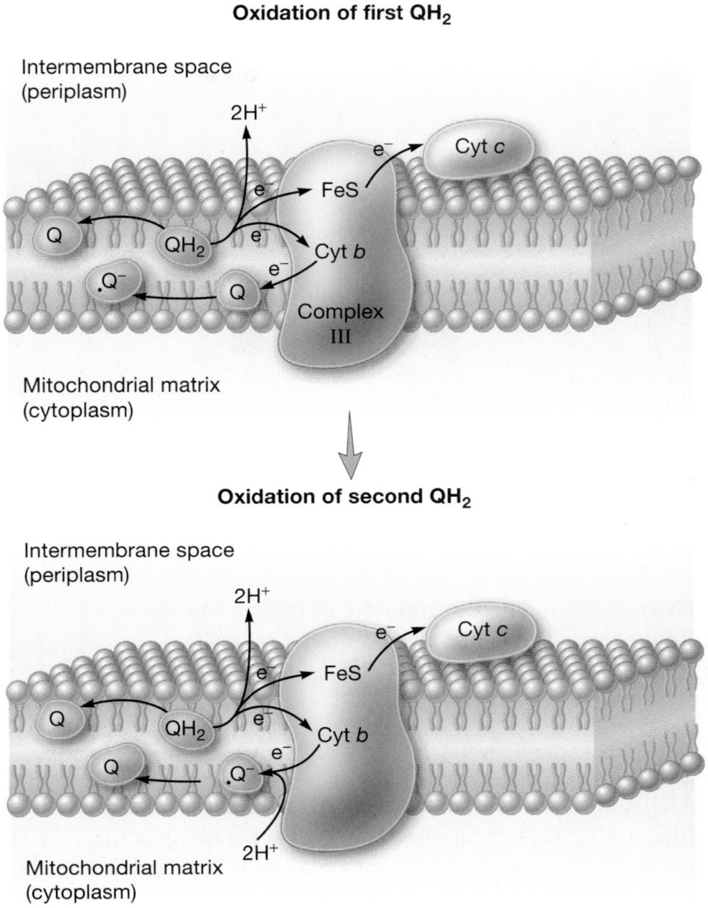

Net reaction:

QH_2 + 2Cyt *c* (oxidized) + $2H^+$ (matrix side)

↓

Q + 2Cyt *c* (reduced) + $4H^+$ (intermembrane space side)

FIGURE 10.13 The Q Cycle. This simplified illustration shows the major events in the Q cycle.

bacterial flagellar motor. Thus the PMF plays a central role in a cell's physiology (**figure 10.14**). ◀◀ *Uptake of nutrients (section 6.6); Flagella (section 3.6)*

The use of PMF for ATP synthesis is catalyzed by **ATP synthase,** a multisubunit enzyme also known as F_1F_0 ATPase because it consists of two components and can catalyze ATP hydrolysis (**figure 10.15**). The mitochondrial F_1 component appears as a spherical structure attached to the mitochondrial inner membrane surface by a stalk. The F_0 component is embedded in the membrane. ATP synthase is on the inner surface of the plasma membrane in bacterial and archaeal cells. F_0 participates in proton movement across the membrane. F_1 is a large complex in which three α subunits alternate with three β subunits. The catalytic sites for ATP synthesis are located on the β subunits. At the center of F_1 is the γ subunit. The γ subunit extends through F_1 and interacts with F_0.

It is now known that ATP synthase functions like a rotary engine, much like the rotary motor of bacterial flagella. It is thought that the flow of protons down the proton gradient through the F_0 subunit causes F_0 and the γ subunit to rotate. As the γ subunit rotates rapidly within the F_1 (much like a car's crankshaft), conformation changes occur in the β subunits (figure 10.15*b*). One conformation change (β_E to β_{HC}) allows entry of

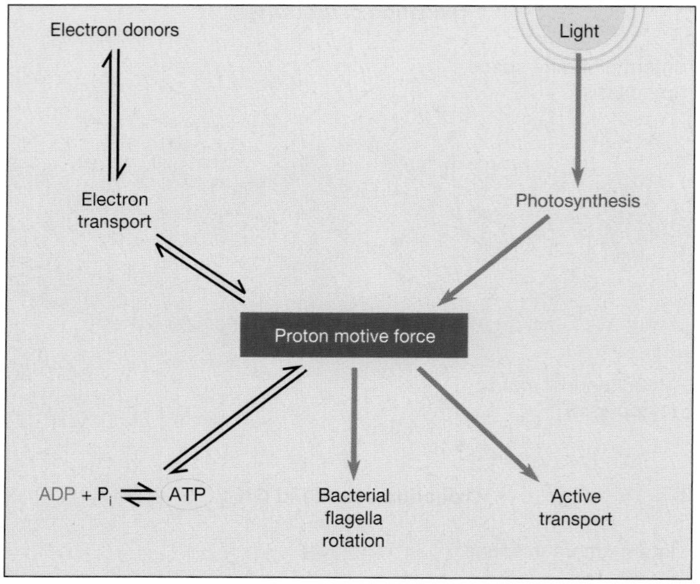

FIGURE 10.14 The Central Role of Proton Motive Force. Active transport is not always driven by PMF.

ADP and P_i into the catalytic site. Another conformation change (β_{HC} to β_{DP}) loosely binds ADP and P_i in the catalytic site. ATP is synthesized when the β_{DP} conformation is changed to the β_{TP} conformation, and ATP is released when β_{TP} changes to the β_E conformation, to start the synthesis cycle anew.

ATP Yield During Aerobic Respiration

It is possible to estimate the number of ATP molecules synthesized per NADH or $FADH_2$ oxidized by an ETC. During aerobic respiration, a pair of electrons from NADH and $FADH_2$ is donated to the ETC and ultimately used to reduce an atom of oxygen to H_2O. This releases enough energy to drive the synthesis of ATP. Prior to the acceptance of the chemiosmotic hypothesis, the concept of the phosphorus to oxygen (P/O) ratio was used as a measure of the number of ATP molecules (phosphorus) generated per oxygen (O) reduced as NADH and $FADH_2$ were oxidized. With the acceptance of the chemiosmotic hypothesis, it was recognized that the important measurement was the number of protons transported across the membrane by NADH oxidation and the number of protons consumed during the synthesis of ATP. Despite this change, the expression of the amount of ATP synthesized as the result of the oxidation of NADH is still often referred to as the P/O ratio. The best estimates of the number of protons moved across the membrane due to NADH oxidation is 10; six electrons are thought to be transferred when $FADH_2$ oxidation initiates electron flow to oxygen. It is currently thought that four protons are consumed during ATP synthesis; three are used by ATP synthase and one is used for transport of ATP, ADP, and P_i. Thus the P/O ratio is estimated to be 2.5 for NADH and 1.5 for $FADH_2$.

Given this information, we can calculate the maximum ATP yield of aerobic respiration in a eukaryote (**figure 10.16**). Substrate-level phosphorylation during glycolysis yields at most two ATP molecules per glucose converted to pyruvate (figures 10.4, 10.5, and 10.6). Two GTP (ATP equivalents) are generated by substrate-level phosphorylation during the two turns of the TCA cycle needed to oxidize two acetyl-CoA molecules (figure 10.7). Thus during aerobic respiration of glucose, at most four ATP molecules are made by substrate-level phosphorylation. Most of the ATP made during aerobic respiration is generated by oxidative phosphorylation. Up to 10 NADH (2 from glycolysis, 2 from pyruvate conversion to acetyl-CoA, and 6 from the TCA cycle) and 2 $FADH_2$ (from the TCA cycle) are generated when glucose is oxidized completely to 6 CO_2. Assuming a P/O ratio of 2.5 for NADH oxidation and 1.5 for $FADH_2$ oxidation, the 10 NADH could theoretically drive the synthesis of 25 ATP, while oxidation of the 2 $FADH_2$ molecules would add another 3 ATP for a maximum of 28 ATP generated via oxidative phosphorylation. Thus oxidative phosphorylation accounts for seven times more ATP than does substrate-level phosphorylation. The maximum total yield of ATP during aerobic respiration by eukaryotes is 32 ATPs.

Bacterial ETCs are often shorter and therefore transport fewer protons across the plasma membrane. Thus their ETCs have lower P/O ratios than eukaryotic chains, and their ATP yields are smaller. For example, *E. coli,* with its truncated ETC, has a P/O ratio around 1.3 when using the cytochrome *bo* path at high oxygen levels and a ratio of only about 0.67 when employing the cytochrome *bd* branch (figure 10.12) at low oxygen concentrations. In this case, ATP production varies with environmental conditions. Perhaps because *E. coli* normally grows in habitats such as the intestinal tract that are very rich in nutrients, it does not have to be particularly efficient in ATP synthesis. Presumably the ETC functions when *E. coli* is in an oxic freshwater environment between hosts.

Two other factors affect the yield of ATP from the catabolism of glucose by aerobic respiration, thus making the theoretical maximum a value that is rarely reached. One is that in bacterial and archaeal cells, the PMF generated by electron transport is used for functions other than ATP synthesis (e.g., bacterial flagella rotation). The second is related to the amphibolic nature of the pathways used to catabolize glucose. Recall that an important function of these pathways is the generation of precursor metabolites for anabolism. For each molecule of glucose degraded, numerous precursor metabolites are made, and as each is made, a microbe must "decide" if that metabolite is needed for anabolism or if it can continue the catabolic process. If the precursor metabolite is used for biosynthesis, fewer NADH and $FADH_2$ molecules may be made and fewer ATP molecules generated. Clearly the flow of the carbons from glucose must be carefully monitored and regulated such that ATP production is appropriately balanced with biosynthesis. The mechanisms for this regulation are described in chapters 9 and 13.

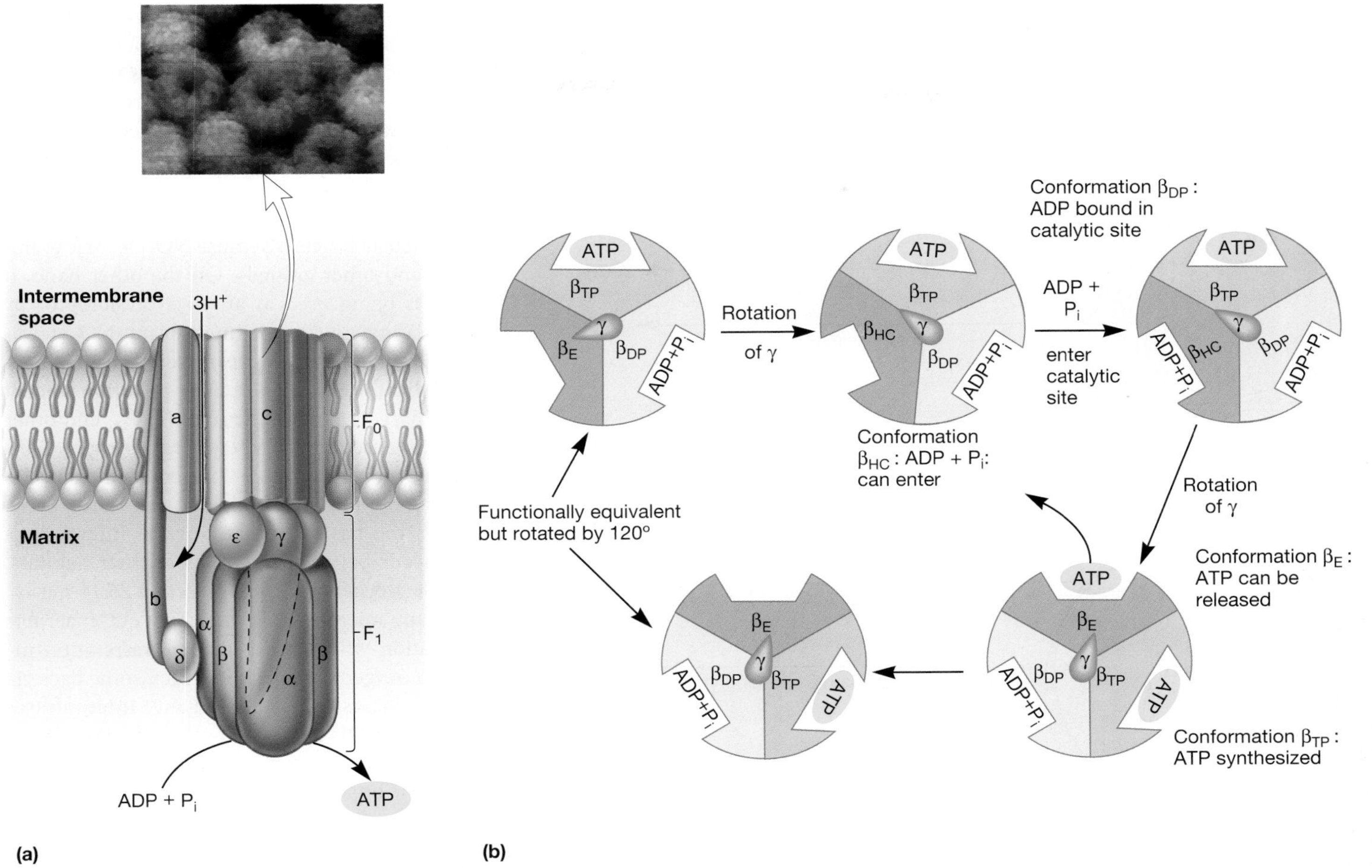

FIGURE 10.15 ATP Synthase Structure and Function. (a) The major structural features of ATP synthase deduced from X-ray crystallography and other studies. F_1 is a spherical structure composed largely of alternating α and β subunits; the three active sites are on the β subunits. The γ subunit extends upward through the center of the sphere and can rotate. The stalk (γ and ε subunits) connects the sphere to F_0, the membrane embedded complex that serves as a proton channel. F_0 contains one a subunit, two b subunits, and 9–12 c subunits. The stator arm is composed of subunit a, two b subunits, and the δ subunit; it is embedded in the membrane and attached to F_1. A ring of c subunits in F_0 is connected to the stalk and may act as a rotor (insert) and move past the a subunit of the stator. As the c subunit ring turns, it rotates the shaft ($\gamma\varepsilon$ subunits). The insert is an atomic force micrograph of the F_0 rotor. (b) The binding change mechanism is a widely accepted model of ATP synthesis. This simplified drawing of the model shows the three catalytic β subunits and the γ subunit, which is located at the center of the F_1 complex. As the γ subunit rotates, it causes conformational changes in each subunit. The β_E (empty) conformation is an open conformation, which does not bind nucleotides. When the γ subunit rotates 30°, β_E is converted to the β_{HC} (half closed) conformation. P_i and ADP can enter the catalytic site when it is in this conformation. The subsequent 90° rotation by the γ subunit is critical because it brings about three significant conformational changes: (1) β_{HC} to β_{DP} (ADP bound), (2) β_{DP} to β_{TP} (ATP bound), and (3) β_{TP} to β_E. Change from β_{DP} to β_{TP} is accompanied by the formation of ATP; change from β_{TP} to β_E allows for release of ATP from ATP synthase.

1. Briefly describe the structure of ETCs and their role in ATP formation. How do mitochondrial chains differ from bacterial and archaeal chains?
2. Describe the current model of oxidative phosphorylation. Briefly describe the structure of ATP synthase and explain how it is thought to function.
3. How do substrate-level phosphorylation and oxidative phosphorylation differ?
4. Calculate the ATP yield when glucose is catabolized completely to six CO_2 by a eukaryotic microbe. How does this value compare to the ATP yield observed for a bacterium? Suppose a bacterium used the Entner-Doudoroff pathway to degrade glucose to pyruvate and then completed the catabolism of glucose via the TCA cycle. How would this affect the total maximum ATP yield? Explain your reasoning.

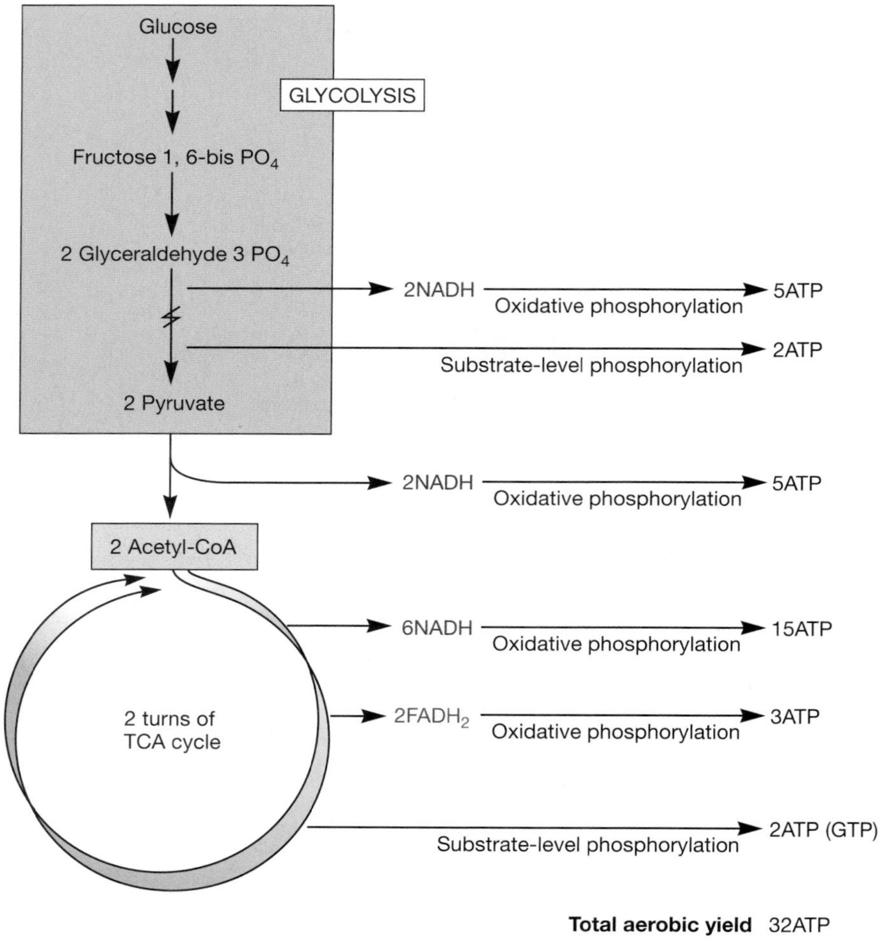

FIGURE 10.16 Maximum Theoretic ATP Yield from Aerobic Respiration. To attain the theoretic maximum yield of ATP, one must assume a P/O ratio of 2.5 for the oxidation of NADH and 1.5 for $FADH_2$.

10.6 Anaerobic Respiration

As we have seen, during aerobic respiration, sugars and other organic molecules are oxidized and their electrons transferred to NAD^+ and FAD to generate NADH and $FADH_2$, respectively. These carriers then donate the electrons to an ETC that uses O_2 as the terminal electron acceptor. However, it is also possible for other terminal electron acceptors to be used for electron transport. **Anaerobic respiration** is the process whereby an exogenous terminal electron acceptor other than O_2 is used for electron transport. It is carried out by many bacteria and archaea, and has recently been identified in a foraminiferan, a eukaryotic microbe. The most common terminal electron acceptors used during anaerobic respiration are nitrate, sulfate, and CO_2, but metals and a few organic molecules can also be reduced (**table 10.1**). ▶▶| Foraminifera *(section 23.4)*

The diversity of alternate electron acceptors has important ecological and practical consequences. For instance, numerous soil microbes use nitrate (NO_3^-) as an alternate electron acceptor in anoxic soils. When this occurs in agricultural soils, it depletes the nitrogen in the soil and decreases the yield from that field. It also forces farmers to use nitrogen-containing fertilizers that can have detrimental consequences such as the contamination of nearby wells, streams, and rivers. The contamination of wells is of particular concern because NO_3^- is toxic to humans and other animals. On the other hand, if NO_3^- is being used as an electron acceptor by bacteria in a sewage treatment plant, NO_3^- levels in the plant effluent are decreased, as is the likelihood that the effluent will cause pollution when it is discharged. Finally, the use of NO_3^- and other electron acceptors by chemoorganotrophs links the carbon cycle to other biogeochemical cycles such as the nitrogen cycle and the sulfur cycle. This is the basis for many of the interactions that occur between microbes in their habitats. ▶▶| *Biogeochemical cycling (section 26.1)*

Many microbes that can carry out anaerobic respiration will perform aerobic respiration instead if oxygen is available. In fact, some bacteria oxidize sugars and other organic molecules to generate NADH and $FADH_2$ using the exact pathways used for aerobic respiration. Thus, depending on the organism, glycolysis and the TCA cycle function in the same way during both aerobic respiration and anaerobic respiration. Furthermore, most of the ATP generated during anaerobic respiration is made by oxidative phosphorylation, as is the case during aerobic respiration.

An example of a bacterium that can carry out both anaerobic respiration and aerobic respiration is *Paracoccus denitrificans.* We have already described the ETC used by this microbe during aerobic respiration (figure 10.11). Under anoxic conditions, *P. denitrificans* uses NO_3^- as its electron acceptor. As shown in **figure 10.17**, the anaerobic ETC is more complex than the aerobic chain. The anaerobic ETC is branched and uses different electron carriers. Electrons are passed from coenzyme Q to cytochrome *b* for the reduction of NO_3^- to nitrite (NO_2^-). Electrons then flow through cytochrome *c* for the sequential reduction of nitrite to gaseous dinitrogen (N_2). Not as many protons are pumped across the membrane during anaerobic growth, but nonetheless, a PMF is established.

The anaerobic reduction of NO_3^- makes it unavailable for assimilation into the cell. Therefore this process is called **dissimilatory nitrate reduction.** As illustrated in figure 10.17, dissimilatory nitrate reduction, also called denitrification, is a multistep process with four enzymes participating: nitrate reductase (Nar), nitrite reductase (Nir), nitric oxide reductase (Nor), and nitrous oxide reductase (Nos).

$$NO_3^- \rightarrow NO_2^- \rightarrow NO \rightarrow N_2O \rightarrow N_2$$

Table 10.1 Some Electron Acceptors Used in Respiration

	Electron Acceptor	Reduced Products	Examples of Microorganisms
Aerobic	O_2	H_2O	All aerobic bacteria, fungi, and protists
Anaerobic	NO_3^-	NO_2^-	Enteric bacteria
	NO_3^-	NO_2^-, N_2O, N_2	*Pseudomonas, Bacillus,* and *Paracoccus*
	SO_4^{2-}	H_2S	*Desulfovibrio* and *Desulfotomaculum*
	CO_2	CH_4	Methanogens
	CO_2	Acetate	Acetogens
	S^0	H_2S	*Desulfuromonas* and *Thermoproteus*
	Fe^{3+}	Fe^{2+}	*Pseudomonas, Bacillus,* and *Geobacter*
	$HAsO_4^{2-}$	$HAsO_2$	*Bacillus, Desulfotomaculum, Sulfurospirillum*
	SeO_4^{2-}	Se, $HSeO_3^-$	*Aeromonas, Bacillus, Thauera*
	Fumarate	Succinate	*Wolinella*

In *P. denitrificans,* the membrane-bound enzyme nitrate reductase catalyzes the reduction of NO_3^- to NO_2^-. Nitrite is reduced to nitric oxide (NO) by the periplasmic enzyme nitrite reductase. Nitric oxide reductase catalyzes the formation of nitrous oxide (N_2O) from NO. It is part of the membrane-bound cytochrome *bc* complex. Finally, the periplasmic enzyme nitrous oxide reductase catalyzes the formation of N_2 from N_2O.

In addition to *P. denitrificans,* some members of the genera *Pseudomonas* and *Bacillus* carry out denitrification. All three genera use denitrification as an alternative to aerobic respiration and are considered facultative anaerobes. Indeed, if O_2 is present, the synthesis of nitrate reductase is repressed and these bacteria use aerobic respiration.

Not all microbes employ anaerobic respiration facultatively. Some are obligate anaerobes that carry out only anaerobic respiration. The methanogens are an example. These archaea use CO_2 or carbonate as a terminal electron acceptor. They are called methanogens because the electron acceptor is reduced to methane (*see figure 18.15*). Bacteria such as *Desulfovibrio* are another example. They donate eight electrons to sulfate (SO_4^{2-}), reducing it to sulfide (S^{2-} or H_2S; *see figure 20.32*).

$$SO_4^{2-} + 8e^- + 8H^+ \rightarrow S^{2-} + 4H_2O$$

It should be noted that both methanogens and *Desulfovibrio* are able to function as chemolithotrophs, using H_2 as an energy source (section 10.11).

Anaerobic respiration is not as efficient in ATP synthesis as is aerobic respiration. Reduction in ATP yield arises from the fact that alternate electron acceptors such as NO_3^- have less positive reduction potentials than O_2. The difference in standard reduction potentials between NADH and NO_3^- is smaller than the difference between NADH and O_2 (figure 10.9). Less energy is available to make ATP in anaerobic respiration because energy yield is directly related to the magnitude of the reduction potential difference. Nevertheless, anaerobic respiration is useful because it allows ATP synthesis by electron transport and oxidative phosphorylation in the absence of O_2.

Anaerobic respiration is prevalent in oxygen-depleted soils and sediments. There, a succession of microorganisms is often observed because of the presence of several electron acceptors. For example, if O_2, nitrate, manganese ion, ferric ion, sulfate, and CO_2 are available, a predictable sequence of electron acceptor use takes place. Oxygen is employed as an electron acceptor first because it inhibits nitrate use by microorganisms capable of respiration with either O_2 or nitrate. While O_2 is available, methanogens are inhibited because they are obligate anaerobes. Once the O_2 and nitrate are exhausted and fermentation products (section 10.7) have accumulated, competition for use of other electron acceptors begins. Manganese and iron are used first, followed by competition between sulfate reducers and methanogens. ▶▶| *Microorganisms in benthic marine environments (section 28.2); The subsurface biosphere (section 29.4)*

1. Describe the process of anaerobic respiration. Does anaerobic respiration yield as much ATP as aerobic respiration? Why or why not?
2. What is denitrification? Why do farmers dislike this process?
3. *E. coli* can use O_2, fumarate, or nitrate as a terminal electron acceptor under different conditions. What is the order of energy yield from highest to lowest for these electron acceptors? Explain your answer in thermodynamic terms.

10.7 Fermentation

Despite the tremendous ATP yield obtained by oxidative phosphorylation, some chemoorganotrophic microbes do not respire. These microbes lack ETCs or they repress the synthesis of ETC components under anoxic conditions, making anaerobic respiration impossible. Some facultative anaerobes may find themselves in an environment that lacks both oxygen and the terminal electron acceptor they use for anaerobic respiration (e.g., nitrate). In each case, NADH produced by the Embden-Meyerhof pathway

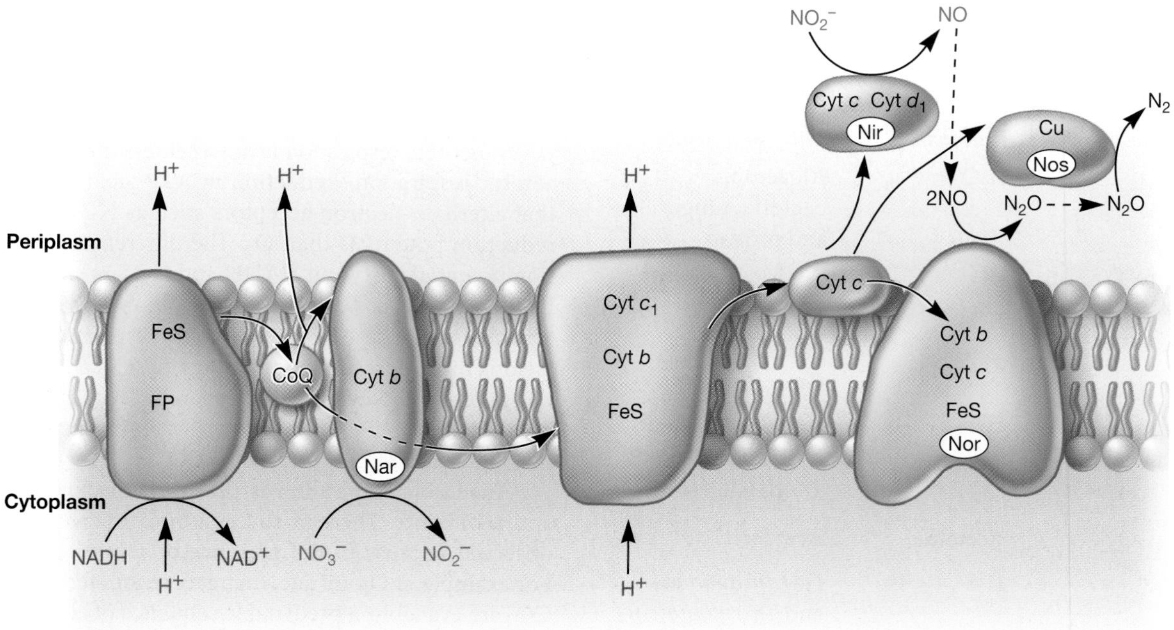

FIGURE 10.17 ***Paracoccus denitrificans*** **Electron Transport Chain Used During Anaerobic Respiration.** This branched ETC is made of both membrane and periplasmic proteins. Nitrate is reduced to diatomic nitrogen by the collective action of four different reductases that receive electrons from CoQ and cytochrome *c*. Locations of proton movement are shown, but the number of protons involved has not been indicated. Abbreviations used: flavoprotein (FP), nitrate reductase (Nar), nitrite reductase (Nir), nitric oxide reductase (Nor), and nitrous oxide reductase (Nos).

reactions during glycolysis (figure 10.4) must still be oxidized back to NAD^+ without the help of an ETC. If NAD^+ is not regenerated, the oxidation of glyceraldehyde 3-phosphate will cease and glycolysis will stop. Many microorganisms solve this problem by slowing or stopping pyruvate dehydrogenase activity and using pyruvate or one of its derivatives as an electron acceptor for the reoxidation of NADH in a fermentation process (**figure 10.18**). There are many kinds of fermentations, and they often are characteristic of particular microbial groups (**figure 10.19**). A few of the more common fermentations are introduced here.

Four unifying themes should be kept in mind when microbial fermentations are examined: (1) NADH is oxidized to NAD^+; (2) O_2 is not needed; (3) the electron acceptor is often either pyruvate or a pyruvate derivative; and (4) an ETC cannot operate, reducing the ATP yield per glucose significantly. Thus in fermentation, the substrate (e.g., glucose) is only partially oxidized and ATP is formed in most organisms exclusively by substrate-level phosphorylation. However, fermenting microbes still need a PMF to do work, in particular to drive transport. To solve the dilemma of how to generate a PMF without an ETC, they use their ATP synthase in the reverse direction. That is, the ATP synthase pumps protons out of the cell, fueling this transport by the energy released when ATP is hydrolyzed to ADP and P_i. Although fermentation yields considerably less ATP, it is an important component of the metabolic repertoire of many microbes, allowing them to adjust to changes in their habitats.

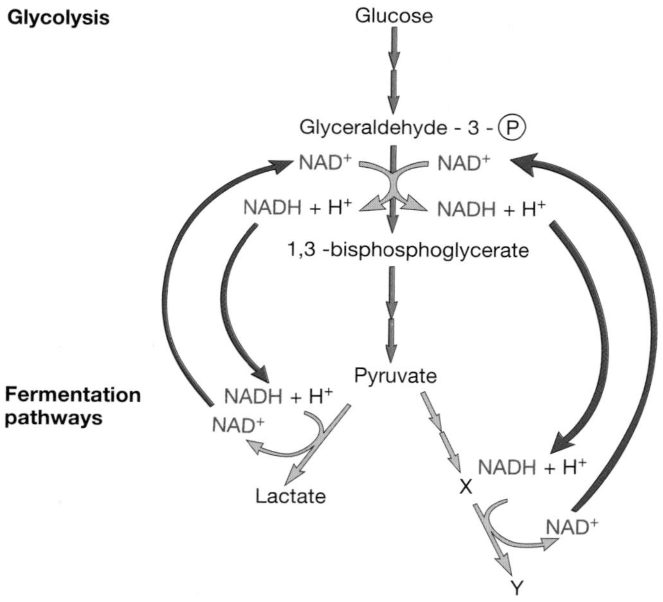

FIGURE 10.18 Reoxidation of NADH During Fermentation. NADH from glycolysis is reoxidized by being used to reduce pyruvate or a pyruvate derivative (X). Either lactate or reduced product Y result.

Figure 10.18 Micro Inquiry

How many NADH are reoxidized to NAD^+ for each glucose catabolized?

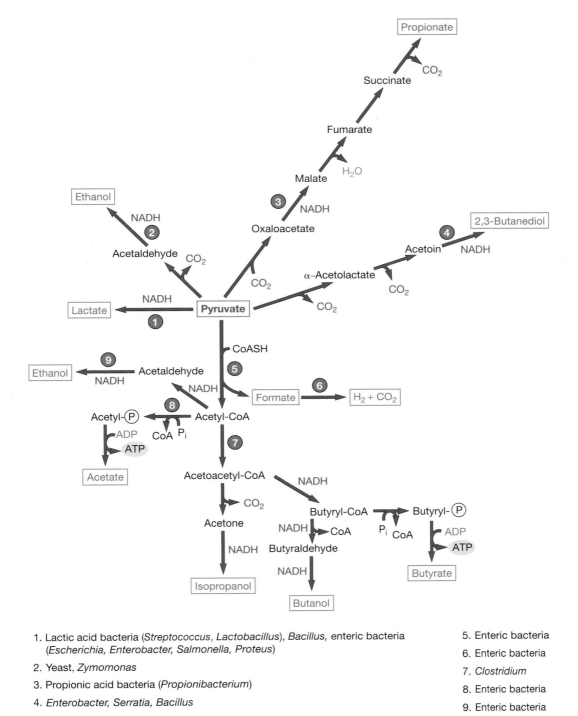

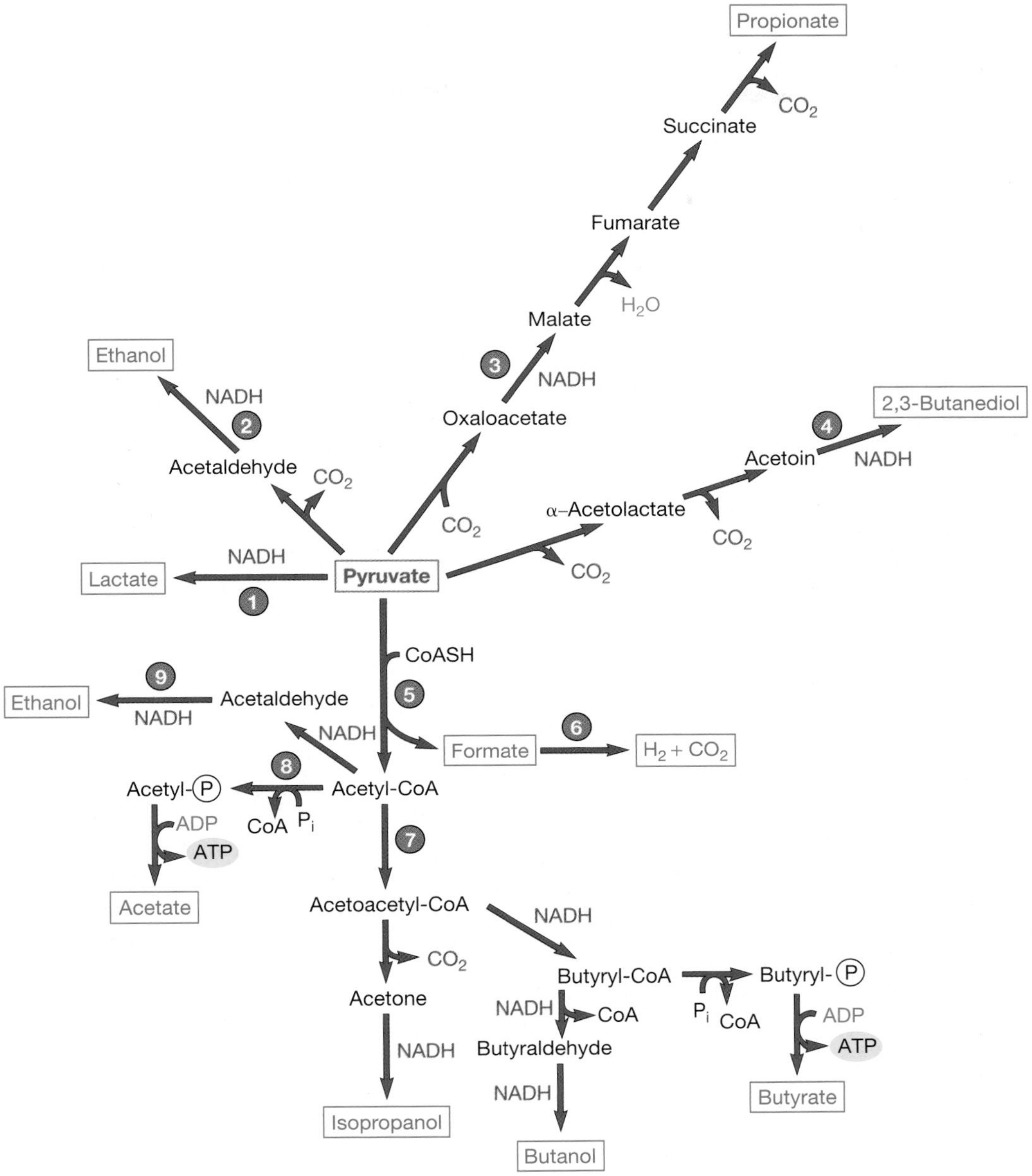

1. Lactic acid bacteria (*Streptococcus, Lactobacillus*), *Bacillus,* enteric bacteria (*Escherichia, Enterobacter, Salmonella, Proteus*)
2. Yeast, *Zymomonas*
3. Propionic acid bacteria (*Propionibacterium*)
4. *Enterobacter, Serratia, Bacillus*
5. Enteric bacteria
6. Enteric bacteria
7. *Clostridium*
8. Enteric bacteria
9. Enteric bacteria

FIGURE 10.19 Some Common Microbial Fermentations. Only pyruvate fermentations are shown for the sake of simplicity; many other organic molecules can be fermented. Most of these pathways have been simplified by deletion of one or more steps and intermediates. Pyruvate and major end products are shown in color.

Fermentation pathways are named after the major acid or alcohol produced by the particular microbe. The most common fermentation is lactic acid (lactate) fermentation, the reduction of pyruvate to lactate (figure 10.19, number 1). It is present in bacteria (lactic acid bacteria, *Bacillus*), protists (*Chlorella* and some water molds), and in animal skeletal muscle. Lactic acid fermenters can be separated into two groups. **Homolactic fermenters** use the Embden-Meyerhof pathway and directly reduce almost all their pyruvate to lactate with the enzyme lactate dehydrogenase. **Heterolactic fermenters** form substantial amounts of products other than lactate; many also produce ethanol and CO_2. Lactic acid bacteria are important to the food industry, where they are used to make a variety of fermented foods (e.g., yogurt). ▶▶| *Microbiology of fermented food (section 40.5)*

Another common fermentation is carried out by many fungi, protists, and some bacteria. These microbes ferment sugars to ethanol and CO_2 in a process called **alcoholic fermentation.** Pyruvate is decarboxylated to acetaldehyde, which is then reduced to ethanol by alcohol dehydrogenase with NADH as the electron donor (figure 10.19, number 2). This fermentation produces the ethanol desired by brewers making beer and ale and by vintners making wine.

Many bacteria, especially members of the family *Enterobacteriaceae,* can metabolize pyruvate to numerous products using several pathways simultaneously. One such complex fermentation is the **mixed acid fermentation,** which results in the excretion of ethanol and a mixture of acids, particularly acetic, lactic, succinic, and formic acids **(table 10.2)**. Members of the genera *Escherichia, Salmonella,* and *Proteus* carry out mixed acid fermentation. They use pathways numbered 1, 5, 8, and 9 in figure 10.19 to make all the fermentation products, except succinate. If the enzyme formic hydrogenlyase is present (figure 10.19, number 6), formate will be degraded to H_2 and CO_2 (*see figure 20.29*). Another complex fermentation is the **butanediol fermentation,** which is characteristic of *Enterobacter, Serratia, Erwinia,* and some species of *Bacillus.* The predominant pathway used during this fermentation process yields butanediol (figure 10.19, number 4). However, large amounts of ethanol also are produced (figure 10.19, number 9), as are smaller amounts of lactic acid (figure 10.19, number 1) and formic acid (figure 10.19, number 5); in some bacteria, the formic acid is further catabolized to H_2 and CO_2 (figure 10.19, number 6). The use of either mixed acid fermentation or butanediol fermentation is important in differentiating members of the *Enterobacteriaceae,* as described in chapter 35. ▶▶| *Order* Enterobacteriales *(section 20.3); Class* Bacilli *(section 21.3)*

Microorganisms carry out a vast array of fermentations using organic substrates other than glucose. Protozoa and fungi ferment a variety of sugars to lactate, ethanol, glycerol, succinate, formate, acetate, butanediol, and additional products. Some members of the genus *Clostridium* ferment mixtures of amino acids. Proteolytic clostridia such as the pathogens *C. sporogenes* and *C. botulinum* carry out the Stickland reaction in which one amino acid is oxidized and a second amino acid acts as the electron acceptor (*see figure 21.8*). Some ATP is formed by substrate-level phosphorylation, and the fermentation is quite useful for growing in anoxic, protein-rich environments. Other bacteria ferment amino acids by different mechanisms. In addition to sugars and amino acids, organic acids such as acetate, lactate, propionate, and citrate can be fermented. Some of these fermentations are of great practical importance. For example, citrate can be converted to diacetyl and give flavor to fermented milk. ▶▶| *Class* Clostridia *(section 21.2); Microbiology of fermented foods (section 40.5)*

Table 10.2 Mixed Acid Fermentation Products of *Escherichia coli*

	Fermentation Balance (μM Product/100 μM Glucose)	
	Acid Growth (pH 6.0)	*Alkaline Growth (pH 8.0)*
Ethanol	50	50
Formic acid	2	86
Acetic acid	36	39
Lactic acid	80	70
Succinic acid	11	15
Carbon dioxide	88	2
Hydrogen gas	75	0.5

1. What are fermentations and why are they useful to many microorganisms?
2. How do the electron acceptors used in fermentation differ from the terminal electron acceptors used during aerobic respiration and anaerobic respiration?
3. Briefly describe alcoholic, lactic acid, mixed acid, and butanediol fermentations. How do homolactic fermenters and heterolactic fermenters differ? How do mixed acid fermenters and butanediol fermenters differ?
4. What is the net yield of ATP during homolactic, acetate, and butyrate fermentations? How do these yields compare to aerobic respiration in terms of both quantity and mechanism of phosphorylation?
5. When bacteria carry out fermentation, only a few reactions of the TCA cycle operate. What purpose do you think these reactions might serve? Why do you think some parts of the cycle are shut down?

10.8 Catabolism of Other Carbohydrates

Thus far our main focus has been on the catabolism of glucose. However, microorganisms can catabolize many other carbohydrates. These carbohydrates may come either from outside the cell or from internal sources generated during normal metabolism. Often the initial steps in the degradation of external carbohydrate polymers differ from those employed with internal reserves.

Carbohydrates

Figure 10.20 outlines some catabolic pathways for the monosaccharides (single sugars) glucose, fructose, mannose, and galactose. The first three are phosphorylated using ATP and easily enter the Embden-Meyerhof pathway. In contrast, galactose must be converted to uridine diphosphate galactose (UDP-gal, *see*

Monosaccharide interconversions

Galactose

Galactose-1-℗

UDP galactose

UDP glucose

Gal-1-℗

UDP Gal

Glucose-1-℗ → Glucose-6-℗

Glucose

Mannose

Mannose-6-℗

Pentose phosphate pathway

Entner-Doudoroff pathway

Fructose-6-℗ Fructose

Embden-Meyerhof pathway

Disaccharide cleavage

1. Maltose + H₂O —Maltase→ 2 glucose

 Maltose + Pᵢ —Maltose phosphorylase→ β-D-glucose-1-℗ + Glucose

2. Sucrose + H₂O —Sucrase→ Glucose + Fructose

 Sucrose + Pᵢ —Sucrose phosphorylase→ α-D-glucose-1-℗ + Fructose

3. Lactose + H₂O —β-galactosidase→ Galactose + Glucose

4. Cellobiose + Pᵢ —Cellobiose phosphorylase→ α-D-glucose-1-℗ + Glucose

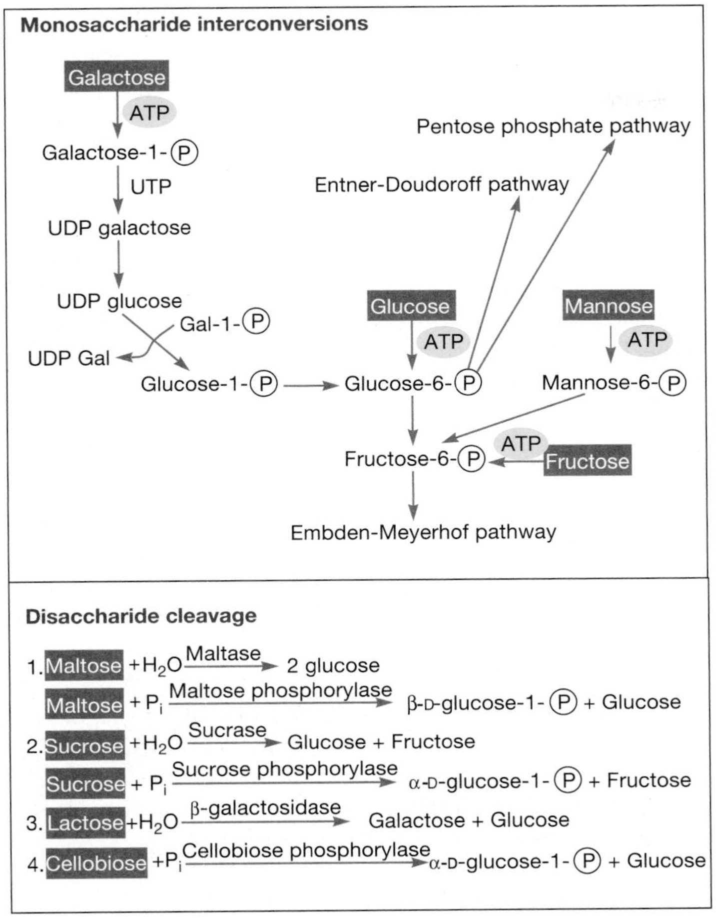

FIGURE 10.20 Carbohydrate Catabolism. Examples of enzymes and pathways used in disaccharide and monosaccharide catabolism. UDP is an abbreviation for uridine diphosphate.

Figure 10.20 Micro Inquiry

What is the difference between a hydrolase and phosphorylase?

figure 11.11) after initial phosphorylation, then converted to glucose 6-phosphate in a three-step process.

The common disaccharides are cleaved to monosaccharides by at least two mechanisms (figure 10.20). Maltose, sucrose, and lactose can be directly hydrolyzed to their constituent sugars. Many disaccharides (e.g., maltose, cellobiose, and sucrose) are also split by a phosphate attack on the bond joining the two sugars, a process called phosphorolysis. This yields the two constituent monosaccharides, one of which is phosphorylated.

Polysaccharides, like disaccharides, are cleaved by both hydrolysis and phosphorolysis. Bacteria, archaea, and fungi degrade external polysaccharides by secreting hydrolytic enzymes. These exoenzymes cleave polysaccharides that are too large to cross the plasma membrane into smaller molecules that can then be assimilated. Starch and glycogen are hydrolyzed by amylases to glucose, maltose, and other products. Cellulose is more difficult to digest; many fungi and a few bacteria (some gliding bacteria, clostridia, and actinomycetes) produce extracellular cellulases that hydrolyze cellulose to cellobiose and glucose. Some actinomycetes and members of the bacterial genus *Cytophaga,* isolated from marine habitats, excrete an agarase that degrades agar. Many soil bacteria and bacterial plant pathogens degrade pectin, a polymer of galacturonic acid (a galactose derivative) that is an important constituent of plant cell walls and tissues. Lignin, another important component of plant cell walls, is usually degraded only by certain fungi that release peroxide-generating enzymes (*see figure 29.2*). ▶▶| *Soil as a microbial habitat (section 29.1)*

Reserve Polymers

Microorganisms often survive for long periods in the absence of exogenous nutrients. Under such circumstances, they catabolize intracellular stores of glycogen, starch, poly-β-hydroxybutyrate, and other carbon and energy reserves. Glycogen and starch are degraded by phosphorylases. Phosphorylases catalyze a phosphorolysis reaction that shortens the polysaccharide chain by one glucose and yields glucose 1-phosphate.

$$(Glucose)_n + P_i \rightarrow (glucose)_{n-1} + glucose\text{-}1\text{-}P$$

Glucose 1-phosphate can enter glycolytic pathways by way of glucose 6-phosphate (figure 10.20).

Poly-β-hydroxybutyrate (PHB) is an important, widespread reserve material. Its catabolism has been studied most thoroughly in the soil bacterium *Azotobacter.* This bacterium hydrolyzes PHB to 3-hydroxybutyrate, then oxidizes the hydroxybutyrate to acetoacetate. Acetoacetate is converted to acetyl-CoA, which can be oxidized in the TCA cycle (figure 10.7).

10.9 Lipid Catabolism

Chemoorganotrophic microorganisms frequently use lipids as energy sources. Triglycerides (also called triacylglycerols) are esters of glycerol and fatty acids that are common energy sources and will serve as our examples (**figure 10.21**). They can be hydrolyzed to glycerol and fatty acids by microbial lipases. The glycerol is then phosphorylated, oxidized to dihydroxyacetone phosphate, and catabolized in the Embden-Meyerhof pathway (figure 10.4).

$$CH_2 - O - \overset{\overset{O}{\|}}{C} - R_1$$
$$CH - O - \overset{\overset{O}{\|}}{C} - R_2$$
$$CH_2 - O - \overset{\overset{O}{\|}}{C} - R_3$$

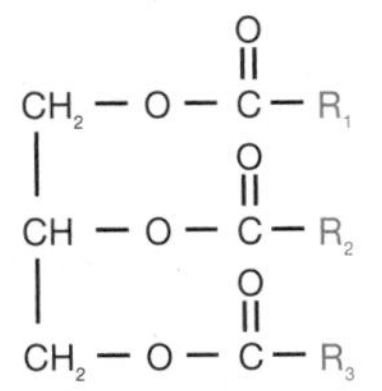

FIGURE 10.21 A Triacylglycerol (Triglyceride). The R groups represent the fatty acid side chains.

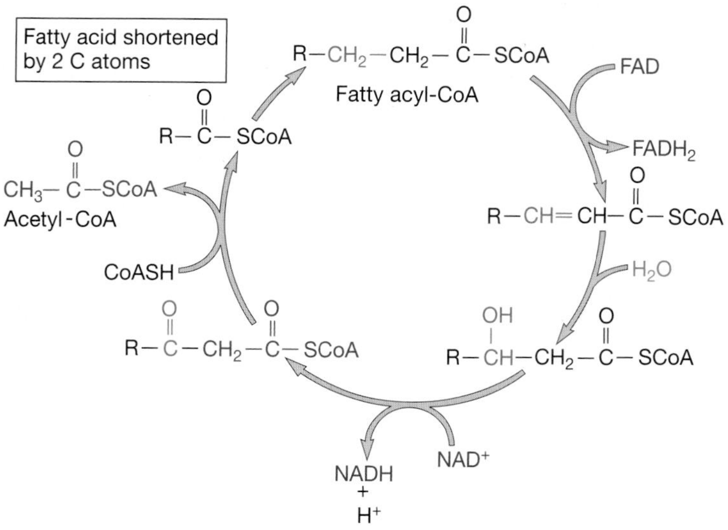

FIGURE 10.22 Fatty Acid β-Oxidation. The portions of the fatty acid being modified are shown in red.

Figure 10.22 Micro Inquiry

Why are fatty acids a rich source of energy even though no ATP is generated when they are degraded by the β-oxidation pathway?

Fatty acids from triacylglycerols and other lipids are often oxidized in the **β-oxidation pathway** after conversion to coenzyme A esters (**figure 10.22**). In this pathway, fatty acids are shortened by two carbons with each turn of the cycle. The two carbon units are released as acetyl-CoA, which can be fed into the TCA cycle or used in biosynthesis. One turn of the cycle produces acetyl-CoA, NADH, and $FADH_2$; NADH and $FADH_2$ can be oxidized by an ETC to provide more ATP. Fatty acids are a rich source of energy for microbial growth. In a similar fashion, some microorganisms grow well on petroleum hydrocarbons under oxic conditions.

10.10 Protein and Amino Acid Catabolism

Some bacteria and fungi—particularly pathogenic, food spoilage, and soil microorganisms—use proteins as their source of carbon and energy. They secrete enzymes called **proteases** that hydrolyze proteins to amino acids, which are transported into the cell and catabolized.

The first step in amino acid catabolism is **deamination,** the removal of the amino group from an amino acid. This is often accomplished by **transamination.** The amino group is transferred from an amino acid to an α-keto acid acceptor (**figure 10.23**). The organic acid resulting from deamination can be converted to

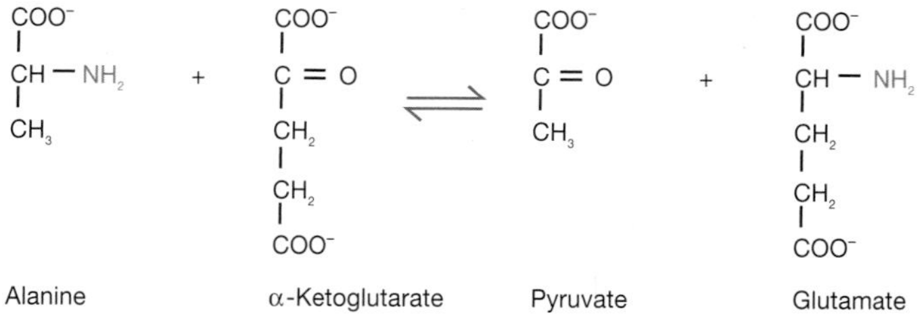

FIGURE 10.23 Transamination. A common example of this process. The α-amino group (blue) of alanine is transferred to the acceptor α-ketoglutarate, forming pyruvate and glutamate. The pyruvate can be fermented, catabolized in the tricarboxylic acid cycle, or used in biosynthesis.

pyruvate, acetyl-CoA, or a TCA cycle intermediate. Depending on the organic acid, it can be fermented or further oxidized in the TCA cycle to release energy. It also can be used as a source of carbon for the synthesis of cell constituents. Excess nitrogen from deamination may be excreted as ammonium ion.

1. Briefly discuss the ways in which microorganisms degrade and use common monosaccharides, disaccharides, and polysaccharides from both external and internal sources.
2. Can members of the genus *Cytophaga* be cultured on standard solidified media? Why or why not?
3. Describe how a microorganism might derive carbon and energy from the lipids and proteins in its diet. What is β-oxidation? Deamination? Transamination?

10.11 Chemolithotrophy

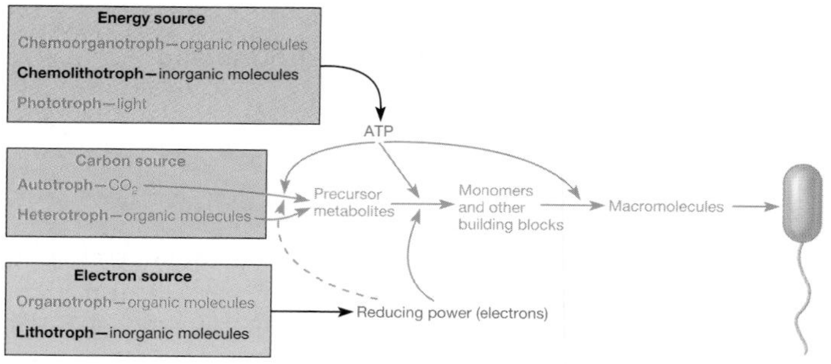

So far, we have considered microbes that synthesize ATP with the energy liberated by the oxidation of organic substrates such as carbohydrates, lipids, and proteins. Certain bacteria and archaea are **chemolithotrophs.** These microbes donate electrons to their ETCs by oxidizing inorganic molecules rather than organic nutrients (**figure 10.24**). Each species is specific in its preferences for electron donors and acceptors (**table 10.3**). The most

Table 10.3 Representative Chemolithotrophs and Their Energy Sources

Bacteria	*Electron Donor*	*Electron Acceptor*	*Products*
Alcaligenes, Hydrogenophaga, and *Pseudomonas* spp.	H_2	O_2	H_2O
Nitrobacter	NO_2^-	O_2	NO_3^-, H_2O
Nitrosomonas	NH_4^+	O_2	NO_2^-, H_2O
Thiobacillus denitrificans	S^0, H_2S	NO_3^-	SO_4^{2-}, N_2
Acidithiobacillus ferrooxidans	Fe^{2+}, S^0, H_2S	O_2	Fe^{3+}, H_2O, H_2SO_4

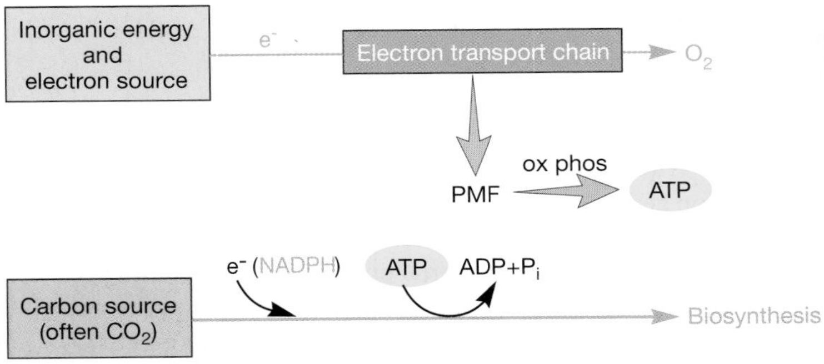

FIGURE 10.24 Chemolithotrophic Fueling Processes. Chemolithotrophic bacteria and archaea oxidize inorganic molecules (e.g., H_2S and NH_3), which serve as energy and electron sources. The electrons released pass through an electron transport chain, generating a proton motive force (PMF). ATP is synthesized by oxidative phosphorylation (ox phos). Most chemolithotrophs use O_2 as the terminal electron acceptor. However, some can use other exogenous molecules as terminal electron acceptors. Note that a molecule other than the energy source provides carbon for biosynthesis. Many chemolithotrophs are autotrophs.

Table 10.4 Energy Yields from Oxidations Used by Chemolithotrophs

Reaction	$\Delta G^{\circ\prime}$ *(kcal/mole)*[a]
$H_2 + 1/2\ O_2 \rightarrow H_2O$	−56.6
$NO_2^- + 1/2\ O_2 \rightarrow NO_3^-$	−17.4
$NH_4^+ + 1\ 1/2\ O_2 \rightarrow NO_2^- + H_2O + 2H^+$	−65.0
$S^0 + 1\ 1/2\ O_2 + H_2O \rightarrow H_2SO_4$	−118.5
$S_2O_3^{2-} + 2O_2 + H_2O \rightarrow 2SO_4^{2-} + 2H^+$	−223.7
$2Fe^{2+} + 2H^+ + 1/2\ O_2 \rightarrow 2Fe^{3+} + H_2O$	−11.2

[a] The $\Delta G^{\circ\prime}$ for complete oxidation of glucose to CO_2 is −686 kcal/mole. A kcal is equivalent to 4.184 kJ.

common energy sources (electron donors) are hydrogen, reduced nitrogen compounds, reduced sulfur compounds, and ferrous iron (Fe^{2+}). The acceptor is usually O_2, but sulfate and nitrate may also be used.

Much less energy is available from the oxidation of inorganic molecules than from the complete oxidation of glucose to CO_2 (**table 10.4**). This is because the reduction potentials of most of the inorganic substrates used by chemolithotrophs are much more positive than the reduction potential of organic molecules such as glucose (figure 10.9). The low yield of ATP means that chemolithotrophs must oxidize a large quantity of inorganic material to grow and reproduce. This is particularly true of chemolithoautotrophs, those chemolithotrophs that fix CO_2 into carbohydrates. CO_2-fixation pathways consume considerable amounts of ATP and reducing power (usually NADPH). Because they must consume a large amount of inorganic material, chemolithotrophs have significant ecological impact. They make important contributions to several biogeochemical cycles, including the nitrogen, sulfur, and iron cycles. ▶▶| *CO_2 fixation (section 11.3); Biogeochemical cycling (section 26.1)*

Several bacterial and archaeal genera can oxidize hydrogen gas using a hydrogenase enzyme (table 10.3).

$$H_2 \rightarrow 2H^+ + 2e^-$$

The $H_2/2H^+$ conjugate redox pair has a very negative standard reduction potential, and its electrons can be donated to either an ETC or NAD^+, depending on the hydrogenase. If NADH is produced, it can be used in ATP synthesis by electron transport and oxidative phosphorylation, with O_2, Fe^{3+}, S^0, and carbon monoxide (CO) as the terminal electron acceptors. Often these hydrogen-oxidizing microorganisms use organic compounds as energy sources when available.

Some bacteria and archaea use the oxidation of nitrogenous compounds as a source of electrons. Among these chemolithotrophs, the **nitrifying bacteria** are best understood. They are soil and aquatic bacteria of considerable ecological significance that carry out **nitrification**—the oxidation of ammonia to nitrate (*see figure 20.13*). Nitrification is a two-step process that depends on the activity of at least two different genera. In the first step, ammonia is oxidized to nitrite by a number of genera, including *Nitrosomonas:*

$$NH_4^+ + 1\ 1/2O_2 \rightarrow NO_2^- + H_2O + 2H^+$$

MICROBIAL DIVERSITY & ECOLOGY

10.1 Acid Mine Drainage

Each year millions of tons of sulfuric acid flow to the Ohio River from the coal mines of the Appalachian Mountains. This sulfuric acid is of microbial origin and leaches enough metals from the mines to make the river reddish and acidic. The primary culprit is *Acidithiobacillus ferrooxidans,* a chemolithotrophic bacterium that derives its energy from oxidizing ferrous ion to ferric ion and sulfide ion to sulfate ion. The combination of these two energy sources is important because of the solubility properties of iron. Ferrous ion is somewhat soluble and can be formed at pH values of 3.0 or less in moderately reducing environments. However, when the pH is greater than 4.0 to 5.0, ferrous ion is spontaneously oxidized to ferric ion by O_2 in the water and precipitates as a hydroxide. If the pH drops below 2.0 to 3.0 because of sulfuric acid production by spontaneous oxidation of sulfur or sulfur oxidation by thiobacilli and other bacteria, the ferrous ion remains reduced, soluble, and available as an energy source. Remarkably, *A. ferrooxidans* grows well at such acidic pHs and actively oxidizes ferrous ion to an insoluble ferric precipitate. The water is rendered toxic for most aquatic life and unfit for human consumption.

The ecological consequences of this metabolic lifestyle arise from the common presence of pyrite (FeS_2) in coal mines. The bacteria oxidize both elemental components of pyrite for their growth and in the process form sulfuric acid, which leaches the remaining minerals.

Autoxidation or bacterial action

$$2FeS_2 + 7O_2 + 2H_2O \rightarrow 2Fe^{2+} + 4SO_4^{2-} + 4H^+$$

A. ferrooxidans

$$2Fe^{2+} + 1/2O_2 + 2H^+ \rightarrow 2Fe^{3+} + H_2O$$

Pyrite oxidation is further accelerated because the ferric ion generated by bacterial activity readily oxidizes more pyrite to sulfuric acid and ferrous ion. In turn the ferrous ion supports further bacterial growth. It is difficult to prevent *A. ferrooxidans* growth as it requires only pyrite and common inorganic salts. Because *A. ferrooxidans* gets its O_2 and CO_2 from the air, the only feasible method of preventing its damaging growth is to seal the mines to render the habitat anoxic.

In the second step, the nitrite is oxidized to nitrate by genera such as *Nitrobacter:*

$$NO_2^- + 1/2O_2 \rightarrow NO_3^-$$

Whereas most chemolithotrophs use O_2 as the terminal electron acceptor, some do not. The anammox bacteria (phylum *Planctomycetes*) are an important example. These bacteria carry out *an*aerobic *amm*onia *ox*idation (anammox) using a membrane-bound organelle called the anammoxosome. In the anammoxosome, ammonia oxidation is coupled with the reduction of nitrite, generating N_2 and H_2O (*see figure 19.11*). Thus this is a type of denitrification. Recall that nitrification differs from denitrification in that nitrification involves the oxidation of inorganic nitrogen compounds to yield nitrate. Nitrification is a form of chemolithotrophy in which the nitrogenous compounds donate electrons to the ETC. By contrast, denitrification is the result of anaerobic respiration in which the electron acceptor is an oxidized nitrogenous compound that is reduced to nitrogen gas (p. 244). The anammox reaction is unusual in that it features both the oxidation and reduction of nitrogenous compounds.

Sulfur-oxidizing microbes are the third major group of chemolithotrophs. The metabolism of *Thiobacillus* and *Acidithiobacillus* has been best studied. These bacteria oxidize sulfur (S^0), hydrogen sulfide (H_2S), thiosulfate ($S_2O_3^{2-}$), and other reduced sulfur compounds to sulfuric acid; therefore they have a significant ecological impact (**Microbial Diversity & Ecology 10.1**). Interestingly, they generate ATP by both oxidative phosphorylation and substrate-level phosphorylation. Substrate-level phosphorylation involves adenosine 5′-phosphosulfate (APS). This high-energy molecule is formed from sulfite and adenosine monophosphate (**figure 10.25**).

Some sulfur-oxidizing microbes are extraordinarily flexible metabolically. For example, *Sulfolobus brierleyi,* an archaeon, and some bacteria can grow aerobically by oxidizing sulfur with oxygen as the electron acceptor (*see figure 18.11*). However, in the absence of O_2, they carry out anaerobic respiration and oxidize organic material with sulfur as the electron acceptor. Furthermore, many sulfur-oxidizing chemolithotrophs use CO_2 as their carbon source but will grow heterotrophically if they are supplied with reduced organic carbon sources such as glucose or amino acids.

Energy released by the oxidation of ammonia, nitrite, and sulfur-containing compounds is used to make ATP by oxidative phosphorylation. As we have already noted, the ATP yields for these chemolithotrophs are low. Complicating this already difficult lifestyle is the fact that many of these microbes are also autotrophs, and autotrophs need NAD(P)H (reducing power) as well as ATP to reduce CO_2 and other molecules (figure 10.24).

(a) Direct oxidation of sulfite

$$SO_3^{2-} \xrightarrow{\text{sulfite oxidase}} SO_4^{2-} + 2e^-$$

(b) Formation of adenosine 5′-phosphosulfate

$$2SO_3^{2-} + 2AMP \longrightarrow 2APS + 4e^-$$

$$2APS + 2P_i \longrightarrow 2ADP + 2SO_4^{2-}$$

$$2ADP \longrightarrow AMP + ATP$$

$$2SO_3^{2-} + AMP + 2P_i \longrightarrow 2SO_4^{2-} + ATP + 4e^-$$

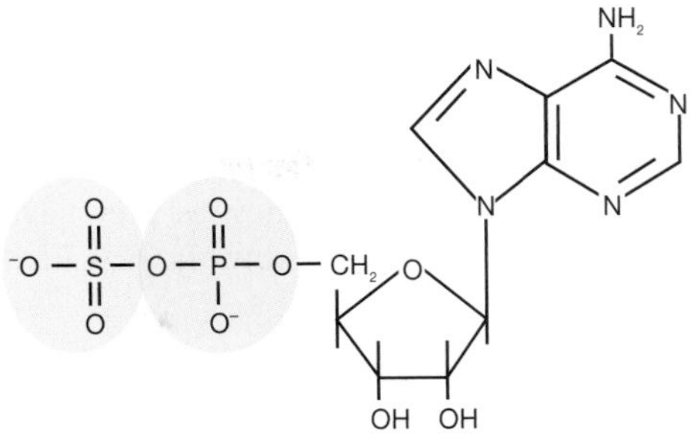

(c) Adenosine 5′-phosphosulfate

FIGURE 10.25 Energy Generation by Sulfur Oxidation. (a) Sulfite can be directly oxidized to provide electrons for electron transport and oxidative phosphorylation. (b) Sulfite can also be oxidized and converted to adenosine 5′-phosphosulfate (APS). This route produces electrons for use in electron transport and ATP by substrate-level phosphorylation with APS. (c) The structure of APS.

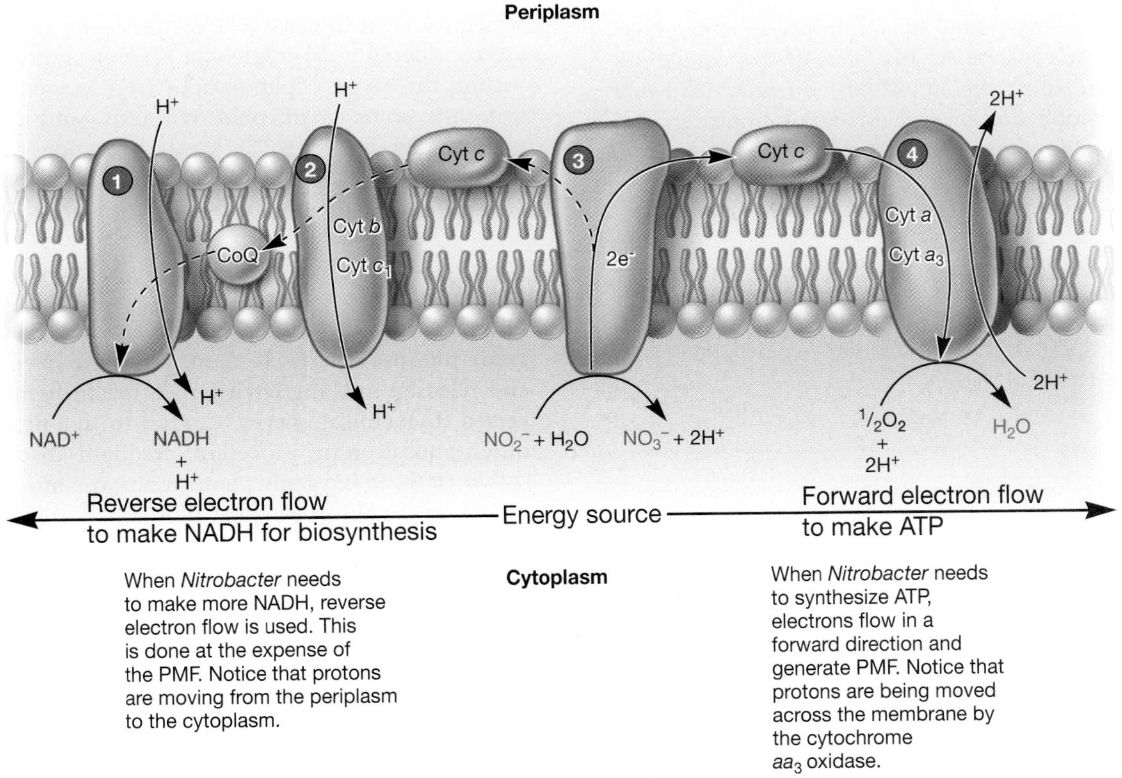

FIGURE 10.26 Electron Flow in *Nitrobacter* Electron Transport Chain. *Nitrobacter* oxidizes nitrite and carries out normal electron transport to generate proton motive force for ATP synthesis. This is the right-hand branch of the diagram. Some of the proton motive force also is used to force electrons to flow up the reduction potential gradient from nitrite to NAD^+ (left-hand branch). Cytochrome *c* and four complexes are involved: NADH-ubiquinone oxidoreductase (1), ubiquinol-cytochrome *c* oxidoreductase (2), nitrite oxidase (3), and cytochrome aa_3 oxidase (4).

Molecules such as ammonia, nitrite, and H_2S have more positive reduction potentials than the $NAD(P)^+/NAD(P)H$ conjugate redox pair (figure 10.9). Therefore, they cannot directly donate their electrons to $NAD(P)^+$. Recall that electrons spontaneously move only from donors with more negative reduction potentials to acceptors with more positive potentials. Thus these chemolithotrophs face a major dilemma: how to make the NAD(P)H they need. They solve this problem by using a process called **reverse electron flow.** During reverse electron flow, electrons derived from the oxidation of inorganic substrates (reduced nitrogen or sulfur compounds) are moved up their ETCs to reduce $NAD(P)^+$ to NAD(P)H (**figure 10.26**). Of course, this is not thermodynamically favorable, so energy in the form of the proton motive force must be diverted from performing other cellular work (e.g., ATP synthesis, transport, motility) to "push" the electrons from molecules of relatively positive reduction potentials to those that are more negative. Chemolithotrophs can afford this inefficiency, as they have no serious competitors for their unique energy sources. ◀◀ *Electron transport chains (section 9.6)*

1. How do chemolithotrophs obtain their ATP and NAD(P)H? What is their most common source of carbon?
2. Describe energy production by hydrogen-oxidizing bacteria, nitrifying bacteria, and sulfur-oxidizing bacteria.
3. Why can hydrogen-oxidizing bacteria and archaea donate electrons to NAD^+, whereas sulfur- and ammonia-oxidizing bacteria and archaea cannot?
4. What is reverse electron flow and why do many chemolithotrophs perform it?
5. Arsenate is a compound that inhibits substrate-level phosphorylation. Compare the effect of this compound on an H_2-oxidizing chemolithotroph, on a sulfite-oxidizing chemolithotroph, and on a chemoorganotroph carrying out fermentation.

10.12 Phototrophy

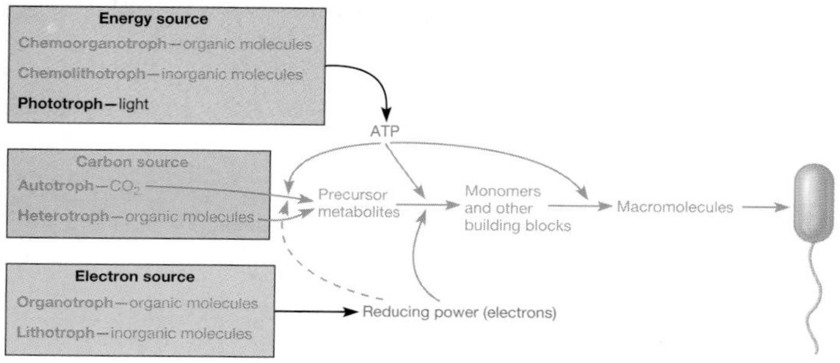

Many microbes capture the energy in light and use it to synthesize ATP and reducing power (e.g., NADPH). When the ATP and reducing power are used to reduce and incorporate CO_2 (CO_2 fixation), the process is called **photosynthesis.** Photosynthesis is one of the most significant metabolic processes on Earth because almost all our energy is ultimately derived from solar energy. Photosynthetic organisms serve as the base of most food chains in the biosphere. One type of photosynthesis is also responsible for replenishing the supply of O_2 in Earth's atmosphere. Although most people associate photosynthesis with plants, over half the photosynthesis on Earth is carried out by microorganisms (**table 10.5**).

Photosynthesis is divided into two parts. In the **light reactions,** light energy is trapped and converted to chemical energy and reducing power. These are then used to fix CO_2 and synthesize cell constituents in the **dark reactions.** The term **phototrophy** refers to the use of light energy to fuel a variety of cellular activities but not necessarily CO_2 fixation. In this sense, photoheterotrophs, which use light to drive ATP synthesis but not carbon fixation, are considered phototrophs but are not photosynthetic. In this section, three types of phototrophy are discussed: oxygenic photosynthesis, anoxygenic photosynthesis, and rhodopsin-based phototrophy (**figure 10.27**). The dark reactions of photosynthesis are reviewed in chapter 11. ▶▶ *CO_2 fixation (section 11.3)*

Table 10.5 Diversity of Phototrophic Organisms

Eukaryotes	Plants; multicellular green, brown and red algae; unicellular protists (e.g., euglenoids, dinoflagellates, diatoms)
Bacteria	Cyanobacteria, green sulfur bacteria, green nonsulfur bacteria, purple sulfur bacteria, purple nonsulfur bacteria, heliobacteria, acidobacteria
Archaea	Halophiles

Light Reactions in Oxygenic Photosynthesis

Photosynthetic eukaryotes and cyanobacteria carry out **oxygenic photosynthesis,** so named because oxygen is generated and released into the environment when light energy is converted to chemical energy. Central to this process, and to all other phototrophic processes, are light-absorbing pigments (**table 10.6**). In oxygenic phototrophs, the most important pigments are the **chlorophylls.** Chlorophylls are large planar molecules composed of four substituted pyrrole rings with a magnesium atom coordinated to the four central nitrogen atoms (**figure 10.28**). Several chlorophylls are found in eukaryotes; the two most important are chlorophyll *a* and chlorophyll *b.* These two molecules differ slightly in their structure and spectral properties. When dissolved in acetone, chlorophyll *a* has a light absorption peak at 665 nm; the corresponding peak for chlorophyll *b* is at 645 nm. In addition to absorbing red light, chlorophylls also absorb blue light strongly (the second absorption peak for chlorophyll *a* is at 430 nm). Because chlorophylls absorb primarily in the red and blue ranges, green light is transmitted, and these organisms appear green. A long hydrophobic tail attached to the chlorophyll ring aids in its attachment to membranes, the site of the light reactions.

Other photosynthetic pigments also trap light energy. The most widespread of these are the carotenoids, long molecules, usually yellowish in color, that possess an extensive conjugated double bond system (**figure 10.29**). β-Carotene is present in cyanobacteria belonging to the genus *Prochloron* and most photosynthetic protists; fucoxanthin is found in protists such as diatoms and dinoflagellates. Red algae and cyanobacteria have

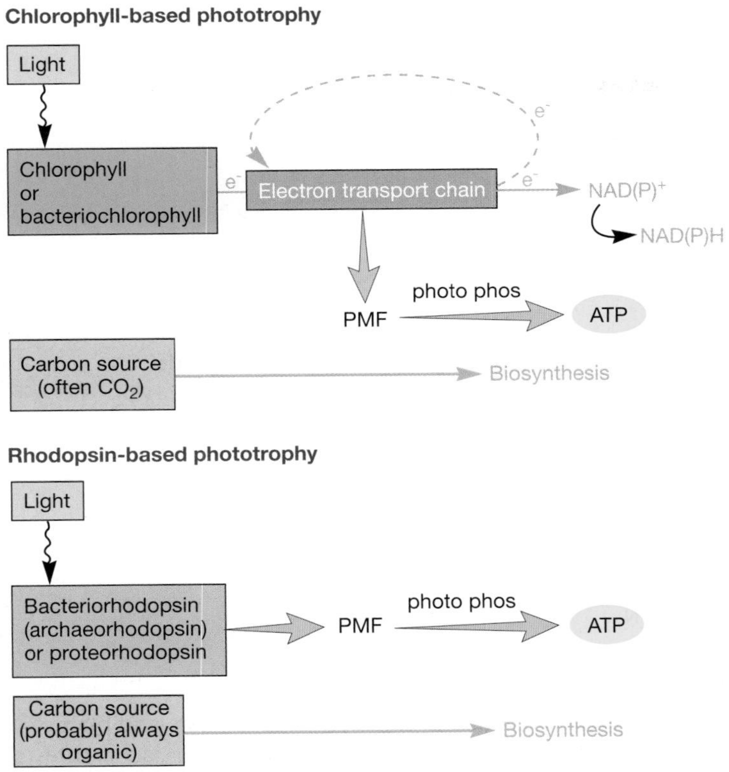

FIGURE 10.27 Phototrophic Fueling Reactions. Phototrophs use light to generate a proton motive force (PMF), which is then used to synthesize ATP by a process called photophosphorylation (photo phos). The process requires light-absorbing pigments. When the pigments are chlorophyll or bacteriochlorophyll, the absorption of light triggers electron flow through an electron transport chain, accompanied by the pumping of protons across a membrane. The electron flow can be either cyclic (dashed line) or noncyclic (solid line), depending on the organism and its needs. Rhodopsin-based phototrophy differs in that the PMF is formed directly by the light-absorbing pigment, which is a light-driven proton pump. Many phototrophs are autotrophs and must use much of the ATP and reducing power they make to fix CO_2.

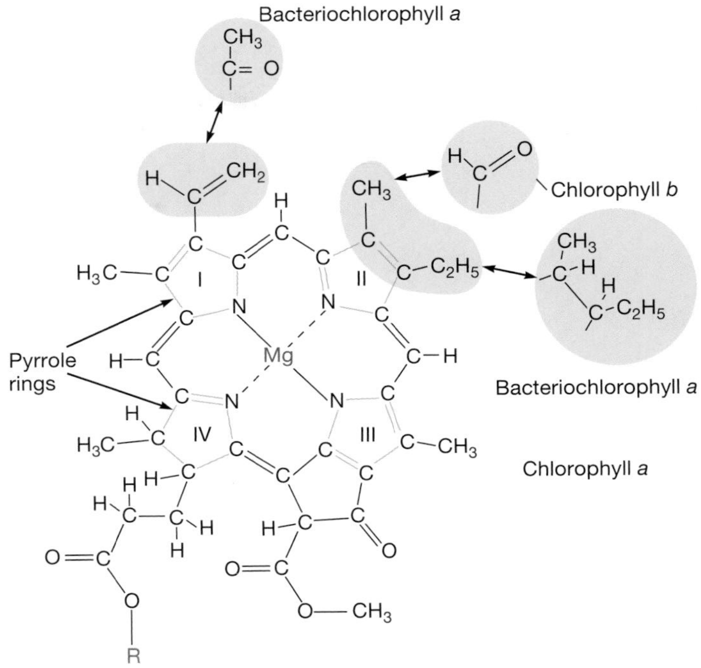

FIGURE 10.28 Chlorophyll Structure. The structures of chlorophyll *a*, chlorophyll *b*, and bacteriochlorophyll *a*. The complete structure of chlorophyll *a* is given. Only one group is altered to produce chlorophyll *b*, and two modifications in the ring system are required to change chlorophyll *a* to bacteriochlorophyll *a*. The side chain (R) of bacteriochlorophyll *a* may be either phytyl (a 20-carbon chain also found in chlorophylls *a* and *b*) or geranylgeranyl (a 20-carbon side chain similar to phytyl but with three more double bonds).

photosynthetic pigments called **phycobiliproteins,** consisting of a protein with a linear tetrapyrrole attached (figure 10.29). Phycoerythrin is a red pigment with a maximum absorption around 550 nm, and phycocyanin is blue (maximum absorption at 620 to 640 nm).

Table 10.6 Properties of Chlorophyll-Based Photosynthetic Systems

Property	*Eukaryotes*	*Cyanobacteria*	*Green Bacteria, Purple Bacteria, Heliobacteria, and Acidobacteria*
Photosynthetic pigment	Chlorophyll *a*	Chlorophyll *a*	Bacteriochlorophyll
Number of photosystems	2	2[a]	1
Photosynthetic electron donors	H_2O	H_2O	H_2, H_2S, S, organic matter
O_2 production pattern	Oxygenic	Oxygenic[b]	Anoxygenic
Primary products of energy conversion	ATP + NADPH	ATP + NADPH	ATP
Carbon source	CO_2	CO_2	Organic or CO_2

[a] A recently discovered cyanobacterium lacks photosystem II.
[b] Some cyanobacteria can function anoxygenically under certain conditions. For example, *Oscillatoria* can use H_2S as an electron donor instead of H_2O.

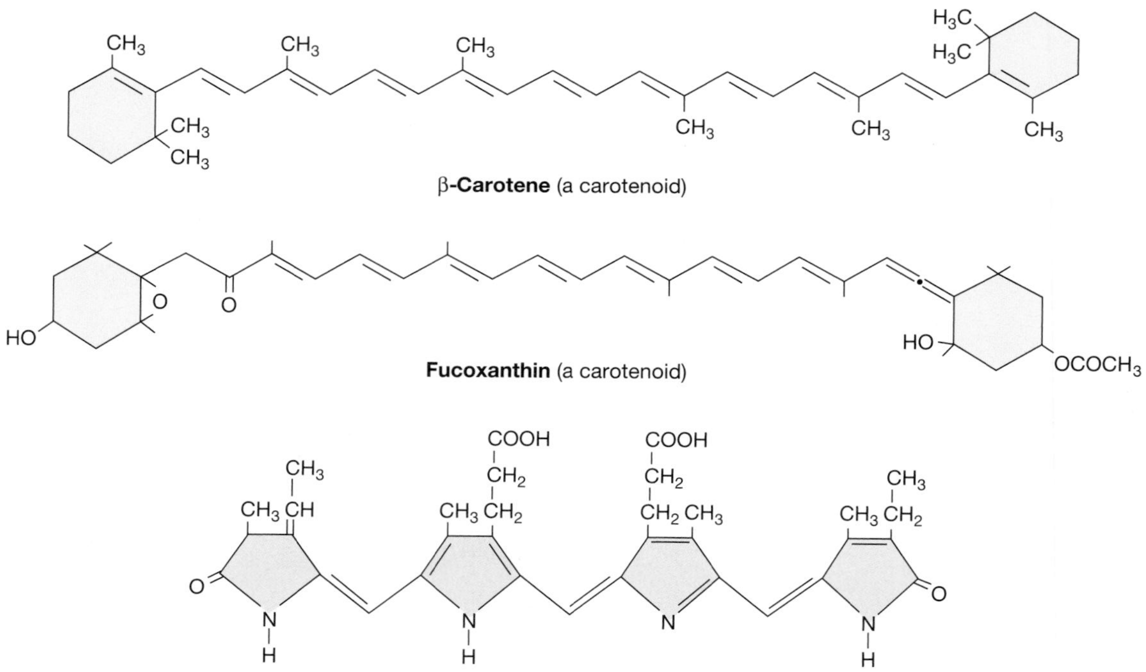

FIGURE 10.29 Representative Accessory Pigments. Beta-carotene is a carotenoid found in photosynthetic protists and plants. Note that it has a long chain of alternating double and single bonds called conjugated double bonds. Fucoxanthin is a carotenoid accessory pigment in several divisions of algae (the dot in the structure represents a carbon atom). Phycocyanobilin is an example of a linear tetrapyrrole that is attached to a protein to form a phycobiliprotein.

Carotenoids and phycobiliproteins are often called **accessory pigments** because of their role in photosynthesis. They are important because they absorb light in the range not absorbed by chlorophylls (the blue-green through yellow range; about 470 to 630 nm) (*see figure 19.4*). This light energy is transferred to chlorophyll. In this way accessory pigments make photosynthesis more efficient over a broader range of wavelengths. In addition, this allows organisms to use light not used by other phototrophs in their habitat. Accessory pigments also protect microorganisms from intense sunlight, which could oxidize and damage the photosynthetic apparatus.

Chlorophylls and accessory pigments are assembled in highly organized arrays called antennas, whose purpose is to create a large surface area to trap as many photons as possible. An antenna has about 300 chlorophyll molecules. Light energy is captured in an antenna and transferred from chlorophyll to chlorophyll until it reaches a **reaction-center chlorophyll pair** that is directly involved in photosynthetic electron transport (*see figure 19.6*). In oxygenic phototrophs, there are two kinds of antennas associated with two different photosystems (**figure 10.30**). **Photosystem I** absorbs longer wavelength light (≥680 nm) and funnels the energy to a reaction center chlorophyll *a* pair called P700. The term P700 signifies that this molecule most effectively absorbs light at a wavelength of 700 nm. **Photosystem II** traps light at shorter wavelengths (≥680 nm) and transfers its energy to the reaction center chlorophyll pair P680.

When the photosystem I antenna transfers light energy to P700, P700 absorbs the energy and is excited, and its reduction potential becomes very negative. This allows P700 to donate its excited, high energy electron to a specific acceptor, probably a special chlorophyll *a* molecule or an iron-sulfur protein. The electron is eventually transferred to ferredoxin and then travels in either of two directions. In the cyclic pathway (the dashed lines in figure 10.30), the electron moves in a cyclic route through a series of electron carriers and back to the oxidized P700. The pathway is termed cyclic because the electron from P700 returns to P700 after traveling through the photosynthetic ETC. PMF is formed during cyclic electron transport and used to synthesize ATP. This process is called **cyclic photophosphorylation** because electrons travel in a circle and ATP is formed. Only photosystem I participates.

Electrons also can travel in a noncyclic pathway involving both photosystems. P700 is excited and donates electrons to ferredoxin as before. In the noncyclic route, however, reduced ferredoxin reduces $NADP^+$ to NADPH (figure 10.30). Because the electrons contributed to $NADP^+$ cannot be used to reduce oxidized P700, photosystem II participation is required. It donates

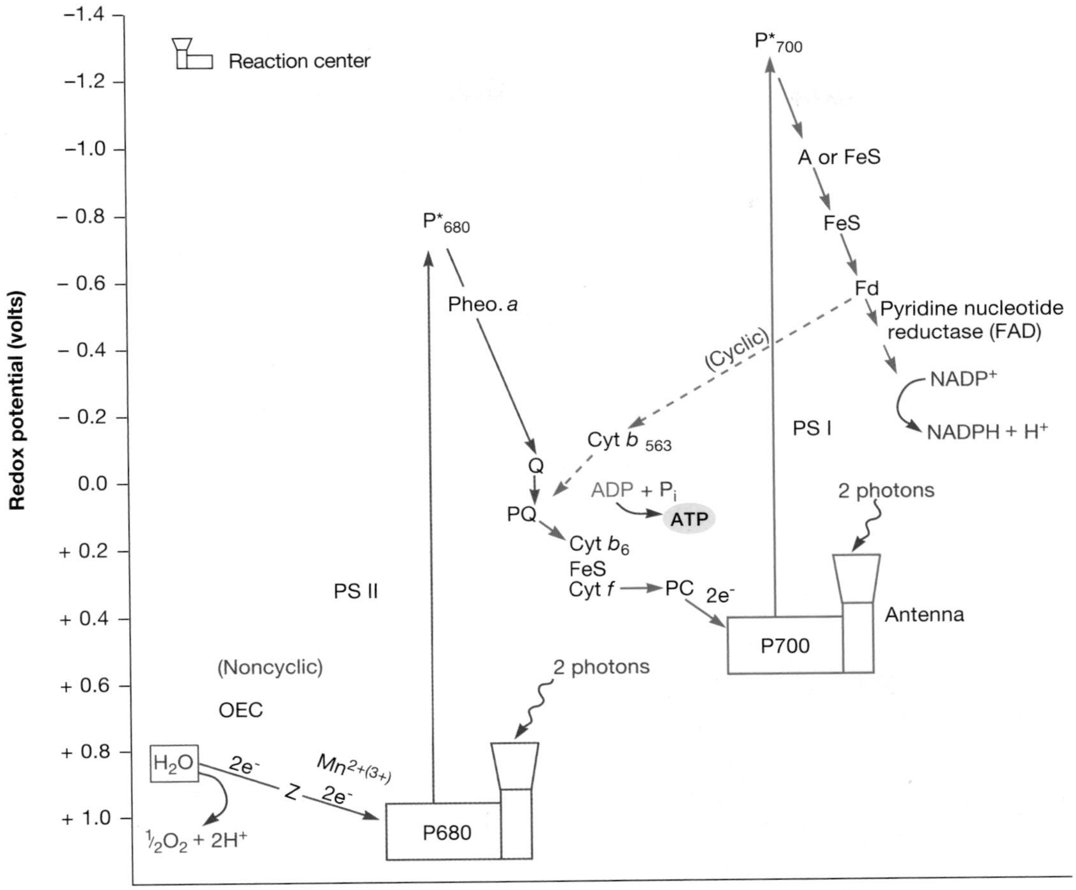

FIGURE 10.30 Oxygenic Photosynthesis. Cyanobacteria and eukaryotic algae are similar in having two photosystems, although they may differ in some details. The carriers involved in electron transport are ferredoxin (Fd) and other FeS proteins; cytochromes b_6, b_{563}, and *f*; plastoquinone (PQ); copper containing plastocyanin (PC); pheophytin *a* (Pheo. *a*); possibly chlorophyll *a* (A); and the unknown quinone Q, which is probably a plastoquinone. Both photosystem I (PS I) and photosystem II (PS II) are involved in noncyclic photophosphorylation; only PS I participates in cyclic photophosphorylation. The oxygen evolving complex (OEC) that extracts electrons from water contains manganese ions and the substance Z, which transfers electrons to the PS II reaction center.

Figure 10.30 Micro Inquiry

When electrons from P700 are used to reduce $NADP^+$, what compound supplies electrons for the re-reduction of P700?

electrons to oxidized P700 and generates ATP in the process. The photosystem II antenna absorbs light energy and excites P680, which then reduces pheophytin *a*. Pheophytin *a* is chlorophyll *a* in which two hydrogen atoms have replaced the central magnesium. Electrons subsequently travel to the plastoquinone pool and down an ETC to reduce P700. Now P680 must also be reduced if it is to accept more light energy. Figure 10.30 indicates that the standard reduction potential of P680 is more positive than that of the $1/2O_2/H_2O$ conjugate redox pair. Thus H_2O can be used to donate electrons to P680, resulting in the release of oxygen. Because electrons flow from water to $NADP^+$ with the aid of energy from two photosystems, ATP is synthesized by **noncyclic photophosphorylation.** It appears that one ATP and one NADPH are formed when two electrons travel through the noncyclic pathway.

It is worth reemphasizing that although light is the source of energy for chlorophyll-based phototrophy, the process used to make ATP is virtually the same as seen for chemotrophs: oxidation-reduction reactions occurring in ETCs generate a PMF that is used by ATP synthase to make ATP. Furthermore, just as is true of mitochondrial electron transport, photosynthetic electron transport takes place within a membrane. Chloroplast granal membranes contain both photosystems and their antennas. **Figure 10.31** shows a thylakoid membrane carrying out noncyclic photophosphorylation by the chemiosmotic mechanism. Protons move to the thylakoid interior during photosynthetic electron transport and return to the stroma when ATP is formed. It is thought that stromal lamellae possess only photosystem I and are involved in cyclic photophosphorylation alone. In cyanobacteria, photosynthetic light reactions are located in thylakoid membranes within the cell. *Photosynthetic Electron Transport and ATP Synthesis*

The dark reactions of oxygenic phototrophs use three ATPs and two NADPHs to reduce one CO_2 to carbohydrate (CH_2O).

$$CO_2 + 3ATP + 2NADPH + 2H^+ + H_2O \rightarrow (CH_2O) + 3ADP + 3P_i + 2NADP^+$$

Oxygenic phototrophs use noncyclic and cyclic electron flow to supply these needs. The noncyclic system generates one NADPH and one ATP per pair of electrons; therefore four electrons passing through the system produce two NADPHs and two ATPs. A total of 8 quanta of light energy (4 quanta for each photosystem) is needed to propel the four electrons from water to $NADP^+$. Because the ratio of ATP to NADPH required for CO_2 fixation is 3:2, at least one more ATP must be supplied. Cyclic photophosphorylation probably operates independently to generate the extra ATP. This requires absorption of another 2 to 4 quanta. It follows that around 10 to 12 quanta of light energy are needed to reduce and incorporate one molecule of CO_2 during photosynthesis.

Light Reactions in Anoxygenic Photosynthesis

Certain bacteria carry out a second type of photosynthesis called **anoxygenic photosynthesis.** This phototrophic process derives its name from the fact that molecules other than water are used as

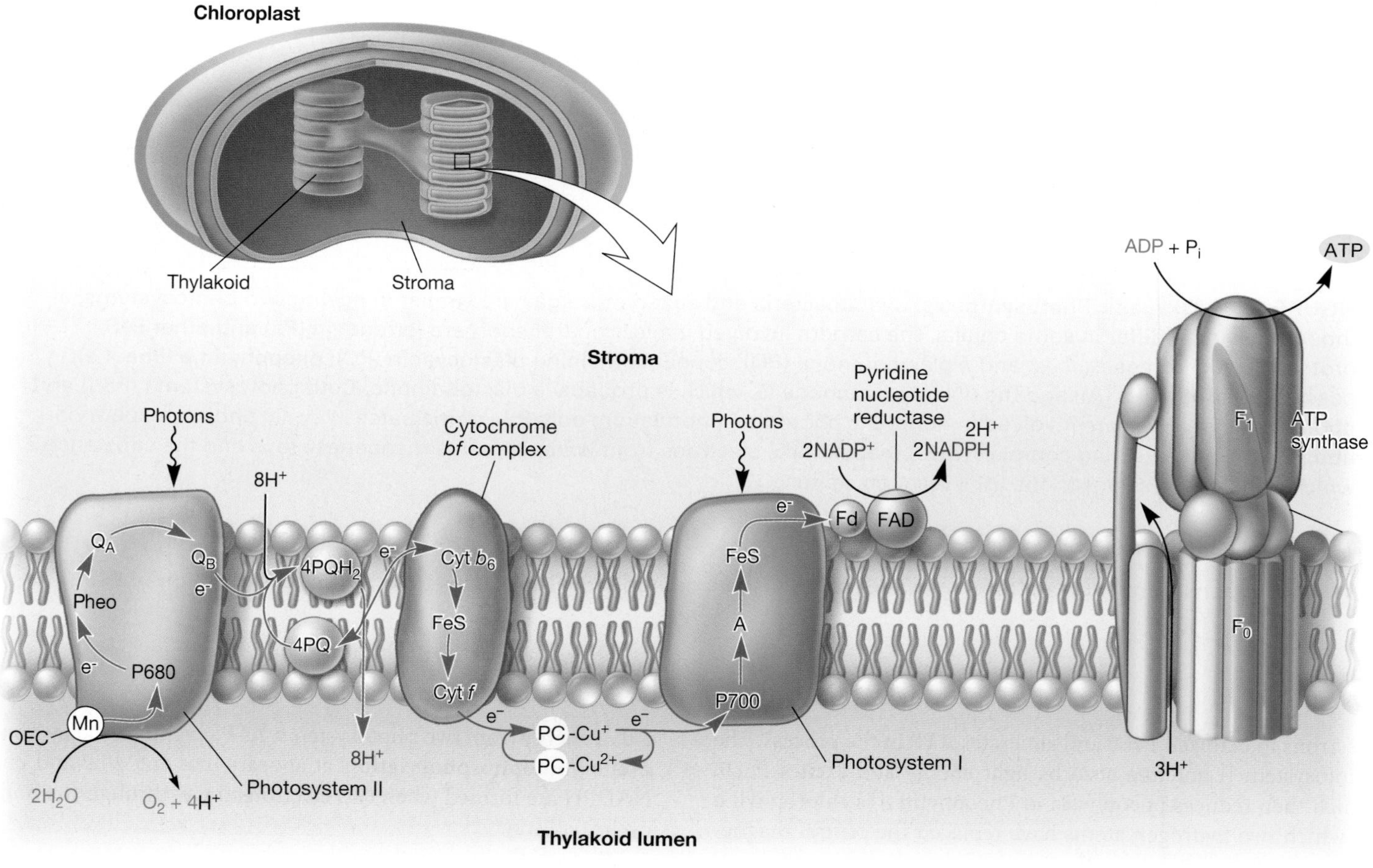

FIGURE 10.31 The Mechanism of Photosynthesis. An illustration of the chloroplast thylakoid membrane showing photosynthetic ETC function and noncyclic photophosphorylation. The chain is composed of three complexes: PS I, the cytochrome *bf* complex, and PS II. Two diffusible electron carriers connect the three complexes. Plastoquinone (PQ) connects PS I with the cytochrome *bf* complex, and plastocyanin (PC) connects the cytochrome *bf* complex with PS II. The light-driven electron flow pumps protons across the thylakoid membrane and generates a PMF, which can then be used to make ATP. Water is the source of electrons and the oxygen-evolving complex (OEC) produces oxygen.

an electron source and therefore O_2 is not produced. The process also differs in terms of the photosynthetic pigments used, the participation of just one photosystem, and the mechanisms used to generate reducing power. Members of five bacterial phyla carry out anoxygenic photosynthesis: *Proteobacteria* (purple sulfur and purple nonsulfur bacteria), *Chlorobi* (green sulfur bacteria) *Chloroflexi* (green nonsulfur bacteria), *Firmicutes* (heliobacteria), and *Acidobacteria*. The biology and ecology of these organisms are described in more detail in chapters 19, 20, and 21.

Anoxygenic phototrophs have light-absorbing pigments called **bacteriochlorophylls** (figure 10.28). In some bacteria, these are located in membranous vesicles called chlorosomes. The absorption maxima of bacteriochlorophylls (Bchl) are at longer wavelengths than those of chlorophylls. Bacteriochlorophylls *a* and *b* have maxima in ether at 775 and 790 nm, respectively. In vivo maxima are about 830 to 890 nm (Bchl *a*) and 1,020 to 1,040 nm (Bchl *b*). This shift of absorption maxima into the infrared region better adapts these bacteria to their ecological niches. ▶▶| *Photosynthetic bacteria (section 19.3)*

Many differences found in anoxygenic phototrophs are because they have a single photosystem. This means that they are restricted to cyclic photophosphorylation and are unable to produce O_2 from H_2O. Indeed, almost all anoxygenic phototrophs are strict anaerobes. A tentative scheme for the photosynthetic ETC of a purple nonsulfur bacterium is given in **figures 10.32** and **10.33**. When the reaction-center bacteriochlorophyll P870 is excited, it donates an electron to bacteriopheophytin. Electrons then flow to quinones and through an ETC back to P870 while generating sufficient PMF to drive ATP synthesis by ATP synthase. Note that although both green and purple bacteria lack two photosystems, the purple bacteria have a photosynthetic apparatus similar to photosystem II of oxygenic phototrophs, whereas the green sulfur bacteria have a system similar to photosystem I. ▶▶| *Purple nonsulfur bacteria (section 20.1); Phylum* Chlorobi *(section 19.3)*

Anoxygenic photoautotrophs face a further problem because they also require reducing power (NAD[P]H or reduced ferredoxin) for CO_2 fixation and other biosynthetic processes. They are able to generate reducing power in at least three ways, depending on the bacterium. (1) Some have hydrogenases that are used to produce NAD(P)H directly from the oxidation of hydrogen gas. This is possible because hydrogen gas has a more negative reduction potential than NAD^+ (figure 10.9). (2) Others, such as the photosynthetic purple bacteria, use reverse electron flow to generate NAD(P)H (figure 10.32). In this mechanism, electrons are drawn off the photosynthetic ETC and "pushed" to $NAD(P)^+$ using PMF. This process is similar to that seen for chemolithotrophs having inorganic energy sources with more positive reduction potentials than that of $NAD(P)^+$/NAD(P)H. (3) Phototrophic green bacteria and heliobacteria also draw off electrons from their ETCs. However, because the reduction potential of the component of the chain where this occurs is more negative than NAD^+ and oxidized ferredoxin, the electrons flow spontaneously to these electron acceptors. Thus these bacteria exhibit a simple form of noncyclic photosynthetic electron flow (**figure 10.34**). The diversion of electrons from the photosynthetic ETC means that continued generation of PMF will cease unless the electrons are replaced. Electrons from electron donors such as hydrogen sulfide, elemental sulfur, and organic compounds replace the electrons removed from the ETC in this way. *Cyclic and Noncyclic Photophosphorylation*

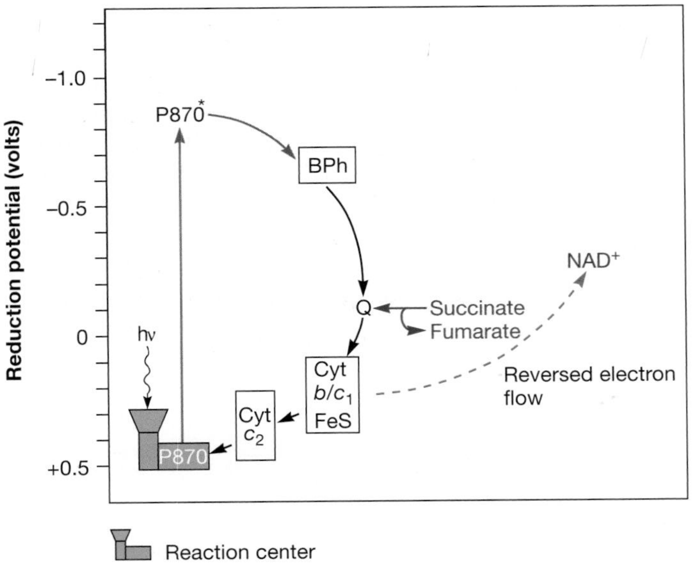

FIGURE 10.32 Purple Nonsulfur Bacterial Photosynthesis. The photosynthetic ETC in the purple nonsulfur bacterium *Rhodobacter sphaeroides*. This scheme is incomplete and tentative. Ubiquinone (Q) is very similar to coenzyme Q. BPh stands for bacteriopheophytin. The electron source succinate is in blue.

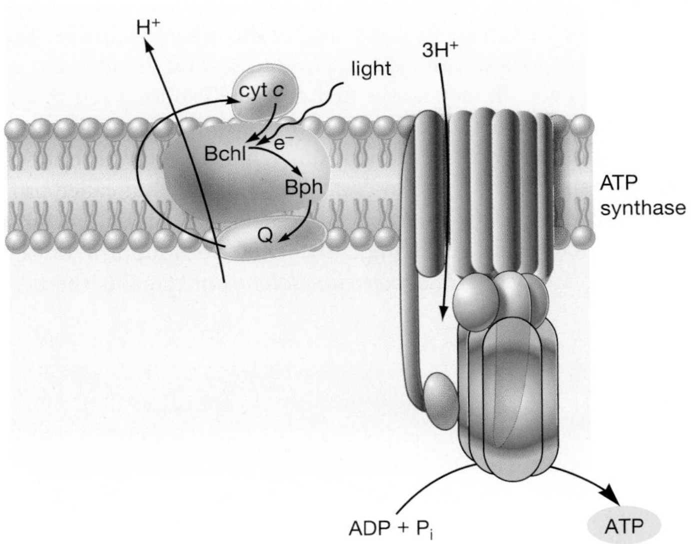

FIGURE 10.33 Cyclic Electron Flow During Anoxygenic Phototrophy. The photosystem of *Rhodopseudomanas viridis* is illustrated. Bchl, bacteriochlorophyll; Bph, bacteriophaeophytin; Q, quinones; and cyt *c*, cytochrome *c*.

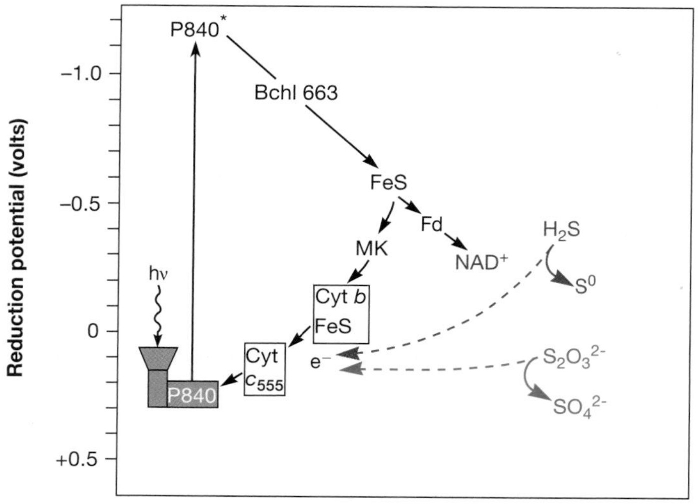

FIGURE 10.34 Green Sulfur Bacterial Photosynthesis. The photosynthetic ETC in the green sulfur bacterium *Chlorobium limicola.* Light energy is used to make ATP by cyclic photophosphorylation and to move electrons from thiosulfate ($S_2O_3^{2-}$) and H_2S (green and blue) to NAD^+. The ETC has a quinone called menaquinone (MK).

Rhodopsin-Based Phototrophy

So far we have discussed chlorophyll-based types of phototrophy—that is, the use of chlorophyll or bacteriochlorophyll to absorb light and initiate the conversion of light energy to chemical energy. Within the last decade, it has become apparent that many bacteria and archaea are capable of chlorophyll-independent phototrophy. These microbes rely on a form of microbial rhodopsin, a molecule similar to that found in the mammalian eye. The first microbial rhodopsin was discovered several decades ago in certain archaea. In fact, it was first described so long ago that it was called **bacteriorhodopsin;** we now know that it should more correctly be called archaeorhodopsin. While a variety of rhodopsins have since been found in *Proteobacteria* (and thus called proteorhodospin), *Flavobacteria,* extremely halophilic bacteria (xanthorhodopsin), and fungi, the bacteriorhodopsin of the halophilic archaeon *Halobacteirum salinarum* remains the best studied and is the focus of our discussion here. ▶▶| *Halobacteria (section 18.3); Microorganisms in the open ocean (section 28.2)*

H. salinarum normally depends on aerobic respiration for the release of energy from an organic energy source. It cannot grow anaerobically by anaerobic respiration or fermentation. However, under conditions of low oxygen and high light intensity, it synthesizes bacteriorhodopsin, a deep-purple pigment that closely resembles the rhodopsin found in the rods and cones of vertebrate eyes. Bacteriorhodopsin's chromophore is retinal, a type of carotenoid. The chromophore is covalently attached to the pigment protein, which is embedded in the plasma membrane in such a way that the retinal is in the center of the membrane.

Bacteriorhodopsin functions as a light-driven proton pump. When retinal absorbs light, a proton is released and the bacteriorhodopsin undergoes a sequence of conformation changes that translocate the proton into the periplasmic space *(see figure 18.17).* The light-driven proton pumping generates a pH gradient that can be used to power the synthesis of ATP by chemiosmosis. This phototrophic capacity is particularly useful to *Halobacterium* because oxygen is not very soluble in concentrated salt solutions and may decrease to an extremely low level in *Halobacterium's* habitat. When the surroundings become temporarily anoxic, the archaeon uses light energy to synthesize sufficient ATP to survive until oxygen levels rise again. Note that this type of phototrophy does not involve electron transport.

1. Define the following terms: light reactions, dark reactions, chlorophyll, carotenoid, phycobiliprotein, antenna, and photosystems I and II.
2. What happens to a reaction center chlorophyll pair, such as P700, when it absorbs light?
3. What is the function of accessory pigments?
4. What is photophosphorylation? What is the difference between cyclic and noncyclic photophosphorylation?
5. Compare and contrast anoxygenic phototrophy and oxygenic photosynthesis. How do these two types of phototrophy differ from rhodopsin-based phototrophy?
6. Suppose you isolated a bacterial strain that carried out oxygenic photosynthesis. What photosystems would it possess and what group of bacteria would it most likely belong to?

Summary

10.1 Chemoorganotrophic Fueling Processes

a. Chemotrophic organisms use chemical sources of electrons and energy.

b. Chemoorganotrophic microorganisms can use three kinds of electron acceptors during energy metabolism (**figure 10.1**). Electrons from the oxidized nutrient can be accepted by an endogenous electron acceptor (fermentation), by oxygen (aerobic respiration), or by another external electron acceptor (anaerobic respiration).

c. Chemoorganotrophs can use a variety of molecules as their energy, electron, and carbon source. These molecules are degraded by pathways that funnel into pathways used for glucose catabolism (**figure 10.2**).

d. Some of the pathways used by chemoorganotrophs are amphibolic, having both catabolic and anabolic functions (**figure 10.3**).

10.2 Aerobic Respiration

a. Aerobic respiration of glucose begins with glycolytic pathways, which catabolize glucose to pyruvate. Pyruvate is fed into the TCA cycle for completion of catabolism.

b. The glycolytic pathways and the TCA cycle generate numerous NADH and $FADH_2$ molecules. These are oxidized by the electron transport chain, using oxygen as the terminal electron acceptor. Electron flow generates a proton motive force (PMF), which is used to synthesize ATP by oxidative phosphorylation.

10.3 Glycolytic Pathways

a. Glycolysis, used in its broadest sense, refers to all pathways used to break down glucose to pyruvate.

b. The Embden-Meyerhof pathway has a net production of two NADHs and two ATPs, the latter being produced by substrate-level phosphorylation. It also produces several precursor metabolites (**figure 10.4**).

c. In the Entner-Doudoroff pathway, glucose is oxidized to 6-phosphogluconate, which is then dehydrated and cleaved to pyruvate and glyceraldehyde 3-phosphate (**figure 10.5**). The latter product can be oxidized by enzymes of the Embden-Meyerhof pathway to provide ATP, NADH, and another molecule of pyruvate.

d. In the pentose phosphate pathway, glucose 6-phosphate is oxidized twice and converted to pentoses and other sugars. It is a source of NADPH, ATP, and several precursor metabolites (**figure 10.6**).

10.4 Tricarboxylic Acid Cycle

a. Pyruvate from the glycolytic pathways is fed into the tricarboxylic acid cycle by a reaction that converts pyruvate to acetyl-CoA. In the process, one of pyruvate's carbons is released in the form of carbon dioxide.

b. The tricarboxylic acid cycle oxidizes acetyl-CoA to CO_2 and forms one GTP, three NADHs, and one $FADH_2$ per acetyl-CoA (**figure 10.7**). It also generates several precursor metabolites.

10.5 Electron Transport and Oxidative Phosphorylation

a. The NADH and $FADH_2$ produced from the oxidation of carbohydrates, fatty acids, and other nutrients can be oxidized in an electron transport chain (ETC). Electrons flow from carriers with more negative reduction potentials to those with more positive potentials (**figures 10.8** to **10.10**), and free energy is released for ATP synthesis by oxidative phosphorylation.

b. Bacterial and archaeal ETCs are often different from eukaryotic chains with respect to such aspects as carriers and branching (**figures 10.11** and **10.12**).

c. ATP synthase catalyzes the synthesis of ATP (**figure 10.15**). In eukaryotes, it is located on the inner surface of the inner mitochondrial membrane. Bacterial and archaeal ATP synthases are on the inner surface of the plasma membrane.

d. The most widely accepted mechanism of oxidative phosphorylation is the chemiosmotic hypothesis in which proton motive force (PMF) drives ATP synthesis (**figure 10.10**).

e. In eukaryotes, the P/O ratio for NADH is about 2.5 and that for $FADH_2$ is around 1.5; P/O ratios are usually much lower in bacterial and archaeal chains. Aerobic respiration in eukaryotes can theoretically yield a maximum of 32 ATPs (**figure 10.16**).

10.6 Anaerobic Respiration

a. Anaerobic respiration is the process of ATP production by electron transport in which the terminal electron acceptor is an exogenous molecule other than O_2. The most common acceptors are nitrate, sulfate, and CO_2 (**figure 10.17**).

b. For some microorganisms, the same pathways used to aerobically respire an organic energy source are also used for anaerobic respiration.

c. Less energy is provided by anaerobic respiration than aerobic respiration because the alternate electron acceptors have reduction potentials that are less positive than the reduction potential of oxygen.

10.7 Fermentation

a. During fermentation, an endogenous electron acceptor is used to reoxidize any NADH generated by the catabolism of glucose to pyruvate (**figure 10.18**).

b. Flow of electrons from the electron donor to the electron acceptor does not involve an ETC, and in most organisms, ATP is synthesized only by substrate-level phosphorylation.

c. There are many different fermentation pathways. These are of practical importance in clinical and industrial settings (**figure 10.19**).

10.8 Catabolism of Other Carbohydrates

a. Microorganisms catabolize many extracellular carbohydrates. Monosaccharides are taken in and phosphorylated; disaccharides may be cleaved to monosaccharides by either hydrolysis or phosphorolysis.

b. External polysaccharides are degraded by hydrolysis and the products are absorbed. Intracellular glycogen and starch are converted to glucose 1-phosphate by phosphorolysis (**figure 10.20**).

10.9 Lipid Catabolism

a. Triglycerides are hydrolyzed to glycerol and fatty acids by enzymes called lipases.

b. Fatty acids are usually oxidized to acetyl-CoA in the β-oxidation pathway (**figure 10.22**).

10.10 Protein and Amino Acid Catabolism

a. Proteins are hydrolyzed to amino acids that are then deaminated (**figure 10.23**).

b. The carbon skeletons produced by deamination can be fermented or fed into the TCA cycle (**figure 10.2**).

10.11 Chemolithotrophy

a. Chemolithotrophs synthesize ATP by oxidizing inorganic compounds—usually hydrogen, reduced nitrogen and sulfur compounds, or ferrous iron—with an ETC. O_2 is the usual electron acceptor (**figure 10.24** and **table 10.3**). The PMF produced is used by ATP synthase to make ATP.

b. Many of the energy sources used by chemolithotrophs have a more positive standard reduction potential than the NAD^+/NADH conjugate redox pair. These chemolithotrophs must expend energy (PMF) to drive reverse electron flow and produce the NADH they need for CO_2 fixation and other processes (**figure 10.26**).

10.12 Phototrophy

a. In oxygenic photosynthesis, eukaryotes and cyanobacteria trap light energy with chlorophyll and accessory pigments, and move electrons through photosystems I and II to make ATP and NADPH (the light reactions). The ATP and NADPH are used in the dark reactions to fix CO_2.

b. Cyclic photophosphorylation involves the activity of photosystem I alone and generates ATP only. In noncyclic photophosphorylation, photosystems I and II operate together to move electrons from water to $NADP^+$, producing ATP, NADPH, and O_2 (**figure 10.30**). In both cases, electron flow generates PMF, which is used by ATP synthase to make ATP.

c. Anoxygenic phototrophs differ from oxygenic phototrophs in possessing bacteriochlorophyll and having only one photosystem (**figures 10.32** to **10.34**). Cyclic electron flow generates a PMF, which is used by ATP synthase to make ATP (i.e., cyclic photophosphorylation). They are anoxygenic because they use molecules other than water as an electron donor for electron flow and the production of reducing power.

d. Some bacteria and archaea use a type of phototrophy that involves a proton-pumping pigment rhodopsin. This type of phototrophy generates PMF but does not involve an ETC.

Critical Thinking Questions

1. Without looking in chapters 19 and 20, predict some characteristics that would describe niches occupied by green and purple photosynthetic bacteria.

2. From an evolutionary perspective, discuss why most microorganisms use aerobic respiration to generate ATP.

3. How would you isolate a thermophilic chemolithotroph that uses sulfur compounds as a source of energy and electrons? What changes in the incubation system would be needed to isolate bacteria using sulfur compounds in anaerobic respiration? How would you tell which process is taking place through an analysis of the sulfur molecules present in the medium?

4. Certain chemicals block ATP synthesis by allowing protons and other ions to "leak across membranes," disrupting the charge and proton gradients established by electron flow through an ETC. Does this observation support the chemiosmosis hypothesis? Explain your reasoning.

5. Two flasks of *E. coli* are grown in batch culture in the same medium (2% glucose and amino acids; no nitrate) and at the same temperature (37°C). Culture #1 is well aerated. Culture #2 is anoxic. After 16 hours the following observations are made:

   - Culture #1 has a high cell density; the cells appear to be in stationary phase, and the glucose level in the medium is reduced to 1.2%.
   - Culture #2 has a low cell density; the cells appear to be in logarithmic phase, although their doubling time is prolonged (over 1 hour). The glucose level is reduced to 0.2%.

   Why does culture #2 have so little glucose remaining relative to culture #1, even though culture #2 displayed slower growth and has less biomass?

6. A recently discovered cyanobacterium has been found to have photosystem I but not photosystem II. It is unable to fix CO_2. How do you think this microbe makes ATP? Reducing power? Why is it to this microbe's advantage to be a heterotroph?

7. The archaeon *Metallosphaera sedula* is of great interest to microbiologists in the field of biomining because it might be used to recover base and precious metals. This aerobic thermoacidophile can use organic carbon, ferrous iron (Fe^{2+}),

and reduced inorganic sulfur compounds including elemental sulfur (S^0) and tetrathionate ($S_4O_6^{2-}$) as a source of electrons. Thus *M. sedula* can grow chemoorganotrophically and chemolithotrophically. Genome sequencing reveals that *M. sedula* has five major terminal oxidase complexes.

Refer to table 9.1 and figure 10.9 to propose the flow of electrons through the electron transport chain when using each of these compounds as the electron source. Which will conserve the most energy and which the least? Why do you think this microbe requires five different terminal oxidases if it always uses O_2 as the terminal electron acceptor?

Read the original paper: Auernik, K. S., and Kelly, R. M. 2008. Identification of components of electron transport chains in the extremely thermoacidophilic crenarchaeon *Metallosphaera sedula* through iron and sulfur oxidation transcriptomes. *App. Environ. Microbiol.* 74:7723.

8. When oil is drilled at sea, seawater is injected during recovery. This essential step has the undesirable consequence of introducing a large population of sulfate-reducing bacteria into the oil. These microbes produce H_2S—a phenomenon known as souring. Although it is possible to chemically remove H_2S postrecovery, it is more efficient to prevent its production. This can be done by introducing nitrate to stimulate the growth of chemolithotrophic nitrate-reducing–sulfide-oxidizing bacteria (NR-SOB) and heterotrophic nitrate-reducing bacteria (hNRB). The hNRB compete with the sulfate-reducing bacteria (SRB) for degradable organic electron donors such as the oil by-products lactate and volatile fatty acids.

Compare the electron transport chains of the NR-SOB and hNRB with that of the SRB. Based on these ETCs, explain why NR-SOBs prevent the accumulation of H_2S. Refer to table 9.1 and figure 10.9, and compare the amounts of energy conserved for each type of ETC; which type of bacteria will be most favored? Finally, what is the most likely fate of the nitrite produced by the nitrate-reducing bacteria? (Hint: Remember that there are many types of bacteria and archaea present in the seawater, although at relatively low densities.)

Read the original paper: Hubert, C., and Voodouw, G. 2007. Oil field souring control by *Sulfurospirillum* spp. that out compete sulfate-reducing bacteria for organic electron donors. *Appl. Environ. Microiol.* 73:2644.

Concept Mapping Challenge

Construct a concept map that describes the metabolic processes used by chemoorganotrophs to catabolize glucose. Use the concepts that follow, any other concepts or terms you need, and your own linking words between each pair of concepts in your map.

aerobic respiration	anaerobic respiration
fermentation	glycolysis
TCA cycle	ETC
oxidative phosphorylation	substrate-level phosphorylation
terminal electron acceptor	PMF

Learn More

Learn more by visiting the text website at www.mhhe.com/willey8, where you will find a complete list of references.

11

Anabolism: The Use of Energy in Biosynthesis

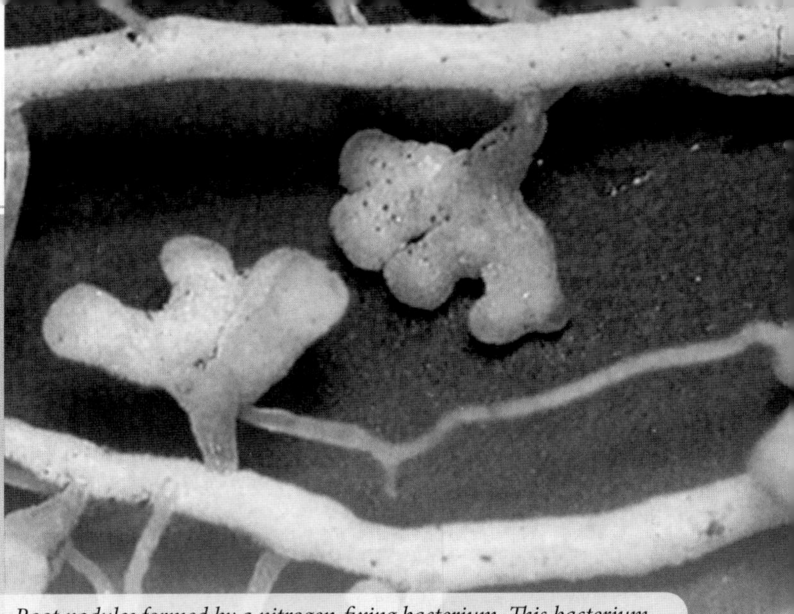

Root nodules formed by a nitrogen-fixing bacterium. This bacterium can use N_2 gas as a source of nitrogen for amino acid and nucleotide synthesis.

CHAPTER GLOSSARY

anaplerotic reactions Reactions that replenish depleted tricarboxylic acid cycle intermediates.

assimilatory nitrate reduction The reduction of nitrate and its incorporation into organic material.

assimilatory sulfate reduction The reduction of sulfate and its incorporation into organic material.

Calvin cycle The main pathway for the fixation (i.e., reduction and incorporation) of CO_2 into organic material by photoautotrophs and chemolithoautotrophs.

gluconeogenesis The synthesis of glucose from noncarbohydrate precursors such as lactate and amino acids.

glyoxylate cycle A modified tricarboxylic acid cycle used to replenish precursor metabolites normally provided by the TCA cycle.

macromolecule A large molecule that is a polymer of smaller units joined together.

nitrogen fixation The metabolic process in which atmospheric molecular nitrogen (N_2) is reduced to ammonia.

nucleoside A combination of ribose or deoxyribose with a purine or pyrimidine nitrogenous base.

nucleotide A nucleoside plus one or more phosphates.

purine A basic, nitrogen-containing molecule consisting of two joined rings; found in nucleic acids and other cell constituents; includes adenine and guanine.

pyrimidine A basic, cyclic, nitrogen-containing molecule found in nucleic acids and other cell constituents; includes cytosine, thymine, and uracil.

ribulose-1,5-bisphosphate carboxylase The enzyme that catalyzes the incorporation of CO_2 in the Calvin cycle.

transaminases Enzymes that catalyze the transfer of an amino group from an amino acid to an α-keto acid.

transpeptidation The reaction that forms the peptide cross-links during peptidoglycan synthesis.

As chapter 10 makes clear, microorganisms can obtain energy in many ways. Much of this energy is used in anabolism. During anabolism, an organism begins with simple inorganic molecules and a carbon source, and constructs ever more complex molecules until new organelles and cells arise (**figure 11.1**). Although there is considerably less diversity in anabolic processes as compared to catabolic processes, anabolism is amazing in its own right. From just 12 precursor metabolites, the cell is able to manufacture the myriad of molecules from which it is constructed. Furthermore, numerous antibiotics exert their control over microbial growth by inhibiting anabolic pathways.

In this chapter, we discuss the synthesis of some of the most important types of cell constituents. We begin with a general introduction to anabolism and the role played by the precursor metabolites in biosynthetic pathways. We then focus on CO_2 fixation and the synthesis of carbohydrates, amino acids, purines and pyrimidines, and lipids. Because protein and nucleic acid synthesis are so significant and complex, the polymerization reactions that yield these macromolecules are described separately in chapter 12.

Anabolism is the creation of order. Because a cell is highly ordered and immensely complex, much energy is required for biosynthesis. This is readily apparent from estimates of the biosynthetic capacity of rapidly growing *Escherichia coli* (**table 11.1**). Although most ATP dedicated to biosynthesis is employed in amino acid and protein synthesis, ATP is also used to make other cell material.

It is intuitively obvious why rapidly growing cells need a large supply of ATP. But even nongrowing cells need energy for the biosynthetic processes they carry out. This is because nongrowing cells continuously degrade and resynthesize cellular molecules during a process known as turnover. Thus cells are never the same from one instant to the next. Clearly metabolism must be carefully regulated if the rate of turnover is to be balanced by the rate of biosynthesis. It must also be regulated in response to a microbe's environment. Some of the mechanisms of metabolic regulation have already been introduced in chapter 9; others are discussed in chapter 13.

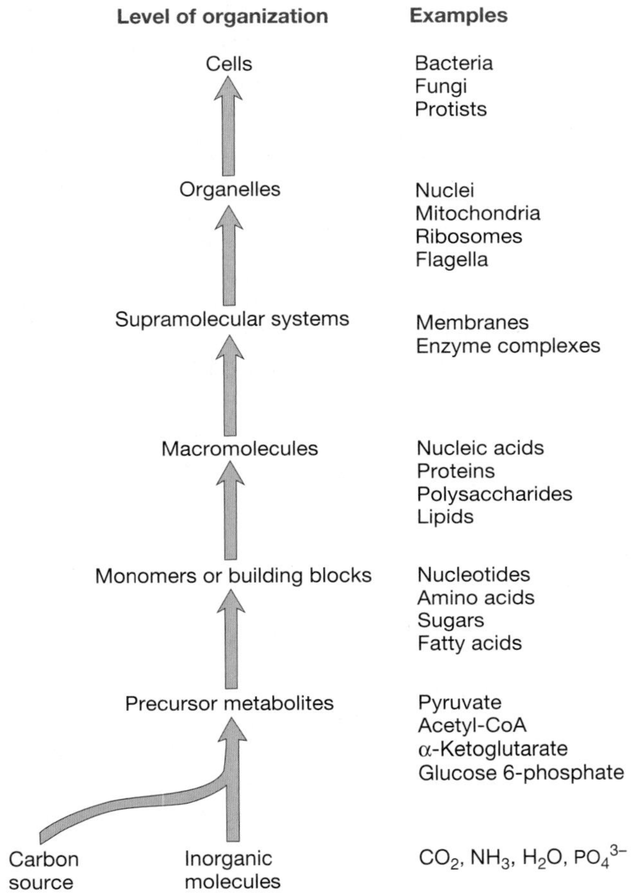

FIGURE 11.1 The Construction of Cells. The biosynthesis of cells and their constituents is organized in levels of ever greater complexity.

11.1 Principles Governing Biosynthesis

The problem faced by all cells is how to make the many molecules they need as efficiently as possible. Cells have solved this problem by carrying out biosynthesis using a few basic principles. Six are now briefly discussed.

1. **Large molecules are made from small molecules.** The construction of large **macromolecules** (complex molecules) from a few simple structural units (monomers) saves much genetic storage capacity, biosynthetic raw material, and energy. A consideration of protein synthesis clarifies this. Proteins—whatever size, shape, or function—are made of only 20 common amino acids joined by peptide bonds. Different proteins simply have different amino acid sequences but not new and dissimilar amino acids. Suppose that proteins were composed of 40 different amino acids instead of 20. The cell would then need the enzymes to manufacture twice as many amino acids (or would have to obtain the extra amino acids in its diet). Genes would be required for the extra enzymes, and the cell would have to invest raw materials and energy in the synthesis of these additional genes, enzymes, and amino acids. Clearly the use of a few monomers linked together by a single type of covalent bond makes the synthesis of macromolecules a highly efficient process. ▶▶| *Proteins (appendix I); Protein structure (section 12.3)*
2. **Many enzymes do double duty.** Many enzymes are used for both catabolic and anabolic processes, saving additional materials and energy. For example, most glycolytic

Table 11.1 Biosynthesis in *Escherichia coli*

Cell Constituent	*Number of Molecules per Cell*[a]	*Molecules Synthesized per Second*	*Molecules of ATP Required per Second for Synthesis*
DNA	1[b]	0.00083	60,000
RNA	15,000	12.5	75,000
Polysaccharides	39,000	32.5	65,000
Lipids	15,000,000	12,500.0	87,000
Proteins	1,700,000	1,400.0	2,120,000

[a] Estimates for a cell with a volume of 2.25 μm^3, a total weight of 1×10^{-12} g, a dry weight of 2.5×10^{-13} g, and a 20-minute cell division cycle.
[b] It should be noted that bacteria can contain multiple copies of their genomic DNA.

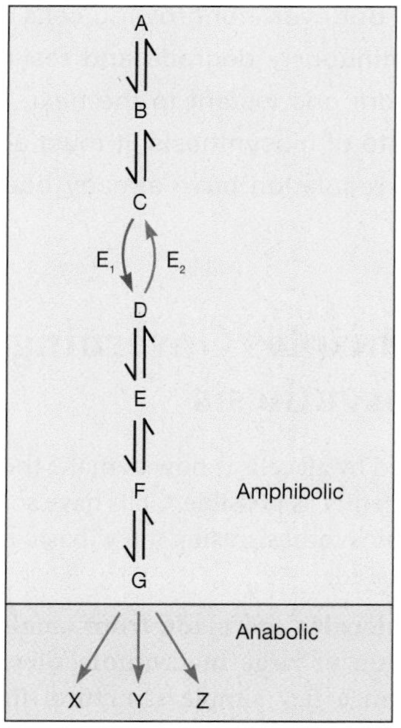

FIGURE 11.2 A Hypothetical Biosynthetic Pathway. The routes connecting G with X, Y, and Z are purely anabolic because they are used only for synthesis of the end products. The pathway from A to G is amphibolic—that is, it has both catabolic and anabolic functions. Most reactions are used in both roles; however, the interconversion of C and D is catalyzed by two separate enzymes, E_1 and E_2.

enzymes are involved in both the synthesis and the degradation of glucose (*see figure 10.3*).

3. **Some enzymes function in one direction only.** Although many steps of amphibolic pathways are catalyzed by enzymes that act reversibly, some are not. These steps require the use of separate enzymes: one to catalyze the catabolic reaction, the other to catalyze the anabolic reaction. The use of two enzymes allows independent regulation of catabolism and anabolism (**figure 11.2**). Thus catabolic and anabolic pathways are never identical, although many enzymes are shared. Although both types of pathways can be regulated by their end products as well as by the concentrations of ATP, ADP, AMP, and NAD^+, end-product regulation generally assumes more importance in anabolic pathways.
4. **Anabolic pathways are irreversible.** To synthesize molecules efficiently, anabolic pathways must operate irreversibly in the direction of biosynthesis. Cells achieve this by connecting some biosynthetic reactions to the breakdown of ATP and other nucleoside triphosphates. When these two processes are coupled, the free energy made available during nucleoside triphosphate breakdown drives the biosynthetic reaction to completion. ◀◀ *ATP (section 9.4)*
5. **Catabolism and anabolism are physically separated.** Catabolic and anabolic pathways can be localized into distinct cellular compartments—a process called compartmentation. In eukaryotes, this is easily done because eukaryotic cells have numerous membrane-bound organelles that carry out specific functions. Compartmentation also occurs in bacterial and archaeal cells. For instance, carboxysomes separate CO_2 fixation from other processes in bacteria. Compartmentation makes it easier for catabolic and anabolic pathways to operate simultaneously yet independently.
6. **Catabolism and anabolism use different cofactors.** Usually catabolic oxidations produce NADH, a substrate for electron transport. In contrast, when an electron donor is needed during biosynthesis, NADPH often serves as the donor.

After macromolecules have been constructed from simpler precursors, they are assembled into larger, more complex structures such as supramolecular systems and organelles (figure 11.1). Macromolecules normally contain the necessary information to form supramolecular systems spontaneously in a process known as **self-assembly.** For example, ribosomes are large assemblages of many proteins and ribonucleic acid molecules, yet they arise by the self-assembly of their components without the involvement of extra factors.

1. Define anabolism, turnover, and self-assembly.
2. Summarize the six principles by which biosynthetic pathways are organized.

11.2 Precursor Metabolites

Generation of the **precursor metabolites** is critical to anabolism, because they give rise to all other molecules made by the cell. Precursor metabolites are carbon skeletons (i.e., carbon chains) used as the starting substrates for the synthesis of monomers and other building blocks needed for the synthesis of macromolecules. Precursor metabolites are referred to as carbon skeletons because they are molecules that lack functional moieties such as amino and sulfhydryl groups; these are added during the biosynthetic process. The precursor metabolites and their use in biosynthesis are shown in **figures 11.3** and **11.4**. Several things should be noted in figure 11.3. First, all the precursor metabolites are intermediates of the glycolytic pathways (Embden-Meyerhof, Entner-Doudoroff, and the pentose phosphate pathways) and the tricarboxylic acid (TCA) cycle. Therefore these pathways play a central role in metabolism and are often referred to as the **central metabolic pathways.** Note, too, that most of the precursor metabolites are used for synthesis of amino acids and nucleotides. ◀◀ *Glycolytic pathways (section 10.3); Tricarboxylic acid cycle (section 10.4)* ▶▶ *Common metabolic pathways (appendix II)*

If an organism is a chemoorganotroph using glucose as its energy, electron, and carbon source (either aerobically or anaerobically), it generates the precursor metabolites as it generates ATP and reducing power. But what if the chemoorganotroph is using an amino acid as its sole source of carbon, electrons, and energy? And what about autotrophs? How do they generate

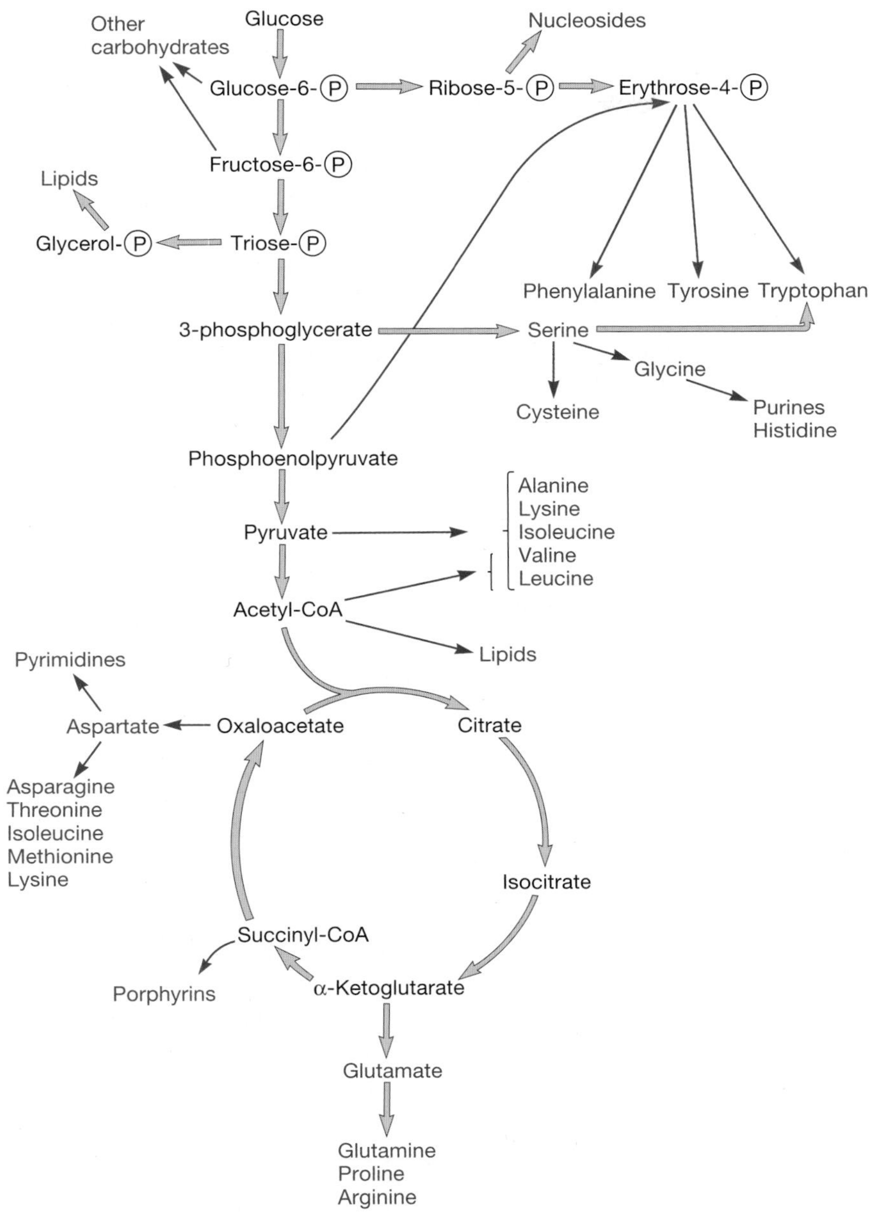

FIGURE 11.3 The Organization of Anabolism. Biosynthetic products (in blue) are derived from precursor metabolites.

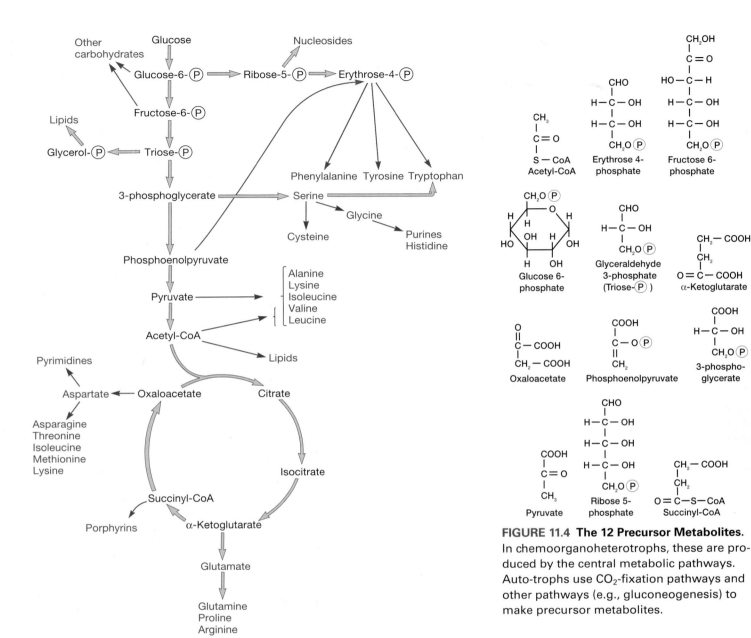

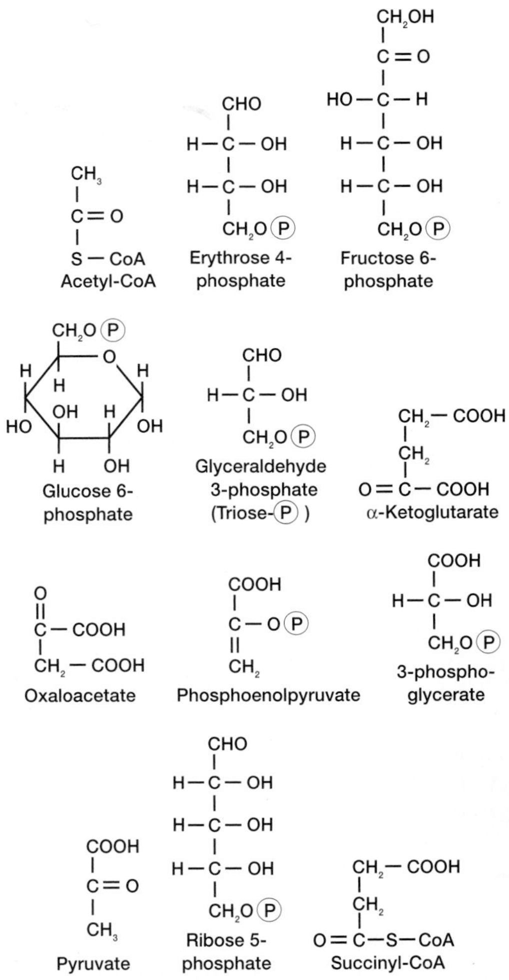

FIGURE 11.4 The 12 Precursor Metabolites. In chemoorganoheterotrophs, these are produced by the central metabolic pathways. Auto-trophs use CO_2-fixation pathways and other pathways (e.g., gluconeogenesis) to make precursor metabolites.

precursor metabolites from CO_2, their carbon source? Heterotrophs growing on something other than glucose degrade that carbon and energy source into one or more intermediates of the central metabolic pathways. From there, they can generate the remaining precursor metabolites. Autotrophs must first convert CO_2 into organic carbon from which they can generate the precursor metabolites. Many of the reactions that autotrophs use to generate the precursor metabolites are reactions of the central metabolic pathways, operating in either the catabolic direction or the anabolic direction. Thus the central metabolic pathways are important to the anabolism of both heterotrophs and autotrophs.

We begin our discussion of anabolism by first considering CO_2 fixation by autotrophs. Once CO_2 is converted to organic carbon, the synthesis of other precursor metabolites, amino acids, nucleotides, and additional building blocks is essentially the same in both autotrophs and heterotrophs. Recall that the precursor metabolites provide the carbon skeletons for the synthesis of other important organic molecules. In the process of transforming a precursor metabolite into an amino acid or a nucleotide, the carbon skeleton is modified in a number of ways, including the addition of nitrogen, phosphorus, and sulfur. Thus as we discuss the synthesis of monomers from precursor metabolites, we also address the assimilation of nitrogen, sulfur, and phosphorus. *A Biochemical Pathway*

11.3 CO_2 Fixation

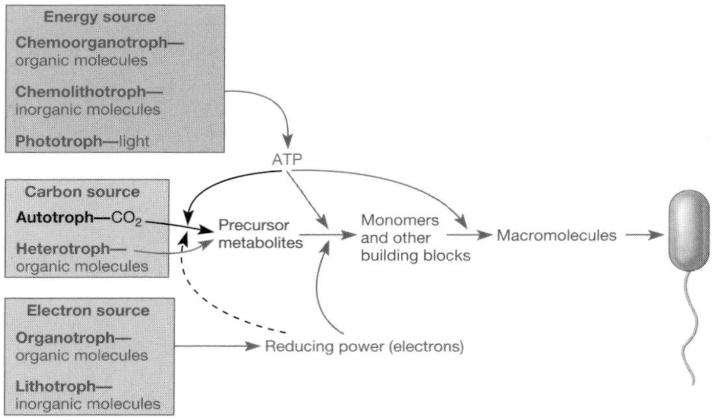

Autotrophs use CO_2 as their sole or principal carbon source, and the reduction and incorporation of CO_2 requires much energy. Many autotrophs obtain energy by trapping light during the light reactions of photosynthesis, but some derive energy from the oxidation of inorganic electron donors. Autotrophic CO_2 fixation is crucial to life on Earth because it provides the organic matter on which heterotrophs depend. |◀◀ *Chemolithotrophy (section 10.11); Phototrophy (section 10.12);* ▶▶| *Carbon cycle (section 26.1)*

Five different CO_2-fixation pathways have been identified in microorganisms. Most autotrophs use the **Calvin cycle,** which is also called the Calvin-Benson cycle or the reductive pentose phosphate cycle. The Calvin cycle is found in photosynthetic eukaryotes and most photosynthetic bacteria. Other pathways are used by some obligatory anaerobic and microaerophilic bacteria. Autotrophic archaea also use alternative pathways for CO_2 fixation. We consider the Calvin cycle first and then briefly introduce the other CO_2-fixation pathways.

Calvin Cycle

The Calvin cycle is also called the reductive pentose phosphate cycle because it is essentially the reverse of the pentose phosphate pathway. Thus many of the reactions are similar, in particular the sugar transformations. The reactions of the Calvin cycle occur in the chloroplast stroma of eukaryotic autotrophs. In cyanobacteria, some nitrifying bacteria, and thiobacilli (sulfur-oxidizing chemolithotrophs), the Calvin cycle is associated with inclusions called **carboxysomes.** These polyhedral structures contain the enzyme critical to the Calvin cycle and are the site of CO_2 fixation. |◀◀ *Pentose phosphate pathway (section 10.3)*

The Calvin cycle is divided into three phases: carboxylation phase, reduction phase, and regeneration phase (**figure 11.5** and *appendix II*). During the carboxylation phase, the enzyme **ribulose 1,5-bisphosphate carboxylase/oxygenase,** (RuBisCO), catalyzes the addition of CO_2 to the five-carbon molecule

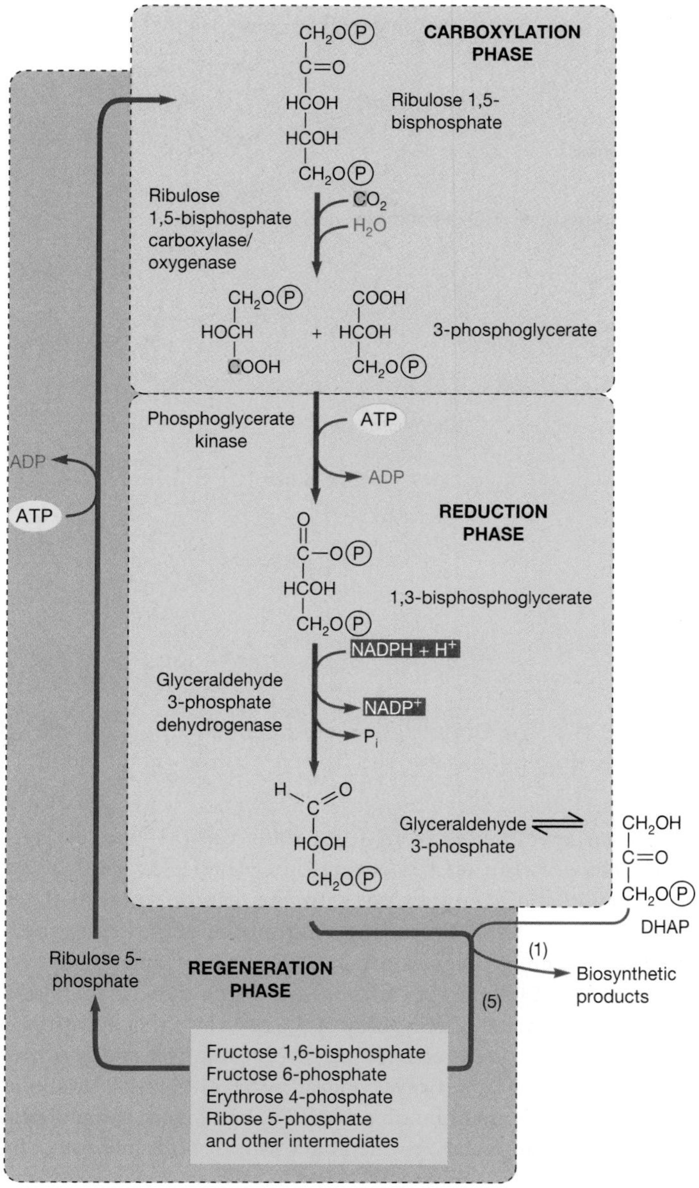

FIGURE 11.5 The Calvin Cycle. This overview of the cycle shows only the carboxylation and reduction phases in detail. Three ribulose 1,5-bisphosphates are carboxylated to give six 3-phosphoglycerates in the carboxylation phase. These are converted to six glyceraldehyde 3-phosphates, which can be converted to dihydroxyacetone phosphate (DHAP). Five of the six trioses (glyceraldehyde phosphate and dihydroxyacetone phosphate) are used to reform three ribulose 1,5-bisphosphates in the regeneration phase. The remaining triose is used in biosynthesis. The numbers in parentheses at the lower right indicate this carbon flow.

ribulose 1,5-bisphosphate (RuBP), forming a six-carbon intermediate that rapidly and spontaneously splits into two molecules of 3-phosphoglycerate (PGA). Note that PGA is an intermediate of the Embden-Meyerhof pathway (EMP), and in the reduction phase, PGA is reduced to glyceraldehyde 3-phosphate by two reactions that are essentially the reverse of two EMP reactions. The difference is that the Calvin cycle enzyme glyceraldehyde 3-phosphate dehydrogenase uses NADP rather than NAD (compare figures 11.5 and 10.4). Finally, in the regeneration phase, RuBP is reformed, so that the cycle can repeat. In addition, this phase produces carbohydrates such as glyceraldehyde 3-phosphate, fructose 6-phosphate, and glucose 6-phosphate, all of which are precursor metabolites (figures 11.3 and 11.4). This portion of the cycle is similar to the pentose phosphate pathway and involves the transketolase and transaldolase reactions. *The Calvin Cycle*

To synthesize fructose 6-phosphate or glucose 6-phosphate from CO_2, the cycle must operate six times to yield the desired hexose and reform the six RuBP molecules.

$$6\text{RuBP} + 6CO_2 \rightarrow 12\text{PGA} \rightarrow 6\text{RuBP} + \text{fructose-6-P}$$

The incorporation of one CO_2 into organic material requires three ATPs and two NADPHs. The formation of glucose from CO_2 may be summarized by the following equation.

$$6CO_2 + 18\text{ATP} + 12\text{NADPH} + 12H^+ + 12H_2O \rightarrow \text{glucose} + 18\text{ADP} + 18P_i + 12\text{NADP}^+$$

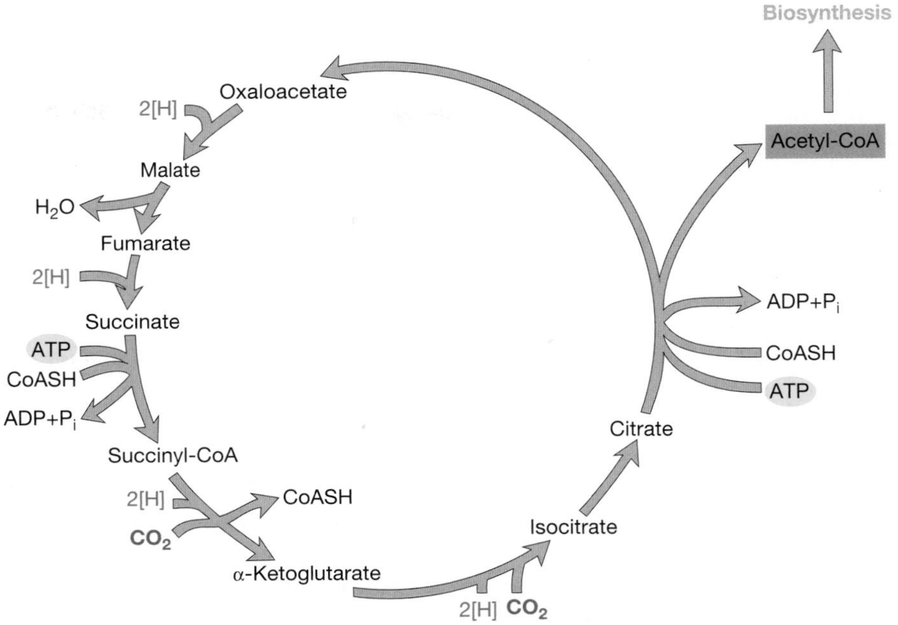

FIGURE 11.6 The Reductive TCA Cycle. This cycle is used by green sulfur bacteria and some chemolithotrophic bacteria and archaea to fix CO_2. The cycle runs in the opposite direction as the TCA cycle. ATP and reducing equivalents [H] power the reversal. In green sulfur bacteria, the reducing equivalents are provided by reduced ferredoxin. The product of this process is acetyl-CoA, which can be used to synthesize other organic molecules and precursor metabolites.

Other CO_2-Fixation Pathways

Certain bacteria and archaea fix CO_2 using the reductive TCA cycle, the 3-hydroxypropionate cycle, the acetyl-CoA pathway, or the 3-hydroxypropionate/4-hydroxybutyrate pathway. The **reductive TCA cycle** (**figure 11.6**) is used by some chemolithoautotrophs (e.g., *Thermoproteus* and *Sulfolobus,* two archaeal genera, the bacterial genus *Aquifex*, and some proteobacteria) and anoxygenic phototrophs such as *Chlorobium*, a green sulfur bacterium. The reductive TCA cycle is so named because it runs in the reverse direction of the normal, oxidative TCA cycle (compare figures 11.6 and 10.7). A few archaeal genera and the green nonsulfur bacteria (another group of anoxygenic phototrophs) use the **3-hydroxypropionate cycle** to fix CO_2. **Figure 11.7** shows the cycle as it is thought to function in the green nonsulfur bacterium *Chloroflexus aurantiacus*. How its product, glyoxylate, is assimilated is unclear. Methanogens use portions of the **acetyl-CoA pathway** for carbon fixation; the pathway as it is used by *Methanobacterium thermoautotrophicum* is illustrated in **figure 11.8.** Acetogens use the pathway in its entirety. Both the acetyl-CoA pathway and methanogenesis involve the activity of a

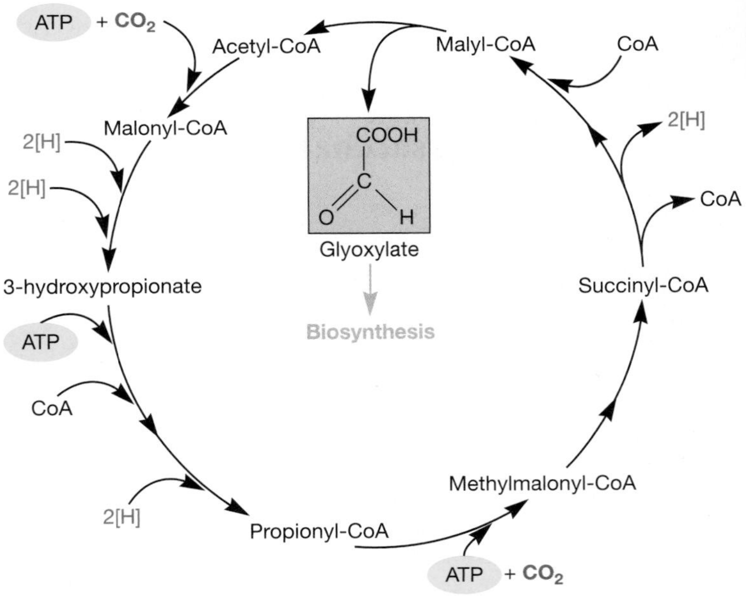

FIGURE 11.7 The 3-Hydroxypropionate Pathway. This pathway functions in green nonsulfur bacteria, a group of anoxygenic phototrophs. The product of the cycle is glyoxylate, which is used in biosynthesis by mechanisms that have not been definitively elucidated.

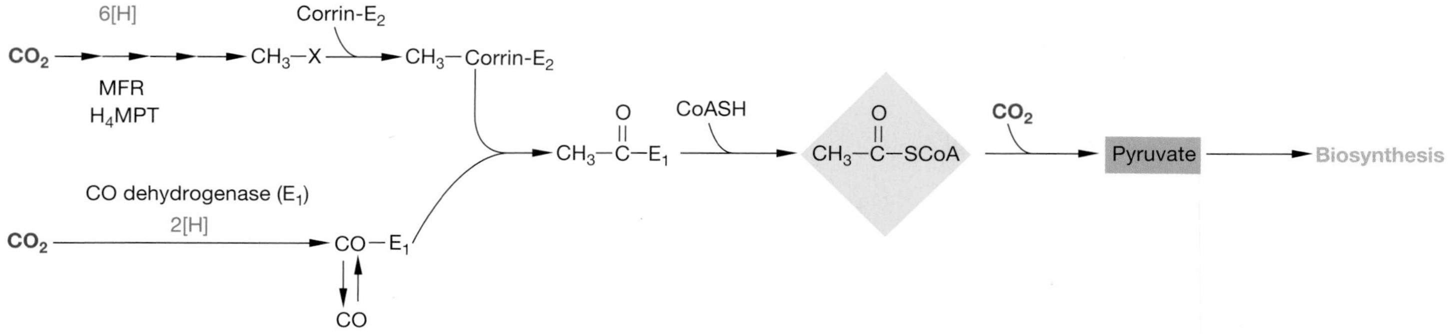

FIGURE 11.8 The Acetyl-CoA Pathway. Methanogens reduce two molecules of CO_2, each by a different mechanism, and combine them to form acetyl and then acetyl-CoA. Acetogenic bacteria use a slightly different version of the pathway.

number of unusual enzymes and coenzymes. These are described in more detail in chapter 18. The final pathway for CO_2-fixation was first described in 2007. This pathway was discovered in an archaeon. It begins using reactions also observed in the 3-hydroxypropionate cycle but then uses a series of unique reactions that produce 4-hydroxybutyrate. The pathway is called the **3-hydroxypropionate/4-hydroxybutyrate pathway** (**figure 11.9**). ▶▶| *Phylum* Crenarchaeota *(section 18.2); Methanogens (section 18.3);* Aquificae *and* Thermotogae *(section 19.1); Photosynthetic bacteria (section 19.3)*

1. Briefly describe the three phases of the Calvin cycle. What other pathways are used to fix CO_2?
2. Which two enzymes are specific to the Calvin cycle?

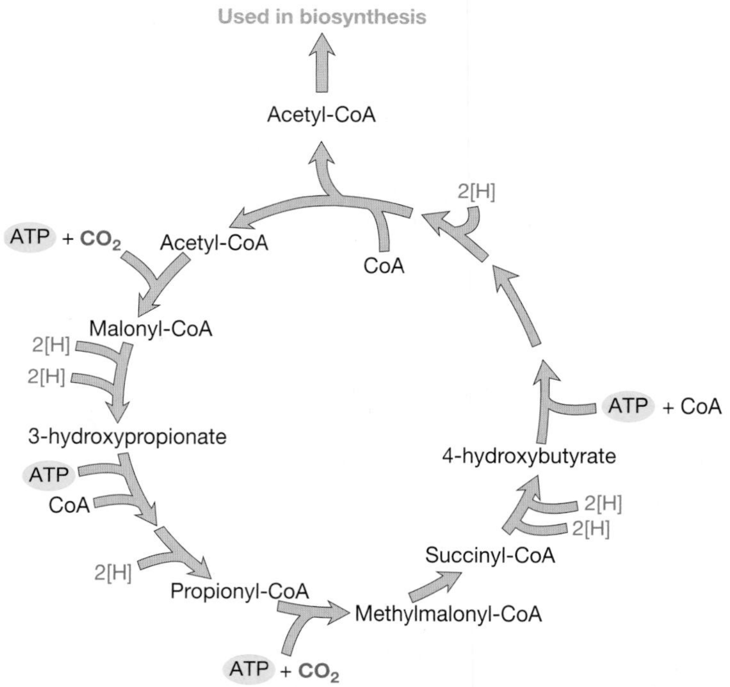

FIGURE 11.9 The 3-Hydroxypropionate/4-Hydroxybutyrate Pathway.

11.4 Synthesis of Sugars and Polysaccharides

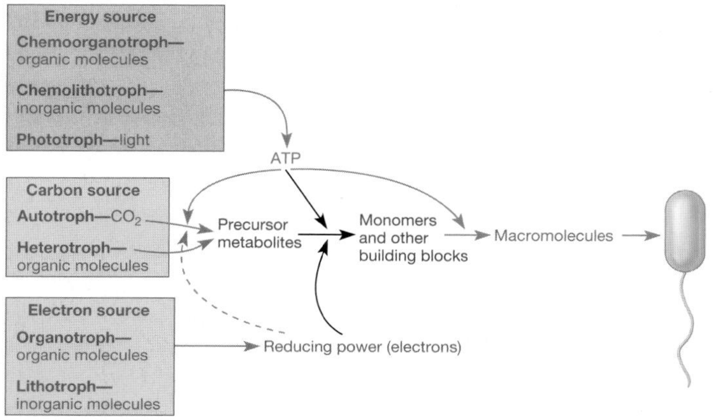

Autotrophs using CO_2-fixation processes other than the Calvin cycle and heterotrophs growing on carbon sources other than sugars must be able to synthesize glucose. The synthesis of glucose from noncarbohydrate precursors is called **gluconeogenesis.** The gluconeogenic pathway shares six enzymes with the Embden-Meyerhof pathway. However, the two pathways are not identical (**figure 11.10**). Four reactions are catalyzed by enzymes that are specific for gluconeogenesis. Two of these enzymes are involved in the conversion of pyruvate to phosphoenolpyruvate. Fructose bisphosphatase catalyzes the formation of fructose 6-phosphate from fructose 1,6-bisphosphate. Finally, glucose 6-phosphatase removes the phosphate from glucose 6-phosphate to generate glucose.

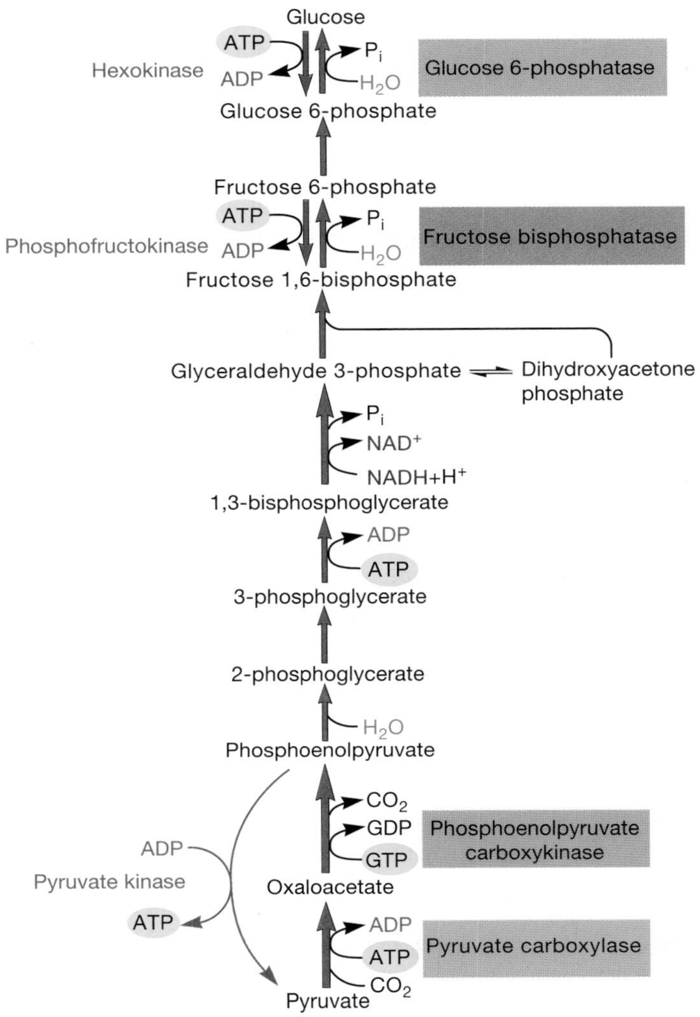

FIGURE 11.10 Gluconeogenesis. The gluconeogenic pathway used in many microorganisms. The names of the four enzymes catalyzing irreversible reactions that are different from those found in the Embden-Meyerhof pathway (EMP) are in shaded boxes. EMP steps are shown in blue for comparison.

Figure 11.10 Micro Inquiry

Why do you think it is important that the enzymes dedicated to gluconeogenesis (rather than those used in both catabolic and anabolic reactions) are found in both the 3-carbon and 6-carbon stages?

Synthesis of Monosaccharides

As can be seen in figure 11.10, gluconeogenesis synthesizes fructose 6-phosphate and glucose 6-phosphate. Once these two precursor metabolites have been formed, other common sugars can be manufactured. For example, mannose comes directly from fructose 6-phosphate by a simple rearrangement of a hydroxyl group *(see figure AI.9)*. Several sugars are synthesized while attached to a nucleoside diphosphate. The most important nucleoside diphosphate sugar is **uridine diphosphate glucose (UDPG),** which is formed when glucose reacts with uridine triphosphate (**figure 11.11**). UDP carries glucose around the cell for participation in enzyme reactions much like ADP bears phosphate in the form of ATP. Other important uridine diphosphate sugars are UDP-galactose and UDP-glucuronic acid.

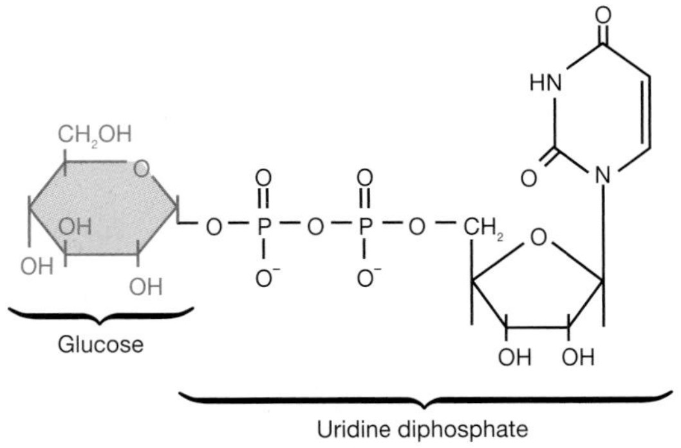

FIGURE 11.11 Uridine Diphosphate Glucose.

Synthesis of Polysaccharides

Nucleoside diphosphate sugars also play a central role in the synthesis of polysaccharides such as starch and glycogen, both of which are long chains of glucose. Again, biosynthesis is not simply a direct reversal of catabolism. For instance, during the synthesis of glycogen and starch in bacteria and protists, adenosine diphosphate glucose (ADP-glucose) is formed from glucose 1-phosphate and ATP. It then donates glucose to the end of growing glycogen and starch chains.

$$\text{ATP} + \text{glucose 1-phosphate} \rightarrow \text{ADP-glucose} + \text{PP}_i$$

$$(\text{Glucose})_n + \text{ADP-glucose} \rightarrow (\text{glucose})_{n+1} + \text{ADP}$$

Synthesis of Peptidoglycan

Nucleoside diphosphate sugars also participate in the synthesis of peptidoglycan. Recall that peptidoglycan is a large, complex molecule consisting of long polysaccharide chains made of alternating *N*-acetylmuramic acid (NAM) and *N*-acetylglucosamine (NAG) residues. Pentapeptide chains are attached to the NAM groups. Peptidoglycan chains are cross-linked by bonds formed between the pentapeptides of adjacent chains. In gram-negative bacteria and many gram-positives, the pentapeptide chains are directly linked. In some gram-positive bacteria, the pentapeptides are linked by an interbridge consisting of one or more amino acids *(see figures 3.11 and 3.16)*. ◀◀ *Bacterial cell walls (section 3.3)*

Not surprisingly, such an intricate structure requires an equally intricate biosynthetic process, especially because some reactions occur in the cytoplasm, others in the membrane, and

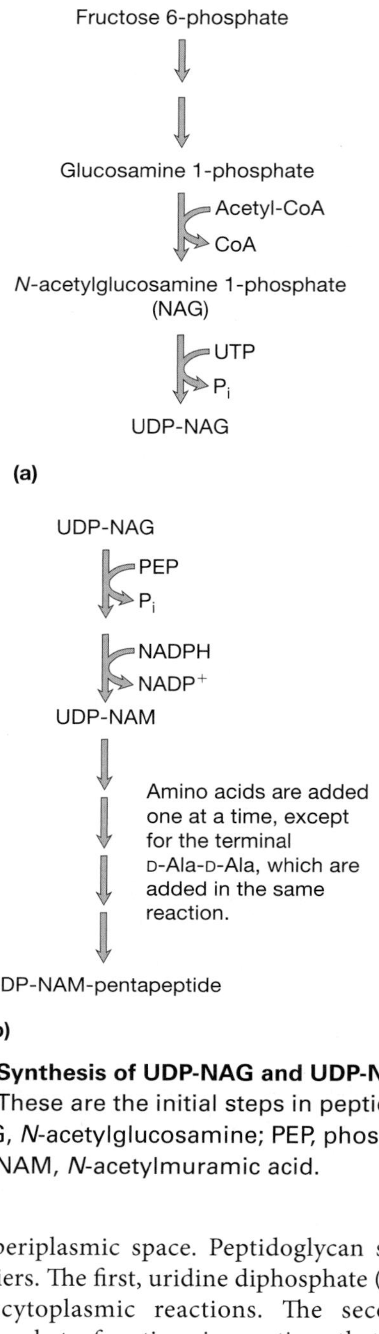

FIGURE 11.12 Synthesis of UDP-NAG and UDP-NAM-Pentapeptide. These are the initial steps in peptidoglycan synthesis. NAG, *N*-acetylglucosamine; PEP, phosphoenolpyruvate; NAM, *N*-acetylmuramic acid.

others in the periplasmic space. Peptidoglycan synthesis involves two carriers. The first, uridine diphosphate (UDP), functions in the cytoplasmic reactions. The second carrier, bactoprenol phosphate, functions in reactions that occur in the plasma membrane.

In the first step of peptidoglycan synthesis, UDP derivatives of NAM and NAG are formed (**figure 11.12**). Amino acids are then added sequentially to UDP-NAM to form the pentapeptide chain. NAM-pentapeptide is then transferred to bactoprenol phosphate (also called undecaprenyl phosphate), which is located at the cytoplasmic side of the plasma membrane (**figure 11.13**). The resulting intermediate is often called Lipid I. Bactoprenol is a 55-carbon, isoprene-derived alcohol that is linked to NAM by a pyrophosphate group (**figure 11.14**). Next, UDP transfers NAG to the bactoprenol-NAM-pentapeptide complex (Lipid I) to generate Lipid II. This creates the peptidoglycan repeat unit. The repeat unit is transferred across the membrane by bactoprenol. If the peptidoglycan unit requires an interbridge, it is added while the repeat unit is within the membrane. Bactoprenol pyrophosphate stays within the membrane and does not enter the periplasmic space. After releasing the peptidoglycan repeat unit into the periplasmic space, bactoprenol pyrophosphate is dephosphorylated to bactoprenol phosphate and returns to the cytoplasmic side of the plasma membrane, where it can function in the next round of synthesis. Meanwhile, the peptidoglycan repeat unit is added to the growing end of a peptidoglycan chain. The final step in peptidoglycan synthesis is **transpeptidation** (**figure 11.15**), which creates the peptide cross-links between the peptidoglycan chains. The enzyme that catalyzes the reaction removes the terminal D-alanine as the cross-link is formed. *Peptidoglycan Biosynthesis*

To grow and divide efficiently, a bacterial cell must add new peptidoglycan to its cell wall in a precise and well-regulated way while maintaining wall shape and integrity in the presence of high osmotic pressure (*see figure 7.9*). Because the cell wall peptidoglycan is essentially a single, enormous network, the growing bacterium must be able to degrade it just enough to provide acceptor ends for the incorporation of new peptidoglycan units. It must also reorganize peptidoglycan structure when necessary. This limited peptidoglycan digestion is accomplished by enzymes known as **autolysins,** some of which attack the polysaccharide chains, while others hydrolyze the peptide cross-links. Autolysin inhibitors are produced to keep the activity of these enzymes under tight control. ◀◀ *Cellular growth and determination of cell shape (section 7.2)*

The synthesis of peptidoglycan is a particularly effective target for antimicrobial agents because of its importance to bacterial cell wall structure and function. Inhibition of any stage of synthesis weakens the cell wall and can lead to lysis. Many commonly used antibiotics interfere with peptidoglycan synthesis. For example, penicillin inhibits the transpeptidation reaction (figure 11.15), and bacitracin blocks the dephosphorylation of bactoprenol pyrophosphate (figure 11.13). ▶▶ *Inhibitors of cell wall synthesis (section 34.4)*

1. What is gluconeogenesis? Why is it important?
2. Describe the formation of mannose, galactose, starch, and glycogen. What are nucleoside diphosphate sugars? How do microorganisms use them?
3. Suppose that a microorganism is growing on a medium that contains amino acids but no sugars. In general terms, how would it synthesize the pentoses and hexoses it needs?
4. Diagram the steps involved in the synthesis of peptidoglycan and show where they occur in the cell. What are the roles of bactoprenol and UDP? What is unusual about the synthesis of the pentapeptide chain?
5. What would happen to a cell if it did not produce any autolysins? What if it produced too many in an unregulated fashion?

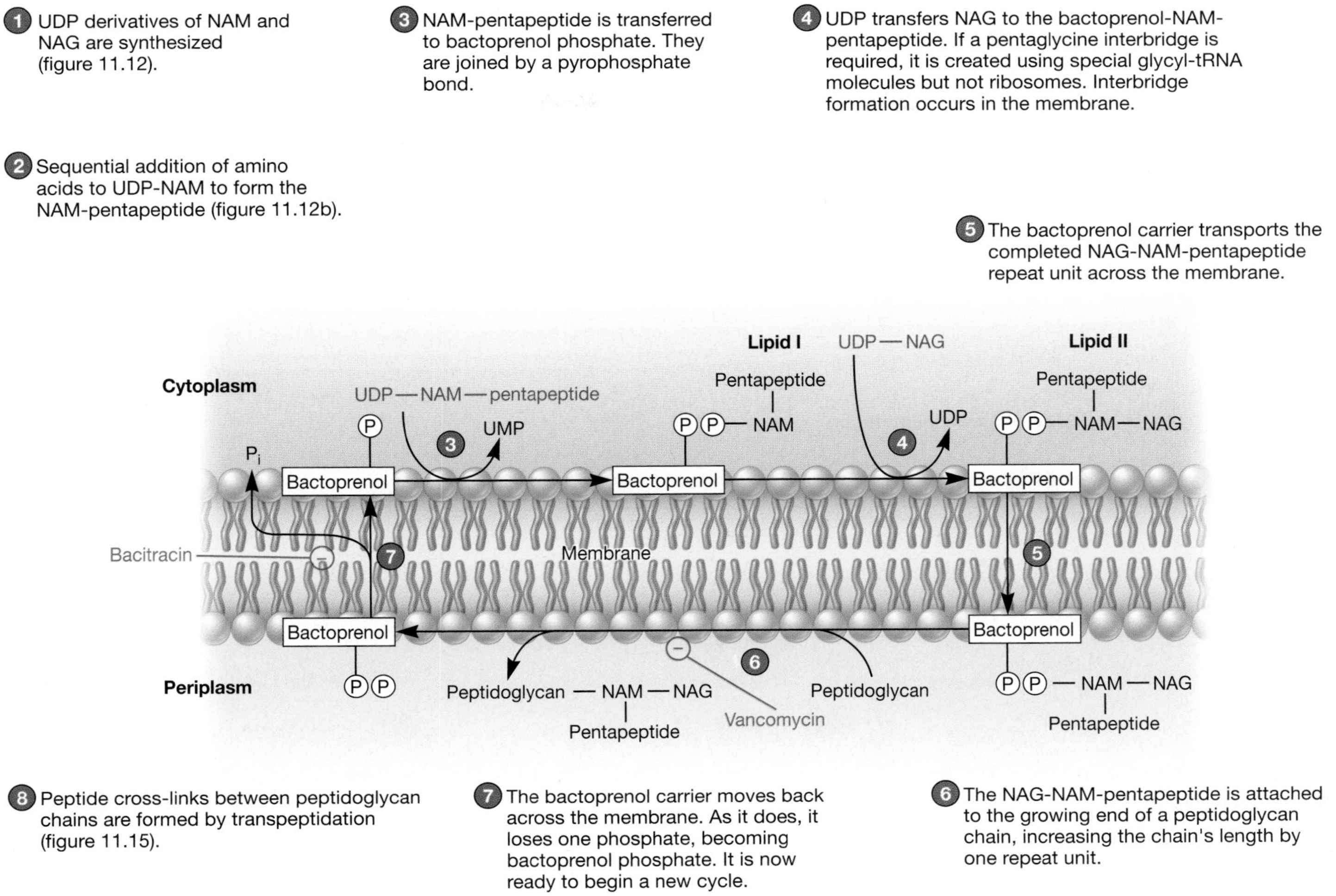
1 UDP derivatives of NAM and NAG are synthesized (figure 11.12).
3 NAM-pentapeptide is transferred to bactoprenol phosphate. They are joined by a pyrophosphate bond.
4 UDP transfers NAG to the bactoprenol-NAM-pentapeptide. If a pentaglycine interbridge is required, it is created using special glycyl-tRNA molecules but not ribosomes. Interbridge formation occurs in the membrane.
2 Sequential addition of amino acids to UDP-NAM to form the NAM-pentapeptide (figure 11.12b).
5 The bactoprenol carrier transports the completed NAG-NAM-pentapeptide repeat unit across the membrane.
Cytoplasm
UDP—NAM—pentapeptide
P
UMP
Lipid I
Pentapeptide
PP— NAM
UDP—NAG
UDP
Lipid II
Pentapeptide
PP— NAM—NAG
Pi
Bactoprenol
Bactoprenol
Bactoprenol
Bacitracin
Membrane
Bactoprenol
Bactoprenol
Periplasm
PP
Peptidoglycan — NAM—NAG
Pentapeptide
Vancomycin
Peptidoglycan
PP — NAM—NAG
Pentapeptide
8 Peptide cross-links between peptidoglycan chains are formed by transpeptidation (figure 11.15).
7 The bactoprenol carrier moves back across the membrane. As it does, it loses one phosphate, becoming bactoprenol phosphate. It is now ready to begin a new cycle.
6 The NAG-NAM-pentapeptide is attached to the growing end of a peptidoglycan chain, increasing the chain's length by one repeat unit.

FIGURE 11.13 Peptidoglycan Synthesis. NAM is *N*-acetylmuramic acid and NAG is *N*-acetylglucosamine. The pentapeptide contains L-lysine in *Staphylococcus aureus* peptidoglycan, and diaminopimelic acid (DAP) in *E. coli.* Inhibition by bacitracin, cycloserine, and vancomycin also is shown.

Figure 11.13 Micro Inquiry

What is the difference between Lipid I and Lipid II? What biosynthetic reaction occurs within the plasma membrane?

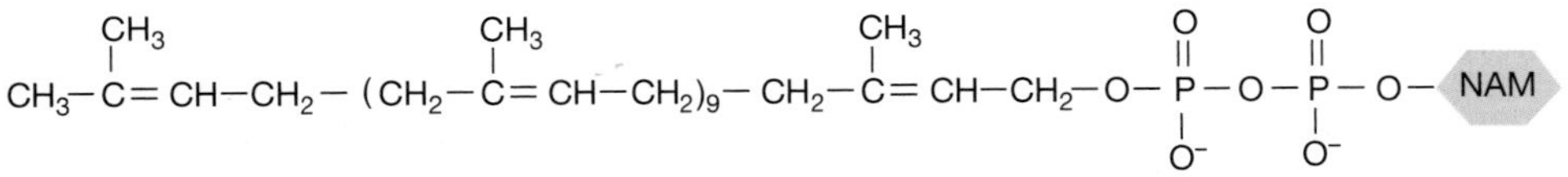
NAM

FIGURE 11.14 Bactoprenol-NAM. Bactoprenol is connected to *N*-acetylmuramic acid (NAM) by pyrophosphate.

11.5 Synthesis of Amino Acids

Many of the precursor metabolites serve as starting substrates for the synthesis of amino acids (figure 11.3). In the amino acid biosynthetic pathways, the carbon skeleton is remodeled and an amino group, and sometimes sulfur, is added. In this section, we first examine the mechanisms by which nitrogen and sulfur are assimilated and incorporated into amino acids. This is followed by a brief consideration of the organization of amino acid biosynthetic pathways.

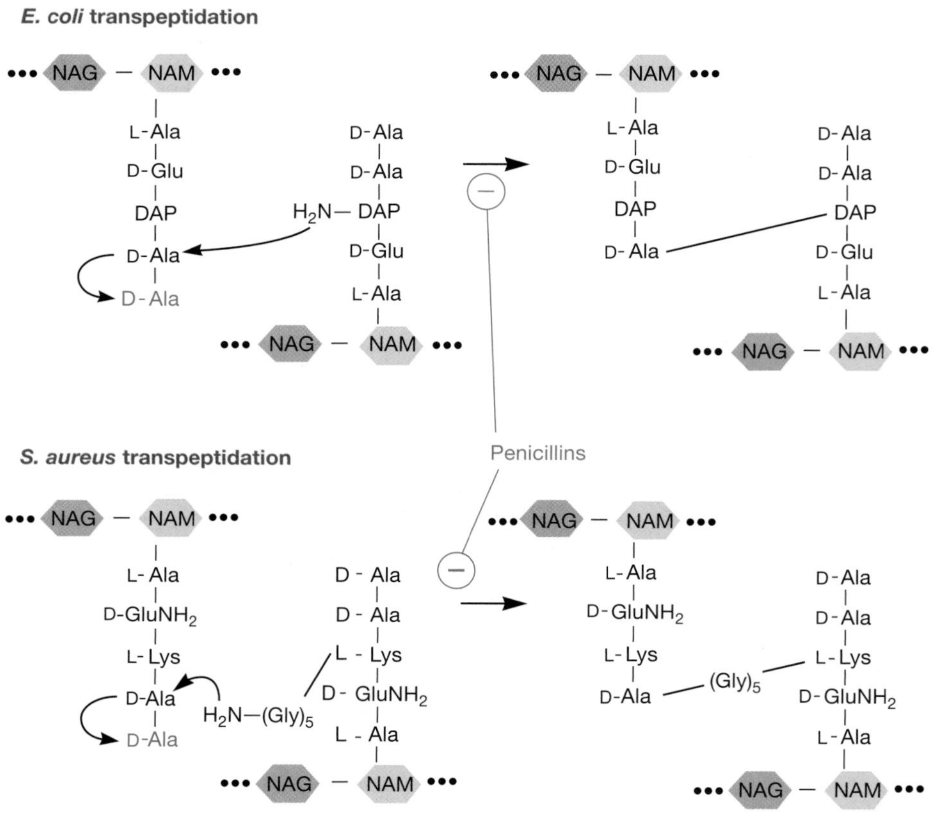

FIGURE 11.15 Transpeptidation. The transpeptidation reactions in the formation of the peptidoglycans of *Escherichia coli* and *Staphylococcus aureus.*

Nitrogen Assimilation

Nitrogen is a major component not only of proteins but also of nucleic acids, coenzymes, and many other cell constituents. Thus the cell's ability to assimilate inorganic nitrogen is exceptionally important. Although nitrogen gas is abundant in the atmosphere, only a few bacteria and archaea can reduce the gas and use it as a nitrogen source. Most must incorporate either ammonia or nitrate. We examine ammonia and nitrate assimilation first and then briefly discuss nitrogen assimilation in microbes that fix N_2.

Ammonia Incorporation

Ammonia can be incorporated into organic material relatively easily and directly because it is more reduced than other forms of inorganic nitrogen. Ammonia is initially incorporated into carbon skeletons by one of two mechanisms: reductive amination or the glutamine synthetase–glutamate synthase system. Once incorporated, the nitrogen can be transferred to other carbon skeletons by enzymes called transaminases.

The major **reductive amination pathway** involves the formation of glutamate from α-ketoglutarate, catalyzed in many bacteria and fungi by glutamate dehydrogenase when the ammonia concentration is high (**figure 11.16**).

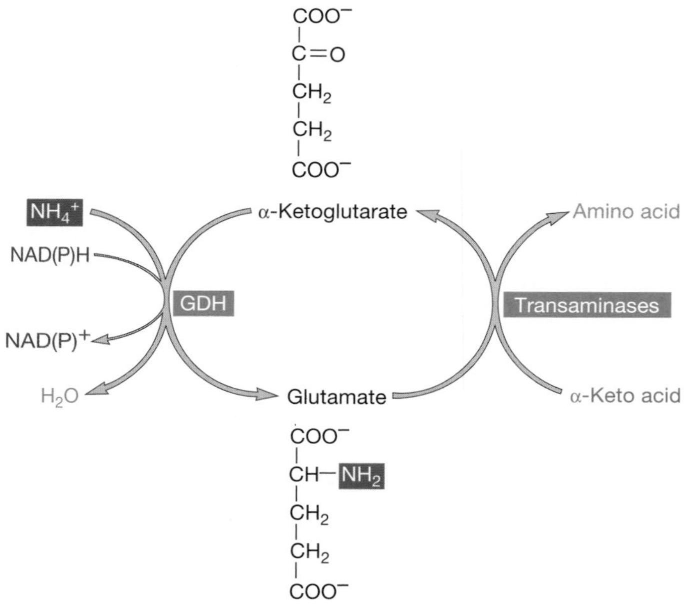

FIGURE 11.16 Ammonia Assimilation by Reductive Amination and Transaminases. Either NADPH- or NADH-dependent glutamate dehydrogenases may be involved. This route is most active at high ammonia concentrations.

$$\alpha\text{-ketoglutarate} + NH_4^+ + NAD(P)H + H^+ \rightleftharpoons \text{glutamate} + NAD(P)^+ + H_2O$$

Once glutamate has been synthesized, the newly formed α-amino group can be transferred to other carbon skeletons by transamination reactions to form different amino acids. **Transaminases** possess the coenzyme pyridoxal phosphate, which is responsible for the amino group transfer. Microorganisms have a number of transaminases, each of which catalyzes the formation of several amino acids using the same amino acid as an amino group donor.

The **glutamine synthetase–glutamate synthase (GS-GOGAT)** system is observed in *E. coli, Bacillus megaterium,* and other bacteria (**figure 11.17**). It functions when ammonia levels are low. Incorporation of ammonia by this system begins when ammonia is used to synthesize glutamine from glutamate in a reaction catalyzed by **glutamine synthetase** (**figure 11.18**). Then the amide nitrogen of glutamine is transferred to α-ketoglutarate to generate a new glutamate molecule. This reaction is catalyzed by **glutamate synthase.** Because glutamate acts as an amino donor in transaminase reactions, ammonia may be used to synthesize all common amino acids when suitable transaminases are present.

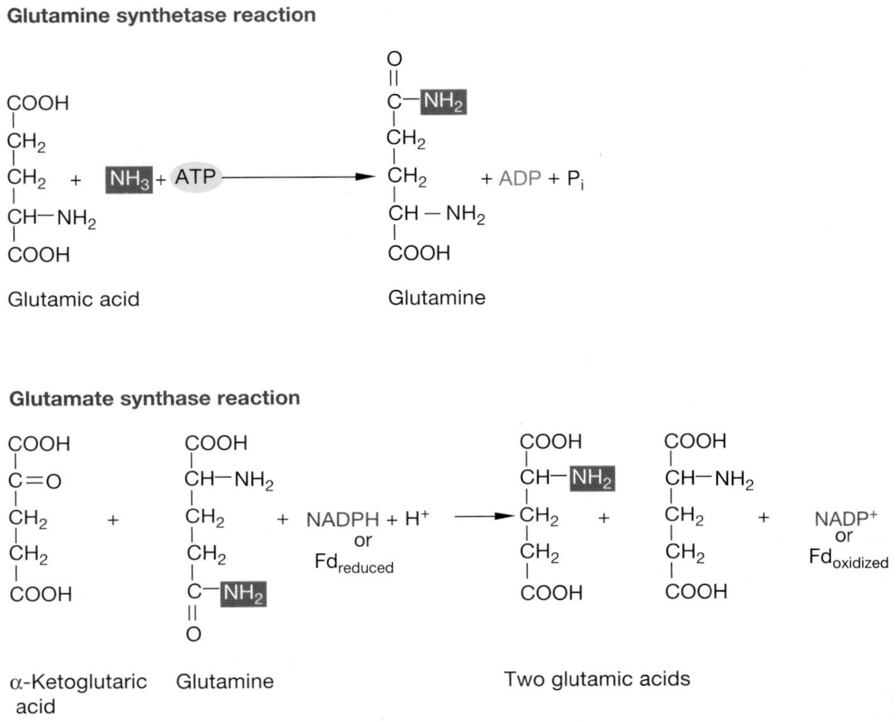

FIGURE 11.17 Glutamine Synthetase and Glutamate Synthase. The glutamine synthetase and glutamate synthase reactions involved in ammonia assimilation. Some glutamine synthases use NADPH as an electron source; others use reduced ferredoxin (Fd).

Figure 11.17 Micro Inquiry

What purpose is served for the cell when these reactions are used to produce two glutamate rather than a single glutamine?

Assimilatory Nitrate Reduction

The nitrogen in nitrate (NO_3^-) is much more oxidized than that in ammonia. Therefore nitrate must first be reduced to ammonia before the nitrogen can be converted to an organic form. This reduction of nitrate is called **assimilatory nitrate reduction,** which is not the same as that which occurs during anaerobic respiration (dissimilatory nitrate reduction). In assimilatory nitrate reduction, nitrate is incorporated into organic material and does not participate in energy conservation. The process is widespread among bacteria, fungi, and photosynthetic protists, and it is an important step in the nitrogen cycle. ◀◀ *Anaerobic respiration (section 10.6);* ▶▶ *Nitrogen cycle (section 26.1)*

Assimilatory nitrate reduction takes place in the cytoplasm in bacteria. The first step in nitrate assimilation is its reduction to nitrite by **nitrate reductase,** an enzyme that contains both

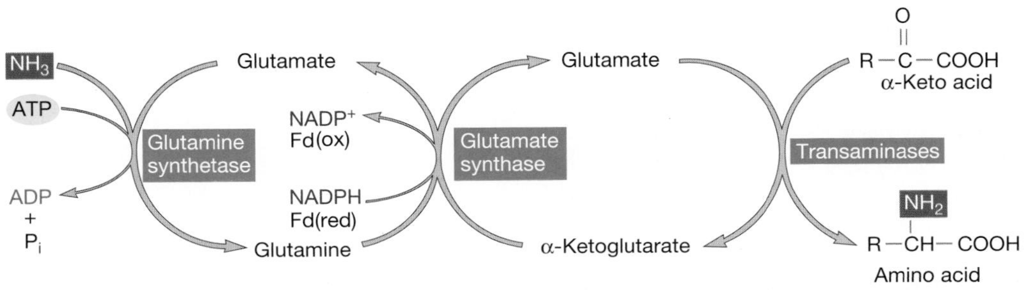

FIGURE 11.18 Ammonia Incorporation Using Glutamine Synthetase, Glutamate Synthase, and Transaminases. This route is effective at low ammonia concentrations.

FAD and molybdenum (**figure 11.19**). NADPH is the electron donor.

$$NO_3^- + NADPH + H^+ \rightarrow NO_2^- + NADP^+ + H_2O$$

Nitrite is next reduced to ammonia with a series of two electron additions catalyzed by nitrite reductase and possibly other enzymes. The ammonia is then incorporated into amino acids by the routes already described.

Nitrogen Fixation

The reduction of atmospheric gaseous nitrogen to ammonia is called **nitrogen fixation.** Because ammonia and nitrate levels often are low and only a few bacteria and archaea can carry out nitrogen fixation (eukaryotic cells completely lack this ability), the rate of this process limits plant growth in many situations. Nitrogen fixation occurs in (1) free-living chemotrophic bacteria and archaea (e.g., *Azotobacter, Klebsiella, Clostridium,* and *Methanococcus*), (2) bacteria living in symbiotic associations with plants such as legumes (e.g., *Rhizobium*), and (3) cyanobacteria (e.g., *Nostoc, Anabaena,* and *Trichodesmium*). These microbes play a critical role in the nitrogen cycle. They complete the cycle from NO_3^- (the most oxidized form of nitrogen) to ammonia (the most reduced form of nitrogen) via N_2, which has an intermediate oxidation state. The biological aspects of nitrogen fixation are discussed in chapters 26 and 29. The biochemistry of nitrogen fixation is the focus of this section.

The reduction of nitrogen to ammonia is catalyzed by the enzyme **nitrogenase.** Although the enzyme-bound intermediates in this process are still unknown, it is thought that nitrogen is reduced by two-electron additions in a way similar to that illustrated in **figure 11.20**. The reduction of molecular nitrogen to ammonia is quite exergonic, but the reaction has a high activation energy because molecular nitrogen is an unreactive gas with a triple bond connecting the two nitrogen atoms. Therefore nitrogen reduction is expensive, requiring a large ATP expenditure—at least 8 electrons and 16 ATP molecules (4 ATPs per pair of electrons). Ferridoxin is used as the electron donor.

$$N_2 + 8H^+ + 8e^- + 16ATP \rightarrow 2NH_3 + H_2 + 16ADP + 16P_i$$

Nitrogenase is a complex enzyme consisting of two major protein components, a MoFe protein (MW 220,000) joined with one or two Fe proteins (MW 64,000). The MoFe protein contains 2 atoms of molybdenum and 28 to 32 atoms of iron; the Fe protein has 4 iron atoms. Fe protein is first reduced by ferredoxin and then binds ATP (**figure 11.21**). ATP binding changes the conformation of the Fe protein and lowers its reduction potential (−0.29 to −0.40 V), enabling it to reduce the MoFe protein. ATP is hydrolyzed when this electron transfer occurs. Finally, reduced MoFe protein donates electrons to atomic nitrogen. Nitrogenase is quite sensitive to O_2 and must be protected from O_2 inactivation within the cell. Microbes use a variety of strategies to protect nitrogenase, as we discuss more fully in chapters 19 and 29.

The reduction of N_2 to NH_3 occurs in three steps, each of which requires an electron pair (figures 11.20 and 11.21). Six electron transfers take place, and this uses a total of 12 ATPs per N_2 reduced. The overall process actually requires at least 8 electrons and 16 ATPs because nitrogenase also catalyzes an additional reaction that essentially "short-circuits" the nitrogen-fixation process. The additional reaction is the reduction of protons to H_2. The H_2 then reacts with diimine (HN=NH) to regenerate N_2 (and also form H_2). Thus nitrogen fixation becomes even more expensive. Symbiotic nitrogen-fixing bacteria can consume almost 20% of the ATP produced by the host plant. Once molecular nitrogen has been reduced to ammonia, the ammonia can be incorporated into organic compounds.

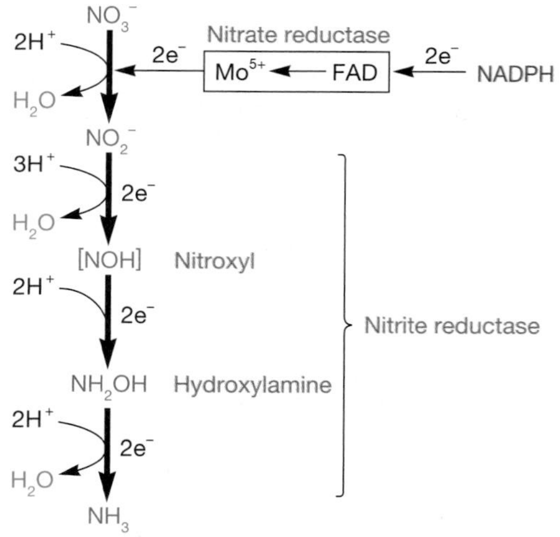

FIGURE 11.19 Assimilatory Nitrate Reduction. This sequence is thought to operate in bacteria that can reduce and assimilate nitrate nitrogen.

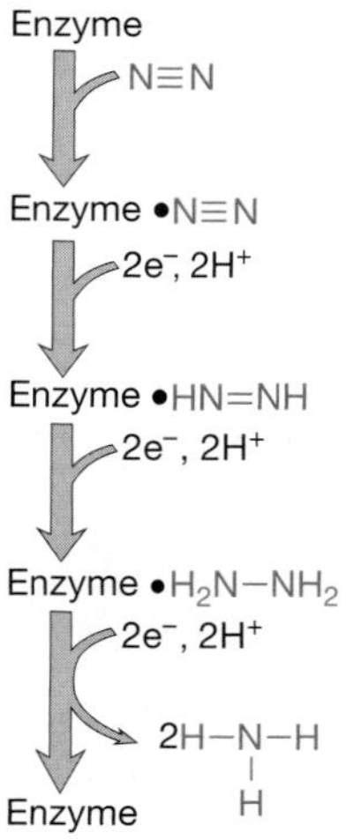

FIGURE 11.20 Nitrogen Reduction. The proposed sequence of nitrogen reduction by nitrogenase.

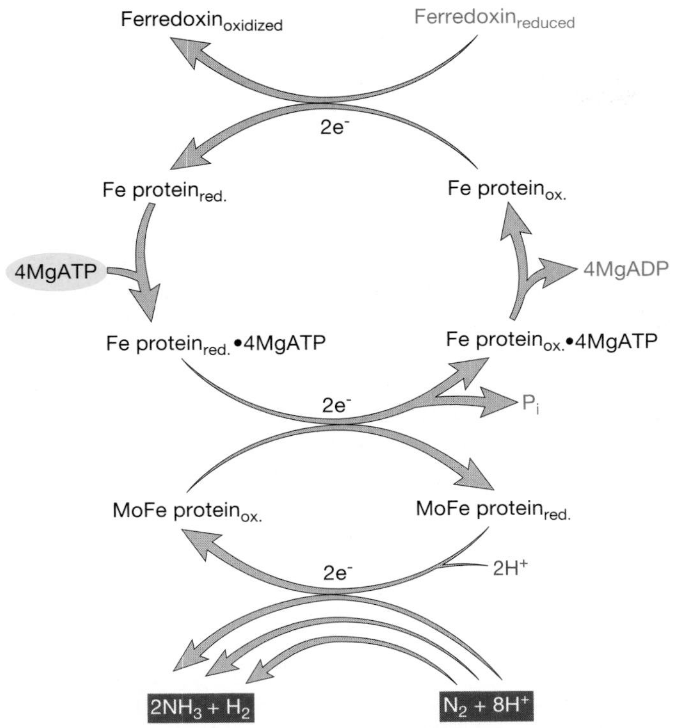

FIGURE 11.21 Mechanism of Nitrogenase Action. The flow of two electrons from ferredoxin to nitrogen is outlined. This process is repeated three times in order to reduce N_2 to two molecules of ammonia. The stoichiometry at the bottom includes proton reduction to H_2.

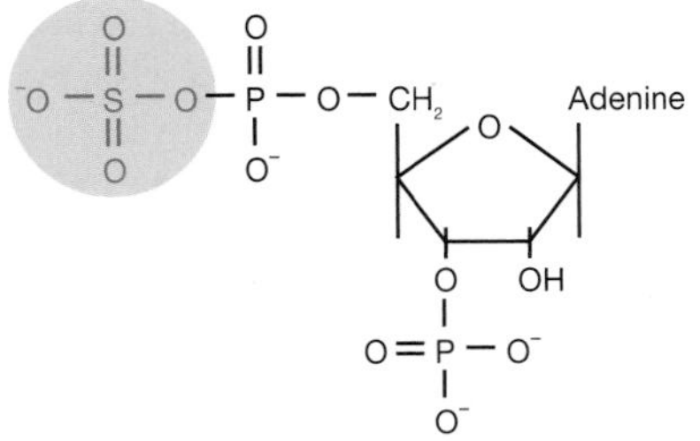

FIGURE 11.22 Phosphoadenosine 5′-phosphosulfate (PAPS). The sulfate group is in color.

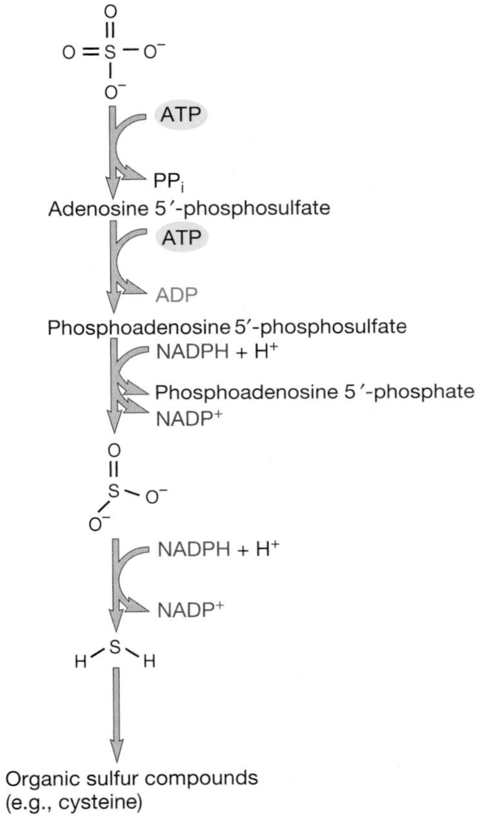

FIGURE 11.23 The Sulfate Reduction Pathway.

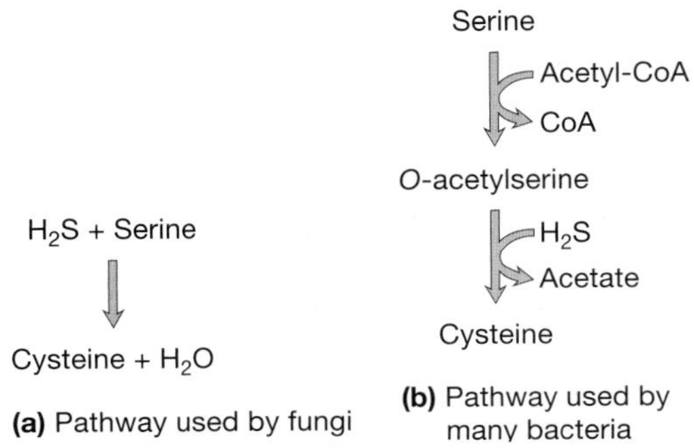

FIGURE 11.24 Incorporation of Sulfur from H_2S. H_2S is the product of sulfate reduction.

Sulfur Assimilation

Sulfur is needed for the synthesis of the amino acids cysteine and methionine. It is also needed for the synthesis of several coenzymes (e.g., coenzyme A and biotin). Sulfur is obtained from two sources: (1) the amino acids cysteine and methionine, and (2) sulfate. Many microorganisms use cysteine and methionine, obtained from either external sources or intracellular amino acid reserves. Because the sulfur is already in a reduced organic form, it is readily incorporated into other organic molecules. The sulfur atom in sulfate is more oxidized than it is in cysteine and methionine; thus sulfate must be reduced before it can be assimilated. This process is known as **assimilatory sulfate reduction** to distinguish it from the dissimilatory sulfate reduction that takes place when sulfate acts as an electron acceptor during anaerobic respiration. ⏮ *Anaerobic respiration (section 10.6);* ⏭ *Sulfur cycle (section 26.1)*

Assimilatory sulfate reduction involves sulfate activation through the formation of phosphoadenosine 5′-phosphosulfate (**figure 11.22**), followed by reduction of the sulfate. As shown in **figure 11.23**, sulfate is first reduced to sulfite (SO_3^{2-}), then to hydrogen sulfide. Cysteine can be synthesized from hydrogen sulfide in two ways (**figure 11.24**). Fungi appear to combine hydrogen sulfide with serine to form cysteine (figure 11.24*a*), whereas many bacteria join hydrogen sulfide with O-acetylserine instead (figure 11.24*b*). Once formed, cysteine can be used in the synthesis of other sulfur-containing organic compounds, including the amino acid methionine.

Amino Acid Biosynthetic Pathways

Some amino acids are made directly by transamination of a precursor metabolite. For example, alanine and aspartate are made directly from pyruvate and oxaloacetate, respectively, using glutamate as the amino group donor. However, most precursor metabolites must be altered by more than just the addition of an amino group. In many cases, the carbon skeleton must be reconfigured, and for cysteine and methionine, the carbon skeleton must be amended by the addition of sulfur. These biosynthetic pathways are more complex. They often involve many steps and are branched. By using branched pathways, a single precursor metabolite can be used for the synthesis of a family of related amino acids. For example, the amino acids lysine, threonine, isoleucine, and methionine are synthesized from oxaloacetate by a branching route (**figure 11.25**). The biosynthetic pathway for the aromatic amino acids phenylalanine, tyrosine, and tryptophan is also branched; it begins with two precursor metabolites, phosphoenolpyruvate and erythrose 4-phosphate (**figure 11.26**). Because of the need to conserve nitrogen, carbon, and energy, amino acid synthetic pathways are usually tightly regulated by allosteric and feedback mechanisms. ◀◀ *Posttranslational regulation of enzyme activity (section 9.10)*

Anaplerotic Reactions and Amino Acid Biosynthesis

When an organism is actively synthesizing amino acids, a heavy demand for precursor metabolites is placed on the central metabolic pathways, especially the TCA cycle. Therefore it is critical that TCA cycle intermediates be readily available. This is especially true for organisms carrying out fermentation, where the TCA cycle does not function in the catabolism of glucose. To ensure an adequate supply of TCA cycle–generated precursor metabolites, microorganisms use **anaplerotic reactions** (Greek *anaplerotic,* filling up) to replenish TCA cycle intermediates.

Most microorganisms can replace TCA cycle intermediates using two reactions that generate oxaloacetate from either phosphoenolpyruvate or pyruvate, both of which are intermediates of the Embden-Meyerhof pathway (**figure 11.27**). These 3-carbon molecules are converted to oxaloacetate by a carboxylation reaction (i.e., CO_2 is added to the molecule, forming a carboxyl group).

The conversion of pyruvate to oxaloacetate is catalyzed by the enzyme pyruvate carboxylase, which requires the cofactor biotin.

$$\text{Pyruvate} + CO_2 + \text{ATP} + H_2O \longrightarrow \text{oxaloacetate} + \text{ADP} + P_i$$

Biotin is often the cofactor for enzymes catalyzing carboxylation reactions. Because of its importance, biotin is a required growth factor for many species. The pyruvate carboxylase reaction is observed in yeasts and some bacteria. Other microorganisms, such as the bacteria *E. coli* and *Salmonella* spp., have the enzyme phosphoenolpyruvate carboxylase, which catalyzes the carboxylation of phosphoenolpyruvate.

$$\text{Phosphoenolpyruvate} + CO_2 \rightarrow \text{oxaloacetate} + P_i$$

Other anaplerotic reactions are part of the **glyoxylate cycle,** which functions in some bacteria, fungi, and protists (**figure 11.28**). This cycle is made possible by two unique enzymes, isocitrate lyase and malate synthase. The glyoxylate cycle is actually a modified TCA cycle. The two decarboxylations of the TCA cycle (the isocitrate dehydrogenase and α-ketoglutarate dehydrogenase steps) are bypassed, making possible the conversion of acetyl-CoA to form oxaloacetate without loss of acetyl-CoA carbon as CO_2. In this fashion, acetate and any molecules that give rise to it can contribute carbon to the cycle and support microbial growth. ◀◀ *Tricarboxylic acid cycle (section 10.4)*

1. Describe the roles of glutamate dehydrogenase, glutamine synthetase, glutamate synthase, and transaminases in ammonia assimilation.
2. How is nitrate assimilated? How does assimilatory nitrate reduction differ from dissimilatory nitrate reduction? What is the fate of nitrate following assimilatory nitrate reduction versus its fate following denitrification?
3. What is nitrogen fixation? Briefly describe the structure and mechanism of action of nitrogenase.
4. How do organisms assimilate sulfur? How does assimilatory sulfate reduction differ from dissimilatory sulfate reduction?
5. Why is using branched pathways an efficient mechanism for synthesizing amino acids?
6. Define an anaplerotic reaction. Give three examples of anaplerotic reactions.
7. Describe the glyoxylate cycle. How is it similar to the TCA cycle? How does it differ?

11.6 Synthesis of Purines, Pyrimidines, and Nucleotides

Purine and pyrimidine biosynthesis is critical for all cells because these molecules are used to synthesize ATP, several cofactors, ribonucleic acid (RNA), deoxyribonucleic acid (DNA), and other important cell components. Nearly all microorganisms can synthesize their own purines and pyrimidines as these are crucial to cell function. ▶▶ *DNA replication (section 12.4); Transcription (section 12.6)*

Purines and **pyrimidines** are cyclic nitrogenous bases with several double bonds. **Adenine** and **guanine** are purines, consisting of two joined rings, whereas pyrimidines (**uracil, cytosine,**

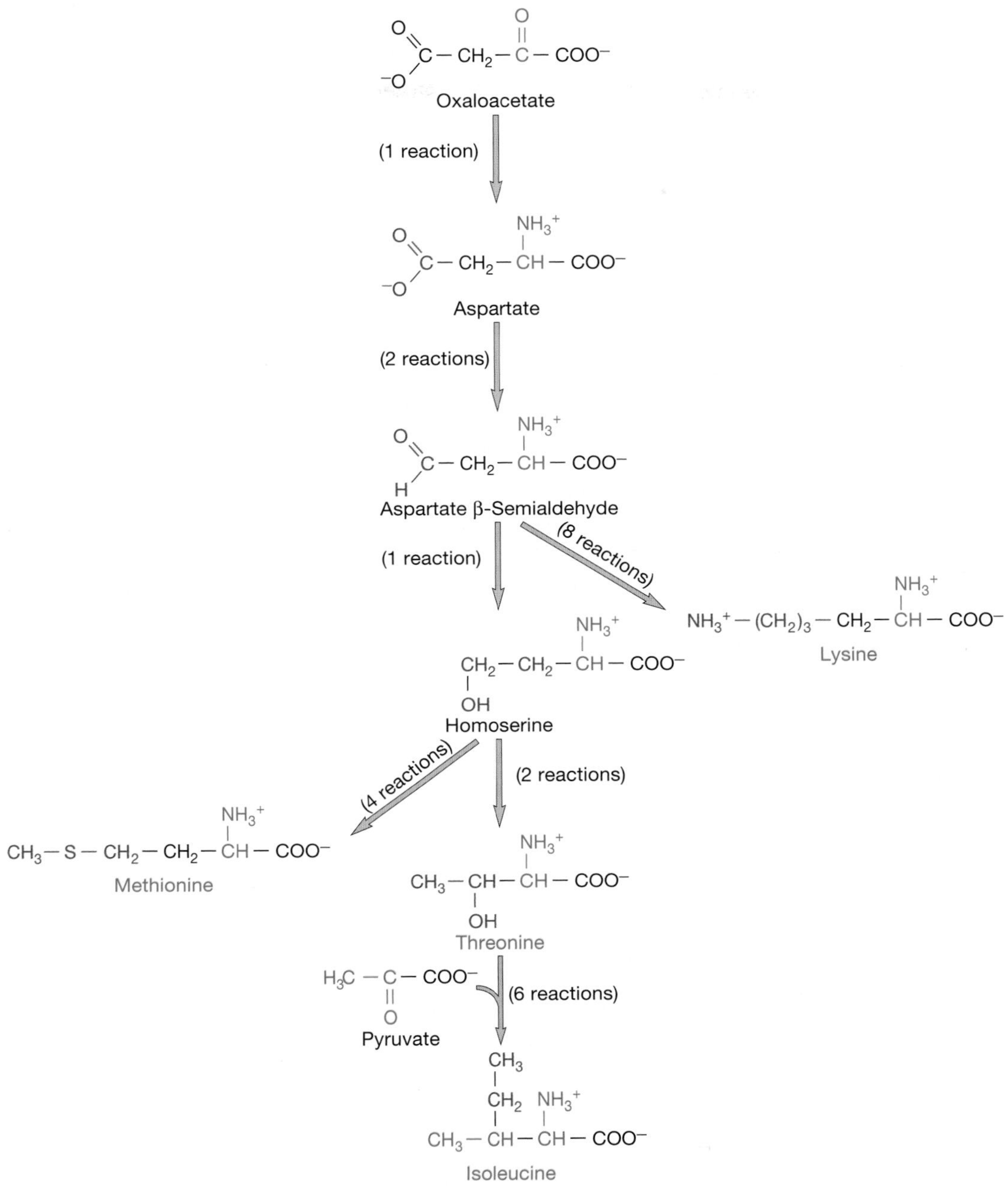

FIGURE 11.25 The Branching Synthetic Pathway to Methionine, Threonine, Isoleucine, and Lysine. Notice that most arrows represent numerous enzyme-catalyzed reactions. Also not shown is the consumption of reducing power and ATP. For instance, the synthesis of isoleucine consumes two ATPs and three NADPHs.

Figure 11.25 Micro Inquiry

Suppose a mutant microbial strain were unable to produce homoserine. What amino acids would it have to obtain from its diet?

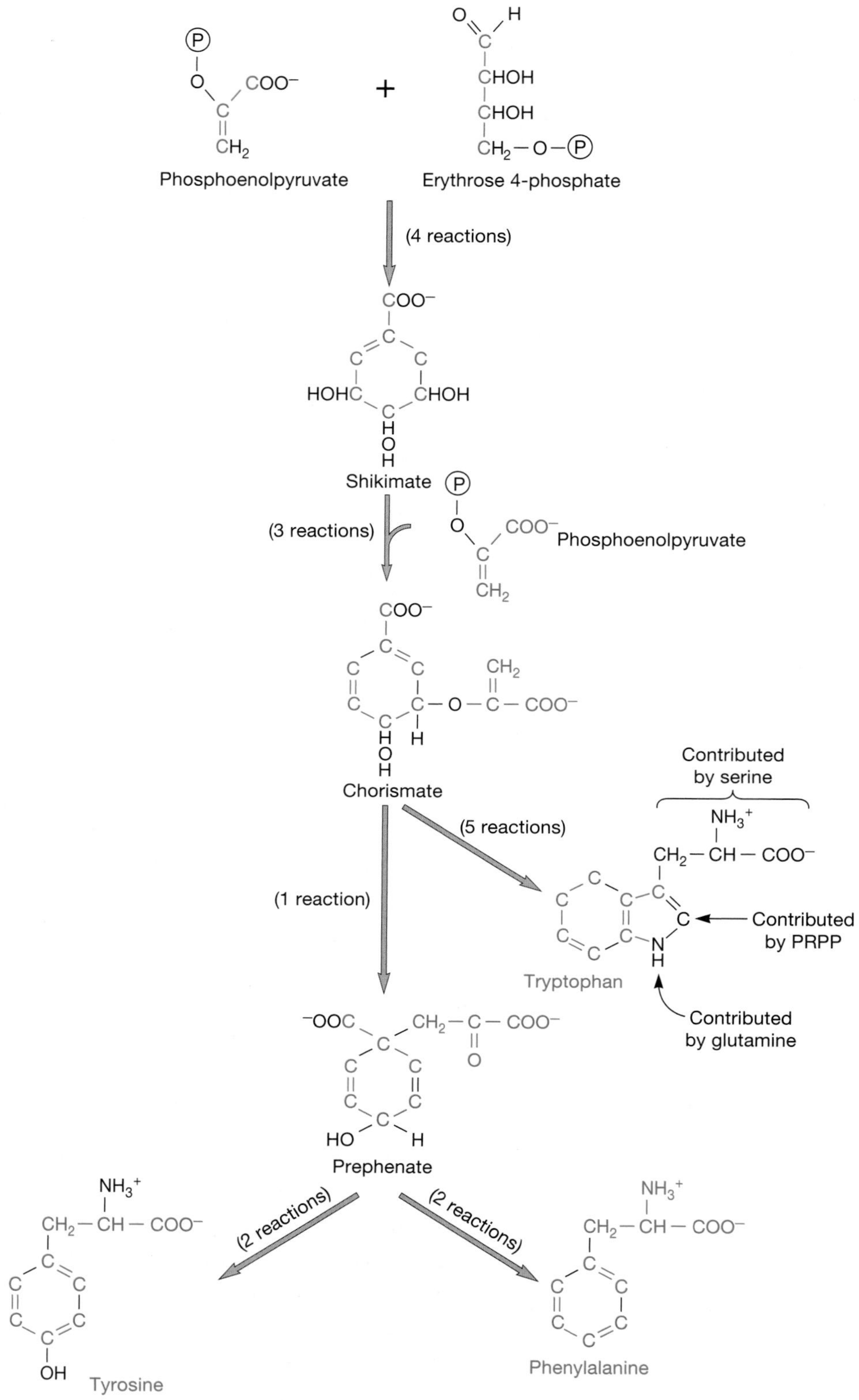

FIGURE 11.26 Aromatic Amino Acid Synthesis. The carbons arising from phosphoenolpyruvate (green) and erythrose 4-phosphate (red) are shown. The remaining carbons present in trytophan are provided by 5-phosphoribosyl-1-pyrophosphate (PRPP) and the amino acid serine. PRPP also is important in purine biosynthesis (*see figure AI1.9*).

Figure 11.26 Micro Inquiry

From which glycolytic pathways are phosphoenolpyruvate and erythrose 4-phosphate derived?

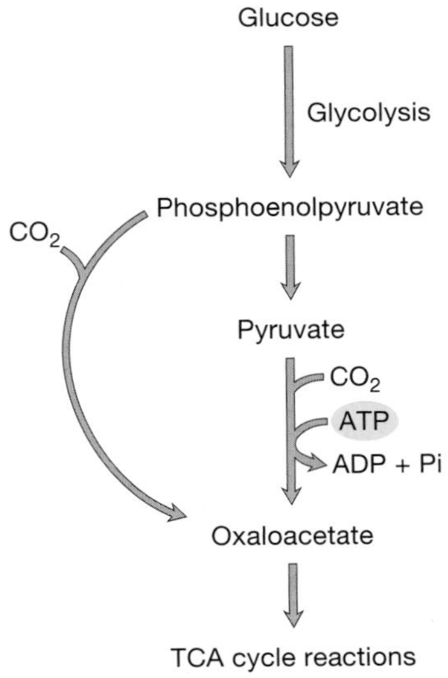

FIGURE 11.27 Anaplerotic Reactions That Replace Oxaloacetate, a TCA Cycle Intermediate.

and **thymine**) have only one ring. A purine or pyrimidine base joined with a pentose sugar, either ribose or deoxyribose, is a **nucleoside.** A **nucleotide** is a nucleoside with one or more phosphate groups attached to the sugar.

Amino acids participate in the synthesis of nitrogenous bases and nucleotides in a number of ways, including providing the nitrogen that is part of all purines and pyrimidines. The phosphorus present in nucleotides is provided by other mechanisms. We begin this section by examining phosphorus assimilation. We then examine the pathways for synthesis of nitrogenous bases and nucleotides.

Phosphorus Assimilation

In addition to nucleic acids, phosphorus is found in proteins (i.e., phosphorylated proteins), phospholipids, and coenzymes such as

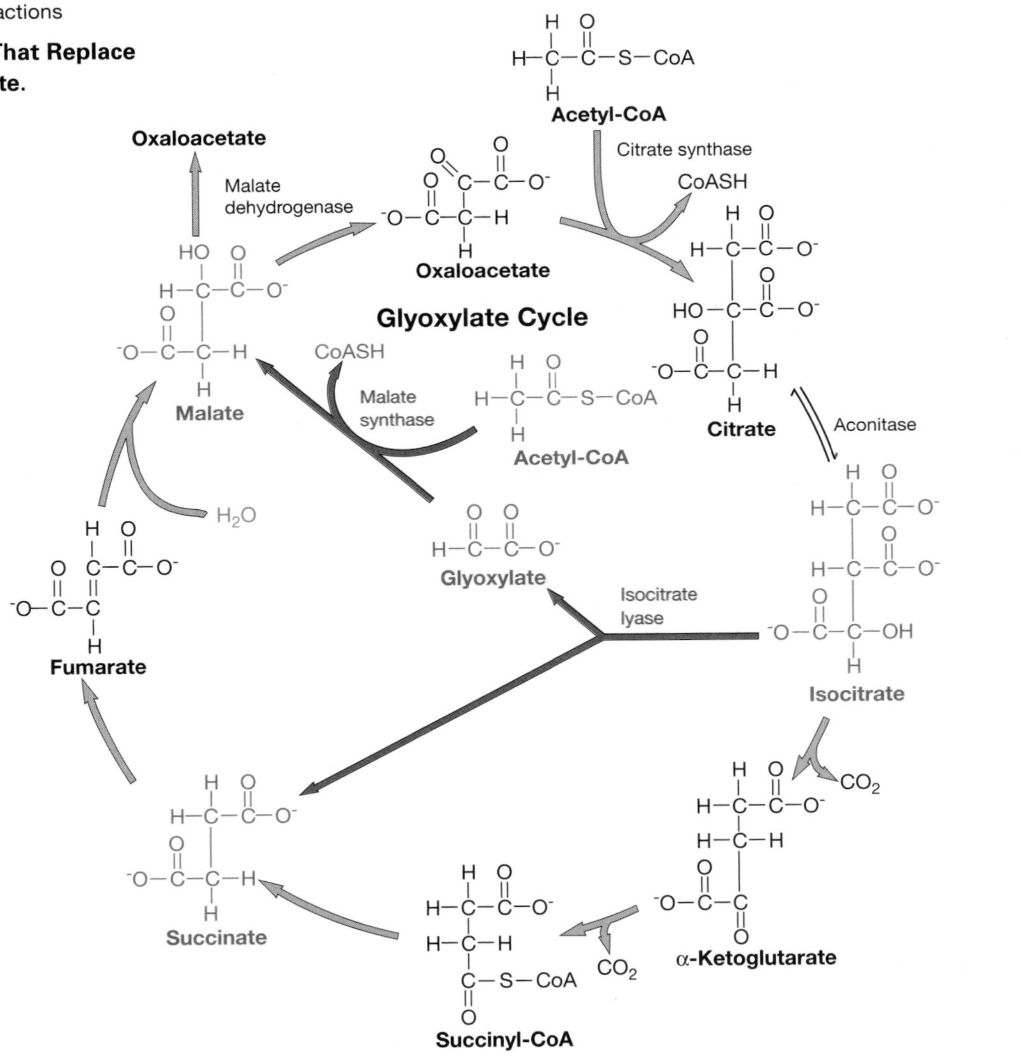

Overall equation:

$$2\text{ Acetyl-CoA} + \text{FAD} + 2\text{NAD}^+ + 3\text{H}_2\text{O} \longrightarrow \text{Oxaloacetate} + 2\text{CoA} + \text{FADH}_2 + 2\text{NADH} + 2\text{H}^+$$

FIGURE 11.28 The Glyoxylate Cycle. The reactions and enzymes unique to the cycle are shown in red and blue, respectively.

$NADP^+$. The most common phosphorus sources are inorganic phosphate and organic phosphate esters. Inorganic phosphate is incorporated through the formation of ATP in one of three ways: (1) photophosphorylation, (2) oxidative phosphorylation, and (3) substrate-level phosphorylation. ◀◀ *Glycolytic pathways (section 10.3); Electron transport and oxidative phosphorylation (section 10.5); Phototrophy (section 10.12)*

Microorganisms may obtain organic phosphates from their surroundings in dissolved or particulate form. **Phosphatases** very often hydrolyze organic phosphate esters to release inorganic phosphate. Gram-negative bacteria have phosphatases in the periplasmic space, which allows phosphate to be taken up immediately after release. On the other hand, protists can directly use organic phosphates after ingestion or hydrolyze them in lysosomes and incorporate the phosphate.

Purine Biosynthesis

The biosynthetic pathway for purines is a complex, 11-step sequence *(see appendix II)* in which seven different molecules contribute parts to the final purine skeleton (**figure 11.29**). The pathway begins with ribose 5-phosphate, and the purine skeleton is constructed on this sugar. Therefore the first purine product of the pathway is the nucleotide inosinic acid, not a free purine base. The cofactor folic acid is very important in purine biosynthesis. Folic acid derivatives contribute carbons two and eight to the purine skeleton.

Once inosinic acid has been formed, relatively short pathways synthesize adenosine monophosphate and guanosine monophosphate (**figure 11.30**), and produce nucleoside diphosphates and triphosphates by phosphate transfers from ATP. DNA contains deoxyribonucleotides (the ribose lacks a hydroxyl group on carbon two) instead of the ribonucleotides found in RNA. Deoxyribonucleotides arise from the reduction of nucleoside diphosphates or nucleoside triphosphates by two different routes. Some microorganisms reduce the triphosphates with a system requiring vitamin B_{12} as a cofactor. Others, such as *E. coli,* reduce the ribose in nucleoside diphosphates. Both systems employ a small, sulfur-containing protein called thioredoxin as their reducing agent.

Pyrimidine Biosynthesis

Pyrimidine biosynthesis begins with aspartic acid and carbamoyl phosphate, a high-energy molecule synthesized from bicarbonate and ammonia provided by the amino acid glutamine (**figure 11.31**). Aspartate carbamoyltransferase catalyzes the condensation of these two substrates to form carbamoylaspartate, which is then converted to the initial pyrimidine product, orotic acid. A nucleotide is produced by the addition of ribose 5-phosphate, using the high-energy intermediate 5-phosphoribosyl 1-pyrophosphate. Thus construction of the pyrimidine ring is completed before ribose is added, in contrast with purine ring synthesis, which begins with ribose 5-phosphate.

Orotidine monophosphate is decarboxylated, yielding uridine monophosphate. This is followed by formation of uridine triphosphate and cytidine triphosphate. These two nucleotides are reduced in the same way that purine nucleotides are to make the deoxy-forms needed for DNA synthesis. Deoxythymidine monophosphate is made when deoxyuridine monophosphate is methylated with a folic acid derivative (**figure 11.32**).

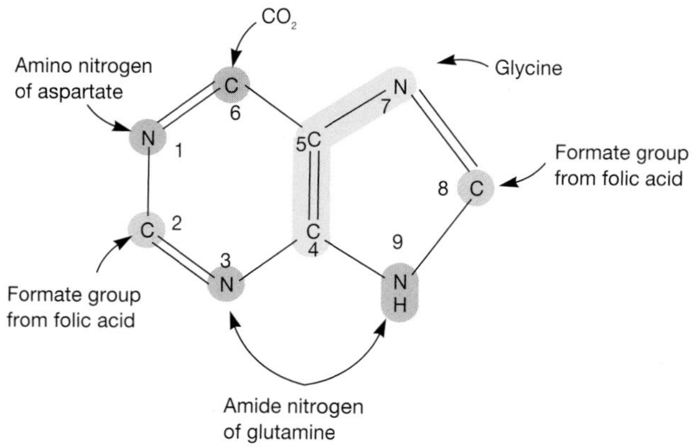

FIGURE 11.29 Purine Biosynthesis. The sources of nitrogen and carbon are indicated.

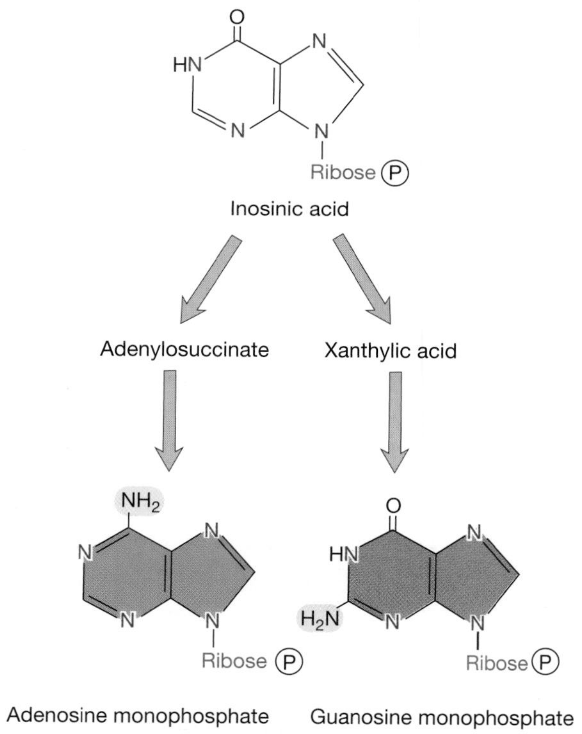

FIGURE 11.30 Synthesis of Adenosine Monophosphate and Guanosine Monophosphate. The highlighted groups differ from those in inosinic acid.

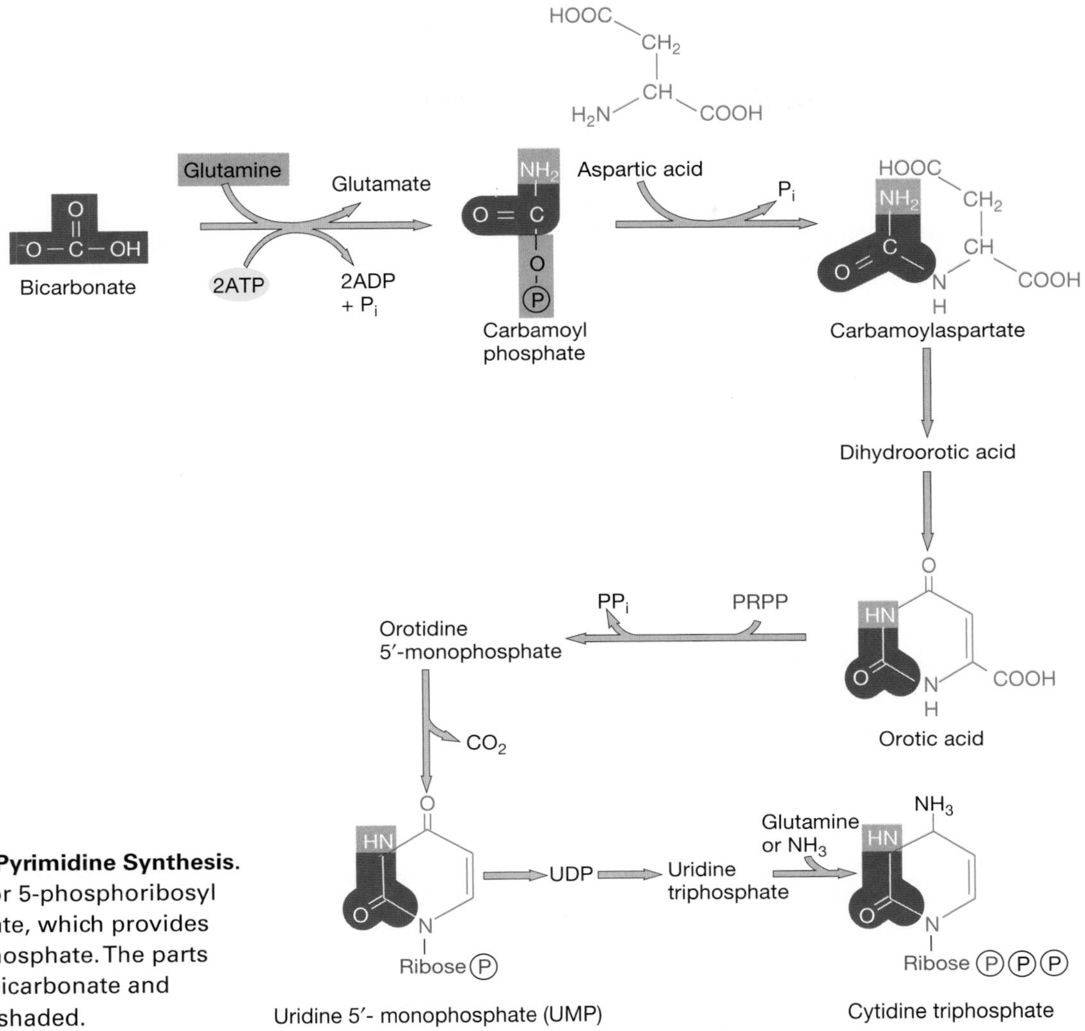

FIGURE 11.31 Pyrimidine Synthesis. PRPP stands for 5-phosphoribosyl 1-pyrophosphate, which provides the ribose 5-phosphate. The parts derived from bicarbonate and glutamine are shaded.

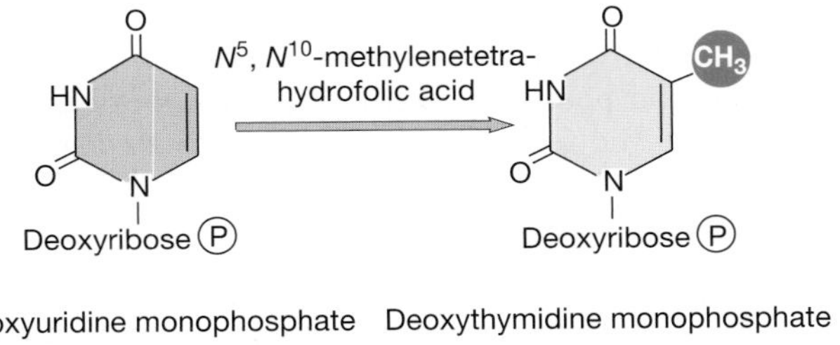

FIGURE 11.32 Deoxythymidine Monophosphate Synthesis. Deoxythymidine differs from deoxyuridine in having the shaded methyl group.

1. How is phosphorus assimilated? What roles do phosphatases play in phosphorus assimilation? Why can phosphate be directly incorporated into cell constituents, whereas nitrate, nitrogen gas, and sulfate cannot?
2. Explain the difference between a purine and a pyrimidine, and between a nucleoside and a nucleotide.
3. Outline the way in which purines and pyrimidines are synthesized. How is the deoxyribose component of deoxyribonucleotides made?

11.7 Lipid Synthesis

Lipids are absolutely required by cells, as they are the major components of cell membranes. Most bacterial and eukaryal lipids contain fatty acids or their derivatives. Fatty acids are monocarboxylic acids with long alkyl chains that usually have an even number of carbons (the average length is 18 carbons). Some may be unsaturated—that is, have one or more double bonds. Most microbial fatty acids are straight chained, but some are branched. Gram-negative bacteria often have cyclopropane fatty acids (fatty acids with one or more cyclopropane rings in their chains). The synthesis of lipids is complex and is the target of some antimicrobial agents used to treat infectious disease caused by eukaryotic pathogens. ▶▶| *Lipids (appendix I); Antifungal drugs (section 34.5)*

Fatty acid synthesis is catalyzed by the **fatty acid synthase** complex with acetyl-CoA and malonyl-CoA as the substrates and NADPH as the electron donor. Malonyl-CoA arises from the ATP-driven carboxylation of acetyl-CoA (**figure 11.33**). Synthesis takes place after acetate and malonate have been transferred from coenzyme A (CoA) to the sulfhydryl group of the acyl carrier protein (ACP), a small protein that carries the growing fatty acid chain during synthesis. Fatty acid synthase adds two carbons at a time to the carboxyl end of the growing fatty acid chain in a two-stage process (figure 11.33). First, malonyl-ACP reacts with the fatty acyl-ACP to yield CO_2 and a fatty acyl-ACP two carbons longer. The loss of CO_2 drives this reaction to completion. Notice that ATP is used to add CO_2 to acetyl-CoA, forming malonyl-CoA. The same CO_2 is lost when malonyl-ACP donates carbons to the chain. Thus carbon dioxide is essential to fatty acid synthesis but is not permanently incorporated. Indeed, some microorganisms require CO_2 for optimum growth, but they can do without it in the presence of a fatty acid such as oleic acid (an 18-carbon unsaturated fatty acid). In the second stage of synthesis, the α-keto group arising from the initial condensation reaction is removed in a three-step process involving two reductions and a dehydration. The fatty acid is then ready for the addition of two more carbon atoms.

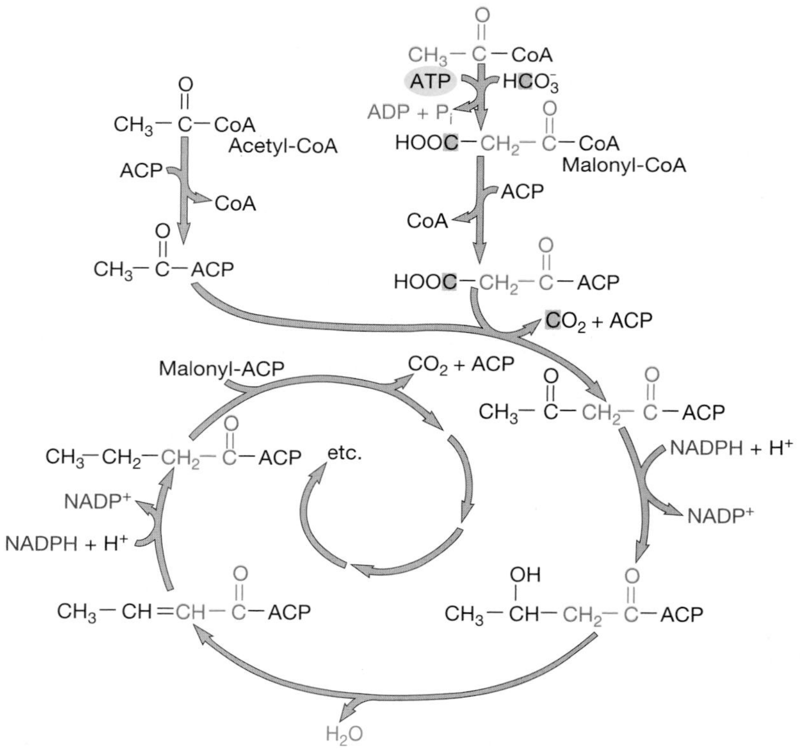

FIGURE 11.33 Fatty Acid Synthesis. The cycle is repeated until the proper chain length has been reached. Carbon dioxide carbon and the remainder of malonyl-CoA are shown in red. ACP stands for acyl carrier protein.

Unsaturated fatty acids are synthesized in two ways. Eukaryotes and aerobic bacteria such as *B. megaterium* employ an aerobic pathway using both NADPH and O_2.

$$\text{R–(CH}_2)_9\text{–}\overset{\overset{\text{O}}{\|}}{\text{C}}\text{–SCoA} + \text{NADPH} + \text{H}^+ + \text{O}_2 \rightarrow$$

$$\text{R–CH=CH(CH}_2)_7\text{–}\overset{\overset{\text{O}}{\|}}{\text{C}}\text{–SCoA} + \text{NADP}^+ + 2\text{H}_2\text{O}$$

A double bond is formed between carbons nine and ten, and O_2 is reduced to water with electrons supplied by both the fatty acid and NADPH. Anaerobic bacteria and some aerobes create double bonds during fatty acid synthesis by dehydrating hydroxy fatty acids. Oxygen is not required for double bond synthesis by this pathway. The anaerobic pathway is present in a number of gram-negative bacteria (e.g., *E. coli, Salmonella* spp., and cyanobacteria), and gram-positive bacteria (e.g., *Lactobacillus plantarum* and *Clostridium pasteurianum*).

Eukaryotic microorganisms and a few gram-positive bacteria can store carbon and energy as **triacylglycerol,** glycerol esterified to three fatty acids. Glycerol arises from the reduction of the precursor metabolite dihydroxyacetone phosphate to glycerol 3-phosphate, which is then esterified with two fatty acids to give phosphatidic acid (**figure 11.34**). Phosphate is hydrolyzed from phosphatidic acid, giving a diacylglycerol, and the third fatty acid is attached to yield a triacylglycerol.

Phospholipids are major components of eukaryotic and bacterial cell membranes. Their synthesis usually proceeds by way of phosphatidic acid and a cytidine diphosphate (CDP) carrier that plays a role similar to that of uridine and adenosine diphosphate carriers in carbohydrate biosynthesis. For example, bacteria synthesize phosphatidylethanolamine, a major cell membrane component, through the initial formation of CDP-diacylglycerol (figure 11.34). This CDP derivative then reacts with serine to form the phospholipid phosphatidylserine, and decarboxylation yields phosphatidylethanolamine. In this way, a complex membrane lipid is constructed from the products of glycolysis, fatty acid biosynthesis, and amino acid biosynthesis.

1. What is a fatty acid? Describe in general terms how fatty acid synthase manufactures a fatty acid.
2. How are unsaturated fatty acids made?
3. Briefly describe the pathways for triacylglycerol and phospholipid synthesis. Of what importance are phosphatidic acid and CDP-diacylglycerol?
4. Activated carriers participate in carbohydrate, peptidoglycan, and lipid synthesis. Briefly describe these carriers and their roles. Are there any features common to all the carriers? Explain your answer.

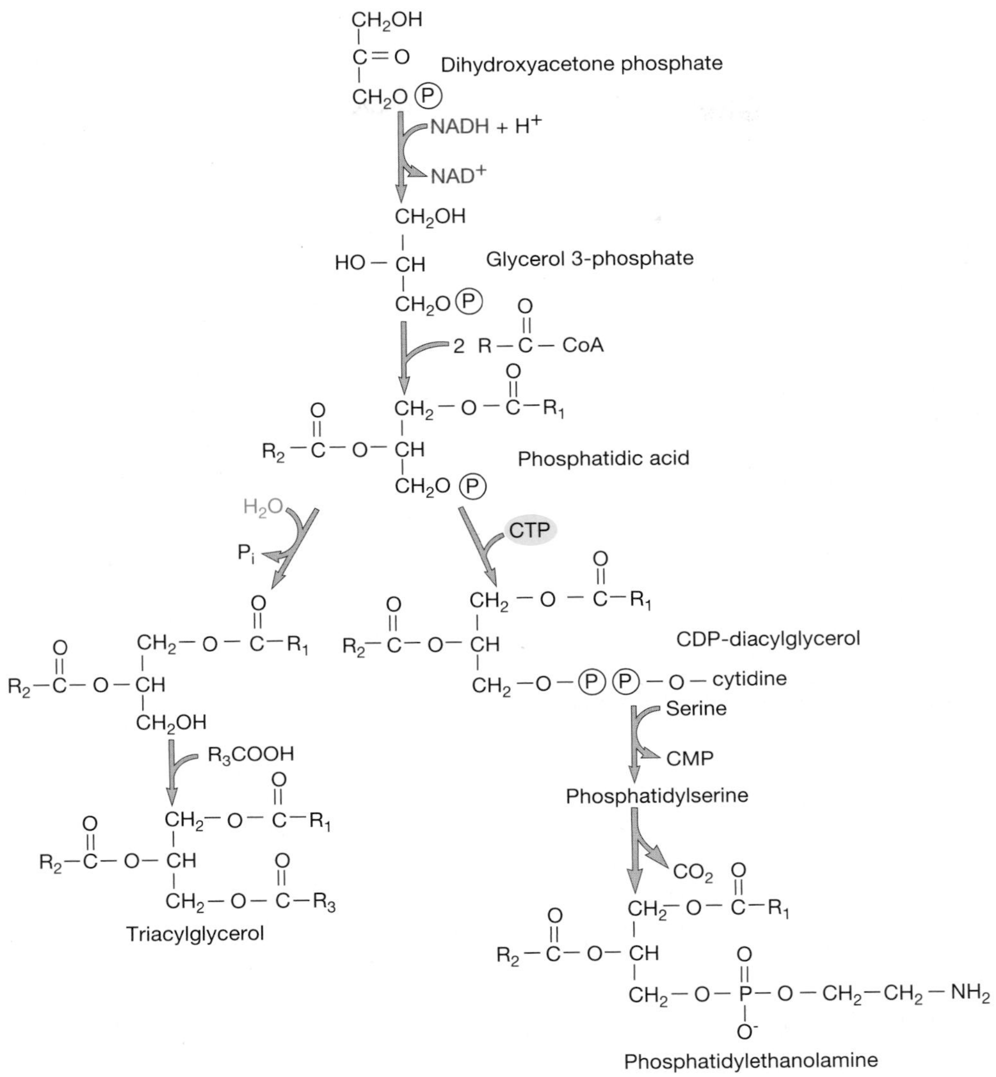

FIGURE 11.34 Triacylglycerol and Phospholipid Synthesis.

Summary

11.1 Principles Governing Biosynthesis

a. Many important cell constituents are macromolecules, large polymers constructed of simple monomers.

b. Although many catabolic and anabolic pathways share enzymes for the sake of efficiency, some of their enzymes are separate and independently regulated.

c. Macromolecular components often undergo self-assembly to form the final molecule or complex.

11.2 Precursor Metabolites

a. Precursor metabolites are carbon skeletons used as the starting substrates for biosynthetic pathways. They are intermediates of glycolytic pathways and the TCA cycle (i.e., the central metabolic pathways) (**figures 11.3** and **11.4**).

b. Most precursor metabolites are used for amino acid biosynthesis; others are used for synthesis of purines, pyrimidines, and lipids.

11.3 CO_2 Fixation

a. Five different CO_2-fixation pathways have been identified in autotrophic microorganisms: the Calvin cycle, the reductive TCA cycle, the acetyl-CoA pathway, the 3-hydroxypropionate cycle, and the 3-hydroxypropionate/4-hydroxybutyrate pathway.

b. The Calvin cycle is used by most autotrophs to fix CO_2. It can be divided into three phases: the carboxylation phase, the reduction phase, and the regeneration phase (**figure 11.5**). Three ATPs and two NADPHs are used during the incorporation of one CO_2.

c. The reductive TCA cycle, acetyl-CoA pathway, 3-hydroxypropionate cycle, and 3-hydroxypropionate/4-hydroxybutyrate pathway are used by many bacteria and archaea to fix CO_2 (**figures 11.6–11.9**).

11.4 Synthesis of Sugars and Polysaccharides

a. Gluconeogenesis is the synthesis of glucose and related sugars from nonglucose precursors.

b. Glucose, fructose, and mannose are gluconeogenic intermediates or are made directly from them (**figure 11.10**); galactose is synthesized with nucleoside diphosphate derivatives (**figure 11.11**). Bacteria and protists synthesize glycogen and starch from adenosine diphosphate glucose.

c. Peptidoglycan synthesis is a complex process involving both UDP derivatives and the lipid carrier bactoprenol, which transports NAG-NAM-pentapeptide units across the cell membrane. Cross-links are formed by transpeptidation (**figures 11.12–11.15**).

11.5 Synthesis of Amino Acids

a. The addition of nitrogen to the carbon chain provided by a precursor metabolite is an important step in amino acid biosynthesis. Ammonia, nitrate, or N_2 can serve as the source of nitrogen.

b. Ammonia can be directly assimilated by the activity of transaminases and either glutamate dehydrogenase or the glutamine synthetase–glutamate synthase system (**figures 11.16–11.18**).

c. Nitrate is incorporated through assimilatory nitrate reduction catalyzed by the enzymes nitrate reductase and nitrite reductase (**figure 11.19**).

d. Nitrogen fixation is catalyzed by nitrogenase. Atmospheric molecular nitrogen is reduced to ammonia, which is then incorporated into amino acids (**figures 11.20** and **11.21**).

e. Microorganisms can use cysteine, methionine, and inorganic sulfate as sulfur sources. Sulfate must be reduced to sulfide before it is assimilated. This occurs during assimilatory sulfate reduction (**figures 11.23** and **11.24**).

f. Some amino acids are made directly by the addition of an amino group to a precursor metabolite, but most amino acids are made by pathways that are more complex. Many amino acid biosynthetic pathways are branched. Thus a single precursor metabolite can give rise to several amino acids (**figures 11.25** and **11.26**).

g. Anaplerotic reactions replace TCA cycle intermediates to keep the cycle in balance while it supplies precursor metabolites. The anaplerotic reactions include the glyoxylate cycle (**figures 11.27** and **11.28**).

11.6 Synthesis of Purines, Pyrimidines, and Nucleotides

a. Purines and pyrimidines are nitrogenous bases found in DNA, RNA, and other molecules. The nitrogen is supplied by certain amino acids that participate in purine and pyrimidine biosynthesis. Phosphorus is provided by either inorganic phosphate or organic phosphate.

b. Phosphorus can be assimilated directly by phosphorylation reactions that form ATP from ADP and P_i. Organic phosphorus sources are the substrates of phosphatases that release phosphate from the organic molecule.

c. The purine skeleton is synthesized beginning with ribose 5-phosphate and initially produces inosinic acid. Pyrimidine biosynthesis starts with carbamoyl phosphate and aspartate, and ribose is added after the skeleton has been constructed (**figures 11.29–11.32**).

11.7 Lipid Synthesis

a. Fatty acids are synthesized from acetyl-CoA, malonyl-CoA, and NADPH by fatty acid synthase. During synthesis, the intermediates are attached to the acyl carrier protein (**figure 11.33**). Double bonds can be added in two different ways.

b. Triacylglycerols are made from fatty acids and glycerol phosphate. Phosphatidic acid is an important intermediate in this pathway (**figure 11.34**).

c. Phospholipids such as phosphatidylethanolamine can be synthesized from phosphatidic acid by forming CDP-diacylglycerol, then adding an amino acid.

Critical Thinking Questions

1. Discuss the relationship between catabolism and anabolism. How does anabolism depend on catabolism?
2. In metabolism, important intermediates are covalently attached to carriers, as if to mark these as important so the cell does not lose track of them. List a few examples of these carriers and indicate whether they are involved primarily in anabolism or catabolism.
3. Intermediary carriers are in a limited supply: When they cannot be recycled because of a metabolic block, serious consequences ensue. Think of some examples of these consequences.
4. Magnetotactic bacteria are morphologically and metabolically complex microbes that produce intracellular, membrane-bound magnetic crystals (*see figure 3.37*). The magnetotactic bacterial isolates called MV-1 and MC-1 both grow chemolithoautotrophically. However, MV-1 uses the Calvin-Benson cycle to fix carbon, whereas MC-1 uses the reductive TCA cycle.

   Refer to figures 11.5 and 11.6 to identify key enzymatic steps in each pathway. How do you think the microbiologists studying these fascinating microbes were able to prove that these two bacterial isolates use two different pathways for carbon fixation?

   Read the original paper: Williams, T. J., et al. 2006. Evidence for autotrophy via the reverse tricarboxylic acid cycle in the marine magnetotactic coccus strain MC-1. *Appl. Environ. Microbiol.* 72:1322.
5. Although triacylglycerols (TAGs) are abundant in eukaryotes, only a few bacteria such as *Streptomyces, Nocardia*, and *Mycobacterium* accumulate this lipid reserve molecule. As shown in figure 11.34, the last step in TAG biosynthesis is the esterification of diacylglycerol (DAG) with a fatty acid. Based on analysis of the complete genome of *S. coelicolor*, three candidate genes were identified that might encode the enzyme that catalyzes this reaction.

   Using both genetic and biochemical analysis, how do you think the microbiologists studying this biosynthetic pathway determined which gene encoded the enzyme that catalyzed this reaction? Be sure to think about what metabolites you would measure to be sure you were examining the reaction in question.

   Read the original paper: Arabolaza, A., et al. 2008. Multiple pathways for triacylglycerol biosynthesis in *Streptomyces coelicolor. Appl. Environ. Microbiol.* 74:2573.

Learn More

Learn more by visiting the text website at www.mhhe.com/willey8, where you will find a complete list of references.

12

Genes: Structure, Replication, and Expression

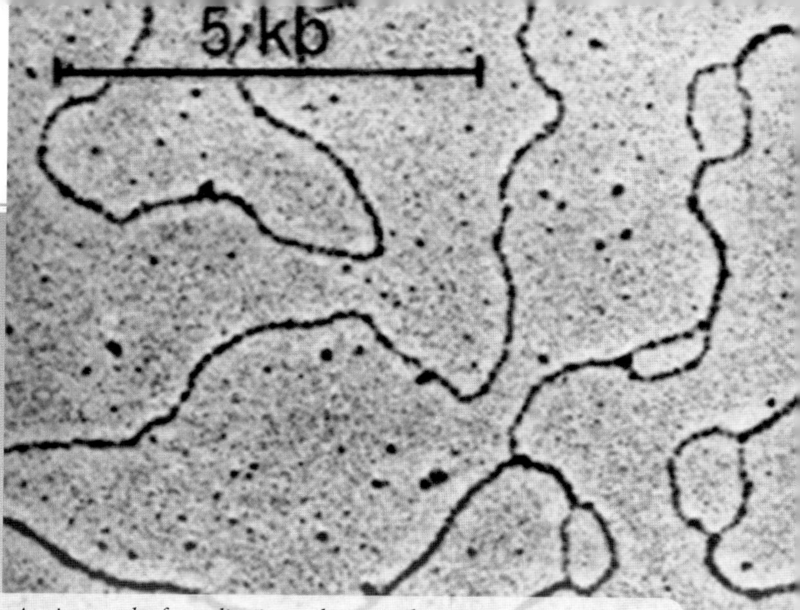

A micrograph of a replicating, eukaryotic chromosome.

CHAPTER GLOSSARY

anticodon The base triplet on a tRNA that is complementary to the triplet codon on mRNA.

codon A sequence of three nucleotides in mRNA that directs the incorporation of an amino acid during protein synthesis or signals the stop of translation.

DNA polymerase An enzyme that synthesizes new DNA using a parental nucleic acid strand as a template.

gene A DNA segment or sequence that codes for a polypeptide, rRNA, or tRNA.

genome The full set of genes (including those on both the chromosome and plasmids) present in a cell or virus.

lagging strand The strand of DNA that is discontinuously synthesized by the formation of short Okazaki fragments.

leader sequence A sequence in a gene that corresponds to a nontranslated region at the 5′ end of mRNA.

leading strand The strand of DNA that is synthesized continuously.

molecular chaperones Proteins that assist in the folding and stabilization of other proteins or in directing newly synthesized proteins to secretion systems or other locations in the cell.

Okazaki fragments Short stretches of polynucleotides produced during discontinuous DNA replication.

peptidyl transferase The rRNA ribozyme (23S rRNA in *Bacteria*) that catalyzes the addition of an amino acid to the growing peptide chain during translation.

promoter The region on DNA at the start of a gene to which RNA polymerase binds before beginning transcription.

reading frame The way in which nucleotides in DNA and mRNA are grouped into codons for reading the message contained in the nucleotide sequence.

replication The process by which an exact copy of a parental nucleic acid (DNA in cells; RNA in some viruses) is made with the parental molecule serving as a template.

replication fork The Y-shaped structure where DNA is replicated; the arms of the Y contain the template strand and a newly synthesized DNA copy.

replicon A unit of the genome that contains an origin of replication and in which DNA is replicated.

replisome A complex of proteins that replicates DNA.

RNA polymerase An enzyme that catalyzes the synthesis of mRNA under the direction of a template.

Shine-Dalgarno sequence A segment in the leader of bacterial and some archaeal mRNA that binds to a sequence on the 16S rRNA of the small ribosomal subunit; it helps properly orient the mRNA on the ribosome.

sigma factor A protein that enables bacterial RNA polymerase core enzyme to recognize a promoter.

stop codon A codon that does not code for an amino acid but is a signal to stop protein synthesis; also called a nonsense codon.

template strand A DNA strand that specifies the base sequence of a new complementary strand of DNA or RNA.

terminator A sequence that marks the end of a gene and stops transcription.

transcription The process by which single-stranded RNA with a base sequence complementary to the template strand of DNA (or RNA in some viruses) is synthesized.

translation The process by which the genetic message carried by mRNA directs the synthesis of polypeptides with the aid of ribosomes and other cell constituents.

In this chapter, we turn our attention to the synthesis of three major macromolecules—DNA, RNA, and proteins—from their constituent monomers. DNA serves as the storage molecule for the genetic instructions that enable organisms to carry out metabolism and reproduction. RNA functions in the expression of genetic information so that enzymes and other proteins can be made. These proteins are used to build cellular structures and to do other cellular work. The study of the synthesis of DNA, RNA, and protein falls into the realms of molecular genetics and biology.

In the mid-1800s the discipline of genetics was born from the work of Gregor Mendel, who studied the inheritance of various traits in pea plants. In the early twentieth century, Mendel's work was rediscovered and furthered by scientists working with fruit flies and plants such as corn. The use of microorganisms as models for genetic studies soon followed. Microorganisms, especially bacteria, have significant advantages as model organisms, in part because of their unique characteristics. One important feature is the nature of their genomes. The term **genome** refers to all DNA present in a cell or virus. Bacterial and archaeal cells normally have one set of genes; that is, they are haploid (1N). In addition, they often carry extrachromosomal genetic elements called plasmids. Eukaryotic organisms, including eukaryotic microorganisms, usually have two sets of genes; that is, they are diploid (2N). Eukaryotes rarely have plasmids. Viral genomes differ significantly from those of cellular organisms; the genetics and molecular biology of viruses are discussed in chapter 25.

◀◀ *Plasmids (section 3.5)*

We now review some of the most basic concepts of molecular genetics: how genetic information is stored and organized in the DNA molecule, the way in which DNA is replicated, gene structure, and how genes function (i.e., gene expression). Based on the foundation provided in this chapter, chapter 13 considers the regulation of gene expression. The regulation of gene expression is important because it links the **genotype** of an organism—the specific set of genes it possesses—to the **phenotype** of an organism—the collection of characteristics that are observable. Not all genes are expressed at the same time or in the same place, and the environment profoundly influences which genes are expressed at any given time. Finally, chapter 14 contains information on the nature of mutation, DNA repair, and genetic recombination. These three chapters provide the background needed to understand recombinant DNA technology (chapter 15) and microbial genomics (chapter 16). Much of the information presented in chapters 12 through 14 will be familiar to those who have taken an introductory genetics course. Because of the importance of bacteria as model organisms, primary emphasis is placed on their genetics. The genetics of the *Archaea* are discussed more fully in chapter 18.

Although modern genetic analysis began with studies of fruit flies and corn, the nature of genetic information, gene structure, the genetic code, and mutation were elucidated by elegant experiments involving bacteria and bacterial viruses. We will first review a few of these early experiments and then summarize the relationship of DNA, RNA, and protein that has guided much of modern research.

12.1 DNA as Genetic Material

Although it is now hard to imagine, it was once thought that DNA was too simple a molecule to store genetic information. It is composed of only four different nucleotides, and it seemed that a molecule of much greater complexity must house the genetic information of a cell. It was argued that proteins, being composed of 20 different amino acids, were the better candidate for this important cellular function.

The early work of Fred Griffith in 1928 on the transfer of virulence in the pathogen *Streptococcus pneumoniae*, commonly called pneumococcus (**figure 12.1**), set the stage for research showing that DNA was indeed the genetic material. Griffith found that if he boiled virulent bacteria and injected them into mice, the mice were not affected and no pneumococci could be recovered from the animals. When he injected a combination of killed virulent bacteria and a living nonvirulent strain, the mice died; moreover, he could recover living virulent bacteria from the dead mice. Griffith called this change of nonvirulent bacteria into virulent pathogens transformation.

Oswald Avery and his colleagues then set out to discover which constituent in the heat-killed virulent pneumococci was responsible for Griffith's transformation. These investigators selectively destroyed constituents in purified extracts of virulent pneumococci (S cells), using enzymes that would hydrolyze DNA, RNA, or protein. They then exposed nonvirulent pneumococcal strains (R strains) to the treated extracts. Transformation of the nonvirulent bacteria was blocked only if the DNA was destroyed, suggesting that DNA was carrying the information required for transformation (**figure 12.2**). The publication of these studies by Avery, C. M. MacLeod, and M. J. McCarty in 1944 provided the first evidence that Griffith's transforming principle was DNA and therefore that DNA carried genetic information.

Some years later (1952), Alfred Hershey and Martha Chase performed several experiments indicating that DNA was the

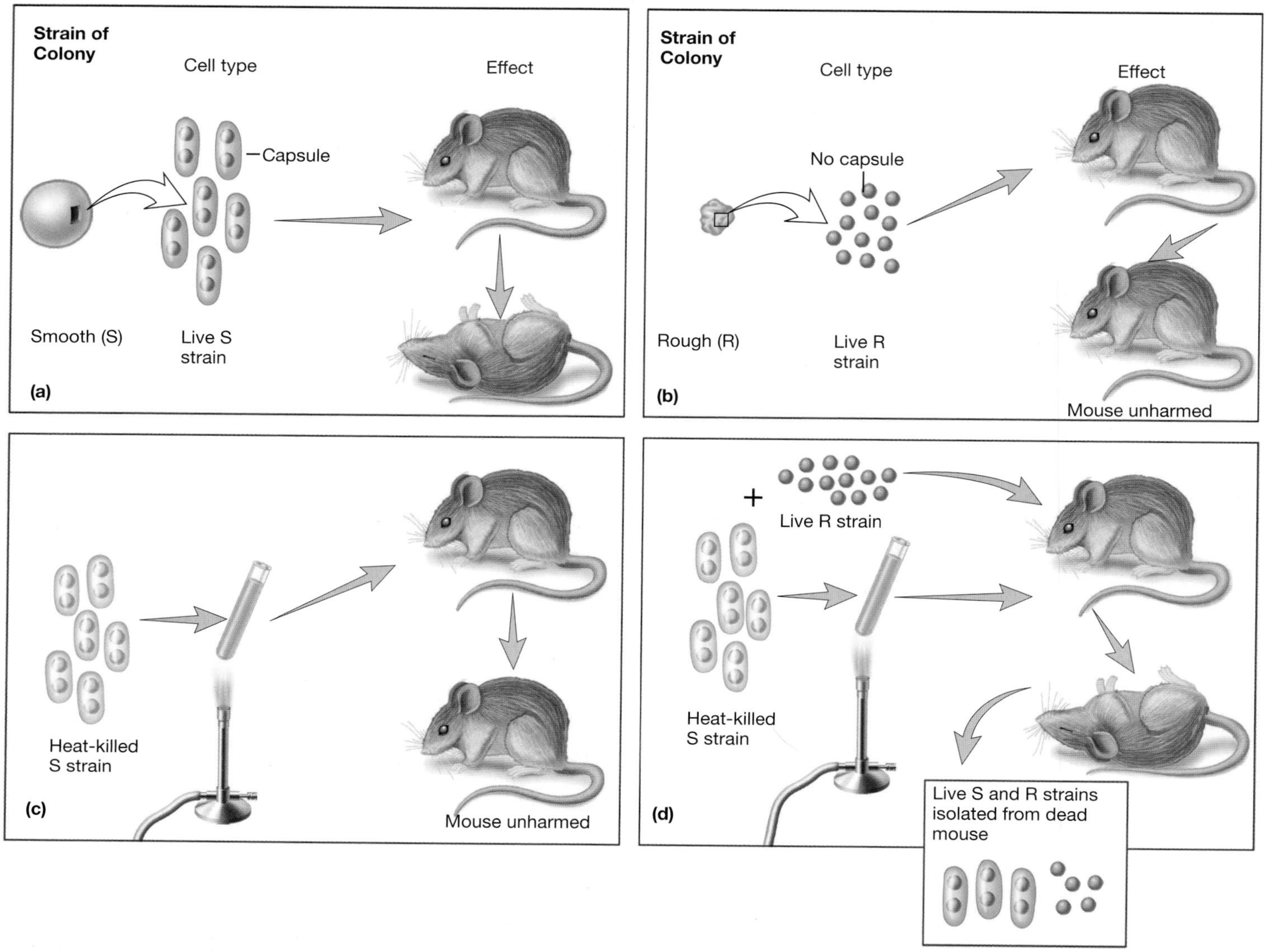

FIGURE 12.1 Griffith's Transformation Experiments. (a) Mice died of pneumonia when injected with pathogenic strains of S pneumococci, which have a capsule and form smooth-looking colonies. (b) Mice survived when injected with a nonpathogenic strain of R pneumococci, which lacks a capsule and forms rough colonies. (c) Injection with heat-killed strains of S pneumococci had no effect. (d) Injection with a live R strain and a heat-killed S strain gave the mice pneumonia, and live S strain pneumococci could be isolated from the dead mice.

Figure 12.1 Micro Inquiry

Based on what we now know about proteins, why can we conclude from this experiment that genetic information was unlikely to be carried by proteins?

genetic material in a bacterial virus called T2 bacteriophage. Some luck was involved in their discovery, for the genetic material of many viruses is RNA and the researchers happened to select a DNA virus for their studies. Imagine the confusion if T2 had been an RNA virus! The controversy surrounding the nature of genetic information might have lasted considerably longer than it did. Hershey and Chase made the virus's DNA radioactive with ^{32}P, or they labeled its protein coat with ^{35}S. They mixed radioactive bacteriophage with *Escherichia coli* and incubated the mixture for a few minutes. The suspension was then agitated

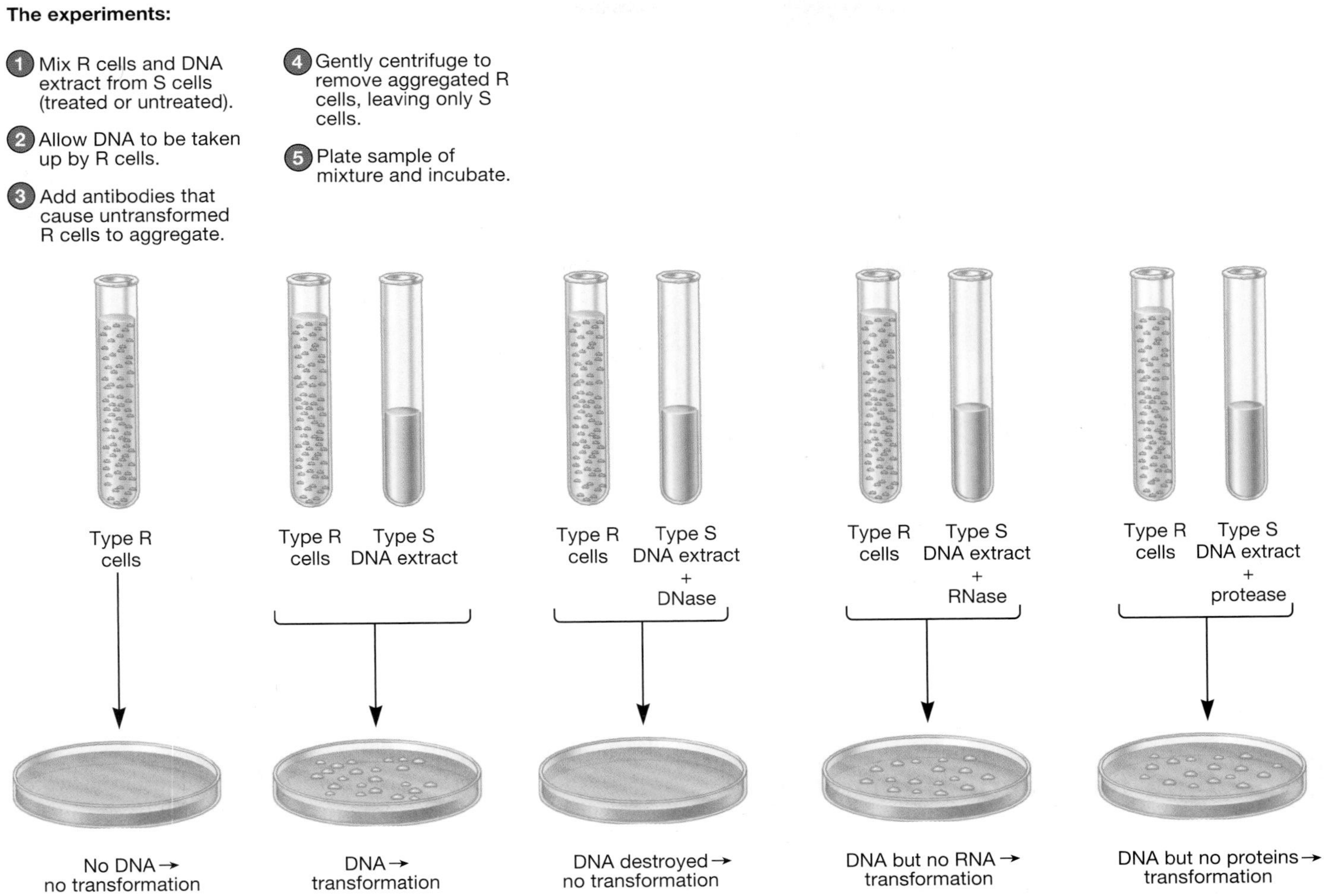
The experiments:
1 Mix R cells and DNA extract from S cells (treated or untreated).
2 Allow DNA to be taken up by R cells.
3 Add antibodies that cause untransformed R cells to aggregate.
4 Gently centrifuge to remove aggregated R cells, leaving only S cells.
5 Plate sample of mixture and incubate.
Type R cells
Type R cells
Type S DNA extract
Type R cells
Type S DNA extract + DNase
Type R cells
Type S DNA extract + RNase
Type R cells
Type S DNA extract + protease
No DNA → no transformation
DNA → transformation
DNA destroyed → no transformation
DNA but no RNA → transformation
DNA but no proteins → transformation

FIGURE 12.2 Some Experiments on the Transforming Principle. Earlier experiments done by Avery, MacLeod, and McCarty had shown that only DNA extracts from S cells caused transformation of R cells to S cells. To demonstrate that contaminating molecules in the DNA extract were not responsible for transformation, the DNA extract from S cells was treated with RNase, DNase, and protease, and then mixed with R cells. Time was allowed for the DNA from S cells to be taken up by the R cells and expressed, transforming R cells into S cells. Then, antibodies (immune system proteins that recognize specific structures) that recognized R cells but not S cells were added to the mixture. The addition of antibodies caused the R cells (i.e., those R cells that had not been transformed) to aggregate. These aggregated R cells were removed from the mixture by gentle centrifugation. Thus the only cells remaining in the mixture were cells that had been transformed and were now S cells. Only treatment of the DNA extract from S cells with DNase destroyed the ability of the extract to transform the R cells.

violently in a blender to shear off any adsorbed bacteriophage particles (**figure 12.3**). After centrifugation, radioactivity in the supernatant (where the virus remained) versus the bacterial cells in the pellet was determined. They found that most radioactive protein was released into the supernatant, whereas ^{32}P DNA remained within the bacteria. Since genetic material was injected and T2 progeny were produced, DNA must have been carrying the genetic information for T2.

Subsequent studies on the genetics of viruses and bacteria were largely responsible for the rapid development of molecular genetics. Furthermore, much of the recombinant DNA technology described in chapter 15 has arisen from studies of bacterial and viral genetics. Research in microbial genetics has had a profound impact on biology as a science and on technology that affects everyday life.

1. Define genome, genotype, and phenotype.
2. Briefly summarize the experiments of Griffith; Avery, MacLeod, and McCarty; and Hershey and Chase. What did each show, and why were these experiments important to the development of microbial genetics?

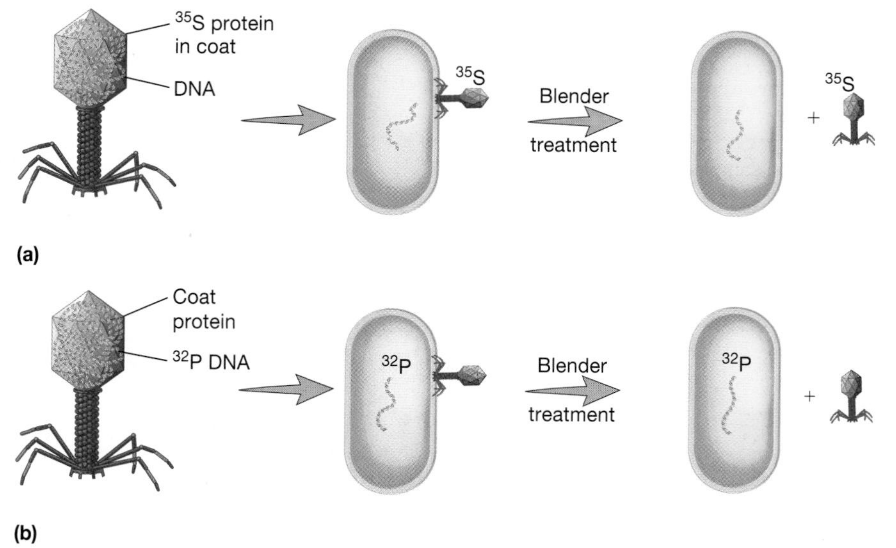

FIGURE 12.3 The Hershey-Chase Experiment. (a) When *E. coli* was infected with a T2 phage containing ^{35}S protein, most of the radioactivity remained outside the host cell. (b) When a T2 phage containing ^{32}P DNA was mixed with the host bacterium, the radioactive DNA was injected into the cell and phages were produced. Thus DNA was carrying the virus's genetic information.

12.2 Flow of Genetic Information

Biologists have long recognized a relationship among DNA, RNA, and protein, and this recognition has guided a vast amount of research over the past decades. The pathway from DNA to RNA and RNA to protein is conserved in all cellular forms of life and is often called the central dogma. **Figure 12.4** illustrates two essential concepts: the flow of genetic information from one generation to the next (replication); and the flow of information within a single cell, a process also called gene expression.

The transmission of genetic information from one generation to the next is shown in figure 12.4*a*. DNA functions as a storage molecule, holding genetic information for the lifetime of a cellular organism and allowing that information to be duplicated and passed on to its progeny. Synthesis of the duplicate DNA is directed by both strands of the parental molecule and is called **replication.** This process is catalyzed by DNA polymerase enzymes.

The genetic information stored in DNA is divided into units called **genes.** For an organism to function properly and reproduce, its genes must be expressed at the appropriate time and place. Gene expression begins with the synthesis of an RNA copy of the gene. This process of DNA-directed RNA synthesis is called **transcription** because the DNA base sequence is rewritten as an RNA base sequence. RNA polymerase enzymes catalyze transcription. Although DNA has two complementary strands, only one strand, the template strand, of a particular gene is transcribed. If both strands of a single gene were transcribed, two different RNA molecules would result in two different products. However, different genes may be encoded on opposite strands, thus both strands of DNA can serve as templates for RNA synthesis, depending on the orientation of the gene on the DNA. Transcription yields three major types of RNA, depending on the gene transcribed. These are messenger RNA (mRNA), transfer RNA (tRNA), and ribosomal RNA (rRNA) (figure 12.4*b*).

During the last phase of gene expression, **translation,** genetic information in the form of an RNA base sequence in a messenger RNA (mRNA) is decoded and used to govern the synthesis of a polypeptide. Thus the amino acid sequence of a protein is a direct reflection of the base sequence in mRNA. In turn, the mRNA nucleotide sequence is complementary to a portion of the DNA genome. In addition to mRNA, translation also requires the activities of transfer RNA and ribosomal RNA. Thus all three types of RNA are involved in the production of protein, which is determined by the code present in the DNA.

1. Describe the general relationship between DNA, RNA, and protein.
2. What are the products of replication, transcription, and translation?
3. Until relatively recently, the "one gene–one protein" hypothesis was used to define the role of genes in organisms. Refer to figure 12.4 and explain why this description of a gene no longer applies.

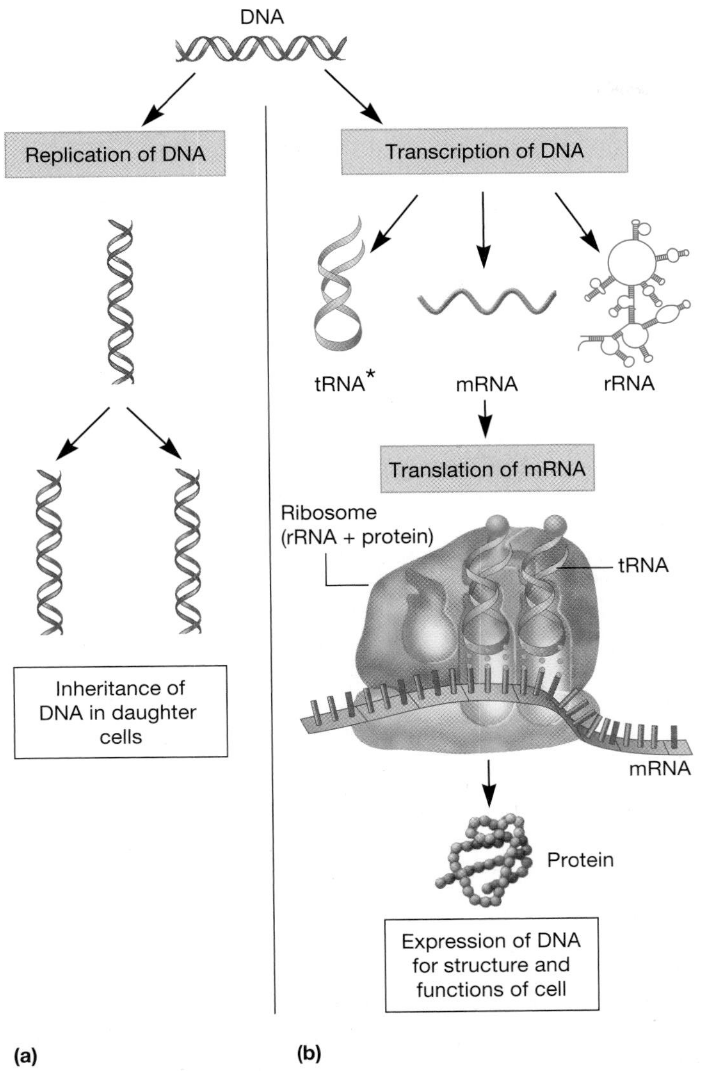

*The sizes of RNA are enlarged to show details.

FIGURE 12.4 Summary of the Flow of Genetic Information in Cells. DNA serves as the storehouse for genetic information. (a) During cellular reproduction, DNA is replicated and passed to progeny cells. (b) To function properly, a cell must express the genetic information stored in DNA. This is accomplished when the genetic code is transcribed into mRNA molecules. The information in the mRNA is then translated into protein.

12.3 Nucleic Acid and Protein Structure

DNA, RNA, and protein are often called informational molecules. The information exists as the sequence of monomers from which they are built. DNA and RNA are polymers of nucleotides (**figure 12.5**) linked together by phosphodiester bonds (**figure 12.6*a***). Proteins are polymers of amino acids (**figure 12.7**) linked by peptide bonds (**figure 12.8**). DNA and RNA differ in terms of the nitrogenous bases they contain, the sugar component of their nucleotides, and whether they are double or single stranded. These differences, as well as an overview of protein structure, are described in this section.

DNA Structure

Deoxyribonucleic acid (DNA) contains the bases adenine, guanine, cytosine, and thymine. The sugar found in the nucleotides is deoxyribose (figure 12.5*b*). DNA molecules are very large and are usually composed of two polynucleotide chains coiled together to form a double helix 2.0 nm in diameter (figure 12.6). Each chain contains purine and pyrimidine deoxyribonucleosides joined by phosphodiester bonds (figure 12.6*a*). That is, a phosphate forms a bridge between a 3′-hydroxyl of one sugar and a 5′-hydroxyl of an adjacent sugar. Purine and pyrimidine bases are attached to the 1′-carbon of the deoxyribose sugars; they extend toward the middle of the cylinder formed by the two chains. (The numbers designating the carbons in the sugars are given a prime to distinguish them from the numbers designating the carbons and nitrogens in the nitrogenous bases.) The bases are stacked on top of each other in the center, one base pair every 0.34 nm. The purine adenine (A) of one strand is always paired with the pyrimidine thymine (T) of the opposite strand by two hydrogen bonds. The purine guanine (G) pairs with cytosine (C) by three hydrogen bonds. This AT and GC base pairing means that the two strands in a DNA double helix are **complementary.** In other words, the bases in one strand match up with those of the other according to specific base pairing rules. Because the sequences of bases in these strands encode genetic information, considerable effort has been devoted to determining the base sequences of DNA and RNA from many organisms, including a variety of microbes. ▶▶| *Microbial genomics (chapter 16)*

The two polynucleotide strands of DNA fit together much like the pieces in a jigsaw puzzle. Inspection of figure 12.6*b,c* shows that the two strands are not positioned directly opposite one another in the helical cylinder. Therefore when the strands twist about one another, a wide major groove and narrower minor groove are formed by the backbone. There are 10.5 base pairs per turn of the helix, and each turn of the helix has a vertical length of 3.4 Å. The helix is right-handed—that is, the chains turn counterclockwise as they approach a viewer looking down the longitudinal axis. The two backbones are antiparallel, which means they run in opposite directions with respect to the orientation of their sugars. One end of each strand has an exposed 5′-hydroxyl group, often with phosphates attached, whereas the other end has a free 3′-hydroxyl group (figure 12.6*a*). In a given direction, one strand is oriented 5′ to 3′ and the other, 3′ to 5′ (figure 12.6*b*).

The structure of DNA just described is that of the B form, the most common form in cells. Two other forms of DNA have been

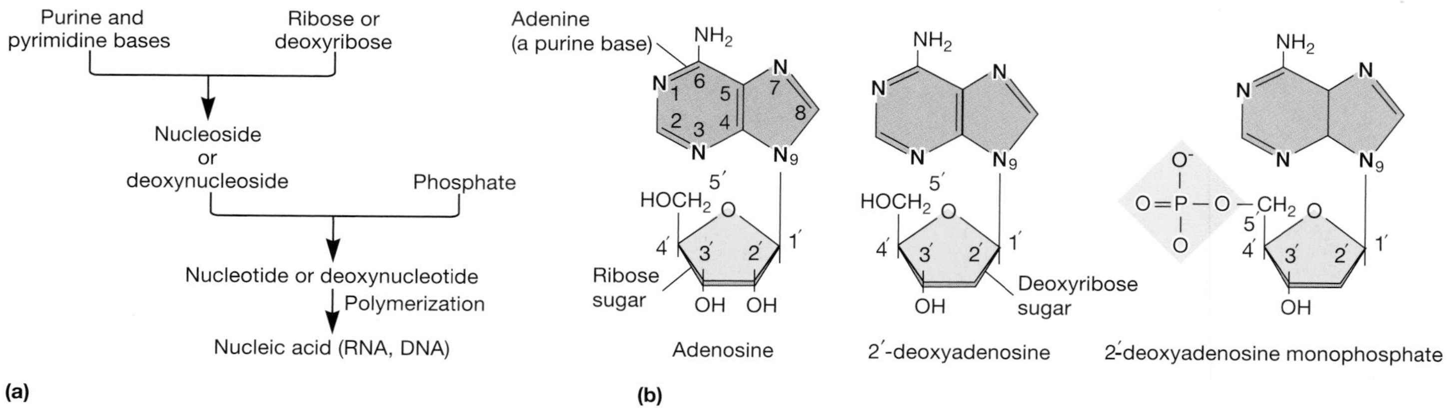

FIGURE 12.5 The Composition of Nucleic Acids. (a) A diagram showing the relationships of various nucleic acid components. Combination of a purine or pyrimidine base with ribose or deoxyribose gives a nucleoside (a ribonucleoside or deoxyribonucleoside). A nucleotide contains a nucleoside and one or more phosphates. Nucleic acids result when nucleotides are connected together in polynucleotide chains. (b) Examples of nucleosides—adenosine and 2′-deoxyadenosine—and the nucleotide 2′-deoxyadenosine monophosphate. The carbons of nucleoside and nucleotide sugars are indicated by numbers with primes to distinguish them from the carbons in the bases.

Figure 12.5 Micro Inquiry

To which carbon of ribose (deoxyribose) is each of the following bonded: adenine and the hydroxyls of adenosine and deoxyadenosine?

identified. The A form primarily differs from the B form in that it has 11 base pairs per helical turn, rather than 10.5, and a vertical length of 2.6 Å, rather than 3.4. Thus it is wider than the B form. The Z form is dramatically different, having a left-handed helical structure, rather than right-handed as seen in the B and A forms. The Z form has 12 base pairs per helical turn and a vertical rise of 3.7 Å. Thus it is more slender than the B form. At this time, it is unclear whether the A form is found in cells. However, evidence exists that small portions of chromosomes can be in the Z form. The role, if any, for these stretches of Z DNA is unknown. *DNA Structure*

RNA Structure

Ribonucleic acid (RNA) contains the bases adenine, guanine, cytosine, and uracil (instead of thymine, although tRNA contains a modified form of thymine). The nucleotides are joined by a phosphodiester bond, just as they are in DNA. The sugar in RNA is ribose (figure 12.5*b*). Most RNA molecules are single stranded. However, an RNA strand can coil back on itself to form a hairpin-shaped structure with complementary base pairing and helical organization (figure 12.4*b* and p. 308, figure 12.21). The formation of double-stranded regions in tRNA and rRNA is critical to their function. Double-stranded regions can also form in some mRNA molecules. These regions are important in regulating gene expression. ▶▶| *Regulation of transcription elongation (section 13.3)*

Protein Structure

Amino acids are the basic building blocks of proteins. An amino acid is defined by the presence of a central carbon (the α carbon) to which is attached a carboxyl group, an amino group, and a side chain. The 20 amino acids normally found in proteins are shown in figure 12.7. As can be seen, the side chains differ in terms of their molecular makeup, which determines if the side chain is nonpolar, polar, or charged. The amino acids are linked by peptide bonds formed by a reaction between the carboxyl group of one amino acid and the amino group of the next amino acid (**figure 12.8**). A polypeptide has polarity just as DNA and RNA do. At one end of the chain is an amino group, and at the other end is a carboxyl group. Thus a polypeptide has an amino or N terminus and a carboxyl or C terminus.

1. What are nucleic acids? How do DNA and RNA differ in structure?
2. What does it mean to say that the two strands of the DNA double helix are complementary and antiparallel? Examine figure 12.6*b* and explain the differences between the minor and major grooves.
3. For each amino acid shown in figure 12.7, identify the α carbon, the carboxyl group, amino group, and side chain. Which amino acids might be found in the transmembrane portion of a polypeptide located in a cell's plasma membrane? Explain your answer.

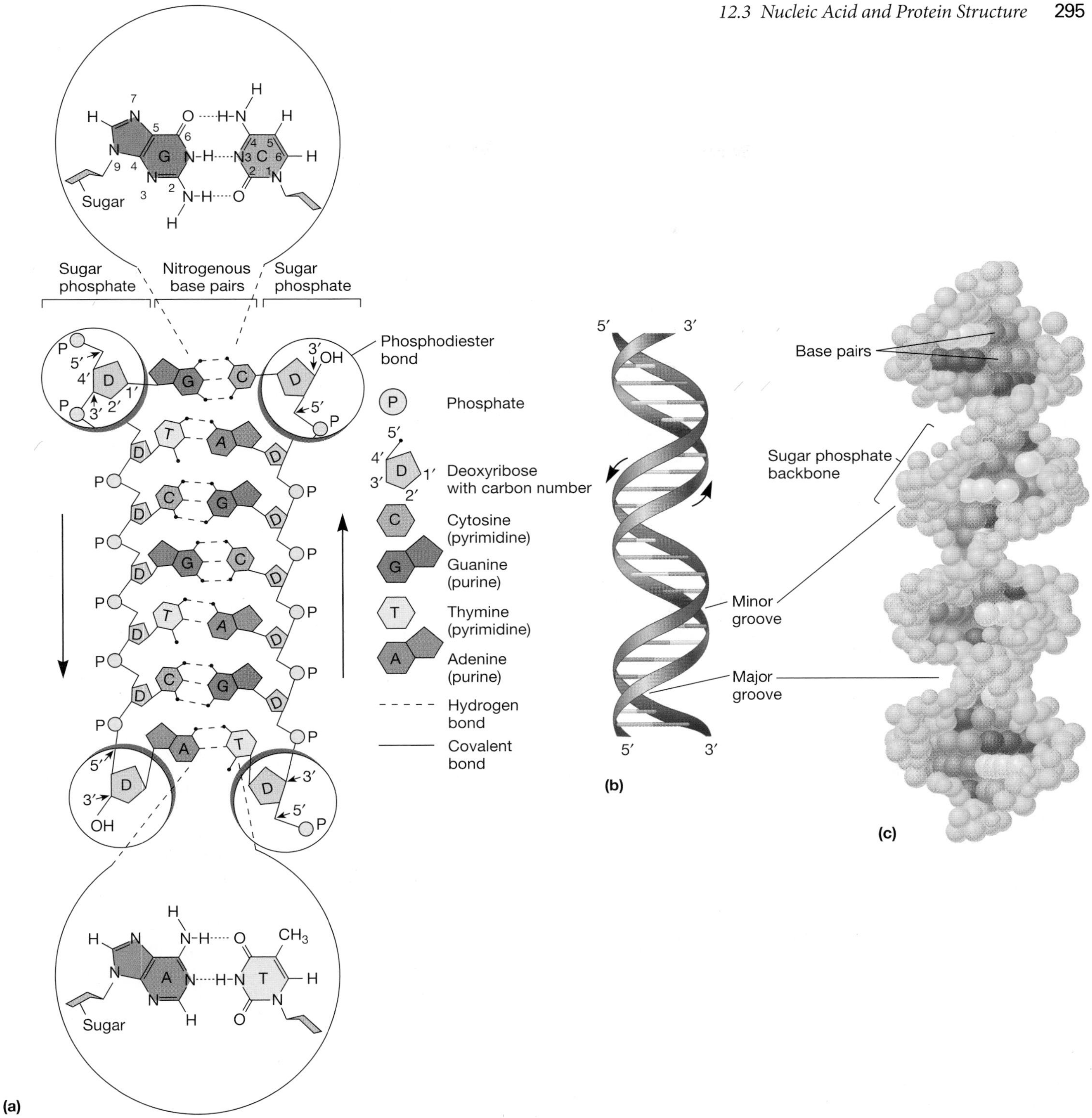

FIGURE 12.6 DNA Structure—B Form of DNA. DNA is usually a double-stranded molecule. (a) A schematic, nonhelical model. In each strand, phosphates are esterified to the 3′-carbon of one deoxyribose sugar (blue) and the 5′-carbon of the adjacent sugar. The two strands are held together by hydrogen bonds (dashed lines). Because of the specific base pairing, the base sequence of one strand determines the sequence of the other. The two strands are antiparallel—that is, the backbones run in opposite directions, as indicated by the two arrows, which point in the 5′ to 3′ direction. (b) A simplified model that highlights the antiparallel arrangement and the major and minor grooves. (c) A space-filling model of the B form of DNA. Note that the sugar-phosphate backbone spirals around the outside of the helix and the base pairs are embedded inside.

***Figure 12.6** Micro Inquiry*

How many H bonds are there between adenine and thymine and between guanine and cytosine?

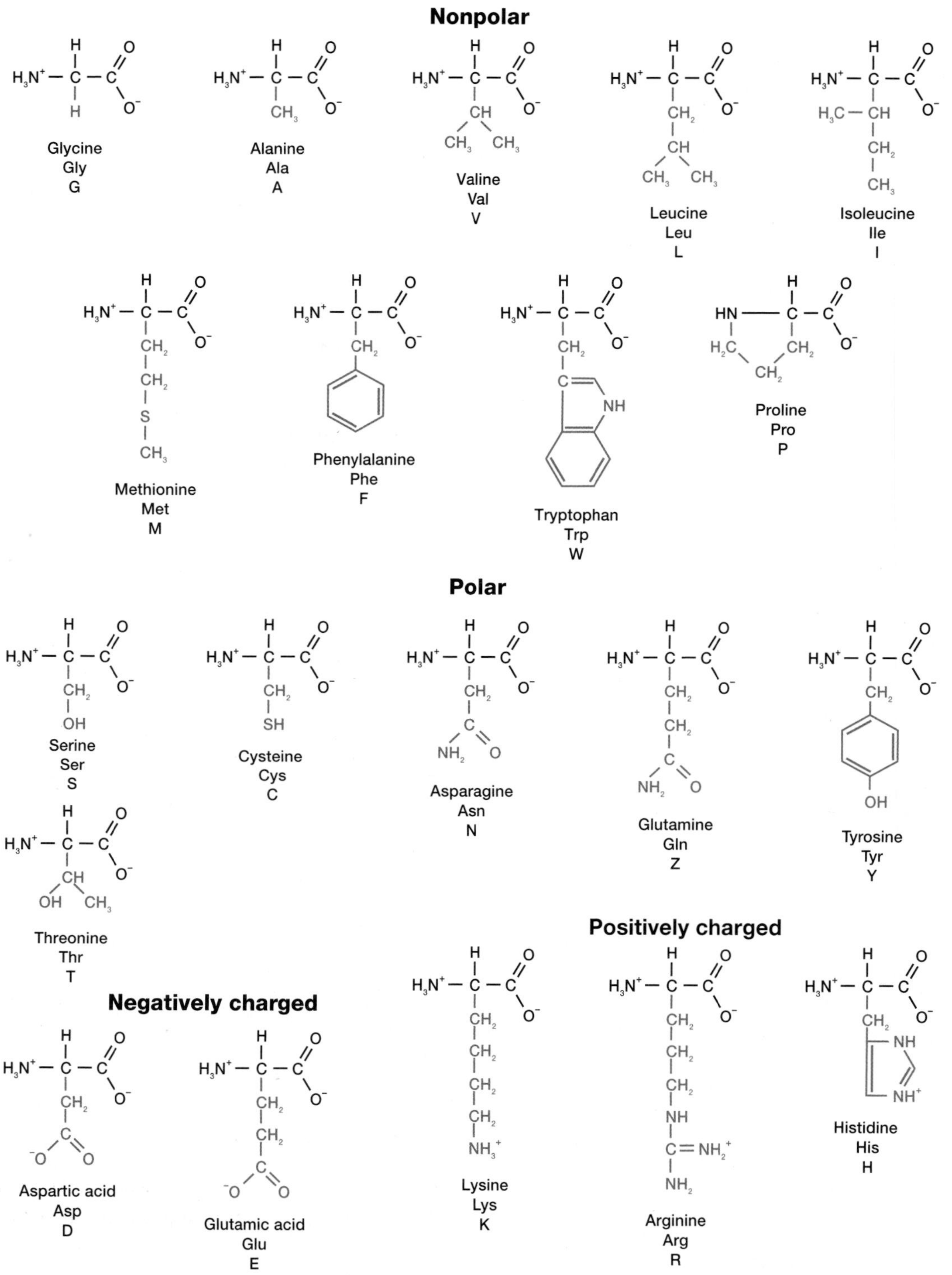

FIGURE 12.7 The Common Amino Acids. The structures of the α-amino acids normally found in proteins. Their side chains are shown in blue, and they are grouped together based on the nature of their side chains—nonpolar, polar, negatively charged (acid), or positively charged (basic). Proline is actually an imino acid, rather than an amino acid. Each amino acid has a 3-letter and 1-letter abbreviation.

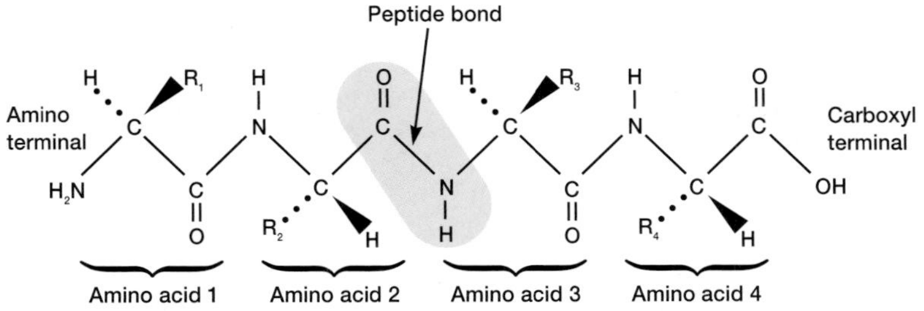

FIGURE 12.8 Peptide Bonds Link Amino Acids Together in Peptide Chains. A tetrapeptide chain is shown. One of the peptide bonds linking the four amino acids together is highlighted in blue. At one end of the peptide is an amino group (amino or N terminal); at the other end is a carboxyl group (carboxyl or C terminal).

Figure 12.8 Micro Inquiry

Identify the two other peptide bonds of the tetrapeptide chain.

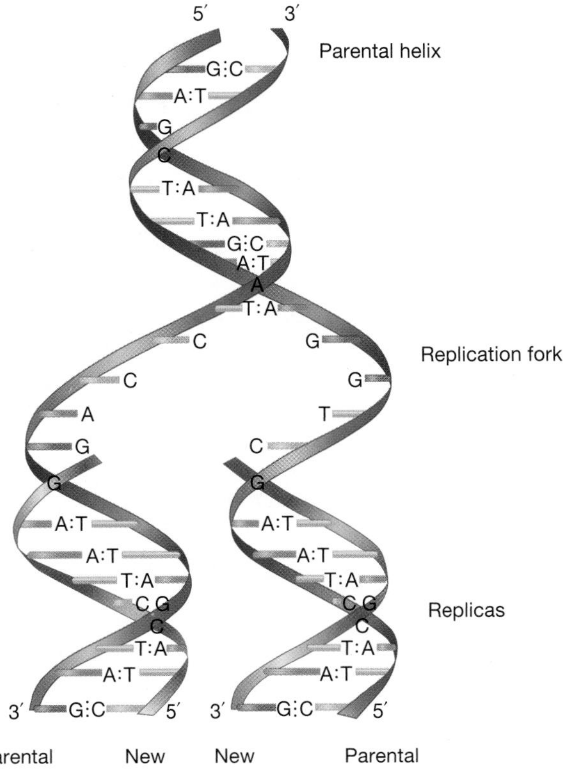

FIGURE 12.9 Semiconservative DNA Replication. The replication fork of DNA showing the synthesis of two progeny strands. Newly synthesized strands are purple. Each copy contains one new and one old strand. This process is called semiconservative replication.

12.4 DNA Replication

DNA replication is an extraordinarily important and complex process upon which all life depends. During DNA replication, the two strands of the double helix are separated; each then serves as a template for the synthesis of a complementary strand according to the base pairing rules. Each of the two progeny DNA molecules consists of one new strand and one old strand. Thus DNA replication is semiconservative (**figure 12.9**). DNA replication is also extremely accurate; *E. coli* makes errors with a frequency of only 10^{-9} or 10^{-10} per base pair replicated (or about one in a million [10^{-6}] per gene per generation). Despite its complexity and accuracy, replication is very rapid. In *Bacteria,* replication rates approach 750 to 1,000 base pairs per second. Eukaryotic replication is slower, about 50 to 100 base pairs per second.

In this section, we first discuss the various patterns of DNA replication observed in cells. We then consider the mechanism of DNA replication in *E. coli,* beginning with an examination of the replication machinery and then events at the replication fork.

Patterns of DNA Synthesis

Replication patterns are somewhat different in *Bacteria, Archaea,* and eukaryotes. Recall from chapter 3 that the DNA of most bacteria is circular. When the circular DNA chromosome is copied, replication begins at a single point, the origin (**figure 12.10**). Synthesis of DNA occurs at the **replication fork,** the place at which the DNA helix is unwound and individual strands are replicated. Two replication forks move outward from the origin until they have copied the whole **replicon**—the portion of the genome that contains an origin and is replicated as a unit. When the replication forks move around the circle, a structure shaped like the Greek letter theta (θ) is formed. Because the bacterial chromosome is a single replicon, the forks meet on the other side and two separate chromosomes are released. Archaeal chromosomes are also circular. Until recently, it was thought that all bacterial and archaeal cells have a single origin of replication. However, two members of the archaeal genus *Sulfolobus* have more than one origin; the reason for this is unknown. Eukaryotic DNA is linear and much longer than bacterial and archaeal DNA. Many replication forks copy eukaryotic DNA simultaneously so that the molecule can be duplicated relatively quickly (**chapter opener**). The ends of the linear chromosomes, both eukaryotic and bacterial, also present a problem to the cell, as we discuss on p. 304. *DNA Replication Fork; Bidirectional DNA Replication*

Replication Machinery

DNA replication is essential to organisms, and a great deal of effort has been devoted to understanding its mechanism. The replication of *E. coli* DNA requires at least 30 proteins

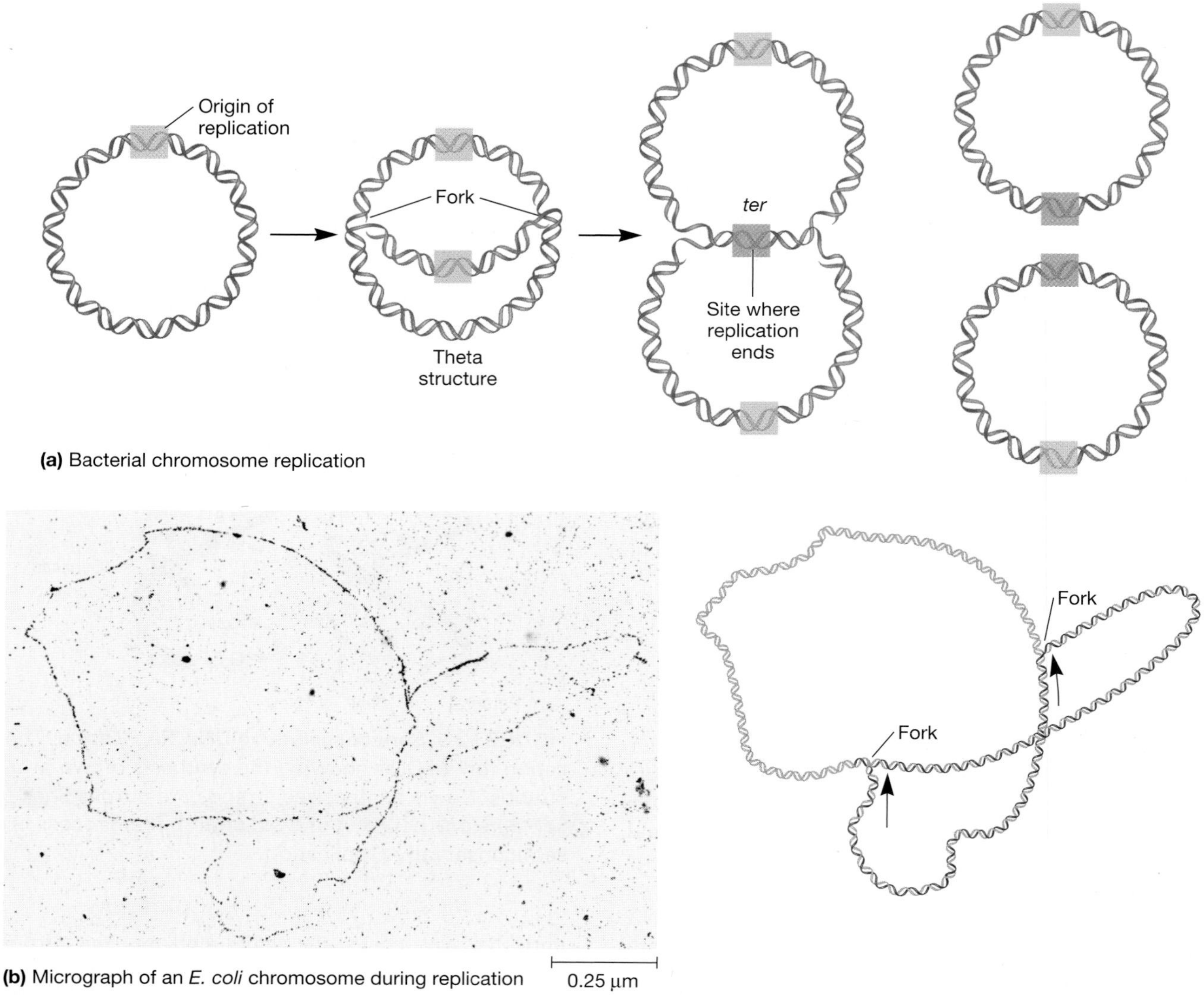

(a) Bacterial chromosome replication

(b) Micrograph of an *E. coli* chromosome during replication

FIGURE 12.10 Bidirectional Replication of the *E. coli* Chromosome. (a) Replication begins at one site on the chromosome, called the origin of replication. Two replication forks proceed in opposite directions from the origin until they meet at a site called the replication termination site (*ter*). The theta structure is a commonly observed intermediate of the process. (b) An autoradiograph of a replicating *E. coli* chromosome; about one-third of the chromosome has been replicated. To the right is a schematic representation of the chromosome. Parental DNA is blue; new DNA strands are purple; arrow represents direction of fork movement.

(**table 12.1**). The archaeal and eukaryotic replication enzymes and proteins differ from those used by bacterial cells. In general, the archaeal replication machinery is more similar to the eukaryotic machinery than it is to the bacterial system. Despite these differences, the overall process is similar in all organisms.

Enzymes called **DNA polymerases** catalyze DNA synthesis. All known DNA polymerases catalyze the synthesis of DNA in the 5′ to 3′ direction. The nucleotide to be added is a deoxynucleoside triphosphate (dNTP). The new deoxynucleotide is added when a phosphodiester bond is formed by a reaction between the hydroxyl group at the 3′ end of the growing DNA strand and the α-phosphate (the phosphate closest to the 5′ carbon) of the in-coming deoxynucleotide (**figure 12.11**). The energy needed to form the phosphodiester bond is provided by the release of the terminal two phosphates as pyrophosphate (PP_i) from the nucleotide that is added. The PP_i is subsequently hydrolyzed to two separate phosphates (P_i). Thus the deoxynucleoside triphosphates dATP, dTTP, dCTP, and dGTP serve as DNA polymerase substrates while deoxynucleoside monophosphates (dNMPs: dAMP, dTMP, dCMP, dGMP) are incorporated into the growing chain. *How Nucleotides Are Added in DNA Replication*

Table 12.1 Components of the *E. coli* Replication Machinery

Protein	*Function*
DnaA protein	Initiation of replication; binds origin of replication (*oriC*)
DnaB protein	Helicase (5′→3′); breaks hydrogen bonds holding two strands of double helix together; promotes DNA primase activity; involved in primosome assembly
DNA gyrase	Relieves supercoiling of DNA produced as DNA strands are separated by helicases; separates daughter molecules in final stages of replication
SSB proteins	Bind single-stranded DNA after strands are separated by helicases
DnaC protein	Helicase loader; helps direct DnaB protein (helicase) to DNA template
n′ protein	Component of primosome; helicase (3′→5′)
n protein	Primosome assembly; component of primosome
n″ protein	Primosome assembly
I protein	Primosome assembly
DNA primase	Synthesis of RNA primer; component of primosome
DNA polymerase III holoenzyme	Complex of about 20 polypeptides; catalyzes most of the DNA synthesis that occurs during DNA replication; has 3′→5′ exonuclease (proofreading) activity
DNA polymerase I	Removes RNA primers; fills gaps in DNA formed by removal of RNA primer
Ribonuclease H	Removes RNA primers
DNA ligase	Seals nicked DNA, joining DNA fragments together
DNA replication terminus site-binding protein	Termination of replication
Topoisomerase IV	Segregation of chromosomes upon completion of DNA replication

For DNA polymerase to catalyze the synthesis of DNA, it needs three things. The first is a template, which is read in the 3′ to 5′ direction and is used to direct the synthesis of a complementary DNA strand. The second is a primer (e.g., an RNA strand or a DNA strand) to provide a free 3′-hydroxyl group to which nucleotides can be added (figure 12.11). The third is a set of dNTPs. *E. coli* has five different DNA polymerases (DNA polymerase I-V). DNA polymerase III plays the major role in replication, although it is assisted by DNA polymerase I.

DNA polymerase III holoenzyme is a complex of 10 different proteins, three of which form two core enzymes (**figure 12.12**). The core enzymes are responsible for catalyzing DNA synthesis and proofreading the product to ensure fidelity of replication. A dimer of another subunit (tau) connects the two core enzymes. Associated with each core enzyme is a subunit called the β clamp. This protein tethers the core enzyme to one strand of the DNA molecule. A complex of proteins called the γ complex is responsible for loading the β clamp onto the DNA. Because there are two core enzymes, both strands of DNA are bound by a single DNA polymerase III holoenzyme.

In *E. coli,* replication begins when a collection of DnaA proteins binds to specific nucleotide sequences (DnaA boxes) within the **origin of replication.** The DnaA proteins use the energy from ATP hydrolysis to break or "melt" the hydrogen bonds between the DNA strands, thus making this localized region single stranded. Although this provides the initial template for replication, DNA polymerase III cannot by itself unwind and maintain the single-stranded DNA. These activities are provided by the action of other proteins found in the **replisome,** a huge complex of proteins that includes DNA polymerase III holoenzyme.

The other proteins found in the replisome include helicases, single-stranded DNA binding proteins, and topoisomerases (**figure 12.13**). **Helicases** are responsible for separating (unwinding) the DNA strands. These enzymes use energy from ATP hydrolysis to unwind short stretches of helix just ahead of the replication fork. **Single-stranded DNA binding proteins (SSBs)** keep the strands apart once they have been separated, and **topoisomerases** relieve the tension generated by the rapid unwinding of the double helix (the replication fork may rotate as rapidly as 75 to 100 revolutions per second). This is important because rapid unwinding can lead to the formation of supercoils in the helix, and these can impede replication if not removed. Topoisomerases change the structure of DNA by transiently breaking one or two strands without altering the nucleotide sequence of the DNA (e.g., a topoisomerase might

DNA polymerase reaction

DNA polymerase reaction:

$$n[\text{dATP, dGTP, dCTP, dTTP}] \xrightarrow[\text{DNA template}]{\text{DNA polymerase}} \text{DNA} + n\text{PP}_i$$

The mechanism of chain growth

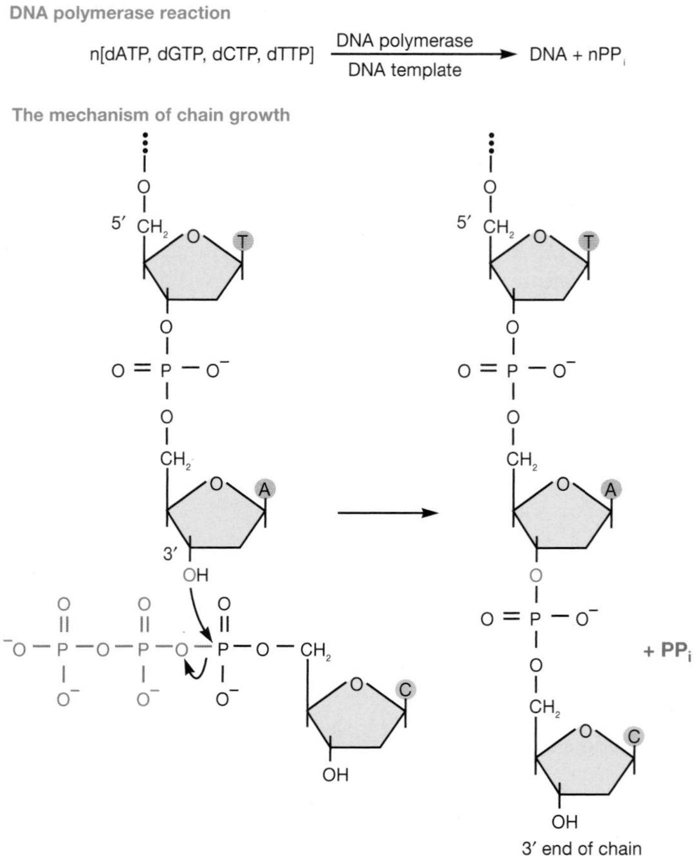

FIGURE 12.11 The DNA Polymerase Reaction and Its Mechanism. The mechanism involves a nucleophilic attack by the hydroxyl of the 3′ terminal deoxyribose on the alpha phosphate of the nucleotide substrate (in this example, adenosine attacks cytidine triphosphate).

Figure 12.11 Micro Inquiry

What provides the energy to fuel this reaction?

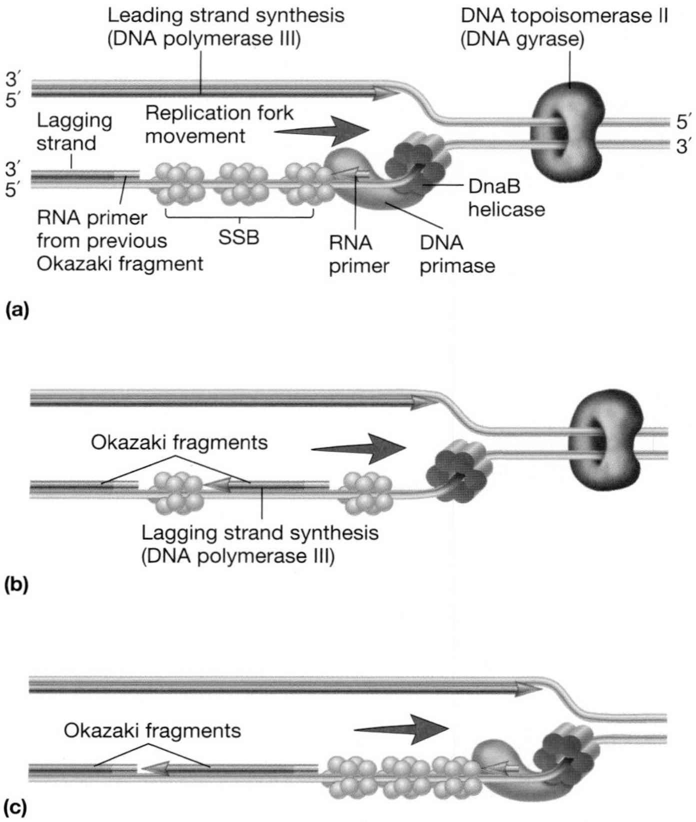

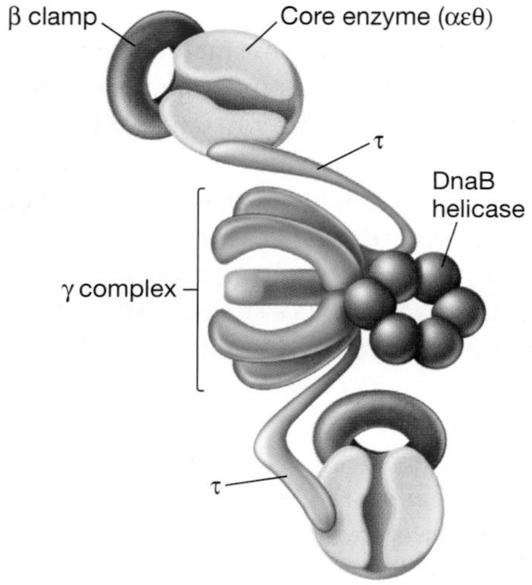

FIGURE 12.12 DNA Polymerase III Holoenzyme. The holoenzyme consists of two core enzymes (three subunits each; α, ε, θ, not shown) and several other subunits. The two tau (τ) subunits connect the two core enzymes to a large complex called the gamma (γ) complex. Each core enzyme is associated with a β clamp, which tethers a DNA template to each core enzyme. Also shown is the DnaB helicase. It is responsible for separating DNA strands ahead of the DNA polymerase holoenzyme.

(a)

(b)

(c)

FIGURE 12.13 Other Replisome Proteins. A simplified diagram of DNA replication in *E. coli.* A single replication fork showing the activity of replisome proteins other than DNA polymerase III (not shown) is illustrated. DnaB helicase is responsible for separating the two strands of parental DNA. The strands are kept apart by single-stranded DNA binding proteins (SSB), which allows for synthesis of an RNA primer by DNA primase. DNA gyrase eases the strain introduced into the DNA double helix by helicase activity. Both leading strand and lagging strand synthesis are illustrated. The lagging strand is synthesized in short fragments called Okazaki fragments. A new RNA primer is required for the synthesis of each Okazaki fragment.

Figure 12.13 Micro Inquiry

What is the difference between helicase and gyrase? Which is a topoisomerase?

tie or untie a knot in a DNA strand). **DNA gyrase** is an important topoisomerase in *E. coli.* It is not only important during DNA replication but also for introducing negative supercoiling in the bacterial chromosome that helps compact it. ◀◀ *Nucleoid (section 3.5)*

Once the template is prepared, the primer needed by DNA polymerase III can be synthesized. An enzyme called **primase** synthesizes a short RNA strand, usually around 10 nucleotides long and complementary to the DNA; this serves as the primer (figure 12.13). RNA is used as the primer because unlike DNA polymerase, RNA polymerases (such as primase) can initiate RNA synthesis without an existing 3′-OH. It appears that primase requires the assistance of several other proteins, and the complex of primase and its accessory proteins is called the **primosome** (table 12.1). The primosome is another important component of the replisome.

As noted previously, DNA polymerase enzymes synthesize DNA in the 5′ to 3′ direction. Therefore only one of the strands, called the **leading strand,** can be synthesized continuously at its 3′ end as the DNA unwinds (figure 12.13). The other strand, called the **lagging strand,** cannot be extended in the same direction as the movement of the replication fork because there is no free 3′-OH to which a nucleotide can be added. As a result, the lagging strand is synthesized discontinuously in the 5′ to 3′ direction (i.e., in the direction opposite of the movement of the replication fork) and produces a series of fragments called **Okazaki fragments,** after their discoverer, Reiji Okazaki. Discontinuous synthesis occurs as primase makes many RNA primers along the template strand. DNA polymerase III then extends these primers with DNA, and eventually the Okazaki fragments are joined to form a complete strand. Thus while the leading strand requires only one RNA primer to initiate synthesis, the lagging strand has many RNA primers that must eventually be removed. Okazaki fragments are about 1,000 to 2,000 nucleotides long in *Bacteria* and approximately 100 nucleotides long in eukaryotic cells. *DNA Replication; Structural Basis of DNA Replication*

Events at the Replication Fork

The details of DNA replication are outlined in **figure 12.14.** In *E. coli,* DNA replication is initiated at specific nucleotides called the *oriC* locus (for *ori*gin of *c*hromosomal replication). Here, we present replication as a series of discrete steps, but in the cell these events occur quickly and simultaneously on both the leading and lagging strands.

1. To initiate replication, as many as 40 DnaA proteins bind *oriC* while hydrolyzing ATP. Binding of DnaA proteins causes the DNA to bend around the protein complex, resulting in separation of the double-stranded DNA at regions within the origin having many AT base pairs. Recall that adenines pair with thymines using only two hydrogen bonds, so AT-rich segments of DNA become single stranded more readily than do GC-rich regions. Once the strands have separated, replication proceeds through the next four stages.
2. Helicases unwind the helix with the aid of topoisomerases such as DNA gyrase (figure 12.14, step 1). DnaB protein is the helicase most actively involved in replication. The single strands are kept separate by SSBs.
3. Primase synthesizes RNA primers as needed (figure 12.14, step 1). A single DNA polymerase III holoenzyme catalyzes both leading strand and lagging strand synthesis from the RNA primers. Lagging strand synthesis is particularly amazing because of the "gymnastic" feats performed by the replisome. It must discard old β clamps (figure 12.14, step 3), load new β clamps (figure 12.14, step 2), and tether the template to the core enzyme with each new round of Okazaki fragment synthesis. All of this occurs as DNA polymerase III is synthesizing DNA. Thus DNA polymerase III is a multifunctional enzyme.
4. After most of the lagging strand has been synthesized by the formation of Okazaki fragments, DNA polymerase I removes the RNA primer. DNA polymerase I does this because, unlike other DNA polymerases, it has the ability to snip off nucleotides one at a time starting at the 5′ end while moving toward the 3′ end of RNA primer. This ability is referred to as 5′ to 3′ exonuclease activity. DNA polymerase I begins its exonuclease activity at the free 5′ end of the RNA primer. With the removal of each ribonucleotide, the adjacent 3′-OH from the deoxynucleotide is used by DNA polymerase I to fill the gap between Okazaki fragments (**figure 12.15**).
5. Finally, the Okazaki fragments are joined by the enzyme **DNA ligase,** which forms a phosphodiester bond between the 3′-OH of the growing strand and the 5′-phosphate of an Okazaki fragment (**figure 12.16**).

Amazingly, DNA polymerase III, like all DNA polymerases, has an additional function that is critically important: **proofreading.** Proofreading is the removal of a mismatched base immediately after it has been added; its removal must occur before the next base is incorporated. One of the protein subunits of the DNA polymerase III core enzyme (the ε subunit) has 3′ to 5′ exonuclease activity. This activity enables the polymerase core enzyme to check each newly incorporated base to see that it forms stable hydrogen bonds. In this way, mismatched bases can be detected. If the wrong base has been mistakenly added, the exonuclease activity is used to remove it. A mismatched base can be removed only as long as it is still at the 3′ end of the growing strand. Once removed, holoenzyme backs up and adds the proper nucleotide in its place. DNA proofreading is not 100% efficient, and as discussed in chapter 14, the mismatch repair system is the cell's second line of defense against the potential harm caused by the incorporation of the incorrect nucleotide. *Proofreading Function of DNA Polymerase*

As we have seen, DNA polymerase III is a remarkable multiprotein complex, with several enzymatic activities. In *E. coli,* the polymerase component is encoded by the *dnaE* gene. *Bacillus*

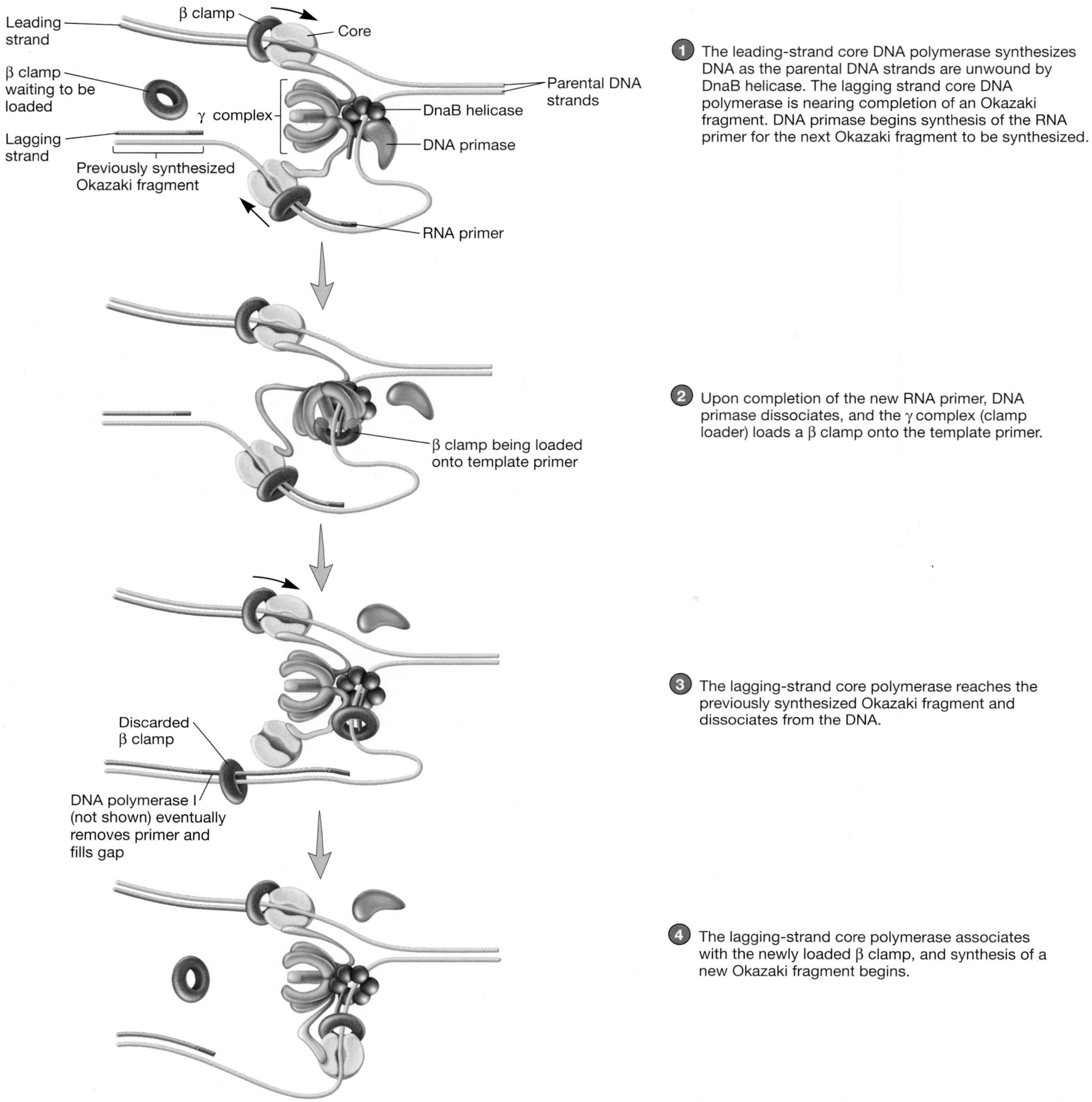
Leading strand
β clamp
Core
β clamp waiting to be loaded
γ complex
Parental DNA strands
DnaB helicase
Lagging strand
DNA primase
Previously synthesized Okazaki fragment
RNA primer
1 The leading-strand core DNA polymerase synthesizes DNA as the parental DNA strands are unwound by DnaB helicase. The lagging strand core DNA polymerase is nearing completion of an Okazaki fragment. DNA primase begins synthesis of the RNA primer for the next Okazaki fragment to be synthesized.
β clamp being loaded onto template primer
2 Upon completion of the new RNA primer, DNA primase dissociates, and the γ complex (clamp loader) loads a β clamp onto the template primer.
Discarded β clamp
DNA polymerase I (not shown) eventually removes primer and fills gap
3 The lagging-strand core polymerase reaches the previously synthesized Okazaki fragment and dissociates from the DNA.
4 The lagging-strand core polymerase associates with the newly loaded β clamp, and synthesis of a new Okazaki fragment begins.

FIGURE 12.14 Events at the Replication Fork. DNA polymerase III holoenzyme and other components of the replisome are responsible for the synthesis of both leading and lagging strands. The arrows show the movement of each DNA core polymerase. After completion of each new Okazaki fragment, the old β clamp is discarded and a new one loaded onto the template DNA (steps 2 and 3). This is achieved by the activity of the γ complex (figure 12.12), which is also known as the clamp loader. Okazaki fragments are eventually joined together (figure 12.15) after removal of the RNA primer and synthesis of DNA to fill the gap, both catalyzed by DNA polymerase I; DNA ligase then seals the nick and joins the two fragments (figure 12.16).

portion of the telomere. The internal RNA template then acts as the template for DNA synthesis and elongates the single strand. (i.e., the 3′-OH of the telomere DNA strand serves as the primer for DNA synthesis). After being lengthened sufficiently, there is room for synthesis of an RNA primer, and the single strand of telomere DNA can serve as the template for synthesis of the complementary strand by DNA polymerase III. Thus the length of the chromosome is maintained.

Telomerase has solved the problem of end replication for eukaryotes, but how do bacteria with linear chromosomes replicate the ends of their chromosomes? Unfortunately little is known about the replication of linear bacterial chromosomes. However, a recent discovery in *Streptomyces*, an important group of soil bacteria, has led to speculation that a telomerase-like process may function in these bacteria. The ends of the linear chromosome of *Streptomyces coelicolor* are associated with a complex of proteins, including one with in vitro reverse transcriptase activity. No RNA has been found in the *Streptomyces* complex. Therefore it is unclear if the protein functions as a reverse transcriptase and what it might use as a template if it does.

1. Define the following: origin of replication, replicon, replication fork, primosome, and replisome.
2. Describe the nature and functions of the following replication components and intermediates: DNA polymerases I and III, topoisomerase, DNA gyrase, helicase, single-stranded DNA binding proteins, Okazaki fragment, DNA ligase, leading strand, lagging strand, primase, and telomerase.
3. Outline the steps involved in DNA synthesis at the replication fork. How do DNA polymerases correct their mistakes?

12.5 Gene Structure

DNA replication allows genetic information to be passed from one generation to the next. But how is the genetic information used? To answer that question, we must first look at how genetic information is organized. The basic unit of genetic information is the gene. The gene has been regarded in several ways. At first, it was thought that a gene contained information for the synthesis of one enzyme—the one gene–one enzyme hypothesis. This was modified to the one gene–one polypeptide hypothesis because of the existence of enzymes and other proteins composed of two or more different polypeptide chains coded for by separate genes. Historically, a segment of DNA that encodes a single polypeptide was termed a cistron; this term is still sometimes used. However, not all genes encode proteins; some code instead for rRNA and tRNA (figure 12.4). In addition, it is now known that some eukaryotic genes encode more than one protein (p. 313). Thus a gene might be defined as a polynucleotide sequence that codes for a functional product (i.e., a polypeptide, tRNA, or rRNA).

When the nucleotide sequences of protein-coding genes are transcribed, the resulting mRNA can be "read" in discrete sets of three nucleotides, each set being a **codon.** Each codon codes for a single amino acid. The sequence of codons is "read" in only one way—the **reading frame** (**figure 12.18**). Each strand of DNA usually consists of gene sequences that do not overlap one another (**figure 12.19*a***). However, there are exceptions to the rule. Some viruses such as the phage ϕX174 have overlapping genes (figure 12.19*b*), and parts of genes overlap in some bacterial genomes.

Bacterial and archaeal gene structure differs greatly from that of eukaryotes. In bacterial and archaeal cells, the coding information within a gene normally is continuous. However, in eukaryotic organisms, many gene sequences that code for polypeptides (exons)

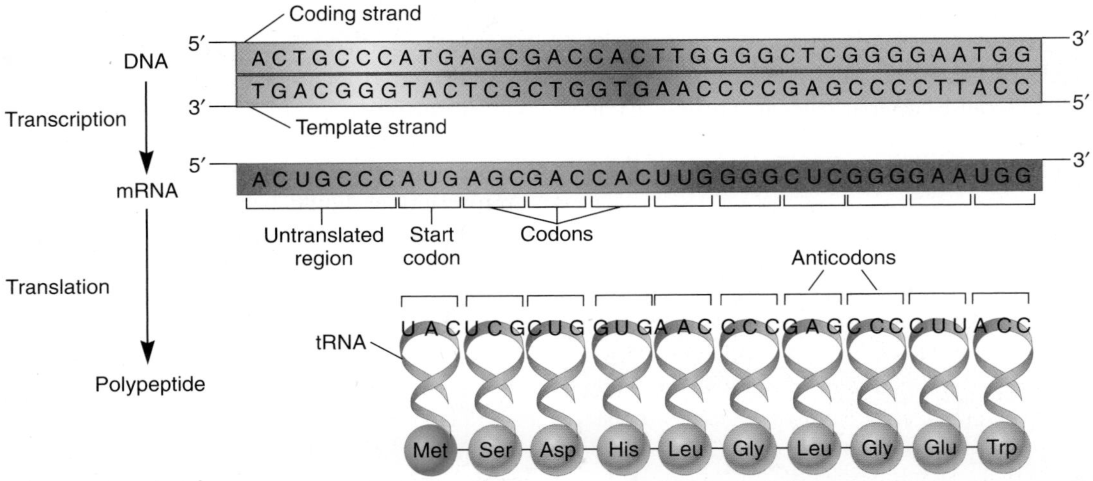

FIGURE 12.18 Reading Frame. During transcription, an mRNA complementary to the template strand of DNA is synthesized. The nucleotides in the mRNA are organized into groups of three, each group being a codon. The first codon translated into protein is the start codon. It establishes the reading frame and therefore the sequence of amino acids in the polypeptide chain that is made from the mRNA.

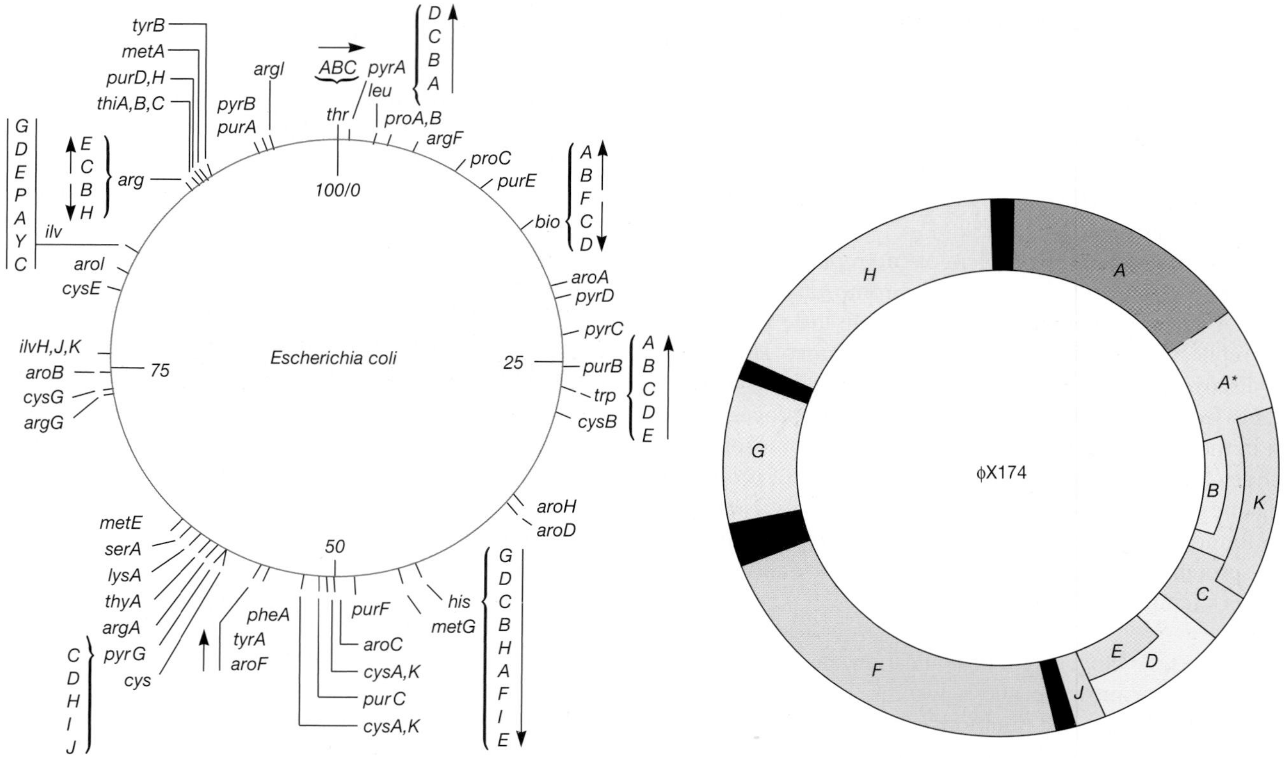

(a) *E.coli* chromosome. Note that genes are separated from each other and don't overlap.

(b) Bacteriophage ϕX174 chromosome. Note that the coding regions for some proteins overlap. This increases the coding efficiency of the phage's genome.

FIGURE 12.19 Chromosomal Organization in *Bacteria* and Viruses. (a) Simplified genetic map of *E. coli.* The *E. coli* map is divided into 100 minutes. (b) The map of phage ϕX174 shows the overlap of gene *B* with *A, K* with *A* and *C,* and *E* with *D.* The solid regions are spaces lying between genes. Protein A* consists of the last part of protein A and arises from reinitiation of translation of gene *A* mRNA.

are interrupted periodically by noncoding sequences (introns). The introns must be spliced out of the mRNA before the protein is made. This affords eukaryotes the ability to cut and paste mRNA molecules so that they can encode more than one polypeptide, a process known as **alternative splicing.** Because bacterial genes are the best characterized, the more detailed description of gene structure that follows focuses on the genes of the model bacterium *E. coli.*

Protein-Coding Genes

For genetic information in DNA to be used, it must first be transcribed to form an RNA molecule. The RNA product of a gene that codes for a protein is messenger RNA (mRNA). Although DNA is double stranded, only one strand of a gene directs RNA synthesis. This strand is called the **template strand,** and the complementary DNA strand is known as the coding strand because it is the same nucleotide sequence as the mRNA, except in DNA bases (**figure 12.20**). Messenger RNA is synthesized from the 5′ to the 3′ end in a manner similar to that seen for DNA synthesis. Therefore the polarity of the DNA template strand is 3′ to 5′. In other words, the beginning of the gene is at the 3′ end of the template strand.

An important site called the **promoter** is located at the start of the gene. The promoter is a recognition/binding site for RNA polymerase, the enzyme that synthesizes RNA. The promoter is neither transcribed nor translated; it functions strictly to orient RNA polymerase a specific distance from the first DNA nucleotide that will serve as a template for RNA synthesis. The promoter thus specifies which strand is to be transcribed and where transcription should begin. As we discuss in chapter 13, the sequences near the promoter are very important in regulating when and where a gene is transcribed.

The transcription start site (labeled +1 in figure 12.20) represents the first nucleotide in the mRNA synthesized from the gene. However, the initially transcribed portion of the gene does not necessarily code for amino acids. Instead, it is a **leader sequence** that is transcribed into mRNA but is not translated into amino acids. In *Bacteria,* the leader sequence includes a region called the **Shine-Dalgarno sequence,** which is important in the initiation of translation. The leader sometimes is also involved in regulation of transcription and translation. ▶▶| *Regulation of transcription elongation (section 13.3); Regulation of translation (section 13.4)*

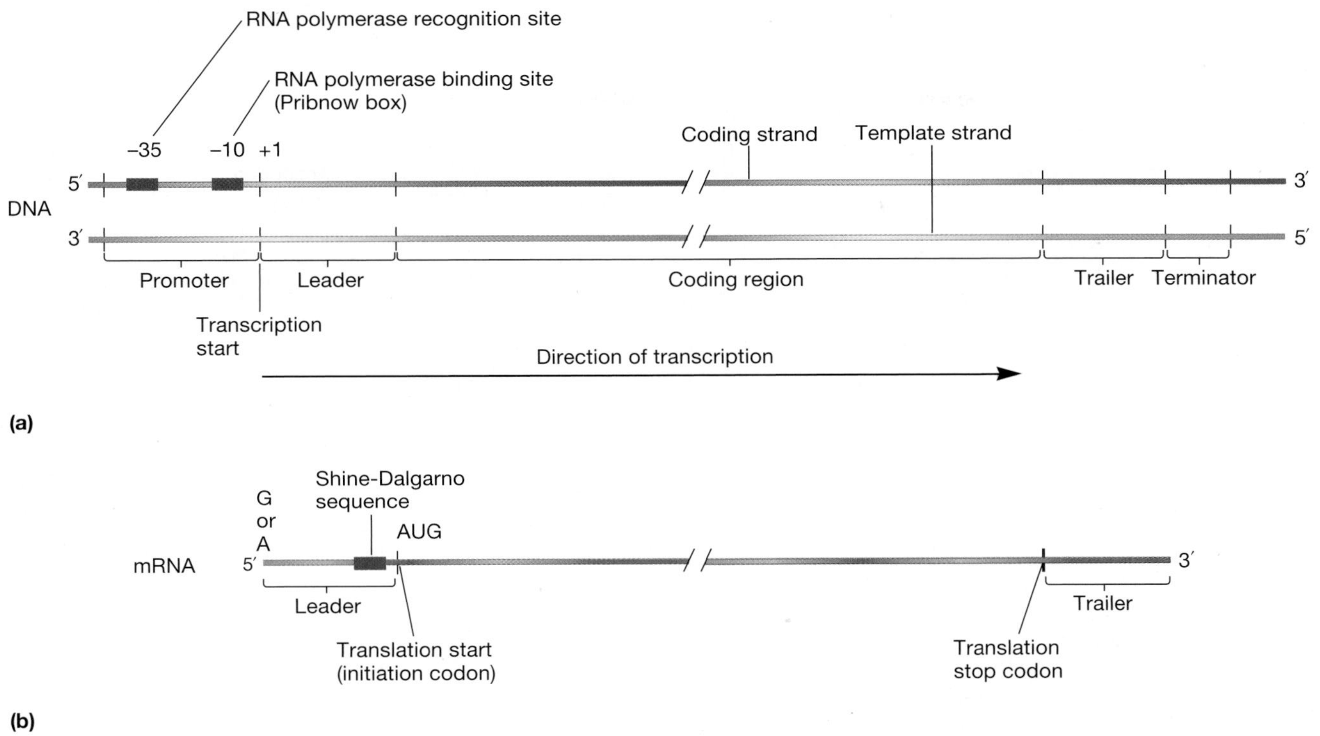

FIGURE 12.20 A Bacterial Structural Gene and Its mRNA Product. (a) The organization of a typical structural gene in a bacterial cell. Leader and trailer sequences are included even though some genes lack one or both. Transcription begins at the +1 position in DNA and proceeds to the right, as shown. The template is read in the 3′ to 5′ direction. (b) Messenger RNA product of the gene shown in part a. The first nucleotide incorporated into mRNA is usually GMP or AMP. Translation of the mRNA begins with the AUG initiation codon. Regulatory sites are not shown.

***Figure 12.20** Micro Inquiry*

Why do you think the nontemplate strand is called the "coding strand"?

Immediately next to (and downstream of) the leader is the most important part of the gene, the **coding region** (figure 12.20). The coding region typically begins with the template DNA sequence 3′-TAC-5′. This is transcribed into the codon 5′-AUG-3′, which in *Bacteria* codes for *N*-formylmethionine, a modified amino acid used to initiate protein synthesis. The remainder of the coding region is transcribed into a sequence of codons that specifies the sequence of amino acids for that particular protein. The coding region ends with a sequence that when transcribed is a **stop codon.** It signals the end of the protein and stops the ribosome during translation. The stop codon is immediately followed by the **trailer sequence,** which is transcribed but not translated. The trailer contains sequences that prepare the RNA polymerase for release from the template strand. Indeed, just beyond the trailer (and sometimes slightly overlapping it) is the **terminator.** The terminator is a sequence that signals the RNA polymerase to stop transcription. It is involved in dislodging the RNA polymerase from the template DNA.

tRNA and rRNA Genes

The DNA segments that code for tRNA and rRNA also are considered genes, although they give rise to RNA rather than protein. In *E. coli,* the genes for tRNA consist of a promoter and transcribed leader and trailer sequences that are removed during the process of tRNA maturation (**figure 12.21*a***). Genes coding for tRNA may code for more than a single tRNA molecule or type of tRNA. The segments coding for tRNAs are separated by short spacer sequences that are removed after transcription by ribonucleases—enzymes (and in some cases ribozymes) that degrade RNA. In addition, many bacterial and archaeal tRNA genes contain introns that must be removed during tRNA maturation. ◀◀ *Ribozymes (section 9.8)*

The genes for rRNA also have promoters, trailers, and terminators (figure 12.21*b*). Interestingly, all the rRNAs are transcribed as a single, large precursor molecule that is cut up by ribonucleases after transcription to yield the final rRNA products. In *E. coli* the trailer regions and the spaces between

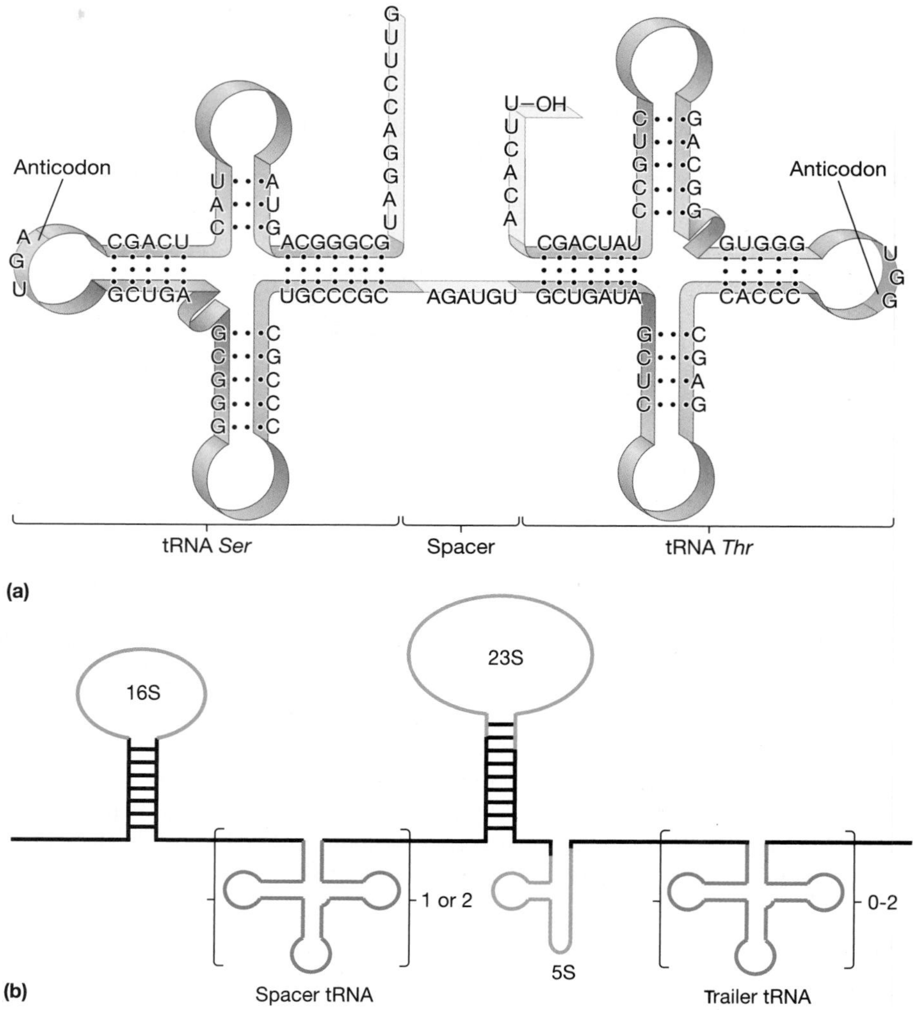

FIGURE 12.21 tRNA and rRNA Genes. (a) A tRNA precursor from *E. coli* that contains two tRNA molecules. The spacer and extra nucleotides at both ends are removed during processing. (b) The *E. coli* ribosomal RNA gene codes for a large transcription product that is cleaved into three rRNAs and one to three tRNAs. The 16S, 23S, and 5S rRNA segments are represented by blue lines, and tRNA sequences are placed in brackets. The seven copies of this gene vary in the number and kind of tRNA sequences.

rRNA-coding regions in the precursor molecule contain tRNA genes. Thus the precursor rRNA encodes for both tRNA and rRNA. The modification of precursor RNA molecules is called posttranscriptional modification.

1. Define or describe the following: gene, template and coding strands, promoter, leader, coding region, reading frame, trailer, and terminator.
2. How do the genes of bacterial, archaeal, and eukaryotic cells usually differ from each other?
3. Briefly discuss the general organization of tRNA and rRNA genes. How does their expression differ from that of protein-coding genes with respect to posttranscriptional modification of the gene product?

12.6 Transcription

Synthesis of RNA under the direction of DNA is called transcription, and the RNA product has a sequence complementary to the DNA template directing its synthesis. Although adenine directs the incorporation of thymine during DNA replication, it usually codes for uracil during RNA synthesis. Transcription generates three major kinds of RNA. **Transfer RNA (tRNA)** carries amino acids during protein synthesis, and **ribosomal RNA (rRNA)** molecules are components of ribosomes. **Messenger RNA (mRNA)** bears the message for protein synthesis. In *Bacteria* and *Archaea,* a single mRNA molecule often bears coding information transcribed from adjacent genes. Therefore it is said to be polycistronic (**figure 12.22*a***). Eukaryotic mRNAs, on the other

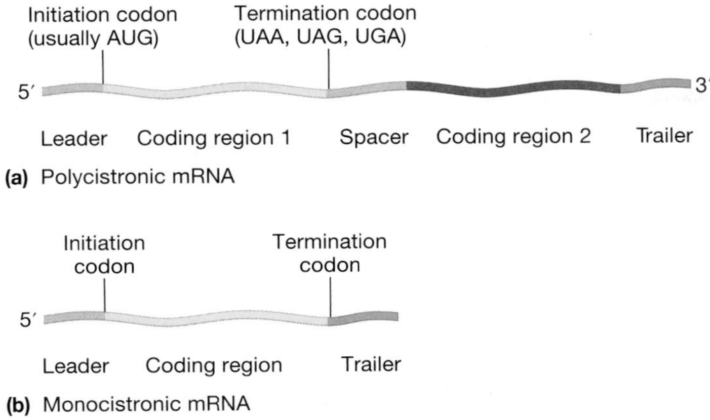

FIGURE 12.22 Polycistronic and Monocistronic mRNAs. Polycistronic mRNAs are commonly observed in *Bacteria* and *Archaea.* Eukaryotic genes give rise to monocistronic mRNAs.

hand, are usually monocistronic (figure 12.22*b*), containing information from a single gene. The synthesis of bacterial mRNA is described first. *Stages of Transcription*

Transcription in *Bacteria*

RNA is synthesized under the direction of DNA by the enzyme **RNA polymerase.** The reaction is quite similar to that catalyzed by DNA polymerase (figure 12.11). ATP, GTP, CTP, and UTP are used to produce RNA complementary to the DNA template, and pyrophosphate is produced as the ribonuleoside monophosphates are incorporated into the growing RNA chain. Pyrophosphate is hydrolyzed to fuel the process. RNA synthesis also proceeds in a 5′ to 3′ direction with new nucleotides being added to the 3′ end of the growing chain (**figure 12.23**).

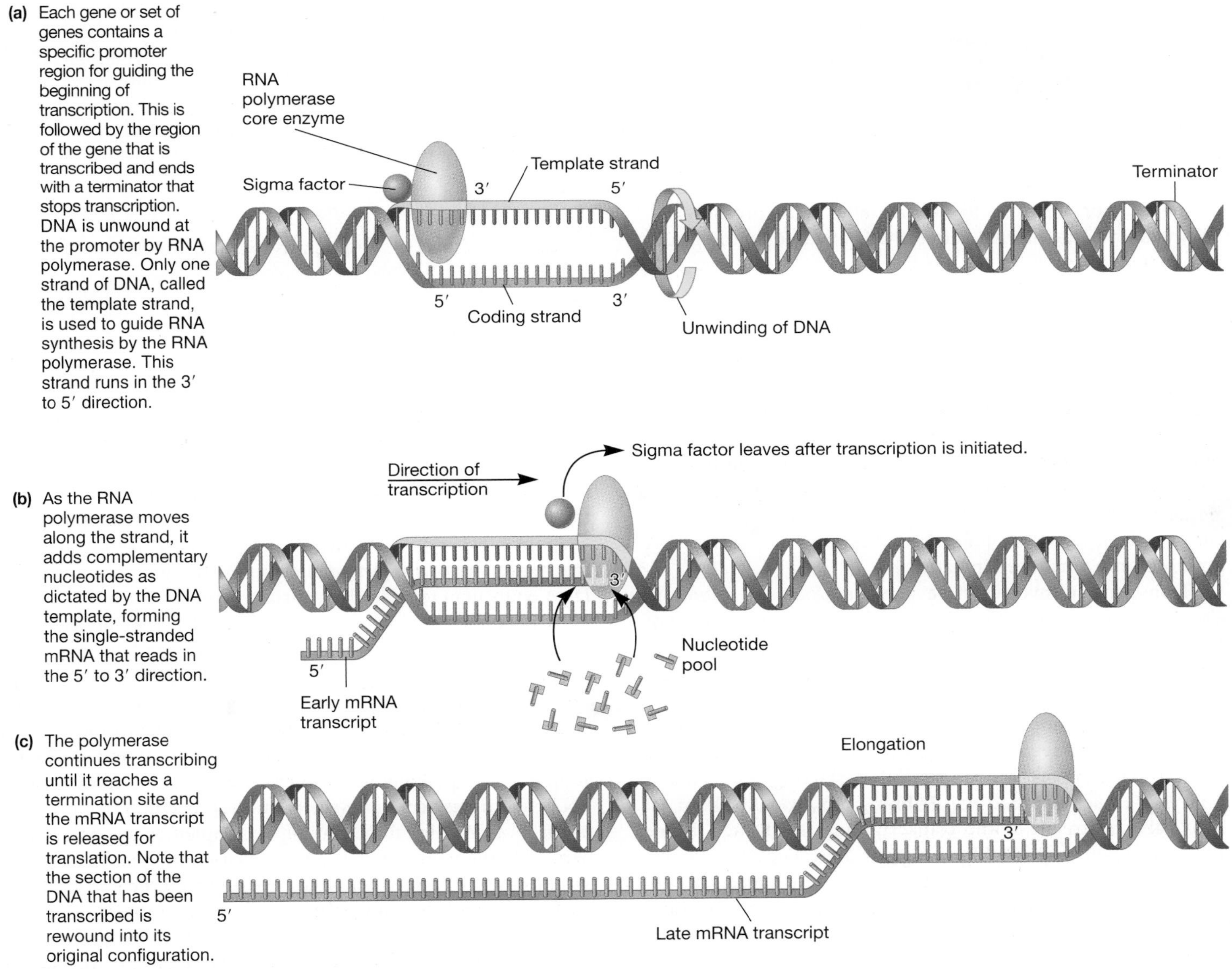

(a) Each gene or set of genes contains a specific promoter region for guiding the beginning of transcription. This is followed by the region of the gene that is transcribed and ends with a terminator that stops transcription. DNA is unwound at the promoter by RNA polymerase. Only one strand of DNA, called the template strand, is used to guide RNA synthesis by the RNA polymerase. This strand runs in the 3′ to 5′ direction.
RNA polymerase core enzyme
Sigma factor
Template strand
3′
5′
Terminator
5′
Coding strand
3′
Unwinding of DNA
(b) As the RNA polymerase moves along the strand, it adds complementary nucleotides as dictated by the DNA template, forming the single-stranded mRNA that reads in the 5′ to 3′ direction.
Direction of transcription
Sigma factor leaves after transcription is initiated.
3′
5′
Early mRNA transcript
Nucleotide pool
(c) The polymerase continues transcribing until it reaches a termination site and the mRNA transcript is released for translation. Note that the section of the DNA that has been transcribed is rewound into its original configuration.
Elongation
3′
5′
Late mRNA transcript

FIGURE 12.23 Transcription in *Bacteria.*

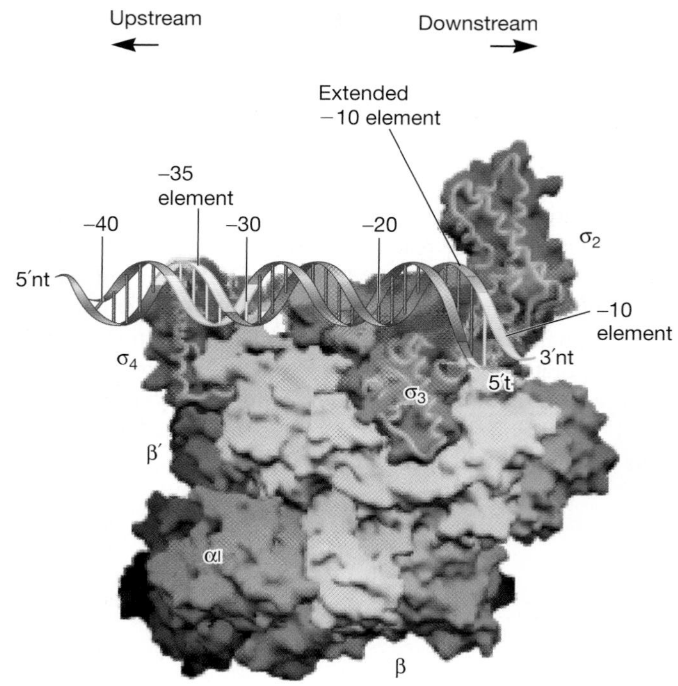

FIGURE 12.24 RNA Polymerase Structure. The holoenzyme-DNA complex of the bacterium *Thermus aquaticus*. Protein surfaces that contact the DNA are in green and are located on the σ factor. The −10 and −35 elements in the promoter are in yellow. The internal active site is covered by the β subunit in this view.

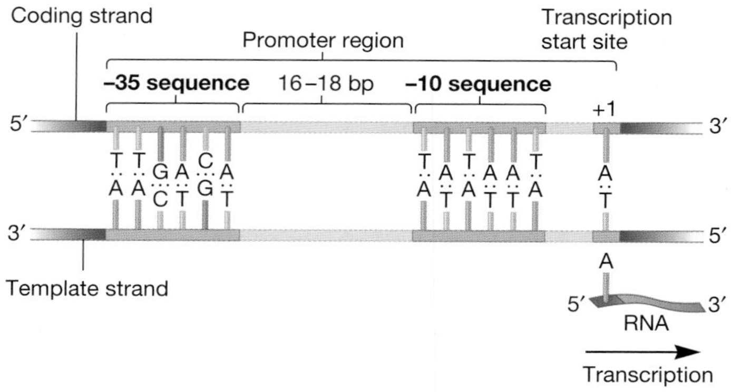

FIGURE 12.25 The Conventional Numbering System of Bacterial Promoters. The first nucleotide that acts as a template for transcription is designated +1. The numbering of nucleotides to the left of this spot is in a negative direction, while the numbering to the right is in a positive direction. For example, the nucleotide that is immediately to the left of the +1 nucleotide is numbered −1, and the nucleotide to the right of the +1 nucleotide is numbered +2. There is no zero nucleotide in this numbering system. In many bacterial promoters, sequence elements at the −35 and −10 regions play a key role in promoting transcription.

Figure 12.25 Micro Inquiry

Are the −35 and −10 regions considered "upstream" or "downstream" of the +1 nucleotide?

Most bacterial RNA polymerases contain five types of polypeptide chains: α, β, β′, ω, and σ (**figure 12.24**). The **RNA polymerase core enzyme** is composed of five polypeptides (α_2, β, β′, and ω) and catalyzes RNA synthesis. The **sigma factor** (σ) has no catalytic activity but helps the core enzyme recognize the promoter. When sigma is bound to core enzyme, the six-subunit complex is termed **RNA polymerase holoenzyme.** Only holoenzyme can begin transcription, but core enzyme completes RNA synthesis once it has been initiated.

Transcription involves three separate processes: initiation, elongation, and termination. Only a relatively short segment of DNA (i.e., a gene) is transcribed, unlike replication in which the entire chromosome must be copied. Sigma factor is critical to the initiation process. It positions the RNA polymerase core enzyme at the promoter. Bacterial promoters have two characteristic features: a sequence of six bases (often TTGACA) about 35 base pairs before the transcription starting point and a TATAAT sequence called the **Pribnow box,** usually about 10 base pairs upstream of the transcriptional start site (**figure 12.25**; also figure 12.20). These regions are called the −35 and −10 sites, respectively, because these are their distances in nucleotides before or "upstream" of the first nucleotide to be transcribed (i.e., the +1 site). Sigma factor recognizes the −10 and −35 sequences, and directs the RNA polymerase core enzyme to them. The −10 and −35 sequences are similar in all promoters and are called **consensus sequences.**

Once bound to the promoter, RNA polymerase unwinds the DNA (**figure 12.26**). The −10 site is rich in adenines and thymines, making it easier to break the hydrogen bonds that keep the DNA double stranded; when the DNA is unwound at this region, it is called an open complex. A region of unwound DNA equivalent to about 16 to 20 base pairs becomes the "transcription bubble," which moves with the RNA polymerase as it synthesizes mRNA from the template DNA strand during elongation (**figure 12.27**). Within the transcription bubble, a temporary RNA:DNA hybrid is formed. As RNA polymerase holoenzyme progresses along the DNA template, the sigma factor usually dissociates from the other subunits and can help another RNA polymerase core enzyme initiate transcription. RNA polymerase core enzyme continues to synthesize mRNA in the 5′ to 3′ direction, making it complementary and antiparallel to the template DNA. As elongation of the mRNA continues, single-stranded mRNA is released, and the two strands of DNA behind the transcription bubble resume their double helical structure. As shown in figure 12.23, RNA polymerase is a remarkable enzyme capable of several activities, including

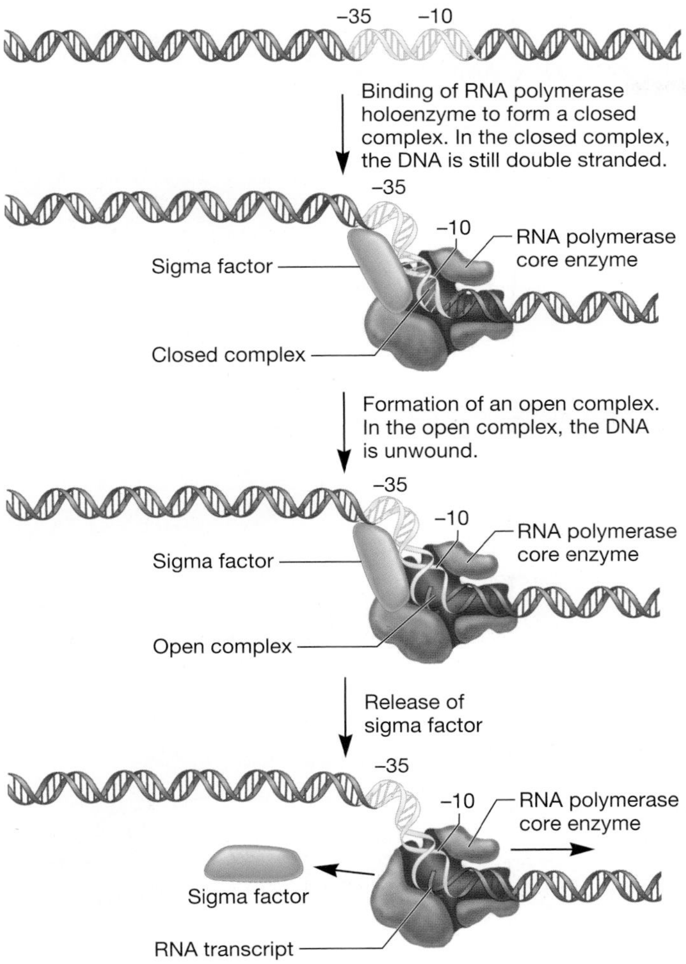

FIGURE 12.26 The Initiation of Transcription in *Bacteria*. The sigma factor of the RNA polymerase holoenzyme is responsible for positioning the core enzyme properly at the promoter. Sigma factor recognizes two regions in the promoter, one centered at –35 and the other centered at –10. Once positioned properly, the DNA at the –10 region unwinds to form an open complex. The sigma factor dissociates from the core enzyme after transcription is initiated.

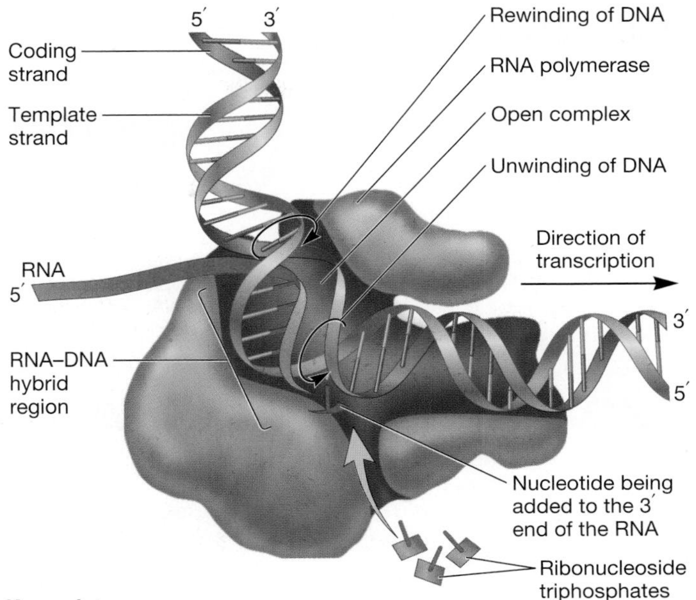

Key points:

- RNA polymerase slides along the DNA, creating an open complex as it moves.
- The template strand is used to make a complementary copy of RNA as an RNA–DNA hybrid.
- The RNA is synthesized in a 5′ to 3′ direction using ribonucleoside triphosphates as precursors. Pyrophosphate is released (not shown).
- The complementarity rule is the same as the AT/GC rule except that U is substituted for T in the RNA.

FIGURE 12.27 The "Transcription Bubble."

unwinding the DNA, moving along the template, and synthesizing RNA.

Termination of transcription occurs when the core RNA polymerase dissociates from the template DNA. This is brought about by the terminator. There are two kinds of terminators. The first type causes intrinsic or rho-independent termination (**figure 12.28**). This terminator follows a nucleotide sequence that, when transcribed into RNA, forms hydrogen bonds within the single-stranded RNA. This intrastrand base pairing creates a hairpin-shaped stem-loop structure, which causes RNA polymerase to stop transcription. Within this terminator is a stretch of about six uracil bases. Once the RNA polymerase pauses after formation of the hairpin loop in the mRNA, the A-U base pairs in the uracil-rich region of the terminator are too weak to hold the RNA:DNA duplex together and RNA polymerase falls off. The second kind of terminator functions in a very different manner. It requires the aid of a protein called rho factor (ρ) and causes rho-dependent termination. It is thought that rho binds to mRNA and moves along the molecule until it reaches the RNA polymerase halted at the terminator by a stem-loop structure that forms in the mRNA (**figure 12.29**). The rho factor, which has hybrid RNA:DNA helicase activity, then causes RNA polymerase to dissociate from the mRNA, probably by unwinding the mRNA-DNA complex. *mRNA Synthesis*

Transcription in Eukaryotes

Transcriptional processes in eukaryotic microorganisms (and in other eukaryotic cells) differ in several ways from bacterial transcription. There are three major RNA polymerases, not one as in *Bacteria.* RNA polymerase II is responsible for mRNA synthesis. Polymerases I and III synthesize rRNA and tRNA, respectively. RNA polymerase II is a large aggregate, at least 500,000 daltons in size, with about 10 or more subunits. Unlike bacterial RNA

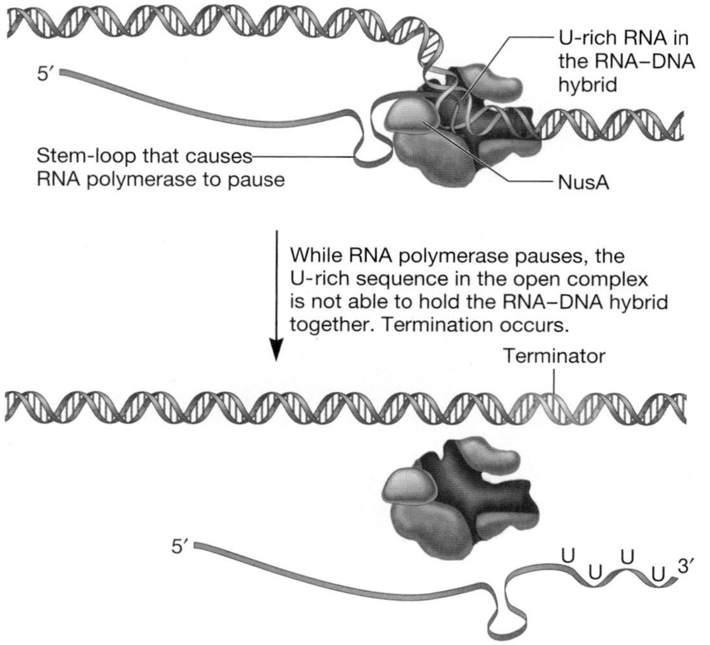

FIGURE 12.28 Intrinsic Termination of Transcription. This type of terminator contains a U-rich sequence downstream from a stretch of nucleotides that can form a stem-loop structure. Formation of the stem and loop in the newly synthesized RNA causes RNA polymerase to pause. This pausing is stabilized by the NusA protein. The U-A bonds in the uracil-rich region are not strong enough to hold the RNA and DNA together. Therefore the RNA, DNA, and RNA polymerase dissociate and transcription stops.

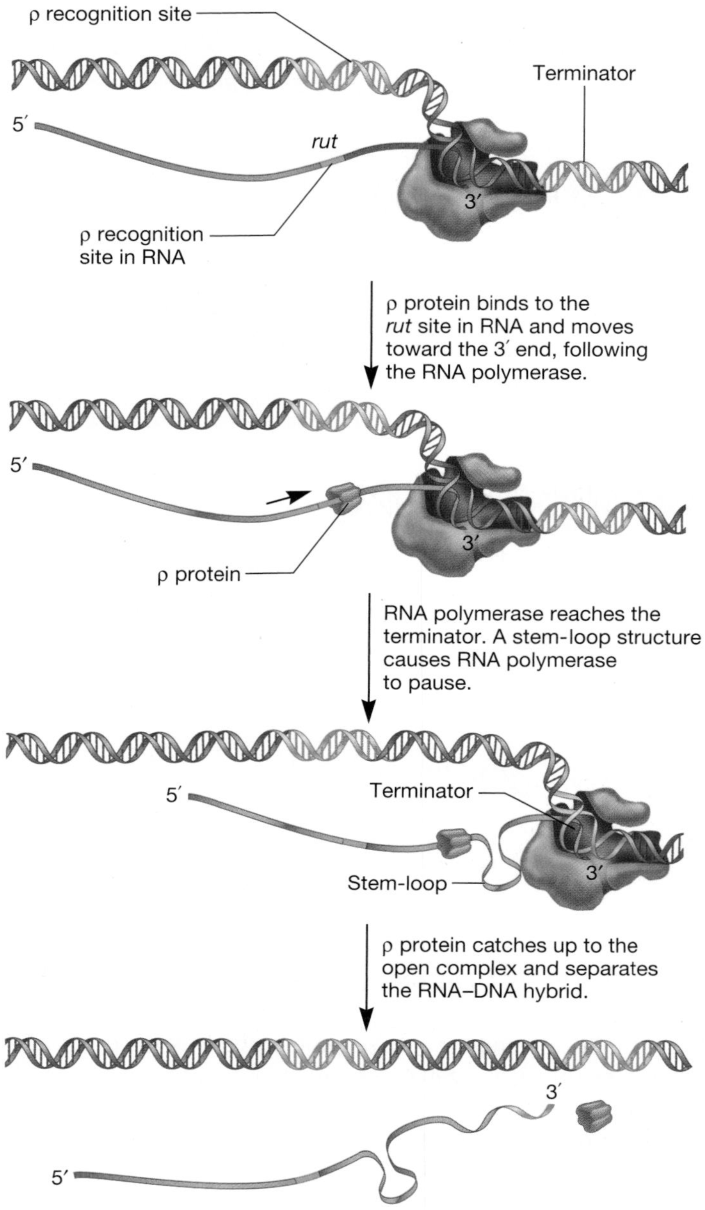

FIGURE 12.29 Rho-Factor (ρ)-Dependent Termination of Transcription. The *rut* site stands for *r*ho *ut*ilization site.

polymerase, RNA polymerase II requires several transcription factors to recognize its promoters (**figure 12.30**). Eukaryotic promoters also differ from those in *Bacteria.* They have combinations of several elements. Three of the most common are the TATA box (located about 30 base pairs upstream of the start point), and the GC and CAAT boxes (located between 50 to 100 base pairs upstream of the start site). The general transcription factor TFIID plays an important role in transcription initiation in eukaryotes (figure 12.30). This multiprotein complex contains the TATA-binding protein (TBP). TBP has been shown to bend the DNA sharply on attachment, making the DNA more accessible to other transcription factors. A variety of other general transcription factors, promoter specific factors, and promoter elements have been discovered in different eukaryotic cells. Each eukaryotic gene seems to be regulated differently, and more research will be required to understand fully the regulation of eukaryotic transcription.

Unlike bacterial mRNA, eukaryotic mRNA arises from **posttranscriptional modification** of large RNA precursors, about 5,000 to 50,000 nucleotides long, sometimes called pre-mRNA (**figure 12.31**). As pre-mRNA is synthesized, a 5′ cap is added. The 5′ cap is the unusual nucleotide 7-methylguanosine. It is attached to the 5′-phosphate of the pre-mRNA by a triphosphate linkage. After synthesis is completed, the pre-mRNA is modified by the addition of a 3′ poly-A tail, about 200 nucleotides long. The functions of the 5′ cap and poly-A tail are not completely clear. The 5′ cap on eukaryotic mRNA may promote the initial binding of ribosomes to the mRNA. It also may protect the mRNA from enzymatic attack. Poly-A protects mRNA from rapid enzymatic degradation. The poly-A tail must be shortened to about 10 nucleotides before mRNA can be degraded. Poly-A also seems to aid in mRNA translation.

As noted earlier, many eukaryotic genes contain **exons** (expressed sequences), which are found in mRNA molecules and

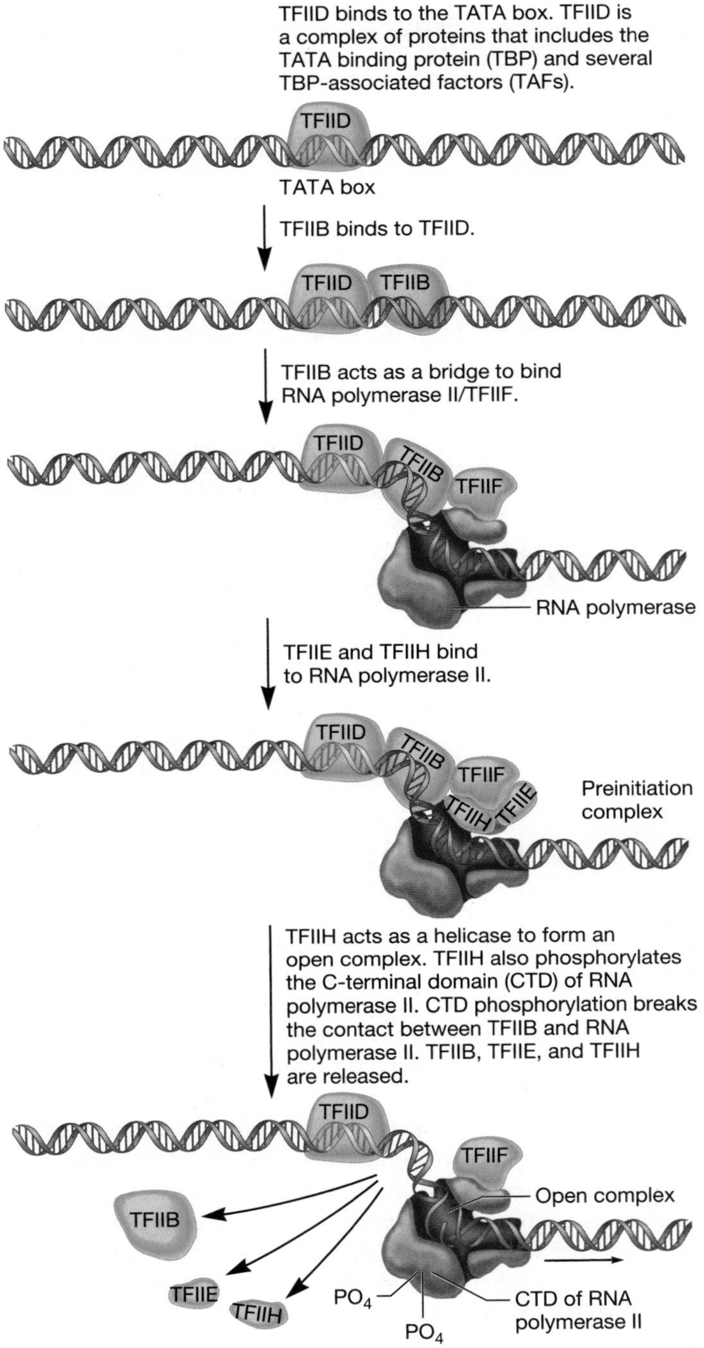

TFIID binds to the TATA box. TFIID is a complex of proteins that includes the TATA binding protein (TBP) and several TBP-associated factors (TAFs).
TFIID
TATA box
TFIIB binds to TFIID.
TFIID
TFIIB
TFIIB acts as a bridge to bind RNA polymerase II/TFIIF.
TFIID
TFIIB
TFIIF
RNA polymerase
TFIIE and TFIIH bind to RNA polymerase II.
TFIID
TFIIB
TFIIF
TFIIH
TFIIE
Preinitiation complex
TFIIH acts as a helicase to form an open complex. TFIIH also phosphorylates the C-terminal domain (CTD) of RNA polymerase II. CTD phosphorylation breaks the contact between TFIIB and RNA polymerase II. TFIIB, TFIIE, and TFIIH are released.
TFIID
TFIIF
Open complex
TFIIB
TFIIE
TFIIH
PO4
PO4
CTD of RNA polymerase II

FIGURE 12.30 Initiation of Transcription in Eukaryotes. The TATA box is a major component of eukaryotic promoters. *TATA binding protein* (TBP), which is a component of a complex of proteins called TFIID (*transcription factor IID*), binds the TATA box. Note that numerous other transcription factors are required for initiation of transcription, unlike *Bacteria,* where only the sigma factor is needed. Initiation of transcription in *Archaea* is similar to that seen in eukaryotes.

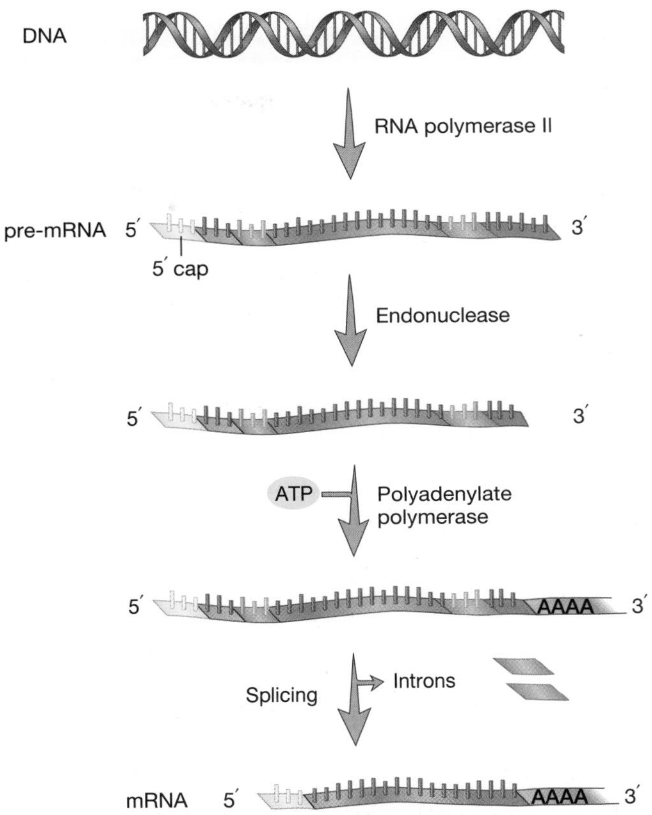

DNA
RNA polymerase II
pre-mRNA 5′
3′
5′ cap
Endonuclease
5′
3′
ATP
Polyadenylate polymerase
5′
AAAA
3′
Splicing
Introns
mRNA 5′
AAAA
3′

FIGURE 12.31 Eukaryotic mRNA Synthesis. The production of eukaryotic messenger RNA. The 5′ cap is added shortly after synthesis of the mRNA begins.

Figure 12.31 Micro Inquiry

What functions may be served by the 5′ cap and the poly-A tail?

are translated into protein, and **introns** (intervening sequences), which are transcribed but not translated. Introns are removed from the pre-mRNA transcript by a process called RNA splicing (**figure 12.32**). Splicing of pre-mRNA occurs in a large complex called a **spliceosome.** It is composed of the pre-mRNA, several small nuclear ribonucleoproteins (snRNPs, pronounced "snurps"), and several non-snRNP splicing factors. The snRNPs consist of small nuclear RNA (snRNA) molecules (about 60 to 300 nucleotides long) associated with proteins. Some snRNPs recognize and bind exon-intron junctions. Exon-intron junctions have a GU sequence at the intron's 5′ boundary and an AG sequence at its 3′ end. Thus the intron's borders are clearly marked for accurate removal.

Sometimes a pre-mRNA is spliced so that different patterns of exons remain. This alternative splicing allows a single gene to code for more than one protein. The splice pattern determines which protein is synthesized. Splice patterns can be cell-type

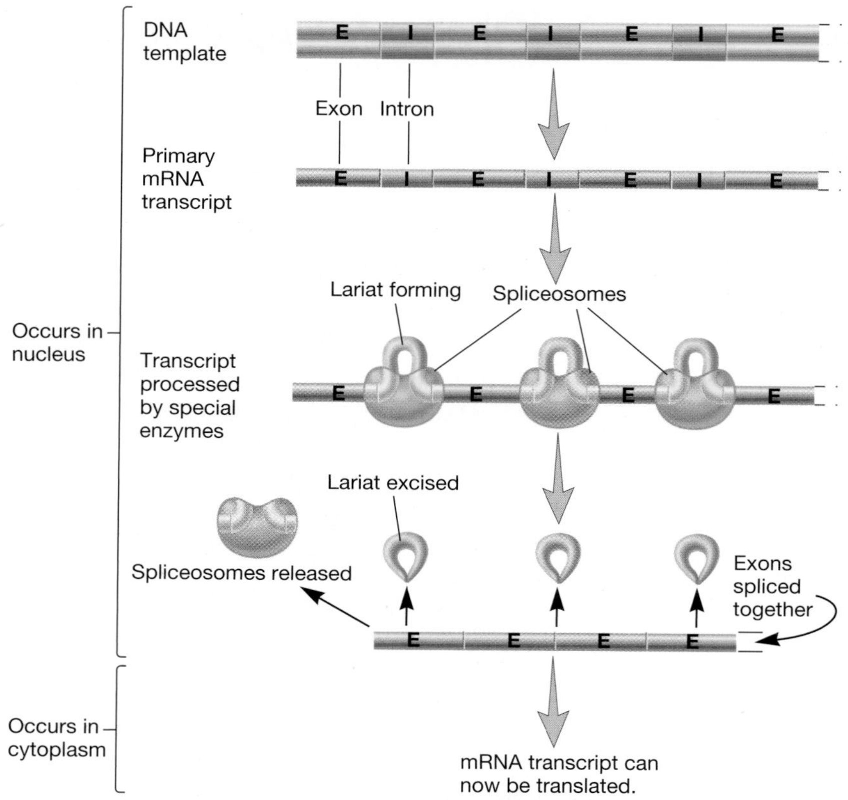

FIGURE 12.32 Splicing of Eukaryotic mRNA Molecules.

specific or determined by the needs of the cell. The importance of alternative splicing in multicellular eukaryotes was clearly demonstrated when it was discovered that the human genome has only about 20,000 genes, rather than the anticipated 100,000. It is thought that alternative splicing is one mechanism by which human cells produce their vast array of proteins using fewer genes. *Processing of Genetic Information: Prokaryotes versus Eukaryotes*

Transcription in *Archaea*

Transcription in the *Archaea* is similar to and distinct from what is observed in *Bacteria* and eukaryotes. Each archaeon has a single RNA polymerase responsible for transcribing all genes in the cell (as in *Bacteria*). However, the RNA polymerase is larger and contains more subunits, many of which are similar to subunits in RNA polymerase II of eukaryotes. The promoters of archaeal genes are similar to those of eukaryotes in having a TATA box; binding of the archaeal RNA polymerase to its promoter requires a TATA-binding protein, just as in eukaryotes. Like the eukaryotic counterpart, the archaeal RNA polymerase also needs several additional transcription factors to function properly. Furthermore, some archaeal genes have introns, which must be removed by posttranscriptional processing. Finally, the mRNA molecules produced by transcription in *Archaea* are usually polycistronic, as in *Bacteria*. This intriguing mixture of bacterial and eukaryotic features in the *Archaea* fuels a great deal of speculation about the evolution of the three domains of life. ▶▶| *Archaeal transcription (section 18.1)*

1. Define the following: polycistronic mRNA, RNA polymerase core enzyme, sigma factor, RNA polymerase holoenzyme, and rho factor.
2. Define or describe posttranscriptional modification, pre-mRNA, 3′ poly-A sequence, 5′ cap, split or interrupted genes, exon, intron, RNA splicing, snRNA, snRNP, and spliceosome.
3. How does bacterial RNA polymerase differ from bacterial DNA polymerase? How does RNA polymerase know when to begin and end transcription?
4. How do bacterial RNA polymerases and promoters differ from those of *Archaea* and eukaryotes?

12.7 The Genetic Code

The final step in the expression of genes that encode proteins is translation. The mRNA nucleotide sequence is translated into the amino acid sequence of a polypeptide chain. Protein synthesis is called translation because it is a decoding process. The information encoded in the language of nucleic acids must be rewritten in the language of proteins. Therefore before we discuss protein synthesis, we examine the nature of the genetic code.

The genetic code, presented in RNA form, is summarized in **table 12.2**. There are several important features of the code. One is that the code words (codons) are three letters (bases) long. Each codon is recognized by an anticodon present on a tRNA molecule. Another feature is that the code has "punctuation." One codon, AUG, is almost always the first codon in the protein-coding portion of mRNA molecules. It is called the **start codon** because it serves as the start site for translation by coding for the initiator tRNA. Three other codons (UGA, UAG, and UAA) are involved in the termination of translation and are called stop or **nonsense codons.** These codons do not encode an amino acid and therefore do not have a tRNA bearing their anticodon. Thus only 61 of the 64 codons in the code, the **sense codons**, direct amino acid incorporation into protein. Finally, the genetic code exhibits **code degeneracy.** That is, there are up to six different codons for a given amino acid.

Despite the existence of 61 sense codons, there are fewer than 61 different tRNAs. It follows that not all codons have a corresponding tRNA. Cells can successfully translate mRNA using

Table 12.2 The Genetic Code

First Position (5′ End)[a]	Second Position: U		C		A		G		Third Position (3′ End)
U	UUU	Phe	UCU	Ser	UAU	Tyr	UGU	Cys	U
	UUC		UCC		UAC		UGC		C
	UUA	Leu	UCA		UAA	STOP	UGA	STOP	A
	UUG		UCG		UAG		UGG	Trp	G
C	CUU	Leu	CCU	Pro	CAU	His	CGU	Arg	U
	CUC		CCC		CAC		CGC		C
	CUA		CCA		CAA	Gln	CGA		A
	CUG		CCG		CAG		CGG		G
A	AUU	Ile	ACU	Thr	AAU	Asn	AGU	Ser	U
	AUC		ACC		AAC		AGC		C
	AUA		ACA		AAA	Lys	AGA	Arg	A
	AUG	Met	ACG		AAG		AGG		G
G	GUU	Val	GCU	Ala	GAU	Asp	GGU	Gly	U
	GUC		GCC		GAC		GGC		C
	GUA		GCA		GAA	Glu	GGA		A
	GUG		GCG		GAG		GGG		G

[a] The code is presented in the RNA form. Codons run in the 5′ to 3′ direction.

fewer tRNAs because loose pairing between the 5′ base in the anticodon and the 3′ base of the codon is tolerated. Thus as long as the first and second bases in the codon correctly base pair with an anticodon, the tRNA bearing the correct amino acid will bind to the mRNA during translation. This is evident on inspection of the code. Note that the codons for a particular amino acid most often differ at the third position (table 12.2). This somewhat loose base pairing is known as **wobble,** and it relieves cells of the need to synthesize so many tRNAs (**figure 12.33**). Wobble also decreases the effects of mutations. ▶▶| *Mutations: their chemical basis and effects (section 14.1)*

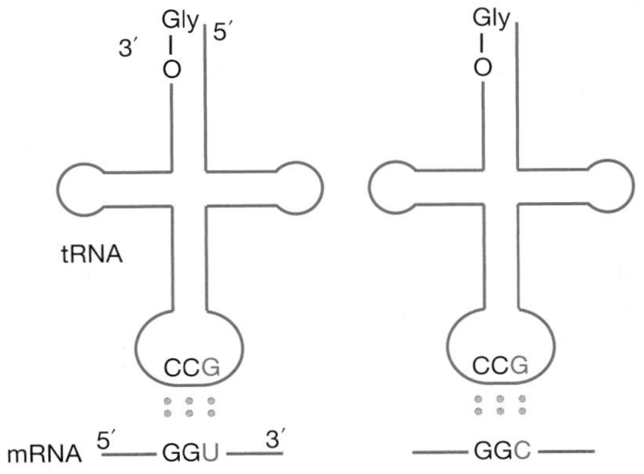

(a) Base pairing of one glycine tRNA with two codons due to wobble

Glycine mRNA codons: GGU, GGC, GGA, GGG (5′ ⟶ 3′)

Glycine tRNA anticodons: CCG, CCU, CCC (3′ ⟶ 5′)

(b) Glycine codons and anticodons

FIGURE 12.33 Wobble and Coding. The use of wobble in coding for the amino acid glycine. (a) Because of wobble, G in the 5′ position of the anticodon can pair with either C or U in the 3′ position of the codon. Thus two codons can be recognized by the same tRNA. (b) Because of wobble, only three tRNA anticodons are needed to translate the four glycine (Gly) codons.

12.8 Translation

Translation involves decoding mRNA and covalently linking amino acids together to form a polypeptide; this occurs at the ribosome. Translation begins when a ribosome binds mRNA and is positioned properly so that translation will yield the correct amino acid sequence in the polypeptide chain. Transfer RNA molecules carry amino acids to the ribosome so that they can be added to the polypeptide chain as the ribosome moves down the mRNA molecule. Just as DNA and RNA synthesis proceeds in one direction, so too does protein synthesis. Polypeptide synthesis begins with the amino acid at the end of the chain with a free amino group (the N-terminal) and moves in the C-terminal direction (figure 12.8). Thus translation is said to occur in the amino terminus to carboxyl terminus direction. Protein synthesis is not only quite accurate but also very rapid. In *E. coli,* synthesis occurs at a rate of at least 900 amino acids added per minute; eukaryotic translation is slower, about 100 amino acids residues per minute.

Cells that grow quickly must use each mRNA with great efficiency to synthesize proteins at a sufficiently rapid rate. The two subunits of the ribosome (the 50S subunit and the 30S subunit in *Bacteria* and *Archaea;* 60S and 40S in eukaryotes) are free in the cytoplasm if protein is not being synthesized. They come together to form the complete ribosome complexed with mRNA only when translation occurs. Frequently mRNAs are simultaneously complexed with several ribosomes, each ribosome reading the mRNA message and synthesizing a polypeptide. At maximal rates of mRNA use, there may be a ribosome every 80 nucleotides along the mRNA or as many as 20 ribosomes simultaneously reading an mRNA that codes for a 50,000 dalton polypeptide. A complex of mRNA with several ribosomes is called a **polyribosome** or polysome (**figure 12.34**). Polysomes are present in all organisms. *Bacteria* and *Archaea* can further increase the efficiency of gene expression by coupling transcription and translation (figure 12.34*b*). While RNA polymerase is synthesizing an mRNA, ribosomes can already be attached to the mRNA so that transcription and translation occur simultaneously. Coupled transcription and translation is possible in bacterial and archaeal cells because a nuclear envelope does not separate the translation machinery from DNA, as it does in eukaryotes. *How Translation Works; Protein Synthesis*

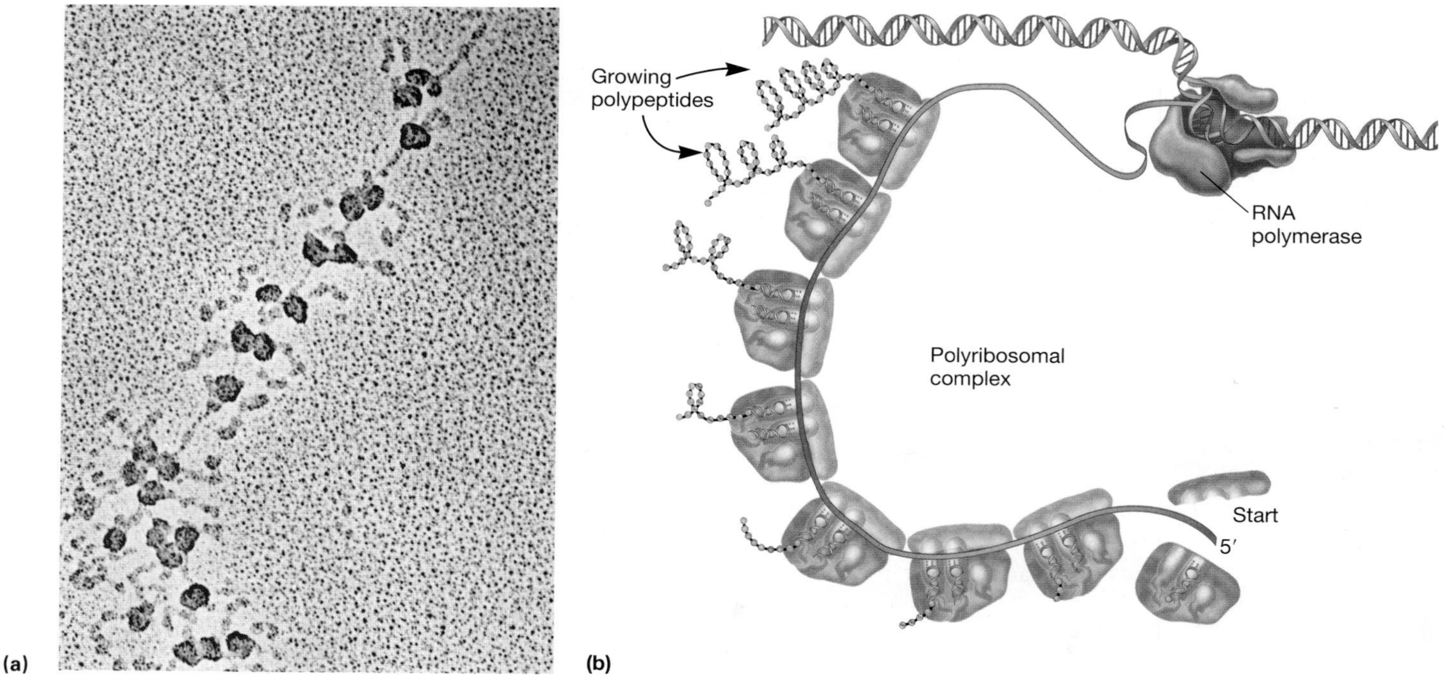

FIGURE 12.34 Coupled Transcription and Translation in *Bacteria* and *Archaea*. (a) A transmission electron micrograph showing a polyribosome. (b) A schematic representation of coupled transcription and translation. As the DNA is transcribed, ribosomes bind the free 5′ end of the mRNA. Thus translation is started before transcription is completed. Note that there are multiple ribosomes bound to the mRNA, forming a polyribosome. With the exception of the ribosome initiating translation at the 5′ end of the mRNA, all other ribosomes show translation just after the formation of the peptide bond. As the bond is formed, the growing peptide chain is transferred to the tRNA in the A site.

Figure 12.34 Micro Inquiry

Why is simultaneous transcription and translation impossible in eukaryotes?

Transfer RNA and Amino Acid Activation

For translation to occur, a ready supply of tRNA molecules bearing the correct amino acid must be available. Thus a preparatory step for protein synthesis is **amino acid activation,** the process in which amino acids are attached to tRNA molecules. Before we discuss this process, it is necessary to examine the structure of tRNA molecules.

Transfer RNA molecules are about 70 to 95 nucleotides long and possess several characteristic structural features. These features become apparent when the tRNA is folded so that base pairing within the tRNA strand is maximized. When represented two-dimensionally, this base pairing causes the tRNA to assume a cloverleaf conformation (**figure 12.35*a***). However, the three-dimensional structure caused by the base pairing looks like the letter L (figure 12.35*b*). One important feature of tRNAs is the acceptor stem, which holds the activated amino acid. The 3′ end of all tRNAs has the same —C—C—A sequence, and in all cases, the amino acid is attached to the terminal adenylic acid. Another important feature of the tRNA is the **anticodon**. The anticodon is complementary to the mRNA codon and is located on the anticodon arm (figure 12.35*a*). Two other large arms are readily observed in the cloverleaf representation: the D or DHU arm and the T or TψC arm. Both are named because of the presence of unusual nucleotides that are unique to tRNA molecules. Finally, as can be seen in the cloverleaf representation, the tRNA has a variable arm whose length changes with the overall length of the tRNA; the other arms are fairly constant in size.

Amino acids are activated for protein synthesis through a reaction catalyzed by enzymes called **aminoacyl-tRNA synthetases** (**figure 12.36**). Just as is true of DNA and RNA synthesis, the reaction is driven to completion when the pyrophosphate product is hydrolyzed to two phosphates. The amino acid is attached to the 3′-hydroxyl of the terminal adenylic acid on the tRNA by a high-energy bond. This bond, when broken during translation, releases the energy needed to attach the amino acid to the growing polypeptide chain. The amino acid is said to be activated because it is ready to be linked to a growing polypeptide chain.

There are at least 20 aminoacyl-tRNA synthetases, each specific for a single amino acid and its tRNAs (cognate tRNAs). It is critical that each tRNA attach the corresponding amino acid because if an incorrect amino acid is attached to a tRNA, it will be incorporated into a polypeptide in place of the correct amino acid. The protein synthetic machinery recognizes only the anticodon of the aminoacyl-tRNA and cannot tell whether the correct amino acid is attached. Some aminoacyl-tRNA synthetases proofread just like DNA polymerases do. If the wrong amino acid is attached to tRNA, the enzyme hydrolyzes the amino acid from the tRNA, rather than release the incorrect product. *Aminoacyl-tRNA Structure*

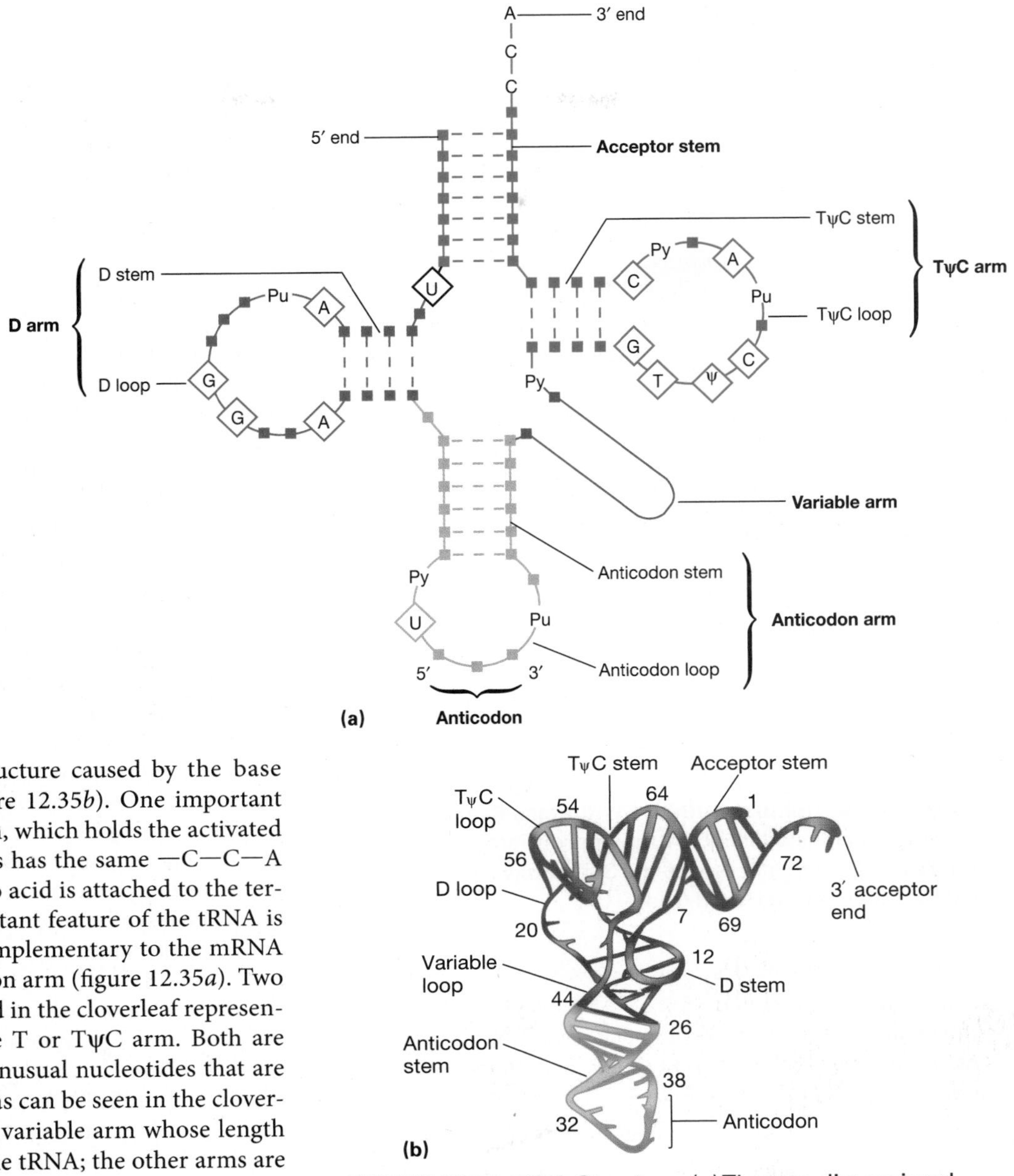

FIGURE 12.35 tRNA Structure. (a) The two-dimensional cloverleaf structure for tRNA. Bases found in all tRNAs are in diamonds; purine and pyrimidine positions in all tRNAs are labeled Pu and Py, respectively. (b) The three-dimensional structure of tRNA. The various regions are distinguished with different colors.

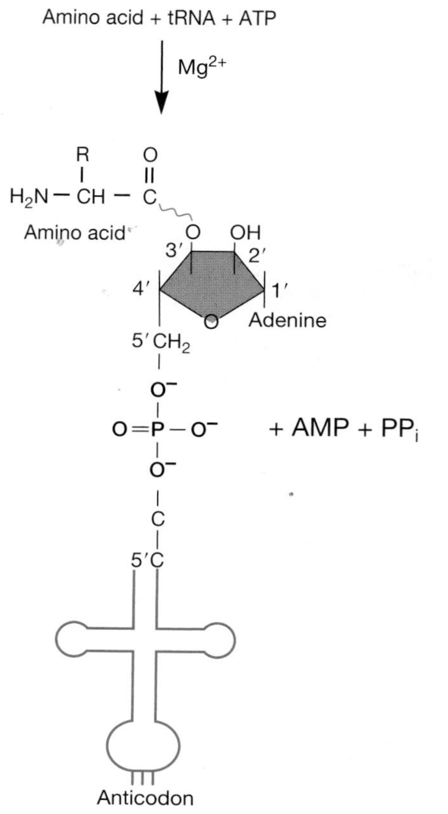

FIGURE 12.36 Aminoacyl-tRNA Synthetase Reaction. The amino acid is attached by the appropriate aminoacyl-tRNA synthetase to the 3′-hydroxyl of adenylic acid by a high-energy bond (red).

Figure 12.36 *Micro Inquiry*

What would be the outcome if an aminoacyl-tRNA synthetase added the wrong amino acid to a tRNA (i.e., the anticodon specified a different amino acid than that added to the 3′ end of the tRNA)?

Ribosome Structure

Protein synthesis takes place on ribosomes that serve as workbenches, with mRNA acting as the blueprint. Recall that ribosomes are formed from two subunits, the large subunit and the small subunit, and each contains one or more rRNA molecules and numerous polypeptide chains. A bacterial ribosome and its components are shown in **figure 12.37**. The region of the ribosome directly responsible for translation is called the translational domain. Both subunits contribute to this domain. The growing peptide chain emerges from the large subunit at the exit domain (figure 12.37*d*).

Ribosomal RNA is thought to have three roles. It obviously contributes to ribosome structure. The 16S rRNA of the 30S subunit is needed for the initiation of protein synthesis in *Bacteria.* The 3′ end of the 16S rRNA binds to a site on the mRNA called the Shine-Dalgarno sequence, which is located in the **ribosome-binding site (RBS).** This helps position the mRNA on the ribosome. The 16S rRNA also binds a protein needed to initiate translation (initiation factor 3) and the 3′ CCA end of aminoacyl-tRNA. Finally, the 23S rRNA is a ribozyme that catalyzes peptide bond formation.

Initiation of Protein Synthesis

Like transcription and DNA replication, protein synthesis is divided into three stages: initiation, elongation, and termination. The initiation of protein synthesis is very elaborate. Apparently the complexity is necessary to ensure that the ribosome does not start synthesizing a polypeptide chain in the middle of a gene—a disastrous error.

E. coli and most other bacteria begin protein synthesis with a modified aminoacyl-tRNA, *N*-formylmethionyl-$tRNA^{fMet}$ (fMet-tRNA), which is coded for by the start codon, AUG (**figure 12.38**). The amino acid of the initiator tRNA has a formyl group covalently bound to the amino group. This aminoacyl-tRNA can be used only for initiation because of the presence of the formyl group. When methionine is to be added to a growing polypeptide chain (i.e., at an AUG codon in the middle of the mRNA), a normal methionyl-$tRNA^{Met}$ is employed. Although most bacteria start protein synthesis with *N*-formylmethionine, the formyl group is not retained but is hydrolytically removed. In fact, one to three amino acids may be removed from the amino terminal end of the polypeptide after synthesis. Eukaryotic protein synthesis (except in mitochondria and chloroplasts) and archaeal protein synthesis begin with a special initiator methionyl-$tRNA^{Met}$.

The initiation stage is crucial for the translation of mRNA into the correct polypeptide (**figure 12.39**). In *Bacteria,* it begins when initiator fMet-tRNA binds to a free 30S ribosomal subunit. The 16S rRNA within the 30S subunit possesses nucleotide sequences at its 3′ end that are complementary to the Shine-Dalgarno sequence in the leader sequence of the mRNA. By aligning the Shine-Dalgarno sequence with the 16S rRNA of the 30S ribosomal subunit, the **initiator codon** (AUG or sometimes GUG, though GUG is a not as good an initiator) specifically binds with the fMet-tRNA anticodon. This ensures that the codon for the initiator fMet-tRNA will be translated first. When the 50S subunit of the ribosome binds to the 30S subunit-mRNA, an active ribosome-mRNA complex is formed, with the fMet-tRNA positioned at the peptidyl or P site (see description of the elongation cycle). This complex is referred to as the initiation complex.

In *Bacteria,* three protein initiation factors (IF-1, IF-2, and IF-3) are required for formation of the initiation complex, and GTP is hydrolyzed during the association of the 50S and 30S subunits (figure 12.39). Eukaryotes and archaeal cells require more initiation factors; otherwise the process is quite similar to that of *Bacteria.*

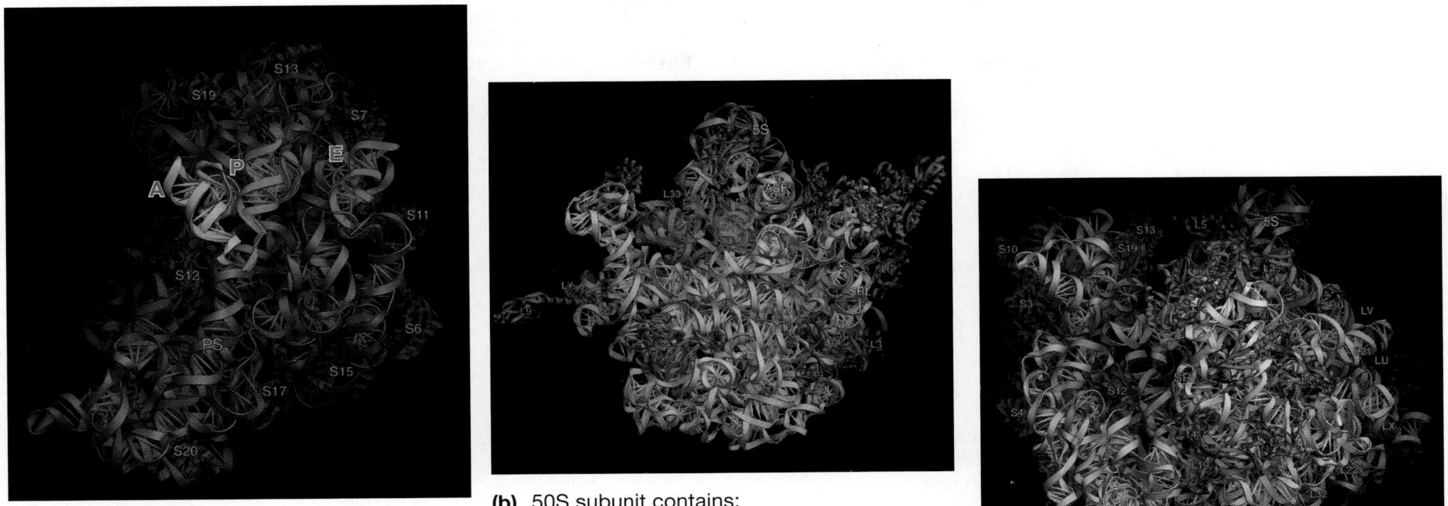

(a) 30S subunit contains:
16S rRNA
21 polypeptides
(0.9×10⁶ daltons)

(b) 50S subunit contains:
5S rRNA
23S rRNA
34 polypeptides
(1.8×10⁶ daltons)

(c) 70S ribosome (2.8×10⁶ daltons)

(d)

EF-Tu

tRNA

EF-G

Messenger RNA

Translational domain

Exit domain

Nascent protein

FIGURE 12.37 Bacterial Ribosome Structure. (a) Interior interface view of the 30S subunit of the *T. thermophilus* 70S ribosome showing the positions of the A, P, and E site tRNAs. (b) Interior interface view of the *T. thermophilus* 50S subunit and portions of its three tRNAs. (c) The complete *T. thermophilus* 70S ribosome viewed from the right-hand side with the 30S subunit on the left and the 50S subunit on the right. The anticodon arm of the A site tRNA is visible in the interface cavity. (d) A diagram of ribosomal structure showing the translational and exit domains. The locations of elongation factor and mRNA binding are indicated. The growing peptide chain probably remains unfolded and extended until it leaves the large subunit. The components in figures (a)–(c) are colored as follows: 16S rRNA, cyan; 23S rRNA, gray; 5S rRNA, light blue; 30S proteins, dark blue; 50S proteins, magenta; and A, P, and E site tRNAs (gold, orange, and red, respectively).

$$CH_3-S-CH_2-CH_2-CH-\overset{\overset{O}{\|}}{C}-tRNA^{fMet}$$

NH
|
C=O
|
H

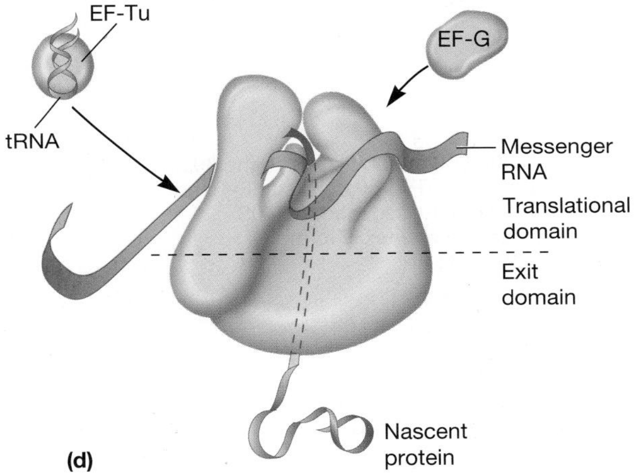

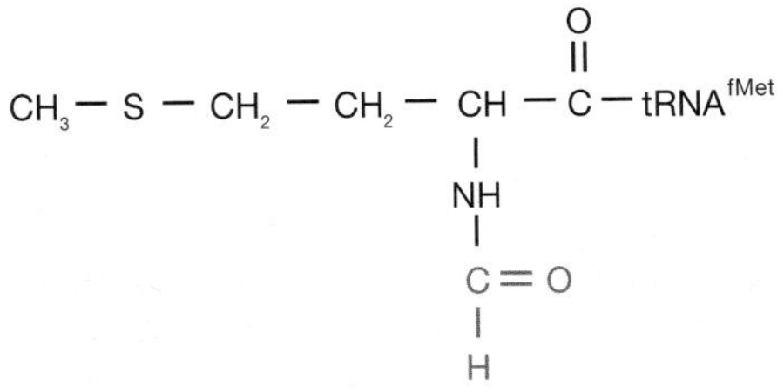

FIGURE 12.38 Bacterial Initiator tRNA. The initiator aminoacyl-tRNA, *N*-formylmethionyl-tRNA^fMet, is used by *Bacteria*. The formyl group is in color. *Archaea* and eukaryotes use methionyl-tRNA for initiation.

Figure 12.38 Micro Inquiry

Why would it be impossible for fMet-tRNA to initiate peptide bond formation with another amino acid? (Hint: Examine figure 12.41 closely.)

Elongation of the Polypeptide Chain

Every addition of an amino acid to a growing polypeptide chain is the result of an elongation cycle composed of three phases: aminoacyl-tRNA binding, the transpeptidation reaction, and translocation. The process is aided by proteins called **elongation factors (EF)**. In each turn of the cycle, an amino acid corresponding to the proper mRNA codon is added to the C-terminal end of the polypeptide chain as the ribosome moves down the mRNA in the 5′ to 3′ direction. The bacterial elongation cycle is described next.

The ribosome has three sites for binding tRNAs: (1) the **peptidyl** or **donor site** (**P site**), (2) the **aminoacyl** or **acceptor site** (**A site**), and (3) the **exit site** (**E site**). At the beginning of an elongation cycle, the P site is filled with either fMet-tRNA or a tRNA

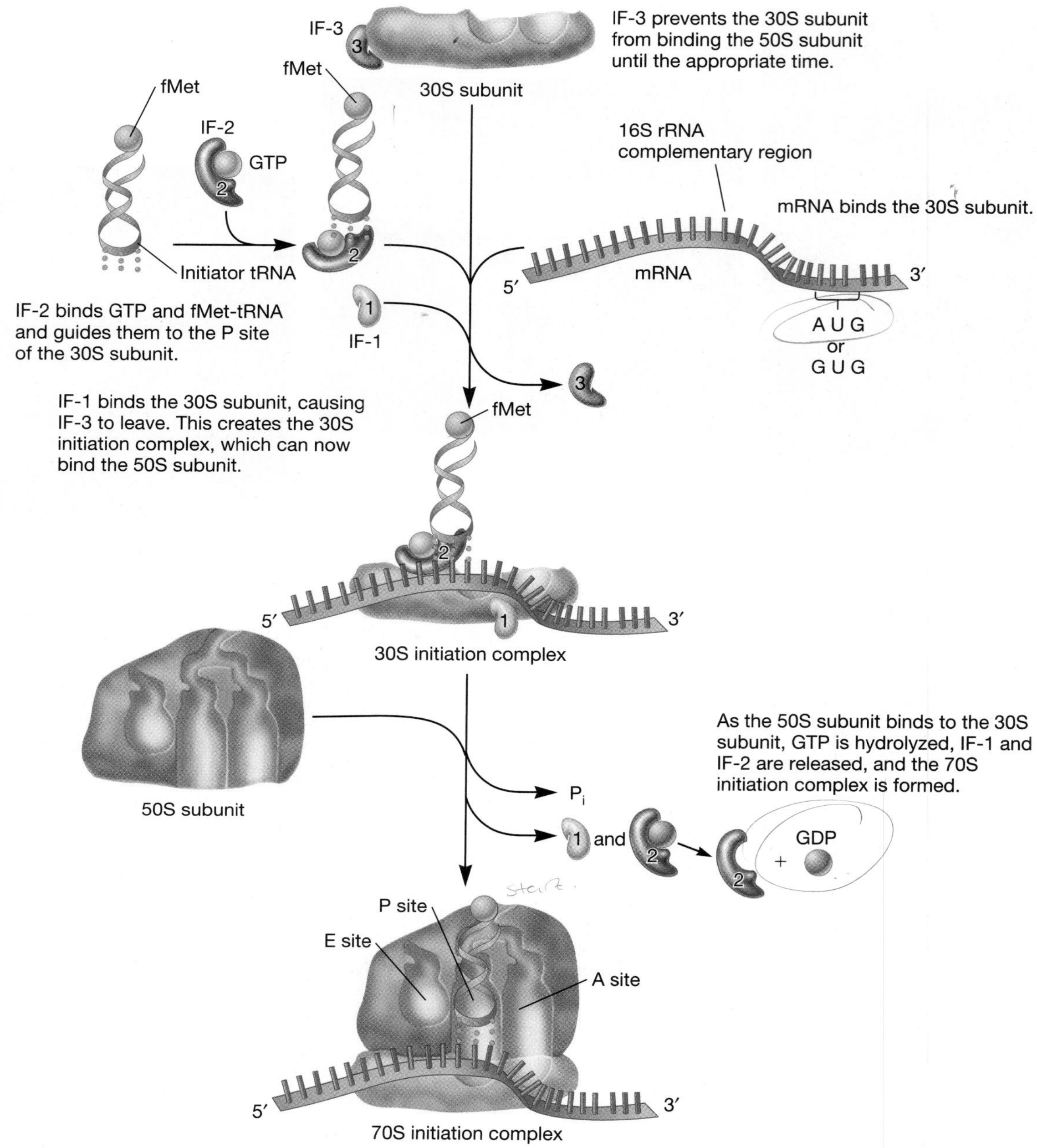

FIGURE 12.39 Initiation of Protein Synthesis. The initiation of protein synthesis in *Bacteria*. The following abbreviations are employed: IF-1, IF-2, and IF-3 stand for initiation factors 1, 2, and 3; initiator tRNA is *N*-formylmethionyl-$tRNA^{fMet}$. The ribosomal locations of initiation factors are depicted for illustration purposes only. They do not represent the actual initiation factor binding sites.

bearing a growing polypeptide chain (peptidyl-tRNA), and the A and E sites are empty (**figure 12.40**). Messenger RNA is bound to the ribosome in such a way that the proper codon interacts with the P site tRNA (e.g., an AUG codon for fMet-tRNA). The next codon is located within the A site and is ready to accept an aminoacyl-tRNA.

The first phase of the elongation cycle is the aminoacyl-tRNA binding phase. The aminoacyl-tRNA corresponding to the codon in the A site is inserted so its anticodon is aligned with the codon on the mRNA. In bacterial cells, this is aided by two elongation factors and requires the expenditure of one GTP (figure 12.40). Aminoacyl-tRNA binding to the A site initiates the second phase of the elongation cycle, the transpeptidation reaction (figure 12.40 and **figure 12.41**). Transpeptidation is catalyzed by the **peptidyl transferase** activity of the 23S rRNA ribozyme, which is part of the 50S ribosomal subunit. In this reaction, the amino group of

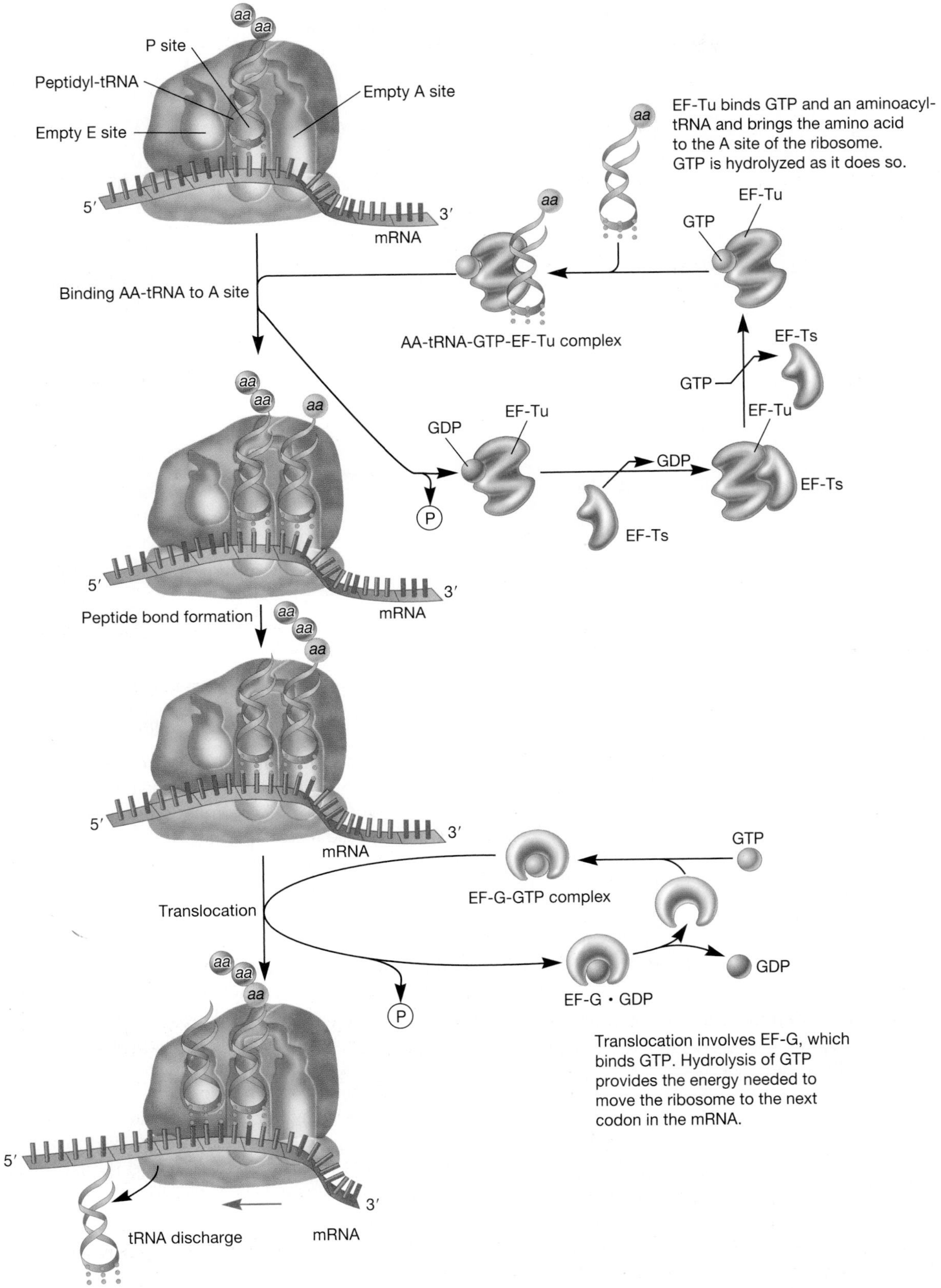

FIGURE 12.40 The Elongation Cycle of Protein Synthesis. The ribosome possesses three sites, a peptidyl or donor site (P site), an aminoacyl or acceptor site (A site), and an exit site (E site). The arrow below the ribosome in the translocation step shows the direction of mRNA movement.

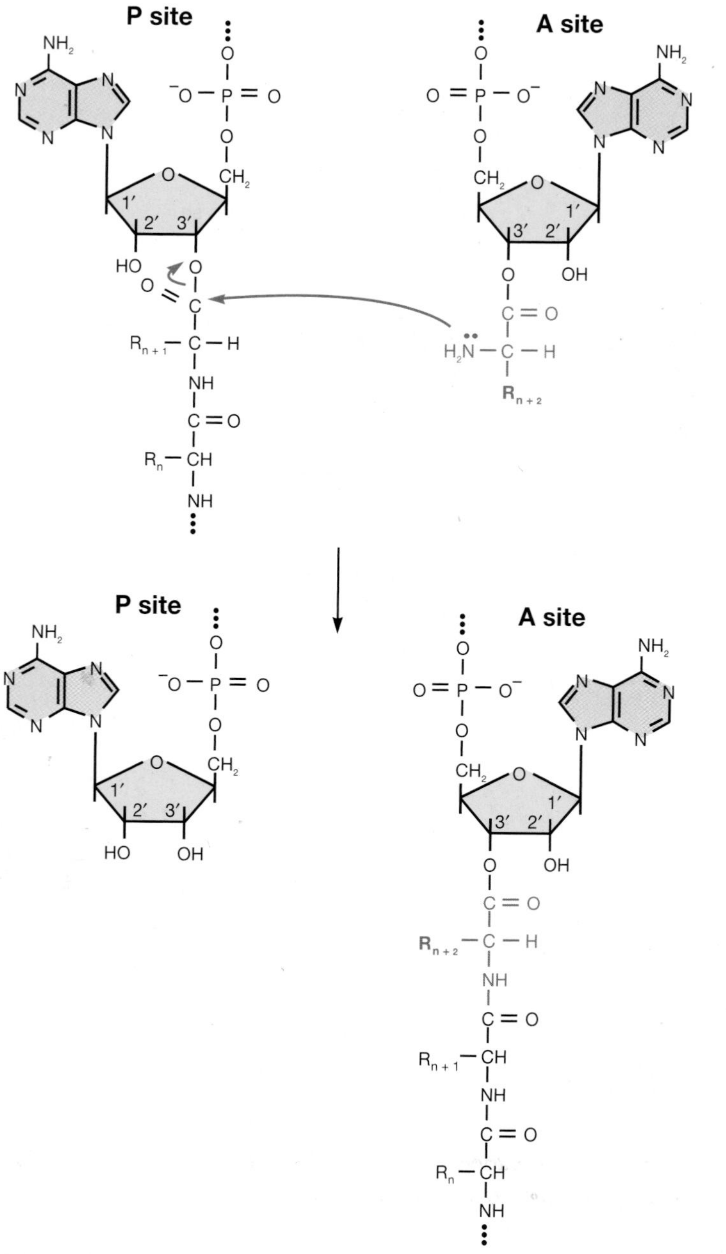

FIGURE 12.41 Transpeptidation. The peptidyl transferase reaction. The peptide grows by one amino acid and is transferred to the A site.

Figure 12.41 Micro Inquiry

What provides the energy to fuel this reaction?

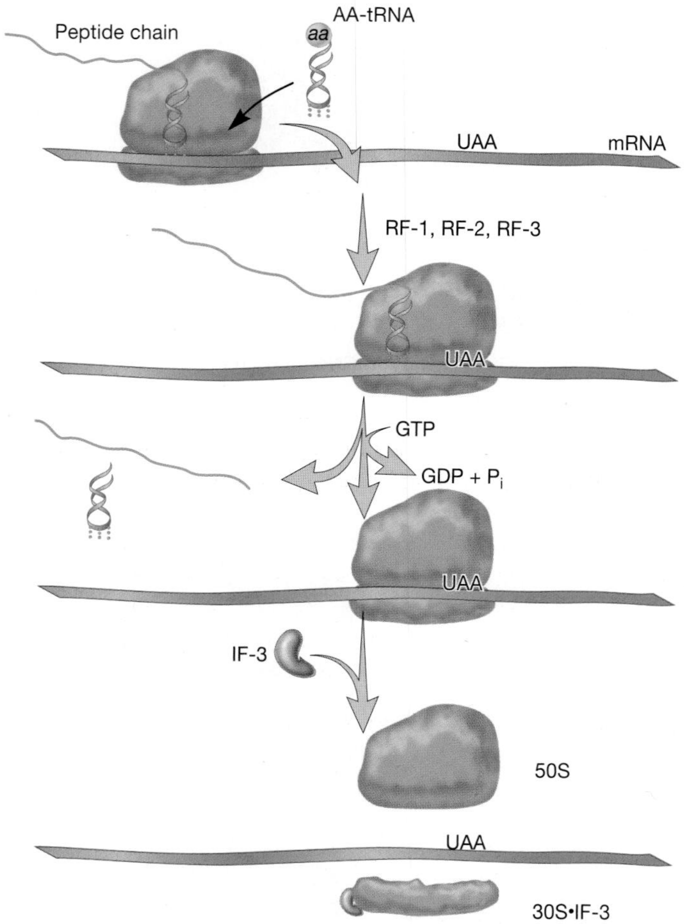

FIGURE 12.42 Termination of Protein Synthesis in *Bacteria*. Although three different nonsense codons can terminate chain elongation, UAA is most often used for this purpose. Three release factors (RF) assist the ribosome in recognizing nonsense codons and terminating translation. GTP hydrolysis is probably involved in termination. Transfer RNAs are in pink.

the A site amino acid reacts with the carboxyl group of the C-terminal amino acid on the P site tRNA (figure 12.41). This results in the transfer of the peptide chain from the tRNA in the P site to the tRNA in the A site, as a peptide bond is formed between the peptide chain and the incoming amino acid. No extra energy source is required for peptide bond formation because the bond linking an amino acid to tRNA is high in energy (figure 12.36).

The final phase in the elongation cycle is **translocation.** Three things happen simultaneously: (1) the peptidyl-tRNA moves from the A site to the P site; (2) the ribosome moves one codon along mRNA so that a new codon is positioned in the A site; and (3) the empty tRNA moves from the P site to the E site and subsequently leaves the ribosome. Ribosomal proteins are involved in these tRNA movements. The intricate process also requires the participation of another elongation factor and GTP hydrolysis. *Translation Elongation*

Termination of Protein Synthesis

Protein synthesis stops when the ribosome reaches one of three nonsense codons—UAA, UAG, and UGA (**figure 12.42**). The nonsense (stop) codon is found on the mRNA immediately before the trailer region. Three release factors (RF-1, RF-2, and RF-3) aid the ribosome in recognizing these codons. Because there is no cognate tRNA for a nonsense codon, the ribosome stops. Peptidyl transferase hydrolyzes the bond linking the polypeptide to the tRNA in the P site, and the polypeptide and the empty tRNA are released. GTP hydrolysis seems to be required during this sequence in some organisms, although it may not be needed for termination in *Bacteria*. Next, the ribosome dissociates from its mRNA and separates into 30S and 50S subunits. Thus ribosomal subunits associate during protein synthesis and separate afterward. The termination of archaeal and eukaryotic protein synthesis is similar except that only one release factor appears to be active. *Translation Termination*

Protein synthesis is a very expensive process. Three GTP molecules probably are used during each elongation cycle, and two ATP high-energy bonds are required for amino acid activation (ATP is converted to AMP, rather than to ADP). Therefore five high-energy bonds are required to add one amino acid to a growing polypeptide chain. GTP also is used in initiation and termination of protein synthesis (figures 12.39 and 12.42). Presumably this large energy expenditure is required to ensure the fidelity of protein synthesis. Fidelity is also ensured by "proofreading" done by the ribosome. Recent studies have shown that the ribosome monitors pairing of tRNA and mRNA. If a mismatch between the anticodon and codon is detected, protein synthesis is halted and the incomplete protein is released. Amazingly, the mismatch can be detected and synthesis halted even after formation of the peptide bond.

1. In which direction are polypeptides synthesized? What is a polyribosome and why is it useful?
2. Briefly describe the structure of transfer RNA and relate this to its function. How are amino acids activated for protein synthesis, and why is the specificity of the aminoacyl-tRNA synthetase reaction so important?
3. What are the translational and exit domains of the ribosome? What roles does ribosomal RNA have?
4. Describe the nature and function of the following: fMet-tRNA, initiator codon, elongation cycle, peptidyl and aminoacyl sites, transpeptidation reaction, peptidyl transferase, translocation, nonsense codon, and release factors.
5. How many ATP and GTP molecules would be hydrolyzed in the synthesis of a 125 amino acid protein? Explain why this is a good argument for careful regulation of gene expression (especially considering that most proteins are larger than 125 amino acids).

12.9 Protein Maturation and Secretion

As a polypeptide emerges from a ribosome, it is not yet ready to assume its cellular functions. Protein function depends on its three-dimensional shape. Proteins must be properly folded and in some cases associated with other protein subunits to generate a functional enzyme (e.g., DNA and RNA polymerases are multimeric proteins). In addition, proteins must be delivered to the proper subcellular or extracellular site. We now discuss these posttranslational events.

Protein Folding and Molecular Chaperones

Although the amino acid sequence of a polypeptide determines its final conformation, helper proteins aid the newly formed or nascent polypeptide in folding to its proper functional shape. These proteins, called **molecular chaperones** or simply chaperones, recognize only unfolded polypeptides or partly denatured proteins and do not bind to normal, functional proteins. Their role is essential because the cytoplasm is filled with new polypeptide chains. Under such conditions, it is possible for polypeptides to fold improperly and aggregate to form nonfunctional complexes. Molecular chaperones suppress incorrect folding and may reverse any incorrect folding that has already taken place. They are so important that chaperones are present in all cells.

Several chaperones and cooperating proteins aid proper protein folding in *Bacteria*. The process has been well studied in *E. coli* and involves at least four chaperones—DnaK, DnaJ, GroEL, and GroES—and the stress protein GrpE. After a sufficient length of nascent polypeptide extends from the ribosome, a series of reactions involving DnaJ and DnaK fold the protein into its native conformation. This requires the expenditure of ATP (**figure 12.43**). Sometimes the polypeptide does not reach its native conformation in one series of reactions. If this occurs, it may repeat the folding process. Alternatively, the partially folded protein may be transferred to chaperones GroEL and GroES, which complete the folding process. This chaperone system also expends ATP as it folds the protein into its proper conformation.

Chaperones were first discovered because they dramatically increase in concentration when cells are exposed to high temperatures, metabolic poisons, and other stressful conditions that cause protein denaturation. Thus many chaperones are called **heat-shock proteins.** When an *E. coli* culture is switched from 30 to 42°C, the concentrations of some 20 different heat-shock proteins increase greatly within about 5 minutes. If the cells are exposed to a lethal temperature, the heat-shock proteins are still synthesized but most other proteins are not. Thus chaperones protect the cell from thermal damage and other stresses as well as promote the proper folding of new polypeptides. For example, DnaK protects *E. coli* RNA polymerase from thermal inactivation in vitro. In addition, DnaK reactivates thermally inactivated RNA polymerase, especially if ATP, DnaJ, and GrpE are present. GroEL and GroES also protect intracellular proteins from aggregation. As expected, large quantities of chaperones are present in hyperthermophiles such as *Pyrodictium occultum*, an archaeon that grows at temperatures as high as 110°C. *P. occultum* has a chaperone similar to GroEL of *E. coli*. This archaeal chaperone hydrolyzes ATP most rapidly at 100°C and makes up almost three-quarters of the cell's soluble protein when *P. occultum* grows at 108°C.

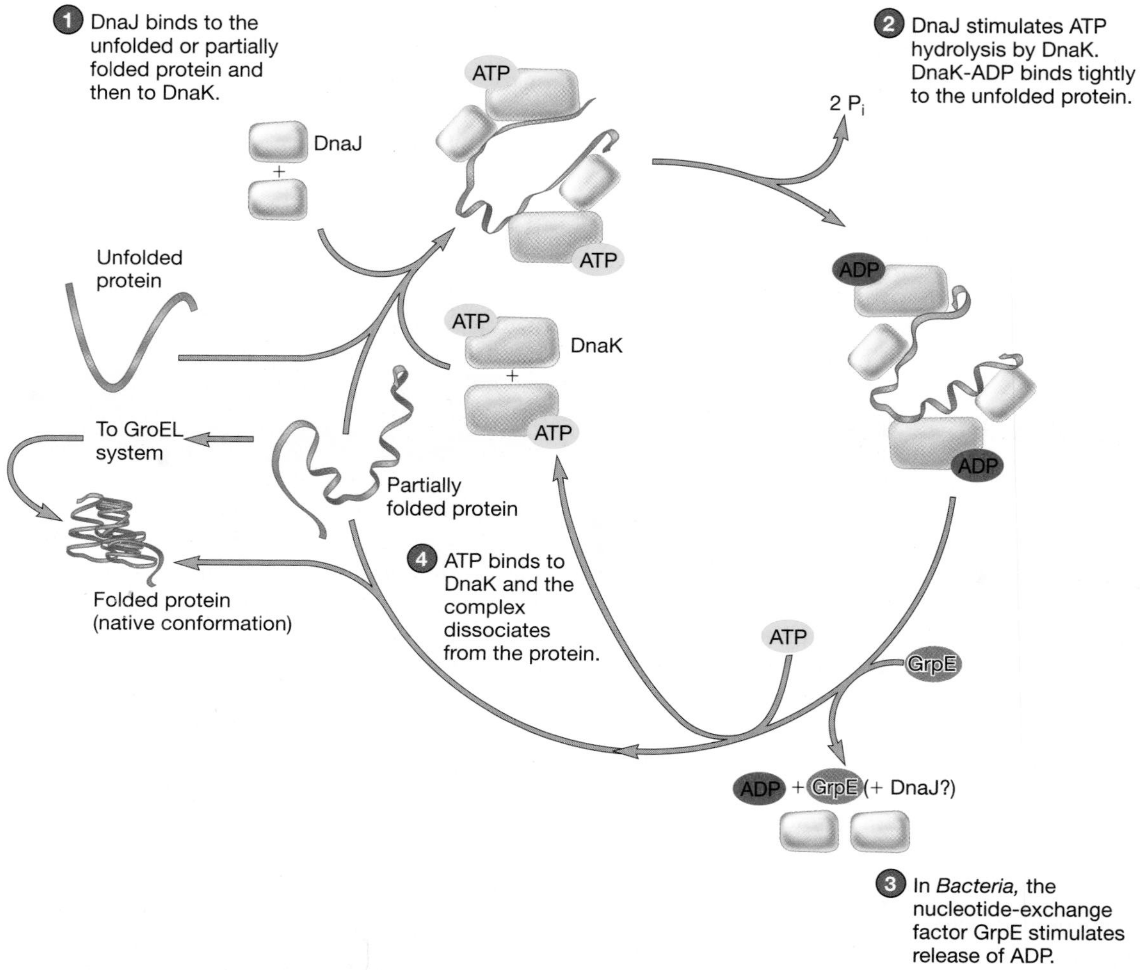

1 DnaJ binds to the unfolded or partially folded protein and then to DnaK.
DnaJ
ATP
ATP
2 P$_i$
2 DnaJ stimulates ATP hydrolysis by DnaK. DnaK-ADP binds tightly to the unfolded protein.
Unfolded protein
ATP
DnaK
ATP
ADP
ADP
To GroEL system
Partially folded protein
Folded protein (native conformation)
4 ATP binds to DnaK and the complex dissociates from the protein.
ATP
GrpE
ADP + GrpE (+ DnaJ?)
3 In *Bacteria*, the nucleotide-exchange factor GrpE stimulates release of ADP.

FIGURE 12.43 Chaperones and Polypeptide Folding. The involvement of bacterial chaperones in the proper folding of a newly synthesized polypeptide chain is depicted in this diagram. Three possible outcomes of a chaperone reaction cycle are shown. A native protein may result, the partially folded polypeptide may bind again to DnaK and DnaJ, or the polypeptide may be transferred to GroEL and GroES to complete folding.

Research indicates that bacterial cells differ from eukaryotes with respect to the timing of protein folding. In terms of conformation, proteins are composed of compact, self-folding, structurally independent regions. These regions, normally around 100 to 300 amino acids in length, are called domains. Larger proteins such as immunoglobulins (important proteins in the immune response) may have two or more domains that are linked by less-structured portions of the polypeptide chain. In eukaryotes, domains fold independently right after being synthesized by the ribosome. It appears that bacterial polypeptides, in contrast, do not fold until after the complete chain has been synthesized. Only then do the individual domains fold. This difference in timing may account for the observation that chaperones seem to be more important in the folding of bacterial proteins. Folding a whole polypeptide is more complex than folding one domain at a time and would require the aid of chaperones. Currently, most archaeal chaperones that have been studied function to protect proteins in stressful conditions. The time of protein folding in archaeal cells is not known. ▶▶| *Antibodies (section 33.7)*

Protein Splicing

A further level of complexity in the formation of proteins has been discovered in microbes belonging to all three domains of life. Some microbial proteins are spliced after translation. In protein splicing, a part of the polypeptide is removed before the polypeptide folds into its final shape. Self-splicing proteins begin as larger precursor proteins composed of an internal intervening sequence called an **intein** (about 130 to 600 amino acids in length) flanked by external sequences called **exteins** (**figure 12.44**). Inteins remove themselves from the precursor protein. When the splicing is completed, two proteins have been formed: the intein protein and the protein formed by splicing the two exteins together. Thus far, more than 130 inteins have been discovered.

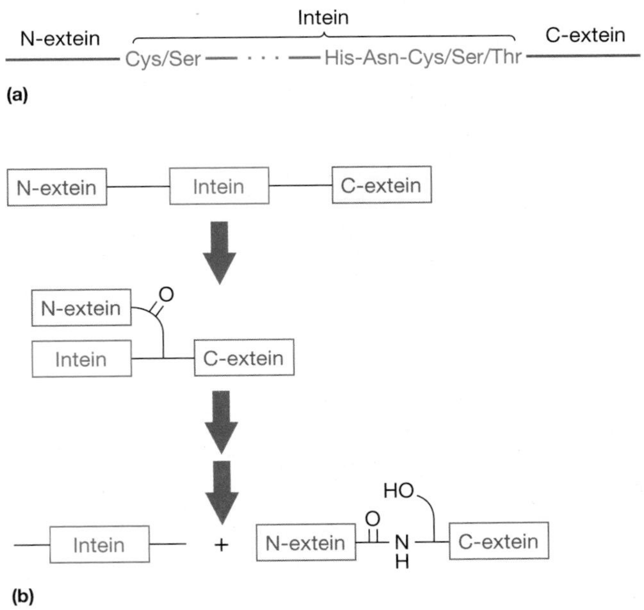

FIGURE 12.44 Protein Splicing. (a) A generalized illustration of intein structure. The amino acids that are commonly present at each end of the inteins are shown. Note that many are thiol- or hydroxyl-containing amino acids. (b) An overview of the proposed pattern or sequence of splicing. The precise mechanism is not yet known but presumably involves the hydroxyls or thiols located at each end of the intein.

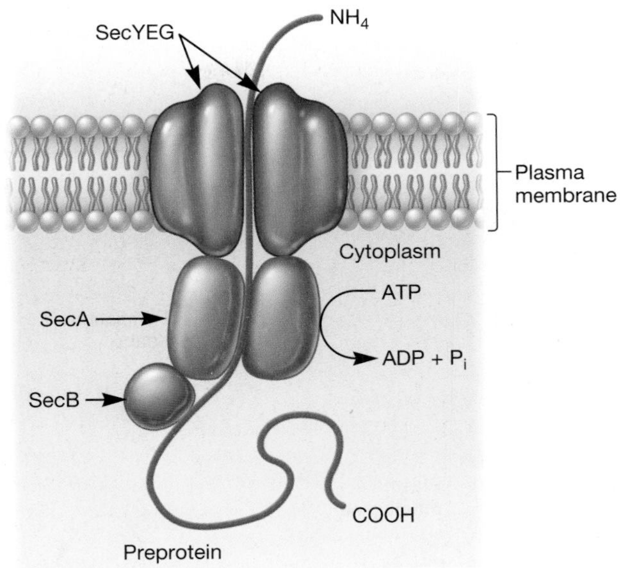

FIGURE 12.45 The Sec System Pathway of *E-Coli*.

Protein Translocation and Secretion in *Bacteria*

It has been estimated that almost one-third of the proteins synthesized by cells leave the cytoplasm to reside in membranes, the periplasmic space of bacterial and archaeal cells, or the external environment. It is not surprising then that over 15 different systems for moving proteins out of the cytoplasm have evolved. Some of these systems are found in all domains of life. Others are unique to bacterial and archaeal cells, and others are observed only in gram-negative bacterial cells. When proteins are moved from the cytoplasm to the membrane or to the periplasmic space, the movement is called translocation. Protein secretion refers to the movement of proteins from the cytoplasm to the external environment. Many of the secretion pathways are designated with numbers (e.g., type I secretion system, type II secretion system, etc.).

Why are so many proteins moved out of the cytoplasm? Many important proteins are located in membranes. These include transport proteins that bring needed materials into the cell and wastes out of the cell. They also include proteins involved in electron transport. In gram-negative bacteria, the periplasmic space is loaded with proteins such as chemotaxis proteins, enzymes involved in cell wall synthesis, and periplasmic components of nutrient uptake systems. Many organisms secrete hydrolytic enzymes into the external environment. These enzymes break down macromolecules into monomers that are more easily brought into the cell. The protein subunits of external structures such as flagella and fimbriae must also be moved out of the cell and assembled on its external surface. Pathogenic microbes often release toxins that are important in the infection process.

Protein secretion poses different difficulties, depending on the structure of the microbial cell envelope. For gram-positive bacteria to secrete proteins, the proteins must be translocated across the plasma membrane. Once across the plasma membrane, the protein either passes through the relatively porous peptidoglycan into the external environment or becomes embedded in or attached to the peptidoglycan. Likewise, secreted archaeal proteins must be moved across or into their unique cell walls. Gram-negative bacteria have more hurdles to jump when they secrete proteins. They, too, must transport the proteins across the plasma membrane, but to complete the secretion process, the proteins must be transported across the outer membrane. Our focus here is on the better-studied bacterial translocation and secretion pathways.

Common Translocation and Secretion Systems

In *Bacteria*, the major pathway for translocating proteins across the plasma membrane is the Sec (*sec*retion) pathway (**figure 12.45**). In gram-negative bacteria, proteins can be transported across the outer membrane by several different mechanisms, some of which bypass the Sec system, moving proteins directly from the cytoplasm to the outside of the cell (**figure 12.46**). All protein translocation and secretion pathways described here require the expenditure of energy at some step in the process. The energy is usually supplied by the hydrolysis of high-energy molecules such as ATP and GTP.

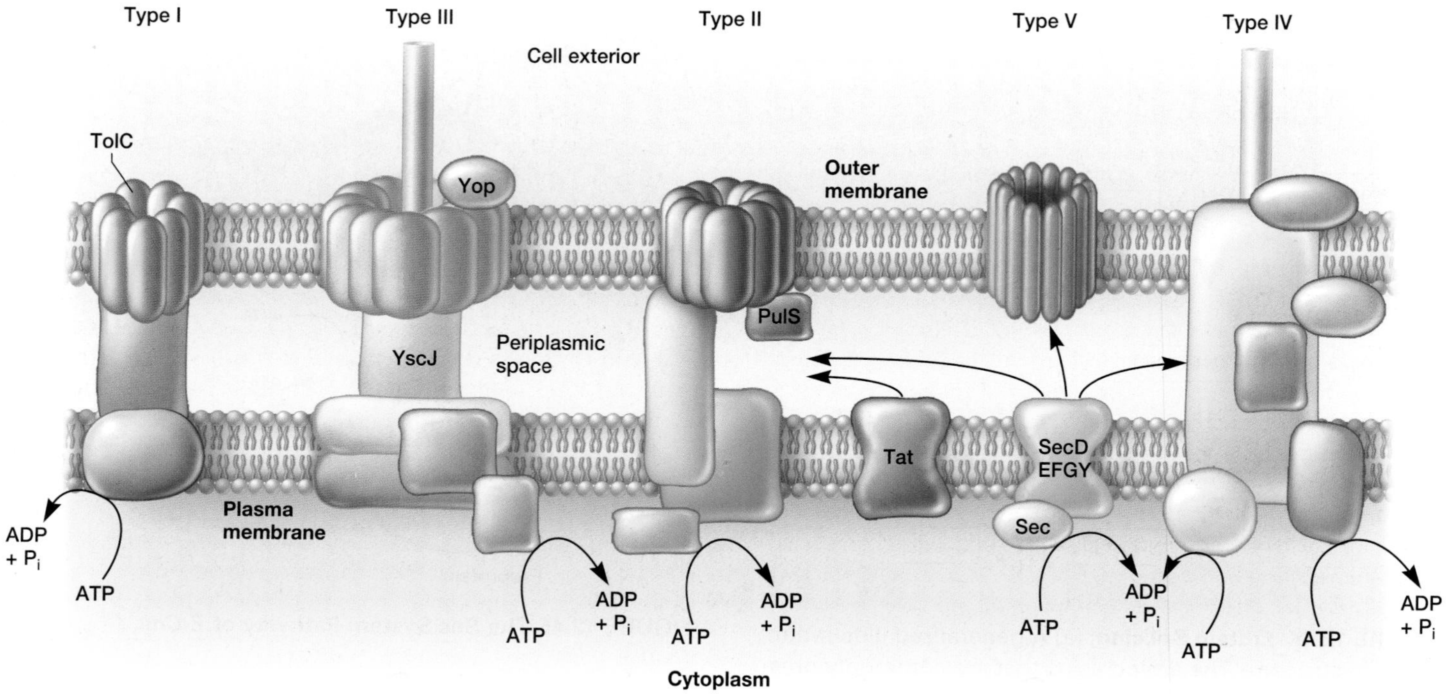

FIGURE 12.46 The Protein Secretion Systems of Gram-Negative Bacteria. Five of the six secretion systems of gram-negative bacteria are shown. The type VI secretion system is not shown because its structure has not been fully elucidated. The Sec and Tat pathways deliver proteins from the cytoplasm to the periplasmic space and also are found in archaea and gram-positive bacteria. The type II, type V, and sometimes type IV secretion systems complete the secretion process begun by the Sec system. The Tat system appears to deliver proteins only to the type II pathway. The type I and type III secretion systems bypass the Sec and Tat pathways, moving proteins directly from the cytoplasm, through the outer membrane, to the extracellular space. The type IV secretion system can work either with the Sec system or can work alone to transport proteins to the extracellular space. Type IV systems are also observed in gram-positive bacteria. Proteins translocated by the Sec system or secreted by the type III secretion system are delivered to those systems by chaperone proteins.

***Figure 12.46* Micro Inquiry**

What are two distinguishing features of protein translocation via the Tat system as compared to that using the Sec system?

Sec pathway translocates unfolded proteins

Tat system translocates folded proteins w/ twin arginine residues

However, the proton motive force also sometimes plays a role. *Active transport (section 6.6); ATP (section 9.4); Electron transport and oxidative phosphorylation (section 10.5)*

The **Sec system,** sometimes called the general secretion pathway, is highly conserved, having been identified in all three domains of life (figure 12.45). It translocates unfolded proteins across the plasma membrane or integrates them into the membrane itself. Proteins to be transported across the plasma membrane by this pathway are synthesized as presecretory proteins called preproteins. The amino-terminus of the preprotein has a **signal peptide,** which is recognized by the Sec machinery. Soon after the signal peptide is synthesized, chaperone proteins (e.g., SecB) bind it. This helps delay protein folding, thereby helping the preprotein reach the Sec transport machinery in the conformation needed for transport. Evidence exists that translocation of proteins can begin before the completion of their synthesis by ribosomes. Certain Sec proteins (SecY, SecE, and SecG) are thought to form a channel in the membrane through which the preprotein passes. Another protein (SecA) binds to the SecYEG proteins and the SecB-preprotein complex. SecA acts as a motor, using the energy released from ATP hydrolysis to translocate the preprotein through the plasma membrane. When the preprotein emerges from the plasma membrane, free from chaperones, an enzyme called signal peptidase removes the signal peptide. The protein then folds into the proper shape.

In *Bacteria* and some archaea, another plasma membrane translocation system called the **Tat system** can move proteins across the plasma membrane. This system is distinguished from the Sec pathway by the nature of the protein translocated. The Sec pathway translocates unfolded proteins; the Tat pathway translocates folded proteins. Furthermore, the Tat pathway only translocates proteins that feature two, or "twin," arginine residues in their signal sequence—in fact, *Tat* stands for *t*win *a*rginine *t*ranslocase. In gram-negative bacteria, proteins translocated by the Tat pathway are delivered to a type II secretion system for transport across the outer membrane.

Type I secretion systems are ubiquitous in gram-positive and gram-negative bacteria as well as *Archaea* (figure 12.46). They are evolutionarily related to ABC transport systems described in chapter 6. In pathogenic gram-negative bacteria, type I secretion systems are involved in the secretion of toxins (α-hemolysin), as well as proteases, lipases, and specific peptides. Secreted proteins usually contain C-terminal secretion signals that help direct the newly synthesized protein to the type I machinery. In gram-negative bacteria, the type I machinery spans the plasma membrane, the periplasmic space, and the outer membrane. These systems move proteins in one step across both membranes, bypassing the Sec system. Gram-positive bacteria use a modified version of a type I system to translocate proteins across the plasma membrane. Analysis of the *Bacillus subtilis* genome has identified 77 type I secretion systems. This may reflect the fact that ABC transporters move a wide variety of solutes in addition to proteins, including sugars and amino acids, as well as exporting antibiotics from the cell interior.

Type IV secretion systems are unique in that they are used to secrete proteins as well as to transfer DNA from a donor bacterium to a recipient during a process called bacterial conjugation. These systems are observed in both gram-positive and gram-negative bacteria. The type IV systems of gram-negative bacteria are best studied. They are composed of many different proteins and are described in more detail in chapter 14.

Protein Secretion in Gram-Negative Bacteria

Currently six protein secretion systems (types I to VI) have been identified in gram-negative bacteria (figure 12.46; the type VI system is not shown because its structure has not been fully elucidated). Some of these systems have already been described as they are present in both gram-negative and gram-positive bacteria (type I and type IV secretion). All others are unique to gram-negative bacteria. Most are used to secrete virulence factors produced by plant and animal pathogens. Gram-negative bacteria use the type II and type V systems to transport proteins across the outer membrane after the protein has first been translocated across the plasma membrane by the Sec system. The type I, III, and VI systems do not transport proteins that are first translocated by the Sec system, so they are said to be Sec-independent. The type IV pathway for secretion sometimes is linked to the Sec pathway but usually functions on its own.

Two of the systems form a needlelike structure that is sometimes called an injectisome (figure 12.46; *also see figure 38.10*). The injectisome extends beyond the outer membrane and can make contact with other cells. The **type III secretion system** is best studied because it injects virulence factors directly into the plant and animal host cells that these pathogens attack. The virulence factors include toxins, phagocytosis inhibitors, stimulators of cytoskeleton reorganization in the host cell, and promoters of host cell suicide (apoptosis). Type III systems also transport other proteins, including (1) some of the proteins from which the system is built, (2) proteins that regulate the secretion process, and (3) proteins that aid in the insertion of secreted proteins into target cells. Important examples of bacteria with type III systems are *Salmonella, Yersinia, Shigella, E. coli, Bordetella, Pseudomonas aeruginosa,* and *Erwinia.* The participation of type III systems in bacterial virulence is further discussed in chapter 38.

Type V secretion systems also warrant comment. They employ proteins called autotransporters because after being translocated across the plasma membrane by the Sec pathway, the proteins are able to transport themselves across the outer membrane. Autotransporters have two domains. One domain is thought to form a pore in the outer membrane through which the other domain (a virulence factor) is transported.

1. What are molecular chaperones and heat-shock proteins? Describe their functions.
2. Give the major characteristics and functions of the protein secretion systems described in this section.
3. Which secretion system is most widespread?
4. What is a signal peptide? Why do you think a protein's signal peptide is not removed until after the protein is translocated across the plasma membrane?

Summary

12.1 DNA as Genetic Material

a. DNA is composed of only four different building blocks (nucleotides). Therefore it was originally thought to be too simple to function as an organism's genetic material.

b. The knowledge that DNA is the genetic material for cells came from studies on transformation by Griffith and Avery and from experiments on T2 phage by Hershey and Chase (**figures 12.1–12.3**).

12.2 Flow of Genetic Information

a. DNA serves as the storage molecule for genetic information. DNA replication is the process by which DNA is duplicated so that it can be passed on to the next generation (**figure 12.4**).

b. During transcription, genetic information in DNA is rewritten as an RNA molecule. The three products of transcription are messenger RNA, ribosomal RNA, and transfer RNA.

c. Translation converts genetic information in the form of a messenger RNA molecule into a polypeptide. Ribosomal RNA and transfer RNA participate in the decoding of genetic information during translation.

12.3 Nucleic Acid and Protein Structure

a. DNA differs in composition from RNA in having deoxyribose and thymine, rather than ribose and uracil.

b. DNA is double stranded, with complementary AT and GC base pairing between the strands. The strands run antiparallel and are twisted into a right-handed double helix (**figure 12.6**).

c. RNA is normally single stranded, although it can coil upon itself and base pair to form hairpin structures.

d. Proteins are polymers of amino acids (**figure 12.7**) linked by peptide bonds (**figure 12.8**).

12.4 DNA Replication

a. Circular bacterial DNAs have a single origin of replication. They are copied by two replication forks moving around the circle to form a theta-shaped (θ) figure (**figure 12.10**).

b. All known archaeal cells have circular DNA. Their chromosomes are thought to be replicated in a manner similar to circular bacterial chromosomes. However, some archaeal species have more than one origin of replication.

c. Eukaryotic DNA has many replicons and replication origins.

d. The replisome is a huge complex of proteins and is responsible for DNA replication.

e. DNA polymerase enzymes catalyze the synthesis of DNA in the 5′ to 3′ direction while reading the DNA template in the 3′ to 5′ direction (**figure 12.13**).

f. The double helix is unwound by helicases with the aid of topoisomerases such as DNA gyrase. DNA binding proteins keep the strands separate.

g. DNA polymerase III holoenzyme synthesizes a complementary DNA copy beginning with a short RNA primer made by the enzyme primase.

h. The leading strand is replicated continuously, whereas DNA synthesis on the lagging strand is discontinuous and forms Okazaki fragments (**figures 12.13** and **12.14**).

i. DNA polymerase I excises the RNA primer and fills in the resulting gap. DNA ligase then joins the fragments together (**figures 12.15** and **12.16**).

j. Telomerase is responsible for replicating the ends of eukaryotic chromosomes (**figure 12.17**).

12.5 Gene Structure

a. A gene may be defined as the nucleic acid sequence that codes for a polypeptide, tRNA, or rRNA.

b. The template strand of DNA carries genetic information and directs the synthesis of the RNA transcript.

c. The gene also contains a promoter, a coding region, and a terminator; it may have a leader and a trailer (**figure 12.20**).

d. The genes for tRNA and rRNA often code for a precursor that is subsequently processed to yield several products (**figure 12.21**).

e. RNA polymerase binds to the promoter region, which contains RNA polymerase recognition and RNA polymerase binding sites (**figures 12.23, 12.25,** and **12.26**).

12.6 Transcription

a. RNA polymerase synthesizes RNA that is complementary to the DNA template strand (**figure 12.23**).

b. The sigma factor helps the bacterial RNA polymerase bind to the promoter region at the start of a gene (**figure 12.26**).

c. A terminator marks the end of a gene. The protein rho is needed for RNA polymerase release from some terminators (**figures 12.28** and **12.29**).

d. In eukaryotes, RNA polymerase II synthesizes pre-mRNA, which then undergoes posttranscriptional modification by RNA cleavage and addition of a 3′ poly-A sequence and a 5′ cap to generate mRNA (**figure 12.31**).

e. Many eukaryotic genes are split or interrupted genes that have exons and introns. Exons are joined by RNA splicing, which is carried out by spliceosomes (**figure 12.32**).

12.7 The Genetic Code

a. Genetic information is carried in the form of 64 nucleotide triplets called codons (**table 12.2**); 61 sense codons direct amino acid incorporation, and three stop or nonsense codons terminate translation.

b. The code is degenerate—that is, there is more than one codon for most amino acids.

12.8 Translation

a. In translation, ribosomes attach to mRNA and synthesize a polypeptide beginning at the N-terminal end. A polysome or polyribosome is a complex of mRNA with several ribosomes (**figure 12.34**).

b. Amino acids are activated for protein synthesis by attachment to the 3′ end of transfer RNAs. Activation requires ATP, and the reaction is catalyzed by aminoacyl-tRNA synthetases (**figure 12.36**).

c. Ribosomes are large, complex organelles composed of rRNAs and many polypeptides. Amino acids are added to a growing peptide chain at the translational domain of the ribosome (**figure 12.37**).

d. Protein synthesis begins with the binding of fMet-tRNA (*Bacteria*) (**figure 12.38**) or an initiator methionyl-$tRNA^{Met}$

(eukaryotes and *Archaea*) to an initiator codon on mRNA and to the two ribosomal subunits. This involves the participation of protein initiation factors (**figure 12.39**).

e. In the elongation cycle, the proper aminoacyl-tRNA binds to the A site (**figure 12.40**). Then the transpeptidation reaction is catalyzed by peptidyl transferase (**figure 12.41**). Finally, during translocation, the peptidyl-tRNA moves to the P site and the ribosome moves down the mRNA by one codon. The empty tRNA leaves the ribosome by way of the exit site.

f. Protein synthesis stops when a nonsense codon is reached. Bacteria require three release factors for codon recognition and ribosome dissociation from the mRNA (**figure 12.42**).

12.9 Protein Maturation and Secretion

a. Bacterial and archaeal proteins may not fold until completely synthesized, whereas eukaryotic protein domains fold as they leave the ribosome. Molecular chaperones are involved in ensuring that proteins fold properly or are delivered to their destination site (**figure 12.43**).

b. A few proteins are self-splicing and excise portions of themselves before folding into their final shape (**figure 12.44**).

c. Many proteins must be transported across the plasma membrane of cells. The most commonly used mechanism is the Sec system, which is found in all microbes (**figure 12.45**). Type I secretion systems are also observed in many bacteria and archaea. Type IV secretion systems are observed in both gram-positive and gram-negative bacteria (**figure 12.46**).

d. Gram-negative bacteria have evolved additional systems for moving proteins across the outer membrane of the cell wall. These include type II, III, V, and VI secretion systems.

Critical Thinking Questions

1. Many scientists say that RNA was the first of the information molecules (i.e., RNA, DNA, protein) to arise during evolution. Given the information in this chapter, what evidence is there to support this hypothesis?

2. *Streptomyces coelicolor* has a linear chromosome. Interestingly, there are no genes that encode essential proteins near the ends of the chromosome in this bacterium. Why do you think this is the case?

3. You have isolated several *E. coli* mutants:

   Mutant #1 has a mutation in the –10 region of the promoter of a structural gene encoding an enzyme needed for synthesis of the amino acid serine.

   Mutant #2 has a mutation in the –35 region in the promoter of the same gene.

   Mutant #3 is a double mutant with mutations in both the –10 and –35 region of the promoter of the same gene.

   Only Mutant #3 is unable to make serine. Why do you think this is so?

4. DNA polymerase I (Pol I) of *E. coli* consists of three functional parts or domains: an N-terminal domain with 5′ to 3′ exonuclease activity required for removal of the RNA primer, a central domain responsible for 3′ to 5′ exonuclease proofreading, and a C-terminal domain with polymerase activity. Pol I is thought to simultaneously remove RNA primers and fill in the gaps that result (figure 12.15). A group of proteins known as RNaseH also have 5′ to 3′ exonuclease activity and can thus remove RNA primers. However, they lack the other two functions observed for Pol I. Predict the ability of the following mutants to replicate DNA: (1) A strain with a mutant gene encoding Pol I such that it no longer has polymerase activity (but retains both types of nuclease activities); (2) A strain without RNaseH proteins; (3) A strain with a mutant gene encoding Pol I such that it no longer has 5′ to 3′ exonuclease activity (but retains 3′ to 5′ nuclease and polymerase activities); (4) A strain with the mutant Pol I described in (3) and lacking all RNaseH proteins. Explain your reasoning for each.

   Read the original paper: Fukushima, S., et al. 2008. Reassessment of the in vivo functions of DNA polymerase I and RNaseH in bacterial cell growth. *J. Bacteriol.* 189:8575.

5. *Helicobacter pylori* is a gram-negative bacterium that causes gastric ulcers and cancer. It has a type IV secretion system that secretes the protein CagA. The function of CagA is unknown, but its export into host cells causes inflammation and is associated with increased virulence. The CagA type IV secretion system consists of at least 14 essential protein components and seven proteins with accessory activities that enable secretion of CagA. Refer to figure 12.46 and predict the specific cell envelope location of the essential CagA type IV secretion system components. Which part of the cell envelope do you think will have the most protein components? What might be the function of the so-called "accessory" proteins, which are not part of the secretion system but are nonetheless required for CagA export?

   Read the original paper: Kutter, S., et al. 2008. Protein subassemblies of the *Helicobacter pylori* CagA type IV secretion system revealed by localization and interaction studies. *J. Bacteriol.* 190:2161.

Concept Mapping Challenge

Construct a concept map that describes transcription, translation, and DNA replication. Use the concepts that follow, any other concepts or terms you need, and your own linking words between each pair of concepts in your map.

template	primer	monomer
peptide bond	ribosome	tRNA
leading strand	promoter	terminator
stop codon	DNA ligase	RNA polymerase
DNA polymerase III	AUG	Okazaki fragment
lagging strand	f-Met	DNA polymerase I

Learn more

Learn more by visiting the text website at www.mhhe.com/willey8, where you will find a complete list of references.

13

Microbial Genetics: Regulation of Gene Expression

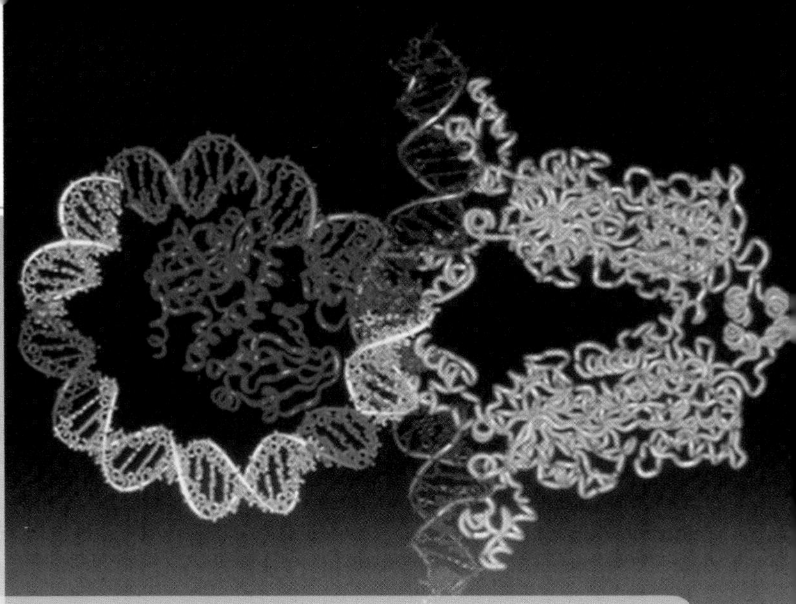

The lac *repressor (lavender) and catabolite activator protein (blue) are bound to the* lac *operon.*

CHAPTER GLOSSARY

activator protein A regulatory protein that can bind a specific site on DNA and promote transcription.

antisense RNA A single-stranded RNA with a base sequence complementary to a segment of another RNA molecule. Antisense RNA specifically binds to the target RNA and alters its activity.

attenuation A mechanism for the regulation of transcription elongation of some bacterial operons by aminoacyl-tRNAs.

catabolite repression Inhibition of the synthesis of several catabolic enzymes in response to the presence of preferred carbon and energy sources such as glucose.

constitutive gene A gene that is consistently expressed at the same level.

corepressor A small molecule that binds and activates a repressor protein, which is then able to bind to and inhibit the transcription of a repressible gene.

global regulatory systems Genetic regulatory mechanisms that control a suite of genes simultaneously.

housekeeping gene A gene that is needed for normal, vegetative growth.

inducer A small molecule that stimulates the transcription of an inducible gene either by binding and inactivating a repressor protein or by binding and activating an activator protein.

inducible gene A gene whose expression can be increased by either positive or negative transcriptional control.

negative transcriptional control A mechanism that uses a repressor protein to alter the level of expression of a gene or genes.

operator The segment of bacterial DNA to which a repressor protein binds.

operon The sequence of bases in bacterial and archaeal DNA that contains two or more structural genes transcribed from a single promoter.

positive transcriptional control A mechanism that uses an activator protein to regulate the level of expression of a gene or genes.

quorum sensing The process in which microbes monitor cell density and regulate expression of density-dependent genes by sensing signal molecules released by microorganisms.

regulon A collection of genes or operons that is controlled by a common regulatory protein.

repressible gene A gene whose level of expression can be decreased by either negative or positive transcriptional control.

repressor protein A regulatory protein that can bind a specific site on DNA and inhibit transcription.

riboswitch A site in the leader of an mRNA molecule that interacts with a metabolite or other small molecule, causing the leader to change its folding pattern; this change alters either transcription levels or translation levels.

structural gene A gene that codes for the synthesis of tRNA, rRNA or a polypeptide with a nonregulatory function.

two-component signal transduction system A signal transduction system that uses the transfer of phosphoryl groups to control gene transcription and protein activity; it has two major components: a **sensor kinase** and a **response regulator**. Some systems include additional transfers and are called **phosphorelay systems**.

The gram-positive soil bacterium *Bacillus subtilis* senses that the nutrient levels in its environment are decreasing, and it must determine if it should initiate sporulation. An *Escherichia coli* cell is in an environment rich in carbon and energy sources, and it must determine which to use and when to use them. A pathogen is transmitted from a stream to the intestinal tract of its animal host, and it must adjust to the warmer temperature, increased nutrient supply, and defenses of the host. These are just a few examples of situations to which microbes must react. To make the most efficient use of the resources in the current environment and their own cellular machinery, microbes must respond to changes by altering physiological and behavioral processes. How is this accomplished?

Cells use two general approaches to regulate cellular processes. (1) They can alter the activity of enzymes and other proteins. This type of regulation occurs after the protein is synthesized and is called posttranslational control. (2) They can change the rate of synthesis of enzymes and other proteins. This type of regulation is often called regulation of gene expression.

The control of cellular processes by regulating the activity of enzymes and other proteins is a fine-tuning mechanism: It acts rapidly to adjust metabolic activity from moment to moment. The control of gene expression occurs over longer intervals. For example, the *E. coli* chromosome can code for about 4,500 polypeptides, yet not all are produced at the same time. Regulation of gene expression conserves energy and raw materials, maintains the balance between the amounts of various cell proteins, and enables microbes to acclimate to long-term environmental change. Thus control of gene expression complements posttranslational control. ◀◀ *Posttranslational regulation of enzyme activity (section 9.10)*

In this chapter, we explore the various mechanisms microorganisms use to regulate cellular processes. We begin with a brief discussion of the many levels at which regulation can occur. We then introduce some important examples of the regulation of transcription initiation, transcription elongation, and translation. In chapter 9, we introduced posttranslational modification of enzymes as a mechanism for controlling enzyme activity. Here, we revisit posttranslational modification as a means for regulating complex cellular behaviors; we use chemotaxis as our example. Next, by describing several examples of global regulation, we examine how cells use these various regulatory mechanisms to control suites of genes in response to changes in their environments. We end the chapter with a brief discussion of regulation in eukaryotic and archaeal cells.

13.1 Levels of Regulation

Figure 13.1 summarizes the steps leading from the information coded in DNA to a functional protein. As we discuss in chapter 12, the process begins with transcription of the gene, followed by translation of the mRNA to yield a protein. Regulation that occurs by controlling the processes of transcription and translation is often called regulation of gene expression. When gene expression is regulated by controlling mRNA synthesis, it is said to be "at the level of transcription" or at the "transcriptional level." Likewise, when gene expression is governed by controlling protein synthesis, it is said to be "at the level of translation" or at the "translational level." Translation yields a protein that may or may not be functional. The activity level of proteins may be altered by posttranslational modification.

Although the overall processes of transcription and translation in the three domains of life are similar, some differences affect gene expression. Thus regulation of gene expression is somewhat different in each domain. Similar mechanisms for posttranslational modifications are observed in all domains. Our focus in this chapter is on well-understood bacterial regulatory processes. Regulation at the level of transcription has been the focus of microbial geneticists for many years and is well understood. Transcription can be regulated by governing its initiation or its elongation, and these topics are covered in sections 13.2 and 13.3, respectively. We begin our discussion by introducing two phenomena: induction of enzyme synthesis and repression of enzyme synthesis. Induction and repression provided the first models for gene regulation. These models involve the action of regulatory proteins, and the notion that gene expression is regulated solely by proteins persisted for many years. Eventually it was clearly demonstrated that RNA molecules also could have regulatory functions.

The mechanisms by which gene expression is controlled at the level of translation have not been studied as long, so our understanding of them is not as fully developed. This is the focus of section 13.4. Finally, posttranslational regulation is discussed in section 13.5.

13.2 Regulation of Transcription Initiation

Induction and repression are historically important, as they were the first regulatory processes for gene expression to be understood in any detail. In this section, we first describe these phenomena and then examine the underlying regulatory events.

Induction and Repression of Enzyme Synthesis

Just as an automobile or a refrigerator does not work forever and eventually must be replaced, so too enzymes only remain functional for a certain time. The "old" enzymes are eventually

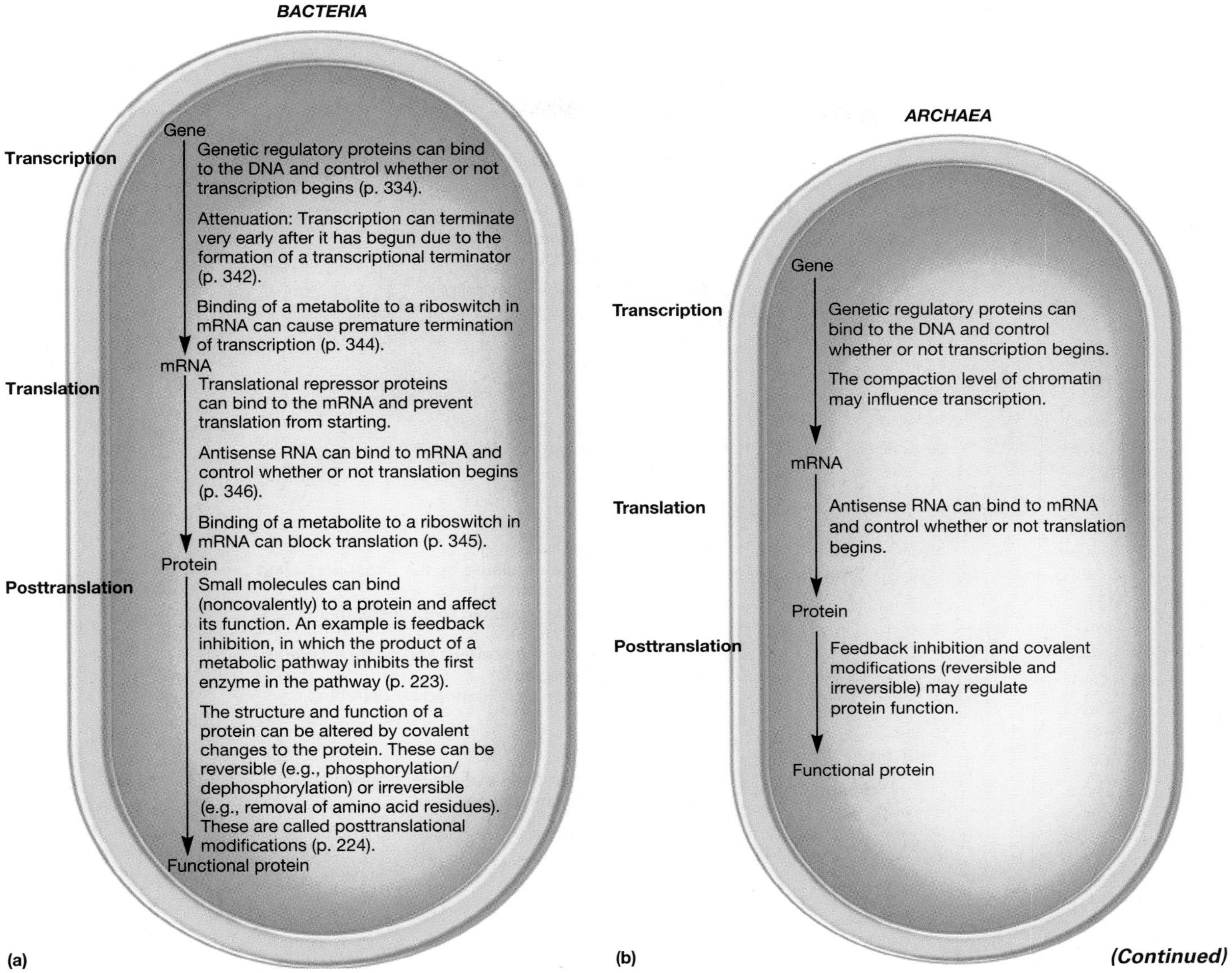
BACTERIA
Transcription
Translation
Posttranslation
Gene
Genetic regulatory proteins can bind to the DNA and control whether or not transcription begins (p. 334).
Attenuation: Transcription can terminate very early after it has begun due to the formation of a transcriptional terminator (p. 342).
Binding of a metabolite to a riboswitch in mRNA can cause premature termination of transcription (p. 344).
mRNA
Translational repressor proteins can bind to the mRNA and prevent translation from starting.
Antisense RNA can bind to mRNA and control whether or not translation begins (p. 346).
Binding of a metabolite to a riboswitch in mRNA can block translation (p. 345).
Protein
Small molecules can bind (noncovalently) to a protein and affect its function. An example is feedback inhibition, in which the product of a metabolic pathway inhibits the first enzyme in the pathway (p. 223).
The structure and function of a protein can be altered by covalent changes to the protein. These can be reversible (e.g., phosphorylation/dephosphorylation) or irreversible (e.g., removal of amino acid residues). These are called posttranslational modifications (p. 224).
Functional protein
(a)
ARCHAEA
Transcription
Translation
Posttranslation
Gene
Genetic regulatory proteins can bind to the DNA and control whether or not transcription begins.
The compaction level of chromatin may influence transcription.
mRNA
Antisense RNA can bind to mRNA and control whether or not translation begins.
Protein
Feedback inhibition and covalent modifications (reversible and irreversible) may regulate protein function.
Functional protein
(b)

(Continued)

FIGURE 13.1 Gene Expression and Common Regulatory Mechanisms in the Three Domains of Life.

degraded by proteasomes and related protein-degrading systems. This degradation not only removes enzymes that may not be functioning properly but also recycles amino acids so that they can be used to synthesize new proteins. Despite the value of this, it also creates a problem for a cell because many enzymes catalyze reactions that are needed almost all the time (e.g., enzymes of the central metabolic pathways). Their functions are often referred to as "housekeeping functions," and the genes that encode them are often called **housekeeping genes.** To maintain the proper amount of housekeeping enzymes, many housekeeping genes are expressed continuously by the cell. Those housekeeping genes that are expressed continuously are said to be **constitutive genes.** Other enzymes are only needed at certain times and in certain environments. The genes encoding these enzymes are expressed only when needed; therefore their expression is regulated. The β-galactosidase gene is an example of a regulated gene.

The enzyme β-galactosidase catalyzes the hydrolysis of the disaccharide sugar lactose to glucose and galactose (**figure 13.2**). When *E. coli* grows with lactose as its only carbon source, each cell contains about 3,000 β-galactosidase molecules, but it has less than three molecules in the absence of lactose. β-galactosidase is an inducible enzyme—that is, its level rises in the presence of a small effector molecule called an **inducer** (in this case, the lactose derivative allolactose). Likewise, the genes that encode inducible enzymes such as β-galactosidase are referred to as **inducible genes.**

β-galactosidase is an enzyme that functions in a catabolic pathway, and many catabolic enzymes are inducible enzymes. On

FIGURE 13.1 ***(Continued)***

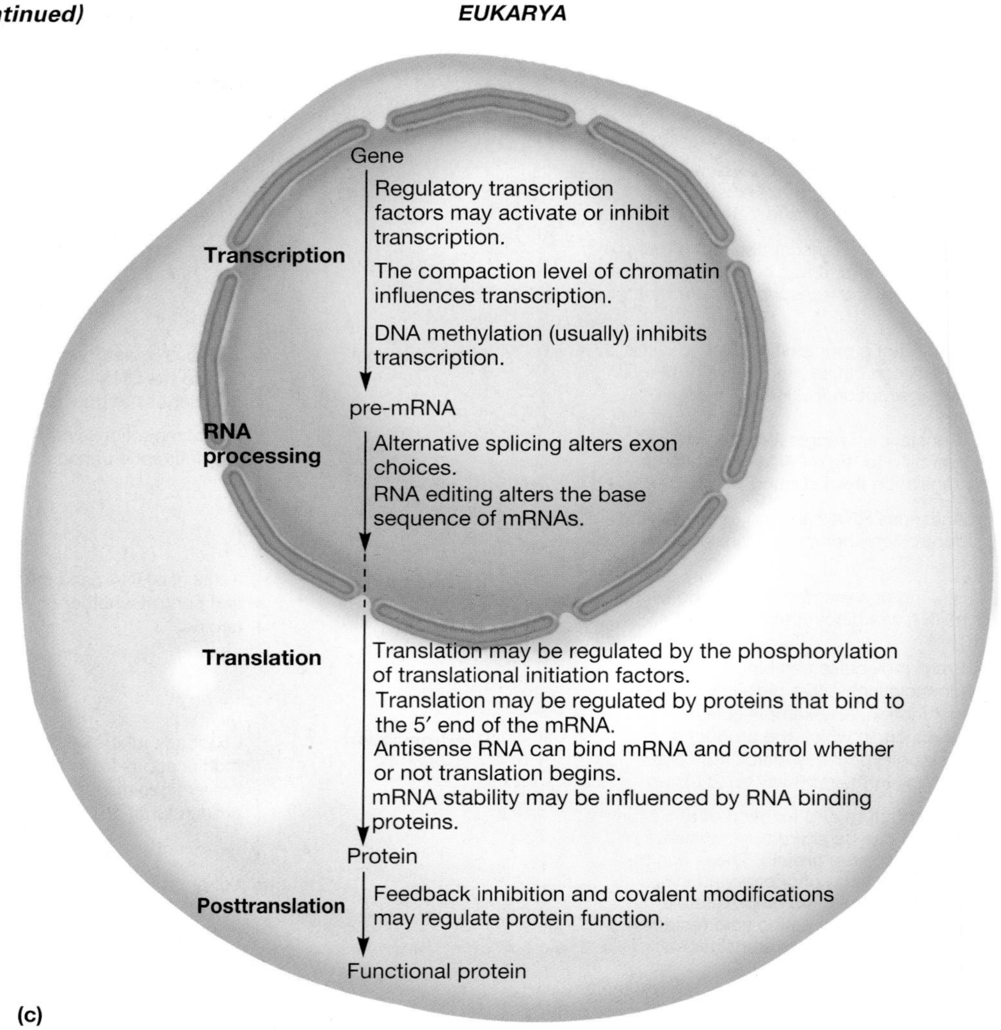

the other hand, the genes for enzymes involved in biosynthetic pathways are often called **repressible genes,** and their products are called repressible enzymes. For instance, an amino acid present in the surroundings may inhibit the formation of enzymes responsible for its biosynthesis. This makes sense because the microorganism does not need the biosynthetic enzymes for a particular substance if it is already available. Generally, repressible enzymes are necessary for synthesis and are present unless the end product of their pathway is available. Inducible enzymes, in contrast, are required only when their substrate is available; they are missing in the absence of the inducer.

EUKARYA
Gene
Transcription
Regulatory transcription factors may activate or inhibit transcription.
The compaction level of chromatin influences transcription.
DNA methylation (usually) inhibits transcription.
pre-mRNA
RNA processing
Alternative splicing alters exon choices.
RNA editing alters the base sequence of mRNAs.
Translation
Translation may be regulated by the phosphorylation of translational initiation factors.
Translation may be regulated by proteins that bind to the 5′ end of the mRNA.
Antisense RNA can bind mRNA and control whether or not translation begins.
mRNA stability may be influenced by RNA binding proteins.
Protein
Posttranslation
Feedback inhibition and covalent modifications may regulate protein function.
Functional protein
(c)

Control of Transcription Initiation by Regulatory Proteins

The action of regulatory proteins is often responsible for induction and repression. Many of these regulatory proteins are DNA-binding proteins that form dimers and bind short, inverted sequences of bases in the DNA called palindromes. Only a small portion of these proteins actually interacts with the palindromes, and these small regions are referred to as DNA-binding domains. Examination of numerous regulatory proteins has led to the recognition of two common motifs in DNA-binding domains: helix-turn-helix and zinc fingers.

Helix-turn-helix DNA-binding domains are often observed in bacterial regulatory proteins (e.g., *lac* repressor, *trp* repressor, and CAP; pp. 338, 339, and 350, respectively). They are generally about 20 amino acids in length and are folded into two α-helices (*see figure A1.14*) separated by a β-turn. One of the helices protrudes from the surface of the protein and is often positioned in the major groove of the DNA. Thus when the regulatory protein forms a dimer, two α-helices, one from each protein subunit, interacts with the major groove of the DNA.

Zinc finger DNA-binding domains have been observed in some bacterial regulatory proteins but are most common in eukaryotic regulatory proteins. Zinc fingers are long loops formed by about 30 amino acids. The loop is stabilized by an interaction

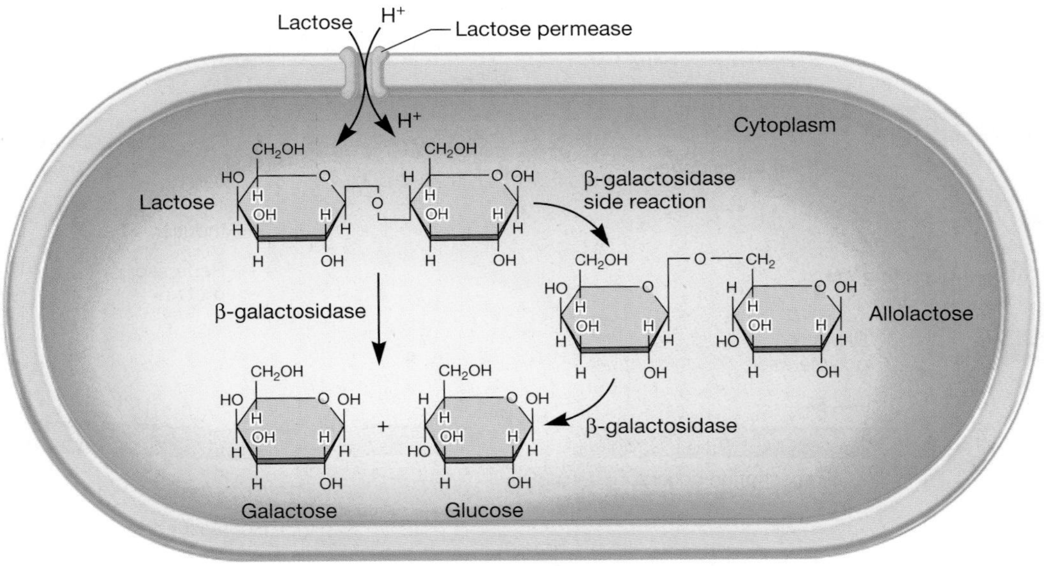

FIGURE 13.2 The Reactions of β-Galactosidase. The main reaction catalyzed by β-galactosidase is the hydrolysis of lactose, a disaccharide, into the monosaccharides galactose and glucose. The enzyme also catalyzes a minor reaction that converts lactose to allolactose. Allolactose acts as the inducer of β-galactosidase synthesis.

between certain amino acids in the loop and Zn^{2+}. Regulatory proteins often have more than one zinc finger, each of which interacts with DNA. However, the function of each finger can differ. For instance, one finger may bind DNA nonspecifically, whereas another is important for recognition of a particular sequence in the DNA.

Regulatory proteins can exert either negative or positive control. **Negative transcriptional control** occurs when the binding of the protein to DNA inhibits initiation of transcription. Regulatory proteins that act in this fashion are called **repressor proteins. Positive transcriptional control** occurs when the binding of the protein to DNA promotes transcription initiation. These proteins are called **activator proteins.** *Regulatory Proteins*

Repressor and activator proteins usually act by binding DNA at specific sites. In *Bacteria,* repressor proteins bind a region called the **operator,** which usually overlaps or is downstream of the promoter (i.e., closer to the coding region) (**figure 13.3*a,b***). When bound, the repressor protein either blocks binding of RNA polymerase to the promoter or prevents its movement. Activator proteins bind **activator-binding sites** (figure 13.3*c,d*). These are often upstream of the promoter (i.e., farther away from the coding region). Binding of an activator to its regulatory site generally promotes RNA polymerase binding.

Repressor and activator proteins must exist in both active and inactive forms if transcription initiation is to be controlled appropriately. The activity of regulatory proteins is modified by small effector molecules, most of which bind the regulatory protein noncovalently (i.e., allosteric regulation). Figure 13.3 shows the four basic ways in which the interactions of an effector and a regulatory protein can affect transcription. (1) For negatively controlled inducible genes (e.g., those encoding enzymes needed for catabolism of a sugar), the repressor protein is active and prevents transcription when the substrate of the pathway is not available (figure 13.3*a*). It is inactivated by binding of the inducer (e.g., the substrate of the pathway). (2) For negatively controlled repressible genes (e.g., those encoding enzymes needed for the synthesis of an amino acid), the repressor protein is initially synthesized in an inactive form called the **aporepressor.** It is activated by binding of the **corepressor** (figure 13.3*b*). For repressible enzymes that function in a biosynthetic pathway, the corepressor is often the product of the pathway (e.g., an amino acid). (3) The activator of a positively regulated inducible gene is activated by the inducer (figure 13.3*c*), whereas (4) the activator of a positively regulated repressible gene is inactivated by an inhibitor (figure 13.3*d*). ◀◀ *Allosteric regulation (section 9.10)*

Before we continue our discussion, we must consider two general aspects of regulation. The first is that gene expression is rarely an all-or-nothing phenomenon; it is a continuum. Inhibition of transcription usually does not mean that genes are "turned off" (though this terminology is frequently used). Rather it means the level of mRNA synthesis is decreased significantly and in most cases is occurring at very low levels. In other words, many promoters of regulated genes and operons are considered "leaky," in that there is always some low, basal level of transcription. The second aspect of regulation is the "decision-making" process used by microbial cells. Although cells do not have thought processes, it is convenient to think of regulation in this way. Consider the

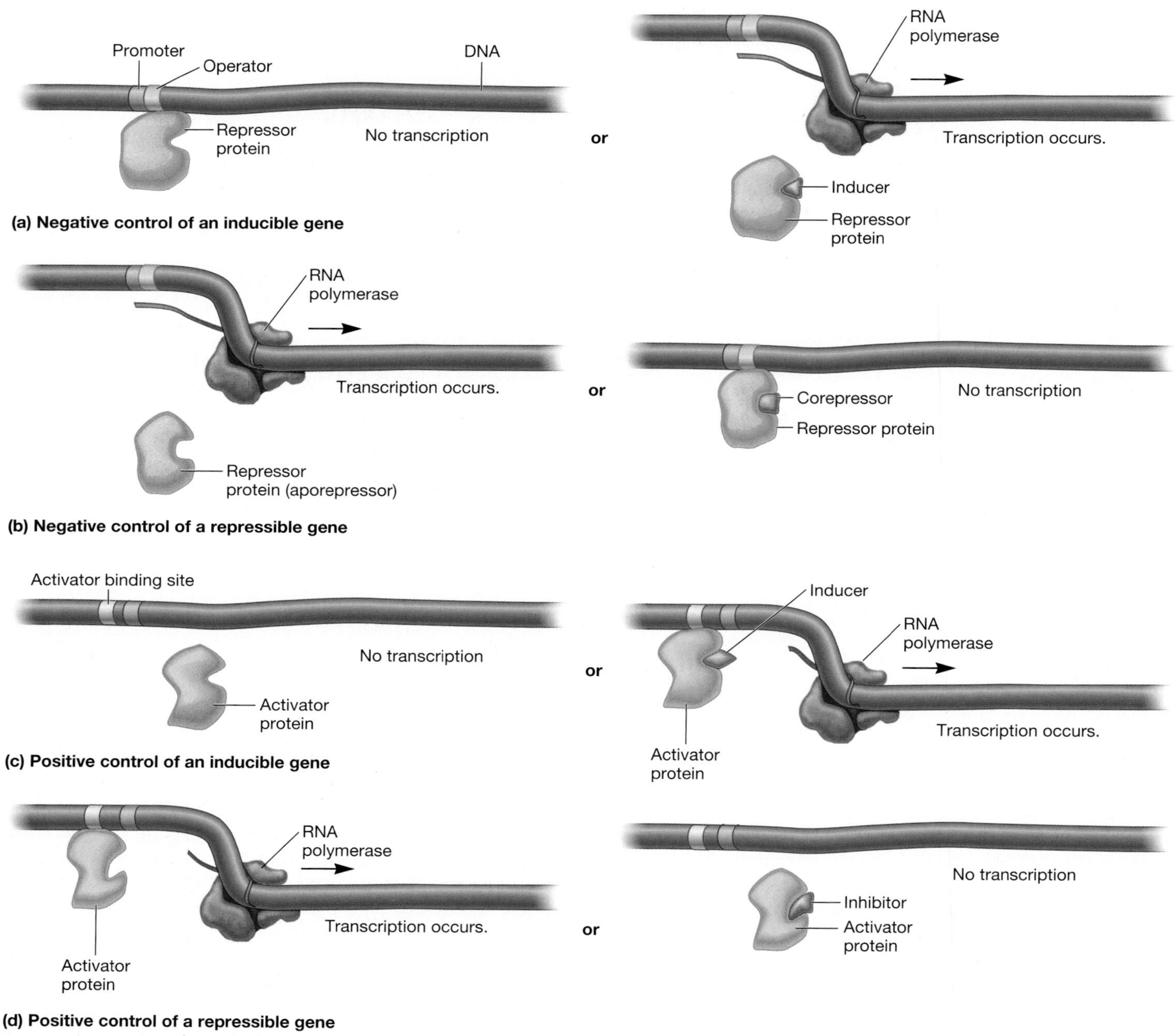

FIGURE 13.3 Action of Bacterial Regulatory Proteins. Bacterial regulatory proteins have two binding sites, one for a small effector molecule and one for DNA. The binding of the effector molecule changes the regulatory protein's ability to bind DNA. (a) In the absence of an inducer, the repressor protein blocks transcription. The presence of an inducer prevents the repressor from binding DNA and transcription occurs. (b) In the absence of a corepressor, the repressor is unable to bind DNA and transcription occurs. When the corepressor is bound to the repressor, the repressor is able to bind DNA and transcription is blocked. (c) The activator protein is only able to bind DNA and activate transcription when it is bound in the inducer. (d) The activator binds DNA and promotes transcription unless the inhibitor is present. When the inhibitor is present, the activator undergoes a conformational change that prevents it from binding DNA; this inhibits transcription.

Figure 13.3 Micro Inquiry

In what way is an inducer molecule that binds a repressor protein similar to an inhibitor molecule that binds an activator protein?

regulatory "decisions" made by an *E. coli* cell. It need only synthesize the enzymes of a specific catabolic pathway if the substrate of the pathway is present in the environment and a preferred carbon and energy source (e.g., glucose) is not (**figure 13.4**; *also see figure 20.4*). Preferred carbon and energy sources usually are more easily catabolized or yield more energy. Thus it is to the cell's advantage to use the preferred source before another source. Conversely, synthesis of the enzymes involved in biosynthetic pathways is inhibited when the end product of the pathway is present.

Recall that functionally related bacterial and archaeal genes are often transcribed from a single promoter. The **structural genes**—the genes coding for nonregulatory polypeptides (e.g., enzymes)—are simply lined up together on the DNA, and a single, polycistronic mRNA carries all the messages (*see figure 12.22*). The sequence of bases coding for one or more polypeptides, together with the promoter and operator or activator-binding sites, is called an **operon.** Many operons have been discovered and studied. Three well-studied operons are discussed next. They demonstrate different ways that regulatory proteins can be used to control gene expression at the level of transcription initiation.

Lactose Operon: Negative Transcriptional Control of Inducible Genes

The best-studied negative control system is the lactose (*lac*) operon of *E. coli.* The *lac* operon contains three structural genes controlled by the *lac* repressor, which is encoded by *lacI* (**figure 13.5**). One gene codes for β-galactosidase; a second gene directs the synthesis of β-galactoside permease, the protein responsible for lactose uptake. The third gene codes for the enzyme β-galactoside transacetylase, whose function still is uncertain. The presence of the first two genes in the same operon ensures that the rates of lactose uptake and breakdown will vary together.

Lactose is one of many organic molecules *E. coli* can use as a carbon and energy source. It is wasteful to synthesize enzymes of the *lac* operon when lactose is not available. Therefore the cell only expresses this operon at high levels when lactose is the only carbon and energy source present in the environment; the *lac* repressor is responsible for inhibiting transcription when there is no lactose.

The *lac* repressor is composed of four identical subunits (i.e., it is a tetramer), each with a helix-turn-helix DNA-binding

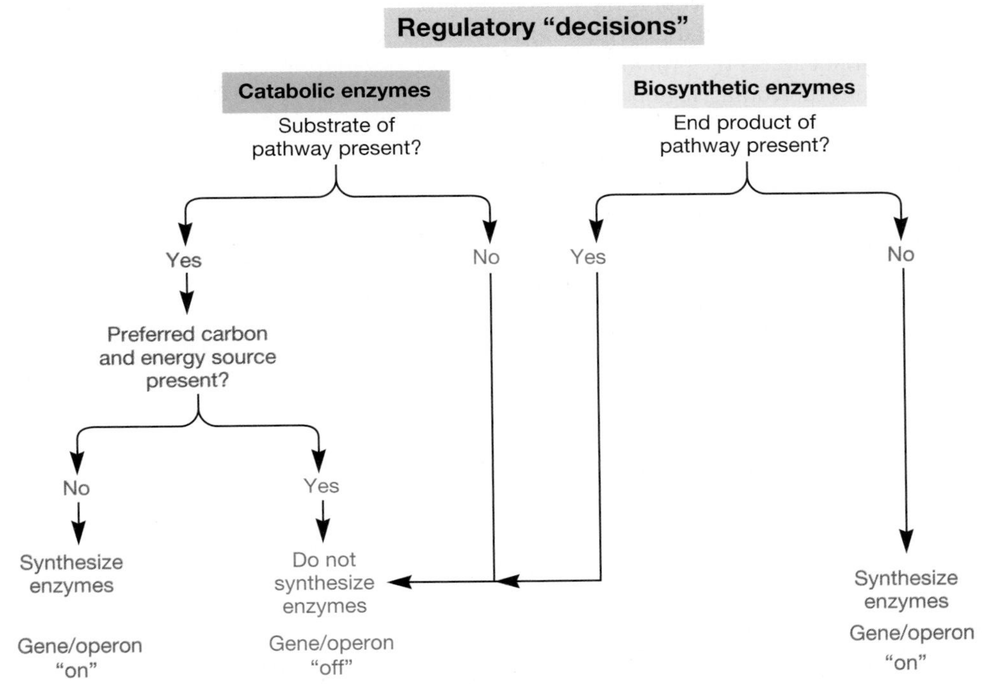

FIGURE 13.4 Examples of Regulatory "Decisions" Made by Cells.

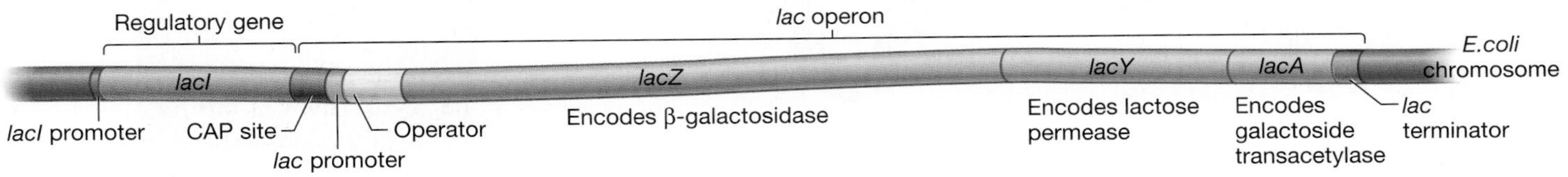

FIGURE 13.5 The *lac* Operon. The *lac* operon consists of three genes—*lacZ, lacY,* and *lacA,* which are transcribed as a single unit from the *lac* promoter. The operon is regulated both negatively and positively. Negative control is brought about by the *lac* repressor, which is the product of the *lacI* gene. The operator is the site of *lac* repressor binding. Positive control results from the action of CAP. CAP binds the CAP site located just upstream from the *lac* promoter. CAP is partly responsible for a phenomenon called catabolite repression, an example of a global control network, in which numerous operons are controlled by a single protein. For simplicity, the operator is represented as a single region. In reality, the *lac* operator consists of three distinct sites, as shown in figure 13.6.

domain. The tetramer is formed when two dimers interact. When lactose catabolism is not required, each dimer recognizes and tightly binds one of three different *lac* operator sites: O_1, O_2, and O_3 (**figure 13.6*a***). O_1 is the main operator site and must be bound by the repressor if transcription is to be inhibited. When one dimer is at O_1 and another is at one of the two other operator sites, the dimers bring the two operator sites close together, with a loop of DNA forming between them. The binding of *lac* repressor is a two-step process. First, the repressor binds nonspecifically to DNA. Then it rapidly slides along the DNA until it reaches an operator site. Two α-helices of the repressor fit into the major groove of operator-site DNA (figure 13.6*b*).

How does the repressor inhibit transcription? The promoter to which RNA polymerase binds is located near the *lac* operator sites. When there is no lactose, the repressor binds O_1 and one of the other operator sites, bending the DNA in the promoter region. This prevents initiation of transcription either because RNA polymerase cannot access the promoter or because it is blocked from moving into the coding region (**figure 13.7*a***). When lactose is available, it is taken up by lactose permease. Once inside the cell, β-galactosidase converts lactose to allolactose, the inducer of the operon (figure 13.2). This occurs because there is always a low level of permease and β-galactosidase synthesis. Allolactose noncovalently binds to the *lac* repressor and causes the repressor to change to an inactive shape that is unable to bind any operator sites. The inactivated repressor leaves the DNA and transcription occurs (figure 13.7*b*).

Close examination of figures 13.5 and 13.6 clearly shows that the regulation of the *lac* operon is not as simple as has just been described. That is because the *lac* operon is regulated by a second regulatory protein called catabolite activator protein (CAP). CAP functions in a global regulatory network that allows *E. coli* to use glucose preferentially over all other carbon and energy sources by a mechanism called catabolite repression. The use of two different regulatory proteins to control the synthesis of an operon illustrates another important point about regulatory processes—that there are often layers of regulation. In the case of the lactose operon, the lactose repressor regulates gene expression in response to the presence or absence of lactose. CAP regulates the operon in response to the presence or absence of glucose. As described in section 13.6, the use of two regulatory proteins generates a continuum of expression levels. The highest levels of transcription occur when lactose is available and glucose is not; the lowest levels occur when lactose is not available and glucose is. For almost all of the examples described in this chapter, regulation occurs by more than one mechanism. *The* lac *Operon*

Tryptophan Operon: Negative Transcriptional Control of Repressible Genes

The tryptophan (*trp*) operon of *E. coli* consists of five structural genes that encode enzymes needed for synthesis of the amino acid tryptophan (**figure 13.8**). It is regulated by the *trp* repressor,

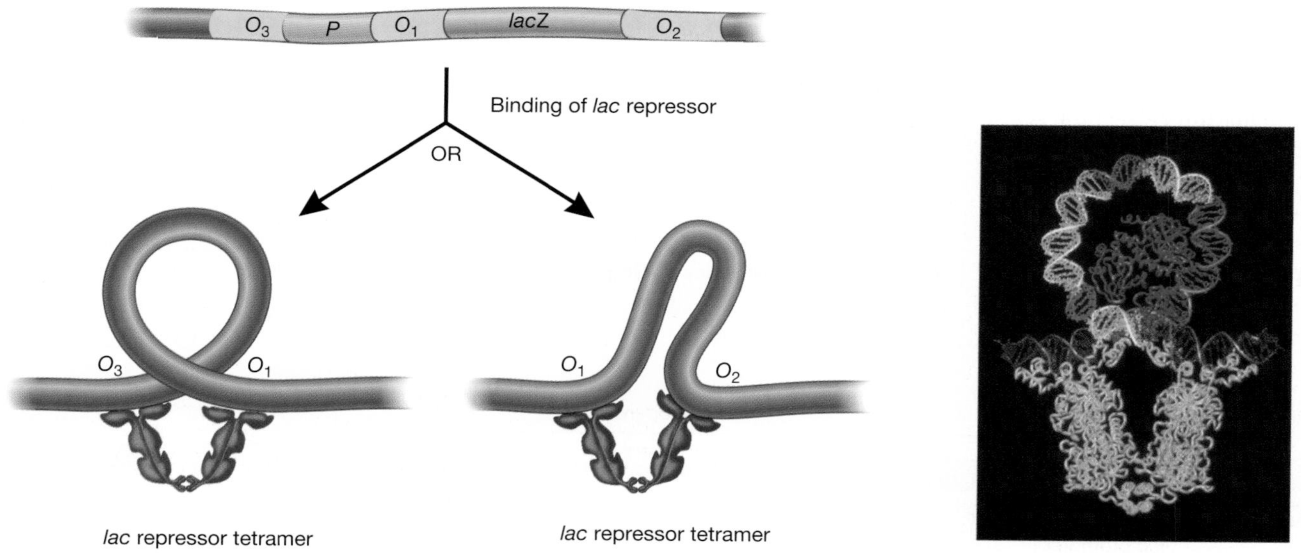

(a) Possible DNA loops caused by the binding of the *lac* repressor

(b) Proposed model of the *lac* repressor binding to O_1 and O_3 (red) based on crystallography studies

FIGURE 13.6 The *lac* Operator Sites. The *lac* operon has three operator sites: O_1, O_2, and O_3 (a). As shown in (a) and (b), the *lac* repressor (violet) binds O_1 and one of the other operator sites, forming a DNA loop. The DNA loop contains the –35 and –10 binding sites (green) recognized by RNA polymerase. Thus these sites are inaccessible and transcription is blocked. The DNA loop also contains the CAP binding site, and CAP (blue) is shown bound to the DNA (b). When the *lac* repressor is bound to the operator, CAP is unable to activate transcription.

(a) No lactose in the environment

lac repressor (active)

lac repressor binds to the operator and inhibits transcription.

(b) Lactose present

The binding of allolactose prevents the *lac* repressor from binding to the operator site.

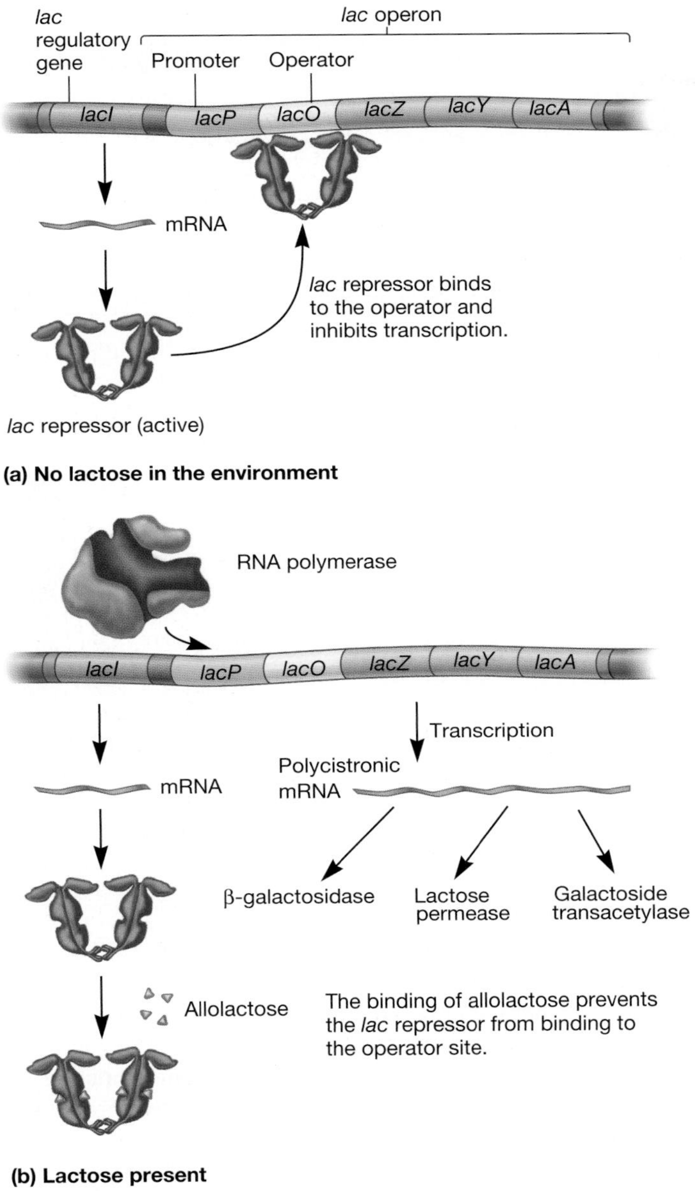

FIGURE 13.7 Regulation of the *lac* Operon by the *lac* Repressor. (a) The *lac* repressor is active and can bind the operator as long as the inducer of the operon, allolactose, is not present. Binding of the repressor to the operator inhibits transcription of the operon by RNA polymerase. (b) When lactose is available, some of it is converted to allolactose by β-galactosidase. When sufficient amounts of allolactose are present, it binds and inactivates the *lac* repressor. The repressor leaves the operator and RNA polymerase is free to initiate transcription.

Figure 13.7 Micro Inquiry

Is allolactose a corepressor or inducer molecule?

which is encoded by the *trpR* gene. Because the enzymes encoded by the *trp* operon function in a biosynthetic pathway, it is wasteful to make the enzymes needed for tryptophan synthesis when tryptophan is readily available. This is especially true because tryptophan is the most difficult amino acid to make, consuming many metabolites and considerable energy. Therefore the operon functions only when tryptophan is not present and must be made de novo from precursor molecules (figure 13.4). To accomplish

(a) Low trytophan levels, transcription of the entire *trp* operon occurs

Corepressor–repressor bind to operator and block transcription.

Tryptophan

Corepressor–repressor form active complex.

(b) High tryptophan levels, repression occurs

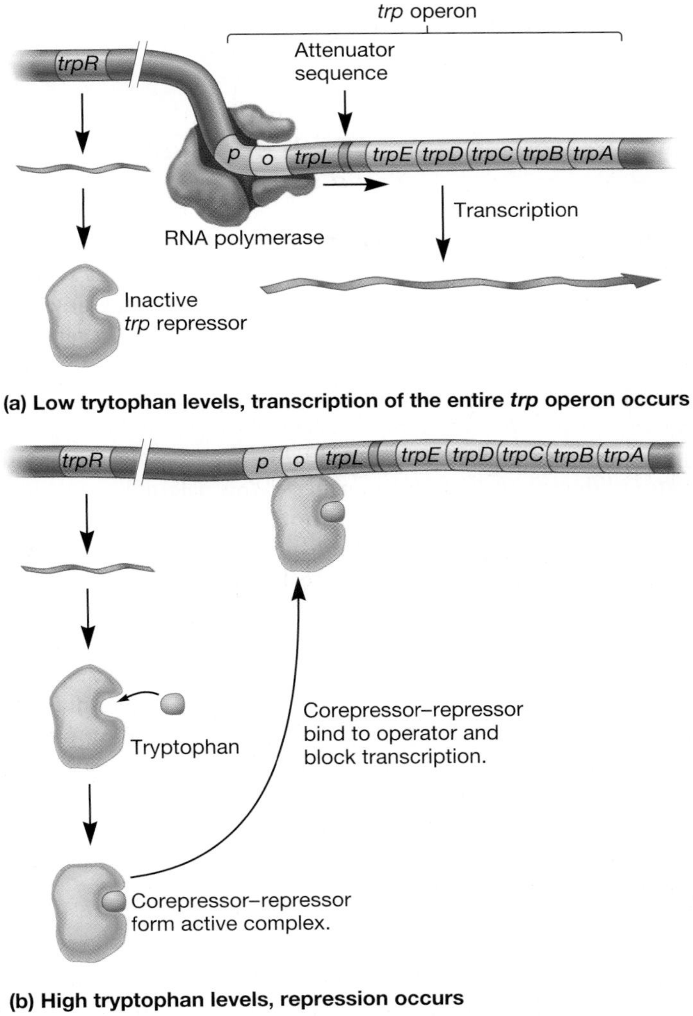

FIGURE 13.8 Regulation of the *trp* Operon by Tryptophan and the *trp* Repressor. The *trp* repressor is inactive when first synthesized and therefore is unable to bind the operator. It is activated by the binding of tryptophan, which serves as the corepressor. (a) When tryptophan levels are low, the repressor is inactive and transcription occurs. The enzymes encoded by the operon catalyze the reactions needed for tryptophan biosynthesis. (b) When tryptophan levels are sufficiently high, it binds the repressor. The repressor–corepressor complex binds the operator and transcription of the operon is inhibited.

this regulatory goal, the *trp* repressor is synthesized in an inactive form that cannot bind the *trp* operator as long as tryptophan levels are low (figure 13.8*a*). When tryptophan levels increase, tryptophan acts as a corepressor, binding the repressor and activating it. The repressor-corepressor complex then binds the operator, blocking transcription initiation (figure 13.8*b*).

Like the *lac* operon, the *trp* operon is subject to another layer of regulation. In addition to being controlled at the level of transcription initiation by the *trp* repressor, expression of the *trp* operon is also controlled at the level of transcription elongation by a process called attenuation. This mode of regulation is discussed in section 13.3. *Tryptophan Repressor*

Arabinose Operon: Transcriptional Control by a Protein That Acts Both Positively and Negatively

Many regulatory proteins are versatile and can function as repressors for one operon and activators for others. The regulation of the *E. coli* arabinose (*ara*) operon illustrates how the same protein can function either positively or negatively, depending on the environmental conditions. The *ara* operon encodes enzymes needed for the catabolism of arabinose to xylulose 5-phosphate, an intermediate of the pentose phosphate pathway. The *ara* operon is regulated by AraC, which can bind three different regulatory sequences: *araO*$_2$, *araO*$_1$, and *araI* (**figure 13.9**). When arabinose is not present, one molecule of AraC binds *araI* and another binds *araO*$_2$. The two AraC proteins interact, causing the DNA to bend. This prevents RNA polymerase from binding to the promoter of the *ara* operon, thereby blocking transcription. In these conditions, AraC acts as a repressor (figure 13.9*a*). However, when arabinose is present, it binds AraC and prevents AraC molecules from interacting. This breaks the DNA loop. Furthermore, binding of two AraC-arabinose complexes to the *araI* site promotes transcription. Thus when arabinose is present, AraC acts as an activator (figure 13.9*b*). The *ara* operon, like the *lac* operon, is also subject to catabolite repression (section 13.5). *Pentose phosphate pathway (section 10.3)*

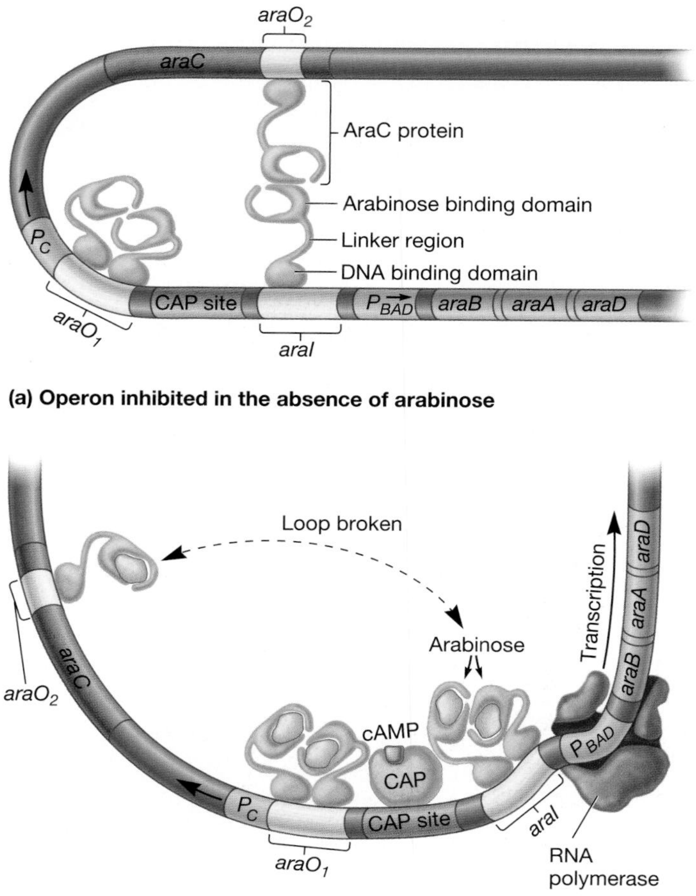

(a) Operon inhibited in the absence of arabinose

(b) Operon activated in the presence of arabinose

FIGURE 13.9 Regulation of the *ara* Operon by the AraC Protein. The AraC protein can act both as a repressor and as an activator, depending on the presence or absence of arabinose. (a) When arabinose is not available, the protein acts as a repressor. Two AraC proteins are involved. One binds the *araI* site and the other binds the *araO*$_2$ site. The two proteins interact in such a way that the DNA between the two operator sites is bent, making it inaccessible to RNA polymerase. (b) When arabinose is present, it binds AraC, disrupting the interaction between the two AraC proteins. Subsequently, two AraC proteins, each bound to arabinose, form a dimer, which binds to the *araI* site. The AraC dimer functions as an activator and transcription occurs.

Two-Component Regulatory Systems and Phosphorelay Systems

The activity levels of the *lac* repressor, *trp* repressor, and AraC protein are controlled by metabolites of those pathways. However, many environmental conditions do not produce a metabolite that can interact directly with a regulatory protein. These include temperature, osmolarity, and oxygen levels. How do organisms sense and respond to such stimuli? Many genes and operons are regulated in response to these types of signals by regulatory proteins that function in **two-component signal transduction systems.** These systems link events occurring outside the cell to gene expression inside the cell.

Two-component signal transduction systems are found in all three domains of life and are named after the two proteins that govern the regulatory pathway. The first is a **sensor kinase protein** that spans the cytoplasmic membrane so that part of it is exposed to the extracellular environment (periplasm, in gram-negative bacteria), while another part is exposed to the cytoplasm (**figure 13.10**). In this way, it can sense specific changes in the environment and communicate information to the cell's interior. The second component is the **response-regulator protein.** The

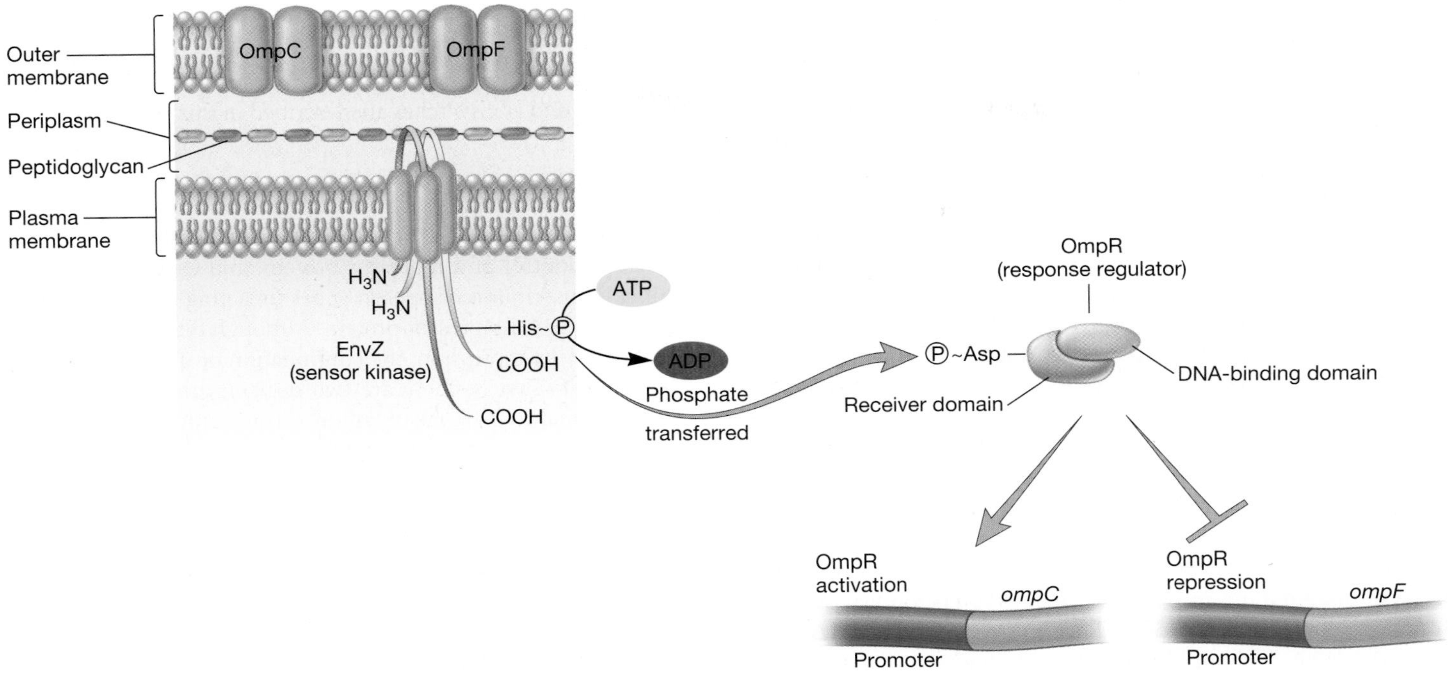

FIGURE 13.10 Two-Component Signal Transduction System and the Regulation of Porin Proteins. In this system, the sensor kinase protein EnvZ loops through the cytoplasmic membrane so that both its C- and N-termini are in the cytoplasm. When EnvZ senses an increase in osmolarity, it autophosphorylates a histidine residue at its C-terminus. EnvZ then passes the phosphoryl group to the response regulator OmpR, which accepts it on an aspartic acid residue located in its N-terminus. This activates OmpR so that it is able to bind DNA and repress *ompF* expression and enhance that of *ompC*.

Figure 13.10 Micro Inquiry

Relative to each promoter, where would you predict phosphorylated OmpR would bind the *ompC* and *ompF* genes?

response regulator is a DNA-binding protein that, when activated by the sensor kinase, can act as either an activator or a repressor. As the former, it promotes transcription of genes or operons whose expression is needed for acclimation to the detected environmental stimulus. As the latter, it inhibits transcription of genes or operons whose products are no longer needed.

The regulation of the ratio of OmpF and OmpC porin proteins in *E. coli* is one of the best-understood two-component signal transduction systems. Recall that the outer membrane of gram-negative bacteria contains channels made of porin proteins *(see figure 3.21)*. The two most important porins in *E. coli* are OmpF and OmpC (Omp for *o*uter *m*embrane *p*rotein). OmpC pores are slightly smaller and are made when the bacterium grows at high osmotic pressures. It is the dominant porin when *E. coli* is in the intestinal tract. The larger OmpF pores are favored when *E. coli* grows in a dilute environment; OmpF allows solutes to diffuse into the cell more readily. The cell must maintain a constant level of porin protein in the membrane, with the relative levels of the two porins corresponding to the osmolarity of the medium. *E. coli* uses two mechanisms: the two-component regulatory system, which we describe here; and regulation at the translational level. The latter regulatory mechanism is described in section 13.4.

The sensor kinase in the OmpF:OmpC two-component regulatory system is the homodimeric protein EnvZ (Env for cell *en*-*v*elope). It is an integral membrane protein anchored to the membrane by two membrane-spanning domains. EnvZ is looped through the membrane such that a central domain protrudes into the periplasm, while the amino and carboxyl termini are exposed to the cytoplasm (figure 13.10).

The response-regulator protein, OmpR, is a soluble, cytoplasmic protein that regulates transcription of the *ompF* and *ompC* structural genes. The N-terminal end of OmpR is called the receiver domain because it possesses a specific aspartic acid residue that accepts the signal (a phosphoryl group) from the sensor kinase. Upon receipt of the signal, the C-terminal end of OmpR is able to regulate transcription by binding DNA. When EnvZ senses high osmolarity, it phosphorylates itself (autophosphorylation) on a specific histidine residue. This phosphoryl group is quickly transferred to the N-terminus of OmpR. Once OmpR is phosphorylated, it is able to regulate transcription of the porin genes so that *ompF* transcription is repressed and *ompC* transcription is activated.

Two-component signal transduction systems are simple in design: The signal recognized by the sensor kinase is directly transduced (sent) to the response regulator that mediates the required changes in gene expression; in many cases, numerous genes and operons may be regulated by the same response regulator. Thus two-component systems often function in global control networks, as we describe in section 13.6. The effectiveness of two-component systems is illustrated by their abundance: Most bacterial and archaeal cells use a variety of two-component signal transduction systems to respond to an array of environmental stresses. For example, members of the soil bacterial genus *Streptomyces* have at least 80 such systems.

Two-component signal transduction systems involve a simple phosphorelay where the sensor kinase transfers a phosphoryl group directly to the response-regulator protein. However, there are instances when more proteins participate in the transfer of phosphoryl groups. These longer pathways are called **phosphorelay systems.** An important and well-studied phosphorelay system functions during sporulation in *Bacillus subtilis* and is described in section 13.6. It should be noted that some phosphorelay systems control protein activity, rather than gene transcription. In this case, the system functions at the posttranslational level. An example of this type of system is chemotaxis in *E. coli,* which is described in section 13.5.

1. Many genes and operons are regulated at the level of transcription initiation. What do you think are the advantages to regulating proteins before they are made, rather than during or after translation?
2. What are induction and repression? How do bacteria use them to respond to changing nutrient supplies?
3. Define repressor protein, activator protein, operator, activator-binding site, inducer, corepressor, structural gene, and operon.
4. Using figure 13.4 as a guide, trace the "decision-making" pathway of an *E. coli* cell that is growing in a medium containing arabinose but lacking tryptophan.
5. *E. coli* has two phosphate uptake systems (one with high affinity for phosphate, the other with a low affinity). Describe how a two-component regulatory system might be used by this microbe to regulate phosphate transport.

13.3 Regulation of Transcription Elongation

Organisms can also regulate transcription by controlling the termination of transcription. In this type of regulation, transcription is initiated but prematurely stopped depending on the environmental conditions and the needs of the organism. Attenuation was the first example of this kind of regulation. It was discovered in the 1970s by studies of the *trp* operon. More recently riboswitches have been discovered. These regulatory sequences in the leader of an mRNA both sense and respond to environmental conditions by either prematurely terminating transcription or blocking translation. Both attenuation and riboswitches are described in this section.

Attenuation

As noted in section 13.2, the tryptophan (*trp*) operon of *E. coli* is under the control of a repressor protein, and excess tryptophan inhibits transcription of operon genes by acting as a corepressor and activating the repressor protein. Although the operon is regulated mainly by repression, the continuation of transcription also is controlled. That is, there are two decision points involved in transcriptional control: the initiation of transcription and the continuation of transcription past the leader region. This additional level of control serves to adjust levels of transcription in a more subtle fashion, such that the two systems of control can decrease transcription levels more than either one alone. When the repressor protein is not active, RNA polymerase begins transcription of the leader region. However, it often does not progress to the first structural gene in the operon. Instead, transcription is terminated within the leader region; this is called **attenuation.**

The ability to attenuate transcription is based on the nucleotide sequences in the leader region and on the fact that transcription is coupled with translation in bacterial and archaeal cells *(see figure 12.34)*. The leader of the *trp* operon mRNA is unusual in that it is translated. The product is called the leader peptide. It has never been isolated, presumably because it is rapidly degraded. In addition to encoding the leader peptide, the leader contains **attenuator** sequences (**figure 13.11**). When transcribed, these sequences form stem-loop secondary structures in the newly formed mRNA. We define these sequences numerically (regions 1, 2, 3, and 4). When regions 1 and 2 pair with one another (1:2; figure 13.11*a*), they form a secondary structure called the pause loop, which causes RNA polymerase to slow down. The pause loop forms just prior to the formation of the terminator loop, which is made when regions 3 and 4 base pair (3:4; figure 13.11*a*). A poly-U sequence follows the 3:4 terminator loop, just as it does in rho-independent transcriptional terminators (*see figure 12.28*). However, in this case, the terminator is in the leader, rather than at the end of the gene. Another stem-loop structure can be formed in the leader region by the pairing of regions 2 and 3 (2:3, figure 13.11*b*). The formation of this antiterminator loop prevents the generation of both the 1:2 pause and 3:4 terminator loops.

How do these various loops control transcription termination? Three scenarios describe the process. In the first, translation is not coupled to transcription because protein synthesis is not occurring. In other words, no ribosome is associated with the mRNA. In this scenario, the pause and terminator loops form, stopping transcription before RNA polymerase reaches the *trpE* gene (figure 13.11*a*).

In the next two scenarios, translation and transcription are coupled; that is, a ribosome associates with the leader mRNA as the rest of the mRNA is being synthesized. The interaction between

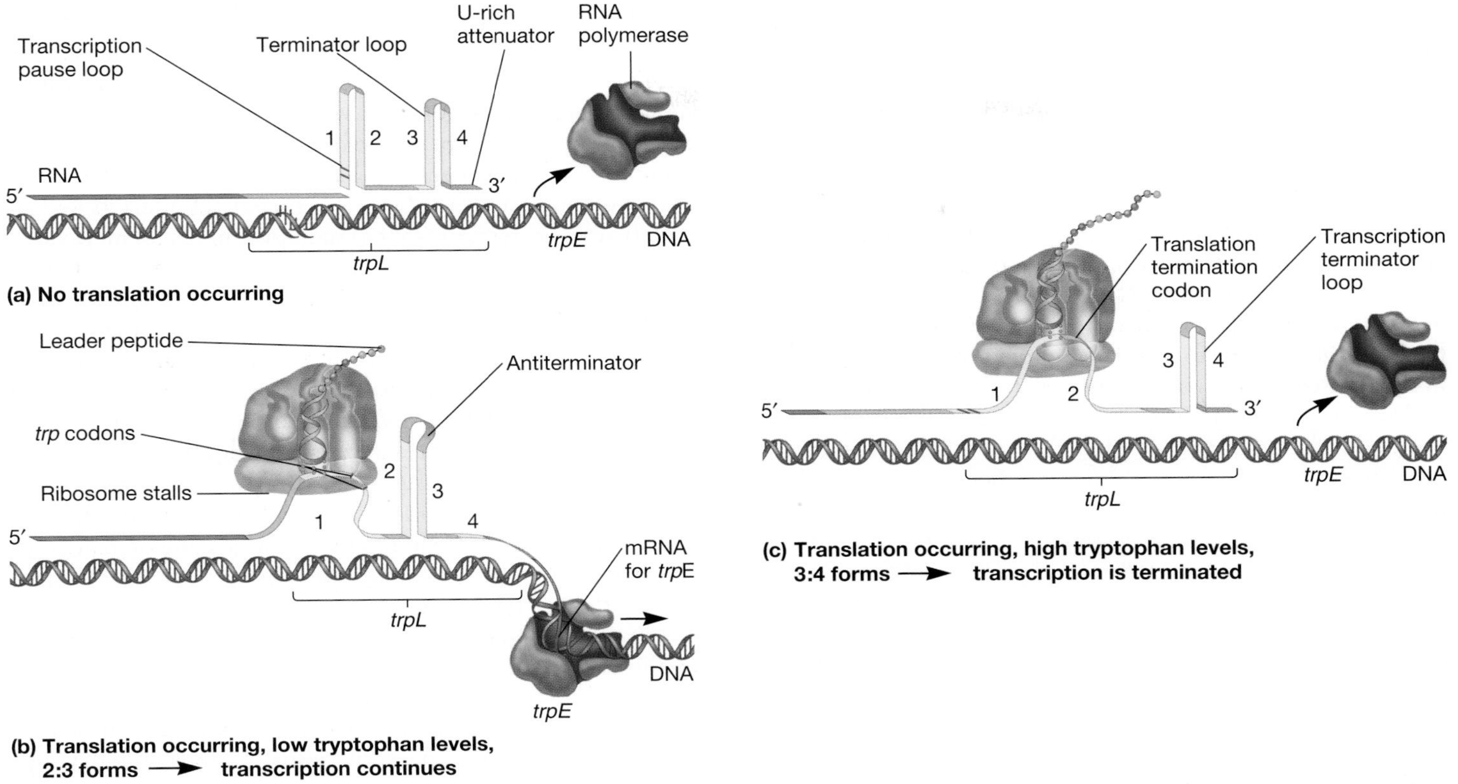

FIGURE 13.11 Attenuation of the *trp* Operon. (a) When protein synthesis has slowed, transcription and translation are not tightly coupled. Under these conditions, the most stable form of the mRNA occurs when region 1 hydrogen bonds to region 2 (RNA polymerase pause loop) and region 3 hydrogen bonds to region 4 (transcription terminator or attenuator loop). The formation of the transcription terminator causes transcription to stop just beyond *trpL* (*trp* leader). (b) When protein synthesis is occurring, transcription and translation are coupled, and the behavior of the ribosome on *trpL* influences transcription. If tryptophan levels are low, the ribosome pauses at the *trp* codons in *trpL* because of insufficient amounts of charged $tRNA^{trp}$. This blocks region 1 of the mRNA, so that region 2 can hydrogen bond only with region 3. Because region 3 is already hydrogen bonded to region 2, the 3:4 terminator loop cannot form. Transcription proceeds and the *trp* biosynthetic enzymes are made. (c) If tryptophan levels are high, translation of *trpL* progresses to the stop codon, blocking region 2. Regions 3 and 4 can hydrogen bond and transcription terminates.

Figure 13.11 Micro Inquiry

How does this attenuation respond to rates of protein synthesis in addition to the intracellular level of tryptophan?

RNA polymerase and the nearest ribosome determines which stem-loop structures are formed. As a ribosome translates the mRNA, it follows the RNA polymerase. Among the first several nucleotides of region 1 are two tryptophan (*trp*) codons; this is unusual because normally there is only one tryptophan residue per 100 amino acids in *E. coli* proteins. If tryptophan levels are low, there will not be enough charged $tRNA^{trp}$ to fill the A site of the ribosome when the ribosome encounters the two *trp* codons, and it will stall (figure 13.11*b*). Meanwhile RNA polymerase continues to transcribe mRNA, moving away from the stalled ribosome. The presence of the ribosome on region 1 prevents region 1 from base pairing with region 2. As RNA polymerase continues, region 3 is transcribed, enabling the formation of the 2:3 antiterminator loop. This prevents the formation of the 3:4 terminator loop. Because the terminator loop is not formed, RNA polymerase is not ejected from the DNA and transcription continues into the *trp* biosynthetic genes. If, on the other hand, there is plenty of tryptophan in the cell, there will be an abundance of charged $tRNA^{trp}$, and the ribosome will translate without hesitation the two *trp* codons in the leader peptide sequence. Thus the ribosome remains close to the RNA polymerase. As RNA polymerase and the ribosome continue through the leader, regions 1 and 2 are transcribed and readily form a pause loop. Then regions 3 and 4 are transcribed, the terminator loop forms, and RNA polymerase is ejected from the DNA template. Finally, the presence of a UGA stop codon between regions 1 and 2 causes early termination of translation (figure 13.11*c*). ◀◀ *The genetic code (section 12.7)*

Attenuation's usefulness is apparent. If the bacterium is deficient in an amino acid other than tryptophan, protein synthesis will slow and tryptophanyl-tRNA will accumulate. Transcription of the tryptophan operon will be inhibited by attenuation. When the bacterium begins to synthesize protein rapidly, tryptophan may be scarce and the concentration of tryptophanyl-tRNA may be low. This would reduce attenuation activity and stimulate operon transcription, resulting in larger quantities of the tryptophan biosynthetic enzymes. Acting together, repression and attenuation can coordinate the rate of synthesis of amino acid biosynthetic enzymes with the availability of amino acid end products and with the overall rate of protein synthesis. When tryptophan is present at high concentrations, any RNA polymerases not blocked by the activated repressor protein probably will not get past the attenuator sequence. Repression decreases transcription about 70-fold and attenuation slows it another eight- to ten-fold; when both mechanisms operate together, transcription can be slowed about 600-fold.

Attenuation is important in regulating at least five other operons that encode amino acid biosynthetic enzymes. In all cases, the leader peptide sequences resemble the tryptophan system in organization. For example, the leader peptide sequence of the histidine operon codes for seven histidines in a row and is followed by an attenuator that is a terminator sequence.

Riboswitches

Regulation by riboswitches (also called sensory RNAs) is a specialized form of transcription attenuation that involves mRNA folding but not ribosome behavior. In this case, if the leader of an mRNA is folded one way, transcription continues; if folded another, transcription is terminated. The leader region is called a **riboswitch** because it is analogous to a light switch turning lights on or off, depending on whether the switch is in the on position (one folding pattern of the mRNA) or the off position (another folding pattern of the mRNA). What makes riboswitches unique and exciting is that the mRNA alters its folding pattern in direct response to the binding of an effector molecule—a capability previously thought to be associated only with proteins.

The riboswitch that regulates the riboflavin (*rib*) biosynthetic operon of *B. subtilis* serves as our example (**figure 13.12**). The production of riboflavin biosynthetic enzymes is repressed by flavin mononucleotide (FMN), which is derived from riboflavin. When transcription of the *rib* operon begins, sequences in the leader region of the mRNA fold into a structure called the RFN-element. This element binds FMN and in doing so alters the folding of the leader region, creating a terminator that stops transcription.

Controlling transcription attenuation with sensory RNAs is an important method used by gram-positive bacteria to regulate amino acid-related genes. As with the *rib* operon, the leader regions of these mRNAs contain a regulatory element. In this case, the region is called the T box. T box sequences give rise to competing terminator and antiterminator loops. The development of either a terminator or an antiterminator is determined by the binding of an uncharged tRNA corresponding to the relevant amino acid. For instance, expression of a tyrosyl-tRNA synthetase gene (i.e., a gene that encodes the enzyme that links tyrosine to a tRNA molecule) is governed by the presence of $tRNA^{Tyr}$. When the level of charged $tRNA^{Tyr}$ falls, the anticodon of an uncharged tRNA binds directly to the "specifier sequence" codon in the leader of the mRNA. At the same time, the antiterminator loop is stabilized by base pairing between sequences in the loop and the acceptor end of the tRNA, which normally binds the amino acid. This prevents formation of the terminator structure, and transcription of the tyrosyl-tRNA synthetase gene continues. Genomic

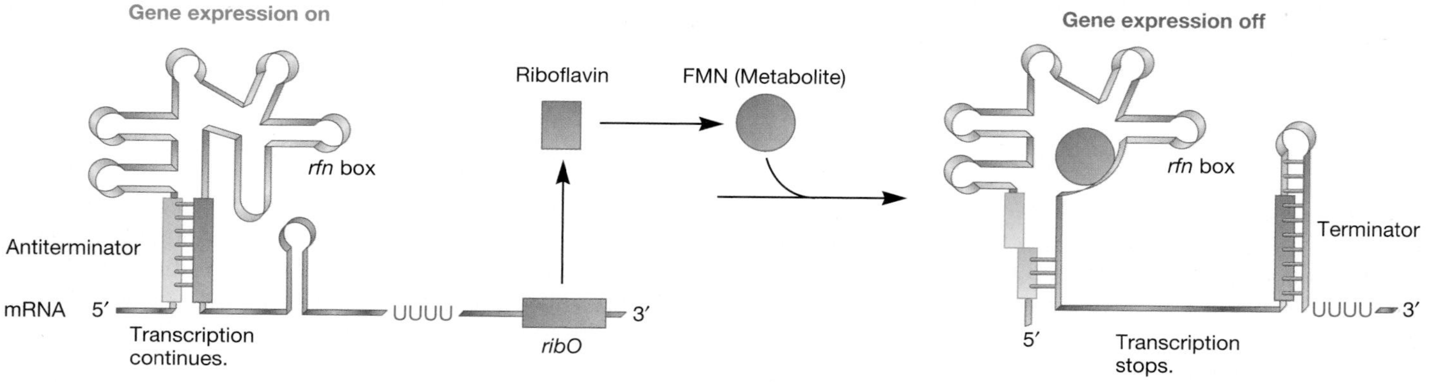

FIGURE 13.12 Riboswitch Control of the Riboflavin (*rib*) Operon of *Bacillus subtilis*. The *rib* operon produces enzymes needed for the synthesis of riboflavin, a component of flavin mononucleotide (FMN). Binding of FMN to the rfn (*ri*fampi*n*) box in the leader of *rib* mRNA causes a change in mRNA folding, which results in the formation of a transcription terminator and cessation of transcription.

Table 13.1 Regulation of Gene Expression by Riboswitches

System	*Microbe(s)*	*Target genes encode:*	*Effector and Regulatory Response*
T box	Many gram-positive bacteria	Amino acid biosynthetic enzymes	Uncharged tRNA; anticodon base pairs to 5′ end of mRNA, preventing formation of transcriptional terminator
Vitamin B_{12} element	*E. coli*	Cobalamine biosynthetic enzymes	Adenosylcobalamine (AdoCbl) binds to *btuB* mRNA and blocks translation
THI box	*Rhizobium etli* *E. coli* *B. subtilis*	Thiamine (Vitamin B_1) biosynthetic and transport proteins	Thiamine pyrophosphate (TPP) causes either premature transcriptional termination (*R. etli, B. subtilis*) or blocks ribosome binding (*E. coli*)
RFN-element	*B. subtilis*	Riboflavin biosynthetic enzymes	Flavin mononucleotide (FMN) cases premature transcriptional termination
S box	Low G + C gram-positive bacteria	Methionine biosynthetic enzymes	S-adenosylmethionine (SAM) causes premature transcriptional termination

analysis suggests that the T box mechanism may be involved in regulating over 300 genes and operons. Some other genes that bear sensory RNA in their leader regions are listed in **table 13.1**. Other riboswitches have been shown to function at the level of translation. They are described in section 13.4.

13.4 Regulation of Translation

It appears that in general, the riboswitches found in gram-positive bacteria function by transcriptional termination, whereas the riboswitches discovered in gram-negative bacteria regulate the translation of mRNA. Translation is usually regulated by blocking its initiation. In addition, some small RNA molecules can control translation initiation. Both are described in this section.

Regulation of Translation by Riboswitches

Similar to the riboswitches described in section 13.3, riboswitches that function at the translational level contain effector-binding elements at the 5′ end of the mRNA. Binding of the effector molecule alters the folding pattern of the mRNA leader, which often results in occlusion of the Shine-Dalgarno sequence and other elements of the ribosome-binding site. This inhibits ribosome binding and initiation of translation (**figure 13.13**). An example of this type of regulation is observed for the thiamine biosynthetic operons of numerous bacteria and some archaea. The leader regions of thiamine operons contain a structure called the THI-element, which can bind thiamine pyrophosphate. Binding of thiamine pyrophosphate to the THI-element causes a conformational change in the leader region that sequesters the Shine-Dalgarno sequence and blocks translation initiation.

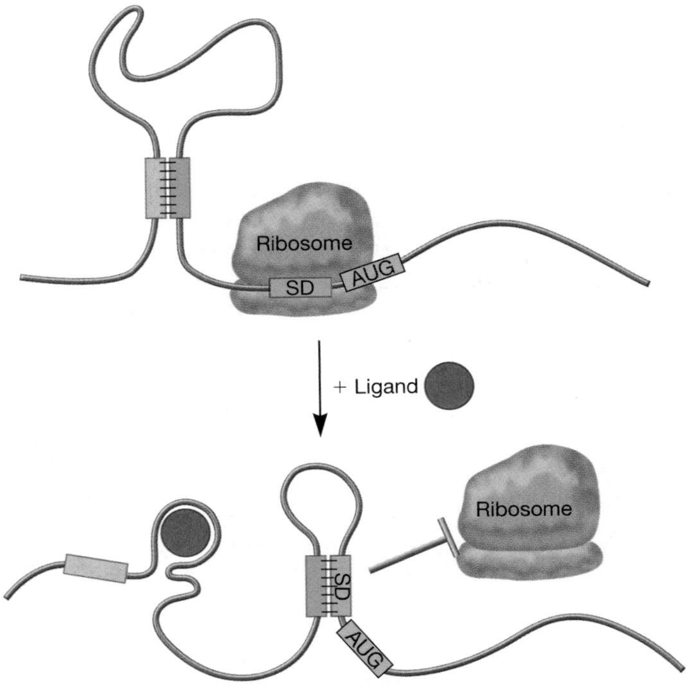

FIGURE 13.13 Regulation of Translation by a Riboswitch. In the absence of a relevant metabolite, an effector binding site is formed in the leader of the mRNA (red) when complementary sequences (orange box and green box) hydrogen bond. This folding pattern exposes important sequences in the ribosome-binding site (e.g., the Shine-Dalgarno sequence; blue box) and translation occurs. When the appropriate effector molecule is present, it binds the leader, disrupting the existing structure and creating a new structure with the ribosome-binding-site sequences. Thus the ribosome-binding site becomes inaccessible and translation is blocked.

Regulation of Translation by Small RNA Molecules

A large number of RNA molecules have been discovered that do not function as mRNAs, tRNAs, or rRNAs. Microbiologists often refer to them as **small RNAs (sRNAs)** or as noncoding RNAs (ncRNAs). In *E. coli,* there are more than 40 sRNAs, ranging in size from around 40 to 400 nucleotides. It is thought that eukaryotes may have hundreds to thousands of sRNAs with lengths from 21 to over 10,000 nucleotides. Although some sRNAs have been implicated in the regulation of DNA replication and transcription, many function at the level of translation.

In *E. coli,* most sRNAs regulate translation by base pairing to the leader region of a target mRNA. Thus they are complementary to the mRNA and are called **antisense RNAs.** It seems intuitive that by binding to the leader, antisense RNAs would block ribosome binding and inhibit translation. Indeed, many antisense RNAs work in this manner. However, some antisense RNAs actually promote translation upon binding to the mRNA. Whether inhibitory or activating, most *E. coli* antisense RNAs work with a protein called Hfq to regulate their target RNAs. The Hfq protein is an RNA chaperone—that is, it interacts with RNA to promote changes in its structure. In addition, the Hfq protein may promote RNA-RNA interactions.

The regulation of synthesis of OmpF and OmpC porin proteins provides an example of translation control by an antisense RNA. In addition to regulation by the OmpR protein described in section 13.2 (figure 13.10), expression of the *ompF* gene is regulated by an antisense RNA called MicF RNA, the product of the *micF* gene (*mic* for *m*RNA-*i*nterfering *c*omplementary RNA). The MicF RNA is complementary to *ompF* at the translation initiation site (**figure 13.14**). It base pairs with *ompF* mRNA and represses translation. MicF RNA is produced under conditions such as high osmotic pressure or the presence of some toxic material, both of which favor *ompC* expression. Production of MicF RNA helps ensure that OmpF protein is not produced at high levels at the same time as OmpC protein. Some other antisense RNAs are listed in **table 13.2**.

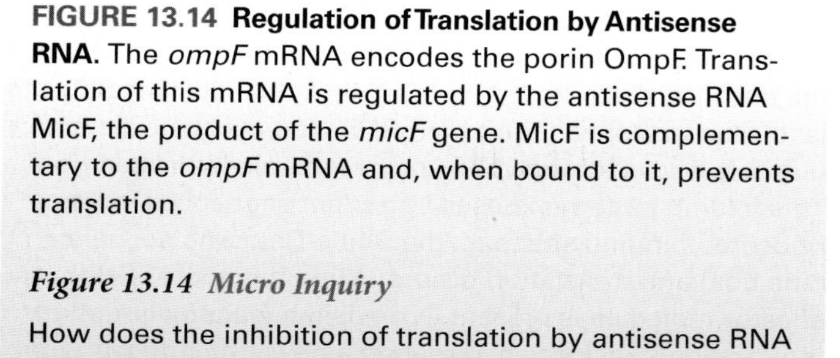

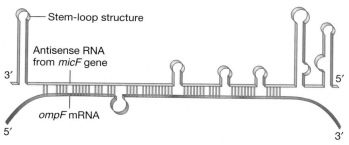

FIGURE 13.14 Regulation of Translation by Antisense RNA. The *ompF* mRNA encodes the porin OmpF. Translation of this mRNA is regulated by the antisense RNA MicF, the product of the *micF* gene. MicF is complementary to the *ompF* mRNA and, when bound to it, prevents translation.

Figure 13.14 Micro Inquiry

How does the inhibition of translation by antisense RNA differ from translational inhibition by riboswitches?

1. Define attenuation. What are the functions of the leader region and ribosome in attenuation?
2. Describe how attenuation activity would vary when *E. coli* is placed in a tryptophan-rich medium, where it rapidly consumes the tryptophan within 8 hours of growth.
3. What are riboswitches? How are they similar to attenuation as described for the *trp* operon? How do they differ? How do riboswitches differ from sRNAs?
4. How do you think *micF* transcription is regulated?

13.5 Posttranslational Regulation

In chapter 9, we introduce the control of metabolic pathways brought about by modulating the activity of certain regulatory enzymes that function in the pathway. This was an example of posttranslational regulation. **Posttranslational regulation** can occur by allosteric control or covalent modification. Covalent modification can be either irreversible (proteolysis) or reversible (e.g., methylation/demethylation and phosphorylation/dephosphorylation). Both kinds of covalent modification are involved in the chemosensory system we describe here.

Recall from chapter 3 that microorganisms are able to sense chemicals in their environment and move either toward them or away from them, depending on whether the chemical is an attractant or a repellent. For simplicity, we only concern ourselves with movement toward an attractant. The best-studied chemotactic system is that of *E. coli,* which, like many other bacteria, exhibits two movement modalities: a smooth swimming motion called a run, interrupted by tumbles. A run occurs when the flagellum rotates in a counterclockwise direction (CCW), and a tumble occurs when the flagellum rotates clockwise (CW) *(see figures 3.45 and 3.50).* The cell alternates between these two types of movements, with the tumble establishing the direction of movement in the run that follows.

When *E. coli* is in an environment that is homogenous—that is, the concentration of all chemicals in the environment is the same throughout its habitat—the cell moves about randomly, with no apparent direction or purpose; this is called a random walk. However, if a chemical gradient exists in its environment, the frequency of tumbles decreases as long as the cell is moving toward the attractant. In other words, the length of time spent moving toward the attractant is increased and eventually the cell gets closer to the attractant. The process is not perfect, and the cell must continually readjust its direction through a trial-and-error process that is mediated by tumbling. When one examines

Table 13.2 Regulation of Gene Expression by Small Regulatory RNAs

Small RNA	*Size*	*Bacterium*	*Function*
RhyB	90 nt[1]	*E. coli*	Represses translation of iron-containing proteins (e.g., *sodB*) when iron availability is low
Spot 42	109 nt	*E. coli*	Inhibits translation of *galK* mRNA (encodes galactokinase)
Rpr A	105 nt	*E. coli*	Promotes translation of *rpoS* mRNA (encodes σ^s, a stationary phase sigma factor); antisense repressor of global negative regulator H-NS (involved in stress responses)
MicF	109 nt	*E. coli*	Inhibits *ompF* mRNA translation
OxyS	109 nt	*E. coli*	Inhibits translation of transcriptional regulator *fhlA* mRNA and *rpoS* mRNA
DsrA	85 nt	*E. coli*	Increases translation of *rpoS* mRNA
CsrB	366 nt	*E. coli*	Inhibits CsrA, a translational regulatory protein that positively regulates flagella synthesis, acetate metabolism, and glycolysis
RNAIII	512 nt	*Staphylococcus aureus*	Activates genes encoding secreted proteins (e.g., α hemolysin); represses genes encoding surface proteins
RNA α	650 nt	*Vibrio anguillarum*	Decreased expression of fat, an iron-uptake protein
RsmB′	259 nt	*Erwinia carotovora subsp. carotovora*	Stabilizes mRNA of virulence proteins (e.g., cellulases, proteases, pectinolytic enzymes)

[1] nt: nucleotides

the path taken by the cell, it is similar to a random walk but is biased toward the attractant.

For over three decades, scientists have been dissecting this complex behavior in order to understand how *E. coli* senses the presence of an attractant, how it switches from a run to a tumble and back again, and how it "knows" it is heading in the correct direction. These studies reveal that the chemotactic response of *E. coli* involves a number of enzymes and other proteins that are regulated by covalent modification. One important component is a phosphorelay system. Recall from section 13.2 that phosphorelay systems are more elaborate versions of two-component regulatory systems. In section 13.2, we describe how such a system can be used to regulate transcription initiation. As we note there, some two-component regulatory systems and phosphorelay systems are used to regulate enzyme activity. This is the case for chemotaxis.

For chemotaxis to occur, *E. coli* must determine if an attractant is present and then modulate the activity of the phosphorelay system that dictates the rotational direction of the flagellum (i.e., either run or tumble). *E. coli* senses chemicals in its environment when they bind to chemoreceptors (**figure 13.15**). Numerous chemoreceptors have been identified. Here, we focus on one class of receptors called methyl-accepting chemotaxis proteins (MCPs). The phosphorelay system that controls direction of flagellar rotation consists of the sensor kinase CheA and the response regulator CheY. When activated, CheA phosphorylates itself using ATP (figure 13.15*c*). The phosphoryl group is then quickly transferred to CheY. Phosphorylated CheY diffuses through the cytoplasm to the flagellar motor. Upon interacting with the motor, the direction of rotation is switched from CCW to CW, and a tumble ensues. When CheA is inactive, the flagellum rotates in its default mode (CCW), and the cell moves forward in a smooth run.

As implied by the preceding discussion, the state of the MCPs must be communicated to the CheA/CheY phosphorelay system. How is this accomplished? The MCPs are embedded in the plasma membrane with different parts exposed on each side of the membrane (figure 13.15*c*). The periplasmic side of each MCP has a binding site for one or more attractant molecules. The cytoplasmic side of an MCP interacts with two proteins, CheW and CheA. The CheW protein binds to the MCP and helps attach the CheA protein. Together with CheW and CheA, the MCP receptors form large receptor clusters at one or both poles of the cell (figure 13.15*b*). It is thought that smaller aggregations of the MCPs, CheA, and CheW function as signaling teams and are the building blocks of the receptor clusters. The number of each of these molecules in the signaling team is not clear, but it has been suggested that each team includes three receptors, often of different types, two CheW molecules, and one CheA dimer (**figure 13.16**). The signaling teams become interconnected by an unknown mechanism to form the receptor clusters visible at the poles of the cell.

No matter what the precise stoichiometry or architecture of the receptor clusters, evidence exists that the MCPs in each signaling team work cooperatively to modulate CheA activity. When any one of the MCPs in the signaling team is bound to an attractant, CheA autophosphorylation is inhibited, the flagellum continues rotating CCW, and the cell continues in its run. Because of this cooperation, the cell can respond to very low concentrations of attractant. Furthermore, it can integrate signals from all receptors in the team (figure 13.16). On the other hand, if attractant levels decrease, so that the level of attractant bound to the MCPs in a signaling team decreases, CheA is stimulated to autophosphorylate, the phosphorelay is set into motion, and the cell begins to tumble. However, tumbling does not continue indefinitely. About 10 seconds after the switch to CW rotation occurs, the phosphoryl group is removed from CheY by the CheZ protein, and CCW rotation is resumed.

But how does *E. coli* measure the concentration of attractant in its environment, and how does it know when it is moving toward the attractant? *E. coli* measures the concentration of an attractant every few seconds and determines if the concentration is increasing or decreasing over time. As long as the concentration increases, the cell continues a run. If the concentration decreases, a tumble is triggered. To compare concentrations of the attractant over time, *E. coli* must have a mechanism for "remembering" the previous concentration. *E. coli* accomplishes this by using another type of covalent modification: methylation. *E. coli* compares the overall methylation level of the MCPs (on the cytoplasmic side) with the overall amount of attractant bound (on the periplasmic face). The cytoplasmic portion of each MCP has four to six glutamic acid residues that can be

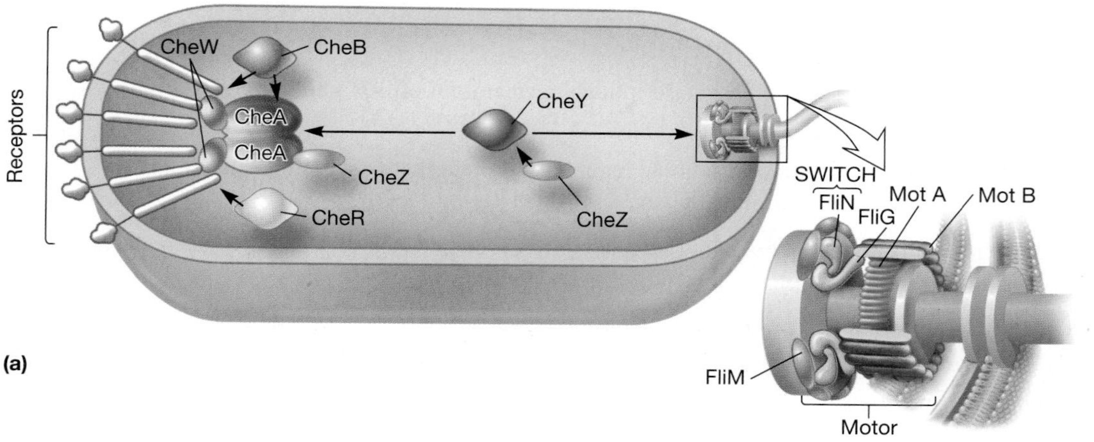

(a)

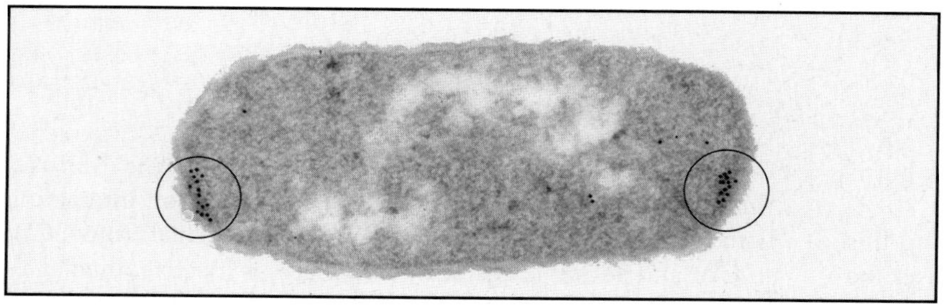

(b)

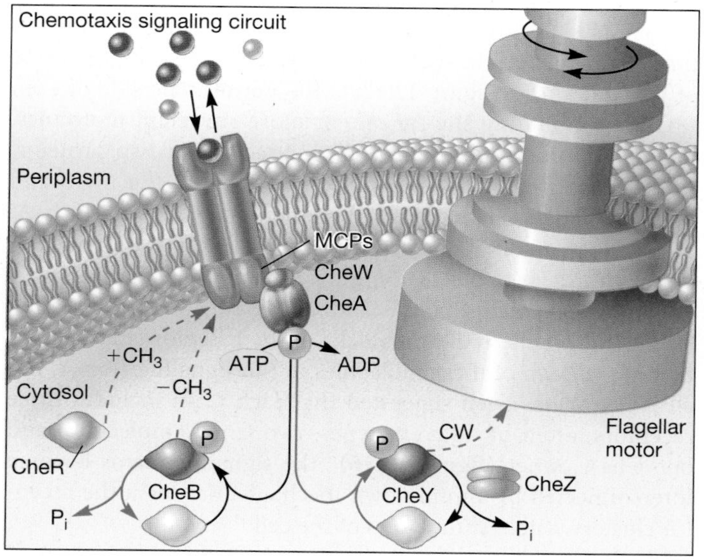

(c)

FIGURE 13.15 Proteins and Signaling Pathways of the Chemotaxis Response in *E. coli*. (a) The methyl-accepting chemotaxis proteins (MCPs) form clusters associated with the CheA and CheW proteins. CheA is a sensor kinase that when activated phosphorylates CheB, a methylesterase, or CheY. Phosphorylated CheY interacts with the FliM protein of the flagellar motor, causing rotation of the flagellum to switch from counterclockwise (CCW) to clockwise (CW). This results in a switch from a run (CCW rotation) to a tumble (CW rotation). (b) MCPs, CheW, and CheA complexes form large clusters of receptors at either end of the cell, as shown in this electron micrograph of *E. coli*. Gold-tagged antibodies were used to label the receptor clusters, which appear as black dots (encircled). (c) The chemotactic signaling pathways of *E. coli*. The pathways that increase the probability of CCW rotation are shown in red. CCW rotation is the default rotation. It is periodically interrupted by CW rotation, which causes tumbling. The pathways that lead to CW rotation are shown in green. Molecules shown in gray are unphosphorylated and inactive. Note that MCP, CheA, and CheZ are homodimers. CheW, CheB, CheY, and CheR are monomers.

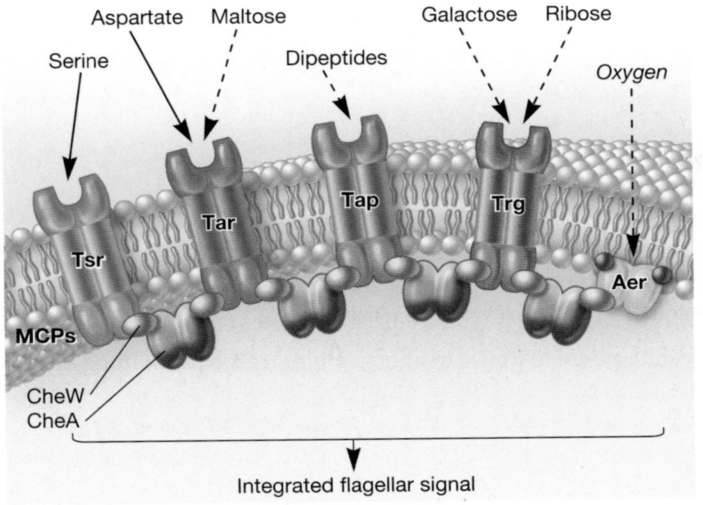

FIGURE 13.16 The Methyl-Accepting Chemotaxis Proteins of *E. coli*. The attractants sensed by each methyl-accepting chemotaxis protein (MCP) are shown. Some are sensed directly, when the attractant binds the MCP (solid lines). Others are sensed indirectly (dashed lines). The attractants maltose, dipeptides, galactose, and ribose are detected by their interaction with periplasmic binding proteins. Oxygen is detected indirectly by the Aer chemoreceptor, which differs from other MCPs in that it lacks a periplasmic sensing domain. Instead, the cytoplasmic domain has a binding site for FAD. FAD is an important electron carrier found in many electron transport systems. The redox state of the MCP-bound FAD molecule is used to monitor the functioning of the electron transport system. This in turn mediates a tactic response to oxygen.

Figure 13.16 Micro Inquiry

Why doesn't the Aer receptor need a periplasmic domain?

methylated. As long as the concentration of the attractant keeps increasing, the number of MCPs bound to attractant remains high, and the MCP methylation level remains high. However, if the attractant concentration decreases, the level of methylation will exceed the level of attractant bound. This disparity in methylation level and MCP-bound attractant stimulates CheA to autophosphorylate. As a result, the phosphorelay signal for CW flagellar rotation is initiated, and the cell tumbles in an attempt to reorient itself in the gradient so that it is moving up the gradient (toward the attractant), rather than down the gradient (away from the attractant). At the same time, some of the methyl groups are removed from the MCPs by the methylesterase CheB, establishing a methylation level that is commensurate with the number of MCPs bound to the attractant. A few seconds later, the number of MCPs bound to attractant will be compared to this new methylation level. Based on the correspondence of the two, the cell will determine if it is again moving up the gradient. If it is, tumbling will be suppressed and the run will continue.

1. What is a phosphorelay system?
2. Describe the MCP-CheW-CheA receptor complex. What two proteins are phosphorylated by CheA? What is the role of each?
3. How does the MCP regulate the rate of CheA autophosphorylation? How does this mediate chemotaxis?

13.6 Global Regulatory Systems

Thus far, we have considered the function of isolated operons. However, organisms must respond rapidly to a wide variety of changing environmental conditions and cope with such stressors as nutrient deprivation, desiccation, and major temperature fluctuations. They also have to compete successfully with other organisms for scarce nutrients and use these nutrients efficiently. These challenges require regulatory systems that can rapidly control many operons and proteins at the same time. Regulatory systems that affect many genes and pathways simultaneously are called **global regulatory systems.**

Although it is usually possible to regulate all the genes of a metabolic pathway in a single operon, there are good reasons for more complex global systems. Some processes involve too many genes to be accommodated in a single operon. For example, the machinery required for protein synthesis is composed of 150 or more gene products, and coordination requires a regulatory network that controls many separate operons. Sometimes two levels of regulation are required because individual operons must be controlled independently but also cooperate with other operons. Regulation of sugar catabolism in *E. coli* is a good example. *E. coli* uses glucose when it is available; in such a case, operons for other catabolic pathways are repressed. If glucose is unavailable and another nutrient is present, only the appropriate operon is activated.

Global regulatory systems are so complex that a specialized nomenclature is used to describe the various kinds. Perhaps the most basic type is the **regulon.** A regulon is a collection of genes or operons that is controlled by a common regulatory protein. Usually the operons are associated with a single pathway or function (e.g., the production of heat-shock proteins or the catabolism of glycerol). A somewhat more complex situation is seen with a **modulon.** This is an operon network under the control of a common global regulatory protein but whose constituent operons also are controlled separately by their own regulators. A good example of a modulon is catabolite repression, which is discussed on p. 350. The most complex global systems are referred to as stimulons. A stimulon is a regulatory system in which all operons respond together in a coordinated

Table 13.3 *E. coli* Sigma Factors

Sigma Factor	*Genes Transcribed*
σ^{70}	Genes needed during exponential growth
σ^{S}	Genes needed during the general stress response and during stationary phase
σ^{E}	Genes needed to restore membrane integrity and the proper folding of membrane proteins
σ^{H} (σ^{32})	Genes needed to protect against heat shock and other stresses, including genes encoding chaperones that help maintain or restore proper folding of cytoplasmic proteins and proteases that degrade damaged proteins
FecI σ	Genes that encode the iron citrate transport machinery in response to iron starvation and the availability of iron citrate
σ^{F} (σ^{28})	Genes involved in flagellum assembly
σ^{60}	Genes involved in nitrogen metabolism

way to an environmental stimulus. It may contain several regulons and modulons, and some of these may not share regulatory proteins. For instance, the genes involved in a response to phosphate limitation are scattered among several regulons and are part of one stimulon.

Mechanisms Used for Global Regulation

Global regulation is complex and often involves more than one regulatory mechanism. Most global regulatory networks are controlled by one or more regulatory proteins. Two-component regulatory systems and phosphorelay systems also play important roles in global control. In *Bacteria,* many global regulatory networks make use of **alternate sigma factors,** which can immediately change expression of many genes as they direct RNA polymerase to specific subsets of a bacterium's genome. This is possible because RNA polymerase core enzyme needs the assistance of a sigma factor to bind a promoter and initiate transcription. Each sigma factor recognizes promoters that differ in sequence, especially at the −10 and −35 positions. The specific sequences recognized by a given sigma factor are called its consensus sequences. When a complex process requires a radical change in transcription or a precisely timed sequence of transcription, it may be regulated by a series of sigma factors. ◀◀ *Transcription (section 12.6)*

E. coli synthesizes several sigma factors (**table 13.3**). Under normal conditions, a sigma factor called σ^{70} directs RNA polymerase activity. (The superscript number or letter indicates the size or function of the sigma factor; 70 stands for 70,000 Da.) When flagella and chemotactic proteins are needed, *E. coli* produces σ^{F} (σ^{28}). σ^{F} then binds its consensus sequences in promoters of genes whose products are needed for flagella biosynthesis and chemotaxis. If the temperature rises too high, σ^{H} (σ^{32}) is produced and stimulates the formation of about 17 heat-shock proteins that protect the cell from thermal destruction. Importantly, each sigma factor has its own set of promoters to which it binds.

In the discussions that follow, we describe four global regulatory networks. The first is the catabolite repression modulon, which involves regulation of transcription by both repressors and activators. As you will see, CAP, the global regulatory protein that controls catabolite repression, is in turn regulated by the small effector molecule cyclic-AMP (cAMP). Cyclic nucleotides and other unusual nucleotides are involved in regulating other cellular processes, and these are described next. A third global regulatory network that is garnering a great deal of attention is quorum sensing, which was introduced in chapter 7. Finally, we examine sporulation in the gram-positive bacterium *B. subtilis.* Regulation of endospore formation involves numerous control mechanisms, including phosphorelay and sequential use of alternative sigma factors.

Catabolite Repression

If *E. coli* grows in a medium containing both glucose and lactose, it uses glucose preferentially until the sugar is exhausted. Then after a short lag, growth resumes at a slower rate with lactose as the carbon source (**figure 13.17**). This allows glucose, the more easily catabolized and more energy-yielding molecule, to be used first, followed by other energy sources that may be more difficult to degrade or yield less energy. This biphasic growth pattern is called **diauxic growth.** The cause of diauxic growth is complex and not completely understood, but **catabolite repression** plays a part. The enzymes for glucose catabolism are constitutive. However, operons that encode enzymes required for the catabolism of carbon sources that must first be modified before entering glycolysis (e.g., the *lac* operon) are regulated by catabolite repression. These include the *ara, mal* (maltose), and *gal* (galactose) operons, as well as the *lac* operon. Collectively, these can be called catabolite operons, and their expression is coordinately (or globally) repressed when glucose is plentiful.

The coordinated regulation of catabolite operons is brought about by **catabolite *a*ctivator *p*rotein (CAP),** which is also

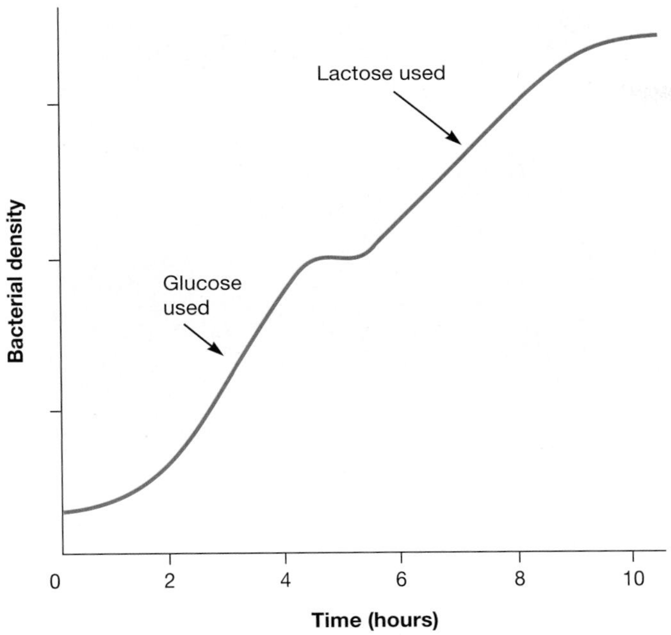

FIGURE 13.17 Diauxic Growth. The diauxic growth curve of *E. coli* grown with a mixture of glucose and lactose. Glucose is first used, then lactose. A short lag in growth is present while the bacteria synthesize the enzymes needed for lactose use.

Figure 13.17 Micro Inquiry

For what other compounds would you also see this kind of diauxic growth?

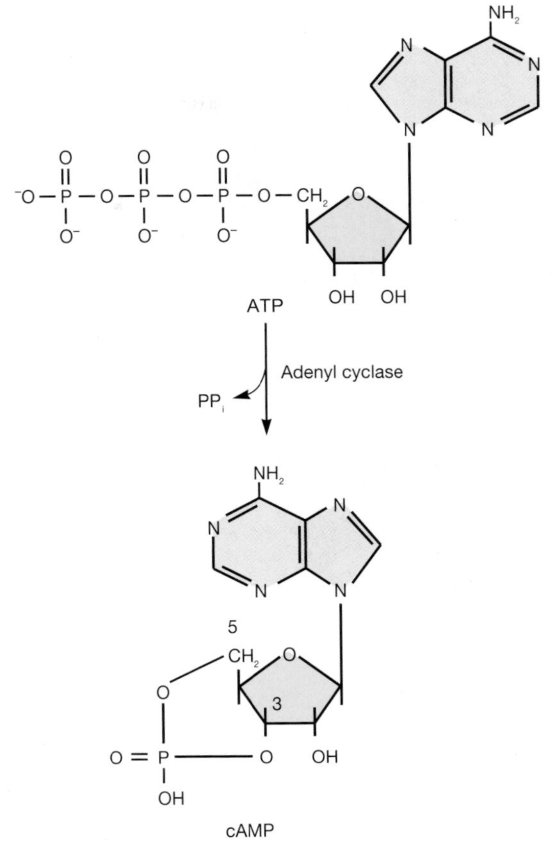

FIGURE 13.18 Cyclic Adenosine Monophosphate (cAMP). The phosphate extends between the 3′ and 5′ hydroxyls of the ribose sugar. The enzyme adenyl cyclase forms cAMP from ATP.

called *c*yclic AMP *r*eceptor *p*rotein (CRP). CAP exists in two states: It is active when the small cyclic nucleotide **3′, 5′-cyclic adenosine monophosphate (cAMP; figure 13.18)** is bound, and it is inactive when it is free of cAMP. The levels of cAMP are controlled by the enzyme adenyl cyclase, which converts ATP to cAMP and PP_i. Adenyl cyclase is active only when little or no glucose is available. Thus the level of cAMP varies inversely with that of glucose: When glucose is unavailable and the catabolism of another sugar might be needed, the amount of cAMP in the cell increases, allowing cAMP to bind to and activate CAP.

All catabolite operons contain a CAP binding site, and CAP must be bound to this site before RNA polymerase can bind the promoter and begin transcription. Upon binding, CAP bends the DNA within two helical turns, which stimulates transcription (figure 13.6*b* and **figure 13.19**). Thus all catabolite operons are controlled by two regulatory proteins: the regulatory protein specific to each operon (e.g., *lac* repressor and AraC protein) and CAP. In the case of the *lac* operon, if glucose is absent and lactose is present, the inducer allolactose will bind to and inactivate the *lac* repressor protein, CAP will be in the active form (with cAMP bound), and transcription will proceed (**figure 13.20*a***). However, if glucose and lactose are both in short supply, even though CAP binds to the *lac* promoter, transcription will be inhibited by the presence of the repressor protein, which remains bound to the operator in the absence of inducer (figure 13.20*c*). Dual control ensures that the *lac* operon is expressed only when lactose catabolic genes are needed.

We have seen how CAP controls catabolite operons; now let us turn our attention to the regulation of the levels of cAMP. The decrease in cAMP levels that occurs when glucose is present is due to the effect of the phosphoenolpyruvate: phosphotransferase system (PTS) on the activity of adenyl cyclase (**figure 13.21**). Recall from chapter 6 that in the PTS, a phosphoryl group is transferred by a series of proteins from phosphoenolpyruvate (PEP) to glucose, which then enters the cell as glucose 6-phosphate *(see figure 6.7)*. When glucose is present, enzyme IIA transfers the phosphoryl group to enzyme IIB, which phosphorylates glucose. However, when glucose is absent, the phosphoryl groups from PEP are transferred to

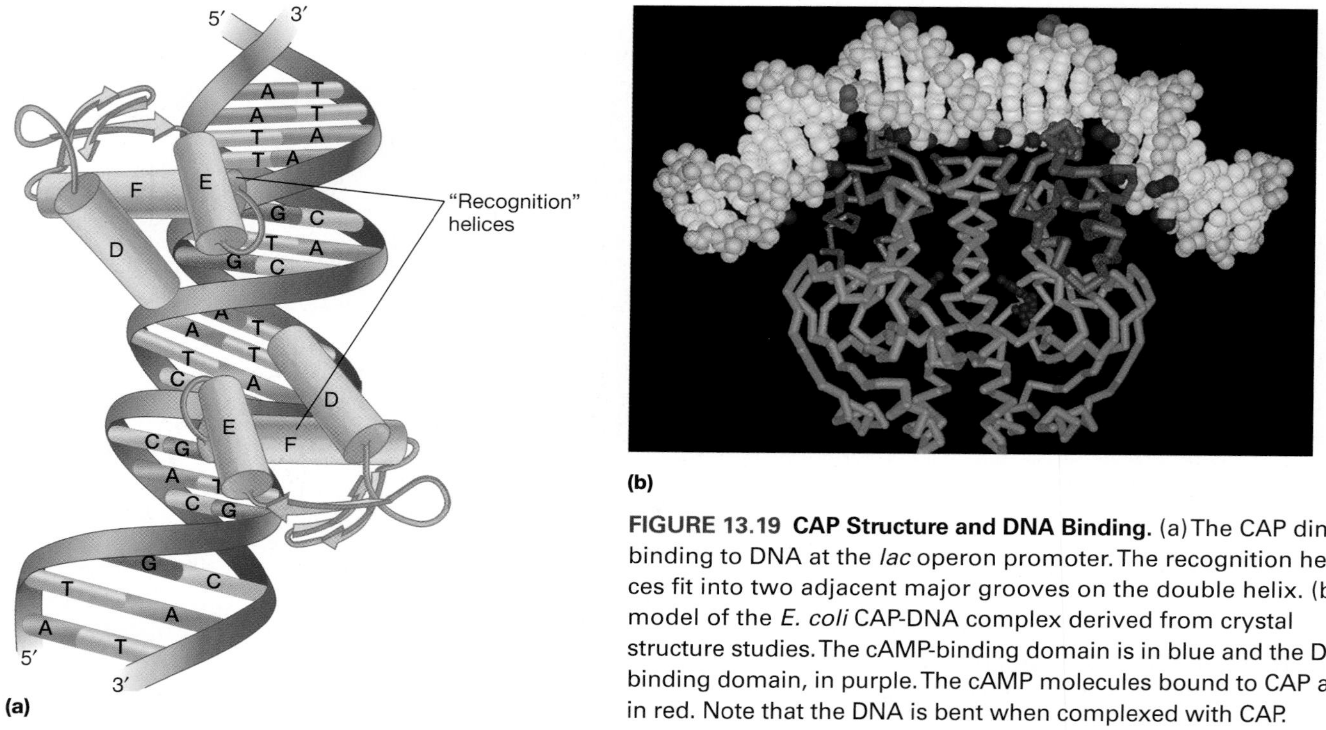

FIGURE 13.19 CAP Structure and DNA Binding. (a) The CAP dimer binding to DNA at the *lac* operon promoter. The recognition helices fit into two adjacent major grooves on the double helix. (b) A model of the *E. coli* CAP-DNA complex derived from crystal structure studies. The cAMP-binding domain is in blue and the DNA-binding domain, in purple. The cAMP molecules bound to CAP are in red. Note that the DNA is bent when complexed with CAP.

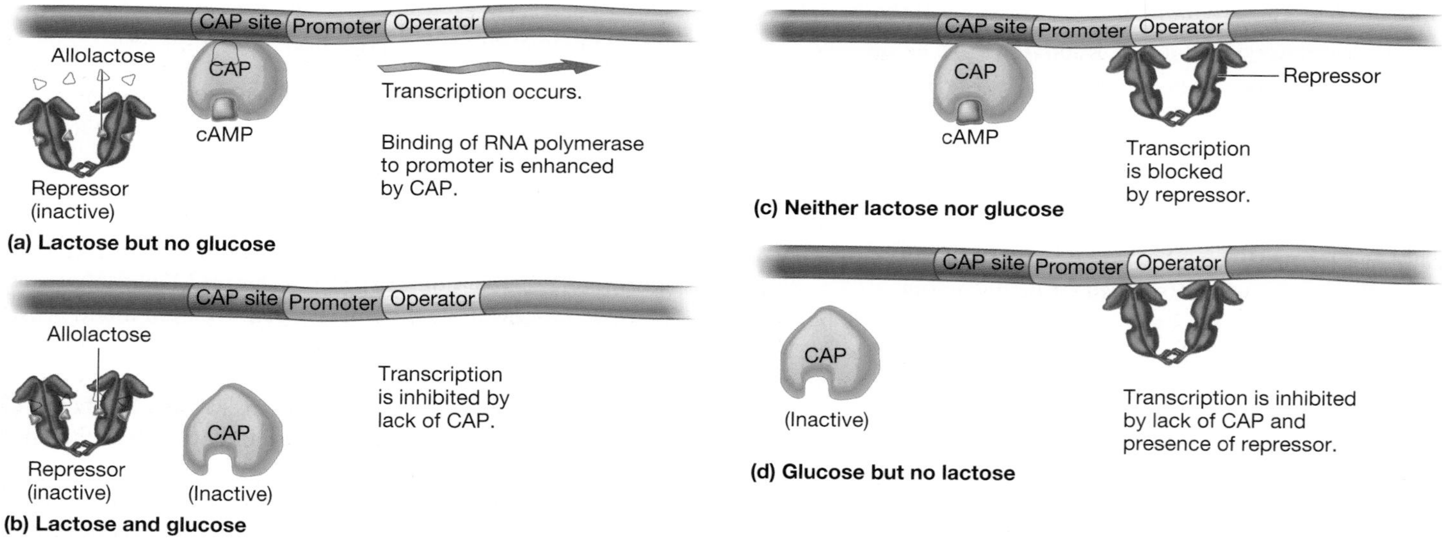

FIGURE 13.20 Regulation of the *lac* Operon by the *lac* Repressor and CAP. A continuum of *lac* mRNA synthesis is brought about by the action of CAP, an activator protein, and the *lac* repressor. (a) When lactose is available and glucose is not, the repressor is inactivated and cAMP levels increase. Cyclic AMP binds CAP, activating it. CAP binds the CAP binding site near the *lac* promoter and facilitates binding of RNA polymerase. Under these conditions, transcription occurs at maximal levels. (b) When both lactose and glucose are available, both CAP and the *lac* repressor are inactive. Because RNA polymerase cannot bind the promoter efficiently without the aid of CAP, transcription levels are low. (c) When neither glucose nor lactose is available, both CAP and the *lac* repressor are active. In this situation, both proteins are bound to their regulatory sites. CAP binding enhances the binding of RNA polymerase to the promoter. However, the repressor blocks transcription. Transcription levels are low. (d) When glucose is available and lactose is not, CAP is inactive and the *lac* repressor is active. Thus RNA polymerase binds inefficiently, and those polymerase molecules that do bind are blocked by the repressor. This condition results in the lowest levels of transcription observed for the *lac* operon.

enzyme IIA but are not transferred to enzyme IIB. The phosphorylated form of enzyme IIA accumulates. This form of the enzyme activates adenyl cyclase, stimulating cAMP production.

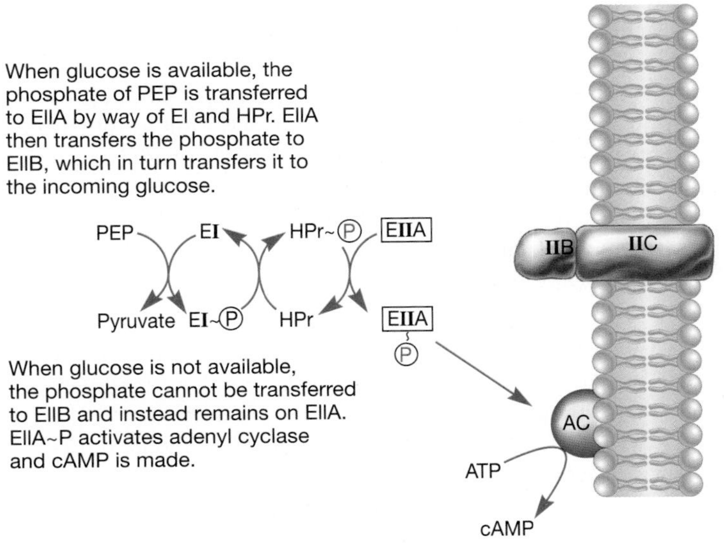

When glucose is available, the phosphate of PEP is transferred to EIIA by way of EI and HPr. EIIA then transfers the phosphate to EIIB, which in turn transfers it to the incoming glucose.
PEP
EI
HPr~P
EIIA
Pyruvate
EI~P
HPr
EIIA
P
When glucose is not available, the phosphate cannot be transferred to EIIB and instead remains on EIIA. EIIA~P activates adenyl cyclase and cAMP is made.
IIB
IIC
AC
ATP
cAMP

FIGURE 13.21 Activation of Adenyl Cyclase by the Phosphoenolpyruate: Sugar Phosphotransferase System (PTS). PEP is phosphoenolpyruvate. EI and EII are enzymes I and II of the PTS, respectively. EII is composed of three subunits: A, B, and C. HPr stands for heat-stable protein. A factor called factor x is also involved in activating the adenyl cyclase enzyme (not shown). It has never been isolated, and its nature is not known.

Group translocation (section 6.6) *Combination of Switches: The* lac *Operon*

Catabolite repression is of considerable advantage to *E. coli.* It will use the most easily catabolized sugar (glucose) first, rather than synthesize the enzymes necessary for catabolism of another carbon and energy source. Catabolite repression is used by a variety of bacteria to regulate numerous metabolic pathways.

Regulation by Other Nucleotides

As we have just seen, cAMP is an important effector molecule in the catabolite repression global regulation network. Two other nucleotides also play important regulatory roles in *Bacteria*: guanosine tetraphosphate (ppGpp) and cyclic dimeric GMP (c-di-GMP). ppGpp was discovered over 40 years ago when it was shown to function in the stringent response of *E. coli* (**figure 13.22**). Cyclic-di-GMP was discovered more recently. It has been shown to regulate virulence genes and the transition from a motile lifestyle to a nonmotile lifestyle. This transition is important in biofilm formation. Our focus here is on the stringent response.

The **stringent response** occurs when cells are starved for amino acids. When *E. coli* cells are growing in a nutrient-rich environment, many of the RNA polymerase molecules in the cell are transcribing tRNA and rRNA genes, providing a ready supply of these molecules for protein synthesis. When cells are starved for amino acids, protein synthesis cannot proceed as it does in the nutrient-rich conditions. Therefore it is wasteful for the cell to keep synthesizing large quantities of tRNA and rRNA. Instead, the cell decreases synthesis of these molecules and increases transcription of the appropriate amino acid biosynthetic genes. Eventually the cell "resets" its level of metabolic activity to one commensurate with the nutrients available in its environment.

The mechanism by which these changes are brought about have become increasingly clear over the 40 years this phenomenon has been studied. When *E. coli* is starved for one or more amino acids, tRNA molecules cannot become attached to their cognate amino acids. Initially the uncharged tRNAs do not enter the A site of the ribosome and the ribosome stalls (**figure 13.23**). Eventually an uncharged tRNA enters the A site of the stalled ribosome. When this occurs, an enzyme called RelA catalyzes the formation of ppGpp from GDP and ATP (figure 13.22*b*). RelA also catalyzes a reaction between GTP and ATP, generating guanosine pentaphosphate (pppGpp), which is quickly converted to ppGpp. ppGpp then interacts with RNA polymerase and downregulates synthesis of tRNA and rRNA, while simultaneously upregulating transcription of amino acid biosynthetic genes. However, ppGpp does not act alone. Instead, it exerts its effects with the help of the protein

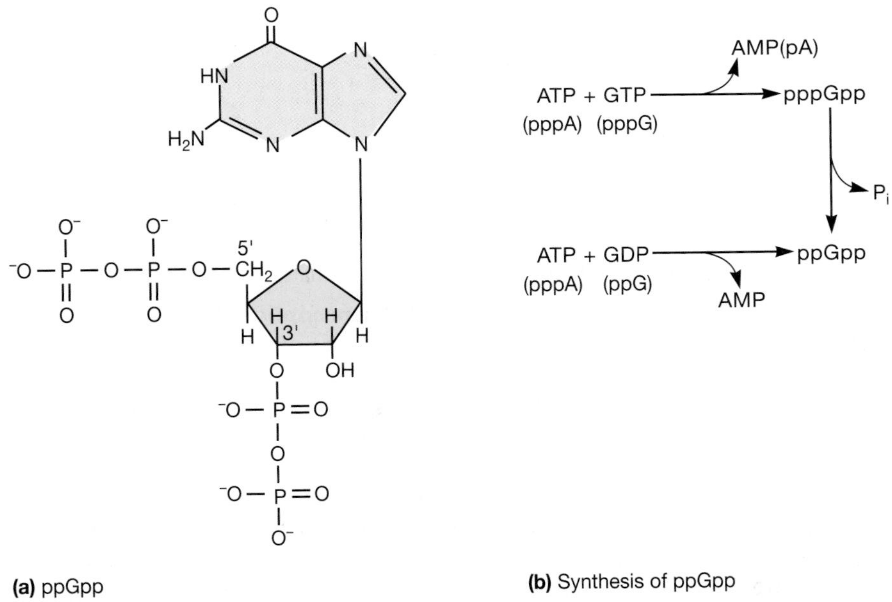

FIGURE 13.22 Guanosine Tetraphosphate. (a) The structure of ppGpp. (b) The two processes that generate ppGpp. In the stringent response, these reactions are catalyzed by the enzyme RelA.

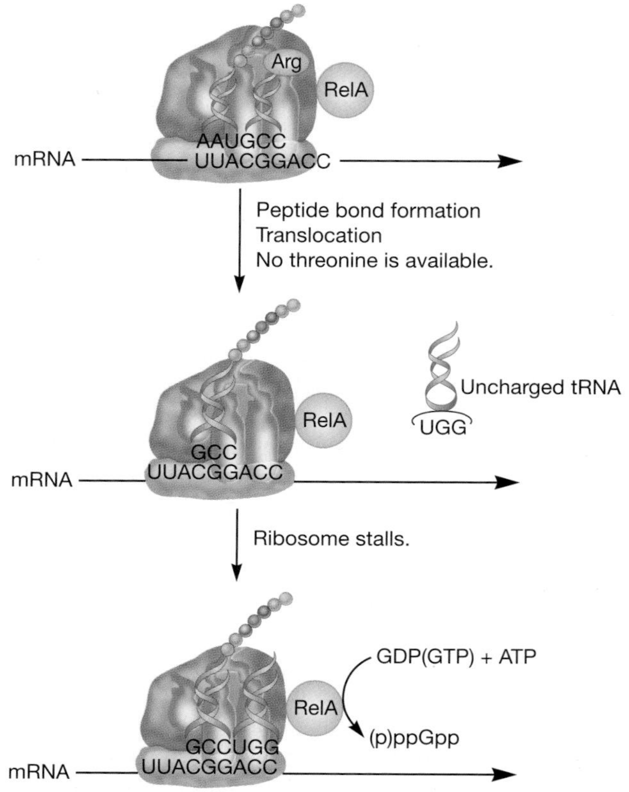

FIGURE 13.23 Activation of RelA in Response to Amino Acid Starvation.

DksA. Together, they destabilize open complexes formed during transcription initiation of the downregulated genes and enhance the rate of open complex formation of the upregulated genes (**figure 13.24**).

An important question for understanding the stringent response is how the cell knows which promoters to downregulate and which to upregulate. Comparisons of the promoters of rRNA genes and amino acid synthetic genes have answered this question. Recall from chapter 12 that sigma factor (σ) positions RNA polymerase core enzyme properly on the promoter. The enzyme then melts the two strands of DNA, forming an open complex (*see figure 12.26*). In ppGpp-regulated genes, there is a region between the −10 site of the promoter and the +1 nucleotide that is important for controlling transcription initiation. In downregulated genes (i.e., rRNA and tRNA genes), this region is GC rich, whereas in amino acid biosynthetic genes it is AT rich (figure 13.24). The GC-rich regions form unstable open complexes during transcription initiation. DksA and ppGpp further destabilize the open complexes, lowering transcription levels. The AT-rich regions of the upregulated genes form much more stable open complexes and are not destabilized by DksA and ppGpp. Instead, these two molecules increase the rate of formation of open complexes, increasing transcription.

More recently ppGpp has been shown to function as an "alarm," signaling that changes in transcription must be made to protect the cell from a number of stresses in addition to amino acid starvation. These include deprivation for other nutrients (e.g., phosphorus, iron, fatty acids). In these situations, production of ppGpp is catalyzed by a different enzyme, called SpoT. Despite being an area of intensive research for many years, many of the details of this global response to stress are still being explored.

1. What are global regulatory systems and why are they necessary? Briefly differentiate between regulons, modulons, and stimulons.
2. What is diauxic growth? Explain how catabolite repression causes diauxic growth.
3. Describe the events that occur with *E. coli* in each of the following growth conditions: in a medium containing glucose but not lactose; in a medium containing both sugars; in a medium containing lactose but no glucose; and in a medium containing neither sugar.
4. Why is it wasteful for a bacterium to synthesize many, many tRNA and rRNA molecules when it is starved for amino acids?

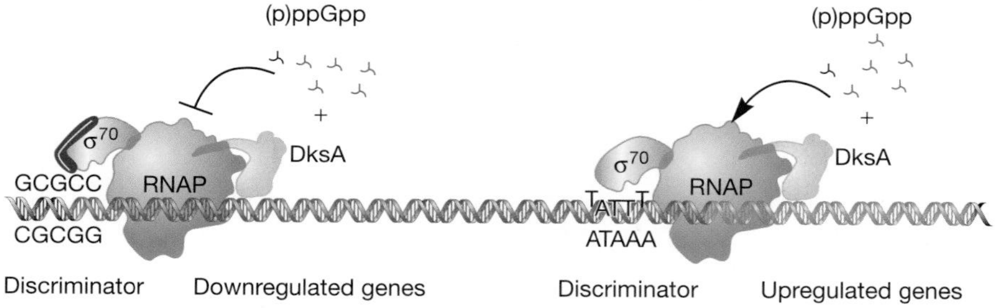

FIGURE 13.24 Effects of ppGpp and DksA on Transcription. Genes encoding tRNA and rRNA have a region between the −10 site of the promoter and the +1 nucleotide of the coding region that is GC rich. This region interacts with sigma factor as it positions the RNA polymerase core enzyme at the promoter. The presence of the GC region destabilizes open complexes formed during initiation. During the stringent response, ppGpp working with DksA further destabilizes the open complexes, thus lowering the rate of transcription. Elsewhere in the genome are genes encoding enzymes needed for amino acid biosynthesis. The promoters for genes are rich for AT base pairs in the same area of the promoter. These regions do not destabilize open complex formation. The interaction of ppGpp and DksA promotes open complex formation, and transcription is increased for these genes.

Quorum Sensing

Cell-to-cell communication among bacterial cells occurs by the exchange of small molecules often termed signals or signaling molecules. The exchange of signaling molecules is essential in the coordination of gene expression in microbial populations. One of the best studied quorum-sensing systems is that of the marine bioluminescent bacterium *Vibrio fischeri,* which produces light only if cells are at high density. It has since been discovered that intercellular communication plays an essential role in the regulation of genes whose products are needed for the establishment of virulence, symbiosis, biofilm production, plasmid transfer, and morphological differentiation in a wide range of bacteria. Here, we describe how signals that are secreted by a group of cells can regulate the genetic expression of that population. Our focus is on the regulation of a single operon. However, it should be kept in mind that **quorum sensing** can regulate multiple genes and operons. ◀◀ *Cell-cell communication within microbial populations (section 7.7)*

Quorum sensing in *V. fischeri* and many other gram-negative bacteria uses an ***N*-acylhomoserine lactone (AHL)** signal (**figure 13.25**). Synthesis of this small molecule is catalyzed by an enzyme called AHL synthase, the product of the *luxI* gene. The *luxI* gene is subject to positive autoregulation. That is to say, transcription of *luxI* increases as AHL accumulates in the cell. This is accomplished through the transcriptional activator LuxR, which is active only when it binds AHL. Thus a simple feedback loop is created. Without AHL-activated LuxR, the *luxI* gene is transcribed only at basal levels. AHL freely diffuses out of the cell and accumulates in the environment. As cell density increases, the concentration of AHL outside the cell eventually exceeds that inside the cell, and the concentration gradient is reversed. As AHL flows back into the cell, it binds and activates LuxR. LuxR can now activate high-level transcription of *luxI* and the genes whose products are needed for bioluminescence *(luxCDABEG)*. Quorum sensing is often called **autoinduction,** and the AHL signal is termed **autoinducer (AI)** to reflect the autoregulatory nature of this system. *Quorum Sensing*

Another kind of quorum sensing depends on an elaborate, two-component signal transduction system. It is found in both gram-negative and gram-positive bacteria, including *Staphylococcus aureus, Ralstonia solanacearum, Salmonella enterica, Vibrio cholerae,* and *E. coli.* It has been best studied in the bioluminescent bacterium *Vibrio harveyi.*

Unlike *V. fischeri, V. harveyi* responds to three autoinducer molecules: HAI-1, AI-2, and CAI-1. HAI-1 (harveyi autoinducer-1) is a homoserine lactone, and its synthesis depends on the *luxM* gene. AI-2 (autoinducer-2) is furanosylborate, a small molecule that contains a boron atom—quite an unusual component in an organic molecule *(see figure 7.35)*. Its synthesis relies on the product of the *luxS* gene. CAI-1 (cholerae autoinducer-1) is the product of the enzyme CqsA. The structure of CAI-1 was recently determined and is 3-hydroxytridecan-4-one. As shown in **figure 13.26**, HAI-1, AI-2, and CAI-1 are secreted by cells, which then use separate proteins called LuxN, LuxPQ, and CqsS, respectively, to detect their presence. LuxN, LuxQ, and CqsS are sensor kinases. At low cell density in the absence of any autoinducer, the three sensor kinases autophosphorylate and converge on a single phosphotransferase protein called LuxU. LuxU accepts phosphates from each

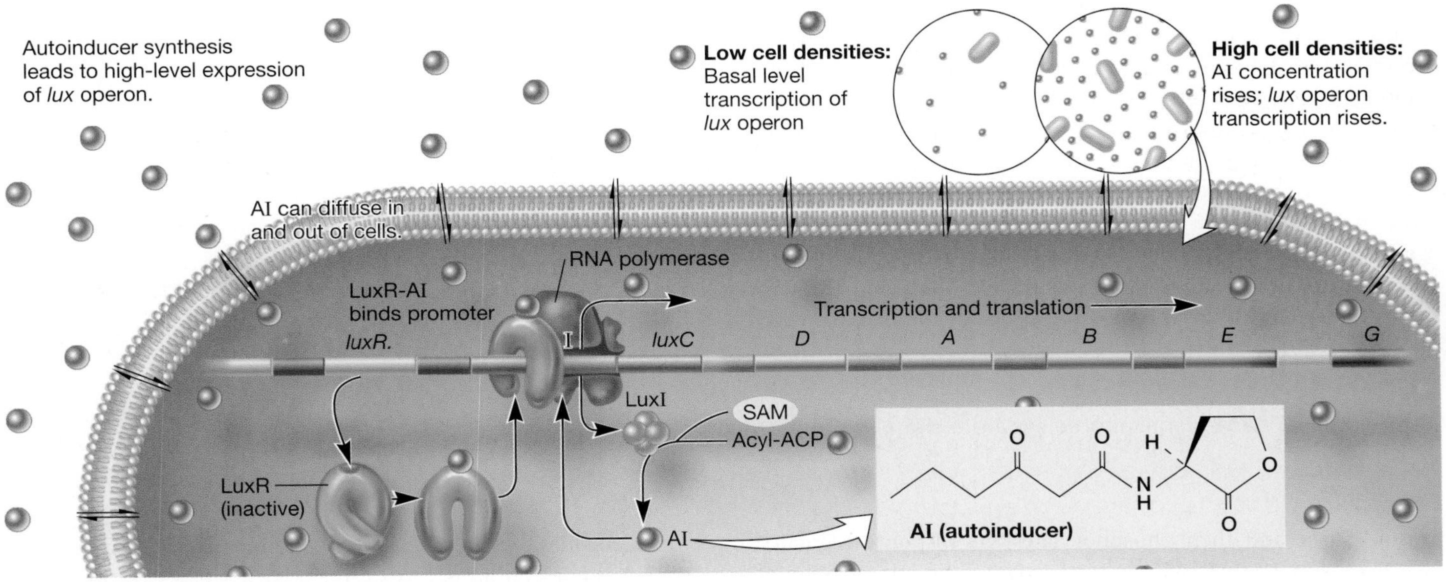

FIGURE 13.25 Quorum Sensing in *V. fischeri.* The AHL signaling molecule (AI) diffuses out of the cell; when cell density is high, AHL diffuses back into the cell, where it binds to and activates the transcriptional regulator LuxR. Active LuxR then stimulates transcription of the gene coding for AHL synthase (*luxI*), as well as the genes encoding proteins needed for light production.

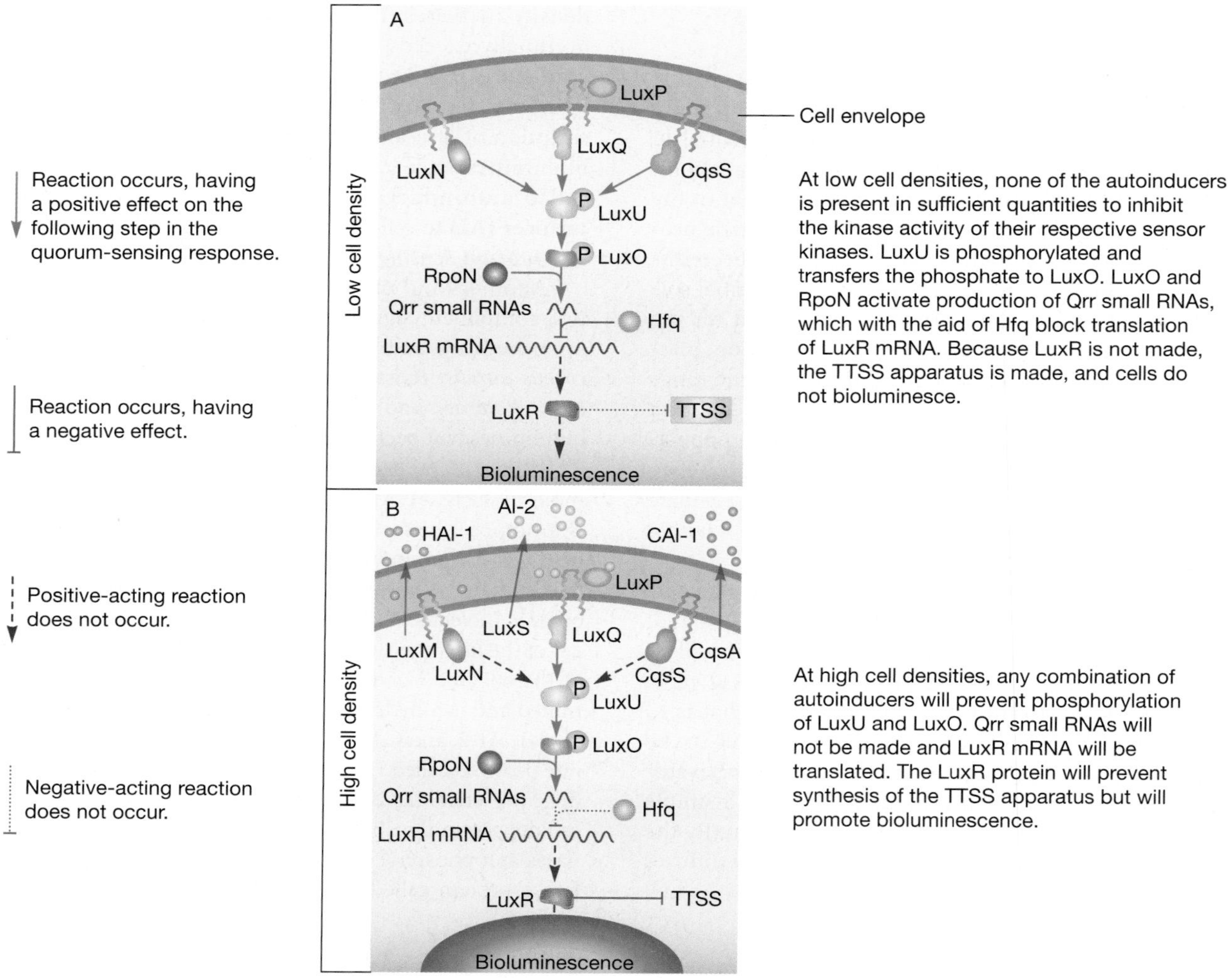
A
LuxP
Cell envelope
LuxQ
LuxN
CqsS
P
LuxU
P LuxO
RpoN
Qrr small RNAs
Hfq
LuxR mRNA
LuxR
TTSS
Bioluminescence
Low cell density
Reaction occurs, having a positive effect on the following step in the quorum-sensing response.
Reaction occurs, having a negative effect.
At low cell densities, none of the autoinducers is present in sufficient quantities to inhibit the kinase activity of their respective sensor kinases. LuxU is phosphorylated and transfers the phosphate to LuxO. LuxO and RpoN activate production of Qrr small RNAs, which with the aid of Hfq block translation of LuxR mRNA. Because LuxR is not made, the TTSS apparatus is made, and cells do not bioluminesce.
B
AI-2
HAI-1
CAI-1
LuxP
LuxS
LuxQ
LuxM
LuxN
CqsA
CqsS
P
LuxU
P LuxO
RpoN
Qrr small RNAs
Hfq
LuxR mRNA
LuxR
TTSS
Bioluminescence
High cell density
Positive-acting reaction does not occur.
Negative-acting reaction does not occur.
At high cell densities, any combination of autoinducers will prevent phosphorylation of LuxU and LuxO. Qrr small RNAs will not be made and LuxR mRNA will be translated. The LuxR protein will prevent synthesis of the TTSS apparatus but will promote bioluminescence.

FIGURE 13.26 Quorum Sensing in *V. harveyi*. Three autoinducing signals HAI-1, AI-2, and CAI-1 are detected. HAI-1 is only produced by *V. harveyi* and one other closely related species. AI-2 is produced by many gram-negative and gram-positive bacteria. CAI-1 is produced by many members of the genus *Vibrio*. Thus *V. harveyi* not only "measures" the density of its own population but that of other bacteria as well. LuxN, LuxA, and CqsA are sensor kinases involved in detecting HAI-1, AI-2, and CAI-1, respectively. Their kinase activity is inhibited by the presence of their respective autoinducer. All three quorum-sensing signals converge at LuxU, which when phosphorylated transfers the phosphate to LuxO. LuxO working with the sigma factor RpoN activates transcription of the *qrr* (quorum regulatory RNA) genes. The products of these genes are small RNAs that block translation of the mRNA encoding the protein LuxR. This is accomplished in conjunction with a small protein, Hfq, which is an RNA chaperone that helps RNA molecules bind each other or aids in their folding. Lux R is a regulatory protein that represses transcription of genes encoding components of a type III secretion system (TTSS) and activates transcription of genes required for bioluminescence.

Figure 13.26 Micro Inquiry

Why does *V. harveyi* make three separate signaling molecules?

sensor kinase and then phosphorylates the response regulator LuxO. Phosphorylated LuxO in turn activates the transcription of genes encoding several small RNAs that destabilize *luxR* mRNA. This is accomplished with the aid of the RNA chaperone Hfq (p. 346). LuxR is a transcriptional activator of the operon *luxCDABE*, which encodes proteins needed for bioluminescence. Because *luxR* mRNA is not translated, LuxR protein is not made. Therefore cells do not make light at low cell density.

An interesting thing happens as the density of any one of these autoinducers increases: LuxN binds HAI-1, LuxPQ binds

AI-2, and CqsS binds CAI-1. When this happens, the proteins switch from functioning as kinases to phosphatases, proteins that dephosphorylate, rather than phosphorylate, their substrates. The flow of phosphates is now reversed; LuxO is inactivated by dephosphorylation, the small RNAs are not made, and *luxR* mRNA is translated. LuxR now activates transcription of *luxCDABE* and light is produced. Careful inspection of figure 13.26 reveals that another set of genes is controlled by the quorum-sensing system of *V. harveyi*. In this microbe, genes for a type III protein secretion system (TTSS) are controlled in the opposite manner as those for bioluminescence.

At this point, we should pause and ask an important question: Why does *V. harveyi* need three different autoinducers? It appears that these molecules allow the bacterium to carry out three different kinds of conversations. HAI-1 is specific to *V. harveyi* (and one other closely related species). Thus it is thought to allow *V. harveyi* to communicate with members of its own species. In essence, this autoinducer conveys the message, "There are many *V. harveyi* nearby." AI-2 is made by many gram-negative and gram-positive bacteria. Thus its message is thought to be, "There are many bacteria nearby." Finally, CAI-1 is produced by other members of the genus *Vibrio*, including *V. cholerae*, for which it is named. Its message is thought to be, "There are many *Vibrio* spp. nearby." As might be expected, each autoinducer has a different signal strength. HAI-1 is the strongest and CAI-1 is the weakest. The presence of all three autoinducers maximizes expression of the bioluminescence genes and fully represses expression of the type III secretion system genes.

Sporulation in *Bacillus subtilis*

As discussed in chapter 3, endospore formation is a complex process that involves asymmetric division of the cytoplasm to yield a large mother cell and a smaller forespore, engulfment of the forespore by the mother cell, and construction of additional layers of spore coverings (**figures 13.27*a*** and *3.53*). Sporulation takes approximately 8 hours. It is controlled by phosphorelay, post-translational modification of proteins, numerous transcription initiation regulatory proteins, and alternate sigma factors. The latter are particularly important. When growing vegetatively, *B. subtilis* RNA polymerase uses sigma factors σ^A and σ^H to recognize genes for normal survival. However, when cells sense a starvation signal, a cascade of events is initiated that results in the production of alternative sigma factors that are differentially expressed in the developing endospore and mother cell.

Initiation of sporulation is controlled by the protein Spo0A, a response-regulator protein that is part of a phosphorelay system (figure 13.27*b*). Sensor kinases associated with this system detect environmental stimuli that trigger sporulation. One of the most important sensor kinases is KinA, which senses nutrient starvation. When *B. subtilis* finds its nutrients are depleted, KinA autophosphorylates a specific histidine residue. The phosphoryl group is then transferred to an aspartic acid residue on Spo0F. However, Spo0F cannot directly regulate gene expression; instead, Spo0F

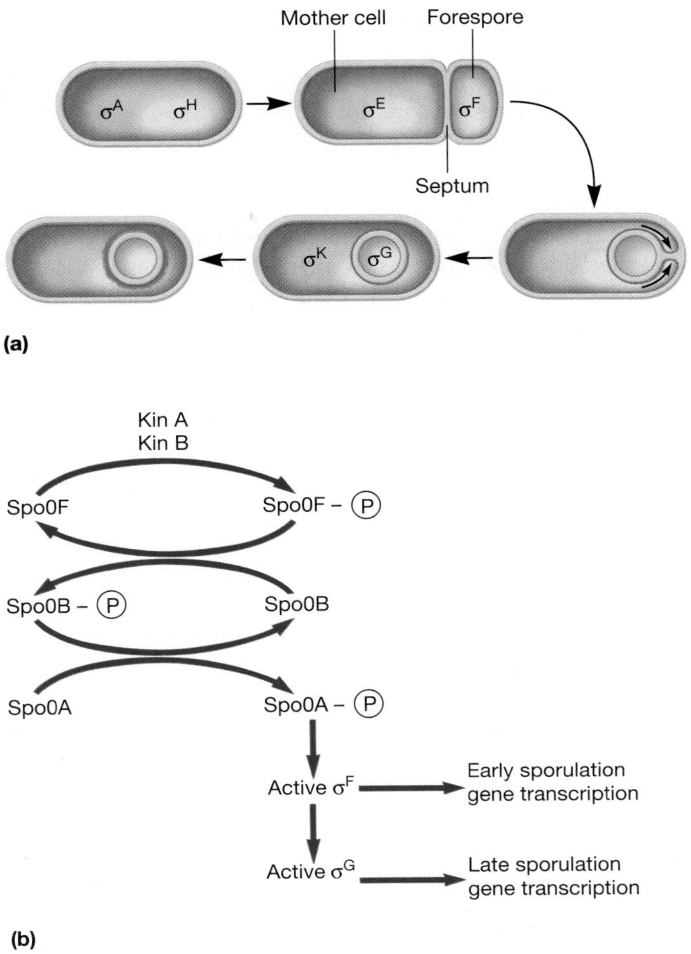

FIGURE 13.27 Genetic Regulation of Sporulation in *Bacillus subtilis*. (a) The initiation of sporulation is governed in part by the activities of two spatially separated sigma factors. σ^F is located in the forespore, while σ^E is confined to the mother cell. These sigma factors direct the initiation of transcription of genes whose products are needed for early events in sporulation. Later, σ^G and σ^K are localized to the developing endospore and mother cell, respectively. They control the expression of genes whose products are involved in the later steps of sporulation. (b) The activation of σ^F is accomplished through a phosphorelay system that is triggered by the activation of the sensor kinase protein KinA. When KinA senses starvation, it autophosphorylates a specific histidine residue. The phosphoryl group is then passed in relay fashion from Spo0F to Spo0B and finally to Spo0A.

Figure 13.27 Micro Inquiry

Compartmentation—the spatial localization of proteins as a means of regulation—is usually considered in the context of eukaryotic cells. How is the process of sporulation an example of compartmentation?

donates the phosphoryl group to a histidine on Spo0B. Spo0B in turn relays the phosphoryl group to Spo0A. Phosphorylated Spo0A positively controls genes needed for sporulation and negatively controls genes that are not needed. In response to Spo0A, the expression of over 500 genes is altered. This has earned Spo0A the name "master regulator." Among the genes whose expression is stimulated by Spo0A is *sigF,* the gene encoding sigma factor σ^F, and *spoIIGB,* the gene encoding an inactive form of σ^E (pro-σ^E).

When sporulation starts, the chromosome has replicated, with one copy remaining in the mother cell and another to be partitioned in the forespore. Shortly after the formation of the spore septum, σ^F is found in the forespore, and pro-σ^E is localized to the mother cell. Pro-σ^E is cleaved by a protease to form active σ^E. The two sigma factors, σ^F and σ^E, bind to the promoters of genes needed in the forespore and mother cell, respectively. There they direct the expression of genes whose products are needed for the early steps of endospore formation. These genes are primarily responsible for the engulfment process. Another gene regulated by σ^F is one that encodes the sigma factor, σ^G, which will replace σ^F in the developing endospore. Likewise, σ^E directs the transcription of a mother-cell–specific sigma factor, σ^K. Like σ^E, σ^K is first produced in an inactive form, pro-σ^K. Upon activation of pro-σ^K by proteolysis, σ^K ensures that genes encoding late-stage sporulation products are transcribed. These include genes for synthesis of the cortex and coat layers of the endospore. Overall, temporal regulation is achieved because σ^F and σ^E direct transcription of genes that are needed early in the sporulation process, whereas σ^G and σ^K are needed for the transcription of genes whose products function later. In addition, spatial control of gene expression is accomplished because σ^F and σ^G are located in the forespore and developing endospore, whereas σ^E and σ^K are found only in the mother cell.

1. What would be the phenotype of a *V. fischeri* mutant that could not regulate *luxI,* so that it was constantly producing autoinducer at high levels?
2. Why do you think bacteria use quorum sensing to regulate genes needed for virulence? How might this reason be related to the rationale behind using quorum sensing to establish a symbiotic relationship?
3. Briefly describe how a phosphorelay system and sigma factors are used to control sporulation in *B. subtilis.* Give one example of posttranslational modification as a means to regulate this process.

13.7 Regulation of Gene Expression in *Eukarya* and *Archaea*

As is the case in *Bacteria,* the regulation of gene expression in *Eukarya* and *Archaea* can occur at transcriptional, translational, and posttranslational levels (figure 13.1). Much of the work on gene regulation in *Eukarya* has focused on transcription initiation. More recently regulation by small RNA molecules has attracted considerable attention. Unfortunately our understanding of the regulation of archaeal gene expression lags considerably behind what we know for *Eukarya* and *Bacteria.* However, some intriguing discoveries are briefly introduced here.

Transcription initiation in *Eukarya* involves numerous transcription factors (*see figure 12.30*). Many transcription factors, such as TFIID, are general transcription factors that are part of the machinery common to transcription initiation of all eukaryotic genes. On the other hand, **regulatory transcription factors** are specific to one or more genes and alter the rate of transcription. Those transcription factors that function as activators bind regulatory sites called **enhancers,** whereas those that function as repressors bind sites called **silencers** (**figure 13.28**). After binding an enhancer or silencer, regulatory transcription factors act indirectly to increase or decrease the rate of transcription. Many regulatory transcription factors control transcription initiation by interacting with general transcription factors, in particular TFIID (figure 13.28*a*) and a multisubunit protein complex called mediator (figure 13.28*b*). *Transcription Factors*

Another regulatory mechanism observed in eukaryotes (and *Bacteria*) is the use of sRNA molecules (p. 346) to control gene expression. Many sRNAs act as antisense RNAs and function at the level of translation, as described in section 13.4. Some eukaryotic antisense RNAs are much smaller than typical bacterial antisense RNAs and are called microRNAs (miRNAs). Some sRNA molecules are important components of the spliceosome, where they contribute to the selection of splice sites used during mRNA processing. In doing so, different proteins can be made at certain times in the life cycle of the organism by combining different exons. *Control of Gene Expression in Eukaryotes*

The regulation of gene expression in the *Archaea* has garnered a great deal of interest. This is because the archaeal transcription and translation machinery is most similar to that of the *Eukarya,* yet it functions in cells with typical, bacteria-like genome organization. The question being asked is whether archaeal regulation of gene expression is more like bacterial regulation or more like eukaryotic regulation. Thus far, the answer is mixed. Most of the archaeal regulatory proteins function much like bacterial activators and repressors—that is, they bind DNA sites near the promoter and enhance or block binding of RNA polymerase, respectively. However, a few seem to function more like eukaryotic regulatory transcription factors in that they bring about their effects by interacting with DNA-binding proteins, such as the TATA-binding protein (*see figure 12.30*). Small RNA molecules also have been identified in some archaea; their role in regulation is still being elucidated. ▶▶| *Archaeal transcription (section 18.1)*

1. How are bacterial regulatory proteins and eukaryotic regulatory transcription factors similar? How do they differ?
2. What regulatory sequences in bacterial genomes are analogous to the enhancers and silencers observed in eukaryotic genomes?

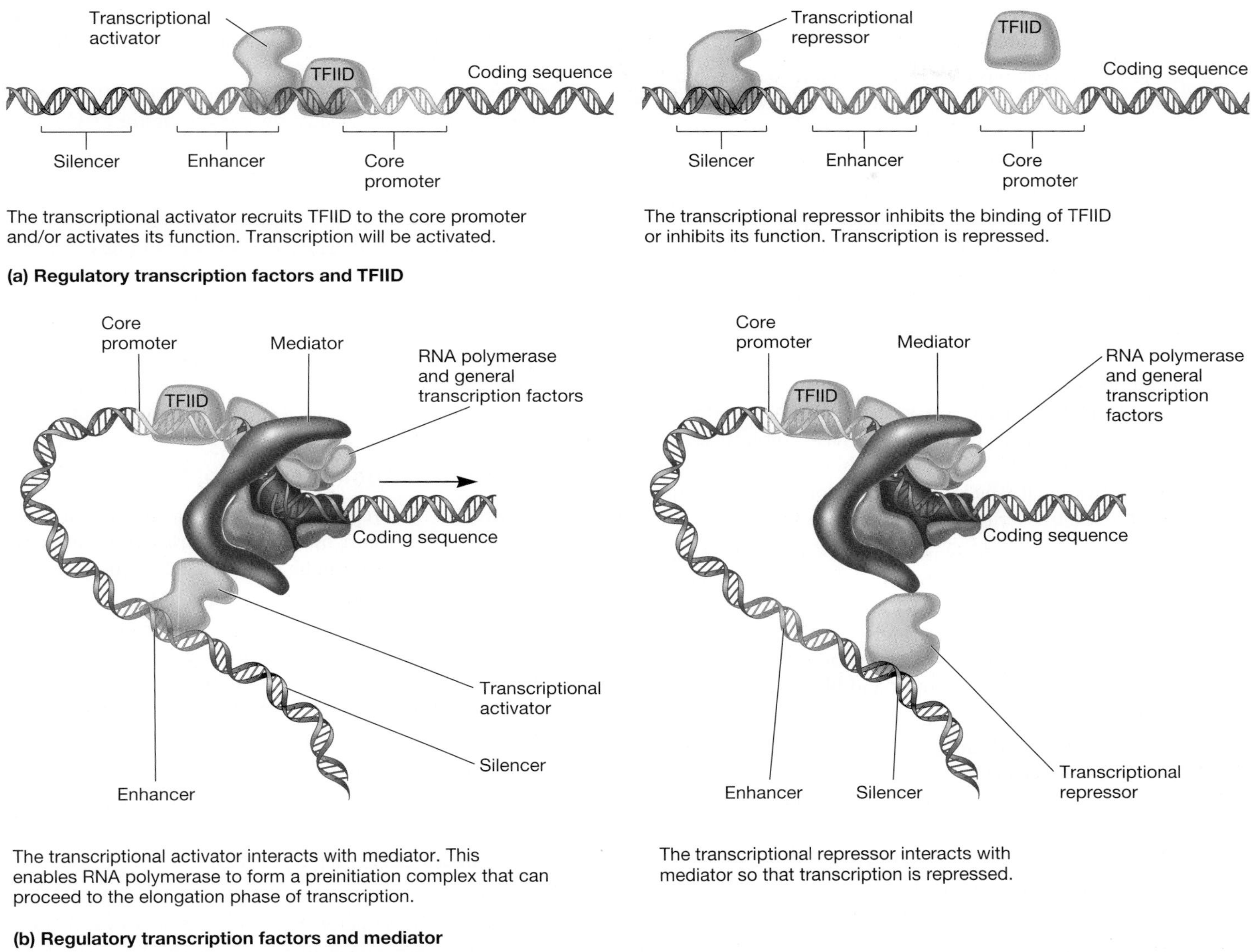

Transcriptional activator
TFIID
Coding sequence
Silencer
Enhancer
Core promoter
The transcriptional activator recruits TFIID to the core promoter and/or activates its function. Transcription will be activated.
Transcriptional repressor
TFIID
Coding sequence
Silencer
Enhancer
Core promoter
The transcriptional repressor inhibits the binding of TFIID or inhibits its function. Transcription is repressed.
(a) Regulatory transcription factors and TFIID
Core promoter
Mediator
RNA polymerase and general transcription factors
TFIID
Coding sequence
Transcriptional activator
Silencer
Enhancer
The transcriptional activator interacts with mediator. This enables RNA polymerase to form a preinitiation complex that can proceed to the elongation phase of transcription.
Core promoter
Mediator
RNA polymerase and general transcription factors
TFIID
Coding sequence
Enhancer
Silencer
Transcriptional repressor
The transcriptional repressor interacts with mediator so that transcription is repressed.
(b) Regulatory transcription factors and mediator

FIGURE 13.28 The Activity of Eukaryotic Regulatory Transcription Factors. Regulatory transcription factors do not exert their effects directly on RNA polymerase. Instead, they act via other proteins, most commonly the general transcription factors TFIID and a protein called mediator. Mediator's role in transcription is to aid RNA polymerase in switching from the initiation stage of transcription to the elongation stage. (a) A regulatory protein acting through TFIID. Activators could influence transcription-enhancing TFIID recruitment of RNA polymerase to the promoter. Repressors would inhibit this ability. (b) A regulatory protein acting through the mediator protein. An activator would stimulate mediator activity; a repressor would decrease mediator activity.

Summary

13.1 Levels of Regulation

a. Regulation of gene expression can be controlled at many levels, including transcription initiation, transcription elongation, translation, and posttranslation (**figure 13.1**).

b. The three domains of life differ in terms of their genome structure and the steps required to complete gene expression. These differences affect the regulatory mechanisms they use.

13.2 Regulation of Transcription Initiation

a. Induction and repression of enzyme levels are two important regulatory phenomena. They usually occur because of the activity of regulatory proteins.

b. Regulatory proteins are DNA-binding proteins. When bound to DNA, they can either inhibit transcription (negative control) or promote transcription (positive control).

Their activity is modulated by small effector molecules called inducers, corepressors, and inhibitors (**figure 13.3**).

c. Repressors are responsible for negative control. They block transcription by binding an operator and interfering with the binding of RNA polymerase to its promoter or by blocking the movement of RNA polymerase after it binds DNA.

d. Activator proteins are responsible for positive control. They bind DNA sequences called activator-binding sites and, in doing so, promote binding of RNA polymerase to its promoter.

e. The *lac* operon of *E. coli* is an example of a negatively controlled inducible operon. When there is no lactose in the surroundings, the *lac* repressor is active and transcription is blocked. When lactose is available, it is converted to allolactose by the enzyme β-galactosidase. Allolactose acts as the inducer of the *lac* operon by binding the repressor and inactivating it. The inactive repressor cannot bind the operator and transcription occurs (**figure 13.7**).

f. The *trp* operon of *E. coli* is an example of a negatively controlled repressible operon. When tryptophan is not available, the *trp* repressor is inactive and transcription occurs. When tryptophan levels are high, tryptophan acts as a corepressor and binds the *trp* repressor, activating it. The *trp* repressor binds the operator and blocks transcription (**figure 13.8**).

g. The *ara* operon of *E. coli* is an example of an inducible operon that is regulated by the dual-function regulatory protein AraC. AraC functions as a repressor when arabinose is not available. It functions as an activator when arabinose, the inducer, is available (**figure 13.9**).

h. Some regulatory proteins are members of two-component signal transduction systems and phosphorelay systems. These systems have a sensor kinase that detects an environmental change. The sensor kinase transduces the environmental signal to the response-regulator protein either directly (two-component system) or indirectly (phosphorelay) by transferring a phosphoryl group to it. The response regulator then activates genes needed to adapt to the new environmental conditions and inhibits expression of those genes that are not needed (**figure 13.10**).

13.3 Regulation of Transcription Elongation

a. In the tryptophan operon, a leader region lies between the operator and the first structural gene (**figure 13.11**). It codes for the synthesis of a leader peptide and contains an attenuator, a rho-independent termination site. Synthesis of the leader peptide by a ribosome while RNA polymerase is transcribing the leader region regulates transcription: Therefore the tryptophan operon is expressed only when insufficient tryptophan is available. This mechanism of transcription control is called attenuation.

b. The leader regions of some mRNA molecules can bind metabolites that act as effector molecules. Binding of the metabolite to the mRNA causes a change in the leader structure, which can terminate transcription. This regulatory mechanism is called a riboswitch (**figure 13.12**).

13.4 Regulation of Translation

a. Some riboswitches regulate gene expression at the level of translation. For these riboswitches, the binding of a small molecule to specific sequences in the leader region of the mRNA alters leader structure and prevents ribosome binding (**figure 13.13**).

b. Translation can also be controlled by antisense RNAs. These small RNA molecules are noncoding. They base pair to the mRNA and usually inhibit translation (**figure 13.14**).

13.5 Posttranslational Regulation

a. Posttranslational regulation occurs after a protein has been synthesized. It occurs by either allosteric control or covalent modification. It can be used to regulate complex processes such as chemotaxis.

b. Phosphorylation of proteins is important in sending signals regarding the level of an attractant in a cell's environment. It is part of a phosphorelay system that controls the direction of rotation of the *E. coli* flagellum (**figure 13.15**).

c. Methylation is used to measure the amount of a chemoattractant encountered over time. In this way, *E. coli* can determine if it is moving toward or away from an attractant.

13.6 Global Regulatory Systems

a. Global regulatory systems control many operons simultaneously and help microbes respond rapidly to a wide variety of environmental challenges.

b. Global regulatory systems often involve many layers of regulation. Mechanisms such as regulatory proteins, alternate sigma factors, two-component signal transduction systems, and phosphorelay systems are often used.

c. Diauxic growth is observed when *E. coli* is cultured in the presence of glucose and another sugar such as lactose (**figure 13.17**). This growth pattern is the result of catabolite repression, where glucose is used preferentially over other sugars. Operons that are part of the catabolite repression system are regulated by the activator protein CAP. CAP activity is modulated by cAMP, which is produced only when glucose is not available (**figure 13.21**). Thus when there is no glucose, CAP is active and promotes transcription of operons needed for the catabolism of other sugars (**figure 13.20**).

d. Unusual nucleotides such as cyclic-dimeric-GMP and guanosine tetraphosphate can be used to regulate global regulatory networks. Guanosine tetraphosphate (ppGpp) functions in the stringent response. This response is made to amino acid starvation. It results in a decrease in transcription of tRNA and rRNA genes and increased transcription of amino acid biosynthetic genes (**figures 13.22–13.24**).

e. Quorum sensing is a type of cell-to-cell communication mediated by small signaling molecules such as *N*-acylhomoserine lactone (AHL). Quorum sensing couples cell density to regulation of transcription. Well-studied quorum-sensing systems include the regulation of bioluminescence in *Vibrio* spp. (**figures 13.25** and **13.26**). Other systems regulate virulence genes and biofilm formation.

f. Endospore formation in *B. subtilis* is another example of a global regulatory system. Two important regulatory mechanisms used during sporulation are a phosphorelay system that is important in initiation of sporulation and the use of alternate sigma factors (**figure 13.27**).

13.7 Regulation of Gene Expression in *Eukarya* and *Archaea*

a. Regulatory transcription factors are used by members of *Eukarya* to control transcription initiation. They can exert either positive or negative control (**figure 13.28**).

b. Antisense RNAs are used by eukaryotes to regulate translation.

c. Microbiologists know relatively little about archaeal regulation of gene expression. Some archaea use regulatory proteins that are similar to bacterial activators and repressors to control initiation of transcription.

Critical Thinking Questions

1. Attenuation affects anabolic pathways, whereas repression affects either anabolic or catabolic pathways. Provide an explanation for this.

2. Describe the phenotype of the following *E. coli* mutants when grown in two different media: glucose only and lactose only. Explain the reasoning behind your answer.

   a. A strain with a mutation in the gene encoding the *lac* repressor; the mutant repressor cannot bind allolactose.

   b. A strain with a mutation in the gene encoding CAP; the mutant form of CAP binds but cannot release cAMP.

   c. A strain in which the Shine-Dalgarno sequence has been deleted from the gene encoding adenyl cyclase.

3. What would be the phenotype of an *E. coli* strain in which the tandem *trp* codons in the leader region were mutated so that they coded for serine instead?

4. What would be the phenotype of a *B. subtilis* strain whose gene for σ^G has been deleted? Consider the ability of the mutant to survive in nutrient-rich versus nutrient-depleted conditions.

5. Propose a mechanism by which a cell might sense and respond to levels of Na^+ in its environment.

6. It has recently been discovered that *E. coli* demonstrates chemotactic behavior when placed in a gradient of thymine and uracil. This is seen in all *E. coli* strains except mutants missing the methyl-accepting chemotactic protein (MCP) Tap, which binds dipeptides. To explore the hypothesis that Tap is mediating a pyrimidine chemotaxis response, microbiologists engineered hybrid MCPs in which the cytoplasmic domain of Tap and Tsr were switched.

   Refer to figure 13.16 and predict the behavior of cells in chemotaxis trials in which cells are exposed to a gradient of dipeptides, serine, pyrimidines, and sugars when they possess the cytoplasmic region of Tap and the periplasmic domain of Tsr and vice versa. Can you think of another experiment the investigators might have performed to test their hypothesis?

   Read the original paper: Liu, X., and Parales, E. 2008. Chemotaxis signaling of *Escherichia coli* to pyrimidines: a new role for the signal transducer Tap. *J. Bacteriol.* 190:972.

7. The two-component signal transduction system consisting of WalK and WalR (signal kinase and response regulator, respectively) is found in a number of gram-positive bacteria, including *B. subtilis, Enterococcus faecalis, Listeria monocytogenes*, and *Streptococcus pneumoniae*. It has been most thoroughly studied in *Staphylococcus aureus*. In all these microbes, the genes *walK* and *walR* are essential; that is, if they are deleted, the cells die. Dubrac and coworkers noticed that when *S. aureus* cells without these genes die, they do not lyse. This suggested that WalK and WalR (WalKR) may regulate cell wall autolysins—enzymes that degrade peptidoglycan. Indeed, they discovered that WalKR positively regulates nine genes involved in different steps of cell wall degradation, including the two major autolysins in *S. aureus*. Furthermore, WalKR positively regulates biofilm formation—an important aspect of *S. aureus* pathogenicity.

   What do you think might be potential WalK ligand (or ligands) that triggers autophosphorylation? How would you test your hypotheses? Why do you think deletion of *walK* and *walR* is lethal? Finally, why do you think cell wall degradation/remodeling and biofilm formation are controlled by the same two-component system?

   Read the original paper: Dubrac, S., et al., 2007. New insights into the WalK/WalR (YycG/YycF) essential signal transduction pathway reveal a major role in controlling cell wall metabolism and biofilm formation in *Staphylococcus aureus*. *J. Bacteriol.* 189:8257.

8. The extracytoplasmic sigma factor (ECF) σ^E is required for full virulence in *Mycobacterium tuberculosis*, which causes tuberculosis. Many stimuli induce σ^E production, including detergent stress to the cell surface, alkaline pH, heat shock, and oxidative stress. Unlike most ECF sigma factors, σ^E does not regulate its own transcription. Because it is so important in *M. tuberculosis* pathogenicity, microbiologists want to understand what controls σ^E production. They have discovered that at least three different types of regulation are in play. (1) A two-component regulatory system consisting of MprA and MprB (signal kinase and response regulator, respectively) activates transcription of the σ^E structural gene *sigE* in response to cell surface stress and increases in pH. (2) The *M. tuberculosis* heat-shock sigma factor, σ^H, activates transcription of *sigE* in response to heat shock and oxidative stress, and (3) an antisigma factor RsrA binds to and inactivates σ^E.

   Sketch this regulatory network using arrows to represent induction and cross-hatched lines to indicate repression. How do you think RsrA might be regulated so that it is not always bound to σ^E? Why do you think *M. tuberculosis* uses three different regulatory mechanisms to control *sigE* expression? Do you think this type of regulatory layering is unusual in bacterial and archaeal cells? Explain your answer.

   Read the original paper: Dona, V., et al., 2008. Evidence of transcriptional, translational, and posttranslational regulation of the extracytoplasmic sigma factor σ^E in *Mycobacterium tuberculosis. J. Bacteriol.* 190: 5963.

Concept Mapping Challenge

Construct a concept map that describes the regulation of the lactose operon. Use the concepts that follow, any other concepts or terms you need, and your own linking words between each pair of concepts in your map.

Operator	Promoter	Inducer
lac repressor	CAP	Catabolite repression
cAMP		

Learn More

Learn more by visiting the text website at www.mhhe.com/willey8, where you will find a complete list of references.

14

Microbial Genetics: Mechanisms of Genetic Variation

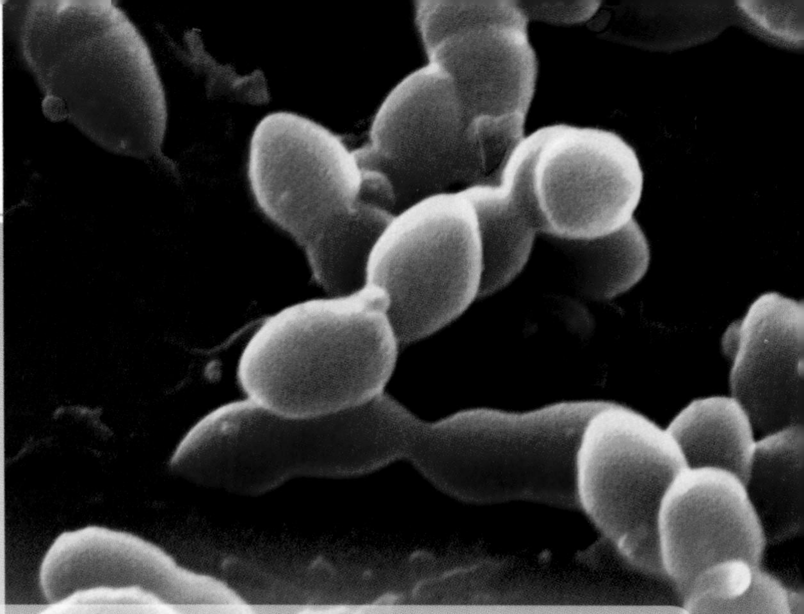

The scanning electron micrograph shows Streptococcus pneumoniae, *the bacterium first used to study transformation and obtain evidence that DNA is the genetic material of organisms.*

CHAPTER GLOSSARY

allele An alternative form of a gene.

auxotroph An organism with a mutation that causes it to lose the ability to synthesize an essential nutrient; the organism must obtain the nutrient or a precursor from its surroundings.

competent cell A cell that can take up free DNA and incorporate it into its genome during transformation.

conjugation The form of gene transfer and recombination in bacterial and archaeal cells that requires direct cell-to-cell contact.

excision repair A type of DNA repair mechanism in which damaged DNA is excised and replaced, using the complementary strand as a template.

F factor (F plasmid) The fertility factor; a plasmid that carries genes for bacterial conjugation and makes its *Escherichia coli* host the gene donor during conjugation.

frameshift mutation Mutations arising from the loss or gain of a base or DNA segment, leading to a change in the codon reading frame.

generalized transduction The transfer of any part of a cell's genome when the DNA fragment is packaged within a virus's capsid by mistake.

Hfr strain An *E. coli* strain that donates its genes with high frequency to a recipient cell during conjugation because the F factor is integrated into the donor's chromosome.

homologous recombination Recombination involving two DNA molecules that are similar in nucleotide sequence.

horizontal (lateral) gene transfer (HGT) The process by which genes are transferred from one mature, independent organism to another.

insertion sequence A simple, mobile genetic element that contains genes only for transposition.

mismatch repair system A type of DNA repair in which a portion of a newly synthesized strand of DNA containing mismatched base pairs is removed and replaced, using the parental strand as a template.

missense mutation A point mutation that changes a codon for one amino acid into a codon for another.

mutagen A chemical or physical agent that causes mutations.

mutation A heritable change in the genetic material.

nonsense mutation A mutation that converts a sense codon to a nonsense (stop) codon.

point mutation A mutation that changes a single base pair.

prototroph A microorganism that requires the same nutrients as most of the members of its species.

recombination The process by which a new recombinant chromosome is formed by combining genetic material from two organisms.

site-specific recombination Recombination of nonhomologous genetic material with a chromosome at a specific site.

SOS response A complex, inducible process that allows bacterial cells with extensive DNA damage to survive, although often in a mutated form.

specialized transduction The process by which only a specific set of bacterial or archaeal genes is carried to a recipient cell by a temperate virus.

suppressor mutation A mutation that overcomes the effect of another mutation and restores the normal phenotype.

transduction The transfer of genes between bacterial or archaeal cells by viruses.

transformation A mode of horizontal gene transfer in bacterial and archaeal cells in which a piece of free DNA is taken up by a cell and stably maintained.

transposition The movement of a piece of DNA around a cell's genome.

transposon A mobile genetic element that carries the genes required for transposition, as well as other genes.

wild type The genotype that is most commonly found in nature; it can refer to either a specific strain of a microbial species or a particular gene.

Chapters 12 and 13 introduce the fundamentals of molecular genetics—the way genetic information is organized, stored, replicated, and expressed. For life to exist with stability, it is essential that the nucleotide sequence of genes is not disturbed to any great extent. However, sequence changes do occur and can result in altered phenotypes. These changes may be detrimental, but those that are not are important in generating new variability in populations and in contributing to the process of evolution.

In this chapter, we focus on processes that contribute to genetic variation in populations of microbes. We begin with an overview of the chemical nature of mutations and the effects of mutations at both the molecular and organismal levels. The chapter continues with a discussion of DNA repair mechanisms. Although these have evolved to prevent the occurrence of mutations, some cellular attempts to correct DNA damage actually generate mutations. Finally, we examine microbial recombination and gene transfer in *Bacteria.* These processes are the basis for microbial evolution and have practical implications in terms of antibiotic resistance and the development of new infectious agents.

14.1 Mutations: Their Chemical Basis and Effects

Mutations (Latin *mutare,* to change) were initially characterized as altered phenotypes, but they are now understood at the molecular level. Several types of mutations exist. Some mutations arise from the alteration of single pairs of nucleotides and from the addition or deletion of one or two nucleotide pairs in the coding regions of a gene. Such small changes in DNA are sometimes called microlesions, and the smallest of these are called **point mutations** because they affect only one base pair in a given location. Larger mutations (macrolesions) are less common. These include large insertions, deletions, inversions, duplications, and translocations of nucleotide sequences.

Mutations occur in one of two ways. (1) **Spontaneous mutations** arise occasionally in all cells and in the absence of any added agent. (2) On the other hand, **induced mutations** are the result of exposure to a **mutagen,** which can be either a physical or a chemical agent. Mutations are characterized according to either the kind of genotypic change that has occurred or their phenotypic consequences. In this section, the molecular basis of mutations and mutagenesis is first considered. Then the phenotypic effects of mutations are discussed.

Spontaneous Mutations

Spontaneous mutations result from errors in DNA replication or from the action of mobile genetic elements such as transposons. A few of the more prevalent mechanisms are described here.

Replication errors can occur when the nitrogenous base of a template nucleotide takes on a tautomeric form. Tautomerism is the relationship between two structural isomers that are in chemical equilibrium and readily change into one another. Nitrogenous bases typically exist in the keto form. However, they can at times take on either an imino or enol form (**figure 14.1*a***). These tautomeric shifts change the hydrogen-bonding characteristics of the bases, allowing purine for purine or pyrimidine for pyrimidine substitutions that can eventually lead to a stable alteration of the nucleotide sequence (figure 14.1*b*). Such substitutions are known as **transition mutations** and are relatively common. On the other hand, **transversion mutations,** mutations where a purine is substituted for a pyrimidine or a pyrimidine for a purine, are rarer due to the steric problems of pairing purines with purines and pyrimidines with pyrimidines.

Replication errors can also result in the insertion and deletion of nucleotides. These mutations generally occur where there is a short stretch of repeated nucleotides. In such a location, the pairing of template and new strand can be displaced by the distance of the repeated sequence, leading to insertions or deletions of bases in the new strand (**figure 14.2**).

Spontaneous mutations can originate from lesions in DNA as well as from replication errors. For example, it is possible for purine nucleotides to be depurinated—that is, to lose their base. This results in the formation of an apurinic site, which does not base pair normally and may cause a transition-type mutation after the next round of replication. Likewise, pyrimidines can be lost, forming an apyrimidinic site. Other lesions are caused by reactive forms of oxygen such as oxygen free radicals and peroxides produced during aerobic metabolism. For example, guanine can be converted to 8-oxo-7,8-dihydrodeoxyguanine, which often pairs with adenine, rather than cytosine, during replication.

Induced Mutations

Virtually any agent that damages DNA, alters its chemistry, or in some way interferes with its functioning will induce mutations. Mutagens can be conveniently classified according to their mode of action. Three common types of chemical mutagens are base analogs, DNA-modifying agents, and intercalating agents. A number of physical agents (e.g., radiation) are mutagens that damage DNA.

Base analogs are structurally similar to normal nitrogenous bases and can be incorporated into the growing polynucleotide chain during replication (**table 14.1**). Once in place, these compounds typically exhibit base pairing properties different from the bases they replace and can eventually cause a stable mutation. A widely used base analog is 5-bromouracil, an analog of thymine. It undergoes a tautomeric shift from the normal keto form to an enol much more frequently than does a normal base. The enol tautomer forms hydrogen bonds like cytosine, pairing with guanine rather than adenine. The mechanism of action of other base analogs is similar to that of 5-bromouracil.

There are many **DNA-modifying agents**—mutagens that change a base's structure and therefore alter its base pairing specificity. Some of these mutagens are selective; they preferentially react with certain bases and produce a particular kind of DNA damage. For example, methyl-nitrosoguanidine is an alkylating agent that adds methyl groups to guanine, causing it to mispair with thymine (**figure 14.3**). A subsequent round of replication can then result in a GC-AT transition. Hydroxylamine is another example of a DNA-modifying agent. It hydroxylates the C-4 nitrogen of cytosine *(see figure 12.6)*, causing it to base pair like thymine.

Intercalating agents distort DNA to induce single nucleotide pair insertions and deletions. These mutagens are planar and insert themselves (intercalate) between the stacked bases of the helix. This results in a mutation, possibly through the formation of a loop in DNA. Intercalating agents include acridines such as proflavin and acridine orange.

Many mutagens, and indeed many carcinogens, damage bases so severely that hydrogen bonding between base pairs is impaired or prevented and the damaged DNA can no longer act as a template for replication. For instance, UV radiation generates cyclobutane dimers, usually thymine dimers, between adjacent pyrimidines (**figure 14.4**). Other examples are ionizing radiation and carcinogens such as the fungal toxin aflatoxin B1 and other benzo(a)pyrene derivatives.

Effects of Mutations

The effects of a mutation can be described at the protein level and in terms of observed phenotypes. In all cases, the impact is readily noticed only if it produces a change in phenotype. In general, the more prevalent form of a gene and its associated phenotype is called the **wild type.** A mutation from wild type to a mutant form is called a **forward mutation** (**table 14.2**). A forward mutation can be reversed by a second mutation that restores the wild-type phenotype. The second mutation can occur at the same site as the original mutation or at another site.

When the second mutation is at the same site as the original mutation, it is called a **reversion mutation.** Some reversions convert the mutant nucleotide sequence back to the wild-type sequence. Others create a new codon that codes for the same amino acid. Reversions can also restore the wild-type phenotype by creating a codon that specifies an amino acid that is similar to the amino acid found at that location in

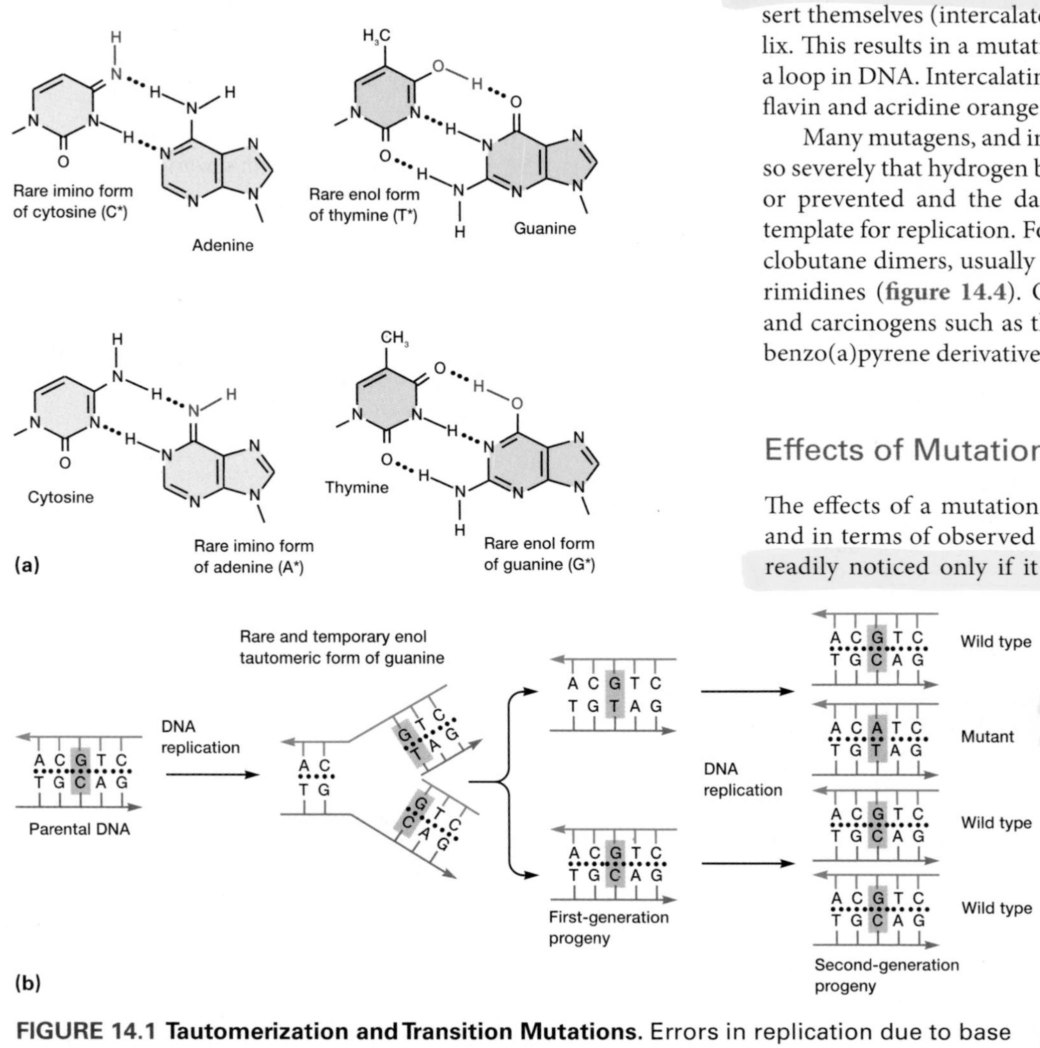

FIGURE 14.1 Tautomerization and Transition Mutations. Errors in replication due to base tautomerization. (a) Normally AT and GC pairs are formed when keto groups participate in hydrogen bonds. In contrast, enol tautomers produce AC and GT base pairs. The alteration in the base is shown in blue. (b) Mutation as a consequence of tautomerization during DNA replication. The temporary enolization of guanine leads to the formation of an AT base pair in the mutant, and a GC–to–AT transition mutation occurs. The process requires two replication cycles. Mutation only occurs if the abnormal first-generation GT base pair is missed by repair mechanisms. Wild type is the form of the gene before mutation occurred.

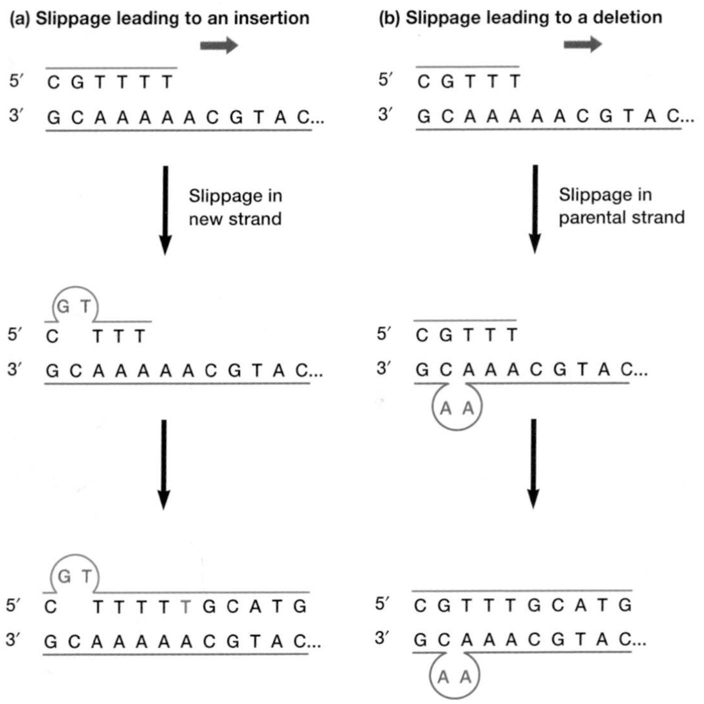

FIGURE 14.2 Insertions and Deletions. A mechanism for the generation of insertions and deletions during replication. The direction of replication is indicated by the blue arrow. In each case, there is strand slippage resulting in the formation of a small loop that is stabilized by hydrogen bonding in the repetitive sequence, the AT stretch in this example. (a) If the new strand slips, an addition of one T results. (b) Slippage of the parental strand yields a deletion (in this case, a loss of two Ts).

the wild-type protein (e.g., both amino acids are nonpolar; *see figure 12.7*).

If the second mutation is at a different site than the original mutation, it is called a **suppressor mutation.** Suppressor mutations may be within the same gene (intragenic suppressor mutation) or in a different gene (extragenic suppressor mutation). Because point mutations are the most common types of mutations, their effects are the focus here.

Mutations in Protein-Coding Genes

Point mutations in protein-coding genes can affect protein structure in a variety of ways. Point mutations are named according to if and how they change the encoded protein. The most common types of point mutations are silent mutations, missense mutations, nonsense mutations, and frameshift mutations. Examples of each are shown in table 14.2.

Silent mutations change the nucleotide sequence of a codon but do not change the amino acid encoded by that codon. This is possible because of the degeneracy of the genetic code. Therefore, when there is more than one codon for a given amino acid, a single base substitution may result in the formation of a new codon for the same amino acid. For example, if the codon CGU were changed to CGC, it would still code for arginine, even though a mutation had occurred. When there is no change in the protein, there is no change in the phenotype of the organism. ◀◀ *The genetic code (section 12.7)*

Table 14.1 Examples of Mutagens

Mutagen	*Effect(s) on DNA Structure*
Chemical	
5-Bromouracil	Base analog
2-Aminopurine	Base analog
Ethyl methanesulfonate	Alkylating agent
Hydroxylamine	Hydroxylates cytosine
Nitrogen mustard	Alkylating agent
Nitrous oxide	Deaminates bases
Proflavin	Intercalating agent
Acridine orange	Intercalating agent
Physical	
UV light	Promotes pyrimidine dimer formation
X rays	Causes base deletions, single-strand nicks, cross-linking, and chromosomal breaks

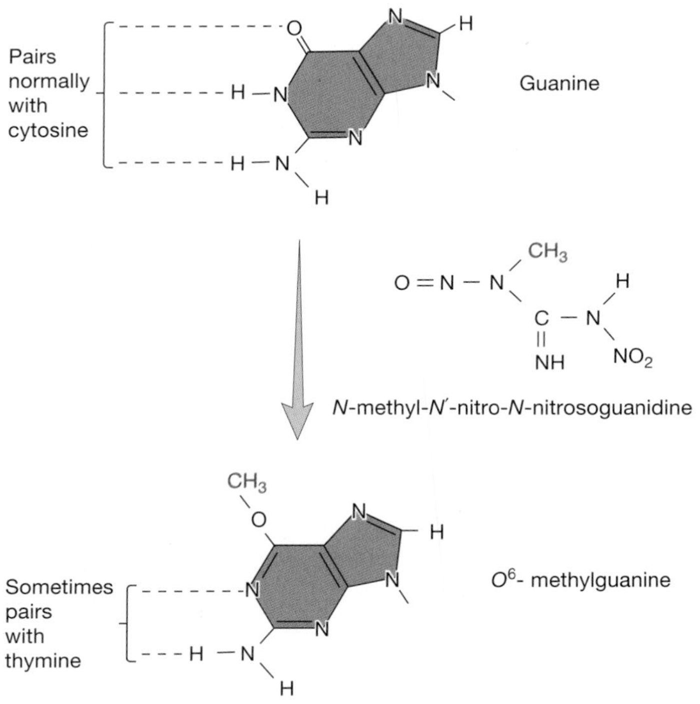

FIGURE 14.3 Methyl-Nitrosoguanidine Mutagenesis. Mutagenesis by methyl-nitrosoguanidine due to the methylation of guanine.

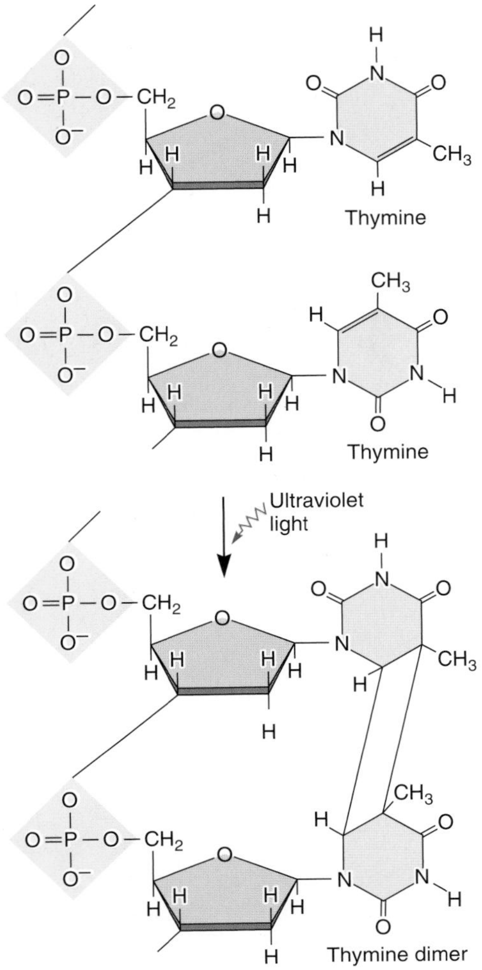

FIGURE 14.4 Thymine Dimer. Thymine dimers are formed by ultraviolet radiation.

Missense mutations involve a single base substitution that changes a codon for one amino acid into a codon for another. For example, the codon GAG, which specifies glutamic acid, could be changed to GUG, which codes for valine. The effects of missense mutations vary. They alter protein structure, but the effect of this change may range from complete loss of activity to no change at all. This is because the effect of missense mutations on protein function depends on the type and location of the amino acid substitution. For instance, replacement of a nonpolar amino acid in the protein's interior with a polar amino acid can drastically alter the protein's three-dimensional structure and therefore its function. Similarly the replacement of a critical amino acid at the active site of an enzyme often destroys its activity. However, the replacement of one polar amino acid with another at the protein surface may have little or no effect. Missense mutations play a very important role in providing new variability to drive evolution because they often are not lethal and therefore remain in the gene pool. ▶▶| *Proteins (appendix I)* *Mutation by Base Substitution*

Nonsense mutations convert a sense codon (i.e., one that codes for an amino acid) to a nonsense codon (i.e., a stop codon: one that does not code for an amino acid). This causes the early termination of translation and therefore results in a shortened polypeptide. Depending on the location of the mutation in the gene, the phenotype may be more or less severely affected. Most proteins retain some function if they are shortened by only one or two amino acids; complete loss of normal function usually results if the mutation occurs closer to the beginning or middle of the gene.

Frameshift mutations arise from the insertion or deletion of one or two base pairs within the coding region of the gene. Since the code consists of a precise sequence of triplet codons, the addition or deletion of fewer than three base pairs causes the reading frame to be shifted for all codons downstream. Frameshift mutations usually are very deleterious and yield mutant phenotypes resulting from the synthesis of nonfunctional proteins. In addition, frameshift mutations often produce a stop codon so that the peptide product is shorter as well as different in sequence. Of course, if the frameshift occurs near the end of the gene or if there is a second frameshift shortly downstream from the first that restores the reading frame, the phenotypic effect might not be as drastic. A second, nearby frameshift that restores the proper reading frame is an example of an intragenic suppressor mutation (table 14.2). *Addition and Deletion Mutations*

As noted, changes in protein structure can lead to changes in protein function, which in turn can alter the phenotype of an organism in several different ways. Morphological mutations change the microorganism's colonial or cellular morphology. Lethal mutations, when expressed, result in the death of the microorganism. Because a microbe must be able to grow to be isolated and studied, lethal mutations are recovered only if they are recessive in diploid organisms or are conditional mutations in haploid organisms. **Conditional mutations** are those that are expressed only under certain environmental conditions. For example, a conditional lethal mutation in *E. coli* might not be expressed under permissive conditions such as low temperature but would be expressed under restrictive conditions such as high temperature. Thus the mutant would grow normally at cooler temperatures but would die at high temperatures.

Biochemical mutations are those causing a change in the biochemistry of the cell. Since these mutations often inactivate a biosynthetic pathway, they frequently eliminate the capacity of the mutant to make an essential molecule such as an amino acid or nucleotide. A strain bearing such a mutation has a conditional phenotype: it is unable to grow on medium lacking that molecule but grows when the molecule is provided. Such mutants are called **auxotrophs,** and they are said to be auxotrophic for the molecule they cannot synthesize. If the wild-type strain from which the mutant arose is a chemoorganotroph able to grow on a minimal medium containing only salts (to supply needed elements such as nitrogen and phosphorus) and a carbon source, it is called a **prototroph.** Another type of biochemical mutant is the resistance mutant. These mutants have acquired resistance to some pathogen, chemical, or antibiotic.

Table 14.2 Types of Point Mutations

Type of Mutation	*Change in DNA*	*Example*
Forward Mutations		
None	None	5′-A-T-G-A-C-C-T-C-C-C-C-G-A-A-A-G-G-G-3′ Met - Thr - Ser - Pro - Lys - Gly
Silent	Base substitution	5′-A-T-G-A-C-A-T-C-C-C-C-G-A-A-A-G-G-G-3′ Met - Thr - Ser - Pro - Lys - Gly
Missense	Base substitution	5′-A-T-G-A-C-C-T-G-C-C-C-G-A-A-A-G-G-G-3′ Met - Thr - Cys - Pro - Lys - Gly
Nonsense		5′-A-T-G-A-C-C-T-C-C-C-C-G-T-A-A-G-G-G-3′ Met - Thr - Ser - Pro - STOP!
Frameshift	Insertion/deletion	5′-A-T-G-A-C-C-T-C-C-G-C-C-G-A-A-A-G-G-G-3′ Met - Thr - Ser - Ala - Glu - Arg
Reverse Mutations		
	Base substitution	5′-A-T-G-A-C-C-T-C-C $\xrightarrow{\text{forward}}$ A-T-G-C-C-C-T-C-C $\xrightarrow{\text{reverse}}$ A-T-G-A-C-C-T-C-C Met - Thr - Ser → Met - Pro - Ser → Met - Thr - Ser
	Base substitution	5′-A-T-G-A-C-C-T-C-C $\xrightarrow{\text{forward}}$ A-T-G-A-C-C-T-G-C $\xrightarrow{\text{reverse}}$ A-T-G-A-C-C-A-G-C Met - Thr - Ser → Met - Thr - Cys → Met - Thr - Ser
	Base substitution	5′-A-T-G-A-C-C-T-C-C $\xrightarrow{\text{forward}}$ A-T-G-C-C-C-T-C-C $\xrightarrow{\text{reverse}}$ A-T-G-C-T-C-T-C-C Met - Thr - Ser (polar amino acid) → Met - Pro - Ser (nonpolar amino acid) → Met - Leu - Ser (polar amino acid) pseudo-wild type
Suppressor Mutations		
Frameshift of opposite sign (intragenic suppressor)	Insertion/deletion	5′-A-T-G-A-C-C-T-C-C-C-C-G-A-A-A-G-G-G-3′ Met - Thr - Ser - Pro - Lys - Gly ↓ Forward mutation 5′-A-T-G-A-C-C-T-C-C-G-C-C-G-A-A-A-G-G-G-3′ Met - Thr - Ser - Ala - Glu - Arg ↓ Suppressor mutation (deletion) 5′-A-T-G-A-C-C-C-C-G-C-C-G-A-A-A-G-G-G-3′ Met - Thr - Pro - Pro - Lys - Gly
Extragenic suppressor		
Nonsense suppressor		Gene (e.g., for tyrosine tRNA) undergoes mutational event in its anticodon region that enables it to recognize and align with a mutant nonsense codon (e.g., UAG) to insert an amino acid (tyrosine) and permit completion of translation.
Physiological suppressor		A defect in one chemical pathway is circumvented by another mutation—for example, one that opens up another chemical pathway to the same product or one that permits more efficient uptake of a compound produced in small quantities because of the original mutation.

Auxotrophic and resistance mutants are quite important in microbial genetics due to the ease of their detection and their relative abundance.

Mutations in Regulatory Sequences

Some of the most interesting and informative mutations studied by microbial geneticists are those that occur in the regulatory sequences responsible for controlling gene expression. Constitutive lactose operon mutants in *E. coli* are excellent examples. Many of these mutations map in the operator site and produce altered operator sequences that are not recognized by the repressor protein. Therefore the operon is continuously transcribed, and β-galactosidase is always synthesized. Mutations in promoters also have been identified. If the mutation renders the promoter sequence nonfunctional, the mutant will be unable to synthesize the product, even though the coding region of the structural gene is completely normal. Without a fully functional promoter, RNA polymerase rarely transcribes a gene as well as wild type. ◀◀ *Regulation of transcription initiation (section 13.2)*

Mutations in tRNA and rRNA Genes

Mutations in tRNA and rRNA alter the phenotype of an organism through disruption of protein synthesis. In fact, these mutants often are initially identified because of their slow growth. On the other hand, a suppressor mutation involving tRNA restores normal (or near normal) growth rates. In these mutations, a base substitution in the anticodon region of a tRNA allows the insertion of the correct amino acid at a mutant codon (table 14.2).

1. List three ways in which spontaneous mutations might arise.
2. How do the mutagens 5-bromouracil, methylnitrosoguanidine, proflavin, and UV radiation induce mutations?
3. Give examples of intragenic and extragenic suppressor mutations.
4. Sometimes a point mutation does not change the phenotype. List all the reasons why this is so.
5. Why might a missense mutation at a protein's surface not affect the phenotype of an organism, whereas the substitution of an internal amino acid does?

14.2 Detection and Isolation of Mutants

To study microbial mutants, they must be readily detected, even when they are rare, and then efficiently isolated from wild-type organisms and other mutants that are not of interest. Microbial geneticists typically increase the likelihood of obtaining mutants by using mutagens to increase the rate of mutation. The rate can increase from the usual one mutant per 10^7 to 10^{11} cells to about one per 10^3 to 10^6 cells. Even at this rate, mutations are rare, and carefully devised means for detecting or selecting a desired mutation must be used. This section describes some techniques used in mutant detection, selection, and isolation.

Mutant Detection

When collecting mutants of a particular organism, the wild-type characteristics must be known so that an altered phenotype can be recognized. A suitable detection system for the mutant phenotype also is needed. The use of such detection systems is called screening. Screening for mutant phenotypes in haploid organisms is straightforward because any mutation usually can be seen immediately. Some screening procedures require only examination of colony morphology. For instance, if albino mutants of a normally pigmented bacterium are being studied, detection simply requires visual observation of colony color. Other screening methods are more complex. For example, the **replica plating** technique is used to screen for auxotrophic mutants. It distinguishes between mutants and the wild-type strain based on their ability to grow in the absence of a particular biosynthetic end product (**figure 14.5**). A lysine auxotroph, for instance, grows on lysine-supplemented media but not on a medium lacking an adequate supply of lysine because it cannot synthesize this amino acid.

Once a screening method is devised, mutants are collected. However, mutant collection can present practical problems. Consider a search for the albino mutants mentioned previously. If the mutation rate were around one in a million, on average a million or more organisms would have to be tested to find one albino mutant. This probably would require several thousand plates. The task of isolating auxotrophic mutants in this way would be even more taxing with the added labor of replica plating. Thus, if possible, it is more efficient to use a selection system employing some environmental factor to separate mutants from wild-type microorganisms. Examples of selection systems are described next.

Mutant Selection

An effective selection technique uses incubation conditions under which the mutant grows because of properties conferred by the mutation, whereas the wild type does not. Selection methods often involve reversion mutations or the development of resistance to an environmental stress. For example, if the intent is to isolate revertants from a lysine auxotroph (Lys$^-$), the approach is quite easy. A large population of lysine auxotrophs is plated on minimal medium lacking lysine, incubated, and examined for colony formation. Only cells that have mutated to restore the ability to manufacture lysine will grow on minimal medium. Several million cells can be plated on a single Petri dish, but only the rare revertant cells will grow. Thus many cells can be tested for mutations by scanning a few Petri dishes for growth. This method has proven very useful in determining the relative mutagenicity of many substances.

Methods for selecting mutants resistant to a particular environmental stress follow a similar approach. Often wild-type cells are susceptible to virus attack or antibiotic treatment, so it is possible to grow the microbe in the presence of the agent and look for surviving organisms. Consider the example of a phage-sensitive wild-type bacterium. When it is cultured in medium lacking the virus and then plated on selective medium containing viruses, any colonies that form are resistant to virus attack and very likely are mutants in this regard. This type of selection can be used for virtually any environmental parameter; resistance to antibiotics and specific temperatures are commonly used.

Substrate utilization mutations also are employed in bacterial selection. Many bacteria use only a few primary carbon sources. With such bacteria, it is possible to select mutants by plating a culture on medium containing an alternate carbon source. Any colonies that appear can use the substrate and are probably mutants.

Mutant screening and selection methods are used for purposes other than understanding more about the nature of genes or the biochemistry of a particular microorganism. One very important role of mutant selection and screening techniques is in the study of carcinogens. Next we briefly describe one of the first and perhaps best known of the carcinogen testing systems.

Mutagens and Carcinogens

An increased understanding of the mechanisms of mutation and their role in cancer has stimulated efforts to identify environmental carcinogens—agents that cause cancer. The observation that many carcinogens also are mutagens was the basis for development of the Ames test by Bruce Ames in the 1970s. This test determines if a substance increases the rate of mutation; that is, if it is a mutagen. If the substance is a mutagen, then it is likely that it will also be carcinogenic if an animal is exposed to it at sufficient levels. Note that the test does not directly test for carcinogenicity. This is because it uses a bacterium as the test organism. Carcinogenicity can only be directly demonstrated with animals. Such testing is extremely expensive and takes much longer to complete than does the Ames test. Thus the Ames test serves as an inexpensive screening procedure to identify chemicals that may be carcinogenic.

The Ames test is a mutational reversion assay employing several strains of *Salmonella enterica* serovar Typhimurium, each of which has a different mutation in the histidine biosynthesis operon; that is to say, they are histidine auxotrophs. The bacteria also have mutational alterations of their cell walls that make them more permeable to test substances. To further increase assay sensitivity, the strains are defective in the ability to repair DNA correctly.

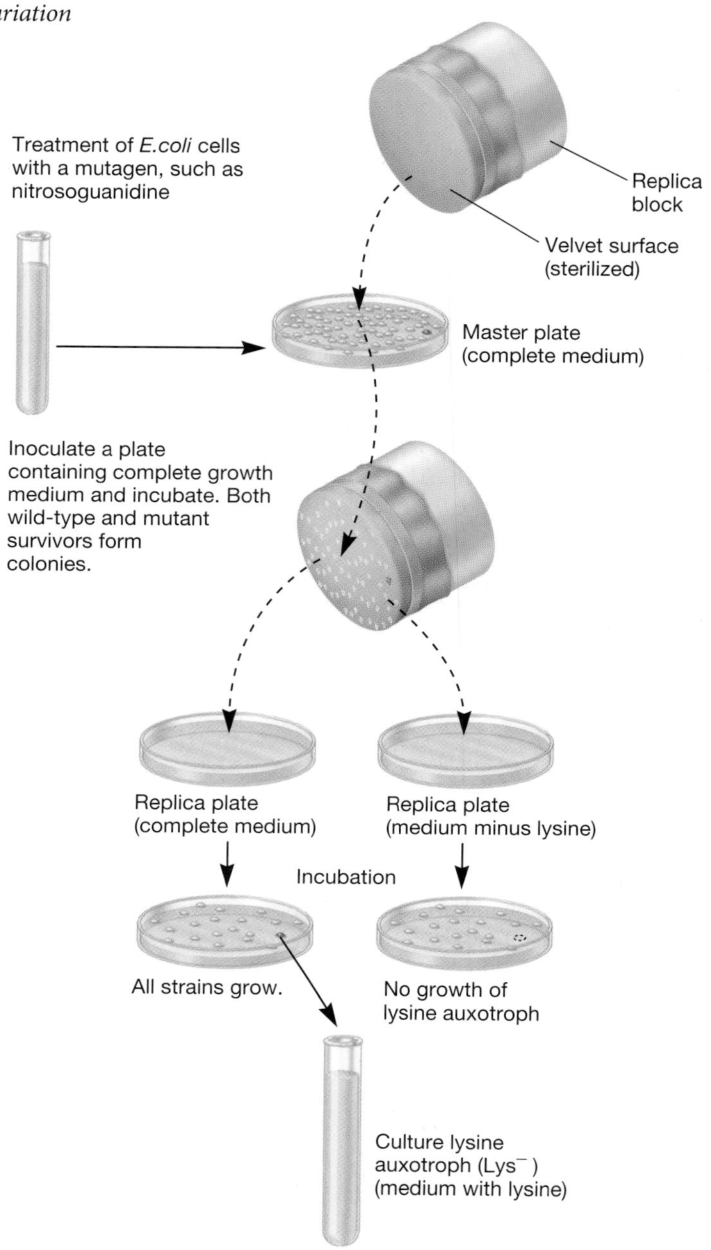

Treatment of *E.coli* cells with a mutagen, such as nitrosoguanidine
Replica block
Velvet surface (sterilized)
Master plate (complete medium)
Inoculate a plate containing complete growth medium and incubate. Both wild-type and mutant survivors form colonies.
Replica plate (complete medium)
Replica plate (medium minus lysine)
Incubation
All strains grow.
No growth of lysine auxotroph
Culture lysine auxotroph (Lys$^-$) (medium with lysine)

FIGURE 14.5 Replica Plating. The use of replica plating to isolate a lysine auxotroph. After growth of a mutagenized culture on a complete medium, a piece of sterile velvet is pressed on the plate surface to pick up bacteria from each colony. Then the velvet is pressed to the surface of other plates, and organisms are transferred to the same position as on the master plate. After the location of Lys$^-$ colonies growing on the replica with complete medium are determined, the auxotrophs can be isolated and cultured.

Figure 14.5 Micro Inquiry

How would you screen for a tryptophan auxotroph? How would you select for a mutant that is resistant to the antibiotic ampicillin but sensitive to tetracycline (assume the parental stain is resistant to both antibiotics)?

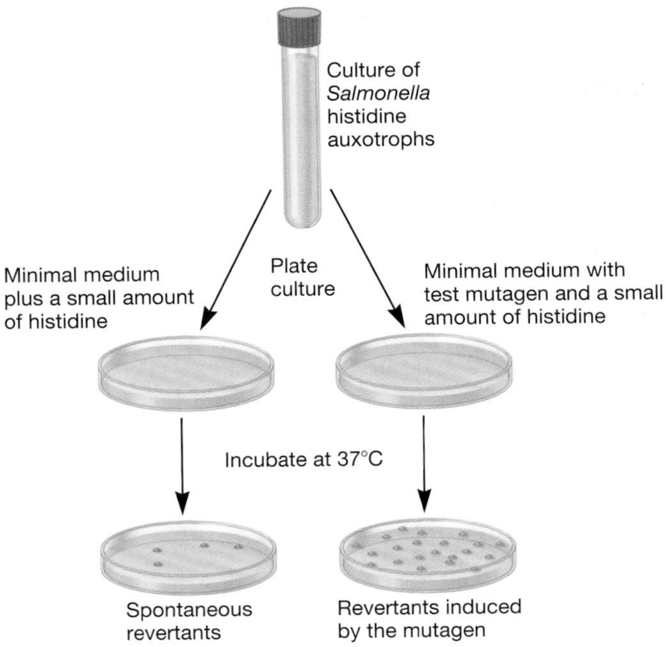

FIGURE 14.6 The Ames Test for Mutagenicity.

In the Ames test, these tester strains of *Salmonella* are plated with the substance being tested and the appearance of visible colonies followed (**figure 14.6**). To ensure that DNA replication can take place in the presence of the potential mutagen, the bacteria and test substance are mixed in dilute molten top agar to which a trace of histidine has been added. This molten mix is then poured on top of minimal agar plates and incubated for 2 to 3 days at 37°C. All of the histidine auxotrophs grow for the first few hours in the presence of the test compound until the histidine is depleted. This is necessary because replication is required for the development of a mutation (figure 14.1). Once the histidine supply is exhausted, only revertants that have mutationally regained the ability to synthesize histidine continue to grow and produce visible colonies. These colonies need only be counted and compared to controls to estimate the relative mutagenicity of the compound: the more colonies, the greater the mutagenicity.

Some chemicals tested may not be mutagenic unless they are transformed into another, more active form. In animals, such transformations occur in the liver. Indeed, many known carcinogens, (e.g., aflatoxins; *see figure 40.6*) are not actually carcinogenic until they are modified in the liver. One function of the liver is to produce enzymes that destroy toxins and other materials that may be circulating in the blood. However, in some cases, these enzymes transform chemicals into more dangerous forms. For this reason, a mammalian liver extract is often added to the molten top agar prior to plating the bacterial cells used in the Ames test. The extract converts potential mutagens into derivatives that readily react with DNA, mimicking the enzymatic transformations that occur in mammals. Carrying out the Ames test with and without the addition of the extract shows which compounds have intrinsic mutagenicity and which need activation after uptake. Despite the use of liver extracts, only about half the potential animal carcinogens are detected by the Ames test.

1. Describe how replica plating is used to detect and isolate auxotrophic mutants.
2. Why are mutant selection techniques generally preferable to screening methods?
3. Briefly discuss how reversion mutations, resistance to an environmental factor, and the ability to use a particular nutrient can be employed in mutant selection.
4. Describe how you would isolate a mutant that required histidine for growth and was resistant to penicillin. The wild type is a prototroph.
5. What is the Ames test and how is it carried out? What assumption concerning mutagenicity and carcinogenicity is it based upon?

14.3 DNA Repair

Because replication errors and mutagens can alter nucleotide sequences, a microorganism must be able to repair any changes that might be lethal. In addition to **proofreading** by DNA polymerases during replication—the removal of an incorrect nucleotide immediately after its addition to the growing end of the chain—several other mechanisms can restore the sequence or repair damaged DNA. Repair in *E. coli* is best understood and is briefly described in this section. ⏮ *DNA replication (section 12.4)*

Excision Repair

Excision repair corrects damage that causes distortions in the double helix. Two types of excision repair systems have been described: nucleotide excision repair and base excision repair. They both use the same approach to repair: Remove the damaged portion of a DNA strand and use the intact complementary strand as the template for synthesis of new DNA. They are distinguished by the enzymes used to correct DNA damage.

In **nucleotide excision repair,** a repair enzyme called UvrABC endonuclease removes damaged nucleotides and a few nucleotides on either side of the lesion. The resulting single-stranded gap is filled by DNA polymerase I, and DNA ligase joins the fragments (**figure 14.7**). This system can remove thymine dimers (figure 14.4) and repair almost any other injury that produces a detectable distortion in DNA.

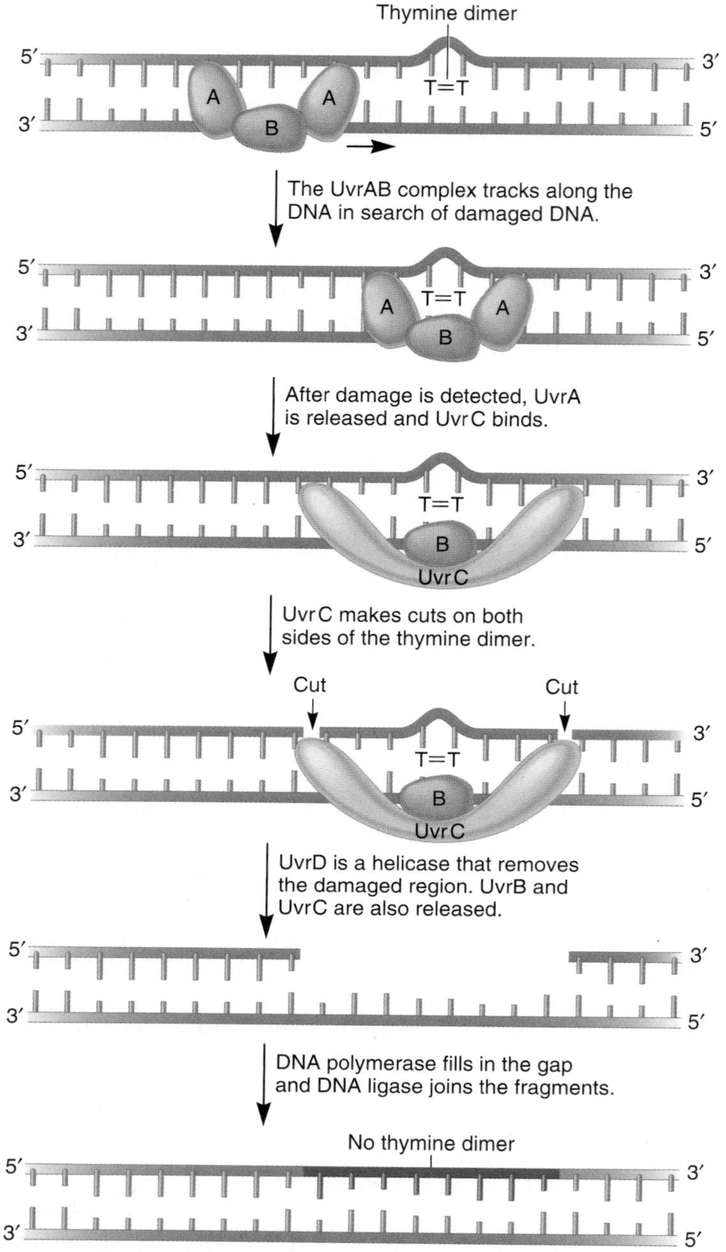

FIGURE 14.7 Nucleotide Excision Repair in *E. coli*.

Figure 14.7 Micro Inquiry

How is damaged DNA recognized by the UvrAB complex?

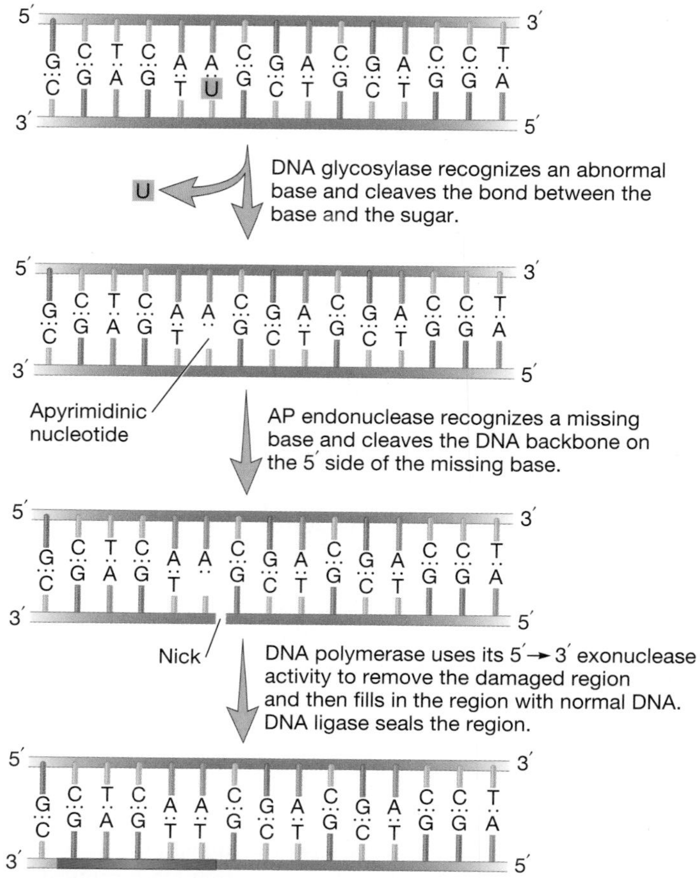

FIGURE 14.8 Base Excision Repair.

Base excision repair employs DNA glycosylases to remove damaged or unnatural bases yielding apurinic or apyrimidinic (AP) sites. Enzymes called AP endonucleases recognize the damaged DNA and nick the backbone at the AP site (**figure 14.8**). DNA polymerase I removes the damaged region, using its 5′ to 3′ exonuclease activity. It then fills in the gap, and DNA ligase joins the DNA fragments.

Direct Repair

Thymine dimers and alkylated bases often are corrected by **direct repair. Photoreactivation** repairs thymine dimers (figure 14.4) by splitting them apart with the help of visible light. This photochemical reaction is catalyzed by the enzyme photolyase. Methyls and some other alkyl groups that have been added to the O^6 position of guanine can be removed with the help of an enzyme known as alkyltransferase or methylguanine methyltransferase. Thus damage to guanine from mutagens such as methylnitrosoguanidine (figure 14.3) can be repaired directly. *Direct Repair*

Mismatch Repair

Despite the accuracy of DNA polymerase and its continual proofreading, errors still are made during DNA replication. Remaining mismatched bases are usually detected and repaired by the **mismatch repair** system in *E. coli* (**figure 14.9**). The mismatch correction enzyme MutS scans the newly replicated DNA for mismatched pairs. Another enzyme, MutH, removes a stretch of newly synthesized DNA around the mismatch. A DNA polymerase then replaces the excised nucleotides, and the resulting

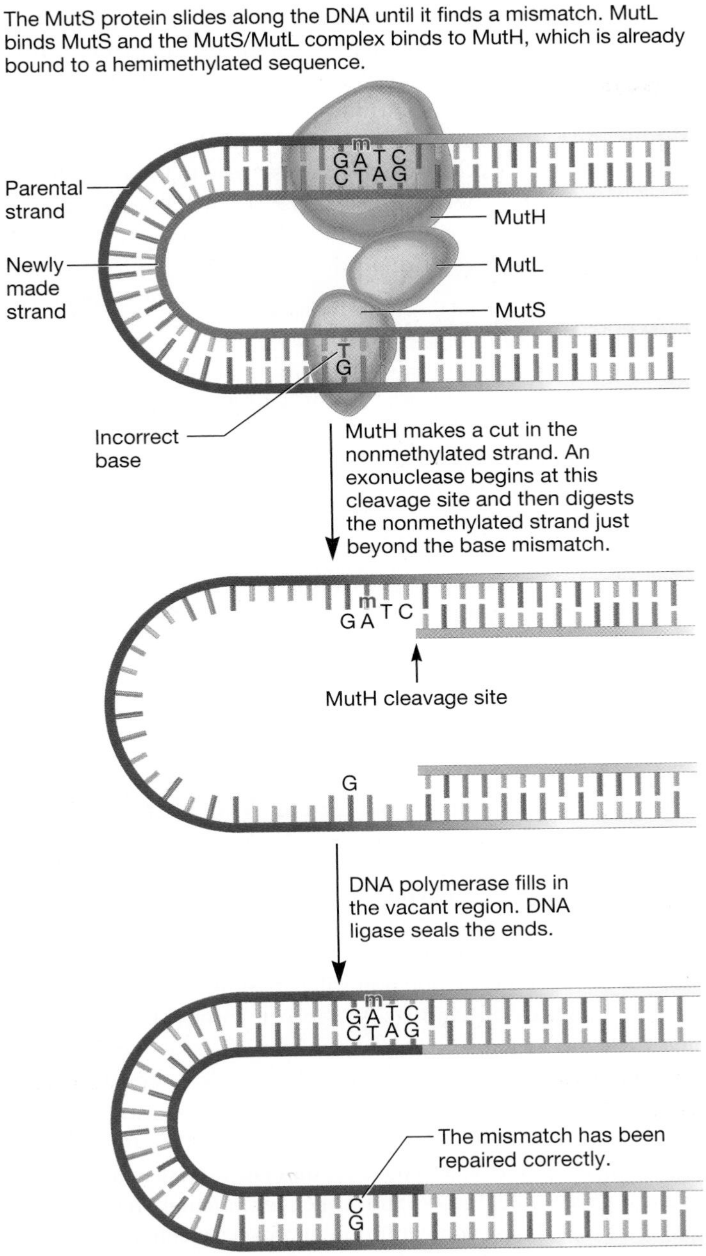

The MutS protein slides along the DNA until it finds a mismatch. MutL binds MutS and the MutS/MutL complex binds to MutH, which is already bound to a hemimethylated sequence.
Parental strand
Newly made strand
m
GATC
CTAG
MutH
MutL
MutS
T
G
Incorrect base
MutH makes a cut in the nonmethylated strand. An exonuclease begins at this cleavage site and then digests the nonmethylated strand just beyond the base mismatch.
GATC
MutH cleavage site
G
DNA polymerase fills in the vacant region. DNA ligase seals the ends.
GATC
CTAG
The mismatch has been repaired correctly.
C
G

FIGURE 14.9 Methyl-Directed Mismatch Repair in *E. coli*. The role of MutH is to identify the methylated strand of DNA, which is the nonmutated parental strand. The methylated adenine is designated by an m.

Figure 14.9 Micro Inquiry

How is mismatch repair similar to DNA polymerase proofreading? How is it different?

nick is sealed by DNA ligase. In this regard, mismatch repair is similar to excision repair.

Successful mismatch repair depends on the ability of enzymes to distinguish between old and newly replicated DNA strands. This distinction is possible because newly replicated DNA strands lack methyl groups on their bases, whereas older DNA has methyl groups on the bases of both strands. **DNA methylation** is catalyzed by DNA methyltransferases and results in three different products: *N*6-methyladenine, 5-methylcytosine, and *N*4-methylcytosine. After strand synthesis, the *E. coli* DNA *a*denine *m*ethyltransferase (DAM) methylates adenine bases in GATC sequences to form *N*6-methyladenine. For a short time after the replication fork has passed, the new strand lacks methyl groups while the template strand is methylated. In other words, the DNA is temporarily hemimethylated. The repair system cuts out the mismatch from the unmethylated strand. *Mismatch Repair*

Recombinational Repair

Recombinational repair corrects damaged DNA in which both bases of a pair are missing or damaged, or where there is a gap opposite a lesion. In this type of repair, the **RecA protein** cuts a piece of template DNA from a sister molecule and puts it into the gap or uses it to replace a damaged strand (**figure 14.10**). Although bacterial cells are haploid, another copy of the damaged segment often is available because either it has recently been replicated or the cell is growing rapidly and has more than one copy of its chromosome. Once the template is in place, the remaining damage can be corrected by another repair system.

SOS Response

Despite having multiple repair systems, sometimes the damage to an organism's DNA is so great that the normal repair mechanisms just described cannot repair all the damage. As a result, DNA synthesis stops completely. In such situations, a global control network called the **SOS response** is activated. In this response, numerous genes are activated when a repressor called LexA is destroyed. LexA negatively controls these genes, and once it is destroyed, they are transcribed and the SOS response ensues.

The SOS response, like recombinational repair, depends on the activity of the RecA protein. RecA binds to single- or double-stranded DNA breaks and gaps generated by cessation of DNA synthesis. RecA binding initiates recombinational repair. Simultaneously RecA takes on coprotease function. It interacts with LexA, causing LexA to destroy itself (autoproteolysis). Destruction of LexA increases transcription of genes for excision repair and recombinational repair, in particular. The first genes to be transcribed are those that encode the Uvr proteins needed for nucleotide excision repair (figure 14.7). Then expression of genes involved in recombinational repair is further increased. To give the cell time to repair its DNA, the protein SfiA is produced; SfiA blocks cell division. Finally, if the DNA has not been fully repaired after about 40 minutes, a process called **translesion DNA synthesis** is triggered. In this process, DNA polymerase IV (also known as DinB) and DNA polymerase V (UmuCD) synthesize DNA across gaps and other lesions (e.g., thymine dimers) that had stopped DNA

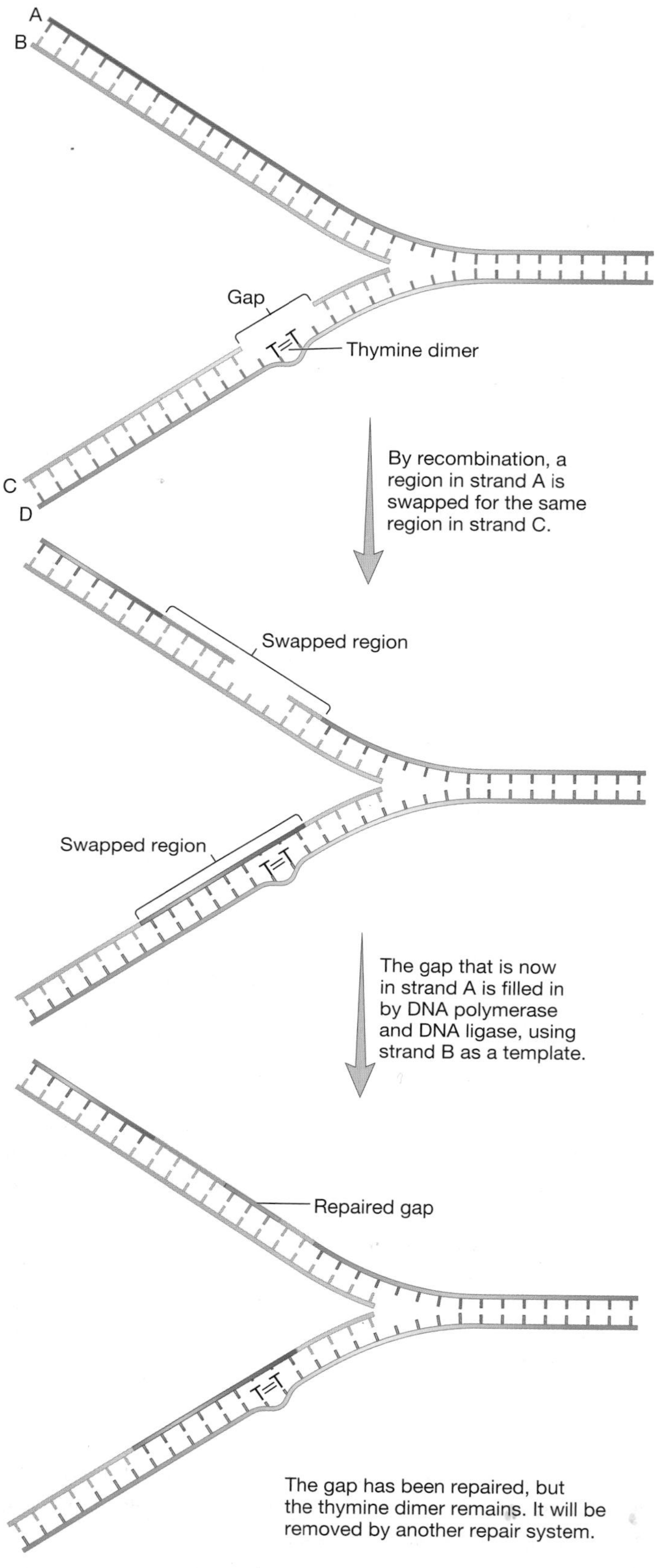
A
B
Gap
T=T
Thymine dimer
C
D
By recombination, a region in strand A is swapped for the same region in strand C.
Swapped region
Swapped region
T=T
The gap that is now in strand A is filled in by DNA polymerase and DNA ligase, using strand B as a template.
Repaired gap
T=T
The gap has been repaired, but the thymine dimer remains. It will be removed by another repair system.

FIGURE 14.10 Recombinational Repair.

polymerase III. However, because an intact template does not exist, these DNA polymerases often insert incorrect bases. Furthermore, they lack proofreading activity. Therefore even though DNA synthesis continues, it is highly error prone and results in the generation of numerous mutations. The SOS response is so named because it is a response made in a life-or-death situation. The response increases the likelihood that some cells will survive by allowing DNA synthesis to continue. For the cell, the risk of dying because of failure to replicate DNA is greater than the risk posed by the mutations generated by this error-prone process.

1. Compare and contrast each of the following repair processes: excision repair, recombinational repair, direct repair, and SOS response.
2. Explain how the following DNA alterations and replication errors would be corrected (there may be more than one way): base addition errors by DNA polymerase III during replication, thymine dimers, AP sites, methylated guanines, and gaps produced during replication.

14.4 Creating Genetic Variability

As discussed in section 14.1, the consequences of mutations can range from no effect to being lethal, depending not only on the nature of the mutation but also on the environment in which the organism lives. Thus all mutations are subject to selective pressure, and this determines if a mutation will persist in a population. Each mutant form is called an **allele,** an alternate form of the gene. Mutant alleles, as well as the wild-type allele, can be combined with other genes, leading to an increase in the genetic variability within a population. Each genotype in a population can be selected for or selected against. Organisms with genotypes, and therefore phenotypes, that are best suited to the environment survive and are most likely to pass on their genes. Shifts in environmental pressures can lead to changes in the population and ultimately result in the evolution of new species. The mechanisms by which new combinations of genes are generated are the topic of this section. All involve **recombination,** the process in which one or more nucleic acid molecules are rearranged or combined to produce a new nucleotide sequence. This is normally accompanied by a phenotypic change. Geneticists refer to organisms produced following a recombination event as recombinant organisms or simply **recombinants.**

Horizontal Gene Transfer in *Bacteria* and *Archaea*

The transfer of genes from parents to progeny is called vertical gene transfer. In eukaryotes, vertical gene transfer also introduces genetic variation into populations. With each new generation, recombinant progeny are formed when gametes from parents fuse during sexual reproduction. However, *Bacteria* and *Archaea* do not reproduce sexually. This suggests that genetic variation in populations of these microbes should be relatively limited, only occurring with the advent of a new mutation and its transfer to the next generation. However, this is not the case. *Bacteria* and *Archaea* have evolved three different mechanisms for creating recombinants. These mechanisms are referred to collectively as **horizontal (lateral) gene transfer (HGT).** HGT is distinctive from vertical gene transfer because genes from one independent, mature organism are transferred to another, often creating a stable recombinant having characteristics of both the donor and the recipient. *Horizontal Gene Transfer*

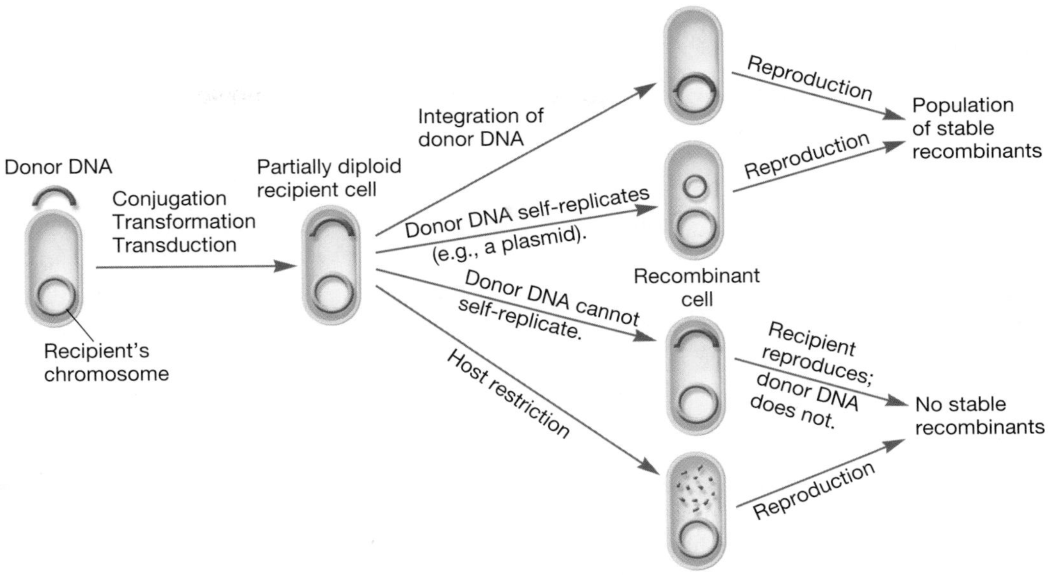

FIGURE 14.11 The Fate of Donor DNA During Horizontal Gene Transfer.

The importance of HGT cannot be overstated. HGT has been important in the evolution of many species, and it continues to be commonplace in many environments. Clear examples are known of DNA transfer between distantly related genera, particularly among bacterial cells sharing an ecological niche (e.g., the human gut). Another important example is the transfer of groups of genes encoding virulence factors from pathogenic bacteria to other bacteria. These groups of genes are called pathogenicity islands, and their discovery has provided valuable insights into the evolution of pathogenic bacteria. Finally, in most cases, the acquisition of new traits by HGT enables microbes to rapidly expand their ecological niche. This is clearly seen when HGT results in the spread of antibiotic-resistance genes among pathogenic bacteria. ▶▶| *Evolutionary processes and the concept of a microbial species (section 17.5)*

Search This: HGT/Antibiotic resistance

During HGT, a piece of donor DNA, sometimes called the exogenote, enters a recipient cell. The transfer can occur in three ways: direct transfer between two cells temporarily in physical contact (conjugation), transfer of a naked DNA fragment (transformation), and transport of DNA by viruses (transduction). If the donor DNA contains genes already present in the recipient, the recipient will become temporarily diploid for those genes. This partially diploid cell is sometimes called a merozygote (**figure 14.11**). The donor DNA has four possible fates in the recipient. First, when the donor DNA has a sequence homologous to one in the recipient's chromosome (sometimes called an endogenote), integration may occur. That is, the donor's DNA may pair with the recipient's DNA and be incorporated to yield a recombinant genome. The recombinant then reproduces, yielding a population of stable recombinants. Second, if the donor DNA is able to replicate itself (e.g., it is a plasmid), it may persist separate from the recipient's chromosome. When the recipient reproduces, the donor DNA replicates and a population of stable recombinants is formed. Third, the donor DNA remains in the cytoplasm but is unable to replicate. When the recipient reproduces, the progeny cells lack the donor DNA and it is eventually lost from the population. Finally, a process called host restriction may occur. In this process, host cell nucleases degrade the donor DNA, thereby preventing the formation of a recombinant cell.

Recombination at the Molecular Level

Although different processes are used by organisms in the three domains of life to create recombinant organisms, the mechanisms of recombination at the molecular level are remarkably similar. Three types of recombination are observed: homologous recombination, site-specific recombination, and transposition.

Homologous recombination, the most common recombination event, usually involves a reciprocal exchange between a pair of DNA molecules with the same nucleotide sequence. It can occur anywhere on the chromosome, and it results from DNA strand breakage and reunion leading to crossing-over. Homologous recombination is carried out by the products of the *rec* genes, including the RecA protein, which is also important for DNA repair (**table 14.3**). The most widely accepted model of homologous recombination is the **double-stranded break model** (**figure 14.12**). It proposes that duplex DNA with a double-stranded break is processed to create DNA with single-stranded ends. RecA promotes the insertion of one single-stranded end into an intact, homologous piece of DNA. This is

Table 14.3 ***E. coli* Homologous Recombination Proteins**

Protein	*Description*
Rec BCD	Recognizes double-stranded breaks and then generates single-stranded regions at the break site that are involved in strand invasion
Single-strand binding protein	Prevents excessive strand degradation by RecBCD
RecA	Promotes strand invasion and displacement of complementary strand to generate D loop
RecG	Helps form Holliday junctions and promotes branch migration
RuvABC	Endonuclease that binds Holliday junctions, promotes branch migration, and cuts strands in the Holliday junction in order to separate chromosomes

called strand invasion. As can be seen in figure 14.12, strand invasion results in the formation of two gaps in the two parent DNA molecules. The gaps are filled, yielding a structure with **heteroduplex DNA;** that is, it contains strands derived from both parent molecules. The two parental DNA molecules are now linked together by two structures called Holliday junctions. These structures move along the DNA molecule during branch migration until they are finally cut and the two DNA molecules are separated. Depending on how this occurs, the resulting DNA molecules will be either recombinant or nonrecombinant. In some cases, a nonreciprocal form of homologous recombination occurs (**figure 14.13**). In the Fox model of nonreciprocal homologous recombination, a piece of genetic material is inserted into the chromosome through the incorporation of a single strand to form a stretch of heteroduplex DNA.

The other two types of recombination do not depend on long regions of sequence homology. **Site-specific recombination** is particularly important in the integration of viral genomes into host chromosomes. The enzymes responsible for this event are often specific for sequences within the particular virus and its host. **Transposition** can occur at many sites in the genome and is discussed in more detail in section 14.5.

1. Distinguish among the three forms of recombination mentioned in this section.
2. What four fates can DNA have after entering a bacterium?

14.5 Transposable Elements

The chromosomes of viruses and cells contain pieces of DNA that can move and integrate into different sites in the chromosomes. Such movement is called transposition, and it plays important roles in the generation of new gene combinations. DNA segments that carry the genes required for transposition are **transposable elements,** sometimes called mobile genetic elements or "jumping genes." Unlike other processes that reorganize DNA, transposition does not require extensive areas of homology between the transposon and its destination site. Mobile genetic elements were first discovered in the 1940s by Barbara McClintock during her studies on maize genetics (a discovery for which she was awarded the Nobel Prize in 1983). They have been most intensely studied in *Bacteria.* *Transposition: Shifting Segments of the Genome*

The simplest transposable elements are **insertion sequences,** or IS elements for short (**figure 14.14*a***). An IS element is a short sequence of DNA (around 750 to 1,600 base pairs [bp] in length). It contains only the gene for the enzyme transposase, and it is bounded at both ends by inverted repeats—identical or very similar sequences of nucleotides in reversed orientation. Inverted repeats are usually about 15 to 25 base pairs long and vary among IS elements so that each type of IS has its own characteristic inverted repeats. **Transposase** is required for transposition and accurately recognizes the ends of the IS. Each IS element is named by giving it the prefix IS followed by a number. IS elements have been observed in a variety of bacteria and some archaea.

Mobile genetic elements that contain genes in addition to those required for transposition (e.g., antibiotic resistance or toxin genes) are called **transposons.** Some transposons consist of a central region containing the extra genes, flanked on both sides by IS elements that are identical or very similar in sequence (figure 14.14*b*). Others are simpler, bounded only by short, inverted repeats, while the coding region contains both transposition genes and the extra genes. It is believed that transposons are formed when two IS elements associate with a central segment containing one or more genes. This association could arise if an IS element replicates and moves only a gene or two down the chromosome. Transposon names begin with the prefix Tn. Some properties of selected transposons are given in **table 14.4**.

The process of transposition in bacterial cells occurs by two basic mechanisms. **Simple transposition,** also called **cut-and-paste transposition,** involves transposase-catalyzed excision of the transposable element, followed by cleavage of a new target site and ligation of the element into this site (**figure 14.15**). Target sites are specific sequences about five to nine base pairs long. When a mobile genetic element inserts at a target site, the

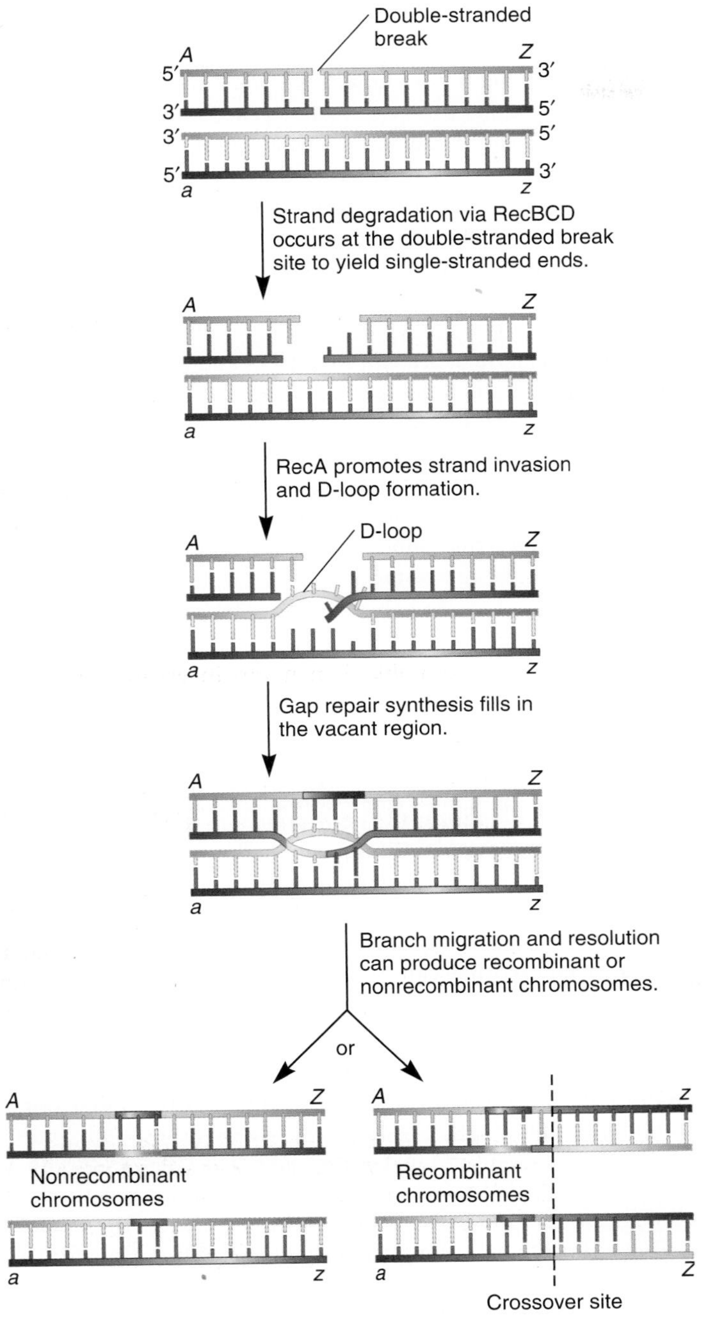

FIGURE 14.12 The Double-Stranded Break Model of Homologous Recombination.

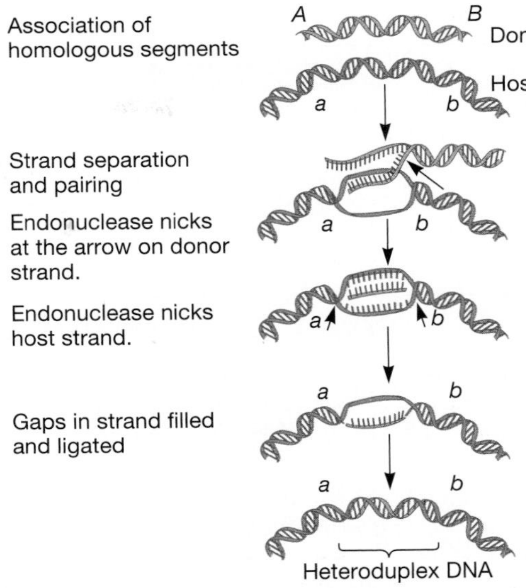

FIGURE 14.13 The Fox Model of Nonreciprocal Homologous Recombination. This mechanism has been proposed for the recombination occurring during transformation in some bacteria.

target sequence is duplicated so that short, direct-sequence repeats flank the element's terminal inverted repeats. *Simple Transposition*

The second transposition mechanism is **replicative transposition.** It is observed for transposons called replicative transposons (figure 14.14*c*). These transposons contain two important genes: one encoding transposase and the other encoding an enzyme called resolvase. In replicative transposition, the original transposon remains at the parental site on the chromosome and a copy is inserted at the target DNA site (**figure 14.16**). The transposition of Tn*3* is a well-studied example of replicative transposition. In the first stage, DNA containing Tn*3* fuses with the target DNA to form a cointegrate molecule (figure 14.16, step 1). This process requires the Tn*3* transposase enzyme coded for by the *tnpA* gene (**figure 14.17**). Note that the cointegrate has two copies of Tn*3*. In the second stage, the cointegrate is resolved to yield two DNA molecules, each with a copy of the transposon (figure 14.16, step 3). Resolution involves a crossover and is catalyzed by a resolvase enzyme coded for by the *tnpR* gene (figure 14.17). *Mechanisms of Transposition*

Transposable elements produce a variety of important effects. They can insert within a gene to cause a mutation or stimulate DNA rearrangement, leading to deletions of genetic material. Because some transposons carry stop codons or termination sequences, when transposed into genes they may block translation or transcription, respectively. Likewise, other transposons carry promoters and can activate genes near the point of insertion. Thus some transposons can turn genes on or off. Transposons also are located in plasmids and participate in such processes as plasmid fusion, insertion of plasmids into chromosomes, and plasmid evolution.

The role of transposons in plasmid evolution is of particular note. Plasmids can contain several different transposon-target sites. Therefore transposons frequently move between plasmids. Of concern is the fact that many transposons contain

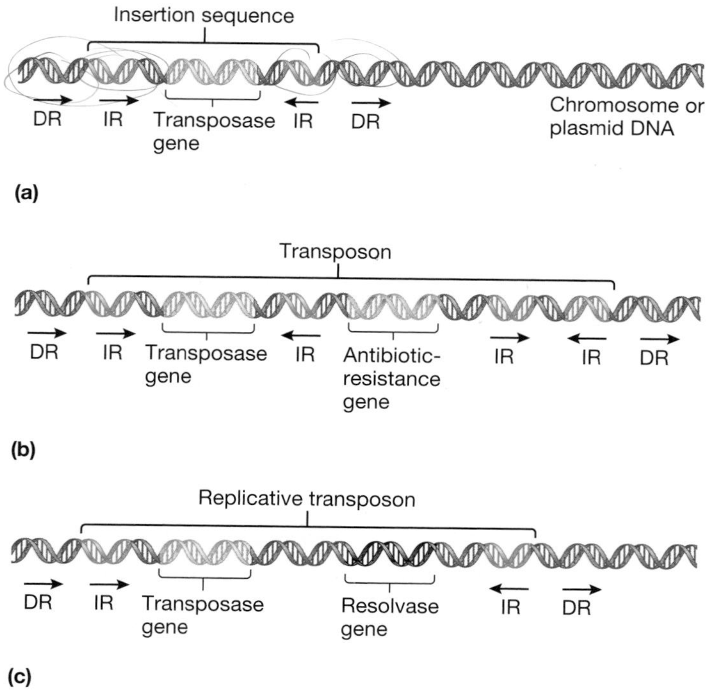

FIGURE 14.14 Transposable Elements. All transposable elements contain common features. These include inverted repeats (IRs) at the ends of the element and a transposase gene. (a) Insertion sequences consist only of IRs on either side of the transposase gene. (b) Transposons and (c) replicative transposons contain additional genes. Insertion sequences and transposons move by simple (cut-and-paste) transposition. Replicative transposons move by making a copy of themselves. One copy remains at the original site and the other moves to the new site. DRs, direct repeats in host DNA, flank a transposable element.

Figure 14.14 Micro Inquiry

What features are common to all types of transposable elements?

antibiotic-resistance genes. Thus as they move from one plasmid to another, resistance genes are introduced into the target plasmid, creating a resistance (R) plasmid. Multiple drug-resistance plasmids can arise from the accumulation of transposons in a plasmid (figure 14.17). Many R plasmids are able to move from one cell to another during conjugation, which spreads the resistance genes throughout a population. Finally, because transposons also move between plasmids and chromosomes, drug-resistance genes can exchange between these two molecules, resulting in the further spread of antibiotic resistance.

Some transposons bear transfer genes and can move between bacteria through the process of conjugation, as discussed in section 14.7. A well-studied example of a **conjugative transposon** is Tn*916* from *Enterococcus faecalis.* Although Tn*916* cannot replicate autonomously, it can transfer itself from *E. faecalis* to a variety of recipients and integrate into their chromosomes. Because it carries a gene for tetracycline resistance, this conjugative transposon also spreads drug resistance.

14.6 Bacterial Plasmids

Conjugation, the transfer of DNA between bacteria involving direct contact, depends on the presence of a plasmid. Recall from chapter 3 that **plasmids** are small, double-stranded DNA molecules that can exist independently of host chromosomes. They have their own replication origins and autonomously replicate and are stably inherited. Some plasmids are **episomes,** plasmids that can exist either with or without being integrated into host chromosomes. Although many plasmid types exist, our concern here is with **conjugative plasmids.** These plasmids can transfer copies of themselves to other bacteria during the process of conjugation, which is discussed in section 14.7.

Perhaps the best-studied conjugative plasmid is **F factor.** It plays a major role in conjugation in *E. coli,* and it was the first conjugative plasmid to be described (**figure 14.18**). The F factor is about 100,000 bases long and bears genes responsible for cell

Table 14.4 The Properties of Selected Transposons

Transposon	*Length (bp)*	*Terminal Repeat Length*	*Terminal Module*	*Genetic Markers*
Tn*3*	4,957	38		Ampicillin resistance
Tn*501*	8,200	38		Mercury resistance
Tn*1681*	2,061		IS*1*	Heat-stable enterotoxin
Tn*2901*	11,000		IS*1*	Arginine biosynthesis

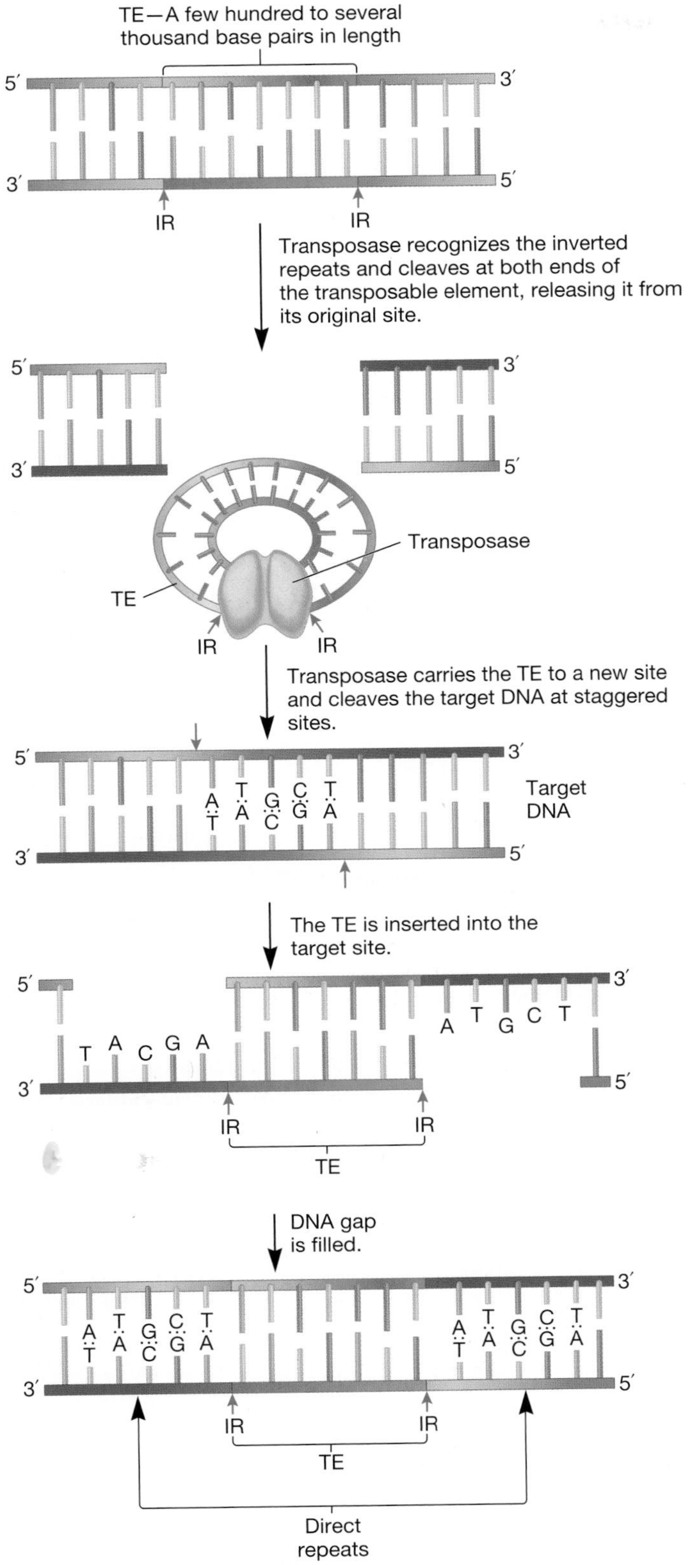

FIGURE 14.15 Simple Transposition. TE, transposable element; IR, inverted repeat.

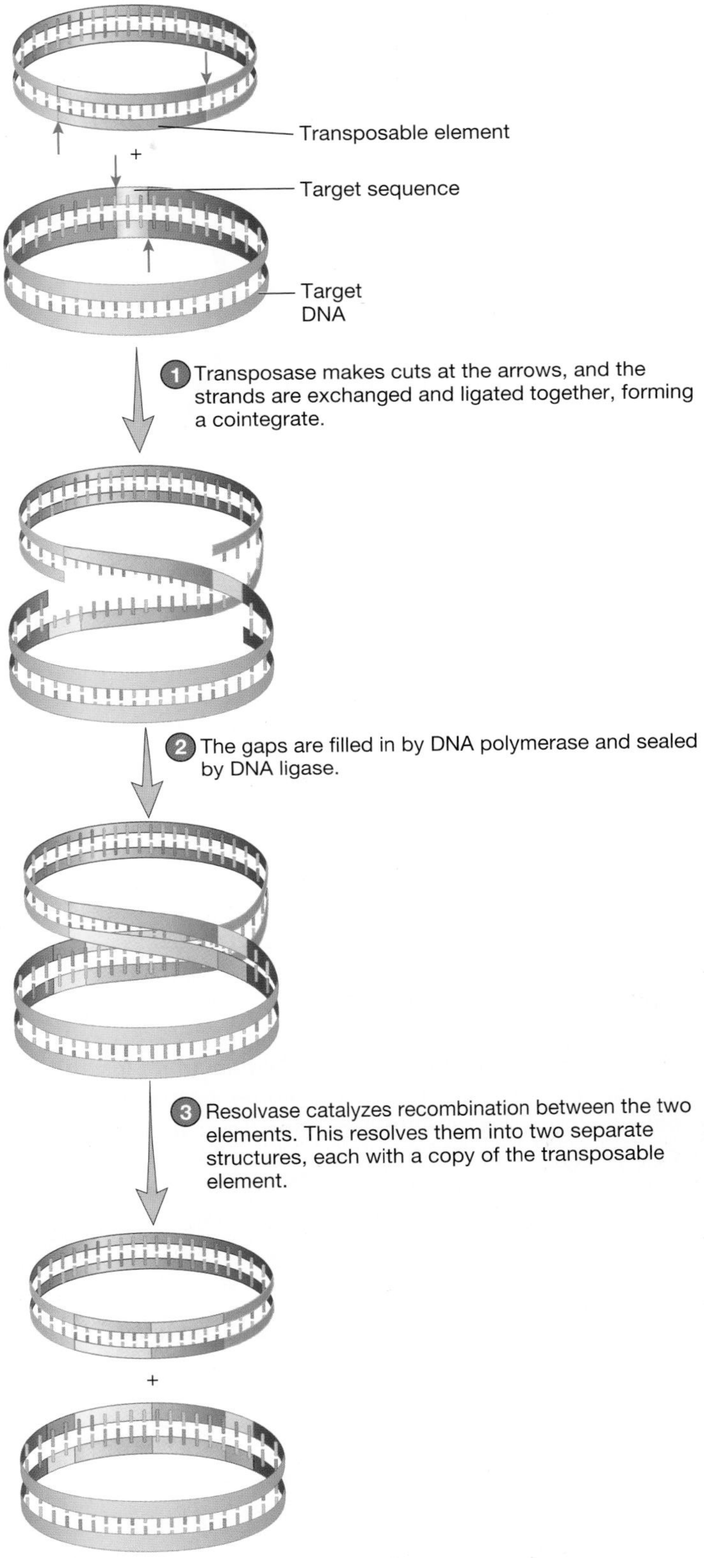

FIGURE 14.16 Replicative Transposition.

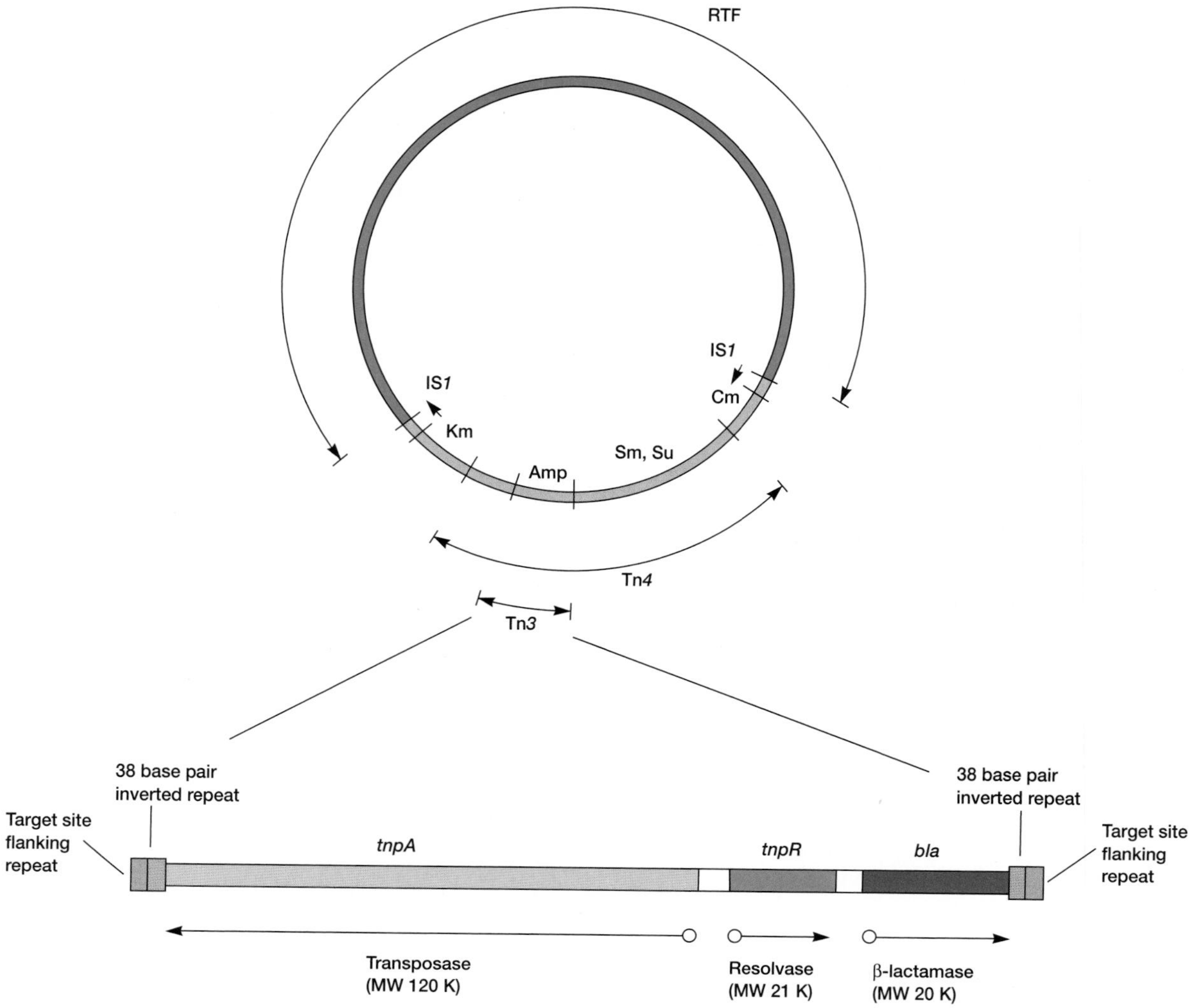

FIGURE 14.17 The Tn*3* Transposon Within an R Plasmid. Tn*3* is a replicative transposon that contains the gene for β-lactamase (*bla*), an enzyme that confers resistance to the antibiotic ampicillin (amp). The arrows below the Tn*3* genes indicate the direction of transcription. Tn*3* can be found in the resistance plasmid R1, where it is inserted into another transposable element, Tn*4*. Tn*4* carries genes that provide resistance to streptomycin (Sm) and sulfonamide (Su). The plasmid also carries resistance genes for kanamycin (Km) and chloramphenicol (Cm). The RTF region of R1 codes for proteins needed for plasmid replication and transfer.

Figure 14.17 Micro Inquiry

As a replicative transposon, what would happen if Tn*3* "hopped" from this R1 plasmid into a different plasmid?

attachment and plasmid transfer between specific *E. coli.* Most of the information required for plasmid transfer is located in the *tra* operon, which contains at least 28 genes. Many of these direct the formation of sex pili that attach the F^+ cell (the donor cell containing an F plasmid) to an F^- cell (**figure 14.19**). Other gene products aid DNA transfer. In addition, the F factor has several IS elements that assist plasmid integration into the host cell's chromosome. Thus the F factor is an episome that can exist outside the bacterial chromosome or can be integrated into it (**figure 14.20**).

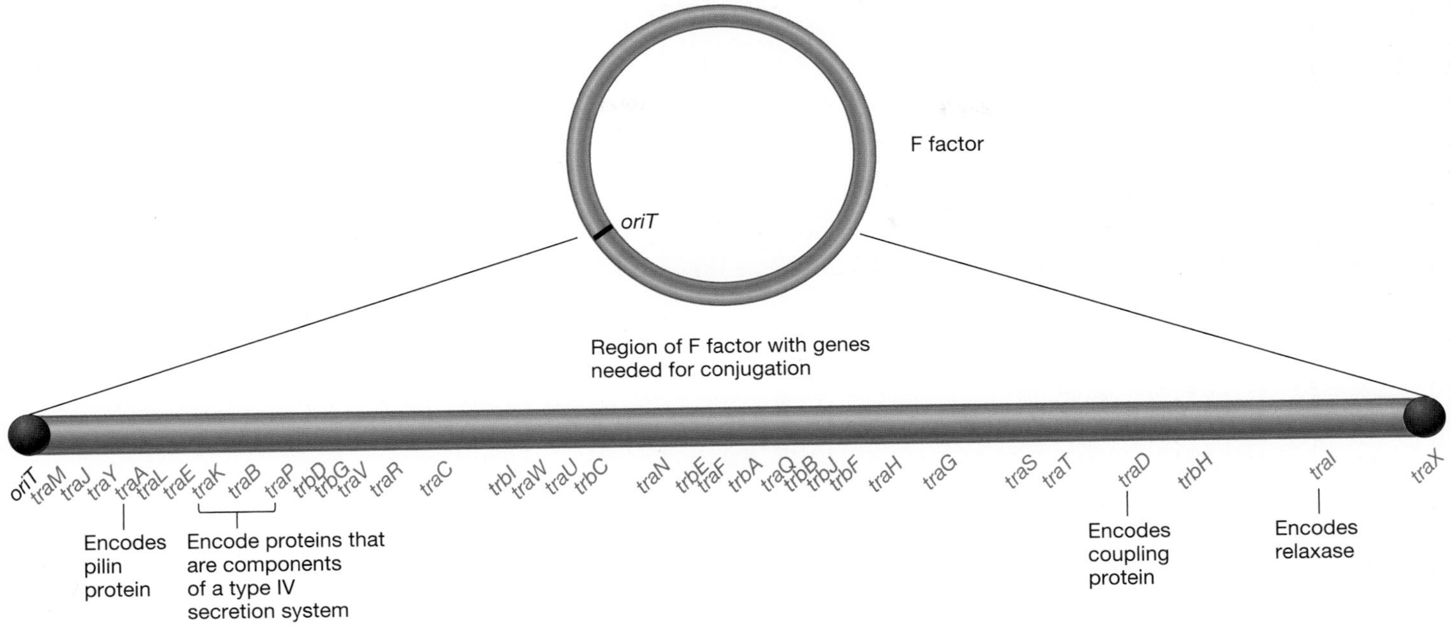

FIGURE 14.18 The F plasmid. Genes that play a role in conjugation are shown, and some of their functions are indicated. The plasmid also contains three insertion sequences and a transposon. The site for initiation of rolling-circle replication and gene transfer during conjugation is *oriT*.

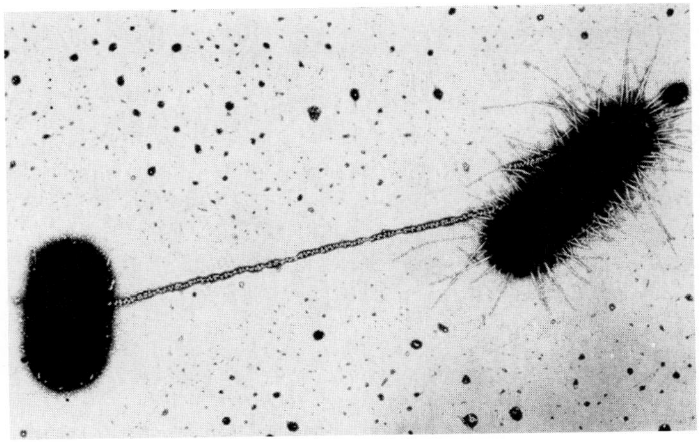

FIGURE 14.19 Bacterial Conjugation. An electron micrograph of two *E. coli* cells in an early stage of conjugation. The F^+ cell to the right is covered with fimbriae, and a sex pilus connects the two cells.

1. Compare and contrast plasmids and transposable elements. Compare and contrast insertion sequences, transposons, and replicative transposons.
2. What is simple (cut-and-paste) transposition? What is replicative transposition? How do the two mechanisms of transposition differ? What happens to the target site during transposition?
3. What effect would you expect the existence of transposable elements and plasmids to have on the rate of microbial evolution? Give your reasoning.
4. How do multiple-drug–resistant plasmids often arise?

14.7 Bacterial Conjugation

The initial evidence for bacterial **conjugation,** the transfer of DNA by direct cell-to-cell contact, came from an elegant experiment performed by Joshua Lederberg and Edward Tatum in 1946. They mixed two auxotrophic strains, incubated the culture for several hours in nutrient medium, and then plated it on minimal medium. To reduce the chance that their results were due to simple reversion, they used double and triple auxotrophs on the assumption that two or three simultaneous reversions would be extremely rare. When recombinant prototrophic colonies appeared on the minimal medium after incubation, they concluded that the two auxotrophs were able to associate and undergo recombination.

Lederberg and Tatum did not directly prove that physical contact of the cells was necessary for gene transfer. This evidence was provided by Bernard Davis (1950), who constructed a U tube

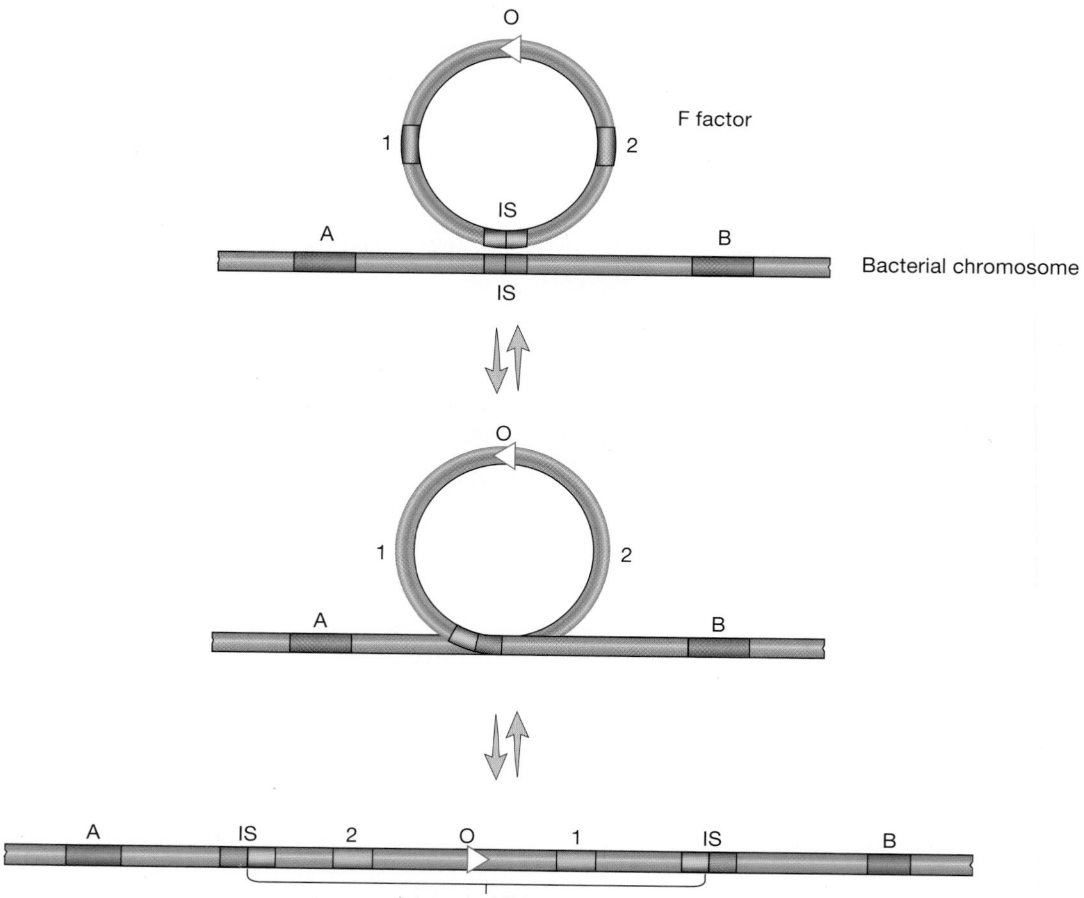

FIGURE 14.20 F Plasmid Integration. The reversible integration of an F plasmid or factor into a host bacterial chromosome. The process begins with association between plasmid and bacterial insertion sequences. The O arrowhead (white) indicates the site at which oriented transfer of chromosome to the recipient cell begins. A, B, 1, and 2 represent genetic markers.

***Figure 14.20* Micro Inquiry**

What does the term episome mean?

consisting of two pieces of curved glass tubing fused at the base to form a U shape with a glass filter between the halves. The filter allowed the passage of media but not bacteria. The U tube was filled with a growth medium and each side inoculated with a different auxotrophic strain of *E. coli.* During incubation, the medium was pumped back and forth through the filter to ensure medium exchange between the halves. When the bacteria were later plated on minimal medium, Davis discovered that if the two auxotrophic strains were separated from each other by the filter, gene transfer did not take place. Therefore direct contact was required for the recombination that Lederberg and Tatum had observed. F factor–mediated conjugation is one of the best-studied conjugation systems. It is the focus of this section.
Bacterial Conjugation

$F^+ \times F^-$ Mating

In 1952 William Hayes demonstrated that the gene transfer observed by Lederberg and Tatum was unidirectional. That is, there were definite donor (F^+, or fertile) and recipient (F^-, or nonfertile) strains, and gene transfer was nonreciprocal. He also found that in $F^+ \times F^-$ mating, the progeny were only rarely changed with regard to auxotrophy (i.e., chromosomal genes usually were not transferred). However, F^- strains frequently became F^+.

These results are readily explained in terms of the F factor described in section 14.6 (figure 14.18). The F^+ strain contains an extrachromosomal F factor carrying the genes for sex pilus formation and plasmid transfer. The **sex pilus** is used to establish contact between the F^+ and F^- cells (**figure 14.21**). Once

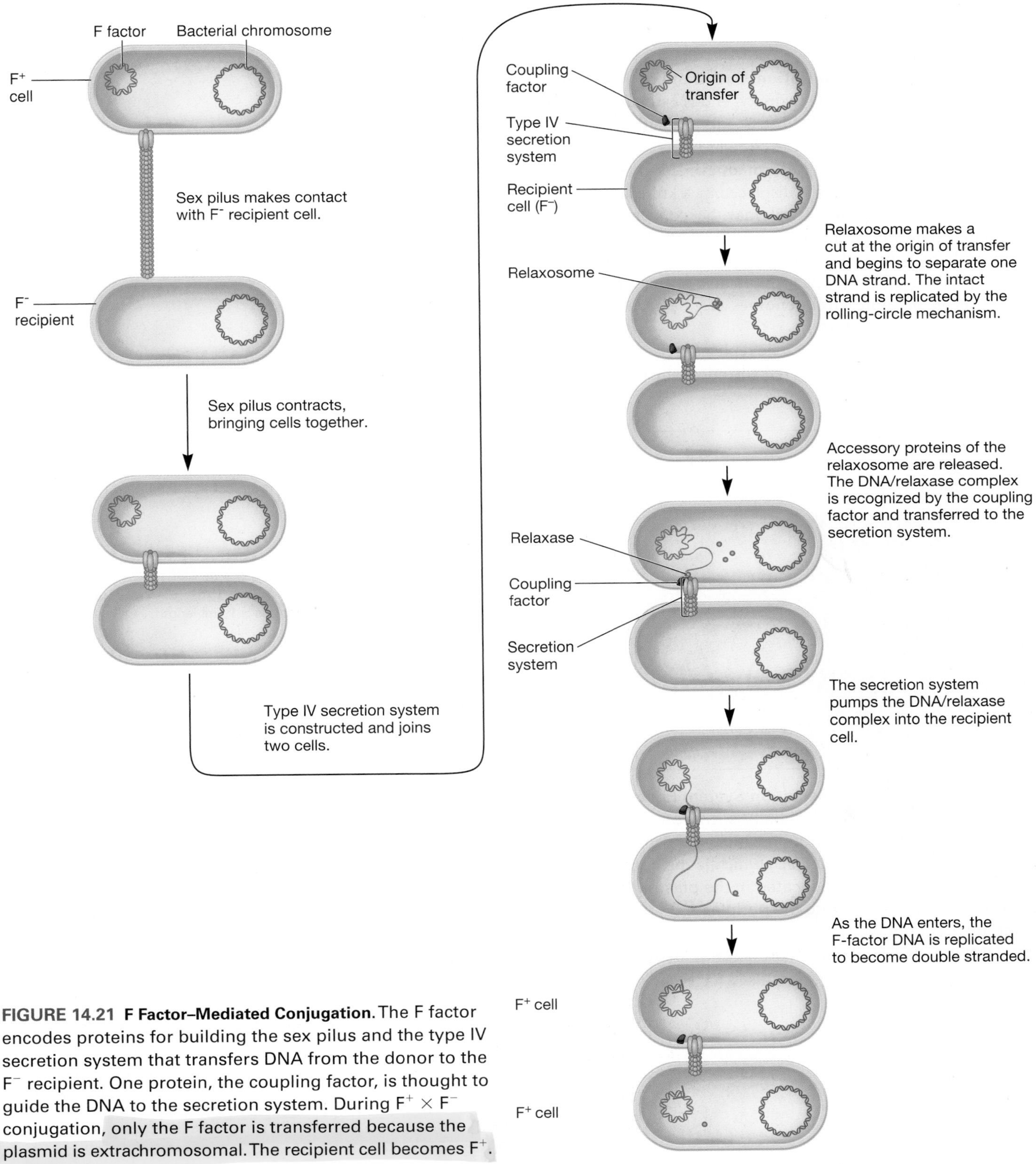

FIGURE 14.21 F Factor–Mediated Conjugation. The F factor encodes proteins for building the sex pilus and the type IV secretion system that transfers DNA from the donor to the F^- recipient. One protein, the coupling factor, is thought to guide the DNA to the secretion system. During $F^+ \times F^-$ conjugation, only the F factor is transferred because the plasmid is extrachromosomal. The recipient cell becomes F^+.

contact is made, the pilus retracts, bringing the cells into close physical contact. The F^+ cell prepares for DNA transfer by assembling a type IV secretion apparatus, using many of the same genes used for sex pilus biogenesis; the sex pilus is embedded in the secretion structure (**figure 14.22**). The F factor then replicates by rolling-circle replication. *Conjugation: Transfer of the F Plasmid*

During **rolling-circle replication,** one strand of the circular DNA is nicked, and the free 3′-hydroxyl end is extended by replication enzymes (**figure 14.23**). The 3′ end is lengthened while the growing point rolls around the circular template and the 5′ end of the strand is displaced to form an ever-lengthening tail, much like the peel of an apple is displaced by a knife as an apple is pared. The single-stranded tail may be converted to the double-stranded form by complementary strand synthesis. We are concerned here with rolling-circle replication of a plasmid. However, rolling-circle replication is also observed during the replication of some viral genomes (e.g., phage lambda). *Rolling-Circle Replication*

During conjugation, rolling-circle replication is initiated by a complex of proteins called the relaxosome, which nicks one

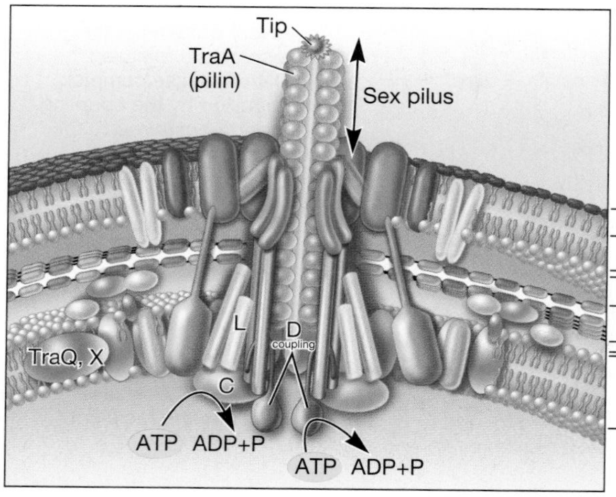

FIGURE 14.22 The Type IV Secretion System Encoded by F Factor. The F factor–encoded type IV secretion system is composed of numerous Tra proteins, including TraA proteins, which form the sex pilus, and TraD, which is the coupling factor. Some Tra proteins are located in the plasma membrane (PM); others extend into the periplasm (P) and pass through the peptidoglycan layer (PG) into the outer membrane (OM) and its lipopolysaccharide (LPS) layer.

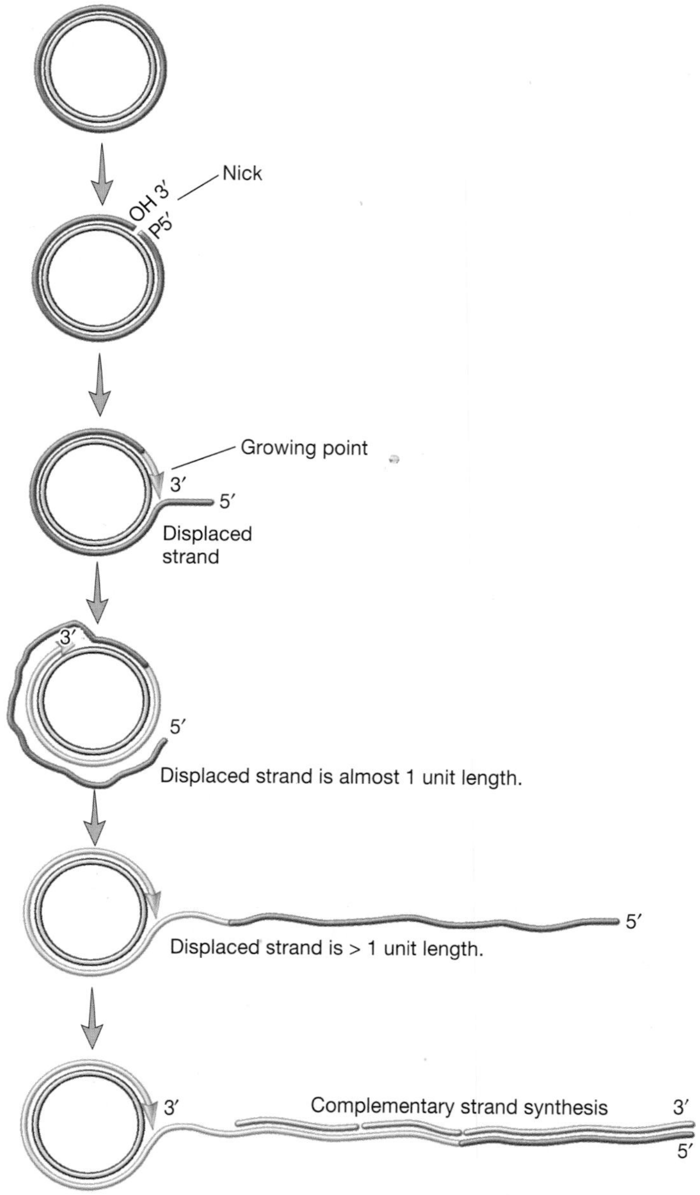

FIGURE 14.23 Rolling-Circle Replication. A single-stranded tail, often composed of more than one genome copy, is generated and can be converted to the double-stranded form by synthesis of a complementary strand. The "free end" of the rolling-circle strand is probably bound to the primosome. OH 3′ is the 3′-hydroxyl and P 5′ is the 5′-phosphate created when the DNA strand is nicked.

strand of the F factor at a site called *oriT* (for origin of transfer). Relaxase, an enzyme associated with the relaxosome, remains attached to the 5′ end of the nicked strand. As F factor is replicated, the displaced strand and the attached relaxase enzyme move through the type IV secretion system to the recipient cell. During plasmid transfer, the entering strand is copied to produce double-stranded DNA. When this is completed, the F^- recipient cell becomes F^+.

Hfr Conjugation

By definition, an F^+ cell has the F factor free from the chromosome, so in an $F^+ \times F^-$ mating, chromosomal DNA is not transferred. However within this population, a few cells have the F plasmid integrated into their chromosomes. This explains why not long after the discovery of $F^+ \times F^-$ mating, a second type of F factor–mediated conjugation was discovered. In this type of conjugation, the donor transfers chromosomal genes with great efficiency but does not change the recipient bacteria into F^+ cells. Because of the *h*igh *f*requency of *r*ecombinants produced by this mating, it is referred to as **Hfr conjugation** and the donor is called an **Hfr strain.** Hfr strains contain the F factor integrated into their chromosome, rather than free in the cytoplasm (**figure 14.24*a***). When integrated, the F plasmid's *tra* operon is still functional; the plasmid can direct the synthesis of pili, carry out rolling-circle replication, and transfer genetic material to an F^- recipient cell. However, rather than transferring itself, the F factor directs the transfer of the host chromosome. DNA transfer begins when the integrated F factor is nicked at its origin of transfer site. As it is replicated, the chromosome moves to the recipient (figure 14.24*b*). Only part of the F factor is initially transferred and the host chromosome follows. Transfer of the entire chromosome with the integrated F factor requires about 100 minutes in *E. coli,* but the connection between the cells usually breaks before this process is finished. Thus a complete F factor is rarely transferred, and the recipient remains F^-.

As mentioned earlier, when an Hfr strain participates in conjugation, bacterial genes are frequently transferred to the recipient. Gene transfer can be in either a clockwise or a counterclockwise direction around the circular chromosome, depending on the orientation of the integrated F factor. After the replicated donor chromosome enters the recipient cell, it may be degraded or incorporated into the F^- genome by recombination. *Transfer of Chromosomal DNA*

F′ Conjugation

Because the F plasmid is an episome, it can leave the bacterial chromosome and resume status as an autonomous F factor. Sometimes during excision, the plasmid makes an error and picks up a portion of the chromosome. Because it is now genotypically distinct from the original F factor, it is called an **F′ plasmid** (**figure 14.25*a***). A cell containing an F′ plasmid retains all of its genes, although some of them are on the plasmid. It mates only with an F^- recipient, and $F' \times F^-$ conjugation is similar to an $F^+ \times F^-$ mating. Once again, the plasmid is transferred as it is copied by rolling-circle replication. Bacterial genes on the chromosome are not transferred (figure 14.25*b*), but bacterial genes on the F′ plasmid are transferred. These genes need not be incorporated into the recipient chromosome to be expressed. The recipient becomes F′ and is partially diploid because the same bacterial genes present on the F′ plasmid are also found on the recipient's chromosome. In this way, specific bacterial genes may spread rapidly throughout a bacterial population.

Other Examples of Bacterial Conjugation

Although most research on plasmids and conjugation has been done using *E. coli* and other gram-negative bacteria, conjugative plasmids are present in gram-positive bacterial genera such as *Bacillus, Streptococcus, Enterococcus, Staphylococcus,* and *Streptomyces.* Much less is known about these systems. It appears that fewer transfer genes are involved, possibly because a sex pilus may not be required for plasmid transfer. For example, *Enterococcus faecalis* recipient cells release short peptide chemical signals that activate transfer genes in donor cells containing the proper plasmid. Donor and recipient cells directly adhere to one another through plasmid-encoded proteins released by the activated donor cell. Plasmid transfer then occurs. *Transfer of a Plasmid*

1. What is bacterial conjugation and how was it discovered?
2. Distinguish between F^+, Hfr, and F^- strains of *E. coli* with respect to their physical nature and role in conjugation.
3. Describe how $F^+ \times F^-$ and Hfr conjugation processes proceed, and distinguish between the two in terms of mechanism and the final results.
4. What is F′ conjugation and how does the F′ plasmid differ from a regular F plasmid?

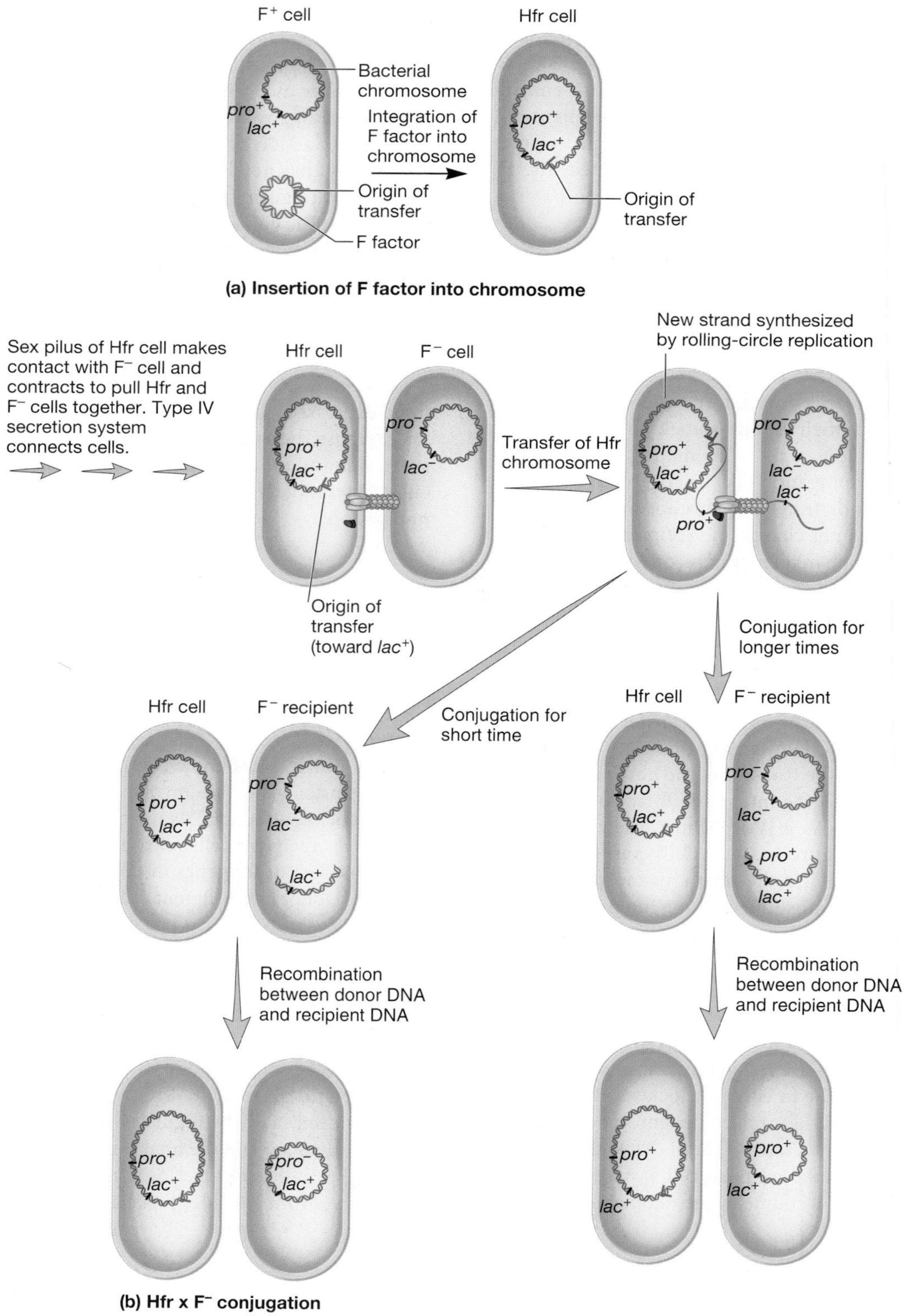

FIGURE 14.24 Creation of Hfr Strains and Hfr × F⁻ Conjugation. (a) Integration of the F factor into the donor cell's chromosome creates an Hfr cell. (b) During Hfr × F⁻ conjugation, some plasmid genes and some chromosomal genes are transferred to the recipient. Note that only a portion of the F factor moves into the recipient. Because the entire plasmid is not transferred, the recipient remains F⁻. In addition, the incoming DNA must recombine into the recipient's chromosome if it is to be stably maintained.

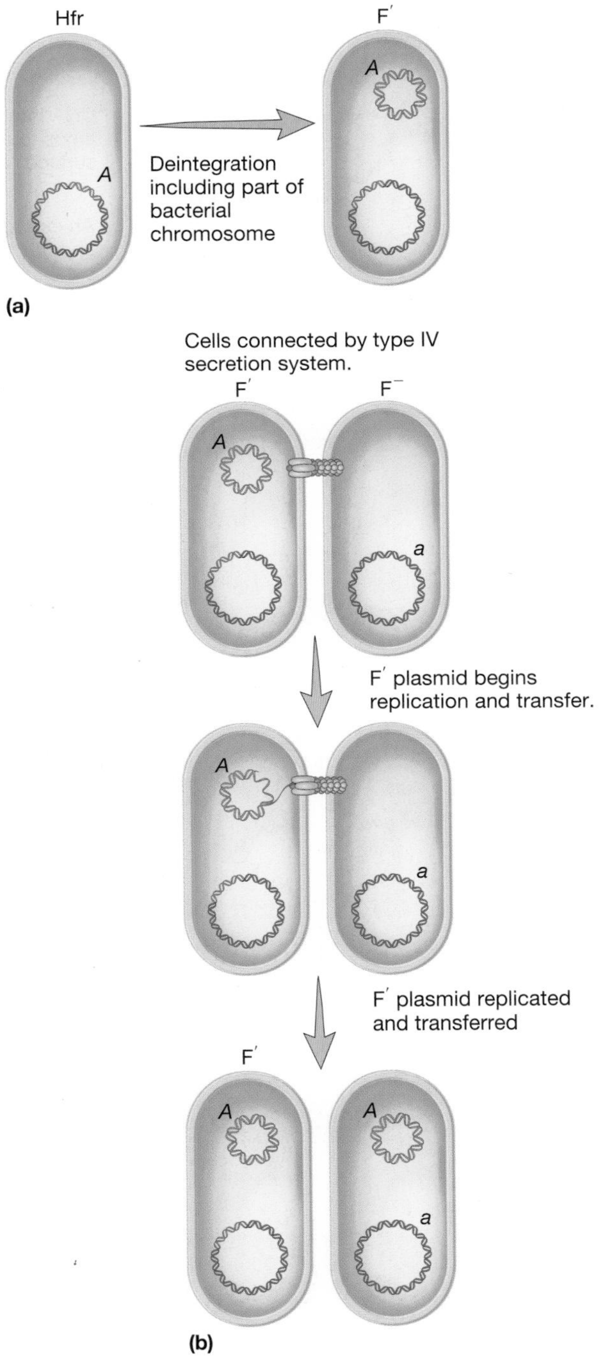

FIGURE 14.25 F′ Conjugation. (a) Due to an error in excision, the *A* gene of an Hfr cell is picked up by the F factor. (b) During conjugation, the *A* gene is transferred to a recipient, which becomes diploid for that gene (i.e., Aa).

***Figure 14.25* Micro Inquiry**

What will determine if the F′ made by deintegration will be capable of transfer to an F^- cell? (Hint: Examine figure 14.18.)

14.8 Bacterial Transformation

The second way DNA can move between bacteria is through transformation, discovered by Fred Griffith in 1928. **Transformation** is the uptake by a cell of DNA, either a plasmid or a fragment of linear DNA, from the medium and the maintenance of the DNA in the recipient in a heritable form. In natural transformation, the DNA comes from a donor bacterium. The process is random, and any portion of the donor's genome may be transferred.

When bacteria lyse, they release considerable amounts of DNA into the surrounding environment. These fragments may be relatively large and contain several genes. If a fragment contacts a **competent cell**—a cell that is able to take up DNA and be transformed—the DNA can be bound to the cell and taken inside (**figure 14.26*a***). The transformation frequency of very competent cells is around 10^{-3} for most genera when an excess of DNA is used. That is, about one cell in every thousand will take up and integrate the gene. Competency is a complex phenomenon and depends on several conditions. Bacteria need to be in a certain stage of growth; for example, *Streptococcus pneumoniae* becomes competent during the exponential phase when the population reaches about 10^7 to 10^8 cells per milliliter. When a population becomes competent, bacteria such as *S. pneumoniae* secrete a small protein called the competence factor that stimulates the production of eight to 10 new proteins required for transformation. Natural transformation has been discovered in some archaea and in several bacterial phyla (e.g., *Deinococcus-Thermus, Cyanobacteria, Chlorobi, Proteobacteria,* and *Firmicutes*). Gene transfer by this process occurs in soil and aquatic ecosystems and may be an important route of genetic exchange in biofilm and other microbial communities.

The mechanism of transformation has been intensively studied in *S. pneumoniae* (**figure 14.27**). A competent cell binds a double-stranded DNA fragment if the fragment is moderately large; the process is random, and donor fragments compete with each other. The DNA then is cleaved by endonucleases to double-stranded fragments about 5,000 to 15,000 base pairs in size. DNA uptake requires energy expenditure. One strand is hydrolyzed by an envelope-associated exonuclease during uptake; the other strand associates with small proteins and moves through the plasma membrane. The single-stranded fragment can then align with a homologous region of the genome and be integrated, probably by a mechanism similar to that depicted in figure 14.13.

Transformation in *Haemophilus influenzae,* a gram-negative bacterium, differs from that in *S. pneumoniae* in several respects. *H. influenzae* does not produce a protein factor to stimulate the development of competence, and it takes up DNA from only closely related species (*S. pneumoniae* is less particular about the source of its DNA). Double-stranded DNA, complexed with proteins, is taken in by membrane vesicles. The specificity of *H. influenzae* transformation is due to an 11 base pair sequence that is repeated over 1,400 times in *H. influenzae*

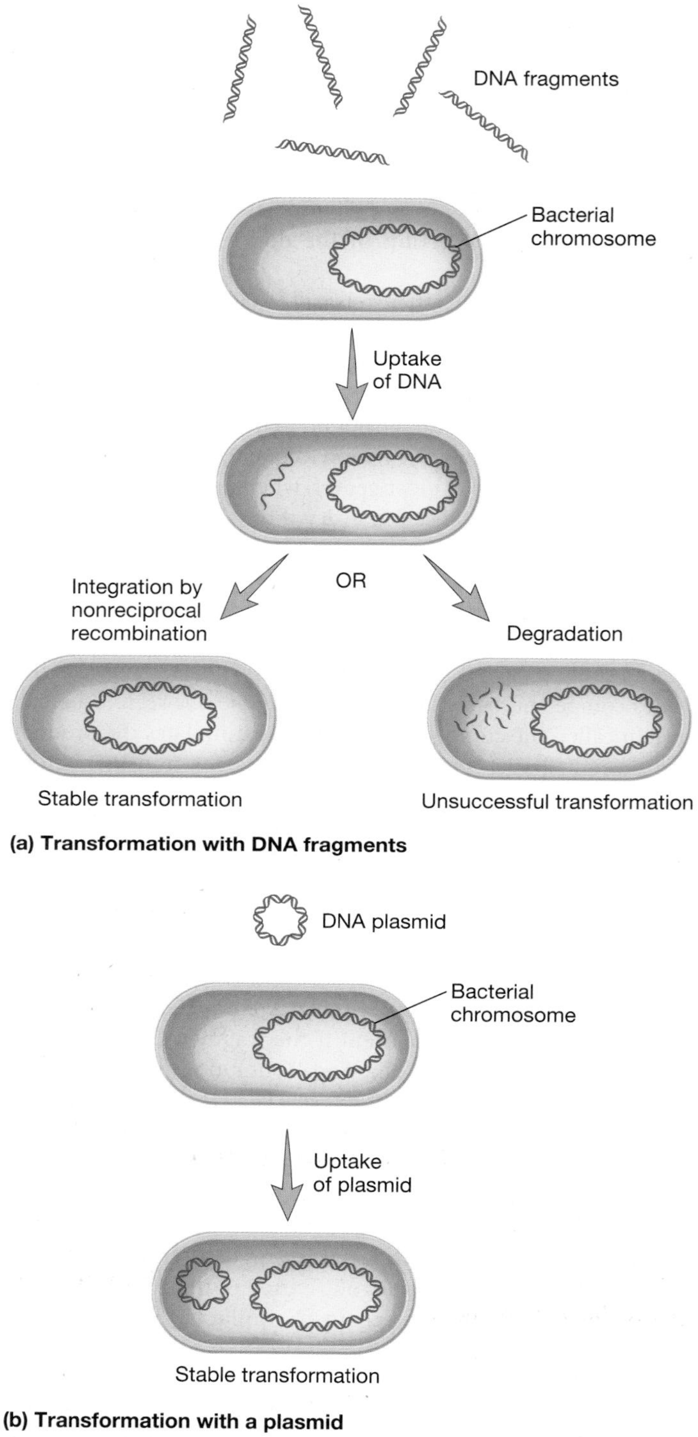

FIGURE 14.26 Bacterial Transformation. Transformation with (a) DNA fragments and (b) plasmids. Transformation with a plasmid often is induced artificially in the laboratory. The transforming DNA is in purple, and integration is at a homologous region of the genome.

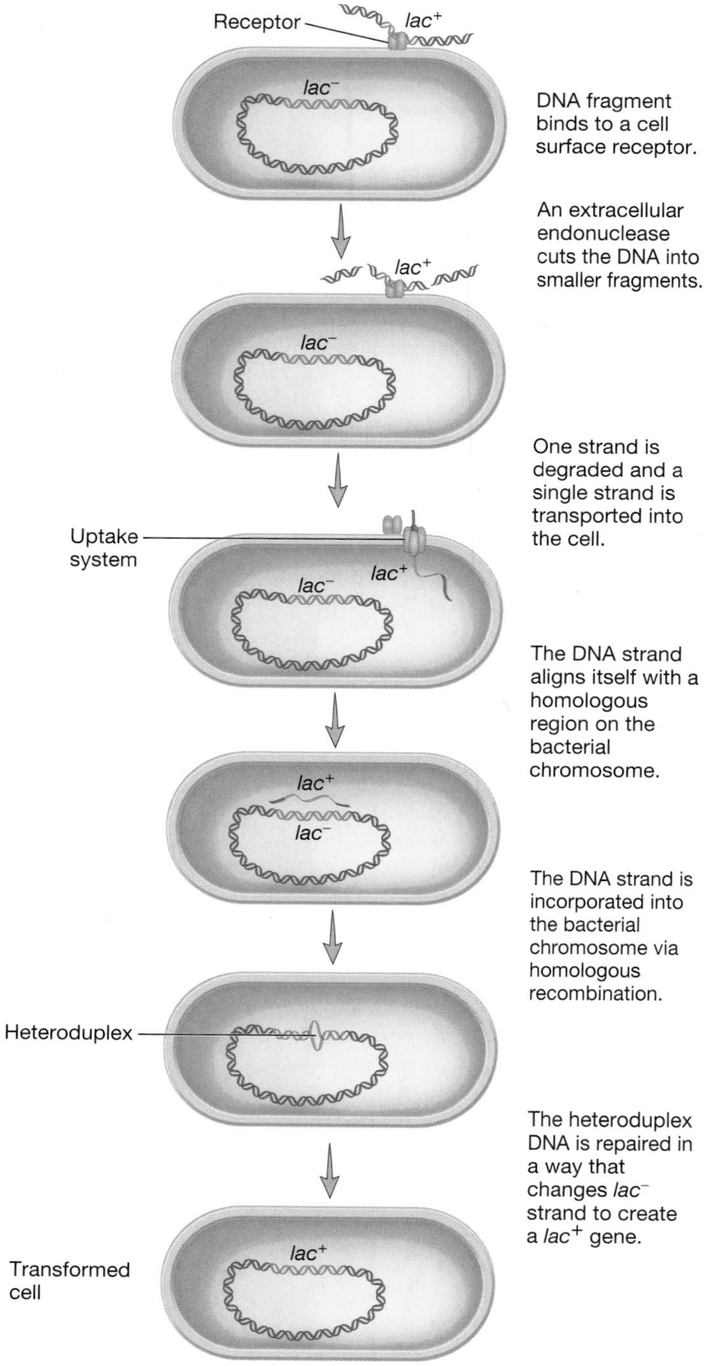

Receptor
lac+
lac−
DNA fragment binds to a cell surface receptor.
An extracellular endonuclease cuts the DNA into smaller fragments.
One strand is degraded and a single strand is transported into the cell.
Uptake system
The DNA strand aligns itself with a homologous region on the bacterial chromosome.
The DNA strand is incorporated into the bacterial chromosome via homologous recombination.
Heteroduplex
The heteroduplex DNA is repaired in a way that changes lac− strand to create a lac+ gene.
Transformed cell

FIGURE 14.27 Bacterial Transformation as Seen in *S. pneumoniae*.

Figure 14.27 Micro Inquiry

According to this model, what would happen if DNA that lacked homology to the *S. pneumoniae* chromosome were taken into the cell?

DNA. DNA must have this sequence to be bound by a competent cell.

The protein complexes that take up free DNA must be able to move it through gram-negative and gram-positive cell walls. As expected, the machinery is quite large and complicated, and appears related to protein secretion systems. **Figure 14.28*a*** illustrates the complex used by the gram-negative bacterium *Neisseria gonorrhoeae.* The protein PilQ aids in the movement across the outer membrane, and the pilin complex (PilE) moves the DNA through the periplasm and peptidoglycan. ComE is a DNA-binding protein; N is the nuclease that degrades one strand before the DNA enters the cytoplasm through the transmembrane channel formed by ComA. The machinery in the gram-positive bacterium *Bacillus subtilis* is depicted in figure 14.28*b*. It is localized to the poles of the cell, and many of the components are similar to those of *N. gonorrhoeae:* the pilin complex (ComGC), DNA-binding protein (ComEA), nuclease (N), and channel protein (ComEC). ComFA is a DNA translocase that moves the DNA into the cytoplasm. A gram-negative equivalent of ComFA has not been identified yet in *N. gonorrhoeae.* *Bacterial Transformation*

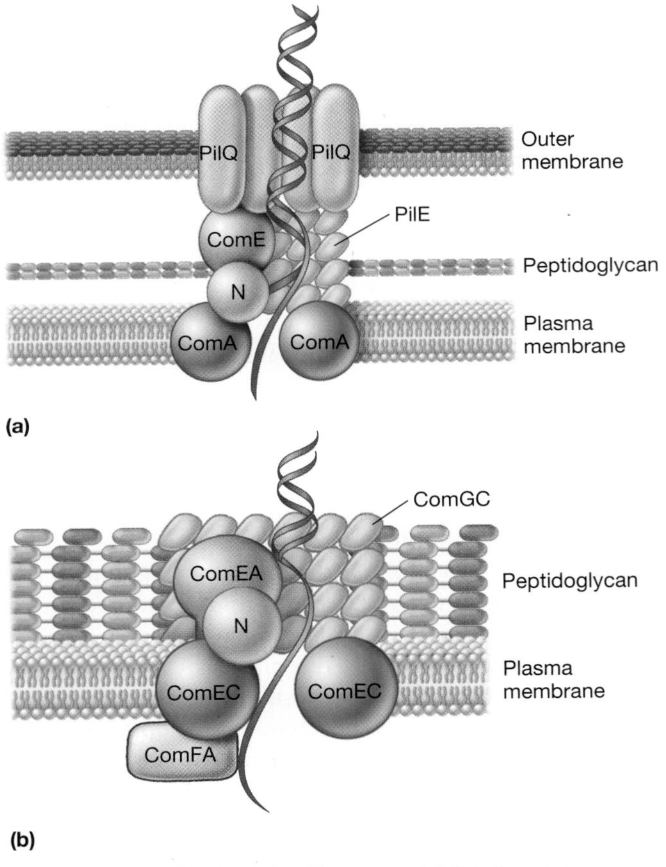

FIGURE 14.28 DNA Uptake Systems. (a) DNA uptake machinery in *N. gonorrhoeae.* (b) Uptake machinery in *B. subtilis.*

Microbial geneticists exploit transformation to move DNA (usually recombinant DNA) into cells. Because many species, including *E. coli,* are not naturally transformation competent, these bacteria must be made artificially competent by certain treatments. Two common techniques are electrical shock and exposure to calcium chloride. Both approaches render the cell membrane temporarily more permeable to DNA, and both are used to make artificially competent *E. coli* cells. To increase the transformation frequency with *E. coli,* strains that lack one or more nucleases are used. These strains are especially important when transforming the cells with linear DNA, which is vulnerable to attack by nucleases. It is easier to transform bacteria with plasmid DNA since plasmids can replicate within the host and are not as easily degraded as are linear fragments (figure 14.26*b*). ▶▶| *Introducing recombinant DNA into host cells (section 15.6)*

1. Define transformation and competence.
2. Describe how transformation occurs in *S. pneumoniae.* How does the process differ in *H. influenzae?*
3. Discuss two ways in which artificial transformation can be used to place functional genes within bacterial cells.

14.9 Transduction

The third mode of bacterial gene transfer is **transduction.** It is a frequent mode of horizontal gene transfer in nature and is mediated by viruses. Recall from chapter 5 that viruses are structurally simple, often composed of just a nucleic acid genome protected by a protein coat called the capsid. They are unable to multiply autonomously. Instead, they infect and take control of a host cell, forcing the host to make many copies of the virus. Viruses that infect bacteria are called bacteriophages, or phages for short. **Virulent bacteriophages** multiply in their bacterial host immediately after entry. After the number of progeny phages reaches a certain number, they cause the host to lyse, so they can be released and infect new host cells (**figure 14.29**). Thus this process is called the **lytic cycle. Temperate bacteriophages,** on the other hand, do not immediately kill their host. Many temperate phages enter the host bacterium and insert their genomes into the bacterial chromosome. The inserted viral genome is called a **prophage.** The host bacterium is unharmed by this, and the phage genome is passively replicated as the host cell's genome is replicated. The relationship between these viruses and their host is called **lysogeny,** and bacteria that have been lysogenized are called **lysogens.** Temperate phages can remain inactive in their hosts for many generations. However, they can be induced to switch to a lytic

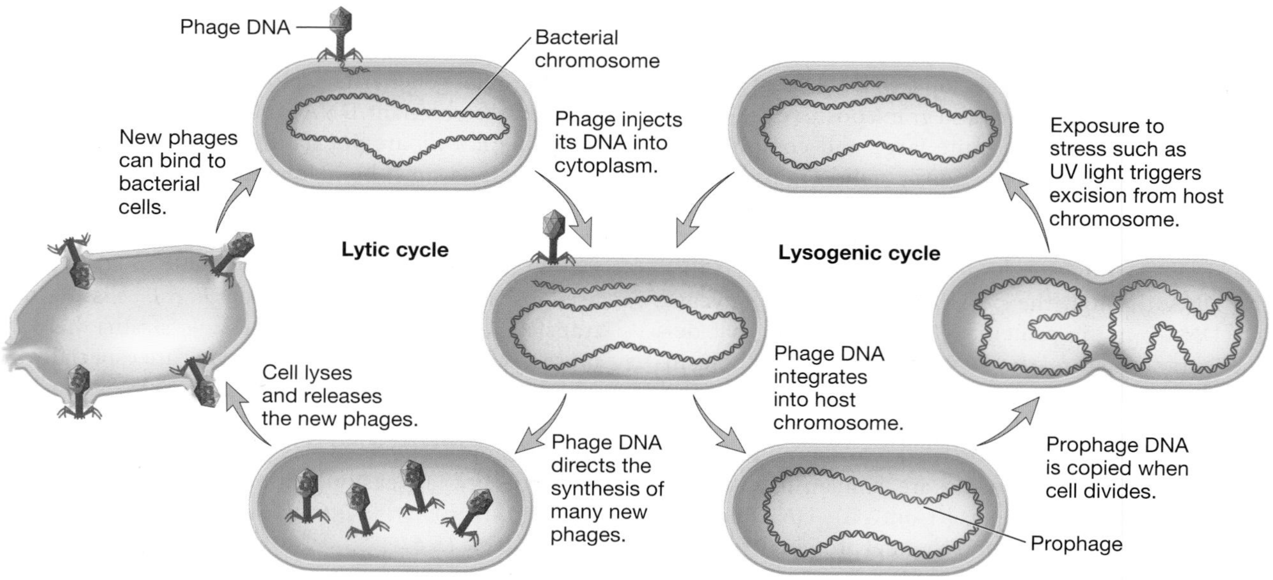

FIGURE 14.29 Lytic and Lysogenic Cycles of Temperate Phages. Virulent phages undergo only the lytic cycle. Temperate phages have two phases to their life cycles. The lysogenic cycle allows the genome of the virus to be replicated passively as the host cell's genome is replicated. Certain environmental factors such as UV light can cause a switch from the lysogenic cycle to the lytic cycle. In the lytic cycle, new virions are made and released when the host cell lyses.

Figure 14.29 Micro Inquiry

What is the term used to describe a lysogenic phage genome when it is integrated into the host genome?

cycle under certain conditions, including UV irradiation. When this occurs, the prophage is excised from the bacterial genome and the lytic cycle proceeds. ▶▶| *Types of viral infections (section 5.4)*

Transduction is the transfer of bacterial or archaeal genes by viruses. The genes are packaged in the virus because of errors made during its life cycle. The virus containing these genes then transfers them to a recipient cell. Two kinds of bacterial transduction have been described: generalized and specialized.

Generalized Transduction

Generalized transduction occurs during the lytic cycle of virulent and some temperate phages. Any part of the bacterial genome can be transferred (**figure 14.30**). During the assembly stage, when the viral chromosomes are packaged into capsids, random fragments of the partially degraded bacterial chromosome may mistakenly be packaged. Because the capsid can contain only a limited quantity of DNA, the viral DNA is left behind. The quantity of bacterial DNA carried depends primarily on the size of the capsid. The P22 phage of *Salmonella enterica* serovar Typhimurium can carry about 1% of the bacterial genome; the P1 phage of *E. coli* and a variety of gram-negative bacteria can package about 2.0 to 2.5% of the genome. The resulting virus often injects the DNA into another bacterial cell but cannot initiate a lytic cycle. This phage is known as a generalized transducing particle and is simply a carrier of genetic information from the original bacterium to another cell. As in transformation, once the DNA fragment has been injected, it must be incorporated into the recipient cell's chromosome to preserve the transferred genes. The DNA remains double stranded during transfer, and both strands are integrated into

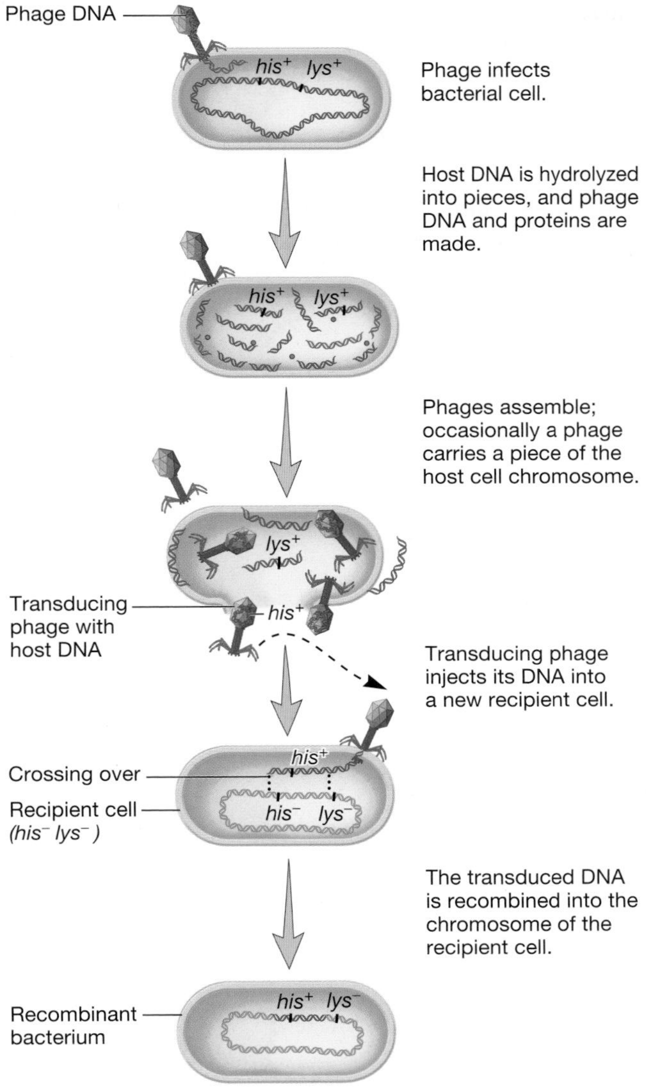
Phage DNA
his^+ lys^+
Phage infects bacterial cell.
Host DNA is hydrolyzed into pieces, and phage DNA and proteins are made.
his^+ lys^+
Phages assemble; occasionally a phage carries a piece of the host cell chromosome.
lys^+
Transducing phage with host DNA
his^+
Transducing phage injects its DNA into a new recipient cell.
Crossing over
his^+
Recipient cell (his^- lys^-)
his^- lys^-
The transduced DNA is recombined into the chromosome of the recipient cell.
Recombinant bacterium
his^+ lys^-

The recombinant bacterium has a genotype (his^+lys^-) that is different from recipient bacterial cell (his^- lys^-).

FIGURE 14.30 Generalized Transduction in Bacteria.

the recipient's chromosome. About 70 to 90% of the transferred DNA is not integrated but often is able to remain intact temporarily and be expressed. Abortive transductants are bacteria that contain this nonintegrated, transduced DNA and are partial diploids.

Specialized Transduction

In **specialized transduction,** only specific portions of the bacterial genome are carried by transducing particles. Specialized transduction is made possible by an error in the lysogenic life cycle of phages that insert their genomes into a specific site in the host chromosome. When a prophage is induced to leave the host chromosome, excision is sometimes carried out improperly. The resulting phage genome contains portions of the bacterial chromosome (about 5 to 10% of the bacterial DNA) next to the integration site, much like the situation with F′ plasmids (**figure 14.31**). A transducing particle genome usually is defective because it lacks some of its genome. However, it will inject bacterial genes into another bacterium, even though the defective phage cannot reproduce without assistance. The bacterial genes may become stably incorporated under the proper circumstances.

The best-studied example of specialized transduction is carried out by the *E. coli* phage lambda. The lambda genome inserts into the host chromosome at specific locations known as attachment or *att* sites (**figure 14.32**). The phage *att* sites and bacterial *att* sites are similar and can complex with each other. The *att* site for lambda is next to the *gal* and *bio* genes on the *E. coli* chromosome; consequently when lambda excises incorrectly to generate a specialized transducing particle, these bacterial genes are most often present. The product of cell lysis (lysate) resulting from the induction of lysogenized *E. coli* contains normal phage and a few defective transducing particles. These particles are called lambda *dgal* if they carry the galactose utilization genes or lambda *dbio* if they carry the *bio* from the other side of the *att* site (figure 14.32). ▶▶| *Bacteriophage lambda: A temperate bacteriophage (section 25.2)* *Specialized Transduction*

1. Describe generalized transduction and how it occurs. What is an abortive transductant?
2. What is specialized transduction and how does it come about?
3. How might one tell whether horizontal gene transfer was mediated by generalized or specialized transduction?
4. Why doesn't a cell lyse after successful transduction with a temperate phage?
5. Describe how conjugation, transformation, and transduction are similar. How are they different?

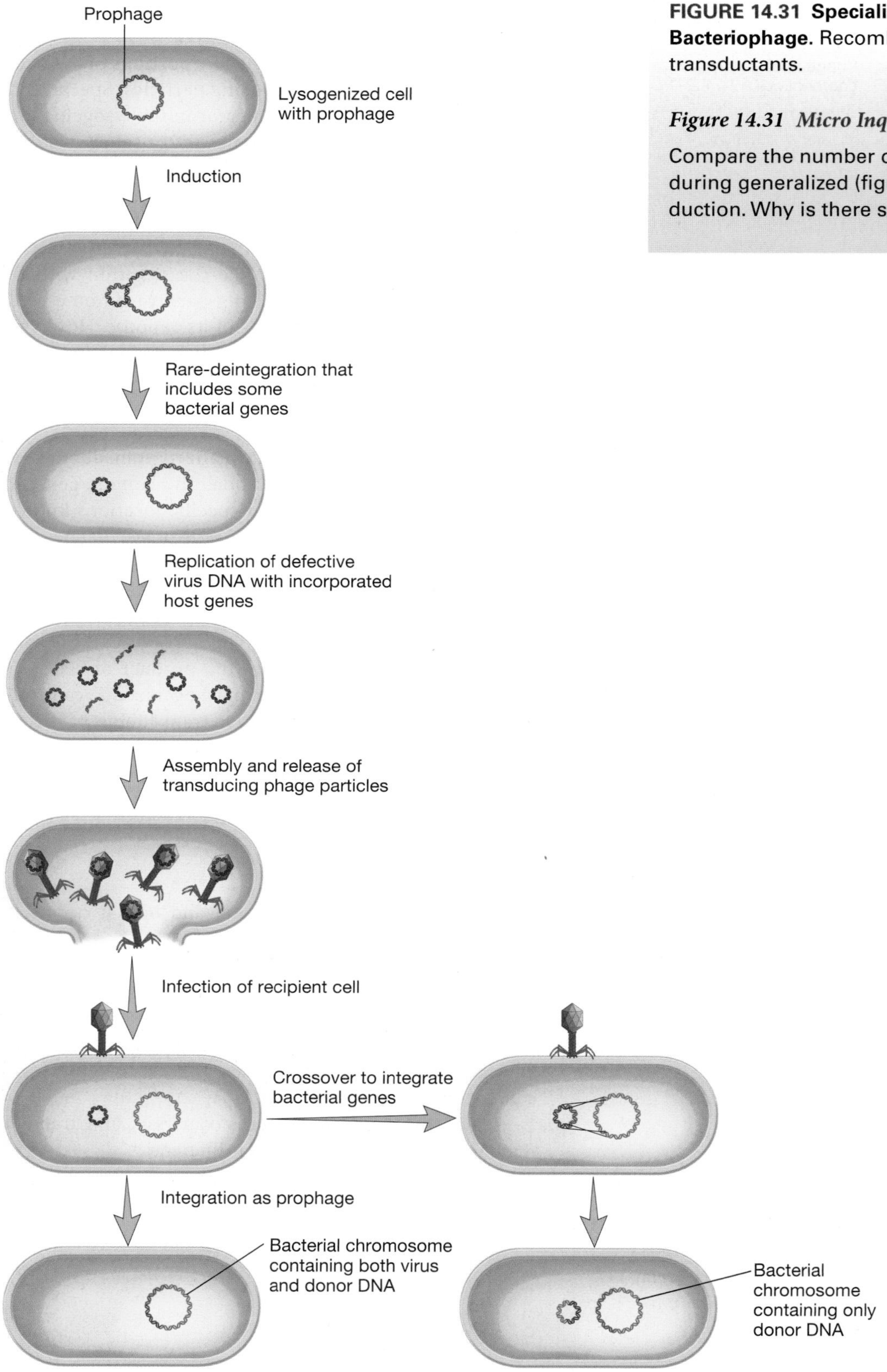

FIGURE 14.31 Specialized Transduction by a Temperate Bacteriophage. Recombination can produce two types of transductants.

Figure 14.31 Micro Inquiry

Compare the number of transducing particles that arise during generalized (figure 14.30) and specialized transduction. Why is there such a big difference?

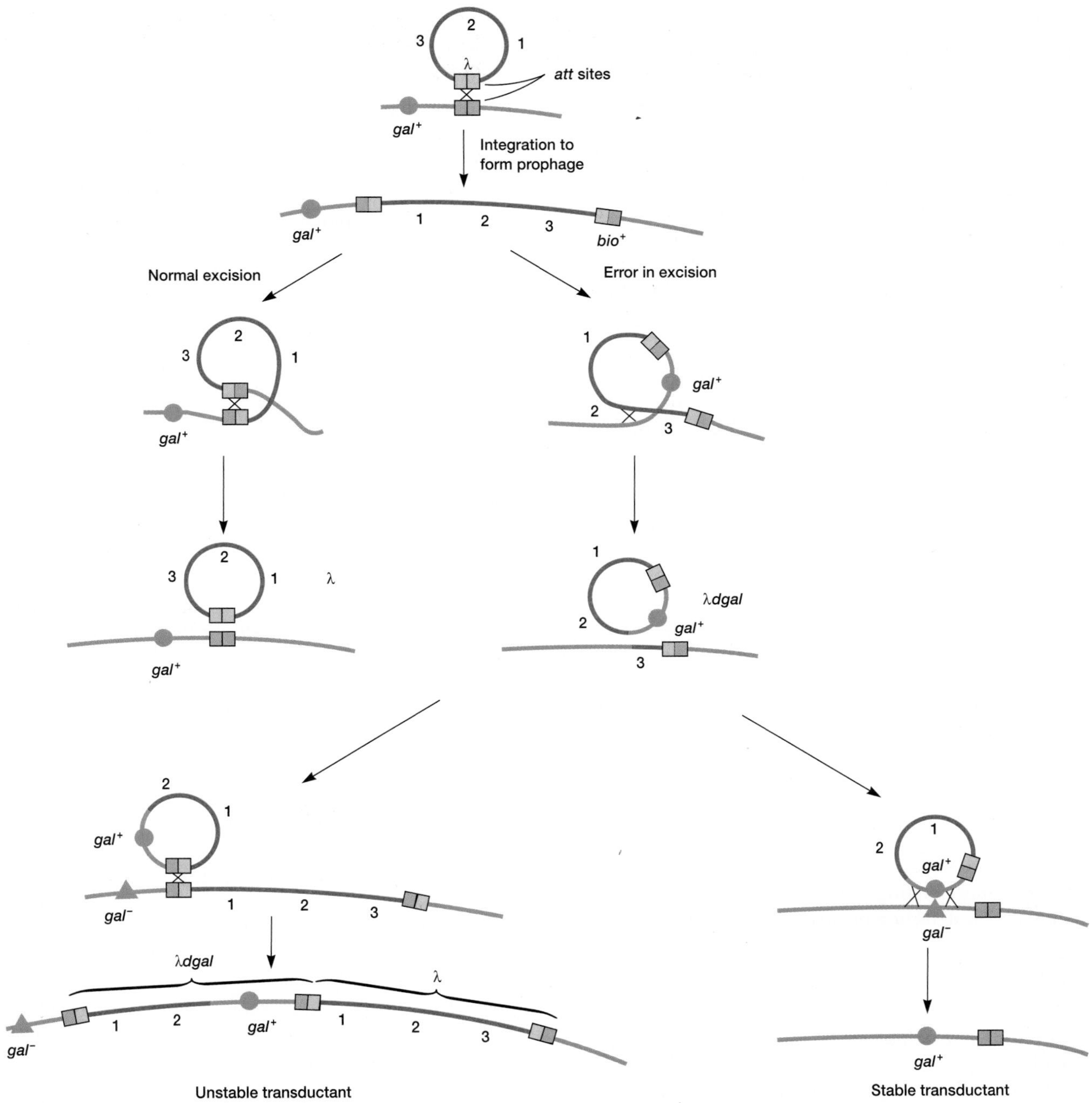

FIGURE 14.32 The Mechanism of Transduction for Phage Lambda and *E. coli.* Integrated lambda phage lies between the *gal* and *bio* genes. When it excises normally (top left), the new phage is complete and contains no bacterial genes. Excision rarely occurs asymmetrically (top right), and either the *gal* or *bio* genes are picked up and some phage genes are lost (only aberrant excision involving the *gal* genes is shown). The result is a defective lambda phage that carries bacterial genes and can transfer them to a new recipient.

Figure 14.32 Micro Inquiry

Why can't the *gal* and *bio* genes be transduced by the same transducing particle?

Summary

14.1 Mutations: Their Chemical Basis and Effects

a. A mutation is a stable, heritable change in the nucleotide sequence of the genetic material.

b. Spontaneous mutations can arise from replication errors (transition, transversion, and insertion and deletion of nucleotides), and from DNA lesions (apurinic sites, apyrimidinic sites, oxidation of DNA) (**figures 14.1** and **14.2**).

c. Induced mutations are caused by mutagens. Mutations may result from the incorporation of base analogs, specific mispairing due to alterations of a base caused by DNA-modifying agents, the presence of intercalating agents, and severe damage to the DNA caused by exposure to radiation (**table 14.1; figures 14.3** and **14.4**).

d. Mutations are usually recognized when they cause a change from the more prevalent wild-type phenotype. A mutant phenotype can be restored to wild type by either reversions or suppressor mutations (**table 14.2**).

e. There are four important types of point mutations: silent mutations, missense mutations, nonsense mutations, and frameshift mutations (**table 14.2**).

f. Mutations can affect phenotype in numerous ways. Some major types of mutations categorized based on their effects on phenotype are morphological, lethal, conditional, biochemical, and resistance mutations.

14.2 Detection and Isolation of Mutants

a. A sensitive and specific screening method is needed for detecting and isolating mutants. An example is replica plating for the detection of auxotrophs (**figure 14.5**).

b. One of the most effective techniques for isolating mutants is to select for mutants by adjusting environmental conditions so that the mutant will grow while the wild type does not.

c. Because many carcinogens are also mutagenic, one can test for mutagenicity with the Ames test and use the results as an indirect indication of carcinogenicity (**figure 14.6**).

14.3 DNA Repair

a. Cells have multiple mechanisms for correcting mispaired and damaged DNA.

b. Excision repair systems remove damaged portions from a single strand of DNA (e.g., thymine dimers) and use the other strand as a template for filling in the gap (**figures 14.7** and **14.8**).

c. Direct repair systems correct damaged DNA without removing damaged regions. For instance, during photoreactivation, thymine dimers are repaired by splitting the two thymines apart. This is catalyzed in the presence of light by the enzyme photolyase.

d. Mismatch repair is similar to excision repair, except that it replaces mismatched base pairs (**figure 14.9**).

e. Recombinational repair removes damaged DNA by recombination of the damaged DNA with a normal DNA strand elsewhere in the cell (**figure 14.10**).

f. When DNA damage is severe, DNA replication is halted. This triggers the SOS response. During the SOS response, genes of the repair systems are transcribed at a higher rate. In addition, special DNA polymerases are produced. These are able to replicate damaged DNA. However, they do so without a proper template and therefore create mutations.

14.4 Creating Genetic Variability

a. In recombination, genetic material from two different DNA molecules is combined to form a new hybrid molecule.

b. Horizontal gene transfer is an important mechanism for creating genetic diversity in *Bacteria* and *Archaea*. It is a one-way process in which donor DNA is transferred from the donor to a recipient. In many transfers, the donor DNA must be integrated into the recipient's chromosome to be stably maintained (**figure 14.11**).

c. There are three types of recombination: homologous recombination, site-specific recombination, and transposition (**figures 14.12** and **14.13**).

14.5 Transposable Elements

a. Mobile genetic elements (transposable elements) are DNA segments that move about the genome in a process known as transposition.

b. There are three important types of transposable elements: insertion sequences, transposons, and replicative transposons (**figure 14.14**).

c. Simple (cut-and-paste) transposition and replicative transposition are two distinct mechanisms of transposition (**figures 14.15** and **14.16**).

d. Transposable elements can cause mutations, turn genes on and off, aid plasmid insertion, and carry antibiotic resistance genes.

14.6 Bacterial Plasmids

a. Plasmids are small, autonomously replicating DNA molecules that can exist independent of the host chromosome.

b. Episomes are plasmids that can be reversibly integrated with the host chromosome.

c. The F factor is one type of conjugative plasmid; that is, it is able to transfer itself from one bacterium to another (**figure 14.18**).

14.7 Bacterial Conjugation

a. Conjugation is the transfer of genes between bacteria that depends upon direct cell-to-cell contact. F factor conjugation in *E. coli* is mediated by a sex pilus and a type IV secretion system.

b. In $F^+ \times F^-$, mating, the F factor remains independent of the chromosome and a copy is transferred to the F^- recipient; donor genes are not usually transferred (**figure 14.21*a***).

c. Hfr strains transfer bacterial genes to recipients because the F factor is integrated into the host chromosome. A complete copy of the F factor is not often transferred (**figure 14.24*a*,*b***).

d. When the F factor leaves an Hfr chromosome, it occasionally picks up some bacterial genes to become an F′ plasmid, which readily transfers these genes to other bacteria (**figure 14.25**).

14.8 Bacterial Transformation

a. Transformation is the uptake of naked DNA by a competent cell and its incorporation into the genome (**figures 14.26** and **14.27**).

b. Only certain bacterial species are naturally transformation competent. Other species can be made competent by artificial means.

14.9 Transduction

a. Bacterial viruses (bacteriophages) can reproduce and destroy the host cell (lytic cycle) or become a latent prophage that remains within the host (lysogenic cycle) (**figure 14.29**).

b. Transduction is the transfer of genes by viruses.

c. In generalized transduction, any host DNA fragment can be packaged in a virus capsid and transferred to a recipient (**figure 14.30**).

d. Certain temperate phages carry out specialized transduction by incorporating bacterial genes during prophage induction and then donating those genes to another bacterium (**figure 14.31**).

Critical Thinking Questions

1. Mutations are often considered harmful. Give an example of a mutation that would be beneficial to a microorganism. What gene would bear the mutation? How would the mutation alter the gene's role in the cell, and what conditions would select for this mutant allele?

2. Mistakes made during transcription affect the cell but are not considered "mutations." Why not?

3. Given what you know about the differences between bacterial and eukaryotic cells, give two reasons why the Ames test detects only about half of potential carcinogens, even when liver extracts are used.

4. Suppose that transduction took place when a U-tube experiment was conducted. How would you confirm that a virus was passed through the filter and transduced the recipient?

5. Suppose that you carried out a U-tube experiment with two auxotrophs and discovered that recombination was not blocked by the filter but was stopped by treatment with deoxyribonuclease. What gene transfer process is responsible? Why would it be best to use double or triple auxotrophs in this experiment?

6. What would be the evolutionary advantage of having a period of natural "competence" in a bacterial life cycle? What would be possible disadvantages?

7. Unlike *E. coli*, *Streptococcus pneumoniae* lacks the capacity to mount an SOS response (p. 373). Prudhomme and colleagues tested to see if transformation might be an alternative response to DNA damage. To do this, they treated *S. pneumoniae* with the drug mitomycin C, which blocks the replication fork, thereby generating a region of single-stranded DNA to which the protein RecA binds. They discovered that transformation in *S. pneumoniae* was indeed induced by mitomycin C exposure. They then tested an assortment of antibiotics that are known to trigger the SOS response in *E. coli* for their ability to induce transformation in *S. pneumoniae.* They found that some but not all of the antibiotics resulted in *S. pneumoniae* transformation.

   Why do you think *S. pneumoniae* responds to certain stresses by genetic transformation? Do you think the discovery that certain antibiotics trigger transformation in this pathogen should be considered when prescribing antibiotics currently used and designing new drugs? If so, explain why and how.

   Read the original paper: Prudhomme, M., et al. 2006. Antibiotic stress induces genetic transformability in the human pathogen *Streptococcus pneumoniae. Science* 313:89.

8. *Helicobacter pylori* is a gastric pathogen that causes ulcers and cancer of the stomach. It is naturally competent and its genome encodes at least four type IV secretion systems. Thus

it is not surprising that isolates of *H. pylori* are able transfer chromosomal DNA using a conjugation-like mechanism. *Campylobacter jejuni* is closely related to *H. pylori* and is a major cause of food-borne illness. Microbiologists were curious to see if *H. pylori* could transfer DNA to *C. jejuni* via conjugation.

How would you design such an experiment? How would you detect successful conjugation without resorting to molecular genetic techniques such as DNA sequencing? What kinds of controls would you need to include so that you could be confident of your experimental results?

Read the original paper: Oyarzabal, O. A., et al. 2007. Conjugative transfer of chromosomally encoded antibiotic resistance from *Helicobacter pylori* to *Campylobacter jejuni*. *J. Clin. Microbiol.* 45:402.

Concept Mapping Challenge

Construct a concept map that describes the types of DNA repair mechanisms employed by organisms to protect their DNA. Use the concepts that follow, any other concepts or terms you need, and your own linking words between each pair of concepts in your map.

Proofreading	Excision repair	Mismatch repair
SOS response	Direct repair	RecA protein
Translesion	DNA synthesis	Error-prone repair

Learn More

Learn more by visiting the text website at www.mhhe.com/willey8, where you will find a complete list of references.

15

Recombinant DNA Technology

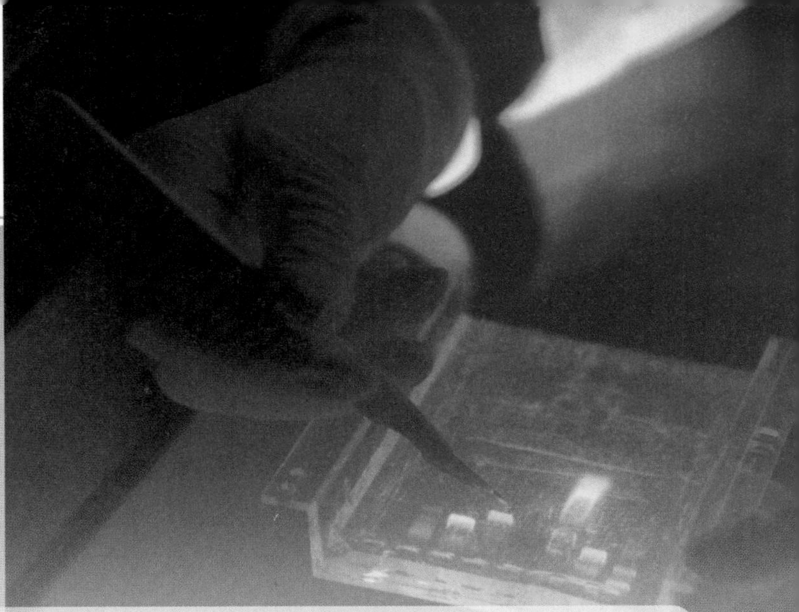

A scientist examines DNA following agarose gel electrophoresis. Each bright band is a fragment of DNA stained with ethidium bromide, so that upon illumination with ultraviolet light, the DNA fluoresces.

CHAPTER GLOSSARY

cloning The generation of a large number of genetically identical DNA molecules.

cloning vector A DNA molecule that can replicate independently of the host chromosome and transport a piece of inserted foreign DNA, such as a gene, into a recipient cell. It may be a plasmid, phage, cosmid, or artificial chromosome.

complementary DNA (cDNA) A DNA copy of an RNA molecule (e.g., a DNA copy of an mRNA).

cosmid A plasmid vector with lambda phage *cos* sites that can be packaged in a phage capsid; it is useful for cloning large DNA fragments.

expression vector A special cloning vector used to express a recombinant gene in host cells; the gene is transcribed and its protein synthesized.

gel electrophoresis A technique that separates molecules according to charge and size.

genetic engineering The deliberate modification of an organism's genetic content by changing its genome.

genomic library A collection of clones that contains fragments that together represent the complete genome of an organism.

polymerase chain reaction (PCR) An in vitro technique used to synthesize large quantities of specific nucleotide sequences from small amounts of genetic material. It employs oligonucleotide primers complementary to specific sequences in the target gene and special heat-stable DNA polymerases.

primer A short piece of DNA or RNA that, when bound to complementary DNA, can start or prime DNA synthesis.

probe A short, labeled segment of RNA or DNA complementary in base sequence to part of another nucleic acid; used to identify or isolate a particular "target" nucleic acid from a mixture through its ability to bind specifically with the target.

recombinant DNA technology The techniques used in carrying out genetic engineering; they involve the identification and isolation of a specific gene, the insertion of the gene into a vector such as a plasmid to form a recombinant molecule, and the production of large quantities of the gene and its product.

restriction enzymes Enzymes that cleave DNA at specific points called recognition sites. They evolved to protect cells from virus infection; they are used in vitro to carry out genetic engineering. Also called restriction endonucleases.

Although human beings have been altering the genetic makeup of organisms for centuries by selective breeding, only recently has the direct manipulation of DNA been possible. The deliberate modification of an organism's genetic information by directly changing the sequence of nucleotides in its genome is called **genetic engineering** and is accomplished by a collection of methods known as **recombinant DNA technology.** The generation of a large number of genetically identical DNA molecules is called **cloning.** The most commonly used steps to clone a gene or other DNA element are outlined in figure 15.1. First, the DNA of interest is identified and isolated (figure 15.1, steps 1 and 2). Once purified, the gene or genes are fused with another piece of DNA called a cloning vector to form recombinant DNA molecules (step 3). These are propagated by insertion into an organism that may not even be in the same domain as the original gene donor but will nonetheless express the gene (step 4). *Steps in Cloning a Gene*

Although the term has several definitions, **biotechnology** refers here to those processes in which living organisms are manipulated, particularly at the molecular genetic level, to form useful products. In this chapter, we introduce the techniques used in biotechnology by briefly discussing each step in the gene cloning process. We then turn our attention to the purification and analysis of cloned gene products, that is, recombinant proteins. Biotechnology is also key to the success of **industrial microbiology**—the use of microbes to manufacture important compounds or the use of microbes as products in their own right. Industrial microbiology is the topic of chapter 41.

15.1 Key Developments in Recombinant DNA Technology

Recombinant DNA technology opened up new areas of research and applied biology. Indeed, the pace of research and discovery in the last half century has been remarkable (**table 15.1**). Many of these early discoveries are essential to biotechnology as it is practiced today; these are now discussed.

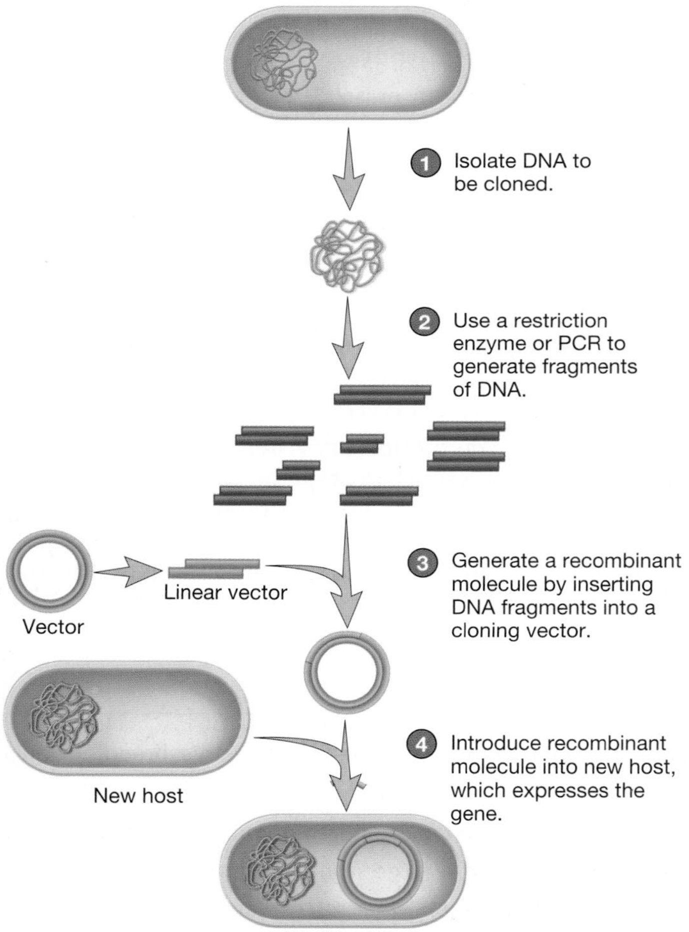

1 Isolate DNA to be cloned.
2 Use a restriction enzyme or PCR to generate fragments of DNA.
Vector
Linear vector
3 Generate a recombinant molecule by inserting DNA fragments into a cloning vector.
New host
4 Introduce recombinant molecule into new host, which expresses the gene.

FIGURE 15.1 Steps in Cloning a Gene. Each step shown in this overview is discussed in more detail in this chapter.

Figure 15.1 Micro Inquiry

Which of the DNA molecules shown are recombinant?

Restriction Enzymes

Recombinant DNA is DNA with a new nucleotide sequence. Such DNA is formed by joining fragments from two or more different sources. One of the first breakthroughs leading to recombinant DNA technology was the discovery by Werner Arber and Hamilton Smith in the late 1960s of bacterial enzymes that cut double-stranded DNA. These enzymes, known as **restriction enzymes** or restriction endonucleases, recognize and cleave specific sequences about four to eight base pairs long (**figure 15.2**). Restriction enzymes recognize specific DNA sequences called **recognition sites**. Each restriction enzyme has its own recognition site. Hundreds of different restriction enzymes have been purified and are commercially available. Type I and type III endonucleases identify their unique recognition sites and then cleave DNA at a defined distance from it. The more common type II endonucleases cut DNA directly at their recognition sites. These enzymes can be used to prepare DNA fragments containing specific genes or portions of genes. For example, the restriction enzyme EcoRI, isolated by Herbert Boyer in 1969 from *Escherichia coli,* cleaves DNA between G and A in the base sequence 5′-GAATTC-3′ (**figure 15.3**). Because DNA is antiparallel, this sequence is reversed on the complementary strand of DNA. When EcoRI cleaves between the G and A residues, unpaired 5′-AATTC-3′ remains at the end of each strand. The complementary bases on two EcoRI-cut fragments can hydrogen bond, thus EcoRI and other endonucleases like it generate cohesive or **sticky ends.** In contrast, cleavage by restriction enzymes such as AluI and HaeIII leaves blunt ends. A few restriction enzymes and their recognition sites are listed in **table 15.2**. Note that each enzyme is named after the bacterium from which it is purified. *Restriction Enzymes*

Genetic Cloning and cDNA Synthesis

An important advance in cloning DNA came in 1972, when David Jackson, Robert Symons, and Paul Berg reported that they had successfully generated recombinant DNA molecules. They allowed the sticky ends of fragments to anneal—that is, to base pair with one another—and then covalently joined the fragments with the enzyme DNA ligase (*see figure 12.16*). Within a year, plasmid vectors that carry foreign DNA fragments during gene cloning had been developed and combined with foreign DNA

Table 15.1 Some Milestones in Biotechnology and Recombinant DNA Technology

1958	DNA polymerase purified
1970	A complete gene synthesized in vitro Discovery of the first sequence-specific restriction endonuclease and the enzyme reverse transcriptase
1972	First recombinant DNA molecules generated
1973	Use of plasmid vectors for gene cloning
1975	Southern blotting technique for detecting specific DNA sequences
1976	First prenatal diagnosis using a gene-specific probe
1977	Methods for rapid DNA sequencing Discovery of introns and exons in eukaryotic genes
1978	Human genomic library constructed
1979	Insulin synthesized using recombinant DNA
1982	Commercial production by *E. coli* of genetically engineered human insulin Isolation, cloning, and characterization of a human cancer gene
1983	Engineered Ti plasmids used to transform plants
1985	Development of the polymerase chain reaction technique
1987	Insertion of a functional gene into a fertilized mouse egg cures the shiverer mutation disease of mice, a normally fatal genetic disease
1988	The first successful production of a genetically engineered staple crop (soybeans)
1989	First field test of a genetically engineered virus (a baculovirus that kills cabbage looper caterpillars)
1990	Production of the first fertile corn transformed with a foreign gene (a gene for resistance to the herbicide bialaphos)
1991	Development of transgenic pigs and goats capable of manufacturing proteins such as human hemoglobin First test of gene therapy on human cancer patients
1994	The Flavr Savr tomato introduced, the first genetically engineered whole food approved for sale Fully human monoclonal antibodies produced in genetically engineered mice
1995	*Haemophilus influenzae* genome sequenced
1997	Human clinical trials of antisense drugs and DNA vaccines begun
1998	First cloned mammal (the sheep Dolly)
2003	Completion of the draft of the human genome
2005	Reconstruction of 1918 influenza virus
2008	First synthetic genome constructed
2007	First complete human genome from a single individual sequenced

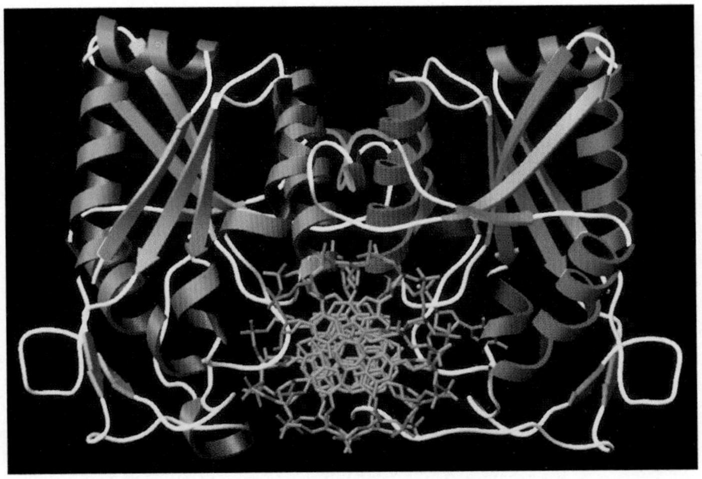

FIGURE 15.2 Restriction Endonuclease Binding to DNA. The structure of BamHI binding to DNA viewed down the DNA axis. The enzyme's two subunits lie on each side of the DNA double helix. The α-helices are in green, the β conformations in purple, and DNA is in orange.

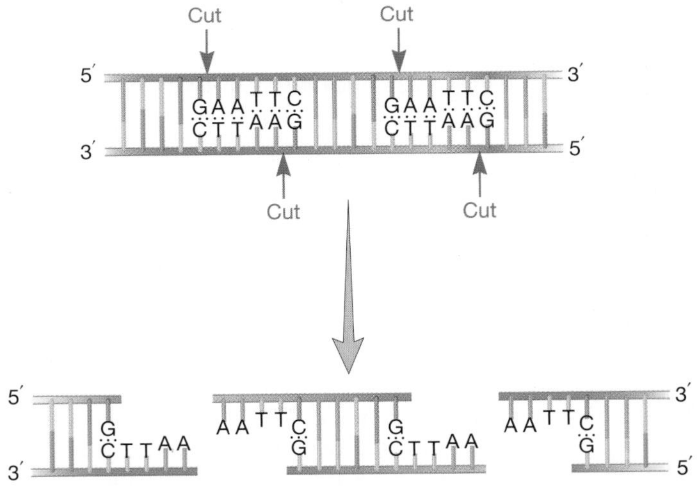

FIGURE 15.3 Restriction Endonuclease Action. The cleavage catalyzed by the restriction endonuclease EcoRI. The enzyme makes staggered cuts on the two DNA strands to form sticky ends.

Figure 15.3 Micro Inquiry

Examine the uncut piece of DNA shown in the upper half of this figure. Where would an exonuclease cleave?

Table 15.2 Some Type II Restriction Endonucleases and Their Recognition Sequences

Enzyme	*Microbial Source*	*Recognition Sequence*[a]	*End Produced*
AluI	*Arthrobacter luteus*	↓ 5′AGCT 3′ 3′ TCGA 5′ ↑	5′ AG CT 3′ 3′ TC GA 5′
BamHI	*Bacillus amyloliquefaciens* H	↓ 5′ GGATCC 3′ 3′ CCTAGG5′ ↑	5′ G GATCC 3′ 3′ CCTAG G 5′
EcoRI	*Escherichia coli*	↓ 5′ GAATTC 3′ 3′ CTTAAG 5′ ↑	5′ G AATTC 3′ 3′ CTTAA G 5′
HaeIII	*Haemophilus aegyptius*	↓ 5′ GGCC 3′ 3′ CCGG 5′ ↑	5′ GG CC 3′ 3′ CC GG 5′
HindIII	*Haemophilus influenzae* d	↓ 5′ AAGCTT 3′ 3′ TTCGAA 5′ ↑	5′ A AGCTT 3′ 3′ TTCGA A 5′
NotI	*Nocardia otitidis-caviarum*	↓ 5′ GCGGCCGC 3′ 3′ CGCCGGCG 5′ ↑	5′ GC GGCCGC 3′ 3′ CGCCGG CG 5′
PstI	*Providencia stuartii*	↓ 5′ CTGCAG 3′ 3′ GACGTC 5′ ↑	5′ CTGCA G3′ 3′ G ACGTC 5′
SalI	*Streptomyces albus*	↓ 5′ GTCGAC 3′ 3′ CAGCTG 5′ ↑	5′ G TCGAC 3′ 3′ CAGCT G 5′

[a]The arrows indicate the sites of cleavage on each strand.

(**figure 15.4**). Recombinant plasmids replicate within a microbial host and maintain the cloned fragment of DNA. *Early Genetic Engineering Experiment*

Once genes could be recombined into cloning vectors, biologists sought to clone specific genes from various organisms. However, it was evident that cloning eukaryotic DNA into bacterial hosts would be problematic. This is because eukaryotic pre-mRNA must be processed (e.g., introns spliced out), and bacteria lack the molecular machinery to perform this task. In 1970, Howard Temin and David Baltimore independently discovered the enzyme that solved this dilemma. They isolated the enzyme **reverse transcriptase (RT)** from retroviruses. These viruses have an RNA genome that is copied into DNA prior to replication. The mechanism by which reverse transcriptase accomplishes this is outlined in **figure 15.5**. Processed mRNA can be used as a template for **complementary DNA (cDNA)** synthesis in vitro. The resulting cDNA can then be cloned, eliminating the need for RNA processing. ▶▶| *Viruses with single-stranded RNA genomes (section 25.7)* *cDNA*

Southern Blotting

Another problem early biotechnologists faced was the inability to distinguish the fragment of DNA possessing the gene of interest from the numerous chromosomal fragments produced by restriction enzyme digestion. In 1975 Edwin Southern solved this problem with his **Southern blotting technique.** This procedure enables the detection of specific DNA fragments from a mixture of DNA molecules (**figure 15.6**). Southern blotting is a three-step process: separate DNA molecules, transfer separated DNA molecules to a membrane, and hybridize to a labeled probe specific for the gene of interest.

In the Southern blotting procedure, DNA fragments are separated by size with agarose gel electrophoresis (see section 15.3). The fragments are then rendered single stranded (i.e., denatured), and the denatured DNA molecules are transferred to a nylon membrane. The transfer occurs when buffer flows through the gel and the membrane, as shown in figure 15.6. Alternatively, the negatively charged DNA fragments can be electrophoresed from

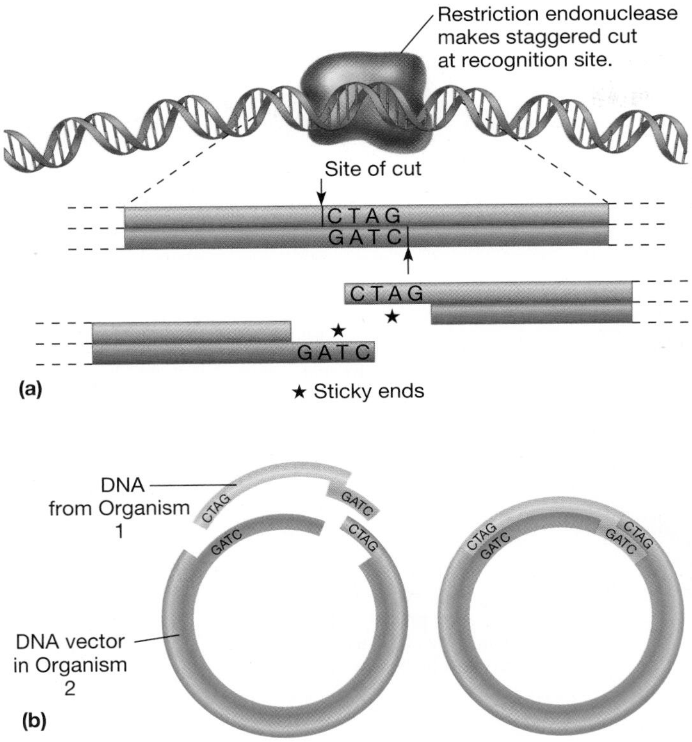

FIGURE 15.4 Recombinant Plasmid Construction. (a) A restriction endonuclease recognizes and cleaves DNA at its specific recognition site. Cleavage produces sticky ends that accept complementary tails for gene splicing. (b) The sticky ends can be used to join DNA from different organisms by cutting it with the same restriction enzyme, ensuring that all fragments have complementary ends.

Figure 15.4 Micro Inquiry

Why might two DNA fragments inadvertently be cloned into a single vector when using this cloning strategy?

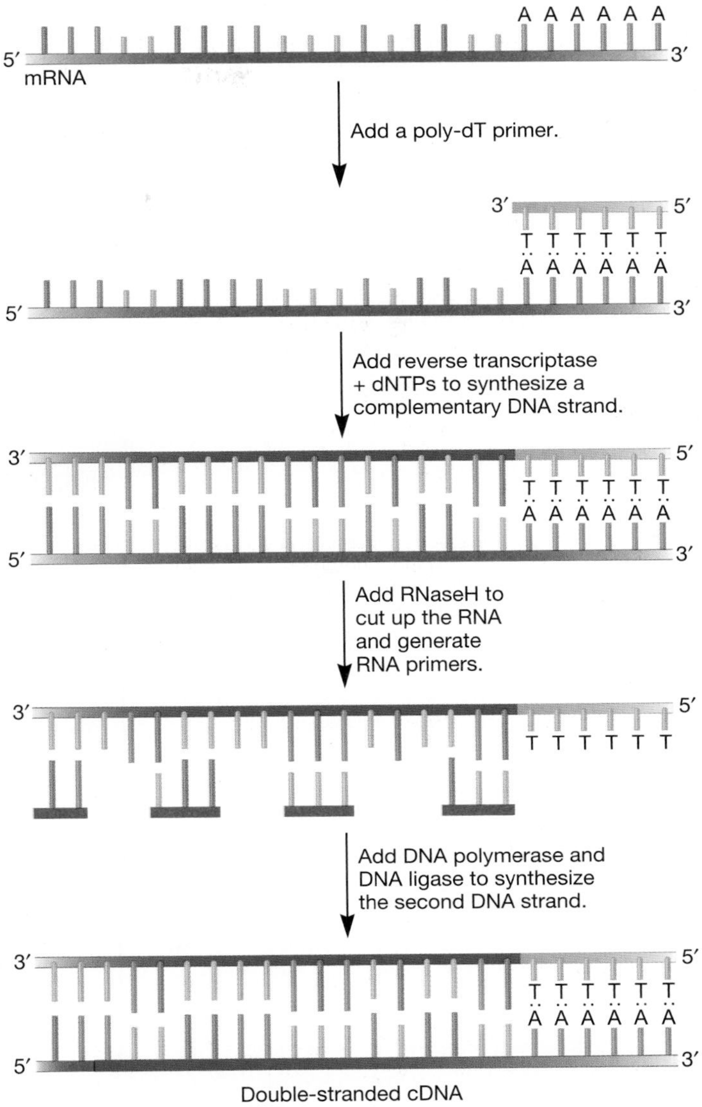

FIGURE 15.5 Synthesis of cDNA. A poly-dT primer anneals to the 3′ end of mRNAs. Reverse transcriptase then catalyzes the synthesis of a complementary DNA strand (cDNA). RNaseH digests the mRNA into short pieces that are used as primers by DNA polymerase to synthesize the second DNA strand. The 5′ to 3′ exonuclease function removes all of the RNA primers except the one at the 5′ end (because there is no primer upstream from this site). This RNA primer can be removed by the subsequent addition of another RNase. After the double-stranded cDNA is made, it can be inserted into vectors, as described in figure 15.4.

Figure 15.5 Micro Inquiry

Why must introns be removed from eukaryotic DNA before it can be expressed in a bacterium?

the gel onto the blotting membrane. After transfer of the DNA, the membrane is treated so that each DNA fragment is firmly bound to the filter at the same position as on the gel. The DNA of interest is identified by bathing the filter in a solution containing a radioactive **probe,** which is a fragment of labeled, single-stranded nucleic acid complementary to the DNA of interest. The DNA to which the probe hydrogen bonds is now radioactive and is readily detected by **autoradiography.** In autoradiography, a sheet of photographic film is placed over the membrane. When the film is developed, bands appear wherever the radioactive probe is bound because the energy released by the isotope causes the formation of dark-silver grains. Nonradioactive probes may also be used to detect specific DNAs. A common label is the small

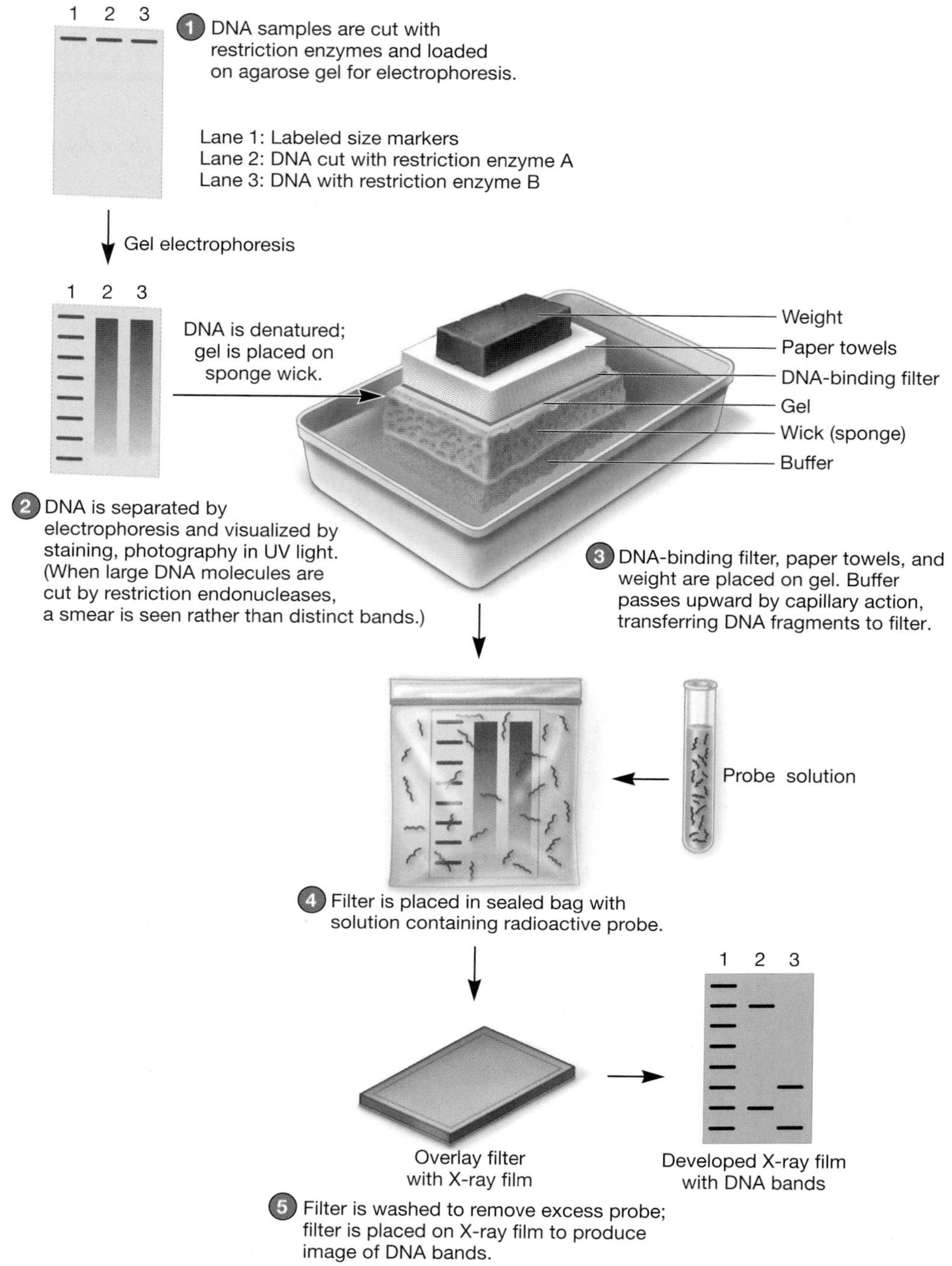
1 2 3
1 DNA samples are cut with restriction enzymes and loaded on agarose gel for electrophoresis.
Lane 1: Labeled size markers
Lane 2: DNA cut with restriction enzyme A
Lane 3: DNA with restriction enzyme B
Gel electrophoresis
1 2 3
DNA is denatured; gel is placed on sponge wick.
Weight
Paper towels
DNA-binding filter
Gel
Wick (sponge)
Buffer
2 DNA is separated by electrophoresis and visualized by staining, photography in UV light. (When large DNA molecules are cut by restriction endonucleases, a smear is seen rather than distinct bands.)
3 DNA-binding filter, paper towels, and weight are placed on gel. Buffer passes upward by capillary action, transferring DNA fragments to filter.
Probe solution
4 Filter is placed in sealed bag with solution containing radioactive probe.
1 2 3
Overlay filter with X-ray film
Developed X-ray film with DNA bands
5 Filter is washed to remove excess probe; filter is placed on X-ray film to produce image of DNA bands.

FIGURE 15.6 The Southern Blotting Technique.

molecule biotin, as discussed in **Techniques & Applications 15.1.** Nonradioactive labels often are more rapidly detected and are safer to use than radioisotopes. *Southern Blotting*

By the late 1970s the techniques for cloning DNA were harnessed to produce recombinant human insulin, and by 1982 commercial production of insulin from genetically engineered *Escherichia coli* began. This was an important development for several reasons: first, diabetic individuals no longer had to depend on insulin from pigs or other animals; second, it demonstrated the commercial feasibility of using recombinant DNA to make a better product.

TECHNIQUES & APPLICATIONS

15.1 Streptavidin-Biotin Binding and Biotechnology

Egg white contains many proteins and glycoproteins with unique properties. One of the most interesting, which binds tenaciously to biotin, was isolated in 1963. This glycoprotein, called avidin due to its "avid" binding of biotin, was suggested to play an important role: making egg white antimicrobial by "tying up" the biotin needed by many microorganisms. Avidin, which functions best under alkaline conditions, has the highest known binding affinity between a protein and a

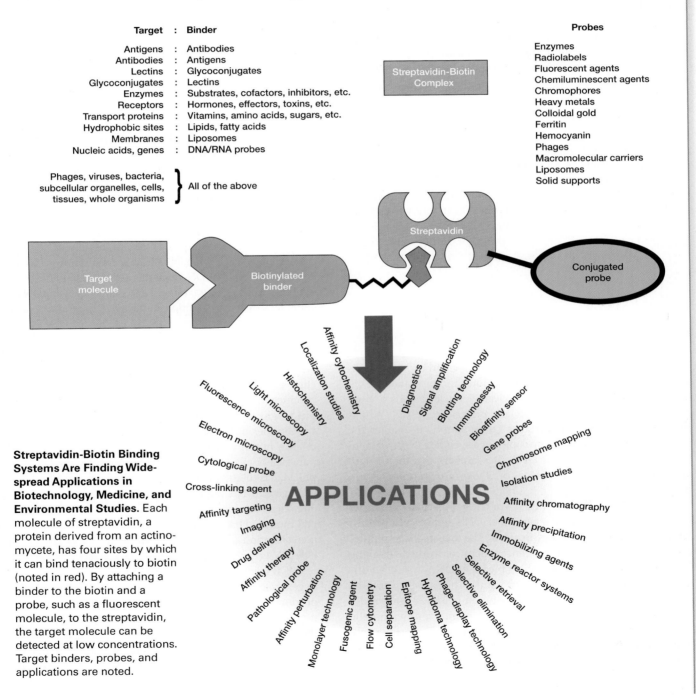

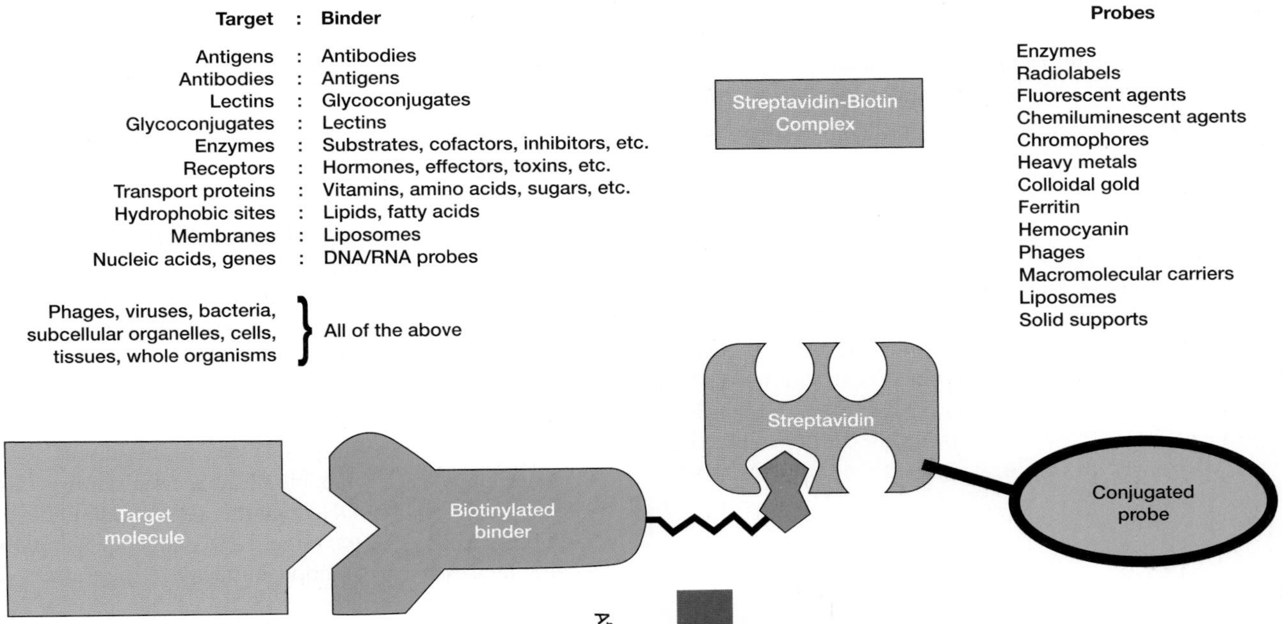

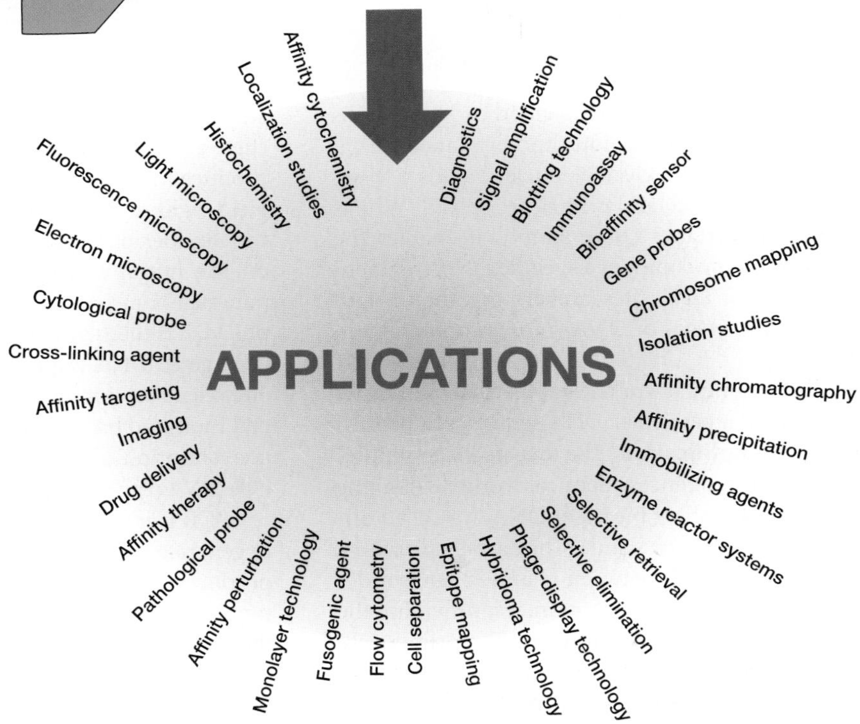

Streptavidin-Biotin Binding Systems Are Finding Widespread Applications in Biotechnology, Medicine, and Environmental Studies. Each molecule of streptavidin, a protein derived from an actinomycete, has four sites by which it can bind tenaciously to biotin (noted in red). By attaching a binder to the biotin and a probe, such as a fluorescent molecule, to the streptavidin, the target molecule can be detected at low concentrations. Target binders, probes, and applications are noted.

TECHNIQUES & APPLICATIONS

(Continued)

ligand. Several years later, scientists at Merck & Co., Inc., discovered a similar protein produced by the actinomycete *Streptomyces avidini,* which binds biotin at a neutral pH and does not contain carbohydrates. These characteristics make this protein, called streptavidin, an ideal binding agent for biotin, and it has been used in an almost unlimited range of applications, as shown in the **box figure**. The streptavidin protein is joined to a probe. When a sample is incubated with the biotinylated binder, the binder attaches to any available target molecules. The presence and location of target molecules can be determined by treating the sample with a streptavidin probe because the streptavidin binds to the biotin on the biotinylated binder, and the probe is then visualized. This detection system is employed in a wide variety of biotechnological applications, including use as a nonradioactive probe in hybridization studies and as a critical component in biosensors for a wide range of environmental monitoring and clinical applications.

1. Describe restriction enzymes, sticky ends, and blunt ends. Can you think of a cloning situation where blunt-ended DNA might be more useful than DNA with sticky ends?
2. What is cDNA? Why is it necessary to generate cDNA before cloning and expressing a eukaryotic gene in a bacterium?
3. What is the purpose of Southern blotting? How is a probe selected? Why do you think the Southern blotting technique was an important breakthrough when it was introduced?

15.2 Polymerase Chain Reaction

The **polymerase chain reaction (PCR),** invented by Kary Mullis in the early 1980s, exploded onto the biotechnology landscape. It has changed the way genes are cloned, nucleic acids are sequenced, diseases are diagnosed, and crimes are solved. Why is PCR so versatile and important? Quite simply, it enables the rapid synthesis of many, many copies of a specific DNA fragment from a complex mixture of DNA. Researchers can thus obtain large quantities of specific pieces of DNA for experimental and diagnostic purposes.

Figure 15.7 outlines how PCR works. Suppose that one wishes to make large quantities of a particular DNA sequence, a process known as gene or **DNA amplification.** The first step is to synthesize DNA fragments with sequences identical to those flanking the targeted sequence. These short pieces of DNA are called **oligonucleotides** (Greek *oligo,* few or scant). They are made with a DNA synthesizer and are generally between 15 and 30 nucleotides long. The oligonucleotides serve as DNA **primers,** providing the 3′-OH needed for DNA synthesis during PCR. The primers are one component of the reaction mixture, which also contains the target, template DNA (often copies of an entire genome), a thermostable DNA polymerase, and each of the four deoxyribonucleoside triphosphates (dNTPs).

PCR requires a series of repeated reactions, called cycles. Each cycle has three steps that are precisely executed in a machine called a thermocycler. In the first step, the DNA containing the sequence to be amplified is denatured by raising the temperature to about 95°C. Next, the temperature is lowered to about 50°C so that the primers can hydrogen bond (anneal) to the DNA on both sides of the target sequence. Because the primers are very small and are present in excess, the targeted DNA strands anneal to the primers, rather than to each other. Finally, the temperature is raised, usually to 68 to 72°C so that DNA polymerase can extend the primers and synthesize copies of the target DNA sequence using dNTPs. Only polymerases able to function at the high temperatures can be used. The most commonly used thermostable enzyme is **Taq polymerase** from the thermophilic bacterium *Thermus aquaticus.*

At the end of one PCR cycle, the targeted sequences on both strands have been copied. When the three-step cycle is repeated (figure 15.7), the two strands from the first cycle are copied to produce four fragments. These are amplified in the third cycle to yield eight double-stranded products. Thus, each cycle increases the number of target DNA molecules exponentially. Depending on the initial concentration of the template DNA and other parameters such as the G + C composition of the DNA to be amplified, it is theoretically possible to produce about 1 million copies of targeted DNA sequence after 20 cycles and over 1 billion after 30 cycles. Pieces ranging in size from less than 100 base pairs to several thousand base pairs in length can be amplified, and the initial concentration of target DNA can be as low as 10^{-20} to 10^{-15} M.

◀◀ *DNA replication (section 12.4)* *Polymerase Chain Reaction*

PCR is most frequently used in one of two ways. If one wants to generate large quantities of a specific piece of DNA, the reaction products are collected and purified at the end of a designated number of cycles. This is sometimes called end-point PCR, and the final number of DNA fragments amplified is not quantitative. This means that the amount of final product does not always reflect the amount of template DNA present. In contrast, **real-time PCR** is quantitative; in fact, it is referred to as qPCR. That is, it allows one to ask how much DNA or RNA template (which is con-

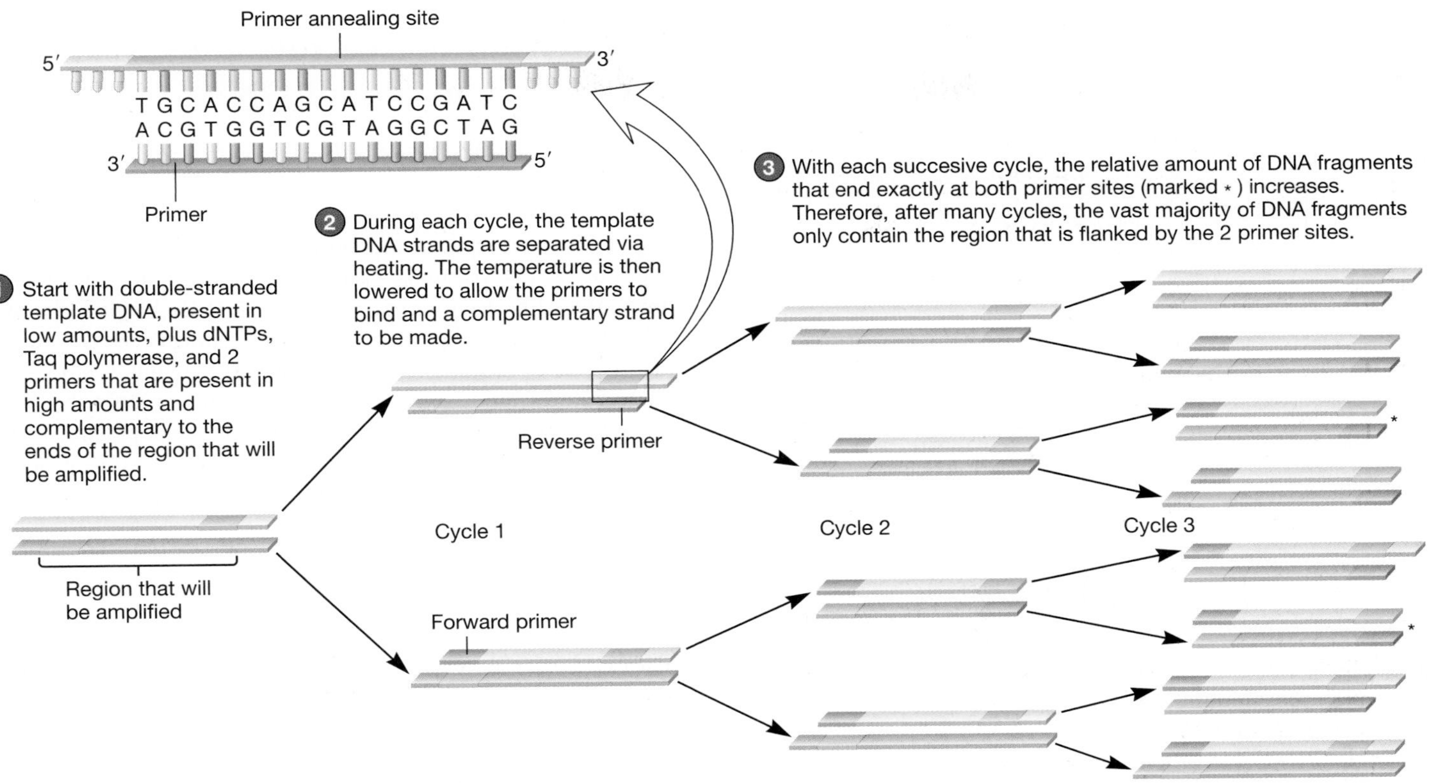

FIGURE 15.7 The Polymerase Chain Reaction (PCR) Technique. During each cycle, oligonucleotides that are complementary to the ends of the targeted DNA sequence bind to the DNA and act as primers for the synthesis of this DNA region. The primers used in actual PCR experiments are usually 15 to 20 nucleotides in length. The region between the two primers is typically hundreds of nucleotides in length, not just several nucleotides as shown here. The net result of PCR is the synthesis of many copies of DNA in the region that is flanked by the two primers.

Figure 15.7 Micro Inquiry

Why, after three cycles, are the vast majority of amplified DNA molecules (i.e., PCR products) the size defined by the distance between the forward and reverse primers?

verted to DNA with reverse transcriptase) is present in a given sample. This is accomplished by adding a fluorescently labeled probe to the reaction mixture and measuring its signal during the initial cycles. This is when the rate of DNA amplification is logarithmic. However, as the PCR cycles continue, substrates are consumed and polymerase efficiency declines. So although the amount of product increases, its rate of synthesis is no longer exponential (this is why end-point collection of PCR products is not quantitative). Thermocyclers specifically designed for real-time PCR record the amount of PCR product generated as it occurs, thus the term real-time PCR. Gene expression studies often rely on real-time PCR, because mRNA transcripts can be copied and amplified by reverse transcriptase. Therefore the procedure monitors the level of transcription of the gene targeted by the primers.

PCR is an essential tool in many areas of molecular biology, medicine, and biotechnology. As shown in **figure 15.8**, when PCR is used to obtain DNA for cloning, a number of steps in traditional cloning procedures are no longer required. PCR is also used to generate DNA for nucleotide sequencing. Because the primers used in PCR target specific DNA, PCR can isolate particular fragments of DNA (e.g., genes) from solutions that contain many different genomes, such as soil, water, and blood. For instance, PCR is used to amplify specific genes from the environment without first culturing members of the microbial community. It has also become an essential part of certain diagnostic tests, including those for AIDS, Lyme disease, chlamydia, tuberculosis, hepatitis, human papillomavirus, and other infectious agents and diseases. The tests are rapid, sensitive, and specific. PCR is also employed in forensic science, where it is used in criminal cases as part of DNA fingerprinting technology. ▶▶| *Metagenomics (section 16.8); Genomic fingerprinting (section 17.3)*

Search This: PCR song

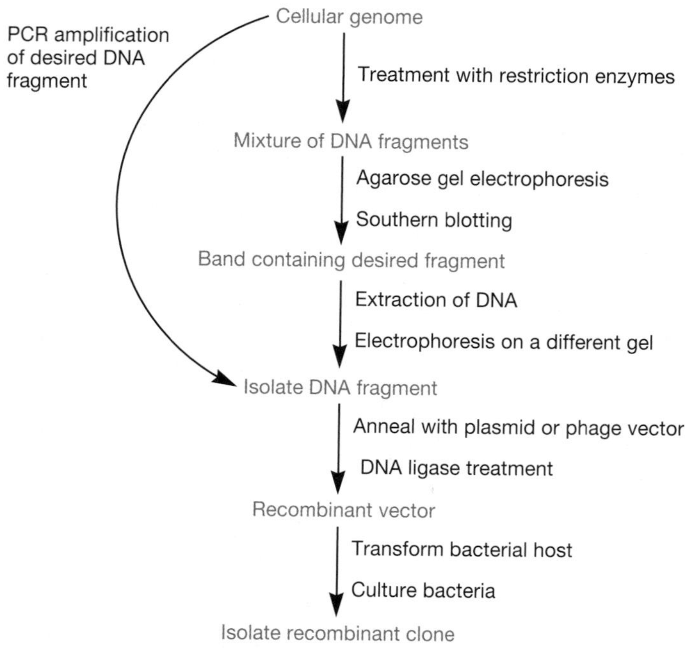

FIGURE 15.8 Cloning Cellular DNA Fragments. The preparation of a recombinant clone from isolated DNA fragments or DNA generated by PCR.

15.3 Gel Electrophoresis

Agarose or polyacrylamide **gel electrophoresis** is routinely used to separate DNA fragments. When DNA molecules are placed at the negative end of an electrical field, they migrate toward the pole with a positive charge (**figure 15.9**). Each fragment's migration rate is determined by its molecular weight so that the smaller a fragment is, the faster it moves through the gel. Migration rate is also a function of gel density. In practice, this means that higher concentrations of gel material (agarose or acrylamide) provide better resolution of small fragments and vice versa. DNA that has not been digested with restriction enzymes is usually supercoiled and does not migrate to a position that corresponds to its molecular weight. For this and other reasons, DNA is usually cut with restriction enzymes prior to electrophoresis. Small DNA molecules usually yield only a few bands because there are few restriction enzyme recognition sites. If the DNA fragment is large or an entire chromosome is digested, many restriction recognition sites are present and the DNA is cut in numerous places. When such DNA is electrophoresed, it produces a smear representing many thousands of DNA fragments of similar sizes that cannot be individually resolved. The region of the gel containing the desired DNA fragment must then be located using the Southern blotting technique (figure 15.6).

Search This: Gel electrophoresis tutorial

15.4 Cloning Vectors and Creating Recombinant DNA

Recombinant DNA technology depends on the propagation of many copies of the nucleotide sequence of choice. To accomplish this, genes or other genetic elements to be cloned are inserted into **cloning vectors** that replicate in a host organism. There are four major types of vectors: plasmids, bacteriophages and other viruses, cosmids, and artificial chromosomes (**table 15.3**). Each type has its own advantages, so the selection of the proper cloning vector is critical to the success of any cloning experiment. Most cloning vectors share three features: an origin of replication; a region of DNA that bears unique restriction sites, called a multicloning site or polylinker; and a selectable marker. These elements are described in this section in the discussion of plasmids, the most frequently used cloning vectors. *Construction of a Plasmid Vector*

Plasmids

Plasmids make excellent cloning vectors because they replicate autonomously (i.e., independently of the chromosome) and are easy to purify. They can be introduced into microbes by conjugation or transformation. Many different plasmids are used in biotechnology, all derived from naturally occurring plasmids that have been genetically engineered (**figure 15.10**). ◀◀ *Plasmids (section 3.5); Bacterial conjugation (section 14.7); Bacterial transformation (section 14.8)*

Origin of Replication

The **origin of replication** (*ori*) allows the plasmid to replicate in the microbial host independently of the chromosome. pUC19, an *E. coli* plasmid, is said to have a high copy number because it replicates about 100 times in the course of one generation. High copy number is often important because it facilitates plasmid purification and can dramatically increase the amount of cloned gene product produced by the cell. Some plasmids have two origins of replication, each recognized by different host organisms. These plasmids are called **shuttle vectors** because they can move or "shuttle" from one host to another. YEp24 is a shuttle vector that can replicate in yeast (*Saccharomyces cerevisiae*) and in *E. coli* because it has the 2μ circle yeast replication element and *E. coli* origin of replication (figure 15.10).

Selectable Marker

Following the uptake of vector by host cells, one must be able to discriminate between cells that successfully obtained vector (transformants) from those that did not (nontransformants). Furthermore, one must be able to continue to select for the presence of plasmid, otherwise the host cell may stop replicating it. This is achieved by the presence of a gene that encodes a protein that is needed for the cell to survive under certain, selective conditions.

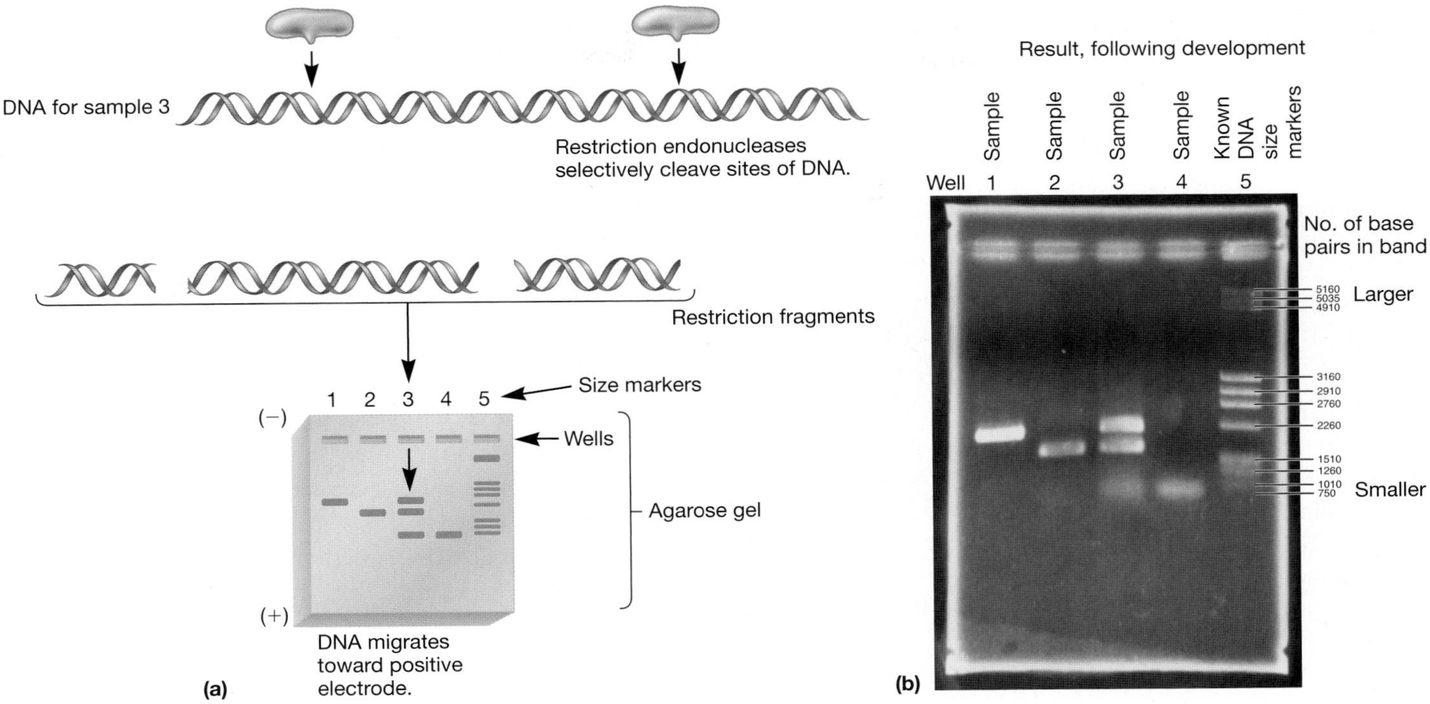

FIGURE 15.9 Gel Electrophoresis of DNA. (a) After cleavage into fragments, DNA is loaded into wells on one end of an agarose gel. When an electrical current is passed through the gel (from the negative pole to the positive pole), the DNA, being negatively charged, migrates toward the positive pole. The larger fragments, measured in numbers of base pairs, migrate more slowly and remain nearer the wells than the smaller (shorter) fragments. (b) An actual developed and stained gel reveals a separation pattern of the fragments of DNA. The size of a given DNA band can be determined by comparing it to a known set of molecular weight markers (lane 5) called a ladder.

Figure 15.9 Micro Inquiry

If a linear piece of DNA is cut with a restriction enzyme for which it has four recognition sites, how many bands would be visible on a gel if the products of digestion were electrophoresed? (Hint: Make a drawing of the DNA and its restriction sites.)

Such a gene is called a **selectable marker**. In the case of pUC19, the selectable marker encodes the ampicillin resistance enzyme (*amp*^{R}, sometimes called *bla,* for β-*la*ctamase). The shuttle vector YEp24 bears both the *amp*^{R} gene for selection in *E. coli* and URA3, which encodes a protein essential for uracil biosynthesis in yeast. Therefore when in *S. cerevisiae,* this plasmid must be maintained in uracil auxotrophs.

Multicloning Site or Polylinker

A region of restriction enzyme cleavage sites found only once in the plasmid is essential for the insertion of foreign DNA. Cleavage at a unique restriction site generates a linear plasmid. Cleavage of the gene to be cloned with the same restriction enzyme results in compatible sticky ends, so that it may be inserted (ligated) into the **multicloning site (MCS).** Alternatively, two different, unique sites within the MCS may be cleaved and the DNA sequence between the two sites replaced with cloned DNA (**figure 15.11**). In either case, the plasmid and the DNA to be inserted are incubated in the presence of the enzyme DNA ligase so that when compatible sticky ends hydrogen bond, phosphodiester bonds can be generated between the cloned DNA fragment and the vector. This requires the input of energy; thus ATP is added to this in vitro ligation reaction (*see figure 12.16*).

pUC19 has a number of unique restriction sites in its MCS (figure 15.10); this provides a number of cleavage options, making it easier to obtain the same, or compatible, sticky ends in both vector and the DNA to be inserted. In pUC19, the MCS is located within the 5′ end of the *lacZ* gene, which encodes β-galactosidase (β-Gal). This enzyme cleaves the disaccharide lactose into galactose and glucose. When DNA has been cloned into the MCS, the *lacZ* gene is no longer intact, so a functional enzyme is not produced. This can be detected by the color of

Table 15.3 Recombinant DNA Cloning Vectors

Vector	*Insert Size (kb, 1 kb = 1,000 bp)*	*Example*	*Features*
Plasmid	<20 kb	pBR322, pUC19	Replicates independently of microbial chromosome so many copies may be maintained in a single cell
Bacteriophage	9–25 kb	λ 1059, λ gt11, M13mp18, EMBL3	Packaged into lambda phage particles; single-stranded DNA viruses such as M13 have been modified (e.g., M13mp18) to generate either double- or single-stranded DNA in the host.
Cosmids	30–47 kb	pJC720, pSupercos	Can be packaged into lambda phage particles for efficient introduction into bacteria, then replicates as a plasmid
PACs (P1 artificial chromosomes)	75–100 kb	pPAC	Based on the bacteriophage P1 packaging mechanism
BACs (bacterial artificial chromosomes)	75–300 kb	pBAC108L	Modified F plasmid that can carry large DNA inserts; very stable within the cell
YACs (yeast artificial chromosomes)	100–1,000 kb	pYAC	Can carry largest DNA inserts; replicates in *Saccharomyces cerevisiae*

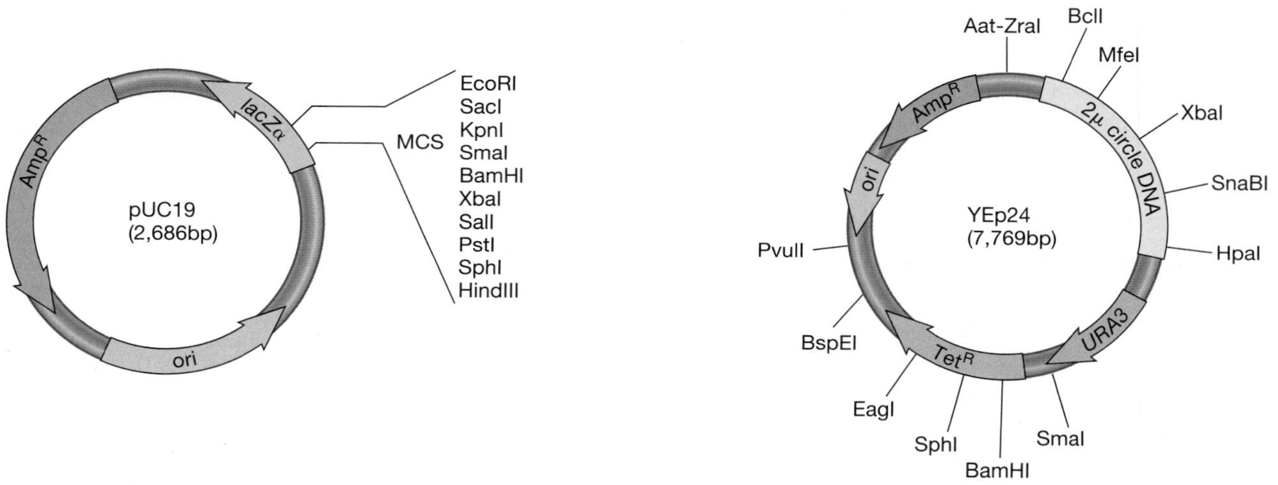

FIGURE 15.10 The Cloning Vectors pUC19 and YEp24. Restriction sites that are present only once in each vector are shown. pUC19 replicates only in *E. coli,* while YEp24 replicates in both *E. coli* and *S. cerevisiae.*

Figure 15.10 Micro Inquiry

Which plasmid is a shuttle vector? Why?

colonies: cells turn blue when β-Gal splits the alternative substrate, X-Gal (5-bromo-4-chloro-3-indolyl-β-D-galactopyranoside), which is included in the medium (figure 15.11*c*). This is important because the ligation of foreign DNA into a vector is never 100% efficient. Thus when the ligation mixture is introduced into host cells, one must be able to distinguish cells that carry just plasmid from those that carry plasmid into which DNA was successfully inserted. In the case of pUC19, all *E. coli* cells that take up plasmid (with or without insert) are selected by their resistance to ampicillin (Amp^R); that is to say, only pUC19

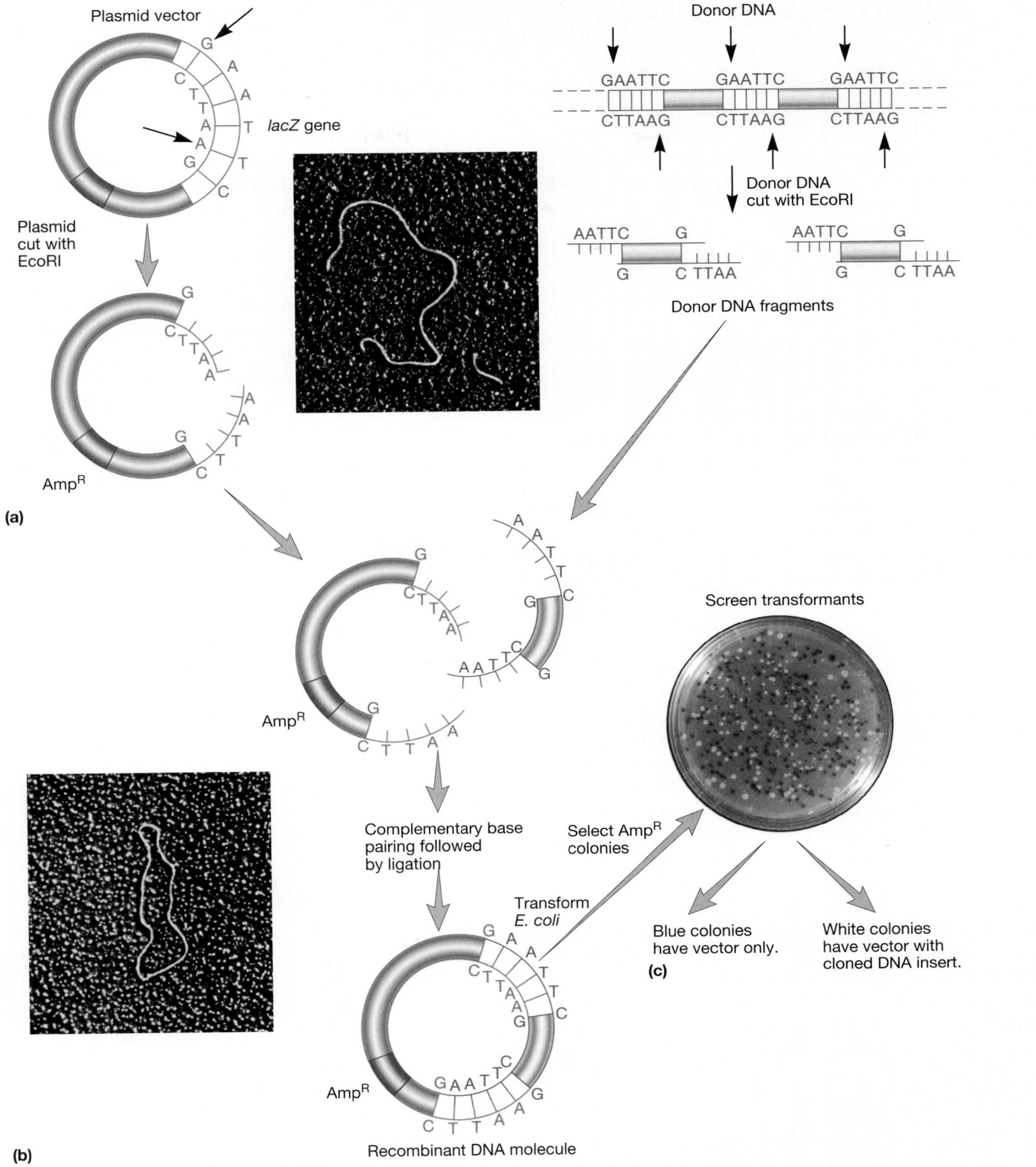

FIGURE 15.11 Recombinant Plasmid Construction and Cloning. The construction and cloning of a recombinant plasmid vector using an antibiotic resistance gene to select for the presence of the plasmid. The interruption of the *lacZ* gene by cloned DNA is used to detect vectors with insert. The scale of the sticky ends of the fragments and plasmid has been enlarged to illustrate complementary base pairing. (a) The electron micrograph shows a plasmid that has been cut by a restriction enzyme and a donor DNA fragment. (b) The micrograph shows a recombinant plasmid. (c) After transformation, *E. coli* cells are plated on medium containing ampicillin and X-Gal so that only ampicillin-resistant transformants grow; X-Gal enables the visualization of colonies that were transformed with recombinant vector (vector + insert, white colonies).

Figure 15.11 Micro Inquiry

What would you conclude if you obtained only blue colonies after cloning into a vector that enables blue/white screening of transformants? Why might such a result occur?

transformants grow. Among these, colonies with plasmid lacking DNA insert will be blue (due to the presence of functional *lacZ* gene), while those with pUC19 into which DNA was successfully cloned will be white. There are a number of other clever ways in which cells with vector versus those with vector plus insert can be differentiated; the detection of blue versus white colonies is a common approach.

Phage Vectors

Phage vectors are engineered phage genomes that have been genetically modified to include useful restriction enzyme recognition sites for the insertion of foreign DNA. Once DNA has been inserted, the recombinant phage genome is packaged into viral capsids and used to infect host cells. The resulting phage lysate consists of thousands of phage particles that carry cloned DNA as well as the genes needed for host lysis. Two commonly used vectors are derived from the bacteriophages T7 and lambda (λ), both of which have double-stranded DNA genomes. Although these phages infect *E. coli,* phage vectors have been engineered for a number of different bacterial host species. ◀◀ *Structure of viruses (section 5.2)*

Cosmids

Cosmids were developed when it became clear that cloning vectors were needed that could tolerate larger fragments of cloned DNA (table 15.3). Unlike phages and plasmids, cosmids do not exist in nature. Instead, these engineered vectors have been constructed to contain features from both. Cosmids have a selectable marker and MCS from plasmids, and a *cos* site from phage. In phage, the *cos* site is where multiple copies of phage genome are linked prior to packaging. Cleavage at the *cos* sites yields single genomes that are the right size for packaging. Cosmids take advantage of the fact that the only requirement for phage heads to package DNA is two *cos* sites on a linear DNA molecule or a single *cos* site on a circular one (*see figure 25.11*). As long as the cosmid with its cloned DNA is the appropriate size (about 37 to 52 kb), it will be packaged. The phage is then used to introduce the recombinant DNA into *E. coli,* where it replicates as a plasmid.

Artificial Chromosomes

Artificial chromosomes are special cloning vectors used when particularly large fragments of DNA must be cloned, as when constructing a genomic library or sequencing an organism's entire genome. In fact, **bacterial artificial chromosomes (BACs)** were crucial to the timely completion of the human genome project and the construction of a synthetic genome (**Techniques & Applications 15.2**). Like natural chromosomes, artificial chromosomes replicate only once per cell cycle. **Yeast artificial chromosomes (YACs)** were developed first and consist of a yeast telomere at each end (TEL), a centromere sequence (CEN), a yeast origin of replication (ARS, *a*utonomously *r*eplicating *s*equence), a selectable marker such as URA3, and an MCS to facilitate the insertion of foreign DNA (**figure 15.12*a***). YACs are used when extraordinarily large DNA pieces (up to 1,000 kb; table 15.3) are to be cloned. BACs were developed, in part, because YACs tend to be unstable and may recombine with host chromosomes, causing mutations and rearrangement of the cloned DNA. Although BACs accept smaller DNA inserts than do YACs (up to 300 kb), they are generally more stable. BACs are based on the F fertility factor of *E. coli (see figure 14.18)*. The example shown in figure 15.12*b* is typical in that it includes genes that ensure a replication complex will be formed (*repE*), as well as proper partitioning of one newly replicated BAC to each daughter cell (*sopA, sopB,* and *sopC*). It also includes features common to many plasmids such as an MCS within the *lacZ* gene for blue/white colony screening and a selectable marker, in this case for resistance to the antibiotic chloramphenicol (Cm^R).

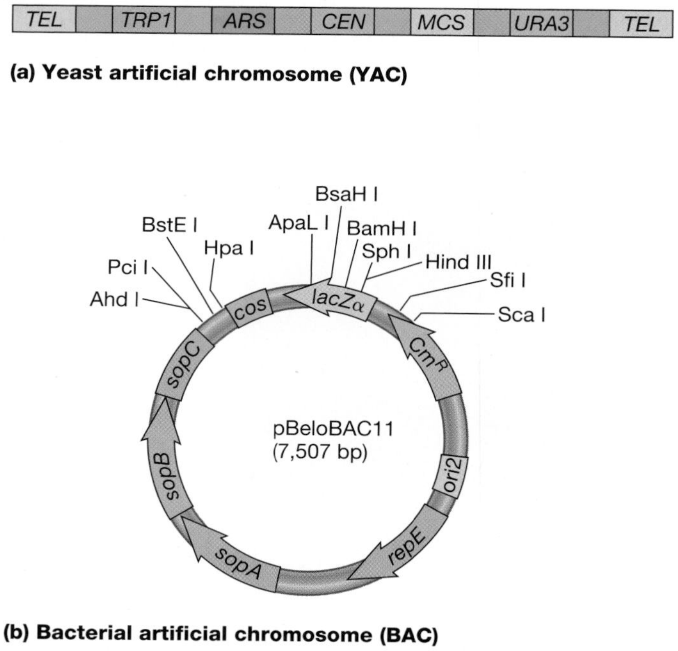

FIGURE 15.12 Artificial Chromosomes Can Be Used as Cloning Vectors. (a) A yeast artificial chromosome and (b) the bacterial artificial chromosome pBeloBAC11.

***Figure 15.12* Micro Inquiry**

In what ways does the BAC shown here differ from the plasmid pUC19 shown in figure 15.10?

TECHNIQUES & APPLICATIONS

15.2 How to Build a Genome

Biotechnology has repeatedly demonstrated how useful the genetic modification of an organism can be. From the production of recombinant human insulin to recombinant indigo (the dye used in blue jeans), it is clear that "gene juggling" can potentially be used for the production of a wide range of products. But what if, rather than taking the genes from one organism and inserting them in another, one could simply design the ideal genome for the expression of the genes that are of interest? One area in which this is being considered is the development of new bacteria that will generate biofuels. ▶▶| *Bioconversion processes (section 41.3)*

The first step in reaching this goal was to determine if building a genome from scratch is really possible. To test this possibility, molecular biologists at the J. Craig Venter Institute (JCVI) started with the one of the smallest bacterial genomes, that of *Mycoplasma genitalium*. This pathogen has a single circular chromosome of only 580,076 base pairs (bp), so it was reasoned this was good place to start.

The genome was first mapped into 101 sections, each about 5,000 to 7,000 bp each. Each segment contained at least one gene, and the ends of the segments were in regions between genes, that is, intergenic regions. The nucleotide sequence in each segment was used as a blueprint to make a cassette of artificially synthesized DNA. In addition, six of the cassettes had special nucleotide sequences, called watermarks, inserted. Just as a watermark is used to identify the origin of a piece of stationary or currency, these nucleotide sequence watermarks were added to identify this genome as artificial. Also, an antibiotic resistance gene was inserted into the middle of a gene that is required for virulence. By inserting the antibiotic resistance gene into an essential virulence gene, pathogenicity was "knocked out," while antibiotic resistance was "knocked in."

Once all the cassettes were constructed, the tricky part was to assemble them in the proper order in a stable molecule of greater than half a million bp. The team used a five-part strategy to accomplish this (**box figure**). First, groups of four neighboring cassettes were ligated together, with a total average size of about 24,000 bp, or 24 kilobp (kb). This was possible thanks to about an 80 bp overlap between cassettes. Specifically, common nucleotide sequences were found at the ends of cassettes 1 and 2, 2 and 3, and 3 and 4, and so on. A nuclease that removes only the 3′ end of DNA was used to make sticky ends, so that the formerly blunt cassette ends could base pair and be ligated together. Groups of four cassettes were then cloned into BACs and introduced into *E. coli*. These became the "A" set of clones. Next, the inserts from three BACS that represented 12 sequential cassettes were excised and ligated together to make 72 kb inserts for a new set of BACs; these were the "B" clones. Two sequential B clones were ligated together and inserts of 144 kb were cloned into BACs, making the "C" clones. However, when the team tried to join two C clones to make a BAC carrying half the *M. genitalium* chromosome, they found that the BACs were unstable and would not accept such large inserts. Thus the half chromosomes (D clones) were assembled on YACs, and finally, the two half-chromosomes were united to form a complete *M. genitalium* chromosome in a yeast host.

The JCVI team was careful to check the nucleotide sequence of the each set of clones along the way, so that when they had assembled the entire genome, they were confident that they had the exact genome they set out to build. This "proof of principle" accomplishment may have raised more questions—both technical and philosophical—than it answered, but by all means, it demonstrates the power of modern molecular biology.

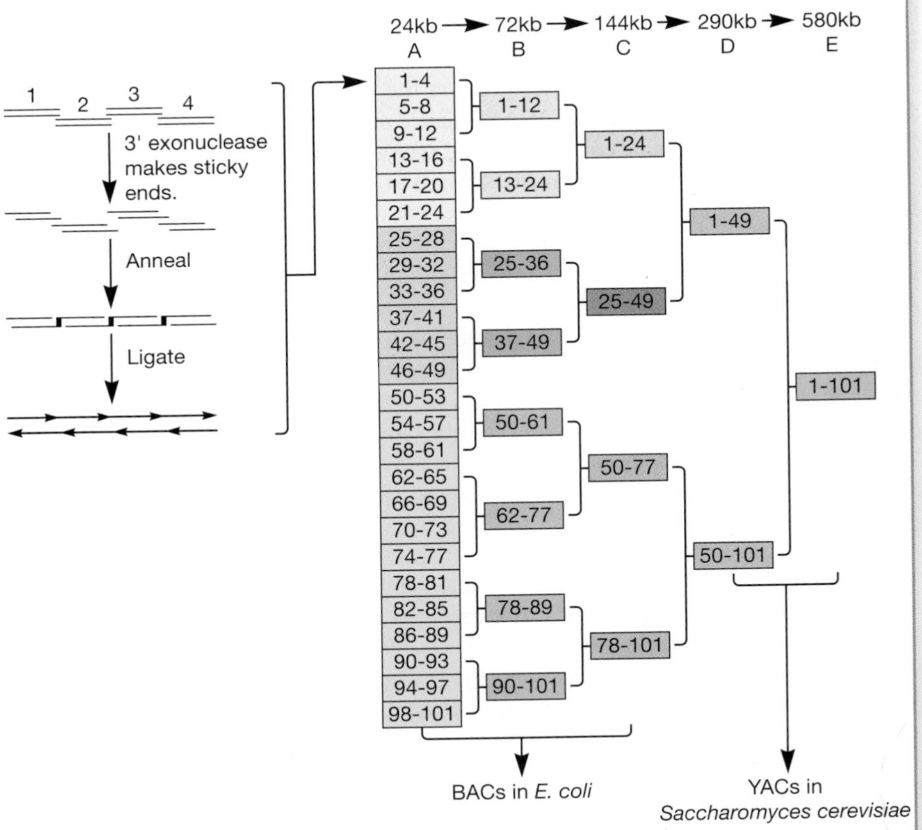

1. Briefly describe the polymerase chain reaction. What is the difference between a step and a cycle?
2. Why can PCR be used to detect very small quantities of a specific microbe, such as the HIV genome, despite the presence of other genomes?
3. Explain the difference between end-point and real-time PCR. Suggest an application for each.
4. What is electrophoresis? How is it used in Southern blotting? Why is it possible to visualize a PCR product on an agarose gel even if the template genome is present at such a low concentration that it cannot be seen?
5. How are plasmids, cosmids, and artificial chromosomes structurally different? What are some of the different purposes served by each?

15.5 Construction of Genomic Libraries

The DNA to be cloned can be obtained in several ways. It can be synthesized by PCR, or it can be located on the chromosome by Southern blotting. However, PCR amplification of a gene requires foreknowledge of its nucleotide sequence (or at least sequences flanking the gene), and a suitable probe must be obtained for Southern blotting. In both cases, once the DNA fragment is purified, it is cloned using a procedure like that described for recombinant plasmids and shown in figure 15.11. However, what if researchers wanted to clone a gene based on the function of its product but had no idea what its DNA sequence might be? A genomic library must then be constructed and screened.

The goal of **genomic library** construction is to have an organism's genome cut into many separate fragments with each fragment cloned into a separate vector. Ideally the entire genome is represented; that is, the sum of the different fragments equals the whole genome. In this way, specific groups of genes can be analyzed and isolated. The construction of a genomic library begins with cleaving the genome into small pieces by a restriction endonuclease (**figure 15.13**). These genomic DNA fragments are then either cloned into vectors and introduced into a microbe or packaged into phage particles that are used to infect the host. In either case, many thousands of different clones—each with a different genomic DNA insert—are created.

To select the desired clone from the library, it is necessary to know something about the function of the target gene or genetic element. If the genomic library has been inserted into a microbe that expresses the foreign gene, it may be possible to assay each clone for a specific protein or phenotype. For example, if one is studying a newly isolated soil bacterium and wants to find genes that encode enzymes needed for the biosynthesis of the amino acid alanine, the library could be expressed in an *E. coli* or *Bacillus subtilis* alanine auxotroph (figure 15.13). Recall that alanine auxotrophs require the addition of this amino acid to the medium. Following introduction of the genomic library into host cells, those that now grow without alanine would be good candidates for the genomic library fragment that possesses the alanine biosynthetic genes. Success with this approach depends on the assumption that the function of the cloned gene product is similar in both organisms. If this is not the case, the host must be the same species from which the library was prepared. In this example, a soil bacterial mutant lacking the gene in question (e.g., an alanine auxotroph) is used as the genomic library host. The genetic complementation of a deficiency in the host cell is sometimes called **phenotypic rescue.** ◀◀ *Detection and isolation of mutants (section 14.2)*

If a genomic library is prepared from a eukaryote in an effort to isolate a structural gene, a cDNA library is usually constructed. In this way, introns are not present in the genomic library. Instead, only the protein-coding regions of the genome are cloned. cDNA is prepared (figure 15.5) and cloned into a suitable vector. After the library is introduced into the host microbe, it may be screened by phenotypic rescue or by hybridization with an oligonucleotide, as described for Southern blotting. In some cases, neither phenotypic rescue nor hybridization with a probe is possible. Then the researcher must develop a novel approach that suits the particular set of circumstances to screen the genomic library.

15.6 Introducing Recombinant DNA into Host Cells

In cloning procedures, the selection of a host organism is as important as the choice of cloning vector. *E. coli* is the most frequent bacterial host and *S. cerevisiae* is most common among eukaryotes. Host microbes that have been engineered to lack restriction enzymes and the recombination enzyme RecA make better hosts because it is less likely that the newly acquired DNA will be degraded or recombined with the host chromosome. There are several ways to introduce recombinant DNA into a host microbe. Transformation and electroporation are two commonly employed techniques. Often the host microbe does not have the capacity to be transformed naturally. This is the case with *E. coli* and most gram-negative bacteria as well as many gram-positive bacteria and archaea. In these cases, the host cells may be rendered competent by treatment with divalent cations and artificially transformed by heat shocking the cells. ◀◀ *Bacterial transformation (section 14.8)*

Electroporation is a simple technique that has wide application to a number of host organisms, including plant and animal cells. In this procedure, cells are mixed with the recombinant DNA and exposed to a brief pulse of high-voltage electricity. The plasma membrane becomes temporarily permeable and DNA molecules are taken up by some of the cells. The cells are then grown on media that select for the presence of the cloning vector, as described in section 15.4.

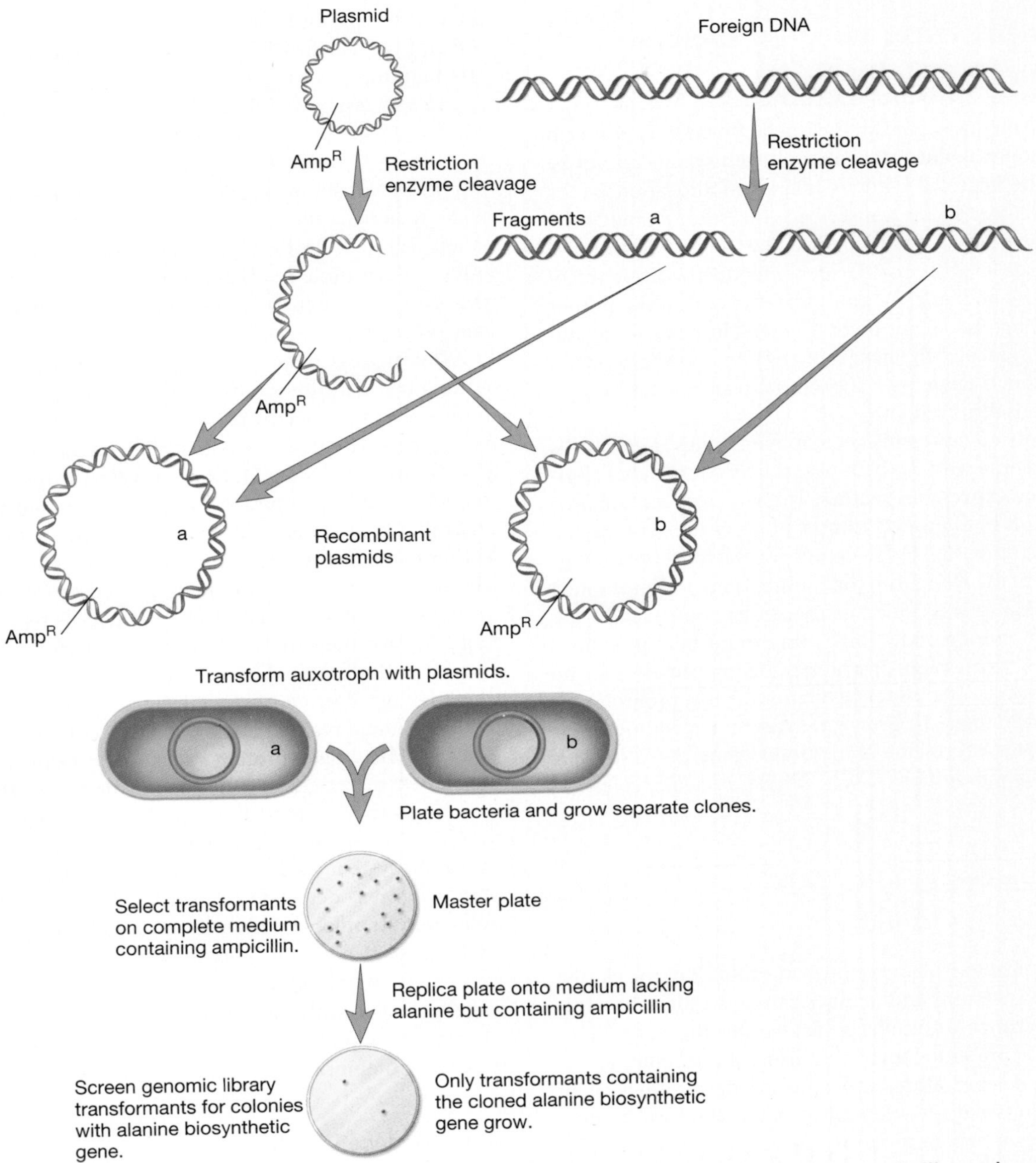

FIGURE 15.13 Construction of a Genomic Library and Screening by Phenotypic Rescue. A genomic library is made by cloning fragments of an organism's entire genome into a vector. For simplicity, only two genomic fragments and recombinant vectors are shown. In reality, a large mixture of vectors with inserts is generated. This mixture is then introduced into a suitable host. Phenotypic rescue is one way to screen the colonies for the gene of interest. It involves using a host with a genetic defect that can be complemented or "rescued" by the expression of a specific gene that has been cloned.

Figure 15.13 Micro Inquiry

Why are long fragments (e.g., 20,000 bp) of genomic DNA often desired when constructing a genomic library?

15.7 Expressing Foreign Genes in Host Cells

When a gene from one organism is cloned into another, it is said to be a **heterologous gene.** Many heterologous genes cannot be expressed in the host cell without further modification of the recombinant vector. To be transcribed, the recombinant gene must have a promoter that is recognized by the host RNA polymerase. Translation of its mRNA depends on the presence of leader sequences and mRNA modifications that allow proper ribosome binding. These are quite different in eukaryotes and bacteria. For instance, if the host is bacterium and the gene has been cloned from a eukaryote, a bacterial leader must be provided and introns removed.

The problems of expressing recombinant genes in host cells are largely overcome with the help of specially constructed cloning vectors called **expression vectors.** These vectors contain promoters that result in high-level transcription of the gene cloned within a multicloning site. Often expression vectors contain regulatory regions of the *lac* operon (or another inducible promoter) so that expression of the cloned genes can be controlled in the same manner as the operon. This is important because sometimes high-level expression of a heterologous protein can be toxic to the host cell. The presence of an inducible promoter allows the biotechnologist to grow the cells to a certain density before inducing the expression of the cloned gene. ⏮ *Regulation of transcription initiation (section 13.2); Catabolite repression (section 13.6)*

Purification and Study of Recombinant Proteins

It is often necessary to isolate the protein product of the cloned gene so that its structure and activity can be studied. In addition, it is sometimes desirable to determine the subcellular localization of a protein. Finally, as discussed in chapter 41, heterologous genes are often expressed for the production of commercially prepared recombinant protein. Here, we discuss some clever ways in which proteins can be purified and visualized in living cells.

Protein Purification

The heterologous expression of a gene is often done with the intention of collecting the protein product and purifying it. For example, one might be very interested in a protein but the microbe that produces it may not grow well in the laboratory. Even if the microbe can be easily grown, the protein of interest may be present at very low levels. It has become standard practice to clone the gene for such a protein into a vector that replicates in *E. coli* or other suitable microbial host. This solves the problem of growing the microbe and producing the protein, but what about obtaining the protein in its pure form? The most common method by which proteins are purified is called **polyhistidine tagging,** or simply **His-tagging.** His-tagging involves adding a series of histidine amino acid residues to either the N- or C-terminus of the protein. Most often six residues are added, and this is called a 6xHis-tag. Because histidine has a high affinity for metal ions, a protein bearing multiple histidine residues will bind preferentially to a solid-phase material, called a resin, which has exposed nickel or cobalt atoms. This permits the separation of the His-tagged protein from other cellular proteins. The tagged protein is then removed from the resin, and the histidine residues may be removed with a specific protease.

The process of purifying a His-tagged protein begins by adding the histidine residues to the protein. This is most commonly accomplished by cloning the protein into an expression vector that already has the histidine-encoding sequence present (**figure 15.14**, step 1). Alternatively, the histidine-encoding sequence can be amplified by PCR and added to the gene of interest. Once the protein-coding sequence is fused to the histidine codons, the vector is introduced into the microbe of choice, usually *E. coli.* As the *E. coli* cells grow, the His-tagged protein is produced (step 3). When the cells reach a sufficiently high density, they are lysed by treatment with detergent or enzymes, and all the soluble protein within the cell—including the His-tagged protein—is collected. These proteins are mixed with the metal resin that is bound to a column (step 4) so that only the His-tagged protein sticks to the resin. The remaining, unwanted material is washed from the resin. The His-tagged protein is then removed by passing the aromatic compound imidazole through the column (step 6). Depending on the purification system used, the His-tag can be cleaved from the protein when it is detached from the resin, or it can be removed following its release. Often the protein is functional with the His-tag, so it may not be removed. His-tagging is used if the protein is soluble and remains in the cytoplasm. While His-tagging is certainly the most popular approach to purifying recombinant proteins, other approaches exist, especially for purifying membrane-associated proteins.

Fluorescence Labeling

What if the goal of cloning a gene is to study the regulation or function of the protein product in vivo? It is possible to visualize the activity of a specific promoter as well as observe the localization of a protein by fluorescence. The development of fluorescent labeling of living cells began when it was discovered that the jellyfish *Aequorea victoria* produces a protein called **green fluorescent protein (GFP).** GFP is encoded by a single gene that, when translated, undergoes self-catalyzed modification to generate a strong, green fluorescence. This means that it is easily cloned and expressed in any organism. In fact, an entire palette of fluorescent labels is now available. These include mutant versions of the GFP gene encoding a variety of proteins that glow throughout the blue-green-yellow spectrum. In addition, new fluorescent peptides

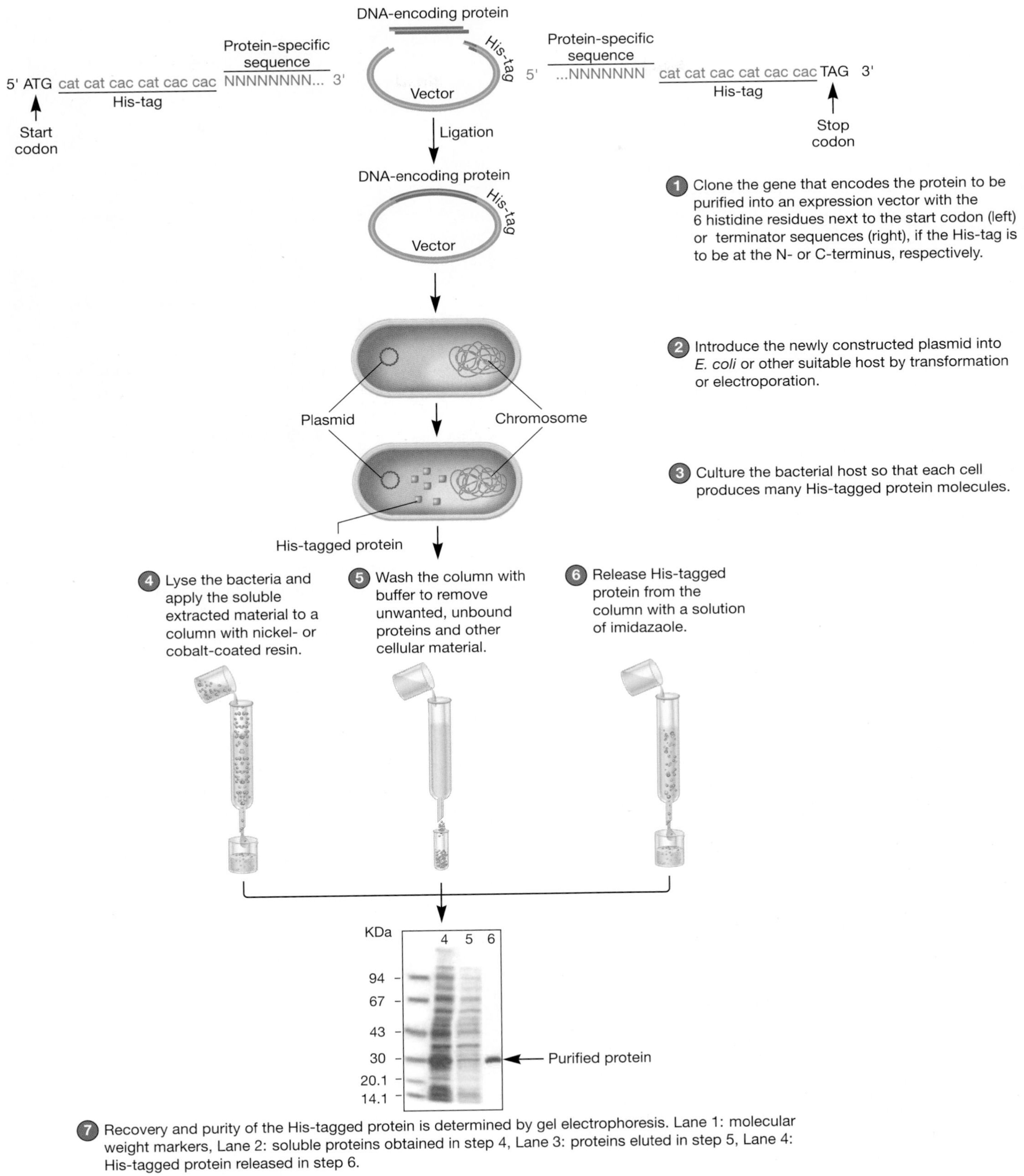

FIGURE 15.14 Construction and Purification of a Polyhistidine-Tagged Protein.

that likewise undergo self-catalyzed light production have been developed.

There are two ways in which the amino acid sequence that encodes fluorescent peptides can be attached to a gene. A transcriptional fusion is used to study gene regulation. Transcriptional fusions replace the entire coding sequence of the gene of interest with the fluorescence-encoding gene (**figure 15.15*a***). Because the expression of the fluorescence gene is now governed by the promoter of the replaced gene, promoter activity can be visualized. That is, when the promoter is turned "on," the cells glow. If the transcriptional fusion has been made in a multicellular organism, then not only is the timing of promoter activity seen but the specific cell type in which the gene is normally expressed is also determined. On the other hand, a translational fusion is used to determine where a particular protein is localized in the cell. Translational fusions add the GFP-encoding sequence to the structural gene of interest. This creates a chimeric protein—a protein that consists of two parts: the protein being studied and GFP. Of course, care must be taken to ensure that the protein fusion still functions like the original protein. One way to do this is to test for phenotypic rescue of a mutant lacking the structural gene of interest. Once it has been determined that the chimeric protein is fully functional, the cellular "address" of the protein can be determined (figure 15.15*b*).

Search This: The GFP site

1. What is a genomic library? Describe how a genomic library might be screened for a gene that confers the production of an extracellular enzyme that degrades casein, a protein found in milk. (Hint: Agar with skim milk loses its opacity when casein is degraded.)
2. Explain selection for antibiotic resistance followed by blue versus white screening of colonies containing recombinant plasmids. Why must both antibiotic selection and color screening be used? What would you conclude if, after transforming a ligation mixture into *E. coli*, only blue colonies were obtained?
3. How can one prevent recombinant DNA from undergoing recombination in a bacterial host cell?
4. List several reasons why a cloned gene might not be expressed in a host cell.
5. Think of a situation in which you might want to keep the His-tag on a recombinant protein and a situation in which it would be important to remove it.
6. You are studying chemotaxis proteins in a newly described bacterium. You have cloned a gene that encodes the CheA protein. You need to be sure that this protein localizes to the inside of the cytoplasmic membrane (*see figure 13.16*). Will you use a transcriptional or translational fusion? Explain your choice.

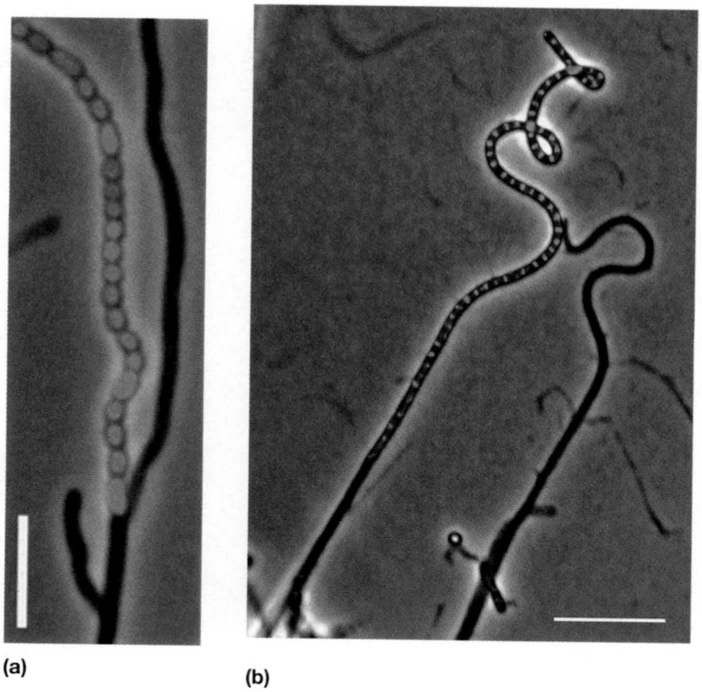

FIGURE 15.15 Fluorescent Protein Labeling. (a) A transcriptional fusion places the gene encoding a fluorescent peptide under the control of a promoter; it replaces the coding sequence of the gene. In this case, the mCherry gene is controlled by the promoter that normally governs transcription of a gene in the actinomycete *Streptomyces coelicolor* that is expressed only when aerial hyphae form cross walls (septa) that give rise to a chain of spores. Thus red fluorescence is seen uniformly throughout the chain of spores but not in the aseptate hyphae. (b) A translational fusion generates a chimeric protein consisting of the normal protein attached to the GFP label. Here, the FtsZ gene has been constructed as a translational fusion so that it is fused to GFP. This shows that FtsZ localizes at the septa. Notice that the filament to the right is not forming septae, so FtsZ is not present.

Figure 15.15 Micro Inquiry

What special considerations are necessary if one is constructing a translational fusion of a protein known to be exported?

Summary

15.1 Key Developments in Recombinant DNA Technology

a. Genetic engineering became possible after the discovery of restriction enzymes and reverse transcriptase, and the development of essential methods in nucleic acid chemistry such as the Southern blotting technique.

b. Restriction enzymes are important because they cut DNA at specific sequences, thereby releasing fragments of DNA that can be cloned or otherwise manipulated (**figure 15.3** and **table 15.2**).

15.2 The Polymerase Chain Reaction

a. The polymerase chain reaction (PCR) allows small amounts of specific DNA sequences to be increased in concentration thousands of times (**figure 15.7**).

b. PCR has dramatically changed molecular biology. For instance, it is used to obtain genes for cloning and in diagnostic and forensic science.

15.3 Gel Electrophoresis

a. Gel electrophoresis is used to separate molecules according to charge and size.

b. DNA fragments are separated on agarose and acrylamide gels. Because DNA is acidic, it migrates from the negative to the positive end of a gel (**figure 15.9**).

15.4 Cloning Vectors and Creating Recombinant DNA

a. There are four types of cloning vectors: plasmids, phages and viruses, cosmids, and artificial chromosomes. Cloning vectors generally have at least three components: an origin of replication, a selectable marker, and a multicloning site or polylinker (**table 15.3; figures 15.10** and **15.12**).

b. The most common approach to cloning is to digest both vector and DNA to be inserted with the same restriction enzyme or enzymes so that compatible sticky ends are generated. The vector and DNA to be cloned are then incubated in vitro in the presence of DNA ligase, which catalyzes the formation of phosphodiester bonds once the DNA fragment inserts into the vector.

c. Once the recombinant plasmid has been introduced into host cells, cells carrying vector must be selected. This is often accomplished by allowing the growth of only antibiotic-resistant cells because the vector bears an antibiotic-resistance gene. Cells that took up vector with inserted DNA can be distinguished from those that contain only vector in several ways. Often a blue-versus-white colony phenotype is used; this is based on the presence or absence of a functional *lacZ* gene, respectively (**figure 15.11**).

15.5 Construction of Genomic Libraries

a. It is sometimes necessary to find a gene without the knowledge of the gene's DNA sequence. A genomic library is constructed by cleaving an organism's genome into many fragments, each of which is cloned into a vector to make a unique recombinant plasmid.

b. Genomic libraries can be screened for the gene of interest by either phenotypic rescue (genetic complementation) or DNA hybridization with an oligonucleotide probe. However, there are instances when a novel approach to screening the library for a specific gene must be devised (**figure 15.13**).

15.6 Introducing Recombinant DNA into Host Cells

a. The bacterium *E. coli* and the yeast *S. cerevisiae* are the most common host species.

b. DNA can be introduced into microbes by transformation or electroporation.

15.7 Expressing Foreign Genes in Host Cells

a. An expression vector has the necessary features to express any recombinant gene it carries.

b. If a eukaryotic gene is to be expressed, cDNA is used because it lacks introns.

c. Purification of recombinant proteins is often accomplished by fusing the coding sequence of a protein to six histidine residue codons found on special expression vectors. When introduced and expressed in bacteria, the His-tagged protein can be selectively purified (**figure 15.14**).

d. Green fluorescent protein can be used to study the regulation of gene expression (transcriptional fusions) and protein localization (translational fusions) (**figure 15.15**).

Critical Thinking Questions

1. Could the Southern blotting technique be applied to RNA? If so, how might this be done?
2. Initial attempts to perform PCR were carried out using the DNA polymerase from *E. coli.* What was the major difficulty?
3. You have cloned a structural gene required for riboflavin synthesis in *E. coli.* You find that an *E. coli* riboflavin auxotroph carrying the cloned gene on a vector makes less riboflavin than does the wild-type strain. Why might this be the case?
4. Suppose that you inserted a plasmid vector carrying a human interferon gene into *E. coli* but none of the transformed bacteria produced interferon. Give as many plausible reasons as possible for this result.
5. You are interested in the activity and regulation of a protease made by the gram-positive microbe *Geobacillus stearothermophilus.* What would be the purpose of constructing each of the following: a His-tagged protease, a transcriptional GFP fusion to the protease gene, and a translational GFP fusion to the protease gene.
6. The deep-sea bacterium *Photobacterium profundum* SS9 is a piezophile; that is, it grows best under high atmospheric pressure. It has a single polar flagellum and also makes lateral flagella under some conditions. Marine microbiologists are interested in comparing this motility system with that of nonpiezophilic bacteria including the well-characterized system of *E. coli* and the closely related nonpiezophile *P. profundum* 3TCK. How could you use heterologous gene expression with phenotypic rescue in the new host to assess the functional similarity between the flagellin proteins encoded by *P. profundum* SS9 and these other bacteria?

   Read the original paper: Eloe, E. A., et al. 2008. The deep-sea bacterium *Photobacterium profundum* SS9 utilizes separate flagellar systems for swimming and swarming under high-pressure conditions. *Appl. Environ. Microbiol.* 74:6298.
7. Rapid and accurate identification of pathogens in clinical samples is paramount for effective patient care. To that end, the identification of microbes in human tissue and blood using PCR has become the norm for some infectious agents. However, before a PCR approach can be adopted by clinical microbiologists, it must be shown that it is specific and accurate. Fungal infections are particularly difficult to treat and have become more widespread as the population of patients with compromised immune systems grows. A PCR protocol for the identification of several important fungal pathogens, including *Aspergillus, Candida,* and *Cryptococcus,* was recently described. This assay is based on the amplification of the 28S large subunit ribosomal RNA gene from each microbe. Recall that as a eukaryote, the human host also has a homologous gene. Outline how you think these researchers performed the PCRs and what they did to control for false positive and false negatives. Why do you think real-time PCR was used, rather than end-point PCR?

   Read the original paper: Vollmer, T., et al. 2008. Evaluation of novel broad-range real-time PCR assay for rapid detection of human pathogenic fungi in various clinical specimens. *J. Clin. Microbiol.* 46:1919.

Concept Mapping Challenge

Use the following words to construct a concept map that describes the process of cloning and expressing a gene. You may need to provide your own linking terms.

Electroporation
Plasmid
DNA insert
Vector
Cosmid
Expression vector
S. cerevisiae
Host cell
E. coli
Restriction endonucleases
Transformation

Learn More

Learn more by visiting the text website at www.mhhe.com/willey8, where you will find a complete list of references.

16

Microbial Genomics

Each dot in the microarray pictured here consists of an oligonucleotide fragment of a single gene bound to a glass slide. Gene expression of two types of cells (e.g., one wild type and one mutant) can be compared by labeling the cDNA from each cell with either a red or green fluorescent label and allowing the cDNAs to bind to homologous sequences attached to the microarray. The color of each spot reveals the relative level of expression of each gene.

CHAPTER GLOSSARY

bioinformatics The interdisciplinary field that manages and analyzes large biological data sets, including genome and protein sequences.

coding sequence (CDS) A nucleotide sequence that encodes or is thought to encode a protein.

comparative genomics The analysis of genomes from different organisms to look for significant differences and similarities.

core genome The common set of genes found in all genomes in a species or taxon.

DNA microarrays Solid supports that have DNA attached in an organized grid pattern; used to evaluate gene expression or microbial population composition.

functional genomics The analysis of genome transcripts and the proteins they encode.

genome annotation The process of determining the location and potential function of genes and genetic elements in a genome sequence.

genomics The study of the molecular organization of genomes, their information content, and the gene products they encode.

in silico analysis The study of physiology and genetics through the examination of nucleic acid and amino acid sequences.

metagenomics The study of genomes recovered from natural samples without first isolating members of the microbial community and growing them in pure cultures.

open reading frame (ORF) A sequence of DNA not interrupted by a stop codon and with an apparent promoter and ribosome binding site at the 5′ end and a terminator at the 3′ end.

orthologue A gene found in the genomes of two or more different organisms that share a common ancestry and are presumed to (or have been demonstrated to) have similar functions.

pan-genome The collection of genes found in all strains of a species or other taxonomic unit.

paralogues Two or more genes in the genome of a single organism that arose through duplication of a common ancestral gene.

phylotype A taxon that is characterized only by its nucleic acid sequence; generally discovered during metagenomic analysis.

proteome The complete collection of proteins that an organism produces.

single-cell genomic sequencing A method whereby the genome of a single microbial cell is copied many times by the **multiple displacement amplification (MDA)** technique. The genome is then sequenced by a post-Sanger technique such as pyrosequencing.

transcriptome All the messenger RNA that is transcribed from the genome of an organism under a given set of circumstances.

whole-genome shotgun sequencing An approach to genome sequencing in which the complete genome is broken into random fragments, which are individually sequenced. The fragments are then placed in the proper order based on overlapping nucleotide sequences.

Genomics is an exciting and growing field that has changed the ways in which key questions in microbial physiology, genetics, ecology, and evolution are pursued. Prior to the advent of genomics, analysis of gene expression was limited to the identification of a small subset of transcripts (i.e., mRNAs) and proteins. As we will see, genomic analysis enables scientists to study the cell in a holistic way by capturing a snapshot of the entire pool of transcripts or proteins. The cell can now be viewed as a network of interconnected circuits, not as a series of individual pathways. Further, genomics provides a window into entire microbial communities: microbial ecologists no longer need to confine their studies to the tiny fraction of microorganisms that have been cultivated. Finally, our understanding of the evolution of all organisms can be illuminated by the insights we gain studying microbial evolution using genomic approaches. In these ways and more, genomics has truly revolutionized biology.

We begin our exploration of genomics with a general overview of the topic. This is followed by an introduction to DNA sequencing techniques, including newer technologies designed to sequence genomes faster and more cheaply. Next, the whole-genome shotgun sequencing method is briefly described. Genome function and the analysis of the transcripts and proteins produced by microbes are then explored. We focus on annotation, DNA microarrays, and proteomics. Finally, comparative and environmental genomics (metagenomics) are reviewed.

16.1 Determining DNA Sequences

Techniques for sequencing DNA were developed in 1977 by Alan Maxam and Walter Gilbert (collaborating on one technique), and Frederick Sanger. Sanger's method is most commonly used and is discussed here. We also introduce three newer sequencing methods designed to make large sequencing projects, such as whole-genome sequencing, more economical.

Sanger DNA Sequencing

The Sanger method involves the synthesis of a new strand of DNA using the DNA to be sequenced as a template. The reaction begins when single strands of template DNA are mixed with primer (a short piece of DNA complementary to the region to be sequenced), DNA polymerase, the four deoxynucleoside triphosphates (dNTPs), and dideoxynucleoside triphosphates (ddNTPs). ddNTPs differ from dNTPs in that the 3′ carbon lacks a hydroxyl group (**figure 16.1**). In such a reaction mixture, DNA synthesis will continue until a ddNTP, rather than a dNTP, is added to the growing chain. Without a 3′-OH group to attack the 5′-PO_4 of the next dNTP to be incorporated, synthesis stops *(see figure 12.11)*. Indeed, Sanger's technique is frequently referred to as the **chain-termination DNA sequencing** method.

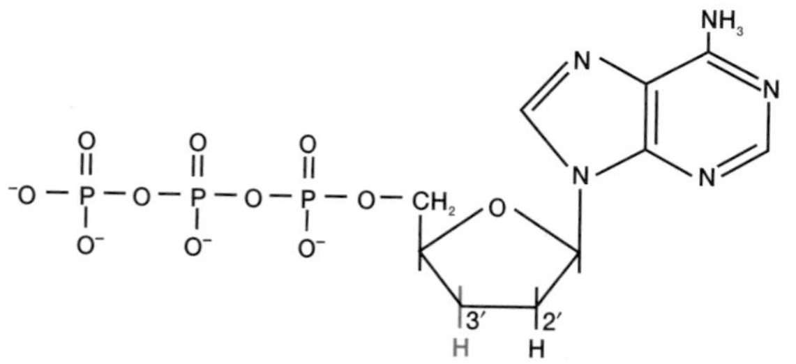

FIGURE 16.1 Dideoxyadenosine Triphosphate (ddATP). Note the lack of a hydroxyl group on the 3′ carbon, which prevents further chain elongation by DNA polymerase.

Figure 16.1 Micro Inquiry

What is the function of the 3′-OH during DNA synthesis?

To obtain sequence information, four separate synthesis reactions must be prepared, one for each ddNTP (**figure 16.2**). When each DNA synthesis reaction is stopped, a collection of DNA fragments of varying lengths has been generated. The reaction prepared with ddATP produces fragments ending with an A, those with ddTTP produce fragments with T termini, and so forth. If the DNA is to be manually sequenced, radioactive dNTPs are used and each reaction is electrophoresed in a separate lane on a polyacrylamide gel. Recall that each fragment's migration rate is inversely proportional to the log of its molecular weight. Simply put, the smaller a fragment is, the faster it moves through the gel. Because synthesis proceeds with the addition of a nucleotide to the 3′-OH of the primer, the dideoxynucleotide at the end of the shortest fragment is assigned as the 5′ end of the DNA sequence, while the largest fragment is the 3′ end. In this way, the DNA sequence can be read directly from the gel from the smallest to the largest fragment. ◀◀ *Gel electrophoresis (section 15.3)* *Sanger Sequencing*

DNA is more often prepared for automated sequencing (**figure 16.3**). Here, the four reaction mixtures can be combined and loaded into a single lane of a gel because each ddNTP is labeled with a different colored fluorescent dye. These fragments are then electrophoresed and a laser beam determines the order in which they exit the bottom of the gel. A chromatogram is generated in which the amplitude of each spike represents the fluorescent intensity of each particular fragment (figure 16.3*b*). The corresponding DNA sequence is listed above the chromatogram.

Post-Sanger DNA Sequencing

Once an organism's genome has been sequenced, the level of inquiry and the pace of research are greatly enhanced. Using Sanger's chain termination method, the cost of sequencing the human genome (completed in 2001) was about \$300 million and took about a decade to finish. One reason for this was sample preparation; as discussed next, when Sanger sequencing is used, a genomic library must be constructed. However, scientists would

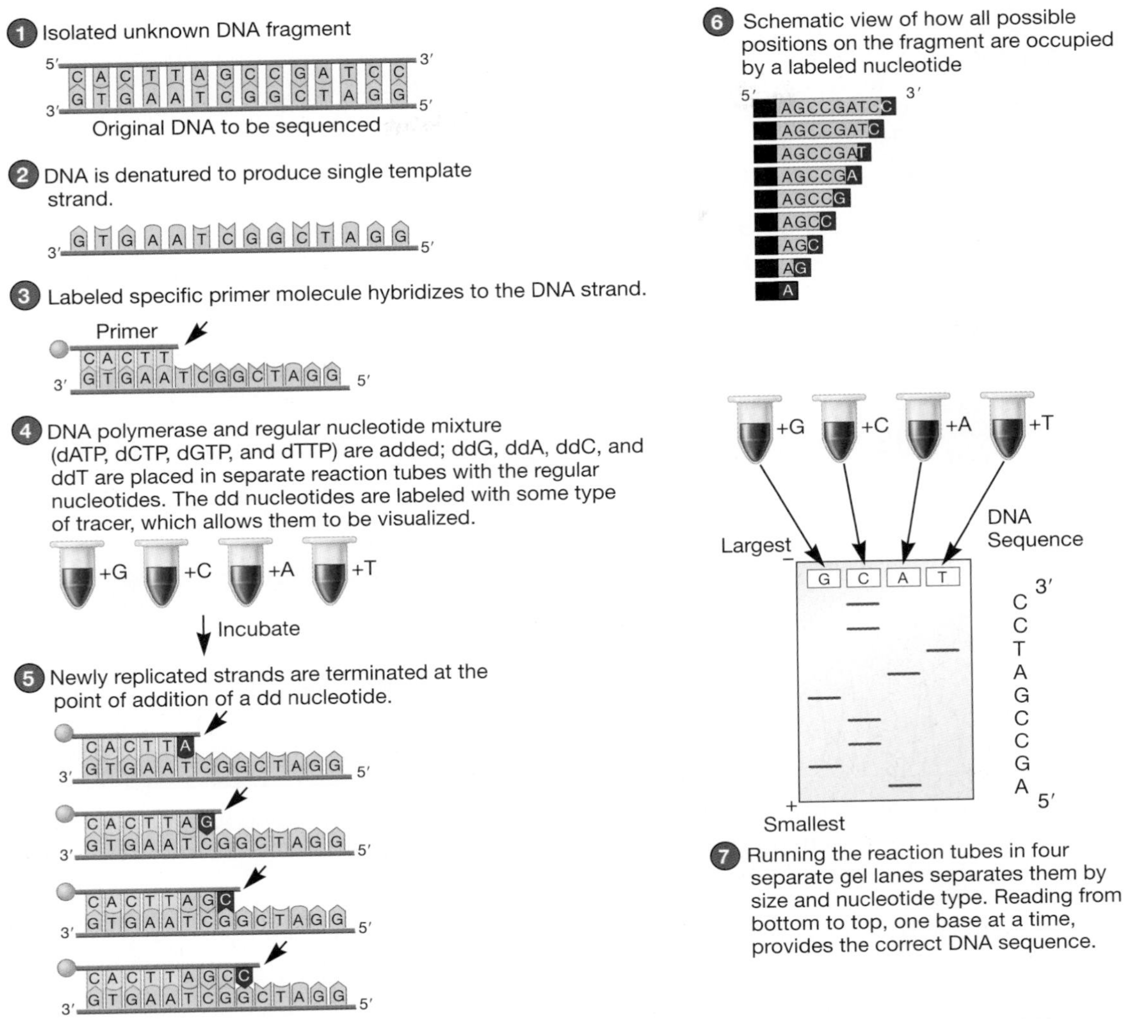

FIGURE 16.2 The Sanger Method of DNA Sequencing. Steps 1–6 are used for both manual and automated sequencing. Step (7) shows preparation of a gel for manual sequencing in which radiolabeled ddNTPs are used.

like to be able to sequence genomes faster and more cheaply. Indeed, the race for the "$1,000 genome" was accelerated by the Archon X Prize challenge, which will award $10 million to the first privately funded company to complete 100 genomes in 10 days for less than $10,000 per genome.

Here, we review three new, "post-Sanger" technologies. All of these approaches avoid cloning, that is, there is no need to construct a genomic library. Instead, these techniques attach the chromosomal fragments to be sequenced to a solid substrate and use the polymerase chain reaction (PCR) to amplify the genomic sequences. What sets each method apart is the technique used to sequence the DNA. A major limitation of all these techniques is that only short nucleotide lengths can be read at a time, so that many, many "reads" need to be assembled to complete an entire genome. Most scientists are confident this problem will be solved, and indeed, significant progress has been made over the last several years. Indeed, these technologies have proven useful in some nucleotide sequencing projects. ◀◀ *Polymerase chain reaction (section 15.2); Construction of genomic libraries (section 15.5)*

Search This: Archon X Prize for Genomics

A widely used emergent technology is that of 454 Life Sciences Corporation. Their approach, called **pyrosequencing,** sequencing by synthesis, or even just 454 sequencing, uses a series of enzymatic reactions that results in the generation of visible light (**figure 16.4*a***). The intensity of the light produced is proportional to the number of nucleotides added to a growing DNA strand. To begin, the genome is cut into 300 to 500 base pair (bp) fragments. Each fragment is made single stranded (denatured) and attached to a microscopic plastic bead. The fragments are then copied by PCR within an oil-water emulsion so as to cover the bead with

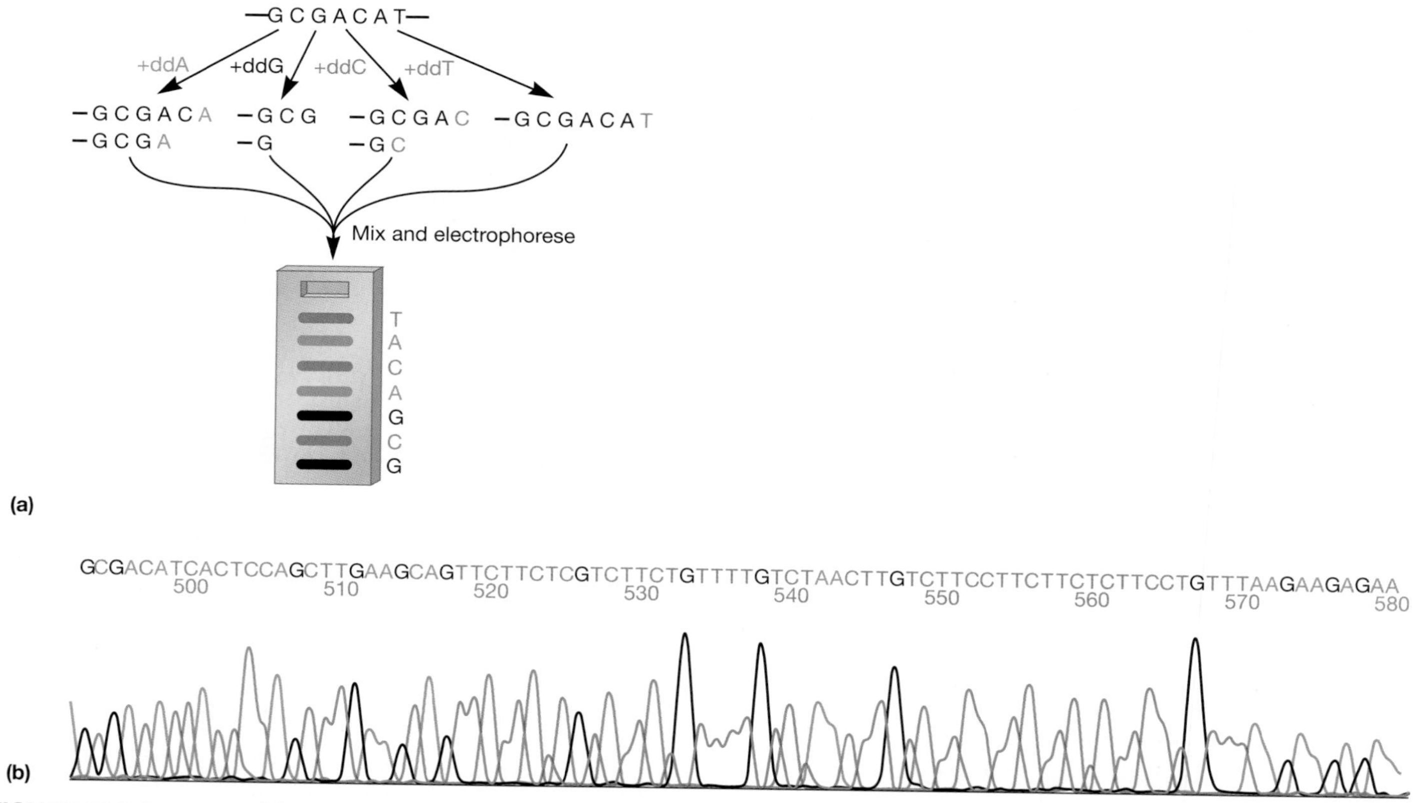

FIGURE 16.3 Automated Sanger DNA Sequencing. (a) Part of an automated DNA sequencing run. Here, the ddNTPs are labeled with fluorescent dyes. (b) Data generated during an automated DNA sequencing run. Bases 493 to 580 are shown.

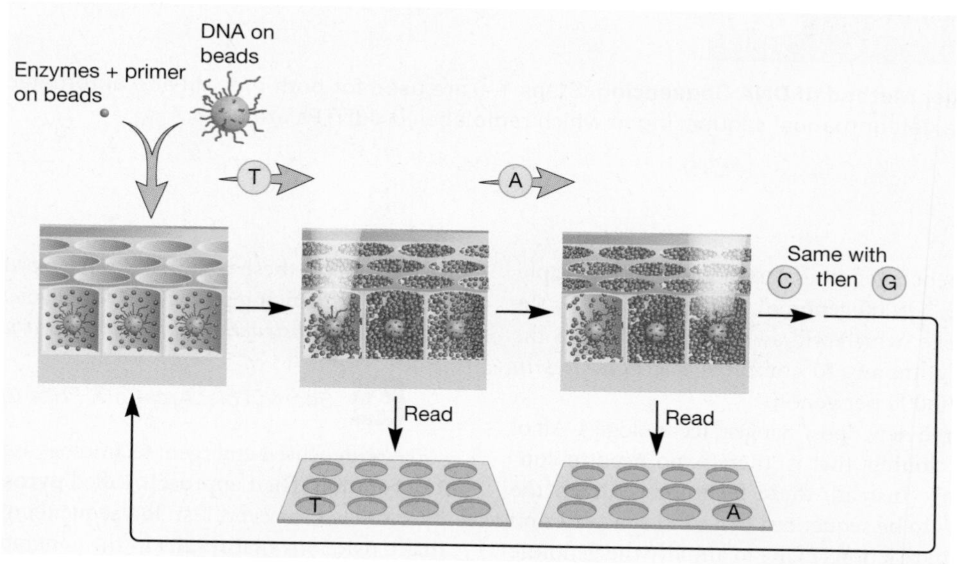

(a) Pyrosequencing

(Continued)

FIGURE 16.4 Post-Sanger DNA Sequencing Methodologies. Each technique attaches chromosomal fragments to a solid substrate such as a bead, and PCR is then used to amplify the fragments (not shown). (a) In pyrosequencing, each bead coated with PCR-amplified chromosomal fragments is placed in a miniature well, and pyrophosphate-based sequencing is performed in parallel on each DNA fragment.

FIGURE 16.4 ***(Continued)***

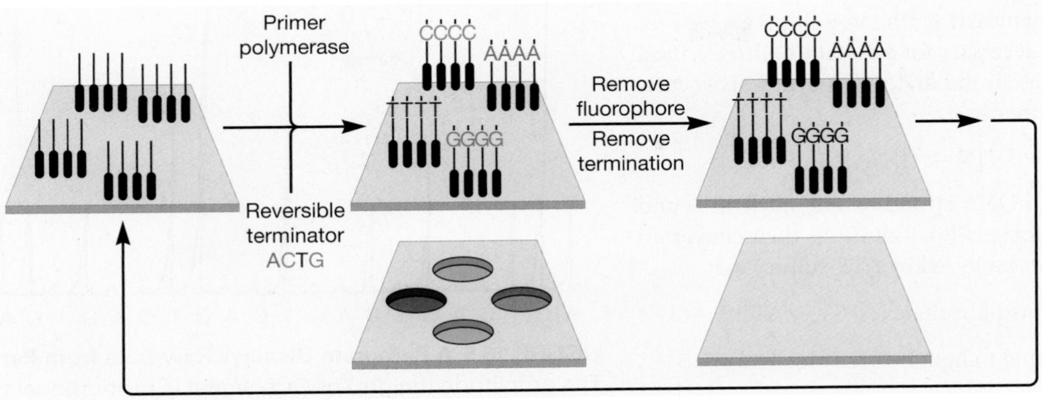

(b) SOLEXA sequencing

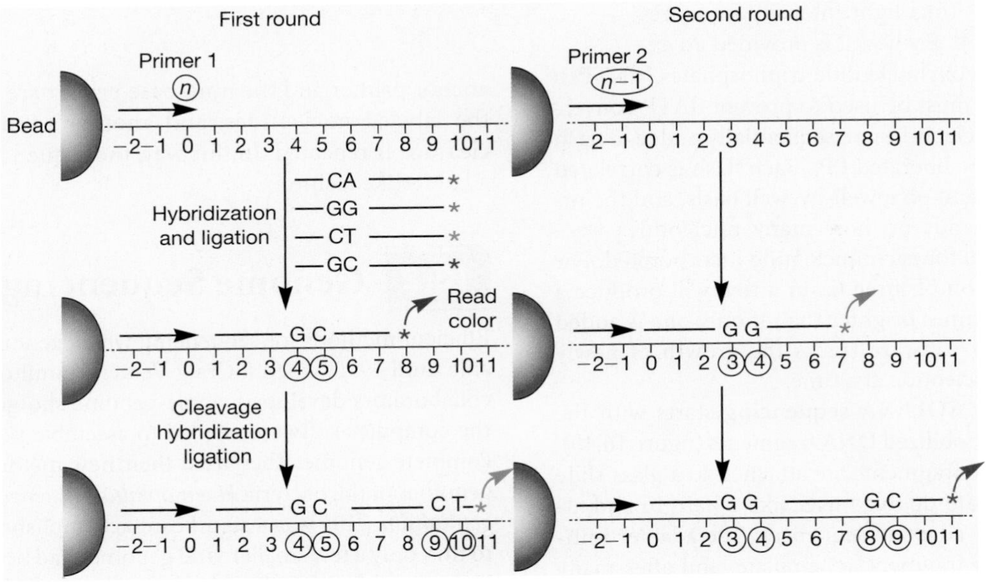

(c) Sequencing by ligation (SOLiD technology)

FIGURE 16.4 Post-Sanger DNA Sequencing Methodologies. (b) The chromosomal fragments in SOLEXA are bent in a hairpin-like fashion, so that each end is attached to an immobilized adapter. PCR amplification and denaturation generates about 1,000 copies of each single-stranded fragment. DNA synthesis reagents are added, and DNA synthesis is recorded one nucleotide at a time for each fragment as the incorporation of a fluorescently labeled nucleotide. The fluorescent label is removed and the process is repeated. (c) DNA sequencing by SOLiD technology involves the sequential hybridization and ligation of oligonucleotides with the central nucleotides fluorescently labeled. Upon hybridization, the color is recorded, the oligonucleotide is cleaved, and the process is repeated. Note that in the first round, the anchoring primer extends to the 0 position, while in the second round, it extends only to −1, thus hybridization in round 1 identifies the nucleotides at positions 4, 5; 9, 10; 14, 15; etc., while in round 2, nucleotides at positions 3, 4; 8, 9; 13, 14; etc., are determined.

Figure 16.4 Micro Inquiry

Why is it important that the DNA to be sequenced is immobilized in all three of these techniques?

about 10 million copies of each fragment. Individual beads are next deposited into picoliter-sized wells; one plate can have as many as 1.6 million wells. Each well receives the reagents necessary for a cascade of three critical enzymatic reactions. In the first, DNA synthesis is catalyzed by DNA polymerase:

$$(DNA)_n + dNTP \rightarrow (DNA)_{n+1} + PP_i$$

This lengthens each DNA strand by one nucleotide and releases pyrophosphate (PP_i). The PP_i is then converted to ATP in a reaction catalyzed by ATP sulfurylase:

$$PP_i + \text{adenosine phosphosulfate (APS)} \rightarrow ATP + SO_4^{2-}$$

ATP is then converted to light by firefly luciferase:

$$ATP + \text{luciferin} + O_2 \rightarrow AMP + PP_i + \text{Oxyluciferin} + CO_2 + \text{Light}$$

The amount of light is directly proportional to the number of ATPs hydrolyzed. Thus light intensity measures the amount of PP_i released. Each well is provided an excess of one of the four deoxynucleotide triphosphates (dNTPs); however, a dATP analog must be used to prevent dATP conversion into light. As the nucleotides are sequentially added, a flash of light is generated by the liberated PP_i. Each flash is correlated with the nucleotides present on a well-by-well basis, and the intensity of the flash depends on how many nucleotides were added; one PP_i is released for each nucleotide incorporated. For instance, the incorporation of three Cs in a row will produce a light signal that is three times brighter than if only one is added (**figure 16.5**). Computer software tracks the growth of newly synthesized DNA one nucleotide at a time.

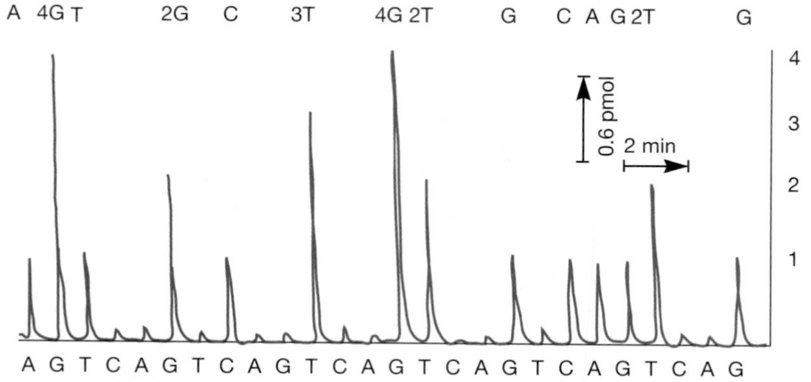

FIGURE 16.5 A Pyrogram Displays Raw Data from Pyrosequencing. The amplitude (height) of each signal is proportional to the number of nucleotides incorporated. The order of nucleotide addition is listed below the pyrogram, and the sequence (indicating the number of each nucleotide incorporated) is listed above the pyrogram.

Like pyrosequencing, **SOLEXA sequencing** starts with the PCR amplification of immobilized DNA fragments (figure 16.4*b*). However, in this case, the fragments are attached to a glass slide such that the fragments are doubled over like a hairpin and attached to primers called "adapter" sequences. The adapter initiates synthesis using each fragment as template, and after many cycles of synthesis and denaturation, about 1,000 copies of single-stranded DNA are spread randomly over the surface of the slide. The DNA is now ready for a Sanger-type sequencing reaction that is performed in a flow cell so that reagents can be flushed out and replaced after each nucleotide is added. By tracking many fragments in parallel, the growth of many fragments of DNA is simultaneously recorded.

A somewhat different approach is called **sequencing by ligation** or **SOLiD technology,** for *s*upported *o*ligo *l*igation *de*tection. DNA fragments attached to a substrate are lengthened by the addition of a short piece of DNA called an anchor primer (figure 16.4*c*). Additional eight-base oligonucleotides are then added. These are made in every possible combination of A-C-T-G, and each has a fluorescently labeled A, T, C, or G in the 4th and 5th positions. When a short primer binds exactly to a DNA fragment, the enzyme ligase covalently attaches it to the anchor primer. By exciting the fluorescent labels with laser light, the identity of the newly bound bases is determined. The anchor primer and the nine-base primer are cleaved to remove the labeled nucleotides, and another round of ligation and cleavage is repeated. In this way, the sequence is determined in a ladderlike fashion.

16.2 Genome Sequencing

Efficient methods for sequencing whole genomes were not available until 1995, when J. Craig Venter, Hamilton Smith, and their collaborators developed whole-genome shotgun sequencing and the computer software needed to assemble sequence data into a complete genome. They used their new method to sequence the genomes of the bacteria *Haemophilus influenzae* and *Mycoplasma genitalium.* This was a significant accomplishment because prior to this, only a few smaller viral genomes had been fully sequenced. Venter and Smith ushered in what has been called the genomic era. Within 15 years, the number of complete genomes published grew from 2 to over 700 with over 2,700 ongoing genome sequencing projects

Search This: Genome Online Database

Whole-Genome Shotgun Sequencing

The process of **whole-genome shotgun sequencing** is fairly complex when considered in detail, but the following summary gives a general idea of the procedure. For simplicity, this approach may be broken into four stages: library construction, random sequencing, fragment alignment and gap closure, and editing.

1. **Library construction.** The DNA molecules are randomly broken into fragments using ultrasonic waves; the fragments are then purified (**figure 16.6**). These fragments

are next inserted into cosmid or bacterial artificial chromosome (BAC) vectors and isolated. *Escherichia coli* strains lacking restriction enzymes are transformed with the cosmids or BACs to produce a library of clones. ◀◀ *Cloning vectors and creating recombinant DNA (section 15.4)*

2. **Random sequencing.** DNA is purified from each clone, and thousands of DNA fragments are sequenced with automated sequencers, employing dye-labeled primers that recognize the plasmid DNA sequences adjacent to the cloned, chromosomal insert. The sequencing machines are very fast and can run for 24 hours without operator attention. As many as 96 samples can be sequenced simultaneously, making it possible to sequence as many as 1 million bases per day, per sequencer. Ideally all stretches of the genome are sequenced multiple times to increase the accuracy of the final results. The number of times a region is sequenced is called "coverage," so for a genome with at least 8× coverage, all regions of the genome have been sequenced a minimum of eight times.
3. **Fragment alignment and gap closure.** Using computer analysis, the DNA sequence of each fragment is assembled into longer stretches of sequence. Two fragments are joined together to form a larger stretch of DNA if the sequences at their ends overlap and match. This comparison process results in a set of larger, contiguous nucleotide sequences called **contigs.** Finally, the contigs are aligned in the proper order to form the complete genome sequence. If gaps exist between two contigs, sometimes fragments with ends in two adjacent contigs are available. These fragments are analyzed and the gaps filled in with their sequences. When this approach is not possible, a variety of other techniques are used to align contigs and fill gaps.
4. **Editing.** The sequence is then carefully proofread to resolve any ambiguities or frameshift mutations in the sequence. Proofreading is accomplished by ensuring that the sequences of the two DNA strands are complementary.

Using this approach, it took less than 4 months to sequence the *M. genitalium* genome (about 500,000 base pairs in size). The shotgun technique was also used to sequence the human genome. Many bacterial and archaeal genomes have since been sequenced

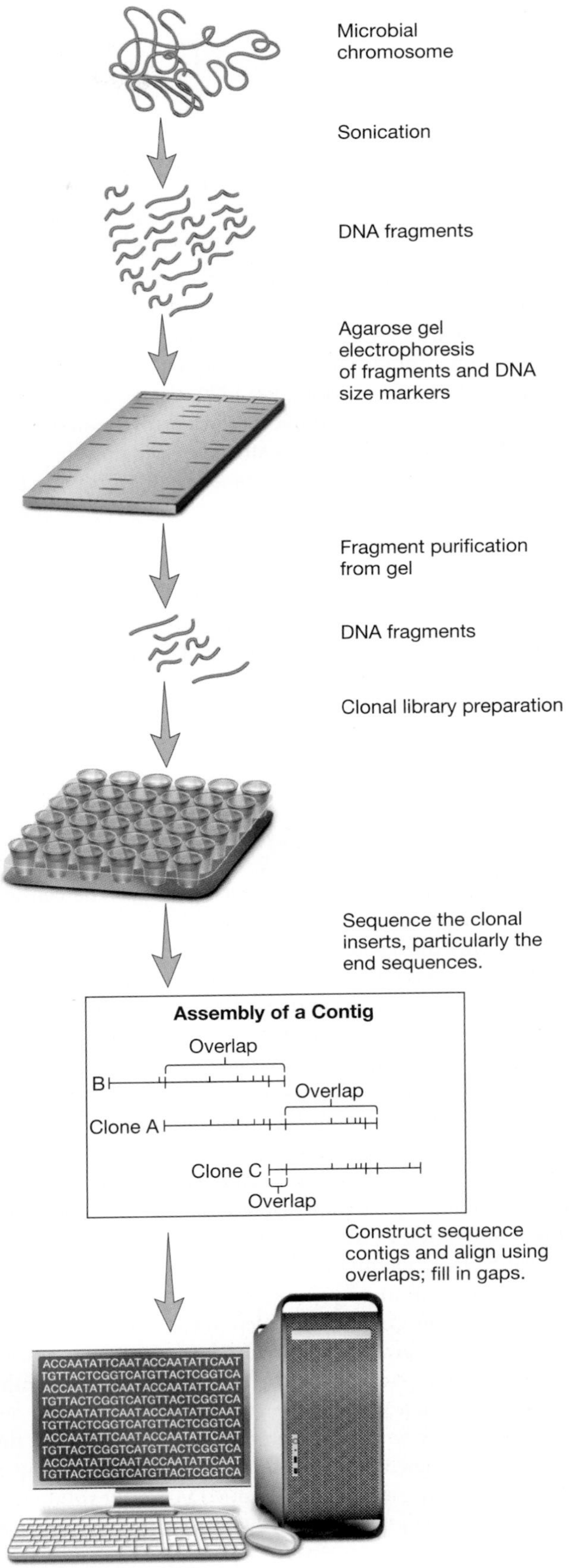

FIGURE 16.6 Whole-Genome Shotgun Sequencing.

Figure 16.6 Micro Inquiry

Which step (or steps) in this process is (are) replaced by PCR amplification and immobilization of fragments to a solid support in the post-Sanger sequencing techniques?

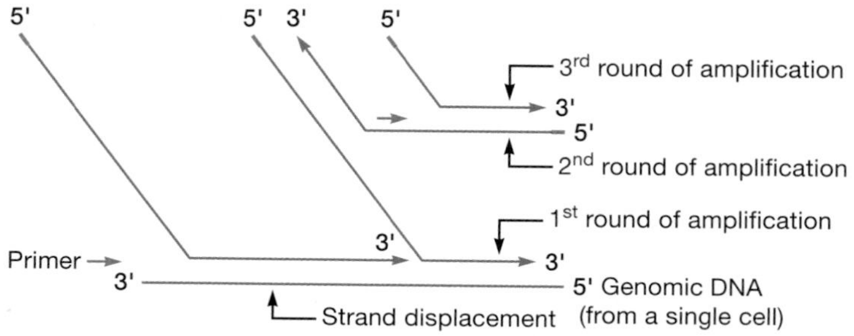

FIGURE 16.7 Single-Cell Genomic Sequencing. Many copies of DNA extracted from a cell are generated by multiple displacement amplification. Random hexamer primers (red) bind to complementary template sequences, and DNA polymerase from bacteriophage phi29 is used to catalyze synthesis in the 5′ to 3′ direction (black arrowheads). When the end of a newly synthesized strand meets double-stranded DNA, one strand is displaced by the growing DNA.

using this general approach. Because this technology requires cloning and the use of multiple automated sequencing machines, genomes have traditionally been sequenced at a limited number of sequencing facilities at a relatively high financial cost. This has spurred the development of the post-Sanger sequencing techniques discussed in section 16.1.

Single-Cell Genomic Sequencing

Only a decade ago, genomic sequencing of a single cell would have seemed a preposterous idea to most. However, it is now possible to amplify the few femtograms (10^{-15} gram) of DNA to the several micrograms (10^{-6} gram) needed for sequencing by either Sanger or post-Sanger techniques. This is an important breakthrough because the vast majority of microbes cannot be cultured axenically and thus cannot be studied from a genomic perspective. The capacity to sequence a single cell extracted from its natural environment is prompting new research strategies in microbial genetics, ecology, and infectious disease.

The process of **single-cell genomic sequencing** is based on a method called **multiple strand displacement (MDA)** (**figure 16.7**). In this reaction, the DNA extracted from a single cell is synthesized using DNA polymerase from the bacteriophage phi29. In addition, a collection of primers of random sequence, all six bases in length (hexamers), is added to the reaction along with dNTPs. The primers hydrogen bond to complementary sequences scattered throughout the genome. As DNA synthesis proceeds from each primer, the growing 3′ end of one newly made strand will eventually bump into and then displace the 5′ end of another newly growing strand. In this way, many new stands are rapidly synthesized. The MDA-synthesized strands have an average length of about 12,000 bases (12 kb) but can be as long as 100 kb. This makes them suitable for cloning and DNA sequencing.

Single cells of *E. coli* and *Myxococcus xanthus* were the first to be sequenced using MDA. Comparison of single-cell genomic sequence with that obtained by whole-genome shotgun sequencing showed that about 70 to 75% of the genome was amplified by MDA. Since that time, MDA has been used to amplify the genomes of other microbes, including a single filament of an uncultured species of *Beggiatoa* and a single spirochete *Borrelia burgdorferi*, the causative agent of Lyme disease, from a tick's midgut. The problem of only 70-to-75% amplification of the genome can be diminished or even eliminated by performing MDA on single genomes from several cells of the same species collected from the same sample. Bacteriophage phi29 DNA polymerase rarely incorporates the wrong base, that is, it has a high fidelity, so the sequence data has fewer mistakes than that obtained from DNA amplified by PCR. ▶▶| *Methods in microbial ecology (chapter 27)*

1. Why is the Sanger technique of DNA sequencing also called the chain-termination method?
2. How would one recognize a gap in the genome sequence following whole-genome shotgun sequencing?
3. Compare whole-genome shotgun sequencing with one of the emergent sequencing approaches. What do you think are the advantages of the new sequencing methodologies besides reduced cost? What are the disadvantages?
4. Suggest a medical and an ecological application of single-cell genomic sequencing.

16.3 Bioinformatics

The analysis of entire genomes generates not only a tremendous amount of nucleotide sequence data but also a rapidly growing volume of information regarding genome content, structure, and arrangement, as well as data detailing protein structure and function. This has led to the development of the field of **bioinformatics,** which combines biology, mathematics, computer science, and statistics. Determining the location and nature of genes or presumed genes on a newly sequenced genome is a complex process called **genome annotation**. Once genes have been identified, bioinformaticists can perform computer or **in silico analysis** to further examine the genome.

Obviously, obtaining nucleotide sequences without any understanding of the location and nature of individual genes would be a pointless exercise. The goal of genome annotation is to identify every potential (putative) protein-coding gene as well as each rRNA and tRNA coding gene. A protein-coding gene is usually first recognized as an **open reading frame (ORF);** to find all ORFs, both strands of DNA must be analyzed in all three reading frames (**figure 16.8**). A bacterial or

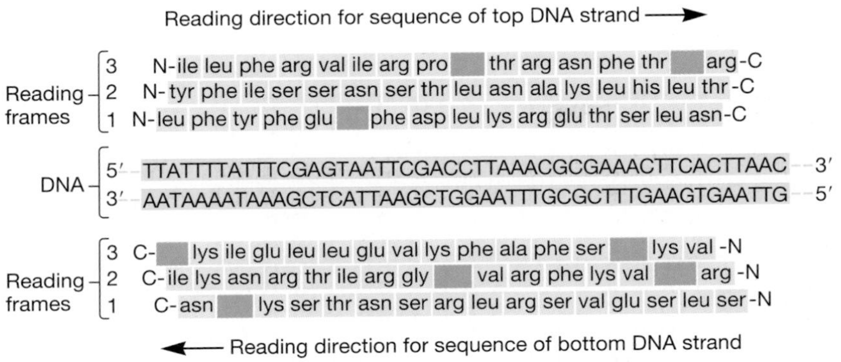

FIGURE 16.8 Finding Potential Protein Coding Genes. Annotation of genomic sequence requires that both strands of DNA be translated from the 5′ to 3′ direction in each of three possible reading frames. Stop codons are shown in green.

archaeal ORF is generally defined as a sequence of at least 100 codons that it is not interrupted by a stop codon and has terminator sequences at the 3′ end. Only if these elements are present is an ORF considered a putative protein-coding gene. This process is performed by gene prediction programs designed to find genes that encode proteins or functional RNA products. In most cases, the computer-called genes are then manually inspected by bioinformaticists to verify the computer-generated assignment. This process is called genome curation.

ORFs presumed to encode proteins are called ***coding sequences* (CDS).** Bioinformaticists have developed algorithms to compare the sequence of predicted CDS with those in large databases containing nucleotide and amino acid sequences of known proteins. The base-by-base comparison of two or more gene sequences is called **alignment.** Scientists most often use **BLAST (*b*asic *l*ocal *a*lignment *s*earch *t*ool)** programs to perform this task. These programs compare the amino acid or nucleotide sequence of interest (the query sequence) to all other sequences entered in the database. The results ("hits") are ranked in order of decreasing similarity. An E-value is assigned to each alignment; this value measures the possibility of obtaining an alignment by chance, thus highly homologous sequences have very low E-values.

Search This: BLAST

The translated amino acid sequences are almost always used to annotate potential genes or ORFs to gain an understanding of potential protein structure and function. Often a short pattern of amino acids, called a motif or domain, will represent a functional unit within a protein, such as the active site of an enzyme. For instance, **figure 16.9** shows the C-terminal domain of the cell division protein MinD from a number of microbes. Because these amino acids are found in such a wide range of organisms, they are considered phylogenetically well conserved. In this case, the conserved region is predicted to form a coil needed for proper localization of the protein to the membrane. Finding this high level of conservation allows the genome curator to assign a function to the domain. ◀◀ *Cytokinesis (section 7.2)*

Indeed, only if an ORF aligns above a certain threshold level with one or more genes in the database is it presumed to encode a protein of similar function. Pairs of such similar ORFs are called **orthologues.** Often an ORF will align with other genes over only a portion of its length, but depending on the degree of similarity and the overall percent of sequence that matches, this may be sufficient to assign the product of that ORF a putative (presumed) function. For instance, a newly discovered gene whose translated amino acid sequence aligns with the MinD sequences shown in figure 16.9 would be called a "putative *minD* gene." The assignment of ORF function is strengthened if the gene product of at least one of the orthologous genes has been experimentally evaluated. ORFs that do not align with known amino acid sequences fall into two classes: (1) **Conserved hypothetical proteins** are encoded by ORFs that have matches in the database but no function has yet been assigned to any of the sequences. (2) **Proteins of unknown function** are the products of ORFs unique to that organism. However, as more genomes are published, future comparisons may reveal a match in another organism. Finally, alignment of nucleotides on the same genome (i.e., from a single organism) may show that the nucleotide sequences of two or more genes are so alike that they most probably arose through gene duplication; such genes are called **paralogues.**

16.4 Functional Genomics

Once the putative identities of the genes in a given genome have been identified and classified, **functional genomics** seeks to place the information in a biological context. A microbial genome is often drawn as a **physical map** (**figure 16.10**). It is helpful to color-code genes according to category. For instance, when the *H. influenzae* genome was first annotated, about one-third of the genes were of unknown function (denoted by white regions on the map) and about 65 genes were assigned regulatory functions (dark blue regions). However, the *H. influenzae* was only the second bacterial genome sequenced. With over 700 complete bacterial and archaeal sequences now available for comparison, the number of genes with unknown function has fallen dramatically and will continue to decline as additional genome sequences are made publicly available.

The careful annotation of a microbe's genome can be used to piece together metabolic pathways, transport systems, and potential regulatory and signal transduction mechanisms. A common outcome of a genome project is to use the genes present to infer the functional metabolic pathways, transport mechanisms,

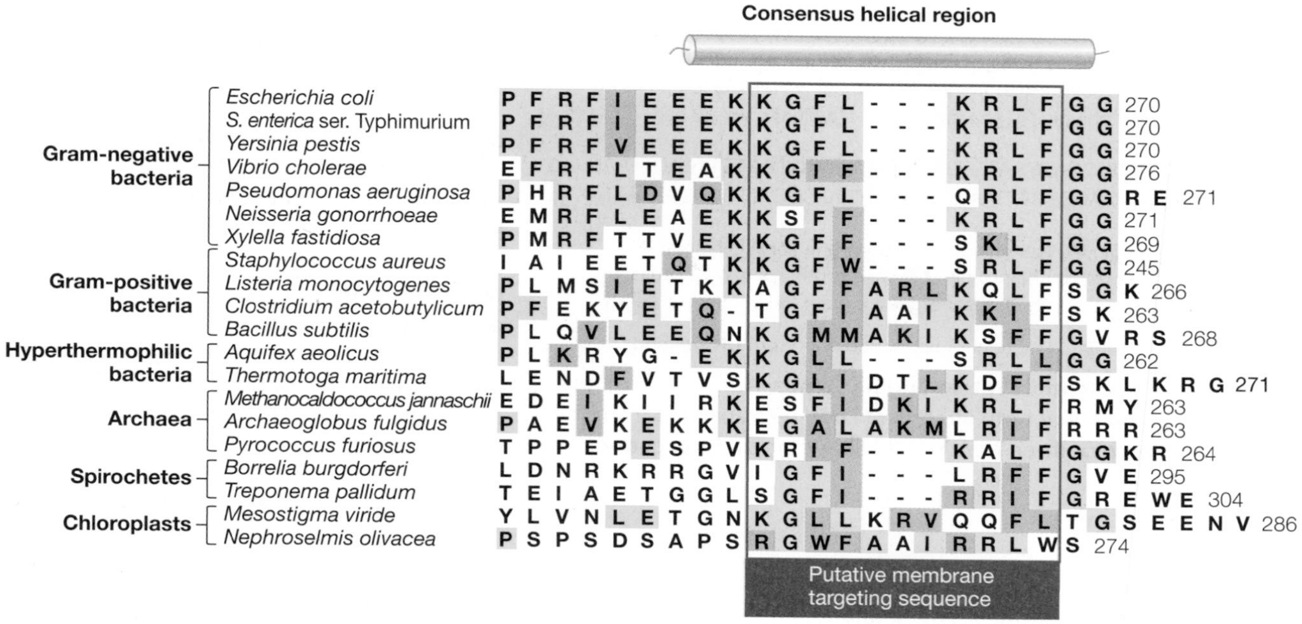

FIGURE 16.9 Analysis of Conserved Regions of Phylogenetically Well-Conserved Proteins. C-terminal amino acid residues of MinD from 20 organisms and chloroplasts are aligned to show strong similarities. Amino acid residues identical to *E. coli* are boxed in yellow, and conservative substitutions (e.g., one hydrophobic residue for another) are boxed in orange. Dashes indicate the absence of amino acids in those positions; such gaps may be included to maintain an alignment. The number of the last residue shown relative to the entire amino acid sequence is shown at the extreme right of each line.

Figure 16.9 Micro Inquiry

Which amino acids are most highly conserved?

and other physiological features of the microbe (**figure 16.11**). This is of special interest for the causative agent of syphilis, *Treponema pallidum*. Because it is not possible to grow *T. pallidum* outside the human body, we know little about its metabolism or the way it avoids host defenses. This has prevented the development of a syphilis vaccine. The sequencing and annotation of the *T. pallidum* genome reveal that *T. pallidum* is metabolically crippled. It can use carbohydrates as an energy source but lacks TCA cycle and oxidative phosphorylation enzymes. *T. pallidum* also lacks many biosynthetic pathways (e.g., for enzyme cofactors, fatty acids, nucleotides, and some electron transport proteins) and must rely on molecules supplied by its host. In fact, about 5% of its genes code for transport proteins. Given the lack of several critical pathways, it is not surprising that the pathogen has not been cultured successfully. The genes for surface proteins are of particular interest. *T. pallidum* has a family of surface protein genes characterized by many repetitive sequences. Some have speculated that these genes might undergo recombination to generate new surface proteins, enabling the organism to avoid attack by the immune system. It may be possible to develop a vaccine for syphilis using some of these surface proteins. It may also be possible to identify strains of *T. pallidum* using these surface proteins, which would be of great importance in syphilis epidemiology. The annotated genome should ultimately help us understand how *T. pallidum* causes syphilis. ▶▶| *Phylum* Spirochaetes *(section 19.6)*

Microarray Analysis

Once the identity and function of the genes that comprise a genome have been analyzed, the key question remains, "Which genes are expressed at any given time?" Prior to the genomic era, researchers could identify only a limited number of genes whose expression was altered under specific circumstances. However, the development of **DNA microarrays** now allows scientists to look at the expression level of a vast collection of genes at once. DNA microarrays are solid supports, usually of glass or silicon, upon which DNA is attached in an organized grid fashion. Each spot of DNA, called a **probe,** represents a single gene or ORF. The location of each gene on the grid is carefully recorded so that when analyzed, the genetic identity of each spot is known.

Commercially prepared microarrays can be purchased or custom arrays can be made in research laboratories. The preparation

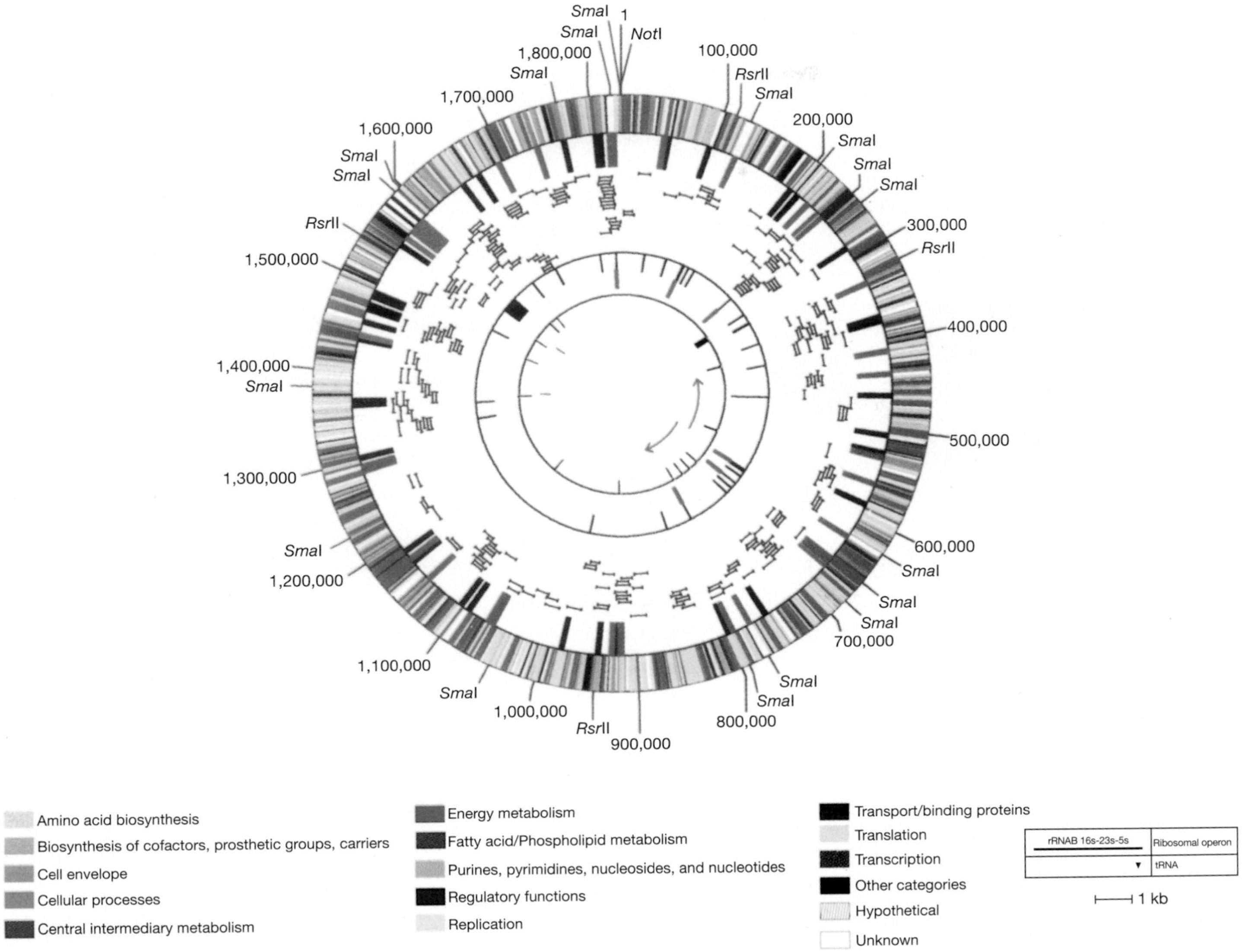

FIGURE 16.10 Physical Map of the *Haemophilus influenzae* Genome. The predicted coding regions in the outer concentric circle are indicated with colors representing their functional roles. The outer perimeter shows the *NotI, RsrII,* and *SmaI* restriction sites. The inner concentric circle shows regions of high G + C content (red and blue) and high A + T content (black and green). The third circle shows the coverage by λ clones (blue). The fourth circle shows the locations of rRNA operons (green), tRNAs (black), and the Mu-like prophage (blue). The fifth circle shows simple tandem repeats and the probable origin of replication (outward pointing green arrows). The red lines are potential termination sequences.

Reprinted with permission from Fleischman, R. D., et al. 1995. Whole-genome random sequencing and assembly of Haemophilus influenzae *Rd.* Science *269:496–512. Figure 1, p. 507, and the Institute of Genomic Research.*

of custom, or spotted, arrays relies on the robotic application of a probe to the microarray. The probe may be a polymerase chain reaction (PCR) product, complementary DNA (cDNA), or a short DNA fragment within the gene or ORF, called an **oligonucleotide.** When eukaryotic genomes are to be analyzed, oligonucleotide probes are called **expressed sequence tags (ESTs)** because each is derived from cDNA. Recall that cDNA lacks introns because it is DNA that is copied from processed mRNA *(see figure 15.5).* Spotted arrays are especially useful in the study of bacteria and archaea that are studied by a relatively small number of researchers. Spotted arrays are also used for eukaryotic microbes, which have much larger genomes than bacteria and archaea. Often eukaryotic genomes are not fully sequenced because of cost constraints. The genes to be represented by DNA on spotted arrays can be carefully selected, enabling the development of custom-made microarrays. Commercial microarrays are available for

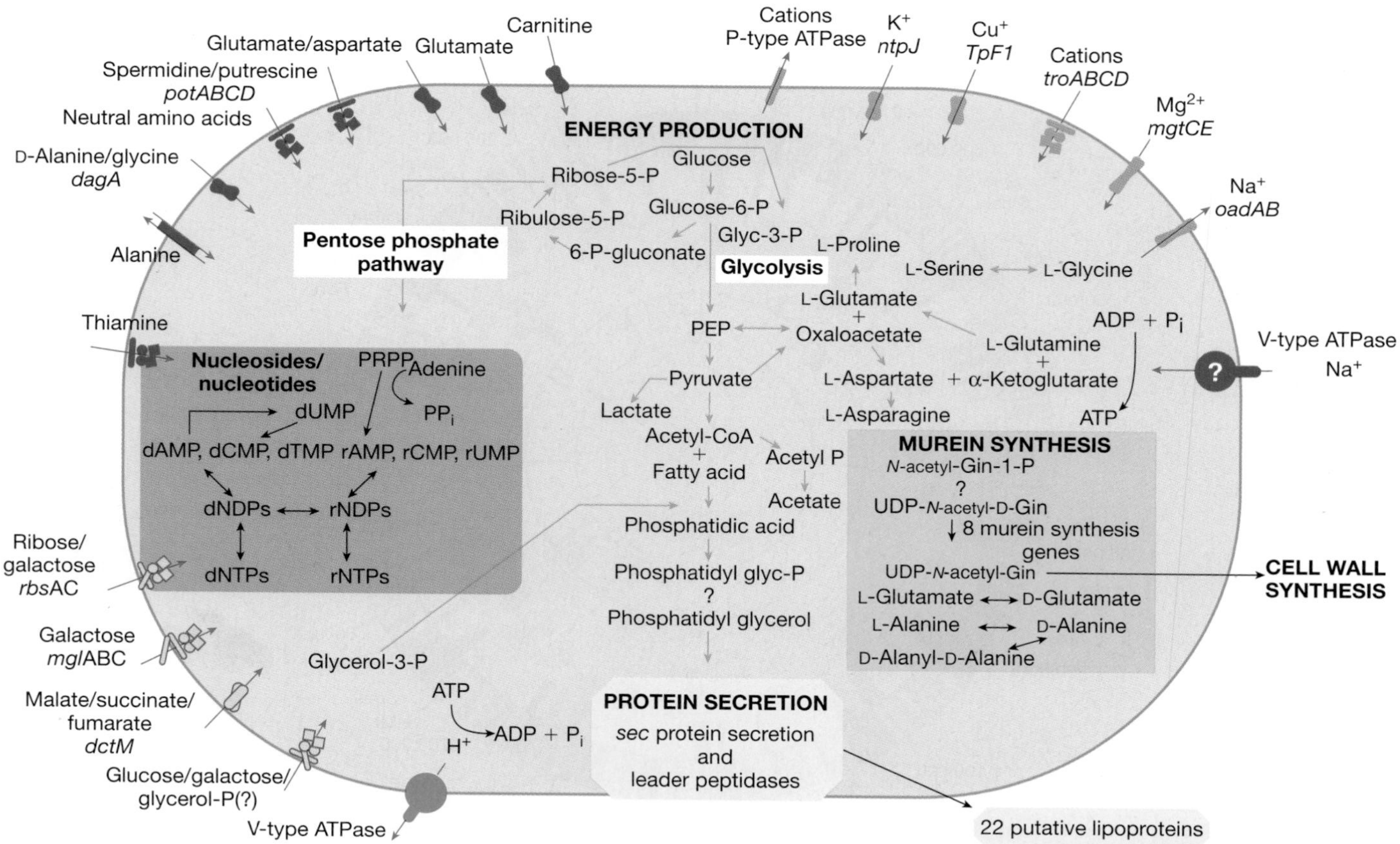

FIGURE 16.11 Metabolic Pathways and Transport Systems of *Treponema pallidum*. This depicts *T. pallidum* metabolism as deduced from genome annotation. Note the limited biosynthetic capabilities and extensive array of transporters. Although glycolysis is present, the TCA cycle and respiratory electron transport are lacking. Question marks indicate where uncertainties exist or expected activities have not been found.

***Figure 16.11* Micro Inquiry**

Based on this genomic reconstruction, can you determine if *Treponema* has a respiratory or fermentative metabolism?

many pathogenic and widely studied model microbes such as *E. coli*, *Mycobacterium tuberculosis*, and the yeast *Saccharomyces cerevisiae*.

The analysis of gene expression using microarray technology, like many other molecular genetic techniques, is based on hybridization between the single-stranded probe DNA and the nucleic acids to be analyzed, often called the targets, which may be either mRNA or single-stranded cDNA (**figure 16.12**). The target nucleotides are labeled with fluorescent dyes and incubated with the chip under conditions that ensure proper binding of target mRNA (or cDNA) to its complementary probe. Unbound target is washed off and the chip is then scanned with laser beams. Fluorescence at each spot or probe indicates that mRNA hybridized. Analysis of the color and intensity of each probe shows which genes were expressed. *DNA Probe*

DNA microarray analysis can be used to determine which genes are expressed during cellular differentiation or as the result of mutation. One common application of microarray technology in microbiology is to determine the genes whose expression is changed (either up- or downregulated) in response to environmental changes (figure 16.12). For instance, the pathogen *Helicobacter pylori* dwells in the stomach, where it causes ulcers. One might want to know which *H. pylori* genes are repressed or induced upon exposure to acidic conditions. To determine this, total cellular mRNA from bacteria grown at a neutral pH is prepared and tagged with a green fluorochrome to serve as a control or reference, while mRNA from cells exposed to acidic medium is labeled red. The green (reference) and red (experimental) mRNA samples are then mixed and hybridized to the same microarray. The red and green mRNA transcripts for any given gene must compete with each other for binding to the target. After the unattached mRNA is washed off, the

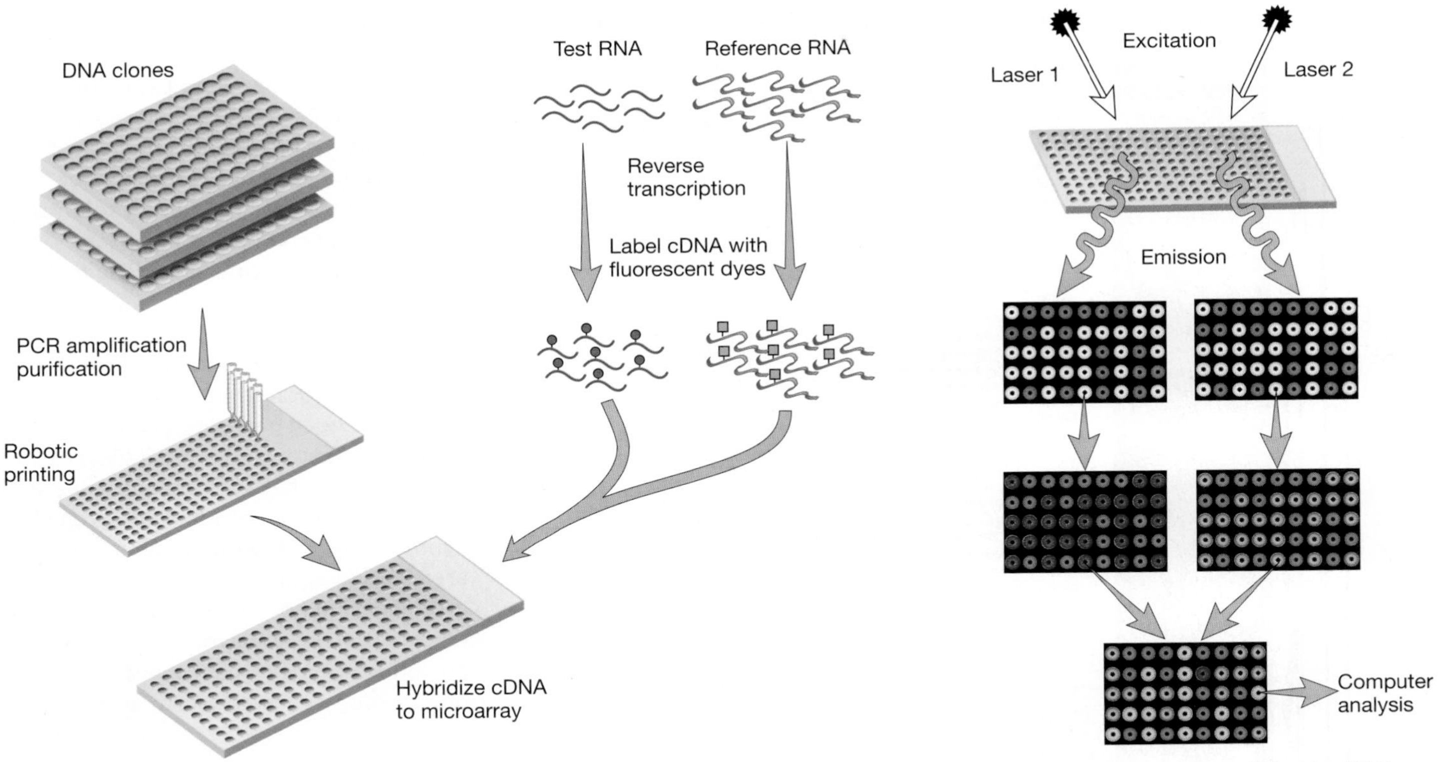

FIGURE 16.12 A Microarray System for Monitoring Gene Expression. Cloned genes from an organism are amplified by PCR, and after purification, samples are applied by a robotic printer to generate a spotted microarray. To monitor enzyme expression, mRNA from test and reference cultures is converted to cDNA by reverse transcriptase and labeled with two different fluorescent dyes. The labeled mixture is hybridized to the microarray and scanned using two lasers with different excitation wavelengths. The fluorescence responses are measured as normalized ratios that show whether the test gene response is higher or lower than that of the reference.

microarray is scanned and the image is computer analyzed. A yellow spot or probe indicates that roughly equal numbers of green and red mRNA molecules were bound, so there was no change in the level of gene expression for that gene. If a target is red, more mRNA from bacteria grown under the experimental acidic conditions was present when the two mRNAs were mixed, thus this gene was induced. Conversely, a green target indicates that the gene was repressed upon exposure to acid stress. Careful image analysis is used to determine the relative intensity of each spot, so that the magnitude of induction or repression of each gene whose expression is altered can be approximated. *Microarrays*

Microarray experiments yield a vast amount of information that must be organized in some meaningful fashion. One common approach is to use **hierarchical cluster analysis,** which groups genes with similar function or patterns of regulation together. As shown in **figure 16.13**, hierarchical cluster analysis was used to group genes based in part on their related functions but primarily in terms of their level of expression: induced genes (red spots) are grouped separately from repressed genes (green spots); genes whose expression remains unaltered are shown in black.

The data shown in figure 16.13 was generated in an experiment on the bacterium *Deinococcus radiodurans*. This microbe has the remarkable ability to survive intense desiccation and γ-radiation at doses many times above that needed to kill humans. Ionizing radiation causes double-stranded breaks in DNA—the most lethal form of DNA damage. *D. radiodurans* is able to reassemble its genome after it has been fragmented into thousands of pieces. Its genome consists of two circular chromosomes (2.6 Mb and 0.4 Mb), a megaplasmid (0.18 Mb), and a small plasmid (45,704 base pairs). It was thought that sequencing this bacterium's genome would reveal that *D. radiodurans* is superb in executing DNA repair. But, surprisingly, *D. radiodurans* was discovered to have fewer DNA repair genes than *E. coli*.

To understand these unexpected findings, microbiologists used microarrays with about 94% of *D. radiodurans* genes represented to examine the **transcriptome** (all the mRNA present) following radiation treatment. These results were then scanned for genes with similar functions and then grouped by relatedness (degree of relatedness is statistically quantified by a correlation coefficient, or "r value"). Such analysis confirmed that the DNA repair gene *recA* and genes involved in DNA replication and recombination are dramatically upregulated following irradiation. In addition, genes whose products direct cell wall metabolism and cellular transport, and many genes whose protein products are unknown

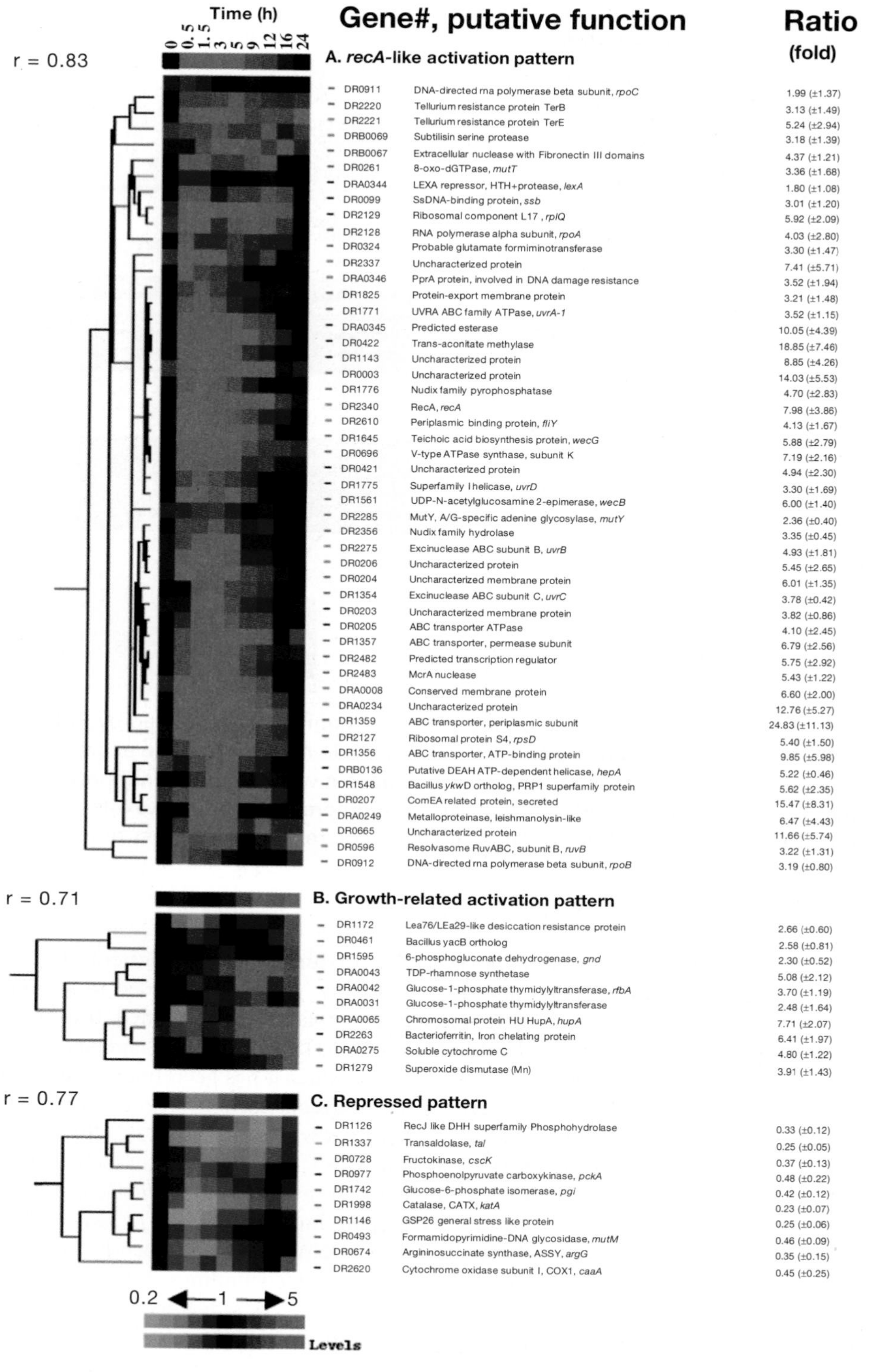

Time (h)
0 0.5 1.5 3 5 9 12 16 24
Gene#, putative function
Ratio (fold)
r = 0.83
A. recA-like activation pattern
DR0911 DNA-directed rna polymerase beta subunit, rpoC 1.99 (±1.37)
DR2220 Tellurium resistance protein TerB 3.13 (±1.49)
DR2221 Tellurium resistance protein TerE 5.24 (±2.94)
DRB0069 Subtilisin serine protease 3.18 (±1.39)
DRB0067 Extracellular nuclease with Fibronectin III domains 4.37 (±1.21)
DR0261 8-oxo-dGTPase, mutT 3.36 (±1.68)
DRA0344 LEXA repressor, HTH+protease, lexA 1.80 (±1.08)
DR0099 SsDNA-binding protein, ssb 3.01 (±1.20)
DR2129 Ribosomal component L17 , rplQ 5.92 (±2.09)
DR2128 RNA polymerase alpha subunit, rpoA 4.03 (±2.80)
DR0324 Probable glutamate formiminotransferase 3.30 (±1.47)
DR2337 Uncharacterized protein 7.41 (±5.71)
DRA0346 PprA protein, involved in DNA damage resistance 3.52 (±1.94)
DR1825 Protein-export membrane protein 3.21 (±1.48)
DR1771 UVRA ABC family ATPase, uvrA-1 3.52 (±1.15)
DRA0345 Predicted esterase 10.05 (±4.39)
DR0422 Trans-aconitate methylase 18.85 (±7.46)
DR1143 Uncharacterized protein 8.85 (±4.26)
DR0003 Uncharacterized protein 14.03 (±5.53)
DR1776 Nudix family pyrophosphatase 4.70 (±2.83)
DR2340 RecA, recA 7.98 (±3.86)
DR2610 Periplasmic binding protein, fliY 4.13 (±1.67)
DR1645 Teichoic acid biosynthesis protein, wecG 5.88 (±2.79)
DR0696 V-type ATPase synthase, subunit K 7.19 (±2.16)
DR0421 Uncharacterized protein 4.94 (±2.30)
DR1775 Superfamily I helicase, uvrD 3.30 (±1.69)
DR1561 UDP-N-acetylglucosamine 2-epimerase, wecB 6.00 (±1.40)
DR2285 MutY, A/G-specific adenine glycosylase, mutY 2.36 (±0.40)
DR2356 Nudix family hydrolase 3.35 (±0.45)
DR2275 Excinuclease ABC subunit B, uvrB 4.93 (±1.81)
DR0206 Uncharacterized protein 5.45 (±2.65)
DR0204 Uncharacterized membrane protein 6.01 (±1.35)
DR1354 Excinuclease ABC subunit C, uvrC 3.78 (±0.42)
DR0203 Uncharacterized membrane protein 3.82 (±0.86)
DR0205 ABC transporter ATPase 4.10 (±2.45)
DR1357 ABC transporter, permease subunit 6.79 (±2.56)
DR2482 Predicted transcription regulator 5.75 (±2.92)
DR2483 McrA nuclease 5.43 (±1.22)
DRA0008 Conserved membrane protein 6.60 (±2.00)
DRA0234 Uncharacterized protein 12.76 (±5.27)
DR1359 ABC transporter, periplasmic subunit 24.83 (±11.13)
DR2127 Ribosomal protein S4, rpsD 5.40 (±1.50)
DR1356 ABC transporter, ATP-binding protein 9.85 (±5.98)
DRB0136 Putative DEAH ATP-dependent helicase, hepA 5.22 (±0.46)
DR1548 Bacillus ykwD ortholog, PRP1 superfamily protein 5.62 (±2.35)
DR0207 ComEA related protein, secreted 15.47 (±8.31)
DRA0249 Metalloproteinase, leishmanolysin-like 6.47 (±4.43)
DR0665 Uncharacterized protein 11.66 (±5.74)
DR0596 Resolvasome RuvABC, subunit B, ruvB 3.22 (±1.31)
DR0912 DNA-directed rna polymerase beta subunit, rpoB 3.19 (±0.80)
r = 0.71
B. Growth-related activation pattern
DR1172 Lea76/LEa29-like desiccation resistance protein 2.66 (±0.60)
DR0461 Bacillus yacB ortholog 2.58 (±0.81)
DR1595 6-phosphogluconate dehydrogenase, gnd 2.30 (±0.52)
DRA0043 TDP-rhamnose synthetase 5.08 (±2.12)
DRA0042 Glucose-1-phosphate thymidylyltransferase, rfbA 3.70 (±1.19)
DRA0031 Glucose-1-phosphate thymidylyltransferase 2.48 (±1.64)
DRA0065 Chromosomal protein HU HupA, hupA 7.71 (±2.07)
DR2263 Bacterioferritin, Iron chelating protein 6.41 (±1.97)
DRA0275 Soluble cytochrome C 4.80 (±1.22)
DR1279 Superoxide dismutase (Mn) 3.91 (±1.43)
r = 0.77
C. Repressed pattern
DR1126 RecJ like DHH superfamily Phosphohydrolase 0.33 (±0.12)
DR1337 Transaldolase, tal 0.25 (±0.05)
DR0728 Fructokinase, cscK 0.37 (±0.13)
DR0977 Phosphoenolpyruvate carboxykinase, pckA 0.48 (±0.22)
DR1742 Glucose-6-phosphate isomerase, pgi 0.42 (±0.12)
DR1998 Catalase, CATX, katA 0.23 (±0.07)
DR1146 GSP26 general stress like protein 0.25 (±0.06)
DR0493 Formamidopyrimidine-DNA glycosidase, mutM 0.46 (±0.09)
DR0674 Argininosuccinate synthase, ASSY, argG 0.35 (±0.15)
DR2620 Cytochrome oxidase subunit I, COX1, caaA 0.45 (±0.25)
0.2 1 5
Levels

FIGURE 16.13 Hierarchical Cluster Analysis of Gene Expression of *D. radiodurans* Following Exposure to γ-Radiation. Each row of colored strips represents a single gene, and the color indicates the level of expression over nine time intervals. The far-left column is the control and thus is black (at control levels of expression). The level of induction or repression relative to the control value is indicated as the ratio (fold). Each group of genes has been scored for relatedness, and a "tree" has been generated on the far left of the clusters, with the indicated correlation coefficient (r value). A large number of genes encoding DNA repair, synthesis, and recombination proteins are induced upon radiation. These are grouped together. Fewer genes that encode proteins involved in metabolism are induced. Finally, genes involved in other aspects of metabolism are repressed.

Figure 16.13 Micro Inquiry

What is the difference in timing of gene induction in the *recA*-like activation pattern compared to the growth-related activation pattern?

are also induced. In 2006 an international team of scientists showed that DNA repair requires extensive DNA synthesis. In contrast to repair found in other cells in which one strand serves as template for synthesis of a new strand, *Deinococcus* uses DNA polymerase A to synthesis two new strands. The newly synthesized fragments are then pieced together in a process that depends on RecA. The newly repaired genome thus becomes a mosaic of old and new DNA fragments. One hypothesis that is consistent with these results is that *D. radiodurans* must have desiccation and radioresistant DNA repair and synthesis proteins. To test this prediction, it is necessary to compare the cellular pool of proteins synthesized under stressful and unstressful conditions. This is accomplished by proteomic analysis, the next topic of discussion. ◀◀ Deinococcus-Thermus *(section 19.2)*

1. What is genome annotation? Why do you think it requires knowledge of mathematics and statistics as well as biology and computer science?
2. How can genomic sequencing be used to ask specific questions about the physiology of a given microbe?

3. Describe how microarrays are constructed and used to analyze gene expression. How might the following scientists use DNA microarrays in their research?
   (a) A microbial ecologist who is interested in how the soil microbe *Rhodopseudomonas palustris* degrades the toxic compound 3-chlorobenzene.
   (b) A medical microbiologist who wants to learn about how the pathogen *Salmonella* survives within a host cell.

16.5 Proteomics

Genome function can be studied at the translation level as well as the transcription level. The entire collection of proteins that an organism produces is called its **proteome.** Thus **proteomics** is the study of the proteome or the array of proteins an organism can produce. Proteomics provides information about genome function that mRNA studies cannot because a direct correlation between mRNA and the pool of cellular proteins does not always exist. This is because mRNA can be unstable (and therefore not detected) and protein activity may be posttranslationally regulated, for example, by phosphorylation. Much of the research in this area is referred to as **functional proteomics.** It is focused on determining the function of different cellular proteins, how they interact with one another, and the ways in which they are regulated.

Although new techniques in proteomics are continuously being developed, we will focus briefly only on the most common approach, **two-dimensional gel electrophoresis.** In this procedure, a mixture of proteins is separated using two different electrophoretic procedures (dimensions). This permits the visualization of thousands of cellular proteins that would not otherwise be separated in a single electrophoretic dimension. As shown in **figure 16.14a**, the first dimension makes use of **isoelectric focusing,** in which proteins move through a pH gradient (e.g., pH 3 to 10). First, the protein mixture is applied to an acrylamide gel in a tube with a stable pH gradient and electrophoresed. Each protein moves along the pH gradient until the protein's net charge is zero and the protein stops moving. The pH at this point is equal to the protein's **isoelectric point.** Thus the first dimension separates proteins based on the content of ionizable amino acids. The second dimension is SDS *p*oly*a*crylamide *g*el *e*lectrophoresis (SDS-PAGE). SDS (sodium dodecyl sulfate) is an anionic detergent that denatures proteins and coats them with a negative charge. After the isoelectric gel has been completed, the tube gel is soaked in SDS buffer and then placed at the edge of an SDS-PAGE gel. A voltage is then applied. Because all the proteins are negatively charged thanks to SDS treatment, they will migrate from the negative pole of the gel to the positive. In this way, polypeptides are separated according to their molecular weight; that is, the smallest polypeptide will travel fastest and farthest. Two-dimensional gel electrophoresis can resolve thousands of proteins; each protein is visualized as a spot of varying intensity, depending on its cellular abundance (figure 16.14*b*). Radiolabeled proteins are often used, enabling greater sensitivity so that newly synthesized proteins can be distinguished. Computer analysis is used to compare two-dimensional gels from microbes grown under different conditions or to compare wild-type and mutant strains. Websites have been developed for the deposition of such images, allowing researchers access to valuable and ever-growing databases.

Search This: Proteomics database

Two-dimensional gel electrophoresis is even more powerful when coupled with **mass spectrometry (MS).** The unknown protein spot is cut from the gel and cleaved into fragments by treatment with proteolytic enzymes. Then the fragments are analyzed by a mass spectrometer and the mass of the fragments is plotted. This mass fingerprint can be used to estimate the probable amino acid composition of each fragment and tentatively identify the protein. Sometimes proteins or collections of fragments are run through two mass spectrometers in sequence, a process known as **tandem MS** (**figure 16.15**). The first spectrometer separates proteins and fragments, which are further fragmented. The second spectrometer then determines the amino acid sequence of each smaller fragment. The sequence of a whole protein often can be determined by analysis of such fragment sequence data. Alternatively, if the genome of the organism has been sequenced, only a partial amino acid sequence is needed. Computer analysis is then used to compare this amino acid sequence with the predicted translated sequences of all the annotated ORFs on the organism's genome. In this way, both the protein and the gene that encodes it can be identified. Further investigation of the protein may rely on one of the large databases of protein sequences that enable comparative analysis. Comparing amino acid sequences can provide information regarding the protein structure, function, and evolution.

A second branch of proteomics is called **structural proteomics.** Here, the focus is on determining the three-dimensional structures of many proteins and using these to predict the structures of other proteins and protein complexes. The assumption is that proteins fold into a limited number of shapes and can be grouped into families of similar structures. When a number of protein structures are determined for a given family, the patterns of protein structure organization or protein-folding rules will be known. Then computational biologists use this information to predict the most likely shape of a newly discovered protein; a process known as **protein modeling.**

Search This: RCSB protein databank

Not surprisingly, a variety of other "-omics" are studied. For instance, **lipidomics** is used to determine a cell's lipid profile at a particular time, and **glycomics** is the systematic study of the cellular pool of carbohydrates. The goal of **metabolomics** is to identify all the small-molecule metabolites, regardless of chemical structure, present in the cell at a given point in time. Because many metabolites are used in multiple pathways, the overview presented by a complete metabolome enables a researcher to discern which pathways are functional under a given set of conditions. Thus

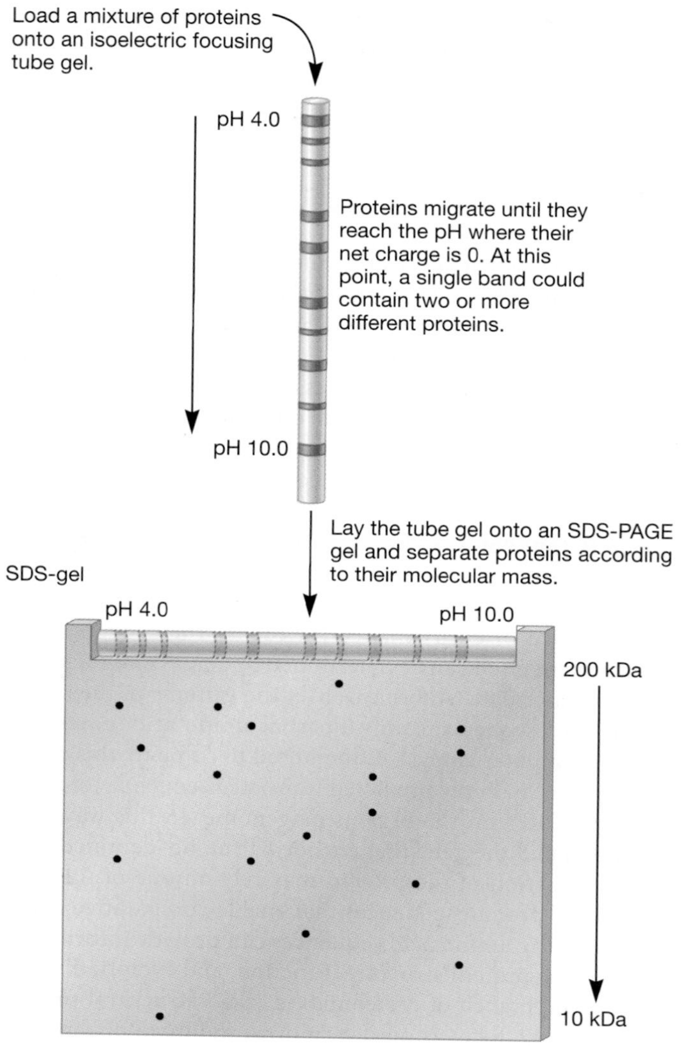

Load a mixture of proteins onto an isoelectric focusing tube gel.
pH 4.0
Proteins migrate until they reach the pH where their net charge is 0. At this point, a single band could contain two or more different proteins.
pH 10.0
Lay the tube gel onto an SDS-PAGE gel and separate proteins according to their molecular mass.
SDS-gel
pH 4.0
pH 10.0
200 kDa
10 kDa

(a) The technique of two-dimensional gel electrophoresis

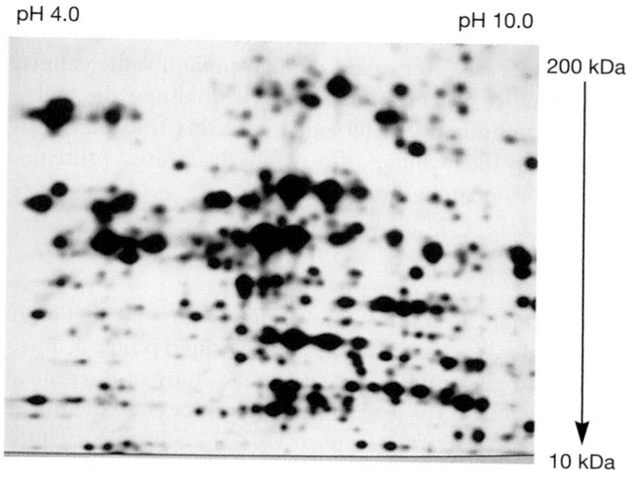

(b) An autoradiograph of a two-dimensional gel. Each protein is a discrete spot.

FIGURE 16.14 Two-Dimensional Gel Electrophoresis.

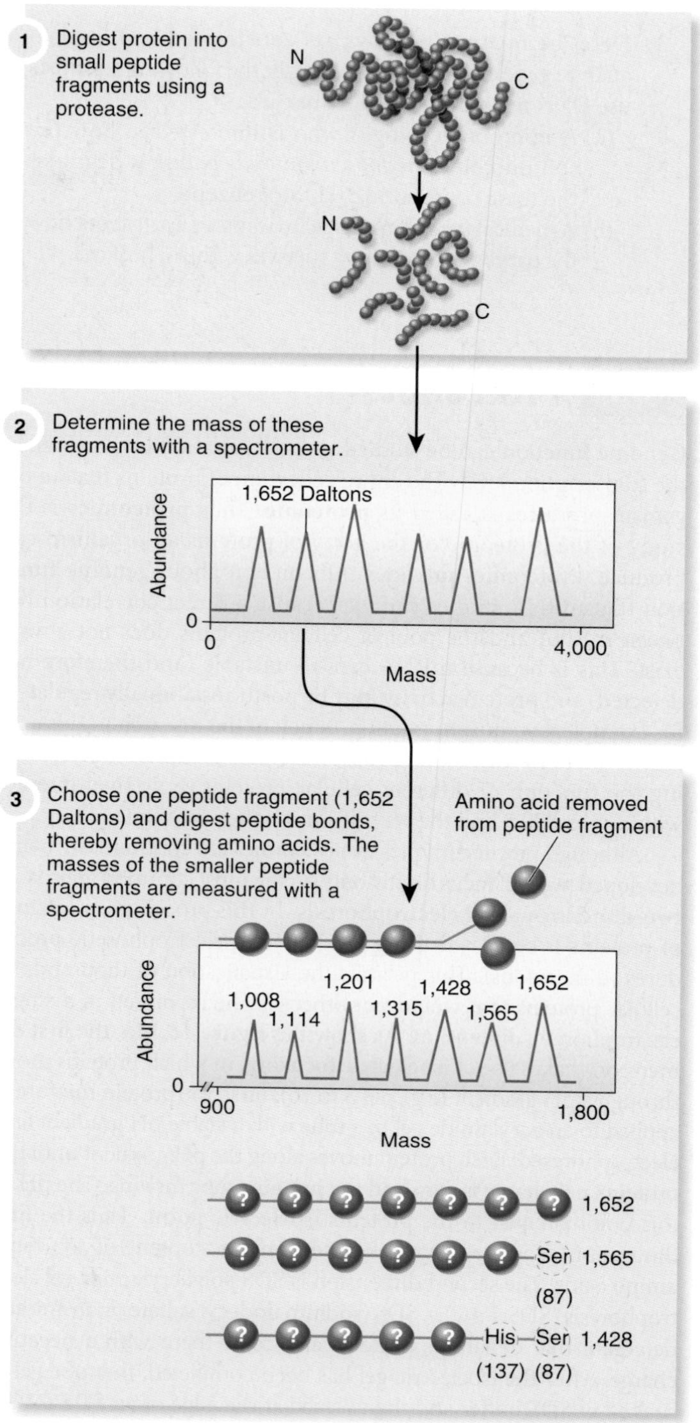

1 Digest protein into small peptide fragments using a protease.

1,652 Daltons

3 Choose one peptide fragment (1,652 Daltons) and digest peptide bonds, thereby removing amino acids. The masses of the smaller peptide fragments are measured with a spectrometer.

His Ser 1,428

FIGURE 16.15 The Use of Tandem Mass Spectrometry to Determine the Amino Acid Sequence of a Peptide.

Figure 16.15 Micro Inquiry

How would determining the amino acid sequence of a 30 amino acid residue peptide differ from determining the amino acid sequence of a 200,000 dalton protein?

metabolomics provides a high-resolution means to evaluate physiological status. These "-omics" rely heavily on the use of chromatography and as such are beyond the scope of this chapter. Nonetheless, the development of these fields illustrates the dynamic and ever-changing nature of microbiology and the need to take a holistic view if one seeks to understand the structure, function, and behavior of cells.

Probing DNA-Protein Interactions

As described in chapters 12 and 13, many proteins directly interact with DNA—for example, regulatory proteins and proteins involved in replication and transcription. Until recently, the most common way to study protein-DNA interaction was through a procedure known as electophoretic mobility shift assays (EMSA). This technique is also called a gel shift assay because when the protein thought to bind DNA is added to a mixture of purified target DNA, the mobility of the DNA is slowed in an agarose gel. When compared to the same DNA without protein, the DNA-protein complex appears to have shifted to a higher molecular weight. The magnitude of the shift is usually related to the protein:DNA ratio such that when sufficient protein is added, the largest shift is seen. To perform EMSA, the target protein must be purified and present in high concentration. But what if the DNA-binding protein being tested binds to a different region of DNA? This will not be detected. ⏮ *Gel electrophoresis (section 15.3)*

Protein-DNA interactions can also be identified through the use of **chromatin immunoprecipitation (ChIP),** which is sometimes coupled with microarray technology, called the ChIP-chip technique. ChIP was developed to identify eukaryotic transcription factors but is now widely used in the study of bacteria and archaea. In contrast to EMSA, which is entirely in vitro and uses one protein and one amplified fragment of DNA, ChiP involves treating living cells with a cross-linking agent such as formaldehyde (**figure 16.16**). This "fixes" any DNA-binding proteins to their genomic targets. The cells are then broken open and the DNA (with any bound proteins) is cut into small pieces. Antibodies, which are molecules that bind to specified proteins, are used to tag the DNA-binding proteins of interest and the antibody-protein-DNA complexes are precipitated. At this point, any region of DNA to which the protein binds can be identified by PCR using primers specific for the DNA regions that the protein of interest is thought to bind. ⏭ *Antibodies (section 33.7)*

In ChIP-chip analysis, the proteins are removed from the precipitated protein-DNA complexes and the DNA is denatured so it is now single stranded. After labeling the DNA strands, they can then be used to hybridize to a microarray that bears all or part of the microbe's genome (figure 16.12). In this way, the whole genome is scanned for sites to which the protein of interest binds. Regardless of whether ChIP is performed with PCR or a microarray, this technology enables the identification of DNA-protein interactions under different conditions. The same microbial strain can be treated with cross-linking agents when grown in different situations or at different phases of growth, so the DNA-binding profile of a protein can be followed.

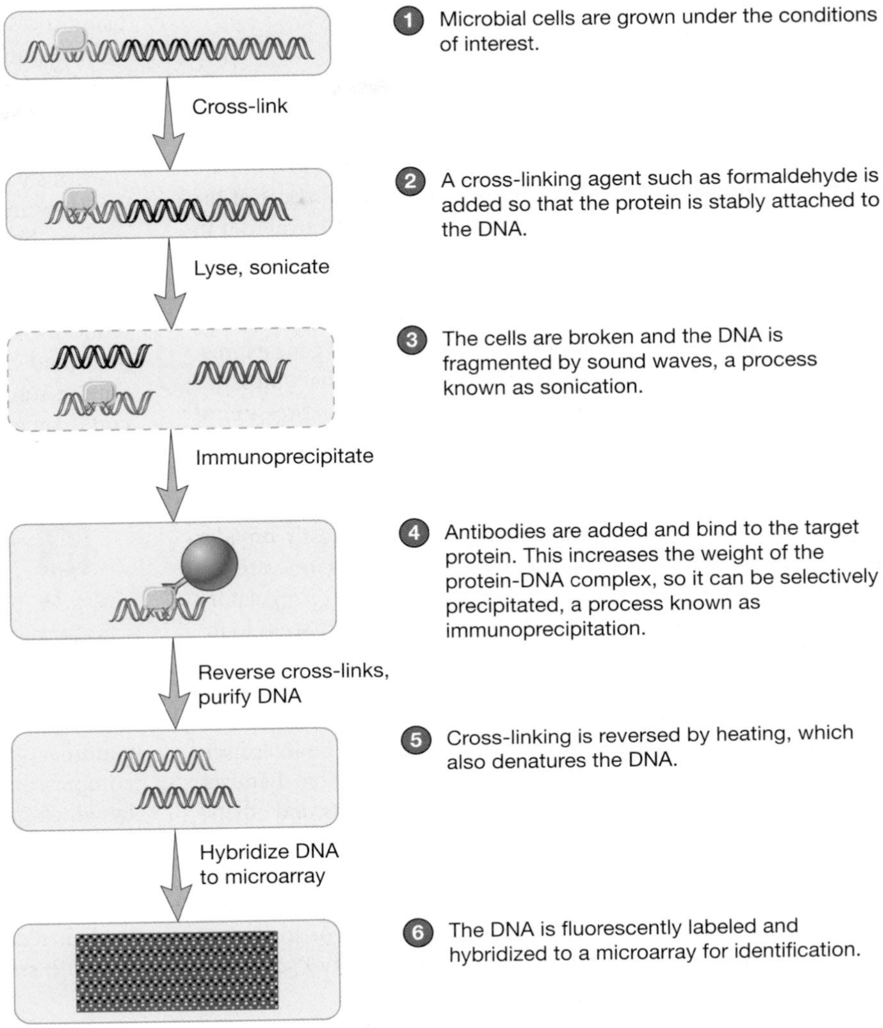

Cross-link
Lyse, sonicate
Immunoprecipitate
Reverse cross-links, purify DNA
Hybridize DNA to microarray
1 Microbial cells are grown under the conditions of interest.
2 A cross-linking agent such as formaldehyde is added so that the protein is stably attached to the DNA.
3 The cells are broken and the DNA is fragmented by sound waves, a process known as sonication.
4 Antibodies are added and bind to the target protein. This increases the weight of the protein-DNA complex, so it can be selectively precipitated, a process known as immunoprecipitation.
5 Cross-linking is reversed by heating, which also denatures the DNA.
6 The DNA is fluorescently labeled and hybridized to a microarray for identification.

FIGURE 16.16 ChIP-Chip Analysis. This technology can be used to identify the DNA to which a protein binds.

1. Why does two-dimensional gel electrophoresis allow the visualization of many more cellular proteins than seen in electrophoresis in a single dimension?
2. What is the role of mass spectrometry in proteomics?
3. What is the difference between functional and structural proteomics? How do you think structural proteomics might be used in vaccine development?
4. Describe a ChIP-chip experiment in which you want to identify the promoters to which the stationary phase sigma factor RpoS binds in *E. coli.* Could you also determine when RpoS binds? Explain your answer.

16.6 Systems Biology

Imagine the following experiment. You have created a microbial strain in which you have deleted, or "knocked out," a single gene; it doesn't really matter what the gene encodes. You want to compare the transcriptome (all the mRNA transcripts) and proteome of the knock-out mutant with its parent strain, which is identical to it in every way except the presence of this particular gene. To evaluate the transcriptome, you perform microarray analysis, and to assess the proteome, you run two-dimensional gels. You find that the loss of just one gene has huge ramifications with regard to gene expression and protein synthesis. Not only is the pathway that you knew your gene product was involved in altered, but the expression of many more genes and proteins has also been affected.

This scenario has been repeated many times for numerous microorganisms. Life scientists now realize that most of the genes and gene products that have been widely investigated have a vastly broader role in cellular physiology than previously thought. Gone are the days of reductionist biology, when a single metabolic or regulatory pathway can be considered in isolation. Systems biology seeks to integrate the "parts list" of cells—mRNA, proteins, small molecules—with the molecular interactions that become pathways for catabolism, anabolism, regulation, behavior, responses to environmental signals, etc. To accomplish this, systems biology must be interdisciplinary, requiring the expertise of physiologists, biochemists, bioinformaticists, mathematicians, geneticists, ecologists, and others.

While almost all life scientists will agree that cells must be viewed holistically, it is hard to find a consensus for a precise definition of systems biology. It has been viewed as a "paradigm shift" from a reductionist to an integrative outlook. For some, it is a term that describes the activities and interactions of scientists using tools that already exist, such as genome sequencing, proteomics, and bioinformatics. For others, it is an emerging field that will develop its own toolbox. Some envision the development of methodologies that are highly predictive, so that, for instance, the results of the mutant versus wild type experiment described previously could be modeled theoretically.

How might systems biology impact microbiology? Because systems biology is so well suited to investigating whole tissues, organs, and organisms, one application is the study of host-microbe interactions. As systems biology develops, it may provide clinically relevant information regarding host responses to pathogens (and vice versa) that would be impossible to glean from tissue culture studies. The potential predictive value of systems biology could then be harnessed to prescreen vaccines, antimicrobials, and other therapeutic agents. This is just one of many examples of impacts that this developing field will no doubt have.

Search This: Institute for Systems Biology

16.7 Comparative Genomics

A natural consequence of the tremendous number of sequenced genomes is the increased ability to make meaningful genome comparisons. **Comparative genomics** is a useful set of analyses by which gene function and evolution can be inferred by studying similar nucleotide and amino acid sequences found among organisms. Perhaps the most fundamental comparison is the relative sizes of microbial genomes. Several generalizations can be deduced from such analysis. Inspection of **figure 16.17** reveals that the smallest genomes published to date, across all three

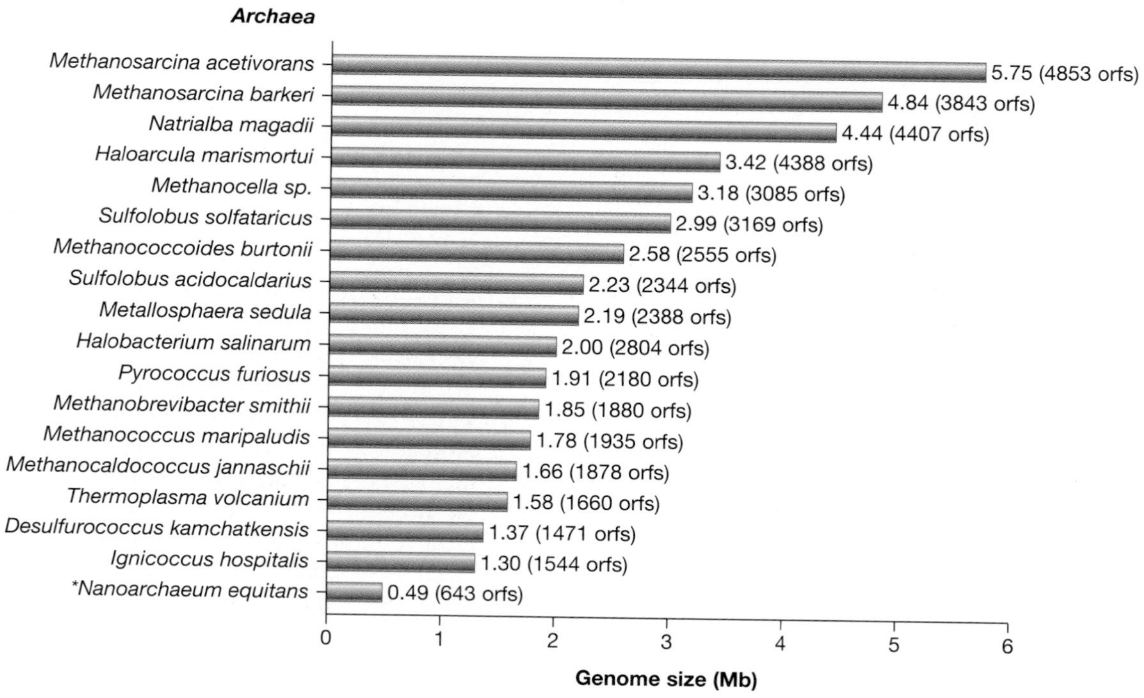

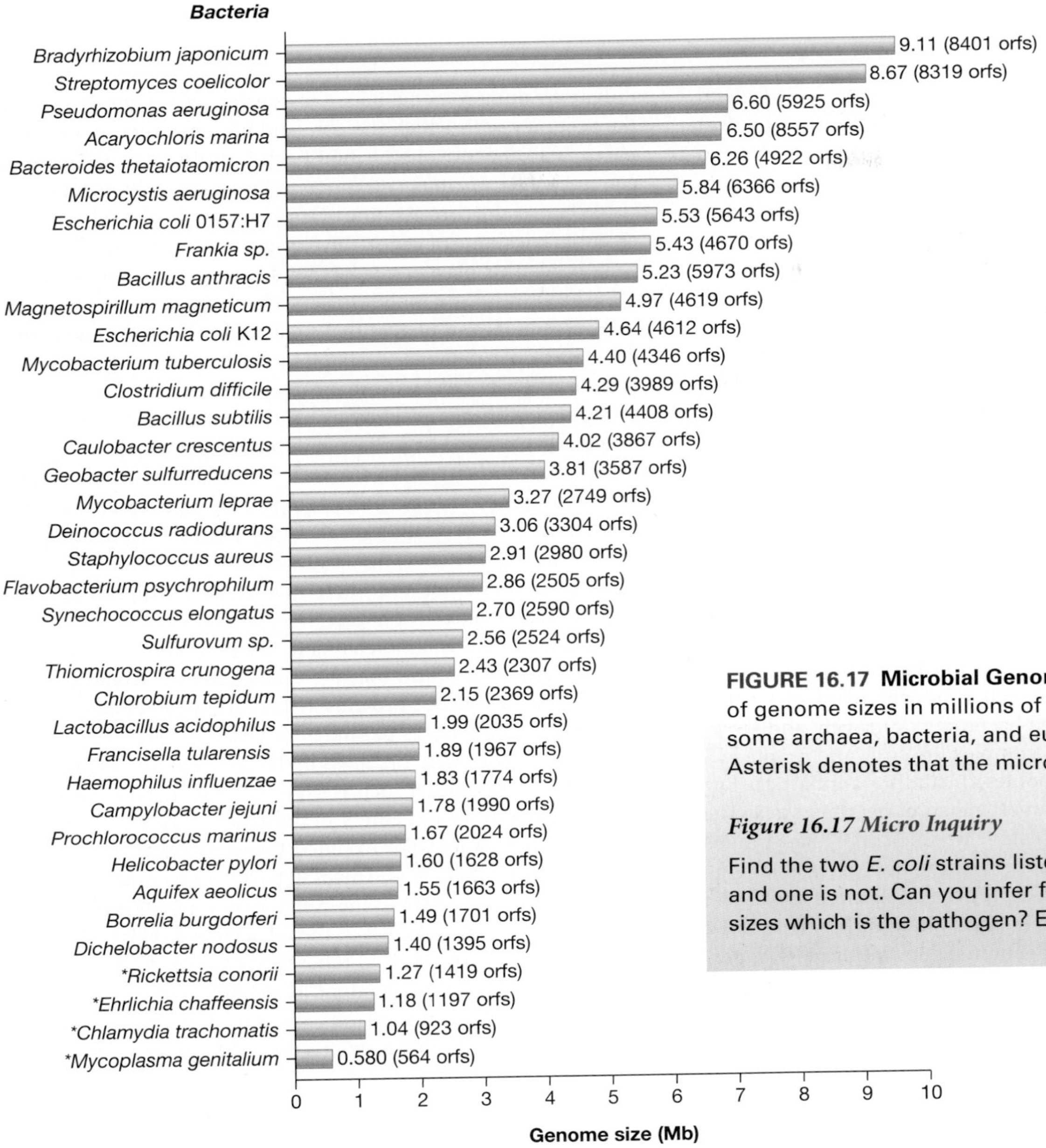

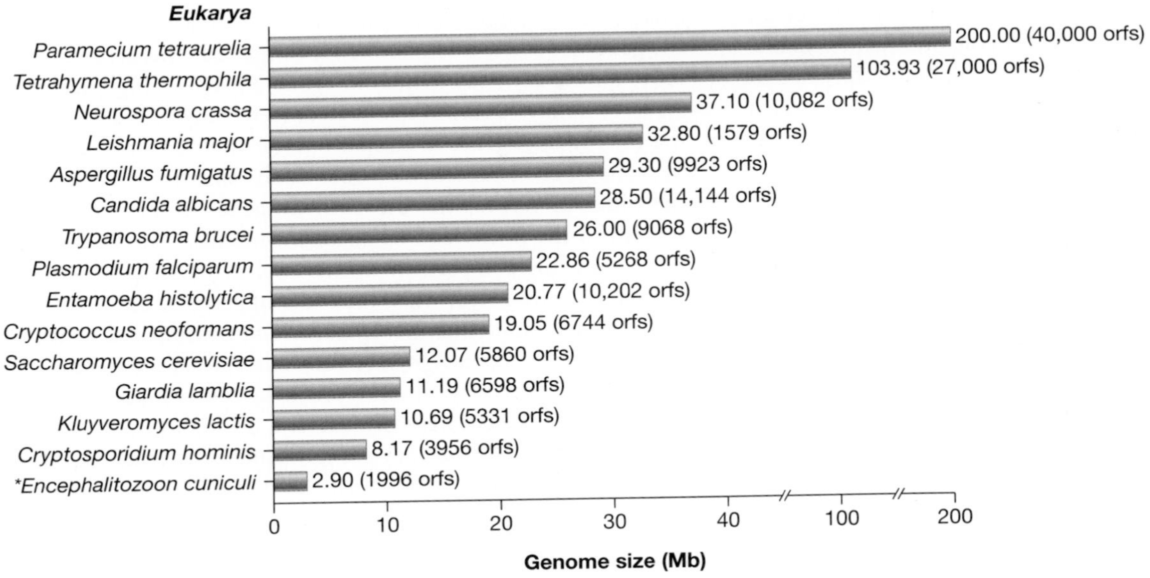

FIGURE 16.17 Microbial Genome Sizes. The distribution of genome sizes in millions of base pairs (Mb) among some archaea, bacteria, and eukaryotic microbes. Asterisk denotes that the microbe is parasitic.

Figure 16.17 Micro Inquiry

Find the two *E. coli* strains listed. One is pathogenic, and one is not. Can you infer from the relative genome sizes which is the pathogen? Explain your answer.

domains, belong to parasitic microbes. These organisms depend on their hosts for many metabolites and have presumably lost the genes encoding the enzymes needed for their assembly. Also note that as the genome size increases, the trend is toward more open reading frames. Bacterial and archaeal genomes tend to have more open reading frames per nucleotide, that is, they have a higher gene density. This is because they lack introns and much of the other noncoding sequences found in eukaryotes. Indeed, some archaea and bacteria have reduced noncoding sequence to a minimum and have many overlapping genes. This is particularly true for those that live in nutrient-limited environments; synthesizing DNA, which requires both phosphate and nitrogen, is costly.

Within domains, we see that archaea tend to have smaller genomes than many bacteria (figure 16.17*a,b*). The largest known archaeal genome, that of the methanogen *Methanosarcina acetivorans,* falls in the middle of the genome size range for bacteria. Bacterial genomes range from the very small (about 500,000 base pairs) to over 9 million base pairs (Mb). In general, genome size reflects metabolic and morphological complexity. For instance, the two largest genomes listed in figure 16.7*b* belong to *Bradyrhizobium japonicum*, which produces plant-associated nodules in which differentiated bacteria fix nitrogen, and *Streptomyces coelicolor,* which has a complex life cycle and produces a number of specialized metabolites, including antibiotics. Eukaryotic microbes have the largest range of genome sizes. The genome of the intracellular pathogenic microsporidian fungus *Encephalitozoon cuniculi* is smaller than many bacterial and archaeal genomes, yet the ciliated protist *Paramecium tetraurelia* has a tremendously large genome. It is thought that this protist underwent multiple rounds of genomic duplication. ▶▶| *Suborder* Streptomycineae *(section 22.7);* Alveolata *(section 23.5); Nitrogen fixation (section 29.3)*

Comparisons of the genomes of strains within a species and among species within a phylum have revealed that microbial genomes consist of an older core genome and a more recently acquired pan-genome (**figure 16.18**). The **core genome** is that set of genes found in all members of a species (or other monophyletic group). Thus it is thought to represent the minimal number of genes needed for the microbes to survive. In general, these genes encode "informational" proteins involved in replication, transcription, and translation. They are thought to be those genes present in the group's common ancestor. By contrast, the **pan-genome** is the combination of all the different genes found in all the strains of a given species. These more recently acquired genes enable the microbe to colonize new niches. A comparison of the core genome size with the actual genome size of a particular strain thus indicates the evolution of new traits. For instance, current values of the core genome and pan-genome of *Bacillus anthracis* differ by only roughly 200 genes (about 3,600 versus 3,800, respectively), while the same estimate for *E. coli* is over 3,000 genes (about 2,800 for the core genome and 6,000 for the pan-genome). The limited genetic diversity among *B. anthracis* strains reflects the fact that *B. anthracis* growth is restricted to only a few habitats. The enormous diversity among strains of *E. coli* has allowed this species to radiate to numerous habitats. Obviously, for any given species, the accuracy with which the core and pan-genome can be compared will depend on the number of strains with sequenced genomes. ▶▶| *Microbial evolutionary processes (section 17.5)*

Whereas the core genome of a species can be thought of as an essential backbone of genes, the pan-genome represents a flexible gene pool. This plasticity is largely due to the fact that many of these genes reside on mobile genetic elements, including transposons, plasmids, and phages that mediate **horizontal gene transfer (HGT).** As detailed in chapter 14, HGT is the exchange of genetic material between organisms that need not be of similar evolutionary lineages. Genome analysis has revealed that HGT is frequently mediated by viruses and that lysogeny may be the rule, rather than the exception. In fact, some bacteria and archaea carry multiple proviruses. Some temperate phages carry lysogenic conversion genes, so called because they change the phenotype of their hosts. For instance, the virulence genes for some important pathogens, including the bacteria that cause diphtheria and cholera, are encoded by phages.

When mobile genetic elements are permanently integrated into a microbial genome, they are called **genomic islands.** When these new genes encode proteins that contribute to or confer virulence, they are known as **pathogenicity islands.** How is it known that certain genes were obtained by HGT? Genomic islands provide evidence of transfer such as a G + C content and a codon bias that differs from the remainder of the genome. This is most easily explained by example. For instance, the genomes of *Streptomyces* spp. are about 72% G + C. This is in part because codons that are AT-rich are rarely used (e.g., the cysteine codon ACG is used much more frequently than ACA). In these bacteria, a region of the genome that has a low G + C ratio coupled with AT-rich codons suggests the presence of a genomic island and the acquisition of the genes by HGT. Genomic islands also sometimes retain inverted repeat sequences or other elements needed for transfer or integration. It has become clear that HGT is a major evolutionary force in short-term microbial evolution and long-term speciation.

It follows that phylogenetic relationships between microbes can be explored by comparative genomics. One approach is to examine how similar the organization of orthologous genes are in the genomes of the microbes being compared. This is called **synteny,** and closely related microbes generally have a high degree of synteny. It may be easiest to understand synteny by first considering the order of genes in similar operons found in two different microbes. For instance, one would not be surprised to learn that the order of tryptophan biosynthetic genes is the same (conserved) in the *E. coli* and *Salmonella* genomes. Analysis of synteny takes this comparison to the level of whole genomes. As shown in **figure 16.19**, when the order of genes in the genome of the plant symbiont *Sinorhizobium meliloti* is compared with that of the plant pathogen *Agrobacterium tumefaciens*, it is clear that the gene order is very similar. This is in contrast to *E. coli* and *S. meliloti.* Although these two bacteria belong to the same phylum, the *Proteobacteria,* they are not sufficiently related to show synteny. ▶▶| Alphaproteobacteria *(section 20.1); Microorganism associations with vascular plants (section 29.3)*

Comparative genomics can also provide great insight into genes and gene products that are associated with virulence when pathogen genomes are analyzed. Such insights are particularly needed in understanding *Mycobacterium tuberculosis,* the causative agent of tuberculosis (TB). About one-third of the human population has TB. After establishing residence in immune cells in the lung, *M. tuberculosis* often remains in a dormant state until the host's immune system is compromised. It then goes on to cause active disease, killing about 2 million people annually. The *M. tuberculosis* genome has been compared to the genomes of two relatives—*M. leprae,* which causes leprosy, and *M. bovis,* the causative agent of TB in a wide range of animals, including cows and humans. The genomes of *M. bovis* and *M. tuberculosis* are most similar—about 99.5% identical at the sequence level. However, the *M. bovis* genome is missing 11 separate regions, making its genome slightly smaller (4.3 Mb versus 4.4. Mb). The sequence dissimilarities involve the inactivation of some genes, leading to major differences in the way the two bacteria respond to environmental conditions. This may account for the host range differences between these two closely related pathogens. ▶▶| *Suborder* Corynebacterineae *(section 22.4); Mycobacterial infections (section 38.1)*

The divergence between *M. tuberculosis* and *M. leprae* is even more striking. The *M. leprae* genome is a third smaller than that of *M. tuberculosis.* About half the genome is devoid of functional genes. Instead, there are over 1,000 degraded, nonfunctional genes called **pseudogenes** (**figure 16.20**). In total, *M. leprae* seems to have lost more that 1,000 genes during its career as an intracellular parasite. *M. leprae* even lacks some of the enzymes required for energy production and DNA replication. This might explain

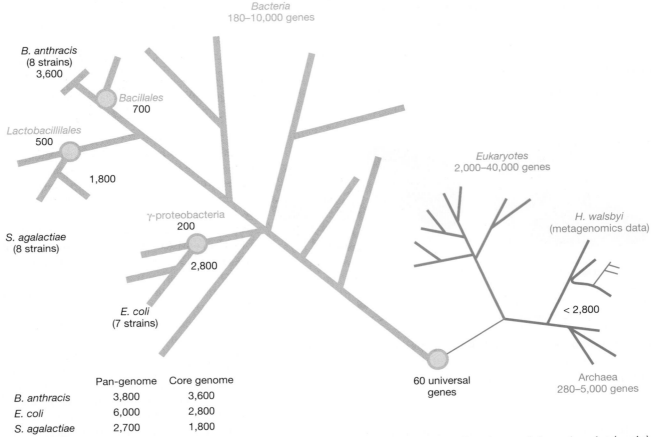

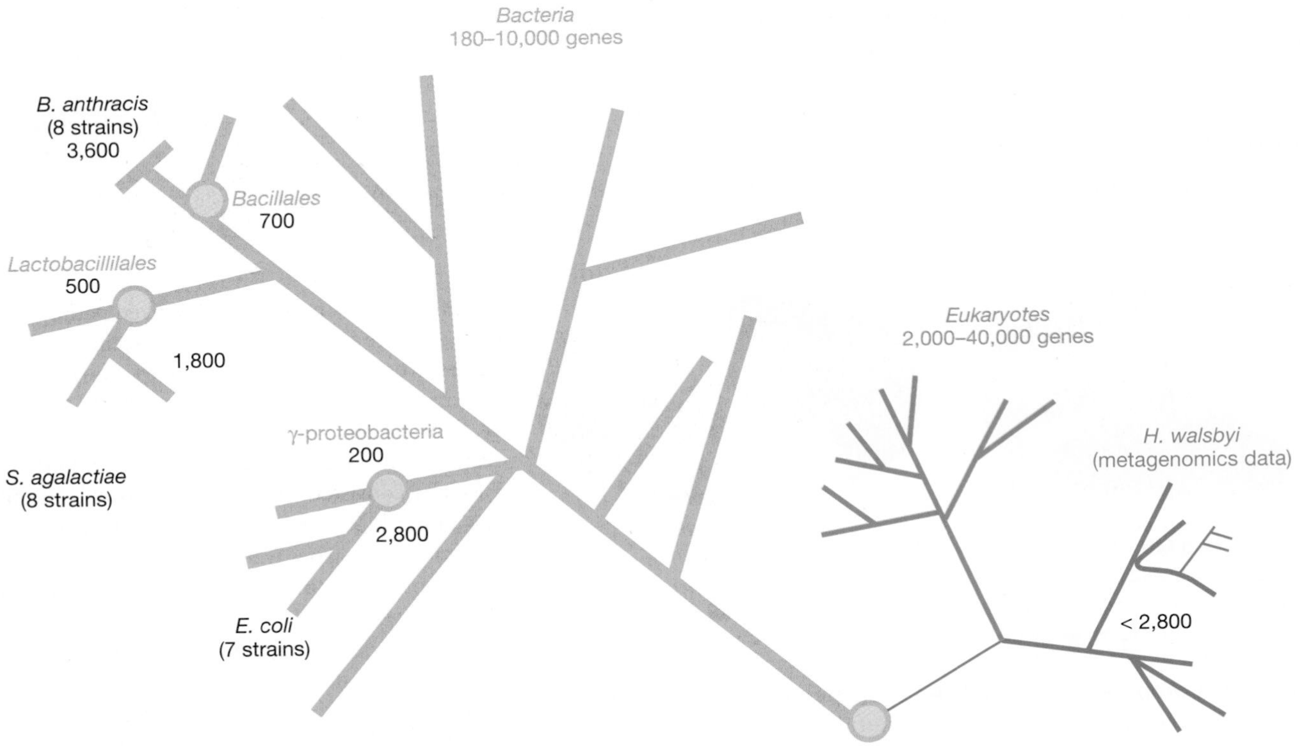

	Pan-genome	Core genome
B. anthracis	3,800	3,600
E. coli	6,000	2,800
S. agalactiae	2,700	1,800

FIGURE 16.18 Core and Pan-genomes. The number of genes in each core genome is listed at each branch point (node). The size of the pan-genomes for *E. coli, Streptococcus agalactiae, B. anthracis,* and *Haloquadratum walsbyi* are noted. A precise pan-genome size for *H. walsbyi* cannot be determined because only one strain has been sequenced, but the pan-genome is assumed to be less than the 2,800 genes found in this strain's genome.

Figure 16.18 Micro Inquiry

What can be inferred about the rate of horizontal gene transfer and the diversity of habitats when the size difference between the core and pan-genomes is compared?

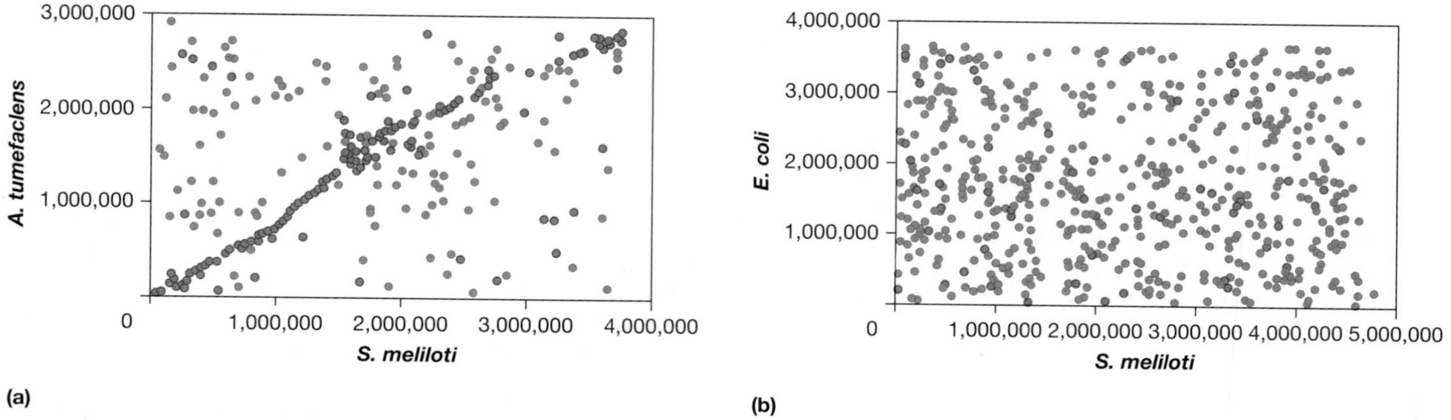

FIGURE 16.19 Synteny. Each axis on the graphs map the genes on each organism's genome. For instance, each number on the X axis in both (a) and (b) indicates the position of genes (blue dots) on the 3.7 Mb *S. meliloti* genome. (a) When the genomes of *S. meliloti* and *A. tumefaciens* are compared, it is evident that the distribution of orthologous genes on the genome is quite similar, as shown by the red line indicating similar placement on the genome (red dots represent *A. tumefaciens* genes). (b) *E. coli* and *S. meliloti* are not closely related and thus their genomes show no synteny (red dots represent *E. coli* genes).

***Figure 16.19** Micro Inquiry*

Do you think genes that show synteny are more likely to be part of the core or the pan-genome?

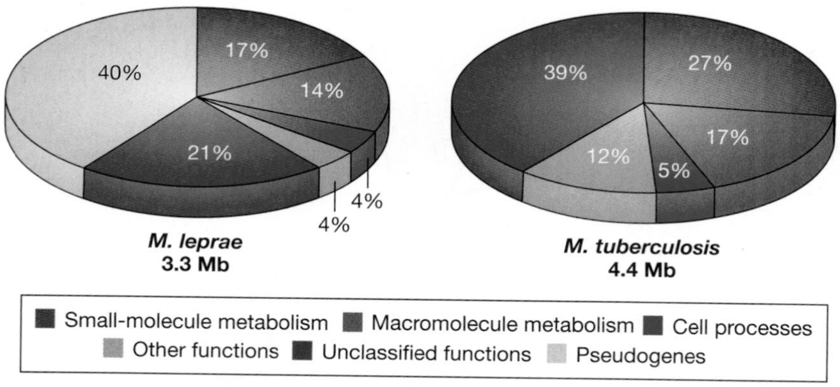

FIGURE 16.20 Comparative Genomic Analysis Between Two *Mycobacterium* species. The genome of *M. leprae* is significantly smaller than that of *M. tuberculosis.* Over one-third of its genes are nonfunctional pseudogenes that will presumably be lost as genome reduction continues.

why the bacterium has such a long doubling time, about two weeks in mice. One hope from genomic analysis is that critical surface proteins can be discovered and used to develop a sensitive test for early detection of leprosy. This would allow immediate treatment of the disease before nerve damage occurs. ▶▶| *Leprosy (section 38.3)*

Search This: TB database

Comparative genomics has given rise to the field of **reverse vaccinology**—the development of new vaccines comprised of only specific proteins produced by a given pathogen. As described more fully in chapter 36, vaccines present molecules from a given pathogen to the host immune system so that if the host encounters the living pathogen, the host's "immunological memory" prevents infection. Throughout most of the twentieth century, vaccines consisted of whole organisms—the microbe was either killed or mutated to lose pathogenicity. Alternative approaches were later introduced that involve the use of a single molecule (or molecules) produced by the pathogen; these are called subunit vaccines. However, for a variety of pathogens, none of these approaches results in an effective vaccine. This is the case if the pathogen cannot be grown in vitro (e.g., hepatitis B and C viruses), it produces molecules that are too similar to host molecules (e.g., serogroup B *Neisseria meningitidis,* also called MenB), or because there is too much genetic diversity among pathogenic strains (e.g., group B *Streptococcus,* also called GBS). GBS causes serious infections in infants so a vaccine is highly desired. In these cases, reverse vaccinology may be the only approach.

A molecule must possess certain features to be targeted by the host immune system. These molecules (antigens) must (1) be expressed by the pathogen during infection; (2) be either secreted or found on the surface of the pathogen; (3) be found in all strains of the pathogen; (4) elicit a host immune response; and (5) be essential for the survival of the pathogen, at least while it is in the host. Considering the thousands of ORFs that are revealed by each annotated genome, finding genes whose products meet all of these criteria might seem next to impossible. However, reverse vaccinology can rapidly create a short list of candidate protein targets. These antigens are then tested in a variety of assays to discover the best molecules for vaccine development.

There are two types of reverse vaccinology: Either a single species' genome is examined for vaccine targets, or a pan-genomic approach is taken so that the genomes of a number of strains are compared. The first approach was used to develop a MenB vaccine. Currently there are two vaccines made of capsular polysaccharides (CPS) available that protect against serogroups A, C, Y, and W135. Because the CPS of MenB is identical to a human polysaccharide, it is not included in these vaccines. By using the complete genome sequence of serogroup B *N. meningitidis*, 600 surface proteins were identified and about 300 were tested in mice. Of these about 25 were worthy of further study, and a MenB vaccine is now in clinical trials. (i.e., being tested on human volunteers).

Search This: Meningitis Research Foundation

A pan-genomic approach was used to develop a GBS vaccine because there are many genetically variable pathogenic strains. Thus the genomes of eight GBS isolates were compared and 312 surface proteins were identified as potential targets. When tested for their ability to protect mice from infection, four were found to be effective. These were combined into a single vaccine capable of protecting animals against infection by all known clinical strains. This vaccine is also in clinical trials. Reverse vaccinology is also being applied to the development of vaccines to protect against a variety of other pathogens including *Staphylococcus aureus, Chlamydophila pneumoniae,* and *Bacillus anthracis.*

Search This: JGI microbial genomics

1. Cite an infectious disease for which you think a systems biology approach would be appropriate. List three specific questions you would like to explore regarding this host-pathogen interaction.
2. Why do you think that, in general, the core genome consists of informational genes whereas the pan-genome consists of genes whose products are involved in metabolism?
3. For what types of microorganisms is extensive gene loss common? What is the most likely explanation for this phenomenon?

16.8 Metagenomics

It is clear that structural, functional, and comparative genomics have revolutionized the study of microbial physiology and genetics. Equally important is the growing field of environmental genomics, also called **metagenomics.** While the dominant role of microorganisms in driving the nutrient cycles that support life on Earth has long been recognized, efforts to comprehensively understand microbial communities have been stymied by the fact that only about 1% of all bacterial and archaeal species have so far been cultured in laboratories. New genomic techniques offer cultivation-independent approaches to studying microbial biodiversity. Environmental genomics can be used to take a census of microbial populations, as well as to discern the presence and abundance of certain classes of genes. That is to say, genomics can ask, "Who is there and what are they doing?" To do this, DNA fragments are extracted directly from the environment and cloned into plasmid vectors. In this way, a library of environmental DNA fragments can be maintained and amplified (**figure 16.21**). Alternatively, certain genes may be obtained by PCR amplification of DNA fragments derived from environmental samples. In either case, a stable source of nucleotide sequences is generated that reflects the diversity of microbes growing in nature, not just those that can be grown in the laboratory. The nucleotides are then sequenced and analyzed, or expressed in a microbial host and screened for a specific function, such as the production of novel antimicrobial compounds. ⏮ *The polymerase chain reaction (section 15.2);* ⏭ *Techniques for determining microbial taxonomy and phylogeny (section 17.3); Assessing microbial community activity (section 27.3)*

Metagenomics has dramatically transformed our understanding of microbial diversity. Metagenomic libraries prepared from DNA extracted from natural samples, such as seawater and soil, have generated billions of nucleotide sequences. Recently pyrosequencing, rather than traditional Sanger-based sequencing, has been used because it is not necessary to first clone all the DNA fragments. This avoids bias that may be introduced by cloning efficiency. When sequences are examined for the presence of genes that indicate taxonomy (e.g., 16S rRNA encoding genes), genomic species, called **phylotypes,** can be determined. As described in chapter 27, microarrays called phylochips have been developed so that community diversity can be assessed. Amazingly, a single gram of pristine soil harbors more phylotypes than the number of currently catalogued bacterial and archaeal species. When analyzed for function, the metabolic activities of a microbial community can be explored.

One field that metagenomics has revolutionized is marine microbiology. An average of 1 million microbial cells can be found per milliliter of seawater. While it has long been recognized that marine microbes account for the majority of the oceans' biomass, where they perform about half of the global photosynthesis, it has been difficult to study their taxonomic and metabolic diversity. Thousands of phylotypes, many of them unrelated to known sequences, have been identified from a number of marine ecosystems. In 2000 metagenomics led to the discovery of a new gene that encodes a protein in the rhodopsin family. Rhodopsins convert light energy directly into a transmembrane proton gradient, generating a proton motive force that fuels the production of ATP. It had long been held that the *Archaea* were the only microbes to produce a protein in the rhodopsin family. When rhodopsin-like genes were amplified from a variety of bacteria found in seawater, they were called proteorhodopsins because they were found in γ-proteobacteria. Since then, genes for proteorhodopsin have been found in other microbes as well. The nature of microbial metabolic diversity in the sea is now being reconsidered as it is estimated that

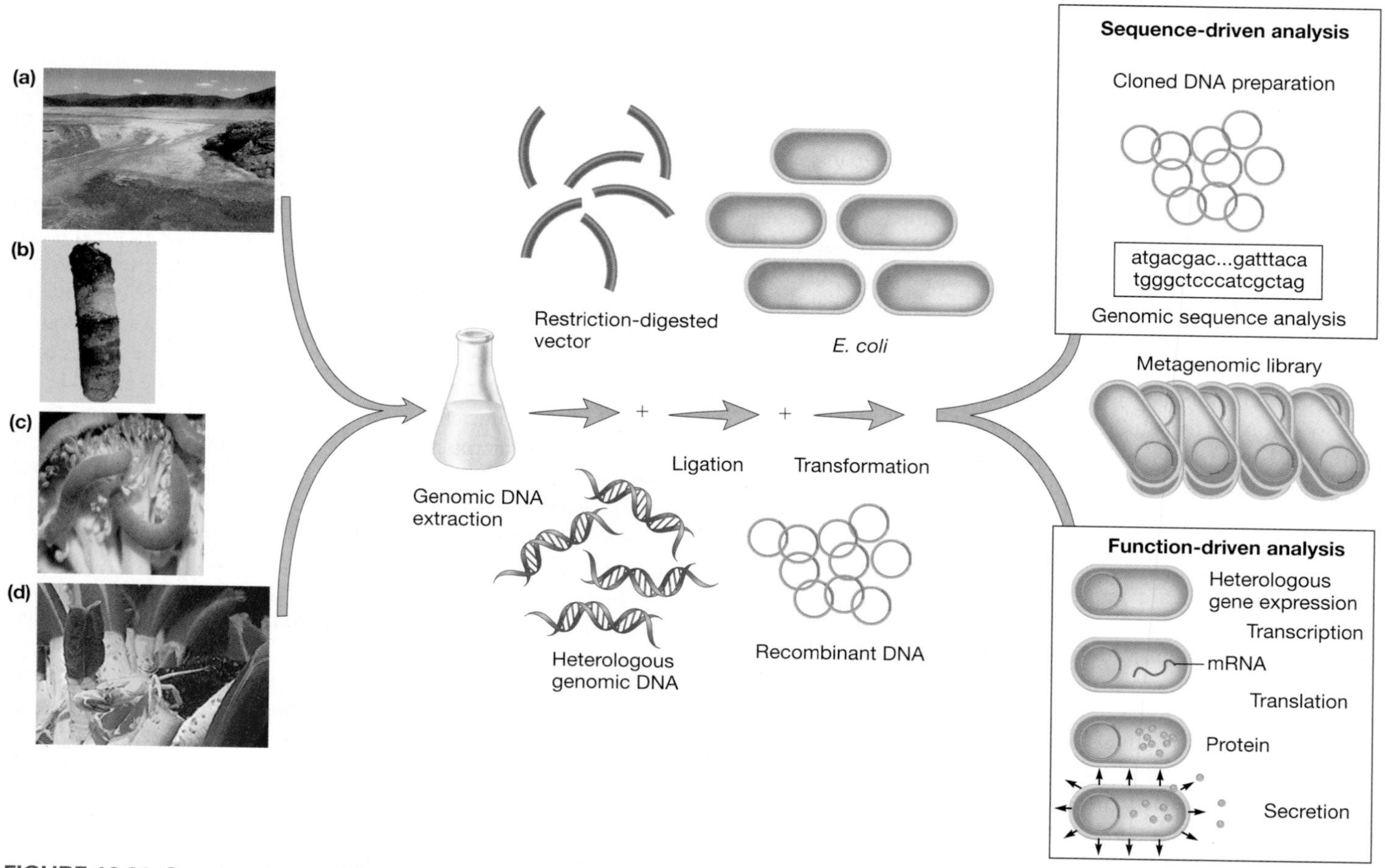

FIGURE 16.21 Construction and Screening of Genomic Libraries Directly from the Environment. DNA has been extracted directly from (a) bacterial mats at Yellowstone National Park, (b) soil samples from Alaska, (c) cabbage white butterfly larvae, and (d) tube worms from hydrothermal vents. The DNA is cloned into suitable vectors and transformed into a bacterial host. Sequences or gene products are then analyzed.

at least 13% of marine microbes may have the genes to encode rhodopsin-based, light-driven proton pumps. Perhaps it is not so surprising that proteorhodopsin genes are so pervasive because they were acquired by HGT. Only seven genes need be transferred to gain a functional proton pump that generates a proton motive force that can drive the synthesis of ATP. Thus these genes confer a phenotype for which there is strong natural selection. ⏮ *Phototrophy (section 10.12);* ⏭ *Phylum* Euryarchaeota *(section 18.3); Microorganisms in the open ocean (section 28.2)*

An ambitious metagenomics project was performed by J. Craig Venter, Hamilton Smith, and colleagues. They wanted to determine the bacterial and archaeal biodiversity of the Sargasso Sea, that portion of the Atlantic Ocean that surrounds Bermuda. They collected seawater and used filtration to exclude viruses (0.2 μm) and eukaryotes (0.8 μm). An environmental genomic library was prepared from DNA extracted from the remaining seawater. After sequencing over 1 billion base pairs, followed by manual and computer analysis to determine sequence relatedness, it was determined that at least 1,800 phylotypes were represented. Among these, about 145 phylotypes were previously unknown and most likely represent new species. Phylogenetic diversity was further evaluated by PCR amplification of genes, such as the recombinase gene *recA*, the 16S rRNA gene, and the gene that encodes RNA polymerase B (*rpoB*). The sequences for these genes are highly conserved, making them good candidates for assessing species diversity (**figure 16.22**). Amazingly, Venter, Smith, and their collaborators report the discovery of 1.2 million previously unknown genes (this number is controversial), including over 700 new proteorhodopsin-like photoreceptors from taxa not previously known to possess light-harvesting capabilities. These results demonstrate that the metagenomic analysis opens the door to more study so that we can ultimately fully appreciate the biological diversity in the world's oceans.

Another interesting metagenomics project is the sequencing of the human microbiome. The human **microbiome** represents all the genes present in the human genome and those present in the trillions of microbes living in and on adults. The microbiome is at least several hundred times larger than our own and encodes a variety of metabolic processes we lack, such as the

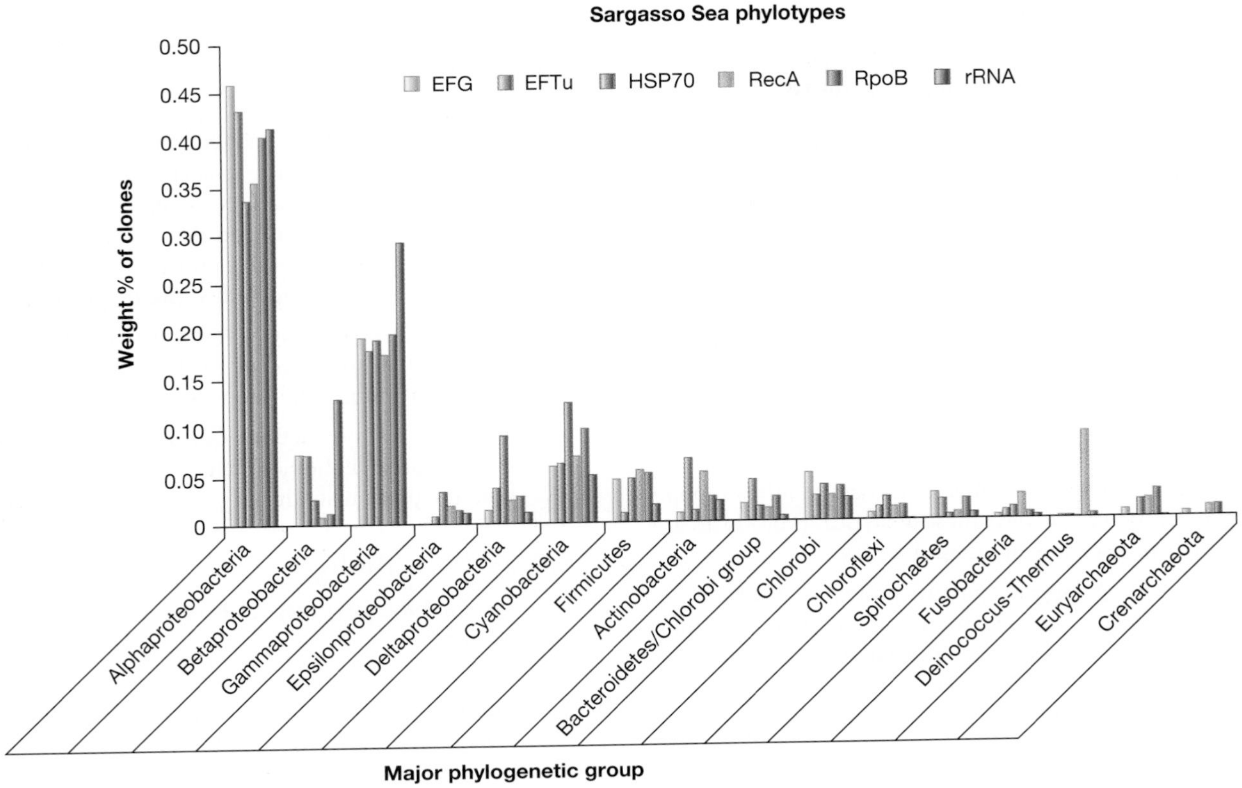

FIGURE 16.22 Phylogenetic Diversity of Sargasso Sea Microbes. The relative abundance (weight % of clones) of each group of microbes is shown according to the specific conserved gene that was used for genetic analysis. The genes used were those encoding elongation factor G (EFG), elongation factor Tu (EFTu), heat-shock protein 70 (HSP70), recombinase A (RecA), RNA polymerase B (RpoB), and the gene that encodes 16S rRNA.

synthesis of vitamin K and the degradation of a variety of otherwise indigestible food items. There are currently a number of microbiome sequencing projects, each focusing on one group of human microbial inhabitants, such as those residing in the colon, skin, or vagina. Clearly metagenomics is a powerful field that will change how we view the world and ourselves. ▶▶| *Normal microbiota of the human body (section 30.3)*

Search This: Human Microbiome Project

1. What is a phylotype?
2. Post-Sanger nucleotide sequencing techniques are considered well suited for metagenomics because a genomic library does not have to be constructed. Apart from convenience, explain why this is important for metagenomic analysis.
3. Examine figure 16.18. How might metagenomics be used to isolate a novel antifungal compound?

Summary

16.1 Determining DNA Sequences

a. Genomics is the study of the molecular organization of genomes, their information content, and the gene products they encode.

b. DNA fragments are normally sequenced using dideoxynucleotides and the Sanger chain termination technique (**figure 16.2**).

c. Recently developed DNA sequencing techniques seek to make DNA sequencing faster and cheaper. Three promising methodologies include pyrosequencing, SOLEXA sequencing, and sequencing by ligation (**figures 16.4** and **16.5**).

16.2 Genome Sequencing

a. Most often microbial genomes are sequenced using the whole-genome shotgun technique of Venter, Smith, and collaborators. Four stages are involved: library construction, sequencing of randomly produced fragments, fragment alignment and gap closure, and editing the final sequence (**figure 16.6**).

b. It is now possible to use the multiple displacement amplification technique to copy the genome of a single microbial cell so there is sufficient DNA for sequencing. In this way, the genomes of uncultivated microbes can be studied (**figure 16.7**).

16.3 Bioinformatics

a. Analysis of vast amounts of genome data requires sophisticated computer software; these analytical procedures are a part of the discipline of bioinformatics.

b. An open reading frame is a putative protein coding gene; the identity of the gene and its product are inferred by aligning the translated amino acid sequence with others like it in the database (**figures 16.8** and **16.9**).

c. Bioinformatics enables the comparison of genes within genomes to identify paralogues and of genes between different organisms to identify orthologues.

16.4 Functional Genomics

a. Functional genomics is used to reveal genome structure and function relationships.

b. The physical map of a microbe's genome is often displayed so that genes of common function are easily identified (**figure 16.10**).

c. The presence of genes whose products are known to function in specific metabolic processes such as catabolic pathways and transport enables the genomic reconstruction of the organism—a type of map that illustrates these processes within the cell (**figure 16.11**).

d. DNA microarrays can be used to assess gene expression as a measure of individual gene transcripts (mRNA). Gene expression can be determined for mutant versus wild-type strains or for a given organism under specific environmental conditions (**figure 16.12**).

16.5 Proteomics

a. The entire collection of proteins that an organism can produce is its proteome, and its study is called proteomics.

b. The proteome is often analyzed by two-dimensional gel electrophoresis, in which the total cellular protein pool can be visualized. In many cases, the amino acid sequence of individual proteins is determined by mass spectrometry; if this is coupled to genomics, both a protein of interest and the gene that encodes it can be identified (**figures 16.14** and **16.15**).

c. Structural proteomics seeks to model the three-dimensional structure of proteins based on computer analysis of amino acid sequence data.

d. Protein-DNA interactions can be analyzed by chromatin immunoprecipitation, which enables the identification of both the protein and its target DNA (**figure 16.16**).

16.6 Systems Biology

a. Systems biology is a relatively new interdisciplinary field of study.

b. Systems biology is the opposite of reductionist biology because it integrates molecular networks with the goal of understanding the function of the cell as a whole.

16.7 Comparative Genomics

a. Comparing genome sequences reveals information about genome structure and evolution, including the importance of lateral gene transfer.

b. The analysis of many microbial genomes has revealed the presence of a core genome and a pan-genome (**figure 16.18**).

c. Comparative genomics is an important tool in discerning how microbes have adapted to particular ecological niches and in developing new therapeutic agents.

16.8 Metagenomics

a. Metagenomics enables the study of the biodiversity and metabolic potential of microbial communities without culturing individual microbes (**figure 16.21**).

b. Metagenomics has been applied to many natural microbial habitats, including the human body.

Critical Thinking Questions

1. Propose an experiment that can be done easily with a DNA microarray that would have required years to do before this new technology.

2. It has been suggested that once the cost of post-Sanger sequencing techniques is further reduced, it may largely replace microarray analysis. What advantage, in terms of scientific inquiry, would genome sequencing have over microarray analysis? Can you think of questions that microarrays would be better suited to address?

3. You are developing a new vaccine for a pathogen. You want your vaccine to recognize specific cell-surface proteins. Explain how you will use genome analysis to identify potential

protein targets. What functional genomics approaches will you use to determine which of these proteins is produced when the pathogen is in its host?

4. The bacterial genus *Borrelia* is the cause of several tick-borne diseases. In addition, *B. recurrentis* causes louse-borne relapsing fever. In 2008 a team of scientists reported that the genome of *B. recurrentis* is a smaller, degraded version of the genome of *B. duttonii,* the cause of tick-borne relapsing fever. What do you think these researchers discovered when they compared these two genomes that enabled them to make this conclusion? In addition, it was found that *B. recurrentis* lacks genes encoding *recA* and *mutS*; refer to section 14.1 and explain why the loss of these genes may have accelerated the decay of this genome.

   Read the original paper: Lescot, M., et al. 2008. The genome of *Borrelia recurrentis,* the agent of deadly louse-borne relapsing fever, is a degraded subset of tick-borne *Borrelia duttonii. PLoS* 4:e1000185.

5. The archaeon *Sulfolobus acidocaldarius* was the third *Sulfolobus* species to have its genome sequenced. It was found to possess 2,292 predicted protein-coding genes, of which 305 are specific to *S. acidocaldarius,* and 866 are specific to the genus *Sulfolobus.* It was also observed that the *S. acidocaldarius* genome has a limited number of mobile genetic elements. What types of genes (informational or metabolic/transport) do you think are common to the genus, and what types of genes do you think are particular to this species? How do you think the small number of mobile genetic elements has influenced the evolution of this archaeon? How could you determine which other archaeal genera (with sequenced genomes) are closely related to *Sulfolobus*?

   Read the original paper: Chen, L., et al. 2005. The genome of *Sulfolobus acidocaldarius,* a model organism of the *Crenarchaeota. J. Bacteriol.* 187:4992.

Concept Mapping Challenge

Use the words listed below to construct a concept map that describes the ways in which a genome might be sequenced and analyzed. Provide your own linking words.

Whole-genome shotgun sequencing
Sanger DNA sequencing
Pyrosequencing
Paralogue
SOLiD DNA microarray
ORF
Post-Sanger DNA sequencing
SOLEXA
Orthologue
Genome annotation

Learn More

Learn more by visiting the text website at www.mhhe.com/willey8, where you will find a complete list of references.

17

Microbial Taxonomy and the Evolution of Diversity

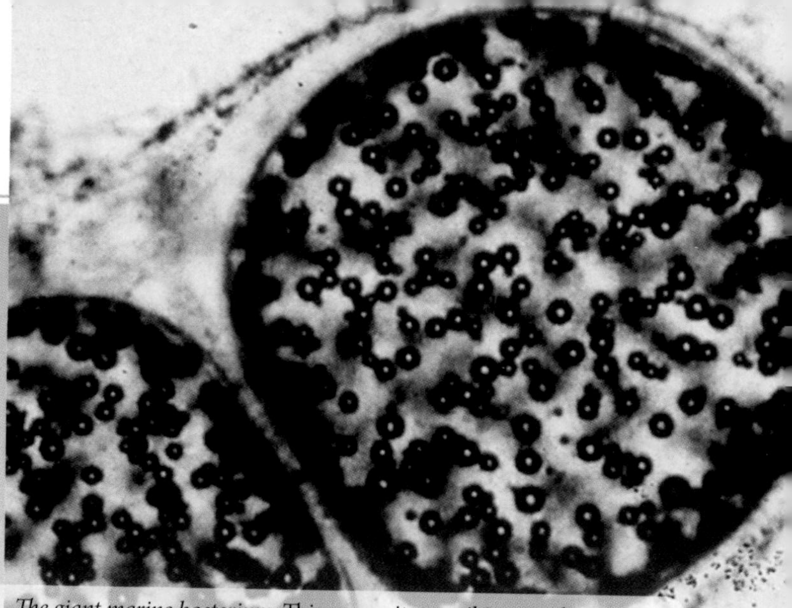

The giant marine bacterium, Thiomargarita namibiensis, *about 100 to 300 μm in diameter, accumulates sulfur from sediments in its refractive sulfur granules and nitrate from overlying waters to support its growth. Its discovery expanded an appreciation of microbial metabolic and morphological diversity.*

CHAPTER GLOSSARY

ecotype A population of microbes that is genetically very similar but ecologically distinct from others of the same taxon.

endosymbiotic hypothesis The hypothesis that mitochondria and chloroplasts arose when bacteria established an endosymbiotic relationship with ancestral cells and then evolved into organelles.

G + C content The percent of an organism's genome that consists of guanine and cytosine (G + C); a parameter that is used in taxonomic analysis.

genotypic classification The use of genetic data to construct a classification scheme for the identification of an unknown species or the phylogeny of a group of microbes.

melting temperature (T_m) The temperature at which double-stranded DNA becomes single stranded; determined by the G + C content of the genome.

monophyletic Refers to organisms that arose from a single common ancestor.

natural classification A classification system that arranges organisms into groups whose members share many characteristics.

oligonucleotide signature sequences Short, conserved nucleotide sequences that are specific for a phylogenetically defined group of organisms.

phenetic system A classification system that groups organisms together based on the similarity of their observable characteristics.

phylogenetic or phyletic classification system A classification system based on evolutionary relationships, rather than the general similarity of characteristics.

phylogenetic tree A graph made of nodes and branches, much like a tree in shape, that shows phylogenetic and evolutionary relationships between groups of organisms.

polyphasic taxonomy An approach in which taxonomic schemes are developed using a wide range of phenotypic and genotypic information.

RNA world The hypothesis that the first replicating entity was an RNA molecule.

strain Microbes that are descendants of a single, pure microbial culture. A single species may have many strains.

systematics The scientific study of organisms with the ultimate objective of characterizing and arranging them in an orderly manner; often considered synonymous with taxonomy.

taxon A group into which related organisms are classified.

taxonomy The science of biological classification; it consists of three parts: classification, nomenclature, and identification.

universal phylogenetic tree A phylogenetic tree that considers the evolutionary relationships among organisms from all three domains of life: *Bacteria, Archaea,* and *Eukarya*.

Perhaps the most fascinating aspect of the microbial world is its extraordinary variety. It seems that almost every possible shape, size, physiology, and lifestyle can be found. In this section of the text, we focus on microbial diversity. The classification of such a vast array of microorganisms has been fraught with challenges, controversies, and discoveries that continue to this day. Because of the unique nature of microbes, their scientific identification requires special consideration. In this chapter, we review the dynamic process of microbial classification. Because evolution is the foundation on which all classification methods currently used rest, we also discuss the evolution of microbes. This is followed by a seven-chapter (18–25) survey of the most important microbial groups.

17.1 Introduction to Microbial Taxonomy

Microbiologists are faced with the daunting task of understanding the diversity of life forms that cannot be seen with the naked eye but can live seemingly anywhere on Earth. One of the first tools needed to survey this level of diversity is a reliable classification system. **Taxonomy** (Greek *taxis,* arrangement or order, and *nomos,* law, or *nemein,* to distribute or govern) is defined as the science of biological classification. In a broader sense, it consists of three separate but interrelated parts: classification, nomenclature, and identification. Once a classification scheme is selected, it is used to arrange organisms into groups called taxa (s., **taxon**) based on mutual similarity. **Nomenclature** is the branch of taxonomy concerned with the assignment of names to taxonomic groups in agreement with published rules. Biologists use the binomial system described in section 17.2. Identification is the practical side of taxonomy—the process of determining if a particular isolate belongs to a recognized taxon and, if so, which one. The term **systematics** is often used for taxonomy. However, many taxonomists define systematics in more general terms as the scientific study of organisms with the ultimate objective of characterizing and arranging them in an orderly manner. Any study of the nature of organisms, when the knowledge gained is used in taxonomy, is a part of systematics. Thus systematics encompasses disciplines such as morphology, ecology, epidemiology, biochemistry, genetics, molecular biology, and physiology.

One of the oldest classification systems, called **natural classification,** arranges organisms into groups whose members share many characteristics. The Swedish botanist Carl von Linné, or Carolus Linnaeus as he often is called, developed the first natural classification in the middle of the eighteenth century. It was based largely on anatomical characteristics and was a great improvement over previously employed artificial systems because knowledge of an organism's position in the scheme provided information about many of its properties. For example, classification of humans as mammals denotes that they have hair, self-regulating body temperature, and milk-producing mammary glands in the female.

When natural classification is applied to higher organisms, evolutionary relationships become apparent simply because the morphology of a given structure (e.g., wings) in a variety of organisms (ducks, songbirds, hawks) suggests how that structure might have been modified to adapt to specific environments or behaviors. However, the traditional taxonomic assignment of microbes was not necessarily rooted in evolutionary relatedness. For instance, bacterial pathogens and microbes of industrial importance were historically given names that described the diseases they cause or the processes they perform (e.g., *Vibrio cholerae, Clostridium tetani,* and *Lactococcus lactis*). Although these labels are of practical use, they do little to guide the taxonomist concerned with the vast majority of microbes that are neither pathogenic nor of industrial consequence. Our present understanding of the evolutionary relationships among microbes now serves as the theoretical underpinning for taxonomic classification.

In practice, determining the genus and species of a newly isolated microbe is based on **polyphasic taxonomy.** This approach includes phenotypic, phylogenetic, and genotypic features. To understand how all of these data are incorporated into a coherent profile of taxonomic criteria, we must first consider the individual components.

Phenetic Classification

For a very long time, microbial taxonomists had to rely exclusively on a **phenetic system,** which classifies organisms according to mutual similarity of their phenotypic characteristics. This system succeeded in bringing order to biological diversity and clarified the function of morphological structures. For example, because motility and flagella are always associated in particular microorganisms, it is reasonable to suppose that flagella are involved in at least some types of motility. Although phenetic studies can reveal possible evolutionary relationships, this is not always the case. Many traits are compared without assuming that some might be more phylogenetically important than others; that is, unweighted traits are employed in estimating general similarity. Obviously the best phenetic classification is one constructed by comparing as many attributes as possible. Organisms sharing many characteristics make up a single phenetic group or taxon.

Phylogenetic Classification

With the publication in 1859 of Charles Darwin's *On the Origin of Species,* biologists began developing **phylogenetic** or **phyletic classification systems** that sought to compare organisms on the basis of evolutionary relationships. The term **phylogeny** (Greek *phylon,* tribe or race, and *genesis,* generation or origin) refers to the evolutionary development of a species. Scientists realized that when they observed differences and similarities between organisms as a result of evolutionary processes, they also gained insight into the history of life on Earth. However, for much of the twentieth century, microbiologists could not effectively employ phylogenetic classification systems, primarily because of the lack of a good fossil record. When Carl Woese and George Fox proposed using small subunit (SSU) rRNA nucleotide sequences to assess evolutionary relationships among microorganisms, the door opened to the resolution of long-standing inquiries regarding the origin and evolution of the majority of life forms on Earth—the microbes. The validity of this approach is now widely accepted, and there are currently over 500,000 different 16S and 18S rRNA sequences in the international databases GenBank and the Ribosomal Database Project (RDP-II). As discussed later (p. 452), the power of rRNA as a phylogenetic and taxonomic tool rests on the features of the rRNA molecule that make it a good indicator of evolutionary history and on the ever-increasing size of the rRNA sequence database.

Search This: Ribosomal Database Project

Genotypic Classification

Currently the genotype of a microbe can be evaluated in taxonomic terms in many ways. In general, **genotypic classification** seeks to compare the genetic similarity between organisms. Individual genes or whole genomes can be compared. Since the 1970s it has been widely accepted that prokaryotes whose genomes are at least 70% homologous belong to the same species. Unfortunately this 70% threshold value was established to avoid disrupting existing species assignments; it is not based on theoretical considerations of species identity. Fortunately genetic data obtained using newer molecular approaches often concur with these older assignments. The means by which microbes are genotypically classified is discussed further in section 17.3.

1. What is a natural classification? What microbial features might have been considered when devising a natural classification scheme?
2. What is polyphasic taxonomy and what three types of data does it consider? Do you think each type of data should be of equal weight?
3. Consider the finding that bacteria capable of anoxygenic photosynthesis belong to several different phylogenetic lineages. How do you think these bacteria were originally classified, and what types of data do you think were key in making the most recent taxonomic assignments?

17.2 Taxonomic Ranks

The definition of a bacterial or archaeal species is widely debated, as discussed in section 17.5. Nonetheless, for practical reasons it is essential that the established rules of taxonomy are followed. Microbes are placed in hierarchical taxonomic levels, with each level or rank sharing a common set of specific features. The ranks are arranged in a nonoverlapping hierarchy so that each level includes not only the traits that define the rank above it but also a new set of more restrictive traits (**figure 17.1**). The highest rank is the domain; *Bacteria* and *Archaea* consist only of microbes, while the *Eukarya* includes both micro- and macroorganisms. Within each domain, each organism is assigned (in descending order) to a phylum, class, order, family, genus, and species epithet. Some microbes are also given a subspecies designation. Microbial groups at each level have a specific suffix indicative of that rank or level. Microbiologists often use informal names in place of formal, hierarchical ones. Typical examples of such names are purple bacteria, spirochetes, methane-oxidizing bacteria, sulfate-reducing bacteria, and lactic acid bacteria. As we shall see, these informal names may not have taxonomic significance as they can include species from several phyla. A good example of this is the sulfur bacteria.

The most fundamental definition of a bacterial or archaeal **species** is a collection of strains that share many stable properties and differ significantly from other groups of strains. A **strain** consists of the descendants of a single, pure microbial culture. Strains within a species may be described in a number of different ways. **Biovars** are variant strains characterized by biochemical or physiological

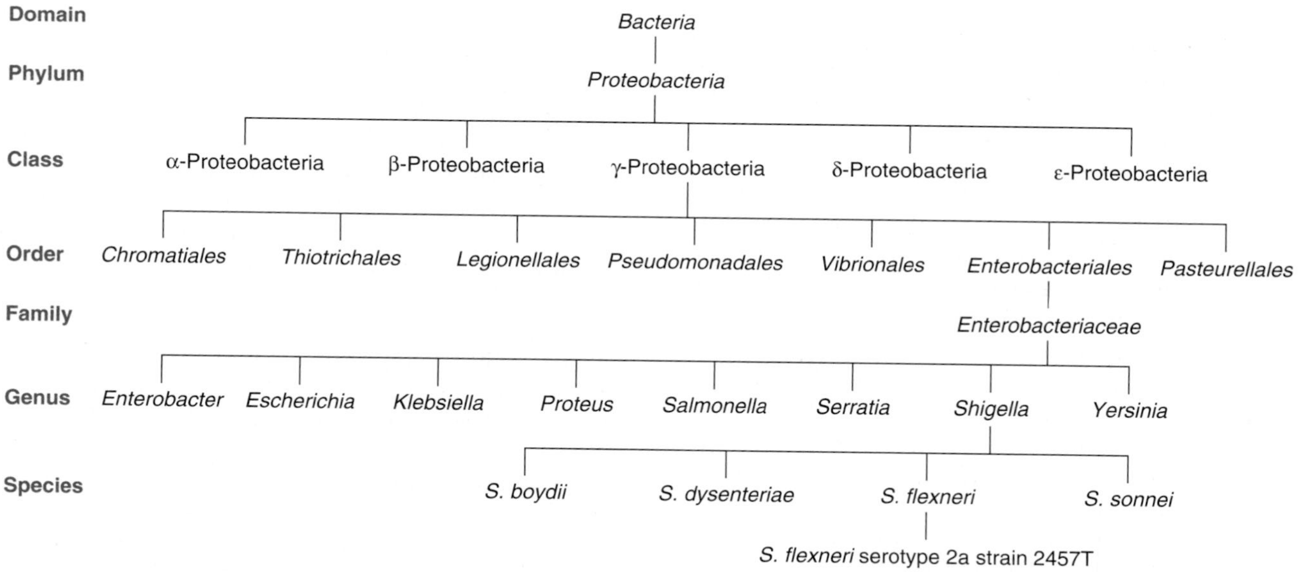

FIGURE 17.1 Hierarchical Arrangement in Taxonomy. In this example, members of the genus *Shigella* are placed within higher taxonomic ranks. Not all classification possibilities are given for each rank to simplify the diagram. Note that *-ales* denotes order and *-ceae* indicates family.

differences, **morphovars** differ morphologically, and **serovars** have distinctive antigenic properties. For each species, one strain is designated as the **type strain.** It is usually one of the first strains studied and often is more fully characterized than other strains; however, it does not have to be the most representative member. The type strain for the species is considered the type species and is the holder of the species name. This ensures permanence of names when nomenclature revisions occur because the type species must remain within the original species. Only those strains very similar to the type strain or type species are included in a species. Each species is assigned to a **genus,** the next rank in the taxonomic hierarchy. A genus is a well-defined group of one or more species that is clearly separate from other genera. In practice, considerable subjectivity occurs in assigning species to a genus, and taxonomists may disagree about the composition of genera.

Microbiologists name microorganisms by using the **binomial system** of Linnaeus. The Latinized, italicized name consists of two parts. The first part is the generic name (i.e., the genus), and the second is species name (e.g., *Yersinia pestis,* the causative agent of plague). The species name is stable; the oldest epithet for a particular organism takes precedence and must be used. In contrast, a generic name can change if the organism is assigned to another genus because of new information. For example, some members of the genus *Streptococcus* were placed into two new genera, *Enterococcus* and *Lactococcus,* based on rRNA analysis and other characteristics. Thus *Streptococcus faecalis* is now *Enterococcus faecalis.* To be recognized as a new species, genomic, metabolic, morphological, reproductive, and ecological data must be accepted and published in the *International Journal of Systematic and Evolutionary Microbiology;* until that time, the new species name will appear in quotation marks. Microbes that have not been grown in pure culture but for which there is sufficient genetic characterization may be given a provisional genus and species name preceded by the term *Candidatus,* meaning candidate. For instance, a novel aerobic phototrophic member of the phylum *Acidobacterium* has been grown only in coculture with another isolate. It has been given the provisional name *Candidatus* Chloracidobacterium thermophilum (note that the genus and species are not italicized). *Bergey's Manual of Systematic Bacteriology* contains only recognized bacterial and archaeal species and is discussed in section 17.6. ▶▶| *Photosynthetic bacteria (section 19.3)*

1. What is the difference between a microbial species and a strain?
2. Define morphovar, serovar, and type strain.
3. The genus *Salmonella* was once thought to contain five species. Most scientists now consider only two species valid: *S. bongori* and *S. enterica.* The latter contains six subspecies, and *S. enterica* subspecies *enterica* is further subdivided into eight serovars. Three of these, Enteritidis, Typhi, and Typhimurium, were once considered *Salmonella* species. (a) Inspect figure 17.1 and construct a similar lineage. (b) What criteria do you think were used to make these taxonomic changes?

17.3 Techniques for Determining Microbial Taxonomy and Phylogeny

Many different approaches are used in classifying and identifying microorganisms that have been isolated and grown in pure culture. For clarity, we divide them into two groups: classical and molecular. The most durable identifications are those that are based on a combination of approaches. Methods often employed in routine laboratory identification of bacteria are covered in chapter 35, Clinical Microbiology and Immunology.

Classical Characteristics

Classical approaches to taxonomy make use of morphological, physiological, biochemical, and ecological characteristics. These characteristics have been employed in microbial taxonomy for many years and form the basis for phenetic classification. When used in combination, they are quite useful in routine identification and may provide phylogenetic information as well.

Morphological Characteristics

Morphological features are important in microbial taxonomy for many reasons. Morphology is easy to study and analyze, particularly in eukaryotic microorganisms and more complex bacteria and archaea. In addition, morphological comparisons are valuable because structural features depend on the expression of many genes, are usually genetically stable, and normally (at least in eukaryotes) do not vary greatly with environmental changes. Thus morphological similarity often is a good indication of phylogenetic relatedness.

Many different morphological features are employed in the classification and identification of microorganisms (**table 17.1**). Although the light microscope has always been a very important tool, its resolution limit of about 0.2 μm reduces its usefulness in viewing smaller microorganisms and structures. The transmission and scanning electron microscopes, with their greater resolution, have immensely aided the study of all microbial groups. |◀◀ *Microscopy (chapter 2)*

Physiological and Metabolic Characteristics

Physiological and metabolic characteristics are very useful because they are directly related to the nature and activity of microbial enzymes and transport proteins. Because proteins are gene products, analysis of these characteristics provides an indirect comparison of microbial genomes. **Table 17.2** lists some of the most important of these properties.

Biochemical Characteristics

Among the more useful biochemical characteristics used in microbial taxonomy is the analysis of bacterial fatty acids using a technique called *f*atty *a*cid *m*ethyl *e*ster (FAME) analysis. A fatty

Table 17.1 Some Morphological Features Used in Classification and Identification

Feature	*Microbial Groups*
Cell shape	All major groups[a]
Cell size	All major groups
Colonial morphology	All major groups
Ultrastructural characteristics	All major groups
Staining behavior	Bacteria, some fungi
Cilia and flagella	All major groups
Mechanism of motility	Gliding bacteria, spirochetes, protists
Endospore shape and location	Endospore-forming bacteria
Spore morphology and location	Bacteria, protists, fungi
Cellular inclusions	All major groups
Colony color	All major groups

[a]Used in classifying and identifying at least some bacteria, archaea, fungi, and protists.

acid profile can reveal specific differences in chain length, degree of saturation, branched chains, and hydroxyl groups. Microbes of the same species will have identical fatty acid profiles, provided they are grown under the same conditions. This is an important caveat: careful attention must be paid to growth conditions because these will alter the fatty acid composition. Obviously this limits FAME analysis to only those microbes that can be grown in pure culture. Finally, because the identification of a species is done by comparing the results of the unknown microbe in question with the FAME profile of other, known microbes, identification is only possible if the species in question has been previously analyzed. Nonetheless, FAME analysis is particularly important in public health, food and water microbiology. In these applications, microbiologists seek to identify members of a specific group of bacterial pathogens. ▶▶| *Microbiology of food (chapter 40); Water purification and sanitary analysis (section 42.1)*

Ecological Characteristics

The ability of a microorganism to colonize a specific environment is of taxonomic value. Some microbes may be very similar in many other respects but inhabit different ecological niches, suggesting they may not be as closely related as first suspected. Some examples of taxonomically important ecological properties are life cycle patterns; the nature of symbiotic relationships; the ability to cause disease in a particular host; and habitat preferences such as requirements for temperature, pH, oxygen, and osmotic concentration. Many growth requirements are considered physiological characteristics as well (table 17.2). |◀◀ *Influences of environmental factors on growth (section 7.6);* ▶▶| *Microbial interactions (section 30.1)*

Table 17.2 Some Physiological and Metabolic Characteristics Used in Classification and Identification

Carbon and nitrogen sources
Cell wall constituents
Energy sources
Fermentation products
General nutritional type
Growth temperature optimum and range
Luminescence
Mechanisms of energy conversion
Motility
Osmotic tolerance
Oxygen relationships
pH optimum and growth range
Photosynthetic pigments
Salt requirements and tolerance
Secondary metabolites formed
Sensitivity to metabolic inhibitors and antibiotics
Storage inclusions

Molecular Characteristics

It is hard to overestimate how the study of DNA, RNA, and proteins has advanced our understanding of microbial evolution and taxonomy. Evolutionary biologists studying plants and animals draw from a rich fossil record to assemble a history of morphological changes; in these cases, molecular approaches supplement such data. In contrast, microorganisms have left almost no fossil record, so molecular analysis is the only feasible means of collecting a large and accurate data set that explores microbial evolution. When scientists are careful to make only valid comparisons, phylogenetic inferences based on molecular approaches provide the most robust analysis of microbial evolution.

Nucleic Acid Base Composition

Microbial genomes can be directly compared, and taxonomic similarity can be estimated in many ways. The first, and possibly the simplest, technique to be employed is the determination of DNA base composition. Recall that DNA contains four purine and pyrimidine bases: adenine (A), guanine (G), cytosine (C), and thymine (T). Base-pairing rules dictate that the (G + C)/(A+ T) ratio or **G + C content**—the percent of G + C in DNA—reflects the base sequence and varies with sequence changes as follows:

$$\text{Mol\% G + C} = \frac{G + C}{G + C + A + T} \times 100$$

The base composition of DNA can be determined in several ways. Although the G + C content can be ascertained after hydrolysis of DNA and analysis of its bases with high-performance liquid chromatography (HPLC), physical methods are easier and more often used. The G + C content often is determined from the **melting temperature (T_m)** of DNA. In double-stranded DNA, three hydrogen bonds join GC base pairs and two bonds connect AT base pairs. As a result, DNA with a greater G + C content has more hydrogen bonds, and its strands separate at higher temperatures; that is, it has a higher melting point. For example, compare *Mycoplasma hominis* with a Mol% G + C of about 29 and a T_m of 65°C with *Micrococcus luteus*, which is almost 79% G + C with a T_m of 85°C.

DNA melting can be easily followed spectrophotometrically because the absorbance of DNA at 260 nm increases during strand separation. When a DNA sample is slowly heated, the absorbance increases as hydrogen bonds are broken and reaches a plateau when all the DNA has become single stranded (**figure 17.2**). The midpoint of the rising curve gives the melting temperature, a direct measure of the G + C content.

The G + C content of thousands of organisms has been determined. The G + C content of DNA from animals and higher plants averages around 40% and ranges between 30 and 50%. In contrast, the G + C content of microorganisms varies greatly. Bacterial and archaeal G + C content ranges from about 25 to almost 80% and is more variable than that of fungi and protists. Despite such a wide range of variation, the G + C content of strains within a particular species is constant and varies very little within a genus (**table 17.3**). If two organisms differ in their G + C content by more than about 10%, their genomes have quite different base sequences, indicating that they are not closely related. On the other hand, it is not safe to assume that organisms with very similar G + C contents also have similar DNA base sequences because two very different base sequences can be constructed from the same proportions of A + T and G + C base pairs. Only if two microorganisms also are alike phenotypically does their similar G + C content suggest close relatedness.

Search This: G + C T_m calculator

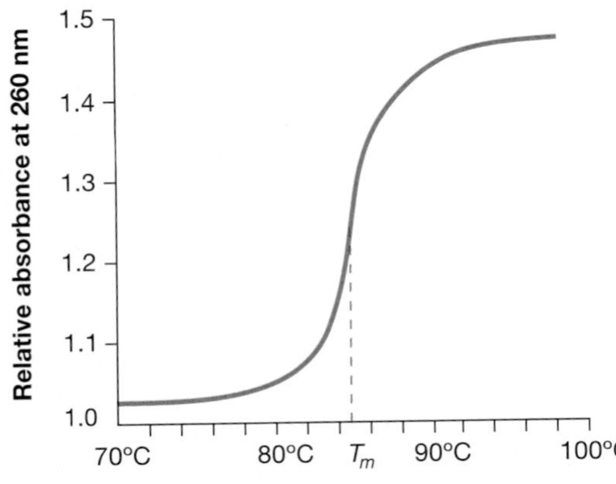

FIGURE 17.2 A DNA Melting Curve. The T_m is indicated.

***Figure 17.2* Micro Inquiry**

Would this curve be shifted to the left or the right for a microbe with an exceptionally low G + C composition? Explain your answer.

Nucleic Acid Hybridization

The similarity between genomes can be compared more directly by use of nucleic acid hybridization studies, also called **DNA-DNA hybridization.** If the genomes of two microbial isolates are heated to become single-stranded (ss) DNA and then cooled and held at a temperature about 25°C below the T_m, strands with complementary base sequences will reassociate to form stable dsDNA. However, noncomplementary strands will remain unpaired. Because strands with similar but not identical sequences associate to form less temperature-stable dsDNA hybrids, incubation of the mixture at 30 to 50°C below the T_m allows hybrids of more diverse ssDNAs to form. Incubation at 10 to 15°C below the T_m permits hybrid formation only with almost identical strands.

In one of the more widely used hybridization techniques, nylon filters with bound nonradioactive DNA strands are incubated at the appropriate temperature with radioactive ssDNA fragments. After the fragments hybridize with the membrane-bound ssDNA, the membrane is washed to remove any nonhybridized ssDNA and its radioactivity is measured. The quantity of radioactivity bound to the filter reflects the amount of hybridization and thus the similarity of the DNA sequences. The degree of similarity or homology is expressed as the percent of experimental DNA radioactivity retained on the filter compared with the percent of homologous DNA radioactivity bound under the same conditions (**table 17.4** provides examples). Two strains whose DNAs show at least 70% relatedness under optimal hybridization conditions and less than a 5% difference in T_m often, but not always, are considered members of the same species.

If DNA molecules are very different in sequence, they will not form a stable, detectable hybrid. Therefore DNA-DNA hybridization is used to study only closely related microorganisms. More distantly related organisms can be compared by carrying out DNA-RNA hybridization experiments using radioactive ribosomal or transfer RNA. Distant relationships can be detected because rRNA and tRNA genes represent only a small portion of the total DNA genome and have not evolved as rapidly as most other microbial genes. The technique is similar to that employed for DNA-DNA hybridization: membrane-bound DNA is incubated with radioactive rRNA, washed, and counted. ◀◀ *tRNA and rRNA genes (section 12.5)* *DNA Probe (DNA-DNA Hybridization)*

Table 17.3 Representative G + C Content of Microorganisms

Organism	Percent G + C	Organism	Percent G + C	Organism	Percent G + C
Bacteria		*Streptococcus*	33–44	*Paramecium* spp.	29–39
Anabaena	39–44	*Streptomyces*	69–73	*Trichomonas*	29–34
Caulobacter	62–65	**Archaea**		*Trypanosoma*	45–59
Chlamydia	41–44	*Halobacterium*	66–68	**Fungi**	
Chlorobium	49–58	*Methanobacterium*	32–50	*Agaricus bisporus*	44
Deinococcus	62–70	*Sulfolobus*	31–37	*Amanita muscaria*	57
Escherichia	48–59	**Protists**		*Aspergillus niger*	52
Mycobacterium	62–70	*Acetabularia mediterranea*	37–53	*Candida albicans*	33–35
Myxococcus	68–71	*Amoeba proteus*	66	*Coprinus lagopus*	52–53
Neisseria	48–56	*Chlamydomonas*	60–68	*Mucor rouxii*	38
Pseudomonas	58–69	*Chlorella*	43–79	*Neurospora crassa*	52–54
Rhodospirillum	62–66	*Dictyostelium*	22–25	*Rhizopus nigricans*	47
Staphylococcus	30–38	*Euglena gracilis*	46–55	*Saccharomyces cerevisiae*	36–42

Table 17.4 Comparison of *Neisseria* Species by DNA Hybridization Experiments

Membrane-Attached DNA[a]	*Percent Homology*[b]
Neisseria meningitidis	100
N. gonorrhoeae	78
N. sicca	45
N. flava	35

Source: Data from Staley, T. E., and Colwell, R. R. 1973. Applications of molecular genetics and numerical taxonomy to the classification of bacteria. Annu. Rev. Ecol. and Systematics, *8:282.*

[a]The experimental membrane-attached nonradioactive DNA from each species was incubated with radioactive *N. meningitidis* DNA, and the amount of radioactivity bound to the membrane was measured. The more radioactivity bound, the greater the homology between DNA sequences.

[b]*N. meningitidis* DNA bound to experimental DNA/Amount bound to membrane attached *N. meningitidis* DNA × 100

Nucleic Acid Sequencing

Despite the usefulness of G + C content determination and nucleic acid hybridization studies, rRNAs from small ribosomal subunits (16S from bacterial and archaeal cells and 18S from eukaryotes) have become the molecules of choice for inferring microbial phylogenies and making taxonomic assignments at the genus level. The **small subunit rRNAs (SSU rRNAs)** are almost ideal for studies of microbial evolution, relatedness, and genus identification because they play the same role in all microorganisms. In addition, because the ribosome is absolutely necessary for survival and the SSU rRNAs are part of the complex ribosome structure, the genes encoding these rRNAs cannot tolerate large mutations. Thus these genes change very slowly with time and do not appear to be subject to horizontal gene transfer, an important factor in comparing sequences for phylogenetic purposes. The utility of SSU rRNAs is extended by the presence of certain sequences that are variable among organisms and other regions that are quite similar. The variable regions enable comparison between closely related microbes, whereas the stable sequences allow the comparison of distantly related microorganisms.

Comparative analysis of SSU rRNA sequences from thousands of organisms has demonstrated the presence of **oligonucleotide signature sequences** (**figure 17.3**). These are short, conserved nucleotide sequences that are specific for a phylogenetically defined group of organisms. Thus the signature sequences found in *Bacteria* are rarely or never found in *Archaea* and vice versa. Likewise, the 18S rRNA of eukaryotes bears signature sequences that are specific to the domain *Eukarya.* Either complete rRNAs or, more often, specific rRNA fragments can be compared.

The proper alignment of SSU rRNA nucleotide sequences and the application of computer algorithms enable sequence comparison between any number of organisms. The ability to amplify regions of rRNA genes (rDNA) by the polymerase chain reaction (PCR) and sequence the DNA using automated sequencing technology has greatly increased the efficiency by which SSU rRNA sequences can be obtained. PCR can be used to amplify rDNA from the genomes of organisms because conserved nucleotide sequences flank the regions of interest. In practice, this means that PCR primers are readily available or can be generated to amplify rDNA from both cultured and uncultured microbes. The validity of SSU rRNA sequences as a taxonomic marker has been evaluated in at least two ways. First, analysis of thousands of

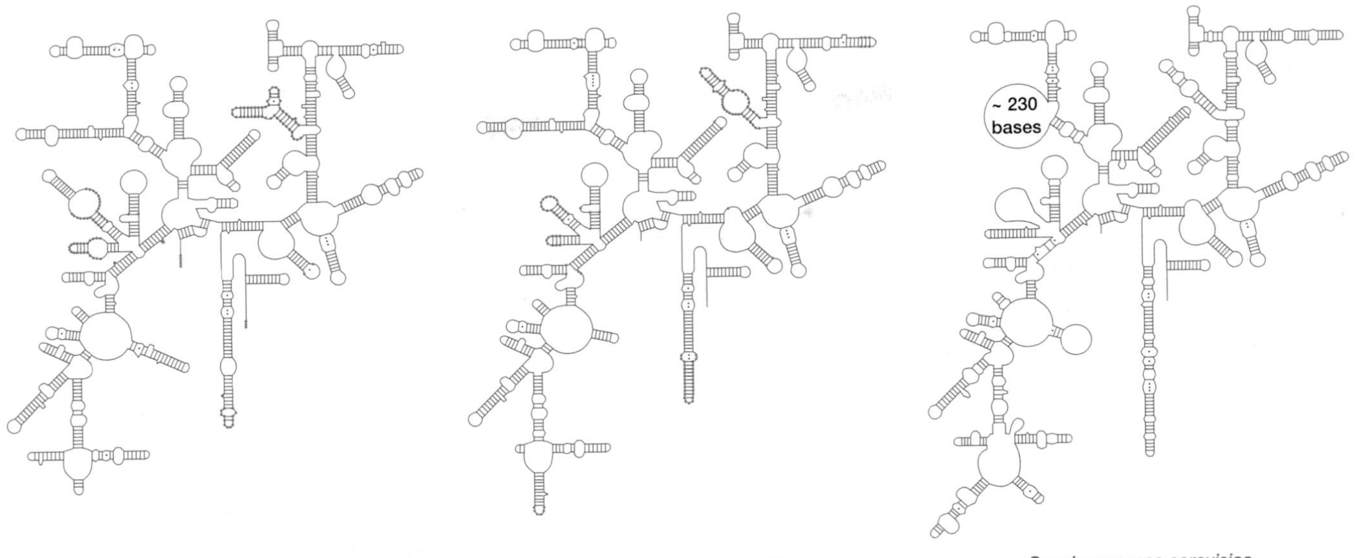

FIGURE 17.3 Small Ribosomal Subunit RNA. Representative examples of rRNA secondary structures from the three domains: *Bacteria (Escherichia coli), Archaea (Methanococcus vannielii)*, and *Eukarya (Saccharomyces cerevisiae)*. The red dots mark positions where *Bacteria* and *Archaea* normally differ.

Source: Data from Woese, C. P. 1987. Microbiological Rev. *51(2):221–27.*

bacterial and archaeal 16S rRNA sequences reveals that no two microbes with 98.5% 16S rRNA sequence identity have been found with less than the 70% DNA-DNA hybridization cutoff value for the operational definition of a microbial species. Second, several genome-based studies have demonstrated that 16S rRNA is a valid and robust measurement of taxonomic relatedness. ◀◀ *Polymerase chain reaction (section 15.2); Determining DNA sequences (section 16.1)*

Signature sequences are present in genes other than those encoding rRNA. Many genes have insertions or deletions of specific lengths and sequences at fixed positions. A particular insertion or deletion may be found exclusively among all members of one or more phyla; these are called conserved **indels** (for *in*sertion/*del*etion). These signature sequences are particularly useful in phylogenetic studies when they are flanked by conserved regions. In such cases, observed changes in the signature sequence cannot be due to sequence misalignments. The signature sequences located in some highly conserved housekeeping genes do not appear greatly affected by horizontal gene transfers and, like SSU rRNA, can be employed in phylogenetic analysis.

Genomic Fingerprinting

The molecular techniques discussed so far can be used to identify a bacterium or archaeon to the genus level. To identify a microbial species, a group of techniques called **genomic fingerprinting** can be used. Genomic fingerprinting requires the evaluation of genes that evolve more quickly than those that encode rRNA. In fact, rather than using a single gene, five to seven conserved housekeeping genes can be sequenced and compared in a technique called **multilocus sequence analysis (MLSA).** Multiple genes are examined to avoid misleading results that can arise through horizontal gene transfer. MLSA compares orthologous genes from a number of strains that may not belong to the same genus. Because many, many different versions, or alleles, of each gene can exist, the finding that two microbial isolates share the same alleles for multiple genes is very strong evidence that the two strains are closely related, perhaps even the same strain. MLSA was derived from **multilocus sequence typing (MLST),** which was originally developed to discriminate among strains belonging to the same pathogenic species. MLSA differs from MLST because its broad application to microbial taxonomy requires the comparison of genes from a more heterogeneous collection of microbes.

Another form of genetic fingerprinting called **restriction fragment length polymorphism (RFLP)** analysis relies on the capacity of restriction endonucleases to recognize specific nucleotide sequences. Specific genes from the microbes in question are amplified by PCR and cut by particular restriction enzymes. The DNA is then examined by gel electrophoresis (*see figure 15.9 and table 15.2*). Thus RFLP analysis uses changes in restriction enzyme recognition sequences as a means to detect differences in nucleotide sequence among microbial strains. Microbes with the same pattern of DNA fragments, called restriction fragments, are probably very closely related. The most common gene amplified and subjected to RFLP analysis is the 16S rRNA gene. When the similarity between rRNA genes is determined by restriction length polymorphisms rather than direct nucleotide sequencing,

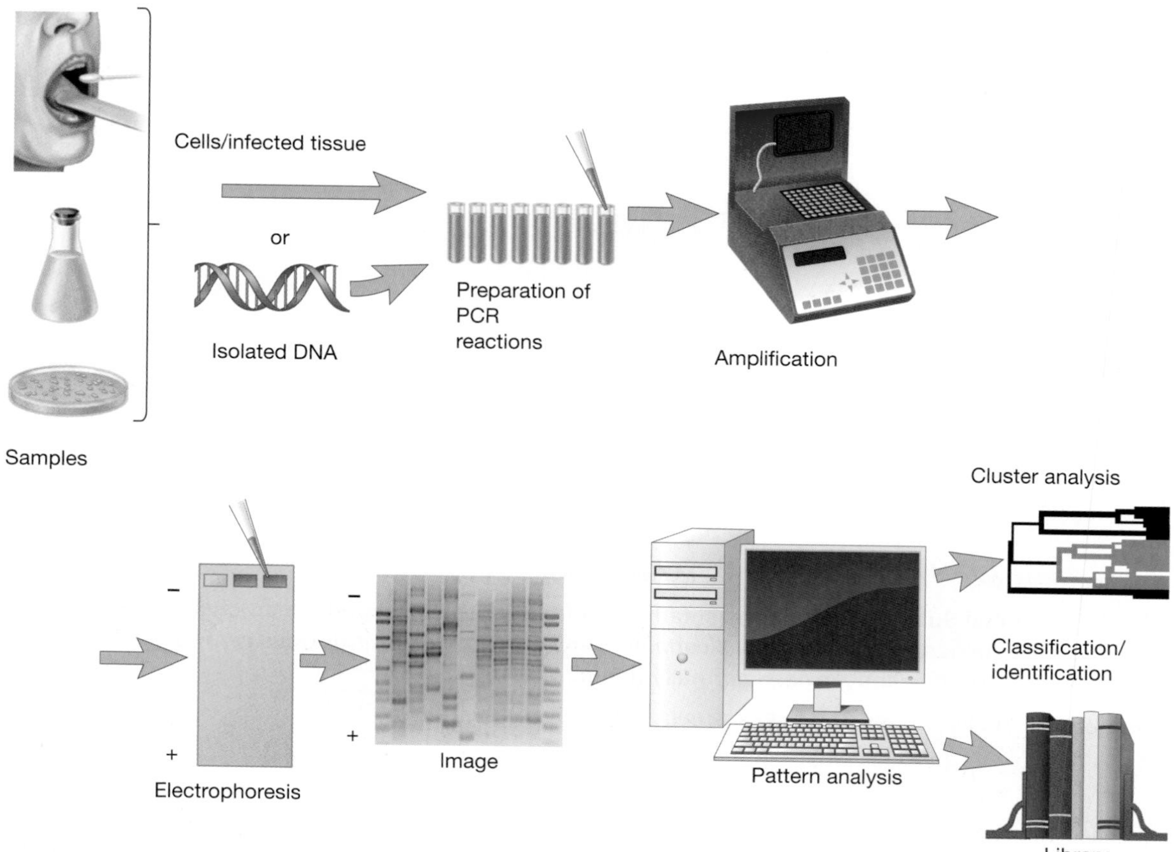

FIGURE 17.4 An Overview of the Genomic Fingerprinting Technique Based on Repetitive Nucleotide Sequences.

it is called **ribotyping.** Sometimes other genes are used instead of or in addition to SSU rRNA genes. *Restriction Fragment Length Polymorphisms*

Highly conserved and repetitive DNA sequences present in many copies in the genomes of most gram-negative and some gram-positive bacteria can also be used to help identify microbes. There are three families of repetitive sequences: the 154 bp BOX elements, the 124–127 bp enterobacterial repetitive intergenic consensus (ERIC) sequence, and 35–40 bp repetitive extragenic palindromic (REP) sequences. These sequences are generally found at distinct sites between genes; that is, they are intergenic. Because they are conserved among genera, oligonucleotide primers can be used to specifically amplify the repetitive sequences by PCR. Different primers are used for each type of repetitive element, and the results are classified as arising from BOX-PCR, ERIC-PCR, or REP-PCR (**figure 17.4**). In each case, the amplified fragments from many microbial samples can be resolved and visualized on an agarose gel. Each lane of the gel corresponds to a single bacterial isolate, and the pattern created by many samples resembles a UPC bar code. The "bar code" is then computer analyzed using pattern recognition software as well as software that calculates phylogenetic relationships. Because DNA fingerprinting enables identification to the level of species, subspecies, and often strain, it is valuable not only in the study of microbial diversity but in the identification of human, animal, and plant pathogens as well.

Search This: Rep-PCR genome fingerprinting

In comparing the molecular methodological requirements for species definition, we see that with the exception of DNA-DNA hybridization, only a small region of the genome is sampled. As shown in **figure 17.5**, 16S rRNA analysis focuses on one specific gene, while MLSA and MLST sample multiple housekeeping genes. While much information is gained using these approaches, several disadvantages have been noted. Novel species of bacteria and archaea can be missed when using so-called universal primers to amplify 16S rRNA sequences from natural samples. Evolutionary changes in housekeeping genes of certain bacteria, particularly pathogens and other microbes that live in association with other organisms, can be driven by interactions with the host. These changes are not necessarily correlated with housekeeping functions and therefore may not be good candidates for MLSA or MLST.

An alternate technique, **single nucleotide polymorphisms (SNP,** pronounced "snip"**),** samples a larger fraction of the genome than either 16S rRNA analysis or MLSA (figure 17.5). Originally designed for use in humans, SNP analysis looks at

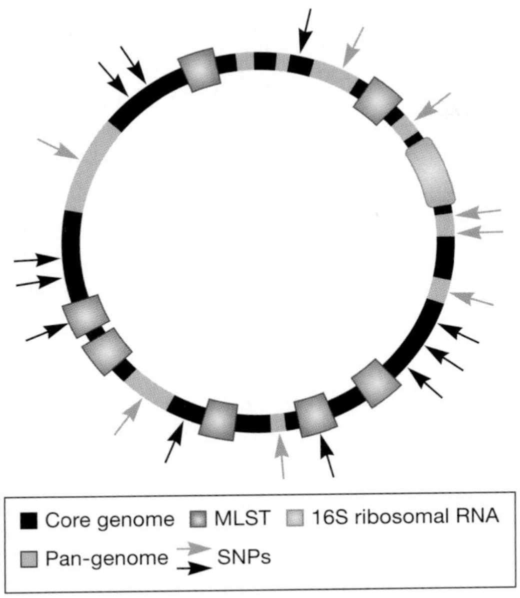

FIGURE 17.5 Genome Coverage of Genetic Taxonomic Approaches. A typical bacterial or archaeal genome contains at least one locus encoding 16S rRNA, which is always in the core genome (black). MLSA samples multiple genes that may be found on both the core genomic and pan-genomic regions. SNP analysis compares more, but shorter, nucleotide sequences that span the entire genome.

single nucleotide changes, or polymorphisms, in specific genes, intergenic regions, or other noncoding regions. These particular regions are targeted because they are normally conserved, so single changes in a base pair reveal evolutionary change. SNP analysis has been used to distinguish between strains of *Bacillus anthracis* and *Mycobacterium tuberculosis*.

Amino Acid Sequencing

The amino acid sequences of proteins directly reflect mRNA sequences and so represent the genes coding for their synthesis. However, the value of a given protein in taxonomic and phylogenetic studies varies. The sequences of proteins with dissimilar functions often change at different rates; some sequences change quite rapidly, whereas others are very stable. Therefore the most direct approach is to determine the amino acid sequence of proteins with the same function. If these sequences are similar, the organisms possessing them may be closely related. The sequences of cytochromes and other electron transport proteins, histones and heat-shock proteins, transcription and translation proteins, and a variety of metabolic enzymes have been used in taxonomic and phylogenetic studies. In contrast, rapidly evolving proteins, such as the outer surface proteins of the syphilis pathogen *Treponema pallidum*, are not appropriate for taxonomic or phylogenetic purposes. Suitable proteins may offer some advantages over rRNA

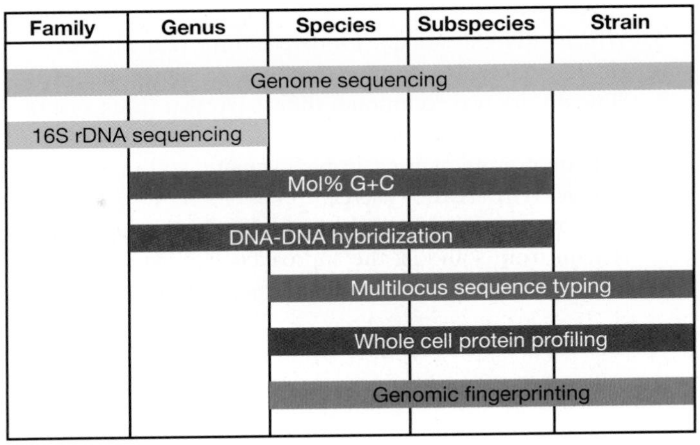

FIGURE 17.6 Relative Taxonomic Resolution of Various Molecular Techniques.

comparisons. A sequence of 20 amino acids has more information per site than a sequence of four nucleotides. Protein sequences are less affected by organism-specific differences in G + C content than are DNA and RNA sequences.

There are several ways to compare proteins. Indirect methods of comparing proteins frequently have been traditionally employed. Specifically, the electrophoretic mobility of proteins has been used to study relationships at the species and subspecies levels. In addition, antibodies can discriminate between very similar proteins, and immunologic techniques can be used to compare proteins from different microorganisms. More recently, the use of mass spectrometry has been adopted for relatively rapid amino acid sequencing, which enables the direct comparison of amino acid sequences among proteins of interest (*see figure 16.15*). **Figure 17.6** shows the taxonomic utility of several kinds of molecular analyses, including protein profiling; with the exception of genome sequencing, it is clear that a combination of approaches is best for identification at the species level or lower. ◀◀ *Proteomics (section 16.5)*

1. What are the advantages of using each major group of characteristics (morphological, physiological and metabolic, biochemical, ecological, genetic, and molecular) in classification and identification? How is each group related to the nature and expression of the genome? Give examples of each type of characteristic.
2. What is the G + C content of DNA, and how can it be determined through melting temperature studies?
3. Why is it not safe to assume that two microorganisms with the same G + C content belong to the same species? In what ways are G + C content data taxonomically valuable?
4. Describe how nucleic acid hybridization studies are carried out using membrane-bound DNA. Why might one wish to vary the incubation temperature during hybridization?

5. Why is rRNA so suitable for determining relatedness?
6. How is ribotyping similar to rRNA sequence analysis? How do the two techniques differ? Do you think one is more accurate than the other?
7. List some proteins used in phylogenetic and taxonomic studies. Why are they useful?
8. You have recently established a pure culture of a new archaeon from soil. List the approaches you would use to identify your new microbe to the species level.

17.4 Phylogenetic Trees

Microbial taxa within *Bacteria* and *Archaea* form discrete, genealogically clustered groups that can be illustrated in phylogenetic trees. **Phylogenetic trees** show inferred evolutionary relationships in the form of multiple branching lineages connected by nodes (**figure 17.7**). The organism whose nucleotide or amino acid sequences have been analyzed is identified at the tip of each branch. Each node (branchpoint) represents a divergence event, and the length of the branches represents the

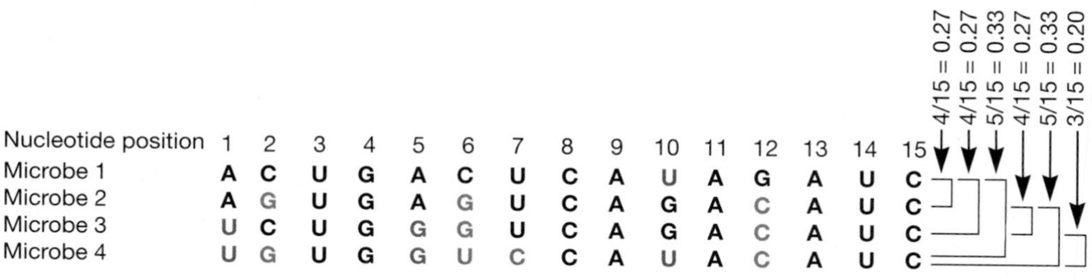

(a) Sequence alignment and analysis

Microbe	1	2	3	4
1	1.0	0.27	0.27	0.33
2		1.0	0.27	0.33
3			1.0	0.20
4				1.0

(b) Calculated evolutionary distance

Microbe	1	2	3	4
1	1.0	0.32	0.32	0.44
2		1.0	0.32	0.44
3			1.0	0.26
4				1.0

(c) Corrected evolutionary distance

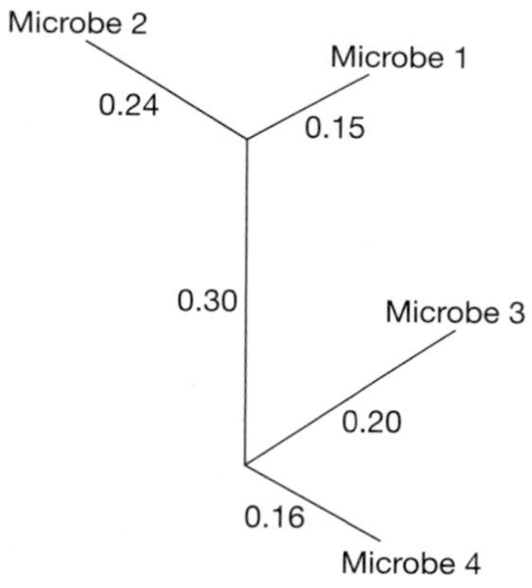

(d) Phylogenetic tree

FIGURE 17.7 Constructing a Phylogenetic Tree Using a Distance Method. (a) Nucleotide sequences are aligned and pairwise comparison is made. The number of nonidentical nucleotides is then scored. For instance, when sequences from microbes 1 and 2 are compared, there are 4 mismatches out of 15 total nucleotides, yielding a calculated evolutionary distance (E_D) of 0.27. (b) The calculated E_D values are corrected to account for back mutation to the original genotype or other forward mutations that could have occurred at the same site before generating the observed genotype. (c) A tree-building method is then selected (in this case, a distance method is used), and computer analysis of the values generates a phylogenetic tree. E_D values are indicated for each branch.

number of molecular changes that have taken place between the two nodes.

Often sequences are obtained from microbes that have been grown in pure culture; however, this is not always the case. SSU rRNA sequences have become particularly important in both identifying microbes in nature and constructing phylogenetic trees to describe their evolutionary relationships. An inclusive term for the biological source of each branch tip is the **operational taxonomic unit (OTU).** We now briefly describe how phylogenetic trees can be built with the intention of increasing an understanding of what they represent.

There are five steps in building a phylogenetic tree. First, the nucleotide or amino acid sequence must be aligned. Alignment is usually done using an online resource like CLUSTAL, although manual inspection of the alignment is also important (figure 17.7*a*). Although SSU rRNA gene sequences are most common, protein-coding genes may also be part of the analysis, in which case the alignment of amino acids is preferred. This is because the genetic code is degenerate, so even if a nucleotide sequence is not conserved, the amino acid sequence may be. Next, the alignment must be examined for a phylogenetic signal; this will determine if it is appropriate to continue with tree building (figure 17.7*b*). There are two extremes in this regard: at one end of the spectrum, the sequences align perfectly; the other extreme is the absence of any matches whatsoever. Phylogenetic analysis can only be performed on those sequences that present a mixture of random and matched positions. The third step is the hardest: one must choose which tree-building method to use. We briefly review some of the more popular methods next. The last two steps involve the application of the method, which is performed by a computer, followed by manual examination of the resulting tree to make sure it makes sense. For example, a tree that places a mammal and an archaeon on the same branch would inspire very little confidence.

As mentioned, there are a number of different approaches to build a phylogenetic tree. These can be divided into two broad categories: a distance-based (phenetic) approach and a character-based (cladistic) approach. Distance-based approaches are the most intuitive. Here the differences between the aligned sequences are counted for each pair and summarized into a single statistic, which is roughly the percent difference between the two sequences (figure 17.7*b,c*). A tree is then generated by serially linking pairs that are ever more most distantly related (i.e., start with those with the least number of sequence differences and move to those with the most). This is called cluster analysis, and it has the unattractive capability of generating trees even in the absence of evolutionary relationships. Neighbor joining is another distance-based method that uses a slightly different matrix that attempts to avoid this problem by modifying the distance between each pair of nodes based on the average divergence from all other nodes.

Character-based methods for phylogenetic tree building are more complicated but generate more robust trees. These methods start with assumptions about the pathway of evolution, infer the ancestor at each node, and choose the best tree according to a specific model of evolutionary change. These methods include maximum parsimony, which assumes that the fewest number of changes occurred between ancestor and extant (living) organisms. Another approach is called maximum likelihood. This requires a large data set because for each possible tree that can be built, its probability (i.e., the likelihood) based on certain evolutionary and molecular information is evaluated. All tree building methods have their advantages and disadvantages, so it is usually advisable to use several methods to analyze the same data set. Similar trees generated by different approaches is the best outcome.

Importantly, a tree may be unrooted or rooted. An unrooted tree (**figure 17.8*a***) simply represents phylogenetic relationships

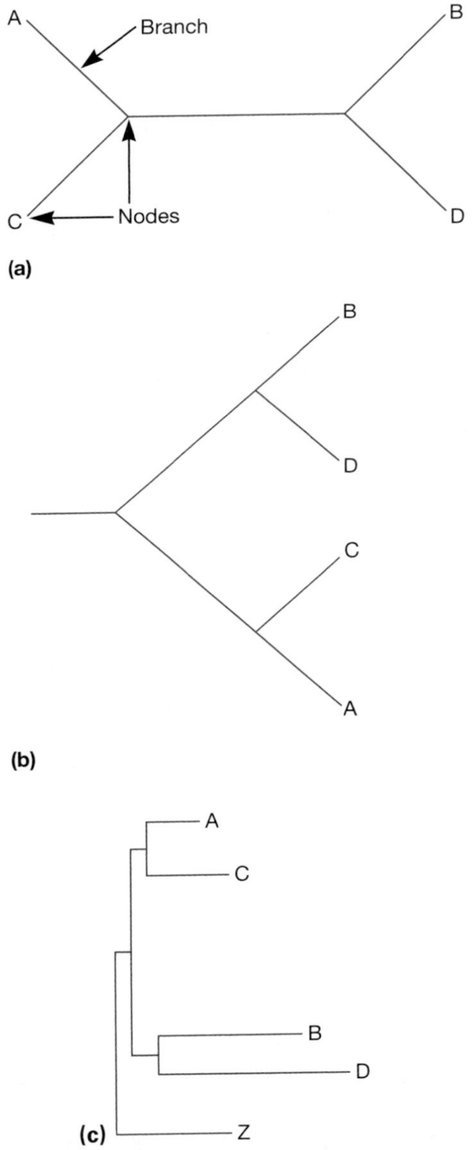

FIGURE 17.8 Phylogenetic Tree Topologies. (a) Unrooted tree joining four taxonomic units. (b) Rooted tree. (c) The tree shown in (a) can be rooted by adding an outgroup, represented by Z.

but does not provide an evolutionary path. Figure 17.8*a* shows that A is more closely related to C than it is to either B or D, but it does not specify the common ancestor for the four species or the direction of change. In contrast, the rooted tree (figure 17.8*b*) gives a node that serves as the common ancestor and shows the development of the four species from this root. It is much more difficult to develop a rooted tree. For example, there are 15 possible rooted trees that connect four species but only three possible unrooted trees.

An unrooted tree can be rooted by adding data from an outgroup—a species known to be very distantly related to all the species in the tree (figure 17.8*c*). The root is determined by the point of the tree where the outgroup joins. This provides a point of reference to identify the oldest node on the tree, which is the node closest to the outgroup.

Once a tree is constructed, it is important to get a sense of whether the placement of branches and nodes is legitimate. There are a variety of methods to assess the "strength" of a tree, but the most common is bootstrapping. Bootstrapping involves phylogenetic analysis of a randomly selected subset of the data presented on the tree. A bootstrap value is the percent of analyses in which that particular branch was found. Typically bootstrap values of 70% or greater are thought to support a tree. Another approach, called Bayesian inference, is becoming common. Rather than looking at a single tree, Bayesian inference analyzes multiple potential trees and calculates the probability that each branch would appear based on this comparison. Although these values are also reported as percentages, they are not directly comparable to bootstrap values. Only values greater than 95% are acceptable when Bayesian inference is used.

An important feature in phylogenetic trees is the scale. Just as a scale bar on a map indicates distance in number of kilometers per centimeter, the scale bar on a phylogenetic tree illustrates the evolutionary distance. This is usually measured in number of mutations per 1,000 nucleotides or amino acid substitutions per 1,000 amino acid residues. This may be expressed as a number without units, e.g., 0.02. To continue our analogy with a road map, just as a map does not reveal how long it takes to get from one point to another (due to traffic, weather, variable road conditions), the branches on a phylogenetic tree do not indicate the length of time it took for an ancestral microbe to give rise to an extant form. As discussed in section 17.5, the theory of punctuated equilibria is one important reason why evolutionary distance, as measured by the similarity of genes or proteins in living organisms, provides little or no information regarding how long ago evolutionary divergence occurred.

One of the biggest challenges in constructing a satisfactory tree is widespread, frequent horizontal gene transfer (HGT). Although microbiologists are careful to exclude from their analysis genes and proteins known to have been subject to HGT, the influence of HGT on phylogeny and evolution cannot be ignored. Indeed, eukaryotes possess genes from both bacteria and archaea, and there has been frequent gene swapping between the bacterial and archaeal domains. Some bacteria even have acquired eukaryotic genes. Clearly the pattern of microbial evolution is not as linear and treelike as previously thought. This has prompted the development of trees that attempt to display HGT (**figure 17.9**). Such trees resemble a web or network with many lateral branches linking various trunks, each branch representing the transfer of one or a few genes. Instead of having a single main trunk or common ancestor at its base, these trees have several trunks or groups of primitive cells that contribute to the original gene pool. Although extensive gene transfer occurs between archaea and bacteria, eukaryotes rarely participated in lateral gene transfer after the formation of fungi, plants, and animals. ◀◀ *Mechanisms of genetic variation (chapter 14)*

Search This: Building phylogenetic trees

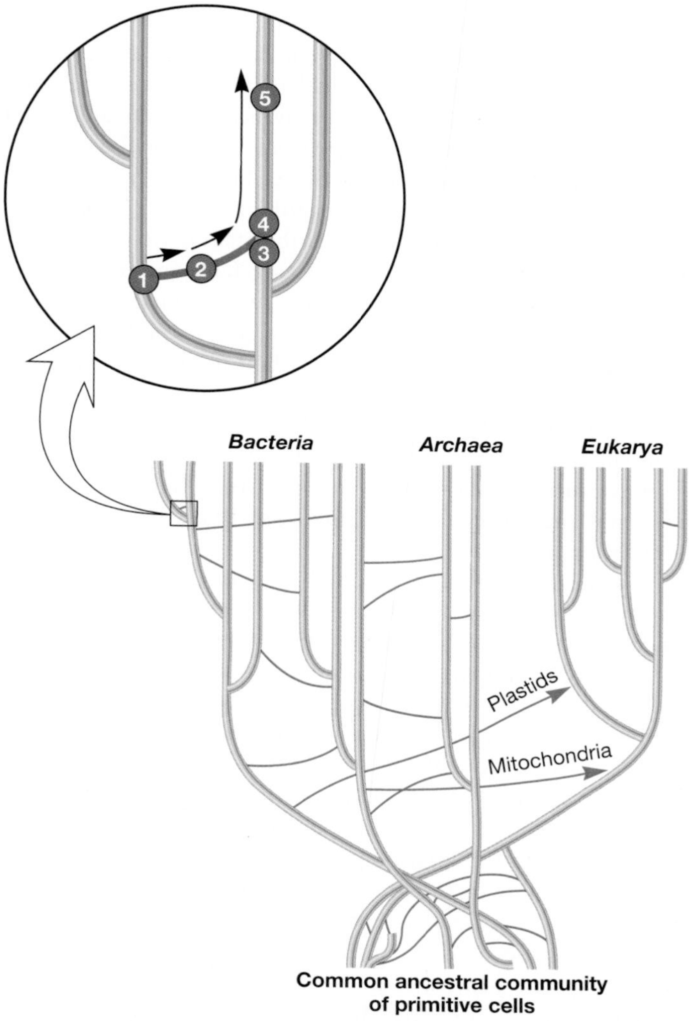

FIGURE 17.9 Universal Phylogenetic Tree with Lateral Gene Transfers. The effect of HGT on the evolution of life results in a tree with weblike interconnections that complicate the emergence of the three domains of life. The insert displays the series of events needed to give rise to the stable inheritance of a gene in a new organism.

1. Could a phylotype be considered an OTU? What about a species?
2. List the differences between distance-based and character-based methods for constructing a phylogenetic tree. Which type is maximum parsimony? Explain your answer.
3. What is the difference between a rooted and unrooted tree? Which provides more information?
4. You are building a tree based on 16S rRNA sequence alignments of a group of spirochetes. Suggest a possible outgroup so that you can build a rooted tree. Refer to chapter 19 if you need more information about these bacteria.

17.5 Evolutionary Processes and the Concept of a Microbial Species

It is our goal here to describe current models that seek to explain the evolution of new microbial species. Before we begin this discussion, however, it is helpful to consider the origin of all microbes—bacterial, archaeal, and eukaryotic. We then need to review the controversy that surrounds the word *species* as applied to bacteria and archaea. Only then will it be possible to understand and appreciate the evolutionary mechanisms that drive the development of new species of microorganisms.

Evolution of the Three Domains of Life

As we present in chapter 1, many believe that the first self-replicating entity was RNA. This is because RNA has the capacity to reproduce itself as well as catalyze chemical reactions. It is thought that when early RNA became enclosed in a lipid sphere, the first primitive cell-like forms were generated (*see figures 1.6 and 1.7*). Considerable evidence indicates that by at least 3.5 billion years ago, such proto-cells (Greek, *protos,* meaning first) had evolved to form the ancestors of our extant microbes. Moreover by 2.5 billion years ago, bacteria and archaea not only abounded, but each had evolved distinct taxonomic lineages. For instance, *Archaea* had diverged into two major phyla, *Crenarchaeota* and *Euryarchaeota.* Similarly the gram-positive bacterial phylum *Firmicutes* and gram-negative phyla *Proteobacteria* and *Cyanobacteria* had developed. Indeed, the ancestors of modern *Cyanobacteria* performed the oxygenic photosynthesis responsible for converting our anoxic planet to an oxygenated one.

A reexamination of the tree of life based on SSU rRNA (**figure 17.10**) shows that the root of the tree is on the earliest region of the bacterial branch. As we discuss in chapter 1, the root is considered the *l*ast *u*niversal *c*ommon *a*ncestor, or LUCA. The placement of LUCA indicates that although *Bacteria* and *Archaea* share similar cellular construction, they are not phylogenetically linked. Because LUCA maps to the bacterial branch of the tree, it

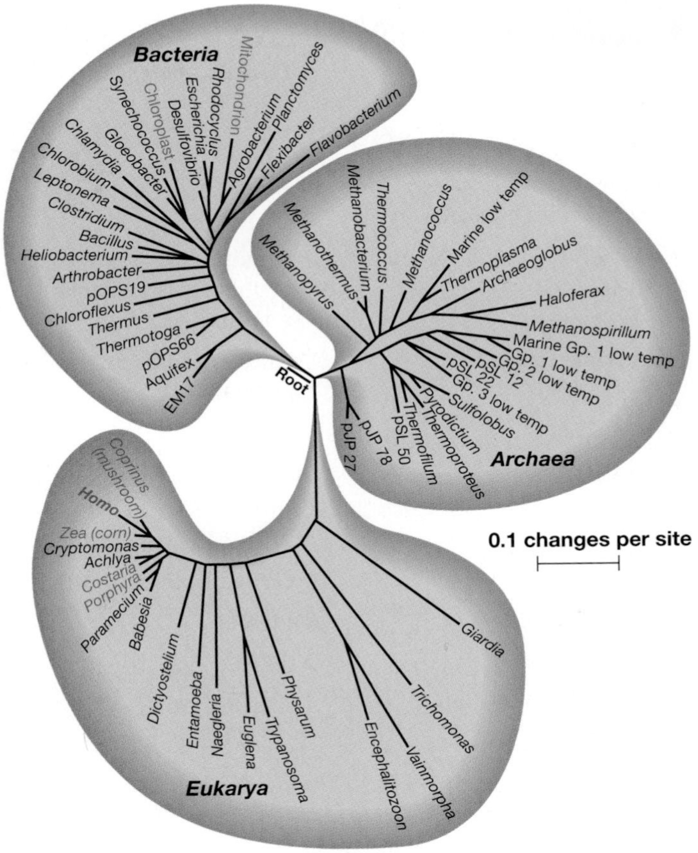

FIGURE 17.10 Universal Phylogenetic Tree. These evolutionary relationships are based on rRNA sequence comparisons. Length of branches indicates evolutionary relationships between organisms but not time. Microbes are printed in black.

Figure 17.10 Micro Inquiry

Why do chloroplasts and mitochondria map to *Synechococcus* (a cyanobacterium) and *Agrobacterium,* respectively?

is thought that *Archaea* and *Eukarya* evolved independently of *Bacteria.* Recall that this was first suggested by Carl Woese and George Fox in the 1970s. Since that time, the differences in evolutionary distance they inferred from bacterial and archaeal SSU rRNA sequences have also been demonstrated by a number of biochemical differences. These include membrane lipids, cell wall construction, and enzymes involved in gene transcription (**table 17.5**). The details of domain specific differences are discussed in chapters 3 and 18.

Notice that the placement of LUCA suggests that *Archaea* and *Eukarya* descend from the same ancestor but diverged to become separate domains. As seen in table 17.5, *Archaea* possess "information processing" proteins and mechanisms that are more

Table 17.5 Comparison of *Bacteria, Archaea,* and *Eukarya*

Property	**Bacteria**	**Archaea**	**Eukarya**
Membrane-Enclosed Nucleus with Nucleolus	Absent	Absent	Present
Complex Internal Membranous Organelles	Absent	Absent	Present
Cell Wall	Almost always have peptidoglycan containing muramic acid	Variety of types, no muramic acid; some have pseudomurein	No muramic acid
Membrane Lipid	Have ester-linked, straight-chained fatty acids	Have ether-linked, branched aliphatic chains	Have ester-linked, straight-chained fatty acids
Gas Vesicles	Present	Present	Absent
Transfer RNA	Thymine present in most tRNAs	No thymine in T or TΨC arm of tRNA	Thymine present
	N-formylmethionine carried by initiator tRNA	Methionine carried by initiator tRNA	Methionine carried by initiator tRNA
Polycistronic mRNA	Present	Present	Absent
mRNA Introns	Absent	Absent	Present
mRNA Splicing, Capping, and Poly A Tailing	Absent	Absent	Present
Ribosomes			
Size	70S	70S	80S (cytoplasmic ribosomes)
Elongation factor 2 reaction with diphtheria toxin	Does not react	Reacts	Reacts
Sensitivity to chloramphenicol and kanamycin	Sensitive	Insensitive	Insensitive
Sensitivity to anisomycin	Insensitive	Sensitive	Sensitive
DNA-Dependent RNA Polymerase			
Number of enzymes	One	One	Three
Structure	Simple subunit pattern (6 subunits)	Complex subunit pattern similar to eukaryotic enzymes (8–12 subunits)	Complex subunit pattern (12–14 subunits)
Rifampicin sensitivity	Sensitive	Insensitive	Insensitive
RNA Polymerase II Type Promoters	Absent	Present	Present
Metabolism			
Similar ATPase	No	Yes	Yes
Methanogenesis	Absent	Present	Absent
Nitrogen fixation	Present	Present	Absent
Chlorophyll-based photosynthesis	Present	Absent	Present[a]
Chemolithotrophy	Present	Present	Absent

[a]Present in chloroplasts (of bacterial origin).

similar to eukaryotes than to bacteria. These proteins include RNA polymerases, and the mechanisms are those used to regulate gene transcription. The genetics and molecular biology of the *Archaea* are discussed more fully in section 18.1.

Although *Archaea* and *Eukarya* share a recent common ancestor, this does not mean that eukaryotes possess only archaeal traits. Indeed, as first suggested by the protistologist F. J. R. (Max) Taylor, the prevailing view among biologists is that the evolution of organelles that chiefly define the domain *Eukarya* arose by incorporating endosymbiotic bacteria. This is known as the **endosymbiotic hypothesis** of the origin of organelles. This theory posits the following series of events. The ancestral eukaryotic cell lost its rigid cell wall. We know that cell walls are found in almost all bacteria and archaea. Those without cell walls are thought to once have had them, but they were later lost. This proto-eukaryote was then able to develop endocytosis (*see figure 4.10*). By this time, the cell had evolved actin (or its precursor) that enabled amoeboid motility. These mobile proto-eukaryotes became predators of other cells, including bacteria. Predation imposed selection for enlargement of the proto-eukaryote and increased motility. Engulfment without digestion of bacterial prey evolved because a smaller bacterial cell provided energy for the larger host cell. The relationship was beneficial for both partners because the host protected and supplied nutrients to the bacterial cell. The energy supplied by the endosymbiont must have conferred a growth advantage to the proto-eukaryote, enabling its dominance over other cells that lacked both cell walls and endosymbionts.

As the endosymbiont became more dependent on its host for nutrients and protection, there was little selective pressure for the retention of genes involved in these processes. Conversely, because the endosymbiont was "permitted" to remain only if it captured and stored energy (presumably as ATP), there was strong selective pressure to retain the genes involved in energy conservation. Thus genes whose products were redundant to the host were eroded and eventually lost. In fact, such genome reduction is the rule, rather than the exception, among obligate intracellular microbes. Good examples include the aphid endosymbiont *Buchnera aphidicola* (*see p. 714*) and the intracellular pathogen *Chlamydia* (*see p. 506*).

Finally, as genome reduction continued, the endosymbiont evolved into an energy-providing organelle. It was originally postulated that the endosymbiont was capable of oxidative phosphorylation and so gave rise to the mitochondrion. However, more recent evidence suggests that the endosymbiont was an anaerobic bacterium with a fermentative metabolism. This endosymbiont then evolved into either hydrogenosomes or mitochondria. Hydrogenosomes are found in some extant protists, where they take up pyruvate that results from glycolysis within the host cytoplasm. Within the hydrogenosome, pyruvate is reduced to acetate, H_2, and CO_2 with an additional ATP generated (*see figure 4.17*). Mitochondria appear to be more highly derived (i.e., took longer to evolve), since these organelles are the site of oxidative phosphorylation. Several lines of evidence support the notion that hydrogenosomes and mitochondria evolved from a single common ancestor, probably related to the α-proteobacterium *Rickettsia prowazekii* (*see p. 518*). The most compelling evidence to support this idea is: (1) The heat-shock proteins found in α-proteobacteria, mitochondria, and hydrogenosomes are closely related; (2) subunits of the mitochondrial enzyme NADH dehydrogenase are active in hydrogenosomes of the protist *Trichomonas vaginalis;* and (3) the primitive genome found in the hydrogenosomes of the protist *Nyctotherus ovalis* encodes components of a mitochondrial electron transport chain. Taken together, these data suggest that mitochondria and hydrogenosomes are aerobic and anaerobic versions, respectively, of the same ancestral organelle.

Later, these new aerobic eukaryotes engulfed a cyanobacterium—probably an ancestor of *Prochlorococcus* (*see p. 503*). Again, this led to the development of a mutualistic relationship. Such endosymbioses exist today in certain protists that retain living cyanobacteria or the functional chloroplasts of their algal prey. Such host cells and their photosynthetic endosymbionts evolved into our extant green plants and algae—organisms that possess both mitochondria and chloroplasts. This picture is a bit complicated by a secondary endosymbiotic event that gave rise to the red algae (e.g., the heterokonts, haptophytes, and cryptomonads; *see chapter 23*). Here it is thought a eukaryotic cell with a mitochondrion-like organelle engulfed a eukaryotic-like cell that possessed both a mitochondrion and chloroplast.

The identification of *Rickettsia prowazekii* and *Prochlorococcus* as the most probable extant relatives of the endosymbionts that gave rise to mitochondria and chloroplasts, respectively, is partly based on homology between the genomes of these microbes and those found in these organelles. In addition, the ribosomes of mitochondria and chloroplasts closely resemble those of their ancestors. Indeed, inspection of figure 17.10 shows that mitochondria and chloroplasts map to the *Bacterial* domain. Note that chloroplasts share a branch with *Synechococcus,* a cyanobacterium, and mitochondria are on the same branch as *Agrobacterium,* a representative α-proteobacterium.

In contrast to the general agreement among biologists regarding the endosymbiotic origin of hydrogenosomes, mitochondria, and chloroplasts, no consensus exists regarding how eukaryotes developed a nucleus. The two most highly regarded hypotheses both account for genes of archaeal and bacterial heritage found on the eukaryotic nuclear genome. One hypothesis asserts that an archaeon and a bacterium living in close association fused (**figure 17.11*a***). It is postulated that the archaeon benefited from the carbon and energy secreted by a fermentative bacterium, which was the progenitor of the hydrogenosome. This hypothesis asserts that the formation of the nuclear membrane occurred concurrently with the evolution of other organelles. The second hypothesis involves a series of endosymbioses, rather than fusion (figure 17.11*b*). Like the fusion hypothesis, this theory starts with an archaeon and bacterium; however, the archaeal cell engulfs the bacterium. This is the first in a series of endosymbiotic events that ultimately gave rise to the nucleus, mitochondria, and chloroplasts.

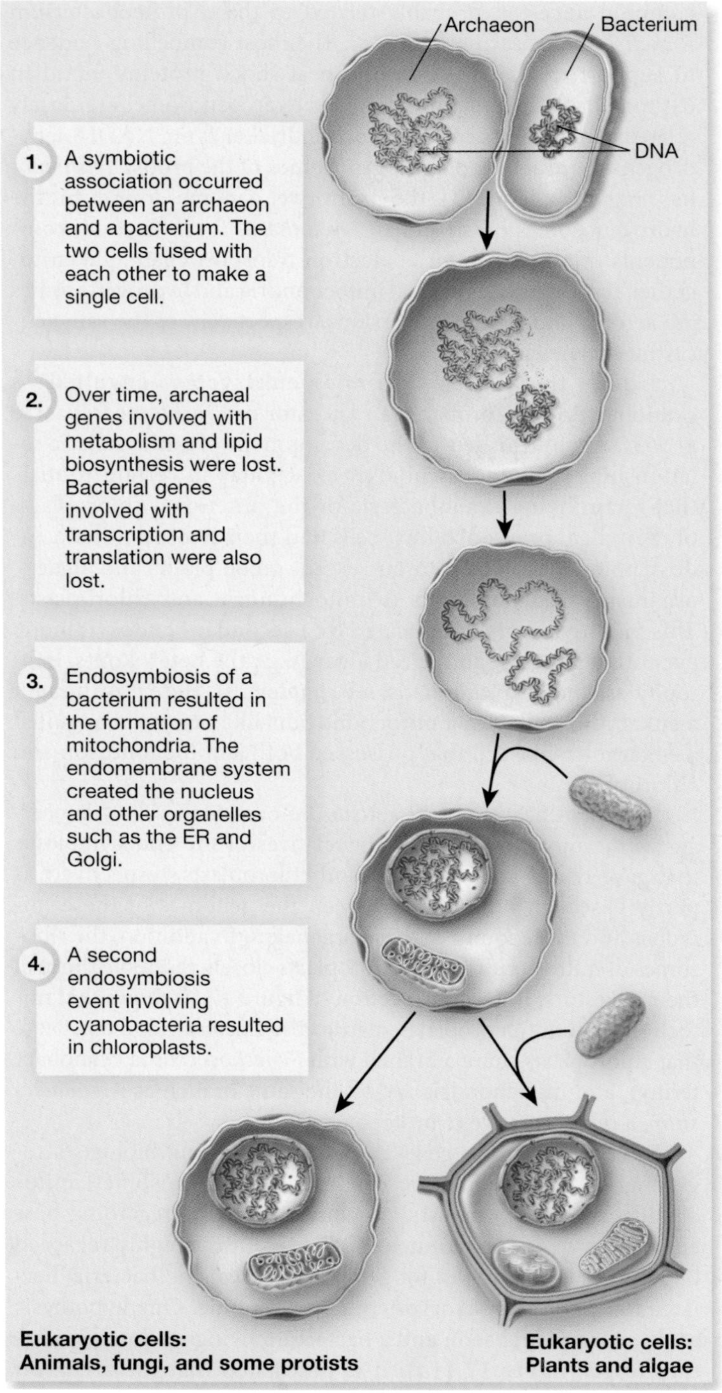
Archaeon
Bacterium
DNA
1. A symbiotic association occurred between an archaeon and a bacterium. The two cells fused with each other to make a single cell.
2. Over time, archaeal genes involved with metabolism and lipid biosynthesis were lost. Bacterial genes involved with transcription and translation were also lost.
3. Endosymbiosis of a bacterium resulted in the formation of mitochondria. The endomembrane system created the nucleus and other organelles such as the ER and Golgi.
4. A second endosymbiosis event involving cyanobacteria resulted in chloroplasts.
Eukaryotic cells: Animals, fungi, and some protists
Eukaryotic cells: Plants and algae

(a) Origin from symbiotic relationship

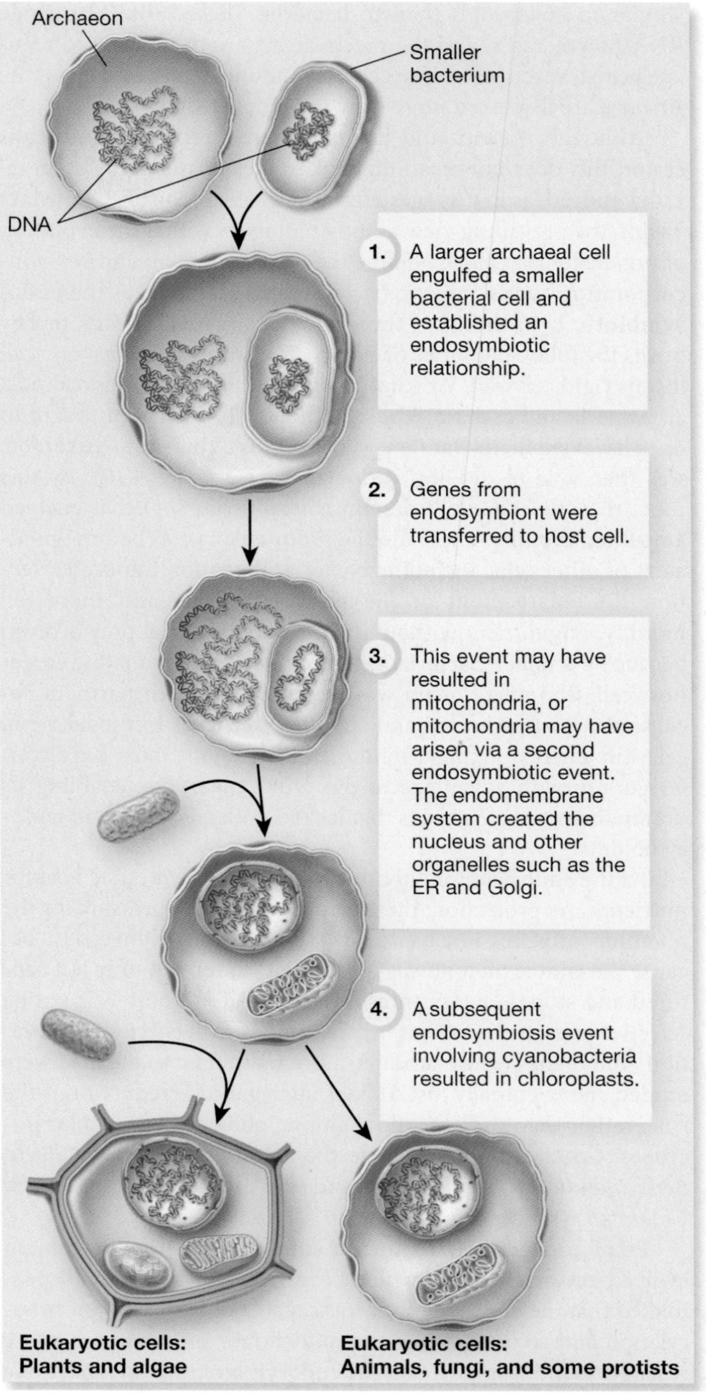
Archaeon
Smaller bacterium
DNA
1. A larger archaeal cell engulfed a smaller bacterial cell and established an endosymbiotic relationship.
2. Genes from endosymbiont were transferred to host cell.
3. This event may have resulted in mitochondria, or mitochondria may have arisen via a second endosymbiotic event. The endomembrane system created the nucleus and other organelles such as the ER and Golgi.
4. A subsequent endosymbiosis event involving cyanobacteria resulted in chloroplasts.
Eukaryotic cells: Plants and algae
Eukaryotic cells: Animals, fungi, and some protists

(b) Origin from endosymbiotic relationship

FIGURE 17.11 The Origin of Eukaryotic Cells.

The Concept and Definition of a Microbial Species

"Species concept" is a theoretical framework used to understand how and why certain organisms can be sorted into discrete taxonomic groups. It describes the biological characteristics that must be common to all members of the same group or taxon. We have already discussed "species definition" in sections 17.2 and 17.3, when we reviewed the criteria that determine to which species a microbe belongs. From this, we can see that species definition is the application of the species concept.

Both species concept and definition have changed over time and continue to be difficult for microbiologists to agree upon. Historically, microbiologists used taxonomic systems designed to identify and classify plants and animals. However, bacteria and archaea lack sexual reproduction, extensive morphological features, and a microbial fossil record. It was therefore clear very early that the phenetic system was limited, even when a polyphasic approach was employed. In the meantime, microbial taxonomists were keen to find new technologies that might advance their efforts to construct valid and robust criteria for the definition of a bacterial species.

The first breakthrough occurred when computer technology was advanced enough to handle large numerical data sets. No longer widely used, numerical taxonomy is truly polyphasic because it takes information about morphological, biochemical, and physiological properties of organisms and converts it into a form suitable for numerical analysis. Matrices are developed and then compared. The resulting classification is based on general similarity as judged by comparison of these many characteristics, each given equal weight.

Numerical taxonomy was next combined with chromosome-based approaches, including G + C ratios and DNA-DNA hybridization. Biochemical methods also gained popularity, and characterization of the cell wall components, lipids (e.g., FAME analysis), and other chemotaxonomic indicators were used. Although Carl Woese introduced SSU rRNA gene sequencing about 30 years ago and its validity has been intensively studied, it was not until 2002 that those charged with determining the necessary criteria for microbial species definition, the International Committee on Systematics of Prokaryotes (ICSP), recommended its use. Importantly, unlike whole genome hybridization, SSU rRNA gene comparisons can be performed on microorganisms that are not in pure culture.

Historically, the application of different criteria in making species assignments has led to confusion in microbial taxonomy. In some cases, a single microbial species is so metabolically and genetically diverse that it seems probable that the group represents multiple species. On the other hand, some species are very narrowly defined, such that two species differ very little. For instance, *Bacillus anthracis* strains are so similar that the species is considered "genetically monomorphic." In fact, many believe that all *B. anthracis* strains are really members of the *B. cereus* species. It is argued that only because *B. anthracis* causes anthrax does it have its own species designation.

In an effort to clarify and standardize microbial taxonomy, the ICSP recommends a "gold standard" for species assignment: phenotypic or morphological similarity to others in the group, whole genome similarity as determined by DNA-DNA hybridization of at least 70%, the melting temperature of the DNA (a reflection of the G + C content) within 5°C, and less than 3% divergence in rRNA gene sequence. Some remain uncomfortable with these criteria, pointing out that two microorganisms with, for instance, only 75% similarity in DNA and 98% rRNA gene sequence identity can be considered the same species. However, if these criteria were applied to eukaryotes, all primates (monkey, apes, you) would be lumped together as a single species! Indeed, it remains unresolved whether or not the species concept can be applied to microbes.

Microbial Evolutionary Processes

While the debate regarding the operational definition of a microbial species continues, most microbiologists agree that the microbial species concept is grounded in natural selection and evolution. As the most ancient life forms on Earth, bacteria and archaea have had the opportunity to evolve and adapt to virtually every habitat. While their diverse metabolic strategies and ability to tolerate extreme conditions explain *why* microbes display such enormous diversity, natural selection explains *how* this diversity came to be.

Recall that genetic diversity in *Archaea* and *Bacteria* must occur asexually. Thus heritable genetic changes in these organisms are introduced principally by two mechanisms: mutation and horizontal gene transfer (HGT), both of which are subject to natural selection. Because diverse individuals arise from a genetically homogeneous population, HGT is not relevant when considering the initial evolution of diversity within a given taxon. Rather genetic variation must arise within the population by mutation and, to a lesser extent, by gene loss and gain, and intragenomic recombination (**figure 17.12*a***).

Anagenesis, also known as **genetic drift,** refers to small, random genetic changes that occur over generations. It might seem that very small genetic differences within a microbial population would be of little evolutionary significance. However, model studies designed to assess competition between microbial populations has led to some surprising observations. When selection is applied, very small genetic differences can result in one population overtaking another. How does this happen when individuals within a population have similar mutation rates, and most of these mutations are neutral and have no phenotypic effect? Only those rare mutations that confer a growth advantage, called **adaptive mutations**, are retained and passed from one generation to the next; that is, the mutation is fixed. The descendents of that individual continue to evolve through mutation and other intraspecific mechanisms.

Given these mechanisms, it is possible to imagine two different evolutionary processes that could give rise to microbial diversity within a population. The first evolutionary process we consider accounts for the gradual development of diversity and is sometimes referred to as the **metapopulation** model. Here groups of microbes live in separate patches that represent slightly different niches (figure 17.12*b*). Individuals within a patch may start out as a clonal population, but as the population grows, genetic drift causes it to become genetically heterogeneous. As resources become scarce, microbes in one patch may move to and populate other patches. As small changes accumulate, the degree of genetic difference between two populations that arose from the same ancestor may be so great that they are no longer considered the same taxon. If the divergence is so large that they no longer compete for the same limited resources, they may continue to survive in the same ecosystem but inhabit different physiological niches.

By contrast, the second model, the ecotype model, involves more abrupt changes. An **ecotype** is a population of microbes that is genetically similar (usually determined by 16S rRNA sequencing)

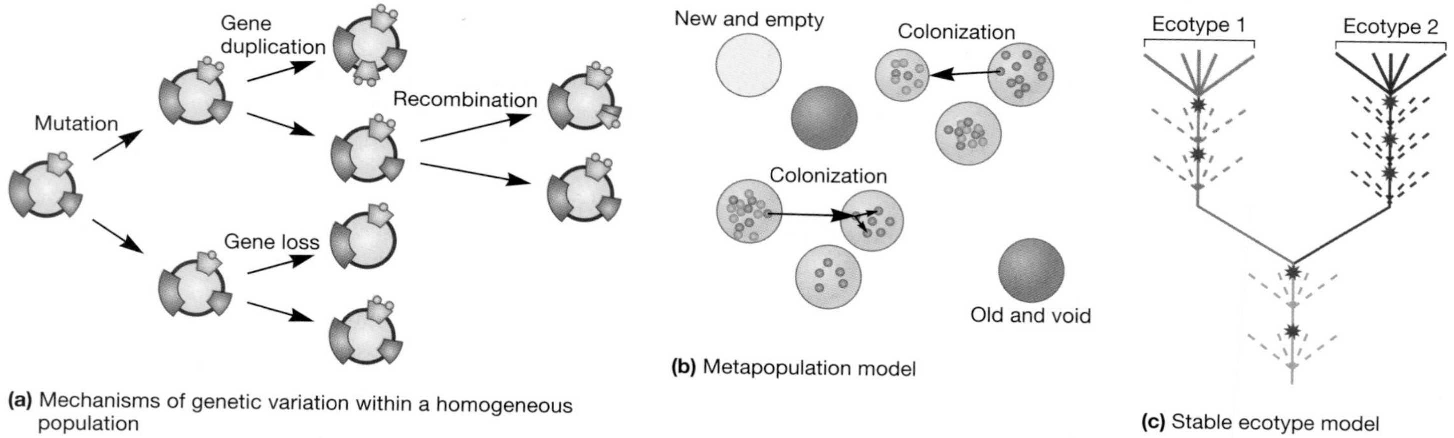

FIGURE 17.12 Evolution of Microbial Diversity. (a) Mechanisms by which a single, genetically homogeneous population of microbes can develop genetic variation include mutation, gene loss and duplication, and recombination. (b) The metapopulation model posits that patches (circles) may be empty or colonized by a genetically heterogeneous population of microbes (dots of different colors). When an empty patch is colonized by an individual (or genetically identical individuals), new populations are established that can give rise to new genetic diversity. Occasionally, a patch can no longer support a population, as indicated by the old and void patch in the lower right. (c) Genetic changes within a population can lead to the development of new ecotypes. Each ecotype is subject to periodic selection events (indicated by stars) that enable cells with adaptive mutations to outcompete other lineages, which are eventually driven to extinction (dotted lines). The solid lines represent successful populations or lineages; those at the top are extant.

Figure 17.12 Micro Inquiry

Construct a scenario in which each of the following factors lead to the establishment of two ecotypes from a single common ancestor, as shown in (c): the availability of carbon and nitrogen sources; terminal electron acceptor; and mean local temperature.

but ecologically distinct. Ecotypes arise when members of a microbial population living in a specific ecosystem undergo a genetic event (or series of events) that enables them to outcompete the remainder of the population. According to the ecotype model, the acquisition of adaptive mutations ultimately drives the remaining members of the population into extinction and reduces the amount of genetic diversity within the surviving population (figure 17.12*c*). The fossil record shows that the pace of evolution does not always occur at a constant rate but is periodically interrupted by rapid bursts of speciation driven by abrupt changes in the environment. Niles Eldredge and Steven Jay Gould coined the term **punctuated equilibria** to describe this phenomenon. Regardless of which model is applied, the 3.5 billion year history of microbial life on Earth affords the accumulation of many, many mutations—that in turn has resulted in speciation. ◀◀ *Mechanisms of genetic variation (chapter 14); Comparative genomics (section 16.7)*

In considering microbial evolution and diversity, we cannot completely ignore horizontal transfer of genetic material. As discussed in chapter 16, the concept of a core genome has emerged from the genome sequences of over 700 bacteria and archaea. A **core genome** is the set of homologous genes found in all genomes of a phylogenetically cohesive group. It tends to consist of informational genes—those that encode proteins involved in DNA replication, transcription, and translation (this would include the genes for SSU rRNAs). These genes are presumed to have been present on the genome of the common ancestor and not to have been acquired by HGT. Therefore any variation found among them is thought to be the result of mutation. Thus metapopulation and ecotype models focus on the evolution of the core genome.

By contrast, the **pan-genome** is the complete gene repertoire of a taxon, so it includes the core genome plus "housekeeping" and dispensable genes. Housekeeping genes are generally defined as those genes whose products are required for normal metabolism and growth. In general, genes within the pan-genome are considered to have been acquired by horizontal gene transfer (**figure 17.13**).

Unlike the variation introduced in the metapopulation and ecotype models, HGT-driven genetic variation requires a genetically heterogeneous group of microbes. This is because HGT does not rely on replication but rather on the exchange of genetic material within a single generation. The rate of HGT is extremely variable. Some microbes have very reduced genomes with no evidence of HGT. These microbes are generally highly adapted to a specific, stable ecological niche. The most extreme examples are obligate

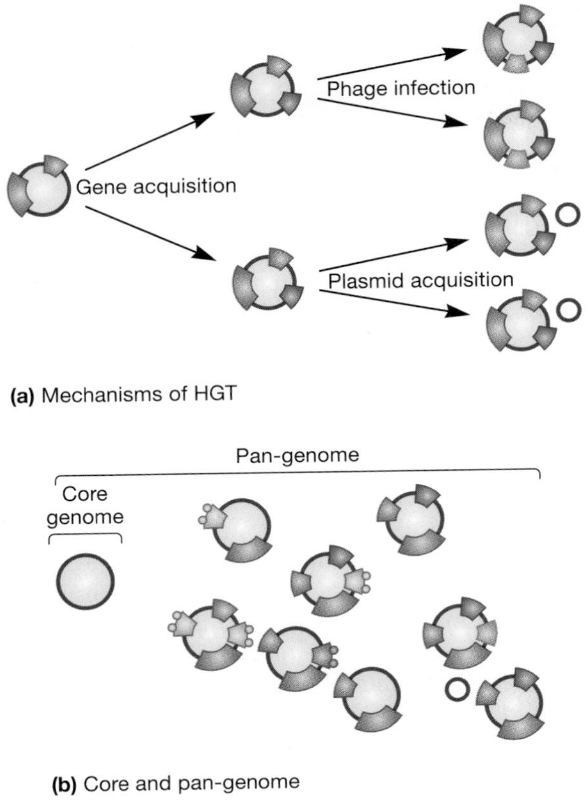

FIGURE 17.13 Horizontal Gene Transfer and the Pan-Genome. (a) The acquisition of genes from other populations of microbes is mediated by phage infection, conjugation, and transformation. During conjugation and transformation, transferred genes can be integrated directly into the chromosome or encoded on a plasmid, as shown here. (b) The prevalence of genes that have been horizontally transferred has given rise to the concept of a core genome—genes found in all members of a taxon, and a pan-genome—the combination of the core genome and all additional genes acquired principally by HGT within the taxon.

Figure 17.13 Micro Inquiry

Which do you think would have a pan-genome more closely related to its core genome: a microbial species whose strains are obligate intracellular symbionts, or a species whose strains constitute the normal flora of the mammalian gut? Explain your answer.

intracellular symbionts that can only grow within their host cells. By contrast, some microbes appear to have a high rate of HGT with almost a third of their genome acquired from genetic exchange. In these bacteria and archaea, genes acquired by horizontal transfer frequently expand metabolic capabilities. It is thought that in general, speciation is driven by mutation, while the rapid adaptation to new environmental challenges is mediated by HGT.

Search This: TalkOrigins Archive

Finally, we pose the question: How many bacterial and archaeal species are there? There are two major obstacles to formulating an answer. First, most microbial species resist growth in the laboratory, so they can only be detected by metagenomic or other culture-independent approaches. Second, as we have seen, microbiologists cannot agree on a biological species concept. So we must resort to the operational definition of a species in terms of G + C content and percent SSU rRNA and DNA homologies, as well as similarities in physiology, morphology, and ecology. Based on these considerations, there are over 7,000 accepted species. But estimates range from 100,000 to 1,000,000 species in nature, with about 10^{30} individual cells. These estimates reveal that there are about a billion more microbes on Earth than stars in the universe.

1. Define ecotype. Do you think it is necessary to obtain microbes in pure culture before assigning different ecotypes?
2. What are the principle differences between the metapopulation and ecotype models of microbial evolution?
3. What is the difference between the core genome and pan-genome? What might you infer if you compare two genera, one in which the size of the core genome and pan-genome are very similar, and one in which the core genome is much smaller than the pan-genome?
4. Of the following genes, which do you think are part of the pan-genome and which are part of the core genome: the genes for lactose catabolism in *E. coli*; the genes for heat stable DNA polymerase in *Thermus aquaticus*; the genes for proteorhodopsin in marine bacteria; the genes for toxin production in *Vibrio cholerae*?
5. Which protein do you think would be best to use in constructing a phylogenetic tree: a protein encoded on the core genome or a one encoded only on the pan-genome? Explain your answer.

17.6 *Bergey's Manual of Systematic Bacteriology*

In 1923 David Bergey, professor of bacteriology at the University of Pennsylvania, and four colleagues published *Bergey's Manual of Determinative Bacteriology*, a classification of bacteria that could be used for the identification of bacterial species. This single-volume manual is now in its ninth edition. It continues to serve as a relatively brief reference guide in the identification of bacteria based on physiological and morphological traits.

In 1984 the first edition of *Bergey's Manual of Systematic Bacteriology* was published. It contained descriptions of all bacterial and archaeal species then identified (**Microbial Diversity & Ecology 17.1**). A more recent second edition consists of five volumes

MICROBIAL DIVERSITY & ECOLOGY

17.1 "Official" Nomenclature Lists—A Letter from *Bergey's*

On a number of occasions lately, the impression has been given that the status of a bacterial taxon in *Bergey's Manual of Systematic Bacteriology* or *Bergey's Manual of Determinative Bacteriology* is in some sense official. Similar impressions are frequently given about the status of names in the Approved List of Bacterial Names and in the Validation Lists of newly proposed names that appear regularly in the *International Journal of Systematic Bacteriology*. It is therefore important to clarify these matters.

There is no such thing as an official classification. *Bergey's Manual* is not "official"—it is merely the best consensus at the time, and although great care has always been taken to obtain a sound and balanced view, there are also always regions in which data are lacking or confusing, resulting in differing opinions and taxonomic instability. When *Bergey's Manual* disavows that it is an official classification, many bacteriologists may feel that the solid earth is trembling. But many areas are in fact reasonably well established. Yet taxonomy is partly a matter of judgment and opinion, as is all science, and until new information is available, different bacteriologists may legitimately hold different views. They cannot be forced to agree to any "official classification." It must be remembered that, as yet, we know only a small percentage of the bacterial species in nature. Advances in technique also shed new light on bacterial relationships. Thus we must expect that existing boundaries of groups will have to be redrawn in the future, and it is expected that molecular biology, in particular, will experience a good deal of change over the next few decades.

The position with the Approved Lists and the Validation Lists is rather similar. When bacteriologists agreed to make a new start in bacteriological nomenclature, they were faced with tens of thousands of names in the literature of the past. The great majority were useless, because, except for about 2,500 names, it was impossible to tell exactly what bacteria they referred to. These 2,500 were therefore retained in the Approved Lists. The names are only approved in the sense that they were approved for retention in the new bacteriological nomenclature. The remainder lost standing in the nomenclature, which means they do not have to be considered when proposing new bacterial names (although names can be individually revived for good cause under special provisions).

The new International Code of Nomenclature of Bacteria requires all new names to be validly published to gain standing in the nomenclature, either by being published in papers in the *International Journal of Systematic Bacteriology* or, if published elsewhere, by being announced in the Validation Lists. The names in the Validation Lists are therefore valid only in the sense of being validly published (and therefore they must be taken account of in bacterial nomenclature). The names do not have to be adopted in all circumstances; if users believe the scientific case for the new taxa and validly published names is not strong enough, they need not adopt the names. For example, *Helicobacter pylori* was immediately accepted as a replacement for *Campylobacter pylori* by the scientific community, whereas *Tatlockia micdadei* had not generally been accepted as a replacement for *Legionella micdadei*. Taxonomy remains a matter of scientific judgment and general agreement.

From Sneath, P. H. A., and Brenner, D. J. 1992. Official nomenclature lists. ASM News. *58(4):175.*

published over a number of years, starting in 2001. Each volume covers a specific group of microbes and is written by experts in that particular field. The morphology, physiology, growth conditions, ecology, and other information is provided, making this a valuable reference for microbiologists.

The second volume of *Bergey's Manual of Systematic Bacteriology* reflects the enormous progress that has been made in microbial taxonomy since the first edition was published. This is particularly true of the molecular approaches to phylogenetic analysis. Thus whereas microbial classification in the first edition was phenetic (based on phenotypic characterization), classification in the second edition of *Bergey's Manual* is largely phylogenetic. Although gram-staining properties are generally considered phenetic characteristics, they also play a role in the phylogenetic classification of microbes. Some of the major differences between gram-negative and gram-positive bacteria are summarized in **table 17.6**.

In addition to the reorganization based on phylogeny, the second edition has more ecological information about individual taxa. It does not group all the clinically important bacteria together as the first edition did. Instead, pathogenic species are placed phylogenetically and thus scattered throughout the following five volumes.

Volume 1, *The Archaea and the Deeply Branching and Phototrophic Bacteria*
Volume 2, *The Proteobacteria*
Volume 3, *The Low G + C Gram-Positive Bacteria*
Volume 4, *The High G + C Gram-Positive Bacteria*
Volume 5, *The Planctomycetes, Spirochaetes, Fibrobacteres, Bacteriodetes, Fusobacteria, Chlamydiae, Acidobacteria, Verrumicrobia, and Dictyoglomi* (Volume 5 also will contain a section that updates descriptions and phylogenetic arrangements that have been revised since publication of volume 1.)

Table 17.7 summarizes the organization of the second edition and indicates where the discussion of a particular group may be found in this textbook. **Figure 17.14** illustrates most of the groups covered in detail in chapters 18–22.

Table 17.6 Some Characteristic Differences Between Gram-Negative and Gram-Positive Bacteria

Property	*Gram-Negative Bacteria*	*Gram-Positive Bacteria*
Cell wall	Gram-negative type wall with inner 2–7 nm peptidoglycan layer and outer membrane (7–8 nm thick) of lipid, protein, and lipopolysaccharide	Gram-positive type wall with a homogeneous, thick cell wall (20–80 nm) composed mainly of peptidoglycan. Other polysaccharides and teichoic acids may be present.
Cell shape	Spheres, ovals, straight or curved rods, helices or filaments; some have sheaths or capsules.	Spheres, rods, or filaments; may show true branching
Reproduction	Binary fission, sometimes budding	Binary fission, filamentous forms grow by tip extension
Metabolism	Phototrophic, chemolithoautotrophic, or chemoorganoheterotrophic	Usually chemoorganoheterotrophic, a few phototrophic
Motility	Motile or nonmotile. Flagella placement can be varied. Motility may also result from the use of axial filaments (spirochetes) or gliding motility.	Most often nonmotile; have peritrichous flagella when motile
Appendages	Can produce several types of appendages—pili and fimbriae, prosthecae, stalks	Usually lack appendages (may have spores on hyphae)
Endospores	Cannot form endospores	Some groups

Table 17.7 Organization of *Bergey's Manual of Systematic Bacteriology*

Taxonomic Rank	*Representative Genera*	*Textbook Coverage*
Volume 1. The Archaea and the Deeply Branching and Phototrophic Bacteria		
Domain *Archaea*		
Phylum *Crenarchaeota*		
Class I. *Thermoprotei*	*Thermoproteus, Pyrodictium, Sulfolobus*	pp. 482–485
Phylum *Euryarchaeota*		
Class I. *Methanobacteria*	*Methanobacterium*	pp. 485–488
Class II. *Methanococci*	*Methanococcus*	
Class III. *Methanomicrobia*	*Methanomicrobium*	
Class IV. *Halobacteria*	*Halobacterium, Halococcus*	pp. 488–490
Class V. *Thermoplasmata*	*Thermoplasma, Picrophilus, Ferroplasma*	p. 490-491
Class VI. *Thermococci*	*Thermococcus, Pyrococcus*	p. 491-492
Class VII. *Archaeoglobi*	*Archaeoglobus*	p. 492-493
Class VIII. *Methanopyri*	*Methanopyrus*	
Domain *Bacteria*		
Phylum *Aquificae*	*Aquifex, Hydrogenobacter*	p. 496
Phylum *Thermotogae*	*Thermotoga, Geotoga*	p. 496
Phylum *Thermodesulfobacteria*	*Thermodesulfobacterium*	
Phylum *Deinococcus-Thermus*	*Deinococcus, Thermus*	pp. 496–497
Phylum *Chrysiogenetes*	*Chrysogenes*	

(*Continued*)

Table 17.7 Organization of *Bergey's Manual of Systematic Bacteriology* (Continued)

Phylum *Chloroflexi*	*Chloroflexus, Herpetosiphon*	pp. 499-500
Phylum *Thermomicrobia*	*Thermomicrobium*	
Phylum *Nitrospira*	*Nitrospira*	
Phylum *Deferribacteres*	*Geovibrio*	
Phylum *Cyanobacteria*	*Prochloron, Synechococcus, Pleurocapsa, Oscillatoria, Anabaena, Nostoc, Stigonema*	pp. 501–505
Phylum *Chlorobi*	*Chlorobium, Pelodictyon*	p. 499
Volume 2. The Proteobacteria		
Phylum *Proteobacteria*		
Class I. *Alphaproteobacteria*	*Rhodospirillum, Rickettsia, Caulobacter, Rhizobium, Brucella, Nitrobacter, Methylobacterium, Beijerinckia, Hyphomicrobium*	pp. 515–523
Class II. *Betaproteobacteria*	*Neisseria, Burkholderia, Alcaligenes, Comamonas, Nitrosomonas, Methylophilus, Thiobacillus*	pp. 524–528
Class III. *Gammaproteobacteria*	*Chromatium, Leucothrix, Legionella, Pseudomonas, Azotobacter, Vibrio, Escherichia, Klebsiella, Proteus, Salmonella, Shigella, Yersinia, Haemophilus*	pp. 528–540
Class IV. *Deltaproteobacteria*	*Desulfovibrio, Bdellovibrio, Myxococcus, Polyangium*	pp. 541–547
Class V. *Epsilonproteobacteria*	*Campylobacter, Helicobacter*	pp. 547–548
Volume 3. The Low G + C Gram-Positive Bacteria		
Phylum *Firmicutes*		
Class I. *Clostridia*	*Clostridium, Peptostreptococcus, Eubacterium, Desulfotomaculum, Heliobacterium, Veillonella*	pp. 555–558
Class II. *Mollicutes*	*Mycoplasma, Ureaplasma, Spiroplasma, Acholeplasma*	pp. 551–555
Class III. *Bacilli*	*Bacillus, Caryophanon, Paenibacillus, Thermoactinomyces, Lactobacillus, Streptococcus, Enterococcus, Listeria, Leuconostoc, Staphylococcus*	pp. 559–566
Volume 4. The High G + C Gram-Positive Bacteria		
Phylum *Actinobacteria*		
Class *Actinobacteria*	*Actinomyces, Micrococcus, Arthrobacter, Corynebacterium, Mycobacterium, Nocardia, Actinoplanes, Propionibacterium, Streptomyces, Thermomonospora, Frankia, Actinomadura, Bifidobacterium*	pp. 568–581
Volume 5. The Planctomycetes, Spirochaetes, Fibrobacteres, Bacteriodetes, Fusobacteria, Chlamydiae, Acidobacteria, Verrucomicrobia, and Dictyoglomi		
Phylum *Planctomycetes*	*Planctomyces, Gemmata*	pp. 505-506
Phylum *Chlamydiae*	*Chlamydia*	pp. 506–507
Phylum *Spirochaetes*	*Spirochaeta, Borrelia, Treponema, Leptospira*	pp. 507–509
Phylum *Fibrobacteres*	*Fibrobacter*	
Phylum *Acidobacteria*	*Acidobacterium*	
Phylum *Bacteroidetes*	*Bacteroides, Porphyromonas, Prevotella, Flavobacterium, Sphingobacterium, Flexibacter, Cytophaga*	pp. 509–511
Phylum *Fusobacteria*	*Fusobacterium, Streptobacillus*	
Phylum *Verrucomicrobia*	*Verrucomicrobium*	p. 511
Phylum *Dictyoglomi*	*Dictyoglomus*	
Phylum *Gemmatimonadetes*	*Gemmatimonas*	

Summary

17.1 Introduction to Microbial Taxonomy

a. Taxonomy, the science of biological classification, is composed of three parts: classification, nomenclature, and identification.

b. A polyphasic approach is used to classify microbes. This incorporates information gleaned from genetic, phenotypic, and phylogenetic analysis.

17.2 Taxonomic Ranks

a. Taxonomic ranks are arranged in a nonoverlapping hierarchy (**figure 17.1**).

b. A bacterial or archaeal species is a collection of strains that have many stable properties in common and differ significantly from other groups of strains.

c. Microorganisms are named according to the binomial system.

17.3 Techniques for Determining Microbial Taxonomy and Phylogeny

a. The classical approach to determining microbial taxonomy and phylogeny includes the use of morphological, physiological and metabolic, and ecological characteristics.

b. The G + C content of DNA is easily determined and taxonomically valuable because it is an indirect reflection of the base sequence.

c. Nucleic acid hybridization studies are used to compare DNA or RNA sequences and thus determine genetic relatedness.

d. Nucleic acid sequencing is the most powerful and direct method for comparing genomes. The sequences of SSU rRNA are used most often in phylogenetic studies of microbes (**figure 17.3**). When complete microbial genomes are available, they offer the most comprehensive phylogenetic analysis.

e. Genomic fingerprinting represents a group of methodologies that can be used to determine microbial phylogeny. They include multilocus sequence analysis and typing (MLSA and MLST, respectively), restriction fragment length polymorphism (RFLP) analysis, the study of repetitive sequences (**figure 17.4**), and single nucleotide polymorphism analysis (**figure 17.5**).

f. Amino acid sequence of some proteins can be taxonomically and phylogenetically relevant, although the value of each protein must be assessed individually.

17.4 Phylogenetic Trees

a. Phylogenetic relationships often are shown in the form of branched diagrams called phylogenetic trees. Trees are based on pairwise comparison of amino acid or nucleotide sequences, followed by computer analysis (**figure 17.7**).

b. Trees may be either rooted or unrooted and are created in several different ways. Unrooted trees can be rooted by including an outgroup when the tree is constructed (**figure 17.8**).

17.5 Evolutionary Processes and the Concept of a Microbial Species

a. *Bacteria* and *Archaea* evolved separately and therefore represent different domains (**figure 17.10**). Most biologists agree that endosymbioses gave rise to mitochondria and chloroplasts, and thus to eukaryotes. There is less agreement with regard to the evolution of the nucleus. Some posit that an archaeon fused with a bacterium, while others view the development of a nucleus as the result of a bacterial endosymbiont within an archaeon (**figure 17.11**).

b. The operational definition of a microbial species is based on criteria that must be met for approval by the International Committee on the Systematics of Prokaryotes. This includes at least 70% whole genome similarity as determined by DNA-DNA hybridization, at least 97% 16S rRNA homology, no more than a 5°C difference in % G + C, as well as physiological, morphological, and ecological similarity.

c. The concept of a microbial species is based on evolution. The metapopulation model explains gradual change and how microbes move from one niche to another. The ecotype model describes the outcome of a periodic natural selection on a genetically homogeneous microbial population. In both cases, individuals that have acquired adaptive mutations are the source of microbial diversity (**figure 17.12**).

d. Horizontal gene transfer is also important in microbial evolution. However, speciation is thought to be the outcome of mutation, while rapid adaptation to new niches is mediated by horizontal gene transfer (**figure 17.13**).

17.6 *Bergey's Manual of Systematic Bacteriology*

a. *Bergey's Manual of Systematic Bacteriology* gives the accepted system of prokaryotic taxonomy.

b. The second edition of *Bergey's Manual* provides phylogenetic classifications. Prokaryotes are divided between two domains and 26 phyla (**table 17.7**). Comparisons of nucleic acid sequences, particularly 16S rRNA sequences, are the foundation of this classification.

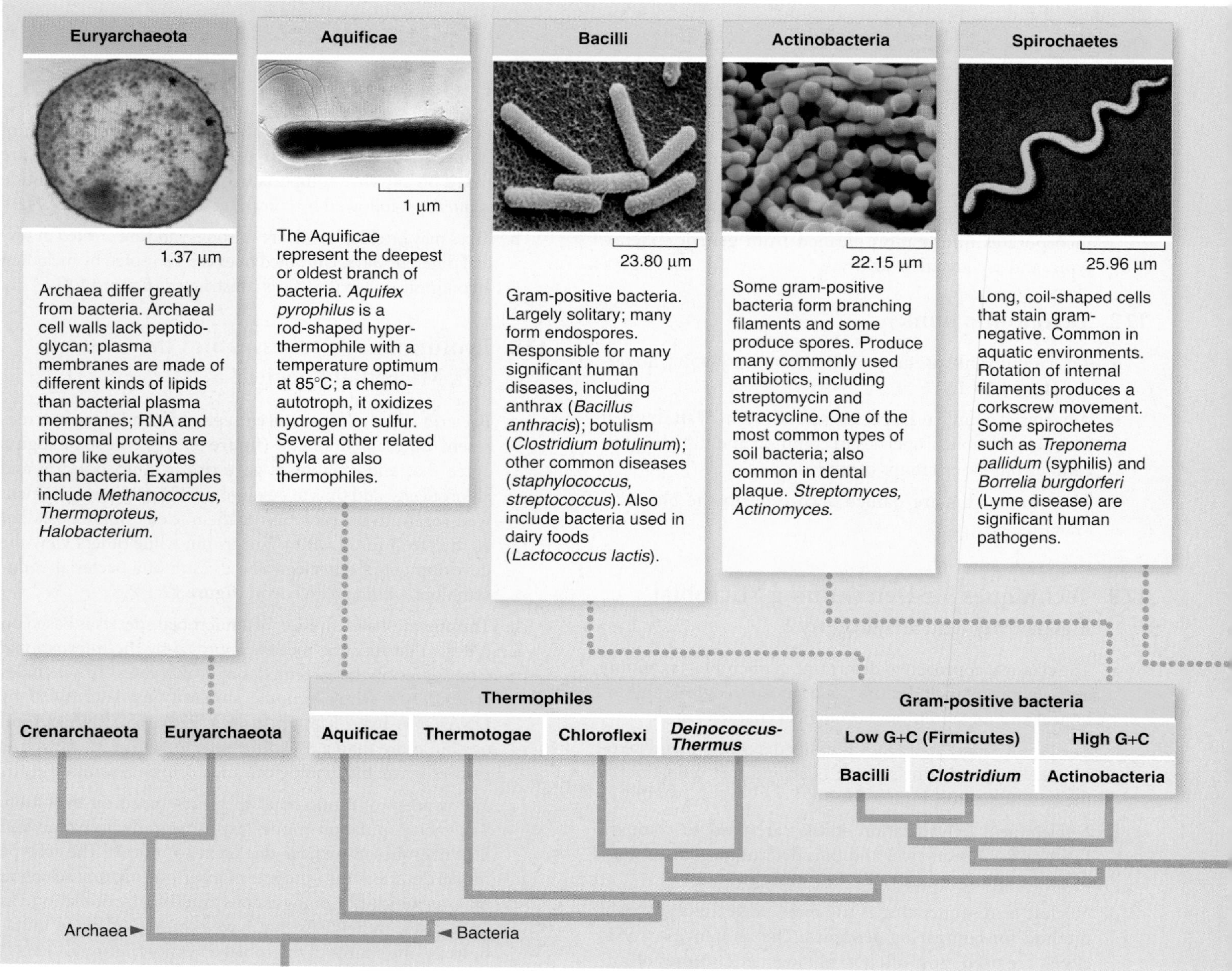

FIGURE 17.14 Some Major Clades of *Bacteria* and *Archaea*. This classification scheme is based on *Bergey's Manual of Systematic Biology, 2d ed.* Phylogenetic relationships between EHEC SNP genotypes (SGs) based on the distance matrix of pairwise difference between SGs. Bootstrap values are shown at each branch.

Critical Thinking Questions

1. Consider the fact that the use of 16S rRNA sequencing as a taxonomic and phylogenetic tool has resulted in tripling the number of bacterial phyla. Why do you think the advent of this genetic technique has expanded the currently accepted number of microbial phyla?

2. *Bacteria* and *Archaea* were classified phenetically in the first edition of *Bergey's Manual of Systematic Bacteriology.* What do you think are the advantages and disadvantages of the phylogenetic classification used in the second edition?

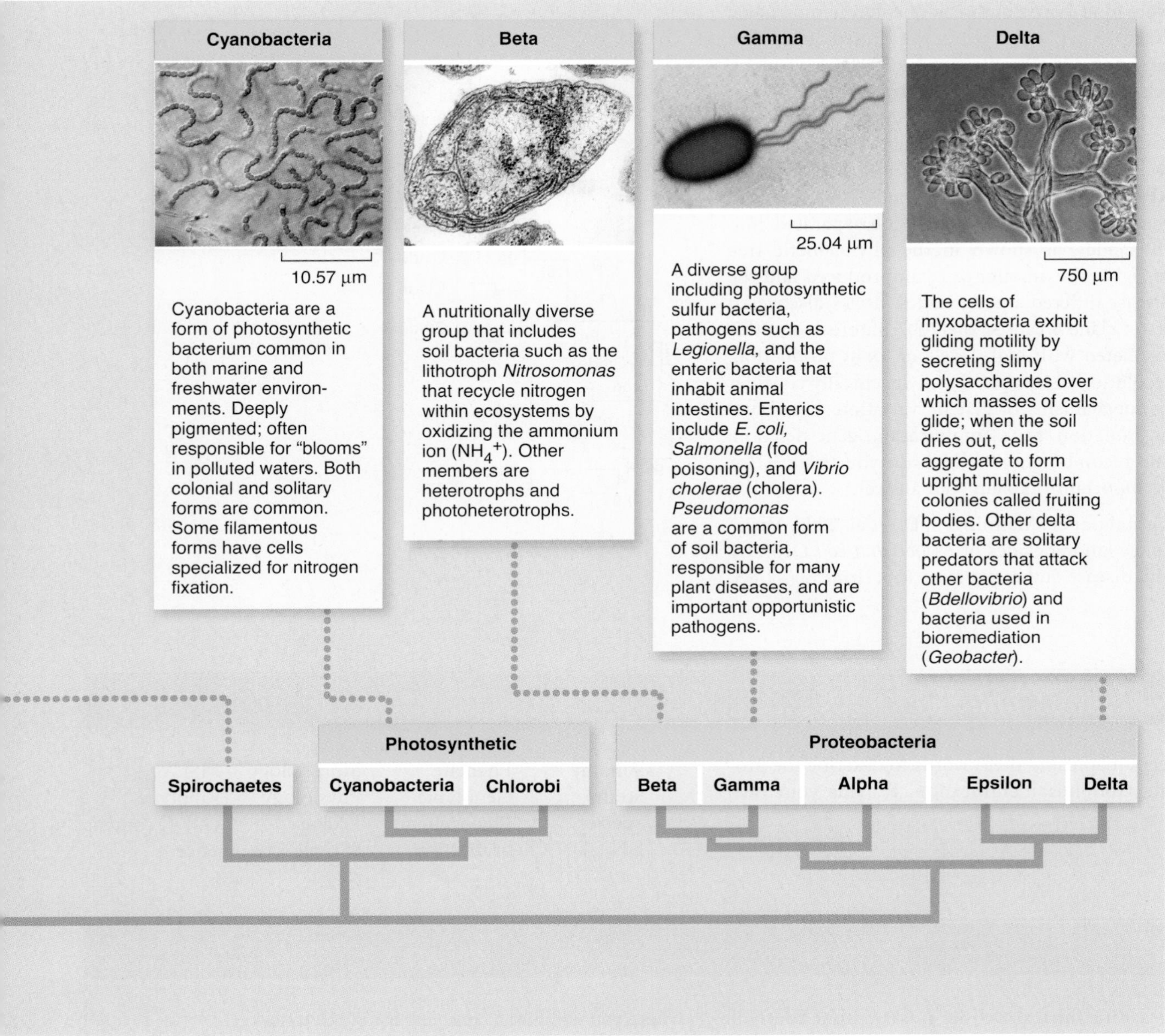

Cyanobacteria
10.57 μm
Cyanobacteria are a form of photosynthetic bacterium common in both marine and freshwater environments. Deeply pigmented; often responsible for "blooms" in polluted waters. Both colonial and solitary forms are common. Some filamentous forms have cells specialized for nitrogen fixation.
Beta
A nutritionally diverse group that includes soil bacteria such as the lithotroph *Nitrosomonas* that recycle nitrogen within ecosystems by oxidizing the ammonium ion (NH_4^+). Other members are heterotrophs and photoheterotrophs.
Gamma
25.04 μm
A diverse group including photosynthetic sulfur bacteria, pathogens such as *Legionella*, and the enteric bacteria that inhabit animal intestines. Enterics include *E. coli*, *Salmonella* (food poisoning), and *Vibrio cholerae* (cholera). *Pseudomonas* are a common form of soil bacteria, responsible for many plant diseases, and are important opportunistic pathogens.
Delta
750 μm
The cells of myxobacteria exhibit gliding motility by secreting slimy polysaccharides over which masses of cells glide; when the soil dries out, cells aggregate to form upright multicellular colonies called fruiting bodies. Other delta bacteria are solitary predators that attack other bacteria (*Bdellovibrio*) and bacteria used in bioremediation (*Geobacter*).
Photosynthetic
Proteobacteria
Spirochaetes
Cyanobacteria
Chlorobi
Beta
Gamma
Alpha
Epsilon
Delta

3. Discuss the problems in developing an accurate phylogenetic tree. Do you think it is possible to create a completely accurate universal phylogenetic tree? Explain your answer.

4. Why is the current classification system for *Bacteria* and *Archaea* likely to change considerably? How would one select the best features to use in the identification of unknown microbes and determination of relatedness?

5. Although the theory of punctuated equilibrium has been studied for many years, it has been difficult to determine the magnitude of its contribution to evolutionary change. In reviewing carefully selected gene alignments, scientists have estimated that for all organisms examined, about 22% of nucleotide substitutions are due to changes that occurred during periods of punctuated equilibrium, while the remainder accumulated gradually. Interestingly, they found that fungi and plants showed more than twice the rate of change attributable to punctuated equilibrium than animals. What do you think accounts for this difference? Note that bacteria and archaea were not examined in this study; why do you think these microbes were excluded? Do you think their evolutionary rates would be more similar to animals or fungi and plants? Explain your rationale.

Read the original paper: Pagel, M., et al., 2006. Large punctuational contribution of speciation to evolutionary divergence at the molecular level. *Science.* 314:119.

6. In 2007 a severe food-borne outbreak of enterohemorrhagic *E. coli* (EHEC) O157:H7 occurred in the United States. The strain that caused this outbreak seemed to be more virulent, causing more severe illness, than previous outbreaks. To test if new, more virulent strains of EHEC might be evolving, SNP analysis was performed on 500 clinical isolates of the bacterium that were collected before, during, and after the 2007 outbreak. SNPs were detected in 96 loci, and 39 discrete SNP genotypes were identified. These could be separated into nine different clades, as shown in the phylogenetic tree shown to the right. The amount of toxin produced and the severity of disease differed among clades. It was discovered that members of clade 8 caused the 2007 outbreak and this clade was associated with high levels of toxin production. Discuss the evolution of this pathogen. Specifically consider each of these mechanisms of genetic variation: horizontal gene transfer, mutation, gene amplification, gene deletion, and intragenic recombination. Which do you think is (or are) the most important? Explain your answer.

   Read the original paper: Manning, S. D., et al. 2008. Variation in virulence among clades of *Escherichia coli O157:H7* associated with disease outbreaks. *Proc. Nat. Acad. Sci., USA.* 105: 4868.

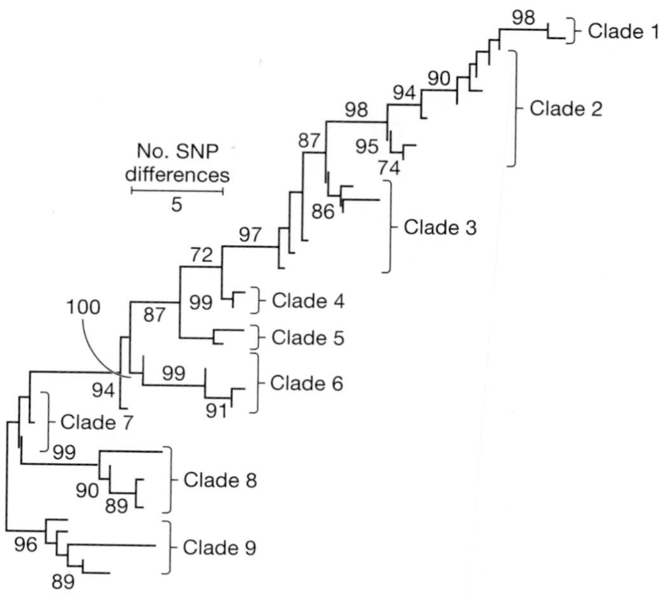

Concept Mapping Challenge

Construct a concept map using these words to describe the process of microbial taxonomic identification. Provide your own linking terms.

Taxonomy
Strain
Species
DNA-DNA hybridization
Phenetic
Genotypic
Phylogenetic
Morphology
Physiology
Ecology
SSU rRNA sequence
MSLA
Ribotype
G + C content

Learn More

Learn more by visiting the text website at www.mhhe.com/willey8, where you will find a complete list of references.

18

The Archaea

Archaea are often found in extreme environments such as this geyser in Yellowstone National Park. The orange color is due to the carotenoid pigments of thermophilic archaea.

CHAPTER GLOSSARY

Archaea The domain that contains microbes with isoprenoid glycerol diether or diglycerol tetraether lipids in their membranes and archaeal rRNA (among many other features).

bacteriorhodopsin A transmembranous protein to which retinal is bound; it functions as a light-driven proton pump, allowing photophosphorylation without chlorophyll or bacteriochlorophyll. Found in the purple membrane of halophilic archaea.

halophile A microorganism that requires high levels of sodium chloride for growth.

hyperthermophile A microorganism that grows optimally at temperatures greater than 85°C.

methanogen Strictly anaerobic archaea that derive energy by converting CO_2, H_2, formate, acetate, and other compounds to either methane or methane and CO_2.

pseudomurein A peptidoglycan-like polymer found in some archaeal cell walls. It is distinguished from peptidoglycan by the presence of L-amino acids, *N*-acetyltalosaminuronic acid (NAT) and $\beta\ (1 \rightarrow 3)$ glycosidic bonds.

sensory rhodopsin A form of rhodopsin found in halobacteria and cyanobacteria that senses the spectral quality of light.

thermoacidophiles Microorganisms that grow best at high temperatures and low pH.

Carl Woese describes his discovery of the *Archaea* in the mid-1970s as "serendipitous but not unexpected." He was examining small subunit (SSU) rRNA sequences in an effort to construct a phylogenetic tree that would represent all life forms (*see figure 1.1*). Fortunately when he began his survey of bacterial SSU rRNAs, the laboratory of Ralph Wolfe, a noted methanogen expert, was nearby. Wolfe provided Woese with 16S rRNA from a methanogen—a microbe that makes methane. When these sequences were determined, Woese knew he had made an exciting discovery: a new life form. However, other scientists could not easily believe that a well-known "bacterium" was unique. In fact, he spent the next 20 years building his case, first convincing fellow microbiologists and then the entire field of life scientists that the *Archaea* represent the third domain of life. It was a difficult task to argue because many felt that the presence of the same cellular organization in *Bacteria* and *Archaea* meant they had to be in the same "kingdom." However, *Archaea* are no more related to *Bacteria* than they are to *Eukarya*.

In this chapter, we survey the *Archaea*, focusing on the two phyla covered in *Bergey's Manual*, with the caveat that the phylogeny and taxonomy of members of this domain are the subject of intense study and debate.

18.1 Overview of the *Archaea*

As discussed in chapter 17, ***Archaea*** have many features in common with *Eukarya*, others in common with *Bacteria*, and still other elements that are uniquely archaeal (*see table 17.5*). In general, archaeal informational genes—those that encode proteins involved in replication, transcription, and translation—share homology with those of *Eukarya*, whereas genes involved in metabolism are similar to bacterial genes. Like all generalizations, this is not uniformly the case; for instance, archaea have a unique tRNA structure and only archaea are capable of methanogenesis. Like *Bacteria, Archaea* are quite diverse, both in morphology and physiology. They may be spherical, rod-shaped, spiral, lobed, cuboidal, triangular, plate-shaped, irregularly shaped, or pleomorphic. Some are single cells, whereas others form filaments or aggregates. They range in diameter from 0.1 to over 15 μm, and some filaments can grow up to 200 μm in length. They can stain either gram positive or gram negative, but they have unique cell walls—different from that of the *Bacteria.* Multiplication is usually by binary fission but may be by budding, fragmentation, or other mechanisms. The *Archaea* are just as diverse physiologically. They can be aerobic, facultatively anaerobic, or strictly anaerobic. Nutritionally, they range from chemolithoautotrophs to organotrophs. They include psychrophiles, mesophiles, and hyperthermophiles that can grow above 100°C.

Archaea inhabit a wide variety of habitats, but for many years they were considered microbes of "extreme environments," or "extremophiles." Indeed, many, but by no means all, archaea inhabit niches that have very high or low temperatures or pH, concentrated salts, or are completely anoxic. However, terms such as extreme and hypersaline reflect a human perspective, meaning that they are situations where humans could not survive. On the contrary, most of Earth (the oceans) is an "extreme environment" where it is very cold (about 4°C), dark, and under high pressure. Many archaea are well adapted to these environments in which they can grow to high numbers. For instance, in some hypersaline environments, their populations become so dense that the brine is red with archaeal pigments. By contrast, archaea also make up about 20% of the prokaryotic biomass of marine plankton and are important members of some soil communities—environments that cannot be considered extreme. In addition, some are symbionts in the digestive tracts of animals, but to date, no pathogenic archaea have been described. Thus the notion that archaea are exclusively "extremophiles" is no longer valid as archaea are known to inhabit temperate and tropical soils and waters.

Archaeal Taxonomy

As shown in **table 18.1**, well-characterized archaea can be divided into five major groups based on physiological and morphological differences. However, we present these microbes as they are phylogenetically classified, as defined by *Bergey's Manual.* Thus, we divide the *Archaea* into the phyla *Euryarchaeota* (Greek *eurus,* wide, and *archaios,* ancient or primitive) and *Crenarchaeota* (Greek *crene,* spring or fount, and *archaios*) (**figure 18.1**). The

Table 18.1 Characteristics of the Major Archaeal Physiological Groups

Group	*General Characteristics*	*Representative Genera*
Methanogenic archaea	Strict anaerobes. Methane is the major metabolic end product. S^0 may be reduced to H_2S without yielding energy production. Cells possess coenzyme M, factors 420 and 430, and methanopterin.	*Methanobacterium* (E)[a] *Methanococcus* (E) *Methanomicrobium* (E) *Methanosarcina* (E)
Archaeal sulfate reducers	Irregular gram-negative staining coccoid cells. H_2S formed from thiosulfate and sulfate. Autotrophic growth with thiosulfate and H_2. Can grow heterotrophically. Traces of methane also formed. Extremely thermophilic and strictly anaerobic. Possess factor 420 and methanopterin but not coenzyme M or factor 430.	*Archaeoglobus* (E)
Extremely halophilic archaea	Rods, cocci, or irregular shaped cells, that may include pyramids or cubes. Stain gram negative or gram positive but like all archaea lack peptidoglycan. Primarily chemoorganoheterotrophs. Most species require sodium chloride ≥1.5 M, but some survive in as little as 0.5 M. Most produce characteristic bright-red colonies; some are unpigmented. Neutrophilic to alkalophilic. Generally mesophilic; however, at least one species is known to grow at 55°C. Possess either bacteriorhodopsin or halorhodopsin and can use light energy to produce ATP.	*Halobacterium* (E) *Halococcus* (E) *Natronobacterium* (E)
Cell wall-less archaea	Pleomorphic cells lacking a cell wall. Thermoacidophilic and chemoorganotrophic. Facultatively anaerobic. Plasma membrane contains a mannose-rich glycoprotein and a lipoglycan.	*Thermoplasma* (E)
Extremely thermophilic S^0-metabolizers	Gram-negative staining rods, filaments, or cocci. Obligately thermophilic (optimum growth temperature between 70–100°C). Usually strict anaerobes but may be aerobic or facutlative. Acidophilic or neutrophilic. Autotrophic or heterotrophic. Most are sulfur metabolizers. S^0 reduced to H_2S anaerobically; H_2S or S^0 oxidized to H_2SO_4 aerobically.	*Desulfurococcus* (C) *Pyrodictium* (C) *Pyrococcus* (E) *Sulfolobus* (C) *Thermococcus* (E) *Thermoproteus* (C)

[a] Indicates phylum; *E, Euryarchaeota, C, Crenarchaeota*

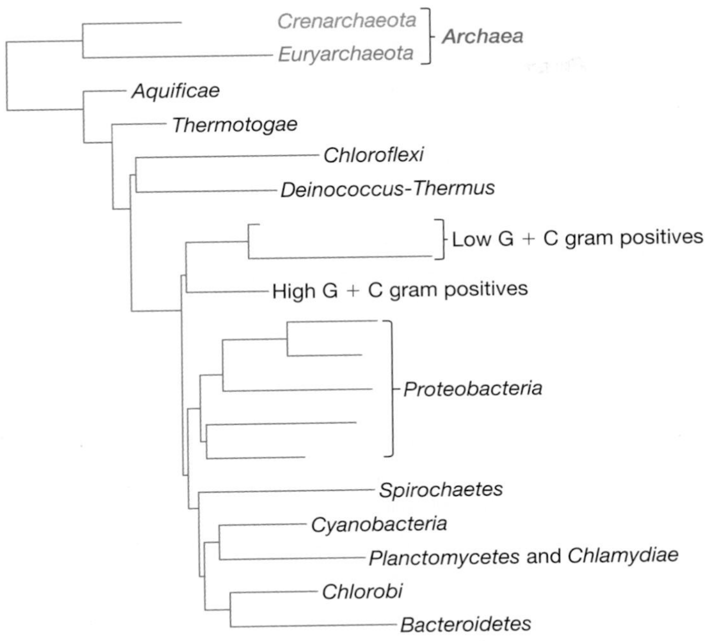

FIGURE 18.1 Phylogenetic Relationships Among *Bacteria* and *Archaea*. The *Archaea* are highlighted.

euryarchaeotes are given this name because they occupy many different ecological niches and have a variety of metabolic patterns. The methanogens, extreme halophiles, sulfate reducers, and many extreme thermophiles with sulfur-dependent metabolism are *Euryarchaeota*. The crenarchaeotes are thought to resemble the ancestral *Archaea*, and almost all the well-characterized species are thermophiles or hyperthermophiles.

Whereas all known archaea currently in pure culture belong to either the *Crenarchaeota* or the *Euryarchaeota*, an ever-growing collection of 16S rRNA nucleotide sequences cloned directly from the environment suggests that archaeal phylogeny is more complicated. Indeed, metagenomic analysis has detected many mesophilic and psychrophilic archaea. These archaeal sequences were originally placed into two groups: group I became known as the "mesophilic crenarchaeota," thereby distinguishing them from the majority of crenarchaeotes in pure culture, which are thermophilic; and group II are members of the *Euryarchaeota*.

It has been particularly difficult to resolve the phylogeny of group I archaea, in part because 16S rRNA sequences from hyperthermophiles have a much higher G + C content than that of the mesophiles; high G + C content can generate artifacts during phylogenetic tree construction. A potential resolution to the taxonomic confusion of the group I archaea occurred when the genome of a group I species, *Crenarchaeum symbiosum*, a marine sponge symbiont, was sequenced. This made it possible to compare the amino acid sequences of all the ribosomal proteins encoded by this microbe to those of 48 other archaea for which complete genomes had been sequenced. Because ribosomal proteins are under the same natural selection constraints as rRNA, they are good phylogenic markers. The resulting maximum likelihood phylogenetic tree is shown in **figure 18.2**. This tree places the group I mesophilic archaea as a separate phylum with the proposed name "*Thaumarchaeota*" (Greek *thaumas*, wonder). This implies that these mesophiles might have diverged before the speciation of the *Euryarchaeota* and the thermophilic crenarchaeota. Interestingly, the tree places the unusual microbe *Nanoarchaeum equitans* in the euryarchaeota lineage. This is controversial because based on 16S rRNA data alone this archaeon appears to constitute its own phylum. Another interesting outcome of this analysis is the placement of methanogenic archaea, which suggests that methane production may not be the ancestral metabolism of the euryarchaeotes, as is often postulated. ⏮ *Metagenomics (section 16.8); Phylogenetic trees (section 17.4)*

Because this tree is based on ribosomal proteins, it does not consider an entire group of archaea that until very recently was known only by 16S rRNA sequences extracted from nature. These microbes, tentatively called the *Korarchaeota* (from the Greek word for "young man"), were placed as a separate phylum based solely on the SSU rRNA data. In 2008 the complete genome of one member of this group, "*Candidatus* Korarchaeum cryptofilium," was sequenced. Note that this archaeon has not been grown in pure culture, so it has candidate (*Candidatus*) species status. Comparative genomic analysis places this archaeon as an ancient lineage with affinity to the *Crenarchaeota* (figure 18.2). However, genes encoding proteins involved in DNA replication and tRNA modification more closely resemble those of the *Euryarchaeota*, suggesting that the *Korarchaeota* may have retained some ancient archaeal cellular features. It is not clear if the separate phylum designation will hold up to more complete analysis when more genomic sequences become available.

Thus although there are currently two accepted archaeal phyla, it is uncertain how many phyla will eventually be recognized. This state of flux in archaeal phylogeny demonstrates how dynamic microbial taxonomy can be. The use of molecular probes to dissect microbial communities, combined with innovative culture techniques, ensures that phylogenetic analysis will continue to evolve.

Archaeal Cell Surfaces and Membranes

Archaeal flagella more closely resemble bacterial type IV pili than the well-studied bacterial flagella (*see figures 3.42 and 3.44*). Archaeal cell walls also differ from those of bacteria. With the exception of the *Mollicutes* (bacteria that lack cell walls), the *Planctomycetes*, and the *Chlamydiae*, all bacteria have peptidoglycan and can be classified as either gram positive or gram negative. By contrast, archaea cannot be classified in this manner. Most archaeal cell envelopes consist of a single protein or glycoprotein surface layer (S-layer) directly attached to the underlying plasma membrane (*see figure 3.28*), whereas bacterial S-layers are fastened to the cell wall. For the *Crenarchaeota*, the distance between the S-layer and the plasma membrane, although variable, is wide—ranging from 20 to 70 nm. This is not the case for *Euryarchaeota* S-layers, which in general are only 10 to 15 nm. Other

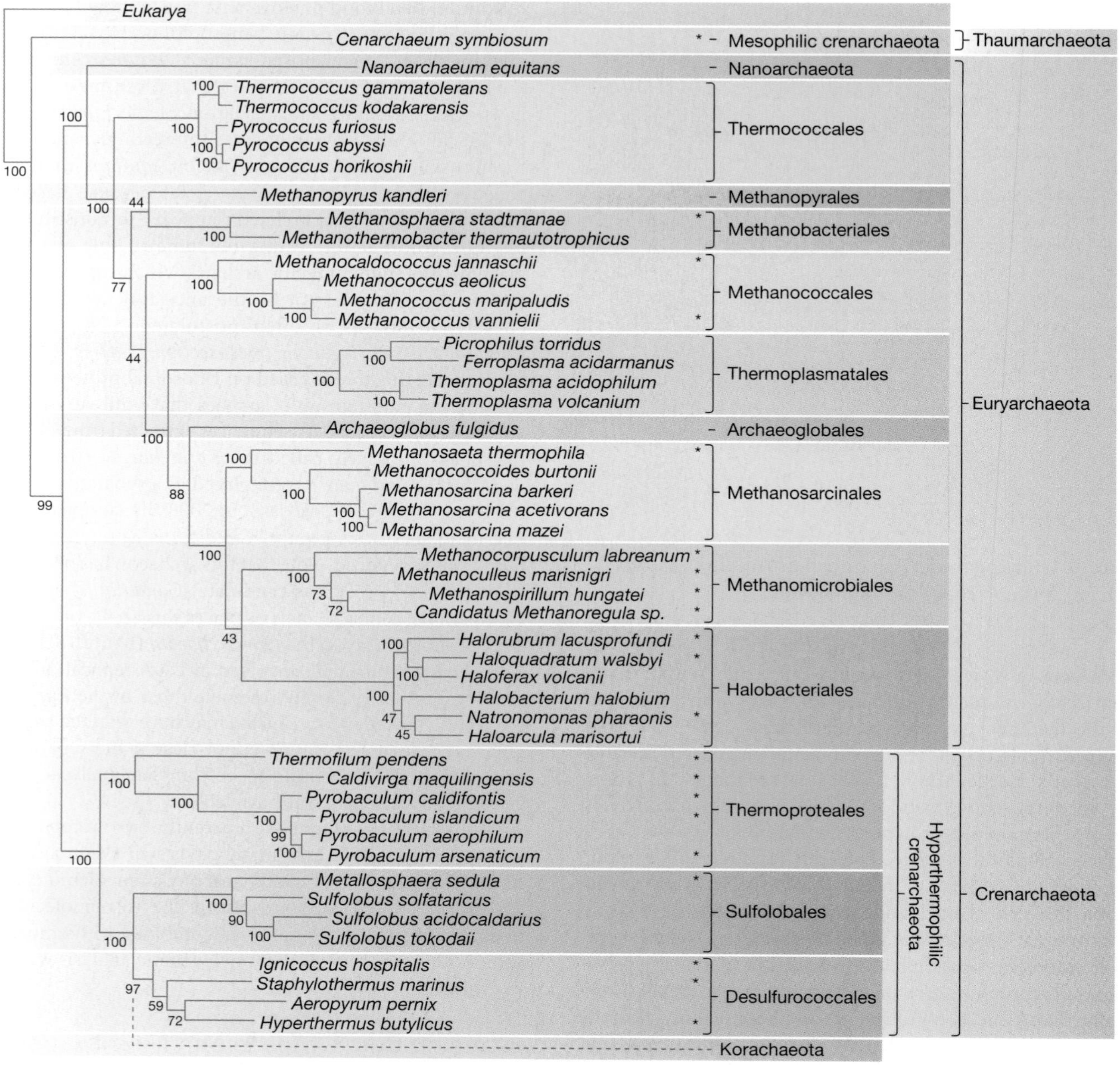

FIGURE 18.2 Maximum Likelihood Phylogenetic Tree of the *Archaea*. This phylogenetic tree was constructed by comparing ribosomal proteins. Note that the newly proposed phylum *Thaumarchaeota* has not been adopted by *Bergey's Manual*. The hypothesized phylum *Korarchaeota* is displayed with a dashed line because its placement is not based on the data used to construct this tree. Numbers at nodes represent bootstrap values. The scale bar represents the number of amino acid substitutions per site.

Source: Brochier-Armanet, C., et al. 2008. Mesophilic crenarchaeota: Proposal for a third archaeal phylum, the Thaumarchaeota. Nature Rev. Microbiol. *6:245.*

archaea, such as some methanogenic euryarchaeota, *Halococcus,* and *Natronococcus,* possess **pseudomurein** (a peptidoglycan-like polymer that is cross-linked with L-amino acids; *see figure 3.29*). *Methanosarcina* contains a complex polysaccharide similar to the chondroitin sulfate of animal connective tissue. Some filamentous methanogens have proteinaceous sheaths. ▶▶| *Phylum* Planctomycetes *(section 19.4); Phylum* Chlamydiae *(section 19.5); Class* Mollicutes *(section 21.1)*

The unusual microbe *Ignicoccus* is currently the only known archaeon to have an outer membrane. This hyperthermophile lives in association with cells of a smaller archaeon, *Nanoarchaeum equitans,* which grow on the surface of the *Ignicoccus*

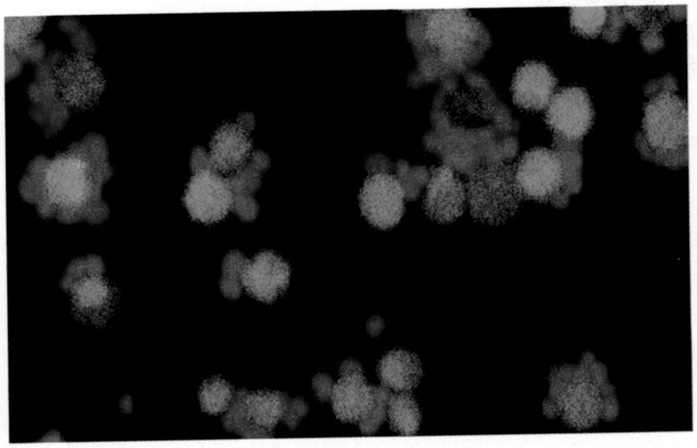

FIGURE 18.3 Parasitic Cells of *Nanoarchaeum equitans* Attached to the Surface of Its Host, *Ignicoccus*. The crenarchaeote *Ignicoccus* is the only archaeon known to have an outer membrane, which may be important for *N. equitans* attachment. In this confocal laser scanning micrograph, *Ignicoccus* is stained green and *N. equitans* is red.

outer membrane (**figure 18.3**). *Ignicoccus* lacks an S-layer but has a large periplasm in which vesicles are found. These vesicles fuse with the outer membrane and appear to be released to the medium. It is not known if these vesicles are used to interact with *N. equitans* cells. ◀◀ *Archaeal cell envelopes (section 3.4); Flagella (section 3.6); Archaeal flagellar movement (section 3.7)*

One of the most distinctive archaeal features is their membrane lipids. Many archaea must maintain membrane integrity at high temperatures and osmolarity. Archaeal membranes are typically less permeable to ions than bacterial membranes. As shown in **figure 18.4**, *Archaea* differ from both *Bacteria* and *Eukarya* in having branched chain hydrocarbons attached to glycerol by ether (rather than ester) linkages. The stereochemistry of the lipid attachment also differs. Halobacteria (which are archaea) and most other mesophilic archaea synthesize C_{20} diether lipid side chains, which are assembled into a bilayer membrane similar to that of bacterial and eukaryotic cells (figure 18.4, membrane lipid 3). Thermophilic archaea sometimes link two glycerol groups to form long tetraethers of 40 carbon atoms (membrane lipids 1 and 2). In addition, cells can adjust chain lengths by cyclizing the chains to form cyclopentane rings (membrane lipid 2). These rings are more densely packed within the membrane, making them more stable at high temperatures. In fact, thermophilic archaea increase the number of cyclopentane rings and the ratio of tetraether to diether side chains as growth temperature increases. Tetraether-based membranes are more impermeable to ions than are diether-based membranes. This is important because it prevents the loss of ions. Cyclopentane ring–containing lipids have also been discovered in nonthermophilic *Crenarchaeota*. These lipids, called **crenarchaeol,** are unique to these organisms, so they are used as a biomarker for the presence of crenarchaeotes in natural environments such as marine plankton. Polar phospholipids, sulfolipids, and glycolipids are also found in archaeal membranes. ◀◀ *Archaeal plasma membranes (section 3.4)*

Genetics and Molecular Biology

Some features of archaeal genetics are similar to those in *Bacteria*, whereas others more closely resemble *Eukarya*. Like many bacteria, all archaeal chromosomes examined so far are circular, and many genomes include plasmids. As we discuss in chapter 12, some archaea possess eukaryotic-like histone proteins, which are absent in *Bacteria*. In thermophiles and hyperthermophiles, histones may help prevent heat denaturation; note that the genomes of these microbes do not have especially high G + C ratios. In fact, the range of G + C ratios—from about 21 to 68%—illustrates archaeal diversity. Genome sizes are also variable. Chemolithotrophic archaea tend to have smaller genomes, between 1.5 to 2.0 Mbp, whereas other, more metabolically complex archaea have genomes up to about 6.0 Mbp (*see figure 16.17a*). One of the smallest genomes belongs to *N. equitans* (0.491 Mbp). Eukaryotic-like self-splicing introns have been observed in a few archaea. ◀◀ *Nucleic acid and protein structure (section 12.3)*

Archaeal DNA Replication and Cell Division

Archaeal DNA replication appears to be a complex mixture of eukaryotic and bacterial features. As in *Bacteria*, replication appears to be bidirectional from a single A-T–rich origin of replication, with the exception of *Sulfolobus*, which has two origins. However, most of the proteins involved in replication are similar to those found in eukaryotes (*see table 17.5*). In archaeal genomes that have been sequenced, the replication origin is flanked by genes encoding the eukaryotic-like initiation proteins Cdc6/Orc1. These proteins bind DNA elements called origin recognition boxes. A eukaryotic-like helicase called MCM then binds the region, and the DNA is locally unwound so that replication can begin. Once denatured, single-stranded DNA is stabilized by single-stranded binding proteins (SSBs). Some archaeal SSBs resemble the eukaryotic SSB, replication protein A (RPA), whereas others are more like bacterial SSBs.

Once the DNA is locally unwound, DNA synthesis can begin. All DNA polymerase enzymes include a sliding clamp—a protein that increases processivity by facilitating the action of proteins with which it interacts. Although the primary structure of the sliding clamp is not well conserved, its three-dimensional structure is. In the sliding clamps of all domains, DNA strands slide through the ring-shaped protein (*see figure 12.12*). The monomeric β-sliding clamp of *E. coli* has been well characterized, as has the homotrimeric eukaryotic sliding clamp, PCNA (*p*roliferating *c*ell *n*uclear *a*ntigen). All the archaea examined to date have a eukaryotic PCNA-like sliding clamp. *Euryarchaeota* have homotrimeric proteins, and *Crenarchaeota* possess heterotrimers, although there is evidence for more complex multimeric complexes. The thermostability of this eukaryotic-like protein in many archaea is thought to arise by a high density of ion pairs within the protein. This leads to increased electrostatic

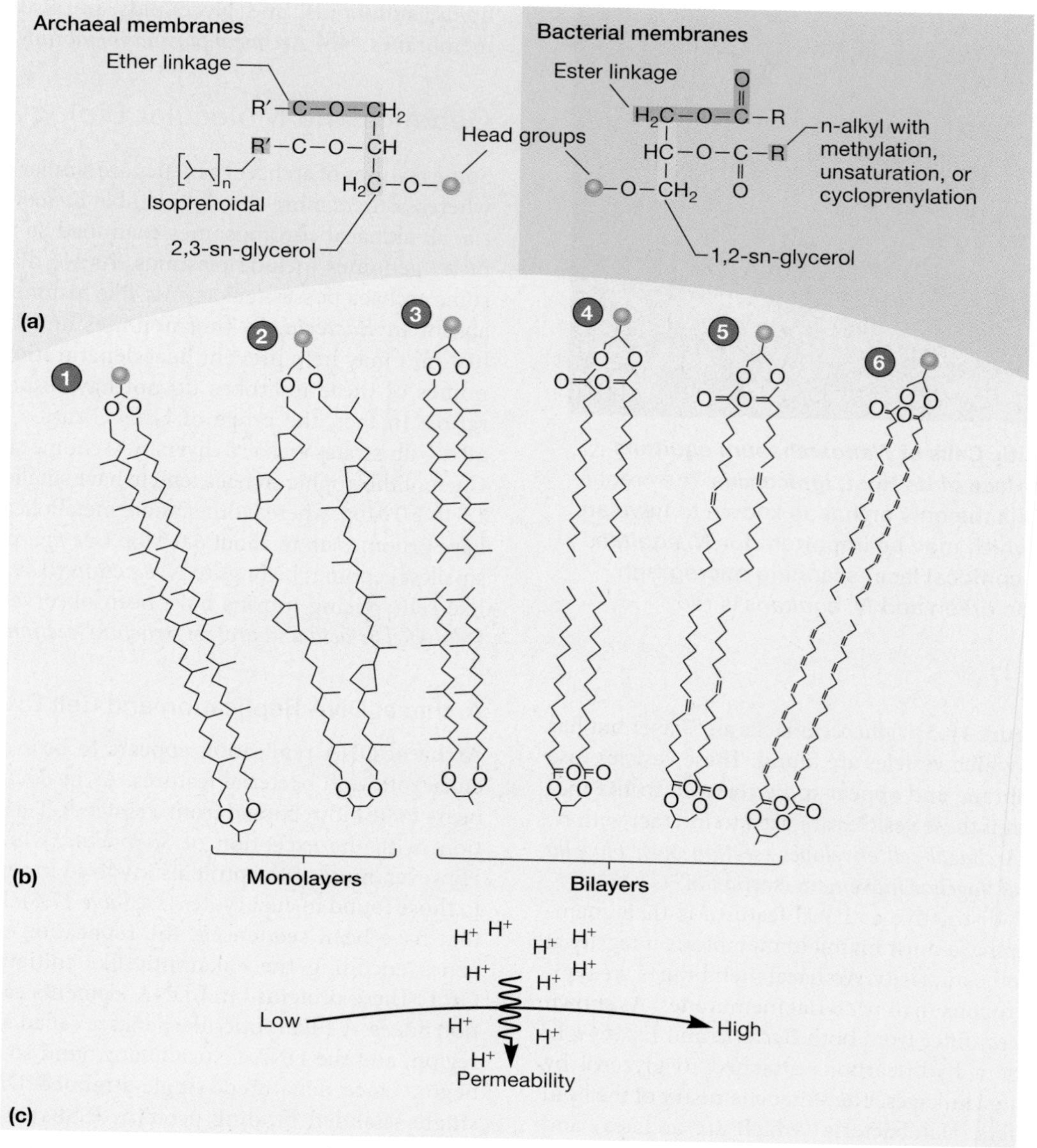

FIGURE 18.4 Comparison of Archaeal and Bacterial Membranes. (a) Archaeal membrane lipids are attached to a glycerol molecule by ether linkages instead of ester linkages, as found in bacteria. The stereochemistry of the attachment is also different, as noted by the position of the head group in the two representative lipids. (b) Membrane lipids 1, 2, and 3 are examples of archaeal membrane lipids, as noted by the branched chain hydrocarbons, the presence of tetraethers 40 carbons in length (1 and 2), and cyclopentane rings (2). Representative bacterial membrane lipids (4, 5, and 6) do not branch, are straight chains, and they do not link glycerol units to form tetraesters. (c) The relative permeability to protons increases as the complexity of the membrane lipid decreases. Because archaea have more complex membrane lipids, they tend to have more impermeable membranes.

Figure 18.4 Micro Inquiry

Considering the fact that some archaea live in very hot, acidic ecosystems, how is the construction of their membranes both in terms of hydrocarbon structure and H^+ permeability advantageous?

interactions that help prevent denaturation. ◀◀ *DNA replication (section 12.4)*

All archaea examined to date possess eukaryotic B-family DNA polymerases. Interestingly, euryarchaeal DNA replication requires both a B-family and an archaeal specific polymerase, PolD. In these microbes, the B polymerase serves to "read ahead" for deaminated cytosines (read as uracils). It is thought this feature may have evolved in these thermophiles in response to the higher rate of deamination under high temperature.

Although very little is known about cell division in the archaea, it is interesting that the sequenced genomes of hyperthermophilic crenarchaea lack genes for FtsZ/tubulin and MreB/actin homologues (*see figure 7.4*). However, unlike other microbes, they encode homologues of eukaryotic endosomal sorting system proteins Vps4 and ESCRT-III. In eukaryotes, these proteins are involved in endosomal trafficking and membrane dynamics during cytokinesis. In *Sulfolobus*, these proteins localize to the midcell during cell division, where they are important for normal cytokinesis. ◀◀ *Endocytic pathway (section 4.4)*

Archaeal Transcription

Transcription in the *Archaea* likewise blends bacterial and eukaryotic features. Archaeal RNA polymerases resemble eukaryotic RNA polymerases II and III, consisting of at least 10 subunits. Also, like eukaryotic nuclear RNA polymerase, archaeal RNA polymerases do not efficiently recognize promoter regions without the aid of additional proteins. Specifically, promoter recognition depends on at least two eukaryotic-like proteins: the *T*ATA-box-*b*inding *p*rotein (TBP) and *t*ranscription *f*actor *B* (TFB). It is therefore not surprising that many archaeal promoters are similar to certain eukaryotic promoters, possessing a TATA box (a 7-bp sequence found about 25 bp before the transcriptional start site) preceded by a purine-rich region called the *B r*esponsive *e*lement (BRE). In eukaryotes, BRE is the site to which transcription factor IIB binds. Likewise, archaeal TFB and TBP bind the BRE region of DNA as a prerequisite for the assembly of RNA polymerase subunits prior to the initiation of transcription (**figure 18.5**). Termination of transcription appears to involve pyrimidine or T-rich stretches. ◀◀ *Transcription in* Archaea *(section 12.6)*

In contrast to the eukaryotic-like transcriptional machinery, regulation of transcription in *Archaea* is very much like that in *Bacteria*. Like bacterial transcripts, some archaeal mRNAs are polycistronic. In addition, many of the transcriptional regulators found in archaea have homologues in bacteria, although some archaeal-specific regulatory proteins have been described. So while some archaeal transcriptional regulatory proteins mediate their effect by interacting directly with RNA polymerase (i.e., to either facilitate or prevent its binding to the promoter, as seen in bacteria), others target the eukaryotic-like transcription factors. For instance, some activators facilitate recruitment of TBP to the TATA box; conversely, repressors have been found that compete with TBP for promoter binding.

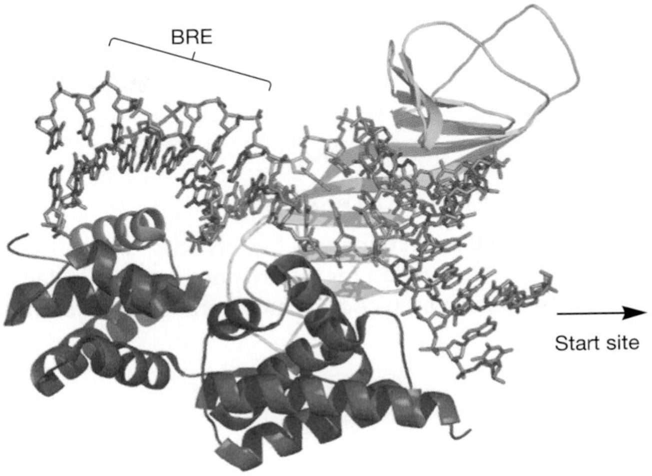

FIGURE 18.5 Archaeal Promoters Resemble Those of Eukaryotes. The crystal structure of the ternary complex between TBP, the carboxyl terminus of TFB, and a region of DNA containing a TATA box and BRE. DNA is shown in gray; TBP is the yellow ribbon structure; and TFB is magenta, with its recognition helix in turquoise. TBP = TATA-box binding protein; TFB = transcription factor B; BRE = B-responsive element.

Archaeal Translation and Protein Secretion

Because translation is such a fundamental cellular process, it is generally agreed that the last universal common ancestor (LUCA) possessed fairly sophisticated translational machinery, with ribosomes much like those in extant cells. Indeed, prior to the discovery that *Archaea* represent a third domain of life, it had been postulated that LUCA had a bacterial-like translational strategy and that the features that set eukaryotic translation apart evolved to address the more complex cellular needs of these organisms. The fact that translation in *Archaea* combines both eukaryotic and bacterial elements has challenged this view. For instance, although *Archaea* have 70S ribosomes, translation is insensitive to the antibiotics chloramphenicol and kanamycin, which bind bacterial ribosomes. By contrast, they are sensitive to the eukaryotic inhibitor anisomycin. Likewise, although many archaeal mRNAs possess ribosome-binding sites homologous to bacterial Shine-Delgarno sequences, many other mRNA transcripts lack a leader like eukaryotic mRNAs. Genome annotation has so far identified six putative initiation factors, two elongation factors (EFs), and a single release factor. Like eukaryotic EF-2s, archaeal EF-2s possess a modified histidine residue, which is ADP-ribosylated by diphtheria toxin (*see table 17.5*). Also like eukaryotes, the archaeal initiator tRNA carries methionine. However, unlike both *Bacteria* and eukaryotes, the TΨC arm of archaeal tRNA lacks thymine and contains pseudouridine or 1-methylpseudouridine. ◀◀ *Translation (section 12.8)*

Most interesting is the incorporation of the 21st amino acid **selenocysteine** in two specific regions of mRNA (**figure 18.6**).

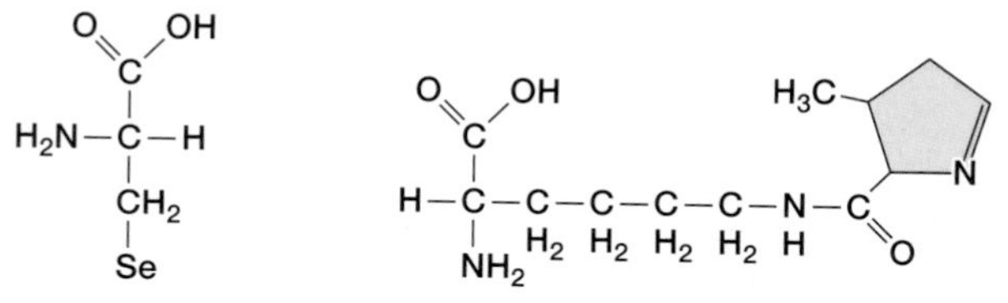

(a) Selenocysteine (b) Pyrrolysine

FIGURE 18.6 Selenocysteine and Pyrrolysine. These unusual amino acids are not coded for by DNA but are inserted in response to specific stop codons and stem-loop structures in mRNA.

Selenocysteine is synthesized from serine in the following way. When certain tRNAs are charged with serine, the enzyme selenocysteine synthase converts the serine to selenocysteine. This novel amino acid is recognized by a specific elongation factor and is incorporated when a UGA stop codon is encountered in association with specific nucleotide sequences called cis-acting *seleno*cysteine *i*nsertion *s*equence elements (SECIS). It is now known that selenocysteine is found in all domains of life. In *Bacteria,* SECIS are found immediately after the UGA stop codon. In *Archaea* and *Eukarya,* SECIS are located in the 3′ untranslated region of the gene. By contrast, **pyrrolysine,** the 22nd amino acid, has been discovered only in several methanogenic archaea and a single bacterium. It is inserted in response to UAG stop codons and a sequence element in the mRNA that forms a specific stem-loop structure called PYLIS (*pyrrolysine insertion sequence*). Like SECIS, PYLIS prevents translation from stopping and results in the insertion of the rare amino acid. It appears that organisms that use pyrrolysine employ UAA or UGA to stop translation instead of UAG. These organisms synthesize an unusual tRNA with a CUA anticodon and an animoacyl-tRNA synthase that charges it with pyrrolysine, which is derived from lysine. So unlike selenocysteine, which is made from an amino acid already attached to a tRNA, pyrrolysine is synthesized before it is added directly to its tRNA. ◀◀ *Translation (section 12.8)*

Protein secretion in all three domains can involve signal recognition particles (SRPs) that target new proteins to translocation sites, but the archaeal SRP differs from those in the other two domains. As in *Bacteria,* the archaeal SRP binds to the signal sequence of a preprotein and can direct it to the Sec protein secretion pathway for transport through the plasma membrane. The archaeal Sec pathway proteins, however, more closely resemble those of the eukaryotic pathway than the bacterial proteins. After the preprotein is moved across the membrane, its signal sequence is removed by a signal peptidase that resembles a subunit of the eukaryotic peptidase. Like *Bacteria, Archaea* can secrete folded proteins through the TAT system. In fact, the signal sequences are identical in both domains. The TAT system appears to be the predominant export mechanism among the halobacteria, and in some it is essential for viability. ◀◀ *Protein maturation and secretion (section 12.9)*

Metabolism

Not surprisingly, in view of the variety of archaeal lifestyles, archaeal metabolism varies greatly among the members of different groups. Some archaea are heterotrophs; others are autotrophic. A few carry out rhodopsin-based phototrophy, as discussed in section 18.3. ◀◀ *Rhodopsin-based phototrophy (section 10.12)*

Archaeal carbohydrate metabolism is best understood. At least three novel pathways have been identified among the *Archaea:* a modified Embden-Meyerhof (EM) pathway and two variations of the Entner-Doudoroff pathway. The modified EM pathway used by hyperthermophilic euryarchaeotes involves several novel enzymes, including an ADP-dependent glucokinase and phosphofructokinase (**figure 18.7**). These kinases each result in the production of AMP; in fact, unlike the classical EM pathway, there is no net ATP produced. Another unique step in this modified EM pathway is the conversion of glyceraldehyde 3-phosphate directly to 3-phosphoglycerate, with the use of ferridoxin as the electron acceptor, rather than NAD^+. By contrast, the thermophilic crenarchaeote *Thermoproteus* employs NAD^+ as the electron acceptor when it uses the modified EM pathway.

Extreme halophiles and *Thermoproteus* catabolize glucose using a phosphorylative form of the Entner-Doudoroff pathway (ED). As shown in **figure 18.8**, glucose is oxidized to gluconate and then dehydrated to 2-keto-3-deoxy-gluconate (KDG). KDG is phosphorylated at the expense of ATP to KDPG, and the remainder of the ED pathway is like that of the *Bacteria.* This results in a net yield of one ATP. A second, nonphosphorylative version of the ED pathway has been discovered in *Sulfolobus, Thermoplasma,* and *Thermoproteus* (note that *Thermoproteus* can utilize all three archaeal-specific pathways). This pathway begins like that of the halophiles, but KDG is cleaved by KDG aldolase to pyruvate and glyceraldehyde. Glyceraldehyde is oxidized to glycerate, which is then phosphorylated to 2-phosphoglycerate. Pyruvate is then generated as in the classical and halophilic ED pathway. Because these microbes bypass the energy-conserving step in which 1,3-bisphosphoglycerate is converted to 3-phosphoglycerate, there is no net yield of ATP. ◀◀ *Glycolytic pathways (section 10.3)*

All archaea that have been studied can oxidize pyruvate to acetyl-CoA. However, they lack the pyruvate dehydrogenase complex present in eukaryotes and respiratory bacteria, and use the enzyme pyruvate oxidoreductase to oxidize pyruvate to acetylCoA. Halophiles and the extreme thermophile *Thermoplasma* seem to have a functional tricarboxylic acid cycle. Methanogens do not catabolize glucose to any significant extent, and so it is not surprising that they lack a complete tricarboxylic acid cycle. However, some store glycogen, which is catabolized via a modified EM pathway (figure 18.7). Evidence for functional respiratory chains has been obtained in halophiles and thermophiles. ◀◀ *Tricarboxylic acid cycle (section 10.4)*

Very little is known in detail about biosynthetic pathways in the *Archaea.* Evidence suggests that the synthetic pathways for amino acids, purines, and pyrimidines are similar to those in other

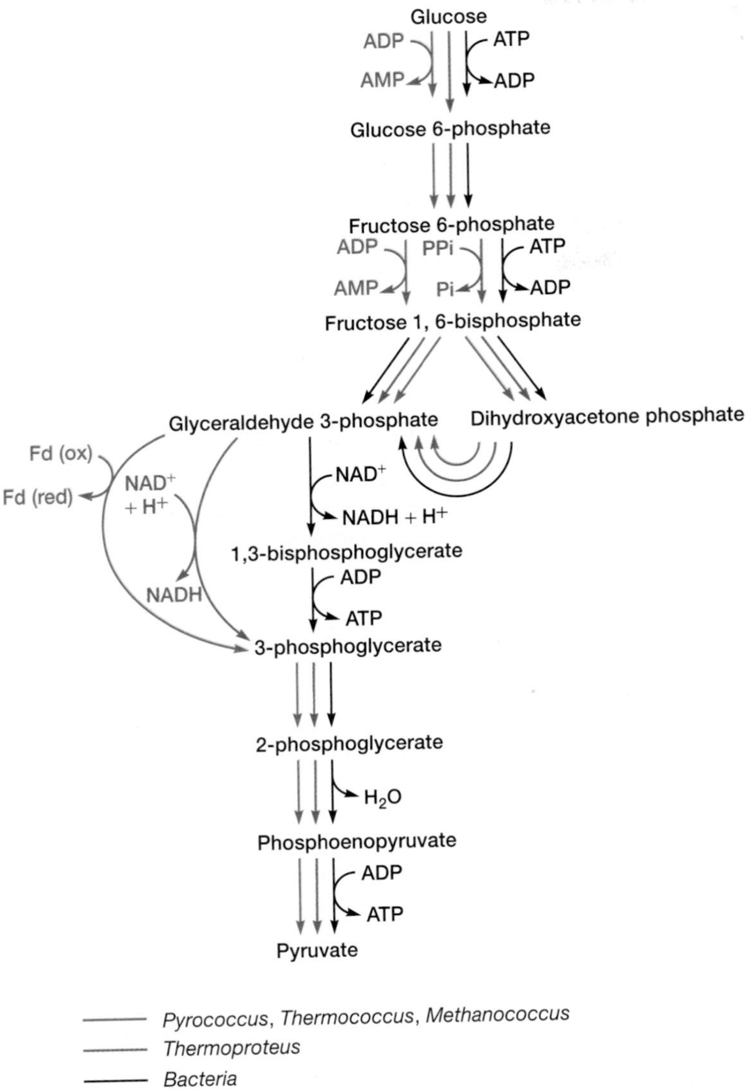

FIGURE 18.7 Modified Embden-Meyerhof Pathway. The pathway used by some archaea compared to that of the *Bacteria*. The only difference between the pathway used by *Thermoproteus* and the other archaea shown is the use of NAD^+ as the electron acceptor rather than ferridoxin when glyceraldehyde 3-phosphate is oxidized to 3-phosphoglycerate. There is no net ATP produced in the archaeal pathway.

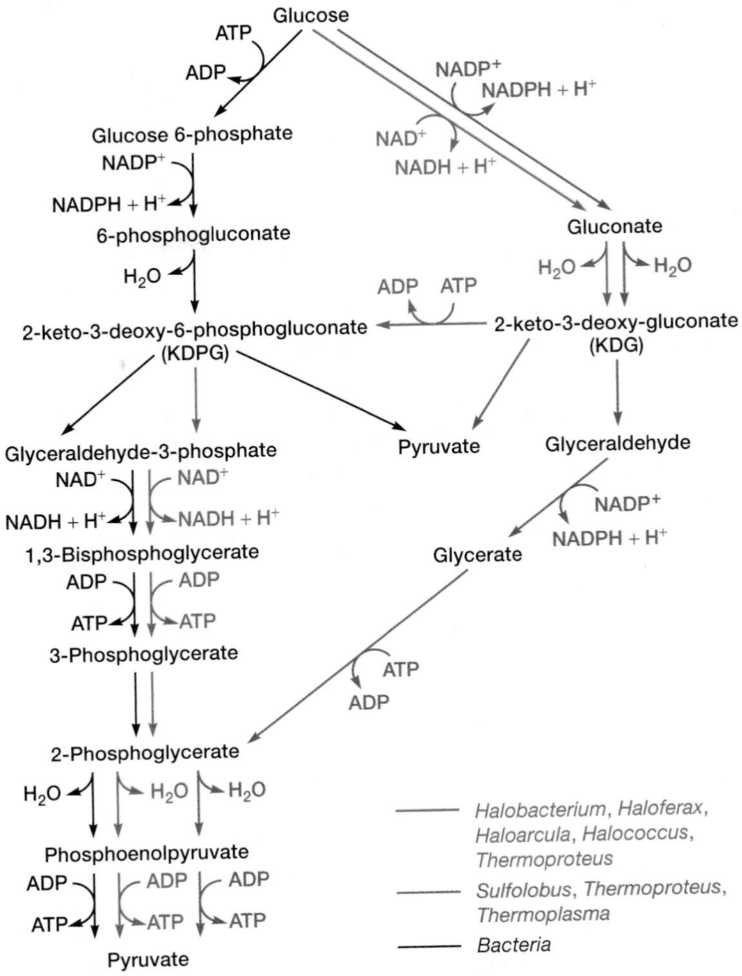

FIGURE 18.8 Two Alternate Versions of the Entner-Doudoroff Pathway. The phosphorylative pathway used by the halophilic archaea involves the ATP-dependent phosphorylation of KDG to KDPG; the remainder of the pathway is like that of the *Bacteria* and an ATP is gained. The nonphosphorylative pathway used by *Sulfolobus* and others joins the bacterial ED pathway when glycerate is converted to 2-phosphoglycerate so there is no net ATP produced.

organisms. Some methanogens can fix atmospheric dinitrogen. Many archaea, including halophiles and methanogens, reverse the EM pathway to synthesize glucose (a process known as gluconeogenesis; *see figure 11.10*), and at least some methanogens and extreme thermophiles employ glycogen as their major reserve material. ◀◀ *Synthesis of sugars and polysaccharides (section 11.4); Synthesis of amino acids (section 11.5)*

Autotrophy is widespread among the methanogens and extreme thermophiles, and CO_2 fixation occurs in more than one way. *Thermoproteus* incorporates CO_2 by the reductive tricarboxylic acid cycle (*see figure 11.6*). This pathway is also present in the green sulfur bacteria and some ε-proteobacteria. Methanogenic archaea and probably most extreme thermophiles incorporate CO_2 by the reductive acetyl-CoA pathway (*see figure 11.8*). A similar pathway also is present in acetogenic bacteria and autotrophic sulfate-reducing bacteria. Recently two novel CO_2 fixation pathways were discovered in the *Crenarchaeota*. Members of the *Sulfolobales* use the 3-hydroxyproprionate/4-hydroxybutyrate cycle, while anaerobic *Ignicoccus* uses the dicarboxylate/4-hydroxybutyrate cycle. In both cycles, acetylCoA and two CO_2 are joined to form succinylCoA, although each cycle uses its own specific enzymes.

Thermostability

The ability of many archaea to grow at very high temperatures has stimulated much research. The means by which archaeal membranes tolerate high temperatures has been discussed (p. 477). We now turn our attention to the thermostability of proteins, which has garnered considerable interest because proteins that function at high temperature can be used in a variety of industrial applications. This is because thermostable enzymes remain folded at temperatures that would denature similar proteins from mesophiles. Several strategies are used by thermophiles to increase protein rigidity. These include increasing the hydrophobicity of the protein core, elevating the number of ionic interactions on the protein surface, increasing atomic packing density, introducing additional hydrogen bonds, and shorter surface loop structures. Relatively minor changes in amino acid sequences can bring about these changes. For instance, thermostable proteins have relatively higher levels of valine, glutamate, and lysine, and fewer glutamine and valine residues. In addition to these changes, hyperthermophiles produce a specific class of chaperones at very high temperatures that function to refold partially denatured proteins. ⏮ *Protein maturation and secretion (section 12.9)*

Hyperthermophiles use more than one mechanism to prevent DNA from denaturing (becoming single stranded). They have a reverse DNA gyrase that induces positive, rather than negative, supercoils into DNA. Positive supercoiling dramatically enhances DNA thermostability; indeed, reverse DNA gyrase is found only in hyperthermophiles. Some hyperthermophiles also increase the solute concentration in their cytoplasm. This helps prevent the loss of purines and pyrimidines that can occur at high temperature. It is also thought that archaeal histones contribute to genome thermostability. ⏮ *Nucleic acid structure (section 12.3); DNA replication (section 12.4)*

The discovery of ever more thermophilic archaea suggests that the upper thermal limit of growth has yet to be discerned. Several years ago, a crenarchaeote provisionally called "strain 121" was isolated from a hydrothermal vent in the Pacific Ocean (*see figure 30.4*). This microbe is the current record holder for thermotolerance, remaining viable even after incubation at 130°C for two hours. This begs the question: What is the upper temperature limit for life on Earth? Best estimates are between 140 and 150°C. Above 150°C, essential macromolecules such as ATP are extremely unstable. This estimate has implications not only for Earth-bound life but also for those who seek life on other planets—if it is presumed that such life has the same macromolecular structure as life on Earth.

1. How are the phyla *Euryarchaeota* and *Crenarchaeota* distinguished?
2. How do archaeal cell walls differ from those of the *Bacteria*? What is pseudomurein?
3. In what ways do archaeal membrane lipids differ from those of *Bacteria* and eukaryotes? How do these differences contribute to the survival of thermophilic and hyperthermophilic archaea?
4. List the differences between *Archaea* and the other domains with respect to DNA replication, transcription, and translation. Do you think *Archaea* resemble *Eukarya* or *Bacteria*, or are they largely unique in this regard?
5. What are the three variations of glucose catabolism used by archaea? Compare the ATP and NADH yield of each of these archaeal pathways with that used by bacteria.
6. In what two unusual ways do archaea incorporate CO_2? How does this compare with other domains?

18.2 Phylum *Crenarchaeota*

Bergey's Manual of Systematic Microbiology lists only the well-characterized thermophilic members of the phylum *Crenarchaeota*. These archaea belong to a single class, *Thermoprotei,* which is divided into four orders and six families. The order *Thermoproteales* contains two families. The family *Thermoproteaceae* includes anaerobic to facultative, hyperthermophilic rods genera. The family *Thermophilaceae* has only one representative, *Thermophilum pendens*, an anaerobic thermoacidophile that is filamentous with cells extending more than 100 μm. Members of the order *Sulfolobales* are coccus-shaped thermoacidophiles. The two families within the order *Desulfurococcales* contain coccoid or disk-shaped hyperthermophiles. They grow chemolithotrophically by hydrogen oxidation or organotrophically by fermentation or respiration with sulfur as the electron acceptor. Members of the family *Desulfurococcaceae* are also capable of heterotrophic growth using sulfur as the electron acceptor. This family includes *Ignicoccus,* the host of *N. equitans* (figure 18.3). The family *Pyrodictiaceae* includes the genus *Pyrodictium,* which forms tubelike structures that connect cells, thereby forming a network called cannulae. Also included in this family is the genus *Pyrolobus. P. fumarii* is one of the most thermophilic microbes isolated to date. Its optimum growth temperature is 106°C and its maximum is 113°C. The order *Caldisphaerales* has only one genus, *Caldisphaera,* whose members are thermoacidophilic, aerobic, and heterotrophic cocci.

Many of these thermophiles are sulfur dependent. The sulfur may be used either as an electron acceptor in anaerobic respiration or as an electron donor by lithotrophs. Many are strict anaerobes. They grow in geothermally heated water or soils that contain elemental sulfur. These environments are scattered all over the world. Familiar terrestrial examples are the sulfur-rich hot springs in Yellowstone National Park (**figure 18.9**). Many hyperthermophiles have been isolated from the waters surrounding areas of submarine volcanic activity. Such habitats are sometimes called solfatara. These archaea can be very thermophilic and often are classified as **hyperthermophiles.** The most extreme example was isolated from an active hydrothermal vent in the northeast Pacific Ocean. This is one of three novel isolates that constitute a new genus in the *Pyrodictiaceae* family. Its optimum growth temperature is about 105°C, but even autoclaving this microbe at 121°C for 1 hour fails to kill it. It is strictly anaerobic, using Fe(III) as a terminal electron acceptor and H_2 or formate as electron donors and energy sources (**figure 18.10**). ⏭ *Sulfide-based mutualisms (section 30.1)*

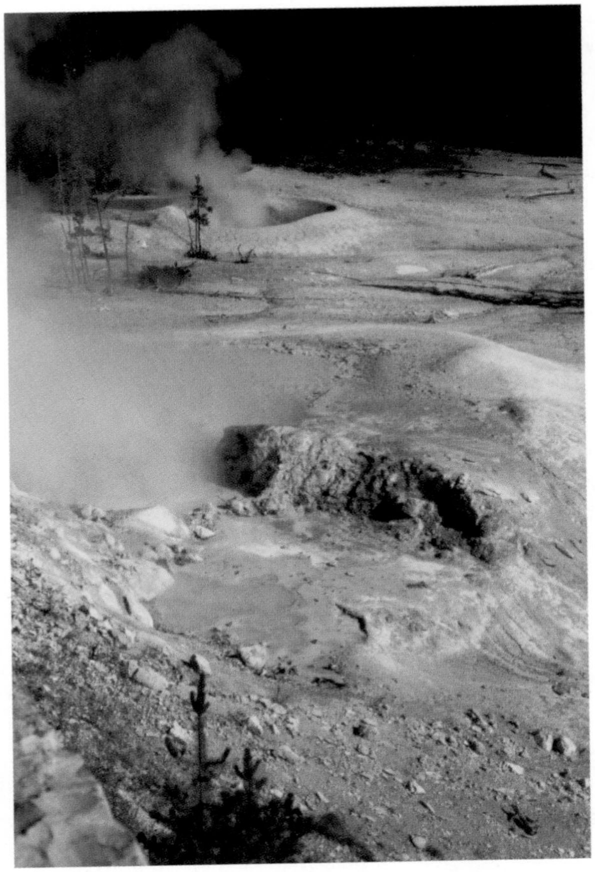

FIGURE 18.9 Habitat for Thermophilic Archaea. The Sulfur Cauldron in Yellowstone National Park. The water is at its boiling point and very rich in sulfur. *Sulfolobus* grows well in such habitats.

At present, *Crenarchaeota* contains 25 genera; two of the better-studied genera are *Sulfolobus* and *Thermoproteus.* Members of the genus *Sulfolobus* stain gram negative and are aerobic, irregularly lobed, spherical archaea with a temperature optimum around 80°C and a pH optimum of 2 to 3. For this reason, they are **thermoacidophiles.** Their cell walls contain lipoprotein and carbohydrate. They grow heterotrophically under oxic conditions, but they can also grow chemolithotrophically, oxidizing H_2, H_2S, and FeS_2 with oxygen as the terminal electron acceptor, although the use of ferric iron has been reported. To date, the genomes of three *Sulfolobus* species have been sequenced and annotated. These genomes show a high level of plasticity; indeed, the *S. solfataricus* genome has 200 integrated insertion sequences. Although *S. solfataricus* grows at pH 2 to 4, it maintains a cytoplasmic pH of about 6.5, thereby generating a large pH gradient across the plasma membrane. This energy is conserved in the formation of ATP by membrane-bound ATP synthases and at least 15 secondary transport systems that couple the transport of organic solutes (e.g., sugars) with the movement of protons (**figure 18.11**). ABC transporters are also used for nutrient uptake; these have high substrate affinities, making them important in low nutrient environments. When growing heterotrophically, sugars are oxidized via the nonphosphorylative version of the Entner-Doudoroff pathway (figure 18.8) and a complete TCA cycle. Unlike most

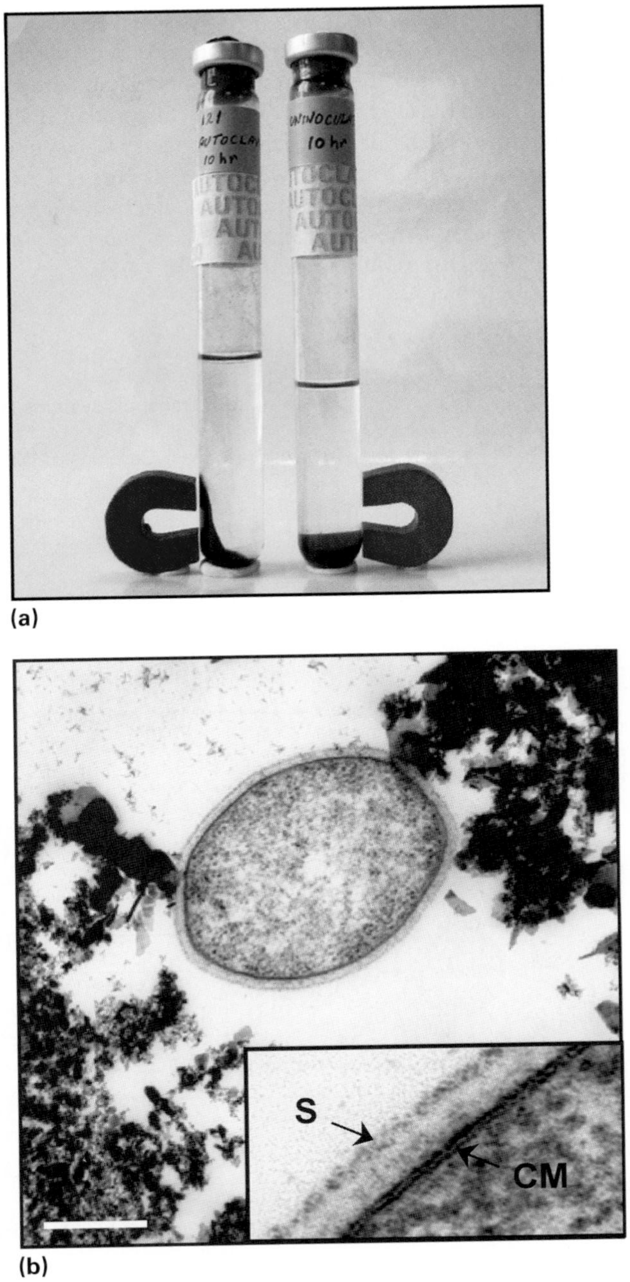

FIGURE 18.10 An Extremely Hyperthermophilic Crenarchaeote. (a) A member of the family *Pyrodictiaceae* (tube on left) grows following autoclaving at 121°C, as shown by its ability to reduce Fe(III) to magnetite when incubated anaerobically (tube on right is sterile control). (b) A transmission electron micrograph shows the single-layer cell envelope (S) and plasma membrane. Scale bar = 1 μm.

organisms, *S. solfataricus* rarely uses NAD^+ as an electron acceptor; instead, it uses $NADP^+$ and ferredoxin (fd)-dependent oxidoreductases. Note that the electron transport chain shown in figure 18.11 uses a ferridoxin (fd) oxidoreductase, rather than the usual respiratory NADH dehydrogenase. In fact, archaea in general seem to lack this enzyme, whereas it is integral to the function of bacterial and mitochondrial electron transport chains. ⏮ *Electron transport and oxidative phosphorylation (section 10.5); Transposable elements (section 14.5)*

Thermoproteus is a long, thin rod that can be bent or branched. Its cell wall is composed of glycoprotein. *Thermoproteus* is an anaerobe that grows at temperatures from 75 to 100°C. Some species are acidophiles, with optimum pH values between 3 and 4, while others are neutrophiles. It is found in hot springs and other hot aquatic habitats rich in sulfur. It can grow organotrophically and oxidize glucose, amino acids, alcohols, and organic acids with elemental sulfur as the electron acceptor during anaerobic respiration. It will also grow chemolithotrophically, oxidizing H_2

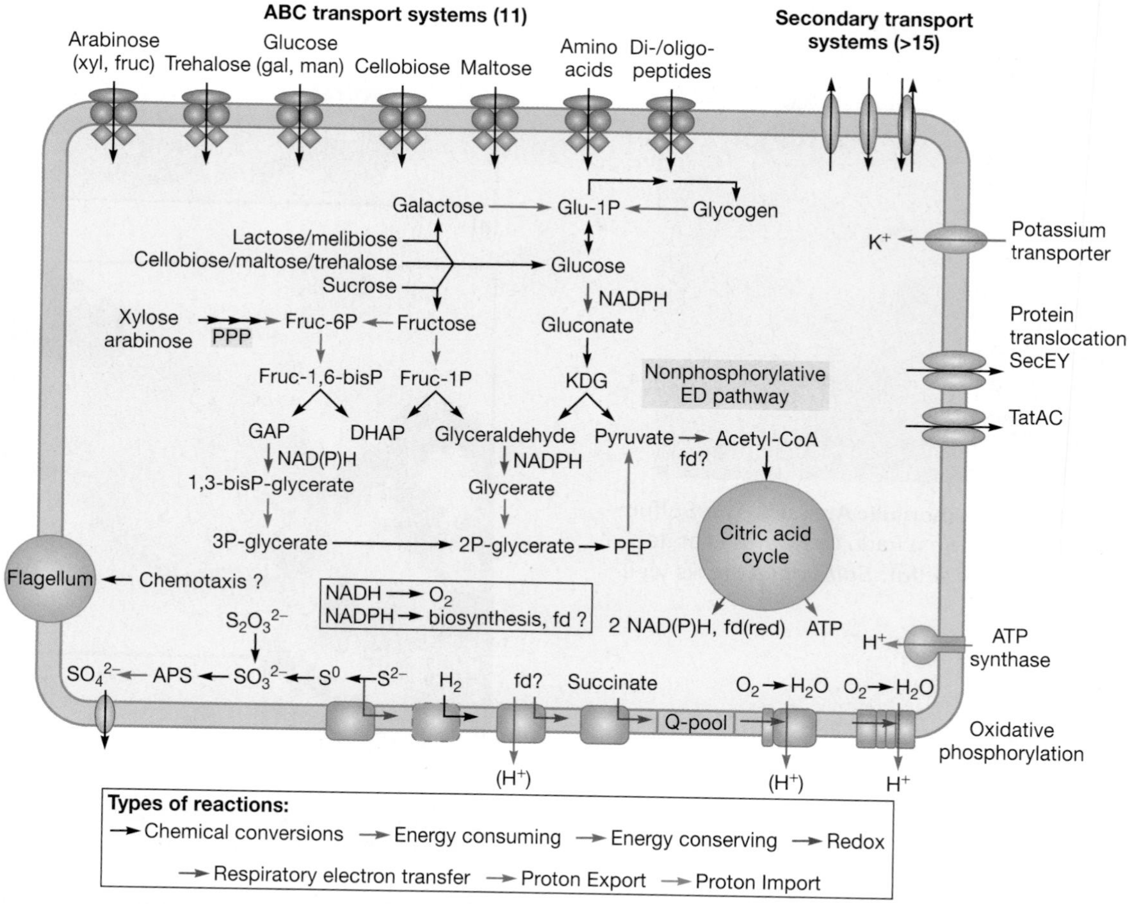

FIGURE 18.11 Genomic reconstruction of *Sulfolobus solfataricus*. Each type of reaction is denoted by a different color arrow, as shown in the key. Components of the aerobic respiratory network include the reduction of a quinone (Q) pool, a putative ferridoxin dehydrogenase (fd?), and succinate dehydrogenase. Alternative electron donors include hydrogen and sulfide, which reduce the quinone pool through hydrogenase and sulfide reductase, respectively. Elemental sulfur and thiosulfate are converted to sulfate. The ABC transport systems import xylose (xyl), fructose (fruc), glucose (glu), galactose (gal), and mannose (man). Transporters for carbohydrates that support growth are listed in blue. The flagellar apparatus more closely resembles type IV pili, as is the case with other archaea, and a gene encoding a methyl-accepting chemotaxis protein is present.

Figure 18.11 Micro Inquiry

Identify the potential electron donors to the electron transport chain. What is the source of energy for the potassium transporter shown on the right-hand side of the cell? Hint: examine arrow color code box.

with S^0 as the electron acceptor. CO_2 can serve as the sole carbon source; it is incorporated through the reductive TCA cycle.

The discovery of the crenarchaeote-specific lipid crenarchaeol and sequence analysis of DNA fragments derived directly from environmental samples revealed that crenarchaeol-containing archaea are widespread in nature (**figure 18.12**). Recall that only a small fraction of microbes have been grown in culture, so the ability to analyze microbial communities using molecular techniques is an important way to truly understand microbial diversity. Such studies demonstrate that significant populations of these archaea dwell in marine plankton from polar, temperate, and tropical waters. They also inhabit rice paddies, soils, and freshwater lake sediments, and at least two symbiotic species have been isolated, one from a cold water sea cucumber and *Cenarchaeum symbiosum* from a marine sponge. Collectively, these microbes are often called the group I archaea or mesophilic *Crenarchaeota*, but as discussed in section 18.1, comparative genome analysis suggests that these microbes may belong to a separate phylum (figure 18.2). ◀◀ *Metagenomics (section 16.8)*

The purification and growth of one of these mesophilic archaea, the marine crenarchaeote *Nitrosopumilus maritimus*, confirmed the notion that archaea are capable of nitrification. This had been suggested by large genomic inserts cloned directly from soil and seawater. *N. maritimus* grows chemolithoautotrophically, capturing energy through the oxidation of ammonia to nitrite using oxygen as the terminal electron acceptor. The first step in ammonia oxidation is its conversion to hydroxylamine (NH_2OH), a reaction catalyzed by ammonia monooxygenase (AMO; *see figure 20.13*). AMO is composed of three subunits, AmoA, AmoB, and AmoC; the archaeal versions of these genes are phylogenetically distinct from bacterial *amo* genes. This has made it possible to assess the presence and, through reverse transcription PCR studies, activity of archaeal nitrification in soils and waters. While it had long been thought that α- and β-proteobacteria were solely responsible for nitrification, it appears that ammonia oxidation by mesophilic archaea may be more important. ▶▶ *Nitrifying bacteria (section 20.1)*

1. What are thermoacidophiles and where do they grow? In what ways do they use sulfur in their metabolism?
2. Briefly describe *Sulfolobus* and *Thermoproteus*.
3. Discuss the role of external pH on the magnitude of the proton motive force generated by *Sulfolobus*.
4. How is crenarchaeol used in discovering habitats for archaea not previously identified?

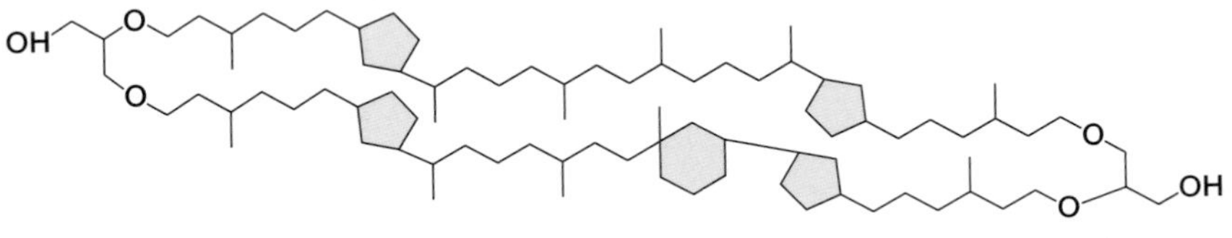

FIGURE 18.12 Crenarchaeol. This membrane lipid is composed of dicyclic biphytane and a tricyclic biphytane.

18.3 Phylum *Euryarchaeota*

Euryarchaeota is a very diverse phylum with many genera (figure 18.2). We discuss five major physiologic groups within the euryarchaeotes.

Methanogens

Late in the eighteenth century, Italian physicist Alessandra Volta discovered that he could ignite gas from anoxic marshes. In this way, Volta discovered the biological production of methane and demonstrated the industrial importance of natural gas. The production of methane—**methanogenesis**—is the last step in the anaerobic degradation of organic compounds. Because the ΔG of methanogenesis compares unfavorably to other forms of respiration, methanogenesis occurs only when O_2 and most other electron acceptors are unavailable, making it a strictly anaerobic process. All methanogenic microbes are archaea and are called **methanogens.** They use H_2 and CO_2 or short-chain organic compounds, such as formate, acetate, and methanol, as substrates. These compounds are usually the fermentation products of other microbes that live in the same community. When using H_2 and CO_2, their growth is autotrophic. These archaea form acetyl-CoA from two molecules of CO_2 and then convert the acetyl-CoA to pyruvate and other products (*see figure 11.8*).

Methanogens are the largest group of cultured archaea. There are five orders (*Methanobacteriales, Methanococcales, Methanomicrobiales, Methanosarcinales,* and *Methanopyrales*) and 26 genera, which differ greatly in overall shape, 16S rRNA sequence, cell wall chemistry and structure, membrane lipids, and other features. For example, methanogens construct three different types of cell walls. Several genera have walls with pseudomurein; other walls contain either proteins or heteropolysaccharides. The morphology of two representative methanogens is shown in **figure 18.13**, and selected properties of representative genera are presented in **table 18.2**.

As might be inferred from the methanogens' ability to produce methane anaerobically, their metabolism is unusual. These euryarchaeotes contain several unique cofactors: tetrahydromethanopterin (H_4MPT), methanofuran (MFR), coenzyme M (2-mercaptoethanesulfonic acid), coenzyme F_{420}, and coenzyme F_{430} (**figure 18.14**). These cofactors are used in the following series of reactions (**figure 18.15**). (1) Gaseous CO_2 is bound to MFR, and as such, it is now activated and part of a formyl group (-HC=O). (2) The formyl group is transferred to H_4MPT and dehydrated. (3) Using F_{420} as the electron donor, the dehydrated formyl group (=HC-) is reduced to the methyl level (-CH_3) in a two-step process. (4) The methyl group is transferred to CoM. This reaction releases energy. (5) The methyl group is now ready for final reduction to methane (CH_4). This is catalyzed by the F_{430}-containing methyl-reductase. H_2 is the electron donor and protons are released.

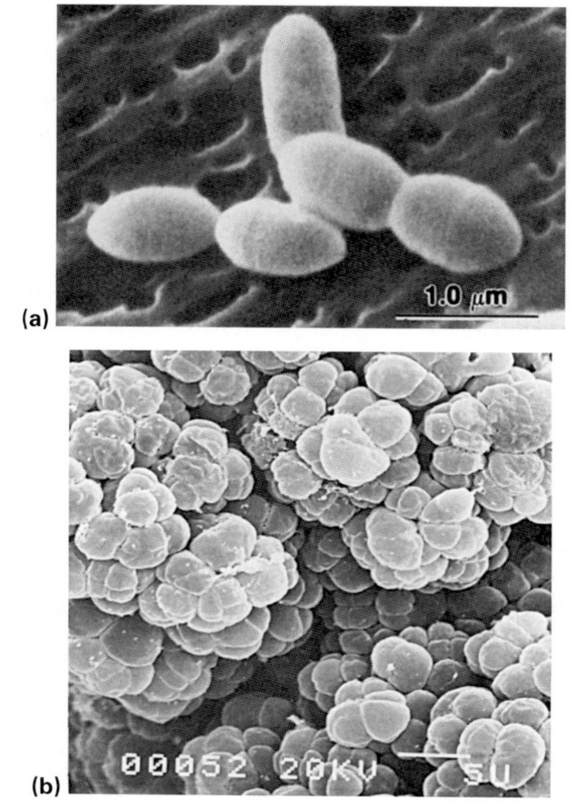

FIGURE 18.13 Selected Methanogens. (a) *Methanobrevibacter smithii.* (b) *Methanosarcina mazei;* SEM. Bar = 5 μm.

It appears that ATP synthesis is linked with methanogenesis by electron transport, proton pumping, and a chemiosmotic mechanism. The precise means by which a proton motive force is generated is not known but may involve the oxidation of H_2 on the outer surface of the membrane (figure 18.15, step 5). In addition, the transfer of the methyl groups from methyl-H_4MPT to HS-CoM releases sufficient energy for the uptake of sodium ions (step 4). This results in a sodium motive force that could also drive ATP synthesis. This may explain why methanogens require about 1 mM Na^+ for growth. However, some microbiologists assert that both Na^+ and H^+ ATP syntheses translocate protons. ◀◀ *Electron transport and oxidative phosphorylation (section 10.5)*

It can be useful to compare methanogens that possess cytochromes with those that lack them. Among the five orders of methanogens, only members of the least ancient order, the *Methanosarcinales*, have cytochromes and methanophenazine, a menaquinone-like molecule. Most of these archaea have a broad substrate range and grow on acetate, methanol, and methylamines, with a few species of *Methanosarcina* also growing on H_2 and CO_2. These archaea tend to have relatively high growth yields of around 7 grams of dried cellular material per mole of CH_4 produced. By contrast, most members of the other four orders convert only H_2 and CO_2 to methane; a few can also use formate as a methanogenic substrate. These archaea have growth yields of between 1.5 to 3 grams per mole CH_4.

Methanogenic archaea are potentially of great practical importance since methane is a clean-burning fuel and an excellent

Table 18.2 Selected Characteristics of Representative Genera of Methanogens

Genus	*Morphology*	*% G + C*	*Wall Composition*	*Gram Reaction*	*Motility*	*Methanogenic Substrates Used*
Order *Methanobacteriales*						
Methanobacterium	Long rods or filaments	32–61	Pseudomurein	+ to variable	–	$H_2 + CO_2$, formate
Methanothermus	Straight to slightly curved rods	33	Pseudomurein with an outer protein S-layer	+	+	$H_2 + CO_2$
Order *Methanococcales*						
Methanococcus	Irregular cocci	29–34	Protein	–	+	$H_2 + CO_2$, formate
Order *Methanomicrobiales*						
Methanomicrobium	Short curved rods	45–49	Protein	–	+	$H_2 + CO_2$, formate
Methanogenium	Irregular cocci	52–61	Protein or glycoprotein	–	–	$H_2 + CO_2$, formate
Methanospirillum	Curved rods or spirilla	47–52	Protein	–	+	$H_2 + CO_2$, formate
Order *Methanosarcinales*						
Methanosarcina	Irregular cocci, packets	36–43	Polysaccharide or protein	+ to variable	–	$H_2 + CO_2$, methanol, methylamines, acetate

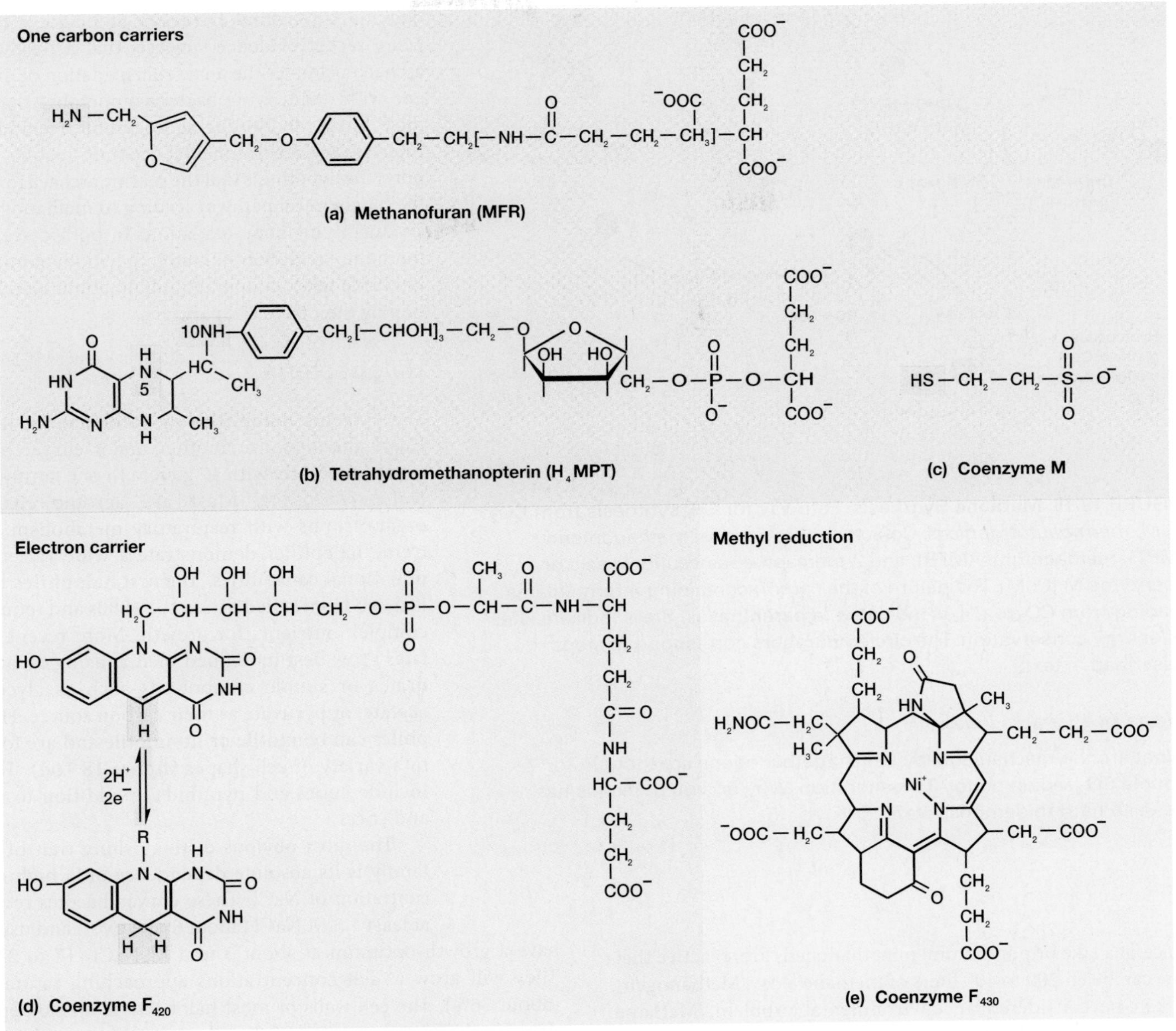

FIGURE 18.14 Methanogen Coenzymes. (a) Coenzyme MFR, (b) H_4MPT, and (c) coenzyme M are used to carry one-carbon units during methanogenesis. MFR and a simpler form of H_4MPT called methanopterin (MPT; not shown) also participate in the synthesis of acetyl-CoA. The portions of the coenzymes that carry the one-carbon units are highlighted. H_4MPT carries carbon units on nitrogens 5 and 10, like the more common enzyme tetrahydrofolate. (d) Coenzyme F_{420} participates in redox reactions. The part of the molecule that is reversibly oxidized and reduced is highlighted. (e) Coenzyme F_{430} participates in reactions catalyzed by the enzyme methyl-CoM methylreductase.

energy source. Anaerobic digester microbes degrade particulate wastes such as sewage sludge to H_2, CO_2, and acetate (*see figure 41.8*). CO_2-reducing methanogens form CH_4 from CO_2 and H_2, while aceticlastic methanogens cleave acetate to CO_2 and CH_4 (about two-thirds of the methane produced by an anaerobic digester comes from acetate). A kilogram of organic matter can yield up to 600 liters of methane. It is quite likely that future research will greatly increase the efficiency of methane production and make methanogenesis an important source of pollution-free energy.

▶▶| *Microbial energy conversion (section 41.3); Wastewater treatment (section 42.2)*

Methanogenic archaea are estimated to produce about 1 billion tons of methane annually. The rate of methane production can be so great that bubbles of methane sometimes rise to the

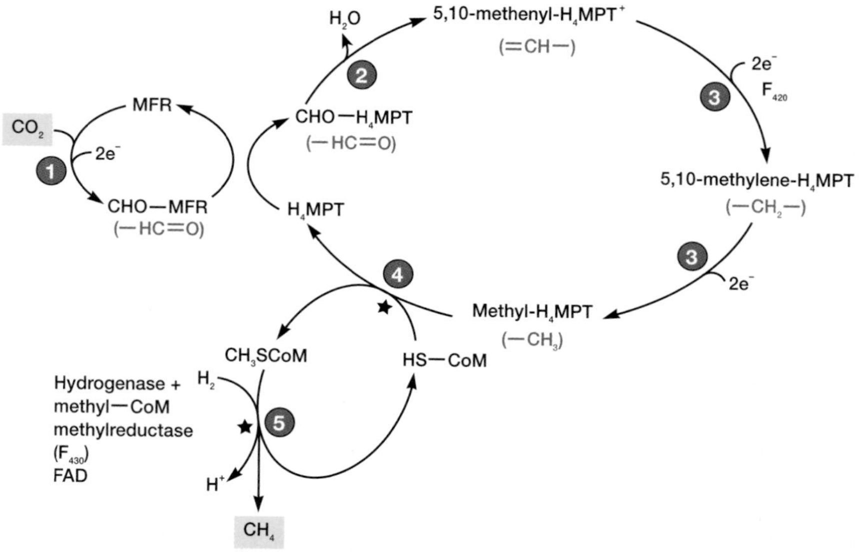

FIGURE 18.15 Methane Synthesis. Pathway for CH_4 synthesis from CO_2 in *M. thermoautotrophicus.* Cofactor abbreviations: methanopterin (MPT), methanofuran (MFR), and 2-mercaptoethanesulfonic acid or coenzyme M (CoM). The nature of the carbon-containing intermediates leading from CO_2 to CH_4 is indicated in parentheses. Stars indicate sites of energy conservation. The circled numbers correspond to steps described in text.

***Figure 18.15** Micro Inquiry*

What are the mechanisms by which methanogens are thought to couple CO_2 reduction to ATP generation. Why do you think this has been so hard to demonstrate?

surface of a lake or pond. Rumen methanogens are so active that a cow can belch 200 to 400 liters of methane a day. Methanogenesis can be considered an environmental problem. Methane absorbs infrared radiation and is a more potent greenhouse gas than CO_2. Atmospheric methane concentrations have been rising over the last 200 years. Methane production may significantly promote global warming; this subject is discussed in chapter 26.
▶▶| *The rumen ecosystem (section 30.1); Global climate change (section 26.2)*

Search This: USGS methanogenesis

No doubt the levels of archaeal-generated methane would render this planet uninhabitable were it not for the oxidation of methane by microbes capable of methylotrophy. For many decades, it was thought that this process—the capacity to use methane as a carbon source—was restricted only to aerobic *Proteobacteria.* However, it is now recognized that certain archaea can oxidize methane anaerobically. This was first discovered in marine archaea that live in close association with sulfate-reducing bacteria (**Microbial Diversity & Ecology 18.1**). More recent evidence suggests that a freshwater archaeon couples the anaerobic oxidation of methane with denitrifying bacteria—microbes that reduce nitrate to N_2 during anaerobic respiration. Interestingly, environmental genomic analysis supports the hypothesis that the marine archaea reverse the biochemical pathway leading to methanogenesis during methane oxidation. In both cases, the methane oxidation is only thermodynamically favorable when coupled to sulfate or nitrate oxidation by their bacterial partners.

Halobacteria

The **extreme halophiles** or **halobacteria,** order *Halobacteriales,* are another major euryarchaeal group, currently with 17 genera in one family, the *Halobacteriaceae.* Most are aerobic chemoorganotrophs with respiratory metabolism. Extreme halophiles demonstrate a wide variety of nutritional capabilities. The first **halophiles** were isolated from salted fish in the 1880s and required complex nutrients for growth. More recent isolates grow best in defined media, using carbohydrates or simple compounds such as glycerol, acetate, or pyruvate as their carbon source. Halophiles can be motile or nonmotile and are found in a variety of cell shapes (**figure 18.16*a***). These include cubes and pyramids in addition to rods and cocci.

The most obvious distinguishing trait of this family is its absolute dependence on a high concentration of NaCl. These euryarchaeotes require at least 1.5 M NaCl (about 8%, wt/vol) and usually have a growth optimum at about 3 to 4 M NaCl (17 to 23%). They will grow at salt concentrations approaching saturation (about 36%). The cell walls of most halobacteria are so dependent on the presence of NaCl that they disintegrate when the NaCl concentration drops below 1.5 M. Thus halobacteria only grow in high-salinity habitats such as marine salterns and salt lakes such as the Dead Sea between Israel and Jordan, and the Great Salt Lake in Utah. Halobacteria often have red-to-yellow pigmentation from carotenoids that are probably used as protection against strong sunlight. They can reach such high population levels that salt lakes, salterns, and salted fish actually turn red (figure 18.16*b*). In fact, halophiles are used in the production of many salted food products.

Halophiles use two strategies to cope with osmotic stress. The first approach is to increase cytoplasmic osmolarity by accumulating small organic molecules called **compatible solutes.** These include glycine, betaine, polyols, ectoine, and amino acids. When this strategy is used, salt adaptation of cytoplasmic proteins is not necessary. Compatible solutes are found in extremely halophilic methanogenic archaea, as well as halophilic bacteria.

MICROBIAL DIVERSITY & ECOLOGY

18.1 Methanotrophic *Archaea*

The marine environment may contain as much as 10,000 billion tons of methane hydrate buried in the ocean floor, around twice the amount of all known fossil fuel reserves. Although some methane rises toward the surface, it often is used before it escapes from the sediments in which it is buried. This is fortunate because methane is a much more powerful greenhouse gas than carbon dioxide. If the atmosphere were flooded with methane, Earth could become too hot to support life as we know it. The reason for this disappearance of methane in sediments has been unclear until a recent discovery.

By using fluorescent probes for specific DNA sequences, an assemblage of archaea and bacteria was discovered in anoxic, methane-rich sediments. These clusters of cells contain a core of about 100 cells from the order *Methanosarcinales* surrounded by a layer of sulfate-reducing bacteria related to the *Desulfosarcina* (**box figure**). These two groups appear to cooperate metabolically in such a way that methane is anaerobically oxidized and sulfate reduced; perhaps the bacteria use waste products of methane oxidation to derive energy from sulfate reduction. Isotope studies show that the archaea feed on methane and the bacteria get much of their carbon from the archaea. These methanotrophs may be crucial contributors to Earth's carbon cycle because it is thought that they oxidize as much as 70 billion kilograms of methane annually.

Since the discovery of this first consortium in 2000, others have described anaerobic methane-oxidizing archaea living in association with sulfate-reducing bacteria. It has been suggested that these methanotrophic archaea evolved the ability to reverse methanogenesis from an ancestral methanogen. Support for this model comes from DNA sequencing of methane-oxidizing archaea from the deep sea. These microbes appear to possess all the genes needed for methanogenesis, implying that anaerobic methane oxidation is accomplished by running methanogenesis backwards.

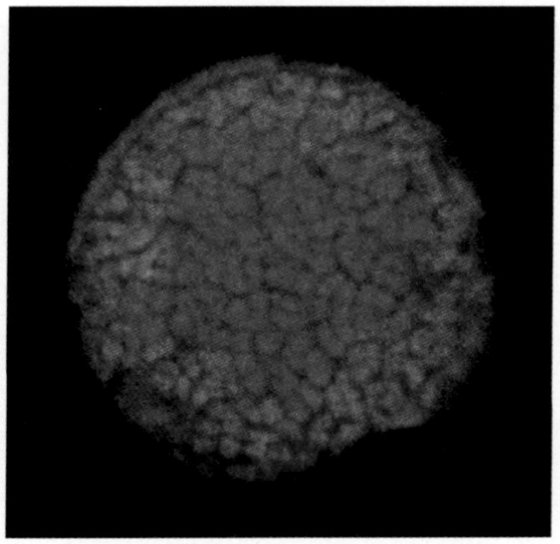

Methane-Consuming Archaea. A cluster of methanotrophic archaea, stained red by a specific fluorescent probe, surrounded by a layer of bacteria labeled by a green fluorescent probe.

Members of the order *Halobacteriales, Haloanaerobiales,* and *Salinibacter* employ a "salt-in" approach. These archaea use Na^+/H^+ antiporters and K^+ symporters to concentrate KCl and NaCl to levels equivalent to the external environment. Thus proteins need to be protected from salt denaturation and dehydration. Proteins from these microbes have evolved to possess a limited number of hydrophobic amino acids and a greater number of acidic residues. The acidic amino acids tend to be located on the surface of the folded protein, where they attract cations, which form a hydrated shell around the protein, thereby maintaining its solubility.

Probably the best-studied member of the family is *Halobacterium salinarium.* This archaeon is unusual because it produces a protein called **bacteriorhodopsin** that can trap light energy without the presence of chlorophyll. Structurally similar to the rhodopsin found in the mammalian eye, bacteriorhodopsin functions as a light-driven proton pump. Like all members of the rhodopsin family, bacteriorhodopsin has two distinct features: (1) a chromophore that is a derivative of retinal (an aldehyde of vitamin A), which is covalently attached to the protein by a Schiff base with the amino group of lysine (**figure 18.17**); and (2) seven membrane-spanning domains connected by loops on either side of the membrane with the retinal resting within the membrane. Bacteriorhodopsin molecules form aggregates in a modified region of the plasma membrane called the **purple membrane.** When retinal absorbs light, the double bond between carbons 13 and 14 changes from a trans to a cis configuration and the Schiff base loses a proton. The proton is accepted by a specific aspartate residue in the protein portion of the molecule. Because this residue sits on the periplasmic face of the

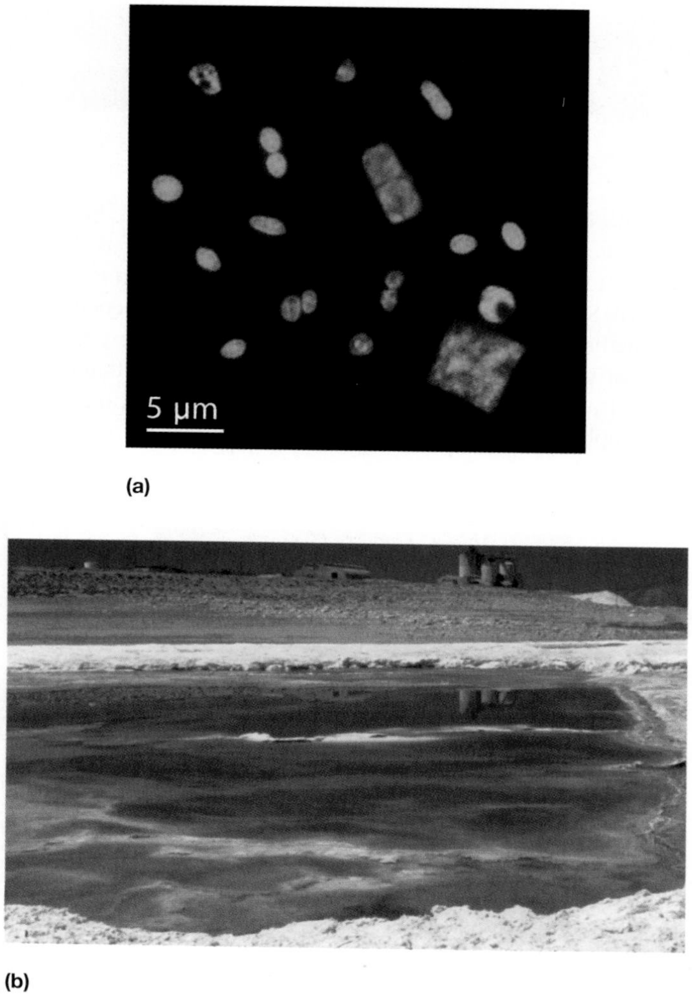

FIGURE 18.16 Halophilic Archaea. (a) A sample taken from a saltern in Australia viewed by fluorescence microscopy. Note the range of cell shapes includes cocci, rods, and cubes. (b) A solar evaporation pond in Owen's Lake, California, is extremely high in salt and mineral content. The archaea that dominate this hot, saline habitat produce brilliant red pigments.

membrane, its subsequent release of the proton moves the proton to the outside of the cell. At the same time, another aspartate residue on the cytoplasmic side of the membrane accepts another proton. This proton is donated to retinal, which isomerizes back to the trans-configuration and the photocycle can begin anew. This light-driven proton pumping generates a pH gradient that can be used to power the synthesis of ATP by a chemiosmotic mechanism. ◀◀ *Electron transport and oxidative phosphorylation (section 10.5); Rhodopsin-based phototrophy (section 10.12)*

Halobacterium has three additional rhodopsins, each with a different function. Halorhodopsin uses light energy to transport chloride ions into the cell and maintain a 4 to 5 M intracellular KCl concentration. The two additional rhodopsins are called sensory rhodopsin I (SRI) and SRII. **Sensory rhodopsins** act as photoreceptors; in this case, one for red light and one for blue. They control flagellar activity to position the organism optimally in the water column. *Halobacterium* moves to a location of high light intensity that lacks the ultraviolet light that would be lethal.

Surprisingly, rhodopsin is widely distributed among bacteria and archaea. DNA sequence analysis of uncultivated marine bacterioplankton reveals the presence of rhodopsin genes among α- and β-proteobacteria and the *Bacteroidetes*. This newly discovered rhodopsin is called **proteorhodopsin.** Cyanobacteria also have sensory rhodopsin proteins, which like SRI and SRII of the halobacteria, sense the spectral quality of light. All these bacterial and archaeal rhodopsins conserve the seven transmembrane helices through the cell membrane and the lysine residue that forms the Schiff base linkage with retinal.

Genomic analysis of *Halobacterium* NRC-1 illustrates many of the strategies discussed (**figure 18.18**). The use of both bacteriorhodopsin-mediated proton motive force and oxidative phosphorylation to drive ATP synthesis is shown, as is the import of cations to increase cytoplasmic osmolarity. The signals from sensory rhodopsins converge on the flagellar apparatus, along with input from 17 methyl-accepting proteins. *Halobacterium* uses a variety of ATP transporters for the uptake of cationic amino acids, peptides, and sugars. Carbohydrates are oxidized via the phosphorylative variant of the Entner-Doudoroff pathway (figure 18.8) and the TCA cycle. Interestingly, genes for exporting the toxic heavy metals arsenite and cadmium are present, as well as genes homologous to the nonspecific multidrug-resistance exporters.

Thermoplasms

Archaea in the class *Thermoplasmata* are thermoacidophiles that lack cell walls. At present, three genera, *Thermoplasma, Picrophilus,* and *Ferroplasma* are known. They are sufficiently different from one another to be placed in separate families: *Thermoplasmataceae, Picrophilaceae,* and *Ferroplasmataceae.*

Members of the genus *Thermoplasma* grow in refuse piles of coal mines. These piles contain large amounts of iron pyrite (FeS), which is oxidized to sulfuric acid by chemolithotrophic bacteria. As a result, the piles become very hot and acidic. This is an ideal habitat for *Thermoplasma* because it grows best at 55 to 59°C and pH 1 to 2. Although it lacks a cell wall, its plasma membrane is strengthened by large quantities of diglycerol tetraethers, lipid-containing polysaccharides, and glycoproteins. The organism's DNA is stabilized by association with archaeal histones that condense the DNA into structures resembling eukaryotic nucleosomes. At 59°C, *Thermoplasma* takes the form of an irregular filament, whereas at lower temperatures, it is spherical. The cells may be flagellated and motile.

Picrophilus is even more unusual than *Thermoplasma.* It originally was isolated from moderately hot solfataric fields in Japan. *Picrophilus* has an S-layer outside its plasma membrane. The cells

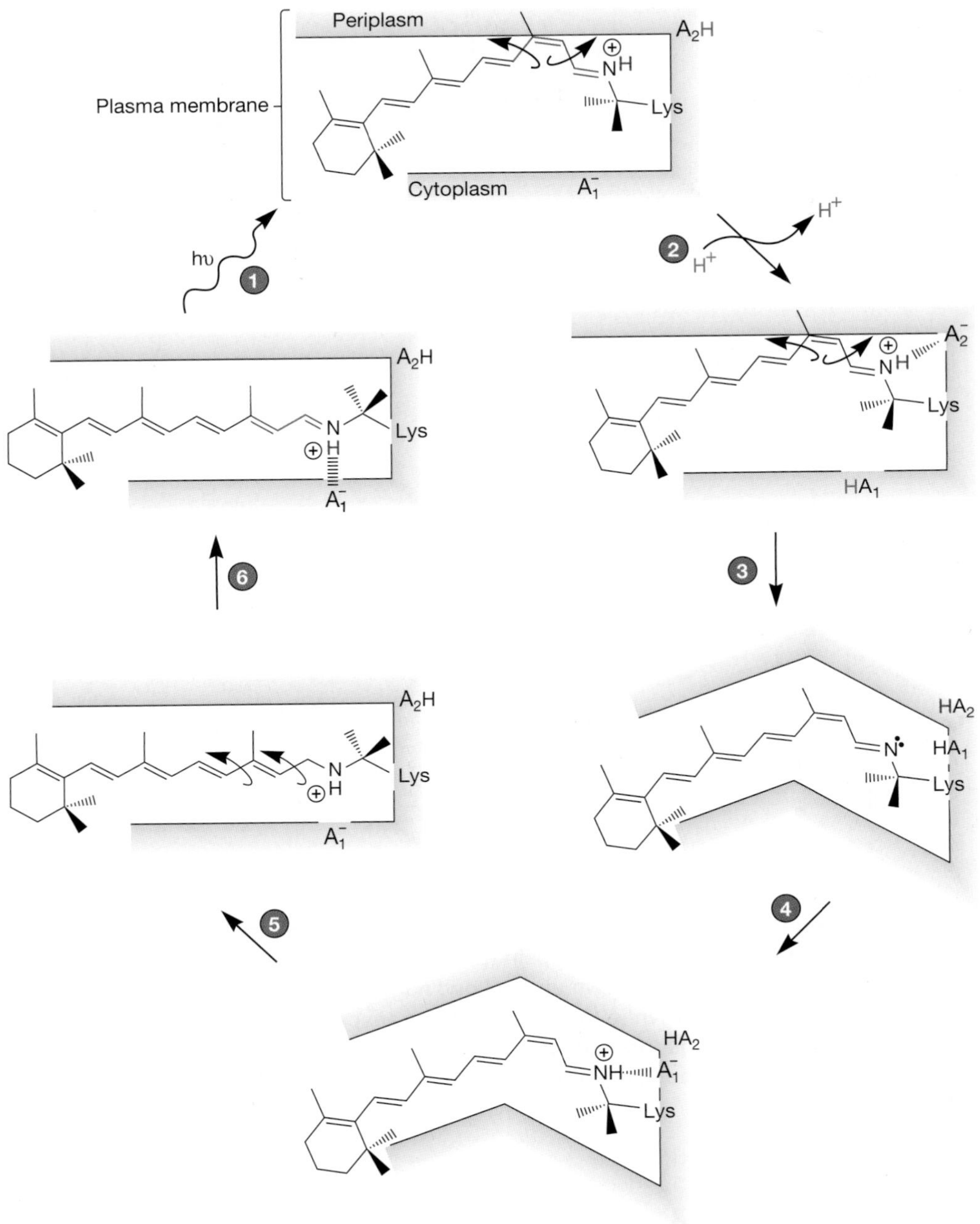

FIGURE 18.17 The Photocycle of Bacteriorhodopsin. In this hypothetical mechanism, the retinal component of bacteriorhodopsin is buried in the membrane and retinal interacts with two amino acids, A_1 and A_2 (aspartates 96 and 85), that can reversibly accept and donate protons. A_2 is connected to the cell exterior, and A_1 is closer to the cell interior. Light absorption by retinal in step 1 triggers an isomerization from 13-*trans*-retinal to 13-*cis*-retinal. The retinal then donates a proton to A_2 in steps 2 and 3, while A_1 is picking up another proton from the interior and A_2 is releasing a proton to the outside. In steps 4 and 5, retinal obtains a proton from A_1 and isomerizes back to the 13-*trans* form. The cycle is ready to begin again after step 6.

grow as irregularly shaped cocci, around 1 to 1.5 μm in diameter, and have large cytoplasmic cavities that are not membrane bounded. *Picrophilus* is aerobic and grows between 47 and 65°C with an optimum of 60°C. It is most remarkable in its pH requirements: it grows only below pH 3.5 and has a growth optimum at pH 0.7. Growth even occurs at about pH 0.

Extremely Thermophilic S^0-Reducers

This physiological group contains the class *Thermococci,* with one order, *Thermococcales.* The *Thermococcales* are strictly anaerobic and can reduce sulfur to sulfide. They are motile by flagella and have optimum growth temperatures around 88 to 100°C. The order contains one family and three genera, *Thermococcus, Paleococcus,* and *Pyrococcus.*

Sulfate-Reducing *Euryarchaeota*

Euryarchaeal sulfate reducers are found in the class *Archaeoglobi* and the order *Archaeoglobales.* This order has only one family and three genera. *Archaeoglobus* contains gram-negative–staining, irregular coccoid cells with cell walls consisting of glycoprotein subunits. It can extract electrons from a variety of electron donors (e.g., H_2, lactate, glucose) and reduce sulfate, sulfite, or thiosulfate

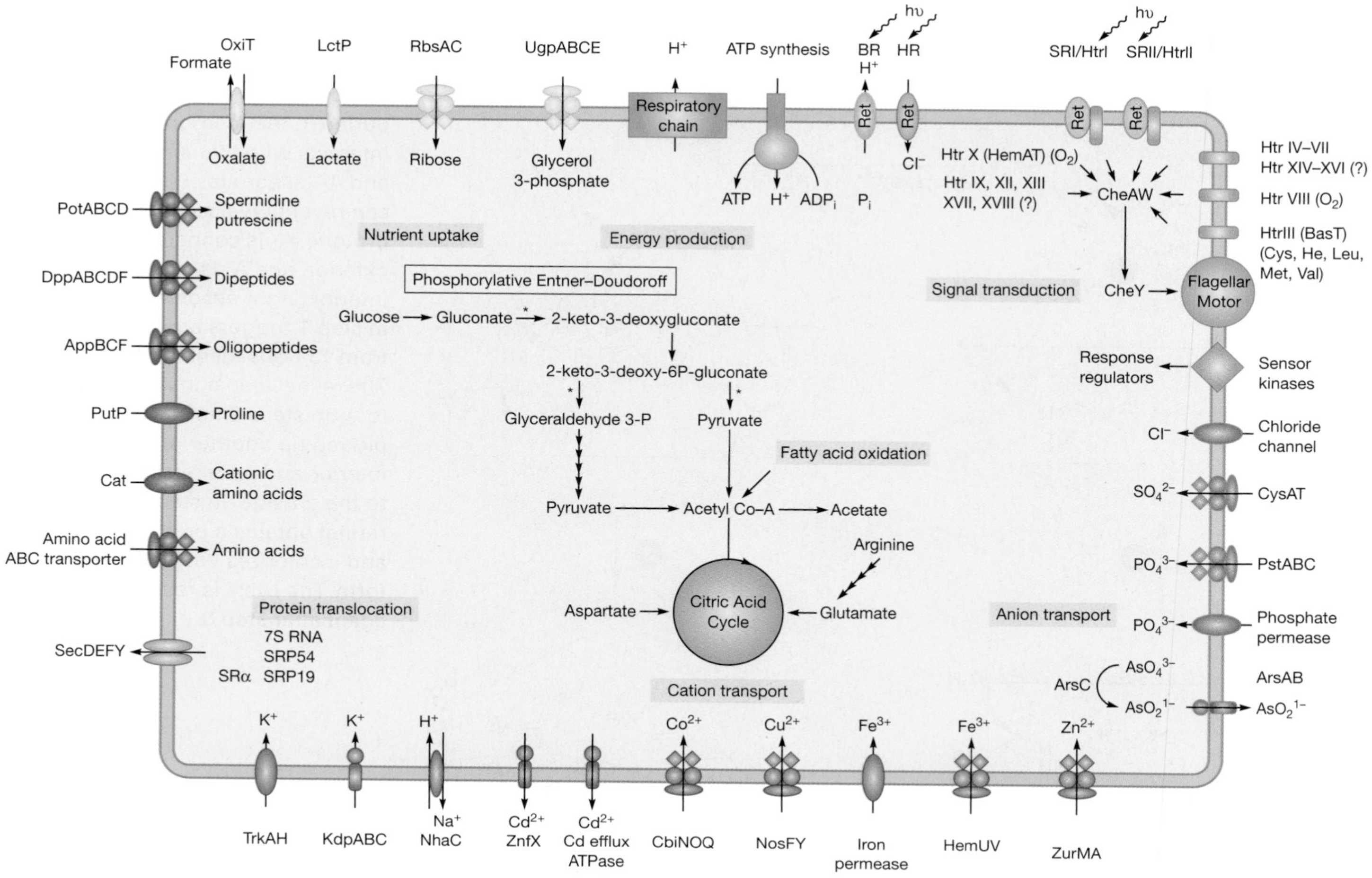

FIGURE 18.18 Genomic Reconstruction of *Halobacterium* NRC-1. ATP synthesis occurs through respiratory dependent chemiosmosis and the ATP synthase, light-driven proton pumping by bacteriorhodopsin (BR, pink oval), and chloride transport by halorhodopsin (HR, blue oval). Nutrient uptake systems are shown in brown and yellow; names of transporters correspond with the gene(s) encoding them; the substrates transported are shown in the inside of the transporter. Cation transporters are shown in purple and anion transporters in red. Sensory rhodopsins (orange) and 17 light, oxygen, and amino acid receptors (Htr proteins) regulate chemotaxis and are thus shown linked to the flagellar motor through the chemotaxis protein CheY.

to sulfide. Elemental sulfur is not used as an acceptor. *Archaeoglobus* is thermophilic (optimum growth is about 83°C) and has been isolated from marine hydrothermal vents. It is not only unusual in being able to reduce sulfate, unlike other archaea, but it also possesses the methanogen coenzymes F_{420} and methanopterin. In fact, it produces tiny amounts of methane, but genome sequencing reveals that it lacks the critical enzyme methyl-CoM reductase (figure 18.15, step 5), so the mechanism by which methane is made remains unclear.

A recently characterized thermophilic euryarchaeote that is not yet listed in *Bergey's Manual* bears mention here. *Aciduliprofundum* is an acidophile that requires a pH of 3.3 to 5.8 and a temperature between 60 and 75°C for growth. Members of this genus are often the dominant archaeal members of microbial communities growing at hydrothermal vents with actively venting sulfide deposits. It is a sulfur- and iron-reducing heterotroph. Its significance is twofold. First, it is the first thermoacidophile (as opposed to acid-tolerant) microbe to be isolated from environments rich in sulfide deposits. Second, its abundance at such sites implies that it is important in sulfur and iron cycling in these ecosystems. ▶▶| *Sulfur cycle (section 26.1)*

1. Why are the growth yields of most methanogens much lower than those of microbes that grow aerobically?
2. Why do you think methanogenesis requires so many cofactors? (Hint: What would happen to CO_2 and CH_4 if they were not bound to larger molecules?)

3. What is the ecological and practical importance of methanogens?
4. Where are extreme halophiles found, and what is unusual about their cell walls and growth requirements?
5. What is the difference between sensory rhodopsin and bacteriorhodopsin? Find examples of each in figure 18.18.
6. How is *Thermoplasma* able to live in acidic, very hot coal refuse piles when it lacks a cell wall? How is its DNA stabilized? What is so remarkable about *Picrophilus?*
7. Characterize *Archaeoglobus.* In what way is it similar to the methanogens, and how does it differ from other extreme thermophiles?

Summary

18.1 Overview of the *Archaea*

a. The *Archaea* are highly diverse with respect to morphology, reproduction, physiology, and ecology. Although best known for their growth in anoxic, hypersaline, and high-temperature habitats, they also inhabit marine Arctic, temperate, and tropical waters.

b. *Archaea* may be divided into five physiological groups: methanogenic archaea, sulfate reducers, extreme halophiles, cell wall-less archaea, and extremely thermophilic S^0-metabolizers (**table 18.1**).

c. The second edition of *Bergey's Manual* divides the *Archaea* into two phyla, the *Crenarchaeota* and *Euryarchaeota*; however, this may eventually be revised to include additional phyla (**figure 18.2**).

d. Archaeal cell walls do not contain peptidoglycan and differ from bacterial walls in structure. They may be composed of pseudomurein, polysaccharides, or glycoproteins and other proteins.

e. Archaeal membrane lipids differ from those of other organisms in having branched chain hydrocarbons connected to glycerol by ether links. Bacterial and eukaryotic lipids have glycerol connected to fatty acids by ester bonds (**figure 18.4**).

f. Their tRNA, ribosomes, elongation factors, RNA polymerases, and other components distinguish *Archaea* from *Bacteria* and eukaryotes.

g. Although much of archaeal metabolism appears similar to that of other organisms, the *Archaea* differ with respect to glucose catabolism, using a modified version of the Embden-Meyerhof and Entner-Doudoroff pathways. They also differ in CO_2 fixation pathways and in the ability of some to synthesize methane (**figures 18.7** and **18.8**).

18.2 Phylum *Crenarchaeota*

a. The extremely thermophilic S^0-metabolizers in the phylum *Crenarchaeota* depend on sulfur for growth and are frequently acidophiles. The sulfur may be used as an electron acceptor in anaerobic respiration or as an electron donor by chemolithotrophs. Many are strict anaerobes and grow in geothermally heated soil and water that is rich in sulfur.

b. More recent metagenomic evidence suggests that group I mesophilic *Crenarchaeota* are not confined to extreme environments. However, these microbes may represent a new phylum (**figure 18.2**).

18.3 Phylum *Euryarchaeota*

a. The phylum *Euryarchaeota* contains five major physiological groups: methanogens, halobacteria, the thermoplasms, extremely thermophilic S^0-reducers, and sulfate-reducing archaea.

b. Methanogenic archaea are strict anaerobes that can obtain energy through the synthesis of methane. They have several unusual cofactors that are involved in methanogenesis (**figures 18.14** and **18.15**).

c. Extreme halophiles or halobacteria are aerobic chemoheterotrophs that require at least 1.5 M NaCl for growth. They are found in habitats such as salterns, salt lakes, and salted fish.

d. *Halobacterium salinarum* can carry out phototrophy without chlorophyll or bacteriochlorophyll by using bacteriorhodopsin, which employs retinal to pump protons across the plasma membrane (**figures 18.17** and **18.18**).

e. The thermophilic archaeon *Thermoplasma* grows in hot, acidic coal refuse piles and survives despite the lack of a cell wall. Another thermoplasm, *Picrophilus*, can grow at pH 0.

f. The class *Thermococci* contains extremely thermophilic organisms that can reduce sulfur to sulfide.

g. Sulfate-reducing archaea are placed in the class *Archaeoglobi*. The extreme thermophile *Archaeoglobus* differs from other archaea in using a variety of electron donors to reduce sulfate. It also contains the methanogen cofactors F_{420} and methanopterin.

Critical Thinking Questions

1. Some believe that the *Archaea* should not be separate from the *Bacteria* because both groups are prokaryotic. Do you agree or disagree? Give your reasoning.
2. Explain why the fixation of CO_2 by *Thermoproteus* using a reductive reversal of the TCA cycle is not photosynthesis. Recently members of the ε-proteobacteria that inhabit hydrothermal vents have been shown to fix CO_2 using the reductive TCA cycle. How could you determine if this was an instance of horizontal gene transfer or convergent evolution?
3. When the temperature increases, some archaea change their shapes from elongated rods into spheres. Suggest one reason for this change.
4. Why would ether linkages be more stable in membranes than ester lipids? How would the presence of tetraether linkages stabilize a thermophile's membrane?
5. Suppose you wished to isolate an archaeon from a hot spring in Yellowstone National Park. How would you go about it?
6. Thermophilic archaea capable of ammonia oxidation to nitrite were discovered in enrichment cultures derived from microbial mats from the Siberian Garga hot spring. To demonstrate the presence of these microbes and their ability to oxidize ammonia, the authors did the following:
   - Screened for α- and β-proteobacterial 16S rRNA genes and the bacterial gene that encodes the key enzyme in ammonia oxidation (*see chapter 22*).
   - Screened for crenarchaeal specific 16S rRNA and crenarchaeal specific genes that encode the ammonia oxidation subunits AmoA and AmoB.
   - Constructed two phylogenetic trees—one based on the crenarchaeal 16S rRNA genes and one based on the AmoA and AmoB amino acid sequences as compared to other archaeal AmoA and AmoB sequences.
   - Performed reverse-transcriptase PCR using mRNA extracted from the enrichment cultures. Primers were specifically targeted crenarchaeal *amoA* nucleotide sequences.

   What was the rationale (i.e., what were the authors testing) for in each procedure? Since the authors concluded that a thermophilic ammonia-oxidizing crenarchaeote was responsible for the conversion of ammonia to nitrite in this system, what was the likely result of each test?

   Read the original paper: Hatzenpichler, R., et al. 2008. A moderately thermophilic ammonia-oxidizing crenarchaeote from a hot spring. *Proc. Nat. Acad. Sci., USA*. 105: 2134–39.

Concept Mapping Challenge

Use the following words as concepts to construct a concept map after providing the linking words.

Eukarya	Pseudomurein	Methanogens
Archaea	S-layer	Crenarchaeol
Crenarchaeota	*Euryarchaeota*	Hyperthermophiles
Cyclopentane ring		

Learn More

Learn more by visiting the text website at www.mhhe.com/willey8, where you will find a complete list of references.

19

Bacteria: *The Deinococci & Nonproteobacteria Gram Negatives*

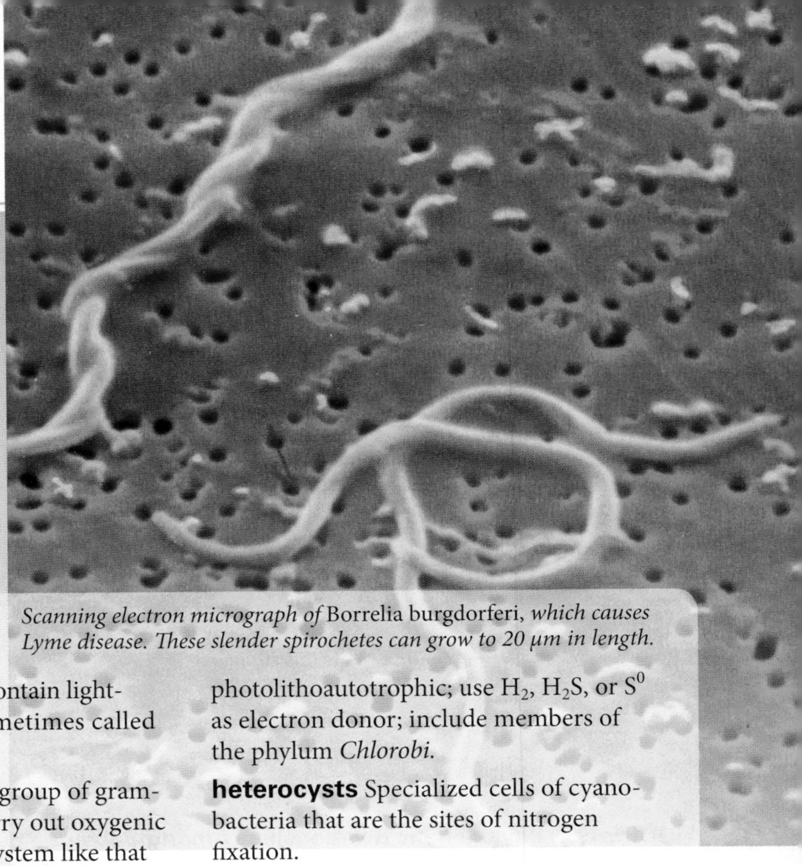

Scanning electron micrograph of Borrelia burgdorferi, *which causes Lyme disease. These slender spirochetes can grow to 20 μm in length.*

CHAPTER GLOSSARY

akinetes Specialized, nonmotile, dormant, thick-walled resting cells formed by some cyanobacteria.

anoxygenic photosynthesis Photosynthesis that does not oxidize water to produce oxygen; a form of photosynthesis characteristic of purple and green photosynthetic bacteria and heliobacteria.

axial fibrils or periplasmic flagella The flagella that lie under the outer sheath and extend from both ends of the spirochete cell to overlap in the middle and form the axial filament.

carboxysomes Polyhedral inclusions that contain the CO_2-fixation enzyme ribulose 1,5-bisphosphate carboxylase oxygenase; found in cyanobacteria, nitrifying bacteria, and thiobacilli.

chlamydiae Obligate intracellular bacteria that have a unique mode of reproduction.

chlorosomes Elongated, intramembranous vesicles found in green sulfur and nonsulfur bacteria; contain light-harvesting pigments. Sometimes called chlorobium vesicles.

cyanobacteria A large group of gram-negative bacteria that carry out oxygenic photosynthesis using a system like that present in photosynthetic eukaryotes.

elementary body (EB) A small, dormant body that serves as the agent of transmission between host cells in the chlamydial life cycle.

gliding motility A type of motility in which a microbial cell moves along a solid surface.

green nonsulfur bacteria Anoxygenic photosynthetic bacteria that contain bacteriochlorophylls *a* and *c;* usually photoheterotrophic and display gliding motility; include members of the phylum *Chloroflexi.*

green sulfur bacteria Anoxygenic photosynthetic bacteria that contain bacteriochlorophylls *a* plus *c, d,* or *e;* photolithoautotrophic; use H_2, H_2S, or S^0 as electron donor; include members of the phylum *Chlorobi.*

heterocysts Specialized cells of cyanobacteria that are the sites of nitrogen fixation.

oxygenic photosynthesis Photosynthesis that oxidizes water to form oxygen; the form of photosynthesis characteristic of plants, protists, and cyanobacteria.

phototaxis The ability of certain phototrophic microbes to move, either by gliding or swimming motility, in response to a light source.

phycobilisomes Particles on the membranes of cyanobacteria that contain photosynthetic pigments and electron transport chains.

reticulate body (RB) The cellular form in the chlamydial life cycle whose role is growth and reproduction within the host cell.

Volumes 1 and 5 of *Bergey's Manual of Systematic Bacteriology* describe a wide variety of microbes that are members of the domain *Bacteria.* This chapter is devoted to 10 of these bacterial phyla. Their phylogenetic locations are depicted in **figure 19.1**. This chapter therefore discusses the major group of gram-negative bacteria that do not belong to the phylum *Proteobacterium.* We follow the general organization and perspective of the second edition of *Bergey's Manual* in most cases. In describing each bacterial phylum, we include aspects such as distinguishing characteristics, morphology, reproduction, physiology, metabolism, and ecology. The taxonomy of each major group is summarized, and representative species are discussed.

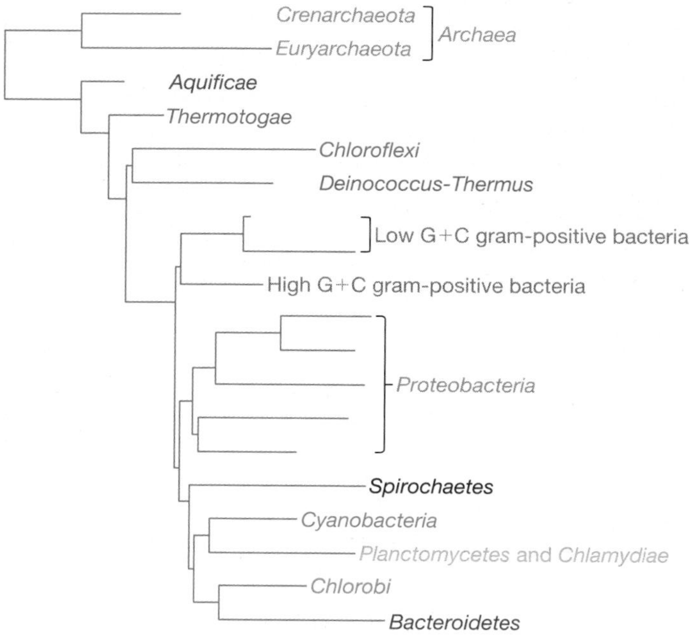

FIGURE 19.1 Phylogenetic Relationships Among *Bacteria* and *Archaea*. The *Deinococcus-Thermus* group and other nonproteobacterial gram-negatives are highlighted.

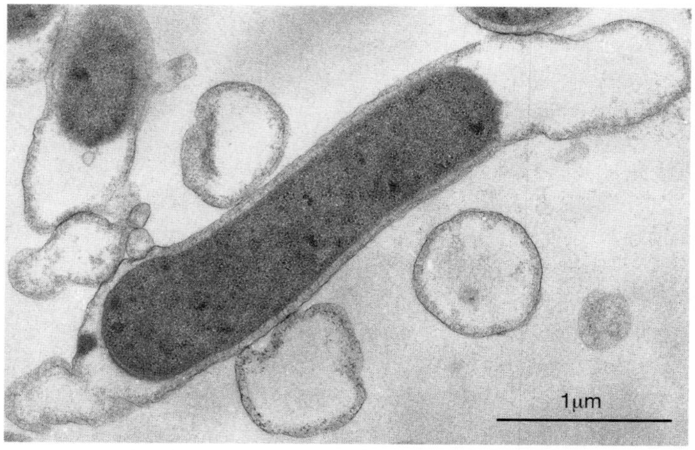

FIGURE 19.2 *Thermotoga maritima*. Note the loose sheath extending from each end of the cell.

19.1 *Aquificae* and *Thermotogae*

Thermophilic microbes are found in both the bacterial and archaeal domains, but to date, almost all hyperthermophilic microbes (those with optimum growth temperatures above 80°C) belong to the *Archaea*. The phyla *Aquificiae* and *Thermotoga* are two examples of bacterial thermophiles.

The phylum *Aquificae* is thought to represent the deepest, oldest branch of *Bacteria* (*see figure 17.10*). It contains one class, one order, and eight genera. Two of the best-studied genera are *Aquifex* and *Hydrogenobacter. Aquifex pyrophilus* is a gram-negative, microaerophilic rod. It is thermophilic with a temperature optimum of 85°C and a maximum of 95°C. *Aquifex* is a chemolithoautotroph that captures energy by oxidizing hydrogen, thiosulfate, and sulfur with oxygen as the terminal electron acceptor. Because *Aquifex* and *Hydrogenobacter* are thermophilic chemolithoautotrophs, it has been suggested that the original bacterial ancestor was probably thermophilic and chemolithoautotrophic. ◀◀ *Chemolithotrophy (section 10.11)*

The second deepest branch is the phylum *Thermotogae,* which has one class, one order, and six genera. The members of the genus *Thermotoga* (Greek *therme,* heat; Latin *toga,* outer garment), like *Aquifex,* are thermophiles with a growth optimum of 80°C and a maximum of 90°C. They are gram-negative rods with an outer sheathlike envelope (like a toga) that can extend or balloon out from the ends of the cell (**figure 19.2**). They grow in active geothermal areas found in marine hydrothermal systems and terrestrial solfataric springs. In contrast to *Aquifex, Thermotoga* is a chemoheterotroph with a functional glycolytic pathway that can grow anaerobically on carbohydrates and protein digests.

The genome of *Aquifex* is about a third the size of the *Escherichia coli* genome and, as expected, contains the genes required for chemolithoautotrophy. The *Thermotoga* genome is somewhat larger and has genes for sugar degradation. About 24% of its coding sequences are similar to archaeal genes; this proportion is greater than that of other bacteria, including *Aquifex* (16% similarity), and may be due to horizontal gene transfer. ◀◀ *Comparative genomics (section 16.7)*

19.2 *Deinococcus-Thermus*

The phylum *Deinococcus-Thermus* contains the class *Deinococci* and the orders *Deinococcales* and *Thermales.* There are only three genera in the phylum. Nine of the 11 species are mesophilic; *Deinococcus geothermalis* and *D. murrayai* are thermophiles with optimum growth between 45°C and 55°C. Deinococci are spherical or rod-shaped and nonmotile. They often are associated in pairs or tetrads (**figure 19.3*a***) and are aerobic and catalase positive; usually they produce acid from only a few sugars. Although they stain gram positive, their cell wall is layered with an outer membrane like gram-negative bacteria (figure 19.3*b*). They also differ from gram-positive cocci in having L-ornithine in their peptidoglycan, lacking teichoic acid, and having a plasma membrane with large amounts of palmitoleic acid rather than phosphatidylglycerol phospholipids. Almost all strains are extraordinarily resistant to both desiccation and radiation; they can survive as much as 3 to 5 million rad of radiation (an exposure of 100 rad can be lethal to humans).

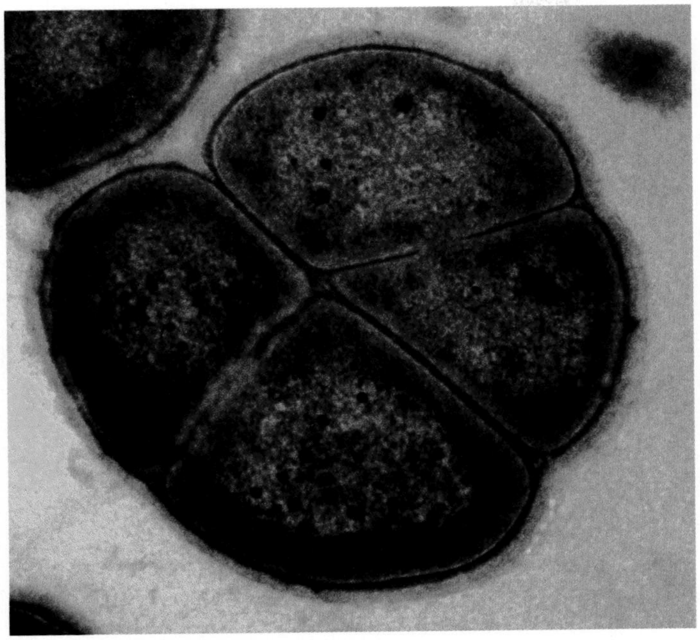
(a)

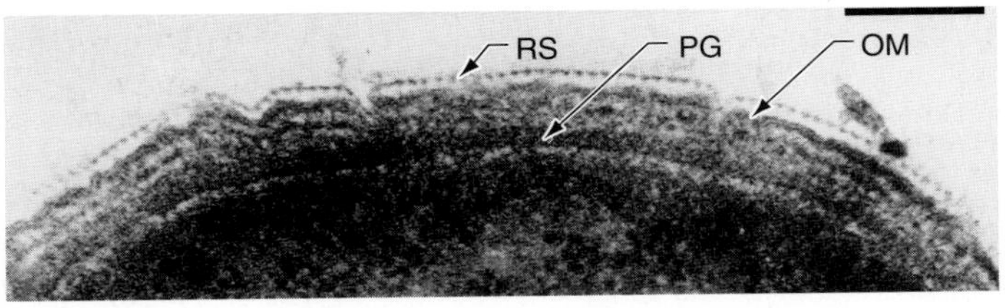

(b)

FIGURE 19.3 The *Deinococci*. (a) A *Deinococcus radiodurans* tetracoccus, or cluster of four cells (average cell diameter 2.5 μm). (b) The cell wall of *D. radiodurans* with a regular surface protein array (RS), a peptidoglycan layer (PG), and an outer membrane (OM). Bar = 100 nm.

Deinococci can be isolated from ground meat, feces, air, freshwater, and other sources, but their natural habitat is not yet known. Their great resistance to radiation results from their ability to repair a severely damaged genome, which consists of a circular chromosome, a megaplasmid, and a small plasmid. When exposed to high levels of radiation, the genome is broken into many, many fragments. Amazingly, within 12 to 24 hours, the genome is pieced back together, ensuring viability. It is unclear how this is accomplished. Genome studies have shown that *D. radiodurans* lacks novel DNA repair genes (*see figure 16.13*). Evidence suggests that while *Deinococcus* DNA is not especially radiation and desiccation resistant, the microbe's proteins are. It appears that the ability to accumulate high levels of Mn(II) may help protect the microbe from damage caused by radiation-induced toxic oxygen species. Interestingly, manganese ions selectively protect proteins, rather than DNA, from hydroxyl radicals generated during radiation. For *Deinococcus* to piece its fragmented genome back together, it needs functional proteins, including DNA repair enzymes. Thus it is hypothesized that one way this genus differs from others is its use of manganese to prevent protein damage.

19.3 Photosynthetic Bacteria

Chlorophyll-based photosynthesis is well established in five bacterial phyla: the gram-negative *Chlorobi, Chloroflexi, Cyanobacteria, Proteobacteria,* and the gram-positive *Firmicutes.* In addition, recent metagenomic research led to the discovery of a new chlorophyll-containing thermophilic bacterium in the phylum *Acidobacteria* from hot springs in Yellowstone National Park, USA. This gram-negative bacterium has not yet been grown in pure culture but is known to contain bacteriochlorophylls *a* and *c* under oxic conditions and to grow photoheterotrophically. Metabolically, the bacterium may be most similar to the aerobic anoxygenic phototrophs (AAnPs) found in marine systems. The discovery of 16S rRNA sequences in Thailand and Tibet similar to those found at Yellowstone suggests a global distribution of this thermophile. The phylum *Acidobacterium* has only one order and one family, but it may be far more diverse than previously imagined. Indeed, the number of photosynthetic microbes may exceed our long-held beliefs.

Other gram-negative photosynthetic bacteria include the purple bacteria, aerobic anoxygenic phototrophs, green bacteria, and cyanobacteria (**table 19.1**). The cyanobacteria differ fundamentally from the other photosynthetic bacteria because they perform **oxygenic photosynthesis.** They have photosystems I and II, use water as an electron donor, and generate oxygen during photosynthesis. In contrast, purple, AAnPs, and green bacteria have only one photosystem and use **anoxygenic photosynthesis.** Because they are unable to use water as an electron source, they employ other reduced molecules as electron donors. AAnPs and purple nonsulfur bacteria are photoorganoheterotrophs, so they use electrons obtained during the oxidation of organic substrates. The purple and green sulfur bacteria use inorganic electron donors such as hydrogen sulfide, sulfur, and hydrogen, as well as organic matter, as their electron source for the reduction of $NAD(P)^+$ to NAD(P)H. Consequently many of these bacteria form sulfur granules. Purple sulfur bacteria accumulate granules within their cells, whereas green sulfur bacteria deposit the sulfur granules outside their cells. Differences also occur in photosynthetic pigments, the organization of photosynthetic membranes, nutritional requirements, and oxygen relationships. ◀◀ *Phototrophy (section 10.12)*

Photosynthetic microbes contribute a significant amount of fixed carbon in a variety of habitats, and the differences in photosynthetic pigments and oxygen requirements among the photosynthetic bacteria have important ecological consequences. As shown

Table 19.1 Characteristics of the Major Groups of Gram-Negative Photosynthetic Bacteria

	Anoxygenic Phototrophic Bacteria					Oxygenic Photosynthetic Bacteria
Characteristic	*Green Sulfur*[a]	*Green Nonsulfur*[b]	*Purple Sulfur*	*Purple Nonsulfur*	*Aerobic Anoxygenic Phototrophs*	*Cyanobacteria*
Major photosynthetic pigments	Bacteriochlorophylls *a* plus *c*, *d*, or *e* (the major pigment)	Bacteriochlorophylls *a* and *c*	Bacteriochlorophyll *a* or *b*	Bacteriochlorophyll *a* or *b*	Bacteriochlorophyll *a*	Chlorophyll *a* plus phycobiliproteins *Prochlorococcus* has divinyl derivatives of chlorophyll *a* and *b*
Morphology of photosynthetic membranes	Photosynthetic system partly in chlorosomes that are independent of the plasma membrane	Chlorosomes present when grown anaerobically	Photosynthetic system contained in spherical or lamellar membrane complexes that are continuous with the plasma membrane	Photosynthetic system contained in spherical or lamellar membrane complexes that are continuous with the plasma membrane	Few, if any, intracytoplasmic membranes	Thylakoid membranes lined with phycobilisomes
Photosynthetic electron donors	H_2, H_2S, S^0	Photoheterotrophic donors—a variety of sugars, amino acids, and organic acids; photoautotrophic donors—H_2S, H_2	H_2, H_2S, S^0	Usually organic molecules: sometimes reduced sulfur compounds or H_2	Photoheterotrophic donors—a variety of sugars, amino acids, and organic acids	H_2O
Sulfur deposition	Outside of the cell	N/A[c]	Inside the cell[d]	Outside of the cell in a few cases	N/A	N/A
Nature of photosynthesis	Anoxygenic	Anoxygenic	Anoxygenic	Anoxygenic	Anoxygenic	Oxygenic (some are also facultatively anoxygenic)
General metabolic type	Obligate anaerobic photolithoautotrophs	Usually photoheterotrophic; sometimes photoautotrophic or chemoheterotrophic (when aerobic and in the dark)	Obligate anaerobic photolithoautotrophs	Usually anaerobic photoorganoheterotrophs; some facultative photolithoautotrophs (in the dark, chemoorganoheterotrophs)	Aerobic photoorganoheterotrophs	Aerobic photolithoautotrophs
Motility	Nonmotile; some have gas vesicles	Gliding	Motile with polar flagella; some are peritrichously flagellated	Motile with polar flagella or nonmotile; some have gas vesicles	Some are motile with a single or a few polar or subpolar flagella	Nonmotile, swimming motility without flagella or gliding motility; some have gas vesicles
Percent G + C	48–58	53–55	45–70	61–72	57–72	35–71
Phylum or class	*Chlorobi*	*Chloroflexi*	γ-proteobacteria	α-proteobacteria, β-proteobacteria (*Rhodocyclus*)	α-proteobacteria, β-proteobacteria, γ-proteobacteria	*Cyanobacteria*

[a] Characteristics of *Chlorobi*.
[b] Characteristics of *Chloroflexus*.
[c] N/A: Not applicable
[d] With the exception of *Ectothiorhodospira*.

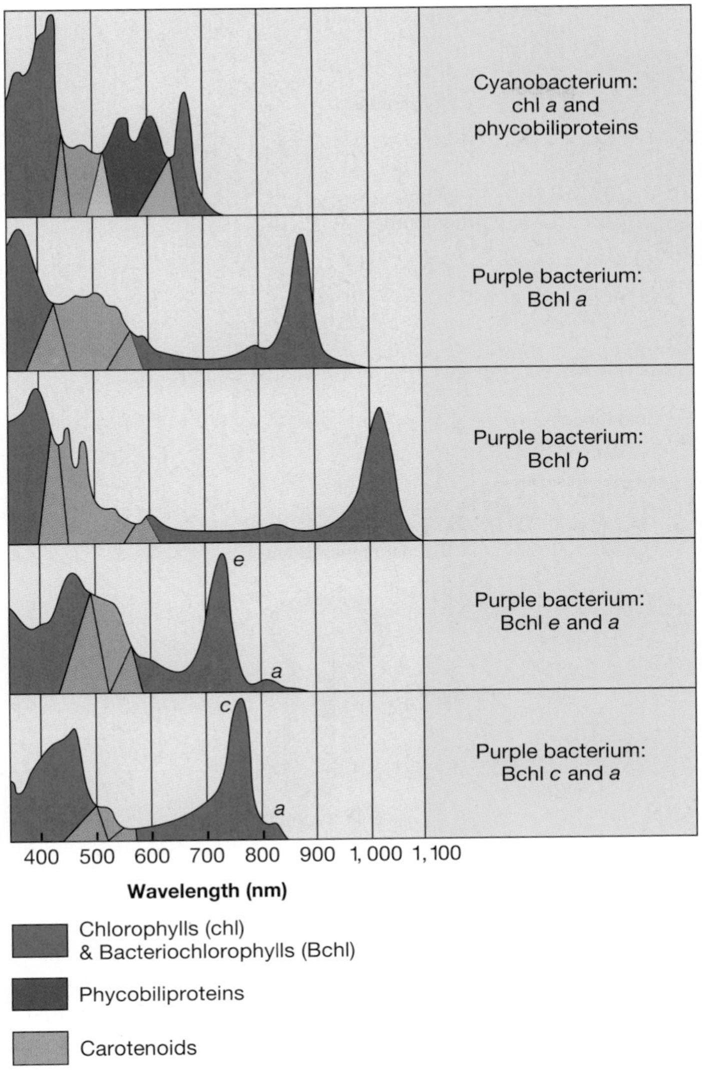

FIGURE 19.4 Photosynthetic Pigments. Absorption spectra of five photosynthetic bacteria showing the differences in absorption maxima and the contributions of various accessory pigments.

in **figure 19.4**, the chlorophylls (Chl), bacteriochlorophylls (Bchl), and their associated accessory pigments have distinct absorption spectra. Accessory pigments collect light of different wavelengths and transfer it to chlorophyll (or Bchl), where it is converted to chemical energy. Oxygenic cyanobacteria and photosynthetic protists dominate the aerated upper layers of freshwater and marine microbial communities, where they absorb large amounts of red and blue light. Below these microbes, the anoxygenic purple and green photosynthetic bacteria inhabit the deeper anoxic zones that are rich in hydrogen sulfide and other reduced compounds that can be used as electron donors. Their bacteriochlorophyll and accessory pigments enable them to use light in the far-red spectrum that is not used by other photosynthetic organisms. In addition, the bacteriochlorophyll absorption peaks at about 350 to 550 nm, allowing them to grow at greater depths because shorter wavelength light can penetrate water farther. As a result, when the water is sufficiently clear, a layer of green and purple bacteria develops in the anoxic, hydrogen sulfide-rich zone.

Bergey's Manual places photosynthetic bacteria into seven major groups distributed between five bacterial phyla. The phylum *Chloroflexi* contains the green nonsulfur bacteria, and the phylum *Chlorobi*, the green sulfur bacteria. The cyanobacteria are placed in their own phylum, *Cyanobacteria*. Purple bacteria are divided between three groups. Purple sulfur bacteria are placed in the γ-proteobacteria, families *Chromatiaceae* and *Ectothiorhodospiraceae*. The purple nonsulfur bacteria are distributed between the α-proteobacteria (five different families) and one family of the β-proteobacteria. Finally, the gram-positive heliobacteria in the phylum *Firmicutes* are also photosynthetic. There appears to have been considerable horizontal transfer of photosynthetic genes among the five phyla. At least 50 genes related to photosynthesis are common to all five. In this chapter, we describe the cyanobacteria and green bacteria; purple bacteria are discussed in chapter 20, while heliobacteria are discussed in chapter 21.

Phylum *Chlorobi*

The phylum *Chlorobi* has only one class (*Chlorobia*), order (*Chlorobiales*), and family (*Chlorobiaceae*). The **green sulfur bacteria** are a small group of obligately anaerobic photolithoautotrophs that use hydrogen sulfide, elemental sulfur, and hydrogen as electron sources. When sulfide is oxidized, elemental sulfur is deposited outside the cell. Their photosynthetic pigments are located in ellipsoidal vesicles called **chlorosomes** or chlorobium vesicles, which are attached to the plasma membrane by a proteinaceous baseplate that contains Bchl *a* molecules (**figure 19.5**). Within chlorosomes, bacteriochlorophyll molecules (Bchl *b, d,* or *e)* are grouped into rodlike structures held together by carotenoids and lipids. The light harvested by these pigments is transferred by baseplate Bchl *a* to reaction centers, which also contain Bchl *a*. Reaction centers are located in the plasma membrane.

These bacteria flourish in the anoxic, sulfide-rich zones of lakes. Although they lack flagella and are nonmotile, some species have gas vesicles to adjust their depth for optimal light and hydrogen sulfide (*see figure 3.36*). Those without gas vesicles are found in sulfide-rich muds at the bottom of lakes and ponds.

The green sulfur bacteria are very diverse morphologically. They may be rods, cocci, or vibrios; some grow singly, and others form chains and clusters. They are either grass-green or chocolate-brown in color. Representative genera are *Chlorobium, Prosthecochloris,* and *Pelodictyon*.

Phylum *Chloroflexi*

The phylum *Chloroflexi* has both photosynthetic and nonphotosynthetic members. *Chloroflexus* is the major representative of the photosynthetic **green nonsulfur bacteria.** However, the term green nonsulfur is a misnomer because not all members of this group are green, and some can use sulfur. *Chloroflexus* is a filamentous, gliding, thermophilic bacterium that often is isolated from neutral to

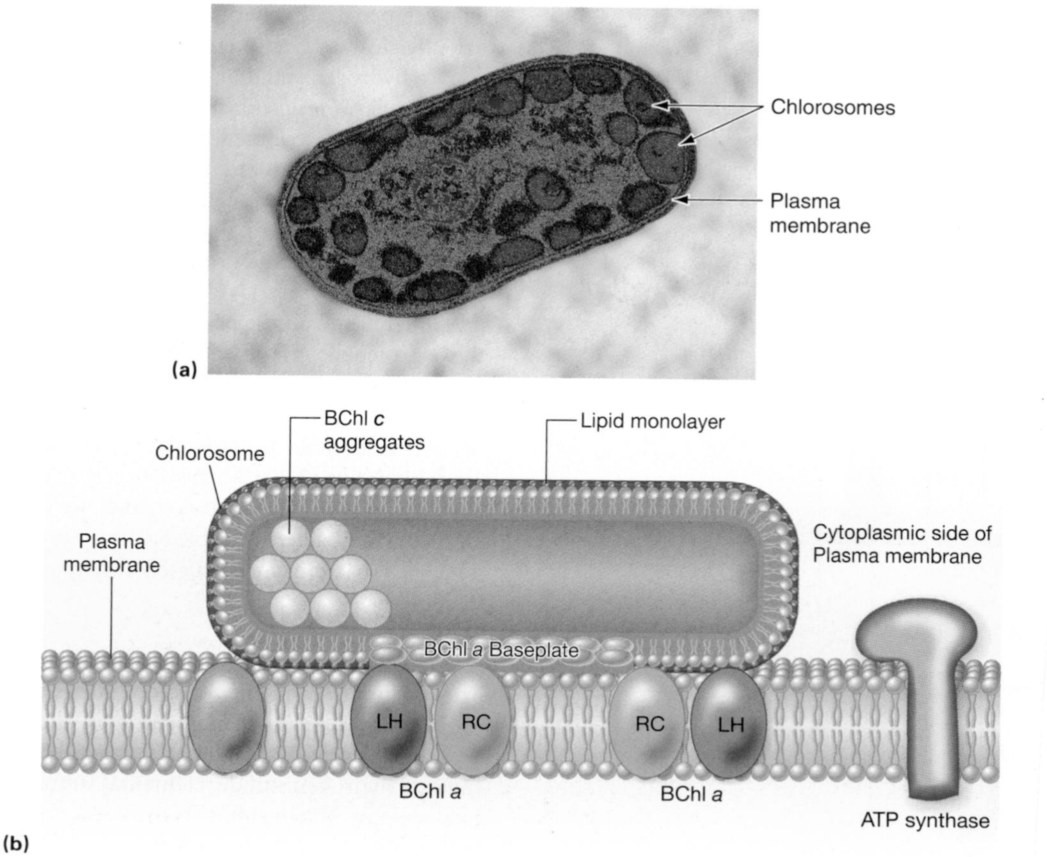

FIGURE 19.5 The Green Sulfur Bacterium *Chlorobium*. (a) Light-harvesting chlorosomes surround the inside perimeter of the cytoplasmic membrane. Antenna bacteriochlorophylls (Bchl *c, d,* or *e*) transfer light energy to reaction center Bchl *a* in the plasma membrane. (b) A chlorosome is attached to the inner surface of the plasma membrane by a baseplate composed of a lipid monolayer and Bchl *a* molecules. The chlorosome absorbs light by Bchl *c, d,* or *e* molecules aligned in rod-shaped complexes (Bchl *c* is shown here). The reaction center (RC) is located in the plasma membrane and contains only Bchl *a*. The light-harvesting complex (LH) and ATP synthase are also located in the plasma membrane.

alkaline hot springs, where it grows in the form of orange-reddish mats, usually in association with cyanobacteria. It possesses small chlorosomes with accessory Bchl *c*. Its light-harvesting complexes contain Bchl *a* and are in the plasma membrane. Metabolically, it is similar to the purple nonsulfur bacteria. *Chloroflexus* can carry out anoxygenic photosynthesis with organic compounds as carbon sources or grow aerobically as a chemoheterotroph. It doesn't appear closely related to any other bacterial group based on 16S rRNA studies and is a deep and ancient branch of the bacterial tree (*see figure 17.10*). Genomic analysis of *Chloroflexus aurantiacus* should help elucidate the origin of photosynthesis.

The nonphotosynthetic, gliding, rod-shaped, or filamentous bacterium *Herpetosiphon* also is included in this phylum. *Herpetosiphon* is an aerobic chemoorganotroph with respiratory metabolism that uses oxygen as the electron acceptor. It can be isolated from freshwater and soil habitats.

Search This: JGI anoxygenic bacteria

1. What are the major distinguishing characteristics of *Aquifex, Thermotoga,* and the deinococci? What is thought to contribute to the radiation resistance of the deinococci?
2. How do oxygenic and anoxygenic photosynthesis differ from each other? What is the ecological significance of these differences?
3. In general terms, give the major characteristics of the following groups: purple sulfur bacteria, purple nonsulfur bacteria, and green sulfur bacteria. How do purple and green sulfur bacteria differ?
4. What are chlorosomes?
5. Compare the green nonsulfur (*Chloroflexi*) and green sulfur bacteria (*Chlorobi*). Which are more metabolically versatile? What are the ecological implications of this metabolic flexibility?

Phylum *Cyanobacteria*

The **cyanobacteria** are the largest and most diverse group of photosynthetic bacteria. There is little agreement about the number of cyanobacterial species. Originally studied by botanists, they were classified as *Chlorophyceae*, commonly known as blue-green algae. The discovery that these microbes are really gram-negative bacteria did not resolve their taxonomic issues. Older classifications list as many as 2,000 or more species. *Bergey's Manual* describes 56 genera. Cyanobacterial diversity is reflected in the G + C content of the group, which ranges from 35 to 71%.

Although cyanobacteria are gram-negative bacteria, their photosynthetic apparatus closely resembles that of eukaryotes, and endosymbiotic cyanobacteria are thought to have evolved into chloroplasts. It follows that all cyanobacteria and most photosynthetic eukaryotes have chlorophyll *a* and photosystems I and II, and thereby perform oxygenic photosynthesis. Like the red algae, cyanobacteria use **phycobiliproteins** as accessory pigments. Photosynthetic pigments and electron transport chain components are located in thylakoid membranes lined with particles called **phycobilisomes** (**figure 19.6**). These contain phycobilin pigments, particularly **phycocyanin** and **phycoerythrin.** The pigments are arranged in stacklike structures around centrally located molecules of allophycocyanin, which transfer energy to photosystem II (figure 19.6*b*). ◀◀ *Evolution of the three domains of life (section 17.5)*

Carbon dioxide is assimilated through the Calvin cycle (*see figure 11.5*). The enzymes needed for this process are found in internal structures called **carboxysomes.** The reserve carbohydrate is glycogen. Sometimes cyanobacteria store extra nitrogen as polymers of arginine and aspartic acid in **cyanophycin** granules. Phosphate is stored in polyphosphate granules. Because cyanobacteria lack the enzyme α-ketoglutarate dehydrogenase, they do not have a fully functional TCA cycle. The pentose phosphate pathway plays a central role in their carbohydrate metabolism. Although most cyanobacteria are obligate photolithoautotrophs, a few can grow slowly in the dark as chemoheterotrophs by oxidizing glucose and a few other sugars. Under anoxic conditions, *Oscillatoria limnetica* oxidizes hydrogen sulfide instead of water and carries out anoxygenic photosynthesis much like the green photosynthetic bacteria. *Oscillatoria* is one of several cyanobacterial genera that produce geosmins, volatile organic compounds that often have an earthy odor. When populations of these cyanobacteria become dominant in lakes and rivers, this can be a public nuisance, even changing the taste of drinking water. ◀◀ *Inclusions (section 3.5);* ▶▶ *Microorganisms in marine and freshwater ecosystems (chapter 28)*

Cyanobacteria also vary greatly in shape and appearance. They range in diameter from about 1 to 10 μm and may be unicellular, exist as colonies of many shapes, or form filaments called trichomes (**figure 19.7*a***). A **trichome** is a row of bacterial cells that are in close contact with one another over a large area. Although many cyanobacteria appear blue-green because of phycocyanin, isolates from the open ocean are red or brown because their phycobilisomes contain the pigment phycoerythrin (figure 19.6*b*). Cyanobacteria modulate the relative amounts of these pigments in a process known as **chromatic adaptation**.

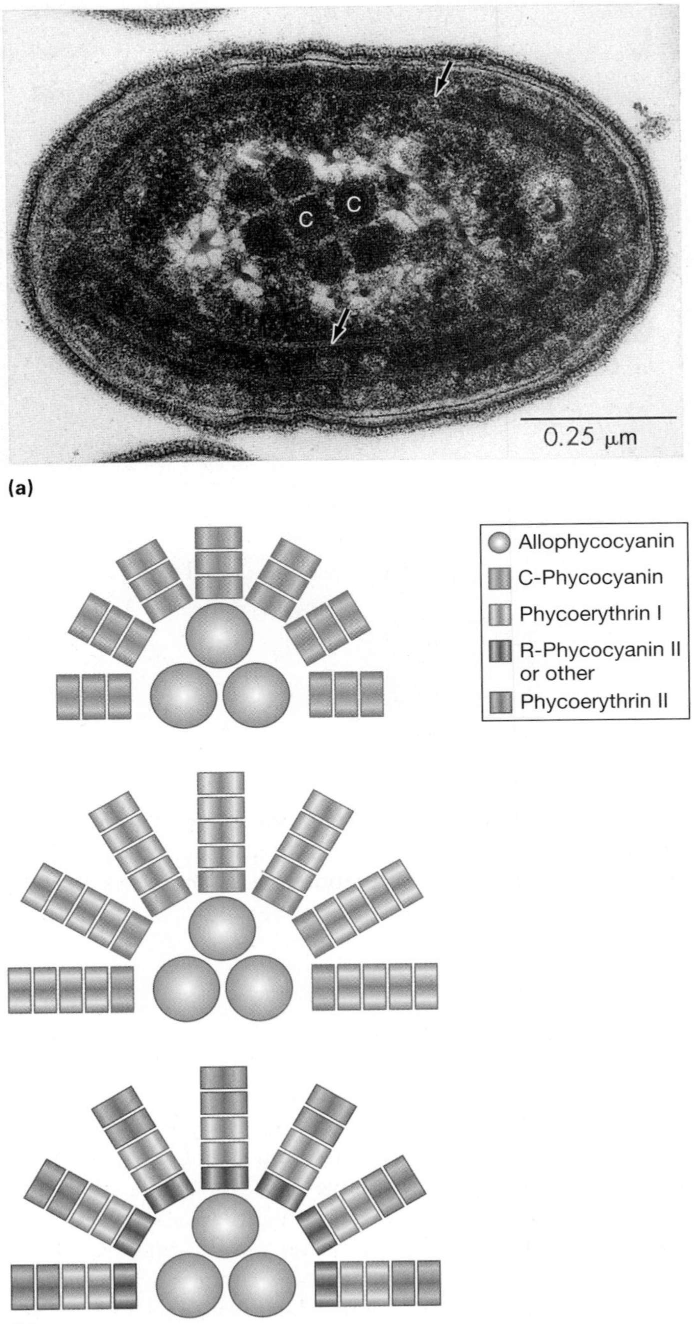

FIGURE 19.6 Cyanobacterial Thylakoids and Phycobilisomes. (a) Marine *Synechococcus* with thylakoids (arrows). Carboxysomes are labeled c. (b) Examples of pigment arrangements in phycobilisomes. Phycobilisomes are found in all cyanobacteria except prochlorophytes.

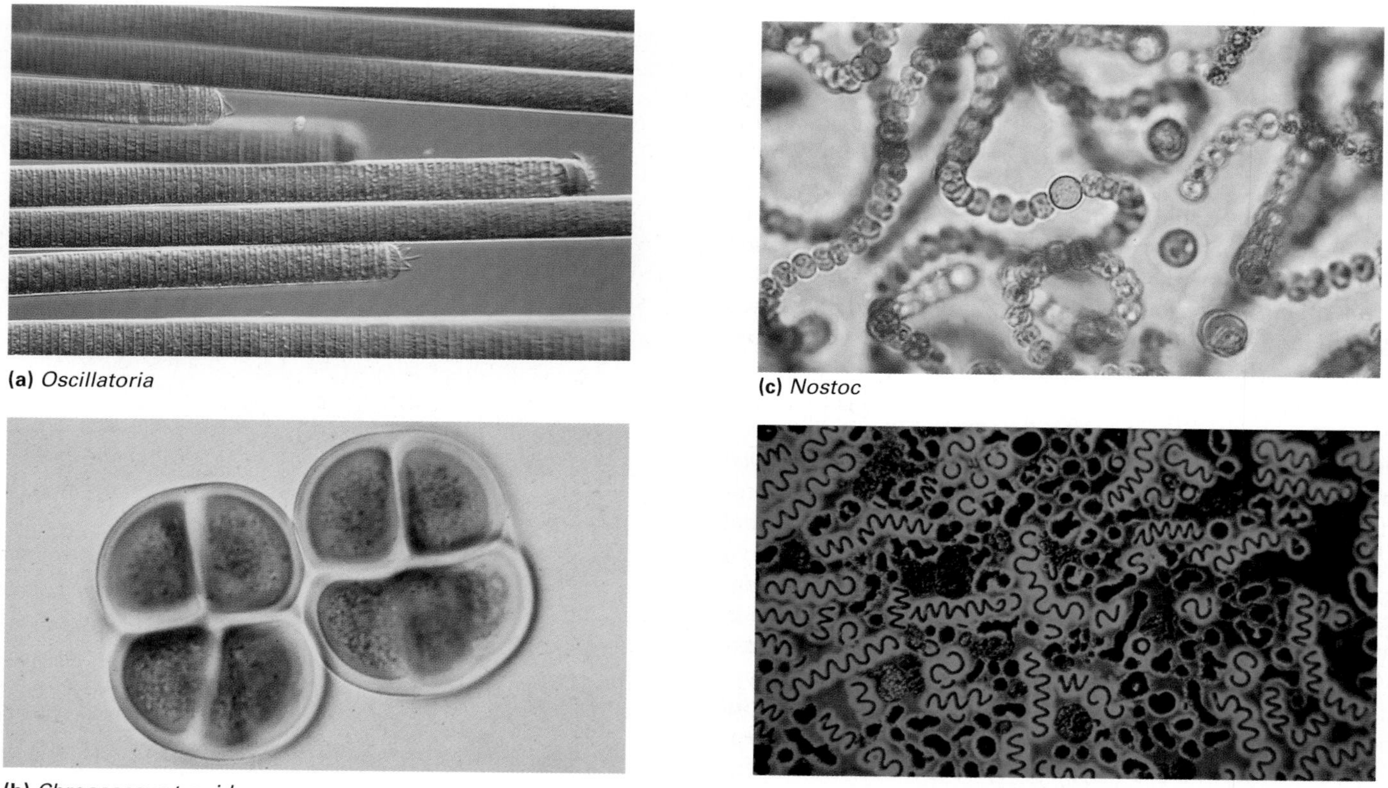

(a) *Oscillatoria*

(c) *Nostoc*

(b) *Chroococcus turgidus*

(d) *Anabaena spiroides* and *Microcystic aeruginosa*

FIGURE 19.7 Oxygenic Photosynthetic Bacteria. Representative cyanobacteria. (a) *Oscillatoria* trichomes seen with Nomarski interference-contrast optics (×250). (b) *Chroococcus turgidus,* two colonies of four cells each (×600). (c) *Nostoc* with heterocysts (×550). (d) The cyanobacteria *Anabaena spiroides* and *Microcystis aeruginosa.* The spiral *A. spiroides* is covered with a thick gelatinous sheath (×1,000).

When orange light is sensed, phycocyanin production is stimulated, whereas blue and blue-green light promote the production of phycoerythrin. Cyanobacteria use a variety of strategies to sense light, including sensory rhodopsins (*see p. 490*), plant-like proteins called phytochromes, and membrane-associated receptors that sense UV and blue light. Many cyanobacterial species use gas vacuoles to position themselves in optimum illumination in the water column—a form of **phototaxis.** Gliding motility is used by other cyanobacteria. Although cyanobacteria lack flagella, about one-third of the marine *Synechococcus* strains swim at rates up to 25 μm per second by an unknown mechanism. Swimming motility is not used for phototaxis; instead, it appears to be used for chemotaxis toward simple nitrogenous compounds such as urea. ⏮ *Motility and chemotaxis (section 3.7)*

Cyanobacteria show great diversity with respect to reproduction and employ a variety of mechanisms: binary fission, budding, fragmentation, and multiple fission. In the last process, a cell enlarges and then divides several times to produce many smaller progeny, which are released upon the rupture of the parental cell. Fragmentation of filamentous cyanobacteria can generate small, motile filaments called **hormogonia.** Some species develop **akinetes,** specialized, dormant, thick-walled resting cells that are resistant to desiccation (**figure 19.8*a***). These later germinate to form new filaments.

Many filamentous cyanobacteria fix atmospheric nitrogen by means of special cells called **heterocysts** (figure 19.8*a,b*). Around 5 to 10% of the cells develop into heterocysts when these cyanobacteria are deprived of both nitrate and ammonia, their preferred nitrogen sources. Within these specialized cells, photosynthetic membranes are reorganized and the proteins that make up photosystem II and phycobilisomes are degraded. Photosystem I remains functional to produce ATP, but no oxygen is generated. Lack of O_2 production is critical because the enzyme nitrogenase is extremely oxygen sensitive. Heterocysts develop thick cell walls, which slow or prevent O_2 diffusion into the cell, and any O_2 present is consumed during respiration. Heterocyst structure and physiology ensure that these specialized cells remain anaerobic; the heterocyst is dedicated to nitrogen fixation and does not replicate. Nutrients are obtained from adjacent vegetative cells, while the heterocysts contribute fixed nitrogen in the form of amino acids. Nitrogen fixation also is carried out by some cyanobacteria that lack heterocysts. Some fix nitrogen under dark, anoxic conditions in microbial mats. Planktonic forms

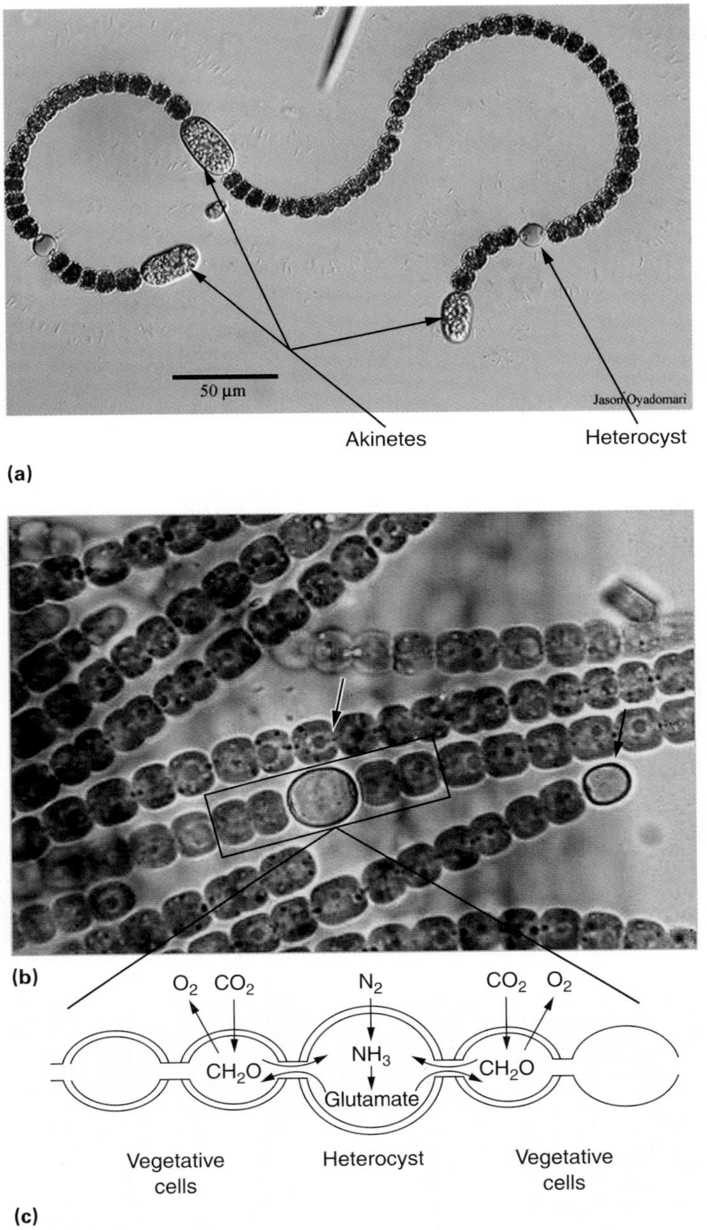

FIGURE 19.8 Cyanobacteria with Akinetes and Heterocysts. (a) The filamentous cyanobacterium *Anabaenea* with terminal and subterminal akinetes and heterocysts. (b) *Anabaena* heterocysts (arrow) and vegetative cells. (c) Heterocysts fix dinitrogen to form glutamate, which is exchanged with adjacent vegetative cells for carbohydrate. Heterocysts use only photosystem I, so photosynthesis is anoxygenic in these differentiated cells.

Figure 19.8 Micro Inquiry

Why do you think NH_3 is converted to the amino acid glutamate in heterocysts before its transfer to neighboring vegetative cells?

such as *Trichodesmium* fix nitrogen and contribute significantly to the marine nitrogen budget. ◀◀ *Nitrogen fixation (section 11.5);* ▶▶ *Nitrogen cycle (section 26.1)*

The cyanobacteria that are collectively referred to as **prochlorophytes** (genera *Prochloron, Prochlorococcus,* and *Prochlorothrix*) are distinguished by the presence of chlorophyll *a* and *b,* and the lack of phycobilins. This pigment arrangement imparts a grass-green color to these microbes. As the only bacteria to possess chlorophyll *b,* the ancestors of unicellular prochlorophytes are considered to be the best candidates as the endosymbionts that gave rise to chloroplasts. Considerable controversy has surrounded their phylogeny, but small subunit rRNA analysis places them with the cyanobacteria. ◀◀ *Endosymbiotic origin of mitochondria, chloroplasts, and hydrogenosomes (section 1.2)*

The three recognized prochlorophyte genera are quite different from one another. *Prochloron* was first discovered as an extracellular symbiont growing either on the surface or within the cloacal cavity of marine colonial ascidian invertebrates. These bacteria are single-celled, spherical, and 8 to 30 μm in diameter. Their mol% of G + C is 31 to 41. *Prochlorothrix* is free living, has cylindrical cells that form filaments, and grows in freshwater. Its DNA has a higher G + C content (53 mol%).

Prochlorococcus marinus, a coccobacillus less than 1 μm in diameter, flourishes in marine plankton. It differs from other prochlorophytes in having divinyl chlorophyll *a* and *b,* and α-carotene, instead of chlorophyll *a* and β-carotene. During the summer, it reaches concentrations of 5×10^5 cells per milliliter. It is one of the most numerous of the marine plankton and may be the most abundant oxygenic photosynthetic organism on Earth.

The classification of cyanobacteria is still unsettled. At present all taxonomic schemes must be considered tentative, and while many genera have been assigned, species names have not yet been designated in most cases. *Bergey's Manual* divides the cyanobacteria into five subsections (**table 19.2**). These are distinguished using cell or filament morphology and reproductive patterns. Some other properties important in cyanobacterial characterization are ultrastructure, genetic characteristics, physiology and biochemistry, and habitat/ecology (preferred habitat and growth habit). Subsection I contains unicellular rods or cocci. Most are nonmotile, and all reproduce by binary fission or budding. Organisms in subsection II are also unicellular, though several individual cells may be held together in an aggregate by an outer wall. Members of this group reproduce by multiple fission to form spherical, very small, reproductive cells called **baeocytes,** which escape when the outer wall ruptures (*see figure 7.2*). Some baeocytes disperse through gliding motility. The other three subsections contain filamentous cyanobacteria. Filaments are often surrounded by a sheath or slime layer. Cyanobacteria in subsection III form unbranched trichomes composed only of vegetative cells, whereas the other two subsections produce heterocysts in the absence of a fixed nitrogen source and also may form akinetes. Heterocystous cyanobacteria are subdivided into those that form unbranched filaments (subsection IV) and those that divide in a second plane to produce branches or aggregates (subsection V).

Table 19.2 Characteristics of the Cyanobacterial Subsections

Subsection	*General Shape*	*Reproduction and Growth*	*Heterocysts*	*% G + C*	*Other Properties*	*Representative Genera*
I	Unicellular rods or cocci; nonfilamentous aggregates	Binary fission, budding	–	31–71	Nonmotile or swim without flagella	*Chroococcus* *Gleocapsa* *Prochlorococcus* *Synechococcus*
II	Unicellular rods or cocci; may be held together in aggregates	Multiple fission to form baeocytes	–	40–46	Some baeocytes are motile.	*Pleurocapsa* *Dermocarpella* *Chroococcidiopsis*
III	Filamentous, unbranched trichome with only vegetative cells	Binary fission in a single plane, fragmentation	–	34–67	Usually motile	*Lyngbya* *Oscillatoria* *Prochlorothrix* *Spirulina* *Pseudanabaena*
IV	Filamentous, unbranched trichome may contain specialized cells	Binary fission in a single plane, fragmentation to form hormogonia	+	38–47	Often motile, may produce akinetes	*Anabaena* *Cylindrospermum* *Nostoc* *Calothrix*
V	Filamentous trichomes either with branches or composed of more than one row of cells	Binary fission in more than one plane, hormogonia formed	+	42–44	May produce akinetes; greatest morphological complexity and differentiation in cyanobacteria	*Fischerella* *Stigonema* *Geitleria*

Another indication of the vast diversity among the cyanobacteria is the wide range of habitats they occupy. Many live in symbiotic association with protists and fungi. Thermophilic species may grow at temperatures up to 75°C in neutral to alkaline hot springs. Because these photoautotrophs are so hardy, they are primary colonizers of soils and surfaces that are devoid of plant growth. Some unicellular forms even grow in the fissures of desert rocks. In nutrient-rich warm ponds and lakes, filamentous cyanobacteria such as *Anacystis* and *Anabaena* can reproduce rapidly to form blooms (**figure 19.9**). The release of large amounts of organic matter upon the death of the bloom microorganisms stimulates the growth of chemoheterotrophic bacteria. These microbes subsequently deplete available oxygen. This kills fish and other organisms. Some species can produce toxins that can kill animals that drink the water. Other cyanobacteria (e.g., *Oscillatoria*) are so pollution resistant and characteristic of freshwater with high organic matter content that they are used as water pollution indicators. ▶▶| *Microorganisms in estuaries and salt marshes (section 28.2)*

FIGURE 19.9 Bloom of Cyanobacteria and Algae in a Eutrophic Pond.

Cyanobacteria have a profound effect on the global carbon cycle. Two marine genera in particular, *Synechococcus* and *Prochlorococcus,* account for about one-third of global CO_2 fixation, yet these unicellular microbes constitute less than 1% of the oceanic photosynthetic biomass. These microbes have coevolved to take maximum advantage of the incident light. *Synechococcus* occupies the upper 25 meters of the open ocean, where white and blue-green light can be harvested. *Prochlorococcus* is divided into two ecotypes: the high-light ecotype lives between about 25 to 100 meters, while the low-light ecotype lives from 80 to 200 meters, where the waters are penetrated by blue-violet light. Thus the two genera do not compete for the essential resource of light, much as heterotrophic organisms coexist best in the absence of overlapping food requirements. ◀◀ *The concept of a microbial species (section 17.5);* ▶▶ *Microorganisms in the open ocean (section 28.2)*

Search This: Cyanobacteria resources

1. What are the major characteristics of the cyanobacteria that distinguish them from other photosynthetic organisms?
2. What is a trichome and how does it differ from a simple chain of cells?
3. Briefly discuss the ways in which cyanobacteria reproduce.
4. Describe how a vegetative cell, a heterocyst, and an akinete are different. How are heterocysts modified to carry out nitrogen fixation? Why must heterocysts use only photosystem I?
5. Compare the prochlorophytes with other cyanobacteria. Why do you think the phylogeny of the prochlorophytes has been so difficult to establish?
6. List some important positive and negative impacts cyanobacteria have on humans and the environment.

19.4 Phylum *Planctomycetes*

The phylum *Planctomycetes* contains one class and one order. Four genera are listed in *Bergey's Manual*, although recently at least four additional genera have been recognized. The planctomycetes are morphologically unique bacteria, having compartmentalized cells (**figure 19.10**). Although each species is unique, all follow a basic cellular organization that includes a plasma membrane closely surrounded by the cell wall, which lacks peptidoglycan. The largest internal compartment, the intracytoplasmic membrane, is separated from the plasma membrane by a peripheral, ribosome-free region called the paryphoplasm. The nucleoid of the planctomycete *Gemmata obscuriglobus* is located in the nuclear body, which is enclosed in a double membrane.

The recently described genera *Brocadia, Kuenenia, Scalindua,* and *Anammoxoglobus* do not localize DNA within a nuclear body. Instead, these microbes possess another compartment, the **anammoxosome.** This is the site of anaerobic ammonia oxidation called the **anammox reaction,** or simply anammox. This unique and

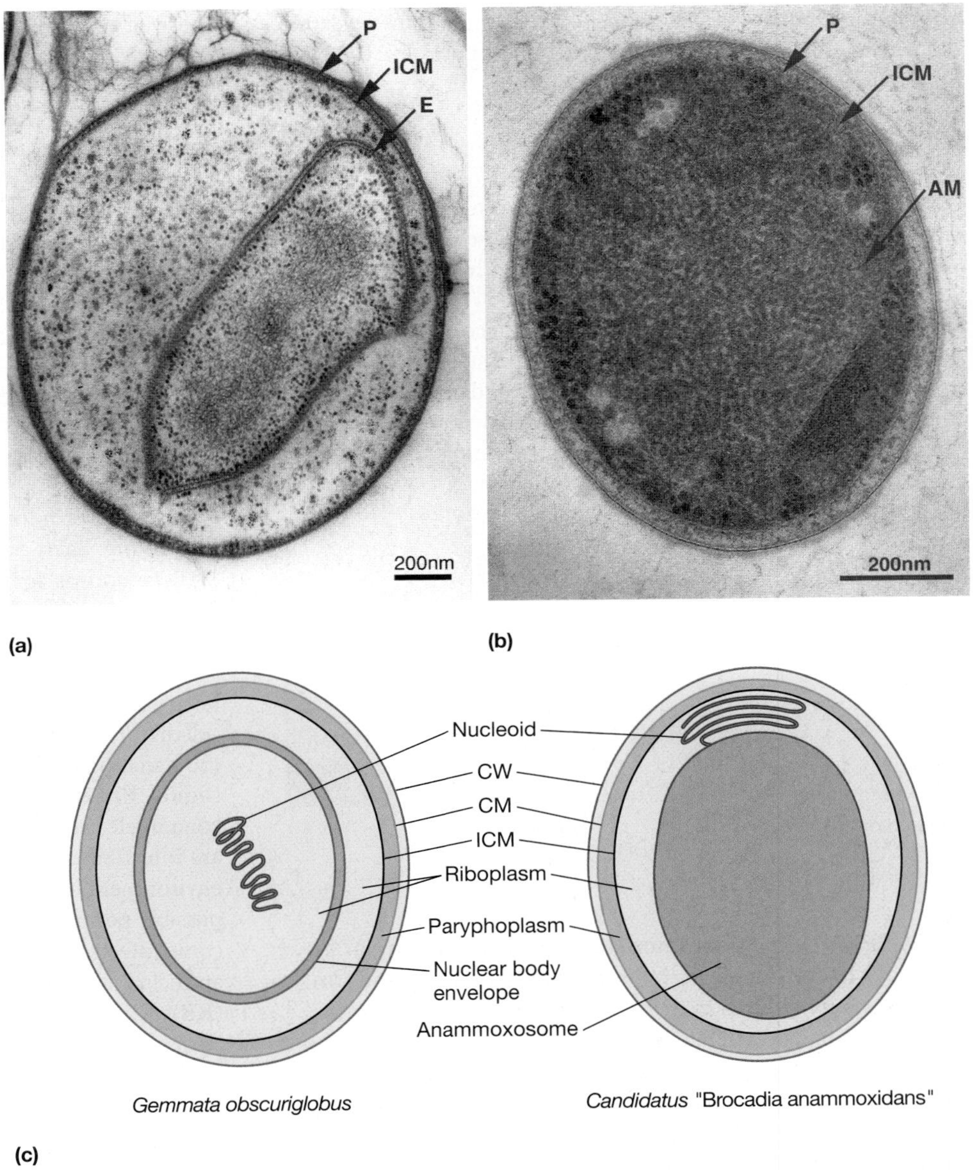

FIGURE 19.10 Planctomycete Cellular Compartmentalization. (a) An electron micrograph of *Gemmata obscuriglobus* showing the nuclear body envelope (E), the intracytoplasmic membrane (ICM), and the paryphoplasm (P). (b) An electron micrograph of the anaerobic ammonia-oxidizing planctomycete *Candidatus* "Brocadia anammoxidans." The anammoxosome is labeled AM. (c) Schematic drawings corresponding to (a) and (b): cell wall (CW), cytoplasmic membrane (CM).

recently discovered form of chemolithotrophy employs ammonium ion (NH_4^+) as the electron donor and nitrite (NO_2^-) as the terminal electron acceptor—it is reduced to nitrogen gas (N_2) (**figure 19.11**). Hydrazine (N_2H_4) is normally a cellular toxin, but in this case, it is an important intermediate. This is because it is sufficiently reduced to donate electrons to ferredoxin. The four high-energy electrons carried by reduced ferredoxin can then be used to produce a proton motive force (PMF) that is able to drive ATP synthesis, thereby conserving energy. Reduced ferredoxin can also be used in CO_2 fixation, which is accomplished using the acetyl Co-A pathway (*see figure 11.8*). Because nitrite must act as both electron donor for biomass production and electron acceptor for ammonium oxidation and generation of a PMF, the overall reaction is in two parts:

An anabolic reaction:

$$2NO_2^- + CO_2 + H_2O \rightarrow CH_2O + 2NO_3^-$$

And a catabolic reaction:

$$NO_2^- + NH_4^+ \rightarrow N_2 + 2H_2O$$

The ecological consequence of the anammox reaction is significant. It is estimated that anammox may contribute as much as 70% to the cycling of nitrogen in the world's oceans. ▶▶| *Nitrogen cycle (section 26.1)*

The genus *Planctomyces* attaches to surfaces through a stalk and holdfast; the other genera in the order lack stalks. Most of these bacteria have life cycles in which sessile cells bud to produce motile swarmer cells. The swarmer cells are flagellated and swim for a while before attaching to a substrate and beginning reproduction.

Search This: Anammox online resource

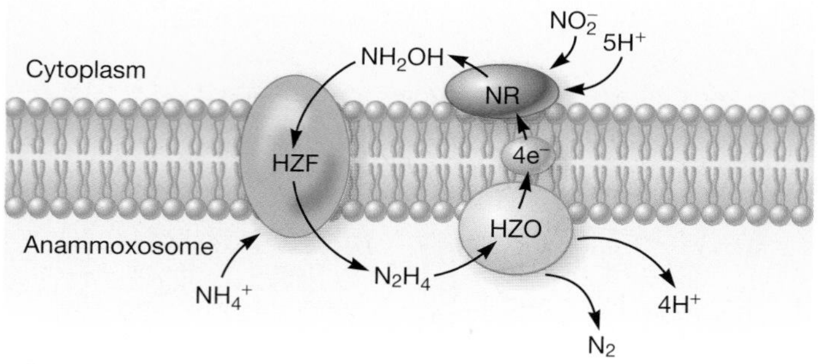

FIGURE 19.11 The Anammox Reaction. Ammonium is thought to combine with hydroxylamine (NH_2OH) to form hydrazine (N_2H_4) in a reaction catalyzed by the hydrazine forming enzyme (HZF). Oxidation of hydrazine by hydrazine oxidase (HZO) results in the formation of one N_2 and 4 protons. The electrons are passed to NO_2^-, which is reduced by nitrite reductase (NR) to form another molecule of hydroxylamine and the cycle continues.

***Figure 19.11* Micro Inquiry**

How does a planctomycete use the anammox reaction to generate a proton motive force?

19.5 Phylum *Chlamydiae*

The gram-negative **chlamydiae** are obligate intracellular parasites. This means they must grow and reproduce within host cells. Although their ability to cause disease is widely recognized, many species grow within protists and animal cells without adverse affects. It is thought that these hosts represent a natural reservoir for the *Chlamydiae.* In the human host, *C. trachomatis* is responsible for the most common sexually transmitted disease, known as nongonococcal urethritis. This microbe is also the most common cause of preventable blindness in the world. These diseases are discussed in chapter 38.

The phylum *Chlamydiae* has one class, one order, four families, and only six genera. The genus *Chlamydia* is by far the most important and best studied; it is the focus of our attention. *Chlamydiae* are nonmotile, coccoid bacteria, ranging in size from 0.2 to 1.5 μm. They reproduce only within cytoplasmic vesicles of host cells by a unique developmental cycle involving the formation of two cell types: elementary bodies and reticulate bodies. Although their envelope resembles that of other gram-negative bacteria, the cell wall differs in lacking muramic acid and a peptidoglycan layer. *Chlamydiae* are extremely limited metabolically, relying on their host cells for key metabolites. This is reflected in the size of their genome. It is relatively small at 1.0 to 1.3 Mb; the G + C content is 41 to 44%. Comparative genomics reveals that some *Chlamydia* genes were acquired by horizontal gene transfer (HGT) from free-living bacteria to a common chlamydial ancestor, and HGT accounts for the transfer of genes from chlamydiae to their protist hosts in the genera *Trypanosoma* and *Leishmania.* ▶▶| Euglenozoa *(section 23.2)*

Chlamydial reproduction is unique. Although they undergo binary fission, *Chlamydia* is one of only a few bacteria that lack the cell division protein FtsZ (*see figure 7.4*). Reproduction begins with the attachment of an **elementary body (EB)** to the host cell surface (**figure 19.12**). Elementary bodies are 0.2 to 0.6 μm in diameter, contain electron-dense nuclear material and a rigid cell wall, and are infectious. They achieve osmotic stability in the extracellular environment by cross-linking their outer membrane proteins, and possibly periplasmic proteins, with disulfide bonds. Host cells become infected with EBs by phagocytosing them. An EB is held in an inclusion body where it reorganizes to form a **reticulate body (RB).** The RB is specialized for reproduction rather than infection. Reticulate bodies are 0.5 to 1.5 μm in diameter and have less dense nuclear material and more ribosomes than EBs; their walls are also more flexible. About 8 to 10 hours after infection, the reticulate body undergoes binary fission, and RB reproduction continues until the host cell dies. Interestingly, a chlamydia-filled inclusion can become large enough to be seen in a light microscope and even fill the host cytoplasm. After 20 to 25 hours, RBs begin to differentiate into infectious EBs and continue this process until the host cell lyses and releases the chlamydiae EBs 48 to 72 hours after infection.

Chlamydial metabolism is very different from that of other gram-negative bacteria. It had been thought that chlamydiae

(a)
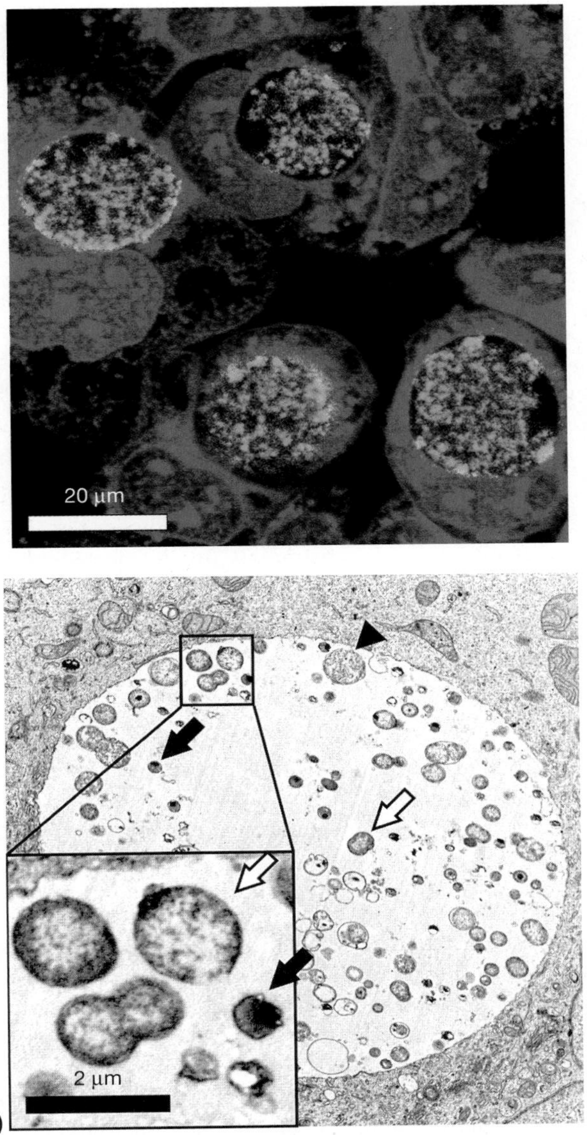

(b)

FIGURE 19.12 The Chlamydial Life Cycle. (a) Fluorescence light micrograph of human cells (red) infected with *C. trachomatis* (green). (b) A transmission electron micrograph of human cells that contain RBs (white arrows), EBs (black arrows), and an intermediate form called "aberrant bodies" (black arrowhead).

Figure 19.12 Micro Inquiry

Why do you think RBs have been so difficult to study?

cannot catabolize carbohydrates or synthesize ATP. *Chlamydia psittaci,* one of the best-studied species, lacks both flavoprotein and cytochrome electron transport chain carriers but has a membrane translocase that acquires host ATP in exchange for ADP. The protist parasite *Chlamydia* UWE25 cannot synthesize NAD^+ but synthesizes a transport protein for the import of host NAD^+. Thus chlamydiae may seem to be energy parasites that are completely dependent on their hosts for ATP. However, the *C. trachomatis* genome sequence indicates that this bacterium may be able to synthesize at least some ATP. Although there are two genes for ATP/ADP translocases, there also are genes for substrate-level phosphorylation, electron transport, and oxidative phosphorylation. When supplied with precursors from the host, RBs can synthesize DNA, RNA, glycogen, lipids, and proteins. Presumably the RBs have porins and active membrane transport proteins, but little is known about them. They also can synthesize at least some amino acids and coenzymes. The EBs have minimal metabolic activity and cannot take in ATP or synthesize proteins. They are designed exclusively for transmission.

1. Describe the *Planctomycetes* and their distinctive morphological and metabolic properties.
2. Explain how some *Planctomycetes* produce both NO_3^- and N_2.
3. Briefly describe the steps in the chlamydial life cycle. Why do you think *Chlamydiae* differentiate into specialized cell types for infection and reproduction?
4. How does chlamydial metabolism differ from that of other bacteria?

19.6 Phylum *Spirochaetes*

The phylum *Spirochaetes* (Greek *spira,* a coil, and *chaete,* hair) contains gram-negative, chemoheterotrophic bacteria distinguished by their structure and mechanism of motility. *Bergey's Manual* divides the phylum *Spirochaetes* into one class, one order (*Spirochaetales*), and three families (*Spirochaetaceae, Serpulinaceae,* and *Leptospiraceae*). At present, there are 13 genera in the phylum. Spirochetes can be anaerobic, facultatively anaerobic, or aerobic. Carbohydrates, amino acids, long-chain fatty acids, and long-chain fatty alcohols may serve as carbon and energy sources. **Table 19.3** summarizes some of the more distinctive properties of selected genera.

Spirochetes are morphologically unique. They are slender, long bacteria (0.1 to 3.0 µm by 5 to 250 µm) with a flexible, helical shape (**figure 19.13** and **chapter opener**). The central protoplasmic cylinder, which contains cytoplasm and the nucleoid, is bounded by a plasma membrane and a gram-negative cell wall (**figure 19.14**). Two to more than a hundred flagella, called **axial fibrils** or **periplasmic flagella,** extend from both ends of the cylinder and often overlap one another in the center third of the cell (*also see figure 3.47*). The whole complex of periplasmic flagella, called the **axial filament,** lies inside a flexible outer sheath. The outer sheath contains lipid, protein, and carbohydrate, and varies in structure between different genera. Its precise function is unknown, but the sheath is essential because spirochetes die if it is damaged or removed. The outer sheath of *Treponema pallidum,* the causative agent of syphilis, has few proteins exposed on its surface. This allows the spirochete to avoid attack

Table 19.3 Characteristics of Spirochete Genera

Genus	*Dimensions (μm) and Flagella*	*G + C Content (mol%)*	*Oxygen Relationship*	*Carbon + Energy Source*	*Habitats*
Spirochaeta	0.2–0.75 × 5–250; 2–40 periplasmic flagella (almost always 2)	51–65	Facultatively anaerobic or anaerobic	Carbohydrates	Aquatic and free living
Cristispira	0.5–3.0 × 30–180; ≥100 periplasmic flagella	N.A.[a]	Thought to be facultatively anaerobic	N.A.	Mollusk digestive tract
Treponema	0.1–0.4 × 5–20; 2–16 periplasmic flagella	25–53	Anaerobic or microaerophilic	Carbohydrates or amino acids	Mouth, intestinal tract, and genital areas of animals; some are pathogenic (syphilis, yaws)
Borrelia	0.2–0.5 × 3–20; 14–60 periplasmic flagella	27–32	Anaerobic or microaerophilic	Carbohydrates	Mammals and arthropods; pathogens (relapsing fever, Lyme disease)
Leptospira	0.1 × 6–24; 2 periplasmic flagella	35–53	Aerobic	Fatty acids and alcohols	Free living or pathogens of mammals, usually located in the kidney (leptospirosis)
Leptonema	0.1 × 6–20; 2 periplasmic flagella	51–53	Aerobic	Fatty acids	Mammals
Brachyspira	0.2 × 1.7–6.0; 8 periplasmic flagella	25–27	Anaerobic	Carbohydrates	Mammalian intestinal tract
Serpulina	0.3–0.4 × 7–9; 16–18 periplasmic flagella	25–26	Anaerobic	Carbohydrates and amino acids	Mammalian intestinal tract

[a]N.A., information not available.

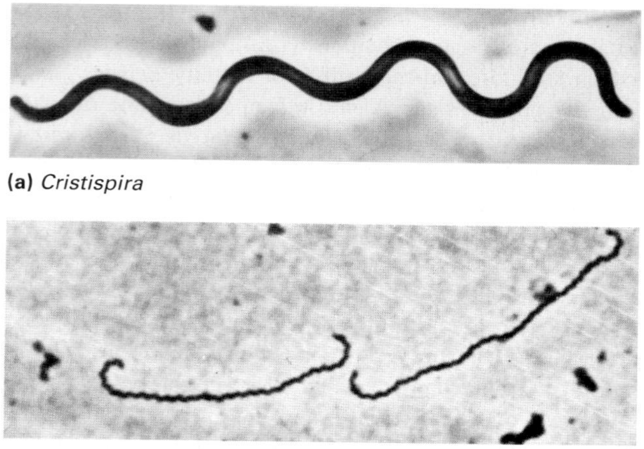

(a) *Cristispira*

(b) *Leptospira interrogans*

FIGURE 19.13 The Spirochetes. Representative examples. (a) *Cristispira* sp. from a clam; phase contrast (×2,200). (b) *Leptospira interrogans* (×2,200).

by host antibodies. *T. pallidum* is also discussed in chapter 16 *(see figure 16.11)*. ▶▶| *Direct contact diseases (section 38.3)*

Motility in the spirochetes is uniquely adapted to movement through viscous solutions. Like the external flagella of other bacteria, the axial fibrils rotate. This results in the rotation of the outer sheath in the opposite direction relative to the periplasmic cylinder, and the cell moves in a corkscrewlike movement through the medium (**figure 19.15**). Fibril rotation can also enable cell flexing and crawling on solid substrates. Thus, unlike other bacteria flagella, axial fibrils mediate movement through liquids and on solid surfaces.

Spirochetes are exceptionally diverse ecologically and grow in habitats ranging from mud to the human mouth. Members of the genus *Spirochaeta* are free-living and often grow in anoxic and sulfide-rich freshwater and marine environments. Some species of the genus *Leptospira* grow in oxic water and moist soil. Some spirochetes form symbiotic associations with other organisms and are found in a variety of locations: the hindguts of termites and wood-eating roaches, the digestive tracts of mollusks (*Cristispira*) and mammals, and the oral cavities of animals (*Treponema denticola, T. oralis*). Spirochetes from termite hindguts and freshwater sediments can fix nitrogen. Evidence exists that they contribute significantly to the nitrogen nutrition of termites. Spirochetes coat the surfaces of many protists from termite and wood-eating roach hindguts. For example, the flagellate *Myxotricha paradoxa* is covered with slender spirochetes (0.15 by 10 μm in length) that are firmly attached and help move the protozoan. ▶▶| *Microorganism-insect interactions (section 30.1)*

Borrelia burgdorferi is one of several species of this genus that cause Lyme disease. It is an unusual bacterium in several respects. Its genome includes a linear chromosome of 0.91 Mbp and 17 linear

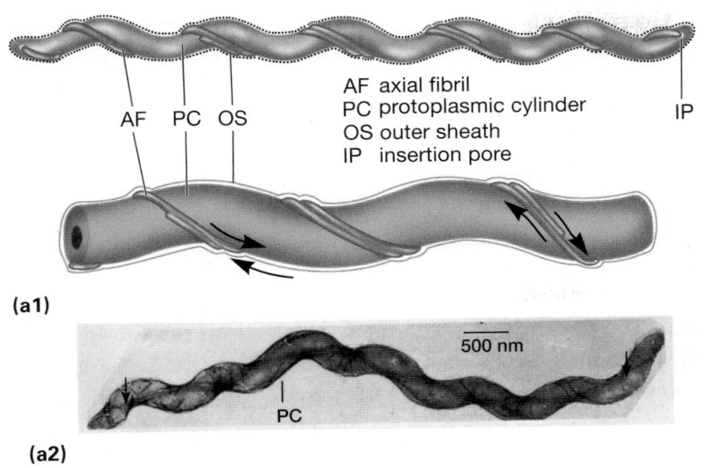

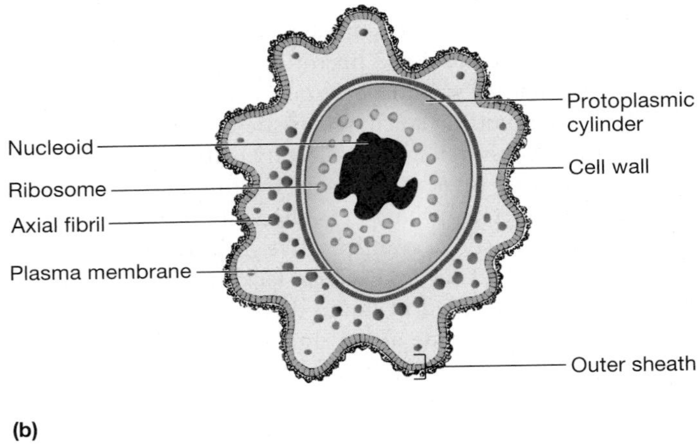

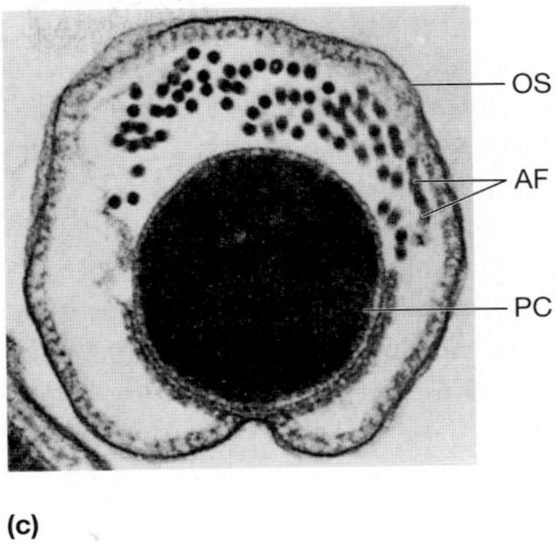

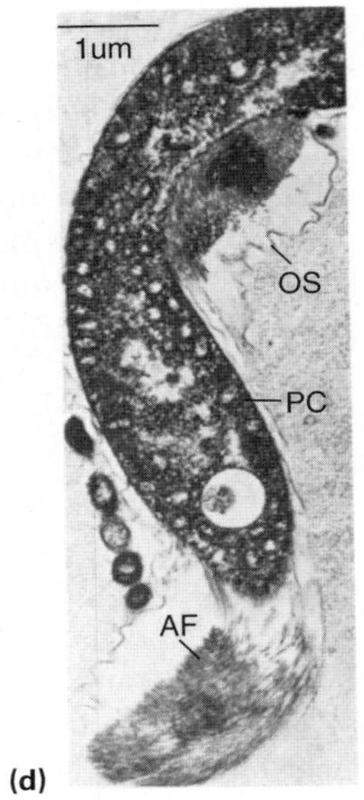

FIGURE 19.14 Spirochete Morphology. (a1) A surface view of spirochete structure as interpreted from electron micrographs. (a2) A longitudinal view of *Treponema zuelzerae* with axial fibrils extending most of the cell length. (b) A cross section of a typical spirochete showing morphological details. (c) Electron micrograph of a cross section of *Clevelandina* from the termite *Reticulitermes flavipes* showing the outer sheath, protoplasmic cylinder, and axial fibrils (×70,000). (d) Longitudinal section of *Cristispira* showing the outer sheath (OS), the protoplasmic cylinder (PC), and the axial fibrils (AF).

and circular plasmids, which together total another 0.53 Mbp. Like the intracellular bacteria *Chlamydiae* and *Mycoplasma genitalium* (another cause of sexually transmitted nongonococcal urethritis), *B. burgdorferi* lacks genes for many cellular biosynthetic pathways. For instance, *N*-acetyl glucosamine (NAG) is required for growth. Recall that NAG is a component of peptidoglycan; *B. burgdorferi* may also use it as an energy source. The microbe uses the Embden-Meyerhof pathway to oxidize glucose monomers to pyruvate, which is then reduced to lactate, consistent with the microaerophilic nature of the bacterium. Reducing power is generated through the pentose phosphate pathway; *B. burgdorferi* lacks a TCA cycle and oxidative phosphorylation. The similarity between *B. burgdorferi* and *M. genitalium* suggests that both bacteria independently underwent genome reduction as they evolved to become more dependent on their host. These diseases are discussed in chapter 38.

1. Define the following: protoplasmic cylinder, axial fibrils (periplasmic flagella), axial filament, and outer sheath. Draw and label a diagram of spirochete morphology, locating these structures.

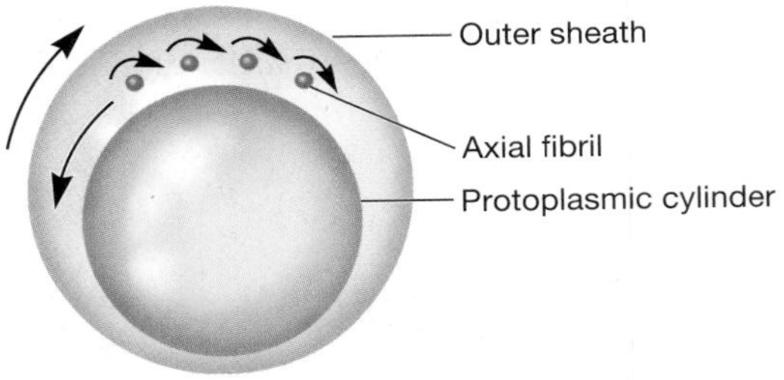

FIGURE 19.15 Spirochete Motility. A hypothetical mechanism for spirochete motility.

2. Compare periplasmic flagella with the flagella of other bacteria. Why do you think spirochete motility is especially well suited for movement through viscous fluids?

19.7 Phylum *Bacteroidetes*

The phylum *Bacteroidetes* is very diverse and seems most closely related to the phylum *Chlorobi.* The phylum has three classes (*Bacteroides, Flavobacteria,* and *Sphingobacteria*), 12 families, and 63 genera. The class *Bacteroides* contains anaerobic, gram-negative, nonspore-forming, motile, or nonmotile rods of various shapes. These bacteria are chemoheterotrophic and usually produce a mixture of organic acids as fermentation end products. They do not reduce sulfate or other sulfur compounds. The genera differ in general shape, motility and flagellation pattern, and fermentation end products. These bacteria grow in habitats such as the oral cavity and intestinal tract of vertebrates and the rumen of ruminants. ▶▶| *The rumen ecosystem (section 30.1); Normal microbiota of the human body (section 30.3)*

Although difficulty culturing these anaerobes has hindered our understanding of them, they are clearly widespread and important. Often they benefit their host. *Bacteroides ruminicola* is a major component of the rumen flora; it ferments starch, pectin, and other carbohydrates. About 30% of the bacteria isolated from human feces are members of the genus *Bacteroides,* and these organisms may provide extra nutrition by degrading cellulose, pectins, and other complex carbohydrates (*see figure 30.17*). The family also is involved in human disease. Members of the genus *Bacteroides* are associated with diseases of major organ systems, ranging from the central nervous system to the skeletal system. *B. fragilis* is a particularly common anaerobic pathogen found in abdominal, pelvic, pulmonary, and blood infections.

Another important taxon in the *Bacteroidetes* is the class *Sphingobacteria.* These bacteria often have sphingolipids in their cell walls. Within this class, the genera *Cytophaga, Sporocytophaga,* and *Flexibacter* differ from each other in morphology, life cycle, and physiology. Bacteria of the genus *Cytophaga* are slender rods, often with pointed ends (**figure 19.16*a***). *Sporocytophaga* is similar to *Cytophaga* but forms spherical resting cells called microcysts (figure 19.16*b*). *Flexibacter* produces long, flexible threadlike cells when young (figure 19.16*c*) and is unable to use complex polysaccharides. Often colonies of these bacteria are yellow to orange because of carotenoid or flexirubin pigments. Some of the flexirubins are chlorinated, which is unusual for biological molecules.

Members of the genera *Cytophaga* and *Sporocytophaga* are aerobes that actively degrade complex polysaccharides. Soil cytophagas digest cellulose; both soil and marine forms attack chitin, pectin, and keratin. Some marine species also degrade agar, a component of seaweed. Cytophagas play a major role in the mineralization of organic matter and can cause great damage to exposed fishing gear and wooden structures. They also are a major component of the bacterial population in sewage treatment plants and presumably contribute significantly to the waste treatment process. Although most cytophagas are free-living, some can be isolated from vertebrate hosts and are pathogenic. *Cytophaga columnaris* and others cause diseases such as columnaris disease, cold water disease, and fin rot in freshwater and marine fish. ▶▶| *Wastewater treatment (section 42.2)*

The gliding motility so characteristic of these organisms is quite different from flagellar motility. **Gliding motility** is present in a wide diversity of taxa: fruiting and nonfruiting aerobic chemoheterotrophs, cyanobacteria, green nonsulfur bacteria, and at least two gram-positive genera (*Heliobacterium* and *Desulfonema*). Gliding bacteria lack flagella and are stationary when suspended in liquid medium. When in contact with a surface, they glide along and some leave a slime trail. Movement can be very rapid; some cytophagas travel 150 μm in a minute, whereas filamentous gliding bacteria may reach speeds of more than 600 μm per minute. Young organisms are the most motile, and motility often is lost with age. Low nutrient levels usually stimulate gliding. Unlike flagellar-mediated swimming motility, the mechanisms that enable gliding motil-

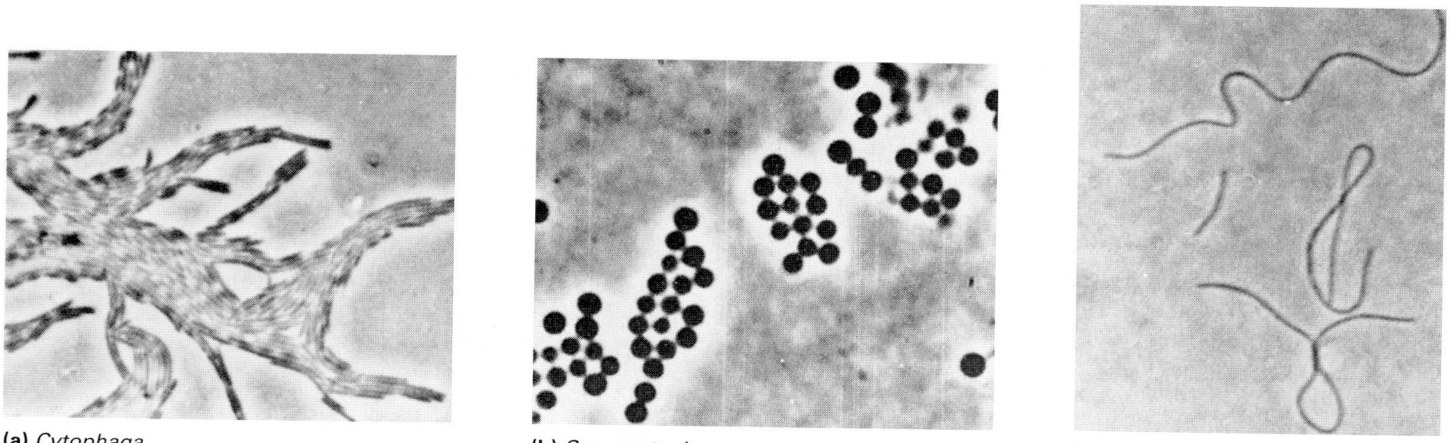

(a) *Cytophaga* (b) *Sporocytophaga myxococcoides* microcysts (c) *Flexibacter elegans*

FIGURE 19.16 Nonphotosynthetic, Nonfruiting, Gliding Bacteria. Representative members of the order *Cytophagales.* (a) *Cytophaga* sp. (×1,150). (b) *Sporocytophaga myxococcoides,* mature microcysts (×1,750). (c) Long thread cells of *Flexibacter elegans* (×1,100).

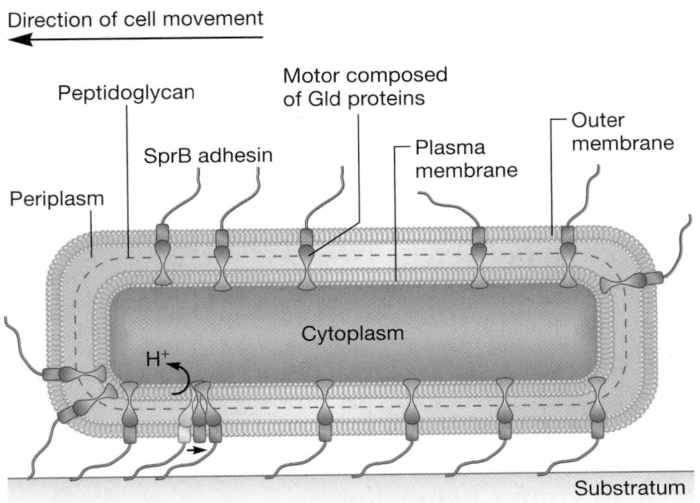

FIGURE 19.17 Proposed Mechanism for *Flavobacterium johnsoniae* Gliding Motility. Gld proteins within the cell envelope are thought to form motors that propel adhesions such as SprB along the cell surface.

Figure 19.17 *Micro Inquiry*

How would you test whether or not Gld proteins are necessary for gliding motility? Explain the experiment you would perform and what your prediction would be.

ity are not highly conserved, and are not well understood. Gliding in *Flavobacterium johnsoniae* depends on a number of proteins that appear to be unique to the *Bacteroidetes*. Some of these are called glide (Gld) proteins and are thought to assemble into motors that cross the plasma membrane and extend into the periplasm (**figure 19.17**). Like the motor proteins in flagellar basal bodies, the Gld proteins probably harvest the energy released from the proton motive force, which then drives movement. The cell surface bears large proteins called SprB adhesions and slender filaments, 5 nm wide, that may be surface adhesins. ◀◀ *Twitching and gliding motility (section 3.7)*

Gliding motility gives a bacterium many advantages. Many aerobic chemoheterotrophic gliding bacteria actively digest insoluble macromolecular substrates such as cellulose and chitin, and gliding motility is ideal for searching these out. Because many of the digestive enzymes are cell-wall associated, the bacteria must be in contact with insoluble nutrient sources; gliding motility makes this possible. Gliding movement is well adapted to drier habitats and to movement within solid masses such as soil, sediments, and rotting wood that are permeated by small channels. Finally, gliding bacteria, like flagellated bacteria, can position themselves at optimal levels of oxygen, hydrogen sulfide, light intensity, temperature, and other factors that influence growth and survival.

19.8 Phylum *Verrucomicrobia*

Bergey's Manual lists a single class and order within the phylum *Verrucomicrobia*. Indeed, this phylum contains just a few cultivated strains, most of which are aerobic or anaerobic heterotrophs that grow in complex medium. In 2007 a new member of this phylum was reported, *Acidimethylosilex fumarolicum*. The discovery and cultivation of this bacterium is noteworthy because it is thermophilic, acidophilic, and methanotrophic; that is, it can grow using methane as a carbon and energy source.

Like any organic energy substrate, methane must be oxidized by the organism using it to support growth. Methane oxidation is significant because methane is an important greenhouse gas. For many years, it was thought that only 13 genera of aerobic α- and β-proteobacteria could oxidize methane. But in 2000 it was discovered that certain archaea anaerobically oxidize methane when growing in association with sulfate reducing bacteria (*see Microbial Diversity & Ecology 18.1, p. 489*). In all cases, methane oxidation ceased in acid environments below pH 4.2. However, it had been noted that methane is produced abiotically in certain geochemically active areas. These sites are also rich in H_2S, which microbes convert to sulfuric acid (H_2SO_4), so the pH can be as low as 1.0. Researchers hypothesized that despite the low pH, aerobic methane oxidation might occur in such environments. They used PCR to amplify a methane monooxygenase gene conserved in proteobacterial methanotrophs directly from the soil. Enrichment cultures were then used to isolate *A. fumarolicum*. This bacterium grows between pH 0.8 and 5.8 with a temperature optimum of 55°C. Whereas *A. fumarolicum* was discovered in Italy, another *Verrucomicrobia* isolate was discovered from a similar acidic, geologically active site in New Zealand. Both grow well on methane and methanol, but simple organic acids such as acetate and succinate completely inhibit growth. These bacteria demonstrate that methanotrophy is more widespread and that this phylum is more diverse than previously thought. Indeed, the discovery of additional novel *Verrucomicrobia* is anticipated as environmental clone libraries from a variety of ecosystems contain SSU rRNA of microbes that may be assigned to this phylum. ▶▶ *Phylum* Methylococcales *(section 20.3)*

1. List the major properties of the class *Bacteroides*.
2. How do these bacteria benefit and harm their hosts?
3. List three advantages of gliding motility.
4. Briefly describe the following genera: *Cytophaga*, *Sporocytophaga*, and *Flexibacter*.
5. Why are the cytophagas ecologically important?
6. What do you think were some of the culture conditions used to isolate *A. fumarolicum*?

Summary

19.1 *Aquificae* and *Thermotogae*

a. *Aquifex* and *Thermotoga* are hyperthermophilic gram-negative rods that represent the two deepest or oldest phylogenetic branches of the *Bacteria*.

b. *Aquifex* spp. are chemolithoautotrophs. *Thermotoga* spp. are chemoheterotrophs.

19.2 *Deinococcus-Thermus*

a. Members of the order *Deinococcales* are aerobic cocci and rods.

b. *Deinococcus radiodurans* is distinctive because of its unusually great resistance to desiccation and radiation.

19.3 Photosynthetic Bacteria

a. Cyanobacteria carry out oxygenic photosynthesis; purple and green bacteria use anoxygenic photosynthesis.

b. The four most important groups of purple and green photosynthetic bacteria are the purple sulfur bacteria, the purple nonsulfur bacteria, the green sulfur bacteria, and the green nonsulfur bacteria (**table 19.1**).

c. The bacteriochlorophyll pigments of purple and green bacteria enable them to live in deeper, anoxic zones of aquatic habitats.

d. The phylum *Chlorobi* includes the green sulfur bacteria—obligately anaerobic photolithoautotrophs that use hydrogen sulfide, elemental sulfur, and hydrogen as electron sources.

e. Green nonsulfur bacteria such as *Chloroflexus* are placed in the phylum *Chloroflexi. Chloroflexus* is a filamentous, gliding thermophilic bacterium that is metabolically similar to the purple nonsulfur bacteria. Both *Chlorobi* and *Chloroflexus* have chlorosomes, which function to harvest light that is transferred to the reaction center in the plasma membrane (**figure 19.5**).

f. Cyanobacteria carry out oxygenic photosynthesis by means of a photosynthetic apparatus similar to that of the eukaryotes. Phycobilisomes contain the light-harvesting pigments phycocyanin and phycoerythrin (**figure 19.6**).

g. Cyanobacteria reproduce by binary fission, budding, multiple fission, and fragmentation by filaments to form hormogonia. Some produce dormant akinetes.

h. Many nitrogen-fixing cyanobacteria form heterocysts, specialized cells in which nitrogen fixation occurs (**figure 19.8**).

i. *Bergey's Manual* divides the cyanobacteria into five subsections and includes the prochlorophytes in the phylum *Cyanobacteria* (**table 19.2**).

19.4 Phylum *Planctomycetes*

a. Members of the phylum *Planctomycetes* lack peptidoglycan in their walls.

b The *Planctomycetes* have unusual cellular compartmentalization, and some perform the anammox reaction (**figures 19.10** and **19.11**).

19.5 Phylum *Chlamydiae*

a. Chlamydiae are nonmotile, coccoid, gram-negative bacteria that lack peptidoglycan and must reproduce within the cytoplasmic vacuoles of host cells by a life cycle involving elementary bodies (EBs) and reticulate bodies (RBs) (**figure 19.12**).

19.6 Phylum *Spirochaetes*

a. The spirochetes are slender, long, helical, gram-negative bacteria.

b. Spirochetes are motile because of the axial filament underlying an outer sheath or outer membrane (**figures 19.13–19.15**).

19.7 Phylum *Bacteroidetes*

a. Members of the class *Bacteroides* are obligately anaerobic, chemoheterotrophic, nonsporing, motile, or nonmotile rods of various shapes. Some are important rumen and intestinal symbionts; others can cause disease.

b. Gliding motility is present in a diverse range of bacteria, including the sphingobacteria and cytophagas (**figure 19.17**).

c. Cytophagas degrade proteins and complex polysaccharides, and are active in the mineralization of organic matter.

19.8 Phylum *Verrucomicrobia*

a. This is a relatively unexplored phylum. However, a new thermoacidophile was discovered recently that is the first microbe known to oxidize methane at very low pH. It is methylotrophic because it can use methane as a carbon and energy source.

Critical Thinking Questions

1. The cyanobacterium *Anabaena* grows well in liquid medium that contains nitrate as the sole nitrogen source. Suppose you transfer some of these filaments to the same medium except it lacks nitrate and other nitrogen sources. Describe the morphological and physiological changes you would observe.
2. Compare the structural and functional differences between chlorosomes and thylakoid membranes. Do you think this is an example of convergent evolution—two structures evolving separately to fulfill the same function, or divergent evolution—two structures that arose from a single ancestor but adapted to meet the particular needs of the organism that bears them? What other data would you need (apart from structure and function) to answer this question?
3. Many types of movement are employed by bacteria in these phyla. Review them and propose mechanisms by which energy (ATP or proton gradients) might drive the locomotion.
4. Propose two experimental approaches you might use to examine the mechanism by which *Cytophaga* glides.
5. Microbiologists using transmission electron microscopy have examined the cellular structure of a thermoacidophilic methylotroph belonging to the phylum *Verrucomicrobia* that was isolated from an acidic hot spring in Kamchatka, Russia. Surprisingly, this microbe has polyhedrals that look much like carboxysomes (figure 19.6). By contrast, proteobacterial methylotrophs possess intracytoplasmic membranes in which particulate methane monooxygenase is embedded. This enzyme uses O_2 to oxidize CH_4. What do you think is the utility of such enzyme compartmentalization? How would you determine if methane monooxygenase is localized to the polyhedral inclusion bodies found in thermoacidiphilic methylotrophs? Assuming the monooxygenase is confined to these structures, why do you think enzyme compartmentalization evolved in both types of bacteria?

   Read the original paper: Islam., R., et al. 2008. Methane oxidation at 55°C and pH 2 by a thermoacidophilic bacterium belonging to the *Verrucomicrobia* phylum. *Proc. Nat. Acad. Sci., USA*. 105:300.

Concept Mapping Challenge

Construct a concept map using the following words; use your own connecting words.

Photosynthesis	Green sulfur bacteria	*Chloroflexus*
Oxygenic	Cyanobacteria	*Chlorobium*
Anoxygenic	Prochlorophytes	Phycobilisomes
Chlorosomes	Phycocyanin	Phycoerythrin
Green nonsulfur bacteria		

Learn More

Learn more by visiting the text website at www.mhhe.com/willey8, where you will find a complete list of references.

20

Bacteria: *The Proteobacteria*

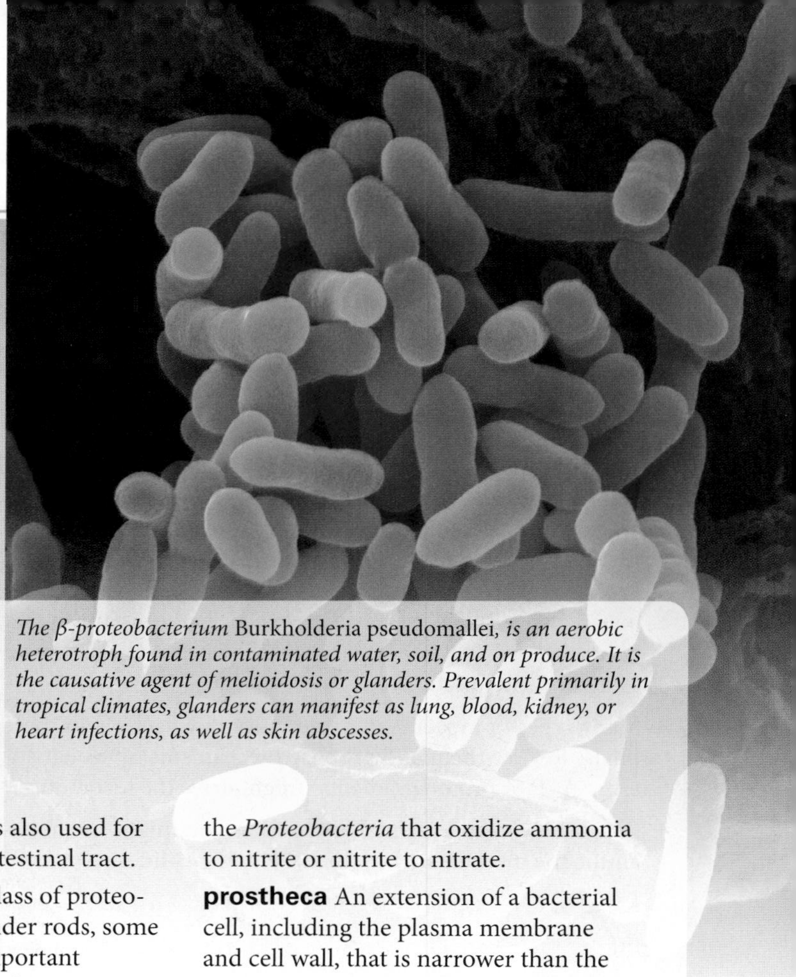

The β-proteobacterium Burkholderia pseudomallei, *is an aerobic heterotroph found in contaminated water, soil, and on produce. It is the causative agent of melioidosis or glanders. Prevalent primarily in tropical climates, glanders can manifest as lung, blood, kidney, or heart infections, as well as skin abscesses.*

CHAPTER GLOSSARY

α-proteobacteria The class of proteobacteria that contains most of the oligotrophic proteobacteria. Some have unusual metabolic modes such as methylotrophy, chemolithotrophy, and nitrogen fixing ability; many have distinctive morphological features.

β-proteobacteria The class of proteobacteria whose members are similar to the α-proteobacteria metabolically but tend to use substances that diffuse from organic matter decomposition in anoxic zones.

budding A means of asexual reproduction in which the daughter cell is smaller than the parent; found in yeast and some bacteria.

δ-proteobacteria The class of proteobacteria that includes chemoorganotrophic bacteria that usually are either predators of other bacteria or anaerobes that generate sulfide from sulfate and sulfite.

enteric bacteria or **enterobacteria** Members of the family *Enterobacteriaceae* (γ-proteobacteria, facultatively anaerobic, straight rods with simple nutritional requirements); the term is also used for bacteria that live in the intestinal tract.

ϵ-proteobacteria The class of proteobacteria that includes slender rods, some of which are medically important (*Campylobacter* and *Helicobacter*).

fruiting body A specialized structure that holds sexually or asexually produced spores; found in fungi and some bacteria (e.g., the myxobacteria).

γ-proteobacteria The largest of the five classes of proteobacteria. This subgroup is very diverse physiologically; many important genera are facultatively anaerobic chemoorganotrophs.

holdfast A structure produced by some bacteria (e.g., *Caulobacter*) that attaches them to a solid object.

methylotroph A microbe that uses reduced one-carbon compounds such as methane and methanol as its sole source of carbon and energy.

nitrification The oxidation of ammonia to nitrate.

nitrifying bacteria Chemolithotrophic bacteria found in several families within the *Proteobacteria* that oxidize ammonia to nitrite or nitrite to nitrate.

prostheca An extension of a bacterial cell, including the plasma membrane and cell wall, that is narrower than the mature cell. Found in the genus *Caulobacter,* an important model bacterium.

Proteobacteria A large phylum of gram-negative bacteria that 16S rRNA sequence comparisons show to be phylogenetically related; *Proteobacteria* contain the purple photosynthetic bacteria and their relatives, and are composed of the α, β, γ, δ, and ϵ classes.

purple nonsulfur bacteria Anoxygenic photosynthetic bacteria belonging to the α- and β-proteobacteria. Most are anaerobic photoorganoheterotrophs; some are facultative photolithoautotrophs (i.e., in the dark, they function as chemoorganoheterotrophs).

purple sulfur bacteria Obligate anaerobic photolithoautotrophic bacteria belonging to the γ-proteobacteria.

sheath A hollow, tubelike structure surrounding a cell or chain of cells.

In this chapter, we introduce the bacteria that are covered in volume 2 of the second edition of *Bergey's Manual of Systematic Bacteriology.* This volume is devoted entirely to the **proteobacteria.** This is the largest group of bacteria; there are over 500 known genera. Although 16S rRNA studies show that they are phylogenetically related, proteobacteria are remarkably diverse. The morphology of these gram-negative bacteria ranges from simple rods and cocci to genera with prosthecae, buds, and even fruiting bodies. Physiologically, they include photoautotrophs, chemolithotrophs, and chemoheterotrophs. No obvious overall pattern in metabolism, morphology, or reproductive strategy characterizes proteobacteria.

Comparison of 16S rRNA sequences has revealed five lineages of descent within the phylum *Proteobacteria: Alphaproteobacteria, Betaproteobacteria, Gammaproteobacteria, Deltaproteobacteria,* and *Epsilonproteobacteria* (**figure 20.1**). However, new sequence data suggest that the separation between the *Betaproteobacteria* and *Gammaproteobacteria* is less distinct than once thought. If further analysis supports this, the *Betaproteobacteria* may be considered a subgroup of the *Gammaproteobacteria.* Members of the purple photosynthetic bacteria are found among the α-, β-, and γ-proteobacteria. This has led to the proposal that the proteobacteria arose from a single photosynthetic ancestor, presumably similar to the purple bacteria. Subsequently photosynthesis would have been lost by various lines and new metabolic capacities acquired as these bacteria adapted to different ecological niches. Although the phylum is currently considered monophyletic, this has recently been questioned. In discussing this extremely diverse group of microbes, we review their morphology and physiology.

20.1 CLASS *ALPHAPROTEOBACTERIA*

The class *Alphaproteobacteria* has seven orders and 20 families. **Figure 20.2** illustrates the phylogenetic relationships among major groups within the **α-proteobacteria** as determined by 16S rRNA sequence and used by *Bergey's Manual.* As more bacterial

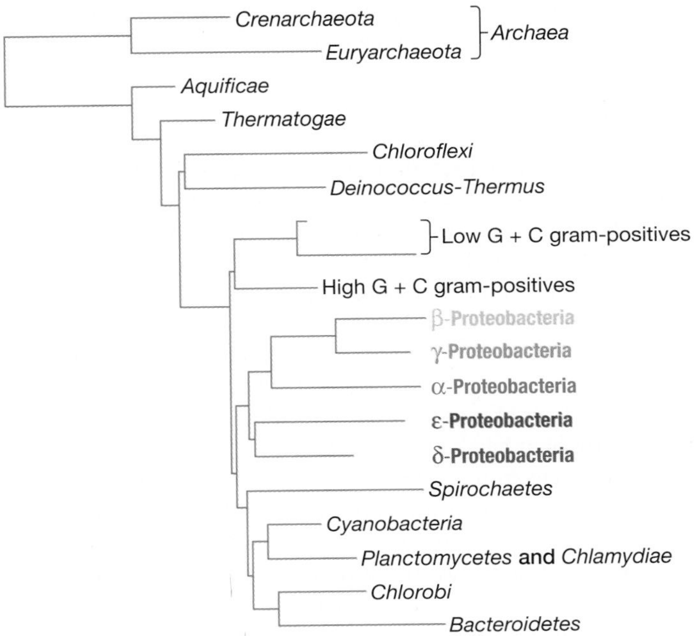

FIGURE 20.1 Phylogenetic Relationships Among *Bacteria* and *Archaea*. The *Proteobacteria* are highlighted.

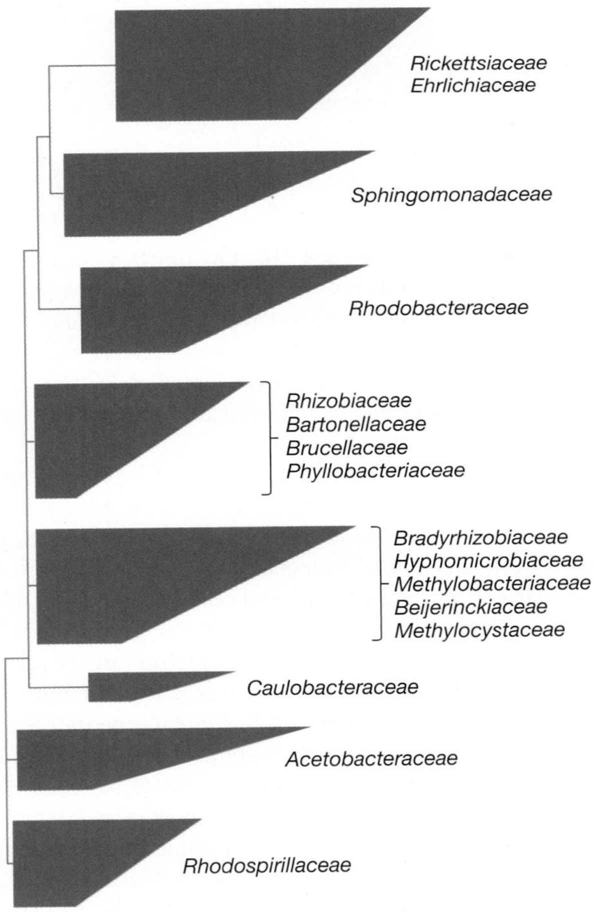

FIGURE 20.2 Phylogenetic Relationships Among Major Families Within the α-Proteobacteria. The relationships are based on 16S rRNA sequence data.

Table 20.1 Characteristics of Selected α-Proteobacteria

Genus	Dimensions (μm) and Morphology	G + C Content (mol%)	Genome Size (Mb)	Oxygen Requirement	Other Distinctive Characteristics
Agrobacterium	0.6–1.0 × 1.5–3.0; motile, nonsporing rods with peritrichous flagella	57–63	2.5	Aerobic	Chemoorganotroph that can invade plants and cause tumors
Caulobacter	0.4–0.6 × 1–2; rod- or vibrioid-shaped with a flagellum and prostheca and holdfast	62–65	4.0	Aerobic	Heterotrophic and oligotrophic; asymmetric cell division
Hyphomicrobium	0.3–1.2 × 1–3; rod-shaped or oval with polar prosthecae	59–65	Nd[a]	Aerobic	Reproduces by budding; methylotrophic
Nitrobacter	0.5–0.9 × 1.0–2.0; rod- or pear-shaped, sometimes motile by flagella	59–62	3.4	Aerobic	Chemolithotroph, oxidizes nitrite to nitrate
Rhizobium	0.5–1.0 × 1.2–3.0; motile rods with flagella	57–66	5.1	Aerobic	Invades leguminous plants to produce nitrogen-fixing root nodules
Rhodospirillum	0.7–1.5 wide; spiral cells with polar flagella	62–64	4.4	Anaerobic, microaerophilic, aerobic	Anoxygenic photoheterotroph under anoxic conditions
Rickettsia	0.3–0.5 × 0.8–2.0; short nonmotile rods	29–33	1.1–1.3	Aerobic	Obligate intracellular parasite

[a] Nd: Not determined; genome not yet sequenced

genomes have been sequenced, this phylum has been reexamined using a large number of protein sequences. In contrast to figure 20.2, these analyses suggest that the *Rickettsiales* are the earliest-branching (most ancient) α-proteobacteria, followed by the *Rhodospirillales* and then the *Sphingomonadales.* In addition, the stalked bacteria *Hyphomonadaceae* and the *Caulobacteriales* may be more closely related.

The *Alphaproteobacteria* include most of the oligotrophic proteobacteria (those capable of growing at low nutrient levels). Indeed, the most abundant bacteria in the world's oceans (and thus the most abundant worldwide) are α-proteobacteria. In addition, many have evolved to live within plant and animal cells. For those that inhabit insect and mammalian cells, this has resulted in genome reduction in some ancient lineages (*see figure 16.17b*). By contrast, the genomes of plant-associated α-proteobacteria have expanded through mechanisms such as gene duplication and horizontal gene transfer. One particularly interesting intracellular α-proteobacteria is *Walbachia pipientis.* This bacterium infects many species of insects and can actually change the male-to-female ratio of host progeny as well as feminize some male insects. This fascinating microbe is discussed in chapter 30. ▶▶| *Microbial Diversity & Ecology 30.1*

Some α-proteobacteria have unusual metabolic modes such as methylotrophy—the ability to grow using methane as a carbon source (*Methylobacterium*), chemolithotrophy (*Nitrobacter*), and the ability to fix nitrogen (*Rhizobium*). Members of genera such as *Rickettsia* and *Brucella* are important pathogens. Many genera are characterized by distinctive morphology such as prosthecae. **Table 20.1** summarizes the general characteristics of many of the bacteria discussed in the following sections.

Purple Nonsulfur Bacteria

All purple nonsulfur bacteria are α-proteobacteria, with the exception of *Rhodocyclus* (β-proteobacteria). These bacteria use anoxygenic photosynthesis and possess bacteriochlorophylls *a* or *b.* Their photosynthetic apparatus is continuous with the plasma membrane. The plasma membrane is invaginated, making many intracellular folds (**figure 20.3*a,b***). These intracytoplasmic membranes (ICMs) increase the surface area, thereby enabling space for more *p*hotosynthetic *u*nits (PSUs). Each PSU includes a reaction center (RC), consisting of a protein-bacteriochlorophyll (Bchl) complex that absorbs light in the infrared range. The RC-Bchl is responsible for donating electrons to the electron transport chain that will result in charge separation and a proton motive force sufficient to fuel ATP synthesis. The energy required for charge separation is supplied by the photons directly absorbed by the RC bacteriochlorophyll or transferred from a light-harvesting complex. All purple bacteria (nonsulfur and sulfur) have a primary light-harvesting complex called LH-I that is closely associated with the RC. Most also have additional light-harvesting complexes, called LH-II.

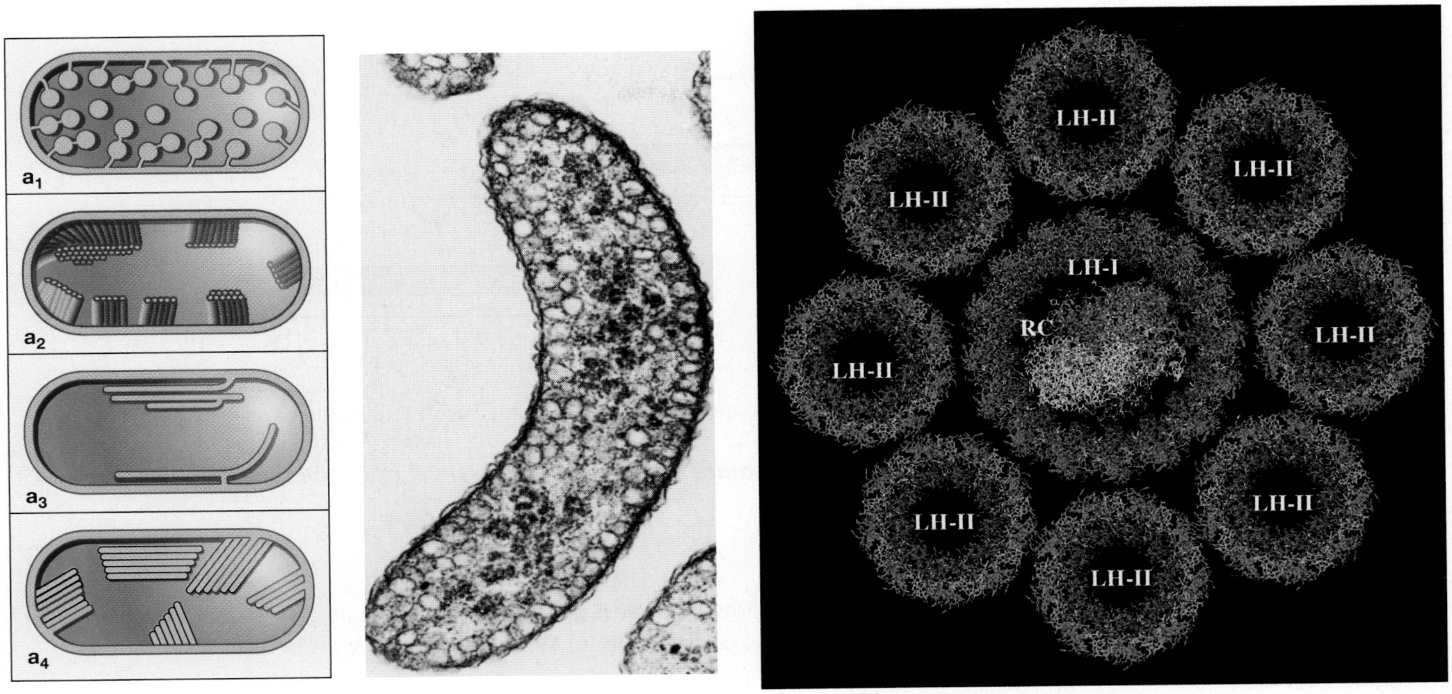

(a) Intracytoplasmic membranes **(b)** *R. rubrum* **(c)** Integral light-harvesting complex

FIGURE 20.3 The Photosynthetic Apparatus of Purple Bacteria. (a) The arrangement of intracytoplasmic membranes (ICMs) varies, but in all cases, they are associated with the plasma membrane. (a_1) ICMs that resemble vesicles are found in the nonsulfur α-proteobacteria *Rhodospirillum rubrum* and *Rhodobacter capsulatus,* as well as the sulfur γ-proteobacteria *Chromatium vinusum* and *Thiocapsa roseopersicina.* (a_2) ICMs can be more tubelike, as found in *T. pfennigii,* or (a_3) may be arranged in small stacks, as seen in *Rhodospirillum molischianum* and the γ-proteobacterium *Ectothiorhodospira mobilis.* (a_4) *Rhodopseudomonas palustris* has larger stacked ICMs. (b) A transmission electron micrograph of *R. rubrum* grown anaerobically in the light shows the vesicle-like ICMs. (c) Crystal structure of light-harvesting complex from a photosynthetic bacterium with the reaction center (RC) and closely associated light-harvesting complex I (LH-I). The RC is encircled by additional light-harvesting complexes (LH-II).

Light-harvesting complexes are composed of Bchl *a* and carotenoid pigments. As shown in figure 20.3*c*, LH-IIs encircle the RC/LH-I complex. This maximizes the efficiency of energy transfer. ◀◀ *Light reactions in anoxygenic photosynthesis (section 10.12); Photosynthetic bacteria (section 19.3)*

The **purple nonsulfur bacteria** are exceptionally flexible in their choice of an energy source. Normally they grow anaerobically as photoorganoheterotrophs, trapping light energy and employing organic molecules as both electron and carbon sources (table 20.1). Although they are called nonsulfur bacteria, some species can oxidize very low, nontoxic levels of sulfide to sulfate, but they do not oxidize elemental sulfur to sulfate. In the absence of light, most purple nonsulfur bacteria grow aerobically as chemoorganoheterotrophs, but some species carry out fermentations anaerobically. Oxygen inhibits bacteriochlorophyll and carotenoid synthesis so that cultures growing aerobically in the dark are colorless.

Rhodospirillum rubrum is perhaps the best-studied purple nonsulfur bacterium. It exemplifies the metabolic flexibility of this group. The "decision-making" process this microbe uses is displayed in **figure 20.4**. When in oxic conditions, the photosynthetic apparatus is inhibited, and it grows chemoorganically. However, when oxygen is not present, it can use fermentation, photoheterotrophy, or photoautotrophy to sustain growth. When growing phototrophically, *R. rubrum* makes photopigments that give it a rich red color. *R. rubrum* has sparked the interest of industrial microbiologists because it produces H_2. In addition, it can produce novel biodegradable plastics when grown on β-hydroxycarboxylic and *n*-alkanoic acids. Finally, it has the rare ability to oxidize carbon monoxide to carbon dioxide. ▶▶ *Industrial microbiology (chapter 41)*

Purple nonsulfur bacteria vary considerably in morphology. Most are motile by polar flagella. They may be spirals (*Rhodospirillum*), rods (*Rhodopseudomonas*), half circles or circles (*Rhodocyclus*), or they may even form prosthecae and buds (*Rhodomicrobium*). Because of their metabolism, they are most prevalent in the mud and water of lakes and ponds with abundant organic matter and low sulfide levels. Marine species also exist.

In addition to the anaerobic anoxygenic photosynthesis displayed by the purple nonsulfur bacteria, some α-proteobacteria are capable of **aerobic anoxygenic photosynthesis (AAnP).**

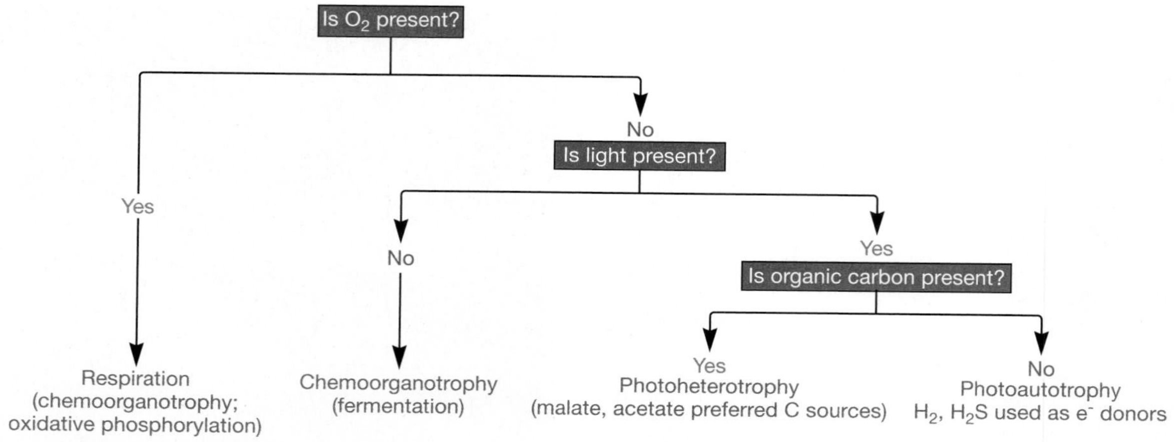

FIGURE 20.4 The Metabolic Flexibility of *Rhodospirillum rubrum*. This microbe can switch from one type of metabolism to another, depending on the availability of light and oxygen.

Figure 20.4 Micro Inquiry

What would *Rhodospirillum rubrum* use for carbon and energy sources if grown (a) anaerobically in the dark in the presence of malate, (b) anaerobically in the light without organic carbon, and (c) aerobically in the light in the presence of malate?

These aerobic bacteria are photoheterotrophs that metabolize carbon and other organic materials using O_2-dependent respiration. They also contain bacteriochlorophyll *a* and carotenoid pigments, and are able to fix CO_2 during photosynthesis, thereby enhancing their growth. They are found in marine and freshwater environments, and are often present in microbial mat communities. Important genera include the α-proteobacteria *Erythromonas, Roseococcus, Porphyrobacter,* and *Roseobacter*, although recently it has been discovered that a wider diversity of bacteria are capable of AAnP (*see table 19.1*). Their contribution to the global carbon cycle is an area of active investigation.

Rhodospirillum and *Azospirillum* (both in the family *Rhodospirillaceae*) are among several bacterial genera capable of forming **cysts** in response to nutrient limitation (**figure 20.5**). These resting cells differ from the well-characterized endospores made by the low G + C gram-positive bacteria *Bacillus* and *Clostridium.* Although cysts are also very resistant to desiccation, they are less tolerant of other environmental stresses such as heat and UV light. They have a thick outer coat and store an abundance of poly-β-hydroxybutyrate (PHB). Cyst-forming bacteria are not limited to α-proteobacteria; for instance, *Azotobacter,* a γ-proteobacterium, also forms cysts. ◀◀ *Bacterial endospores (section 3.8)*

Rickettsia

The genus *Rickettsia* is placed in the order *Rickettsiales* and family *Rickettsiaceae* of the α-proteobacteria. These bacteria are rod-shaped, coccoid, or pleomorphic with typical gram-negative walls and no flagella. Although their size varies, they tend to be very small. For example, members of the genus *Rickettsia* are 0.3 to 0.5 μm in diameter and 0.8 to 2.0 μm long. All species are parasitic or mutualistic. The parasitic forms grow in vertebrate erythrocytes, macrophages, and vascular endothelial cells. Some also live in blood-sucking arthropods such as fleas, ticks, mites, or lice, which serve as vectors or primary hosts. ▶▶ *Microbial interactions (section 30.1)*

The rickettsias are believed to be descended from a free-living bacterium, and a series of events have been proposed to account for this. First, a free-living, aerobic bacterium became an intracellular parasite of a proto-eukaryotic cell, that is, an ancestral eukaryotic cell that lacked organelles. In this way, the bacterium was able to assimilate many host substrates. This made the genes for the biosynthesis of these substrates unnecessary. Thus if they were mutated, there was no selection to prevent their change in function or their conversion to pseudogenes (genes that do not encode a product). To increase cellular efficiency, many pseudogenes were eventually lost. This gene reduction gave rise to a small genome that prevented the bacterium from living outside the evolving host cell.

Rickettsias are very different from most other bacteria in physiology and metabolism. They lack glycolytic pathways and do not use glucose as an energy source but rather oxidize glutamate and tricarboxylic acid cycle intermediates such as succinate. The rickettsial plasma membrane has carrier-mediated transport systems, and host cell nutrients and coenzymes are absorbed and directly used. For example, *R. prowazekii* takes up both NAD^+ and uridine diphosphate glucose (UDPG). Because *R. prowazekii*

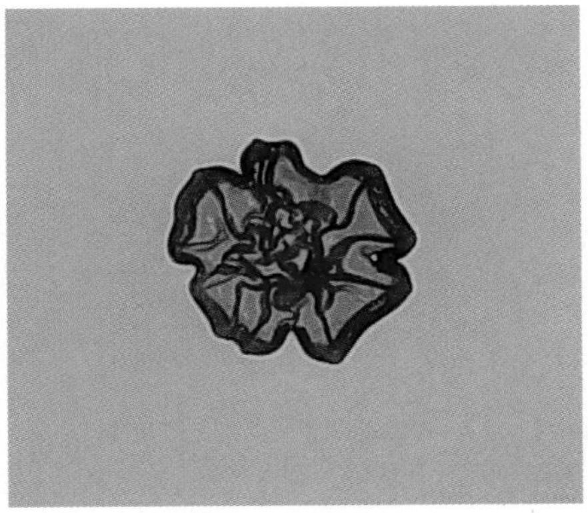

(a)

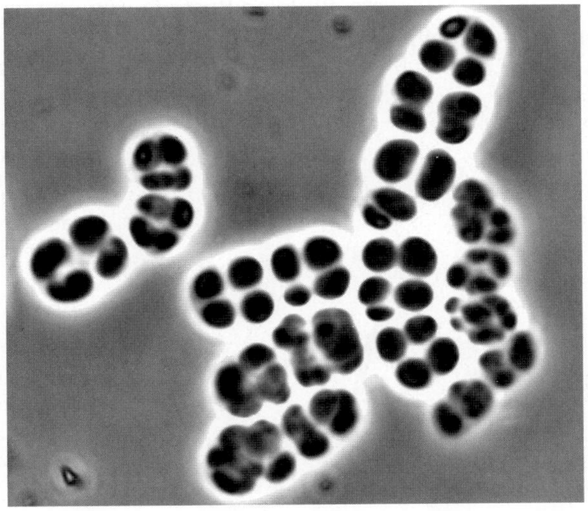

(b)

FIGURE 20.5 Cyst formation in *Rhodospirillum centrum*. (a) Colony morphology when grown on a medium that limits nutrients. (b) The cells within the colony (a) have formed cysts. In this species, cysts are most commonly found in clusters of four surrounded by a thick outer coat consisting of lipopolysaccharides and lipoproteins. Cysts are nonmotile and more spherical than growing cells.

***Figure 20.5* Micro Inquiry**

How are bacterial cysts different from the endospores of certain gram-positive bacteria?

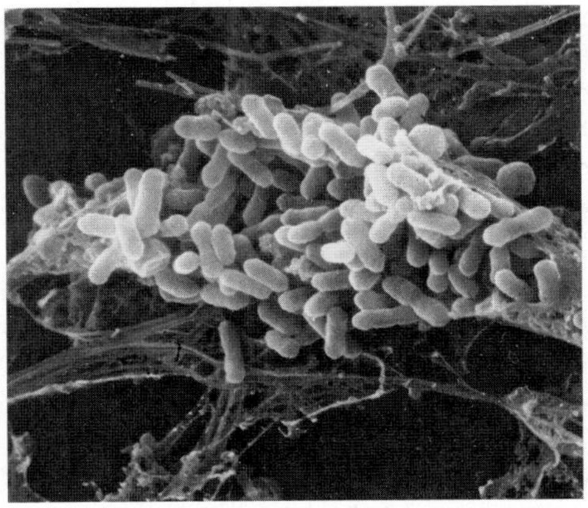

(a)

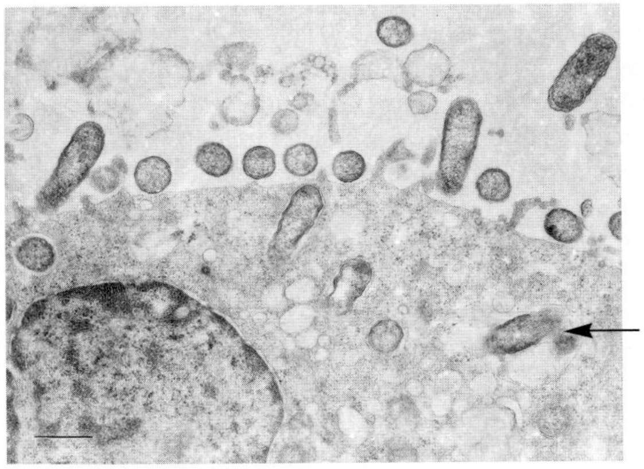

(b)

FIGURE 20.6 *Rickettsia*. Rickettsial morphology and reproduction. (a) A human fibroblast filled with *Rickettsia prowazekii*. (b) *Rickettsiae* attachment to endothelial cell and subsequent phagocytosis. *Rickettsiae* leave a disrupted phagosome (arrow) and enter the cytoplasmic matrix.

cannot catabolize glucose, it is believed to use host UDPG for the synthesis of its slime layer, lipopolysaccharide, and peptidoglycan. Its membrane also has a carrier that exchanges ADP for external ATP. Thus host ATP may provide some of the energy needed for growth. *R. prowazekii* possesses a complete TCA cycle and a membrane-bound ATP synthase, distinguishing it from the *Chlamydia*, which are also obligate intracellular pathogens. ◀◀ *Phylum* Chlamydiae *(section 19.5)*

Genome sequencing shows that *R. prowazekii* is similar in many ways to mitochondria, suggesting that these organelles arose from an endosymbiotic association with an ancestor of *Rickettsia*. In fact, of the 834 protein-coding genes found on the *R. prowazekii* genome, about one-fifth of them share significant homology to mitochondrial genes. ◀◀ *Endosymbiotic origin of mitochondria, chloroplasts, and hydrogenosomes (section 1.2)*

This genus includes important human pathogens including *R. prowazekii* and *R. typhi*, the causative agents of louse-borne and murine typhus, respectively, as well as *R. rickettsii*, which causes Rocky Mountain spotted fever. This has prompted intensive study of their reproduction and metabolism. They enter the host cell by inducing phagocytosis but immediately escape the phagocytic vacuole and reproduce by binary fission in the cytoplasm (**figure 20.6**). Eventually the host cell bursts, releasing new organisms. Besides incurring damage from cell lysis, the host is harmed by the toxic effects of bacterial cell walls (wall toxicity appears related to the mechanism of penetration into host cells). ▶▶| *Phagocytosis (section 32.5); Arthropod-borne diseases (section 38.2)*

Caulobacteraceae and *Hyphomicrobiaceae*

A number of the proteobacteria are not simple rods or cocci but have some sort of appendage. These bacteria have interesting life cycles that feature a prostheca or reproduction by budding. A **prostheca** (pl., prosthecae), also called a **stalk,** is an extension of the cell, including the plasma membrane and cell wall, that is narrower than the mature cell. **Budding** is distinctly different from the binary fission normally used by bacteria. The bud first appears as a small protrusion at a single point and enlarges to form a mature cell. Most or all of the bud's cell envelope is newly synthesized. In contrast, portions of the parental cell envelope are shared with the progeny cells during binary fission *(see figure 7.9)*. Finally, the parental cell retains its identity during budding, and the new cell is often smaller than its parent. In binary fission, the parental cell disappears as it forms progeny of equal size. The families *Caulobacteraceae* and *Hyphomicrobiaceae* of the α-proteobacteria contain two of the best studied prosthecate genera: *Caulobacter* and *Hyphomicrobium.* |◀◀ *Bacterial cell cycle (section 7.2)* *Appendaged Bacteria*

The genus *Hyphomicrobium* contains chemoheterotrophic, aerobic, budding bacteria that frequently attach to solid objects in freshwater, marine, and terrestrial environments. The vegetative cell measures about 0.5 to 1.0 by 1 to 3 μm (**figure 20.7**). At the beginning of the reproductive cycle, the mature cell produces a prostheca (also called a hypha) 0.2 to 0.3 μm in diameter that grows to several μm in length (**figure 20.8**). The nucleoid divides, and a copy moves into the hypha while a bud forms at its end. As the bud matures, it produces one to three flagella, and a septum divides the bud from the hypha. The bud is finally released as an oval- to pear-shaped swarmer cell, which swims off, then settles down and begins budding. The mother cell may bud several times at the tip of its hypha.

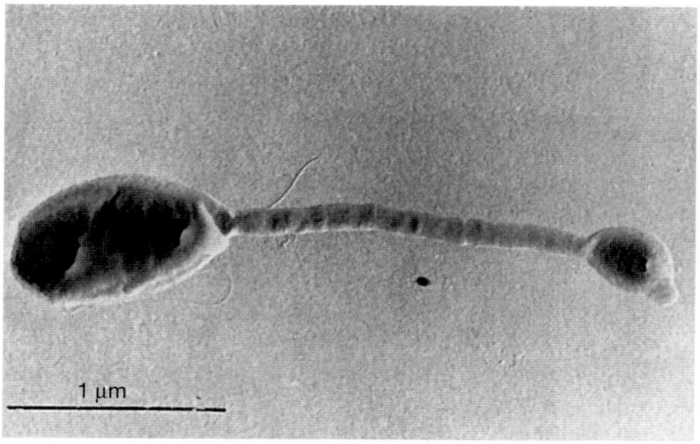

FIGURE 20.7 Prosthecate, Budding Bacteria. *Hyphomicrobium facilis* with hypha and young bud.

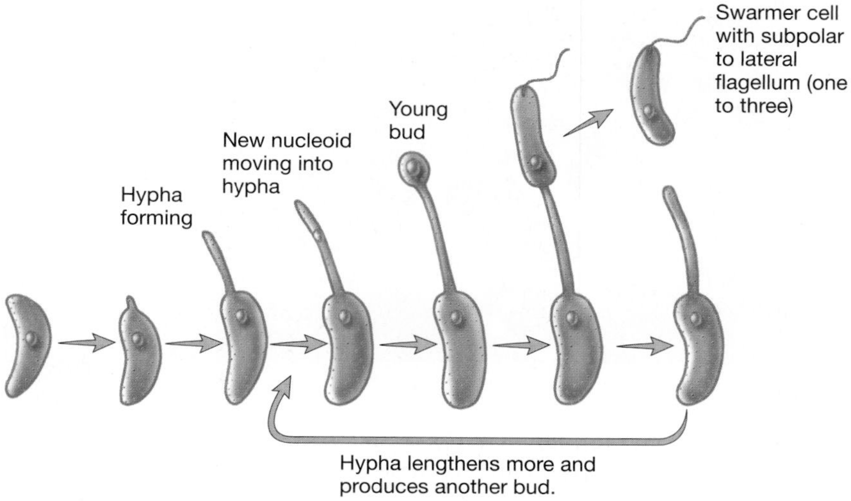

FIGURE 20.8 The Life Cycle of *Hyphomicrobium.*

Hyphomicrobium also has distinctive physiology and nutrition. Sugars and most amino acids do not support abundant growth; instead, *Hyphomicrobium* grows on ethanol and acetate, and flourishes with one-carbon compounds such as methanol, formate, and formaldehyde. That is, it is a facultative **methylotroph** and can derive both energy and carbon from reduced one-carbon compounds. It is so efficient at acquiring one-carbon molecules that it can grow in a medium without an added carbon source (presumably the medium absorbs sufficient atmospheric carbon compounds). *Hyphomicrobium* may comprise up to 25% of the total bacterial population in oligotrophic or nutrient-poor freshwater habitats.

Bacteria in the genus *Caulobacter* alternate between polarly flagellated rods and cells that possess a prostheca and a **holdfast,** by which they attach to solid substrata (**figure 20.9**). Incredibly, the material secreted at the end of the *Caulobacter crescentus* holdfast is the strongest biological adhesion molecule known—a sort of bacterial superglue. Caulobacters are usually isolated from freshwater and marine habitats with low nutrient levels, but they also are present in the soil. They often adhere to bacteria, photosynthetic protists, and other microorganisms, and may absorb nutrients released by their hosts. The prostheca differs from that of *Hyphomicrobium* in that it lacks cytoplasmic components and is composed almost totally of the plasma membrane and cell wall. It grows longer in nutrient-poor media and can reach more than 10 times the length of the cell body. The prostheca may improve the efficiency of nutrient uptake from dilute habitats by increasing surface area; it also gives the cell extra buoyancy.

The life cycle of *Caulobacter* is unusual (**figure 20.10**). When ready to reproduce, the cell elongates and a single polar flagellum forms at the end opposite the prostheca. The cell then undergoes

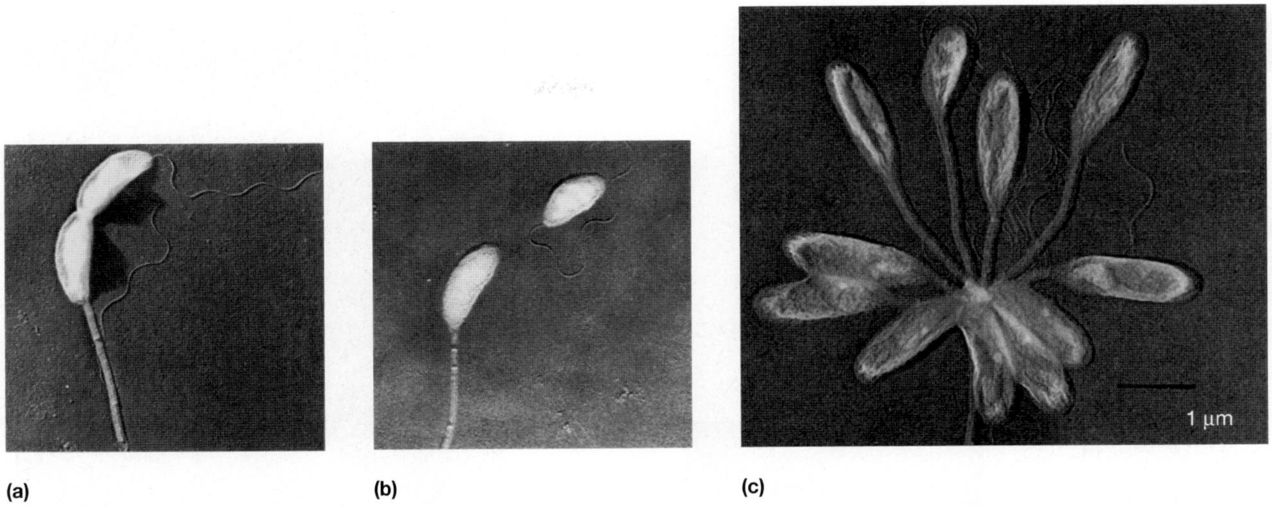

FIGURE 20.9 ***Caulobacter* Morphology and Reproduction.** (a) A cell dividing to produce a swarmer (×6,030). Note prostheca and flagellum. (b) A stalked cell and a flagellated swarmer cell (×6,030). (c) A rosette of cells as seen in the electron microscope.

Figure 20.9 Micro Inquiry

If *Caulobacter* were grown in a phosphate-limited broth, what would happen to the length of the stalks?

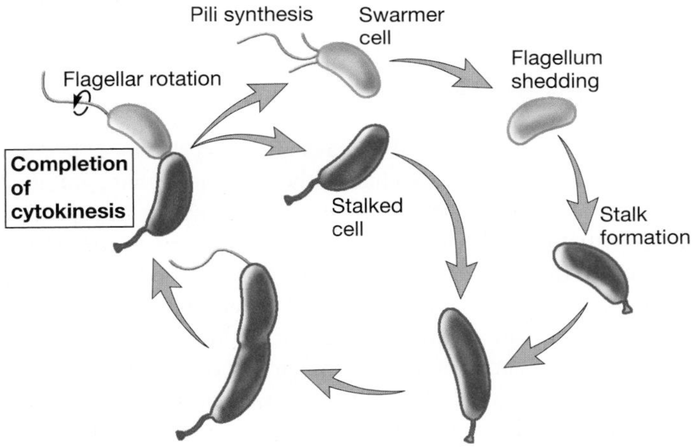

FIGURE 20.10 ***Caulobacter* Life Cycle.** Stalked cells attached to a substrate undergo asymmetric binary fission, producing a stalked and a flagellated cell, called a swarmer cell. The swarmer cell swims freely and makes pili until it settles, ejects its flagella, and forms a stalk. Only stalked cells can divide.

asymmetric transverse binary fission to produce a flagellated swarmer cell that swims away. The swarmer, which cannot reproduce, comes to rest, ejects its flagellum, and forms a new prostheca on the formerly flagellated end. The new stalked cell then starts the cycle anew. This process takes about 2 hours to complete. The species *C. crescentus* has become an important model organism in the study of microbial development and the bacterial cell cycle.

Search This: Caulobacter.org

Order *Rhizobiales*

The order *Rhizobiales* of the α-proteobacteria contains 11 families with a great variety of phenotypes. This includes the family *Hyphomicrobiaceae,* which has already been discussed. Another important family in this order is *Rhizobiaceae,* which includes the aerobic genera *Rhizobium* and *Agrobacterium.*

Members of the genus *Rhizobium* are 0.5 to 0.9 by 1.2 to 3.0 µm motile rods that become pleomorphic (having many shapes) under adverse conditions (**figure 20.11**). Cells often contain poly-β-hydroxybutyrate inclusions. They grow symbiotically within root nodule cells of legumes as nitrogen-fixing bacteroids (figure 20.11*b*; *see also figure 29.9*). In fact, *Leguminosae* is the most successful plant family on Earth, with over 18,000 species. Their proliferation reflects their capacity to establish symbiotic relationships with bacteria that form nodules on their roots. Within the nodules, the microbes reduce or fix atmospheric nitrogen to ammonia, making it directly available to the plant host. The process by which bacteria perform this fascinating and important symbiosis is discussed in chapter 29. ◀◀ *Nitrogen fixation (section 11.5);* ▶▶ *The Rhizobia (section 29.3)*

The genus *Agrobacterium* is placed in the family *Rhizobiaceae* but differs from *Rhizobium* because it is a plant pathogen.

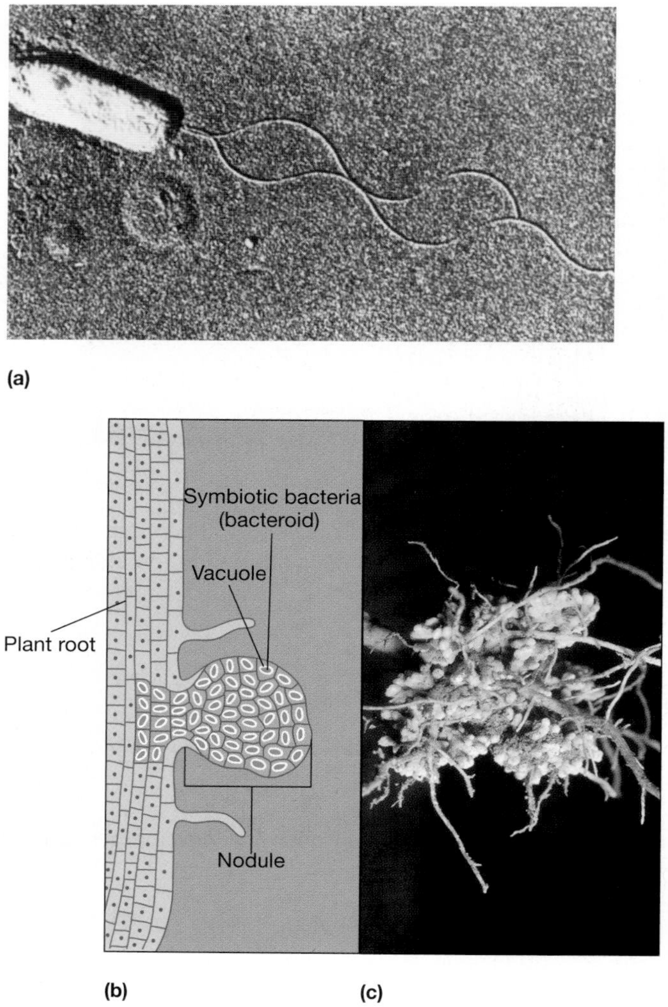

FIGURE 20.11 ***Rhizobium.*** (a) *Rhizobium leguminosarum* with two polar flagella (×14,000). (b) Diagram of differentiated bacteria, called bacteroids, in the nodule of a legume root. (c) Nodules on a legume root.

FIGURE 20.12 ***Agrobacterium.*** Crown gall tumor of a tomato plant caused by *Agrobacterium tumefaciens.*

Agrobacteria invade the crown, roots, and stems of many plants, and transform plant cells into autonomously proliferating tumor cells (*see figures 29.12 and 29.13*). The best-studied species is *A. tumefaciens,* which enters many broad-leaved plants through wounds and causes crown gall disease (**figure 20.12**). The ability to produce tumors depends on the presence of a large Ti (for *tu*mor *i*nducing) plasmid. Tumor production by *Agrobacterium* is discussed in greater detail in chapter 29, while the use of the Ti plasmid in the genetic engineering of plants is covered in chapter 41 (*see figure 41.10*).

The order *Rhizobiales* also includes the family *Brucellaceae.* The genus *Brucella* is an important human and animal pathogen. Brucellosis, also called undulant fever, is a zoonotic disease—a disease transmitted from animals to humans. It is usually caused by tiny, faintly staining coccobacilli of the species *B. abortus, B. melitensis, B. suis,* or *B. canis.* An ongoing controversy in the northwestern United States centers on the possible transmission of *Brucella,* endemic in the wild bison and elk populations, to otherwise *Brucella*-free cattle. ▶▶| *Brucellosis (section 38.5)*

Nitrifying Bacteria

The **nitrifying bacteria** and **archaea** are a very diverse collection of microbes. Nitrifiers are chemolithoautotrophs that gain electrons from the oxidation of ammonium or nitrite in a process called **nitrification**. Electrons are donated to the electron transport chain from either ammonia, which is oxidized to nitrite, or from nitrite, which is then oxidized to nitrate (**figure 20.13**). Oxygen serves as the terminal electron acceptor. No microbe can perform both reactions, and as shown in **table 20.2**, some *Proteobacteria* are classified as ammonia-oxidizing bacteria, while others are listed as nitrite-oxidizing bacteria. To date, there are no known archaea capable of nitrite oxidation, and few ammonia-oxidizing archaea have been grown in pure culture. Typically, nitrifying bacteria use CO_2 as their carbon source and thus are chemolithoautotrophs, but some can function as chemolithoheterotrophs and use reduced organic carbon sources. |◀◀ *Chemolithotrophy (section 10.11); Phylum Crenarchaeota (section 18.2)*

Bergey's Manual places nitrifying bacteria in several families. All terrestrial ammonia-oxidizing bacteria belong to the β-proteobacteria, including *Nitrosomonas* and *Nitrosospira* in the *Nitrosomonadaceae* family, while marine ammonia oxidizers include both β-proteobacteria and γ-proteobacteria such as *Nitrosococcus* in the *Chromatiaceae* family. The recently identified ammonia oxidizing archaea (not listed in *Bergey's Manual*) are

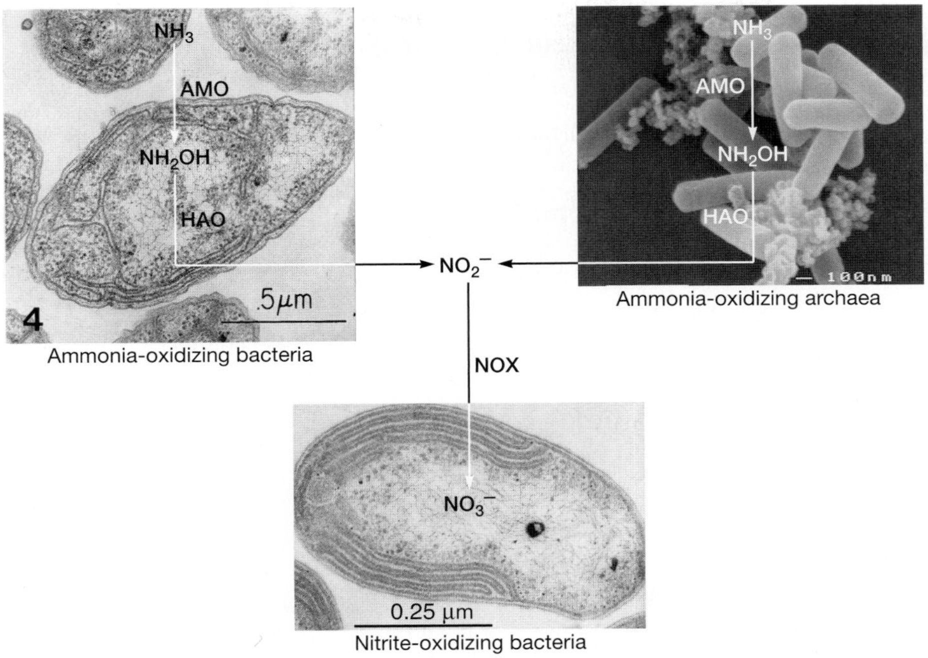

FIGURE 20.13 Nitrification. The process of nitrification involves three different groups of microorganisms. Ammonia-oxidizing bacteria (e.g., *Nitrosomonas europaea*) and archaea ("*Nitrosopumilus maritimus*") use the enzyme ammonia monooxygenase (AMO) to convert ammonia to the intermediate hydroxylamine. Hydroxylamine oxidoreductase (HAO) then catalyzes the production of nitrite, which is released into the environment. Nitrite-oxidizing bacteria such as *Nitrobacter winogradskyi* use nitrite oxidoreductase (NOX) to catalyze the oxidation of nitrite to nitrate.

Figure 20.13 Micro Inquiry

Why do nitrifying microbes possess intracellular membranes?

found in both soils and marine ecosystems. Nitrite oxidation is carried out by bacteria in the α-proteobacterial class, *Nitrobacter* (*Bradyrhizobiaceae* family), and the γ-proteobacteria class, *Nitrococcus* (*Ectothiorhodospiraceae* family). These microbes differ considerably in morphology and may be rod-shaped, ellipsoidal, spherical, spirillar, or lobate. They may possess either polar or peritrichous flagella. Often they have extensive membrane complexes in their cytoplasm (figure 20.13). Just as is the case with the purple bacteria, these intracellular membranes increase the surface area available for electron transport. Chemolithotrophy yields very little free energy, so reduced inorganic compounds such as ammonia or nitrite must be oxidized at a great rate in order to conserve enough energy to support growth.

Nitrifying microorganisms make important contributions to the nitrogen cycle. Nitrification occurs rapidly in oxic soil treated with fertilizers containing ammonium salts. Nitrification increases the availability of nitrogen to plants because nitrate is readily used by plants. On the other hand, it is also rapidly lost through leaching of water-soluble nitrates and by denitrification to nitrogen gas, so the benefits gained from nitrification can be fleeting. ▶▶| *Nitrogen cycle (section 26.1)*

1. Describe the general properties of the α-proteobacteria.
2. Discuss the characteristics and physiology of the purple nonsulfur bacteria. Where would one expect to find them growing?
3. Briefly describe the characteristics and life cycle of the genus *Rickettsia.* Why are these bacteria thought to be the closest extant relative to the endosymbiont that evolved into mitochondria?
4. What is unusual about the physiology of *Hyphomicrobium?* How does this influence its ecological distribution?
5. What unique features of the *Caulobacter* life cycle do you think make this microbe such an attractive model system?
6. How do *Agrobacterium* and *Rhizobium* differ in lifestyle? What effect does *Agrobacterium* have on plant hosts?
7. Give the major characteristics of the nitrifying bacteria and discuss their ecological importance. How does the metabolism of *Nitrobacter* differ from that of *Nitrosomonas?*

Table 20.2 Selected Characteristics of Representative Nitrifying Bacteria

Species	*Cell Morphology and Size (μm)*	*Reproduction*	*Motility*	*Cytomembranes*	*G + C Content (mol%)*	*Habitat*
Ammonia-Oxidizing Bacteria						
Nitrosomonas europaea (β-proteobacteria)	Rod; 0.8–1.1 × 1.0–1.7	Binary fission	–	Peripheral, lamellar	50.6–51.4	Soil, sewage, freshwater, marine
Nitrosococcus oceani (γ-proteobacteria)	Coccoid; 1.8–2.2 in diameter	Binary fission	+; 1 or more subpolar flagella	Centrally located parallel bundle, lamellar	50.5	Obligately marine
Nitrosospira briensis (β-proteobacteria)	Spiral; 0.3–0.4 in diameter	Binary fission	+ or –; 1 to 6 peritrichous flagella	Rare	53.8–54.1	Soil
Nitrite-Oxidizing Bacteria						
Nitrobacter winogradskyi (α-proteobacteria)	Rod, often pear-shaped; 0.5–0.9 × 1.0–2.0	Budding	+ or –; 1 polar flagellum	Polar cap of flattened vesicles in peripheral region of the cell	61.7	Soil, freshwater, marine
Nitrococcus mobilis (γ-proteobacteria)	Coccoid; 1.5–1.8 in diameter	Binary fission	+; 1 or 2 subpolar flagella	Tubular cytomembranes randomly arranged in cytoplasm	61.3 (1 strain)	Marine

From Brenner, D. J., et al., eds. 2005. Bergey's Manual to Systemic Bacteriology, *2d ed. Vol. 2: The Proteobacteria. Garrity, G. M. Ed-in-Chief. New York: Springer.*

20.2 CLASS *BETAPROTEOBACTERIA*

The **β-proteobacteria** overlap the α-proteobacteria metabolically but tend to use substances that diffuse from organic decomposition in anoxic habitats. Some of these bacteria use hydrogen, ammonia, methane, volatile fatty acids, and similar substances. As with the α-proteobacteria, there is considerable metabolic diversity; the β-proteobacteria may be chemoheterotrophs (including methylotrophs), photolithotrophs, and chemolithotrophs.

The class *Betaproteobacteria* has seven orders and 12 families. **Figure 20.14** shows the phylogenetic relationships among major groups within the β-proteobacteria, and **table 20.3** summarizes the general characteristics of many of the bacteria discussed in this section.

Order *Neisseriales*

The order *Neisseriales* has one family, *Neisseriaceae,* with 15 genera. The best-known and most intensely studied genus is *Neisseria.* Members of this genus are nonmotile, aerobic, gram-negative cocci that most often occur in pairs with adjacent sides flattened. They may have capsules and fimbriae (often referred to

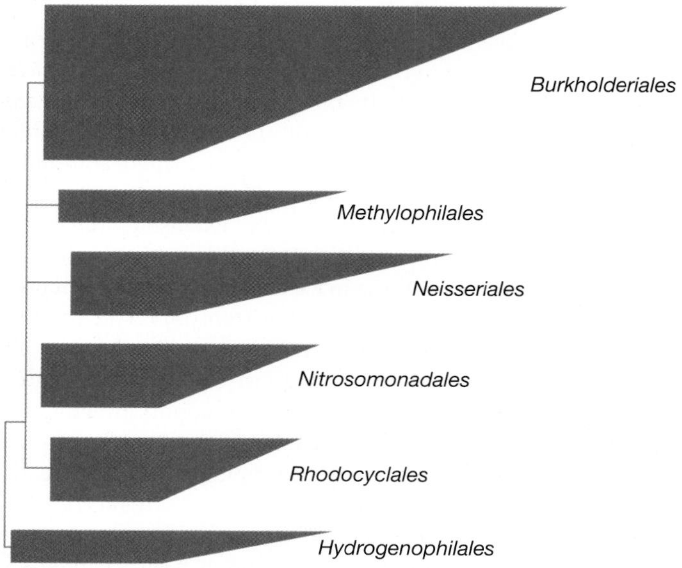

FIGURE 20.14 Phylogenetic Relationships Among Major Groups Within the β-Proteobacteria. The relationships are based on 16S rRNA sequence data.

Table 20.3 Characteristics of Selected β-Proteobacteria

Genus	*Dimensions (μm) and Morphology*	*G + C Content (mol%)*	*Genome Size (Mb)*	*Oxygen Requirement*	*Other Distinctive Characteristics*
Bordetella	0.2–0.5 × 0.5–2.0; nonmotile coccobacillus	66–70	3.7–5.3	Aerobic	Requires organic sulfur and nitrogen; mammalian parasite
Burkholderia	0.5–1.0 × 1.5–4; straight rods with single flagella or a tuft at the pole	59–69.5	4.1–7.2	Aerobic, some capable of anaerobic respiration with NO_3^-	Poly-β-hydroxybutyrate as reserve; can be pathogenic
Leptothrix	0.6–1.5 × 2.5–15; straight rods in chains with sheath, free cells flagellated	68–71	Nd[a]	Aerobic	Sheaths encrusted with iron and manganese oxides
Neisseria	0.6–1.9; cocci in pairs with flattened adjacent sides	48–56	2.2–2.3	Aerobic	Inhabitant of mucous membranes of mammals
Nitrosomonas	Size varies with strain; spherical to ellipsoidal cells with intracytoplasmic membranes	45–54	2.8	Aerobic	Chemolithotroph that oxidizes ammonia to nitrite
Sphaerotilus	1.2–2.5 × 2–10; single chains of cells with sheaths, may have holdfasts	70	Nd	Aerobic	Sheaths not encrusted with iron and manganese oxides
Thiobacillus	0.3–0.5 × 0.9–4; rods, often with polar flagella	52–68	Nd	Aerobic	All chemolithotrophic, oxidizes reduced sulfur compounds to sulfate; some also chemoorganotrophic

[a] Nd: Not determined; genome not yet sequenced

as pili). The genus is chemoorganotrophic and produces the enzymes oxidase and catalase. They are inhabitants of the mucous membranes of mammals, and some are human pathogens. *Neisseria gonorrhoeae* is the causative agent of gonorrhea; *Neisseria meningitidis* is responsible for some cases of bacterial meningitis. ▶▶| *Meningitis (section 38.1); Gonorrhea (section 38.3)*

Order *Burkholderiales*

The order *Burkholderiales* contains four families, three of them with well-known genera. The genus *Burkholderia* is placed in the family *Burkholderiaceae.* This genus was established when *Pseudomonas* was divided into at least seven genera based on rRNA data: *Acidovorax, Aminobacter, Burkholderia, Comamonas, Deleya, Hydrogenophaga,* and *Methylobacterium.* Members of the genus *Burkholderia* are gram-negative, aerobic, nonfermentative, nonsporing, mesophilic straight rods (**chapter opener**). With the exception of one species, all are motile with a single polar flagellum or a tuft of polar flagella. Catalase is produced, and they often are oxidase positive. Most species use poly-β-hydroxybutyrate as their carbon reserve. One of the most important species is *B. cepacia,* which can degrade over 100 different organic molecules and is very active in recycling organic materials in nature. Originally described as the plant pathogen that causes onion rot, it has emerged in the last 20 years as a major nosocomial (hospital acquired) pathogen. It is a particular problem for cystic fibrosis patients. Two other species, *B. mallei* and *B. pseudomallei,* are human pathogens that could be misused as bioterrorism agents. ▶▶| *Bioterrorism preparedness (section 36.8)*

Surprisingly, two genera within the *Burkholderiaceae* family are capable of forming nitrogen-fixing symbioses with legumes much like the rhizobia that belong to the α-proteobacteria. Genome analysis of nitrogen-fixing *Burkholderia* and *Ralstonia* isolates reveals the presence of nodulation (*nod*) genes that are very similar to those of the rhizobia. This suggests a common origin. It is thought these β-proteobacteria gained the capacity to form symbiotic, nitrogen-fixing nodules with legumes through horizontal gene transfer.

Some members of the order *Burkholderiales* have a **sheath**—a hollow, tubelike structure surrounding a chain of cells. Sheaths often are close fitting, but they are never in intimate contact with the cells they enclose. Some contain ferric or manganic oxides. They have at least two functions. Sheaths help bacteria attach to solid surfaces and acquire nutrients from slowly running water as it flows past, even if it is nutrient-poor. Sheaths also protect against predators such as protozoa.

Two well-studied sheathed genera are in the family *Comamonadaceae*. *Sphaerotilus* forms long, sheathed chains of rods, 0.7 to 2.4 μm by 3 to 10 μm, attached to submerged plants, rocks, and other solid objects, often by a holdfast. Single swarmer cells with a bundle of subpolar flagella escape the filament and form a new chain after attaching to a solid object at another site. *Sphaerotilus* grows best in slowly running freshwater polluted with sewage or industrial waste. It grows so well in activated sewage sludge that it sometimes forms tangled masses of filaments and interferes with the proper settling of sludge (*see figure 42.5*). *Leptothrix* characteristically deposits large amounts of iron and manganese oxides in its sheath (**figure 20.15**). This seems to protect it and allow *Leptothrix* to grow in the presence of high concentrations of soluble iron compounds.

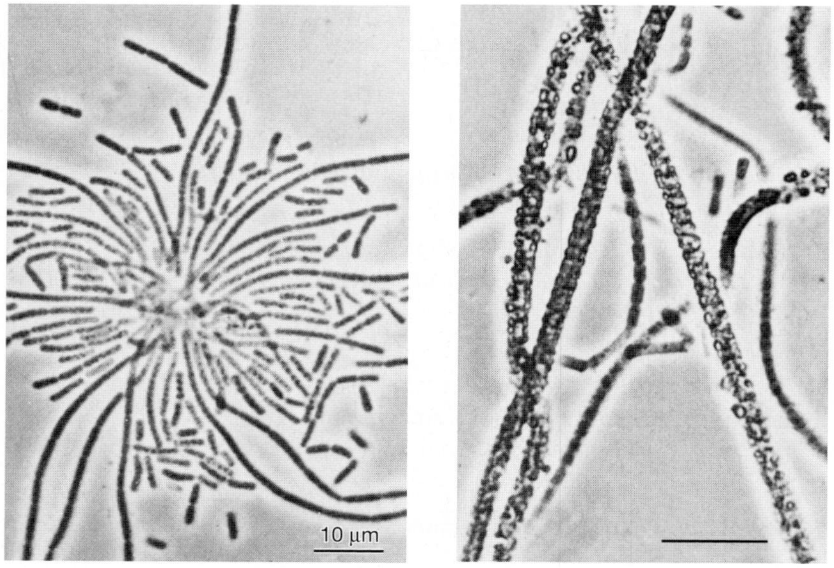

(a) (b)

FIGURE 20.15 Sheathed Bacteria, *Leptothrix* Morphology. (a) *L. lopholea* trichomes radiating from a collection of holdfasts. (b) *L. cholodnii* sheaths encrusted with MnO_2.

The family *Alcaligenaceae* contains the genus *Bordetella*. This genus is composed of gram-negative, aerobic coccobacilli about 0.2 to 0.5 μm by 0.5 to 2.0 μm in size. *Bordetella* is a chemoorganotroph with respiratory metabolism that requires organic sulfur and nitrogen (amino acids) for growth. It is a mammalian parasite that multiplies in respiratory epithelial cells. *B. bronchiseptica* is the causative agent of the canine condition kennel cough, and *B. pertussis* is a nonmotile, encapsulated species that causes pertussis (whooping cough), although *B. parapertussis* is a closely related species that causes a milder form of the disease. ▶▶| *Pertussis (section 38.1)*

Order *Nitrosomonadales*

A number of chemolithotrophs are found in the order *Nitrosomonadales*. The stalked chemolithotroph *Gallionella* is in this order. The family *Spirillaceae* has one genus, *Spirillum*. Two genera of nitrifying bacteria (*Nitrosomonas* and *Nitrosospira*) are members of the family *Nitrosomonadaceae* (figure 20.13). Both of these bacteria oxidize ammonia to nitrite. A metabolic and transport reconstruction based on the genome of *Nitrosomonas europaea* is shown in **figure 20.16**. This microbe uses the enzyme ammonia monooxygenase (AMO) and hydroxylamine oxidoreductase to oxidize ammonia to nitrite in the reaction:

$$NH_3 + O_2 + 2H^+ + 2e^- \rightarrow NH_2OH + H_2O \rightarrow$$

$$NO_2 + 5\,H^+ + 4e^-$$

Of the four electrons released, two must return to AMO to continue ammonia oxidation, while the other two either can be used as reductants for biosynthesis or be donated to the electron transport chain to reduce the terminal electron acceptor, oxygen. Figure 20.16 also illustrates the autotrophic metabolism of this microbe, which uses the Calvin cycle to fix CO_2.

Order *Hydrogenophilales*

This small order contains *Thiobacillus*, one of the best-studied chemolithotrophs and most prominent of the **colorless sulfur bacteria**. Many are unicellular rod-shaped or spiral, sulfur-oxidizing bacteria that can be either nonmotile or flagellated (**table 20.4**). *Bergey's Manual* divides the colorless sulfur bacteria between two classes; for example, *Thiobacillus* and *Macromonas* are β-proteobacteria, whereas *Thiomicrospira, Thiobacterium, Thiospira, Thiothrix, Beggiatoa,* and others are γ-proteobacteria. Only some of these bacteria have been isolated and studied in pure culture.

The metabolism of *Thiobacillus* has been intensely studied. It grows aerobically by oxidizing a variety of inorganic sulfur compounds (elemental sulfur, hydrogen sulfide, thiosulfate, thiocyanate) to sulfate. Most species temporarily store the intermediates of sulfide oxidation (sulfur, sulfite, polythionates). The genus includes obligate aerobes and facultative denitrifiers, as discussed next. ATP is produced by a combination of oxidative phosphorylation and substrate-level phosphorylation by means of adenosine 5′-phosphosulfate (APS) (**figure 20.17**). Although all *Thiobacillus* spp. are autotrophic, fixing CO_2 via the Calvin-Bensen cycle (*see figure 11.5*), *T. novellus* and a few other species can also grow heterotrophically. |◀◀ *Chemolithotrophy (section 10.11)*

The metabolic flexibility of *Thiobacillus denitrificans* has made it the focus of much investigation. It uses reduced inorganic sulfur compounds such as hydrogen sulfide and thiosulfate to donate electrons to an electron transport chain that uses either oxygen or nitrate as the terminal electron acceptor. *T. denitrificans* can sequentially reduce nitrate to nitrogen gas—a process called **denitrification**. Recently, it was discovered that *T. denitrificans* can also

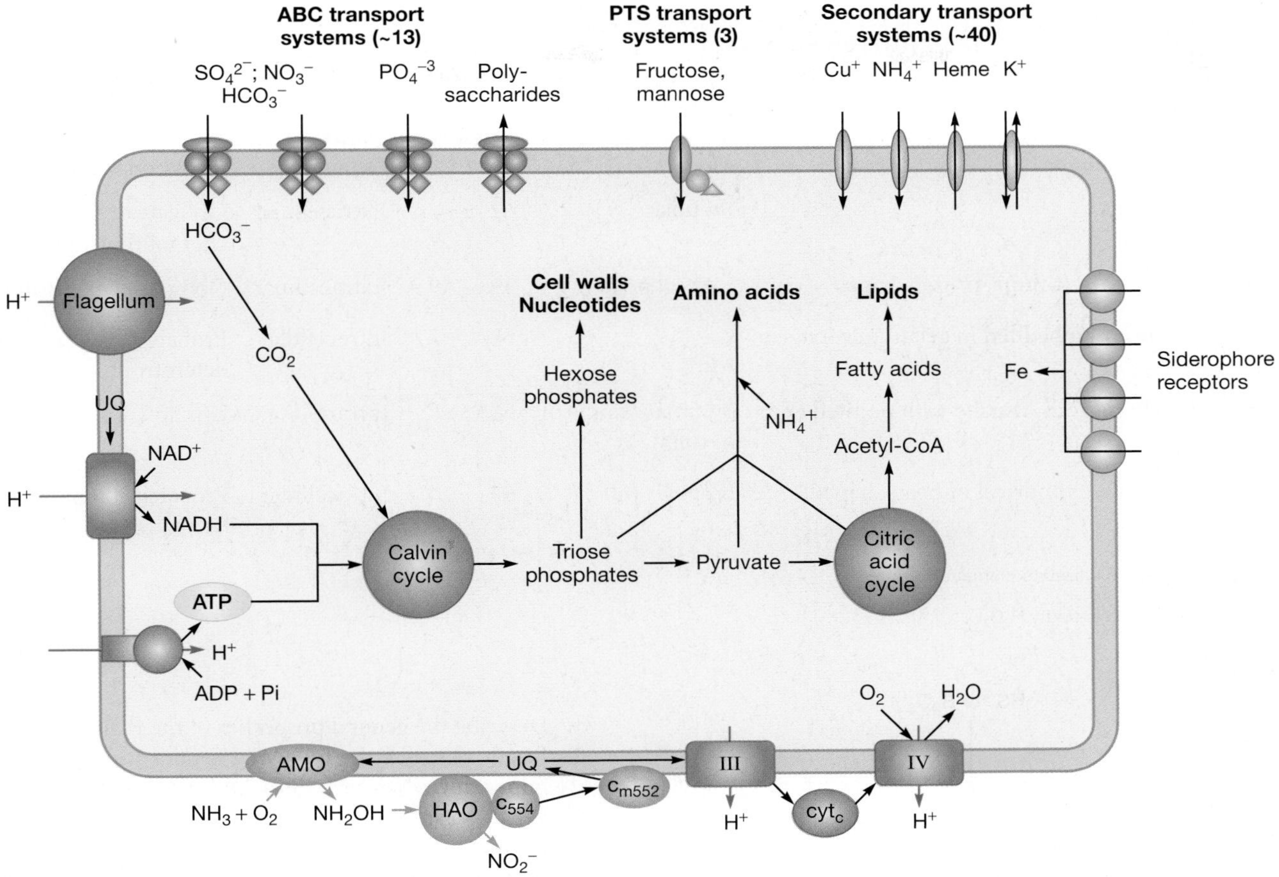

FIGURE 20.16 Genomic Reconstruction of *Nitrosomonas europaea*. This chemolithoautotroph uses the Calvin cycle to fix CO_2. Ammonia oxidation is catalyzed by a membrane-bound ammonia monooxygenase (AMO) and hydroxylamine oxidoreductase (HAO). Electrons from HAO flow through two cytochromes (c_{554} and cm_{552}) into an electron transport chain at the level of ubiquinone-cytochrome c oxidoreductase. O_2 is the terminal electron acceptor. Electrons from ammonia oxidation must also flow in reverse to supply reductant for NAD^+ and $NADP^+$ reduction. These enzymes, the flagellar motor, the ATP synthase, and secondary transport systems are fueled by the proton motive force established by ammonia oxidation.

Chain, P., et al. 2003. Complete genome sequence of the ammonia-oxidizing bacterium and obligate chemolithoautotroph Nitrosomonas europaea. *J. Bacteriol. 185:2759–73.*

Figure 20.16 Micro Inquiry

Why do you think *N. europaea* has PTS transport systems for the sugars fructose and mannose if it is autotrophic?

use ferrous iron (Fe^{2+}) as an electron donor; this form of chemolithotrophy has been well documented in *T. ferrooxidans*.

By contrast, the other sulfur-oxidizing β-proteobacterium, *Macromonas,* grows heterotrophically and probably does not derive energy from sulfur oxidation. It may use the process to detoxify metabolically produced hydrogen peroxide.

Sulfur-oxidizing bacteria have a wide distribution and great practical importance. *Thiobacillus* spp. are ubiquitous in soil, freshwater, and marine habitats. Many *Thiobacillus* spp. demonstrate great acid tolerance (e.g., *T. thiooxidans* grows at pH 0.5 and cannot grow above pH 6), and these bacteria prosper in habitats they have acidified by sulfuric acid production, even though most other organisms cannot. Sulfuric acid and ferric iron produced by *T. ferrooxidans* corrodes concrete and pipe structures. Thiobacilli often cause extensive acid and metal pollution when they release metals from mine wastes. However, sulfur-oxidizing bacteria also are beneficial. They may increase soil fertility by oxidizing elemental sulfur to sulfate. Thiobacilli are used in processing low-grade metal ores because of their ability to leach metals from ore. There is much interest in using *T. denitrificans* to decontaminate water with high levels of nitrate. ▶▶| *Sulfur cycle (section 26.1)*

Table 20.4 Colorless Sulfur-Oxidizing Genera

Genus	*Cell Shape*	*Motility; Location of Flagella*	*G + C Content (mol%)*	*Sulfur Deposit*[a]	*Nutritional Type*
Thiobacillus	Rods	+; polar	62–67	Extracellular	Obligate or facultative chemolithotroph
Thiomicrospira	Spirals, comma, or rod shaped	− or +; polar	39.6–49.9	Extracellular	Obligate chemolithotroph
Thiobacterium	Rods embedded in gelatinous masses	−	N.A.[b]	Intracellular[c]	Probably chemoorgano-heterotroph
Thiospira	Spiral rods, usually with pointed ends	+; polar (single or in tufts)	N.A.	Intracellular	Unknown
Macromonas	Rods, cylindrical or bean shaped	+; polar tuft	67	Intracellular[c]	Probably chemoorgano-heterotroph

[a] When hydrogen sulfide is oxidized to elemental sulfur.
[b] N.A., data not available.
[c] May use sulfur oxidation to detoxify H_2O_2.

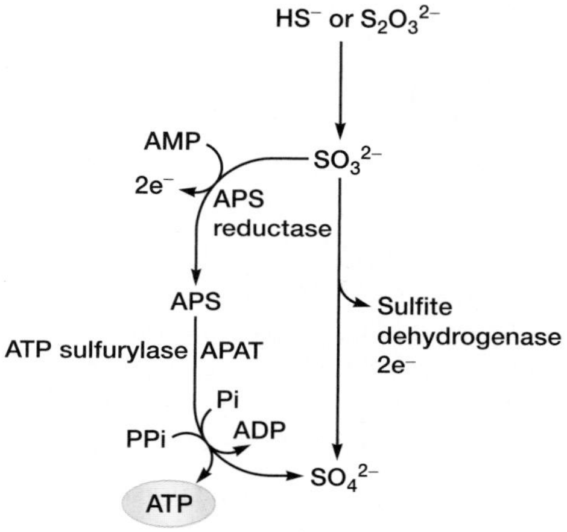

FIGURE 20.17 Sulfur oxidation in *Thiobacillus ferrooxidans*. Reduced sulfur compounds such as sulfide (HS^-) or thiosulfate ($S_2O_3^{2-}$) are oxidized to sulfite (SO_3^{2-}). Electrons from sulfite can be donated to an electron transport chain via sulfite dehydrogenase and used to generate a proton motive force. Alternatively, the enzyme APS reductase catalyzes the oxidation of sulfite to sulfate, which is used to convert adenosine monophosphate (AMP) to APS. ATP sulfurylase and APS:phosphate adenylyltransferase (APAT) then catalyze the exchange of the sulfate on APS with PPi to generate ATP via substrate-level phosphorylation.

Figure 20.17* *Micro Inquiry

Identify the reactions that contribute to substrate-level phosphorylation and those that might contribute to ATP synthesis by a membrane-bound ATP synthase.

1. Describe the general properties of the β-proteobacteria.
2. Briefly describe the following genera and their practical importance: *Neisseria, Burkholderia,* and *Thiobacillus.*
3. What is a sheath and of what advantage is it?
4. Examine figure 20.16. How does *N. europea* generate a proton motive force and in what processes does it use this PMF to energize?
5. How do colorless sulfur bacteria obtain energy by oxidizing sulfur compounds?
6. What role does iron play in the redox metabolism of *Thiobacillus ferrooxidans*?
7. List several positive and negative impacts sulfur-oxidizing bacteria have on the environment and human activities.

20.3 CLASS *GAMMAPROTEOBACTERIA*

The **γ-proteobacteria** constitute the largest subgroup of proteobacteria with an extraordinary variety of physiological types. Many important genera are chemoorganotrophic and facultatively anaerobic. Other genera contain aerobic chemoorganotrophs, photolithotrophs, chemolithotrophs, or methylotrophs. According to some DNA-rRNA hybridization studies, the γ-proteobacteria are composed of several deeply branching groups. One consists of the purple sulfur bacteria; a second includes the intracellular parasites *Legionella* and *Coxiella.* The two largest groups contain a wide variety of nonphotosynthetic genera. These microbes can be grouped into rRNA superfamilies based on the degree of sequence conservation among rRNA (*see figure 17.3*). Ribosomal RNA superfamily I includes the families *Vibrionaceae, Enterobacteriaceae,* and *Pasteurellaceae.* These bacteria use the Embden-Meyerhof and pentose phosphate pathways to catabolize carbohydrates. Most are facultative anaerobes. Ribosomal RNA superfamily II contains mostly aerobes that often use the Entner-Doudoroff and pentose phosphate pathways to catabolize many different kinds of organic molecules. The genera *Pseudomonas, Azotobacter, Moraxella,* and *Acinetobacter* belong to this superfamily. ◀◀ *Techniques for determining microbial taxonomy and phylogeny (section 17.3)*

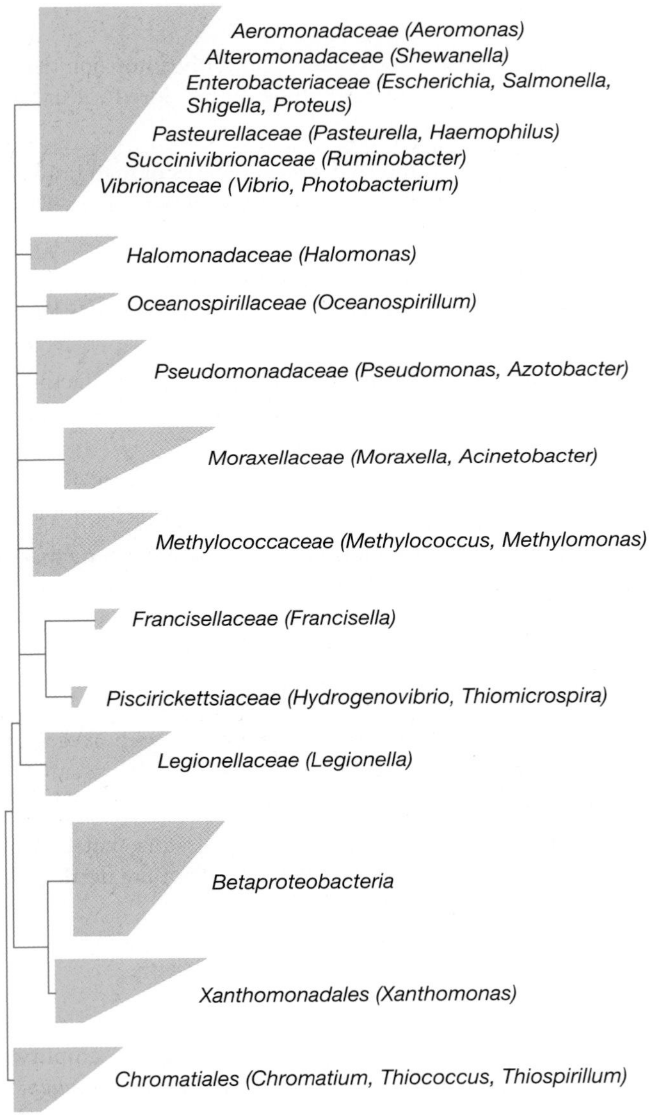

FIGURE 20.18 Phylogenetic Relationships Among γ-Proteobacteria. The major phylogenetic groups based on 16S rRNA sequence comparisons. Representative genera are given in parentheses. Each tetrahedron in the tree represents a group of related organisms; its horizontal edges show the shortest and longest branches in the group. Multiple branching at the same level indicates that the relative branching order of the groups cannot be determined from the data.
Source: The Ribosomal Database Project.

The exceptional diversity of these bacteria is evident from the fact that *Bergey's Manual* divides the class γ-proteobacteria into 14 orders and 28 families. **Figure 20.18** illustrates the phylogenetic relationships among major groups and selected γ-proteobacteria, and **table 20.5** outlines the general characteristics of some of the bacteria discussed in this section.

Purple Sulfur Bacteria

We have seen that the purple photosynthetic bacteria are distributed between three subgroups of the proteobacteria. Most of the purple nonsulfur bacteria are α-proteobacteria and were discussed in section 20.1. Because the purple sulfur bacteria are γ-proteobacteria, they are described here.

The purple sulfur bacteria are strict anaerobes and usually photolithoautotrophs. *Bergey's Manual* divides the **purple sulfur bacteria** into two families: the *Chromatiaceae* and *Ectothiorhodospiraceae* in the order *Chromatiales.* The family *Ectothiorhodospiraceae* contains eight genera. *Ectothiorhodospira* has red, spiral-shaped, polarly flagellated cells that deposit sulfur globules externally (**figure 20.19**). Internal photosynthetic membranes are organized as lamellar stacks (figure 20.3). The majority of purple sulfur bacteria are in the family *Chromatiaceae,* which contains 26 genera.

Chromatiaceae oxidize hydrogen sulfide to sulfur and deposit it internally as sulfur granules (usually within invaginated pockets of the plasma membrane); often they eventually oxidize the sulfur to sulfate (**figure 20.20**). Hydrogen also may serve as an electron donor. *Thiospirillum, Thiocapsa,* and *Chromatium* are typical purple sulfur bacteria, although it was recently discovered that a *Thiocapsa* isolate is able to use nitrite as an electron donor during anoxygenic photosynthesis. These bacteria are found in anoxic, sulfide-rich zones of lakes, marshes, and lagoons where large blooms can occur under certain conditions (**figure 20.21**).

Search This: JGI purple sulfur bacteria

Order *Thiotrichales*

The order *Thiotrichales* contains three families; the largest is the family *Thiotrichaceae.* This family has several genera that oxidize sulfur compounds. Morphologically both rods and filamentous forms are present.

Two of the best-studied gliding genera in this family are *Beggiatoa* and *Leucothrix. Beggiatoa* is microaerophilic and grows in sulfide-rich habitats such as sulfur springs, freshwater with decaying plant material, rice paddies, salt marshes, and marine sediments. Its filaments contain short, disklike cells and lack a sheath (**figure 20.22**). *Beggiatoa* is very versatile metabolically. It oxidizes hydrogen sulfide to form large sulfur grains located in pockets formed by invaginations of the plasma membrane. *Beggiatoa* can subsequently oxidize the sulfur to sulfate and donate electrons to the electron transport chain in energy production. Many strains also can grow heterotrophically with acetate as a carbon source, and some incorporate CO_2 autotrophically.

Table 20.5 Characteristics of Selected γ-Proteobacteria

Genus	*Dimensions (μm) and Morphology*	*G + C Content (mol%)*	*Oxygen Requirement*	*Other Distinctive Characteristics*
Azotobacter	1.5–2.0; ovoid cells, pleomorphic, peritrichous flagella or nonmotile	63.2–67.5	Aerobic	Can form cysts, fix nitrogen nonsymbiotically
Beggiatoa	1–200 × 2–10; colorless cells form filaments, either single or in colonies	35–39	Aerobic or microaerophilic	Gliding motility; can form sulfur inclusions with hydrogen sulfide present
Chromatium	1–6 × 1.5–16; rod-shaped or ovoid, straight or slightly curved, polar flagella	48–50	Anaerobic	Anoxygenic photolithoautotroph that can use sulfide; sulfur stored within the cell
Ectothiorhodospira	0.7–1.5 in diameter; vibrioid- or rod-shaped, polar flagella	61.4–68.4	Anaerobic, some aerobic or microaerophilic	Internal lamellar stacks of membranes; deposits sulfur granules outside cells
Escherichia	1.1–1.5 × 2–6; straight rods, peritrichous flagella or nonmotile	48–59	Facultatively anaerobic	Mixed acid fermenter; formic acid converted to H_2 and CO_2, lactose fermented, citrate not used
Haemophilus	<1.0 in width, variable lengths; coccobacilli or rods, nonmotile	37–44	Aerobic or facultatively anaerobic	Fermentative; requires growth factors present in blood; parasites on mucous membranes
Leucothrix	Long filaments of short cylindrical cells, usually holdfast is present	46–51	Aerobic	Dispersal by gonidia, filaments don't glide; rosettes formed; heterotrophic
Methylococcus	0.8–1.5 × 1.0–1.5; cocci with capsules, nonmotile	59–65	Aerobic	Can form cysts; uses methane, methanol, and formaldehyde as sole carbon and energy sources
Photobacterium	0.8–1.3 × 1.8–2.4; straight, plump rods with polar flagella	39–44	Facultatively anaerobic	Two species can emit blue-green light; Na^+ needed for growth
Pseudomonas	0.5–1.0 × 1.5–5.0; straight or slightly curved rods, polar flagella	58–69	Aerobic or facultatively anaerobic	Respiratory metabolism with oxygen or nitrate as acceptor; some use H_2 or CO as energy source
Vibrio	0.5–0.8 × 1.4–2.6; straight or curved rods with sheathed polar flagella	38–51	Facultatively anaerobic	Fermentative or respiratory metabolism; sodium ions stimulate or are needed for growth; oxidase positive

Three genera of *Thiotrichales*—*Beggiatoa, Thioploca*, and *Thiomargarita*—are notable because they form large mats on the seafloor near nutrient-rich coasts, regions where methane seeps out of Earth's crust, and at hydrothermal vents. These filamentous microbes are among the largest known bacteria. For example, *Thiomargarita* can be over 100 μm in diameter and hundreds of centimeters long (**figure 20.23**). All three genera grow in bundles, making them even more conspicuous. These bacteria appear hollow because they form large intracellular vacuoles where high concentrations of nitrate (up to 500 mM) are stored. Sulfur inclusions are found in a thin layer of cytoplasm around the vacuoles. Freshwater *Beggiatoa* is much smaller than marine representatives, which have not been yet been grown in pure culture. However, it has been possible to sequence the genome of a giant, marine *Beggiatoa* sp. by extracting and amplifying DNA from single filaments. ◀◀ *Single-cell genome sequencing (section 16.2)*

Leucothrix is an aerobic chemoorganotroph that forms filaments or trichomes up to 400 μm long (**figure 20.24**). It is usually marine and is attached to solid substrates by a holdfast. *Leucothrix* has a complex life cycle in which it is dispersed by the formation of gonidia. Rosette formation often is seen in culture. *Thiothrix* is a related genus that forms sheathed filaments and releases gonidia from the open end of the sheath. In contrast to *Leucothrix, Thiothrix* is a chemolithotroph that oxidizes hydrogen sulfide and deposits sulfur granules internally. It also requires

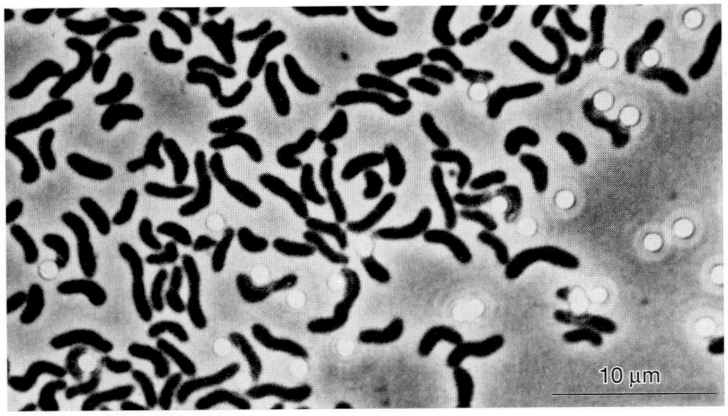

FIGURE 20.19 Purple Bacteria. *Ectothiorhodospira mobilis*; light micrograph. Note external sulfur globules.

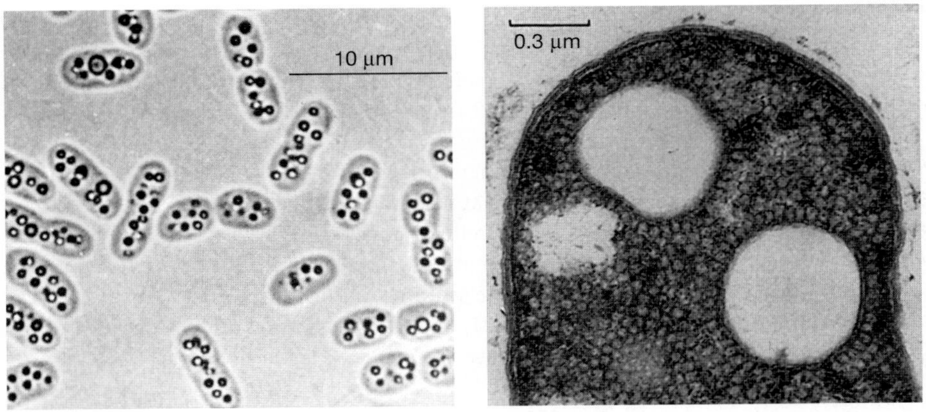

(a) *Chromatium vinosum* (b) *C. vinosum*

FIGURE 20.20 Typical Purple Sulfur Bacteria. (a) *Chromatium vinosum* with intracellular sulfur granules. (b) Electron micrograph of *C. vinosum*. Note the intracytoplasmic vesicular membrane system. The large white areas are the former sites of sulfur globules.

organic carbon for growth. *Thiothrix* grows in sulfide-rich flowing water and activated sludge sewage systems.

The genus *Thiomicrospira*, family *Piscirickettsiaceae*, has an interesting history. Originally it included all marine, spiral-shaped, sulfur-oxidizing bacteria. Later 16S rRNA analysis showed that this group was not monophyletic, and members are now split between the γ- and ϵ-proteobacterial classes. For instance *Thiomicrospira denitrificans* is now known to be a member of ϵ-proteobacteria, but a new genus name has not yet been established.

Figure 20.25 shows a cell model based on the annotated genome of *Thiomicrospira crunogen* XCL-2. This deep-sea microbe was isolated from a hydrothermal vent (*see figure 30.4*). This unique environment is characterized by the mixture of hot, anoxic, highly reduced vent fluid with cold, oxic bottom water. Of importance to this autotroph is the difference in dissolved CO_2 concentrations that it encounters: vent fluid is rich in CO_2, whereas bottom water is not. This has led to the evolution of a high cellular affinity for both HCO_3^- and CO_2 under low CO_2 conditions. Also of note is the multienzyme complex capable of oxidizing a number of reduced sulfur compounds to sulfate. This "Sox" complex (for *s*ulfur *ox*idation) is conserved in a number of sulfur-dependent chemolithoautotrophs such as the α-proteobacterium *Thiobacillus*. Although *Thiomicrospira* can use a variety of reduced sulfur compounds as electron donors, it can use only oxygen as an electron acceptor. Interestingly, genome sequencing reveals the presence of what appears to be a complete prophage within the chromosome.

Order *Methylococcales*

The single family in this order is *Methylococcaceae*. It contains rods, vibrios, and cocci that use methane, methanol, and other reduced

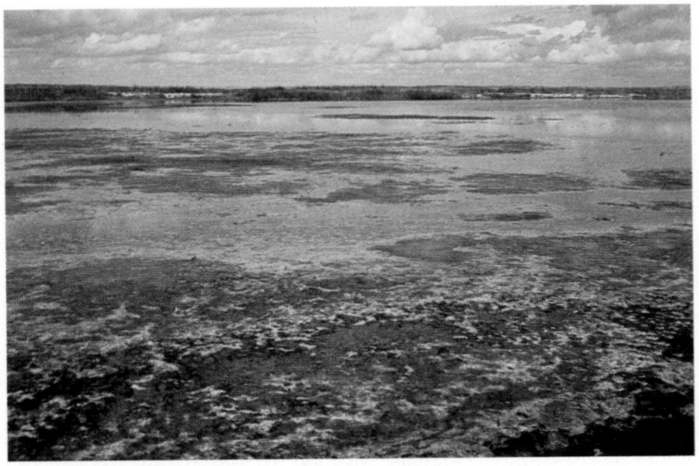

FIGURE 20.21 Purple Photosynthetic Sulfur Bacteria. Purple photosynthetic sulfur bacteria growing in a marsh.

FIGURE 20.22 ***Beggiatoa alba.*** A light micrograph showing part of a colony (×400). Note the dark sulfur granules within many of the filaments.

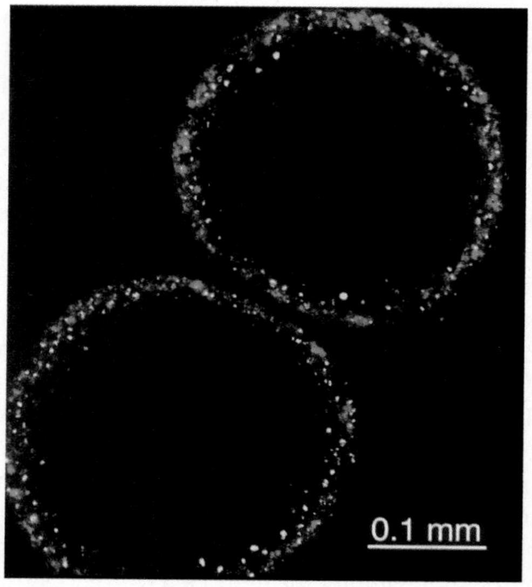

FIGURE 20.23 ***Thiomargarita namibiensis,* the World's Largest Known Bacterium.** This bacterium, usually 100–300 μm in diameter occasionally reaches a size of 750 μm (larger than a period on this page), 100 times the size of a common bacterium. *T. namibiensis* uses sulfide from bottom sediments as an energy source and nitrate, which is found in the overlying waters, as an electron acceptor. It appears hollow because it stores high concentrations of nitrate in a central vacuole.

Figure 20.23 *Micro Inquiry*

Given that bacteria must depend on membrane transport and intracellular diffusion of nutrients, why is it important that a microbe the size of *T. namibiensis* contain a central vacuole rather than cytoplasm?

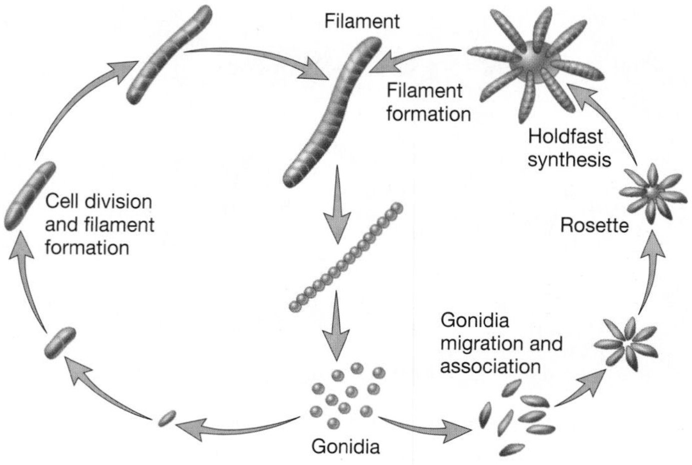

FIGURE 20.24 Life cycle and morphology of *Leucothrix mucor.*

one-carbon compounds as their sole carbon and energy sources under aerobic or microaerobic (low oxygen) conditions. They are methylotrophs, different from bacteria that use methane exclusively as their carbon and energy source, which are called **methanotrophs.** The family contains seven genera, including *Methylococcus* (spherical, nonmotile cells) and *Methylomonas* (straight, curved, or branched rods with a single, polar flagellum). When oxidizing methane, the bacteria contain complex arrays of intracellular membranes. Almost all are capable of forming cysts. Methanogenesis from archaea using substrates such as H_2 and CO_2 is widespread in anoxic soil and water, and methylotrophic microbes grow above these habitats all over the world. ⏮ *Phylum* Euryarchaeota *(section 18.3)*

Methane-oxidizing bacteria use methane as a source of both energy and carbon. This can be simplified to a three-step process (**figure 20.26**). First, methane (or other single-carbon compound) is oxidized to formaldehyde. Next, formaldehyde is either converted to CO_2 or assimilated to produce biomass. Thus formaldehyde is the central intermediate that can be either oxidized to conserve energy or serve as the building block for the synthesis of other metabolites. It is paradoxical that formaldehyde is the key intermediate because it is a poisonous molecule on which the survival of these bacteria depends. The key to their viability appears to be the rate at which the formaldehyde is converted to either CO_2 or other cellular constituents. It is assimilated into cell material by the activity of either of two pathways, one involving the formation of the amino acid serine and the other proceeding through the synthesis of sugars such as fructose 6-phosphate and ribulose 5-phosphate.

Search This: C1 metabolism

Methylotrophy is not confined to the γ-proteobacteria. In fact, one of the best-studied methylotrophs is the α-proteobacterium *Methylobacterium extorquens*. Methylotrophic bacteria are also found among the β-proteobacteria, the *Verrucomicrobia,* the *Firmicutes*, and the *Archaea* that live in association with sulfate-reducing bacteria. ⏮ *Microbial Diversity & Ecology 18.1; Verrucomicrobia (section 19.8)*

Order *Legionellales*

Two families make up the order *Legionellales*. The first is *Legionellaceae*, with its single genus, *Legionella*. The second family is the *Coxiellaceae*, which has two genera, *Coxiella* and *Rickettsiella* (not to be confused with the α-proteobacterium *Rickettsia*). All these microbes are intracellular pathogens that display a dimorphic (i.e., two cell types) lifestyle within the host that is reminiscent of the *Chlamydiae*. Here we discuss *Legionella* and *Coxiella*, the two genera that include human pathogens. ⏮ *Phylum* Chlamydiae *(section 19.5)*

Although there are over 40 species of *Legionella, L. pneumophila* has been most intensely studied because it causes a specific type of pneumonia called Legionnaire's disease, as discussed in chapter 38. In nature, *L. pneumophilia* is an intracellular parasite

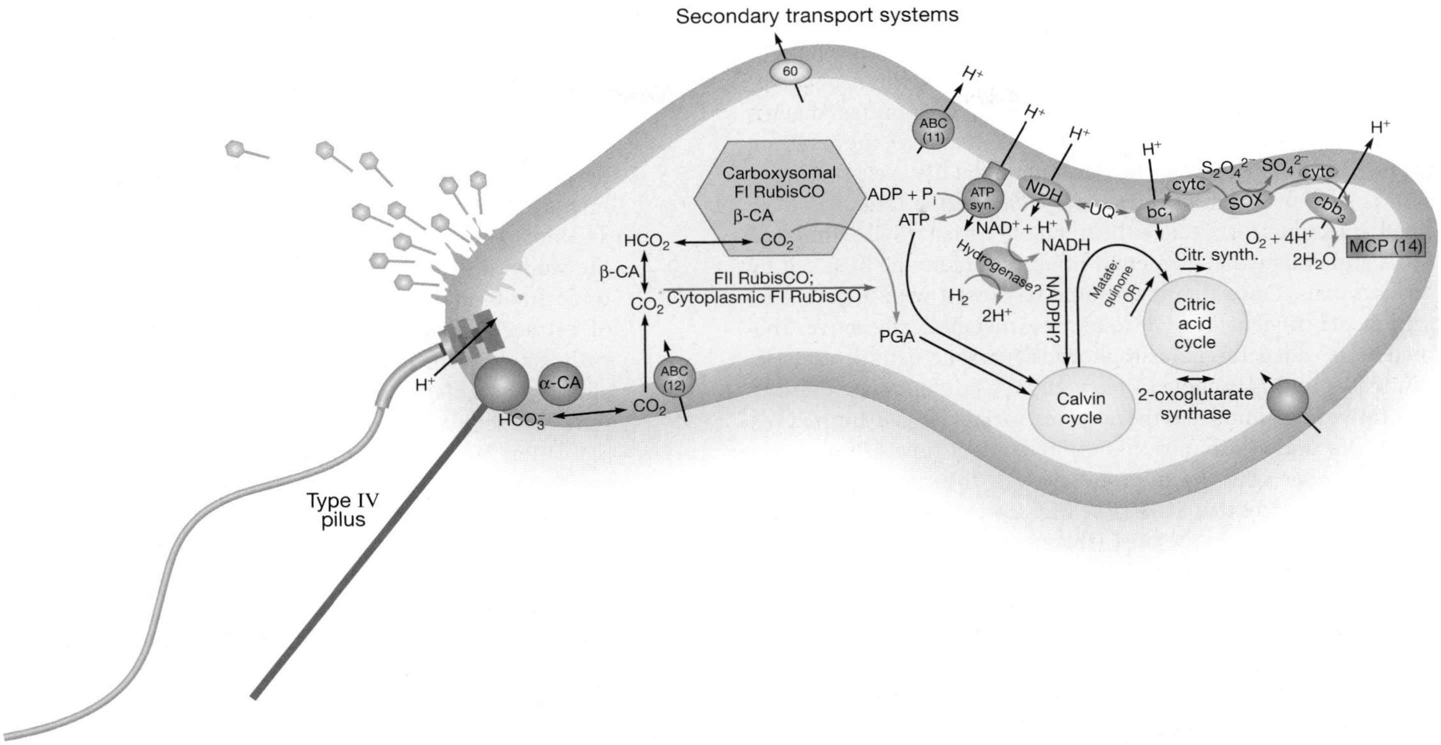

FIGURE 20.25 Model *Thiomicrospira crunogena* XCL-2 cell. This chemolithoautotroph has a high affinity carbonic anhydrase (α-CA and β-CA), enabling the accumulation of CO_2 100 times higher than extracellular concentrations. Three ribulose 1,5-bisphosphate carboxylase-oxygenase (RubisCO) genes are present, and it is hypothesized that one RubisCO protein is localized in carboxysomes. Energy and reducing power for carbon fixation is captured through Sox-dependent oxidation of reduced sulfur compounds. The electron transport chain includes cytochromes c (cytc), bc_1, and cbb_3 with O_2 as the terminal electron acceptor. A prophage is present within the genome; a suite of genes encoding lytic and lysogeny proteins suggests that the prophage can enter the lytic cycle, as indicated by the release of phage on the left side of the image. Transporters include over 60 secondary transporters (light blue) for metal, inorganic, and organic compounds, as well as a dozen ATP-binding cassette (ABC) uptake proteins and 11 ABC exporters (red). Other membrane proteins include 14 methyl-accepting proteins (MCP, purple) that help mediate chemotaxis and 5 unclassified transporters (lavender).

Source: Scott, K. M., et al. 2006. PLoS Biology. *4:2196.*

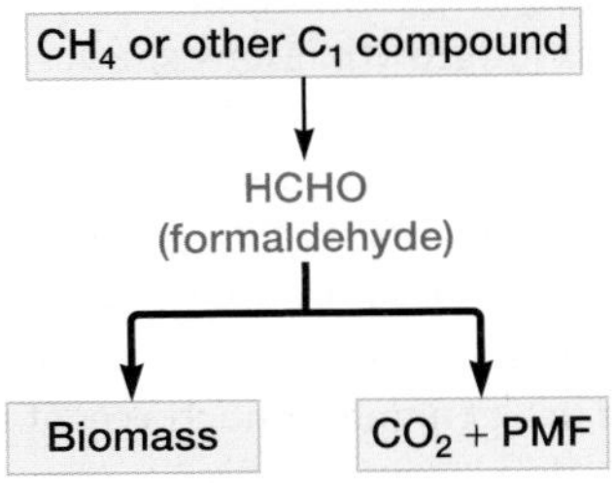

FIGURE 20.26 Methylotrophy. Methane or another reduced single-carbon compound (i.e., not CO_2) is converted to the toxic intermediate formaldehyde. This is tolerated only because it is rapidly incorporated into larger molecules to increase biomass, or it is quickly oxidized to generate a proton motive force (PMF). Thick arrows indicate the conversion of formaldehyde must occur faster than it can accumulate in the cell.

of protozoa that live in moist soil and aquatic environments that include cooling towers, air conditioning units, and hot tubs. In its protozoan hosts, as well as in some human cell cultures, *L. pneumophilia* replicates by binary fission to produce slender, rod-shaped cells with a typical gram-negative cell wall. The replicative forms (RFs) reside in a special host vacuole called the replicative endosome. Late during infection, RFs differentiate into cystlike cells called *m*ature *i*ntracellular *f*orms (MIFs), which when released represent a nonreplicative, highly infectious form. MIFs are chubby rods and metabolically dormant with an invaginated plasma membrane that forms intracellular folds. They also have a distinct cell wall structure that includes a high concentration of the protein Hsp60. This is important because this protein acts as an invasin, that is, a protein required for host cell invasion. In addition, MIFs are heat tolerant and resistant to a variety of antibiotics. These features may help explain why *L. pneumophila* becomes more infective to human cells after it has grown within a protist.

▶▶| *Legionnaire's disease and Pontiac fever (section 38.1)*

Coxiella also displays two cell types within its host, although it has a much more impressive range of hosts, including insects, fish, birds, rodents, sheep, goats, and humans. As with *L. pneumophilia,* humans become infected by inhaling contaminated aerosols, although in the case of the pathogen *C. burnetti,* the resulting Q fever is a flulike illness. The developmental life cycle of *C. burnetti* is similar to that of *L. pneumophila* and the *Chlamydiae.* A small, infectious form called the *s*mall *c*ell *v*ariant (SCV, similar to MIFs and chlamydial elementary bodies) enters the host cell by phagocytosis. Once the phagosome has fused with a lysosome, the low pH triggers the SCV to become metabolically active. This vacuole is called the parasitophorous (parasite eating) vacuole. While it is in this vacuole *C. burnetti* SCVs differentiate into the replicative form known as the *l*arge *c*ell *v*ariant (LCV). Both LCVs and SCVs undergo binary fission and infect host cells, so one might wonder why the microbe evolved the capacity to form SCVs. It is thought that SCVs, like MIFs, are responsible for long-term survival outside the host (this makes these pathogens distinct from the *Chlamydiae* and *Rickettsia,* which cannot survive outside the host). In addition, it has been suggested that the compacted and protected genome of SCVs makes these forms better suited for the initial respiratory burst and release of degradative enzymes when the phagosomes and lysosomes initially fuse. ◀◀ Chlamydiae *(section 19.5)* ▶▶ *Phagocytosis (section 32.5); Q fever (section 38.5)*

Order *Pseudomonadales*

Pseudomonas is the most important genus in the order *Pseudomonadales,* the family *Pseudomonaceae.* These bacteria are straight or slightly curved rods, 0.5 to 1.0 μm by 1.5 to 5.0 μm in length, and are motile by one or several polar flagella. They are chemoheterotrophs that usually carry out aerobic respiration. Sometimes nitrate is used as the terminal electron acceptor in anaerobic respiration. All pseudomonads have a functional tricarboxylic acid cycle and can oxidize substrates completely to CO_2. Most hexoses are degraded by the Entner-Doudoroff pathway, rather than the Embden-Meyerhof pathway. ◀◀ *Glycolytic pathways (section 10.3); Tricarboxylic acid cycle (section 10.4)*

The genus *Pseudomonas* is an exceptionally heterogeneous taxon currently composed of about 60 species. Many can be placed in one of seven rRNA homology groups. The three best-characterized groups are subdivided according to properties such as the presence of poly-β-hydroxybutyrate (PHB), the production of a fluorescent pigment, pathogenicity, the presence of arginine dihydrolase, and glucose utilization. For example, the fluorescent subgroup does not accumulate PHB and produces a diffusible, water-soluble, yellow-green pigment that fluoresces under UV radiation. *P. aeruginosa, P. fluorescens, P. putida,* and *P. syringae* are members of this group.

The pseudomonads have a number of practical impacts, including:

1. Many can degrade an exceptionally wide variety of organic molecules. Thus they are very important in mineralization processes (the microbial breakdown of organic materials to inorganic substances) in nature and in sewage treatment. The fluorescent pseudomonads can use approximately 80 different substances as their carbon and energy sources. ▶▶ *Microorganisms in the soil environment (section 29.2)*
2. Several species (e.g., *P. aeruginosa*) are important experimental subjects. Many advances in microbial physiology and biochemistry have come from their study. For example, the study of *P. aeruginosa* has significantly advanced our understanding of how bacteria form biofilms and the role of extracellular signaling in bacterial communities and pathogenesis. The genome of *P. aeruginosa* has an unusually large number of genes for catabolism, nutrient transport, the efflux of organic molecules, and metabolic regulation. This may explain its ability to grow in many environments and resist antibiotics. ◀◀ *Biofilms (section 7.7); Quorum sensing (section 13.6)*
3. Some pseudomonads are major animal and plant pathogens. *P. aeruginosa* infects people with low resistance, such as cystic fibrosis patients. It also invades burns and causes urinary tract infections. *P. syringae* is an important plant pathogen.
4. Pseudomonads such as *P. fluorescens* are involved in the spoilage of refrigerated milk, meat, eggs, and seafood because they grow at 4°C and degrade lipids and proteins.

The genus *Azotobacter* also is in the family *Pseudomonadaceae.* The genus contains ovoid bacteria, 1.5 to 2.0 μm in diameter, that may be motile by peritrichous flagella. The cells are often pleomorphic, ranging from rods to coccoid shapes, and form cysts as the culture ages. The genus is aerobic, catalase positive, and fixes nitrogen nonsymbiotically. *Azotobacter* is widespread in soil and water.

Order *Alteromonadales*

The genus *Altermonas* in the order *Aleromonadales* includes strictly aerobic, nonspore-forming, straight or curved rods that are motile with a single polar flagella. They are mesophilic and require sodium ions for growth. Many species originally classified in this genus have been reclassified to different genera within the order, including *Marinomonas, Pseudoalteromonas,* and *Shewanella. Alteromonas, Marinomonas,* and *Pseudoalteromonas* are marine bacteria, while *Shewanella* spp. have been isolated from diverse habitats, including seawater, lake sediments, and salted foods. *Shewanella* has generated much recent interest and is the focus of our discussion.

The genus *Shewanella* includes 48 facultatively anaerobic species that form straight or curve rods. It was first noted that *S. oneidensis* could use over 10 different electron acceptors and reduce thiosulfate and elemental sulfur to sulfide, a property usually seen only in strict anaerobes. Later it was discovered that as a group, *Shewanella* spp. show similar metabolic flexibility and can use a variety of metal electron acceptors, including uranium, chromium, iodate, technetium, neptunium, plutonium, selnite, tellurite, and vanadate, as well as nitroaromatic compounds. This wide range of metal electron acceptors makes *Shewanella* spp. excellent candidates as agents for bioremediation of environments contaminated with radionuclides. In addition, these bacteria can use a number of carbon substrates, and some can use H_2 as an electron source. ▶▶| *Biodegradation and bioremediation by natural communities (section 42.3)*

The use of metals as terminal electron acceptors is considered dissimilatory metal reduction because the metals serve no other physiological role. In other words, they are not assimilated into biomass. Dissimilatory metal reduction presents a problem for the microbe because many metals, such as iron and manganese, are insoluble in their oxidized forms, which is the form needed by the microbe as an electron acceptor. Dissimilatory metal-reducing bacteria have evolved several strategies that enable the use of insoluble metals such as Fe(III) or Mn(IV) as electron acceptors. The most direct approach is to localize cytochromes in the outer membrane, rather than the plasma membrane (**figure 20.27*a***). This enables the direct transfer of electrons from the cytochromes to the extracellular metal. This approach is used by *Shewanella*, as well as by a number of gram-negative dissimilatory metal-reducing bacteria such as the δ-proteobacterium *Geobacter sulfurreducens* (p. 543). Another strategy used by *Shewanella*, *Pseudomonas*, and others is to transfer electrons to the metal via external, intermediary compounds such as humic acids or secreted metabolites. These compounds are collectively known as **electron shuttles** because they pass electrons from the last point on the electron transport chain (ETC) through the growth substrate to the mineral surface. *Shewanella, Geobacter,* and some cyanobacteria have evolved a third strategy: the production of electrically conductive **nanowires.** These pililike appendages, about 100 nm in diameter and tens of μm in length, transfer electrons from the terminal point in the ETC to a metal surface (figure 20.27*b,c*). Nanowires enable the bacteria to transport electrons to solid-phase acceptors that are physically distant from the cell. This has made these microbes of particular interest to those developing microbial fuel cells (*see figure 41.9*). ▶▶| *Bioconversion processes (section 41.3)*

Search This: Shewanella *blogger*

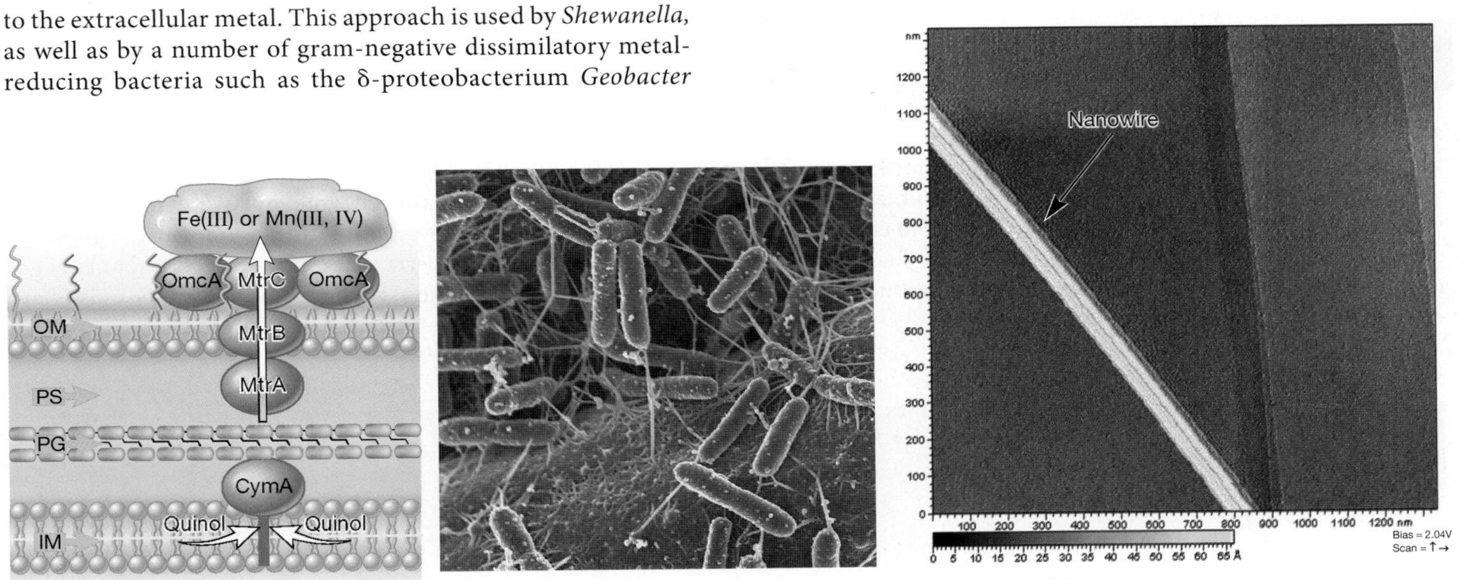

(a) **(b)** Nanowire, SEM **(c)** Nanowire, STM

FIGURE 20.27 External terminal electron acceptors. (a) *Shewanella oneidensis* MR1 transfers electrons from the plasma membrane, through the outer membrane, to an insoluble metal. CymA is a quinol dehydrogenase that oxidizes quinol and passes the electrons to the c-type cytochrome MtrA. MtrB is not a cytochrome but helps move the electrons through the outer membrane to the surface-exposed c-type cytochromes, MtrC and OmcA. These redox proteins donate electrons to the oxidized, insoluble metals iron and manganese, which thereby serve as external, terminal electron acceptors. (b) When electron acceptors are in short supply, *S. oneidensis* MR1 makes nanowires along which electrons can travel to a terminal electron acceptor. This scanning electron microgram (SEM) shows the production of nanowires that occurs when cells are oxygen limited. (c) A scanning tunneling micrograph (STM) of a *S. oneidensis* MR1 nanowire with a diameter of 100 nm. Note the ridges and troughs that run along the axis of the nanowire. IM, inner or plasma membrane; PG, peptidoglycan; PS, periplasmic space; OM, outer membrane.

Figure 20.27* *Micro Inquiry

Do you think a gram-positive microbe could use an insoluble metal as an terminal electron acceptor in the same way *Shewanella* does in (a)?

Order *Vibrionales*

Three closely related orders of the γ-proteobacteria contain a number of important bacterial genera. Each order has only one family of facultatively anaerobic gram-negative rods. **Table 20.6** summarizes the distinguishing properties of the families *Enterobacteriaceae, Vibrionaceae,* and *Pasteurellaceae.*

The order *Vibrionales* contains only one family, the *Vibrionaceae.* Members of the family *Vibrionaceae* are straight or curved, flagellated rods. Most are oxidase positive, and all use D-glucose as their sole or primary carbon and energy source (table 20.6). The majority are aquatic microorganisms, widespread in freshwater and the sea. The family has eight genera: *Vibrio, Photobacterium, Salinivibrio, Listonella, Allomonas, Enterovibrio, Catencoccus,* and *Grimontia.*

Several vibrios are important pathogens. *Vibrio cholerae* causes cholera, and *V. parahaemolyticus* can cause gastroenteritis in humans following consumption of contaminated seafood. *V. anguillarum* and others are responsible for fish diseases, which can be especially problematic in fish farms. ▶▶| *Cholera (section 38.4)*

The *V. cholerae* genome contains about 3,800 open reading frames distributed between two circular chromosomes, chromosome 1 (2.96 million base pairs) and chromosome 2 (1.07 million bp). The larger chromosome primarily has genes for essential cell functions such as DNA replication, transcription, and protein synthesis. It also has most of the virulence genes. For example, the cholera toxin gene is located in an integrated CTX phage on chromosome 1. Chromosome 2 also has essential genes such as transport genes and ribosomal protein genes. Copies of some genes are present on both chromosomes.

Some members of the family are unusual in being bioluminescent. *Vibrio fischeri, V. harveyi,* and at least two species of *Photobacterium* are marine bacteria capable of bioluminescence. They emit a blue-green light because of the activity of the enzyme luciferase (**Microbial Diversity & Ecology 20.1**). The light is usually blue-green in color (472 to 505 nm), but one strain of *V. fischeri* emits yellow light with a major peak at 545 nm. Although many of these bacteria are free-living, *V. fischeri, V. harveyi, P. phosphoreum,* and *P. leiognathi* live symbiotically in the luminous organs of fish (**figure 20.28**) and squid (*see figure 7.36*).

Order *Enterobacteriales*

The family *Enterobacteriaceae* is the largest of the families listed in table 20.6. It contains peritrichously flagellated or nonmotile, facultatively anaerobic, straight rods with simple nutritional requirements. The order *Enterobacteriales* has only one family, *Enterobacteriaceae,* with 44 genera.

The metabolic properties of the *Enterobacteriaceae* are very useful in characterizing its constituent genera. Members of the family, often called **enterobacteria** or **enteric bacteria** (Greek *enterikos,* pertaining to the intestine), all degrade sugars by means of the Embden-Meyerhof pathway. Under microaerobic or anoxic conditions, the enzyme **pyruvate formate-lyase (PFL)** catalyzes the cleavage of pyruvate to formate and acetyl CoA. Those enteric bacteria that produce large amounts of gas during sugar fermentation, such as *Escherichia* spp., also have

Table 20.6 Characteristics of Families of Facultatively Anaerobic Gram-Negative Rods

Characteristics	*Enterobacteriaceae*	*Vibrionaceae*	*Pasteurellaceae*
Cell dimensions	0.3–1.0 × 1.0–6.0 μm	0.3–1.3 × 1.0–3.5 μm	0.2–0.4 × 0.4–2.0 μm
Morphology	Straight rods; peritrichous flagella or nonmotile	Straight or curved rods; polar flagella; lateral flagella may be produced on solid media	Coccoid to rod-shaped cells, sometimes pleomorphic; nonmotile
Physiology	Oxidase negative	Oxidase positive; all can use D-glucose as sole or principal carbon source	Oxidase positive; heme and/or NAD^+ often required for growth; organic nitrogen source required
G + C content	38–60%	38–51%	38–47%
Symbiotic relationships	Some parasitic on mammals and birds; some species are plant pathogens	Most not pathogens; several inhabit light organs of marine organisms	Parasites of mammals and birds
Representative genera	*Escherichia, Shigella, Salmonella, Citrobacter, Klebsiella, Enterobacter, Erwinia, Serratia, Proteus, Yersinia*	*Vibrio, Photobacterium*	*Pasteurella, Haemophilus*

MICROBIAL DIVERSITY & ECOLOGY

20.1 Bacterial Bioluminescence

Several species in the genera *Vibrio* and *Photobacterium* can emit light of a blue-green color. The enzyme luciferase catalyzes the reaction and uses reduced flavin mononucleotide, molecular oxygen, and a long-chain aldehyde as substrates.

$$FMNH_2 + O_2 + RCHO \xrightarrow{\text{luciferase}} FMN + H_2O + RCOOH + \text{light}$$

Evidence suggests that an enzyme-bound, excited flavin intermediate is the direct source of luminescence. Because the electrons used in light generation are probably diverted from the electron transport chain and ATP synthesis, the bacteria expend considerable energy on luminescence. It follows that luminescence is regulated and can be turned off or on under the proper conditions.

Much speculation arises about the role of bacterial luminescence and its value to bacteria, particularly because it is such an energetically expensive process. Luminescent bacteria occupying the luminous organs of fish do not emit light when they grow as free-living organisms in the seawater. Free-living luminescent bacteria can reproduce and infect young fish. Once settled in a fish's luminous organ, the quorum-sensing molecule autoinducer produced by the bacteria stimulates the emission of light. Other luminescent bacteria growing on potential food items such as small crustacea may use light to attract fish to the food source. After ingestion, they could establish a symbiotic relationship in the host's gut.

The mechanism by which autoinducer regulates light production in these marine bacteria is an important model for understanding quorum sensing in many gram-negative bacteria, including a number of pathogens. ◀◀ *Cell-cell communication within microbial populations (section 7.7); Quorum sensing (section 13.6)*

(a)

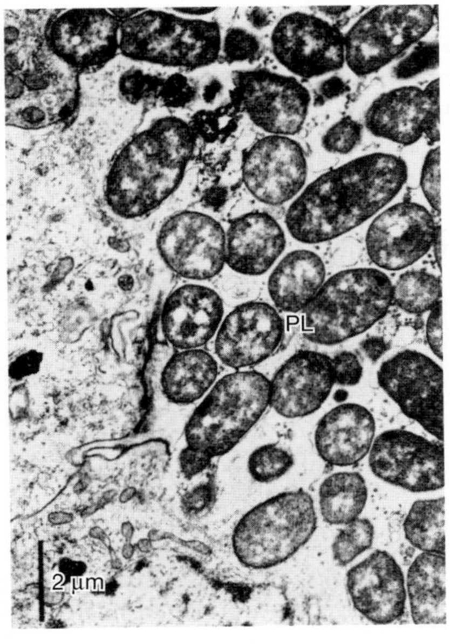

(b)

FIGURE 20.28 Bioluminescence. (a) A photograph of the Atlantic flashlight fish *Kryptophanaron alfredi.* The light area under the eye is the fish's luminous organ, which can be covered by a lid of tissue. (b) Ultrathin section of the luminous organ of a fish, *Equulities novaehollandiae*, with the bioluminescent bacterium *Photobacterium leiognathi*, PL.

the enzyme **formate dehydrogenase (FDH)** that then degrades formic acid to H_2 and CO_2. This enzyme complex is crucial to the survival of these enterics during fermentation, as one-third of the carbon derived from glucose is converted to formate. If this intermediate were allowed to accumulate within the cell, the internal pH would become far too acidic to support viability. In fact, *Escherichia coli* has three different FDH enzymes (**figure 20.29**). FDH-H (H for hydrogen production) is synthesized only during fermentative conditions. FDH-N is produced under anoxic conditions when nitrate serves as the terminal electron acceptor. FDH-O is produced under microaerobic conditions as well as when nitrate is the terminal electron acceptor.

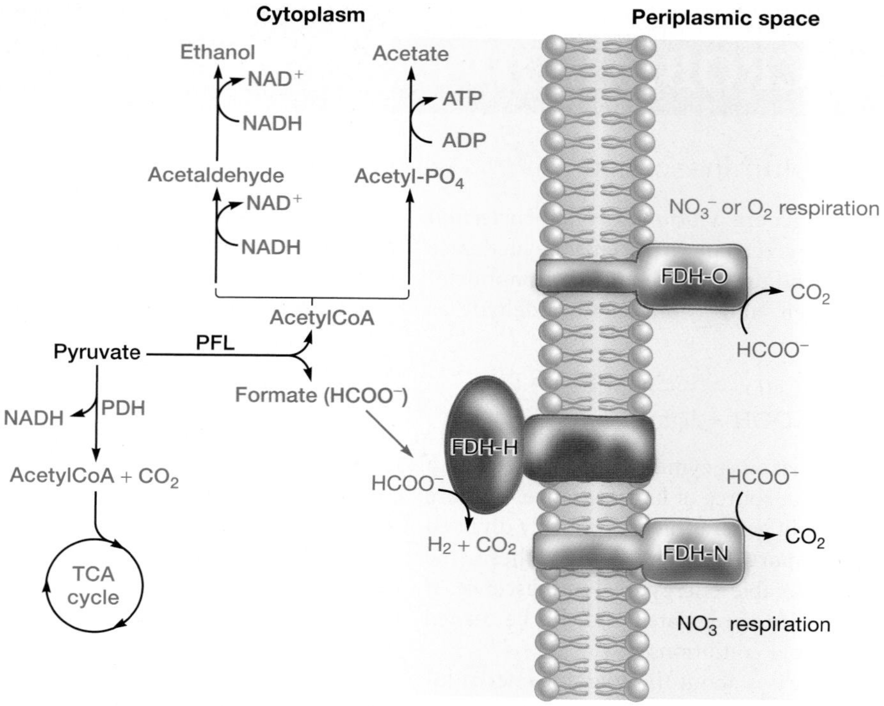

FIGURE 20.29 Metabolism of Pyruvate and Formate in *E. coli*. During aerobic growth, pyruvate is converted to CO_2 and acetyl-CoA by the pyruvate dehydrogenase complex (PDH) so that acetyl-CoA can enter the TCA cycle. Under low oxygen conditions, some pyruvate is also cleaved to formate and acetyl-CoA by pryruvate formate-lyase (PFL). Formate must then be converted into CO_2 and H_2 by one of three formate dehydrogenase (FDH) enzymes. FDH-O is active during aerobic respiration when oxygen levels are low. If respiration continues under anoxic conditions with nitrate as the terminal electron acceptor, both FDH-N and FDH-O split formate. These enzymes are located on the periplasmic face of the plasma membrane. During fermentative conditions, FDH-H is made, resulting in mixed acid fermentation: formate is cleaved to yield CO_2 and H_2, whereas acetyl-CoA is converted to ethanol and acetate. FDH-H is a peripheral membrane protein, with its active site exposed to the cytoplasm. Aerobic metabolism is shown in pink, fermentative metabolism in green.

Figure 20.29 Micro Inquiry

Why do you think the FDH-O and FDH-N are on the periplasmic face of the plasma membrane while FDH-H is on the cytoplasmic face?

The family can be divided into two groups based on their fermentation products. The majority (e.g., the genera *Escherichia, Proteus, Salmonella,* and *Shigella*) carry out mixed acid fermentation and produce mainly lactate, acetate, succinate, formate (or H_2 and CO_2), and ethanol. In contrast, *Enterobacter, Serratia, Erwinia,* and *Klebsiella* are butanediol fermenters. The major products of butanediol fermentation are butanediol, ethanol, and carbon dioxide. The two types of fermentations are distinguished by the methyl red and Voges-Proskauer tests, respectively. ◀◀ *Fermentation (section 10.7)*

Because the enteric bacteria are so similar in morphology, biochemical tests are normally used to identify them after a preliminary examination of their morphology, motility, and growth responses (**figure 20.30** provides a simple example). Some more commonly used tests are those for the type of fermentation, lactose and citrate utilization, indole production from tryptophan, urea hydrolysis, and hydrogen sulfide production. For example, lactose fermentation occurs in *Escherichia* and *Enterobacter* but not in *Shigella, Salmonella,* or *Proteus.* **Table 20.7** summarizes a few of the biochemical properties useful in distinguishing between genera of enteric bacteria. The mixed acid fermenters are located on the left in this table and the butanediol fermenters on the right. The usefulness of biochemical tests in identifying enteric bacteria is shown by the popularity of commercial identification systems, such as the Enterotube and API 20-E systems, that are based on these tests. ▶▶ *Identification of microorganisms from specimens (section 35.2)*

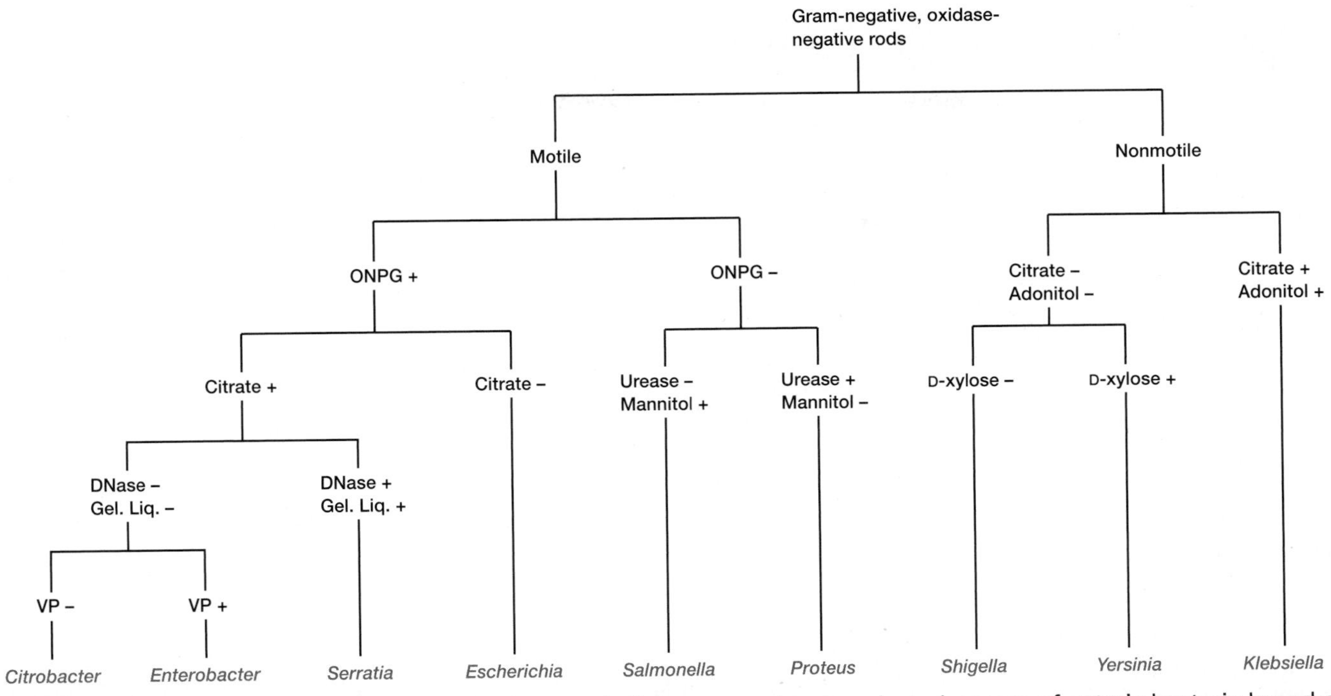

FIGURE 20.30 Identification of Enterobacterial Genera. A dichotomous key to selected genera of enteric bacteria based on motility and biochemical characteristics. The following abbreviations are used: ONPG, o-nitrophenyl-β-D-galactopyranoside (a test for β-galactosidase); DNase, deoxyribonuclease; Gel. Liq., gelatin liquefaction; and VP, Voges-Proskauer (a test for butanediol fermentation).

***Figure 20.30** Micro Inquiry*

What would be the identity of a motile microbe that synthesizes the enzyme β-galactosidase, assimilates citrate, produces DNase, and liquefies gelatin?

Members of the *Enterobacteriaceae* are so common, widespread, and important that they are probably seen more often in laboratories than any other bacteria. *E. coli* is undoubtedly the best-studied bacterium and the experimental organism of choice for many microbiologists. It is an inhabitant of the colon of humans and other warm-blooded animals, and it is quite useful in the analysis of water for fecal contamination. Some strains cause gastroenteritis or urinary tract infections. Several genera contain very important human pathogens responsible for a variety of diseases: *Salmonella,* typhoid fever and gastroenteritis; *Shigella,* bacillary dysentery; *Klebsiella,* pneumonia; and *Yersinia,* plague. Members of the genus *Erwinia* are major pathogens of crop plants and cause blights, wilts, and several other plant diseases.

Order *Pasteurellales*

The family *Pasteurellaceae* in the order *Pasteurellales* differs from the *Vibrionales* and the *Enterobacteriales* in several ways (table 20.6). Most notably, they are small (0.2 to 0.4 μm in diameter) and nonmotile, normally oxidase positive, have complex nutritional requirements, and are parasitic in vertebrates. The family contains seven genera: *Pasteurella, Haemophilus, Actinobacillus, Lonepinella, Mannheimia, Phocoenobacter,* and *Gallibacterium.*

As might be expected, members of this family are best known for the diseases they cause in humans and many animals. *Pasteurella multocida* and *P. haemolytica* are important animal pathogens. *P. multocida* is responsible for fowl cholera, which kills many chickens, turkeys, ducks, and geese each year. *P. haemolytica* is at least partly responsible for pneumonia in cattle, sheep, and goats (e.g., "shipping fever" in cattle). *H. influenzae* serotype b is a major human pathogen that causes a variety of diseases, including meningitis, sinusitis, pneumonia, and bronchitis. Fortunately, a sharp reduction in the incidence of *H. influenzae* serotype b infections began in the mid-1980s due to administration of the *H. influenzae* type b ("Hib") vaccine. However, *H. influenzae* serotype b still causes at least 3 million cases of serious disease and several hundreds of thousands of deaths each year globally. ▶▶| *Meningitis (section 38.1)*

Table 20.7 Some Characteristics of Selected Genera in the *Enterobacteriaceae*

Characteristics	*Escherichia*	*Shigella*	*Salmonella*	*Citrobacter*	*Proteus*
Methyl red	+	+	+	+	+
Voges-Proskauer	–	–	–	–	d
Indole production	(+)	d	–	d	d
Citrate use	–	–	(+)	+	d
H_2S production	–	–	(+)	d	(+)
Urease	–	–	–	(+)	+
β-galactosidase	(+)	d	d	+	–
Gas from glucose	+	–	(+)	+	+
Acid from lactose	+	–	(–)	d	–
Phenylalanine deaminase	–	–	–	–	+
Lysine decarboxylase	(+)	–	(+)	–	–
Ornithine decarboxylase	(+)	d	(+)	(+)	d
Motility	d	–	(+)	+	+
Gelatin liquifaction (22°C)	–	–	–	–	+
% G + C	48–59	49–53	50–53	50–52	38–41
Genome size (Mb)	4.6–5.5	4.6	4.5–4.9	Nd[d]	Nd
Other characteristics	1.1–1.5 × 2.0–6.0 μm; peritrichous flagella when motile	No gas from sugars	0.7–1.5 × 2–5 μm; peritrichous flagella	1.0 × 2.0–6.0 μm; peritrichous flagella	0.4–0.8 × 1.0–3.0 μm; peritrichous flagella

[a] (+) usually present
[b] (–) usually absent
[c] d, strains or species vary in possession of characteristic
[d] Nd: Not determined; genome not yet sequenced

1. Describe the general properties of the γ-proteobacteria.
2. What are the major characteristics of the purple sulfur bacteria? Contrast the families *Chromatiaceae* and *Ectothiorhodospiraceae.*
3. Describe the genera *Beggiatoa, Leucothrix,* and *Thiothrix.* Why do *Thiomargarita* cells appear hollow?
4. In what habitats would one expect to see the *Methylococcaceae* growing and why?
5. What is a methylotroph? How do methane-oxidizing bacteria use methane as both an energy source and a carbon source? How do these microbes avoid poisoning themselves with the formaldehyde they produce as an intermediate in C1 metabolism?
6. Give the major distinctive properties of the genera *Pseudomonas* and *Azotobacter.*
7. Why are the pseudomonads such important bacteria? What is mineralization?
8. Why is metal reduction by *Shewanella* considered dissimilatory?
9. List the major distinguishing traits of the families *Vibrionaceae, Enterobacteriaceae,* and *Pasteurellaceae.*
10. Briefly describe bioluminescence and the way it is produced.
11. Into what two groups can the enteric bacteria be placed based on their fermentation patterns?

	Yersinia	*Klebsiella*	*Enterobacter*	*Erwinia*	*Serratia*
	+	(+)[a]	(−)[b]	+	d[c]
	− (37°C)	(+)	+	(+)	+
	d	d	−	(−)	(−)
	(−)	(+)	+	(+)	+
	−	−	−	(+)	−
	d	(+)	(−)	−	−
	+	(+)	+	+	+
	(−)	(+)	(+)	(−)	d
	(−)	(+)	(+)	d	d
	−	−	(−)	(−)	−
	(−)	(+)	d	−	d
	d	−	(+)	−	d
	− (37°C)	−	+	+	+
	(−)	−	d	d	(+)
	46–50	53–58	52–60	50–54	52–60
	4.6	Nd	Nd	5.1	5.1
	0.5–0.8 × 1.0–3.0 μm; peritrichous flagella when motile	0.3–1.0 × 0.6–6.0 μm; capsulated	0.6–1.0 × 1.2–3.0 μm; peritrichous flagella	0.5–1.0 × 1.0–3.0 μm; peritrichous flagella; plant pathogens and saprophytes	0.5–0.8 × 0.9–2.0 μm; peritrichous flagella; colonies often pigmented

20.4 CLASS DELTAPROTEOBACTERIA

Although the **δ-proteobacteria** are not a large assemblage of genera, they show considerable morphological and physiological diversity. These bacteria can be divided into two general groups, all of them chemoorganotrophs. Some genera are predators, such as the bdellovibrios and myxobacteria. Others are anaerobes that use sulfate and sulfur as terminal electron acceptors while oxidizing organic nutrients. The class has eight orders and 20 families. **Figure 20.31** illustrates the phylogenetic relationships among major groups within the δ-proteobacteria, and **table 20.8** summarizes the general properties of some representative genera.

Orders *Desulfovibrionales*, *Desulfobacterales*, and *Desulfuromonadales*

Desulfovibrionales, Desulfobacterales, and *Desulfuromonadales* are a diverse group of **sulfate-** or **sulfur-reducing bacteria (SRB)** that are united by their anaerobic nature and ability to reduce elemental sulfur or sulfate and other oxidized sulfur compounds to hydrogen sulfide during anaerobic respiration. The use of sulfate as a terminal electron acceptor is a multistep process. Sulfate cannot be directly reduced. Because sulfate is a stable molecule, it must first be activated by reacting with ATP to form adenosine phosphosulfate, or APS (**figure 20.32**). During anaerobic respiration, APS accepts two electrons, which results in the production of sulfite (SO_3^{2-}) and adenosine monophosphate, or AMP. Six additional electrons are needed to reduce sulfite completely

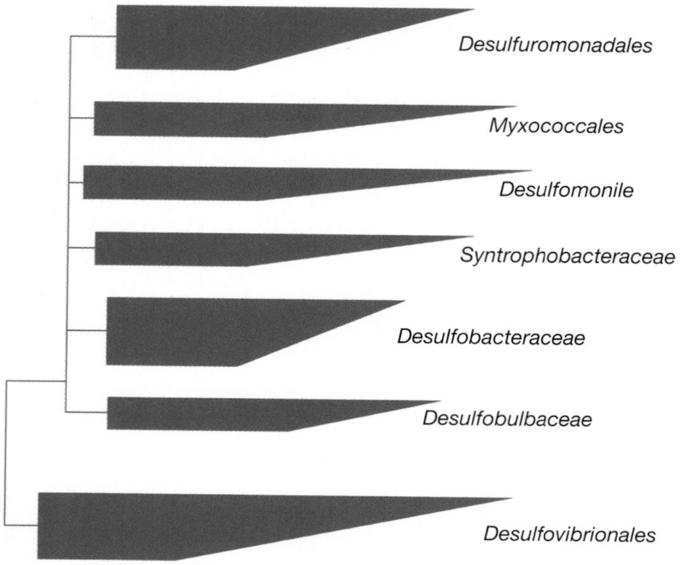

FIGURE 20.31 Phylogenetic Relationships Among Major Groups Within the δ-Proteobacteria. The relationships are based on 16S rRNA sequence.

to hydrogen sulfide. These are captured by a hydrogenase enzyme located on the periplasmic face of the plasma membrane. The hydrogenase donates the electrons from molecular hydrogen to cytochrome c_3, which in turn transfers them to the integral membrane protein Hmc. Finally an iron-sulfur protein delivers the electrons to sulfite reductase, and hydrogen sulfide is produced. Note that the hydrogenase releases protons in the periplasm; the resulting proton motive force is sufficient to drive the synthesis of ATP.

Recall that anaerobic respiration is coupled with the oxidation of reduced organic compounds and in general the SRB can be grouped according to carbon metabolism. Many SRB, including all but one member of the order *Desulfovibrionales* (*Desulfothermus naphthae*), oxidize lactate, formate, butyrate, proprionate, pyrvate, and aromatic compounds to acetate, which is not further metabolized. Other SRBs can oxidize acetate completely to CO_2. The best-studied sulfate-reducing genus is *Desulfovibrio; Desulfuromonas* uses only elemental sulfur as an acceptor. ◀◀ *Anaerobic respiration (section 10.6)*

The SRB are very important in the cycling of sulfur within the ecosystem. Because significant amounts of sulfate are present

Table 20.8 Characteristics of Selected δ- and ε-Proteobacteria

Class Genus	*Dimensions (μm) and Morphology*	*G + C Content (mol%)*	*Oxygen Requirement*	*Other Distinctive Characteristics*
δ-Proteobacteria				
Bdellovibrio	0.2–0.5 × 0.5–1.4; comma-shaped rods with a sheathed polar flagellum	49.5–51	Aerobic	Preys on other gram-negative bacteria where it grows in the periplasm; alternates between predatory and intracellular reproductive phases
Desulfovibrio	0.5–1.5 × 2.5–10; curved or sometimes straight rods, motile by polar flagella	46.1–61.2	Anaerobic	Oxidizes organic compounds to acetate and reduces sulfate or sulfur to H_2S
Desulfuromonas	0.4–0.9 × 1.0–4.0; straight or slightly curved or ovoid rods, lateral or subpolar flagella	54–62	Anaerobic	Reduces sulfur to H_2S, oxidizes acetate to CO_2; forms pink or peach-colored colonies
Myxococcus	0.4–0.7 × 2–8; slender rods with tapering ends, gliding motility	68–71	Aerobic	Forms fruiting bodies with microcysts not enclosed in a sporangium
Stigmatella	0.7–0.8 × 4–8; straight rods with tapered ends, gliding motility	67–68	Aerobic	Stalked fruiting bodies with sporangioles containing myxospores (0.9–1.2 × 2–4 μm)
ε-Proteobacteria				
Campylobacter	0.2–0.8 × 0.5–5; spirally curved cells with a single polar flagellum at one or both ends	29–47	Microaerophilic	Carbohydrates not fermented or oxidized; oxidase positive and urease negative; found in intestinal tract, reproductive organs, and oral cavity of animals
Helicobacter	0.2–1.2 × 1.5–10; helical, curved, or straight cells with rounded ends; multiple, sheathed flagella	24–48	Microaerophilic	Catalase and oxidase positive; urea rapidly hydrolyzed; found in the gastric mucosa of humans and other animals

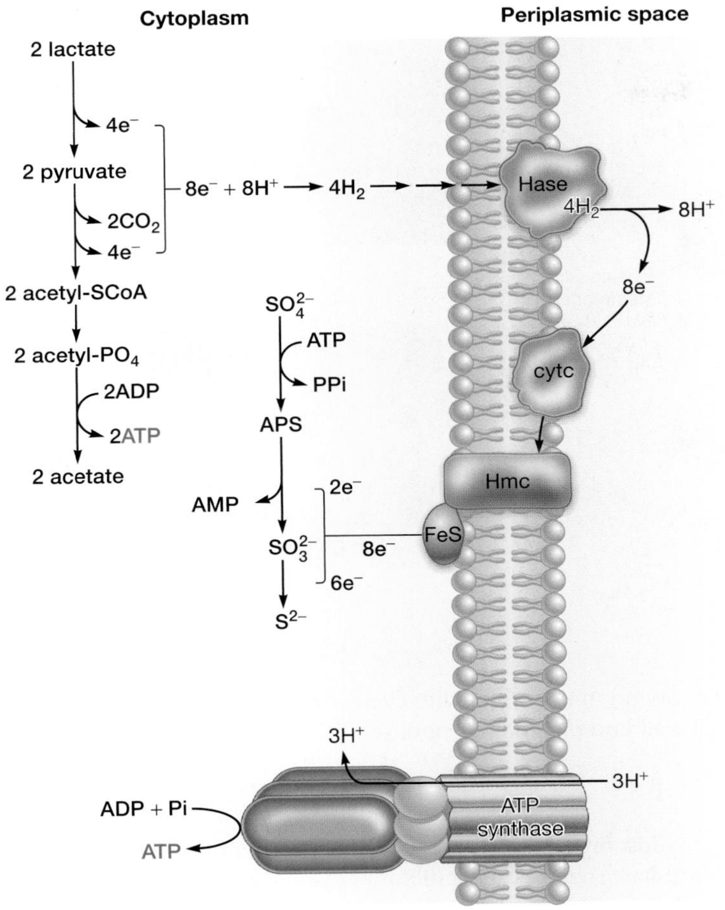

FIGURE 20.32 Dissimilatory Sulfate Reduction by *Desulfovibrio*. These microbes make ATP by substrate-level phosphorylation during acetate fermentation and by ATP synthase. Synthesis of ATP by ATP synthase is driven by a proton motive force generated by a periplasmic hydrogenase, which oxidizes H_2 to protons and electrons. The electrons are passed back into the cell, where they are used to reduce sulfate completely to sulfide. As detailed in the text, the oxidation of sulfate to sulfide requires 8 electrons and the activation of sulfate by its covalent attachment to ADP to form APS.

in almost all aquatic and terrestrial habitats, SRB are widespread and active in locations made anoxic by microbial digestion of organic materials. *Desulfovibrio* and other SRB thrive in habitats such as muds and sediments of polluted lakes and streams, sewage lagoons and digesters, and waterlogged soils. *Desulfuromonas* is most prevalent in anoxic marine and estuarine sediments. It also can be isolated from methane digesters and anoxic hydrogen sulfide–rich muds of freshwater habitats. Often sulfate and sulfur reduction are apparent from the smell of hydrogen sulfide and the blackening of water and sediment by iron sulfide. Hydrogen sulfide production in waterlogged soils can kill animals, plants, and microorganisms. SRB negatively impact industry because of their primary role in the anaerobic corrosion of iron in pipelines, heating systems, and other structures. ▶▶| *Sulfur cycle (section 26.1)*

Order *Desulfuromonales*

The order *Desulfuromonales* combines three families, the *Desulfuromonaceae, Geobacteraceae,* and *Pelobacteraceae.* All are strictly anaerobic with respiratory or fermentative metabolism. They can be chemolithoheterotrophs, obtaining electrons from reduced inorganic compounds, or chemoorganotrophs. They are mesophilic and have been isolated from anoxic marine and freshwater environments.

Geobacter metallireducens was the first microbe described to couple the oxidation of organic substrates such as acetate with the use of Fe(III) as a terminal electron acceptor. The capacity of *Geobacter* spp. to conserve energy from dissimilatory metal reduction is of great interest to environmental microbiologists. They have discovered that these bacteria, like the γ-proteobacterium *Shewanella,* can also reduce a number of toxic and radioactive metals. In addition, when *Geobacter* spp. oxidize organic substrates, they can transfer the electrons directly to an electrode. This makes it possible to harvest electricity during the degradation of organic waste. *Geobacter* spp. use two strategies to reduce insoluble metals, which they use as terminal electron acceptors. As discussed on p. 535, they have an outer membrane reductase that transfers the electrons they receive from the plasma membrane-bound ETC to an external, insoluble acceptor. They also synthesize nanowires. They do not use external electron shuttles, as does *Shewanella.* However, much as with *Shewanella,* there is great interest in the use of *Geobacter* spp. in bioremediation and electricity generated by microbial fuel cells. ▶▶| *Microbial energy conversion (section 41.3); Biodegradation and bioremediation in natural communities (section 42.3)*

Order *Bdellovibrionales*

The order *Bdellovibrionales* has only the family *Bdellovibrionaceae* and four genera. The genus *Bdellovibrio* (Greek *bdella,* leech) contains aerobic gram-negative, curved rods with polar flagella. The flagellum is unusually thick due to the presence of a sheath that is continuous with the cell wall. *Bdellovibrio* has a distinctive lifestyle: it preys on other gram-negative bacteria and alternates between a nongrowing predatory phase and an intracellular reproductive phase.

The life cycle of *Bdellovibrio* is complex, although it requires only 1 to 3 hours for completion (**figure 20.33**). The free bacterium swims along very rapidly (about 100 cell lengths per second) until it collides violently with its prey. It attaches to the bacterial surface, begins to rotate as fast as 100 revolutions per second, and bores a hole through the host cell wall in 5 to 20 minutes by releasing several hydrolytic enzymes. Its flagellum is lost during penetration of the cell.

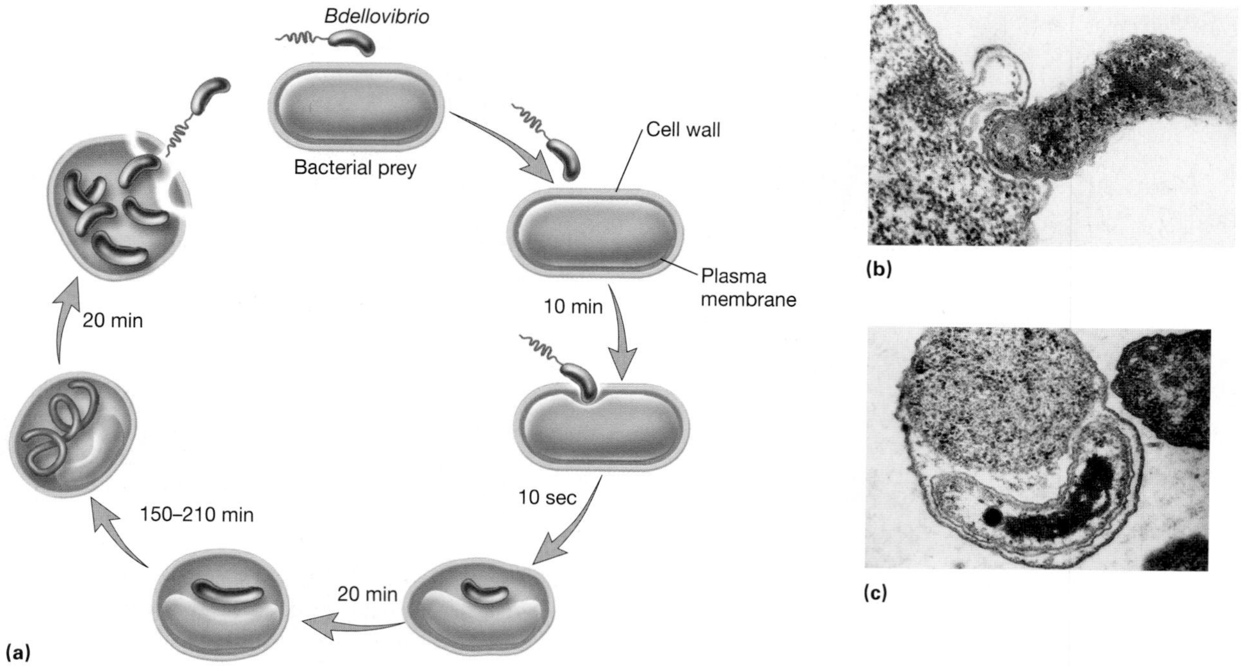

FIGURE 20.33 The Life Cycle of *Bdellovibrio*. (a) A general diagram showing the complete life cycle. (b) Penetrating the cell wall of *E. coli* (×55,000). (c) A *Bdellovibrio* encapsulated between the cell wall and plasma membrane of *E. coli* (×60,800).

After entry, *Bdellovibrio* takes control of the host cell and grows between the cell wall and plasma membrane while the host cell loses its shape and rounds up. The predator quickly inhibits host DNA, RNA, and protein synthesis, and disrupts the host's plasma membrane so that cytoplasmic constituents leak out of the cell. The growing bacterium uses host amino acids as its carbon, nitrogen, and energy source. It employs host fatty acids and nucleotides directly in biosynthesis, thus saving carbon and energy. The bacterium rapidly grows into a long filament under the cell wall and then divides into many smaller, flagellated progeny, which escape upon host cell lysis. Such multiple fission is rare in *Bacteria* and *Archaea*. Bdellovibrio

The *Bdellovibrio* life cycle resembles that of bacteriophages in many ways. Not surprisingly, when *Bdellovibrio* is plated on agar with host bacteria, plaques will form in the bacterial lawn. This technique is used to isolate pure strains and count the number of viable organisms just as with phages.

Order *Myxococcales*

The **myxobacteria** are gram-negative, aerobic soil bacteria characterized by gliding motility, a complex life cycle that includes production of multicellular structures called fruiting bodies and the formation of spores called myxospores. Myxobacterial cells are rods, about 0.4 to 0.7 μm by 2 to 8 μm long, and may be either slender with tapered ends or stout with rounded, blunt ends. The order *Myxococcales* is divided into six families based on the shape of vegetative cells, myxospores, and sporangia.

Most myxobacteria are predators of other microbes. They secrete lytic enzymes and antibiotics to kill their prey. Unlike *Bdellovibrio*, replication is not dependent on predation, and they can be grown in the absence of prey. The digestion products, primarily small peptides, are absorbed. Most myxobacteria use amino acids as their major source of carbon, nitrogen, and energy. All are aerobic chemoheterotrophs with respiratory metabolism.

The myxobacterial life cycle is quite distinctive and in many ways resembles that of the cellular slime molds (**figure 20.34**). In the presence of a food supply, myxobacteria glide along a solid surface, feeding and leaving slime trails. During this stage, the cells often form a swarm and move in a coordinated fashion. Some species congregate to produce a sheet of cells that moves rhythmically to generate waves or ripples. When their nutrient supply is exhausted, the myxobacteria aggregate and differentiate into a fruiting body, a structure that is largely made up of myxospores. The life cycle of the species *Myxococcus xanthus* has been well studied. Development in this microbe is induced by nutrient limitation and involves the exchange of at least five different extracellular signaling molecules that allow the cells to communicate with one another. ◀◀ *Motility and chemotaxis (section 3.7)*

Fruiting bodies range in height from 50 to 500 μm and often are colored red, yellow, or brown by carotenoid pigments. Each species forms a characteristic fruiting body. They vary in complexity from simple globular objects made of about 100,000 cells (*Myxococcus*) to the elaborate, branching, treelike structures formed by *Stigmatella* and *Chondromyces* (**figure 20.35**). Some

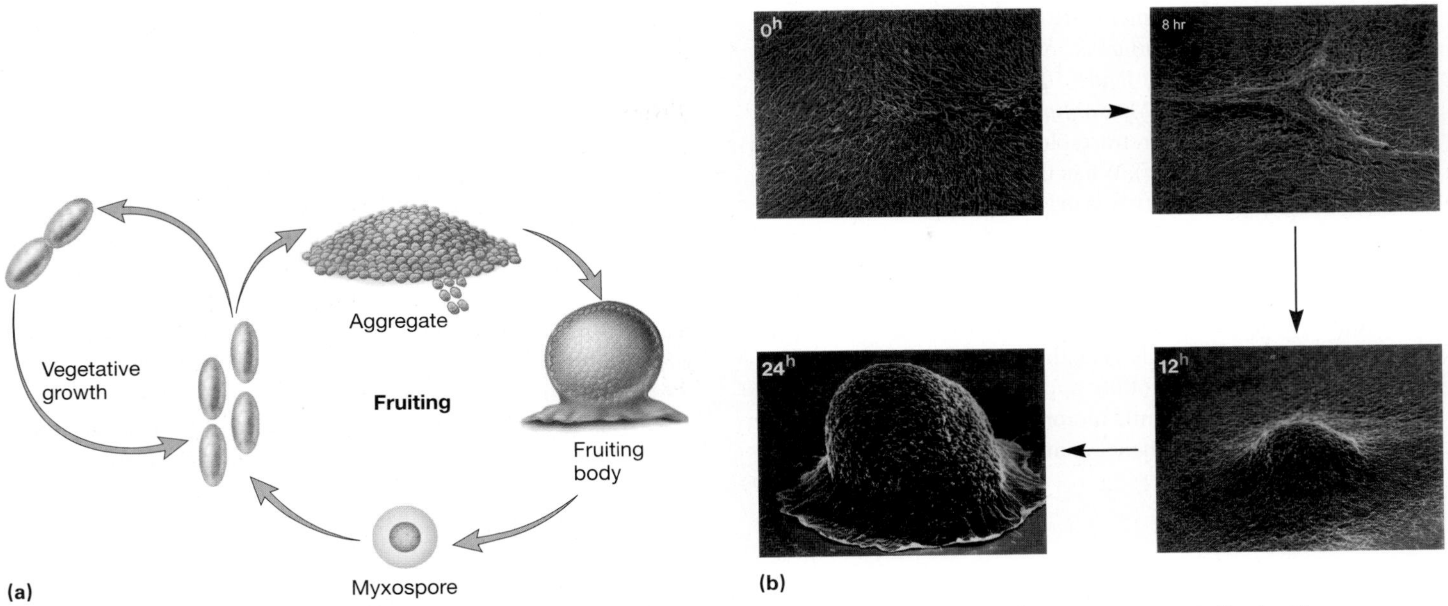

FIGURE 20.34 Life cycle of *Myxococcus xanthus*. (a) When nutrients are plentiful, *M. xanthus* grows vegetatively. However, when nutrients are depleted, a complex exchange of extracellular signaling molecules triggers the cells to aggregate and form fruiting bodies. Most of the cells within a fruiting body will become resting myxospores that will not germinate until nutrients are available. (b) Scanning electron micrographs taken during aggregate (0–12 hours) and fruiting body (24 hours) formation.

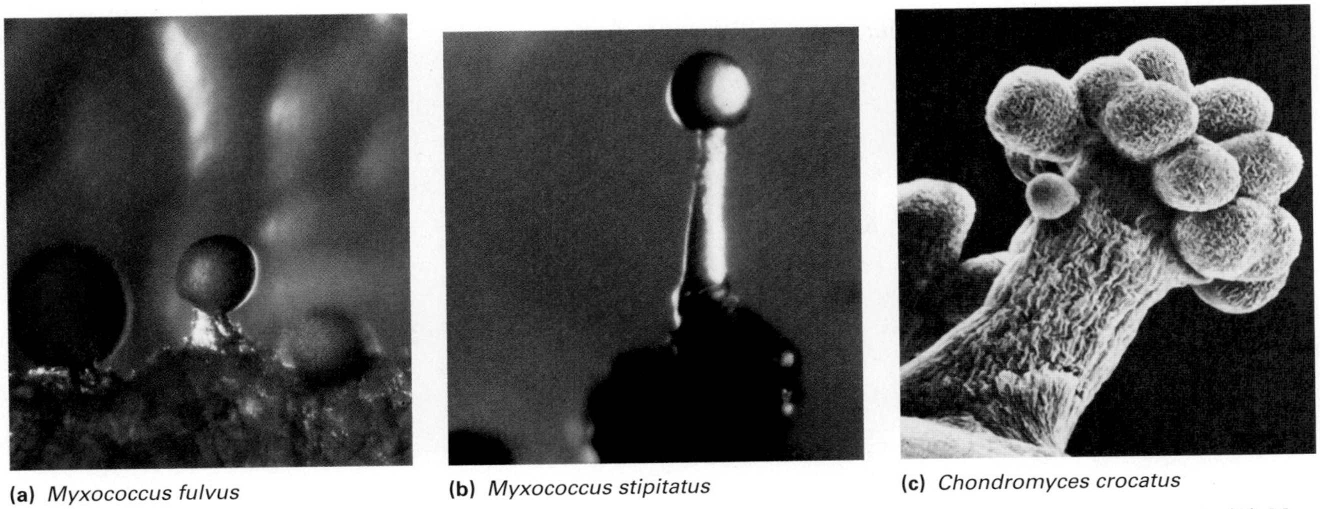

FIGURE 20.35 Myxobacterial Fruiting Bodies. (a) *Myxococcus fulvus.* Fruiting bodies are about 150–400 μm high. (b) *Myxococcus stipitatus.* The stalk is as tall as 200 μm. (c) *Chondromyces crocatus* viewed with the SEM. The stalk may reach 700 μm or more in height.

cells develop into dormant myxospores that frequently are enclosed in walled structures called sporangioles or sporangia. As might be expected for such complex structures, a variety of new proteins are synthesized during fruiting body formation.

Myxospores are dormant and desiccation-resistant; they may survive up to 10 years under adverse conditions. The use of fruiting bodies provides further protection for the myxospores and assists in their dispersal. (The myxospores often are suspended above the soil surface.) Because myxospores are kept together within the fruiting body, a colony of myxobacteria automatically develops when the myxospores are released and germinate. This communal organization may be advantageous because myxobacteria obtain nutrients by secreting hydrolytic enzymes and absorbing soluble digestive products. A mass of myxobacteria can produce enzyme concentrations sufficient to digest their prey more easily than can an individual cell. Extracellular enzymes diffuse away from their source, and an individual cell has more difficulty overcoming diffusional losses than a swarm of cells.

Fruiting body development also requires gliding motility. Gliding motility in *M. xanthus* is much slower (2–4 mm/min) than in *Flavobacterium (see figure 19.17)*. Two types of motility have been characterized in *M. xanthus*. Social (S) motility is governed by the production of retractable type IV pili from the front end of the cell (**figure 20.36*a***). When the pili retract, the cell creeps forward. This type of motility was originally called social motility because it is only observed in cells that are close together. It is now known that cell-to-cell contact is required for S motility because cells share outer membrane lipoproteins involved in pili secretion. The second type of motility is called adventurous (A) motility because it is exhibited by single cells that leave the group, perhaps scouting for prey. A-type motility appears to involve slime secretion and clusters of cytoplasmic motor proteins (AlgZ) that marks regions where the cell wall makes contact with the substrate. AlgZ protein complexes are stationary as the rod-shaped cells rotate forward. However, when an AlgZ complex reaches the lagging pole of the cell, it appears to disperse and reassemble at the leading pole (figure 20.36*b*). It is hypothesized that AlgZ complexes may travel along a helical cytoskeletal fiber that runs the length of the cell. It is intriguing that a single microbe has evolved two motility systems, and while S motility appears to be similar to twitching motility in *Pseudomonas aeruginosa*, neither system seems to be similar to the mechanism by which *Bacteroidetes*, cyanobacteria, or *Mycoplasma* species glide. This suggests that gliding motility has evolved independently multiple times. |◀◀ *Twitching and gliding motility (section 3.7); Phylum* Bacteriodetes *(section 19.7)*; ▶▶| *Class* Mollicutes *(The mycoplasmas) (section 21.1)*

Myxobacteria are found in soils worldwide. They are most commonly isolated from neutral soils or decaying plant material such as leaves and tree bark, and from animal dung. Although they grow in habitats as diverse as tropical rain forests and the Arctic tundra, they are most abundant in warm areas.

Search This: Xanthus wikimods

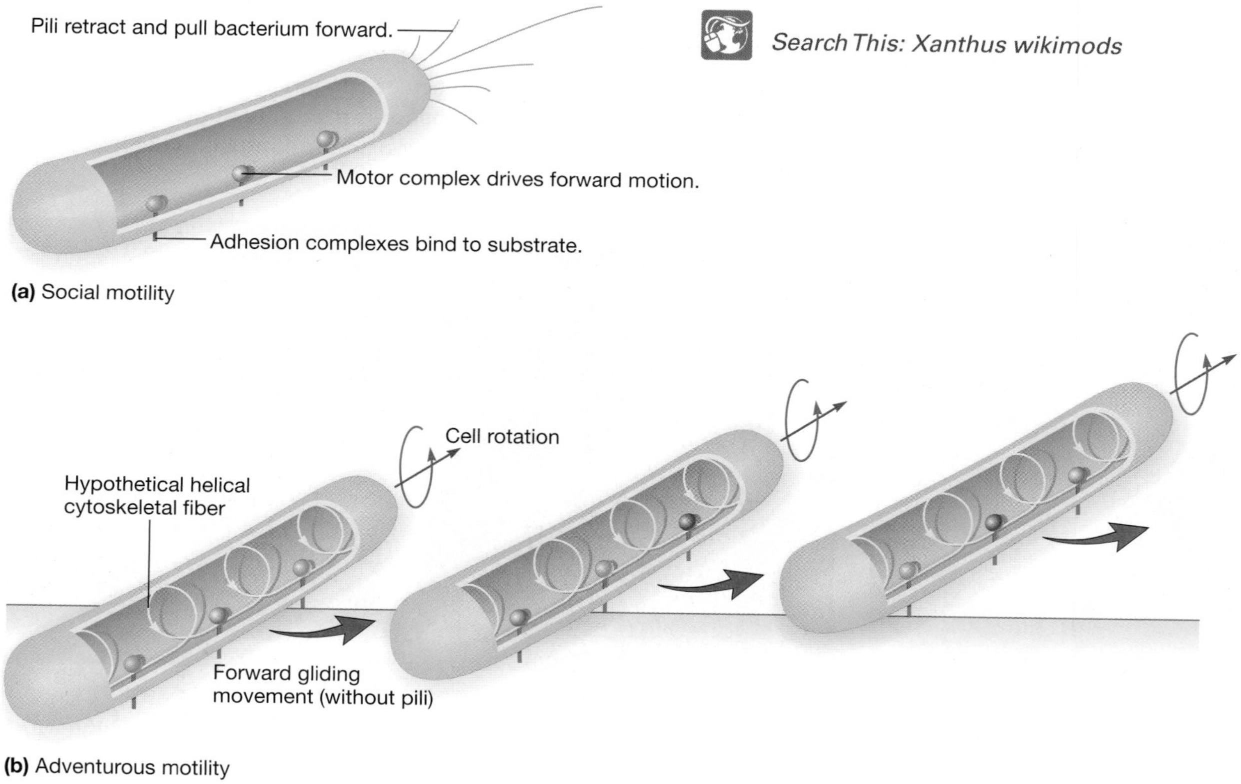

FIGURE 20.36 Model for Gliding Motility in *Myxococcus xanthus*. *M. xanthus* displays two types of surface-associated motility. During social motility, type IV pili extend and retract from the leading pole, much like grappling hooks. When cells reverse direction, the pili disassemble and reassemble at the opposite pole. (b) Adventurous motility is thought to be mediated by stationary motor proteins (AlgZ clusters) that make contact with the substrate. When an AlgZ cluster reaches the lagging end of the pole, it disperses and is reassembled at the leading pole. It is hypothesized that AlgZ clusters may move along a helical cytoskeletal filament.

Figure 20.36 Micro Inquiry

How could the analysis of motility mutants be used to determine if the adhesion complexes are used in both S and A motility?

1. Briefly characterize the δ-proteobacteria.
2. Describe the metabolic specialization of the dissimilatory sulfate- or sulfur-reducing bacteria. How do SRB make ATP?
3. In what ways is *Geobacter* similar to *Shewanella*? How do they differ?
4. Characterize the genus *Bdellovibrio* and outline its life cycle in detail. Why do you think it preys only on gram-negative bacteria?
5. Briefly describe the myxobacterial life cycle. What are fruiting bodies and myxospores? Why do you think extracellular signaling molecules are required for fruiting body formation?

20.5 CLASS *EPSILONPROTEOBACTERIA*

The **ϵ-proteobacteria** are the smallest of the five proteobacterial classes. They all are slender gram-negative rods, which can be straight, curved, or helical. *Bergey's Manual* places the ϵ-proteobacteria in one order, *Campylobacterales,* and three families: *Campylobacteraceae, Helicobacteraceae,* and the recently added *Nautiliaceae.* However, recent phylogenetic analysis that includes SSU rRNA sequences from newly cultured and uncultured organisms identifies two orders: the *Campylobacterales* (families *Campylobacteraceae, Helicobacteraceae,* and *Hydrogenimonaceae*) and *Nautiliales* (genera *Nautilia, Caminobacter,* and *Lebetimonas*). Two pathogenic genera, *Campylobacter* and *Helicobacter,* are microaerophilic, motile, helical or vibrioid, gram-negative rods. Table 20.8 summarizes some of the characteristics of these two genera.

The genus *Campylobacter* contains both nonpathogens and species pathogenic for humans and other animals. *C. foetus* causes reproductive disease and abortions in cattle and sheep. It is associated with a variety of conditions in humans ranging from septicemia (pathogens or their toxins in the blood) to enteritis (inflammation of the intestinal tract). It is also the most frequent infection that precedes the development of Gullain-Barre syndrome (GBS). GBS is the most common form of flaccid paralysis. It is thought that *Campylobacter* triggers GBS by a phenomenon known as molecular mimicry. In this case, the structure of *Campylobacter* lipopolysaccharide resembles components found on nerve cells. This causes the immune system to mistakenly attack the host's peripheral nervous system even after infection has cleared.

C. jejuni is a slender, gram-negative, motile, curved rod found in the intestinal tract of animals. It causes an estimated 2 million human cases of *Campylobacter* gastroenteritis—inflammation of the intestine—or campylobacteriosis and subsequent diarrhea in the United States each year. Studies with chickens, turkeys, and cattle have shown that as much as 50 to 100% of a flock or herd of these birds or animals excrete *C. jejuni.* These bacteria also can be isolated in high numbers from surface waters. They are transmitted to humans by contaminated food and water, contact with infected animals, or anal-oral sexual activity.

The other pathogenic ϵ-proteobacterial genus is *Helicobacter.* There are at least 23 species of *Helicobacter,* all isolated from the stomachs and upper intestines of humans, dogs, cats, and other mammals. In developing countries, 70 to 90% of the population is infected; the rate in developed countries ranges from 25 to 50%. The major human pathogen is *Helicobacter pylori,* which causes gastritis and peptic ulcer disease. *H. pylori* is an obligate microaerophile. The only carbohydrate the bacterium can use is glucose, which it metabolizes by both respiratory and fermentative pathways. It uses the pentose-phosphate pathways, which provides NADPH for biosynthesis, and the Entner-Doudoroff pathway to produce pyruvate. The microbe does not have a complete TCA cycle and appears to run the cycle "forwards" (i.e., oxidatively) from oxaloacetate to α-ketoglutarate and "backwards" (i.e., in the reductive direction) from oxaloacetate to succinate. Fumarate may act as the terminal electron acceptor during anaerobic respiration. It can also carry out aerobic respiration. Pyruvate is fermented to lactate, acetate, and formate during fermentation. As a resident of the gastric mucosa, it would seem that *H. pylori* is exposed to very acidic conditions, yet in the laboratory, it cannot be cultured below pH 4.5. The presence of urease enzymes in the cytoplasm and the cell surface help explain this paradox. *H. pylori* burrows within the gastric mucosa, which is not as acidic as the lumen. There it depends on urease activity to convert urea to CO_2 and NH_3. The ammonia drives the local pH up while the microbe uses chemotaxis to maintain its position within the mucous membranes. Indeed, motility is required for initial colonization of the host.

The ϵ-proteobacteria are now recognized to be more metabolically and ecologically diverse than previously thought. For instance, filamentous microbial mats in anoxic, sulfide-rich cave springs are dominated by members of the ϵ-proteobacteria (**figure 20.37**). Many have been cultured from marine hydrothermal vents; terrestrial forms have been isolated from groundwater, oil-field brines, limestone caves, and sulfidic springs. Most are thermophilic and chemolithoautotrophic; however, a few are heterotrophic or chemolithoheterotrophic—that is, they use organic carbon for biosynthesis but inorganic compounds, such as sulfide and molecular hydrogen, as electron donors for energy conservation (**table 20.9**). The autotrophic species use the reductive TCA cycle for CO_2 fixation (*see figure 11.6*). Because these bacteria generate a lot of biomass (as demonstrated in figure 20.37) and have relatively fast growth rates, they are thought to make a significant contribution to the elemental cycling in the unusual habitats in which they thrive.

▶▶| *Sulfide-based mutualisms (section 30.1)*

Search This: Cave bacteria

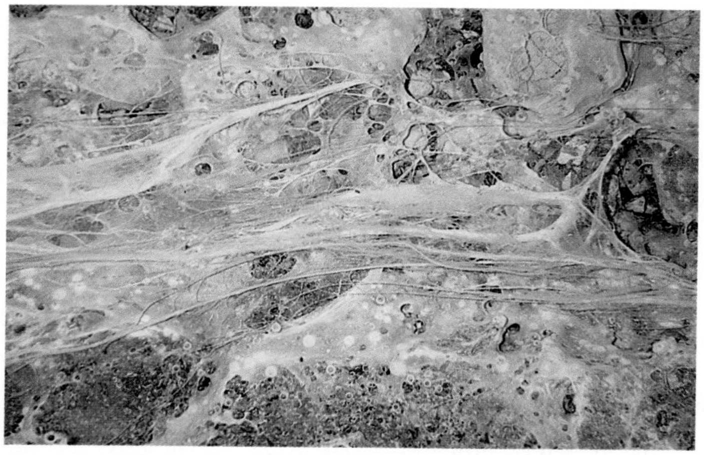

FIGURE 20.37 **ε-Proteobacteria Dominate Filamentous Microbial Mats in a Wyoming Sulfidic Cave Spring.**

1. Briefly describe the properties of the ϵ-proteobacteria.
2. How does *H. pylori* survive the acidic conditions of the stomach?
3. Compare the sulfur metabolism described for the ϵ-proteobacteria with that of the sulfate-reducing γ-proteobacteria and the purple sulfur bacteria.

Table 20.9 Some Recently Isolated ε-Proteobacteria

Species	*Isolation Site*	*Optimum Growth Temperature*	*Carbon Metabolism*	*Electron Donor*	*Electron Acceptor*	*Sulfur/Nitrogen Reduction Product*
Nautilia lithotrophica	Hydrothermal vent	52°C	Heterotroph	H_2, formate	SO_3^{2-}, S^0	H_2S
Caminibacter hydrogeniphilus	Hydrothermal vent	60°C	Heterotroph	H_2, complex organic compounds	NO_3^-, S^0	H_2S, NH_3
Nitratiruptor tergarcus	Hydrothermal vent	55°C	Autotroph	H_2	O_2, (microaerobic), NO_3^-, S^0	H_2S, N_2
Sulfurospirillum sp. str. Am-N	Hydrothermal vent	41°C	Heterotroph	Formate, fumarate	S^0	H_2S
Arcobacter sp. str. FWKO B	Oil-field production water	30°C	Autotroph	H_2, formate, HS^-	O_2, (microaerobic), NO_3^-, S^0	H_2S, N_2O^-
Sulfuricurvum kujiense	Oil-field production water	25°C	Autotroph	H_2, HS^-, $S_2O_3^{2-}$, S^0	O_2, (microaerobic), NO_3^-	NO_2^-

From Campbell, B. J., et al. 2006. The versatile ε-proteobacteria: Key players in sulphidic habitats. Nature Rev. Microbiol. *4:458–67.*

Summary

20.1 Class *Alphaproteobacteria*

a. The purple nonsulfur bacteria can grow anaerobically as photoorganoheterotrophs and often aerobically as chemoorganoheterotrophs. When growing photosynthetically, purple bacteria localize the light-harvesting proteins and reactions center on intracytoplasmic membranes (**figure 20.3**).

b. Rickettsias are obligate intracellular parasites. They have numerous transport proteins in their plasma membranes and make extensive use of host cell nutrients, coenzymes, and ATP.

c. Many proteobacteria have prosthecae, stalks, or reproduction by budding. Most of these bacteria are placed among the α-proteobacteria.

d. Two examples of budding or appendaged bacteria are *Hyphomicrobium* (budding bacteria that produce swarmer cells) and *Caulobacter* (bacteria with prosthecae and holdfasts) (**figures 20.7–20.10**).

e. *Rhizobium* carries out nitrogen fixation, whereas *Agrobacterium* causes the development of plant tumors. Both are in the family *Rhizobiaceae* (**figures 20.11** and **20.12**).

f. Chemolithotrophic bacteria derive energy and electrons from reduced inorganic compounds. Nitrifying bacteria are aerobes that oxidize either ammonia to nitrite or nitrite to nitrate and are responsible for nitrification (**figure 20.13** and **table 20.2**).

20.2 Class *Betaproteobacteria*

a. The genus *Neisseria* contains nonmotile, aerobic, gram-negative cocci that usually occur in pairs. They colonize mucous membranes and cause several human diseases, including meningitis and gonorrhea.

b. *Sphaerotilus, Leptothrix,* and several other genera have sheaths, hollow tubelike structures that surround chains of cells without being in intimate contact with the cells.

c. The order *Nitrosomonadales* includes two genera of nitrifying bacteria, including the ammonia-oxidizing microbe *Nitrosomonas europaea* (**figure 20.16**).

d. The colorless sulfur bacteria such as *Thiobacillus* oxidize elemental sulfur, hydrogen sulfide, and thiosulfate to sulfate while generating energy chemolithotrophically (**figure 20.17**).

20.3 Class *Gammaproteobacteria*

a. The γ-proteobacteria are the largest subgroup of proteobacteria and have great variety in physiological types (**table 20.5** and **figure 20.18**).

b. The purple sulfur bacteria are anaerobes and usually photolithoautotrophs. They oxidize hydrogen sulfide to sulfur and deposit the granules internally (**figure 20.20**).

c. Bacteria such as *Beggiatoa* and *Leucothrix* grow in long filaments or trichomes. Both genera have gliding motility. *Beggiatoa* is primarily a chemolithotroph and *Leucothrix,* a chemoorganotroph.

d. Marine *Beggiatoa, Thioploca,* and *Thiomargarita* are among the largest microbes. *Thiomargarita* form large intracellular vacuoles in which nitrate is stored (**figure 20.23**).

e. The genus *Thiomicrospira* can be found in both the γ- and ϵ-proteobacteria. The marine bacterium *T. crunogena* is a model γ-proteobacterial *Thiomicrospira.* It is a chemolithoautotroph that uses reduced sulfur compounds as a source of energy and electrons and O_2 as the terminal electron acceptor (**figure 20.25**).

f. The *Methylococcaceae* are methylotrophs; they use methane, methanol, and other reduced one-carbon compounds as their sole carbon and energy sources (**figure 20.26**).

g. The genus *Pseudomonas* contains straight or slightly curved, gram-negative, aerobic rods that are motile by one or several polar flagella and do not have prosthecae or sheaths.

h. The pseudomonads participate in natural mineralization processes, are major experimental subjects, cause many diseases, and often spoil refrigerated food.

i. The genus *Shewanella* is notable for its metabolic flexibility and ability to use extracellular, insoluble metals as terminal electron acceptors. It is one of several microbes that has evolved several strategies to accomplish this, including the use of electron shuttles and formation of nanowires (**figure 20.27**).

j. The most important facultatively anaerobic, gram-negative rods are found in three families: *Vibrionaceae, Enterobacteriaceae,* and *Pasteurellaceae* (**table 20.6**).

k. The *Enterobacteriaceae,* often called enterobacteria or enteric bacteria, are gram-negative, peritrichously flagellated or nonmotile, facultatively anaerobic, straight rods with simple nutritional requirements.

l. The enteric bacteria are usually identified by a variety of physiological tests. *E. coli* performs a mixed acid fermentation, which depends on the enzymes pyruvate formate-lyase and formate dehydrogenase (**figure 20.29**).

20.4 Class *Deltaproteobacteria*

a. The δ-proteobacteria contain chemoorganotrophic, gram-negative bacteria that are anaerobic and can use elemental sulfur and oxidized sulfur compounds as electron acceptors in anaerobic respiration. Other δ-proteobacteria are predatory aerobes (**table 20.8**).

b. The sulfate-reducing bacteria are very important in sulfur cycling in the ecosystem. *Desulfovibrio* conserves energy by fermentation and a proton motive force (**figure 20.32**).

c. *Bdellovibrio* is an aerobic curved rod with sheathed polar flagellum that preys on other gram-negative bacteria and grows within their periplasmic space (**figure 20.33**).

d. *Myxobacteria* are gram-negative, aerobic soil bacteria with gliding motility and a complex life cycle that leads to the production of dormant myxospores held within fruiting bodies (**figures 20.34–20.36**).

20.5 Class *Epsilonproteobacteria*

a. The ϵ-proteobacteria are the smallest of the proteobacterial classes and contain two important pathogenic genera: *Campylobacter* and *Helicobacter.* These are microaerophilic, motile, helical or vibrioid, gram-negative rods (**table 20.8**).

b. Recently a new family, the *Nautiliaceae,* has been added. Many of these bacteria are chemolithoautotrophs from deep-sea hydrothermal vent ecosystems (**table 20.9**).

Critical Thinking Questions

1. The advantages of metabolic flexibility in *Rhodospirillum rubrum* must offset the energetic burden of carrying all the structural and regulatory genes that make this possible. What factors make this trade-off possible?
2. Why do you think no microbe is able to oxidize ammonia completely to nitrate?
3. Why might the ability to form dormant cysts be of great advantage to *Agrobacterium* but not as much to *Rhizobium?*
4. Examine figures 20.17 and 20.32. Note that in both cases APS is formed, but in the case of *T. ferrooxidans* (figure 20.17), it is involved in sulfide oxidation, whereas for *Desulfovibrio* (figure 20.32), it is used during sulfate reduction. Why must APS be used in both cases?
5. How might electron shuttles facilitate growth of bacteria in a biofilm that is limited in the availability of a soluble terminal electron acceptor?
6. Intercellular communication has been studied in detail in *Myxococcus xanthus.* Review the life cycle of this developmentally complex microbe and indicate at what points during this process intercelllar signaling molecules might be exchanged and the message they would convey.
7. The application of cyro-electron microscopy and cryo-electron tomography (*see chapter 2*) to *Bdellovibrio bacteriovorus* shows that the cellular flexibility of these microbes enables rapid, localized deformation without loss of cell integrity. Discuss how the ability of these cells to truly bend is of great advantage. Also, considering what is known about the structure of the gram-negative cell envelope, what structural adaptations do you think *Bdellovibrio* might need to survive such microbial gymnastics?

   Read the original paper: Bargnia, M., et al., 2008. Three-dimensional imaging of the highly bent architecture of *Bdellovibrio bacteriovorus* by using cryo-electron tomography. *J. Bacteriol.* 190:2588.
8. The γ-proteobacterium *Congregibacter litoralis* is one of several cultured aerobic anoxygenic phototrophs (AAnPs). AAnPs are also found among the α- and β-proteobacteria. These microbes are photoheterotrophs that contain Bchl*a*. They cannot grow photoautotrophically or anaerobically; *C. litoralis* is a microaerophile. It possesses the genes needed for the oxidation of reduced sulfur compounds, but the addition of reduced, inorganic sulfur to the media does not enhance growth. Compare the distribution of the AAnPs among the proteobacterial classes with that of the purple sulfur and nonsulfur bacteria. How would you determine if the AAnPs and the purple bacteria evolved from a single photosynthetic ancestor or have acquired the capacity to grow phototrophically by horizontal gene transfer? Finally, if *C. litoralis* does not use reduced sulfur compounds as a source of electrons, what do you think is the electron donor?

   Read the original paper: Fuchs, B. M., et al. 2007. Characterization of a marine gammaproteobacterium capable of aerobic anoxygenic photosynthesis. *Proc. Nat. Acad. Sci., USA.* 104:2891.

Learn More

Learn more by visiting the text website at www.mhhe.com/willey8, where you will find a complete list of references.

21

Bacteria: *The Low* G + C *Gram Positives*

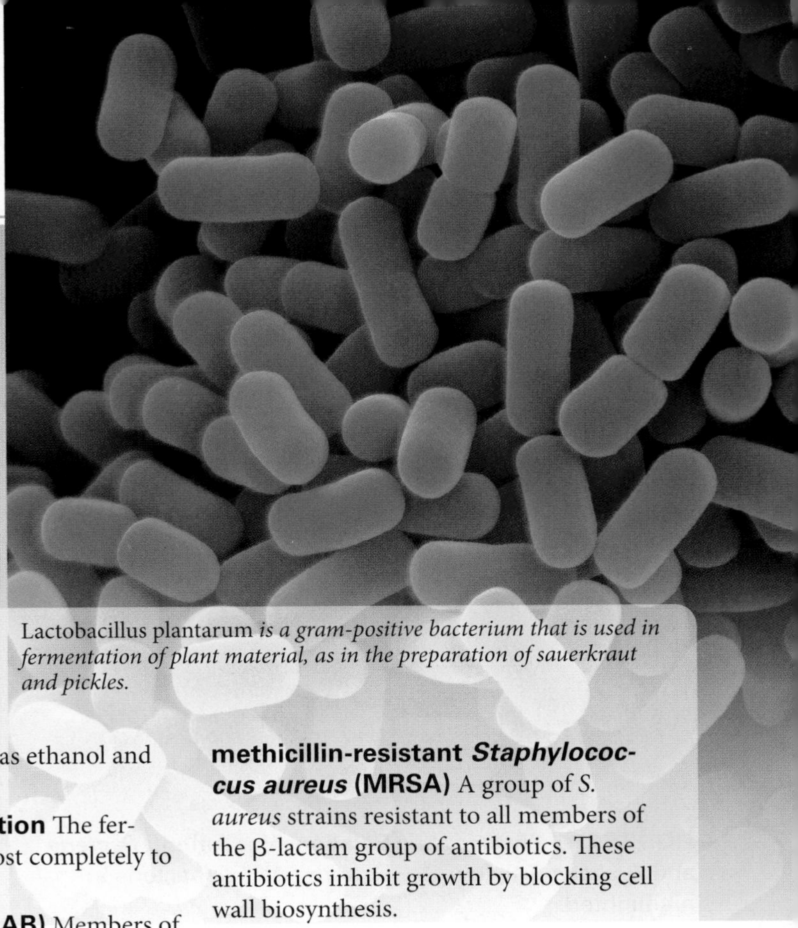

Lactobacillus plantarum is a gram-positive bacterium that is used in fermentation of plant material, as in the preparation of sauerkraut and pickles.

CHAPTER GLOSSARY

α-hemolysis A greenish zone of partial clearing around a bacterial colony growing on blood agar, resulting from incomplete destruction of red blood cells and hemoglobin breakdown.

β-hemolysis A zone of complete clearing around a bacterial colony growing on blood agar, resulting from complete destruction of red blood cells and hemoglobin breakdown.

heterolactic fermentation The fermentation of sugars to form lactate and other products such as ethanol and CO_2.

homolactic fermentation The fermentation of sugars almost completely to lactic acid.

lactic acid bacteria (LAB) Members of the order *Lactobacillales* that generate lactic acid as their major or sole fermentation product.

Lancefield grouping system A method by which streptococci can be placed in serologically distinguishable groups.

methicillin-resistant *Staphylococcus aureus* (MRSA) A group of *S. aureus* strains resistant to all members of the β-lactam group of antibiotics. These antibiotics inhibit growth by blocking cell wall biosynthesis.

mycoplasmas Bacteria that are members of the class *Mollicutes* that lack cell walls and cannot synthesize peptidoglycan precursors; most require sterols for growth.

Chapters 21 and 22 introduce the gram-positive bacteria. These bacteria were historically grouped on the basis of their general shape (e.g., rods, cocci, or irregular) and their ability to form endospores. However, analysis of phylogenetic relationships within the gram-positive bacteria shows that they are better divided into a low G + C group and a high G + C, or actinobacterial, group (**figure 21.1**). *Bergey's Manual of Systematic Bacteriology* places the low G + C gram-positive bacteria in volume 3. This volume describes over 1,300 species placed in 255 genera. ◀◀ *Gram-positive cell walls (section 3.3)*

The low G + C gram-positive bacteria are placed in the phylum *Firmicutes* and divided into three classes: *Mollicutes, Clostridia,* and *Bacilli.* The phylum *Firmicutes* is large and complex; it has 10 orders and 34 families. The mycoplasmas, class *Mollicutes*, are also considered low G + C gram positives despite their lack of a cell wall. Ribosomal RNA data indicate that the mycoplasmas are closely related to the lactobacilli. In this chapter, we focus on the mycoplasmas, *Clostridium* and its relatives, and the bacilli and lactobacilli. **Figure 21.2** shows the phylogenetic relationships among some of the bacteria reviewed in this chapter.

21.1 Class *Mollicutes* (The Mycoplasmas)

Members of the class *Mollicutes*, commonly called **mycoplasmas,** are unusual because they lack cell walls and have small genomes and simplified metabolic pathways. When they were first discovered, they were thought to be quite primitive, but molecular analysis has revealed that they are descendents of a gram-positive bacterial ancestor. Their small genome appears to be the result of genome reduction such that they now lack a variety of metabolic capabilities including the ability to synthesize peptidoglycan precursors (*see figure 16.17*).

The class *Mollicutes* has five orders and six families. The best-studied genera are found in the orders *Mycoplasmatales* (*Mycoplasma, Ureaplasma*), *Entomoplasmatales* (*Entomoplasma, Mesoplasma, Spiroplasma*), *Acholeplasmatales* (*Acholeplasma*),

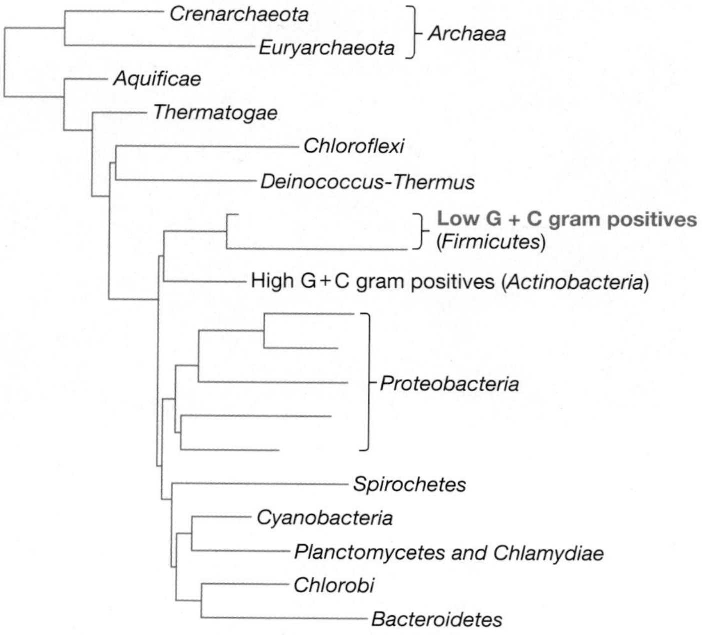

FIGURE 21.1 Phylogenetic Relationships Among *Bacteria* and *Archaea*. The low G + C gram-positive bacteria are highlighted.

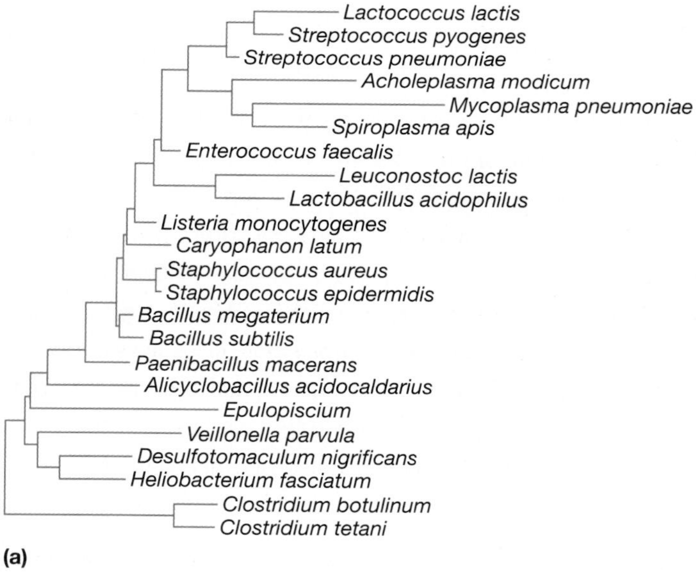
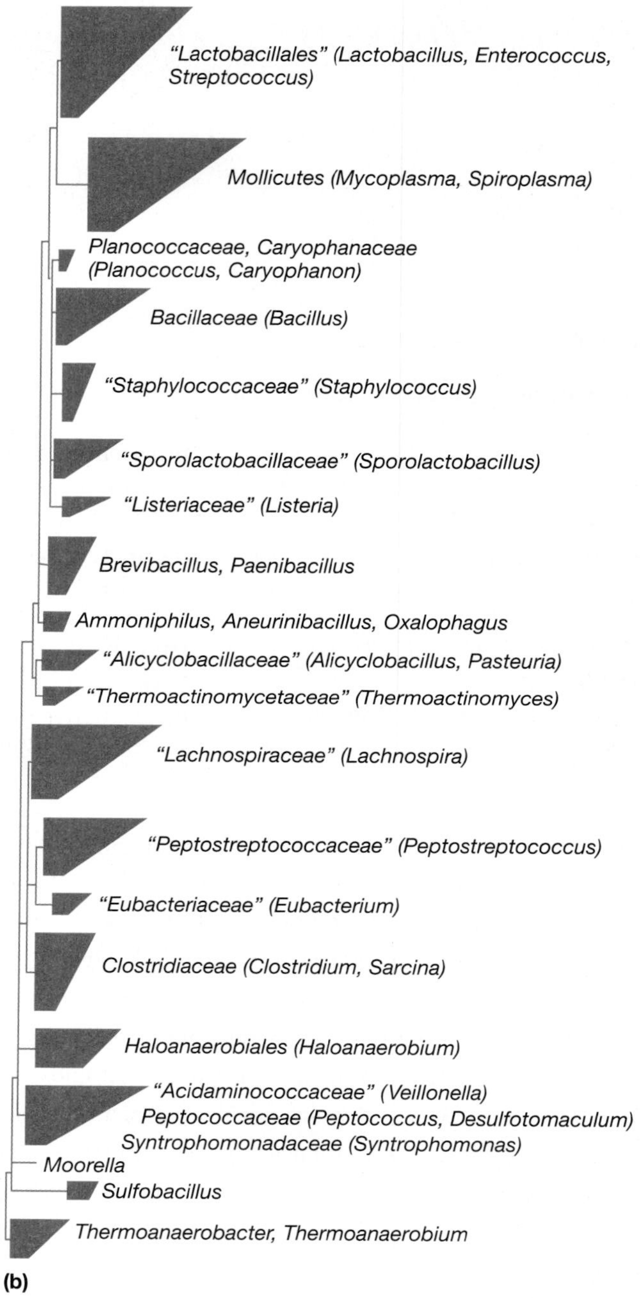

FIGURE 21.2 Phylogenetic Relationships in the Phylum *Firmicutes* (Low G + C Gram Positives). (a) The relationships of a few species based on 16S rRNA sequence date. (b) The major phylogenetic groups with representative genera in parentheses. Each tetrahedron in the tree represents a group of related organisms; its horizontal edges show the shortest and longest branches in the group. Multiple branching at the same level indicates that the relative branching order of the groups cannot be determined from the data.

Source: The Ribosomal Database Project.

Table 21.1 Properties of Some Members of the Class *Mollicutes*

Genus	*No. of Recognized Species*	*G + C Content (mol%)*	*Genome Size (Mb)*	*Sterol Requirement*	*Habitat*	*Other Distinctive Features*
Acholeplasma	13	26–36	1.50–1.65	No	Vertebrates, some plants and insects	Optimum growth 30–37°C
Anaeroplasma	4	29–34	1.50–1.60	Yes	Bovine or ovine rumen	Oxygen-sensitive anaerobes
Asteroleplasma	1	40	1.50	No	Bovine or ovine rumen	Oxygen-sensitive anaerobes
Entomoplasma	5	27–29	0.79–1.14	Yes	Insects, plants	Optimum growth 30°C
Mesoplasma	12	27–30	0.87–1.10	No	Insects, plants	Optimum growth 30°C; sustained growth in serum-free medium only with 0.04% detergent (Tween 80)
Mycoplasma	104	23–40	0.60–1.35	Yes	Humans, animals	Optimum growth usually 37°C
Spiroplasma	22	25–30	0.94–2.20	Yes	Insects, plants	Helical filaments; optimum growth at 30–37°C
Ureaplasma	6	27–30	0.75–1.20	Yes	Humans, animals	Urea hydrolysis

Adapted from J. G. Tully, et al., "Revised Taxonomy of the Class Mollicutes" in International Journal of Systematic Bacteriology, *43(2):378–85. Copyright © 1993 American Society for Microbiology, Washington, D.C. Reprinted by permission.*

and *Anaeroplasmatales (Anaeroplasma, Asteroleplasma).* **Table 21.1** summarizes some of the major characteristics of these genera. Because they are bounded only by a plasma membrane, these bacteria are pleomorphic and vary in shape from spherical or pear-shaped organisms, about 0.3 to 0.8 μm in diameter, to branched or helical filaments (**figure 21.3**). Some mycoplasmas (e.g., *M. genitalium*) have a specialized terminal structure that projects from the cell and gives them a flask or pear shape. This structure aids in attachment to eukaryotic cells. They are among the smallest bacteria capable of self-reproduction. Most species are facultative anaerobes, but a few are obligate anaerobes. When growing on agar, most form colonies with a "fried egg" appearance because they grow into the agar surface at the center while spreading outward on the surface at the colony edges (**figure 21.4**).

Mollicute genomes are among the smallest found among bacteria, ranging from 0.7 to 1.7 Mb (table 21.1). The sequenced genomes of the human pathogens *Mycoplasma genitalium, M. pneumoniae,* and *Ureaplasma urealyticum* have fewer than 1,000 genes, suggesting a minimal genome size for a free-living existence. As suggested by their limited number of genes, the mycoplasmas are incapable of synthesizing a number of macromolecules and thus require complex growth media. For instance, because *M. genitalium* relies so heavily on its host's biosynthetic capacity, it has lost the genes that encode the enzymes needed to synthesize amino acids, purines, pyrimidines, and fatty acids. In addition, most species require sterols for growth, which they obtain from the host as cholesterol. Sterols are an essential component of the mycoplasma plasma membrane, where they may facilitate osmotic stability in the absence of a cell wall. Some produce ATP by the Embden-Meyerhof pathway and

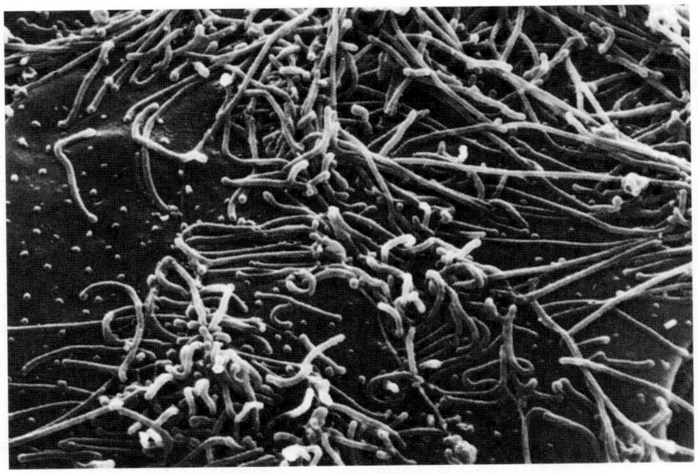

FIGURE 21.3 The Mycoplasmas. A scanning electron micrograph of *Mycoplasma pneumoniae* shows its pleomorphic nature. (×26,000).

lactic acid fermentation. The pentose phosphate pathway seems to be functional in at least some mycoplasmas; none appears to have the complete tricarboxylic acid cycle. Arginine can be catabolized to generate ATP in the following series of reactions:

$$\text{Arginine} + H_2O \rightarrow \text{citrulline} + NH_3$$
$$\text{Citrulline} + P_i \rightarrow \text{ornithine} + \text{carbamyl-P}$$
$$\text{Carbamyl-P} + \text{ADP} \rightarrow \text{ATP} + CO_2 + NH_3$$

The first reaction is catalyzed by the enzyme arginine deaminase, which is found in several other low G + C gram positives, including *Bacillus* and *Clostridium* species.

Ureaplasma urealyticum, a causative agent of urinary tract infections, has a unique means of conserving energy: it hydrolyzes urea to generate an electrochemical gradient via the accumulation of ammonia/ammonium (NH_3/NH_4^+). This gradient is responsible for the chemiosmotic potential that drives ATP synthesis (**figure 21.5**).

Some mycoplasmas are capable of gliding motility. *Mycoplasma mobile* moves at a steady pace of 2 to 5 μm/second. This appears to be mediated by cell surface proteins that surround the "neck" of the cell (**figure 21.6**). In *M. mobile,* these proteins are thought to attach to the cytoskeleton and function like microscopic legs. Unlike flagellar motility, gliding in this microbe is powered by ATP hydrolysis. By contrast, most other motile mycoplasmas move more like inchworms, at a tenth of the speed of *M. mobile.* For instance the terminal structure of *M. pneumoniae* is covered with cytoskeletal proteins that may function as motors to cause the contraction and extension of the cell, moving it forward.

Mycoplasmas are remarkably widespread and can be isolated from animals, plants, the soil, and even compost piles. Although their complex growth requirements can make their growth in pure (axenic) cultures difficult, at least 10% of the mammalian cell cultures in use are probably contaminated with mycoplasmas. This seriously interferes with tissue culture experiments and has resulted in the development of commercially available rapid tests for the detection of mycoplasmas in cell culture.

Search This: Mycoplasma contamination

In animals, mycoplasmas colonize mucous membranes and joints, and often are associated with diseases of the respiratory and urogenital tracts. Mycoplasmas cause several major diseases in livestock; for example, contagious bovine pleuropneumonia in cattle (*M. mycoides*), chronic respiratory disease in chickens (*M. gallisepticum*), and pneumonia in swine (*M. hyopneumoniae*). Spiroplasmas have been isolated from insects, ticks, and a variety of plants. They cause disease in citrus plants, cabbage, broccoli, corn, honeybees, and other hosts. Arthropods often act as vectors and carry the spiroplasmas between plants. In humans, *U. urealyticum* and *M. hominis* are common parasitic microorganisms of the genital tract, and their transmission is related to sexual activity. Both mycoplasmas can opportunistically cause inflammation of the reproductive organs of males and females. In addition, *Ureaplasma urealyticum* is associated with premature delivery of newborns, as well as neonatal meningitis and pneumonia. *M. pneumoniae* causes primary atypical pneumonia in humans.

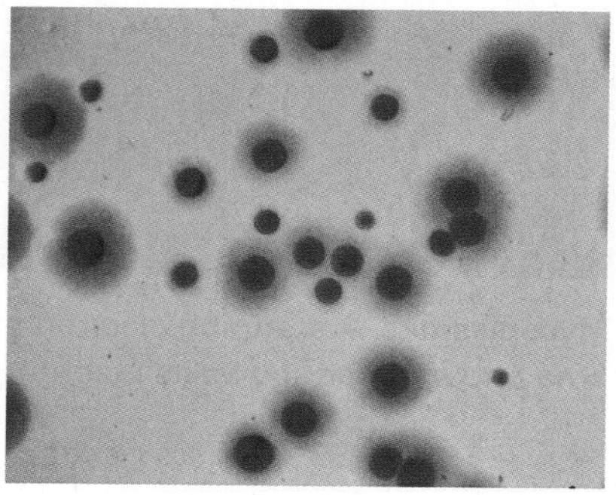

FIGURE 21.4 Mycoplasma Colonies. Note the "fried egg" appearance; colonies stained before photographing (×100).

Figure 21.4 Micro Inquiry

How does the "fried egg" appearance come about?

Search This: Mycoplasma resources

1. What morphological feature distinguishes the mycoplasmas? In what class are they found? Why have they been placed with the low G + C gram-positive bacteria?
2. What might mycoplasmas use sterols for?
3. Explain the relationship between *Mycoplasma* genome size and growth requirements. Why is genome reduction common in intracellular microbes?
4. List several animal and human diseases caused by mycoplasmas. What kinds of organisms do spiroplasmas usually infect?

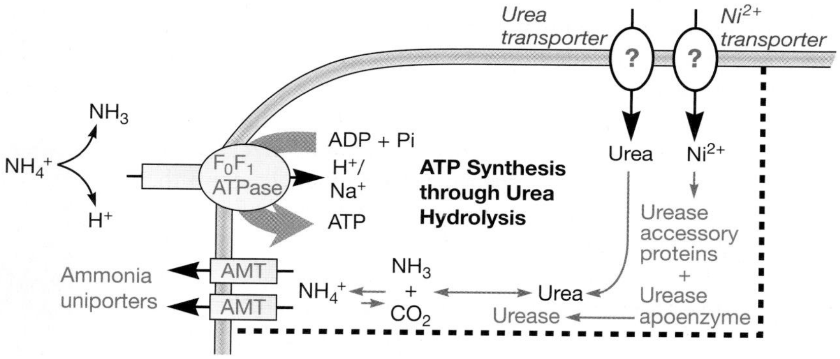

FIGURE 21.5 Energy Conservation in *Ureaplasma urealyticum.* To generate an electrochemical gradient through the accumulation of NH_3/NH_4^+, *U. urealyticum* is thought to import urea through specific transporters. Once inside the cell, urease (a nickel-binding enzyme) catalyzes the hydrolysis of urea to NH_3 and CO_2. Ammonia accepts a proton to become NH_4^+, which is exported through NH_4^+ uniporters (AMT). Once outside the cell, NH_4^+ is converted to NH_3 and H^+, providing the protons to drive ATP synthesis via the membrane bound ATP synthase.

Adapted from Glass, J. I., et al. 2000. The complete genome of the mucosal pathogen Ureaplasma urealyticum, Nature. *407:757–62.*

Figure 21.5 Micro Inquiry

How does the generation of PMF and use of the ATP synthase by *U. urealyticum* differ from the usual fermentative microbe?

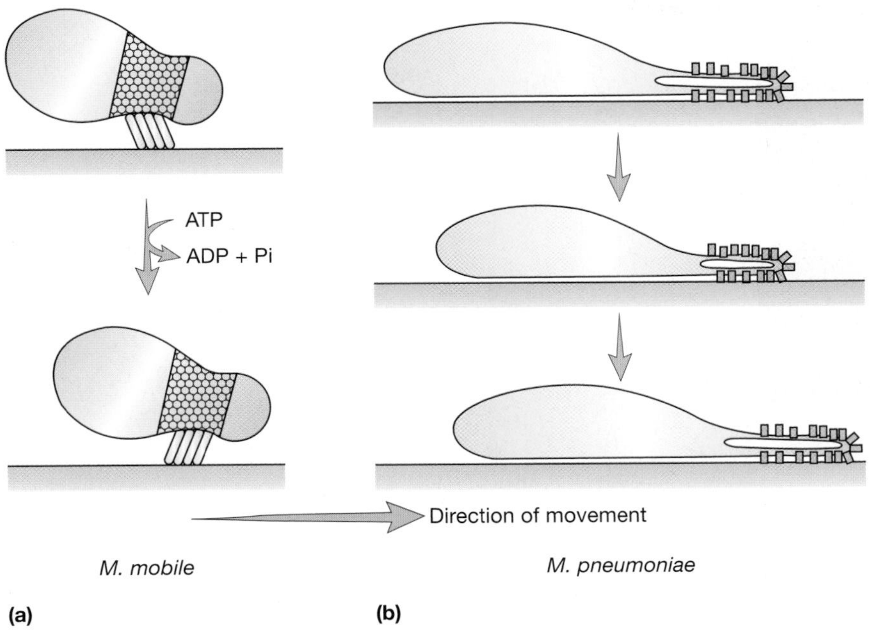

FIGURE 21.6 Two Forms of Gliding Motility in the Mycoplasmas. (a) *M. mobile* uses cytoskeletal "leg" proteins to move the microbe at a constant speed. (b) *M. pneumoniae* has cytoplasmic motor proteins that cause the cell to expand and contract, so that it moves more slowly and at an uneven, inchwormlike pace.

21.2 Class *Clostridia*

There are two classes of low G + C endospore-forming bacteria: *Clostridia* and *Bacilli*. Recall that endospores are made within the mother cell upon nutrient deprivation. Once released, these spores are remarkably resistant to environmental stresses, surviving thousands, if not millions, of years. The class *Clostridia* has a very wide variety of gram-positive bacteria distributed into three orders and 11 families. The characteristics of some of the more important genera are summarized in **table 21.2**. Phylogenetic relationships are shown in figure 21.2. ◀◀ *Bacterial endospores (section 3.8)*

The largest genus in the class *Clostridia* is *Clostridium*. It includes obligately anaerobic (although some can withstand O_2 exposure), fermentative, gram-positive bacteria that form endospores (**figure 21.7**). The genus contains well over 100 species in several distinct phylogenetic clusters and may be subdivided into several genera in the future. Members of the genus *Clostridium* have great practical impact. Because they are anaerobic and form heat-resistant endospores, they are responsible for many cases of food spoilage, even in canned foods. Clostridia often can ferment amino acids to produce ATP by oxidizing one amino acid while using another as an electron

Table 21.2 Characteristics of Selected Members of the Class *Clostridia*

Genus	*Dimensions (µm), Morphology, and Motility*	*G + C Content (mol%)*	*Oxygen Relationship*	*Other Distinctive Characteristics*
Clostridium	0.3–2.0 × 1.5–20; rod-shaped, often pleomorphic, nonmotile or peritrichous flagella	22–55	Anaerobic	Does not carry out dissimilatory sulfate reduction; usually chemoorganotrophic, fermentative, and catalase negative; forms oval or spherical endospores
Desulfotomaculum	0.3–1.5 × 3–9; straight or curved rods, peritrichous or polar flagella	37–50	Anaerobic	Reduces sulfate to H_2S, forms subterminal to terminal endospores; stains gram negative but has gram-positive wall; catalase negative
Heliobacterium	1.0 × 4–10; rods that are frequently bent, gliding motility	52–55	Anaerobic	Photoheterotrophic with bacteriochlorophyll *g*; stains gram negative but has gram-positive wall; some form endospores
Veillonella	0.3–0.5; cocci in pairs, short chains, and masses; nonmotile	36–43	Anaerobic	Stains gram negative; pyruvate and lactate fermented but not carbohydrates; acetate, propionate, CO_2, and H_2 produced from lactate; parasitic in mouths, intestines, and respiratory tracts of animals

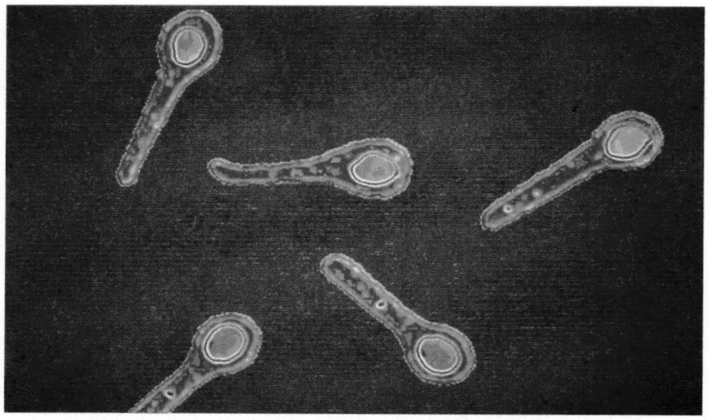

FIGURE 21.7 ***Clostridium tetani.*** This pathogen makes endospores that are round and terminal (×3,000).

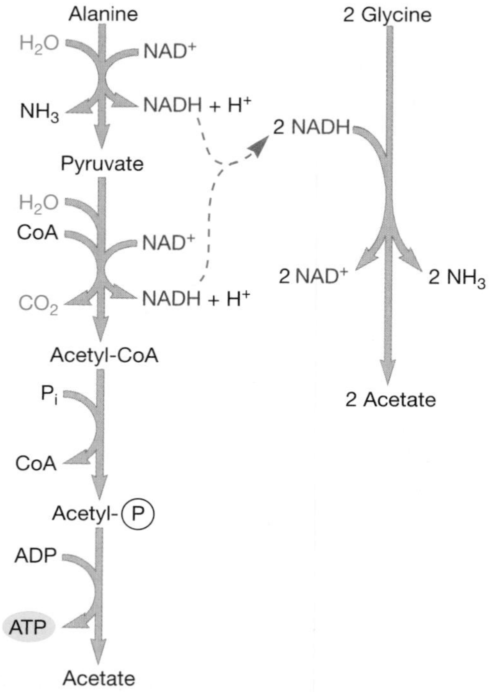

FIGURE 21.8 The Stickland Reaction. Alanine is oxidized to acetate, and glycine is used to reoxidize the NADH generated during alanine degradation. The fermentation also produces some ATP.

acceptor in a process called the **Stickland reaction** (**figure 21.8**). This reaction generates ammonia, hydrogen sulfide, fatty acids during the anaerobic decomposition of proteins. These products are responsible for many unpleasant odors arising during putrefaction.

Although clostridia are industrially valuable (for example, *C. acetobutylicum* is used to manufacture butanol), the pathogenic species that produce toxins are most well known. For instance, *C. perfringens* causes gas gangrene and food poisoning. Although it lacks a TCA cycle and makes ATP only by fermentation, it has an extraordinary doubling time of only 8 to 10 minutes when in the human host. This is presumably due to the ready supply and assimilation of host macromolecules. *C. botulinum* is the causative agent of botulism, and *C. tetani* causes tetanus. ▶▶| *Botulism (section 38.4); Tetanus (section 38.3)*

C. tetani ferments a number of amino acids to lactate, butyrate, and H_2 (**figure 21.9**). It lacks the genes to catabolize sugars. This is in contrast to *C. acetobutylicum* and *C. perfringens,* which can catabolize both carbohydrates and amino acids. *C. tetani* uses a V-type ATPase and an ATP-binding cassette transporter to couple ATP hydrolysis to the flow of sodium ions out of the cell. This means that some of the ATP made by substrate-level phosphorylation during fermentation is consumed to expel sodium ions. This is not a futile exercise, because it creates a sodium motive force, which in turn drives the amino acid uptake systems. Although *C. tetani* also exports protons via a pyrophosphatase, external protons seem to serve only as the counter ion in Na^+-H^+ antiport. Surprisingly, *C. tetani* possesses genes for a membrane-bound electron transport chain; microbes that make ATP exclusively by fermentation do not generally possess electron transport chains. In this case, it appears that this chain links Na^+-extrusion with oxidation of NADH generated during fermentation. Finally, *C. tetani* possess two Na^+-dependent multidrug efflux systems.

Desulfotomaculum is an endospore-forming genus that reduces sulfate and sulfite to hydrogen sulfide during anaerobic respiration (**figure 21.10**). Although it stains gram negative, electron microscopic studies have shown that *Desulfotomaculum* has a gram-positive–type cell wall. This concurs with phylogenetic studies that place it with the low G + C gram positives.

The heliobacteria are an excellent example of the diversity in this class. The genera *Heliobacterium* and *Heliophilum* are a group of unusual anaerobic, photosynthetic bacteria characterized by the presence of bacteriochlorophyll *g*. They have a photosystem I–type reaction center like the green sulfur bacteria but have no intracytoplasmic photosynthetic membranes; pigments are contained in the plasma membrane (**figure 21.11**). They also differ from most other anoxygenic photosynthetic bacteria because they are unable to grow autotrophically. Pyruvate is used during photoheterotrophic growth in light; it is used during fermentation under dark, anoxic conditions. Like the clostridia, heliobacteria are capable of nitrogen fixation. Although they have a gram-positive cell wall, they have a low peptidoglycan content, and they stain gram negative. Some heliobacteria form endospores. ▶▶| *Light reactions in anoxygenic photosynthesis (section 10.12)*

The phylogenetic placement of the genus *Veillonella* bears mentioning. Although these bacteria stain gram negative, *Bergey's Manual* places them in the family *Acidominococcaceae,* in the order *Clostridiales.* Members of the genus *Veillonella* are anaerobic, chemoheterotrophic cocci ranging in diameter from about 0.3 to 2.5 μm. Usually they are diplococci (often with their adjacent sides flattened), but they may exist as single cells, clusters, or chains. All have complex nutritional requirements and ferment substances such as carbohydrates, lactate and other organic acids, and amino acids to

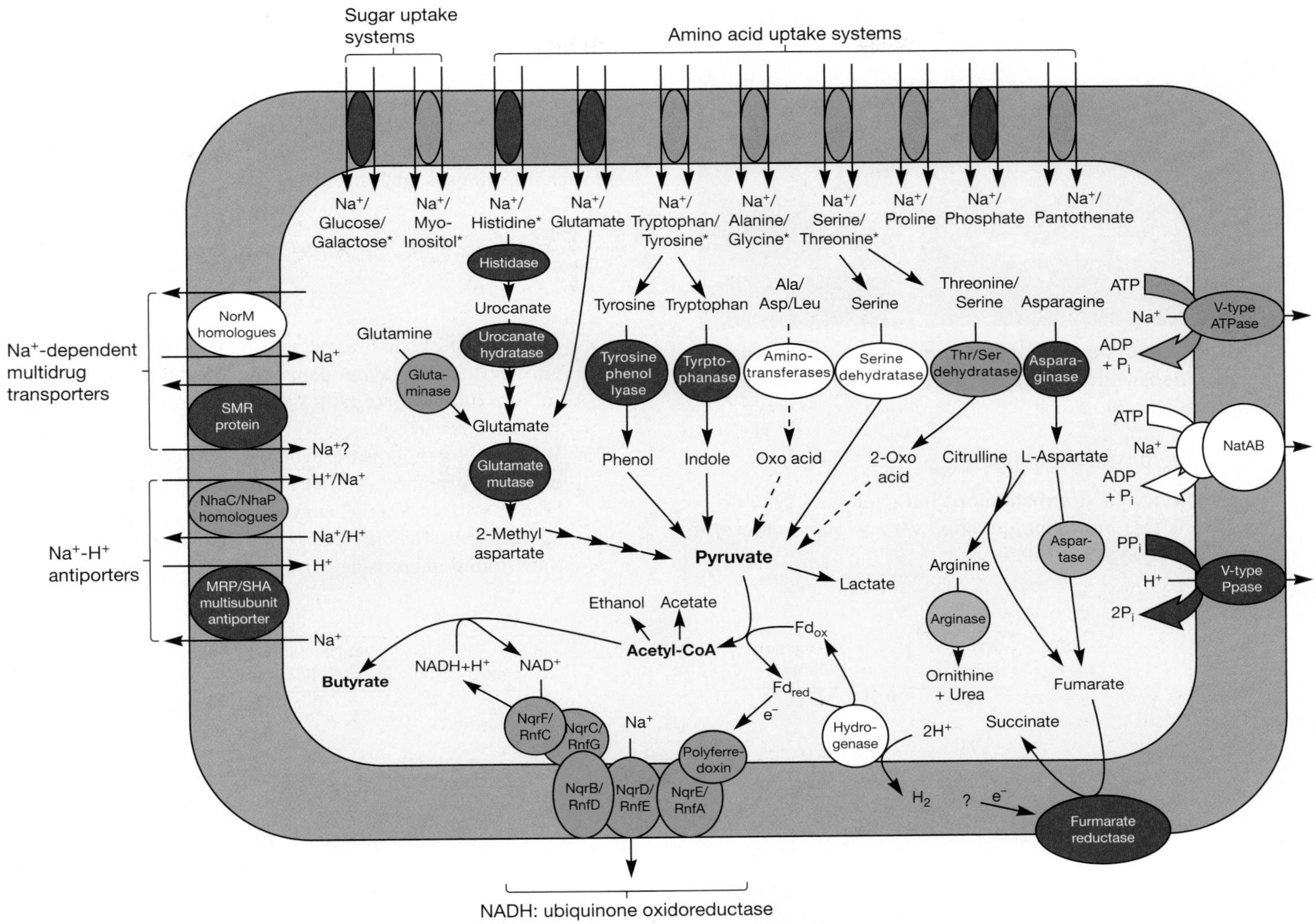

FIGURE 21.9 Amino Acid Fermentation and Na^+ Flux in *Clostridium tetani*. The fermentation of amino acids and the generation of a Na^+ motive force is central to survival of this pathogen. Note that amino acids are transported into the cell by Na^+-dependent uptake systems. Sugars are also transported in this manner but are used in anabolic reactions only. Several Na^+-dependent transporters with unknown substrates are present. A V-type ATPase and an ATP-binding cassette transporter (NatAB) pump Na^+ out of the cell at the expense of ATP hydrolysis. A proton translocating pyrophosphatase (V-type Ppase) is used to export H^+. Components of a membrane-bound electron transport chain are present but, it is thought that the NADH dehydrogenase participates in amino acid fermentation. The lack of quinones suggests that the ETC is not used to establish a proton motive force.

Adapted from Bruggemann, H., et al., 2003. The genome sequence of Clostridium tetani, *the causative agent of tetanus disease.* Proc. Nat. Acad. Sci., USA. *100:1316–21.*

Figure 21.9 Micro Inquiry

The presence of an electron transport chain in a fermentative microbe is unusual. Describe how this microbe uses its ETC and how this differs from a respiratory microbe.

produce gas (CO_2 and often H_2) plus a mixture of volatile fatty acids. They are parasites of homeothermic (warm-blooded) animals.

Like many groups of anaerobic bacteria, members of this genus have not been thoroughly studied. Some species are part of the normal biota of the mouth, the gastrointestinal tract, and the urogenital tract of humans and other animals. For example, *Veillonella* is plentiful on the tongue and dental plaque of humans, and it can be isolated from the vagina. *Veillonella* is unusual in

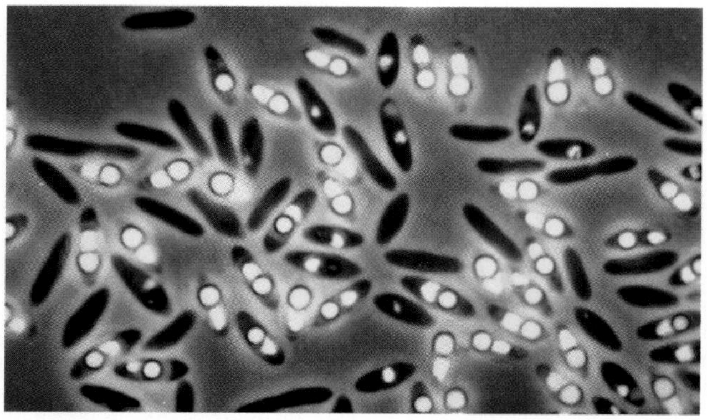

FIGURE 21.10 *Desulfotomaculum.* *Desulfotomaculum acetoxidans* with endospores (bright spheres); phase contrast (×2,000).

Figure 21.10* *Micro Inquiry

In addition to its gram staining characteristics, why do you think *Desulfomatculum* spp. were originally thought to be gram negative microbes?

growing well on organic acids such as lactate, pyruvate, and malate while being unable to ferment glucose and other carbohydrates. It is well adapted to the oral environment because it can use the lactic acid produced from carbohydrates by the streptococci and other oral bacteria. *Veillonella* are found in infections of the head, lungs, and the female genital tract, but their precise role in such infections is unclear.

Search This: Pathema Clostridium

1. List three species of *Clostridium* and why they are notable.
2. Do you think *C. tetani* gains any advantage by using a sodium motive force rather than a proton motive force? Explain your answer.
3. What do you think is the evolutionary significance of the discovery of a photosynthetic genus within the low G + C gram-positive bacteria? Do you think the photosynthetic apparatus in *Heliobacterium* evolved independently of those found in gram-negative anoxygenic phototrophs?

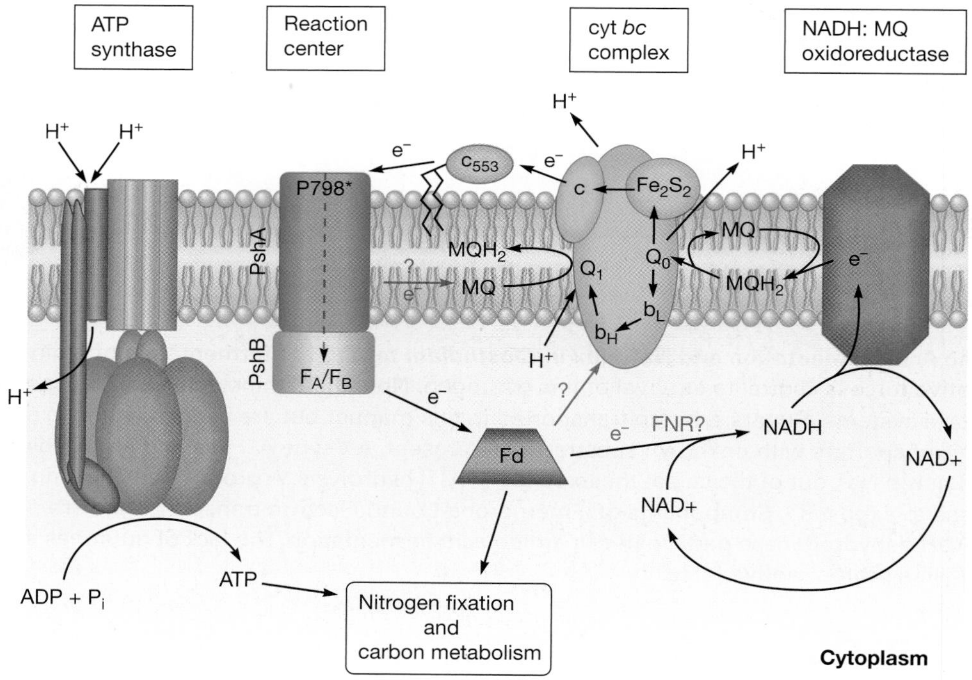

FIGURE 21.11 Electron Transfer in *Heliobacterium modesticaldum.* Genes encoding all subunits of NADH:quinone oxidoreductase are present on the *H. modesticaldum* genome. This suggests that electrons flow from NADH to the NADH:menaquinonone (MQ) oxidoreductase, to cytochrome (cyt) *bc* complex, which in turn reduces a 2Fe-2S subunit and cytochrome b_L. Electrons from cyt *bc* can be transferred via cyt c_{553} to the reaction center (RC) primary electron donor P798. Based on similarities between the heliobacterial RC and photosystem I in other anoxygenic photosynthetic bacteria, it is hypothesized that a proton motive force is generated by cyclic electron flow.

Figure 21.11* *Micro Inquiry

What are the two sources of electrons for the *H. modesticaldum* electron transport chain?

21.3 Class *Bacilli*

The second edition of *Bergey's Manual* gathers a large variety of gram-positive bacteria into one class, *Bacilli,* and two orders, *Bacillales* and *Lactobacillales.* These orders contain 17 families and over 70 gram-positive genera representing cocci, endospore-forming rods and cocci, and nonsporing rods. The biology of some members of the order *Bacillales* is described first, then important representatives of the order *Lactobacillales* are considered. The phylogenetic relationships between some of these organisms are shown in figure 21.2, and the characteristics of selected genera are summarized in **table 21.3**.

Table 21.3 Characteristics of Members of the Class *Bacilli*

Genus	*Dimensions (μm), Morphology, and Motility*	*G + C Content (mol%)*	*Genome Size (Mb)*	*Oxygen Relationship*	*Other Distinctive Characteristics*
Bacillus	0.5–2.5 × 1.2–10; straight rods, peritrichous flagella, spore-forming	32–69	4.2–5.4	Aerobic or facultative	Catalase positive; chemoorganotrophic
Caryophanon	1.5–3.0 × 10–20; multicellular rods with rounded ends, peritrichous flagella, nonsporing	41–46	Nd[a]	Aerobic	Acetate only major carbon source; catalase positive; trichome cells have greater width than length; trichomes can be in short chains
Enterococcus	0.6–2.0 × 0.6–2.5; spherical or ovoid cells in pairs or short chains, nonsporing, sometimes motile	34–42	3.2	Facultative	Ferments carbohydrates to lactate with no gas; complex nutritional requirements; catalase negative; occurs widely, particularly in fecal material
Lactobacillus	0.5–1.2 × 1.0–10; usually long, regular rods, nonsporing, rarely motile	32–53	1.9–3.3	Facultative or microaerophilic	Fermentative, at least half the end product is lactate; requires rich, complex media; catalase and cytochrome negative
Lactococcus	0.5–1.2 × 0.5–1.5; spherical or ovoid cells in pairs or short chains, nonsporing, nonmotile	38–40	2.4	Facultative	Chemoorganotrophic with fermentative metabolism; lactate without gas produced; catalase negative; complex nutritional requirements; in dairy and plant products
Leuconostoc	0.5–0.7 × 0.7–1.2; cells spherical or ovoid, in pairs or chains; nonmotile and nonsporing	38–44	Nd	Facultative	Requires fermentable carbohydrate and nutritionally rich medium for growth; fermentation produces lactate, ethanol, and gas; catalase and cytochrome negative
Staphylococcus	0.9–1.3; spherical cells occurring singly and in irregular clusters, nonmotile and nonsporing	30–39	2.5–2.8	Facultative	Chemoorganotrophic with both respiratory and fermentative metabolism; usually catalase positive; associated with skin and mucous membranes of vertebrates
Streptococcus	0.5–2.0; spherical or ovoid cells in pairs or chains, nonmotile and nonsporing	34–46	1.8–2.2	Facultative	Fermentative, producing mainly lactate and no gas; catalase negative; commonly attacks red blood cells (α- or β-hemolysis); complex nutritional requirements; commensals or parasites on animals
Thermoactinomyces	0.4–1.0 in diameter; branched, septate mycelium resembles those of actinomycetes	52–54.8	Nd	Aerobic	Usually thermophilic; true endospores form singly on hyphae; numerous in decaying hay, vegetable matter, and compost

[a] Nd: Not determined; genome not yet sequenced.

Order *Bacillales*

The genus *Bacillus,* family *Bacillaceae,* is the largest in the order *Bacillales.* The genus contains endospore-forming, chemoheterotrophic rods that are usually motile with peritrichous flagella (**figure 21.12**). It is aerobic, or sometimes facultative, and catalase positive. Many species once included in this genus have been placed in other families and genera based on rRNA sequence data. Some examples of organisms that were formerly in the genus *Bacillus* are *Paenibacillus alvei, P. macerans,* and *P. polymyxa.*

Bacillus subtilis, the type species for the genus, is the most well-studied gram-positive bacterium. It is a facultative anaerobe that can use nitrate as a terminal electron acceptor or perform mixed acid fermentation with lactate, acetate, and acetoin as major end products. It is nonpathogenic and a terrific model organism for the study of gene regulation, cell division, quorum sensing, and cellular differentiation. Its 4.2-Mb genome was one of the first genomes to be completely sequenced and reveals a number of interesting elements. For instance, several families of genes have been expanded by gene duplication; the largest such family encodes ABC transporters—the most frequent class of protein in *B. subtilis.* There are 18 genes that encode sigma factors. Recall that the use of alternative sigma subunits of RNA polymerase is one way in which bacteria regulate gene expression. In this case, many of the sigma factors govern spore formation and other responses to stressful conditions. The genome contains genes for the catabolism of many diverse carbon sources, antibiotic synthesis, and natural transformation competence. There are at least 10 integrated prophages or remnants of prophages. ◀◀ *Bacterial endospores (section 3.8); Protein maturation and secretion (section 12.9); Sporulation in* Bacillus subtilis *(section 13.6); Comparative genomics (section 16.7)*

B. subtilis is a soil-dwelling microbe whose spores are easily extracted from a diverse collection of soils. Interestingly, it is often found in its vegetative state (i.e., not as a spore) when growing in association with actively decaying material and in close association with plant roots. In addition to differentiating into spores, *B. subtilis* also develops biofilms (**figure 21.13**). The vast wealth of genetic information regarding *B. subtilis* has made it a model for the study of these important structures.

Many species of *Bacillus* are of considerable importance. Some produce the antibiotics bacitracin, gramicidin, and polymyxin. *B. cereus* (figure 21.12*b*) causes some forms of food poisoning. Several species are used as insecticides. For example, *B. thuringiensis* and *B. sphaericus* form a solid protein crystal, the **parasporal body,** next to their endospores during spore formation (**figure 21.14**). As discussed in chapter 41, the *B. thuringiensis* parasporal body contains protein toxins that kill over 100 species of moths by dissolving in the alkaline gut of caterpillars and destroying the epithelium. The *B. sphaericus* parasporal body contains proteins toxic for mosquito larvae and may be useful in controlling the mosquitoes that carry the malaria parasite *Plasmodium. B. anthracis* is the causative agent of the disease anthrax, which can affect both farm animals and humans. ▶▶ *Anthrax (section 38.5); Microbes as products (section 41.5)*

The genus *Thermoactinomyces* has historically been classified as an actinomycete because its soil-associated substrate hyphae diffentiate into upwardly growing aerial hyphae. However, phylogenetic analysis places it with the low G + C microbes in the order *Bacillales,* family *Thermoactinomycetaceae.* Its G + C content (52–55 mol%) is considerably lower than that of the *Actinobacteria.* Also unlike the *Actinobacteria, Thermoactinomyces* species form true endospores within both the aerial and substrate hyphae (**figure 21.15**). These spores have a typical endospore structure, including the presence of calcium and dipicolinic acid, and can survive at 90°C for 30 minutes (figure 21.15*b*). This genus is thermophilic and grows between 45 and 60°C. *Thermoactinomyces* is commonly found in damp haystacks, compost piles, and other high-temperature habitats. *Thermoactinomyces vulgaris* from haystacks, grain storage silos, and compost piles is a causative agent of farmer's lung, an allergic disease of the respiratory system. Spores from *Thermoactinomyces vulgaris* were recovered from the

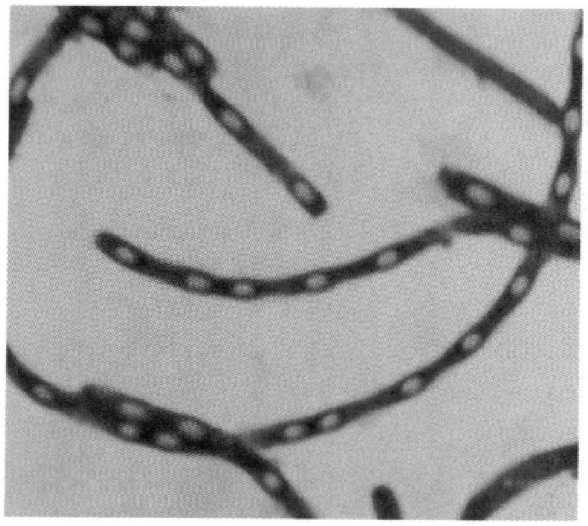

(a) *Bacillus anthracis*

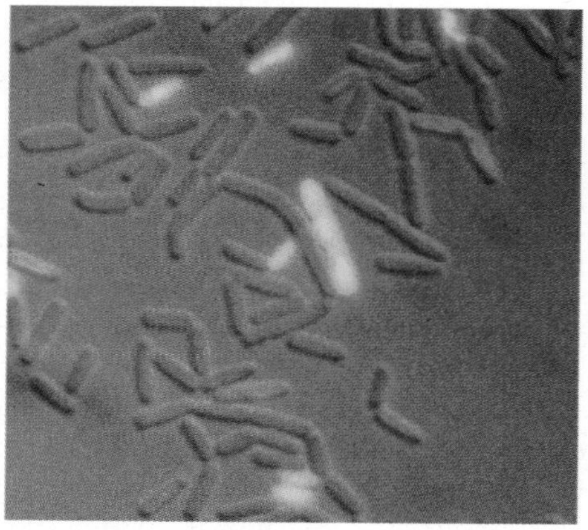

(b) *B. cereus*

FIGURE 21.12 ***Bacillus.*** (a) *B. anthracis* endospores are elliptical and central (×1,600). (b) *B. cereus* stained with SYTOX Green nucleic acid stain and viewed by epifluorescence and differential interference contrast microscopy. The cells that glow green are dead.

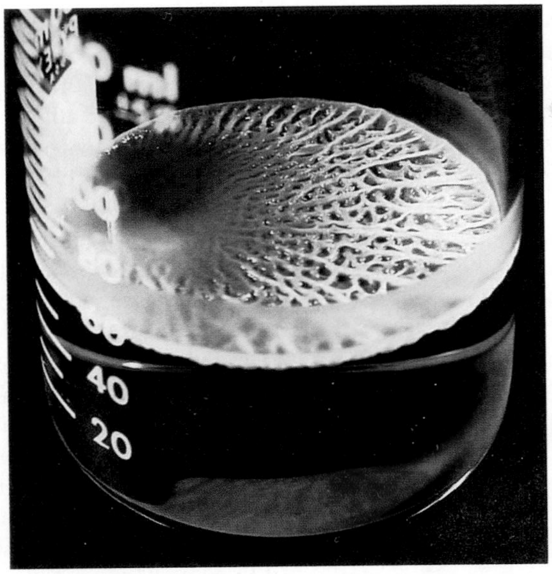

(a) Biofilm formation

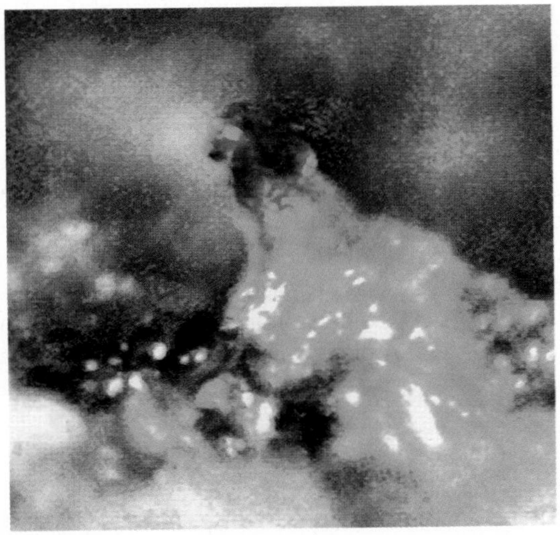

(b) Fruiting body formation

FIGURE 21.13 Biofilm Formation by *Bacillus subtilis.* (a) When grown for several days without agitation in nutrient limiting medium, *B. subtilis* forms a hydrophobic biofilm or pellicle at the aqueous-air interface. (b) *B. subtilis* forms multicellular structures, or fruiting bodies, when grown on minimal salts agar for several days. These cells are labeled so that spores turn blue, demonstrating that spores are abundant at the tips of these structures. Only *B. subtilis* isolates freshly obtained from soils produce fruiting bodies, suggesting that prolonged growth in the laboratory selects for mutations that lead to the loss of this developmental process.

Figure 21.13 Micro Inquiry

What growth or survival advantage might this type of multicellular behavior confer to *B. subtilis?*

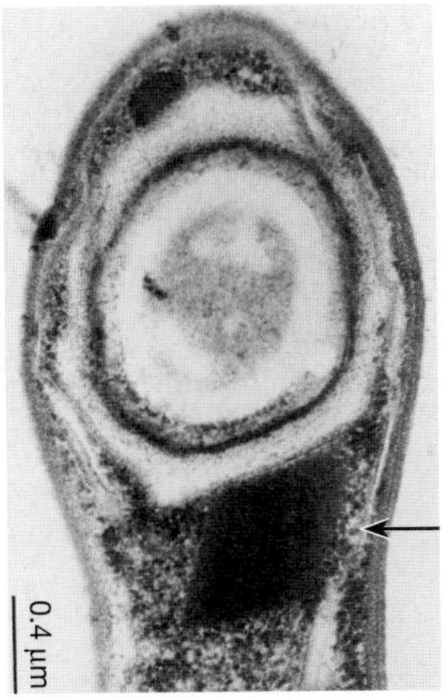

(a)

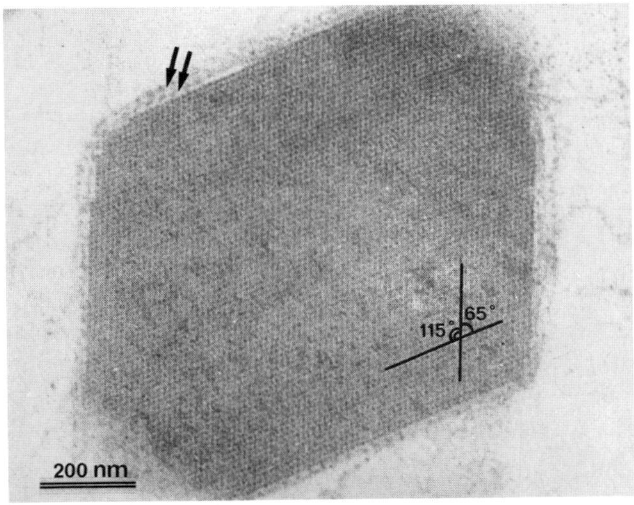

(b)

FIGURE 21.14 The Parasporal Body. (a) An electron micrograph of a *B. sphaericus* sporulating cell containing a parasporal body (arrow) just beneath the endospore. (b) The crystalline parasporal body at a higher magnification. The crystal is surrounded by a two-layered envelope (arrows).

mud of a Minnesota lake and found to be viable after about 7,500 years of dormancy.

Search This: Bacillus research community

One other member within the order *Bacillales* bears mentioning: *Sporosarcina.* This microbe is one of five genera in the family *Panococcaceae* and the only known endospore-forming bacterium that has a coccoid rather than rod shape. *S. ureae* is particularly interesting because it is tolerant of very alkaline conditions (up to pH 10), which it creates when it degrades urea to CO_2 and NH_3. Interestingly, it is easily isolated from agricultural soils on which animals frequently urinate.

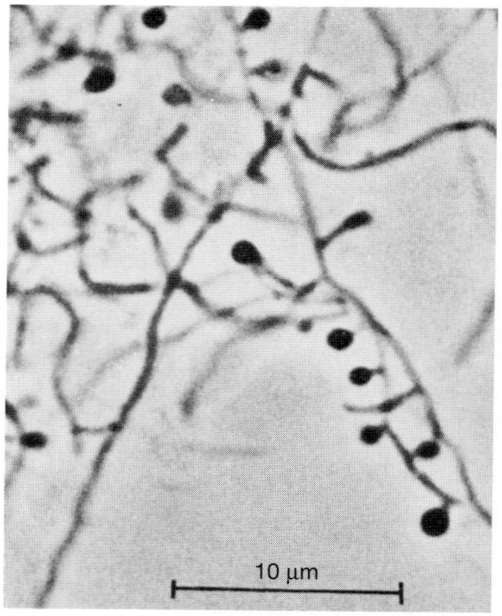

(a)

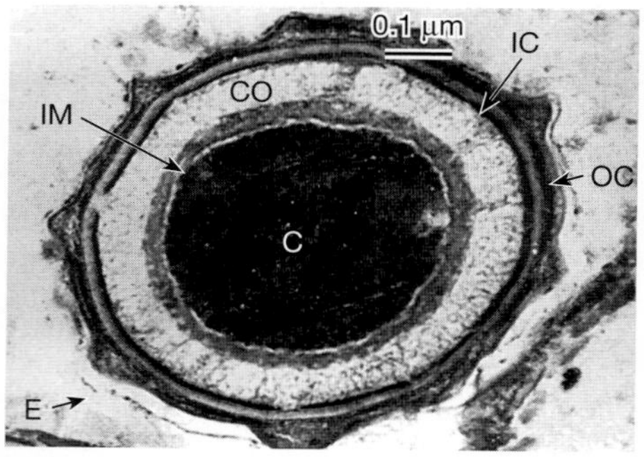

(b)

FIGURE 21.15 *Thermoactinomyces.* (a) *Thermoactinomyces vulgaris* aerial mycelium with developing endospores at tips of hyphae. (b) Thin section of a *T. sacchari* endospore. E, exosporium; OC, outer spore coat; IC, inner spore coat; CO, cortex; IM, inner forespore membrane; C, core.

Family *Staphylococcaceae*

The family *Staphylococcaceae* contains five genera, the most important of which is the genus *Staphylococcus.* Members of this genus are facultatively anaerobic, nonmotile, gram-positive cocci, 0.5 to 1.5 μm in diameter, occurring singly, in pairs, and in tetrads, and characteristically dividing in more than one plane to form irregular clusters (**figure 21.16**). They are usually catalase positive and oxidase negative. They respire using oxygen as the terminal electron acceptor, although some can reduce nitrate to nitrite. They are also capable of fermentation and convert glucose principally to lactate.

Staphylococci are normally associated with the skin, skin glands, and mucous membranes of warm-blooded animals. They are responsible for many human diseases. *S. epidermidis* is a common skin resident that is sometimes responsible for endocarditis and infections of patients with lowered resistance (e.g., wound, surgical, urinary tract, and body piercing infections). *S. aureus* is the most important human staphylococcal pathogen and causes boils, abscesses, wound infections, pneumonia, toxic shock syndrome, and other diseases. Strains of **methicillin-resistant *Staphylococcus aureus* (MRSA)** and vancomycin-resistant *S. aureus* (VRSA) are among the most threatening antibiotic-resistant pathogens known. Vancomycin is considered the "drug of last resort," and infections caused by vancomycin-resistant *S. aureus* generally cannot be treated by antibiotic therapy. How did *S. aureus* evolve the capacity to defeat a diverse array of antibiotics in just 50 years or so? Comparative genome analysis of MRSA strains with antibiotic-sensitive *S. aureus* strains shows that this microbe is very adept at acquiring genetic elements from other bacteria. In fact, large, mobile genetic elements appear to encode both antibiotic-resistance factors and proteins that increase virulence.

In addition to causing skin and wound infections, *S. aureus* is a major cause of food poisoning. For instance, several

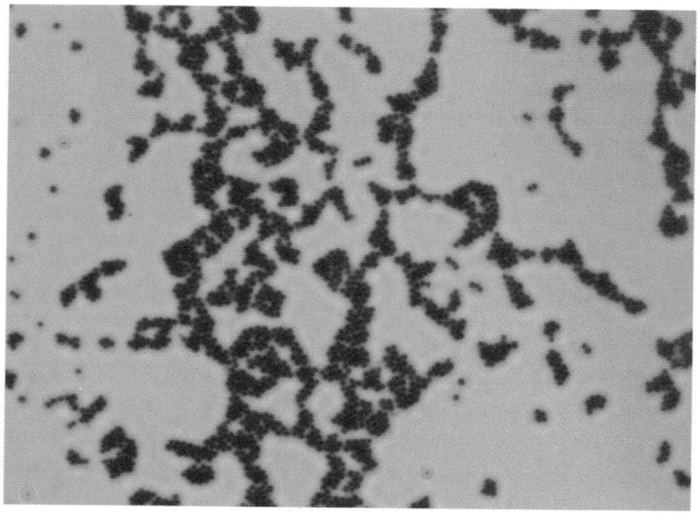
(a)

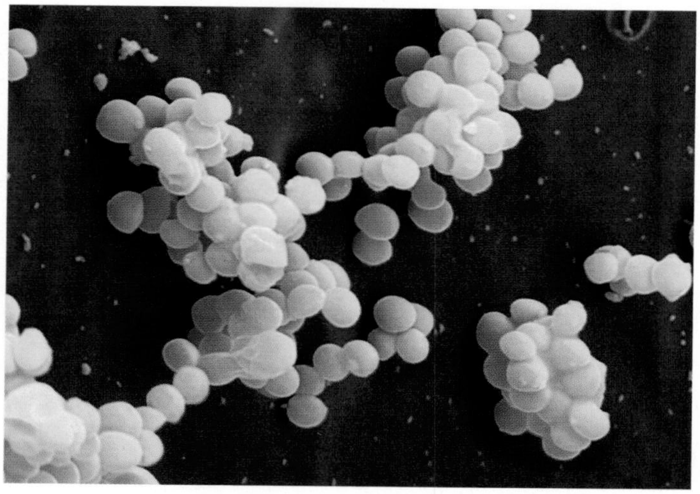
(b)

FIGURE 21.16 *Staphylococcus.* (a) *Staphylococcus aureus,* gram-stained smear (×1,500). (b) *S. aureus* cocci arranged like clusters of grapes; color-enhanced scanning electron micrograph. Each cell is about 1 μm in diameter.

years ago in Texas, 1,364 elementary school children were sickened by tainted chicken. A food service worker responsible for deboning the chicken was the most likely source of contamination. After the chicken was deboned, it was cooled to room temperature before further processing and refrigeration. This provided sufficient time for the bacteria to grow and produce toxins. This case is not uncommon. Poultry accounts for about one-quarter of all bacterial food poisoning cases in which the source of poisoning is known, and it must be cooked and handled carefully. ▶▶| *Controlling food spoilage (section 40.2)*

One of the virulence factors produced by *S. aureus* is the enzyme coagulase, which causes blood plasma to clot. Growth and hemolysis patterns on blood agar are also useful in identifying these staphylococci. The structure of staphylococcal β-hemolysin has been determined. The toxin lyses a cell by forming solvent-filled channels in its plasma membrane. Water-soluble toxin monomers bind to the cell surface and associate with each other to form pores. The hydrophilic channels then allow free passage of water, ions, and small molecules. ▶▶| *Staphylococcal diseases (section 38.3)*

Family *Listeriaceae*

The family *Listeriaceae* includes two genera: *Brochothrix* and *Listeria. Brochothrix* is most commonly found in meat but is not pathogenic. *Listeria* is the medically important genus in this family. The genus contains short rods that are facultatively anaerobic, catalase positive, and motile by peritrichous flagella. In addition to aerobic respiration, these microbes ferment glucose mainly to lactate. *Listeria* is widely distributed in nature, particularly in decaying matter. *L. monocytogenes* is a pathogen of humans and other animals and causes listeriosis, an important food infection. ▶▶| *Types of food-borne disease (section 40.3)*

Order *Lactobacillales*

Many members of the order *Lactobacillales* produce lactic acid as their major or sole fermentation product and are sometimes collectively called **lactic acid bacteria (LAB).** *Streptococcus, Enterococcus, Lactococcus, Lactobacillus,* and *Leuconostoc* are all members of this group. Lactic acid bacteria do not form endospores and are usually nonmotile. They depend on sugar fermentation for energy. They lack cytochromes and obtain energy by substrate-level phosphorylation, rather than by electron transport and oxidative phosphorylation. Nutritionally, they are fastidious, and many vitamins, amino acids, purines, and pyrimidines must be supplied because of their limited biosynthetic capabilities. Lactic acid bacteria usually are categorized as facultative anaerobes, but some classify them as aerotolerant anaerobes. |◀◀ *Oxygen concentration (section 7.6); Fermentation (section 10.7)*

The largest genus in this order is *Lactobacillus,* with about 100 species. *Lactobacillus* includes rods and some coccobacilli. All lack catalase and cytochromes, and produce lactic acid as their main or sole fermentation product (*see figure 10.19*). Lactobacilli carry out either **homolactic fermentation** using the Embden-Meyerhof pathway or **heterolactic fermentation** with the phosphoketolase pathway (**figure 21.17**). They grow optimally under slightly acidic conditions, at a pH between 4.5 and 6.4. The genus is found on plant surfaces and in dairy products, meat, water, sewage, beer, fruits, and many other materials.

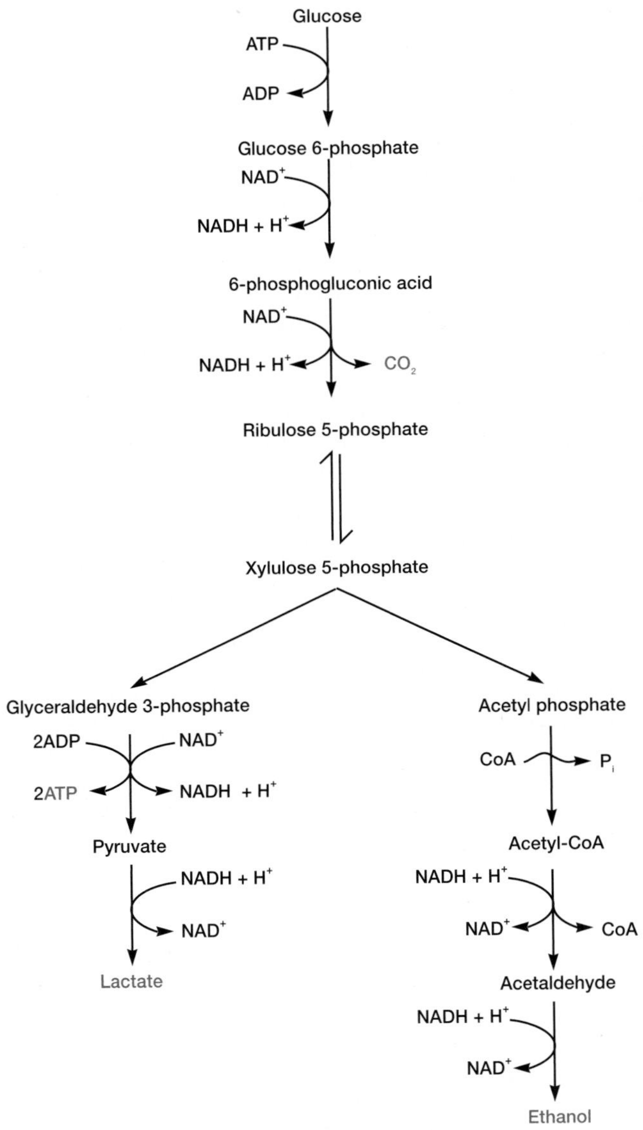

FIGURE 21.17 Heterolactic Fermentation and the Phosphoketolase Pathway. The phosphoketolase pathway converts glucose to lactate, ethanol, and CO_2.

Figure 21.17 Micro Inquiry

What is the net yield of ATP for each glucose that enters the phosphoketolase pathway?

Lactobacilli also are part of the normal flora of the human body in the mouth, intestinal tract, and vagina. They are rarely pathogenic.

Lactobacillus is indispensable to the food and dairy industry. Lactobacilli are used in the production of fermented vegetable foods (e.g., sauerkraut, pickles, silage), sourdough bread, Swiss and other hard cheeses, yogurt, and sausage. Yogurt is probably the most popular fermented milk product in the United States and, as described in chapter 40, *Streptococcus thermophilus* and *Lactobacillus delbrueckii* subspecies *bulgaricus* are used in its production. ▶▶| *Fermented milks (section 40.5)*

At least one species, *L. acidophilus,* is sold commercially as a probiotic agent that may provide some health benefits for the consumer. On the other hand, some lactobacilli also create problems. They sometimes are responsible for spoilage of beer, wine, milk, and meat because their metabolic end products contribute undesirable flavors and odors.

Leuconostoc, family *Leuconostocaceae,* contains facultative gram-positive cocci, which may be elongated or elliptical and arranged in pairs or chains (**figure 21.18**). Leuconostocs lack catalase and cytochromes and carry out heterolactic fermentation by converting glucose to D-lactate and ethanol or acetic acid by means of the phosphoketolase pathway (figure 21.17). They can be isolated from plants, silage, and milk. The genus is used in wine production, in the fermentation of vegetables such as cabbage (sauerkraut) and cucumbers (pickles), and in the manufacture of buttermilk, butter, and cheese. *L. mesenteroides* synthesizes dextrans from sucrose and is important in industrial dextran production. *Leuconostoc* species are involved in food spoilage and tolerate high sugar concentrations so well that they grow in syrup and are a major problem in sugar refineries.

Enterococcaceae and *Streptococcaceae* are important families of chemoheterotrophic, mesophilic, nonsporing cocci. They occur in pairs or chains when grown in liquid media (**figure 21.19** and **table 21.4**) and usually are nonmotile. They are unable to respire and have a strictly fermentative metabolism. They ferment sugars to produce lactic acid but no gas; that is, they carry out homolactic fermentation. A few species are anaerobic rather than aerotolerant. The enterococci such as *E. faecalis* are normal residents of the intestinal tracts of humans and most other animals. *E. faecalis* is an opportunistic pathogen that can cause urinary tract infections and endocarditis. Enterococci are major agents in the horizontal transfer of antibiotic-resistance genes.

The family *Streptococcaceae* includes only two genera: *Lactococcus* and *Streptococcus. Lactococcus* spp. ferment sugars to lactic acid and can grow at 10°C but not at 45°C. *L. lactis* is widely used in the production of buttermilk and cheese because it can curdle milk and add flavor through the synthesis of diacetyl and other products.

The genus *Streptococcus* is large and complex. All species in the genus are facultatively anaerobic and catalase negative. Many characteristics are used to identify these cocci. One of their most important taxonomic characteristics is the ability to lyse erythrocytes when growing on blood agar, an agar medium containing 5% sheep or horse blood (**figure 21.20**). In **α-hemolysis,** a 1-to-3 mm greenish zone of incomplete hemolysis forms around the colony; **β-hemolysis** is characterized by a clear zone of complete lysis. In addition, other hemolytic patterns are sometimes seen. Serological studies are also very important in identification because these genera often have distinctive cell wall antigens. Polysaccharide and teichoic acid antigens found in the cell wall or between the wall and the plasma membrane are used to identify these cocci, particularly pathogenic β-hemolytic streptococci, by the **Lancefield grouping system.** Biochemical and

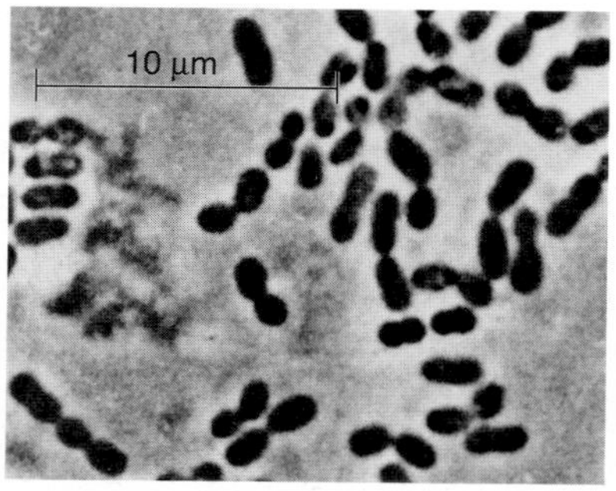

FIGURE 21.18 ***Leuconostoc.*** *Leuconostoc mesenteroides*; phase-contrast micrograph.

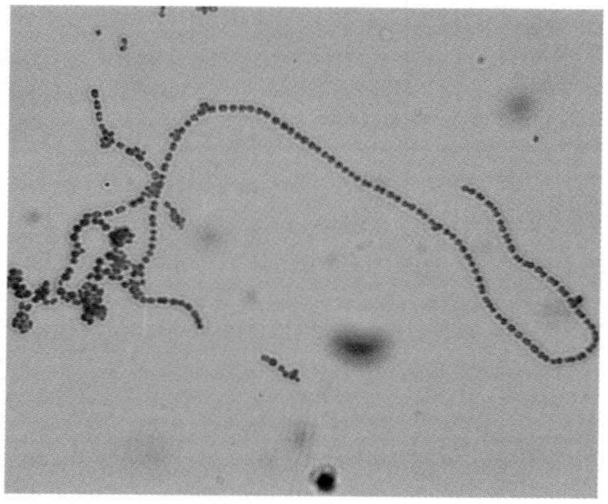

(a) *Streptococcus pyogenes*

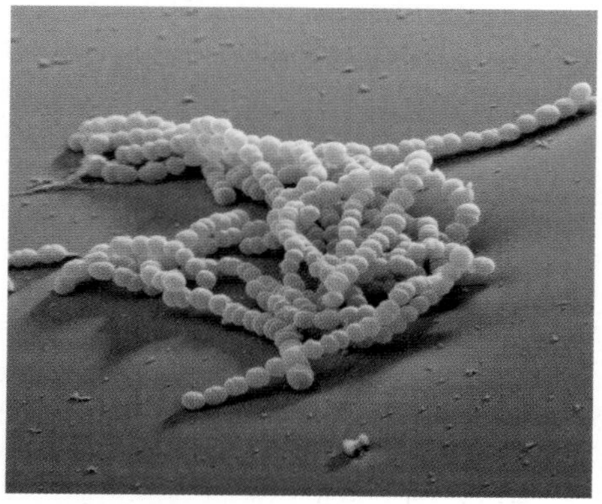

(b) *S. agalactiae*

FIGURE 21.19 ***Streptococcus.*** (a) *Streptococcus pyogenes* (×900). (b) *Streptococcus agalactiae,* the cause of Group B streptococcal infections. Note the long chains of cells; color–enhanced scanning electron micrograph (×4,800).

Table 21.4 Classification of the Streptococci, Entercocci, and Lactococci

Characteristics	*Streptococcus*	*Enterococcus*	*Lactococcus*
Predominant arrangement (most common first)	Chains, pairs	Pairs, chains	Pairs, short chains
Capsule/slime layer	+	–	–
Habitat	Mouth, respiratory tract	Gastrointestinal tract	Dairy products
Growth at 45°C	Variable	+	–
Growth at 10°C	Variable	Usually +	+
Growth at 6.5% NaCl broth	Variable	+	–
Growth at pH 9.6	Variable	+	–
Hemolysis	Usually β (pyogenic) or α (oral)	α, β, –	Usually –
Serological group (Lancefield)	Variable (A–O)	Usually D	Usually N
Mol% G + C	34–46	34–42	38–40
Representative species	Pyogenic streptococci	*E. faecalis*	*L. lactis*
	S. agalactiae, S. pyogenes, S. equi, S. dysgalactiae	*E. faecium*	*L. raffinolactis*
	Oral streptococci	*E. avium*	*L. plantarum*
	S. gordonii, S. salivarius, S. sanguis, S. oralis	*E. durans*	
	S. pneumoniae, S. mitis, S. mutans	*E. gallinarum*	
	Other streptococci		
	S. bovis, S. thermophilus		

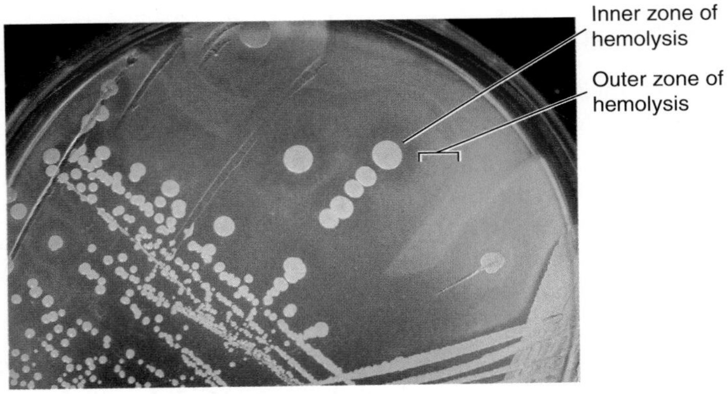

(a)

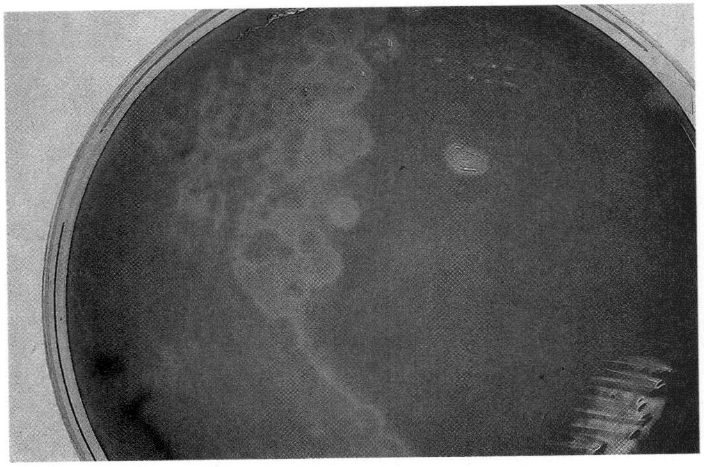

(b)

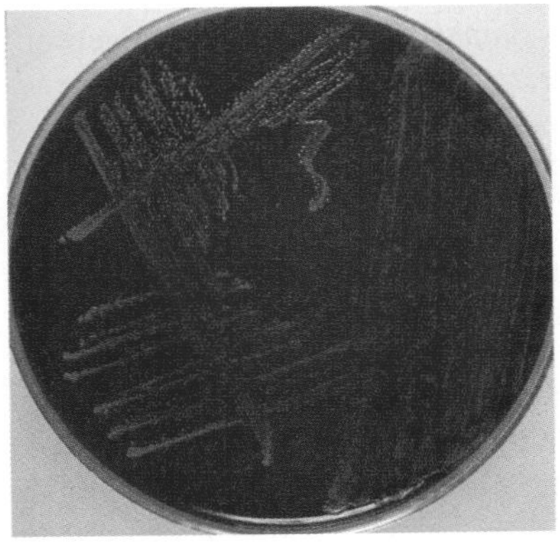

(c)

FIGURE 21.20 Staphylococcal and Streptococcal Hemolytic Patterns. (a) *Staphylococcus aureus* on blood agar, illustrating β-hemolysis. (b) *Streptococcus pneumoniae* on blood agar, illustrating α-hemolysis. (c) *Staphylococcus epidermidis* on blood agar with no hemolysis.

Figure 21.20 Micro Inquiry

Why is the cleared zone green following α-hemolysis?

physiological tests are essential in identification (e.g., growth temperature preferences, carbohydrate fermentation patterns, acetoin production, reduction of litmus milk, sodium chloride and bile salt tolerance, and the ability to hydrolyze arginine, esculin, hippurate, and starch). Sensitivity to bacitracin, sulfa drugs, and optochin (ethylhydrocuprein) also are used to identify particular species. Some of these techniques are being replaced by molecular genetic approaches such as multilocus sequence typing (MLST). ▶▶| *Clinical microbiology and immunology (chapter 35)* |◀◀ *Techniques for determining microbial taxonomy and phylogeny (section 17.3)*

Many bacteria originally placed within the genus have been moved to two other genera, *Enterococcus* and *Lactococcus.* The streptococci have been divided into three groups: pyogenic streptococci, oral streptococci, and other streptococci. Pyogenic streptococci are pathogens associated with pus formation (*pyogenic* means pus producing). Most species are β-hemolytic on blood agar and form chains of cells. The major human pathogen in this group is *S. pyogenes*, which causes streptococcal sore throat, acute glomerulonephritis, and rheumatic fever. The normal habitat of oral streptococci is the oral cavity and upper respiratory tract of humans and other animals. In other respects, oral streptococci are not necessarily similar. *S. mutans* is associated with the formation of dental caries. *S. pneumoniae* is an example of the "other" group. It is α-hemolytic and grows as pairs of cocci. It is associated with lobar pneumonia and otitis media (inflammation of the middle ear) in young children. ▶▶| *Streptococcal diseases (section 38.3)*

1. List the major properties of the genus *Bacillus.* Why do you think this bacterium has become a model organism?
2. Briefly describe the genus *Thermoactinomyces,* with particular emphasis on its unique features. What disease does it cause?
3. Describe the genus *Staphylococcus.* On what basis are pathogenic and nonpathogenic species and strains distinguished?
4. List the major properties of the genus *Lactobacillus.* In what ways is it important in the food and dairy industries?
5. Describe the major distinguishing characteristics of the following taxa: *Streptococcus, Enterococcus, Lactococcus,* and *Leuconostoc.*
6. Of what practical importance is *Leuconostoc?* What are lactic acid bacteria?
7. What are α-hemolysis, β-hemolysis, and the Lancefield grouping system?

Summary

21.1 Class *Mollicutes* (The Mycoplasmas)

a. Mycoplasmas stain gram-negative because they lack cell walls and cannot synthesize peptidoglycan precursors; however, they are phylogenetically related to the low G + C gram-positive bacteria (**figures 21.2** and **21.3**).

b. Many mycoplasmas require sterols for growth. They are some of the smallest bacteria capable of self-reproduction and usually grow on agar to give colonies a "fried egg" appearance (**figure 21.4**).

21.2 Class *Clostridia*

a. Members of the genus *Clostridium* are anaerobic gram-positive rods that form endospores and do not carry out dissimilatory sulfate reduction. They are responsible for food spoilage, botulism, tetanus, and gas gangrene. They have a fermentative metabolism (**figures 21.7** to **21.9**).

b. *Desulfotomaculum* is an anaerobic, endospore-forming genus that reduces sulfate to sulfide during anaerobic respiration.

c. The heliobacteria are anaerobic, photosynthetic bacteria with bacteriochlorophyll *g*. Some form endospores (**figure 21.11**).

d. The family *Veillonellaceae* contains anaerobic cocci that stain gram negative. Some are parasites of warm-blooded animals.

21.3 Class *Bacilli*

a. The class *Bacilli* is divided into two orders: *Bacillales* and *Lactobacillales.*

b. The genus *Bacillus* contains aerobic and facultative, catalase-positive, endospore-forming, chemoheterotrophic rods that are usually motile with peritrichous flagella. *Bacillus subtilis* is a model organism that forms endospores and multicellular structures such as biofilms and fruiting bodies (**figure 21.12**).

c. *Thermoactinomyces* is a thermophile that forms a mycelium and true endospores (**figure 21.15**). It causes allergic reactions and leads to farmer's lung.

d. Members of the genus *Staphylococcus* are facultatively anaerobic, nonmotile, gram-positive cocci that form irregular clusters. They carry out aerobic respiration; some also ferment and use nitrate as terminal electron acceptor (**figure 21.16**).

e. Several important genera such as *Lactobacillus* and *Listeria* contain regular, nonsporing, rods. *Lactobacillus* carries out lactic acid fermentation and is extensively used in the food and dairy industries. *Listeria* is an important agent in food poisoning. *Leuconostoc* carries out heterolactic fermentation using the phosphoketolase pathway (**figure 21.17**) and is involved in the production of fermented vegetable products, buttermilk, butter, and cheese.

f. The genera *Streptococcus, Enterococcus,* and *Lactococcus* contain cocci arranged in pairs and chains that are usually facultative and carry out homolactic fermentation (**table 21.4**). Some important species are the pyogenic coccus *S. pyogenes, S. pneumoniae,* the enterococcus *E. faecalis,* and the lactococcus *L. lactis.*

Critical Thinking Questions

1. Many low G + C bacteria are parasitic. The dependence on a host might be a consequence of the low G + C content. Elaborate on this concept.
2. How might one go about determining whether the genome of *M. genitalium* is the smallest compatible with a parasitic existence?
3. Account for the ease with which anaerobic clostridia can be isolated from soil and other generally aerobic niches.
4. Cells of *Bacillus subtilis* that are committed to sporulate secrete a signaling protein that prevents sister cells from entering sporulation and a killing factor that causes them to lyse. The nutrients released by these cells are consumed by the "killer" cells, providing the nutrients needed to complete spore formation. It has been proposed that this provides evidence that sporulation is a "last resort" for survival by this microbe. Do you agree that the capacity to cannibalize sister cells is evidence that the *B. subtilis* uses spore formation as last resort? Explain your answer, being careful not to project human terms onto the microbes (i.e., anthropomorphize).

   Read the original papers: Gonzalez-Paster, E., et al. 2003. Cannibalism by sporulating bacteria. *Science* 301:510; and Ellermeier, C. D., et al. 2006. A three-protein signaling pathway governing immunity to a bacterial cannibalism toxin. *Cell* 124:549.
5. *Staphylococcus aureus* is an important source of nosocomial (hospital acquired) infections. Recent evidence suggests that O_2 levels in the site of infection influences expression of virulence genes. For instance, abscesses can be completely anaerobic. These observations prompted Fuchs and his colleagues to explore the anaerobic physiology of *S. aureus* using a proteomics approach (*see p. 433*). They determined that in the absence of O_2 or NO_3^-, TCA cycle enzymes were diminished, while the following enzymes were upregulated: glycolytic enzymes, lactate dehydrogenase, alcohol dehydrogenases, acetolactate synthase, and acetoin reductase. Refer to chapter 10 (*see figures 10.5 and 10.17*) to determine how *S. aureus* makes ATP and reoxidizes NADH under anaerobic conditions in the absence of nitrate.

   Read the original paper: Fuchs, S., et al. 2007. Anaerobic gene expression in *Staphylococcus aureus. J. Bacteriol.* 189:4275.

Learn More

Learn more by visiting the text website at www.mhhe.com/willey8, where you will find a complete list of references.

22

Bacteria: *The High G + C Gram Positives*

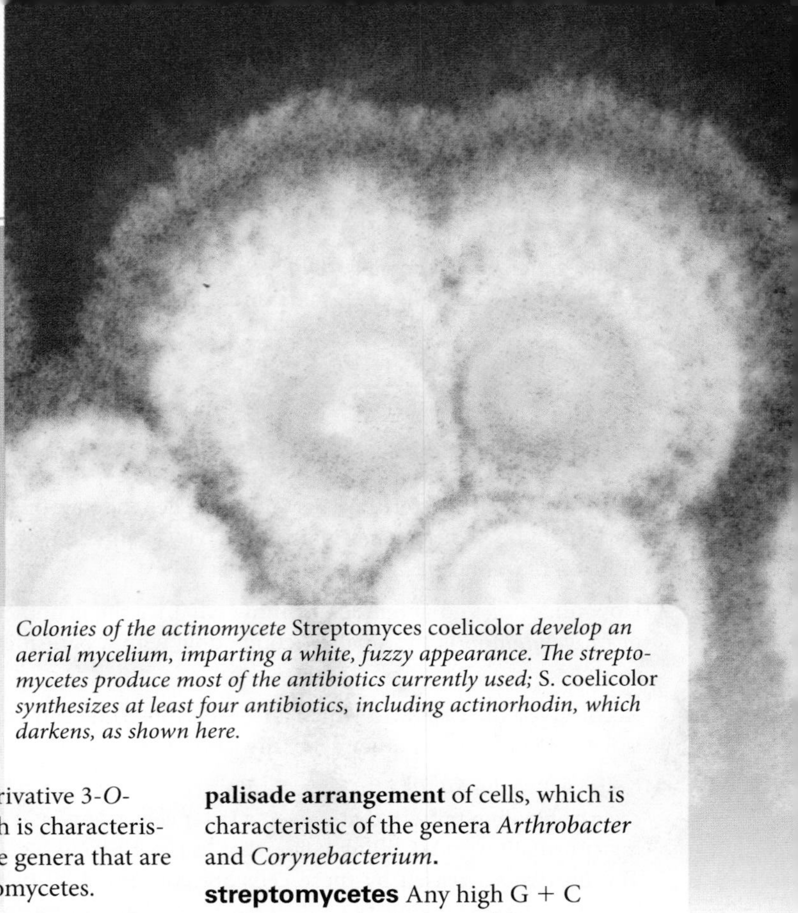

Colonies of the actinomycete Streptomyces coelicolor *develop an aerial mycelium, imparting a white, fuzzy appearance. The streptomycetes produce most of the antibiotics currently used;* S. coelicolor *synthesizes at least four antibiotics, including actinorhodin, which darkens, as shown here.*

CHAPTER GLOSSARY

actinobacteria A general term used to refer to all high G + C gram-positive bacteria that are placed in the phylum *Actinobacteria.*

actinomycetes High G + C gram-positive, aerobic bacteria that produce filamentous hyphae and differentiate to produce asexual spores.

aerial mycelium A mat of hyphae formed by actinomycetes that grows above the substrate, imparting a fuzzy appearance to colonies.

hyphae Tubular, filamentous cellular structures formed by many actinomycetes and most fungi.

madurose The sugar derivative 3-*O*-methyl-D-galactose, which is characteristic of several actinomycete genera that are collectively called maduromycetes.

mycolic acids Complex 60 to 90 carbon fatty acids with a hydroxyl on the β-carbon and an aliphatic chain on the α-carbon; found in the cell walls of mycobacteria.

nocardioforms Bacteria resembling members of the genus *Nocardia;* they develop a substrate mycelium that readily fragments into rods and coccoid elements.

snapping division A distinctive type of binary fission resulting in an angular or a palisade arrangement of cells, which is characteristic of the genera *Arthrobacter* and *Corynebacterium.*

streptomycetes Any high G + C gram-positive bacterium of the genera *Kitasatospora, Streptomyces,* and *Streptoverticillium.* The term is also often used to refer to other closely related families, including *Streptosporangiaceae* and *Nocardiopsaceae.*

substrate mycelium In the actinomycetes, hyphae that are on the surface and may penetrate into the solid medium on which they are growing.

Chapter 22, the last of the survey chapters on bacteria, describes the high G + C gram-positive bacteria (**figure 22.1**). They are found in volume 4 of *Bergey's Manual of Systematic Bacteriology.* Many of these bacteria are called actinomycetes. **Actinomycetes** are gram-positive, aerobic bacteria that are distinctive because most have filamentous cells called hyphae that differentiate to produce asexual spores. Many closely resemble fungi in overall morphology. Presumably this resemblance results partly from adaptation to the same habitats. In this chapter, we first summarize the general characteristics of the actinomycetes. Representatives are described next, with emphasis on morphology, taxonomy, reproduction, and general importance. The actinomycetes (s., actinomycete) are a diverse group, but they share many properties.

22.1 General Properties of the Actinomycetes

The actinomycetes are a fascinating group of microorganisms. They are the source of most of the antibiotics used in medicine today. They also produce metabolites that are used as anticancer drugs, antihelminthics (e.g., ivermectin given to dogs to prevent heartworm), and drugs that suppress the immune system in patients who have received organ transplants. This practical aspect of the actinomycetes is linked very closely to their mode of growth. Like the myxobacteria, the prosthecate bacteria, and several other microbes described in previous chapters, many actinomycetes have a complex life cycle. The life cycle of these actinomycetes includes the

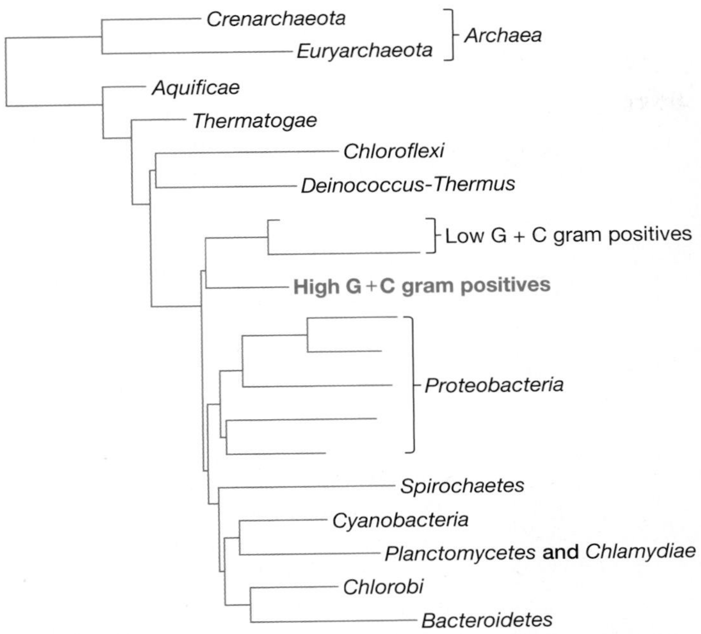

FIGURE 22.1 Phylogenetic Relationships Among *Bacteria* and *Archaea*. The high G + C gram-positive bacteria are highlighted.

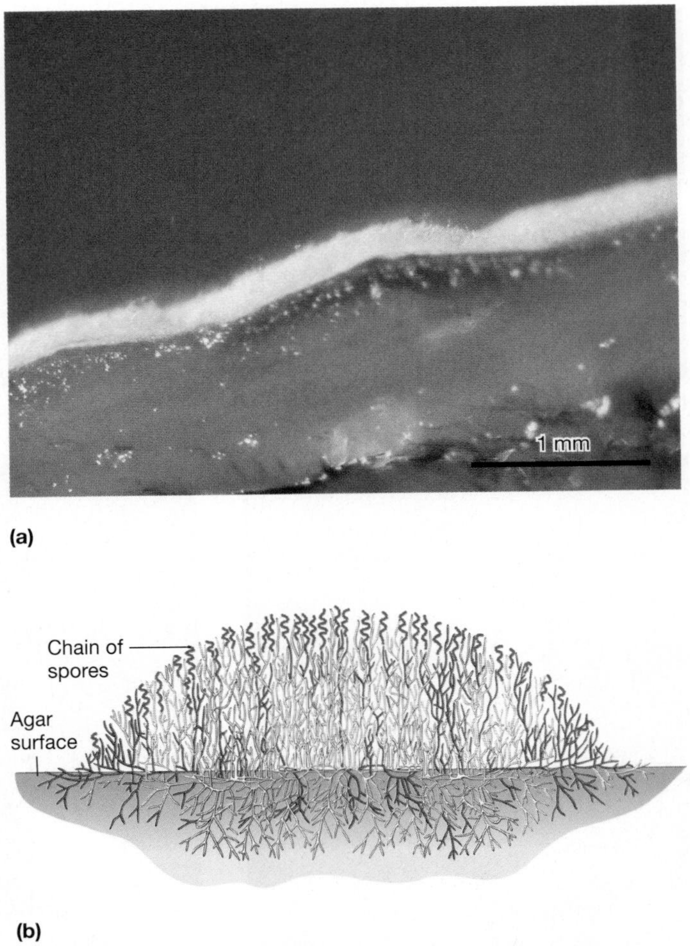

FIGURE 22.2 Cross Section of an Actinomycete Colony. (a) This photomicrograph shows *Streptomyces griseus* substrate hyphae (yellow) growing into the agar and the white aerial hyphae growing away from the colony surface. (b) The colony consists of actively growing aerial hyphae that cannibalize the substrate hyphae to obtain nutrients. Live hyphae are blue, dead are white.

development of filamentous cells, called **hyphae,** and spores. When growing on a solid substratum such as soil or agar, the actinomycetes develop a branching network of hyphae (**figure 22.2**). The hyphae grow both on the surface of the substratum and into it to form a dense hyphal mat called a **substrate mycelium.** Septae usually divide the hyphae into long cells (20 μm and longer) each containing several nucleoids. In many actinomycetes, substrate hyphae differentiate into upwardly growing hyphae to form an **aerial mycelium** that extends above the substratum. It is at this time that medically useful compounds are formed. Because the physiology of the actinomycete has switched from actively growing vegetative cells into this special cell type, these compounds are often called **secondary metabolites.**

Over time, aerial hyphae septate to form thin-walled spores. These spores are considered **exospores** because they do not develop within a mother cell like the endospores of *Bacillus* and *Clostridium.* If the spores are located in a sporangium, they may be called **sporangiospores.** The spores can vary greatly in shape (**figure 22.3**). Like spore formation in other bacteria, actinomycete sporulation is usually in response to nutrient deprivation. In general, actinomycete spores are not particularly heat resistant but withstand desiccation well, so they have considerable adaptive value. Most actinomycetes are not motile, and spores dispersed by wind or animals may find a new habitat to provide needed nutrients. In the few motile genera, movement is confined to flagellated spores.

Actinomycetes also have great ecological significance. They are primarily soil inhabitants and are widely distributed. They can degrade an enormous variety of organic compounds and are extremely important in the mineralization of organic matter. Although most actinomycetes are free-living microorganisms, a few are pathogens of humans, other animals, and some plants.

Actinomycete cell wall composition varies greatly among different groups and is of considerable taxonomic importance. Four major cell wall types can be distinguished based on the structure of the peptidoglycan and the cell wall sugar content (**figure 22.4** and **table 22.1**, *also see figure 3.16*). Two aspects of peptidoglycan structure are used: the amino acid in position 3 of the tetrapeptide side chain and the presence of glycine in interpeptide bridges. Cell walls contain other sugars beside the *N*-acetylmuramic acid and *N*-acetylglucosamine found in peptidoglycan, and such sugars have historically been used to categorize the actinomycetes. Whole cell extracts are used and the sugar content is easily determined by paper chromatography. Some other taxonomically valuable properties are the morphology and color of the mycelium and sporangia, the surface features and arrangement of spores, the percent G + C in DNA, the phospholipid composition of cell membranes, and spore heat resistance. |◀◀ *Peptidoglycan structure (section 3.3)*

Although 16S rRNA sequences are most valuable, it has been clear for some time that some of the phenotypic traits traditionally used to determine actinomycete taxonomy do not always fit with 16S rRNA sequence data. By definition, high G + C gram-positive bacteria have a DNA base composition above approximately 50 mol% G + C. Based on 16S rRNA sequence data, all are placed in the phylum *Actinobacteria* and classified as shown in **figure 22.5**. The phylum is large and very complex; it contains 1 class (*Actinobacteria*), 5 subclasses, 6 orders, 14 suborders, and 44 families. In this system, the **actinobacteria** are composed of the actinomycetes and their high G + C relatives. **Table 22.2** summarizes the characteristics of some of the genera discussed in this chapter. |◀◀ *Techniques for determining microbial taxonomy and phylogeny (section 17.3)*

Most of the genera discussed in the following survey are in the subclass *Actinobacteridae* and order *Actinomycetales*, which is divided into 10 suborders. We focus on several of these suborders. The order *Bifidobacteriales* also is briefly described.

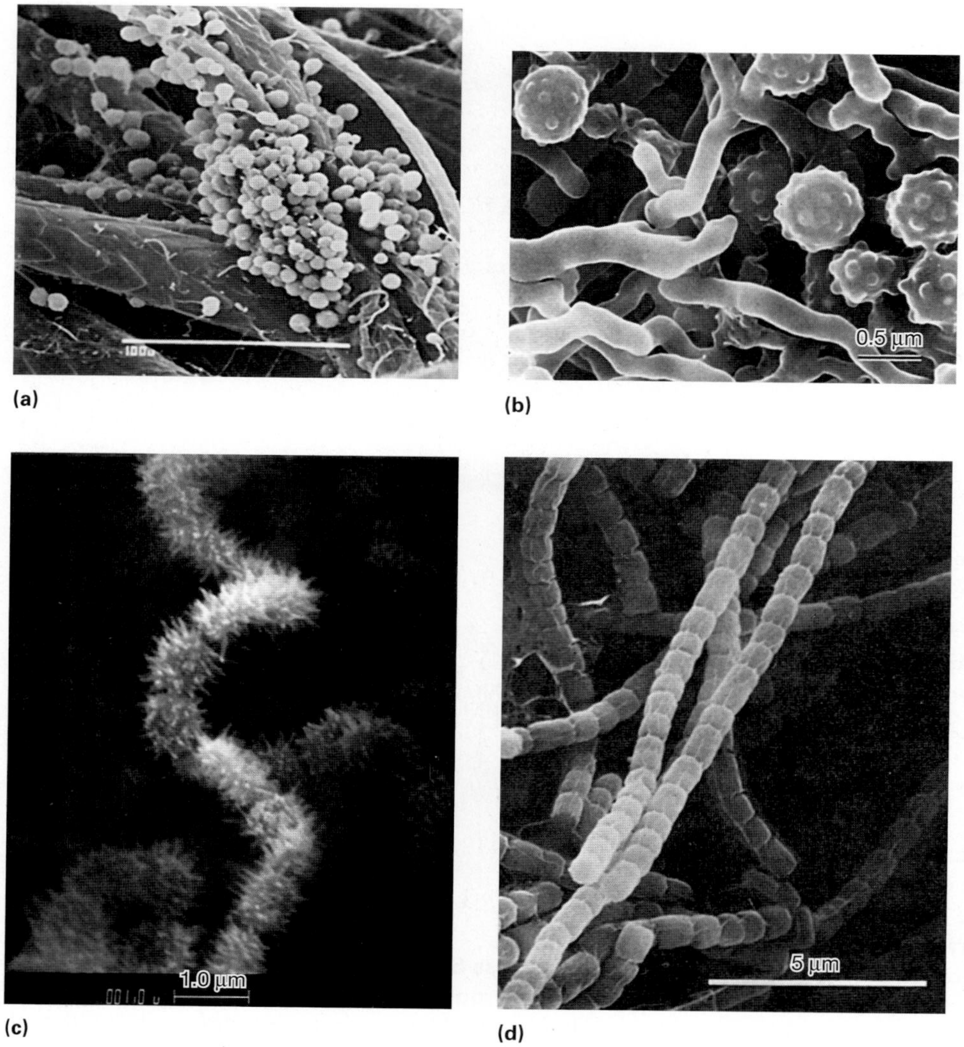

FIGURE 22.3 Examples of Actinomycete Spores as Seen in the Scanning Electron Microscope. (a) Spores of *Pilimelia columellifera* on mouse hair (×520). (b) *Micromonospora echinospora.* (c) A chain of hairy streptomycete spores. (d) Spore chain of *Kitasatosporia setae.*

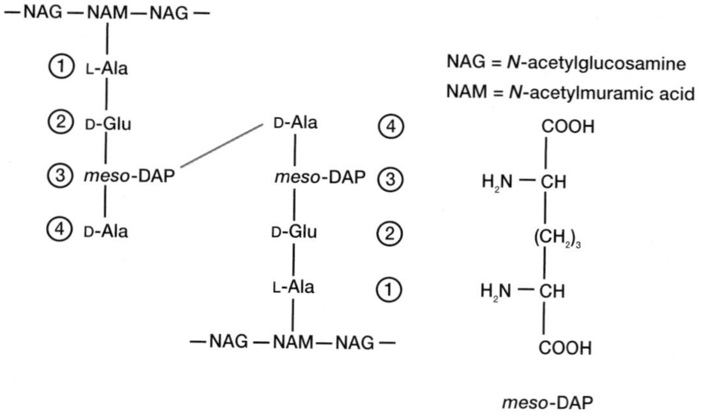

FIGURE 22.4 Actinobacterial Peptidoglycan Structure. Peptidoglycan with meso-diaminopimelic acid in position 3 and a direct cross-linkage between positions 3 and 4 of the peptide subunits is shown here. This is found in a variety of bacteria, including actinobacteria containing type III and IV cell walls (e.g., *Frankia* and *Nocardia*). Other actinobacteria have a glycine residue linking tetrapeptides.

Figure 22.4 Micro Inquiry

Refer to table 22.1 and determine how the cell walls of *Streptomyces* and *Actinoplanes* would differ from that shown here.

Table 22.1 Actinomycete Cell Wall Types and Whole Cell Sugar Patterns

Cell Wall Type	Diaminopimelic Acid Isomer	Glycine in Interpeptide Bridge	Characteristic Sugars	Representative Genera
I	L, L	+	NA[a]	*Nocardioides, Streptomyces*
II	*meso*	+	NA	*Micromonospora, Pilimelia, Actinoplanes*
III	*meso*	−	NA	*Actinomadura, Frankia*
IV	*meso*	−	Arabinose, galactose	*Saccharomonospora, Nocardia*

Whole Cell Sugar Patterns[b]	Characteristic Sugars	Representative Genera
A	Arabinose, galactose	*Nocardia, Rhodococcus, Saccharomonospora*
B	Madurose[c]	*Actinomadura, Streptosporangium, Dermatophilus*
C	None	*Thermomonospora, Actinosynnema, Geodermatophilus*
D	Arabinose, xylose	*Micromonospora, Actinoplanes*

[a] NA, either not applicable or no diagnostic sugars.
[b] Characteristic sugar patterns are present only in wall types II–IV, those actinomycetes with *meso*-diaminopimelic acid.
[c] Madurose is 3-*O*-methyl-D-galactose.

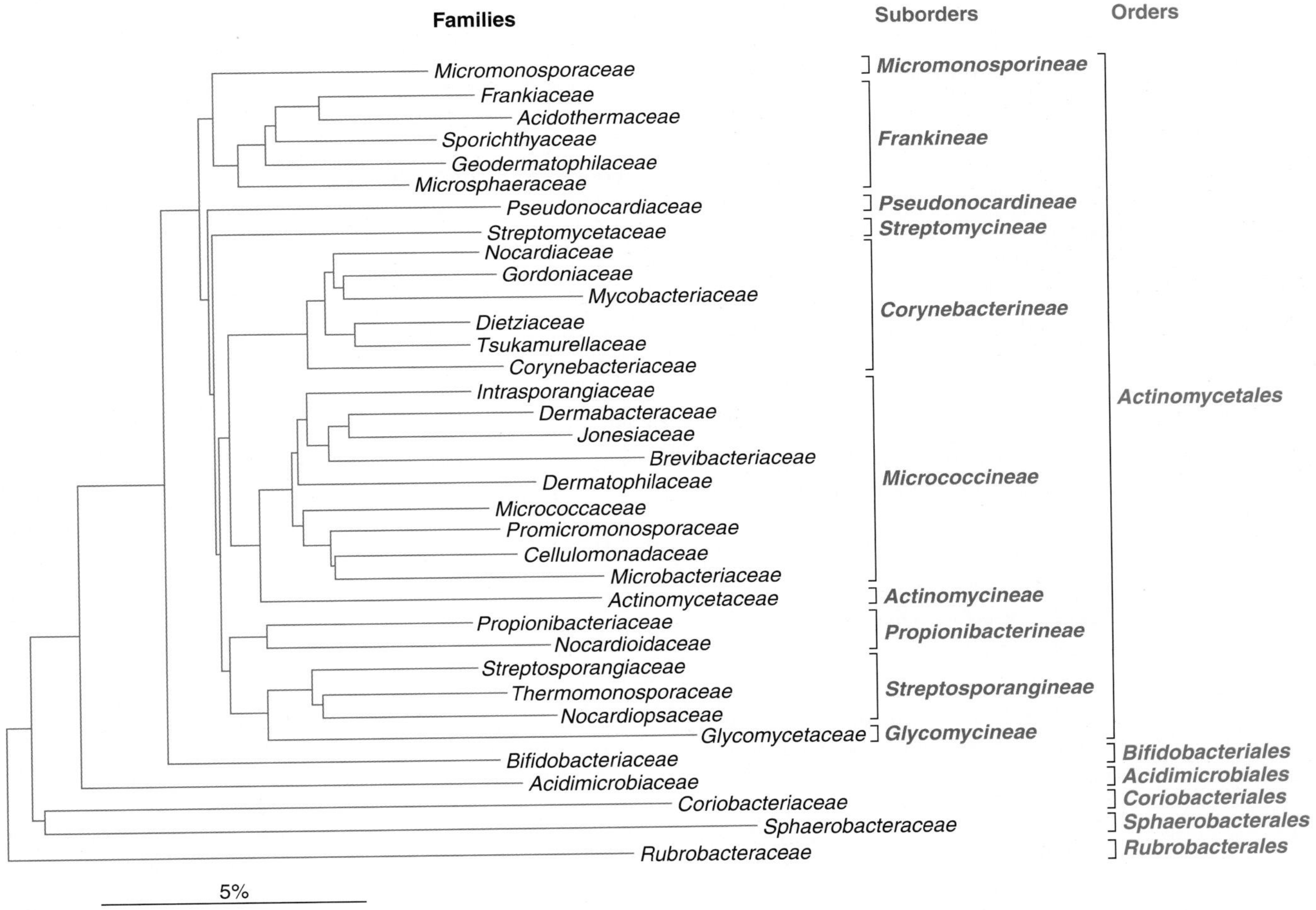

FIGURE 22.5 Classification of the Phylum *Actinobacteria*. The phylogenetic relationships between orders, suborders, and families based on 16S rRNA data are shown. The bar represents 5 nucleotide substitutions per 100 nucleotides.

Source: (a) Stackebrandt, E., et al. 1997. Proposal for a new hierarchic classification system, Actinobacteria, *classis nov. Fig. 3, p. 482.* Int. J. Syst. Bacteriol. *47(2):479–91.*

Figure 22.5 Micro Inquiry

What microbe is used as the outgroup in this tree? Which two genera appear to be most evolutionarily derived?

Table 22.2 Characteristics of Actinobacteria

Genus	*Dimensions (μm), Morphology and Motility*	*G + C Content (mol%)*	*Oxygen Relationship*	*Other Distinctive Characteristics*
Actinoplanes	Nonfragmenting, branching mycelium with little aerial growth; sporangia formed; motile spores with polar flagella	72–73	Aerobic	Hyphae often in palisade arrangement; highly colored; type II cell walls; found in soil and decaying plant material
Arthrobacter	0.8–1.2 × 1.0–8.0; young cells are irregular rods, older cells are small cocci; usually nonmotile	59–70	Aerobic	Has rod-coccus growth cycle; metabolism respiratory; catalase positive; mainly in soil
Bifidobacterium	0.5–1.3 × 1.5–8; rods of varied shape, usually curved; nonmotile	55–67	Anaerobic	Cells can be clubbed or branched, pairs often in V arrangement; ferment carbohydrates to acetate and lactate but no CO_2; catalase negative
Corynebacterium	0.3–0.8 × 1.5–8.0; straight or slightly curved rods with tapered or clubbed ends; nonmotile	51–63	Facultatively anaerobic	Cells often arranged in a V formation or in palisades of parallel cells; catalase positive and fermentative; metachromatic granules
Frankia	0.5–2.0 in diameter; vegetative hyphae with limited-to-extensive branching and no aerial mycelium; multilocular sporangia formed	66–71	Aerobic to microaerophilic	Sporangiospores nonmotile; usually fixes nitrogen; type III cell walls; most strains are symbiotic with angiosperm plants and induce nodules
Micrococcus	0.5–2.0 diameter; cocci in pairs, tetrads, or irregular clusters; usually nonmotile	64–75	Aerobic	Colonies usually yellow or red; catalase positive with respiratory metabolism; primarily on mammalian skin and in soil
Mycobacterium	0.2–0.6 × 1.0–10; straight or slightly curved rods, sometimes branched; acid-fast; nonmotile and nonsporing	62–70	Aerobic	Catalase positive; can form filaments that are readily fragmented; walls have high lipid content; in soil and water; some parasitic
Nocardia	0.5–1.2 in diameter; rudimentary to extensive vegetative hyphae that can fragment into rod-shaped and coccoid forms	64–72	Aerobic	Aerial hyphae formed; catalase positive; type IV cell wall; widely distributed in soil
Propionibacterium	0.5–0.8 × 1–5; pleomorphic nonmotile rods, may be forked or branched; nonsporing	53–67	Anaerobic to aerotolerant	Fermentation produces propionate and acetate, and often gas; catalase positive
Streptomyces	0.5–2.0 in diameter; vegetative mycelium extensively branched; aerial mycelium forms chains of three to many spores	69–78	Aerobic	Forms discrete lichenoid or leathery colonies that often are pigmented; respiratory metabolism; uses many organic compounds as nutrients; soil organisms

22.2 Suborder *Actinomycineae*

There is one family with five genera in the suborder *Actinomycineae.* These genera are *Actinomyces, Actinobaculum, Arcanobacterium, Mobiluncus,* and *Varibaculum.* Most are irregularly shaped, nonsporing, gram-positive rods with aerobic or facultative metabolism. The rods may be straight or slightly curved and usually have swellings, club shapes, or other deviations from normal rod-shape morphology.

Members of the genus *Actinomyces* are either straight or slightly curved rods, or slender filaments with true branching

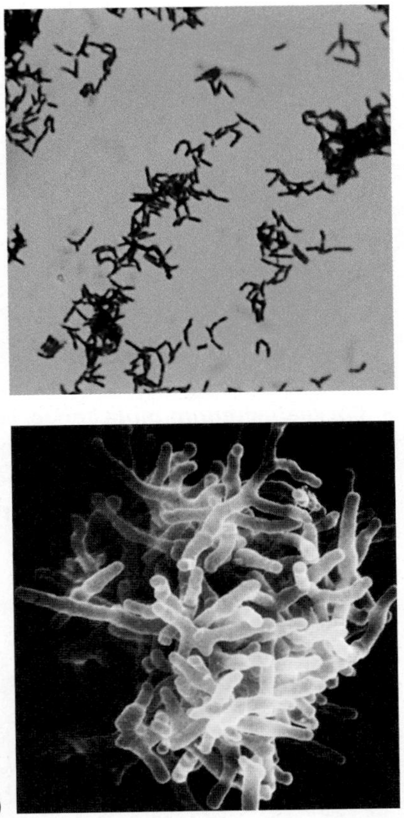

(a)

(b)

FIGURE 22.6 Representatives of the Genus *Actinomyces*. (a) *A. naeslundii*; Gram stain (×1,000). (b) *Actinomyces*; scanning electron micrograph (×18,000). Note the filamentous nature of the colony.

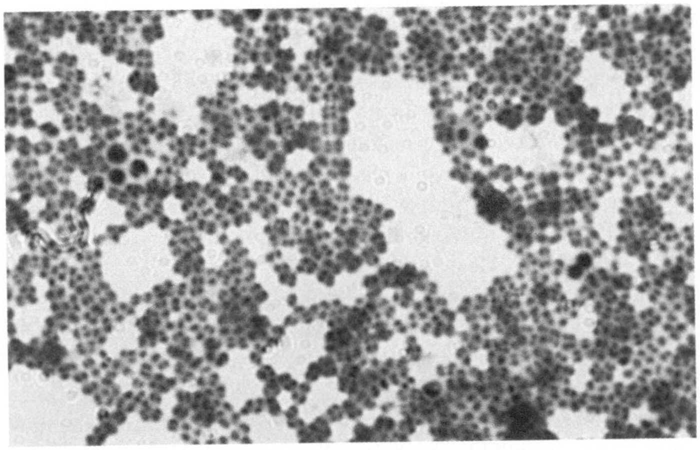

FIGURE 22.7 *Micrococcus*. *Micrococcus luteus*, methylene blue stain (×1,000).

Figure 22.7 Micro Inquiry

What makes the morphology of *M. luteus* distinct from most other actinomycetes discussed in this chapter?

(**figure 22.6**). The rods and filaments may have swollen or clubbed ends. They are either facultative or strict anaerobes that require CO_2 for optimal growth. The peptidoglycan contains lysine but not diaminopimelic acid or glycine. *Actinomyces* species are normal inhabitants of mucosal surfaces of humans and other warm-blooded animals; the oral cavity is their preferred habitat. *A. bovis* causes lumpy jaw in cattle. *Actinomyces* is responsible for actinomycoses, ocular infections, and periodontal disease in humans.

22.3 Suborder *Micrococcineae*

The suborder *Micrococcineae* has 14 families and a wide variety of genera. Two of the best-known genera are *Micrococcus* and *Arthrobacter*. The genus *Micrococcus* contains aerobic, catalase-positive cocci that occur mainly in pairs, tetrads, or irregular clusters and are usually nonmotile (**figure 22.7**). Unlike many other actinomycetes, *Micrococcus* does not undergo morphological differentiation. Micrococci colonies often are yellow, orange, or red. They are widespread in soil, water, and on mammalian skin, which may be their normal habitat.

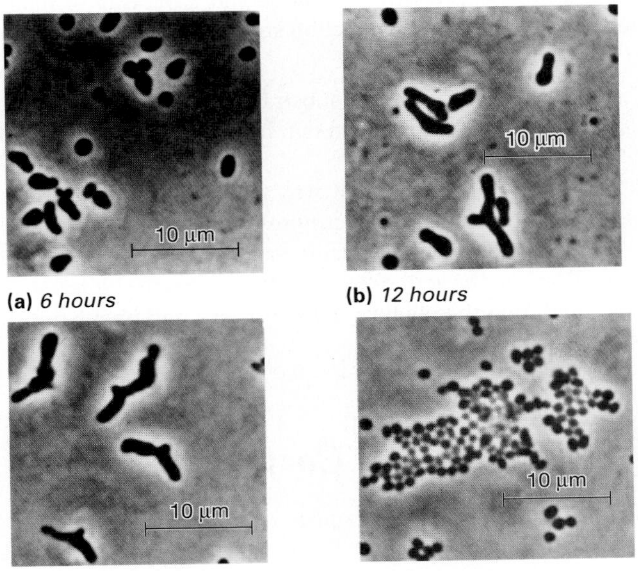

(a) *6 hours* (b) *12 hours* (c) *24 hours* (d) *72 hours*

FIGURE 22.8 The Rod-Coccus Growth Cycle. The rod-coccus cycle of *Arthrobacter globiformis* when grown at 25°C. The cells used for inoculation resembled these stationary-phase cocci; the duration of time passed since inoculation is noted.

The genus *Arthrobacter* contains aerobic, catalase-positive rods with respiratory metabolism. They possess peptidoglycan with lysine. Its most distinctive feature is a rod-coccus growth cycle (**figure 22.8**). When *Arthrobacter* grows in exponential phase, the bacteria are irregular, branched rods. As they enter stationary phase, the cells change to a coccoid form. Upon transfer to fresh medium, the coccoid cells differentiate to form actively growing rods. Although arthrobacters often are isolated from fish, sewage, and plant surfaces, their most important habitat is the soil, where they constitute a significant component of the culturable microbial community. They are well adapted to this niche because they are very resistant to desiccation and nutrient deprivation, even though they do not form spores. This genus is unusually flexible nutritionally and can even degrade some herbicides and pesticides.

Both *Arthrobacter* and *Corynebacteria* (section 22.4) undergo a particular kind of binary fission called **snapping division.** The mechanism of snapping division has been studied in *Arthrobacter.* These bacteria have a two-layered cell wall, and only the inner layer grows inward to generate a transverse wall dividing the new cells. The completed transverse wall or septum thickens and puts tension on the outer wall layer, which still holds the two cells together. Eventually increasing tension ruptures the outer layer at its weakest point, and a snapping movement tears the outer layer apart around most of its circumference. The new cells now rest at an angle to each other and are held together by the remaining portion of the outer layer, which acts as a hinge.

A third genus in this suborder is *Dermatophilus. Dermatophilus* (type IIIB) also forms packets of motile spores with tufts of flagella. It is a facultative anaerobe and a parasite of mammals responsible for the skin infection streptothricosis.

1. Define actinomycete, substrate mycelium, aerial mycelium, and exospore. Explain how these structures confer a survival advantage.
2. Describe how cell wall structure and sugar content are used to classify the actinomycetes. Include a brief description of the four major wall types.
3. Why are the actinomycetes of such practical interest?
4. Describe the phylum *Actinobacteria* and its relationship to the actinomycetes.

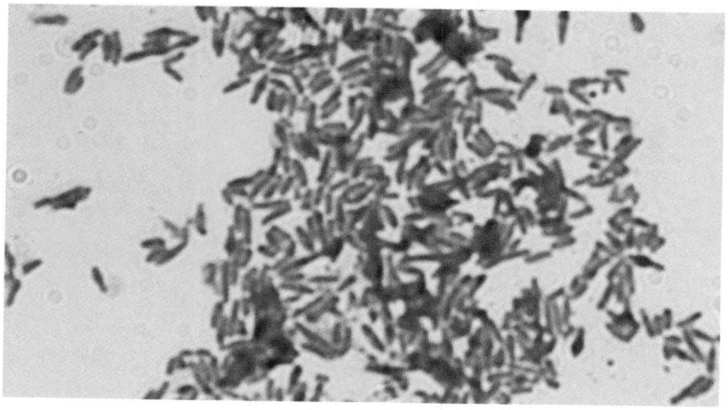

FIGURE 22.9 ***Corynebacterium diphtheriae.*** Note the irregular shapes of individual cells, the angular associations of pairs of cells, and palisade arrangements (×1,000). These gram-positive rods do not form endospores.

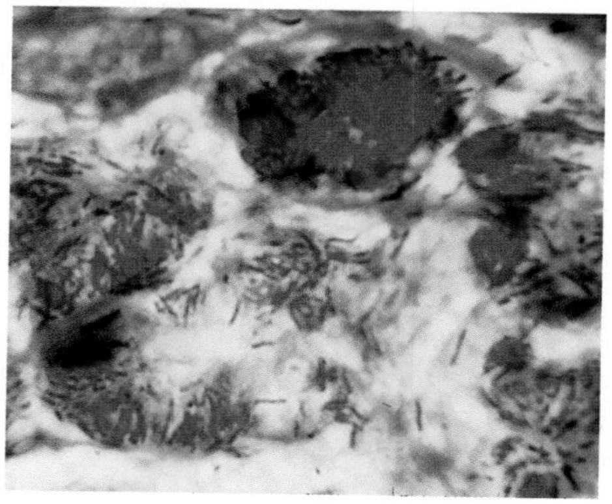

FIGURE 22.10 The Mycobacteria. *Mycobacterium leprae.* Acid-fast stain (×400). Note the masses of red mycobacteria within blue-green host cells.

22.4 Suborder *Corynebacterineae*

The suborder *Corynebacterineae* contains seven families with several well-known genera. Three of the most important genera are *Corynebacterium, Mycobacterium,* and *Nocardia.*

The family *Corynebacteriaceae* has one genus, *Corynebacterium,* which includes aerobic and facultative, catalase-positive, straight to slightly curved rods, often with tapered ends. Club-shaped forms are also seen. The bacteria often remain partially attached after snapping division, resulting in angular arrangements of the cells or a **palisade arrangement** in which rows of cells are lined up side by side (**figure 22.9**). *Corynebacteria* form metachromatic granules, and their walls have meso-diaminopimelic acid. Although some species are harmless saprophytes, many corynebacteria are plant or animal pathogens. For example, *C. diphtheriae* is the causative agent of diphtheria in humans. ▶▶| *Airborne diseases (section 38.1)*

The family *Mycobacteriaceae* contains the genus *Mycobacterium,* which is composed of slightly curved or straight rods that sometimes branch or form filaments (**figure 22.10**). Mycobacterial filaments differ from those of actinomycetes because they readily fragment into rods and coccoid bodies. They are aerobic and catalase positive. Mycobacteria grow very slowly and must be incubated for 2 to 40 days after inoculation on a solidified complex medium to form a visible colony. Their cell walls have a very high lipid content due to the presence of **mycolic acids.** These are complex fatty acids that feature an invariant C26 fatty acid that is attached to a longer, variable fatty acid (**figure 22.11**). Mycolic acids are attached to the underlying peptidoglycan by a layer of the polysaccharide arabinoglycan. The surface of the cell wall consists of the glycolipid trehalose dimycolate, once called cord factor. This combination of lipids and polysaccharides makes the cell wall extremely hydrophobic and impenetrable to most organic molecules, including antibiotics. Cryo-electron microscopy has revealed that unlike other gram-positive bacteria, mycobacteria

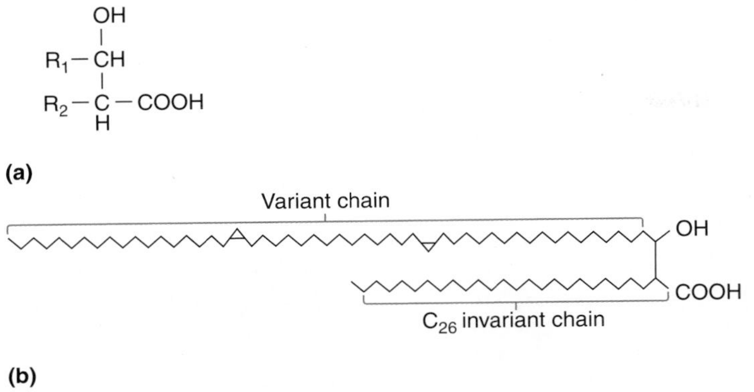

FIGURE 22.11 Mycolic Acid Structure. (a) The generic structure of mycolic acids, a family that includes over 500 different types. (b) An example mycolic acid with two cyclopropane rings.

Figure 22.11 Micro Inquiry

Porins are usually associated with the outer membrane of gram-negative bacteria. Why do *Mycobacterium* spp. also have porins?

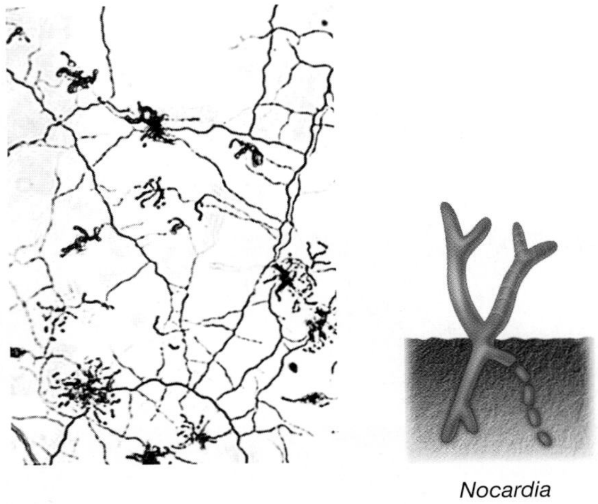

FIGURE 22.12 *Nocardia*. *Nocardia asteroides,* substrate mycelium and aerial mycelia with conidia; illustration and light micrograph (×1,250).

possess a true outer membrane. Mycolic acids appear to be required for the formation of this outer membrane. It follows that mycobacteria have evolved porins that appear to be unique to these bacteria. The cell wall of mycobacteria is **acid-fast** (basic fuchsin dye cannot be removed from the cell by acid alcohol treatment). ◀◀ *Differential staining (section 2.3)*

Although some mycobacteria are free-living saprophytes, they are best known as animal pathogens. *M. bovis* causes tuberculosis in cattle, other ruminants, and primates. Because this bacterium can produce tuberculosis in humans, dairy cattle are tested for the disease yearly; milk pasteurization kills the pathogen. Prior to widespread milk pasteurization, contaminated milk was a source of transmission. Currently, *M. tuberculosis* is the chief source of tuberculosis in humans. It is estimated that about one-third of the world's population is infected with *M. tuberculosis* (*see figure 38.4*). Another major mycobacterial human disease is leprosy, caused by *M. leprae.* Finally, *M. avium* and the related species *M. intracellulare* comprise *M. avium* complex (MAC), which has become a common mycobacterial disease in developed countries. ◀◀ *Comparative genomics (section 16.7)* ▶▶ Mycobacterium *infections (section 38.1)*

Search This: Mycobacterium biohealth

The family *Nocardiaceae* is composed of two genera, *Nocardia* and *Rhodococcus.* These bacteria develop a substrate mycelium that readily breaks into rods and coccoid elements (**figure 22.12**). They also form an aerial mycelium that rises above the substratum and may produce conidia. Almost all are strict aerobes. Most species have peptidoglycan with meso-diaminopimelic acid and no peptide interbridge. The wall usually contains a carbohydrate composed of arabinose and galactose, and mycolic acids are present. These and related genera that resemble members of the genus *Nocardia* (named after Edmond Nocard [1850–1903], French bacteriologist and veterinary pathologist) are collectively called **nocardioforms.**

Nocardia is distributed worldwide in soil and aquatic habitats. They are involved in the degradation of hydrocarbons and waxes, and can contribute to the biodeterioration of rubber joints in water and sewage pipes. Although most are free-living saprophytes, some species, particularly *N. asteroides,* are opportunistic pathogens that cause nocardiosis in humans and other animals. People with low resistance due to other health problems, such as individuals with HIV-AIDS, are most at risk. The lungs are typically infected; other organs and the central nervous system may be invaded as well.

Rhodococcus is widely distributed in soils and aquatic habitats. It is of considerable interest because members of the genus can degrade an enormous variety of molecules such as petroleum hydrocarbons, detergents, benzene, polychlorinated biphenyls (PCBs), and various pesticides. It may be possible to use rhodococci to remove sulfur from fuels, thus reducing air pollution from sulfur oxide emissions.

1. Describe the major characteristics of *Corynebacterium* and *Mycobacterium.* Include comments on their normal habitat and importance.
2. What is snapping division?
3. What implications, if any, do you think the complex hydrophobic cell wall of *Mycobacterium* has in terms of treating diseases caused by this genera?
4. What is a nocardioform, and how can the group be distinguished from other actinomycetes?

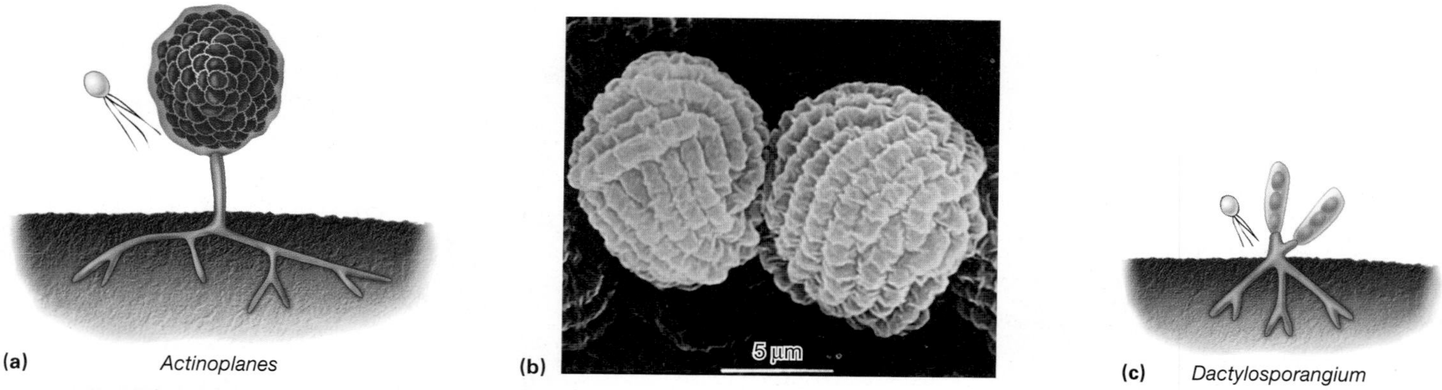

(a) *Actinoplanes* (b) (c) *Dactylosporangium*

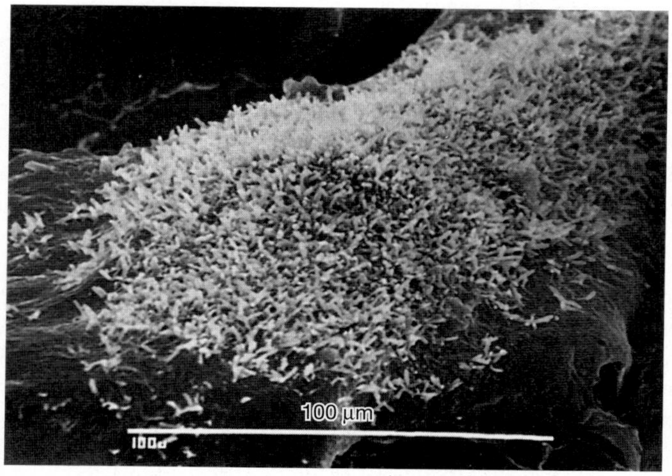

(d)

FIGURE 22.13 Family *Micromonosporaceae*. Actinoplanete morphology. (a) *Actinoplanes* structure. (b) A scanning electron micrograph of mature *Actinoplanes* sporangia. (c) *Dactylosporangium* structure. (d) A *Dactylosporangium* colony covered with sporangia.

22.5 Suborder *Micromonosporineae*

The suborder *Micromonosporineae* contains only one family, *Micromonosporaceae*. Genera include *Micromonospora, Dactylosporangium, Pilimelia,* and *Actinoplanes*. Often the family is collectively referred to as the actinoplanetes (Greek *actinos,* a ray or beam, and *planes,* a wanderer). They have an extensive substrate mycelium and are cell wall type IID cells (tables 22.1 and 22.2). Often the hyphae are highly colored, and diffusible pigments may be produced. Normally an aerial mycelium is absent or rudimentary. Spores are usually formed within a sporangium raised above the surface of the substratum at the end of a special hypha called a sporangiophore. The spores can be either motile or nonmotile. These bacteria vary in the arrangement and development of their spores. Some genera (*Actinoplanes, Pilimelia*) have spherical, cylindrical, or irregular sporangia with a few to several thousand spores per sporangium (figure 22.3*a* and **figure 22.13**). The spores are arranged in coiled or parallel chains (figure 22.13*b*). *Dactylosporangium* forms club-shaped, fingerlike, or pyriform sporangia with one to six spores (figure 22.13*c,d*). *Micromonospora* bears single spores, which often occur in branched clusters of sporophores (figure 22.3*b*).

Actinoplanetes grow in almost all soil habitats, ranging from forest litter to beach sand. They also flourish in freshwater, particularly in streams and rivers (probably because of abundant oxygen and plant debris). Some have been isolated from the ocean. The soil-dwelling species may have an important role in the decomposition of plant and animal material. *Pilimelia* grows in association with keratin. *Micromonospora* actively degrades chitin and cellulose, and produces antibiotics such as gentamicin.

22.6 Suborder *Propionibacterineae*

The suborder *Propionibacterineae* contains two families and 14 genera. The genus *Propionibacterium* contains pleomorphic, nonmotile, nonsporing rods that are often club-shaped with one end tapered and the other end rounded. Cells also may be coccoid or branched. They can be single, in short chains, or in clumps. The genus is facultatively anaerobic or aerotolerant; lactate and sugars are fermented to produce large quantities of propionic and acetic acids, and often carbon dioxide. *Propionibacterium* is usually catalase positive. The genus is found growing on the skin and in the digestive tract of animals, and in dairy products such as cheese. *Propionibacterium* fermentation gives Swiss cheese its characteristic flavor. *P. acnes* is involved with the development of body odor and acne vulgaris.

22.7 Suborder *Streptomycineae*

The suborder *Streptomycineae* has only one family, *Streptomycetaceae,* and three genera, the most important of which is *Streptomyces.* These bacteria have aerial hyphae that divide in a single plane to form chains of multiple nonmotile spores with surface texture ranging from smooth to spiny and warty (figure 22.3*c*; **figures 22.14** and **22.15**). All have a type I cell wall and a G + C content of 69 to 78%. Filaments grow by tip extension, rather than by fragmentation. Members of this family and similar bacteria are often called **streptomycetes** (Greek *streptos,* bent or twisted, and *myces,* fungus).

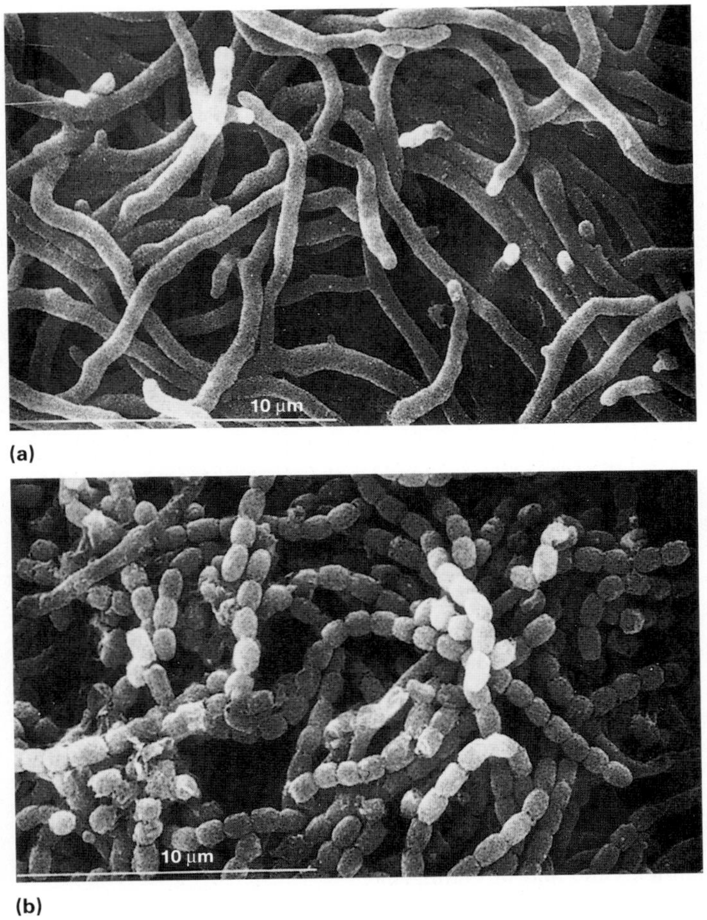

FIGURE 22.14 Streptomyces Development. (a) *S. coelicolor* vegetative hyphae are straight with branches. (b) Chains of *S. coelicolor* spores that will eventually pinch off and be released into the environment.

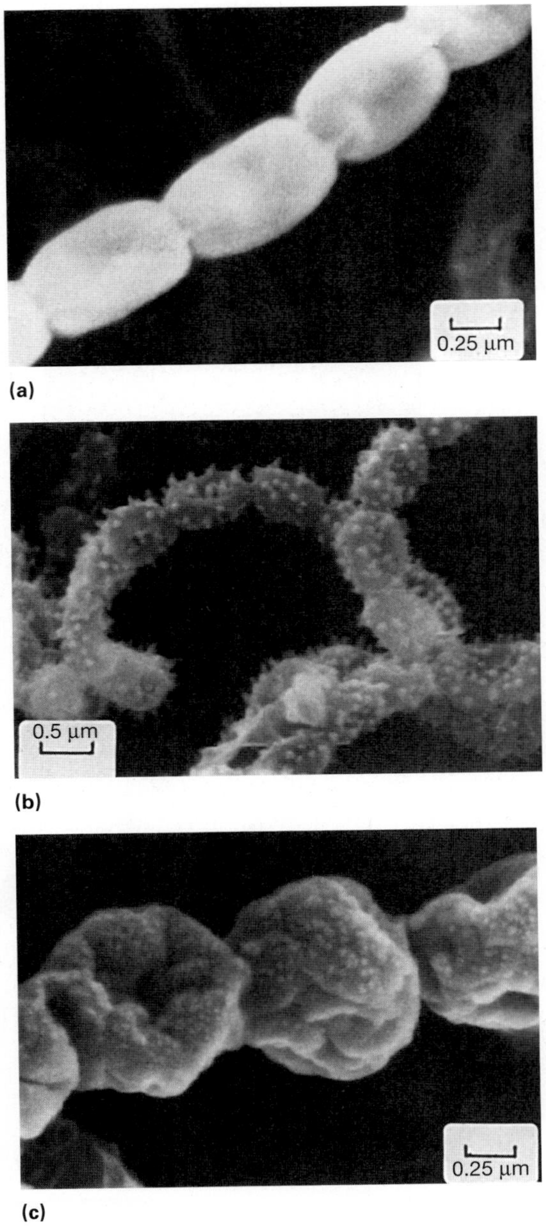

FIGURE 22.15 Streptomycete Spore Chains. (a) Smooth spores of *S. niveus.* (b) Spiney spores of *S. viridochromogenes.* (c) Warty spores of *S. pulcher.* Scanning electron micrographs.

Streptomyces is a large genus; there are around 150 species. Members of the genus are strict aerobes. *Streptomyces* species are determined by means of a mixture of morphological and physiological characteristics, including the following: the color of the aerial and substrate mycelia, spore arrangement, surface features of individual spores, carbohydrate use, antibiotic production, melanin synthesis, nitrate reduction, and the hydrolysis of urea and hippuric acid.

Streptomycetes are very important both ecologically and medically. The natural habitat of most streptomycetes is the soil, where they may constitute from 1 to 20% of the culturable population. In fact, the odor of moist earth is largely the result of streptomycete production of volatile substances such as geosmin. Streptomycetes play a major role in mineralization. They are flexible nutritionally and can degrade recalcitrant (resistant) substances such as pectin, lignin, chitin, keratin, latex, agar, and aromatic compounds. Streptomycetes are best known for their synthesis of a vast array of antibiotics.

Selman Waksman's discovery that *S. griseus* (**figure 22.16*a*** and figure 22.2) produces streptomycin was an enormously important contribution to science and public health. Streptomycin was the first drug to effectively combat tuberculosis, and in 1952 Waksman earned the Nobel Prize. In addition, this discovery set off a massive search resulting in the isolation of new *Streptomyces* species that produce other compounds of medicinal importance. In fact, since that time, the streptomycetes have been found to produce over 10,000 bioactive compounds. Hundreds of these natural products are now used in medicine and industry; about two-thirds of the antimicrobial agents used in human and veterinary medicine are derived from the streptomycetes. Examples include amphotericin B, chloramphenicol, neomycin,

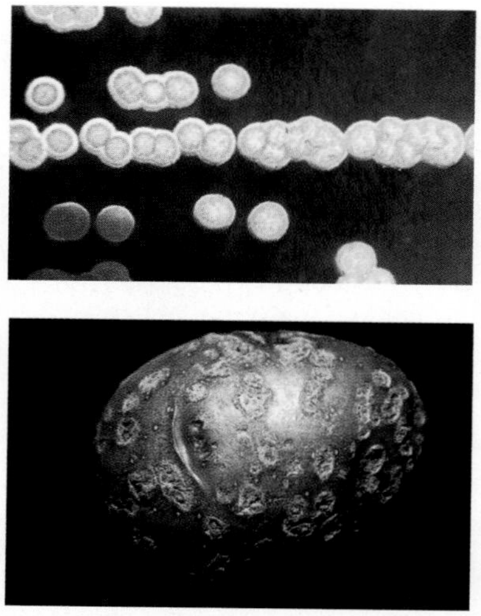

(a)

(b)

FIGURE 22.16 Streptomycetes of Practical Importance. (a) *Streptomyces griseus*. Colonies of the actinomycete that produces streptomycin. (b) *S. scabies* growing on a potato.

nystatin, and tetracycline. Some *Streptomyces* species produce more than one antibiotic. Antibiotic-producing bacteria have genes that encode proteins (or antisense RNA) that make them resistant to the antibiotics they synthesize. ▶▶| *Antimicrobial chemotherapy (chapter 34)*

The genome of *Streptomyces coelicolor*, which produces four antibiotics and serves as a model species for research, is one of the largest bacterial genomes (8.67 Mbp). Its large number of genes (7,825) no doubt reflects the number of proteins required to undergo a complex life cycle. Many genes are devoted to regulation, with an astonishing 65 predicted RNA polymerase sigma subunits and about 80 two-component regulatory systems. The ability to exploit a variety of soil nutrients is also demonstrated by the presence of a large number of ABC transporters, the Sec and two Tat protein translocation systems, and secreted degradative enzymes. Interestingly, genes were discovered that are thought to encode enzymes for the production of an additional 18 secondary metabolites. |◀◀ *Protein maturation and secretion (section 12.9); Two-component regulatory systems and phosphorelay systems (section 13.2); Global regulatory systems (section 13.6)*

Although most streptomycetes are nonpathogenic saprophytes, a few are associated with plant and animal diseases. *Streptomyces scabies* causes scab disease in potatoes and beets (figure 22.16*b*). *S. somaliensis* is the only streptomycete known to be pathogenic to humans. It is associated with actinomycetoma, an unusual infection of subcutaneous tissues that produces lesions that lead to swelling, abscesses, and even bone destruction if untreated. *S. albus* and other species have been isolated from patients with various ailments and may be pathogenic.

Search This: Streptomyces coelicolor

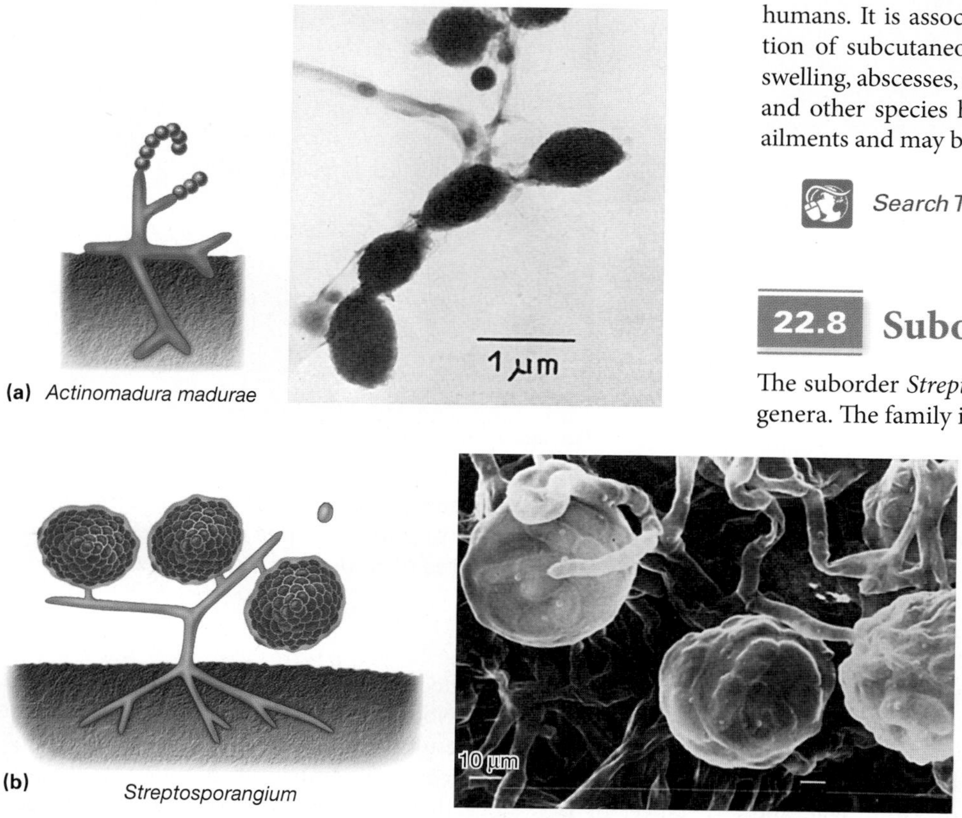

(a) *Actinomadura madurae*

(b) *Streptosporangium*

FIGURE 22.17 Maduromycetes. (a) *Actinomadura madurae* morphology; illustration and electron micrograph of a spore chain (×16,500). (b) *Streptosporangium* morphology; illustration and micrograph of *S. album* on oatmeal agar with sporangia and hyphae; SEM.

22.8 Suborder *Streptosporangineae*

The suborder *Streptosporangineae* contains three families and 16 genera. The family includes the maduromycetes, which have type III cell walls and the sugar derivative **madurose** (3-*O*-methyl-D-galactose) in whole cell homogenates. Their G + C content is 64 to 74 mol%. Aerial hyphae bear pairs or short chains of spores, and the substrate hyphae are branched (**figure 22.17**). Some genera form sporangia; spores are not heat resistant. Like *S. somaliensis*, *Actinomadura* is another actinomycete associated with the disease actinomycetoma. *Thermomonospora* produces single spores on the aerial mycelium or on both the aerial and substrate mycelia. It has been isolated from moderately high-temperature habitats such as compost piles and hay; it can grow at 40 to 48°C.

22.9 Suborder *Frankineae*

The suborder *Frankineae* includes the genera *Frankia* and *Geodermatophilus.* Both form multilocular sporangia, characterized by clusters of spores when a hypha divides both transversely and longitudinally. (Multilocular means having many cells or compartments.) They have type III cell walls, although the cell extract sugar patterns differ. The G + C content varies from 57 to 75 mol%. *Geodermatophilus* (type IIIC) has motile spores and is an aerobic soil organism. *Frankia* (type IIID) forms nonmotile spores in a sporogenous body (**figure 22.18**). It grows in symbiotic association with the roots of at least eight families of higher nonleguminous plants (e.g., alder trees) and is a microaerophile that can fix atmospheric nitrogen.

The roots of infected plants develop nodules containing *Frankia* that fix nitrogen so efficiently that a plant such as an alder can grow in the absence of combined nitrogen (e.g., NO_3^-) (figure 22.18*b*). Within the nodule cells, *Frankia* forms branching hyphae with globular vesicles at their ends, which may be the sites of nitrogen fixation. The nitrogen-fixation process resembles that of *Rhizobium* in that it is oxygen sensitive and requires molybdenum and cobalt. ▶▶| *Nitrogen-fixing bacteria (section 29.3)*

Another genus in this suborder, *Sporichthya,* is one of the strangest of the actinomycetes. It lacks a substrate mycelium. The hyphae remain attached to the substratum by holdfasts and grow upward to form aerial mycelia that release motile, flagellate spores in the presence of water.

(a)

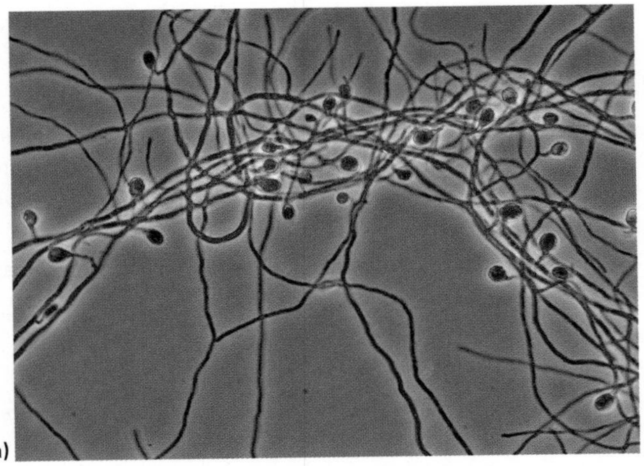

(b)

FIGURE 22.18 ***Frankia.*** (a) A phase contrast micrograph showing hyphae, multilocular sporangia, and spores. (b) Nodules of the alder *Alnus rubra.*

22.10 Order *Bifidobacteriales*

The order *Bifidobacteriales* has one family, *Bifidobacteriaceae,* and 10 genera (five of which have unknown affiliation). *Falcivibrio* and *Gardnerella* are found in the human genital/urinary tract; *Gardnerella* is thought to be a major cause of bacterial vaginitis. *Bifidobacterium* probably is the best-studied genus. Bifidobacteria are nonmotile, nonsporing, gram-positive rods of varied shapes that are slightly curved and clubbed; often they are branched (**figure 22.19**). The rods can be single or in clusters and V-shaped pairs. *Bifidobacterium* is anaerobic and actively ferments carbohydrates to produce acetic and lactic acids but not carbon dioxide. It is found in the mouth and intestinal tract of warm-blooded vertebrates, in sewage, and in insects. *B. bifidum* is a pioneer colonizer of the human intestinal tract, particularly when babies are breast fed. A few *Bifidobacterium* infections have been reported in humans, but the genus does not appear to be a major cause of disease. In fact, several species are sold as probiotic agents, which are thought to impart a health benefit (*see figure 40.11*). ▶▶| *Microbiology of fermented foods (section 40.5)*

Search This: Bifidobacterium *probiotics*

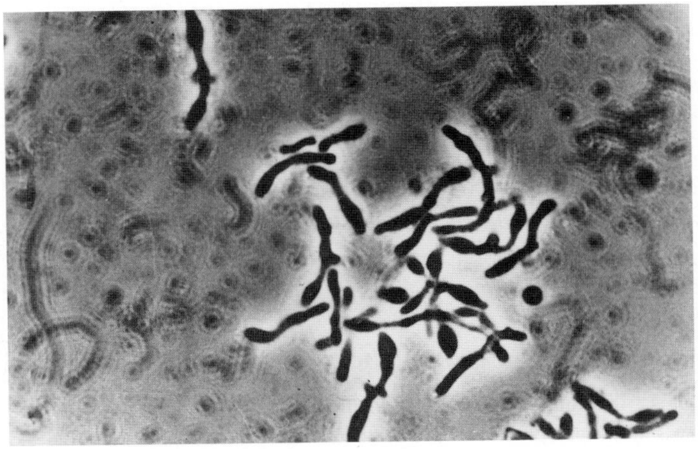

FIGURE 22.19 ***Bifidobacterium.*** *Bifidobacterium bifidum;* phase-contrast photomicrograph (×1,500).

1. Give the distinguishing properties of the actinoplanetes. Why are they morphologically distinct from other actinomycetes?
2. Describe the genus *Propionibacterium* and comment on its practical importance.
3. Describe the major properties of the genus *Streptomyces.*
4. Describe three ways in which *Streptomyces* is of ecological importance. Why do you think *Streptomyces* produce antibiotics?
5. Briefly describe the genera of the suborder *Streptosporangineae.* What is madurose? Why is *Actinomadura* important?
6. Describe *Frankia* and discuss its importance.
7. Characterize the genus *Bifidobacterium.* Where is it found and why is it significant?

Summary

22.1 General Properties of the Actinomycetes

a. Actinomycetes are aerobic, gram-positive bacteria that form branching, usually nonfragmenting hyphae and asexual spores (**figure 22.2**).

b. The asexual spores borne on aerial mycelia are called exospores because they do not develop within a mother cell. Some produce sporangiospores.

c. Actinomycetes have several distinctively different types of cell walls and often also vary in terms of the sugars present in cell extracts. Properties such as color and morphology are also taxonomically useful (**tables 22.1** and **22.2; figure 22.4**).

d. *Bergey's Manual* classifies the high G + C bacteria phylogenetically using 16S rRNA data. The phylum *Actinobacteria* contains the actinomycetes and their high G + C relatives (**figure 22.5**).

22.2 Suborder *Actinomycineae*

a. The suborder *Actinomycineae* contains the genus *Actinomyces,* members of which are irregularly shaped, nonsporing rods that can cause disease in cattle and humans.

22.3 Suborder *Micrococcineae*

a. The suborder *Micrococcineae* includes the genera *Micrococcus, Arthrobacter,* and *Dermatophilus.*

b. *Arthrobacter* has an unusual rod-coccus growth cycle and carries out snapping division (**figure 22.8**).

22.4 Suborder *Corynebacterineae*

a. The genera *Corynebacterium, Mycobacterium,* and *Nocardia* are placed in the suborder *Corynebacterineae.*

b. Mycobacteria form either rods or filaments that readily fragment. Their cell walls have a high lipid content and mycolic acids; the presence of these lipids makes them acid-fast (**figures 22.10** and **22.11**).

c. Nocardioform actinomycetes have hyphae that readily fragment into rods and coccoid elements, and often form aerial mycelia with spores (**figure 22.12**).

22.5 Suborder *Micromonosporineae*

a. The suborder *Micromonosporineae* has genera that include *Micromonospora* and *Actinoplanes.* These actinomycetes have an extensive substrate mycelium and form aerial sporangia (**figure 22.13**).

b. These bacteria are present in soil, freshwater, and the ocean. The soil forms are probably important in decomposition.

22.6 Suborder *Propionibacterineae*

a. The genus *Propionibacterium* is in the suborder *Propionibacterineae.* Members of this genus are common skin and intestinal inhabitants, and are important in cheese manufacture and the development of acne vulgaris.

22.7 Suborder *Streptomycineae*

a. The suborder *Streptomycineae* includes the genus *Streptomyces.* Members of this genus have type I cell walls and aerial hyphae bearing chains of 3 to 50 or more nonmotile spores (**figures 22.14** and **22.15**).

b. Streptomycetes are important in the degradation of resistant organic material in the soil and produce many useful antibiotics. A few cause diseases in plants and animals. However, they are most notable for the number of antibiotics and other medically and industrially important compounds they produce.

22.8 Suborder *Streptosporangineae*

a. Many genera in suborder *Streptosporangineae* have the sugar derivative madurose and type III cell walls.

22.9 Suborder *Frankineae*

a. The genera *Frankia* and *Geodermatophilus* are placed in the suborder *Frankineae*. They produce clusters of spores at hyphal tips and have type III cell walls.

b. *Frankia* grows in symbiotic association with nonleguminous plants and fixes nitrogen (**figure 22.18**).

22.10 Order *Bifidobacteriales*

a. The genus *Bifidobacterium* is placed in the order *Bifidobacteriales*. This irregular, anaerobic rod is one of the first colonizers of the intestinal tract in nursing babies.

Critical Thinking Questions

1. Even though these are high G + C organisms, there are regions of the genome that are AT-rich. Suggest a few such regions and explain why they must be more AT-rich.
2. Choose two different species in the phylum *Actinobacteria* and investigate their physiology and ecology. Compare and contrast these two organisms. Can you determine why having a high G + C genomic content might confer an evolutionary advantage?
3. *Streptomyces coelicolor* is studied as a model system for cellular differentiation. Some of the genes involved in sporulation contain an A + T rich leucine codon not used in vegetative genes. Suggest how *Streptomyces* might use this rare codon to regulate sporulation.
4. Suppose that you discovered a nodulated plant that could fix atmospheric nitrogen. How might you show that a bacterial symbiont was involved and that *Frankia*, rather than *Rhizobium*, was responsible?
5. Most of the actinobacteria form either rod-shaped or filamentous cells. This requires localized peptidoglycan synthesis at the poles of rod-shaped cells or at the ends of hyphae in filamentous forms. In *Corynebacterium glutamicum, Mycobacteria smegmatis*, and *Streptomyces coelicolor*, the protein DivIVA localizes to these regions of growth. When this protein is depleted in *C. glutamicum*, cells become coccoid; deletion of the gene encoding DivIVA causes cell death. How would you determine if this protein functions in a similar way in other actinobacteria? How could you test if DivIVA interacts directly with peptidoglycan or with components of peptidoglycan synthesis (*see figures 7.8 and 7.9*)? Why do you think this gene is essential in *C. glutamicum* and how would you test this hypothesis? Finally, refer to chapter 7 to determine why the authors of this research comment that *C. glutamicum* lacks the protein MreB.

   Read the original paper and see terrific images of fluorescently labeled proteins in vivo: Letek, M., et al. 2008. DivIVA is required for polar growth in the MreB-lacking rod-shaped actinomycete *Corynebacterium glutamicum*. *J. Bacteriol.* 190:3283.

6. It has long been held that soil microbes such as the streptomycetes produce antibiotics so that they can kill neighboring microbes and use the nutrients released. However, it has been found that each of 480 soil bacterial isolates were resistant to 6 to 20 different antibiotics. Separate research demonstrates that hundreds of soil bacteria can use assorted antibiotics as their sole carbon source. These data, together with the ability of sublethal concentrations of antibiotics to trigger diverse responses in nontarget microbes, has stimulated the hypothesis that in nature, antibiotics are used as signaling molecules rather than weapons. Do you think this argument has any merit and how would you test it?

   Read the original papers: D'Costa, V. M., et al. 2006. Sampling the antibiotic resistome. *Science*. 311:374; Dantas, G., et al. 2008. Bacteria subsisting on antibiotics. *Science*. 320:100.

Learn More

Learn more by visiting the text website at www.mhhe.com/willey8, where you will find a complete list of references.

23

The Protists

The protist Acanthoplegma *spp. is a radiolarian. It uses long, slender pseudopodia (filopodia) to catch and consume its prey.*

CHAPTER GLOSSARY

apical complex A collection of organelles, fibrils, microtubules, and vacuoles found in apicoplexan protists.

apicomplexans Protists that lack special locomotor organelles but have an apical complex and a spore-forming stage. All are intra- or extracellular parasites of animals.

apicoplast A plastid found in apicomplexans that arose by the endosymbiosis of a cyanobacterium but is not a site of photosynthesis in these chemoorganotrophic protists.

ciliates (*Ciliophora*) Protists that move by means of rapidly beating cilia and belong to the *Alveolata* (supergroup *Chromalveolata*).

cyst The inactive or resting stage of a protozoan.

diatoms (*Bacillariophyta*) Photosynthetic protists with siliceous cell walls called frustules. They constitute a substantial fraction of the phytoplankton in marine and freshwater ecosystems.

dinoflagellates (*Dinoflagellata*) Photosynthetic protists characterized by two flagella used to swim in a spinning pattern.

filopodia Long, narrow pseudopodia found in certain protists, for example, the *Rhizaria.*

holozoic nutrition A nutritional strategy in which food items (such as bacteria) are endocytosed with the subsequent formation of a food vacuole or phagosome.

lobopodia Rounded pseudopodia found in some amoeboid protists.

mitosome An organelle found in certain protists that lack mitochondria or hydrogenosomes. The presence of proteins with iron-sulfur clusters that resemble those found in mitochondria suggests that these protists may have once had mitochondria.

osmotrophy A form of nutrition in which soluble nutrients are absorbed through the cytoplasmic membrane; found in prokaryotes, fungi, and some protists.

pseudopodia (s., pseudopodium) Dynamic cytoplasmic extensions of the cell that enable amoeboid protists to move and feed.

reticulopodia Netlike pseudopodia found in certain protists, for example, *Foraminifera.*

saprophyte An organism that takes up nonliving organic nutrients in dissolved form and usually grows on decomposing organic matter.

sporozoite The motile, infective stage of apicomplexan protists, including *Plasmodium,* the causative agent of malaria.

trophozoite The active, motile feeding stage of a protozoan.

So far, our discussion of microbial diversity has focused on the *Bacteria* and *Archaea.* In this chapter, we turn our attention to protists. We follow the higher-level classification scheme of eukaryotes as proposed by the International Society of Protistologists in 2005. Our goal is to survey the vast diversity in the eukaryotic microbial world with an emphasis on the adaptations that make these microorganisms successful competitors. The general structure and reproduction of eukaryotic microbes are reviewed in chapter 4.

Eukaryotic microorganisms suffer from a confused taxonomic history. The kingdom Protista, as presented in many introductory biology texts, is an artificial grouping of over 64,000 different single-celled life forms that lack common evolutionary heritage; that is to say, they are not monophyletic (**figure 23.1**). In this text, we refer to these organisms simply as **protists.** The protists are unified only by what they lack: absent is the level of tissue organization found in the more evolved fungi, plants, and animals. The term **protozoa** (s., protozoan; Greek *protos,* first, and *zoon,* animal) traditionally

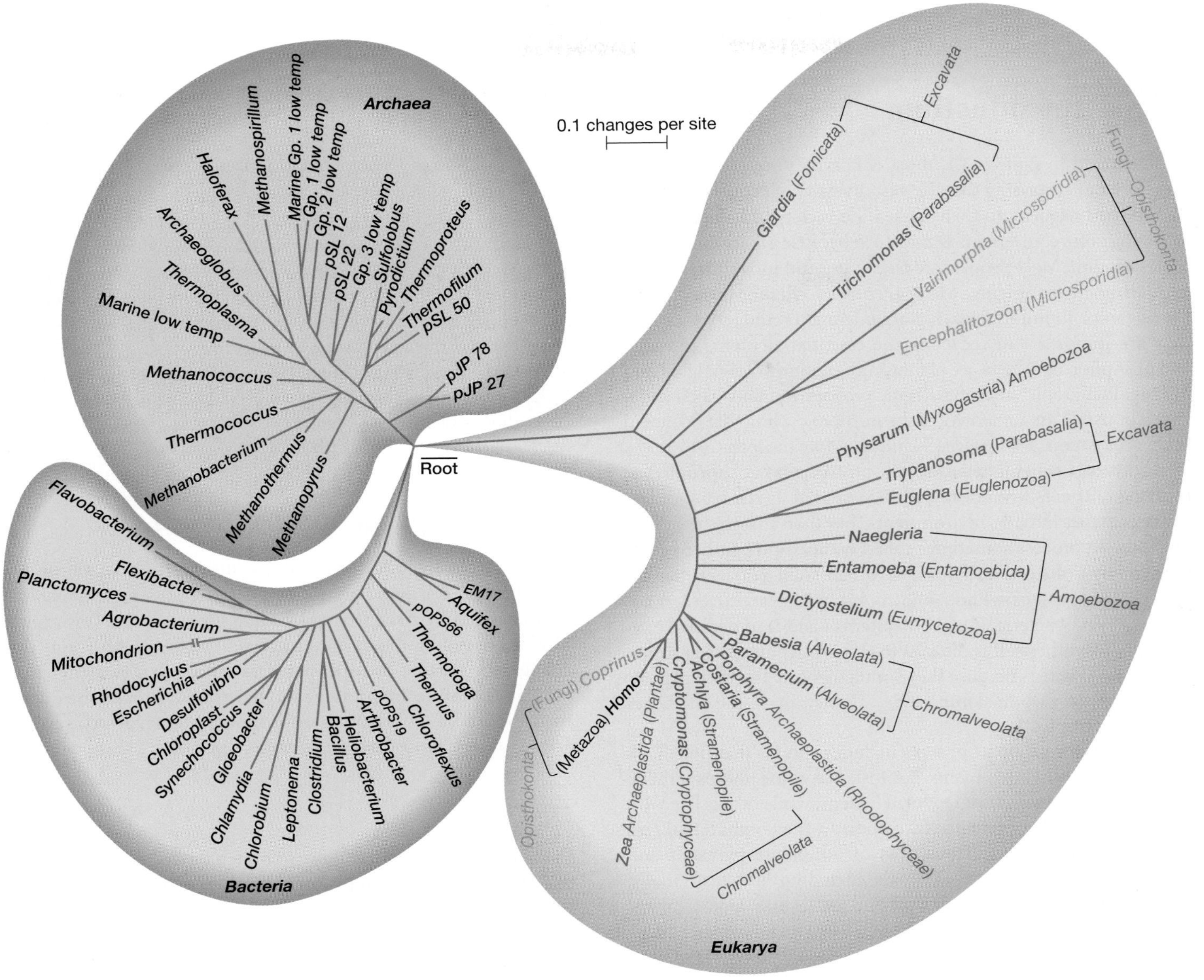

FIGURE 23.1 Universal Phylogenetic Tree, Highlighting Eukaryotic Microorganisms. Recent molecular phylogenetic evidence in combination with morphological and biochemical data has resulted in the establishment of supergroups among the *Eukarya* (each supergroup is highlighted with a different color). This new, higher-order classification scheme is not entirely congruent with the placement of the protists on the universal tree of life; the latter is based entirely on analysis of SSU rRNA. Nonetheless, it demonstrates that the protists are a highly paraphyletic group.

Figure 23.1 Micro Inquiry

How many branches within the *Eukarya* do not represent protists? What does this imply about protist evolution?

refers to chemoorganotrophic protists; protozoology is the study of protozoa. Like animal cells, these microbes usually lack a cell wall. The term **algae** can be used to describe photosynthetic protists that, like plant cells, possess a cell wall. It was originally used to refer to all "simple aquatic plants," but this term has no phylogenetic utility. The study of photosynthetic protists (algae) is often referred to as **phycology** and is the realm of both botanists and protistologists. The study of all protists, regardless of their metabolic type, is called **protistology**.

23.1 Introduction

As illustrated in figure 23.1, most eukaryotes are microbes. It is therefore not surprising that the vast diversity of protists is a function of their capacity to thrive in a wide variety of habitats. Their one common requirement is moisture because all are susceptible to desiccation. Most protists are free living and inhabit freshwater or marine environments. Many terrestrial chemoorganotrophic forms can be found in decaying organic matter and in soil, where they are important in recycling the essential elements nitrogen and phosphorus. Others are planktonic—floating free in lakes and oceans. Planktonic microbes (both prokaryotic and eukaryotic) are responsible for a majority of the nutrient cycling that occurs in these ecosystems. ▶▶| *Biogeochemical cycling (chapter 26)*

Protozoa, or chemoheterotrophic protists, may be **saprophytes,** securing nutrients from dead organic material by releasing degradative enzymes into the environment. They then absorb the soluble products—a process sometimes called **osmotrophy.** Other protozoa employ **holozoic nutrition,** in which solid nutrients are acquired by phagocytosis. Photoautotrophic protists are strict aerobes and, like cyanobacteria, use photosystems I and II to perform oxygenic photosynthesis. It is difficult to classify the nutritional strategies of some protists because they simultaneously use both organic and inorganic carbon compounds. This strategy is sometimes called **mixotrophy.**

Every major group of protists includes species that live in association with other organisms. For instance, some photosynthetic protists associate with fungi to form lichens, while others live with corals, where they provide fixed carbon to the coral animal (*see figures 30.3 and 30.12*). Thousands of others are parasites and cause important diseases in animals, including humans. Finally, protists are useful in biochemical and molecular biological research. Not only do they display an amazing array of unique adaptations, but many biochemical pathways used by protists are found in other eukaryotes, making them useful model organisms.

Ever since Antony van Leeuwenhoek described the first protozoan "animalcule" in 1674, the taxonomic classification of the protists has remained in flux. During the twentieth century, classification schemes were based on morphology rather than evolutionary relationships. Protists were often classified into four major groups based on their means of locomotion: flagellates (*Mastigophora*), ciliates (*Infusoria* or *Ciliophora*), amoebae (*Sarcodina*), and stationary forms (*Sporozoa*). Although nonprotistologists still use these terms, these divisions have no bearing on evolutionary relationships and should be avoided. It is now agreed that the old classification system is best abandoned, but little agreement exists on what should take its place. However, in 2005 the International Society of Protistologists proposed a higher-level classification system for the eukaryotes based on more recent morphological, biochemical, and phylogenetic analyses. This scheme does not use formal hierarchical rank designations such as class and order, reflecting the fact that protist taxonomy remains an area of active research. This chapter is organized in accordance with this classification scheme. The **Classification of the Protists** as proposed by the International Society of Protistologists, is presented in a table on the text website *(www.mhhe.com/willey8.)*

Search This: International Society of Protistologists

1. Why are protists not considered monophyletic?
2. What is the difference between a protozoan and an alga?
3. Describe the ways in which protists take up nutrients. How does this differ from bacteria and archaea?
4. What is the distribution of these microbes?
5. What does the fact that photosynthetic protists use only oxygenic photosynthesis suggest about the evolution of these organisms?

23.2 Supergroup *Excavata*

The *Excavata* includes some of the most primitive, or deeply branching, eukaryotes (figure 23.1). Most possess a cytostome characterized by a suspension-feeding groove with a posteriorly directed flagellum that is used to generate a feeding current. This enables the capture of small particles. Those that lack this morphological feature are presumed to have had it at one time during their evolution—that is to say, it is thought to have been secondarily lost. |◀◀ *Protist morphology (section 4.9)*

Fornicata

Ever curious, Antony van Leeuwenhoek described *Giardia lamblia* (**figure 23.2*a***) from his own diarrheic feces. Over 300 years later, this species continues to be a public health concern. Today this microaerophilic protist most often infects campers and other individuals who unwittingly consume contaminated water. It is the most common cause of epidemic waterborne diarrheal disease (about 30,000 cases yearly). Members of the *Fornicata* bear flagella and lack mitochondria. However, mitochondria-like double-bounded membrane organelles called **mitosomes** have been reported in *Giardia.* Because these organelles have proteins that have iron-sulfur clusters resembling those found in mitochondria, it has been suggested that these primitive organelles share ancestry with mitochondria and hydrogenosomes. This would suggest that *Giardia* and other mitosome-containing protists once had mitochondria, but the organelles were lost as the microbes evolved. Unlike *Giardia,* most *Fornicata* are harmless symbionts. The few free-living forms are most often found in waters that are heavily polluted with organic nutrients (a condition known as eutrophication). Only asexual reproduction by binary fission has been observed (*see figure 4.25*). Other pathogenic species include *Hexamida salmonis,* a troublesome fish parasite found in hatcheries and fish farms, and *H. meleagridis,* a turkey pathogen that is responsible for the annual loss of millions of dollars in poultry revenue. |◀◀ *Organelles involved in energy conservation (section 4.6)* ▶▶| *Giardiasis (section 39.5)*

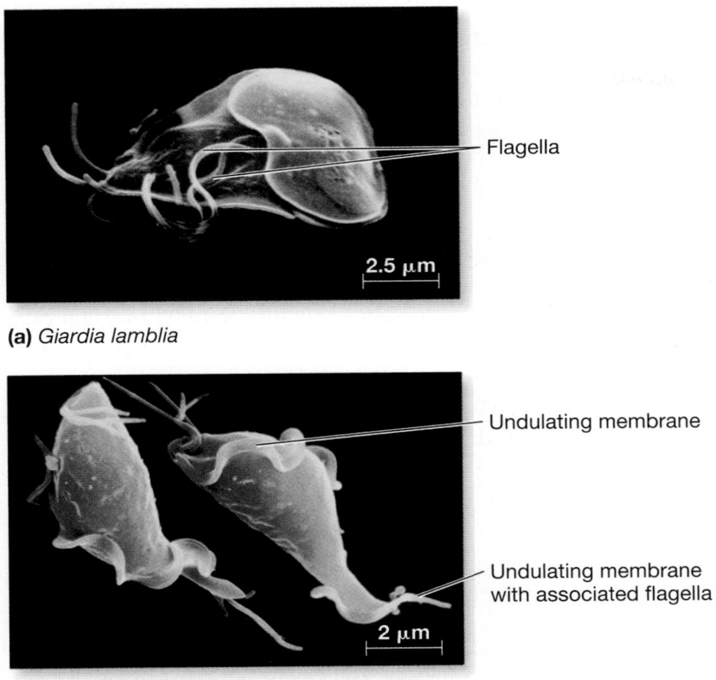

(a) *Giardia lamblia*

(b) *Trichomonas vaginalis*

FIGURE 23.2 Parasitic Members of the Supergroup *Excavata*. (a) *Giardia lamblia.* (b) *Trichomonas vaginalis.* These specialized parasitic flagellates absorb nutrients from living hosts.

Figure 23.2 Micro Inquiry

How are these two protists morphologically similar but physiologically different?

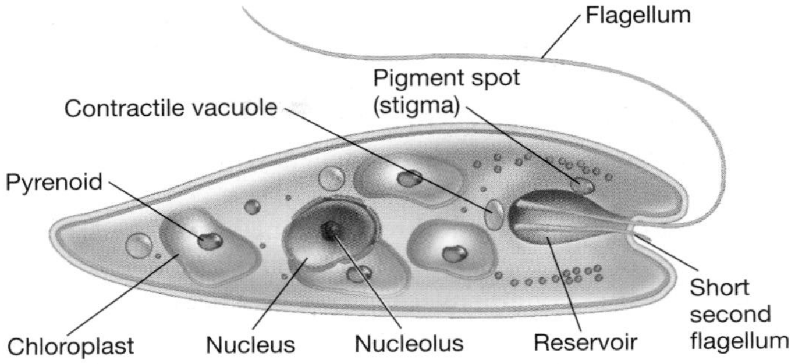

FIGURE 23.3 *Euglena:* Principle Structures Found in This Euglenoid. Notice that a short second flagellum does not emerge from the anterior invagination. In some euglenoids, both flagella are emergent.

Figure 23.3 Micro Inquiry

What are the functions of the stigma and the reservoir?

Parabasalia

Members of the *Parabasalia* are flagellated; most are endosymbionts of animals. They use phagocytosis to engulf food items, although they lack a distinct cytostome. Here, we consider two subgroups: the *Trichonymphida* and the *Trichomonadida*. *Trichonymphida* are obligate mutualists in the digestive tracts of wood-eating insects such as termites and wood roaches. They secrete the enzyme cellulase needed for the digestion of wood particles, which they entrap with pseudopodia. One species, *Trichonympha campanula,* can account for up to one-third of the biomass of an individual termite (*see figure 30.2*). This species is particularly large for a protist (several hundred micrometers) and can bear several thousand flagella. Although asexual reproduction is the norm, a hormone called ecdysone produced by the host when molting triggers sexual reproduction. ▶▶| *Microorganism-insect mutualisms (section 30.1)*

Trichomonadida, or simply the trichomonads, do not require oxygen and possess hydrogenosomes rather than mitochondria (*see figure 4.17*). They undergo asexual reproduction only. They are symbionts of the digestive, reproductive, and respiratory tracts of many vertebrates. *Tritrichomonas foetus* is a cattle parasite and an important cause of spontaneous abortion in these animals. Four species infect humans: *Dientamoeba fragilis, Pentatrichomonas hominis, Trichomonas tenax,* and *T. vaginalis* (figure 23.2*b*). *D. fragilis* has recently been recognized as a cause of diarrhea, while *T. vaginalis* has long been known to be pathogenic. Found in the genitourinary tract of both men and women, most *T. vaginalis* strains are either not pathogenic or only mildly so. Nonetheless, the sexual transmission of pathogenic strains accounts for an estimated 7 million cases of trichomoniasis annually in the United States and 180 million cases annually worldwide. ▶▶| *Trichomoniasis (section 39.4)*

Euglenozoa

The *Euglenozoa* are commonly found in freshwater, although a few species are marine. About one-third of euglenids are photoautotrophic; the remaining are free-living chemoorganotrophs. Most chemoorganotrophic forms are saprophytic, although a few parasitic species have been described. The representative genus is the photoautotroph *Euglena*. A typical *Euglena* cell (**figure 23.3**) is elongated and bounded by a plasmalemma. The pellicle consists of proteinaceous strips and microtubules; it is elastic enough to enable turning and flexing of the cell, yet rigid enough to prevent excessive alterations in shape. *Euglena* contains chlorophylls *a* and *b* together with carotenoids. The large nucleus contains a prominent nucleolus. The primary storage product is paramylon (a polysaccharide composed of $\beta(1\rightarrow3)$ linked glucose molecules), which is unique to euglenoids. A red eyespot called a **stigma** helps the organism orient to light and is located near an anterior reservoir. A contractile vacuole near the reservoir continuously collects water from the cell and empties it into the reservoir, thus regulating the osmotic pressure within the organism. Two flagella arise from the base of the reservoir, although only one emerges from the canal and actively beats to move the cell. Reproduction in euglenoids is by longitudinal mitotic cell division.

Several protists of medical relevance belong to the *Euglenozoa.* These include members of the genus *Leishmania,* which cause a group of conditions, collectively termed leishmaniasis, that include systemic and skin/mucous membrane afflictions affecting some 12 million people. Chagas' disease is caused by *Trypanosoma cruzi,* which is transmitted by "kissing bugs" (*Triatominae*), so called because they bite the face of sleeping victims (*see figure 39.10*). Two to 3 million citizens of South and Central America show the central and peripheral nervous system dysfunction that is characteristic of this disease; of these, about 45,000 die of the disease each year. ▶▶| *Leishmaniasis and trypanosomiasis (section 39.3)*

The **trypanosomes** are also of great importance. These microbes exist only as parasites of plants and animals, and have global significance. *Trypanosoma gambiense* and *T. rhodesiense* (often considered a subspecies of *T. brucei*) cause African sleeping sickness (**figure 23.4**). Ingestion of these parasites by the blood-sucking tsetse fly triggers a complex cycle of development and reproduction, first in the fly's gut and then in its salivary glands (*see figure 39.11*). From there the parasite is easily transferred to a vertebrate host, where it often causes a fatal infection. It is estimated that about 65,000 people die annually of sleeping sickness. The presence of this dangerous parasite prevents the use of about 11 million square kilometers of African grazing land.

Trypanosomes have a thick glycoprotein layer coating the cell wall surface. The chemical composition of the glycoprotein layer is switched cyclically, expressing only one of 1,000 to 2,000 variable antigens at any given time. This process, known as **antigenic variation,** enables the parasite's escape from host immune surveillance. It is therefore not surprising that there are no vaccines for either Chagas' disease or African sleeping sickness and that the few drugs available for treatment are not particularly effective. Interestingly, it was recently shown that when African trypanosomes are blocked in the production of one of the flagellar axoneme proteins (*see figure 4.22*), nonviable "monster" aggregates of cells are formed. It is thought that the protists use their flagella to mark cell polarity when dividing. Because these proteins are specific to trypanosomes, it opens the door to the development of agents that specifically target the axonemal proteins. |◀◀ *External structures (section 4.7)*

Search This: CDC Division of Parasitic Diseases

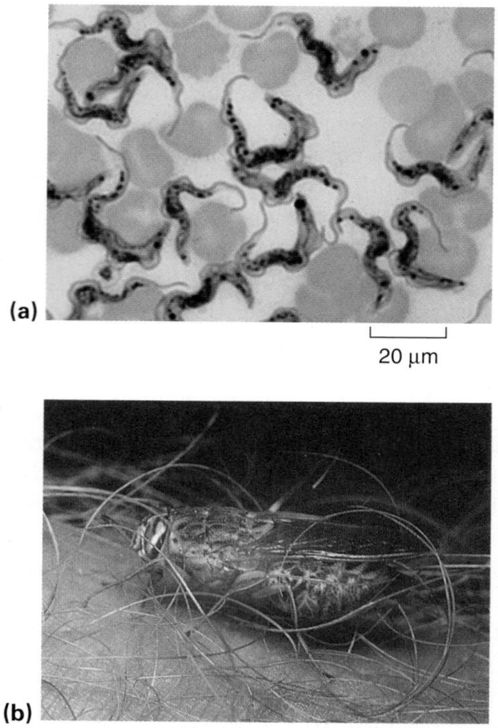

FIGURE 23.4 The Euglenozoan *Trypanosoma* and Its Insect Host. (a) *Trypanosoma* among red blood cells. Note the dark-staining nuclei, anterior flagella, and undulating changeable shape (×500). (b) The tsetse fly, shown here sucking blood from a human arm, is an important vector of the *Trypanosoma* species that cause African sleeping sickness.

1. What are some features that distinguish the *Parabasalia* from the *Euglenozoa?*
2. What is the function of the plasmalemma and pellicle in *Euglena?* How are these structures similar to the bacterial cell wall?
3. What *Euglenozoa* genera cause disease? What adaptations make these protists successful pathogens?

23.3 Supergroup *Amoebozoa*

It is clear that the amoeboid form arose independently numerous times from various flagellated ancestors. Thus some amoebae are placed in the supergroup *Amoebozoa,* while others are in the *Rhizaria.* One of the morphological hallmarks of amoeboid motility is the use of **pseudopodia** (meaning "false feet") for both locomotion and feeding (**figure 23.5**). Pseudopodia can be rounded (**lobopodia**), long and narrow (**filopodia**), or form a netlike mesh (**reticulopodia**). Amoebae that lack a cell wall or other supporting structures and are surrounded only by a plasma membrane are called **naked amoebae.** In contrast, the plasma membrane of a **testate amoeba** is covered by material that is either made by the protist itself or collected by the organism from the environment. Binary fission is the usual means of asexual division (*see figure 4.25*), although some amoebae form cysts that undergo multiple fission.

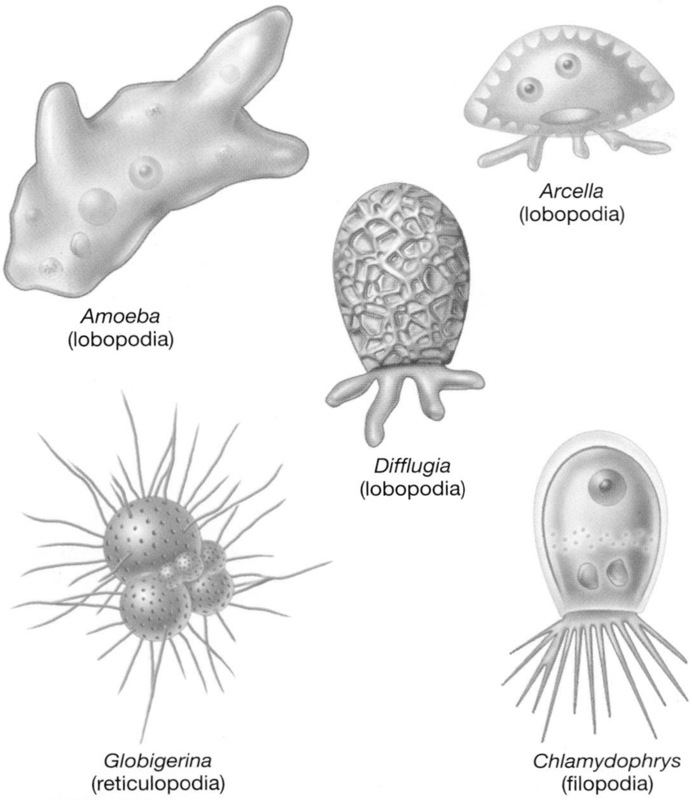

FIGURE 23.5 Pseudopodia.

Tubulinea

The *Tubulinea* inhabit almost any environment where they will remain moist; this includes glacial meltwater, marine plankton, tide pools, lakes, and streams. Free-living forms are known to dwell in ventilation ducts and cooling towers, where they feed on microbial biofilms. Others are endosymbionts, commensals, or parasites of invertebrates, fishes, and mammals. Some harbor intracellular symbionts, including algae, bacteria, and viruses, but the nature of these relationships is not well understood. *Amoeba proteus,* a favorite among introductory biology laboratory instructors, is included in this group.

Entamoebida

Like the *Excavata,* these protists lack mitochondria and hydrogenosomes but contain mitosomes. However, the identification of about 20 putative mitochondrial proteins in the free-living amoeba *Mastigamoeba balamuthi* suggests that these are more advanced than the mitosomes found in *Giarida. M. balamuthi* is a relative of *Entamoeba histolytica,* the cause of amoebic dysentery—the third-leading cause of parasitic death worldwide. Individuals acquire this pathogen by eating feces-contaminated food or by drinking water contaminated by *E. histolytica* cysts. These pass unharmed through the stomach and undergo multiple fission when introduced to the alkaline conditions in the intestines (*see figure 39.19*). There they graze on bacteria and produce a suite of digestive enzymes that degrade gut epithelial cells. *E. histolytica* can penetrate into the bloodstream and migrate to the liver, lungs, or skin. Cysts in feces remain viable for weeks but are killed by heat greater than 40°C. ▶▶| *Amebiasis (section 39.5)*

Eumycetozoa

First described in the 1880s, the *Eumycetozoa* or "slime molds" have been classified as plants, animals, and fungi. As we examine their morphology and behavior, the source of this confusion should become apparent. Recent analysis of certain proteins (e.g., elongation factor EF-1, α-tubulin, and actin) as well as physiological, behavioral, biochemical, and developmental data point to a monophyletic group (figure 23.1). The *Eumycetozoa* include the *Myxogastria* and *Dictyostelia.* The **acellular slime mold,** or ***Myxogastria,*** life cycle includes a distinctive stage when the organisms exist as streaming masses of colorful protoplasm that creep along in amoeboid fashion over moist, rotting logs, leaves, and other organic matter, which they degrade (**figure 23.6**). Their name derives from the presence of a large, multinucleate mass that lacks cell membranes. This is called a plasmodium, and there can be as many as 10,000 synchronously dividing nuclei within a single plasmodium (figure 23.6*b*). Feeding is by endocytosis. When starved or dried, the plasmodium develops ornate fruiting bodies. As these mature, they form stalks with cellulose walls that are resistant to environmental stressors (figure 23.6*c,d*). When conditions improve, spores germinate and release haploid amoeboflagellates. These fuse, and as the resulting zygotes feed, nuclear division and synchronous mitotic divisions give rise to the multinucleate plasmodium.

The **cellular slime molds** (*Dictyostelia*) are strictly amoeboid and use endocytosis to feed on bacteria and yeasts. Their complex life cycle involves true multicellularity, despite their primitive evolutionary status (**figure 23.7*a***). The species *Dictyostelium discoideum* is an attractive model organism. The vegetative cells move as a mass, sometimes called a pseudoplasmodium, because individual cells retain their cell membranes. Starved cells release cyclic AMP and a specific glycoprotein, which serve as molecular signals. Other cells sense these compounds and respond by forming an aggregate around the signal-producing cells (figure 23.7*b*). In this way large, motile, multicellular slugs develop and serve as precursors to fruiting body formation (figure 23.7*c*). Fruiting body morphogenesis commences when the slug stops and cells pile on top of each other. Cells at the bottom of this vertically oriented structure form a stalk by secreting cellulose, while cells at the tip differentiate into spores (figure 23.7*d,e*). Germinated spores become vegetative amoebae to start this asexual cycle anew.

It was recently discovered that *Dictyostelium* cells can also differentiate such that they resemble primitive immune cells. During slug formation, some cells become "sentinel cells" and vanquish harmful bacteria. Sentinel cells accomplish this by

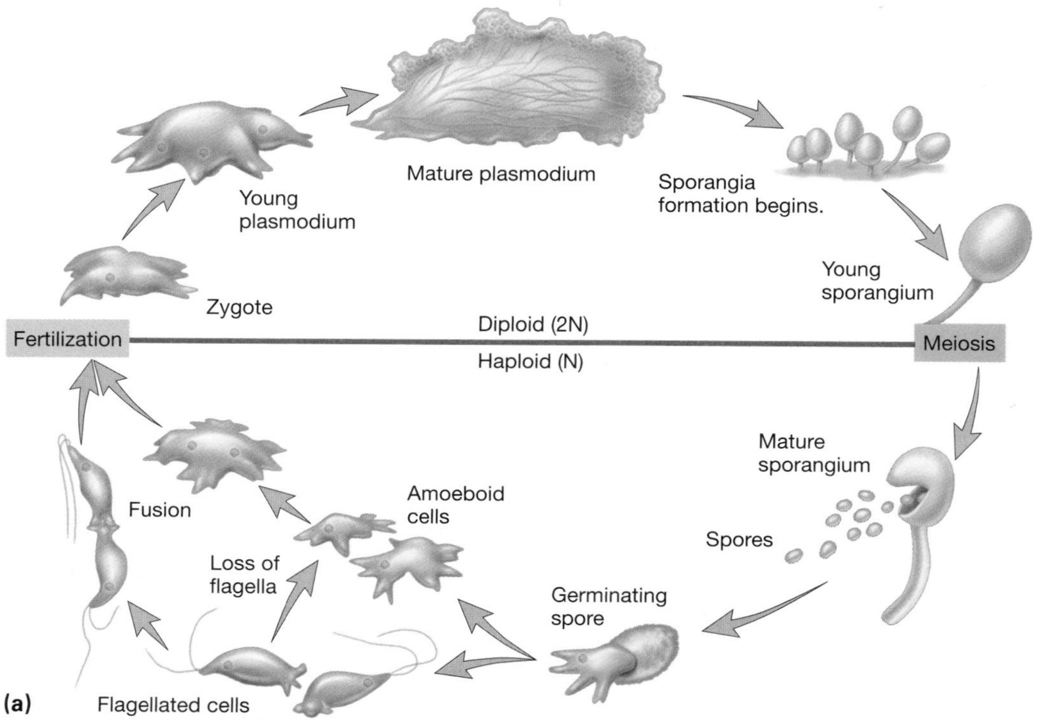

(a)

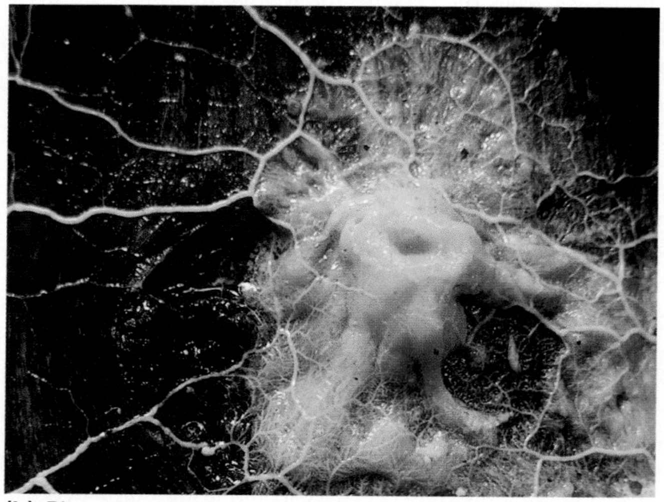
(b) *Physarum* sp.

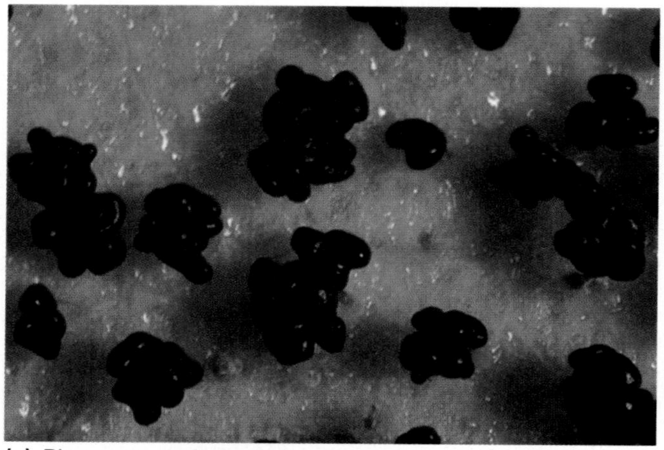
(c) *Physarum polycephalum*

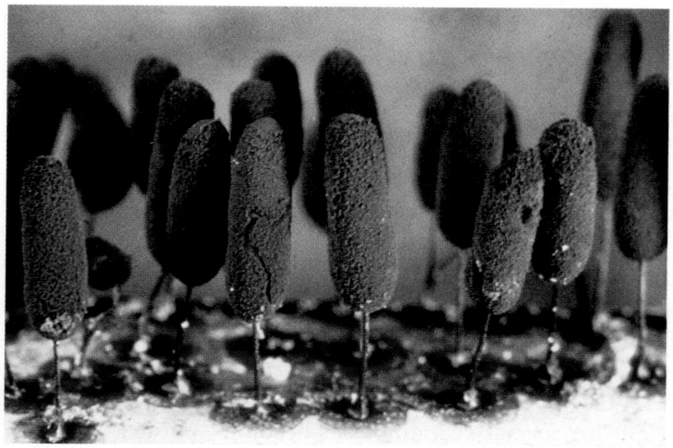
(d) *Stemonitis*

FIGURE 23.6 Acellular Slime Molds. (a) The life cycle of a plasmodial slime mold includes sexual reproduction; when conditions are favorable for growth, the adult diploid forms sporangia. Following meiosis, the haploid spores germinate, releasing haploid amoeboid or flagellated cells that fuse. (b) Plasmodium of the slime mold *Physarum* sp. (×175). Sporangia of (c) *Physarum polycephalum,* and (d) *Stemonitis.*

***Figure 23.6* Micro Inquiry**

How does a plasmodium differ from a single cell?

producing proteins that are similar to those involved in immune responses in higher organisms. These cells roam within the slug as if patrolling for pathogenic bacteria such as *Legionella pneumophila,* which are known to infect *Dictyostelium.* Sentinel cells have been found in several species related to *Dictyostelium*; immunologists are not too surprised, noting that all multicellular organisms need protection against bacterial pathogens.

Sexual reproduction in *D. discoideum* involves the formation of special spores call macrocysts. These arise by a form of conjugation that has some unusual features. First, a group of amoebae become enclosed within a wall of cellulose. Following conjugation, a single, large amoeba forms and cannibalizes the remaining amoebae. The now giant amoeba matures into a macrocyst. Macrocysts can remain dormant within their cellulose

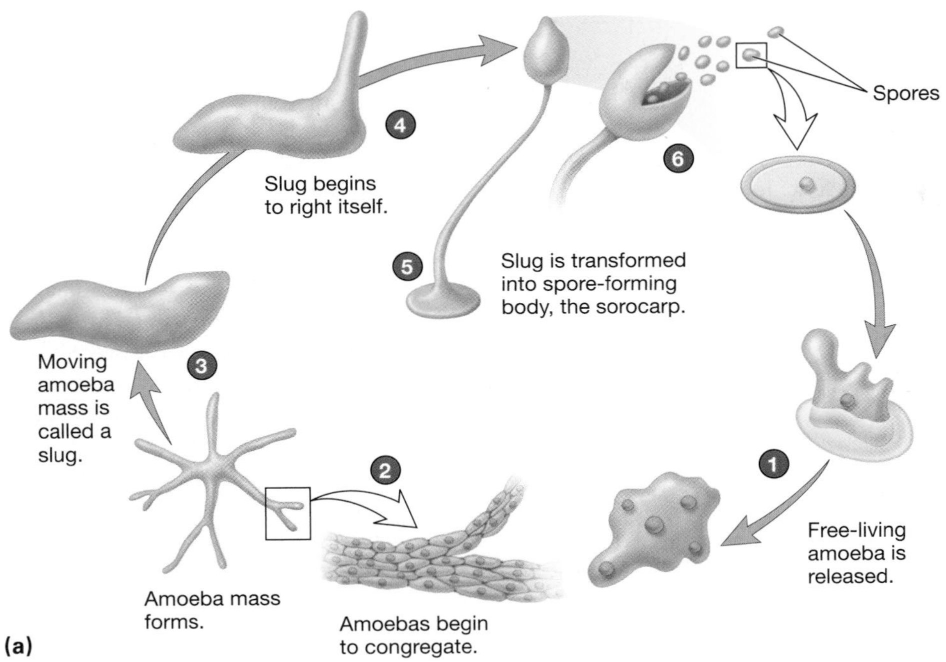
Spores
4
Slug begins
to right itself.
6
5
Slug is transformed
into spore-forming
body, the sorocarp.
Moving
amoeba
mass is
called a
slug.
3
2
1
Free-living
amoeba is
released.
Amoeba mass
forms.
Amoebas begin
to congregate.
(a)

walls for extended periods of time. Vegetative growth resumes after the diploid nucleus undergoes meiosis to generate haploid amoebae.

The 33.8-Mb genome of *D. discoideum* supports the notion that these soil-dwelling microbes are more primitive than the fungi. For instance, *D. discoideum* has 14 different histidine kinase receptor proteins; these proteins are typically found in bacteria and archaea. Also of note is the presence of 40 genes that appear to be involved in cellulose biosynthesis or degradation. These genes could be involved in producing the cellulose that the microbe needs during morphological differentiation or for degrading cellulose-containing microbes.

Search This: Dictyostelium movies

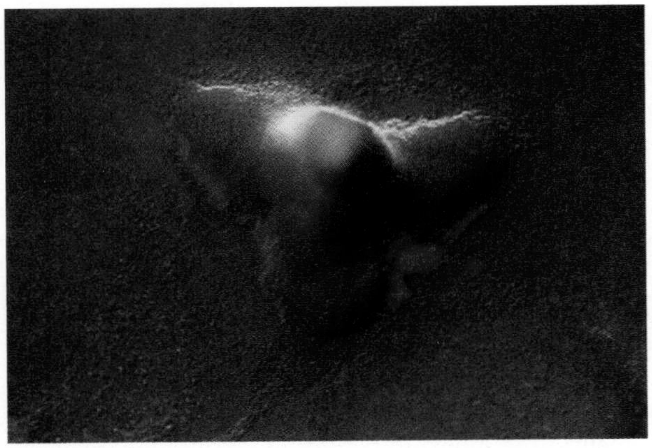
(b)

(c)

(d)

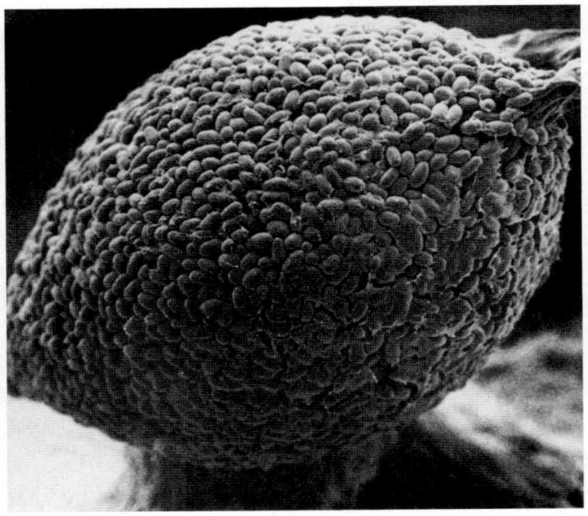
(e)

FIGURE 23.7 Development of *Dictyostelium discoideum,* a Cellular Slime Mold. (a) Life cycle. (b) Aggregating *D. discoideum* become polar and begin to move in an oriented direction in response to the molecular signal cAMP. (c) Slug begins to right itself and (d) forms a spore-forming body called a sorocarp. (e) Electron micrograph of a sorocarp showing individual spores (×1,800).

Figure 23.7 Micro Inquiry

How would a mutant strain unable to secrete cAMP behave if mixed with a wild type strain? What about if it were mixed with another cAMP mutant?

23.4 Supergroup *Rhizaria*

The *Rhizaria* are amoeboid in morphology and thus were historically grouped with the members of what we now call the *Amoebozoa.* However, molecular phylogenetic analysis makes it clear that the *Amoebozoa* and *Rhizaria* are not monophyletic. In fact, more recent evidence strongly suggests that the *Rhizaria* form a monophyletic clade with the *Stramenopila* and *Alveolata,* which we place with the *Chromalveolata* in agreement with the International Society of Protistologists. Morphologically, the *Rhizaria* can be distinguished by their filopodia, which can be simple, branched, or connected. Filopodia supported by microtubules are known as **axopodia** (s., axopodium). Axopodia protrude from a central region of the cell called the axoplast and are primarily used in feeding (**figure 23.8*a***).

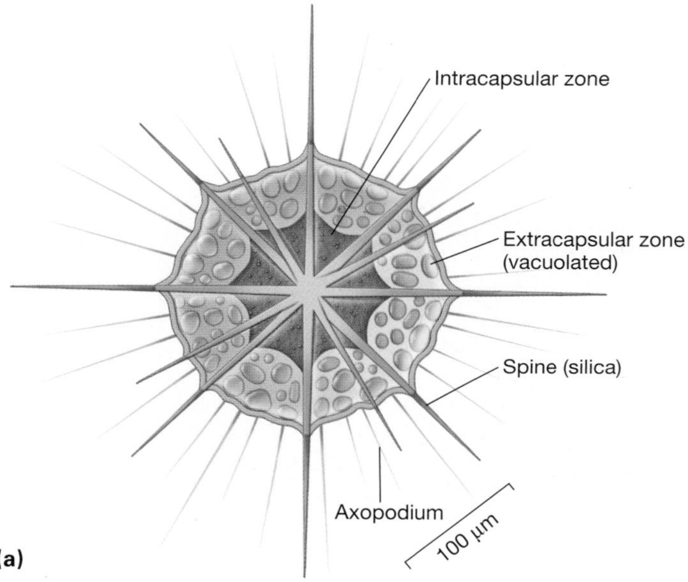

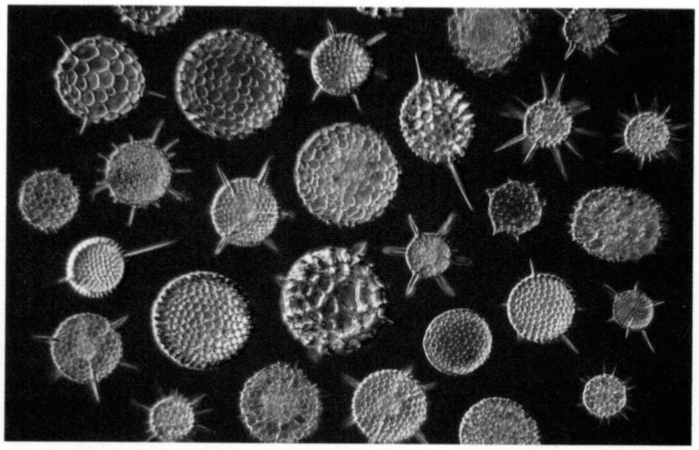

(b) Radiolarian shells

FIGURE 23.8 Radiolaria. (a) The radiolarian *Acanthometra elasticum* demonstrates the internal skeleton. (b) Radiolarian shells made of silica.

Radiolaria

Most *Radiolaria* have an internal skeleton made of siliceous material; however, members of the subgroup *Acantharia* have endoskeletons consisting of strontium sulfate. A few genera have an exoskeleton of siliceous spines or scales, while some lack a skeleton completely. Skeletal morphology is highly variable and often includes radiating spines that help the organisms float, as does the storage of oils and other low-density fluids (figure 23.8*b*).

The *Radiolaria* feed by endocytosis, using mucus-coated filopodia to entrap prey including bacteria, other protists, and even small invertebrates. Large prey items are partially digested extracellularly before becoming encased in a food vacuole. Many surface-dwelling radiolarians have algal symbionts thought to enhance their net carbon assimilation. The acantharians reproduce only sexually by consecutive mitotic and meiotic divisions that release hundreds of biciliated cells. Asexual reproduction (binary or multiple fission, or budding) is most common in other *Rhizaria,* but sexual reproduction can be triggered by nutrient limitation or a heavy feeding. In this case, two haploid nuclei fuse to form a diploid zygote encased in a cyst from which it is released when survival conditions improve.

Foraminifera

The *Foraminifera* (also called forams) range in size from roughly 20 μm to several centimeters. Their filopodia are arranged in a branching network called reticulopodia. These bear vesicles at their tips that secrete a sticky substance used to trap prey. Many species harbor endosymbiotic algae that migrate out of reticulopodia (without being eaten) to expose themselves to more sunlight. These algae contribute significantly to foram nutrition. Forams have characteristic tests arranged in multiple chambers that are sequentially added as the protist grows (**figure 23.9**).

Some forams are free-floating planktonic forms; others are benthic, living within the sediments. Recently benthic forams inhabiting anoxic sediments were shown to perform denitrification, reducing nitrate completely to dinitrogen gas. Foraminifera are thus the first and only example of a eukaryote capable of this process. How this impacts the global nitrogen cycle awaits further study. ◀◀ *Anaerobic respiration (section 10.6)* ▶▶ *Nitrogen cycle (section 26.1)*

Foraminiferan life cycles can be complex. While some smaller species reproduce only asexually by budding or multiple fission, larger forms frequently alternate between sexual and asexual phases. During the sexual phase, flagellated gametes pair, fuse, and generate asexual individuals (agamonts). Meiotic division of the agamonts gives rise to haploid gamonts. There are several mechanisms by which gamonts return to the diploid condition. For instance, a variety of forams release flagellated gametes that become fertilized in the open water. In others, two or more gamonts attach to one another, enabling gametes to fuse within the chambers of the paired tests. When the shells separate, newly formed agamonts are released.

Foraminifera are found in marine and estuarine habitats. Some are planktonic, but most are benthic. Foraminiferan tests accumulate on the seafloor, where they create a fossil record dating back to the Early Cambrian (543 million years ago),

FIGURE 23.9 A Foraminiferan. The reticulopodia are seen projecting through pores in the calcareous test, or shell, of this protist.

FIGURE 23.10 White Cliffs of Dover. The limestone that forms these cliffs is composed almost entirely of fossil shells of protists, including foraminifera.

which is helpful in oil exploration. Their remains, or ooze, can be up to hundreds of meters deep in some tropical regions. Foram tests make up most modern-day chalk, limestone, and marble, and are familiar to most as the white cliffs of Dover in England (**figure 23.10**). They also formed the stones used to build the great pyramids of Egypt.

Search This: Cushman Foundation/forminifera

1. Describe filopodia, lobopodia, and reticulopodia form and function.
2. Why do you think the slime molds have been so hard to classify?
3. What is a plasmodium? How does it differ between the acellular and cellular slime molds?
4. Describe the life cycle of *Dictyostelium discoideum.* Why is this organism a good model for the study of cellular differentiation and coordinated cell movement?
5. What adaptations do the planktonic radiolaria have to help them float?
6. Compare the means by which radiolaria use axopodia with the way foraminifera use reticulopodia.
7. Describe the forms of sexual reproduction in the forams.

23.5 Supergroup *Chromalveolata*

The supergroup *Chromalveolata* is very diverse as it includes autotrophic, mixotrophic, and heterotrophic protists. These protists are united in plastid origin, which appears to have been acquired by endosymbiosis with an ancestral archaeplastid (i.e., green or red alga). This view has been challenged, and it has been suggested that two of the "first rank" groups, the *Haptophyta* and *Cryptophyceae* form a clade that is separate from the other two first order groups, the *Alveolata* and *Stramenopila.* Here, we introduce three of these groups, the *Alveolata, Stramenopila,* and *Haptophyta;* we discuss the physiology and ecology of these microbes and stress that their phylogeny is unsettled.

Alveolata

The *Alveolata* is a large group that includes the *Dinoflagellata* (dinoflagellates), *Ciliophora,* and *Apicomplexa.* We begin our discussion with the **dinoflagellates**—a large group most commonly found in marine plankton, where some species are responsible for the phosphorescence sometimes seen in seawater. Their nutrition is complex; photoautotrophy, heterotrophy, and mixotrophy are all observed. Most of the heterotrophic forms are saprophytic (either entirely or as facultative chemoorganotrophs), but some also use endocytosis. Each cell bears two distinctively placed flagella: One is wrapped around a transverse groove (the girdle), and the other is draped in a longitudinal groove (the sulcus; **figure 23.11*a,b***). The orientation and beating patterns of these flagella cause the cell to spin as it is propelled forward; the name *dinoflagellate* is derived from the Greek *dinein,* "to whirl." Many dinoflagellates are covered with cellulose plates that are secreted within the alveolar sacs that lie just under the plasma membrane. These forms are said to be thecate or armored; those with empty alveoli are called athecate or naked and include the luminescent genus *Noctiluca.* Like some other protists, dinoflagellates have specialized compressed proteins called **trichocysts** (figure 23.11*c*). When attacked, trichocysts shoot out from the cell as a means of defense.

Most dinoflagellates are free living, although some form important associations with other organisms. Endosymbiotic dinoflagellates that live as undifferentiated cells occasionally send out motile cells called **zooxanthellae.** The most well-known zooxanthellae belong to the genus *Symbiodinium.* These are photosynthetic endosymbionts of reef-building coral (*see figure 30.3*). They provide fixed carbon to the coral animal and help maintain the internal chemical environment needed for the coral to secrete its calcium carbonate exoskeleton. Dinoflagellates are also responsible for toxic "red tides" that harm other organisms, including humans. Common red tide dinoflagellates include members of the genera *Alexandrium* and *Lingulodinium* (formerly called *Gonyaulax*). ▶▶| *Microorganisms in estuaries and saltmarshes (section 28.2); Zooxanthellae (section 30.1)*

The **ciliates (*Ciliophora*)** include about 12,000 species. All are chemoorganotrophic and range from about 10 μm to 4.5 mm long. They inhabit both benthic and planktonic communities in marine and freshwater systems, as well as moist soils. As their name implies, *Ciliophora* employ many cilia for locomotion and feeding. The cilia are generally arranged either in longitudinal rows (**figure 23.12**) or in spirals around the body of the organism. They beat with an oblique stroke, causing the protist to revolve as it swims. Ciliary beating is so precisely coordinated that ciliates can go both forward and backward. There is great variation in shape, and most ciliates do not look like the slipper-shaped *Paramecium.* Some species, including *Vorticella,* attach to substrates by a long stalk. *Stentor* attaches to substrates and stretches out in a trumpet shape to feed. A few species have tentacles for the capture of prey. Some can discharge toxic, threadlike darts called toxicysts, which are used in capturing prey. A striking feature of the *Ciliophora* is their ability to quickly entrap many particles by the action of the cilia around the buccal cavity. Food first enters the cytostome and passes into phagocytic vacuoles that fuse with lysosomes after detachment from the cytostome and the vacuole's contents are digested. After the digested material has been absorbed into the cytoplasm, the vacuole fuses with the cytoproct and waste material is expelled.

Most ciliates have two types of nuclei: a large **macronucleus** and a smaller **micronucleus.** The micronucleus is diploid and contains

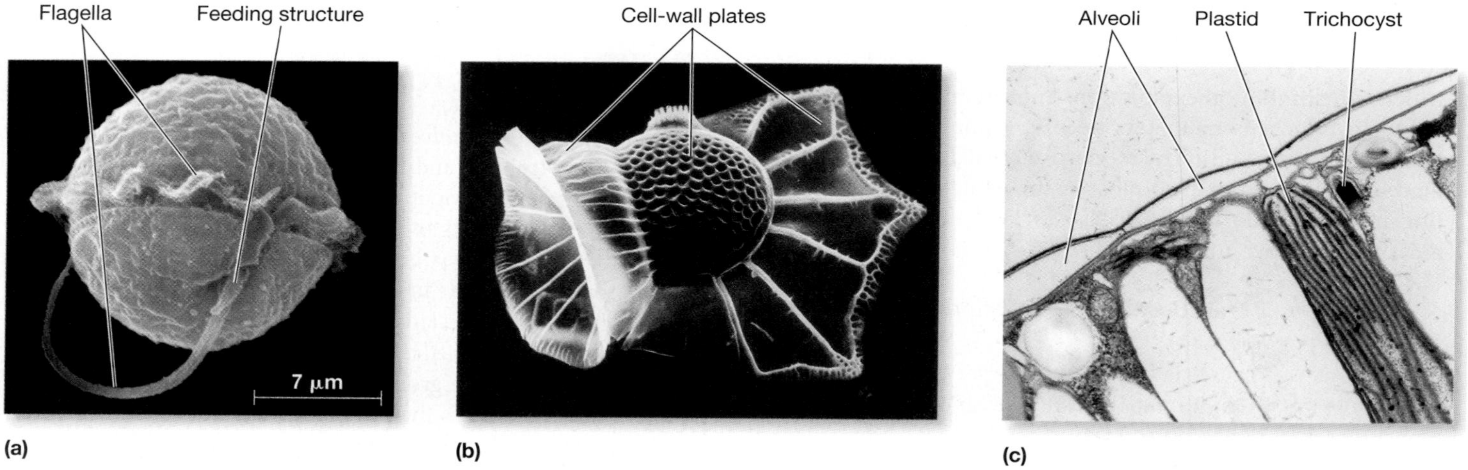

FIGURE 23.11 Dinoflagellates. (a) The surface of *Peridiniopsis berolinensis* is smooth because the alveoli appear empty. Two types of flagella can be seen in this freshwater dinoflagellate. One flagellum coils around the girdle; as it moves, this flagellum causes the cell to spin. The straight flagellum that extends from the cell acts as a rudder to determine the direction of movement. (b) The alveoli of the marine dinoflagellate genus *Ornithocercus* contains cellulose cell-wall plates. The cell wall extends to form sail-like structures. (c) Alveoli appear as sac-shaped membranous vesicles below the plasma membrane in this transmission electron micrograph of a dinoflagellate cell. Note the trichocysts ready for discharge and membrane within the plastid.

Figure 23.11 Micro Inquiry

What do you think is the function of the "sails" in *Ornithocercus*?

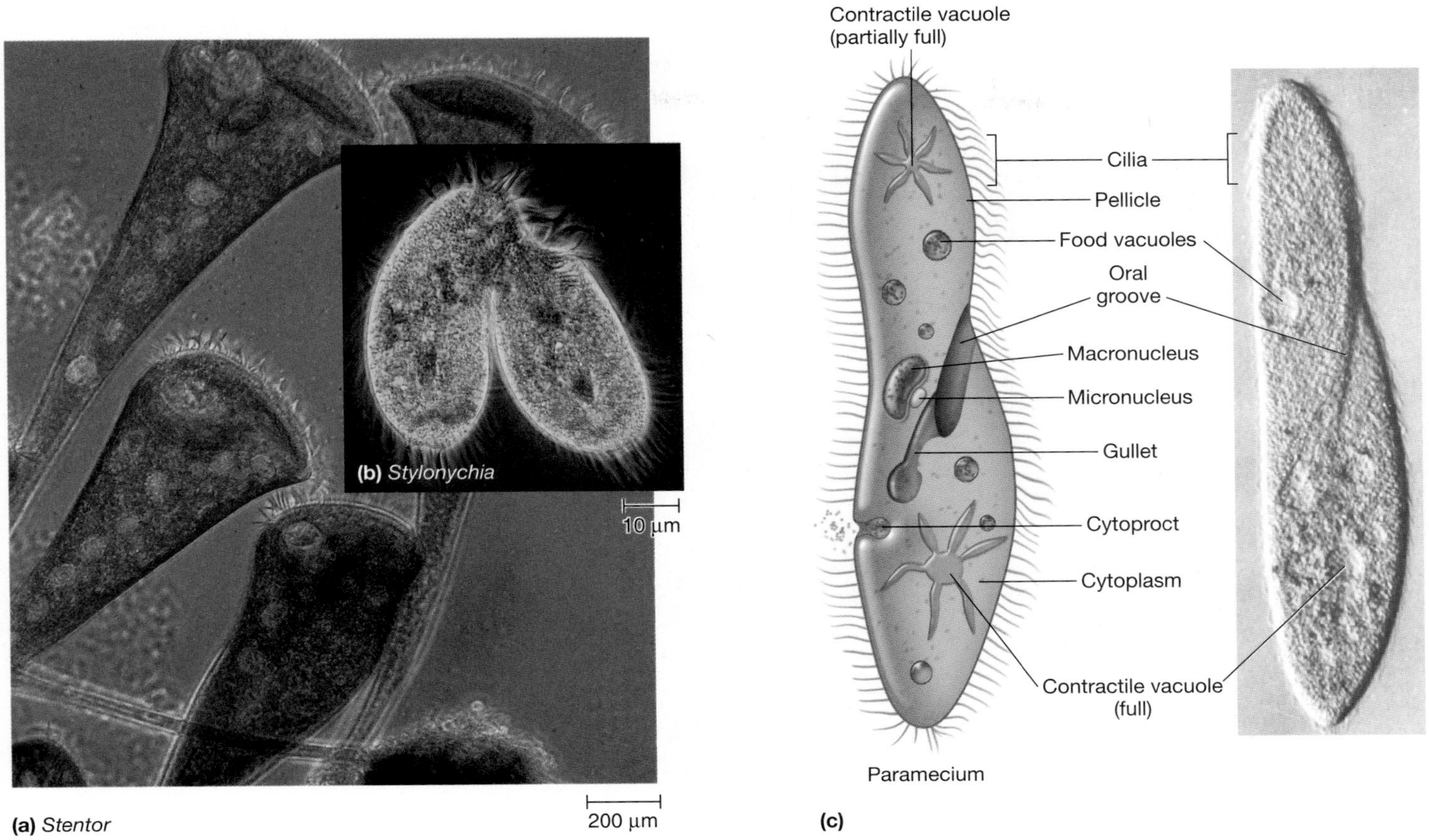

FIGURE 23.12 The *Ciliophora*. (a) *Stentor*, a large, vase-shaped, freshwater protozoan. (b) Two *Stylonychia* conjugating. (c) Structure of *Paramecium*, adjacent to an electron micrograph.

the normal somatic chromosomes. It divides by mitosis and transmits genetic information through meiosis and sexual reproduction. Macronuclei are derived from micronuclei by a complex series of steps. Within the macronucleus are many chromatin bodies, each containing many copies of only one or two genes. Macronuclei are thus polyploid and divide by elongating and then constricting. They produce mRNA to direct protein synthesis, maintain routine cellular functions, and control normal cell metabolism.

The fate of the macro- and micronuclei can be followed during conjugation of *Paramecium caudatum* (**figure 23.13**). In this process, there is an exchange of gametes between paired cells of complementary mating types (conjugants). At the beginning of conjugation, two ciliates unite, fusing their pellicles at the contact point. The macronucleus in each is degraded. The individual micronuclei undergo meiosis to form four haploid pronuclei, three of which disintegrate. The remaining pronucleus divides again mitotically to form two gametic nuclei—a stationary one and a migratory one. The migratory nuclei pass into the respective conjugants. Then the ciliates separate, the gametic nuclei fuse, and the resulting diploid zygote nucleus undergoes three rounds of mitosis. The eight resulting nuclei have different fates: one nucleus is retained as a micronucleus; three others are destroyed; and the four remaining nuclei develop into macronuclei. Each separated conjugant now undergoes cell division. Eventually progeny with one macronucleus and one micronucleus are formed.

Genomic sequencing has revealed that the macronuclear chromosomes of *Paramecium tetraurelia* have undergone at least three whole genome duplications. It is thought that the 40,000 genes found in the extant microbe arose from duplication of about 19,552 genes on the ancestral genome. Thus present-day *P. tetraurelia* has multiple copies (paralogues) of many genes; this seems to be particularly true of genes that are highly expressed.

Although most ciliates are free living, symbiotic forms also exist. Some live as harmless commensals; for example, *Entodinium* is found in the rumen of cattle and *Nyctotherus* occurs in the colon of frogs. Other ciliates are strict parasites; for example, *Balantidium coli* lives in the intestines of mammals, including humans, where it can cause dysentery. *Ichthyophthirius* lives in freshwater, where it can attack many species of fish, producing a disease known as "ick."

All **apicomplexans** are either intra- or intercellular parasites of animals and are distinguished by a unique arrangement of fibrils, microtubules, vacuoles, and other organelles, collectively called the **apical complex**, which is located at one end of the cell

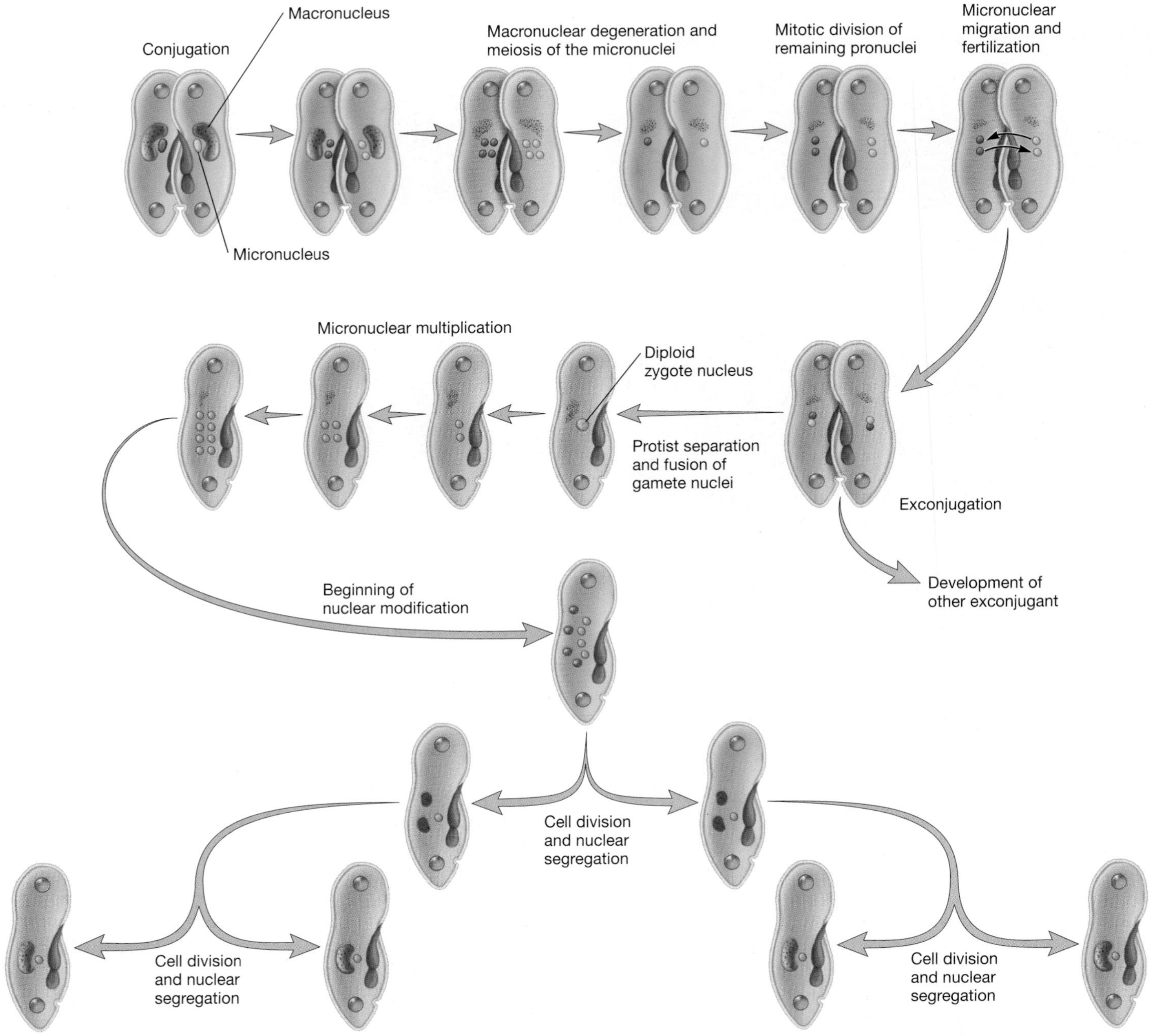

FIGURE 23.13 Conjugation in *Paramecium caudatum*. After the conjugants separate, only one of the exconjugants is shown; however, a total of eight new protists result from each conjugation.

Figure 23.13 Micro Inquiry

What are the functions of the micronucleus and macronucleus?

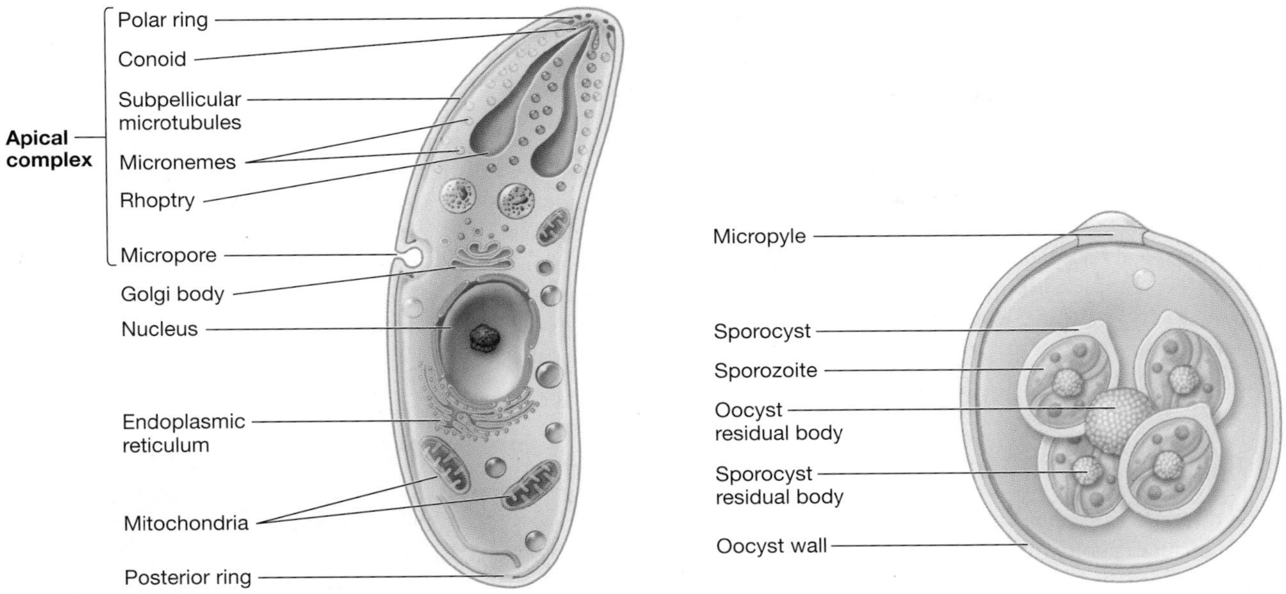

(a) Merozoite **(b)** Oocyst

FIGURE 23.14 The Apicomplexan Cell. (a) The vegetative cell, or merozoite, illustrating the apical complex, which consists of the polar ring, conoid, rhoptries, subpellicular microtubules, and micropore. (b) The infective oocyst of *Eimeria.* The oocyst is the resistant stage and has undergone multiple fission after zygote formation (sporogony).

(**figure 23.14*a***). There are three secretory organelles, the rhoptries, micronemes, and dense granules, that release enzymes and calcium needed to penetrate host cells. In addition, apicomplexans (with the exception of *Cryptosporidium*) contain plastids, called **apicoplasts.** Although these arose by endosymbiosis of an ancient cyanobacterium, unlike chloroplasts, apicoplast are not the site of photosynthesis. However, in all apicomplexans studied to date, the apicoplast is essential. Apicoplast genome sequencing suggest that it synthesizes fatty acids, isoprenoids, and heme. The fact that apicoplasts are essential for viability and are specific to apicoplexans has made them attractive targets for the development of drugs to inhibit the growth of pathogenic apicomplexans. Interestingly, a recently identified protist closely related to the apicoplexans, called *Chromera velia*, was discovered to have pigmented, functional, apicoplast. The discovery of this protist will facilitate study of apicomplexan evolution.

Apicomplexans have complex life cycles in which certain stages sometimes occur in one host and other stages occur in a different host. The life cycle has both asexual (clonal) and sexual phases, and is characterized by an alternation of haploid and diploid generations. The clonal and sexual stages are haploid, except for the zygote. The motile, infective stage is called the **sporozoite.** When this haploid form infects a host, it differentiates into a **gamont;** male and female gamonts pair and undergo multiple fission, which produces many gametes. Released gametes pair, fuse, and form zygotes. Each zygote secretes a protective covering and is then considered a spore. Within the spore, the nucleus undergoes meiosis (restoring the haploid condition) followed by mitosis to generate eight sporozoites ready to infect a new host (figure 23.14*b*). Motility (flagellated or amoeboid) is confined to the gametes and zygotes of a few species.

A number of apicomplexans are important infectious agents. *Eimera* is the causative agent of cecal coccidiosis in chickens, a condition that costs hundreds of millions of dollars in lost animals each year in the United States. *Theilaria parva* and *T. annulata* are tick-borne parasites that cause a fatal disease in cattle called "East Coast fever." This kills over a million animals annually in sub-Saharan Africa, costing over $200 million and affecting farmers who can least afford such losses. Toxoplasmosis, caused by members of the genus *Toxoplasma,* is transmitted either by consumption of undercooked meat or by fecal contamination from a cat's litter box. Cryptosporidia are responsible for cryptosporidiosis, an infection that begins in the intestines but can disseminate to other parts of the body. Cryptosporidiosis and another apicomplexan parasite, *Cyclospora,* have become problematic for AIDS patients and other immunocompromised individuals. The most important human pathogen among the protists is *Plasmodium,* the causative agent of malaria. Human malaria is caused by four species of *Plasmodium: P. falciparum, P. malariae, P. vivax,* and *P. ovale.* The life cycle of *P. vivax* is shown in figure 39.6. About 500 million new cases of malaria develop each year, and about 1 to 3 million people die of the disease annually.

▶▶| *Malaria (section 39.3); Toxoplasmosis (section 39.4); Cryptosporidiosis (section 39.5)*

Search This: UCMP protista

1. What are alveoli? Describe the difference between a naked and an armored dinoflagellate.
2. What is the morphology of typical *Ciliophora?* Why do you think these are the fastest-moving protists?
3. Describe conjugation as it occurs in the *Ciliophora.* What is the fate of the micronucleus and the macronucleus during this process?
4. What is the apical complex seen in apicomplexans?
5. Consider the complex life cycle of the apicomplexans. Which stage do you think would be most vulnerable to drug treatment in an effort to treat a human disease caused by an apicomplexan?

Stramenopila

The large and diverse *Stramenopila* group (informally known as the stramenopiles) includes photosynthetic protists such as the diatoms, brown and golden algae (the *Chrysophyceae*), as well as chemoorganotrophic (saprophytic) genera such as the öomycetes (*Peronosporomycetes*), labyrinthulids (slime nets), and the *Hyphochytriales.* The *Stramenopila* also include brown seaweeds and kelp that form large, rigid structures and macroscopic forms that were once considered fungi and plants. One unifying feature of this very diverse taxon is the possession of **heterokont flagella** at some point in the life cycle. This is characterized by two flagella—one extending anteriorly and the other posteriorly. These flagella bear small hairs with a unique, three-part morphology; stramenopila means "straw hair" (**figure 23.15**).

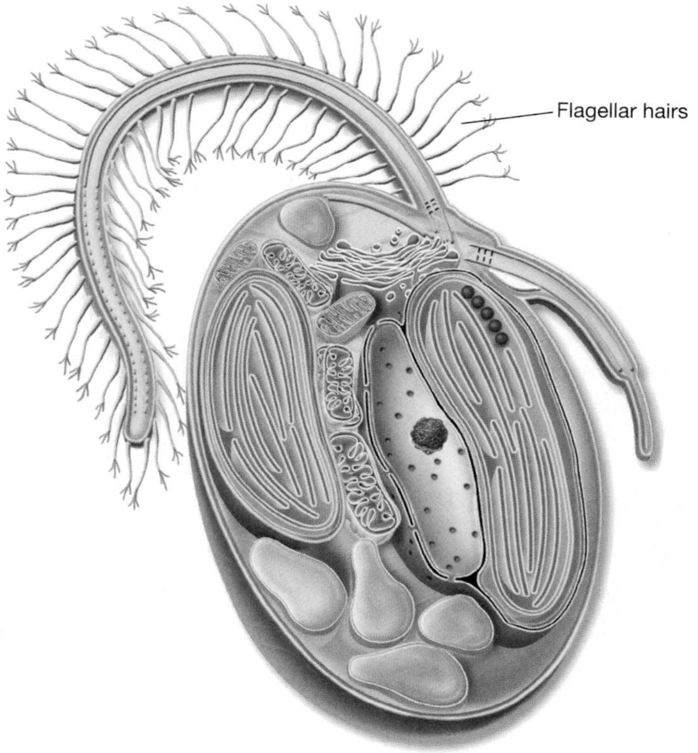

FIGURE 23.15 A *Stramenopila* Cell. Note the two flagella, one with "straw hairlike" projections.

The **diatoms (*Bacillariophyta*)** possess chlorophylls *a* and c_1/c_2, and the carotenoid fucoxanthin. When fucoxanthin is the dominant pigment, the cells have a golden-brown color. Their major carbohydrate reserve is chrysolaminarin (a polysaccharide storage product composed principally of $\beta(1\rightarrow3)$-linked glucose residues). Diatoms have a distinctive, two-piece cell wall of silica called a **frustule.** Diatom frustules are composed of two halves or thecae that overlap like those of a Petri dish (**figure 23.16*a***). The larger half is the epitheca, and the smaller half is the hypotheca. Diatom frustules are composed of crystallized silica [$Si(OH)_4$] with very fine markings (figure 23.16*b*). They have distinctive, and often exceptionally beautiful, patterns that are unique for each species. Frustule morphology

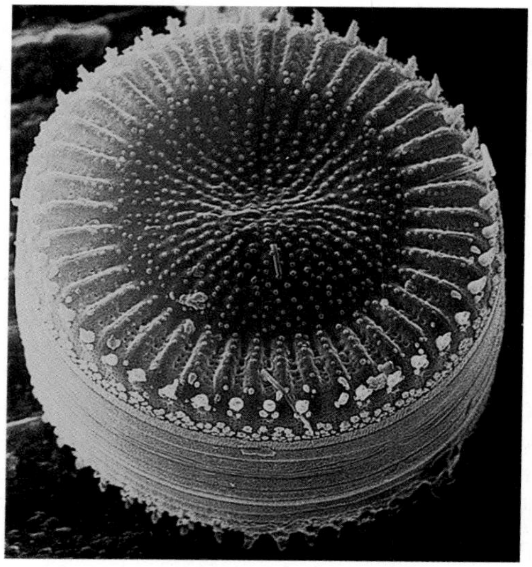

(a) *Cyclotella meneghiniana*

(b)

FIGURE 23.16 Diatoms. (a) The silaceous epitheca and hypotheca of the diatom *Cyclotella meneghiniana* fit together like halves of a Petri dish. (b) A variety of diatoms show the intricate structure of the silica cell wall.

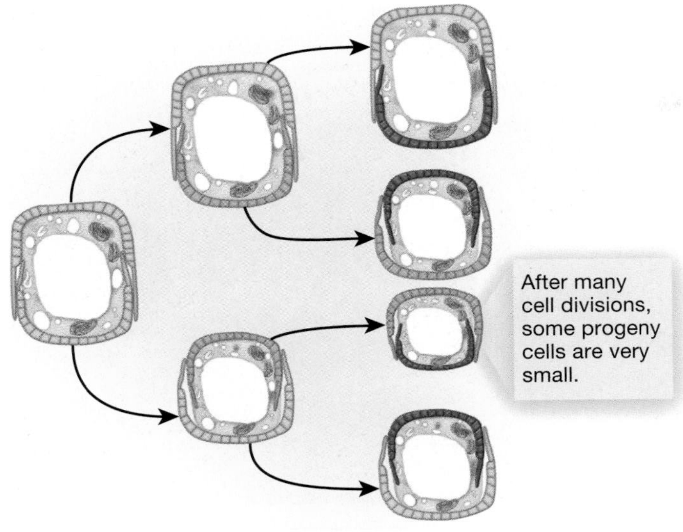

After many cell divisions, some progeny cells are very small.

(a) Asexual reproduction in diatoms

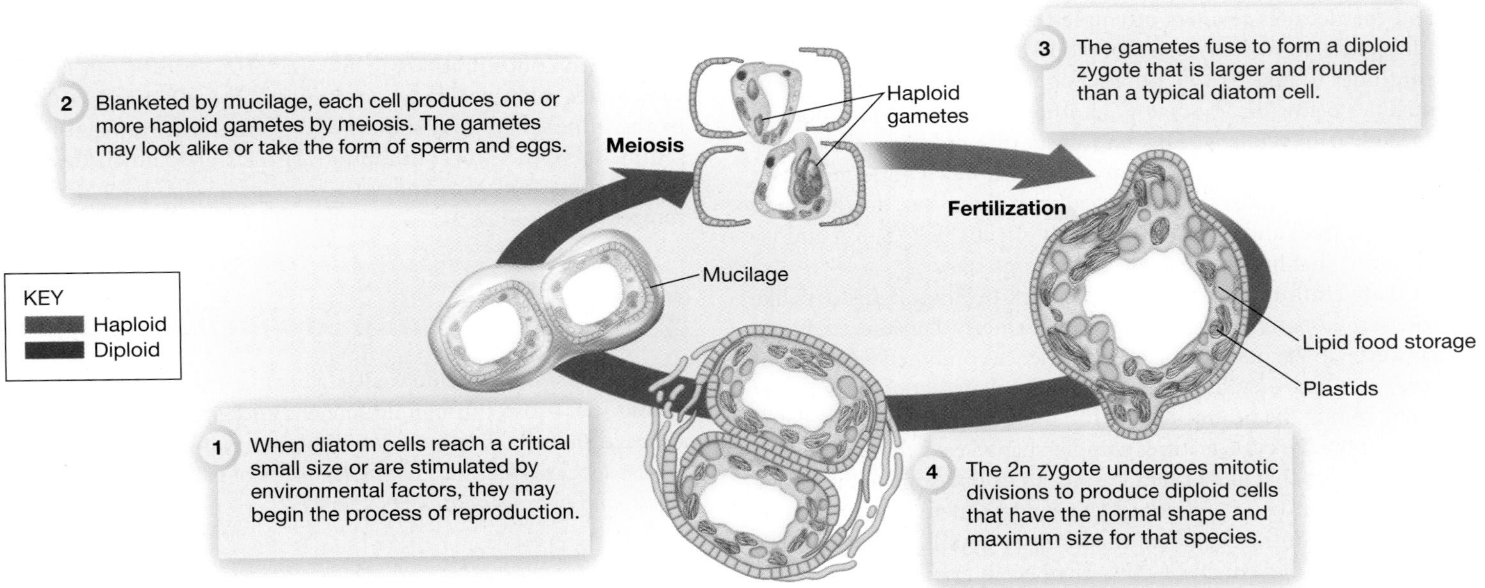

2 Blanketed by mucilage, each cell produces one or more haploid gametes by meiosis. The gametes may look alike or take the form of sperm and eggs.
Meiosis
Haploid gametes
3 The gametes fuse to form a diploid zygote that is larger and rounder than a typical diatom cell.
Fertilization
KEY
Haploid
Diploid
Mucilage
Lipid food storage
Plastids
1 When diatom cells reach a critical small size or are stimulated by environmental factors, they may begin the process of reproduction.
4 The 2n zygote undergoes mitotic divisions to produce diploid cells that have the normal shape and maximum size for that species.

(b) Sexual reproduction in diatoms

FIGURE 23.17 Gametic Life Cycle as Illustrated by Diatoms. (a) Diatom asexual reproduction involves repeated mitotic divisions. Because a new lower cell-wall piece is always synthesized, asexual reproduction may eventually cause the mean cell size to decline in a diatom population. (b) Small cell size may trigger sexual reproduction, which regenerates maximal cell size.

Figure 23.17 Micro Inquiry

How does the trigger for sexual reproduction (i.e., reduced size) compare with more common stimuli that induce sexual reproduction? (Hint: compare to *Radiolaria*.)

is very useful in diatom identification, and diatom frustules have a number of practical applications. The fine detail and precise morphology of these frustules have made them attractive for nanotechnology. ▶▶| *Microbes as products (section 41.5)*

Although the majority of diatoms are strictly photoautotrophic, some are facultative chemoorganotrophs, absorbing carbon-containing molecules through the holes in their walls. The vegetative cells of diatoms are diploid and can be unicellular, colonial, or filamentous. They lack flagella and have a single, large nucleus and smaller plastids. Reproduction consists of the organism dividing asexually, with each half then constructing a new hypotheca (**figure 23.17*a***). Because the epitheca and hypotheca are of different sizes, each time the hypotheca is used as a template to construct a new hypotheca, the diatom gets smaller. However, when a cell has diminished to about 30% of its original size, sexual reproduction is usually triggered. The diploid vegetative cells undergo meiosis to form gametes, which then fuse to produce a zygote. The zygote develops into an auxospore, which increases in size again and forms a new wall. The mature auxospore eventually divides mitotically to produce vegetative cells with frustules of the original size (figure 23.17*b*).

Diatoms are found in freshwater lakes, ponds, streams, and throughout the world's oceans. Marine planktonic diatoms produce 40 to 50% of the organic carbon in the ocean; they are therefore very important in global carbon cycling. In fact, marine diatoms are thought to contribute as much fixed carbon as all rain forests combined. ▶▶| *Carbon cycle (section 26.1); Microorganisms in the open ocean (section 28.2)*

Search This: International Society for Diatom Research

A group of protists once considered fungi and traditionally called **öomycetes,** meaning "egg fungi," were recently assigned the name **peronosporomycetes.** They differ from fungi in a number of ways, including their cell wall composition (cellulose and β-glucan instead of chitin) and the fact that they are diploid throughout their life cycle. When undergoing sexual reproduction, they form a relatively large egg cell (öogonium) that is fertilized by either a sperm cell or a smaller gametic cell (called an antheridium) to produce a zygote. When the zygote germinates, the asexual zoospores display heterokont flagellation.

Peronosporomycetes such as *Saprolegnia* and *Achlya* are saprophytes that grow as cottony masses on dead algae and animals, mainly in freshwater environments. Some öomycetes are parasitic on the gills of fish. *Peronospora hyoscyami* is responsible for "blue mold" on tobacco plants, and grape downy mildew is caused by *Plasmopara viticola.* Certainly the most famous öomycete is *Phytophthora infestans,* which attacked the European potato crop in the mid-1840s, spawning the Irish famine. The original classification of *P. infestans* as a fungus was misleading, and for decades farmers attempted to control its growth with fungicide, to which it is (of course) resistant. This protist continues to take its toll; potato blight costs some $5 billion annually worldwide. *P. sojae* is another agriculturally important species; it infects soybeans and costs this industry millions of dollars per year. Finally, *P. ramorum* is a newly emerged plant pathogen that causes a disease called sudden oak death, which kills oak trees and a variety of woody shrubs (e.g., laurel and viburnum) that live in the same ecosystems.

Labyrinthulids also have a complex taxonomic history: like the *Peronosporomycetes,* they were formerly considered fungi. However, molecular phylogenetic evidence combined with the observation that they form heterokont flagellated zoospores places them among the *Stramenopila.* The more familiar, nonflagellated stage of the life cycle features spindle-shaped cells that form complex colonies that glide rapidly along an ectoplasmic net made by the organism. This net is actually an external network of calcium-dependent contractile fibers made up of actinlike proteins that facilitate the movement of cells. Their feeding mechanism is like that of fungi: osmotrophy aided by the production of extracellular degradative enzymes. In marine habitats, the genus *Labyrinthula* grows on plants and algae, and is thought to play a role in the "wasting disease" of eelgrass, an important intertidal plant.

Haptophyta

An important subgroup of the *Haptophyta* is the *Coccolithales.* These photosynthetic protists bear ornate calcite scales called coccoliths (**figure 23.18**). Together with the *Foraminifera,* the **coccolithophores** precipitate calcium carbonate ($CaCO_3$) in the ocean, thereby influencing Earth's carbon budget. Cells are usually biflagellated and possess a unique organelle called a haptonema, which is somewhat similar to a flagellum but differs in microtubule arrangement. One species, *Emiliania huxleyi,* has been studied extensively. Like all coccolithophores, it is planktonic and can cause massive blooms in the open ocean (*see figure 28.11*).

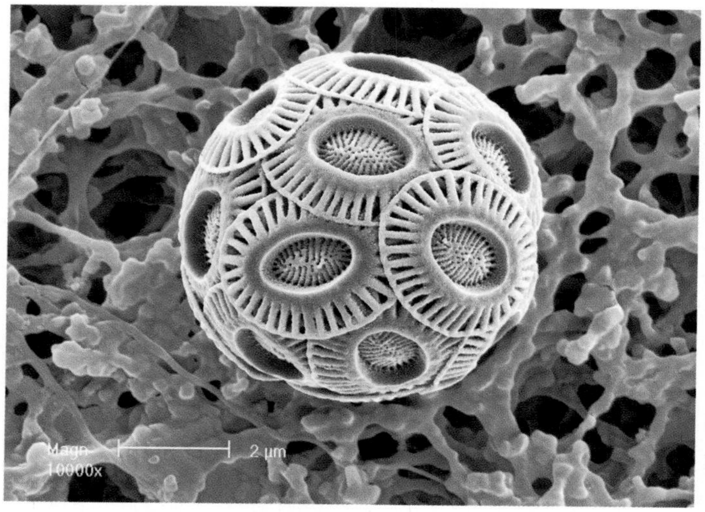

FIGURE 23.18 The Haptophyte *Emiliana huxleyi.* Note the ornamental scales made of calcite.

These high concentrations or blooms of *E. huxleyi* can significantly alter nutrient flux by emitting sulfur (as dimethyl sulfide) to the atmosphere and sequestering calcium carbonate in the sediments. Other coccolithophore species are known to cause toxic blooms. ▶▶| *Nutrient cycling in marine and freshwater environments (section 28.1)*

23.6 Supergroup *Archaeplastida*

The *Archaeplastida* includes all organisms with a photosynthetic plastid that arose through an ancient endosymbiosis with a cyanobacterium. It thus includes all higher plants as well as many protist species. |◀◀ *Endosymbiotic origin of mitochondria, chloroplasts, and hydrogenosomes (section 1.2)*

Chloroplastida

The *Chloroplastida* are often referred to as green algae (Greek *chloros,* green). These phototrophs grow in fresh and salt water, in soil, and on and within other organisms. They have chlorophylls *a* and *b* along with specific carotenoids, and they store carbohydrates such as starch. Many have cell walls made of cellulose. They exhibit a wide diversity of body forms, ranging from unicellular to colonial, filamentous, membranous or sheetlike, and tubular types (**figure 23.19**). The smallest known eukaryote, *Ostreococcus tauri,* is a member of the *Chloroplastida* (subgroup *Prasinophyceae*). This marine, planktonic microbe is smaller than most bacteria and archaea, with an average size of 0.8 µm. Some species have a holdfast structure that anchors them to a substratum. Both asexual and sexual reproduction are observed.

Chlamydomonas is a member of the subgroup *Chlorophyta* (**figure 23.20**). Individuals have two flagella of equal length at the anterior end by which they move rapidly in water. Each cell has a

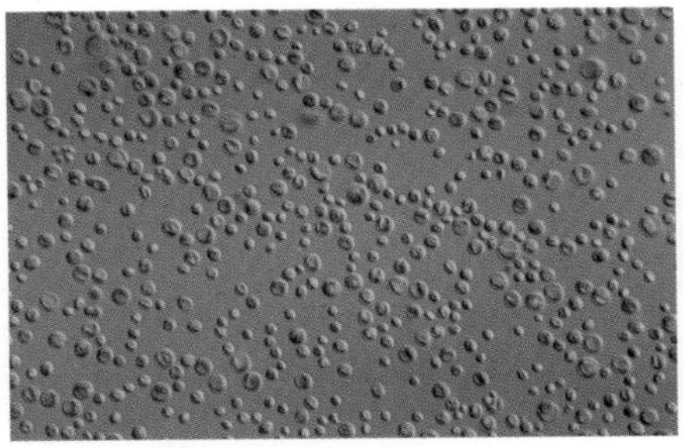
(a) *Chlorella*

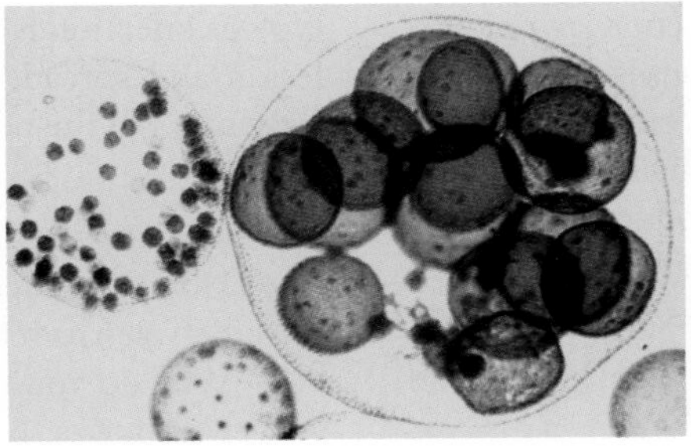
(b) *Volvox*

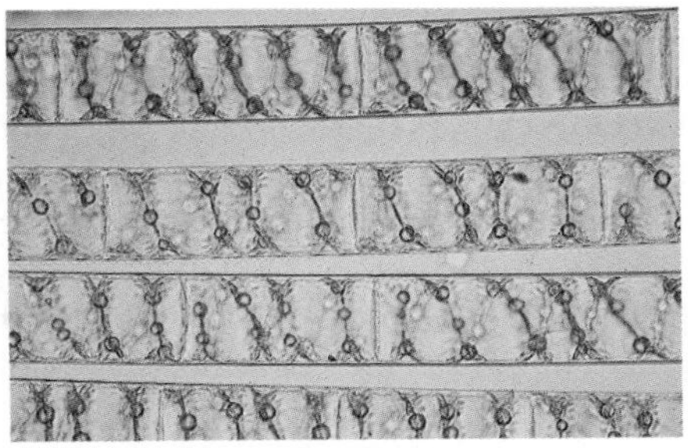
(c) *Spirogyra*

(d) *Acetabularia*

FIGURE 23.19 *Chlorophyta* (Green Algae); Light Micrographs. (a) *Chlorella*, a unicellular nonmotile *Chlorophyte* (×160). (b) *Volvox*, which demonstrates colonial growth (×450). (c) *Spirogyra* (×100). Four filaments are shown. Note the ribbonlike, spiral chloroplasts within each filament. (d) *Acetabularia*, the mermaid's wine goblet.

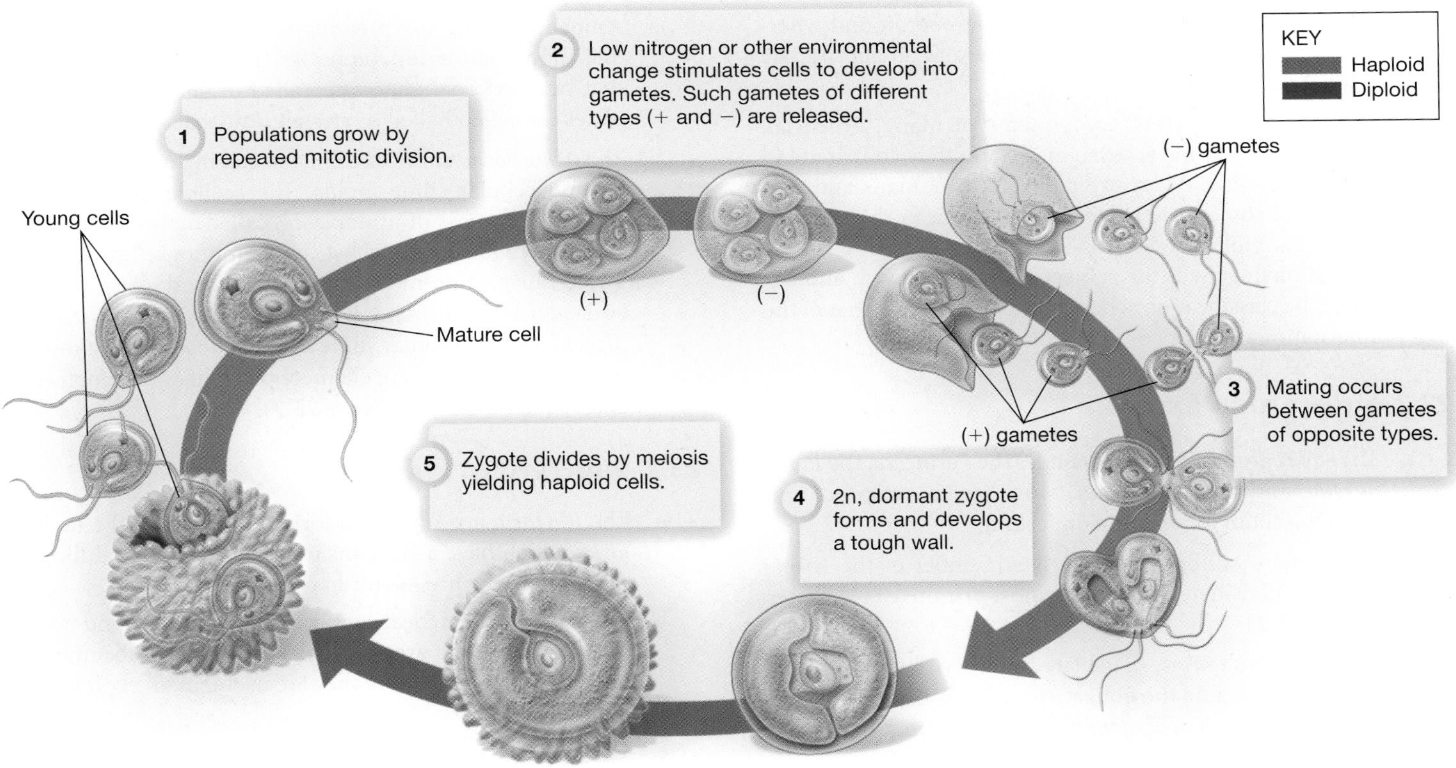

FIGURE 23.20 *Chlamydomonas:* The Structure and Life Cycle of This Motile Green Alga. During asexual reproduction, all structures are haploid; during reproduction, only the zygote is diploid.

***Figure 23.20** Micro Inquiry*

Which of the cells shown above are haploid and which are diploid?

single haploid nucleus, a large chloroplast, a conspicuous pyrenoid, and a stigma that aids the cell in phototactic responses. Two small contractile vacuoles at the base of the flagella function as osmoregulatory organelles. *Chlamydomonas* reproduces asexually by producing zoospores through cell division. Sexual reproduction occurs when some products of cell division act as gametes and fuse to form a four-flagellated, diploid zygote that ultimately loses its flagella and enters a resting phase. Meiosis occurs at the end of this resting phase and produces four haploid cells that give rise to adults.

The chlorophyte *Prototheca moriformis* causes the disease protothecosis in humans and animals. *Prototheca* cells are fairly common in the soil, and it is from this site that most infections occur. Severe systemic infections, such as massive invasion of the bloodstream, have been reported in animals. The subcutaneous type of infection is more common in humans. It starts as a small lesion and spreads slowly through the lymph glands, covering large areas of the body.

1. Explain the unique structural features of the diatoms. How does their morphology play a role in the alternation between asexual and sexual reproduction?
2. Why do you think the öomycetes and the labyrinthulids were formerly considered fungi?
3. What is the ecological importance of the coccolithophores?
4. Compare the morphology of *Chlamydomonas* with that of *Euglena* (figure 23.3).

Summary

23.1 Introduction

a. Protists are widespread in the environment, found wherever water, suitable organic nutrients, and an appropriate temperature occur.

b. Protists are important components of many terrestrial, aquatic, and marine ecosystems, where they contribute to nutrient cycling. Many are parasitic in humans and animals, and some have become very useful in the study of molecular biology.

c. Protist phylogeny is the subject of active research and debate. The classification scheme presented here is that of the International Society of Protistologists.

23.2 Supergroup *Excavata*

a. The super group *Excavata* includes the *Fornicata*, the *Parabasalia*, and the *Euglenozoa*. Most have a cytoproct and use a flagellum for suspension feeding.

b. The human pathogen *Giardia* is a member of the *Fornicata*. It is considered one of the most primitive eukaryotes (**figure 23.2**).

c. Most *Parabasalia* are flagellated endosymbionts of animals. They include the obligate mutualists of wood-eating insects such as *Trichonympha* and the human pathogen *Trichomonas.*

d. Many of the members of the *Euglenozoa* are photoautotrophic (**figure 23.3**). The remainder are chemoorganotrophs, of which most are saprophytic. Important human pathogens include members of the genus *Trypanosoma.* Trypanosomes cause a number of important human diseases including leishmaniasis, Chagas' disease, and African sleeping sickness.

23.3 Supergroup *Amoebozoa*

a. Amoeboid forms use pseudopodia, which can be lobopodia, filopodia, or reticulopodia (**figure 23.5**). Amoebae that bear external plates are called testate; those without plates are called naked or atestate amoebae.

b. The *Amoebozoa* subclass *Eumycetozoa* includes the acellular and cellular slime molds. The acellular slime molds form a large mass of protoplasm, called a plasmodium, in which individual cells lack a cell membrane (**figure 23.6**). The cellular slime molds produce a pseudoplasmodium, and each cell within has a cell membrane. *Dictyostelium discoideum* is a cellular slime mold that is used as a model organism in the study of chemotaxis, cellular development, and cellular behavior (**figure 23.7**).

23.4 Supergroup *Rhizaria*

a. The *Rhizaria* are amoeboid forms that include the *Radiolaria,* which have a siliceous internal skeleton and filopodia, which they use to entrap prey (**figure 23.8**).

b. The *Foraminifera* bear netlike reticulopodia and tests. Most foraminifera are benthic and their tests accumulate on the ocean floor, where they are useful in oil exploration (**figure 23.9**).

c. The *Rhizaria* and *Foraminifera* are structurally ornate.

23.5 Supergroup *Chromalveolata*

a. The super group *Chromalveolata* is diverse. It includes the *Alveolata,* which consists of the dinoflagellates, stramenopiles, ciliates, and the apicomplexans.

b. The dinoflagellates are a large group of nutritionally complex protists. Most are marine and planktonic. They are known for their phosphorescence and for causing toxic blooms (**figure 23.11**).

c. The *Ciliophora* are chemoorganotrophic protists that use cilia for locomotion and feeding. In addition to asexual reproduction, conjugation is used in sexual reproduction (**figures 23.12** and **23.13**).

d. Apicomplexans are parasitic with complex life cycles. The motile, infective stage is called the sporozoite (**figure 23.14**).

e. The *Stramenopila* are extremely diverse and include diatoms, golden and brown algae, the öomycetes, and labyrinthulids. Diatoms are found in fresh- and saltwater, and are important components of marine plankton (**figure 23.16**). The öomycetes and labyrinthulids were once thought to be fungi.

f. The haptophytes include the coccolithophores, planktonic photosynthetic protists that contribute to the global carbon budget by precipitating calcium carbonate for their ornate scales (**figure 23.18**).

23.6 Supergroup *Archaeplastida*

a. The *Archaeplastida* include the *Chlorophyta*, also known as green algae. All are photosynthetic with chlorophylls *a* and *b* along with specific carotenoids. They exhibit a wide range of morphologies (**figure 23.19**).

b. *Chlamydomonas* is a model protist, which undergoes sexual and asexual reproduction (**figure 23.20**).

Critical Thinking Questions

1. Why do you think our knowledge of the biology of protists has lagged so far behind that of bacteria, archaea, fungi, and higher eukaryotes?

2. Which of the protists discussed in this chapter do you think are the most evolutionarily advanced or derived? Explain your reasoning.

3. Vaccine development for diseases caused by protists (e.g., malaria, Chagas' disease) has been much less successful than for bacterial diseases. Discuss one biological reason and one geopolitical reason for this fact.

4. Some protists reproduce asexually when nutrients are plentiful and conditions are favorable for growth but reproduce sexually when environmental or nutrient conditions are not favorable. Why is this an evolutionarily important and successful strategy?

5. Most apicomplexans bear the unpigmented chloroplast remnant called the apicoplast (p. 595). Only the recently described apicomplexan *Chomera velia* has been shown to use the apicoplastid as the site of photosynthesis. What genetic and physiological features do you think were characterized to show that this organism is truly photo-synthetic? How would you determine if this protist is photoautotrophic, photoheterotrophic, or mixotrophic?

   Read the original paper: Moore, R. B., et al. 2008. A photosynthetic alveolate closely related to apicomplexan parasites. *Nature.* 451:959.

6. Benthic foraminifera inhabit marine sediments. It was discovered that the benthic foram *Globobulimina pseudospinescens* stores intracellular nitrate, which it reduces to nitrogen gas. This is the first and only case of eukaryotic denitrification. How would you go about designing a comparative analysis of eukaryotic and bacterial denitrification? What physical factors and physiological and enzymatic features would you characterize?

   Read the original paper: Risgaard-Peterson, N., et al. 2006. Evidence for complete denitrification in a benthic foraminifera. *Nature.* 443:93.

Learn More

Learn more by visiting the text website at www.mhhe.com/willey8, where you will find a complete list of references.

24

The Fungi (Eumycota)

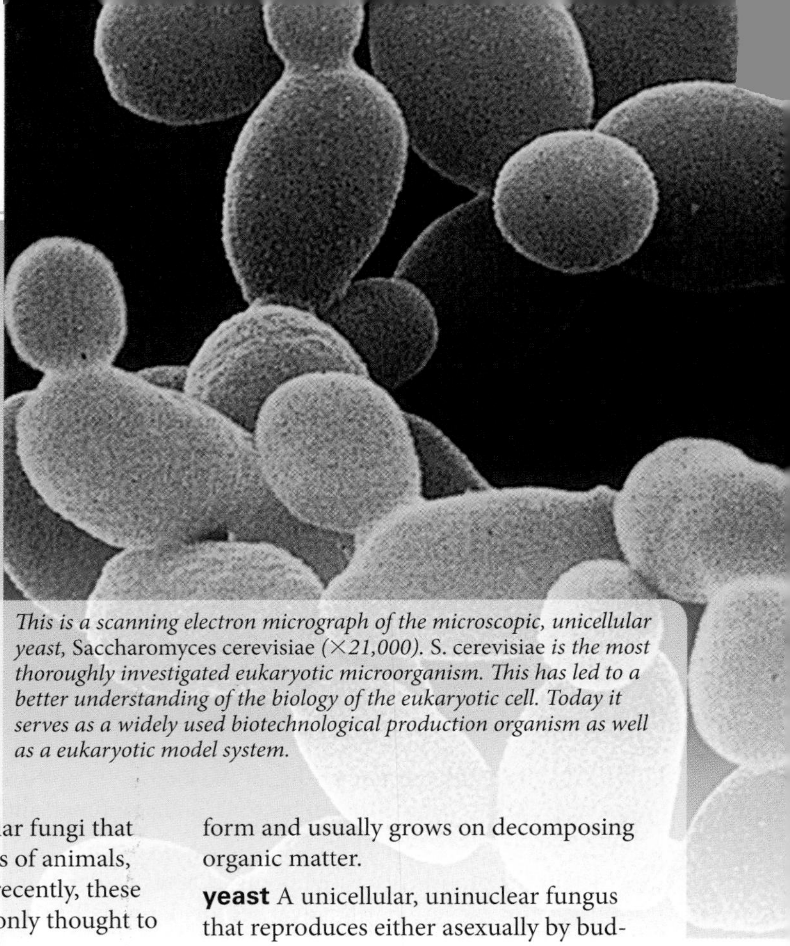

This is a scanning electron micrograph of the microscopic, unicellular yeast, Saccharomyces cerevisiae *(×21,000).* S. cerevisiae *is the most thoroughly investigated eukaryotic microorganism. This has led to a better understanding of the biology of the eukaryotic cell. Today it serves as a widely used biotechnological production organism as well as a eukaryotic model system.*

CHAPTER GLOSSARY

ascomycetes Fungi that form spores in a saclike structure, the **ascus**.

basidiomycetes Fungi in which the spores are borne on club-shaped organs called basidia.

chytrids Members of *Chytridiomycota.* They are simple terrestrial and aquatic fungi that produce motile zoospores with single, posterior, whiplash flagella.

glomeromycetes Fungi that include the mycorrhizal fungi, which form important mutualistic relationships with the roots of most plants.

microsporidia Unicellular fungi that are intracellular pathogens of animals, including humans. Until recently, these unique fungi were commonly thought to be protists.

mycosis (pl., mycoses) Any disease caused by a fungus.

osmotrophy A form of nutrition in which soluble nutrients are absorbed through the cytoplasmic membrane; found in prokaryotes, fungi, and some protists.

saprophyte An organism that takes up nonliving organic nutrients in dissolved form and usually grows on decomposing organic matter.

yeast A unicellular, uninuclear fungus that reproduces either asexually by budding or fission, or sexually through spore formation.

zygomycetes Fungi that usually have a coenocytic mycelium with chitinous cell walls and lack motile spores. Sexual reproduction normally involves the formation of zygospores.

In this chapter, we introduce the ***Fungi,*** sometimes referred to as the true fungi or ***Eumycota*** (Greek *eu*, true, and *mykes*, fungus). Microbiologists use the term fungus (pl., fungi; Latin *fungus*, mushroom) to describe eukaryotic organisms that are spore-bearing, have absorptive nutrition, lack chlorophyll, and reproduce both sexually and asexually. Scientists who study fungi are **mycologists,** and the scientific discipline devoted to fungi is called **mycology.** The study of fungal toxins and their effects is called **mycotoxicology,** and the diseases caused by fungi in animals are known as **mycoses** (s., mycosis).

Fungi is an enormous group of organisms; about 90,000 fungal species have been described, and some estimates suggest that 1.5 million species exist. It is thus not surprising that the taxonomy of these organisms has been revised numerous times. Most recently the application of molecular techniques combined with morphological and ecological considerations has led to a better understanding of the phylogenetic relationships between various fungal groups (**figure 24.1**). In its 2005 classification of the eukaryotes, the International Society of Protistologists places fungi in the supergroup *Opisthokonta,* which includes all metazoa. Here, we present six major fungal groups: the *Chytridiomycota, Zygomycota, Glomeromycota, Ascomycota, Basidiomycota,* and *Microsporidia* (**figure 24.2**). The *Basidiomycota* and *Ascomycota* are both dikarya, meaning that during sexual reproduction, the two parental nuclei are initially paired. These haploid nuclei eventually fuse and undergo meiosis

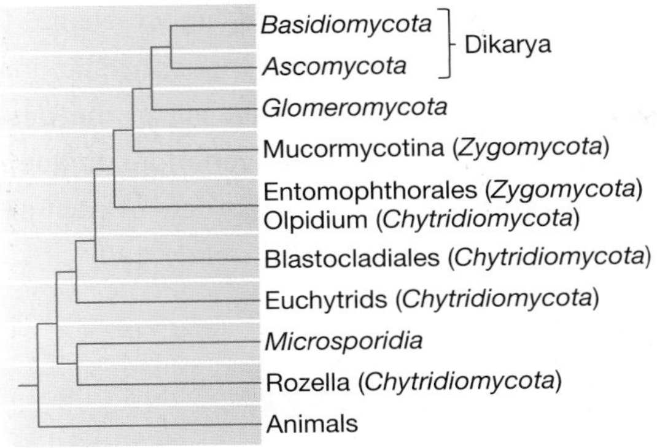

FIGURE 24.1 The Major Branches of the *Fungi.* Phylogenetic analysis of the six fungal groups shows that the *Zygomycota* and the *Chytridiomycota* are paraphyletic.

Figure 24.1 Micro Inquiry

Compare the nodes and branches that lead to the monophyletic *Basidiomycota* and *Ascomycota.* How do these differ from those leading to the *Zygomycota* and *Chytridiomycota*? What can be concluded from this?

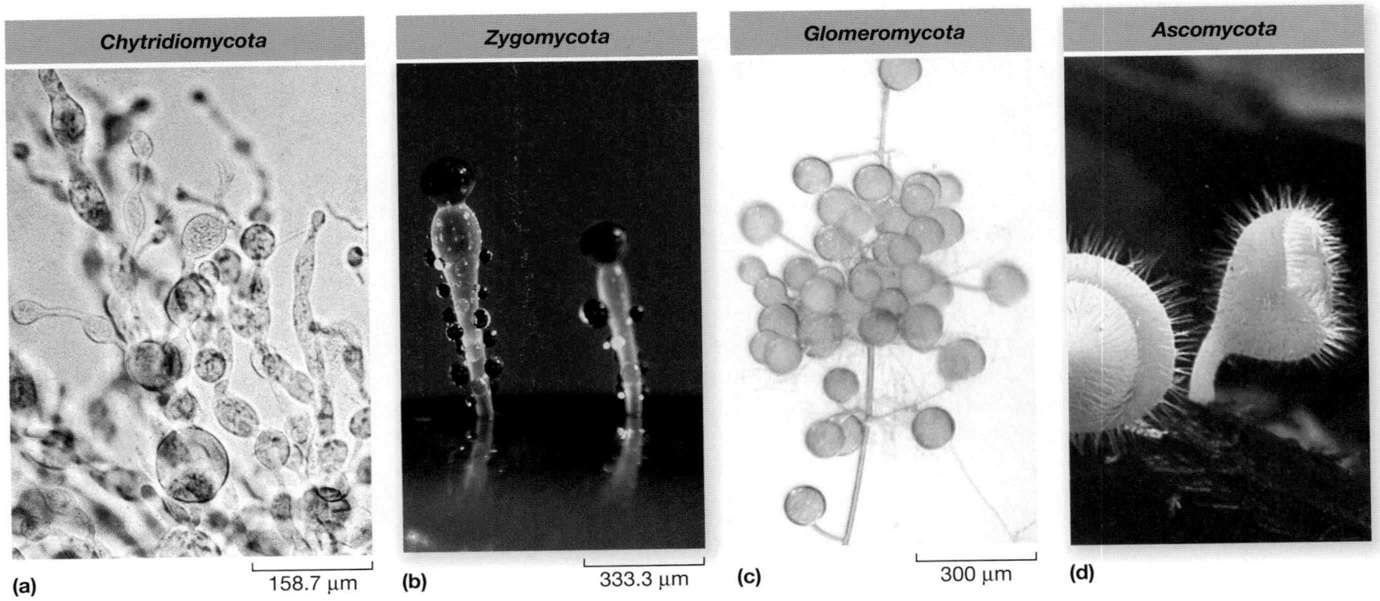

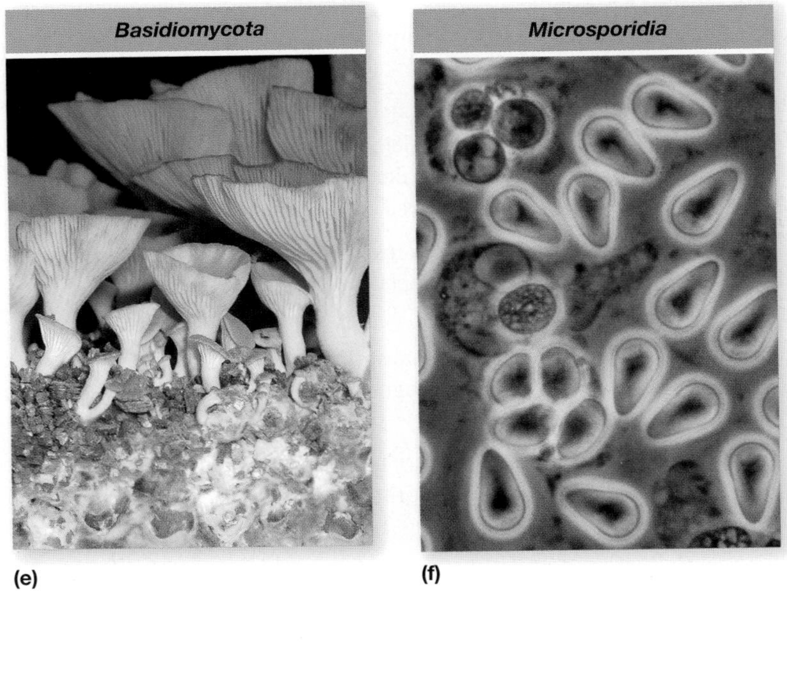

FIGURE 24.2 Representatives of the Fungal Groups. (a) Some chytrids, including members of the genus *Hypochytrium*, are plant parasites, whereas others are free living. (b) *Pilobolus*, a zygomycete, grows on animal dung and also on culture medium. Stalks about 10 mm long contain dark, spore-bearing sacs. (c) Spores of *Glomus intraradices*, a glomeromycete associated with roots. (d) The cup fungus *Cookeina tricholoma* is an ascomycete from the rain forest of Costa Rica. In the cup fungi, the spore-producing structures line the cup; in basidiomycetes that form mushrooms such as *Amanita*, they line the gills beneath the cap of the mushroom. (e) *Pleurotus ostreatus,* the oyster mushroom, is a basidiomycete. All visible structures of fungi arise from an extensive network of filamentous hyphae that penetrates and is interwoven with the substrate as they grow. (f) By contrast, the unusual fungi that belong to the *Microsporidia* form spores that germinate in host cells. Shown here is the mosquito pathogen *Edhazardia aedis.*

to produce haploid progeny (*see figure 4.28*). The *Glomeromycota* live only as symbionts of higher plants. They are most closely related to the dikaryotic fungi. Neither the *Zygomycota* nor the *Chytridiomycota* are monophyletic; instead they are considered paraphyletic (figure 24.1). A **paraphyletic** taxonomic group includes some, but not all, of the descendents of a single common ancestor. The most primitive are *Rozella*, an unusual *Chytridiomycota,* and the even more unique *Microsporidia.* **Table 24.1** surveys these groups; we focus most of our discussion on the general features of members of each of these groups. Fungal cellular structure is covered in chapter 4. ◀◀ *Overview of fungal structure and function (section 4.10)*

Table 24.1 Abbreviated Classification of the *Fungi* Proposed by the International Society of Protistologists[a]

Subclass	*Characteristics*	*Examples*
Chytridiomycota	Flagellated cells in at least one stage of life cycle; may have one or more flagella. Cell walls with chitin and β-1,3-1,6-glucan; glycogen is used as a storage carbohydrate. Sexual reproduction often results in a zygote that becomes a resting spore or sporangium; saprophytic or parasitic. Chytrid subdivisions include *Blastocladiales, Monoblepharidales, Neocallimastigaceae, Spizellomycetales,* and *Chytridiales.*	*Allomyces* *Blastocladiella* *Coelomomyces* *Physoderma* *Synchytrium*
Zygomycota	Thalli usually filamentous and nonseptate, without cilia; sexual reproduction gives rise to thick-walled zygospores that are often ornamented. Includes seven subdivisions: *Basidiobolus, Dimargaritales, Endogonales, Entomophthorales, Harpellales, Kickxellales, Mucorales,* and *Zoopagales.* Human pathogens found among the *Mucorales* and *Entomophthorales.*	*Amoebophilus* *Mucor* *Phycomyces* *Rhizopus* *Thamnidium*
Ascomycota	Sexual reproduction involves meiosis of a diploid nucleus in an ascus, giving rise to haploid ascospores; most also undergo asexual reproduction with the formation of conidiospores with specialized aerial hyphae called conidiophores. Many produce asci within complex fruiting bodies called ascocarps. Includes saprophytic, parasitic forms; many form mutualisms with phototrophic microbes to form lichens. Four monophyletic subdivisions: *Saccharomycetes, Pezizomycotina, Taphrinomycotina,* and *Neolecta.*	*Ascobolus* *Aspergillis* *Candida* *Crinula* *Neurospora* *Penicillium* *Pneumocystis* *Saccharomyces*
Basidiomycota	Includes many common mushrooms and shelf fungi. Sexual reproduction involves formation of a basidium (small, club-shaped structure that typically forms spores at the ends of tiny projections) within which haploid basidiospores are formed. Usually 4 spores per basidium but can range from 1 to 8. Sexual reproduction involves fusion with opposite mating type resulting in a dikaryotic mycelium with parental nuclei paired but not initially fused. No subdivisions recognized.	*Agaricus* *Boletes* *Dacrymyces* *Lycoperdon* *Polyporus* *Russula* *Tremella*
Urediniomycota	Mycelial or yeast forms. Sexual reproduction involves fusion of parental nuclei in probasidium followed by meiosis in a separate compartment. Many are plant pathogens called rusts, animal pathogens, nonpathogenic endophytes, and rhizosphere species. Considered a basidiomycete.	*Caeoma* *Melampsora* *Uromyces*
Ustilaginomycota	Plant parasites that cause rusts and smuts. Mycelial in parasitic phase; meiospores formed on septate or aseptate basidia; cell wall principally composed of glucose polymers. No subdivisions recognized. Considered a basidiomycete.	*Malassezia* *Tilletia* *Ustilago*
Glomeromycota	Filamentous, most are endomycorrhizal, arbuscular; lack cilium; form asexual spores outside of host plant; lack centrioles, conidia, and aerial spores. No subdivisions recognized.	*Acaulospora* *Entrophospora* *Glomus*
Microsporidia	Obligate intracellular parasites usually of animals. Lack mitochondria, peroxisomes, kinetosomes, cilia, and centrioles; spores have an inner chitin wall and outer wall of protein; produce a tube for host penetration. Subdivisions currently uncertain.	*Amblyospora* *Encephalitozoon* *Enterocytozoon* *Nosema*

[a] *Adapted from: Adl, S. M., et al. 2005. The new higher level classification of Eukaryotes with emphasis on the taxonomy of protists.* J. Eukaryot. Microbiol. *52:399–451.*

24.1 Fungal Distribution and Importance

Unlike protists, fungi are primarily terrestrial organisms, although a few are found in aquatic ecosystems. They have a global distribution from polar to tropical regions. Fungi are **saprophytes,** securing nutrients from dead organic material by releasing degradative enzymes into the environment. They then absorb the soluble products—a process sometimes called **osmotrophy.** Together with many bacteria and archaea and a few other groups of chemoorganotrophic organisms, fungi act as decomposers, a role of enormous significance. They degrade complex organic materials in the environment to simple organic compounds and inorganic molecules. In this way, carbon, nitrogen, phosphorus, and other critical constituents of dead organisms are released and made available for living organisms. Many fungi are pathogenic and infect plants and animals. Over 5,000 species attack economically valuable crops, garden plants, and many wild plants, and about 20 new human fungal pathogens are documented each year. Fungi also form beneficial relationships with other organisms. For example, the vast majority of vascular plant roots form associations with fungi, called mycorrhizae. ▶▶| *Mycorrhizae (section 29.3); Microbial interactions (section 30.1)*

Fungi, especially the yeasts, are essential to many industrial processes involving fermentation. Examples include the making of bread, wine, beer, cheeses, and soy sauce. They are also important in the commercial production of many organic acids (citric, gallic) and certain drugs (ergometrine, cortisone), and in the manufacture of many antibiotics (penicillin, griseofulvin) and the immunosuppressive drug cyclosporin. In addition, fungi are important research tools in the study of fundamental biological processes. Cytologists, geneticists, biochemists, biophysicists, and microbiologists regularly use fungi in their research. The yeast *Saccharomyces cerevisiae* is the best understood eukaryotic cell. It has been a valuable model organism in the study of cell biology, genetics, and cancer. ▶▶| *Microbiology of food (chapter 40)*

Search This: Fungus of the month

1. How can fungi be defined?
2. What is the distribution of these microbes?
3. What does the term paraphyletic mean? How is it relevant to fungi?
4. Why do you think the fungi are such important model organisms?
5. Why do you think the mycelial morphology of the fungi makes them especially effective saprophytes?

24.2 *Chytridiomycota*

The simplest of the fungi belong to the *Chytridiomycota,* or **chytrids.** Free-living members of this taxon are saprophytic, living on plant or animal matter in freshwater, mud, or soil. Parasitic forms infect aquatic plants and animals, including insects (**figure 24.3**). A few are found in the anoxic rumen of herbivores. Recently evidence suggests that chytrids may be responsible for large-scale mortality of amphibians and that this may be

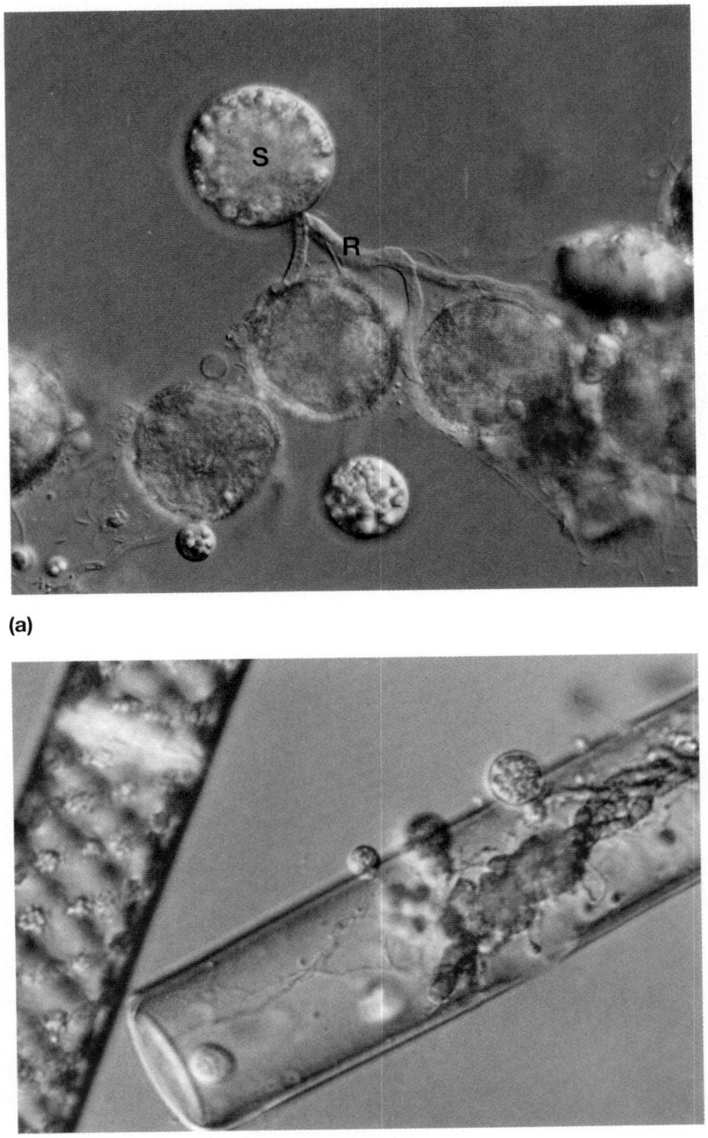

(a)

(b)

FIGURE 24.3 The *Chytridiomycota.* (a) A chytrid with a single round sporangium (S) that is about 50 μm in diameter. In this image, the fungus is growing between brown pollen grains, which it will eventually penetrate with its branched rhizoids (R). (b) Parasitic chytrids attached to the surface of a photosynthetic protist (a green alga).

Figure 24.3 Micro Inquiry

What is the function of the chytrid rhizoids?

linked to climate change. The interaction between chytrid growth, amphibians, and a warming planet is discussed in chapter 26. ▶▶| *Global climate change (section 26.2)*

Chytrids are unique among fungi in the production of a zoospore with a single, posterior, whiplash flagellum. It has long been hypothesized that the chytrids are the most primitive division of fungi and that a single event resulted in the loss of flagella in the more advanced fungi. However, recent analysis suggests at least four independent losses of flagella. The loss of flagella is thought to have coincided with the development of other means of spore dispersal, such as aerial dispersal from mycelial mats. The chytrids are not considered monophyletic (note four clades of *Chytridiomycota* in figure 24.1). Instead, they derive from early diverging lineages that bear zoospores. The largest clade, the *Euchytrids,* unites several orders of chytrids. Still somewhat unresolved is the placement of the endoparasitic chytrids *Rozella* with the primitive fungi the microsporidia, and the genus *Olpidium* with the *Zygomycota.*

Chytridiomycota display a variety of life cycles involving both asexual and sexual reproduction. Members of this group are microscopic in size and may consist of a single cell, a small multinucleate mass, or a true **mycelium**—a mat of hyphae capable of penetrating porous substrates. Many are capable of degrading cellulose and even keratin, which enables the degradation of crustacean exoskeletons. The genus *Allomyces* is used to study morphogenesis. This model organism has a complex life cycle that includes four types of sporangia and five developmentally distinct spore types. In addition, a population's growth and development can be synchronized by manipulating culture conditions.

24.3 *Zygomycota*

The *Zygomycota* contain fungi informally called **zygomycetes.** Most live on decaying plant and animal matter in the soil; a few are parasites of plants, insects, other animals, and humans. The hyphae of zygomycetes are coenocytic, with many haploid nuclei. Asexual spores develop in sporangia at the tips of aerial hyphae and are usually wind dispersed. Sexual reproduction produces tough, thick-walled zygotes called zygospores that can remain dormant when the environment is too harsh for growth of the fungus.

The mold *Rhizopus stolonifer* is a common member of this division. This fungus grows on the surface of moist, carbohydrate-rich foods, such as breads, fruits, and vegetables. On breads, for example, *Rhizopus* hyphae can rapidly cover the surface. Hyphae called rhizoids extend into the bread and absorb nutrients (**figure 24.4**). Other hyphae (stolons) become erect, then arch back into the substratum, forming new rhizoids.

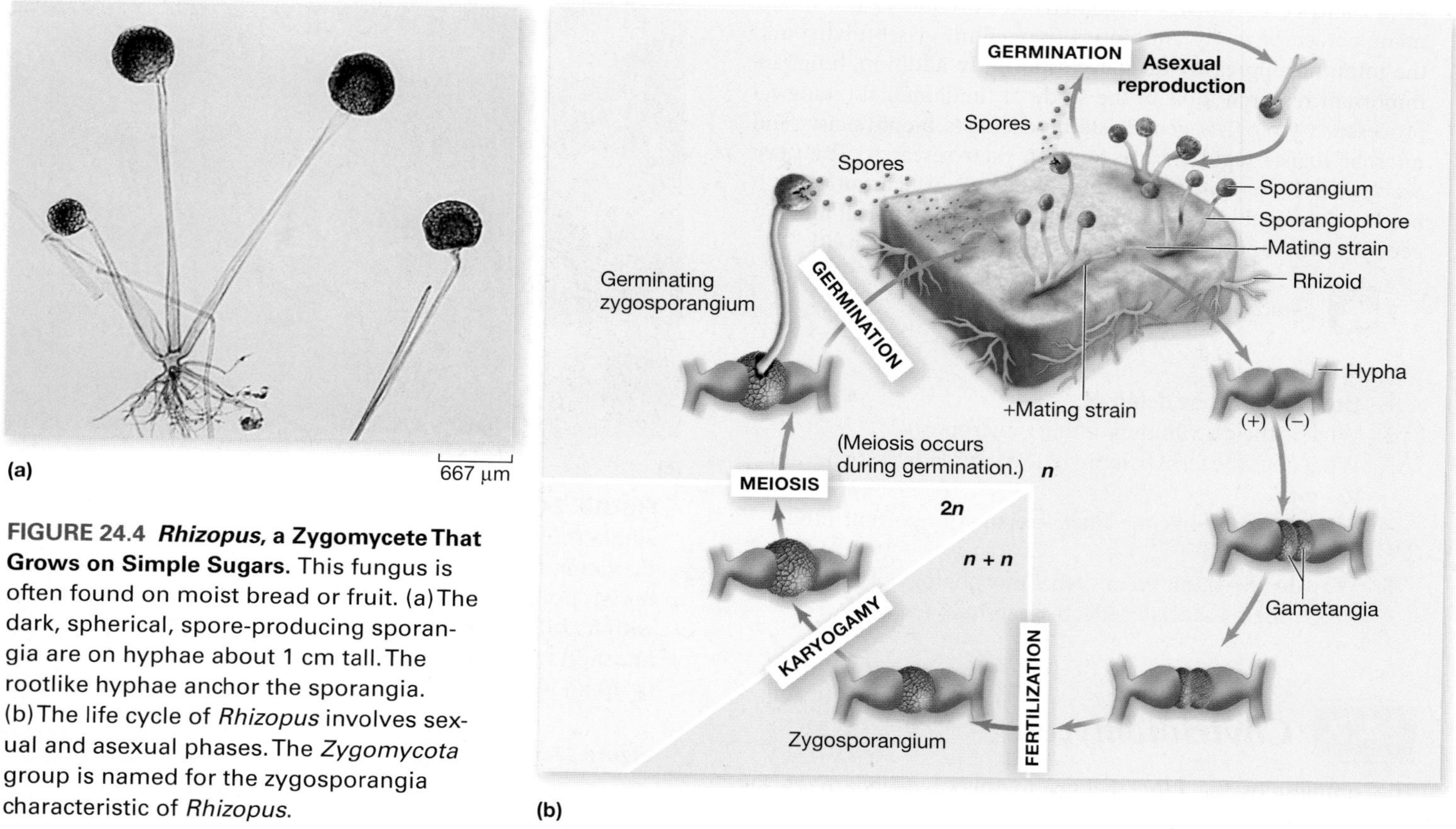

FIGURE 24.4 *Rhizopus,* a Zygomycete That Grows on Simple Sugars. This fungus is often found on moist bread or fruit. (a) The dark, spherical, spore-producing sporangia are on hyphae about 1 cm tall. The rootlike hyphae anchor the sporangia. (b) The life cycle of *Rhizopus* involves sexual and asexual phases. The *Zygomycota* group is named for the zygosporangia characteristic of *Rhizopus.*

Still others remain erect and produce at their tips asexual sporangia filled with black spores, giving the mold its characteristic color. Each spore, when liberated, can germinate to start a new mycelium.

Rhizopus usually reproduces asexually, but if food becomes scarce or environmental conditions unfavorable, sexual reproduction occurs (figure 24.4*b*). Sexual reproduction requires compatible strains of opposite mating types. When the two mating strains are close, hormones, called pheromones, are produced that cause their hyphae to form projections called progametangia that mature into gametangia. After fusion of the gametangia, the nuclei of the two gametes fuse, forming a zygote. The zygote develops a thick, rough, black coat and becomes a dormant zygospore. Meiosis often occurs at the time of germination; the zygospore then splits open and produces a hypha that bears an asexual sporangium to begin the cycle again.

The genus *Rhizopus* is also important because it is involved in the rice disease known as seedling blight. If one considers that rice feeds more people on Earth than any other crop, the implications of this disease are obvious. It was thought that *Rhizopus* secreted a toxin that kills rice seedlings, so scientists set about isolating the toxin and the genes that produce it. Much to everyone's surprise, an α-proteobacterium, *Burkholderia,* found growing within the fungus produces the toxin. Although the evolutionary history and nature of the *Rhizopus-Burkholderia* symbiosis remains to be clarified, there is at least one interesting twist to this story: the same toxin has been shown to stop cell division in some human cancer cells and is now being investigated as an antitumor agent.

The zygomycetes also contribute to human welfare. For example, one species of *Rhizopus* is used in Indonesia to produce a food called tempeh from boiled, skinless soybeans. Another zygomycete (*Mucor* spp.) is used with soybeans in Asia to make a curd called sufu. Others are employed in the commercial preparation of some anesthetics, birth control agents, industrial alcohols, meat tenderizers, and the yellow coloring used in margarine and butter substitutes. ▶▶| *Microbiology of fermented foods (section 40.5)*

24.4 *Glomeromycota*

The **glomeromycetes** are of critical ecological importance because they are mycorrhizal symbionts of vascular plants. **Mycorrhizal fungi** form important associations with the roots of almost all herbaceous plants and tropical trees. This is considered a mutualistic relationship because both the host plant and the fungus benefit: the fungus helps protect its host from stress and delivers soil nutrients to the plant, which in turn provides carbohydrate to the fungus. As discussed in chapter 29, most mycorrhizal fungi belong to one of two groups. The ectomycorrhizae do not penetrate host root cells; instead, the fungal hyphae grow between and around cells. By contrast, arbuscular mycorrhizal hyphae penetrate the host root cell wall but not the plasma membrane (**figure 24.5**). ▶▶| *Mycorrhizae (section 29.3)*

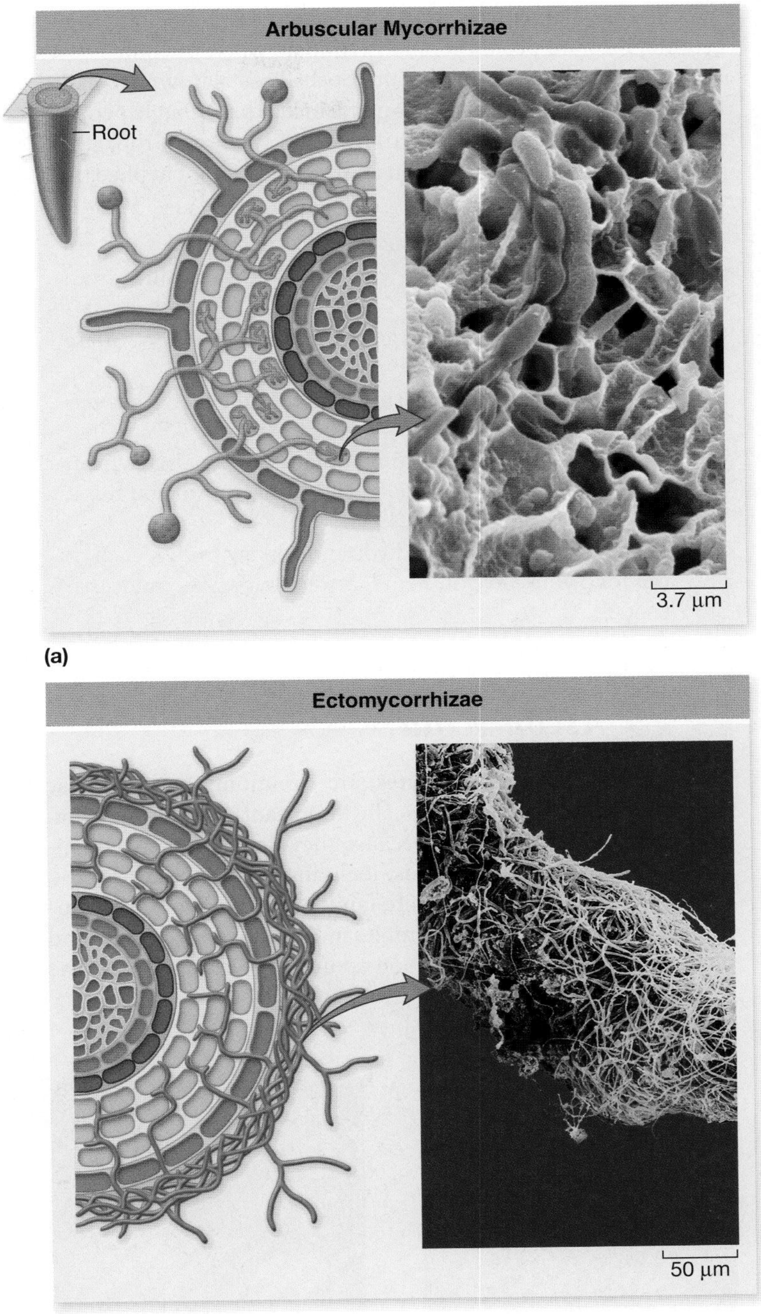

FIGURE 24.5 The *Glomeromycota:* Arbuscular Mycorrhizae and Ectomycorrhizae. (a) In arbuscular mycorrhizae, fungal hyphae penetrate the root cell wall of plants but not the plant membranes. (b) Ectomycorrhizae on the roots of a *Eucalyptus* tree do not penetrate root cells but grow around and extend between the cells.

Figure 24.5 Micro Inquiry

Why do you think neither the arbuscular mycorrhizae nor the ectomycorrhizae enter the host cell cytoplasm?

Only asexual reproduction is known to occur in glomeromycetes. Spores are produced and germinate when in contact with the roots of a suitable host plant. Specialized flat hyphae called **appressoria** (s., appressorium) are formed. These enable penetration and subsequent reproduction within the host. Propagation can also occur by fragmentation and colonization of hyphae from the soil or a nearby plant.

1. What are the *Chytridiomyceta?* How do they differ from other fungi?
2. Describe how a typical zygomycete reproduces. What are some beneficial uses for zygomycetes?
3. Compare the function of rhizoids and stolons in *Rhizopus stolonifer.*
4. Given that mycorrhizal fungi obtain carbohydrates from their plant partners, what specific nutrients do you think the fungi provide to plants?
5. Which fungal class would you most likely find in each of the following environments: 2-week-old cake, mud, tree roots?

24.5 *Ascomycota*

The *Ascomycota,* or **ascomycetes,** are commonly known as **sac fungi.** Ascomycetes are ecologically important in freshwater, marine, and terrestrial habitats because they degrade many chemically stable organic compounds, including lignin, cellulose, and collagen. Many species are quite familiar and economically important (**figure 24.6**). For example, most of the red, brown, and blue-green molds that cause food spoilage are ascomycetes. The powdery mildews that attack plant leaves and the fungi that cause chestnut blight and Dutch elm disease are ascomycetes. Many yeasts as well as edible morels and truffles are also ascomycetes. The pink bread mold *Neurospora crassa* is an important research tool in genetics and biochemistry.

The ascomycetes are named for their characteristic reproductive structure, the saclike **ascus** (pl., asci; Greek *askos,* sac). Many ascomycetes are yeasts. The term **yeast** refers to unicellular fungi that reproduce asexually by either budding or binary fission (**figure 24.7*a***); the life cycle of the yeast *Saccharomyces cerevisiae,* commonly known as brewer's yeast, is well understood. *S. cerevisiae* alternates between haploid and diploid states (figure 24.7*b*). As long as nutrients remain plentiful, haploid and diploid cells undergo mitosis to produce haploid and diploid daughter cells, respectively. Each daughter cell leaves a scar on the mother cell as it separates, and daughter cells bud only from unscarred regions of the cell wall. When a mother cell has no more unscarred cell wall remaining, it can no longer reproduce and will senesce (die). When nutrients are limited, diploid *S. cerevisiae* cells undergo meiosis to produce four haploid cells that remain bound within a common cell wall, the ascus. Upon the addition of nutrients, two haploid cells of opposite mating types (a and α) come into contact and fuse to create a diploid. Typically only cells of opposite mating types can fuse; this process is tightly regulated by the action of pheromones—chemical signals that are exchanged by the haploid cells.

S. cerevisiae is a valuable model organism. Research on this organism has revealed the importance of many cellular processes. For instance, it is a favorite model system for studying cell cycling and the events during mitosis. This research is critical not only for our understanding of normal cell division, but it also has been invaluable for understanding the loss of cell cycle control that occurs in cancerous cells.

Search This: Fungal genomes

Filamentous ascomycetes form septate hyphae. Asexual reproduction is common in these ascomycetes and is associated with the production of **conidia** (**figure 24.8**). Sexual reproduction involves ascus formation, with each ascus usually bearing eight haploid ascospores, although some species can produce over 1,000. Such hyphae are said to be ascogenous. Mating starts when two strains of opposite mating types form ascogenous hyphae into which pairs of nuclei migrate (**figure 24.9**). One nucleus of each pair originates from a "male" mycelium (antheridium) or cell and the other from a "female" organ or cell (ascogonium) that has fused with it. As the ascogenous hyphae grow, the paired nuclei divide so that there is one pair of nuclei in each cell. After the ascogenous hyphae have matured, nuclear fusion occurs at the hyphal tips in the ascus mother cells. The diploid zygote nucleus then undergoes meiosis, and the resulting four haploid nuclei divide mitotically again to produce a row of eight nuclei in each developing ascus. These nuclei are walled off from one another. Thousands of asci may be packed together in a cup- or flask-shaped fruiting body called an ascocarp. When the ascospores mature, they often are released from the asci with great force. If the mature ascocarp

(a) *Morchella esculenta* (b) *Tuber brumale*

FIGURE 24.6 The *Ascomycota.* (a) The common morel, *Morchella esculenta,* is one of the choicest edible fungi. It fruits in the spring. (b) The black truffle, *Tuber brumale,* is highly prized for its flavor by gourmet cooks. Truffles are mycorrhizal associations on oak trees.

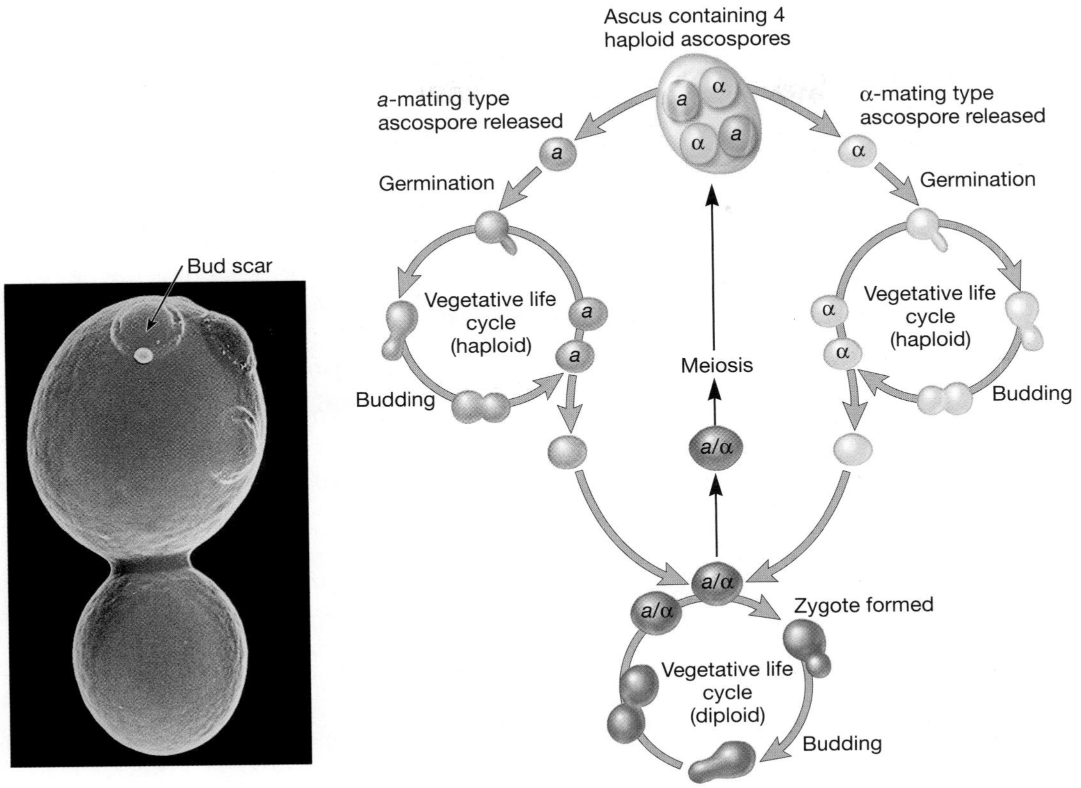

(a) *Saccharomyces cerevisiae:* budding division

(b) *S. cerevisiae* life cycle

FIGURE 24.7 The Life Cycle of the Yeast *Saccharomyces cerevisiae.* (a) Budding division results in asymmetric septation and the formation of a smaller daughter cell. (b) When nutrients are abundant, haploid and diploid cells undergo mitosis and grow vegetatively. When nutrients are limited, diploid *S. cerevisiae* cells undergo meiosis to produce four haploid cells that remain bound within a common cell wall, the ascus. Upon the addition of nutrients, two haploid cells of opposite mating types (a and α) fuse to create a diploid cell.

Figure 24.7 Micro Inquiry

What determines when a yeast cell can no longer bud?

FIGURE 24.8 Asexual Reproduction in *Ascomyota.* Characteristic conidiospores of *Aspergillus* as viewed with the scanning electron microscope (×1,200).

is jarred, it may appear to belch puffs of "smoke" consisting of thousands of ascospores. Upon reaching a suitable environment, the ascospores germinate and start the cycle anew.

Although conidia are the major form of dissemination, some filamentous fungi also produce sclerotia. **Sclerotia** are compact masses of hyphae that can survive the winter. In the spring they germinate to produce more hyphae or conidia. These structures confer a competitive advantage to the fungi that produce them. For instance, some species of *Aspergillus* that infect plants form sclerotia to remain viable in the soil, where they can take advantage of nutrient resources when the temperature warms.

Comparative genomic analysis of three important *Aspergillus* species is noteworthy. *A. fumigatus* is ubiquitous in the environment, commonly found tracked into homes and the workplace. It is known to trigger allergic responses and is implicated in the increased incidence in severe asthma and sinusitis. It is also pathogenic, infecting immunocompromised individuals with a mortality

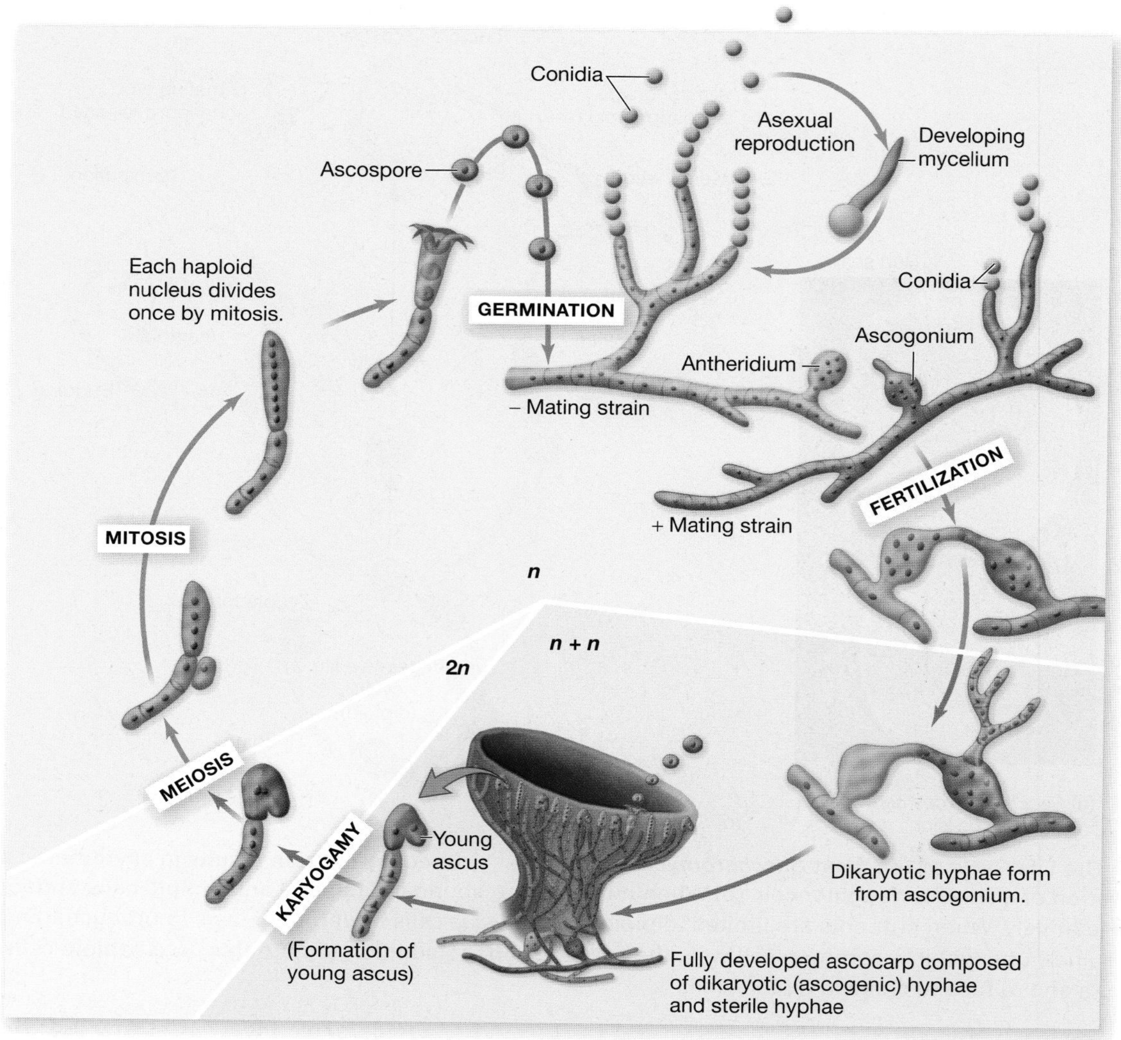

FIGURE 24.9 The Typical Life Cycle of a Filamentous Ascomycete. Sexual reproduction involves the formation of asci and ascospores. Within the ascus, karyogamy is followed by meiosis to produce the ascospores.

Figure 24.9 *Micro Inquiry*

In what ways is the life cycle of the filamentous ascomycetes similar to that of the zygomycetes shown in figure 24.4b? How do the two life cycles differ?

rate of nearly 50%. Its 29.4-Mb genome consists of eight chromosomes and about 10,000 genes. *A. nidulans* is a model organism that is used to study questions of eukaryotic cell and developmental biology. *A. oryzae* is used in the production of traditional fermented foods and beverages in Japan, including saki and soy sauce. Because it secretes many industrially useful proteins and can be genetically manipulated, it has become an important organism in biotechnology. Its metabolic and genetic versatility is reflected in the size of its genome—at 37 Mb, it is considerably larger than either *A. fumigatus* or *A. nidulans.* Much of this additional sequence is involved in the production of secreted hydrolytic enzymes, nutrient transport systems, and secondary metabolites. There are over 5,000 nonprotein-coding regions that are conserved in all three species. A variety of regulatory elements are present, including a riboswitch and other forms of translational control. These genome sequences will be useful to scientists seeking to understand the interaction between *Aspergillus* and the immune system, its role in food and industrial microbiology, and eukaryotic evolution. ◀◀ *Riboswitches (section 13.4); Comparative genomics (section 16.7);* ▶▶ *Microbiology of fermented foods (section 40.5)*

Many ascomycetes are parasites on higher plants. *Claviceps purpurea* parasitizes rye and other grasses, causing the plant disease ergot (**figure 24.10**). Ergotism, the toxic condition in humans and animals that eat grain infected with the fungus, is often accompanied by gangrene, psychotic delusions, nervous spasms, abortion, and convulsions. During the Middle Ages ergotism, then known as St. Anthony's fire, killed thousands of people. For example, over 40,000 deaths from ergot poisoning were recorded in France in the year 943. It has been suggested that the widespread accusations of witchcraft in Salem Village and other New England communities in the 1690s may have resulted from outbreaks of ergotism. The pharmacological activities are due to an active ingredient, lysergic acid diethylamide (LSD). In controlled dosages, other active compounds can be used to induce labor, lower blood pressure, and ease migraine headaches.

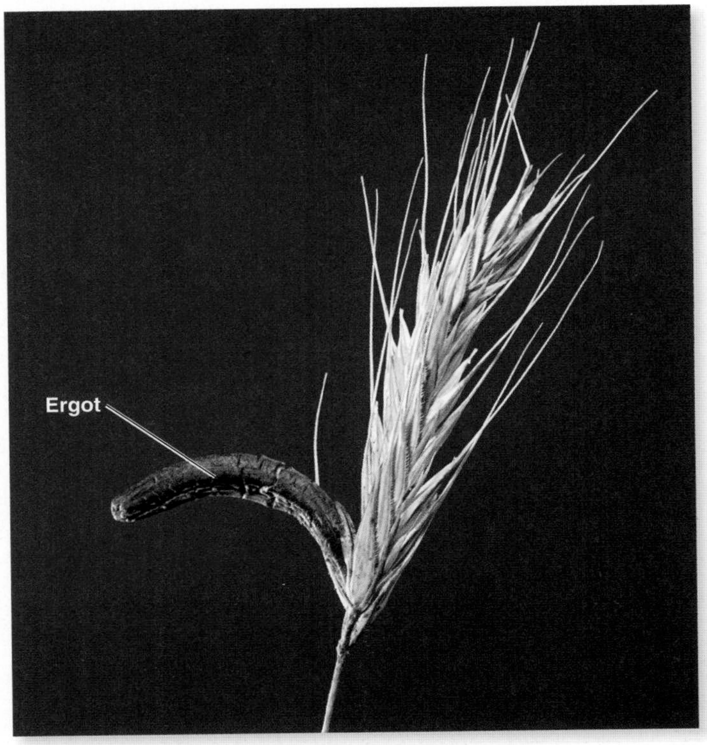

FIGURE 24.10 Ergot of Rye. The ascomycete *Claviceps purpurea* infects rye and other grasses, producing hard masses of hyphae known as ergots in place of some of the grains (fruits). Ergots produce alkaloids related to LSD and thus cause psychotic delusions in humans and animals that consume products made with infected grain. Ergots were used to treat migraine headaches and hasten childbirth.

Most fungal pathogens that infect humans are ascomycetes. Many are opportunistic pathogens such as *Candida, Blastomyces,* and *Histoplasma.* In addition, the cause of "sick building syndrome," *Stachybotrys,* is also an ascomycete (**figure 24.11**). Finally, the *Aspergillus* toxins known as aflatoxins are an important cause of food contamination. Exposure to aflatoxins can result in liver cancer. These and other fungal pathogens are discussed in chapter 39, while aflatoxins are covered in both chapters 39 and 40. Although ascomycete infections can have catastrophic consequences for humans, the use of an ascomycete that infects destructive and pathogen-bearing insects is under development as a biocontrol measure for malaria.

Search This: Dr. Fungus

1. Describe the ascomycete life cycle. How are the ascomycetes useful to humans?
2. How do yeasts reproduce sexually? Asexually? Why do you think *Saccharomyces cerevisiae* has become such an important model organism?
3. Review the life cycle of a filamentous ascomycete. Compare the dikaryotic stage with the haploid stage. Why do you think these fungi delay nuclear fusion?

(a) (b)

FIGURE 24.11 Fungal Growth in a Building. Fungal growth on sheet rock removed from a water-damaged building. (a) Stereo microscopic view showing black discoloration. Bar = 500 mm. (b) Scanning electron micrograph of dense mycelia and condiophores characteristic of *Stachybotrys*. Bar = 10 mm.

24.6 *Basidiomycota*

The *Basidiomycota* includes the **basidiomycetes,** commonly known as **club fungi.** Examples include jelly fungi, rusts, shelf fungi, stinkhorns, puffballs, toadstools, mushrooms, and bird's nest fungi. Basidiomycetes are named for their characteristic structure or cell, the **basidium,** which is involved in sexual reproduction (**figure 24.12**). A basidium (Greek *basidion,* small base) is produced at the tip of hyphae and normally is club shaped. Two or more **basidiospores** are produced by the basidium, and basidia may be held within fruiting bodies called **basidiocarps.**

The basidiomycetes affect humans in many ways. Most are saprophytes that decompose plant debris, especially cellulose and lignin. For example, the common fungus *Polyporus squamosus* forms large, shelflike structures that project from the lower portion of dead trees, which they help decompose. The fruiting body can reach 60 cm in diameter and has many pores, each lined with basidia that produce basidiospores. Thus a single fruiting body can produce millions of spores. Many mushrooms are used as food throughout the world. The cultivation of the mushroom *Agaricus campestris* is a multimillion-dollar business (*see figure 40.16*).

Of course not all mushrooms are edible. Many mushrooms produce alkaloids that act as either poisons or hallucinogens. One such example is the "death angel" mushroom, *Amanita phalloides* (figure 24.12*a*). Two toxins isolated from this species are phalloidin and α-amanitin. Phalloidin primarily attacks liver cells, where it binds to plasma membranes, causing them to rupture and leak their contents. Alpha-amanitin attacks the cells lining the stomach and small intestine, causing severe gastrointestinal symptoms associated with mushroom poisoning. *A. muscaria* is both poisonous and hallucinogenic. It produces a toxin that inhibits eukaryotic RNA polymerase II.

Search This: Mushroom hunting

The basidiomycete *Cryptococcus neoformans* is an important human and animal pathogen. It produces the disease called

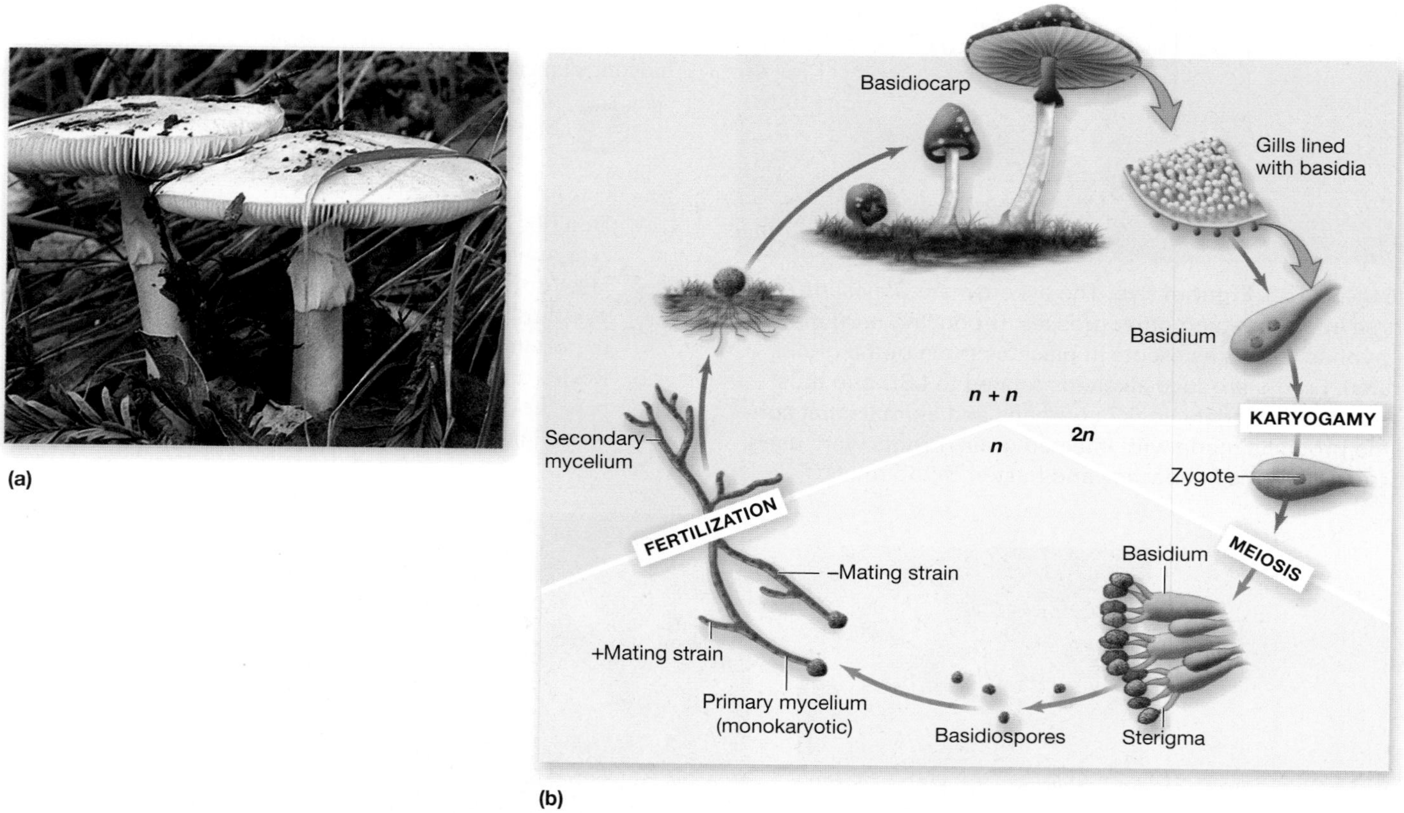

FIGURE 24.12 The *Basidiomycota*. (a) The death cap mushroom *Amanita phalloides* is usually fatal when eaten. (b) The life cycle of a typical soil basidiomycete starts with a basidiospore germinating to produce a monokaryotic mycelium (a single nucleus in each septate cell). The mycelium grows and spreads throughout the soil. When this primary mycelium meets another monokaryotic mycelium of a different mating type, the two fuse to initiate a dikaryotic secondary mycelium. The secondary mycelium is divided by septa into cells, each of which contains two nuclei, one of each mating type. This dikaryotic mycelium is eventually stimulated to produce basidiocarps. A solid mass of hyphae forms a button that pushes through the soil, elongates, and develops a cap. The cap contains many platelike gills, each of which is coated with basidia. The two nuclei in the tip of each basidium fuse to form a diploid zygote nucleus, which immediately undergoes meiosis to form four haploid nuclei. These nuclei push their way into the developing basidiospores, which are then released at maturity.

cryptococcosis, a systemic infection primarily involving the lungs and central nervous system. This fungus always grows as a large, budding yeast, and the production of an elaborate capsule is an important virulence factor for the microbe. In the environment, *C. neoformans* is a saprophyte with a worldwide distribution. Aged, dried pigeon droppings are an apparent source of infection, although it can also grow on plant surfaces. Cryptococcosis is found in approximately 15% of AIDS patients. The fungus enters the body by the respiratory tract, causing a minor pulmonary infection that is usually transitory. However, some pulmonary infections spread to the skin, bones, viscera, and the central nervous system. Once the nervous system is involved, cryptococcal meningitis usually results. Cryptococcosis must be treated with systemic antifungal agents. ▶▶| *Antifungal drugs (section 34.5); Airborne diseases (section 39.2)*

Urediniomycota and *Ustilaginomycota*

The *Basidiomycota* that belong to the urediniomycetes and the ustilaginomycetes include important plant pathogens causing "rusts" and "smuts." In addition, some urediniomycetes include human pathogens, and some are virulent plant pathogens that cause extensive damage to cereal crops, destroying millions of dollars worth of crops annually. These fungi do not form large basidiocarps. Instead, small basidia arise from hyphae at the surface of the host plant. The hyphae grow either intra- or extracellularly in plant tissue.

The ustilaginomycete *Ustilago maydis* is a common corn pathogen that has become a model organism for plant smuts (**figure 24.13**). It is dimorphic; plant-associated fungi grow in the mycelial form but are yeastlike in the external environment. Yeastlike saprophytic *U. maydis* cells can be easily grown in the laboratory. In nature, the yeast form (haploid sporidia) must mate to produce infectious, filamentous dikaryons that depend on the host plant for continued development (**figure 24.14**). Once a plant is infected, *U. maydis* forms appressoria within the host. This triggers the plant to form tumors in which the fungus proliferates and eventually produces diploid spores called **teliospores.** Upon germination, cells undergo meiosis and haploid sporidia are released. These sporidia can spread from plant to plant, where they start the growth and infection cycle again.

FIGURE 24.13 *Ustilago maydis.* This pathogen causes tumor formation in corn. Note the enlarged tumors releasing dark teliospores in place of normal corn kernels.

24.7 *Microsporidia*

Of all the fungi, the ***Microsporidia*** have had the most confused taxonomic history. These obligately endoparasitic microbes have been considered protists and are sometimes still cited as such. However, molecular analysis of ribosomal RNA and specific proteins such as α- and β-tubulin shows that they are related to fungi. Inspection of figure 24.1 shows that they are believed to have evolved from an endocytic chytrid ancestor similar to *Rozella*. However, unlike other fungi, they lack mitochondria, peroxisomes, and centrioles but possess mitosomes (*see page 584*). Their genomes are highly reduced in size and complexity. But like all fungi, microsporidian spores contain chitin and trehalose.

Microsporidia morphology is also unique among eukaryotes. It is distinct in the possession of an organelle called the polar tube that is essential for host invasion. Small spores of 1 to 40 μm are viable outside the host. Depending on the species, spores may be spherical, rodlike, or egg- or crescent-shaped. Spore germination is triggered by a signal from the host and results in the expulsion of

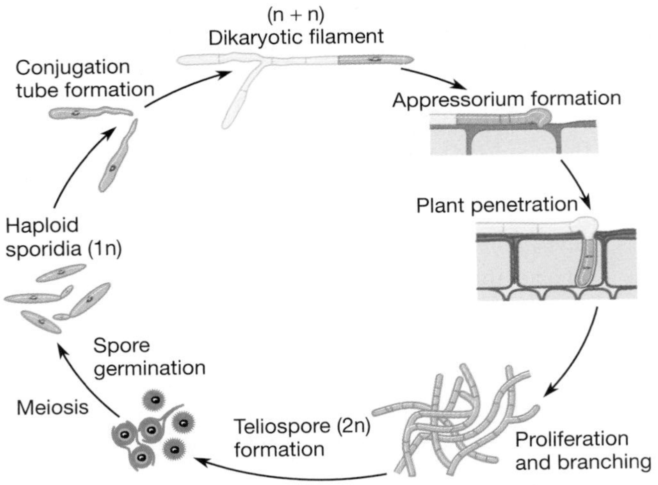

FIGURE 24.14 Life Cycle of the Plant Pathogen *Ustilago maydis.* Haploid sporidia must mate to form infectious dikaryons, which in turn depend on the host plant for continued development. The formation of fungal appressoria within the host induces tumor formation.

Figure 24.14 Micro Inquiry

What adaptations do you think *Ustilago* has evolved that has led to its pathogenicity? Contrast this with the mycorrhizae.

the tightly packed polar tube or filament (**figure 24.15**). The polar tube is ejected with enough force to pierce the host cell membrane, which permits parasite entry. Once inside the host cell, the microsporidian undergoes a developmental cycle that differs among the various microsporidian species. However, in all cases, more spores are produced and eventually take over the host cell.

Microsporidia are important obligate intracellular parasites that infect insects, fish, and humans. In particular, they are particularly problematic for immunocompromised individuals, especially those with HIV/AIDS. Common human pathogens include *Enterocystozoon bieneusi,* which causes diarrhea and pneumonia (depending on whether it was acquired through ingestion or inhalation, respectively), and *Encephalitozoon cuniculi,* which causes encephalitis and nephritis (kidney disease). Aquatic birds are an important source of pathogenic microsporidia in nature; indeed, a single flock of waterfowl can deposit over 10^8 microsporidian spores in surface waters after one visit. ▶▶| *Microsporidia (section 39.6)*

1. Describe the life cycle of a typical basidiomycete. Discuss their importance.
2. How does *Ustilago maydis* trigger tumor formation? How is tumor formation advantageous to the fungi?
3. Why do you think *Microsporidia* are still sometimes considered protists? Describe the germination of a microsporidian spore.

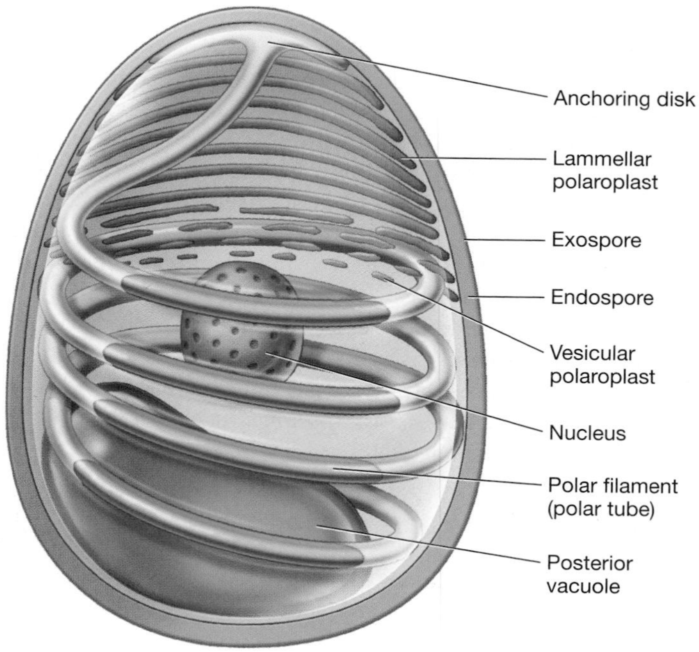

FIGURE 24.15 The Unique Structure of a Microsporidian Spore. Upon germination in a host cell, the polar tube is ejected with enough force to pierce the host cell membrane, allowing the fungus to gain entry.

Summary

24.1 Fungal Distribution and Importance

a. There are six major divisions within the *Fungi:* the *Ascomycota* and *Basidiomycota* are dikaryons, the *Glomeromycota* are important plant symbionts, *Zygomycota* and *Chytridiomycota* are not monophyletic, and the *Microsporidia* are the most ancient (**figures 24.1** and **24.2**).

b. Fungi are important decomposers that break down organic matter; live as parasites on animals, humans, and plants; play a role in many industrial processes; and are used as research tools in the study of fundamental biological processes. Fungi secrete enzymes outside their body structure and absorb the digested food.

24.2 *Chytridiomycota*

a. *Chytridiomycota* produce motile spores. Most are saprophytic; some reside in the rumen of herbivores (**figure 24.3**).

b. These fungi are microscopic and display both sexual and asexual reproduction.

24.3 *Zygomycota*

a. Zygomycetes are coenocytic. Most are saprophytic. One example is the common bread mold, *Rhizopus stolonifer.* Sexual reproduction occurs through conjugation between strains of opposite mating type (**figure 24.4**).

b. Zygomycetes are used in food and pharmaceutical manufacturing.

24.4 *Glomeromycota*

a. *Glomeromycota* include the mycorrhizal fungi that grow in association with plant roots. They serve to increase plant nutrient uptake.

b. Most mycorrhizal fungi are either ectomycorrhizae, which do not penetrate the host plant cells, or arbuscular mycorrhizae, which grow between the plant cell wall and plasma membrane (**figure 24.5**).

24.5 *Ascomycota*

a. *Ascomycota* are known as the sac fungi because they form a sac-shaped reproductive structure called an ascus. Ascomycetes can have either a yeast morphology (**figure 24.7**) or a mold morphology (**figure 24.8**).

b. Important ascomycetes include the yeast *Saccharomyces cerevisiae, Aspergillus* spp., *Claviceps purpurea*—the cause of ergotism, and the indoor mold *Stachybotrys.*

24.6 *Basidiomycota*

a. *Basidiomycota* are the club fungi. They are named after their basidium that produces basidiospores (**figure 24.12**).

b. Urediniomycetes and ustilaginomycetes are often considered *Basidiomycota*. Genera from both groups include important plant pathogens (**figure 24.13**).

24.7 *Microsporidia*

a. *Microsporidia* have a unique morphology and are still sometimes considered protists.

b. They include virulent human pathogens; some infect other vertebrates and insects (**figure 24.15**).

Critical Thinking Questions

1. What are some logical physiological or morphological targets to exploit in treating animals or plants suffering from fungal infections? Compare these with the targets you would use when treating infections caused by bacteria.

2. Some fungi can be viewed as coenocytic organisms that exhibit little differentiation. When differentiation does occur, such as in the formation of reproductive structures, it is preceded by septum formation. Why does this occur?

3. Both bacteria and fungi are major environmental decomposers. Obviously competition exists in a given environment, but fungi usually have an advantage. What characteristics specific to the fungi provide this advantage?

4. *Neurospora crassa* and several other filamentous fungi demonstrate a circadian rhythm such that conidiation occurs in roughly 21.5 hour cycles, depending on temperature and the light/dark regime. This can be observed in nature as bands of conidia alternating with bands of undifferentiated fungi. Why do you think these fungi evolved a circadian rhythm? If you were to study this phenomenon, what questions would you want to explore?

5. A study designed to explore climate change impact on ecosystem function used historical records of 315 fungal species in 1,400 locations in Britain. The records listed the dates at which fruiting bodies first emerged in the spring and the last date they were visible in the autumn. It was concluded that in 2005 the average first fruiting date was about 8 days earlier and the last date in the autumn in which fungi could be observed was about a week later than in 1950. Why were the scientists interested in fungal growth as part of ecosystem function? How do you think the scientists were able to link these changes to climate change? What other factors do you think they might have needed to consider?

   Read the original paper: Gange, A. C., et al. 2007. Rapid and recent changes in fungal fruiting patterns. *Science.* 316:71.

6. Analysis of sclerotia and conidia formation in the filamentous ascomycete *Aspergillus flavus* suggests that the formation of these structures is cell-density dependent. Specifically, high cell density cultures yield conidia, whereas low cell density triggers sclerotia formation. Review the structure and function of conidia and sclerotia and formulate a hypothesis to explain why their development would be regulated in a density-dependent manner. List several specific testable predictions based on your hypothesis.

   Read the original paper: Horowitz-Brown, S., et al. 2008. Morphological transitions governed by density dependence and lipooxygenase activity in *Aspergillis flavus. Appl. Environ. Microbiol.* 74:5674.

Learn More

Learn more by visiting the text website at www.mhhe.com/willey8, where you will find a complete list of references.

25

The Viruses

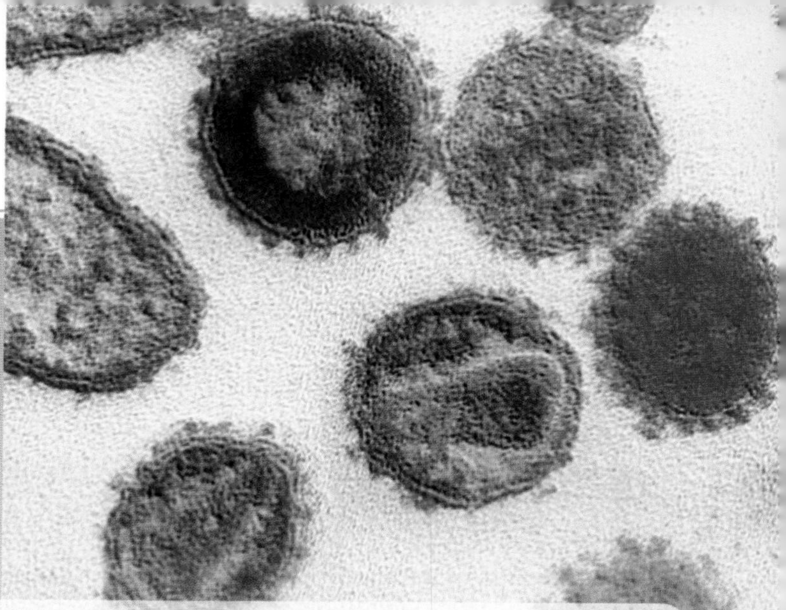

The human immunodeficiency virus (HIV) is an enveloped retrovirus. Protein spikes are embedded in the membrane, and the ssRNA genome is enclosed within a core.

CHAPTER GLOSSARY

concatemers Long DNA molecules consisting of several genomes linked together in a row.

DNA-dependent DNA polymerase An enzyme that catalyzes the synthesis of DNA using a DNA molecule as template.

DNA-dependent RNA polymerase An enzyme that catalyzes the synthesis of RNA using a DNA molecule as template.

integrase A viral enzyme that catalyzes integration of a provirus into the host cell's chromosome.

minus-strand (negative-strand) A viral nucleic acid with a sequence of nucleotides that is complementary to the mRNA produced by the virus.

plus-strand (positive-strand) A viral nucleic acid with a nucleotide sequence identical to that in the mRNA produced by the virus, except when the nucleic acid is DNA. Then DNA thymine (T) replaces uracil (U) in the sequence.

polyprotein A large polypeptide that contains several proteins, which are released when the polyprotein is cut by proteases. Polyproteins are often produced by viruses having positive-strand RNA genomes.

RecA A protein present in all domains of life that is important in recombination and DNA repair. It is also involved in the induction process that occurs when bacteriophage lambda switches from lysogeny to a lytic cycle.

replicase The term used to describe the activity of an RNA-dependent RNA polymerase when it is functioning to replicate the RNA genome of a virus.

replicative form (RF) A double-stranded DNA or RNA molecule that is generated by some single-stranded DNA and RNA viruses during their life cycles. It is used to synthesize new viral genomes and sometimes to synthesize viral mRNA molecules.

restriction A defense mechanism used by some bacteria to protect themselves from viral infection. It involves degradation of the viral DNA by restriction enzymes.

reverse transcriptase A multifunctional enzyme used by retroviruses and hepadnaviruses to generate double-stranded DNA from an RNA template. The enzyme has RNA-dependent DNA polymerase activity, DNA-dependent DNA polymerase activity, and RNaseH activity.

RNA-dependent DNA polymerase An enzyme that synthesizes DNA using an RNA template.

RNA-dependent RNA polymerase An enzyme that synthesizes RNA using an RNA template.

subgenomic mRNA An mRNA molecule that is shorter than the genomic RNA of an RNA virus.

Over 100 years ago, the concept of a virus was developed by Dmitrii Ivanowski and Martinus Beijerinck (studying *Tobacco mosaic virus*) and by Friedrich Loeffler and Paul Frosch (studying the virus responsible for foot-and-mouth disease). Within about 20 years, the concept was applied to an agent that infected bacteria, when Félix d'Herelle coined the term bacteriophage. Since then viruses have captured the imagination of countless virologists because of their amazing diversity and ability to affect humans, other organisms, and ecosystems. Viruses also have captured the imagination of novelists, who weave apocalyptic stories of the demise of human civilization and of great plagues caused by these tiniest of microbes. The fascination with viruses by the general population is based on the long history humans have had with plagues caused by viruses, and it is reinvigorated on a regular basis by reports of viruses such as Ebola that kill swiftly and horrifically.

As we outline in chapter 5, the diversity of viruses is in part due to their structural features, especially the nature of their genomes. A virus's genome dictates important aspects of a virus's life cycle and is in part the basis of a scheme (the Baltimore System) that greatly simplifies the discussion of viral life cycles. The Baltimore system is not the classification system used by virologists to identify and name viruses. That role belongs to the International Committee on the Taxonomy of Viruses. The Baltimore system will serve to organize the discussion of viral diversity in this chapter. For each Baltimore group, we present a brief overview of how viruses in the group multiply, as well as the complete life cycle of at least one virus.

Another aspect of viral diversity results from the variety of strategies they use to regulate events in their life cycles and the activities of their host cells. Thus for each virus, we will examine some of these strategies. In many ways, this discussion is a continuation of topics introduced in chapter 13, where we considered the regulation of cellular processes. For a virus to reproduce, it must "know" its host intimately so that it can take advantage of cellular processes as needed. Indeed, viruses are the masters of regulation.

25.1 Virus Taxonomy and Phylogeny

The classification of viruses is in a much less satisfactory state than that of cellular microorganisms. In part, this is due to a lack of knowledge of their origin and evolutionary history. In 1971 the International Committee on Taxonomy of Viruses (ICTV) developed a uniform classification system. Since then the number of viruses and taxa has continued to expand. In its eighth report, the ICTV described almost 2,000 virus species and placed them in 3 orders, 73 families, 9 subfamilies, and 287 genera. (See the text website at *www.mhhe.com/willey8* for a table showing **Some Common Virus Groups and Their Characteristics.**) The committee places greatest weight on specific properties to define families: nucleic acid type, presence or absence of an envelope, symmetry of the capsid, and dimensions of the virion and capsid. Virus order names end in *-virales;* virus family names in *-viridae;* subfamily names, in *-virinae;* and genus names, in *-virus.* An example of this nomenclature scheme is shown in **figure 25.1**, and the diversity of viral morphology and genome structure among viruses is illustrated in **figure 25.2.** ◀◀ *Structure of viruses (section 5.2)*

Although the ICTV reports are the official authority on viral taxonomy, many virologists find it useful to group viruses using a scheme devised by Nobel laureate David Baltimore. The Baltimore system complements the ICTV system but focuses on the viral genome and the process used to synthesize viral mRNA. Recall from chapter 5 that all four nucleic acid types can be found in viruses: double-stranded (ds) DNA, single-stranded (ss) DNA, dsRNA, and ssRNA. The characterization of the genome of a ssRNA virus is further complicated by the sense of the ssRNA. Some ssRNA viruses have an RNA genome that is identical in base sequence to that of mRNA produced by the virus. Such viruses are said to have **plus-strand** or **positive-strand RNA** (**figure 25.3**). Other ssRNA viruses have genomes that are complementary to the mRNA they produce. These viruses are said to have **minus-strand** or **negative-strand RNA.** Baltimore's system organizes viruses into seven groups (**table 25.1**). This system helps virologists (and microbiology students) simplify the vast array of viral life cycles into a relatively small number of basic types; thus we use it here.

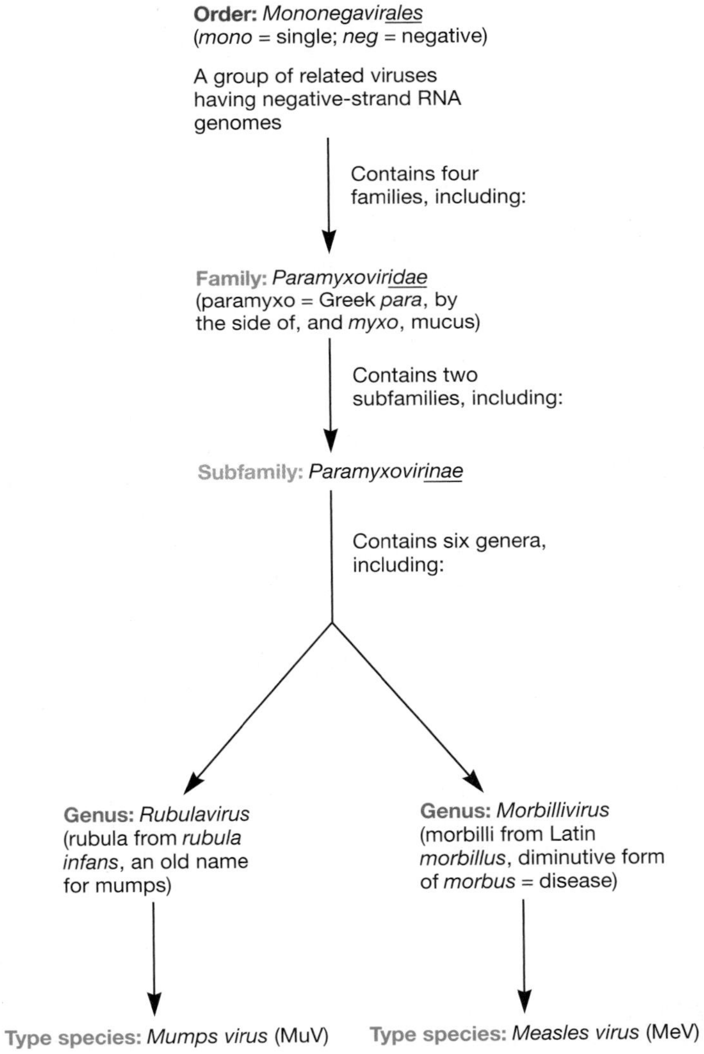
Order: Mononegavirales
(mono = single; neg = negative)
A group of related viruses having negative-strand RNA genomes
Contains four families, including:
Family: Paramyxoviridae
(paramyxo = Greek para, by the side of, and myxo, mucus)
Contains two subfamilies, including:
Subfamily: Paramyxovirinae
Contains six genera, including:
Genus: Rubulavirus
(rubula from rubula infans, an old name for mumps)
Genus: Morbillivirus
(morbilli from Latin morbillus, diminutive form of morbus = disease)
Type species: Mumps virus (MuV)
Type species: Measles virus (MeV)

FIGURE 25.1 The Naming of Viruses. Because of the difficulty in establishing evolutionary relationships, most virus families have not been placed into an order. Virus names are derived from various aspects of their biology and history, including the features of their structure, diseases they cause, and locations where they were first identified or recognized.

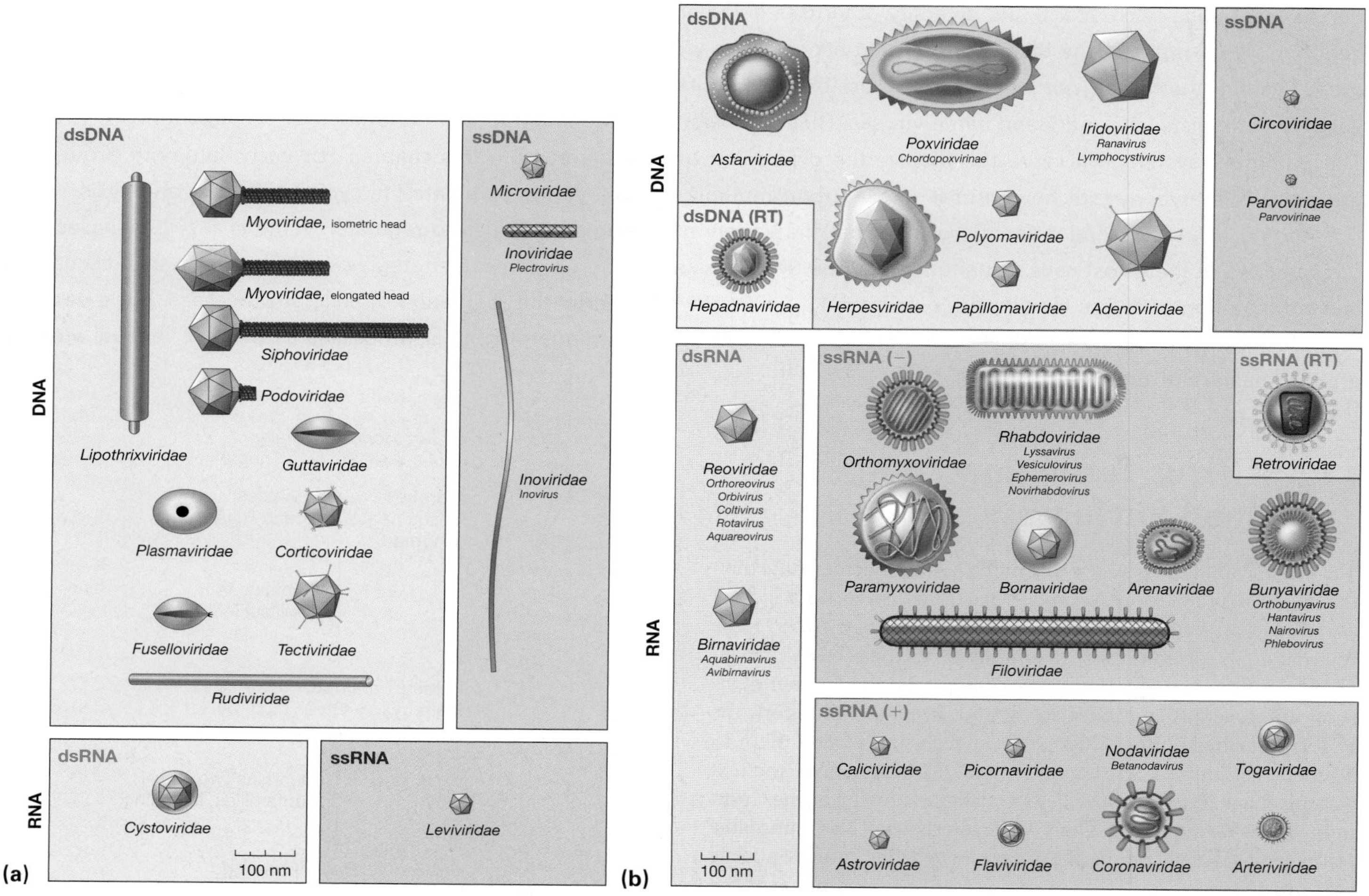

FIGURE 25.2 Important Families (*-idae*), Subfamilies (*-inae*), and Genera of Viruses. (a) Viruses that infect *Bacteria* and *Archaea*. The *Myoviridae* have contractile tails. *Plasmaviridae* are pleomorphic. *Tectiviridae* have distinctive double capsids. *Corticoviridae* have lipid-containing capids. (b) Viruses that infect vertebrates. Many of these viruses are enveloped viruses.

As is the case with cellular life forms, the genomes of more and more viruses have been sequenced. Virologists are now carrying out comparative analyses of these genomes to try to discern evolutionary relationships. These studies suggest that viruses in Baltimore groups VI (retroviruses) and VII (e.g., *Hepatitis B virus*) share a common evolutionary history. All of these viruses use an enzyme called reverse transcriptase in their life cycles. Another group of related viruses have large dsDNA genomes and share other characteristics. These viruses are referred to as nucleo-cytoplasmic large DNA (NCLD) viruses and have garnered considerable interest in part because they seem to blur the distinctions between cells and viruses. These viruses are considered in section 25.2.

1. List some characteristics used in classifying viruses. Which seem to be the most important?
2. What are the endings for the names of virus families, subfamilies, and genera?

25.2 Viruses with Double-Stranded DNA Genomes (Group I)

Perhaps the largest group of known viruses is the double-stranded (ds) DNA viruses; most bacteriophages have dsDNA genomes, as do several insect viruses and a number of important vertebrate viruses, including the herpesviruses and poxviruses. The pattern of multiplication for dsDNA viruses is shown in **figure 25.4**. The synthesis of DNA and RNA is similar to what occurs in cellular organisms; therefore some dsDNA viruses can rely entirely on their host's DNA and RNA polymerases.

Bacteriophage T4: A Virulent Bacteriophage

The life cycle of T4 bacteriophage (family *Myoviridae*) serves as our example of a virulent dsDNA phage. Virulent (lytic) bacteriophages are capable only of the lytic cycle. That is, their infection of a host always ends with cell lysis. As with any virus, the first step of viral

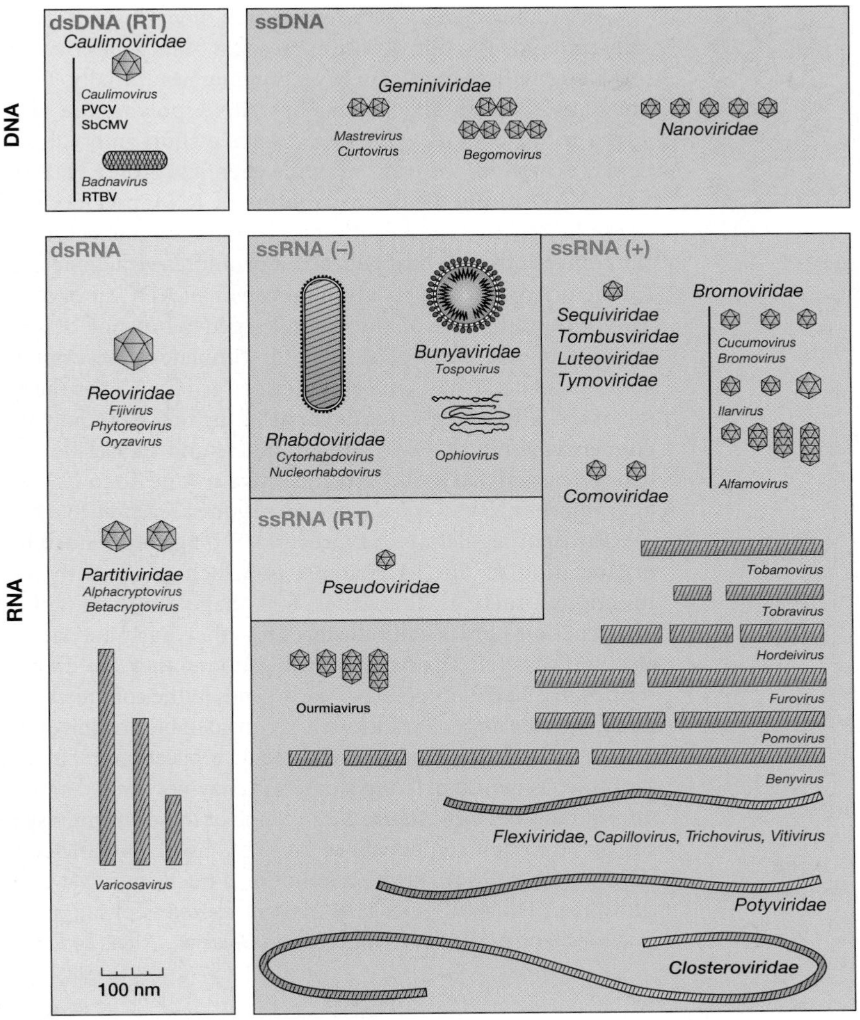

(c)

FIGURE 25.2 ***(continued)*****.** (c) Viruses that infect plants. RT stands for reverse transcriptase, an enzyme able to synthesize dsDNA from ssRNA.

infection is attachment (adsorption) to the host cell surface. T4 phage attachment begins when a tail fiber contacts either the lipopolysaccharide or certain proteins in the outer membrane of its *Escherichia coli* host (**figure 25.5*a***). As more tail fibers make contact, the baseplate settles down on the surface (figure 25.5*b*). After the baseplate is seated firmly on the cell surface, the baseplate and sheath change shape, and the tail sheath reorganizes so that it shortens from a cylinder 24 rings long to one of 12 rings (figure 25.5*c,d*). As the sheath becomes shorter and wider, the central tube is pushed through the bacterial cell wall. The baseplate contains the protein gp5, which has lysozyme activity. Lysozyme is an enzyme that breaks the bonds linking the sugars of peptidoglycan together (*see figure 32.3*). Thus this protein aids in the penetration of the tube through the peptidoglycan layer. Finally, the linear DNA is extruded from the head, probably through the tail tube, and into the host cell (figure 25.5*e,f*). The tube may interact with the plasma membrane to form a pore through which DNA passes.

Within 2 minutes after injection of T4 DNA into an *E. coli* cell, the *E. coli* RNA polymerase starts synthesizing T4 mRNA (**figure 25.6**). This mRNA is called early mRNA because it is made before viral DNA is made. Within 5 minutes, viral DNA synthesis commences, catalyzed by a virus-encoded **DNA-dependent DNA polymerase.** DNA replication is initiated from several origins of replication and proceeds bidirectionally from each. Viral DNA replication is followed by the synthesis of late mRNAs, which are important in later stages of the infection.

Viral mRNA	5′...GAC UCG AGC...3′
Plus-strand RNA	5′...GAC UCG AGC...3′
Negative-strand RNA	3′...CUG AGC UCG...5′
Plus-strand DNA	5′...GAC TCG AGC...3′
Negative-strand DNA	3′...CTG AGC TCG...5′

FIGURE 25.3 Plus-Strand and Negative-Strand Viral Genomes. The genomes and replication intermediates of viral genomes can be either plus strand or negative strand. This designation is relative to the sequence of nucleotides in the mRNA of the virus. Plus-strand genomes have the same sequence as the mRNA, either using DNA nucleotides if a DNA genome or RNA nucleotides if an RNA genome. Negative-strand genomes are complementary to the viral mRNA.

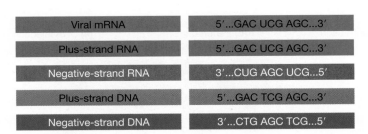

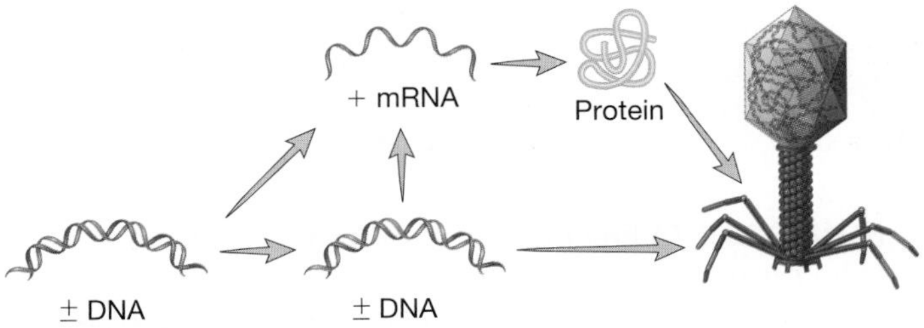

FIGURE 25.4 Multiplication Strategy of Double-Stranded DNA Viruses. Because the genome of double-stranded DNA viruses is similar to the host, genome replication and mRNA synthesis closely resemble that of the host cell and can involve the use of host polymerases, viral polymerases, or both. The DNA serves as the template for DNA replication and mRNA synthesis. Translation of the mRNA by the host cell's translation machinery yields viral proteins, which are assembled with the viral DNA to make mature virions. These are eventually released from the host.

Table 25.1 The Baltimore System

Group	*Description*
I	Double-stranded DNA genome *genome replication: dsDNA* $\rightarrow$ *dsDNA* *mRNA synthesis: dsDNA* $\rightarrow$ *mRNA*
II	Single-stranded DNA genome *genome replication: ssDNA* $\rightarrow$ *dsDNA* $\rightarrow$ *ssDNA* *mRNA synthesis: ssDNA* $\rightarrow$ *dsDNA* $\rightarrow$ *mRNA*
III	Double-stranded RNA genome *replication: dsRNA* $\rightarrow$ *ssRNA* $\rightarrow$ *dsRNA* *mRNA synthesis: dsRNA* $\rightarrow$ *mRNA*
IV	Plus-strand RNA genome *replication:* $+RNA \rightarrow -RNA \rightarrow +RNA$ *mRNA synthesis:* $+RNA = mRNA$
V	Negative-strand RNA genome *replication:* $-RNA \rightarrow +RNA \rightarrow -RNA$ *mRNA synthesis:* $-RNA \rightarrow mRNA$
VI	Single-stranded RNA genome *replication: ssRNA* $\rightarrow$ *dsDNA* $\rightarrow$ *ssRNA* *mRNA synthesis: ssRNA* $\rightarrow$ *dsDNA* $\rightarrow$ *mRNA*
VII	Double-stranded gapped DNA genome *replication: gapped dsDNA* $\rightarrow$ *dsDNA* $\rightarrow$ $+RNA \rightarrow$ $-DNA \rightarrow$ *gapped dsDNA* *mRNA synthesis: gapped dsDNA* $\rightarrow$ *dsDNA* $\rightarrow$ *mRNA*

As with most viruses, the expression of viral genes is temporally ordered. T4 controls the expression of its genes by regulating the activity of the *E. coli* RNA polymerase. Initially, T4 genes are transcribed by the regular host RNA polymerase and the sigma factor σ^{70} (*see table 13.3*). After a short interval, a viral enzyme catalyzes the transfer of the chemical group ADP-ribose from NAD to one of the α-subunits of RNA polymerase (*see figure 12.24*). This modification of the host enzyme helps inhibit the transcription of host genes and promotes viral gene expression. Later the second α-subunit receives an ADP-ribosyl group. This turns off some of the early T4 genes but not before the product of one early gene (*motA*) stimulates transcription of somewhat later genes. One of these later genes encodes the sigma factor gp55. This viral sigma factor helps the cell's RNA polymerase core enzyme bind to viral late promoters and transcribe the late genes. The viral late genes become active around 10 to 12 minutes after infection. ◀◀ *Global regulatory systems (section 13.6)*

The tight regulation of expression of T4 genes is aided by the organization of the T4 genome, in which genes with related functions—such as the genes for phage head or tail fiber construction—are usually clustered together. Early and late genes also are clustered separately on the genome; they are even transcribed in different directions—early genes in the counterclockwise direction and late genes, clockwise. A considerable portion of the T4 genome encodes products needed for its replication, including all the protein subunits of its replisome and enzymes needed to prepare for synthesis of DNA (**figure 25.7**). Some of these enzymes synthesize an important component of T4 DNA, hydroxymethylcytosine (HMC) (**figure 25.8**). HMC is a modified nucleotide that replaces cytosine in T4 DNA. Once HMC is synthesized, replication ensues by a mechanism similar to that seen in *Bacteria.* After T4 DNA has

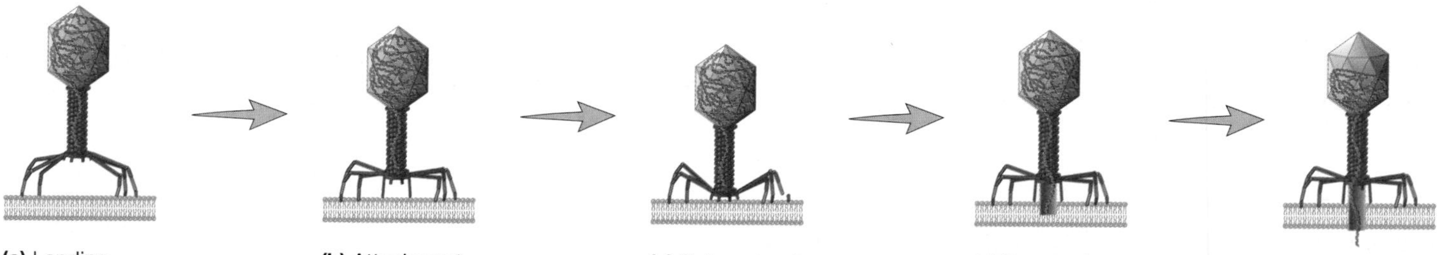

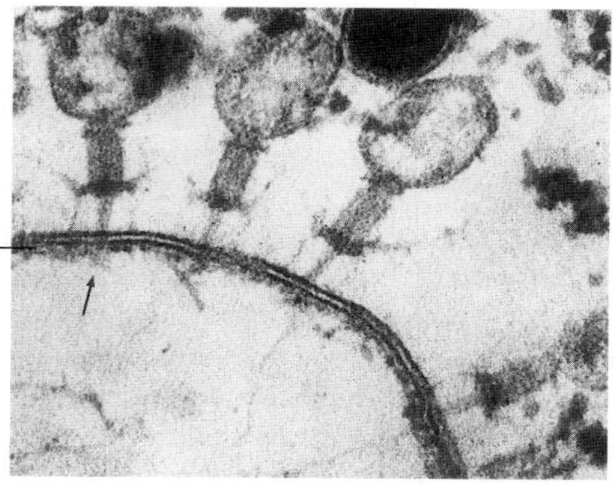

FIGURE 25.5 T4 Phage Adsorption and DNA Injection. (a–e) Adsorption and DNA injection is mediated by the phage's tail fibers and base plate, as shown here. (f) An electron micrograph of an *E. coli* cell being infected by "T-even" phages. These phages have injected their nucleic acid through the cell wall and now have empty capsids.

Figure 25.5 Micro Inquiry

What enzyme found in the T4 baseplate facilitates penetration through the cell wall?

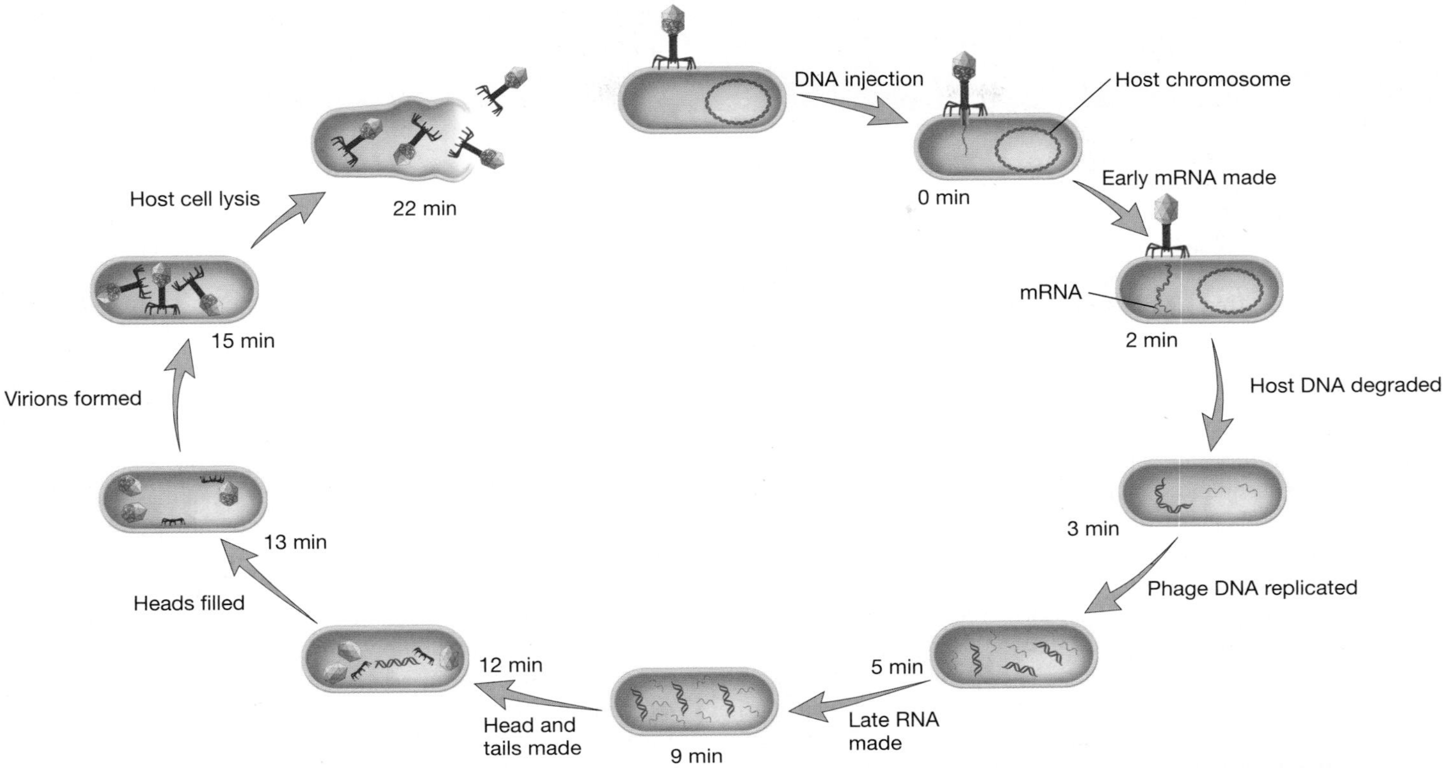

(a)

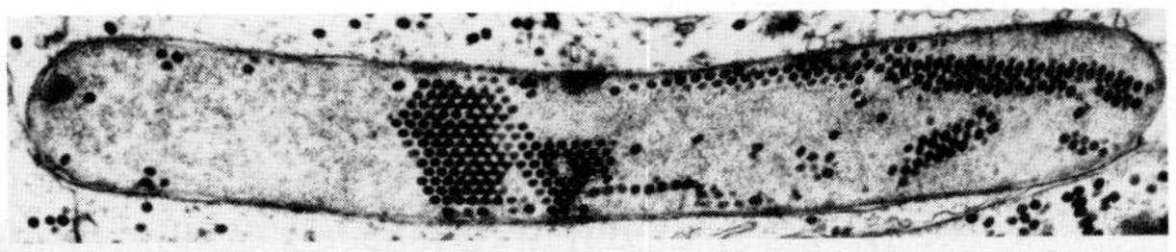

(b)

FIGURE 25.6 The Life Cycle of Bacteriophage T4. (a) A diagram depicting the life cycle with the minutes after DNA injection given for each stage. (b) Electron micrograph showing a bacterium that contains numerous completed phages, about 30 minutes after injection.

Figure 25.6 Micro Inquiry

Why do you think T4 evolved to initiate DNA replication from multiple origins, rather than from a single origin of replication as seen in its host cell?

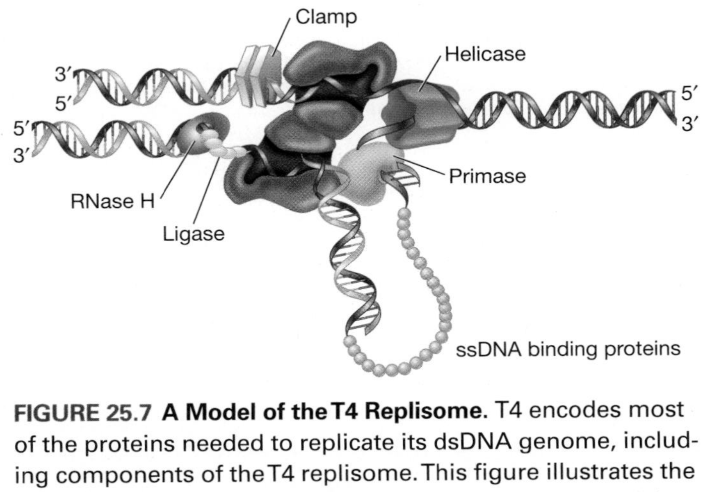

FIGURE 25.7 A Model of the T4 Replisome. T4 encodes most of the proteins needed to replicate its dsDNA genome, including components of the T4 replisome. This figure illustrates the viral replisome at the replication fork. Compare this model with the bacterial replisome shown in figure 12.14.

FIGURE 25.8 5-Hydroxymethylcytosine (HMC). In T4 DNA, the HMC often has glucose attached to its hydroxyl.

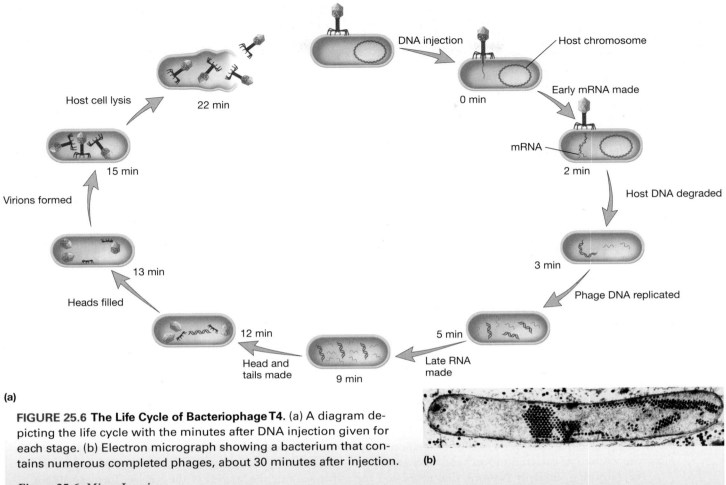

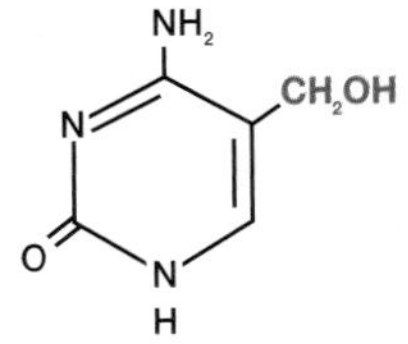

Figure 25.8 Micro Inquiry

What function does HMC glucosylation serve?

been synthesized, it is glucosylated by the addition of glucose to the HMC residues. Glucosylated HMC residues protect T4 DNA from attack by *E. coli* endonucleases called **restriction enzymes,** which would otherwise cleave the viral DNA at specific points and destroy it. This bacterial defense mechanism is called **restriction.** Restriction is just one mechanism used by *E. coli* to protect itself from infection by a bacteriophage (**Microbial Diversity & Ecology 25.1**). Other phages also chemically modify their DNA to protect against host restriction. ◀◀ *Restriction enzymes (section 15.1)*

MICROBIAL DIVERSITY & ECOLOGY

25.1 *Bacteria* (and *Archaea*) Fight Back

It has been recognized for several decades that bacteria have mechanisms for protecting themselves from foreign DNA such as that injected into a bacterium by a bacteriophage. This simple defense system was discovered in the 1960s and took the form of enzymes called restriction endonucleases, or simply restriction enzymes. These enzymes were named this because they restrict infection by a bacteriophage when they degrade the phage's genome.

As more and more bacterial and archaeal genomes have been sequenced, a new protective mechanism has been found. In approximately 40% of the sequenced bacterial genomes and about 90% of archaeal genomes a similar motif of sequences was found. This motif is called the CRISPR system, which stands for *c*lustered, *r*egularly *i*nterspaced *s*hort *p*alindromic *r*epeats. Each CRISPR system consists of a set of repeated sequences separated by spacers (**box figure**). Upstream of the repeats and spacers is a set of genes called CRISPR-asssociated sequences (CAS). By definition, the repeats are identical (or nearly so) within each CRISPR system. The spacers, however, differ considerably within a system. Surprisingly, when the spacer sequences were used to probe nucleic acid databases that included viral genomes, it was discovered that the spacers exhibited significant similarity to a variety of bacterial and archaeal virus genomes.

The surprising similarity between viral genes and the CRISPR spacers led scientists to question how the spacers came to exist. Numerous studies support the hypothesis that when a bacterial or archaeal cell is infected by a virus, a new set of sequences is added to the CRISPR region. This occurs at the end of the region closest to the CAS genes, and each new set consists of a new repeat and a spacer consisting of DNA from the infecting virus. Thus, the CRISPR sequences are akin to the growth rings on trees. Just as the history of the yearly growth of a tree can be ascertained by examining its growth rings, so too can the history of viral infections be determined. The oldest infection is located at the 3′ end of the CRISPR region and the most recent closest to the CAS genes.

But how do these sequences protect the bacterial cell? It has been demonstrated that the CRISPR region is transcribed to yield an RNA containing all repeats and spacers. The protein products of the CAS genes are thought to associate with the full-length RNA and process it into smaller RNA molecules called crRNAs. Each of these consists of one repeat and one spacer. CAS proteins are thought to remain associated with the crRNAs. During an infection by an appropriate virus (i.e., one that has previously infected the cell), the CAS-crRNAs associate with either the viruses DNA or mRNA, leading to destruction of the molecule. Because of this, the virus cannot multiply, and infection is thwarted.

Another interesting aspect of the CRISPR protective mechanism is that it in some ways resembles a defense mechanism observed in eukaryotic cells. This defense mechanism is called RNA silencing or RNA interference (RNAi). In the eukaryotic system, small double-stranded (ds) RNA molecules are generated from large dsRNA molecules formed during the replication of an RNA virus within its eukaryotic host. The resulting RNAs are called silencing RNAs (siRNAs). The siRNAs associate with a complex of proteins that separate the strands of the siRNA. Then the siRNA-protein complex binds to viral mRNAs, which are degraded. In the CRISPR system, the CAS proteins are thought to function in a manner analogous to the proteins that function in RNA silencing. Considerable work needs to be done to support this hypothesis and to work out the details of how the CRISPR system works.

Sources: Barrangou, R., et al. 2007. CRISPR provides acquired resistance against viruses in prokaryotes. Science *315:1709–12; Sorek, R., et al. 2008. CRISPR—a widespread system that provides acquired resistance against phages in bacteria and archaea.* Nature Rev. Microbiol. *6:181–6.*

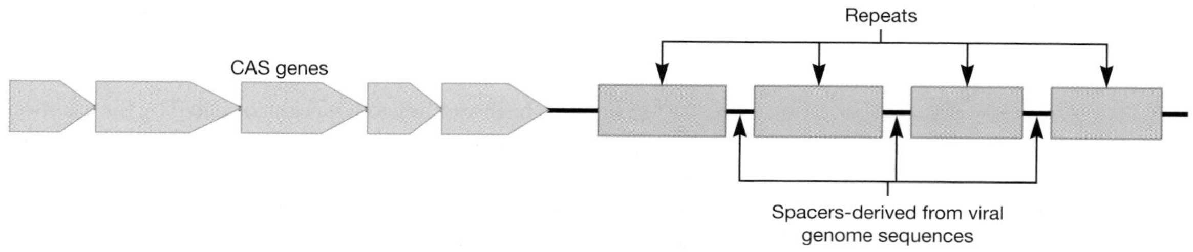

The linear dsDNA genome of T4 shows what is called terminal redundancy—that is, a base sequence is repeated at each end of the molecule. These two characteristics contribute to the formation of long DNA molecules called **concatemers,** which are composed of several genome units linked together in the same orientation (**figure 25.9**). Why does this occur? As we discuss in chapter 12, the ends of most linear DNA molecules cannot be replicated without mechanisms such as the enzyme telomerase. T4 and its *E. coli* host do not have telomerase activity. Therefore, each progeny DNA molecule has single-stranded 3′ ends. These ends participate in homologous recombination with double-stranded regions of other progeny DNA molecules, generating the concatemers. During assembly, concatemers are cleaved such that the genome packaged in the capsid is slightly longer than the T4 gene set. Thus each progeny virus has a genome unit that begins with a different gene. However, if each genome of the progeny viruses were circularized, the sequence of genes in each virion would be the same. Therefore the T4 genome is said to be circularly permuted, and the genetic map of T4 is drawn as a circular molecule.

The formation of new T4 phages is an exceptionally complex self-assembly process that involves viral proteins and some host cell factors (*see figure 5.11*). Late mRNA transcription begins about 9 minutes after injection of T4 DNA into *E. coli* (figure 25.6). Late mRNA products include phage structural proteins, proteins that help with phage assembly without becoming part of the virion, and proteins involved in cell lysis and phage release. These proteins are used in four fairly independent subassembly lines that ultimately converge to generate a mature T4 virion.

DNA packaging within T4 is accomplished by a complex of proteins sometimes called the "packasome." The packasome includes a set of proteins called the terminase complex, which generates double-stranded ends at the ends of the concatemers created during viral genome replication. These double-stranded ends are needed for packaging the T4 genome. The packasome then moves the viral genome into the head, and the concatemer is cut when the phage head is filled with DNA—a DNA molecule roughly 3% longer than the length of one set of T4 genes.

Finally, the virions are released so that they can infect new cells and begin the cycle anew. T4 lyses *E. coli* when about 150 virus

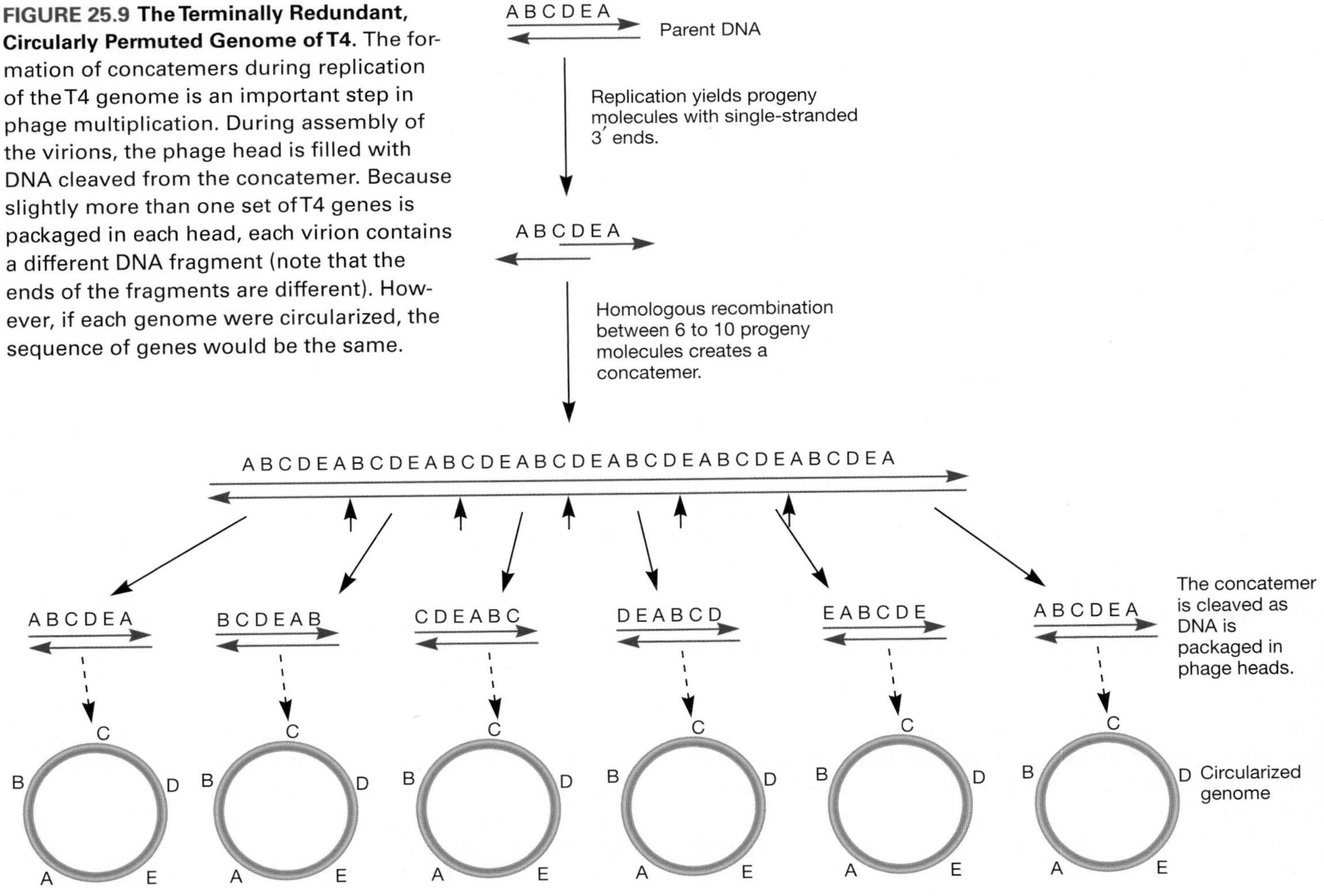

FIGURE 25.9 The Terminally Redundant, Circularly Permuted Genome of T4. The formation of concatemers during replication of the T4 genome is an important step in phage multiplication. During assembly of the virions, the phage head is filled with DNA cleaved from the concatemer. Because slightly more than one set of T4 genes is packaged in each head, each virion contains a different DNA fragment (note that the ends of the fragments are different). However, if each genome were circularized, the sequence of genes would be the same.

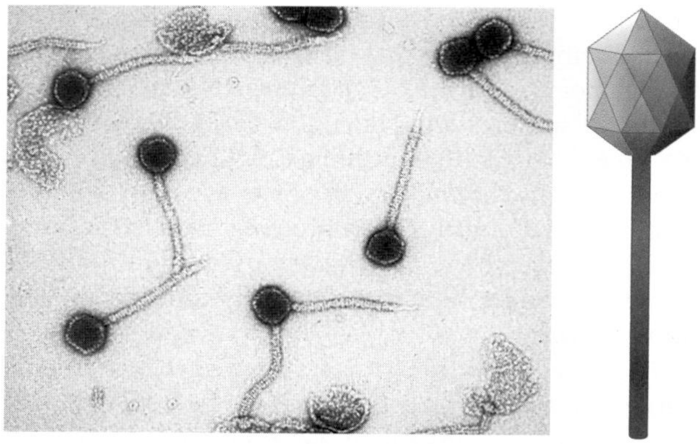

FIGURE 25.10 Bacteriophage Lambda (λ).

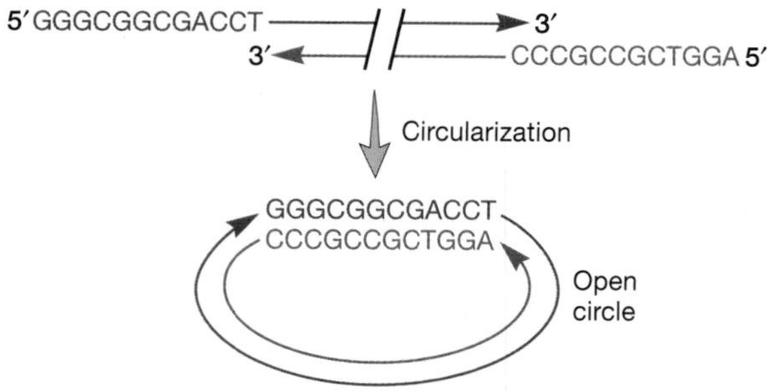

FIGURE 25.11 Lambda Phage DNA. A diagram of λ phage DNA showing its 12 base, single-stranded cohesive ends, and the circularization their complementary base sequences make possible.

particles have accumulated in the host cell. T4 encodes two proteins to accomplish this. The first, called holin, creates holes in the *E. coli* plasma membrane. The second enzyme, T4 lysozyme, degrades peptidoglycan in the host's cell wall. Thus the activity of holin enables T4 lysozyme to move from the cytoplasm to the peptidoglycan so that both the plasma membrane and the cell wall are destroyed.

Bacteriophage Lambda: A Temperate Bacteriophage

Unlike bacteriophage T4 and other "T-even" phages, temperate bacteriophages, such as phage lambda (family *Siphoviridae*), can enter either the lytic or lysogenic cycle upon infecting a host cell. If a temperate virus enters the lysogenic cycle, its dsDNA genome often is integrated into the host's chromosome, where it resides as a prophage until conditions for induction occur (*see figure 5.15*). Upon induction, the viral genome is excised from the host genome and directs the initiation of the lytic cycle. ◀◀ *Infections of bacterial and archaeal cells: Lysis and lysogeny (section 5.4); Transduction (section 14.9)*

How does a temperate phage "decide" which cycle to follow? The process by which phages make this decision is best illustrated by bacteriophage lambda (λ), which infects *E. coli*. As shown in **figure 25.10**, λ has an icosahedral head 55 nm in diameter and a noncontractile tail with a thin tail fiber at its end. Its DNA genome is a linear molecule with cohesive ends—single-stranded stretches, 12 nucleotides long, that are complementary to each other and can base pair.

Like most bacteriophages, λ attaches to its host and then injects its genome into the cytoplasm, leaving the capsid outside. Once inside the cell, the linear genome is circularized when the two cohesive ends base pair with each other; the breaks in the strands are sealed by the host cell's DNA ligase (**figure 25.11**). The λ genome has been carefully mapped, and over 40 genes have been located (**figure 25.12**). Most genes are clustered according to their function, with separate groups involved in head synthesis, tail synthesis, lysogeny, DNA replication, and cell lysis. This organization is important because once the genome is circularized, a cascade of regulatory events occurs that determine if the phage pursues a lytic cycle or establishes lysogeny. Regulation of appropriate genes is facilitated by clustering and coordinated transcription from the same promoters.

The cascade of events leading to either lysogeny or the lytic cycle serves as a model for complex regulatory processes. It involves the action of several regulatory proteins that function as repressors or activators or both, proteins that regulate transcription termination, and antisense RNA molecules (**figure 25.13**). The protein cII is an activator protein that plays a pivotal role in determining if λ will establish lysogeny or follow a lytic pathway. If the cII protein reaches high enough levels early in the infection process, lysogeny will occur; if it does not reach a critical level, the lytic cycle will occur.

Transcription of the λ genome is catalyzed by the host cell's **DNA-dependent RNA polymerase,** and the cII protein is synthesized relatively early in the infection. If the cII protein levels are high enough, it will increase transcription of the *int* gene, which encodes the enzyme integrase. **Integrase** catalyzes the integration of the λ genome into the host cell's chromosome, thus establishing lysogeny. The cII protein also increases transcription of the *cI* gene. This gene encodes a regulatory protein that is often called the λ repressor because it represses the transcription of all genes (except its own). This repression maintains the lysogenic state. As just noted, the cI protein (λ repressor) allows transcription of its own gene. This is because cI functions as an activator protein when it binds to the P_{RM} promoter from which the *cI* gene can be transcribed (**table 25.2**).

Integration of the λ genome into the host chromosome takes place at a site in the host chromosome called the attachment site (*att*). A homologous site is present on the phage genome, so the phage and bacterial *att* sites can base pair with each other. The bacterial site is located between the galactose (*gal*) and biotin (*bio*) operons, and as a result of integration, the circular λ genome becomes a linear stretch of DNA located between these two host operons (*see figure 14.32*). The prophage can remain integrated indefinitely, being replicated as the bacterial genome is replicated.

Because the cII protein is made early in the infectious process, it might seem that it would accumulate quickly and ensure

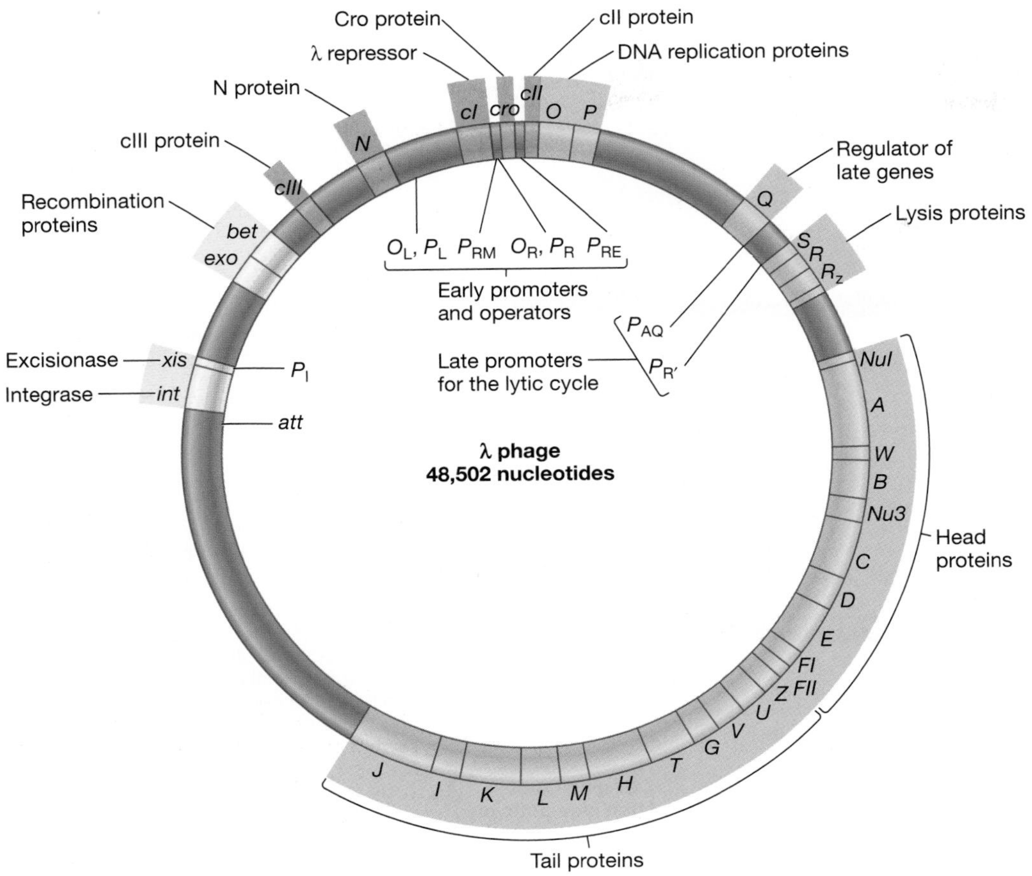

FIGURE 25.12 The Genome of Phage λ. The genes are color-coded according to function. Those genes shaded in orange encode regulatory proteins that determine if the lytic or lysogenic cycle will be followed. Genes involved in establishing lysogeny are shaded in yellow. Those required for the lytic cycle are shaded in tan. Important promoters and operators are also noted.

that lysogeny occurs. However, cII is degraded by a host enzyme (HflB) unless it is protected by a viral protein called cIII. The cIII protein is synthesized at the same time as cII, and as long as its levels remain high enough, cII will be protected. However, if cII is not protected sufficiently from the host degradative enzyme, the level of a protein called Cro (product of the *cro* gene) will increase. Cro protein is both a repressor protein and an activator protein. It inhibits transcription of the *cIII* gene and the *cI* gene, further decreasing the amount of cII and λ repressor. However, it increases its own synthesis as well as the synthesis of another regulatory protein called Q. When Q protein accumulates at a high enough level, it activates transcription of genes required for the lytic cycle. When that occurs, the infection process is committed to the lytic cycle. Ultimately the host is lysed and the new virions are released.

We have now considered the regulatory processes that dictate whether lysogeny is established or the lytic cycle is pursued. However, how does induction reverse lysogeny? Induction usually occurs in response to environmental factors such as UV light or chemical mutagens that damage DNA. This damage alters the activity of the host cell's **RecA protein.** As we describe in chapter 14, RecA plays important roles in recombination and DNA repair processes. When activated by DNA damage, RecA interacts with λ repressor, causing the repressor to cleave itself. As more and more repressor proteins destroy themselves, transcription of the *cI* gene is decreased, further lowering the amount of λ repressor in the cell. Eventually the level becomes so low that transcription of the *xis, int,* and *cro* genes begins. The *xis* gene encodes the protein excisionase. It binds integrase, causing it to reverse the integration process, and the prophage is freed from the host chromosome. As λ repressor levels decline, the Cro protein levels increase. Eventually synthesis of λ repressor is completely blocked, protein Q levels become high, and the lytic cycle proceeds to completion.

Our attention has been on λ phage, but there are many other temperate phages. Most, like λ, exist as integrated prophages in the lysogen. However, not all temperate phages integrate into the host chromosome at specific sites. Bacteriophage Mu uses a transposition mechanism to integrate randomly into the genome. It then expresses a repressor protein that inhibits lytic growth. Furthermore, integration is not an absolute requirement for lysogeny. The *E. coli* phage P1 is similar to λ in that it circularizes after infection and begins to manufacture repressor. However, it remains as an independent circular DNA molecule in the lysogen and is replicated at the same time as the host chromosome. When *E. coli* divides, P1 DNA is apportioned between the daughter cells so that all lysogens contain one or two copies of the phage genome.

1. Explain why the T4 phage genome is circularly permuted.
2. Precisely how, in molecular terms, is a bacterial cell made lysogenic by a temperate phage such as phage λ?
3. How is a prophage induced to become active again?
4. Describe the roles of the cII protein, CIII protein, λ repressor, Cro protein, Q protein, RecA protein, integrase, and excisionase in lysogeny and induction.
5. How do the temperate phages Mu and P1 differ from λ phage?

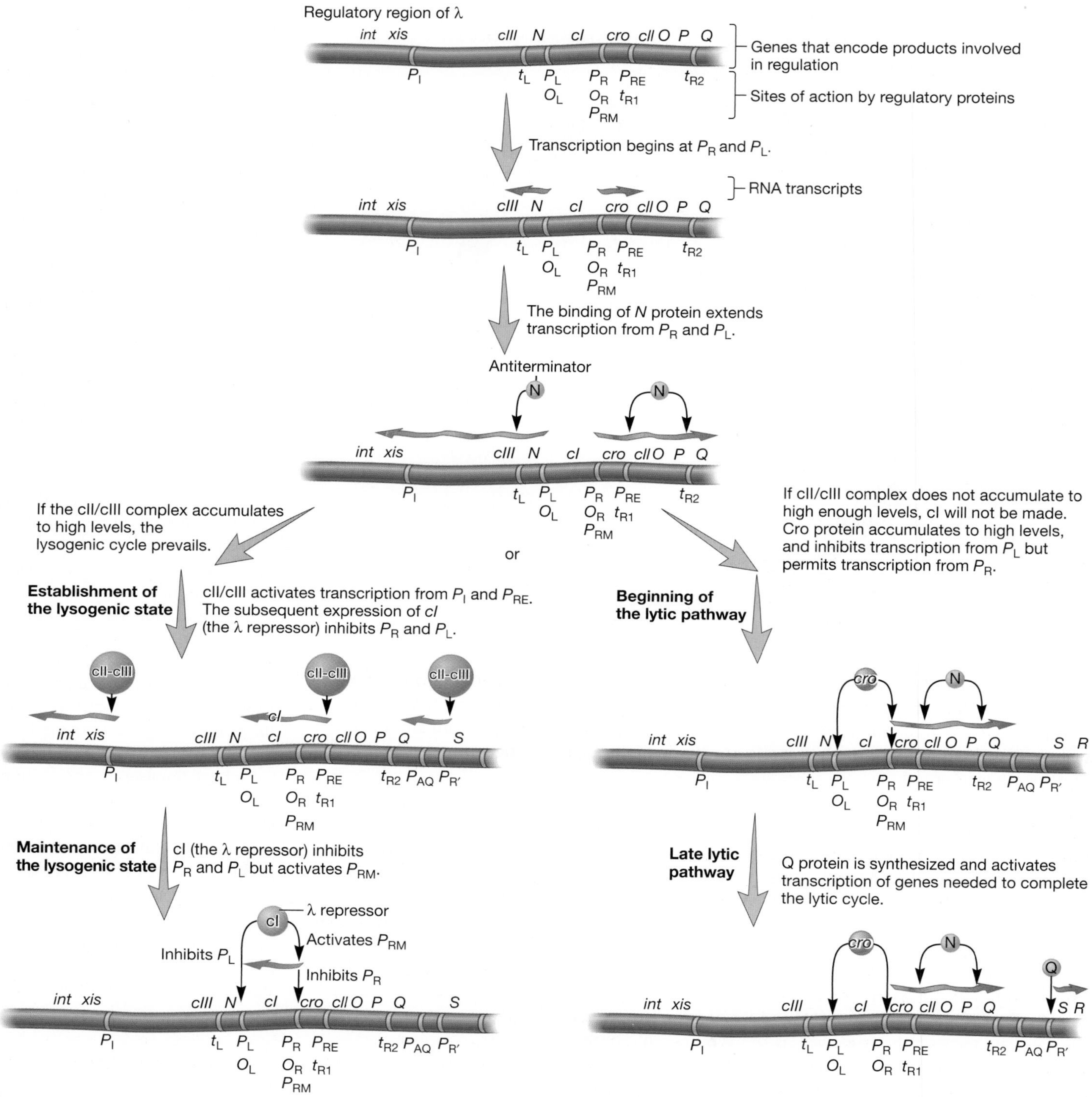

FIGURE 25.13 The Decision-Making Process for Establishing Lysogeny or the Lytic Pathway. The initial transcripts are synthesized by the host RNA polymerase. These encode the N protein and the Cro protein. The N protein is an antiterminator that allows transcription to proceed past the terminator sequences t_L, t_{R1}, and t_{R2}. This allows transcription of other regulatory genes as well as the *xis* and *int* genes. The latter genes encode the enzymes excisionase and integrase, respectively. The left side of the figure illustrates what occurs if lysogeny is established. The right side of the figure shows the lytic pathway.

Table 25.2 Functions of Lambda Promoters and Operators

Promoter or Operator	Name Derivation	Function
P_L	**P**romoter **L**eftward	Promoter for transcription of *N, cIII, xis,* and *int* genes; important in establishing lysogeny
O_L	**O**perator **L**eftward	Binding site for lambda repressor and Cro protein; binding by lambda repressor maintains lysogenic state; binding by Cro protein prevents establishment of lysogeny
P_R	**P**romoter **R**ightward	Promoter for transcription of *cro, cII, O, P,* and *Q* genes; Cro, O, P, and Q proteins are needed for lytic cycle; CII protein helps establish lysogeny
O_R	**O**perator **R**ightward	Binding site for lambda repressor and Cro protein; binding by lambda repressor maintains lysogenic state; binding by Cro allows transcription to occur
P_{RE}	**P**romoter for Lambda **R**epressor **E**stablishment	Promoter for *cI* gene (lambda repressor gene); recognized by CII protein, a transcriptional activator; important in establishing lysogeny
P_I	**P**romoter for **I**ntegrase Gene	Transcription from P_I generates mRNA for integrase protein but not excisionase; recognized by the transcriptional activator CII; important for establishing lysogeny
P_{AQ}	**P**romoter for **A**nti-***Q*** mRNA	Transcription from P_{AQ} generates an antisense RNA that binds *Q* mRNA, preventing its translation; recognized by the transcriptional activator CII; important for establishing lysogeny
P_{RM}	**P**romoter for **R**epressor **M**aintenance	Promoter for transcription of lambda repressor gene (cI); activated by lambda repressor; important in maintaining lysogeny
$P_{R'}$	**P**romoter **R**ightward′	Promoter for transcription of viral structural genes; activated by Q protein; important in lytic cycle

Archaeal Viruses

Our understanding of archaeal viruses lags significantly behind that of bacterial viruses and animal viruses. However, as more archaeal viruses are isolated and studied, virologists are beginning to develop a clearer picture of the nature of these interesting biological entities. All known archaeal viruses have dsDNA genomes and so are discussed here.

Although some archaeal viruses exhibit morphologies similar to that of the tailed bacteriophages, many others have unusual morphologies, such as bottle-shaped, droplet-shaped, and spindle-shaped. Some of these are shown in figure 25.2. These unusually shaped viruses have defined new virus families. Many archaeal viruses are enveloped, and their genomes can be either linear or circular. Relatively little is known about their life cycles. However, some are clearly virulent viruses that always cause lytic infections and others have been shown to establish lysogeny. In some cases, the provirus integrates into the host cell's chromosome, but in other cases, it remains free in the cytoplasm. Because of the evolutionary history of *Archaea* and *Eukarya,* it is assumed that the life cycles of these viruses and the regulatory mechanisms they use will have unique features. ◀◀ *Microbial Diversity & Ecology 5.1*

Herpesviruses

So far our discussion of dsDNA viruses has focused on bacteriophages and archaeal viruses. Bacteriophages have served as important model organisms for studying DNA replication, regulatory processes, and assembly of viruses. We now turn our attention to some important dsDNA viruses of eukaryotes. We begin our discussion with herpesviruses. As you will see, the life cycle of these viruses shares many features with bacteriophages such as T4. However, important distinctions in the life cycles exist because of the nature of their respective host cells.

Herpesviruses are members of the family *Herpesviridae,* which is subdivided into three subfamilies (alpha-, beta-, and gammaherpesviruses) and a group of unclassified herpesviruses. Alphaherpesviruses include herpes simplex viruses type 1 and type 2, which cause cold sores and genital herpes, respectively. The varicella-zoster virus, which causes chickenpox and shingles, is also an alphaherpesvirus. An important betaherpesvirus is cytomegalovirus, and an important gammaherpesvirus is the Epstein-Barr virus, which causes infectious mononucleosis and has been implicated in some human cancers.

Herpesvirus virions are 120 to 200 nm diameter, somewhat pleomorphic, and enveloped (**figure 25.14**). The envelope

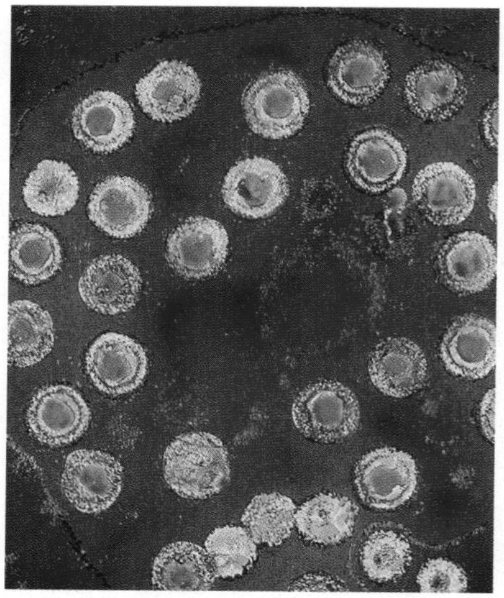

FIGURE 25.14 HSV-2. Herpes simplex virus type 2 inside an infected cell.

contains distinct viral projections (called spikes) that are regularly dispersed over the surface. The envelope surrounds a layer of proteins called the tegument, which in turn surrounds the nucleocapsid. Herpesvirus genomes are linear, about 160,000 base pairs long, and contain 50 to 100 genes. The nucleocapsid is icosahedral with DNA wrapped around a spool-like fibrillar core. When herpesviruses target cells of vertebrate hosts, some bind to epithelial cells, others to neurons. Host cell binding is mediated by specific interaction with the envelope spikes.

Herpesviruses cause both productive infections and latent infections. In the productive infection, the virus multiplies explosively; between 50,000 and 200,000 new virions are produced from each infected cell. As it multiplies, the host cell's metabolism is inhibited and the host's DNA is degraded. As a result, the cell dies. The first exposure to a herpesvirus usually causes this type of infection. Some of the cells infected in the initial infection are neurons. It is within neurons that a latent infection develops. During the latent infection, infectious virions cannot be detected. However, the virus can be reactivated in the neurons, leading to a productive infection. The viral genome remains in the neuron after reactivation; thus once infected, the host can experience repeated productive infections.

The productive infection begins with receptor-mediated attachment followed by fusion of the viral envelope with the host cell membrane (**figure 25.15**). The initial association is between proteoglycans of the epithelial cell surface and viral glycoproteins. This is followed by a specific interaction with one of several cellular receptors collectively termed HVEMs for *h*erpes*v*irus *e*n-try *m*ediators. The nucleocapsid is released into the cytoplasm and is transported by the host cell's microtubules to the nucleus. The linear dsDNA is then released from the nucleocapsid and enters the nucleus by way of a nuclear pore complex. Immediately upon release of the viral DNA into the host nucleus, the DNA circularizes and is transcribed by host DNA-dependent RNA polymerase to form mRNAs, which direct the synthesis of several immediate early and early proteins. These are mostly regulatory proteins and the enzymes required for replication of viral DNA (figure 25.15, steps 1 and 2). Replication of the genome with a virus-specific DNA-dependent DNA polymerase begins in the cell nucleus within 4 hours after infection (figure 25.15, step 3). Viral structural proteins are the products of the late genes. These proteins enter the nucleus, where they are assembled into nucleocapsids. Capsid formation and the genome packaging process are similar to that seen for the bacteriophage T4.

The acquisition of the tegument and envelope, and exit from the host cell are interesting processes that require several steps to complete. Once the nucleocapsid is assembled, it makes contact with the inner membrane of the nucleus. It then buds into the space between the two nuclear membranes. This generates an envelope that is called the primary viral envelope. The primary envelope is lost when it fuses with the outer nuclear membrane, releasing the herpesvirus nucleocapsid into the cytoplasm. At this point in the life cycle, some tegument proteins associate with the nucleocapsid. Additional tegument proteins are added when the developing virion is enveloped by membranes of the trans-Golgi network. This event yields a mature virion enclosed within a membrane vesicle. The vesicle transports the virus to the plasma membrane, fuses with it, and releases the mature virus from the host cell. Thus, unlike many other enveloped viruses, the source of the virus's envelope arises from the Golgi, rather than the plasma membrane.

Just as the establishment of lysogeny by temperate phages such as λ has been of interest as a model of regulatory processes, so too has the establishment of latency by herpesviruses. It appears that both viral and host proteins play a role in determining if a productive infection or latent infection will occur. Several studies have demonstrated that in epithelial cells, a viral protein called VP16 and a host protein called simply host cell factor (HCF) enter the nucleus with the viral genome. VP16 and HCF are needed for full expression of the immediate early genes. In neurons, VP16 and HCF do not enter the nucleus and the expression of many immediate early genes is decreased. In addition, small noncoding RNAs (microRNAs) produced by the virus further decrease expression of immediate early genes needed for a lytic infection.

Nucleo-Cytoplasmic Large DNA Viruses

The nucleo-cytoplasmic large DNA (NCLD) viruses are a group of eukaryotic viruses that are thought to have arisen from a common ancestor. NCLD viruses have similar life cycles in which most, if not all, of the events occur in the cytoplasm. They all have icosahedral capsids that enclose a lipid membrane. At the core of their virions is a large dsDNA genome. Thus both the virion itself and the genome are large, often rivaling the sizes of small bacteria and their genomes. NCLD viruses are found in a number of virus families, including *Poxviridae* (e.g., *Vaccinia virus* and *Variola virus*, the cause of smallpox), *Iridoviridae* (viruses that infect fish and amphibians), *Asfarviridae* (a family consisting of only one member, which infects swine), *Phycodnaviridae* (viruses that

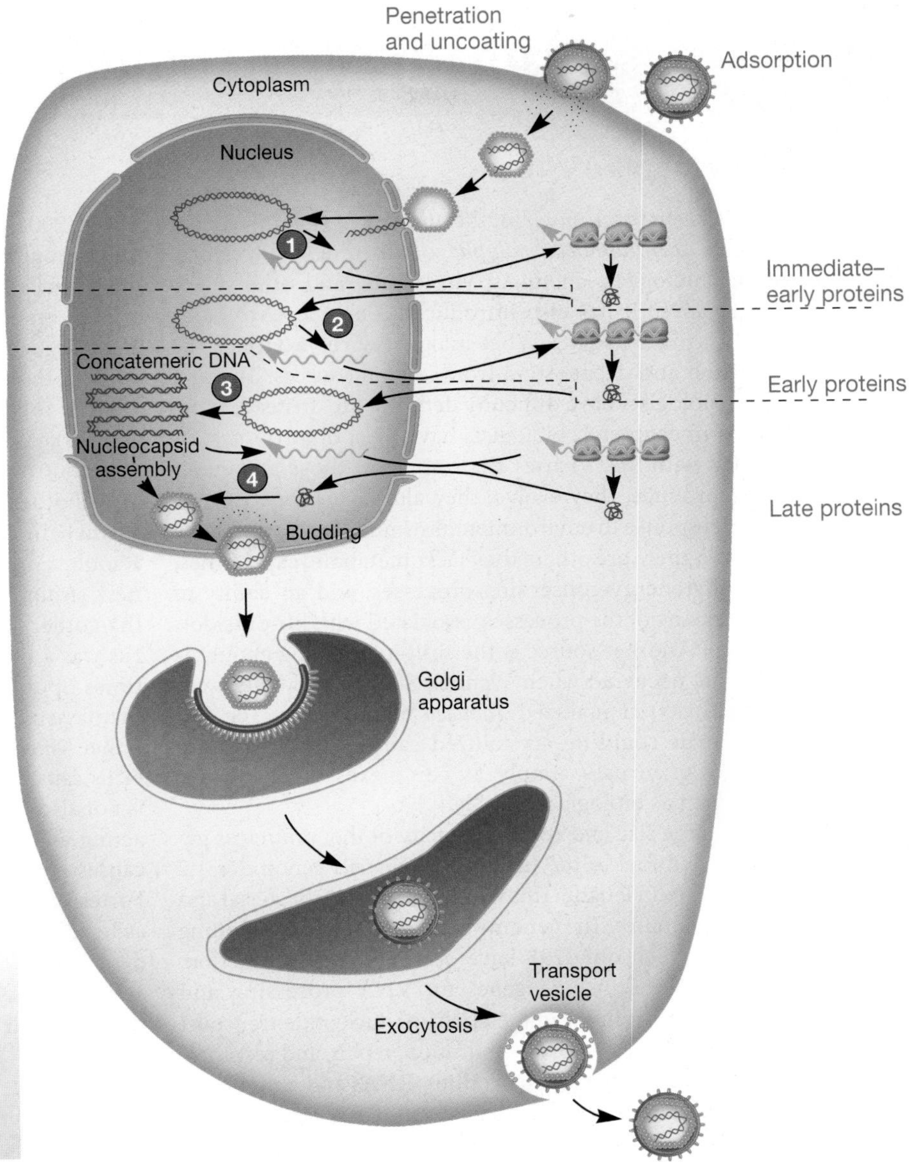

1 Circularization of genome and transcription of immediate-early genes
2 Immediate-early proteins (products of immediate-early genes) stimulate transcription of early genes.
3 Early proteins (products of early genes) function in DNA replication, yielding concatemeric DNA. Late genes are transcribed.
4 Late proteins (products of late genes) participate in virion assembly.
Penetration and uncoating
Adsorption
Cytoplasm
Nucleus
Immediate–early proteins
Concatemeric DNA
Early proteins
Nucleocapsid assembly
Late proteins
Budding
Golgi apparatus
Transport vesicle
Exocytosis

FIGURE 25.15 Generalized Life Cycle for Herpes Simplex Virus Type 1.

Figure 25.15 Micro Inquiry

How is the envelope of this virus formed? How does this differ from most enveloped viruses?

infect algae), and *Mimiviridae. Mimiviridae* consists only of the virus *Mimivirus*. It is the largest NCLD virus and infects the amoeba *Acanthamoeba polyphaga. Mimivirus* is 400 nm in diameter and has a genome that is 1.2 million bases long. This is as large as many bacterial genomes.

The large genomes of the NCLD viruses enable them to encode many proteins. Thus NCLD viruses are very self-sufficient and rely on their hosts for much less than do other viruses. They generally encode all the proteins needed for DNA replication, enzymes involved in recombination, RNA polymerases and associated transcription factors, and chaperones. In addition, they encode some components of the translational machinery (tRNAs and aminoacyl-tRNA synthetases). Indeed, their coding capacity and size have led many virologists to rethink the definition of viruses and cells (**Microbial Diversity & Ecology 25.2**).

Poxviruses are the best-studied NCLD viruses. Their virions are large, brick-shaped, and contain a dumbbell-shaped core. The *Variola virus* (300 by 250 to 200 nm) is slightly larger than cells of the bacterial genus *Chlamydia.* Its genome contains over 200 genes and is replicated in the host cell's cytoplasm. Poxviruses enter through receptor-mediated endocytosis; the central core escapes from the endosome and enters the cytoplasm. The core contains a viral DNA-dependent RNA polymerase. The polymerase synthesizes early mRNAs, one of which directs the production of an enzyme that completes virus uncoating. DNA polymerase and other enzymes needed for DNA replication are also synthesized early in the life cycle. About the time DNA replication starts, transcription of late genes is initiated. Many late proteins are structural proteins used in capsid construction. The complete life cycle of poxviruses takes about 24 hours.

MICROBIAL DIVERSITY & ECOLOGY

25.2 What Is a Virus?

The discovery of *Mimivirus*, the giant virus that infects the amoeba *Acanthamoeba polyphaga*, has renewed a debate among biologists, chemists, virologists, and others: Are viruses alive? It has also introduced a new level of complexity to this debate: What makes a virus a virus? The confusion and debate stem from several sources. The first is that biologists have difficulty defining life. Instead, as we discuss in chapter 1, biologists have a list of attributes associated with life. Viruses share some of those attributes: they reproduce, they evolve, they alter their gene expression in response to environmental stimuli (e.g., induction). However, there are others they lack: metabolic capabilities, especially energy-conserving processes, and an ability to carry out any of the processes associated with life outside a host cell. Another source is the shift in thinking about viruses that occurred when Wendell Stanley and colleagues (all chemists) crystallized *Tobacco mosaic virus*. The fact that a virus could be crystallized caused many to begin thinking of viruses simply as complexes of chemicals, rather than as biological entities.

The large size and coding capacity of the *Mimivirus* genome added fuel to the fire. The *Mimivirus* genome is 1.2 million bases in length. This is as large as many bacterial and archaeal genomes. The genome contains 911 protein-coding genes, of which only 298 have been assigned a function. Their functions include genes for RNA processing and modification; amino acid metabolism; nucleotide transport and metabolism; and DNA replication, repair, and recombination. In addition, genes encoding tRNAs and aminoacyl-tRNA synthetases have been identified. These genes encode products needed for translation, a surprising discovery since viruses usually rely totally on their hosts for translation.

In 2008 Didier Raoult and Patrick Forterre formalized their thoughts about what makes a virus a virus in a paper they published in *Nature Reviews Microbiology*. They proposed, first of all, that viruses are living organisms. They then argued that organisms should be divided into two groups: those that synthesize translational machinery (i.e., cellular organisms) and those that synthesize capsid (i.e., viruses). As you might imagine, this has caused considerable controversy.

Another recent discovery has clouded the water even further. In 2008 a new *Mimivirus* strain was isolated by Raoult, Forterre, and their co-workers. They refer to the new strain as mamavirus, because it seemed even larger than other *Mimivirus* strains. Associated with the mamavirus was a small icosahedral virus they call Sputnik. Sometimes Sputnik was observed in and near the area where mamavirus virions are synthesized (the viral factory). Other times Sputnik particles were observed inside mamavirus capsids. Sputnik cannot reproduce in a host cell that is not also infected with the mamavirus; thus mamavirus is acting as a helper virus. However, Sputnik's reproduction causes abnormalities in mamavirus virions. Thus Raoult, Forterre, and their colleagues consider Sputnik to be a parasite of mamavirus and have coined the term virophage to describe it. The arguments against this designation have already begun, and so the debate continues. What do you think? Is a virus alive?

Sources: LaScola, B., et al. 2003. A giant virus in amoebae. Science *299:2033; Raoult, D., and Forterre, P. 2008. Redefining viruses: Lessons from* Mimivirus. Nature Rev. Microbiol. *6:315–19; LaScola, B., et al. 2008. The virophage as a unique parasite of the giant* Mimivirus. Nature *455:100–4.*

1. Why do cold sores reoccur throughout the lifetime of an HSV-1–infected individual?
2. In what part of the host cell does a herpesvirus genome replicate? Where does the viral genome reside during a latent infection?
3. Many small DNA viruses rely on host enzymes for replication and transcription. Why are NCLD viruses able to use their own DNA-dependent RNA polymerases and DNA polymerases?

25.3 Viruses with Single-Stranded DNA Genomes (Group II)

Most DNA viruses are double-stranded, but several important viruses with single-stranded (ss) DNA genomes have been described. The life cycles of ssDNA viruses are similar to those of dsDNA viruses with one major exception. An additional step must occur in the synthesis stage because the ssDNA genome needs to be converted to a dsDNA molecule. A few ssDNA viruses are discussed next.

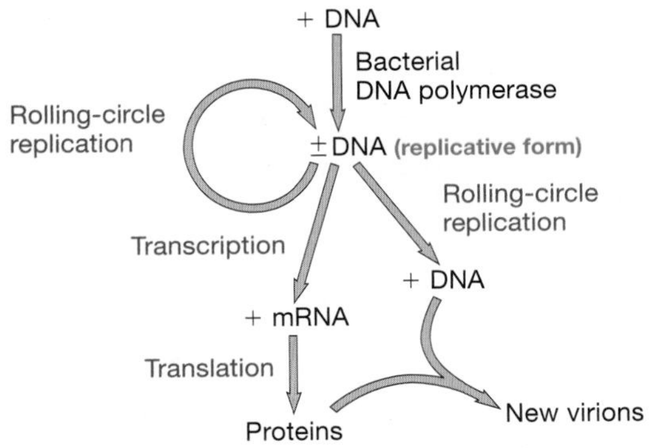

FIGURE 25.16 The Multiplication Strategy of ϕX174, a Plus-Strand DNA Phage.

Figure 25.16 Micro Inquiry

Why is the ϕX174 genome considered plus-stranded?

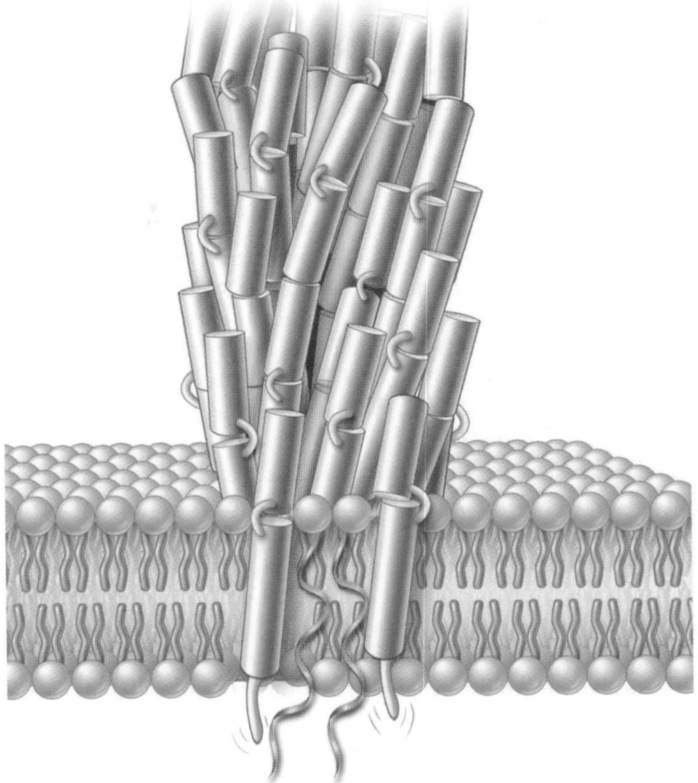

FIGURE 25.17 Release of Pf1 Phage. The Pf1 phage is a filamentous bacteriophage that is released from *Pseudomonas aeruginosa* without lysis. In this illustration, the blue cylinders are hydrophobic α-helices that span the plasma membrane, and the red cylinders are amphipathic helices that lie on the membrane surface before virus assembly. In each protomer, the two helices are connected by a short, flexible peptide loop (yellow). It is thought that the blue helix binds with circular, single-stranded viral DNA (green) as it is extruded through the membrane. The red helix simultaneously attaches to the growing virus coat that projects from the membrane surface. Eventually the blue helix leaves the membrane and also becomes part of the capsid.

Figure 25.17 Micro Inquiry

What is the fate of the host cell infected with Pf1 phage?

Bacteriophages ϕX174 and fd

The life cycle of ϕX174 (family *Microviridae*) begins with its attachment to the cell wall of its *E. coli* host. The circular ssDNA genome is injected into the cell, while the protein capsid remains outside the cell. The ϕX174 ssDNA genome has the same base sequence as viral mRNA and is therefore **plus-strand DNA.** For either transcription or genome replication to occur, the phage DNA must be converted to a double-stranded **replicative form (RF)** (**figure 25.16**). This is catalyzed by the host's DNA polymerase. The RF directs the synthesis of more RF copies and plus-strand DNA, both by rolling-circle replication (*see figure 14.23*). After assembly of virions, the host is lysed by a viral enzyme (enzyme E) that blocks peptidoglycan synthesis. Enzyme E inhibits the activity of the bacterial protein MraY, which catalyzes the transfer of peptidoglycan precursors to lipid carriers (*see figure 11.13*). Blocking cell wall synthesis weakens the host cell wall, causing the cell to lyse and release the progeny virions.

Although the fd phage (family *Inoviridae*) also has a circular, positive-strand DNA genome, it behaves quite differently from ϕX174 in many respects. It is shaped like a long fiber about 6 nm in diameter by 900 to 1,900 nm in length. Its ssDNA lies in the center of the filament and is surrounded by a tube made of a coat protein organized in a helical arrangement. The virus infects F^+, Hfr, and F′ *E. coli* cells by attaching to the tip of the sex pilus; the DNA enters the host along or possibly through the F factor–encoded sex pilus with the aid of an adsorption protein. As with ϕX174, an RF is first synthesized and then transcribed. A phage-coded protein then aids in replication of the phage DNA by rolling-circle replication. A distinguishing feature of fd and other filamentous phages (e.g., Pf1 phage of *Pseudomonas aeruginosa*) is that they do not kill their host cell. Instead, they establish a relationship in which new virions are continually released by a secretory process. Filamentous phage coat proteins are first inserted into the membrane. The coat then assembles around the viral DNA as it is secreted through the host plasma membrane (**figure 25.17**). Although the host cell is not lysed, it grows and divides at a slightly reduced rate.

Parvoviruses

Parvoviruses (family *Parvoviridae*) are viruses of eukaryotic cells. They infect numerous animal hosts, including crustaceans, dogs, cats, mice, and humans. Since its discovery in 1974, *Human parvovirus B19* (genus *Erythrovirus*) has emerged as a significant human pathogen. Parvovirus virions are uniform, icosahedral, nonenveloped particles approximately 26 nm in diameter. Their genomes are composed of one ssDNA molecule of about 5,000 bases. Most of the genomes are **negative-strand DNA** molecules. That is, their sequence of nucleotides is complementary to that of the viral mRNA (figure 25.3). Parvoviruses are among the simplest of the DNA viruses. The genome is so small that it directs the synthesis of only three proteins and some smaller polypeptides. None has enzymatic activity.

Each parvovirus uses a particular host cell molecule for attachment. For instance, the B19 virus uses a molecule only found on the surface of progenitors of red blood cells. The virus then enters by receptor-mediated endocytosis. The mechanism by which the nucleocapsid is released from the endosome is still being studied. Once it escapes the endosome, it is thought to be transported to the nucleus by the host cell's microtubules. It is not clear whether the virus or just its DNA enters the nucleus.

Since the parvovirus genome does not code for any enzymes, the virus must use host cell enzymes for all biosynthetic processes. Thus viral DNA can only be replicated in the nucleus during the S phase of the cell cycle, when the host cell replicates its own DNA. Because the genome is negative-strand DNA, it serves as the template for mRNA synthesis. Some of the RNA products encode polypeptides that are required for the interesting mechanism used during replication of the virus's genome. The ends of the parvovirus genome are palindromic sequences that can fold back on themselves. Formation of the hairpin at the 3′ end of the genome provides the primer needed for replication (**figure 25.18**). This is recognized by the host DNA polymerase, and DNA replication ensues by a process that is somewhat similar to rolling-circle replication. The parvovirus version of this replication method is often called rolling-hairpin replication because DNA polymerase seems to shuttle back and forth as it synthesizes genomes. The process involves a dsDNA intermediate, much like ϕX174. ◀◀ *DNA replication (section 12.4)*

1. Why is it necessary for some ssDNA viruses to manufacture a replicative form?
2. From the point of view of the virus, compare the advantages and disadvantages of host cell lysis, as seen in ϕX174, with the continuous release of phage without lysis, as with fd phage.
3. How do parvoviruses "trick" the host DNA polymerase into replicating its genome? Why must this occur in the nucleus only during the S phase of the host's cell cycle?

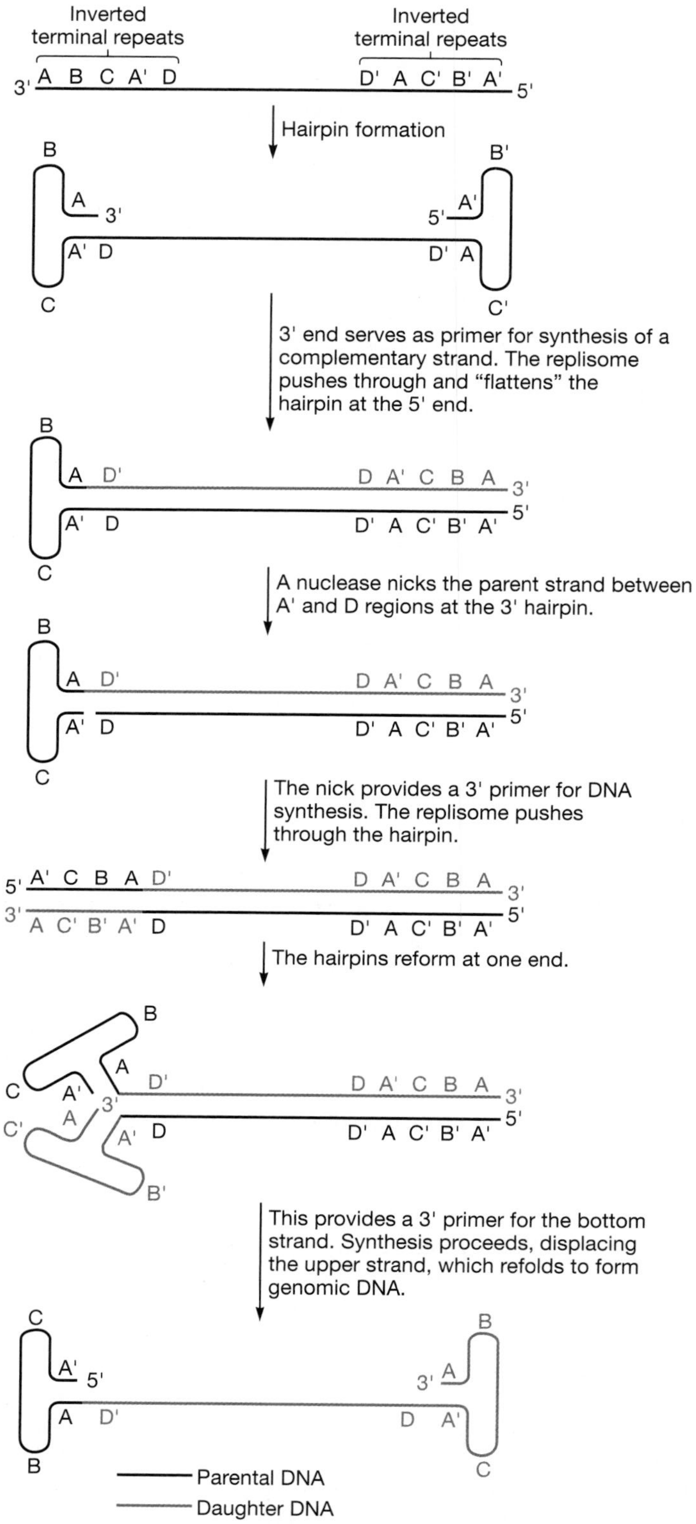

FIGURE 25.18 The Rolling-Hairpin Replication Used by Parvoviruses to Replicate Their DNA.

25.4 Viruses with Double-Stranded RNA Genomes (Group III)

Because their host cells have dsDNA genomes, viruses with RNA genomes cannot rely on host cell enzymes for genome replication or mRNA synthesis. Instead, RNA viruses in groups III, IV, and V use a viral enzyme called **RNA-dependent RNA polymerase** to complete their life cycles. When an RNA-dependent RNA polymerase is used to replicate the RNA genome, it is often referred to as a replicase. When it is used to synthesize mRNA, it is often said to have transcriptase activity. Most RNA viruses use the identical enzyme to carry out both functions.

Among viruses with RNA genomes, those with dsRNA appear to be least abundant. These viruses share a common multiplication strategy (**figure 25.19**). Here, we discuss two representative dsRNA viruses: a bacteriophage and a vertebrate virus.

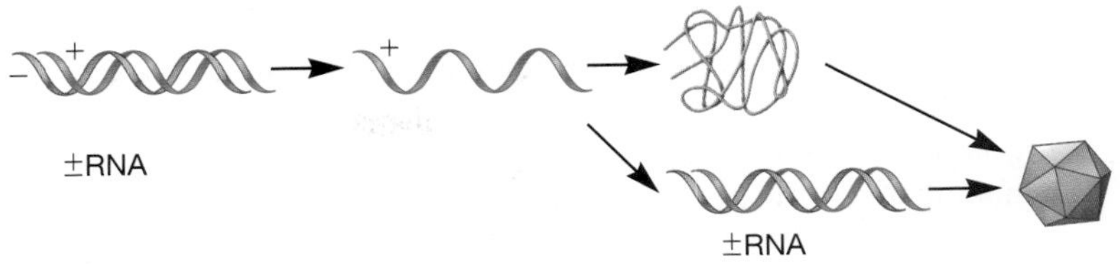

FIGURE 25.19 Multiplication Strategy of Double-Stranded RNA Viruses. Both strands can serve as templates for synthesis of mRNA by transcriptase. The negative strand of the viral genome also serves as the template for synthesis of positive strands by replicase. The positive strands are used by replicase to synthesize negative strands, thus replicating the dsRNA genome.

Bacteriophage ϕ6

Several dsRNA phages have been discovered and form the family *Cystoviridae*. The bacteriophage ϕ6 of *Pseudomonas syringae* pathovar Phaseolicola (previously called *Pseudomonas phaseolicola*), a plant pathogen, is the best studied. It is an unusual bacteriophage for several reasons. One is that it is an enveloped phage. Enclosed by the envelope is a nucleocapsid containing an RNA-dependent RNA polymerase and a segmented genome. Segmented genomes are composed of multiple linear RNA molecules, each encoding a distinct set of products. The segmented genome of ϕ6 consists of three dsRNA molecules. The life cycle of the virus also has unusual features. Like some other phages, ϕ6 attaches to the side of a pilus. However, ϕ6 uses an envelope protein to facilitate adsorption. Retraction of the pilus brings the phage into contact with the outer membrane of its gram-negative host. The viral envelope then fuses with the cell's outer membrane, a process mediated by another envelope protein. Fusion of the two membranes delivers the nucleocapsid into the periplasmic space. A protein associated with the nucleocapsid digests the peptidoglycan, allowing the nucleocapsid to cross this layer of the cell wall. Finally, the intact nucleocapsid enters the host cell by a process that resembles endocytosis. Because bacterial cells do not have the proteins and other factors needed for endocytosis, it is thought that viral proteins mediate this mechanism of entry.

Once inside the host, the viral RNA polymerase acts as a **transcriptase,** catalyzing synthesis of viral mRNA from each dsRNA segment. The enzyme also acts as a **replicase**, synthesizing plus-strand RNA from each segment. These are enclosed within newly formed capsid proteins, where they serve as templates for the synthesis of the complementary negative strand, regenerating the dsRNA genome. Once the nucleocapsid is completed, a nonstructural viral protein called P12 functions in surrounding the nucleocapsid with a plasma membrane-derived envelope, within the host cytoplasm. Finally, additional viral proteins are added to the envelope and the host cell is lysed, releasing the mature virions.

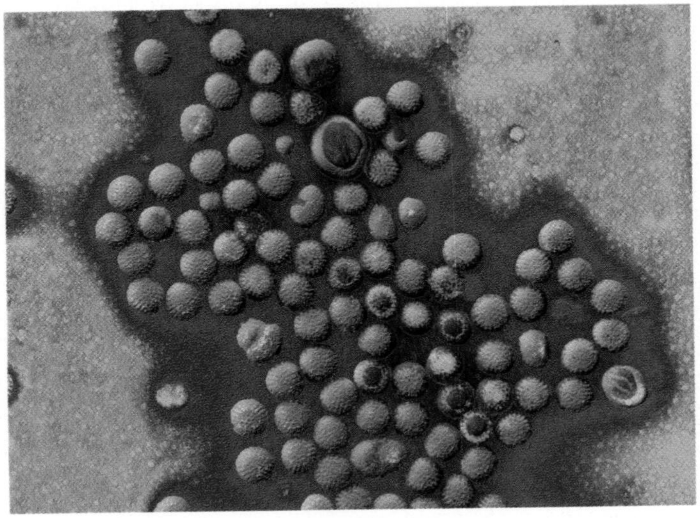

FIGURE 25.20 Rotaviruses. Electron micrograph of rotaviruses in a human gastroenteritis stool filtrate (×90,000). Note the wheel-like appearance of the icosahedral capsids that surround double-stranded RNA within each virion.

Rotaviruses

Human rotaviruses are responsible for the deaths of over 600,000 children worldwide each year. They cause severe diarrhea, which rapidly causes dehydration and death if appropriate therapy is not provided. Because of their impact on humans, rotaviruses have been studied since their discovery in an attempt to better understand their life cycles and pathogenesis.

Viewed by electron microscopy, rotaviruses (family *Reoviridae*) have a characteristic wheel-like appearance (rotavirus is derived from the Latin *rota,* meaning wheel). They are nonenveloped viruses with a genome composed of 11 segments of dsRNA surrounded by three concentric layers of proteins (**figure 25.20**). The RNA segments code for six structural and six nonstructural proteins. Rotavirus virions are very stable in the environment, often remaining infective for several days. This is an important characteristic because rotaviruses are often transmitted by ingestion of fecal material that is present in water or food, or on contaminated surfaces.

When a rotavirus enters a host cell, it loses the outermost protein layer and is then referred to by virologists as a double-layered particle (DLP). Once in the cytoplasm, the genome is transcribed by the viral transcriptase while still inside the DLP. The mRNA then passes through channels in the DLP and is released into the cytoplasm of the host cell. There, the mRNAs are translated by the host cell's protein synthesis machinery. The newly formed proteins cluster together, forming an inclusion in the host cell's cytoplasm called a viroplasm. It is within the viroplasm that new DLPs are formed. Initially the DLPs contain plus-strand RNA, but this is soon used as a template for synthesis of the negative strand. Thus the dsRNA molecule is synthesized within the developing DLP.

DLPs containing dsRNA eventually leave the viroplasm and are enveloped by membranes of the endoplasmic reticulum. While in the endoplasmic reticulum, the outermost layer is added to the DLP, converting it to the mature triple-layered virion. The mature virion is released by an unknown mechanism, but it is thought that it is transported to the surface of the cell in a membrane vesicle. The vesicle then fuses with the plasma membrane of the host cell, releasing the virion to the extracellular space.

1. Describe the life cycle of ϕ6 phage. What makes this phage unusual when compared with other bacteriophage?
2. How does the rotavirus structure give rise to its name?
3. In what ways are the life cycles of ϕ6 and rotaviruses similar? How do they differ?

25.5 Viruses with Plus-Strand RNA Genomes (Group IV)

Plus-strand RNA viruses have nonsegmented genomes that can act as mRNA and be translated upon entry into the host cell. All of these viruses replicate in the host cytoplasm and synthesize an RNA-dependent RNA polymerase. For some viruses, the polymerase is used to synthesize negative-strand RNAs, which are then used to make more plus-strand RNAs (**figure 25.21**). In some cases, this occurs by way of a double-stranded RF. In animal viruses, the synthesis of the viral genomes and assembly of the progeny virions occur in a structure called a replication complex. These are membrane-bound structures formed in response to factors produced by the virus, and they are derived from different cell organelles, depending on the virus. There are a number of important positive-strand animal viruses (e.g., the viruses that cause polio, SARS, and hepatitis A), and most plant viruses have plus-stranded RNA genomes. Although plus-strand RNA bacteriophages are a rarity, two have served as model systems for understanding the life cycles of plus-strand RNA viruses. They are described next.

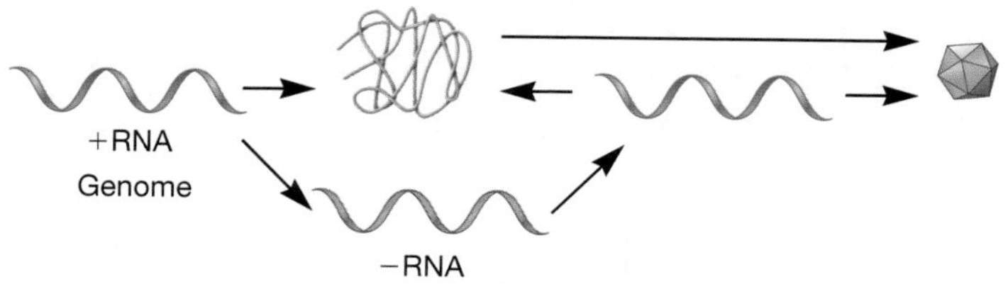

FIGURE 25.21 Multiplication Strategy of Plus-Strand RNA Viruses. The plus-strand RNA genome can serve directly as mRNA. One of the first viral proteins synthesized is RNA replicase, which replicates the genome, sometimes via a double-stranded replicative form. Transcriptase is responsible for synthesizing additional mRNA molecules.

Figure 25.21 Micro Inquiry

Where in the host does the plus-strand RNA genome replicate?

Bacteriophages MS2 and Qβ

Bacteriophages MS2 and Qβ (family *Leviviridae*) are small, tailless, icosahedral viruses. MS2 and Qβ have only three or four genes and are genetically the simplest phages known. In MS2, one protein is involved in phage adsorption to the host cell (and possibly also in virion construction or maturation). The other three genes code for a coat protein, RNA-dependent RNA polymerase, and a protein needed for host cell lysis.

MS2 and Qβ attach to the side of the F pilus of their *E. coli* host. Retraction of the pilus brings the virions close to the outer membrane of the cell, from which they gain entry. As with many bacteriophages, the capsids of these viruses remain outside the cell and only the RNA genome enters. Once the viral RNA polymerase is synthesized, it uses the plus-strand genome to synthesize an RF. The replicase then uses the RF to synthesize thousands of copies of plus-strand RNA. Some are used to make more RF in order to accelerate replication of the genome. Other plus-strand RNAs act as mRNA. Eventually plus-strand RNAs are incorporated into the capsid and mature virions are released by lysis.

Poliovirus

Poliovirus, the causative agent of poliomyelitis, has caused disease in humans for centuries. Polio was first recognized in 1789. It primarily targets children, with paralysis being the tragic result in many cases. The disturbing images of crippled children and individuals in iron lungs in the early part of the twentieth century prompted the development of a vaccine in the 1950s. Currently the vaccine is being used to eradicate the disease worldwide. This attempt has been largely successful, but pockets of disease still remain in certain developing countries.

The biology of *Poliovirus* (family *Picornaviridae*) has been intensely studied for decades. But many questions still linger. This nonenveloped virus makes its way into a human host by ingestion. It attaches to a cell surface molecule called human PV receptor. Current evidence suggests that the nucleocapsid enters the host cell, but the mechanism by which it does so has not been determined. Once inside the host cell, the plus-strand RNA genome is released into the cytoplasm. The genome acts as mRNA and is translated by host cell ribosomes. This process has garnered considerable attention because the virus's RNA does not have the 5′ cap found on eukaryotic mRNAs. This cap is important in ribosome binding. *Poliovirus* "tricks" its host into translating its capless RNA using a 5′ region on the RNA called the internal ribosome binding site (IRES). This region has numerous elements and forms extensive secondary structures, which are important for recognition of the RNA by the ribosome (**figure 25.22**).

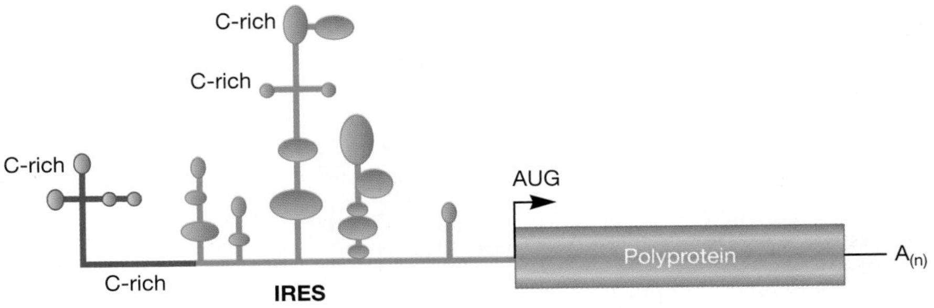

FIGURE 25.22 The *Poliovirus* Genome Showing the IRES Region Used by the Ribosome to Bind the Plus-Strand RNA and Synthesize the Polyprotein. Note the extensive secondary structure.

Translation of the *Poliovirus* genome yields a large protein called a **polyprotein** (**figure 25.23**). The polyprotein has protease activity and cleaves itself into three smaller proteins. Additional cuts in these proteins eventually yield all the structural proteins needed for capsid formation, as well as the virus's RNA-dependent RNA polymerase. The polymerase is used to generate negative-strand RNA molecules that serve as templates for synthesis of plus-strand RNAs. Some of these are translated, but eventually most will be used as new genomes and incorporated into capsids. All of the synthetic activity occurs in a membrane-bound structure. Mature particles are released from the cell by lysis.

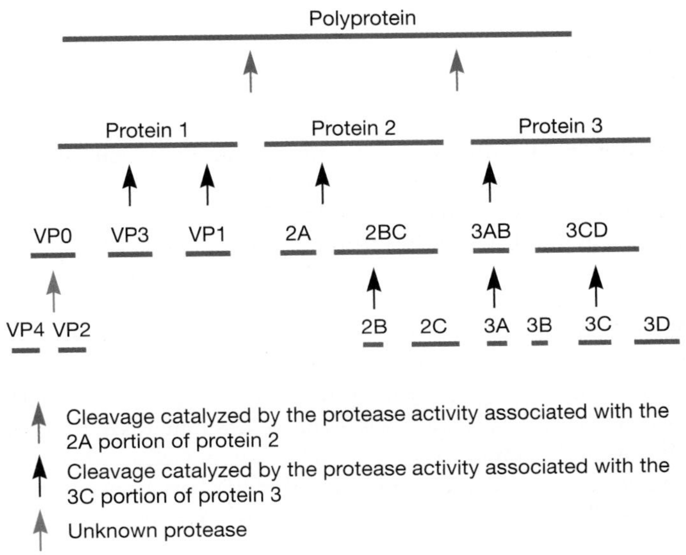

FIGURE 25.23 Cleavage of the *Poliovirus* Polyprotein. Proteins VP1-4 are structural proteins. Protein 3D is the RNA-dependent RNA polymerase.

***Figure 25.23* Micro Inquiry**

What are two other strategies used by plus-strand RNA viruses to generate multiple proteins from a single mRNA?

Poliovirus illustrates one method used by plus-strand RNA viruses of vertebrates to generate multiple proteins from a single RNA: Make a polyprotein and cut it into smaller proteins. Other viruses use other mechanisms. For instance, after synthesizing RNA-dependent RNA polymerase from the genomic plus-strand RNA, some viruses use the polymerase later in the life cycle to synthesize mRNAs that are smaller than the genomic RNA. These **subgenomic mRNA** molecules are translated to yield their encoded proteins. Other viruses use a process called **ribosomal frame-shifting.** This allows overlapping coding regions to be translated. When one ribosome reads the RNA, it stops at a stop codon in one reading frame, thus generating one protein. Another ribosome, however, may reach the stop codon and shift the reading frame at that point in the message. This generates a longer protein that is then processed to yield the protein encoded in the second reading frame. Finally, some viruses have **readthrough** mechanisms that allow the plus-strand RNA to be translated into two different proteins: a shorter protein, if the stop codon at the end of the open reading frame for the shorter protein is heeded by the ribosome; or a longer protein, if the stop codon is ignored. In all of these alternate routes for generating multiple proteins from a single plus-strand RNA, polyproteins are produced and cleaved to yield smaller functional proteins.

Tobacco Mosaic Virus

Most plant viruses are RNA viruses and, of these, plus-strand RNA viruses are most common. *Tobacco mosaic virus* (TMV) is the best-studied plus-strand RNA plant virus. TMV virions are filamentous with coat proteins arranged in a helical pattern. Plant viruses usually enter the host through an abrasion or wound on the plant; biting insects are often involved in transmission of the virus. Following entry into its host, the TMV RNA genome is translated. Two proteins are produced. One protein is the 126 kilodalton (kD) protein; the other is the 183 kD protein. The larger 183 kD protein is produced when a stop codon at the end of the coding region for the

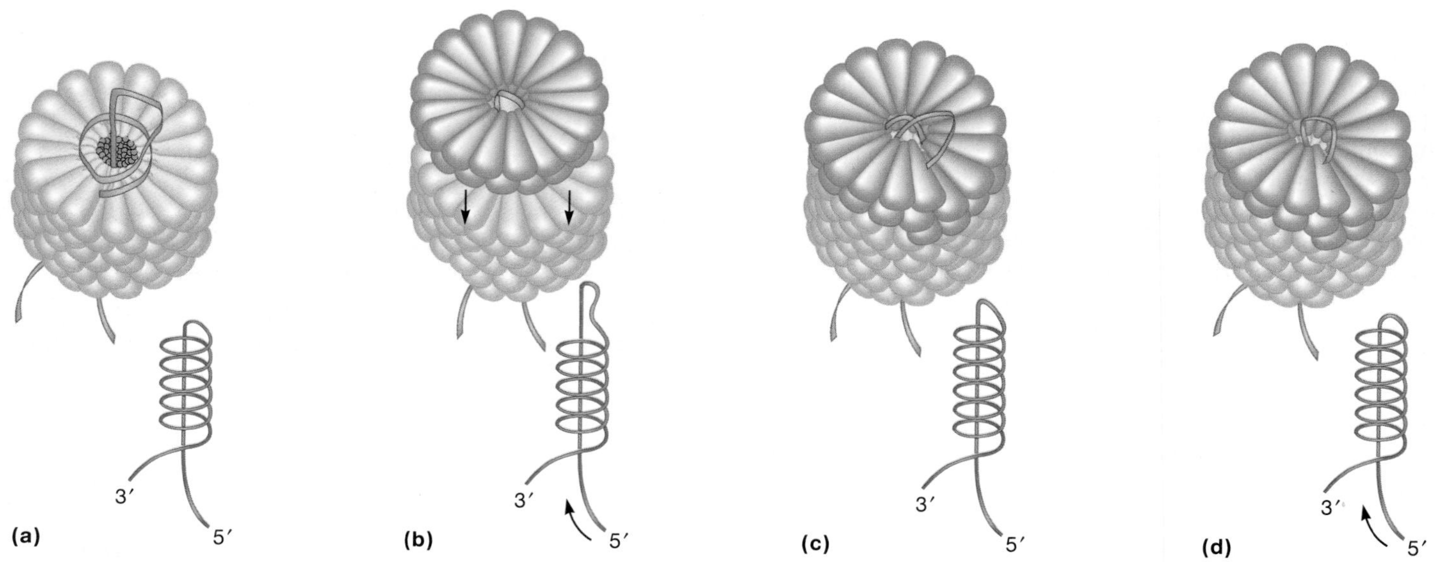

FIGURE 25.24 TMV Assembly. The elongation phase of *Tobacco mosaic virus* nucleocapsid construction. The lengthening of the helical capsid through the addition of a protein disk to its end is shown in a sequence of four illustrations; line drawings depicting RNA behavior are included. The RNA genome inserts itself through the hole of an approaching disk and then binds to the groove in the disk as it locks into place at the end of the cylinder.

126 kD protein is ignored (readthrough) by the ribosome. The 183 kD protein is thought to provide RNA-dependent RNA polymerase activity. The enzyme functions as both a transcriptase and a replicase. As the replicase, it synthesizes negative-strand RNA using the plus-strand genome as the template. It is not clear whether a double-stranded RF is created in vivo, but RFs have been observed in vitro. The negative-strand RNA serves as template for synthesis of new genomes. It also serves as the template for synthesis of subgenomic mRNAs that are translated into a variety of proteins needed by the virus.

After the coat protein and RNA genome of TMV have been synthesized, they spontaneously assemble into complete TMV virions in a highly organized process (**figure 25.24**). The protomers come together to form disks composed of two layers of protomers arranged in a helical spiral. Association of coat protein with TMV RNA begins at a specific assembly initiation site close to the 3′ end of the genome. The helical capsid grows by the addition of protomers, probably as disks, to the end of the rod. As the rod lengthens, the RNA passes through a channel in its center and forms a loop at the growing end. In this way, the RNA easily fits as a spiral into the interior of the helical capsid.

Multiplication of plant viruses within their host depends on the virus's ability to spread throughout the plant. Viruses can move long distances through the plant vasculature; usually they travel in the phloem. The spread of plant viruses in nonvascular tissue is hindered by the presence of tough cell walls. TMV thus spreads slowly, about 1 mm per day or less, moving from cell to cell through the plasmodesmata. These slender cytoplasmic strands extend through holes in adjacent cell walls and join plant cells by narrow bridges. Viral "movement proteins" are required for transfer from cell to cell. TMV movement proteins accumulate in the plasmodesmata, where they cause the plasmodesmata to increase in diameter. They are thought to coat the viral RNA, causing the RNA to assume a linear conformation, rather than the helical conformation found in TMV virions. This linear conformation is the proper size to fit through the plasmodesmata. Evidence exists that movement proteins associate with microtubules. However, the role of the interaction in movement of viral RNA through the plasmodesmata is still being studied.

Several cytological changes can take place in TMV-infected cells. TMV infections of plants produce microscopically visible intracellular inclusions that are similar to the replication complexes observed in animal cells infected with plus-strand viruses. Hexagonal crystals of almost pure TMV virions sometimes develop in TMV-infected cells. In addition, host cell chloroplasts become abnormal and often degenerate, while new chloroplast synthesis is inhibited.

1. What is a polyprotein? How is it used by many plus-strand RNA viruses to complete their life cycles?
2. What is an IRES? Why is it important?
3. How does TMV spread from one host cell to another?

25.6 Viruses with Minus-Strand RNA Genomes (Group V)

Negative-strand RNA viruses seem to be a more recent evolutionary development, sharing a basic genome arrangement and significant sequence similarities. In general, the virions are spherical to pleomorphic, measure 80 to 120 nm in diameter (influenza viruses can also assume a filamentous shape, 200 to 300 nm long and 20 nm in diameter), and are enveloped. The grouping of negative-strand RNA viruses includes those that have segmented and unsegmented genomes. Recall that segmented genomes are composed of multiple linear pieces of RNA. All minus-strand RNA genomes have self-complementary 3′ and 5′ ends, and the transcripts have a 5′ cap and a 3′ poly (A) tail. Segmented genomes may have evolved from unsegmented ones; they may have been created by the reduction of redundant genetic regions. The unsegmented, negative-strand RNA genome is arranged with genes in a highly conserved order. The genes are tandemly linked and typically separated by nontranscribed intergenic sequences. The negative-strand viruses include the families *Rhabdoviridae* (e.g., *Rabies virus*), the *Filoviridae* (e.g., Marburg and Ebola viruses), and the *Paramyxoviridae* (e.g., *Measles virus*). Of the segmented negative-strand RNA genomes, the *Bunyaviridae* (e.g., hantaviruses) have three segments transcribing six proteins, and the *Orthomyxoviridae* (e.g., influenza viruses) have seven or eight segments transcribing 10 proteins. The *Bunyaviridae* replicate in the host cell cytoplasm, whereas the *Orthomyxoviridae* are the only RNA viruses that replicate in the host nucleus.

The genomes of negative-strand RNA viruses cannot function as mRNA. Therefore these viruses must bring at least one RNA-dependent RNA polymerase into the host cell during entry. Initially the viral genome serves as the template for mRNA synthesis (**figure 25.25**). Later the virus switches from mRNA synthesis to genome replication, as the RNA-dependent RNA polymerase synthesizes a distinct plus-strand RNA for replication. During this phase of the life cycle, the plus-strand RNA molecules synthesized from the minus-strand genome serve as templates for the manufacture of new negative-strand RNA genomes.

Influenza (Italian, to be influenced by the stars—*un influenza di freddo*), or the flu, is a respiratory system disease caused by viruses that belong to the family *Orthomyxoviridae*. There are three types of influenza virus: influenza viruses A, B, and C. They contain seven to eight segments of linear RNA, with a genome length from 12,000 to 15,000 nucleotides (*see figure 5.4*). The virus is acquired by inhalation or ingestion of virus-infected respiratory secretions. The virus adheres to the epithelium of the respiratory system with the aid of two viral proteins. The **neuraminidase (NA)** present in envelope spikes may hydrolyze the mucus that covers the epithelium. This allows the **hemagglutinin (HA)** spike protein to interact with receptors on the surface of epithelial cells, and the virion enters by receptor-mediated endocytosis. This encloses the virus in an endosome (**figure 25.26**). The hemagglutinin molecule in the virus's envelope undergoes a dramatic conformational change when the endosomal pH decreases. The hydrophobic ends of the hemagglutinin spring outward and extend toward the endosomal membrane. After they contact the membrane, fusion occurs and the nucleocapsids are released into the cytoplasm.

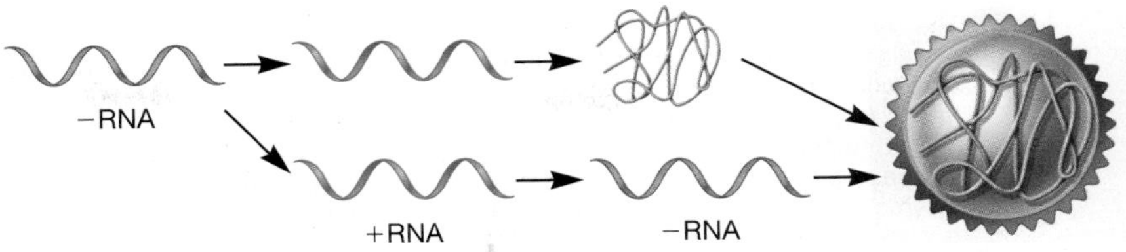

FIGURE 25.25 Multiplication Strategy of Negative-Strand RNA Viruses. An RNA-dependent RNA polymerase enters the host cell at the same time the negative-strand RNA genome enters. The genome serves as a template for synthesis of mRNA. Later in the infection, the negative-strand genome is used for plus-strand synthesis. These plus strands then act as templates for replication of the negative-strand genomes.

Once the genome and its associated RNA-dependent RNA polymerase enter the host cell, the genome serves as the template for mRNA synthesis (figure 25.26, step 1). Later, the virus switches from mRNA synthesis to genome replication. During this phase of the life cycle, the plus-strand RNA molecules synthesized from the minus-strand genome segments serve as templates for new negative-strand RNA genomes (figure 25.26, step 4). The viruses exit the host cell by budding and thus acquire their envelope (figure 25.26, step 6; *see also figure 5.14*).

1. What function do the influenza virus neuraminidase and hemagglutinin glycoproteins serve?
2. Trace the reproduction of an influenza virus starting with host cell attachment and ending with the exit of virions.

25.7 Viruses with Single-Stranded RNA Genomes (Group VI—Retroviruses)

Retroviruses have positive-strand RNA genomes. However, their genomes do not function as mRNA. Instead, retroviruses first convert their ssRNA genomes into dsDNA using an enzyme called **reverse transcriptase** (**figure 25.27**). The dsDNA then integrates into the host's DNA, where it can serve as a template for mRNA synthesis. The dsDNA also serves as the template for synthesis of the plus-strand RNA genome. The synthesis of both is catalyzed by the host cell's DNA-dependent RNA polymerase.

Numerous retroviruses have been identified and studied. However, the human immunodeficiency virus (HIV), the cause of AIDS (acquired immune deficiency syndrome), is of particular interest. AIDS is now recognized as the greatest pandemic of the second half of the twentieth century. Because of its global importance, we focus exclusively on HIV in this section, although there are a number of other retroviruses whose molecular biology is far less complicated than that of HIV.

1. The endonuclease activity of the PB1 protein cleaves the cap and about 10 nucleotides from the 5′ end of host mRNA (cap snatching). The fragment is used to prime viral mRNA synthesis by the RNA-dependent RNA polymerase activity of the PB1 protein.
2. Viral mRNA is translated. Early products include more NP and PB1 proteins.
3. RNA polymerase activity of the PB1 protein synthesizes +ssRNA from genomic −ssRNA molecules.
4. RNA polymerase activity of the PB1 protein synthesizes new copies of the genome using +ssRNA made in step 3 as templates. Some of these new genome segments serve as templates for the synthesis of more viral mRNA. Later in the infection, they will become progeny genomes.
5. Viral mRNA molecules transcribed from other genome segments encode structural proteins such as hemagglutinin (HA) and neuraminidase (NA). These messages are translated by ER-associated ribosomes and delivered to the cell membrane.
6. Viral genome segments are packaged as progeny virions bud from the host cell.

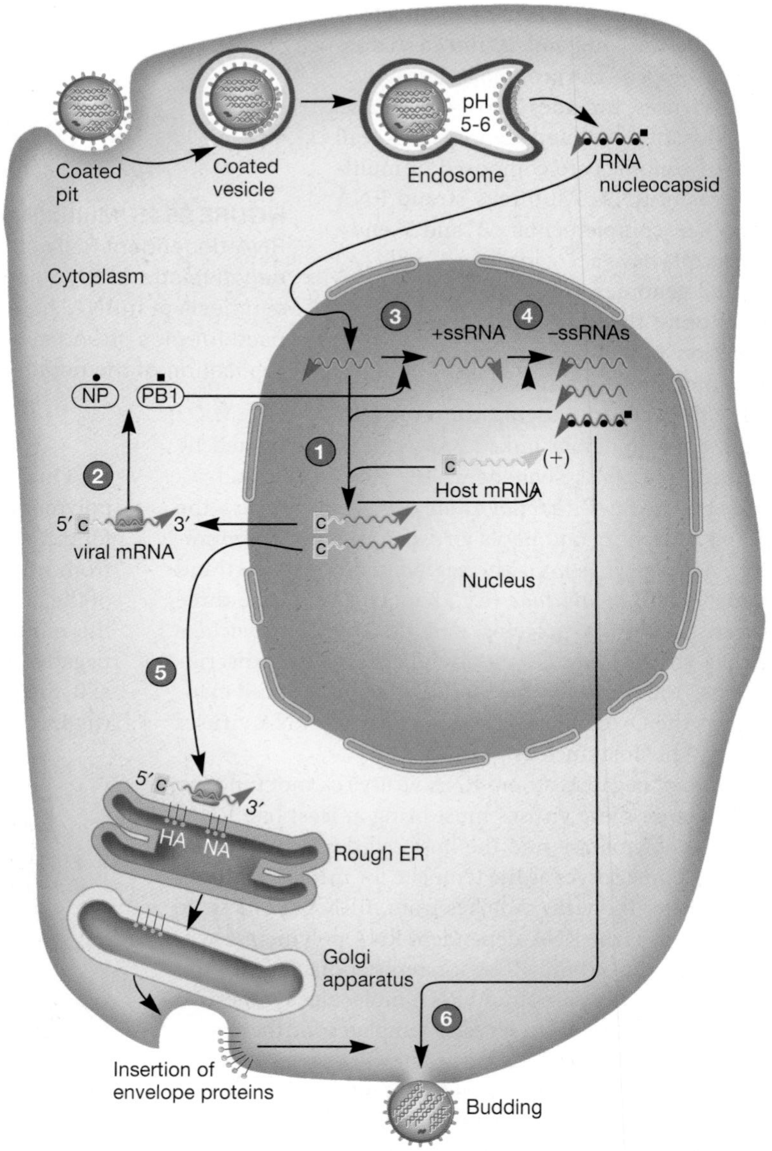

FIGURE 25.26 Simplified Life Cycle of an Influenza Virus. Several steps have been eliminated for simplicity and clarity. After entry by receptor-mediated endocytosis, the virus envelope fuses with the endosome membrane, releasing the nucleocapsids into the cytoplasm (each genome segment is associated with nucleocapsid proteins [NP] and PB1 to form the nucleocapsid). The nucleocapsids enter the nucleus, where synthesis of viral mRNA and genomes occurs. Critical to the production of viral mRNA and genomes is the enzyme RNA-dependent RNA polymerase. This is one activity of the PB1 enzyme. The other is endonuclease activity, which is used to cleave the 5′ ends from host mRNA. Steps 1 through 6 illustrate the remaining steps of the virus's life cycle.

First described in 1981, AIDS is the result of an infection by HIV, a member of the genus *Lentivirus* within the family *Retroviridae.* In the United States, AIDS is caused primarily by HIV-1 (some cases result from an HIV-2 infection). HIV-1 is an enveloped virus. The envelope surrounds an outer shell, which encloses a somewhat cone-shaped core (**figure 25.28**). The core contains two copies of the HIV RNA genome and several enzymes, including the enzymes reverse transcriptase and integrase. Thus far 10 virus-specific proteins have been discovered in the HIV virion.

Once inside the body, the gp120 viral envelope protein binds to host cells that have a surface glycoprotein called CD4. These cells include $CD4^+$ T cells (also called T-helper cells), macrophages, dendritic cells, and monocytes—all cells with key roles in host defenses. Dendritic cells are present throughout the

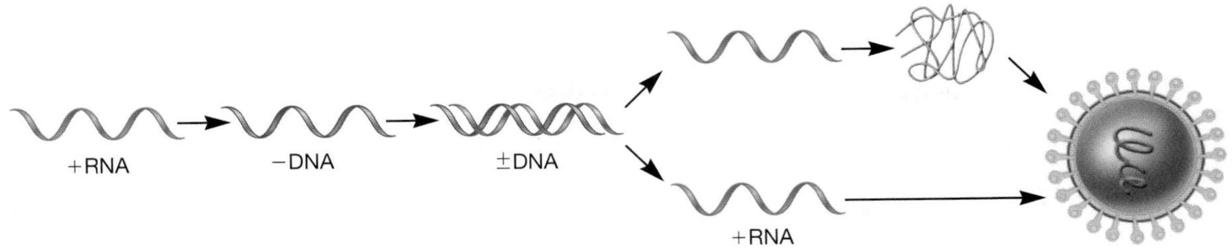

FIGURE 25.27 Multiplication Strategy of Retroviruses. Retroviruses have a plus-strand RNA genome that is first converted into dsDNA by the enzyme reverse transcriptase. The viral dsDNA integrates into the host chromosome, where it serves as the template for synthesis of viral mRNA and viral genomes. Both are synthesized using the host cell's DNA-dependent RNA polymerase.

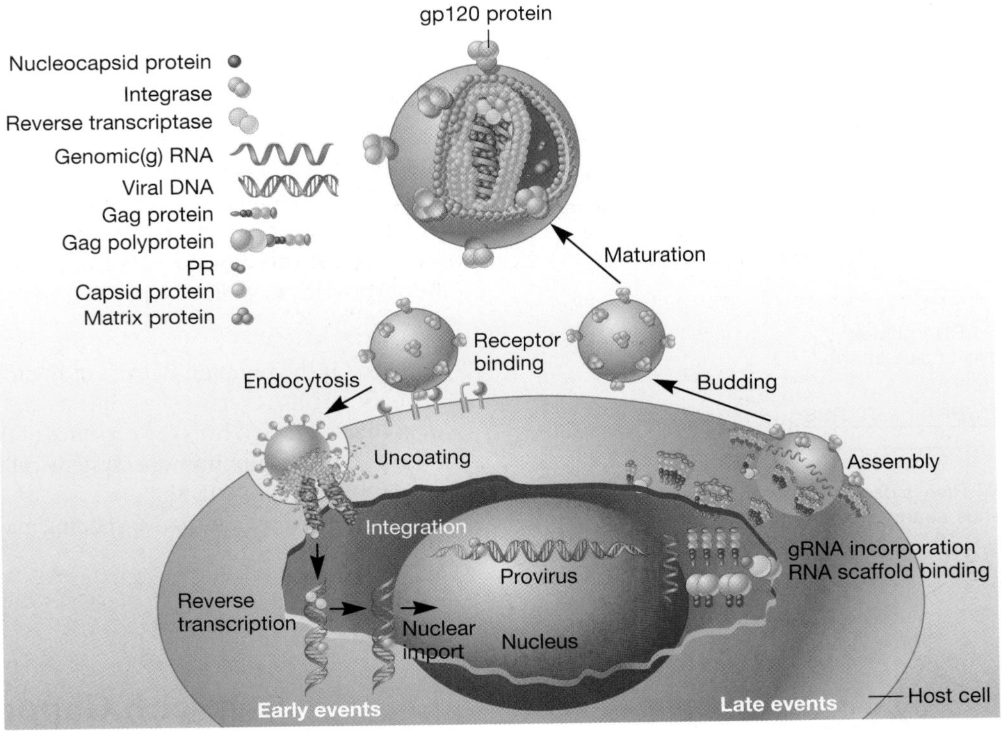

FIGURE 25.28 The HIV-1 Life Cycle.

Figure 25.28 Micro Inquiry

What three enzymatic activities does reverse transcriptase display? When during the HIV life cycle is each activity used?

body's mucosal surfaces, and it is possible that these are the first cells infected by HIV. The virus requires a coreceptor in addition to CD4, and this varies depending on the host cells infected. ▶▶| *Cells, tissues, and organs of the immune system (section 32.4); T-cell biology (section 33.5)*

For many years, entry into the host cell was thought to begin when the HIV envelope fused with the cell's plasma membrane and the virus released its core into the cytoplasm. However, a study reported in 2009 clearly demonstrated that the virus enters by receptor-mediated endocytosis (figure 25.28). Inside the infected cell, the core protein dissociates from the RNA, and the RNA is copied into a single strand of DNA by the reverse transcriptase enzyme. The RNA is next degraded by reverse transcriptase, and the DNA strand is duplicated to form a double-stranded DNA copy of the original RNA genome (figure 25.27).

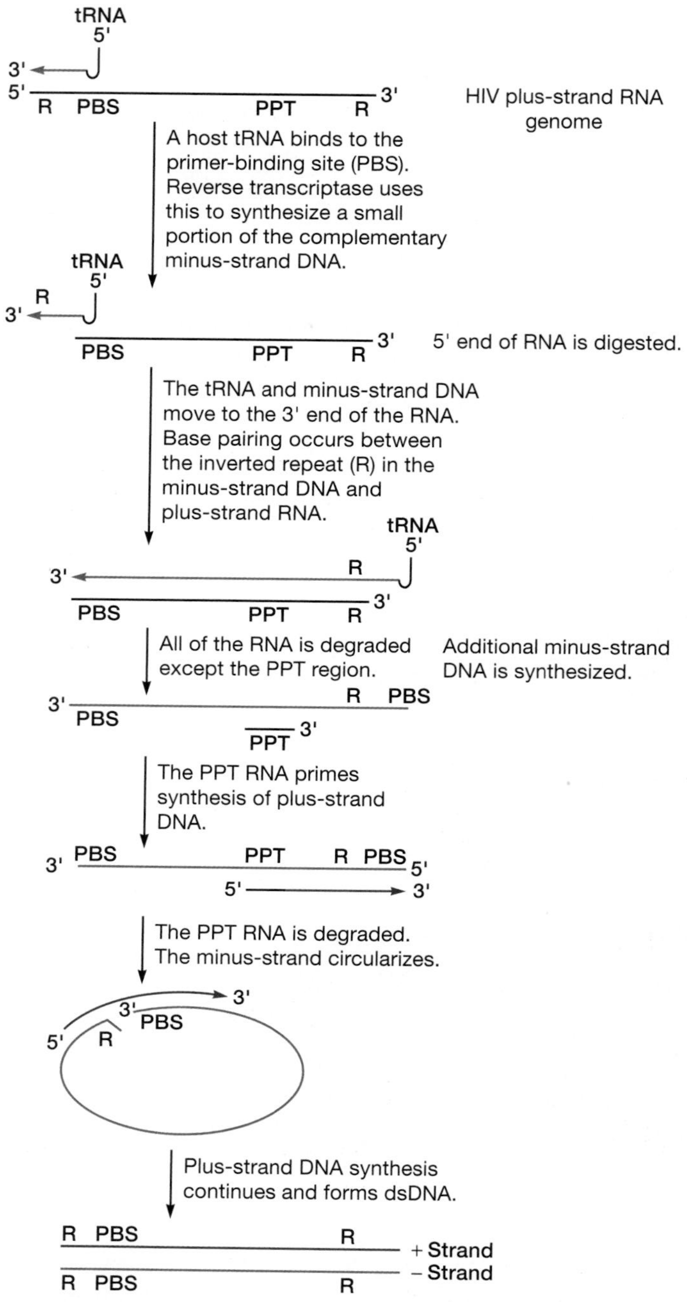

FIGURE 25.29 Reverse Transcription.

Reverse transcription is a critical step in the life cycle of HIV. As we just described, reverse transcriptase is a remarkable enzyme with multiple activities. It is an RNA-dependent DNA polymerase, a DNA-dependent DNA polymerase, and a ribonuclease. This latter function is referred to as RNaseH activity. Despite its versatility, reverse transcriptase lacks a function observed in other DNA polymerases: proofreading. Thus it makes many errors as it synthesizes DNA. Nonetheless, reverse transcription is an amazing process that involves the use of a host tRNA molecule as primer for initial steps in DNA synthesis (**figure 25.29**). In the process, a newly synthesized, small negative-strand DNA molecule is transferred from one end of the RNA template to the other to prime the synthesis of the rest of the minus-strand DNA. Later the full-length minus-strand DNA circularizes to allow completion of the plus-strand DNA.

Once the double-stranded (ds) DNA is formed, a complex of the dsDNA (the provirus), integrase enzyme, and other factors (including some host molecules) moves into the nucleus. Then the proviral DNA is integrated into the cell's DNA through a sequence of reactions catalyzed by integrase. Once integrated, the provirus can force the cell to synthesize viral mRNA (figure 25.28). Both the full-length mRNA and shorter mRNAs formed by alternative splicing are translated into the 10 proteins needed to form HIV virions. All mRNAs yield polyproteins that are cleaved to give rise to the needed proteins. In addition, some early proteins synthesized are involved in regulating cellular processes so that HIV genes are preferentially expressed. Eventually viral proteins and the complete HIV-1 RNA genome are assembled into new nucleocapsids that bud from the infected host cell (figure 25.28). After some time, the host cell dies, in part from repeated budding but by other processes as well.

1. What is the function of each of the following HIV products: gp120, reverse transcriptase, and integrase?
2. What is CD4? Why is the presence of this molecule on the surface of many immune system cells important to the development of AIDS?
3. What role does alternative splicing play in the life cycle of HIV-1?

25.8 Viruses with Gapped DNA Genomes (Group VII)

The hepadnaviruses such as *Hepatitis B virus* (HBV) are quite different from other DNA viruses with respect to genome replication. HBV is classified as an *Orthohepadnavirus* within the family *Hepadnaviridae.* The infecting virus is a 42-nm spherical particle that contains the viral genome. The HBV genome is 3.2 kb in length, consisting of four partially overlapping, open reading frames that encode viral proteins. Production of new viruses takes place predominantly in liver cells (hepatocytes).

The HBV genome is a circular dsDNA molecule that consists of one complete but nicked strand and a complementary strand that has a large gap—that is, it is incomplete (**figure 25.30**). After infecting the cell, the virus's gapped DNA is released into the nucleus. There, host repair enzymes fill the gap and seal the nick, yielding a covalently closed, circular DNA. Transcription of viral

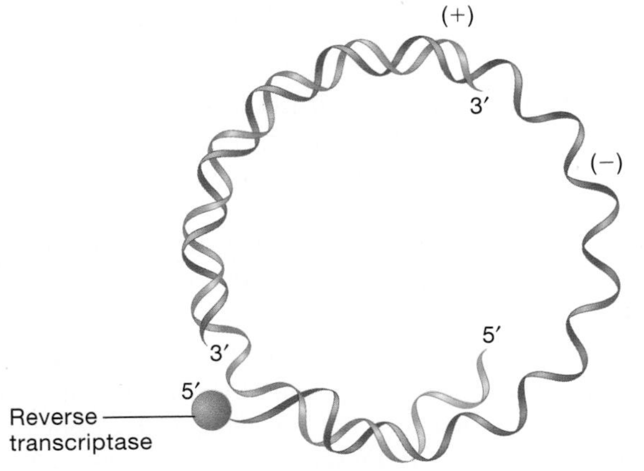

FIGURE 25.30 The Gapped Genome of Hepadnaviruses. The genomes of hepadnaviruses are unusual in several respects. The negative strand of the dsDNA molecule is complete but nicked. The enzyme reverse transcriptase is attached to its 5′ end. The positive strand is gapped; that is, it is incomplete (shown in purple). A short stretch of RNA is attached to its 5′ end (shown in green). Upon entry into the host nucleus, the nick in the negative strand is sealed and the gap in the positive strand is filled, yielding a covalently closed, circular dsDNA molecule.

genes occurs in the nucleus using host RNA polymerase and yields several mRNAs, including a large 3.4-kb RNA known as the pregenome (a plus-strand RNA). The RNAs move to the cytoplasm, and the mRNAs are translated to produce viral proteins, including core protein and reverse transcriptase. After synthesis of reverse transcriptase, it associates with the plus-strand RNA pregenome and core protein to form an immature core particle. Reverse transcriptase subsequently reverse transcribes the RNA using a protein primer to form a minus-strand DNA from the pregenome RNA. After almost all the pregenome RNA has been degraded by the RNaseH activity of the reverse transcriptase, the remaining RNA fragment serves as a primer for synthesis of the gapped dsDNA genome, using the minus-strand DNA as template. Finally, the nucleocapsid is completed and the progeny virions are released.

1. Describe the HBV genome. How is it converted to covalently closed, circular DNA in the host?
2. Trace the HBV reproductive cycle, paying particular attention to localization within the host cell during the biosynthetic stage.

Summary

25.1 Virus Taxonomy and Phylogeny

a. Currently viruses are classified with a taxonomic system placing primary emphasis on the type and strandedness of viral nucleic acids, and on the presence or absence of an envelope.

b. The Baltimore system is used by many virologists to organize viruses based on their genome type and the mechanisms used to synthesize mRNA and replicate their genomes (**table 25.1**).

25.2 Viruses with Double-Stranded DNA Genomes (Group I)

a. T4 phage is a virulent bacteriophage that causes lytic infections of *E. coli*. After attachment to a specific receptor site on the bacterial surface, T4 injects its dsDNA into the cell (**figures 25.5** and **25.6**). T4 DNA contains hydroxymethylcytosine (HMC) in place of cytosine, and glucose is often added to the HMC to protect the phage DNA from attack by host restriction enzymes (**figure 25.8**). T4 DNA replication produces concatemers, long strands of several genome copies linked together (**figure 25.9**).

b. Lambda (λ) phage is a temperate bacteriophage. It can establish lysogeny, rather than pursuing a lytic infection. During lysogeny, the viral DNA, called a prophage or provirus, is replicated as the cell's genome is replicated. Lysogeny is reversible, and the prophage can be induced to become active again and lyse its host. This highly regulated process is an important model system for regulatory processes. The protein cII plays a central role in regulating the choice between lysogeny and a lytic cycle. If cII protein levels are high enough, lysogeny is established. If not, the lytic cycle is initiated (**figures 25.12** and **25.13**).

c. Herpesviruses are a large group of dsDNA viruses. They cause acute infections such as cold sores, genital herpes, chickenpox, and mononucleosis. This is followed by a lifelong latent infection in which the virus genome resides within neurons. The virus can be reactivated at later times to cause a productive infection.

25.3 Viruses with Single-Stranded DNA Genomes (Group II)

a. ϕX174 is an example of a ssDNA bacteriophage. Its replication involves the formation of a dsDNA replicative form (RF) (**figure 25.16**).

b. fd phage is filamentous phage that upon infection is continuously released by the host without causing lysis.

c. The parvoviruses cause a spectrum of diseases in a wide variety of animals. The parvovirus genome is replicated by host DNA polymerase in the host cell's nucleus using a process that is similar to rolling-circle replication.

25.4 Viruses with Double-Stranded RNA Genomes (Group III)

a. Double-stranded RNA viruses use a viral enzyme called RNA-dependent RNA polymerase to synthesize mRNA (transcriptase activity) and replicate their genomes (replicase activity) (**figure 25.19**).

b. The bacteriophage ϕ6 is an unusual phage in that it is enveloped and enters the host bacterium through a process that resembles endocytosis.

c. Rotaviruses are one cause of viral diarrhea. They are a major cause of diarrhea in children and are responsible for a large number of deaths in developing countries. They multiply in an inclusion called the viroplasm.

25.5 Viruses with Plus-Strand RNA Genomes (Group IV)

a. The genomes of plus-strand RNA viruses serve directly as mRNA molecules. Among the first viral proteins synthesized is an RNA-dependent RNA polymerase that replicates the plus-strand RNA genome, sometimes by forming a dsRNA replicative intermediate. The negative RNA strands produced by the RNA-dependent RNA polymerase can be used to make either more genomes or mRNA (**figure 25.21**).

b. Bacteriophages MS2 and Qβ are small phages with only a few genes. They enter the host cell by attaching to the F pilus of their *E. coli* host.

c. *Poliovirus* genomic RNA is translated into a polyprotein that is cleaved to form all the proteins needed by the virus during its life cycle (**figure 25.23**).

d. TMV is like most known plant viruses in that it has a plus-strand RNA genome. The TMV nucleocapsid forms spontaneously when disks of coat protein protomers complex with the RNA (**figure 25.24**).

25.6 Viruses with Minus-Strand RNA Genomes (Group V)

a. For minus-strand RNA viruses to synthesize mRNA and replicate their genomes, their virions must carry an RNA-dependent RNA polymerase, which enters the host cells as the genome does. The polymerase first synthesizes mRNA from the negative-strand genome. Later it is used to replicate the genome by way of a plus-strand intermediate (**figure 25.25**).

b. Influenza viruses have segmented genomes. Host cell entry involves both the NA and HA glycoproteins. The viruses bring their own RNA-dependent RNA polymerase into the host cell for transcription and genome replication; they exit by budding (**figure 25.26**).

25.7 Viruses with Single-Stranded RNA Genomes (Group VI—Retroviruses)

a. Retroviruses replicate their genomes and synthesize mRNA via a dsDNA intermediate. The dsDNA is formed by an enzyme called reverse transcriptase (**figure 25.27**).

b. HIV is an enveloped retrovirus with a cone-shaped core that contains two copies of its genome and several enzymes, including reverse transcriptase. Upon infection, its ssRNA genome is converted to dsDNA, which is then integrated into the host genome (**figure 25.28**).

25.8 Viruses with Gapped DNA Genomes (Group VII)

a. Hepadnaviruses, including *Hepatitis B virus,* have dsDNA genomes that consist of one complete but nicked strand and an incomplete (i.e., gapped) complementary strand.

b. Upon infection, the host cell repairs the gap and seals the nick to generate a covalently closed, circular viral genome (**figure 25.30**). This serves as the template for the synthesis of pregenome RNA, which is the template for reverse transcription. Reverse transcription produces the dsDNA, gapped genome.

Critical Thinking Questions

1. No temperate RNA phages have yet been discovered. How might this absence be explained?
2. The choice between lysogeny and lysis is influenced by many factors. How would external conditions such as starvation or crowding be "sensed" and communicated to the transcriptional machinery and influence this choice?
3. The most straightforward explanation as to why the endolysin of T4 is expressed so late in infection is that its promoter is recognized by the gp55 alternative sigma factor. Propose a different explanation.
4. Several characteristics of AIDS render it particularly difficult to detect, prevent, and treat effectively. Discuss two of them. Contrast the disease with polio and smallpox.
5. In terms of molecular genetics, why is influenza such a prevalent viral infection in humans?

6. Will it be possible to eradicate many viral diseases in the same way as smallpox? Why or why not?
7. White spot syndrome virus (WSSV) is an enveloped dsDNA virus that infects crustaceans. It is particularly problematic for shrimp farmers as white spot disease can cause 100% shrimp mortality within a week. During infection, WSSV produces a large amount of a protein known as ICP11. This protein is a DNA mimic; that is, its folded structure resembles the acidic double helical structure of DNA. ICP11 binds directly to the DNA-binding site of host histones (*see figure 4.11*).

   How can a protein look like DNA, given that each is a different macromolecule composed of distinct monomers? Do you think ICP11 alone could cause host cell death? If so, explain the specific steps that would lead to host cell mortality. If not, what other virus-mediated activities would be necessary?

   Read the original paper: Wang, H. C., et al. 2008. White spot syndrome virus protein ICP11: A histone-binding DNA mimic that disrupts nucleosome assembly. *Proc. Nat. Acad. Sci., USA.* 105:20758.
8. Upon infection of host epithelial cells, papillomavirus (family *Papillomaviridae*) genomes are stably maintained in the nuclei for many years. These viral genomes are extrachromosomal and lack the capacity to segregate during host cell division. To prevent their loss during host cell division, the viral genomes use a protein known as E2 to attach to host chromosomes. For instance, the E2 protein of bovine papillomavirus type 1 (BPV-1) binds the host protein Brd4, which helps the viral genome attach to host chromosomes during mitosis. By contrast, the E2 protein of human papillomavirus 8 (HPV-8) does not require other host factors and binds directly to specific regions of host DNA during mitosis.

   How do you think the viral genome is released from a host chromosome following cell division? How do you think it was established that the HPV-8 E2 functions independently of host proteins whereas the E2 of BPV-1 requires Brd-4? Do you think the E2 proteins could be exploited to develop antiviral drugs? If so, why? How would you determine that a drug that targets E2 is specific to the viral protein and does not interfere with host cell function?

   Read the original paper: Poddar, A., et al. 2009. The human papillomavirus type 8 E2 tethering protein targets the ribosomal DNA loci of host mitotic chromosomes. *J. Virol.* 83:640.

Concept Mapping Challenge

Construct a concept map that illustrates which enzymes are involved in the life cycles of each of the Baltimore groups of viruses and how those enzymes are used. Use the concepts that follow, any other concepts or terms you need, and your own linking words between each pair of concepts in your map. Also provide specific examples of viruses that use these enzymes.

DNA-dependent DNA polymerase
DNA-dependent RNA polymerase
RNA-dependent RNA polymerase
RNA-dependent DNA polymerase
Excisionase
Reverse transcriptase
Transcriptase
Replicase
Integrase
Protease
RNAseH

Learn More

Learn more by visiting the text website at www.mhhe.com/willey8, where you will find a complete list of references.

26

Biogeochemical Cycling

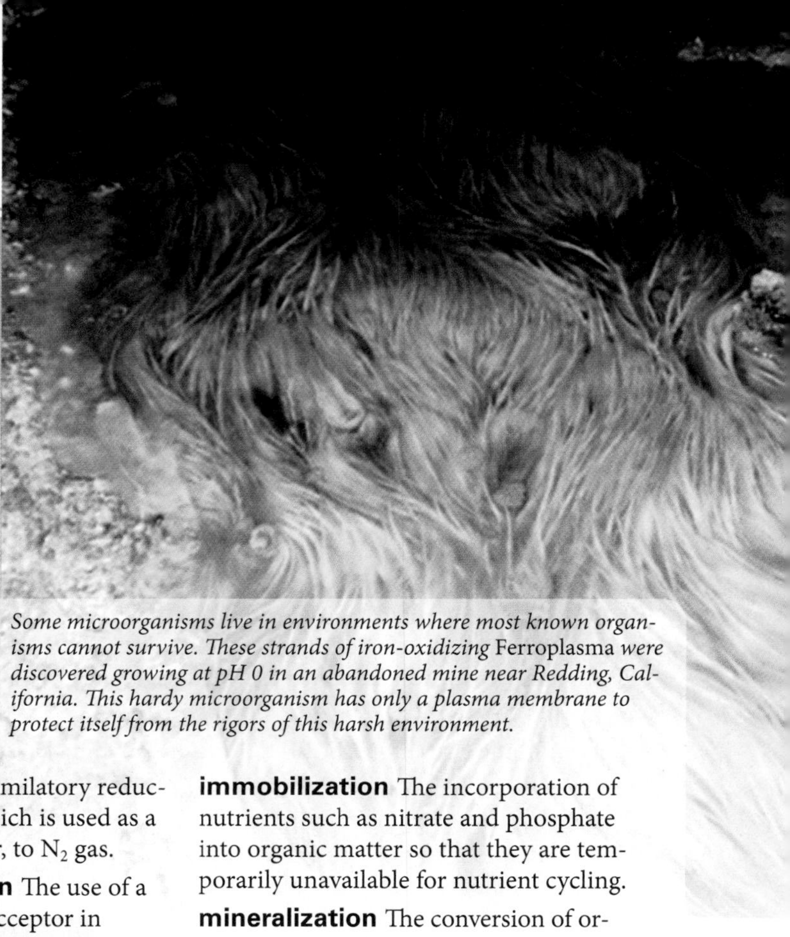

Some microorganisms live in environments where most known organisms cannot survive. These strands of iron-oxidizing Ferroplasma *were discovered growing at pH 0 in an abandoned mine near Redding, California. This hardy microorganism has only a plasma membrane to protect itself from the rigors of this harsh environment.*

CHAPTER GLOSSARY

anammox reaction The chemolithotrophic use of ammonium ion (NH_4^+) as electron donor and nitrite (NO_2^-) as terminal electron acceptor, which is reduced completely to N_2 gas.

assimilatory reduction The reduction of an inorganic molecule to incorporate it into organic material. No energy is made available during this process.

biogeochemical cycling The oxidation and reduction of substances carried out by living organisms and abiotic processes that results in the cycling of elements within and between different parts of ecosystems (the soil, aquatic environments, and atmosphere).

denitrification The dissimilatory reduction of nitrate (NO_3^-), which is used as a terminal electron acceptor, to N_2 gas.

dissimilatory reduction The use of a substance as an electron acceptor in energy conservation. The acceptor (e.g., sulfate or nitrate) is reduced but not incorporated into organic matter during biosynthetic processes.

greenhouse gas Gases such as CO_2, CH_4, and nitrogen oxides that absorb and emit radiation within Earth's atmosphere. Greenhouse gases are needed to maintain Earth's temperature, but their atmospheric accumulation within the last century has resulted in an increase in the mean global surface temperature.

immobilization The incorporation of nutrients such as nitrate and phosphate into organic matter so that they are temporarily unavailable for nutrient cycling.

mineralization The conversion of organic nutrients into inorganic material during microbial growth and metabolism.

nitrification The oxidation of ammonia to nitrate.

nitrogen fixation The metabolic process in which atmospheric molecular nitrogen (N_2) is reduced to ammonia; carried out by cyanobacteria, *Rhizobium*, methanogens, and other nitrogen-fixing bacteria and archaea.

Microbes have been on Earth for 3.5 billion years. During that time, they have made profound changes to our planet. They have changed Earth's atmosphere through oxygenic photosynthesis and nitrogen fixation, making animal life possible. Their interaction with minerals has altered the composition of freshwater and marine systems, where their primary production sustains complex ecosystems. Microbial activity in soils has enabled the development of the rich soils in which we grow crops.

Previous chapters have emphasized the capacity of bacteria and archaea to use enzyme catalyzed redox reactions to assemble macromolecules from the elements C, N, H, O, and S. We now turn our attention to microbial interactions with the environment. These are dynamic activities that balance the abiotic physical and chemical constraints with changes microbes introduce to the environment itself. The sum of the microbial and chemical processes that drive the flow of elements between sediments, waters, and the atmosphere is known as **biogeochemical cycling**. In the nineteenth century, Beijerinck, Winogradsky, and Pasteur recognized that all life depends on microbes. Indeed, in the twenty-first century, we continue to explore microorganisms as biogeochemical engineers that can still surprise us with their capacity to thrive in environments once thought too harsh to sustain even microbial growth.

In this chapter, we explore biogeochemical cycling by focusing on the flux of the major elements carbon, nitrogen, sulfur, phosphorus, iron, and magnesium. We discuss how these cycles can be linked and then turn our attention to global climate change. Collectively, this is the realm of **environmental microbiology,** which is concerned with microbial processes that occur in ecosystems, rather than the individual microorganisms responsible for these essential activities. One can study these biologically mediated activities and their global impact without dissecting the specific microenvironment and its microorganisms. However, it is critical to understand that bacteria, archaea, fungi, and protists are the agents of change. With less than 1% of these microbes currently in culture, the identities and specific physiological activities of most microbes await discovery. Chapter 27 describes some approaches microbial ecologists employ to make these discoveries.

26.1 Biogeochemical Cycling

The long success and vast diversity of microbes are in part because Earth has always presented habitats that feature complex variations in temperature, pressure, salinity, pH, nutrient availability, and perhaps most importantly, redox potential. **Redox potential,** or more formally, reduction/oxidation potential, is a measure of the tendency of a chemical compound to accept electrons, thereby becoming reduced. The term redox potential is sometimes used to collectively describe microbial habitats. An environment dominated with compounds of a high redox potential (more positive) will be more likely to accept electrons (i.e., be reduced) from any new compound added to the habitat. It follows that the newly added compound that donates electrons becomes oxidized. Conversely, compounds within a habitat with a low redox potential (more negative) are more likely to give up electrons, and thereby become oxidized, when a new compound is introduced. These newly added compounds thus accept electrons and become reduced. The oxidation state of the elements within any given habitat is therefore a function of the redox state of the environment (**table 26.1**). As we will see, the redox state of the environment plays a critical role determining the biogeochemical flux of elements. ◄◄ *Oxidation-reduction reactions (section 9.5)*

Carbon Cycle

A simplified carbon cycle is shown in **figure 26.1**. Carbon in the environment is present in reduced forms, such as methane (CH_4) and organic matter, and in more oxidized forms, such as carbon monoxide (CO) and carbon dioxide (CO_2). Although carbon is continuously transformed from one form to another, for the sake of clarity, we shall say that the cycle "begins" with carbon fixation—the conversion of CO_2 into organic matter. Plants such as trees and crops are often regarded as the principal CO_2-fixing organisms, but at least half the carbon on Earth is fixed by

Table 26.1 The Major Forms of Carbon, Nitrogen, Sulfur, and Iron Important in Biogeochemical Cycling

		Major Forms and Valences				
Cycle	*Significant Gaseous Component Present?*	*Reduced Forms*	*Intermediate Oxidation State Forms*			*Oxidized Forms*
C	Yes	Methane: CH_4 (−4)	Carbon monoxide: CO (+2)			CO_2 (+4)
N	Yes	Ammonium: NH_4^+; organic N (−3)	Nitrogen gas: N_2 (0)	Nitrous oxide N_2O (+1)	Nitrite: NO_2^- (+3)	Nitrate: NO_3^- (+5)
S	Yes	Hydrogen sulfide: H_2S; SH groups in organic matter (−2)	Elemental sulfur: S^0 (0)	Thiosulfate: $S_2O_3^{2-}$ (+2)	Sulfite: SO_3^{2-} (+4)	Sulfate: SO_4^{2-} (+6)
Fe	No	Ferrous iron: Fe^{2+} (+2)				Ferric Iron: Fe^{3+} (+3)

Note: The carbon, nitrogen, and sulfur cycles have significant gaseous components, and these are described as gaseous nutrient cycles. The iron cycle does not have a gaseous component, and this is described as a sedimentary nutrient cycle. Major reduced, intermediate oxidation state, and oxidized forms are noted, together with valences.

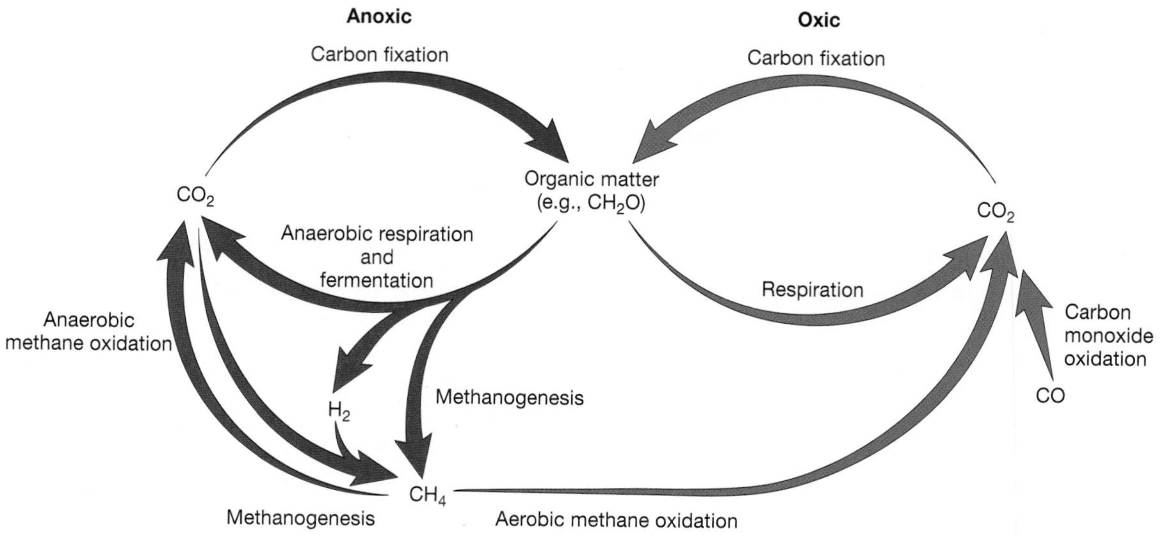

FIGURE 26.1 The Basic Carbon Cycle in the Environment. Carbon fixation can occur through the activities of photoautotrophic and chemoautotrophic microorganisms. Methane can be produced from inorganic substrates ($CO_2 + H_2$) or from organic matter. Carbon monoxide (CO)—produced by sources such as automobiles and industry—is returned to the carbon cycle by CO-oxidizing bacteria. Aerobic processes are noted with blue arrows, and anaerobic processes are shown with red arrows.

Figure 26.1 ***Micro Inquiry***

What microbes are capable of methanogenesis and methane oxidation?

microbes, particularly marine photosynthetic bacteria and protists (e.g., the cyanobacteria *Prochlorococcus* and *Synechococcus,* and diatoms, respectively). Importantly, microbes also fix carbon in anoxic environments using anoxygenic photosynthesis as well as by chemolithoautotrophy in the absence of light.

All fixed carbon enters a common pool of organic matter that can then be oxidized back to CO_2 through aerobic or anaerobic respiration and fermentation. Alternatively, inorganic carbon (CO_2) and organic carbon can be reduced anaerobically to methane (CH_4). It is thought that about 2% of the net CO_2 fixed annually is converted to methane. Recall that only archaea can form methane. They use either $H_2 + CO_2$ or acetate, as denoted by the production of methane from both CO_2 and organic matter in figure 26.1. The anoxic zones where methane is produced are usually below an oxic zone. However, once produced, methane moves up the water column and is rapidly oxidized by methylotrophic α- and γ-proteobacteria or by the newly discovered gram-negative acidophiles in the phylum *Verrucomicrobia*. Thus in these regions, little methane escapes to the atmosphere. By contrast, methane is usually released directly to the atmosphere in environments without an overlying oxic water zone. However, in some specialized anoxic environments, anaerobic oxidation of methane occurs. These habitats include marine sediments and freshwater ecosystems rich in methane, nitrate, and nitrite. In marine sediments, methane-oxidizing archaea form consortia with sulfate-reducing bacteria (*see p. 489*). In freshwater ecosystems, methane oxidation is coupled to denitrification. This is also accomplished by a consortium of an archaeon and a bacterium that represents a new phylum. Globally, rice paddies, ruminant animals, coal mines, sewage treatment plants, landfills, and marshes are important sources of methane. Even archaea such as *Methanobrevibacter* in the guts of termites contribute to global methane production. ◀◀ *Methanogens (section 18.3); Phylum* Verrucomicrobia *(section 19.8)*

In the carbon cycle depicted in figure 26.1, no distinction is made between the different types of organic matter formed and degraded. This is a marked oversimplification because organic matter varies widely in physical characteristics and in the biochemistry of its synthesis and degradation. Organic matter differs in terms of elemental composition, structure of basic repeating units, linkages between repeating units, and physical and chemical characteristics. Its degradation is influenced by a series of factors. These include (1) oxidation-reduction potential as determined by the relative abundance of electron donors (e.g., hydrogen) and electron acceptors (e.g., oxygen) in the environment; (2) nutrient availability; (3) abiotic conditions such as pH, temperature, O_2, and osmotic conditions; and (4) the microbial community present.

Many of the complex organic substrates used by microorganisms are summarized in **table 26.2**. Because microorganisms require each macronutrient in specific relative amounts, if an environment is enriched in one nutrient but relatively deficient in another, the nutrients may not be completely recycled into living biomass. For instance, inspection of table 26.2 reveals that protein and chitin (present in insect exoskeletons and fungal cell walls) contain carbon, hydrogen, oxygen, and nitrogen, whereas lipids and nucleic acids contain phosphorus as well as these elements. These compounds may provide more nitrogen and phosphorus than can be used given the amount of carbon available. Thus if

Table 26.2 Complex Organic Substrate Characteristics That Influence Decomposition and Degradability

			Elements Present in Large Quantity					Degradation	
Substrate	*Basic Subunit*	*Linkages (if Critical)*	*C*	*H*	*O*	*N*	*P*	*With O_2*	*Without O_2*
Starch	Glucose	α(1→4) α (1→6)	+	+	+	–	–	+	+
Cellulose	Glucose	β(1→4)	+	+	+	–	–	+	+
Hemicellulose	C6 and C5 monosaccharides	β(1→4), β(1→3), β(1→6)	+	+	+	–	–	+	+
Lignin	Phenylpropene	C–C, C–O bonds	+	+	+	–	–	+	+/–
Chitin	*N*-acetylglucosamine	β(1→4)	+	+	+	+	–	+	+
Protein	Amino acids	Peptide bonds	+	+	+	+	–	+	+
Hydrocarbon	Aliphatic, cyclic, aromatic		+	+	–	–	–	+	+/–
Lipids	Glycerol, fatty acids; some contain phosphate and nitrogen	Esters, ethers	+	+	+	+	+	+	+
Nucleic acids	Purine and pyrimidine bases, sugars, phosphate	Phosphodiester and *N*-glycosidic bonds	+	+	+	+	+	+	+

these substrates are used for growth, these minerals may be released to the environment in the process of **mineralization.** This is the process by which organic matter is decomposed to release simpler, inorganic compounds (e.g., CO_2, NH_3, CH_4, H_2).

The other complex substrates listed in table 26.2 contain only carbon, hydrogen, and oxygen. If microorganisms are to grow using these substrates, they must acquire the remaining nutrients they need for biomass synthesis from the environment. This is often very difficult, as the concentration of nitrogen, phosphorus, and iron may be very low. The inability to assimilate sufficient levels of one or more macronutrients may then limit the growth of a given population. For instance, in open-ocean microbial communities, growth of many autotrophic microbes is often nitrogen limited. In other words, if higher concentrations of usable nitrogen (NO_3^-, NH_4^+) were available, the rate of growth of individual microbes would increase, as would their overall population size. Those nutrients that are converted into biomass become temporarily "tied up" and are unavailable for nutrient cycling; this is sometimes called nutrient **immobilization.**

Most carbon substrates can be degraded easily regardless of whether or not oxygen is present, but this is not always the case. Notable exceptions are hydrocarbons and lignin. Hydrocarbons are unique in that microbial degradation most often involves the initial addition of molecular O_2. When anaerobic degradation of hydrocarbons does occur, sulfate or nitrate are used as electron acceptors. This proceeds more slowly and only in microbial communities that have been exposed to these compounds for extended periods. Nonetheless, under oxic or anoxic conditions, the relative insolubility of hydrocarbons reduces their availability for microbial degradation. ▶▶| *Biodegradation and bioremediation by natural communities (section 42.3)*

Lignin, an important structural component in mature plant materials, is a family of complex amorphous polymers linked by carbon-carbon and carbon-ether bonds (*see figure 29.2*). Because lignin is structurally very heterogeneous (i.e., there is no single lignin molecule), no single degradative enzyme can attack lignin using a uniform strategy. Fungi and the streptomycetes degrade lignin by oxidative depolymerization, a process that requires oxygen. A few microbes, such as the purple bacterium *Rhodopseudomonas palustris,* can degrade lignin anaerobically but at a very slow rate. Lignin's diminished biodegradability under anoxic conditions results in accumulation of lignified materials, including the formation of peat bogs and muck soils.

Oxygen availability also affects the final products that accumulate when organic substrates have been processed and mineralized by microorganisms. Under oxic conditions, oxidized products such as nitrate, sulfate, and carbon dioxide will result from microbial degradation of complex organic matter (**figure 26.2**). In comparison, under anoxic conditions, reduced end products tend to accumulate, including ammonium ion, sulfide, and methane.

1. What is biogeochemical cycling?
2. Describe the consortia of microorganisms that oxidize methane anaerobically. Why are two physiologically distinct microbes required for this reaction?
3. Define mineralization and immobilization, and give examples.
4. What is unique about lignin and its degradation?
5. What C, N, and S forms will accumulate after anaerobic degradation of organic matter? Compare these with the forms that accumulate after aerobic degradation.

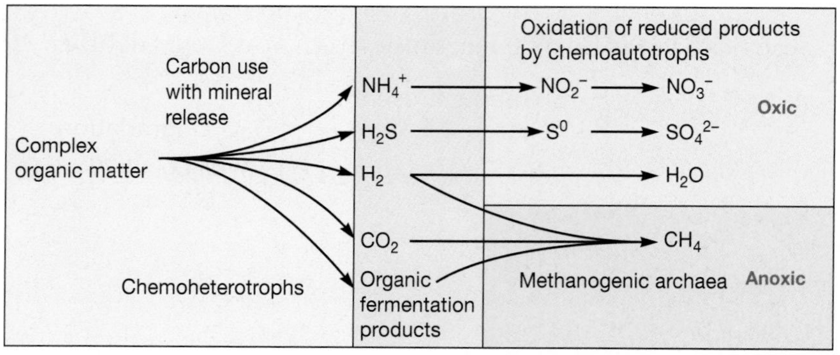

FIGURE 26.2 The Influence of Oxygen on Organic Matter Decomposition. When organic carbon is mineralized to CO_2 by chemoheterotrophs, autotrophic organisms can recycle that carbon back into the ecosystem. Under oxic conditions, chemolithoautotrophs couple the reduction of CO_2 to the oxidation of inorganic molecules, which results in the accumulation of oxidized products. Conversely, in anoxic environments, reduced products accumulate.

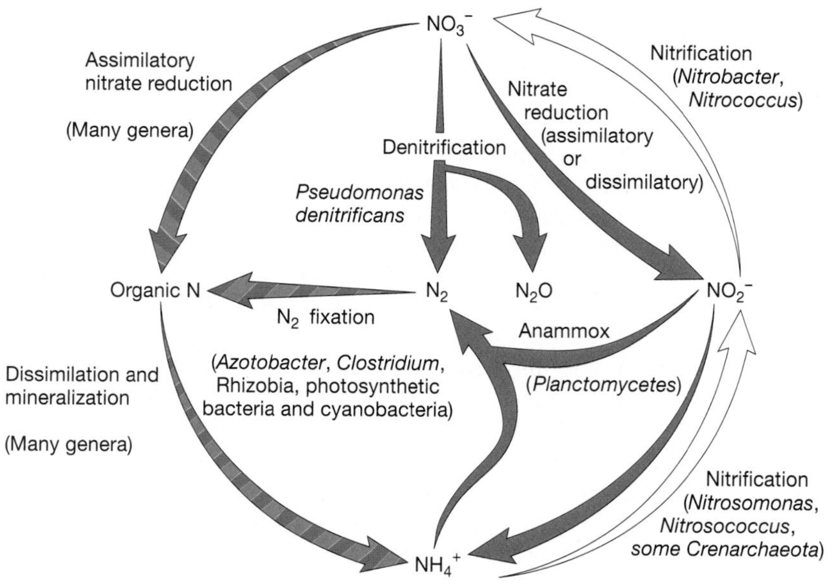

FIGURE 26.3 A Simplified Nitrogen Cycle. Reactions that occur predominantly under oxic conditions are noted with open arrows. Anaerobic processes are noted with solid purple arrows. Processes occurring under both oxic and anoxic conditions are marked with striped arrows. Important genera contributing to the nitrogen cycle are noted. The reduction of NO_3^- to NH_4^+ can be either assimilatory, if the NH_4^+ is incorporated into organic substrates (e.g., amino acids), or dissimilatory, as occurs during NO_3^- or NO_2^- respiration (i.e., the use of these compounds as terminal electron acceptors).

Figure 26.3 Micro Inquiry

What are the products of denitrification, nitrification, dissimilatory nitrate reduction, and assimilatory nitrate reduction?

Nitrogen Cycle

Like the carbon cycle, cycling of nitrogenous materials makes life on Earth possible. A simplified nitrogen cycle is presented in **figure 26.3**. Again, we begin our discussion of this cycle with the fixation of the inorganic gaseous element (N_2) to its organic form (e.g., amino acids, purines, and pyrimidines). **Nitrogen fixation** is performed only by some bacteria and archaea; apart from a limited amount of nitrogen fixation that occurs during lightning strikes and from the manufacture of fertilizer, all naturally produced organic nitrogen is of bacterial and archaeal origin. Although the nitrogenase enzyme is sensitive to oxygen, nitrogen fixation can be carried out under both oxic and anoxic conditions. Microbes such as *Azotobacter* and the cyanobacterium *Trichodesmium* fix nitrogen aerobically, while free-living anaerobes such as members of the genus *Clostridium* fix nitrogen anaerobically. Perhaps the best-studied nitrogen-fixing microbes are the bacterial symbionts of leguminous plants, including *Rhizobium,* its α-proteobacterial relatives, and some recently discovered β-proteobacteria (e.g., *Burkholderia* and *Ralstonia*). Other bacterial symbionts fix nitrogen as well. For instance, the actinomycete *Frankia* fixes nitrogen while colonizing many types of woody shrubs, and the heterocystous cyanobacterium *Anabaenea* fixes nitrogen when in association with the water fern *Azolla.* ◀◀ *Nitrogen fixation (section 11.5); Phylum* Cyanobacteria *(section 19.3);* ▶▶ *Nitrogen-fixing bacteria (section 29.3)*

The product of N_2 fixation is ammonia (NH_3), which is immediately incorporated into organic matter as an amine. These amine N-atoms are eventually introduced into proteins, nucleic acids, and other biomolecules. The N cycle continues when these organic molecules are degraded (dissimilated) and mineralized, and ammonium (NH_4^+) is released. Many microbial genera are capable of dissimilation of organic nitrogen substrates, but complete mineralization requires an assemblage of microbes.

One important fate of ammonium is its conversion to nitrate (NO_3^-), a process called **nitrification.** This is a two-step chemolithotrophic process whereby ammonium is first oxidized to nitrite (NO_2^-), which is then oxidized to nitrate. No single microbial genus can perform both steps of nitrification. For example, some archaea and the bacteria *Nitrosomonas* and *Nitrosococcus* play important roles in ammonia oxidation, and *Nitrobacter* and related bacteria carry out nitrite oxidation. In addition, *Nitrosomonas eutropha* has been found to oxidize ammonium ion anaerobically to nitrite and nitric oxide (NO) using nitrogen dioxide (NO_2) as an acceptor in a denitrification-related reaction (*see figure 20.13*). The diversity of nitrifying microbes and the means by which nitrification occurs appear to be more complex than previously thought.

The production of nitrate is important because it can be reduced and incorporated into microbial and plant cell biomass; this process is known as **assimilatory nitrate reduction.** Alternatively, some microorganisms use nitrate as a terminal electron acceptor during anaerobic respiration. Because the nitrate-nitrogen is not incorporated into cellular material, this is called **dissimilatory nitrate reduction.** A variety of microbes, including *Geobacter metallireducens* and *Desulfovibrio* spp. are responsible for dissimilatory nitrate reduction. When nitrate is reduced to dinitrogen gas (N_2), nitrogen is removed from the ecosystem and returned to the atmosphere through a series of reactions collectively known as **denitrification.** This dissimilatory process is performed by a variety of heterotrophic bacteria, such as *Pseudomonas denitrificans.* The major products of denitrification include nitrogen gas (N_2) and nitrous oxide (N_2O), although nitrite (NO_2^-) also can accumulate. ◀◀ *Anaerobic respiration (section 10.6); Nitrogen assimilation (section 11.5)*

A recently identified form of nitrogen conversion is called the **anammox reaction** (*an*oxic *amm*onium *ox*idation). In this anaerobic reaction, chemolithotrophs use ammonium ion (NH_4^+) as the electron donor and nitrite (NO_2^-) as the terminal electron acceptor; it is reduced to nitrogen gas (N_2). In effect, the anammox reaction is a shortcut to N_2, proceeding directly from ammonium and nitrite, without having to cycle first through nitrate (figure 26.3). Although this reaction was known to be energetically possible, microbes capable of performing the anammox reaction were only recently documented. The discovery that marine bacteria perform the anammox reaction in the anoxic waters just below oxygenated regions in the open ocean solved a long-standing mystery. For many years, microbiologists wondered where the "missing" NH_4^+ could be—mass calculations did not agree with experimentally derived nitrogen measurements. The discovery that planctomycete bacteria oxidize measurable amounts of NH_4^+ to N_2, thereby removing it from the marine ecosystem, has necessitated a reevaluation of nitrogen cycling in the open ocean. ◀◀ *Phylum* Planctomycetes *(section 19.4)*

Search This: Nutrient cycling visionlearning

1. Why is nitrogen fixation important and under what circumstances does it occur?
2. Describe the two-step process that makes up nitrification. Why do you think nitrification requires two different types of microbes?
3. What is the difference between assimilatory nitrate reduction and denitrification? Which reaction is performed by most microbes, and which is a more specialized metabolic capability?
4. Describe the anammox reaction. Why do you think it was so difficult for microbiologists to discover the microbes that perform this reaction?

Phosphorus Cycle

Biogeochemical cycling of phosphorus is important for a number of reasons. All living cells require phosphorus for nucleic acids, lipids, and some polysaccharides. However, unlike the carbon and nitrogen cycles, the phosphorus cycle has no gaseous component (**figure 26.4**). Because most environmental phosphorus is present in low concentrations, locked within Earth's crust, phosphorus frequently limits growth.

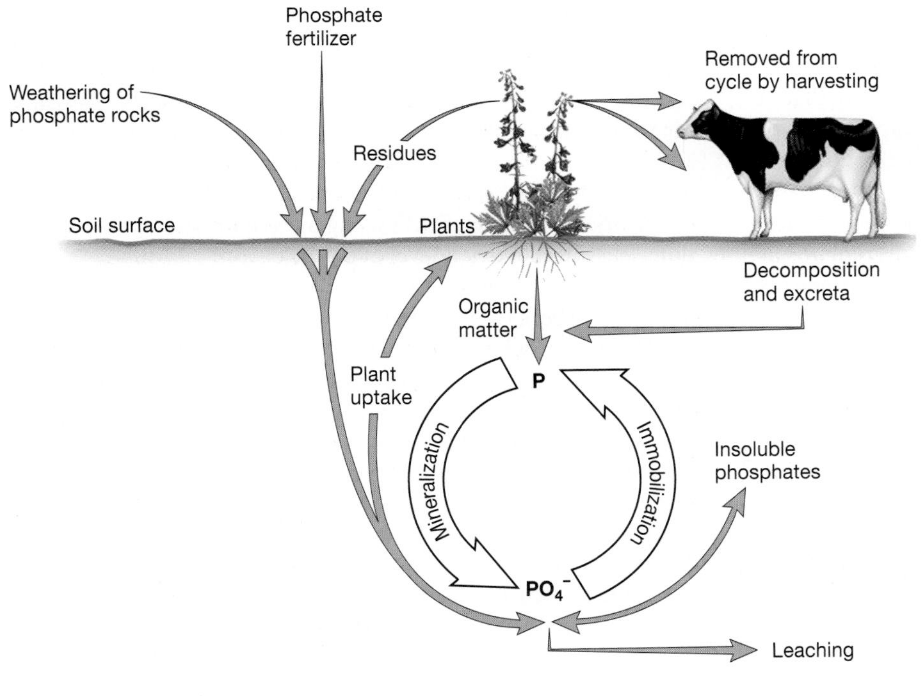

FIGURE 26.4 A Simplified Phosphorus Cycle. Phosphorus enters soil and water through the weathering of rocks, phosphate fertilizer, and surface residue of plant degradation. Plants and microbes rapidly convert inorganic phosphorus to its organic form, causing immobilization. However, much of the soil phosphorus can leach great distances or complex with cations to form relatively insoluble compounds.

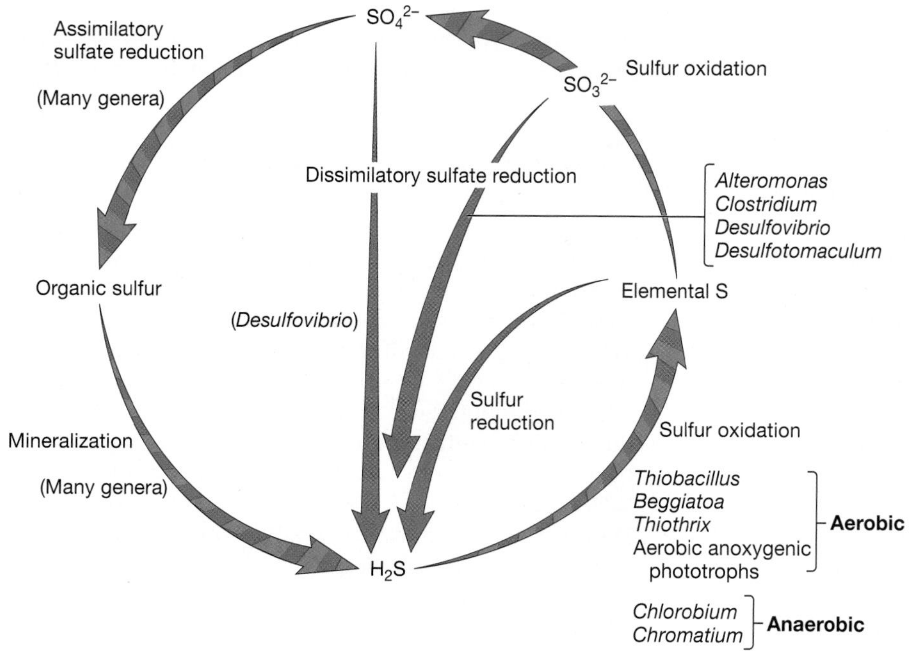

FIGURE 26.5 A Simplified Sulfur Cycle. Photosynthetic and chemosynthetic microorganisms contribute to the environmental sulfur cycle. Sulfate and sulfite reductions carried out by *Desulfovibrio* and related microorganisms, noted with purple arrows, are dissimilatory processes. Elemental sulfur reduction to sulfide is carried out by *Desulfuromonas,* thermophilic archaea, or cyanobacteria in hypersaline sediments. Sulfate reduction also can occur in assimilatory reactions, resulting in organic sulfur forms. Sulfur oxidation can be carried out by a wide range of aerobic chemotrophs and by aerobic and anaerobic phototrophs.

Figure 26.5 Micro Inquiry

During what process does dissimilatory sulfate reduction take place?

Unlike carbon and nitrogen, which can be obtained from the atmosphere, phosphorus is derived solely from the weathering of phosphate-containing rocks. In soil, phosphorus exists in both inorganic and organic forms. Organic phosphorus includes not only that found in biomass but also phosphorus in materials such as humus and other organic compounds. The phosphorus in these organic materials is recycled by microbial activity. Inorganic phosphorus is negatively charged, so it complexes readily with cations in the environment, such as iron, aluminum, and calcium. These compounds are relatively insoluble, and their dissolution is pH dependent such that it is most available to plants and microbes between pH 6 and 7. Under such conditions, these organisms rapidly convert phosphate to its organic form so that it becomes available to animals. The microbial transformation of phosphorus features the transformation of simple orthophosphate (PO_4^{3-}) to more complex forms. These include the polyphosphates present as inclusions as well as more familiar macromolecules. Note that the phosphorus in all of these organic forms remains in the +5 valence state. ◀◀ *Inclusions (section 3.5); Phosphorus assimilation (section 11.6)*

Sulfur Cycle

Microorganisms contribute greatly to the sulfur cycle; a simplified version is shown in **figure 26.5**. Sulfate is reduced by plants and microbes for use in amino acid and protein biosynthesis; this is described as assimilatory sulfate reduction. By contrast, when sulfate diffuses into anoxic habitats, it provides an opportunity for different groups of microorganisms to carry out **dissimilatory sulfate reduction.** For example, when a usable organic electron donor is present, sulfate serves as a terminal electron acceptor during anaerobic respiration by a variety of microbes, including δ-proteobacteria such as *Desulfovibrio, Desulfonema,* and *Desulfuromonas*, and archaea such as *Archaeoglobus.* This results in sulfide accumulation. Recall that sulfide can serve as an electron source for anoxygenic photosynthetic microorganisms and chemolithoautotrophs such as *Chlorobi* and *Thiobacillus,* respectively. These microbes convert sulfide to elemental sulfur and sulfate.

Other microorganisms have been found to carry out dissimilatory elemental sulfur (S^0) reduction. These include *Desulfuromonas,* thermophilic archaea, and cyanobacteria in hypersaline sediments. Sulfite (SO_3^{2-}) is another critical intermediate that can be reduced to sulfide by a wide variety of microorganisms, including *Alteromonas* and *Clostridium,* as well as *Desulfovibrio* and *Desulfotomaculum. Desulfovibrio* is usually considered an obligate anaerobe. Research, however, has shown that this interesting organism also respires using oxygen under hypoxic conditions (dissolved oxygen level of 0.04%).

Minor compounds in the sulfur cycle play major roles in biology. An excellent example is dimethylsulfonopropionate, which is used by bacterioplankton (floating bacteria) as a sulfur source for protein synthesis and is transformed to dimethylsulfide, a volatile sulfur form that can affect atmospheric processes, including the formation of clouds.

When pH and oxidation-reduction conditions are favorable, several key transformations in the sulfur cycle also occur as the result of chemical reactions in the absence of microorganisms. An important example of such an abiotic process is the oxidation of sulfide to elemental sulfur. This takes place rapidly at neutral pH, with a half-life of approximately 10 minutes for sulfide at room temperature.

Iron Cycle

The iron cycle principally features the interchange of ferrous iron (Fe^{2+}) to ferric iron (Fe^{3+}) (**figure 26.6**). In environments that are fully aerated with a neutral pH, iron is present primarily as insoluble minerals of either oxidation state. The solubility of both reduced, ferrous iron (Fe^{2+}) and oxidized, ferric iron (Fe^{3+}) increases as the pH decreases, so that below pH 4, Fe^{2+} is found in the aqueous form. The assimilation of iron presents two challenges for most aerobic microorganisms. First, while iron is an essential element, free iron is usually present in very small quantities. Second, Fe^{3+} dominates in oxic environments, but cells generally incorporate Fe^{2+} into biomolecules such as enzymes involved in redox reactions. The use of siderophores resolves both problems. Siderophores are low molecular weight organic molecules that bind Fe^{3+}, which is transported into the cell, where it is reduced to Fe^{2+} (*see figure 6.8*). ◀◀ *Iron uptake (section 6.6)*

Dissimilatory reduction occurs when ferric iron serves as a terminal electron acceptor during anaerobic respiration. In most environments, Fe^{3+} is found chiefly in a crystalline phase (e.g., hematite and magnetite) and as a component of sediment clays. As described more fully in chapter 20, recent studies have shown that microbes can donate electrons from the electron transport chain to these solid phase forms of Fe^{3+} outside the cell. Different microbes appear to use different strategies to transfer electrons to these external electron acceptors. For instance, the δ-proteobacterium *Geobacter* and the γ-proteobacterium *Shewanella* use electrically conductive piluslike structures called nanowires to transfer electrons from a reductase in the outer membrane to particulate Fe^{3+}(*see figure 20.27*). *Shewanella* also produces redox-active flavins that shuttle electrons from the outer membrane to crystalline Fe^{3+} oxides. ◀◀ *Anaerobic respiration (section 10.6); Order* Alteromonadales *(section 20.3)*

A wide range of bacteria and archaea are capable of dissimilatory Fe^{3+} reduction. This includes genera from the *Euryarchaeota* and the *Crenarchaeota,* all five classes of *Proteobacteria,* the *Firmicutes, Deferribacteraceae, Acidobacteria, Thermotoga,* and *Thermus*. This phylogenetic diversity may reflect the antiquity of Fe^{3+} reduction. It is believed that early life began 3.5 to 3.8 billion years ago in an environment that was hot and rich in Fe^{2+}. The photooxidation of Fe^{2+} to Fe^{3+} and H_2 would have provided an electron acceptor and energy source, respectively, to early cellular forms. Banded iron formation that occurred when atmospheric oxygen levels were beginning to increase at the end of the Precambrian era may be evidence of increased bacterial iron metabolism. ◀◀ *Microbial evolution (section 1.2)*

In addition to ferric iron (Fe^{3+}) as a terminal electron acceptor, some magnetotactic bacteria such as *Aquaspirillum magnetotacticum* transform extracellular iron to the mixed valence iron oxide mineral magnetite (Fe_3O_4) and construct intracellular magnetic compasses. Magnetotactic bacteria may be described as magneto-aerotactic bacteria because they are thought to use magnetic fields to migrate to the position in a bog or swamp where the oxygen level best meets their needs. Furthermore, some dissimilatory iron-reducing bacteria accumulate magnetite as an extracellular product. ◀◀ *Inclusions (section 3.5)*

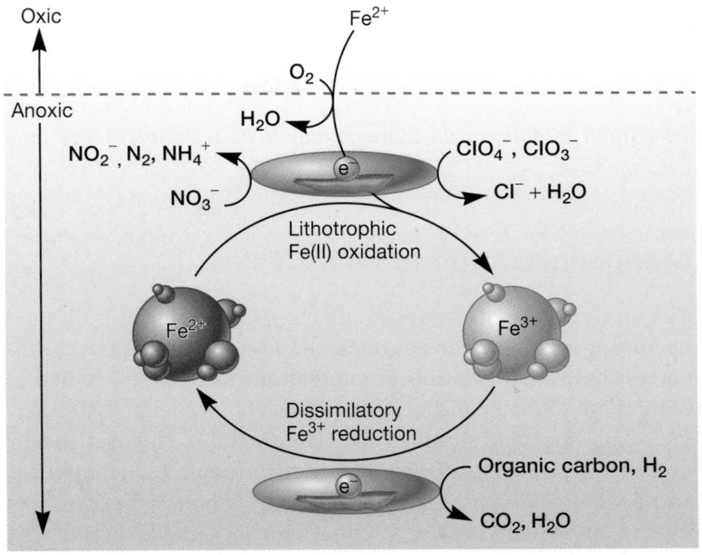

FIGURE 26.6 The Iron Cycle. Dissimilatory reduction of ferrous iron takes place when Fe^{3+} is used as a terminal electron acceptor during anaerobic respiration. This occurs only in anoxic environments. By contrast, oxidation of Fe^{2+} can occur both under both oxic and anoxic conditions. Most well-known is the use of Fe^{2+} as an electron donor and O_2 as electron acceptor. Fe^{2+} can also be oxidized by lithotrophs that use NO_3^- or even the environmental contaminants chlorate (ClO_3^-) and perchlorate (ClO_4^-) as electron acceptors. Green ovals represent microbial communities.

Figure 26.6 Micro Inquiry

Why does the use of Fe^{2+} as an electron donor occur in chiefly in acidic environments?

The use of ferrous iron (Fe^{2+}) as an electron donor by lithotrophic microbes in acidic, oxic environments where Fe^{2+} is soluble and oxygen can serve as the terminal electron acceptor was first documented over a century ago. It has been well characterized in the β-proteobacterium *Acidithiobacillus ferrooxidans* and the thermophilic crenarchaeote *Sulfolobus*. In addition, it is now known that a number of microbes oxidize Fe^{2+} at neutral pH under oxic conditions. Best studied is the γ-proteobacterium *Marinobacter* and the β-proteobacteria *Leptothrix* and *Gallionella.* Ferrous iron can also be oxidized under anoxic conditions with nitrate as the electron acceptor. One interesting anaerobic microbe is *Dechlorosoma suillum,* which oxidizes Fe^{2+} using perchlorate (ClO_4^-) and chlorate (ClO_3^-) as electron acceptors. Because perchlorate is a major component of explosives and rocket propellents, it is a frequent contaminant at retired munitions facilities and military bases. Thus *D. suillum* may be used in the bioremediation (biological cleanup) of such sites. This process also occurs in aquatic sediments with depressed levels of oxygen and may be another route by which large zones of oxidized iron have accumulated in environments with lower oxygen levels. *Biodegradation and bioremediation by natural communities (section 42.3)*

Manganese and Mercury Cycles

The importance of microorganisms in manganese cycling is becoming much better appreciated. The manganese cycle involves the transformation of manganous ion (Mn^{2+}) to MnO_2 (equivalent to manganic ion [Mn^{4+}]), which occurs in hydrothermal vents, bogs, and stratified lakes (**figure 26.7**). Phylogenetically diverse bacteria such as *Leptothrix*, the actinomycete *Arthrobacter*, and *Pedomicrobium*, a member of the *Hyphomicrobiaceae* family, are important in Mn^{2+} oxidation. *Shewanella, Geobacter*, and other chemoorganotrophs can carry out the complementary manganese reduction process. *Microorganisms in lakes (section 28.3)*

The mercury cycle illustrates many characteristics of those metals that can be methylated. Mercury compounds were widely used in industrial processes over the centuries. A devastating situation developed in southwestern Japan when large-scale mercury poisoning occurred in the Minamata Bay region because of industrial mercury released into the marine environment. Inorganic mercury that accumulated in bottom muds of the bay was methylated by anaerobic bacteria of the genus *Desulfovibrio* (**figure 26.8**). Such methylated mercury forms are volatile and lipid soluble, and the mercury concentrations increased in the food chain by the process of **biomagnification.** Biomagnification is the progressive accumulation of refractile compounds by successive trophic levels. Fish containing high levels of mercury were ultimately ingested by humans—the "top consumers"—leading to severe neurological disorders, particularly among children. Clearly microbial interactions with metals such as manganese, and mercury can have far-reaching consequences.

Interaction Between Elemental Cycles

So far we have presented biogeochemical cycling as a discrete series of nutrient fluxes. However, it is important to understand that biogeochemical cycling involves dynamic, interconnected processes that over a geologic time scale keep the biosphere in a self-sustaining steady state. In considering links between cycles, one need look no further than anaerobic respiration. For instance, when a saprophytic microbe uses decaying biomass as its source of carbon and electrons, and nitrate as its terminal electron acceptor, it contributes to both the flux of carbon and nitrogen. Likewise when sulfate is used as the terminal electron acceptor following the oxidation of organic carbon, the carbon and sulfur cycles are linked.

Global biogeochemical cycling frequently entails the production or consumption of one-carbon compounds such as methane and CO_2. Recall that chemolithoautotrophs fix CO_2 and use reduced inorganic compounds such as ammonium or ferrous iron as a source of electrons. Because a relatively small amount of energy is released during the oxidation of inorganic chemicals, chemolithotrophs must oxidize large amounts of these compounds to generate enough ATP and reducing equivalents needed for carbon fixation. Thus chemolithotrophs not only link the carbon cycle to either the nitrogen, sulfur, or iron (or other metal) cycles, they significantly impact the global flux of these elements. Recent advances in microbial ecology have revealed that methane oxidation can link the carbon cycle to the nitrogen and sulfur cycles. For many years, it was believed that methane had to be oxidized aerobically. However, we now know that methane is oxidized by microbial consortia, consisting of methane-oxidizing archaea and bacteria in anoxic environments. Two types of consortia have been described. In 2000 clusters of methane-oxidizing archaea surrounded by sulfate reducing δ-proteobacteria were discovered. More recently (2006) consortia of methanotrophic archaea and bacteria that reduced nitrate or, to a lesser extent, nitrite completely to gaseous N_2 were reported. In both cases, the oxidation of methane is made thermodynamically favorable by the redox activity of the bacterial partners. *Phylum* Euryarchaeota *(section 18.3)*

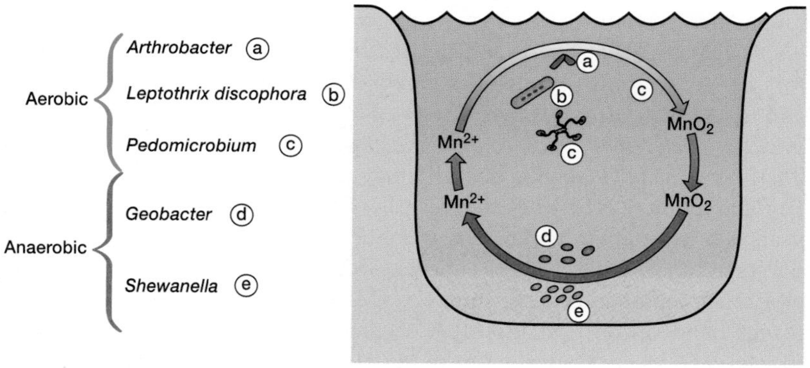

FIGURE 26.7 The Manganese Cycle, Illustrated in a Stratified Lake. Microorganisms make many important contributions to the manganese cycle. After diffusing from anoxic (pink) to oxic (blue) zones, manganous ion (Mn^{2+}) is oxidized chemically and by many morphologically distinct microorganisms in the oxic water column to manganic oxide—$MnO_2^{(IV)}$, valence equivalent to 4+. When the $MnO_2^{(IV)}$ diffuses into the anoxic zone, bacteria such as *Geobacter* and *Shewanella* carry out the complementary reduction process. Similar processes occur across oxic/anoxic transitions in soils, muds, and other environments.

1. Trace the fate of a single phosphorus atom from a rock to a stream and back to the earth.
2. Describe the difference between assimilatory and dissimilatory sulfate reduction. Compare these to assimilatory and dissimilatory nitrate reduction.
3. How might iron have been important in the evolution of early life?
4. What are some important microbial genera that contribute to manganese cycling?
5. How can microbial activity render some metals more or less toxic to warm-blooded animals?
6. Suggest how a chemolithoautotroph might link both the carbon and sulfur cycles.

26.2 Global Climate Change

Microbial activity is critical in maintaining the dynamic equilibrium that defines our biosphere. The capacity of microbes to thrive in seemingly every niche on Earth is the result of 3.5 billion years of coevolution between microbes and their habitats. Importantly, changes in the physical and chemical environment to which microbes have had to adapt have generally occurred over geological time scales. However, since the beginning of the twentieth century, the rate at which CO_2 and other so-called greenhouse gases have entered the atmosphere has been faster than at any other time in the known history of life on Earth. Atmospheric gases such as CO_2, CH_4, and nitrogen oxides are called **greenhouse gases** because they trap the heat that is reflected from Earth's surface in the atmosphere, rather than allowing it to radiate into space. The rate at which these gases are entering the atmosphere exceeds the rate by which the natural carbon and nitrogen cycles can recycle them, thus they accumulate. It is the ever-increasing atmospheric concentration of these gases that has resulted in what is often called global warming. However, the term global climate change more accurately reflects the changes in patterns of wind, precipitation, and ocean temperatures that we are now experiencing.

Search This: EIA—Greenhouse gases

The most abundant greenhouse gas is CO_2. Since the onset of the industrial era, about 150 years ago, CO_2 levels have risen from 278 ppm (parts per million) to a current level of about 380 ppm (**figure 26.9**). Most of this can be attributed to the burning of fossil fuel, which releases about 7 billion tons of CO_2 each year. Changes in land use management, principally through deforestation, account for another 1.6 billion tons of CO_2 annually. To understand why these activities result in the release of CO_2, consider that fossil fuel (e.g., oil) is formed from decaying organic matter. This organic matter arose through millions of years of CO_2 fixation. When it is burned, the energy of combustion

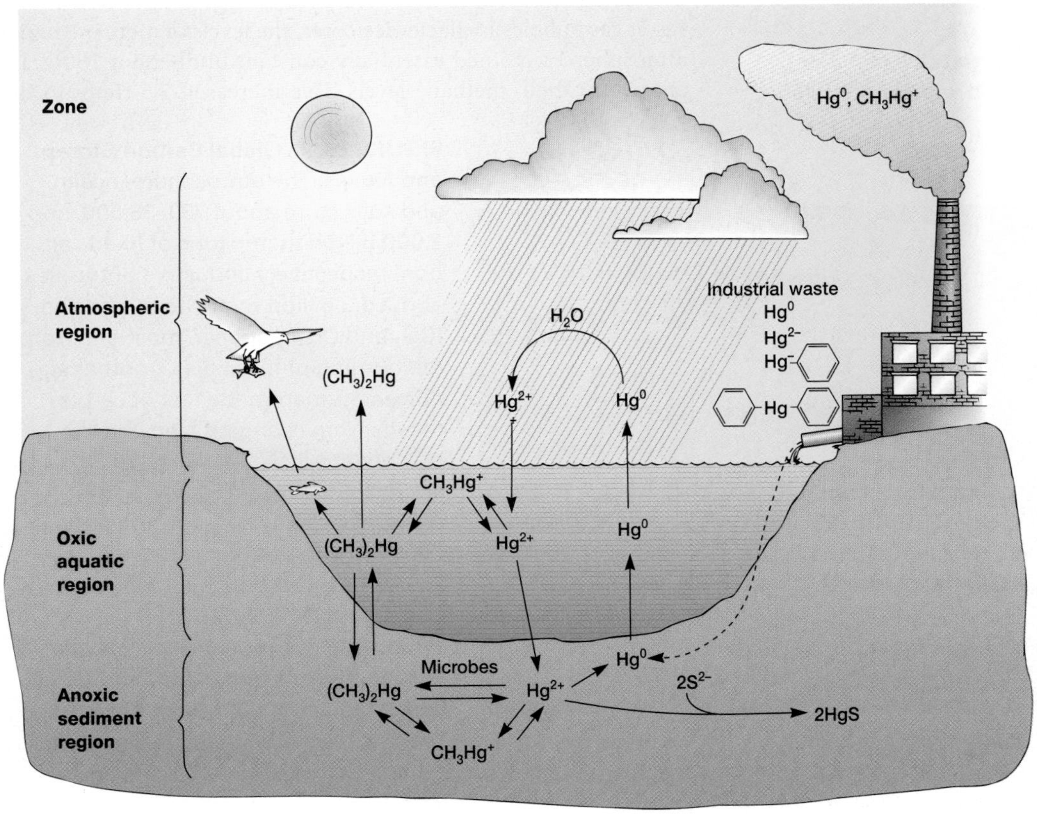

FIGURE 26.8 The Mercury Cycle. Interactions between the atmosphere, oxic water, and anoxic sediment are critical. Microorganisms in anoxic sediments, primarily *Desulfovibrio,* can transform mercury to methylated forms that can be transported to water and the atmosphere. These methylated forms also undergo biomagnification. The production of volatile elemental mercury (Hg^0) releases this metal to waters and the atmosphere. Sulfide, if present in the anoxic sediment, can react with ionic mercury to produce less soluble HgS.

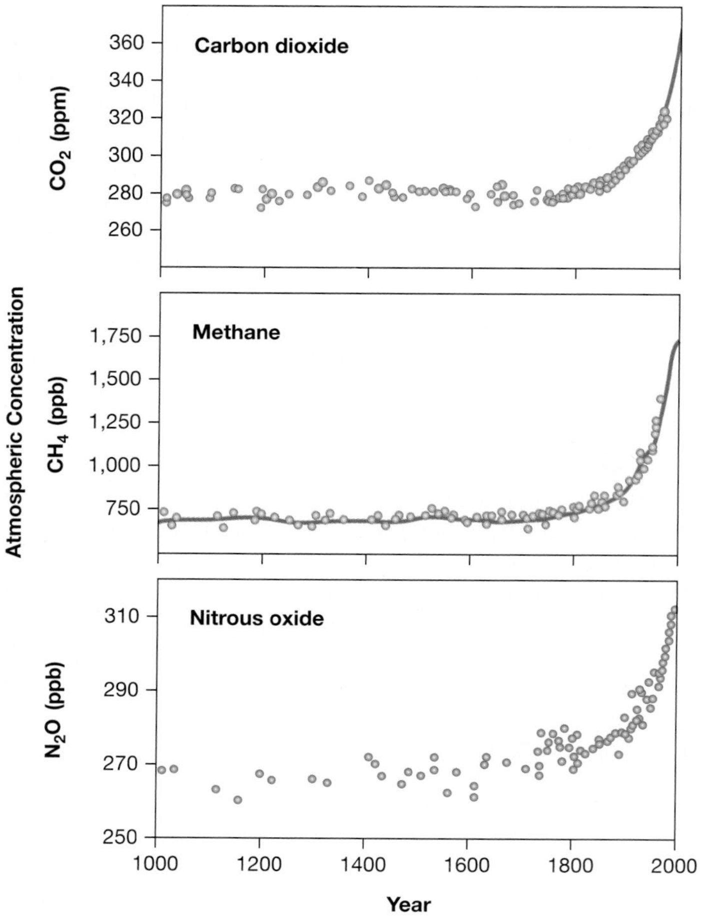

FIGURE 26.9 Global Atmospheric Concentrations of Greenhouse Gases. ppm = parts per million, ppb = parts per billion.

oxidizes this material back to CO_2 in a tiny fraction of the time it took to accumulate (**figure 26.10**).

The situation is different with deforestation. Forests are generally regarded as CO_2 "sinks" because the abundance of photoautotrophic plants pulls a great deal of CO_2 out of the atmosphere and converts it to biomass. Indeed, during the 1990s terrestrial ecosystems sequestered about 3 gigatons of carbon per year (a gigaton is 1 billion metric tons). Thus when forests are cut down for firewood or agricultural reasons, the Earth loses a carbon sink. While this increases the amount of atmospheric CO_2, some believe that other, remaining plants will assimilate the extra CO_2 and simply grow faster. An international team of scientists recently addressed this question by growing a variety of trees for several years under higher concentrations of CO_2. Indeed, they detected plant growth stimulation, but it was coupled to an increase in respiration by soil microbes. Recall that as respiration rates increase, so does the quantity of CO_2 released. Similarly, recent analysis shows that warming permafrost in subarctic regions has dramatically increased soil microbial respiration. A variety of geoengineering strategies to bury or otherwise remove CO_2 from the atmosphere have been proposed, but no clear solution is currently in sight.

Methane is a greenhouse gas of increasing concern because it has about 23 times the global warming potential than CO_2 will over the next century. This means that a single molecule of CH_4 released into the atmosphere has the same thermal retention capacity as 23 CO_2 molecules over the next 100 years. Sources of methane include ruminants, rice paddies, landfills, and even the methanogenic archaea that inhabit termite guts. Based on analyses of gas bubbles in glacier ice cores, the levels of methane in the atmosphere remained essentially constant until about 150 years ago. Since then, methane levels have increased 2.5 times to the

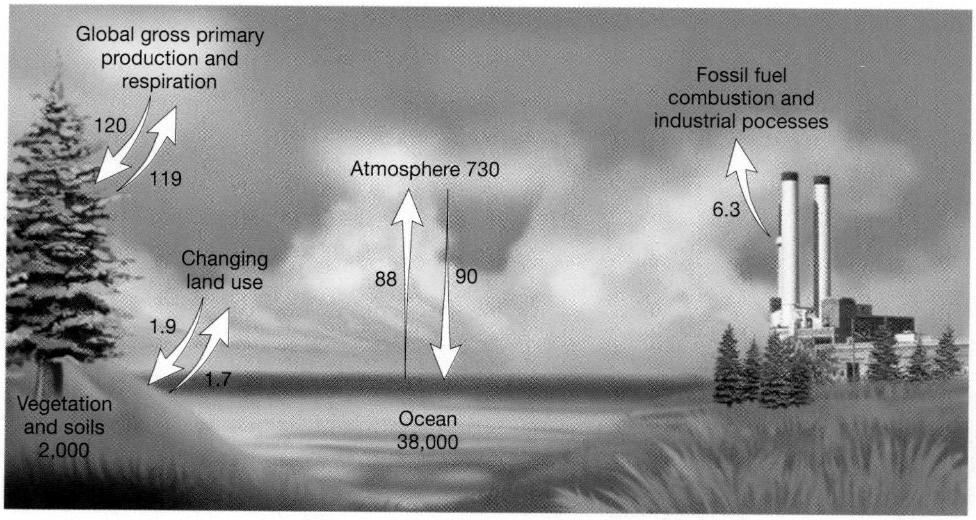

FIGURE 26.10 Global Carbon Stores and Fluxes. The atmosphere, ocean, and soils store about 730, 38,000, and 2,000 billion metric tons of fixed carbon, respectively. Industry contributes about 6.3 billion metric tons of carbon (CO_2 and CH_4) per year, most of it from the burning of fossil fuel. Another significant human-made flux of carbon results from changing land use (e.g., deforestation). Natural carbon fluxes are indicated by broad white arrows. All carbon fluxes are measured in billion metric tons of carbon per year.

Figure 26.10 Micro Inquiry

What is the largest reservoir of carbon on Earth? What is the net flux of carbon (i.e., the difference between the flux to and from the atmosphere) that can be attributed to vegetation?

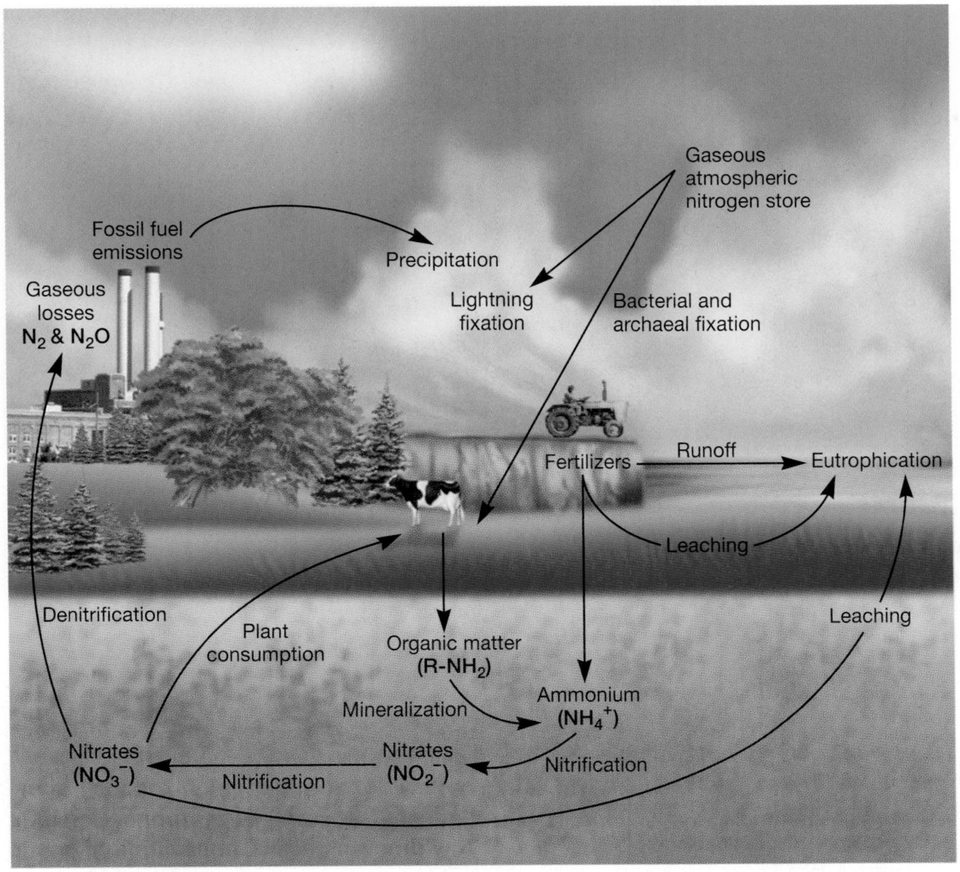

FIGURE 26.11 Natural and Human-Made Influences on the Nitrogen Cycle.

Figure 26.11 Micro Inquiry

What organisms benefit from nitrification?

present level of 1,730 parts per billion (ppb) (figure 26.9). Considering these trends, there is a worldwide interest in understanding the factors that control methane synthesis and use by microorganisms.

Search This: What's my carbon footprint?

To understand the chief source of the nitrogen oxide greenhouse gases NO and N_2O, collectively referred to as "Nox," we need to examine food production. The "green revolution" of the mid-twentieth century ensured that agricultural output, boosted by the application of human-made fertilizer, could keep pace with population growth. The manufacture of fertilizer is an energy-intensive process that uses hydrogen gas to reduce N_2 to NH_4^+ at high temperature and pressure. This process was first described by Fritz Haber and Karl Bosch in 1913, but it was not until the middle of the twentieth century that the Haber-Bosch reaction was used at an industrial level. Current fertilizer production is about 1.21×10^{14} g per year, or about 400 times greater now than it was in 1940. In contrast to the positive outcome of increased agricultural output, the addition of so much ammonium to the soil has altered the balance of the nitrogen cycle. What becomes of the extra NH_4^+? That which is not taken up by plants generally has one of two fates: runoff or nitrification followed by denitrification (**figure 26.11**). Ammonium runoff leaches into lakes and streams, frequently causing eutrophication—an increase in nutrient levels that stimulates the growth of a limited number of organisms, thereby disturbing the ecology of these aquatic ecosystems. By contrast, microbial nitrification can result in the oxidation of ammonium to more nitrate than can be immobilized by plants and microbes (organisms need a specific ratio of C:N:P). The process of denitrification converts this extra nitrate to N_2 and the reactive greenhouse nitrogen oxides, collectively known as Nox. This cycle of nitrification/denitrification fueled by NH_4^+ introduced as fertilizer is responsible for the highest N_2O levels in 650,000 years (figure 26.9).

What are the consequences of disrupting the carbon and nitrogen cycles? Global warming is the most obvious example. It is important to keep in mind that weather is not the same as climate. While North America has suffered some of the hottest summers on record in the last decade, a single day or week in July that is particularly hot is not, by itself, evidence of global climate change. Global climate change is measured over decades and includes many parameters such as surface temperature on land, sea, the atmosphere and troposphere, rates of precipitation, and frequency of extreme weather. Based on these analyses, the average global temperature has increased 0.74°C, and this rise is directly correlated with fossil fuel combustion to CO_2 (**figure 26.12**).

Search This: International Panel on Climate Change (IPCC)

In addition to generating and consuming greenhouse gases, microbes also respond to them. In fact, scientists question how global warming may change patterns of infectious disease outbreaks in humans and other animals. Just as weather is not the same as climate, a single outbreak of a tropical infectious disease in a temperate climate cannot be considered evidence of the effects of global climate change on infectious disease epidemiology.

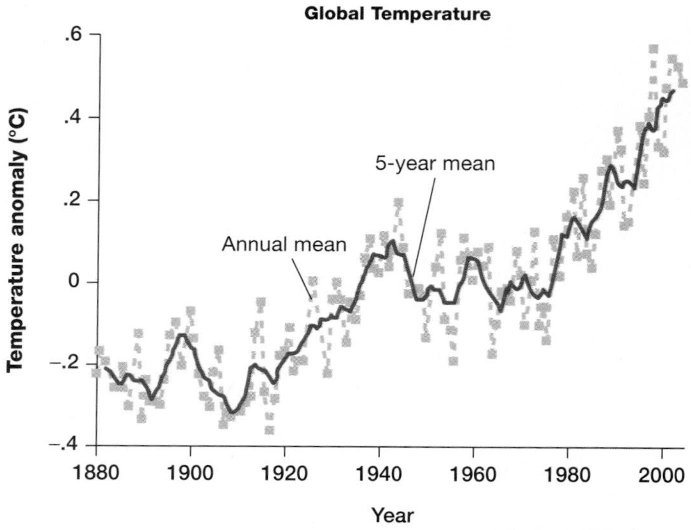

FIGURE 26.12 Global Annual-Mean Surface Air Temperature Change. Data is derived from the meteorological station network, Goddard Institute for Space Sciences, http://data.giss.nasa.gov/gistemp/graphs/.

FIGURE 26.13 Chytrid Fungi Appear to Be Responsible for the Extinction of Many Species of Harlequin Frogs. This species of Panamanian golden frog can still be seen, but many tree frog species have been eliminated as a result of fungal infection. The fungus (*Batrachochytrium dendrobatidis*) has expanded its range in response to warmer temperatures.

Figure 26.13 Micro Inquiry

Can you think of other vector-borne infectious microbes whose range may expand due to an increase in mean global temperatures?

Rather, patterns must be analyzed and the biology of the pathogen examined. For instance, ecologists have documented the extinction over the last 20 years of 67% of the 110 species of harlequin frogs (*Atelopus*) native to tropical America (**figure 26.13**). Some believe that the pathogenic chytrid fungus (*Batrachochytrium dendrobatidis*) found to be responsible for many of these deaths has expanded its range in response to warmer temperatures. This is not the first account of a pathogen thought to be responding to climate change. Pine blister rust, caused by the fungus *Cronartium ribicola*, is on the rise in mountainous regions of North America because its vector, the mountain pine beetle (*Dendroctonus poderosae*), is now able to complete its life cycle in one, rather than two, years because of warmer temperatures. Epidemiologists are currently monitoring the geography of infectious disease outbreaks in an effort to model potential patterns on a warmer planet. ◂◂ Chytridiomycota *(section 24.2)*

1. List some greenhouse gases. Discuss their origins.
2. Discuss the possible role of forests in the control of CO_2.
3. How do you think changes in the nitrogen cycle caused by fertilization influence the carbon cycle?
4. What tropical diseases that you are familiar with might increase their range due to global climate change? What other factors, besides climate change, do you think might be responsible for the increased incidence of tropical diseases in temperate climates?

Summary

26.1 Biogeochemical Cycling

a. Microorganisms—functioning with plants, animals, and the environment—play important roles in nutrient cycling, which is also termed biogeochemical cycling. Assimilatory processes involve incorporation of nutrients into the organism's biomass during metabolism; dissimilatory processes involve the release of nutrients to the environment after metabolism.

b. Biogeochemical cycling involves oxidation and reduction processes, and changes in the concentrations of gaseous cycle components, such as carbon, nitrogen, phosphorus, and sulfur, can result from microbial activity (**figures 26.1–26.8**).

c. Major organic compounds used by microorganisms differ in structure, linkage, elemental composition, and susceptibility to degradation under oxic and anoxic conditions. Lignin is degraded only under oxic conditions, a fact that has important implications in terms of carbon retention in the biosphere.

26.2 Global Climate Change

a. Microbes have evolved slowly over time; this has given rise to the biogeochemical cycles that sustain life on Earth. However, since the beginning of the twentieth century, the rate at which CO_2, CH_4, and nitrogen oxides have been released into the atmosphere has outpaced the rate at which they can be recycled. Thus these greenhouse gases are accumulating in Earth's atmosphere (**figures 26.9–26.11**).

b. The increase in greenhouse gases is correlated with an increase in the global annual-mean surface air temperature (**figure 26.12**).

c. Global climate change has far-reaching implications for infectious disease ecology. This has already been documented for diseases that affect plants and animals.

Critical Thinking Questions

1. Examine the carbon cycle shown in figure 26.1. What do you think are some major biogenic sources of CO_2 emission into the atmosphere? What are the major biogenic sinks of atmospheric CO_2?

2. Examine figure 26.11. Farmers in Argentina use a crop/cattle rotation system in which cattle graze on pastures for about 5 years, then are moved to a different area and crops are planted. Crops are grown for 3 years, and then the cattle are rotated back and the field reverts to a pasture. This dramatically reduces the amount of nitrogen fertilizer that must be added when crops are grown. Why is this the case? Why is it argued that this practice also minimizes fossil fuel emissions?

3. A bacterium isolated from sewage sludge was recently found to be capable of anoxygenic photosynthesis using nitrite as the electron donor and converting it to nitrate. Compare this form of nitrification to that which is well characterized. By removing nitrite from the environment for reductant, what other processes within the nitrogen cycle might this bacterium be influencing?

   Read the original paper: Griffen, B. M., et al. 2007. Nitrite, an electron donor for anoxygenic photosynthesis. *Science*. 316:1870.

4. Heterotrophic microbes contribute two forms of organic matter to ecosystems: the degradation products of the substrates they consume and themselves. That is, bacteria, archaea, and fungi are consumed by particle-ingesting consumers. While the notion is well established that heterotrophic microbes are food items in aquatic ecosystems, this has been more difficult to document in soil, marsh, and estuarine environments. A recent study found that bacteria, rather than fungi, were the most actively growing microbes in a marsh characterized by submerged, decaying plant material. These scientists were able to measure how much carbon was immobilized by bacteria and fungi, and the amount of CO_2 each type of microbe emitted.

   Why do you think the researchers were a bit surprised that bacteria were more dominant than fungi in this environment? If a complete carbon budget for this ecosystem were to be developed, what other microbes would have to be analyzed? Explain their potential contributions to the carbon budget.

   Read the original paper: Buesing, N., and Gessner, M. O. 2006. Benthic bacterial and fungal productivity and carbon turnover in a freshwater marsh. *Appl. Environ. Microbiol.* 72:596.

Concept Mapping Challenge

Construct a concept map using the words listed below and using your own linking words.

C cycle	S cycle	N cycle
Dissimilatory reduction	Sulfur respiration	N_2 fixation
Aerobic	Anaerobic	Anammox
Assimilatory reduction	Methanogenesis	Denitrification
Fermentation	Nitrification	

Learn More

Learn more by visiting the text website at www.mhhe.com/willey8, where you will find a complete list of references.

27

Methods in Microbial Ecology

This hot spring appears green due to the abundant growth of cyanobacteria. Microbial ecologists seek to determine the species composition of such a community, how members of the community interact, and the physical, chemical, and biological factors that influence growth.

CHAPTER GLOSSARY

axenic culture A pure culture, consisting of only one strain of microorganism.

community An assemblage of different types of organisms or a mixture of different populations in the same ecosystem.

enrichment culture The growth of a specific microbe or group of physiologically similar microbes from natural samples by including selective features (light, temperature, nutrients) to promote the growth of the desired microorganisms while counter-selecting other microbes that may be found in the same environment.

fluorescent in situ hybridization (FISH) A technique for identifying certain genes or organisms in which specific DNA fragments are labeled with fluorescent dye and hybridized to the chromosomes of interest.

great plate count anomaly The discrepancy between the number of viable microbial cells and the number of colonies that can be cultivated from the same natural sample.

microcosm A simplified, model ecosystem designed to simulate biological activities at a larger scale.

microelectrodes Thin electrodes that can measure a variety of parameters (e.g., pH, O_2, H_2) at very high spatial resolution.

most probable number A serial dilution test designed to estimate the number of target microbes in a sample.

multiplex PCR The identification of more than one type of PCR product in a single reaction.

phylotype A term applied to a microorganism that has been classified by its genetic uniqueness, usually by SSU rRNA sequences, although sometimes specific protein-coding gene sequences are used.

population An assemblage of organisms of the same type within a given ecosystem.

In previous chapters, microorganisms usually have been considered as isolated entities. However, all microorganisms exist in **communities** consisting of some combination of viruses, bacteria, archaea, molds, and protists. Recent estimates suggest that most microbial communities have between 10^{10} to 10^{17} individuals representing at least 10^7 different taxa. How can such huge groups of organisms exist and, moreover, survive together in a productive fashion? To answer this question, one must know which microbes are present and how they interact—that is, one must study microbial ecology. Mary Ann Moran, a microbial ecologist at the University of Georgia, has described the current approach to studying microbial ecology as an interdependent series of inquiries that involves traditional, laboratory-based analyses, metagenomics, and in situ (Latin, in place) biogeochemical assessments (**figure 27.1**). The application of these techniques has brought an explosion of recent advances and generated a new appreciation of the vast diversity of microbes and their role in biogeochemical cycling. ◀◀ *Metagenomics (section 16.8)*

In this chapter, we begin our consideration of this multidisciplinary field by presenting some of the advances that have been developed to coax previously uncultured microbes into pure culture in the laboratory. This underscores the fact that although culture-independent techniques are tremendously important, the "gold standard" remains the capacity to study a microbe in the laboratory. We next present an overview of some of the more important tools and techniques used to assess microbial diversity. We then turn our attention to the methods used to measure the activities of microbes in nature. This chapter thus provides the foundation for a more detailed review of microbial communities in nature (chapters 28 and 29).

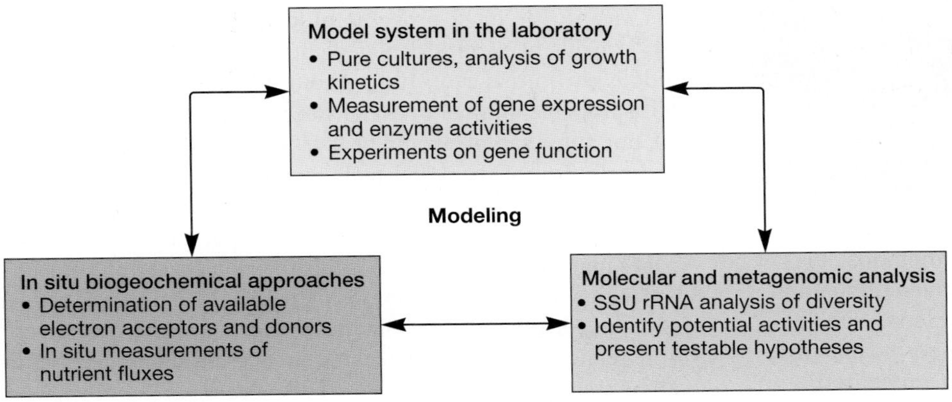

FIGURE 27.1 Microbial Ecology Is an Interdisciplinary Science. It relies on the interplay between a variety of approaches, which can be broadly classified as shown here.

Table 27.1 Estimates of the Percent Cultured Microorganisms from Various Environments

Environment	*Estimated Percent Cultured*
Seawater	0.001–0.100
Freshwater	0.25
Mesotrophic lake	0.1–1.0
Unpolluted estuarine waters	0.1–3.0
Activated sludge	1–15
Sediments	0.25
Soil	0.3

Source: Cowan, D. A. 2000. Microbial genomes—the untapped resource. *Tibtech* 18:14–16, table 2, p. 15.

27.1 Culturing Techniques

As we have emphasized throughout the text, it is now well understood that the vast majority of microbes have not been grown under laboratory conditions (**table 27.1**). The discrepancy between the number of microbial cells observed by microscopic examination and the number of colonies that can be cultivated from the same natural sample has been called the **great plate count anomaly (GPCA).** At least two reasons account for the GPCA. First, some of the organisms observed upon microscopic examination of natural samples may be truly nonviable. Alternatively, the right conditions for their growth in the lab have not yet been created. This has led to the description of such potentially viable microbes as being "nonculturable." A microbe is deemed **viable but nonculturable (VBNC)** if it fails to be cultivated but shows motility or the presence of dividing cells when directly observed, is known to grow in a natural environment, or is determined to be viable when stained with dyes that discriminate between live and dead cells (**figure 27.2**).

When starting with a mixed assemblage of microbes, it is often best to increase, or enrich for, the microbe of interest. This is the case if one wants to isolate specific groups of microbes or to attempt to search for organisms with new capabilities. **Enrichment culture** techniques are based on the expansion of the microenvironment to allow abundant growth of a microorganism formerly restricted to a small ecological niche. At the same time, the growth of other microbial types is inhibited. This approach plays a central role in finding new microbes. The success of an enrichment culture depends on a good understanding of the specific niche the microbe of interest inhabits and the physiological features that set that microbe apart from others. A simple example would be the isolation of anoxygenic photoautotrophs such as green sulfur bacteria. One might incubate a natural sample thought to harbor these bacteria under anoxic conditions with a source of reduced sulfur, carbonate (for CO_2), and light of a long wavelength. The absence of organic carbon selects against heterotrophs, while the anoxic conditions limit the growth of cyanobacteria and eukaryotic oxygenic photosynthetic microorganisms. ◀◀ *Photosynthetic bacteria (section 19.3)*

One approach to estimating the number of microbes in a natural sample is the **most probable number (MPN)** technique. This can be used to quantify specific types of microbes in natural samples, enrichment cultures, or as is often the case, in food or water samples. The microbe of interest must be capable of growth in the laboratory as the MPN technique involves establishing serial, tenfold dilutions of the sample. Triplicate tubes of each dilution are usually established, although sometimes five or 10 replicates of each dilution are used for more accurate results. The theoretical basis of the MPN approach is that even if only one cell is inoculated into a tube, growth will occur. As shown in **figure 27.3**, after incubation at the appropriate temperature and for a specific duration, the tubes are examined for growth. Ideally, beyond a certain dilution, no cells are introduced into the medium and the tubes fail to show growth. This is seen in the 10^{-5} and 10^{-6} triplicates in figure 27.3. In this case, the first dilution without growth would be marked as the last dilution in a set of three that includes the two dilutions that precede it. The number of tubes in the first set (10^{-3}, in this case) and the middle set

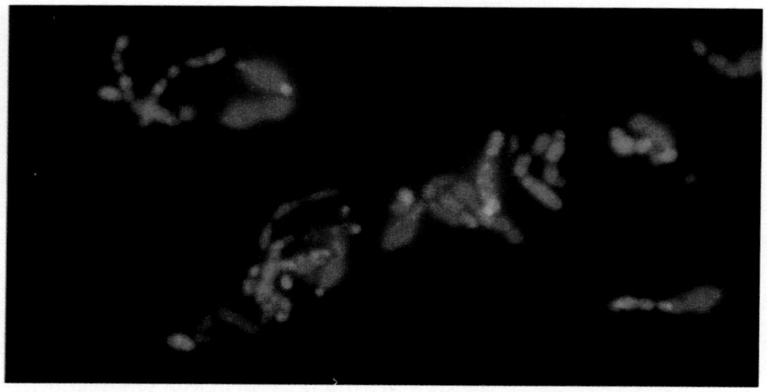

FIGURE 27.2 Assessment of Microbial Viability by Use of Direct Staining. By using differential staining methods, it is possible to estimate the portion of cells in a given population that are viable. In the LIVE/DEAD *Bac*Light Bacterial Viability procedure, two stains are used: a membrane-permeable green fluorescent, nucleic acid stain, and propidium iodide (red), which penetrates only cells with damaged membranes. In this *Bifidobacterium* culture, living cells stain green, while dead and dying cells stain red.

is recorded. In our example, we see that this gives us a pattern of three tubes with growth in the 10^{-3} dilution tubes, one tube with growth in 10^{-4}, and zero tubes in 10^{-5}. A table based on statistical and theoretical considerations is consulted (figure 27.3*b*), and we see that the pattern 3-1-0 gives a MPN of 0.43. What does this mean? The most probable outcome is that an average of 0.43 organisms were inoculated into each of the tubes of the middle set (in this case, 10^{-4}). From this, we conclude that the most-probable number of organisms per 1 milliliter of the original, undiluted sample would be 0.43×10^4 or 4.3×10^3.

Inspection of the MPN table in figure 27.3 reveals that the results do not always show the expected lack of growth only in the lowest dilution tubes. Why would this be the case? The microbes may not have been randomly distributed within the sample, or they may grow in clumps or chains. When this happens, less than ideal results are obtained, and it may be best to use more replicates or a plating technique so that individual colonies can be counted.

Search This: MPN/FDA

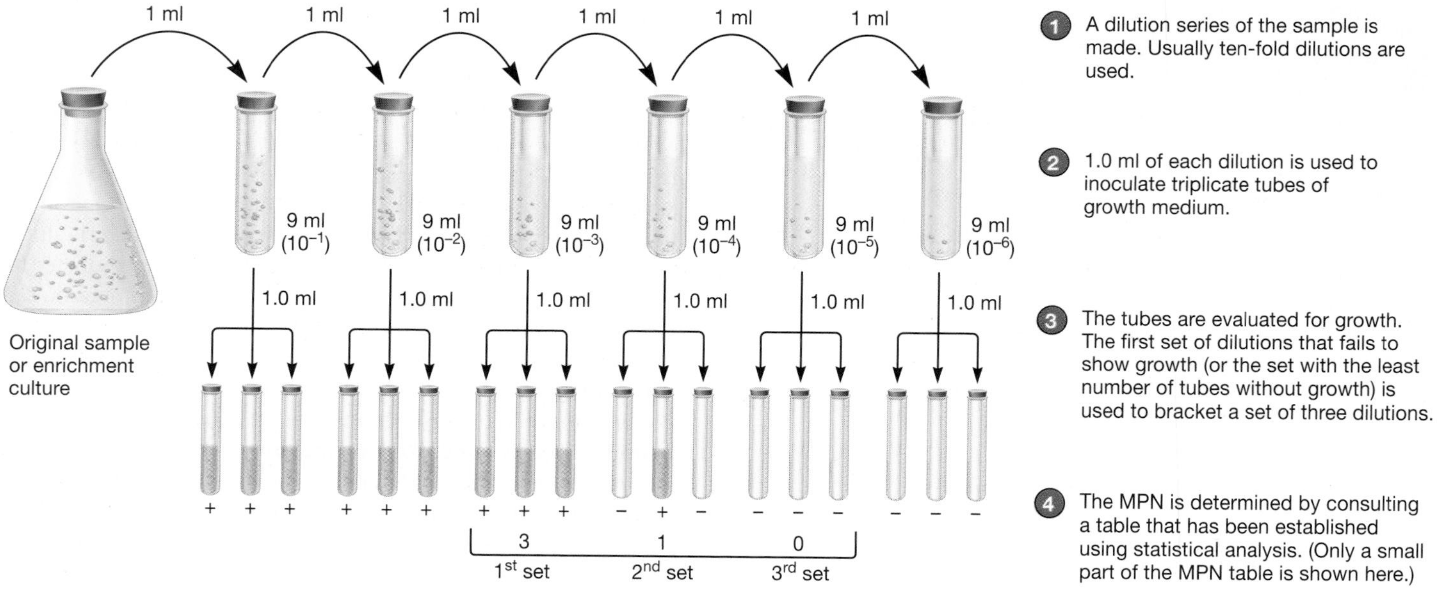
1 ml
9 ml (10⁻¹)
9 ml (10⁻²)
9 ml (10⁻³)
9 ml (10⁻⁴)
9 ml (10⁻⁵)
9 ml (10⁻⁶)
1.0 ml
Original sample or enrichment culture
+ + + + + + + + + − + − − − − − − −
3
1
0
1st set
2nd set
3rd set
1 A dilution series of the sample is made. Usually ten-fold dilutions are used.
2 1.0 ml of each dilution is used to inoculate triplicate tubes of growth medium.
3 The tubes are evaluated for growth. The first set of dilutions that fails to show growth (or the set with the least number of tubes without growth) is used to bracket a set of three dilutions.
4 The MPN is determined by consulting a table that has been established using statistical analysis. (Only a small part of the MPN table is shown here.)

(a) Procedure for MPN analysis

Number of positive tubes in:			MPN
1st set	2nd set	3rd set	
3	0	3	0.95
3	1	0	0.43
3	1	1	0.75

MPN equals 4.3×10^3 microbes per 1.0 ml of sample

(b) MPN table

FIGURE 27.3 Most Probable Number Technique.

Clearly, the MPN technique is only valid if the microbe of interest can be grown in the laboratory. Although this is not true of most microorganisms, it would be premature to claim that microbes currently unculturable will never be cultivated. Traditional growth media are now giving way to novel media formulations. One approach is to use unusual electron donors and acceptors. For instance, a novel isolate from Mono Lake (California) sediments was cultivated only when anoxic conditions with arsenite as the electron donor and nitrate as the electron acceptor were provided. It has also become clear that growing "unculturable" microbes requires patience. Microbiologists are most familiar with doubling times in the order of hours or, at the most, just a few days, when in fact, this is not always the case. For example, newly described anaerobic soil isolates were cultured only when solid cultures were incubated over a three-month period. Indeed, 20% of the colonies did not appear until after the second month of incubation. Furthermore, microbiologists have been trained to equate turbidity with growth. This is not always true, as exemplified by the cultivation of *Pelagibacter ubique*—a marine α-proteobacterium that is arguably the world's most abundant microbe (*see p. 681*). After years of unsuccessful attempts to culture the bacterium, it was finally coaxed into growing in dilute, sterilized seawater supplemented with a small amount of phosphate and ammonium. Unlike most cultured microbes, which grow to densities exceeding 10^8 cells per milliliter, *P. ubique* stops replicating after reaching about 10^6 cells per milliliter. Curiously, this is the density at which they are generally found in nature, suggesting that natural factors in seawater somehow control population growth.

A variety of new culture techniques have been applied that address specific needs of microorganisms growing in diverse environments. We discuss two of these "high-throughput" methods. Interestingly, both techniques use water collected from the natural environment, rather than laboratory media formulations. The **extinction culture technique** is so named because natural samples are first microscopically examined and then diluted to a density of 1 to 10 cells, that is, diluted to extinction, much like in the MPN assay. Multiple 1-milliliter cultures are established in 48 well microtiter dishes. After incubating for the desired time under appropriate conditions, the samples are stained and examined microscopically to screen for growth. In many cases, cultures that do not appear turbid will nonetheless show signs of viable microbial growth and can be transferred to fresh medium. It is important that the initial culture has fewer than 10 cells; additional transfers and dilutions can ensure that a culture arising from a single cell (i.e., clonal growth) eventually is obtained. This technique and variations of it can be used in any microbiology lab because it does not require special, high-tech equipment.

In the second example, the **microdroplet culture** method, cells from mixed assemblages are encapsulated in a gel matrix. The matrix is then emulsified to generate gel "microdroplets;" each microdroplet contains a single cell. The gel matrix is sufficiently porous so that when microdroplets are incubated together in continuous cultures, nutrients are delivered to the encapsulated microbes. In addition, waste material is removed, and metabolites can be exchanged. The fact that microbes can exchange small molecules is an important feature, as this may be needed for growth. After incubation, the microdroplets are examined by flow cytometry, a method in which fluorescently labeled cells are sorted and enumerated by laser beams. In this way, microbeads containing colonies due to clonal growth of the initial inoculum are separated from free-living cells and empty microdroplets. Microbial cells within microbeads can then be isolated for further culture, and microscopic and phylogenetic analysis. ⏮ *Continuous culture of microorganisms (section 7.5);* ⏭ *Flow cytometry (section 35.3)*

What if one knows very little about the physiological requirement of a microbe but has obtained a 16S rRNA sequence from analysis of a microbial community? In this case, the microbe could be living in one of many microniches, necessitating the use of multiple culture and incubation conditions to discover what it requires. A group of microbiologists at Michigan State University has developed a technique called "**plate wash polymerase chain reaction**" to address this needle-in-a-haystack problem (**figure 27.4**). In this method, a natural sample is used to inoculate multiple Petri plates, each with different growth conditions (e.g., different electron donors or acceptors, temperature, levels of aeration, even concentrations of CO_2). After growth has occurred, colonies from each plate incubated under the same condition are pooled, and the polymerase chain reaction (PCR) is performed using primers designed to amplify DNA from the desired microbe. Although this does not identify the individual colony that generated the nucleotide sequence, it does identify the conditions under which the microbe grew. Additional cultures can then be established using these newly discovered growth parameters. These cultures may be done in microtiter plates so that each well, which contains medium, can be inoculated with a single colony. ⏮ *Polymerase chain reaction (section 15.2)*

It is often not possible to obtain an **axenic culture** (i.e., a pure culture), but single-cell isolations enable the examination of uncontaminated specimens. This is accomplished by **optical tweezers**—a laser beam that drags a microbe away from its neighbors—coupled with micromanipulation. With a micromanipulator, a desired cell or cellular organelle is drawn up into a micropipette after direct observation (**figure 27.5**). Once the microbe is isolated, PCR amplification of the DNA from the individual cell provides sequence data for phylogenetic analysis. In addition, it is now possible to sequence almost the entire genome with DNA extracted from a single microorganism. ⏮ *Single-cell genome sequencing (section 16.2)*

Consideration of microbial ecology on the scale of the individual cell has led to important ecological insights. It is now evident that there is surprising heterogeneity in what have been assumed to be homogenous microbial populations. Cells of a genetically uniform population do not always have similar phenotypic attributes. This gives rise to the phenomenon of phenotypic or population heterogeneity. For instance, in a population of starving *Bacillus subtilis* cells, some cells will secrete a toxin to kill other, genetically identical cells prior to committing to endospore formation. Microbial ecology on the scale of the individual cell is important in understanding microbial processes in complex environments and disease processes.

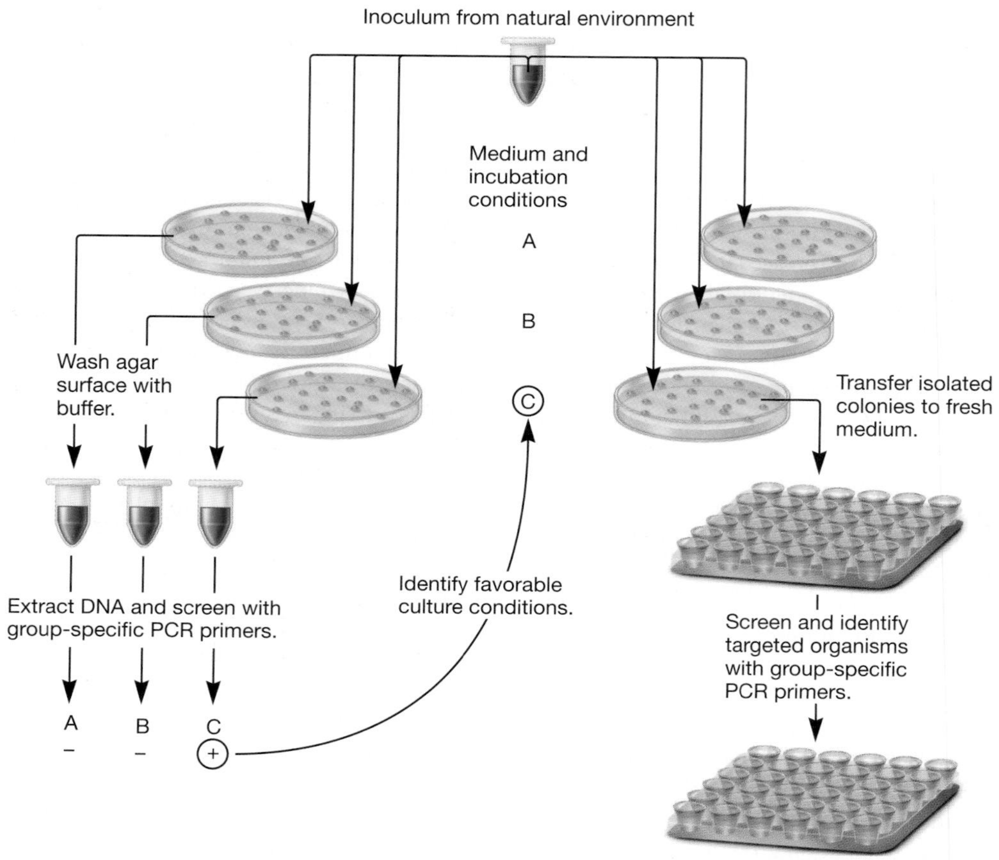

Inoculum from natural environment
Medium and incubation conditions
A
B
C
Wash agar surface with buffer.
Transfer isolated colonies to fresh medium.
Extract DNA and screen with group-specific PCR primers.
Identify favorable culture conditions.
Screen and identify targeted organisms with group-specific PCR primers.
A
–
B
–
C
+

FIGURE 27.4 Plate Wash Polymerase Chain Reaction. The method identifies the culture conditions that support the growth of a desired microbe.

Figure 27.4 Micro Inquiry

What are the advantages of a pure culture of microbes as compared to culture-independent analysis?

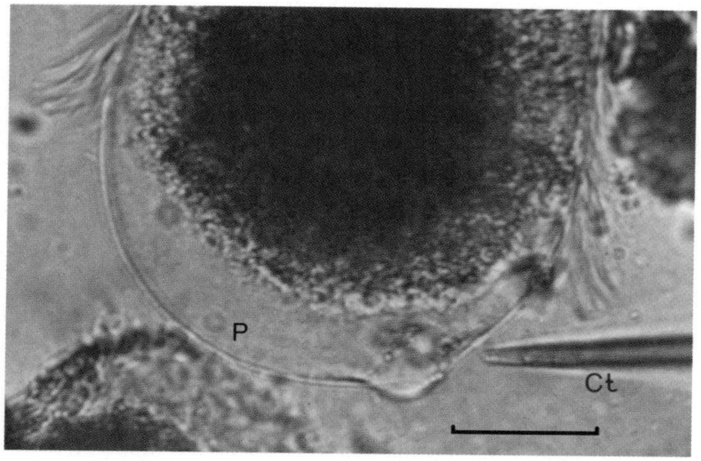

P
Ct

FIGURE 27.5 Micromanipulation for Isolation of Single Cells. Recovery of an endosymbiotic mycoplasma from a single cell of the protist *Koruga bonita* by micromanipulation (bar = 10 μm). P, protist; Ct, capillary tube.

1. Why might a microbe be considered VBNC?
2. What is an enrichment culture? Describe an enrichment culture designed to isolate nitrogen-fixing soil bacteria.
3. Compare the rationale upon which the extinction culture technique is based with that of the microdroplet approach.

27.2 Assessing Microbial Diversity

Microbial ecologists study natural microbial communities that exist in soils or waters, or in association with other organisms, including humans. Regardless of the habitat they study, these scientists seek to assess microbial diversity: How many microbes are present? What genera, species, or ecotypes are represented in the ecosystem? Community dynamics are also explored: How do these microorganisms make a living? How are carbon sources, electron donors and acceptors allocated among microbes? A microbial **population** is defined as a group of microorganisms within an ecosystem that are similar. They may be the same species, but a microbial population may also be defined by physiological activity. For instance, a microbial ecologist might be interested in the population of microbes that performs denitrification in a soil. This would most likely include a diverse collection of bacterial species.

Traditionally, microbial populations have been studied by obtaining isolates in axenic culture. This has generated a great deal of insight into the microbial morphology, physiology, and genetics of a variety of organisms. Unfortunately, until it was realized that only a small fraction (about 1%) of microbes have been cultured in the laboratory, misunderstandings arose in terms of the degree of community diversity and the relative importance of specific genera. While it is still widely recognized that an axenic culture is the "gold standard" for microbial analysis, most microbiologists agree that to understand natural microbial populations, one must also consider the diversity of life forms found within each specific niche. Thus as we as discuss the examination of microbial populations, we review both culture- and nonculture-based approaches.

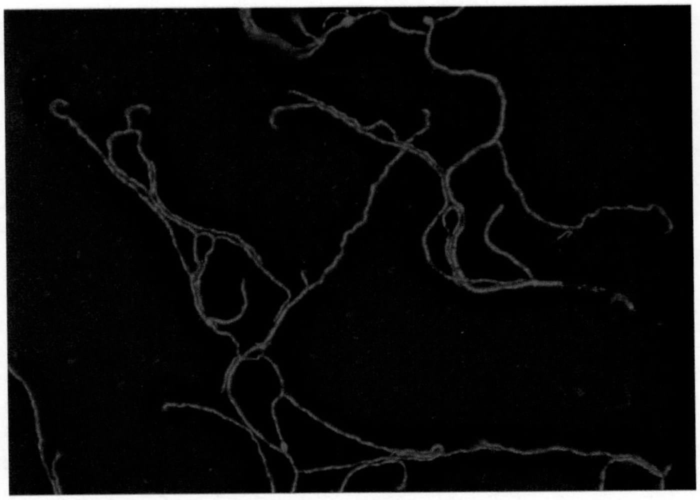

FIGURE 27.6 DAPI Stains Nucleic Acids. This image shows filaments of the actinomycete *Streptomyces coelicolor.* Because this microbe makes hyphae with incomplete septae, the chromosomes (and thus DAPI staining) are found throughout the filaments. Scale bar = 5 μm.

Staining Techniques

The most direct way to assess microbial populations and community structure is to observe microbes in nature. This can be carried out in situ using immersed slides or electron microscope grids placed in a location of interest, which are then recovered later for observation. Samples taken from the environment often are examined in the laboratory using classical cellular stains, fluorescent stains, or fluorescent molecular probes. The fluorescent stain DAPI (4′, 6-*dia*mido-2-*p*henyl*i*ndole) is commonly used to visualize microbes in environmental, food, and clinical samples (**figure 27.6**). It specifically labels nucleic acids (i.e., both DNA and RNA), and little sample preparation is required prior to its application. Thus DAPI staining is a convenient way to enumerate all the microbes in a specimen.

However, often a microbiologist will want to enumerate a specific genus or type of microbe. In these cases, a technique called ***f*luorescent *in s*itu *h*ybridization (FISH)** can be used to identify microbes. Here, fluorescently labeled oligonucleotides (small pieces of DNA) that are known to be specific to the microbe of interest are used to label natural samples (**figure 27.7**). The oligonucleotides are also called **probes**. Sequences within the gene encoding small subunit (SSU) ribosomal RNA (16S for bacteria and archaea, 18S for eukaryotes) are commonly used probes. As discussed in chapter 17, certain sequences within rRNA are unique to different microbial genera (*see figure 17.3*). When hybridized to the rRNA of the target organisms, the probe fluoresces. This can be detected by **epifluorescence microscopy,** in which the microscope is fitted with light filters that enable the excitation of the specimen with a specific wavelength. The fluorescent tag then emits light of a longer wavelength. Alternatively, individual cells can be separated and counted (but not visualized) by flow cytometry. ⏮ *Fluorescence microscope (section 2.2);* ⏭ *Flow cytometry (section 35.3)*

Depending on the probe used, FISH can be specific for species, strain, or ecotype. This has made it a popular tool in clinical diagnostics and food microbiology, as well as microbial ecology. For example, FISH was used to discover the association between the archaea *Nanoarchaeum equitans* and *Ignicoccus* (**figure 27.8**). Two fluorescent tags—one green, the other red—were attached to 16S rRNA probes, each having a different genus-specific nucleotide sequence. In this way, it was found that *N. equitans,* which is only 0.4 μm in diameter, grows on the larger archaeon as an obligate symbiont. Thus the combination of direct observation and molecular probes made it possible to document the lifestyle of this unusual archaeal symbiont and extend our understanding of archaeal diversity. ⏮ *Archaeal taxonomy (section 18.1);* ⏭ *Identification of microorganisms from specimens (section 35.2)*

The application of FISH to a variety of natural habitats has shown that the fluorescent signal is sometimes not bright enough to be detected microscopically. This is because microbes living in these environments often grow quite slowly, so each cell has relatively few ribosomes. A clever modification of the FISH technique has been developed to amplify the signal produced by each cell. The trick is to label the fluorescent probe (i.e., label the label) with an enzyme that, when exposed to substrate, makes lots of fluorescent product (**figure 27.9**). This is called *ca*talyzed *r*eported *d*eposition-FISH, or simply **CARD-FISH.** The enzyme attached to the oligonucleotide is usually horseradish peroxidase (HRP). After the oligonucleotides are hybridized to the sample, the HRP substrate tyramide is added. Tyramide is oxidized by HRP and the level of fluorescence is amplified. In this way, a single oligonucleotide bound to just one molecule of rRNA can generate hundreds of fluorescent signals.

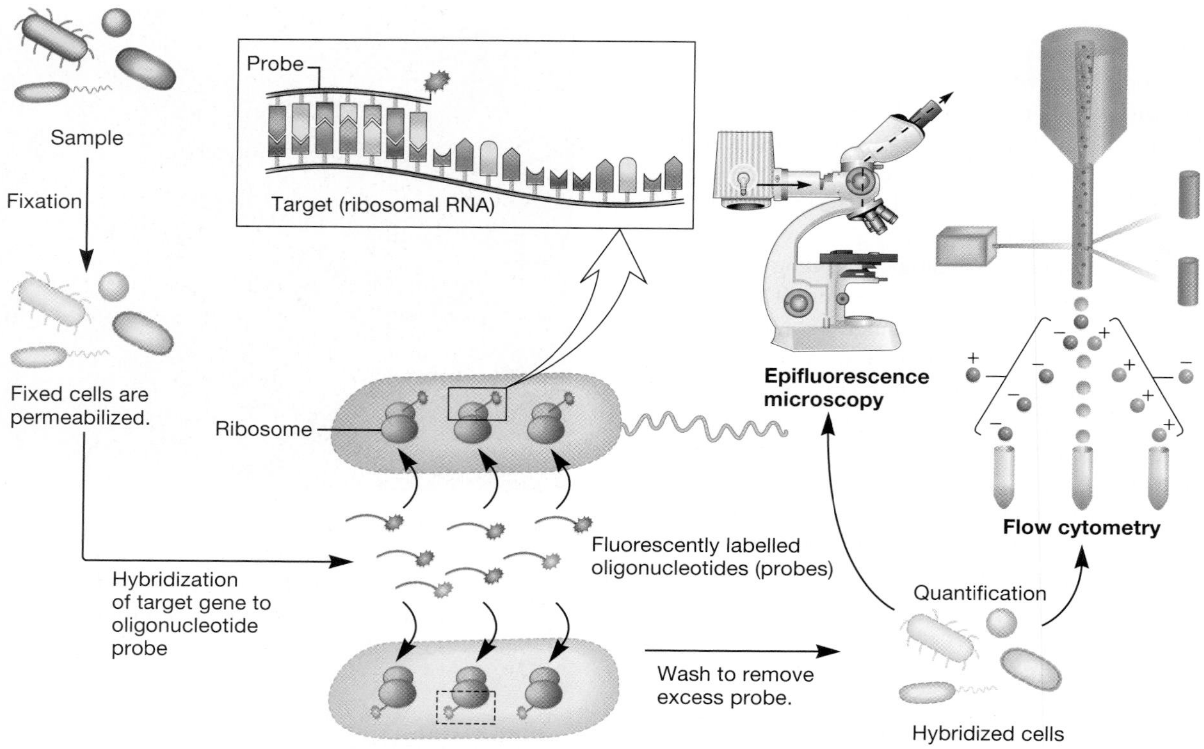

FIGURE 27.7 Fluorescence in Situ Hybridization (FISH).

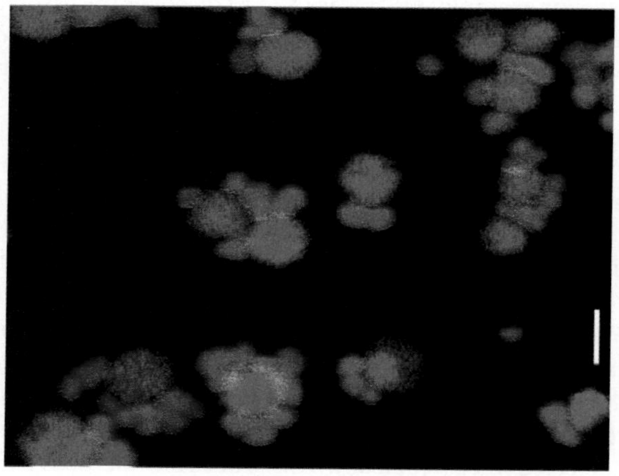

FIGURE 27.8 The Use of FISH to Study Microbial Populations. In this study, fluorescently tagged 16S rRNA probes were used to study the physical relationships between two archaea: a larger host, a member of the genus *Ignicoccus* (*green*), and a nanosized (~400 nm) hyperthermophilic symbiont named *Nanoarchaeum equitans* (*red*). Bar = 1 µm.

Figure 27.8 Micro Inquiry

Under what circumstances would one use epifluorescence microscopy, rather than flow cytometry, and vice versa?

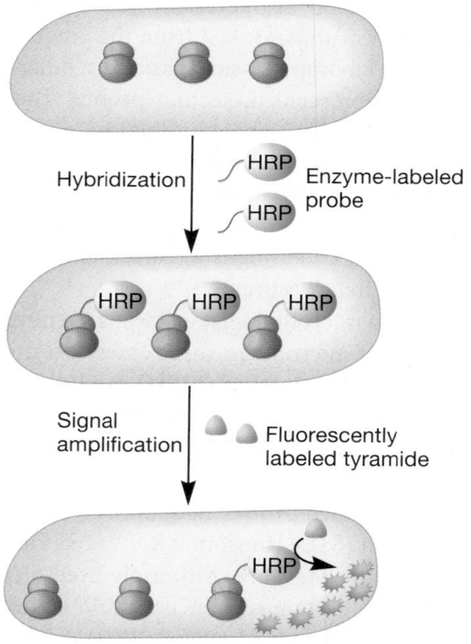

FIGURE 27.9 Catalyzed Reported Deposition-Fluorescence in Situ Hybridization (CARD-FISH). This technique is designed to intensify the fluorescent signal produced during FISH. This is accomplished by attaching an enzyme (horseradish peroxidase, HRP) to the oligonucleotide probe. Once the probe is bound, the HRP substrate tyramide is added. Oxidation of tyramide by HRP amplifies the fluorescent signal so that even if only a few probes have bound per cell, they can be detected.

Molecular Techniques

Because so few microbes have so far been grown in culture, the importance of molecular techniques to identify and quantify microorganisms is paramount. Nucleic acids are routinely recovered from soils and waters by direct extraction techniques. However, it has been found that the quality and quantity of DNA obtained can vary, depending on the method employed. A second problem, especially with muds and soils, is that one does not know the source of the DNA. Some of the DNA recovered by this approach may not be derived from living organisms. Third, the DNA may have been recovered from nonfunctional propagules such as fungal or bacterial spores, or other resting structures.

With these caveats in mind, microbial ecologists regularly use SSU rRNA analysis to identify the microbes that populate a community. SSU rRNA can be amplified by PCR directly from samples of soil, water, or other natural material (e.g., sputum or blood in a clinical setting). The use of specific primers that target either archaeal, bacterial, or eukaryotic SSU rRNA genes (or other conserved genes) enables researchers to use PCR to obtain a sufficient number of nucleic acid fragments for DNA sequencing. More recently, the use of quantitative or real-time PCR of the **internal transcribed spacer region (ITS)** between the 16S and 23S rRNA genes has also been used (**figure 27.10**). ◀◀ *Polymerase chain reaction (section 15.2)*

Despite the power of PCR, its use under these circumstances can be problematic because certain SSU rRNA genes are more readily amplified than others. The most obvious reason for this is relative abundance of a microbe—a microbe present in great numbers should generate more PCR product than a rare microbe—but this is not the only reason. Other factors include (but are not limited to) the G + C content of a microbe's genome, the amount of nucleic acid degradation that might have occurred, and the relative ease with which some microbes can be lysed and thus their nucleic acids extracted. It is widely understood that PCR amplification of SSU rRNA genes introduces a level of bias that is unquantifiable; that is, it cannot be measured. Thus while amplification and analysis of SSU rRNA genes from microbial assemblages can identify the genera present, it is not quantitative, so it cannot be used to show their relative abundance.

The observation that certain nucleotide templates are more readily amplified than others is termed **PCR bias.** It is very desirable to avoid PCR bias and detect rRNA molecules directly from the environment. This is especially true when one considers that rRNA molecules are transcribed only by metabolically active organisms and thus can indicate which microbes have a functional role in a given ecosystem. The isolation of rRNA molecules from natural samples is technically challenging, and most progress has been made in the direct extraction of these nucleic acids from aquatic habitats (e.g., lakes, rivers, oceans, blood). ◀◀ *Metagenomics (section 16.8); Techniques for determining microbial taxonomy and phylogeny (section 17.3)*

New variations of SSU rRNA analysis are continuously being introduced. One such modification enables the detection of multiple SSU rRNA genes—thus multiple microbial genera—in a single sample. This technique, called **hierarchical oligonucleotide primer extension (HOPE),** uses PCR primers representing the different 16S rRNA genes of interest. Each primer includes a string of adenine nucleotides (a poly-A tail) at the 5′ end, ensuring that each primer will amplify a product of a different size. This mixture of primers is added to the sample containing DNA extracted from the mixed assemblage of microbes. Once annealed to the target genes, synthesis from the 3′ end of the primer occurs as it would in any PCR, but fluorescently labeled nucleotides that will terminate the reaction are used. The unique size of each product enables their discrimination. For example in a soil sample, members of the genus *Bacillus* will be identified with a PCR product of one size, while *Streptomyces* spp. will be recognized by a PCR product of another. The identification of more than one type of PCR product in a single reaction is called **multiplex PCR.** HOPE was developed for use after the DNA in a sample has been amplified by PCR (i.e., for relatively high levels of template DNA) but is currently being modified so that it can be used for the direct detection of rRNA from the environment, thereby avoiding PCR bias.

Once nucleotide sequences are obtained, they can be compared with sequences from the SSU rRNA genes isolated from other microbes by using several different databases. The Ribosomal Database II is most frequently used as it currently has over 500,000 rRNA sequences. In this way, microbial ecologists can get a reasonable idea of the identity of the microbes that occupy a specific niche. Because whole organisms are not isolated and studied, it is said that specific **phylotypes** or unique SSU rRNA genes have been identified. Sometimes other genes besides SSU rRNA analysis are used. When protein-coding genes are used to define phylotypes, the genes must meet three important criteria: they need to be found in all known microbes, they need to have only one copy in the genome, and they need to rarely show evidence of horizontal gene transfer.

Search This: Ribosomal Database Project

As discussed in chapter 17, it is sometimes desirable to avoid DNA sequencing and obtain a DNA fingerprint instead. DNA fingerprints generate a barcodelike image of the population by separating the amplified rRNA gene fragments on an agarose gel. One microbial population can then be compared with another (differing in either time or space). However, sometimes a single pair of primers is used to amplify the same region of a SSU rRNA gene from a population of genomes. Despite different nucleotide sequences, the PCR products will have very similar molecular weights

FIGURE 27.10 Internal Transcribed Spacer Region Between rRNA Genes. The intergenic region between rRNA genes is sometimes amplified as a means of helping to quantify the relative abundance of rRNA obtained from microbes in natural samples.

and thus appear as a single band on an agarose gel. How can one separate such a collection of PCR products to obtain specific patterns of bands on a gel? **Denaturing gradient gel electrophoresis (DGGE)** can be used (**figure 27.11**).

DGGE relies on the fact that although the PCR fragments are about the same size, they differ in nucleotide sequence. Recall that the temperature at which double-stranded DNA becomes single stranded varies with the G + C content; that is, it varies with DNA sequence. DGGE takes advantage of the observation that DNA of different nucleotide sequences will denature at varying rates, although a gradient of chemicals that denature the DNA (usually urea and formamide), rather than temperature, is used. In this technique, a mixture of DNA fragments is placed in a single well in a gradient gel. As electrophoresis proceeds, fragments will migrate until they become denatured. What appeared to be a single DNA fragment (band) on a nongradient gel will resolve into separate fragments (multiple bands) by DGGE. Other techniques for generating DNA fingerprints of microbial populations include temperature gradient gel electrophoresis (TGGE; much like DGGE, except a temperature gradient is used), single-strand conformation polymorphism (SSCP), and terminal restriction fragment length polymorphism (T-RFLP). SSCP separates fragments based on the altered melting temperature that arises from differing secondary structure (i.e., internal base pairing) in single-stranded DNA. Because SSCP and other gel-based methods can be limited by poor gel-to-gel reproducibility, multiple banding of a single nucleic acid molecule that has more than one stable conformation, and comigration of several nucleic acid molecules, their popularity among microbial ecologists has declined in recent years. ◀◀ *Gel electrophoresis (section 15.3); Determining DNA sequences (section 16.1)*

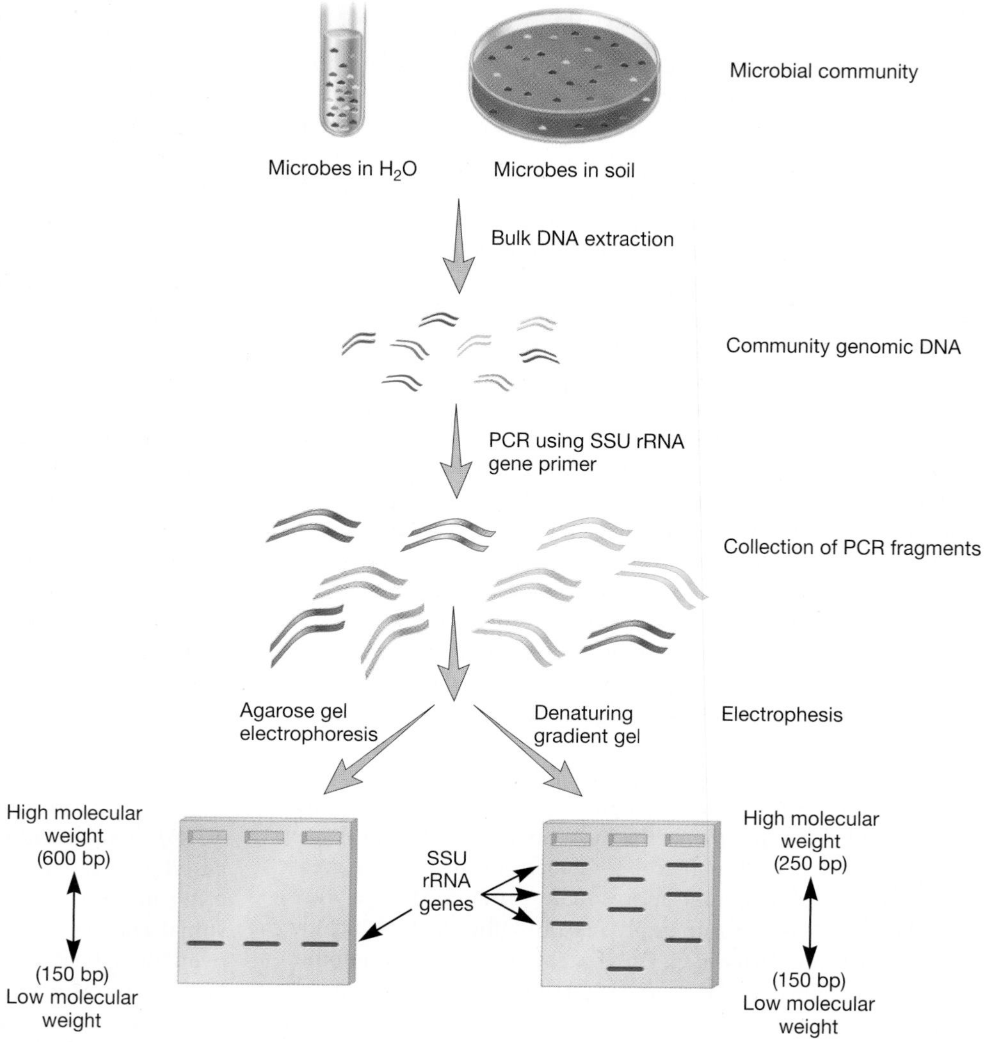

FIGURE 27.11 Denaturing Gradient Gel Electrophoresis (DGGE). The identification of phylotypes starts with the extraction of DNA from a microbial community and the PCR amplification of the gene of choice, usually that which encodes SSU rRNA. Because the majority of amplified DNA fragments have about the same molecular weight, when visualized by agarose gel electrophoresis, they appear identical, as shown in the gel on the left. However, DGGE uses a gradient of DNA denaturing agents to separate the fragments based on the condition under which they become single stranded. When a fragment is denatured, it stops migrating through the gel matrix (gel on right). Individual DNA fragments can then be cut out of the gel and cloned, and the nucleotide sequence determined.

Figure 27.11 Micro Inquiry

Compare the molecular weight range for the agarose gel (left) with that of the denaturing gradient gel. Why do they differ?

The importance of microbial communities has led to vigorous debate in recent years regarding the magnitude of microbial diversity. Additional techniques used to validate, or sometimes replace, assessments based on SSU rRNA studies have been developed. One such approach is to "count" the number of genomes. This is accomplished through **DNA reassociation.** When DNA is rendered single stranded by heating, it will spontaneously reanneal (become double stranded again) when cooled. The rate at which

DNA reanneals depends on its size: the larger the fragment (or chromosome), the longer it takes. Determining microbial diversity based on rates of DNA reassociation rests on the notion that DNA extracted from the environment can be viewed as a giant genome; the length of time it takes to reanneal can be divided by the length of time it takes for the average genome to reanneal. This then gives researchers a general idea of how many genomes are present. The use of this technique has revealed that on the whole, microbial communities in soil are more diverse than those in aquatic ecosystems and that unspoiled environments have more microbial species than those that are polluted (**table 27.2**). Most recently the application of mathematical models to analyze new and previously reported data suggests that microbial diversity may be even greater than previously imagined, with over a million different species in a single pristine soil community.

Table 27.2 Bacterial and Archaeal Diversity

DNA Source	*Abundance (cells/cm³)*[a]	*Genome Equivalents*[b]
Forest soil	4.8×10^9	6,000
Pasture soil	1.8×10^{10}	3,500–8,800
Arable soil	2.1×10^{10}	140–350
Marine fish farm	7.7×10^9	50
Hypersaline pond (22% salinity)	6.0×10^9	7

[a] Direct cell counts using fluorescence microscopy.
[b] DNA reassociation rate of DNA; genome equivalents are based on the assumption that the average bacterial genome is about the size of the *E. coli* genome (4.1 Mb).

From Torsvik, V., et al. 2002. Prokaryotic diversity—magnitude, dynamics, and controlling factors. Science. *296:1064–66.*

Another emerging technology in the assessment of microbial diversity is the use of a type of microarray called a **phylochip** (**figure 27.12**). As described more fully in chapter 16, microarrays are prepared robotically with single-stranded DNA "probes" attached in a gridlike pattern on a solid matrix, often a glass slide. Single-stranded nucleotides from the sample in question are then allowed to hybridize to complementary sequences. The level of hybridization is measured by laser beams.

Several types of microarrays have been developed for evaluating microbial diversity. **Phylogenetic oligonucleotide arrays (POAs)** contain rRNA sequences as probes and are used as a means to detect complementary rRNA sequences from the environment. The design of the oligonucleotide probes is critical because whereas many regions of the rRNA molecule (and its gene) are highly conserved, others are not. To obtain the necessary level of specificity, both highly conserved and hypervariable regions that are genus-specific can be used, along with a variety of control sequences. POAs can also be used to assess changes in microbial community structure. In one study, alterations in rumen populations were monitored in response to different cattle feed. **Community genome arrays (CGAs)** are also used to assess the diversity of a microbial population. Rather than using only SSU rRNA genes as probes, CGAs include probes that represent certain genes involved in specific functions, such as nitrogen fixation or sulfate reduction. Thus potential microbial activity can be explored. In addition, strain specific sequences can be used as probes. This can be useful when a sample may contain multiple strains of a single genus. For instance, CGAs were used to identify 15 *Helicobacter pylori* strains that differ in their capacity to cause gastric ulcers.

Search This: Phylochip Project

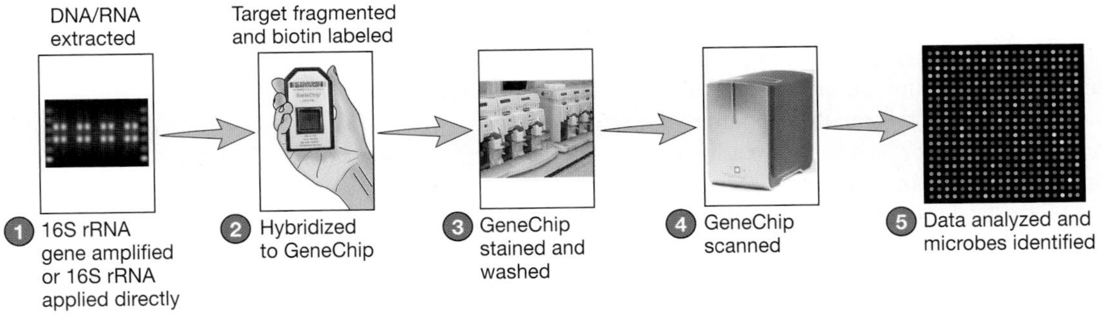
DNA/RNA extracted
Target fragmented and biotin labeled
1 16S rRNA gene amplified or 16S rRNA applied directly
2 Hybridized to GeneChip
3 GeneChip stained and washed
4 GeneChip scanned
5 Data analyzed and microbes identified

FIGURE 27.12 Phylochip Analysis. This form of microarray analysis enables the rapid evaluation of microbial diversity. (1) 16S rRNA genes or 16S rRNA molecules are obtained from the natural sample and labeled. (2) The 16S nucleic acids are hybridized to probes on a microarray called a phylochip. The probes consist of short regions of 16S rRNA genes obtained from preselected microbes. (3) The chip is washed to remove labeled 16S rRNA nucleic acids that could not hybridize because their complement is not found among the probes. (4) Laser beams scan the phylochip and (5) generate a series of spots of various colors and intensities that indicate the presence, absences, and to a lesser extent, relative abundance of each microbe represented with an rRNA probe. This data is computer analyzed.

Figure 27.12 Micro Inquiry

How do phylochips differ from microarrays designed to examine a single microbial species (*see chapter 16*)?

1. What macromolecule does DAPI stain? What are its advantages and disadvantages?
2. What is FISH? Describe an example in which the application of CARD-FISH might yield better results than traditional FISH.
3. What is PCR bias? Why is it undesirable? List one approach that attempts to eliminate PCR bias.
4. Compare advantages and disadvantages of phylochip analysis with DNA fingerprinting and DGGE to assess community diversity. How might phylochip analysis be useful in food safety investigations?

27.3 Assessing Microbial Community Activity

Microbial ecologists also want to understand how microbes drive the biogeochemical cycles that are essential for life on Earth. In any given biome, they seek to identify and measure the flux of nutrients, the interactions between microbes that make these fluxes possible, and the influences of abiotic factors. Here, we divide the methods into biogeochemical and molecular techniques, but it should become clear that the results of one type of approach are often used to formulate hypotheses that will be tested by another (figure 27.1). This illustrates why microbial ecology might be considered the most multidisciplinary branch of microbiology.

Biogeochemical Approaches

Microbial growth rates in complex systems can be measured directly. Changes in microbial numbers are followed over time, and the frequency of dividing cells is also used to estimate production. Alternatively, the incorporation of radiolabeled components such as thymidine (a DNA constituent, usually labeled with ^{3}H) into microbial biomass provides information about growth rates and microbial turnover. This approach is most easily applied to studies of aquatic microorganisms; it is more difficult to directly observe and extract terrestrial microbial communities.

Aquatic microorganisms can be recovered directly by filtration; the volume, dry weight, or chemical content of the microorganisms can then be measured. It is sometimes not possible to directly measure the cell density. In that case, carbon, nitrogen, phosphorus, or an organic constituent of the cells (such as lipids or protein) can be determined. This will give a single-value estimate of the microbial community. Such chemical measurements may be used either directly or expressed as microbial biomass.

To determine community activity, it is necessary to have a good understanding of the microenvironment inhabited by the microbial community in question. In many ecosystems, these microhabitats are studied by **microelectrodes**—electrodes capable of measuring pH, oxygen tension, H_2, H_2S, or N_2O. The tips of these electrodes range from 2 to 100 μm, enabling measurements in increments of a millimeter or less. Because these parameters include electron donor and acceptors, microbial activities can be inferred. Microelectrodes have been widely used in the study of microbial mats (**figure 27.13**). These highly striated communities of photoautotrophs and chemoheterotrophs develop in saline lakes, marine intertidal zones, and hot springs.

In determining community function, it is often instructive to explore the cycling of a particular elemental species. In this case, **stable isotope analysis** may be used. Isotopes are those forms of elements that differ in atomic weight because they bear different numbers of neutrons. Isotopes can be stable or unstable; only radioactive isotopes are unstable and decay. As shown in **table 27.3**, stable isotopes are rare when compared to their lighter counterparts. For example, both ^{14}N and ^{15}N occur in nature, but ^{14}N is found in great abundance, whereas ^{15}N is rare. Organisms discriminate between stable isotopes, and for any given element, the lighter (in this case, ^{14}N) is preferentially incorporated into biomass, a phenomenon known as **isotopic fractionation.** This discrimination can be used to follow the fate of a specific element. For instance, one might want to explore the fate of NO_3^- in a soil ecosystem. Samples of the soil would be incubated in the presence of $^{15}NO_3^-$. One would then sample the soil microorganisms and measure the amount of ^{15}N incorporated into biomass, released as $^{15}N_2$, and reduced to $^{15}NH_4^+$. Because the absolute value of ^{15}N is so small, it cannot be directly measured. Instead, the ratio of ^{15}N to ^{14}N in the sample is compared to the standard ratio found in inorganic material. An organism that assimilates the $^{15}NO_3^-$ would contain more ^{15}N when compared to the standard. This difference, called delta (δ), is calculated as follows:

$$[(R_{sample}\text{-}R_{standard})/(R_{standard})] \times 1000 = \delta_{sample\text{-}standard}$$

where R_{sample} is the ratio of heavy to light isotope in the sample and $R_{standard}$ is the ratio of heavy to light isotope in the standard. For instance, if our hypothetical soil community is found to have a $^{15}N/^{14}N$ ratio greater than the standard by 3 parts per thousand, this value is reported as $\delta\ ^{15}N = +3\ \delta$ ‰.

Search This: Stable isotope overview

Another technique called **stable isotope probing** can be used to examine nutrient cycling in microbial communities as well as identify the microbes taking up the element of interest. As an example, this technique was employed to explore methanogenesis in a rice paddy soil ecosystem. Researchers built **microcosms**—small incubation chambers with conditions that mimic the natural setting. In this case, they simulated natural rice paddies and introduced $^{13}CO_2$. Gaseous $^{13}CH_4$ was collected and RNA was isolated from the soil. The ^{12}C-containing RNA was separated from the ^{13}C-RNA on the basis of differing densities. The ^{13}C-containing rRNA, which could only be synthesized by those bacteria that assimilated the $^{13}CO_2$, was then used to identify the microbes. Recall that methanogenic microbes are archaea; the archaea found in this study have not yet been cultured but were identified by phylotype as members of "Rice Cluster-I" (RC-I). These

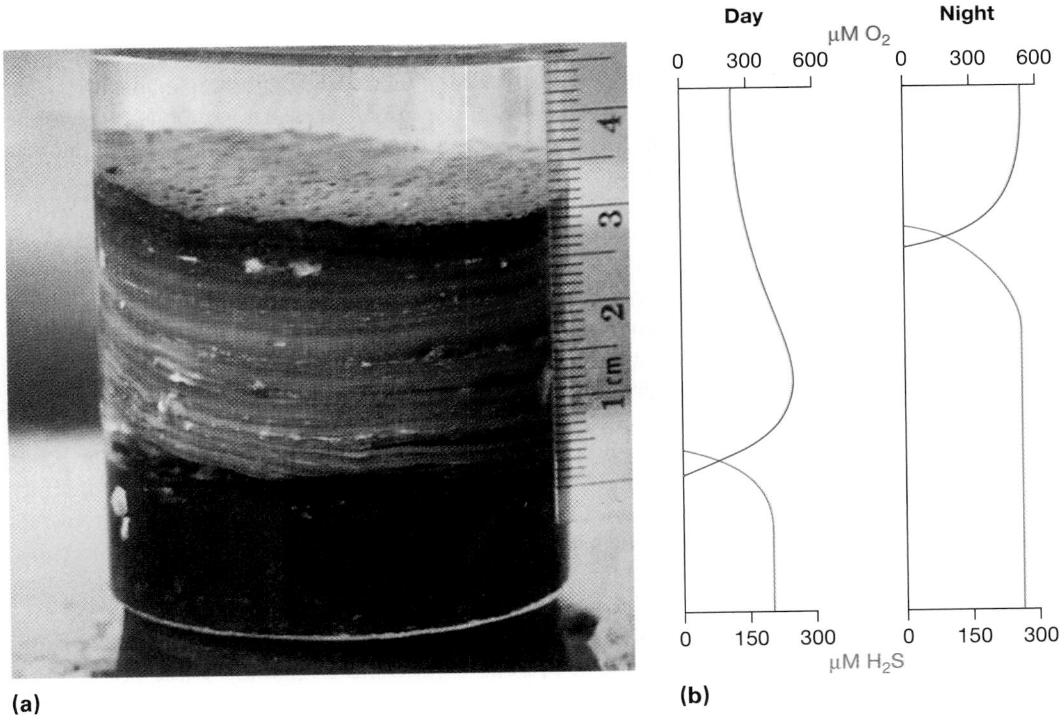

FIGURE 27.13 Microelectrode Analysis of Microbial Mats. (a) Microorganisms, through their metabolic activities, can create environmental gradients resulting in layered ecosystems. A vertical section of a hot spring (55°C) microbial mat, showing the various layers of microorganisms. (b) The measurement of oxygen and H_2S with microelectrodes demonstrates the diurnal shift in gas production: during the day, oxygen produced by cyanobacteria in the upper layers inhibits anaerobic respiration, so sulfate reduction is confined to the lower (dark) strata. At night, atmospheric oxygen has a limited range of diffusion into the mat, and H_2S diffuses upward, where it can be used as a reductant by purple photosynthetic bacteria.

Table 27.3 Natural Abundances of Some Biologically Relevant Isotopes

Hydrogen	*Carbon*	*Nitrogen*	*Oxygen*	*Sulfur*
^{1}H–99.984%	^{12}C–98.89%	^{14}N–99.64%	^{16}O–99.763%	^{32}S–95.02%
^{2}D–0.0156% (deuterium)	^{13}C–1.11%	^{15}N–0.36%	^{17}O–0.0375%	^{33}S–0.75%
			^{18}O–0.1995%	^{34}S–4.21%
				^{36}S–0.02%

methanogens rely on other microbes to ferment photosynthetically derived compounds excreted from plant roots. The H_2 produced by fermentation is then used by the RC-I archaea to reduce CO_2 to CH_4. This study demonstrates the value and importance of culture-independent approaches that combine biogeochemistry and molecular biology to understand microbial community ecology and physiology. ◀◀ *Methanogens (section 18.3)*

Molecular Approaches

While molecular techniques have been applied to microbial taxonomy for at least two decades, their use in community ecology is relatively new. The introduction of metagenomics and microarray technologies to the investigation of mixed populations has revolutionized the depth to which community dynamics can be assessed. The fundamentals of these approaches are presented in chapter 16; here, we focus on their application to microbial community ecology.

Metagenomics applied to diverse natural habitats can have several outcomes, including the identification of new genes and gene products from uncultured microbes, the assembly of whole genomes, and comparisons of community gene content from microbial assemblages of different origin. Metagenomic data can also be used to validate the measurement of gene expression when mRNA is extracted directly from the environment. Researchers are still finding large numbers of new genes from metagenomic analysis, and it is agreed that the discovery of new genes is not complete and many new genes and gene products are yet to be sampled (*see figure 29.4*). It is likely that the number of metagenomic sequences obtained from uncultivated microbes in mixed assemblages will soon outnumber the data sets obtained from microbes grown in pure culture. Here, we discuss a few metagenomic approaches that are most commonly used by microbial ecologists. ◀◀ *Metagenomics (section 16.8)*

One approach to avoiding PCR bias (p. 667) is to clone DNA fragments extracted directly from the environment—in other words, a metagenomic approach. As described in section 16.8, metagenomics involves the insertion of genomic DNA into cloning vectors to obtain a stable source of community DNA fragments that can be amplified by growth in a host bacterium. The nucleotide sequences of DNA fragments can then be determined and the sequence data used to infer community function. In a recent study, the comparison of DNA fragments cloned from a variety of biomes (e.g., mines, hypersaline ponds, marine, freshwater, etc.) showed that it was possible to predict the biogeochemical conditions of a habitat based on its metagenome. For instance, the community metagenome from a well-oxygenated coral reef was rich in genes encoding respiration proteins, while the metagenomes from the colons of terrestrial animals have a high number of genes coding for proteins involved in fermentation.

Most recently mRNA present in the environment has been monitored. Here, mRNA extracted from the natural environment is reverse transcribed to cDNA (*see figure 15.5*). To avoid PCR bias, cDNAs can be analyzed by pyrosequencing (*see figure 16.4*). This approach is generally considered more accurate than community metagenomic surveys. In metagenomic surveys, the number of times a gene is captured is considered a function of the population size of a specific taxon. However, mRNA analysis can identify genes previously undetected in metagenomic data sets. Furthermore, the fact that mRNA is extracted demonstrates that a gene is actively transcribed and the gene product is active. Such a study was used to show that the level of nitrogen fixation by marine unicellular cyanobacteria was much higher than imagined. This was determined by recovering mRNAs of nitrogenase structural genes (*nif* genes) from seawater. ◀◀ *Determining DNA sequences (section 16.1)*

Another way the level of specific mRNA, and therefore the expression of a given gene, can be evaluated uses FISH (figure 27.8). This approach is called **in situ reverse transcriptase (ISRT)-FISH**. Recall that in situ means "in place," so this technique allows the microbial ecologist to examine specimens and determine if a gene of interest is being expressed, as well as visualize the microbes expressing it. Let's say that one is interested in recently identified archaea capable of nitrification (*see figure 20.13*). These microbes must express the *amo* (ammonia monooxygenase) gene to oxidize ammonium to nitrite. Thus one could design an *amo* DNA probe—a short piece of DNA that is complementary to the *amo* mRNA. Once the DNA probe has hybridized to the *amo* mRNA in the natural sample, reverse transcriptase is used to generate a cDNA, which is then amplified by PCR. One now has an abundance of nucleotide fragments that can be fluorescently tagged in FISH. In our example of the archaeal nitrifier, we might want to determine where these archaea are metabolically active in the water column of an oceanic sampling station. Water samples of varying depths would be collected and filtered on board a research vessel. The filters would most likely be fixed with preservative for later analysis (it is difficult to perform microscopy on a moving boat). The data collected by ISRT-FISH would indicate where archaeal nitrification was most active. This could then be correlated with physical parameters such as light penetration, oxygen concentration, and levels of nitrate, nitrite, ammonia, and dissolved organic nitrogen. In this way, the microbial ecologist can begin to build a hypothesis about the importance and distribution of these interesting microbes.

But what if one wanted to identify the microbe responsible for a specific metabolic activity? Recall that ISRT-FISH enables the visualization of the microbes expressing a specific gene but does not identify the microbe. To simultaneously assess metabolic activity and phylogenetic identity, a relatively new technique that combines an older method, **microautoradiography (MAR)**, with FISH can be employed. This method is thus called MAR-FISH. To understand MAR-FISH, it is necessary to first explain MAR. Here, a radioactive substrate is mixed with the sample and incubated under specific conditions, usually closely resembling those in the field. At the completion of the incubation period, the sample is washed to remove any radioactive substrate not taken up by the cells. A sample attached to a microscope slide is treated with a photographic emulsion. The slide is carefully wrapped and stored in complete darkness. As the incorporated radioactive material decays, the silver grains in the photographic emulsion are exposed. When developed, this leaves a series of black dots around the cells that assimilated the radioactive substrate (**figure 27.14**). When combined with FISH, one can analyze the phylogenetic identity and the specific substrate uptake patterns of microbes of interest at the level of a single cell in complex microbial communities.

Microarray analysis can also be used to assess community function. **Functional gene arrays** (FGAs) are designed to assess microbial activity, rather than identity. These arrays are prepared with sequences of genes known to be important in biogeochemical cycling processes (e.g., *nif* genes). However, the use of microarrays to assess microbial communities is inherently problematic: only known genes can be used to make oligonucleotide probes. This can generate an information bias such that no novel genes in the

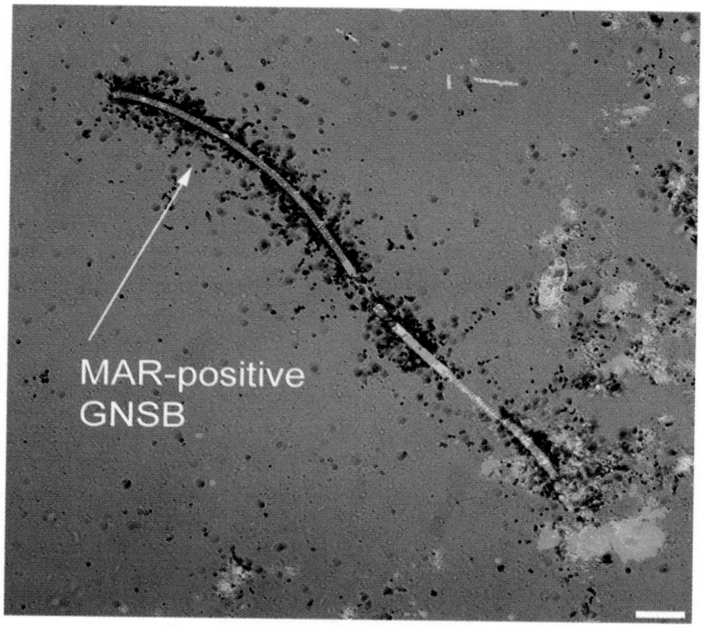

FIGURE 27.14 MAR-FISH. This technique combines microautoradiography with FISH. ^{14}C-labeled amino acids were incubated with samples to detect heterotrophic microbes. MAR-positive indicates that the microbe assimilated the amino acids, as seen by the black dots that surround the filamentous microbe. The microorganism was identified using FISH as an uncultured *g*reen *n*onsulfur *b*acterium (GNSB) by using a red fluorescent probe specific to the phylum *Chloroflexi.* The green fluorescence is the result of hybridization to a probe that detects SSU of most bacteria. Scale bar = 10 μm.

community can be detected. In addition, the interpretation of community-based microarrays presents challenges because the mRNA pool is derived from multiple genera. This results in varying levels of homology to different probes. In turn, this can generate high background levels of fluorescence and obscure valid results. These technical issues will no doubt be minimized by further development. Environmental microarrays hold great promise because they can yield a great deal of genetic information without nucleotide sequencing. ◀◀ *Functional genomics (section 16.4)*

Search This: Functional gene arrays/microbial ecology

Microbial ecologists also use **reporter microbes** to characterize the physical microenvironment on the scale of an individual bacterium (around 1 to 3 μm). This is done by constructing cells with reporter genes, often based on green fluorescent protein (GFP), that change their fluorescence in response to environmental and physiological alterations. Such microbes are used to measure oxygen availability, UV radiation dose, pollutant or toxic chemical effects, and stress. For example, when microbes containing a moisture stress reporter gene have less available water, an increase occurs in GFP-based fluorescence. ◀◀ *Fluorescence labeling (section 15.7)*

Just as metagenomics seeks to capture each gene present in a community, the goal of **metaproteomics** is to identify each protein present in a given microenvironment at the time of sampling. Importantly while metagenomics informs the scientist of the potential activities, the ideal metaproteomics study identifies those metabolic processes active at the time the community was assessed. Thus whereas metagenomics paints a general picture about what might be possible under a variety of conditions, metaproteomics generates a snapshot of the events at a given time.

In contrast to metagenomics, which benefits from a relatively long history of development, metaproteomics is an emerging technology. There are two general approaches by which all the proteins in a community can be sampled and identified: two-dimensional polyacrylamide gel electronphoresis and two-dimensional nano-liquid chromatography.

Two-dimensional polyacrylamide gel electrophoresis (2D-PAGE) is the more labor-intensive technique because each protein recovered from the environment is visualized (*see figure 16.14*). Individual proteins of interest can be eluted from the gel and analyzed by mass spectroscopy (MS). Total peptide mass and amino acid sequence information is then used to search databases such as BLAST to identify the proteins. This approach is best if the relative concentration of individual proteins is considered important.

Two-dimensional nano-liquid chromatography (2D nano-LC) is used when qualitative data is most important. It is a more automated, high-throughput approach. Community proteins are extracted and digested with a protease (usually trypsin) to generate protein fragments called peptides. This complex mixture of peptides is separated by two different chromatographic methods: peptides are first separated by charge and then by hydrophobicity. The peptide masses are determined by advanced mass spectrometry. This is followed by tandem MS (MS-MS) to obtain amino acid sequence information, which can then be used to query databases such as BLAST (*see figure 16.15*). In both metaproteomics methods, the protein sequences obtained are compared to the metagenomic data obtained from the same environment.

Metaproteomics is limited by the complexity of the environment being assessed. Complex environments such as a pristine soil may have as many as a billion or more different proteins per gram. Even the most advanced and automated MS approaches are not yet able to resolve such a large and diverse group of proteins. On the other hand, metaproteomics has been successfully applied to environments of limited microbial diversity, such as biofilms present in acid mine drainage.

1. What are stable isotopes? Design an experiment in which you plan on determining whether or not denitrification is taking place in a soil ecosystem.
2. How might microelectrodes be used to gather data in preparation for a stable isotope study?
3. How do functional gene arrays differ from phylochips? When would one use a functional gene array, rather than a phylochip?
4. What is metaproteomics? How is it similar to metagenomics? How does it differ?

Summary

27.1 Culturing Techniques

a. Enrichment cultures are used to promote the growth of one type of microbe, while suppressing the growth of others. This is accomplished by designing media and culture conditions that favor the growth of the desired microorganisms but limit the growth of others.

b. Although most microbes cannot be grown in pure culture, novel approaches have been developed that have resulted in the isolation of previously uncultured microorganisms.

27.2 Assessing Microbial Diversity

a. A variety of staining techniques are used to observe microbes in natural environments. Fluorescent in situ hybridization (FISH) labels specific microbes that possess a specific nucleotide sequence, usually a region of the SSU rRNA gene (**figures 27.7–27.9**).

b. A survey of SSU rRNA is the most common approach for assessing microbial diversity. There are several approaches by which SSU rRNA can be recovered; those that avoid the use of PCR to amplify rRNA genes are gaining popularity because they avoid PCR bias.

c. DNA reassociation is also used to evaluate the size and complexity of microbial communities. This technique considers the community as a collection of genomes.

d. Microarrays that have SSU rRNAs as probes, called phylochips, can quickly evaluate population and community diversity (**figure 27.12**).

27.3 Assessing Microbial Community Activity

a. Microelectrodes can be used to determine physical and biological parameters such as pH, O_2, and H_2S concentrations in microniches (**figure 27.13**).

b. Stable isotope analysis is based on the fact that organisms discriminate between heavy and light naturally occurring isotopes. It is often used to monitor nutrient flux through ecosystems (**table 27.3**).

c. Functional gene arrays can assess community activity because the probes represent genes that are involved in nutrient cycling.

d. Metaproteomics can be used to identify the proteins that are expressed by a microbial community. Because the number of proteins is so large, traditional 2D gel analysis is usually not practical and high-throughput chromatographic techniques are used instead.

Critical Thinking Questions

1. How might you attempt to grow in the laboratory a chemolithoautotroph that uses ferrous iron as an electron donor and oxygen as an electron acceptor?

2. What genes would you use as probes on a functional array designed to assess microbial activity in the plankton of a nitrogen-limited aquatic ecosystem?

3. Both acetate and CO_2 can be used by methanogenic archaea to generate methane (CH_4). How would you determine the source of carbon for methanogenesis in a waterlogged peat?

4. Kindaichi et al. used FISH and microelectrodes to investigate the spatial organization of nitrifying and anammox microbes in a biofilm. Recall that nitrification is an aerobic process that involves ammonia oxidation to nitrite by both archaea and bacteria followed by nitrite oxidation to nitrate by bacteria (*see p. 522*). Gram-negative bacteria in the phylum *Planctomycetes* perform the anammox reaction—the anaerobic oxidation of ammonium with nitrite as the electron acceptor and N_2 as the major end product (*see figure 19.11*). Explain the purpose of each technique used by these investigators. How do you think these approaches would contribute to an understanding of the spatial distribution of each type of microbe as well as help document the metabolic process occurring in each microregion of the biofilm?

Read the original paper: Kindaichi, T., et al. 2007. In situ activity and spatial organization of anaerobic, ammonium-oxidizing (anammox) bacteria in biofilms. *Appl. Environ. Microbiol.* 73:4931.

5. A PCR approach was used to explore the diversity of aerobic anoxygenic phototrophs (AAnPs; *see p. 497*). The *pufM* gene encodes a protein unique to the photosynthetic reaction center in AAnPs. When *pufM* was amplified from a genomic library from the Delaware River, it was determined that most of the AAnPs belonged to the β- and γ-proteobacteria. On the other hand, quantitative PCR (*p. 405*) showed that the Delaware estuary was dominated by α-proteobacterial AAnPs, although β- and γ-proteobacteria were also present. Why do you think these two type of molecular approaches resulted in different conclusions? Which class of microbes do you think were numerically dominant? Which were most likely responsible for most of the aerobic anoxygenic phototrophy? Explain your reasoning.

Read the original paper: Waidner, L. A., and Kirchman, D. L. 2008. Diversity and distribution of ecotypes of the aerobic anoxygenic phototrophy gene *pufM* in the Delaware Estuary. *Appl. Environ. Microbiol.* 74:4021.

Concept Mapping Challenge

Construct a concept map using the words below; provide your own linking terms.

Enrichment culture	Community activity	MAR-FISH
In situ analysis	Axenic culture	Microelectrodes
Microdroplet culture	Stable isotope analysis	FISH

Learn More

Learn more by visiting the text website at www.mhhe.com/willey8, where you will find a complete list of references.

28

Microorganisms in Marine and Freshwater Ecosystems

Oceanographers sampling water on an expedition to investigate phytoplankton blooms in the open ocean.

CHAPTER GLOSSARY

allochthonous Material in a freshwater system that originates from an external source, such as land or groundwater.

autochthonous Material in a freshwater system that originates from within that body of water.

carbonate equilibrium system The balance of CO_2, bicarbonate, and carbonate that keeps the ocean buffered between pH 7.6 and 8.2.

dissolved organic matter (DOM) Nutrients that are available in the soluble, or dissolved, state.

eutrophic Pertaining to a nutrient-enriched environment.

marine snow Flocculent suspended particles in the ocean's water column; consist primarily of fecal pellets, diatom frustules, and other materials that are not easily degraded.

microbial loop The cycling of organic matter synthesized by photosynthetic microorganisms through the activity of other microbes, such as bacteria and protozoa. This process "loops" nutrients back for reuse by primary producers and makes some organic matter unavailable to higher consumers.

oligotrophic environment An environment containing low levels of nutrients, particularly those that support microbial growth.

particulate organic matter (POM) Nutrients that are not dissolved or soluble, including microorganisms or their cellular debris after senescence or viral lysis.

photosynthate Nutrient material that leaks from phototrophic organisms. It contributes to the pool of dissolved organic matter that is available to microorganisms.

phytoplankton Phototrophic bacteria and protists that drift in the upper, sunlit regions of marine and freshwater ecosystems.

primary production The incorporation of carbon dioxide into organic matter by photosynthetic and chemoautotrophic organisms (i.e., primary producers).

SAR11 The most abundant microbe on Earth, this lineage of α-proteobacteria has been found in almost all marine ecosystems studied; includes the isolate *Pelagibacter ubique*.

virioplankton Viruses that drift in marine and freshwater environments; such viruses are very abundant.

In this chapter, we turn our attention to the microbial ecology of aquatic environments. The two main disciplines of aquatic biology are oceanography and limnology. **Oceanography** is the study of marine systems and the biological, physical, geological, and chemical factors that impact biogeochemical cycling, water circulation, and climate. Microbial oceanography seeks to understand marine microorganisms at the level of the individual taxon and at the same time elucidate how they influence biogeochemical cycling at a global level. **Limnology** is the investigation of aquatic systems within continental boundaries, including glaciers, groundwater, rivers, streams, and wetlands. Although these ecosystems do not have a tremendous impact on global fluxes, all terrestrial life depends on the sustained provision of clean freshwater. Thus like microbial oceanography, freshwater microbiology is an essential discipline because the physiology of individual microbes can profoundly impact life on Earth.

It is an exciting time for aquatic microbiology. The development of molecular microbial ecology methods, novel culturing techniques, remote sensing, and deep-sea exploration has propelled this discipline to a new age of discovery. Recent reports have revealed a level of microbial diversity not previously imagined as well as the impact of microbes in maintaining a balanced ecosystem. We now realize the importance of microbes in addressing problems such as global warming and pollution. In addition, new technologies have advanced our understanding of the food webs that govern the world's fisheries.

We open this chapter by describing water as a habitat that sustains microbial growth. We next introduce marine microbiology. The oceans occupy about 71% of Earth's surface and are on average about 4 km deep. Although it may seem like a single continuous ecosystem, the ocean is actually a collection of smaller, interconnecting habitats maintained by global-scale interactions between the ocean and the atmosphere. We thus discuss different microbial habitats within the marine environment. Likewise, freshwater systems can be divided into discrete habitats and these are also reviewed.

28.1 Water as a Microbial Habitat

The nature of water as a microbial habitat depends on a number of physical factors such as temperature, pH, and light penetration. One of the most important of these is dissolved oxygen content. The flux rate of oxygen through water is about 10,000 times less than its rate through air. However, in some aquatic habitats, the limits to oxygen diffusion can be offset by the increased solubility of oxygen at colder temperatures and increasing atmospheric pressures. For instance, in the very deep ocean, the dissolved oxygen concentration actually increases with depth, even though the air-water interface can be literally miles away. On the other hand, tropical lakes and summertime temperate lakes may become oxygen limited only meters below the surface. In these cases, aerobic microbes consume the surface-associated oxygen faster than it can be replenished. This leads to the formation of hypoxic or anoxic zones in these environments. This enables anaerobic microbes, both chemotrophic and phototrophic, to grow in the lower regions of lakes where light can penetrate.

The second major gas in water, CO_2, plays many important roles in chemical and biological processes. The pH of unbuffered distilled water is determined by dissolved CO_2 in equilibrium with the air and is approximately 5.0 to 5.5. The pH of freshwater systems such as lakes and streams, which are usually only weakly buffered, is therefore controlled by terrestrial input (e.g., minerals that may be either acidic or alkaline) and the rate at which CO_2 is removed by photosynthesis. When autotrophic organisms such as diatoms use CO_2, the pH of the water is often increased.

In contrast, seawater is strongly buffered by the balance of CO_2, bicarbonate (HCO_3^-), and carbonate (CO_3^{2-}). Atmospheric CO_2 enters the oceans and either is converted to organic carbon by photosynthesis or reacts with seawater to form carbonic acid (H_2CO_3), which quickly dissociates to form bicarbonate and carbonate (**figure 28.1**):

$$CO_2 + H_2O \rightleftharpoons H_2CO_3 \rightleftharpoons H^+ + HCO_3^- \rightleftharpoons 2H^+ + CO_3^{2-}$$

The oceans are effectively buffered between pH 7.6 and 8.2 by this **carbonate equilibrium system.** Much like the buffer one might use in a chemistry experiment, the pH of seawater is determined by the relative concentrations of the weak acids bicarbonate and carbonate.

The reactions of the carbonate equilibrium system have taken on new importance. Many oceanographers predict that the pH of the ocean will drop by 0.35 to 0.50 units by 2100, unless effective means of limiting greenhouse gas emissions (in particular, CO_2) are implemented. Indeed, compared to preindustrial levels, the average surface ocean pH has declined by 0.1 unit. Recall that the pH scale is logarithmic; it is unclear what the implications of this change in carbonate equilibrium will mean for the marine environment and indeed for all life on Earth. However, investigation of natural acidification by underwater volcanic activity in waters near Sicily may be instructive. Following a decline in pH from 8.2 to 7.8, researchers found decreased abundance of calcareous taxa and a community shift away from coralline species toward brown algae.

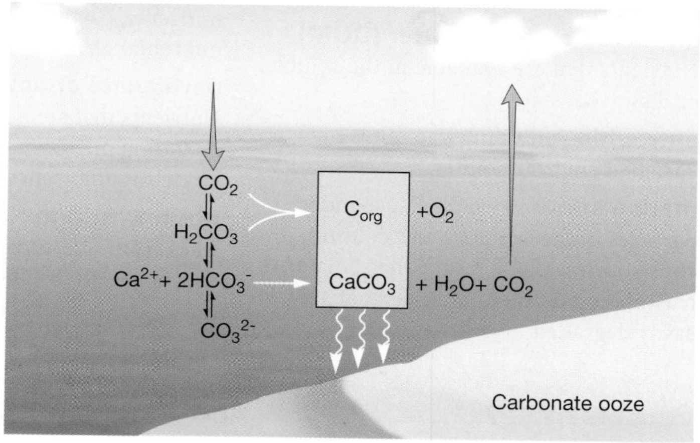

FIGURE 28.1 The Carbonate Equilibrium System. Atmospheric CO_2 enters seawater and is converted to organic carbon (C_{org}) or is converted to carbonic acid (H_2CO_3) that rapidly dissociates into the weak acids bicarbonate (HCO_3^-) and carbonate (CO_3^{2-}). Calcium carbonate ($CaCO_3$), a solid, precipitates to the seafloor, where it helps form a carbonate ooze. This system keeps seawater buffered at about pH 8.0.

Figure 28.1 Micro Inquiry

Which carbon species, CO_2 or CO_3^-, acidifies seawater?

Other gases also are important in aquatic environments. These include N_2, used as a nitrogen source by nitrogen fixers; H_2, which is both a waste product and a vital electron donor; and methane (CH_4). These gases vary in their water solubility; methane is the least soluble of the three. Methane produced under anoxic conditions either accumulates in benthic (bottom) environments (e.g., natural gas deposits) or diffuses up through the water column, where it is oxidized by methanotrophs or released to the atmosphere. This eliminates the problem of waste accumulation that occurs with many microbial metabolic products, such as organic acids and ammonium ion.

Light is also critical for the health of marine and freshwater ecosystems. Like all life on Earth, organisms in these environments depend on **primary producers**—autotrophic organisms—to provide organic carbon. In streams, lakes, and coastal marine systems, macroscopic algae and plants are the chief primary producers. Organic carbon also enters these systems in terrestrial runoff. The situation is very different in the open ocean, where all organic carbon is the product of microbial autotrophy. In fact, about half of all **primary production** on Earth is the result of this microbial carbon fixation. The **photic zone** can be defined as the water from the surface to the depth to which light penetrates with sufficient intensity so that the rate of photosynthesis by microscopic autotrophs exceeds the collective rate of respiration. We see a marked difference in the depth of the photic zone when we compare nearshore waters with the open ocean. In lakes and estuaries where the water is turbid, the photic zone may be only a meter or two deep. This is in sharp contrast to nutrient-depleted areas such as the open ocean and many tropical areas where the water seems "crystal clear." In these regions, the photic zone ranges from 150 to 200 m.

Solar radiation warms the water and this can lead to thermal stratification. Warm water is less dense than cool water, so as the sun heats the surface in tropical and temperate waters, a **thermocline** develops. A thermocline can be thought of as a mass of warmer water "floating" on top of cooler water. These two water masses remain separate until there is either a substantial mixing event, such as a severe storm, or in temperate climates, the onset of autumn. As the weather cools, the upper layer of warm water becomes cooled and the two water masses mix. This is often associated with a pulse of nutrients from the lower, darker waters to the surface. This pulse of nutrients can trigger a sudden and rapid increase in the population of certain microbes and a bloom may develop. This is considered more fully in the discussion of microorganisms in coastal marine systems (*p. 677*) and lakes (*p. 689*).

1. What factors influence oxygen solubility? How is this important in considering aquatic environments?
2. Describe the buffering system that regulates the pH of seawater. What might be the implications of this stable buffering on microbial evolution?
3. What is the photic zone? How and why does it differ in lakes and coastal ecosystems versus the open ocean?
4. What is a thermocline?

Nutrient Cycling in Marine and Freshwater Environments

Obviously many differences exist between nearshore and open-ocean environments. From a microbial point of view, lakes, estuaries, and other coastal regions have relatively high rates of primary production. They therefore must have a higher influx and turnover (reuse) of essential nutrients. In these regions, nearby agricultural and urban activities frequently generate runoff that provides substantial nutrients (especially nitrogen and phosphate). In contrast, nutrient levels are very low in the open ocean, which is unaffected by rivers, streams, and terrestrial runoff. Here, nitrogen, phosphorus, iron, and even silica, which diatoms need to construct their frustules, can be limiting. ◀◀ Stramenopila *(section 23.5)*

The major source of organic matter in illuminated surface waters of large lakes and the open ocean is photosynthetic activity, primarily from **phytoplankton** (Greek *phyto,* plant, and *planktos,* wandering)—autotrophic organisms that float in the photic zone. Common marine planktonic cyanobacterial genera are *Prochlorococcus* and *Synechococcus.* These unicellular microbes can reach densities of 10^4 to 10^5 cells per milliliter in the photic zone. **Picoplankton** (planktonic microbes between 0.2 and 2.0 μm in size) can represent 20 to 80% of the total phytoplankton biomass. Larger eukaryotic autotrophs, especially diatoms, also contribute a significant fraction of fixed carbon to these ecosystems. Lakes have a more diverse assemblage of planktonic primary producers. The picoplankton often include freshwater species of *Synechococcus,* while filamentous cyanobacteria such as *Anabaena* and *Microcystis* constitute the macroplankton (microbes larger than 200 μm). The microplankton (between 20 and 200 μm) consist of diatoms and dinoflagellates in both marine and freshwater ecosystems. Collectively these planktonic autotrophic microbes make up the phytoplankton community. ◀◀ *Phylum* Cyanobacteria *(section 19.3)*

As they grow and fix CO_2 to form organic matter, phytoplankton acquire needed nitrogen and phosphorus from the surrounding water. In marine systems, the nutrient composition of the water affects the final carbon-nitrogen-phosphorus (C:N:P) ratio of the phytoplankton, which is termed the **Redfield ratio,** named for the oceanographer Alfred Redfield. This ratio is 106 parts C, 16 parts N, and 1 part P. Although this ratio is an average, its value is surprisingly constant even when diverse oceanic phytoplankton populations are measured. Changes in the Redfield ratio are generally associated with non-steady state growth conditions. It can therefore be used as an important marker when following nutrient dynamics, especially mineralization and immobilization processes, and for studying the sensitivity of oceanic photosynthesis to changes in atmospheric levels of CO_2, nitrogen, sulfur, and iron.

As primary producers, microbes play an essential role in cycling other nutrients. Traditionally the interaction of organisms at different trophic levels has been depicted as a food chain in which primary producers are most numerous, followed by herbivores, which are then consumed by carnivores. Such diagrams generally show microorganisms functioning strictly as decomposers, mineralizing most of the waste products in the ecosystem. However, this simplified version of trophic interactions does not adequately describe the important and varied roles of microbes. Consider that microbial ecologists estimate at least 6×10^{30} microbial cells reside on the planet at any given time. This unseen biomass far exceeds that of all other organismal groups combined. If microbes functioned only to degrade and mineralize organic material to their inorganic forms, these essential elements would be in danger of being irreversibly removed from the ecosystem. Instead, microbes interact with several trophic levels, serving to recycle nutrients many times within the community before any given element is either mineralized or sinks.

The **microbial loop** describes the many roles that microbes serve. As depicted in **figure 28.2**, it is most applicable to planktonic ecosystems such as that found in the pelagic, or deep, region of lakes and oceans. In these systems, autotrophic microbes are the primary producers and heterotrophic microbes consume **dissolved organic matter (DOM)** released by larger organisms. Exudates of larger plants and animals, the liquid waste of zooplankton, and material that leaks from the phytoplankton, sometimes called **photosynthate,** all contribute to a common pool of DOM. Viruses are also a source of DOM. Marine viruses can be present at concentrations up to 10^8 per milliliter; the lysis of their host cells contributes significantly to the return of nutrients into the microbial loop (p. 683). Protists, including flagellates and ciliates, consume smaller microbes (figure 28.2*b,c*), which can be thought of as **particulate organic matter (POM).** Because protists are then consumed by zooplankton, both DOM and POM are recycled for use at a number of trophic levels. The net effect of

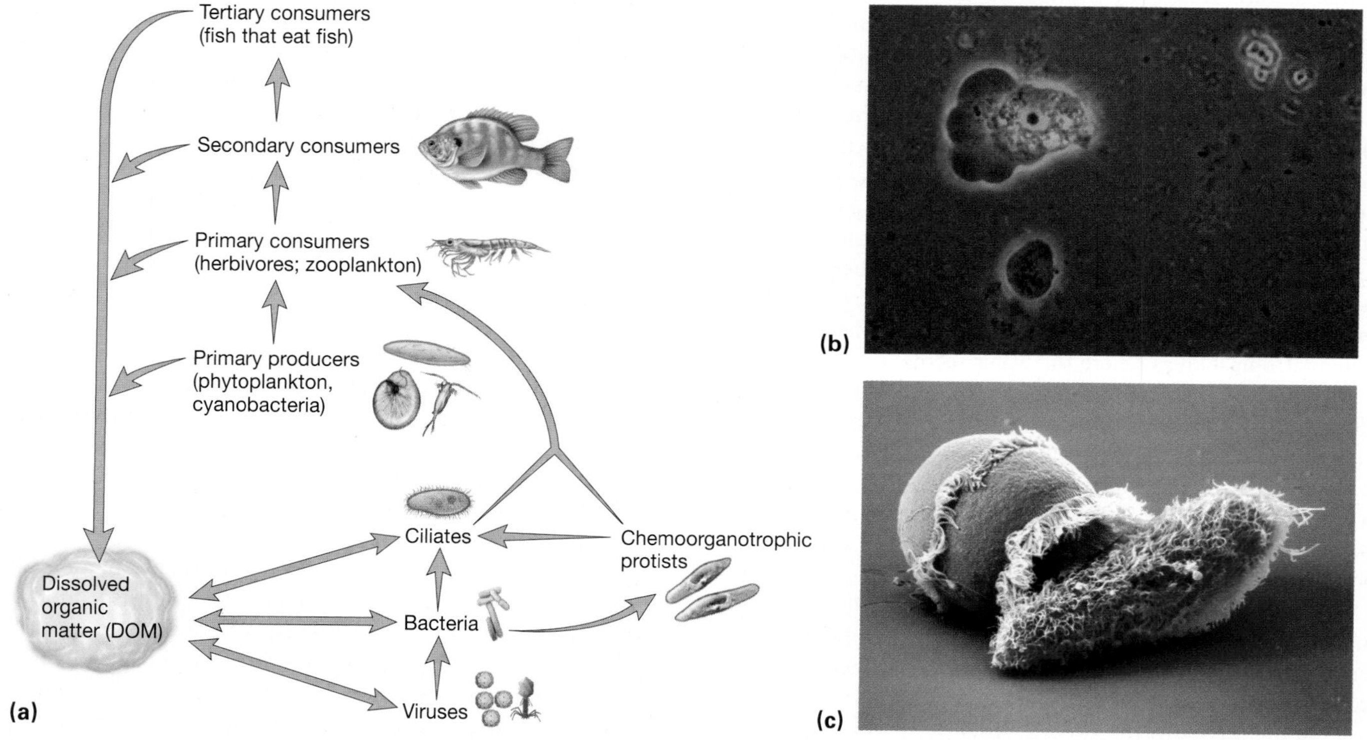

FIGURE 28.2 The Microbial Loop. (a) Microorganisms play vital roles in ecosystems as primary producers, decomposers, and primary consumers. All organisms contribute to a common pool of dissolved organic matter (DOM) that is consumed by microbes. Viruses contribute DOM by lysing their hosts, and bacterial and archaeal cells are consumed by protists, which also consume other protists. These microbes are then consumed by herbivores that often select their food items by size, thereby ingesting both heterotrophic and autotrophic microbes. Thus nutrient cycling is a complex system driven in large part by microbes. (b) Protists consume bacteria; in this case, a naked amoeba is consuming the cyanobacterium *Synechococcus,* which fluoresces red. (c) Protists consume protists; here, the ciliate *Didinium* (gold) is preying upon another ciliate, *Paramecium.*

Figure 28.2 Micro Inquiry

How do heterotrophic microbes contribute to the pool of DOM? How does this compare with the manner by which microbial autotrophs contribute to the DOM pool?

the microbial loop is the reuse of essential nutrients such as organic carbon, nitrate, and phosphate within the photic zone. Only a small fraction of this material sinks to the waters below, and that which does resists degradation.

Search This: Cycling through the food web

1. Describe the differences in nutrient input in coastal ecosystems as compared to that in the open ocean.
2. What is picoplankton and why is it important?
3. How does the microbial loop differ from a food chain?

28.2 Microorganisms in Marine Ecosystems

As terrestrial organisms, we must remind ourselves that about 96% of Earth's water is in marine environments. Although much of this is in the deep sea, from a microbiological perspective, the surface waters have been most intensely studied. Only recently have scientists had the ability to probe the benthos—the deep-sea sediments and the subsurface. Investigations of this kind are revealing many surprises. We begin our discussion of marine ecosystems with estuaries, then discuss the microbial communities that inhabit the open ocean, and finally, the dark, cold, high-pressure benthos.

Microorganisms in Estuaries and Salt Marshes

An estuary is a semienclosed coastal region where a river meets the sea. Estuaries are defined by tidal mixing between freshwater and saltwater. They feature a characteristic salinity profile called a **salt wedge** (**figure 28.3**). Salt wedges are formed because saltwater is denser than freshwater, so seawater sinks below overlying freshwater. As the contribution from the incoming river increases and that of the ocean decreases, the relative amount of seawater declines with the estuary's distance from the sea.

The distance the salt wedge intrudes up the estuary is not static. Most estuaries undergo large-scale tidal flushing; this forces organisms to adapt to changing salt concentrations on a daily basis. Microbes that live under such conditions combat the resulting osmotic stress by adjusting their intracellular osmolarity to limit the difference with that of the surrounding water. Most protists and fungi produce osmotically active carbohydrates for this purpose, whereas bacteria and archaea regulate internal concentrations of potassium or special amino acids such as ecoine and betaine. Thus most microbes that inhabit estuaries are halotolerant, which is distinct from halophilic. **Halotolerant** microbes can withstand significant changes in salinity; halophilic microorganisms have an absolute requirement for high salt concentrations. ◀◀ *Solutes and water activity (section 7.6)*

Estuaries are unique in many respects. Their calm, nutrient-rich waters serve as nurseries for juvenile forms of many commercially important fish and invertebrates. However, despite their importance to the commercial fishing industry, estuaries are among the most polluted marine environments. Estuaries and the rivers that feed them are the receptacles of pollutants discharged during industrial processes as well as agricultural and street runoff. Recall that from the seventeenth century through most of the twentieth century, industries dumped wastes without fear of punitive consequences. The cleanup of rivers and estuaries contaminated with industrial wastes such as polychlorinated biphenyls (PCBs) continues to this day. Agricultural runoff and industrial pollution includes organic materials that can be used as nutrients. This further magnifies the problem because chemoorganotrophic microbes consume available oxygen, forming anoxic dead zones. Such anoxic regions, which are devoid of almost all macroscopic life, are now present in the Gulf of Mexico, Chesapeake Bay, and fjords entering the North Sea. ▶▶ *Biodegradation and bioremediation by natural communities (section 42.3)*

Pollution can also create the opposite problem: too much growth. However, in this case, a single microbial species, either algal or cyanobacterial, grows at the expense of all other organisms in the community. This phenomenon, called a **bloom,** often results from the introduction of nutrients combined with changes in light, temperature, and mixing. If the microbes produce a toxic product

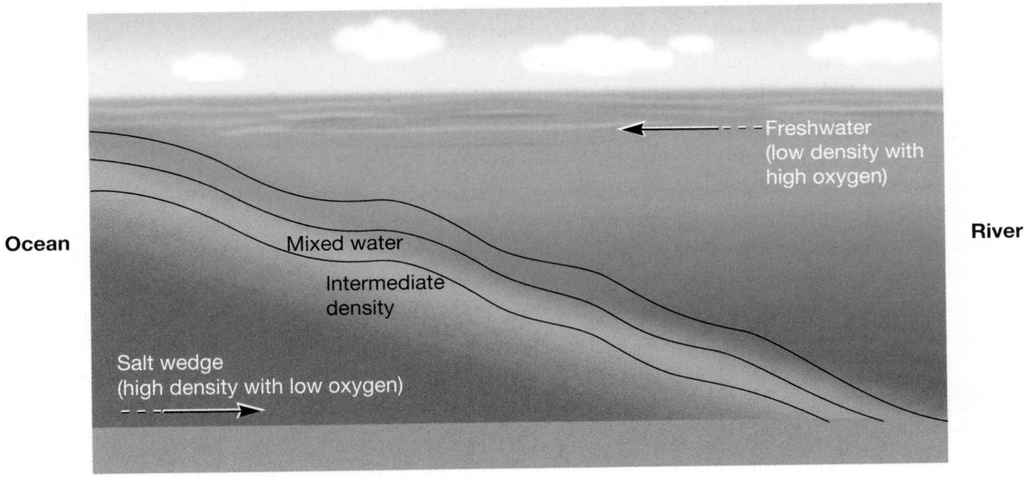

FIGURE 28.3 A Salt Wedge. An estuary contains both freshwater and saltwater. Because seawater is denser than freshwater, the water masses do not mix. Rather, the seawater remains below the freshwater with the relative amount of seawater decreasing in the upper reaches of the estuary.

FIGURE 28.4 *Pfiesteria* Lesions. Lesions on menhaden resulting from parasitism by the dinoflagellate *Pfiesteria piscicida.*

or are in themselves toxic to other organisms such as shellfish or fish, the term **harmful algal bloom (HAB)** is used. Some HABs are called **red tides** because the microbial density is so great that the water becomes red or pink (the color of the algae).

The number of HABs has dramatically increased in the last decade or so, although the reasons for this are unclear. HABs are sometimes responsible for killing large numbers of fish or marine mammals. For instance, off the coast of California, an HAB was responsible for sudden, large-scale deaths of sea lions. In this case, the bloom species was a diatom of the genus *Pseudonitzschia.* Anchovies consumed the diatoms and a potent neurotoxin, domoic acid, accumulated in the fish. The mammals were poisoned after they ingested large quantities of the fish, an important component of the sea lion diet. Periodic deaths of sea lions in this region due to domoic acid poisoning continues to this day. ⏮ Stramenopila *(section 23.5)*

HABs are often caused by dinoflagellates. Some bloom-causing dinoflagellates also produce potent neurotoxins. Dinoflagellates of the genus *Alexandrium* produce a toxin responsible for most of the paralytic shellfish poisoning (PSP), which affects humans as well as other animals in coastal, temperate North America. The toxin produced by the dinoflagellate *Karenia brevis* killed a number of endangered manatees and bottlenose dolphins in 2002 in a bloom in Florida. Another dinoflagellate, *Pfiesteria piscicida,* has become a problem in the Chesapeake Bay and regions south. This protist produces lethal lesions in fish (**figure 28.4**) and has had a devastating effect on the local fishing industry. Exposure to this microbe may also cause neurological damage to humans. ⏮ Alveolata *(section 23.5)*

Search This: Harmful algal blooms/noaa

Salt marshes generally differ from estuaries because their freshwater input is from multiple sources, rather than a single river. Smaller streams enter salt marshes, which are flatter and have a wider expanse of sediment and plant life exposed at low tide. The microbial communities within salt marsh sediments are

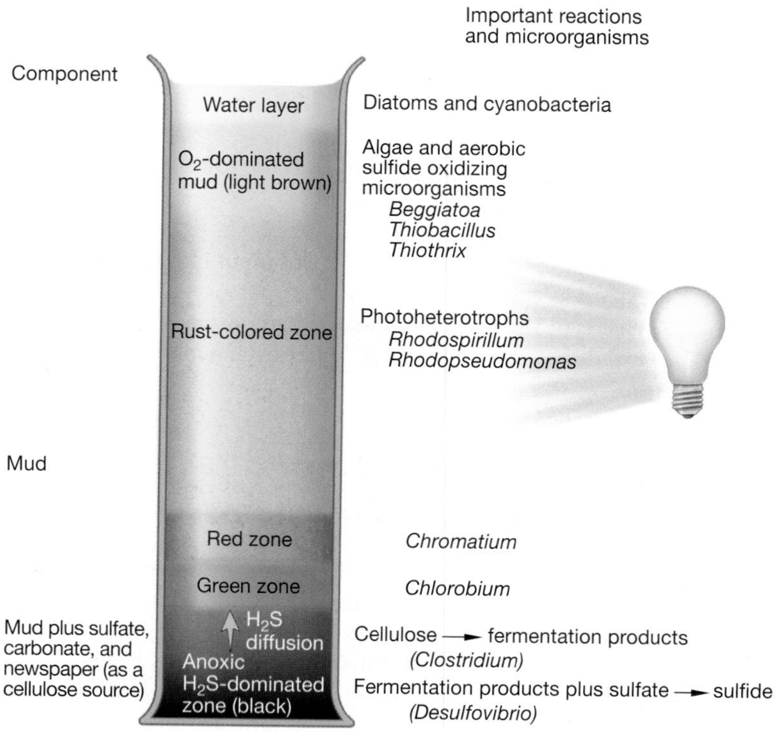

FIGURE 28.5 The Winogradsky Column. A microcosm in which microorganisms and nutrients interact over a vertical gradient. Fermentation products and sulfide migrate up from the reduced lower zone, and oxygen penetrates from the surface. This creates conditions similar to those in a lake or salt marsh with nutrient-rich sediments. Light is provided to simulate the penetration of sunlight into the anoxic lower region, which allows photosynthetic microorganisms to develop.

Figure 28.5 Micro Inquiry

How is sulfur cycled between the anoxygenic photosynthetic microbes *Chlorobium* and *Chromatium*, and anaerobic heterotrophic *Desulfovibrio*?

very dynamic. These ecosystems can be modeled in **Winogradsky columns** (**figure 28.5**), named after the pioneering microbial ecologist Sergei Winogradsky (1856–1953).

A Winogradsky column is easily constructed using a glass cylinder into which sediment is placed and then overlaid with saltwater. When the top of the column is sealed, much of the cylinder becomes anoxic. The addition of shredded newspaper introduces a source of cellulose that is degraded to fermentation products by members of the genus *Clostridium.* With these fermentation products available as electron donors, *Desulfovibrio* and other microbes use sulfate as a terminal electron acceptor to produce hydrogen sulfide (H_2S). The H_2S diffuses upward toward the oxygenated zone, creating a stable H_2S gradient. In this gradient, the photoautotrophs *Chlorobium* and *Chromatium* develop as visible olive green and purple zones, respectively. These lithoautotrophs use

H_2S as an electron source and CO_2, from sodium carbonate, as a carbon source. Above this region, the purple nonsulfur bacteria of the genera *Rhodospirillum* and *Rhodopseudomonas* can grow. These photoorganotrophs use organic matter as an electron donor under anoxic conditions and function in a zone where the sulfide level is lower. Both O_2 and H_2S may be present higher in the column, allowing specially adapted microorganisms to function. These include the chemolithotrophs *Beggiatoa* and *Thiothrix*, which use reduced sulfur compounds as electron donors and O_2 as an acceptor. In the upper portion of the column, diatoms and cyanobacteria may be visible. ◀◀ *Anaerobic respiration (section 10.6); Phototrophy (section 10.12); Photosynthetic bacteria (section 19.3); Purple nonsulfur bacteria (section 20.1); Class* Gammaproteobacteria *(section 20.3); Class* Clostridia *(section 21.2)*

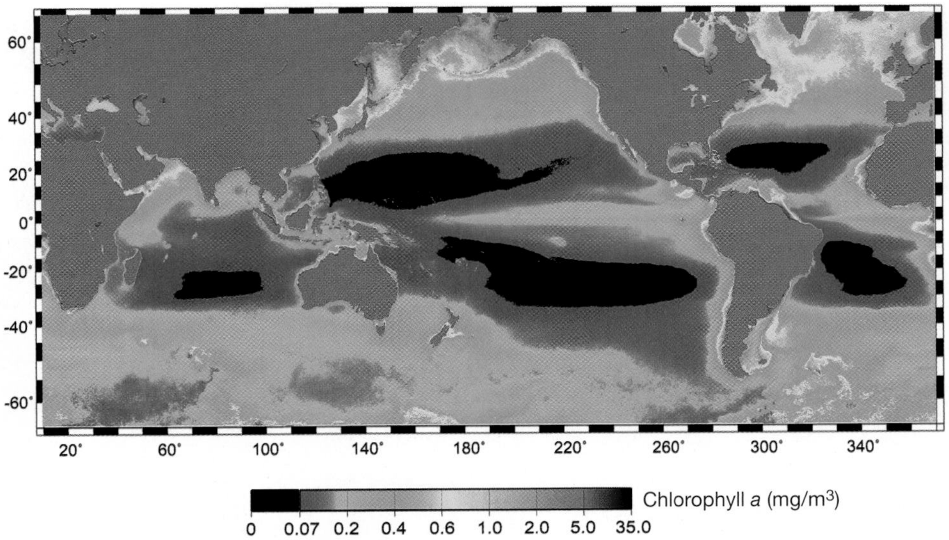

FIGURE 28.6 Mean Annual Surface Chlorophyll Levels. Global chlorophyll as measured by ocean satellites. The black regions correspond to subtropical gyres (massive spirals of water) where nutrient levels are especially low; note scale below image.

Image courtesy of NASA-Goddard Space Flight Center and the Orbital Sciences Corporation.

1. Define salt wedge and explain its influence on estuarine microbial communities.
2. Consider the fact that during droughts, rivers that normally flow into an estuary have a significantly diminished flow. Examine figure 28.3 and explain how this could result in salt intrusion into a public water supply. What do you think the consequences would be on estuarine microbial communities?
3. Name two groups of protists known to cause HABs in marine ecosystems. What hazards do HABs pose for other marine organisms and humans?
4. Describe the ecosystem that develops within a Winogradsky column. Trace the flow of electron donors and acceptors.

Microorganisms in the Open Ocean

Sometimes called "the invisible rain forest," the upper 200 to 300 m of the open ocean are home to a diverse collection of microbes. Nutrient cycling in the ocean is dominated by microbes, which outpace larger, eukaryotic marine organisms in terms of number, diversity, and metabolic activity. This can be observed in a single microliter of surface seawater, which is a trophic mishmash: among phototrophs, this drop of water would contain about 100 to 200 cyanobacteria and 10 to 20 photosynthetic protists. These microbes secrete fixed carbon in the form of dissolved organic carbon (DOC), which would be consumed by the 1,000 heterotrophic bacteria in this drop. All of these microbes would be subject by predation because this same tiny drop of water would include about 10,000 viruses and at least 10 protozoa (figure 28.2).

One of the major goals of microbial oceanography is to figure out the fate of all the carbon fixed by marine microbes. The use of satellite imagery to measure surface chlorophyll *a* concentrations (**figure 28.6**) shows that coastal areas are most productive and regions of subtropical gyres—huge vortexes of water—are least productive. The open ocean is an **oligotrophic environment**—that is to say, nutrient levels are very low. Because the influx of new nutrients is limited, primary productivity (and thus the whole ecosystem) depends on rapid recycling of nutrients. In contrast to terrestrial ecosystems, at least 90% of the nutrients required by the microbial community are recycled within the microbial loop.

Global climate change has had a measurable impact on phytoplankton communities, as satellite imaging of chlorophyll *a* concentrations shows (figure 28.6). For instance, in 2009 it was reported that summertime surface chlorophyll *a* declined by 12% over the preceding 30 years along the western shelf of the Antarctic Peninsula. This decrease corresponds to observed changes in krill (zooplankton) and penguin populations in this region, demonstrating the effect this has on higher trophic levels.

Global climate change has also focused intense scrutiny on determining how much CO_2 phytoplankton can remove from the atmosphere and sequester in the benthos. As shown in **figure 28.7**, there is a constant exchange of CO_2 at the ocean surface, as well as export of carbon to the seafloor. Organic matter escapes the photic zone and falls through the depths as **marine snow** (**figure 28.8*a***). This material gets its name from its appearance, sometimes seen in video images of mid- and deep-ocean water, where drifting flocculent particles look much like snow. Marine snow consists primarily of fecal pellets, diatom frustules, and other materials that are not easily degraded. During its fall to the bottom, marine snow is colonized by microbes and mineralization continues. Bacterial population densities within and on marine snow can reach 10^8 to 10^9 per milliliter, which is orders of

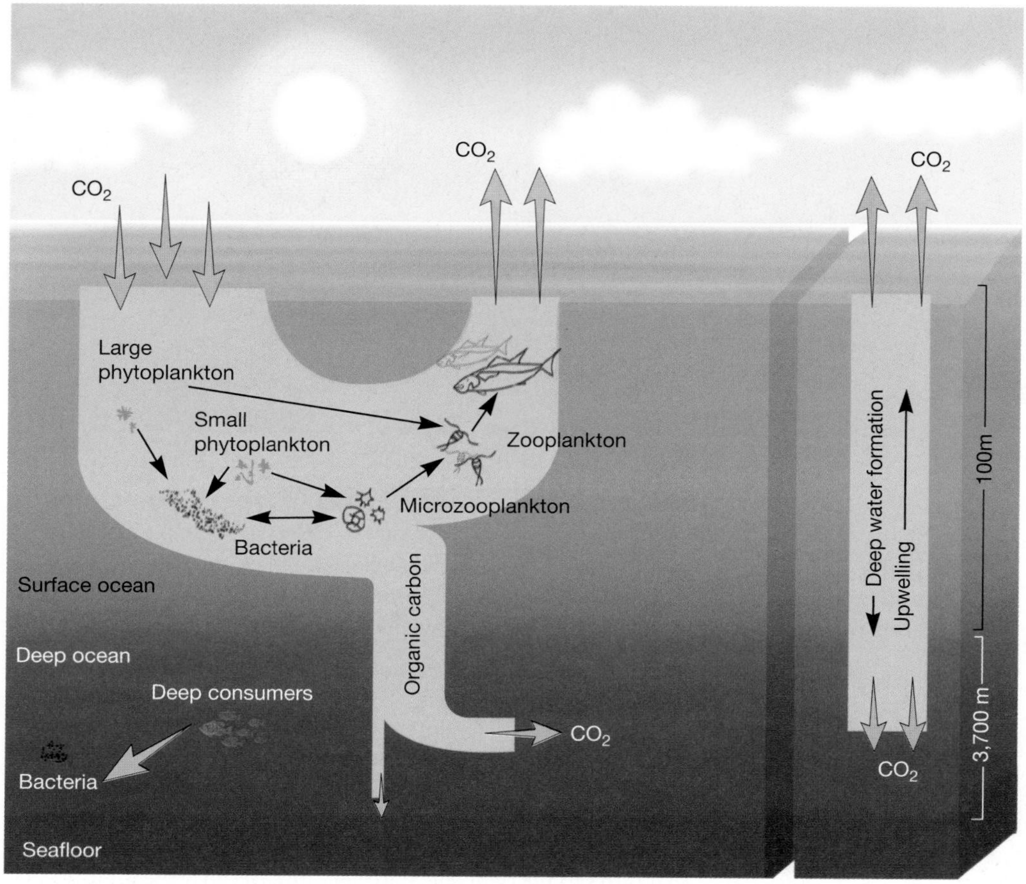

FIGURE 28.7 The Biological Carbon Pump. The vast majority of the carbon fixed by microbial autotrophs in the open ocean remains in the photic zone. However, a small fraction is "pumped" to the seafloor, which in turn returns much of it to the surface in regions where deep and surface waters mix (upwelling regions). The flux of CO_2 in and out of the world's oceans is in equilibrium, but this equilibrium may be upset by the increase in atmospheric CO_2, the hallmark of global warming. Note depth scale on right.

magnitudes higher than the surrounding deep water. They secrete a variety of exoenzymes to break down organic polymers and assimilate the resulting monomers. However, it appears that microbial degradation and assimilation are not tightly linked, as sinking marine snow leaves a plume of DOM, nitrogen, phosphorus, and iron during sinking (figure 28.8*b*). This material does not go to waste: it is consumed by free-swimming microbes. By the time a piece of marine snow reaches the seafloor, less than 1% of photosynthetically derived material remains unaltered. Once there, only a tiny fraction will be buried in the sediments; most will return to the surface in upwelling regions (figure 28.7).

An urgent question is whether or not the oceans can draw down (remove and bury) more CO_2 as atmospheric concentrations continue to increase. Presently it is estimated that this "biological pump" buries (sequesters) about 3 gigatons of carbon per year in deep-ocean sediments. Theoretically, increases in atmospheric CO_2 could be offset by increasing the rate and volume of carbon sequestration. This in turn would slow global warming. Some have attempted to accomplish this by fertilizing specific regions of the oceans. These regions, known as **high-nitrate, low-chlorophyll (HNLC)** areas, are limited by iron. A number of experiments have been performed wherein large transects of the southern Pacific have been fertilized with iron to trigger diatom blooms. However, as shown in **figure 28.9**, it is not apparent that this will be effective in diminishing atmospheric CO_2.

Search This: Ocean iron fertilization/whoi

Unlike the HNLC areas, most of the remainder of the open ocean is limited by nitrogen, not iron. Because there is no terrestrial runoff, the only source of "new" organic nitrogen at sea is biological nitrogen fixation. There appear to be at least four sources of biological N_2 fixation. For many years, it was thought that the filamentous cyanobacteria *Richelia intracellularis* was the only microbe that fixed significant amount of N_2. This microbe lives in association (either on the surface or intracellularly) with certain diatoms. Then in the 1990s it was discovered that the filamentous cyanobacterium *Trichodesmium* fixes far more N_2 than previously thought. A short time later, it was reported that marine unicellular cyanobacteria also fix N_2. These cyanobacteria protect their oxygen-sensitive nitrogenase enzymes by fixing nitrogen at night, when they respire, rather than use photosystems I and II, which results in the evolution of oxygen. This leads us to the fourth type of N_2-fixing microbe. In 2008 unicellular cyanobacteria were discovered that fix N_2 but lack photosystem II. This enables N_2 fixation during the day because these microbes do not evolve oxygen. ◀◀ *Nitrogen fixation (section 11.5); Phylum* Cyanobacteria *(section 19.3)*

The other recent discovery that has revolutionized our understanding of nitrogen cycling is the presence of bacteria that perform the anammox reaction below the photic zone, where oxygen concentrations reach a minimum (*see figure 19.11*). In nitrogen-limited areas, it is generally understood that there is a net loss of ammonium, nitrate, and nitrite. A large fraction of this loss has been attributed to denitrification (the anaerobic reduction of nitrate to N_2) occurring at the oxygen-minimum zone just below the photic zone. It now appears that bacteria capable of the anammox reaction—the

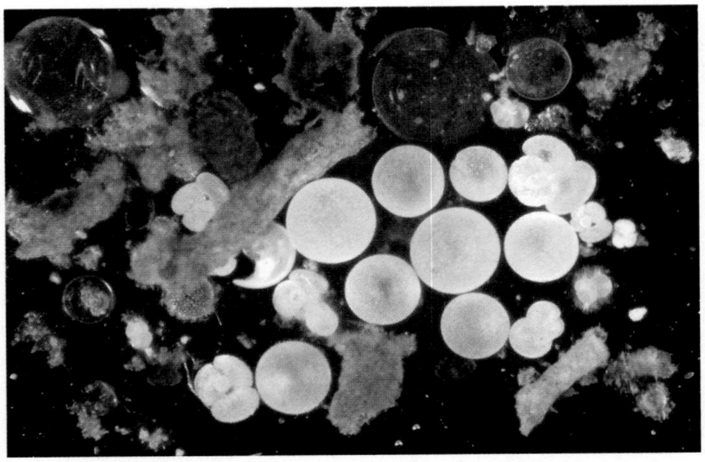

(a)

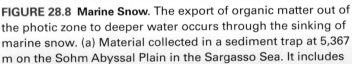

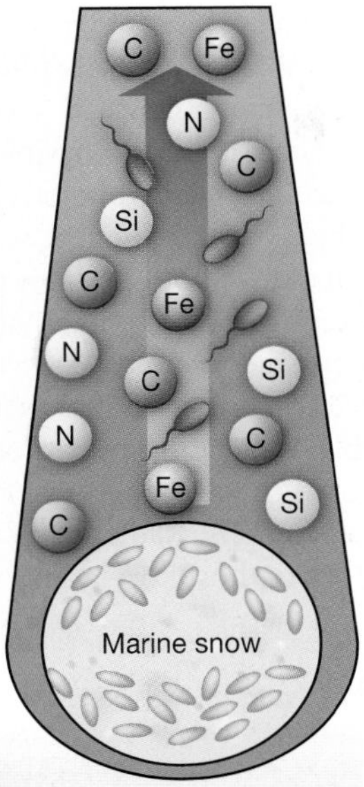

(b)

FIGURE 28.8 Marine Snow. The export of organic matter out of the photic zone to deeper water occurs through the sinking of marine snow. (a) Material collected in a sediment trap at 5,367 m on the Sohm Abyssal Plain in the Sargasso Sea. It includes cylindrical fecal pellets, planktonic tests (round white objects), transparent snail-like pteropod shells, radiolarians, and diatoms. (b) The hydrolytic activity of the exoenzymes secreted by the microbes that colonize marine snow generates a plume of nutrients during sinking. This in turn attracts more microbes.

Figure 28.8 Micro Inquiry

How does marine snow contribute to the biological carbon pump shown in figure 28.7?

anaerobic oxidation of NH_4^+ to N_2—are responsible for much of the loss of nitrogen that could otherwise support life. These bacteria are members of the interesting phylum *Planctomycetes.* Nitrogen that escapes denitrification and sinks further is therefore dominated by NH_3/NH_4^+, yet most of the nitrogen that accumulates at depth is nitrate (NO_3^-). Recall that nitrification is the oxidation of NH_3 to NO_3^-. Surprisingly much (if not most) of this nitrification is accomplished by crenarchaea. These mesophilic archaea oxidize NH_3 to NO_2^-, which is then oxidized to NO_3^- by bacteria. Once again, the urgency to understand the global oceanic nitrogen budget reflects concern about increasing levels of atmospheric CO_2, global warming, and the capacity of the world's oceans to remove more CO_2 while maintaining overall ecosystem equilibrium. ◀◀ *Phylum* Crenarchaeota *(section 18.2); Phylum* Planctomycetes *(section 19.4); Nitrogen cycle (section 26.1)*

It is clear that when researchers consider microbial activities in the open ocean, they are assessing global processes. So perhaps it is not surprising that the most abundant group of monophyletic organisms on Earth is marine. Members of the α-proteobacterial clade called **SAR11** (named after the *Sar*gasso Sea, where they were first detected) have been detected by their rRNA genes from almost all open-ocean samples taken worldwide. In addition, they have been found at depths of 3,000 m, in coastal waters, and even in some freshwater lakes. Using fluorescence in situ hybridization (FISH; *see figure 27.7*), SAR11 bacteria have been found to constitute 25 to 50% of the total bacteria and archaea in the surface waters in both nearshore and open-ocean samples. Indeed, SAR11 bacteria are estimated to constitute about 25% of all microbial life on the planet. ◀◀ *Culturing techniques (section 27.1)*

The genome of the SAR11 isolate *Pelagibacter ubique* is 1.31 Mb—among the smallest genomes of independently replicating cells sequenced to date. This is partly because the SAR11 genome is devoid of genomic waste. There are no pseudogenes, phage genes, or recent gene duplications, and the number of nucleotides between coding regions is very limited. Nonetheless, *P. ubique* has the genes necessary for the Entner-Doudoroff pathway, the TCA cycle, and a complete electron transport chain. It has adapted to life in the oligotrophic ocean by encoding a number of high-affinity nutrient transport systems and maintaining its small size, thereby optimizing its surface area-to-volume ratio. It uses respiration to capture energy, but it also has a **proteorhodopsin** proton pump to contribute to the generation of a proton motive force.

The physiology of SAR11 illustrates a recurring metabolic theme that is emerging in the study of marine microorganisms: many have evolved strategies to supplement their energy reserves. Since the first discovery of rhodopsin-containing γ-proteobacteria in 2000, the presence of similar genes have been found in two bacterial taxa, α-proteobacteria and *Bacteroidetes*, and the archaeal phylum *Euryarchaeota.* In all cases, the rhodopsin is distinct from the bacteriorhodopsin found in haloarchaea (recall that bacteriorhodopsin was named before the *Archaea* were recognized as a distinct domain). These newly discovered rhodopsins are called proteorhodopsin, even though it is now known that some microbes possessing this form of rhodopsin are not *Proteobacteria.* None of these microorganisms is autotrophic, instead it is hypothesized that they use the light-driven rhodopsin pump to supplement ATP pools in nutrient-depleted waters. This predicts that proteorhodopsin-containing bacteria and archaea would have a growth advantage. This can most easily be tested in pure culture because these microbes should grow faster in the light than the dark. While this has been shown for at least one bacterium, it has been difficult to demonstrate in others. This is in part because many of these microbes are currently difficult to grow in the laboratory. Nonetheless, metagenomic analysis indicates that about 13% of heterotrophic marine bacteria and archaea have the genes needed for proteorhodopsin biosynthesis. ◀◀ *Rhodopsin-based phototrophy (section 10.12); Phylum* Euryarchaeota *(section 18.3)*

Another group of marine microbes that appears to be supplementing their energy reserves are the **aerobic anoxygenic phototrophs** (AAnPs). Although these bacteria were discovered several decades ago, for a very long time they were considered a physiological curiosity. Bacteria capable of aerobic anoxygenic phototrophy are known to possess bacteriochlorophyll *a* (Bchl*a*), but it was a great surprise when it was discovered that this photosynthetic pigment is widely distributed in surface ocean waters. A combination of cultivation and metagenomic techniques demonstrates that members of the α-proteobacteria (the *Roseobacter* clade), as well as the β- and γ-proteobacteria, are capable of this form of phototrophy (*see table 19.1*). It would be incorrect to call these bacteria photosynthetic because they are not autotrophs. Rather, they use light to drive the synthesis of ATP (photophosphorylation) but meet their carbon needs by the assimilation of organic carbon; that is, they are photoheterotrophs. For instance, the γ-proteobacterium AAnP *Congregibacter litoralis* requires carboxylic acids, fatty acids, or oligopeptides for growth. By meeting some of their ATP demands phototrophically, AAnPs can divert more organic carbon to biosynthesis and less to catabolic processes. Bacteria that possess Bchl*a* are much less abundant than those that produce proteorhodopsin. This may reflect the fact that genes for proteorhodopsin have been widely subject to horizontal gene transfer, as only six genes are required for a functional light-driven proton pump. This is in contrast to the 50 genes needed to assemble the anoxygenic light-harvesting complex, reaction center, Bchl*a*, and carotenoids.

The α-proteobacterium *Silicobacter pomeroyi* appears to be capable of a third interesting metabolic strategy to increase its ATP yield: **lithoheterotrophy.** Like photoheterotrophy, in which light is used for energy and organic carbon is consumed, lithoheterotrophy uses inorganic chemicals as a source of energy but requires organic carbon as its carbon source. Within the last several years, a number of carbon monoxide (CO) oxidizing heterotrophs have been cultured from coastal systems. These bacteria produce CO dehydrogenase, thereby oxidizing CO to CO_2, but unlike CO-oxidizing bacteria isolated in previous decades, they do not have the enzymes necessary for the conversion of this CO_2 into biomass. As has been shown with *S. pomeroyi*, they seem to use CO as a supplemental electron donor. Thus they need not oxidize as much organic carbon to generate NADH for this purpose. Instead, more organic carbon can be used for biosynthesis. This hypothesis is supported by thermodynamic considerations: the oxidation of one molecule of CO should release enough energy to drive the production of one ATP. Measurements of CO lithoheterotrophs show that they make up about 7% of the bacterioplankton in open ocean and coastal waters sampled.

It is not difficult to hypothesize that these alternatives to "traditional" organotrophy would be beneficial to individual cells in a nutrient-stressed environment. What is more problematic is estimating how much these processes contribute to the carbon budget of marine habitats. Data in both regards is sparse. Theoretically, the advantage these processes confer must more than offset the energetic cost of carrying the genes required for their expression. But it remains to be seen if these processes actually make these microbes better competitors. Likewise, the impact these metabolic

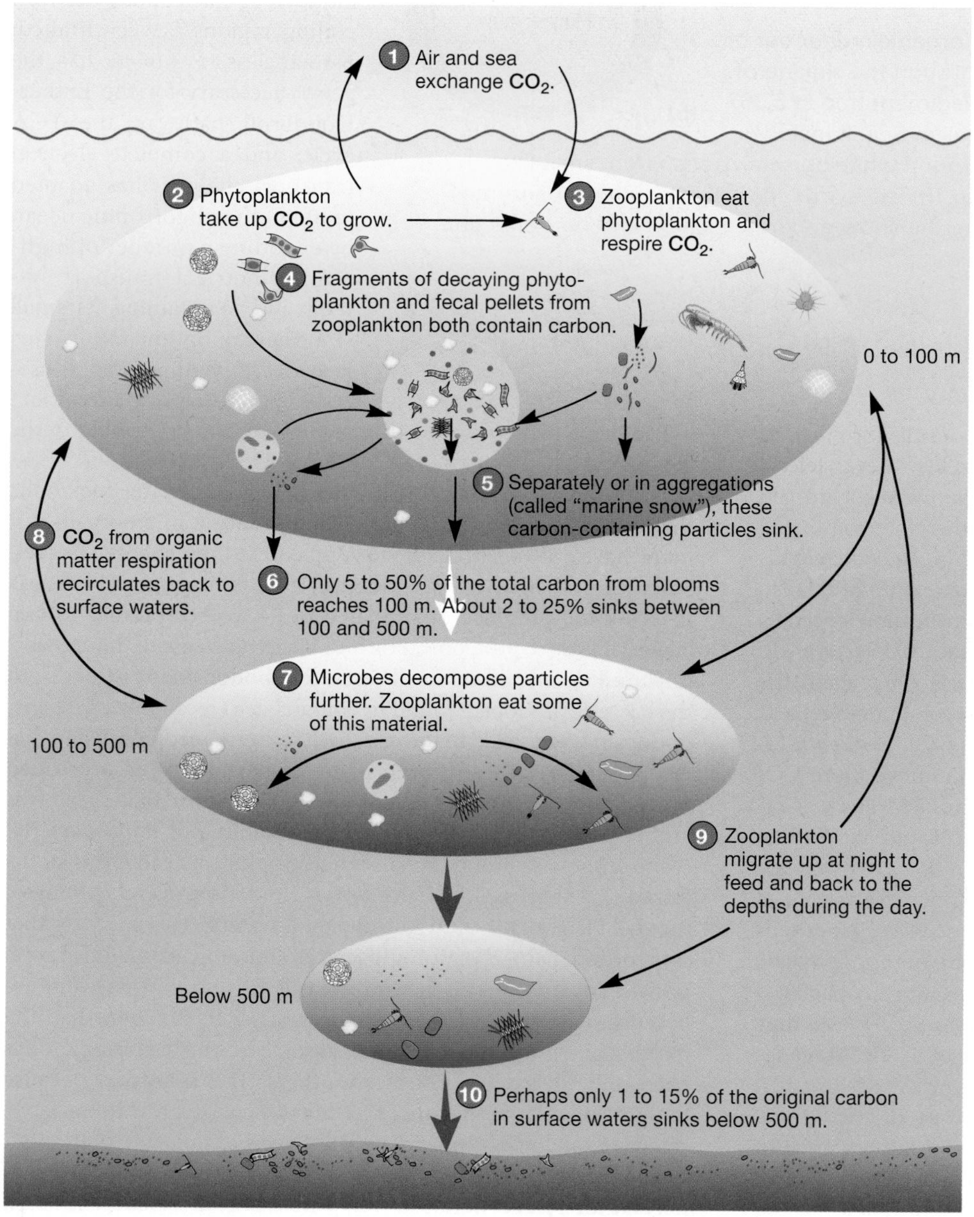
1 Air and sea exchange CO_2.
2 Phytoplankton take up CO_2 to grow.
3 Zooplankton eat phytoplankton and respire CO_2.
4 Fragments of decaying phytoplankton and fecal pellets from zooplankton both contain carbon.
0 to 100 m
5 Separately or in aggregations (called "marine snow"), these carbon-containing particles sink.
8 CO_2 from organic matter respiration recirculates back to surface waters.
6 Only 5 to 50% of the total carbon from blooms reaches 100 m. About 2 to 25% sinks between 100 and 500 m.
7 Microbes decompose particles further. Zooplankton eat some of this material.
100 to 500 m
9 Zooplankton migrate up at night to feed and back to the depths during the day.
Below 500 m
10 Perhaps only 1 to 15% of the original carbon in surface waters sinks below 500 m.

FIGURE 28.9 The Fate of Carbon from the Photic Zone to the Benthos.

strategies have on carbon flux at the community, ecosystem, and global level awaits more investigation. It is questions like these that make microbial oceanography such a dynamic and exciting field.

Aquatic Viruses

As mentioned in our discussion of the microbial loop (figure 28.2), viruses are important members of marine and freshwater microbial communities. In fact, **virioplankton** are the most numerous members of marine ecosystems. However, quantifying viruses is tricky: The traditional method of plaque formation requires knowledge of both the virus and its host, and the ability to grow the host in the laboratory. Because so few microbes have been cultured, this prevents the measurement of virus diversity by examining actual virus infection. Instead, virus particles may be visualized directly. There are two approaches for this. Electron microscopy is the most rigorous but requires the concentration of many tens of liters of water. Epifluorescence microscopy is a common method used to examine viruses in aquatic systems. Viruses are collected on a filter with a pore size that is less than 0.2 μm, and their nucleic acids are stained with a fluorescent dye, such as YO-PRO and SYBR Green. Neither electron or epifluorescence microscopy prove that any given virus can actually infect a host cell, so viruses enumerated in this way are called ***virus*l*ike* *p*articles (VLPs).** Using this approach, the average VLP density in seawater is between 10^6 to 10^7 per milliliter (although in some cases, it may be closer to 10^8 per milliliter); their numbers decline to roughly 10^6 below about 250 m. Marine viruses are so abundant that VLPs are now recognized as the most abundant microbes on Earth (**figure 28.10**).

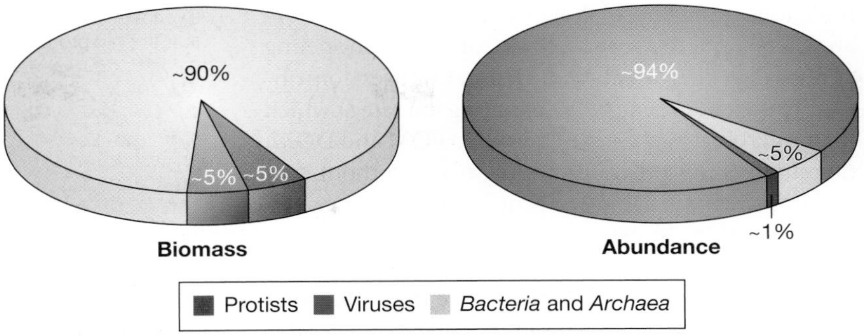

FIGURE 28.10 Relative Biomass and Abundance of Marine Microbes. (a) The biomass of bacteria and archaea far exceeds that of both protists and viruses. (b) When individuals are counted, however, the vast majority of marine microbes are viruses.

Viral diversity is vast, including single- and double-stranded RNA and DNA viruses that infect archaea, bacteria, and protists. Metagenomic analysis of cloned viral genomes has been used to explore viral diversity. Viruses can be effectively separated from other biomass by filtration, so the construction of metagenomic libraries that contain only viral sequences is possible. In one such study, viral communities were sampled from the Arctic Ocean, North Pacific coastal waters, the Gulf of Mexico, and the Sargasso Sea. Of the roughly 1.8 million nucleotides that were obtained, 90% had no recognizable match on any database. Thus not only is viral genetic diversity immense, it is largely unexplored. ◀◀ *The viruses (chapter 25)*

Clearly viruses must be major agents of mortality in the sea. Indeed, viruses are thought to kill on average about 20% of the marine microbial biomass daily. Measurement of virus-induced microbial mortality shows that it is highly variable. Viral abundance frequently corresponds to the microbial host that is most active in the community. Virus-mediated cell lysis can significantly impact community structure. Models predict that as one microbial species (or strain) becomes numerically dominant, it soon will be targeted by lytic viruses and thus its population will decline. This permits another bacterial or archaeal species (or strain) to flourish, which then becomes subject to intense viral lysis, and so on. This "kill the winner" model has garnered much attention, but more experimental evidence is needed before it is widely accepted. Perhaps the most compelling data to date follows the fate of blooms of the coccolithophore *Emiliania huxleyi*. Such blooms are so intense they can be imaged from space, yet their collapse is thought to be (at least in part) the result of viral lysis (**figure 28.11**). It should be noted, however, that in most cases, viral lysis does not result in the complete collapse of a host population. This may be due partly to strains of the host species that are resistant to virus

FIGURE 28.11 Viral Lysis and the Termination of a Massive Plankton Bloom. This satellite image shows the demise of a 500 km-long bloom (seen as light blue smear) of the coccolithophorid *Emiliania huxleyi* (inset). While the exhaustion of available nutrients also contributes to the death of such an enormous bloom, research has demonstrated that viruses specific to this protist are principle agents in the bloom's demise. The green land mass is Ireland.

infection, as has been shown with the abundant cyanobacterium *Synechococcus*. Computer modeling and model experiments indicate that viruses contribute to nutrient cycling by accelerating the rate at which their microbial hosts are converted to POM and DOM, thereby feeding other microorganisms without first making them available for protists and other bacteriovores. This "short-circuits" the microbial loop (**figure 28.12**).

It is not surprising that phages are important vectors for horizontal gene transfer in marine ecosystems. In fact, it has been calculated that in the oceans, phage-mediated gene transfer occurs at an astounding rate of 20 billion times per second. Recently the importance of phage-mediated horizontal gene transfer was demonstrated when it was discovered that cyanophages that infect the cyanobacteria *Synechococcus* and *Prochlorococcus* carry structural genes for photosynthetic reaction center proteins. These phages shuttle these genes between strains of cyanobacteria and may thereby play a critical role in the evolution of these important primary producers.

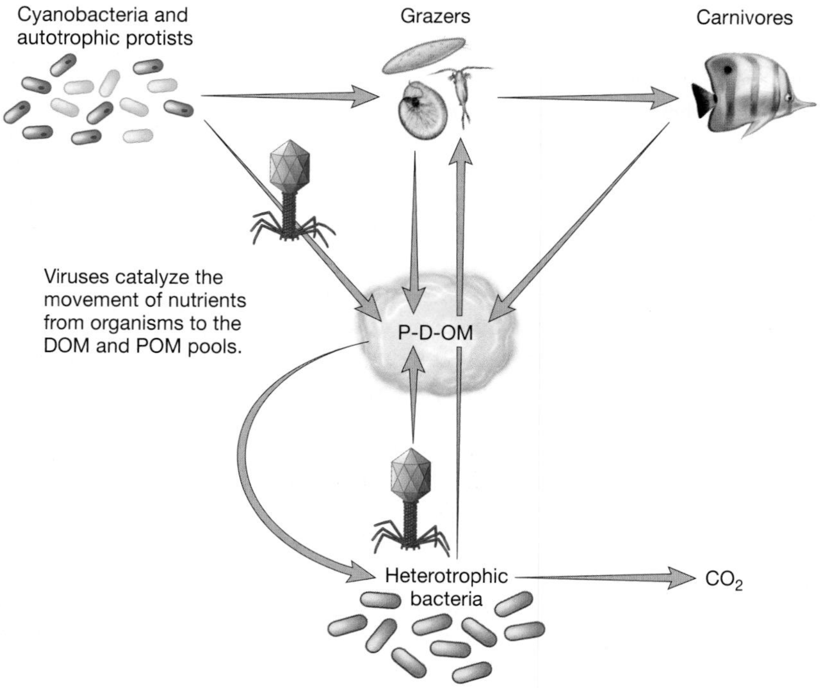

FIGURE 28.12 The Role of Viruses in the Microbial Loop. Viral lysis of autotrophic and heterotrophic microbes accelerates the rate at which these microbes are converted to particulate and dissolved organic matter (P-D-OM). This is thought to increase net community respiration and decrease the efficiency of nutrient transfer to higher trophic levels.

Figure 28.12 Micro Inquiry

When might viruses choose lysogeny rather than lysis and how would this affect carbon flow (also see figure 28.2)?

Microorganisms in Benthic Marine Environments

Sixty-five percent of the Earth's crust is under the sea, which means that of all the world's microbial ecosystems, we know the least about the largest. However, the combination of deep-ocean drilling projects and the exploration of geologically active sites, such as submarine volcanoes and hydrothermal vents, has revealed that the study of ocean sediments, or benthos, can be rewarding and surprising. Marine sediments range from the very shallow to the deepest trenches, from dimly illuminated to completely dark, and from the newest sediment on Earth to material that is millions of years old. The temperature and age of such sediments depend on their proximity to geologically active areas. Hydrothermal vent communities with large and diverse invertebrates, some of which depend on endosymbiotic chemolithotrophic bacteria, have been intensely investigated since their exciting discovery in the late 1970s. These microbes are discussed in chapter 30. Because the vast majority of Earth's crust lies at great depth far from geothermally active regions, most benthic marine microbes live under high pressure, without light, and at temperatures between 1°C to 4°C.

Deep-ocean sediments were once thought to be devoid of all life and were therefore not thought worth the considerable effort it takes to study them. In fact, it took an international consortium of scientists to organize the Ocean Drilling Project in 1985, which has been expanded to the Integrated Ocean Drilling Program (ODP) through 2013. Researchers aboard the research vessel *JOIDES Resolution* drill cores of sediments from water depths up to 8,200 m (at its deepest, the ocean is about 11,000 m). Microbiologists now know that far from being sterile, seafloor sediments contain 10 to 10,000 more cells per unit volume than productive surface waters. Reasons for this difference are complicated but include high energy availability and limited grazing pressure. This result is especially surprising because the majority of the ocean floor receives only about 1 gram of organic carbon per square meter per year.

Global extrapolation of measured subsurface microbial cell populations results in the amazing conclusion that the marine deep biosphere (up to 0.6 km below the sediment surface) is the "hidden majority" of all microbial biomass. Estimates range from half to five-sixths of the Earth's total microbial biomass and between one-tenth to one-third of Earth's total living biomass. As might be predicted, most of these microbes are not amenable to laboratory culture. SSU rRNA sequence analysis of DNA extracted from ODP sites has been surprising in the lack of genes with homology to known sulfate reducers and methanogens, with the notable exception of the archaeon *Methanocaldococcus*. In fact, recent evidence suggests that the majority of subsurface microbes are archaea. Furthermore, the sediment surface and subsurface appear to be teeming with viruses; it is estimated that each year viral lysis in the benthos releases up to 630 million tons of carbon that had been sequestered by marine snow and other falling particulates. This exciting finding suggests that a significant fraction of Earth's biomass is amazingly active yet largely uncharacterized.

Subsurface microbes are astonishing in another regard. Their calculated growth rates—years to thousands of years—seem almost impossibly slow. To survive at these depths, microorganisms must also be able to tolerate atmospheric pressures up to 1,100 atm (pressure increases about 1 atm per 10 m depth). Such microbes are said to be **piezophilic** (**barophilic;** Greek, *baro,* weight, and *philein,* to love). Some are obligate barophiles and must be cultured in hyperbaric incubation chambers. In fact, scientists have yet to find the subterranean depth limit of microbial growth. It appears that this value will not be governed by pressure; rather, it will be determined by temperature. It seems unlikely that the maximum temperature at which life can be sustained has been identified. ⏮ *Thermostability (section 18.1)*

Interestingly, recent deep-sea sediment drilling has turned our understanding of bacterial energetics literally upside down. As discussed in chapter 10, it is generally understood that anaerobic respiration occurs such that there is preferential use of available terminal electron acceptors. Acceptors that yield the most energy (more negative ΔG) from the oxidation of NADH or an inorganic reduced compound (e.g., H_2, H_2S) will be used before those that produce a less negative ΔG (*see table 9.1 and figure 10.9*). Thus following oxygen depletion, nitrate will be reduced, then manganese, iron, sulfate, and finally, carbon dioxide. When sediment cores up to 420 m deep were collected off the coast of Peru, this predictable profile of electron acceptors and their microbial-derived reduced products was observed within the upper strata of the sediments (**figure 28.13**). However, when these signature chemical compounds were measured at great depth, the profile was reversed. This suggests the presence of unknown sources of these electron acceptors at subsurface depths of more than 420 m. In addition, contrary to our long-held notion of thermodynamic limits, methane formation (methanogenesis) and iron and manganese reduction co-occur in sediments with high sulfate concentrations. Although the identity of the microbes that make up these communities awaits further study, it is clear that with densities of 10^8 cells per gram of sediment at the seafloor surface and 10^4 cells per gram in deep subsurface sediments, these communities are important. ⏮ *Free energy and reactions (section 9.3); Oxidation-reduction reactions (section 9.5); Anaerobic respiration (section 10.6)*

Another outcome of deep-ocean sediment exploration has been the discovery of a variety of hydrocarbon-fueled microbial communities on continental margins at depths between 200 and 3,500 m. Depending on the surface topology and the rate at which hydrocarbons are emitted, they are called pockmarks, gas chimneys, mud volcanoes, brine ponds, and oil and asphalt seeps. Although diverse microbes can use hydrocarbons as a sole carbon source, oxidizing them with oxygen or sulfate as the electron acceptor, oxygen is rapidly depleted, making conditions right for sulfate-reducing δ-proteobacteria. In addition, methane-oxidizing archaea grow in consortia with sulfate-reducing bacteria of the *Desulfococcus* and *Desulfobulbus* groups, as discussed in chapter 18. Some cold seeps are the sites of microbial mats, which can sometimes cover several hundreds of meters. These mats are

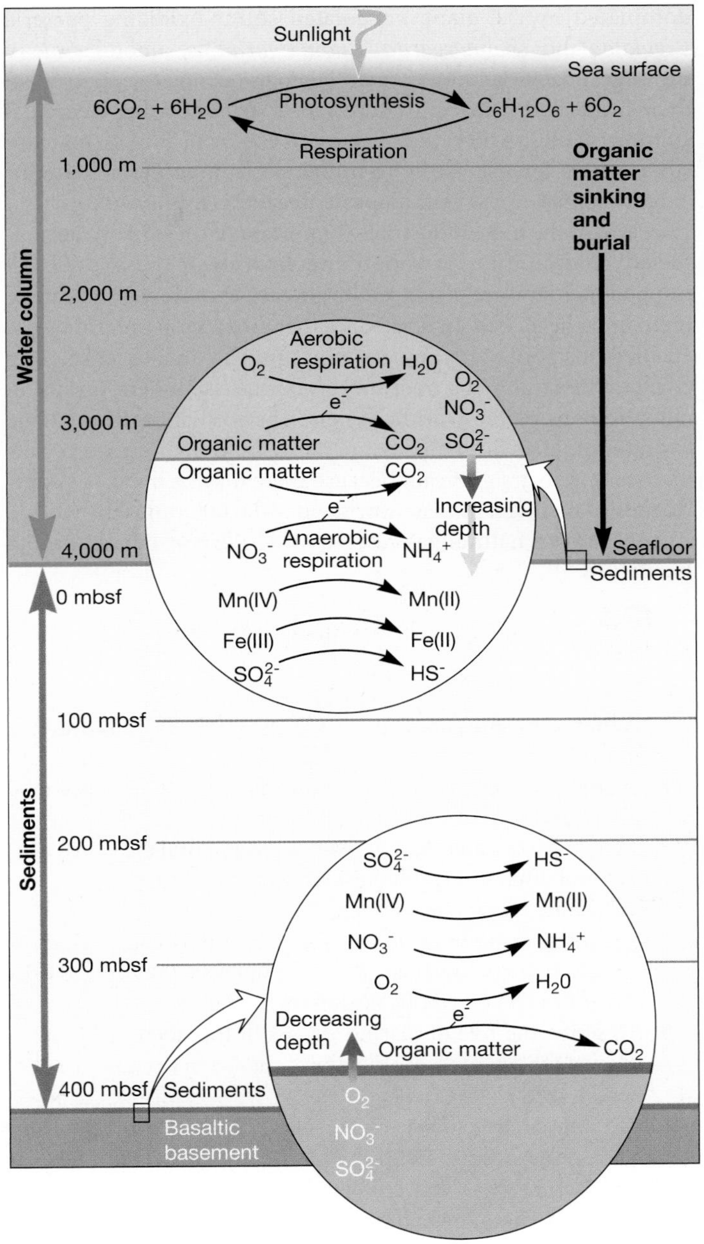

FIGURE 28.13 Microbial Activity in Deep-Ocean Sediments. At the surface of the seafloor, reduction of oxidized substrates that serve as electron acceptors in anaerobic respiration follows a predictable stratification based on thermodynamic considerations. This sequence is just the opposite in very deep subsurface sediments, suggesting a source of electron acceptors from deep within Earth's crust. Meters below seafloor, mbsf.

dominated by the giant, vacuolated sulfur-oxidizing bacteria *Beggiatoa* and *Thiomargarita*. *Thiomargarita* are among some of the largest bacteria known; their vacuoles occupy the majority of their intracellular space. In addition to accumulating elemental sulfur granules, which they use as an electron donor, they also store nitrate for use as the terminal electron acceptor in sulfur oxidation. ◀◀ *Class* Gammaproteobacteria *(section 20.3)*

Perhaps the most industrially important of these hydrocarbon-fueled communities are **methane hydrates.** These pools of trapped methane are produced by archaea that convert acetate to methane. The lack of associated methane-oxidizing microbes has enabled this pool of natural gas to accumulate in latticelike cages of crystalline water 500 m or more below the sediment surface in many regions of the world's oceans. The formation of methane hydrates requires both cold temperatures and high pressure. This discovery is very significant because there may be up to 10^{13} metric tons of methane hydrate worldwide—80,000 times the world's current known natural gas reserve. ◀◀ *Phylum* Euryarchaeota *(section 18.3).*

Search This: JOIDES Resolution/odp

1. What is marine snow? Why is it important in CO_2 draw-down?
2. Name two sources of organic nitrogen in the open ocean that have only recently been recognized.
3. Why do you think that, despite its great abundance, SAR11 was not discovered until the late twentieth century?
4. List some metabolic strategies that have evolved to enable microbial survival in oligotrophic marine habitats. Which do you think is the most successful in terms of numbers of microbes and metabolic flexibility? Explain your answer.
5. Describe the role of marine viruses in the microbial loop. How might community dynamics change in the absence of viral lysis?
6. Explain what is meant by "upside-down microbial energetics," as described for some deep-subsurface ocean sediments.
7. What are methane hydrates? Why are they important?

28.3 Microorganisms in Freshwater Ecosystems

While the vast majority of water on Earth is in marine environments, freshwater is crucial to our terrestrial existence. As shown in **figure 28.14**, of the 3.9% of Earth's water that is not saline, only a tiny fraction represents the familiar streams, rivers and wetlands that we associate with the term freshwater. We begin our discussion with the largest fraction of freshwater—that which is frozen at the Earth's polar regions. We then move to streams, rivers, and lakes.

Microorganisms in Glaciers and Permanently Frozen Lakes

Much of the ice on Earth has remained frozen for millions of years. Indeed, it is important to note that a majority of the Earth's surface never exceeds a temperature of 5°C. This includes polar regions, the deep ocean, and high-altitude terrestrial locations throughout the world. Surprisingly, microbes within glaciers are not dormant. Rather, evidence that active microbial communities exist in these environments has been growing over the last decade. In fact, this is an exciting time in glacial microbiology; determining the diversity in these systems and assessing the role of these microorganisms in biogeochemical cycling can involve novel and creative techniques. The results may be of great consequence because glaciers have traditionally been regarded as areas that do not contribute to the global carbon budget. In addition, since the discovery of ice on Mars and on Jupiter's moon Europa, astrobiologists have become very interested in ice-dwelling psychrophilic microbes. ◀◀ *Temperature (section 7.6)*

Among the frozen landscapes of interest to microbiologists are permanently frozen lakes, such as Antarctica's McMurdo Dry Valley Lakes, where the ice is 3 to 6 m deep. Life in these ecosystems depends on the photosynthetic activity of microbial psychrophiles. In contrast, lakes that lie below glaciers are blocked from solar radiation. These communities are driven by chemosynthesis. One of the most intriguing and well-studied Antarctic habitats is Lake Vostok, one of 68 lakes located 3 to 4 km below the East Antarctic Ice Sheet (**figure 28.15**). Geothermal heating, pressure, and the insulation of the overlying ice keep these lakes in a liquid state. It is thought that Lake Vostok was formed approximately 420,000 years ago. This stable environment supports a number of microbes, including proteobacteria and actinomycetes. The overlying ice also harbors similar microbes, although at lower population densities. To find out if these microbial populations are active, radiolabeled substrates, including ^{14}C-acetate and ^{14}C-glucose, were added to samples incubated at an Antarctic laboratory. Indeed, the respiration of these compounds, measured as ^{14}C-CO_2, demonstrates that these interesting communities are dynamic.

Search This: Subglacial Lake Vostok

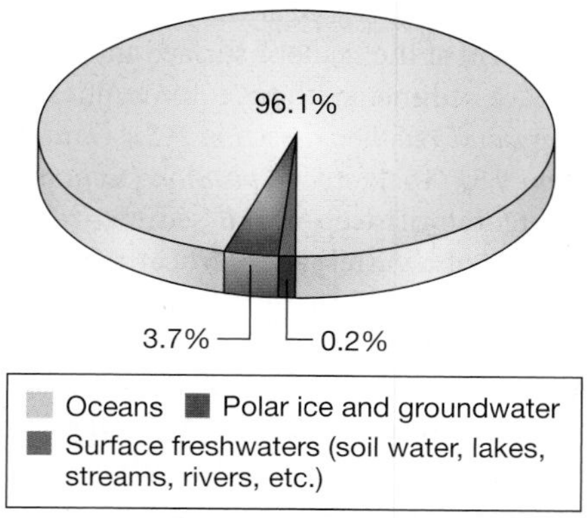

FIGURE 28.14 Relative Distribution of Freshwater on Earth.

Microorganisms in Streams and Rivers

As freshwater glaciers melt, their waters enter streams and rivers. This marks a departure from an environment that is stable on a geologic time scale to one that is extremely changeable.

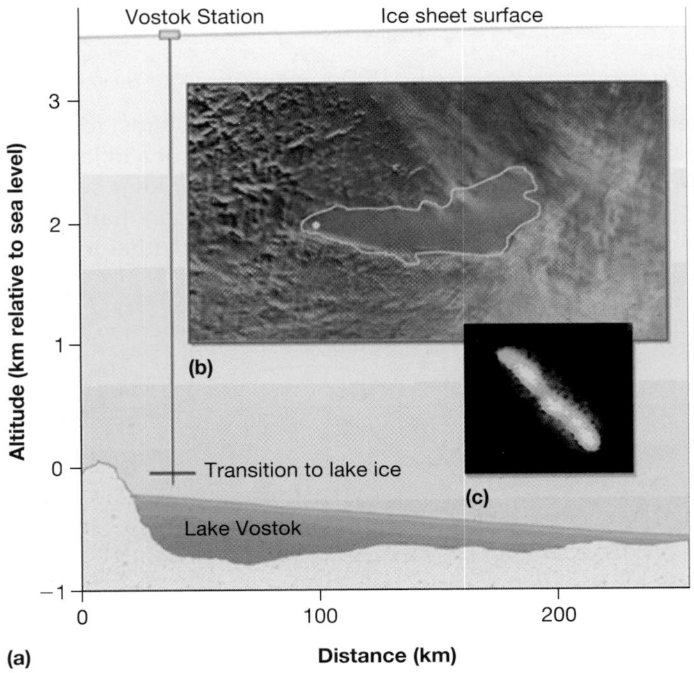

FIGURE 28.15 The East Antarctic Ice Sheet and Lake Vostok. (a) Lake Vostok lies beneath several kilometers of the ice sheet. (b) The deep drilling station from which ice and lake water samples were obtained is indicated by the yellow dot. The yellow outline identifies a smooth plateau of snow and ice that floats on top of Lake Vostok. The surrounding rough ice is formed when the ice sheet moves over bedrock rather than liquid water. (c) Microbes have been found in both the overlying ice and the lake water, as seen in this epifluorescent image.

Some of the physical and biological factors that help regulate the microbial communities of these freshwater systems are listed in **table 28.1**.

Lakes, rivers, and streams can be broadly classified as either **lotic** systems with free-running waters (streams, most rivers, canals) or **lentic** systems with free-standing waters (e.g., ponds; lakes; marshes; very broad, slow-moving rivers). In large, lentic rivers and lakes, phytoplankton thrive and are responsible for most of the fixed carbon. The biologically available carbon that is produced within the system is called **autochthonous.** By contrast, the constant flow of water in streams and all but the largest rivers prevents the development of significant planktonic communities. In addition, phytoplankton growth can be further limited by the amount of available light due to overhanging foliage, turbidity, and rapid mixing of the water. In most streams and rivers, the source of nutrients therefore comes from the surrounding land. Such nutrients are called **allochthonous.** This distinction is important, as lentic systems often have a net autotrophic metabolism, whereas lotic systems are generally heterotrophic.

The allochthonous carbon that enters rivers and streams includes both dissolved and particulate organic carbon (DOC and POC, respectively). Although we encountered these terms when discussing marine habitats, because the origin of the carbon is so different in freshwater systems, we need to revisit organic carbon in this context. The amount of carbon that enters freshwater systems depends on surrounding vegetation, proximity to human activity, and local geology. In addition, hydrological factors such as the duration water stays in the soil before entering the river or stream also influence the nature of the organic carbon. Thus DOC ranges from material that can be readily assimilated to refractory carbon that only a few microbial groups can slowly degrade. Much of the particulate organic carbon includes large items such as leaf litter, which provide a substrate for fungi and bacteria to colonize. Initially, the activity of these microbes results in a pulse of water-soluble DOC; this is followed by the slow leaching of nutrients that may not be as easy to assimilate.

Table 28.1 Physical and Biological Factors Influencing Microbial Communities in Freshwater Habitats

Habitat	*Major Microbial Community*	*Major Physical Characteristic*	*Major Biological Constraints*
Lake	Plankton	Stratification, wind-generated turbulence	Nutrient competition, grazing, parasitism
Floodplain	Plankton, biofilms attached to plants	Periodic desiccation	Competition between algae and macrophytes
Rivers and streams	Benthic microbes	Flow-generated turbulence	Colonization competition, biofilm grazing
Estuary—mudflats	Biofilms	Desiccation, high light, salt exposure	Competition and grazing at mud surface
Estuary—outflow	Plankton	Mixing with saltwater, turbidity	Grazing in water column

Reprinted with permission from: Sigee, D. C. 2005. Freshwater Microbiology. *Hoboken, NJ: Wiley & Sons.*

Most rivers and streams have healthy benthic communities of attached microbes. In healthy rivers and streams, these biofilms are consumed by invertebrates and other grazing organisms. In shallow rivers and streams, some biofilms are dominated by diatoms and other photosynthetic protists, which contribute to the total primary production of the system. Biofilms that consist primarily of heterotrophic microbes are important sources of degradation and mineralization. The role of benthic microbes in rivers and streams can be generally characterized by the microbial loop shown in **figure 28.16**.

As in any microbial community, when river and stream chemoorganotrophic microorganisms metabolize the available organic material, they recycle nutrients within the ecosystem. Autotrophic microorganisms grow using the minerals released from organic matter. This leads to the production of O_2 during daylight hours, when the rate of photosynthesis and respiration are in dynamic equilibrium. When the amount of organic matter added to streams and rivers does not exceed the system's oxidative capacity, productive streams and rivers are maintained. However, the capacity of streams and rivers to process added organic matter is limited. If too much organic matter is added, the system is said to be **eutrophic,** and the rate of respiration greatly exceeds that of photosynthesis. Thus the water may become anoxic. This sometimes occurs in streams and rivers adjacent to urban and agricultural areas. The release of inadequately treated municipal wastes and other materials from a specific location along a river or stream represents point source pollution. Such additions of organic matter can produce distinct and predictable changes in the microbial community and available oxygen, creating an oxygen sag curve in which oxygen is depleted just downstream of the pollution source (**figure 28.17**). Runoff from agriculturally active fields and feedlots is an example of nonpoint source pollution. This can also cause disequilibrium in the microbial community, leading to algal or cyanobacterial blooms.

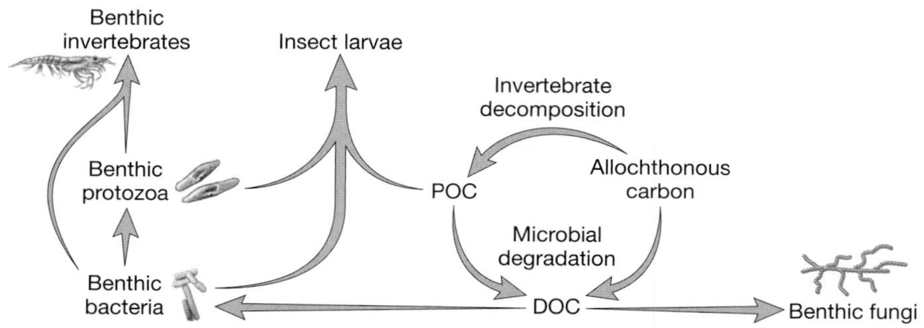

FIGURE 28.16 A Benthic Microbial Loop in Lotic Systems. In contrast to the microbial loop in planktonic systems (figure 28.2), the benthos of a lotic system, such as a fast-moving river or stream, features small eukaryotes such as invertebrates and insect larvae. These organisms play an important part in nutrient cycling. At the same time, macrofauna such as fish and other top carnivores may be absent or play relatively minor roles.

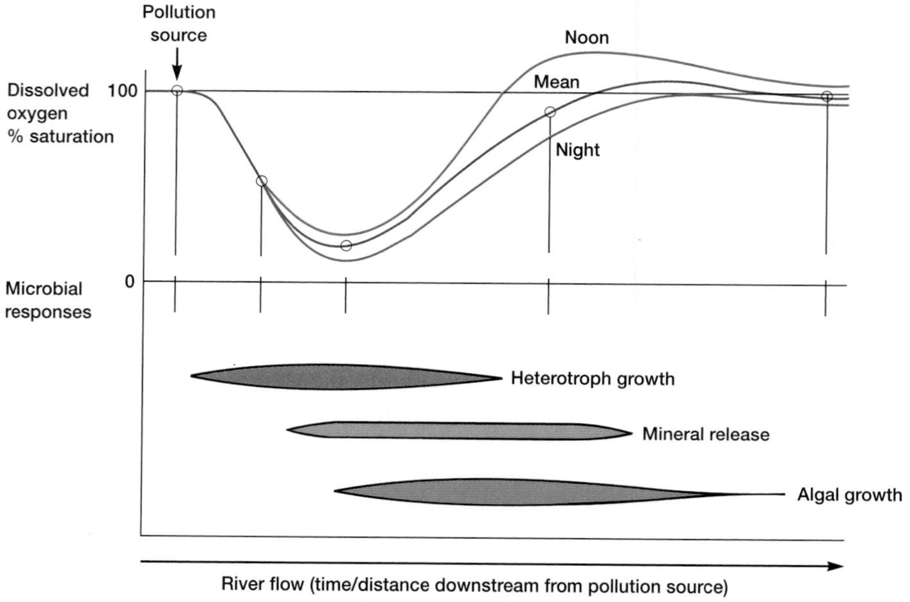

FIGURE 28.17 The Dissolved Oxygen Sag Curve. Microorganisms and their activities can create gradients over distance and time when nutrients are added to rivers. An excellent example is the dissolved oxygen sag curve, caused when organic wastes are added to a clean river system. During the later stages of self-purification, the phototrophic community will again become dominant, resulting in diurnal changes in river oxygen levels.

1. Describe the Lake Vostok ecosystem. Why is this a chemosynthesis-based, rather than a photosynthesis-based, microbial community?
2. What is an oxygen sag curve? What changes in a river cause this effect?
3. What are point and nonpoint source pollution? Can you think of examples in your community?

Microorganisms in Lakes

Lakes differ from rivers and streams because they are lentic systems, thus dominated by planktonic microbes and invertebrates. The topology and hydrology of lakes are key factors in the development of these communities. Lake hydrology incorporates all aspects of water flow: the water in the lake itself, plus the inflow and outflow. Because much of the water in a lake comes from the surrounding land, the geology and use of the land that feeds the lake largely determine the chemistry of the inflow water. For instance, mountain lakes are fed water from relatively infertile land, resulting in oligotrophic

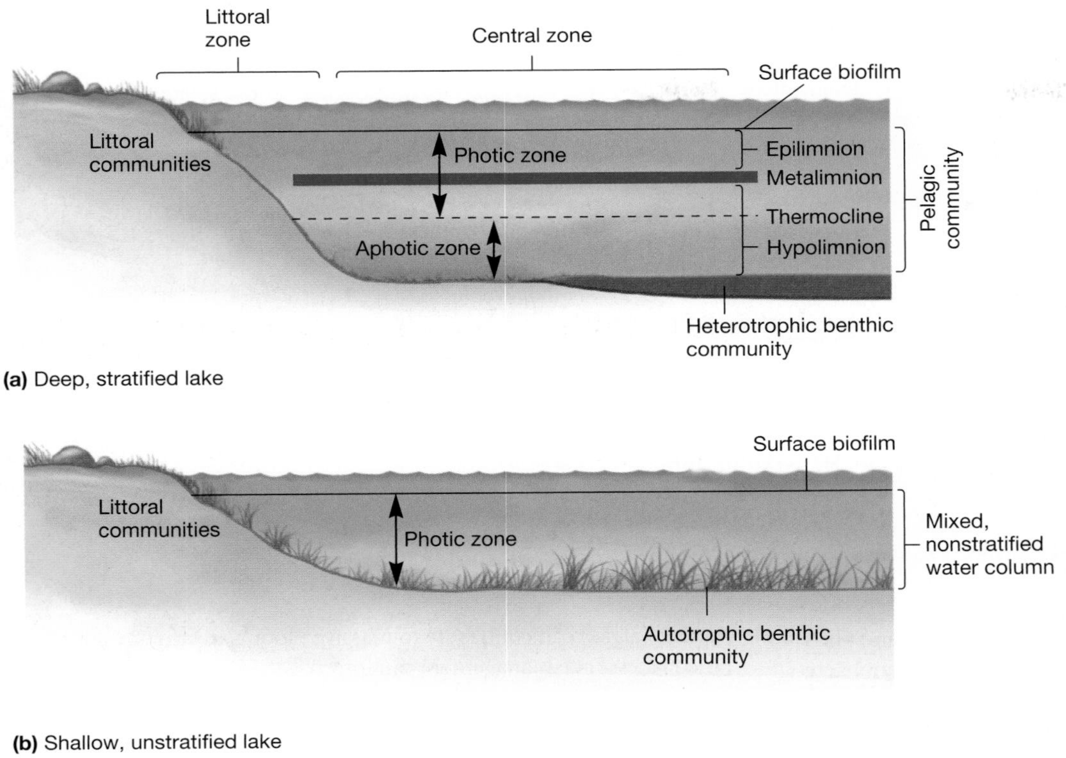

FIGURE 28.18 The Depth of a Lake Is a Key Factor in Shaping the Microbial Community. (a) Deep lakes have autotroph-dominated littoral and epilimnion microbial communities. Waters below the thermocline and photic zone form the hypolimnion, which can become oxygen depleted. Here, a heterotrophic benthic community thrives. (b) Shallow lakes have little mixing and the photic zone reaches the lake bottom. Thus autotrophs dominate the water column and benthic regions throughout the lake.

Figure 28.18 Micro Inquiry

How does the contribution of macrophytes differ in deep lakes as opposed to shallow lakes? How does this affect the relative contribution of phytoplankton in these two aquatic systems?

conditions within the lake. This explains why mountain lakes tend to be so clear—microbial growth is limited to those species that can tolerate low nutrient levels. Such microbes tend to have long generation times. On the other hand, low-lying lakes generally catch water from fertile, cultivated soils and so are nutrient rich. This supports a high level of planktonic growth, and the lake may appear turbid.

Deep lakes can be divided into two main regions: the **littoral zone,** or shoreline, and the central, **pelagic zone** (these terms also apply to marine systems). As shown in **figure 28.18*a*,** the littoral region is also defined by the depth of the photic zone. The two zones have distinct microbial communities. In the littoral zone, allochthonous DOC introduces organic substrates and nutrients such as phosphate and nitrate. Because the water is shallow, light penetrates throughout the water column, enabling the development of larger plants. Biofilms frequently form on these plants and in the sediments.

The microbial community within the pelagic zone is partly determined by the depth of the lake. Light penetrates the entire water column in shallow lakes (less than about 10 m), so benthic autotrophs, including diatoms and cyanobacteria, may be the principle primary producers (figure 28.18*b*) In deeper lakes, the central, pelagic region can become thermally stratified. This is the case in temperate regions (**figure 28.19**). In the summer, the upper layer, called the **epilimnion,** becomes warmed and oxygen is exchanged with the atmosphere. The thermocline spans a region called the **metalimnion;** this region acts as a barrier to mixing of upper and lower water masses. The lower, colder region, known as the **hypolimnion,** can become anoxic. Because these two water masses do not mix (or mix very little), the upper layer supports a rich diversity of primary producers and consumers, which can deplete nutrients. Just the opposite is true for the hypolimnion, where dark, anoxic, cold conditions limit growth. Nutrients accumulate and the microbial community primarily consists of benthic heterotrophs. When autumn weather cools the surface and storms physically mix the two layers, a short-lived bloom can occur as nutrients from the bottom of the lake are mixed into the photic zone. Much longer blooms may occur in the spring, when thermal stratification occurs but the epilimnion has not yet been depleted of nutrients.

Tropical lakes lack significant temperature changes during the year but have seasonal differences in rainfall. Although many aspects of tropical lakes are similar to those of a summer temperate lake, there are several key differences. The warmer, year-round temperatures support higher productivity and plankton blooms may occur at any time during the year. Cyanobacteria capable of nitrogen fixation may dominate, reflecting the low N:P ratios frequently found in tropical lakes. Nutrient cycling in temperate lakes is largely driven by physical mixing, whereas biological processes are more important in tropical lakes. In this regard, the microbial loop shown in figure 28.2 can be applied to the epilimnion of larger tropical lakes that receive few allochthonous nutrients.

Lakes vary in nutrient status. Some are oligotrophic, others are eutrophic. Nutrient-poor lakes remain oxic throughout the year, and seasonal temperature shifts usually do not result in distinct oxygen stratification. In contrast, eutrophic lakes usually have bottom sediments that are rich in organic matter. In oligotrophic lakes, cyanobacteria that are capable of nitrogen fixation may bloom. Several genera, notably *Anabaena, Nostoc,* and *Cylindrospermum,* can fix nitrogen under oxic conditions. The genus *Oscillatoria,* using hydrogen sulfide as an electron donor for photosynthesis,

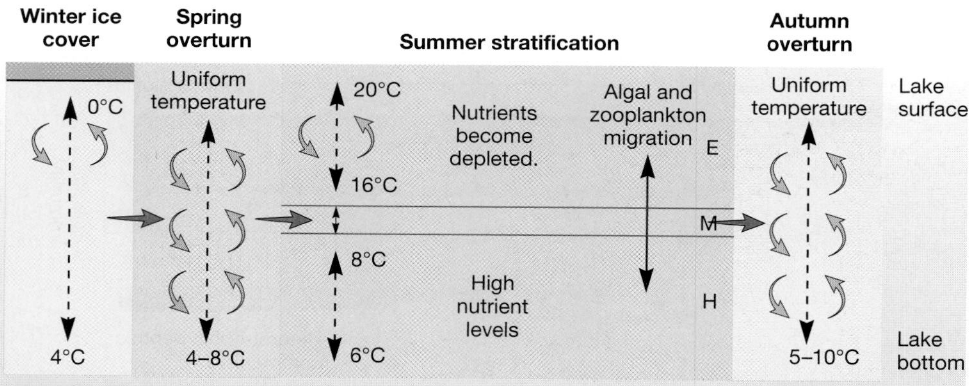

FIGURE 28.19 Seasonal Changes in a Temperate Lake Are Associated with Blooms. In the spring, the water begins to warm, and lack of thermal stratification ensures ample nutrients for microbes that can grow in the relatively warmer surface waters. During the summer as the surface waters warm, thermal stratification separates the cool, deep, nutrient-replete waters from the warm, illuminated waters. Because microbial growth is favored in the warmer surface water, this water becomes depleted of nutrients. During the autumn, waters cool and storms mix the layers, bringing the bottom nutrients to the surface. This pulse of nutrients can support a bloom of microbes or algae. Water turbulence is indicated by arrows; E, epilimnion; M, metalimnion; H, hypolimnion.

Figure 28.19* *Micro Inquiry

Why does water turbulence play only a minor role in mixing the water column in the summer?

can fix nitrogen under anoxic conditions. If both nitrogen and phosphorus are present, cyanobacteria compete with algae. Cyanobacteria function more efficiently in alkaline waters (pH 8.5 to 9.5) and higher temperatures (30 to 35°C). By using CO_2 at rapid rates, cyanobacteria also increase the pH, making the environment less suitable for protists. Cyanobacteria have additional competitive advantages. Many produce hydroxamates, which bind iron, making this important trace nutrient less available for protists. Some cyanobacteria also resist predation because they produce toxins. In addition, some synthesize odor-producing compounds that affect the quality of drinking water. However, both cyanobacteria and protists (algae) can contribute to massive blooms in strongly eutrophied lakes. Lake management can improve the situation by removing or sealing bottom sediments or adding coagulating agents to speed sedimentation. |◀◀ *Phylum* Cyanobacteria *(section 19.3)*

1. What terms can be used to describe the different parts of a lake?
2. What are some important effects of eutrophication on lakes?
3. Why do cyanobacteria often dominate waters that have been polluted by the addition of phosphorus?

Summary

28.1 Water as a Microbial Habitat

a. Most water on the Earth is marine (97%). The majority of this is cold (2 to 5°C) and at high pressure. Freshwater is a minor but important part of Earth's biosphere.

b. Oxygen solubility and diffusion rates in surface waters are limited; water has a low oxygen diffusion rate, in comparison with soils. Carbon dioxide, nitrogen, hydrogen, and methane are also important gases for microbial activity in waters.

c. The carbonate equilibrium system keeps the oceans buffered at pH 7.6 to 8.2 (**figure 28.1**).

d. The penetration of light into surface waters determines the depth of the photic zone. Warming the surface waters can lead to the development of a thermocline.

e. In coastal aquatic systems, much of the fixed carbon comes from large plants, algae, and terrestrial runoff. In large lakes and open oceans, microorganisms provide all the organic carbon on which the ecosystem depends.

f. The microbial loop describes the transfer of nutrients between trophic levels while taking into account the multiple contributions of microbes to recycling nutrients. Nutrients are recycled so efficiently, the majority remain in the photic zone (**figure 28.2**).

28.2 Microorganisms in Marine Ecosystems

a. Tidal mixing in estuaries, as characterized by a salt wedge, is osmotically stressful to microbes in this habitat. Thus they have evolved mechanisms to cope with rapid changes in salinity (**figure 28.3**).

b. Coastal regions such as estuaries and salt marshes can be the sites of harmful algal blooms.

c. Autotrophic microbes in the photic zone within the open ocean account for about one-half of all the carbon fixation on Earth.

d. The carbon and nitrogen budgets of the open ocean are intensely studied because of their implications for controlling global warming (**figures 28.6, 28.7,** and **28.9**).

e. The α-proteobacteria clade SAR11 are the most abundant organisms on Earth and demonstrate unique adaptations to life in the oligotrophic open ocean.

f. Viruses are present at high concentrations in many waters and occur at tenfold higher levels than bacteria. In marine systems, they may play a major role in nutrient turnover (**figures 28.10–28.12**).

g. Sediments deep beneath the ocean's surface are home to perhaps one-half of the world's bacterial and archaeal biomass.

h. Methane hydrates, the result of psychrophilic archaeal methanogenesis under extreme atmospheric pressure, may contain more natural gas than is currently found in known reserves.

28.3 Microorganisms in Freshwater Ecosystems

a. Glaciers and permanently frozen lakes are sites of active microbial communities. The East Antarctic Ice Sheet and Lake Vostok, which lies beneath it, are productive study sites (**figure 28.15**).

b. Nutrient sources for streams and rivers may be autochthonous or allochthonous. Often allochthonous inputs include urban, industrial, and agricultural runoff.

c. The depth of a lake determines the microbial communities it will support. Deep lakes will have a planktonic, autotrophic community in the epilimnion, and shallow lakes support large autotrophic communities (**figure 28.18**).

d. Seasonal mixing of temperate lakes can introduce conditions that support microbial blooms in the spring and fall (**figure 28.19**).

Critical Thinking Questions

1. The unicellular cyanobacterium *Prochlorococcus* is the most abundant photosynthetic microbe in tropical and subtropical oceans. At least two ecotypes exist: one is adapted to high light and the other to lower light intensities. How does the presence of these two ecotypes contribute to their physiological success and their numerical success? How would you determine if, in addition to being numerically dominant, they contribute the most fixed carbon to these open-ocean ecosystems?
2. How might it be possible to cleanse an aging eutrophic lake? Consider chemical, biological, and physical approaches as you formulate your plan.
3. *Clostridium botulinum,* the causative agent of botulism, sometimes causes fish kills in lakes where people swim. Currently there are few, if any, monitoring procedures for this potential source of disease transmission to humans. Do you think a monitoring program is needed, and, if so, how would you implement such a program?
4. Scientists have been surprised by the diversity of previously unidentified eukaryotic microorganisms that are part of the plankton community in the open ocean. Recent nucleotide sequencing of the 18S rRNA genes from a collection of picoplankton reveals the presence of an entirely new, independent phylogenetic group among major eukaryotic taxa. These microbes have been named picobiliphytes because they possess a plastidlike organelle that fluoresces orange, indicative of phycobilins (*see figure 19.6*). Phylogenetic analysis indicates that they are a diverse group, composed of at least three clades, and analysis of SSU rRNA sequences that have been reported by other marine microbiologists suggests these microbes have wide distribution in cold ocean waters. List three questions you would explore, were you to design an oceanographic cruise to sample and analyze these microbes. Explain why you chose each question and the approaches you would use to address them.

   Read the original paper: Not, F., et al. 2007. Picobiliphytes: A marine picoplanktonic algal group with unknown affinities to other eukaryotes. *Science*. 315:253.
5. It is well known that bacterivory (the consumption of bacteria) supports the growth of marine protists that lack plastids. However, measurements of bacterivory among plastid-containing planktonic algae were surprising in showing that small (<5 μm) algae carry out between about 40 to 95% of the bacterivory in the photic zone of the North Atlantic. Discuss how this level of mixotrophy impacts the concept of the microbial loop (figure 28.2) as well as global carbon cycling.

   Read the original paper: Zubkov, M. V., and Tarran, G. A. 2008. High bacterivory by the smallest phytoplankton in the North Atlantic Ocean. *Nature*. 455:224.

Concept Mapping Challenge

Construct a concept map that features aspects of marine microbes that are specific to growth in the open ocean. You will need to provide your own linking terms.

Oligotrophic adaptations	Picoplankton	Bacteria
Archaea	Proteorhodopsin	Protists
AAnP	Lithoheterotrophy	Viruses
Microbial loop	N_2 fixation	

Learn More

Learn more by visiting the text website at www.mhhe.com/willey8, where you will find a complete list of references.

29

Microorganisms in Terrestrial Ecosystems

The plant pathogenic fungus Puccinia graminis *grows within the tissues of wheat plants, using plant nutrients and producing rusty streaks of red spores that erupt at the stem and leaf surface, where spores can be dispersed. By contrast, many plants rely on mycorrhizal fungi and nitrogen-fixing bacteria to promote growth.*

CHAPTER GLOSSARY

bacteroid A modified bacterial cell that carries out nitrogen fixation within the root nodule cells of legumes.

dissolved organic matter (DOM) Nutrients that are available in the soluble, or dissolved, state.

endophyte An organism that grows within a plant.

epiphyte An organism that grows on the surface of a plant.

mycorrhizal fungi Fungi that form stable, mutualistic relationships on (ectomycorrhizal) or in (endomycorrhizal) the root cells of vascular plants. The plants provide carbohydrate for the fungi, while the fungal hyphae extend into the soil and bring nutrients to the plants.

phyllosphere The parts of a plant above ground (e.g., stems and leaves) that supports a microbial community.

rhizobia Any one of a number of α- and β-proteobacteria that form symbiotic nitrogen-fixing nodules on the roots of leguminous plants.

rhizoplane The surface of a plant root.

rhizosphere A region around the plant root where materials released from the root promote microbial growth.

root nodule A knoblike structure on roots that contains endosymbiotic nitrogen-fixing bacteria (e.g., *Rhizobium* or *Bradyrhizobium).*

Soil is a hugely underappreciated resource. Apart from its role in agriculture, it is a vast reservoir of nutrients, wastes, energy, and even pharmaceutical products. Although much is known about the geology and chemistry of soil, our understanding of the microbial communities that drive the development and continued health of soils is far from complete.

Early in the twentieth century, microbiologists believed they could catalogue soil microbes with relative ease. By the end of the century, this optimism had been replaced with the conclusion that soil microbial diversity might be so vast that such a task was utterly beyond reach. However, the development of new, innovative culture techniques combined with culture-independent approaches has proved that far from a hopeless endeavor, the study of terrestrial microbiology is an exciting and dynamic field.

We begin this chapter by describing the soil habitat and how microbes contribute to the development of soils. This is followed by a discussion of specific soil microbial communities and the interaction of microbes with vascular plants. We pay particular attention to two very important relationships: that between fungi and plants and that between nitrogen-fixing bacteria and leguminous plants. Finally, the new and fascinating field of deep subsurface microbiology is introduced.

29.1 Soils as a Microbial Habitat

A soil scientist would describe soil as weathered rock combined with organic matter and nutrients. An agronomist would point out that soil supports plant life. However, a microbial ecologist knows that the formation of organic matter and the growth of plants depend on the microbial community within the soil. The level of microbial diversity in soil exceeds that of any other habitat on Earth. This variety is supported by the complexity of the physical and chemical environments, which provide a vast array of microhabitats. These include soil particles and the pore space between them, which is also critical for the movement of water and gases (**figure 29.1**). Total pore space, and thus gas diffusion, is determined by the texture of the soil. For instance, sandy soils

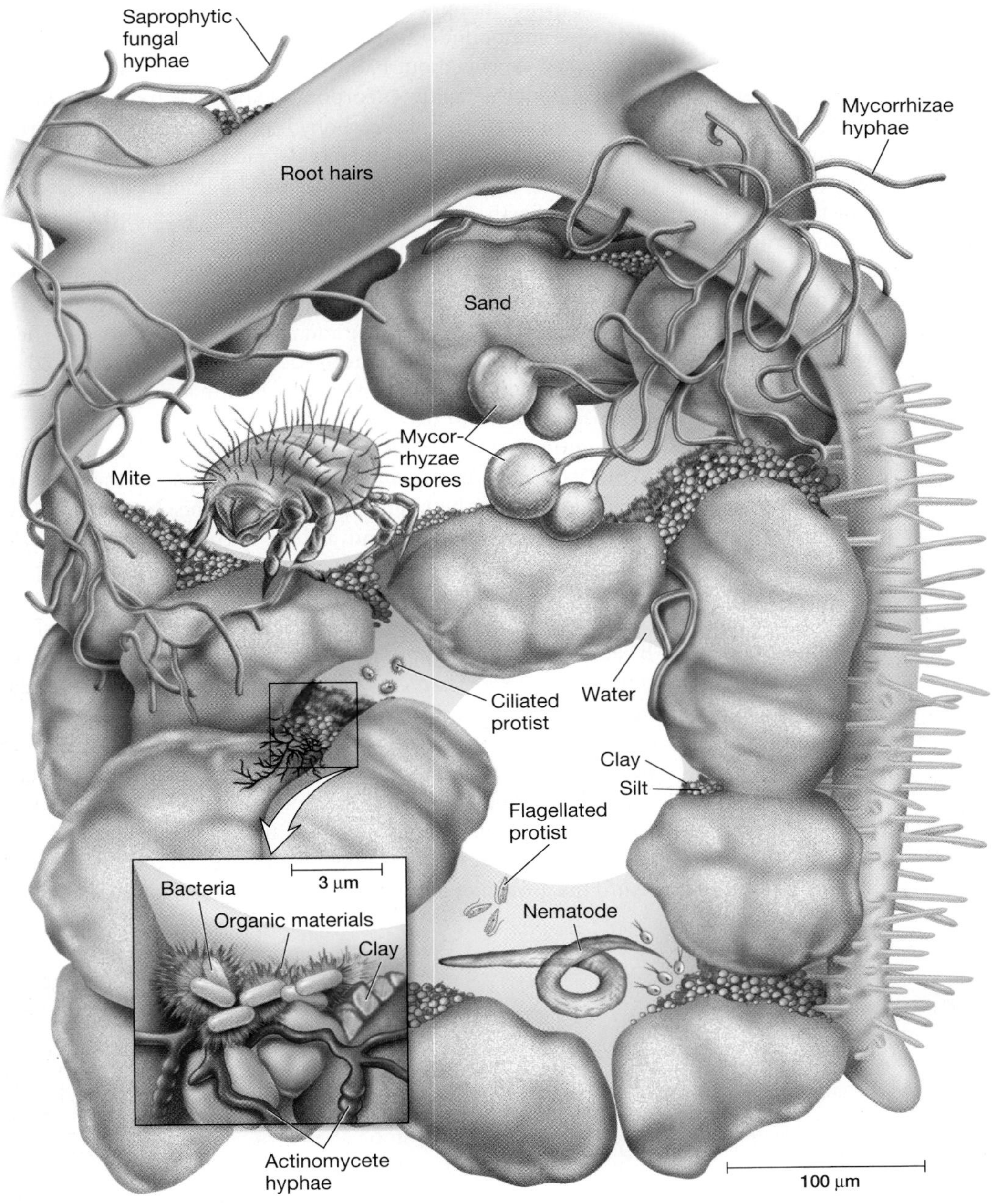

FIGURE 29.1 The Soil Habitat. A typical soil habitat contains a mixture of clay, silt, and sand along with soil organic matter. Roots, animals (e.g., nematodes and mites), as well as chemoorganotrophic bacteria consume oxygen, which is rapidly replaced by diffusion within the soil pores where the microbes live. Note that two types of fungi are present: mycorrhizal fungi, which derive their organic carbon from their symbiotic partners, plant roots; and saprophytic fungi, which contribute to the degradation of organic material.

Figure 29.1 Micro Inquiry

How are filamentous microbes especially well suited to the soil ecosystem?

have larger pore spaces than do clay soils, so sandy soils tend to drain more quickly. Pores are also critical because they provide the optimum environment for microbial growth. Here, microbes are within thin water films on particle surfaces where oxygen is present at high levels and can be easily replenished by diffusion. The oxygen concentrations and flux rates in pores and channels are high, whereas within water-filled zones, the oxygen flux rate is much lower. For example, particles as small as about 2.0 μm can be oxic on the outside and anoxic on the inside.

Soils are dynamic environments. For instance, rainfall or irrigation may rapidly change a well-aerated soil to an environment with isolated pockets of water, generating "miniaquatic" habitats. If flooding continues, a waterlogged soil can be created that is more like lake sediment. If oxygen consumption exceeds that of oxygen diffusion, waterlogged soils can become anoxic. Shifts in water content and gas fluxes also affect the concentrations of CO_2 and other gases. These changes are accentuated in the smaller pores where many bacteria are found. The roots of plants growing in aerated soils also consume oxygen and release CO_2, influencing the concentrations of these gases in the root environment.

Soils can be divided into two general categories: A **mineral soil** contains less than 20% organic carbon, whereas an **organic soil** possesses at least this amount. By this definition, the vast majority of Earth's soils are mineral. The importance of organic matter within soils cannot be underestimated. **Soil organic matter (SOM)** helps to retain nutrients, maintain soil structure, and hold water for plant use. SOM is subject to gains and losses, depending on changes in environmental conditions and agricultural management practices. Plowing and other disturbances expose SOM to more oxygen, leading to extensive microbiological degradation of organic matter. Irrigation causes periodic wetting and drying, which can also lead to increased degradation of SOM, especially at higher temperatures.

Microbial degradation of plant material results in the release of CO_2 and the incorporation of plant carbon into additional microbial biomass. However, a small fraction of the decomposed plant material remains in the soil as SOM. When considering this material, it is convenient to divide the SOM into humic and nonhumic fractions (**table 29.1**). Nonhumic SOM is free from significant biochemical degradation. It can represent up to about 20% of the soil organic matter. Humic SOM, or humus, results when the products of microbial metabolism have undergone chemical transformation within the soil. Although it has no precise chemical composition, humus can be described as a complex blend of phenolic compounds, polysaccharides, and proteins. The recalcitrant nature of SOM to degradation is evident by ^{14}C dating: the average age of most SOM ranges from 150 to 1,500 years.

The degradation of plant material and the development of SOM can be thought of as a three-step process. First, easily degraded compounds such as soluble carbohydrates and proteins are broken down. About half the carbon is respired as CO_2 and the remainder is rapidly incorporated into new biomass. During the second stage, complex carbohydrates, such as the plant structural polysaccharide cellulose, are degraded. Cellulose represents a vast store of carbon, as it is the most abundant organic compound on Earth. Fungi and bacterial genera such as the filamentous *Streptomyces* and *Micromonospora,* and nonfilamentous forms such as *Cytophaga* and *Bacillus* produce extracellular cellulase enzymes that break down cellulose into two to three glucose units called cellobiose and cellotriose, respectively. These smaller compounds are readily degraded and assimilated as glucose monomers. Finally, very resistant material, in particular lignin, is attacked.

Lignin is an important structural component of woody plants. While its exact structure differs among plant species, the common building block is the phenylpropene unit. This consists of a hydroxylated benzene ring and a three-carbon linear side chain (**figure 29.2**). A single lignin molecule can consist of up to 600 cross-linked phenylpropene units. It is therefore not surprising that degradation of lignin is much slower than that of cellulose. Basidiomycete fungi commonly termed "white rot fungi" and actinomycetes (e.g., *Streptomyces* spp.) are capable of extracellular lignin degradation. These microbes produce extracellular phenoloxidase enzymes needed for aerobic lignin degradation. Lignin decomposition is also limited by the physical nature of the material. For example, healthy woody plants are saturated with sap, which limits oxygen diffusion. In addition, high ethylene and CO_2 levels and the presence of phenolic and terpenoid compounds retard the growth of lignin-degrading microbes. It follows that no

Table 29.1 Fractions of Soil Organic Matter

SOM Fraction	*Definition*	*Physical Appearance*
Humic substances	High molecular-weight organic material produced by secondary synthesis reactions	Dark brown to black
Nonhumic substances	Unaltered remains of plants, animals, and microbes from which macromolecules have not yet been extracted	Light brown
Humic acid	Organic matter extracted from soils by various reagents (often dilute alkali treatment) that is then precipitated by acidification	Dark brown to black
Fulvic acid	Soluble organic matter that remains after humic acid extraction	Yellow
Humin	SOM that cannot be extracted from soil with dilute alkali	Variable

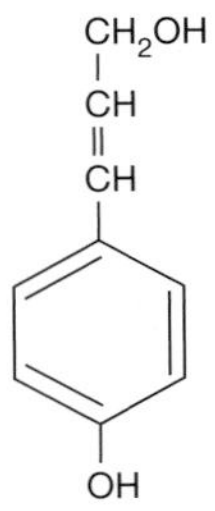

Coumaryl alcohol

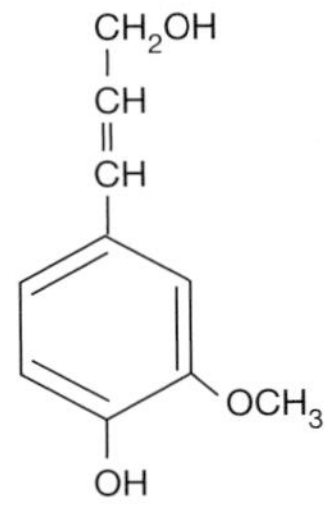

Coniferyl alcohol

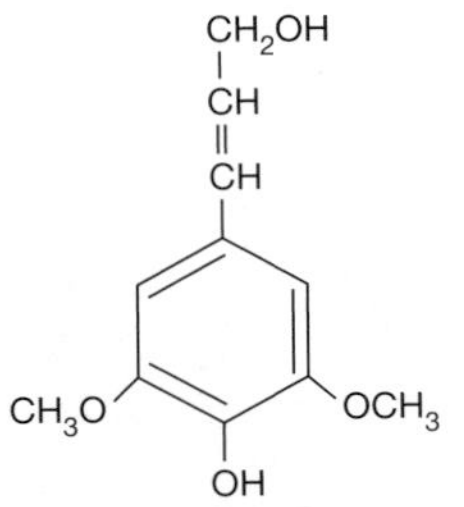

Sinapyl alcohol

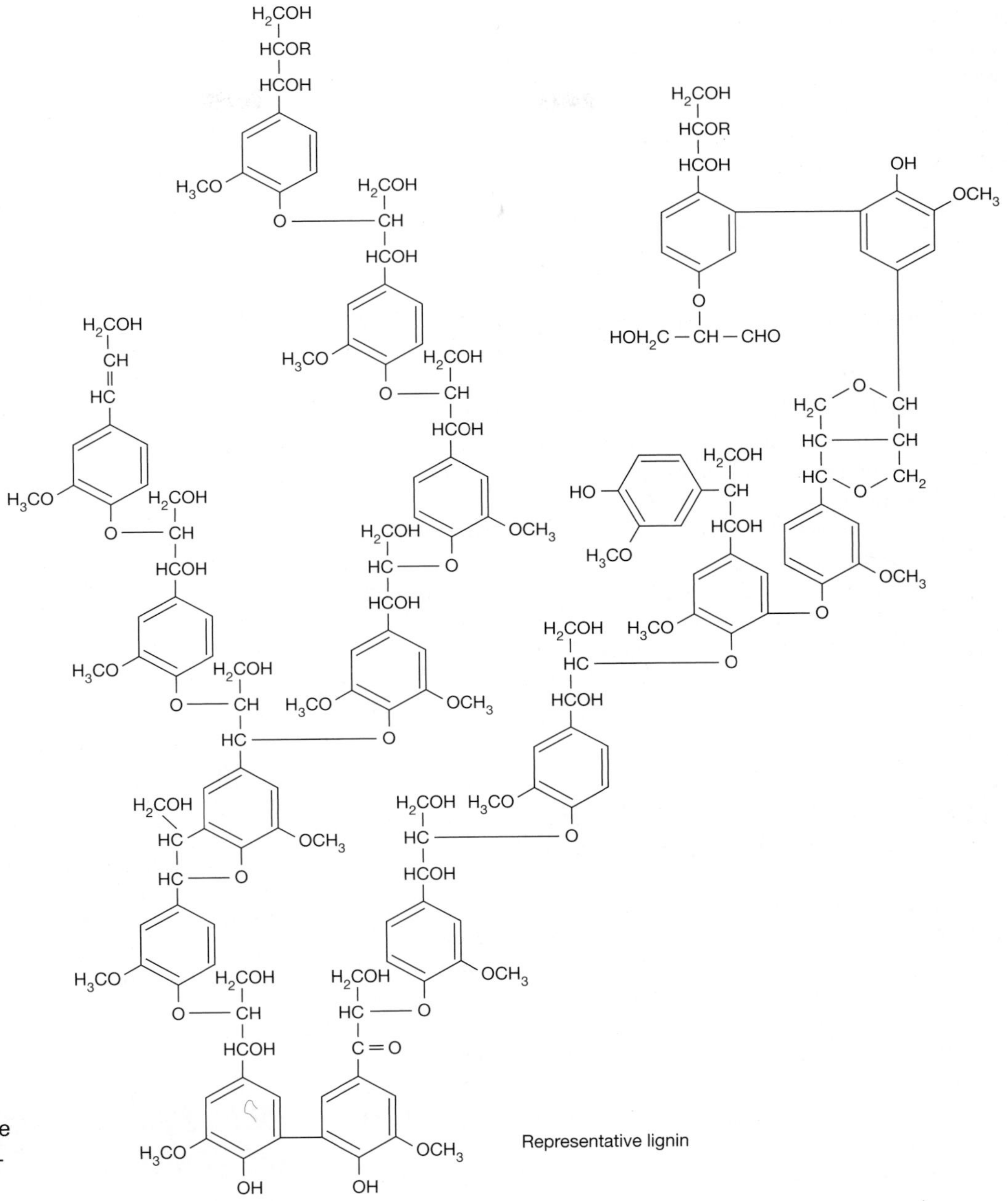

Representative lignin

FIGURE 29.2 Example Phenylpropene Units. These molecules are polymerized to form lignin, an irregular branched polymer.

more than 10% of the carbon found in lignin is recycled into new microbial biomass. Lignin can be degraded anaerobically, but this process is very slow, so lignin tends to accumulate in wet, poorly oxygenated soils, such as peat bogs.

Nitrogen is another important element in soil ecosystems. Nitrogen in soil is often considered in relation to the soil carbon content as the organic **carbon to nitrogen ratio (C/N ratio).** As a general rule, decomposition is thought to be maximal at a C/N ratio of 30 (i.e., 30 times more carbon than nitrogen). A C/N ratio of less than 30 can result in the loss of soluble nitrogen from the system. Conversely ratios above 30 may cause the microbial community to be nitrogen limited. In fact, many soils are nitrogen limited. This is why each year tons of nitrogen fertilizer are added to agricultural soils. The major types of nitrogen fertilizers used in agriculture are liquid ammonia and ammonium nitrate. Ammonium ion usually is added because it will be attracted to the negatively charged clays in a soil and be retained on the clay surfaces until used as a nutrient by the plants. However, the nitrifier populations in a soil can oxidize the ammonium ion to nitrite and nitrate, and these anions can be leached from the plant environment and pollute surface waters and groundwaters. ◀◀ *Nitrogen cycle (section 26.1); Global climate change (section 26.2)*

Phosphorus in fertilizers also is critical. The binding of this anionic fertilizer component to soils depends on the cation exchange capacity and soil pH. As the soil's phosphorus sorption capacity is reached, the excess, together with phosphorus that moves to lakes, streams, and estuaries with soil erosion, can stimulate the growth of freshwater organisms, particularly nitrogen-fixing cyanobacteria, in the process of eutrophication. ◀◀ *Microorganisms in streams and rivers (section 28.3)*

Search This: Soil health

1. What is the importance of soil pores?
2. List three reasons why soil organic matter (SOM) is important.
3. What is the difference between humic and nonhumic SOM? Which is more abundant in most soils? Why?
4. Describe the three phases of plant degradation and SOM formation.
5. What are possible effects of nitrogen-containing fertilizers on microbial communities?
6. Most nitrogen fertilizer is added as ammonium ion. Why is this preferred over nitrate?
7. Why might enrichment of freshwater with phosphorus be even more critical than nitrogen enrichment?

29.2 Microorganisms in the Soil Environment

An average gram of forest soil contains about 4×10^7 bacteria and archaea, while a gram of grassland or cultivated soil is thought to possess 2×10^9 cells. It is difficult to determine how many different taxa are represented, but estimates based on DNA reassociation range from 2,000 to 18,000 distinct genomes per gram of soil. These numbers suggest that the microbial diversity in just one gram of soil may exceed the combined number of known archaeal and bacterial species—just over 16,000. ◀◀ *Assessing microbial diversity (section 27.2)*

Small subunit rRNA genes have been extracted and cloned from a variety of soils in an effort to assess soil microbial diversity. In most cases, primers specific for members of the domain *Bacteria* were used. These data give mixed results for a variety of reasons, as discussed in chapter 27 (*p. 666*). Nonetheless, these data show some major trends. Whereas the majority (79 to 89%) of 16S rRNA gene sequences were from novel genera that have no known representatives in existing databases, most could be assigned to well-studied bacterial phyla. Overall, bacteria belonging to 32 different phyla have been identified, with over 90% of soil library 16S rRNA assigned to the nine phyla displayed in **figure 29.3**. Note that this includes the *Acidobacteria* and *Verrucomicrobia*, phyla about which little is known. Other important phyla not shown in figure 29.3 include *Chlamydiae, Chlorobi, Cyanobacteria, Deinococcus-Thermus,* and *Nitrospirae*. The coryneforms, nocardioforms, and streptomycetes (**table 29.2**)

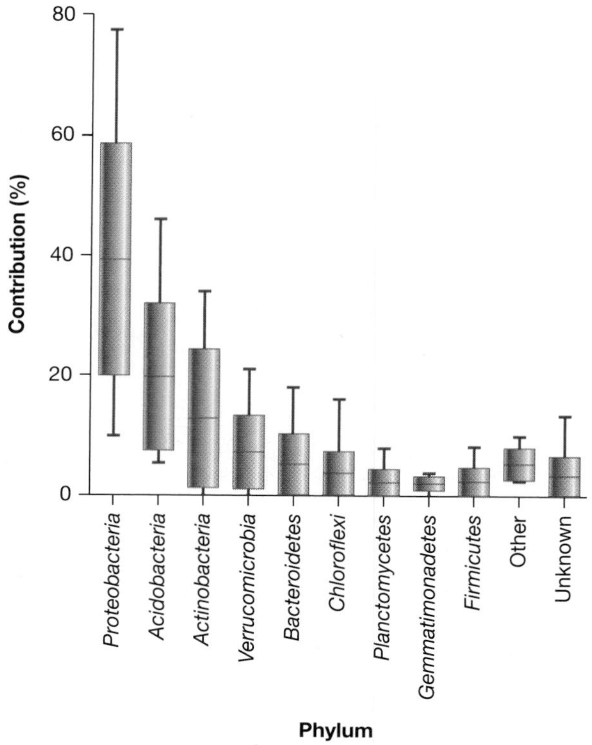

FIGURE 29.3 Bacterial Phyla Identified from 16S rRNA and 16SrRNA Genes Extracted from Soil. The data from 21 soil metagenomic libraries was analyzed to more completely assess soil bacterial diversity. The horizontal line in the middle of each block represents the mean. The block represents one standard deviation around the mean. The vertical lines extending above and below indicate the maximum and minimum values for each phylum. Note that very few of the phyla were unknown.

Table 29.2 Easily Cultured Gram-Positive Bacteria in Soils

Bacterial Group	*Representative Genera*	*Comments and Characteristics*
Coryneforms	*Arthrobacter*	Rod-coccus cycle
	Cellulomonas	Important in degradation of cellulose
	Corynebacterium	Club-shaped cells
Mycobacteria	*Mycobacterium*	Acid-fast
Nocardioforms	*Nocardia*	Rudimentary branching
Streptomycetes	*Streptomyces*	Aerobic filamentous bacteria
Bacilli	*Thermoactinomyces*	Higher temperature growth

are easily cultured and therefore well-characterized members of the soil community. These high G + C gram-positive bacteria play a major role in the degradation of hydrocarbons, older plant materials, and soil humus. In addition, some members of these groups actively degrade pesticides. The filamentous actinomycetes, primarily of the genus *Streptomyces,* produce an odor-causing compound called geosmin, which gives soils their characteristic earthy odor. ◀◀ *High G + C gram-positive bacteria: The* Actinobacteria *(chapter 22)*

Of course bacteria are not the only microbes present in soils. The only feasible way to compare the diversity of *Bacteria, Archaea,* and *Fungi* from soil ecosystems is to use a SSU rRNA approach. Such analysis identifies sequences that are 97% similar to "operational taxonomic units" (OTUs), rather than specific genera. OTU is a means by which microbial ecologists can refer to microbial taxa that have not been cultured and about which little else is known apart from nucleotide and amino acid sequences. **Figure 29.4** illustrates two important findings. First, the number of OTUs discovered is proportional to the number of SSU rRNA sequences cloned. This is common for almost all natural SSU rRNA libraries. Importantly, this implies that fully representative SSU rRNA libraries have yet to be constructed. In other words, so far it appears that no one has been able to construct a metagenomic library that samples each microbe from the given soil habitat. The second result is surprising: archaeal and fungal OTU richness is generally equal to or greater than bacterial diversity. In fact, recent studies indicate that ammonia-oxidizing soil crenarchaeota outnumber the α and β-proteobacteria that are typically associated with this first step in nitrification (*see figure 20.13*). These results were found in a variety of soil types, and together with the newly discovered ammonia-oxidizing marine archaea, it may be that crenarchaea are the most abundant nitrifying microbes on Earth.

If we look at the soil habitat in greater detail (figure 29.1), we find that some of this incredible diversity can be explained by the presence of bacteria, archaea, fungi, and protists that use different functional strategies to take advantage of this complex physical matrix. Most soil bacteria and archaea are located on the surfaces of soil particles, where water and nutrients are in their immediate vicinity. These microbes are found most frequently on surfaces within smaller soil pores (2 to 6 μm in diameter). Here, they are probably less liable to be eaten by protozoa, unlike those located on the exposed outer surface of a sand grain or organic matter particle.

Fungi can extend up to several hundred meters of hyphae per gram of soil. We tend to think of soil fungi as small structures such as the mushrooms sprouting on our lawns. However, the vast majority of fungal biomass is below ground, where filaments bridge open areas between soil particles or aggregates and are exposed to high levels of oxygen. These fungi tend to darken and form oxygen-impermeable structures called sclerotia and hyphal cords. This is particularly important for basidiomycetes, which form such structures as an oxygen-sealing mechanism. Within these structures, the filamentous fungi move nutrients and water over great distances, including across air spaces, a unique part of their functional strategy. Fungal biomass can be enormous; an individual clone of the fungus *Armillaria bulbosa,* which lives associated with tree roots in hardwood forests, covers about 30 acres in the Upper Peninsula of Michigan. It is estimated to weigh a minimum of 100 tons (an adult blue whale weighs about 150 tons) and be at least 1,500 years old. Thus some fungal mycelia are among the largest and most ancient living organisms on Earth. ◀◀ Basidiomycota *(section 24.6)*

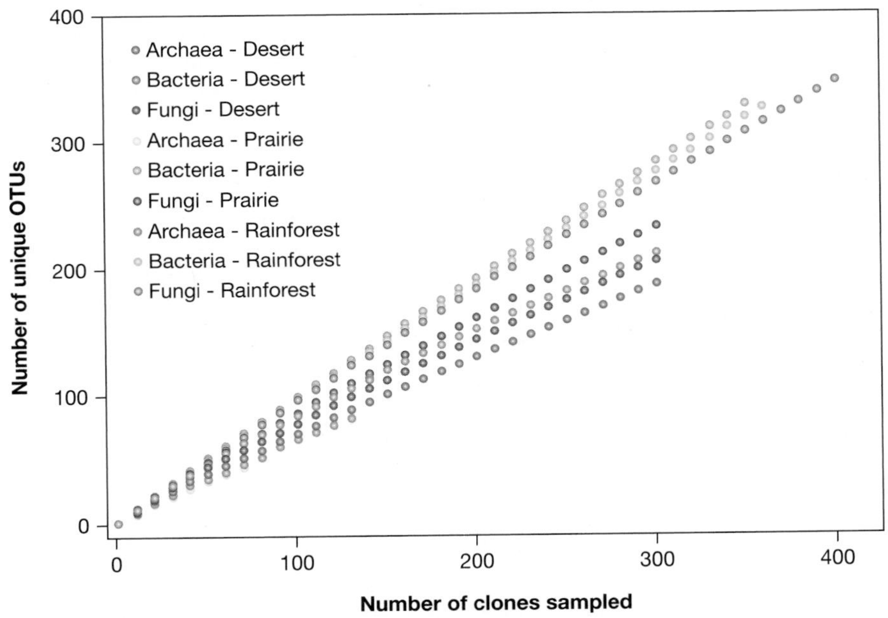

FIGURE 29.4 Soil Metagenomic Libraries and Species Richness. The number of operational taxonomic units (OTUs) identified in nine metagenomic libraries that include fungal, archaeal, and bacterial clones is graphed. In each case, as the number of clones increase, so do the number of OTUs. Because the number of OTUs has not been saturated, that is, the curve continues to climb instead of level off, it appears that none of these metagenomic libraries represents an entire soil microbial community.

Figure 29.4 Micro Inquiry

What soil ecosystems are not represented in this analysis? Why do you think metagenomic libraries cannot sample each microbial taxon present?

Search This: Soil fungi/ars/usda

Nutrients are regenerated in soils through a microbial loop that differs from that which operates in aquatic ecosystems. A major distinction is that plants, rather than microbes, account for most primary production in nearly all terrestrial systems. But much like the microbial loop in pelagic waters, microbes in soils rapidly recycle the organic material derived from plants and animals, including the many nematodes and insects, which function to reduce the size of solid organic matter. In turn, microbes themselves are preyed upon by viruses, other bacteria (e.g., species within the genera *Lysobacter* and *Myxococcus*), and soil protists, whose numbers can reach 100,000 per gram of soil. This makes microbial organic matter available to other trophic levels. Another difference between aquatic and terrestrial microbial loops reflects the physical and biological properties of soil. Degradative enzymes released by microbes, plants, insects, and other animals do not rapidly diffuse away. Instead, they represent a significant contribution to the biological activity in soil ecosystems by contributing to the many hydrolytic degradation reactions needed for the assimilation and recycling of nutrients.

1. What are the differences in preferred soil microhabitats between bacteria and filamentous fungi?
2. What types of archaea have been detected in soils?
3. How does the microbial loop function in soils? Compare this with its function in aquatic systems.

29.3 Microorganism Associations with Vascular Plants

The vast majority of soil microbes are heterotrophic, so it should come as no surprise that many have evolved close relationships with plants, the major source of terrestrial primary production. Many microbe-plant interactions are commensalistic—they do no harm to the plant, whereas the microbe gains some advantage. Many other important interactions are beneficial to both the microorganism and the plant (i.e., they are mutualistic). Finally, other microbes are plant pathogens and parasitize their plant hosts. In all cases, the microbe and the plant have established the capacity to communicate. The microbe detects and responds to plant-produced chemical signaling molecules. This generally triggers the release of microbial compounds that are in turn recognized by the plant, thereby beginning a two-way "conversation" that employs a molecular lexicon. Once a microbe-plant relationship is initiated, microbes and plants continue to monitor the physiology of their partner and adjust their own activities accordingly. The nature of the signaling molecules and the mechanisms by which both plants and microbes respond have provoked exciting multidisciplinary research in soil microbiology, ecology, molecular biology, genetics, and biochemistry.

◀◀ *Cell-cell communication within microbial populations (section 7.7);* ▶▶ *Microbial interactions (section 30.1)*

Because the rhizosphere offers an assortment of plant exudates that are more easily degraded than lignin and cellulose, it is home to a variety of bacteria that must compete with saprotrophic and plant pathogenic fungi also in the rhizosphere. These include fungi whose rapid growth is supported by small organic compounds. Bacteria respond to this with a variety of antifungal strategies, such as the production of antifungal agents, lytic enzymes, and even hydrogen cyanide. Fungi in turn counteract these inhibitors by efflux pumps, detoxification enzymes, and secretion of molecules that alter bacterial gene expression.

Microbe-plant interactions can be broadly divided into two classes: microbes that live on the surface of plants, called **epiphytes,** and those that colonize internal plant tissues, called **endophytes.** Furthermore, we can consider microbes that live in the aboveground, or aerial, surfaces of plants separately from those that inhabit below-ground plant tissues. We begin our discussion of microbe-plant interactions by first introducing the microbial communities associated with aerial regions of plants. We then turn our attention to two important microbe-root symbioses—the mycorrhizal fungi and the nitrogen-fixing rhizobia. Finally, we consider several microbial plant pathogens.

Phyllosphere Microorganisms

The environment of the aerial portion of a plant (e.g., stems and leaves), called the **phyllosphere,** undergoes frequent and rapid changes in humidity, UV exposure, and temperature. This in turn results in fluctuations of the leaching of organic material (primarily simple sugars) that supports a diverse assortment of microbes, including bacteria, filamentous fungi, yeasts, and photosynthetic and heterotrophic protists. Numerically, it appears that the γ-proteobacteria *Pseudomonas syringae, Erwinia,* and *Pantoea* are most important. Plant pectin methylesterase enzymes produce methanol, which in turn supports the growth of methylotrophic α-proteobacteria such as *Methylobacterium.* Another abundant bacterial genus, *Sphingomonas,* produces pigments that function like sunscreen so it can survive the high levels of UV irradiation striking plant surfaces. This bacterium, also common in soils and waters, can reach 10^8 cells per gram of plant tissue. It often represents a majority of the culturable species.

A recent survey of the bacterial diversity of the phyllosphere of the Atlantic Forest in Brazil revealed some intriguing findings. This forest is thought to be the oldest on Earth, with over 20,000 vascular plant species. SSU rRNA analysis of the bacteria in the phyllosphere of three plant species showed amazing species richness—from 95 to 671 bacterial species, and all but 3% are members of undescribed taxa. The diversity is further extended by the observation that each plant species had an almost unique phyllosphere bacterial community. Extrapolation of these results to the entire forest indicates that there are millions of new bacterial species yet to be discovered. These are the types of results that drive the search for new biota.

Search This: Encyclopedia of life

Rhizosphere and Rhizoplane Microorganisms

Plant roots receive between 30 to 60% of the net photosynthesized carbon. Of this, an estimated 40 to 90% enters the soil as a wide variety of materials, including alcohols, ethylene, sugars, amino and organic acids, vitamins, nucleotides, polysaccharides, and enzymes. These materials create a unique environment for soil microorganisms called the **rhizosphere.** The plant root surface, termed the **rhizoplane,** also provides an exceptional environment for microorganisms, as these gaseous, soluble, and particulate materials move from plant to soil. Rhizosphere and rhizoplane microbial community composition and function change when these substrates become available. In addition, rhizosphere and rhizoplane microorganisms serve as labile sources of nutrients for other organisms, creating a soil microbial loop and thereby playing critical roles in organic matter synthesis and degradation.

A wide range of microbes in the rhizosphere can promote plant growth, orchestrated by their ability to communicate with plants using chemical signals. Some of these signaling compounds include auxins, gibberellins, glycolipids, and cytokinins. The use of these compounds in agriculture is an active area of research. For instance, plant growth-promoting rhizobacteria include the genera *Pseudomonas* and *Achromobacter,* which can be added to the plant even in the seed stage, if the bacteria produce the required surface attachment proteins. The genes that control the expression of these attachment proteins are of great interest to agricultural biotechnologists.

A critical process that occurs on the surface of the plant, and particularly in the root zone, is associative nitrogen fixation, in which nitrogen-fixing microorganisms are on the surface of the plant root, the rhizoplane, as well as in the rhizosphere. This process is carried out by members of the genera *Azotobacter, Azospirillum,* and *Acetobacter.* These bacteria contribute to nitrogen accumulation by tropical grasses. However, evidence suggests that their major contribution may not be nitrogen fixation but the production of growth-promoting hormones that increase root hair development, thereby enhancing plant nutrient uptake. This is an area of research that is particularly important to tropical agriculture.

1. Define rhizosphere, rhizoplane, and associative nitrogen fixation.
2. What unique stresses does a microorganism on a leaf contend with that a microbe in the soil does not?
3. What is the importance of plant growth-promoting bacteria?
4. What important genera are involved in associative nitrogen fixation?

Mycorrhizae

Mycorrhizae (Greek, "fungus root") are mutualistic relationships that develop between about 80% of all land plants and a limited number of filamentous fungal species. Mycorrhizae are so abundant that up to 100 m of mycorrhizal filaments can be found in a gram of soil. Both partners in a mutualistic relationship depend on the activities of the other and thus have coevolved. Unlike most fungi, **mycorrhizal fungi** are not saprophytic—that is, they do not obtain organic carbon from the degradation of organic material. Instead, they use photosynthetically derived carbohydrate from their host. In return, they provide a number of services for their plant hosts. In wet environments, they increase the availability of nutrients, especially phosphorus. In arid environments, where nutrients do not limit plant functioning to the same degree, the mycorrhizae aid in water uptake, allowing increased transpiration rates in comparison with nonmycorrhizal plants. ▶▶| *Mutualism (section 30.1)*

Mycorrhizae can be broadly classified as **endomycorrhizae**—those with fungi that enter the root cells, or as **ectomycorrhizae**—those that remain extracellular, forming a sheath of interconnecting filaments (hyphae) around the roots. Although six types of mycorrhizae are detailed in **table 29.3** and **figure 29.5**, we focus primarily on the two most important types: the ectomycorrhizae and the endomycorrhizae called arbuscular mycorrhizae (*see figure 24.5*). |◀◀ Glomeromycota *(section 24.4)*

The ectomycorrhizae (ECM) are formed by both ascomycete and basidiomycete fungi. ECM colonize almost all trees in cooler climates. Their importance arises from their ability to transfer essential nutrients, especially phosphorus and nitrogen, to the root. The development of ECM starts with the growth of a fungal mycelium around the root. As the mycelium thickens, it forms a sheath or mantle so that the entire root may be covered by the fungal mycelium (**figure 29.6**). Most ECM produce signaling molecules that limit the growth of root hairs; thus ECM-colonized roots often appear blunt and covered in fungi. From the root surface, the fungi extend hyphae into the soil; these filaments may aggregate to form **rhizomorphs**—dense mats of hyphae that are often visible to the naked eye. Hyphae on the inner side of the sheath penetrate between (but not within) the cortical root cells, forming a characteristic meshwork of hyphae called the **Hartig net** (figure 29.5*d*). Soil nutrients taken up by rhizomorphs must first pass through the hyphal sheath and then into the Hartig net filaments, which form numerous contacts with root cells. This results in efficient, two-way transfer of soil nutrients to the plant and carbohydrates to the fungus. This relationship has evolved to the point that some plants synthesize sugars such as mannitol and trehelose that cannot be used by the plants and can only be assimilated by their fungal symbionts.

Arbuscular mycorrhizae (AM) are the most common type of mycorrhizae. They can be found in association with many tropical plants and, importantly, with most crop plants. AM fungi belong to the division *Glomeromycota* and have not yet been grown in pure culture without their plant hosts. These microbes enter root cells between the plant cell wall and invaginations in the plasma membrane (figure 29.5*a*). So, although AM are endomycorrhizae, they do not breach the root cell membrane. Instead,

Table 29.3 Mycorrhizal Associations

Mycorrhizal Classification	*Fungi Involved*	*Plants Colonized*	*Fungal Structural Features*	*Fungal Function*
Ectomycorrhizae	Basidiomycetes, including those with large fruiting bodies (e.g., toadstools); some ascomycetes	~90% of trees and woody plants in temperate regions; fungal-plant colonization is often species specific	Hartig net, mantle or sheath; rhizomorphs; root hair development is usually limited.	Nutrient (N and P) uptake and transfer
Arbuscular	Glomeromycetes, in particular six genera of the taxon *Glomeromycota*	Wild and crop plants, tropical trees; fungal-plant colonization is not highly specific	Arbuscules: hyphae-filled invaginations of cortical root cell	Nutrient (N and P) uptake and transfer; facilitate soil aggregation; promote seed production; reduce pest and nematode infection; increase drought and disease resistance
Ericaceous	Ascomycetes Basidiomycetes	Low evergreen shrubs, heathers	Some intracellular, some extracellular	Mineralization of organic matter
Orchidaceous	Basidiomycetes	Orchids	Hyphal coils called pelotons within host tissue	Some orchids are non-photosynthetic and others produce chlorophyll when mature; these organisms are almost completely dependent on mycorrhizae for organic carbon and nutrients.
Ectendomycorrhizae	Ascomycetes	Conifers	Hartig net with some intracellular hyphae	Nutrient uptake and mineralization of organic matter
Monotropoid mycorrhizae	Ascomycetes Basidiomycetes	Flowering plants that lack chlorophyll (*Monotropaceae;* e.g., Indian pipe)	Hartig net one cell deep in the root cortex	Nutrient uptake and transfer

treelike hyphal networks called **arbuscules** develop within the folds of the plasma membrane. Individual arbuscules are transient; they last at most two weeks. Spores released by nearby root-associated fungi germinate and colonize roots. This requires a number of specific steps, each of which is regulated by the exchange of molecular signals between the fungus and its host plant (**figure 29.7**). AM can be vigorous colonizers: a 5 cm segment of root can support the growth of as many as eight species, and hyphae from a single germinated spore can simultaneously colonize multiple roots from unrelated plant species.

AM are believed to provide a number of services to their plant hosts, including protection from disease, drought, nematodes, and other pests. Their capacity to transfer phosphorus to roots has been well documented, and recently the nature of their transfer of nitrogen has been explored in detail. Stable isotope experiments were performed in which AM-colonized plants (and the appropriate negative controls lacking mycorrhizae) were treated with $^{15}NO_3^-$ and $^{15}NH_4^+$ (**figure 29.8**). These forms of nitrogen are incorporated into fungal tissue through the glutamine synthetase–glutamate synthase pathway (*see figure 11.18*). Thus ^{15}N-containing glutamine (which is converted to arginine) can be followed. However, it was discovered that prior to transferring nitrogen to host cells, intracellular fungal hyphae degrade the amino acids, transferring only the $^{15}NH_4^+$. Thus while the fungus provides its host with much needed ammonium, it retains the carbon skeleton it needs for nitrogen uptake. In contrast, little is known about how mycorrhizal fungi take up carbon from their hosts. Recently the carbohydrate transporter from an AM that colonizes a cyanobacterium (the only AM known to do so) was identified. This integral membrane protein appears to function as a proton-sugar cotransporter. Its highest affinity is for glucose, followed by mannose, galactose, then fructose. Interestingly, it appears that this transporter may be unique to AM fungi.

⏮ *Nitrogen assimilation (section 11.5)*

In addition to these two most abundant mycorrhizal types, the orchidaceous mycorrhizae are of particular interest (figure 29.5*b*).

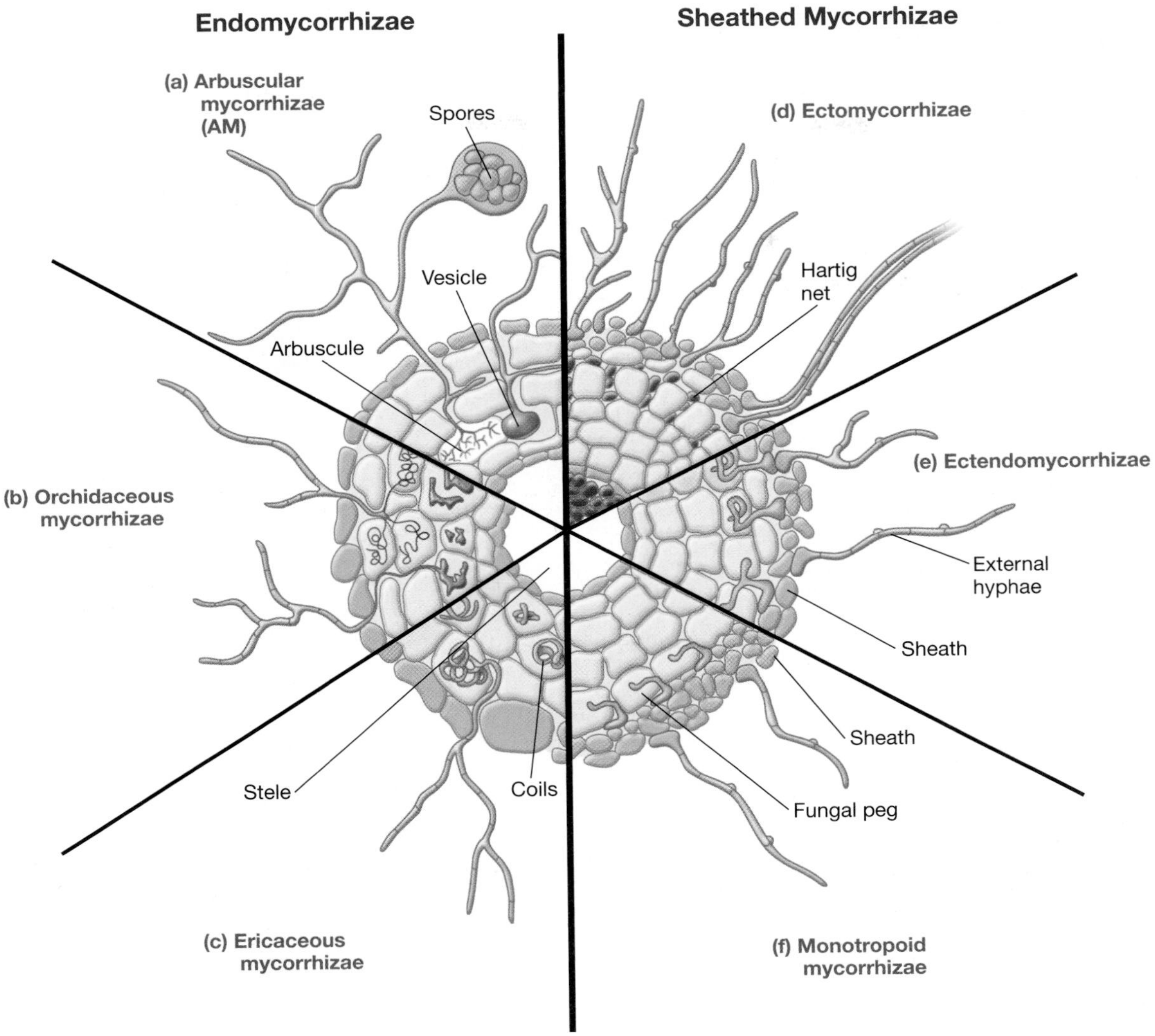

FIGURE 29.5 Mycorrhizae. Fungi can establish mutually beneficial relationships with plant roots, called mycorrhizae. Root cross sections illustrate different mycorrhizal relationships.

Orchids are unusual plants because many never produce chlorophyll, while others only do so after they have matured past the seedling stage. Therefore all orchids have an absolute dependence on their endomycorrhizal partners for at least part of their lives. Indeed, orchid seeds will not germinate unless first colonized by a basidiomycete orchid mycorrhiza. Because the host orchid cannot produce photosynthetically derived organic carbon (or produces very little), orchid mycorrhizal fungi are saprophytic. They must degrade organic matter to obtain carbon, which the orchids then also consume. In this case, the orchid functions as a parasite.

The ecological relationship between mycorrhizae and host is currently being reevaluated. It is now known that in addition to orchids, some green plants such as the serrated wintergreen (*Orthilia secunda*), also parasitize their mycorrhizal partners. Likewise, the growth of the nuisance plant spotted knapweed in the United States may owe its success to its mycorrhizal fungi,

FIGURE 29.6 Ectomycorrhizae as Found on Roots of a Pine Tree. Typical irregular branching of the white, smooth mycorrhizae is evident.

FIGURE 29.7 The Process of Root Colonization by Arbuscular Mycorrhizal Fungi.

Figure 29.7 Micro Inquiry

Which hyphae are growing saprotrophically in this diagram? How does the proportion of the organism growing saprotrophically change during host colonization?

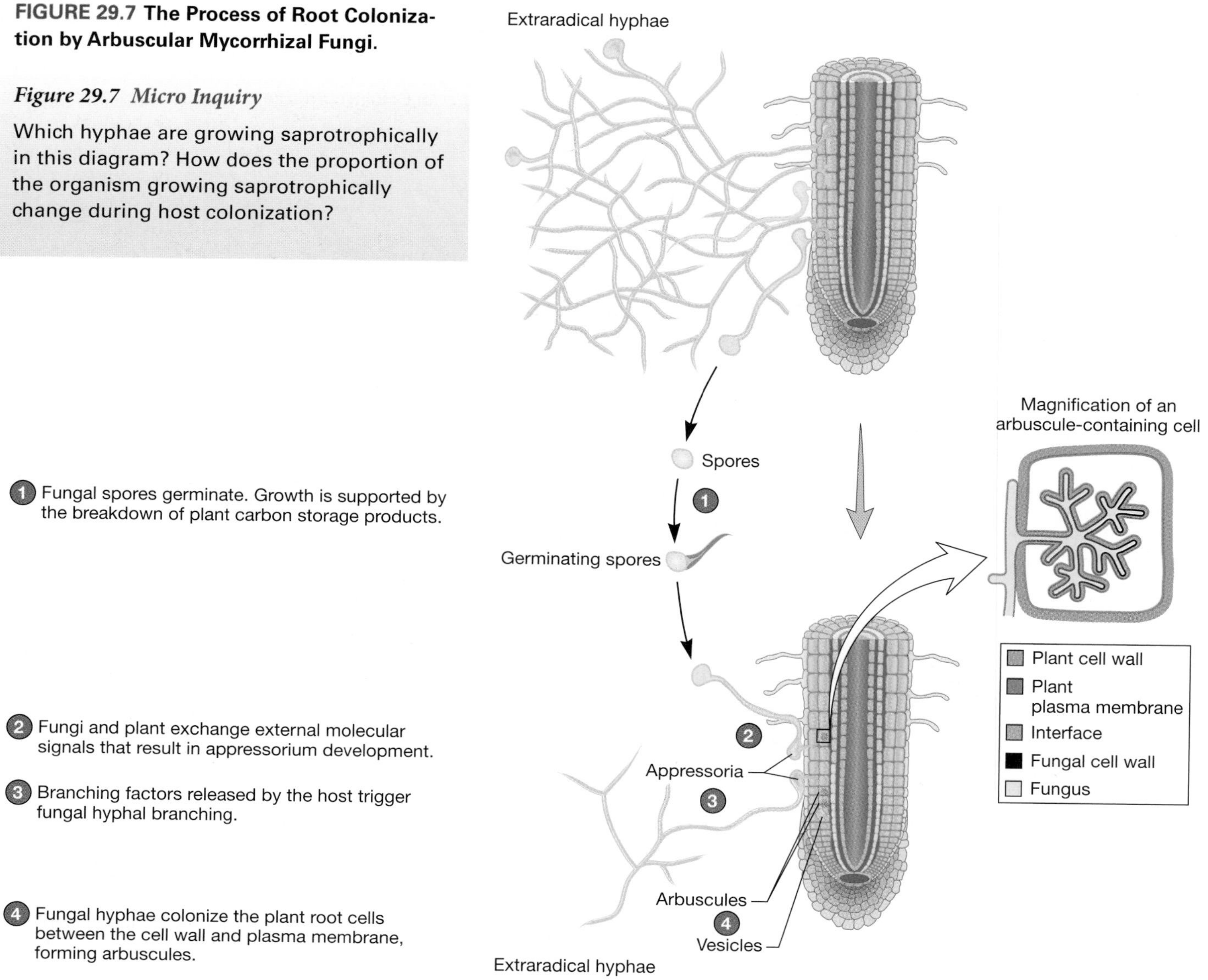

which provide up to 15% of the knapweed's organic carbon, which the fungi obtains from neighboring grasses. Finally, it has also been suggested mycorrhizae may limit plant species diversity in some ecosystems. This is driven by the colonization and growth promotion of seedlings by mycorrhizae that will only associate with a particular plant species.

Extraradical hyphae
Magnification of an arbuscule-containing cell
Spores
1
1 Fungal spores germinate. Growth is supported by the breakdown of plant carbon storage products.
Germinating spores
Plant cell wall
Plant plasma membrane
Interface
Fungal cell wall
Fungus
2
2 Fungi and plant exchange external molecular signals that result in appressorium development.
Appressoria
3
3 Branching factors released by the host trigger fungal hyphal branching.
Arbuscules
4
Vesicles
4 Fungal hyphae colonize the plant root cells between the cell wall and plasma membrane, forming arbuscules.
Extraradical hyphae

Bacteria are also associated with the mycorrhizal fungi. As the external hyphal network radiates out into the soil, a mycorrhizosphere is formed due to the flow of carbon from the plant into the mycorrhizal hyphal network and then into the surrounding soil. These bacteria are called mycorrhization helper bacteria because they may play a role in the development of mycorrhizal relationships involving ectomycorrhizal fungi; they include species of *Pseudomonas, Bacillus,* and *Burkholderia*. Bacterial symbionts have also been found in the cytoplasm of AM fungi. These organisms appear to form a distinct lineage within the β-proteobacteria, most closely related to *Burkholderia, Pandoraea,* and *Ralstonia*. Their role within the fungi remains unclear.

Search This: Mycorrhiza/cropsoil

1. Describe the two-way relationship between mycorrhizal fungi and the plant host.
2. List three major differences between arbuscular mycorrhizae and ectomycorrhizae.
3. What is the function of the rhizomorph and the Hartig net?
4. Describe the uptake and transfer of ammonium by arbuscular mycorrhizae to the plant host. Why do you think only the ammonium is transferred?
5. Propose two potential functions for mycorrhization helper bacteria.

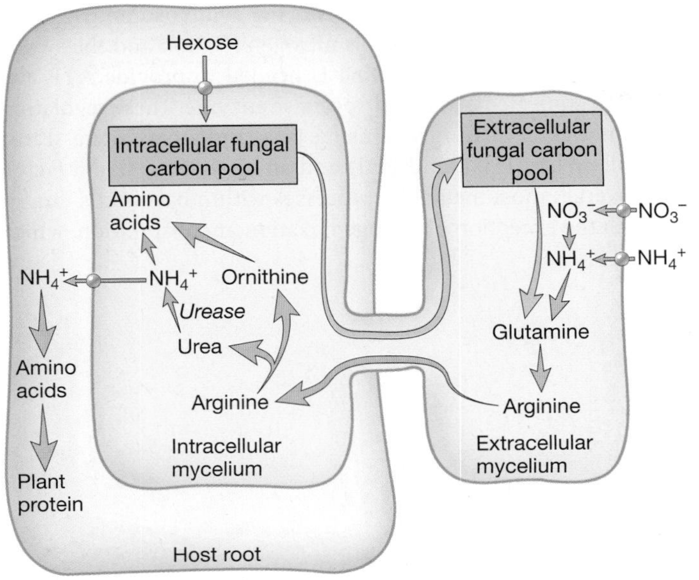

FIGURE 29.8 Nitrogen Exchange Between Arbuscular Mycorrhizal Fungi and Host Plant. Nitrate and ammonium are taken up by the fungal mycelium that is outside the host plant cell (extracellular mycelium) and converted to arginine. This amino acid is transferred to the mycelium within the host plant cell and broken down so that only the ammonium enters the plant.

Nitrogen-Fixing Bacteria

The enzymatic conversion of gaseous nitrogen (N_2) to ammonia (NH_3) often occurs as part of a symbiotic relationship between bacteria and plants. These symbioses produce more than 100 million metric tons of fixed nitrogen annually and are a vital part of the global nitrogen cycle. In addition, symbiotic nitrogen fixation accounts for more than half of the nitrogen used in agriculture (the remainder is applied as fertilizer). The provision of fixed nitrogen enables the growth of host plants in soils that would otherwise be nitrogen limiting. At the same time it reduces loss of nitrogen by denitrification and leaching. It follows that nitrogen fixation, particularly that by members of the genus *Rhizobium* and related α-proteobacteria in association with their leguminous host plants, has been the subject of intense investigation.

The Rhizobia

Several bacterial genera are able to form nitrogen-fixing nodules with legumes. These include the α-proteobacteria *Allorhizobium, Azorhizobium, Bradyrhizobium, Mesorhizobium, Sinorhizobium,* and *Rhizobium.* Collectively these genera are often called the **rhizobia.** Recently the phylogenetic diversity of the rhizobia has been extended by the discovery that the β-proteobacteria *Burkholderia caribensis* and *Ralstonia taiwanensis* also form nitrogen-fixing nodules on legumes. Here, we discuss the general process of nodulation, including some molecular details that have been revealed largely through studies of the genus *Rhizobium.* ◀◀ *Class* Alphaproteobacteria *(section 20.1); Class* Betaproteobacteria *(section 20.2)*

Rhizobia live freely in the soil. In nitrogen-sufficient soils, plants secrete ethylene-mediated compounds that inhibit nodulation. But even in nitrogen-starved soils, when rhizobia approach the plant root, they are assumed to be alien invaders. The plant responds with an oxidative burst, producing a mixture of compounds that can contain superoxide radicals, hydrogen peroxide, and N_2O. This oxidative burst, involving glutathione and homoglutathione, is critical in determining the fate of the infection process. The rhizobia, if they are to be effective colonizers, must use antioxidant defenses to survive and become symbionts. Only rhizobia and related genera with sufficient antioxidant abilities are able to proceed to the next step in the infection process.

The infection is initiated by the exchange of signaling molecules between the plant and rhizobia in the rhizosphere. Plant roots release flavonoid inducer molecules (2-phenyl-1,4-benzopyrone derivatives) that stimulate rhizobial colonization of the root surfaces (**figure 29.9*a***). Flavonoids bind the bacterial protein NodD, which functions as a transcriptional regulator. NodD activates transcription of *nod* genes, which encode the biosynthetic enzymes needed for the production of rhizobial signaling compounds called **Nod factors.** Conversely flavonoids from nonhost plants inhibit the production of Nod factors. The precise structure of individual Nod factors depends on the bacterial species, but all consist of four to five units of β–1,4 linked *N*-acetyl-*D*-glucosamine bearing a species-specific acyl chain at the nonreducing terminal residue and a sulfate attached to the reducing end (figure 29.9*c*; note highlighted moieties). Upon receipt of the Nod factor signal, gene expression in the outer (epidermal) cells of the roots is altered, resulting in oscillations in the levels of intracellular calcium. This "calcium spiking" triggers the root hairs to curl, resembling a shepherd's crook, thereby entrapping bacteria (figure 29.9*c,d*). Nod factors trigger root cortex cells to reinitiate cell division, and these cells will eventually form the nodule primordium to generate cells that will accept the invading rhizobium.

The bacteria now produce Nod factors and a symbiotically active exopolysaccharide to induce changes in the plant cell wall so that the plant plasma membrane invaginates, and new plant material is laid down. These modifications lead to the development of a bacteria-filled, tubelike structure called the **infection thread** (figure 29.9*e,f*). Plant cell division drives the growth of the infection thread. Upon penetrating the base of a root hair cell, the bacteria must stimulate the growth of the infection thread through additional plant cell walls and membranes. This is dependent on the plant hormone cytokinin in concert with Nod-factor stimulation of plant cell mitosis. Finally, the infection thread filled with bacteria reaches the inner plant cortex, where each bacterial cell is endocytosed by a plant cell in an discrete, unwalled membrane compartment that arises from the infection thread. This unit, consisting of an individual bacterium and the surrounding endocytic membrane, is called the **symbiosome** (figure 29.9*h*). It is here that each bacterial cell differentiates into the nitrogen-fixing form called a **bacteroid.** Bacteroids are terminally differentiated—they can neither divide nor revert back to the nondifferentiated state.

The low levels of oxygen within the symbiosome triggers a regulatory cascade that controls nitrogen fixation and the switch to microaerobic respiration that is needed to provide ATP and reducing equivalents to the nitrogenase enzyme. These regulatory proteins include an oxygen-sensing, two-component signal transduction system (FixL and FixJ) and an alternative sigma factor, σ^{54}. Overall, most metabolic processes within bacteroids diminish with the exception of nitrogen fixation and respiration, which

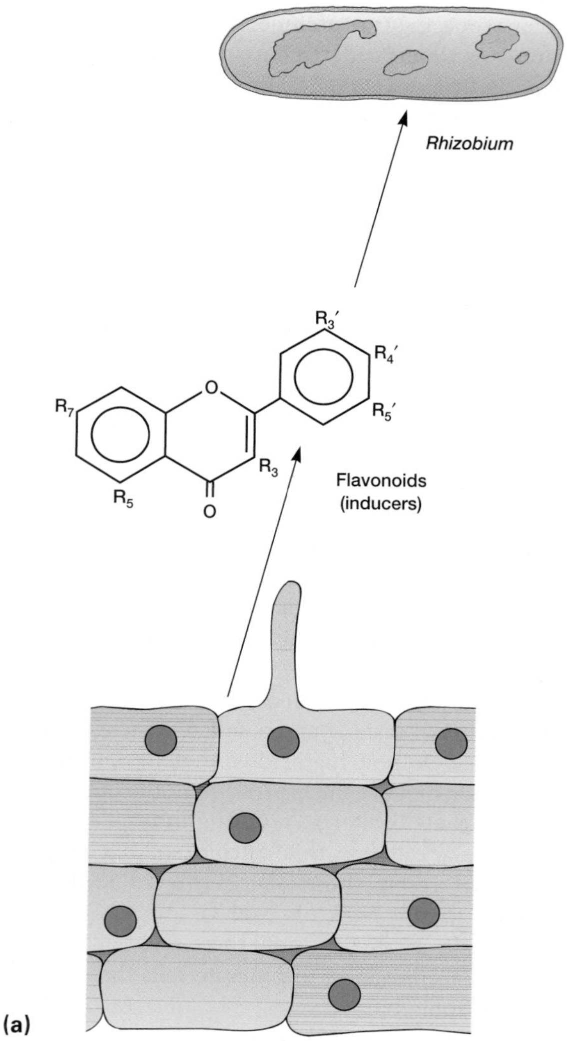

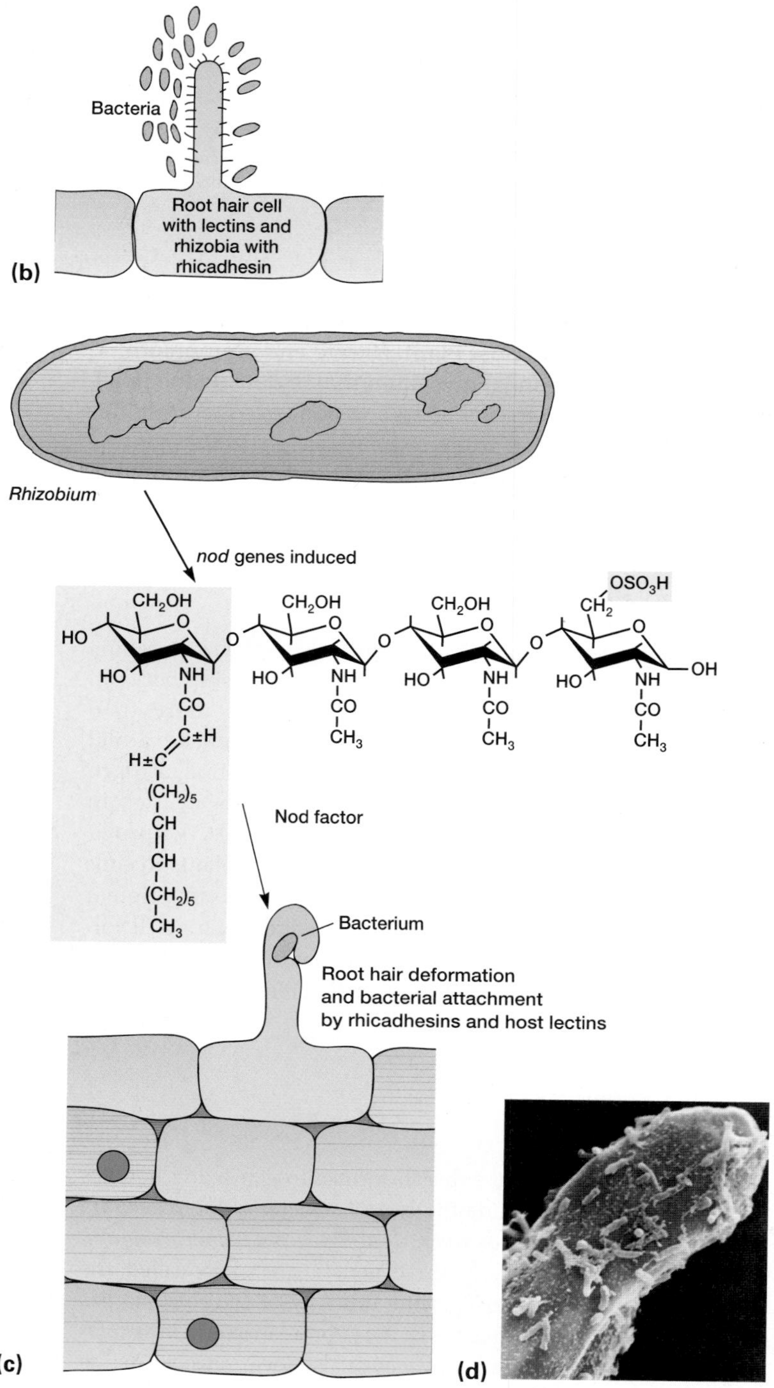

FIGURE 29.9 Root Nodule Formation by *Rhizobium*. Root nodule formation on legumes by *Rhizobium* is a complex process that produces the nitrogen-fixing symbiosis. (a) The plant root releases flavonoids that stimulate the production of various Nod metabolites by *Rhizobium.* Many different Nod factors control infection specificity. (b) Attachment of *Rhizobium* to root hairs involves specific bacterial proteins called rhicadhesins and host plant lectins that affect the pattern of attachment and *nod* gene expression. (c) Structure of a typical Nod factor that promotes root hair curling and plant cortical cell division. The bioactive portion (nonreducing *N*-fatty acyl glucosamine) is highlighted. These Nod factors enter root hairs and migrate to their nuclei. (d) A plant root hair covered with *Rhizobium* and undergoing curling.

continued

are upregulated. The assembly of many symbiosomes gives rise to the **root nodule** (figure 29.9*i,j*). ◀◀ *Nitrogen fixation (section 11.5); Regulation of transcription initiation (section 13.2)*

The reduction of atmospheric N_2 to ammonia by the differentiated bacteroids depends on the production of a protein called **leghemoglobin.** Recall that the nitrogenase enzyme is disabled by oxygen. To help protect the nitrogenase, leghemoglobin binds to oxygen and helps maintain microaerobic conditions within the mature nodule. Leghemoglobin is similar in structure to the hemoglobin found in animals; however, it has a higher affinity for oxygen. Interestingly, the protein moiety is encoded by plant genes, whereas the heme group is a bacterial product.

It had long been thought that rhizobium symbionts transfer ammonia to the host plant, but it is now known that a more complex cycling of amino acids occurs. Apparently the plant provides certain amino acids to the bacteroids so that they do not need to assimilate ammonia. In return, the bacteroids shuttle amino acids (which bear the newly fixed nitrogen) back to the plant. This creates an interdependent relationship, providing selective pressure for the evolution of mutualism. The bacteria also receive carbon and energy in the form of dicarboxylic acids from their host legume.

The molecular mechanisms by which both the legume host and the rhizobial symbionts establish productive nitrogen-fixing bacteroids within nodules continues to be an intense area of

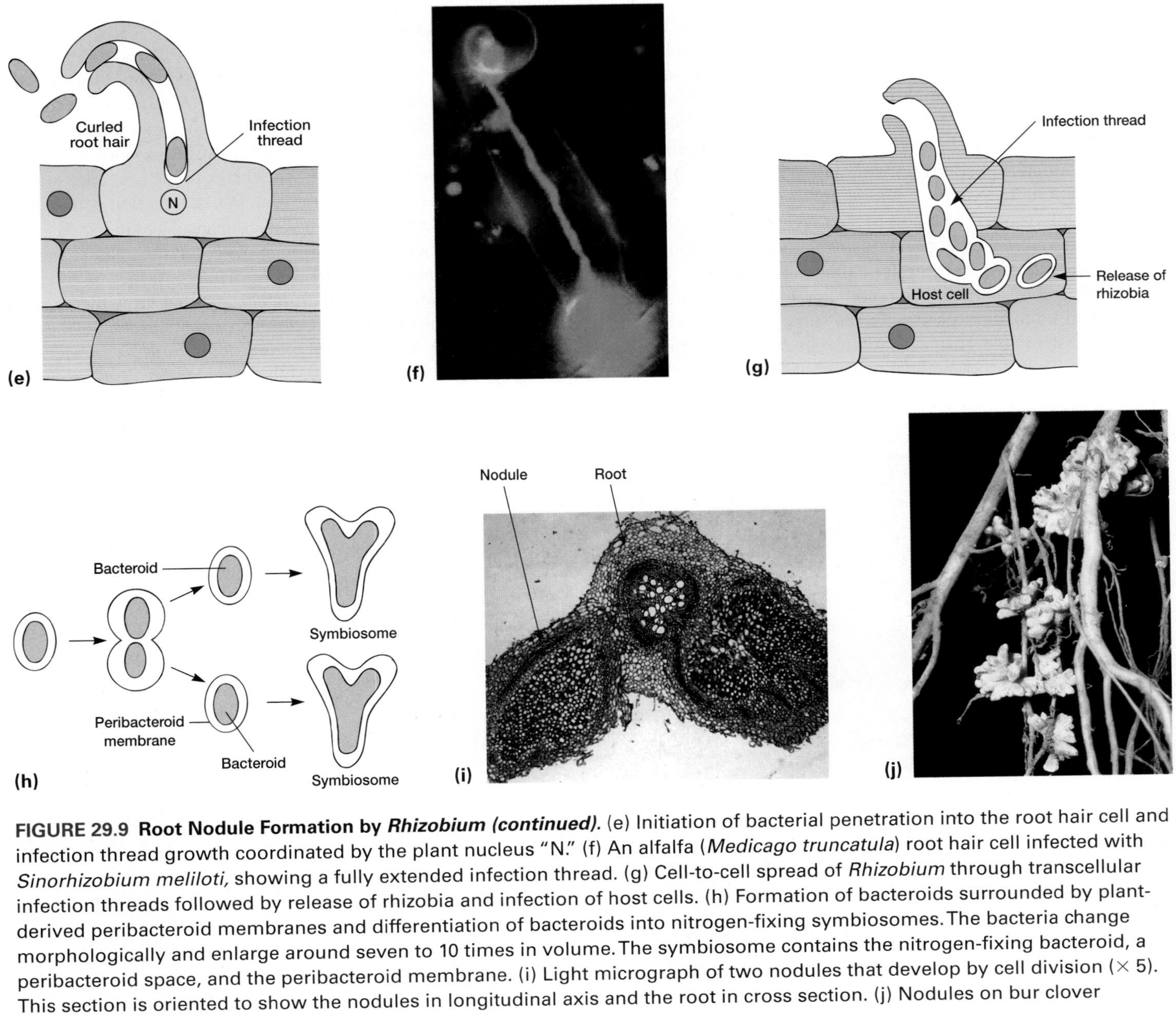

FIGURE 29.9 Root Nodule Formation by *Rhizobium* (continued). (e) Initiation of bacterial penetration into the root hair cell and infection thread growth coordinated by the plant nucleus "N." (f) An alfalfa (*Medicago truncatula*) root hair cell infected with *Sinorhizobium meliloti,* showing a fully extended infection thread. (g) Cell-to-cell spread of *Rhizobium* through transcellular infection threads followed by release of rhizobia and infection of host cells. (h) Formation of bacteroids surrounded by plant-derived peribacteroid membranes and differentiation of bacteroids into nitrogen-fixing symbiosomes. The bacteria change morphologically and enlarge around seven to 10 times in volume. The symbiosome contains the nitrogen-fixing bacteroid, a peribacteroid space, and the peribacteroid membrane. (i) Light micrograph of two nodules that develop by cell division (× 5). This section is oriented to show the nodules in longitudinal axis and the root in cross section. (j) Nodules on bur clover (*Medicago* sp.).

research. A major goal of biotechnology is to introduce nitrogen-fixation genes into plants that do not normally form such associations. It has been possible to produce modified lateral roots on nonlegumes such as rice, wheat, and oilseed rape (the source of canola oil), which are invaded by nitrogen-fixing bacteria. Although these modified root structures have not yet been found to fix useful amounts of nitrogen, they enhance rice production, and work is expected to continue in this area.

Stem-Nodulating Rhizobia

A few leguminous plants support rhizobia that nodulate stems, rather than roots. Nodules form at the base of adventitious roots branching out of the stem just above the soil surface (**figure 29.10*a***). These plants are generally found in waterlogged soils and riverbanks. For instance, the plant *Sesbania rostrata* is nodulated by the α-proteobacterium *Azorhizobium caulinodans,* and the vetch *Aeschynomene sensitiva* is nodulated by a strain of *Bradyrhizobium* that is a photoheterotroph. This symbiosis has been studied in some detail. *Bradyrhizobium* strain BTAi synthesizes its photosynthetic apparatus only under aerobic day-night cycles of illumination, and optimal nodulation and nitrogen fixation occurs only if the microbe is able to photosynthesize. It is thought that light-driven cyclic electron transfer generates a proton gradient used for either ATP synthesis or substrate transfer.

The process of stem nodulation differs from that of root nodulation. Stem nodulation starts with the entry of bacteria through ruptures in stem epidermal cells that occur during emergence of the root tip. Unlike root nodulation, the infection of *Bradyrhizobium* strain BTAi does not involve the production of Nod factors or plant calcium spikes (figure 29.10*b*). Nonetheless, as with root nodulation, bacteria within stem nodules are protected from the external environment and can obtain plant-derived energy substrates.

Actinorhizae

Another example of symbiotic nitrogen fixation occurs between the actinomycete *Frankia* and eight nonleguminous host plant families. These bacterial associations with plant roots are called **actinorhizae** or actinorhizal relationships (**figure 29.11**). *Frankia* fixes nitrogen and is important particularly in trees and shrubs. For example, these associations occur in areas where Douglas fir forests have been clear-cut and in bog and heath environments where bayberries and alders are dominant. The nodules of some plants (*Alnus, Ceanothus*) are as large as baseballs. The nodules of *Casuarina* (Australian pine) approach soccer ball size.

Depending on the host plant, the infection process proceeds either by root hair infection or by direct penetration of the matrix between root cells. When infecting through root hairs, the bacteria induce plant cell division, leading to the development of a prenodule. *Frankia* then enters the cortex of the root and forms a nodule that resembles a lateral root. When infection is independent of root hair infection, *Frankia* is encapsulated in a shell made of pectin that then extends through the plant intercellular spaces. Although *Frankia* nodules vary in morphology, they all bear the general appearance of lateral root clusters that grow by repeated branching of the tips. Some nodules are found just below the surface of the soil, whereas others—particularly those forming on plants that grow in dry, rocky soil—develop very deep below the ground. ◀◀ *Suborder* Frankinae *(section 22.9)*

(a)

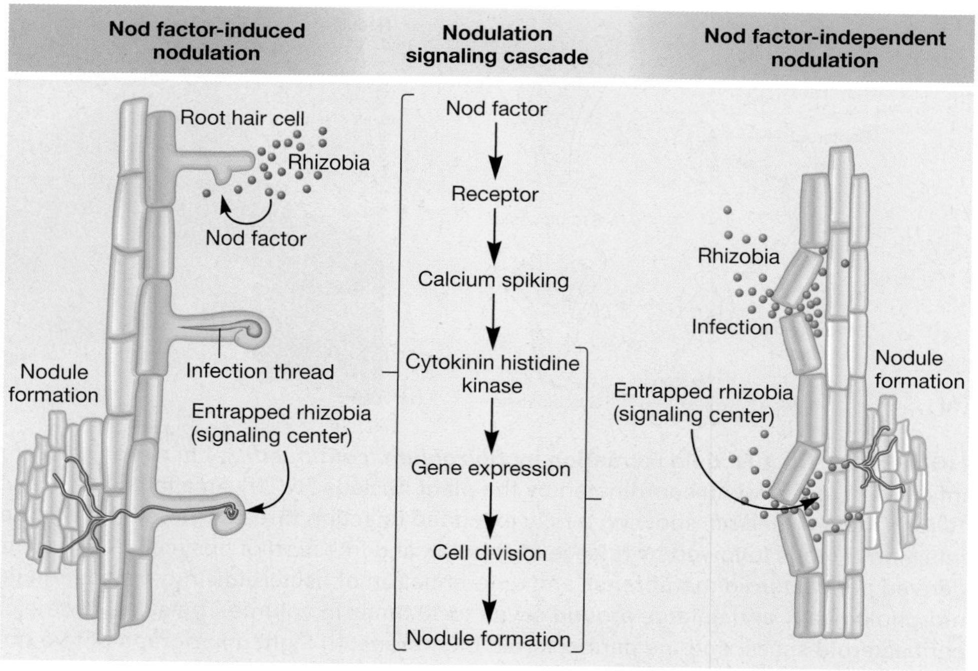

(b)

FIGURE 29.10 Nodulation. (a) Nitrogen-fixing rhizobia can also form nodules on stems of some tropical legumes, as shown in here on *Sesbania rostrata*. (b) Root nodulation (left) involves the production of Nod factors by rhizobium and signaling molecule cytokinin by the host plant. By contrast, Nod factor and plant cytokinin release is not necessary for stem nodulation by at least two rhizobia (right). It is not yet known if this is generally the case.

1. List several bacteria that are considered rhizobia.
2. Describe the communication system between a rhizobial bacterium and its legume host.
3. What is the function of the infection thread? Why do you think it is important that the bacteria do not enter plant cell cytoplasm until they reach the primordium?
4. What does the term terminally differentiated mean? Can you think of other cells that are also terminally differentiated?
5. How is leghemoglobin made and what is its function?
6. What is the major contribution of *Frankia* to its plant hosts? Which types of plants are infected?

Agrobacterium

Clearly some microbe-plant interactions are beneficial for both partners. However, others involve microbial pathogens that harm or even kill their host. *Agrobacterium tumefaciens* is an α-proteobacterium that has been studied intensely for several decades. Initially research focused on how this microbe causes crown gall disease, which results in the formation of tumorlike growths in a wide variety of plants (**figure 29.12**). Within the last 20 years or so, however, *A. tumefaciens* has become one of biotechnology's most important tools. The molecular genetics by which this pathogen infects its host is the basis for plant genetic engineering. ⏮ *Class* Alphaproteobacteria *(section 20.1)*; ⏭ *Agricultural microbiology (section 41.4)*

The genes for plant infection and virulence are encoded on an *A. tumefaciens* plasmid called the **Ti (tumor-inducing) plasmid.** These genes include 21 *vir* genes (*vir* stands for virulence), which are found in six separate operons. Two of these genes, *virD1* and *virD2,* encode proteins that excise a separate region of the Ti plasmid, called **T DNA.** After excision, this T DNA fragment is integrated into the host plant's genome. Once incorporated into a plant cell's genome, T DNA directs the overproduction of phytohormones that cause unregulated growth and reproduction of plant cells, thereby generating a tumor or gall in the plant.

The *vir* genes are not expressed when *A. tumefaciens* is living saprophytically in the soil. Instead, they are induced by the presence of plant phenolics and monosaccharides present in an acidic (pH 5.2–5.7) and cool (below 30°C) environment (**figure 29.13*a***). The microbe usually infects its host through a wound. Upon reception of the plant signal, a two-component signal transduction system is activated: VirA is a sensor kinase that, in the presence of a phenolic signal, phosphorylates the response regulator VirG. Activated VirG then induces transcription of the other *vir* genes. This enables the bacterial cell to become adequately positioned relative to the plant cell, at which point the *virB* operon expresses the apparatus that will transfer the T DNA. This transfer is similar to bacterial conjugation and involves a type IV secretion system. After VirD1 and VirD2 excise the T DNA from the Ti plasmid, the T DNA, with the VirD2 protein attached to the 5′ end, is delivered to the plant cell cytoplasm. The protein VirE2 is also transferred, and together with VirD2, the T DNA is shepherded to the plant cell nucleus, where it is integrated into the host's genome. Here, the T DNA has two specific functions. First, it directs the host cell to overproduce phytohormones that cause tumor formation. Second, it stimulates the plant to produce special amino acid and sugar derivatives called opines (figure 29.13*b*). Opines are not metabolized by the plant but *A. tumefaciens* is attracted to opines; chemotaxis of bacteria from the surrounding soil population will further advance the infection because the bacterium can use opines as sources of carbon, energy, nitrogen, and, in some cases, phosphorus. ⏮ *Two-component*

FIGURE 29.11 Actinorhizae. *Frankia*-induced actinorhizal nodules in *Ceanothus* (buckbrush).

***Figure 29.11* Micro Inquiry**

How do you think *Frankia* protects its nitrogenase from oxygen damage?

FIGURE 29.12 *Agrobacterium.* *Agrobacterium*-caused tumor on a *Kalanchoe* sp. plant.

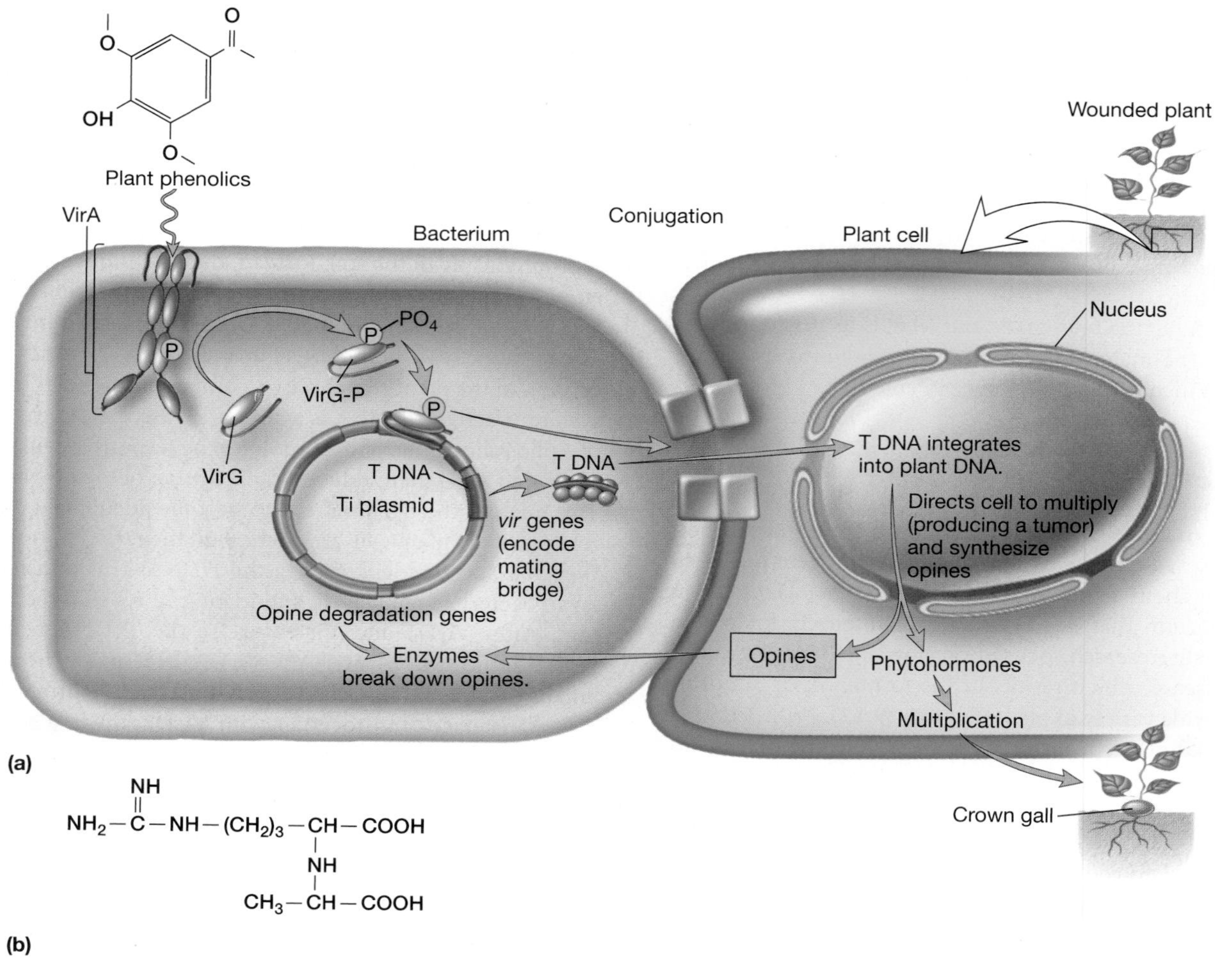

FIGURE 29.13 Functions of Genes Carried on the *Agrobacterium* Ti Plasmid. (a) Genes carried on the Ti plasmid of *Agrobacterium* control tumor formation by a two-component regulatory system that stimulates formation of the mating bridge and excision of the T DNA. The T DNA is moved into the plant cell, where it integrates into the plant DNA. T DNA induces the production of plant hormones that cause the plant cells to divide, producing the tumor. (b) The tumor cells produce opines that can serve as a carbon source for the infecting *Agrobacterium.* Ultimately a crown gall is formed on the stem of the wounded plant above the soil surface.

Figure 29.13 Micro Inquiry

How does the production of opines by the plant accelerate and amplify the *Agrobacterium* infection process?

regulatory systems and phosphorelay systems (section 13.2); Bacterial conjugation (section 14.7)

Search This: Agrobacterium tumefaciens/ *microbial world*

Other Plant Pathogens

In addition to *Agrobacterium,* many other bacteria cause an array of spots, blights, wilts, rots, cankers, and galls, as shown in **table 29.4** and the chapter opener. The soft rots caused by the enterobacteria *Erwinia chrysanthemi* and *E. carotovora* have significant economic impact. These bacteria digest plant tissue by producing extracellular enzymes that degrade pectin, cellulose, and proteins. These exoenzymes are secreted by type II and type I secretion systems; mutants lacking these secretion systems are no longer pathogenic. Similarly, proteobacteria belonging to the genera *Ralstonia, Pseudomonas, Pantoea,* and *Xanthomonas* rely on type III secretion systems to deliver virulence proteins. Although these microbes cause a diverse collection of plant diseases, all colonize the spaces between plant cells, rather than invading the cells themselves. Another group of important plant pathogens includes

Table 29.4 Major Plant Diseases Caused by Bacteria

Symptoms	*Examples*	*Pathogen*
Spots and blights	Wildfire (tobacco)	*Pseudomonas syringae* pv.[a] *tabaci*
	Haloblight (bean)	*P. syringae* pv. *phaseolica*
	Citrus blast	*P. syringae* pv. *syringae*
	Leaf spot (bean)	*P. syringae* pv. *syringae*
	Blight (rice)	*Xanthomonas campestris* pv. *oryzae*
	Blight (cereals)	*X. campestris* pv. *translucens*
	Spot (tomato, pepper)	*X. campestris* pv. *vesicatoria*
	Ring rot (potato)	*Clavibacter michiganensis* pv. *sepedonicum*
Vascular wilts	Wilt (tomato)	*C. michiganensis* pv. *michiganensis*
	Stewart's wilt (corn)	*Erwinia stewartii*
	Fire blight (apples)	*E. amylovora*
	Moko disease (banana)	*P. solanacearum*
Soft rots	Black rot (crucifers)	*X. campestris* pv. *campestris*
	Soft rots (numerous)	*E. carotovora* pv. *carotovora*
	Black leg (potato)	*E. carotovora* pv. *atroseptica*
	Pink eye (potato)	*P. marginalis*
	Sour skin (onion)	*P. cepacia*
Canker	Canker (stone fruit)	*P. syringae* pv. *syringae*
	Canker (citrus)	*X. campestris* pv. *citri*
Galls	Crown galls (numerous)	*Agrobacterium tumefaciens*
	Hairy root	*A. rhizogenes*
	Olive knot	*P. syringae* pv. *savastonoi*

[a] pv., pathovar, a variety of microorganisms with phytopathogenic properties.

Source: From Lengler, J. W., et al. 1999. Biology of the prokaryotes. *Malden, MA: Blackwell Science, table 34.4.*

the wall-less phytoplasms that infect vegetable and fruit crops such as sweet potatoes, corn, and citrus. ◀◀ *Protein maturation and secretion (section 12.9)*

As discussed in chapters 23, 24, and 25 protists, fungi, viruses, and viroids can be devastating plant pathogens. Examples are the fungus *Puccinia graminis,* which causes wheat rust; the öomycete *Phytophthora infestans,* which was responsible for the Irish potato famine; and *Tobacco mosaic virus* (TMV), the first virus to be characterized. A virus of particular interest in terms of plant-pathogen interactions is a hypovirus (family *Hypoviridae*) that infects the fungus *Cryphonectria parasitica,* the cause of chestnut blight. Based on pioneering studies carried out in Italy and France, workers in Connecticut and West Virginia noted that if they infected the fungus with this hypovirus, the rate and occurrence of blight were decreased. They are hoping to treat trees with the less lethal virus strains and eventually transform the indigenous lethal strains of *Cryphonectria* into more benign fungi.

1. Discuss the nature and importance of the Ti plasmid.
2. What functions do the members of the two-component system play in infection of a plant by *Agrobacterium?* What are the roles of phenolics and opines in this infection process?
3. What kinds of exoenzymes are produced by some plant pathogens?
4. How are plant pathologists attempting to control chestnut blight?

29.4 The Subsurface Biosphere

For many years it was thought that life could exist only in the thin veneer of Earth's surface and that any microbes recovered from sediments hundreds of meters deep were contaminants obtained during sampling. This view was drastically altered in the 1980s when the U.S. Department of Energy started looking for novel ways to store toxic waste. The agency began funding studies that applied modern technologies to sample the deep subsurface biosphere. Subsequent reports of microbes at great depth gained credibility, and international teams of geologists and microbiologists have since recovered microbes from thousands of meters below Earth's surface. The application of culture-independent techniques to quantify the numbers and diversity of microbes has revealed that subsurface microbes constitute about one-third of Earth's living biomass. This realization has made deep subsurface microbiology an exciting and active field.

Microbial processes take place in different subsurface regions, including (1) the shallow subsurface where water flowing from the surface moves below the plant root zone; (2) regions where organic matter, originating from the Earth's surface in times past, has been transformed by chemical and biological processes to yield coal (from land plants), kerogens (from marine and freshwater microorganisms), and oil and gas; and (3) zones where methane is being synthesized as a result of microbial activity.

In the shallow subsurface, surface waters often move through aquifers—porous geological structures below the plant root zone. In a pristine system with an oxic surface zone, the electron acceptors used in catabolism are distributed from the most oxidized and energetically favorable (oxygen) near the surface to the least favorable (in which CO_2 is used in methanogenesis) in lower zones (**figure 29.14**).

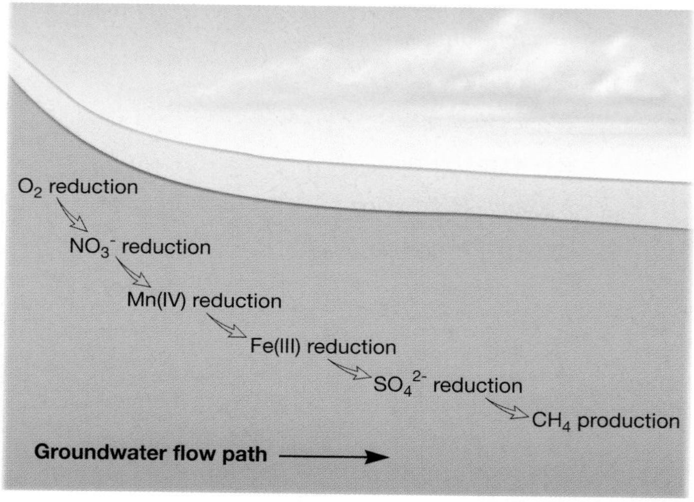

FIGURE 29.14 The Shallow Subsurface Biosphere. The shallow subsurface, as in a stable sediment, showing the distribution with depth of electron acceptors that can occur in an oxic pristine aquifer. In oxic sediments, the electron acceptors will be distributed with the most energetically favorable (oxygen) near the surface and the least energetically favorable at the lower zones of the geological structure.

Source: Lovey, D. K., 1991. Dissimilatory Fe (III) and Mn (IV) reduction. Microbiol. Rev. *55:259–87.*

Figure 29.14 Micro Inquiry

Which electron acceptor has the most positive reduction potential?

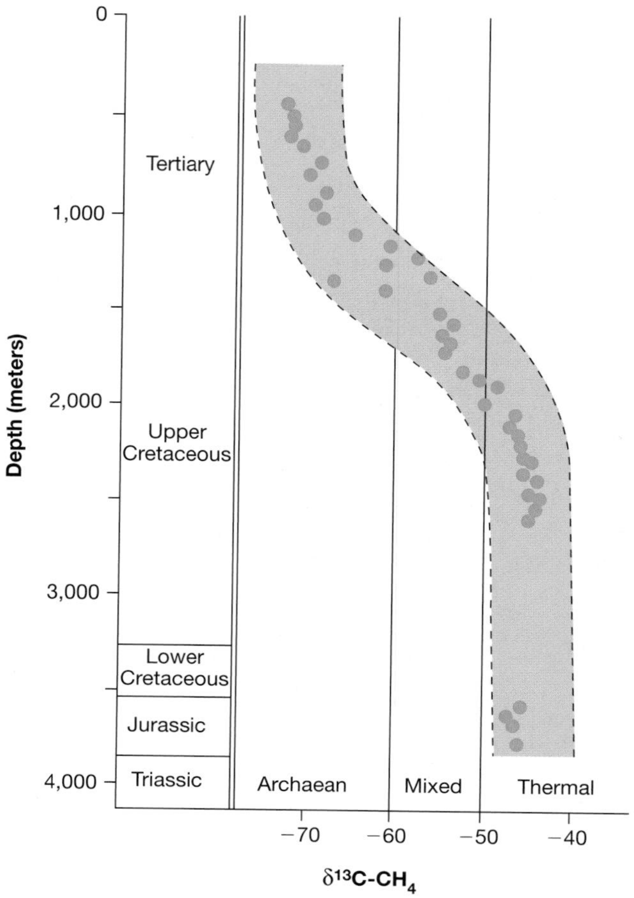

FIGURE 29.15 Methane Synthesis by Archaea in the Subsurface. Stable isotope techniques show that microbially mediated production of methane occurs in the subsurface. The decrease in the occurrence of the ^{13}C isotope of carbon indicates that the methane is produced by microorganisms down to a depth of 1,500 m under the floor of the North Sea. Below 2,000 m, the methane does not have a lower frequency of the ^{13}C isotope of carbon, indicating that it is formed by abiotic processes. The $\delta^{13}C$ value gives an indication of the relative proportion of ^{13}C to ^{12}C in the sample. The more negative scale values signify a decreased presence of the heavier isotope in the upper zone at this location.

Figure 29.15 Micro Inquiry

What would this figure look like if archaea were responsible for methane production to a depth of 2,500 m?

In subsurface regions where organic matter from the Earth's surface has been buried and processed by thermal and possibly biological processes, kerogens and coals break down to yield gas and oil. After their generation, these mobile products, predominantly hydrocarbons, move upward into the more porous geological structures where microorganisms can be active. Chemical signature molecules from plant and microbial biomass are present in these petroleum hydrocarbons.

Below these zones lie vast regions where methane is present in geological structures; this methane is continuously being released to the overlying strata. Based on stable carbon isotope analysis (*see p. 668*), methane made by archaea using H_2 as an energy source and CO_2 as both electron acceptor and carbon source will have less ^{13}C isotope than geologically produced methane. As shown in **figure 29.15**, methane is not depleted of the heavier carbon isotope in the underlying, hotter abiogenic zone. This indicates that archaea produce methane in the upper 1,500 m. Below this depth, the gas is of chemical and thermal origin.

Microorganisms also appear to be growing in intermediate-depth, oil-bearing structures. Recent studies indicate that active bacterial and archaeal assemblages are present in high-temperature (60 to 90°C) oil reservoirs, including such genera as *Thermotoga, Thermoanaerobacter,* and *Thermococcus.* The archaeal genera are dominated by methanogens.

Microbial activity within the "deep hot biosphere," a term suggested by Thomas Gold, has been demonstrated in 2.8 km-deep, high-pressure water within a crustal fracture that was discovered in a South African gold mine while drilling ahead of a tunnel. Microbial diversity was assessed by a combination of 16S rRNA clone libraries and microarrays. Surprisingly, at least 99% of the microbes belong to a single *Firmicute* species. Enough DNA was collected from this unusual habitat so that sequenced nucleotide fragments could be assembled into a complete genome. The bacterium is provisionally called *Candidatus* Desulforudis audaxviator. It appears to be a motile, spore-forming, sulfate-reducing chemoautotroph that is moderately thermophilic. This bacterium most likely infiltrated the fracture zone 3 million to 25 million years ago, and its growth has been sustained by nonphotosynthetically derived sulfate and hydrogen of geologic origin. Indeed, microbial activity from other deep subsurface biomes reveals that chemolithoautotrophy is the predominant, if not only, metabolic strategy in these environments.
◄◄ *Class* Clostridia *(section 21.2)*

The discovery of deep subsurface microbes has a variety of implications. As noted, evidence suggests that these microbes have been trapped in these environments for many years, perhaps as long as 160 million years. They have evolved to exist in a stable environment that is anoxic and without sunlight—some chemolithoautotrophs survive only on H_2, CO_2, and water. The presence of primary production in the absence of sunlight suggests that autotrophy might also be possible in the subsurface sediments of Mars. Exploring these possibilities, along with learning how metabolically active Earth's deep subsurface microbes are, remains the next challenge.
◄◄ *Microbiology in benthic marine environments (section 28.2)*

Search This: Subsurface microbiology/usgs

1. What types of microbial activities have been observed in the deep subsurface?
2. What happens in terms of microbiological processes when organic matter leaches from the surface into the subsurface?
3. Why are stable isotope analyses so important in studies of microbe-geological interactions?
4. What microbial genera have been observed in oil field materials?

Summary

29.1 Soils as a Microbial Habitat

a. In an ideal soil, microorganisms function in thin water films that have close contact with air. Miniaquatic environments can form within soils (**figure 29.1**).

b. Soil organic matter (SOM) helps retain nutrients and water, and maintain soil structure. It can be divided into humic and nonhumic material (**table 29.1**).

29.2 Microorganisms in the Soil Environment

a. Soils support the highest levels of microbial diversity measured.

b. Metagenomic evidence suggests that archaea and fungi are just as abundant as bacteria in soils (**figure 29.4**).

29.3 Microorganism Associations with Vascular Plants

a. The phyllosphere includes all parts of a plant that are aboveground. These structures offer a unique but changeable environment for microbial growth.

b. The rhizosphere is the region around plant roots into which plant exudates are released. The rhizoplane is the plant root surface. A variety of microbes growing in these regions promote plant growth.

c. Mycorrhizal relationships (plant-fungal associations) are varied and complex. Six basic types can be observed, including endomycorrhizal and sheathed/ectomycorrhizal types. The hyphal network of the mycobiont can lead to the formation of a mycorrhizosphere (**figures 29.5–29.8**; **table 29.3**).

d. The rhizobium-legume symbiosis is one of the best-studied examples of plant-microorganism interactions. Rhizobia form nodules on the roots of host plants in which specialized bacterial cells (bacteroids) fix nitrogen (**figure 29.9**).

e. The interaction between rhizobia and host plants is mediated by complex chemicals that serve as communication signals (**figure 29.10*b***).

f. The actinomycete *Frankia* forms nitrogen-fixing symbioses with some trees and shrubs.

g. *Agrobacterium* establishes a complex communication system with its plant host into which it transfers a fragment of DNA. Genes on this DNA encode proteins that result in the formation of plant tumors or galls (**figures 29.12** and **29.13**).

29.4 The Subsurface Biosphere

a. The subsurface includes at least three zones: the shallow subsurface; the zone where gas, oil, and coal have accumulated; and the deep subsurface, where methane synthesis occurs (**figures 29.14** and **29.15**).

b. The "deep hot biosphere" has been sampled to a depth of 2.8 km, and clone libraries suggest that microbial diversity is very limited in these ancient ecosystems that must depend entirely on chemolithoautotrophy.

Critical Thinking Questions

1. Tropical soils throughout the world are under intense pressure in terms of agricultural development. What land use and microbial approaches might be employed to better maintain this valuable resource?

2. Why might vascular plants have developed relationships with so many types of microorganisms? These molecular-level interactions show many similarities when microbe-plant and microbe-human interactions are considered. What does this suggest concerning possible common evolutionary relationships?

3. How might you maintain organisms from the deep hot subsurface under their in situ conditions when trying to culture them? Compare this problem with that of working with microorganisms from deep marine environments.

4. Soil bacteria such as *Streptomyces* produce the bulk of known antibiotics. Look up the competitors for *Streptomyces,* the types of antibiotics these bacteria produce, and how the compounds are effective against competitors (what are the physiological targets?). Would you expect aquatic/marine bacteria to be major producers of antibiotics? Why or why not?

5. It has been reported that the Nod-factor signals produced by the root nodulating bacterium *Sinorhizobium meliloti* are essential for biofilm formation as well as N_2 fixation. In what ways do you think nodulation and biofilm formation are similar, and why do you think this microbe evolved to use the same signal for the development of both?

   Read the original paper: Fujishige, N., et al. 2008. Rhizobium common nod genes are required for biofilm formation. *Mol. Microbiol.* 67:504.

6. In 2008 the annotated genome sequence of the ectomycorrhizal basidiomycete *Laccaria bicolor* was published. Surprisingly, *L. bicolor* produces a large number of small, secreted proteins (SSPs), some of which are only expressed in hyphae that colonize the host root. Another interesting finding was the presence of genes encoding enzymes for the degradation of cell wall polysaccharide only found in organisms other than plants. Formulate an hypothesis regarding the role of the SSPs during host colonization and discuss the metabolic significance of the types of cell walls *L. bicolor* is able to degrade.

   Read the original paper: Martin, F. 2008. The genome of *Laccaria bicolor* provides insights into mycorrhizal symbiosis. *Nature.* 452:88.

Concept Mapping Challenge

Construct a concept map using the words below; provide your own linking terms.

Symbionts	Mycorrhizae	Root nodule
Phosphate	N_2 fixation	Rhizomorphs
Rhizobia	Actinorhizae	Ectomychorrhizae

Learn More

Learn more by visiting the text website at www.mhhe.com/willey8, where you will find a complete list of references.

30

Microbial Interactions

*A "garden" of tube worms (*Riftia pachyptila*) at the Galápagos Rift hydrothermal vent site (depth 2,550 m). Each worm grows to more than a meter in length thanks to endosymbiotic chemolithoautotrophic bacteria, which provide carbohydrate to these gutless worms. The bacteria use the Calvin cycle to fix CO_2 and H_2S as an electron donor.*

CHAPTER GLOSSARY

amensalism A relationship in which the product of one organism has a negative effect on another organism.

commensalism Living on or within another organism without injuring or benefiting the other organism. Usually the smaller of the two organisms is the **commensal.**

competition An interaction between two organisms attempting to use the same resource (nutrients, space, etc.).

competitive exclusion principle When two competing organisms overlap in resource use, one of the organisms will be more successful than the other.

consortium A physical association of two different organisms, usually beneficial to both organisms.

cooperation A positive but not obligatory interaction between two different organisms.

ectosymbiont An organism that lives on or outside the body of another organism in a symbiotic association.

endosymbiont An organism that lives within the body of another organism in a symbiotic association.

gnotobiotic Animals that are germfree (microorganism free) or live in association with one or more known microorganisms.

microbial flora or microbiota The typical or normal microorganisms found on and in a eukaryotic host.

microbiome All the genes of the host, including those of its microbiota; the microbiome is the combined genetic background of a host and its microbiota.

mutualism A type of symbiosis in which both partners gain from the association and are unable to survive without it. The **mutualist** and the host are metabolically dependent on each other.

normal microbial flora or microbiota The microorganisms normally associated with a particular tissue or structure.

opportunistic microorganism or pathogen A microorganism that is usually free living or a part of the host's normal microbiota but that may become pathogenic under certain circumstances, such as when the immune system is compromised.

parasitism A type of symbiosis in which one organism benefits from the other and the host is usually harmed.

pathogen Any virus, bacterium, or other infectious agent that causes disease.

symbiosis The living together or close association of two dissimilar organisms; each of these organisms is known as a **symbiont.**

syntrophism The association in which the growth of one organism either depends on or is improved by the provision of one or more growth factors or nutrients by a neighboring organism. Sometimes both organisms benefit.

Our discussion of microbial ecology has so far considered microbial communities in complex ecosystems. However, the ecology of microorganisms also involves the physiology and behavior of microbes as they interact with one another and with higher organisms. In this chapter, we define types of microbial interactions and present a number of illustrative examples. We then apply these interactions to examine perhaps the most intimate, yet still relatively unexplored, microbial habitat: the human body.

We begin our discussion by defining the term **symbiosis.** Although symbiosis is often used in a nonscientific sense to mean a mutually beneficial relationship, here we use the term in its original broadest sense, as an association of two or more different species of organisms, as suggested by H. A. deBary in 1879.

30.1 Microbial Interactions

Microorganisms can associate physically with other organisms in a variety of ways. One organism can be located on the surface of another, as an **ectosymbiont,** or one organism can be located within another organism, as an **endosymbiont.** The simplest microbial interactions involve two members, a **symbiont** and its host. In such a two-membered relationship, the smaller organism is considered the symbiont and the larger is the host. Sometimes an organism hosts more than one symbiont. The term **consortium** can be used to describe this physical relationship. These physical associations can be intermittent and cyclic or permanent. Important human diseases, including listeriosis, malaria, leptospirosis, legionellosis, and vaginosis, also involve such intermittent and cyclic symbioses. Interesting permanent relationships also occur between bacteria and animals. In these cases, an important characteristic of the host animal is conferred by the permanent bacterial symbiont. The provision of vitamin K by *Escherichia coli* to its human host is a particularly relevant example.

Although it is possible to observe microorganisms in these varied physical associations with other organisms, the fact that there is some type of physical contact provides no information about the nature of the interactions that might be occurring. These interactions include mutualism, cooperation, commensalism, predation, parasitism, amensalism, and competition (**figure 30.1**). These interactions are now discussed.

Mutualism

Mutualism (Latin *mutuus,* borrowed or reciprocal) defines the relationship in which some reciprocal benefit accrues to both partners. This is an obligatory relationship in which the **mutualist** and the host are dependent on each other. In many cases, the individual organisms will not survive when separated. Several examples of mutualism are presented next.

Microorganism-Insect Mutualisms

Mutualistic associations are common between microbes and about 10% of all known insects. This is related to the foods they use, which often include plant sap or animal fluids lacking in essential vitamins and amino acids. The required vitamins and amino acids are provided by bacterial symbionts in exchange for a secure habitat and ample nutrients (**Microbial Diversity & Ecology 30.1**). The aphid is an excellent example of this mutualistic relationship. Cells of this insect harbor the γ-proteobacterium *Buchnera aphidicola,* and a mature insect contains literally millions of these bacteria in its body. It is estimated that the *B. aphidicola*–aphid endosymbiosis was established about 150 million years ago. Strains of *B. aphidicola* provide their hosts with all essential amino acids except tryptophan. If the insect is treated with antibiotics, it dies. Likewise, *B. aphidicola* is an obligate mutualistic symbiont. The inability of either partner to grow without the other indicates that the two organisms underwent **coevolution**—that is, they have evolved together. In fact, the pea aphid (*Acyrthosiphon pisum*) and

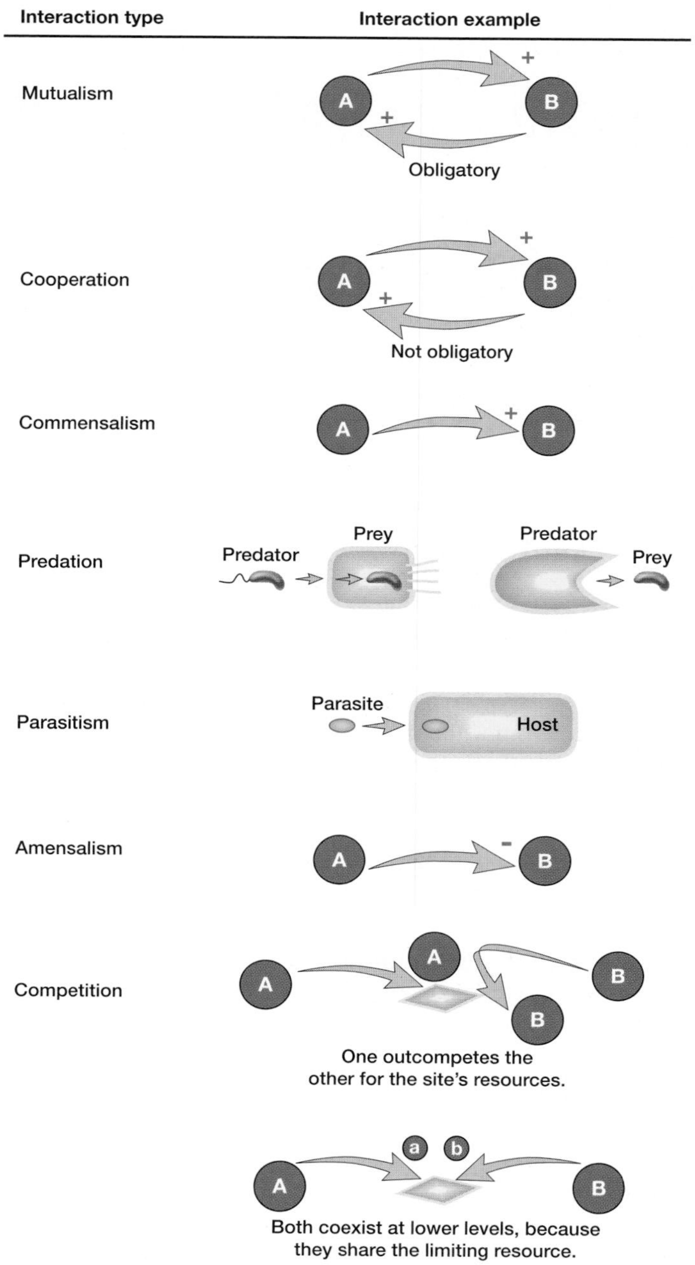

FIGURE 30.1 Microbial Interactions. Basic characteristics of symbiotic interactions that can occur between different organisms.

B. aphidicola symbiont share the genes for the biosynthesis of amino acids, such that some steps occur only in one organism.

The genomic sequences of endosymbiotic *B. aphidicola* strains reveal extreme genomic stability. These strains diverged 50 million to 70 million years ago, and since that time, approximately 75% of the ancestral *B. aphidicola* genome has been eliminated. These genomes are among the smallest known to date: in most cases, only about 0.62 Mb each, although the genome of the bacterium *Candidatus* Carsonella rudii, a symbiont of the sap-sucking insect *Pachypsylla venusta* is a record-setting 0.422 Mb. More than 90% of these genes are common among the symbionts, and the genomes are very stable showing no evidence of gene duplication, translocation, inversion, or horizontal transfer. This implies that although the initial acquisition of the endosymbiont by ancestral aphids enabled their use of an otherwise deficient food source (sap), the bacteria have not continued to expand the ecological niche of their insect host through the acquisition of new traits that might be advantageous elsewhere. ◀◀ *Bioinformatics (section 16.3); Evolutionary processes and the concept of a microbial species (section 17.5)*

Search This: Buchnera *Aphid Symbiosis*

Termites have long been considered a terrific model system for the study of endosymbiosis. There are two classes of termites. Lower termites harbor protist, bacterial, and archaeal endosymbionts and eat only wood. By contrast, higher termites rely solely on bacteria and archaea and forage for a more varied diet. Here, we discuss lower termites. The main structural polysaccharides of wood are cellulose (an unbranched glucose polymer) and hemicellulose (a smaller, branched glucose polymer). This diet poses two problems for the termite: how to degrade polysaccharides that may possess as many as 15,000 glucose monomers and where to get organic nitrogen to support growth. Although lower termites produce a cellulolytic enzyme, they need protists to complete lignocellulose degradation. Nitrogen-fixing bacteria that live in the termite gut solve the problem inherent to consuming a substrate that lacks nitrogen.

Most termite-associated protists are members of the *Excavata* supergroup and as such are quite ancient, possessing hydrogenosomes rather than mitochondria. These protists ferment cellulose to acetate, CO_2, and H_2. Acetate is the termite's preferred carbon source, and termites rely on bacterial symbionts to convert the CO_2 and H_2 released by the protists to acetate through a pathway known as reductive acetogenesis or the acetyl-CoA pathway (*see figure 11.8*). This is not 100% efficient, and termites are known to be prolific H_2 producers. In addition, both lower and higher termites harbor methanogenic archaea that use these substrates to form CH_4. ◀◀ *Organelles involved in energy conservation (section 4.6); Supergroup* Excavata *(section 23.2)*

Some protists also harbor endosymbionts (i.e., bacterial endosymbionts within protist endosymbionts). For instance, the protist *Trichonympha* relies on a bacterial endosymbiont to convert glutamine to other amino acids and nitrogenous compounds (**figure 30.2**). These bacteria, which have not yet

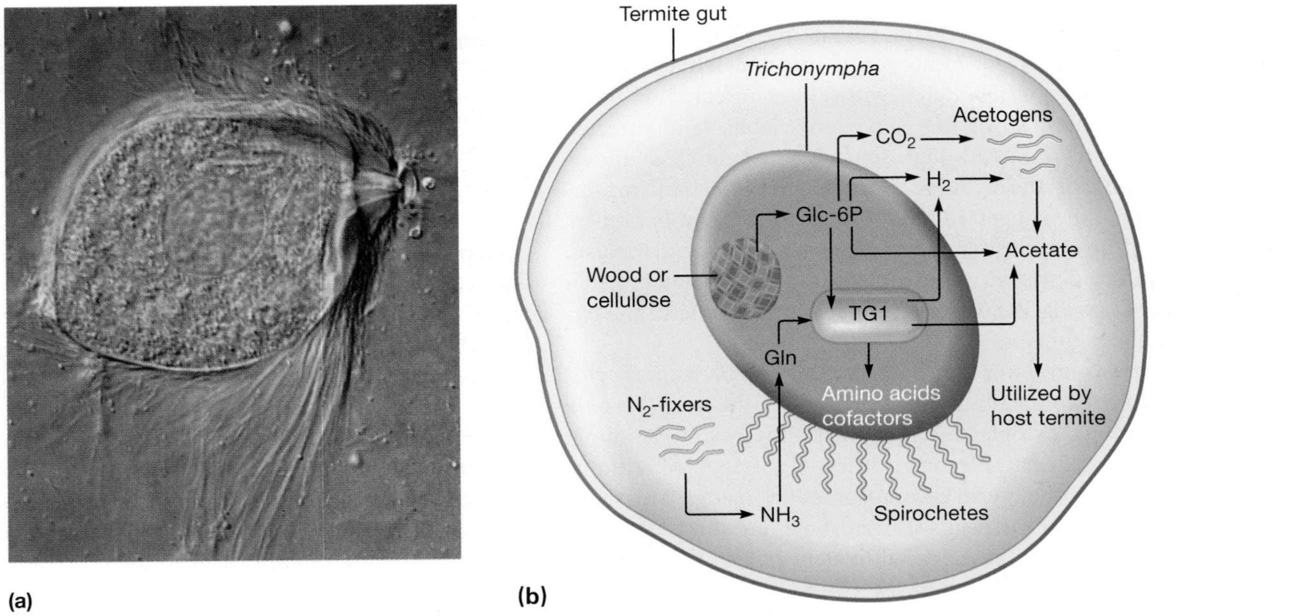

(a) (b)

FIGURE 30.2 Mutualism. (a) Light micrograph of *Trichonympha,* a multiflagellated protozoan from the termite's gut (×135). The ability of *Trichonympha* to break down cellulose allows termites to use wood as a food source. (b) Proposed interaction between the termite endosymbiontic protist *Trichonympha* and its bacterial endosymbiont TG1. Within the termite gut, bacteria convert N_2 to ammonia, which is taken up and converted to glutamine (Gln) by Trichonympha. The protist relies on TG1 to convert Gln to other amino acids, which are released with other cofactors that the protist cannot synthesize. Both *Trichonympha* and TG1 ferment glucose to acetate, H_2, and CO_2. These products are exported by the protist and made available to the termite, which also relies on free-swimming gut bacteria to boost acetate levels through reductive acetogenesis.

Figure 30.2 Micro Inquiry

What methanogenic substrates are also available in the termite gut?

MICROBIAL DIVERSITY & ECOLOGY

30.1 *Wolbachia pipientis*: The World's Most Infectious Microbe?

Most people have never heard of the bacterium *Wolbachia pipientis*, but this rickettsia infects more organisms than does any other microbe. It is known to infect a variety of crustaceans, spiders, mites, millipedes, and parasitic worms, and may infect more than a million insect species worldwide. *Wolbachia* inhabits the cytoplasm of these animals, where it apparently does no harm. To what does *Wolbachia* owe its extraordinary success? Quite simply, this endosymbiont is a master at manipulating its hosts' reproductive biology. In some cases, it can even change the sex of the infected organism.

Wolbachia is transferred from one generation of host to the next through the eggs of infected female hosts. So to survive, this microbe must ensure the fertilization and viability of infected eggs while decreasing the likelihood that uninfected eggs survive. In the 1970s scientists noticed that if a male wasp infected with *Wolbachia* mated with an uninfected female, few, if any, of the uninfected offspring survived. However, if infected females of the same wasp species mated with either infected or uninfected males, all of the eggs were viable—and infected with *Wolbachia*. Although scientists had no clear understanding of the cellular or molecular mechanisms involved, they suggested that "cytoplasmic incompatibility" might be responsible. They proposed that the cytoplasm of infected sperm was toxic to uninfected eggs but that eggs carrying *Wolbachia* produced an antidote to the hypothetical poison.

It was not until the 1990s that the mechanism of cytoplasmic incompatibility became clear. In the wasp *Nasonia vitripennis*, cytoplasmic incompatibility involves timing, not toxins. Specifically, when infected sperm fertilize uninfected eggs, the sperm chromosomes try to align with those of the egg while the egg's chromosomes are still confined to the pronucleus. These eggs ultimately divide as if never fertilized and develop into males. However, chromosomes behave normally when a male and infected female mate. This yields a normal sex distribution, and all progeny are infected with *Wolbachia*.

In other infected species, scientists discovered that in addition to cytoplasmic incompatibility, *Wolbachia* has developed other means to ensure its endurance (**box figure**). In some insect species, the endosymbiont simply kills all the male offspring and induces parthenogenesis of infected females—that is, the mothers simply clone themselves. This limits genetic diversity but allows 100% transmission of *Wolbachia* to the next generation. In still other insects, the microbe allows the birth of males but then modifies their hormones so that the males become feminized and produce eggs.

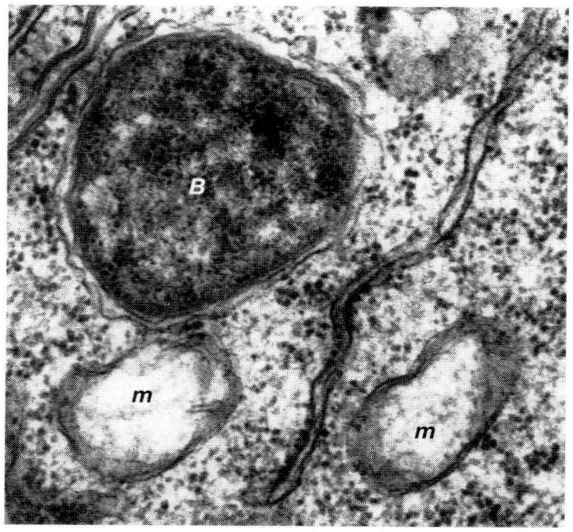

***Wolbachia pipientis* Within the Egg Cytoplasm of the Ant *Gnamptogenys menadensis*.** In this insect, *Wolbachia* is maternally transmitted, so the bacterium has evolved mechanisms to manipulate the sex distribution of the offspring so that the host produces mostly females. The *Wolbachia* cell is indicated by B, and host mitochondria are labeled m.

Wolbachia's ability to manipulate the reproduction of its hosts has led some to question the bacterium's potential role in speciation. It was noticed that two North American wasp species (*N. giraulti* and *N. longicornis*), each carrying a different strain of *Wolbachia*, appeared to be morphologically, behaviorally, and, most importantly, genetically similar. Predictably, when the two species mate with each other, there are no viable offspring. But when wasps are treated with an antibiotic to cure them of their *Wolbachia* infections, the two wasp species mate and produce viable, fertile offspring. Could it be that while there are no genetic barriers to reproduction, the presence of two different *Wolbachia* endosymbionts forms a reproductive wall that could ultimately drive the evolution of new species? The role of *Wolbachia* in generating genetic diversity among its hosts is further illustrated by the discovery of horizontal gene transfer (HGT) from the bacterium to its host's genome. This has been confirmed in at least four insect and four nematode species that harbor *Wolbachia* symbionts. In fact, the fruit fly *Drosophila ananassae* contains the entire *Wolbachia* genome within its own.

Continued

MICROBIAL DIVERSITY & ECOLOGY

30.1 *(Continued)*

Finally, *Wolbachia* may be driving more than speciation. They are required for the embryogenesis of filarial nematodes, which cause diseases such as elephantitis and river blindness. *Onchocerca volvulus* is the filarial nematode that causes river blindness. It is transmitted by blackflies in Africa, Latin America, and Yemen, with a worldwide incidence of about 18 million people. When a fly bites a person, the nematode establishes its home in small nodules beneath the skin, where it can survive for as long as 14 years. During that time, it releases millions of larvae, many of which migrate to the eye. Eventually the host mounts an inflammatory response that results in progressive vision loss (**box figure**). It is now recognized that this inflammatory response is principally directed at the *Wolbachia* infecting the nematodes, not the worms themselves. German researchers discovered that in patients treated with a single course of antibiotic to kill the endosymbiont, nematode reproduction stopped. While inflammatory damage cannot be reversed, disease progression is halted. This may prove a more effective and cost-efficient means of treatment than the current antiparasitic method that must be repeated every six months.

Our understanding of the distribution of *Wolbachia* and the mechanistically clever ways it has evolved to ensure its survival will continue to grow. The microbe may ultimately become a valuable tool in investigating the complexities of speciation, as well as the key to curing a devastating disease.

River blindness. This is the second-leading cause of blindness worldwide. Evidence suggests that it is not the nematode but its endosymbiont, *Wolbachia pipientis*, that causes the severe inflammatory response that leaves many, like the man shown here, blind.

Sources: St. Andre, A., et al. 2002. The role of endosymbiotic Wolbachia *bacteria in the pathogenesis of river blindness.* Science *295:1892–95; Zimmer, C. 2001.* Wolbachia: *A tale of sex and survival.* Science *292:1093–95.*

been grown in pure culture, are called simply TG1, although the genus name "Endomicrobia" has been proposed. In exchange, *Trichonympha* supplies TG1 with glucose-6-phosphate, which can directly enter glycolysis. Finally, motility is often provided by spirochetes that cover the protist surface. Motility is essential to prevent protist expulsion by the termite gut and to acquire food.

Zooxanthellae

Many marine invertebrates (sponges, jellyfish, sea anemones, corals, ciliated protists) harbor endosymbiotic dinoflagellates called zooxanthellae within their tissue. Because the degree of host dependency on the dinoflagellate is variable, only one well-known example is presented. ◀◀ Alveolata *(section 23.5)*

The hermatypic (reef-building) corals satisfy most of their energy requirements using their zooxanthellae, which include members of the dinoflagellate genus *Symbiodinium* (**figure 30.3*a***). These protists line the coral gastrodermal tissue at densities between 5×10^5 and 5×10^6 cells per square centimeter of coral animal. In exchange for up to 95% of their photosynthate (fixed carbon), zooxanthellae receive nitrogenous compounds, phosphates, CO_2, and protection from UV light from their hosts. In addition to *Symbiodinium*, corals host distinct bacterial communities in three habitats (figure 30.3*b*). The surface mucus layer supports the growth of nitrogen-fixing and chitin-degrading

(a)

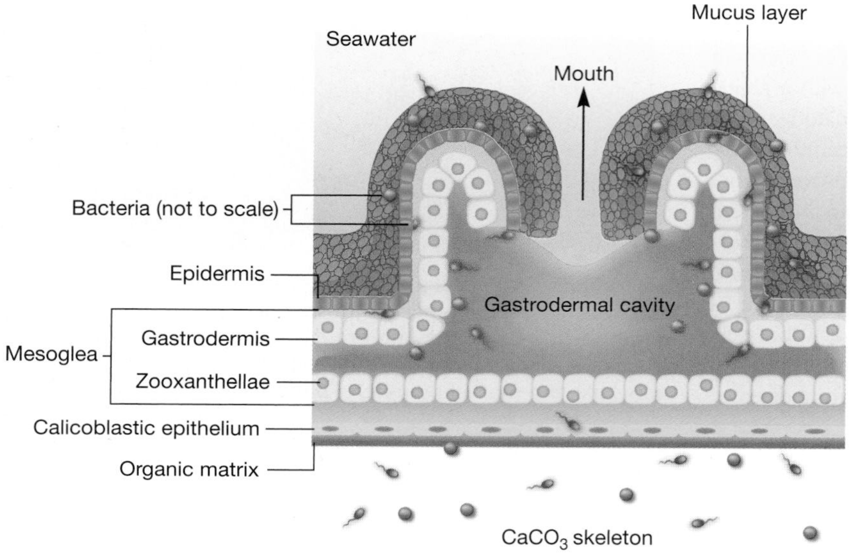

(b)

FIGURE 30.3 Zooxanthellae. (a) The green color of this rose coral (*Manilina*) is due to the abundant zooxanthellae within its tissues. (b) Corals provide three distinct habitats for bacterial growth: the mucus layer at the surface, the gastrodermal cavity, and the coral skeleton. Calicoblastic epithelial cells precipitate calcium carbonate ($CaCO_3$) from the water to build exoskeleton. Zooxanthellae inhabit gastrodermal cells and are shown in green.

bacteria, the animal tissue within the gastrodermal cavity features another population of bacteria (although little is known about these microbes), and the coral skeleton has yet another group of bacteria. Like mucus-associated microbes, these bacteria are important in meeting the animal's nitrogen needs. Together these micro-organisms are responsible for extremely efficient nutrient cycling and tight coupling of trophic levels, which accounts for the stunning success of reef-building corals in developing vibrant ecosystems.

During the past several decades, the number of coral bleaching events has increased dramatically. **Coral bleaching** is defined as a loss of either the photosynthetic pigments from the zooxanthellae or the expulsion of the protists by the coral. It has been determined that damage to photosystem II of the zooxanthellae generates reactive oxygen species (ROS); it is these ROS that appear to be the direct cause of damage (recall that photosystem II uses water as the electron source, resulting in the evolution of oxygen). Coral bleaching appears to be caused by a variety of stressors, but one important factor is temperature. Temperature increases as small as 2°C above the average summer maxima can trigger coral bleaching. Ocean acidification caused by increased CO_2 dissolution in seawater also threatens coral health. As the pH declines, even slightly, the carbonate equilibrium is shifted such that less carbonate is available for coral skeleton production (*see figure 28.1*). Sadly, evidence suggests that as many as one-third of corals and their zooxanthellae will be unable to evolve quickly enough to keep pace with predicted ocean acidification and warming if global climate change continues unchecked. ◀◀ *Oxygen concentration (section 7.6); Light reactions in oxygenic photosynthesis (section 10.12); Water as a microbial habitat (section 28.1); Global climate change (section 26.2)*

Search This: Coral bleaching/noaa

Sulfide-Based Mutualisms

Tube worm–bacterial relationships exist several thousand meters below the surface of the ocean, where the Earth's crustal plates are spreading apart (**figure 30.4**). Vent fluids are anoxic, contain high concentrations of hydrogen sulfide, and can reach a temperature of 350°C. However, because of increased atmospheric pressure, the water does not boil. The seawater surrounding these vents has sulfide concentrations around 250 μM and temperatures 10 to 20°C above the ambient seawater temperature of about 2°C.

Giant (>1 m in length), red, gutless tube worms (*Riftia* spp., **figure 30.5*a***) near these hydrothermal vents provide an example of a remarkably successful form of mutualism in which chemolithotrophic bacterial endosymbionts are maintained within specialized cells of the tube worm host (figure 30.5*b,c,d*). *Riftia* tube worms live at the interface between hot, anoxic, sulfide-containing fluids of the vents and cold, oxygenated seawater. To provide both reduced sulfur and oxygen to their bacterial endosymbionts, the worm's blood contains a unique kind of hemoglobin, which accounts for the bright-red plume extending out of their tubes. Hydrogen sulfide (H_2S) and O_2 are removed from the seawater by the worm's hemoglobin and delivered to an organ called the trophosome. The trophosome is

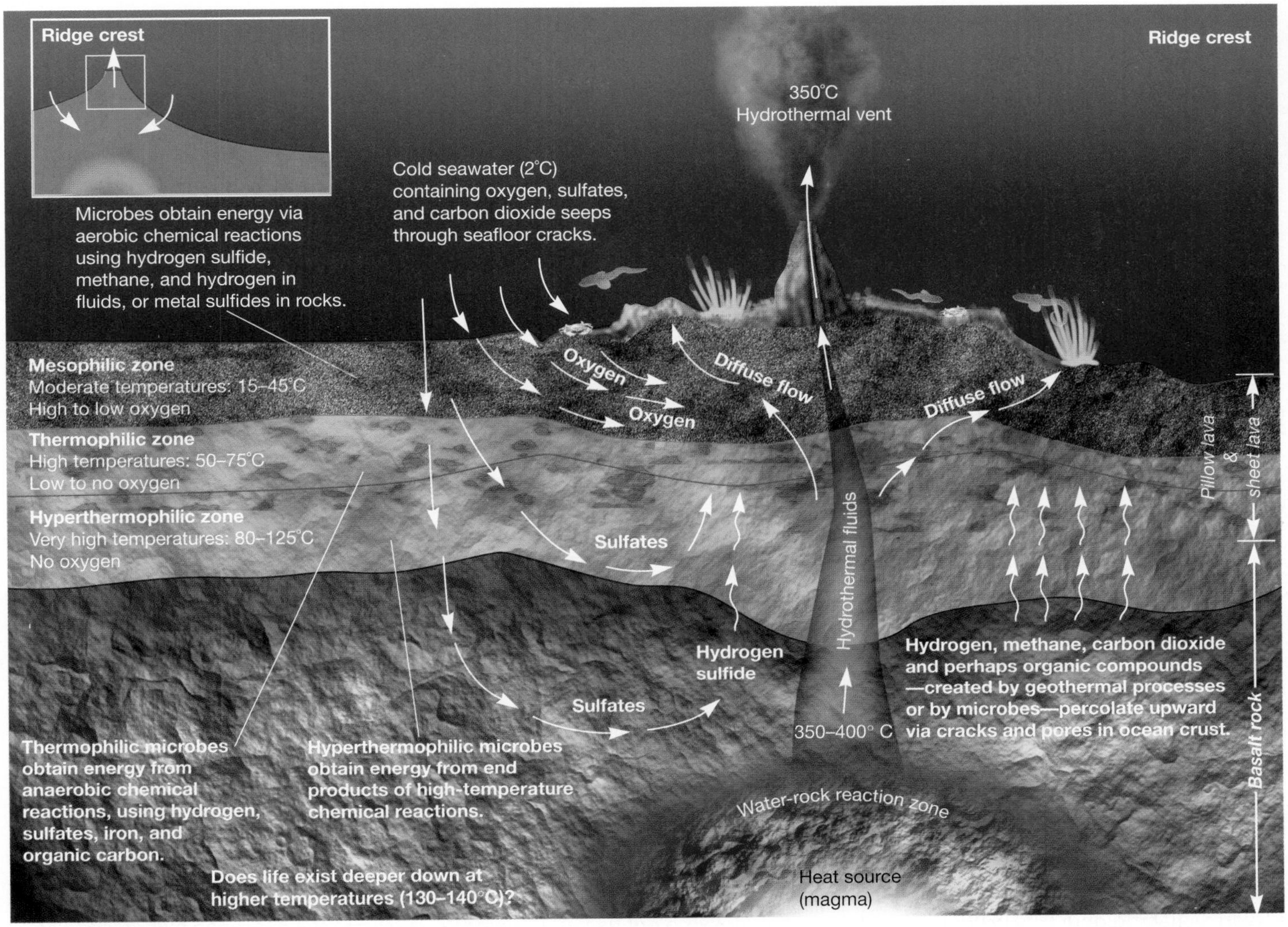
Ridge crest
Ridge crest
350°C
Hydrothermal vent
Cold seawater (2°C)
containing oxygen, sulfates,
and carbon dioxide seeps
through seafloor cracks.
Microbes obtain energy via
aerobic chemical reactions
using hydrogen sulfide,
methane, and hydrogen in
fluids, or metal sulfides in rocks.
Mesophilic zone
Moderate temperatures: 15–45°C
High to low oxygen
Thermophilic zone
High temperatures: 50–75°C
Low to no oxygen
Hyperthermophilic zone
Very high temperatures: 80–125°C
No oxygen
Oxygen
Oxygen
Diffuse flow
Diffuse flow
Pillow lava & sheet lava
Sulfates
Hydrothermal fluids
Hydrogen sulfide
Sulfates
Hydrogen, methane, carbon dioxide
and perhaps organic compounds
—created by geothermal processes
or by microbes—percolate upward
via cracks and pores in ocean crust.
350–400° C
Basalt rock
Thermophilic microbes
obtain energy from
anaerobic chemical
reactions, using hydrogen,
sulfates, iron, and
organic carbon.
Hyperthermophilic microbes
obtain energy from end
products of high-temperature
chemical reactions.
Water-rock reaction zone
Does life exist deeper down at
higher temperatures (130–140°C)?
Heat source
(magma)

FIGURE 30.4 Hydrothermal Vents and Related Geological Activity. The chemical reactions between seawater and rocks that occur over a range of temperatures on the seafloor supply the carbon and energy that support a diverse collection of microbial communities in specific niches within the vent system.

packed with chemolithotrophic bacterial endosymbionts (which have not yet been cultured in the laboratory) that reach densities of up to 10^{11} cells per gram of worm tissue. These endosymbionts fix CO_2 using the Calvin cycle (*see figure 11.5*) with electrons provided by H_2S. The CO_2 is carried to the endosymbionts in three ways: (1) freely in the bloodstream, (2) bound to hemoglobin, and (3) as organic acids such as malate and succinate. When these acids are decarboxylated, they release CO_2. This CO_2 is then fixed using the same pathway used by plants and cyanobacteria, but it occurs in the deepest, darkest reaches of the ocean. This mutualism enables *Riftia* to grow to an astounding size in densely packed communities.

Search This: Hydrothermal vents

An interesting, more complex mutualism occurs between the gutless marine oligochaete *Olavius algarvensis* and bacterial endosymbionts that reside just below the outer surface (cuticle) of this worm. Like *Riftia*, these worms lack any digestive tract so they obtain organic carbon from their endosymbionts. However unlike *Riftia*, *O. algarvensis* has a very primitive, inefficient excretory system, thus the endosymbionts also consume the worm's waste products. Metagenomic analysis reveals that this bacterial endosymbiont community is dominated by two δ-proteobacterial and two γ-proteobacterial species. The δ-proteobacteria are metabolically flexible. They use sulfate as their terminal electron acceptor and are heterotrophic, with genes for the uptake and oxidation of a variety of carbohydrates. In addition, they are capable of carbon fixation through both the acetyl-CoA pathway

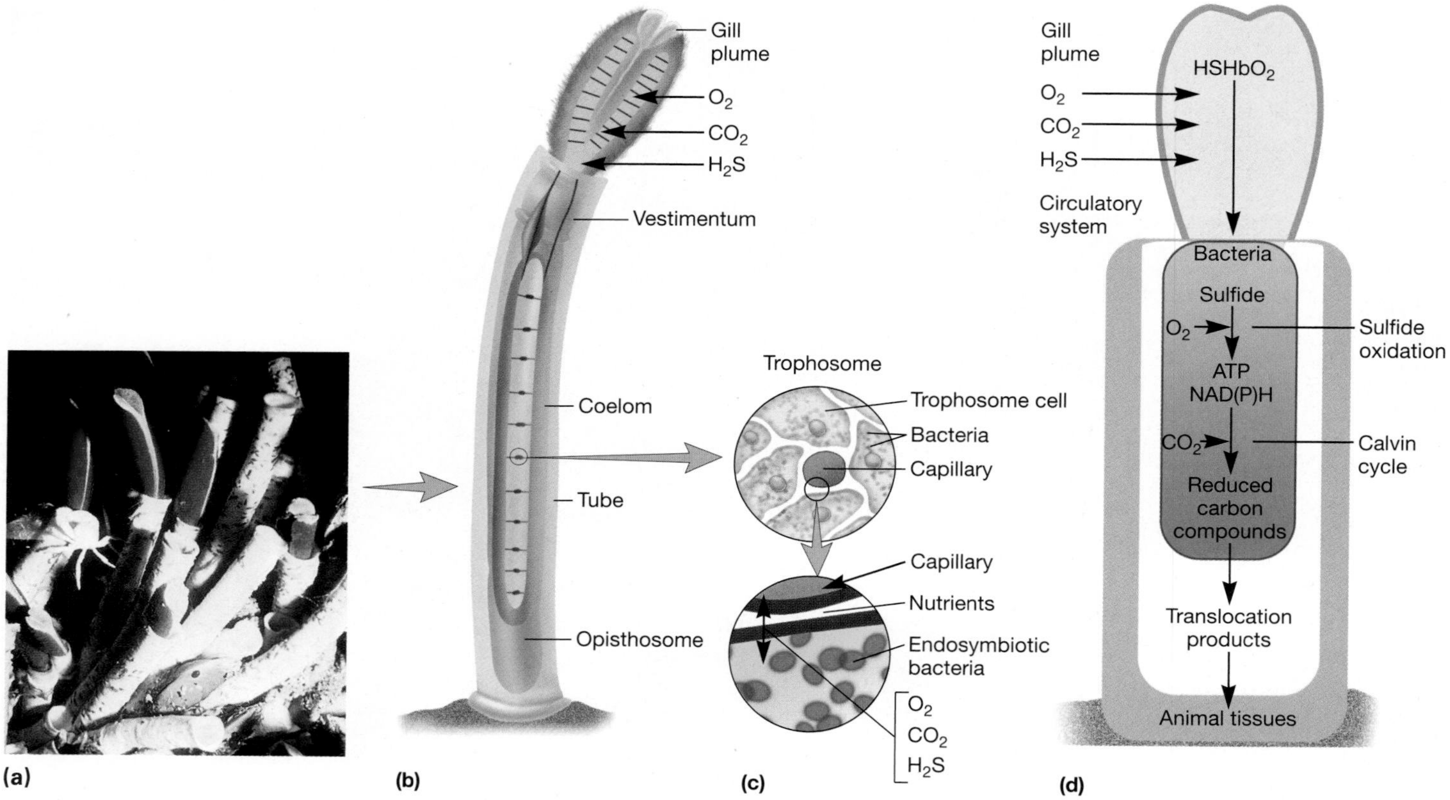

FIGURE 30.5 The Tube Worm–Bacterial Relationship. (a) A community of tube worms (*Riftia pachyptila*) at the Galápagos Rift hydrothermal vent site (depth 2,550 m). Each worm is more than a meter in length and has a 20-cm gill plume. (b, c) Schematic illustration of the anatomical and physiological organization of the tube worm. The animal is anchored inside its protective tube by the vestimentum. At its anterior end is a respiratory gill plume. Inside the trunk of the worm is a trophosome consisting primarily of endosymbiotic bacteria, associated cells, and blood vessels. At the posterior end of the animal is the opisthosome, which anchors the worm in its tube. (d) Oxygen, carbon dioxide, and hydrogen sulfide are absorbed through the gill plume and transported to the blood cells of the trophosome. Hydrogen sulfide is bound to the worm's hemoglobin ($HSHbO_2$) and carried to the endosymbiont bacteria. The bacteria oxidize the hydrogen sulfide and use some of the released energy to fix CO_2 in the Calvin cycle. Some of the reduced carbon compounds synthesized by the endosymbiont are translocated to the animal's tissues.

Figure 30.5 Micro Inquiry

What is the nutritional type of the endosymbiotic bacteria?

and reductive TCA cycle; they can also use H_2 as an electron donor. The γ-proteobacteria are chemoautolithotrophs, using reduced sulfur compounds as electron donors and O_2, nitrate, or organic substrates such as fumarate as electron acceptors. ◀◀ *Anaerobic respiration (section 10.6); Chemolithotrophy (section 10.11); Other CO_2-fixation pathways (section 11.3)*

Why does *O. algarvensis* have four endosymbiotic species? Harboring symbionts is energetically expensive, so it can be argued that all four are needed for the worm to survive. The metabolic flexibility of these bacteria coupled with the worm's habitat in the sediment suggests that the endosymbionts interact with each other, as well as with the host. For instance, as the worm migrates vertically, three different microhabitats are encountered: oxic, anoxic with nitrate and sulfate available, and anoxic with H_2 and reduced sulfur present (**figure 30.6**). This requires that the bacterial endosymbionts adjust their metabolism to fit the prevailing electron acceptors and donors. Thus at the surface, the γ-proteobacteria use O_2 as the terminal electron acceptor and the δ-proteobacteria rely on heterotrophy. Because there is little sulfur at the surface, the bacteria rely on each other to provide reduced and oxidized sulfur as electron donor and acceptor, respectively. In the inter-

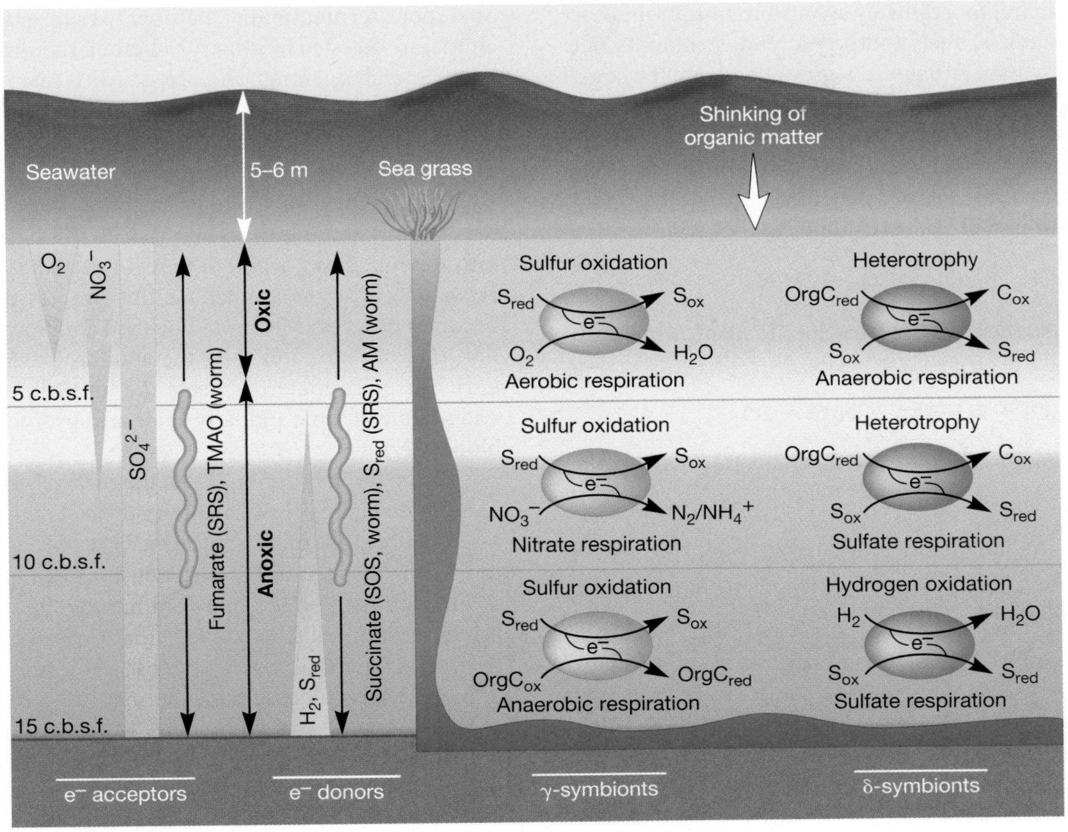

FIGURE 30.6 Proposed Energy Metabolism of Bacterial Endosymbionts of the Mediterranean Worm *Olavius algarvensis*. Electron donors and acceptors change with increasing sediment depth, as measured in centimeters below seafloor (c.b.s.f.). As the worm moves vertically through the sediments, the δ- and γ-proteobacterial endosymbionts switch their metabolism to match the available electron donors and acceptors. OrgC, organic carbon; S_{ox}, oxidized sulfur compounds; S_{red}, reduced sulfur compounds; SOS, sulfur-oxidizing symbionts; SRS, sulfate-reducing symbionts; TMAO, trimethylamine *N*-oxide—an organic electron acceptor produced by the host.

***Figure 30.6** Micro Inquiry*

Why do the δ-proteobacterial symbionts switch to H_2 as an electron donor only in the deepest sediments inhabited by *O. algarvensis*?

mediate zone, the γ-proteobacteria use nitrate as the terminal electron acceptor and may rely on a combination of externally and internally produced reduced sulfur as electron donor. Meanwhile the δ-proteobacteria find plenty of external sulfate to respire. Finally, in the deepest zone, the γ-proteobacteria have access to lots of reduced sulfur in the sediments but must use reduced organics provided by the host or the environment as the terminal electron acceptor. Meanwhile, the δ-proteobacteria may switch to chemotrophy and use H_2 as the electron donor and the oxidized sulfur generated by the γ-proteobacteria as electron acceptor. This complex interplay between the bacteria and host has resulted from millions of years of coevolution.

Methane-Based Mutualisms

Other unique food chains involve methane-oxidizing microorganisms. By converting methane to carbohydrate, these bacteria perform the first step in providing organic matter for consumers. **Methanotrophic bacteria** are capable of using methane as a sole carbon source. Some occur as intracellular symbionts of methane-vent mussels. In these mussels, the thick, fleshy gills are filled with bacteria. In the Barbados Trench, methanotrophic carnivorous sponges have been discovered in a mud volcano at a depth of 4,943 m. Abundant methanotrophic symbionts were confirmed by the presence of enzymes related to methane oxidation in sponge tissues. These sponges are not satisfied with just bacterial symbionts; they also trap swimming prey.

Methanotrophic microorganisms are also important in other ecosystems. For example, methanotrophic endosymbionts help reduce the flux of methane from peat bogs; wetlands are the largest natural source of this greenhouse gas. Sphagnum moss, the principal plant in peat bogs (and a favorite among florists), can grow when submerged in water. Methanotrophic α-proteobacteria living within the outer cortex cells of sphagnum stems oxidize methane as it diffuses through the water column:

$$CH_4 + 2O_2 \rightleftharpoons CO_2 + 2H_2O$$

The resulting CO_2 is then readily fixed by the plant, which uses the Calvin cycle:

$$2CO_2 + 2H_2O \rightleftharpoons 2CH_2O + 2O_2$$

This enables extremely efficient carbon recycling within this ecosystem:

$$CH_4 + CO_2 \rightleftharpoons 2CH_2O$$

The Rumen Ecosystem

Ruminants are the most successful and diverse group of mammals on Earth today. Examples include cattle, deer, elk, bison, water buffalo, camels, sheep, goats, giraffes, and caribou. These animals spend vast amounts of time chewing their cud—a small ball of partially digested grasses that the animal has consumed but not yet completely digested. It is thought that ruminants evolved an "eat now, digest later" strategy because their grazing can often be interrupted by predator attacks.

These herbivorous animals have stomachs that are divided into four chambers (**figure 30.7**). The upper part of the ruminant stomach is expanded to form a large pouch called the rumen and a smaller, honeycomb-like region, the reticulum. The lower portion is divided into an antechamber, the omasum, followed by the "true" stomach, the abomasum. The rumen is a highly muscular, anaerobic fermentation chamber where grasses eaten by the animal are digested by a diverse microbial community that includes bacteria, archaea, fungi, and protists. This microbial community is large—about 10^{12} organisms per milliliter of digestive fluid. When the animal eats plant material, it is mixed with saliva and swallowed without chewing to enter the rumen. Here, microbial attack and further mixing coats the grass with microbes, reducing it to a pulpy, partially digested mass. At this point, the mass moves into the reticulum, where it is regurgitated as cud, chewed, and reswallowed by the animal. As this process proceeds, the grass becomes progressively more liquefied and flows out of the rumen into the omasum and then the abomasum. Here, the nutrient-enriched grass material meets the animal's digestive enzymes, and soluble organic and fatty acids are absorbed into the animal's bloodstream.

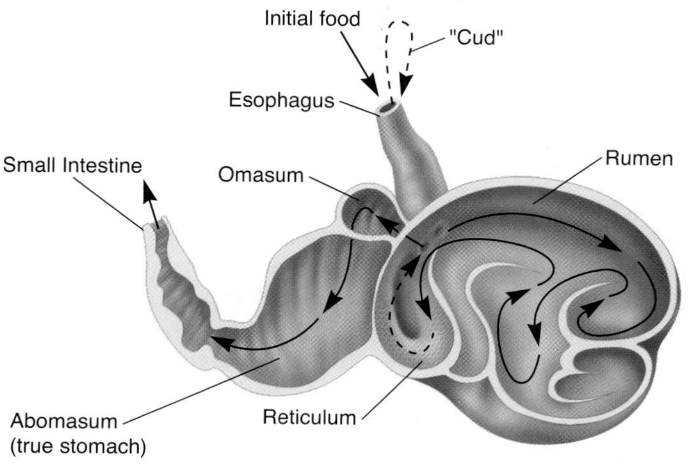

FIGURE 30.7 Ruminant Stomach. The stomach compartments of a cow. The microorganisms are active mainly in the rumen. Arrows indicate direction of food movement.

The microbial community in the rumen is extremely dynamic. The rumen is slightly warmer than the rest of the animal, and with a redox potential of about −30 mV, all resident microorganisms must carry out anaerobic metabolism. Very specific interactions occur within the microbial community. One group of bacteria produces extracellular cellulases that cleave the $\beta(1\rightarrow4)$ linkages between the successive D-glucose molecules that form plant cellulose. The D-glucose is then fermented to organic acids such as acetate, butyrate, and propionate. These organic acids, as well as fatty acids, are the true energy source for the animal. In some ruminants, the processing of organic matter stops at this stage. In others, such as cows, CO_2, H_2, and to a lesser extent, acetate are used by methanogenic archaea to generate methane (CH_4), a greenhouse gas. Methanogens provide two services to the ruminant host: they synthesize most of the vitamins needed by the animal and efficiently remove H_2. This alters the nutritional patterns of fermentative bacteria so that more organic acids are produced. Although methanogens make up probably less than 5% of the microbial community, they are very active. In fact, a single cow can produce as much as 200 to 400 liters of CH_4 per day. The animal releases this CH_4 by a process called eructation (Latin *eructare,* to belch)—and although it may seem preposterous, ruminants contribute a significant fraction of global CH_4. ◀◀ *Fermentation (section 10.7); Global climate change (section 26.2)*

1. What is the critical characteristic of a mutualistic relationship?
2. How could one test to see if an insect-microbe relationship is mutualistic?
3. What is the role of *Riftia* hemoglobin in the success of the tube worm-endosymbiont mutualistic relationship?
4. How is the *Riftia* endosymbiont similar to cyanobacteria? How is it different?
5. Describe how the sphagnum-methanotroph mutualism results in efficient carbon cycling.
6. What structural features of the rumen make it suitable for an herbivorous diet?
7. Why is it important that the rumen is a reducing environment?

Cooperation

Cooperation and commensalism are two positive but not obligatory types of symbioses found widely in the microbial world (figure 30.1). For most microbial ecologists, this nonobligatory aspect differentiates cooperation from mutualism. Unfortunately, it is often difficult to distinguish obligatory from nonobligatory because that which is obligatory in one habitat may not be in another (e.g., the laboratory). Nonetheless, the most useful distinction between cooperation and mutualism is the observation that cooperating organisms can be separated from one another and remain viable, although they may not function as well. Two examples of a cooperative relationship include the association between *Desulfovibrio* and *Chromatium,* in which the carbon and sulfur cycles are linked (**figure 30.8*a***), and the interaction of a nitrogen-fixing microorganism with a cellulolytic organism such as *Cellulomonas* (figure 30.8*b*). In the second example, the cellulose-degrading microorganism liberates glucose from the cellulose, which can be used by nitrogen-fixing microbes.

Cooperative interactions are also found between microorganisms and eukaryotic hosts. In some cases, the carbon fixed by sulfide-dependent autotrophic filamentous microbes serves as the carbon and energy source for its host. Some of the most interesting examples include the polychaete worms *Alvinella pompejana* (**figure 30.9**), the Pompeii worm, and *Paralvinella palmiformis,* the palm worm. Both have filamentous bacteria on their dorsal surfaces. These filamentous bacteria tolerate high levels of metals such as arsenic, cadmium, and copper. When growing on the surface of the animal, they may provide protection from these toxic metals, as well as thermal protection; in addition, they appear to be used as a food source.

Cooperation may also involve syntrophic relationships. **Syntrophism** (Greek *syn,* together, and *trophe,* nourishment) is an association in which the growth of one organism either depends on or is improved by growth factors, nutrients, or substrates provided by another organism growing nearby. Sometimes both organisms benefit.

Important syntrophic relationships occur in anoxic methanogenic ecosystems such as sludge digesters, anoxic freshwater aquatic sediments, flooded soils, and the rumen ecosystem. Obligate proton-reducing acetogens live in these environments. These microbes oxidize ethanol, butyrate, propionate, and other fermentation end products to H_2, CO_2, and acetate (hence the name "acetogen"). Although the enzymology is not well understood, it is

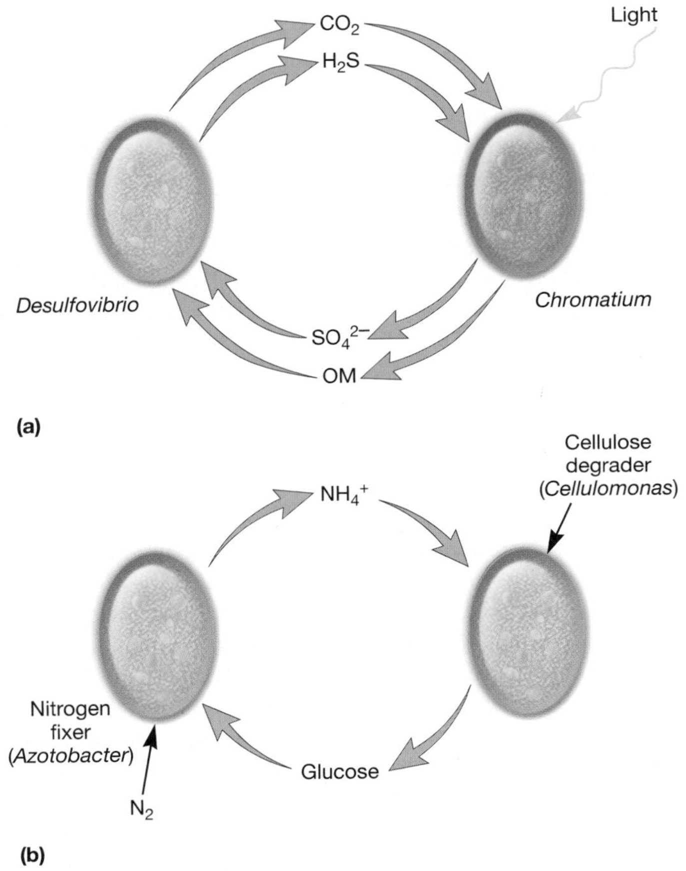

FIGURE 30.8 Examples of Cooperative Symbiotic Processes. (a) The organic matter (OM) and sulfate required by *Desulfovibrio* are produced by the *Chromatium* in its photosynthesis-driven reduction of CO_2 to organic matter and oxidation of sulfide to sulfate. (b) *Azotobacter* uses glucose provided by a cellulose-degrading microorganism such as *Cellulomonas,* which uses the nitrogen fixed by *Azotobacter.*

FIGURE 30.9 A Marine Worm–Bacterial Cooperative Relationship. *Alvinella pompejana,* a 10-cm-long worm, forms a cooperative relationship with bacteria that grow as long threads on the worm's surface. The bacteria and *Alvinella* are found near the black smoker-heated water fonts.

clear that these reactions are thermodynamically unfavorable under standard conditions, that is, they have a positive ΔG. However, because methanogens consume the H_2 as it is being generated, the reaction becomes thermodynamically possible. This is called **interspecies hydrogen transfer.** Various fermentative bacteria produce low molecular-weight fatty acids that can be degraded by anaerobic bacteria such as *Syntrophobacter* to produce H_2 as follows:

$$\text{Propionic acid} \rightarrow \text{acetate} + CO_2 + 3H_2\ (\Delta G^\circ = +76.1\ \text{kJ/mol})$$

Syntrophobacter uses protons ($2H^+ + 2e^- \rightarrow H_2$) as terminal electron acceptors in ATP synthesis. The products H_2 and CO_2 are then used by methanogenic archaea such as *Methanospirillum*:

$$4H_2 + CO_2 \rightarrow CH_4 + 2H_2O\ (\Delta G^\circ = -25.6\ \text{kJ/mol})$$

By synthesizing methane, *Methanospirillum* maintains a low H_2 concentration in the immediate environment of both microbes. Continuous removal of H_2 promotes further fatty acid fermentation and H_2 production. Because increased H_2 production and consumption stimulate the growth of *Syntrophobacter* and *Methanospirillum*, both participants in the relationship benefit. Interestingly, it was long held that these reactions were performed by a single microbe. The syntrophic nature of this relationship was recently further illustrated when it was discovered that the proprionate-oxidizing bacterium *Pelotomaculum thermopropionicum* tethers itself to the methanogen *Methanothermobacter thermautotrophicus*. Remarkably the flagellar tip protein FliD serves as an interspecies signal to increase the rate of methanogenesis by *M. thermoautotrophicus*. ◀◀ *Methanogens (section 18.3)*

1. How does cooperation differ from mutualism? What might be some of the evolutionary implications of both types of symbioses?
2. What is syntrophism? Is physical contact required for this relationship? Why or why not?
3. Why is *Alvinella* a good example of cooperative microorganism-animal interactions?

Commensalism

Commensalism (Latin *com,* together, and *mensa,* table) is a relationship in which one symbiont, the **commensal,** benefits, while the other (sometimes called the host) is neither harmed nor helped, as shown in figure 30.1. This is a unidirectional process. The spatial proximity of the two partners permits the commensal to feed on substances captured or ingested by the host, and the commensal often obtains shelter by living either on or in the host. The commensal is not directly dependent on the host metabolically, so when it is separated from the host experimentally, it can survive without the addition of factors of host origin.

Commensalistic relationships between microorganisms include situations in which the waste product of one microorganism is a substrate for another species. One good example is nitrification—the oxidation of ammonium ion to nitrate. Nitrification occurs in two steps: first, microorganisms such as *Nitrosomonas* and certain crenarchaeotes oxidize ammonium to nitrite, and second, nitrite is oxidized to nitrate by *Nitrobacter* and similar bacteria. *Nitrobacter* benefits from its association with *Nitrosomonas* because it uses nitrite to obtain energy for growth.

Commensalistic associations also occur when one microbial group modifies the environment to make it better suited for another organism. The synthesis of acidic waste products during fermentation stimulates the proliferation of more acid-tolerant microorganisms, which may be only a minor part of the microbial community at neutral pH. A good example is the succession of microorganisms during milk spoilage (*see figure 40.4*). Biofilm formation provides another example. The colonization of a newly exposed surface by one type of microorganism (an initial colonizer) makes it possible for other microorganisms to attach to the microbially modified surface. ◀◀ *Biofilms (section 7.7)*

Commensalism is also important in the colonization of the human body and the surfaces of other animals and plants. The microorganisms associated with an animal's skin and body orifices can use volatile, soluble, and particulate organic compounds from the host as nutrients. Under most conditions, these microbes do not cause harm. However, if the host organism is stressed or the skin is punctured, these normally commensal microorganisms may become pathogenic by entering a different environment. These interactions are discussed in more detail in section 30.3.

1. How does commensalism differ from cooperation?
2. Why is nitrification a good example of a commensalistic process?
3. What is interspecies hydrogen transfer, and why is this beneficial to both producers and consumers of hydrogen?

Predation

As is the case with larger organisms, predation among microbes involves a predator species that attacks and usually kills its prey. Over the last several decades, microbiologists have discovered a number of fascinating bacteria that survive by their ability to prey upon other microbes. Several of the best examples are *Bdellovibrio, Vampirococcus,* and *Daptobacter* (**figure 30.10**).

Bdellovibrio is an active hunter that is vigorously motile, swimming about looking for susceptible gram-negative bacterial prey. Upon sensing such a cell, *Bdellovibrio* swims faster until it collides with the prey cell. It then bores a hole through the outer membrane of its prey and enters the periplasmic space. As it grows, it forms a long filament that eventually septates to produce progeny bacteria. Lysis of the prey cell releases new *Bdellovibrio* cells (*see figure 20.33*). *Bdellovibrio* will not attack mammalian cells, and gram-negative prey bacteria have never been observed to acquire resistance to *Bdellovibrio*. This has raised interest in the use of *Bdellovibrio* as a "probiotic" to treat infected wounds. Although this has not yet been tried, one can imagine that with the rise in antibiotic-resistant

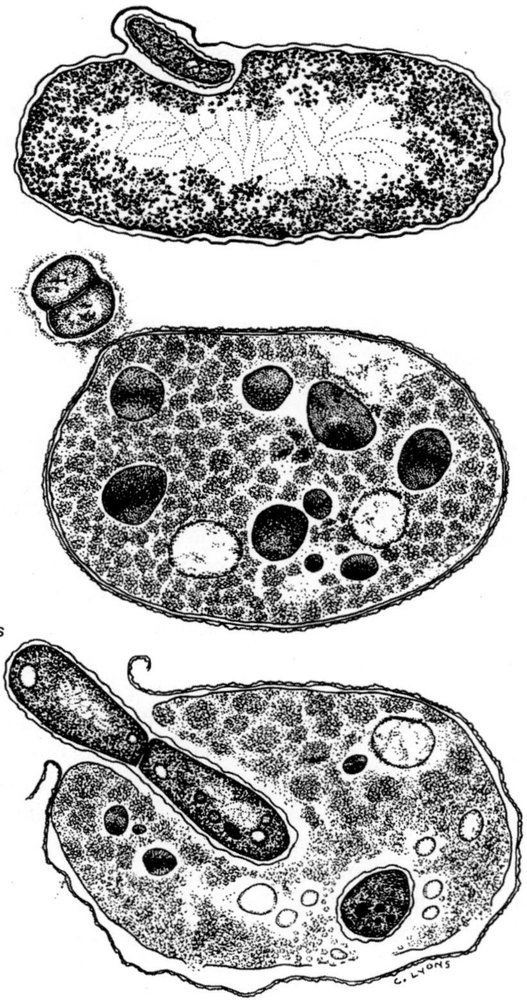

FIGURE 30.10 Examples of Predatory Bacteria Found in Nature. (a) *Bdellovibrio,* a periplasmic predator that penetrates the cell wall and grows outside the plasma membrane, (b) *Vampirococcus* has a unique epibiotic mode of attacking a prey bacterium, and (c) *Daptobacter* showing its cytoplasmic location as it attacks a susceptible bacterium.

pathogens, such forms of treatments may be considered viable alternatives. ◀◀ *Class* Deltaproteobacteria *(section 20.4)*

Search This: Bdellovibrio *video*

Although *Vampirococcus* and *Daptobacter* also kill their prey, they gain entry in a less-dramatic fashion. *Vampirococcus* attaches to the outer membrane of its prey (figure 30.10*b*). It then secretes degradative enzymes that result in the release of the prey's cytoplasmic contents. In contrast, *Daptobacter* penetrates the prey cell and consumes the cytoplasmic contents directly (figure 30.10*c*).

Some bacteria are facultative predators. Sometimes they consume organic matter that is released from dead organisms, whereas at other times, they actively prey on other microbes. Two good examples are the γ-proteobacterium *Lysobacter* and the δ-proteobacterium *Myxococcus.* Both these microbes display what has been called "wolf pack" predation whereby populations of cells use gliding motility to creep toward and over their prey as they release an arsenal of degradative enzymes. *M. xanthus* fruiting body formation can be triggered when its prey has been exhausted (**figure 30.11**).

A surprising finding is that predation has many beneficial effects, especially when one considers interactive populations of predators and prey. Simple ingestion and assimilation of a prey bacterium can lead to increased rates of nutrient cycling, critical for the functioning of the microbial loop (*see figure 28.2*). Ingestion and short-term retention of bacteria also are critical for ciliated protists functioning in the rumen, where methanogenic bacteria contribute to the health of the ciliates by decreasing toxic hydrogen levels. They do so by using H_2 to produce methane, which then is passed from the rumen.

Clearly predation in the microbial world is not straightforward. It often has a fatal and final outcome for an individual prey organism, but it can have a wide range of beneficial effects on prey populations. Regardless of the outcome, predation is critical in the functioning of natural environments.

Parasitism

Parasitism is one of the most complex microbial interactions; the line between parasitism and predation is difficult to define (figure 30.1). This is a relationship between two organisms in which one benefits from the other and the host is usually harmed. It can involve nutrient acquisition, physical maintenance in or on the host, or both. In parasitism, there is always some coexistence between host and **parasite.** This is because a host that dies immediately after parasite invasion may prevent the microbe from reproducing to sufficient numbers to ensure colonization of a new host. But what happens if the host-parasite equilibrium is upset? If the balance favors the host (perhaps by a strong immune defense or antimicrobial therapy), the parasite loses its habitat and may be unable to survive. On the other hand, if the equilibrium is shifted to favor the parasite, the host becomes ill and, depending on the specific host-parasite relationship, may die. One good example is the disease typhus. This disease is caused by *Rickettsia typhi*, which is harbored in fleas that live on rats. It is transmitted to humans who are bitten by such fleas. Humans often live in association with rats, and in such communities, there is always a small number of people with typhus—that is to say, typhus is endemic. However, during times of war or when people are forced to become refugees, lack of sanitation and overcrowding result in an increased number of rat-human interactions. Typhus can then reach epidemic proportions. During the Crimean War (1853–1856), about 213,000 men were killed or wounded in combat, while over 850,000 were sickened or killed by typhus.

On the other hand, a controlled parasite-host relationship can be maintained for long periods of time. For example, lichens (**figure 30.12**) are the association between specific ascomycetes (a fungus) and certain genera of either green algae or cyanobacteria.

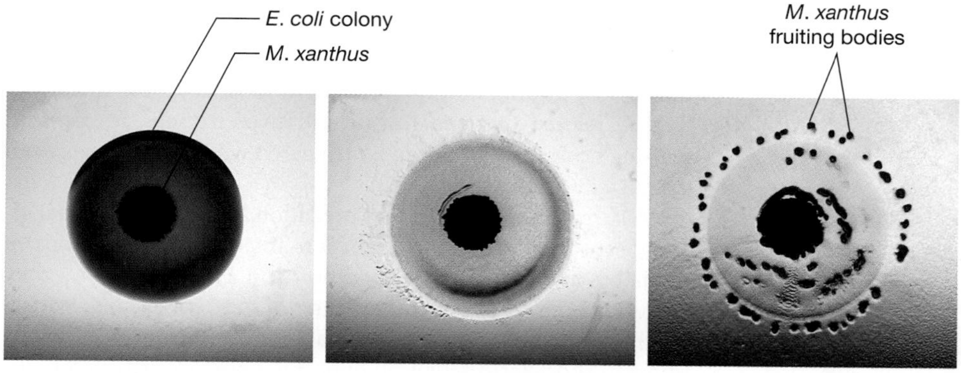

(a) (b) (c)

FIGURE 30.11 Wolf Pack Predation by *Myxococcus xanthus*. (a) In this demonstration, a drop of *M. xanthus* cells was colored with India ink and dropped into the center of a colony of prey bacteria (*Escherichia coli*). (b) The *E. coli* cells lyse as *M. xanthus* glides over and consumes them; thus the colony becomes clear. (c) The *M. xanthus* cells that swarmed over the prey bacteria now form fruiting bodies around the periphery. Note that just a small subpopulation of *M. xanthus* cells swarmed over the prey.

Figure 30.11 Micro Inquiry

How does myxobacterial predation differ from that of *Bdellovibrio*?

(a) (b)

FIGURE 30.12 Lichens. (a) The flattened leaf-shaped genus *Umbilicaria* is a common foliose lichen. (b) A thin slice of *Umbilicaria* viewed with a compound microscope shows that the green algae occur in a thin upper layer and fungal hyphae make up the rest of the lichen.

The fungal partner is termed the **mycobiont** and the algal or cyanobacterial partner, the **phycobiont.** The fungus obtains nutrients from its partner by projections of fungal hyphae called haustoria, which penetrate the phycobiont cell wall. It also uses O_2 produced by the phycobiont for respiration. In turn, the fungus protects the phycobiont from high light intensities, provides water and minerals, and creates a firm substratum within which the phycobiont can grow protected from environmental stress. This is an ancient association; 600 million-year-old lichenlike fossils from South China indicate that the relationship between fungi and their photoautotrophic partners developed prior to the evolution of vascular plants. In the past, the lichen symbiosis was considered to be a mutualistic interaction. However, this appears to be a controlled parasitism between the fungi and phycobiont, as some phytobiotic cyanobacteria and algae grow more quickly when cultured alone. ◀◀ Ascomycota *(section 24.5); Phylum* Cyanobacteria *(section 19.3)*

Search This: Lichen identification

An important aspect of many symbiotic relationships, including parasitism, is that over time, the symbiont, once it has established a relationship with the host, will discard excess, unused genomic information, a process called **genomic reduction** (*see figure 16.17*). This is clearly the case with the aphid endosymbiont *Buchnera aphidicola*, and it has also occurred with the parasitic bacterium *Mycobacterium leprae* and the microsporidium *Encephalitozoon cuniculi.* The latter organism, which parasitizes a wide range of animals, including humans, now can only survive inside the host cell. ◀◀ *Comparative geneomics (section 16.7); Suborder* Corynebacterineae: *(section 22.4);* Microsporidia *(section 24.7)*

1. Define predation and parasitism. How are these similar and different?
2. How can a predator confer positive benefits on its prey? Think of the responses of individual organisms versus populations as you consider this question.
3. What are examples of parasites that are important in microbiology?
4. What is a lichen? Discuss the benefits the phycobiont and mycobiont provide each other.

Amensalism

Amensalism describes the adverse effect that one organism has on another organism (figure 30.1). This is a unidirectional process based on the release of a specific compound by one organism that has a negative effect on another organism. A classic example of amensalism is the microbial production of antibiotics that can inhibit or kill another, susceptible microorganism (**figure 30.13*a***).

Another interesting amensalism demonstrates how complex interactions between organisms can be. Attine ants (ants belonging to a New World tribe) are able to take advantage of an amensalistic relationship between an actinomycete and the parasitic

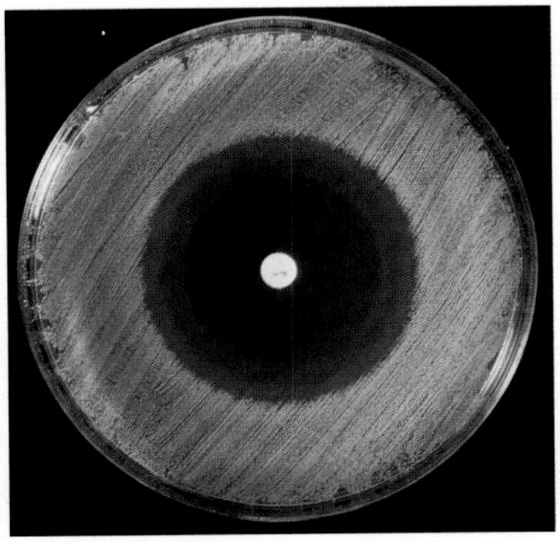

(a)

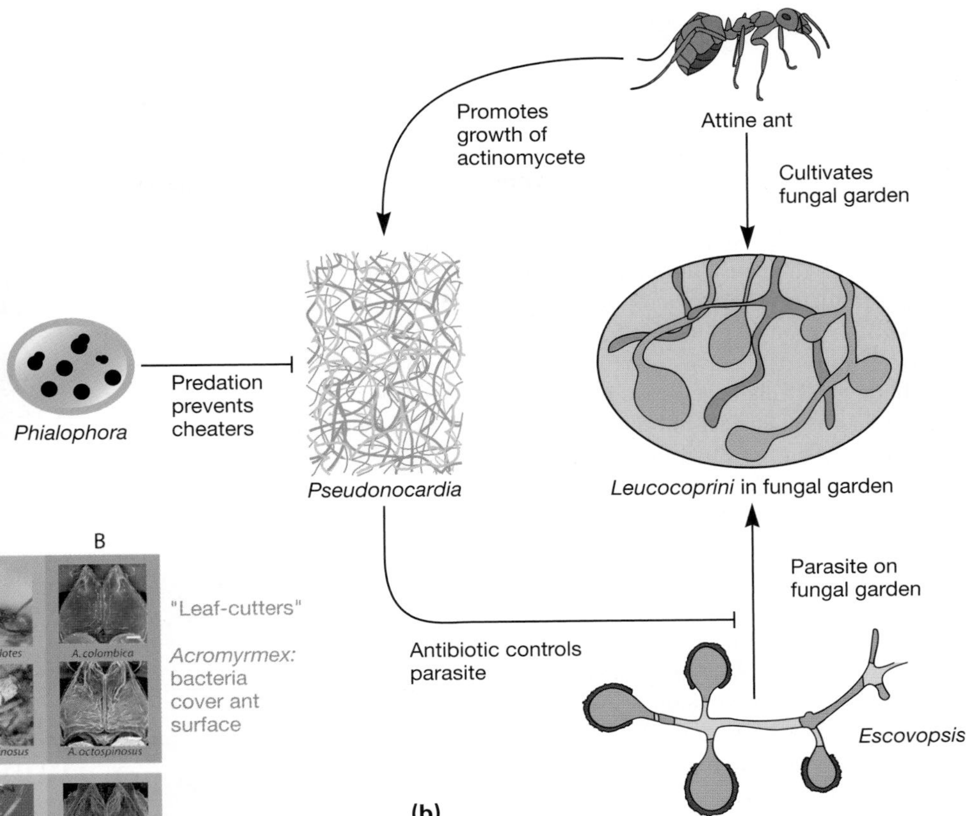

(b)

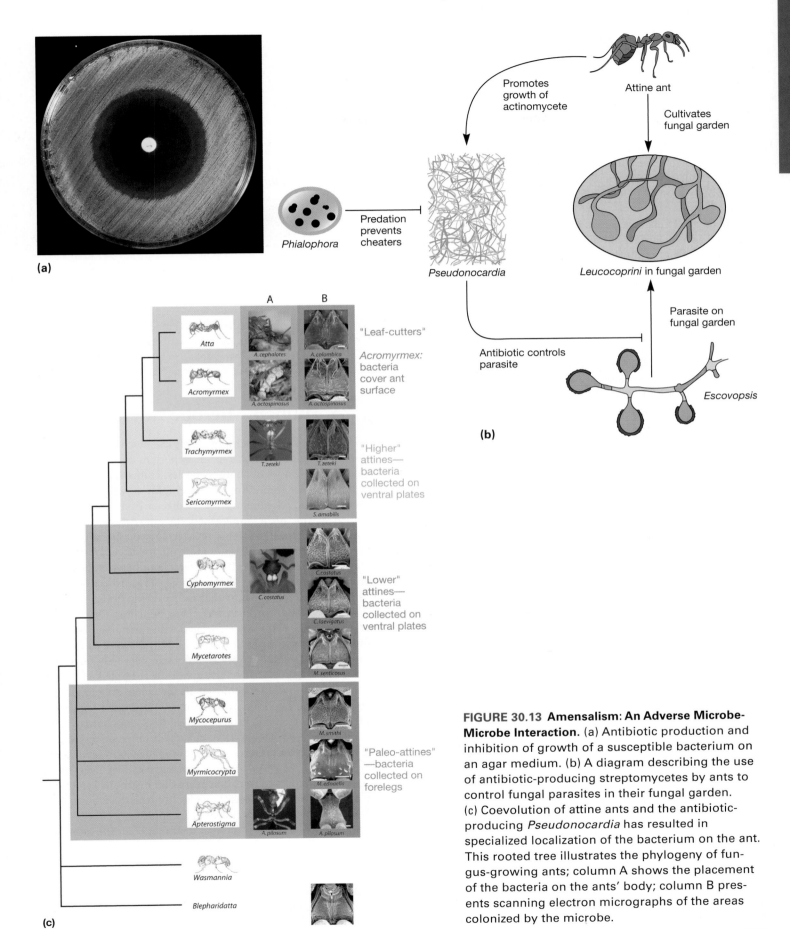

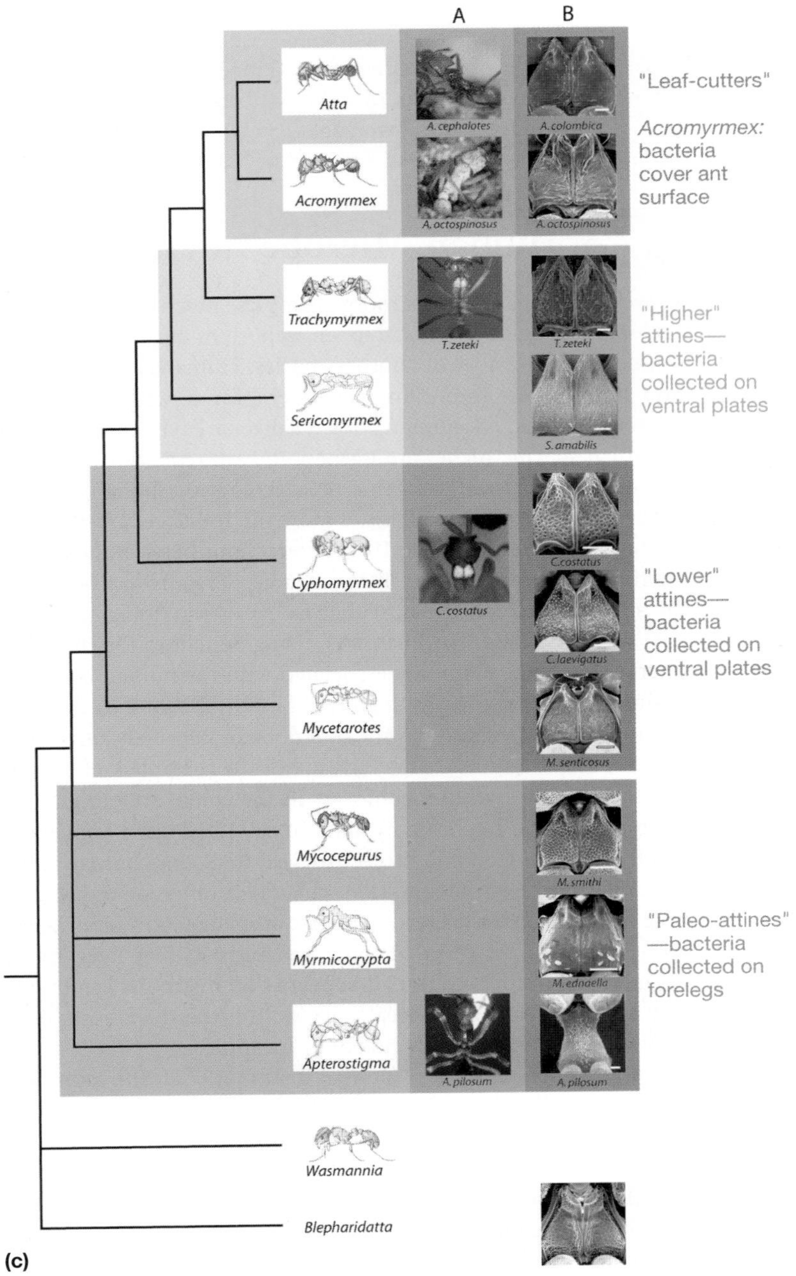

(c)

FIGURE 30.13 Amensalism: An Adverse Microbe-Microbe Interaction. (a) Antibiotic production and inhibition of growth of a susceptible bacterium on an agar medium. (b) A diagram describing the use of antibiotic-producing streptomycetes by ants to control fungal parasites in their fungal garden. (c) Coevolution of attine ants and the antibiotic-producing *Pseudonocardia* has resulted in specialized localization of the bacterium on the ant. This rooted tree illustrates the phylogeny of fungus-growing ants; column A shows the placement of the bacteria on the ants' body; column B presents scanning electron micrographs of the areas colonized by the microbe.

fungi *Escovopsis*. This amensalistic relationship enables the ant to maintain a mutualism with another fungal species, *Leucocoprini*. Amazingly, these ants cultivate a garden of *Leucocoprini* for their own nourishment (figure 30.13*b*). To prevent the parasitic fungus *Escovopsis* from decimating their fungal garden, the ants also promote the growth of an actinomycete of the genus *Pseudonocardia*, which produces an antimicrobial compound that inhibits the growth of *Escovopsis*.

This unique amensalistic process appears to have evolved 50 million to 65 million years ago in South America. Thus this relationship has been subject to millions of years of coevolution, such that particular groups of ants cultivate specific strains of fungi that are then subject to different groups of *Escovopsis* parasites. In addition, the ants have developed intricate crypts within their exoskeletons for the growth of the antibiotic-producing *Pseudonocardia*. As shown in figure 30.13*c*, these crypts have been modified throughout the ants' evolutionary history. The most primitive "paleo-attine" ants carry the bacterium on their forelegs, "lower" and "higher" attines have evolved special plates on their ventral surfaces, while the entire surface of the most recent attines, leaf-cutter ants of the genus *Acromyrmex*, are covered with the bacterium. Related ants that do not cultivate fungal gardens (e.g., *Atta* sp.) do not host *Pseudonocardia*.

Surprisingly, this four-membered symbiosis involving attine ants is now known to be even more complicated, at least in some cases. In 2007 a fifth member of the symbiosis was reported. This member is a black ascomycete yeast of the genus *Phialophora*. It grows in association with the *Pseudonocardia* and on which it also preys. But why would attine ants have evolved the capacity to tolerate a fungus that eats their antibiotic-producing actinomycete? The answer has to do with cheating. Briefly, it is energetically expensive for *Pseudonocardia* to produce an antibiotic, but antibiotic production ensures it of a home on the ant. However, some members of the *Pseudonocardia* population could get a "free ride" by letting other cells in the population produce the antibiotic. It is thought that by applying a second form of selective pressure, these cheaters cannot thrive. This unique multipartner relationship has enabled scientists to explore the evolutionary, behavioral, physiological, and structural aspects of attine ants, which also serve as a model system for symbiotic interactions in general.

Search This: Attine ant fungal garden

Competition

Competition arises when different organisms within a population or community try to acquire the same resource, whether this is a physical location or a particular limiting nutrient (figure 30.1). If one of the two competing organisms can dominate the environment, whether by occupying the physical habitat or by consuming a limiting nutrient, it will overtake the other organism. This phenomenon was studied by E. F. Gause, who in 1934 described it as the **competitive exclusion principle.** He found that if two competing ciliates overlapped too much in terms of their resource use, one of the two protist populations was excluded. In chemostats, competition for a limiting nutrient may occur among microorganisms with transport systems of differing affinity. This can lead to the exclusion of the slower-growing population under a particular set of conditions. If the dilution rate is changed, the previously slower-growing population may become dominant. Often two microbial populations that appear to be similar nevertheless coexist. In this case, they share the limiting resource (space, a limiting nutrient) and coexist while surviving at lower population levels. ⏮ *Continuous culture of microorganisms (section 7.5)*

1. Describe the amensalism observed in attine ant communities. What other types of interactions are present in these communities?
2. The production of antimicrobial agents and their effects on target species have been called a "microbial arms race." Explain what you think this phrase means and its implications for the attine ant system.
3. What is the competitive exclusion principle? Can you think of another example where this principle is demonstrated in the natural world?

30.2 Human-Microbe Interactions

As we have seen, many microorganisms live much of their lives in a special ecological relationship: an important part of their environment is a member of another species. Humans are no exception and are an important ecosystem for a wide variety of microorganisms. The human body is a diverse environment in and on which specific niches are formed, thus the microbial inhabitants may be discussed as the microbial ecology of a human. In fact, the average adult carries 10 times more microbial cells (10^{14}) than human cells (average about 10^{13}). Interactions between hosts and microbes are dynamic, permitting the microorganisms to colonize the specific place on the host that meets its physiological needs (i.e., niche fulfillment) while maximizing benefit to the microbe. Microorganisms commonly associated with the human body are traditionally referred to as the normal microbial flora or the normal microbiota because specific microbes are routinely associated with specific locations on the human host (e.g., staphylococci and diphtheroids on the skin, coliforms in the colon, streptococci in the mouth, etc.). Interestingly, data from Washington University and the Human Genome Project suggest that some human traits (e.g., obesity) result in conditions that favor one type of bacteria over another. Since bacteria colonize humans soon after being born, these data suggest that some human traits may have arisen by microbial influence. Thus it appears that microbial symbionts may have a substantially greater impact on the their hosts than passive association with specific body locations.

In December 2007 the National Institutes of Health launched the Human Microbiome Project to better understand the symbiotic relationships that result in human health and disease. The term **microbiome** was coined to define all the genes of the host, including those of its microbiota; the microbiome reflects the combined or composite genetic background of a host and its microbiota. One goal of the Human Microbiome Project is to sort out the impact that microbial gene function has on human health

MICROBIAL DIVERSITY & ECOLOGY

30.2 Do Bacteria Make People Fat?

The prevalence toward human obesity has increased steadily since 1960. In fact, in the United States, the incidence of obesity increased from 13% to 34% between 1960 and 2006. According to the National Center for Health Statistics, approximately 35 million women and 29 million men 20 years of age and older are obese. These numbers represent approximately one-third of the adults in the United States. As a result, the U.S. Public Health Service has labeled obesity an epidemic. While a sedentary lifestyle coupled with poor nutritional choices probably drive the prevalence of obesity, new evidence suggests that obesity may also be related to the bacteria in your gut.

Studies from Washington University, Arkansas State University, the University of Arkansas, and the Mayo Clinic have implicated two phyla of bacteria that impact body weight. Using metagenomic techniques examining intestinal bacteria, higher relative concentrations of *Firmicutes* (low G + C bacteria) to *Bacteroidetes* (gram-negative, nonendospore-forming bacteria) are correlated with obesity. Conversely, lean people have much fewer *Firmicutes* in their intestines as compared to *Bacteroidetes.* Interestingly, obese people who lost at least 25 pounds over one year by eating diets low in fat or carbohydrates altered their gut microbiota and subsequently had bacterial profiles that more closely resembled those of lean people.

Critics of this research suggest that this observation is not causal, just coincidence. To address this, investigators used germfree mice (p. 730) as test subjects. They discovered that such mice are inherently resistant to weight gain despite a high fat content in their diet. In fact, in an eight-week study, germfree mice consuming a diet containing 40% fat gained 50% less weight than conventional (having their normal microbiota) mice fed the same amount and type of food. However, when investigators transplanted intestinal bacteria from obese mice into germfree mice, the germfree mice gained more weight than germfree mice controls that received bacteria from lean mice. Thus it appears that unique bacterial profiles occur in obese mice and that specific bacteria were highly correlated with obesity. But what could be the mechanism by which intestinal bacteria drive obesity?

In evaluating the microbiome (the composite effect of both host and bacterial genomes), investigators have determined that the bacteria from obese mice had greater gene activation for proteins that catabolize complex carbohydrates. In other words, the firmicute bacteria were more active in catabolism of otherwise indigestible starches and sugars, releasing more energy from dietary nutrients than the bacteroidete bacteria. Furthermore, the firmicute bacteria were able to trigger host genes that slowed fat catabolism and increased fat accrual in various host tissues. Thus the increased energy released from nutrients and their subsequent storage in host tissue may help lead to obesity.

While these recent data suggest a simplistic view of weight gain, decades of other data support the absolute requirement for aerobic exercise and proper nutrition as part of achieving and maintaining a healthy weight. Bacteria may ultimately be proven to play a strong role in weight management, but it seems that part of that story is our role in creating a selective medium for bacterial growth. Remember, a high fat, high carbohydrate environment favors firmicute overgrowth, while a low fat, low carbohydrate environment favors the bacteroidetes. Which bacteria do you want deciding your weight?

and disease. In other words, since our 46 chromosomes do not appear to encode the products necessary for all biological functions of the human body, the project seeks to correlate the contribution of microbial genes to human biological function.

Another way of thinking about the intimate symbiotic associations that humans have with their microbiota is to consider them as "superorganisms." In this context, a superorganism emerges when the gene-encoded metabolic processes of the host become integrated with those of its microbiota. Thus superorganism metabolism can be thought of as a blend of host (human in this case) and microbial traits, where host and microbial cells cometabolize various substrates, resulting in unique products that give rise to host responses that would otherwise not occur. In other words, transgenomic metabolic regulation causes the host to function differently than it would without the contribution of symbionts. A number of provocative studies promoting this concept correlate human microbiome composition with states of health and disease. These studies report differences in human microbiome composition between healthy people, people on various antibiotics, lean and obese people, and even people altering their diets to lose weight. This is certainly an interesting field of science to watch (**Microbial Diversity & Ecology 30.2**).

Search This: Human Microbiome Project

Tolerating a normal microbiota likewise suggests that the host derives some benefit, such as the activation of the human immune system by early bacterial colonizers or the acquisition of vitamin K produced by fecal coliforms. Acquisition of a normal microbiota represents a selective process, where a niche may be defined by cellular receptors, surface properties, or secreted products. In humans, microbial niche variations are also related to age, gender, diet, nutrition, and developmental stage.

The survival of a host, such as a human, depends on an elaborate network of defenses that keeps harmful microorganisms and other foreign material from entering the body. Should they gain access, additional host defenses are summoned to prevent them from establishing another type of relationship, one of parasitism or pathogenicity. **Pathogenicity** (Greek *pathos,* emotion or suffering, and *gennan,* to produce) is the ability to produce pathologic changes or disease. A **pathogen** is any disease-producing microorganism. Here, we introduce the normal human microbiota, which function not as pathogens but as symbionts that are part of the host's first line of defense against harmful infectious agents.

Gnotobiotic Animals

To determine the role of the normal microorganisms associated with a host and evaluate the consequences of colonization, it is possible to deliver an animal by cesarean section and raise that animal in the absence of microorganisms—that is, germfree. These microorganism-free animals provide suitable experimental models for investigating the interactions of animals and their microbial flora. Comparing animals possessing normal microbiota (conventional animals) with germfree animals permits the elucidation of many complex relationships between microorganisms, hosts, and specific environments. Germfree experiments also extend and challenge the microbiologist's "pure culture concept" to in vivo research.

The term **gnotobiotic** (Greek *gnotos,* known, and *biota,* the flora and fauna of a region) has been defined in two ways. Some think of a gnotobiotic environment or animal as one in which all the microbiota are known; they distinguish it from one that is truly germfree. We shall use the term in a more inclusive sense. Gnotobiotic refers to a microbiologically monitored environment or animal that is germfree (axenic [*a,* without, and Greek *xenos,* a stranger]) or in which the identities of all microbiota are known.

Gnotobiotic animals and systems have become commonplace in research laboratories (**figure 30.14**). Germfree animals are usually more susceptible to pathogens because without normal commensal microbiota, pathogenic microorganisms establish themselves very easily. The number of microorganisms necessary to infect a germfree animal and produce a diseased state is much smaller. Conversely, germfree animals are almost completely resistant to the intestinal protozoan *Entamoeba histolytica,* the cause of amebic dysentery, because it is deprived of its bacterial food source. Germfree animals also do not show any dental caries or plaque formation, unless inoculated with cariogenic (caries or cavity-causing) streptococci and fed a high-sucrose diet. ◀◀ *Supergroup* Amoebozoa *(section 23.3)*

1. How might a microbial symbiont effect the uptake and metabolism of amino acids needed by its host?
2. Compare a germfree mouse to a normal one with regard to overall susceptibility to pathogens. What benefits does an animal gain from its microbiota?

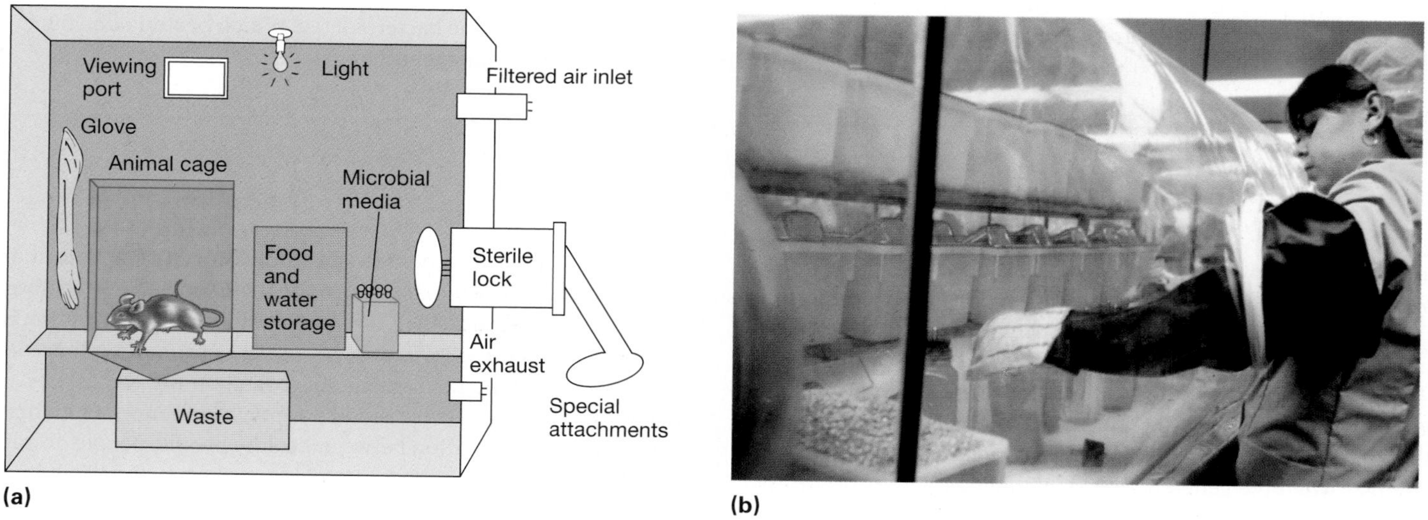

FIGURE 30.14 Raising Gnotobiotic Animals. (a) Schematic of a gnotobiotic isolator. The microbiological culture media monitor the sterile environment. (b) Gnotobiotic isolators for rearing colonies of small mammals.

30.3 Normal Microbiota of the Human Body

Development of a lifelong symbiotic relationship with microbes begins during birth. The human fetus is usually free of microorganisms. During birth, the infant's exposure to vaginal mucosa, skin, hair, food, and other nonsterile objects quickly results in the acquisition of a normal microbiota. Colonization of the newborn varies with respect to its environment. Newborns acquire external microbiota from those who provide their care, while internal microbiota are acquired through their diet. Bifidobacteria represent more than 90% of the culturable intestinal bacteria in breast-fed infants, with *Enterobacteriaceae* and enterococci in smaller proportions. These data also suggest that human milk may act as a selective medium for nonpathogenic bacteria, as bottle-fed babies appear to have a much smaller proportion of intestinal bifidobacteria. Bifidobacteria genomics suggests that it is prototrophic. That is, it is able to synthesize all amino acids and any other growth factors it needs from simple nutrients such as glucose and inorganic sources of nitrogen, sulfur, and phosphorus. Thus, it is well adapted for growth in the colonic environment, which has very low concentrations of these nutrients. Bifidobacteria may even use novel strategies for sugar import when in this environment (**figure 30.15**). Operons encoding enzymes for the catabolism of starch, pullulan, and amylopectin (necessary for weaning from milk onto complex carbohydrates) are also found. Switching to cow's milk or solid food (mostly polysaccharide) appears to result in the loss of bifidobacteria predominance, as *Enterobacteriaceae*, enterococci, *Bacteroides*, lactobacilli, and clostridia increase in number.

In a healthy human, regardless of age, the internal tissues (e.g., brain, blood, cerebrospinal fluid, muscles) are normally free of microorganisms. Conversely, the surface tissues (e.g., skin and mucous membranes) are constantly in contact with environmental microorganisms and become readily colonized by various microbial species. The mixture of microorganisms regularly found at any anatomical site is referred to as the **normal microbiota,** the indigenous microbial population, the **microflora,** or the normal flora. An overview of the microbiota typically native to different regions of the body is presented next (**figure 30.16**). Because bacterial and archaeal species make up most of the normal microbiota, they are emphasized over the fungi (mainly yeasts) and protists.

There are many reasons to understand the normal human microbiota. Three specific examples include:

- An understanding of the different microorganisms at particular locations provides greater insight into the possible infections that might result from injury to these body sites.
- Knowledge of the normal microbiota helps in understanding the causes and consequences of colonization and growth by microorganisms normally absent at a specific body site.
- An increased awareness of the role that these normal microbiota play in stimulating the host immune response can be gained. This is important because the immune system provides protection against potential pathogens.

Skin

The adult human is covered with approximately 2 m^2 of skin. It has been estimated that this surface area supports about 10^{12} bacteria. Microorganisms living on or in the skin can be either resident (normal) or transient microbiota. Resident organisms normally grow on or in the skin. Their presence becomes fixed in well-defined distribution patterns. Those that are temporarily present are transient microorganisms. Transients usually do not become firmly entrenched and are typically unable to multiply.

The anatomy and physiology of the skin vary from one part of the body to another, and the normal resident microbiota reflect these variations. The skin surface or epidermis has a slightly acidic pH, a high concentration of sodium chloride, a lack of moisture in many areas, and certain inhibitory substances (bactericidal or bacteriostatic). For example, the sweat glands release **lysozyme** (muramidase), an enzyme that hydrolyzes the $\beta(1\rightarrow4)$ glycosidic bond connecting *N*-acetylmuramic acid and *N*-acetylglucosamine in the bacterial cell wall peptidoglycan. Sweat glands also produce antimicrobial

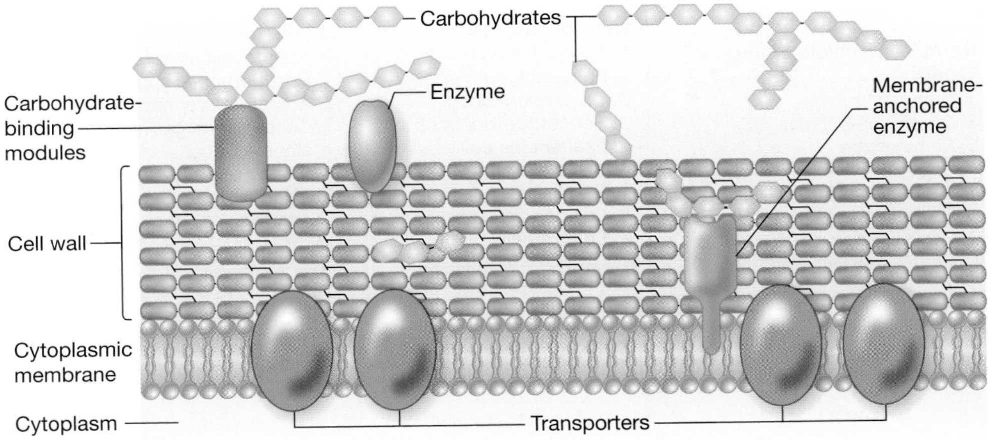

FIGURE 30.15 Sugar Acquisition Strategy of Bifidobacteria. Carbohydrates are bound to the bifidobacterial cell by use of an enzyme "docking station." Anchored to the cell surface, modular glycanases degrade complex carbohydrates to oligosaccharides, which are then transported across the cell membranes.

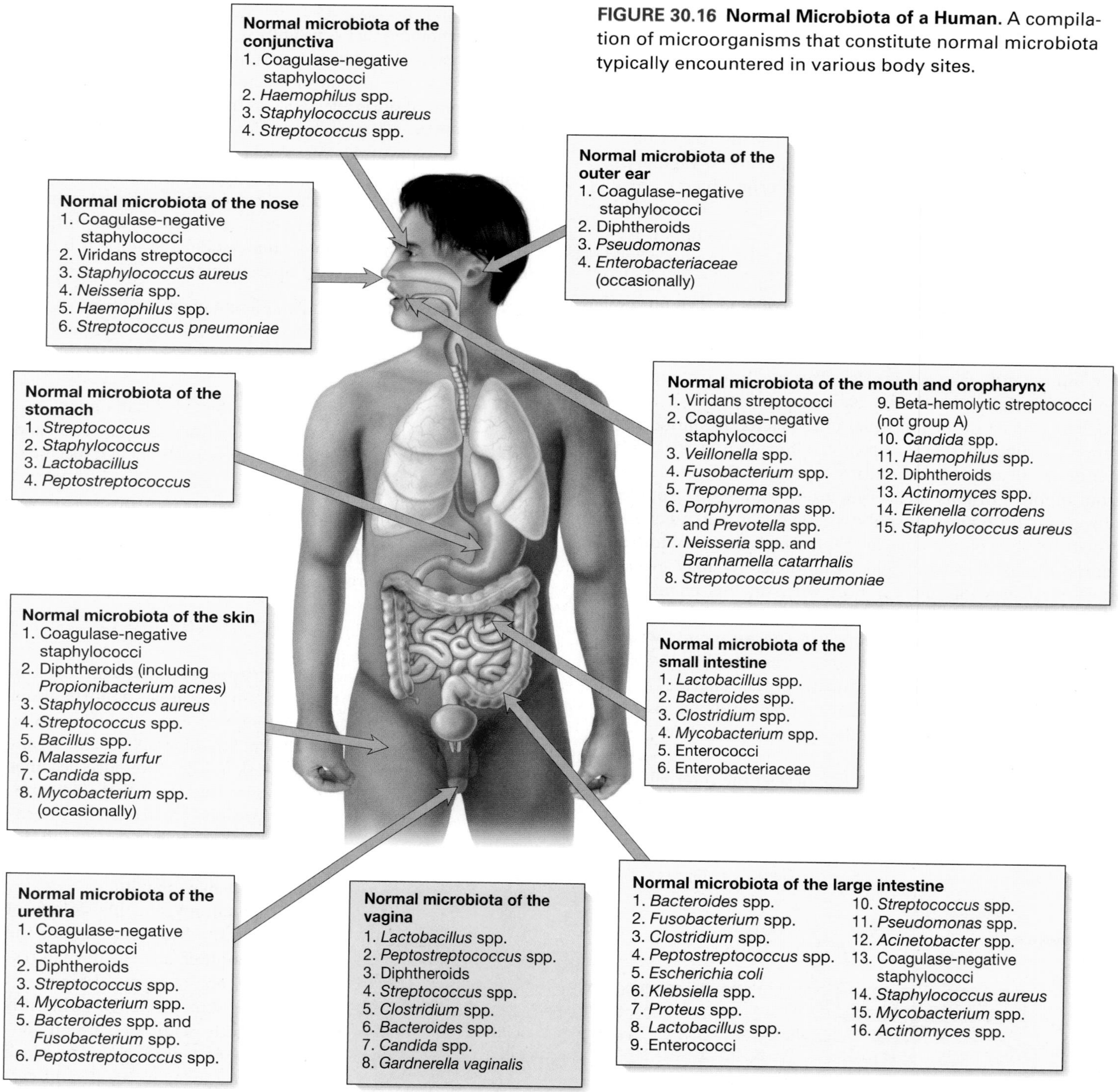

Normal microbiota of the conjunctiva
1. Coagulase-negative staphylococci
2. Haemophilus spp.
3. Staphylococcus aureus
4. Streptococcus spp.
Normal microbiota of the nose
1. Coagulase-negative staphylococci
2. Viridans streptococci
3. Staphylococcus aureus
4. Neisseria spp.
5. Haemophilus spp.
6. Streptococcus pneumoniae
Normal microbiota of the outer ear
1. Coagulase-negative staphylococci
2. Diphtheroids
3. Pseudomonas
4. Enterobacteriaceae (occasionally)
Normal microbiota of the stomach
1. Streptococcus
2. Staphylococcus
3. Lactobacillus
4. Peptostreptococcus
Normal microbiota of the mouth and oropharynx
1. Viridans streptococci
2. Coagulase-negative staphylococci
3. Veillonella spp.
4. Fusobacterium spp.
5. Treponema spp.
6. Porphyromonas spp. and Prevotella spp.
7. Neisseria spp. and Branhamella catarrhalis
8. Streptococcus pneumoniae
9. Beta-hemolytic streptococci (not group A)
10. Candida spp.
11. Haemophilus spp.
12. Diphtheroids
13. Actinomyces spp.
14. Eikenella corrodens
15. Staphylococcus aureus
Normal microbiota of the skin
1. Coagulase-negative staphylococci
2. Diphtheroids (including Propionibacterium acnes)
3. Staphylococcus aureus
4. Streptococcus spp.
5. Bacillus spp.
6. Malassezia furfur
7. Candida spp.
8. Mycobacterium spp. (occasionally)
Normal microbiota of the small intestine
1. Lactobacillus spp.
2. Bacteroides spp.
3. Clostridium spp.
4. Mycobacterium spp.
5. Enterococci
6. Enterobacteriaceae
Normal microbiota of the urethra
1. Coagulase-negative staphylococci
2. Diphtheroids
3. Streptococcus spp.
4. Mycobacterium spp.
5. Bacteroides spp. and Fusobacterium spp.
6. Peptostreptococcus spp.
Normal microbiota of the vagina
1. Lactobacillus spp.
2. Peptostreptococcus spp.
3. Diphtheroids
4. Streptococcus spp.
5. Clostridium spp.
6. Bacteroides spp.
7. Candida spp.
8. Gardnerella vaginalis
Normal microbiota of the large intestine
1. Bacteroides spp.
2. Fusobacterium spp.
3. Clostridium spp.
4. Peptostreptococcus spp.
5. Escherichia coli
6. Klebsiella spp.
7. Proteus spp.
8. Lactobacillus spp.
9. Enterococci
10. Streptococcus spp.
11. Pseudomonas spp.
12. Acinetobacter spp.
13. Coagulase-negative staphylococci
14. Staphylococcus aureus
15. Mycobacterium spp.
16. Actinomyces spp.

FIGURE 30.16 Normal Microbiota of a Human. A compilation of microorganisms that constitute normal microbiota typically encountered in various body sites.

peptides called cathelicidins (Latin *catharticus,* to purge, and *cida,* to kill) that form pores in bacterial plasma membranes. Thus, the normal flora of the skin must be able to tolerate and even thrive in these conditions. ◀◀ *Bacterial cell walls (section 3.3);* ▶▶ *Chemical mediators in nonspecific (innate) resistance (section 32.3)*

Most skin bacteria are found on superficial cells, colonizing dead cells, or closely associated with the oil and sweat glands. Secretions from these glands provide the water, amino acids, urea, electrolytes, and specific fatty acids that serve as nutrients primarily for *S. epidermidis* and aerobic corynebacteria. Gram-negative bacteria generally are found in the moister regions of the skin. The yeasts *Pityrosporum ovale* and *P. orbiculare* normally occur on the scalp. Using 16S rRNA sequence analyses, skin samples from the antecubital area (inside the elbow) of the forearm suggest that the majority of skin bacteria are from the *Proteobacteria, Actinobacteria, Firmicutes, Bacteroidetes, Cyanobacteria,* and *Acidobacteria;* this confirms results obtained from culture-based studies to identify the skin microbiota. However, the 16S data indicate that *S. epidermidis* and *Propionibacterium acnes* represent less than 5% of the skin microbiota, contrary to the commonly held notion that they predominate.

The oil glands secrete complex lipids that may be partially degraded by the enzymes from certain gram-positive bacteria (e.g., *P. acnes*). These bacteria usually are harmless; however, they are associated with the skin disease acne vulgaris. They can change the lipids secreted by the oil glands to unsaturated fatty acids such as oleic acid that have strong antimicrobial activity against gram-negative bacteria and some fungi. Some of these fatty acids are volatile and may have a strong odor. This is why many deodorants contain antibacterial substances that act selectively against gram-positive bacteria to reduce the production of volatile unsaturated fatty acids and body odor.

1. Why is it important to understand the normal human microbiota?
2. Why is the skin not usually a favorable microenvironment for colonization by bacteria?
3. How do microorganisms contribute to body odor?
4. What physiological role does *Propionibacterium acnes* play in the establishment of acne vulgaris?

Nose and Nasopharynx

The normal microbiota of the nose is found just inside the nostrils. *Staphylococcus aureus* and *S. epidermidis* are the predominant culturable bacteria present and are found in approximately the same numbers as on the skin of the face.

The nasopharynx, that part of the pharynx lying above the level of the soft palate, may contain small numbers of potentially pathogenic bacteria such as *Streptococcus pneumoniae, Neisseria meningitidis,* and *Haemophilus influenzae.* Diphtheroids, a large group of nonpathogenic gram-positive bacteria that resemble *Corynebacterium,* are commonly found in both the nose and nasopharynx.

Oropharynx

The oropharynx is that division of the pharynx lying between the soft palate and the upper edge of the epiglottis. The most important bacteria found in the oropharynx are the various α-hemolytic streptococci (*S. oralis, S. milleri, S. gordonii, S. salivarius*); large numbers of diphtheroids; *Branhamella catarrhalis*; and small gram-negative cocci related to *N. meningitidis.* The palatine and pharyngeal tonsils harbor a similar microbiota, except within the tonsillar crypts, where there is an increase in *Micrococcus* and the anaerobes *Porphyromonas, Prevotella,* and *Fusobacterium.*

Respiratory Tract

The upper and lower respiratory tracts (trachea, bronchi, bronchioles, alveoli) do not have a normal microbiota. This is because microorganisms are removed in at least three ways. First, a continuous stream of mucus is generated by the goblet cells. This entraps microorganisms, and the ciliated epithelial cells continually move the entrapped microorganisms out of the respiratory tract (*see figure 32.4*). Second, alveolar macrophages phagocytize and destroy microorganisms. Finally, a bactericidal effect is exerted by the enzyme lysozyme, present in the nasal mucus.

Eye and External Ear

At birth and throughout human life, a small number of bacteria are found on the conjunctiva of the eye. The predominant bacterium is *S. epidermidis,* followed by *S. aureus, Haemophilus* spp., and *S. pneumoniae.*

The normal microbiota of the external ear resembles that of the skin, with coagulase-negative staphylococci and *Corynebacterium* predominating. Mycological studies show the following fungi to be normal microbiota: *Aspergillus, Alternaria, Penicillium, Candida,* and *Saccharomyces.*

Mouth

In many ways, the mouth provides an ideal habitat for microbes. It has a ready supply of water and nutrients. It also has a neutral pH and moderate temperature. However, it subjects the normal microbiota to mechanical perturbations that can dislodge microbes from the mouth. The normal microbiota of the mouth consists of organisms that resist mechanical removal by adhering to surfaces such as the gums and teeth. These removal mechanisms include flushing of the oral cavity contents to the stomach, where they are destroyed by hydrochloric acid, and the continuous desquamation (shedding) of epithelial cells. ◀◀ *Biofilms (section 7.7)*

Soon after an infant is born, the mouth is colonized by microorganisms from the surrounding environment. Initially the microbiota consists mostly of the genera *Streptococcus, Neisseria, Actinomyces, Veillonella, Lactobacillus,* and some yeasts. Most microorganisms that initially invade the oral cavity are aerobes and obligate anaerobes. When the first teeth erupt, anaerobes (*Porphyromonas, Prevotella,* and *Fusobacterium*) become dominant due to the anoxic nature of the space between the teeth and gums. As the teeth grow, *Streptococcus parasanguis* and *S. mutans* attach to their enamel surfaces; *S. salivarius* attaches to the buccal and gingival epithelial surfaces and colonizes the saliva. These streptococci produce a glycocalyx and various other adherence factors that enable them to attach to oral surfaces. The presence of these bacteria contributes to the eventual formation of dental plaque, caries, gingivitis, and periodontal disease. ▶▶ *Periodontal disease (section 38.6)*

Stomach

The very acidic pH (2 to 3) of the gastric contents kills most microorganisms. As a result, the stomach usually contains less than 10 viable bacteria per milliliter of gastric fluid. These are mainly *Streptococcus, Staphylococcus, Lactobacillus, Peptostreptococcus,* and yeasts such as *Candida* spp. Microorganisms may survive if they pass rapidly through the stomach or if the organisms ingested with food are particularly resistant to gastric pH (e.g., mycobacteria).

Small Intestine

The small intestine is divided into three anatomical areas: duodenum, jejunum, and ileum. The duodenum (the first 25 cm of the small intestine) contains few microorganisms because of the combined influence of the stomach's acidic juices and the inhibitory action of bile and pancreatic secretions that are added here. Of the bacteria present, gram-positive cocci and rods comprise most of the microbiota. *Enterococcus faecalis,* lactobacilli, diphtheroids, and the yeast *Candida albicans* are occasionally found in the jejunum. In the distal portion of the small intestine (ileum), the microbiota begins to take on the characteristics of the colon microbiota. It is within the ileum that the pH becomes more alkaline. As a result, anaerobic gram-negative bacteria and members of the family *Enterobacteriaceae* become established.

Large Intestine (Colon)

The large intestine or colon has the largest microbial community in the body. Microscopic counts of feces approach 10^{12} organisms per gram wet weight. Over 400 different bacterial species have been isolated from human feces. These microorganisms consist primarily of anaerobic, gram-negative bacteria and gram-positive rods. Not only are the vast majority of microorganisms anaerobic, but many different species are present in large numbers. Several studies have shown that the ratio of anaerobic to facultative anaerobic bacteria is approximately 300 to 1. Besides the many bacteria in the large intestine, the yeast *Candida albicans* and certain protozoa may occur. The protists *Trichomonas hominis, Entamoeba hartmanni, Endolimax nana,* and *Iodamoeba butschlii* are common inhabitants.

The importance of the microbes living within the human colon, which can be likened to an anaerobic bioreactor (*see figure 41.8*), has prompted a number of investigations using culture-independent molecular approaches. Recent 16S rRNA analysis of microbes shed in feces, as well as microbes collected from gut epithelium, reveals that the majority of colonic bacteria and archaea are currently uncultivated. Metagenomic evaluations of intestinal bacteria collected from infants and their caregivers suggest that bacteria colonize the gastrointestinal tract in a variable and chaotic manner, with a few species appearing, disappearing, and reappearing during the first few months of life. These studies also confirm culture evaluations reporting initial colonization arising from caregivers. The metagenomic profile of gastrointestinal bacteria includes species of *Actinobacteria, Bacteroidetes, Firmicutes, Proteobacteria,* and *Verrucomicrobia.*

One bacterium, *Bacteroides thetaiotaomicron,* has been the focus of recent interest. This microbe is well suited for survival in the gut, where it is able to degrade complex dietary polysaccharides. Genome analysis reveals that *B. thetaiotaomicron* has a large collection of genes that encode proteins needed for the acquisition and metabolism of carbohydrates. It resides in a specific microenvironment: rather than adhering to the intestinal epithelium, it produces substrate-specific binding proteins that allow it to colonize exfoliated host cells, food particles, and even sloughed mucus (**figure 30.17**). It is thought that such attachment helps retain the microbes in the gut, and once they are bound, the induced expression of extracellular hydrolases enables efficient digestion. Of course, the diversity and density of microbes within the colon suggest that such "nutrient rafts" are colonized by a community of bacteria. For example, methanogenic archaea are thought to remove the products of fermentation by converting H_2 and CO_2 to methane, just as they do in the rumen microbial community.

Various physiological processes move the microbiota through the colon so an adult eliminates about 3×10^{13} microorganisms daily. These processes include peristalsis and desquamation of the surface epithelial cells to which microorganisms are attached, and continuous flow of mucus that carries adher-

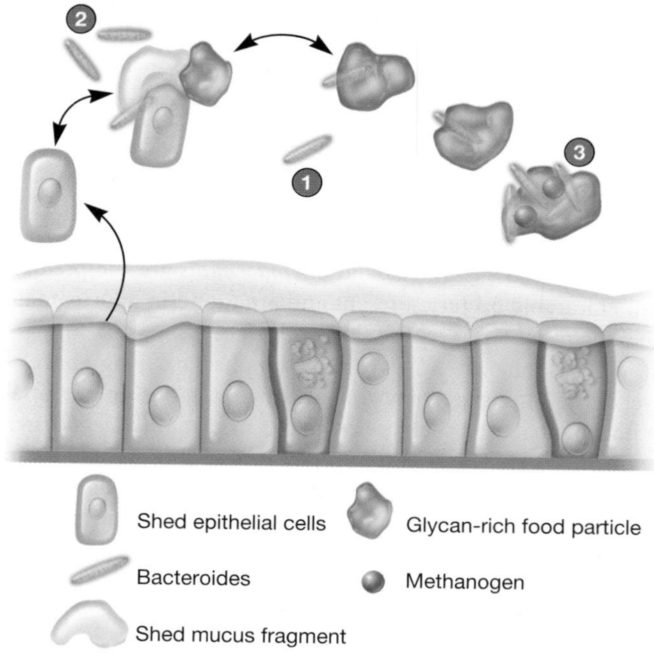

FIGURE 30.17 *Bacteroides thetaiotaomicron* as a Model for Colon Microbial Physiology and Community Dynamics. (1) *B. thetaiotaomicron* is rapidly eliminated if it remains planktonic; however, (2) it efficiently adheres to substrates within the lumen of the gut rather than the gut itself (3) where, together with other microbes including methanogenic archaea, it degrades complex carbohydrates.

TECHNIQUES & APPLICATIONS

30.3 Probiotics for Humans and Animals

There are 10,000 more microbes in the human gut than people that populate Earth. It is thus no surprise that it requires a complex series of interactions to maintain colonic health. Under healthy circumstances, the vast majority of these microorganisms help prevent infection and have a positive effect on nutrition. However, an abrupt change in diet, stress, or antibiotic therapy can upset this microbial balance, making the host susceptible to disease and decreasing the efficiency of food use. The practice of consuming certain foods to help maintain intestinal well-being has a long history. However, the commercialization of such foods has recently swelled, almost tripling in number in the United States and Europe. The term probiotic (Greek *pro*, for, and *bios*, life), which is frequently used by physicians, consumers, as well as marketing campaigns, was clarified by the Food and Agricultural Organization of the United Nations-World Health Organization in 2002 to mean "live microorganisms, which when administered in adequate amounts, confer a health benefit on the host."

Probiotic microorganisms are host specific; thus a strain selected as a probiotic in one animal may not be suitable in another species. Furthermore, microorganisms selected for probiotic use should exhibit these characteristics:

1. Adhere to the intestinal mucosa of the host
2. Be easily cultured
3. Be nontoxic and nonpathogenic to the host
4. Exert a beneficial effect on the host
5. Produce useful enzymes or physiological end products that the host can use
6. Remain viable for a long time
7. Withstand HCl in the host's stomach and bile salts in the small intestine

There are several possible explanations of how probiotic microorganisms displace pathogens and enhance the development and stability of the microbial balance in the large intestine. These include:

1. Competition with pathogens for nutrients and adhesion sites
2. Inactivation of pathogenic bacterial toxins or metabolites
3. Production of substances that inhibit pathogen growth
4. Stimulation of nonspecific immunity

In addition to the growing market of human probiotics, a wide variety of probiotic preparations have been patented for cattle, goats, horses, pigs, poultry, sheep, and domestic animals. Most of these preparations contain lactobacilli or streptococci or both, and bifidobacteria. Evidence has accumulated that certain probiotic microorganisms also offer considerable health benefits for humans. Potential benefits include:

1. Anticarcinogenic activity
2. Control of intestinal pathogens
3. Improvement of lactose digestion in individuals who have lactose intolerance
4. Reduction in the serum cholesterol concentration

In addition to the popularity of yogurt and other fermented milk products, some probiotics are currently used in the treatment of infectious gastroenteritis or in the prevention and cure of antibiotic-induced diarrhea. The introduction of probiotics into medical practice comes as the number of scientific publications describing probiotic use has more than doubled over the last several years. Many of these papers present the results of well-controlled clinical studies. No doubt as our understanding of the human ecology of the normal microflora increases, both consumers and medical practitioners will be able to make more informed choices regarding the nature and efficacy of probiotic foods and formulations.
▶▶| *Probiotics (section 40.5)*

ing microorganisms with it. To maintain homeostasis of the microbiota, the body must continually replace lost microorganisms. The bacterial population in the human colon usually doubles once or twice a day. Under normal conditions, the resident microbial community is self-regulating. Competition and mutualism between different microorganisms and between the microorganisms and their host serve to maintain a status quo. However, if the intestinal environment is disturbed, the normal microbiota may change greatly. Disruptive factors include stress, altitude changes, starvation, parasitic organisms, diarrhea, and use of antibiotics. Finally, it should be emphasized that the actual proportions of the individual bacterial populations within the indigenous microbiota depend largely on a person's diet. The consumption of probiotic foods, which contain live cultures of bacteria thought to impart health emphasizes our growing understanding of the important role of colonic microflora (**Techniques & Application 30.3**).

Genitourinary Tract

The upper urinary tract (kidneys, ureters, and urinary bladder) is usually free of microorganisms. In both males and females, a few bacteria (*S. epidermidis, E. faecalis,* and *Corynebacterium* spp.) usually are present in the distal portion of the urethra.

In contrast, the adult female genital tract, because of its large surface area and mucous secretions, has a complex microbiota that constantly changes with a woman's menstrual cycle. The major microorganisms are the acid-tolerant lactobacilli, primarily *Lactobacillus acidophilus*, often called Döderlein's bacillus. They ferment the glycogen produced by the vaginal epithelium, forming lactic acid. As a result, the pH of the vagina and cervix is maintained between 4.4 and 4.6, inhibiting other microorganisms.

1. What are the most common microorganisms found in the nose? The oropharynx? The nasopharynx? The lower respiratory tract? The mouth? The eye? The external ear? The stomach? The small intestine? The colon? The genitourinary tract?
2. Draw a time line indicating the colonization of the human large intestine. What is the significance of the temporal pattern?
3. What physiological processes move the microbiota through the gastrointestinal tract?
4. Describe the microbiota of the upper and lower female genitourinary tract.

The Relationship Between Normal Microbiota and the Host

The interaction between a host and a microorganism is a dynamic process in which each partner acts to maximize its survival. In some instances, after a microorganism enters or contacts a host, a positive mutually beneficial relationship occurs that becomes integral to the health of the host. These microorganisms become the normal microbiota. In other instances, the microorganism causes deleterious effects on the host; the end result may be disease or even death of the host. ▶▶| *Infection and pathogenicity (chapter 31)*

Our environment is teeming with microorganisms and we come in contact with many of them every day. Products made by colonic bacteria (such as vitamins B and K) demonstrate how these bacteria benefit us. However, some of these microorganisms are pathogenic—that is, they cause disease. Yet these pathogens are at times prevented from causing disease by competition provided by the normal microbiota. In general, the normal microbiota uses space, resources, and nutrients needed by pathogens. In addition, its members may produce chemicals that repel invading pathogens. For instance, the lactobacilli in the female genital tract maintain a low pH and inhibit colonization by pathogenic bacteria and yeast, and the corynebacteria on the skin produce fatty acids that inhibit colonization by pathogenic bacteria. These are excellent examples of amensalism.

Interestingly, studies using germfree animals suggest a strong correlation between the establishment of a stable microbial flora and the induction of immune competency. For example, the introduction of normal fecal flora to germfree rodents stimulates the production and secretion of angiogenin-4, an antimicrobial peptide of intestinal Paneth cells. Furthermore, the reconstitution of germfree rodents with flora from conventionally raised siblings causes the abnormal gut-associated lymphoid tissue and intestinal lamina propria to resemble that of the conventional animals (i.e., their lymphoid tissues and immunity are normalized). Even cell wall fragments from gram-positive bacteria can induce these changes. This normalization also includes an increase in the local lymphocyte populations and increased mucosal antibody production. ▶▶| *Cells, tissues, and organs of the immune system (section 32.4)*

Although normal microbiota offer some protection from invading pathogens, its members may themselves become pathogenic and produce disease under certain circumstances; they then are termed **opportunistic microorganisms** or **pathogens.** Opportunistic microorganisms are adapted to the noninvasive mode of life defined by the limitations of the environment in which they are living. If they are removed from these environmental restrictions and introduced into the bloodstream or tissues, disease can result. For example, streptococci of the viridans group are the most common resident bacteria of the mouth and oropharynx. If they are introduced into the bloodstream in large numbers (e.g., following tooth extraction or a tonsillectomy), they may settle on deformed or prosthetic heart valves and cause endocarditis.

Opportunistic microorganisms often cause disease in compromised hosts. A **compromised host** is seriously debilitated and has a lowered resistance to infection. There are many causes of this condition, including malnutrition, alcoholism, cancer, diabetes, leukemia, another infectious disease, trauma from surgery or an injury, an altered normal microbiota from the prolonged use of antibiotics, and immunosuppression by various factors (e.g., drugs, viruses [HIV], hormones, and genetic deficiencies). For example, *Bacteroides* species are one of the most common residents in the large intestine (figure 30.16) and are quite harmless in that location. If introduced into the peritoneal cavity or into the pelvic tissues as a result of trauma, they cause suppuration (the formation of pus) and bacteremia (the presence of bacteria in the blood). Importantly, the normal microbiota are harmless and are often beneficial in their normal location in the host and in the absence of coincident abnormalities. However, they can produce disease if introduced into foreign locations or compromised hosts.

1. Describe two examples of the normal microbiota benefiting a host.
2. Explain how the principle of competitive exclusion is used by normal host microbiota in preventing the establishment of pathogens.
3. How would you define an opportunistic microorganism or pathogen? A compromised host?

Summary

30.1 Microbial Interactions

a. Symbiotic interactions include mutualism (mutually beneficial and obligatory), cooperation (mutually beneficial, not obligatory), and commensalism (product of one organism can be used beneficially by another organism). Predation involves one organism (the predator) ingesting or killing a larger or smaller prey, parasitism (a longer-term internal maintenance of another organism or acellular infectious agent), and amensalism (a microbial product inhibiting another organism). Competition involves organisms competing for space or a limiting nutrient. This can lead to dominance of one organism or coexistence of both at lower populations (**figure 30.1**).

b. A consortium is a physical association of organisms that have a mutually beneficial relationship based on positive interactions.

c. Mutual advantage is central to many organism-organism interactions. These interactions can be based on material transfers related to energetics or the creation of physical environmental changes that offer protection. With several important mutualistic interactions, chemolithotrophic microorganisms play a critical role in making organic matter available for use by an associated organism (e.g., endosymbionts in *Riftia*) (**figure 30.5**).

d. The rumen is an excellent example of a mutualistic interaction between an animal and a complex microbial community. In this microbial community, complex plant materials are broken down to simple organic compounds that can be absorbed by the ruminant, as well as forming waste gases such as methane that are released to the environment.

e. Cooperative interactions are beneficial for both organisms but are not obligatory. Important examples are marine animals, including *Alvinella* and *Paralvinella*, that involve interactions with hydrogen sulfide-oxidizing chemotrophs.

f. Commensalism or syntrophism simply means growth together. It does not require physical contact and involves a mutually positive transfer of materials, such as interspecies hydrogen transfer.

g. Predation and parasitism are closely related. Predation has many beneficial effects on populations of predators and prey. These include the returning minerals immobilized in organic matter to mineral forms for reuse by chemotrophic and photosynthetic primary producers, protection of prey from heat and damaging chemicals, and possibly aiding pathogenicity.

30.2 Human-Microbe Interactions

a. Animals and environments that are germfree or have one or more known microorganisms are termed gnotobiotic. Gnotobiotic animals and techniques provide good experimental systems with which to investigate the interactions of animals and specific species or microorganisms (**figure 30.14**).

b. The Human Microbiome Project is evaluating the impact of microbial genes as their products interact with human gene products. Early indications suggest a much greater role for coevolution in defining microbial-human relationships.

30.3 Normal Microbiota of the Human Body

a. Various microbes have adapted to specific niches found on the human host. These niches are uniquely able to support microbe growth by maintaining a relatively constant environment (**figure 30.16**).

b. Microorganisms living on or in the skin can be characterized as either transients or residents. The normal microbiota have evolved the ability to tolerate conditions that predominate on skin (e.g., somewhat acidic, salty, and dry)

c. The oral cavity provides a nutrient-rich habitat but subjects microbes to physical processes that can dislodge them. The normal microbiota of the oral cavity is composed of those microorganisms able to resist this mechanical removal.

d. The stomach contains very few microorganisms due to its acidic pH.

e. The distal portion of the small intestine and the entire large intestine have the largest microbial community in the body. Over 400 species have been identified, the vast majority of them anaerobic.

f. The upper urinary tract is usually free of microorganisms. In contrast, the adult female genital tract has a complex microbiota.

g. In some instances, after a microorganism contacts or enters a host, a positive mutually beneficial relationship occurs and becomes integral to the health of the host. In other instances, the microorganism may produce disease or even death of the host.

h. Many of the normal host microbiota compete with pathogenic microorganisms.

i. An opportunistic microorganism is generally harmless in its normal environment but may become pathogenic when moved to a different body location or in a compromised host.

Critical Thinking Questions

1. Describe an experimental approach to determine if a plant-associated microbe is a commensal or a mutualist.
2. Some patients who take antibiotics for acne develop yeast infections of the mouth or genital-urinary tract. Explain.
3. How does knowing the anatomical location of commensal flora help clinicians diagnose infection?
4. Compare and contrast the microbial communities that reside in a ruminant with those in the human gut.
5. The fruit fly *Drosophila melanogaster* is an excellent model system. *D. melanogaster* hosts the ubiquitous insect symbiont *Wolbachia,* but unlike most other insect hosts, it is not subject to reproductive manipulation by the bacterium. Surprisingly, infection with *Wolbachia* protects *D. melanogaster* from infection by three, unrelated viruses that otherwise cause 100% mortality. Discuss the differences in this *Wolbachia*-insect symbiosis with that described in Microbial Diversity & Ecology 30.1. Do you think it is likely that *Wolbachia* also protects wasps against other pathogens, such as viruses? Explain your answer from the point of view of the bacterium and from the insect's point of view.

Read the original paper: Hedges, L. M. 2008. *Wolbachia* and virus protection in insects. *Science* 322:722.

Concept Mapping Challenge

Construct a concept map using the words below; provide your own linking terms.

Symbioses	Phycobiont	Mycobiont	Lichen	Parasitism	Predation
Mutualism	Corals	*Syntrophobacter*	Termites	*Bdellovibrio*	Tube worms
			Commensalism	Methanogens	Human microflora
			Parasitism		

Learn More

Learn more by visiting the text website at www.mhhe.com/willey8, where you will find a complete list of references.

31

Infection and Pathogenicity

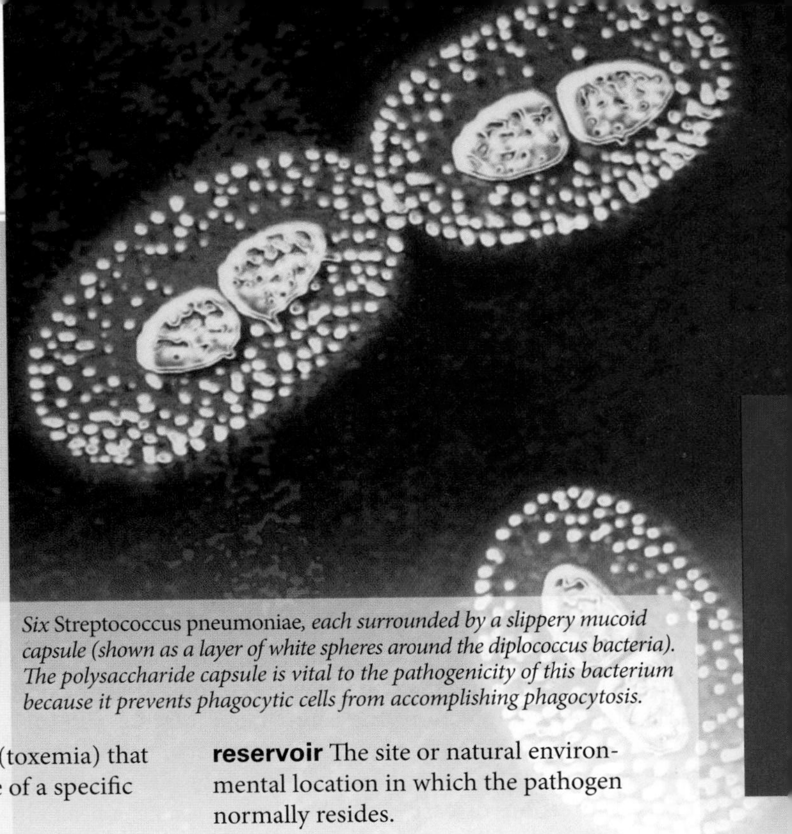

Six Streptococcus pneumoniae, *each surrounded by a slippery mucoid capsule (shown as a layer of white spheres around the diplococcus bacteria). The polysaccharide capsule is vital to the pathogenicity of this bacterium because it prevents phagocytic cells from accomplishing phagocytosis.*

CHAPTER GLOSSARY

AB toxins The structure and activity of many exotoxins; the B subunit binds the toxin to a cell, and the A subunit then enters the cell, where it disrupts cellular functions.

bacteremia The presence of viable bacteria in the blood.

colonization Colonization is the establishment of a site of microbial reproduction on or in a host.

cytotoxin A toxin that acts upon specific cells; cytotoxins are named according to the cell for which they are specific (e.g., neurotoxin).

endotoxin The lipid A component of gram-negative bacterial cell wall lipopolysaccharide. Systemic effects of endotoxin are sometimes referred to as septic shock.

exotoxin A heat-labile, toxic protein produced by a bacterium and usually released into the bacterium's surroundings.

infectious dose 50 (ID_{50}) and lethal dose 50 (LD_{50}) The number of microorganisms or dose that will infect or kill, respectively, 50% of an experimental group of organisms within a specified time period.

intoxication A disease (toxemia) that results from the presence of a specific toxin in the host.

intracellular parasite Any infectious agent that resides and reproduces within the host cell; however, it is often used to refer specifically to protozoan pathogens.

leukocidin A microbial toxin that can damage or kill leukocytes.

mycotoxin A toxin produced as a secondary metabolite by fungi.

opportunistic pathogen An organism that is part of the host's normal microbiota but is able to cause disease when the host is immunocompromised or when the organism has gained access to tissue sites outside its normal niche.

pathogen Any organism or agent that produces a disease.

pathogenicity island A 10 to 200 Kb segment of DNA in some pathogens that contains the genes responsible for virulence.

phospholipase An enzyme that hydrolyzes a specific ester bond in phospholipids.

reservoir The site or natural environmental location in which the pathogen normally resides.

septicemia A disease associated with the presence of pathogens or bacterial toxins in the blood.

toxin A microbial product (e.g., a protein) or component (e.g., lipid A) that injures another cell or organism.

toxoid A bacterial exotoxin that has been modified so that it is no longer toxic but is still immunogenic, stimulating antitoxin formation in a host.

vector Living transmitters (e.g., certain insects and animals) of a pathogen.

vehicle Inanimate material involved in pathogen transmission; also called a **fomite.**

viremia The presence of viruses in the blood stream.

virulence factor A microbial product or structure that contributes to virulence or pathogenicity.

zoonoses Human infectious disease acquired from an animal host.

Chapter 30 introduces the concept of symbiosis and deals with several of its subordinate categories, including commensalism and mutualism. In this chapter, the process of parasitism is presented along with one of its possible consequences—pathogenicity. The parasitic way of life is so successful, it has evolved independently in nearly all groups of organisms. Understanding host-parasite relationships requires an interdisciplinary approach, drawing on knowledge of cell biology, microbiology, entomology, immunology, ecology, and zoology. This chapter examines the parasitic way of life in terms of health and disease with an emphasis on viral and bacterial disease mechanisms.

31.1 Host-Parasite Relationships

Relationships between two organisms can be very complex. A larger organism that supports the survival and growth of a smaller organism is called the **host.** Technically, **parasites** are those organisms that live on or within a host organism and are metabolically dependent on the host. Unfortunately, the term parasite has other meanings. It is often used to specifically refer to a protozoan or helminth living within a host. However, any organism that causes disease is a parasite. Even normal microbiota, such as those associated with the gut, can become parasites when they are present in a location within the host other than the site they normally colonize.

Microbiologists can define infectious disease by the host-parasite relationship, which is complex and dynamic. When a parasite is growing and multiplying within or on a host, the host is said to have an **infection.** The nature of an infection can vary widely with respect to severity, location, and number of organisms involved. An infection may or may not result in overt disease. An **infectious disease** is any change from a state of health in which part or all of the host body is not capable of carrying on its normal functions due to the presence of a parasite or its products. Any organism or agent that produces such a disease is also known as a **pathogen** (Greek *patho,* disease, and *gennan,* to produce). Its ability to cause disease is called **pathogenicity.** A primary pathogen is any organism that causes disease in a healthy host by direct interaction. Conversely, an **opportunistic pathogen** refers to an organism that infects a host having a weakened immune system (a compromised host) and may even be part of the host's normal microbiota.

At times an infectious organism can enter a **latent state** in which no transmission of the organism occurs (i.e., the organism is not infectious at that time) and no symptoms are present within the host. This latency can be either intermittent or quiescent. Intermittent latency is exemplified by the herpesvirus that causes cold sores (fever blisters). After an initial infection, the symptoms subside. However, the virus remains in nerve tissue and can be cyclically activated weeks or years later by factors such as stress or sunlight. In quiescent latency, the organism persists but remains inactive for long periods of time, usually for years. For example, mycobacteria that cause tuberculosis and the varicella-zoster virus that causes chickenpox in children remain after initial infection. In adulthood, under certain conditions, the same microorganism may cause disease, such as reactivation tuberculosis or shingles, respectively.

The outcome of most host-parasite relationships depends on three main factors: (1) the number of microorganisms infecting the host, (2) the virulence of the organism, and (3) the host's defenses or degree of resistance (**figure 31.1**). The term **virulence** (Latin *virulentia,* from virus, poison) refers to the degree or intensity of pathogenicity. Usually the more pathogenic an organism is within a given host, the greater the likelihood that it will overcome or evade the host immune defenses and cause disease. However, a small number of organisms may cause disease if they are extremely virulent or if the host's resistance is low. Such infections can be a serious problem among patients with very low resistance.

$$\text{Infection (infectious disease)} = \frac{\text{No. of organisms} \times \text{Virulence}}{\text{Host resistance}}$$

FIGURE 31.1 Mathematical Expression of Infection. Infection or infectious disease can be evaluated by determining the relative contributions of the number of organisms, their virulence, and the host resistance. Organism number reflects the infectious dose and the rate at which the organism can reproduce. Virulence can be modeled on the total number of virulence factors expressed in the host. Likewise, host resistance is a function of immune status (immunizations, nutrition, previous exposure, etc.) that may be combined with the effects of chemotherapeutic intervention.

1. Define parasitic organism, infection, infectious disease, pathogenicity, virulence, and opportunistic pathogen.
2. What factors determine the outcome of most host-parasite relationships?

31.2 The Infectious Disease Process

An infectious disease results from an infection by microbial agents such as viruses, bacteria, fungi, protozoa, and helminths. Often infectious diseases have characteristic signs and symptoms. **Signs** are objective changes in the body, such as a fever or rash, that can be directly observed. **Symptoms** are subjective changes, such as pain and loss of appetite, that are experienced by the patient. The term symptom is often used in a broader scope to include the clinical signs. A **disease syndrome** is a set of signs and symptoms that are characteristic of the disease. Frequently additional laboratory tests are required for an accurate diagnosis because symptoms and readily observable signs may not be sufficient for diagnosis.

Clinically, the course of an infectious disease usually has a characteristic pattern and can be divided into several phases (**figure 31.2**). The **incubation period** is the time between pathogen entry and the development of signs and symptoms. The pathogen is spreading but has not reached a sufficient level to cause clinical manifestations. This period's length varies with disease. Next, the **prodromal stage** occurs with an onset of signs and symptoms that are not yet specific enough to make a diagnosis. However, the patient often is contagious. This is followed by the

illness period, when the disease is most severe and displays characteristic signs and symptoms. The host immune response is typically triggered at this stage. Finally, during the period of decline, the signs and symptoms begin to disappear. The recovery stage often is referred to as **convalescence.**

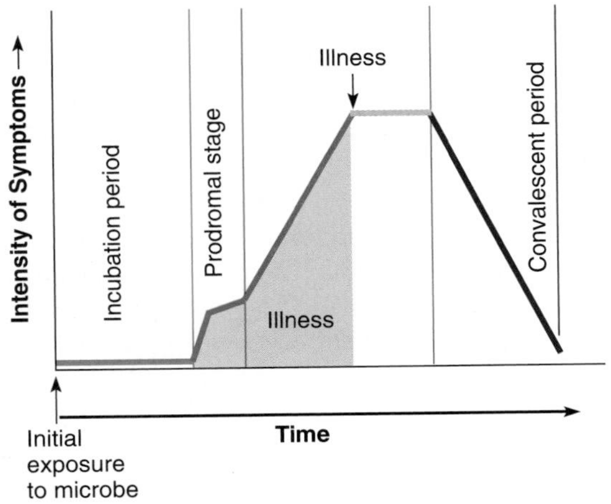

FIGURE 31.2 The Course of an Infectious Disease. Most infectious diseases occur in four stages. The duration of each stage is a characteristic feature of each disease. The shaded area represents when a disease is typically communicable.

Figure 31.2 Micro Inquiry

During which stages does the host exhibit signs and symptoms? How does this correlate to when the disease is communicable?

Biologically, the fundamental process of infection is essentially a competition for resources. The host represents an opportunity for the infecting agent to obtain protection, nutrients, and energy to use for its own survival. The infecting agent must therefore develop mechanisms to access the host and exploit it. Furthermore, to continue surviving, the pathogen must also devise methods to move on to better environments once the immediate one declines in value. Dissemination to another point in one host or into another host must then occur. The infectious disease cycle or chain of infection represents these events in the form of an intriguing mystery, where understanding the disease process is only really revealed when all of the links of the chain are known (**figure 31.3**).

The Source or Reservoir

The **source** of a pathogen is the first link in the infectious disease cycle. The most common reservoirs of human pathogens are humans and other animals. A source is the location from which the pathogen is immediately transmitted to the host, either directly through the environment or indirectly through an intermediate agent. The source can be either animate (e.g., humans or animals) or inanimate (e.g., water, soil, or food). The period of infectivity is the time during which the source is infectious or is disseminating the pathogen. If the source of the infection can be eliminated or controlled, the infectious disease cycle itself will be interrupted and transmission of the pathogen will be prevented.

A **reservoir** is the site or natural environmental location in which the pathogen normally resides. It is also the site from which a source acquires the pathogen or where direct infection of the host can occur. Thus a reservoir sometimes functions as a source. Reservoirs also can be animate or inanimate. The increasing impingement of humans on the environment and increased exposure to antibiotics and mutagens have played a

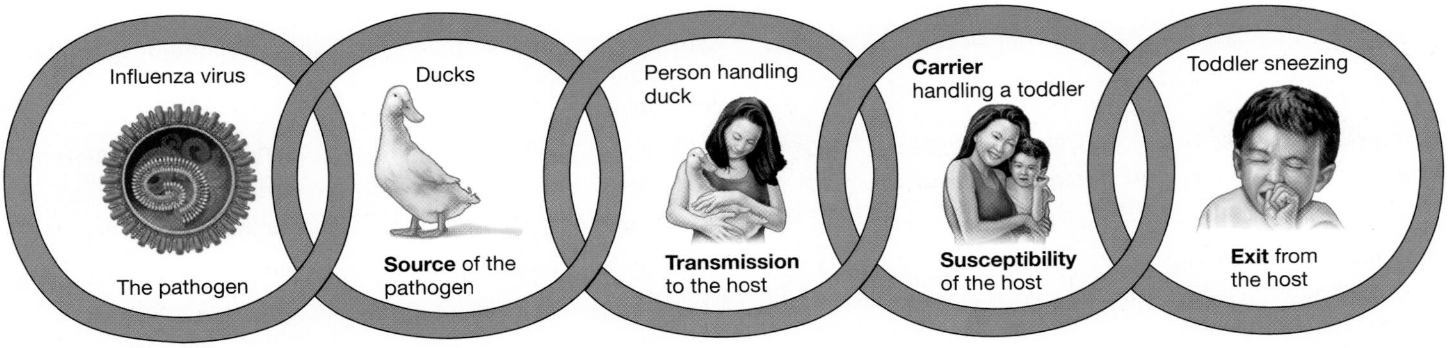

FIGURE 31.3 Proposed Chain of Avian Influenza Infection.

Figure 31.3 Micro Inquiry

What is the term used to describe a disease that can be transmitted from an animal to a human host?

significant role in the substantive change in reservoirs over the last 100 years.

Much of the time, human hosts are the most important animate sources of the pathogen and are called carriers. A **carrier** is an infected individual who is a potential source of infection for others. Carriers play an important role in the epidemiology of disease. Convalescent, healthy, and incubatory carriers may harbor the pathogen for only a brief period (hours, days, or weeks) and are called casual, acute, or transient carriers. If they harbor the pathogen for long periods (months, years, or life), they are called chronic carriers.

Infectious diseases that can be transmitted from animals to humans are termed **zoonoses** (Greek *zoon,* animal, and *nosos,* disease); thus animals also can serve as reservoirs. Humans contract the pathogen by several mechanisms: coming into direct contact with diseased animal flesh (e.g., anthrax); drinking contaminated cow's milk (e.g., tuberculosis and brucellosis); inhaling dust particles contaminated by animal excreta or products (e.g., Q fever, hantavirus pulmonary infection); or eating insufficiently cooked infected flesh (e.g., anthrax, trichinosis). In addition, being bitten by arthropod **vectors** (organisms that spread disease from one host to another) such as mosquitoes, ticks, fleas, mites, or biting flies (e.g., equine encephalomyelitis, malaria, Lyme disease, Rocky Mountain spotted fever, and plague), or being bitten by a diseased animal (e.g., rabies) can lead to infection.

Table 31.1 lists some common zoonoses found in the Western Hemisphere. This table is not inclusive in scope; it merely abbreviates the enormous spectrum of zoonotic diseases that are relevant to human infectious disease. Domestic animals are the most common source of zoonoses because they live in greater proximity to humans than do wild animals. Diseases of wild animals that are transmitted to humans tend to occur sporadically because close contact is infrequent. Other major reservoirs of pathogens are water, soil, and food. These reservoirs are discussed in detail in chapters 37 through 39.

1. What are some important characteristics of a pathogen? What is an infectious disease?
2. Define source, reservoir, and carrier. How are they related?
3. What types of infectious disease carriers are there?
4. Which phase of an infectious disease usually coincides with spread of the disease? Suppose an individual was contagious in the incubation period or in the early prodromal stage. How would that alter the ease with which the pathogen is spread to new hosts?

Pathogen Transmission

An essential feature in the development of an infectious disease is the initial transport of the pathogen to the host; the pathogen must be transmitted from one host or source to another (figure 31.3). Evidence exists that a pathogen's virulence may be strongly influenced by its mode of transmission and ability to live outside

Table 31.1 Infectious Organisms in Nonhuman Reservoirs That May Be Transmitted to Humans

Disease	*Etiologic Agent*	*Usual or Suspected Nonhuman Host*	*Usual Method of Human Infection*
Anthrax	*Bacillus anthracis*	Cattle, horses, sheep, swine, goats, dogs, cats, wild animals, birds	Inhalation or ingestion of spores; direct contact
Babesiosis	*Babesia bovis, B. divergens, B. microti, B. equi*	*Ixodes* ticks of various species	Bite of infected tick
Brucellosis (undulant fever)	*Brucella melitensis, B. abortus, B. suis*	Cattle, goats, swine, sheep, horses, mules, dogs, cats, fowl, deer, rabbits	Milk; direct or indirect contact
Campylobacteriosis	*Campylobacter fetus, C. jejuni*	Cattle, sheep, poultry, swine, pets	Contaminated water and food
Cat-scratch disease	*Bartonella henselae*	Cats, dogs	Cat or dog scratch
Cryptosporidiosis	*Cryptosporidium* spp.	Farm animals, pets	Contaminated water
Encephalitis (California)	Arbovirus	Rats, squirrels, horses, deer, hares, cows	Mosquito
Encephalitis (St. Louis)	Arbovirus	Birds	Mosquito
Encephalomyelitis (Eastern equine)	Arbovirus	Birds, ducks, fowl, horses	Mosquito
Encephalomyelitis (Venezuelan equine)	Arbovirus	Rodents, horses	Mosquito

(Continued)

Table 31.1 Infectious Organisms in Nonhuman Reservoirs That May Be Transmitted to Humans *(Continued)*

Disease	*Etiologic Agent*	*Usual or Suspected Nonhuman Host*	*Usual Method of Human Infection*
Encephalomyelitis (Western equine)	Arbovirus	Birds, snakes, squirrels, horses	Mosquito
Giardiasis	*Giardia intestinalis*	Rodents, deer, cattle, dogs, cats	Contaminated water
Hantavirus pulmonary syndrome	Pulmonary syndrome hantavirus	Deer mice	Contact with the saliva, urine, or feces of deer mice; aerosolized viruses
Herpes B viral encephalitis	Herpesvirus simiae	Monkeys	Monkey bite; contact with material from monkeys
Influenza	Influenza virus	Water fowl, pigs	Direct contact or inhalation
Listeriosis	*Listeria monocytogenes*	Sheep, cattle, goats, guinea pigs, chickens, horses, rodents, birds, crustaceans	Food-borne
Lyme disease	*Borrelia burgdorferi*	Ticks (*Ixodes scapularis* or related ticks)	Bite of infected tick
Lymphocytic choriomeningitis	*Arenavirus*	Mice, rats, dogs, monkeys, guinea pigs	Inhalation of contaminated dust; ingestion of contaminated food
Pasteurellosis	*Pasteurella multocida*	Fowl, cattle, sheep, swine, goats, mice, rats, rabbits	Animal bite
Plague (bubonic)	*Yersinia pestis*	Domestic rats, many wild rodents	Flea bite
Psittacosis	*Chlamydia psittaci*	Birds	Direct contact, respiratory aerosols
Q fever	*Coxiella burnetii*	Cattle, sheep, goats	Inhalation of contaminated soil and dust
Rabies	*Rabies virus*	Dogs, bats, opposums, skunks, raccoons, foxes, cats, cattle	Bite of rabid animal
Relapsing fever (borreliosis)	*Borrelia* spp.	Rodents, porcupines, opposums, armadillos, ticks, lice	Tick or louse bite
Rocky Mountain spotted fever	*Rickettsia rickettsii*	Rabbits, squirrels, rats, mice, groundhogs	Tick bite
Salmonellosis	*Salmonella* spp. (except *S. typhosa*)	Fowl, swine, sheep, cattle, horses, dogs, cats, rodents, reptiles, birds, turtles	Direct contact; food
SARS	SARS coronavirus	Bats, civits	Contact with infected animal or person
Tuberculosis	*Mycobacterium bovis, M. tuberculosis*	Cattle, horses, cats, dogs	Milk; direct contact
Tularemia	*Francisella tularensis*	Wild rabbits, most other wild and domestic animals	Direct contact with infected carcass, usually rabbit; tick bite, biting flies
Typhus fever (endemic)	*Rickettsia mooseri*	Rats	Flea bite
Yellow fever (jungle)	Yellow fever virus	Monkeys, marmosets, lemurs, mosquitoes	Mosquito

Modified from Guy Youmans, et al., The Biologic and Clinical Basis of Infectious Diseases. *Copyright © 1985 W. B. Saunders, Philadelphia, PA. Reprinted by permission.*

HISTORICAL HIGHLIGHTS

31.1 The First Indications of Person-to-Person Spread of an Infectious Disease

In 1773 Charles White, an English surgeon and obstetrician, published his "Treatise on the Management of Pregnant and Lying-in Women." In it, he appealed for surgical cleanliness to combat childbed or puerperal fever. (Puerperal fever is an acute febrile condition that can follow childbirth and is caused by streptococcal infection of the uterus or adjacent regions.) In 1795 Alexander Gordon, a Scottish obstetrician, published his "Treatise on the Epidemic Puerperal Fever of Aberdeen," which demonstrated for the first time the contagiousness of the disease. In 1843 Oliver Wendell Holmes, a noted physician and anatomist in the United States, published a paper entitled "On the Contagiousness of Puerperal Fever" and also appealed for surgical cleanliness to combat this disease.

However, the first person to realize that a pathogen could be transmitted from one person to another was the Hungarian physician Ignaz Phillip Semmelweis. Between 1847 and 1849, Semmelweis observed that women who had their babies at the hospital with the help of medical students and physicians were four times as likely to contract puerperal fever as those who gave birth with the help of midwives. He concluded that the physicians and students were infecting women with material remaining on their hands after autopsies and other activities. Semmelweis thus began washing his hands with a calcium chloride solution before examining patients or delivering babies. This simple procedure led to a dramatic decrease in the number of cases of puerperal fever and saved the lives of many women. As a result, Semmelweis is credited with being the pioneer of antisepsis in obstetrics. Unfortunately, in his own time, most of the medical establishment refused to acknowledge his contribution and adopt his procedures. After years of rejection, Semmelweis had a nervous breakdown in 1865. He died a short time later of a wound infection. It is very probable that it was a streptococcal infection, arising from the same pathogen he had struggled against his whole professional life.

its host (**Historical Highlights 31.1**). When the pathogen uses a mode of transmission such as direct contact, it cannot afford to make the host so ill that it will not be transmitted effectively. This is the case with the common cold, which is caused by rhinoviruses and several other respiratory viruses. If the virus reproduced too rapidly and damaged its host extensively, the person would be bedridden and not contact others. The efficiency of transmission would drop because rhinoviruses shed from the cold sufferer could not contact new hosts and would be inactivated by exposure. Cold sufferers must be able to move about and directly contact others. Thus virulence is low and people are not incapacitated by the common cold.

On the other hand, if a pathogen uses a mode of transmission not dependent on host health and mobility, then the person's health will not be a critical matter. The pathogen might be quite successful—that is, transmitted to many new hosts even though it kills its host relatively quickly. Host death means the end of any resident pathogens, but the species as a whole can spread and flourish as long as the increased transmission rate outbalances the loss due to host death. This situation may arise in several ways. In general, such pathogens are highly contagious. Thus transmission is a critical component of the infection process and is the next link in the infectious disease cycle, occurring by four main routes: airborne, contact, vehicle, and vector-borne (**figure 31.4**).

Transmission alone is not sufficient for infection to occur. Rather, the pathogen must also make contact with the appropriate host tissue. For instance, rhinoviruses are spread by airborne transmission from one host to another. Once the virus makes contact with the upper respiratory tract epithelia, the virus invades to cause disease only in the upper respiratory tract; rhinoviruses do not usually infect anywhere else in the host. This specificity is called a **tropism** (Greek *trope,* turning). Many pathogens exhibit cell, tissue, and organ specificities. A tropism by a specific microbe usually reflects the presence of specific cell surface receptors on the host cell for that microbe.

Airborne Transmission

Because air is not a suitable medium for the growth of pathogens, any pathogen that is airborne must have originated from a source such as humans, other animals, plants, soil, food, or water. In **airborne transmission,** the pathogen is suspended in the air in either droplets or dust, which travel over a meter or more from the source to the host. Typically, this results from host-to-host interaction (coughing, sneezing).

Droplet nuclei can be small particles, 1 to 4 μm in diameter, that result from the evaporation of larger particles (5 μm or more in diameter) called droplets. The route is through the air for a very short distance. As a result, droplet transmission of a pathogen depends on the proximity of the source and the host. In contrast, the much smaller droplet nuclei can remain airborne for hours or days and travel long distances. Contact with oral secretions may result when droplet nuclei contaminate body surfaces that touch mucous membranes (e.g., respiratory secretions on hands that contact eyes). Chicken pox and measles are examples of droplet-spread diseases. Many systemic mycoses are examples of transmission in dust.

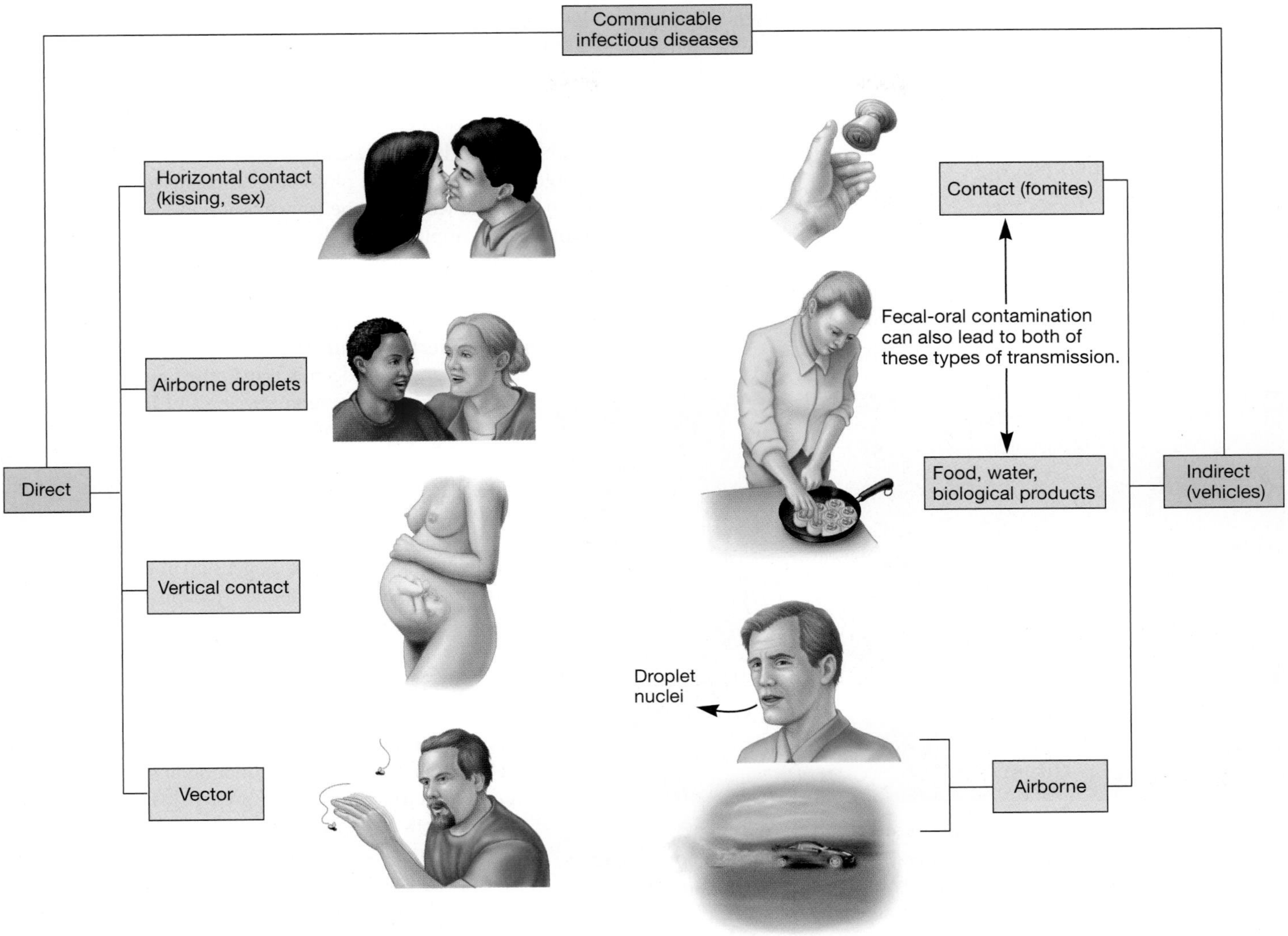

FIGURE 31.4 Transmission of Infectious Disease. Infectious diseases can be transmitted by various direct and indirect methods.

When animals or humans are the source of the airborne pathogen, it usually is propelled from the respiratory tract into the air by coughing, sneezing, or vocalization. For example, enormous numbers of moisture droplets are aerosolized during a typical sneeze (**figure 31.5**). Each droplet is about 10 μm in diameter and initially moves about 100 m/sec, or more than 200 mi/hr!

Dust also is an important route of airborne transmission. At times a pathogen adheres to dust particles and contributes to the number of airborne pathogens when the dust is resuspended by some disturbance. A pathogen that can survive for relatively long periods in or on dust creates an epidemiological problem, particularly in hospitals, where dust can be the source of hospital-acquired (nosocomial) infections.

Contact Transmission

Contact transmission implies the coming together or touching of the source or reservoir of the pathogen and the host. Contact can be direct or indirect. Direct contact implies an actual physical interaction with the infectious source (figure 31.4). This route is frequently called person-to-person contact. Person-to-person transmission occurs primarily by touching, kissing, or sexual contact; by contact with oral secretions or body lesions (e.g., herpes and boils); by nursing mothers (e.g., staphylococcal infections); and through the placenta (e.g., AIDS, syphilis). Some infectious pathogens also can be transmitted by direct contact with animals or animal products (e.g., *Salmonella* and *Campylobacter*).

Vehicle Transmission

Inanimate materials involved in pathogen transmission are called **vehicles** (**figure 31.6*a***). In vehicle transmission, a single inanimate vehicle or source serves to spread the pathogen to multiple hosts. Examples include surgical instruments, drinking vessels, stethoscopes, bedding, eating utensils, and neckties. These common vehicles are called fomites (s., fomes or fomite). A single source containing pathogens (e.g., blood, drugs, IV fluids) can contaminate a common vehicle that causes multiple infections. Food and water are important common vehicles for many human diseases. They often support pathogen reproduction.

Vector-Borne Transmission

Living transmitters of a pathogen are called vectors. Most vectors are arthropods (e.g., insects, ticks, mites, fleas) or vertebrates (e.g., dogs, cats, skunks, bats). **Vector-borne transmission** can be either external or internal. In external (mechanical) transmission, the pathogen is carried on the body surface of a vector. Carriage is passive, with no growth of the pathogen during transmission. An example would be flies carrying *Shigella* organisms on their feet from a fecal source to a plate of food that a person is eating.

By contrast, the pathogen is carried within the vector during internal transmission, which can be either harborage or biologic transmission. In harborage transmission, the pathogen does not undergo morphological or physiological changes within the vector. An example would be the transmission of *Yersinia pestis* (the etiologic agent of plague) by the rat flea from rat to human. Biologic transmission implies that the pathogen undergoes a morphological or physiological change within the vector. An example would be the developmental sequence of the malarial parasite inside its mosquito vector (figure 31.6*b*).

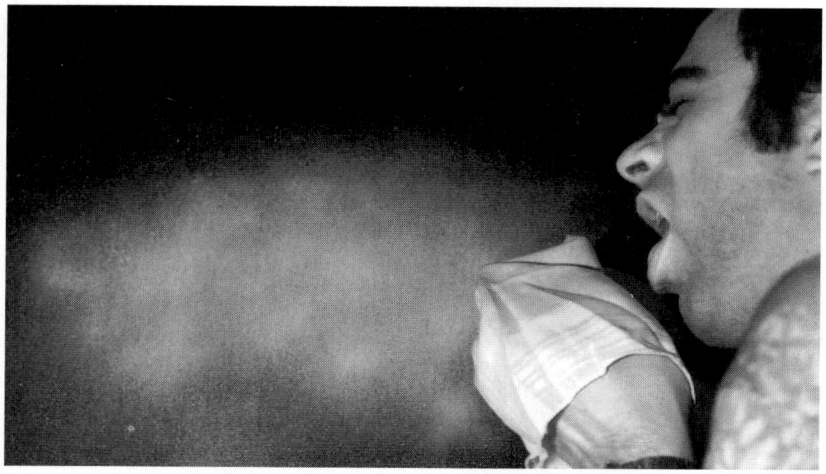

FIGURE 31.5 A Sneeze. High-speed photograph of an aerosol generated by an unstifled sneeze. The expelled particles are comprised of saliva and mucus laden with microorganisms. These airborne particles may be infectious when inhaled by a susceptible host. Even a surgical mask will not prevent the spread of all particles.

When a pathogen is transmitted by a vector, it will benefit by extensive reproduction and spread within the host. If pathogen levels are very high in the host, a vector such as a biting insect has a better chance of picking up the pathogen and transferring it to a new host. Indeed, pathogens transmitted by biting arthropods often are very virulent (e.g., malaria, typhus, sleeping sickness). It is important that such pathogens do not harm their vectors and the vector generally remains healthy, at least long enough for pathogen transmission.

Infectious Dose

As we mentioned, the rate at which an infection proceeds and its severity are a function of the initial inoculum of microorganisms and their virulence (in addition to the host's resistance). Some microorganisms are sufficiently adept at entering a host that very few are required to establish an infection. Conversely, other microorganisms are required in high numbers to establish an infection. The inocula can be measured experimentally to determine the **infectious dose 50 (ID_{50})**, that is, the number of microorganisms required to cause

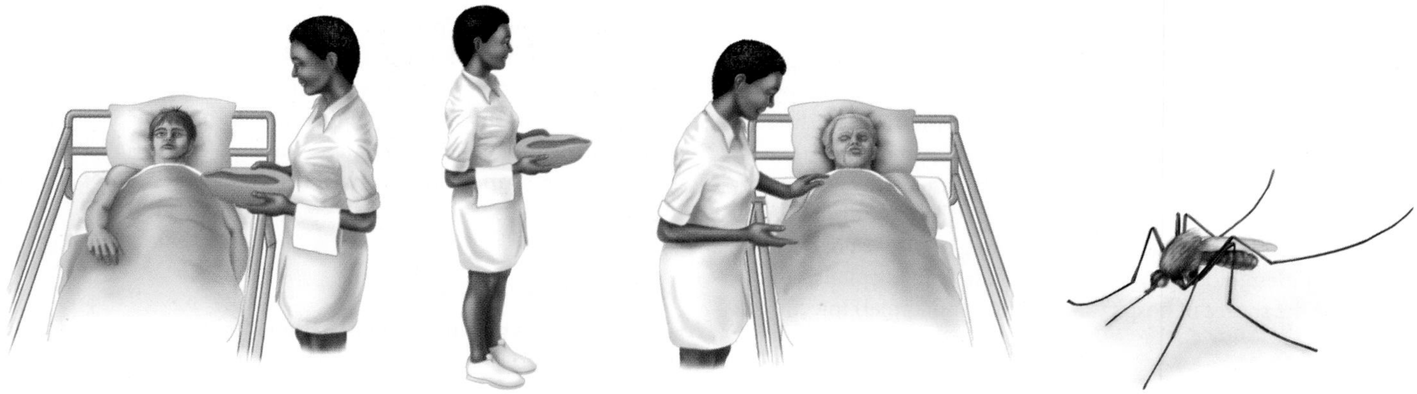

(a) (b)

FIGURE 31.6 Infectious Disease Transmission. (a) Inanimate vehicles transmit pathogens between people. (b) Living vectors, such as this mosquito, also transmit pathogens.

clinical disease in 50% of the inoculated hosts (**figure 31.7**). One example of this is found in the oral transmission of *Salmonella*. *Salmonella* are shed in the feces of infected hosts and often contaminate fingers and hands. During food preparation, the bacteria are transferred to foods, where they can then be ingested. The ID_{50} for *Salmonella* ingested in contaminated food is approximately 1×10^5 bacteria. Suffice it to say, the reason for thorough hand washing as a means of maintaining public health is to decrease the number of pathogens that can be transmitted. For the most part, the lower the microbial dose, the lower the risk of infection.

Growth Rate

For a bacterial pathogen to be successful in growth and reproduction (colonization), it must find an appropriate environment (e.g., nutrients, pH, temperature, redox potential) within the host. Those areas of the host's body that provide the most favorable conditions will harbor the pathogen and allow it to grow and multiply to produce an infection. Infectious microorganisms are considered extracellular pathogens if, during the course of disease, they remain in tissues and fluids but never enter host cells. For instance, some bacteria and fungi can actively grow and multiply in the blood or tissue spaces. *Y. pestis*, and *Aspergillus* (the cause of a variety of fungal infections) are good examples of extremely virulent extracellular pathogens.

Some microbes are able to grow and multiply within various cells of a host and are called **intracellular pathogens;** they can be further subdivided into two groups. **Facultative intracellular pathogens** are those organisms that can reside within the cells of the host or in the environment. A bacterial example is *Brucella abortus*, which is capable of growth and replication within macrophages, neutrophils, and trophoblast cells (cells that surround the developing embryo). A protozoal example is *Histoplasma capsulatum*, which grows within phagocytes. However, facultative intracellular pathogens can also be grown in pure culture without host cell support. In contrast, **obligate intracellular pathogens** are incapable of growth and multiplication outside a host cell. By definition, all viruses are obligate intracellular pathogens in that they require a host cell for multiplication to the detriment of that cell. Influenza and HIV are two examples of pathogens in this group. Examples of obligate intracellular bacterial pathogens include *Chlamydia* and the rickettsia. These microbes cannot be grown in the laboratory outside of their host cells. Malarial parasites are examples of protozoa that require host cells for growth and differentiation.

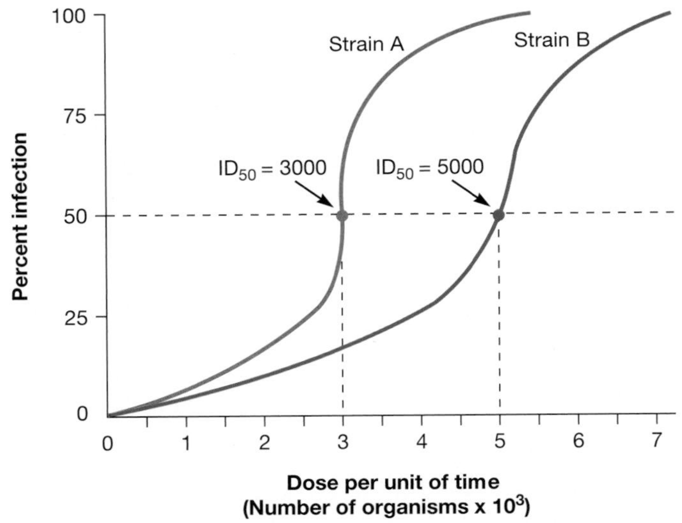

FIGURE 31.7 Determination of the ID_{50} of a Pathogenic Microorganism. Various doses of a specific pathogen are introduced into experimental host animals. Infections are recorded and a graph constructed. In this example, the graph represents the susceptibility of host animals to two different strains of a pathogen—strain A and strain B. For strain A, the ID_{50} is 3000, and for strain B, it is 5000. Hence strain A is more virulent than strain B.

Host Susceptibility

The next link in the infectious disease cycle is the host (figure 31.3). The susceptibility of the host to a pathogen depends on both the pathogenicity of the organism and the nonspecific and specific defense mechanisms of the host. Host immunity is the subject of chapters 32 and 33 and will not be discussed in this chapter. In addition to host defense mechanisms, nutrition, genetic predisposition, cleanliness, and stress also influence host susceptibility to infection.

Exiting the Host

The last link in the infectious disease cycle is release or exit of the pathogen from the host (figure 31.3). From the point of view of the pathogen, successful escape from the host is just as important as its initial entry. Unless a successful escape occurs, the disease cycle will be interrupted and the pathogen will not be perpetuated. Escape can be active or passive, although sometimes a combination of the two occurs. Active escape takes place when a pathogen actively moves to a portal of exit and leaves the host. Examples include parasitic helminths that migrate through the body of their host, eventually reaching the surface and exiting. Passive escape occurs when a pathogen or its progeny leaves the host in feces, urine, droplets, saliva, or desquamated cells. Microorganisms usually employ passive escape mechanisms. Mechanisms of bacterial and viral spread vary, but the most common routes are the bloodstream and the lymphatic system. The presence of viruses in the blood is called **viremia.** In some instances, spread is by way of nerves (e.g., rabies, herpes simplex, and varicella-zoster viruses).

1. Describe the four main types of infectious disease transmission and give examples of each.
2. Define droplet nuclei, vehicle, fomite, and vector.
3. Why are pathogens that are transmitted person-to-person often less virulent than those transmitted by arthropod vectors?

31.3 Virulence

As mentioned in section 31.1, pathogenicity is a general term that refers to an organism's potential to cause disease. Virulence refers more specifically to the magnitude of harm (pathogenicity) caused by a particular microorganism. Various physical and chemical characteristics (such as structures that facilitate attachment and molecules that bypass host defenses) contribute to pathogenicity. These intrinsic characteristics are called **virulence factors.** Virulence is determined by the degree to which the pathogen causes damage, including invasiveness and infectivity. Virulence also is often directly correlated with a pathogen's ability to survive in the external environment. If a pathogen cannot survive well outside its host and does not use a vector, it depends on host survival and will tend to be less virulent. When a pathogen can survive for long periods outside its host, it can afford to leave the host and simply wait for a new one to come along. This seems to promote increased virulence. Host health is not critical, but extensive multiplication within the host will increase the efficiency of transmission. Good examples are tuberculosis, diphtheria, and measles. *Mycobacterium tuberculosis, Corynebacterium diphtheriae,* and *Measles virus* survive for a relatively long time, from hours to days and weeks, outside human hosts.

Human cultural patterns and behavior affect pathogen virulence. Waterborne pathogens such as *Vibrio cholerae* and *Poliovirus* are transmitted through drinking water systems. They can be virulent because immobile hosts still shed pathogens, which frequently reach the water. This is why the establishment of an uncontaminated drinking water supply is critical in limiting a cholera or polio outbreak. The same appears to be true of *Shigella* and dysentery. Often one of the best ways to reduce virulence may be to reduce the frequency of transmission. ▶▶| *Water purification and sanitary analysis (section 42.1)*

Search This: Virulence database

Pathogenicity Islands

The horizontal transfer of virulence genes to bacteria appears to have a rich evolutionary history, as made clear by the presence of genes that encode major virulence factors on large segments of bacterial chromosomal and plasmid DNA called **pathogenicity islands.** Many bacteria (e.g., *Yersinia* spp., *Pseudomonas aeruginosa, Shigella flexneri, Salmonella,* enteropathogenic *Escherichia coli*) carry at least one pathogenicity island. Pathogenicity islands generally increase bacterial virulence and are absent in nonpathogenic members of the same genus or species. Pathogenicity islands can be recognized by several common sequence characteristics: (1) The 3′ and 5′ ends of the islands contain insertion-like elements, suggesting their promiscuity as mobile genetic elements. (2) The G + C nucleotide content of pathogenicity islands differs significantly from that of the remaining bacterial genome. (3) The pathogenicity island DNA also exhibits several open reading frames, suggesting other putative genes. Interestingly, pathogenicity islands are typically associated with genes that encode tRNA. An excellent example of virulence genes carried on pathogenicity islands is seen in protein secretion systems (*see figure 38.10*).

Virulence Factors

In most cases, a single virulence factor may be necessary but by itself is not sufficient for pathogenicity. The role specific virulence factors play in establishing and maintaining infection is an intense area of research. Often, mutant strains of a pathogen lacking the capacity to produce the virulence factor in question are constructed. Its ability to cause disease is then evaluated in an animal. The use of animals that mimic the disease as it is manifested in humans is called an **animal model system.** For instance, guinea pigs are commonly used to study tuberculosis, whereas a mouse (or murine) system is used to study cholera.

Table 31.2 Adherence Properties of Microbes

Microbe	*Disease*	*Adhesion Mechanism*
Neisseria gonorrhoeae	Gonorrhea	Fimbriae attach to genital epithelium
Escherichia coli	Diarrhea	Well-developed fimbrial adhesin
Shigella	Dysentery	Fimbriae can attach to intestinal epithelium
Vibrio	Cholera	Glycocalyx anchors microbe to intestinal epithelium
Treponema	Syphilis	Tapered hook embeds in host cell
Mycoplasma	Pneumonia	Specialized tip at ends of bacteria fuse tightly to lung epithelium
Pseudomonas aeruginosa	Burn, lung infections	Fimbriae and slime layer
Streptococcus pyogenes	Pharyngitis, impetigo	Lipotechoic acid and capsule anchor cocci to epithelium
Streptococcus mutants, S. sobrinus	Dental caries	Dextran slime layer glues cocci to tooth surface
Influenza virus	Influenza	Viral spikes react with receptor on cell surface
Poliovirus	Polio	Capsid proteins attach to receptors on susceptible cells
HIV	AIDS	Viral spikes adhere to white blood cell receptors
Giardia lamblia (protozoan)	Giardiasis	Small suction disc on underside attaches to intestinal surface

Adherence and Colonization

The first step in the infectious disease process is the entrance and attachment of the microorganism to a susceptible host. Entrance may be accomplished through one of the body surfaces: skin, respiratory system, gastrointestinal system, urogenital system, or the conjunctiva of the eye. Some pathogens enter the host by sexual contact, needle sticks, blood transfusions and organ transplants, or insect vectors. Bacteria, fungi, and protozoa require a portal of entry into the host so that they have access to nutrients. Recall that viruses are obligate intracellular parasites and must enter the host's cells to replicate. Some bacteria enter the host's cell (i.e., intracellular pathogens), but most remain in the tissue spaces around the cells (extracellular pathogens).

After being transmitted to an appropriate host, the pathogen must be able to adhere to and colonize host cells or tissues. In this context, colonization means the establishment of a site of microbial replication on or within a host. It does not necessarily result in tissue invasion or damage. Colonization depends on the ability of the pathogen to survive in the new (host) environment and to compete successfully with host cells and the host's normal microbiota for essential nutrients. Specialized structures that allow microorganisms to compete for surface attachment sites also are necessary for colonization.

Pathogens adhere with a high degree of specificity to particular tissues. Adherence structures such as pili and fimbriae (**table 31.2**), membrane and capsular materials, and specialized adhesion molecules on the invading cell surface bind to host sites especially if they have complementary receptor sites on the cell surface (**figure 31.8**), facilitating attachment to host cells. These microbial products and structural components contribute to infectivity, thus they are one type of virulence factor. Specific examples of pathogen adherence are discussed in the context of the disease that is established (chapters 37–39).

Invasion

Pathogens can be described in terms of their infectivity and invasiveness. **Infectivity** is the ability of the organism to establish a discrete, focal point of infection. **Invasiveness** is the ability of the organism to spread to adjacent or other tissues. For some pathogens, a localized infection is sufficient to cause disease. However, most pathogens invade other tissues. Entry into tissues is a specialized strategy used by many bacterial and fungal pathogens for survival and multiplication.

Pathogens can either actively or passively penetrate the host's mucous membranes and epithelium after attachment to the epithelial surface. Active penetration may be accomplished through production of lytic substances that alter the host tissue by (1) attacking the extracellular matrix and basement membranes of integuments and intestinal linings, (2) degrading carbohydrate-protein complexes between cells or on the cell surface (the glycocalyx), or (3) disrupting the host cell surface. Passive mechanisms of penetration are not related to the pathogen itself. Examples include (1) small breaks, lesions, or ulcers in a mucous membrane that permit initial entry; (2) wounds, abrasions, or burns on the skin's surface; (3) arthropod vectors that create small wounds while feeding; and (4) tissue damage caused by other organisms (e.g., a dog bite).

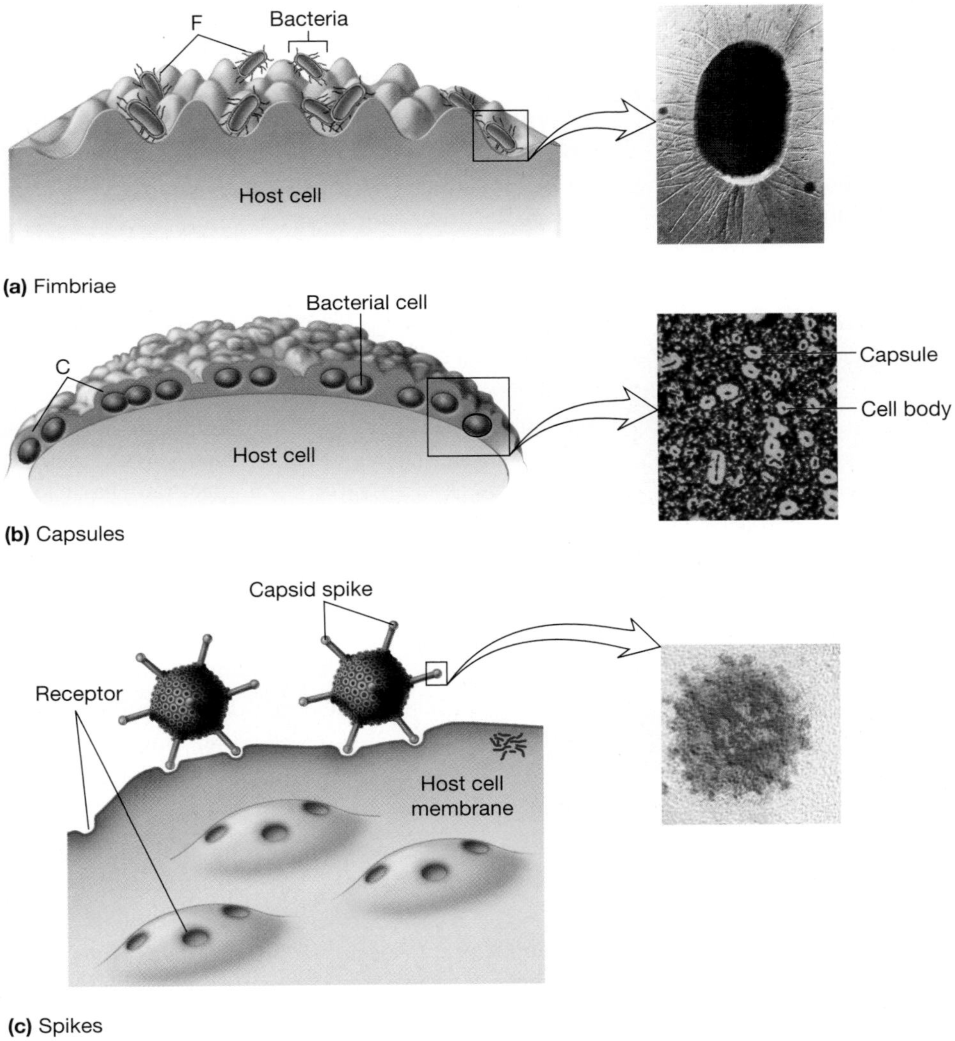

FIGURE 31.8 Microbial Adherence Mechanisms. (a) Bacterial fimbriae. (b) Bacterial capsule. (c) Viral protein spikes.

Once under the mucous membrane, a pathogen may penetrate to deeper tissues and continue disseminating throughout the body of the host. One way the pathogen accomplishes this is by producing specific structures or enzymes that promote spreading (**table 31.3**). These products represent other types of virulence factors. Bacteria may also enter the small terminal lymphatic capillaries that surround epithelial cells. These capillaries merge into large lymphatic vessels that eventually drain into the circulatory system. Once the circulatory system is reached, the bacteria have access to all organs and systems of the host. The presence of viable bacteria in the bloodstream is called **bacteremia.** The infectious disease process caused by bacterial or fungal toxins in the blood is termed **septicemia** (Greek *septikos,* produced by putrefaction, and *haima,* blood).

Invasiveness varies greatly among pathogens. For example, *Clostridium tetani* (the cause of tetanus) produces a variety of virulence factors (e.g., toxins and proteolytic enzymes) but is considered noninvasive because it does not spread from one tissue to another. *Bacillus anthracis* (the cause of anthrax) and *Y. pestis* (the cause of plague) also produce substantial virulence factors (e.g., capsules and toxins) but are also highly invasive. Members of the genus *Streptococcus* span the spectrum of virulence factors and invasiveness. *Microsporidia* use a novel polar tubule to bore into host cells (*see figure 24.15*), whereas many other fungi use hydrolytic enzymes to invade cells and tissues. These virulence factors are discussed in more detail in chapters 38 and 39.

Resisting Host Defenses

Most microorganisms do not cause disease, in part because they are eliminated before infection begins by the host's normal flora or its immune system. However, pathogens have evolved many

Table 31.3 Microbial Virulence Factors Involved in Bacterial Pathogen Invasion and Dissemination

Product	*Organism Involved*	*Mechanism of Action*
Coagulase	*Staphylococcus aureus*	Coagulates (clots) the fibrinogen in plasma. The clot protects the pathogen from phagocytosis and isolates it from other host defenses.
Collagenase	*Clostridium* spp.	Breaks down collagen that forms the framework of connective tissues; allows the pathogen to spread
Deoxyribonuclease (along with calcium and magnesium)	Group A streptococci, staphylococci, *Clostridium perfringens*	Lowers viscosity of exudates, giving the pathogen more mobility
Elastase and alkaline protease	*Pseudomonas aeruginosa*	Cleaves laminin associated with basement membranes
Hemolysins	Staphylococci, streptococci, *Escherichia coli, Clostridium perfringens*	Lyse erythrocytes; make iron available for microbial growth
Hyaluronidase	Groups A, B, C, and G streptococci, staphylococci, clostridia	Hydrolyzes hyaluronic acid, a constituent of the extracellular matrix that cements cells together and renders the intercellular spaces amenable to passage by the pathogen
Hydrogen peroxide (H_2O_2) and ammonia (NH_3)	*Mycoplasma* spp., *Ureaplasma* spp.	Are produced as metabolic wastes. These are toxic and damage epithelia in respiratory and urogenital systems.
Immunoglobulin A protease	*Streptococcus pneumoniae*	Cleaves immunoglobulin A into Fab and Fc fragments
Lecithinase or phospholipase	*Clostridium* spp.	Destroys the lecithin (phosphatidylcholine) component of plasma membranes, allowing pathogen to spread
Leukocidins	Staphylococci, pneumococci, streptococci	Pore-forming exotoxins that kill leukocytes; cause degranulation of lysosomes within leukocytes, which decreases host resistance
Porins	*Salmonella enterica* serovar Typhimurium	Inhibit leukocyte phagocytosis by activating the adenylate cyclase system
Protein A Protein G	*Staphylococcus aureus* *Streptococcus pyogenes*	Located on cell wall. Immunoglobulin G (IgG) binds to protein A by its Fc end, thereby preventing complement from interacting with bound IgG.
Pyrogenic exotoxin B (cysteine protease)	Group A streptococci, *(Streptococcus pyogenes)*	Degrades proteins
Streptokinase (fibrinolysin, staphylokinase)	Group A, C, and G streptococci, staphylococci	A protein that binds to plasminogen and activates the production of plasmin, thus digesting fibrin clots; this allows the pathogen to move from the clotted area.

mechanisms to evade host defenses. Although host innate defense mechanisms come into play upon initial host exposure, specific immune responses normally become paramount in preventing further proliferation of infectious agents. Thus successful pathogens have evolved mechanisms that enable them to elude both the initial host responses as well as those governed by the adaptive immune system. The evasion of host defenses begins when the pathogen first infects the host. However, for the pathogen to cause a successful infection (from the microbe's point of view), it must be able to avoid host immunity so it can increase to a sufficient number; this is when disease ensues.

Some viruses, most notably HIV, infect cells of the immune system and diminish their function. HIV as well as *Measles virus* and cytomegalovirus cause fusion of host cells. This allows these viruses to move from an infected cell to an uninfected cell without exposure to the antimicrobial fluids of the host. Some bacteria such as *Streptococcus pneumoniae, Neisseria meningitidis,* and *Haemophilus influenzae* can produce a slippery mucoid capsule that prevents the host immune cells from effectively capturing the bacterium. Other bacteria evade phagocytosis by producing specialized surface proteins such as the M protein on *S. pyogenes* *(see figure 38.5).*

Many pathogens have devised methods to prevent detection by antimicrobial proteins that either directly or indirectly cause membrane (envelope) lysis. *Hepatitis B virus*–infected cells produce large amounts of proteins not associated with the complete virus. These proteins serve as decoys, binding the available antimicrobial proteins so that virus goes unnoticed in the midst of all the viral protein. To evade antimicrobial activity, some bacteria have capsules (**chapter opener**) that prevent binding of bacteria to host immune cells. Some gram-negative bacteria can lengthen the O side chains in their lipopolysaccharide to prevent host detection. Others such as *Neisseria gonorrhoeae* have modified the lipo-oligosaccharides on their surface that interfere with host mechanisms that normally lyse *Neisseria.*

Some bacteria have evolved the ability to survive inside the host cells that otherwise would destroy bacteria. These microorganisms are very pathogenic because they are impervious to an important innate resistance mechanism of the host (*see p. 783*). One method of evasion is to escape by ejecting themselves from the host cell, as seen with *Listeria monocytogenes, Shigella,* and *Rickettsia.* These bacteria use actin-based motility to move within mammalian host cells and spread between them. They activate the assembly of an actin tail using host cell actin and other cytoskeletal proteins (**figure 31.9*a***). The actin tails propel the bacteria through the cytoplasm of the infected cell to its surface, where they push out against the plasma membrane and form protrusions (figure 31.9*b*). The protrusions are engulfed by adjacent cells, and the bacteria enter the unsuspecting cell neighbor. In this way, the infection spreads to adjacent cells.

Pathogens have evolved a variety of ways to suppress or evade the host's immune response. For instance, some change their surface proteins, by mutation or recombination (e.g., the influenza virus), or downregulate the level of expression of cell surface proteins. To further evade the specific immune response, some bacteria (e.g., *S. pyogenes*) produce capsules that resemble host tissue components and thus remain covert. *N. gonorrhoeae* can evade the specific immune response by two mechanisms: (1) it makes genetic variations in its pili through a process called phase variation, resulting in altered pilin protein sequence and expression.

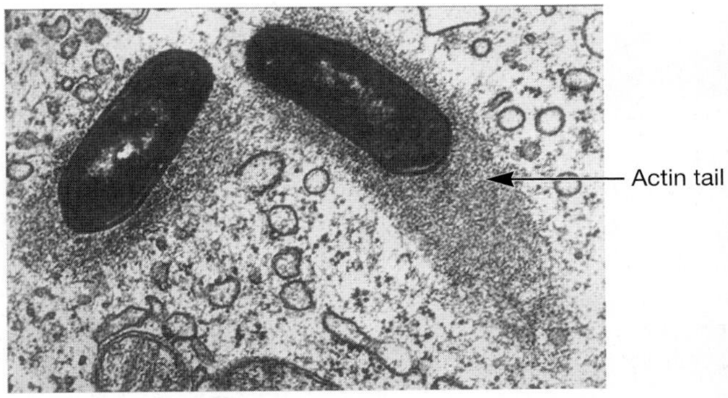

(a)

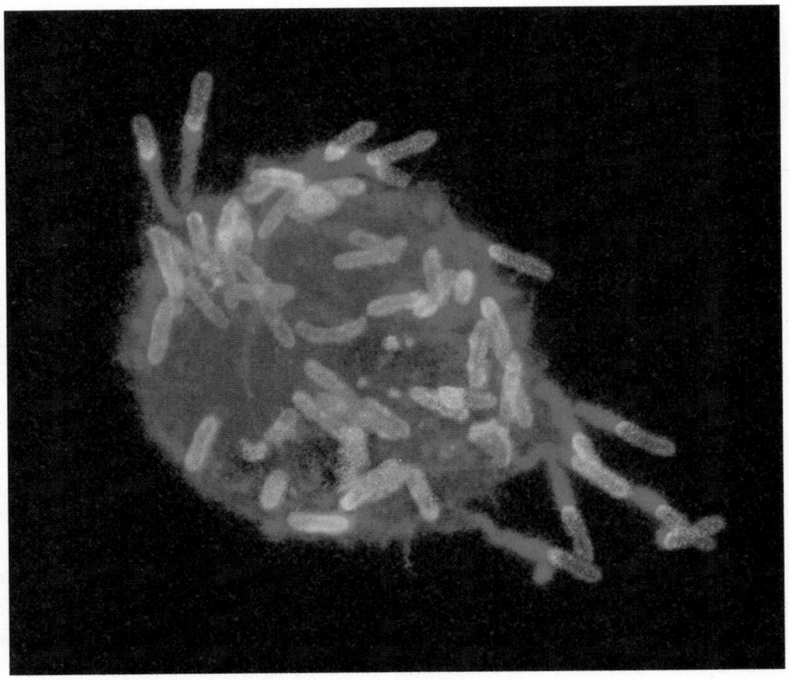

(b)

FIGURE 31.9 Formation of Actin Tails by Intracellular Bacterial Pathogens. (a) Transmission electron micrograph of *Listeria monocytogenes* in a host macrophage. The bacterium has polymerized host actin into a long tail at only one pole that it uses for intracellular propulsion and to move from one host cell to another. (b) *Burkholderia pseudomallei* (stained red) also forms actin tails, as shown in this confocal micrograph. Note that the actin tails enable the bacterial cells to be propelled out of the host cell.

This renders specific host responses useless against the new pili, and adherence to host tissue occurs; and (2) it produces proteases that degrade host proteins that bind the bacteria. This allows adherence. Finally, some bacteria produce special proteins (such as staphylococcal protein A and protein G of *S. pyogenes*) that interfere with the host's ability to detect and remove them.

1. Describe several specific adhesins by which bacterial pathogens attach to host cells.
2. What are virulence factors?
3. If you were annotating the genomic sequence of a bacterium, how would you recognize a pathogenicity island?

Safety in Numbers

One of the long-standing tenants of microbiology is the pure culture. Dating to the early 1880s, Robert Koch set a paradigm that insisted bacteria be studied in pure culture to evaluate their virulence and, by extension, their ability to be controlled by antimicrobial agents. Today the paradigm has been rephrased to mean that pure culture suspensions of individual and unattached (planktonic) cells are to be examined for virulence and susceptibility to antibiotics. This has guided countless experiments in the evaluation of environmentally and medically derived bacteria. However, modern techniques evaluating bacteria in their natural settings suggest that many types of bacteria are only planktonic when dispersing from more stable and heterogeneous communities known as biofilms (*see figures 7.33 and 7.34*).

Within their natural environment, complex (biofilm) communities likely form to provide a means of nutrient acquisition and safety from predators. The role of the biofilm now seems much more complex. This is especially true for biofilms that cause chronic infections in humans. Using high throughput sequencing, fluorescent in-situ hybridization, and global gene analysis, studies have revealed that some pathogenic bacteria in biofilms are physiologically different from their planktonic cousins; bacteria within biofilms exchange plasmids, nutrients, and quorum-sensing molecules so as to behave differently than planktonic forms. The differences are striking in that the bacteria within biofilms coordinate gene expression to upregulate mechanisms that make the biofilm community less sensitive to antibiotics and more resistant to host defense mechanisms. For example, studies on *Pseudomonas aeruginosa* biofilms reveal that shared communication signals activate global response regulators, which increase the activity of multidrug efflux pumps, increase starvation or stress responses, stimulate exopolysaccharide secretion, increase resistance to host cell attack, suppress host cell killing mechanisms, and change rapidly growing, antibiotic-sensitive cells into slower-growing, antibiotic-resistant persister cells.

Importantly, one remarkable activity of biofilms made of pathogenic *P. aeruginosa* is the ability to suppress host immune responses (**figure 31.10**). This is evident in studies evaluating

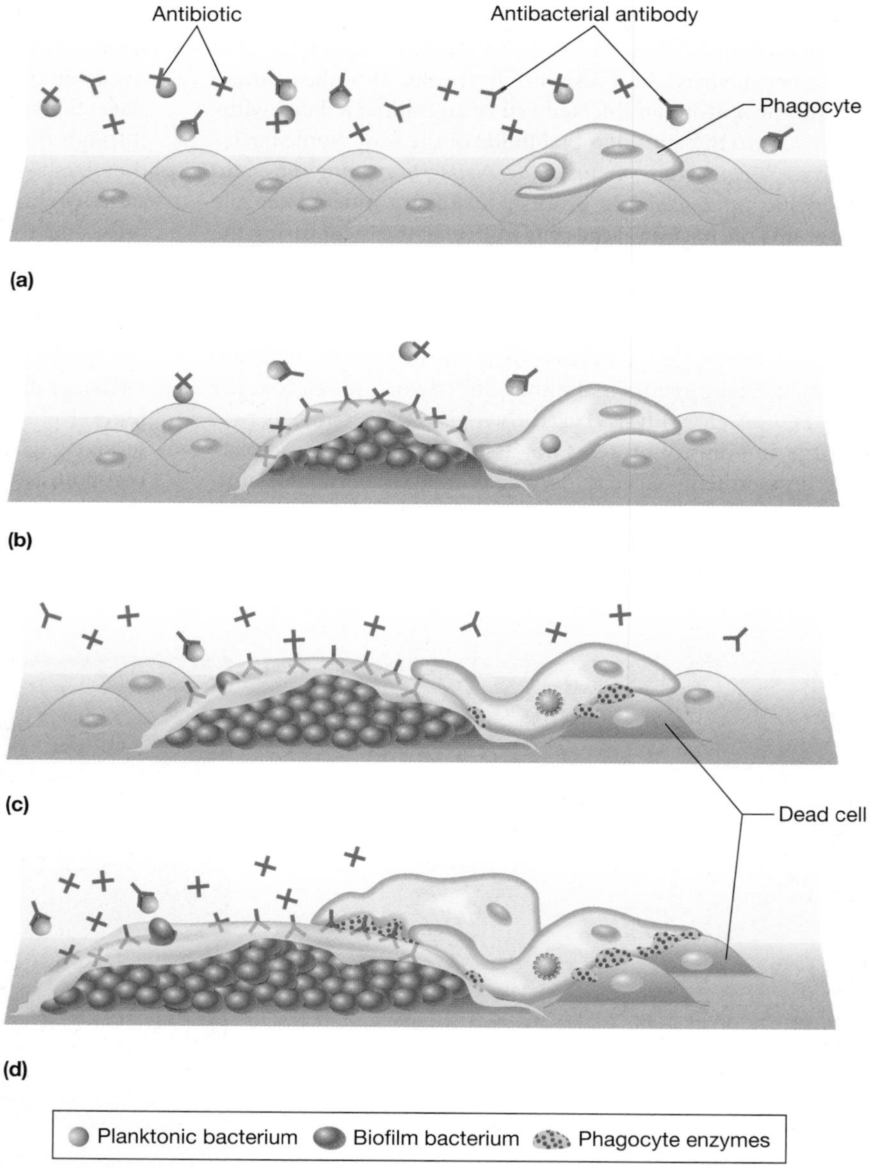

FIGURE 31.10 Frustrated Phagocytosis. (a) Planktonic cells are sensitive to antibiotics and are detected by host phagocytes and antibodies. (b) Planktonic cells settle to form a biofilm and become biofilm cells. (c) Biofilm cells resist antibiotics and antibody detection. Host phagocytes detect and attempt to destroy biofilm cells. (d) Unable to capture biofilm cells, phagocytes release antimicrobial products that kill host cells but not biofilm cells.

P. aeruginosa mixed with antibiotics, antibacterial proteins known as antibodies (*see figure 31.10*), and phagocytic white blood cells known as macrophages and neutrophils. Antibiotics, antibodies, and phagocytes readily attack and kill planktonic *P. aeruginosa* using hydrolytic enzymes (lysozyme, lactoferrin, etc.) and free radicals of oxygen and nitrogen. However, antibiotics, antibodies, and phagocytes are ineffective at killing *P. aeruginosa* in biofilms. *P. aeruginosa* in biofilms appears to prevent penetration of antibiotics and antibodies, and to cause phagocytes to downregulate killing mechanisms. In addition, hydrolytic enzymes are decreased, while enzymes that create super- and nitric oxides are inhibited. In essence, the *P. aeruginosa* biofilms "frustrate" phagocytes by reducing the magnitude of some antibacterial responses and preventing others altogether. Thus it appears that another bacterial virulence factor is the ability to form biofilms.

Toxins

Another major aspect of pathogenic potential is toxigenicity. A toxin (Latin *toxicum,* poison) is a substance, such as a metabolic product, that alters the normal metabolism of host cells with deleterious effects on the host. The term toxemia refers to the condition caused by toxins that have entered the blood of the host. Some toxins are so potent, such as the botulinum toxin of *Clostridium botulinum*, that even if the pathogens that produced them are eliminated, the disease conditions persist. **Toxigenicity** is the pathogen's ability to produce toxins. **Intoxications** are diseases that result from a specific toxin (e.g., botulinum toxin, aflatoxin) produced by the pathogen. Intoxication diseases do not require active infections, as in the case of botulism, where the premade toxin contaminates poorly canned foods. Intoxication diseases can, and often do, start from an active infection, as seen in tetanus and diphtheria. Host damage in an infection results primarily from the pathogen's growth and replication (or invasiveness). Host damage due to intoxication is a function of the specific mechanism of action of the toxin (e.g., inhibition of protein synthesis by ADP-ribosylation of eukaryotic elongation factor-2). Toxins produced by bacteria can be divided into two main categories: exotoxins and endotoxins. The primary characteristics of the two groups are compared in **table 31.4**.

Search This: Agency for Toxic Substances and Disease Registry

Exotoxins

Exotoxins are soluble, heat-labile proteins (inactivated at 60 to 80°C) that usually are released into the surroundings as the bacterial pathogen grows. Most exotoxins are produced by gram-positive bacteria, although some gram-negative bacteria also make exotoxins. Often exotoxins travel from the site of infection to other body tissues or target cells, where they exert their effects.

Exotoxins are usually synthesized by specific bacteria that have plasmids or prophages bearing the toxin genes. They are associated with specific diseases and often are named for the disease they produce (e.g., the diphtheria toxin). Exotoxins are among the most lethal substances known; they are toxic in nanogram-per-kilogram concentrations (e.g., botulinum toxin).

Table 31.4 Characteristics of Exotoxins and Endotoxins

Characteristic	*Exotoxins*	*Endotoxins*
Chemical composition	Protein, often with two components (A and B)	Lipopolysaccharide complex on outer membrane; lipid A portion is toxic
Disease examples	Botulism, diphtheria, tetanus	Gram-negative infections, meningococcemia
Effect on host	Highly variable between different toxins	Similar for all endotoxins
Fever	Usually do not produce fever	Produce fever by induction of interleukin-1 and TNF
Genetics	Frequently carried by plasmids	Synthesized directly by chromosomal genes
Heat stability	Most are heat sensitive and inactivated at 60–80°C	Heat stable to 250°C
Immune response	Antitoxins provide host immunity; highly antigenic	Weakly immunogenic; immunogenicity associated with polysaccharide
Location	Usually excreted outside the bacterial cell	Part of outer membrane of gram-negative bacteria
Production	Produced by gram-positive and negative bacteria	Found only in gram-negative bacteria; released on bacterial death and some liberated during growth
Toxicity	Highly toxic and fatal in nanogram quantities	Less potent and less specific than exotoxin; cause septic shock
Toxoid production	Converted to antigenic, nontoxic toxoids used to immunize (e.g., tetanus toxoid)	Toxoids cannot be made.

Exotoxins exert their biological activity by specific mechanisms. As proteins, the toxins are easily recognized by the host immune system, which produces protein antitoxins—antibodies that recognize and bind to toxins. The toxins can also be inactivated by formaldehyde, iodine, and other chemicals to form immunogenic **toxoids** (e.g., tetanus toxoid). In fact, the tetanus vaccine is a solution of tetanus toxoid.

Exotoxins can be grouped into four types based on their structure and physiological activities. (1) One type is the **AB toxin,** which gets its name from the fact that the B portion of the toxin binds to a host cell receptor and is separate from the A portion, which enters the cell and has enzyme activity that causes the toxicity (**figure 31.11*a***). Thus the B subunit determines the cell type the toxin will affect, whereas the A subunits are enzymatically active. AB toxins act on cells by different mechanisms. Some of the mechanisms include protein synthesis inhibition, ribonuclease activity, signal transduction interference, and blocking neurotransmitters. One mechanism shared by a number of AB toxins is the addition of the small molecule ADP-ribose to the host target protein or structure. This is known as ADP-ribosylation. The general properties of some AB exotoxins are presented in **table 31.5.** *AB Exotoxins*

(2) A second type of exotoxin consists of those toxins that affect a specific host site (nervous tissue, **neurotoxins;** the intestines, **enterotoxins;** general tissues, **cytotoxins**). Some of the bacterial pathogens that produce these exotoxins are presented in table 31.5, including neurotoxins (botulinum toxin and tetanus toxin), enterotoxins (cholera toxin, *E. coli* heat-labile toxins), and cytotoxins (diphtheria toxin, *Shiga* toxin). Note that many AB toxins are also host-site specific, thus categories 1 and 2 are not mutually exclusive.

(3) The third type of exotoxin does not have separable A and B portions and acts by disorganizing host cell membranes. It lyses host cells by disrupting the integrity of the plasma membrane. There are two subtypes of membrane-disrupting exotoxins. The first is a protein that binds to the cholesterol portion

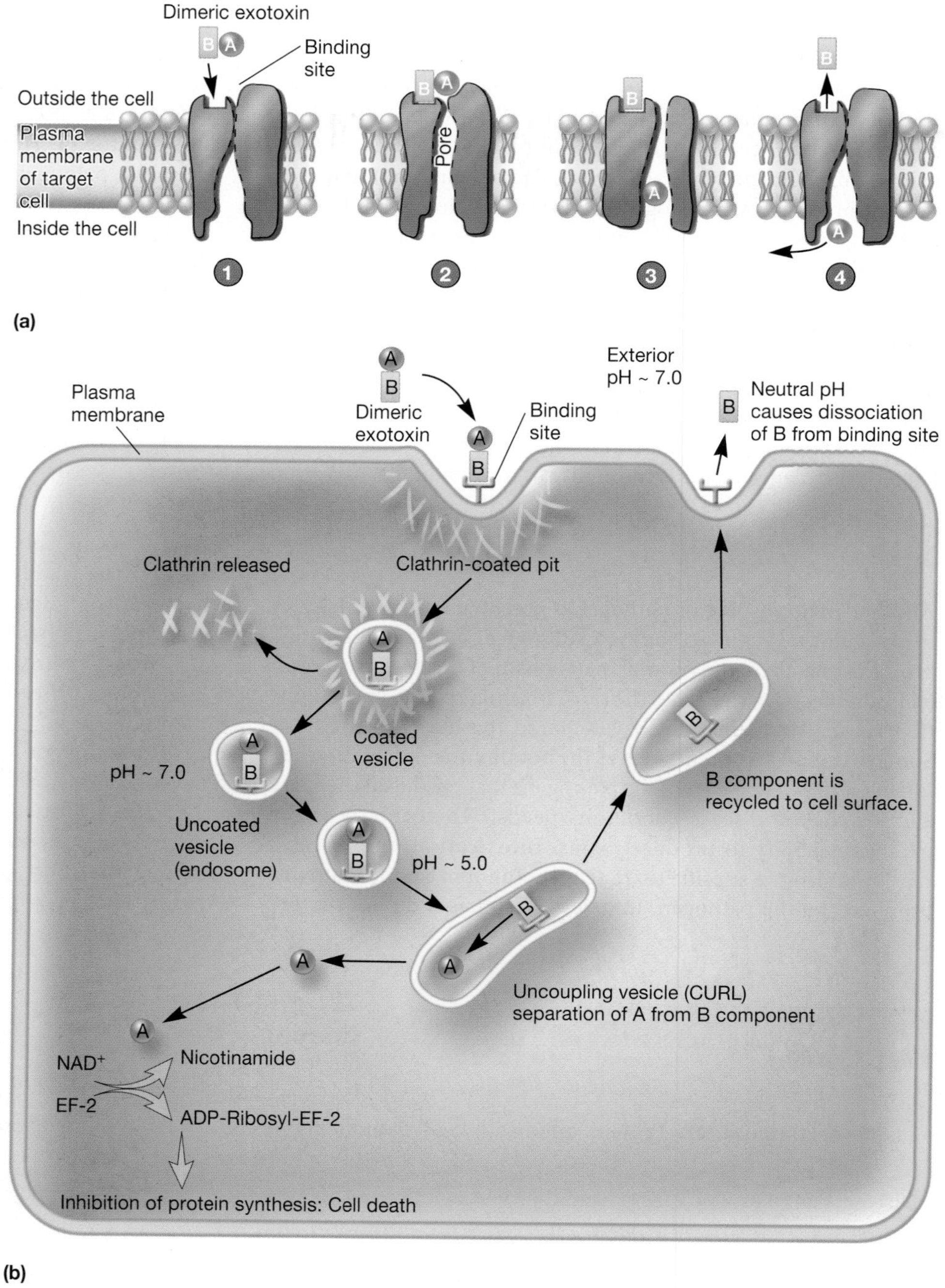

FIGURE 31.11 Two AB Exotoxin Transport Mechanisms. (a) Subunit B of the dimeric exotoxin (AB) binds to a specific membrane receptor of a target cell [1]. A conformational change [2] generates a pore [3] through which the A subunit crosses the membrane and enters the cytosol, followed by re-creation [4] of the binding site. (b) Receptor-mediated endocytosis of the diphtheria toxin involves the dimeric exotoxin binding to a receptor-ligand complex that is internalized in a clathrin-coated pit that pinches off to become a coated vesicle. The clathrin coat depolymerizes, resulting in an uncoated endosome vesicle. The pH in the endosome decreases due to H^+-ATPase activity. The low pH causes A and B components to separate. An endosome in which this separation occurs is sometimes called a CURL (*c*ompartment of *u*ncoupling of *r*eceptor and *l*igand). The B subunit is then recycled to the cell surface. The A subunit moves through the cytosol, catalyzes the ADP-ribosylation of EF-2 (elongation factor 2) and inhibits protein synthesis, leading to cell death.

Table 31.5 Properties of Some AB Model Bacterial Exotoxins

Toxin	Organism	Gene Location	Subunit Structure	Target Cell Receptor	Enzymatic Activity	Biologic Effects
Anthrax toxins	*Bacillus anthracis*	Plasmid	Three separate proteins (EF, LF, PA)[a]	Capillary morphogenesis protein 2 (CMP-2) and tumor endothelium marker 8 (TEM8)	EF is a calmodulin-dependent adenylate cyclase; LF is a zinc-dependent protease that cleaves a host signal transduction molecule (MAPKK).	EF + PA: increase in target cell cAMP level, localized edema; LF + PA: altered cell signaling; death of target cells
Bordetella adenylate cyclase toxin	*Bordetella* spp.	Chromosomal	A-B[b]	CR3 integrin (CD11–CD18)	Calmodulin-activated adenylate cyclase	Increase in target cell cAMP level; decrease in ATP production; modified cell function or cell death
Botulinum toxin	*Clostridium botulinum*	Phage	A-B[c]	Synaptic vesicle 2 (SV2)	Zinc-dependent endoprotease cleavage of presynaptic protein (SNARE)	Decrease in peripheral, presynaptic acetylcholine release; flaccid paralysis
Cholera toxin	*Vibro cholerae*	Phage	A-5B[d]	Ganglioside (GM_1)	ADP ribosylation of adenylate cyclase regulatory protein, G_S	Activation of adenylate cyclase, increase in cAMP level; secretory diarrhea
Diphtheria toxin	*Corynebacterium diphtheriae*	Phage	A-B[e]	Heparin-binding, EGF-like growth factor precursor	ADP ribosylation of elongation factor 2	Inhibition of protein synthesis; cell death
Heat-labile enterotoxins[f]	*E. coli*	Plasmid	——— Similar or Identical to Cholera Toxin ———			
Pertussis toxin	*Bordetella pertussis*	Chromosomal	A-5B[g]	Asparagine-linked oligosaccharide and lactosylceramide sequences	ADP ribosylation of signal-transducing G proteins	Block of signal transduction mediated by target G proteins
Pseudomonas exotoxin A	*P. aeruginosa*	Chromosomal	A-B	α_2-Macroglobulin/LDL receptor	— Similar or Identical to Diphtheria Toxin —	
Shiga toxin	*Shigella dysenteriae*	Chromosomal	A-5B[h]	Globotriaosylceramide (Gb_3)	RNA *N*-glycosidase	Inhibition of protein synthesis, cell death
Shiga-like toxin 1	*Shigella* spp., *E. coli*	Phage	——— Similar or Identical to Shiga Toxin ———			
Tetanus toxin	*C. tetani*	Plasmid	A-B[c]	Ganglioside (GT_1 and/or GD_{1b})	Zinc-dependent endopeptidase cleavage of synaptobrevin	Decrease in neurotransmitter release from inhibitory neurons; spastic paralysis

Adapted from G. L. Mandell, et al., *Principles and Practice of Infectious Diseases,* 3d edition. Copyright © 1990 Churchill-Livingstone, Inc., Medical Publishers, New York, NY. Reprinted by permission.

[a]The binding component (known as protective antigen [PA]) catalyzes/facilitates the entry of either edema factor (EF) or lethal factor (LF).

[b]Apparently synthesized as a single polypeptide with binding and catalytic (adenylate cyclase) domains

[c]Holotoxin is apparently synthesized as a single polypeptide and cleaved proteolytically as diphtheria toxin; subunits are referred to as L: light chain, A equivalent; H: heavy chain, B equivalent.

[d]The A subunit is proteolytically cleaved into A1 and A2, with A1 possessing the ADP-ribosyl transferase activity; the binding component is made up of five identical B units.

[e]Holotoxin is synthesized as a single polypeptide and cleaved proteolytically into A and B components held together by disulfide bonds.

[f]The heat-labile enterotoxins of *E. coli* are now recognized to be a family of related molecules with identical mechanisms of action.

[g]The binding portion is made up of two dissimilar heterodimers labeled S2-S3 and S2-S4 that are held together by a bridging peptide, SS.

[h]Subunit composition and structure similar to cholera toxin

of the host cell plasma membrane, inserts itself into the membrane, and forms a channel (pore) (**figure 31.12*a***). This causes the cytoplasmic contents to leak out. Also, because the osmolality of the cytoplasm is higher than that of the extracellular fluid, there is a sudden influx of water into the cell, causing it to swell and rupture. This subtype of exotoxin is exemplified by **leukocidins** (leukocyte and Latin *caedere*, to kill) and **hemolysins** (Greek *haima*, blood, and *lysis*, dissolution), which characteristically lyse white blood cells and erythrocytes, respectively.

The second subtype of membrane-disrupting toxins are the **phospholipase** enzymes. Phospholipases remove the charged head group (figure 31.11*b*) from the lipid portion of the phospholipids in

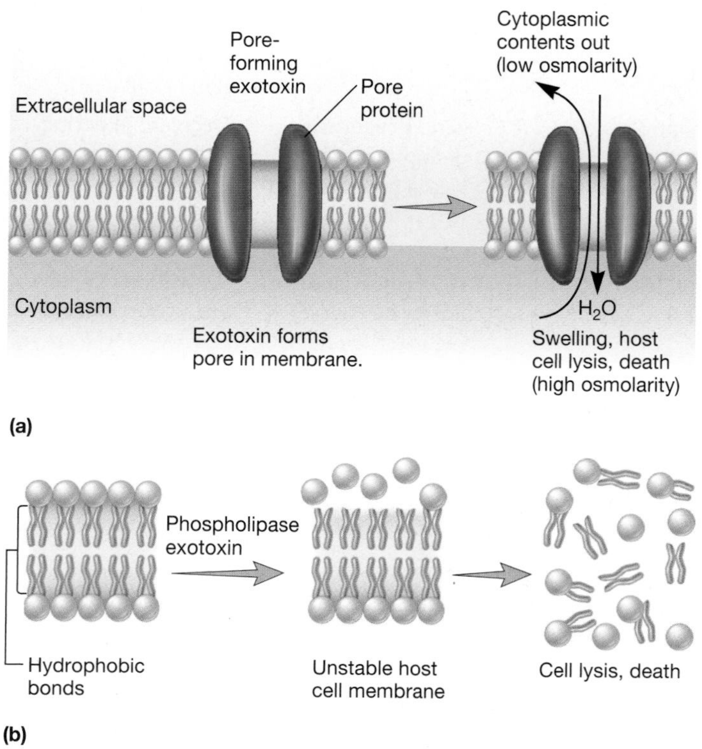

FIGURE 31.12 Two Subtypes of Membrane-Disrupting Exotoxins. (a) A channel-forming (pore-forming) type of exotoxin inserts itself into the normal host cell membrane and makes an open channel (pore). Formation of multiple pores causes cytoplasmic contents to leave the cell and water to move in, leading to cellular lysis and death of the host cell. (b) A phospholipid-hydrolyzing phospholipase exotoxin destroys membrane integrity. The exotoxin removes the charged polar head groups from the phospholipid part of the host cell membrane. This destabilizes the membrane and causes the host cell to lyse.

Figure 31.12 Micro Inquiry

Which mechanism is used by leukocidins and hemolysins?

the host-cell plasma membrane. This destabilizes the membrane so that the cell lyses and dies. One disease to which phospholipases contribute is gas gangrene caused by *Clostridium perfringens*. In this disease, the phospholipase activity of α-toxin almost completely destroys the local population of white blood cells that are drawn in by inflammation to fight the infection.

(4) Finally, exotoxins called **superantigens** act by stimulating immune system cells called T cells, causing the T cells to overexpress and release massive amounts of signalling molecules called cytokines from other host immune cells. The excessive concentration of cytokines causes multiple host organs to fail, giving the pathogen time to disseminate. The mechanism by which superantigens provoke as many as 30% of a person's T cells to release cytokines is discussed in chapter 33 (*see p. 801*). The best-studied superantigen is the staphylococcal enterotoxin B, which also exhibits superantigen activity at nanogram concentrations. It is so potent, it is classified as a select agent because it has the potential to be misused as a bioterror agent. Examples of each specific toxin type are discussed as part of the disease process in chapters 37 through 39.

1. What is the difference between an infection and an intoxication? Define toxemia.
2. Describe some general characteristics of exotoxins.
3. What is the mode of action of a leukocidin? Of a hemolysin?

Endotoxins

Gram-negative bacteria have lipopolysaccharide (LPS) in the outer membrane of their cell wall that, under certain circumstances, is toxic to specific hosts. This LPS is called an **endotoxin** because it is bound to the bacterium and is released when the microorganism lyses. Some is also released during bacterial multiplication. The toxic component of the LPS is the lipid portion, called lipid A. **Lipid A** is not a single macromolecular structure but rather it is a complex array of lipid residues (*see figure 3.20*). Lipid A is heat stable and toxic in nanogram amounts but only weakly immunogenic. ◀◀ *Gram-negative cell walls (section 3.3)*

Unlike the structural and functional diversity of exotoxins, the lipid A of various gram-negative bacteria produces similar systemic effects regardless of the microbe from which it is derived. These include fever (i.e., endotoxin is pyrogenic), shock, blood coagulation, weakness, diarrhea, inflammation, intestinal hemorrhage, and fibrinolysis (enzymatic breakdown of fibrin, the major protein component of blood clots) (**figure 31.13**).

The characteristics of endotoxins and exotoxins are contrasted in table 31.4. The main biological effect of lipid A is an indirect one, mediated by host molecules and systems, rather than by lipid A itself. For example, endotoxins can initially activate Hageman factor (blood clotting factor XII), which in turn activates and overstimulates up to four humoral systems: coagulation, complement, fibrinolytic, and kininogen systems; the sum of which is unregulated blood clotting within capillaries (disseminated intravascular coagulation) and multiorgan failure (figure 31.13). Endotoxins also indirectly induce a fever in the host by causing macrophages to release endogenous pyrogens that reset the hypothalamic thermostat. One important endogenous pyrogen is the cytokine interleukin-1 (IL-1). Other cytokines released by macrophages, such as the tumor necrosis factor, also produce fever. The net effect is often called **septic shock** and can also be induced by certain pathogenic fungi and gram-positive bacteria. All drugs, especially those given intravenously or intramuscularly, must be free of endotoxin. ◀◀ *Cytokines (section 32.3)*

Mycotoxins

Mycotoxins are toxins produced as secondary metabolites by fungi. For example, *Aspergillus flavus* and *A. parasiticus* produce aflatoxins, and *Stachybotrys* produces satratoxins, also known as trichothecene mycotoxins. These fungi commonly contaminate food crops and water-damaged buildings, respectively. An estimated 4.5 billion people in developing countries may be exposed chronically to aflatoxins through their diet. Exposure to aflatoxins is known to cause both chronic and acute liver disease and liver cancer. Aflatoxins are extremely carcinogenic, mutagenic, and immunosuppressive. Approximately 18 different types of aflatoxins exist. Aflatoxins are difuranocoumarins and classified in two broad groups according to their chemical structure (*see figure 40.6*). The *Stachybotrys trichothecene* mycotoxins are potent inhibitors of DNA, RNA, and protein synthesis. They induce inflammation, disrupt surfactant phospholipids in the lungs, and may lead to pathological changes in tissues.

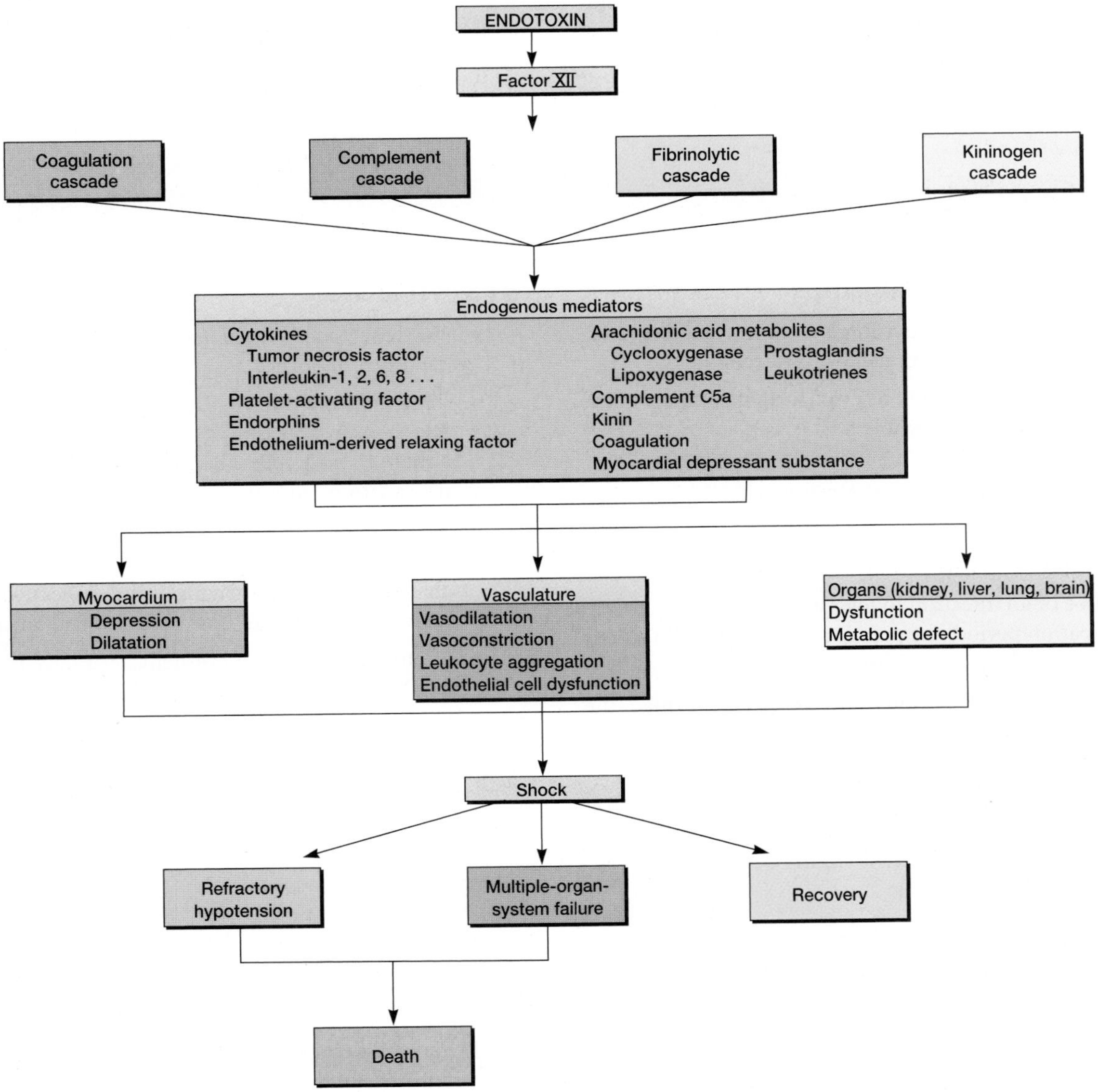

FIGURE 31.13 The Septic Shock Cascade. Gram-negative bacterial endotoxin triggers the biochemical events that lead to such serious complications as shock, adult respiratory distress syndrome, and disseminated intravascular coagulation.

The fungus *Claviceps purpurea* also produces toxic substances. The products are generically referred to as ergots, reflecting the name of the tuberlike structure of the fungi (*see figure 24.10*). The ergot is a fungal resting stage and is composed of a compact mass of hyphae. The ergots from various *Claviceps* spp. produce alkaloids that have varying physiological effects on humans. One such alkaloid is lysergic acid, a psychotropic hallucinogen. The ergot alkaloids have long been suspected as the cause of St. Anthony's fire in the eighth to sixteenth centuries in Europe and the hallucinations associated with the Salem witch trials of early American infamy.

Measuring Virulence

Virulence is often measured at the host level by determining the **lethal dose 50 (LD_{50})**. This value refers to the dose or number of pathogens that kills 50% of an experimental group of hosts within a specified period. At the cellular level, virulence is measured by cytopathology (cellular changes), including death of the cell. The cytopathology can be unique to a specific microorganism.

Other methods are often used to specifically measure virulence by examining virulence factors. For example, a number of experimental systems are used to measure bacterial motility and viral trophism, bacterial and viral adherence to host cells, bacterial and fungal invasiveness, the rates of bacterial, fungal, viral and protozoal replication, and the release of biochemicals that facilitate survival strategies (viral receptor docking, bacterial and fungal toxin production, and parasite fecundity). Chapter 35 presents a number of methods by which virulence can be measured.

1. Describe the chemical structure of the LPS endotoxin.
2. List some general characteristics of endotoxins.
3. How do gram-negative endotoxins induce fever in a mammalian host?
4. Describe some of the physiological effects associated with mycotoxin poisoning.

Summary

31.1 Host-Parasite Relationships

a. Infections occur when microorganisms enter a larger organism, the host, and compete for resources.

b. Infections can be measured by evaluating the impact of the microbes on cells or the whole host.

31.2 The Infectious Disease Process

a. Signs of an infectious disease are objective changes in the body. Symptoms are subjective.

b. The course of an infection follows a characteristic pattern: the incubation period, the prodromal stage, the illness period, and convalescence. **(figure 31.2)**

c. Pathogens reside in various environments, termed the reservoir. They infect the host from the reservoir and can return to it.

d. Pathogens are transmitted to and between hosts by several common routes, including: air, direct contact, arthropod vectors, and ingestible vehicles. **(figure 31.4)**

31.3 Virulence

a. Virulence refers to the magnitude of harm that a pathogen can exert on its host.

b. Pathogens often have numerous virulence factors that assist them in accessing a host. These include the ability to adhere to host cells, chemicals that permit cell and tissue degradation, and processes and mediators that overcome host defenses. **(table 31.3)**

c. Biofilms provide a unique environment for bacteria to exchange genetic information and collectively resist antimicrobial agents and host defense mechanisms.

d. Exotoxins are secreted proteins that have biological activity. Endotoxins are the lipid A components of gram-negative bacterial envelopes. **(figures 31.11–31.13)**

e. Mycotoxins are secondary metabolites of fungi. Their effects include hallucinations, cancer, and immunosuppression.

Critical Thinking Questions

1. Why does a parasitic organism not have to be a parasite?
2. In general, infectious diseases that are commonly fatal are newly evolved relationships between the parasitic organism and the host. Why is this so?
3. Explain the observation that different pathogens infect different parts of the host.
4. Intracellular bacterial infections present a particular difficulty for the host. Why is it harder to defend against these

infections than against viral infections and extracellular bacterial infections?

5. Among the many diseases caused by *Staphylococcus aureus,* necrotizing pneumonia is particularly aggressive, causing lung inflammation, tissue death (necrosis), and hemorrhage. Only *S. aureus* strains that produce the toxin Panton-Valentine leukocidin (PVL) cause necrotizing pneumonia. PVL is a leukocidin because it forms pores in cell membranes of white blood cells (leukocytes) that would otherwise engulf and phagocytose the pathogen. When *S. aureus* infects the respiratory tract, it relies on adhesins to mediate attachment to airway cells. One of the most important adhesins is called staphylococcal protein A (Spa). Spa also blocks the engulfment activity of host phagocytes. *S. aureus* strains that produce PVL also produce high levels of Spa.

   How could you assess the contributions of each protein independently to pathogenesis and their combined effect (note that a mouse model system has been developed for this disease)? Your answer should state your hypotheses (predictions) about each protein tested alone and together, followed by the approach you would take to test each hypothesis.

   Read the original paper: Labandeira-Rey, M., et al., 2007. *Staphylococcus aureus* Panton-Valentine leukocidin causes necrotizing pneumonia. *Science* 315:1130.

6. Understanding how some pathogens survive phagocytosis is an important step in developing strategies to prevent and treat infection. *Bacillus anthracis,* which causes anthrax, not only survives phagocytosis, but its spores actually germinate in phagocytes, where the bacterium then proliferates. The release of reactive oxygen and nitrogen species during phagocytosis is one of the principle mechanisms by which phagocytes kill engulfed pathogens. This prompted Shatalin and co-workers to investigate how *B. anthracis* survives such oxidative stress. Amazingly, this microbe has its own nitric oxide synthase (NOS) that paradoxically protects *B. anthracis* from the oxidative stress (*see p. 180*). The production of NO by the bacterium triggers high-level expression of catalase. It is thus hypothesized that reactive oxygen and nitrogen species produced by immune cells are detoxified upon entry into newly germinated *B. anthracis* cells.

   How do you think *B. anthracis* controls the production of its own NOS? When during the course of germination do you think this NOS is produced? Do you think the *B. anthracis* NOS could be considered a virulence factor? Explain your answer.

   Read the original paper: Shatalin, K., et al. 2008. *Bacillus anthracis*–derived nitric oxide is essential for pathogen virulence and survival in macrophages. *Proc. Nat. Acad. Sci., USA.* 105:1009–13.

Concept Mapping Challenge

Construct a concept map using the following words and your own linking terms.

Infection	Mycotoxins	Carrier	Transmission	AB toxins	Biofilm
Zoonoses	Source	Vehicle	LPS	Exit	Airborne
			Vector	Contact	Endotoxin
			Virulence	Susceptibility	Exotoxins

Learn More

Learn more by visiting the text website at www.mhhe.com/willey8, where you will find a complete list of references.

32

Nonspecific (Innate) Host Resistance

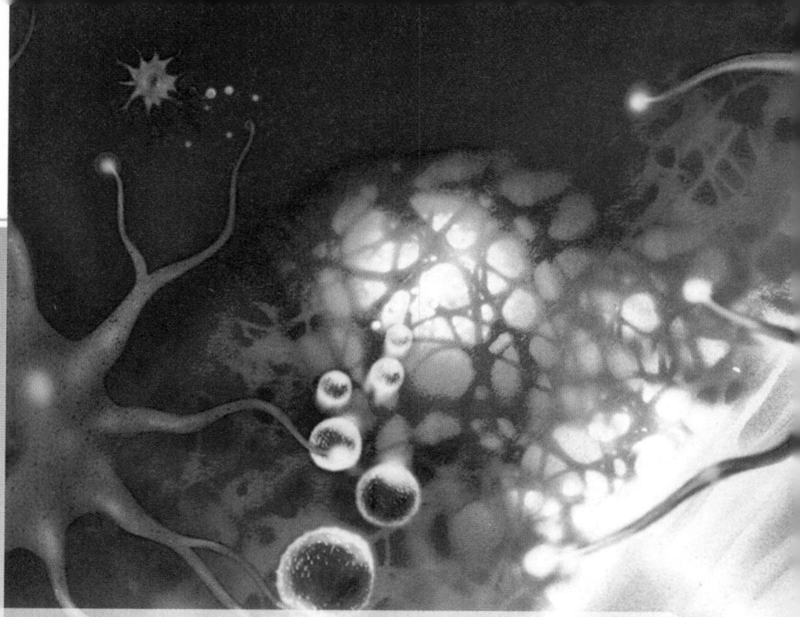

The immune system is a myriad of cells, tissues, and soluble factors that cooperate to defend against foreign invaders.

CHAPTER GLOSSARY

antigen A foreign substance (such as a protein, nucleoprotein, polysaccharide, or sometimes a glycolipid) that induces an immune response.

basophil A white blood cell in the granulocyte lineage that is weakly phagocytic. It synthesizes and stores vasoactive molecules (e.g., histamine) that are released in response to external triggers.

B lymphocyte or B cell A type of lymphocyte derived from bone marrow stem cells that matures into an immunologically competent cell.

chemokine A glycoprotein having potent leukocyte activation and chemotactic activity.

complement system Proteins found in blood plasma that when activated, bind to foreign materials to remove them from the host.

cytokine A general term for proteins released by cells of the immune system in response to specific stimuli. Cytokines influence the activities of other cells.

dendritic cell An antigen-presenting cell that has long membrane extensions resembling the dendrites of neurons. These cells are found in the lymph nodes, spleen, and thymus (interdigitating dendritic cells); skin (Langerhans cells); and other tissues (interstitial dendritic cells).

interferon (IFN) A glycoprotein that has nonspecific antiviral activity by stimulating cells to produce antiviral proteins, which inhibit the synthesis of viral RNA and proteins. Interferons also regulate the growth, differentiation, and function of a variety of immune system cells.

interleukin A cytokine produced by macrophages and T cells that regulates cellular growth and differentiation, particularly of lymphocytes.

lymph node A small secondary lymphoid organ that contains lymphocytes, macrophages, and dendritic cells. It serves as a site for filtration and removal of foreign antigens and for the activation and proliferation of lymphocytes.

lymphokine A biologically active glycoprotein secreted by activated lymphocytes, especially sensitized T cells. It acts as an intercellular mediator of the immune response and transmits signals affecting cell growth, differentiation, and behavior.

macrophage A large, mononuclear phagocytic cell, present in various tissues. Macrophages are derived from **monocytes.** They phagocytose and destroy pathogens; macrophages also activate B cells and T cells.

membrane attack complex (MAC) Complement components (C5b–C9) that assemble to create a pore in the membrane of a target cell, leading to target cell lysis.

monokine A protein (cytokine) produced by mononuclear phagocytes (monocytes or macrophages) that mediates immune responses.

natural killer (NK) cell A type of white blood cell that has a lineage independent of the granulocyte, B-cell, and T-cell lineages. NK cells are capable of killing virus-infected and malignant cells.

neutrophil A mature white blood cell in the granulocyte lineage formed in bone marrow. It has a nucleus with three to five lobes and is phagocytic.

nonspecific immune response General defense mechanisms that are inherited as part of the innate structure and function of each animal; also known as nonspecific, innate, or natural immunity.

opsonization The coating of foreign substances by antibody proteins, complement proteins, or fibronectin to make the substances more readily recognized by phagocytic cells.

pathogen-associated molecular patterns (PAMPs) Common components of pathogens such as lipopolysaccharide, peptidoglycan, fungal cell wall components called zymosan, viral nucleic acids, and foreign DNA.

pattern recognition receptors (PRRs) Surface molecules that function as receptors to recognize common components of pathogens.

phagocytosis The endocytotic process in which a cell encloses large particles in a membrane-delimited phagocytic vacuole or **phagosome.** The material within is destroyed after the phagosome fuses with lysosomes, which deliver degradative and toxin-producing enzymes.

specific immune response (acquired, adaptive, or specific immunity) Refers to the type of specific (adaptive) immunity that develops after exposure to an antigen.

T lymphocyte or T cell A type of lymphocyte derived from bone marrow stem cells that matures into an immunologically competent cell under the influence of the thymus. T cells are involved in a variety of cell-mediated immune reactions and support B cell growth and development.

toll-like receptor (TLR) A type of pattern recognition receptor on phagocytes that triggers the proper response to different classes of pathogens. It signals the production of transcription factor NFκB, which stimulates formation of cytokines, chemokines, and other defense molecules.

The integrity of any eukaryotic organism depends not only on the proper expression of its genes but also on its freedom from invading microorganisms. Commensal relationships aside, when microorganisms inhabit a multicellular host, competition for resources at the cellular level occurs. In this chapter, we explore the mechanisms by which humans (and other mammalian hosts) defend themselves against such microbial invasion.

32.1 Host Resistance Overview

To establish an infection, an invading microorganism must first overcome many surface barriers, such as skin, degradative enzymes, and mucus, that have either direct antimicrobial activity or inhibit attachment of the microorganism to the host. Because neither the surface of the skin nor the mucus-lined body cavities are ideal environments for the vast majority of microorganisms, most **pathogens,** or disease-causing microorganisms, must breach these barriers to reach underlying tissues. Any microorganism that penetrates these barriers encounters two levels of resistance: other nonspecific resistance mechanisms and the specific immune response.

Vertebrates (including humans) are continuously exposed to microorganisms that can cause disease. Fortunately these animals are equipped with an immune system that usually protects against adverse consequences of this exposure. The **immune system** is composed of widely distributed cells, tissues, and organs that recognize foreign substances, including microorganisms. Together they act to neutralize or destroy them.

Immunity

The term immunity (Latin *immunis,* free of burden) refers to the general ability of a host to resist infection or disease. **Immunology** is the science that is concerned with immune responses to foreign challenge and how these responses are used to resist infection. It includes the distinction between "self" and "nonself" and all the biological, chemical, and physical aspects of the immune response.

There are two fundamentally different yet complementary components of the immune response to an invading microorganism or foreign material. The **nonspecific immune response** is also known as **nonspecific resistance** and **innate** or **natural immunity;** it is the first line of defense against any microorganism or foreign material encountered by the vertebrate host. It includes general mechanisms inherited as part of the innate structure and function of each animal (such as skin, mucus, and constitutively produced antimicrobial mediators such as lysozyme). The nonspecific immune response defends against foreign invaders equally and lacks immunological memory—that is, nonspecific responses occur to the same extent each time a microorganism or foreign body is encountered.

In contrast, the **specific immune response,** also known as **acquired, adaptive,** or **specific immunity,** resists a particular foreign agent. The effectiveness of specific immune responses increases on repeated exposure to foreign agents such as viruses, bacteria, or toxins; that is to say, specific responses have "memory." Substances that are recognized as foreign and provoke immune responses are called immunogens (immunity generators) or, more frequently, **antigens.** The presence of foreign antigens causes specific cells to replicate and manufacture a variety of proteins that function to protect the host. One such cell, the B cell, produces and secretes glycoproteins called antibodies. **Antibodies** bind to specific antigens and inactivate them or contribute to their elimination. Other immune cells become activated to destroy host cells harboring intracellular pathogens, such as viruses. The nonspecific and specific responses work together to eliminate pathogenic microorganisms and other foreign agents.

The distinction between the innate and adaptive systems is, to a degree, artificial. Although innate systems predominate immediately upon initial exposure to foreign substances, multiple bridges occur between innate and adaptive immune system components (**figure 32.1**). Importantly, a variety of cells function in both innate and adaptive immunity. These cells are known as the **white blood cells,** or leukocytes. Blood cell development occurs in the bone marrow of mammals during the process of **hematopoesis.** Some leukocytes function in the innate system, whereas others are part of an adaptive immune response. Some are important because they link the nonspecific arm of the immune system to the specific. The leukocytes form the basis for immune responses to invading microbes and foreign substances. Many of these cells reside in specific tissues and organs. Some tissues and organs provide supportive functions in nurturing the cells so that they can mature and respond correctly to antigens.

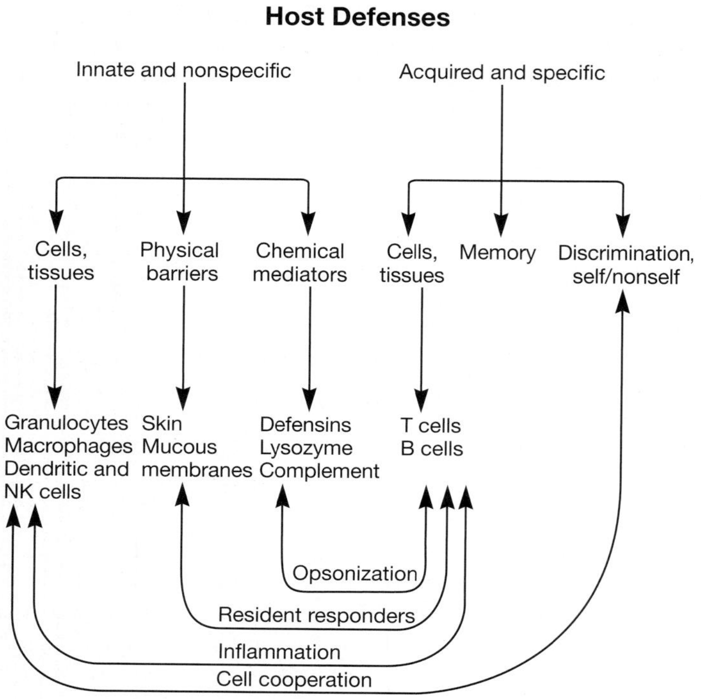

FIGURE 32.1 Some of the Major Components That Make Up the Mammalian Immune System. Double-headed arrows indicate potential bridging events that unite innate and acquired forms of immunity.

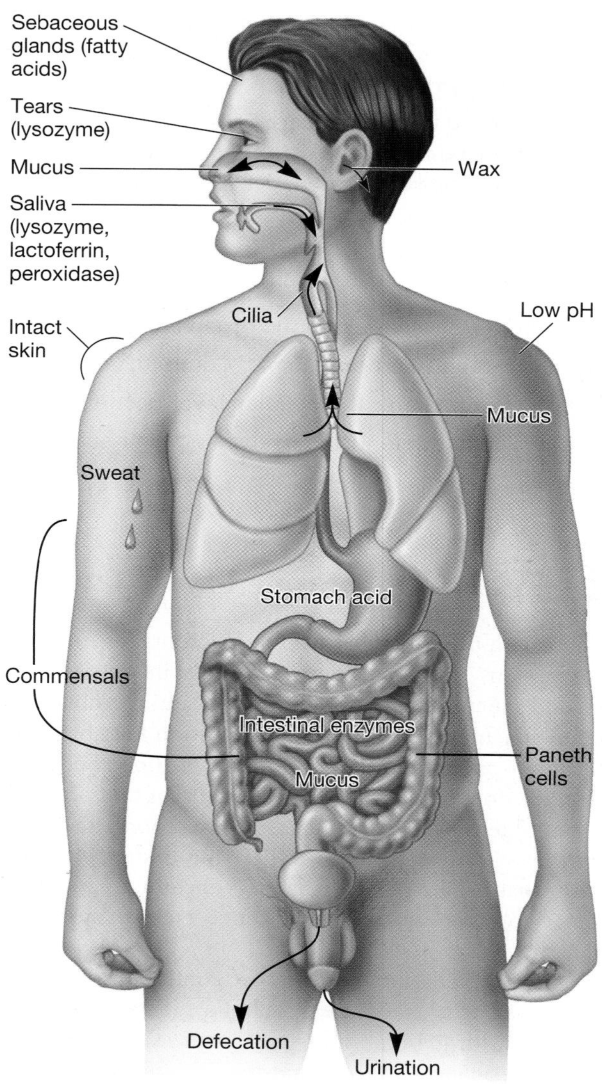

FIGURE 32.2 Host Defenses. Some nonspecific host defense mechanisms that help prevent entry of microorganisms into the host's tissues.

The bridges that interconnect the innate and specific immune responses are numerous and present challenges for any author trying to describe immunity. We begin our discussion with an introduction to the physical and chemical barriers that face microbial entry into a host. We then present the various cells, tissues, and organs of the innate defenses. In this discussion, reference will be made to processes and chemical mediators that are described in detail either later in this chapter or in chapter 33 (specific defenses). For your convenience, cross-references to the detailed discussion are provided.

1. Define each of the following: immune system, immunity, immunology, antigen, antibody.
2. Compare and contrast the specific and nonspecific immune responses.

32.2 Physical Barriers in Nonspecific (Innate) Resistance

With few exceptions, a potential microbial pathogen invading a human host immediately confronts a vast array of nonspecific defense mechanisms (**figure 32.2**). Although the effectiveness of some individual mechanisms is not great, collectively their defense is formidable. Many direct factors (nutrition, physiology, fever, age, genetics) and equally as many indirect factors (personal hygiene, socioeconomic status, living conditions) influence all host-microbe relationships. In addition to these direct and indirect factors, a vertebrate host has some specific physical and mechanical barriers.

Physical and Mechanical Barriers

Physical and mechanical barriers, along with the host's secretions (flushing mechanisms), are the first line of defense against microorganisms. Protection of the most important body surfaces by these mechanisms is discussed next.

Skin

Intact skin contributes greatly to nonspecific host resistance. It forms a very effective mechanical barrier to microbial invasion. Its outer layer consists of thick, closely packed cells called

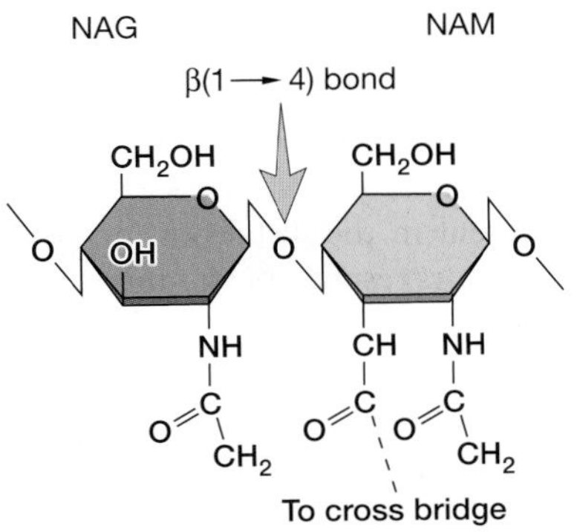

FIGURE 32.3 Action of Lysozyme on the Cell Wall of Gram-Positive Bacteria. This enzyme degrades the cell wall peptidoglycan by hydrolysing the β(1→4) bonds that connect alternating *N*-acetylglucosamine (NAG) with *N*-acetylmuramic acid (NAM) residues.

Figure 32.3 Micro Inquiry

Why is this enzyme not very effective against gram-negative bacteria?

keratinocytes, which produce keratins. Keratins are scleroproteins (i.e., insoluble proteins) and make up the main components of hair, nails, and the outer skin cells. These outer skin cells shed continuously, removing microorganisms that manage to adhere to their surface. The skin is slightly acidic (around pH 5 to 6) due to skin oil, secretions from sweat glands, and organic acids produced by commensal staphylococci. It also contains a high concentration of sodium chloride and is subject to periodic drying.

Mucous Membranes

The mucous membranes of the eye (conjunctiva) and the respiratory, digestive, and urogenital systems withstand microbial invasion because the intact stratified squamous epithelium and mucus secretions form a protective covering that resists penetration and traps many microorganisms. This mechanism contributes to nonspecific immunity. Furthermore, many mucosal surfaces are bathed in specific antimicrobial secretions. For example, cervical mucus, prostatic fluid, and tears are toxic to many bacteria. One antibacterial substance in these secretions is **lysozyme** (muramidase), an enzyme that lyses bacteria by hydrolyzing the β(1→4) bond connecting *N*-acetylmuramic acid and *N*-acetylglucosamine of the bacterial cell wall peptidoglycan—especially in gram-positive bacteria (**figure 32.3**). These mucous secretions also contain specific immune proteins that help prevent the attachment of microorganisms. In addition, they possess significant amounts of the iron-binding protein lactoferrin. **Lactoferrin** is released by activated phagocytic cells called macrophages and polymorphonuclear leukocytes (PMNs). It sequesters iron from the plasma, reducing the amount of iron available to invading microbial pathogens and limiting their ability to multiply. Finally, mucous membranes produce **lactoperoxidase,** an enzyme that catalyzes the production of superoxide radicals, a reactive oxygen intermediate that is toxic to many microorganisms. ◀◀ *Peptidoglycan structure (section 3.3)*

Respiratory System

The mammalian respiratory system has formidable defense mechanisms. The average person inhales at least eight microorganisms a minute, or 10,000 each day. Once inhaled, a microorganism must first survive and penetrate the air-filtration system of the upper and lower respiratory tracts. Because the airflow in these tracts is very turbulent, microorganisms are deposited on the moist, sticky mucosal surfaces. Microbes larger than 10 μm usually are trapped by hairs and cilia lining the nasal cavity. The cilia in the nasal cavity beat toward the pharynx, so that mucus with its trapped microorganisms is moved toward the mouth and expelled (**figure 32.4**). Humidification of the air within the nasal cavity causes many hygroscopic (attracting moisture from the air) microorganisms to swell, and this aids phagocytosis. Microbes smaller than 10 μm often pass through the nasal cavity and are trapped by the **mucociliary blanket** that coats the mucosal surfaces of lower portions of the respiratory system. The trapped microbes are transported by ciliary action—the mucociliary escalator—that moves them away from the lungs (figure 32.4). Coughing and sneezing reflexes clear the respiratory system of microorganisms by expelling air forcefully from the lungs through the mouth and nose, respectively. Salivation also washes microorganisms from the mouth and nasopharyngeal areas into the stomach. Microorganisms that succeed in reaching the alveoli of the lungs encounter a population of fixed phagocytic cells called **alveolar macrophages.** These cells can ingest and kill most bacteria by phagocytosis.

Gastrointestinal Tract

Most microorganisms that reach the stomach are killed by gastric juice (a mixture of hydrochloric acid, proteolytic enzymes, and mucus). The very acidic gastric juice (pH 2 to 3) is sufficient to destroy most organisms and their toxins, although exceptions exist (protozoan cysts, *Helicobacter pylori, Clostridium,* and *Staphylococcus* toxins). However, organisms embedded in food particles are protected from gastric juice and reach the small intestine. There, microorganisms often are damaged by various pancreatic enzymes, bile, enzymes in intestinal secretions, and the GALT system (p. 780). **Peristalsis** (Greek *peri,* around, and *stalsis,* contraction) and the normal loss of columnar epithelial cells act in concert to purge intestinal microorganisms. In addition, the normal microbiota of the large intestine (*see figure 30.16*) is extremely important in preventing the establishment of pathogenic organisms. For example, the metabolic products (e.g., fatty acids) of many normal commensals in the intestinal tract prevent unwanted microorganisms from becoming established. Other normal microbiota outcompete potential pathogens for attachment sites and nutrients.

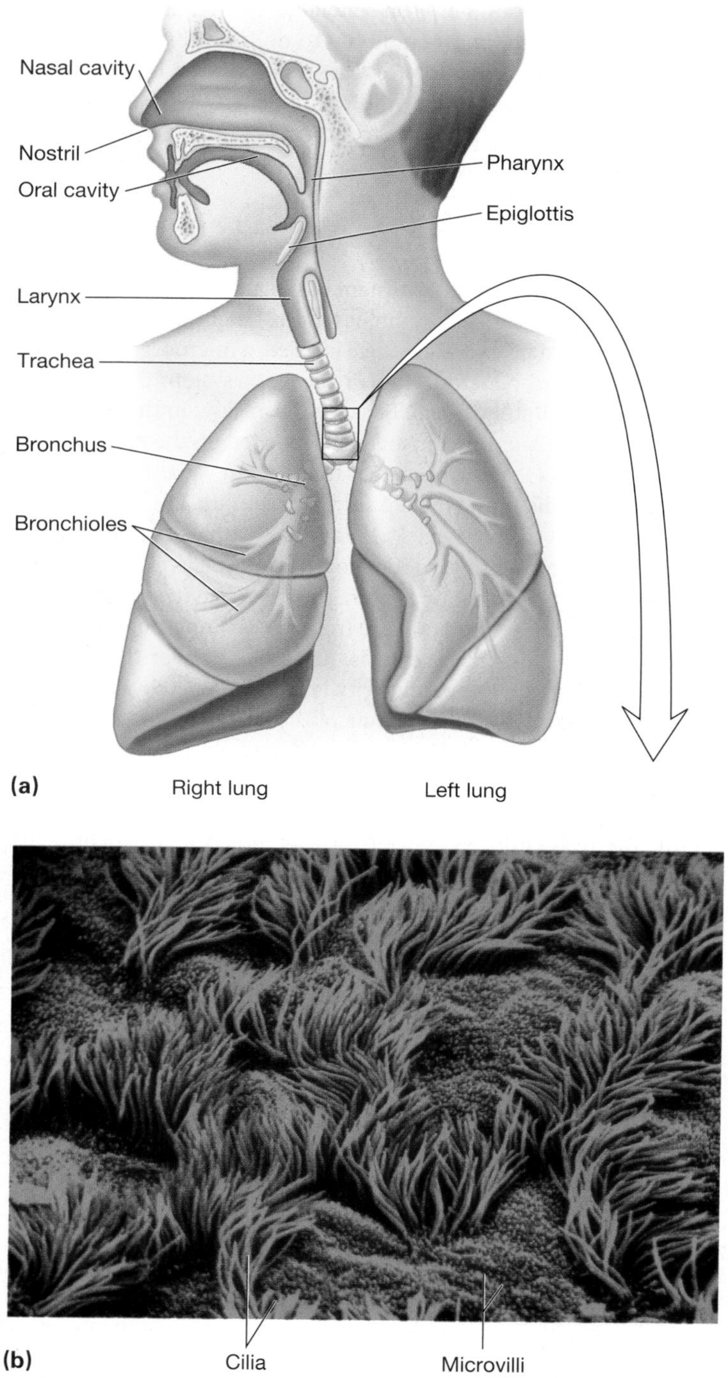

FIGURE 32.4 Ciliated Epithelial Cells of the Respiratory Tract. (a) The respiratory tract is lined with a mucous membrane of ciliated epithelial cells. (b) Cilia (×5,000) sweep particles upward toward the throat to be expectorated.

The mucous membranes of the intestinal tract contain cells called **Paneth cells.** These cells produce lysozyme (figure 32.3) and peptides called **cryptins,** a type of defensin (p. 765). Cryptins are toxic for some bacteria, although their mode of action is not known.

Genitourinary Tract

Under normal circumstances, the kidneys, ureters, and urinary bladder of mammals are sterile. Urine within the urinary bladder also is sterile. However, in both the male and female, a few bacteria are usually present in the distal portion of the urethra (*see figure 30.16*). The factors responsible for this sterility are complex. In addition to removing microbes by flushing action, urine kills some bacteria due to its low pH and the presence of urea and other metabolic end products (uric acid, hippuric acid, indican, fatty acids, mucin, enzymes). The kidney medulla is so hypertonic that few organisms can survive there. In males, the anatomical length of the urethra (20 cm) provides a distance barrier that excludes microorganisms from the urinary bladder. Conversely, the short urethra (5 cm) in females is more readily traversed by microorganisms; this explains why urinary tract infections are 14 times more common in females than in males.

The vagina has another unique defense. Under the influence of estrogens, the vaginal epithelium produces increased amounts of glycogen that acid-tolerant *Lactobacillus acidophilus* bacteria degrade to form lactic acid. Normal vaginal secretions contain up to 10^8 of these bacteria per milliliter. Thus an acidic environment (pH 3 to 5) unfavorable to most organisms is established. Cervical mucus also has some antibacterial activity.

The Eye

The conjunctiva is a specialized, mucus-secreting epithelial membrane that lines the interior surface of each eyelid and the exposed surface of the eyeball. It is kept moist by the continuous flushing action of tears (lacrimal fluid) from the lacrimal glands. Tears contain large amounts of lysozyme, lactoferrin, and antibody, and thus provide chemical as well as physical protection.

1. Why is the skin such a good first line of defense against pathogenic microorganisms?
2. How do intact mucous membranes resist microbial invasion of the host?
3. Describe the different antimicrobial defense mechanisms that operate within the respiratory system of mammals.
4. What factors operate within the gastrointestinal system that help prevent the establishment of pathogenic microorganisms?
5. Except for the anterior portion of the urethra, why is the genitourinary tract a sterile environment?

32.3 Chemical Mediators in Nonspecific (Innate) Resistance

Mammalian hosts have a chemical arsenal with which to combat the continuous onslaught of microorganisms. Some of these chemicals (gastric juices, salivary glycoproteins, lysozyme, oleic acid on the skin, urea) have already been discussed with respect to the specific body sites they protect. In addition, blood, lymph, and other body fluids contain a potpourri of defensive chemicals such as defensins and other polypeptides.

Antimicrobial Peptides

Antimicrobial peptides are low molecular weight proteins that exhibit broad-spectrum antimicrobial activity toward bacteria, viruses, and fungi. They are evolutionarily conserved peptides, many of which are positively charged. Antimicrobial peptides are amphipathic (having hydrophobic and hydrophilic components) so as to be soluble in aqueous and lipid-rich environments. Antimicrobial peptides kill target cells through diverse mechanisms after interacting with their membranes.

Cationic Peptides

Antimicrobial cationic peptides appear to be highly conserved through evolution. We will only discuss those peptides found in humans. There are three generic classes of cationic peptides whose biological activity is related to their ability to damage bacterial plasma membranes. This is accomplished by electrostatic interactions with membranes—the formation of ionic pores or transient gaps, thereby altering membrane permeability.

The first group of cationic peptides includes those that are linear, α-helical peptides that lack cysteine amino acid residues. An important example is **cathelicidin,** a peptide that arises from a precursor protein having a C-terminus bearing the mature peptide of some 12 to 80 amino acids. Cathelicidins are produced by a variety of cells (e.g., neutrophils, respiratory epithelial cells, and alveolar macrophages), and substantial heterogeneity exists between cathelicidins made by various cells. ▶▶| *Protein structure (appendix I)*

A second group, the **defensins,** are peptides that are rich in arginine and cysteine, and disulfide linked. The group is composed of various structural motifs with an approximate average molecular weight of 4,000 Daltons. In mammals, defensins have antiparallel beta sheet structures with beta hairpin loops containing cationic amino acids. Two types of defensins have been reported in humans—α and β. α defensins tend to be peptides of 29 to 35 amino acid residues, whereas β defensins are usually 36 to 42 amino acids in length and are found in the primary granules of neutrophils, intestinal Paneth cells, and intestinal and respiratory epithelial cells.

A third group contains larger peptides that are enriched for specific amino acids and exhibit regular structural repeats. **Histatin,** one such peptide isolated from human saliva, has antifungal activity. Histatin is a 24 to 38 amino acid peptide, heavily enriched with histidine, that translocates to the fungal cytoplasm, where it targets mitochondria.

Other natural antimicrobial products include fragments from (1) histone proteins, (2) lactoferrin, and (3) chemokines. A number of antibacterial peptides are produced by bacteria as well. The most notable of these are the bacteriocins.

Bacteriocins

As noted previously, the first line of defense against microorganisms is the host's anatomical barrier, consisting of the skin and mucous membranes. These surfaces are colonized by normal microbiota, which by themselves provide a biological barrier against uncontrolled proliferation of foreign microorganisms. Many of the bacteria that are part of the normal microbiota synthesize and release toxic proteins called bacteriocins that are lethal to other strains of the same species as well as other closely related bacterial species. Bacteriocin peptides range from about 900 to 5,800 Daltons and can be cationic, neutral, or anionic. Bacteriocins may give their producers, which are naturally immune to the antibacterial products they make, an adaptive advantage against other bacteria. Ironically, they sometimes increase bacterial virulence by damaging host cells such as mononuclear phagocytes. Bacteriocins are produced by gram-negative and gram-positive bacteria. For example, *Escherichia coli* synthesizes bacteriocins called colicins, which are encoded by genes on several different plasmids (ColB, ColE1, ColE2, ColI, and ColV). Some colicins bind to specific receptors on the cell envelope of sensitive target bacteria and cause cell lysis, attack specific intracellular sites such as ribosomes, or disrupt energy production. Other examples include the lantibiotics produced by genera such as *Streptococcus, Bacillus, Lactococcus,* and *Staphylococcus.* These antimicrobial peptides act as defensive effector molecules protecting the bacterial flora and its human host. |◀◀ *Normal microbiota of the human body (section 30.3)*

1. How do cationic peptides function against gram-positive bacteria?
2. How do bacteriocins function?

Complement

Complement was discovered many years ago as a heat-labile component of human blood plasma that augments phagocytosis. This activity was said to "complement" the antibacterial activity of antibody; hence the name complement. It is now known that the **complement system** is composed of over 30 serum proteins that have a complex (and somewhat confusing) nomenclature. This system has three major physiological activities: (1) defending against bacterial infections by facilitating and enhancing phagocytosis through opsonization, chemotaxis, activation of leukocytes, and lysis of bacteria; (2) bridging innate and adaptive immunity by enhancing antibody responses and immunologic memory; and (3) disposing of wastes such as dead host cells and immune complexes, the products of inflammatory injury.

An outcome of complement activation is opsonization. **Opsonization** (Greek *opson,* to prepare victims for) is a process in which microorganisms or other particles are coated by serum components, thereby preparing them for recognition and ingestion by phagocytic cells. Molecules that function in this capacity are collectively known as opsonins. In the opsonin-dependent recognition mechanism, the host serum components act as a

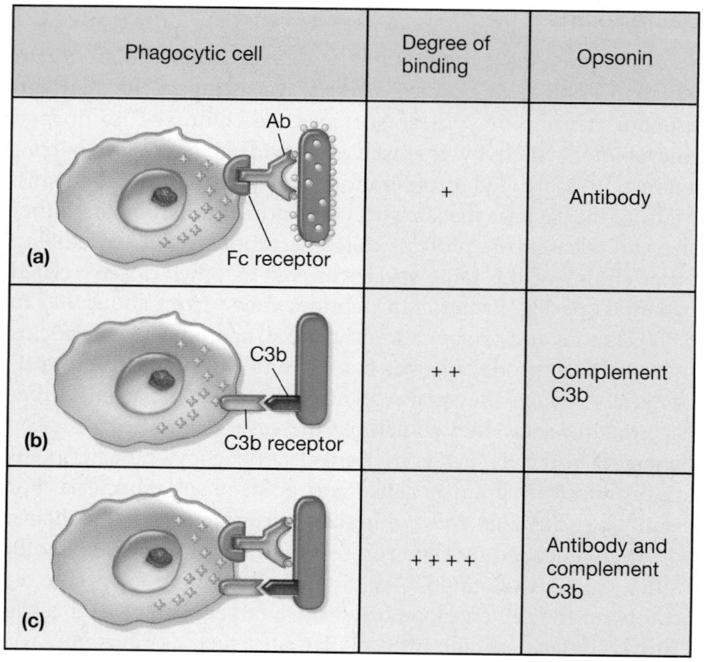

FIGURE 32.5 Opsonization. (a) The intrinsic ability of a phagocyte to bind to a microorganism is enhanced if the microorganism elicits the formation of antibodies (Ab) that act as a bridge to attach the microorganism to the Fc antibody receptor on the phagocytic cell. (b) If the activated complement (C3b) is bound to the microbe, the degree of binding is further increased by the C3b receptor. (c) If both antibody and C3b opsonize, binding is maximal.

bridge between the microorganism and the phagocyte. They connect the surface of the microorganism to specific receptors on the phagocyte surface at the other (**figure 32.5**). These complement proteins are opsonins because when they bind microbial cells, the cells are more easily recognized by phagocytes (figure 32.5*b*).

In addition to opsonization, other complement proteins help protect the host in two ways. Some are strong chemotactic signals that recruit phagocytes to the site of their activation. Still other complement proteins puncture cell membranes to cause lysis. Interestingly, one of the several triggers that can activate the complement process is the recognition of specific antibody on a target cell. All together, the complement activities unite the nonspecific and specific arms of the immune system to assist in the killing and removal of invading pathogens. *Complement Function*

Complement proteins are produced in an inactive form; they become active following enzymatic cleavage. There are three pathways of complement activation: the alternative, lectin, and classical pathways (**figure 32.6**). Although they employ similar mechanisms, specific proteins are unique to the first part of each pathway (**table 32.1**). Each complement pathway is activated in a cascade fashion: the activation of one component results in the activation of the next. Thus complement proteins are poised for immediate activity when the host is challenged by an infectious agent. *Complement Activation*

The **alternative complement pathway** (figure 32.6) plays an important role in the nonspecific immune defense against intravascular invasion by bacteria and some fungi. The alternative pathway is initiated in response to bacterial molecules with repetitive structures such as lipopolysaccharide (LPS). It begins with cleavage of complement protein C3 into fragments C3a and C3b by a blood enzyme. These fragments are initially produced at a slow rate, and free C3b is rapidly cleaved into inactive fragments by another protein called Factor I. However, C3b becomes stable when it binds to the LPS of gram-negative bacterial cell walls or to aggregates of antibodies. In binding to the LPS, C3b acts as an opsonin and thus assists in bacterial capture by phagocytes that have receptors for C3b. A protein in blood termed Factor B adsorbs to bound C3b and is cleaved into two fragments by Factor D, leading to the formation of active enzyme $\overline{\text{C3bBb}}$ (the bar indicates an activated enzyme complex). This complex is called the C3 convertase of the alternative pathway because it cleaves more C3 to C3a and C3b, thereby increasing the rate at which C3 is converted. $\overline{\text{C3bBb}}$ is further stabilized by a second blood protein, properdin, which allows another addition of C3b, forming C5 convertase ($\overline{\text{C3bBb3b}}$). This convertase then cleaves C5 to C5a and C5b. The two proteins C6 and C7 rapidly bind to C5b, forming a $\overline{\text{C5b67}}$ complex that is stabilized by binding to a membrane. C8 and C9 then bind, forming the **membrane attack complex (MAC)** ($\overline{\text{C5b6789}}$), which creates a pore in the plasma membrane or outer membrane of the target cell (**figure 32.7**). If the cell is eukaryotic, Na^+ and H_2O enter through the pore and the cell lyses. Lysozyme can pass through pores in the outer membrane of gram-negative cell walls and digest the peptidoglycan, thus weakening the cell wall and aiding lysis. In contrast, gram-positive bacteria resist the cytolytic action of the membrane attack complex because they lack an exposed outer membrane and have a thick peptidoglycan protecting the plasma membrane. Unfortunately, eukaryotic host cell membranes are also susceptible to attack by complement proteins, and bystander lysis (i.e., lysis of nearby host cells) is a potential consequence of complement activation.

The generation of complement fragments C3a and C5a leads to several important inflammatory effects. For example, binding of C3a and C5a to their cellular receptors induces some cells to release other biological mediators. These amplify the inflammatory signals of C3a and C5a by dilating vessels, increasing permeability, stimulating nerves, and recruiting phagocytic cells. C5a induces a directed, chemotactic migration of neutrophils to the site of complement activation, where they are among the first cells to phagocytose the invading microbe. Other phagocytes, called macrophages, that may be in the area synthesize even more complement components to interact with the invading microbe. All of these defensive events promote the ingestion and ultimate destruction of the microbe by neutrophils and macrophages.

The **lectin complement pathway** (also called the mannan-binding lectin pathway) also begins with the activation of C3 convertase. However, in this case, a lectin, a protein that binds to specific carbohydrates, initiates the proteolytic cascade. When macrophages ingest viruses, bacteria, or other foreign material,

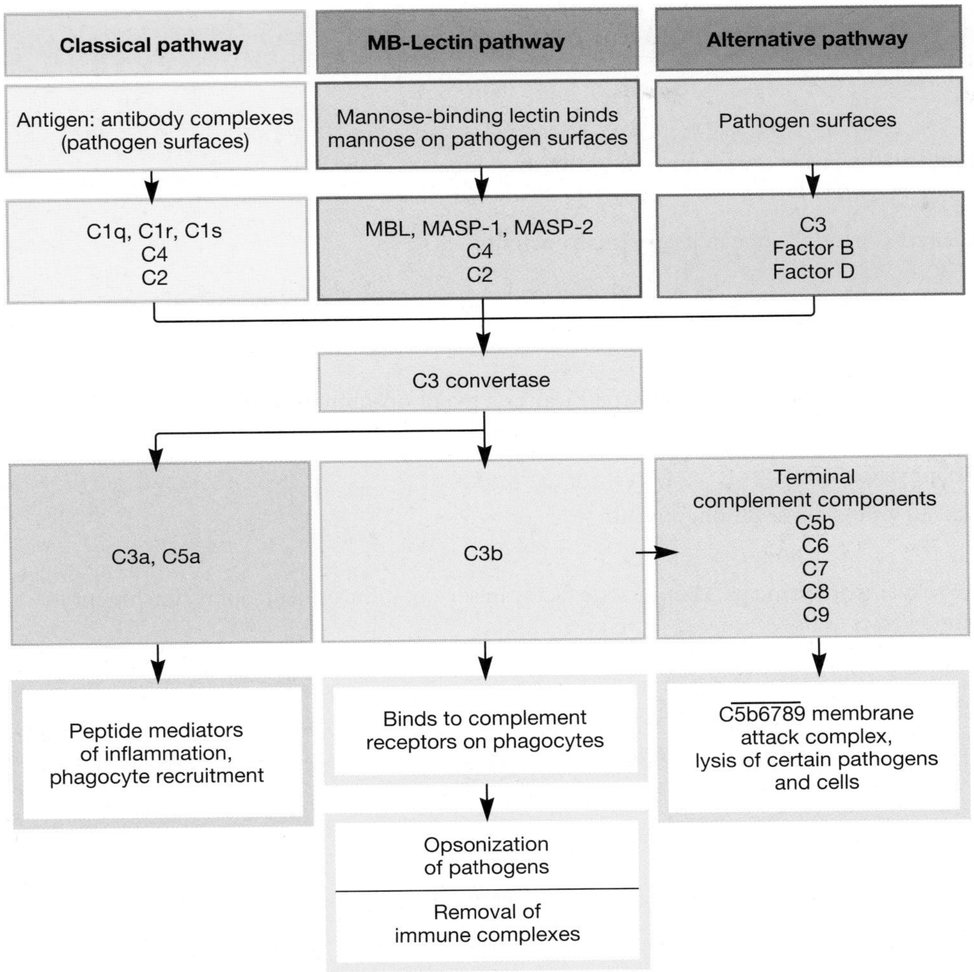

FIGURE 32.6 The Main Components and Actions of Complement. Complement activation involves a series of enzymatic reactions that culminate in the formation of C3 convertase, which cleaves complement component C3 into C3b and C3a. The production of the C3 convertase is where the three pathways converge. C3a is a peptide mediator of local inflammation. C3b binds covalently to the bacterial cell and opsonizes the bacteria, enabling phagocytes to internalize them. C5a and C5b are generated by the cleavage of C5 by a C5 convertase. C5a is also a powerful peptide mediator of inflammation. C5b promotes the terminal components of complement to assemble into a membrane attack complex.

they induce an acute inflammatory response. Part of the acute inflammatory response is the release of macrophage chemicals that stimulate liver cells to secrete acute-phase proteins. One important acute-phase protein is **mannose-binding protein (MBP).** Because mannose, in certain three-dimensional configurations, is a major component of bacterial cell walls and of some virus envelopes and antigen-antibody complexes, MBP binds to these components. MBP enhances phagocytosis and is therefore an opsonin. When MBP is bound to an esterase enzyme, it is called *m*annose-*a*ssociated *s*erine *p*rotease (MASP) and it activates the same C3 convertase found in the alternative complement pathway. Thus the lectin pathway activates the same complement cascade that the classical and alternative pathways do. However, it uses a mechanism that is independent of antibody-antigen interactions (the classical pathway), and it does not require interaction of complement with pathogen surfaces (alternative pathway).

This overview of the alternative and lectin complement pathways (figure 32.6) provides a basis for consideration of the function of complement as an integrated system during an animal's defensive effort. Bacteria arriving at a local tissue site interact with components of the alternative pathway, resulting in the generation of biologically active fragments, opsonization of the bacteria, and initiation of the lytic sequence. If the bacteria persist or if they invade the animal a second time, antibody responses also will activate the classical complement pathway.

Activation of the **classical complement pathway** can occur in response to some microbial products (e.g., lipid A, staphylococcal protein A). However, it is usually initiated by the interaction of antibodies with an antigen (figure 32.6). Antibody secretion is part of an acquired immune response (and is discussed in chapter 33). It is important to emphasize that antibodies are glycoproteins that bind to specific antigens. This binding triggers the C1 complement component, composed of three proteins (q, r, and s), to attach to the antibody through its C1q subcomponent. In the presence of calcium ions, a trimolecular complex (C1qrs + antigen + antibody) with esterase activity is rapidly formed. The activated C1s subcomponent attacks and cleaves its natural substrates in serum (C2 and C4). This leads to binding of a portion of each molecule (C2a and C4b) to the antigen-antibody-complement complex and the release of C4a and C2b fragments. With the binding of C2a to C4b, an enzyme with trypsinlike proteolytic activity is generated. The natural substrate for this enzyme is C3; thus C2a4b is a C3 convertase. Just as we saw with the lectin and alternative pathways, the C3 convertase cleaves C3 into a bound subcomponent C3b and a C3a soluble component. This sets in motion the activation of the complement cascade, which leads to the formation of the membrane attack complex, opsonins, and the release of mediators that influence inflammation. Thus the three complement pathways have three different initiating processes. However, their common outcomes of opsonization, stimulation of inflammatory

Table 32.1 Some Important Proteins of the Complement Cascade

Protein	*Fragment*	*Function*
Recognition Unit		
C1	q	Binds to the Fc portion of antigen-antibody complexes
	r	Activates C1s
	s	Cleaves C4 and C2 due to its enzymatic activity
Activation Unit		
C2		Causes viral neutralization
C3	a	Anaphylatoxin, immunoregulatory
	b	Key component of the alternative pathway and major opsonin in serum
	e	Induces leukocytosis
C4	a	Anaphylatoxin
	b	Causes viral neutralization; opsonin
Membrane Attack Unit		
C5	a	Anaphylatoxin; principal chemotactic factor in serum; induces neutrophil attachment to blood vessel walls
	b	Initiates membrane attack
C6, C7, C8, C9		Participate with C5b in formation of the membrane attack complex that lyses targeted cells
Alternative Pathway		
Factor B		Causes macrophage spreading on surfaces; precursor of C3 convertase
Factor $\overline{D}$		Cleaves Factor B to form active $\overline{C3bBb}$ in alternative pathway
Properdin		Stabilizes alternative pathway C3 convertase
Regulatory Proteins		
Factor H		Promotes C3b breakdown and regulates alternative pathway
Factor I		Degrades C3b and regulates alternative pathway
C4b binding protein		Inhibits assembly and accelerates decay of $\overline{C4bC2a}$
C1 INH complex		Binds to and dissociates C1r and C1s from C1
S protein		Binds fluid-phase $\overline{C5b67}$; prevents membrane attachment

mediators, and lysis of membrane-bound microorganisms achieve the goal of innate host defense against foreign invaders.

1. What effect does the formation of the membrane attack complex have on eukaryotic cells? Bacterial cells? Host cells?
2. How is the alternative pathway activated? The lectin pathway?
3. What role do complement fragments C3a and C5a play in an animal's defense against gram-negative bacteria?
4. How is the classical complement pathway activated?

Cytokines

Defense against viruses, microorganisms and their products, parasites, and cancer cells is mediated by both nonspecific immunity and specific immunity. Cytokines are required for regulation of both of these immune responses. **Cytokine** (Greek *cyto*, cell, and *kinesis*, movement) is a generic term for any soluble protein or glycoprotein released by one cell population that acts as an intercellular (between cells) mediator or signaling molecule. When released from mononuclear phagocytes, these proteins are called **monokines;** when released from T lymphocytes, they are called **lymphokines;** when produced by a leukocyte and the action is on another leukocyte, they are **interleukins;** and if their effect is to

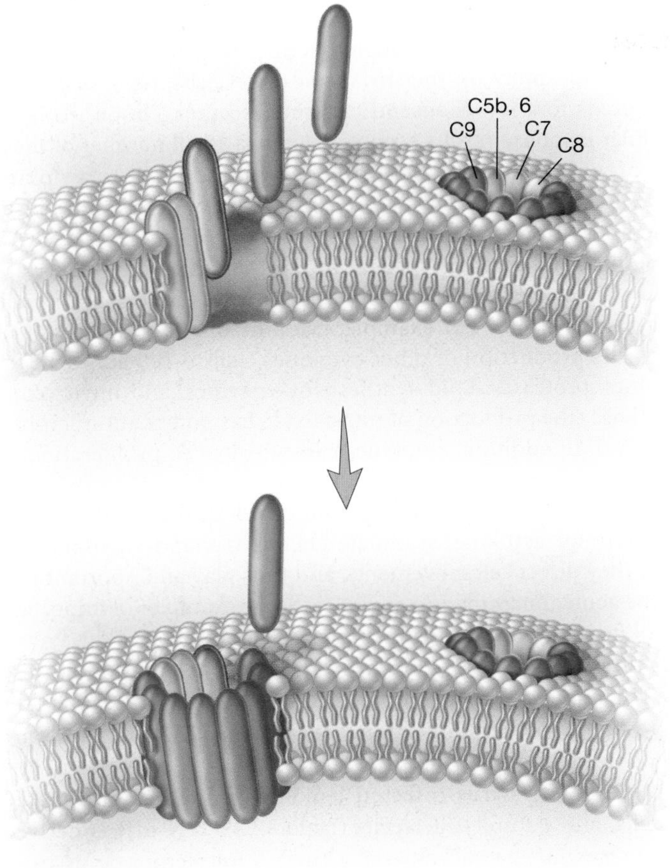

(a)

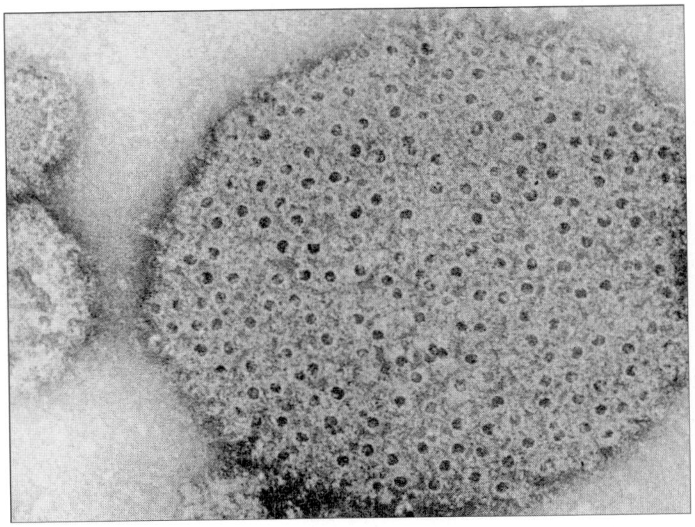

(b)

FIGURE 32.7 The Membrane Attack Complex. The membrane attack complex (MAC) is a tubular structure that forms a transmembrane pore in the target cell's membrane. (a) This representation shows the subunit architecture of the membrane attack complex. The transmembrane channel is formed by a C5b678 complex and 10 to 16 polymerized molecules of C9. (b) MAC pores appear as craters or donuts by electron microscopy.

Figure 32.7 Micro Inquiry

What bacteria are most sensitive to the membrane attack complex? Why?

Table 32.2 The Four Cytokine Families

Family	*Examples*	*Functions*
Chemokines	IL-8, RANTES[a], MIP (macrophage inflammatory protein)	Cytokines that are chemotactic and chemokinetic for leukocytes. They stimulate cell migration and attract phagocytic cells and lymphocytes. Chemokines play a central role in the inflammatory response.
Hematopoietins	Epo (erythropoietin), various colony-stimulating factors	Cytokines that stimulate and regulate the growth and differentiation processes involved in blood cell formation (hematopoiesis)
Interleukins	IL-1 to IL-18	Cytokines produced by lymphocytes and monocytes that regulate the growth and differentiation of other cells, primarily lymphocytes and hematopoietic stem cells. They often also have other biological effects.
Tumor necrosis factor (TNF) family	TNF-α, TNF-β, Fas ligand	Cytokines that are cytotoxic for tumor cells and have many other effects such as promoting inflammation, fever, and shock; some can induce apoptosis.

[a] RANTES: *R*egulated on *a*ctivation, *n*ormal *T* *e*xpressed and *s*ecreted; also called CCL5; member of the IL-8 cytokine superfamily.

stimulate the growth and differentiation of immature leukocytes in the bone marrow, they are called **colony-stimulating factors (CSFs).** Cytokines have been grouped into the following categories or families: chemokines, hematopoietins, interleukins, and members of the tumor necrosis factor (TNF) family. Some examples of these cytokine families are listed in **table 32.2**. Cytokines can affect the same cell responsible for their production (an autocrine function) or nearby cells (a paracrine function), or they can be distributed by the circulatory system to distant target cells (an endocrine function). Their production is induced by nonspecific

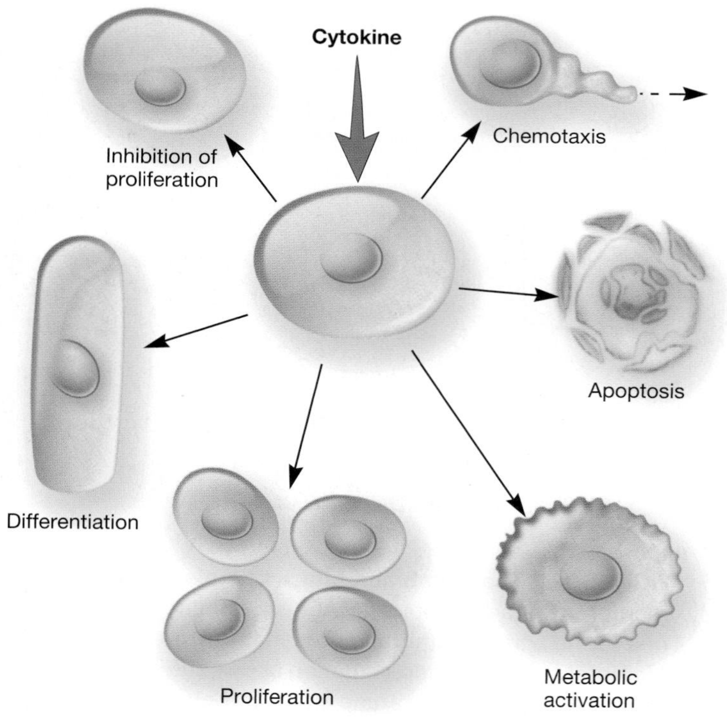

FIGURE 32.8 Range of Biological Actions That Cytokines Have on Eukaryotic Cells. Chemokines are one family of cytokines that induce leukocyte chemotaxis and migration. Other cytokines activate cell metabolism and synthesis. This can lead to the synthesis of a wide range of proteins including cyclooxygenase II, proteolytic enzymes, NO synthase, and various adhesion receptors. In addition, other cytokines can cause proliferation, inhibition of cell proliferation, or apoptosis.

stimuli such as a viral, bacterial, or parasitic infection; cancer; inflammation; or the interaction between a T cell and antigen. Some cytokines also can induce the production of other cytokines.

Cytokines produce biological actions only when they bind to specific receptors on the surface of target cells. Most cells have hundreds to a few thousand cytokine receptors, but a maximal cellular response results even when only a small number of these are occupied by a cytokine. This is because the affinity of cytokine receptors for their specific cytokines is very high, and consequently cytokines are effective at very low concentrations. Cytokine binding activates specific intracellular signaling pathways that switch on genes encoding proteins essential for appropriate cellular functions. For example, cytokine binding may result in the cell's production of other cytokines, cell-to-cell adhesion receptors, proteases, lipid-synthesizing enzymes, and nitric oxide synthase (the production of nitric oxide has potent antimicrobial activity). In addition, cytokines can activate cell proliferation or cell differentiation (**figure 32.8**). They also can inhibit cell division and cause apoptosis (programmed cell death). **Chemokines,** one type of cytokine, stimulate chemotaxis and chemokinesis (i.e., they direct cell movement), and thus play an important role in the acute inflammatory response. Some examples of important cytokines and their functions are given in **table 32.3**.

Interferons (IFNs) are a group of related low molecular weight, regulatory cytokines produced by certain eukaryotic cells in response to a viral infection. Several classes of interferons are recognized: IFN-γ is a family of 20 different molecules that can be synthesized by virus-infected leukocytes, antigen-stimulated T cells, and natural killer cells (table 32.3). IFN-α/β is derived from virus-infected fibroblasts. Although interferons do not prevent virus entry into host cells, they prevent viral replication and assembly, thereby limiting viral infection (**figure 32.9**). *Antiviral Activity of Interferon*

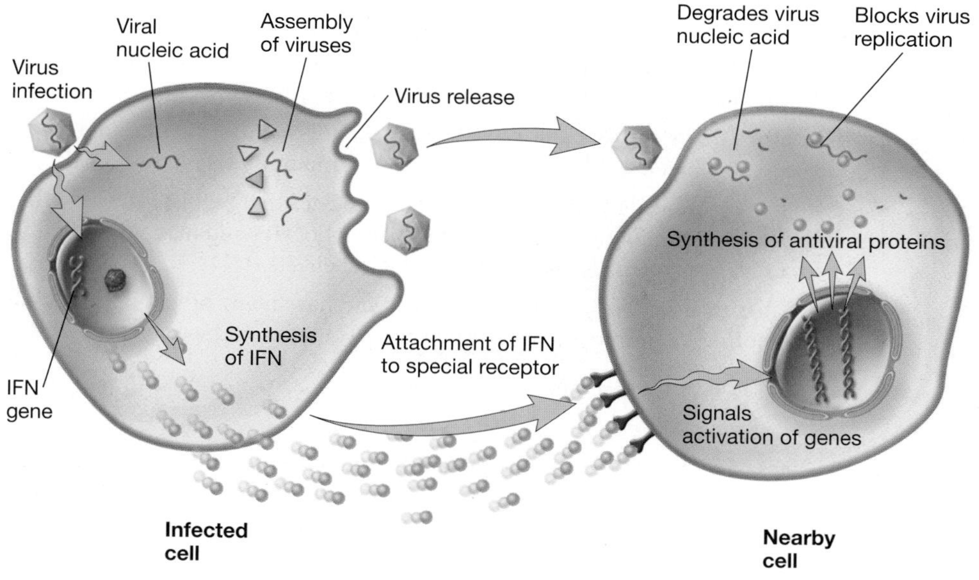

FIGURE 32.9 The Antiviral Action of Interferon. Interferon (IFN) synthesis and release is often induced by a virus infection or double-stranded RNA (dsRNA). Interferon binds to a ganglioside receptor on the plasma membrane of a second cell and triggers the production of enzymes that render the cell resistant to virus infection. The two most important such enzymes are oligo(A) synthetase and a protein kinase. When an interferon-stimulated cell is infected, viral protein synthesis is inhibited by an active endoribonuclease that degrades viral RNA. An active protein kinase phosphorylates and inactivates the initiation factor eIF-2 required for viral protein synthesis.

Table 32.3 Some of the Cytokines That Mediate Immune Responses

Cytokine	*Cell Source*	*Functions*
IL-1 (interleukin-1)	Monocytes/macrophages, endothelial cells, fibroblasts, neuronal cells, glial cells, keratinocytes, epithelial cells	Produces a wide variety of effects on the differentiation and function of cells involved in inflammatory and immune effector responses; also affects central nervous and endocrine systems. It is an endogenous pyrogen.
IL-2 (interleukin-2, T-cell growth factor)	T cells (T_H1)	Stimulates T-cell proliferation and differentiation; enhances cytolytic activity of NK cells; promotes proliferation and immunoglobin secretion of activated B cells
IL-3 (interleukin-3)	T cells, keratinocytes, neuronal cells, mast cells	Stimulates the production and differentiation of macrophages, neutrophils, eosinophils, basophils, and mast cells
IL-4 (interleukin-4, B-cell growth factor-1 [BCGF-1], B-cell stimulatory factor-1 [BCSF-1])	T cells (T_H2), macrophages, mast cells, basophils, B cells	Induces the differentiation of naive $CD4^+$ T cells into T-helper cells; induces the proliferation and differentiation of B cells; exhibits diverse effects on T cells, monocytes, granulocytes, fibroblasts, and endothelial cells
IL-5 (interleukin-5)	T cells (T_H2)	Growth and activation of B cells and eosinophils; chemotactic for eosinophils
IL-6 (interleukin-6, cytotoxic T-cell differentiation factor, B-cell differentiation factor)	T_H2 cells, monocytes/macrophages, fibroblasts, hepatocytes, endothelial cells, neuronal cells	Activates hematopoietic cells; induces growth of T cells, B cells, hepatocytes, keratinocytes, and nerve cells; stimulates the production of acute-phase proteins
IL-8 (interleukin-8)	Monocytes, endothelial cells, fibroblasts, alveolar epithelium, T cells, keratinocytes, neutrophils, hepatocytes	Chemoattractant for PMNs and T cells; causes PMN degranulation and expression of receptors; inhibits adhesion of PMNs to cytokine-activated endothelium; promotes migration of PMNs through nonactivated endothelium
IL-10 (interleukin-10)	T cells (T_H2), B cells, macrophages, keratinocytes	Reduces the production of IFN-γ, IL-1, TNF-α, and IL-6 by macrophages; in combination with IL-3 and IL-4, causes mast cell growth; in combination with IL-2, causes growth of cytotoxic T cells and differentiation of $CD8^+$ cells
IFNs α/β (interferons α/β)	T cells, B cells, monocytes/macrophages, fibroblasts	Antiviral activity, antiproliferative; stimulates macrophage activity; increases MHC class I protein expression on cells; regulates the development of the specific immune response
IFN-γ (interferon-γ)	T cells (T_H1, CTLs), NK cells	Activation of T cells, macrophages, neutrophils, and NK cells; antiviral and antiproliferative activities; increases class I and II MHC molecule expression on various cells
TNF-α (tumor necrosis factor-α [cachectin])	T cells, macrophages, and NK cells	A wide variety of effects due to its ability to mediate expression of genes for growth factors and cytokines, transcription factors, receptors, inflammatory mediators, and acute-phase proteins; plays a role in host resistance to infection by serving as an immunostimulant and mediator of the inflammatory response; cytotoxic for tumor cells
TNF-β (tumor necrosis factor-β [lymphotoxin])	T cells, B cells	Same as TNF-α
G-CSF (granulocyte colony-stimulating factor)	T cells, macrophages, neutrophils	Enhances the differentiation and activation of neutrophils
M-CSF (macrophage colony-stimulating factor)	T cells, neutrophils, macrophages, fibroblasts, endothelial cells	Stimulates various functions of monocytes and macrophages, promotes the growth and development of macrophage colonies from undifferentiated precursors

Another group of noteworthy cytokines are **endogenous pyrogens,** which elicit fever in the host. From a physiological point of view, fever results from disturbances in hypothalamic thermoregulatory activity, leading to an increase of the thermal "set point." In adult humans, fever is defined as an oral temperature above 37°C (98.6°F). The most common cause of a fever is a viral or bacterial infection (or bacterial toxins). Examples of endogenous pyrogens include interleukin-1 (IL-1), IL-6, and tissue necrosis factor (TNF); all are produced by host macrophages in response to pathogenic microorganisms. After their release, these pyrogens circulate to the hypothalamus and induce neurons to secrete prostaglandins. Prostaglandins reset the hypothalamic thermostat to a higher temperature, and temperature-regulating reflex mechanisms then act to bring the core body temperature up to this new setting.

The fever induced by a microorganism augments the host's defenses in three ways: (1) it stimulates leukocytes so that they can destroy the microorganism; (2) it enhances the specific activity of the immune system; and (3) it enhances microbiostasis (growth inhibition) by decreasing available iron to the microorganism. Evidence suggests that some hosts are able to redistribute the iron during a fever in an attempt to withhold it from the microorganism (hypoferremia). Conversely, the virulence of many microorganisms is enhanced with increased iron availability (hyperferremia). For example, gonococci, the causative agent of gonorrhea, spread most often during menstruation, a time in which an increased concentration of free iron is available to these bacteria.

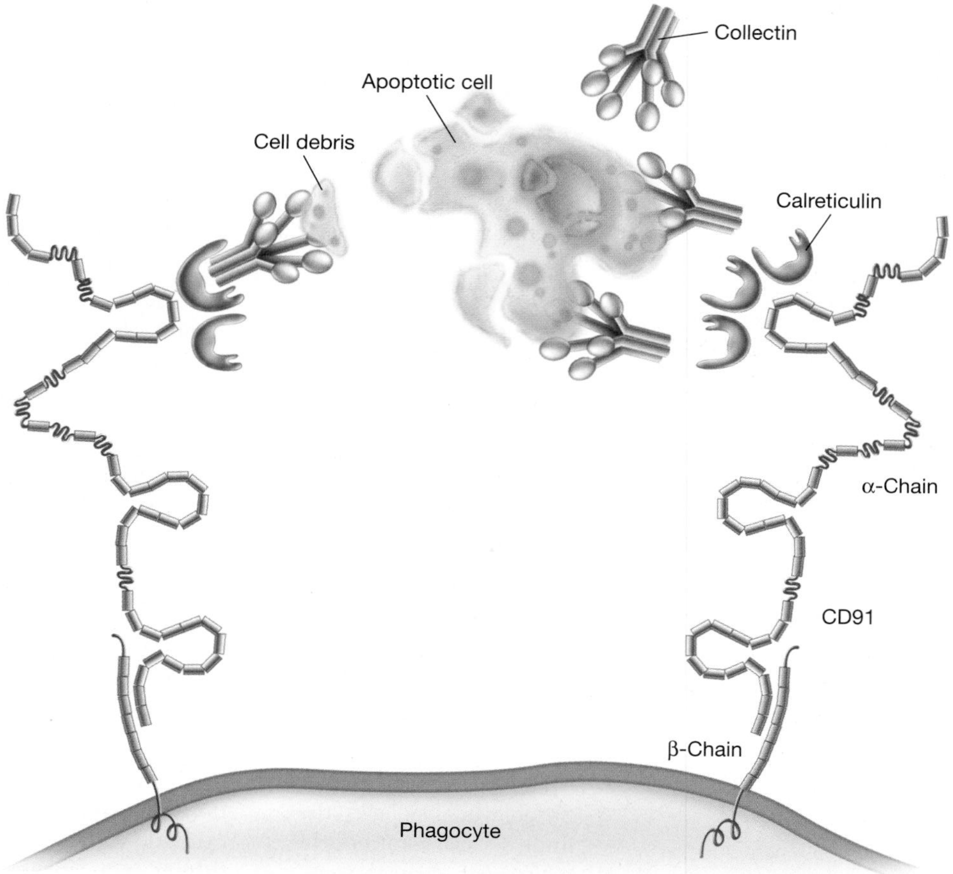

FIGURE 32.10 Collectins Are Molecular Scavengers. Collectins (also known as defense collagens) are a family of similar proteins that bind cellular debris and dying cells through their globular head groups. Their collagenous tails are then recognized by calreticulin associated with α-2 macroglobulin (CD91) on the surface of phagocytes.

Acute-Phase Proteins

Macrophages release cytokines (IL-1, IL-6, IL-8, TNF-α, etc.) upon activation by bacteria, which stimulate the liver to rapidly produce acute-phase proteins. These include C-reactive protein (CRP), mannose-binding lectin (MBL), and surfactant proteins A (SP-A) and D (SP-D), all of which can bind bacterial surfaces and act as opsonins. CRP can interact with C1q to activate the classical complement pathway. MBL activates the alternative complement pathway. SP-A, SP-D, and C1q are collectins—proteins composed of a collagen-like motif connected by α-helices to globular binding sites (**figure 32.10**). These proteins (along with others) police host tissues by binding to and assisting in the removal of bacteria.

1. Describe the role of cytokines and interferons in innate immunity.
2. How do interferons render cells resistant to viruses?
3. How might acute-phase reactants assist in pathogen removal?
4. How can a fever be beneficial to a host?
5. What is the role of collectins in innate immunity?

32.4 Cells, Tissues, and Organs of the Immune System

The immune system is an organization of molecules, cells, tissues, and organs, each with a specialized role in defending against viruses and other microorganisms, cancer cells, and other foreign entities. Immune system cells and tissue are now considered.

Cells of the Immune System

The cells responsible for both nonspecific and specific immunity are the **leukocytes** (Greek *leukos,* white, and *kytos,* cell). All leukocytes originate from pluripotent stem cells in the fetal liver and in the bone marrow of the animal host (**figure 32.11**). Pluripotent stem cells have not yet committed to differentiating into one specific cell type. When they migrate to other body sites, some differentiate into hematopoietic precursor cells that are destined to become blood cells. When stimulated to undergo further development, some leukocytes become residents within tissues, where they respond to local trauma. These cells may sound the alarm that signals invasion by foreign organisms. Other leukocytes circulate in body fluids and are recruited to the sites of infection after the alarm has been raised. The average adult has approximately 7,400 leukocytes per cubic millimeter of blood (**table 32.4**). This average value shifts substantially during an immune response. In defending the host against pathogenic microorganisms, leukocytes cooperate with each other first to recognize the pathogen as an invader and then to destroy it. The different types of leukocytes are now briefly examined.

Granulocytes

Granulocytes have irregularly shaped nuclei with two to five lobes. Their cytoplasm has granules that contain reactive substances that kill microorganisms and enhance inflammation (figure 32.11). Three types of granulocytes exist: basophils, eosinophils, and neutrophils. Because of the irregularly shaped nuclei, neutrophils are also called **polymorphonuclear neutrophils,** or **PMNs.**

Basophils (Greek *basis,* base, and *philein,* to love) have irregularly shaped nuclei with two lobes and granules that stain bluish-black with basic dyes (figure 32.11). Basophils are nonphagocytic cells that release specific compounds from their cytoplasmic granules in response to certain types of stimulation. These molecules include histamine, prostaglandins, serotonin, and leukotrienes. Because these compounds influence the tone and diameter of blood vessels, they are termed **vasoactive mediators.** Basophils (and mast cells) possess high-affinity receptors for the type of antibody associated with allergic responses. When these cells become coated with this type of antibody, binding of antigen to the antibody triggers the secretion of vasoactive mediators. As discussed in chapter 33, vasoactive mediators play a major role in certain allergic responses such as eczema, hay fever, and asthma. ▶▶| *Antibodies (section 33.7); Hypersensitivities (section 33.10)*

Eosinophils (Greek *eos,* dawn, and *philien*) have a two-lobed nucleus connected by a slender thread of chromatin and granules that stain red with acidic dyes (figure 32.11). Unlike basophils, eosinophils migrate from the bloodstream into tissue spaces, especially mucous membranes. They are important in the defense against protozoan and helminth parasites, mainly by releasing cationic peptides (p. 765) and reactive oxygen intermediates (p. 784) into the extracellular fluid. These molecules damage the parasite plasma membrane, killing it. Eosinophils also play a role in allergic reactions, as they have granules containing histaminase and aryl sulphatase, downregulators of the inflammatory mediators histamine and leukotrienes, respectively. Thus their numbers often increase during allergic reactions, especially type 1 hypersensitivities (*see p. 816*).

Neutrophils (Latin *neuter,* neither, and *philien*) are phagocytic cells with a nucleus that has three to five lobes connected by slender threads of chromatin. Neutrophils have inconspicuous organelles known as primary and secondary granules, which contain lytic enzymes and bactericidal substances. Primary granules contain peroxidase, lysozyme, defensins, and various hydrolytic enzymes, whereas the smaller secondary granules have collagenase, lactoferrin, cathelicidins, and lysozyme. These granules help digest foreign material after it is phagocytosed (section 32.5). Neutrophils also use oxygen-dependent and oxygen-independent pathways that generate additional antimicrobial substances to kill ingested microorganisms. Like macrophages, neutrophils have receptors for antibodies and complement proteins (figure 32.5) and are highly phagocytic. However, unlike macrophages, neutrophils do not reside in healthy tissue but circulate in blood so they can rapidly migrate to the site of tissue damage and infection, where they become the principal phagocytic and microbicidal cells. Neutrophils and their antimicrobial compounds are described in more detail in the context of phagocytosis (section 32.5) and the inflammatory response (section 32.6).

Table 32.4 Normal Adult Blood Count

Cell Type	*Cells/mm^3*	*Percent WBC*
Red blood cells	5,000,000	
Platelets	250,000	
White blood cells	7,400	100
Neutrophils	4,320	60
Lymphocytes	2,160	30
Monocytes	430	6
Eosinophils	215	3
Basophils	70	1

Mast Cells

Mast cells are bone marrow–derived cells that differentiate in the blood and connective tissue. Although they contain granules with histamine and other pharmacologically active substances similar to those in basophils, they arise from a different cellular lineage (figure 32.11). Mast cells, along with basophils, are important in the development of allergies and hypersensitivities.

Monocytes and Macrophages

Monocytes (Greek *monos,* single, and *cyte,* cell) are mononuclear leukocytes with an ovoid- or kidney-shaped nucleus and granules in the cytoplasm (figure 32.11). They are produced in the bone

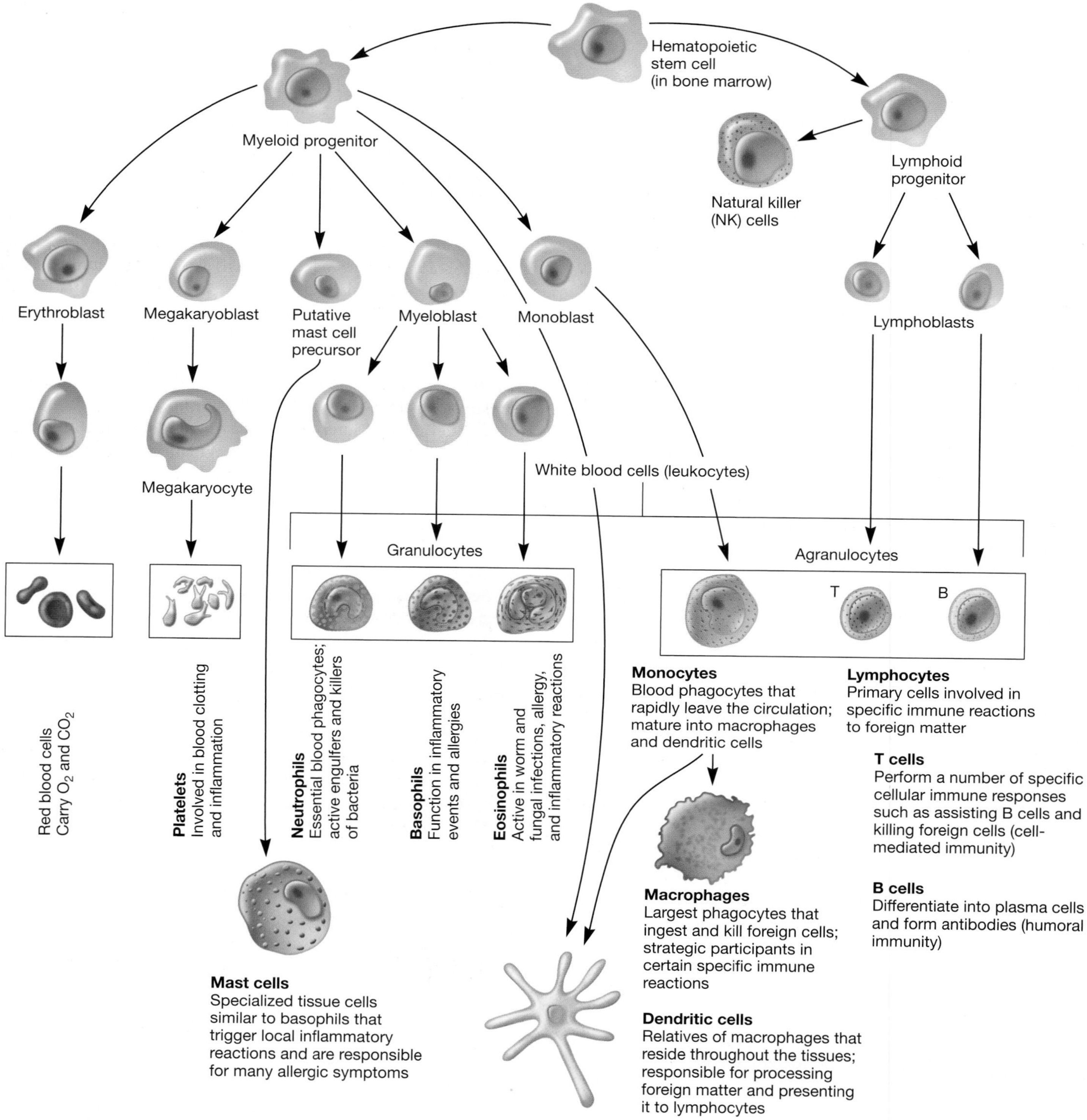
Hematopoietic stem cell (in bone marrow)
Myeloid progenitor
Lymphoid progenitor
Natural killer (NK) cells
Erythroblast
Megakaryoblast
Putative mast cell precursor
Myeloblast
Monoblast
Lymphoblasts
Megakaryocyte
White blood cells (leukocytes)
Granulocytes
Agranulocytes
T
B
Red blood cells
Carry O_2 and CO_2
Platelets
Involved in blood clotting and inflammation
Neutrophils
Essential blood phagocytes; active engulfers and killers of bacteria
Basophils
Function in inflammatory events and allergies
Eosinophils
Active in worm and fungal infections, allergy, and inflammatory reactions
Monocytes
Blood phagocytes that rapidly leave the circulation; mature into macrophages and dendritic cells
Lymphocytes
Primary cells involved in specific immune reactions to foreign matter
T cells
Perform a number of specific cellular immune responses such as assisting B cells and killing foreign cells (cell-mediated immunity)
B cells
Differentiate into plasma cells and form antibodies (humoral immunity)
Macrophages
Largest phagocytes that ingest and kill foreign cells; strategic participants in certain specific immune reactions
Mast cells
Specialized tissue cells similar to basophils that trigger local inflammatory reactions and are responsible for many allergic symptoms
Dendritic cells
Relatives of macrophages that reside throughout the tissues; responsible for processing foreign matter and presenting it to lymphocytes

FIGURE 32.11 The Different Types of Human Blood Cells. Pluripotent stem cells in the bone marrow divide to form two blood cell lineages: (1) the lymphoid progenitor cell gives rise to B cells that become antibody-secreting plasma cells, T cells that become activated T cells, and natural killer cells; and (2) the common myeloid progenitor cell gives rise to dendritic cells (DCs) and the granulocytes (neutrophils, eosinophils, basophils), monocytes that give rise to macrophages and DCs, an unknown precursor that gives rise to mast cells, megakaryocytes that produce platelets, and the erythroblast that produces erythrocytes (red blood cells). DCs can arise directly from myeloid progenitor cells differentiating into early and late DC precursors and from nonproliferating monocytes.

Figure 32.11 Micro Inquiry

What is the difference between pluripotent and hematopoietic stem cells? Are they both found in bone marrow?

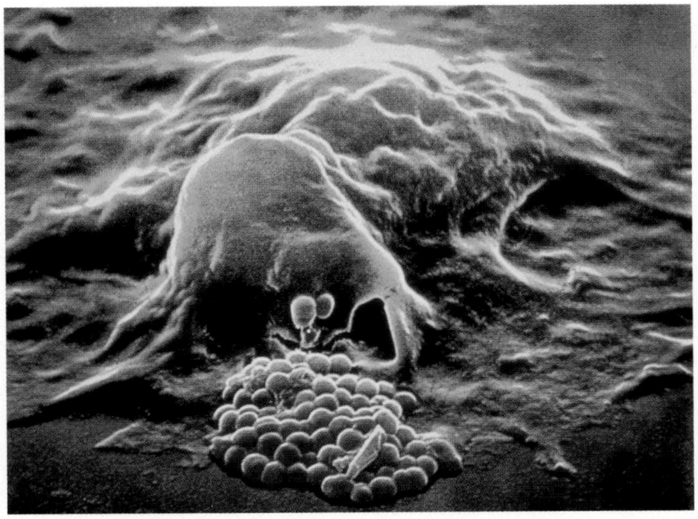

FIGURE 32.12 Phagocytosis by a Macrophage. One type of nonspecific host resistance involves white blood cells called macrophages and the process of phagocytosis. This scanning electron micrograph (×3,000) shows a macrophage devouring a colony of bacteria. Phagocytosis is one of many nonspecific defenses humans and other animals use to combat microbial pathogens.

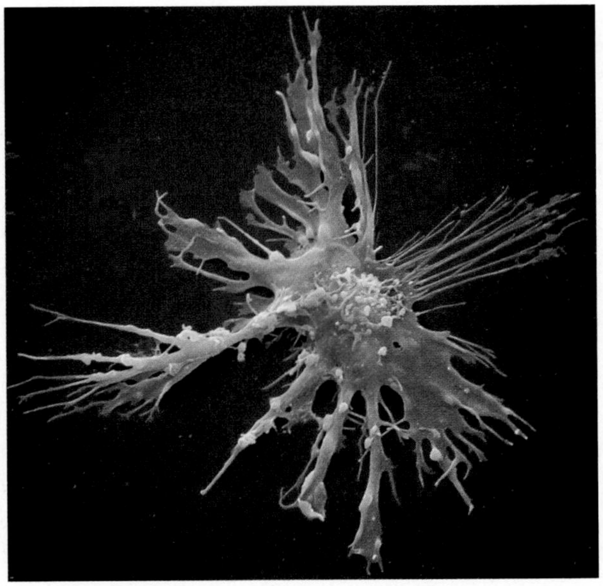

FIGURE 32.13 The Dendritic Cell. The dendritic cell was named for its cellular extensions, which resemble the dendrites of nerve cells. Dendritic cells reside in most tissue sites, where they survey their local environments for pathogens and altered host cells.

Figure 32.13 *Micro Inquiry*

Where are dendritic cells most commonly found?

marrow and enter the blood, circulate for about eight hours, enlarge, migrate to the tissues, and mature into macrophages or dendritic cells.

Macrophages (Greek *macros,* large, and *phagein,* to eat) are derived from monocytes and are classified as mononuclear phagocytic leukocytes. However, they are larger than monocytes, contain more organelles that are critical for phagocytosis, and have a plasma membrane covered with microvilli (**figure 32.12**). Macrophages have surface molecules that function as receptors to recognize common components of pathogens. These receptors include mannose and fucose receptors, and a special class of molecules called toll-like receptors (p. 781), which bind lipopolysaccharide (LPS), peptidoglycan, fungal cell wall components called zymosan, viral nucleic acids, and foreign DNA. These microbial molecules are examples of ***p*****athogen-*a*****ssociated *m*****olecular *p*****atterns (PAMPs)**. PAMPS enable macrophages to distinguish between potentially harmful microbes and other host molecules. After the pathogen is recognized, the macrophage's ***p*****attern *r*****ecognition *r*****eceptors (PRRs)** bind the pathogen and phagocytose it. PAMPs and PRRs are discussed further in section 32.5. Macrophages also have receptors for antibodies and complement glycoproteins (figure 32.5). Both antibody and complement proteins can coat microorganisms or foreign material and enhance their phagocytosis. Recall that this enhancement is termed opsonization and is discussed in detail in section 32.3. Macrophages spread throughout the animal body and take up residence in specific tissues. Because macrophages are highly phagocytic, their function in nonspecific resistance is discussed in more detail in the context of phagocytosis (section 32.5). Although the specifics of phagocytosis will be discussed shortly, here we emphasize that phagocytosis involves the engulfment of particles and microorganisms that are then enclosed in a phagocytic vacuole. Macrophages digest larger particles into small fragments that become attached to surface proteins so as to be used as activation signals for T cells. Recent evidence suggests that the macrophage can also activate B cells by presenting antigen on its membrane to the B-cell receptor. ▶▶| *Antigens (section 33.2); Antibodies (section 33.7)*

Dendritic Cells

Dendritic cells are not a single cell type; they are a heterogeneous group of cells so named because of their dendrite (neuron)-like appendages (**figure 32.13**). They arise from various hematopoietic cell lineages; lymphoid and myeloid dendritic cells are documented. Immature dendritic cells constitute about 0.2% of peripheral blood leukocytes and can be stimulated to maturity by specific cytokines. Most dendritic cells are tissue-bound, where they play an important role in bridging nonspecific resistance and specific immunity. They are present in the skin and mucous membranes of the nose, lungs, and intestines, where they readily contact invading pathogens, phagocytose and process antigens, and display foreign antigens on their surface. This process is known as **antigen presentation** and is discussed in more detail in section 33.4.

Dendritic cells are similar to macrophages in their ability to recognize specific PAMPs on microorganisms. They also possess pattern recognition receptors (PRRs) to bind and phagocytose the pathogen. The dendritic cells then migrate to lymphoid tissues, where, as activated cells, they present antigen to T cells. Antigen presentation triggers the activation of T cells, which is critical for the initiation and regulation of an effective specific immune response. However, unlike macrophages, dendritic cells are capable of eliciting specific immune responses from naïve (i.e., those that have never encountered the specific antigen they recognize) T and B cells. Thus not only do dendritic cells destroy invading pathogens as part of the innate response, but they also help trigger specific immune responses from naïve lymphocytes. ▶▶| *T-cell biology (section 33.5)*

1. Describe the structure and function of each of the following blood cells: basophil, eosinophil, neutrophil, monocyte, macrophage, and dendritic cell. Which cells are phagocytic?
2. What is the significance of the respective blood cell percentages in blood?
3. How does a dendritic cell differ from a macrophage when resisting microbial invasion?

Search This: IL-7 and dendritic cells

Lymphocytes

Lymphocytes (Latin *lympha,* water, and *cyte,* cell) are the major cells of the specific immune system. Lymphocytes can be divided into three populations: T cells, B cells, and natural killer, or NK, cells. Lymphocytes leave the bone marrow in a kind of cellular stasis—not actively replicating like other somatic cells. B and T lymphocytes differentiate from their respective lymphoid precursor cells. Once activated by specific antigens, lymphocytes differentiate into mature forms. Furthermore, subsets of activated lymphocyte clones then enter cellular stasis only to respond more vigorously on reactivation. These cells are called **memory cells.**

After **B lymphocytes** or **B cells** reach maturity within the bone marrow, they circulate in the blood and disperse into various lymphoid organs, where they become activated. The activated B cell becomes more ovoid. Its nuclear chromatin condenses, and numerous folds of endoplasmic reticulum become more visible. A mature, activated B cell is called a **plasma cell,** which secretes large quantities of antibodies (**figure 32.14**). ▶▶| *Actions of antibodies (section 33.8)*

Lymphocytes destined to become **T lymphocytes** or **T cells** leave the bone marrow and mature in the thymus gland. They can remain in the thymus, circulate in the blood, or reside in lymphoid organs such as the lymph nodes and spleen, as B cells do. Also like B cells, T cells require a specific antigen to bind to their receptor to signal the continuation of replication. Unlike B cells, however, T cells do not secrete antibodies. Activated T cells differentiate into helper T cells (T_H) and cytotoxic lymphocytes (CTLs) that produce and secrete **cytokines** (figure 32.14 and p. 799).

Natural killer (NK) cells are a small population of large, nonphagocytic granular lymphocytes that play an important role in innate immunity (figure 32.11). The major NK cell function is to destroy malignant cells and cells infected with microorganisms. They recognize their targets in one of two ways. They can bind to antibodies that coat infected or malignant cells; thus the antibody bridges the two cell types. This process is called **antibody-dependent cell-mediated cytotoxicity (ADCC) (figure 32.15)** and can result in the death of the target cell. The second way that NK cells recognize infected cells and cancer cells relies on the presence of specialized proteins on the surface of all nucleated host cells, known as the class I major histocompatibility complex (MHC) antigen. If a host cell loses this MHC protein, as when some viruses or cancers overtake the cell, the NK cell kills it by releasing pore-forming proteins and cytotoxic enzymes called **granzymes.** Together the pore-forming proteins and the granzymes cause the target cell to die (**figure 32.16**). ▶▶| *Recognition of foreignness (section 33.4)*

1. What is a plasma cell?
2. What is the purpose of T-cell secretion of cytokines?
3. Discuss the role of NK cells in protecting the host. What are the two mechanisms NK cells use to kill other cells?

Organs and Tissues of the Immune System

Based on function, the organs and tissues of the immune system can be divided into primary or secondary lymphoid organs and tissues (**figure 32.17**). The primary organs and tissues are where immature lymphocytes mature and differentiate into antigen-sensitive B and T cells. The thymus is the primary lymphoid organ for T cells, and the bone marrow is the primary lymphoid tissue for B cells. The secondary organs and tissues serve as areas where lymphocytes may encounter and bind antigen, whereupon they proliferate and differentiate into fully active, antigen-specific effector cells. The spleen is a secondary lymphoid organ, and lymph nodes and mucosal-associated tissues (GALT—gut-associated lymphoid tissue, and SALT—skin-associated lymphoid tissues) are secondary lymphoid tissues. These are now discussed in more detail.

Primary Lymphoid Organs and Tissues

The **thymus** is a highly organized lymphoid organ located above the heart. Precursor cells from the bone marrow migrate into the outer cortex of the thymus, where they proliferate. As they mature, about 98% die. This is due to a process known as **thymic**

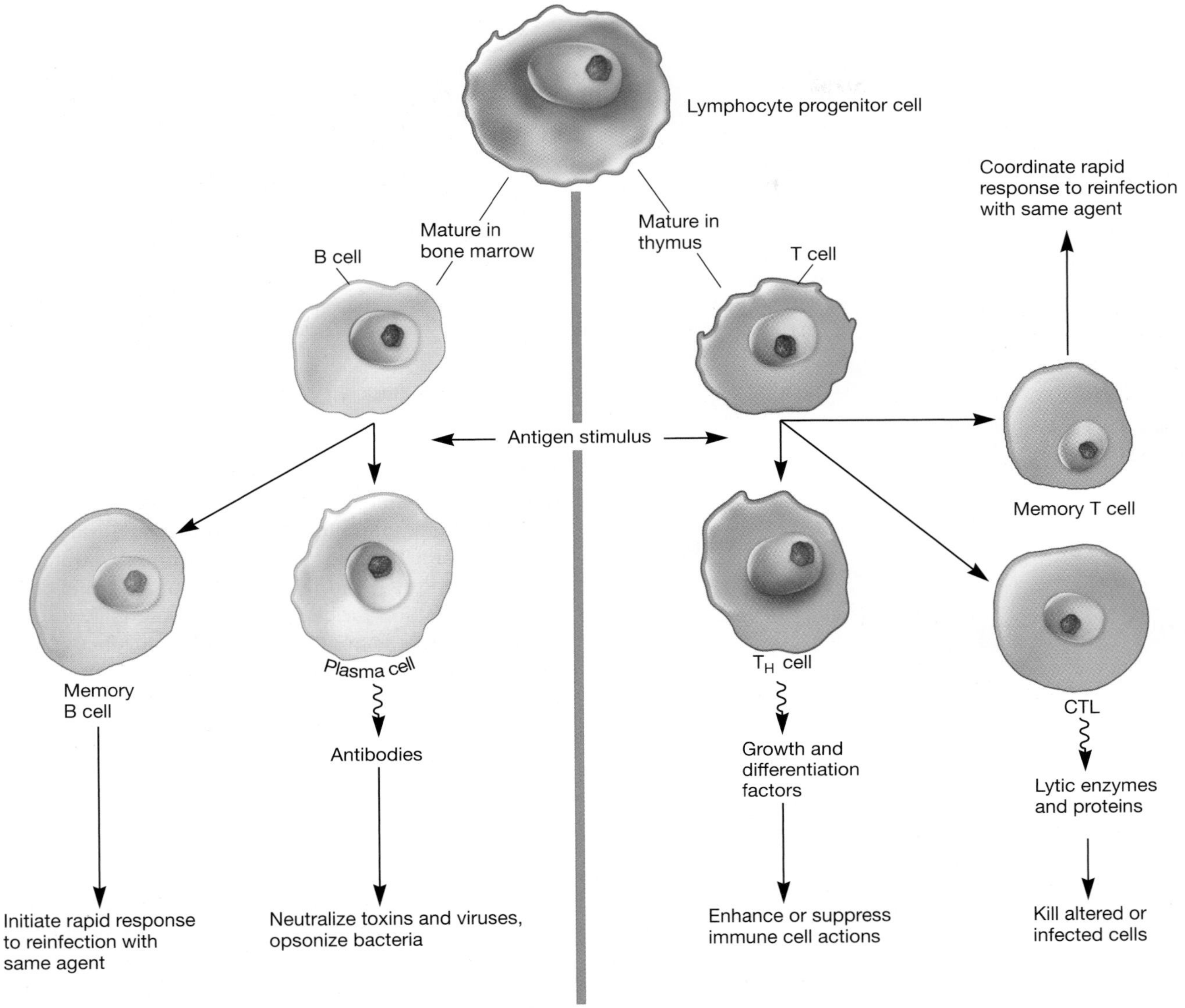

FIGURE 32.14 The Development and Function of B and T Lymphocytes. B cells and T cells arise from the same cell lineage but diverge into two different functional types. Immature B cells and T cells are indistinguishable by histological staining. However, they express different proteins on their surfaces that can be detected by immunohistochemistry. Additionally, the final secreted products of mature B and T cells can be used to identify the cell type.

Figure 32.14 Micro Inquiry

What molecules are secreted by mature B cells? What are the two types of mature T cells?

selection in which T cells that recognize host (self) antigens are destroyed. The remaining 2% move into the medulla of the thymus (figure 32.17*a*), become mature T cells, and subsequently enter the bloodstream. These T cells recognize nonhost (nonself) antigens.

In mammals, the **bone marrow** (figure 32.17*b*) is the site of B-cell maturation. Like thymic selection during T-cell maturation, a selection process within the bone marrow eliminates nonfunctioning B cells and those bearing self-reactive antigen receptors. In birds, undifferentiated lymphocytes move from the bone marrow to the bursa of Fabricius, where B cells mature; this is where B cells were first identified and how they came to be known as "B" (for bursa) cells. ▶▶| *B-cell biology (section 33.6)*

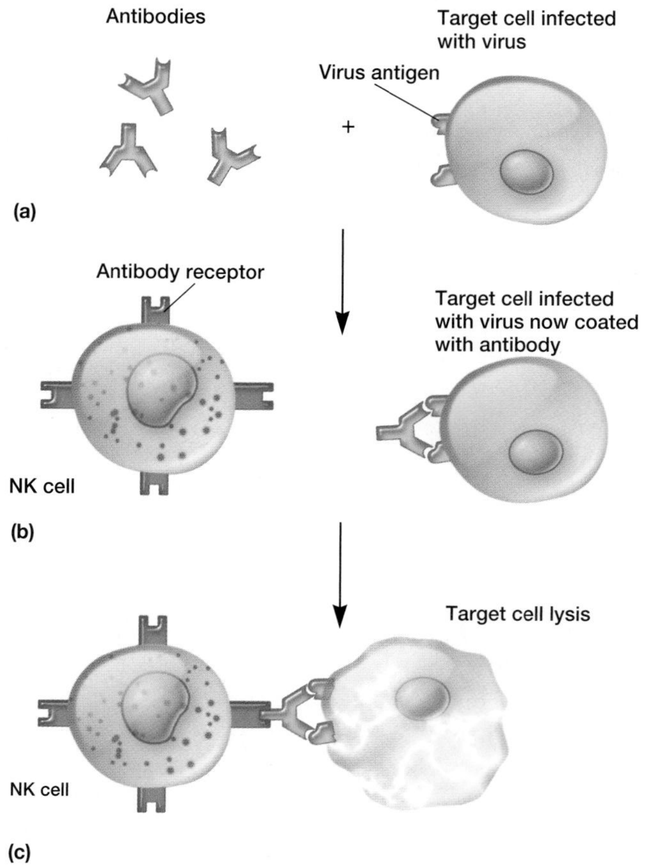

FIGURE 32.15 Antibody-Dependent Cell-Mediated Cytotoxicity. (a) In this mechanism, antibodies bind to a target cell infected with a virus. (b) NK cells have specific antibody receptors on their surface. (c) When the NK cells encounter virus-infected cells coated with antibody, they kill the target cell.

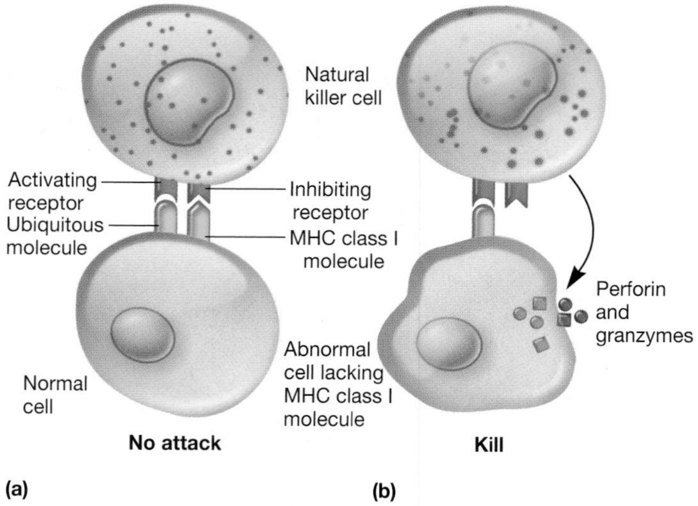

FIGURE 32.16 The System Used by Natural Killer Cells to Recognize Normal Cells and Abnormal Cells That Lack the Major Histocompatibility Complex Class I Surface Molecule. (a) The killer-activating receptor recognizes a ubiquitous molecule on the plasma membrane of a normal cell. Since the killer-inhibitory receptor recognizes the MHC class I molecule, there is no attack. (b) In the absence of the inhibitory signal, the receptor issues an order to the NK cell to attack and kill the abnormal cell. The cytotoxic granules of the NK cell contain perforin and granzymes. With no inhibitory signal, the granules release their contents, killing the abnormal cell.

Secondary Lymphoid Organs and Tissues

The **spleen** is the most highly organized secondary lymphoid organ. The spleen is a large organ located in the abdominal cavity (figure 32.17). It filters the blood and traps blood-borne microorganisms and antigens. Once trapped by splenic macrophages or dendritic cells, the pathogen is phagocytosed, killed, and digested. The resulting peptides are about 50 amino acids or less in length and are referred to as peptide antigens. The peptide antigens are delivered to the macrophage or dendritic cell surface within a protein receptor, where they are presented. This is the most common way B and T lymphocytes become activated to carry out their immune functions.

Lymph nodes lie at the junctions of lymphatic vessels, where phagocytic macrophages and dendritic cells trap pathogens and antigens (figure 32.17*c*). They then phagocytose the foreign material and present antigen to lymphocytes. Antigen presented to the class of T lymphocytes called T-helper cells activates these cells to release cytokines needed for B-cell activation. Thus lymph nodes represent one environment where B cells differentiate into memory cells and antibody-secreting plasma cells.

Lymphoid tissues are found throughout the body and act as regional centers of antigen sampling and processing (figure 32.17). Lymphoid tissues are found as highly organized or loosely associated cellular complexes. Some lymphoid cells are closely associated with specific tissues such as skin (skin-associated lymphoid tissue, or SALT) and mucous membranes (mucosal-associated lymphoid tissue, or MALT). SALT and MALT are good examples of highly organized lymphoid tissues, typically seen histologically as macrophages surrounded by specific areas of B and T lymphocytes and sometimes dendritic cells. Loosely associated lymphoid tissue is best represented by the bronchial-associated lymphoid tissue, or BALT, characterized by the lack of cellular partitioning (figure 32.17). The primary role of these lymphoid tissues is to efficiently organize leukocytes to increase interaction between the innate and the acquired arms of the immune response. As we now discuss, the lymphoid tissues serve as the interface between the nonspecific (innate) and specific (acquired) immunity of a host.

Despite the skin's defenses, at times pathogenic microorganisms gain access to the tissue under the skin surface. Here,

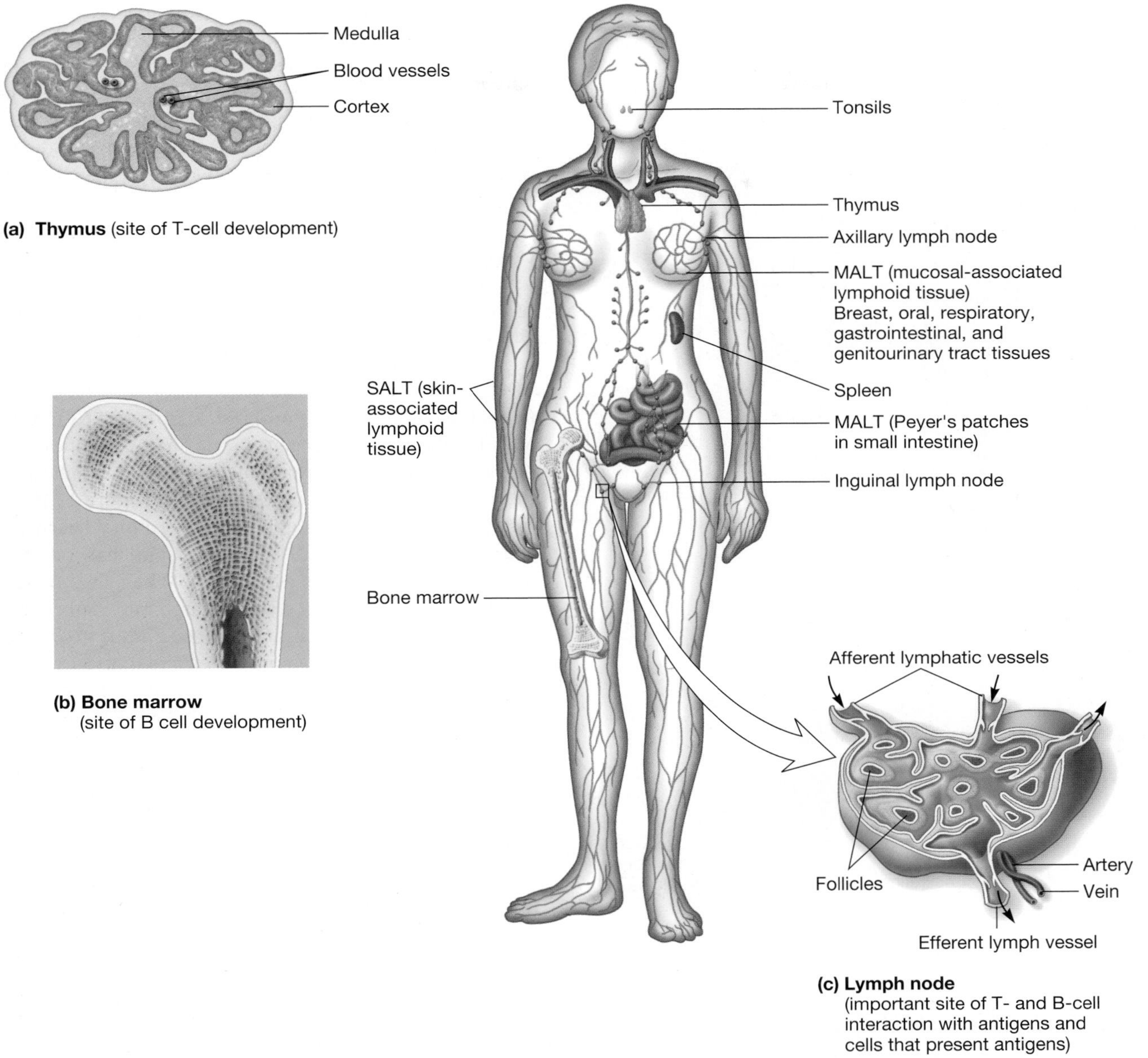

FIGURE 32.17 Anatomy of the Lymphoid System. Lymph is distributed through a system of lymphatic vessels, passing through many lymph nodes and lymphoid tissues. For example, (a) the thymus is involved in T-cell development and atrophies with age; (b) the bone marrow is the site of B-cell development; and (c) lymph enters a lymph node through the afferent lymph vessels, percolates through and around the follicles in the node, and leaves through the efferent lymphatic vessels. The lymphoid follicles are the site of cellular interactions and extensive immunologic activity.

***Figure 32.17** Micro Inquiry*

What is the major difference between primary and secondary lymphoid organs and tissues?

they encounter a specialized set of cells called the **skin-associated lymphoid tissue (SALT)** (**figure 32.18**). The major function of SALT is to confine microbial invaders to the area immediately underlying the epidermis and to prevent them from gaining access to the bloodstream. One type of SALT cell is the **Langerhans cell,** a specialized myeloid cell that can phagocytose antigens. Once the Langerhans cell has internalized the antigen, it migrates from the epidermis to nearby lymph nodes, where it

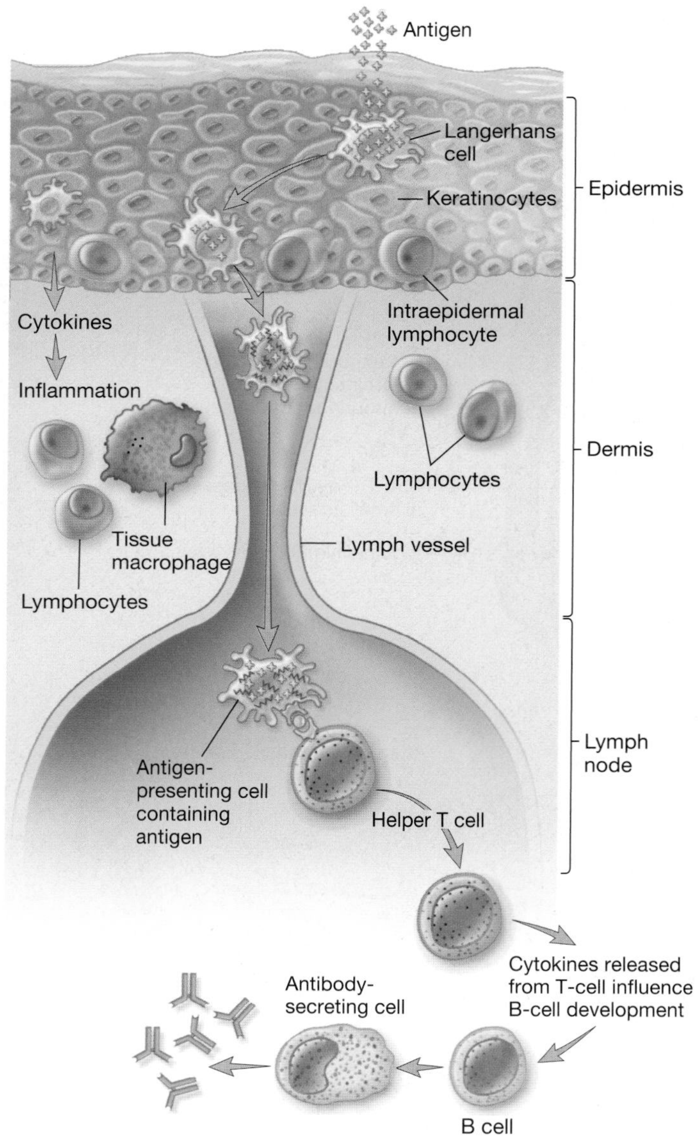

FIGURE 32.18 Skin-Associated Lymphoid Tissue (SALT). Keratinocytes make up 90% of the epidermis. They are capable of secreting cytokines that cause an inflammatory response to invading pathogens. Langerhans cells internalize antigen and move to a lymph node, where they differentiate into dendritic cells that present antigen to helper T cells. The intraepidermal lymphocytes may function as T cells that can activate B cells to induce an antibody response.

differentiates into a mature dendritic cell. Dendritic cells can engulf and degrade antigens into small fragments that are used to activate nearby lymphocytes to induce the acquired immune system. This dendritic cell–lymphocyte interaction illustrates another bridge between the innate and acquired immune systems.

The epidermis also contains another type of SALT cell called the **intraepidermal lymphocyte** (figure 32.18), a specialized T cell having potent cytolytic and immunoregulatory responses to antigen. These cells are strategically located in the skin so that they can intercept any antigens that breach the first line of defense. Most of these specialized SALT cells have limited receptor diversity and have likely evolved to recognize common skin pathogen patterns.

The specialized lymphoid tissue in mucous membranes is called ***m**ucosal-**a**ssociated **l**ymphoid **t**issue* **(MALT).** There are several types of MALT. The system most studied is the ***g**ut-**a**ssociated **l**ymphoid **t**issue* **(GALT).** GALT includes the tonsils, adenoids, diffuse lymphoid areas along the gut, and specialized regions in the intestine called Peyer's patches. Less well-organized MALT also occurs in the respiratory system and is called ***b**ronchial-**a**ssociated **l**ymphoid **t**issue* **(BALT);** the diffuse MALT in the urogenital system does not have a specific name. MALT operates by two basic mechanisms. First, when an antigen arrives at the mucosal surface, it contacts an **M cell** (**figure 32.19*a***). M cells lack the brush border of microvilli found on adjacent columnar epithelial cells. They reside above large epithelial pockets containing B cells, T cells, and macrophages. When an antigen contacts an M cell, it is endocytosed and released into the pocket. Macrophages engulf the antigen or pathogen and try to destroy it. An M cell also can endocytose an antigen and transport it to a cluster of cells called an organized lymphoid follicle (figure 32.19*b*). The B cells within this follicle recognize the antigen and mature into antibody-producing plasma cells. The plasma cells leave the follicle and secrete mucous membrane–associated antibody. The antibody is then transported into the lumen of the gut, where it interacts with the antigen that caused its production. Similar to SALT, GALT intra- and interepithelial lymphocytes are strategically distributed so the likelihood of antigen detection is increased, should the intestinal membrane be breached. ▶▶| *Antibodies (section 33.7)*

Thus a microbe attempting to invade a potential host is greeted by nonspecific, physical, chemical, and granulocyte barriers that are designed to kill the invader, digest the carcass into small antigens, and assist the lymphocytes in formulating long-term protection against the next invasion. We now examine the phagocytic processes in more detail and then consider how the host integrates many of the innate immune activities into a substantial barrier to microbial invasion, known as the inflammatory response.

1. Briefly describe each of the primary lymphoid organs and tissues.
2. What is the function of the spleen? A lymph node? The thymus? What is the importance of thymic selection?
3. Injury to the spleen can lead to its removal. What impact would this have on host defenses?
4. Describe SALT function in the immune response.
5. How do M cells function in MALT?

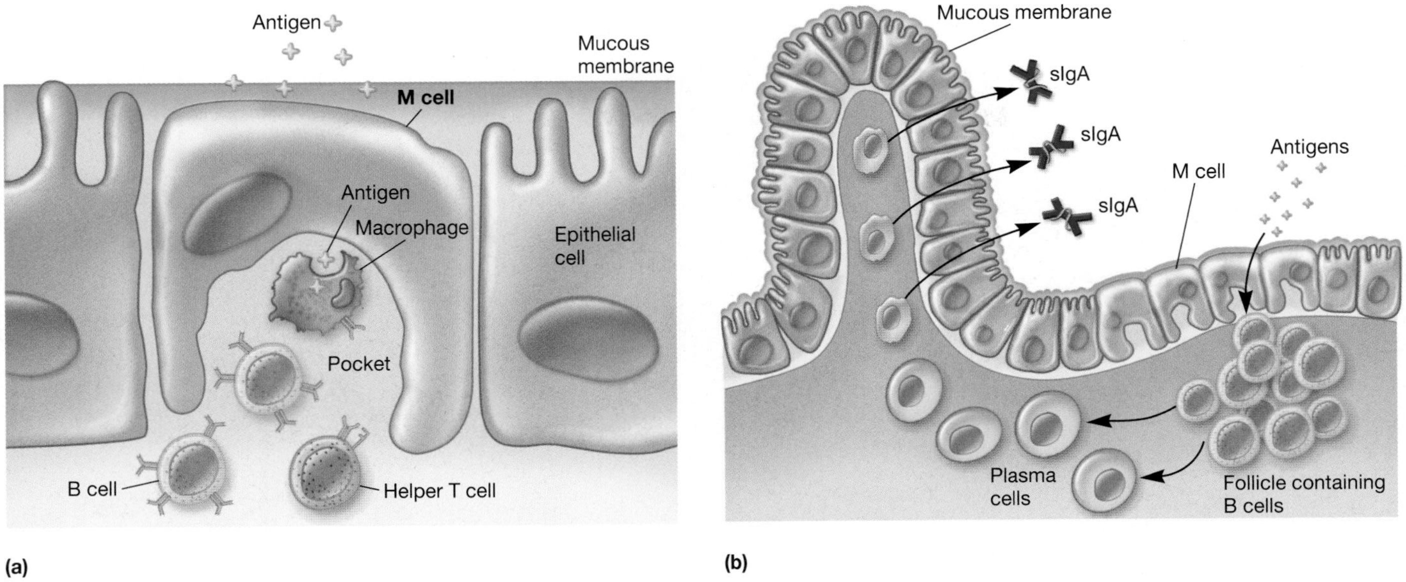

FIGURE 32.19 Function of M Cells in Mucosal-Associated Immunity. (a) Structure of an M cell located between two epithelial cells in a mucous membrane. The M cell endocytoses the pathogen and releases it into the pocket containing helper T cells, B cells, and macrophages. It is within the pocket that the pathogen often is destroyed. (b) The antigen is transported by the M cell to the organized lymphoid follicle containing B cells. The activated B cells mature into plasma cells, which produce antibodies that are released into the lumen, where they react with antigen that caused their production.

32.5 Phagocytosis

During their lifetimes, humans and other vertebrates encounter many microbial species, but only a few of these can grow and cause serious disease in otherwise healthy hosts. Phagocytic cells (monocytes, tissue macrophages, dendritic cells, and neutrophils) are an important early defense against invading microorganisms. These phagocytic cells recognize, ingest, and kill many extracellular microbial species by the process called **phagocytosis** (Greek *phagein,* to eat; *cyte,* cell; and *osis,* a process) (**figure 32.20**). Phagocytic cells use two basic molecular mechanisms for the recognition of microorganisms: (1) opsonin-independent (nonopsonic) recognition and (2) opsonin-dependent (opsonic) recognition. The phagocytic process can be greatly enhanced by opsonization. Recall that we discussed opsonin recognition in section 32.3. We now discuss nonopsonic recognition. *Phagocytosis*

Pathogen Recognition

The **opsonin-independent mechanism** is a receptor-based system wherein components common to many different pathogens are recognized to activate phagocytes (figure 32.20*a*). Phagocytic cells recognize pathogens by several means but appear to exploit a common signaling system to respond. One recognition mode, termed lectin phagocytosis, is based on the binding of a microbial lectin (Latin *legere,* to select or choose), a protein that specifically binds or cross-links carbohydrates to a carbohydrate moiety of a cell receptor (**figure 32.21**). A second mode results from protein-protein interactions between the peptide sequence arginine-glycine-aspartic acid (RGD) on the cell surface of microorganisms and RGD receptors found on all phagocytes. Third, hydrophobic interactions between bacteria and phagocytic cells also promote phagocytosis. A particular microbial species can express multiple binding sites, each recognized by a distinct receptor present on phagocytic cells.

A fourth type of interaction also involves the recognition of microbial antigen and plays a crucial role in nonspecific host resistance. This recognition strategy is based on the detection of conserved molecular structures called PAMPs, which were introduced on page 779. PAMPs are unique to microorganisms and are invariant among microorganisms of a given class. The most well-known examples of PAMPs are the lipopolysaccharide (LPS) of gram-negative bacteria and the peptidoglycan of gram-positive bacteria. These and other PAMPs are recognized by receptors on phagocytic cells called PRRs. Because PAMPs are produced only by microorganisms, they are perceived by the phagocytic cells of the innate immune system as molecular signatures of infection.

Toll-like Receptors

Several structurally and functionally distinct classes of PRRs evolved in phagocytic cells to induce various host defensive pathways. For example, secreted PRRs bind to microbial cells and mark them for destruction. Another class of PRRs function exclusively as signaling receptors. These receptors are known as

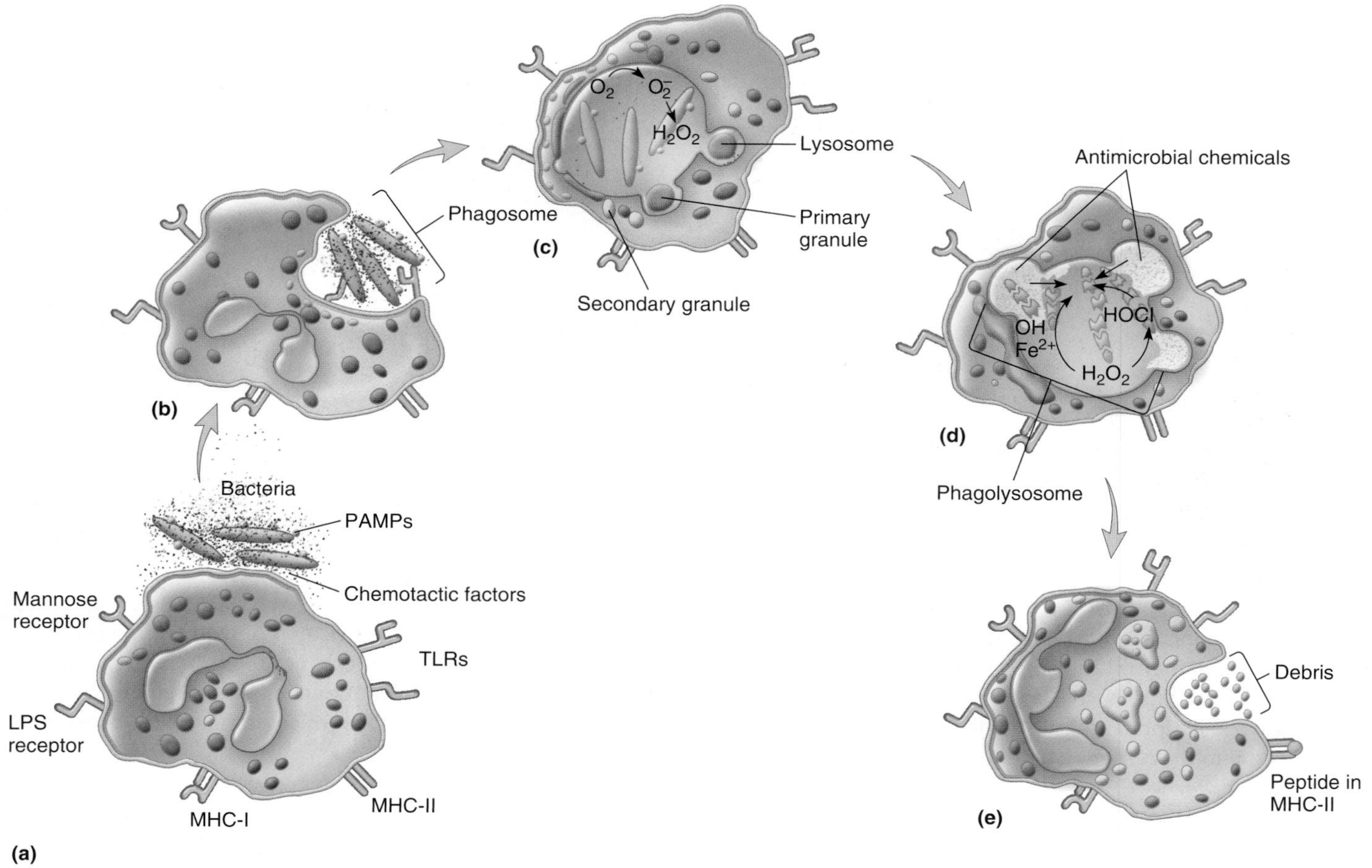

FIGURE 32.20 Phagocytosis. (a) Receptors on a phagocytic cell, such as a macrophage, and the corresponding PAMPs participating in phagocytosis. The process of phagocytosis includes (b) ingestion, (c) participation of primary and secondary granules, and O_2-dependent killing events, (d) intracellular digestion, and (e) exocytosis. LPS receptor: lipopolysaccharide receptor; TLRs: toll-like receptors; MHC-I: class I major histocompatibility protein; MHC-II: class II major histocompatibility protein; PAMPs: pathogen-associated molecular patterns.

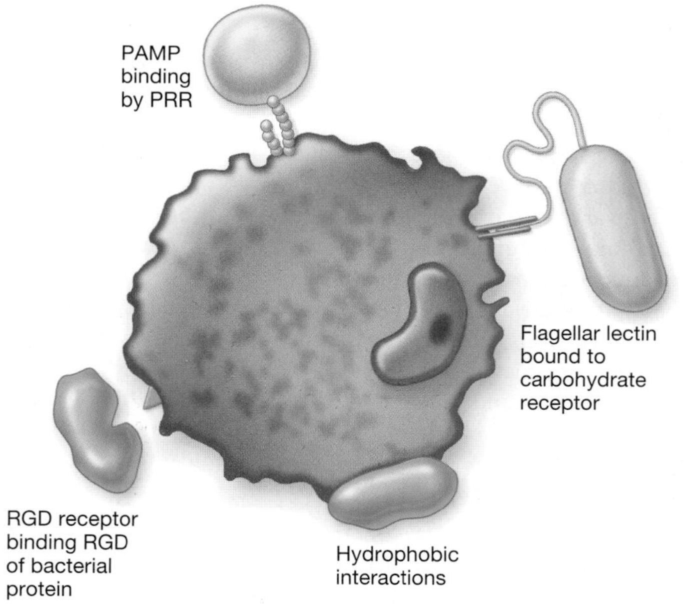

FIGURE 32.21 Some of the Possible Mechanisms by Which a Macrophage Can Recognize and Capture Microbes.

***toll-like receptors* (TLRs)** (**figure 32.22**). TLRs recognize and bind unique PAMPs of different classes of pathogens (viruses, bacteria, or fungi) and subsequently communicate that binding to the host cell nucleus to initiate appropriate gene expression and host response. There are at least 10 distinct proteins in this family of mammalian receptors. For example, TLR-4 signals the presence of bacterial lipopolysaccharide and heat-shock proteins. TLR-9 signals the dinucleotide CpG motif present on DNA released by dying bacteria. TLR-2 signals the presence of bacterial lipoproteins and peptidoglycans.

Binding of TLRs triggers an evolutionarily ancient signaling pathway that activates transcription factor NFκB through the degradation of its inhibitor, IκB. NFκB then induces expression of a variety of genes, including genes for cytokines, chemokines, and co-stimulatory molecules that play essential roles in calling forth and directing the adaptive immune response later in an infection. Thus binding of specific microbial components to phagocyte receptors is an important first step in phagocytosis. Once bound, the microbe or its components can be internalized as part of a phagosome that is then united with a lysosome to facilitate microbial killing and digestion. *Toll-like Receptors*

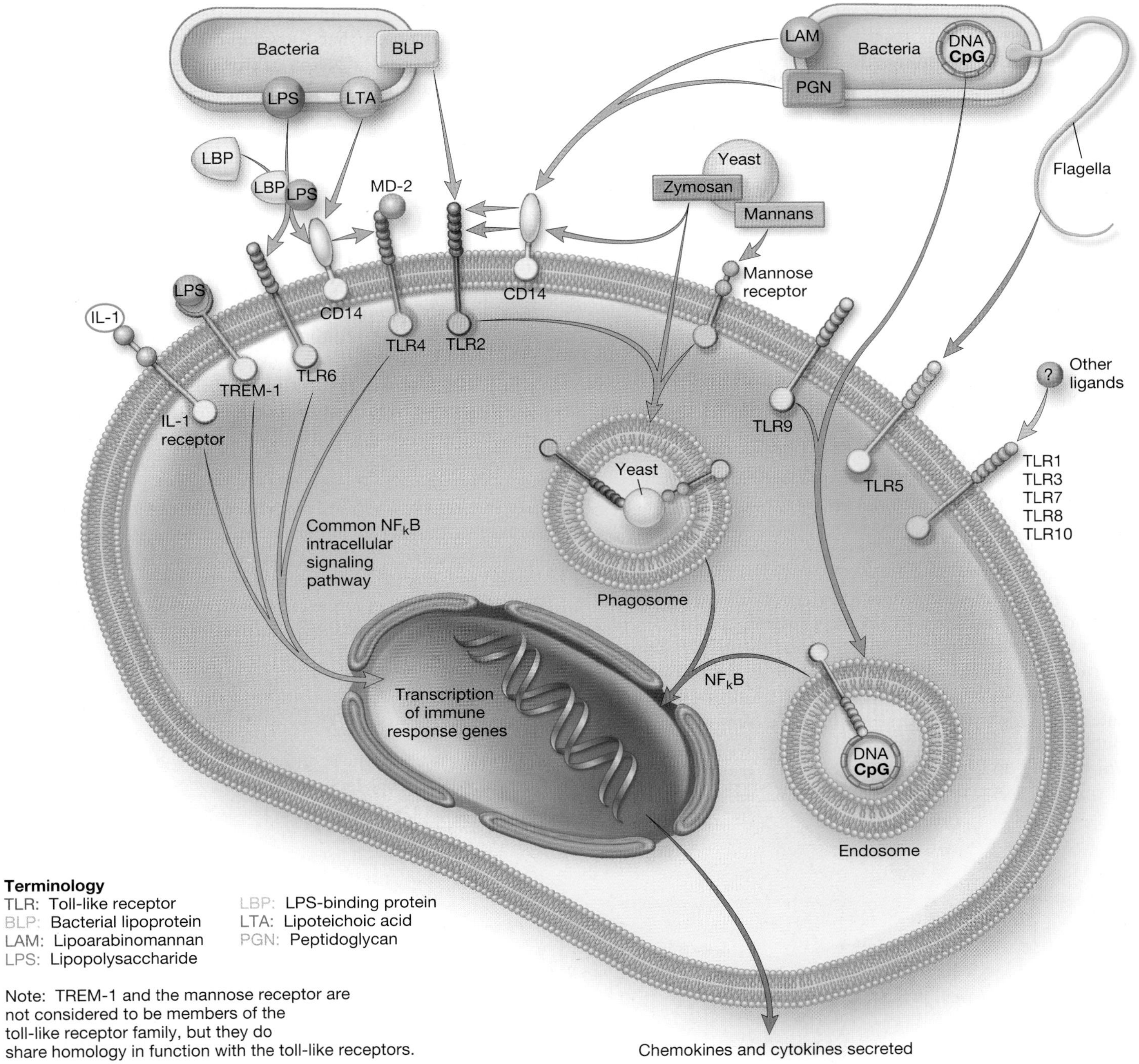

FIGURE 32.22 Recognition of Pathogen-Associated Molecular Patterns (PAMPs) by Toll-like Receptors (TLRs). PAMP binding of TLR results in a signaling process that upregulates gene expression. A common NFκB signal transduction pathway is used.

Intracellular Digestion

Once ingested by phagocytosis, microorganisms in membrane-enclosed vesicles are delivered to a **lysosome** by fusion of the phagocytic vesicle, called a **phagosome,** with the lysosome membrane, forming a new vacuole called a **phagolysosome** (figure 32.20*c,d*). Lysosomes deliver a variety of hydrolases such as lysozyme, phospholipase A_2, ribonuclease, deoxyribonuclease, and proteases. The activity of these degradative enzymes is enhanced by the acidic vacuolar pH. Collectively these enzymes participate in the destruction of the entrapped microorganisms. In addition to these oxygen-independent lysosomal hydrolases, macrophage and neutrophil lysosomes contain oxygen-dependent enzymes that produce toxic **reactive oxygen intermediates (ROIs)** such as the superoxide radical (O_2^-.), hydrogen peroxide (H_2O_2), singlet oxygen ($^1O_2$), and hydroxyl radical (OH^-). Neutrophils also contain the heme-protein myeloperoxidase, which catalyzes the production of hypochlorous acid. Some reactions that form ROIs are shown in **table 32.5**. These reactions result from the **respiratory burst** that accompanies the increased oxygen consumption and ATP generation needed for phagocytosis. These reactions occur within the phagosome as soon as it is formed; lysosome fusion is not necessary for the respiratory burst. In this way the ROIs are immediately effective in killing microorganisms as the phagosome forms. ◀◀ *Oxygen concentration (section 7.6)*

Macrophages, neutrophils, and mast cells have also been shown to form **reactive nitrogen intermediates (RNIs).** These molecules include nitric oxide (NO) and its oxidized forms, nitrite (NO_2^-) and nitrate (NO_3^-). The RNIs are very potent cytotoxic agents and may be either released from cells or generated within cell vacuoles. Nitric oxide is probably the most effective RNI. Nitric oxide can block cellular respiration by complexing with the iron in electron transport proteins. Macrophages use RNIs in the destruction of a variety of infectious agents as well as to kill tumor cells.

Neutrophil granules contain a variety of other microbicidal substances such as several cationic peptides, the bactericidal permeability-increasing protein (BPI), and the family of broad-spectrum antimicrobial peptides including defensins (section 32.3). These substances are compartmentalized strategically so as to locate them for extracellular secretion or delivery to phagocytic vacuoles. Susceptible microbial targets include a variety of gram-positive and gram-negative bacteria, yeasts and molds, and some viruses. Defensins act against bacteria and fungi by permeabilizing cell membranes. They form voltage-dependent membrane channels that allow ionic efflux. Antiviral activity involves direct neutralization of enveloped viruses, so they can no longer bind host cell receptors; nonenveloped viruses are not affected by defensins.

Table 32.5 Formation of Reactive Oxygen Intermediates

Oxygen Intermediate	*Reaction*
Superoxide ($O_2^-\bullet$)	$NADPH + 2O_2 \xrightarrow{\text{NADPH oxidase}} 2O_2^-\bullet + H^+ + NADP^+$
Hydrogen peroxide (H_2O_2)	$2O_2^-\bullet + 2H^+ \xrightarrow{\text{Superoxide dismutase}} H_2O_2 + O_2$
Hypochlorous acid (HOCl)	$H_2O_2 + Cl^- \xrightarrow{\text{Myeloperoxidase}} HOCl + OH^+$
Singlet oxygen ($^1O_2$)	$ClO^- + H_2O_2 \xrightarrow{\text{Peroxidase}} {}^1O_2 + Cl^- + H_2O$
Hydroxyl radical (OH^-)	$O_2^-\bullet + H_2O_2 \xrightarrow{\text{Peroxidase}} 2OH^- + O_2$

Exocytosis

Once the microbial invaders have been killed and digested into small antigenic fragments, the phagocyte may do one of two things. Neutrophils tend to expel the microbial fragments by the process of **exocytosis** (figure 32.20*e*). This is essentially a reverse of the phagocytic process whereby the phagolysosome unites with the cell membrane, resulting in the extracellular release of the microbial fragments. By contrast, macrophages and dendritic cells become **antigen-presenting cells.** This is accomplished by passing some of the microbial fragments from the phagolysosome to the endoplasmic reticulum. Here, the peptide components of the fragments are united with glycoproteins destined for the cell membrane. The glycoproteins bind the peptides so that they are presented outward from the cell once the glycoprotein is secured in the cell membrane. Antigen presentation is critical because it enables wandering lymphocytes to evaluate killed microbes (as antigens) and be activated. Thus antigen presentation links a nonspecific immune response (phagocytosis) to a specific immune response (lymphocyte activation). ▶▶| *Recognition of foreignness (section 33.4)*

1. Once a phagolysosome forms, how is the entrapped microorganism destroyed?
2. What is the purpose of the respiratory burst that occurs within macrophages and other phagocytic cells? Describe the nature and function of reactive oxygen and nitrogen intermediates.
3. How do macrophages and dendritic cells become antigen-presenting cells?

32.6 Inflammation

We next turn our attention to how innate immune cells perceive an impending invasion by pathogens so they can be recruited for host defense. In this regard, the process of inflammation is key. Inflammation (Latin, *inflammatio,* to set on fire) is an important nonspecific defense reaction to tissue injury, such as that caused by a pathogen or wound. Acute inflammation is the immediate response of the body to injury or cell death. The gross features were described over 2,000 years ago and are still known as the cardinal signs of inflammation: redness (*rubor*), warmth (*calor*), pain (*dolor*), swelling (*tumor*), and altered function (*functio laesa*). *Inflammatory Response*

The **acute inflammatory response** begins when injured tissue cells release chemical signals (chemokines) that activate the inner lining (endothelium) of nearby capillaries (**figure 32.23**). Within the capillaries, **selectins** (a family of cell adhesion molecules) are displayed on the activated endothelial cells. These adhesion molecules attract and attach wandering neutrophils to the endothelial cells. This slows the neutrophils and causes them to roll along the endothelium, where they encounter the inflammatory chemicals that act as activating signals (figure 32.23*b*). These signals activate **integrins** (adhesion receptors) on the neutrophils. The integrins then attach tightly to the selectins, causing the neutrophils to stick to the endothelium and stop rolling (margination). The neutrophils now undergo dramatic shape changes, squeeze through the endothelial wall (diapedesis) into the interstitial tissue fluid, migrate to the site of injury (extravasation), and attack the pathogen or other cause of the tissue damage. Neutrophils and other leukocytes are attracted to the infection site by chemotactic factors, which are also called **chemotaxins.** They include substances released by bacteria, endothelial cells, mast cells, and tissue breakdown products. Depending on the severity and nature of tissue damage, other types of leukocytes (e.g., lymphocytes, monocytes, and macrophages) may follow the neutrophils.

The release of inflammatory mediators from injured tissue cells sets into motion a cascade of events that results in the development of the signs of inflammation. One response that ensues is the stimulation of local macrophages and distant liver cells to release antimicrobial and other proteins. The local response is for these mediators to increase the acidity in the surrounding extracellular fluid, which activates the extracellular enzyme **kallikrein** (**figure 32.24**). Kallikrein cleavage releases the peptide bradykinin from its long precursor chain. Bradykinin then binds to receptors on the capillary wall, opening the junctions between cells and allowing fluid, red blood cells, and infection-fighting leukocytes to leave the capillary and enter the infected tissue. Simultaneously, bradykinin binds to mast cells in the connective tissue associated with most small blood vessels. This activates mast cells by causing an influx of calcium ions, which leads to degranulation and release of preformed mediators such as histamine. If nerves in the infected area are damaged, they release substance P, which also binds to mast cells, boosting preformed-mediator release. Histamine in turn makes the intercellular junctions in the capillary wall wider so that more fluid, leukocytes, kallikrein, and bradykinin move out, causing swelling

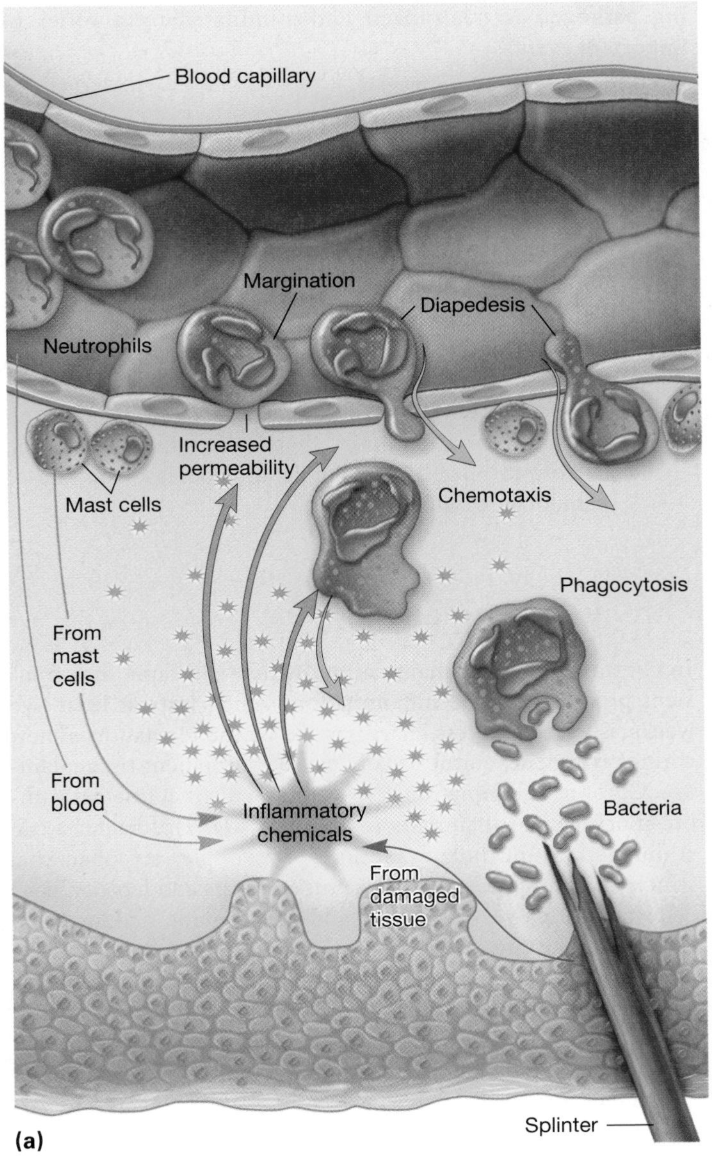

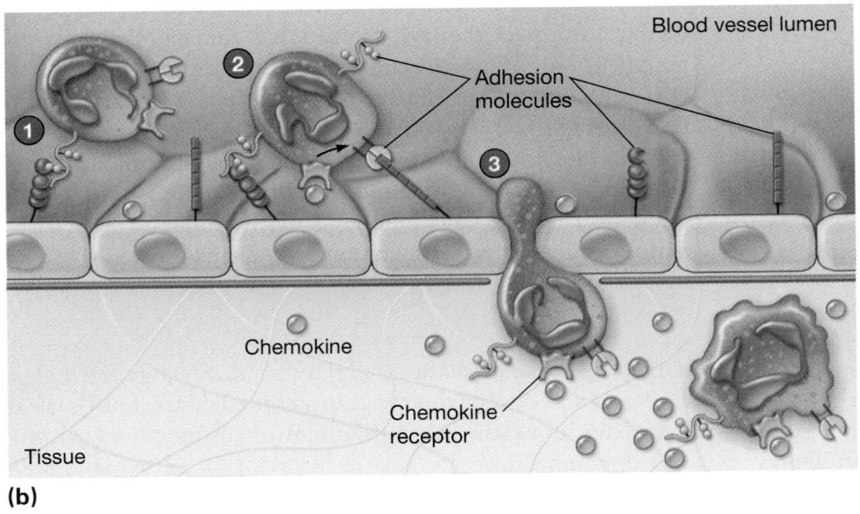

FIGURE 32.23 Physiological Events of the Acute Inflammatory Response. (a) At the site of injury (splinter), chemical messengers are released from the damaged tissue, mast cells, and the blood plasma. These inflammatory chemicals stimulate neutrophil migration, diapedesis, chemotaxis, and phagocytosis. (b) Neutrophil integrins interact with endothelial selectins (1) to facilitate margination (2) and diapedesis (3).

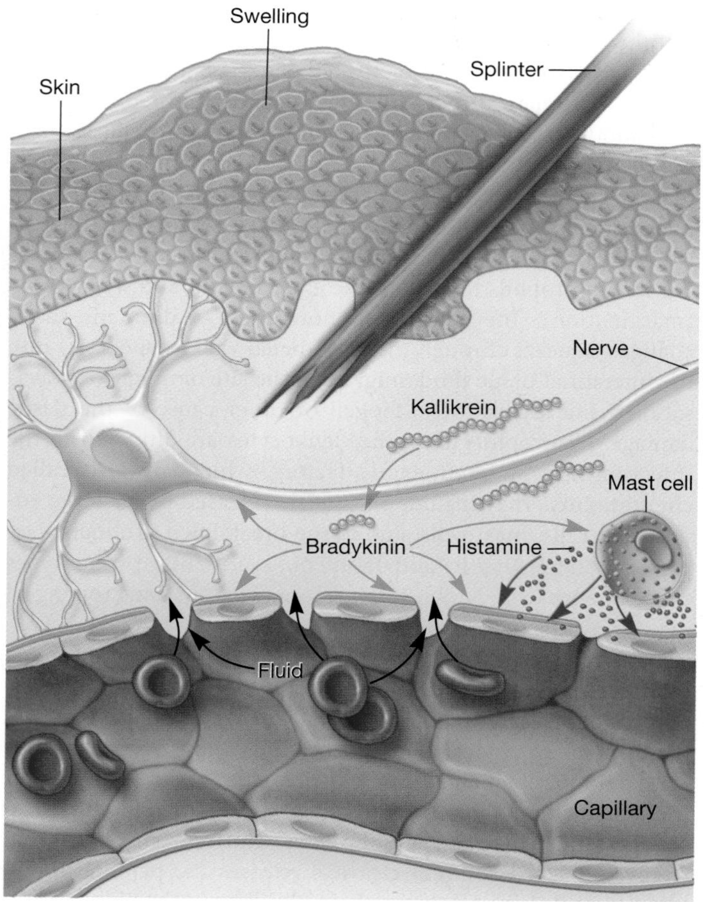

FIGURE 32.24 Tissue Injury Results in the Recruitment of Kallikrein, from Which Bradykinin Is Released. Bradykinin acts on endothelial and nerve cells, resulting in edema and pain, respectively. It also stimulates mast cells to release histamine. Histamine acts on endothelial cells, further increasing blood leakage into injured tissue sites.

or edema. Bradykinin then binds to nearby capillary cells and stimulates the production of prostaglandins (PGE_2 and PGF_2) to promote tissue swelling in the infected area. Prostaglandins also bind to free nerve endings, making them fire and start a pain impulse. At the same time, liver cells release complement proteins, the iron-binding glycoprotein lactoferrin, and various scavenging proteins called collectins.

Activated mast cells also release a small molecule called arachidonic acid, the product of a reaction catalyzed by phospholipase A_2. Arachidonic acid is metabolized by mast cell to form potent mediators, including prostaglandins E_2 and F_2, thromboxane, slow-reacting substance (SRS), and leukotrienes (LTC_4 and LTD_4). These mediators play specific roles in the inflammatory response. During acute inflammation, the offending pathogen is neutralized and eliminated by a series of important events:

1. The increase in blood flow and capillary dilation bring into the area more antimicrobial factors and leukocytes that destroy the pathogen. Dead host cells also release antimicrobial factors.
2. Blood leakage into tissue spaces increases the temperature and further stimulates the inflammatory response and may inhibit microbial growth.
3. A fibrin clot often forms and may limit the spread of the invaders.
4. Phagocytes collect in the inflamed area and phagocytose the pathogen. In addition, chemicals stimulate the bone marrow to release neutrophils and increase the rate of granulocyte production.

Chronic Inflammation

In contrast to acute inflammation, which is a rapid and transient process, chronic inflammation, which lasts at least two weeks, is a slow process characterized by the formation of new connective tissue, and it usually causes permanent tissue damage. Chronic inflammation can occur as a distinct process without much acute inflammation. The persistence of bacteria can stimulate chronic inflammation. For example, mycobacteria, which include species that cause tuberculosis and leprosy, have cell walls with a very high lipid and wax content, making them relatively resistant to phagocytosis and intracellular killing. These bacteria and a number of other microbial pathogens, including some protozoa such as *Leishmania*, can survive within macrophages. In addition, some bacteria produce toxins that stimulate tissue-damaging reactions even after bacterial death. ◀◀ *Suborder* Corynebacterineae *(section 22.4); Supergroup Excavata (section 23.2)*

Chronic inflammation is characterized by a dense infiltration of lymphocytes and macrophages. If the macrophages are unable to protect the host from tissue damage, the body attempts to wall off and isolate the site by forming a **granuloma** (Latin, *granulum,* a small particle; Greek, *oma,* to form). Granulomas are formed when neutrophils and macrophages are unable to destroy the microorganism during inflammation. Infections caused by some microbial pathogens as well as helminth parasites and large antibody-antigen complexes (as in rheumatoid arthritis) result in granuloma formation and chronic inflammation.

1. What major events occur during an inflammatory reaction, and how do they contribute to pathogen destruction?
2. How does chronic inflammation differ from acute inflammation?

Summary

32.1 Host Resistance Overview

a. There are two interdependent components of the immune response to invading microorganisms and foreign material.

b. The nonspecific (innate) response offers resistance to any microorganism or foreign material. It includes general mechanisms that are a part of the animal's innate structure and function. The nonspecific system has no immunological memory—that is, nonspecific responses occur to the same extent each time.

c. The specific (adaptive) response resists a particular foreign agent; moreover, specific immune responses improve on repeated exposure to the agent.

32.2 Physical Barriers in Nonspecific (Innate) Resistance

a. Many direct factors (age, nutrition) or general barriers contribute in some degree to all host-microbe relationships. At times they favor the establishment of the microorganism; at other times they provide some measure of general defense to the host.

b. Physical and mechanical barriers along with host secretions are the host's first line of defense against pathogens. Examples include the skin and mucous membranes and the epithelia of the respiratory, gastrointestinal, and genitourinary systems (**figures 32.2–32.4**).

32.3 Chemical Mediators in Nonspecific (Innate) Resistance

a. Mammalian hosts have specific chemical barriers that help combat the continuous onslaught of pathogens. Examples include cationic peptides, bacteriocins, cytokines, interferons, endogenous pyrogens, acute-phase proteins, and complement.

b. The complement system is composed of a large number of serum proteins that play a major role in an animal's defensive immune response. There are three pathways of complement activation: the classical, alternative, and lectin pathways (**figure 32.6** and **table 32.1**).

c. Cytokines are required for regulation of both the nonspecific and specific immune responses. Cytokines have a broad range of actions on eukaryotic cells (**figure 32.8** and **tables 32.2** and **32.3**).

d. Interferons are a group of cytokines that respond in a defensive way to viral infections, double-stranded RNA, and many pathogens capable of intracellular growth (**figure 32.9**).

e. Acute phase proteins are released from liver cells to bind microbial surfaces. They act as opsonins.

32.4 Cells, Tissues, and Organs of the Immune System

a. The cells responsible for both nonspecific and specific immunity are the white blood cells called leukocytes (**figure 32.11**). Examples include monocytes and macrophages, dendritic cells, granulocytes, and mast cells.

b. Immature undifferentiated lymphocytes are generated in the bone marrow, mature, and become committed to a particular antigenic specificity within the primary lymphoid organs and tissues. In mammals, T cells mature in the thymus and B cells in the bone marrow. The thymus is the primary lymphoid organ; the bone marrow is the primary lymphoid tissue (**figure 32.14**).

c. Natural killer cells are a small population of large, nonphagocytic lymphocytes that destroy cancer cells and cells infected with microorganisms (**figures 32.15** and **32.16**).

d. The secondary lymphoid organs and tissues serve as areas where lymphocytes may encounter and bind antigens, then proliferate and differentiate into fully mature, antigen-specific effector cells. The spleen is a secondary lymphoid organ, and the lymph nodes and mucosal-associated tissues (GALT and SALT) are the secondary lymphoid tissue (**figures 32.17–32.19**).

32.5 Phagocytosis

a. Phagocytosis involves the recognition, ingestion, and destruction of pathogens by lysosomal enzymes, superoxide radicals, hydrogen peroxide, defensins, RNIs, and metallic ions.

b. Phagocytic cells use two basic mechanisms for the recognition of microorganisms: opsonin-dependent and opsonin-independent (**figure 32.20**).

c. Phagocytes use pathogen recognition receptors to detect pathogen-associated molecular patterns on microorganisms. Toll-like receptors are a distinct class of pathogen recognition receptors (**figures 32.21** and **32.22**).

32.6 Inflammation

a. Inflammation is one of the host's nonspecific defense mechanisms to tissue injury that may be caused by a pathogen.

b. Inflammation can either be acute or chronic (**figures 32.23** and **32.24**).

Critical Thinking Questions

1. Some pathogens invade cells, others invade tissue spaces. Explain how the nonspecific immune response differs for both types of pathogens.
2. How might the various antimicrobial chemical factors be developed into new methods to control infectious disease?
3. How might a scientist use selective gene "knock-outs" in bacteria to test the role of the toll-like receptor proteins?
4. Some infectious microbes have evolved to become complement resistant. What modifications to the microbial cell would be needed to be confer complement resistance?
5. For viruses to complete their life cycle in human cells, they must subvert the host's interferon (IFN) response (figure 32.9). The Epstein-Barr virus (EBV), which causes mononucleosis among other diseases, uses a protein known as LF2 to inhibit IFN. Analysis of the EBV genome sequence led microbiologists to suspect that LF2 might be able to block IFN activity, which was then shown experimentally.

   What criteria do you think were used to infer that LF2 might interact with IFN? How do you think it was shown that LF2 blocks IFN activity? (Hint: Scientists often start with an in vitro approach using purified components before moving to in vivo analysis).

   Read the original paper: Wu, L., et al. 2009. Epstein-Barr Virus LF2: An antagonist to Type 1 interferon. *J. Virol.* 83:1140–46.
6. Once monocytes leave the bone marrow, they must differentiate into mature immune cells. This process is triggered by specific cytokines. For instance, in vitro, macrophage colony-stimulating factor (M-CSF) results in the differentiation of monocytes into macrophages, and the combination of GM-CSF and IL-4 results in differentiation into dendritic cells. IL-32 is a cytokine that is produced in response to bacterial infection. It was recently shown that Il-32 induces differentiation of monocytes into macrophage-like cells. It also results in the transformation of GM-CSF/IL-4 induced dendritic cells into macrophage-like cells.

   Why do you think IL-32 would trigger the transformation of certain dendritic cells into cells that behave like macrophages? What do you think might be the difference between a macrophage and a macrophage-like cell? Why do you think scientists are careful to make this distinction?

   Read the original paper: Netea, M. G., et al. 2008. Interleukin-32 induces the differentiation of monocytes into macrophage-like cells. *Proc. Nat. Acad. Sci., USA.* 105:3515–20.

Concept Mapping Challenge

Use the following words to construct a concept map by providing your own linking words:

Phagocytosis	Opsonin	Mucous membrane
Defensins	Cytokines	Bone marrow
Physical barriers	MAC	Neutrophils
Lymphocyte	Leukocytes	Macrophages

Dendritic cells	Cytokines	Lymphoid tissue
Complement system	Opsonization	Skin
Endogenous pyrogen	SALT	IFNs
GALT	MALT	T cells
B cell	Thymus	Basophils

Learn More

Learn more by visiting the text website at www.mhhe.com/willey8, where you will find a complete list of references.

33

Specific (Adaptive) Immunity

Nude (athymic) mice have a genetic defect (nu mutation) that affects thymus gland development. Thus T cells do not form. However, they do have a B-cell component. This unique deficiency provides animals in which to study B/T-cell dichotomy and environmental influences on the maturation and differentiation of T cells, as well as many different immune disorders.

CHAPTER GLOSSARY

acquired immune tolerance The ability to produce antibodies against nonself antigens while "tolerating" (not producing antibodies against) self antigens.

allergen An antigen that induces an allergic response.

antibody or immunoglobulin (Ig) A glycoprotein made by plasma cells (mature B cells) in response to an antigen. The antigen-binding region of an antibody molecule is the three-dimensional mirror image of the antigen that stimulated its synthesis. Thus the antibody can bind to the antigen with exact specificity.

antigen A substance (such as a protein, nucleoprotein, polysaccharide, or glycolipid) to which lymphocytes respond; also known as an immunogen because it induces the immune response.

antigen-presenting cell (APC) Cells that take in protein antigens, process them, and present antigen fragments to other cells, activating them. Macrophages, B cells, and dendritic cells can act as APCs.

antigen processing The hydrolytic digestion of antigens to produce antigen fragments. Antigen fragments can be collected by class I or class II MHC molecules and presented on the surface of a cell. Antigen processing that occurs by proteasome action on antigens that entered a cell by means other than endocytosis (e.g., viral infection) is known as **endogenous antigen processing.** Processed antigens are presented in a class I MHC molecule. Antigen processing that occurs during endocytosis is known as **exogenous antigen processing.** Antigen is presented in a class II MHC molecule.

antitoxin An antibody to a microbial toxin that binds specifically to a toxin, thereby neutralizing it.

autoimmune disease A disease in which self-reactive T and B cells are activated, causing the immune system to attack self antigens and leading to tissue or organ damage.

B-cell receptor (BCR) A transmembrane Ig complex on the surface of a B cell that binds antigen stimulating the B cell. It is composed of a membrane-bound Ig, complexed with the Ig-α/Ig-β heterodimer accessory protein.

cellular (cell-mediated) immunity The type of immunity mediated by T cells, including T-helper (T_H) cells, cytotoxic T lymphocytes (CTLs), and regulatory T (Treg) cells.

class switching The change in Ig isotype (or class) secretion that results during B-cell and then plasma-cell differentiation.

clonal selection The process by which an antigen, when bound to the best-fitting B-cell receptor, activates that B cell, resulting in the synthesis of antibody against that antigen and clonal expansion of the B cells.

cluster of differentiation molecules (CDs) or antigens Functional cell-surface proteins or receptors that can be used to identify cells.

cytotoxic T lymphocyte (CTL) A type of T cell that recognizes antigen in class I MHC molecules, destroying the cell on which the antigen is displayed. Also called **$CD8^+$ T cell.**

epitope An area of an antigen that stimulates the production of and combines with specific antibodies; also known as the antigenic determinant site.

humoral (antibody-mediated) immunity The type of immunity that results from the presence of antibodies in blood and lymph.

major histocompatibility complex (MHC) A chromosome locus encoding the histocompatibility antigens. **Class I MHC** molecules are cell-surface glycoproteins present on all nucleated cells and present endogenous antigens to $CD8^+$ T cells; **class II MHC** glycoproteins are on antigen-presenting cells and present exogenous antigens to $CD4^+$ T cells.

memory cell An inactive lymphocyte clone derived from a sensitized B or T cell capable of a heightened response to a subsequent antigen exposure.

negative selection The process by which lymphocytes that recognize host (self) antigens undergo apoptosis or become anergic (inactive).

perforin pathway The secretion of perforin protein by CTLs and NK cells, which polymerizes to form membrane pores in target cells to help destroy them during cell-mediated cytotoxicity.

plasma cell A mature, differentiated B lymphocyte that synthesizes and secretes antibody.

Regulatory T (Treg) cells $CD4^+$ and $CD8^+$ T cells that suppress host reactions against itself.

superantigen A toxic microbial protein that stimulates T cells to proliferate and release cytokine much more extensively than do normal antigens (e.g., streptococcal scarlet fever toxins and staphylococcal toxic shock syndrome toxin-1).

T-cell receptor (TCR) The receptor on the T-cell surface consisting of two antigen-binding peptide chains associated with a number of other glycoproteins.

T-dependent antigen An antigen that effectively stimulates a B-cell response only with the aid of T-helper cells that produce interleukin-2 and B-cell growth factor.

T-helper (T_H) cell A cell that is needed for T cell–dependent antigens to be effectively presented to B cells. It also promotes cell-mediated immune responses. There are three classes of functional T_H cells: **T_H1** cells interact with cytotoxic T lymphocytes, **T_H2** cells typically interact with B cells, and **T_H17** cells recruit neutrophils.

T-independent antigen An antigen that triggers B cell antibody production without T-cell cooperation.

toxin neutralization The inactivation of toxins by specific antibodies, called antitoxins, that react with them.

type I hypersensitivity A form of immediate hypersensitivity arising from the binding of antigen to IgE attached to mast cells, which then release anaphylaxis mediators such as histamine.

type II hypersensitivity A form of immediate hypersensitivity involving the binding of antibodies to antigens on cell surfaces, followed by destruction of the target cells.

type III hypersensitivity A form of immediate hypersensitivity resulting from the exposure to excessive amounts of antigens to which antibodies bind. These antibody-antigen complexes activate complement and trigger an acute inflammatory response with subsequent tissue damage.

type IV hypersensitivity A delayed hypersensitivity response (appears 24 to 48 hours after antigen exposure) that results from the binding of antigen to activated T lymphocytes, which then release cytokines and trigger inflammation and macrophage attacks that damage tissue.

viral neutralization An antibody-mediated process in which IgG, IgM, and IgA antibodies bind to some viruses during their extracellular phase and inactivate or neutralize them.

Chapter 32 discusses nonspecific host resistance and the innate mechanisms by which the host is protected from invading microorganisms. Recall that the innate resistance system responds to a foreign substance in the same manner and to the same magnitude each time and that its activation can assist in the formation of specific immune responses. In chapter 33, we continue our discussion of the immune response by describing the specific (adaptive) responses used to protect the host. Although all vertebrates are born with the capacity to acquire specific immunity, it requires sufficient time to fully develop. Unlike innate immunity, upon subsequent exposure to the same substance (antigen), activation of a specific immune response is significantly faster and stronger than that of the initial response. A fully mature immune response will therefore involve cooperation between the host's ever-present innate resistance mechanisms and inducible specific responses.

After a brief overview of how the specific immune responses work, we launch into chapter 33 by first defining the biochemical nature of molecules (antigens) that elicit specific immune reactions, including the methods by which specific immunity can be induced, and the role recognition of "self" plays in a host's detection of invaders. We then elaborate on the structure and function of T and B cells, along with their effector products. Finally, we end the chapter with a brief overview of how the host tolerates itself and the disorders that can occur with immunologic tolerance.

33.1 Overview of Specific (Adaptive) Immunity

The specific (adaptive) immune system of vertebrates has three major functions: (1) to recognize anything that is foreign to the body ("nonself"); (2) to respond to this foreign material; and (3) to remember the foreign invader. The recognition response is highly specific. The immune system is able to distinguish one pathogen from another, to identify cancer cells, and to discriminate the body's own "self" proteins and cells as different from "nonself" proteins, cells, tissues, and organs. After recognition of an invader has occurred, the specific immune system responds by activating and amplifying specific lymphocytes to attack it. This is called an

effector response. A successful effector response either eliminates the foreign material or renders it harmless to the host. If the same invader is encountered at a later time, the immune system is prepared to mount a more intense and rapid **memory response** that eliminates the invader once again and protects the host from disease. Four characteristics distinguish specific immunity from nonspecific (innate) resistance:

1. **Discrimination between self and nonself.** The specific immune system almost always responds selectively to nonself and produces specific responses against the stimulus. This is possible because host cells express a unique protein on their surface, marking them as residents of that host, or as self. Thus the introduction of materials lacking that unique self marker results in their attack by the host.
2. **Diversity.** The system is able to generate an enormous diversity of molecules such as cellular receptors and soluble proteins, including antibodies, that recognize trillions of different foreign substances.
3. **Specificity.** Immunity is highly selective in that it can be directed against one particular pathogen or foreign substance (among trillions); the immunity to one pathogen or substance usually does not confer immunity to others.
4. **Memory.** When reexposed to the same pathogen or substance, the host reacts so quickly that there is usually no noticeable pathogenesis. By contrast, the reaction time for nonspecific defenses is just as long for later exposures to a given antigen as it was for the initial one.

Two branches or arms of specific immunity are recognized (**figure 33.1**): humoral immunity and cellular (cell-mediated) immunity. **Humoral (antibody-mediated) immunity,** named for the fluids or "humors" of the body, is based on the action of soluble glycoproteins called antibodies that occur in body fluids and on the plasma membranes of B lymphocytes. Circulating antibodies bind to microorganisms, toxins, and extracellular viruses, neutralizing them or "tagging" them for destruction by phagocytes and other mechanisms described in section 33.8. **Cellular (cell-mediated) immunity** is based on the action of specific kinds of T lymphocytes that directly attack target cells infected with viruses or parasites, transplanted cells or organs, and cancer cells. T cells can induce target cell suicide (apoptosis), lyse target cells, or release chemicals (cytokines) that enhance specific immunity and nonspecific defenses such as phagocytosis and inflammation. Because the activity of the acquired immune response is so potent, it is imperative that T and B cells consistently discriminate between self and nonself with great accuracy. How they accomplish this is discussed next.

33.2 Antigens

The immune system distinguishes between self and nonself through an elaborate recognition process. During their development, B and T cells that would recognize components of their host (self-determinants) are induced to undergo apoptosis (programmed cell death). This ensures that the host will have only lymphocytes that produce specific immunologic reactions against foreign materials and organisms. Self and nonself substances that elicit an immune response and react with the products of that response are called **antigens.** Antigens include molecules such as proteins, nucleoproteins, polysaccharides, and some glycolipids. While the term immunogen (*immun*ity *gen*erator) is a more precise descriptor for a substance that elicits a specific immune response, antigen is used more frequently. Most antigens are large, complex molecules with a molecular weight generally greater than 10,000 Daltons (Da). The ability of a molecule to function as an antigen depends on its size, structural complexity, chemical nature, and degree of foreignness to the host. ◀◀ *Lymphocytes (section 32.4)*

Each antigen can have several **antigenic determinant sites,** or **epitopes** (**figure 33.2**). Epitopes are the regions or sites of the antigen that bind to a specific antibody or T-cell receptor. Chemically, epitopes include sugars, organic acids and bases, amino acid side chains, hydrocarbons, and aromatic groups. The number of epitopes on the surface of an antigen is its **valence.** The valence determines the number of antibody molecules that can combine with the antigen at one time. If one determinant site is present, the antigen is monovalent. Most antigens, however, have more than one copy of the same epitope and are termed multivalent. Multivalent antigens generally elicit a stronger immune response than do monovalent antigens. As we will see in section 33.7, each antibody molecule has at least two antigen-binding sites, so multivalent antigens can be "cross-linked" by antibodies, a phenomenon that can result in precipitation or agglutination of antigen. **Antibody affinity** relates to the strength with which an antibody binds to its antigen at a given antigen-binding site. Affinity tends to increase during the course of an immune response and is discussed in section 33.7. The **avidity** of an antibody relates to its overall ability to bind antigen at all antigen-binding sites.

Haptens

Many small organic molecules are not antigenic by themselves but become antigenic if they bond to a larger carrier molecule such as a protein. These small antigens are called **haptens** (Latin *haptein,* to grasp). When lymphocytes are stimulated by combined hapten-carrier molecules, they can react to either the hapten or the larger carrier molecule. This occurs because the hapten functions as one epitope of the carrier. When the carrier is processed and presented to T cells, recognition of both the hapten and the carrier protein can occur. As a result, both hapten-specific and carrier-specific responses can be made. One example of a hapten is penicillin. By itself, penicillin is a small molecule that is not antigenic. However, when it is combined with certain serum proteins in sensitive individuals, the resulting molecule becomes immunogenic, activates lymphocytes, and initiates a severe and sometimes fatal allergic immune reaction. In these instances, the hapten is acting as an antigenic determinant or epitope on the carrier molecule.

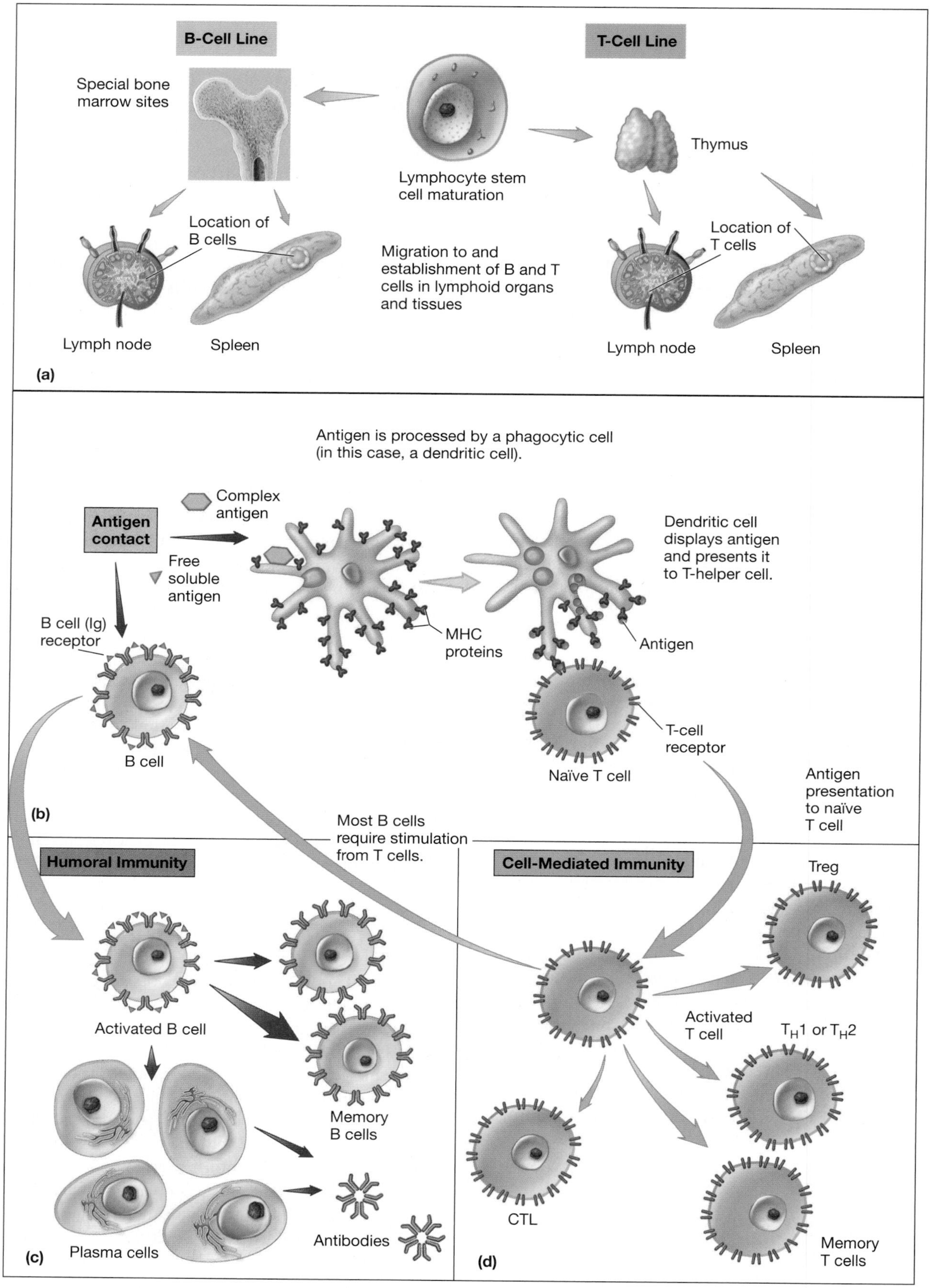

FIGURE 33.1 Acquired Immune System Development. (a) Lymphocyte progenitor cells develop into B- and T-cell precursors that migrate to the bone marrow or thymus, respectively. Mature B and T cells seed secondary lymphoid tissues. (b) Lymphocyte receptor binding of antigen activates B and T cells to become effector cells. (c) B lymphocytes develop into memory cells and antibody-secreting plasma cells. (d) T cells develop into memory cells, helper T cells, cytotoxic T cells, and regulatory T cells.

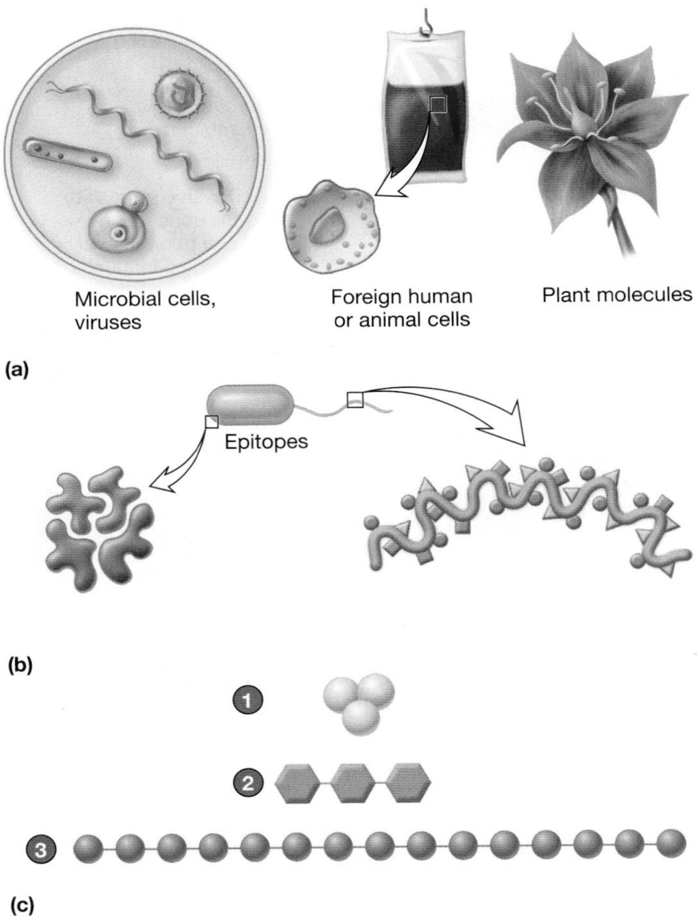

FIGURE 33.2 Antigen Characteristics Are Numerous and Diverse. (a) Whole cells and viruses make good immunogens. (b) Complex molecules with several epitopes also make good immunogens. (c) Poor immunogens include small molecules not attached to a carrier molecule (1), simple molecules (2), and large but repetitive molecules (3).

Figure 33.2 Micro Inquiry

What does the term valence mean and how does an antigen's valence influence an immune response?

1. Distinguish between self and nonself substances.
2. Define and give several examples of an antigen. What is an antigenic determinant site or epitope?

33.3 Types of Specific Immunity

Acquired immunity refers to the type of specific immunity a host develops after exposure to foreign substances or after transfer of antibodies or lymphocytes from an immune donor. Acquired immunity can be obtained actively or passively by natural or artificial means (**figure 33.3**).

Naturally Acquired Immunity

Naturally acquired active immunity occurs when an individual's immune system contacts a foreign stimulus (antigen) such as a pathogen that causes an infection. The immune system responds by producing antibodies and activated lymphocytes that inactivate or destroy the pathogen. The immunity produced can be either lifelong, as with measles or chickenpox, or last for only a few years, as with influenza. **Naturally acquired passive immunity** involves the transfer of antibodies from one host to another. For example, some of a pregnant woman's antibodies pass across the placenta to her fetus. If the woman is immune to diseases such as polio or diphtheria, this placental transfer also gives her fetus and newborn temporary immunity to these diseases. Certain other antibodies can pass from a mother to her offspring in the first secretions (called colostrum) from the mammary glands. These maternal antibodies are essential for providing immunity to the newborn for the first few weeks or months of life, as the child's own immune system matures. Naturally acquired passive immunity generally lasts only a short time (weeks to months, at most).

Artificially Acquired Immunity

Artificially acquired active immunity results when an animal is vaccinated, that is, intentionally exposed to a foreign material and induced to form antibodies and activated lymphocytes. A vaccine may consist of a preparation of killed microorganisms; living, weakened (attenuated) microorganisms; genetically engineered organisms or their products; or inactivated bacterial toxins (toxoids) that are administered to induce immunity artificially. The vaccine material induces production of antibodies that bind to the whole bacterium, virus, toxin, and so forth, to inactivate or help remove them from the host (figure 33.18). Vaccines and immunizations are discussed in detail in chapter 36.

Artificially acquired passive immunity results when antibodies or lymphocytes that have been produced by one host are introduced into another. Although this type of immunity is immediate, it is short-lived, lasting only a few weeks to a few months. An example of artificially acquired passive immunity would be botulinum antitoxin produced in a horse and given to a human suffering from botulism food poisoning, or a bone marrow transplant given to a patient with genetic immunodeficiency. (In this case the immunity then becomes natural and active with time as the bone marrow cells are adopted by the host to reconstitute the immune system.)

1. What are the three related activities mediated by the specific immune systems?
2. What distinguishes specific immunity from nonspecific resistance?
3. What are the two arms of specific immunity?
4. Of the four types of acquired immunity, which do you think your immune system has undergone?

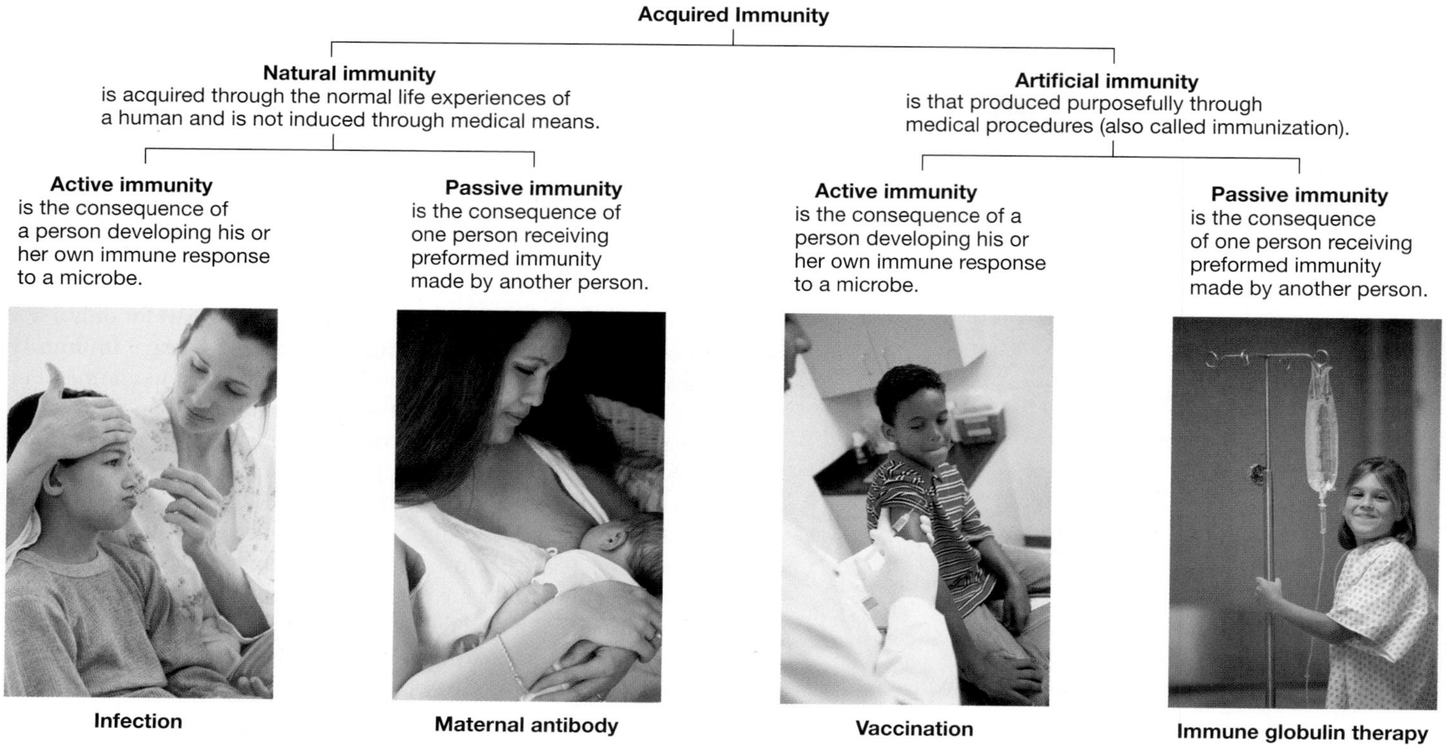

Acquired Immunity
Natural immunity
is acquired through the normal life experiences of a human and is not induced through medical means.
Artificial immunity
is that produced purposefully through medical procedures (also called immunization).
Active immunity
is the consequence of a person developing his or her own immune response to a microbe.
Passive immunity
is the consequence of one person receiving preformed immunity made by another person.
Active immunity
is the consequence of a person developing his or her own immune response to a microbe.
Passive immunity
is the consequence of one person receiving preformed immunity made by another person.
Infection
Maternal antibody
Vaccination
Immune globulin therapy

FIGURE 33.3 Immunity Can Be Acquired by Various Means. Naturally acquired immunity as well as artificially acquired immunity can be either active or passive.

33.4 Recognition of Foreignness

Distinguishing between self and nonself is essential in maintaining host integrity. This distinction must be highly specific and selective so invading pathogens are eliminated but host tissue is not destroyed. We note in chapter 32 that nonspecific (innate) resistance does not readily distinguish between invading foreign particles; innate resistance can detect generic molecular components of microorganisms, but it cannot distinguish between microorganisms. Rather, an important extension of this exquisite survival mechanism, where self is distinguished from nonself, is seen in the modern use of tissue transplantation, where organs, tissues, and cells from unrelated persons are carefully matched to the self-recognition markers on the recipient's cells. While entire textbooks are devoted to this subject, we only discuss the aspects of foreignness recognition that assist us in understanding why and how lymphocytes respond. Without the ability to recognize foreign materials, lymphocytes have no reason to differentiate into cells that seek out and respond to materials that do not belong to the host.

Recall that each cell of a particular host needs to be identified as a member of that host so it can be distinguished from foreign invaders. To accomplish this, each cell must express proteins that mark it as a resident of that host. Furthermore, it is not enough to simply identify resident (self) cells; in addition, effective cooperation between cells must occur so that efficient information sharing and selective effector activities occur. Such a system has evolved in mammals and is encoded in the major histocompatibility gene complex.

The Major Histocompatibility Complex

The ***m**ajor **h**istocompatibility **c**omplex* **(MHC)** is a collection of genes that code for the self/nonself recognition potential of a vertebrate. The term histocompatibility is derived from the Greek word for tissue (*histo*) and the ability to get along (*compatibility*). The human version of MHC is located on chromosome 6 and is called the ***h**uman **l**eukocyte **a**ntigen* **(HLA)** complex. HLA molecules can be divided into three classes: class I molecules are found on all types of nucleated body cells; class II molecules appear only on cells that can process nonself materials and present antigens to other cells (i.e., macrophages, dendritic cells, and B cells); and class III molecules include various secreted proteins that have immune functions. Class III molecules are mostly secreted products that are not required for the discrimination between self and nonself. Furthermore, the class III HLA molecules are not membrane proteins, are not related to class I or II molecules, and have no role in antigen presentation. We will not discuss them further.

Each individual has two sets of MHC genes—one from each parent—and both are expressed (i.e., they are codominant). Thus a person expresses many different HLA products. The HLA proteins differ among individuals; the closer two people are related, the more similar are their HLA molecules. In addition,

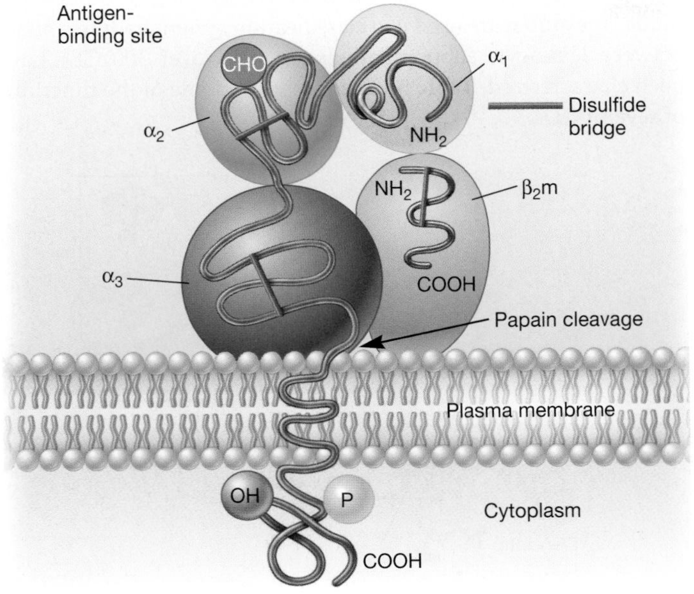

(a) Class I MHC

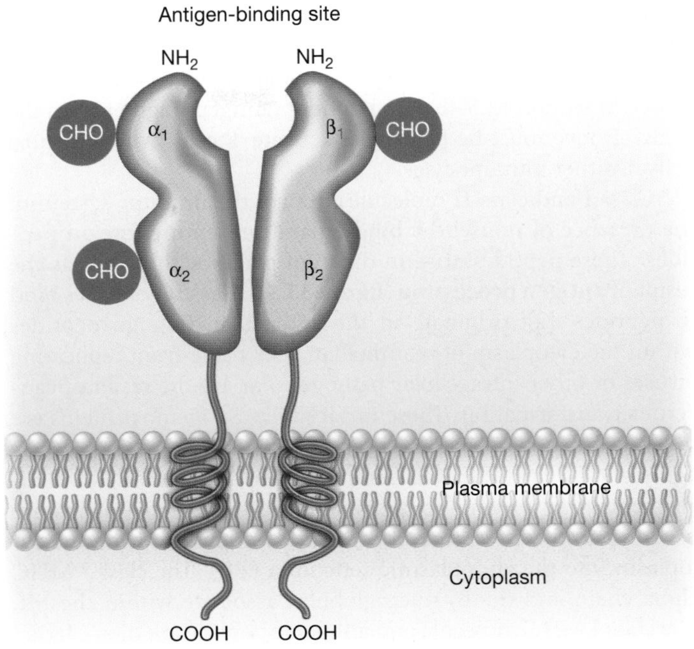

(b) Class II MHC

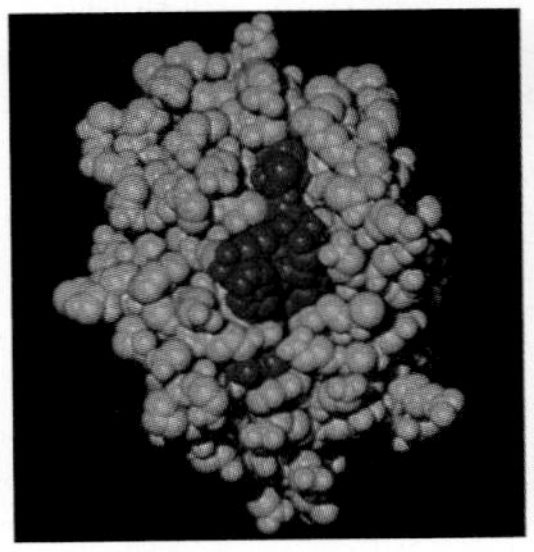

(c)

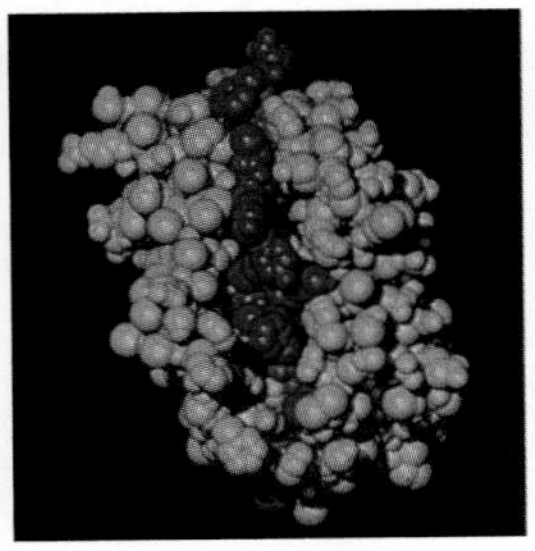

(d)

FIGURE 33.4 The Membrane-Bound Class I and Class II Major Histocompatibility Complex Molecules. (a) The class I molecule is a heterodimer composed of the alpha protein, which is divided into three domains: α_1, α_2, and α_3, and the protein β_2 microglobulin (β_2m). (b) The class II molecule is a heterodimer composed of two distinct proteins called alpha and beta. Each is divided into two domains α_1, α_2 and β_1, β_2, respectively. (c) This space-filling model of a class I MHC protein illustrates that it holds shorter peptide antigens (blue) than does a class II MHC. (d) The difference is because the peptide binding site of the class I molecule is closed off, whereas the binding site of the class II molecules is open on both ends.

Figure 33.4 Micro Inquiry

On what types of cells are MHC class I molecules found? MHC class II?

many forms of each HLA gene exist. This is because multiple alleles of each gene have arisen by high gene mutation rates, gene recombination, and other mechanisms (i.e., each gene locus is polymorphic).

Class I MHC molecules comprise HLA types A, B, and C, and serve to identify almost all cells of the body as "self." They consist of a complex of two protein chains, one with a mass of 45,000 Daltons (Da), known as the alpha chain, and the other with a mass of 12,000 Da called β_2-microglobulin (**figure 33.4*a***). Only the alpha chain spans the plasma membrane, but both chains interact to form an antigen-binding site. Because MHC class I proteins are found on all nucleated cells (i.e., only red blood cells lack them), they stimulate an immune response when cells from one host are introduced into another host with different class I molecules. This is the basis for MHC typing when a patient is being prepared for an organ transplant.

Class II MHC molecules are produced only by certain white blood cells, such as activated macrophages, dendritic cells, mature B cells, some T cells, and certain cells of other tissues. Importantly, class II molecules are required for T-cell communication with macrophages, dendritic cells, and B cells. Class II MHC molecules are also transmembrane proteins consisting of α and β chains of mass 34,000 Da and 28,000 Da, respectively (figure 33.4*b*). Similar to MHC class I receptor molecules, the protein chains expressed from MHC class II genes combine to form a three-dimensional pocket, the antigen-binding pocket, into which a nonself peptide fragment can be captured for presentation to other cells of the immune system (immunocytes). Although MHC class I and class II molecules are structurally distinct, both fold into similar shapes. Each MHC molecule has a deep groove into which a short peptide derived from a foreign substance can bind (figure 33.4*c,d*). As discussed in section 33.5, foreign peptides (antigen fragments) in the MHC groove must be present to activate T cells, which in turn activate other immunocytes.

Class I and class II molecules inform the immune system of the presence of nonself by binding and presenting foreign peptides. These peptides arise in different places within cells as the result of **antigen processing** (**figure 33.5**). Class I molecules bind to peptides that originate in the cytoplasm. Foreign peptides within the cytoplasm of mammalian cells come from replicating viruses or other intracellular pathogens, or are the result of cancerous transformation. These intracellular antigenic proteins are digested by a cytoplasmic structure called the proteasome (*see figure 4.9*) as part of the natural process by which a cell continually renews its protein contents. Specific transport proteins are used to pump the resulting short peptide fragments from the cytoplasm into the endoplasmic reticulum (ER). The class I MHC alpha chain and the b_2-microglobulin associate within the ER. The class I MHC molecule and antigenic peptide are then carried to and anchored in the plasma membrane. This process, known as **endogenous antigen processing,** enables the host cell to present the antigen to a subset of T cells called $CD8^+$, or cytotoxic T lymphocytes. $CD8^+$ T cells bear a receptor that is specific for class I MHC molecules that are presenting antigen; as will be discussed in section 33.5, these T cells bind and ultimately kill host cells presenting this foreign protein fragment.

Class II MHC molecules bind to fragments that initially come from antigens outside the cell, thus they undergo **exogenous antigen processing.** This pathway functions with particles (e.g., bacteria, viruses, toxins) that have been taken up by endocytosis. An **antigen-presenting cell (APC),** such as a macrophage, dendritic cell, or B cell, takes in the antigen or pathogen by receptor-mediated endocytosis or phagocytosis and produces antigen fragments by digestion in the phagolysosome. Fragments then combine with preformed class II MHC molecules and are delivered to the cell surface. The peptide within the MHC class II antigen-binding site can now be recognized by $CD4^+$ T-helper cells. Dendritic cells are particularly adept at presenting foreign peptides to T cells and stimulating them to become activated T cells. Unlike $CD8^+$ T cells, $CD4^+$ T cells do not directly kill target cells. Instead, they respond in two distinct ways. One is to proliferate, thereby increasing the number of $CD4^+$ cells that can react to the antigen. Some of these cells will become memory T cells that can respond to subsequent exposures to the same antigen. The second response is to secrete cytokines that either directly inhibit the pathogen that produced the antigen or recruit and stimulate other cells to join in the immune response. ◀◀ *Phagocytosis (section 32.5); Cytokines (section 32.3)*

Cluster of Differentiation Molecules

Lymphocytes and other immune cells bear cell-surface proteins that have specific roles in intercellular communication and are called **cluster of *d*ifferentiation (CDs) molecules** or **antigens.** CDs are cell-surface proteins and many are receptors. CDs have both biological and diagnostic significance. They can be measured in situ and from peripheral blood, biopsy samples, or other body fluids. They often are used in a classification system to differentiate between leukocyte subpopulations. To date, over 300 CDs have been characterized. **Table 33.1** summarizes some of the functions of several CDs.

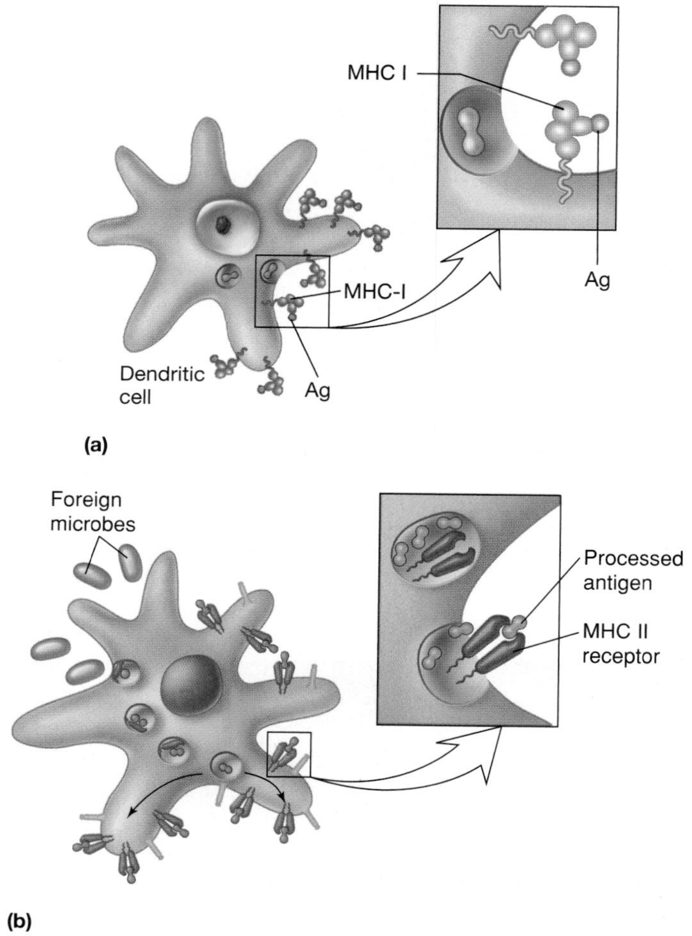

FIGURE 33.5 Antigen Presentation. (a) Antigens arising from within a cell (including intracellular parasites) are degraded by a proteasome and integrated into the antigen-binding site of the MHC class I receptors for presentation on the cell surface. (b) Antigens arising from outside a phagocytic cell are captured by endocytosis and degraded by the lysosomal process and integrated into the antigen-binding site of MHC class II receptors for presentation on the cell surface.

Figure 33.5 Micro Inquiry

What cells recognize and bind to MHC class I molecules that carry antigens? What about MHC class II molecules with antigen?

Table 33.1 Functions of Some Cluster of Differentiation (CD) Molecules

Molecule	*Function*
CD1 a, b, c	MHC class I-like receptor used for lipid antigen presentation
CD3 δ, ε, γ	T-cell antigen receptor
CD4	MHC class II coreceptor on T cells, monocytes, and macrophages; HIV-1 and HIV-2 (gp120) receptor
CD8	MHC class I coreceptor on cytotoxic T cells
CD11 a, b, c, d	α-subunits of integrin found on various myeloid and lymphoid cells; used for binding to cell adhesion molecules
CD19	B-cell antigen coreceptor
CD34	Stem cell protein that binds to sialic acid residues
CD45	Tyrosine phosphatase common to all hematopoietic cells
CD56	NK cell and neural cell adhesion molecule

The presence of various CDs on the cell's surface can be used to determine the cell's identity. For example, it has been established that the CD4 molecule is a cell-surface receptor for human immunodeficiency virus (HIV; the virus that causes AIDS), and CD34 is the cell-surface indicator of stem cells. As we will see, using the CD antigen system to name cells is more efficient than describing all of a cell's functions. We also use this approach in naming specific cell types as we discuss their relative functions in immunity.

1. What are MHCs and HLAs? Describe the roles of the three MHC classes.
2. Define hapten and CD antigens.
3. Give some examples of the biological significance of cluster of differentiation molecules (CDs).
4. How are foreign peptides produced so as to activate CD4+ T helper cells and CD8+ cytotoxic T cells, respectively?

33.5 T-Cell Biology

For acquired immunity to develop, T cells and B cells must be activated. T cells are major players in the cell-mediated immune response (figure 33.1) and have a major role in B-cell activation. They are immunologically specific, can carry a vast repertoire of immunologic memory, and can function in a variety of regulatory and effector ways. Because of their paramount importance, we discuss them first.

T-Cell Receptors

T cells respond to antigen fragments presented in the MHC molecules through specific **T-cell receptor (TCR) complexes** on their plasma membrane surface. Each receptor complex is composed of two parts: a heterodimeric polypeptide receptor composed of α and β polypeptide chains, and six accessory polypeptides, collectively referred to as CD3 (**figure 33.6*a***). The α/β heterodimer forms a transmembrane receptor stabilized by disulfide bonds. The antigen recognition sites of the receptors form when the extracellular amino termini of the α and β chains interact and create a three-dimensional "pocket" having terminal variable sections complementary to antigen fragments (similar to MHC antigen-binding sites). The cytoplasmic tails of the receptors do not contribute to signal transduction but rather associate with the CD3 accessory polypeptides. The accessory molecules transduce antigen binding events into intracellular signals. Each CD3 complex is composed of invariant polypeptides (one CD3δ, one CD3γ, two CD3ε, and two ζ chains). Each CD3 complex is further defined as heterodimers of CD3δ/CD3ε and CD3γ/CD3ε, and a homodimer of ζ chains.

Types of T Cells

T cells originate from CD34+ stem cells in the bone marrow, but T-cell precursors migrate to the thymus for further differentiation. This includes destruction of cells that recognize self antigens (so-called self-reactive T cells). Like all lymphocytes, T cells that have survived the process of development are called mature cells, but they are also considered to be "naïve" cells because they have not yet been activated by a specific MHC-antigen peptide combination. This activation of T cells involves specific molecular signaling events inside the cell, which is discussed in the section on T-cell activation (p. 800). Once activation occurs, T cells proliferate to form activated or **effector cells,** as well as **memory cells** (**figure 33.7**). Effector T cells carry out specific functions to protect the host against the foreign antigen. The three major types of T cells—the T-helper (T_H) cells, the cytotoxic T cells (Tcs)—that mature into cytotoxic T lymphocytes (CTLs), and the elusive regulatory T cells—are discussed first, then some of the details of T-cell activation are examined (**table 33.2**).

T-Helper Cells

T-helper (T_H) cells, also known as $CD4^+$ T cells, are activated by antigen presented by class II MHC molecules on APCs (e.g., macrophages and dendritic cells). They can be further subdivided into

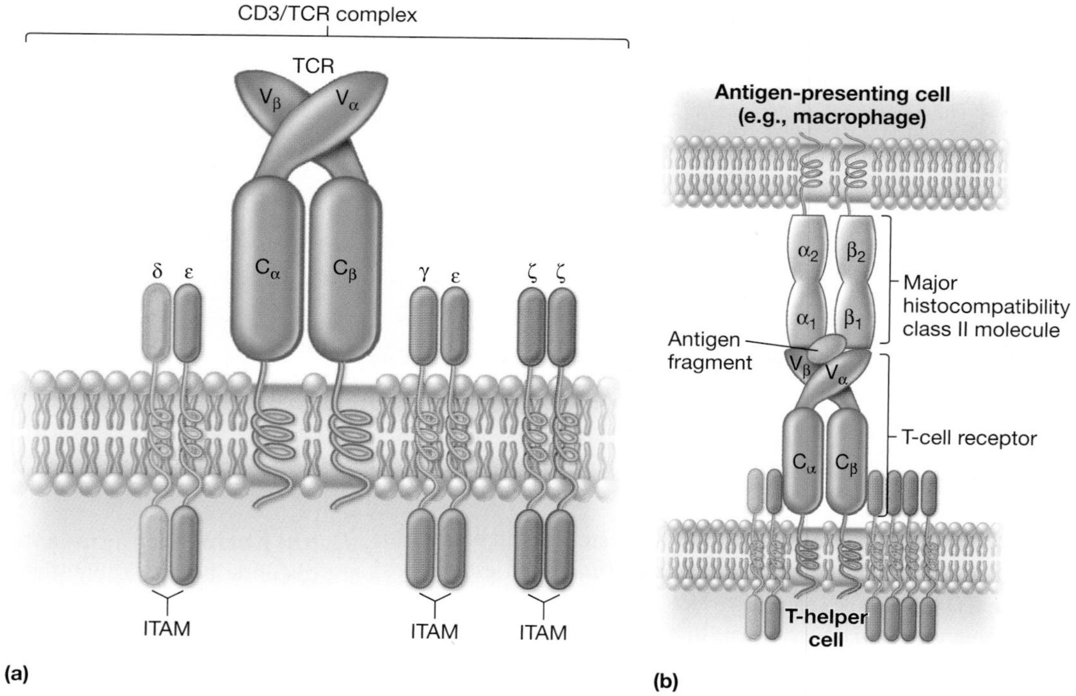

FIGURE 33.6 The T-Cell Receptor Protein Complex in T-Helper Cell Activation. (a) The overall structure of the antigen receptor complex on a T-cell plasma membrane. Notice the CD3 accessory proteins that convey the signal of antigen capture intracellularly. Extracellular antigen binding is communicated internally through the immunoreceptor tyrosine-based activation motifs (ITAMs) on the CD3 proteins. (b) An antigen-presenting cell begins the activation process by displaying an antigen fragment (e.g., peptide) within histocompatibility molecules. A T-helper cell is activated after the variable region of its receptor (designated V_α and V_β) reacts with the antigen fragment in a class II MHC molecule on the presenting cell surface.

T_H0 cells, T_H1 cells, T_H2 cells, T_H17, and Treg (p. 800) cells. **T_H0 cells** are simply undifferentiated precursors of T_H1, T_H2, and T_H17 cells, whereas T_H1, T_H2, and T_H17 cells are distinguished by the types of transcription factors and cytokines they produce, and the cells with which they interact. Activated **T_H1 cells** promote cytotoxic T lymphocyte (CTL) activity, activate macrophages, and mediate inflammation by producing interleukin (IL)-2, interferon (IFN)-γ, and tumor necrosis factor (TNF)-β when the transcription factor T-bet is active. These cytokines are also responsible for delayed-type (type IV) hypersensitivity reactions, in which host cells and tissues are damaged nonspecifically by activated T cells (p. 820). **T_H2 cells** tend to stimulate antibody responses in general and defend against helminth parasites by producing IL-5, IL-6, IL-10, and IL-13. An over abundance of T_H2-type responses may also be involved in promoting allergic reactions. T_H17 cells are found predominantly in the skin and intestinal epithelia, where they respond to bacterial invasion by producing IL-17 and IL-22. T_H17 cells respond to invading bacteria by secreting defensins, recruiting neutrophils, and inducing a strong inflammatory response. In addition, T_H17 cells have been implicated in several autoimmune diseases. Allergic and hypersensitivity reactions are discussed in section 33.10.

Under most circumstances, B cells depend on T-helper cell interaction to produce antibody (p. 802). While this was well understood, it was unclear how these B cells orient to lymph node follicles, where they receive T-cell help. It seems obvious that migration to lymph nodes is through the lymphatic channels. (*see figure 32.17*). However, the mechanism by which migration into the follicles occurs is not clear. A newly discovered kind of T cell called the follicular helper T cell (TFH) has been reported to fulfill this function. These cells lose their ability to migrate toward T-cell sites (homing) and instead acquire the capacity to home to B cell–rich follicles of secondary lymphoid organs. Once in the follicle, TFH cells provide specific signals that aid B-cell recruitment and maturation. The report of TFH cells suggests that new technologies that assess genome and proteome expression of single cells may reveal a growing list of T-cell subsets. ⏮ *Single-cell genome sequencing (section 16.2); Proteomics (section 16.5); Cytokines (section 32.3)*

Cytotoxic T Cells

Cytotoxic T cells (Tcs) are naïve or inactive $CD8^+$ T cells that express antigen-specific T-cell receptor (TCRs). Activation by its specific antigen causes a Tc cell to mature into a cytotoxic lymphocyte (CTL) that functions to destroy host cells that have been infected by intracellular pathogens, such as a virus, or have altered MHC surface proteins. Tcs can be divided into subsets, like T_H cells, based upon cytokine expression; subset 1 expresses high levels of IFN-γ, and subset 2 expresses high levels of IL-2, IL-4, and perforin but no IFN-γ. Activation of Tcs into CTLs can be thought of as a two-step process. First, naïve Tc cells must interact with a dendritic cell that has processed the antigen and presents it to the immature Tc on its class I MHC molecule (figure 33.7). Note that unlike T-helper cells, Tc cells interact with APCs through their class I MHCs. This is important because Tc cells mature into CTLs that can respond to the same antigen presented in the class I MHC of any host cells infected by the same intracellular pathogen. All host cells that present the same antigen are thus targeted for destruction. Once activated, these CTLs kill target cells in at least two ways: the perforin pathway and the CD95 pathway.

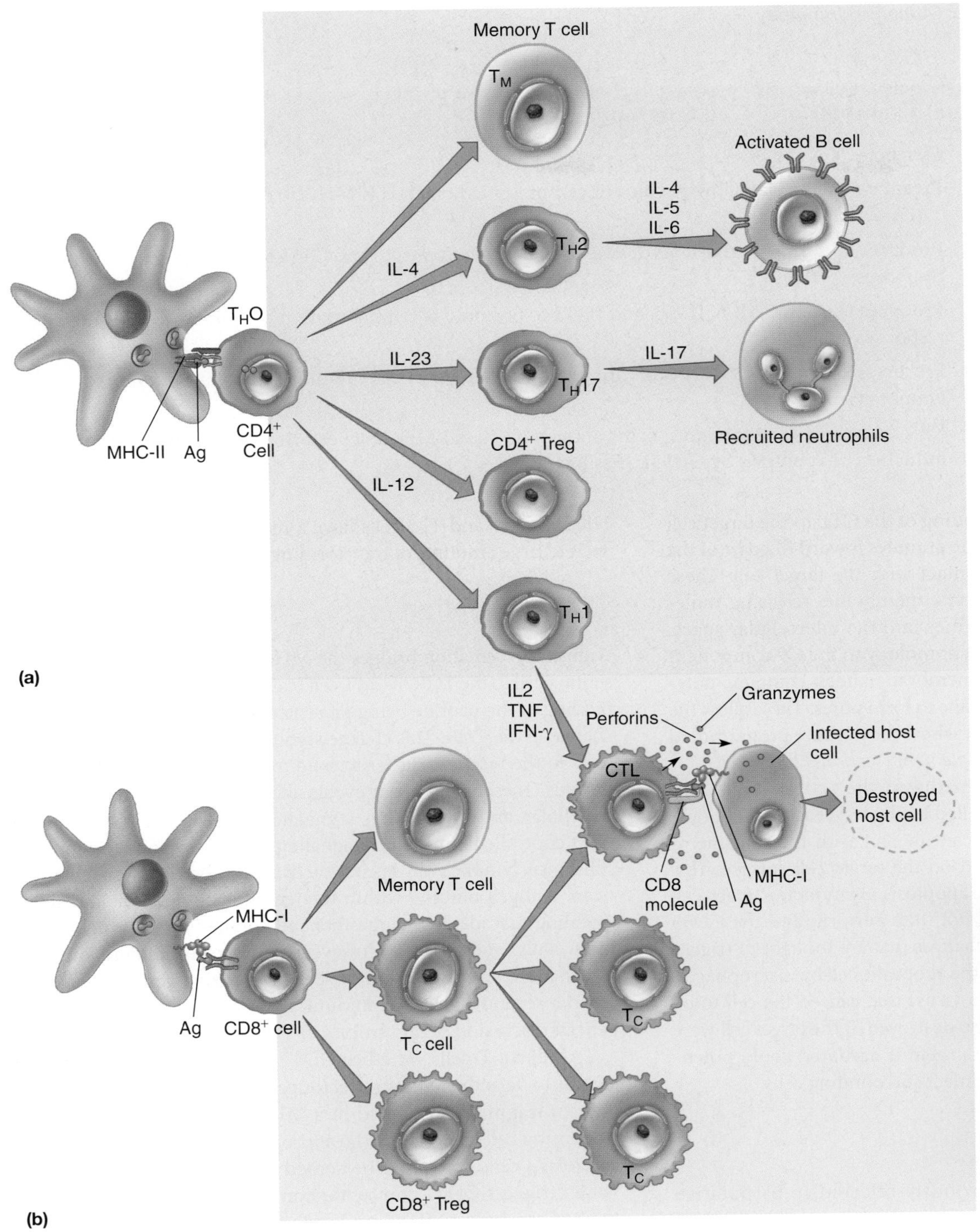

(a)

(b)

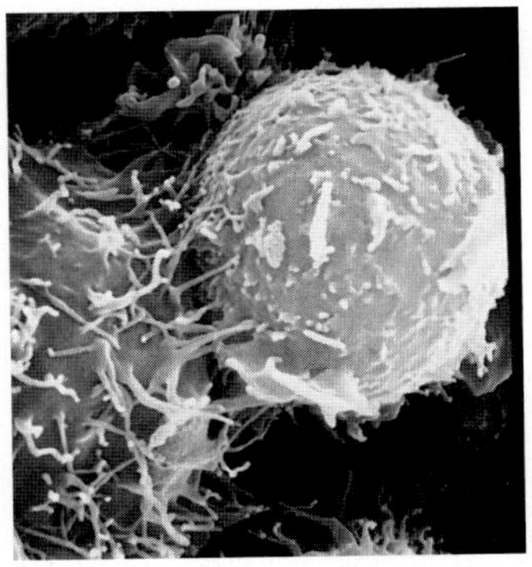

(c)

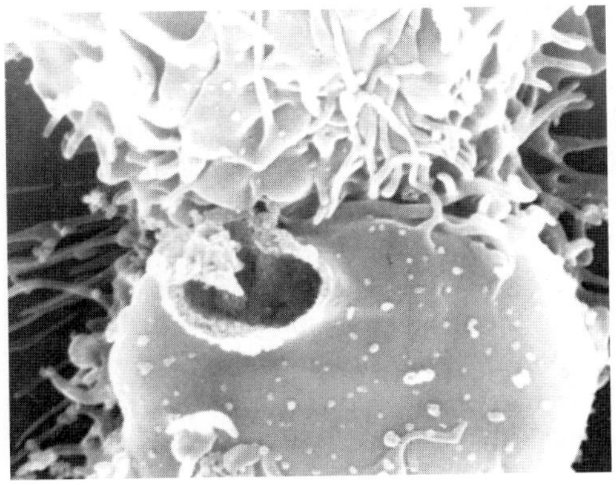

(d)

FIGURE 33.7 T-Cell Development. (a) Once activated, the T_H0 cell may differentiate into T_H1, T_H2, T_H17, or Treg cells, depending on the growth and differentiation factors in its environment (IL-12, IL-4, or IL-23, for example). T_H1 cells secrete IL-2, IFN-γ, and TNF-α, assisting in the activation of Tc cells into cytotoxic T lymphocytess (CTLs). T_H2 cells secrete IL-4, IL-5, and IL-6, followed by IL-10 and IL-13 (not shown), which causes B-cell proliferation and differentiation into antibody secreting plasma cells, secreting specific antiviral antibodies. T_H17 cells secrete IL-17, which recruits neutrophils. Neutrophils phagocytose circulating virus and upregulate inflammation to assist in virus removal. (b) Once the Tc differentiates into an activated CTL, it lysis or causes programmed cell death of a virus-infected cell by either the perforin or Fas pathways. (c) A cytotoxic T cell (left) contacts a target cell (right) ($\times$5,700). (d) The CTL secretes perforin that forms pores in the target cell's plasma membrane. These pores allow the contents of the target cell to leak out and granzymes to enter and induce apoptosis.

Table 33.2 Some Functional Examples of T-Cell Subpopulations

Type	*Function*
$CD4^+$ T-helper (T_H0) cell	Precursor cell activated by specific antigen presented on class II MHC to differentiate into T_H1, T_H2, T_H17, or T_{reg} cells
T_H1 cell	Produces IL-2, IFN-γ, and TNF-α, which activate macrophages and CTLs to promote cellular immune responses
T_H2 cell	Produces IL-4, IL-5, IL-6, IL-10, and IL-13 to promote B-cell maturation and humoral immune responses
$CD8^+$ T cell	Precursor cell activated by specific antigen presented on MHC-I to differentiate into cytotoxic lymphocyte
Cytotoxic T lymphocyte (CTL), derived from cytotoxic T cell (Tc)	Kills cells expressing foreign specific antigen on class I MHC by perforin and granzyme release or induction of apoptosis by CD95L (Fas ligand)

In the **perforin pathway,** binding of the CTL to the target cell triggers movement of cytoplasmic granules toward the part of the plasma membrane that is in contact with the target cell. These CTL granules fuse with the plasma membrane, releasing molecules called perforin and granzymes into the intercellular space. Perforin, which has considerable homology to the C9 component of complement that forms the membrane attack complex, polymerizes in the target cell's membrane to form pores. These allow the granzymes to enter the target cell, where they induce programmed cell death (apoptosis). ◀◀ *Complement (section 32.3)*

In the **Fas-FasL,** or **CD95, pathway,** the activated CTL increases expression of a protein called Fas ligand (FasL; also known as CD95L) on its surface. FasL can interact with the transmembrane Fas protein receptor found on the target cell surface. This induces the target cell to undergo apoptosis. By inducing target-cell apoptosis, rather than cell lysis, both the perforin and the CD95 pathways stimulate membrane changes that are thought to trigger phagocytosis and destruction of the apoptotic cell by macrophages. Thus any infectious agent (e.g., a virus) that caused the cell to be initially attacked by the CTL is also destroyed. If the target cell were simply to be lysed, any infectious agent it harbored could potentially be released unharmed and infect surrounding cells.

Regulatory T Cells

Regulatory T (Treg) cells (previously referred to as putative suppressor T cells) are derived from approximately 10% of $CD4^+$ T cells and 2% of $CD8^+$ T cells. Their regulatory function is induced by the IL-10 induction of transcription factor Foxp3. Foxp3 upregulates expression of CD25 and cytotoxic lymphocyte antigen CTLA-4. CTLA-4 binds to B7 (CD80), blocking the second signal required for lymphocyte activation (p. 801). Treg cells have functional α/β TCRs that recognize self antigens to prevent other T cells from reacting to self. Treg cells secrete IL-10 upon activation, which inhibits T_H1 and T_H17 cells from upregulating inflammatory responses and T_H2 cells from assisting B cells in making antibody and upregulating allergic responses. There are three subsets of Treg cells: NKT, Tr1, and T_H3 cells. NKT cells secrete large amounts of IFN-γ or IL-4 and share several surface molecules with NK cells. Tr1 and T_H3 cells have a reduced CD25 expression and secrete large amounts of transforming growth factor (TGF)-β.

T-Cell Activation

Antigen presentation bridges the MHC class II of the antigen presenting cell to the TCR of the T cell (called receptor clustering). TCR binding of the peptide being presented causes a three-dimensional change in the TCR. This change stimulates the CD3 accessory proteins in the receptor cluster to send molecular signals to the T-cell nucleus. These intracellular events alter T-cell behavior; that is, extracellular antigen binding stimulates T-cell activation through CD3-controlled intracellular signaling. Thus to respond to a foreign substance, lymphocytes must be activated by binding a specific antigen. Antigen binding within the lymphocyte receptor initiates a signaling cascade involving other membrane-bound proteins and intracellular messengers. However, lymphocyte proliferation, differentiation, and expression of specific cytokine genes occur only when a second signal is communicated along with the antigen. A general discussion of this process in T cells now follows.

All naïve T cells, whether $CD4^+$ or $CD8^+$ cells, require two signals to be activated into effector cells. **Signal 1** occurs when an antigen fragment, presented in a MHC molecule of an antigen-presenting cell (APC), fills the appropriate T-cell receptor. In the case of T_H cells, antigen is presented by class II MHC molecules, which triggers $CD4^+$ coreceptors on the T_H cell to interact with the antigen-bound MHC molecule (**figure 33.8**). For Tcs to be activated, an endogenous (cytoplasmic) antigen is presented in class I MHC molecules on the APC, and the $CD8^+$ coreceptor on the Tc interacts with the antigen-bound MHC molecule on the APC to induce CTL function. In both cases, the T-cell coreceptor assists with signal 1 recognition. ◀◀ *Organelles of the secretory and endocytic pathways (section 4.4)*

In addition to signal 1, both naïve cell types require a second, co-stimulatory **signal 2** to become activated. A T cell receiving only signal 1 will often become **anergic,** or unresponsive to that antigen. More than one factor may contribute to signal 2, but the most important seems to be the **B7 (CD80) protein** on the surface of an APC, which binds to the CD28 receptor on the T cell (figure 33.8).

One type of APC that is particularly good at stimulating naïve T cells is the dendritic cell (p. 775). This type of phagocytic cell expresses high levels of B7 constitutively (at all times). Thus the combination of signals 1 and 2 provided by a mature dendritic cell presenting the antigen fragment stimulates molecular events inside the T cell, which causes it to proliferate and differentiate into a fully functional effector cell. However, other APCs may not produce sufficient B7 to communicate signal 2. Poor B7 stimulation of $CD8^+$ cells requires activated T_H cells to stimulate the APC to generate a different signal 2. In this way, T_H cells help $CD8^+$ cells become activated. The activated $CD8^+$ cell then synthesizes and secretes IL-2 to drive its own proliferation and differentiation. Overall, once $CD8^+$ cells have been activated by two signals, they differentiate into CTLs, which can rapidly respond to host cells infected with an intracellular pathogen (i.e., a target cell) by simply recognizing foreign antigen fragments within the target cell's MHC class I protein and docking onto the MHC receptor.

Superantigens

Several bacterial and viral proteins can provoke a drastic and harmful response when they are exposed to T cells. These protein antigens are known as **superantigens** because they "trick" a huge number of T cells into activation when no specific antigen has triggered them. Superantigens accomplish this by bridging class II MHC molecules on APCs to T-cell receptors (TCRs) in the absence of a specific antigen in the MHC-binding site. This interaction results in the activation of many different T cells with different antigen specificities. The consequence of this nonspecific activation is the release of massive quantities of cytokines from $CD4^+$ T cells, leading to organ failure and suppression of specific immune responses. Thus numerous T cells (nearly 30%) can be activated by superantigen to overproduce cytokines such as TNF-α and interleukins 1 and 6, resulting in endothelial damage, circulatory shock, and multiorgan failure. Superantigens can be considered virulence factors whose effects contribute to microbial pathogenicity. Examples of superantigens include the staphylococcal enterotoxins (which can cause food poisoning) and the toxin that causes toxic shock syndrome. Because of these activities, staphylococcal enterotoxin B has been added to the U.S. government's select agent list as a potential agent of terrorism. The devastating effects superantigens have on the host serve to emphasize the importance of tightly regulating a normal immune response.

1. What is the function of an antigen-presenting cell? What is a T-cell receptor and how is it involved in T-cell activation?
2. Describe antigen processing. How does this process differ for endogenous and exogenous antigens?
3. Briefly describe the cytotoxic T cell, its general role, how it is activated into the CTL, and the two ways in which it destroys target cells.
4. Outline the functions of a T-helper cell. How do T_H1, T_H2, and T_H17 cells differ in function? Briefly describe how T_H cells are activated by co-stimulation versus superantigen.

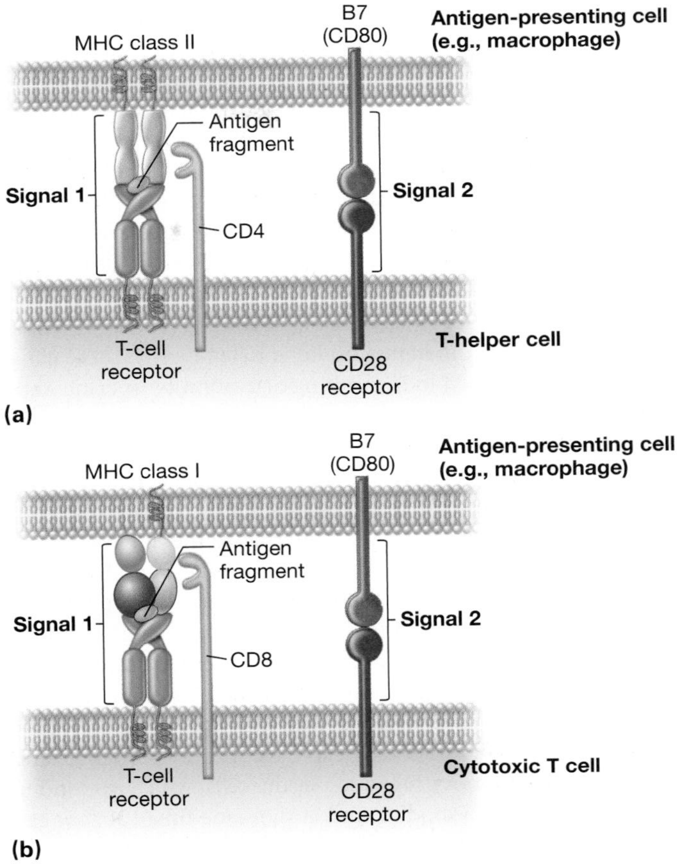

FIGURE 33.8 Two Signals (Co-stimulation) Are Essential for T-Helper Cell Activation. The first signal is the presentation of the antigen fragment by a macrophage or other antigen-presenting cell (a) in the MHC class II molecule to activate the T-helper cell. (b) The first signal for Tc cells is presentation of the antigen fragment in the MHC class I molecule. The second signal occurs when the macrophage presents the B7 (CD80) protein to the T-helper or cytotoxic T cell with its CD28 protein receptor.

Figure 33.8 Micro Inquiry

How can T-helper cells facilitate $CD8^+$ cell activation?

33.6 B-Cell Biology

Stem cells in the bone marrow produce B-cell precursors (figure 33.1). As with T cells, a specific antigen must activate B cells to proliferate into mature, antibody-secreting plasma cells. Prior to antigen stimulation, B cells produce a form of antibody that is attached to their cell membrane and oriented so that the part of the antibody that binds to antigen is facing outward, away from the cell (**figure 33.9**). These cell-surface, transmembrane antibodies (also known as immunoglobulins) act as receptors for the one specific antigen that will activate that particular B cell. When an antigen is captured by the immunoglobulin receptor, the receptor communicates this capture to the nucleus through a signal transduction pathway similar to that described for T cells. On a molecular level, the immunoglobulin receptor molecules on the B-cell surface associate with other proteins known as the Ig-α/Ig-β heterodimer proteins (similar to how the T-cell receptor interacts with CD3). Together the transmembrane immunoglobulin and the heterodimer protein complexes are called **B-cell receptors (BCRs)**.

Each B cell may have as many as 50,000 BCRs on its surface. While each individual human carries BCRs specific for as many as 10^{13} different antigens, each individual B cell possesses BCRs specific for only one particular epitope on an antigen. Therefore a host produces at least 10^{13} different, undifferentiated B cells. These naïve B cells circulate in the blood awaiting activation by specific antigenic epitopes. Upon activation, B cells differentiate into antibody-producing **plasma cells** and memory cells (figure 33.1).

So far we have introduced the mechanism that activates B cells and the fact that activated B cells secrete antibody. However, it is important to note that B cells also internalize the antigen-bound receptor to present its three-dimensional configuration to other cells—that is, BCRs that have captured their antigenic epitope are able to trigger endocytosis of that antigen, leading to antigen processing inside the B cell. As is the case with macrophages and dendritic cells, a small antigen fragment is then presented on the surface of the B cell in association with class II MHC molecules. Thus B cells have two immunological roles: (1) they proliferate and differentiate into memory cells and plasma cells, which respond to antigens by making antibodies, and at the same time (2) they can act as antigen-presenting cells, stimulating T cells to also respond.

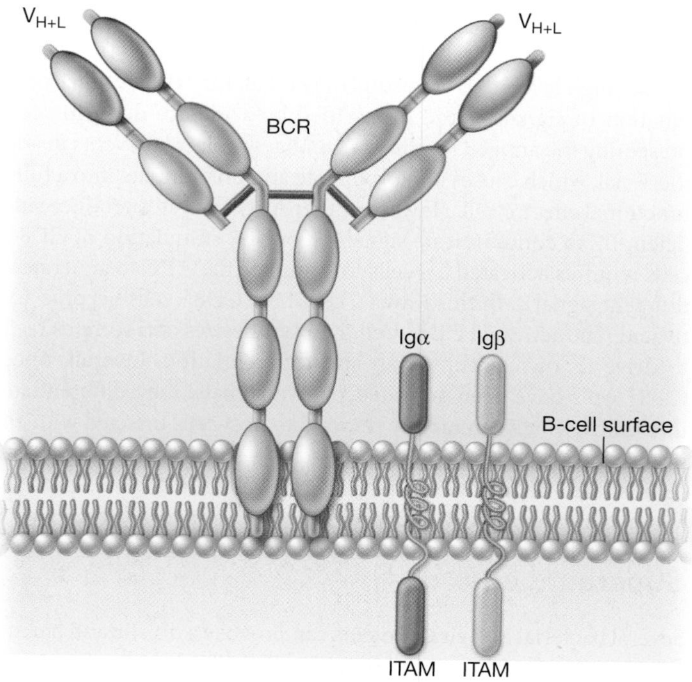

FIGURE 33.9 The Membrane-Bound B-Cell Receptor (BCR). The BCR is composed of monomeric IgM antibody and the coreceptors Igα and Igβ. A B cell is activated after the variable region of one receptor (designated V_{H+L}) binds an antigen to process and present as an antigenic fragment attached to the class II MHC molecule, so as to be recognized by a T_H2 cell (T-dependent activation). Activation can also occur after the variable regions of (two or more) receptors are bridged by an antigen (T-independent activation).

B-Cell Activation

In general, an activated B cell requires growth and differentiation factors supplied by other cells. Recall that T-helper cells produce cytokines, some of which act on B cells to assist in their growth and differentiation. Although activation of the B cell is typically antigen-specific, it can additionally be T-cell dependent or T-cell independent. This distinction reflects additional supportive activity provided by T-helper cells beyond the release of initial cytokine growth factors.

T-Dependent Antigen Triggering

In most cases, B cells that are specific for a given epitope on an antigen (e.g., epitope X) cannot develop into plasma cells that secrete antibody (anti-X) without the collaboration of T-helper cells, a process called **T-dependent antigen triggering.** In other words, binding of epitope X to the B cell may be necessary, but it is not usually sufficient for B-cell activation. Antigens that elicit a response with the aid of T-helper cells are called **T-dependent antigens.** Examples include bacteria, foreign red blood cells, certain proteins, and hapten-carrier combinations.

The basic mechanism for T-dependent antigen triggering of a B cell occurs when an activated T_H2 cell directly associates with the B cell displaying the same antigen-MHC complex that was presented to it on a dendritic cell. B-cell growth factors are then secreted by the activated T_H2 cell (**figure 33.10**). These interactions cause the B cell to proliferate and differentiate into a plasma cell, which starts producing antibodies. Thus B cells, like T cells, require two signals: antigen-BCR interaction (signal 1) and T-cell cytokines (signal 2). This is a very effective process: one plasma cell can synthesize more than 10 million antibody molecules per hour! In addition, once the antigen binds to surface BCRs, B cells become more effective antigen presenters than macrophages, especially at low antigen concentrations. Here, B cells bind antigen, take it up by receptor-mediated endocytosis, and present it to T-helper cells to activate them. In this situation, B cells and T-helper cells activate each other.

T-Independent Antigen Triggering

A few specific antigens can trigger B cells into antibody production without T-cell cooperation. These are called **T-independent antigens,** and their stimulation of B cells is known as **T-independent antigen triggering.** Examples include bacterial lipopolysaccharides, certain tumor-promoting agents, antibodies against other antibodies, and antibodies to certain B-cell differentiation antigens. The T-independent antigens tend to be polymeric; that is, they are

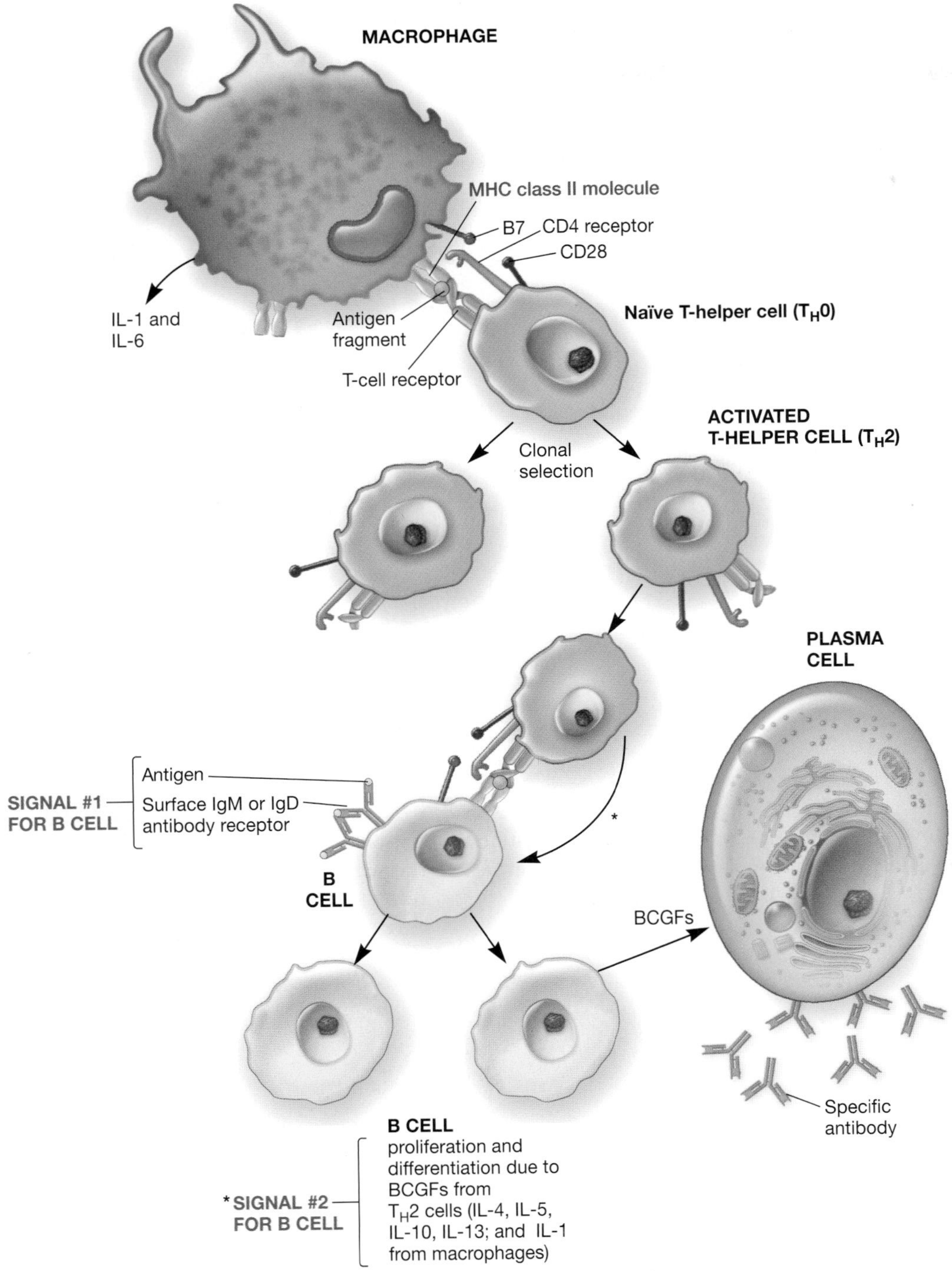

FIGURE 33.10 T-Dependent Antigen Triggering of a B Cell. Schematic diagram of the events occurring in the interactions between macrophages, T-helper cells, and B cells that produce humoral immunity. Many cytokines (e.g., IL-1, IL-4, IL-5, IL-6, IL-10, IL-13) stimulate B-cell proliferation. Cytokines such as IL-2, IL-4, IL-6, and IL-13 stimulate B-cell differentiation into plasma cells.

Figure 33.10 Micro Inquiry

Which cells are functioning as APCs in this figure?

composed of repeating polysaccharide or protein subunits. The resulting antibody generally has a relatively low affinity for antigen.

The mechanism for activation by T-independent antigens probably depends on their polymeric structure. Large molecules present a wide array of identical epitopes to a B cell specific for that determinant. The repeating epitopes cross-link membrane-bound BCRs such that cell activation occurs and antibody is secreted. However, without T-cell help, the B cell cannot alter its antibody production, and no memory B cells are formed. Thus T-independent B-cell activation is less effective than T-dependent B-cell activation: The antibodies produced have a low affinity for antigen and no immunologic memory is formed.

1. What are B-cell receptors? How are they involved in B-cell activation?
2. Briefly compare and contrast B cells and T cells with respect to their formation, structure, and roles in the immune response.
3. How does antigen-antibody binding occur? What is the basis for antibody specificity?
4. How does T-independent antigen triggering of B cells differ from T-dependent triggering?

33.7 Antibodies

So far we have discussed antibody function and how B cells are activated to become antibody-secreting plasma cells. We now turn our attention to antibody structure and how a seemingly infinite diversity of antibodies can be synthesized by a single individual. An **antibody** or **immunoglobulin (Ig)** is a glycoprotein. There are five classes of human antibodies; all share a basic structure that varies in order to accomplish specific biological functions. **Table 33.3** summarizes some of the more important physicochemical properties of the human immunoglobulin classes.

Table 33.3 Physicochemical Properties of Human Immunoglobulin Classes

	Immunoglobulin Classes				
Property	*IgG*[a]	*IgM*	*IgA*[b]	*IgD*	*IgE*
Heavy chain	γ_1	μ	α_1	δ	ε
Mean serum concentration (mg/ml)	9	1.5	3.0	0.03	0.00005
Percent of total serum antibody	80–85	5–10	5–15	<1	<1
Valency	2	5(10)	2(4)	2	2
Mass of heavy chain (kDa)	51	65	56	70	72
Mass of entire molecule (kDa)	146	970	160[c]	184	188
Placental transfer	+	–	–	–	–
Half-life in serum (days)[d]	23	5	6	3	2
Complement activation					
Classical pathway	++	+++	–	–	–
Alternative pathway	–	–	+	–	–
Induces mast cell degranulation	–	–	–	–	+
% carbohydrate	3	7–10	7	12	11
Major characteristics	Most abundant Ig in body fluids; neutralizes toxins; opsonizes bacteria	First to appear after antigen stimulation; very effective agglutinator; expressed as membrane-bound antibody on B cells	Secretory antibody; protects external surfaces	Present on B-cell surface; B-cell recognition of antigen	Anaphylactic-mediating antibody; resistance to helminths

[a] Properties of IgG subclass 1.
[b] Properties of IgA subclass 1.
[c] sIgA = 360 – 400 kDa
[d] Time required for half of the antibodies to disappear.

Immunoglobulin Structure

All immunoglobulin molecules have a basic structure composed of four polypeptide chains: two identical heavy and two identical light chains connected to each other by disulfide bonds (**figure 33.11**). Each light chain polypeptide usually consists of about 220 amino acids and has a mass of approximately 25,000 Da. Each heavy chain consists of about 440 amino acids and has a mass of about 50,000 to 70,000 Da. The heavy chains are structurally distinct for each immunoglobulin class or subclass. Both light (L) and heavy (H) chains contain two different regions. The **constant (C) regions** (**C_L** and **C_H**) have amino acid sequences that do not vary significantly between antibodies of the same class. The **variable (V) regions** (**V_L** and **V_H**) have different amino acid sequences, and these regions fold together to form the antigen-binding sites.

The four chains are arranged in the form of a flexible Y with a hinge region. This hinge allows the antibody molecule to be more flexible, adjusting to the different spatial arrangements of epitopes on antigens. The stalk of the Y is termed the **crystallizable fragment (Fc)** and can bind to a host cell by interacting with the cell-surface Fc receptor. The top of the Y consists of two **antigen-binding fragments (Fab)** that bind with compatible epitopes. The Fc fragments are composed only of constant regions, whereas the Fab fragments have both constant and variable regions. Both the heavy and light chains contain several homologous units of about 100 to 110 amino acids. Within each unit, called a domain, disulfide bonds form a loop of approximately 60 amino acids (figure 33.11). Interchain disulfide bonds also link heavy and light chains together.

The light chain may be either of two distinct forms called kappa (κ) and lambda (λ). These can be distinguished by the

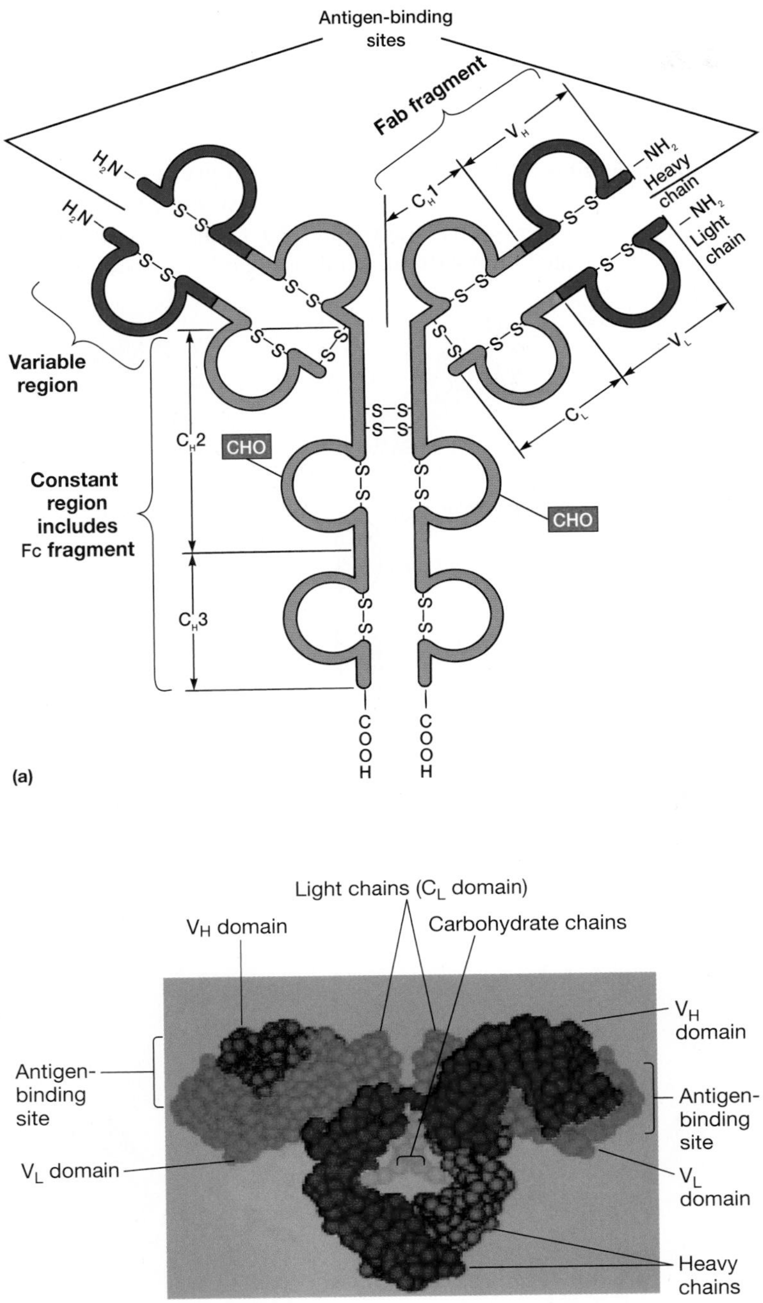

FIGURE 33.11 Immunoglobulin (Antibody) Structure. (a) An immunoglobulin molecule consists of two identical light chains and two identical heavy chains held together by disulfide bonds, which form domains. All light chains contain a single variable domain (V_L) and a single constant domain (C_L). Heavy chains contain a variable domain (V_H) and either three or four constant domains (C_H1, C_H2, C_H3, and C_H4). The variable regions (V_H,V_L), when folded together in three-dimensions, form the antigen-binding sites. (b) A computer-generated model of antibody structure showing the arrangement of the four polypeptide chains.

amino acid sequence of the constant (C) portion of the chain. In humans, the constant regions of all chains are identical. Each antibody molecule produced by a sole B cell will contain either κ or λ light chains but never both. The light-chain variable (V) domain contains hypervariable regions, or complementarity-determining regions (CDRs), that differ in amino acid sequence more frequently than the rest of the variable domain. These figure prominently in determining antigen specificity.

The amino-terminal domain of the heavy chain has a pattern of variability similar to that of the variable (V) region of the light chain domains and is termed the V_H domain. The other domains of the heavy chains are termed constant (C) domains. The constant domains of the heavy chain form the constant (C_H) region. The amino acid sequence of this region determines the classes of heavy chains. In humans, there are five classes of heavy chains designated by lowercase Greek letters: gamma (γ), alpha (α), mu (μ), delta (δ), and epsilon (ε), and usually written as G, A, M, D, or E. The properties of these heavy chains determine, respectively, the five immunoglobulin (Ig) classes—IgG, IgA, IgM, IgD, and IgE (table 33.3). Each immunoglobulin class differs in its general properties, half-life, distribution in the body, and interaction with other components of the host's defensive systems.

1. What is the variable region of an antibody? The hypervariable or complementarity-determining region? The constant region?
2. What is the function of the Fc region of an antibody? The Fab region?
3. Name the two types of antibody light chains.
4. What determines the class of heavy chain of an antibody? Name the five immunoglobulin classes.

Immunoglobulin Function

Each end of the immunoglobulin molecule has a unique role. The Fab region is concerned with binding to antigen, whereas the Fc region mediates binding to receptors (called Fc receptors) found on various cells of the immune system or the first component of the classical complement system. The binding of an antibody to an antigen usually does not destroy the antigen or the microorganism, cell, or agent to which it is attached. Rather, the antibody serves to mark and identify the nonself agent as a target for immunological attack and to activate nonspecific immune responses that can destroy the target.

An antigen binds to an antibody at the antigen-binding site within the Fab region of the antibody. More specifically, a pocket is formed by the folding of the V_H and V_L regions (figure 33.11*b*). At this site, specific amino acids contact the antigen's epitope and form multiple noncovalent bonds between the antigen and amino acids of the binding site. Because binding is due to weak, noncovalent bonds such as hydrogen bonds and electrostatic attractions, the antigen's shape must exactly match that of the antigen-binding site. If the shapes of the epitope and binding site are not truly complementary, the antibody will not effectively bind the antigen. Thus a lock-and-key mechanism operates; however, in at least one case, the antigen-binding site changes shape when it complexes with the antigen (an induced-fit mechanism). In either case, antibody specificity results from the nature of antibody-antigen binding.

Phagocytes have Fc receptors for immunoglobulin on their surface, so bacteria that are covered with antibodies are better targets for phagocytosis by neutrophils and macrophages. This is termed **opsonization.** Other cells, such as natural killer cells, destroy antibody-coated cells through a process called antibody-dependent cell-mediated cytotoxicity (*see figure 32.15*). Immune destruction also is promoted by antibody-induced activation of the classical complement system. ◀◀ *Complement (section 32.3)*

Immunoglobulin Classes

Immunoglobulin γ, or **IgG,** is the major immunoglobulin in human serum, accounting for 80% of the immunoglobulin pool (**figure 33.12*a***). IgG is present in blood plasma and tissue fluids. The IgG class acts against bacteria and viruses by opsonizing the invaders and neutralizing toxins and viruses (p. 812). It is also one of the two immunoglobulin classes that activate complement by the classical pathway. IgG is the only immunoglobulin molecule able to cross the placenta and provide natural immunity in utero and to the neonate at birth.

There are four human IgG isotypes (IgG1, IgG2, IgG3, and IgG4) that vary chemically in their heavy-chain composition and the number and arrangement of interchain disulfide bonds (figure 33.12*b*). About 80–85% of the total serum immunoglobulin is IgG, about 60% is IgG1, and 23% is IgG2. Differences in biological function have been noted in these isotypes. For example, IgG2 antibodies are opsonic and develop in response to toxins. IgG1 and IgG3, upon recognition of their specific antigens, bind to Fc receptors expressed on neutrophils and macrophages. This increases phagocytosis by these cells. The IgG4 antibodies function as skin-sensitizing immunoglobulins.

Immunoglobulin μ, or **IgM,** accounts for about 5–10% of the serum immunoglobulin pool. It is usually a polymer of five monomeric units (pentamer), each composed of two heavy chains and two light chains (**figure 33.13*a***). The monomers are arranged in a pinwheel array with the Fc ends in the center, held together by disulfide bonds and a special J (joining) chain. IgM is the first immunoglobulin made during B-cell maturation, and individual IgM monomers are expressed on B cells, serving as the antibody component of the BCR. Pentameric IgM is secreted into serum during a primary antibody response (p. 808). IgM tends to remain in the bloodstream, where it agglutinates (or clumps) bacteria, activates complement by the classical pathway, and enhances the ingestion of pathogens by phagocytic cells.

Although most IgM appears to be pentameric (5–10% in serum), around 5% or less of human serum IgM exists in a hexameric form. This molecule contains six monomeric units but seems to lack a J chain. Hexameric IgM activates complement up to 20 times more effectively than does the pentameric form. It has been suggested that bacterial cell-wall antigens such as gram-negative lipopolysaccharides may directly stimulate B cells to form hexameric IgM without a J chain. If this is the case, the immunoglobulins formed during primary immune responses are less homogeneous than previously thought.

Immunoglobulin α, or **IgA,** accounts for about 12% of serum immunoglobulin. Some IgA is present in the serum as a monomer. However, IgA is most abundant in mucous secretions, where it is a dimer held together by a J chain (figure 33.13*b*). IgA has features that are associated with secretory mucosal surfaces. IgA, when

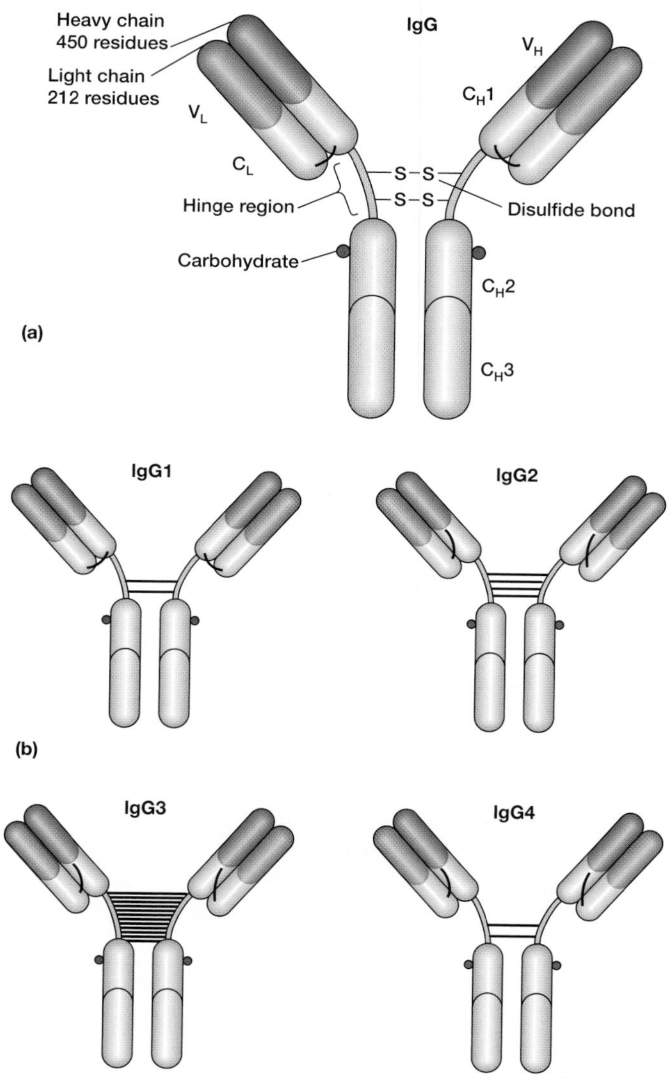

FIGURE 33.12 Immunoglobulin G. (a) The basic structure of human IgG. (b) The structure of the four human IgG subclasses. Note the arrangement and numbers of disulfide bonds (shown as thin black lines). Carbohydrate side chains are shown in red.

Figure 33.12 Micro Inquiry

What makes this class of immunoglobulin unique?

transported from the mucosal-associated lymphoid tissue (MALT) to mucosal surfaces, acquires a protein called the secretory component (*see figure 32.19*). **Secretory IgA (sIgA),** as the modified molecule is called, is the primary immunoglobulin of MALT. Secretory IgA is also found in saliva, tears, and breast milk. In these fluids and related body areas, sIgA plays a major role in protecting surface tissues against infectious microorganisms by the formation of an immune barrier. For example, in the intestine, sIgA attaches to viruses, bacteria, and protozoan parasites. This prevents pathogen adherence to mucosal surfaces and invasion of host tissues, a phenomenon known as **immune exclusion.** In addition, sIgA binds to antigens within the mucosal layer of the small intestine; subsequently the antigen-sIgA complexes are excreted through the adjacent epithelium into the gut lumen. This rids the body of locally formed immune complexes and decreases their access to the circulatory system. Secretory IgA also plays a role in the alternative complement pathway. ◀◀ *Complement (section 32.3); Secondary lymphoid organs and tissues (section 32.4)*

Immunoglobulin δ, or **IgD,** is an immunoglobulin found in trace amounts in blood serum. It has a monomeric structure (figure 33.13*c*) similar to that of IgG. IgD antibodies are abundant in combination with IgM on the surface of B cells and thus are part of the B-cell receptor complex. Therefore their function is to signal the B cell to start antibody production upon initial antigen binding.

Immunoglobulin ε, or **IgE** (figure 33.13*d*), makes up only a small percent of the total immunoglobulin pool. The skin-sensitizing and anaphylactic antibodies belong to this class. It is important to note that a number of different cell types, especially dendritic cells and macrophages, have receptors on their surface that bind antibody or fragments of complement proteins. These receptors bind to their respective ligand after the ligand has bound to its antigenic target. Binding of the target by the antibody or complement fragment is called opsonization (p. 812). Thus opsonized targets are more readily phagocytosed because they are brought to the phagoycte by a receptor-mediated process. Additionally the Fc portion of IgE can bind to Fc receptors specific for IgE that are found on mast cells, eosinophils, and basophils. Thus these cells can become coated with IgE molecules. When two cell-bound IgE molecules are cross-linked by binding to the same antigen, the cells degranulate. This degranulation releases histamine and other mediators of inflammation. It also stimulates production of an excessive number of eosinophils in the blood (eosinophilia) and increased rate of movement of the intestinal contents (gut hypermotility), which aid in the elimination of helminthic parasites. Thus although IgE is present in small amounts, this class of antibodies has potent biological capabilities, as is discussed in section 33.10.

1. Explain the different functions of IgM when it is bound to B cells versus when it is soluble in serum.
2. Describe the major functions of each immunoglobulin class.
3. Why is the structure of IgG considered the model for all five immunoglobulin classes?
4. Which immunoglobulin can cross the placenta?
5. Which immunoglobulin is most prevalent in the immunoglobulin pool? The least prevalent?

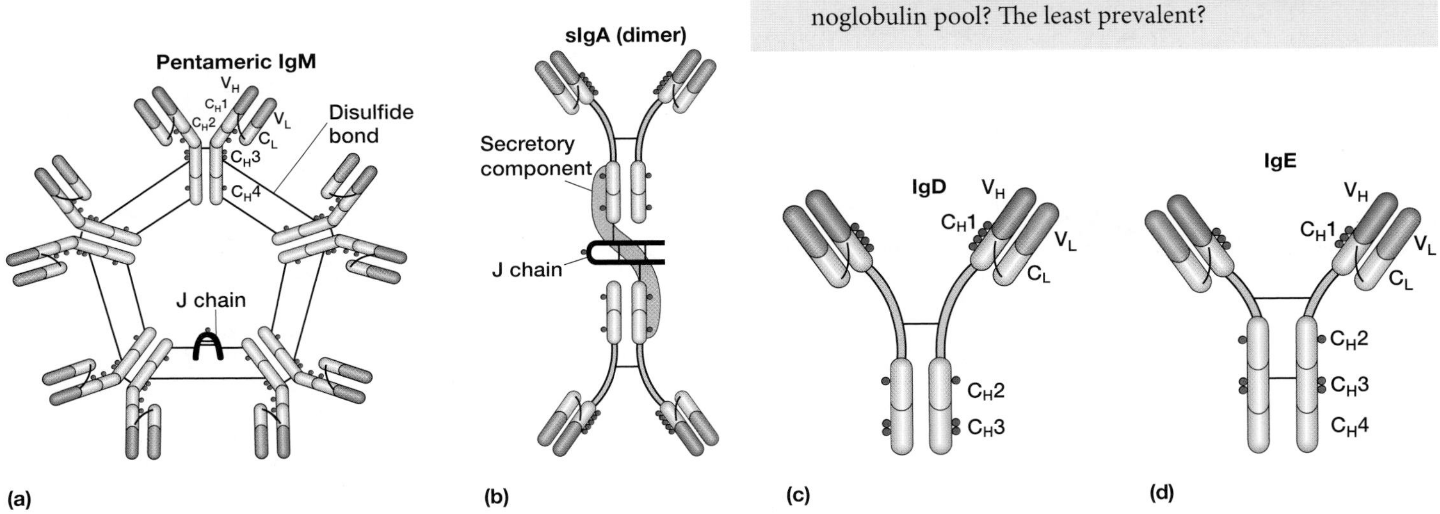

FIGURE 33.13 Immunoglobulins M, A, D, and E. (a) The pentameric structure of human IgM. The disulfide bonds linking peptide chains are shown in black. Note that 10 antigen-binding sites are present. (b) The dimeric structure of human secretory IgA. Notice the secretory component (tan) wound around the IgA dimer and attached to the constant domain of each IgA monomer. (c) The structure of human IgD showing the disulfide bonds that link protein chains (shown in black). (d) The structure of human IgE. Carbohydrate side chains are in red.

***Figure 33.13** Micro Inquiry*

Where is monomeric IgM commonly found?

Antibody Kinetics

The synthesis and secretion of antibody can also be evaluated with respect to time. Monomeric IgM serves as the B-cell receptor for antigen, and pentameric IgM is secreted after B-cell activation. Furthermore, under the influence of T-helper cells, IgM-secreting plasma cells may stop producing and secreting IgM in favor of another antibody class (e.g., IgG, IgA, or IgE). This is known as **class switching** (p. 810). These events take time to unfold.

The Primary Antibody Response

When an individual is exposed to an antigen (e.g., an infection or vaccine), there is an initial lag phase, or latent period, of several days to weeks before an antibody response is mounted. During this latent period, no antigen-specific antibody can be detected in the blood (**figure 33.14**). Once B cells have differentiated into plasma cells, antibody is secreted and can be detected. This explains why antibody-based HIV tests, for example, are not accurate until weeks

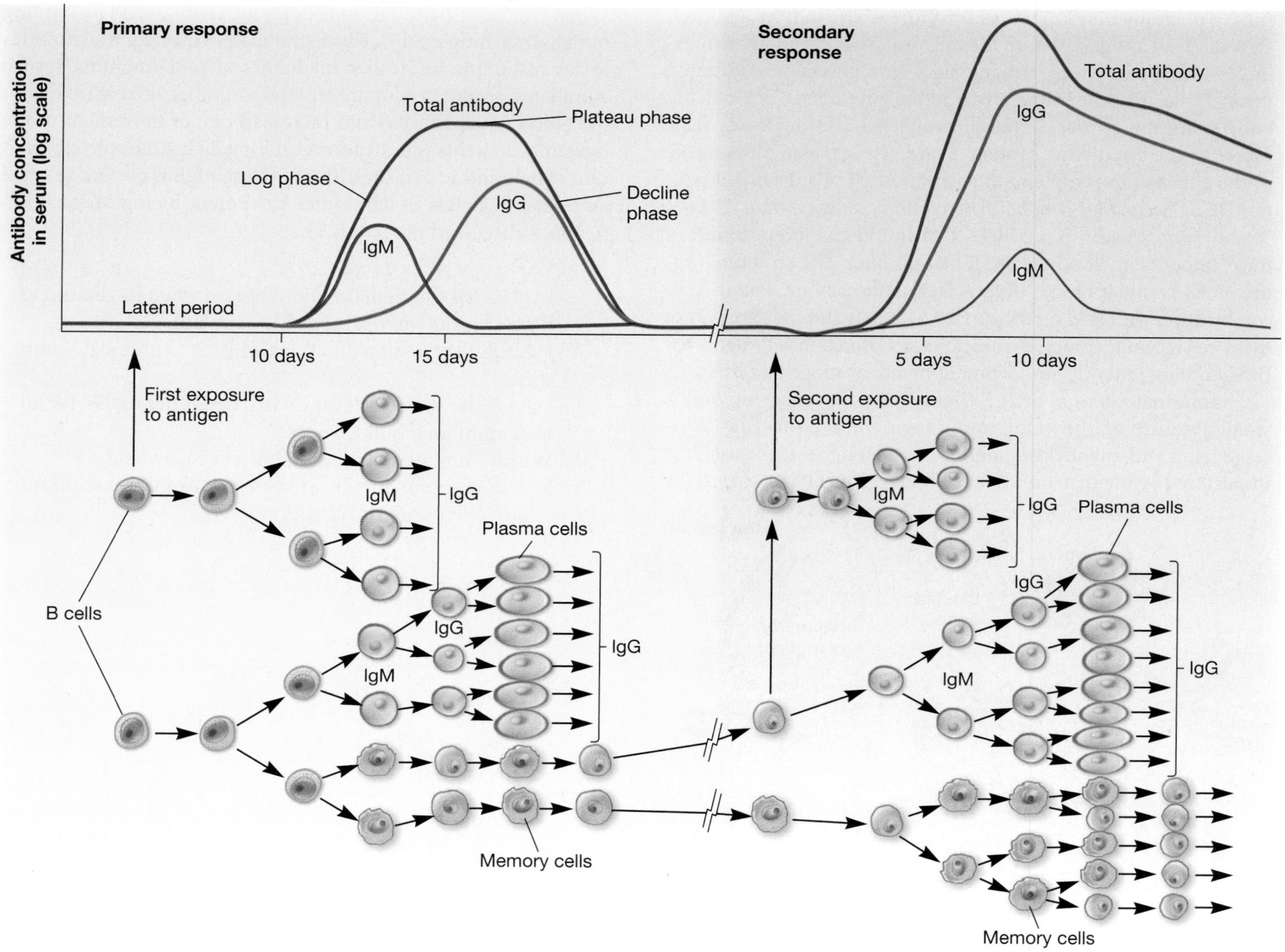

FIGURE 33.14 Antibody Production and Kinetics. The four phases of a primary antibody response correlate to the clonal expansion of the activated B cell, differentiation into plasma cells, and secretion of the antibody protein. The secondary response is much more rapid, and total antibody production is nearly 1,000 times greater than that of the primary response.

Figure 33.14 Micro Inquiry

How does the lag in IgM production compare during a primary and secondary response? Contrast this with the lag seen in IgG production during a primary and secondary response. What does this difference imply about the nature of the secondary response?

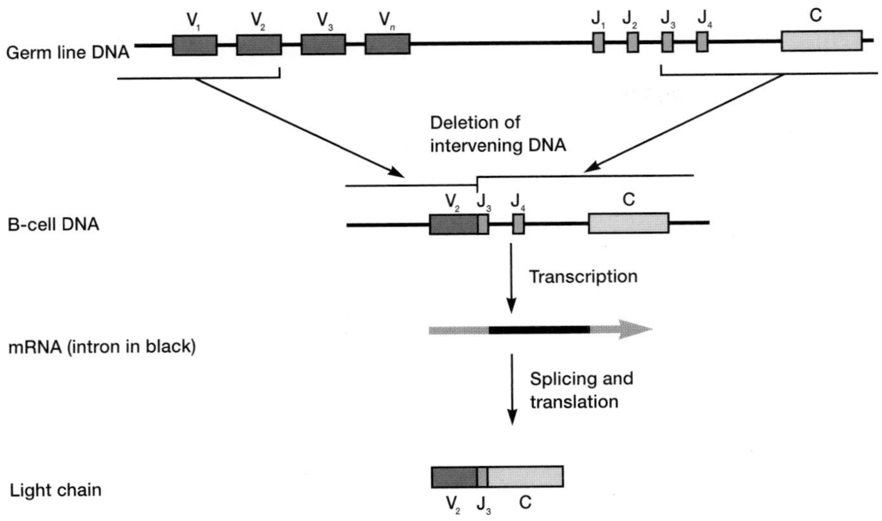

FIGURE 33.15 Light-Chain Production in a Mouse. One V segment is randomly joined with one J-C region by deletion of the intervening DNA. The remaining J segments are eliminated from the RNA transcript during RNA processing. An intron is a segment of DNA occurring between expressed regions of genes.

Table 33.4 Theoretical Antibody Diversity Resulting from Combinatorial Joining of Germ Line Genes[a]

λ light chains	V regions = 2 J regions = 3 Combinations = $2 \times 3 = 6$
κ light chains	V_κ regions = 250 − 350 J_κ regions = 4 Combinations = $250 \times 4 = 1{,}000$ = $350 \times 4 = 1{,}400$
Heavy chains	V_H = 250 − 1,000 D = 10–30 J_H = 4 Combinations = $250 \times 10 \times 4 = 10{,}000$ = $1{,}000 \times 30 \times 4 = 120{,}000$
Diversity of antibodies	κ-containing: $1{,}000 \times 10{,}000 = 10^7$ $1{,}400 \times 120{,}000 = 2 \times 10^8$ λ-containing: $6 \times 10{,}000 = 6 \times 10^4$ $6 \times 120{,}000 = 7 \times 10^5$

[a] Approximate values.

after exposure. The **antibody titer,** which is a measurement of serum antibody concentration (the reciprocal of the highest dilution of an antiserum that gives a positive reaction in the test being used), then rises logarithmically to a plateau during the second, or log, phase. In the plateau phase, the antibody titer stabilizes. This is followed by a decline phase, during which antibodies are naturally metabolized or bound to the antigen and cleared from circulation. During the primary antibody response, IgM appears first, then switches to another antibody class, usually IgG. The affinity of the antibodies for the antigen's determinants is low to moderate during the primary antibody response.

The Secondary Antibody Response

The primary antibody response primes the immune system so that it possesses specific immunological memory through its clones of memory B cells. Upon secondary antigen challenge, as occurs when an individual is reexposed to a pathogen or receives a vaccine booster, B cells mount a heightened secondary response to the same antigen (figure 33.14). Compared to the primary antibody response, the secondary antibody response has a much shorter lag phase and a more rapid log phase, persists for a longer plateau period, attains a higher IgG titer, and produces antibodies with a higher affinity for the antigen.

Diversity of Antibodies

One unique property of antibodies is their remarkable diversity. According to current estimates, each human can synthesize antibodies that can bind to more than 10^{13} (10 trillion) different epitopes. How is this diversity generated? The answer is threefold: (1) rearrangement of antibody gene segments, called combinatorial joining; (2) generation of different codons during antibody gene splicing; and (3) somatic mutations.

Combinatorial joining of immunoglobulin loci occurs because these genes are split into many gene segments. The genes that encode antibody proteins in precursor B cells contain a small number of exons, close together on the same chromosome, that determine the constant (C) region of the light chains. Separated from them but still on the same chromosome is a larger cluster of segments that determines the variable (V) region of the light chains. During B-cell differentiation, exons for the constant region are joined to one segment of the variable region. This occurs by recombination and is mediated by specific enzymes called RAG-1 and RAG-2. This splicing process joins the coding regions for constant and variable regions of light chains to produce a complete light chain of an antibody. A similar splicing produces a complete heavy-chain antibody gene.

Because the light-chain genes actually consist of three parts and the heavy-chain genes consist of four, the formation of a finished antibody molecule is slightly more complicated than previously outlined. The germ line DNA for the light-chain gene contains multiple coding sequences called V and J (joining) regions (**figure 33.15**). During the development of a B cell in the bone marrow, the RAG enzymes join one V gene segment with one J segment. This DNA joining process is termed combinatorial joining because it can create many combinations of the V and J regions (**table 33.4**). In addition, an enzyme called *t*erminal *d*eoxynucleotidyl *t*ransferase (tdt) inserts nucleotides at the V-J junction, creating additional diversity. When the light-chain gene is transcribed, transcription continues through the DNA region that encodes the constant portion of the gene. RNA splicing subsequently joins the V, J, and C regions, creating mRNA.

Combinatorial joining in the formation of a heavy-chain gene occurs by means of DNA splicing of the heavy-chain counterparts of V and J along with a third set of D (diversity) sequences (**figure 33.16*a***). Initially, all heavy chains have the μ type of constant region. This corresponds to antibody class IgM (figure 33.16*b*). If a particular B cell is reexposed to its antigen, another DNA splice joins the VDJ region with a different constant region that can subsequently change the class of antibody produced by the B cell (figure 33.16*c*)—the phenomenon of **antibody class switching.** This process is regulated by antigen-activated helper T cells.

Antibody class switching solves a problem faced by an organism during the immune response. It must be able to bind lots of antigen, but it must also be able to clear the antigen-antibody complexes (immune clearance). Recall that the primary antibody response to antigen is IgM. Because secreted IgM is pentameric, its 10 antigen-binding sites allow it to bind strongly to foreign materials that have multiple repetitive epitopes. However, IgM is not the best antibody class for clearing the antigen from the host. In fact, depending on the antigen and the body site, IgA or IgG provide better effector mechanisms for immune clearance. So how does an antigen-activated B cell change antibody class without altering antigen-binding affinity? Proliferating B cells are able to use combinatorial joining to attach the rearranged variable region coding sequence flanking the μ constant gene to another heavy-chain gene sequence (figure 33.16). For instance, the switch from an IgM to IgG results when the μ and γ constant genes are transcribed with their flanking switch regions.

The switch from IgM to another antibody class results from a unique editing process that occurs when the proliferating B cell synthesizes the enzyme *a*ctivation-*i*nduced cytidine *d*eaminase (AID). AID replaces cytosine residues with uracil when the complementary DNA strands separate during transcription. This causes DNA repair enzymes to remove the inappropriate uracil (now on a DNA backbone). The abasic nucleotide is then excised, nicking one DNA strand. When nicking occurs in the respective μ and γ switch regions, recombination occurs; the μ constant region is excised, and the γ constant region is joined to the VDJ region.

Remember that combinatorial joining is only one of three mechanisms by which antibody diversity is generated. The other two processes are:

1. **Splice-site variability:** The junction for either VJ or VDJ splicing in combinatorial joining can occur between different nucleotides and thus generate different codons in the spliced gene. In addition, the activation of tdt can greatly increase the variability of the nucleotide sequence at the VJ or VDJ junctions during the splicing process. For example, one VJ splicing event can join the V sequence CCTCCC with the J sequence TGGTGG in two ways: CCTCCC + TGGTGG = CCGTGG, which codes for proline and tryptophan. Alternatively, the VJ splicing event can give rise to the sequence CCTCGG, which codes for proline and arginine. Thus the same VJ joining could produce polypeptides differing in a single amino acid.

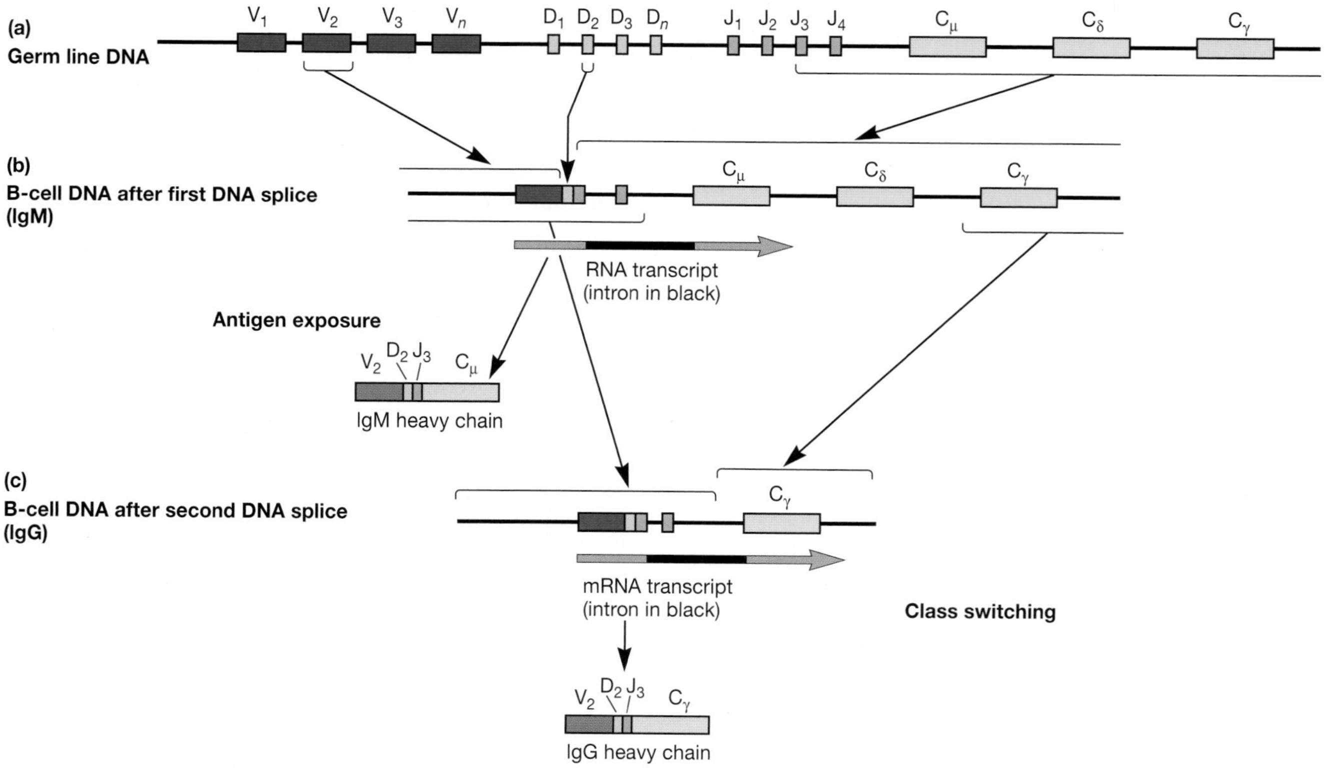

FIGURE 33.16 The Formation of a Gene for the Heavy Chain of an Antibody Molecule.

2. **Somatic hypermutation of V regions:** The V regions of germ-line DNA are susceptible to a high rate of somatic mutation during B-cell development in response to an antigen challenge. To accomplish this, the rearranged variable heavy and light genes encode "hotspots" of four to five nucleotides that are more likely to be altered by point mutation than other areas of the variable regions. The resulting point mutations are recognized by DNA repair enzymes that attempt to correct the altered bases. AID (the same enzyme that mediates antibody class switching) is one of the enzymes that is involved. AID deaminates DNA at cytosine resulting in base pair mismatches and transition-type mutations. Additionally, repair of the base pair mismatch is error-prone and results in additional sequence diversity. Accumulation of the altered bases in the variable regions of B cell immunoglobulin genes during antigen challenge promotes significant antibody diversity and thus appropriate antibody affinity. Adding the impact of antibody class switching to the above two processes should begin to explain how it is possible for unique antibodies to respond to the trillions of potential antigens in the environment.

1. What is the name of each part of the gene that encodes for the different regions of antibody chains?
2. Describe what is meant by combinatorial joining of V, D, and J gene segments.
3. In addition to combinatorial joining, what other two processes play a role in antibody diversity?

Clonal Selection

As noted previously, combinatorial joinings, somatic mutations, and variations in the splicing process generate the great variety of antibodies produced by mature B cells. From a large, diverse B-cell pool, specific cells are stimulated by antigens to reproduce and form B-cell clones containing the same genetic information. This is known as **clonal selection.** Both B and T cells undergo clonal selection. It is important because it accounts for immunological specificity and memory.

The clonal selection theory has four components or tenets. The first tenet is that there exists a pool of lymphocytes that can bind to a

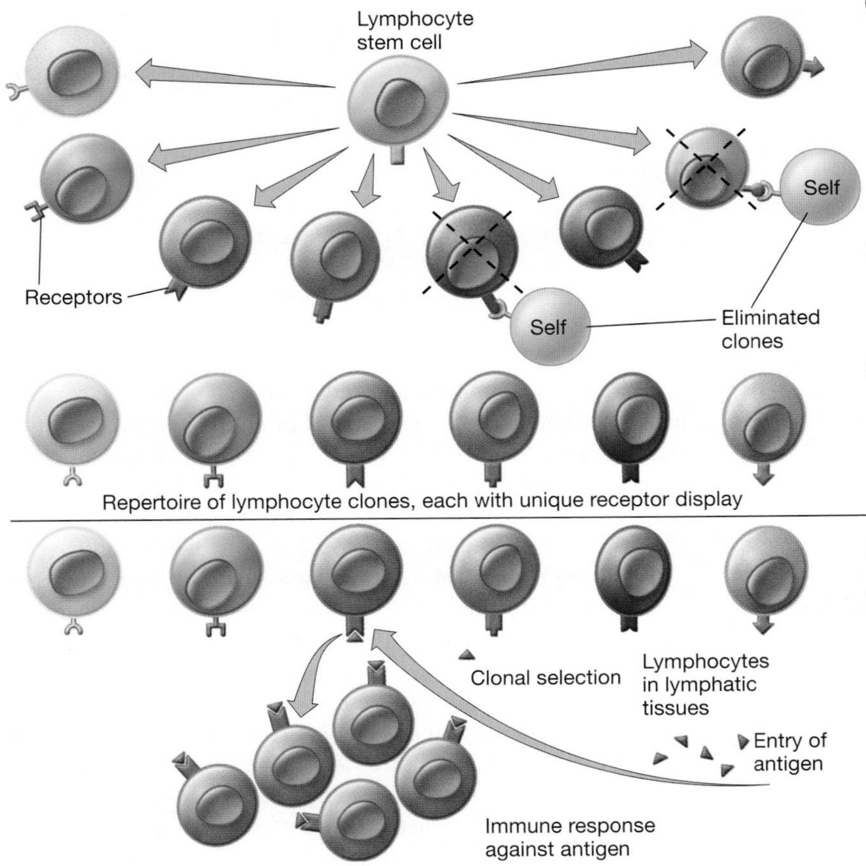

(a) Antigen-Independent Period

1. During development of early lymphocytes from stem cells, a given stem cell undergoes rapid cell division to form numerous progeny.

   During this period of cell differentiation, random rearrangements of the genes that code for cell surface protein receptors occur. The result is a large array of genetically distinct cells, called clones, each clone bearing a different receptor that is specific to react with only a single type of foreign molecule or antigen.

2. At the same time, any lymphocyte clones that have a specificity for self molecules and could be harmful are eliminated from the pool of diversity. This is called immune tolerance.

3. The specificity for a single antigen molecule is programmed into the lymphocyte and is set for the life of a given clone. The end result is an enormous pool of mature but naïve lymphocytes that are ready to further differentiate under the influence of certain organs and immune stimuli.

(b) Antigen-Dependent Period

4. Lymphocytes come to populate the lymphatic organs, where they will finally encounter antigens. These antigens will become the stimulus for the lymphocytes' final activation and immune function. Entry of a specific antigen selects only the lymphocyte clone or clones that carry matching surface receptors. This will trigger an immune response, which varies according to the type of lymphocyte involved.

FIGURE 33.17 **Lymphocyte Clonal Expansion.** (a) Cell populations expand and are restricted based on their MHC expression. (b) They are further expanded when activated by a specific antigen.

tremendous range of epitopes (**figure 33.17**). Because some of the B cells generated by this process will produce antibodies that can react with self-epitopes, the second tenet of the theory is that these self-reactive cells are eliminated at an early stage of development. Indeed, this has been shown to be true for developing B cells (in the bone marrow) and self-reacting T cells (in the thymus). The third tenet is that once a lymphocyte has been released into the body and is exposed to its specific antigen, it proliferates to form a **clone** (a population of identical cells derived from a single parent cell). Note that this clone has been "selected" by exposure to specific antigen, hence the name of the theory. The final tenet states that all clonal cells react with the same antigenic epitope that stimulated its formation. However, the cells may differentiate to have somewhat different functions. Figure 33.17 shows this process for a B cell, which, after proliferating in response to antigen exposure, forms two different cell populations: antibody-producing plasma cells and memory cells.

Plasma cells are literally protein factories that produce about 2,000 antibodies per second in their brief 5- to 7-day lifespan. Memory cells circulate more actively from blood to lymph and live much longer (years or even decades) than do plasma cells. Memory cells are responsible for the immune system's rapid secondary antibody response (figure 33.14) to the same antigen. Finally, memory B cells and plasma cells are usually not produced unless the B cell has interacted with and received cytokine signals from activated T-helper cells (figure 33.10). Clonal selection theory has also led to the development of **monoclonal antibody (mAb)** technology (**Techniques & Applications 33.1**).

33.8 Action of Antibodies

The antigen-antibody interaction is a bimolecular association that exhibits exquisite specificity. The in vivo interactions that occur in vertebrate animals are absolutely essential in protecting the animal against the continuous onslaught of microorganisms and their products, as well as cancer cells. This occurs partly because the antibody coats the invading foreign material, marking it for enhanced recognition by other components of the innate and adaptive immune systems. The mechanisms by which antibodies achieve this are now discussed.

Neutralization

Binding of antibody to biologically active materials (such as bacteria, toxins, or viruses) causes their inactivation or **neutralization** (**figure 33.18**). The capacity of bacteria to colonize the mucosal surfaces of mammalian hosts depends in part on their ability to adhere to mucosal epithelial cells. Secretory IgA (sIgA) antibodies block certain bacterial adherence-promoting factors from attaching to host cells. Thus sIgA can protect the host against infection by some pathogenic bacteria and perhaps other microorganisms on mucosal surfaces by neutralizing (preventing) their adherence to host cells.

Toxins must enter cells to exert their bioactivity. Antibody binding of toxins prevents toxin-mediated injury to cells: The toxin-antibody complex either is unable to attach to receptor sites on host target cells and is unable to enter the cell, or it is ingested by macrophages. For example, diphtheria toxin inhibits protein synthesis after binding to the cell surface by its B subunit and subsequent passage of the toxin into the cytoplasm of the target cell. The antibody blocks the toxic effect by inhibiting the entry of the A subunit by binding to the B fragment. An antibody capable of neutralizing a toxin or antiserum containing neutralizing antibody against a toxin is called **antitoxin.** ⏮ *Toxins (section 31.3)*

IgG, IgM, and IgA antibodies can bind to some viruses during their extracellular phase and inactivate them. This antibody-mediated viral inactivation is called **viral neutralization.** Viral neutralization prevents a viral infection because the antibody prevents the virus from binding to and entering its target cell.

Opsonization

Phagocytes have an intrinsic ability to bind directly to microorganisms by pattern recognition and nonspecific cell-surface receptors, engulf the microorganisms, form phagosomes, and digest the microorganisms. This phagocytic process can be greatly enhanced by opsonization. As noted in section 32.3, opsonization is the process by which microorganisms or other foreign particles are coated with antibody or complement and thus prepared for "recognition" and ingestion by phagocytic cells. Opsonizing antibodies, especially IgG1 and IgG3, bind to Fc receptors on the surface of dendritic cells, macrophages, and neutrophils. This binding provides the phagocyte with a method for the specific capture of antigens. In other words, the antibody forms a bridge between the phagocyte and the antigen, thereby increasing the likelihood of its phagocytosis (*see figure 32.5*).

Immune Complex Formation

Because antibodies have at least two antigen-binding sites and most antigens have at least two antigenic determinants, cross-linking can occur, producing large aggregates termed **immune complexes** (figure 33.18). If the antigens are soluble molecules and the complex becomes large enough to settle out of solution, a **precipitation** (Latin *praecipitare,* to cast down) or **precipitin reaction** occurs. When the immune complex involves the cross-linking of cells or particles, an **agglutination reaction** occurs. These immune complexes are more rapidly phagocytosed in vivo than free antigens.

The extent of immune complex formation depends on the relative concentrations of the antibody and antigen. If there is a large excess of antibody, separate antibody molecules usually bind to each antigenic determinant and the antigen remains soluble. When antigen is present in excess, two separate antigen molecules tend to bind to each antibody and network development or cross-linking is inhibited. The ratio of antibody and antigen is said to be in the **equivalence zone** when their concentration is optimal for the formation of a large network of interconnected antibody and antigen molecules (*see figures 35.10 and 35.14*). When in the equivalence zone, all antibody and antigen molecules precipitate or agglutinate as an insoluble complex.

TECHNIQUES & APPLICATIONS

33.1 Monoclonal Antibody Technology

The value of antibodies as tools for locating or identifying antigens is well established. For many years, antiserum extracted from human or animal blood was the main source of antibodies for tests and therapy, but most antiserum is problematic. It contains polyclonal antibodies, meaning it is a mixture of different antibodies because it reflects dozens of immune reactions from a wide variety of B-cell clones. This characteristic is to be expected, because several immune reactions may be occurring simultaneously, and even a single species of microbe can stimulate several different types of antibodies. Certain applications in immunology require a pure preparation of monoclonal antibodies (mAbs) that originate from a single clone and have a single specificity for antigen.

The technology for producing monoclonal antibodies is possible by hybridizing cancer cells and activated B cells in vitro. This technique began with the discovery that tumors isolated from multiple myelomas (a blood cancer) in mice consist of identical plasma cells. These monoclonal plasma cells secrete a strikingly pure form of antibody with a single specificity and continue to divide indefinitely. Immunologists recognized the potential in these plasma cells and devised a hybridoma approach to creating mAb. The basic idea behind this approach is to hybridize or fuse a myeloma cell with a normal plasma cell from a mouse spleen to create an immortal cell that secretes a supply of functional antibodies with a single specificity.

The introduction of this technology has the potential for numerous biomedical applications. Monoclonal antibodies have provided immunologists with excellent standardized tools for studying the immune system and for expanding disease diagnosis and treatment. Most of the successful applications thus far use mAbs in in vitro diagnostic testing and research. Although injecting monoclonal antibodies to treat human disease is an exciting prospect, this therapy has been stymied because most mAbs are of mouse origin, and many humans will develop hypersensitivity to them. However, using genetic engineering, human antibody constant regions are cloned to mouse antibody-binding regions to create a hybrid antibody that is highly specific but less likely to cause hypersensitivity reactions.

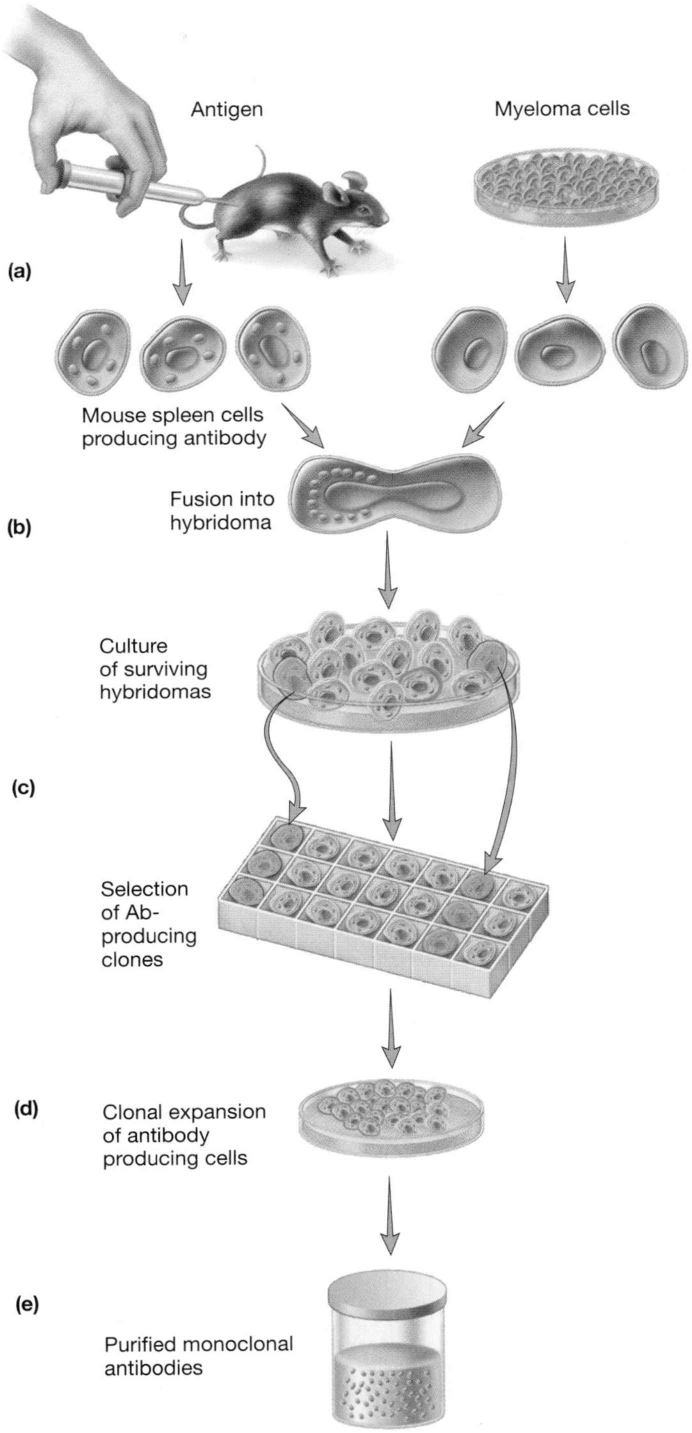

Monoclonal Antibody Formation. (a) A mouse is inoculated with an antigen having the desired specificity, and activated cells are isolated from its spleen. A special strain of mouse provides the myeloma cells. (b) The two cell populations are mixed with polyethylene glycol, which causes some cells in the mixture to fuse and form hybridomas. (c) Surviving cells are cultured and separated into individual wells. (d) Tests are performed on each hybridoma to determine specificity of the antibody (Ab) it secretes. (e) A hybridoma with the desired specificity is grown in tissue culture; antibody is then isolated and purified.

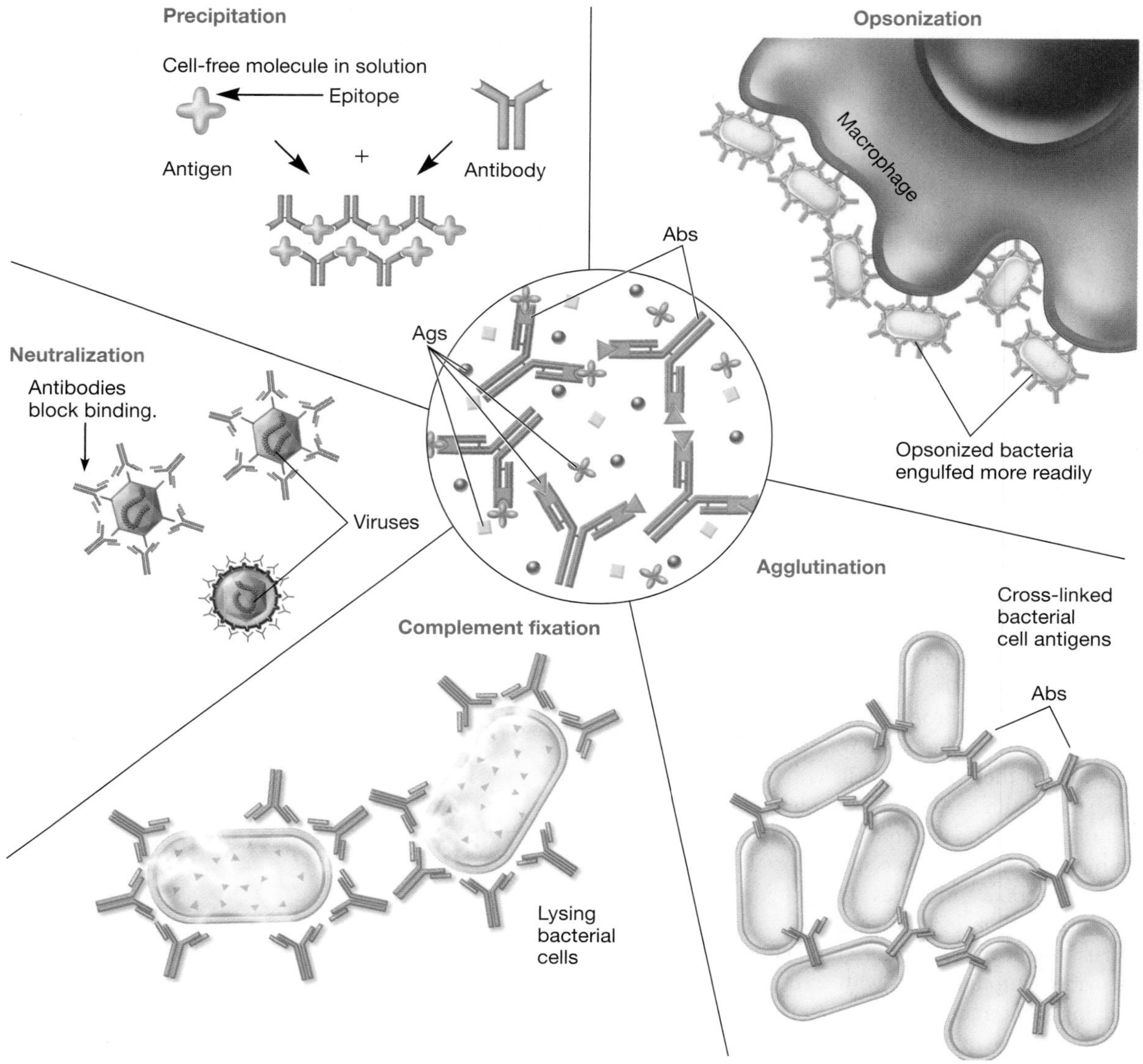

FIGURE 33.18 Consequences of Antigen-Antibody Binding. Immune complexes form when soluble antigens (Ag) bind soluble antibody (Ab), resulting in precipitation. Opsonization occurs when antibody binds to antigens on larger molecules or cells to be recognized by phagocytic cells. Agglutination results when insoluble antigens (such as viral or bacterial cells) are cross-linked by antibody. The classical complement cascade can be activated by immune complexes; this is called complement fixation. Neutralization results when antibody binds to antigens, preventing the antigen from binding to host cells.

Figure 33.18 Micro Inquiry

What is the difference between a precipitation and an agglutination reaction?

1. How does toxin neutralization occur? Viral neutralization?
2. How does opsonization inhibit microbial adherence to eukaryotic cells?
3. Describe an immune complex. What is the equivalence zone?

33.9 Acquired Immune Tolerance

Acquired immune tolerance is the body's ability to produce T cells and antibodies against nonself antigens such as microbial antigens, while "tolerating" (not responding to) self antigens. Some of this tolerance arises early in embryonic life, when immunologic competence is being established. Three general tolerance mechanisms have been proposed: (1) negative selection by clonal deletion, (2) the induction of anergy, and (3) inhibition of the immune response by T cells with suppressor/regulatory function (Treg cells).

Negative selection by clonal deletion removes lymphocytes that recognize any self antigens that are present. These cells are eliminated by apoptosis. T-cell tolerance induced in the thymus and B-cell tolerance in the bone marrow is called **central tolerance.** However, another mechanism is needed to prevent immune reactions against self antigens, termed autoimmunity, because many antigens are tissue-specific and are not present in the thymus or bone marrow.

Mechanisms occurring elsewhere in the body are collectively referred to as **peripheral tolerance.** These supplement central tolerance. Peripheral tolerance is thought to be based largely on incomplete activation signals given to the lymphocyte when it encounters self antigens in the periphery of the body. This mechanism leads to a state of unresponsiveness called **anergy,** which is associated with impaired intracellular signaling and apoptosis. Many autoreactive B cells undergo clonal deletion or become anergic as they mature in the bone marrow. The deletion of self-reactive B cells also takes place in secondary lymphoid tissue, such as the spleen and lymph nodes. Since B cells can recognize unprocessed antigen, there is no need for the participation of MHC molecules in these processes. For those self antigens present at relatively low concentrations, immunologic tolerance is often maintained only within the T-cell population. This is sufficient to sustain tolerance because it denies the help essential for antibody production by self-reactive B cells. T cells with suppressor activity have been defined as cells that can specifically inhibit responses of other T cells in an antigen-specific manner, although their existence has not been conclusively proven.

1. Describe the three ways acquired immune tolerance develops in the vertebrate host.
2. How would you define anergy?

33.10 Immune Disorders

As in any system in a vertebrate animal, disorders also occur in the immune system. Immune disorders can be categorized as hypersensitivities, autoimmune diseases, transplantation (tissue) rejection, and immunodeficiencies. Each of these is now discussed.

Hypersensitivities

Hypersensitivity is an exaggerated specific immune response that results in tissue damage and is manifested in the individual on a second or subsequent contact with an antigen. Hypersensitivity reactions can be classified as either immediate or delayed. Obviously, immediate reactions appear faster than delayed ones, but the main difference between them is the nature of the immune response to the antigen. Realizing this fact in 1963, Peter Gell and Robert Coombs developed a classification system for reactions responsible for hypersensitivities. Their system correlates clinical symptoms with information about immunologic events that occur during hypersensitivity reactions. The **Gell-Coombs classification** system divides hypersensitivity into four types: I, II, III, and IV.

Type I Hypersensitivity

An **allergy** (Greek *allos,* other, and *ergon,* work) is one kind of **type I hypersensitivity** reaction. Allergic reactions occur when an individual who has produced IgE antibody in response to the initial exposure to an antigen (**allergen**) subsequently encounters the same allergen. Upon initial exposure to a soluble allergen, B cells are stimulated to differentiate into plasma cells and produce specific IgE antibodies with the help of T_H cells (**figure 33.19**). This IgE is sometimes called a **reagin,** and the individual has a hereditary predisposition for its production. Once synthesized, IgE binds to the Fc receptors of mast cells (basophils and eosinophils can also be bound) and sensitizes these cells, making the individual sensitized to the allergen. When a subsequent exposure to the allergen occurs, the allergen attaches to the surface-bound IgE on the sensitized mast cells, causing mast cell degranulation.

Degranulation releases physiological mediators such as histamine, leukotrienes, heparin, prostaglandins, PAF (*p*latelet-*a*ctivating *f*actor), ECF-A (*e*osinophil *c*hemotactic *f*actor of *a*naphylaxis), and proteolytic enzymes. These mediators trigger smooth muscle contractions, vasodilation, increased vascular permeability, and mucus secretion (figure 33.19). The inclusive term for these responses is **anaphylaxis** (Greek *ana,* up, back again, and *phylaxis,* protection). Anaphylaxis can be divided into systemic and localized reactions.

Systemic anaphylaxis is a generalized response that is immediate due to a sudden burst of mast cell mediators. Usually there is respiratory impairment caused by smooth muscle constriction in the bronchioles. The arterioles dilate, which greatly reduces arterial blood pressure and increases capillary permeability with rapid loss of fluid into the tissue spaces. These physiological changes can be rapid and severe enough to be fatal within a few minutes from reduced venous return, asphyxiation, reduced blood pressure, and circulatory shock. Common examples of allergens that can produce systemic anaphylaxis include drugs (penicillin), passively administered antisera, peanuts, and insect venom from the stings or bites of wasps, hornets, or bees.

Localized anaphylaxis is called an **atopic** ("out of place") **reaction.** The symptoms that develop depend primarily on the route by which the allergen enters the body. Hay fever (allergic rhinitis) is a good example of atopy involving the upper respiratory tract. Initial exposure involves airborne allergens—such as plant pollen, fungal spores, animal hair and dander, and house dust mites—that sensitize mast cells located within the mucous membranes of the respiratory tract. Reexposure to the allergen causes a localized anaphylactic response: itchy and tearing eyes, congested nasal passages, coughing, and sneezing. Antihistamine drugs are used to help alleviate these symptoms.

Skin testing can be used to identify the antigen responsible for allergies. These tests involve inoculating a small amount of the

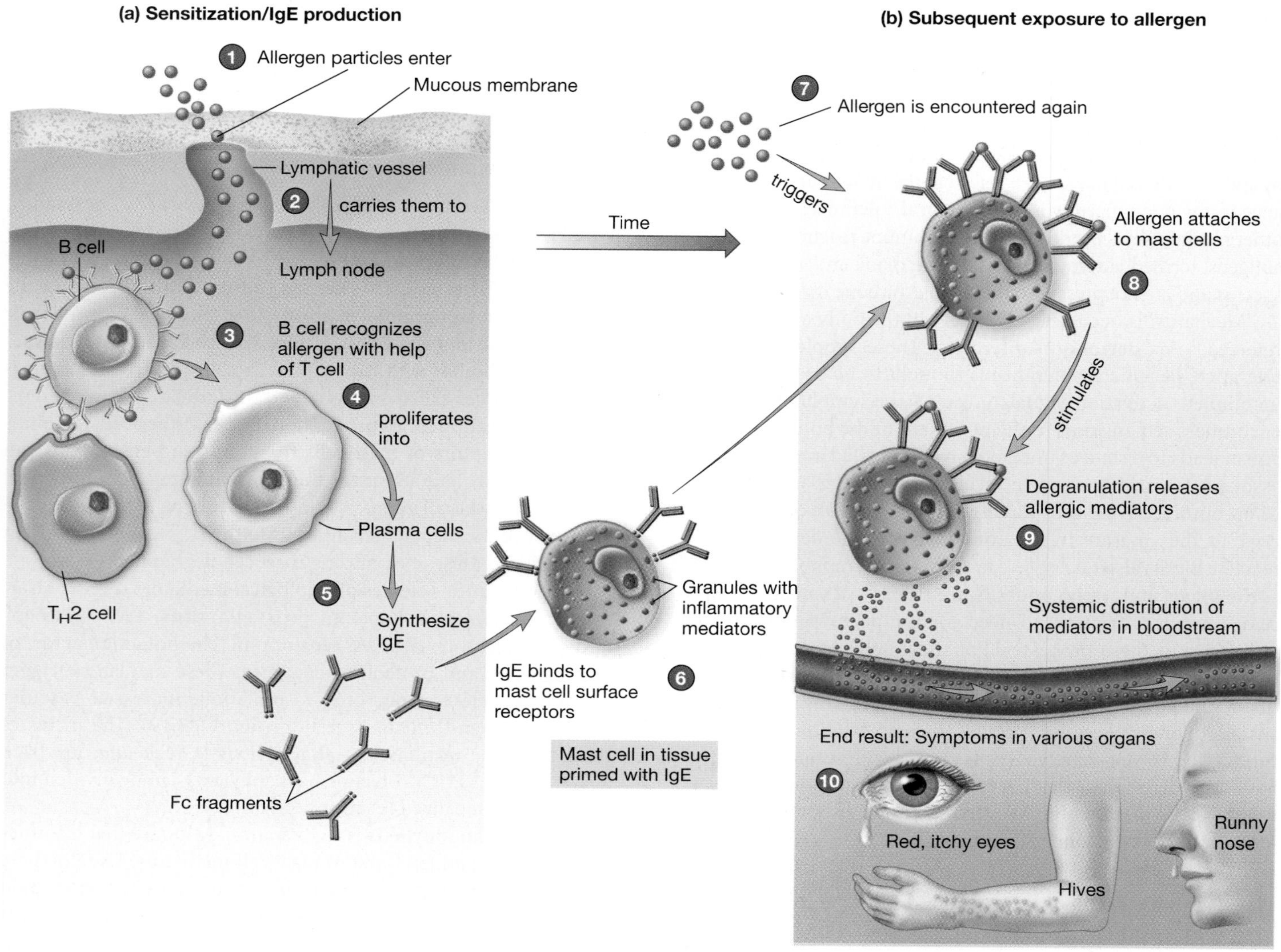

FIGURE 33.19 Type I Hypersensitivity (Allergic Response). (a) The initial contact (sensitization) of lymphocytes by small-protein allergens at mucous membranes results in T_H2 cell–assisted antibody class switching; plasma cells secrete IgE antibody. The IgE binds to its receptor on tissue mast cells (1–6). (b) Subsequent exposure to the same allergens results in their capture by the cell-bound IgE (7), triggering mast cell degranulation (8, 9). Characteristic signs and symptoms (hives, swelling, itching, etc.) of allergy ensue (10).

Figure 33.19 Micro Inquiry

What are some of the allergic mediators released during mast cell degranulation?

suspect allergens into the skin. Sensitivity to an antigen is shown by a rapid inflammatory reaction characterized by redness, swelling, and itching at the site of inoculation (**figure 33.20**). The affected area in which the allergen–mast cell reaction takes place is called a wheal and flare reaction site. Once the responsible allergen has been identified, the individual should avoid contact with it. If this is not possible, **desensitization** is warranted. This procedure consists of a series of allergen doses injected beneath the skin to stimulate the production of IgG antibodies rather than IgE antibodies. The circulating IgG antibodies can then act to intercept and neutralize allergens before they have time to react with mast cell–bound IgE. Desensitizations are about 65 to 75% effective in individuals whose allergies are caused by inhaled allergens.

Type II Hypersensitivity

Type II hypersensitivity is generally called a **cytolytic** or **cytotoxic reaction** because it results in the destruction of host cells, either by lysis or toxic mediators. In type II hypersensitivity, IgG or IgM antibodies are inappropriately directed against cell-surface or

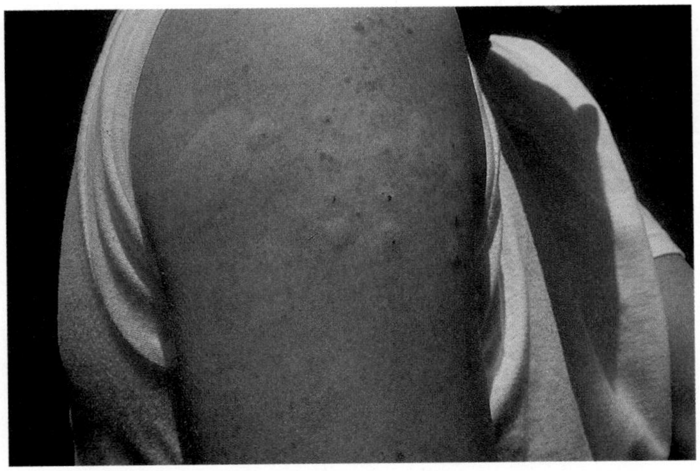

FIGURE 33.20 In Vivo Skin Testing. Skin prick tests with grass pollen in a person with summer hay fever. Notice the various reactions with increasing dosages (from top to bottom).

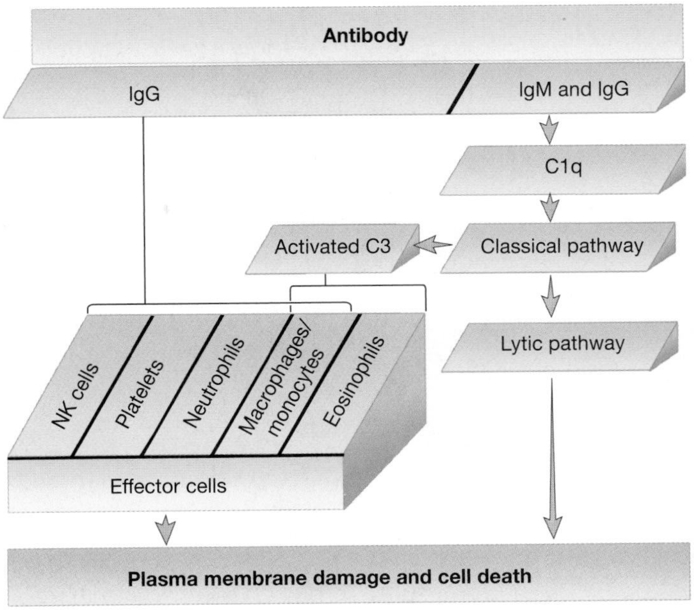

FIGURE 33.21 Type II Hypersensitivity. The action of antibody occurs through effector cells or the membrane attack complex, which damages target cell plasma membranes, causing cell destruction.

tissue-associated antigens. They usually stimulate the classical complement pathway and a variety of effector cells (**figure 33.21**). The antibodies interact with complement (Clq) and the effector cells through their Fc regions. The damage mechanisms are a reflection of the normal physiological processes involved in interaction of the immune system with pathogens. Classical examples of type II hypersensitivity reactions are the response exhibited by a person who receives a transfusion with blood from a donor with a different blood group and erythroblastosis fetalis, which we now discuss.

Blood transfusion was often fatal prior to the discovery of distinct blood types by Karl Landsteiner in 1904. Landsteiner's observation that sera from one person could agglutinate the blood cells of another person led to his identification of four distinct types of human blood. The red blood cell types were subsequently determined to result from cell-surface glycoproteins, now called the **ABO blood groups** (**figure 33.22*a***). Blood types are genetically inherited as the A, B, or O alleles respectively encoding the A- or B-type glycoprotein or no glycoprotein at all. Thus homozygous expression of A or B alleles results in type A or type B blood, respectively. Heterozygous expression of the codominant A and B alleles results in type AB blood. Heterozygous expression of the dominant A or B alleles with the O allele results in type A or type B blood, respectively. Homozygous expression of the O allele results in type O blood. AB glycoproteins are self antigens. Thus AB reactive lymphocytes of the developing host are destroyed during the negative selection process. However, the lymphocytes specific for AB glycoproteins not expressed by the host remain to be activated upon exposure to those specific antigens (which are ubiquitously distributed throughout nature). Consequently, type A hosts produce anti-B antibodies, type B hosts produce anti-A antibodies, and type O hosts produce both anti-A and anti-B antibodies. Type O individuals are considered "universal donors" because their red blood cells lack A and B surface antigens. Conversely, type AB hosts produce neither anti-A nor anti-B antibodies, so such individuals are called "universal recipients." The type II hypersensitivity reaction seen in blood transfusion occurs as complement is activated by cross-linking antibodies (figure 33.22*b*).

Blood typing can be accomplished by a slightly more sophisticated method of Landsteiner's process whereby blood from one host is mixed with antibodies specific for type A or type B blood. Agglutination of red cells by antibody (figure 33.22*c*) is used as a diagnostic tool to determine blood type. Another red blood cell antigen often reported with the ABO type was discovered during experiments with rhesus monkeys. The so-called Rh factor (or D antigen) is determined by the expression of two alleles, one dominant (coding for the factor) and one recessive (not coding for the factor). Thus homozygous or heterozygous expression of the Rh allele confers the antigen (indicated as Rh^+). Expression of two recessive alleles results in the designation of Rh^- (no Rh factor). Incompatibility between Rh^- mothers and their Rh^+ fetus can result in maternal anti-Rh antibodies destroying fetal blood cells (**figure 33.23**). This type II hypersensitivity is called **erythroblastosis fetalis.** Control of this potentially fatal hemolytic disease of the newborn can be mitigated if the mother is passively immunized with anti-Rh factor antibodies, or RhoGAM.

Type III Hypersensitivity

Type III hypersensitivity involves the formation of immune complexes (**figure 33.24**). Normally these complexes are phagocytosed effectively by dendritic cells and macrophages. In the presence of excess amounts of some soluble antigens, the antigen-antibody complexes may not be efficiently removed. Their accumulation can lead to a hypersensitivity reaction from complement that triggers a variety of inflammatory processes. The antibodies of type III

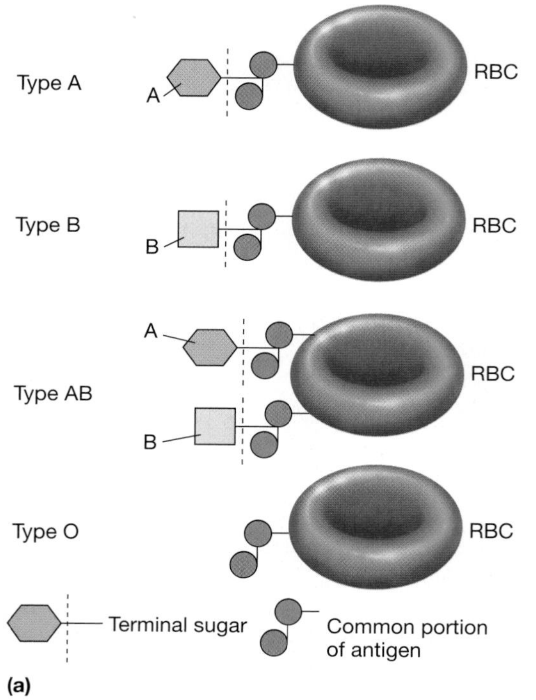

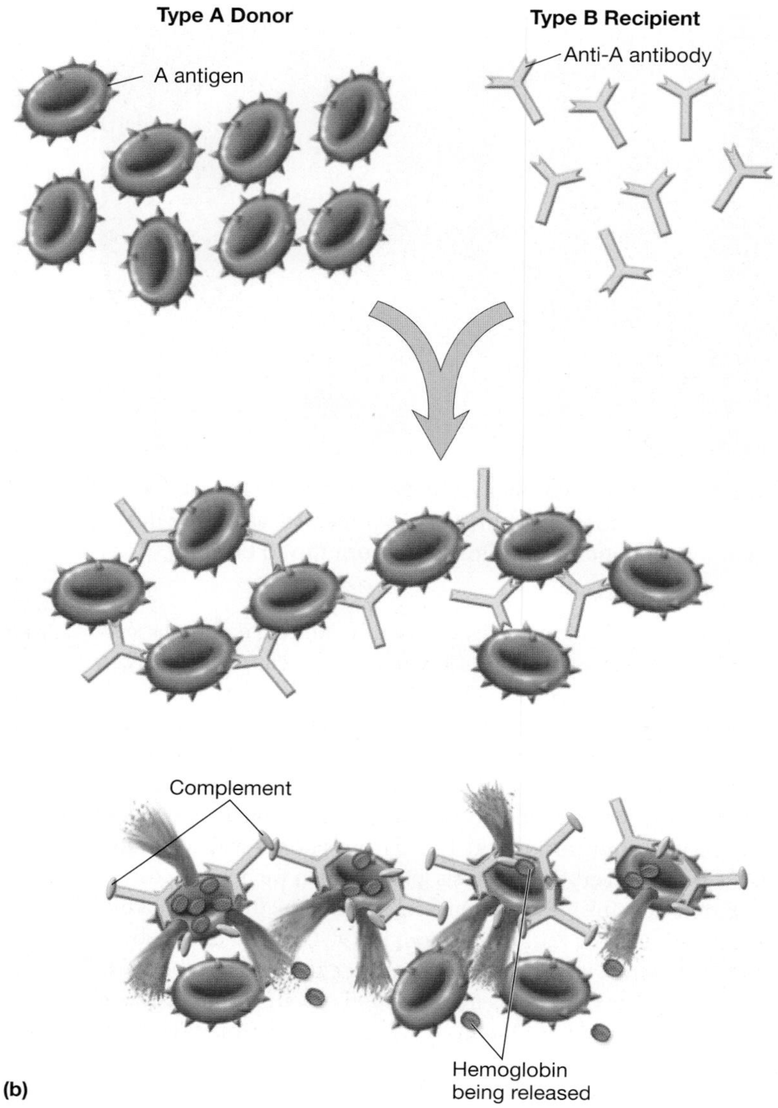

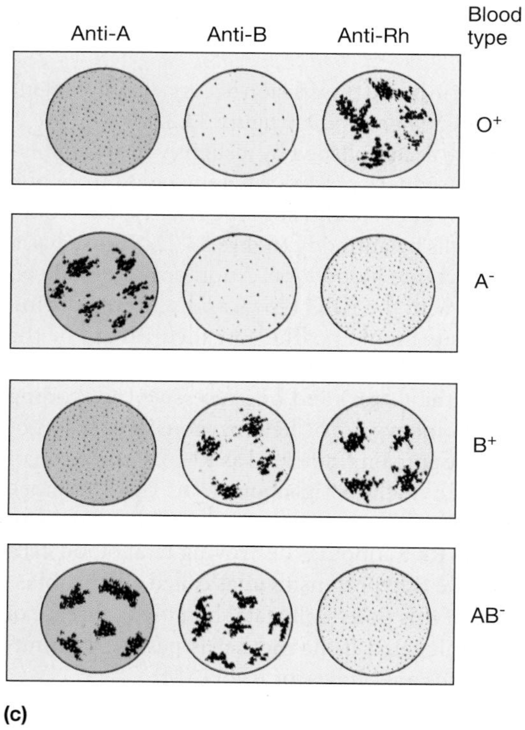

FIGURE 33.22 Immunohematology Is the Study of Immune Reactions Associated with Blood. (a) Red blood cells (RBCs) can have genetically inherited carbohydrate antigens (two possible sugar residues) on their surface. The presence of the antigen(s) determines the blood type. Some individuals may have one, both, or no antigen, resulting in the A or B, AB, or O blood types, respectively. (b) A host does not make antibodies to its own blood antigen(s); antibodies to the blood antigens not found in a host are made. Exposure of blood to antibody specific for its carbohydrate type results in RBC agglutination. RBC lysis can then occur if complement is activated by the antibody-agglutinated cells. (c) RBC agglutination with specific antibody is the basis for blood typing. Another molecule, the rhesus (Rh) factor, is another major RBC antigen that is typed to determine blood compatibility.

Figure 33.22 Micro Inquiry

What would be the genotype of an individual with type O, Rh^- blood?

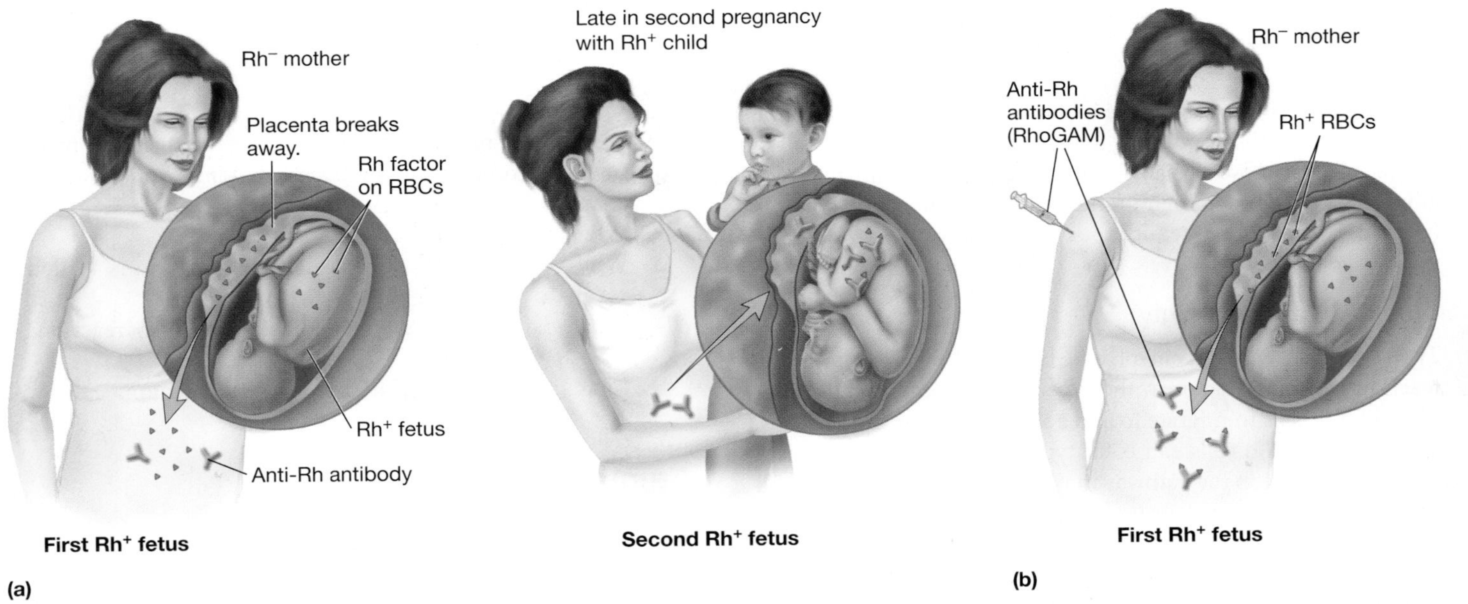

Rh⁻ mother
Placenta breaks away.
Rh factor on RBCs
Rh⁺ fetus
Anti-Rh antibody
First Rh⁺ fetus
(a)
Late in second pregnancy with Rh⁺ child
Second Rh⁺ fetus
Anti-Rh antibodies (RhoGAM)
Rh⁻ mother
Rh⁺ RBCs
First Rh⁺ fetus
(b)

FIGURE 33.23 Rh Factor Incompatibility Can Result in RBC Lysis. (a) A naturally occurring blood cell incompatibility results when a Rh^+ fetus develops within a Rh^- mother. Initial sensitization of the maternal immune system occurs when fetal blood passes the placental barrier. In most cases, the fetus develops normally. However, a subsequent pregnancy with a Rh^+ fetus results in a severe, fetal hemolysis. (b) Anti-Rh antibody (RhoGAM) can be administered to Rh^- mothers during pregnancy to help bind, inactivate, and remove any Rh factor that may be transferred from the fetus. In some cases, RhoGAM is administered before sensitization occurs.

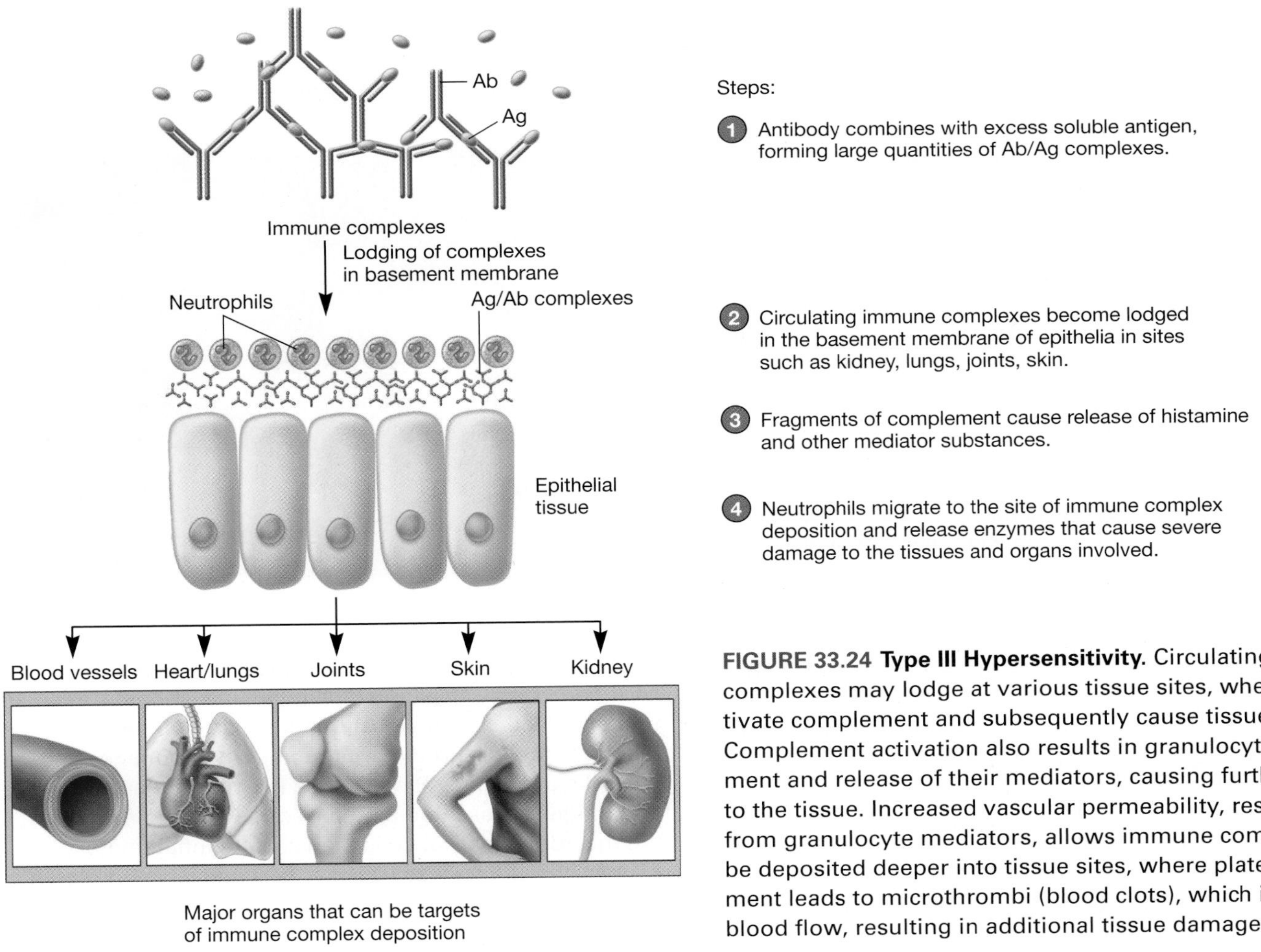

Ab
Ag
Immune complexes
Lodging of complexes in basement membrane
Neutrophils
Ag/Ab complexes
Epithelial tissue
Blood vessels
Heart/lungs
Joints
Skin
Kidney
Major organs that can be targets of immune complex deposition
Steps:
1 Antibody combines with excess soluble antigen, forming large quantities of Ab/Ag complexes.
2 Circulating immune complexes become lodged in the basement membrane of epithelia in sites such as kidney, lungs, joints, skin.
3 Fragments of complement cause release of histamine and other mediator substances.
4 Neutrophils migrate to the site of immune complex deposition and release enzymes that cause severe damage to the tissues and organs involved.

FIGURE 33.24 Type III Hypersensitivity. Circulating immune complexes may lodge at various tissue sites, where they activate complement and subsequently cause tissue cell lysis. Complement activation also results in granulocyte recruitment and release of their mediators, causing further injury to the tissue. Increased vascular permeability, resulting from granulocyte mediators, allows immune complexes to be deposited deeper into tissue sites, where platelet recruitment leads to microthrombi (blood clots), which impair blood flow, resulting in additional tissue damage.

reactions are primarily IgG. The inflammation caused by immune complexes and cells responding to such inflammation can result in significant damage, especially of blood vessels (vasculitis), kidney glomerular basement membranes (glomerulonephritis), joints (arthritis), and skin (systemic lupus erythematosus).

Type IV Hypersensitivity

Type IV hypersensitivity involves delayed, cell-mediated immune reactions. A major factor in the type IV reaction is the time required for T cells to migrate to and accumulate near the antigens. Both T_H and CTL cells can elicit type IV hypersensitivity reactions, depending on the pathway in which the antigen is processed and presented. These events usually take a day or more to plateau.

In general, type IV reactions occur when antigens, especially those binding to serum proteins or tissue cells, are processed and presented to T cells. If the antigen is phagocytosed, it will be presented to T_H cells by the class II MHC molecules on the APC. This activates the T_H1 cell, causing it to proliferate and secrete cytokines such as IFN-γ and TNF-α. If the antigen is lipid soluble, it can cross the cell membrane to be processed within the cytosol and be presented to CTL cells by class I MHC molecules. CTLs secrete cytokines and kill the cell that is presenting the antigen. Regardless of the cytokine source, they stimulate the expression of adhesion molecules on the local endothelium and increase vascular permeability, allowing fluid and cells to enter the tissue space. Cytokines also stimulate keratinocytes of the skin to release their own cytokines. Together, the cytokines attract lymphocytes, macrophages, and basophils to the affected tissue, exacerbating the inflammation. Extensive tissue damage may result. Examples of type IV hypersensitivities include tuberculin hypersensitivity (the TB skin test; **figure 33.25**), allergic contact dermatitis, some autoimmune diseases, transplantation rejection, and killing of cancer cells.

In **tuberculin hypersensitivity,** a partially purified protein called tuberculin, which is obtained from the bacterium that causes tuberculosis, is injected into the skin of the forearm (figure 33.25). The response in a tuberculin-positive individual begins in about 8 hours, and a reddened area surrounding the injection site becomes indurated (firm and hard) within 12 to 24 hours. The T_H1 cells that migrate to the injection site are responsible for the induration. The reaction reaches its peak in 48 hours and then subsides. The size of the induration is directly related to the amount of antigen that was introduced and to the degree of hypersensitivity of the tested individual. Other examples of type IV hypersensitivity reactions include the cell and tissue destruction caused by leprosy, tuberculosis, leishmaniasis, candidiasis, and herpes simplex lesions. These infectious diseases are discussed in chapters 37–39.

Allergic contact dermatitis is a type IV reaction caused by haptens that combine with proteins in the skin to form the allergen that elicits the immune response (**figure 33.26**). The haptens are the antigenic determinants, and the skin proteins are the carrier molecules for the haptens. Examples of these haptens include cosmetics, plant materials (catechol molecules from poison ivy and poison oak), topical chemotherapeutic agents, metals, and jewelry (especially jewelry containing nickel).

1. Discuss the mechanism of type I hypersensitivity reactions and how these can lead to systemic and localized anaphylaxis.
2. What causes a wheal and flare reaction site?
3. Why are type II hypersensitivity reactions called cytolytic or cytotoxic?
4. What characterizes a type III hypersensitivity reaction? Give an example.
5. Characterize a type IV hypersensitivity reaction.

Autoimmune Diseases

As discussed in section 33.1, the body is normally able to distinguish its own self antigens from foreign nonself antigens and usually does not mount an immunologic attack against itself. This phenomenon is called immune tolerance. At times the body loses tolerance and mounts an abnormal immune attack, either with antibodies or T cells, against its own self antigens.

It is important to distinguish between autoimmunity and autoimmune disease. Autoimmunity often is benign, whereas autoimmune disease often is fatal. **Autoimmunity** is characterized by the presence of serum antibodies that react with self antigens. These antibodies are called autoantibodies. The formation of autoantibodies is a normal consequence of aging; is readily inducible by infectious agents or drugs; and is potentially reversible (it disappears

FIGURE 33.25 Type IV (or Delayed-Type) Hypersensitivity. The mechanism of type IV hypersensitivity is illustrated by the tuberculin skin test, used to determine exposure to *M. tuberculosis.* Injection of the tuberculin antigens into the skin of individuals previously sensitized by *M. tuberculosis* results in the localized recruitment of macrophages and T_H1 cells over 12 to 48 hours. The T_H1 cells are activated by the antigens presented by the class II MHC molecules on the macrophages. T_H1 cells then secrete inflammatory cytokines, which increase vascular permeability and recruit other immune cells, resulting in a visible swelling at the injection site.

when the offending "agent" is removed or eradicated). **Autoimmune disease** results from the activation of self-reactive T and B cells that, following stimulation by genetic or environmental triggers, cause actual tissue damage (**table 33.5**). Examples include rheumatoid arthritis and type I diabetes mellitus.

Some investigators believe that the release of abnormally large quantities of antigens may occur when the infectious agent causes tissue damage. The same agents also may cause body proteins to change into forms that stimulate antibody production or T-cell activation. Simultaneously, the capacity of T cells to limit this type of reaction seems to be repressed. Many autoimmune diseases have a genetic component. For example, there is a well-documented association between an individual's susceptibility to Graves' disease (which causes hyperthyroidism) and the neurodegenerative disease multiple sclerosis and specific determinants on the major histocompatibility complex.

Transplantation (Tissue) Rejection

Transplants between genetically different individuals within a species are termed **allografts** (Greek *allos*, other). Some transplanted tissues do not stimulate an immune response. For example, a transplanted cornea is rarely rejected because lymphocytes do not circulate into the anterior chamber of the eye. This site is considered an immunologically privileged site. Another example of a privileged tissue is the heart valve, which in fact can be transplanted from a pig to a human without stimulating an immune response. Such a graft between different species is termed a **xenograft** (Greek *xenos*, strayed).

Transplanting tissue that is not immunologically privileged generates the possibility that the recipient's cells will recognize the donor's tissues as foreign. This triggers the recipient's immune mechanisms, which may destroy the donor tissue. Such a response is called a **tissue rejection reaction.** A tissue rejection reaction can occur by two different mechanisms. First, foreign class II MHC molecules on transplanted tissue, or the "graft," are recognized by host T-helper cells, which aid cytotoxic T cells in graft destruction (**figure 33.27**). Cytotoxic T cells then recognize the graft because it bears foreign class I MHC molecules. This response is much like the activation of CTLs by virally infected host cells. A second mechanism involves the T-helper cells reacting to the graft (transplanted tissue) and releasing cytokines. The cytokines stimulate macrophages to enter, accumulate within the graft, and destroy it. The MHC molecules play a dominant role in

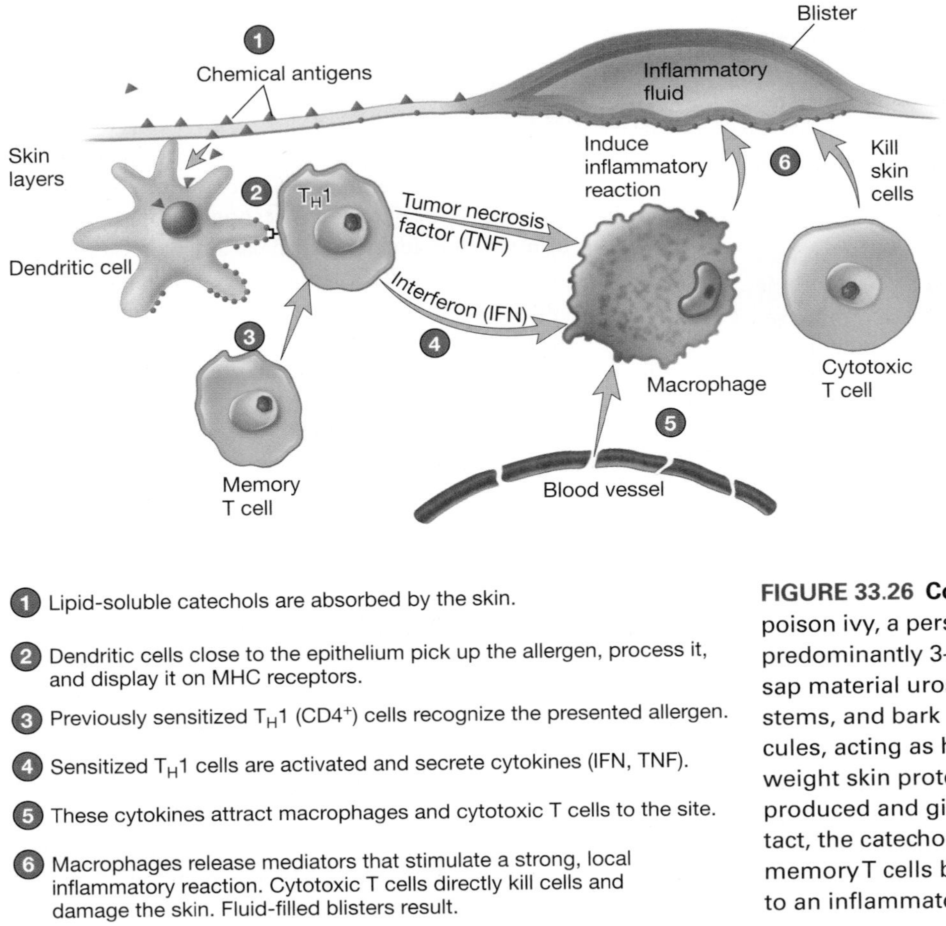

1
Chemical antigens
Blister
Inflammatory fluid
Skin layers
Induce inflammatory reaction
6
Kill skin cells
2
T_H1
Tumor necrosis factor (TNF)
Dendritic cell
Interferon (IFN)
3
4
Macrophage
Cytotoxic T cell
5
Memory T cell
Blood vessel
1 Lipid-soluble catechols are absorbed by the skin.
2 Dendritic cells close to the epithelium pick up the allergen, process it, and display it on MHC receptors.
3 Previously sensitized T_H1 ($CD4^+$) cells recognize the presented allergen.
4 Sensitized T_H1 cells are activated and secrete cytokines (IFN, TNF).
5 These cytokines attract macrophages and cytotoxic T cells to the site.
6 Macrophages release mediators that stimulate a strong, local inflammatory reaction. Cytotoxic T cells directly kill cells and damage the skin. Fluid-filled blisters result.

FIGURE 33.26 Contact Dermatitis. In contact dermatitis to poison ivy, a person initially becomes exposed to the allergen, predominantly 3-n-pentadecyl-catechol found in the resinous sap material uroshiol, which is produced by the leaves, fruit, stems, and bark of the poison ivy plant. The catechol molecules, acting as haptens, combine with high molecular weight skin proteins. After 7 to 10 days, sensitized T cells are produced and give rise to memory T cells. Upon second contact, the catechols bind to the same skin proteins, and the memory T cells become activated in only 1 to 2 days, leading to an inflammatory reaction (contact dermatitis).

Table 33.5 Some Autoimmune Diseases in Humans

Disease	*Autoantigen*	*Pathophysiology*
Acute rheumatic fever	Streptococcal cell wall antigens; antibodies cross-react with cardiomyocytes	Arthritis, scarring of heart valves, myocarditis
Autoimmune hemolytic anemia	Rh blood group, I antigen	Red blood cells are destroyed by complement and phagocytosis, anemia
Autoimmune thrombocytopenia purpura	Platelet integrin	Perfuse bleeding
Goodpasture's syndrome	Basement membrane collagen	Glomerulonephritis, pulmonary hemorrhage
Graves' disease	Thyroid-stimulating hormone receptor	Hyperthyroidism
Multiple sclerosis	Myelin basic protein	Demyelination of axons
Myasthenia gravis	Acetycholine receptor	Progressive muscular weakness
Pemphigus vulgaris	Cadherin in epidermis	Skin blisters
Rheumatoid arthritis	Unknown synovial joint antigen	Joint inflammation and destruction
Systemic lupus erythematosus	DNA, histones, ribosomes	Arthritis, glomerulonephritis, vasculitis, rash
Type 1 diabetes mellitus	Pancreatic beta cell antigen	Beta cell destruction

tissue rejection reactions because of their unique association with the recognition system of T cells. Unlike antibodies, T cells cannot recognize or react directly with non-MHC molecules (viruses, allergens). They recognize these molecules only in association with, or complexed to, an MHC molecule.

Because class I MHC molecules are present on every nucleated cell in the body, they are important targets of the rejection reaction. The greater the antigenic difference between class I molecules of the recipient and donor tissues, the more rapid and severe the rejection reaction is likely to be. Class II MHC molecule mismatch can also result in rejection and may be even more severe than class I mismatch reactions. However, the reaction can sometimes be minimized if recipient and donor tissues are matched as closely as possible. Most recipients are not 100% matched to their donors, so immunosuppressing drugs are used to prevent host-mediated rejection of the graft.

Organ transplant recipients also can develop **graft-versus-host disease.** This occurs when the transplanted tissue contains immune cells that recognize host antigens and attack the host. The immunosuppressed recipient cannot control the response of the grafted tissue. Graft-versus-host disease is a common problem in bone marrow transplants. The transplanted bone marrow contains many mature, postthymic T cells. These cells recognize the host MHC antigens and attack the immunosuppressed recipient's normal tissue cells. Currently one way to prevent graft-versus-host disease is to deplete the bone marrow of mature T cells by using immunosuppressive techniques. Examples include drugs that attack T cells (azathioprine, methotrexate, and cyclophosphamide), immunosuppressive drugs (cyclosporin, tacrolimus, and rapamycin), anti-inflammatory drugs (corticosteroids), irradiation of the lymphoid tissue, and antibodies directed against T-cell antigens.

Immunodeficiencies

Defects in one or more components of the immune system can result in its failing to recognize and respond properly to antigens. Such **immunodeficiencies** can make a person more prone to infection than those people capable of a complete and active immune response. Despite the increase in knowledge of functional derangements and cellular abnormalities in the various immunodeficiency disorders, the fundamental biological errors responsible for them remain largely unknown. To date, most genetic errors associated with these immunodeficiencies are located on the X chromosome and produce primary or congenital immunodeficiencies (**table 33.6**). Other immunodeficiencies can be acquired because of infections by immunosuppressive microorganisms, such as HIV.

1. What is an autoimmune disease and how might it develop?
2. What is an immunologically privileged site and how is it related to transplantation success?
3. How does a tissue rejection reaction occur? How might a patient about to receive a bone marrow transplant be prepared? Explain why this is necessary.
4. Describe an immunodeficiency. How might immunodeficiencies arise?

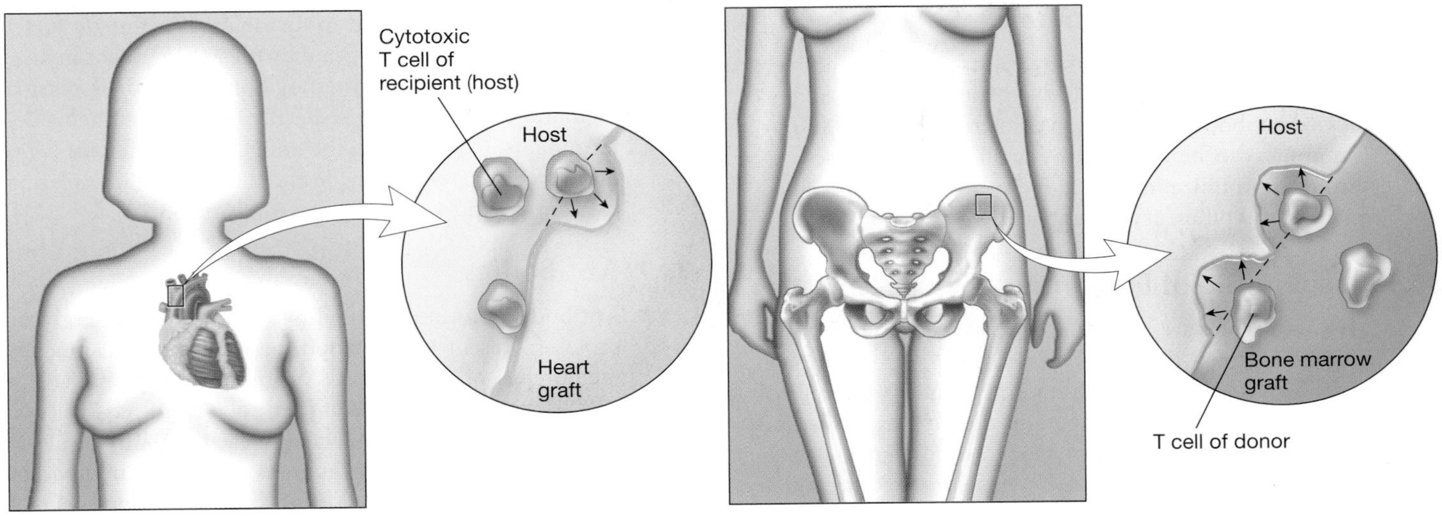

(a) (b)

FIGURE 33.27 Potential Transplantation Reactions. (a) Donated tissues that are not from an identical twin contain cellular MHC proteins that are recognized as foreign by the recipient host (host-versus-graft disease). The tissue is then attacked by host CTLs, resulting in its damage and rejection. (b) Donated tissue may also contain immune cells that react against host antigens. Recognition of a foreign host by donor CTLs results in graft-versus-host disease.

Table 33.6 Some Congenital Immune Deficiencies in Humans

Condition	*Symptoms*	*Cause*
Chronic granulomatous disease	Defective monocytes and neutrophils leading to recurrent bacterial and fungal infections	Failure to produce reactive oxygen intermediates due to defective NADPH oxidase
X-linked agammaglobulinemia	Plasma cell or B-cell deficiency and inability to produce adequate specific antibodies	Defective B-cell differentiation due to loss of tyrosine kinase
DiGeorge syndrome	T-cell deficiency and very poor cell-mediated immunity	Lack of thymus or a poorly developed thymus
Severe combined immunodeficiency disease (SCID)	Both antibody production and cell-mediated immunity impaired due to a great reduction of B- and T-cell levels	Various mechanisms (e.g., defective B- and T-cell maturation because of X-linked gene mutation; absence of adenosine deaminase in lymphocytes)

Summary

33.1 Overview of Specific (Adaptive) Immunity

a. The specific (adaptive) immune response system consists of lymphocytes that can recognize foreign molecules (antigens) and respond to them. Two branches or arms of immunity are recognized: humoral (antibody-mediated) immunity and cellular (cell-mediated) immunity (**figure 33.1**).

b. Acquired immunity refers to the type of specific (adaptive) immunity that a host develops after exposure to a suitable antigen.

33.2 Antigens

a. An antigen is a substance that stimulates an immune response and reacts with the products of that response. Each antigen can have several antigenic determinant sites called epitopes that stimulate production of and combine with specific antibodies (**figure 33.2**).

b. Haptens are small organic molecules that are not antigenic by themselves but can become antigenic if bound to a larger carrier molecule.

33.3 Types of Specific Immunity

a. Immunity can be acquired by natural means—actively through infection or passively through receipt of preformed antibodies, as through colostrum (**figure 33.3**).
b. Immunity can be acquired by artificial means—actively through immunization or passively through receipt of preformed antibodies, as with antisera.

33.4 Recognition of Foreignness

a. MHC molecules are cell-surface proteins coded by a group of genes termed the major histocompatibility complex. Class I MHC proteins are found on all nucleated cells of mammals. Class II MHC proteins are only expressed on B cells and cells that can phagocytose foreign materials and organisms. The human MHC gene products are called the human leukocyte antigens (HLA) (**figure 33.4**).
b. Class I MHC proteins collect foreign peptides processed by the proteasome and present them to cytotoxic T cells (**figure 33.5**).
c. Class II MHC proteins collect foreign peptides processed by the phagosome and present them to T-helper cells.

33.5 T-Cell Biology

a. T cells are pivotal elements of the immune response. T cells have antigen-specific receptor proteins (**figure 33.6**).
b. Antigen-presenting cells—macrophages, dendritic cells, and B cells—take in foreign antigens or pathogens, process them, and present antigenic fragments complexed with MHC molecules to T-helper cells.
c. Cytotoxic T lymphocytes recognize target cells such as virus-infected cells that have foreign antigens and class I MHC molecules on their surface. The CTLs then attack and destroy the target cells using the CD95 pathway, the perforin pathway, or both.
d. T cells control the development of other cells, including effector B and T cells. T-helper cells ($CD4^+$) regulate cell behavior; cytotoxic T cells ($CD8^+$) also regulate cell behavior, but in addition, they can kill altered host cells directly. There are four subsets of T-helper cells: T_H1, T_H2, T_H17, and T_H0. T_H1 cells produce various cytokines and are involved in cellular immunity. The T_H2 cells also produce various cytokines but are involved in humoral immunity. T_H17 cells primarily produce IL-17 and induce inflammation by recruiting neutrophils. T_H0 cells are simply undifferentiated precursors of T_H1, T_H2, and T_H17 cells (**figure 33.7**).
e. Regulatory T (Treg) cells can arise from both $CD4^+$ and $CD8^+$ T cells and can downregulate specific immune reactions, especially those rare reactions attacking self.

33.6 B-Cell Biology

a. B cells defend against antigens by differentiating into plasma cells that secrete antibodies into the blood and lymph, providing humoral or antibody-mediated immunity.
b. B cells can be stimulated to divide or differentiate to secrete antibody when triggered by the appropriate signals.
c. B cells have receptor immunoglobulins on their plasma membrane surface that are specific for given antigenic determinants. Contact with the antigenic determinant is required for the B cell to divide and differentiate into plasma cells and memory cells (**figures 33.9** and **33.10**).

33.7 Antibodies

a. Antibodies (immunoglobulins) are a group of glycoproteins present in the blood, tissue fluids, and mucous membranes of vertebrates. All immunoglobulins have a basic structure composed of four polypeptide chains (two light and two heavy) connected to each other by disulfide bonds (**figure 33.11**). In humans, five immunoglobulin classes exist: IgG, IgA, IgM, IgD, and IgE (**figures 33.12** and **33.13** and **table 33.3**).
b. The primary antibody response in a host occurs following initial exposure to the antigen. This response has lag, log, plateau, and decline phases. Upon secondary antigen challenge, the B cells mount a heightened and accelerated response (**figure 33.14**).
c. Antibody diversity results from the rearrangement and splicing of the individual gene segments on the antibody-coding chromosomes, somatic mutations, the generation of different codons during splicing, and the independent assortment of light- and heavy-chain genes (**figures 33.15** and **33.16**).
d. Immunologic specificity and memory are partly explained by the clonal selection theory (**figure 33.17**).
e. Hybridomas result from the fusion of lymphocytes with myeloma cells. These cells produce a single monoclonal antibody. Monoclonal antibodies have many uses (**Techniques & Applications 33.1**).

33.8 Action of Antibodies

a. Various types of antigen-antibody reactions occur in vertebrates and initiate the participation of other body processes that determine the ultimate fate of the antigen.
b. The complement system can be activated, leading to cell lysis, phagocytosis, chemotaxis, or stimulation of the inflammatory response.
c. Other defensive antigen-antibody interactions include toxin neutralization, viral neutralization, adherence inhibition, opsonization, and immune complex formation (**figure 33.18**).

33.9 Acquired Immune Tolerance

a. Acquired immune tolerance is the ability of a host to react against nonself antigens while tolerating self antigens. It can be induced in several ways.
b. Anergy is a state of immune unresponsiveness.

33.10 Immune Disorders

a. When the immune response occurs in an exaggerated form and results in tissue damage to the individual, the term hypersensitivity is applied. There are four types of hypersensitivity reactions, designated as types I through IV (**figures 33.19–33.26**).

b. Autoimmune diseases result when self-reactive T and B cells attack the body and cause tissue damage. A variety of factors can influence the development of autoimmune disease.

c. The immune system can act detrimentally and reject tissue transplants. There are different types of transplants. Xenografts involve transplants of privileged tissue between different species, and allografts are transplants between genetically different individuals of the same species (**figure 33.27**).

d. Immunodeficiency diseases are a diverse group of conditions in which an individual's susceptibility to various infections is increased; several severe diseases can arise because of one or more defects in the specific (adaptive) or nonspecific (innate) immune response.

Critical Thinking Questions

1. Why do you think antibodies are proteins rather than polysaccharides or lipids? List all properties of proteins that make them suitable molecules from which to make antibodies.

2. How did the clonal selection theory inspire the development of monoclonal antibody techniques?

3. Why do MHC, TCR, and BCR molecules require accessory proteins or coreceptors for a signal to be sent within the cell?

4. Why do you think two signals are required for B- and T-cell activation but only one signal is required for activation of an APC?

5. Most immunizations require multiple exposures to the vaccine (i.e., boosters). Why is this the case?

6. In an effort to gain a better understanding of the mechanisms by which hosts tolerate their normal gut flora, researchers introduced gnotobiotic mice with the gram-negative bacterium *Bacteriodes thetaiotaomicron* (a normal inhabitant of the guts of both mice and humans). They discovered that the more sIgA the mice produced, the better the microbe was tolerated, as monitored by the level of intestinal inflammation. We usually think of antibody production and inflammation as partners in the fight against infectious agents.

   Why do you think that in this case, the production of antibody was not coupled with inflammation and, to the contrary, protected the host against inflammation? What does this suggest about the mechanism by which normal flora is "permitted" to make a specific host its home?

   Read the original paper: Peterson, D. A., et al. 2007. IgA response to symbiotic bacteria as a mediator of gut homeostasis. *Cell Host Microbe* 2:328–39.

7. While the distribution of memory T cells is well understood, the distribution of memory B cells after an infection has not been as well studied. To address this, Joo and coworkers used a mouse influenza pneumonia model to track the distribution of IgG and IgA memory B cells. The influenza infection was limited to the lung; no virus was detected outside the lung. During and after infection, memory B cells were widely dispersed to secondary lymphoid tissues as well as into the lung itself.

   Why do you think memory B cells localize to secondary lymphoid tissues if this virus is specific to lung tissue? What might the implications of this study be for the distribution of memory B cells following vaccination, particularly for a subunit vaccine that does not localize to the normal site of infection (*see p. 889*)?

   Read the original paper: Joo, H. M. 2008. Broad dispersion and lung localization of virus-specific memory B cells induced by influenza pneumonia. *Proc. Nat. Acad. Sci., USA* 105:3485–90.

Concept Mapping Challenge

Use the following words to construct a concept map by providing your own linking words:

Cellular immunity, Humoral immunity, T cells, B cells, Acquired immunity, Memory, Antibodies, T-helper cells, Specificity, Plasma cells, Cytotoxic T cells, MHC proteins

Learn More

Learn more by visiting the text website at www.mhhe.com/willey8, where you will find a complete list of references.

34

Antimicrobial Chemotherapy

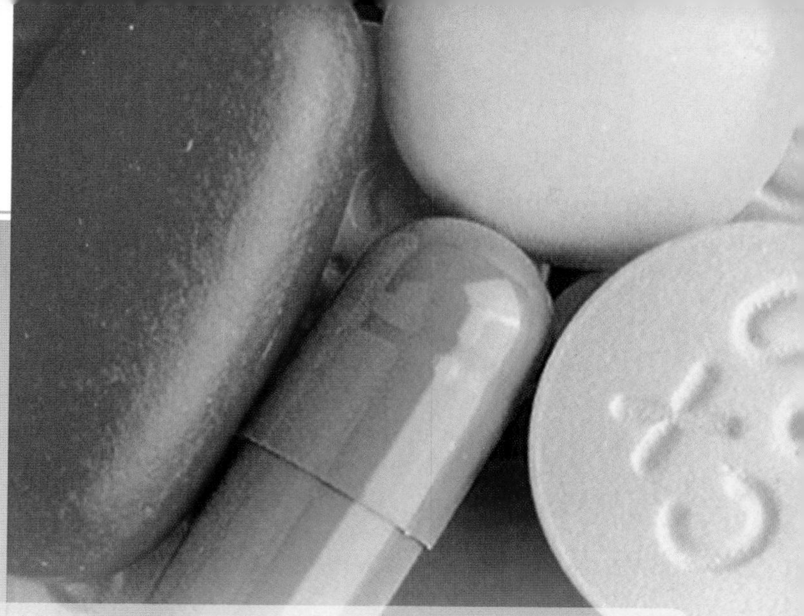

Many antimicrobial medications are available to combat infections. Nonetheless, they fall into a limited number of classes based on their modes of action.

CHAPTER GLOSSARY

aminoglycoside antibiotic An antibiotic that contains a cyclohexane ring and amino sugars; this antibiotic binds to the small ribosomal subunit and inhibits protein synthesis.

antimetabolite A compound that blocks metabolic pathway function by competitively inhibiting a key enzyme's use of a metabolite because it closely resembles the normal enzyme substrate.

broad-spectrum drugs Chemotherapeutic agents that are effective against many different kinds of pathogens.

chemotherapeutic agent Compounds used in the treatment of disease that destroy pathogens or inhibit their growth at concentrations low enough to avoid doing undesirable damage to the host. Includes **antibiotics,** which are natural microbial products, and **antimicrobials,** which may be naturally or chemically synthesized.

cidal Having the capacity to cause cell death.

integron A genetic element with an attachment site for site-specific recombination and an integrase gene. It can capture genes and gene cassettes.

Kirby-Bauer method A disk diffusion test to determine the susceptibility of a microorganism to chemotherapeutic agents.

macrolide antibiotic An antibiotic containing a macrolide ring, a large lactone ring with multiple keto and hydroxyl groups, linked to one or more sugars; inhibits bacterial growth by binding to the 50S ribosomal subunit, thereby blocking protein synthesis.

minimal inhibitory concentration (MIC) The lowest concentration of a drug that prevents the growth of a particular microorganism.

minimal lethal concentration (MLC) The lowest concentration of a drug that kills a particular microorganism.

narrow-spectrum drugs Chemotherapeutic agents that are effective only against a limited variety of microorganisms.

quinolones A class of broad-spectrum antibiotics, derived from nalidixic acid, that bind to bacterial DNA gyrase, inhibiting DNA replication. This group of antibiotics is bacteriocidal.

R plasmids Plasmids bearing one or more drug-resistance genes.

selective toxicity The activity of a drug that kills or inhibits the microbial pathogen while damaging the host as little as possible.

static Having the capacity to limit microbial growth but not cause microbial death.

tetracyclines A family of antibiotics with a common four-ring structure. They are isolated from the genus *Streptomyces* or produced semisynthetically; all are related to chlortetracycline or oxytetracycline.

therapeutic index The ratio between the toxic dose and the therapeutic dose of a drug, used as a measure of the drug's relative safety.

Modern medicine depends on chemotherapeutic agents—chemical agents that are used to treat disease. Ideally chemotherapeutic agents used to treat infectious disease destroy pathogenic microorganisms or inhibit their growth at concentrations low enough to avoid undesirable damage to the host. Most of these agents are **antibiotics** (Greek *anti,* against, and *bios,* life), microbial products or their derivatives that kill susceptible microorganisms or inhibit their growth. Drugs such as the sulfonamides are sometimes called antibiotics although they are synthetic chemotherapeutic agents, not microbially synthesized. This chapter introduces the principles of **antimicrobial** chemotherapy and briefly reviews the characteristics of selected antibacterial, antifungal, antiprotozoan, and antiviral drugs.

34.1 The Development of Chemotherapy

The modern era of chemotherapy began with the work of the German physician Paul Ehrlich (1854–1915). Ehrlich was fascinated with dyes that specifically bind to and stain microbial cells. He reasoned that one of the dyes could be a chemical that would selectively destroy pathogens without harming human cells—a "magic bullet." By 1904 Ehrlich found that the dye trypan red was active against the trypanosome that causes African sleeping sickness (*see figure 23.4*) and could be used therapeutically. Subsequently Ehrlich and a young Japanese scientist named Sahachiro Hata tested a variety of arsenic-based chemicals on syphilis-infected rabbits and found that arsphenamine was active against the syphilis spirochete. Arsphenamine was made available in 1910 under the trade name Salvarsan and paved the way to the testing of hundreds of compounds for their selective toxicity and therapeutic potential.

In 1927 the German chemical industry giant I. G. Farbenindustrie began a long-term search for chemotherapeutic agents under the direction of Gerhard Domagk. Domagk had screened a vast number of chemicals for other "magic bullets" and discovered that Prontosil red, a new dye for staining leather, protected mice completely against pathogenic streptococci and staphylococci without apparent toxicity. Jacques and Therese Trefouel later showed that the body metabolized the dye to sulfanilamide. Domagk received the 1939 Nobel Prize in Physiology or Medicine for his discovery of sulfonamides, or sulfa drugs.

Penicillin, the first true antibiotic, was initially discovered in 1896 by a twenty-one-year-old French medical student named Ernest Duchesne. His work was forgotten until Alexander Fleming accidentally rediscovered penicillin in September 1928. After returning from a weekend vacation, Fleming noticed that a Petri plate of *Staphylococcus* also had mold growing on it and there were no bacterial colonies surrounding it (**figure 34.1**). Although the precise events are still unclear, it has been suggested that a *Penicillium notatum* spore had contaminated the Petri dish before it had been inoculated with the staphylococci. The mold apparently grew before the bacteria and produced penicillin. The bacteria nearest the fungus were lysed. Fleming correctly deduced that the mold produced a diffusible substance, which he called penicillin. Unfortunately, Fleming could not demonstrate that penicillin remained active in vivo long enough to destroy pathogens and thus dropped the research.

In 1939 Howard Florey, a professor of pathology at Oxford University, was in the midst of testing the bactericidal activity of many substances. After reading Fleming's paper on penicillin, one of Florey's coworkers, Ernst Chain, obtained the *Penicillium* culture from Fleming and set about purifying the antibiotic. Norman Heatley, a biochemist, was enlisted to help. He devised the original assay, culture, and purification techniques needed to produce crude penicillin for further experimentation. When purified penicillin was injected into mice infected with streptococci or staphylococci, almost all the mice survived. Florey and Chain's success was reported in 1940, and subsequent human trials were equally successful. Fleming, Florey, and Chain received the Nobel Prize in 1945 for the discovery and production of penicillin.

The discovery of penicillin stimulated the search for other antibiotics. Selman Waksman, while at Rutgers University, announced in 1944 that he and his associates had found a new antibiotic, streptomycin, produced by the actinomycete *Streptomyces griseus (see figure 22.16)*. This discovery arose from the careful screening of about 10,000 strains of soil bacteria and fungi. The importance of streptomycin cannot be understated, as it was the first drug that could successfully treat tuberculosis. Waksman received the Nobel Prize in 1952, and his success led to a worldwide search for other antibiotic-producing soil microorganisms. Microorganisms producing chloramphenicol, neomycin, terramycin, and tetracycline were isolated by 1953.

Search This: Waksman Foundation

The discovery of chemotherapeutic agents and the development of newer, more powerful drugs have transformed modern medicine and greatly alleviated human suffering. Furthermore, antibiotics have proven exceptionally useful in microbiological research.

1. What are chemotherapeutic agents? Antibiotics?
2. Louis Pasteur is often credited with saying, "Chance favors the prepared mind." How does this apply to Fleming?

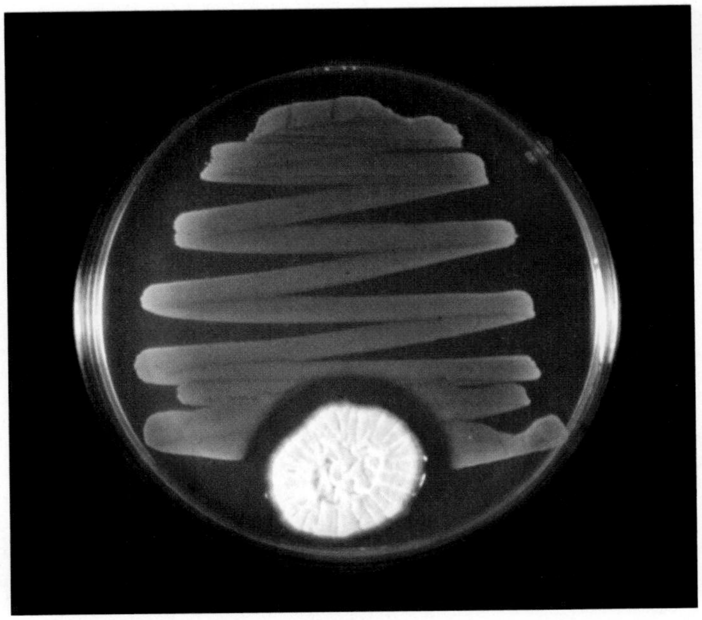

FIGURE 34.1 Bacteriocidal Action of Penicillin. The *Penicillium* mold colony secretes penicillin that kills *Staphylococcus aureus* that was streaked nearby.

34.2 General Characteristics of Antimicrobial Drugs

As Ehrlich so clearly saw, to be successful, a chemotherapeutic agent must have selective toxicity: it must kill or inhibit the microbial pathogen while damaging the host as little as possible. The degree of **selective toxicity** may be expressed in terms of (1) the therapeutic dose—the drug level required for clinical treatment of a particular infection, and (2) the toxic dose—the drug level at which the agent becomes too toxic for the host. The **therapeutic index** is the ratio of the toxic dose to the therapeutic dose. The larger the therapeutic index, the better the chemotherapeutic agent (all other things being equal).

A drug that disrupts a microbial function not found in host animal cells often has a greater selective toxicity and a higher therapeutic index. For example, penicillin inhibits bacterial cell wall peptidoglycan synthesis but has little effect on host cells because they lack cell walls; therefore penicillin's therapeutic index is high. A drug may have a low therapeutic index because it inhibits the same process in host cells or damages the host in other ways. The undesirable effects on the host, or side effects, are of many kinds and may involve almost any organ system. Because side effects can be severe, chemotherapeutic agents must be administered with great care.

Some bacteria and fungi are able to naturally produce many of the commonly employed antibiotics (**table 34.1**). In contrast, several important chemotherapeutic agents, such as sulfonamides, trimethoprim, ciprofloxacin, isoniazid, and dapsone, are synthetic—manufactured by chemical procedures independent of microbial activity. Some antibiotics are semisynthetic—natural antibiotics that have been structurally modified by the addition of chemical groups to make them less susceptible to inactivation by pathogens (e.g., ampicillin and methicillin). In addition, many semisynthetic drugs have a broader spectrum of antibiotic activity than does their parent molecule. This is particularly true of the semisynthetic penicillins (e.g., ampicillin, amoxycillin) as compared to the naturally produced penicillin G and penicillin V.

Drugs vary considerably in their range of effectiveness. Many are **narrow-spectrum drugs**—that is, they are effective only against a limited variety of pathogens (table 34.1). Others are **broad-spectrum drugs** that attack many different kinds of pathogens. Drugs may also be classified based on the general microbial group they act against: antibacterial, antifungal, antiprotozoan, and antiviral. A few agents can be used against more than one group; for example, sulfonamides are active against bacteria and

Table 34.1 Properties of Some Common Antibacterial Drugs

Antibiotic Group	*Primary Effect*	*Mechanism of Action*	*Members*	*Spectrum*	*Common Side Effects*
Cell Wall Synthesis Inhibition					
Penicillins	Cidal	Inhibit transpeptidation enzymes involved in cross-linking the polysaccharide chains of the bacterial cell wall peptidoglycan Activate cell wall lytic enzymes	Penicillin G, penicillin V, methicillin Ampicillin, carbenicillin	Narrow (gram-positive) Broad (gram-positive, some gram-negative)	Allergic responses (diarrhea, anemia, hives, nausea, renal toxicity)
Cephalosporins	Cidal	Same as above	Cephalothin, cefoxitin, cefaperazone, ceftriaxone	Broad (gram-positive, some gram-negative)	Allergic responses, thrombophlebitis, renal injury
Vancomycin	Cidal	Prevent transpeptidation of peptidoglycan subunits by binding to D-Ala-D-Ala amino acids at the end of peptide side chains. Thus it has a different binding site than that of the penicillins.	Vancomycin	Narrow (gram-positive)	Ototoxic (tinnitus and deafness), nephrotoxic, allergic reactions
Protein Synthesis Inhibition					
Aminoglycosides	Cidal	Bind to small ribosomal subunit (30S) and interfere with protein synthesis by directly inhibiting synthesis and causing misreading of mRNA	Neomycin, kanamycin, gentamicin Streptomycin	Broad (gram-negative, mycobacteria) Narrow (aerobic gram-negative)	Ototoxic, renal damage, loss of balance, nausea, allergic responses

Continued

Table 34.1 Properties of Some Common Antibacterial Drugs *(continued)*

Tetracyclines	Static	Same as aminoglycosides	Oxytetracycline, chlortetracycline	Broad (including rickettsia and chlamydia)	Gastrointestinal upset, teeth discoloration, renal and hepatic injury
Macrolides	Static	Bind to 23S rRNA of large ribosomal subunit (50S) to inhibit peptide chain elongation during protein synthesis	Erythromycin, clindamycin	Broad (aerobic and anaerobic gram-positive, some gram-negative)	Gastrointestinal upset, hepatic injury, anemia, allergic responses
Chloramphenicol	Static	Same as above	Chloramphenicol	Broad (gram-positive and -negative, rickettsia and chlamydia)	Depressed bone marrow function, allergic reactions
Nucleic Acid Synthesis Inhibition					
Quinolones and Fluoroquinolones	Cidal	Inhibit DNA gyrase and topoisomerase II, thereby blocking DNA replication	Norfloxacin, ciprofloxacin, Levofloxacin	Narrow (gram-negatives better than gram-positives) Broad spectrum	Tendonitis, headache, lightheadedness, convulsions, allergic reactions
Rifampin	Cidal	Inhibits bacterial DNA-dependent RNA polymerase	R-Cin, rifacilin, rifamycin, rimactane, rimpin, siticox	*Mycobacterium* infections and some gram-negative such as *Neisseria meningitidis* and *Haemophilus influenzae* b	Nausea, vomiting, diarrhea, fatigue, anemia, drowsiness, headache, mouth ulceration, liver damage
Cell Membrane Disruption					
Polymyxin B	Cidal	Bind to plasma membrane and disrupts its structure and permeability properties	Polymyxin B, polymyxin topical ointment	Narrow—mycobacterial infections, principally leprosy	Can cause severe kidney damage, drowsiness, dizziness
Antimetabolites					
Sulfonamides	Static	Inhibits folic acid synthesis by competing with *p*-aminobenzoic acid (PABA)	Silver sulfadiazine, sodium sulfacetamide, sulfamethoxazole, sulfanilamide, sulfasalazine, sulfisoxazole	Broad spectrum	Nausea, vomiting, and diarrhea; hypersensitivity reactions such as rashes, photosensitivity
Trimethoprim	Static	Blocks folic acid synthesis by inhibiting the enzyme tetrahydrofolate reductase	Trimethoprim (in combination with a sulfamethoxazole [1:5])	Broad spectrum	Same as sulfonamides but less frequent
Dapsone	Static	Thought to interfere with folic acid synthesis	Dapsone	Narrow—mycobacterial infections, principally leprosy	Back, leg, or stomach pains; discolored fingernails, lips, or skin; breathing difficulties, fever, loss of appetite, skin rash, fatigue
Isoniazid	Cidal if bacteria are actively growing, static if bacteria are dormant	Exact mechanism is unclear, but it is thought to inhibit lipid synthesis (especially mycolic acid); putative enoyl-reductase inhibitor	Isoniazid	Narrow—mycobacterial infections, principally tuberculosis	Nausea, vomiting, liver damage, seizures, "pins and needles" in extremities (peripheral neuropathy)

some protozoa. Finally, chemotherapeutic agents can be either **cidal** or **static.** Static agents reversibly inhibit growth; if the agent is removed, the microorganisms will recover and grow again. ◀◀ *Senescence and death (section 7.3)*

Although a cidal agent kills the target pathogen, it may be static at low levels. The effect of an agent also varies with the target species: an agent may be cidal for one species and static for another. Because static agents do not directly destroy the pathogen, elimination of the infection depends on the host's own immunity mechanisms. A static agent may not be effective if the host is immunosuppressed. Some idea of the effectiveness of a chemotherapeutic agent against a pathogen can be obtained from the **minimal inhibitory concentration (MIC).** The MIC is the lowest concentration of a drug that prevents growth of a particular pathogen. On the other hand, the **minimal lethal concentration (MLC)** is the lowest drug concentration that kills the pathogen. A cidal drug generally kills pathogens at levels only two to four times the MIC, whereas a static agent kills at much higher concentrations, if at all.

1. Define the following: selective toxicity, therapeutic index, side effect, narrow-spectrum drug, broad-spectrum drug, synthetic and semisynthetic antibiotics, cidal and static agents, minimal inhibitory concentration, and minimal lethal concentration.
2. How do semisynthetic antibiotics commonly differ from their parent molecules?
3. Use the MIC and MLC concepts to distinguish between cidal and static agents.

34.3 Determining the Level of Antimicrobial Activity

Determination of antimicrobial effectiveness against specific pathogens is essential for proper therapy. Testing can show which agents are most effective against a pathogen and give an estimate of the proper therapeutic dose.

Dilution Susceptibility Tests

Dilution susceptibility tests can be used to determine MIC and MLC values. Antibiotic dilution tests can be done in both agar and broth. In the broth dilution test, a series of broth tubes (usually Mueller-Hinton broth) containing antibiotic concentrations in the range of 0.1 to 128 μg per milliliter (two-fold dilutions) is prepared and inoculated with a standard density of the test organism. The lowest concentration of the antibiotic resulting in no growth after 16 to 20 hours of incubation is the MIC. The MLC can be ascertained if the tubes showing no growth are then cultured into fresh medium lacking antibiotic. The lowest antibiotic concentration from which the microorganisms do not grow when transferred to fresh medium is the MLC. The agar dilution test is very similar to the broth dilution test. Plates containing Mueller-Hinton agar and various amounts of antibiotic are inoculated and examined for growth. Several automated systems for susceptibility testing and MIC determination with broth or agar cultures have been developed.

Disk Diffusion Tests

If a rapidly growing microbe such as *Staphylococcus* or *Pseudomonas* is being tested, a disk diffusion technique may be used to save time and media. The principle behind this assay is fairly simple. When an antibiotic-impregnated disk is placed on agar previously inoculated with the test bacterium, the antibiotic diffuses radially outward through the agar, producing an antibiotic concentration gradient. The antibiotic is present at high concentrations near the disk and affects even minimally susceptible microorganisms (resistant organisms will grow up to the disk). As the distance from the disk increases, the antibiotic concentration decreases and only more susceptible pathogens are harmed. A clear zone or ring is present around an antibiotic disk after incubation if the agent inhibits bacterial growth. The wider the zone surrounding a disk, the more susceptible the pathogen is. Zone width also is a function of the antibiotic's initial concentration, its solubility, and its diffusion rate through agar. Thus zone width cannot be used to compare directly the effectiveness of different antibiotics.

Currently the disk diffusion test most often used is the **Kirby-Bauer method,** which was developed in the early 1960s at the University of Washington Medical School by William Kirby, A. W. Bauer, and their colleagues. Freshly grown bacteria are used to inoculate the entire surface of a Mueller-Hinton agar plate. After the agar surface has dried for about 5 minutes, the appropriate antibiotic test disks are placed on it, either with sterilized forceps or with a multiple applicator device (**figure 34.2**). The plate is immediately placed at 35°C. After 16 to 18 hours of incubation, the diameters of the zones of inhibition are measured to the nearest millimeter.

Kirby-Bauer test results are interpreted using a table that relates zone diameter to the degree of microbial resistance (**table 34.2**). The values in table 34.2 were derived by finding the MIC values and zone diameters for many different microbial strains. A plot of MIC (on a logarithmic scale) versus zone inhibition diameter (arithmetic scale) is prepared for each antibiotic (**figure 34.3**). These plots are then used to find the zone diameters corresponding to the drug concentrations actually reached in the body. If the zone diameter for the lowest level reached in the body is smaller than that seen with the test pathogen, the pathogen should have an MIC value low enough to be destroyed by the drug. A pathogen with too high an MIC value (too small a zone diameter) is resistant to the agent at normal body concentrations.

The Etest®

The Etest® from bioMérieux S.A. may be used in sensitivity testing under a majority of conditions. It is particularly convenient for use with anaerobic pathogens. Bacteria are used to inoculate the entire surface of agar medium with the test organism and

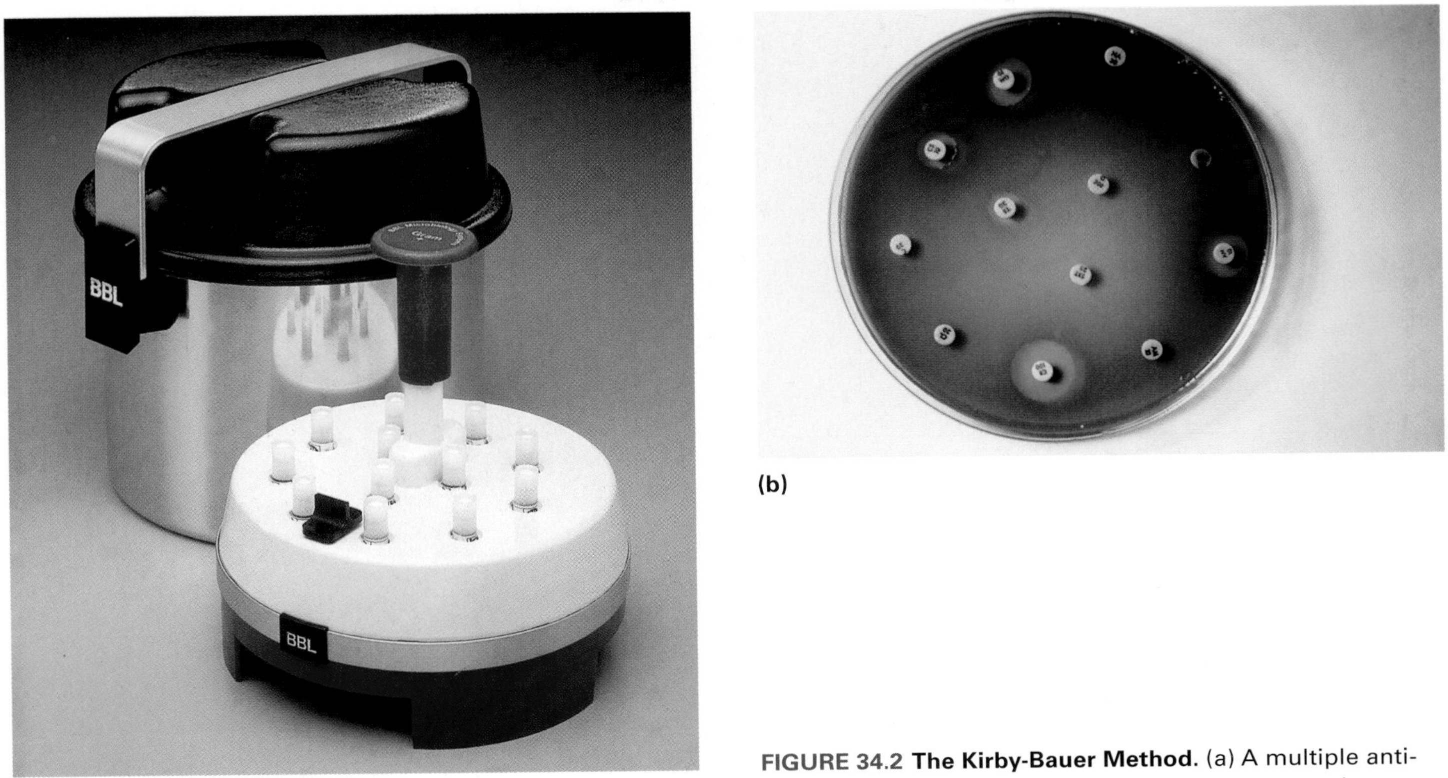

(a)

(b)

FIGURE 34.2 The Kirby-Bauer Method. (a) A multiple antibiotic disk dispenser and (b) disk diffusion test results.

Table 34.2 Inhibition Zone Diameter of Selected Chemotherapeutic Drugs

		Zone Diameter (Nearest mm)		
Chemotherapeutic Drug	*Disk Content*	*Resistant*	*Intermediate*	*Susceptible*
Carbenicillin (with *Proteus* spp. and *E. coli*)	100 μg	≤17	18–22	≥23
Carbenicillin (with *Pseudomonas aeruginosa*)	100 μg	≤13	14–16	≥17
Erythromycin	15 μg	≤13	14–17	≥18
Penicillin G (with staphylococci)	10 U[a]	≤20	21–28	≥29
Penicillin G (with other microorganisms)	10 U	≤11	12–21	≥22
Streptomycin	10 μg	≤11	12–14	≥15
Sulfonamides	250 or 300 μg	≤12	13–16	≥17

[a] One milligram of penicillin G sodium = 1,600 units (U).

then plastic Etest® strips are placed on the surface so that they extend out radially from the center (**figure 34.4**). Each strip contains a gradient of an antibiotic and is labeled with a scale of MIC values. The lowest concentration in the strip lies at the center of the plate. After 24 to 48 hours of incubation, an elliptical zone of inhibition appears. As shown in figure 34.4, MICs are determined from the point of intersection between the inhibition zone and the strip's scale of MIC values.

1. How can dilution susceptibility tests and disk diffusion tests be used to determine microbial drug sensitivity?
2. Briefly describe the Kirby-Bauer test and its purpose.
3. What would you surmise if you examined a Kirby-Bauer assay and found individual colonies of the plated microbe growing within the zone of inhibition?
4. How is the Etest® carried out?

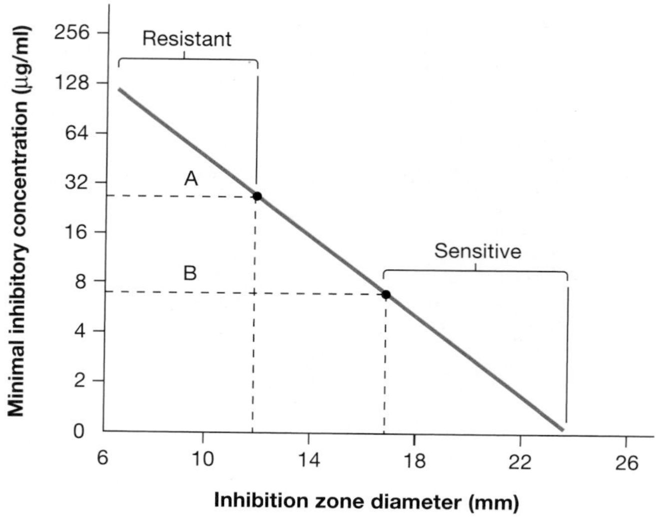

FIGURE 34.3 Interpretation of Kirby-Bauer Test Results. The relationship between the minimal inhibitory concentrations of a hypothetical drug and the size of the zone around a disk in which microbial growth is inhibited. As the sensitivity of microorganisms to the drug increases, the MIC value decreases and the inhibition zone grows larger. Suppose that this drug varies from 7–28 µg/mL in the body during treatment. Dashed line A shows that any pathogen with a zone of inhibition less than 12 mm in diameter will have an MIC value greater than 28 µg/mL and will be resistant to drug treatment. A pathogen with a zone diameter greater than 17 mm will have an MIC less than 7 µg/mL and will be sensitive to the drug (see line B). Zone diameters between 12 and 17 mm indicate intermediate sensitivity and usually signify resistance.

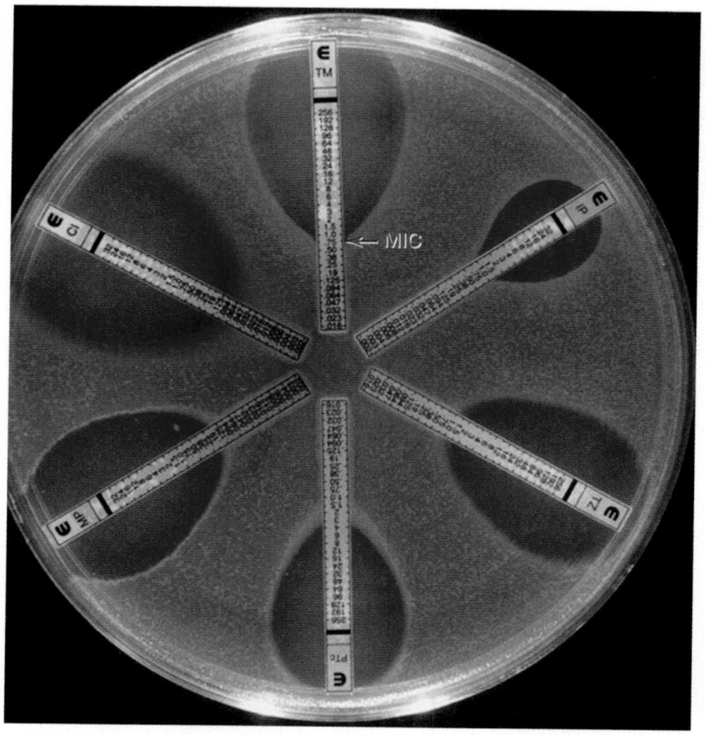

FIGURE 34.4 Etest®. An example of a bacterial culture plate with Etest® strips arranged in a radial pattern with the concentration minimum at the center of the plate. The MIC value is read at the point where the inhibition ellipse intersects the scale on the strip, as shown by the arrow in this example. Etest® is a registered trademark of bioMérieux S.A. or one of its subsidiaries.

Figure 34.4 Micro Inquiry

Based on these zones of diffusion, which Etest® strip contains the most effective antibiotic against this microbe? Which is the least effective?

34.4 Antibacterial Drugs

Since Fleming's discovery of penicillin, many antibiotics have been found that can damage pathogens in several ways. A few antibacterial drugs are described here and summarized in table 34.1, with emphasis on their mechanisms of action.

Inhibitors of Cell Wall Synthesis

The most selective antibiotics are those that interfere with bacterial cell wall synthesis. Drugs such as penicillins, cephalosporins, vancomycin, and bacitracin have a high therapeutic index because they target structures not found in eukaryotic cells. ◀◀ *Bacterial cell walls (section 3.3)*

Penicillins

Most **penicillins** (e.g., penicillin G or benzylpenicillin) are derivatives of 6-aminopenicillanic acid and differ from one another with respect to the side chain attached to the amino group (**figure 34.5**). The most crucial feature of the molecule is the **β-lactam ring,** which is essential for bioactivity. Many penicillin-resistant bacteria produce **penicillinase** (also called **β-lactamase**), an enzyme that inactivates the antibiotic by hydrolyzing a bond in the β-lactam ring.

The structure of the penicillins resembles the terminal D-alanyl-D-alanine found on the peptide side chain of the peptidoglycan subunit. It is thought that this structural similarity blocks the enzyme catalyzing the transpeptidation reaction that forms the peptidoglycan cross-links (*see figure 11.13*). Thus formation of a complete cell wall is blocked, leading to osmotic lysis. This mechanism is consistent with the observation that penicillins act only on growing bacteria that are synthesizing new peptidoglycan. However, the mechanism of penicillin action is actually more complex. Penicillins also bind to several periplasmic proteins (penicillin-binding proteins, or PBPs) and may also destroy bacteria by activating their own autolytic enzymes. However, penicillin may kill bacteria even in the absence of autolysins or murein hydrolases. Lysis could occur after bacterial viability has

Penicillin G

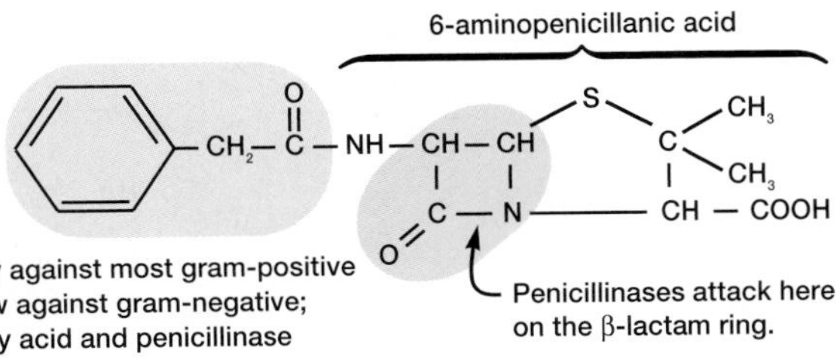

High activity against most gram-positive bacteria, low against gram-negative; destroyed by acid and penicillinase

Penicillin V

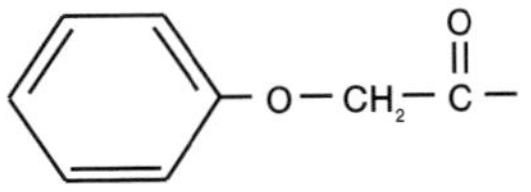

Same spectrum but more acid resistant than penicillin G

Ampicillin

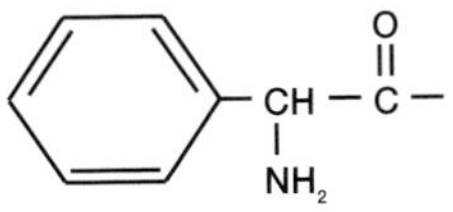

Active against gram-positive and gram-negative bacteria; acid stable

Carbenicillin

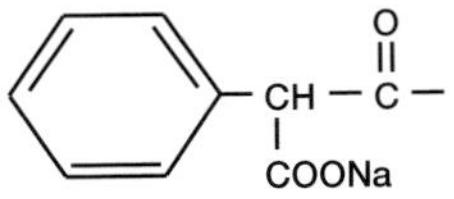

Active against gram-negative bacteria such as *Pseudomonas* and *Proteus;* acid stable; not well absorbed by small intestine

Methicillin

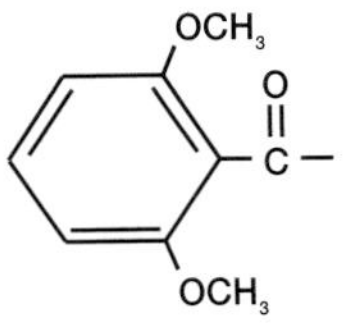

Penicillinase-resistant but less active than penicillin G; acid-labile

Ticarcillin

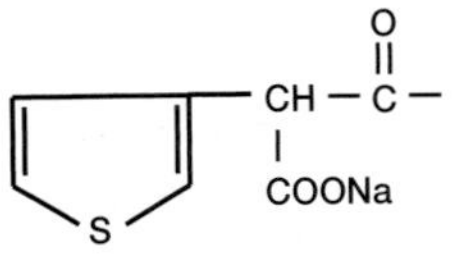

Similar to carbenicillin but more active against *Pseudomonas*

FIGURE 34.5 Penicillins. The structures and characteristics of representative penicillins. All are derivatives of 6-aminopenicillanic acid; in each case, the shaded portion of penicillin G is replaced by the side chain indicated. The β-lactam ring is also shaded (blue).

Figure 34.5 Micro Inquiry

What is the difference between penicillin G and penicillin V? How do the semisynthetic penicillins differ from their parent compounds?

already been lost. Penicillin may stimulate proteins called bacterial holins to form holes or lesions in the plasma membrane, leading directly to membrane leakage and death. Murein hydrolases also could move through the holes, disrupt the peptidoglycan, and lyse the cell.

Penicillins differ from each other in several ways. The two naturally occurring penicillins, penicillin G and penicillin V, are narrow-spectrum drugs (figure 34.5). Penicillin G is effective against gonococci, meningococci, and several gram-positive pathogens such as streptococci and staphylococci. However, it must be administered by injection (parenterally) because it is destroyed by stomach acid. Penicillin V is similar to penicillin G in spectrum of activity but can be given orally because it is more resistant to acid. The semisynthetic penicillins, on the other hand, have a broader spectrum of activity. Ampicillin can be administered orally and is effective against gram-negative bacteria such as *Haemophilus, Salmonella,* and *Shigella.* Carbenicillin and ticarcillin are potent against *Pseudomonas* and *Proteus.*

An increasing number of bacteria have become resistant to natural penicillins and many of the semisynthetic analogs. Physicians frequently employ specific semisynthetic penicillins that are not destroyed by β-lactamases to combat antibiotic-resistant pathogens. These include methicillin (figure 34.5), nafcillin, and oxacillin. However, this practice has been confounded by the emergence of methicillin-resistant bacteria.

Although penicillins are the least toxic of the antibiotics, about 1 to 5% of the adults in the United States develop allergies to them. Occasionally, a person will die of a violent allergic response; therefore patients should be questioned about penicillin allergies before treatment is begun. ⏮ *Hypersensitivities (section 33.10)*

Cephalosporins

Cephalosporins are a family of antibiotics originally isolated in 1948 from the fungus *Cephalosporium.* They contain a β-lactam structure that is very similar to that of the penicillins (**figure 34.6**). As might be expected from their structural similarities to penicillins, cephalosporins also inhibit the transpeptidation reaction during peptidoglycan synthesis. They are broad-spectrum drugs frequently given to patients with penicillin allergies (although about 10% of patients allergic to penicillin are also allergic to cephalosporins).

Cephalosporins are broadly categorized into four generations (groups of drugs that were sequentially developed) based on their spectrum of activity. First-generation cephalosporins are more effective against gram-positive pathogens than gram negatives. Second-generation drugs, developed after the first generation, have improved effects on gram-negative bacteria with some anaerobe coverage. Third-generation drugs are particularly effective against gram-negative pathogens, and some reach the central nervous system. This is of particular note because many antimicrobial agents do not cross the blood-brain barrier. Finally, fourth-generation cephalosporins are broad spectrum with excellent gram-positive and gram-negative coverage and, like their third-generation predecessors, inhibit the growth of the difficult opportunistic pathogen *Pseudomonas aeruginosa.*

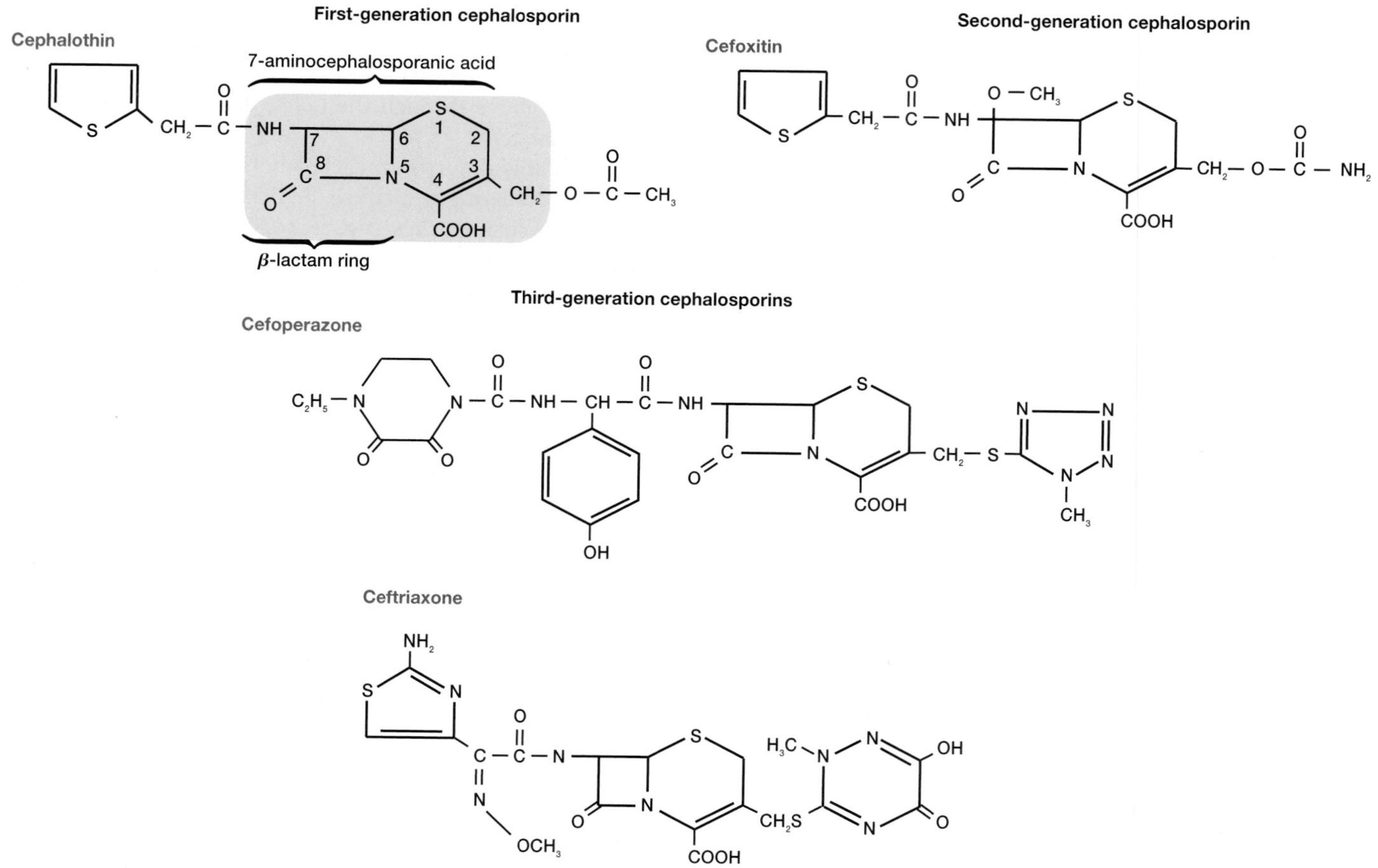

FIGURE 34.6 Cephalosporin Antibiotics. These drugs are derivatives of 7-aminocephalosporanic acid and contain a β-lactam ring.

Figure 34.6 Micro Inquiry

Do you think cephalosporin antibiotics are susceptible to degradation by β-lactamase enzymes? Explain.

Vancomycin and Teicoplanin

Vancomycin is a glycopeptide antibiotic produced by *Streptomyces orientalis.* It is a cup-shaped molecule composed of a peptide linked to a disaccharide. The peptide portion blocks the transpeptidation reaction by binding specifically to the D-alanyl-D-alanine terminal sequence on the pentapeptide portion of peptidoglycan (**figure 34.7**). The antibiotic is bactericidal for *Staphylococcus* and some members of the genera *Clostridium, Bacillus, Streptococcus,* and *Enterococcus.* It is given both orally and intravenously, and has been particularly important in the treatment of antibiotic-resistant staphylococcal and enterococcal infections. However, vancomycin-resistant strains of *Enterococcus* have become widespread and cases of resistant *Staphylococcus aureus* have appeared. In this case, resistance is conferred when bacteria change the terminal D-alanine to either D-lactate or a D-serine residue. Vancomycin resistance poses a serious public health threat: vancomycin has been considered the "drug of last resort" in cases of antibiotic-resistant *S. aureus.* Clearly new drugs must be developed.

Teicoplanin, another glycopeptide antibiotic, is produced by the actinomycete *Actinoplanes teichomyceticus.* It is similar in structure and mechanism of action to vancomycin but has fewer side effects. It is active against staphylococci, enterococci, streptococci, clostridia, *Listeria,* and many gram-positive pathogens.

Protein Synthesis Inhibitors

Many antibiotics inhibit protein synthesis by binding with the bacterial ribosome and other components of protein synthesis. Because these drugs discriminate between bacterial and eukaryotic ribosomes, their therapeutic index is fairly high but not as high as that of cell wall inhibitors. Several different steps in protein synthesis can

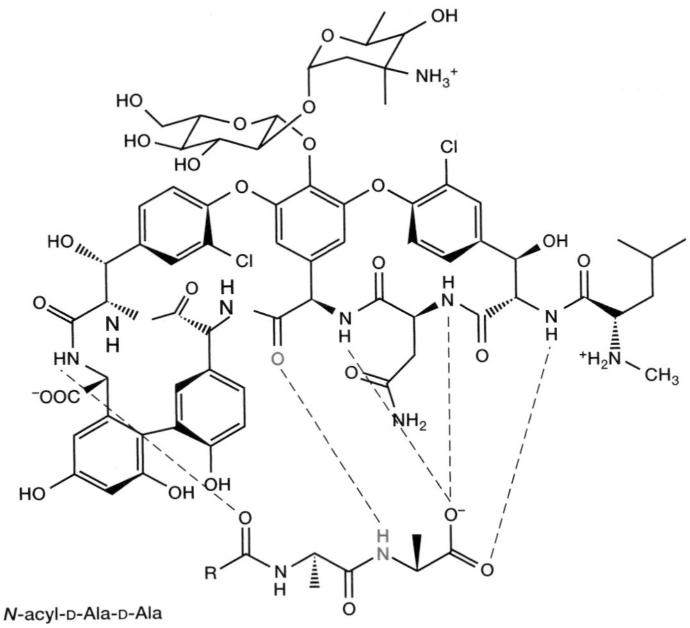

FIGURE 34.7 Vancomycin. The cup-shaped vancomycin molecule binds to the D-alanyl-D-alanine terminal sequence of peptidoglycan.

be affected: aminoacyl-tRNA binding, peptide bond formation, mRNA reading, and translocation. ⏮ *Translation (section 12.8)*

Aminoglycosides

Although considerable variation in structure occurs among several important **aminoglycoside antibiotics,** all contain a cyclohexane ring and amino sugars (**figure 34.8*a***). Streptomycin, kanamycin, neomycin, and tobramycin are synthesized by different species of the genus *Streptomyces,* whereas gentamicin comes from another actinomycete, *Micromonospora purpurea.* Streptomycin's usefulness has decreased greatly due to widespread drug resistance but may still be effective when other aminoglycosides are not tolerated or are contraindicated due to interactions with other drugs (e.g., HIV protease inhibitors). Gentamicin is used to treat *Proteus, Escherichia, Klebsiella,* and *Serratia* infections. Aminoglycosides can be quite toxic, however, and can cause deafness, renal damage, loss of balance, nausea, and allergic responses.

Aminoglycosides bind to the 30S (small) ribosomal subunit to interfere with protein synthesis. These antibiotics are bactericidal and tend to be most effective against gram-negative pathogens. Recently scientists at Boston University revealed the specific mechanisms by which aminoglycosides kill gram-negative bacteria. Using gene expression microarrays comparing the effects of gentamicin and kanamicin, *E. coli* mutant strains that lack envelope stress response were evaluated for effects on protein synthesis, protein translocation, and oxidative stress. By binding to bacterial ribosomes, it appears that these aminoglycosides cause tRNA mismatching and thus protein mistranslation. The mistranslated proteins, along with correctly translated proteins, move into the periplasm by the Sec-dependent translocation system. Once in the periplasm, the mistranslated proteins are degraded but not before some of them are inserted into the plasma membrane, activating the envelope stress-response system (figure 34.8*b*). Upregulation of the stress-response system induces change in the metabolic and respiratory pathways that results in hydroxyl radical formation. The increased oxygen radical production causes cell death. ⏮ *Oxygen concentration (section 7.6); Microarray analysis (section 16.4)*

Tetracyclines

The **tetracyclines** are a family of antibiotics with a common four-ring structure to which a variety of side chains are attached (**figure 34.9**). Oxytetracycline and chlortetracycline are produced naturally by *Streptomyces* species, whereas others are semisynthetic drugs. These antibiotics are similar to the aminoglycosides and combine with the 30S subunit of the ribosome. This inhibits the binding of aminoacyl-tRNA molecules to the A site of the ribosome. Their action is only bacteriostatic.

Tetracyclines are broad-spectrum antibiotics that are active against most bacteria, including rickettsias, chlamydiae, and mycoplasmas. Although their use has declined in recent years, they are still sometimes used to treat acne.

Macrolides

The **macrolide antibiotics** contain 12- to 22-carbon lactone rings linked to one or more sugars (**figure 34.10**). Erythromycin binds to the 23S rRNA of the 50S ribosomal subunit to inhibit peptide chain elongation during protein synthesis. Erythromycin is a relatively broad-spectrum antibiotic effective against gram-positive bacteria, mycoplasmas, and a few gram-negative bacteria, but is usually only bacteriostatic. It is used with patients who are allergic to penicillins and in the treatment of whooping cough, diphtheria, diarrhea caused by *Campylobacter,* and pneumonia from *Legionella* or *Mycoplasma* infections. Clindamycin is effective against a variety of bacteria, including staphylococci and anaerobes such as *Bacteroides.* Azithromycin, which has surpassed erythromycin in use, is particularly effective against many bacteria, including *Chlamydia trachomatis.*

Chloramphenicol

Chloramphenicol was first produced from cultures of *Streptomyces venezuelae* but is now synthesized chemically. Like erythromycin, this antibiotic binds to 23S rRNA on the 50S ribosomal subunit to inhibit the peptidyl transferase reaction. It has a very broad spectrum of activity but, unfortunately, is quite toxic. The most common side effect is depression of bone marrow function, leading to aplastic anemia and a decreased number of white blood cells. Consequently this antibiotic is used only in life-threatening situations when no other drug is adequate.

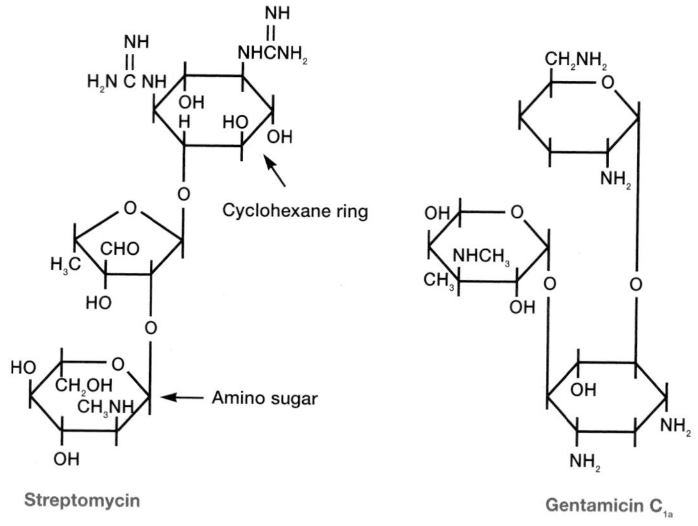

(a)

Metabolic Antagonists

Several valuable drugs act as **antimetabolites:** they antagonize, or block, the functioning of metabolic pathways by competitively inhibiting the use of metabolites by key enzymes. These drugs act as structural analogs, molecules that are structurally similar to naturally occurring metabolic intermediates. These analogs compete with intermediates in metabolic processes because of their similarity but are just different enough that they prevent normal cellular metabolism. By preventing metabolism, they are bacteriostatic but broad spectrum; their removal reestablishes the metabolic activity.

Sulfonamides or Sulfa Drugs

Sulfonamides, or sulfa drugs, are structurally related to sulfanilamide, an analog of *p*-aminobenzoic acid, or PABA (**figure 34.11**). PABA is used in the synthesis of the cofactor folic acid (folate).

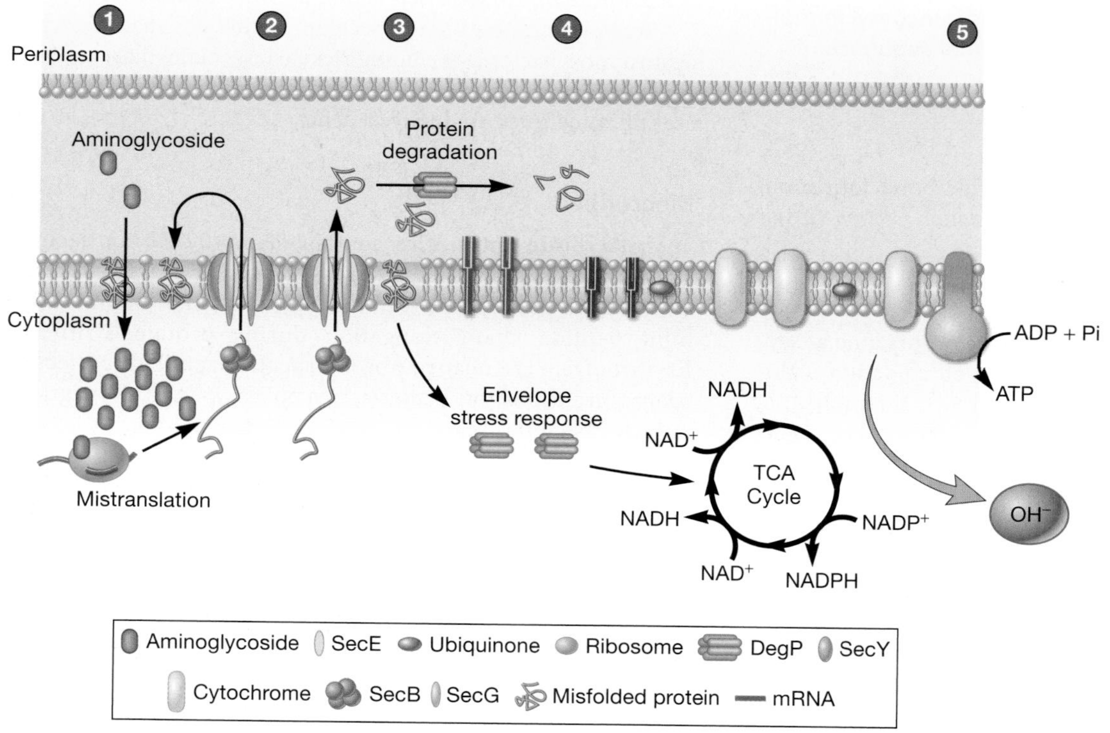

(b)

FIGURE 34.8 Representative Aminoglycoside Antibiotics. (a) Identifying cyclohexane ring and amino sugar. (b) Proposed aminoglycoside killing mechanism. Aminoglycoside (1) enters bacterium causing mistranslation of bacterial proteins. (2) Mistranslated proteins translocate into periplasm and (3) integrate into the plasma membrane, causing (4) envelope stress that (5) upregulates hydroxyl radical formation.

Figure 34.8 Micro Inquiry

How do these drugs inhibit protein synthesis?

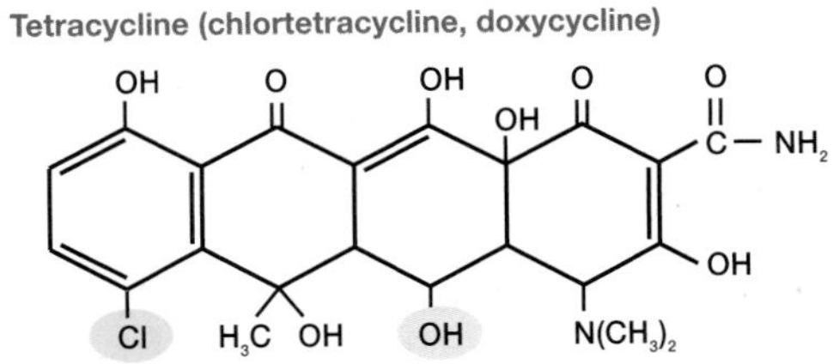

FIGURE 34.9 Tetracyclines. Three members of the tetracycline family. Tetracycline lacks both of the groups that are shaded. Chlortetracycline (aureomycin) differs from tetracycline in having a chlorine atom (blue); doxycycline consists of tetracycline with an extra hydroxyl (light blue).

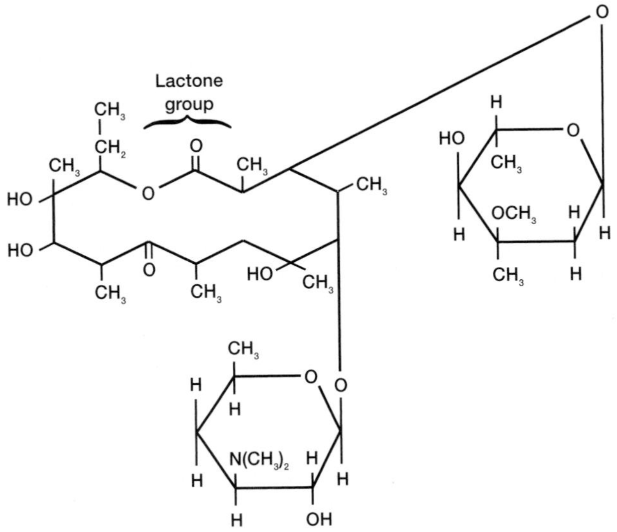

FIGURE 34.10 Erythromycin, a Macrolide Antibiotic. The 14-member lactone ring is connected to two sugars.

Figure 34.10 Micro Inquiry

How is the mechanism by which macrolides block protein synthesis similar to that of the tetracyclines? How is it different?

When sulfanilamide or another sulfonamide enters a bacterial cell, it competes with PABA for the active site of an enzyme involved in folic acid synthesis, causing a decline in folate concentration (*see figure 9.19*). This decline is detrimental to the bacterium because folic acid is a precursor of purines and pyrimidines, the bases used in the construction of DNA, RNA, and other important cell constituents. The resulting inhibition of purine and pyrimidine synthesis leads to cessation of protein synthesis and DNA replication. Sulfonamides are selectively toxic for many bacterial and protozoan pathogens because these microbes manufacture their own folate and cannot effectively take up this cofactor, whereas humans do not synthesize folate (we must obtain it in our diet). Sulfonamides thus have a high therapeutic index.

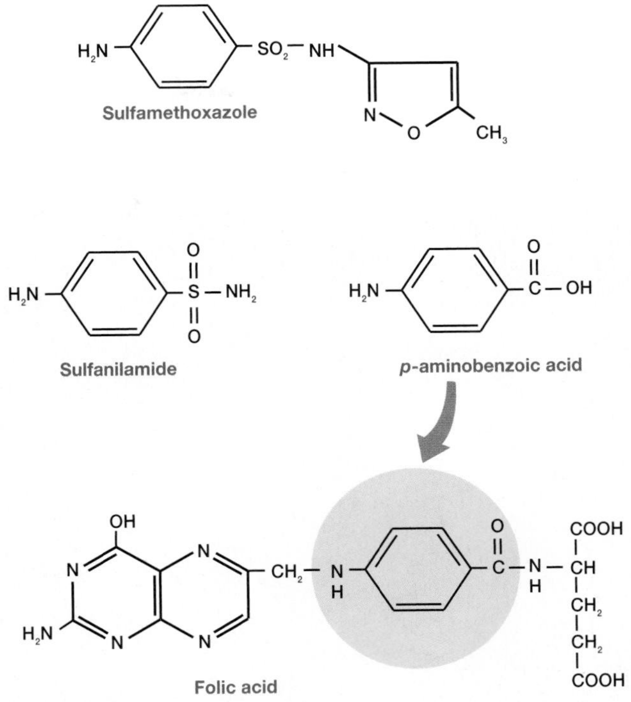

FIGURE 34.11 Sulfa Drugs. Both sulfanilamide and sulfamethoxazole compete with *p*-aminobenzoic acid to block folic acid synthesis.

Figure 34.11 Micro Inquiry

Why do sulfa drugs have a high therapeutic index?

The increasing resistance of many bacteria to sulfa drugs limits their effectiveness. Furthermore, as many as 5% of the patients receiving sulfa drugs experience adverse side effects, chiefly allergic responses such as fever, hives, and rashes.

Trimethoprim

Trimethoprim is a synthetic antibiotic that also interferes with the production of folic acid. It does so by binding to dihydrofolate reductase (DHFR), the enzyme responsible for converting dihydrofolic acid to tetrahydrofolic acid, competing against the dihydrofolic acid substrate (**figure 34.12**). It is a broad-spectrum antibiotic often used to treat respiratory and middle ear infections, urinary tract infections, and traveler's diarrhea. It can be combined with sulfa drugs to increase efficacy of treatment by blocking two key steps in the folic acid pathway. The inhibition of two successive steps in a single biochemical pathway means that less of each drug is needed in combination than when used alone. This is termed a synergistic drug interaction.

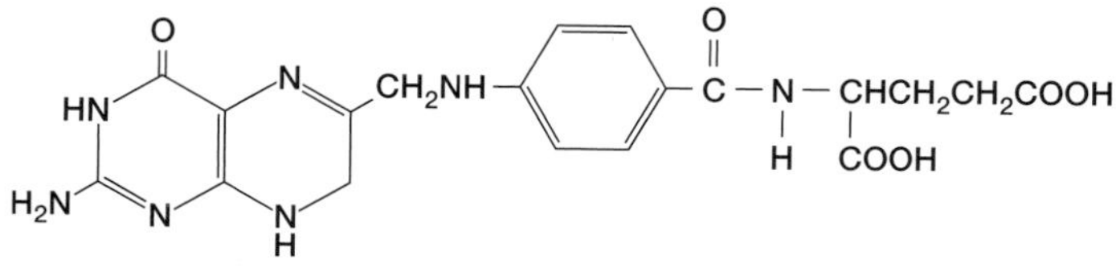

(a) Dihydrofolic acid (DFA)

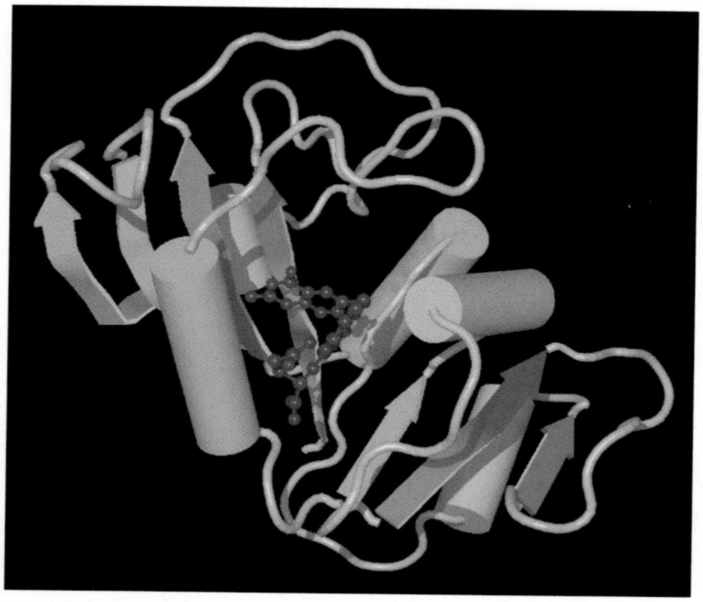

(b) Dihydrofolate reductase

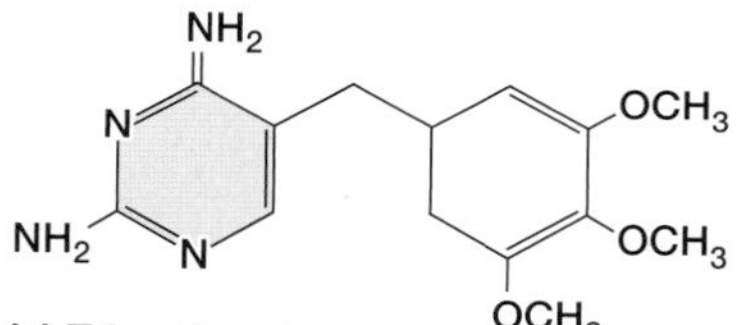

(c) Trimethoprim

FIGURE 34.12 Competitive Inhibition of Dihydrofolate Reductase (DHFR) by Trimethoprim. (a) Dihydrofolic acid (DFA) is the natural substrate for the DHFR enzyme of the folic acid pathway. (b) DHFR structure and its interaction with DFA (red), its natural substrate. Note the chemical structure and how it fits into the active site of the enzyme. (c) Trimethoprim mimics the structural orientation of the DFA and thus competes for the active site of the enzyme. This causes delayed or absent folic acid synthesis because the DFA cannot be converted to tetrahydrofolic acid when trimethoprim occupies the DHFR active site.

Nucleic Acid Synthesis Inhibition

The antibacterial drugs that inhibit nucleic acid synthesis function by inhibiting (1) DNA polymerase and DNA helicase or (2) RNA polymerase to block replication or transcription, respectively. These drugs are not as selectively toxic as other antibiotics because bacteria and eukaryotes do not differ greatly with respect to nucleic acid synthesis.

Quinolones

The **quinolones** are synthetic drugs that contain the 4-quinolone ring. The quinolones are important antimicrobial agents that inhibit nucleic acid synthesis. They are increasingly used to treat a wide variety of infections. The first quinolone, nalidixic acid (**figure 34.13**), was synthesized in 1962. Since that time, generations of fluoroquinolones have been produced. Three of these—ciprofloxacin, norfloxacin, and ofloxacin—are currently used in the United States, and more fluoroquinolones are being synthesized and tested.

Quinolones act by inhibiting the bacterial DNA gyrase and topoisomerase II. DNA gyrase introduces negative twist in DNA and helps separate its strands. Inhibition of DNA gyrase disrupts DNA replication and repair, bacterial chromosome separation during division, and other processes involving DNA. Fluoroquinolones also inhibit topoisomerase II, another enzyme that untangles DNA during replication. It is not surprising that quinolones are bactericidal. ◀◀ *DNA replication (section 12.4)*

The quinolones are broad-spectrum antibiotics. They are highly effective against enteric bacteria such as *Escherichia coli* and *Klebsiella pneumoniae*. They can be used with *Haemophilus*, *Neisseria*, *P. aeruginosa*, and other gram-negative pathogens. The quinolones also are active against gram-positive bacteria such as *S. aureus*, *Streptococcus pyogenes*, and *Mycobacterium tuberculosis*. Thus they are used in treating a wide range of infections.

1. Explain five ways in which chemotherapeutic agents kill or damage bacterial pathogens.
2. Why do penicillins and cephalosporins have a higher therapeutic index than most other antibiotics?
3. Would there be any advantage to administering a bacteriostatic agent along with penicillins? Any disadvantage?
4. What are antimetabolites? Why do you think these are effective against protozoan pathogens while the other drugs presented here generally are not?

34.5 Antifungal Drugs

Treatment of fungal infections generally has been less successful than that of bacterial infections largely because as eukaryotes, fungal cells are much more similar to human cells than are bacterial cells. Many drugs that inhibit or kill fungi are therefore quite toxic for humans and thus have a low therapeutic index. In addition, most fungi have a detoxification system that modifies many antifungal agents. Therefore antibiotics are fungistatic only as long as repeated application maintains high levels of unmodified antibiotic. Nonetheless, a few drugs are useful in treating many major fungal diseases. Effective antifungal agents frequently either extract membrane sterols or prevent their synthesis. Similarly, because fungal cell walls contain chitin, the enzyme chitin synthase is the target for antifungals such as polyoxin D and nikkomycin.

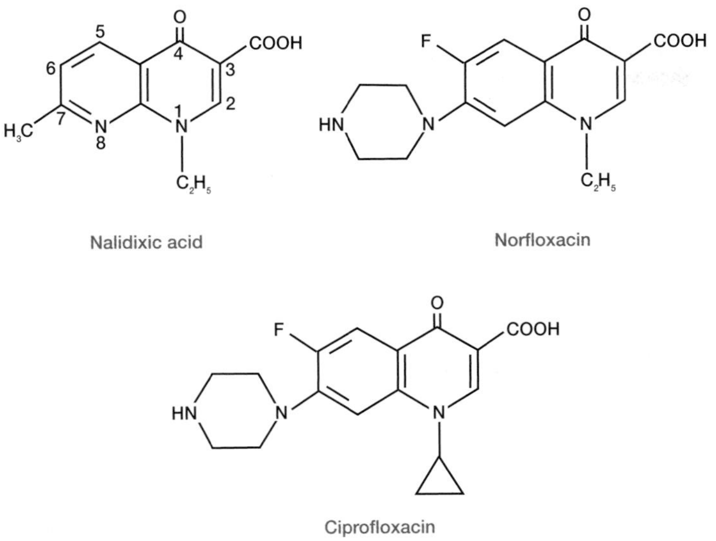

FIGURE 34.13 Quinolone Antimicrobial Agents. Ciprofloxacin and norfloxacin are newer-generation fluoroquinolones. The 4-quinolone ring in nalidixic acid has been numbered.

Fungal infections are often subdivided into infections called superficial mycoses, subcutaneous mycoses, and systemic mycoses. Treatment for these types of disease is very different. Several drugs are used to treat superficial mycoses. Three drugs containing imidazole—miconazole, ketoconazole (**figure 34.14**), and clotrimazole—are broad-spectrum agents available as creams and solutions for the treatment of dermatophyte infections such as athlete's foot, and oral and vaginal candidiasis. They are thought to disrupt fungal membrane permeability and inhibit sterol synthesis. Nystatin (figure 34.14), a polyene antibiotic from *Streptomyces,* is used to control *Candida* infections of the skin, vagina, or alimentary tract. It binds to sterols and damages the membrane, leading to fungal membrane leakage. Griseofulvin (figure 34.14), an antibiotic formed by *Penicillium,* is given orally to treat chronic dermatophyte infections. It is thought to disrupt the mitotic spindle and inhibit cell division; it also may inhibit protein and nucleic acid synthesis. Side effects of griseofulvin include headaches, gastrointestinal upset, and allergic reactions. ▶▶| *Human diseases caused by fungi and protists (chapter 39)*

Systemic infections are very difficult to control and can be fatal. Three drugs commonly used against systemic mycoses are amphotericin B, 5-flucytosine, and fluconazole (figure 34.14). Amphotericin B from *Streptomyces* spp. binds to the sterols in fungal membranes, disrupting membrane permeability and causing leakage of cell constituents. It is quite toxic to humans and used only for serious, life-threatening infections. The synthetic oral antimycotic agent 5-flucytosine (5-fluorocytosine) is effective against most systemic fungi, although drug resistance often develops rapidly. The drug is converted to 5-fluorouracil by fungi, incorporated into RNA in place of uracil, and disrupts RNA function. Its side effects include skin rashes, diarrhea, nausea, aplastic anemia, and liver damage. Atovaquone and pentamidine are used to treat *Pneumocystis jiroveci* (formerly called *P. carinii*). Some reports indicate that pentamidine interferes with *P. jiroveci* metabolism, although the drug only moderately inhibits glucose metabolism, protein synthesis, RNA synthesis, and intracellular amino acid transport in vitro. Fluconazole is used in the treatment of candidiasis, cryptococcal meningitis, and coccidioidal meningitis. Because adverse effects of fluconazole are relatively uncommon, it is used prophylactically to prevent life-threatening fungal infections in AIDS patients and other individuals who are severely immunosuppressed. Posaconazole is a broad-spectrum, antifungal agent in the azole class of antifungals. It is approved for the prophylaxis and treatment of invasive fungal infections. Posaconazole is fungistatic against most pathogenic yeast species, including *Aspergillus, Candida, Cryptococcus,* and *Trichosporon,* and it has good coverage against species of *Aspergillus* and *Candida* that are resistant to other azole antifungal agents. Posaconazole is administered orally and is less toxic to kidneys than other azoles and the polyenes.

Subcutaneous mycoses, such as mycetoma and sporotrichosis, are typically treated with combinations of therapies that would be used for superficial and systemic mycoses. As with systemic mycoses, strict attention to potential toxic side effects is warranted.

1. Summarize the mechanism of action and the therapeutic use of the following antifungal drugs: miconazole, nystatin, griseofulvin, amphotericin B, and 5-flucytosine.
2. Why are immunosuppressed individuals given antifungal agents?

34.6 Antiviral Drugs

Because viruses enter host cells and make use of host cell enzymes and constituents, it was long thought that a drug that blocked virus multiplication would be toxic for the host. However, inhibitors of virus-specific enzymes and life cycle processes have been discovered, and several drugs are used therapeutically. Some important examples are shown in **figure 34.15.** |◀◀ *Viral multiplication (section 5.3)*

Most antiviral drugs disrupt either critical stages in the virus life cycle or the synthesis of virus-specific nucleic acids. **Amantadine** and **rimantadine** can be used to prevent influenza A infections. When given early in the infection (in the first 48 hours), they reduce the incidence of influenza by 50 to 70% in an exposed population. Amantadine blocks the penetration and uncoating of influenza virus particles. Adenine arabinoside (vidarabine) disrupts the activity of DNA polymerase and several other enzymes involved in DNA and RNA synthesis and function. It is given intravenously or applied as an ointment to treat herpes infections. A third drug, acyclovir, is also used in the treatment of herpes infections. Upon phosphorylation,

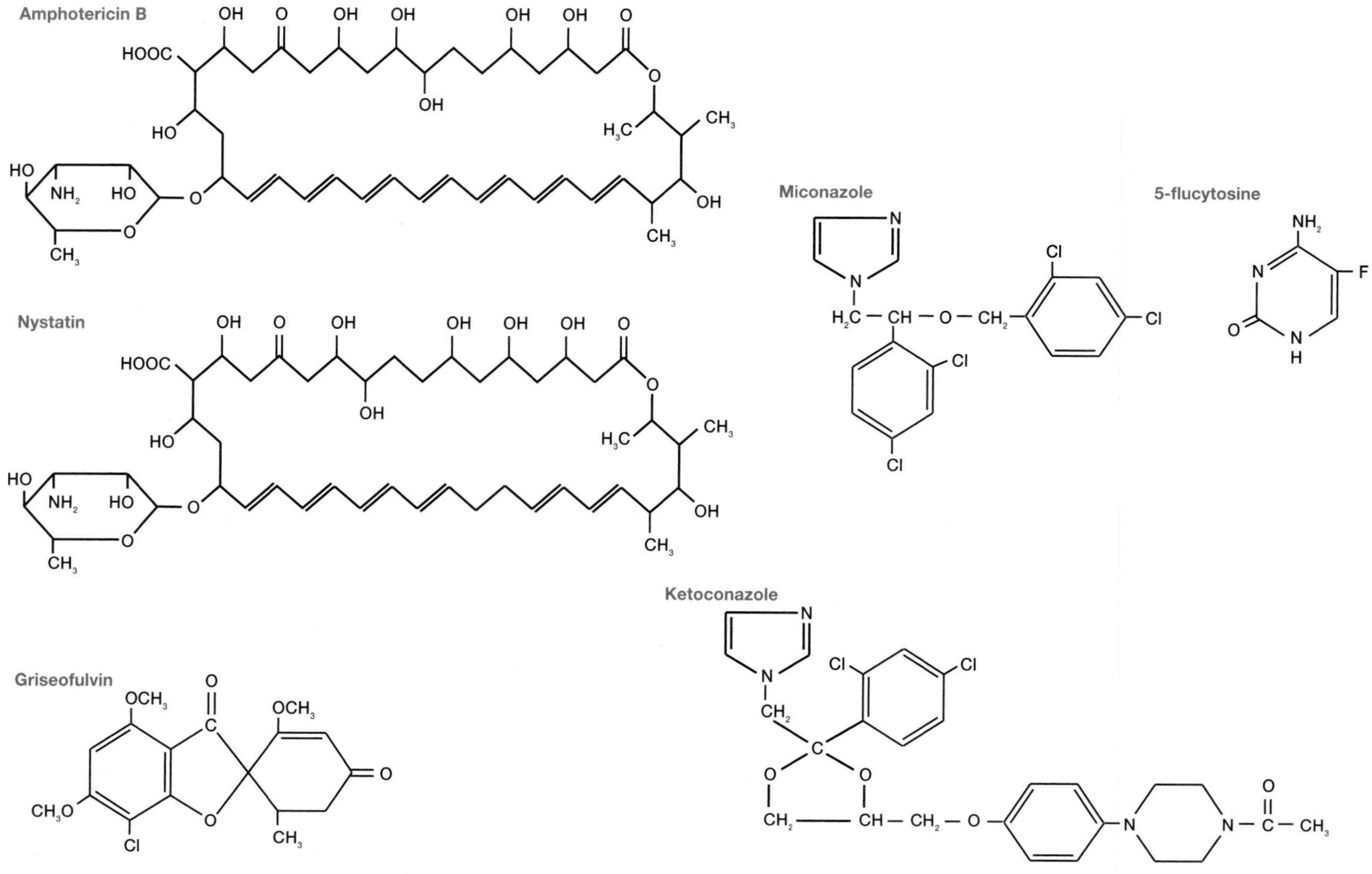

FIGURE 34.14 Antifungal Drugs. Six commonly used drugs are shown.

Figure 34.14 Micro Inquiry

What is the mechanism by which nystatin inhibits growth? How does this compare to that of amphotericin B? Do you think nystatin is less toxic than amphotericin? (Hint: Think about how the two drugs are delivered.)

acyclovir resembles deoxyGTP and inhibits the viral DNA polymerase. Unfortunately, acyclovir-resistant strains of herpes have developed. Effective acyclovir derivatives and relatives are now available. Valacyclovir is an orally administered prodrug form of acyclovir. Prodrugs are inactive until metabolized. Ganciclovir, penciclovir, and penciclovir's oral form, famciclovir, are effective in treatment of herpesviruses. Another kind of drug, foscarnet, inhibits the virus DNA polymerase in a different way. Foscarnet is an organic analog of pyrophosphate (figure 34.15) that binds to the polymerase active site and blocks the cleavage of pyrophosphate from nucleoside triphosphate substrates. It is used in treating herpes and cytomegalovirus infections. ▶▶| *Human diseases caused by viruses and prions (chapter 37)*

Several broad-spectrum anti-DNA virus drugs have been developed. A good example is the drug HPMPC, also known as cidofovir (figure 34.15). It is effective against papovaviruses, adenoviruses, herpesviruses, iridoviruses, and poxviruses. The drug acts on the viral DNA polymerase as a competitive inhibitor and alternative substrate of dCTP. It has been used primarily against cytomegalovirus but also against herpes simplex and human papillomavirus infections.

Research on anti-HIV drugs has been particularly active. Many of the first drugs to be developed were reverse transcriptase inhibitors such as **azidothymidine (AZT)** or zidovudine, lamivudine (3TC), didanosine (ddI), zalcitabine (ddC), and stavudine (d4T) (figure 34.15). These interfere with reverse transcriptase activity. The antiviral drugs currently approved for use in HIV disease are of four types: (1) nucleoside reverse transcriptase inhibitors (NRTIs), which are nucleoside analogues that inhibit the enzyme reverse transcriptase as it synthesizes viral DNA; (2) nonnucleoside reverse transcriptase inhibitors (NNRTIs); (3) protease inhibitors (PIs), which work by blocking the activity of the HIV protease and thus interfere

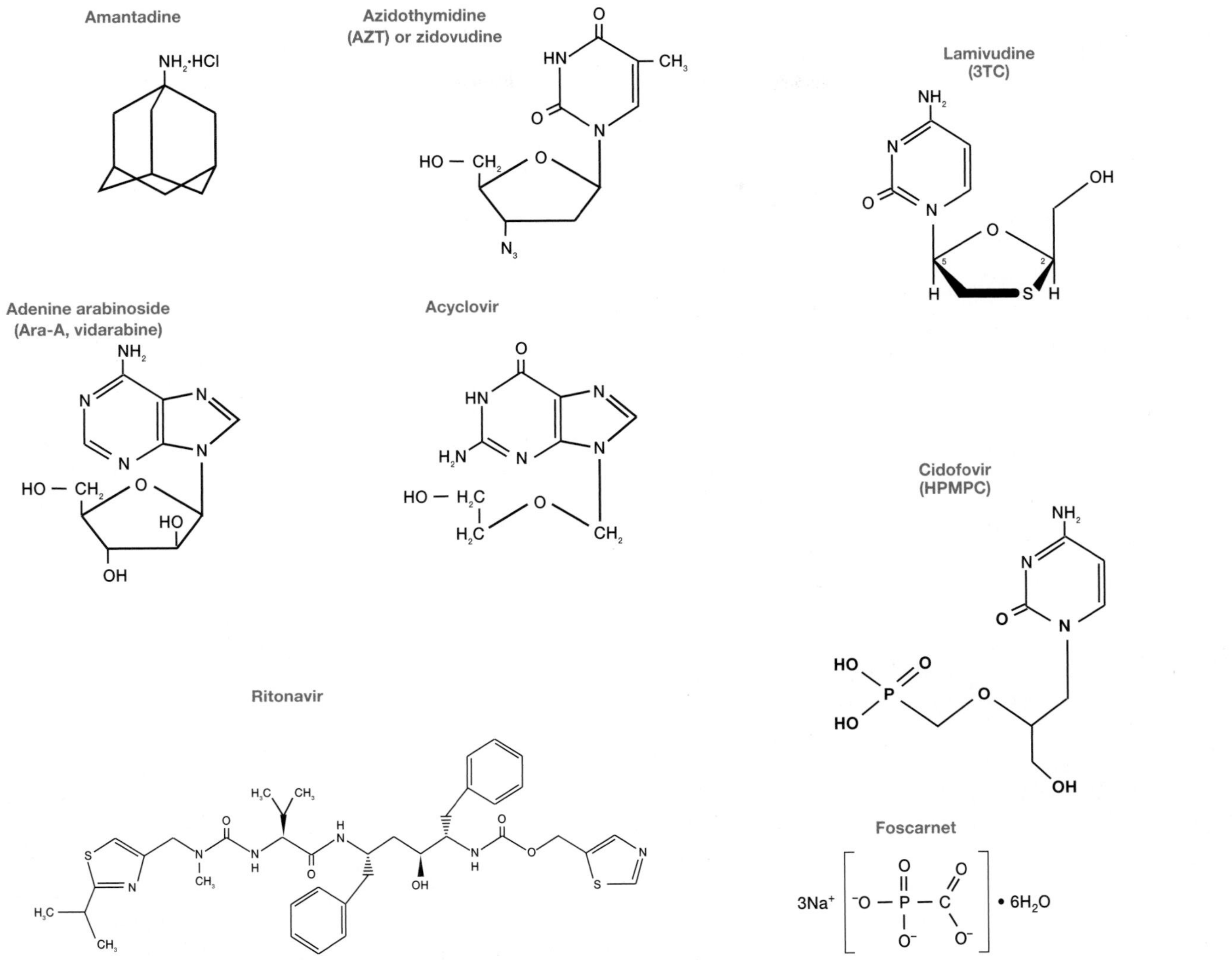

FIGURE 34.15 Antiviral Drugs Target the Viral Replication Cycle. Representative antiviral drugs.

with virion assembly; and (4) fusion inhibitors (FIs), a relatively newer category of drugs that prevent HIV entry into cells (**figure 34.16**). Inhibition of reverse transcription blocks viral DNA synthesis and halts HIV replication. Protease inhibitors are effective because HIV, like many RNA viruses, translates multiple proteins as a single large polyprotein. This polyprotein must then be cleaved into individual proteins required for virus replication. Protease inhibitors mimic the peptide bond that is normally attacked by the protease. Three of the most used PIs are saquinvir, indinavir, and ritonavir. Fusion inhibitors are particularly interesting as an effective blockade to viral entry into host cells, essentially preventing disease. ◀◀ *Viruses with plus-strand RNA genomes (group IV) (section 25.5); Viruses with single-stranded RNA genomes (group VI-retroviruses) (section 25.7)*

The most successful treatment approach to date in combating HIV/AIDS is to use drug combinations. The most successful HIV treatment regimen has been a cocktail of agents given at high dosages to prevent the development of drug resistance. For example, the combination of AZT, 3TC, and ritonavir is very effective in reducing HIV plasma concentrations almost to zero. However, the treatment does not eliminate proviral HIV DNA that still resides in memory T cells and possibly elsewhere. In many patients, the virus disappears from the patient's blood with proper treatment and drug-resistant strains do not seem to arise. HIV can remain dormant in memory T cells, survive drug cocktails, and reactivate. Thus patients are not completely cured with drug treatment. It should be noted that side effects can be very severe, and treatment is prohibitively expensive for those without medical insurance. Globally, the vast majority of HIV-positive individuals do not have access to effective combination therapy. ▶▶ *Acquired immunodeficiency syndrome (AIDS) (section 37.3)*

Probably the most publicized antiviral agent has been **Tamiflu** (generically, oseltamivir phosphate). Tamiflu is a neuraminidase inhibitor that has received much attention in light of the

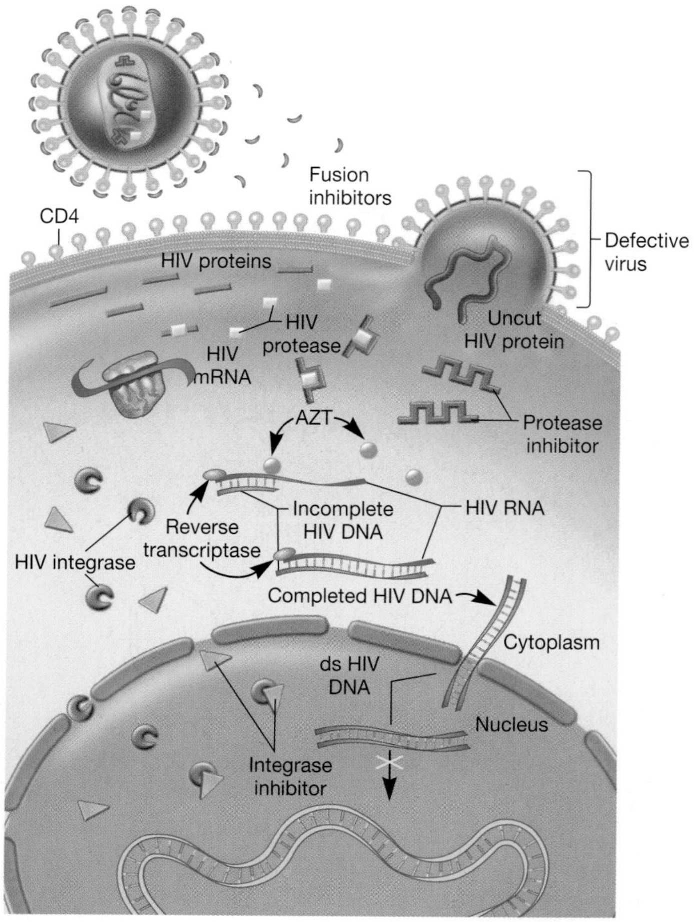

FIGURE 34.16 Examples of how anti-HIV agents block HIV replication.

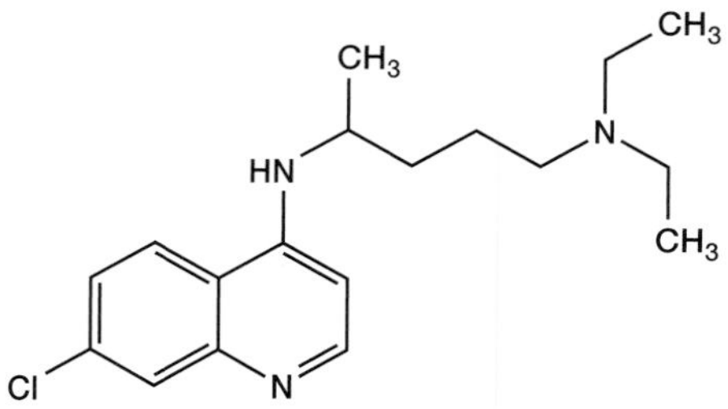

FIGURE 34.17 Chloroquine.

H1N1 influenza pandemic, including novel H1N1 influenza ("swine flu"). While Tamiflu is not a cure for neurominidase-expressing viruses, clinical trials show that patients who take Tamiflu are relieved of flu symptoms 1.3 days faster than patients who do not take Tamiflu. However, prophylactic use has resulted in viral resistance to Tamiflu. Tamiflu is not a substitute for yearly flu vaccination and frequent hand-washing.

34.7 Antiprotozoan Drugs

The mechanism of action for most antiprotozoan drugs is not completely understood. Drugs such as chloroquine, atovaquone, mefloquine, iodoquinol, metronidazole, and nitazoxanide, for example, have potent antiprotozoan action, but a clear mechanism by which protozoan growth is inhibited is unknown. However, most antiprotozoan drugs appear to act on protozoan nucleic acid or some metabolic event.

Chloroquine, qualaquin, and other **quinine drugs** are used to treat malaria (**figure 34.17**). These drugs suppress protozoan reproduction and are effective in eradicating erythrocytic asexual stages. Several mechanisms of action have been reported. They can raise the internal pH, clump the plasmodial pigment, and intercalate into plasmodial DNA. Chloroquine also inhibits heme polymerase, an enzyme that converts toxic heme into nontoxic hemazoin. Inhibition of this enzyme leads to a buildup of toxic heme. It is worth noting that individuals who are traveling to areas where malaria is endemic should receive chemoprophylactic treatment with chloroquine. ▶▶| *Malaria (section 39.3)*

Mefloquine has been found to swell the *Plasmodium falciparum* food vacuoles, where it may act by forming toxic complexes that damage membranes and other plasmodial components. Primaquine is active against a dormant form of the protists (hypnozoites) that are found in the liver; it prevents relapses. However, primaquine should not be taken by pregnant women or those who have a glucose-6-phosphate dehydrogenase deficiency. However, because resistance to these drugs occurs rapidly, more expensive drug combinations are now required. One example is Fansidar, a combination of pyrimethamine and sulfadoxine.

Although not approved for use in the United States, **artemisinin** and its derivatives are becoming more widely used to treat malaria. Having its origins as a treatment in traditional Chinese medicine, artemisinin is a chemical found in the sweet wormwood plant that grows in Africa and Asia. The purified plant compound has low potency. However, a semisynthetic derivative used as a component in artemisinin combination therapies (artemisinin plus a Western antimalarial drug) is an effective treatment of malaria and a truly cost-effective treatment in places where drug resistance and health care access are problems. The mechanism of artemisinin action is not well understood; it appears to form reactive oxygen intermediates inside *Plasmodium*-infected red blood cells, leading to altered hemoglobin catabolism and plasmodium electron transport chain damage.

Metronidazole is used to treat *Entamoeba* infections. Anaerobic organisms readily reduce it to the active metabolite within the cytoplasm. Aerobic organisms appear to reduce it using ferrodoxin (a protein of the electron transport system). Reduced metronidazole interacts with DNA, altering its helical structure and causing DNA fragmentation; it prevents normal nucleic acid synthesis, resulting in cell death.

A number of antibiotics that inhibit bacterial protein synthesis are also used to treat protozoan infection. These include the aminoglycosides clindamycin and paromomycin. Aminoglycosides can be considered polycationic molecules that have a high affinity for nucleic acids. Specifically, aminoglycosides possess high affinities for RNAs. Different aminoglycoside antibiotics bind to different sites on RNAs. RNA binding interferes with the normal expression and function of the RNA, resulting in cell death.

Interference with eukaryotic electron transport is the mechanism of action for atovaquone, which is used to treat *Toxoplasma gondii*. It is an analog of ubiquinone, an integral component of the eukaryotic electron transport system. As an analog of ubiquinone, atovaquone can act as a competitive inhibitor and thus suppress electron transport. The ultimate metabolic effects of electron transport blockade include inhibited or delayed synthesis of nucleic acids and ATP. Another drug that interferes with electron transport is nitazoxanide, which is used to treat cryptosporidiosis. It appears to exert its effect through interference with the pyruvate:ferredoxin oxidoreductase. It has also been reported to form toxic free radicals once the nitro group is reduced intracellularly. Pyrimethamine and dapsone, used to treat *Toxoplasma* infections, appear to act in the same way as trimethoprim—interfering with folic acid synthesis by inhibition of dihydrofolate reductase.

As with other antimicrobial therapies, traditional drug development starts by identifying a unique target to which a drug can bind and thus prevent some vital function. A second consideration is often related to drug spectrum: how many different species have that target so that the proposed drug can be used broadly as a chemotherapeutic agent? This is also true for use of agents needed to remove protozoan parasites from their hosts. However, because protozoa are eukaryotes, the potential for drug action on host cells and tissues is greater than it is when targeting bacteria. Most of the drugs used to treat protozoan infection have significant side effects; nonetheless, the side effects are usually acceptable when weighed against the parasitic burden.

1. Why do you think drugs that inhibit bacterial protein synthesis are also effective against some protists?
2. Why do you think malaria, like tuberculosis, is now treated with several drugs simultaneously?
3. What special considerations must be taken into account when treating infections caused by protozoan parasites?

34.8 Factors Influencing Antimicrobial Drug Effectiveness

It is crucial to recognize that effective drug therapy is not a simple matter. Drugs can be administered in several different ways, and they do not always spread rapidly throughout the body or immediately kill all invading pathogens. A complex array of factors influences the effectiveness of drugs.

First, the drug must actually be able to reach the site of infection. Understanding the factors that control drug activity, stability, and metabolism in vivo are essential in drug formulation. For example, the mode of administration plays an important role. A drug such as penicillin G is not suitable for oral administration because it is relatively unstable in stomach acid. Some antibiotics—for example, gentamicin and other aminoglycosides—are not well absorbed from the intestinal tract and must be injected intramuscularly or given intravenously. Other antibiotics (neomycin, bacitracin) are so toxic that they can only be applied topically to skin lesions. Nonoral routes of administration are called **parenteral routes.** Even when an agent is administered properly, it may be excluded from the site of infection. For example, blood clots, necrotic tissue, or biofilms can protect bacteria from a drug, either because body fluids containing the agent may not easily reach the pathogens or because the agent is absorbed by materials surrounding them.

Second, the pathogen must be susceptible to the drug. Bacteria in biofilms or abscesses may be replicating very slowly and are therefore resistant to chemotherapy because many agents affect pathogens only if they are actively growing and dividing. A pathogen, even though growing, may simply not be susceptible to a particular agent. To control resistance, drug cocktails can be used to treat some infections. A notable example of this is the use of clavulonic acid (to inactivate penicillinase) combined with ampicillin, to treat penicillin-resistant bacteria.

Third, the chemotherapeutic agent must reach levels in the body that exceed the pathogen's MIC value if it is going to be effective. The concentration reached will depend on the amount of drug administered, the route of administration and speed of uptake, and the rate at which the drug is cleared or eliminated from the body. It makes sense that a drug will remain at high concentrations longer if it is absorbed over an extended period and excreted slowly.

Finally, chemotherapy has been rendered less effective and much more complex by the spread of drug-resistance genes and prevention of drug access by biofilm components.

1. What factors do you think must be considered when treating an infection present in a biofilm on a medical implant (e.g., an artificial hip) versus a skin infection caused by the same microbe?
2. What is parenteral administration of a drug? Why is it used?

34.9 Drug Resistance

The spread of drug-resistant pathogens is one of the most serious threats to public health in the twenty-first century. This section describes the ways in which bacteria acquire drug resistance and how resistance spreads within a bacterial population.

Mechanisms of Drug Resistance

The long-awaited "superbug" arrived in the summer of 2002. *S. aureus,* a common but sometimes deadly bacterium, had acquired a new antibiotic-resistance gene. The new strain was isolated from foot ulcers on a diabetic patient in Detroit, Michigan. **Methicillin-resistant *S. aureus* (MRSA)** had been well-known as the bane of hospitals. This newer strain had developed resistance to vancomycin, one of the few antibiotics thought to still control *S. aureus.* This new vancomycin-resistant *S. aureus* (VRSA) strain also resisted most other antibiotics, including ciprofloxacin, methicillin, and penicillin. Isolated from the same patient was another dread of hospitals—**vancomycin-resistant enterococci (VRE).** Genetic analyses revealed that the patient's own vancomycin-sensitive *S. aureus* had acquired the vancomycin-resistance gene, *vanA,* from VRE through conjugation. So was born a new threat to the health of the human race. ◀◀ *Bacterial conjugation (section 14.7); Bacterial plasmids (section 14.6)*

Several factors contribute to the spread of antibiotic resistance in a population of bacteria (**figure 34.18**). Furthermore, a particular resistance mechanism is not confined to a single class of drugs (**figure 34.19**). Two bacteria may use different resistance mechanisms to withstand the same chemotherapeutic agent. In addition, resistant mutants may arise spontaneously and are then selected for in the presence of the drug.

Pathogens often become resistant simply by preventing entrance of the drug. Many gram-negative bacteria are unaffected by penicillin G because it cannot penetrate the bacterial outer membrane. A decrease in permeability can lead to sulfonamide resistance. Mycobacteria resist many drugs because of the high content of mycolic acids (*see figure 22.11*) in a complex lipid layer outside their peptidoglycan. This layer is impermeable to most water-soluble drugs. ◀◀ *Suborder* Corynebacterineae *(section 22.4)*

A second resistance strategy is to pump the drug out of the cell after it has entered. Some pathogens have plasma membrane translocases, often called **efflux pumps,** that expel drugs. Because they are relatively nonspecific and pump many different drugs, these transport proteins often are multidrug-resistance pumps. Many are drug/proton antiporters—that is, protons enter the cell as the drug leaves. Such systems are present in *E. coli, P. aeruginosa,* and *S. aureus,* to name a few.

Many bacterial pathogens resist attack by **drug inactivation** through chemical modification. The best-known example is the hydrolysis of the β-lactam ring of penicillins by the enzyme penicillinase. Drugs also are inactivated by the addition of chemical groups. For example, chloramphenicol contains two hydroxyl groups that can be modified by the addition of acetyl-CoA, a reaction catalyzed by the enzyme chloramphenicol acetyltransferase. Aminoglycosides (figure 34.8a) can be modified and inactivated in several ways. For instance, acetyltransferases catalyze the acetylation of amino groups. Some aminoglycoside-modifying enzymes catalyze the addition to hydroxyl groups of either phosphates (phosphotransferases) or adenyl groups (adenyltransferases).

Because each chemotherapeutic agent acts on a specific target enzyme or cellular structure, resistance arises through **target modification.** We have already discussed this in the case of vancomycin resistance (p. 834). The affinity of ribosomes for erythromycin and chloramphenicol also can be decreased by a change in the 23S rRNA to which they bind. This drastically reduces antibiotic binding. Antimetabolite action may be resisted through alteration of susceptible enzymes. In sulfonamide-resistant bacte-

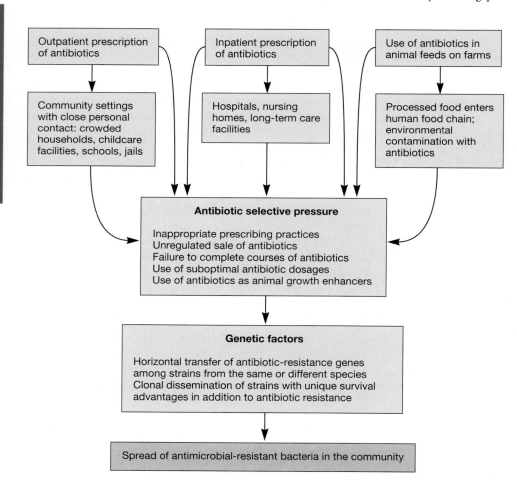

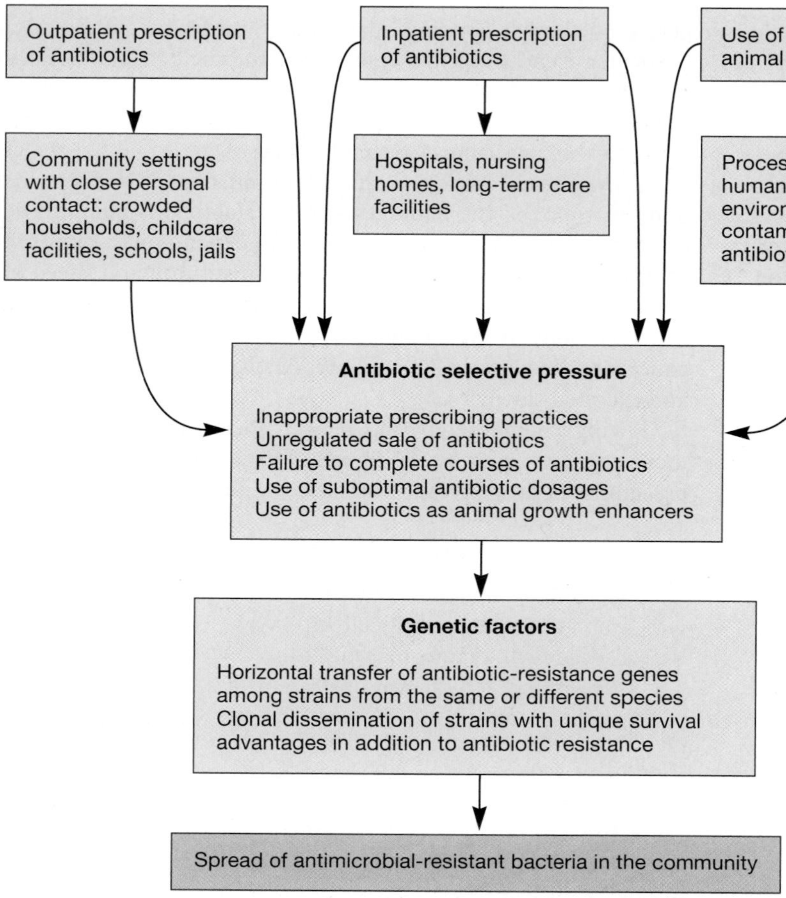

FIGURE 34.18 Antibiotic Resistance Has Many Sources. Incomplete and indiscriminant use of antibiotics in people and animals leads to increased selective pressure on bacteria. Bacteria capable of resisting antibiotics survive and spread these traits by horizontal gene transfer.

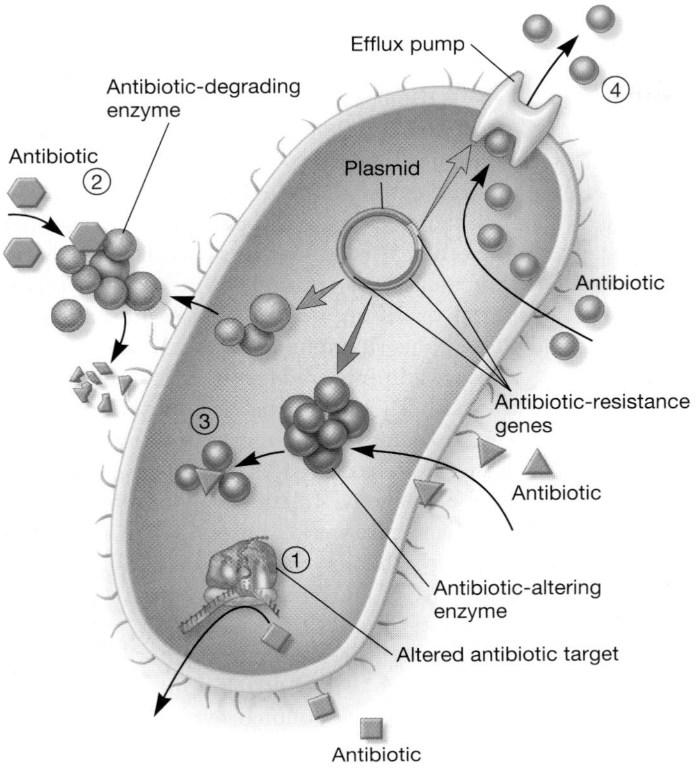

FIGURE 34.19 Antibiotic Resistance Mechanisms. Bacteria can resist the action of antibiotics by (1) preventing access to (or altering) the target of the antibiotic, (2) degrading the antibiotic, (3) altering the antibiotic, or (4) rapidly extruding the antibiotic.

Figure 34.19 Micro Inquiry

What is a common example of drug inactivation by enzyme degradation?

ria, the enzyme that uses *p*-aminobenzoic acid during folic acid synthesis (the dihydropteroic acid synthetase; figure 34.12b) often has a much lower affinity for sulfonamides.

Finally, resistant bacteria may either use an **alternate pathway** to bypass the sequence inhibited by the agent or increase the production of the target metabolite. For example, some bacteria are resistant to sulfonamides simply because they use preformed folic acid from their surroundings, rather than synthesize it themselves. Other strains increase their rate of folic acid production and thus counteract sulfonamide inhibition.

The Origin and Transmission of Drug Resistance

Within three years after the widespread use of penicillin began, penicillin-resistant bacteria were found in clinical specimens. To understand the origin of drug resistance, it is important to recall that in nature, antibiotic-producing microbes must also protect themselves from the antibiotics they secrete. In other words, gaining the function of antibiotic production requires that the same microbe also gain resistance, otherwise they would "commit suicide" by producing the antibiotic. In antibiotic-producing microorganisms, the genes that encode resistance proteins are often referred to as immunity genes. Immunity genes are usually coordinately regulated with genes that code for antibiotic biosynthetic enzymes. It is believed that many genes that encode antibiotic resistance in bacteria were "captured" by means of horizontal gene transfer from producer to nonproducer, thus giving rise to a large pool of resistance-encoding genes outside the producing microorganisms.

It is therefore not surprising that in nonproducing bacteria, genes for drug resistance may be present on bacterial chromosomes, plasmids, transposons, and other mobile genetic elements. Because they are often found on mobile genetic elements, they can freely exchange between bacteria. Spontaneous mutations in the bacterial chromosome can also make bacteria drug resistant, although they do not occur very often (with the exception of *M. tuberculosis*). Usually such mutations result in a change in the drug target; therefore the antibiotic cannot bind and inhibit growth. If a patient fails to take prescribed antibiotics as directed (e.g., does not complete the course of treatment), resistant mutants survive and flourish because of their competitive advantage over nonresistant strains. Preventing the growth of such mutants is the rationale behind the direct observation of TB patients when taking each antibiotic dose. ▶▶| *Mycobacterium infections (section 38.1)*

Frequently a bacterial pathogen is drug resistant because it has a plasmid bearing one or more resistance genes; such plasmids are called **R plasmids** (resistance plasmids; *see figure 14.17*). Plasmid resistance genes often code for enzymes that destroy or modify drugs; for example, the hydrolysis of penicillin or the acetylation of chloramphenicol and aminoglycoside drugs. Plasmid-associated genes have been implicated in resistance to the aminoglycosides, chloramphenicol, penicillins and cephalosporins, erythromycin, tetracyclines, sulfonamides, and others. Once a bacterial cell possesses an R plasmid, the plasmid (or its genes) may be transferred to other cells quite rapidly through normal gene exchange processes such as conjugation, transduction, and transformation (**figure 34.20**). Because a single plasmid may carry genes for resistance to several drugs, a pathogen population can become resistant to several antibiotics simultaneously, even though the infected patient is being treated with only one drug. |◀◀ *Bacterial plasmids (section 14.6); Bacterial conjugation (section 14.7); Bacterial transformation (section 14.8); Transduction (section 14.9).*

Antibiotic resistance genes can be located on genetic elements other than plasmids. Many transposons contain genes for antibiotic resistance and can move rapidly between plasmids and through a bacterial population. They are found in both gram-negative and gram-positive bacteria. Some examples and their resistance markers are Tn*5* (kanamycin, bleomycin, streptomycin), Tn*21* (streptomycin, spectinomycin, sulfonamide), Tn*551*

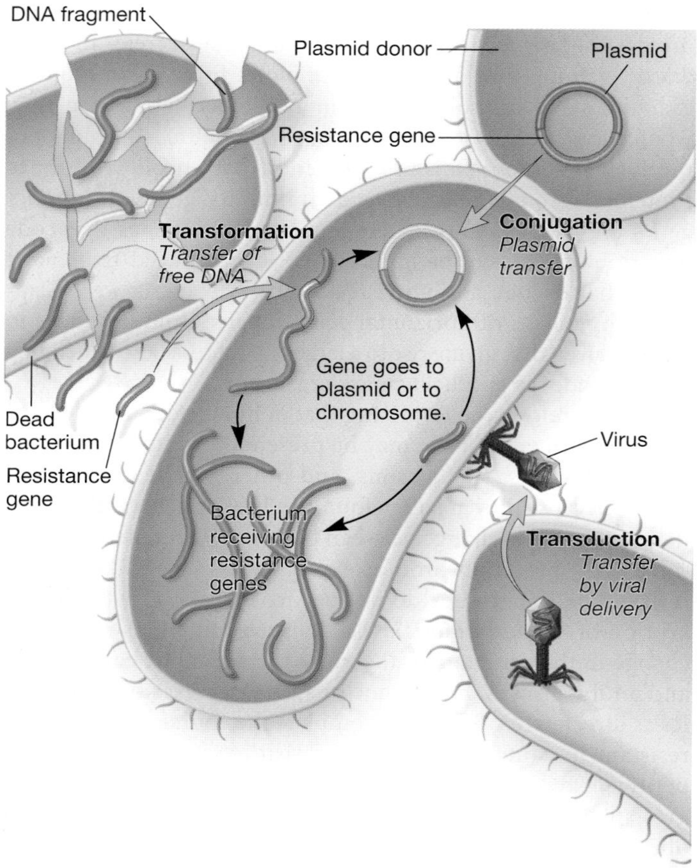

FIGURE 34.20 Horizontal Gene Transfer. Bacteria exchange genetic information, such as antibiotic resistance genes, through conjugation, transformation, or transduction.

Figure 34.20 Micro Inquiry

What is an R plasmid?

(erythromycin), and Tn*4001* (gentamicin, tobramycin, kanamycin). Often several resistance genes are carried together as gene cassettes in association with a genetic element known as an integron. **Gene cassettes** are genetic elements that may exist as circular, nonreplicating DNA when moving from one site to another but normally are a linear part of a transposon, plasmid, or bacterial chromosome. An **integron** is composed of an integrase gene and sequences for site-specific recombination. Thus integrons can capture genes and gene cassettes. Several cassettes can be integrated sequentially in an integron. Thus integrons also are important in spreading resistance genes. Finally, conjugative transposons can carry resistance genes. Because they are capable of moving between bacteria by conjugation, they are also effective in spreading resistance. ⏮ *Transposable elements (section 14.5)*

Overcoming Drug Resistance

Several strategies can be employed to discourage the emergence of drug resistance. The drug can be given in a high enough concentration to destroy susceptible bacteria and most spontaneous mutants that might arise during treatment. Sometimes two or even three different drugs can be administered simultaneously with the hope that each drug will prevent the emergence of resistance to the other. This approach is used in treating tuberculosis, HIV, and malaria, for example. When treating tuberculosis (TB), several drugs are administered simultaneously (e.g., isoniazid [INH], plus rifampin, ethambutol, and pyrazinamide). These drugs are administered for 6 to 9 months as a way of decreasing the possibility that the bacterium develops drug resistance. Inadequate therapy is the most common means by which resistant bacteria are acquired. "Directly observed therapy" (DOT) is the most effective treatment management strategy. During DOT, a health care worker observes each TB patient take each antibiotic dose, thereby ensuring full antibiotic regimen compliance. The use of DOT increased from 21% in 1993 to 57% in 2005, the latest year with available data, effectively contributing to the substantial 50% decline in reported TB cases in the United States between 1993 and 2006. Unfortunately, the number of worldwide TB cases doubled over that same time period. There were 9.2 million cases of TB reported to the WHO in 2008. Of these, 500,000 were *m*ulti*d*rug-*r*esistant strains of tuberculosis (MDR-TB). A multidrug-resistant strain is defined as *M. tuberculosis* that is resistant to INH and rifampin, with or without resistance to other drugs.

Recently strains of *M. tuberculosis* have emerged that are *ex*tremely *d*rug-*r*esistant (XDR); they are resistant to INH, rifampin, and any fluoroquinolone and at least one of three injectable second-line drugs (i.e., amikacin, kanamycin, or capreomycin). Importantly, MDR-TB and XDR-TB can be fatal. The number of drug-resistant TB cases steadily increases in the world, with 1.6 million MDR-TB and XDR-TB cases expected by 2010. In the United States, the total number of MDR-TB cases has fluctuated between 91 and 126 per year, with 98 cases reported in 2007; the number of U.S. XDR-TB cases reported to the Centers for Disease Control was 2.

Search This: WHO/mdr TB

Other strategies used to prevent drug resistance include the strict control on use of chemotherapeutic drugs, particularly broad-spectrum drugs, which should be used only when absolutely necessary. If possible, the pathogen should be identified, drug sensitivity tests performed, and the proper narrow-spectrum drug employed. Patient compliance is just as important because completing a full course of antimicrobial therapy often prevents full mutation to resistant phenotypes.

Despite efforts to control the emergence and spread of drug resistance, the situation continues to worsen. Thus an urgent need exists for new antibiotics that microorganisms have never encountered. Pharmaceutical and biotechnology companies collect and analyze samples from around the world in a search for completely new antimicrobial agents. Both culture-based and metagenomics approaches are used (*see figure 16.21*) Structure-based or rational drug design is another option. If the three-dimensional structure of a susceptible target molecule such as an enzyme essential to microbial function is known, computer programs can be used to design drugs that precisely fit the target molecule. These drugs might be able to bind to the target and disrupt its function sufficiently to destroy the pathogen. Pharmaceutical companies are using these approaches to develop drugs for the treatment of AIDS, cancer, septicemia caused by lipopolysaccharide (LPS), and the common cold. ▶▶| *Biotechnology and industrial microbiology (chapter 41)*

Search This: Preventing antibiotic resistance

Information derived from the sequencing and analysis of pathogen genomes is also useful in identifying new targets for antimicrobial drugs. For example, genomics studies are providing data for research on inhibitors of both aminoacyl-tRNA synthetases and the enzyme that removes the formyl group from the N-terminal methionine during bacterial protein synthesis. The drug susceptibility of enzymes required for fatty acid synthesis is also being analyzed. |◀◀ *Microbial genomics (chapter 16)*

A most interesting response to the current crisis is the renewed interest in an idea first proposed early in the twentieth century by Felix d'Herelle, one of the discoverers of bacterial viruses (bacteriophages). d'Herelle proposed that bacteriophages could be used to treat bacterial diseases. Although most microbiologists did not pursue his proposal actively due to technical difficulties and the advent of antibiotics, Russian scientists developed the medical use of bacteriophages. Currently Russian physicians use bacteriophages to treat many bacterial infections. Bandages are saturated with phage solutions, phage mixtures are administered orally, and phage preparations are given intravenously to treat *Staphylococcus* infections. Several American companies are actively conducting research on phage therapy and preparing to carry out clinical trials. |◀◀ *Viruses and other acellular infectious agents (chapter 5)*

1. Briefly describe the five major ways in which bacteria become resistant to drugs and give an example of each.
2. Define plasmid, R plasmid, integron, and gene cassette. How are R plasmids involved in the spread of drug resistance?
3. List several ways in which the development of antibiotic-resistant pathogens can be slowed or prevented.

Summary

34.1 The Development of Chemotherapy

a. The modern era of chemotherapy began with Paul Ehrlich's work on drugs against African sleeping sickness and syphilis.
b. Other early pioneers were Gerhard Domagk, Alexander Fleming, Howard Florey, Ernst Chain, Norman Heatley, and Selman Waksman.

34.2 General Characteristics of Antimicrobial Drugs

a. An effective chemotherapeutic agent must have selective toxicity. A drug with great selective toxicity has a high therapeutic index and usually disrupts a structure or process unique to the pathogen. It has fewer side effects.
b. Antibiotics can be classified in terms of the range of target microorganisms (narrow spectrum versus broad spectrum); their source (natural, semisynthetic, or synthetic); and their general effect (static versus cidal) (**table 34.1**).

34.3 Determining the Level of Antimicrobial Activity

a. Antibiotic effectiveness can be estimated through the determination of the minimal inhibitory concentration and the minimal lethal concentration with dilution susceptibility tests.
b. Tests such as the Kirby-Bauer test (a disk diffusion test) and the Etest® are often used to estimate a pathogen's susceptibility to drugs quickly (**figures 34.2–34.4**).

34.4 Antibacterial Drugs

a. Members of the penicillin family contain a β-lactam ring and disrupt bacterial cell wall synthesis, resulting in cell lysis (**figure 34.5**). Some, such as penicillin G, are usually administered by injection and are most effective against gram-positive bacteria. Others can be given orally (penicillin V), are broad spectrum (ampicillin, carbenicillin), or are usually penicillinase resistant (methicillin).
b. Cephalosporins are similar to penicillins but can be given to patients with penicillin allergies (**figure 34.6**).
c. Vancomycin is a glycopeptide antibiotic that inhibits transpeptidation during peptidoglycan synthesis (**figure 34.7**). It is used against drug-resistant staphylococci, enterococci, and clostridia.
d. Aminoglycoside antibiotics such as streptomycin and gentamicin bind to the small ribosomal subunit, inhibit protein synthesis, and are bactericidal (**figure 34.8**).
e. Tetracyclines are broad-spectrum antibiotics having a four-ring nucleus with attached groups (**figure 34.9**). They bind to the small ribosomal subunit and inhibit protein synthesis.
f. Erythromycin is a bacteriostatic macrolide antibiotic that binds to the large ribosomal subunit and inhibits protein synthesis (**figure 34.10**).

g. Chloramphenicol is a broad-spectrum, bacteriostatic antibiotic that inhibits protein synthesis. It is quite toxic and used only for very serious infections.
h. Sulfonamides or sulfa drugs resemble *p*-aminobenzoic acid and competitively inhibit folic acid synthesis (**figure 34.11**).
i. Trimethoprim is a synthetic antibiotic that inhibits dihydrofolate reductase, which is required by organisms in the manufacture of folic acid (**figure 34.12**).
j. Quinolones are a family of bactericidal synthetic drugs that inhibit DNA gyrase and thus inhibit DNA replication (**figure 34.13**).

34.5 Antifungal Drugs

a. Because fungi are more similar to human cells than bacteria, antifungal drugs generally have lower therapeutic indexes than antibacterial agents and produce more side effects.
b. Superficial mycoses can be treated with miconazole, ketoconazole, clotrimazole, tolnaftate, nystatin, and griseofulvin (**figure 34.14**). Amphotericin B, 5-flucytosine, and fluconazole are used for systemic mycoses.

34.6 Antiviral Drugs

a. Antiviral drugs interfere with critical stages in the virus life cycle (amantadine, rimantadine, ritonavir) or inhibit the synthesis of virus-specific nucleic acids (zidovudine, adenine arabinoside, acyclovir) (**figure 34.15**).
b. Drug combinations (cocktails) appear to be more effective than monotherapies.

34.7 Antiprotozoan Drugs

a. The mechanisms of action of most drugs used to treat protozoan infection are unknown.
b. Some antiprotozoan drugs interfere with critical steps in nucleic acid synthesis, protein synthesis, electron transport, or folic acid synthesis.

34.8 Factors Influencing Antimicrobial Drug Effectiveness

a. A variety of factors can greatly influence the effectiveness of antimicrobial drugs during use. These include route, effective concentration, and pathogen sensitivity.

34.9 Drug Resistance

a. Bacteria can become resistant to a drug by excluding it from the cell, pumping the drug out of the cell, enzymatically altering it, or modifying the target enzyme or structure so it is no longer affected by the drug. The genes for drug resistance may be found on the bacterial chromosome, a plasmid called an R plasmid, or other genetic elements such as transposons (**figures 34.19** and **34.20**).
b. Chemotherapeutic agent misuse fosters the increase and spread of drug resistance.

Critical Thinking Questions

1. What advantage might soil bacteria and fungi gain from the synthesis of antibiotics?
2. You are the CEO of a biotechnology company that seeks to develop new antibiotics. Given that only 1% of all microbes have so far been cultured, which of the following approaches is your company going to use to find new drugs: (1) the development of new culture techniques; (2) metagenomics (*see section 16.8 and chapter 41*); or (3) the chemical synthesis of novel compounds based on specific microbial metabolic or structural targets? Be able to defend your choice.
3. Some advocate stockpiling the drug Tamiflu in the event of an influenza pandemic. Others point out that wealthy, Western nations would have an unfair advantage because developing nations (where the pandemic is most likely to start) would not have access to this expensive antiviral. Furthermore, some fear that indiscriminate use of the drug would promote the evolution of resistant flu strains. Given these caveats, do you think developed nations should stockpile Tamiflu for the protection and treatment of their citizens? Explain your answer.
4. How might the use of antibiotics as growth promoters in livestock contribute to antibiotic resistance among human pathogens?
5. A recent study found that 480 bacterial strains freshly isolated from the soil are resistant to at least six different antibiotics. In fact, some isolates are resistant to 20 different antibiotic drugs. Why do you think these bacteria, which are neither pathogenic nor exposed to human use of antibiotics, are resistant to so many drugs? What might be the implications for human bacterial pathogens?
6. You are a pediatrician treating a child with an upper respiratory infection that is clearly caused by a virus. The child's mother insists that you prescribe antibiotics—she's not leaving without them! How do you convince the child's mother that antibiotics will do more harm than good?
7. The mainstay of antimicrobial therapy has traditionally been the use of agents that specifically target a pathogen structure or process not found in the host. This approach works best when the pathogen is actively growing. This is one reason *Mycobacterium tuberculosis* is so hard to treat: *M. tuberculo-*

sis replicates very slowly or, in the case of latent infection, not at all. This is also one reason treatment consists of several antibacterial drugs taken for 6 to 24 months. A new approach to treating tuberculosis is suggested by the development of a new agent, called PA-824, which mimics host immunity by stimulating the production of reactive nitrogen species within the pathogen. PA-824 is a nitroimidazole; this class of compounds is inactive until modified by a bacterial enzyme. When PA-824 is converted to its active form, des-nitroimidazole (des-nitro), nitric oxide, and other reactive nitrogen species are generated. These then damage bacterial DNA and enzymes and are bacteriocidal, even in nonreplicating cells.

What host immune process does PA-824 mimic? How does it differ from this process? There are several factors that make PA-824 a good candidate for further development. What do you think they are?

Read the original paper: Singh, R. 2008. PA-824 kills nonreplicating *Mycoplasma tuberculosis* by intracellular NO release. *Science* 322:1392.

8. Antibiotics that target bacterial molecules not previously exploited are desperately needed. One such target is the protein FtsZ. The small molecule 3-methoxybenzamide (3-MBA) is known to inhibit FtsZ in *Bacillus subtilis* but is not bacteriocidal. Nonetheless, researchers reasoned that 3-MBA offered a good starting point for the synthesis of a molecule that might be a potential drug candidate. Over 500 3-MBA analogs were synthesized and screened; one called PA190723 was extremely potent in its capacity to bind FtsZ and inhibit bacterial growth. In fact, when used in a mouse model, PA190723 was bacteriocidal against methicillin- and multidrug-resistant *Staphylococcus aureus*.

   What makes FtsZ a good drug target (*see chapter 7*)? What preliminary information about 3-MBA would be helpful if you were designing the 3-MBA analogs? As these researchers move forward with clinical (human) testing, what other parameters and outcomes must be assessed besides the bacteriocidal activity of PA190723?

   Read the original paper: Haydon, D. J., et al. 2008. An inhibitor of FtsZ with potent and selective anti-staphylococcal activity. *Science* 321:1673.

Concept Mapping Challenge

Map the following concepts to show your understanding of their relatedness.

Antimicrobial drug
Broad spectrum
Kirby-Bauer test
Mobile genetic element
Drug penetration
Drug resistance
Selective toxicity
Sensitivity testing

Learn More

Learn more by visiting the text website at www.mhhe.com/willey8, where you will find a complete list of references.

35

Clinical Microbiology and Immunology

The major objective of the clinical microbiologist is to isolate and identify pathogens from clinical specimens rapidly. In this illustration, a clinical microbiologist is picking up suspect bacterial colonies for biochemical, immunologic, or molecular testing.

CHAPTER GLOSSARY

agglutination reaction The formation of an insoluble immune complex by the cross-linking of cells or particles.

bacteriophage (phage) typing Identification of bacterial strains based on their susceptibility to specific phages.

direct immunofluorescence Visualization technique using fluorescently labeled antibodies to detect antigens.

enzyme-linked immunosorbent assay (ELISA) A serological assay in which bound antigen or antibody is detected by an antibody that is conjugated to an enzyme. The enzyme converts a colorless substrate to a colored product reporting the antibody capture of the antigen.

flow cytometry A tool for defining and enumerating cells. Fluorescently labeled cells pass single file through a capillary tube. Laser light detectors connected to a computer analyze the light patterns to identify cells.

hemadsorption The adherence of red blood cells to the surface of something, such as another cell or a virus.

immunoblotting The electrophoretic transfer of proteins from polyacrylamide gels to filters to demonstrate the presence of specific proteins through reaction with labeled antibodies.

immunodiffusion A technique involving the diffusion of antigen and antibody within a semisolid gel to produce a precipitin reaction in which they meet in proper proportions.

immunoelectrophoresis The electrophoretic separation of protein antigens followed by diffusion and precipitation in gels using antibodies against the separated proteins.

immunofluorescence A technique used to identify particular antigens microscopically in cells or tissues by the binding of a fluorescent antibody conjugate.

immunoprecipitation A reaction involving soluble antigens reacting with antibodies to form a large aggregate that precipitates out of solution.

indirect immunofluorescence A visualization technique using fluorescently labeled antibodies to detect the presence of other antibodies made in response to a specific antigen.

monoclonal antibody (mAb) An antibody of a single type that is produced by a population of genetically identical plasma cells.

plasmid fingerprinting A technique used to identify microbial isolates as belonging to the same strain because they contain the same plasmids as demonstrated by endonuclease digestion, which generates fragments of the same molecular weight.

Quellung reaction The increase in visibility or the swelling of the capsule of a microorganism in the presence of antibodies against capsular antigens.

radioimmunoassay (RIA) A very sensitive assay that uses a purified radioisotope-labeled antigen or antibody to compete for antibody or antigen with unlabeled standard antigen or test antigen in experimental samples to determine the concentration of a substance in the samples.

ribotyping The identification of bacterial strains based on the nucleotide sequence of the rRNA.

serotyping A technique used to differentiate between strains (serovars or serotypes) of microorganisms that have differences in antigenic composition.

viral hemagglutination The clumping or agglutination of red blood cells caused by some viruses.

Pathogens, particularly bacteria and yeasts, coexist with harmless microorganisms on or in the host. The challenge for clinical microbiologists is to identify these pathogens as the cause of infectious diseases. A variety of approaches may be used based on morphological, biochemical, immunologic, and molecular procedures. Time is a significant factor, especially in life-threatening situations, and advances in technology continue to increase the speed at which definitive identifications can be made. In the absence of a culture, polymerase chain reaction (PCR) and immunologic tests can be used to detect pathogens in specimens such as blood and sputum. In the final analysis, the patient's well-being and health can benefit significantly from information provided by the clinical laboratory—the subject of this chapter.

35.1 Overview of the Clinical Microbiology Laboratory

The major goal of the **clinical microbiologist** is to isolate and rapidly identify pathogenic microorganisms from clinical specimens. The purpose of the clinical laboratory is to provide the physician with information concerning the presence or absence of microorganisms that may be involved in the infectious disease process. Clinical microbiology is interdisciplinary, and the clinical microbiologist must have a working knowledge of microbial biochemistry and physiology, immunology, molecular biology, genomics, and the dynamics of host-parasite relationships. Importantly, tests developed to exploit the antigen-antibody binding capabilities, the focus of clinical immunology, can often detect microorganisms in specimens by identifying microbial antigens and quantifying the type and amount of responding antibody (**figure 35.1**).

Many different types of laboratories evaluate clinical specimens. Clinical laboratories range in size and function. A laboratory can be as small as the cubicle in a doctor's office or clinic where urine is checked for protein using a commercial "dip stick" or as large as the reference laboratories often found in state health departments or "teaching" hospitals. Most clinical laboratories are relatively mid-sized, sufficient to support three to eight medical technologists and most of the equipment for basic tests and procedures. Regardless of the laboratory size, the tests and procedures that we describe next are not used by all clinical laboratories. They are nonetheless important in the diagnostic process, regardless of which laboratory uses them.

In clinical microbiology, a clinical specimen represents a portion or quantity of human material that is tested, examined, or studied to determine the presence or absence of particular microorganisms (**figure 35.2**). Safety for the patients, hospital, and laboratory staff is of utmost importance. The guidelines (also known as universal precautions) presented in **Techniques & Applications 35.1** were established by the Centers for Disease Control and Prevention (CDC) to address areas of personal protection and specimen handling. Other important concerns regarding specimens need emphasis:

1. The specimen selected should adequately represent the diseased area and also may include additional sites (e.g., urine and blood specimens) to isolate and identify potential agents of the particular disease process.
2. A quantity of specimen adequate to allow a variety of diagnostic testing should be obtained.
3. Attention must be given to specimen collection to avoid contamination from the many varieties of microorganisms indigenous to the skin, mucous membranes, and environment (*see figure 30.16*).
4. The specimen should be collected in appropriate containers and forwarded promptly to the clinical laboratory.
5. If possible, the specimen should be obtained before antimicrobial agents have been administered to the patient.

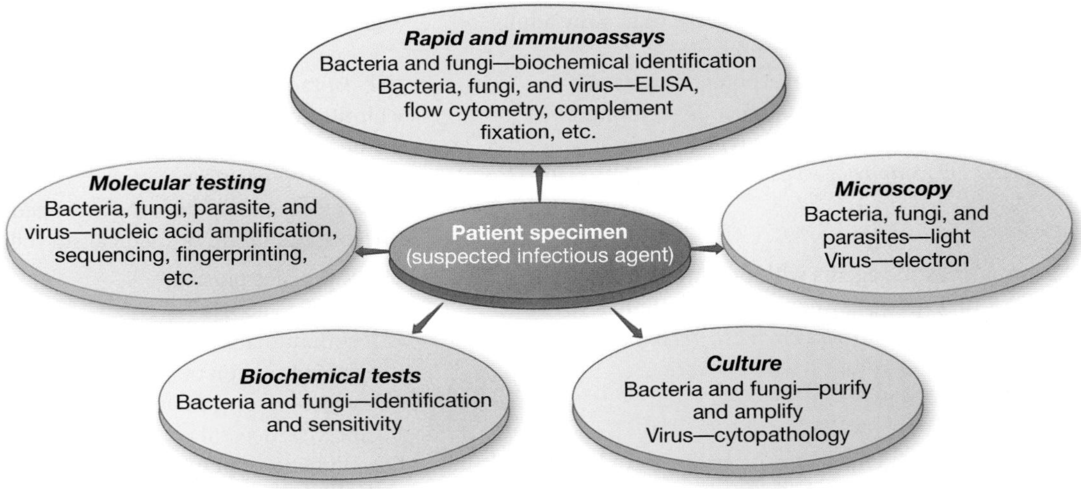

FIGURE 35.1 The Patient Specimen Is Evaluated by Various Techniques. The specimen source and the patient history guide the decision for use of tests and techniques. Not all tests are performed on all specimens, but all laboratories need to be able to perform these tests or refer them to other laboratories.

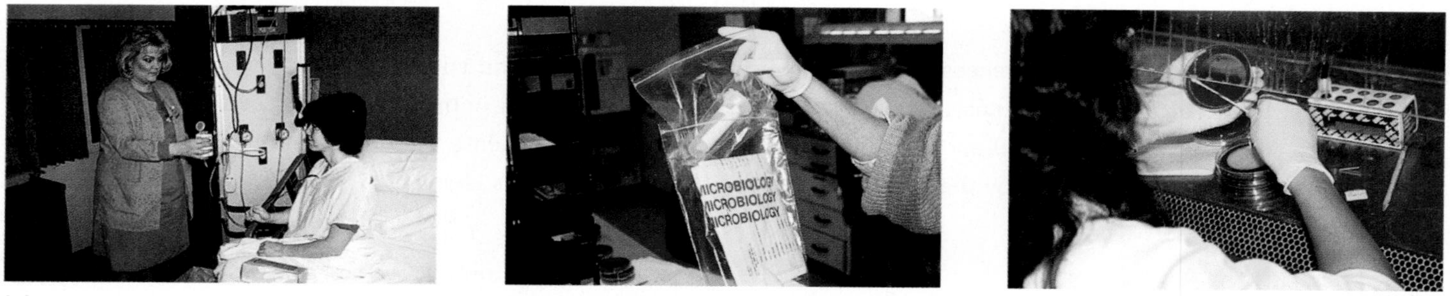

(a) (b) (c)

FIGURE 35.2 Collection of Patient Specimen for Microorganism Testing in a Clinical Laboratory. (a) The identification of the microorganism begins at the patient's beside. The nurse is giving instructions to the patient on how to obtain a sputum specimen. (b) The specimen is sent to the laboratory to be processed. Notice that the specimen and worksheet are in different bag compartments. (c) Patient specimens are tested in the laboratory to identify infectious agents and determine their sensitivity to antimicrobial drugs.

TECHNIQUES & APPLICATIONS

35.1 Standard Microbiological Practices

The identification of potentially fatal, blood-borne infectious agents (HIV, *Hepatitis B and viruses*, and others) spurred the codification of standard microbiological practices to limit exposure to such agents. These standard microbiological practices are minimum guidelines that should be supplemented with other precautions based on the potential exposure risks and biosafety level regulations for the lab. Briefly:

1. Eating, drinking, manipulation of contact lenses, and the use of cosmetics, gum, and tobacco products are strictly prohibited in the lab.
2. Hair longer than shoulder length should be tied back. Hands should be kept away from face at all times. Items (e.g., pencils) should not be placed in the mouth while in the lab. Protective clothing (lab coat, smock, etc.) is recommended while in the lab. Exposed wounds should be covered and protected.
3. Lab personnel should know how to use the emergency eyewash and shower stations.
4. Work space should be disinfected at the beginning and completion of lab time. Hands should be washed thoroughly after any exposure and before leaving the lab.
5. Precautions should be taken to prevent injuries caused by sharp objects (needles, scalpels, etc.). Sharp instruments should be discarded for disposal in specially marked containers.

Recommended guidelines for additional precautions should reflect the laboratory's biosafety level (BSL). The following table defines the BSL for the four categories of biological agents and suggested practices.

BSL	Agents	Practices
1	Not known to consistently cause disease in healthy adults (e.g., *Lactobacillus casei, Vibrio fischeri*)	Standard Microbiological Practices
2	Associated with human disease, potential hazard if percutaneous injury, ingestion, mucous membrane exposure occurs (e.g., *Salmonella typhi, E. coli* O157:H7, *Staphylococcus aureus*)	BSL-1 practice plus: • Limited access • Biohazard warning signs • "Sharps" precautions • Biosafety manual defining any needed waste decontamination or medical surveillance policies
3	Indigenous or exotic agents with potential for aerosol transmission; disease may have serious or lethal consequences (e.g., *Coxiella burnetti, Yersinia pestis,* herpesviruses)	BSL-2 practice plus: • Controlled access • Decontamination of all waste • Decontamination of lab clothing before laundering • Baseline serum values determined in workers using BSL-3 agents
4	Dangerous/exotic agents that pose high risk of life-threatening disease, aerosol-transmitted lab infections; or related agents with unknown risk of transmission (e.g., variola major (smallpox virus), Ebola virus, hemorrhagic fever viruses)	BSL-3 practices plus: • Clothing change before entering • Shower on exit • All material decontaminated on exit from facility

35.2 Identification of Microorganisms from Specimens

The clinical specimen is an unknown, and the microbiology laboratory scientists are the detectives who can identify the microorganism(s) in the specimen; they determine the cause of a patient's infection. The clinical microbiology laboratory provides preliminary or definitive identification of microorganisms using various tests and procedures that have the highest probability of rapid identification based on (1) microscopic examination of specimens, (2) study of the growth and biochemical characteristics of isolated microorganisms (pure cultures), (3) rapid and automated detection of unique microbial signatures, (4) bacteriophage typing (restricted to research settings and the CDC), and (5) molecular methods. Choosing the appropriate test or procedure is determined by the specimen and what is generically expected to be in the specimen based on the patient history.

Microscopy

Light microscopy is used to image organisms that are typically larger than 0.5 μm, for example, protozoa, fungi, and bacteria. Electron microscopy is used to see things that are much smaller, such as viruses. Some morphological and genetic features used in classification and identification of microorganisms are presented in section 17.3 and tables 17.1 and 17.2. Standard references, such as the *Manual of Clinical Microbiology* published by the American Society for Microbiology, provide details about reagents and staining procedures.

Bacteria

Wet-mount, heat-fixed, or chemically fixed specimens can be examined with an ordinary bright-field microscope. Examination can be enhanced with either phase-contrast or dark-field microscopy. The latter is the procedure of choice for the detection of spirochetes in skin lesions associated with early syphilis or Lyme disease. The fluorescence microscope can be used to identify any microorganism after it is stained with fluorochromes such as acridine orange, which stains nucleic acids, or any fluorochrome-labeled antibody that binds to specific microbial antigens. Many stains that can be used to examine specimens for specific microorganisms have been described. Two of the more widely used bacterial stains are the Gram stain and the acid-fast stain (*see figure 2.18*). Because these stains are based on the chemical composition of cell walls, they are not useful in identifying bacteria without cell walls (e.g., mycoplasmas).

The use of fluorescently labeled monoclonal antibodies in microscopic techniques has been an important breakthrough in diagnostic microscopy. In the 1980s immunlogists created hybrid cells (hybridomas) that live a very long time and secrete antibodies (*see Techniques & Applications 33.1*). Recall that each hybridoma cell and its progeny normally produce a **monoclonal antibody (mAb)** of a single specificity that can bind or capture the antigen to which it was produced. These antigen-capture antibodies recognize a single epitope and are therefore used for diagnostics. One such cross-cutting method, known as **immunofluorescence** or **immunohistochemistry**, is used to detect a variety of microorganisms and results from the chemical attachment of fluorescent dyes called fluorochromes to mAbs; the mAb then binds to a single epitope and the fluorescent molecule "reports" that binding. This technique can be used to tag specific microorganisms in a clinical specimen. Examples of commonly used fluorochromes include rhodamine B and fluorescein isothiocyanate (FITC), which can be coupled to antibody molecules without changing the antibody's capacity to bind to a specific antigen. In the clinical microbiology laboratory, fluorescently labeled mAbs to viral or bacterial antigens have replaced polyclonal antisera for use in culture confirmation when accurate, rapid identification is required. With the use of sensitive techniques such as fluorescence microscopy, it is possible to perform antibody-based microbial identifications with improved accuracy, speed, and fewer organisms. ◀◀ *Antibodies (section 33.7)*

Two main kinds of fluorescent antibody assays are used: direct and indirect. **Direct immunofluorescence** involves fixing the specimen (cell or microorganism) containing the antigen of interest onto a slide (**figure 35.3*a***). Fluorochrome-labeled antibodies are then added to the slide and incubated. The slide is washed to remove any unbound antibody and examined with the fluorescence microscope (*see figure 2.13*) for fluorescence. The pattern of fluorescence reveals the antigen's location. Direct immunofluorescence is used to identify antigens such as those found on the surface of group A streptococci and to diagnose enteropathogenic *Escherichia coli, Neisseria meningitidis, Salmonella* spp., *Shigella sonnei, Listeria monocytogenes,* and *Haemophilus influenzae* type b.

Indirect immunofluorescence (figure 35.3*b*) is used to detect the presence of antibodies in serum following an individual's exposure to microorganisms. In this technique, a known antigen (e.g., a virus) is fixed onto a slide. The patient serum containing antibodies is then added, and if the specific antibody is present, it reacts with antigen to form a complex. When a second, fluorescein-labeled antibody (not from the patients' serum) is added, it reacts with the fixed antibody. After incubation and washing, the slide is examined with the fluorescence microscope. The occurrence of fluorescence shows that the antibody specific to the (viral) antigen is present in the serum and that its presence is reported by the fluorescence of the secondary

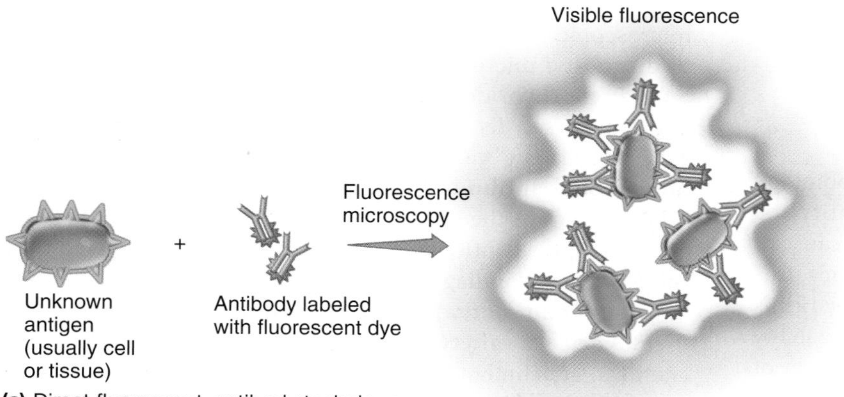

(a) Direct fluorescent–antibody technique

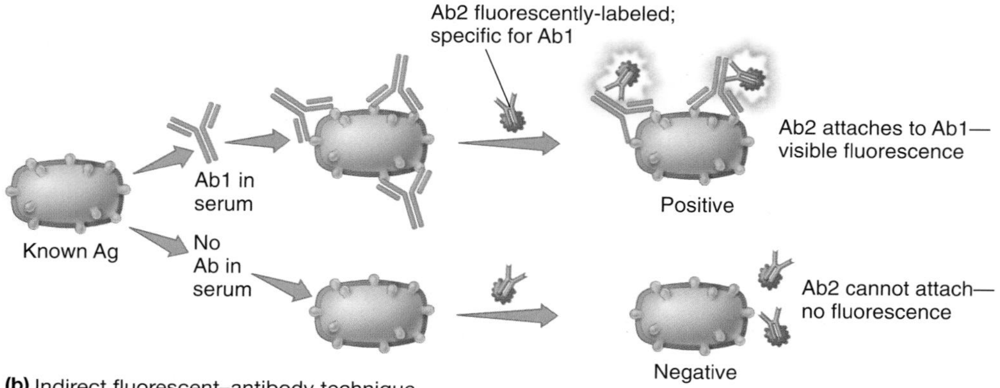

(b) Indirect fluorescent–antibody technique

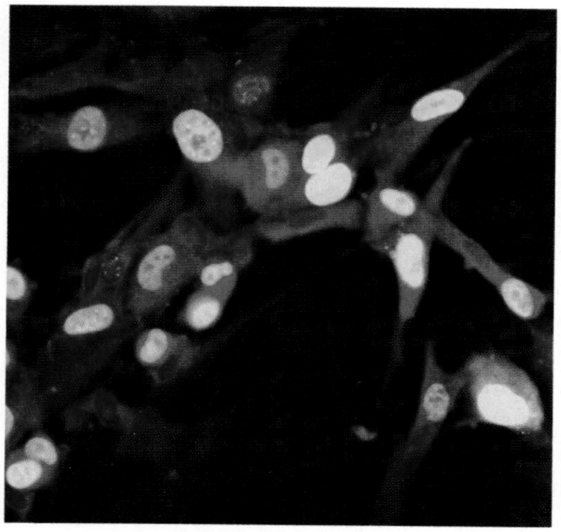

(c)

FIGURE 35.3 Direct and Indirect Immunofluorescence. (a) In the direct fluorescent antibody (DFA) technique, the specimen containing the antigen is fixed to a slide. Fluorescently labeled antibodies that recognize the antigen are then added, and the specimen is examined with a fluorescence microscope for fluorescence. (b) The indirect fluorescent antibody technique (IFA) detects antigen on a slide as it reacts with an antibody directed against it. The antigen-antibody complex is located with a fluorescent antibody that recognizes antibodies. (c) A photomicrograph of cells infected with cytomegalovirus (green fluorescence) and adenovirus (yellow fluorescence) using an indirect immunofluorescent staining procedure.

Figure 35.3 Micro Inquiry

Which type of immunofluorescence would most likely be used to test a clinical sample for the presence of an intracellular pathogen such as *Hepatitis C virus*?

antibody that has attached to the serum antibody (figure 35.3c). A common application of indirect immunofluorescence is the identification of *Treponema pallidum* antibodies in the diagnosis of syphilis.

Chlamydiae can be demonstrated in tissues and cell scrapings with Giemsa staining, which detects the characteristic intracellular inclusion bodies (*see figure 19.12*). Immunofluorescent staining of tissues and cells with monoclonal antibody reagents is a more sensitive and specific means of diagnosis of chlamydial disease.

Fungi

Direct microscopic examination of most specimens suspected of containing fungi can be made by light microscopy as well. Identification of hyphae in clinical specimens is a presumptive positive result for fungal infection. Definitive identification of most fungi is based on the morphology of reproductive structures (spores). Lactophenol aniline (cotton) blue is typically used to stain fungi from cultures. Fungal infections (i.e., mold and yeast infections) often are diagnosed by direct microscopic examination of specimens using fluorescence. For example, the identification of molds often can be made if a portion of the specimen is mixed with a drop of 10% Calcofluor White stain on a glass slide. ◀◀ *Light microscopes (section 2.2)*

Parasites

Identification and characterization of ova, trophozoites, and cysts in the specimen result in the definitive diagnosis of protozoa infection. This is typically accomplished by direct microscopic evaluation of the clinical specimen. Typical histological staining of blood, negative staining of other body fluids, and immunofluorescence staining are routinely used in the identification of parasites. Concentrated wet mounts of blood, stool, or urine specimens can be examined microscopically for the presence of eggs, cysts, larvae, or vegetative cells of parasites. D'Antoni's iodine (1%) is often used to stain internal structures of parasites. Blood smears for apicomplexan (malaria) and flagellate (trypanosome) parasites are stained with Giemsa.

Viruses

Most viruses are too small to be visualized by light microscopy alone. Fluorescence microscopy using fluorescently labeled antibodies is used for some virus detection. For high magnification and resolution, viruses must be imaged by electron microscopy. Specialized fixing and staining procedures are also used to prepare viruses for electron microscopy. However, very few clinical laboratories have access to electron microscopes, except those associated with large hospitals or research departments.

Growth and Biochemical Characteristics

For those microorganisms that can be grown in culture and identified by particular growth patterns and biochemical characteristics, specific tests are used. These tests vary, depending on whether the clinical microbiologist is dealing with viruses, fungi (yeasts, molds), parasites (protozoa, helminths), common gram-positive or gram-negative bacteria, rickettsias, chlamydiae, or mycoplasmas.

Bacteria

Isolation and growth of bacteria are required before many diagnostic tests can be used to confirm the identification of the pathogen. The presence of bacterial growth usually can be recognized by the development of colonies on solid media or turbidity in liquid media. The time needed for visible growth to occur is an important variable in the clinical laboratory. For example, most pathogenic bacteria require only a few hours to produce visible growth, whereas it may take weeks for colonies of mycobacteria or mycoplasmas to become evident. The clinical microbiologist as well as the clinician should be aware of reasonable reporting times for various cultures.

The initial identity of a bacterial organism may be suggested by (1) the source of the culture specimen; (2) its microscopic appearance and gram reaction; (3) its pattern of growth on selective, differential, or metabolism-determining media; and (4) its hemolytic, metabolic, and fermentative properties on the various media (**table 35.1**; *see also table 20.7*). For example, methylene blue is often used to inhibit the growth of gram-positive bacteria, whereas phenylethyl alcohol is often used to inhibit gram-negative bacteria. Sheep blood–supplemented agars can be used to determine hemolytic capabilities (*see figure 21.20*).

After the microscopic and growth characteristics of a pure culture of bacteria are examined, specific biochemical tests can be performed. Some of the most common biochemical tests used to identify bacterial isolates are listed in **table 35.2**. Classic dichotomous keys are coupled with the biochemical tests for the identification of bacteria from specimens. Generally,

Table 35.1 Isolation of Pure Bacterial Cultures from Specimens

Selective Media
A selective medium is prepared by the addition of specific substances to a culture medium that will permit growth of one group of bacteria while inhibiting growth of some other groups. These are examples:
Salmonella-Shigella agar (SS) is used to isolate *Salmonella* and *Shigella* species. Its bile salt mixture inhibits many groups of coliforms. Both *Salmonella* and *Shigella* species produce colorless colonies because they are unable to ferment lactose. Lactose-fermenting bacteria will produce pink colonies.
Mannitol salt agar (MS) is used for the isolation of staphylococci. The selectivity is obtained by the high (7.5%) salt concentration that inhibits growth of many groups of bacteria. The mannitol in this medium helps in differentiating the pathogenic from the nonpathogenic staphylococci, as the former ferment mannitol to form acid while the latter do not. Thus this medium is also differential.
Bismuth sulfite agar (BS) is used for the isolation of *Salmonella enterica* serovar Typhi, especially from stool and food specimens. *S. enterica* serovar Typhi reduces the sulfite to sulfide, resulting in black colonies with a metallic sheen.
Hektoen enteric agar is used to increase the yield of *Salmonella* and *Shigella* species relative to other microbiota. The high bile salt concentration inhibits the growth of gram-positive bacteria and retards the growth of many coliform strains.
Differential Media
The incorporation of certain chemicals into a medium may result in diagnostically useful growth or visible change in the medium after incubation. These are examples:
Eosin methylene blue agar (EMB) differentiates between lactose fermenters and nonlactose fermenters. EMB contains lactose, salts, and two dyes—eosin and methylene blue. *E. coli,* which is a lactose fermenter, will produce a dark colony or one that has a metallic sheen. *S. enterica* serovar Typhi, a nonlactose fermenter, will appear colorless.
MacConkey agar is used for the selection and recovery of *Enterobacteriaceae* and related gram-negative rods. The bile salts and crystal violet in this medium inhibit the growth of gram-positive bacteria and some fastidious gram-negative bacteria. Because lactose is the sole carbohydrate, lactose-fermenting bacteria produce colonies that are various shades of red, whereas nonlactose fermenters produce colorless colonies.
Blood agar: addition of citrated blood to tryptic soy agar makes possible variable hemolysis, which permits differentiation of some species of bacteria. Three hemolytic patterns can be observed on blood agar. 1. α-hemolysis—greenish to brownish halo around the colony (e.g., *Streptococcus gordonii, Streptococcus pneumoniae*) 2. β-hemolysis—complete lysis of blood cells resulting in a clearing effect around the colony (e.g., *Staphylococcus aureus* and *Streptococcus pyogenes*) 3. Nonhemolytic—no change in medium (e.g., *Staphylococcus epidermidis* and *Staphylococcus saprophyticus*)
Media to Determine Biochemical Reactions
Some media are used to test bacteria for particular metabolic activities, products, or requirements. These are examples:
Urea broth is used to detect the enzyme urease. Some enteric bacteria are able to break down urea, using urease, into ammonia and CO_2. Media turns pink if urease is present.
Triple sugar iron (TSI) agar contains lactose, sucrose, and glucose plus ferrous ammonium sulfate and sodium thiosulfate. TSI is used for the identification of enteric organisms based on their ability to metabolize glucose, lactose, or sucrose, and to liberate sulfides from ammonium sulfate or sodium thiosulfate. Acid reactions turn media yellow. Basic reactions turn media orange to red.
Citrate agar contains sodium citrate, which serves as the sole source of carbon, and ammonium phosphate, the sole source of nitrogen. Citrate agar is used to differentiate enteric bacteria on the basis of citrate utilization. Citrate utilization turns media blue.
Lysine iron agar (LIA) is used to differentiate bacteria that can either deaminate or decarboxylate the amino acid lysine. LIA contains lysine, which permits enzyme detection, and ferric ammonium citrate for the detection of H_2S production. Lysine metabolism is indicated by red-purple media.
Sulfide, indole, motility (SIM) medium is used for three different tests. One can observe the production of sulfides, formation of indole (a metabolic product from tryptophan utilization), and motility. This medium is generally used for the differentiation of enteric organisms. Sulfide production turns media black. Indole formation results in a red color on top of media upon addition of Kovacs reagent.

Table 35.2 Some Common Biochemical Tests Used by Clinical Microbiologists in the Diagnosis of Bacteria from a Patient's Specimen

Biochemical Test	*Description*	*Laboratory Application*
Carbohydrate fermentation	Acid and/or gas are produced during fermentative growth with sugars or sugar alcohols.	Fermentation of specific sugars used to differentiate enteric bacteria as well as other genera or species
Casein hydrolysis	Detects the presence of caseinase, an enzyme able to hydrolyze milk protein casein. Bacteria that use casein appear as colonies surrounded by a clear zone.	Used to cultivate and differentiate aerobic actinomycetes based on casein utilization. For example, *Streptomyces* uses casein and *Nocardia* does not.
Catalase	Detects the presence of catalase, which converts hydrogen peroxide to water and O_2	Used to differentiate *Streptococcus* (−) from *Staphylococcus* (+) and *Bacillus* (+) from *Clostridium* (−)
Citrate utilization	When citrate is used as the sole carbon source, this results in alkalinization of the medium.	Used in the identification of enteric bacteria. *Klebsiella* (+), *Enterobacter* (+), *Salmonella* (often +); *Escherichia* (−), *Edwardsiella* (−)
Coagulase	Detects the presence of coagulase. Coagulase causes plasma to clot.	This is an important test to differentiate *Staphylococcus aureus* (+) from *S. epidermidis* (−).
Decarboxylases (arginine, lysine, ornithine)	The decarboxylation of amino acids releases CO_2 and amine.	Used in the identification of enteric bacteria
Esculin hydrolysis	Tests for the cleavage of a glycoside	Used in the differentiation of *Staphylococcus aureus*, *Streptococcus mitis*, and others (−) from *S. bovis*, *S. mutans*, and enterococci (+)
β-galactosidase (ONPG) test	Demonstrates the presence of an enzyme that cleaves lactose to glucose and galactose	Used to separate enterics (*Citrobacter* +, *Salmonella* −) and to identify pseudomonads
Gelatin liquefaction	Detects whether or not a bacterium can produce proteases that hydrolyze gelatin and liquify solid gelatin medium	Used in the identification of *Clostridium*, *Serratia*, *Pseudomonas*, and *Flavobacterium*
Hydrogen sulfide (H_2S)	Detects the formation of hydrogen sulfide from the amino acid cysteine due to cysteine desulfurase	Important in the identification of *Edwardsiella*, *Proteus*, and *Salmonella*
IMViC (indole; methyl red; Voges-Proskauer; citrate)	The indole test detects the production of indole from the amino acid tryptophan. Methyl red is a pH indicator to determine whether the bacterium carries out mixed acid fermentation. VP (Voges-Proskauer) detects the production of acetoin. The citrate test determines whether or not the bacterium can use sodium citrate as a sole source of carbon.	Used to separate *Escherichia* (MR+, VP−, indole +) from *Enterobacter* (MR−, VP+, indole−) and *Klebsiella pneumoniae* (MR−, VP+, indole−); also used to characterize members of the genus *Bacillus*
Lipid hydrolysis	Detects the presence of lipase, which breaks down lipids into simple fatty acids and glycerol	Used in the separation of clostridia
Nitrate reduction	Detects whether a bacterium can use nitrate as an electron acceptor	Used in the identification of enteric bacteria, which are usually +
Oxidase	Detects the presence of cytochrome *c* oxidase that is able to reduce O_2 and artificial electron acceptors	Important in distinguishing *Neisseria* and *Moraxella* spp. (+) from *Acinetobacter* (−) and enterics (all −) from pseudomonads (+)
Phenylalanine deaminase	Deamination of phenylalanine produces phenylpyruvic acid, which can be detected colorimetrically.	Used in the characterization of the genera *Proteus* and *Providencia*
Starch hydrolysis	Detects the presence of the enzyme amylase, which hydrolyzes starch	Used to identify typical starch hydrolyzers such as *Bacillus* spp.
Urease	Detects the enzyme that splits urea to NH_3 and CO_2	Used to distinguish *Proteus*, *Providencia rettgeri*, and *Klebsiella pneumoniae* (+) from *Salmonella*, *Shigella* and *Escherichia* (−)

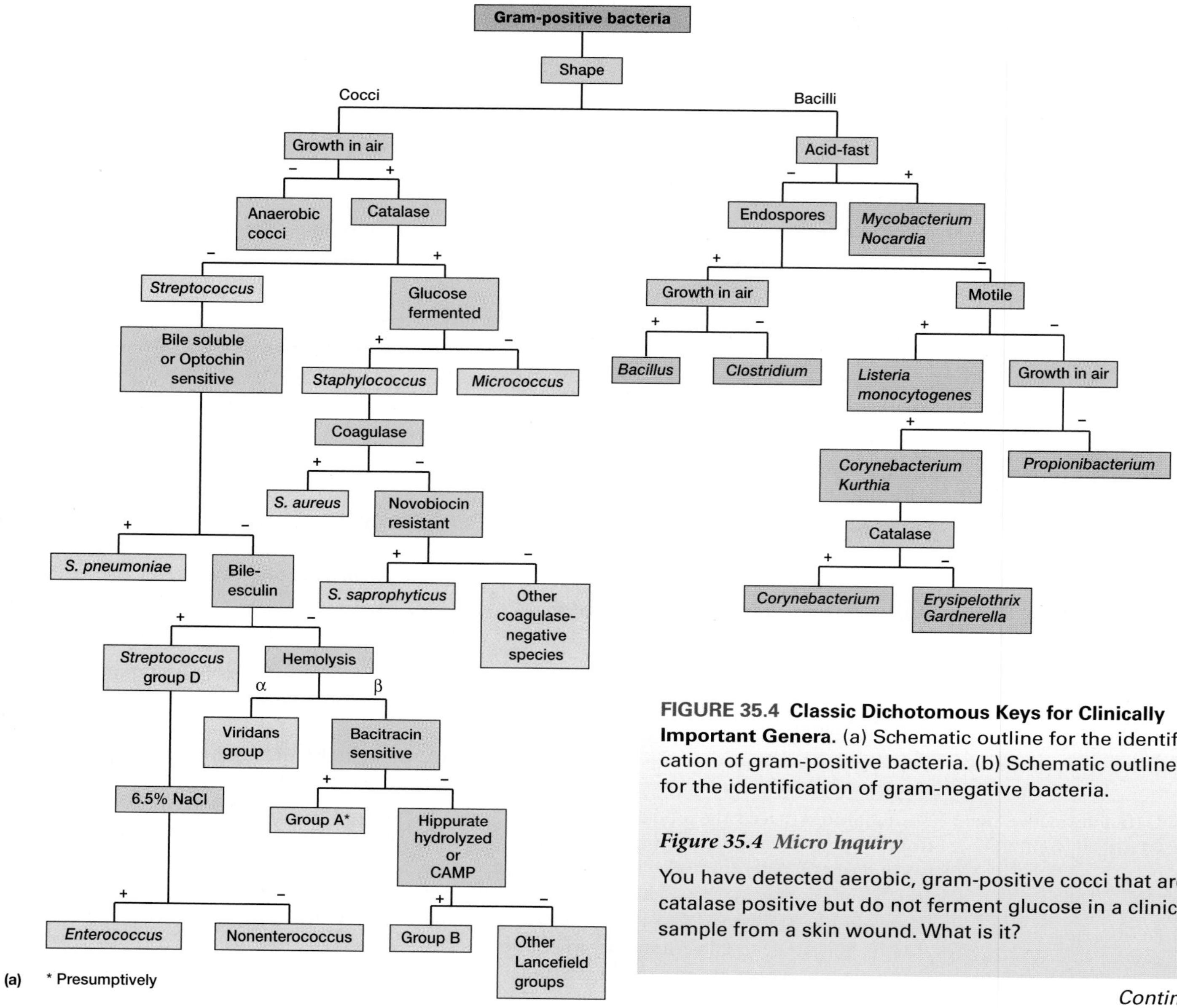

FIGURE 35.4 Classic Dichotomous Keys for Clinically Important Genera. (a) Schematic outline for the identification of gram-positive bacteria. (b) Schematic outline for the identification of gram-negative bacteria.

Figure 35.4 Micro Inquiry

You have detected aerobic, gram-positive cocci that are catalase positive but do not ferment glucose in a clinical sample from a skin wound. What is it?

Continued

fewer than 20 tests are required to identify clinical bacterial isolates to the species level (**figure 35.4**).

Certain bacteria require special considerations. For instance, the rickettsias, chlamydiae, and mycoplasmas differ from other bacterial pathogens in a variety of ways. They can be diagnosed by immunoassays or by isolation of the microorganism. Because isolation is both hazardous and expensive, immunological methods are preferred. Isolation of rickettsias and diagnosis of rickettsial diseases are generally confined to reference and specialized research laboratories.

Fungi

Fungal cultures remain the standard for the recovery of fungi from patient specimens; however, the time needed to culture fungi varies anywhere from a few days to several weeks, depending on the organism. For this reason, fungal cultures demonstrating no growth should be maintained for a minimum of 30 days before they are discarded as a negative result. Cultures should be evaluated for rate and appearance of growth on at least one selective and one nonselective agar medium, with careful examination of colonial morphology, color, and dimorphism. Typically, the isolation

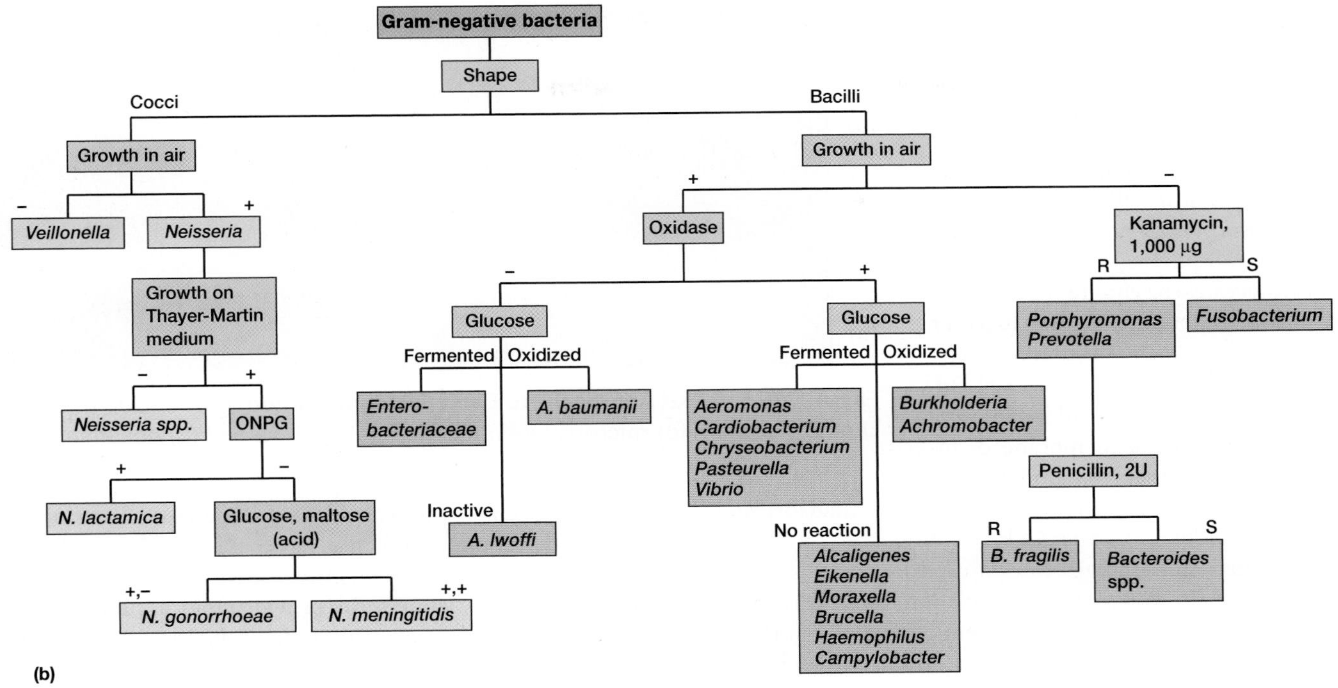

(b)

FIGURE 35.4 ***Continued***

Table 35.3 Examples of Media Used to Isolate Fungi

Culture Medium	*Target Fungi*
Antibiotic agar	Fungi in polymicrobial specimens
Caffeine agar	*Aspergillus, Rhizopus,* and others
Chlamydospore agar	Dimorphic fungi
Cornmeal agar	Most fungi including pathogens
Malt agar	Ascomycetes
Malt extract agar	Basidiomycetes
Potato dextrose-yeast extract agar	Mushrooms
Sabouraud dextrose agar (SAB)	Most fungi
SAB + chloramphenicol + cyclohexamide	Dimorphic fungi and dermatophytes

of fungi is accomplished by concurrent culture of the specimen on media that is respectively supplemented and unsupplemented with antibiotics and cycloheximide. Antibiotics inhibit bacteria that may be in the specimen and cycloheximide inhibits saprophytic (living on decaying matter) molds. However, a number of media formulations are routinely used to culture specific fungi (**table 35.3**).

Parasites

Culture of protozoa from clinical specimens is not routine. Thus detection and identification of these microorganisms is by microscopy, molecular techniques, or both.

Viruses

It is important to note that the culture of virus is not typically done in hospital labs. Viruses are grown and identified by isolation in conventional cell (tissue) culture in specialized labs equipped to provide the biosafety and biocontainment required for the specific virus. Hopsital labs, rather, identify virus in clinical specimens by immunodiagnosis (fluorescent antibody, enzyme immunoassay, radioimmunoassay, latex agglutination, and immunoperoxidase) and by molecular detection methods such as nucleic acid probes and PCR amplification assays. Several types of systems are available for virus cultivation: cell cultures, embryonated hen's eggs, and experimental animals; these are discussed shortly.

Cell cultures are divided into three general classes: (1) **Primary cultures** consist of cells derived directly from tissues such as monkey kidney and mink lung cells that have undergone one or two passages (subcultures) since harvesting. (2) **Semicontinuous cell cultures** or low-passage cell lines are obtained from subcultures of a primary culture and usually consist of diploid fibroblasts that undergo a finite number of divisions. (3) **Continuous or immortalized cell cultures,** such as HEp-2 cells, are derived from transformed cells that are generally epithelial in origin. These cultures grow rapidly, are heteroploid (having a chromosome number that is not a simple multiple of the haploid number), and can be subcultured indefinitely.

Each type of cell culture favors the growth of a different array of viruses, just as bacterial culture media have differing selective and restrictive properties for growth of bacteria. Viral replication in cell cultures is detected in two ways: (1) by observing the presence or absence of cytopathic effects (CPEs) and (2) by hemadsorption. A cytopathic effect is an observable morphological change that occurs in cells because of viral replication. Examples include ballooning, binding together, clustering, or even death of the culture cells. During the incubation period of a cell culture, red blood cells can be added. Several viruses alter the plasma membrane of infected culture cells so that red blood cells adhere firmly to them. This phenomenon is called **hemadsorption.** ⏮ *Viral multiplication (section 5.3)*

Embryonated chicken eggs can be used for virus isolation. There are three main routes of egg inoculation for virus isolation as different viruses grow best on different cell types: (1) the allantoic cavity, (2) the amniotic cavity, and (3) the chorioallantoic membrane. Egg tissues are inoculated with clinical specimens to determine the presence of virus; virus is revealed by the development of pocks on the chorioallantoic membrane, by the development of hemagglutinins in the allantoic and amniotic fluid, and by death of the embryo.

Laboratory animals, especially suckling mice, also may be used for virus isolation. Inoculated animals are observed for specific signs of disease or death. Several serological tests for viral identification make use of mAb-based immunofluorescence. These tests detect viruses such as herpes simplex virus in tissue cultures.

(a)
(b)

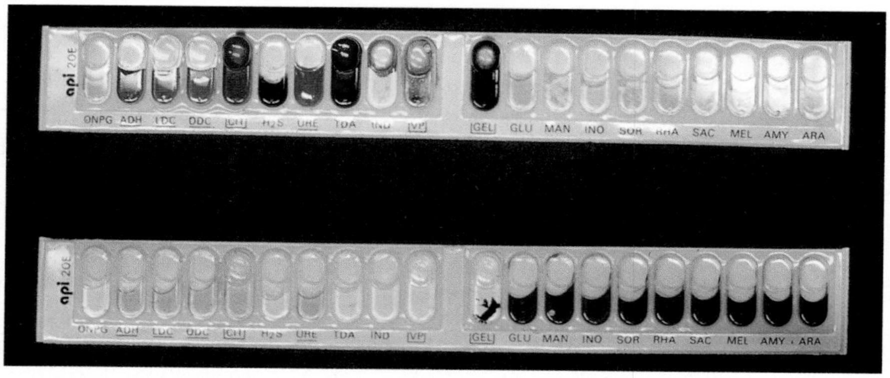

FIGURE 35.5 A "Kit Approach" to Bacterial Identification. The API 20E manual biochemical system for microbial identification. (a) Positive and (b) negative results.

1. Name two specimens for which microscopy would be used in the initial diagnosis of an infectious disease.
2. Name three general classes of cell cultures.
3. Explain two ways by which the presence of viral replication is detected in cell culture.
4. What are the advantages of using monoclonal antibody (mAb) immunofluorescence in the identification of viruses?
5. How can a clinical microbiologist determine the initial identity of a bacterium?
6. Describe a dichotomous key that could be used to identify a bacterium.
7. How can fungi and protozoa be detected in a clinical specimen? Rickettsias? Chlamydiae? Mycoplasmas?

Rapid Methods of Identification

Clinical laboratory scientists (medical technologists) are the trained and certified workforce that is the front line in laboratory-based disease detection. They staff the sentinel laboratories that receive patient specimens. The production of faster and more specific detection technologies has enabled medical technologists to rapidly and accurately identify disease agents. In this way, clinical microbiology laboratories provide rapid, accurate, and timely microbial identification and antimicrobial susceptibility results that assist clinicians in the diagnosis and treatment of infectious disease. Clinical microbiology has also benefited greatly from technological advances in equipment, computer software and databases, molecular biology, and immunochemistry. With new technology, it has been possible to shift from the multistep methods previously discussed to unitary procedures and systems that incorporate standardization, speed, reproducibility, miniaturization, mechanization, and automation. These rapid identification methods can be divided into three categories: (1) manual biochemical "kit" systems, (2) mechanized/automated systems, and (3) immunologic systems.

One example of a "kit approach" biochemical system for the identification of members of the family *Enterobacteriaceae* and other gram-negative bacteria is the API 20E system. It consists of a plastic strip with 20 microtubes containing dehydrated biochemical substrates that can detect certain biochemical characteristics (**figure 35.5**). The biochemical substrates in the 20 microtubes are inoculated with a pure culture of bacteria evenly suspended in sterile physiological saline. After 5 to 12 hours of incubation, the 20 test results are converted to a seven- or nine-digit profile. This profile number can be used with a computer or a book called the *API Profile Index* to identify the bacterium.

In addition to traditional techniques and kits, laboratories also use semiautomated microbial identification and susceptibility instruments. These instruments automate the incubation, reagent addition, and evaluation of test results from miniaturized versions of classical bench tests. Examples of these identification and susceptibility instruments include the Phoenix™ automated microbiology system, the VersaTREK®, and the MicroScan® WalkAway. These systems use carriers, reagents, and reporting components that are uniquely made for them. The typical identification and susceptibility test begins after the primary isolation of the organisms in pure culture. A saline suspension of the pure culture is adjusted to a specific density and is coded so as to be tracked throughout the various stages of evaluation. Aliquots of the bacterial suspension are then injected into various, miniaturized test chambers containing biochemical reagents. Colorimetric or fluorometric reactions resulting from each test are then evaluated against controls to report on the specific capabilities of the organism. The results of the individual reactions are evaluated together to predict organism identification, within a calculated confidence interval. Antimicrobial susceptibility is determined by evaluating organism growth when exposed to various antibiotics and reported by similar means. Thus the organism's identity and its sensitivity to a panel of antibiotics (a profile called its antibiogram) are determined, often within hours.

The rapidly growing discipline of immunology has greatly aided the clinical microbiologist. Numerous technologies now exist that exploit the specificity and sensitivity of monoclonal antibodies to detect and identify microorganisms. These are briefly discussed here and expanded upon in section 35.3. Monoclonal antibodies (mAbs) have many applications. For example, they are routinely used in the typing of tissue; in the identification and epidemiological study of infectious microorganisms, tumors, and other surface antigens; in the classification of leukemias; in the identification of functional populations of different types of T cells; and in the identification and mapping of antigenic determinants (epitopes) on proteins. Importantly, mAbs can be conjugated with molecules that provide colorimetric, fluorometric, or enzymatic activity to report the binding of the mAb to specific microbial antigens. Numerous microbial detection kits are available to screen clinical specimens for the presence of specific microorganisms (**table 35.4**). mAbs in these kits have been produced against a wide variety of bacteria, viruses, fungi, and protozoans, respectively. The mAbs are produced as cross-species or cross-genus isotypes, that is, they have reactivity against epitopes that are common to a number of species within a genus or a number of genera within a class, so as to be used as an adjunct method in the taxonomic identification of the microorganisms being detected. Importantly, those mAbs that define species-specific antigens are extremely valuable in diagnostic reagents, as they can detect the unique species of microorganism causing disease. mAbs that exhibit more restrictive specificity can be used to identify strains or biotypes within a species and in epidemiological studies involving the matching of microbial strains. Coupling sensitive visualization technologies such as fluorescence or scanning tunneling microscopy to mAb detection systems makes it possible to perform microbial identifications with improved accuracy, speed, and fewer organisms.

Importantly, the anthrax attacks in 2001 in the United States spawned a renewed demand for "better, faster, and smarter" microbial detection and identification technologies. While nucleic acid-based detection systems, such as PCR, have garnered much attention as the basis of newer detection systems, antibody-based identification technologies are still considered more flexible and easier to modify. Traditional antibody-based detection technologies are being linked to sophisticated reporting systems that provide "med techs" with an ever-increasing array of cutting-edge technology. Examples of more recent microbial identification technologies include **biosensors** based on: (1) microfluidic antigen sensors, (2) real-time (20-minute) PCR, (3) highly sensitive spectroscopy systems, and (4) liquid crystal amplification of microbial immune complexes. Some of these technologies are being used as part of military sentinel detection programs; others are awaiting approval by various licensing agencies before being deployed in clinical laboratories. Additional technologies are expected as the demand for immediate, highly sensitive microbial detection increases globally.

Table 35.4 Some Common Rapid Immunologic Test Kits for the Detection of Bacteria and Viruses in Clinical Specimens

Culturette Group A Strep ID Kit (Marion Scientific, Kansas City, Mo.)
The Culturette kit is used for the detection of group A streptococci from throat swabs.
Directigen (Hynson, Wescott, and Dunning, Baltimore, Md.)
The Directigen Meningitis Test kit is used to detect *H. influenzae* type b, *S. pneumoniae,* and *N. meningitidis* groups A and C.
The Directigen Group A Strep Test kit is used for the direct detection of group A streptococci from throat swabs.
Gono Gen (Micro-Media Systems, San Jose, Calif.)
The Gono Gen kit detects *Neisseria gonorrhoeae.*
OraQuick (OraSure Technologies, Bethlehem, Pa.)
Detects HIV antibodies in saliva in 10 minutes.
Staphaurex (Wellcome Diagnostics, Research Triangle Park, N.C.)
Staphaurex screens and confirms *Staphylococcus aureus* in 30 seconds.
SureCell Herpes (HSV) Test (Kodak, Rochester, N.Y.)
Detects the herpes (HSV) 1 and 2 viruses in minutes.

1. Describe in general how biochemical tests are used in the API 20E system to identify bacteria.
2. Why might cultures for some microorganisms be unavailable?
3. What are some of the benefits of monoclonal antibodies as diagnostic reagents?

Bacteriophage Typing

Bacteriophages are viruses that attack members of a particular bacterial species or strain within a species. Bacteriophage (phage) typing is based on the specificity of phage surface proteins for cell surface receptors. Only those bacteriophages that can attach to these surface receptors can infect bacteria and cause lysis. On a Petri dish culture, lytic bacteriophages cause plaques on lawns of sensitive bacteria. These plaques represent infection by the virus (*see figure 5.20*). ◀◀ *Viruses and other acellular infectious agents (chapter 5)*

In **bacteriophage typing,** the clinical microbiologist inoculates the bacterium to be tested onto a Petri plate. The plate is heavily and uniformly inoculated so that the bacteria will grow to form a solid lawn of cells. The plate is then marked off into squares (15 to 20 mm per side), and each square is inoculated with a drop of suspension from the different phages available for typing. After the plate is incubated for 24 hours, it is observed for plaques. The phage type is reported as a specific genus and species followed by the types that can infect the bacterium. For example, the series 10/16/24 indicates that this bacterium is sensitive to phages 10, 16, and 24, and belongs to a collection of strains, called a phagovar, that have this particular phage sensitivity. Bacteriophage typing remains a tool of research and reference laboratories.

Molecular Genetic Methods

Some of the most accurate approaches to microbial identification are through the analysis of proteins and nucleic acids. Examples include comparison of proteins; physical, kinetic, and regulatory properties of microbial enzymes; nucleic acid–base composition; nucleic acid hybridization; and nucleic acid sequencing (*see figures 15.6 and 16.2 to 16.6*). Other molecular methods being widely used are nucleic acid probes, real-time PCR amplification of DNA, and DNA fingerprinting. ◀◀ *Techniques for determining microbial taxonomy and phylogeny (section 17.3)*

Nucleic acid–based diagnostic methods for the detection and identification of microorganisms have become routine in clinical microbiology laboratories. For example, DNA hybridization technology can identify a microorganism by probing its genetic composition. The use of cloned DNA as a probe is based on the capacity of single-stranded DNA to bind (hybridize) with a complementary nucleic acid sequence present in test specimens to form a double-stranded DNA hybrid. These hybrids are more sensitive than conventional microbiological techniques, give results in 2 hours or less, and require the presence of fewer microorganisms. DNA probe sensitivity can be increased by over 1 million times if the target DNA is first amplified using PCR. PCR can also be monitored in "real time" so as to detect microbial DNA amplification after each replicative cycle, identifying the microorganisms after only 25 to 30 amplification cycles (**figure 35.6**). In other words, DNA amplification can be measured continuously using real-time (rt) PCR (*see figure 15.7*). The most sensitive methods for demonstrating chlamydiae in clinical specimens involve nucleic acid sequencing and PCR-based methods. ◀◀ *Polymerase chain reaction (section 15.2); Molecular characteristics (section 17.3)* *DNA Hybridization*

The nucleotide sequence of small subunit ribosomal RNA (rRNA) can be used to identify bacterial genera (*see figure 17.3*). Usually the rRNA encoding gene or gene fragment is amplified by PCR. After nucleotide sequencing, the rRNA gene nucleotide sequence is compared with those on international databases. This method of bacterial identification, called **ribotyping,** is based on the high level of 16s rRNA conservation among bacteria. As discussed in chapter 17, multiple "housekeeping" genes, instead of 16S rRNA genes, may be sequenced and analyzed in a technique called **multilocus sequence typing (MLST)** (*see p. 453*).

Genomic fingerprinting is also used in identifying pathogens. This does not involve nucleotide sequencing; rather, it compares the similarity of specific DNA fragments generated by restriction endonuclease digestion. BOX-, ERIC-, and REP-PCR are described in section 17.3 (*see figure 17.4*). Plasmids—autonomously replicating extrachromosomal molecules of DNA—can also be used in DNA fingerprinting. **Plasmid fingerprinting** identifies microbial isolates of the same or similar strains; related strains often contain the same plasmids (**figure 35.7*a***). In contrast, microbial isolates that are phenotypically distinct have different

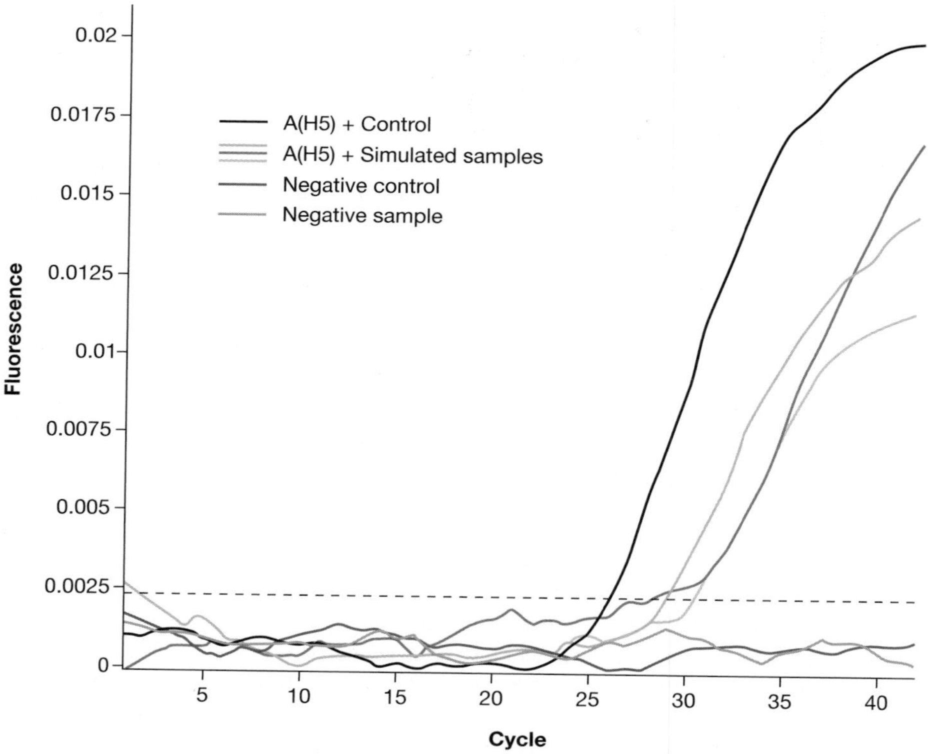

FIGURE 35.6 PCR Kit Results. Real-time detection of avian influenza is possible using a PCR kit. Note that sufficient DNA amplification occurs after 25 to 30 cycles.

plasmid fingerprints. Plasmid fingerprinting of many *E. coli*, *Salmonella*, *Campylobacter*, and *Pseudomonas* strains and species has demonstrated that this method often is more accurate than phenotyping methods such as biotyping, antibiotic resistance patterns, phage typing, and serotyping. Just as in genomic fingerprinting, isolated plasmid DNA is cut with specific restriction endonucleases (*see table 15.2*). The DNA fragments are then separated by gel electrophoresis to yield a pattern of fragments, which appear as bands in the gel. The molecular weight of each plasmid species can then be determined and patterns of restriction fragments compared. ◀◀ *Gel electrophoresis (section 15.3)*

1. What is the basis for bacteriophage typing?
2. How can nucleic acid–based detection methods be used by the clinical microbiologist?
3. How can a suspect bacterium be plasmid fingerprinted?

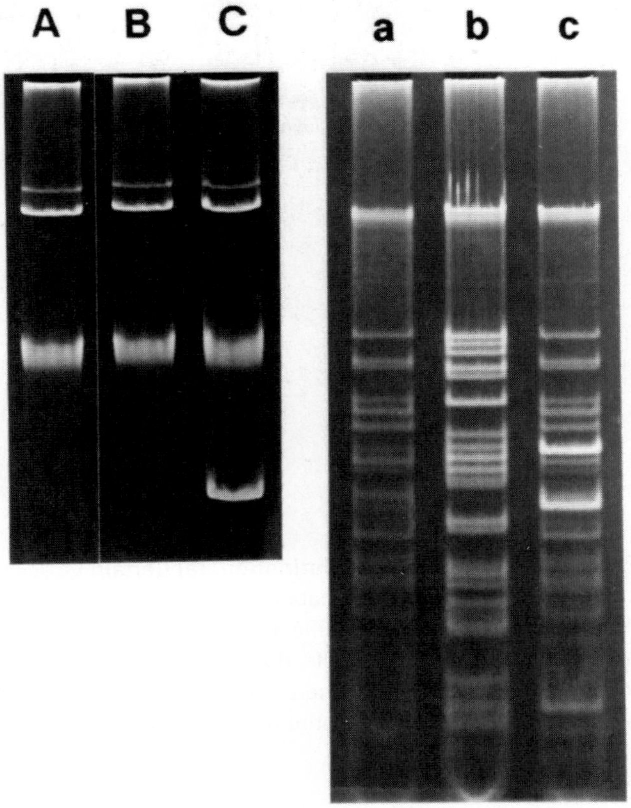

FIGURE 35.7 Plasmid Fingerprinting. Agarose gel electrophoresis of plasmid DNA. A, B, C: plasmids that have not been digested by endonucleases; a, b, c: the same plasmids following restriction enzyme digestion.

Figure 35.7* *Micro Inquiry

How many different plasmids are present in this analysis?

35.3 Clinical Immunology

The culturing of certain viruses, bacteria, fungi, and protozoa from clinical specimens may not be possible because the methodology remains undeveloped (e.g., *Treponema pallidum;* Hepatitis A, B, and C viruses; and Epstein-Barr virus), is unsafe (e.g., rickettsias and HIV), or is impractical for all but a few microbiology laboratories (e.g., mycobacteria, strict anaerobes, *Borrelia*). Cultures also may be negative because of prior antimicrobial therapy. Under these circumstances, detection of antibodies or antigens can be quite valuable diagnostically.

Immunologic systems for the detection and identification of pathogens from clinical specimens are easy to use, give relatively rapid reaction end points, and are sensitive and specific with a low percentage of false positives and negatives. Some of the more popular immunologic rapid test kits for viruses and bacteria are presented in table 35.4.

Dramatic advances in clinical immunology have given rise to a marked increase in the number, sensitivity, and specificity of serological tests. This increase reflects a better understanding of (1) immune cell surface antigens (CD antigens), (2) lymphocyte biology, (3) the production of monoclonal antibodies, and (4) the development of sensitive antibody-binding reporter systems. For a number of reasons, the utility of these tests depends on proper test selection and timing of specimen collection. For instance, each individual's immunologic response to a microorganism is quite variable, making the interpretation of immunologic tests potentially difficult. For example, a single, elevated IgG titer does not distinguish between active and past infections. Rather, an elevated IgM titer typically indicates an active infection, especially when subsidence of symptoms correlates with a fourfold (or greater) decrease in antibody titer. Furthermore, a lack of a measurable antibody titer may reflect an organism's lack of immunogenicity or insufficient time for an antibody response to develop following the onset of the infectious disease. Some patients are also immunosuppressed due to other disease processes or treatment procedures (e.g., cancer and AIDS patients) and therefore do not respond. In this section, some of the more common antibody-based techniques employed in the diagnosis of microbial and immunological diseases are discussed.

Serotyping

Serum is the liquid portion of blood (devoid of clotting factors) that contains many different components, especially the immunoglobulins or antibodies. **Serotyping** refers to the use of serum (antibodies) to specifically detect and identify other molecules. Serotyping can be used to identify specific white blood cells or the proteins on cell surfaces. Serotyping can also be used to differentiate strains (serovars or serotypes) of microorganisms that differ in the antigenic composition of a structure or product. The serological identification of a pathogenic strain has diagnostic value. Often the symptoms of infection depend on the nature of the cell products released by the pathogen. Therefore it is sometimes possible

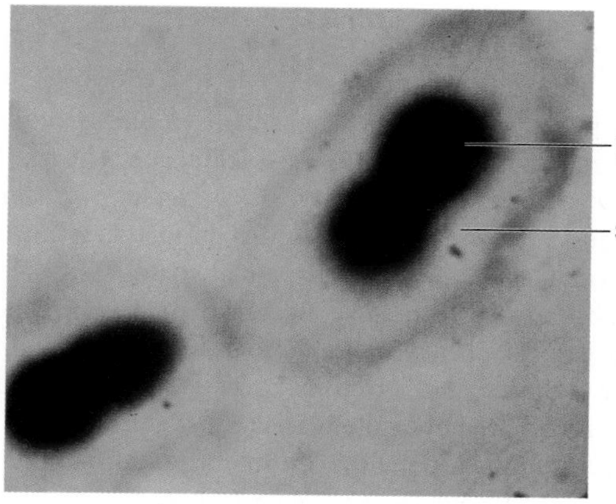

FIGURE 35.8 Serotyping. *Streptococcus pneumoniae* has reacted with a specific pneumococcal antiserum, leading to capsular swelling (the Quellung reaction). The capsules seen around the bacteria indicate potential virulence.

to identify a pathogen serologically by testing for cell antigens. For example, there are 90 different strains of *Streptococcus pneumoniae,* each unique in the nature of its capsular material. These differences can be detected by antibody-induced capsular swelling, termed the **Quellung reaction,** when the appropriate antiserum for a specific capsular type is used (**figure 35.8**).

Agglutination

An **agglutination reaction** occurs when an immune complex is formed by cross-linking cells or particles with specific antibodies (*see figure 33.18*). Agglutination reactions usually form visible aggregates or clumps, called **agglutinates,** that can be seen with the unaided eye. Direct agglutination reactions are very useful in the diagnosis of certain diseases. For example, the **Widal test** is a reaction involving the agglutination of typhoid bacilli when they are mixed with serum containing typhoid antibodies from an individual who has typhoid fever. ◀◀ *Immune complex formation (section 33.8)*

Techniques have also been developed that employ microscopic synthetic latex spheres coated with antigens. These coated microspheres can bind antibodies in a patient's serum specimen to identify viral disease rapidly and when cultures are not feasible (HIV, for example). Latex agglutination tests are also used to detect antibodies that develop during certain mycotic, helminthic, and bacterial infections, as well as in drug testing. Microspheres can also be coated with monoclonal antibodies so as to capture antigens from patient specimens.

Hemagglutination usually results from antibodies cross-linking red blood cells through attachment to surface antigens and is routinely used in blood typing. In addition, some viruses can accomplish **viral hemagglutination.** For example, if a person has a certain viral disease, such as measles, antibodies will be present in the serum to react with the measles viruses and neutralize them. Normally hemagglutination occurs when measles viruses and red blood cells are mixed. However, a person's serum may be mixed first with virions, followed by the addition of red blood cells. If no hemagglutination occurs, the serum antibodies have neutralized the measles viruses. This is considered a positive test result for the presence of virus-specific antibodies (**figure 35.9**). Hemagglutination inhibition tests are widely used to diagnose influenza,

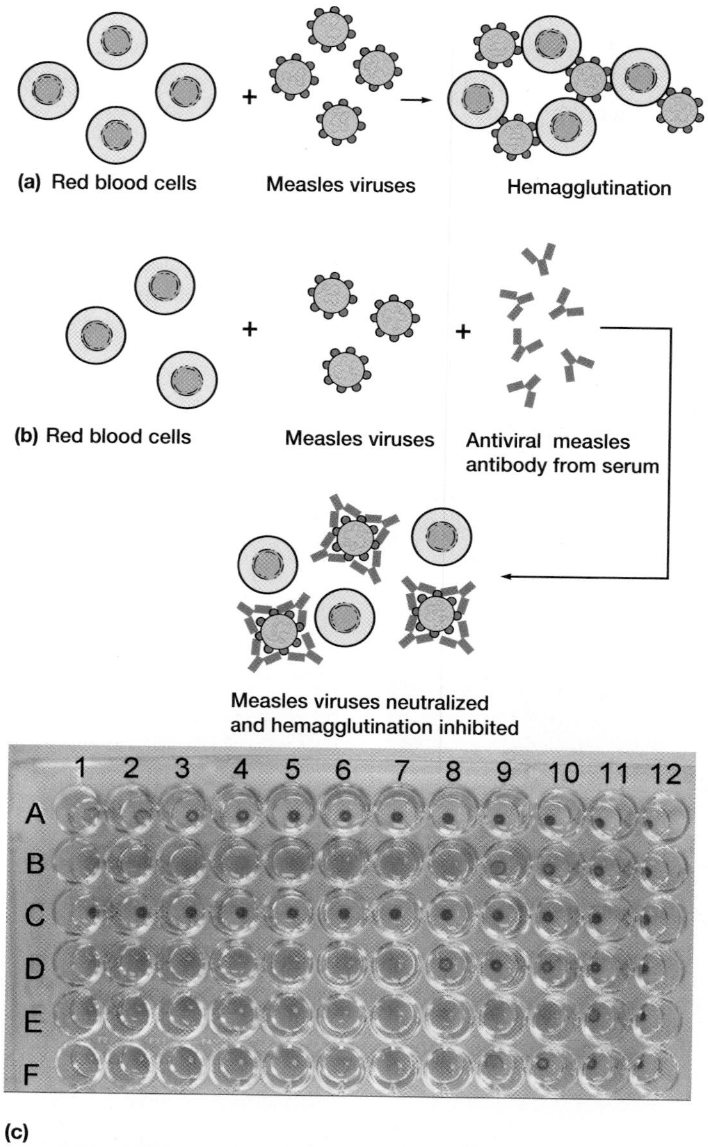

FIGURE 35.9 Viral Hemagglutination. (a) Certain viruses can bind to red blood cells causing hemagglutination. (b) If serum containing specific antibodies to the virus is mixed with the red blood cells, the antibodies will neutralize the virus and inhibit hemagglutination (a positive test). (c) Reovirus hemagglutination test results.

Figure 35.9 Micro Inquiry

Which wells show hemagglutination as characterized by cellular clumping?

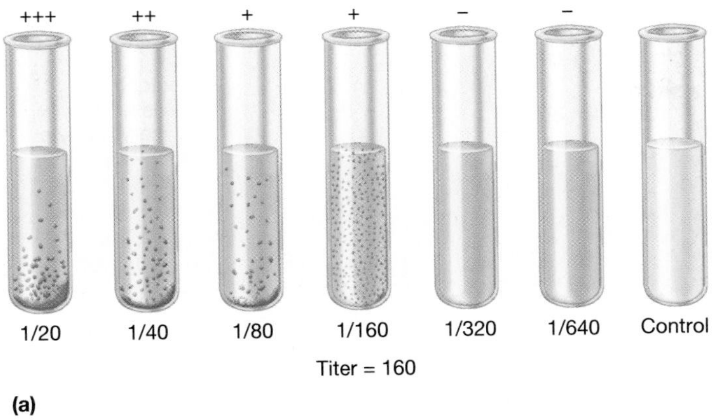

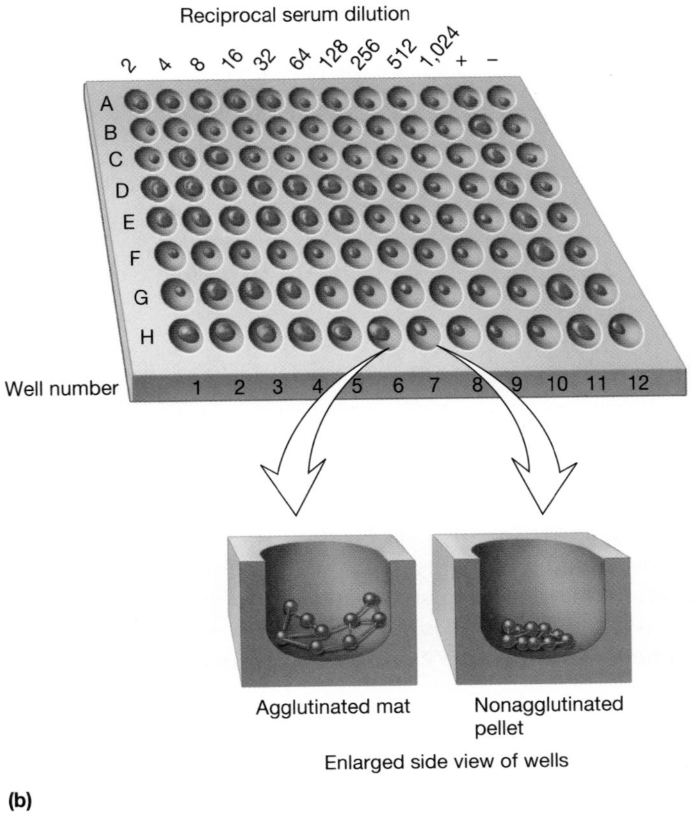

FIGURE 35.10 Agglutination Tests. (a) Tube agglutination test for determining antibody titer. The titer in this example is 160 because there is no agglutination in the next tube in the dilution series (1/320). The blue in the dilution tubes indicates the presence of the patient's serum. (b) A microtiter plate illustrating hemagglutination. The antibody is placed in the wells (rows 1–10). Positive controls (row 11) and negative controls (row 12) are included. Red blood cells are added to each well. If sufficient antibody is present to agglutinate the cells, they sink as a mat to the bottom of the well. If insufficient antibody is present, they form a pellet at the bottom.

Figure 35.10 Micro Inquiry

What is the titer for the antibody tested in row E?

measles, mumps, mononucleosis, and other viral infections. The most routinely used techniques for identification of the mycoplasmas are hemagglutinin reactions, antigen-antibody reactions using the patient's sera, and PCR. These microorganisms are slow growing; therefore positive results from isolation procedures are rarely available before 30 days—a long delay with an approach that offers little advantage over standard techniques. Fungal serology is also designed to detect serum antibody but is limited to a few fungi. The cryptococcal latex antigen test is routinely used for the direct detection of *Cryptococcus neoformans* in serum and cerebrospinal fluid. In the clinical laboratory, nonautomated and automated methods for rapid identification (minutes to hours) are used to detect most yeasts. Any serological method used to detect fungi should always be accompanied by morphological studies examining for pseudohyphae, yeast cell structure, chlamydospores, and so on.

Agglutination tests are also used to measure antibody titer. In tube or well agglutination tests, a specific amount of antigen is added to a series of tubes or shallow wells in a microtiter plate (**figure 35.10**). Serial dilutions of serum (1/20, 1/40, 1/80, 1/160, etc.) containing the antibody are then added to each tube or well. The greatest dilution of serum showing an agglutination reaction is determined, and the reciprocal of this dilution is the serum antibody titer.

1. What is serology?
2. When would you use the Widal test?
3. Why does hemagglutination occur and how can it be used in the clinical laboratory?

Complement Fixation

When complement binds to an antigen-antibody complex, it becomes "fixed" and "used up." Complement fixation tests are very sensitive and can be used to detect extremely small amounts of an antibody for a suspect microorganism in an individual's serum. A known antigen is mixed with test serum lacking complement (**figure 35.11*a***). When immune complexes have had time to form, complement is added (figure 35.11*b*) to the mixture. If immune complexes are present, they will fix and consume complement. Afterward, sensitized indicator cells, usually sheep red blood cells previously coated with complement-fixing antibodies, are added to the mixture. If specific antibodies are present in the test serum and complement is consumed by the immune complexes, insufficient amounts of complement will be available to lyse the indicator cells. On the other hand, in the absence of antibodies, complement remains and lyses the indicator cells

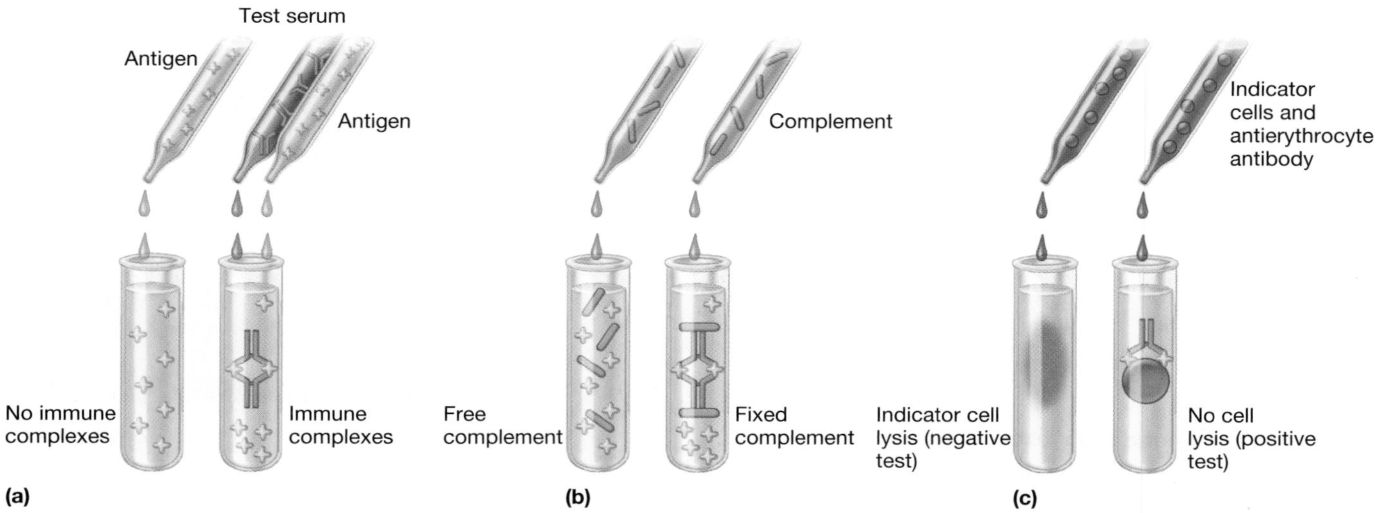

FIGURE 35.11 Complement Fixation. (a) Test serum is added to one test tube. A fixed amount of antigen is then added to both tubes. If antibody is present in the test serum, immune complexes form. (b) When complement is added, if complexes are present, they fix complement and consume it. (c) Indicator cells and a small amount of antierythrocyte antibody are added to the two tubes. If there is complement present, the indicator cells will lyse (a negative test): if the complement has been consumed by the immune complex, no lysis will occur (a positive test).

(figure 35.11*c*). Thus absence of lysis shows that specific antibodies are present in the test serum. Complement fixation was once used in the diagnosis of syphilis (the Wassermann test) and is still used as a rapid, inexpensive screening method in the diagnosis of certain viral, fungal, rickettsial, chlamydial, and protozoan diseases.

Enzyme-Linked Immunosorbent Assay

The ***e**nzyme-**l**inked **i**mmuno**s**orbent **a**ssay* **(ELISA)** has become one of the most widely used serological tests for antibody or antigen detection. This test involves the linking of various "reporting" enzymes to either antigens or antibodies. Two basic methods are used: the indirect immunosorbent assay and the direct (also called the double antibody sandwich assay).

The indirect immunosorbent assay detects antibodies, rather than antigens. In this assay, antigen in appropriate buffer is incubated in the wells of a microtiter plate (**figure 35.12*a***) and is adsorbed onto the walls of the wells. Free antigen is washed away. Test antiserum is added, and if specific antibody is present, it binds to the antigen. Unbound antibody is washed away. Alternatively the test sample can be incubated with a suspension of latex beads that have the desired antigen attached to their surface. After allowing time for antibody-antigen complex formation, the beads are trapped on a filter and unbound antibody is washed away. An anti-antibody that has been covalently coupled to an enzyme, such as horseradish peroxides, is added next. The antibody-enzyme complex (the conjugate) binds to the test antibody, and after unbound conjugate is washed away, the attached ligand is visualized by the addition of a **chromogen.** A chromogen is a colorless substrate acted on by the enzyme portion of the ligand to produce a colored product. The amount of test antibody is quantitated in the same way as an antigen is in the double antibody sandwich method. The indirect immunosorbent assay currently is used to test for antibodies to HIV and *Rubella virus* (German measles), and to detect certain drugs in serum. For example, antigen-coated nitrocellulose membranes are used in the OraQuick HIV-1 test to detect HIV serum antibodies in about 10 minutes.

The double antibody sandwich assay is used for the detection of antigens (figure 35.12*c*). In this assay, specific antibody is placed in wells of a microtiter plate (or it may be attached to a membrane). The antibody is absorbed onto the walls, coating the plate. A test antigen (in serum, urine, etc.) is then added to each well. If the antigen reacts with the antibody, the antigen is retained when the well is washed to remove unbound antigen. A commercially prepared antibody-enzyme conjugate specific for the antigen is then added to each well. The final complex formed is an outer antibody-enzyme, middle antigen, and inner antibody—that is, it is a layered (Ab-Ag-Ab) sandwich. A substrate that the enzyme will convert to a colored product is then added, and any resulting product is quantitatively measured by optical density scanning of the plate. If the antigen has reacted with the absorbed antibodies in the first step, the ELISA test is positive (i.e., it is colored). If the antigen is not recognized by the absorbed antibody, the ELISA test is negative because the unattached antigen has been washed away and no antibody-enzyme is bound (it is colorless). This assay is currently used for the detection of *Helicobacter pylori* infections, brucellosis, salmonellosis, and cholera. Many other antigens also can be detected by the sandwich method. For example, some ELISA kits on the market can test for many different food allergens. *ELISA: Enzyme-Linked Immunosorbent Assay*

Search This: National HIV testing

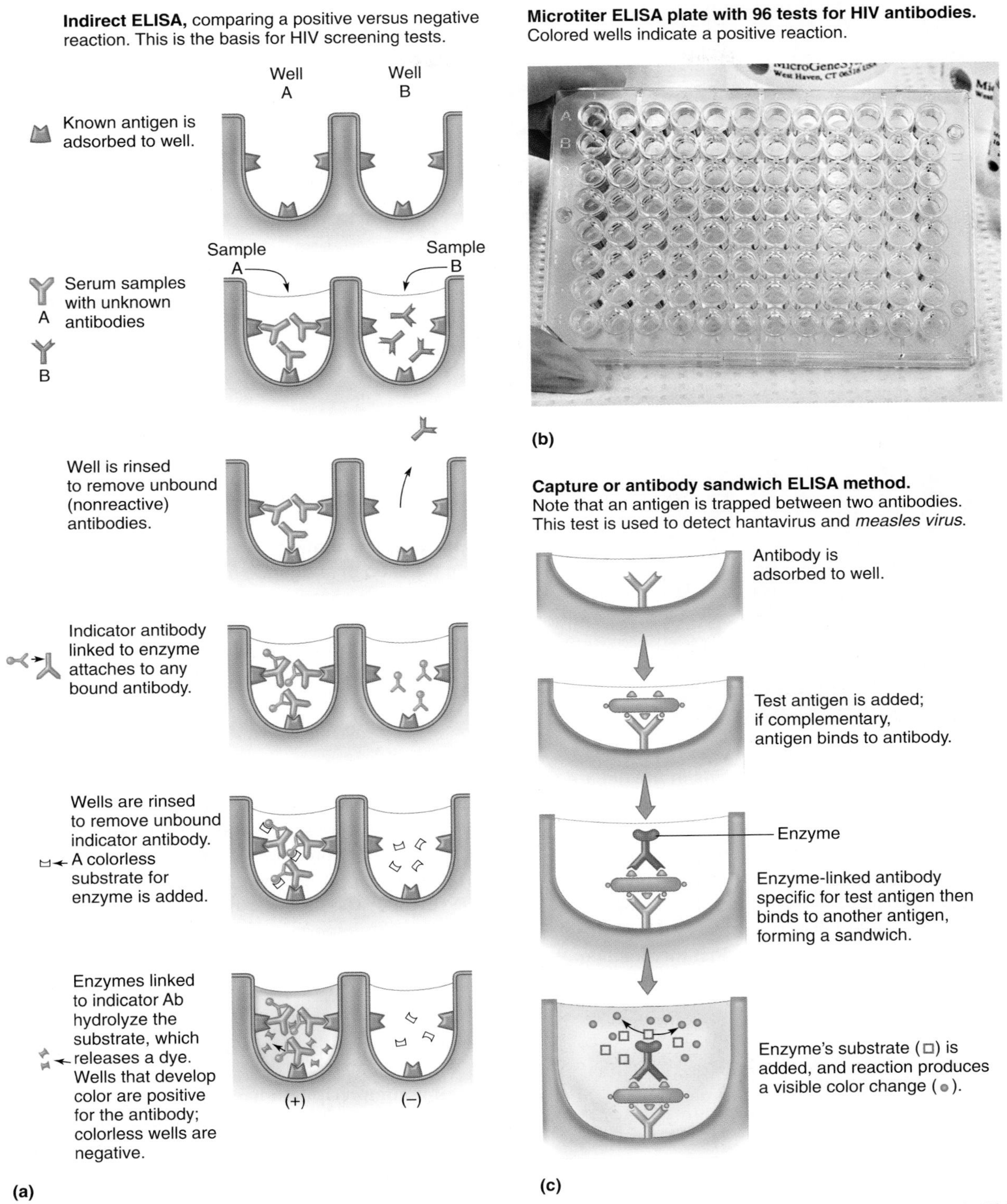
Indirect ELISA, comparing a positive versus negative reaction. This is the basis for HIV screening tests.
Well A
Well B
Known antigen is adsorbed to well.
Sample A
Sample B
Serum samples with unknown antibodies
A
B
Well is rinsed to remove unbound (nonreactive) antibodies.
Indicator antibody linked to enzyme attaches to any bound antibody.
Wells are rinsed to remove unbound indicator antibody. A colorless substrate for enzyme is added.
Enzymes linked to indicator Ab hydrolyze the substrate, which releases a dye. Wells that develop color are positive for the antibody; colorless wells are negative.
(+)
(–)
(a)
Microtiter ELISA plate with 96 tests for HIV antibodies. Colored wells indicate a positive reaction.
(b)
Capture or antibody sandwich ELISA method. Note that an antigen is trapped between two antibodies. This test is used to detect hantavirus and *measles virus*.
Antibody is adsorbed to well.
Test antigen is added; if complementary, antigen binds to antibody.
Enzyme
Enzyme-linked antibody specific for test antigen then binds to another antigen, forming a sandwich.
Enzyme's substrate (□) is added, and reaction produces a visible color change (●).
(c)

FIGURE 35.12 The ELISA Test. (a) The indirect immunosorbent assay for detecting antibodies. (b) HIV antibody ELISA test results. (c) The direct or capture or sandwich method for the detection of antigens.

Figure 35.12 Micro Inquiry

Why do you think an indirect ELISA is used to test for HIV?

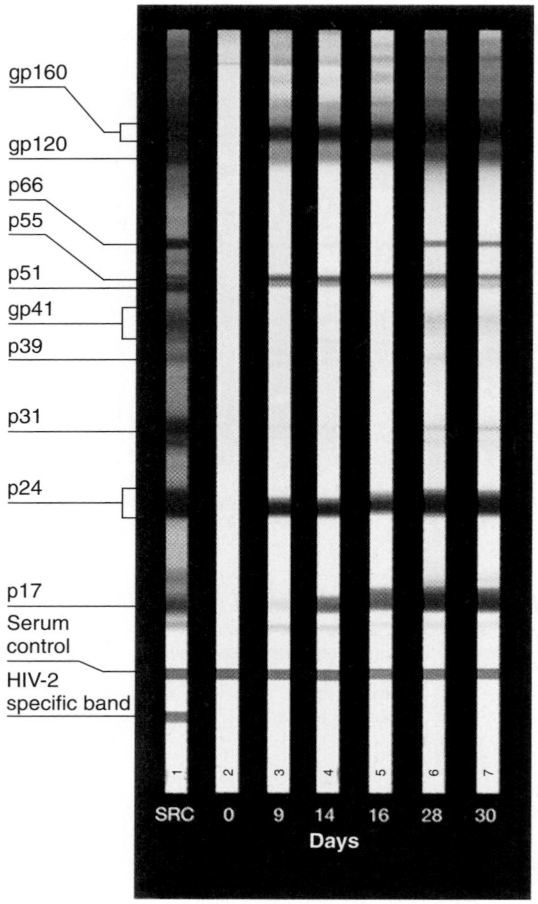

FIGURE 35.13 The Western Blot. HIV proteins are separated by gel electrophoresis. The proteins in the gel are then blotted onto nitrocellulose filter strips. The nitrocellulose strips are incubated with patient serum samples (collected over 30 days) to permit serum antibody to bind to HIV proteins. The strips are developed by probing for antibody using radioactive or colorimetric methods. Identification of HIV antibodies is determined by comparing the patient serum samples against the positive control (SRC).

Figure 35.13 Micro Inquiry

Why do you think the positive control strip (SRC) has more proteins on it than any of the sample strips?

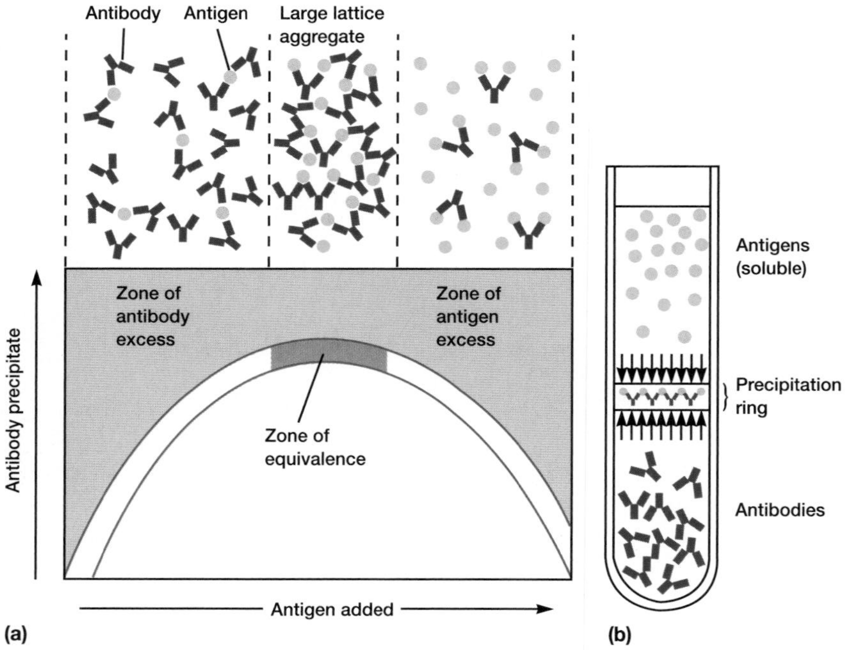

FIGURE 35.14 Immunoprecipitation. (a) Graph showing that a precipitation curve is based on the ratio of antigen to antibody. The zone of equivalence represents the optimal ratio for precipitation. (b) A precipitation ring test. Antibodies and antigens diffuse toward each other in a test tube. A precipitation ring is formed at the zone of equivalence.

Immunoblotting (Western Blotting)

Another immunologic technique used in the clinical microbiology laboratory is immunoblotting, also known as Western blotting (**figure 35.13**). **Immunoblotting** involves polyacrylamide gel electrophoresis of a protein specimen followed by transfer of the separated proteins to sheets of nitrocellulose or polyvinyl difluoride. Protein bands are then visualized by treating the nitrocellulose sheets with solutions of enzyme-tagged antibodies. This procedure demonstrates the presence of common and specific proteins among different strains of microorganisms. Immunoblotting also can be used to show strain-specific immune responses to microorganisms, to serve as an important diagnostic indicator of a recent infection with a particular strain of microorganism, and to allow for prognostic implications with severe infectious diseases.

Immunoprecipitation

The **immunoprecipitation** technique detects soluble antigens that react with antibodies called **precipitins.** The precipitin reaction occurs when bivalent or multivalent antibodies and antigens are mixed in the proper proportions. The antibodies link the antigen to form a large antibody-antigen network or lattice that settles out of solution when it becomes sufficiently large (**figure 35.14*a***). Immunoprecipitation reactions occur only at the equivalence zone when there is an optimal ratio of antigen to antibody so that an insoluble lattice forms. If the precipitin reaction takes place in a test tube (figure 35.14*b*), a precipitation ring forms in the area in which the optimal ratio or equivalence zone develops. ◂◂ *Immune complex formation (section 33.8)* *Complement Fixation Test*

(a)

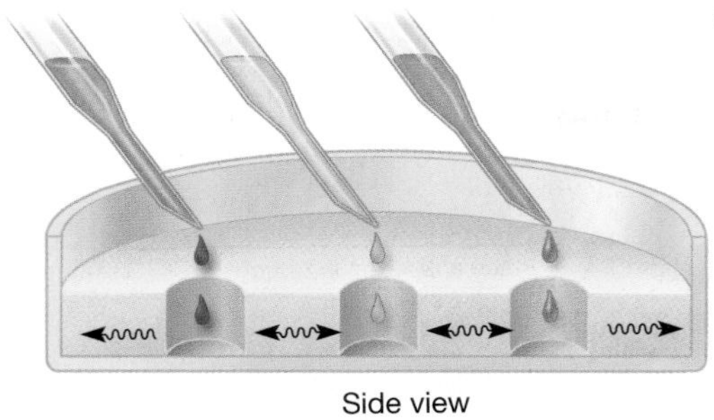

a. In one method of setting up a double-diffusion test, wells are punctured in soft agar, and antibodies (Ab) and antigens (Ag) are added in a pattern. As the contents of the wells diffuse toward each other, a number of reactions can result, depending on whether antibodies meet and precipitate antigens.

(b)

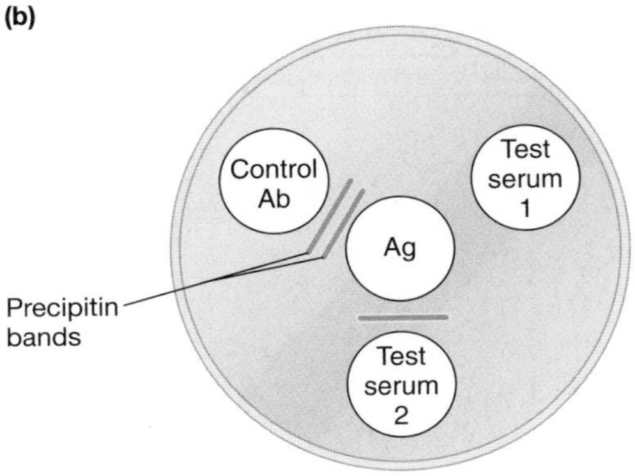

b. Example of test pattern and results. Antigen (Ag) is placed in the center well and antibody (Ab) samples are placed in outer wells. The control contains known Abs to the test Ag. Note bands that form where Ab and Ag meet. The other wells (1, 2) contain unknown test sera. One is positive and the other is negative. Double bands indicate more than one antigen and antibody that can react.

(c)

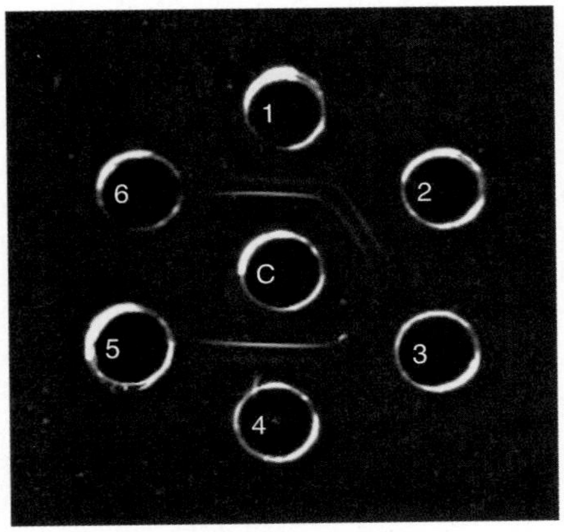

c. Actual test results for detecting infection with the fungal pathogen *Histoplasma*. Numbers 1 and 4 are controls, and 2, 3, 5, and 6 are patient test sera. Can you determine which patients have the infection and which do not?

FIGURE 35.15 Immunodiffusion. Double diffusion agar assay showing characteristics of identity (test serum 2) and a reaction of nonidentity (test serum 1).

Immunodiffusion

Immunodiffusion refers to a precipitation reaction that occurs between an antibody and antigen in an agar gel medium. Two techniques are routinely used: single radial immunodiffusion and double diffusion in agar.

The **single *r*adial *i*mmuno*d*iffusion (RID) assay** or Mancini technique quantitates antigens. Antibody is added to agar, then the mixture is poured onto slides and allowed to set. Wells are cut in the agar and known amounts of standard antigen added. The unknown test antigen is added to a separate well. The slide is left for 24 hours or until equilibrium has been reached, during which time the antigen diffuses out of the wells to form insoluble complexes with antibodies. The size of the resulting precipitation ring surrounding various dilutions of antigen is proportional to the amount of antigen in the well (the wider the ring, the greater the antigen concentration). This is because the antigen's concentration drops as it diffuses farther out into the agar. The antigen forms a precipitin ring in the agar when its level has decreased sufficiently to reach equivalence and combine with the antibody to produce a large, insoluble network. This method is commonly used to quantitate serum immunoglobulins, complement proteins, and other substances.

The **double diffusion agar assay (Öuchterlony technique)** is based on the principle that diffusion of both antibody and antigen (hence double diffusion) through agar can form stable and easily observable immune complexes. Test solutions of antigen and antibody are added to the separate wells punched in agar. The solutions diffuse outward, and when antigen and the appropriate antibody meet, they combine and precipitate at the equivalence zone, producing an indicator line (or lines) (**figure 35.15**). The visible line of precipitation permits a comparison of antigens for identity (same antigenic determinants), partial identity, or nonidentity against a given selected antibody.

Immunoelectrophoresis

Some antigen mixtures are too complex to be resolved by simple diffusion and precipitation. Greater resolution is obtained by the technique of **immunoelectrophoresis** in which antigens are first separated based on their electrical charge, then visualized by the precipitation reaction. In this procedure, antigens are separated by electrophoresis in an agar gel. Positively charged proteins move to the negative electrode, and negatively charged proteins move to the positive electrode (**figure 35.16a**). A trough is then cut next to the wells (figure 35.16b) and filled with antibody. When the plate is incubated, the antibodies and antigens diffuse and form precipitation bands or arcs (figure 35.16c) that can be better visualized by staining (figure 35.16d). This assay is used to separate the major blood proteins in serum for certain diagnostic tests. ◀◀ *Gel electrophoresis (section 15.3)*

1. What does a negative complement fixation test show? A positive test?
2. What are the two types of ELISA methods and how do they work? What is a chromogen?
3. Specifically, when do immunoprecipitation reactions occur?
4. Name two types of immunodiffusion tests and describe how they operate.
5. Describe the immunoelectrophoresis technique.

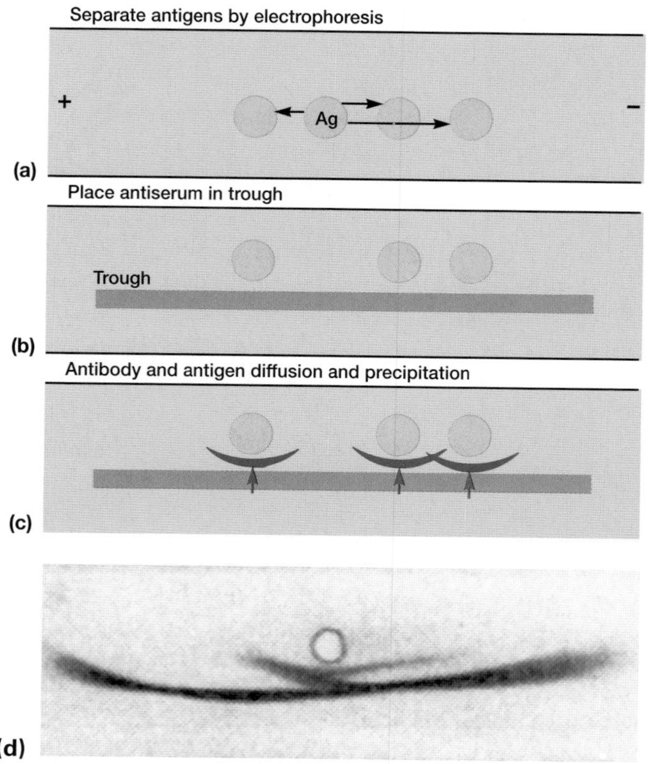

FIGURE 35.16 Classical Immunoelectrophoresis. (a) Antigens are separated in an agar gel by an electrical charge. (b) Antibody (antiserum) is then placed in a trough cut parallel to the direction of the antigen migration. (c) The antigens and antibodies diffuse through the agar and form precipitin arcs. (d) After staining, better visualization is possible.

Flow Cytometry

Flow cytometry allows detection of single- or multiple-microorganisms in clinical samples in an easy, reliable, fast way. In flow cytometry, microorganisms are identified on the basis of their unique cytometric parameters or by means of fluorochromes that can be used either independently or bound to specific antibodies or oligonucleotides. The flow cytometer forces a suspension of cells through a laser beam and measures the light they scatter or the florescence the cells emit as they pass through the beam. For example, cells can be tagged with a fluorescent antibody directed against a specific surface antigen. As the stream of cells flows past the laser beam, each fluorescent cell can be detected, counted, and even separated from the other cells in the suspension. The cytometer also can measure a cell's shape, size, and content of DNA and RNA. This technique has enabled the development of quantitative methods to assess antimicrobial susceptibility and drug cytotoxicity in a rapid, accurate, and highly reproducible way. One outstanding contribution of this technique is the ability to detect the presence of heterogeneous microbial populations with different responses to antimicrobial treatments.

Radioimmunoassay

The **radioimmunoassay (RIA)** technique has become an extremely important tool in biomedical research and clinical practice (e.g., in cardiology, blood banking, diagnosis of allergies, and endocrinology). Indeed, Rosalyn Yalow won the 1977 Nobel Prize in Physiology or Medicine for its development. RIA uses a purified antigen that is radioisotope-labeled and competes for antibody with unlabeled standard antigen or test antigen in experimental samples. The radioactivity associated with the antibody is then detected by means of radioisotope analyzers and autoradiography (photographic emulsions that show areas of radioactivity). If there is a lot of antigen in an experimental sample, it will compete with the radioisotope-labeled antigen for antigen-binding sites on the antibody, and little radioactivity will be bound. A large amount of bound radioactivity indicates that little antigen is present in the experimental sample.

1. Explain how flow cytometry is both qualitative and quantitative (i.e., how it can determine both the identity and the number of a specific cell type).
2. Describe the RIA technique.

Summary

35.1 Overview of the Clinical Microbiology Laboratory

a. The major focus of the clinical microbiologist is identifying microorganisms from clinical specimens accurately and rapidly.

b. A variety of laboratory tests and techniques are used (**figure 35.1**).

c. Because each laboratory reflects the clinical environment in which it is housed, some laboratories are relatively small and may only perform a few common or rapid tests. Other laboratories may be associated with large hospitals or public health departments.

35.2 Identification of Microorganisms from Specimens

a. The clinical microbiology laboratory can provide preliminary or definitive identification of microorganisms based on (1) microscopic examination of specimens; (2) growth and biochemical characteristics of microorganisms isolated from cultures; (3) immunologic techniques that detect antibodies or microbial antigens; (4) bacteriophage typing; and (5) molecular amplification and hybridization techniques.

b. The type of suspected microorganisms dictates the tests that are performed (**figure 35.2**). Bacteria and fungi are often grown in culture to purify and amplify the pathogen for identification. Viruses are identified by isolation in living cells or immunologic tests. Several types of living cells are available: cell culture, embryonated hen's eggs, and experimental animals.

c. The culturing of many types of bacteria is routine; however, a few require special consideration. Rickettsial disease can be diagnosed immunologically. Chlamydiae can be demonstrated in tissue and cell scrapings with Giemsa stain, which detects the characteristic intracellular inclusion bodies. The most routinely used techniques for identification of the mycoplasmas are immunologic.

d. Identification of fungi often can be made if a portion of the specimen is mixed with a drop of 10% Calcofluor White stain. Wet mounts of stool specimens or urine can be examined microscopically for the presence of parasites.

e. Immunofluorescence is a process in which fluorochromes are irradiated with UV, violet, or blue light to make them fluoresce. These dyes can be coupled to an antibody. There are two main kinds of fluorescent antibody assays: direct and indirect (**figure 35.3**).

f. The initial identity of a bacterial organism may be suggested by (1) the source of the culture specimen; (2) its microscopic appearance; (3) its pattern of growth on selective, differential, enrichment, or characteristic media; and (4) its hemolytic, metabolic, and fermentative properties. A simple dichotomous key is used to facilitate the identification (**figure 35.4**).

g. Rapid methods for microbial identification can be divided into three categories: (1) manual biochemical systems, (2) mechanized/automated systems, and (3) immunologic systems.

h. Bacteriophage typing for bacterial identification is based on the fact that phage surface receptors bind to specific cell surface receptors. On a Petri plate culture, bacteriophages cause plaques on lawns of bacteria possessing the proper receptors.

i. Various molecular methods also can be used to identify microorganisms. Examples include nucleic acid-based detection, real-time PCR, and genome and plasmid fingerprinting.

35.3 Clinical Immunology

a. Serotyping refers to serological procedures used to differentiate strains (serovars or serotypes) of microorganisms that have differences in the antigenic composition of a structure or product (**figure 35.8**).

b. In vitro agglutination reactions usually form aggregates or clumps (agglutinates) that are visible with the naked eye. Tests have been developed, such as the Widal test, latex microsphere agglutination reaction, hemagglutination, and viral hemagglutination, to detect antigen as well as to determine antibody titer (**figures 35.9** and **35.10**).

c. The complement fixation test can be used to detect a specific antibody for a suspect microorganism in an individual's serum (**figure 35.11**).

d. The enzyme-linked immunosorbent assay (ELISA) involves linking various enzymes to either antigens or antibodies. Two basic methods are involved: the double antibody sandwich method and the indirect immunosorbent assay (**figure 35.12**). The first method detects antigens and the latter, antibodies.

e. Immunoblotting involves polyacrylamide gel electrophoresis of a protein specimen followed by transfer of the separated proteins to nitrocellulose sheets and identification of specific bands by labeled antibodies (**figure 35.13**).

f. Immunoprecipitation reactions occur only when there is an optimal ratio of antigen and antibody to produce a lattice at the zone of equivalence, which is evidenced by a visible precipitate (**figure 35.14**).

g. Immunodiffusion refers to a precipitation reaction that occurs between antibody and antigen in an agar gel medium. Two techniques are routinely used: single radial diffusion and double diffusion in agar (**figure 35.15**).

h. In immunoelectrophoresis, antigens are separated based on their electrical charge, then visualized by precipitation and staining (**figure 35.16**).

i. Flow cytometry and fluorescence allow detection of single- or multiple-microorganisms based on their cytometric parameters or by means of certain dyes called fluorochromes.

Critical Thinking Questions

1. As more ways of identifying the characteristics of microorganisms emerge, the number of distinguishable microbial strains also seems to increase. Why do you think this is the case?
2. Why are miniaturized identification systems used in clinical microbiology? Describe one such system and its advantage over classic dichotomous keys.
3. It has been speculated that as good as nucleic acid tests are in identifying viruses, antibody-based identification tests will be just as specific but easier and faster to develop and get to the marketplace. Do you agree or disagree? Defend your answer.
4. ELISA tests usually use a primary and secondary antibody. Why? What are the necessary controls one would need to perform to ensure that the antibody specificities are valid (i.e., no false-positive or false-negative reactions)?
5. *Legionella* is a bacterium that is often found in water systems (e.g., showerheads, air-cooling towers). Their dispersal can lead to the development of Legionnaires' disease, a particularly virulent pneumonia in the elderly and immunocompromised. Water systems are routinely monitored for *Legionella* contamination using GVPC medium, which contains glycine and the antibiotics vancomycin, polymyxin B, and cycloheximide. However, this method is known to underestimate the number of *Legionella* cells in the environment. Flow cytometry (FC) is one approach to count cells. Detection by FC is accomplished by staining the cells with a fluorescent dye. When two specific types of dyes are used, viable cells can be distinguished from nonviable cells.

   How would you go about determining if FC is a good way to monitor *Legionella* in the environment? Specifically, how would you collect your samples and compare your results to those obtained by the currently accepted approach (i.e., plating on GVPC medium)? What controls would you need to perform? Based on the information given above, how do you think the results from FC would compare to those of GVPC cultures obtained from the same samples?

   Read the original paper: Allegra, S. 2008. Use of flow cytometry to monitor *Legionella* viability. *Appl. Environ. Microbiol.* 74:7813.
6. Rotaviruses are double-stranded RNA viruses that cause severe diarrhea in children, resulting in over 600,000 deaths annually, most in developing countries. There are seven serotypes of rotaviruses (A through G; A is the most common). Ideally the diagnosis of rotavirus includes the serotype. A PCR-ELISA approach may be the most efficient way to accomplish this. PCR-ELISA uses a 96-well microtiter plate format, thereby maximizing efficiency. PCR products are labeled, usually with a chemical known as digoxigenin, during amplification. A special oligonucleotide called a "capture probe" complementary to the PCR product is then added. The capture probe immobilizes the PCR product to the well and ELISA using antibody against digoxigenin is performed.

   Why do you think it is important to report the serotype of a rotavirus infection? How does PCR-ELISA differ from PCR? How does it differ from a regular ELISA? Why do you think PCR-ELISA might be better than the currently used typing PCR protocol? (Hint: Think about how PCR-ELISA could detect more than one PCR product).

   Read the original paper: Santos, N., et al. 2008. Development of a microtiter plate hybridization-based PCR-enzyme linked-linked immunosorbent assay for identification of clinically relevant human group A rotavirus G and P genotypes. *J. Clin. Microbiol.* 46:462.

Concept Mapping Challenge

Map the following concepts to demonstrate your understanding of their relationships.

Bacteria
Culture
Fungi
Molecular techniques
Parasites
PCR
ELISA
Identification
Microorganism
Microscopy
Plasmid fingerprinting
Rapid testing
Serotyping
Specimen

Learn More

Learn more by visiting the text website at www.mhhe.com/willey8, where you will find a complete list of references.

36

Epidemiology and Public Health Microbiology

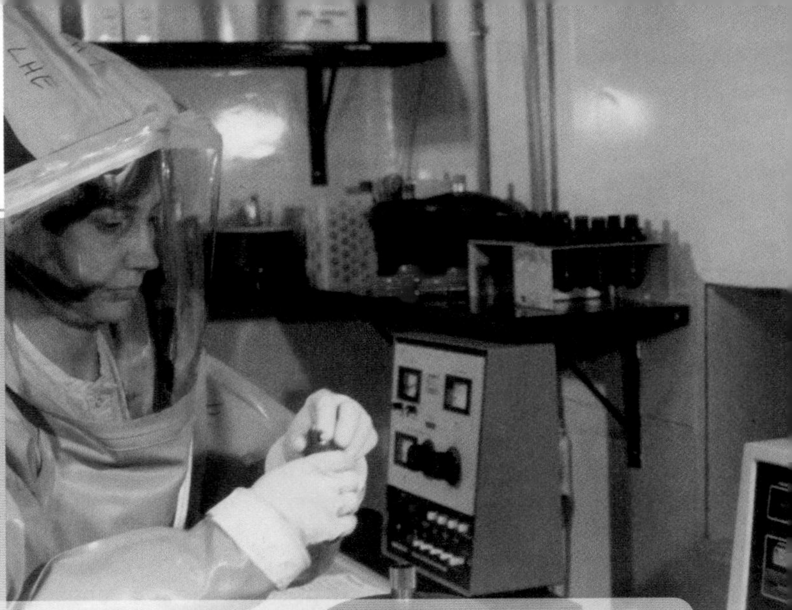

This laboratory worker at the Centers for Disease Control and Prevention is in the highest level of isolation (level 4) to avoid contact with microorganisms and to prevent their escape into the environment. Extensive training is required to work in these labs.

CHAPTER GLOSSARY

antigenic drift A small change in the immunogenic character of an organism that allows it to evade immune system attack.

antigenic shift A major change in the immunogenic character of an organism; it is unrecognized by immune mechanisms.

attack rate The proportional number of cases that develop in a population that was exposed to an infectious agent.

common-source epidemic An epidemic characterized by a sharp rise to a peak and then a less rapid decline in the number of individuals infected; it usually involves a single infectious source.

communicable disease A disease associated with a pathogen that can be transmitted from one host to another.

endemic disease A disease that is constantly present in a population, usually at a steady low frequency.

epidemic A disease that suddenly increases in occurrence above the normal level in a given population.

epidemiology The science that evaluates the occurrence, determinants, distribution, and control of health and disease in a defined human population.

herd immunity The resistance of a population to infection and spread of an infectious agent due to the immunity of a high percentage of the population.

incidence The number of new cases of a disease in a population at risk during a specified time period.

incubation period The period after pathogen entry into a host and before signs and symptoms appear.

morbidity rate The number of individuals who become ill as a result of a particular disease within a susceptible population during a specific time period.

mortality rate The ratio of the number of deaths from a given disease to the total number of cases of the disease.

nosocomial infection An infection that is acquired while a patient is in a hospital or other clinical care facility.

pandemic An increase in the occurrence of a disease within a large and geographically widespread population (often refers to a worldwide epidemic).

public health surveillance The systematic collection, analysis, and interpretation of outcome-based health data so as to control or prevent disease or injury.

toxoid A bacterial exotoxin that has been modified so that it is no longer toxic but will still stimulate antitoxin formation when injected into a person or animal.

vaccine A preparation given to induce an immune response and protect the individual against a pathogen or a toxin.

vector A living organism, usually an arthropod or other animal, that transfers an infective agent between hosts.

vehicle An inanimate substance or medium that transmits a pathogen.

zoonosis A disease of animals that can be transmitted to humans.

In this chapter, we describe the practical goal of epidemiology: to establish effective disease recognition, control, prevention, and eradication measures within a given population. Because emerging and reemerging diseases and pathogens, hospital-acquired (nosocomial) infections, and bioterrorism are major concerns for public health, these topics are also covered here.

36.1 Epidemiology

By definition, **epidemiology** (Greek *epi,* upon; *demos,* people or population; and *logy,* study) is the science that evaluates the occurrence, determinants, distribution, and control of health and disease in a defined human population (**figure 36.1**). An individual who practices epidemiology is an **epidemiologist.** Epidemiologists are, in effect, disease detectives. Their major concerns are the discovery of the factors essential to disease occurrence and the development of methods for disease prevention. In the United States, the **Centers for Disease Control and Prevention** (CDC), headquartered in Atlanta, Georgia, serves as the national agency for developing and carrying out disease prevention and control, environmental health, and health promotion and education activities (**Historical Highlights 36.1**). Its worldwide counterpart is the **World Health Organization** (WHO), located in Geneva, Switzerland.

The science of epidemiology originated and evolved in response to the great epidemic diseases such as cholera, typhoid fever, smallpox, influenza, and yellow fever (**Historical Highlights 36.2**). More recent epidemics of Ebola, HIV/AIDS, cryptosporidiosis, enteropathogenic *Escherichia coli*, SARS, *Salmonella*, and novel H1N1 influenza have underscored the importance of epidemiology in preventing global catastrophes caused by infectious diseases. Today its scope encompasses all public health issues: infectious diseases, genetic abnormalities, metabolic dysfunction, malnutrition, neoplasms, psychiatric disorders, and aging. This chapter emphasizes only infectious disease epidemiology.

Epidemiology is the study of the factors determining and influencing the frequency and distribution of health-related events. When a disease occurs occasionally and at irregular intervals in a human population, it is a sporadic disease (e.g., bacterial meningitis). When it maintains a steady, low-level frequency at a moderately regular interval, it is an **endemic** (Greek *endemos,* dwelling in the same people) **disease** (e.g., the common cold). Hyperendemic diseases gradually increase in occurrence frequency beyond the endemic level but not to the epidemic level (e.g., the common cold during winter months). An **incidence** reflects the number of new cases of a disease in a population at risk during a specified time period. An **outbreak** is the sudden, unexpected occurrence of a disease, usually focally or in a limited segment of a population (e.g., Legionnaires' disease). The **attack rate** is the proportional number of cases that develop in a population that was exposed to an infectious agent. An **epidemic** (Greek *epidemios,* upon the people), on the

HISTORICAL HIGHLIGHTS

36.1 The Birth of Public Health in the United States

The U.S. Public Health Service was born in 1798 out of the need to keep sailors from becoming sick while at sea. As defenders of the new republic, sailors, or seamen as they were called, could not find adequate health care at port cities of the day. Public hospitals were few in number and often did not have the resources to care for the large numbers of sick seamen that could arrive. As the seamen came from all over the new states and former colonies, their health care was determined to be a national responsibility. Thus the Marine Hospital Act was approved in 1798 and established a network of federal hospitals, called the Marine Hospital Service (MHS), to care for sick and disabled seamen. The oversight responsibility was assigned to the Revenue Marine Division of the Treasury Department. The act permitted the taxation (20 cents per month) of the seamen, creating the first medical insurance program in the United States.

In 1870 the MHS was reorganized by congressional action from individually operated hospitals into a centrally controlled, national agency headquartered in Washington, D.C. The new federal agency became a separate bureau of the Treasury Department overseen by a central administrator, the "Supervising Surgeon," who was appointed by the Secretary of the Treasury. The administrative title of Supervising Surgeon was changed to Supervising Surgeon General in 1875. In 1902 the MHS became the Public Health and Marine Hospital Services, and the administrator's title was changed to Surgeon General. A national "hygienic" laboratory was established in 1887 and was located on Staten Island, New York. The Pure Food and Drugs Act was established in 1906. In 1912 the Public Health and Marine Hospital Services was renamed the Public Health Service and authorized to investigate communicable disease.

In 1930 the Hygienic Lab moved to Washington to become the National Institute of Health, but it was not until 1939 that all health, education, and welfare agencies and services created by Congress were combined, under the Federal Security Agency. The Public Health Service opened the "Communicable Disease Center" in Atlanta, Georgia, in 1946. In 1953 the Federal Security Agency became the Department of Health, Education, and Welfare, and in 1970 the Communicable Disease Center became the Center for Disease Control. Ten years later in 1980, the Department of Health, Education, and Welfare became the Department of Health and Human Services, and the Center for Disease Control was renamed the Centers for Disease Control. It was renamed again in 1992 to become the Centers for Disease Control and Prevention (although it is still abbreviated as the CDC).

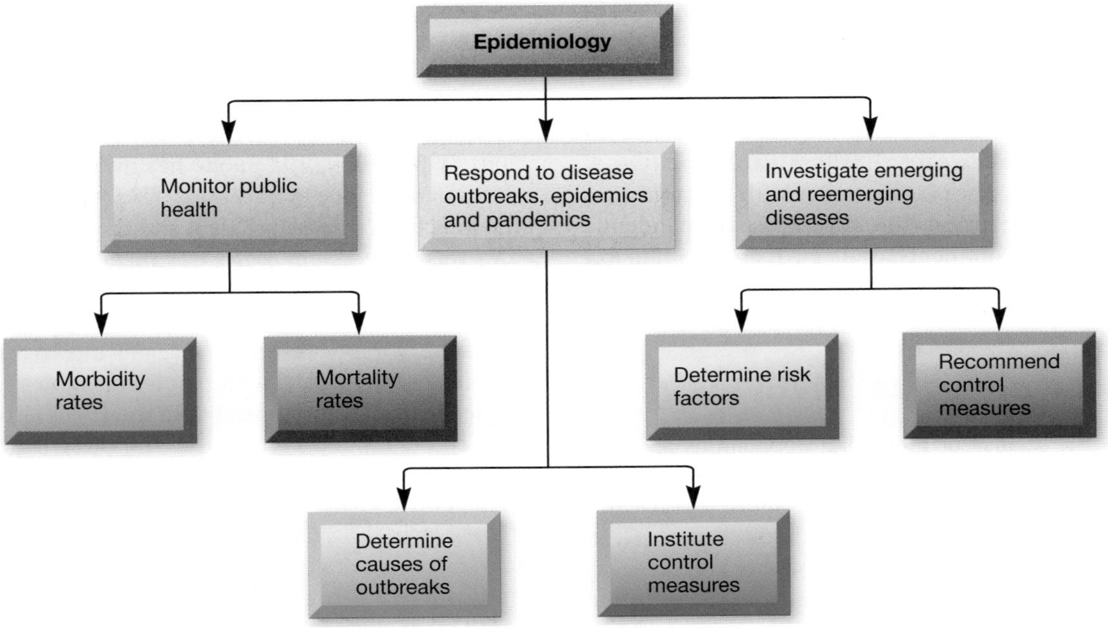

FIGURE 36.1 Epidemiology. Epidemiology is a multifaceted science that investigates diseases to discover their origin, evaluates diseases to assess their risk, and controls diseases to prevent future outbreaks.

HISTORICAL HIGHLIGHTS

36.2 John Snow—The First Epidemiologist

Much of what we know today about the epidemiology of cholera is based on the classic studies conducted between 1849 and 1854 by the British physician John Snow. During this period, a series of cholera outbreaks occurred in London, England, and Snow set out to find the source of the disease. Some years earlier when he was still a medical apprentice, Snow had been sent to help during a cholera outbreak among coal miners. His observations convinced him that the disease was usually spread by unwashed hands and shared food, not by "bad" air or casual direct contact.

Thus when the outbreak of 1849 occurred, Snow believed that cholera was spread among the poor in the same way as among the coal miners. He suspected that water, and not unwashed hands and shared food, was the source of the cholera infection among the wealthier residents. Snow examined official death records and discovered that most of the victims in the Broad Street area had lived close to the Broad Street water pump or had been in the habit of drinking from it. He concluded that cholera was spread by drinking water from the Broad Street pump, which was contaminated with raw sewage containing the disease agent. When the pump handle was removed, the number of cholera cases dropped dramatically.

In 1854 another cholera outbreak struck London. Part of the city's water supply came from two different suppliers: the Southwark and Vauxhall Company, and the Lambeth Company. Snow interviewed cholera patients and found that most of them purchased their drinking water from the Southwark and Vauxhall Company. He also discovered that this company obtained its water from the Thames River below locations where Londoners discharged their sewage. In contrast, the Lambeth Company took its water from the Thames before the river reached the city. The death rate from cholera was over eightfold lower in households supplied with Lambeth Company water. Water contaminated by sewage was transmitting the disease. Finally, Snow concluded that the cause of the disease must be able to multiply in water. Thus he nearly recognized that cholera was caused by a microorganism, though Robert Koch did not discover the causative bacterium (*Vibrio cholerae*) until 1883.

To commemorate these achievements, the John Snow Pub now stands at the site of the old Broad Street pump. Those who complete the Epidemiologic Intelligence Program at the Centers for Disease Control and Prevention receive an emblem bearing a replica of a barrel of Whatney's Ale—the brew dispensed at the John Snow Pub.

other hand, is an outbreak affecting many people at once (i.e., there is a sudden increase in the occurrence of a disease above the expected level). Influenza is an example of a disease that may occur suddenly and unexpectedly in a family and often achieves epidemic status in a community. The first case in an epidemic is called the **index case**. Finally, a **pandemic** (Greek *pan*, all) is an increase in disease occurrence within a large population over a very wide region (the world). Usually pandemic diseases spread among continents. The global H1N1 influenza outbreaks of 1918 and 2009 are good examples.

36.2 Epidemiological Methods

Public health is the science of protecting populations and improving the health of human communities through education, promotion of healthy lifestyles, and prevention of disease and injury. Public health practitioners use a methodical approach to identify population health issues so as to prevent or correct negative outcomes. The methodical approach mirrors the scientific method in that it identifies a population health problem, determines the cause, proposes preventative or corrective action, implements that action, and assesses the outcomes of implementing the action (**figure 36.2**).

Public Health Surveillance

The identification of a population health issue is rooted in public health surveillance. Public health surveillance is the proactive evaluation of genetic background, environmental conditions, human behaviors and lifestyle choices, emerging infectious agents, and microbial responses to chemotherapeutic agents, for example, to monitor the health of a population. In other words, public health practitioners look for cause-and-effect relationships to determine relative risk. The impact of public health surveillance is obvious when one considers the change in the leading causes of death in the United States from 1900 to 2001 (**table 36.1**). The public health landscape of 1900 was a valley of death due to infectious disease. Surveillance identified the infectious disease problem and the risks associated with it. This was remedied by water treatment, strict sanitation guidelines, and later by the use of antimicrobial agents. The result changed that death-by-infection landscape of the early twentieth century to one of very few deaths due to infectious disease by the 1960s. Ironically, the modernization that helped solve the leading causes of mortality in the 1900s also led to a longer life expectancy and a more sedentary lifestyle in the 2000s. This changed the landscape of 2001 to one more like a mountain of metabolic disease.

Public health surveillance really began in the fourteenth century as a means to control bubonic plague. However, it was not until after the germ theory of disease was widely accepted (late 1800s) that a scientifically based monitoring process was used as a means to track communicable disease. This was not the confidential monitoring that we expect today. The first disease

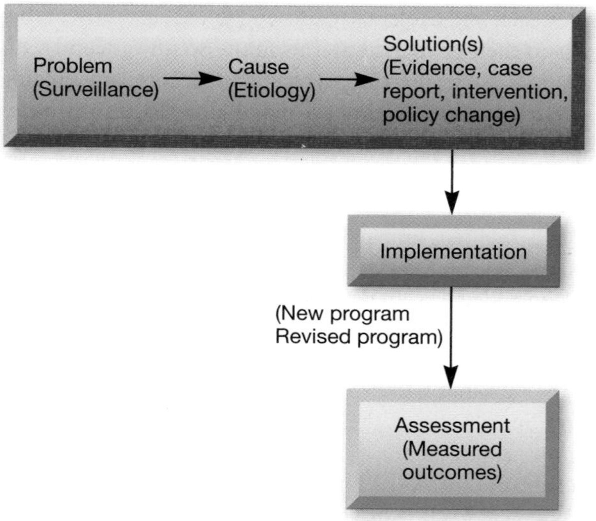

FIGURE 36.2 Evidence-Based Method for Understanding and Preventing Public Health Problems. The first three components (red shade) focus on epidemiology.

Table 36.1 Percentage of Total U.S. Citizen Deaths for the 10 Leading Causes of Death in 1900* and 2001**

Cause	Percentage 1900	Percentage 2001
Heart diseases	8.0	29.0
Cancer	3.7	22.9
Pneumonia and influenza	11.8	2.6
Tuberculosis	11.3	ND
Gastrointestinal diseases	8.3	ND
Cerebrovascular diseases	6.2	6.8
Senility (Alzheimer's diseases)	6.8	2.2
Pulmonary diseases	ND	5.1
Kidney diseases	4.7	1.6
Accidents	4.2	4.2
Diabetes mellitus	ND	3.0
Diphtheria	2.3	ND
Septicemia	ND	1.3
All other causes	32.6	21.3
TOTAL	100.0	100.0

*Source: National Office of Vital Statistics, 1954.
**Source: Anderson, R. N., and Smith, B. L. Deaths: Leading causes for 2001. National Vital Statistics Reports, vol. 5(9). Hyattsville, MD: National Center for Health Statistics; Nov. 7, 2003.
ND: Not determined

HISTORICAL HIGHLIGHTS

36.3 "Typhoid Mary"

In the early 1900s there were thousands of typhoid fever cases, and many people died of the disease. Most of these cases arose when people drank water contaminated with sewage or ate food handled by or prepared by individuals who were shedding the typhoid fever bacterium (*Salmonella enterica* serovar Typhi). The most famous carrier of the typhoid bacterium was Mary Mallon.

Between 1896 and 1906, Mary Mallon worked as a cook in seven homes in New York City. Twenty-eight cases of typhoid fever occurred in these homes while she worked in them. As a result, the New York City Health Department had Mallon arrested and admitted to an isolation hospital on North Brother Island in New York's East River. Examination of her stools showed that she was shedding large numbers of typhoid bacteria, though she exhibited no external symptoms of the disease. An article published in 1908 in the *Journal of the American Medical Association* referred to Mallon as "Typhoid Mary," an epithet by which she is still known today. She was released when she pledged not to cook for others or serve food to them. Mallon changed her name and began to work as a cook again. For five years she managed to avoid capture while continuing to spread typhoid fever. Eventually the authorities tracked her down. She was held in custody for 23 years, until she died in 1938. As a lifetime carrier, Mary Mallon was positively linked with 10 outbreaks of typhoid fever, 53 cases, and 3 deaths.

tracking practices were intrusive, as people were inspected for signs of clinical disease; those found to be infected were quarantined. The change from inspecting people to monitoring disease occurred around 1940. The emphasis on watching disease progress in a population and disseminating information in advance of the disease radically improved the impact on public health by changing the role of public health practitioners from disease informants to health-care allies.

Effective use of public health surveillance data stems from high-integrity data collection (i.e., documented first-hand information of the data origin and the methods used to collect them). Surveillance of health issues is typically accomplished by two methods: population surveys and case reporting. Population surveys tend to be more of a sentinel surveillance method in that they look for trends and outliers within the given population (e.g., immunization records revealing infection control success). Population surveillance may overestimate the number of affected people in that the specific details that define the disease are lacking; that is, they are too generic and thus false positive error can be introduced. Case reports tend to be retrospective comparisons of specific conditions that assign people to risk groups (e.g., influenza test result to rule out SARS). The specific conditions are termed the case definition and may actually underestimate the true relationships as the disease progresses with respect to time, while evaluation of the case may occur outside the disease time frame. Thus false negative data can be introduced into the population analysis. This is why unambiguous definitions must be agreed upon when data are collected and communicated. Various sources of public health surveillance data are used to identify problems and assess effectiveness of solutions. Some examples of surveillance data sources are vital statistics, immunization registries, cancer registries, hospital charts, lab reports, and notifiable disease reports.

Remote Sensing and Geographic Information Systems: Charting Infectious Disease Data

Remote sensing and geographic information systems are map-based tools that can be used to study the distribution, dynamics, and environmental correlates of microbial diseases. **Remote sensing (RS)** is the gathering of data—digital images of Earth's surface from satellites and output from biological sensors, for example—and transformation of the data into maps. A **geographic information system (GIS)** is a data management system that organizes and displays digital map data from RS and facilitates the analysis of relationships between mapped features. Statistical relationships often exist between mapped features and diseases in natural host or human populations. Examples include the location of the habitats of the malaria parasite and mosquito vectors in Mexico and Asia, Lyme disease in the United States, and African trypanosomiasis in both humans and livestock in the southeastern United States. RS and GIS may also permit the assessment of human risk from pathogens such as *Sin Nombre virus* (the virus that causes hantavirus pulmonary syndrome in North America). RS and GIS are most useful if disease dynamics and distributions are clearly related to mapped environmental variables. For example, if a microbial disease is associated with certain vegetation types or physical characteristics (e.g., elevation, precipitation), RS and GIS can identify regions in which risk is relatively high.

After an infectious disease has been recognized in a population, epidemiologists correlate the disease outbreak with a specific organism; its exact cause must be discovered (**Historical Highlights 36.3**). At this point, the clinical or diagnostic microbiology laboratory enters the investigation. Its purpose is to isolate and identify the organism responsible for the disease.

36.3 Measuring Infectious Disease Frequency

To determine if an outbreak, epidemic, or pandemic is occurring, epidemiologists measure disease frequency at single time points and over time. They then use statistics to analyze the data and determine risk factors and other factors associated with disease. Statistics is the branch of mathematics dealing with the collection, organization, and interpretation of numerical data. As a science particularly concerned with rates and the comparison of rates, epidemiology was the first medical field in which statistical methods were extensively used.

To comment on the frequency or rate of an event, an accurate count of the total population, the population exposed, and the number of affected people needs to be made. Disease surveillance practices enable the tracking of infections. Many infectious diseases (e.g., food- and water-borne diseases) must, by law, be reported within a specific time frame. This permits public health agencies to act swiftly to contain the outbreak and mobilize control measures. Timely notification of "reportable" diseases to public health agencies has stemmed numerous epidemics such as cholera and typhoid.

Measures of frequency usually are expressed as fractions. The numerator is the number of individuals experiencing the event—infection—and the denominator is the number of individuals in whom the event could have occurred, that is, the population at risk. The fraction is a proportion or ratio but is commonly called a rate because a time period is specified. (A rate also can be expressed as a percentage.) In population statistics, rates usually are stated per 1,000 individuals, although other powers of 10 may be used for particular diseases (e.g., per 100 for very common diseases and per 10,000 or 100,000 for uncommon diseases). For example, disease incidence is a measure of the number of diseased people during a defined time period, as compared to the total (healthy) population. The incidence of a disease reports not only the rate of occurrence but the relative risk as well. Since disease reflects a change in health status over time, a **morbidity rate** is used. The rate is commonly determined when the number of new cases of illness in the general population is known from clinical reports. It is calculated as follows:

$$\text{Morbidity rate} = \frac{\text{Number of new cases of a disease during a specified period}}{\text{Number of individuals in the population}}$$

For example, if in one month, 700 new cases of influenza per 100,000 individuals occur, the morbidity rate would be expressed as 700 per 100,000, or 0.7%.

The **prevalence rate** refers to the total number of individuals infected in a population at any one time no matter when the disease began. The prevalence rate depends on both the incidence rate and the duration of the illness. Prevalence is calculated as follows:

$$\text{Prevalence} = \frac{\text{Total number of cases in population}}{\text{Total population}} \times 100$$

The **mortality rate** is the relationship between the number of deaths from a given disease and the total number of cases of the disease. The mortality rate is a simple statement of the proportion of all deaths that are assigned to a single cause. It is calculated as follows:

$$\text{Mortality rate} = \frac{\text{Number of deaths due to a given disease}}{\text{Size of the total population with the same disease}}$$

For example, if 15,000 deaths occurred due to AIDS in a year and the total number of people infected was 30,000, the mortality rate would be 15,000 per 30,000, or 1 per 2, or 50%.

The determination of morbidity, prevalence, and mortality rates helps public health personnel in directing health-care efforts to control the spread of infectious diseases. For example, a sudden increase in the morbidity rate of a particular disease may indicate a need for the implementation of preventive measures designed to reduce disease.

Search This: Disease surveillance

1. What is epidemiology?
2. What terms are used to describe the occurrence of a disease in a human population?
3. What types of surveillance data are most useful in determining infectious disease penetration into a population?
4. How might remote sensing be used to track Rocky Mountain spotted fever (*see p. 948*)?
5. Define morbidity rate, prevalence rate, and mortality rate.

36.4 Patterns of Infectious Disease in a Population

An **infectious disease** is a disease resulting from an infection by microbial agents such as viruses, bacteria, fungi, protozoa, and helminths. A **communicable disease** is an infectious disease that can be transmitted from person to person (not all infectious diseases are communicable; e.g., rabies is an infectious disease acquired only through contact with a rabid animal). The manifestations of an infectious or communicable disease can range from mild to deadly, depending on the agent and host. An epidemiologist studying an infectious disease is concerned with the causative agent, the source or reservoir of the disease agent, how it is transmitted, what host and environmental factors can aid development of the disease within a defined population, and how best to control or eliminate the disease. These factors describe the natural history or cycle of an infectious disease. ◀◀ *The infectious disease process (section 31.2)*

Two major types of epidemics are recognized: common source (noncommunicable) and propagated (communicable). A **common-source epidemic** is characterized as having reached a peak level within a short period of time (1 to 2 weeks), followed by a moderately rapid decline in the number of infected patients (**figure 36.3*a***). This type of epidemic usually results from a single, common contaminated source such as food (food poisoning) or water (Legionnaires' disease). By contrast, a **propagated epidemic** is characterized by a relatively slow and prolonged rise, and then a gradual decline in the number of individuals infected (figure 36.3*b*). This type of epidemic usually results from the introduction of a single infected individual into a susceptible population. The initial infection is then propagated to others in a gradual fashion until many individuals within the population are infected. An example is the increase in strep throat cases that coincides with new populations of sensitive children who arrive in classrooms. While one infected child is sufficient to initiate the epidemic, it is the person-to-person transmission of the streptococci that extends the epidemic.

As mentioned, epidemiologists recognize an infectious disease in a population by using various surveillance methods. Surveillance is a dynamic activity that includes gathering information on the development and occurrence of a disease, collating and analyzing the data, summarizing the findings, and using the information to select control methods (**figure 36.4**). Some combination of these surveillance methods is used most often for the:

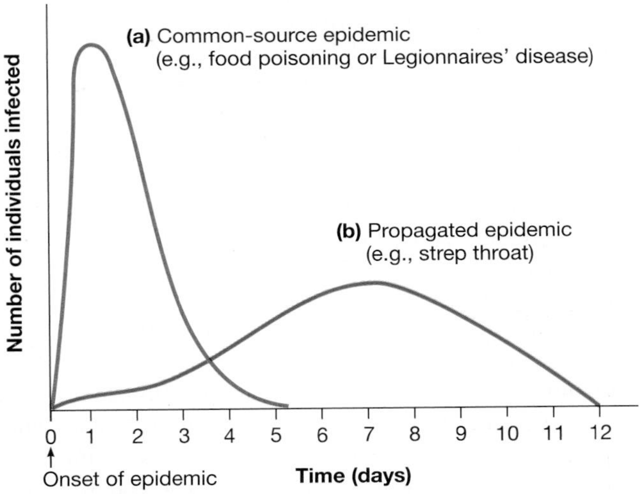

FIGURE 36.3 Epidemic Curves. (a) In a common-source epidemic, there is a rapid increase up to a peak in the number of individuals infected and then a rapid but more gradual decline. Cases usually are reported for a period that equals approximately one incubation period of the disease. (b) In a propagated epidemic, the curve has a gradual rise and then a gradual decline. Cases usually are reported over a time interval equivalent to several incubation periods of the disease.

Figure 36.3 Micro Inquiry

Do you think a flu epidemic is an example of a common-source or a propagated epidemic? Explain your answer.

1. Generation of morbidity data from case reports
2. Collection of mortality data from death certificates
3. Investigation of actual cases
4. Collection of data from reported epidemics
5. Field investigation of epidemics
6. Review of laboratory results: for example, surveys of a population for antibodies against the agent and specific microbial serotypes, skin tests, cultures, and stool analyses
7. Population surveys using statistical sampling to determine who has the disease
8. Use of animal and vector disease data
9. Collection of information on the use of specific biologics—antibiotics, antitoxins, vaccines, and other prophylactic measures
10. Use of demographic data on population characteristics such as human movements during a specific time of the year
11. Use of remote sensing and geographic information systems

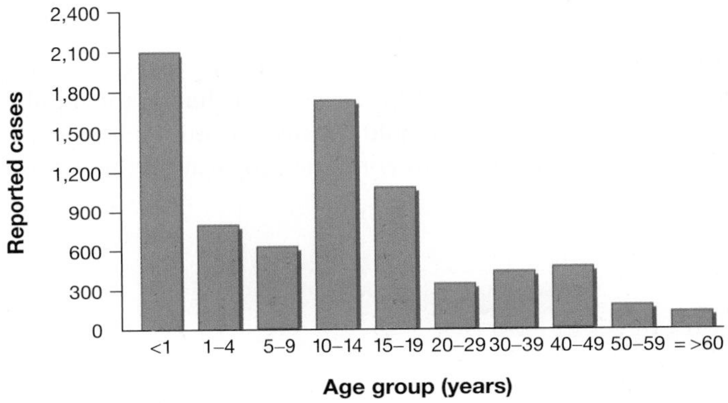

(a) Pertussis—Reported cases by age group, United States, 2000

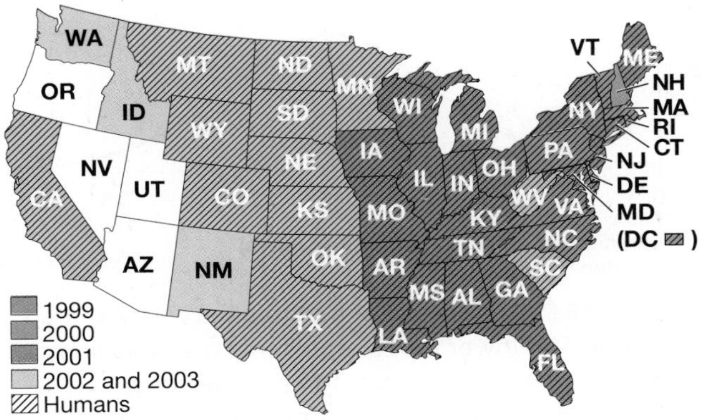

(b) West Nile virus cases were first reported in 1999 among a variety of animals and humans. Colors track its rapid progress across the United States in just 4 years.

FIGURE 36.4 Graphic Representation of Epidemiological Data. The Centers for Disease Control and Prevention collects and evaluates a number of parameters related to human health and disease. The data are then presented in various formats including (a) age and (b) geographic region.

As noted, surveillance may not always require the direct examination of cases. However, to accurately interpret surveillance data and study the course of a disease in individuals, epidemiologists and other medical professionals must be aware of the pattern of infectious diseases. Recall that the course of an infectious disease usually is identified by a characteristic pattern that reflects how the disease spreads through a population or at least to one person (*see figure 31.2*). Epidemiologists study disease patterns so they can rapidly identify new outbreaks before substantial morbidity and mortality occur. As we learn in chapters 37–39, people with infectious diseases demonstrate or present specific signs and symptoms that become clues as to the type of infecting agent, its route of transmission, and its virulence. ⏮ *The infectious disease process (section 31.2)*

To understand how epidemics are propagated, consider **figure 36.5**. At time 0, all individuals in this population are susceptible to a hypothetical pathogen. The introduction of an infected individual initiates the epidemic outbreak (lower curve), which spreads to reach a peak by day 15. As individuals recover from the disease, they become immune and no longer transmit the pathogen (upper curve). The number of susceptible individuals therefore decreases. The decline in the number of susceptibles to the threshold density (the minimum number of individuals necessary to continue propagating the disease) coincides with the peak of the epidemic wave, and the incidence of new cases declines because the pathogen cannot propagate itself.

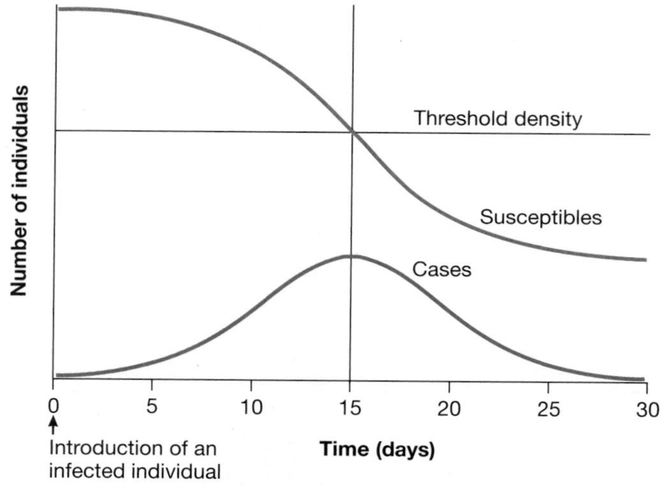

FIGURE 36.5 The Spread of an Imaginary Propagated Epidemic. The lower curve represents the number of cases, and the upper curve, the number of susceptible individuals. Notice the coincidence of the peak of the epidemic wave with the threshold density of susceptible people.

Figure 36.5 Micro Inquiry

Why does the number of susceptible individuals decline after the epidemic has reached its peak?

Herd immunity is the resistance of a population to infection and pathogen spread because of the immunity of a large percentage of the population. The larger the proportion of those immune, the smaller the probability of effective contact between infective and susceptible individuals; that is, many contacts will be immune, and thus the population will exhibit a group resistance. A susceptible member of such an immune population enjoys an immunity that is not of his or her own making but instead arises because of membership in the group.

At times public health officials immunize large portions of the susceptible population in an attempt to maintain a high level of herd immunity. Any increase in the number of susceptible individuals may result in an endemic disease becoming epidemic. The proportion of immune to susceptible individuals must be constantly monitored because new susceptible individuals continually enter a population through migration and birth.

Pathogens cause endemic diseases because infected humans continually transfer them to others (e.g., sexually transmitted diseases) or because they continually reenter the human population from animal reservoirs (e.g., rabies). Other pathogens continue to evolve and may produce epidemics (e.g., AIDS, influenza virus [A strain], and *Legionella* bacteria). One way in which a pathogen changes is by **antigenic shift,** a major genetically determined change in the antigenic character of a pathogen (**figure 36.6**). An antigenic shift can be so extensive that the pathogen is no longer recognized by the host's immune system. For example, influenza viruses frequently change by recombination from one antigenic type to another. Antigenic shift also occurs through the hybridization of different influenza virus serovars; two serovars of a virus intermingle to form a new antigenic type. Hybridization may occur between an animal strain and a human strain of the virus. Even though resistance in the human population becomes so high that the virus can no longer spread (herd immunity), it can be transmitted to animals, where the hybridization takes place. Smaller antigenic changes also can take place by mutations in pathogenic strains that help the pathogen avoid host immune responses. These smaller changes are called **antigenic drift.** ⏮ *Viruses with minus-strand RNA genomes (section 25.6)*

Whenever antigenic shift or drift occurs, the population of susceptible individuals increases because the immune system does not recognize the new mutant strain. If the percentage of susceptible people is above the threshold density (figure 36.5), the level of protection provided by herd immunity will decrease and the morbidity rate will increase. For example, the morbidity rates of influenza among schoolchildren may reach epidemic levels if the number of susceptible people rises above 30% for the whole population. One of the primary goals of public health is to ensure that the general public is sufficiently protected to withstand epidemics. This has resulted in the recommendation that at least 70% of the population be immunized against common infectious diseases, so as to provide the herd immunity necessary for the protection of those who are not immunized. In other words, by keeping the population sufficiently immunized, the chain of infectious disease transmission is broken, limiting the reach of the disease. This is especially true for infectious diseases that quickly arise by antigenic shift, the 2009 novel H1N1 pandemic influenza being a great example.

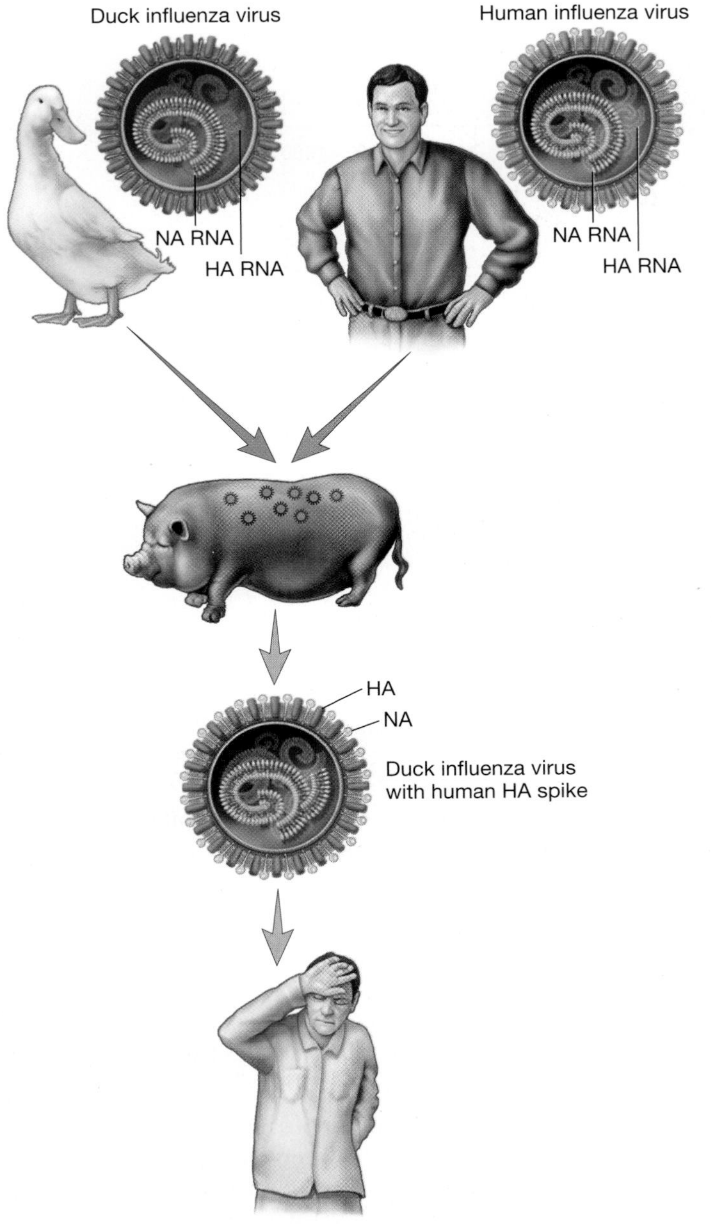

FIGURE 36.6 Antigenic Shift in Influenza Viruses. Close habitation of humans, pigs, and ducks permits the multiple infection of the pigs with both human and duck influenza viruses. Exchange of genetic information then results in a new strain of influenza that is new to humans; humans have no immunity to the new strain.

1. How can epidemiologists recognize an infectious disease in a population?
2. Differentiate between common-source and propagated epidemics.
3. Explain herd immunity. How does this protect the community?
4. What is the significance of antigenic shift and drift in epidemiology?

36.5 Emerging and Reemerging Infectious Diseases and Pathogens

Only a few decades ago, a grateful public trusted that science had triumphed over infectious diseases by building a fortress of health protection. Antibiotics, vaccines, and aggressive public health campaigns had yielded a string of victories over old enemies such as whooping cough, pneumonia, polio, and smallpox. In developed countries, people were lulled into believing that microbial threats were a thing of the past. Trends in the number of deaths caused by infectious diseases in the United States from 1900 through 1982 supported this conclusion (**figure 36.7**). However, this downward trend ended in 1982, and the death rate has since risen. The reality is that while there was a general downtrend of death due to infectious disease prior to 1982, the incidence of infectious disease resulting from emerging microbial populations has increased steadily since 1940, peaking in the late 1980s (**figure 36.8**). The world has seen the global spread of AIDS, the resurgence of tuberculosis, pandemic emergence of the H1N1 influenza virus, and the appearance of new enemies such as

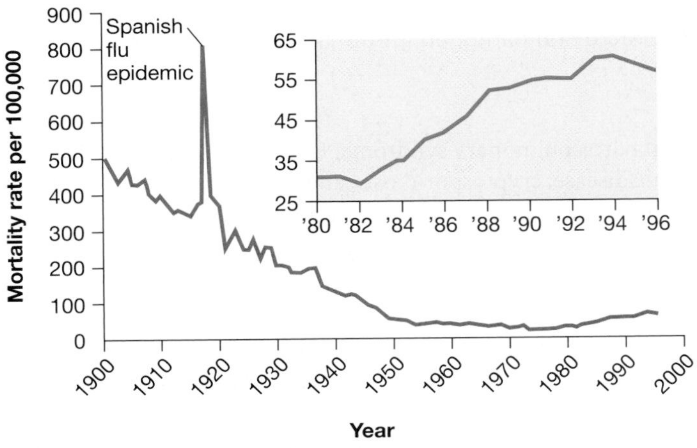

FIGURE 36.7 Infectious Disease Mortality in the United States Decreased Greatly During Most of the Twentieth Century. The insert is an enlargement of the right-hand portion of the graph and shows that the death rate from infectious diseases increased between 1980 and 1994.

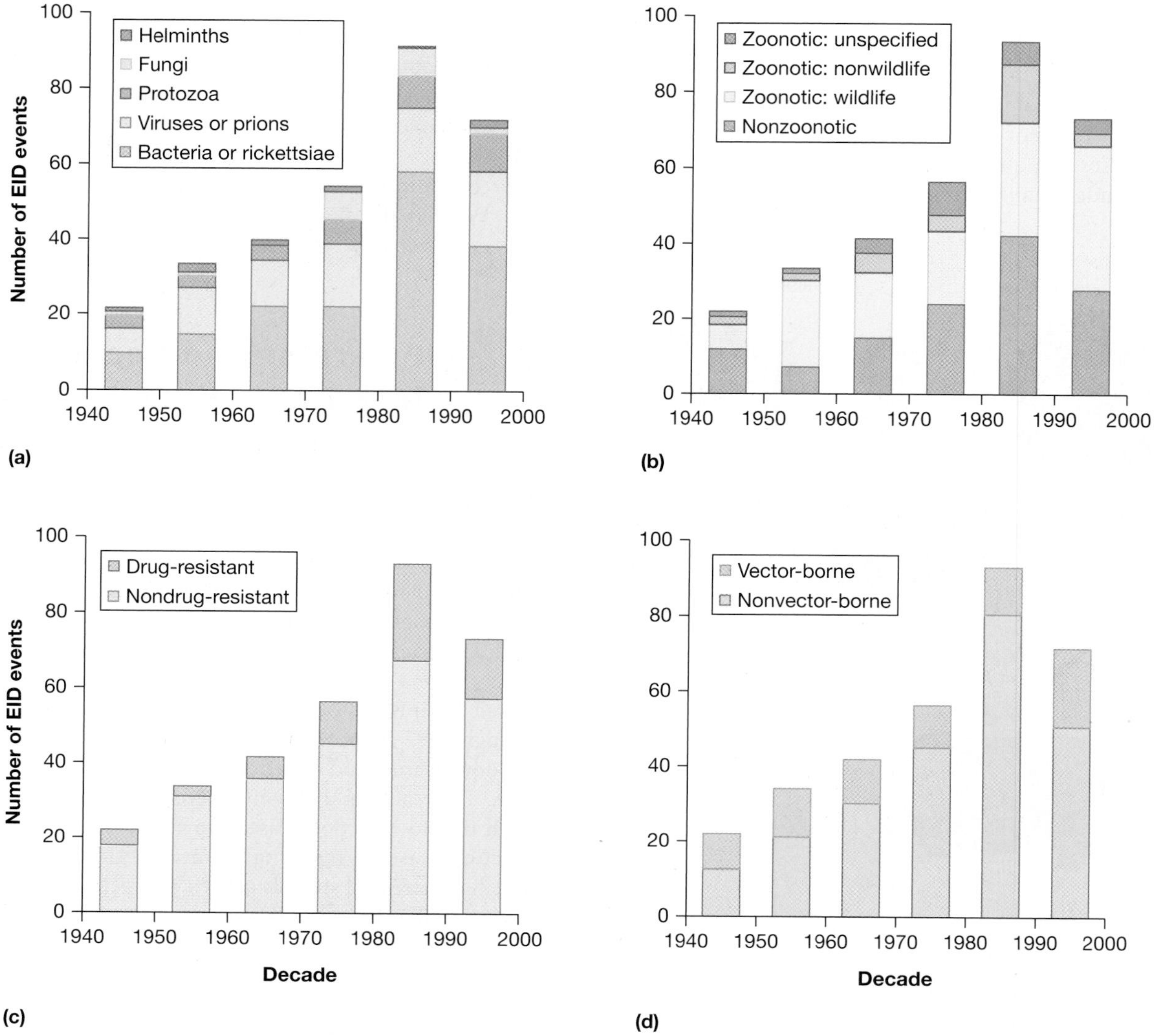

FIGURE 36.8 Emerging Infectious Diseases. Data represent the number of emerging infectious diseases per decade from 1940 to 2000 by (a) pathogen group, (b) transmission type, (c) drug resistance, and (d) mode of transmission.

hantavirus pulmonary syndrome, Hepatitis C and E, Ebola virus, Lyme disease, cryptosporidiosis, and *Escherichia coli* O157:H7. In addition, during this same time period:

1. A "bird flu" virus that had never before attacked humans began to kill people in southeast Asia.
2. A new variant of a fatal prion disease of the brain, Creutzfeldt-Jakob disease, was identified in the United Kingdom, transmitted by beef from animals with "mad cow disease."
3. *Staphylococcus* bacteria with resistance to methicillin and vancomycin, long the antibiotics of first choice and last resort, respectively, were seen for the first time.
4. Several major multistate food-borne outbreaks occurred in the United States, including those caused by *Salmonella* in peanut butter, protists on raspberries, viruses on strawberries, and various bacteria in produce, ground beef, cold cuts, and breakfast cereal.
5. A new strain of tuberculosis that is resistant to many drugs and occurs most often in people infected with HIV arose in New York and other large cities.

By the 1990s the idea that infectious diseases no longer posed a serious threat to human health was obsolete. In the twenty-first century, it is clear that humans will continually be faced with both new infectious diseases and the reemergence of older diseases once thought to be conquered (e.g., tuberculosis, dengue hemorrhagic fever, yellow fever). The Centers for Disease Control and Prevention has defined these diseases as "new, reemerging, or drug-resistant infections whose incidence in humans has increased within the past three decades or whose incidence threatens to increase in the near future." Of note are data that suggest

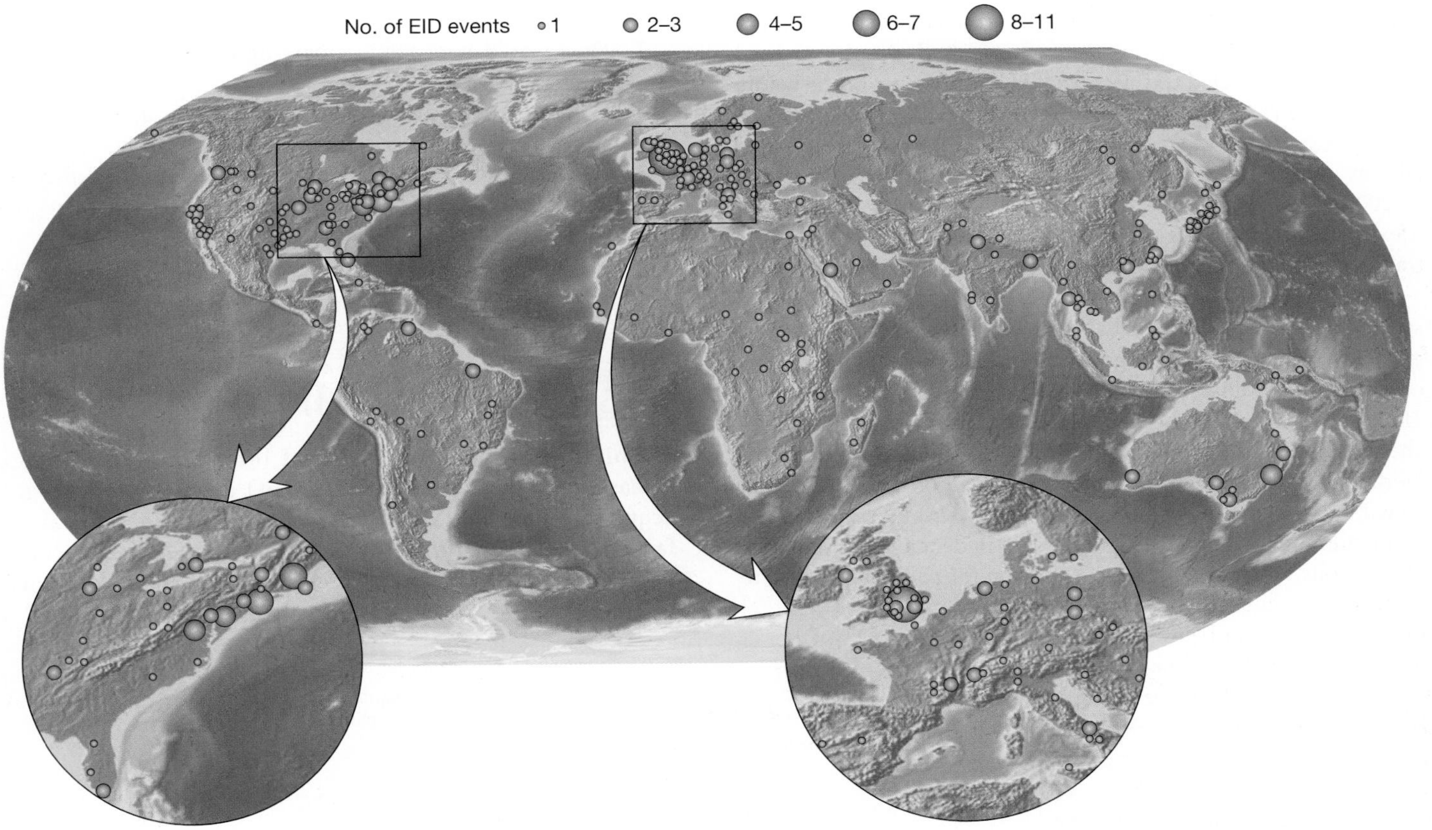

FIGURE 36.9 Map of Geographic Origins of Emerging Infectious Diseases (EID) from 1940 to 2004. Circles represent sites of EID with circle size proportional to the number of events.

that the highest concentration of emerging infectious diseases, per million square kilometers of land, is between 30 and 60 degrees north and between 30 and 40 degrees south; the hot spots of emerging infectious diseases from 1940 to 2004 are in the northeastern United States, western Europe, Japan, and southeastern Australia (**figure 36.9**). The increased importance of emerging and reemerging infectious diseases has stimulated the establishment of a field called **systematic epidemiology,** which focuses on the ecological and social factors that influence the development of these diseases. Coupled with systematic epidemiology and remote sensing techniques, the growing field of public health informatics integrates surveillance activities, defines data elements, and disseminates information via the Internet so as to better track emerging and reemerging infectious diseases in real time.

Why are viruses, bacteria, fungi, and parasites posing such a problem, despite dramatic advances in medical research, drug discovery, technology development, and sanitation? Many factors characteristic of the modern world undoubtedly favor the development and spread of these microorganisms and their diseases. Examples include:

1. Unprecedented worldwide population growth, population shifts, and urbanization
2. Increased international travel, transport (commerce), migration, and relocation of animals and food products
3. Changes in food processing, handling, and agricultural practices
4. Changes in human behavior, technology, and industry
5. Human encroachment on wilderness habitats that are reservoirs for insects and animals that harbor infectious agents
6. Microbial evolution (e.g., selective pressure) and the development of resistance to antibiotics and other antimicrobial drugs
7. Changes in ecology and climate
8. Modern medicine (e.g., immunosuppression)
9. Inadequacy of public infrastructure and vaccination programs
10. Social unrest, civil wars, and bioterrorism (section 36.8)

As population density increases in cities, the dynamics of microbial exposure and evolution increase in humans. Urbanization often crowds humans and increases exposure to microorganisms. Crowding leads to unsanitary conditions and hinders the effective implementation of adequate medical care, enabling more widespread transmission and propagation of pathogens. In modern societies, crowded workplaces, communal-living settings, day-care centers, large hospitals, and public transportation all facilitate microbial transmission. In this new millennium, the speed and volume of international travel are major factors contributing to the global emergence of infectious diseases. The spread of a new disease often used to be limited by the travel time needed to reach a new host population. If the travel time was sufficiently long, as when a ship crossed the ocean, the infected travelers would either recover or die before reaching a new population. Because travel by air has obliterated time between exposure and disease outbreak, a traveler can spread virtually any infectious disease in a matter of hours. The H1N1 pandemic of 2009 is one example of how one virus-infected individual (in Mexico) can lead to a global epidemic as asymptomatic carriers transport the virus to others during normal activities (see p. 899).

Furthermore, land development and the exploration and destruction of natural habitats have increased the likelihood of human exposure to new pathogens and may put selective pressures on pathogens to adapt to new hosts and changing environments. The introduction of pathogens to a new environment or host can alter transmission and exposure patterns, leading to sudden proliferation of disease. For example, the spread of Lyme disease in New England probably is due partly to ecological disruption that eliminates predators of deer. An increase in deer and the deer tick populations provides a favorable situation for pathogen spread to humans. Whenever the environment is altered and new environments are created, this may not only confer a survival advantage but may also increase a pathogen's virulence and alter its drug susceptibility profile. When changes in climate or ecology occur, it should not be surprising to find changes in both beneficial and detrimental microorganisms. Global warming also affects microorganism selection and survival. Finally, mass migrations of refugees, workers, and displaced persons have led to a steady growth of urban centers at the expense of rural areas.

Emerging and reemerging pathogens and their diseases are therefore the outcome of many different factors. Because the world is now so interconnected, we cannot isolate ourselves from other countries and continents. Changes in the disease status of one part of the world or the misuse of antibiotics in another part may well affect health around the planet. As Nobel laureate Joshua Lederberg so eloquently stated, "The microbe that felled one child in a distant continent yesterday can reach your child today and seed a global pandemic tomorrow."

Search This: Emerging infectious disease

1. How would you define emerging or reemerging infectious diseases?
2. What are some of the factors responsible for the emergence or reemergence of pathogens?
3. Describe how virulence and the mode of transmission may be related. What might cause the development of new human diseases?

36.6 Nosocomial Infections

Hospital-acquired or **nosocomial infections** (Greek *nosos,* disease, and *komeion,* to take care of) result from pathogens acquired by patients while in a hospital or other clinical care facility (**figure 36.10**). Besides harming patients, nosocomial infections can affect nurses, physicians, aides, visitors, sales-

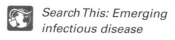

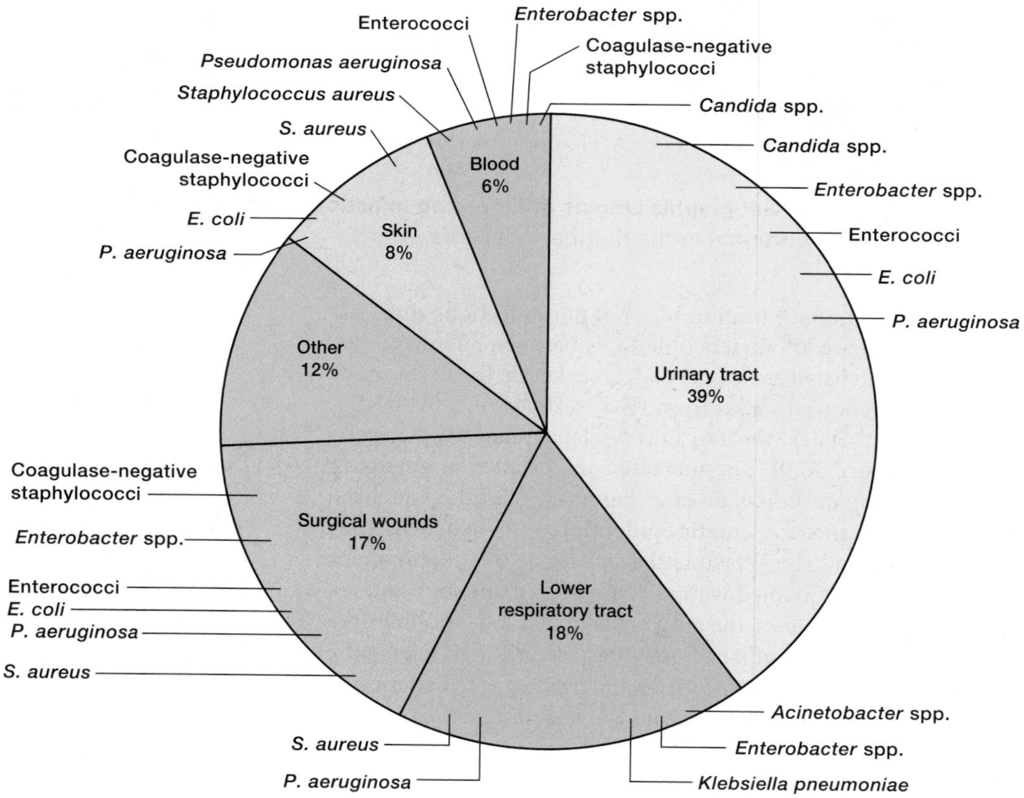

FIGURE 36.10 Nosocomial Infections. Relative frequency by body site. These data are from the National Nosocomial Infections Surveillance, which is conducted by the Centers for Disease Control and Prevention.

Figure 36.10 Micro Inquiry

What is the source of most pathogens that cause nosocomial infections?

people, delivery personnel, custodians, and anyone else who has contact with the hospital. Most nosocomial infections become clinically apparent while patients are still hospitalized; however, disease onset can occur after patients have been discharged. Infections that are incubating when patients are admitted to a hospital are not nosocomial; they are community acquired. However, because such infections can serve as a ready source or reservoir of pathogens for other patients or personnel, they are also considered in the total epidemiology of nosocomial infections. The CDC estimates that about 10% of all hospital patients (up to 4 million people) acquire some type of nosocomial infection. Thus nosocomial infections represent a significant proportion of all infectious diseases acquired by humans.

Nosocomial diseases are usually caused by bacteria, most of which are noninvasive and part of the normal microbiota; viruses, protozoa, and fungi are rarely involved. Figure 36.10 summarizes the most common types of nosocomial infections and the most common nosocomial pathogens. Interestingly, throughout most of the twentieth century, hospital-acquired infections were dominated early on by staphylococci, which initially responded to penicillin. During subsequent years, the emergence of methicillin-resistant *Staphylococcus aureus* (MRSA) increased dramatically; similar patterns are emerging for penicillin-resistant *Streptococcus pneumoniae.* Vancomycin-resistant *S. aureus* infections are now a major nosocomial threat. First reported in the late 1980s, vancomycin-resistant enterococci (VRE) are now common in U.S. hospitals. Recently recognized nosocomial, gram-positive species include *Corynebacterium jeikeium* and *Rhodococcus equi.* The incidences of infections by the gram-negative pathogens *Pseudomonas aeruginosa, Acinetobacter* ssp., *Burkholderia cepacia* and *Stenotrophomonas maltophilia* have increased. With respect to the extended-spectrum β-lactam-resistant gram-negative bacilli, bacteria such as *Klebsiella pneumoniae, E. coli,* other *Klebsiella* spp., *Proteus* spp., *Morganella* spp., *Citrobacter* spp., *Salmonella* spp., and *Serratia marcescens* are resistant to penicillins; first-generation cephalosporins; and some third-generation cephalosporins such as cefotaxime (Claforan), ceftriaxone (Rocephin), ceftazidime, and aztreonam (Azactam).

Sources

The nosocomial pathogens that cause diseases come from either endogenous or exogenous sources. Endogenous sources are the patient's own microbiota; exogenous sources are microbiota other than the patient's. Endogenous pathogens are either brought into the hospital by the patient or are acquired when the patient becomes colonized after admission. In either case, the pathogen colonizing the patient may subsequently cause a nosocomial disease (e.g., when the pathogen is transported to another part of the body or when the host's resistance drops). If it cannot be determined that the specific pathogen responsible for a nosocomial disease is exogenous or endogenous, then the term autogenous is used. An autogenous infection is one that is caused by an agent derived from the microbiota of the patient, despite whether it became part of the patient's microbiota following his or her admission to the hospital.

Many potential exogenous sources exist in a hospital. Animate sources are the hospital staff, other patients, and visitors. Some examples of inanimate exogenous sources are food, computer keyboards, urinary catheters, intravenous and respiratory therapy equipment, and water systems (e.g., softeners, dialysis units, and hydrotherapy equipment).

Control, Prevention, and Surveillance

In the United States, nosocomial infections prolong hospital stays by 4 to 13 days, result in over $4.5 billion a year in direct hospital charges, and lead to over 20,000 direct and 60,000 indirect deaths annually. The enormity of this problem has led most hospitals to allocate substantial resources to developing methods and programs for the surveillance, prevention, and control of nosocomial infections.

All personnel involved in the care of patients should be familiar with basic infection control measures such as isolation policies of the hospital; aseptic techniques; proper handling of equipment, supplies, food, and excreta; and surgical wound care and dressings. To adequately protect their patients, hospital personnel must practice proper aseptic technique and hand-washing procedures, and must wear gloves when contacting blood, mucous membranes, and secretions. Patients should be monitored with respect to the frequency, distribution, symptomatology, and other characteristics common to nosocomial infections. A dynamic control and surveillance program can be invaluable in preventing many nosocomial infections, patient discomfort, extended stays, and further expense. All hospitals desiring accreditation by the Joint Commission on Accreditation of Healthcare Organizations must have a designated individual directly responsible for developing and implementing policies governing control of infections and communicable diseases. This infection control practitioner periodically evaluates laboratory reports, patients' charts, and surveys to determine whether any increase has occurred in the frequency of particular infectious diseases or potential pathogens.

Without a doubt, the key factors responsible for the rise in drug-resistant pathogens have been the excessive or inappropriate use of antimicrobial therapy and the indiscriminate use of broad-spectrum antibiotics. Although the CDC acknowledges that it is too late to solve the resistance problem simply by using antimicrobial agents more prudently, it is no less true that the problem of drug resistance will continue to worsen if this does not occur. Also needed (especially in underdeveloped countries) is a renewed emphasis on alternate prevention and control strategies that prevailed in the years before antimicrobial chemotherapy. These include improved sanitation and hygiene, isolation of infected persons, antisepsis, and vaccination. ⏮ *Drug resistance (section 34.9)*

1. Describe a nosocomial infection and explain why such infections are important.
2. What two general sources are responsible for nosocomial infections? Give some specific examples of each general source.
3. What does an infection control practitioner do to control nosocomial infections?
4. What are some key factors responsible for the rise in drug-resistant bacteria?

36.7 Control of Epidemics

The development of an infectious disease is a complex process involving many factors, as is the design of specific epidemiological control measures. Epidemiologists must consider available resources and time constraints, adverse effects of potential control measures, and human activities that might influence the spread of the infection. Many times control activities reflect compromises among alternatives. To proceed intelligently, one must identify components of the infectious disease cycle that are primarily responsible for a particular epidemic. Control measures should be directed toward that part of the cycle that is most susceptible to control—the weakest link in the chain (*see figure 31.3*).

There are three types of control measures. The first type of control measure is directed toward reducing or eliminating the source or reservoir of infection: Social distancing and isolation of carriers, Destruction of an animal reservoir of infection, Treatment of water and sewage to reduce contamination (*see chapter 42*), Therapy that reduces or eliminates infectivity of the individual.

The second type of control measure is designed to break the connection between the source of the infection and susceptible individuals. Examples include general sanitation measures: Chlorination of water supplies, Pasteurization of milk and other beverages, Supervision and inspection of food and food handlers, Destruction of vectors (e.g., spraying to eliminate mosquitoes).

The third type of control measure reduces the number of susceptible individuals and raises the general level of herd immunity by immunization. Examples include: Passive immunization (*see figure 33.3*) to give a temporary immunity following exposure to a pathogen or when a disease threatens to become an epidemic, Active immunization to protect the individual from the pathogen and the host population from the epidemic, Prophylactic treatment to prevent infection (e.g., use of chloroquine when traveling to malaria-endemic countries), We will discuss the use of vaccines and the subsequent immunization in the next section. Antimicrobial chemotherapy is discussed in chapter 34.

Vaccines and Immunization

A **vaccine** (Latin *vacca,* cow) is a preparation of one or more microbial antigens used to induce protective immunity. **Immunization** is the result achieved by the successful delivery of vaccines; it stimulates immunity. Vaccination attempts to induce antibodies and activated T cells to protect a host from future infection. Many epidemics have been stayed by mass prophylactic immunization (**table 36.2**). Vaccines have eradicated smallpox, pushed polio to the brink of extinction, and spared countless individuals from hepatitis A and B, influenza, measles, rotavirus disease, tetanus, typhus, and other dangerous diseases. **Vaccinomics,** the application of genomics and bioinformatics to vaccine development, is bringing a fresh approach to the Herculean problem of making vaccines against various microorganisms and helminths. ⏮ *Comparative genomics (section 16.7)*

To promote a more efficient immune response, antigens in vaccines can be mixed with an **adjuvant** (Latin *adjuvans,* aiding), which enhances the rate and degree of immunization. Adjuvants can be any nontoxic material that prolongs antigen interaction with immune cells, assists in processing of antigens by antigen-presenting cell (APC), or otherwise nonspecifically stimulates the immune response to the antigen. Several types of adjuvants can be used. Oil in water emulsions (Freund's incomplete adjuvant), aluminum hydroxide salts (alum), beeswax, and various combinations of bacteria (live or killed) are used as vaccine adjuvants. In most cases, the adjuvant materials trap the antigen, thereby promoting a sustained release as APCs digest and degrade the preparation. In other cases, the adjuvant activates APCs so that antigen recognition, processing, and presentation are more efficient.

The modern era of vaccines and immunization began in 1798 with Edward Jenner's use of cowpox as a vaccine against smallpox (**Historical Highlights 36.4**) and in 1881 with Louis Pasteur's rabies vaccine. Vaccines for other diseases did not emerge until later in the nineteenth century, when largely through a process of trial and error, methods for inactivating and attenuating microorganisms were improved and vaccines were produced. Vaccines were eventually developed against most of the epidemic diseases that plagued Western Europe and North America (e.g., diphtheria, measles, mumps, pertussis, German measles, and polio). Indeed, by the late twentieth century, it seemed that the combination of vaccines and antibiotics had solved the problem of microbial infections in developed nations. Such optimism was cut short by the emergence of new and previously unrecognized pathogens and antibiotic resistance among existing pathogens. Nevertheless, vaccination is still one of the most cost-effective weapons for preventing microbial disease.

Vaccination of most children should begin at about age two months (**table 36.3**). Before that age, they are protected by passive natural immunity from maternal antibodies. Further vaccination of teens and adults depends on their relative risk for exposure to infectious disease. Individuals living in close quarters (e.g., college students in residence halls, military personnel), the elderly, and individuals with reduced immunity (e.g., those with chronic and metabolic diseases) should receive vaccines for influenza, meningitis, and pneumonia. International travelers may be immunized against cholera, hepatitis A, plague,

Table 36.2 Examples of Vaccines to Prevent Viral and Bacterial Diseases in Humans

Disease	*Vaccine*	*Booster*	*Recommendation*
Viral Diseases			
Chickenpox	Attenuated Oka strain (Varivax)	None	Children 12–18 months: older children who have not had chickenpox
Hepatitis A	Inactivated virus (Havrix)	6–12 months	International travelers
Hepatitis B	HB viral antigen (Engerix-B, Recombivax HB)	1–4 months 6–18 months	High-risk medical personnel: children, birth to 18 months and 11–12 years of age
Influenza A	Inactivated virus or live attenuated	Yearly	All persons
Measles, Mumps, Rubella	Attenuated viruses (combination MMR vaccine)	None	First dose 12–15 months, 2nd dose 4–6 years
Poliomyelitis	Attenuated (oral poliomyelitis vaccine, OPV) or inactivated virus	Adults as needed	First dose at 2 months, 2nd at 4 months, 3rd at 16–18 months, 4th at 4–6 years
Rabies	Inactivated virus	None	For individuals in contact with wildlife, animal control personnel, veterinarians
Respiratory disease	Live attenuated adenovirus	None	Military personnel
Smallpox	Live attenuated *Vaccinia virus*	None	Laboratory, health-care, and military personnel
Yellow fever	Attenuated virus	10 years	Military personnel and individuals traveling to endemic areas
Bacterial Diseases			
Anthrax	Extracellular components of unencapsulated *B. anthracis*	None	Agricultural workers, veterinary, and military personnel
Cholera	Fraction of *Vibrio cholerae*	6 months	Individuals in endemic areas, travelers
Diphtheria, Pertussis, Tetanus	Diphtheria toxoid, killed *Bordetella pertussis,* tetanus toxoid (DPT vaccine) or with acellular *pertussis* (DtaP); or tetanus toxoid, reduced diphtheria toxoid, and acellular pertussis vaccine, adsorbed (Tdap)	10 years	Children from 2–3 months old to 12 years, and adults; children 10–18 years, at least 5 years after DPT series, should receive Tdap
Haemophilus influenzae type b	Polysaccharide-protein conjugate (HbCV) or bacterial polysaccharide (HbPV)	None	First dose at 2 months, 2nd at 4 months, 3rd at 6 months, 4th at 12–15 months
Meningococcal infections	*Neisseria meningitidis* polysaccharides of serotypes A/C/Y/W-135	None	Military; high-risk individuals; college students living in dormatories; elderly in nursing homes
Plague	Fraction of *Yersinia pestis*	Yearly	Individuals in contact with rodents in endemic areas
Pneumococcal pneumonia	Purified *S. pneumoniae* polysaccharide of 23 pneumococcal types	None	Adults over 50 with chronic disease
Q fever	Inactivated *Coxiella burnetii*	None	Workers in slaughterhouses and meat-processing plants
Tuberculosis	Attenuated *Mycobacterium bovis* (BCG vaccine)	3–4 years	Individuals exposed to TB for prolonged periods of time; used in some countries, not licensed in the U.S.
Typhoid fever	*Salmonella enterica* Typhi Ty21a (live attenuated or polysaccharide)	None	Residents of and travelers to areas of endemic disease
Typhus fever	Killed *Rickettsia prowazekii*	Yearly	Scientists and medical personnel in areas where typhus is endemic

HISTORICAL HIGHLIGHTS

36.4 The First Immunizations

Since the time of the ancient Greeks, it has been recognized that people who have recovered from plague, smallpox, yellow fever, and various other infectious diseases rarely contract the diseases again. The first scientific attempts at artificial immunizations were made in the late eighteenth century by Edward Jenner (1749–1823), who was a country doctor from Berkley, Gloucestershire, England. Jenner investigated the basis for the widespread belief of the English peasants that anyone who had vaccinia (cowpox) never contracted smallpox. Smallpox was often fatal—10 to 40% of the victims died—and those who recovered had disfiguring pockmarks. Yet most English milkmaids, who were readily infected with cowpox, had clear skin because cowpox was a relatively mild infection that left no scars.

It was on May 14, 1796, that Jenner extracted the contents of a pustule from the arm of a cowpox-infected milkmaid, Sarah Nelmes, and injected it into the arm of eight-year-old James Phipps. As Jenner expected, immunization with the cowpox virus caused only mild symptoms in the boy. When he subsequently inoculated the boy with smallpox virus (an act now considered completely unethical), the boy showed no symptoms of the disease. Jenner then inoculated large numbers of his patients with cowpox pus, as did other physicians in England and on the European continent (**box figure**). By 1800 the practice known as variolation had begun in America, and by 1805 Napoleon Bonaparte had ordered all French soldiers to be vaccinated.

Further work on immunization was carried out by Louis Pasteur (1822–1895). Pasteur discovered that if cultures of chicken cholera bacteria were allowed to age for 2 or 3 months, the bacteria produced only a mild attack of cholera when inoculated into chickens. Somehow the old cultures had become less pathogenic (attenuated) for the chickens. He then found that fresh cultures of the bacteria failed to produce cholera in chickens that had been previously inoculated with old, attenuated cultures. To honor Jenner's work with cowpox, Pasteur gave the name *vaccine* to any preparation of a weakened pathogen that was used (as was Jenner's "vaccine virus") to immunize against infectious disease.

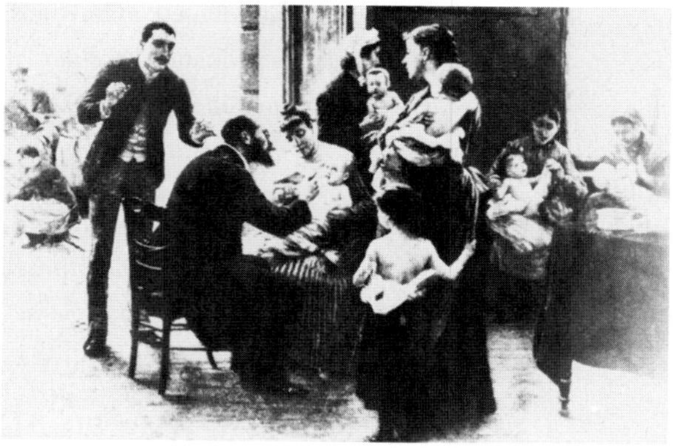

Nineteenth-Century Physicians Performing Vaccinations on Children.

Table 36.3 Recommended Childhood and Adolescent Immunization Schedule, United States, 2009

Vaccine ▼ Age ▶	Birth	*1 month*	*2 months*	*4 months*	*6 months*	*12 months*	*15 months*	*18 months*	*19–23 months*	*2–3 years*	*4–6 years*
Hepatitis B	HepB	HepB				HepB					
Rotavirus			RV	RV	RV						
Diphtheria/Tetanus/Pertussis			DTaP	DTaP	DTaP		DTaP				DTaP
Haemophilus influenzae type b			Hib	Hib	Hib	Hib					
Pneumococcal			PCV	PCV	PCV	PCV				PPSV	
Inactivated poliovirus			IPV	IPV		IPV					IPV
Influenza								Influenza (Yearly)			
Measles, mumps, rubella						MMR					MMR
Varicella						Varicella					Varicella
Hepatitis A							HepA (2 doses)			HepA Series	
Meningococcal										MCV	

Source: http://www.cdc.gov

Table 36.4 A Comparison of Inactivated (Killed) and Attenuated (Live) Vaccines

Major Characteristic	*Inactivated Vaccine*	*Attenuated Vaccine*
Booster shots	Multiple boosters required	Only a single booster, if any, required
Production	Virulent microorganism inactivated by chemicals or irradiation	Virulent microorganism grown under adverse conditions or passed through different hosts until avirulent
Reversion tendency	None	May revert to a virulent form
Stability	Very stable, even where refrigeration is unavailable	Less stable
Type of immunity induced	Humoral	Humoral and cell-mediated

Source: Adapted from Goldsby, R. A.; Kindt, T. J.; and Osborne, B. A. 2003. Kuby Immunology. *New York: W. H. Freeman.*

polio, typhoid, typhus, and yellow fever, depending on the country visited. Veterinarians, forest rangers, and others whose jobs involve contact with animals may be vaccinated against rabies, plague, and anthrax. Health-care workers are typically immunized against *Hepatitis B virus*. The role of immunization as a protective therapy cannot be overstated; immunizations save lives.

Whole-Cell Vaccines

Many of the current vaccines used for humans that are effective against viral and bacterial diseases consist of whole microorganisms, termed **whole-cell vaccines.** These are either inactivated (killed) or attenuated (live but avirulent). The major characteristics of these vaccines are compared in **table 36.4**. **Inactivated vaccines** are effective, but they are less immunogenic, so they often require several boosters and normally do not adequately stimulate cell-mediated immunity or secretory IgA production. In contrast, **attenuated vaccines** usually are given in a single dose and stimulate both humoral and cell-mediated immunity.

Even though whole-cell vaccines are considered the "gold standard" of vaccines, they can be problematic. For example, whole-organism vaccines fail to shield against some diseases. Attenuated vaccines can also cause full-blown illness in an individual whose immune system is compromised (e.g., AIDS patients, cancer patients undergoing chemotherapy, the elderly). These same individuals may also contract the disease from healthy people who have been vaccinated recently. Moreover, attenuated viruses can at times mutate in ways that restore virulence, as has happened in some monkeys given an attenuated simian form of the AIDS virus.

Acellular or Subunit Vaccines

A few of the common risks associated with whole-cell vaccines can be avoided by using only specific, purified macromolecules derived from pathogenic microorganisms. There are three general forms of **subunit vaccines:** (1) capsular polysaccharides, (2) recombinant surface antigens, and (3) inactivated exotoxins called **toxoids.** The purified microbial subunits or their secreted products can be prepared as nontoxic antigens to be used in the formulation of vaccines (**table 36.5**).

Table 36.5 Subunit Vaccines Currently Available for Human Use

Microorganism or Toxin	*Vaccine Subunit*
Capsular polysaccharide	
Haemophilus influenzae type b	Polysaccharide-protein conjugate (HbCV) or bacterial polysaccharide (HbPV)
Neisseria meningitidis	Polysaccharides of serotypes A/C/Y/W-135
Streptococcus pneumoniae	23 distinct capsular polysaccharides
Surface antigen	
Hepatitis B virus	Recombinant surface antigen (HbsAg)
Toxoids	
Corynebacterium diphtheriae toxin	Inactivated exotoxin
Clostridium tetani toxin	Inactivated exotoxin

Recombinant-Vector and DNA Vaccines

Genes isolated from a pathogen that encode major antigens can be inserted into nonvirulent viruses or bacteria. Such recombinant microorganisms serve as vectors, replicating within the host and expressing the gene product of the pathogen-encoded antigenic proteins. The antigens elicit humoral immunity (i.e., antibody production) when they escape from the vector, and they also elicit cellular immunity when they are broken down and properly displayed on the cell surface (just as occurs when host cells harbor an active pathogen). Several microorganisms, such as adenovirus and attenuated *Salmonella*, have been used in the production of these **recombinant-vector vaccines.**

On the other hand, **DNA vaccines** introduce DNA directly into the host cell. When injected into muscle cells, the DNA is taken into the nucleus and the pathogen's DNA fragment is transiently expressed, generating foreign proteins to which the host's immune system responds (**figure 36.11**). DNA vaccines are very stable; refrigeration is often unnecessary. At present, human trials are underway with several different DNA vaccines against malaria, AIDS, influenza, hepatitis B, and herpesvirus. DNA vaccines against a number of cancers (such as lymphomas, prostate, colon) are also being tested. *Constructing Vaccines*

The Role of the Public Health System: Epidemiological Guardian

The control of an infectious disease relies heavily on a well-defined network of clinical microbiologists, nurses, physicians, epidemiologists, and infection control personnel who supply epidemiological information to a network of local, state, national, and international organizations. These individuals and organizations comprise the public health system. For example, each state has a public health laboratory that is involved in infection surveillance and control. The communicable disease section of a state laboratory includes specialized laboratory services for the examination of specimens or cultures submitted by physicians, local health departments, hospital laboratories, sanitarians, epidemiologists, and others. These groups share their findings with other health agencies in the state, the Centers for Disease Control and Prevention, and the World Health Organization.

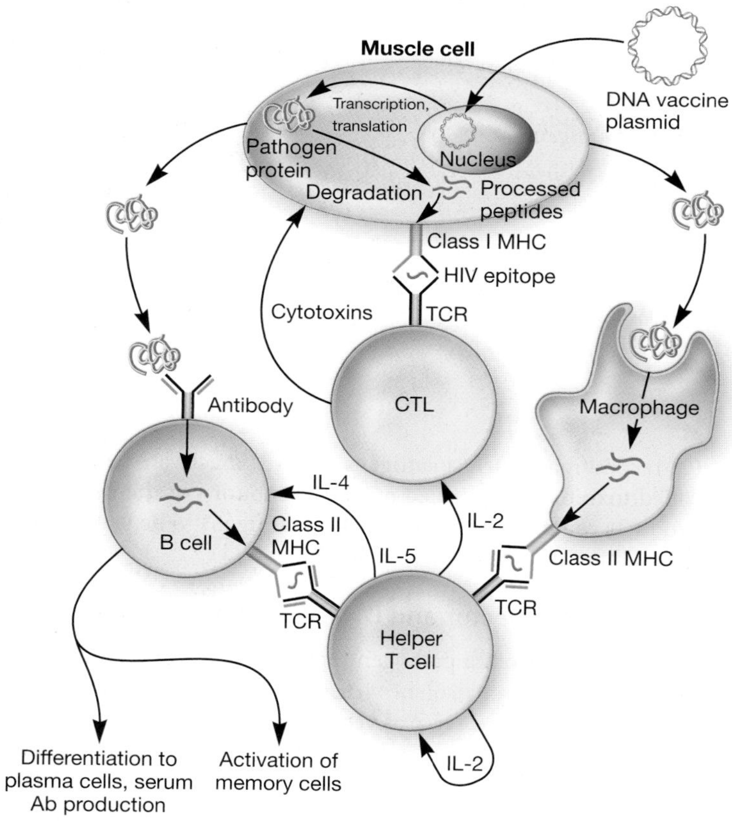

FIGURE 36.11 DNA Vaccine. DNA coding for pathogen proteins is inserted into host cells and expressed. The foreign proteins elicit both cell-mediated and humoral responses.

36.8 Bioterrorism Preparedness

Bioterrorism is defined as "the intentional or threatened use of viruses, bacteria, fungi, or toxins from living organisms to produce death or disease in humans, animals, and plants." The use of biological agents to effect personal or political outcome is not new, and the modern use of biological agents is a reality (**Historical Highlights 36.5**). The most notable intentional uses of biological agents for criminal or terror intent are (1) the use of *Salmonella enterica* serovar Typhimurium in 10 restaurant salad bars (by the Rajneeshee religious cult in The Dalles, Oregon, 1984); (2) the intentional release of *Shigella dysenteriae* in a hospital laboratory break room (perpetrator[s] unknown; Texas, 1996); and (3) the use of *Bacillus anthracis* spores delivered through the U.S. postal system (perpetrator[s] unknown, although the FBI named Bruce Ivins [now deceased] as the likely suspect; five eastern U.S. states, 2001). The *Salmonella*-contaminated salads resulted in 751 documented cases and 45 hospitalizations due to salmonellosis. The *Shigella* release resulted in eight confirmed cases and four hospitalizations for shigellosis. The *Bacillus* spores infected 22 people (11 cases of inhalation anthrax and 11 cases of cutaneous anthrax) and caused five deaths. The list of biological agents that could pose the greatest public health risk in the event of a bioterrorist attack is relatively short and includes viruses, bacteria, parasites, and toxins that, if acquired and properly disseminated, could become a difficult public health challenge in terms of limiting the numbers of casualties and controlling panic (**table 36.6**).

Among weapons of mass destruction, biological weapons can be more destructive than chemical weapons, including nerve gas. In certain circumstances, biological weapons can be as devastating as a nuclear explosion—a few kilograms of anthrax could kill as many people as a Hiroshima-size nuclear bomb. Biological agents are likely to be chosen as a means of localized attack (biocrime) or mass casualty (bioterrorism) for several reasons. They are mostly invisible, odorless, tasteless, and difficult to detect. Use of biological agents for terrorism also means that perpetrators may escape undetected as it may take hours to days before signs and symptoms of their use become evident. Additionally, the general public is not likely to be protected immunologically against agents that might be used in bioterrorism. Ultimately the use of biological agents in terrorism results in fear, panic, and chaos.

In 1998 the U.S. government launched the first national effort to create a biological weapons defense. The initiatives included (1) the first-ever procurement of specialized vaccines and medicines for a national civilian protection stockpile; (2) invigoration of research and development in the science of biodefense;

HISTORICAL HIGHLIGHTS

36.5 1346—The First Recorded Biological Warfare Attack

The Black Death (*see p. 945*), which swept through Europe, Asia, and North Africa in the mid-fourteenth century, was probably the greatest public health disaster in recorded history. Europe, for example, lost an estimated quarter to a third of its population. This is not only of great historical interest but also relevant to current efforts to evaluate the threat of military or terrorist use of biological weapons.

Some believe that evidence for the origin of the Black Death in Europe is found in a memoir by the Genoese Gabriele de' Mussi. According to this fourteenth-century memoir, the Black Death reached Europe from the Crimea (a region of the Ukraine) in 1346 as a result of a biological warfare attack. The Mongol army hurled plague-infected cadavers into the besieged Crimean city of Caffa (now Feodosija, Ukraine), thereby transmitting the disease to the inhabitants; fleeing survivors then spread the plague from Caffa to the Mediterranean Basin. Such transmission was especially likely at Caffa, where cadavers would have been badly mangled by being hurled and the defenders probably often had cut or abraded hands from coping with the bombardment. Because many cadavers were involved, the opportunity for disease transmission was greatly increased. Disposal of victims' bodies in a major disease outbreak is always a problem, and the Mongol army used their hurling machines as a solution to limited mortuary facilities. It is possible that thousands of cadavers were disposed of this way; de' Mussi's description of "mountains of dead" might have been quite literally true. Indeed, Caffa could be the site of the most spectacular incident of biological warfare ever, with the Black Death as its disastrous consequence. It is a powerful reminder of the horrific consequences that can result when disease is successfully used as a weapon.

Table 36.6 Pathogens and Toxins Defined by the CDC as Select Agents

Category	*Definition*	*Disease (Agent)*
A	Easily disseminated or transmitted from person to person; high mortality rates; potential for major public health impact; cause public panic and social disruption; require special action for public health preparedness	**Anthrax** (*Bacillus anthracis*) **Botulism** (*Clostridium botulinum* toxin) **Plague** (*Yersinia pestis*) **Smallpox** (Variola major) **Tularemia** (*Francisella tularensis*) **Viral hemorrhagic fever** (filoviruses and arenaviruses)
B	Moderately easy to disseminate, moderate morbidity and mortality rates; require specific enhancements of CDC's diagnostic capacity and enhanced disease surveillance	**Brucellosis** (*Brucella* species) **Glanders** (*Burkholderia mallei*) **Melioidosis** (*Burkholderia pseudomallei*) **Psittacosis** (*Chlamydia psittaci*) **Q fever** (*Coxiella burnetii*) **Typhus fever** (*Rickettsia prowazekii*) **Viral encephalitis** (alphaviruses) **Toxemia** Ricin from castor beans Staphylococcal enterotoxin B Epsilon toxin *Clostridium perfringens* **Other** Water safety threats (e.g., *Vibrio cholerae, Cryptosporidium parvum*) Food safety threats (e.g., *Salmonella* spp., *E. coli* O157:H7, *Shigella* spp.)
C	Emerging pathogens that could be engineered for mass dissemination; potential for high morbidity and mortality rates; major health impact potential	*Nipah virus* Hantaviruses Tick-borne hemorrhagic fever viruses Tick-borne encephalitis viruses *Yellow fever virus* Multidrug-resistant and extremely drug-resistant *Mycobacterium tuberculosis*

(3) investment of more time and money in genome sequencing, new vaccine research, and new therapeutic research; (4) development of improved detection and diagnostic systems; and (5) preparation of clinical microbiologists and clinical microbiology laboratories as members of first-responder teams to respond in a timely manner to acts of bioterrorism. In 2002 the U.S. Congress enacted the Public Health Security and Bioterrorism Preparedness and Response Act, which identified "select" agents whose use is now tightly regulated. The agents are categorized (A, B, or C) based on (1) ease of dissemination, (2) communicability, and (3) morbidity and mortality. A final rule implementing the provisions of the act that govern the possession, use, and transport of biological agents that are considered likely to be used for biocrimes or bioterrorism was issued in 2005.

In 2003 Congress established the Department of Homeland Security to coordinate the defense of the United States against terrorist attacks. As one of many duties, the secretary of Homeland Security is responsible for maintaining a National Incident Management System to monitor large-scale hazardous events. Bioterrorism and other public health incidents are managed within this system. The Department of Health and Human Services has the initial responsibility for the national public health and will deploy assets as needed within the areas of its statutory responsibility (e.g., the Public Health Service Act and the Federal Food, Drug, and Cosmetic Act) while keeping the secretary of Homeland Security apprised during an incident of the nature of the response. The secretary of Health and Human Services directs the CDC to effect the necessary integration of public health activities.

The events of September and October 2001 in the United States changed the world. Global efforts to prevent terrorism, especially using biological agents, are evolving from cautious planning to proactive preparedness. In the United States, the CDC has partnered with academic institutions across the country to educate, train, and drill public health employees, first responders, and numerous environmental and health-care providers (**chapter opener**). Centers for Public Health Preparedness were established to bolster the overall response capability to bioterrorism. Another CDC-managed program that began in 1999, the Laboratory Response Network (LRN), serves to ensure an effective laboratory response to bioterrorism by helping to improve the nation's public health laboratory infrastructure through its partnership with the FBI and the Association of Public Health Laboratories (APHL). The LRN maintains an integrated network that links state and local public health, federal, military, and international laboratories so that a rapid and coordinated response to bioterrorism or other public health emergencies (including veterinary, agriculture, military, and water- and food-related) can occur.

In the absence of overt terrorist threats and without the ability to rapidly detect bioterrorism agents, it is likely that a bioterrorism act will be defined by a sudden spike in an unusual (nonendemic) disease reported to the public health system. Also, sudden increased numbers of zoonoses, diseased animals, or vehicle-borne illnesses may indicate bioterrorism. Important guidelines and standardized protocols prepared for all sentinel (local hospital, contract, clinic, etc.) laboratories to assist in the management of clinical specimens containing select agents have been established by the American Society for Microbiology in coordination with the CDC and the APHL. A summary of the rule-out tests for six bacterial agents is presented in **table 36.7**. The diseases and

Table 36.7 Criteria for Presumptive Identification of Six Bacterial Select Agents

Pathogen	*Gram Morphology*	*Colonial Morphology*	*Biochemical Results*
Bacillus anthracis	Gram-positive endospore-forming rod	Grey, flat, "Medusa head" irregularity, nonhemolytic on sheep's blood agar	Catalase positive, oxidase positive, urea negative, Voges-Proskauer (VP) positive, phenylalanine (Phe) deaminase negative, NO_3^- to NO_2^- positive
Brucella suis	Gram-negative rod (tiny)	Nonpigmented, convex-raised, pin-point after 48 hr, nonhemolytic	Catalase positive, oxidase variable, urea positive, VP negative
Burkholderia mallei	Gram-negative straight or slightly curved cocobacilli, bundles	Grey, smooth, translucent after 48 hr, nonhemolytic	Catalase positive, oxidase variable, indole negative, Arginine dihydrolase positive, NO_3^- to NO_2^- positive
Clostridium botulinum	Gram-positive endospore-forming rod	Creamy, irregular, rough, broad, nonhemolytic	Catalase negative, urea negative, gelatinase positive, indole negative, VP negative, Phe deaminase negative, NO_3^- to NO_2^- negative
Francisella tularensis	Gram-negative rod (tiny)	Grey-white, shiny, convex, pin-point after 72 hr, nonhemolytic	Catalase positive (weak), oxidase negative, β-lactamase positive, urea negative
Yersinia pestis	Gram-negative rod (bipolar staining)	Grey-white, "fried-egg" irregularity, nonhemolytic, grows faster and larger at 28°C	Catalase positive, oxidase negative, urea negative, indole negative, VP negative, Phe deaminase negative

microbiology associated with specific select agents are discussed in chapters 37–39.

Search This: Bioterrorism defense

1. In what three general ways can epidemics be controlled? Give one or two specific examples of each type of control measure.
2. Name some of the microorganisms that can be used to commit biocrimes. From this list, which pose the greatest risk for causing large numbers of casualties?
3. Why are biological weapons more destructive than chemical weapons?
4. What is the Public Health Security and Bioterrorism Preparedness and Response Act designed to do?

36.9 Global Health Considerations

From a global health perspective, developed countries such as Australia, the European countries, Israel, New Zealand, and the United States have highly effective public health systems. About 25% of the over 6 billion people on Earth live in these countries. As a result, of the approximately 12 million deaths in these countries per year, only about 500,000 (about 4%) are due to infectious diseases. Less-developed areas such as Africa, Central and South America, India, parts of eastern Europe, and Asia have less-developed public health systems and represent 75% of the human population. It is in these regions that infectious diseases are the major cause of death; for example, of approximately 38.5 million deaths per year, about 18 million (about 47%) are attributed to infectious microbial diseases. Despite the efforts of many governmental and nongovernmental agencies, it will take many years before all people have access to clean water, decent sanitation, and a basic health-care infrastructure.

The high incidence of infectious diseases in less-developed countries must be of great concern for people traveling to these destinations. Each year 1 billion passengers travel by air and over 50 million people from developed countries visit less-developed countries. Furthermore, the time required to circumnavigate the globe has decreased from 365 days to fewer than 2 days. Several kinds of precautions can be taken by individuals to prevent travel-related infectious diseases. Examples include:

1. Wash hands with soap and water frequently, especially before each meal.
2. Get or update vaccinations appropriate for specific destinations. Check the CDC travel advisory website (www.cdc.gov/travel) for precautions regarding specific locations.
3. Avoid uncooked food, unbottled water and beverages, and unpasteurized dairy products. Use bottled water for drinking, making ice cubes, and brushing teeth.
4. Use barrier protection if engaging in sexual activity.
5. Minimize skin exposure and use repellents to prevent arthropod-borne illnesses (e.g., malaria, dengue, yellow fever, Japanese encephalitis).
6. Avoid skin-perforating procedures (e.g., acupuncture, body piercing, tattooing, venipuncture, sharing of razors).
7. Do not pet or feed animals, especially dogs and monkeys.
8. Avoid swimming or wading in nonchlorinated fresh water.

Vaccinations are one of the most important strategies of prophylaxis in travel medicine. A medical consultation before travel is an excellent opportunity to update routine immunizations. Selection of immunizations should be based on requirements and risk of infection at the travel destination. According to International Health Regulations, many countries require proof of yellow fever vaccination on the International Certificate of Vaccination. Additionally, a few countries still require proof of vaccination against cholera, diphtheria, and meningococcal disease. The basic CDC immunization recommendations for those traveling abroad are presented in **table 36.8**. In the not-too-distant future, it is hoped that travelers will be offered a variety of oral vaccines against the microorganisms causing dengue fever and travelers' diarrhea (e.g., enterotoxigenic *E. coli*, *Campylobacter*,

Table 36.8 Vaccine Recommendations for Travelers*

Category	*Vaccine*
Routine recommended vaccination**	Diphtheria/Tetanus/Pertussis (DPT)#† Hepatitis B (HBV)# *Haemophilus influenzae* type b (Hib)# Influenza Measles (MMR)# Poliomyelitis (IPV)# Varicella#
Selective vaccination based on exposure risk	Cholera Hepatitis A (HAV) Japanese encephalitis Meningococcal (polysaccharide) Pneumococcal (polysaccharide) Rabies Tick-borne encephalitis Typhoid fever Yellow fever
Mandatory vaccination for entry§	Meningococcal (polysaccharide) Yellow fever

*Based on the CDC and WHO 2005 recommendations
** For travelers aged 2 years or older
#Normally administered during childhood; should be updated if indicated for travel
†Adult booster of tetanus/diphtheria (Td) vaccine should be every 10 years
§Meningococcal vaccination to enter Saudi Arabia; yellow fever vaccination to enter various countries in the endemic zone of South America and Africa

Shigella); vaccines against malaria and AIDS, the infections causing most deaths in travelers, are much further in the future. In addition, a traveler should:

1. Read carefully CDC information about the destination and follow recommendations.
2. Begin the vaccination process early.
3. Find a travel clinic for information and specialized immunizations.
4. Plan ahead when traveling with children or if there are special needs.
5. Learn about safe food and water (contaminated food and water are the major sources of stomach or intestinal illness while traveling), protection against insects, and other precautions.

1. Give some examples of how population movements affect microbial disease transmission.
2. In addition to vaccinations, what are some additional precautions global travelers should take?

Summary

36.1 Epidemiology

a. Epidemiology is the science that evaluates the determinants, occurrence, distribution, and control of health and disease in a defined population.

b. Specific epidemiological terminology is used to communicate disease incidence in a given population. Frequently used terms include sporadic disease, endemic disease, hyperendemic disease, epidemic, index case, outbreak, and pandemic.

36.2 Epidemiological Methods

a. Public health surveillance is necessary for recognizing a specific infectious disease within a given population. This consists of gathering data on the occurrence of the disease, collating and analyzing the data, summarizing the findings, and applying the information to control measures.

b. Population-based and case-based surveillance data are used to track infections within a population.

c. Remote sensing and geographic information systems are used to digitally gather data from natural environments.

36.3 Measuring Infectious Disease Frequency

a. Statistics is an important tool used in the study of modern epidemiology.

b. Epidemiological data can be obtained from such factors as morbidity, prevalence, and mortality rates.

36.4 Patterns of Infectious Disease in a Population

a. A common-source epidemic is characterized by a sharp rise to a peak and then a rapid but not as pronounced decline in the number of individuals infected (**figure 36.3**). A propagated epidemic is characterized by a relatively slow and prolonged rise and then a gradual decline in the number of individuals infected.

b. Herd immunity is the resistance of a population to infection and pathogen spread because of the immunity of a large percentage of the individuals within the population.

36.5 Emerging and Reemerging Infectious Diseases and Pathogens

a. Humans will continually be faced with both new infectious diseases and the reemergence of older diseases once thought to be conquered.

b. The CDC has defined these diseases as "new, reemerging, or drug-resistant infections whose incidence in humans has increased within the past two decades or whose incidence threatens to increase in the near future" (**figure 36.8**).

c. Many factors characteristic of the modern world undoubtedly favor the development and spread of these microorganisms and their diseases.

36.6 Nosocomial Infections

a. Nosocomial infections are infections acquired within a hospital or other clinical care facility and are produced by a pathogen acquired during a patient's stay. These infections come from either endogenous or exogenous sources (**figure 36.10**).

b. Hospitals must designate an individual to be responsible for identifying and controlling nosocomial infections. This person is known as the infection control practitioner.

36.7 Control of Epidemics

a. The public health system consists of individuals and organizations that function in the control of infectious diseases and epidemics.

b. Vaccination is one of the most cost-effective weapons for microbial disease prevention, and vaccines constitute one of the greatest achievements of modern medicine.

c. Many of the current vaccines in use for humans (**table 36.2**) consist of whole organisms that are either inactivated (killed) or attenuated (live but avirulent).

d. Some of the risks associated with whole-cell vaccines can be avoided by using only specific purified macromolecules derived from pathogenic microorganisms. Currently there are three general forms of subunit or acellular vaccines: capsular polysaccharides, recombinant surface antigens, and inactivated exotoxins (toxoids) (**table 36.5**).
e. A number of microorganisms have been used for recombinant-vector vaccines. The attenuated microorganism serves as a vector, replicating within the host and expressing the gene product of the pathogen-encoded antigenic proteins. The proteins can elicit humoral immunity when the proteins escape from the cells and cellular immunity when they are broken down and properly displayed on the cell surface.
f. DNA vaccines elicit protective immunity against a pathogen by activating both branches of the immune system: humoral and cellular.
g. Epidemiological control measures can be directed toward reducing or eliminating infection sources, breaking the connection between sources and susceptible individuals, or isolating the susceptible individuals and raising the general level of herd immunity by immunization.

36.8 Bioterrorism Preparedness

a. Among weapons of mass destruction, biological weapons are more destructive than chemical weapons. The list of biological agents that could pose the greatest public health risk in the event of a bioterrorist attack is short and includes viruses, bacteria, parasites, and toxins (**table 36.6**).
b. Specific laboratory tests define the process for hospital labs to rule out a select agent.

36.9 Global Health Considerations

a. It will take many years before all people have access to clean water, decent sanitation, and a basic health-care infrastructure.
b. Basic hand hygiene and vaccinations are very effective in preventing infectious disease while traveling.

Critical Thinking Questions

1. Why is international cooperation a necessity in the field of epidemiology? What specific problems can you envision if there were no such cooperation?
2. What common sources of infectious disease are found in your community? How can the etiologic agents of such infectious diseases spread from their source or reservoir to members of your community?
3. How could you prove that an epidemic of a given infectious disease was occurring?
4. How can changes in herd immunity contribute to an outbreak of a disease on an island?
5. College dormitories are notorious for outbreaks of flu and other infectious diseases. These are particularly prevalent during final exam weeks. Using your knowledge of the immune response and epidemiology, suggest practices that could be adopted to minimize the risks at such a critical time.
6. Why does an inactivated vaccine induce only a humoral response, whereas an attenuated vaccine induces both humoral and cell-mediated responses?
7. Why is a DNA vaccine delivered intramuscularly and not by intravenous or oral routes?
8. While it is generally assumed that emerging infectious diseases (EIDs) are increasing, it was not until 2008 that a careful, quantitative investigation of EIDs was reported. Jones and her colleagues analyzed a database of 335 EID events between 1940 and 2004. They define an EID event as the first report of a given disease, including the first time drug-resistant forms of old diseases, such as chloroquine-resistant malaria, are detected. Not surprisingly, they found that distribution of EID events was not random. Suggest a reason for each of the following observations:

   The incidence of EID events has increased steadily since 1940, reached a maximum in the 1980s, and has since declined but is still greater than that measured in the 1970s.

   The majority of EIDs were zoonotic in origin (can you cite some examples?). Insect-borne EIDs increased significantly in the last decade; this increase correlates to global climate anomalies in rainfall, temperature, and severe storms.

   EID events were most common in higher latitudes (e.g., Europe, Japan, and North America).

   Read the original paper: Jones, K. E. 2008. Global trends in emerging infectious diseases. *Nature* 451:990.
9. Community-associated methicillin-resistant *Staphylococcus aureus* (CA-MRSA) was an unexpected development. In the United States, CA-MRSA isolates are classified as USA300. It was unknown whether USA300 strains are clonal in origin (i.e., arising from a single, common parental strain), represent convergent evolution of multiple strains, or are some combination of the two. To differentiate between these possibilities, the entire genome sequence of 10 USA300 isolates from eight states was compared. There were very few single nucleotide polymorphisms in eight of the 10 isolates (*see figure 17.5* and related discussion). Furthermore, the same eight were extremely virulent in a mouse model and produced the same exotoxins, whereas the other two were only modestly virulent and produced fewer exotoxins.

Why do you think it was important to determine if USA300 was clonal in origin or if the strains arose from multiple parental strains? What would be the public health implication of each?

Based on this evidence, what do you conclude about the origin of these 10 USA300 strains? Explain your answer.

Read the original paper: Kennedy, A. D. 2008. Epidemic community-associated methicillin-resistant *Staphylococcus aureus*: Recent clonal expansion and diversification. *Proc. Nat. Acad. Sci., USA*. 105:1327.

10. A rigorous study of women infected with the subtype of HIV called HIV-1 revealed that about 20% of the women were also infected with a second strain of HIV. In some cases, it was clear that the second, superinfection occurred a year or more after the initial HIV-1 infection.

How does the phenomenon of HIV superinfection compare with our understanding of acquired active immunity? What are the implications for successful vaccine development?

Read the original paper: Piantadosi, A., et al., Chronic HIV-1 infection frequently fails to protect against superinfection. *PLoS Pathog.* 3:e177 (2007).

Concept Mapping Challenge

Map the following concepts to demonstrate your understanding of their relatedness.

Bioterrorism
Communicable
Epidemiology
Population
Statistics
Vaccine
Immunity
Incidence
Infectious disease
Mortality
Vector
Vehicle
Zoonosis

Learn More

Learn more by visiting the text website at www.mhhe.com/willey8, where you will find a complete list of references.

37

Human Diseases Caused by Viruses and Prions

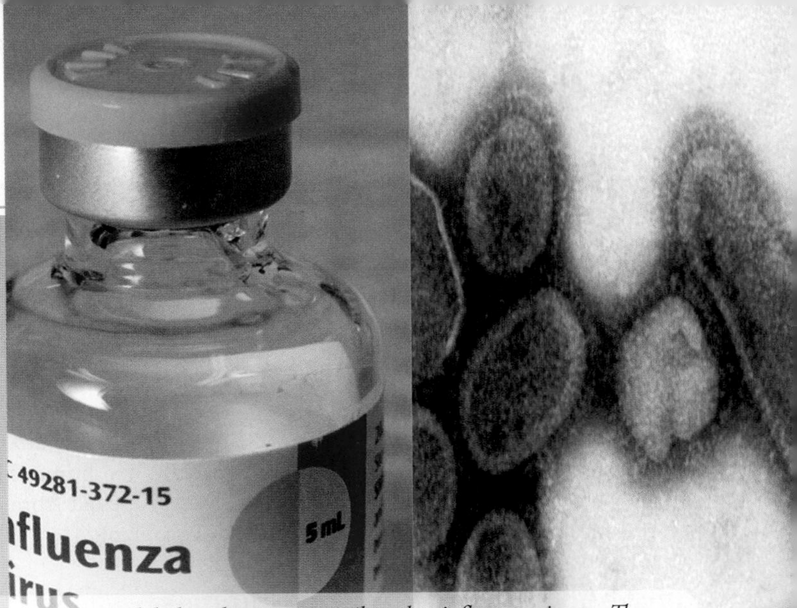

Numerous global epidemics are attributed to influenza viruses. The rapid mutation of their surface antigens results in sudden antigenic shifts that lead to widespread human infections. More frequent mutations leading to antigenic drifting can be anticipated such that yearly immunization with influenza vaccine may prevent infection. The H5N1 strain may be the agent of the next influenza pandemic.

CHAPTER GLOSSARY

Epstein-Barr virus A dsDNA virus in the family *Herpesviridae*; causes infectious mononucleosis (mono).

Hepatitis A virus A naked, icosahedral, plus-strand RNA virus of the genus *Hepatovirus* in the family *Picornaviridae*; causes a type of hepatitis transmitted by fecal-oral contamination.

Hepatitis B virus A dsDNA virus with a gapped genome in the family *Hepadnaviridae*; infections may be asymptomatic or associated with cirrhosis and liver cancer.

Hepatitis C virus An enveloped, plus-strand RNA virus in the family *Flaviviridae* that causes serious liver disease.

herpes simplex virus type 1 (HSV-1) A dsDNA virus that is a member of the family *Herpesviridae*; causes cold sores.

herpes simplex virus type 2 (HSV-2) A dsDNA virus that is a member of the family *Herpesviridae*; causes genital herpes.

human immunodeficiency virus (HIV) The single-stranded (ss) RNA virus of the family *Retroviridae* that causes AIDS.

Human parvovirus B19 A small, ssDNA virus that causes a suite of diseases including fifth disease in young children.

influenza viruses A group of minus-strand RNA viruses with segmented genomes that belong to the family *Orthomyxoviridae*; cause influenza (the flu).

Measles virus An enveloped minus-strand RNA virus in the genus *Morbillivirus* and the family *Paramyxoviridae*; causes measles.

Poliovirus A plus-strand RNA virus in the family *Picornaviridae*; causes poliomyelitis.

Rabies virus A minus-strand RNA virus belonging to the genus *Lyssavirus* in the family *Rhabdoviridae*; causes rabies.

rotaviruses dsRNA viruses belonging to the family *Reoviridae* that are one cause of acute viral gastroenteritis.

Rubella virus A plus-strand RNA virus in the family *Togaviridae*; causes rubella (German measles).

SARS coronavirus A plus-strand RNA virus that causes severe acute respiratory syndrome (SARS).

varicella-zoster virus (VZV) A dsDNA virus that is a member of the family *Herpesviridae*; causes chickenpox and shingles.

Variola virus A large dsDNA virus in the family *Poxviridae*; causes smallpox.

West Nile virus A plus-strand RNA flavivirus that causes West Nile fever (encephalitis).

Chapters 5 and 25 review the general biology of viruses and introduce basic virology. In chapter 37, we continue this coverage by discussing some of the most important viruses that are pathogenic to humans. We group viral diseases according to their mode of acquisition and transmission; viral diseases that occur in the United States are emphasized. Diseases caused by viruses that are listed as select agents are identified within the chapter by two asterisks (**).

More than 400 different viruses can infect humans. Human diseases caused by viruses are particularly interesting, considering the small amount of genetic information introduced into a host cell. This apparent simplicity belies the severe pathological features and clinical consequences that result from many viral diseases. With few exceptions, only prophylactic or supportive treatment is available. Collectively these diseases are some of the most common and yet most puzzling of all infectious diseases. The resulting frustration is compounded when year after year, familiar diseases of unknown etiology or new diseases become linked to virus infections.

37.1 Airborne Diseases

Because air does not support virus growth, any virus that is airborne must have originated from a living source. When humans are the source of the airborne virus, it usually is propelled from the respiratory tract by coughing, sneezing, or vocalizing.

Chickenpox (Varicella) and Shingles (Herpes Zoster)

Chickenpox (varicella) is a highly contagious skin disease primarily of children two to seven years of age. Humans are the reservoir and the source for this virus, which is acquired by droplet inhalation into the respiratory system. The virus is highly infectious, with secondary infection rates in susceptible household contacts of 65% to 86%. In the prevaccine era, about 4 million cases of chickenpox occurred annually in the United States, resulting in approximately 11,000 hospitalizations and 100 deaths.

The causative agent is the enveloped, DNA varicella-zoster virus (VZV), a member of the family *Herpesviridae.* The virus produces at least six glycoproteins that play a role in viral attachment to specific receptors on respiratory epithelial cells. Following an incubation period of 10 to 23 days, small vesicles erupt on the face or upper trunk, fill with pus, rupture, and become covered by scabs (**figure 37.1**). Healing of the vesicles occurs in about 10 days. During this time, intense itching often occurs.

Laboratory testing for VZV is not normally required, as the diagnosis of chickenpox is typically made by clinical assessment. Laboratory confirmation is recommended, though, to confirm the diagnosis of severe or unusual cases of chickenpox. As the incidence of chickenpox continues to decline due to vaccination, fewer cases are seen clinically, resulting in the likelihood of misdiagnosis. Furthermore, in persons who have previously received varicella vaccination, the disease is usually mild or atypical and can pose particular challenges for clinical diagnosis. Chickenpox can be prevented or the infection shortened with an attenuated varicella vaccine (Varivax; see *table 36.2*) or the drug acyclovir (Zovirax or Valtrex). It should be noted that Valtrex (valacyclovir) is an orally administered prodrug of Zovirax or acyclovir (*see figure 34.16*). Valtrex is the valyl ester of acyclovir and is rapidly hydrolyzed to acyclovir in the body.

Individuals who recover from chickenpox are subsequently immune to this disease; however, they are not free of the virus, as viral DNA resides in a dormant (latent) state within the nuclei of cranial nerves and sensory neurons in the dorsal root ganglia.

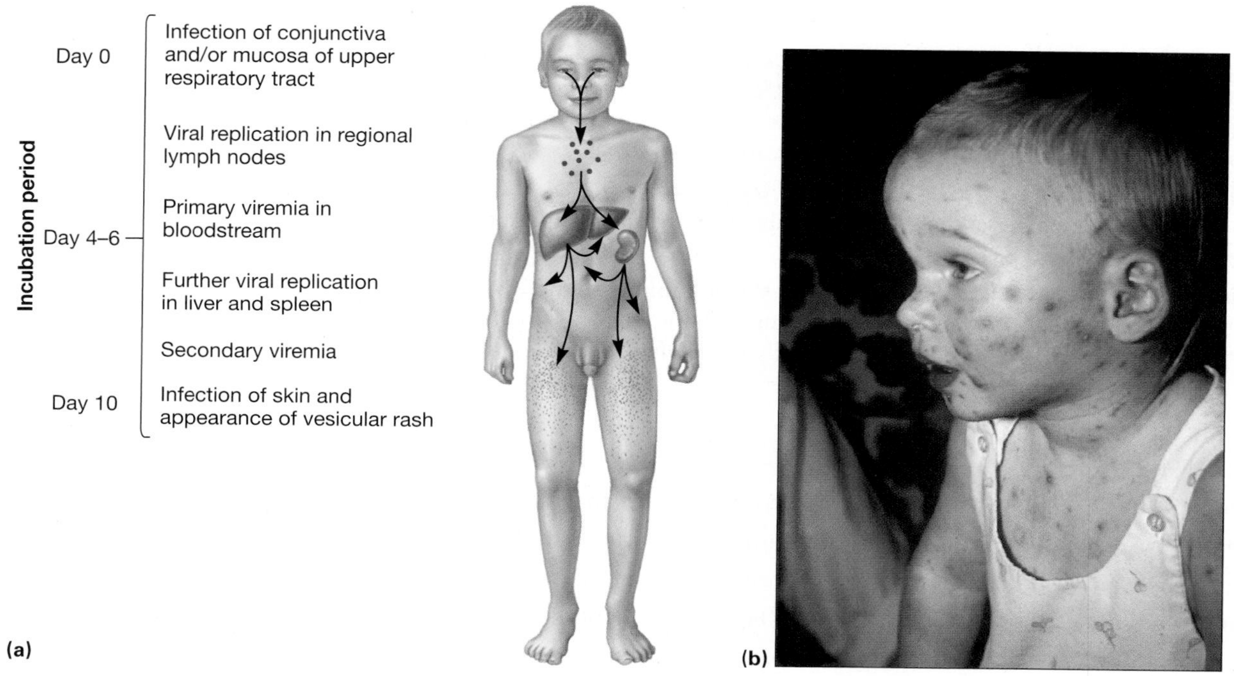

FIGURE 37.1 Chickenpox (Varicella). (a) Course of infection. (b) Typical vesicular skin rash. This rash occurs all over the body but is heaviest on the trunk and diminishes in intensity toward the periphery.

This viral DNA is maintained in infected cells, but virions cannot be detected (**figure 37.2*a***). When the infected person becomes immunocompromised by such factors as age, cancers, organ transplants, AIDS, or psychological or physiological stress, the viruses may become activated (figure 37.2*b*). They migrate down sensory nerves, initiate viral replication, and produce painful vesicles because of sensory nerve damage (figure 37.2*c*). This syndrome is called **postherpetic neuralgia.** To manage the intense pain, corticosteroids or the drug gabapentin (Neurontin) can be prescribed. The reactivated form of chickenpox is called **shingles (herpes zoster).** Most cases occur in people over 50 years of age. Except for the pain of postherpetic neuralgia, shingles does not require specific therapy; however, in immunocompromised individuals, acyclovir, valacyclovir, vidarabine (Vira-A), or famciclovir (Famvir) are recommended. More than 500,000 cases of herpes zoster occur annually in the United States.

Influenza (Flu)

Influenza (Italian, *un influenza di freddo*—to be influenced by the cold), or the flu, was described by Hippocrates in 412 B.C. The first well-documented global epidemic of influenza-like disease occurred in 1580. Since then, 31 possible influenza pandemics have been documented, with four occurring in the twentieth century. The worst pandemic on record occurred in 1918 and killed approximately 50 million people around the world. This disaster, traced to the Spanish influenza virus (*see figure 36.7*), was followed by pandemics of Asian flu (1957), Hong Kong flu (1968), and Russian flu (1977). (The names reflect popular impressions of where the episodes began, although all are now thought to have originated in China.)

Influenza is a respiratory system disease caused by negative-strand RNA viruses that belong to the family *Orthomyxoviridae.*

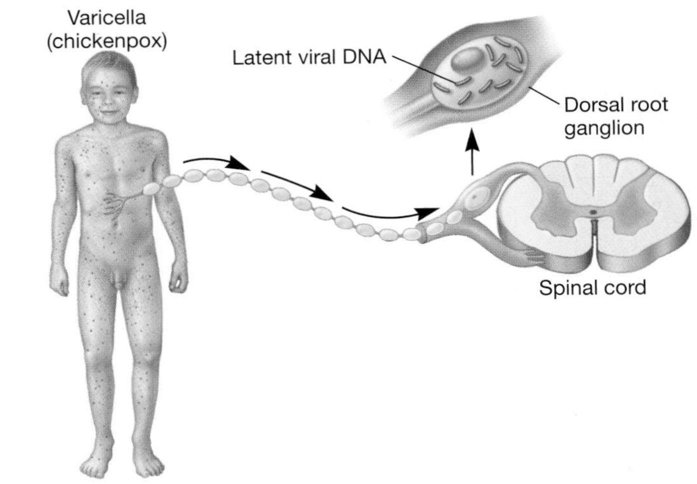

(a) Primary infection—Chickenpox

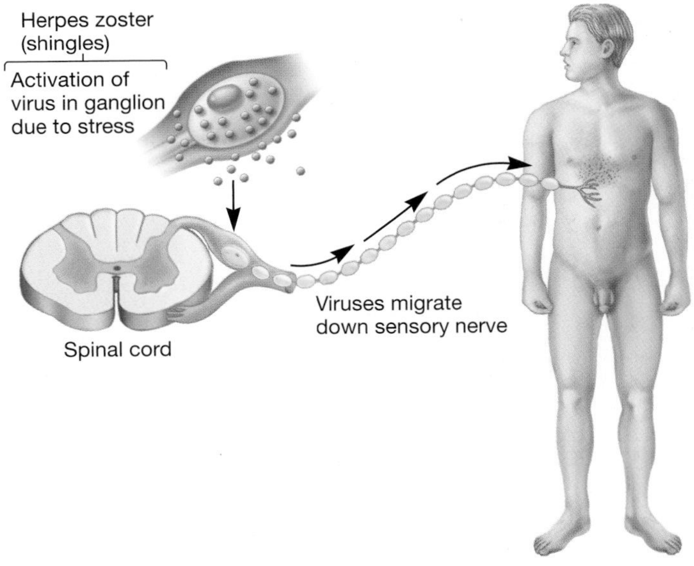

(b) Recurrence—Shingles

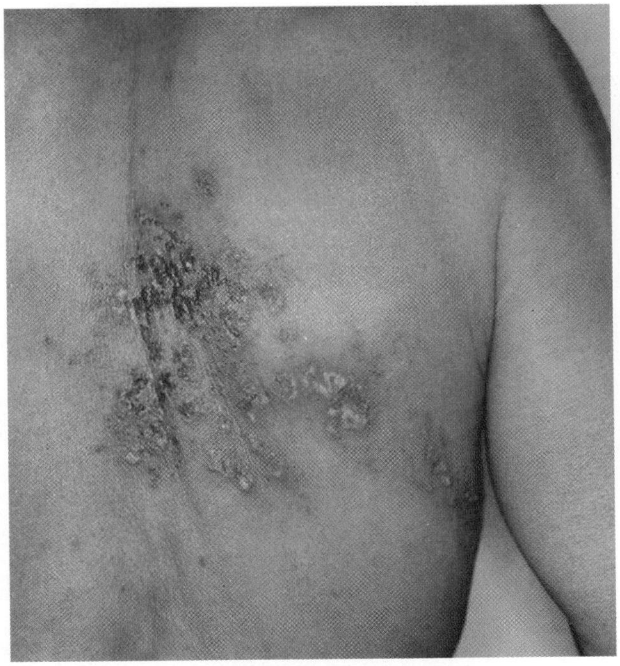

(c)

FIGURE 37.2 Pathogenesis of the Varicella-Zoster Virus. (a) After an initial infection with varicella (chickenpox), the viruses migrate up sensory peripheral nerves to their dorsal root ganglia, producing a latent infection. (b) When a person becomes immunocompromised or is under psychological or physiological stress, the viruses may be activated. (c) They migrate down sensory nerve axons, initiate viral replication, and produce painful vesicles. Since these vesicles usually appear around the trunk of the body, the name *zoster* (Greek for girdle) was used.

***Figure 37.2* Micro Inquiry**

Where does viral DNA reside during latent infection? Can virions be detected at this time?

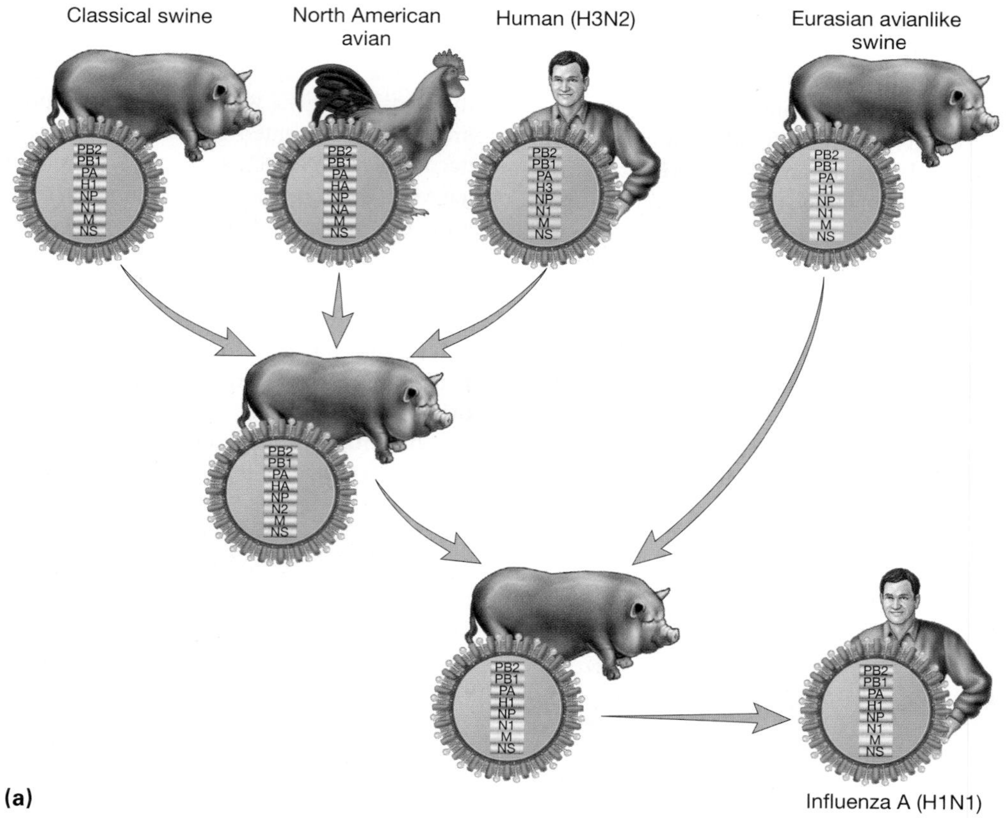

(a)

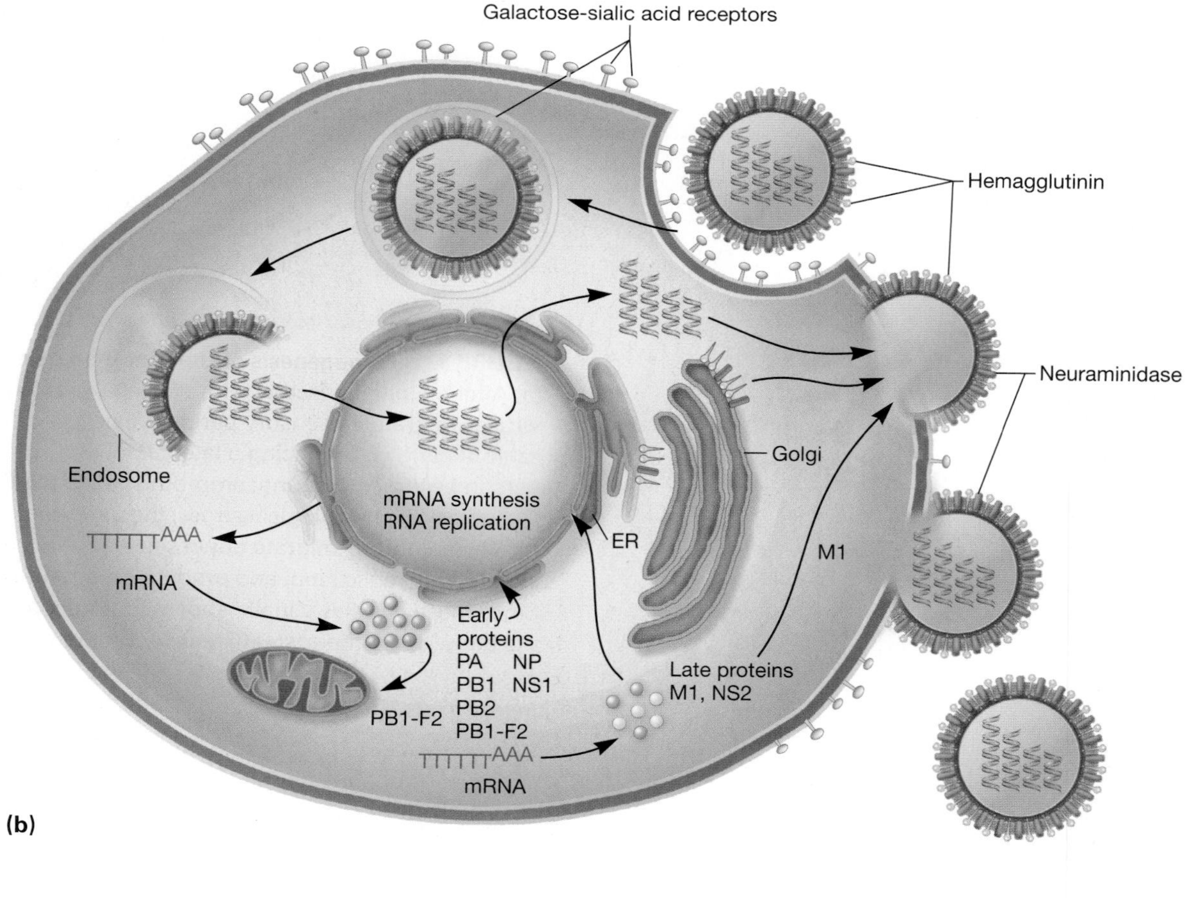

(b)

Continued

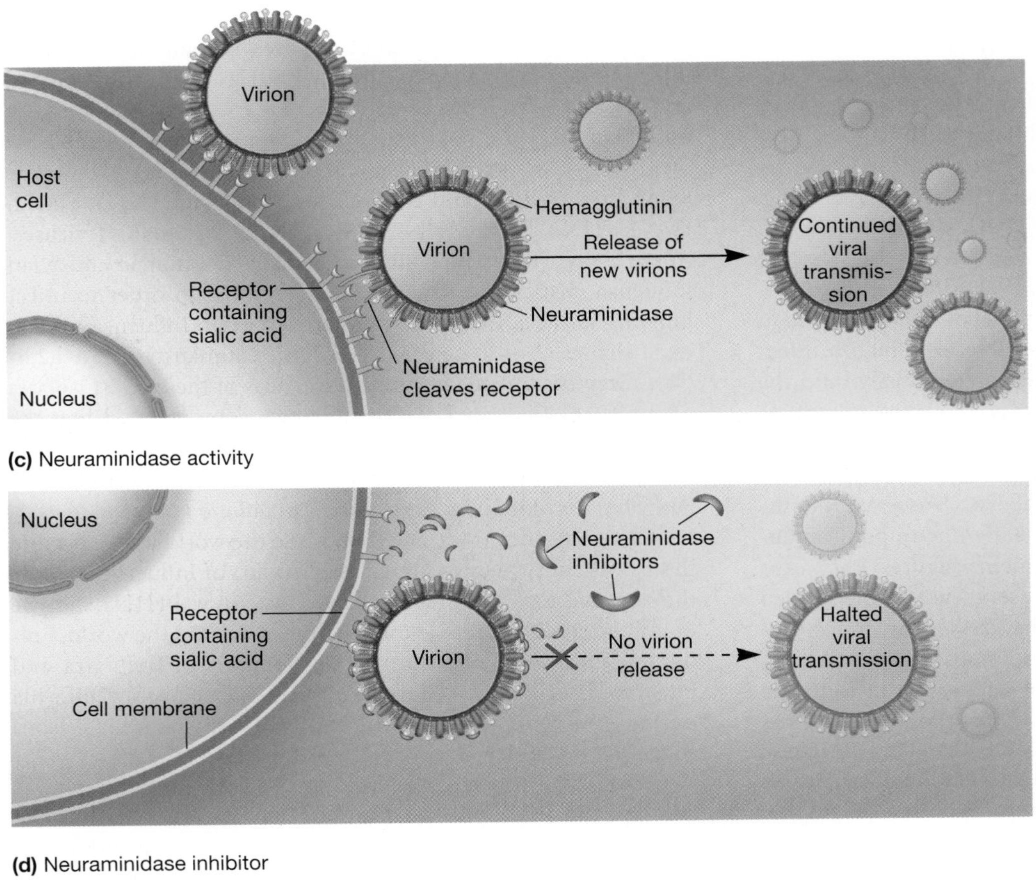

FIGURE 37.3 Influenza Infection. (a) Influenza viruses infect birds, pigs, and humans. Reassortment of viral RNA from more than one viral subtype results in novel subtypes, such as the H1N1 cause of the 2009–2010 influenza pandemic. (b) Influenza virions enter the host cell by receptor-mediated endocytosis. Attachment of viral hemagglutinin proteins to galactose-sialic acid receptors results in endocytosis, RNA release, and subsequent replication and transcription. Viral hemagglutinin and neuraminidase proteins coat the host membrane that will become the viral envelope. (c) Virus shedding from infected cells is also mediated by virion attachment to membrane sialic acid–containing receptors so as to facilitate budding from the host cell. Neuraminidase cleaves the receptors to release the virions from the host. (d) Neuraminidase inhibitors block virion release and prevent transmission of virions to new host cells.

Figure 37.3 Micro Inquiry

What is the function of hemagglutinin following endocytosis?

There are four groups: influenza A, influenza B, influenza C, and Thogoto viruses. Influenza viruses are widely distributed in nature infecting a variety of mammal and bird hosts (**figure 37.3*a***). All influenza viruses (*see figure 5.4*) are acquired by inhalation or ingestion of virus-contaminated respiratory secretions (thus the public health recommendation to "cover your cough and wash your hands frequently"). During an incubation period of 1 to 2 days, the virus adheres to the epithelium of the respiratory system (the neuraminidase present in envelope spikes may hydrolyze the mucus that covers the epithelium). The virus attaches to the epithelial cell by its hemagglutinin spike protein, causing part of the cell's plasma membrane to bulge inward, seal off, and form a vesicle (receptor-mediated endocytosis). This encloses the virus in an endosome (*see figure 4.10*). The hemagglutinin molecule in the virus envelope undergoes a dramatic conformational change when the endosomal pH decreases. The hydrophobic ends of the hemagglutinin spring outward and extend toward the endosomal membrane. After they contact the membrane, fusion occurs and the RNA nucleocapsid is released into the cytoplasm.

Influenza viruses contain seven to eight segments (pieces) of linear RNA, with a total genome length between 12,000 to 15,000 nucleotides. Each RNA segment encodes one or two proteins. Replication and transcription of viral RNAs is accomplished in the host cell nucleus by three polymerase subunits and the nucleoprotein (figure 37.3*b*). Newly synthesized viral ribonucleoproteins are then exported from the nucleus to the cytoplasm, where they assemble into virions at the host cell plasma membrane. The mature virions are released from the host cell, taking host membrane as the viral envelope. The enveloped virion can be spherical (50 to 120 nm in diameter) or filamentous (200 to 300 nm long, 20 nm in diameter) (*see figure 5.4*).

Influenza viruses are classified into subtypes (or strains) based on their membrane surface glycoproteins, hemagglutinin (HA) and neuraminidase (NA). There are 16 HA (H1-H16) and 9 NA (N1-N9) antigenic forms of influenza known; they recombine to produce the various HA/NA subtypes. As mentioned previously, HA and NA function in viral attachment and virulence. Specifically, the HA protein is required for binding of individual virions to host cell receptors and the subsequent fusion of the viral envelope with the host endosomal membrane, initiating infection (figure 37.3*b*). The NA protein facilitates release of newly assembled influenza virions from infected cells by cleaving sialic acid residues between virus and host membranes (figure 37.3*c*).

Together HA and NA protein efficiency define the virulence of influenza by facilitating virus attachment to susceptible host cells and mediating virus shedding, respectively. Host specificity can be explained in part by the difference in receptor binding. Human influenza viruses preferentially attach by way of their HA protein to receptors composed of sialic acid (SA) residues that are bound to galactose (Gal) sugars by an α2,6-linkage (SA-α2,6-Gal). Membranes of human tracheal epithelial cells that contain the SA-α2,6-Gal moiety are thus susceptible to influenza infection. In contrast, avian influenza viruses preferentially bind the SA-α2,3-Gal typically found on intestinal epithelial cells of waterfowl (the main viral replication site of avian influenza viruses), thus restricting infection to birds.

However, the genomes of the influenza viruses are significantly plastic in that they change frequently due to point mutations and reassortment of the RNA segments (figure 37.3*b*). This is noteworthy, as we discuss later, because it leads to the sudden appearance of new influenza viruses that cause pandemics. Genome plasticity is one of the most important features of influenza viruses in that it leads to both subtle and sometimes drastic changes in the viral proteins that are synthesized, altering virus antigenicity. If the protein variation is small, it is called **antigenic drift**. Antigenic drift results from the accumulation of mutations of HA and NA in a single strain of flu virus within a geographic region. This usually occurs every 2 to 3 years, causing local increases in the number of flu infections. **Antigenic shift** is a large antigenic change resulting from the reassortment of genomes when two different strains of flu viruses (from both animals and humans) infect the same host cell and are incorporated into a single new capsid (*see figures 36.6* and 37.3*a*). Because there is a greater change with antigenic shift than antigenic drift, antigenic shifts can yield major epidemics and pandemics. Antigenic variation occurs almost yearly with influenza A virus, less frequently with the B virus, and has not been demonstrated in the C virus, thus the need for yearly influenza immunization. ◀◀ *Patterns of infectious disease in a population (section 36.4)*

Animal reservoirs are critical to the epidemiology of human influenza. For example, rural China is one region of the world in which chickens, pigs, and humans live in close, crowded conditions. Influenza is widespread in birds; although birds can't usually transmit the virus to humans, they can transfer it to pigs. Pigs can transfer it to humans, and humans back to pigs. Recombination between human and avian strains thus occurs in pigs, leading to novel HA/NA combinations causing major antigenic shifts. To track the various influenza subtypes, a standard nomenclature system was developed. It includes the following information: group (A, B, or C), host of origin, geographic location, strain number, and year of original isolation. Antigenic descriptions of the HA and NA surface proteins are given in parentheses for type A viruses only. The host of origin is indicated for viruses obtained from nonhuman hosts; for example, A/swine/Iowa/15/30 (H1N1). However, the host origin is not given for human isolates; for example, A/Hong Kong/03/68 (H3N2).

Influenza A viruses having H1, H2, and H3 HA antigens, along with N1 and N2 NA antigens, are predominant in nature, infecting humans since the early 1900s. H1N1 viruses appeared in 1918 and were replaced in 1957 by H2N2 viruses as the predominant subtype. The H2N2 viruses were subsequently replaced by H3N2 as the principle subtype in 1968. The H1N1 subtype reappeared in 1977 and cocirculates today with H2N1, H3N2, H5N2, H7N2, H7N3, H7N7, H9N2, H10N7, and H5N1 viruses. Importantly, the receptor-binding specificity of human and avian influenza viruses just discussed suggests that in order to infect humans surface HA and NA proteins of avian influenza viruses must change so as to be able to recognize human-type (SA-α2,6-Gal) receptors. Indeed, genome sequences of the earliest isolates of the 1918, 1957, and 1968 pandemics demonstrated that the otherwise avian influenza A viruses possessed HAs that recognized SA-α2,6-Gal receptors usually seen in human influenza. ◀◀ *Patterns of infectious disease in a population (section 36.4)*

Of significance is the fact that, while the world was expecting the H5N1 subtype (also known as bird flu) of influenza A to be the next cause of an influenza pandemic, a novel H1N1 subtype of influenza A evolved and quickly spread around the world. Epidemiological data evaluating large numbers of influenza and pneumonia cases in La Gloria, Veracruz, Mexico, and California during mid-February 2009 (the end of the seasonal flu season) suggested that a new strain of flu was the cause.

An H1N1 (previously referred to as swine flu) outbreak was quickly confirmed by Mexican and U.S. public health officials. On April 26, 2009, the United States declared a public health emergency, implementing the nation's pandemic response plan. By the end of April 2009, the World Health Organization (WHO) increased the global pandemic alert from phase 3 to phase 4 and, shortly thereafter, to phase 5, reflecting international spread of H1N1 with clusters of human-to-human transmission (in at least two countries). Forty-one countries reported 11,034 cases, including 85 deaths by May 21, 2009. On June 11, 2009, the WHO raised the pandemic alert to level 6, indicating that a global pandemic was in progress with more than 70 countries reporting cases of H1N1 infection with ongoing community-level outbreaks in multiple parts of the world. The WHO reported 99,103 cases of H1N1 with 476 fatalities on July 6, 2009. Ten days later, officials at the WHO indicated that the counting of individual cases was no longer essential for monitoring either the level or nature of the risk posed by the pandemic virus or to guide implementation of the most appropriate response measures. As a result, routine counting and reporting of novel H1N1 cases was halted except for newly affected areas.

By June 19, 2009, all 50 states, the District of Columbia, Puerto Rico, and the U.S. Virgin Islands reported novel H1N1 cases. The commencement of the regular influenza season in the Southern Hemisphere complicated the detection and spread of novel H1N1, making accurate counts of H1N1 infection impossible. So althought the Centers for Disease Control and Prevention (CDC) reported 43,771 confirmed cases and 302 deaths on July 24, 2009, officials believed that more than a million Americans had been sickened by the virus. At this time, the majority of H1N1 cases were still occurring in younger people, with the median age reported to be twelve to seventeen years, based on data from Canada, Chile, Japan, the United Kingdom, and the United States. Some reports suggest that patients requiring hospitalization and those with fatal illness may be slightly older and have underlying disease, such as asthma. Cardiovascular disease, respiratory disease, diabetes, and cancer were also considered risk factors for serious H1N1 disease with fatal complications. H1N1 vaccines became available in October of 2009 with the goal to use them in addition to the 2009 seasonal flu vaccine. Limited supply of the H1N1 vaccine by the end of Novemeber 2009 prevented this. CDC estimates that between April and October 2009 there were as many as 34 million H1N1 infections, with an approximate 153,000 hospitalizations and upward of 6,100 deaths.

Search This: CDC Swine flu update

The novel H1N1 (swine) flu pandemic of 2009 diverted attention from the continuing threat of H5N1 (bird) flu. The H5N1 subtype was responsible for six deaths in 1997, although it had been sporadically detected prior to that. Throughout 2003 to 2008, however, H5N1 was responsible for a substantial number of bird infections, resulting in the culling of millions of birds, even though no significant human-to-human transmission had occurred. The birds were destroyed to help prevent transmission of the virus to susceptible humans. Yet by the end of June 2009, H5N1 was responsible for 436 confirmed cases of influenza and 262 deaths in 15 countries.

Widespread outbreaks of avian H5N1 continued from late 2003 and were reported in Southeast Asia in 2004 through 2008, predominantly among poultry. In a number of Asian countries, though, these outbreaks were associated with severe human illnesses and deaths. Because of potentially severe consequences of pandemic H5N1 infection of humans, an international effort coordinated by the WHO monitors virus surveillance, epidemiology, and control efforts for this virus subtype as well as for H1N1. The National Institutes of Health have contracted for the manufacture of a vaccine to H5N1. It began clinical trials in 2005.

Search This: WHO avian influenza data

As with many other viral diseases, only the symptoms of influenza usually are treated. Amantadine (Symmetrel) (*see figure 34.15*), rimantadine (Flumadine), zanamivir (Relenza), and oseltamivir (Tamiflu) have been shown to reduce the duration and symptoms of type A influenza if administered during the first two days of illness. Amantadine and rimantadine are chemically related, antiviral drugs known as adamantanes. These usually have activity against influenza A viruses but not influenza B viruses. Adamantanes are thought to interfere with influenza A virus M2 protein, a membrane ion channel protein. They also inhibit virus uncoating, which inhibits virus replication, resulting in decreased viral shedding. As of January 2009, 99% of influenza A (H1N1) viruses were sensitive to adamantanes. However, all H3N2 viruses were resistant.

Zanamivir and oseltamivir are chemically related antiviral drugs known as neuraminidase inhibitors that have activity against both influenza A and B viruses. Neuraminidase inhibitors prevent the release of newly formed virions from their host cell by blocking the catalytic site of the enzyme neuraminidase. Newly formed virions exit their host cell by budding. However, the final step of release occurs when neuraminidase cleaves the sialic acid residue on a host membrane receptor that is attached to the virus's hemagglutinin spike. With the neuraminidase blocked, virions are not detached from the host cell membrane and therefore can't travel to infect another cell (figure 37.3*d*). Oseltamivir resistance among influenza A viruses (in the United States) increased from 0.7% during the 2006–2007 season to 11.9% in 2007–2008. (The worldwide resistance rate was 16% at the end of the 2007–2008 season.) Importantly, aspirin (salicylic acid) should be avoided in children younger than fourteen years to reduce the risk of Reye's syndrome. ⏮ *Antiviral drugs (section 34.6)*

The mainstay for prevention of influenza since the late 1940s has been inactivated virus vaccines (*see table 36.2*), especially for the chronically ill, individuals over age sixty-five, residents of nursing homes, and health-care workers in close contact with people at risk. Clinical disease in these patients is most likely to be severe. Because of influenza's high genetic variability, efforts are made each year to incorporate new virus subtypes into the vaccine. Even when no new subtypes are identified in a given year, annual immunization is still recommended because immunity using the inactivated virus vaccine typically lasts only 1 to 2 years. ⏮ *Vaccines and immunization (section 36.7)*

Measles (Rubeola)

Measles (**rubeola**: Latin *rubeus*, red) is an extremely contagious skin disease that is endemic throughout most of the world. In March 2000 a group of experts convened by the Centers for Disease Control and Prevention (CDC) concluded that fortunately measles is no longer endemic in the United States. It seems that most U.S. cases of measles (fewer than 200 per year since 1998) were imported from other countries, usually Europe and Asia. However, a number of sporadic cases of measles are reported in teens and young adults whose vaccine protection has lapsed. We discuss measles because it remains of global importance; there were 454,000 deaths due to measles in 2004 outside the United States.

Measles virus is a negative-strand, enveloped RNA virus in the genus *Morbillivirus* and the family *Paramyxoviridae* (**figure 37.4*a***). *Measles virus* is monotypic; this is, it is represented as a single species, but small variations at the epitope level have been described. The variations are based on genetic variability in the virus's genes; the measles genome encodes eight proteins, six of which are structural (figure 37.4*b*). Such variations, however, have no effect on protective function since a measles infection, as well as immunization with measles vaccine, still provides a lifelong immunity against reinfection.

The virus enters the body through the respiratory tract or the conjunctiva of the eyes. The receptors for the measles virus are (1) the complement regulator CD46, also known as membrane cofactor

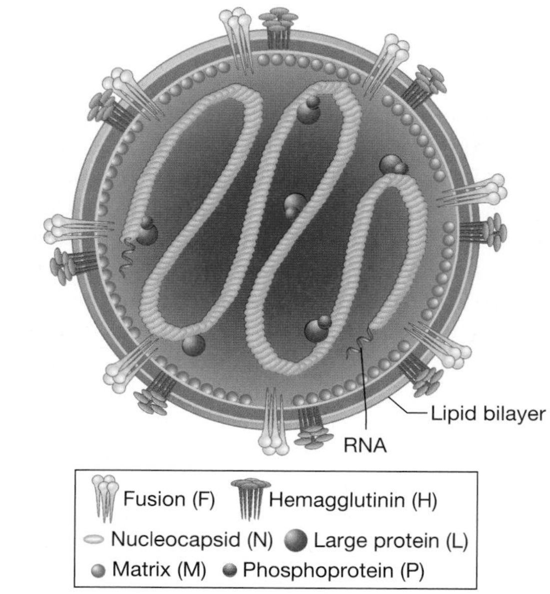

(a)

(b)

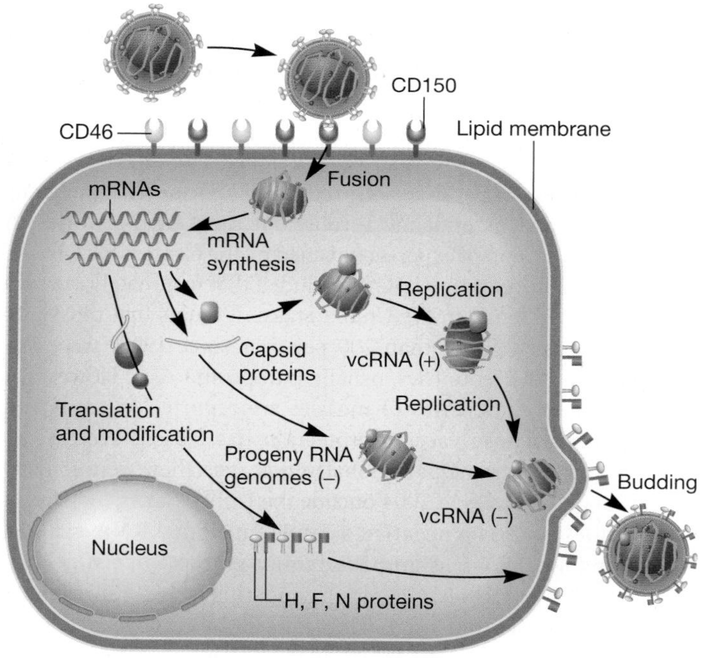

(c)

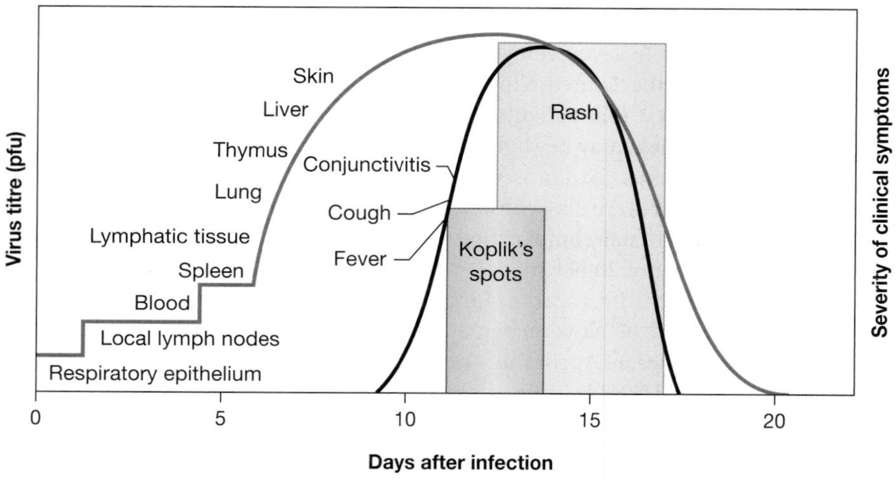

(d)

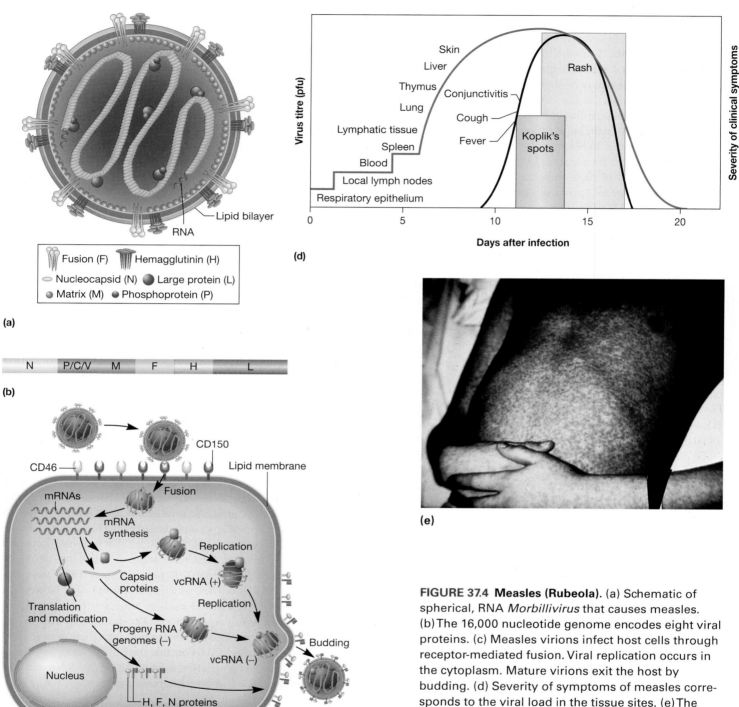

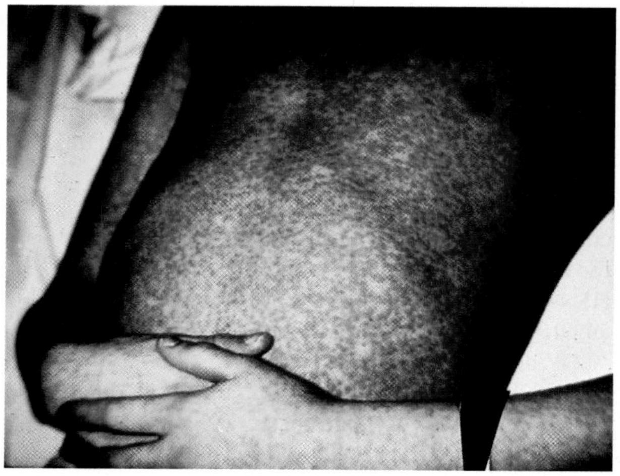

(e)

FIGURE 37.4 Measles (Rubeola). (a) Schematic of spherical, RNA *Morbillivirus* that causes measles. (b) The 16,000 nucleotide genome encodes eight viral proteins. (c) Measles virions infect host cells through receptor-mediated fusion. Viral replication occurs in the cytoplasm. Mature virions exit the host by budding. (d) Severity of symptoms of measles corresponds to the viral load in the tissue sites. (e) The rash of small, raised spots is typical of measles. The rash usually begins on the face and moves downward to the trunk.

protein, and (2) CD150 (the signaling lymphocyte activation molecule found on activated B and T cells, as well as antigen-presenting cells). The hemagglutinin (H) and fusion (F) proteins mediate attachment of the viral envelope to host cell membrane, facilitating entry. Two proteins, L and N, form the nucleocapsid. The remaining proteins (P, V, C and M) are involved in viral replication (figure 37.4*c*).

The incubation period for measles is usually 10 to 14 days, and visible symptoms begin about the tenth day (figure 37.4*d*). The measles prodrome is characterized by nasal discharge, cough, fever, headache, and conjunctivitis, which intensify several days prior to the onset of rash. Within 3 to 5 days, skin eruptions occur as erythematous (red) maculopapular (discolored area of small bumps) lesions that are at first discrete but gradually become confluent (figure 37.4*e*). The rash normally lasts about 5 to 7 days. Lesions of the oral cavity include the diagnostically useful bright-red **Koplik's spots** with a bluish-white speck in the center of each. Koplik's spots represent a viral exanthem (a skin eruption) occurring in the form of macules or papules on the mucosal lining of the mouth. Recovery from measles begins soon after the rash is reported. Very infrequently a progressive degeneration of the central nervous system called **subacute sclerosing panencephalitis** occurs.

No specific treatment is available for measles. The use of attenuated measles vaccine (Attenuvax), or in combination with mumps and rubella (MMR vaccine) or as measles, mumps, rubella, and varicella vaccine (MMRV, ProQuad®) is recommended for all children (*see table 36.3*). Since public health immunization programs began in 1963, measles nearly has been eradicated in the United States. In less well-developed countries, however, morbidity and mortality in young children from measles infection remain high. It has been estimated that *Measles virus* infects 50 million people and kills about 4 million a year worldwide. Serious outbreaks of measles are still reported in North America and Europe, especially among college students.

Mumps

Mumps is an acute, highly contagious disease that has decreased 99% in the United States since the widespread use of the MMR vaccine, with fewer than 300 cases reported annually. *Mumps virus* is a member of the genus *Rubulavirus* in the family *Paramyxoviridae*. This virus is a pleomorphic, enveloped virus that contains a helical nucleocapsid containing negative-strand RNA. The virus is transmitted in saliva and respiratory droplets. Mumps is about as contagious as influenza and rubella but less so than measles or chickenpox. The most prominent manifestations of mumps are swelling and tenderness of the salivary (parotid) glands 16 to 18 days after infection of the host by the virus. The swelling usually lasts for 1 to 2 weeks and is accompanied by a low-grade fever. Severe complications of mumps are rare; however, meningitis, enchephalitis, and inflammation of the epididymis and testes (**orchitis**) leading to sterility can be important complications associated with this disease, especially in the postpubescent male. Therapy of mumps is limited to symptomatic and supportive measures. A live, attenuated mumps virus vaccine is available. It usually is given as part of the trivalent MMR or tetravalent MMRV vaccine (*see table 36.3*).

Respiratory Syndromes and Viral Pneumonia

Acute viral infections of the respiratory system are among the most common causes of human disease. The infectious agents are called acute respiratory viruses, and they collectively produce a variety of clinical manifestations, including rhinitis (inflammation of the mucous membranes of the nose), tonsillitis, laryngitis, and bronchitis. The adenoviruses, coxsackieviruses A and B, echoviruses, influenza viruses, parainfluenza viruses, *Poliovirus*, respiratory syncytial virus, and reoviruses are thought to be responsible. It should be emphasized that for most of these viruses, there is a lack of specific correlation between the agent and the clinical manifestation; hence, the term syndrome. Immunity is not complete, and reinfection is common. The best treatment is rest. ⏮ *Virus taxonomy and phylogeny (section 25.1)*

In those cases of pneumonia for which no cause can be identified, viral pneumonia may be assumed if mycoplasmal pneumonia has been ruled out. The clinical picture is nonspecific. Symptoms may be mild, or there may be severe illness and death.

Respiratory syncytial virus (RSV) often is described as the most dangerous cause of lower respiratory infections in young children. In the United States, over 90,000 infants are hospitalized each year and at least 4,000 die. RSV is a member of the negative-strand RNA virus family *Paramyxoviridae*. The virion is variable in shape and size (average diameter of between 120 and 300 nm). It is enveloped with two virus-specific glycoproteins as part of the structure. One of the two glycoproteins, G, is responsible for the binding of the virus to the host cell; the other, the fusion or F protein, permits fusion of the viral envelope with the host cell plasma membrane, leading to entry of the virus. The F protein also induces the fusion of the plasma membranes of infected cells. RSV thus gets its name from the resulting formation of a syncytium or multinucleated mass of fused cells. The multinucleated syncytia are responsible for inflammation, alveolar thickening, and the filling of alveolar spaces with fluid. The source of the RSV is direct contact with respiratory secretions of humans. The virus is unstable in the environment (surviving only a few hours on environmental surfaces) and is readily inactivated with soap and water, and disinfectants.

Clinical manifestations of RSV infection consist of an acute onset of fever, cough, rhinitis, and nasal congestion. In infants and young children, this often progresses to severe bronchitis and viral pneumonia. Diagnosis is by either Directigen RSV or Test-Pack RSV rapid test kits. The virus is found worldwide and causes seasonal (November to March) outbreaks lasting several months. Treatment is with inhaled ribavirin (Virazole). A series of antibody (RSV-immune globulin) injections has been shown to reduce the severity of this disease in infants by 75%. Prevention and control consist of isolation of RSV-infected individuals, use of nursing barrier protection, and strict attention to good hand-washing practices.

Rubella (German Measles)

Rubella (Latin *rubellus*, reddish) was first described in Germany in the 1800s and was subsequently called German measles. It is a moderately contagious disease that occurs primarily in children five to nine years of age. It is caused by *Rubella virus*, an enveloped, positive-strand RNA virus that is a member of the family *Togaviridae*. Rubella is worldwide in distribution, being spread in droplets that are shed from the respiratory secretions of infected individuals. Once the virus is inside the body, the incubation period ranges from 12 to 23 days. A rash of small red spots (**figure 37.5**), usually lasting no more than 3 days, and a light fever are the normal symptoms. The rash appears as immunity develops and the virus disappears from the blood, suggesting that the rash is immunologically mediated and not caused by the virus infecting skin cells.

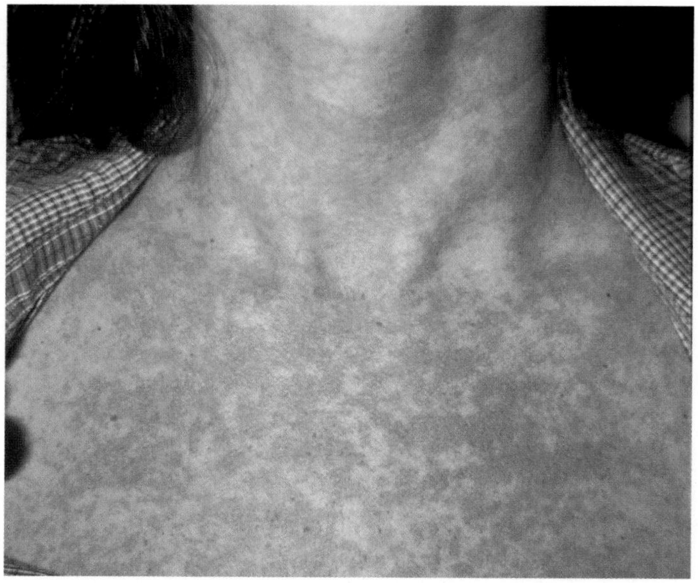

FIGURE 37.5 German Measles (Rubella). This disease is characterized by a rash of red spots. Notice that the spots are not raised above the surrounding skin as in measles (rubeola; see figure 37.4*e*).

Figure 37.5 Micro Inquiry

Why is it thought that the characteristic rash is not caused by viral infection of skin cells?

Rubella can be a disastrous disease in the first trimester of pregnancy and can lead to fetal death, premature delivery, or a wide array of congenital defects that affect the heart, eyes, and ears (**congenital rubella syndrome**). However, because rubella is usually such a mild infection, no treatment is indicated. All children and women of childbearing age who have not been previously exposed to rubella should be vaccinated. The live attenuated rubella vaccine (part of MMR, *see table 36.3*) is recommended. Fewer than 1,000 cases of rubella and 10 cases of congenital rubella currently occur annually.

Severe Acute Respiratory Syndrome

Severe acute respiratory syndrome (SARS) is a highly contagious viral disease caused by a novel coronavirus known as the SARS coronavirus (SARS-CoV). Coronaviruses are positive-strand RNA viruses. Coronaviruses are relatively large (120–150 nm), composed of RNA within a helical nucleocapsid, surrounded by an envelope. Large peplomers (spikes) protrude from the envelope to aid in attachment and entry into host cells. The protruding peplomers extend from the oval-to-spherical virion to give the illusion of a halo, or corona, around the virus (**figure 37.6*a***). The peplomer protein mediates viral attachment and fusion to the host cell membrane. SARS-CoV virions use the angiotensin-converting enzyme 2 (ACE2) on host cells as the virus receptor. A receptor-binding domain of the peplomer is complementary to the ACE2 receptor creating a binding interface (figure 37.6*b*).

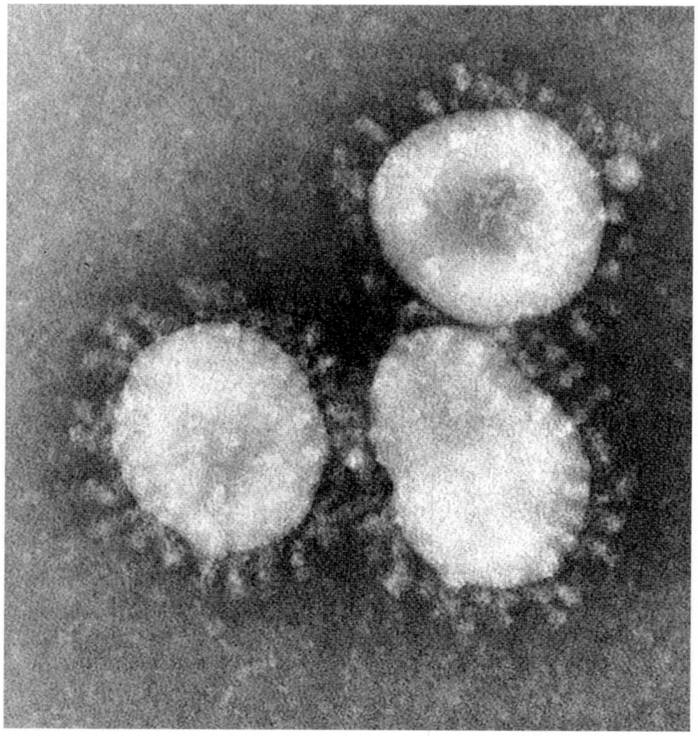

(a)

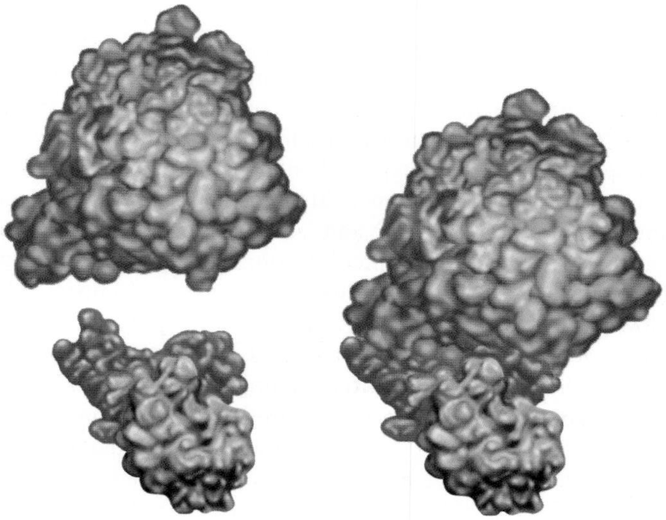

(b)

FIGURE 37.6 SARS Coronavirus (SARS-CoV). (a) Coronaviruses are named for the crownlike (corona) appearance seen by electron microscopy. (b) Space-filling model of Sars-CoV receptor-binding domain (red and cyan) as it interfaces its ACE2 receptor (green).

The virus causes a febrile (100.4°F, 38°C) lower respiratory tract illness. Sudden, severe illness in otherwise healthy individuals is a hallmark of the disease. Other symptoms may include headache; mild, flulike discomfort; and body aches. SARS patients may develop a dry cough after a few days, and most will develop pneumonia. About 10 to 20% of patients have diarrhea. If not detected early, this disease can be fatal even with supportive care. SARS is transmitted by close contact with respiratory secretions (droplet spread).

The initial outbreak of SARS originated in China in late 2002, and it spread rapidly to at least 29 other countries by summer 2003. The outbreak resulted in 8,098 persons with possible SARS, including 744 deaths being reported by the World Health Organization. There were 373 possible SARS cases in the United States; however, SARS-CoV identification has been confirmed in only 8 of them. Seven of the eight cases were likely due to exposure during international travel, and the eighth case was probably due to exposure to one of the other seven. The 2003 SARS epidemic demonstrated to the world the ease with which a virus can spread. Rapid detection and prevention measures were pursued during and after the outbreak. Diligent screening for signs of fever or respiratory disease at airports and the initiation of SARS-CoV vaccine trials are two examples of protective measures. No specific treatment is currently approved.

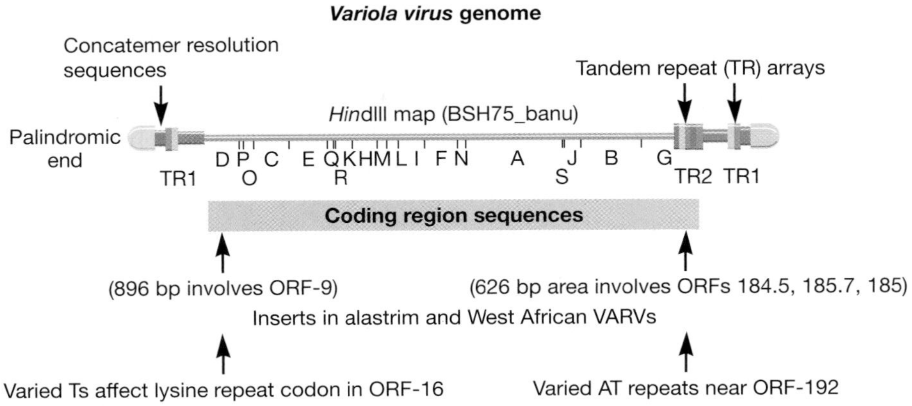

(a)

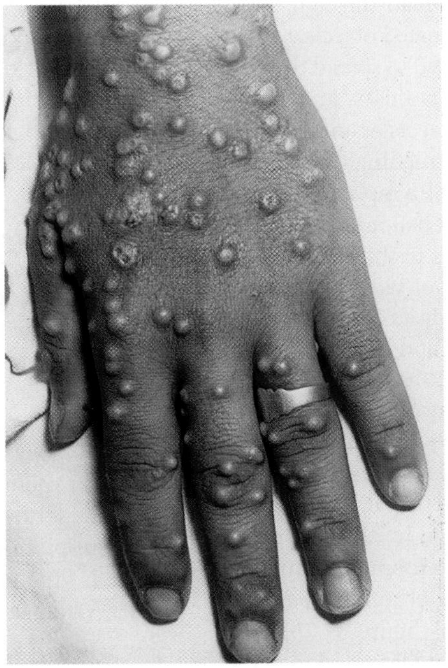

(b)

FIGURE 37.7 *Variola Virus* Causes Smallpox. (a) Genome architecture of *Variola virus* showing its linear, 186 kb pair, covalently closed DNA. (b) Back of hand showing single crop of smallpox vesicles.

Figure 37.7 Micro Inquiry

What specific clinical and biological features of smallpox made its eradication possible?

**Smallpox (Variola)

Smallpox (variola) is a highly contagious illness of humans caused by the orthopoxvirus variola major. *Variola virus* belongs to the family *Poxviridae*, which includes *Monkeypox virus*, *Molluscum contagiosum virus*, and *Vaccinia virus*, the virus that causes cowpox and serves as the antigen in the smallpox vaccine. The *Variola virus* virion is large, brick-shaped, and contains a dumbbell-shaped core. The genome inside the core consists of a single, linear molecule of double-stranded DNA that replicates in the host cell's cytoplasm. The genome is composed of approximately 186 kb pairs with covalently closed ends. The ends adjoin small, inverted terminal repeat (ITR) regions that flank the coding sequences (**figure 37.7*a***). Variola isolates have approximately 200 open reading frames (except in the ITRs) that are closely spaced and nonoverlapping. Variola does not splice its mRNA. Interestingly, the size of the smallpox virus (300 by 250 to 200 nm) is slightly larger than that of some of the smallest bacteria, for example, *Chlamydia*.

Smallpox was once one of the most prevalent of all diseases. It was a universally dreaded scourge for more than 3 millennia, with case fatality rates of 20 to 50%; it disabled and disfigured those that survived. The disease begins with a prodrome of systemic aches. Characteristic symptoms of infection also include acute onset of fever 101°F (38.3°C) followed by a rash that features firm, deep-seated vesicles or pustules in the same stage of development without other apparent cause (figure 37.7*b*). Humans are the only natural hosts of variola. First subjected to some control by variolation in tenth-century India and China, it was gradually suppressed in the industrialized world after Edward Jenner's 1796 landmark discovery that infection with the harmless cowpox (vaccinia) virus renders humans immune to the smallpox virus. ◀◀ *Historical highlights 36.4*

Since the advent of immunization with the vaccinia virus and because of concerted efforts by the World Health Organization, smallpox has been eradicated throughout the world—the greatest public health achievement ever. (The last case from a natural infection occurred in Somalia in 1977.) Eradication was possible because smallpox has obvious clinical features, virtually no asymptomatic carriers, only human hosts as reservoirs, and a short period of infectivity (3 to 4 weeks). The disease was successfully eliminated by a global immunization effort to prevent the spread of smallpox until no new cases developed.

There are two clinical forms of smallpox. Disease caused by variola major is more severe and the most common form of smallpox, with a more extensive rash and higher fever. Historically, this form of smallpox had an overall fatality rate of about 33%, with significant morbidity in those who did not die. Variola minor is a less common form of smallpox, with much less severe disease and death rates of 1% or less. Variola is generally transmitted by direct and fairly prolonged face-to-face contact. It also can be spread through direct contact with infected bodily fluids or contaminated objects such as bedding or clothing. Smallpox has been reported to spread through the air in enclosed settings such as buildings, buses, and trains. Smallpox is not known to be transmitted by insects or animals.

Variola virus enters the respiratory tract, seeding the mucous membranes and passing rapidly into regional lymph nodes. The average incubation period is 12 to 14 days but can range from 7 to 17 days. During this time, the virus multiplies in the host, but the host is not contagious. Another brief period of viremia precedes the prodromal phase. The prodromal phase lasts for 2 to 4 days and is characterized by malaise, severe head and body aches, occasional vomiting, and fever (over 104°F, 40°C), all beginning abruptly. During the prodromal phase, the mucous membranes in the mouth and pharynx become infected, resulting in a rash of small, red spots. These spots develop into open sores that spread large amounts of the virus into the mouth and throat. At this time, the person is highly contagious. The virus then invades the capillary epithelium of the skin, leading to the development of the following sequence of lesions in or on the skin: eruptions, papules, vesicles, pustules, crusts, and desquamation (figure 37.7*b*). The rash appears on the face, spreads to the arms and legs, and then the hands and feet. Usually the rash spreads over the body within 24 hours. By the third day of the rash, it forms raised bumps. By the fourth day, the bumps fill with a thick, opaque fluid and often have a depression in the center that looks like a bellybutton. (This is a major distinguishing characteristic of smallpox, as compared to other diseases that exhibit rashes.) The fever usually declines as the rash appears but rises and is sustained from the time the vesicles form until they crust over. Oropharyngeal and skin lesions contain abundant virus particles, particularly early in the disease process. Death from smallpox is due to toxemia associated with immune-mediated blood clots and elevated blood pressure.

Protection from smallpox is through vaccination (once referred to as variolation). The smallpox vaccine is a live virus immunization using the related *Vaccina virus*. The vaccine is given using a bifurcated (two-pronged) needle that is dipped into the vaccine. The bifurcated needle retains a droplet of the vaccine so that pricking the skin allows vaccinia entry into the skin. This immunization practice causes a sore spot and one or two droplets of blood to form. If the vaccination is successful, a red and itchy bump develops at the vaccine site that becomes a large blister, filled with pus. The blister will dry and form a scab that falls off, leaving a small scar. People vaccinated for the first time have a stronger reaction than those who are revaccinated. Routine immunization for smallpox was discontinued in the United States once global eradication was confirmed. Today smallpox vaccination is controversial in light of its unknown efficacy in bioterrorism prevention and potential side effects. There is no FDA-approved treatment for smallpox, although several antiviral agents have been suggested as adjunct therapies.

There is great concern that the smallpox virus could be used as a bioweapon by terrorists. An accidental or deliberate release of smallpox virus would be catastrophic in an unimmunized population and could cause a major pandemic. Because smallpox vaccination has not been performed routinely since about 1972, there is now a large population of susceptible persons. Currently, less than half the world's population has been exposed to either smallpox (*Variola virus*) or to the vaccine. Thus if an outbreak occurred, prompt recognition and institution of control measures would be paramount. ◀◀ *Bioterrorism preparednesss (section 36.8)*

1. Why are chickenpox and shingles discussed together? What is their relationship?
2. Briefly describe the course of an influenza infection and how the virus causes the symptoms associated with the flu. Why has it been difficult to develop a single flu vaccine?
3. What are some common symptoms of measles? What are Koplik's spots?
4. What is one side effect that mumps can cause in a young postpubescent male?
5. Describe some clinical manifestations caused by the acute respiratory viruses.
6. Is viral pneumonia a specific disease? Explain.
7. When is a German measles infection most dangerous and why?
8. Why is there a concern that *Variola virus* might be used by terrorists?

37.2 Arthropod-Borne Diseases

The arthropod-borne viruses (arboviruses) are transmitted by bloodsucking arthropods from one vertebrate host to another. They multiply in the tissues of the arthropod without producing disease, and the arthropod vector acquires a lifelong infection. Approximately 150 of the recognized arboviruses cause illness in humans. Diseases produced by the arboviruses can be divided into three clinical syndromes: (1) fevers of an undifferentiated type with or without a rash; (2) encephalitis (inflammation of the brain), often with a high fatality rate; and (3) hemorrhagic fevers, also frequently severe and fatal. No vaccines are available for most human arthropod-borne diseases, although supportive treatment is beneficial.

**Equine Encephalitis

Equine encephalitis is caused by viruses in the genus *Alphavirus*, family *Togaviridae*. They are positive-strand, enveloped RNA viruses. In humans, the disease can present as a spectrum from fever and headache to (aseptic) meningitis and encephalitis. However, the disease can be fatal to horses. The human disease can progress to include seizures, paralysis, coma, and death. The virus is transmitted to humans by *Aedes* and *Culex* spp. mosquitoes. Various geographic descriptors are used to define the disease caused by genetically distinct strains of these arboviruses: eastern equine encephalitis (EEE) occurs along the eastern Atlantic coast from Canada to South America; western equine encephalitis (WEE) occurs from Canada to South America along the western coast; and Venezuelan equine encephalitis occurs in central and southern parts of the United States into South America. Reservoir hosts are important in the replication, maintenance, and dissemination of these arboviruses. Treatment consists of the supportive care of symptoms. The equine hosts generally show little or no disease after infection. Currently no vaccine is available to prevent disease. Preventative measures rely on common mosquito precautions.

West Nile Fever (Encephalitis)

West Nile fever (encephalitis) is caused by a positive-strand RNA flavivirus that occurred primarily in the Middle East, Africa, and Southwest Asia. The disease was first discovered in 1937 in the West Nile district of Uganda. In 1999 the virus appeared unexpectedly in the United States (New York), causing seven deaths among 62 confirmed human encephalitis cases and extensive mortality in a variety of domestic and exotic birds. It probably crossed the Atlantic in an infected bird, mosquito, or human traveler.

By 2003, 46 U.S. states reported *West Nile virus* (WNV) infections in over 9,800 people, resulting in 264 deaths. By the start of 2006, every state in the continental United States reported WNV in either animals or humans. In 2008 there were 1,338 human cases of West Nile disease reported to the CDC, of which there were 43 deaths. Over 15,000 people in the United States have tested positive for WNV, according to the CDC. WNV is transmitted predominately to humans by *Culex* spp. mosquitoes that feed on infected birds (crows and sparrows). Mosquitoes harbor the greatest concentration of virus in the early fall; there is a peak of disease in late August to early September. The risk of disease then decreases as the mosquitoes die when the weather becomes colder. Although many people are bitten by WNV-infected mosquitoes, most do not know they have been exposed; most infected individuals remain asymptomatic or exhibit only mild, flulike symptoms. Data from the outbreak in Queens, New York, suggests that 2.6% of the population was infected, 20% of the infected people developed mild illness, and only 0.7% of the infected people developed meningitis or encephalitis.

Human-to-human transmission has been reported through blood and organ donation. However, the risk of acquiring WNV infection from donated blood or organs has greatly diminished since the introduction of a PCR-based detection assay in 2003. There are no data to suggest that WNV transmission to humans occurs from handling infected birds (live or dead), but barrier protection is suggested in handling potentially infected animals. The virus can be recovered from *Culex* mosquitoes, birds, and blood taken in the acute stage of a human infection. Diagnosis is by a rise in neutralizing antibody in a patient's serum. Only one antigenic type exists and immunity is presumed permanent. An enzyme-linked immunosorbent assay (ELISA) test for IgM anti-WNV antibody is the FDA-approved diagnostic test. There is no treatment other than hospitalization and intravenous fluids. There is no human vaccine to prevent WNV. Mosquito abatement and the use of repellents such as DEET appear to be the only control measures. ◀◀ *Clinical immunology (section 35.3)*

37.3 Direct Contact Diseases

Recall that each microorganism has an optimal portal of entry into the host. In addition to air- and vector-borne transmission, virus can be exchanged between people. Transmission of disease from one person to another often requires close personal contact to gain entry. For example, transfer of microorganisms between people occurs by direct contact with the infected person through touching, kissing, sexual contact, contact with body fluids and secretions, or contact with open wounds. We now consider several diseases that are transmitted by direct contact.

Acquired Immune Deficiency Syndrome (AIDS)

It is now recognized that **AIDS (acquired immune deficiency syndrome)** was the great pandemic of the second half of the twentieth century. Fortunately, the 2009 report issued by the Joint United Nations Programme on HIV/AIDS and the WHO indicates that the number of new HIV infections has been decreasing since 1996, when 3.5 million people were infected worldwide. Down by almost 30%, 2.7 million people were infected in 2008. First described in 1981, AIDS is the result of an infection by the **human immunodeficiency virus (HIV),** a positive-strand, enveloped RNA virus within the family *Retroviridae*. Molecular epidemiology data indicate that HIV-1 arose from the simian immunodeficiency virus (SIV) harbored by the chimpanzee (SIVcpz), *Pan troglodytes troglodytes* (Ptt). The deduced evolutionary sequence suggests that SIVcpz ancestors recombined when they crossed between several species of nonhuman primates, then into chimpanzees, and finally into humans. Stable viral infections in humans occurred on at least three occasions, with several groups of mutated SIVcpz viruses adapting to humans residing within chimp habitats. Full-length genome analyses indicate that HIV-1 groups M, N, and O are most similar to SIVcpz viruses of Ptt, evolving from separate SIVcpz lineages. However, only the SIVcpz strain now referred to as the group M HIV-1 gave rise to the virus causing the global AIDS pandemic. Notably, group M HIV-1 has diverged into several subtypes (clades) indicated as A-K. Subtype B was the first to appear in the United States, and it remains the predominant type (80%) through the Americas.

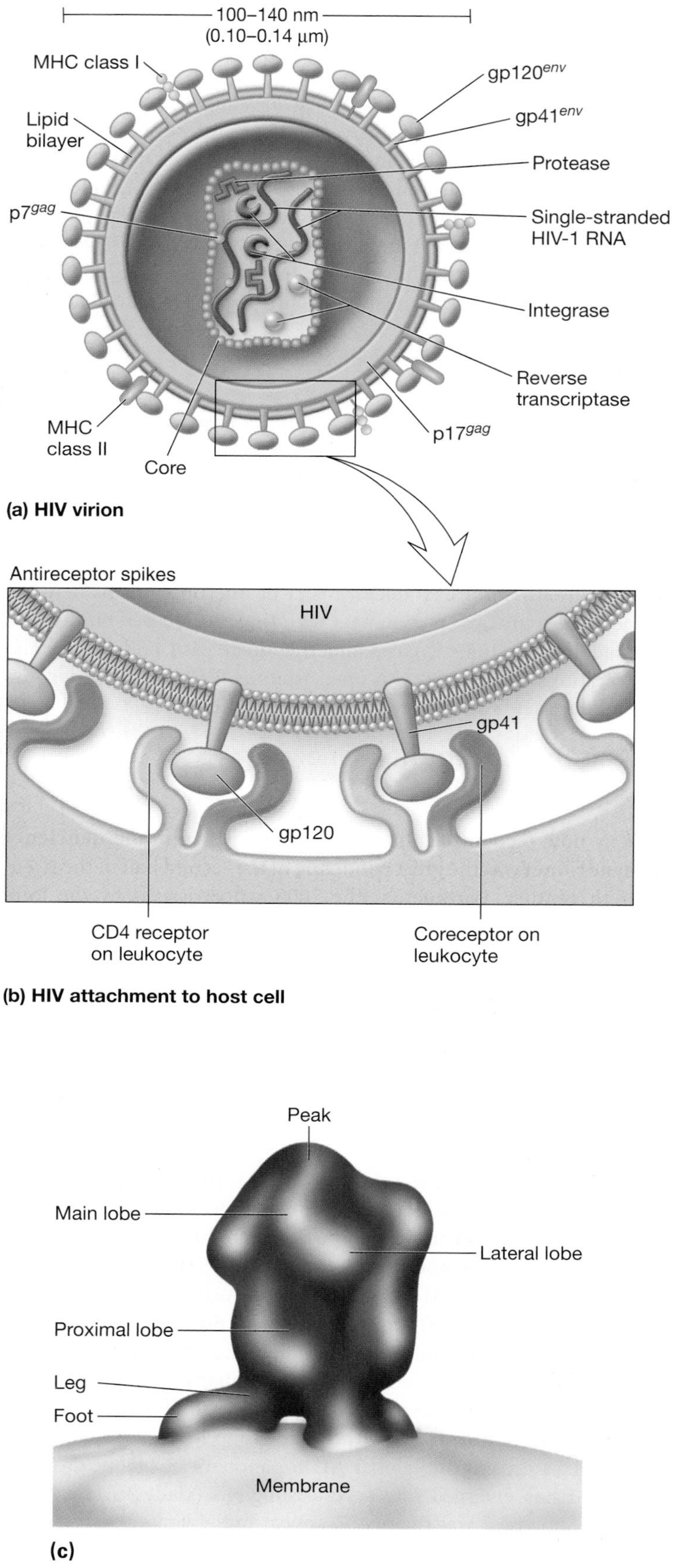

HIV/AIDS occurs worldwide with 33.4 million people living with the virus. In the United States, AIDS is caused primarily by HIV-1 (some cases result from HIV-2 infection). HIV-1 virions are approximately 110 nm in diameter, have a cylindrical core inside their capsids, and their membrane envelopes are peppered with viral spike proteins (**figure 37.8**). The core contains two copies of its RNA genome and several enzymes. Thus far 10 virus-specific proteins have been discovered. One of them, the gp120 envelope protein, participates in HIV-1 attachment to CD4 cells (e.g., T-helper cells; figure 37.8*b*). Electron cryotomography imaging reveals virion surface features. The gp120 spikes are approximately 14 nm long and appear to have no periodic spacing; they are randomly distributed. What is clear, though, is that the gp120 proteins form trimeric spikes; that is, three lobes form the final three-dimensional spike structure, and the final structure sits on three membrane-bound feet (figure 37.8*c*). ⏮ *Electron cryotomography (section 2.4)*

Although rare in the United States, the HIV-2 strain predominately causes AIDS in sub-Saharan Africa, where adult prevalence rates approach 30% (**figure 37.9**). HIV is acquired and may be passed from one person to another when infected blood, semen, or vaginal secretions come in contact with an uninfected person's broken skin or mucous membranes. In the developing world, AIDS affects men and women alike, with many women getting AIDS from men who have multiple sex partners.

FIGURE 37.8 Schematic Diagram of the HIV-1 Virion. (a) The HIV-1 virion is an enveloped structure containing 72 external spikes. These spikes are formed by the two major viral-envelope proteins, gp120 and gp41 (gp stands for glycoprotein—proteins linked to sugars—and the number refers to the mass of the protein, in thousands of Daltons). The HIV-1 lipid bilayer is also studded with various host proteins, including class I and class II major histocompatibility complex molecules, acquired during virion budding. The cone-shaped core of HIV-1 contains four nucleocapsid proteins (p24, p9, p7) each of which is proteolytically cleaved from a 53 kDa *gag* precursor by the HIV-1 protease. The phosphorylated p24 polypeptide forms the chief component of the inner shell of the nucleocapsid, whereas the p17 protein is associated with the inner surface of the lipid bilayer and stabilizes the exterior and interior components of the virion. The p7 protein binds directly to the genomic RNA through a zinc finger structural motif and together with p9 forms the nucleoid core. The retroviral core contains two copies of the single-stranded HIV-1 genomic RNA that is associated with the various preformed viral enzymes, including the reverse transcriptase, integrase, ribonuclease, and protease. (b) The snug attachment of HIV glycoprotein molecules (gp41 and gp120) to their specific receptors on a human cell membrane. These receptors are CD4 and a coreceptor called CXCR-4 (fusin) that permit docking with the host cell and fusion with the cell membrane. (c) Surface-rendered model of the gp120 ENV protein spike of human and simian immunodeficiency viruses.

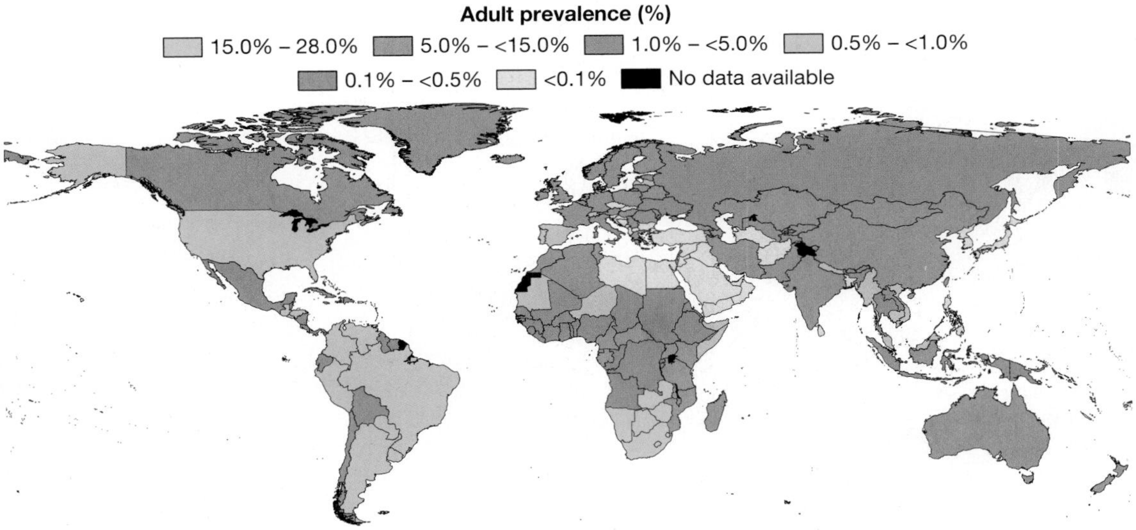

FIGURE 37.9 The Global HIV/AIDS View. The figure shows data from a 2008 United Nations report. According to UN estimates, the number of HIV/AIDS cases may be over 33 million.

Source of data: UNAIDS.

In the United States, the groups most at risk for acquiring AIDS are (in descending order) men who have unprotected sex with other men; intravenous drug users; heterosexuals who have unprotected sex with infected partners; children born of infected mothers, as well as their breast-fed infants; transfusion patients; and transplant recipients. Transmission through semen is now recognized as making HIV up to 100,000 times more virulent than by other routes. This results from the shuttling of HIV particles as they get trapped in prostatic acid phosphatase fragments to target cells. Transmission of HIV in the latter two risk groups is exceedingly rare due to extensive testing of blood products before use. The mortality rate from AIDS is almost 100% if it is not treated. The use of combined antiviral medications has significantly reduced the morbidity and mortality of AIDS in developed nations.

Once inside the body, the virus's gp120 envelope (ENV) protein (figure 37.8*b*) binds to the CD4 glycoprotein plasma membrane receptor on $CD4^+$ T cells, macrophages, dendritic cells, and monocytes. Dendritic cells are present throughout the body's mucosal surfaces and bear the CD4 protein. Thus it is possible that these are the first cells infected by HIV in sexual transmission. The virus requires a coreceptor in addition to CD4. Macrophage-tropic strains, which seem to predominate early in the disease and infect both macrophages and T cells, require the CCR5 (CC-CKR-5) chemokine receptor protein as well as CD4. A second chemokine coreceptor, called CXCR-4 or fusin, is used by T-cell–tropic strains that are active at later stages of infection. These strains induce the formation of syncytia (multinucleated masses of fused cells). There are rare individuals with two defective copies of the CCR5 gene do not seem to develop AIDS; apparently the virus cannot infect their T cells. People with one good copy of the CCR5 gene do get AIDS but survive several years longer than those with no mutation. ◀◀ *Viruses with single-stranded RNA genomes (group VI–retroviruses) (section 25.7); T-cell biology (section 33.5)*

Infection of the host cell begins when the gp120 protein binds to the CD4 receptor and chemokine co-receptor. The virus is taken into the host cell by endocytosis. Eloquent studies using membrane tracking dyes have now demonstrated that the virus uncoats from within the endocytic compartment as the viral envelope fuses with the endosome membrane, releasing the viral contents into the cytoplasm. (**figure 37.10*a***). Inside the infected cell, the core protein remains associated with the RNA as it is copied into a single strand of DNA by the RNA-dependent DNA polymerase activity of the reverse transcriptase enzyme. The RNA is next degraded by another reverse transcriptase component, ribonuclease H, and the DNA strand is duplicated to form a double-stranded DNA copy of the original RNA genome. A complex of the double-stranded DNA (the provirus) and the integrase enzyme moves into the nucleus. Then the proviral DNA is integrated into the cell's DNA through a complex sequence of reactions catalyzed by the integrase. The integrated provirus can remain latent, giving no clinical sign of its presence. Alternatively the provirus can force the cell to synthesize viral mRNA (figure 37.10*b*). Some of the RNA is translated to produce viral proteins by the cell's ribosomes. Some of the proteins have been shown to affect host cell function; for example, HIV NEF decreases MHC class I expression and may prevent apoptosis at some infection stages. Viral proteins and the complete HIV-1 RNA genome are then assembled into new virions that bud from the infected host cell (figure 37.10*c*). Eventually the host cell lyses. ◀◀ *Recognition of foreignness (section 33.4)* *Replication Cycle of a Retrovirus; Replication of HIV; How the HIV Infections Cycle Works*

Once a person becomes infected with HIV, the course of disease may vary greatly. Some rapid progressors may develop clinical AIDS and die within 2 to 3 years. A small percentage of long-term nonprogressors remain relatively healthy for at least 10 years after infection. For the majority of HIV-infected individuals, HIV

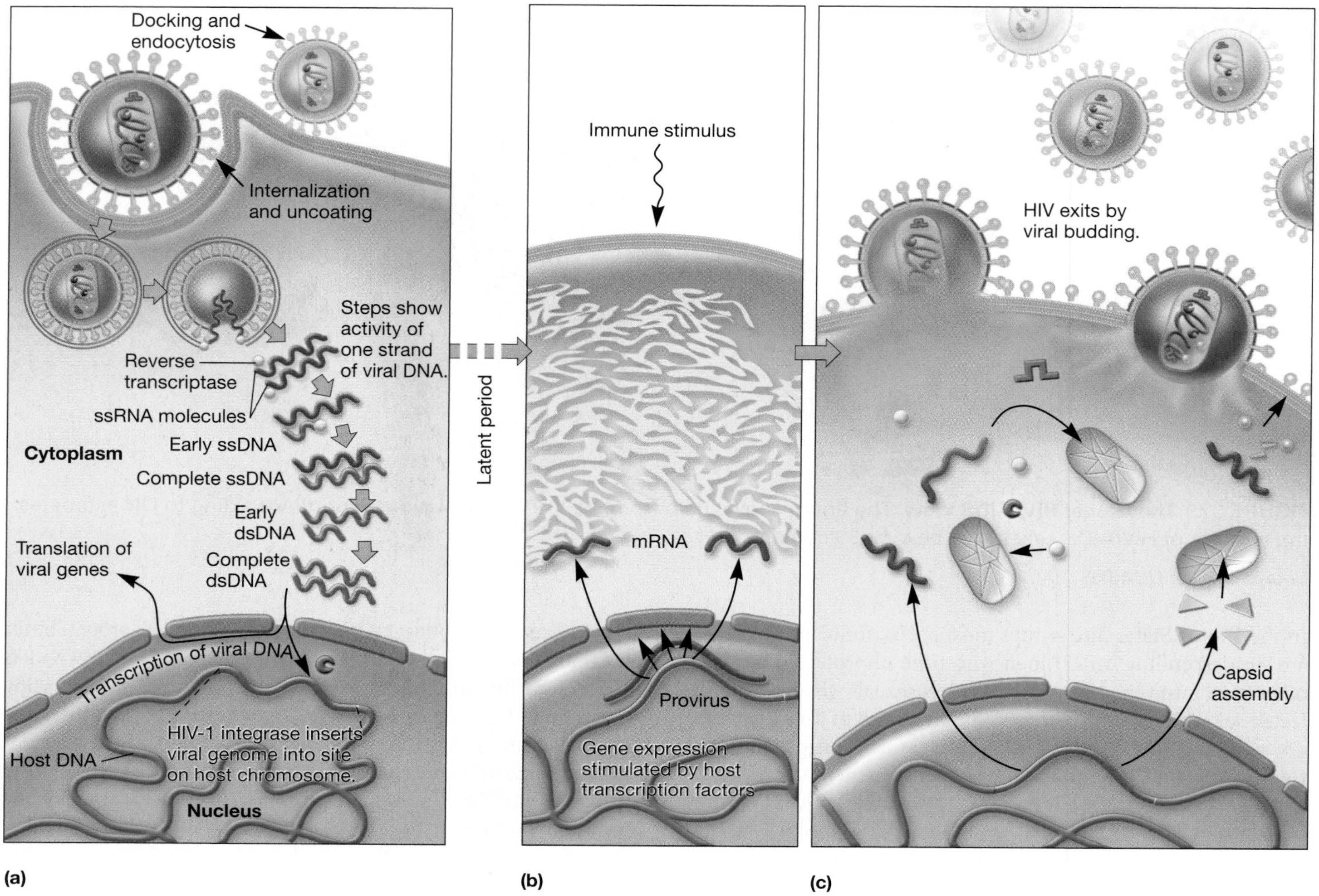

FIGURE 37.10 HIV Life Cycle. (a) The virus binds to the $CD4^+$ host cell and enters the host cell by endocytosis. Membrane labeling studies show that the viral envelope fuses with the endocytic compartment releasing the viral contents into the cytoplasm. The twin RNAs and the reverse transcriptase catalyzes the synthesis of a single complementary strand of DNA. This DNA serves as a template for synthesis of double-stranded (ds) DNA. The dsDNA can be inserted into the host chromosome as a provirus (latency). (b) The provirus genes are transcribed. (c) Viral mRNA is translated into virus components (capsid, reverse transcriptase, spike proteins), and the virus is assembled. Mature virus particles bud from the host cell, taking host membrane as their envelope.

Figure 37.10 Micro Inquiry

What is the function of integrase and how its activity related to latency?

infection progresses to AIDS in 8 to 10 years. The CDC has developed a classification system for the stages of HIV-related conditions: acute, asymptomatic, chronic symptomatic, and AIDS.

The acute infection stage occurs 2 to 8 weeks after HIV infection. About 70% of individuals in this stage experience a brief illness referred to as acute retroviral syndrome, with symptoms that may include fever, malaise, headache, macular (small, red, spotty) rash, weight loss, lymph node enlargement (lymphadenopathy), and oral candidiasis. During this stage, the virus multiplies rapidly and disseminates to lymphoid tissues throughout the body, until an acquired immune response (antibodies and cytotoxic T cells) can be generated to bring virus multiplication under control (**figure 37.11**). During the acute infection stage, levels of HIV may reach 10^5 to 10^6 copies of viral RNA per milliliter of plasma. It is believed that the extent to which the immune response is able to control this initial burst of virus multiplication may determine the amount of time required for progression to the next clinical stage.

The asymptomatic stage of HIV infection may last from 6 months to 10 years or more in some individuals. During this stage, the levels of detectable HIV in the blood decrease, but the virus continues to replicate, particularly in lymphoid tissues. Even before any changes in $CD4^+$ T cells can be detected, the virus may affect certain immune functions, e.g., memory cell responses to common antigens such as tetanus toxoid or *Candida albicans*.

During the chronic symptomatic stage, which can last for months to years, virus multiplication continues and the number of

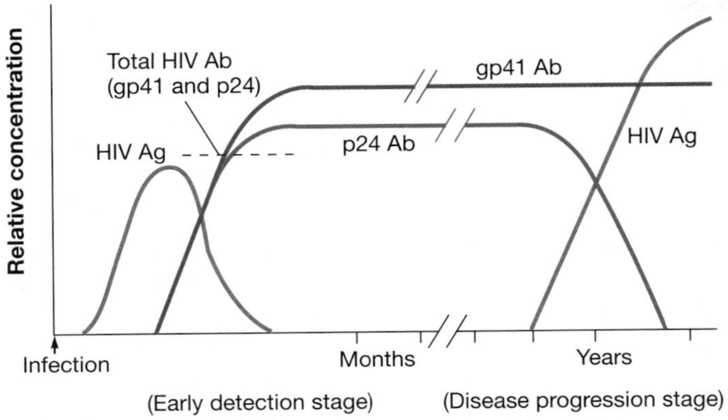

FIGURE 37.11 The Typical Serological Pattern in an HIV-1 Infection. HIV-1 antigen (HIV Ag) is detectable as early as 2 weeks after infection and typically declines as antibody to HIV proteins increases. Seroconversion occurs when HIV-1 antibodies have risen to detectable levels. This usually takes place several weeks to months after the HIV-1 infection. The period between HIV-1 infection and seroconversion often is associated with an acute illness. Whether or not the individual has flulike symptoms, the appearance of circulating HIV-1 antigens typically occurs before IgG antibodies against gp41 and p24 develop. HIV-1 antigen then usually disappears following seroconversion but reappears in the latter stages of the disease. The reappearance of antigen usually indicates impending clinical deterioration. An asymptomatic HIV-1 antigen-positive individual is six times more likely to develop AIDS within 3 years than a similar individual who is HIV-1 antigen negative. Thus testing for the presence of the HIV-1 antigen assists clinicians in monitoring the progression of the disease.

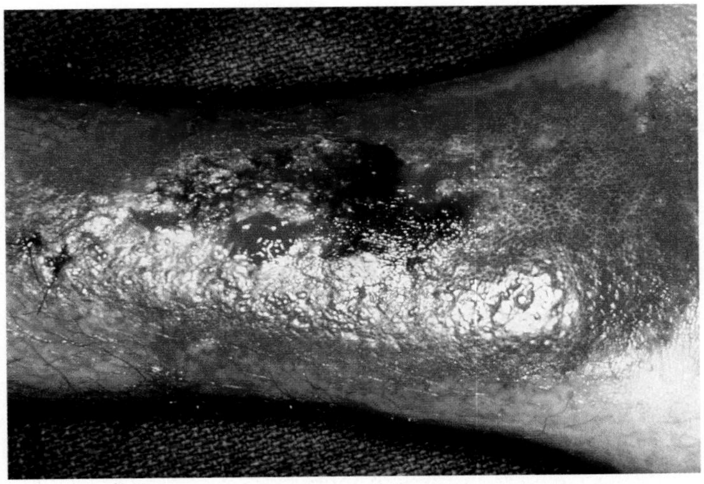

FIGURE 37.12 Kaposi's Sarcoma on the Arm of an AIDS Patient. The flat purple tumors can occur in almost any tissue.

$CD4^+$ T cells in the blood begins to significantly decrease. Because these T-helper cells are critically important in the generation of acquired immunity, individuals at this stage develop a variety of symptoms including fever, weight loss, malaise, fatigue, anorexia, abdominal pain, diarrhea, headaches, and lymphadenopathy. Paradoxically, some patients develop increased serum antibody production during this stage, perhaps as a result of generalized immune dysfunction. These antibodies, however, do little to protect the host from infection. As $CD4^+$ T cell numbers continue to decline, some patients develop opportunistic infections, such as oral candidiasis, or *Pneumocystis* pneumonia(**figure 39.25**).

Many considerations may be incorporated into a diagnosis of AIDS, the fourth stage of HIV infection. At this point, the host immune system is no longer able to defend against the virus, as it has largely depleted the potential effector cells. In 1993 the CDC revised the definition of AIDS to include all HIV-infected individuals who have fewer than 200 $CD4^+$ T cells per microliter of blood (or a CD4 T^+ cell percentage of total lymphocytes of less than 14). The reason for this definition is that the development of particular opportunistic infections is related to the $CD4^+$ T cell concentration. Healthy persons have about 1,000 $CD4^+$ T cells per microliter of blood. This number declines by an average of 40 to 80 cells per microliter per year in HIV-infected individuals. The first opportunistic infections and disease processes typically occur once the $CD4^+$ T cell count declines to 200 to 400 per microliter, despite antimicrobial therapies (**table 37.1**). Examples of these infections and diseases include *Pneumocystis* pneumonia, *Mycobacterium avium-intracellulare* pneumonia, toxoplasmosis, herpes zoster infection, chronic diarrhea caused by *Cyclospora*, cryptococcal meningitis, and *Histoplasma capsulatum* infection.

In addition to its devastating effects on the immune system, HIV infection can also lead to disease of the central nervous system because virus-infected macrophages can cross the blood-brain barrier. The classical symptoms of central nervous system disease in AIDS patients are headaches, fevers, subtle cognitive changes, abnormal reflexes, and ataxia (irregularity of muscular action). Dementia and severe sensory and motor changes characterize more advanced stages of the disease. Another potential complication of HIV infection is cancer. Individuals infected with HIV-1 have an increased risk of three types of tumors: (1) Kaposi's sarcoma (figure 37.12), (2) carcinomas of the mouth and rectum, and (3) B-cell lymphomas or lymphoproliferative disorders. It seems likely that the depression of the initial immune response enables other tumor-causing agents to initiate the cancers.

The laboratory diagnosis of HIV infection can be by viral isolation and culture or by using assays for viral reverse transcriptase activity or viral antigens (figure 37.11). However, diagnosis is most commonly accomplished through the detection of specific anti-HIV antibodies in the blood. For routine screening purposes, an ELISA is commonly used because it is sensitive and relatively inexpensive. However, false-positive results can occur with this method, requiring positive samples to be retested using a more specific Western blot technique. The most sensitive HIV assay employs the polymerase chain reaction (PCR). PCR can be used to amplify and detect tiny amounts of viral RNA and cDNA in infected host cells. Quantitative PCR assays provide an estimate of a patient's viral load. This is particularly significant because the level of virions in the blood, as well as the concentration of $CD4^+$

Table 37.1 Disease Processes Associated with AIDS
Candidiasis of bronchi, trachea, or lungs
Candidiasis, esophageal
Cervical cancer, invasive
Coccidioidomycosis, disseminated or extrapulmonary
Cryptosporidiosis, chronic intestinal (>1 month's duration)
Cyclospora, diarrheal disease
Cytomegalovirus disease (other than liver, spleen, or lymph nodes)
Cytomegalovirus retinitis (with loss of vision)
Encephalopathy, HIV-related
Herpes simplex: chronic ulcer(s) (>1 month's duration); or bronchitis, pneumonitis, or esophagitis
Histoplasmosis, disseminated or extrapulmonary
Isosporiasis, chronic intestinal (>1 month's duration)
Kaposi's sarcoma
Lymphoma, Burkitt's
Lymphoma, immunoblastic
Lymphoma, primary, of brain
Mycobacterium avium complex or *M. kansasii*
Mycobacterium tuberculosis, any site
Mycobacterium, other species or unidentified species
Pneumocystis pneumonia
Pneumonia, recurrent
Progressive multifocal leukoencephalopathy
Salmonella septicemia, recurrent
Toxoplasmosis of brain
Wasting syndrome due to AIDS

Source: Data from MMWR *41 (No. RR17). 1993 Revised Classification System for HIV Infection and Expanded Surveillance Case Definition for AIDS Among Adolescents and Adults.*

cells, is very predictive of the clinical course of the infection. The probable time to development of AIDS can be estimated from the patient's blood virion level and $CD4^+$ cell count. ◀◀ *Polymerase chain reaction (section 15.2); Clinical immunology (section 35.3)*

At present there is no cure for AIDS. Primary treatment is directed at reducing the viral load and disease symptoms, and treating opportunistic infections and malignancies. The antiviral drugs currently approved for use in HIV disease are described in detail in chapter 34. An overview of their mechanisms of action are presented in figure 34.16*b*. Briefly, there are four types. (1) Nucleoside reverse transcriptase inhibitors (NRTIs) are nucleoside analogues that inhibit the enzyme reverse transcriptase as it synthesizes viral DNA. Examples include zidovudine (AZT or Retrovir), didanosine (Videx), zalcitabine (ddC or HIVID), tenofovir (Viread), emtricitabine (Emtriva or Coviracil), stavudine (Zerit), and lamivudine (Epivir or 3TC). (2) The nonnucleoside reverse transcriptase inhibitors (NNRTIs) include delavirdine (Rescriptor), efavirenz (Sustiva), and nevirapine (Viramune). (3) The protease inhibitors (PIs) work by blocking the activity of the HIV protease, which is responsible for generating the mature proteins, and thus interfere with virion assembly. Examples include indinavir (Crixivan), ritonavir (Norvir), nelfinavir (Viracept), lopinavir (Kaletra), Fosamprenavir (Lexiva), atazanavir (Reyataz), and saquinavir (Invirase). (4) The fusion inhibitors (FIs) are a newer category of drugs that prevent HIV entry into cells. This category is represented by enfuvirtide (Fuzeon). The most successful treatment approach in combating HIV/AIDS is to use drug combinations. An effective combination is a cocktail of various NRTIs, NNRTIs, PIs, and the FI. Such drug combination use is referred to as HAART (highly active antiretroviral therapy). *Treatment of HIV Infection*

The development of a vaccine for AIDS has been a long-sought research goal. Such a vaccine would ideally (1) stimulate the production of neutralizing antibodies, which can bind to the viral envelope and prevent the virus from entering host cells; and (2) promote the formation of cytotoxic T cells (CTLs), which can destroy cells infected with virus. Among the many problems encountered in developing an HIV vaccine is the fact that the envelope proteins of the virus continually change their antigenic properties; HIV has a high mutation rate because reverse transcriptase lacks proofreading capabilities.

Many HIV researchers continue to take great interest in HIV-infected persons who are long-term nonprogressors. These individuals maintain $CD4^+$ T cell counts of at least 600 per microliter of blood, have less than 5,000 copies of HIV RNA per milliliter of blood, and have remained this way for more than 10 years after documented infection even in the absence of antiviral agents. Recent data suggest that there are two subpopulations of long-term nonprogressors, so called "viremic controllers" and "elite controllers." The former are defined as having only 50–2000 HIV copies per milliter of blood after 10 years; the latter have less than 50 HIV copies per milliter of blood. Interestingly it seems that immunity in these hosts puts selective pressure on the virus such that mutations resulting in enhanced viral escape from immune cells also results in loss of viral fitness (decreased HIV entry into new cells and decreased integration of HIV DNA into host cell DNA).

Prevention and control of AIDS is achieved primarily through education. Understanding risk factors and practicing strategies to reduce risk are essential in the fight against AIDS. Barrier protection from blood and body fluids greatly limits risk of HIV infection. Education to prevent the sharing of intravenous needles and syringes is also very important. Additionally prevention includes the continued screening of blood and blood products.

Cold Sores

The term herpes is derived from the Greek word meaning "to creep." Clinical descriptions of herpes lesions date back to the time of Hippocrates (circa 400 B.C.). Herpes lesions are caused by the double-stranded DNA virus herpes simplex. Herpes viruses have icosahedral capsids and are enveloped. The genome encodes approximately 100 proteins. The herpesvirus genome must first enter an epithelial cell for the initiation of infection. Transmission is through direct contact of epithelial tissue surfaces with the virus. Herpes simplex virus type 1 (HSV-1) was initially thought to infect oral mucosal epithelium to cause cold sores or fever blisters and herpes simplex virus type 2 (HSV-2) was thought to infect genital epithelium to cause genital herpes. However, both viruses can infect either tissue site. The initial association is between proteoglycans of the epithelial cell surface and viral glycoproteins. This is followed by a specific interaction with one of several cellular receptors collectively termed HVEMs for *h*erpes*v*irus *e*ntry *m*ediators. The capsid, along with some associated proteins, then migrates along the cellular microtubule transport machinery to nuclear envelope pores. This "docking" is thought to result in the viral DNA being injected through the nuclear envelope pores while the capsid remains in the cytoplasm.

Search This: Dr. Wagner's herpes virus research

Active and latent phases have been identified within an infected host. After an incubation period of about a week, the active phase begins. During the active phase, the virus multiplies explosively; between 50,000 and 200,000 new virions are produced from each infected cell. During this replication cycle, herpesvirus inhibits its host cell's metabolism and degrades host DNA, inducing apoptosis. As a result, the cell dies, releasing viral progeny to infect other cells. Such an active infection may be symptom-free, or painful blisters in the infected tissue may occur. Indeed, apparently healthy people can transmit HSV to other hosts or their newborns. This is especially true of transmission with HSV-2; it can be passed to sexual contacts even when there are no clinical signs of infection. Blisters involving the epidermis and surface mucous membranes of the lips, mouth, and gums (**gingivostomatitis**) are referred to as herpes labialis (**figure 37.13**). Primary and recurring HSV infections also may occur in the eyes, causing **herpetic keratitis** (inflammation of the cornea)—currently a major cause of blindness in the United States. Blisters involving the epidermis and surface mucous membranes of the genitals and perianal region are referred to as genital herpes. In all cases, the blisters are the result of cell lysis and the development of a local inflammatory response; they contain fluid and infectious virions. Fever, headache, muscle aches and pains, a burning sensation, and general soreness are frequently present during the active phase.

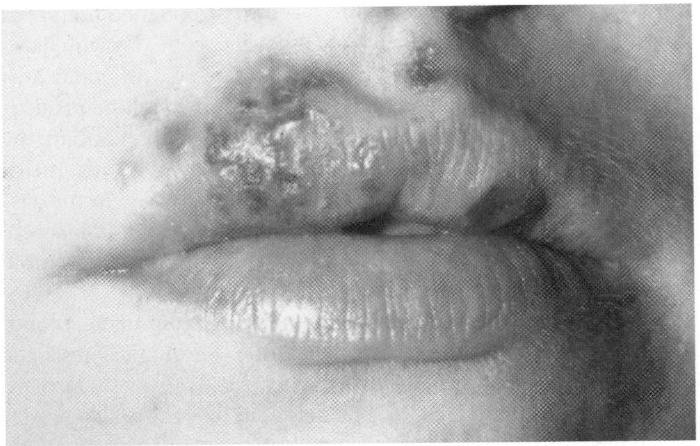

FIGURE 37.13 Cold Sores. Herpes simplex fever blisters on the lip, caused by herpes simplex type 1 virus.

Although blisters usually heal within a week, after a primary infection, a very interesting virulence factor initiates the long-term survival of HSV-1–infected neurons, inducing the latent phase of infection. As the active infection is curtailed by a vigorous cellular immune response, virus released from infected oral epithelia travel to the trigeminal nerve ganglion. The retreat to the nerve ganglion (cells) results in a latent state for the lifetime of the infected host, apparently to prevent further detection by the host immune system. With time, however, HSV-1 occasionally emerges from latency to grow productively and kill its host neuron, potentially attracting the wrath of the immune system. Stressful stimuli such as excessive sunlight, fever, trauma, chilling, emotional stress, and hormonal changes can reactivate the virus. Once reactivated, the virus moves from the nerve ganglion down a peripheral nerve and back to epithelial cells to produce the active phase. Interestingly, not all infected neurons die from a productive infection. Latently infected neurons and some neurons with productive HSV infection survive. How is this? It turns out that a small HSV gene has been identified that can protect the infected neuron from host immune cells. The gene encodes the latency-associated transcript, or LAT, an 8.5 kb RNA. LAT is spliced as an mRNA precursor, but no consistently expressed protein has been found to be translated from it. Instead, a long-lived 2 kb microRNA (miR-LAT) remaining in the nucleus appears to promote degradation of two proteins that inhibit cell proliferation and initiate apoptosis. In other words, miR-LAT prevents the infected cell from self-destruction, ensures neuron survival, and provides a hiding place for the virus. ◀◀ *Regulation of translation by small RNA molecules (section 13.4)*

The drugs vidarabine (Vira-A) and acyclovir (Zovirax) are effective against cold sores. Idoxuridine and trifluridine are used to treat herpes infections of the eye. By adulthood, 70 to 90% of all people in the United States have been infected and have type 1 herpes antibodies. Diagnosis of HSV-1 infection is by ELISA and direct fluorescent antibody screening of tissue. Diagnosis may also be made through the recovery of viral DNA by PCR. These tests are especially useful in individuals who are particularly susceptible to severe infections.

Common Cold

The **common cold** (coryza: Greek *koryza*, discharge from the nostrils) is one of the most frequent infections experienced by humans of all ages. The incidence of infection is greater during the winter months, likely due to increased population density (indoors), the effect of dry winter air on mucous membranes, and the decreased immune function that results from the direct effect of cold temperatures. About 50% of the cases are caused by rhinoviruses (Greek *rhinos*, nose), which are nonenveloped, positive-strand RNA viruses in the family *Picornaviridae* (*see section 25.5*). There are over 115 distinct serotypes, and each of these antigenic types has a varying capacity to infect the nasal mucosa and cause a cold. In addition, immunity to many of them is transitory. Several other respiratory viruses are also associated with colds (e.g., coronaviruses and parainfluenza viruses). Thus colds are common because of the diversity of rhinoviruses, the involvement of other respiratory viruses, and the lack of a durable immunity.

Rhinoviruses provide an excellent example of the medical relevance of research on virus morphology (**figure 37.14**). The complete rhinovirus capsid structure has been elucidated with the use of X-ray diffraction techniques. The results help explain rhinovirus resistance to human immune defenses. The capsid protein that recognizes and binds to cell surface molecules during infection lies at the bottom of a surface cleft (sometimes called a "canyon") about 12 Å deep and 15 Å wide. Thus the binding site is well protected from the immune system while it carries out its functions. Moreover, with greater than 100 serotypes of human rhinoviruses, immunity to one strain may not protect against another strain, making vaccine development problematic. Possibly drugs that could fit in the cleft and interfere with virus attachment can be designed.

Viral invasion of the upper respiratory tract is the basic mechanism in the pathogenesis of a cold. The virus enters the body's cells by binding to cellular adhesion molecules. The clinical manifestations include the familiar nasal stuffiness, sneezing, scratchy throat, and a watery discharge from the nose. The discharge becomes thicker and assumes a yellowish appearance over several days. General malaise is commonly present. Fever is usually absent in uncomplicated colds, although a low-grade (100–102°F, 37.8–38.9°C) fever may occur in infants and children. The disease usually runs its course in about a week. Diagnosis of the common cold is made from observations of clinical symptoms. There are no procedures for direct examination of clinical specimens or for serological diagnosis.

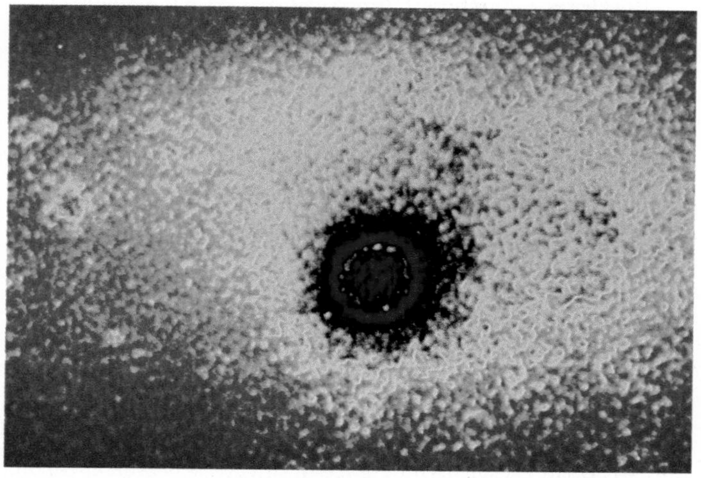

FIGURE 37.14 Rhinovirus. One of over 100 types of virus that cause the common cold.

Sources of the cold viruses include infected individuals excreting viruses in nasal secretions, airborne transmission over short distances by way of moisture droplets, and transmission on contaminated hands or fomites. Epidemiological studies of rhinovirus colds have shown that the familiar explosive, noncontained sneeze (*see figure 31.5*) may not play an important role in virus spread. Rather, hand-to-hand contact between a rhinovirus "donor" and a susceptible "recipient" is more likely. The common cold occurs worldwide with two main seasonal peaks, spring and early autumn. Infection is most common early in life and generally decreases with an increase in age. Treatment for the common cold is mainly rest, extra fluids, and the use of anti-inflammatory agents for alleviating local and systemic discomfort.

Cytomegalovirus Inclusion Disease

Cytomegalovirus inclusion disease is caused by the human cytomegalovirus (HCMV), a member of the family *Herpesviridae*. HCMV is an enveloped, double-stranded DNA virus with an icosahedral capsid. Most people become infected with this virus at some time during their life; in the United States, as many as 80% of individuals older than thirty-five years have been exposed to this virus and carry a lifelong infection. Although most HCMV infections are asymptomatic, certain patient groups are at risk to develop serious illness and long-term effects. For example, this virus remains the leading cause of congenital virus infection in the United States, a significant cause of transfusion-acquired infections, and a frequent contributor to morbidity and mortality among organ transplant recipients and immunocompromised individuals (especially AIDS patients). Because the virus persists in the body, it is shed for several years in saliva, urine, semen, and cervical secretions.

HCMV can infect any cell of the body, where it multiplies slowly and causes the host cell to swell in size; hence the prefix "cytomegalo-", which means "an enlarged cell." Cytomegaloviruses are well-known for their ability to interfere with many host immune functions, such as MHC presentation, cytokine production, and natural killer cell activity. Infected cells contain the unique **intranuclear inclusion bodies** and cytoplasmic inclusions (**figure 37.15**). In fatal cases, cell damage is seen in the gastrointestinal tract, lungs, liver, spleen, and kidneys. In less-severe cases, cytomegalovirus inclusion disease symptoms resemble those of infectious mononucleosis.

Laboratory diagnosis is by virus isolation from urine, blood, semen, and lung or other infected tissue. Serological tests for anti-HCMV IgM and IgG or by rapid test kit (CMV-vue) also are available. Detection of HCMV nucleic acid by PCR is also used.

The virus has a worldwide distribution, especially in developing countries, where infection is universal by childhood. The prevalence of this disease increases with a lowering of socioeconomic status and hygienic practices. The only drugs available, ganciclovir (Cytovene-IV) and cidofovir (Vistide), are used only for high-risk patients. Infection can be prevented by avoiding close personal contact (including sexual) with an actively infected individual. Transmission by blood transfusion or organ transplantation can be avoided by using blood or organs from seronegative donors.

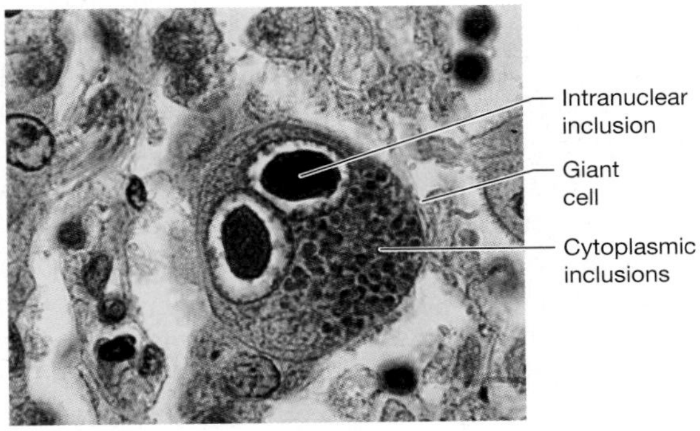

FIGURE 37.15 Cytomegalovirus Inclusion Disease. Light micrograph of a giant cell in a lung section infected with the cytomegalovirus (×480). The intranuclear inclusion body has a typical "owl-eyed" appearance because of its surrounding clear halo. Sometimes virus inclusions also are visible in the cytoplasm.

Genital Herpes

Genital herpes is a life-long infection caused by herpes simplex virus, predominantly type 2 (HSV-2). HSV-2 is classified in the alphaherpes subfamily of the family *Herpesviridae* (**figure 37.16*a***). All members of this subfamily have a very short replication cycle. The core DNA is linear and double stranded. The HSV-2 envelope contains at least eight glycoproteins. HSV-2 is most frequently transmitted by sexual contact. Infection begins when the virus is introduced into a break in the skin or mucous membranes. The virus infects the epithelial cells of the external genitalia, the urethra, and the cervix. Rectal and pharyngeal herpes are also transmitted by sexual contact. The details of HSV-2 infection are very similar to those of HSV-1.

In the case of HSV-2, the viruses retreat to nerve cells in the sacral plexus of the spinal cord, where they remain in a latent form. As described for HSV-1, during the latent phase, the host cell does not die. Because viral genes are not expressed, the infected person is symptom-free. It should be noted that both primary infection and reactivation can occur without any symptoms and apparently healthy people can transmit HSV-2 to their sexual partners or their newborns during vaginal delivery.

Primary and recurring HSV infections also may occur in infants during vaginal delivery, leading to **congenital (neonatal) herpes.** Congenital herpes is one of the most life-threatening of all infections in newborns, affecting approximately 1,500 to 2,200 babies per year in the United States. It can result in neurological involvement

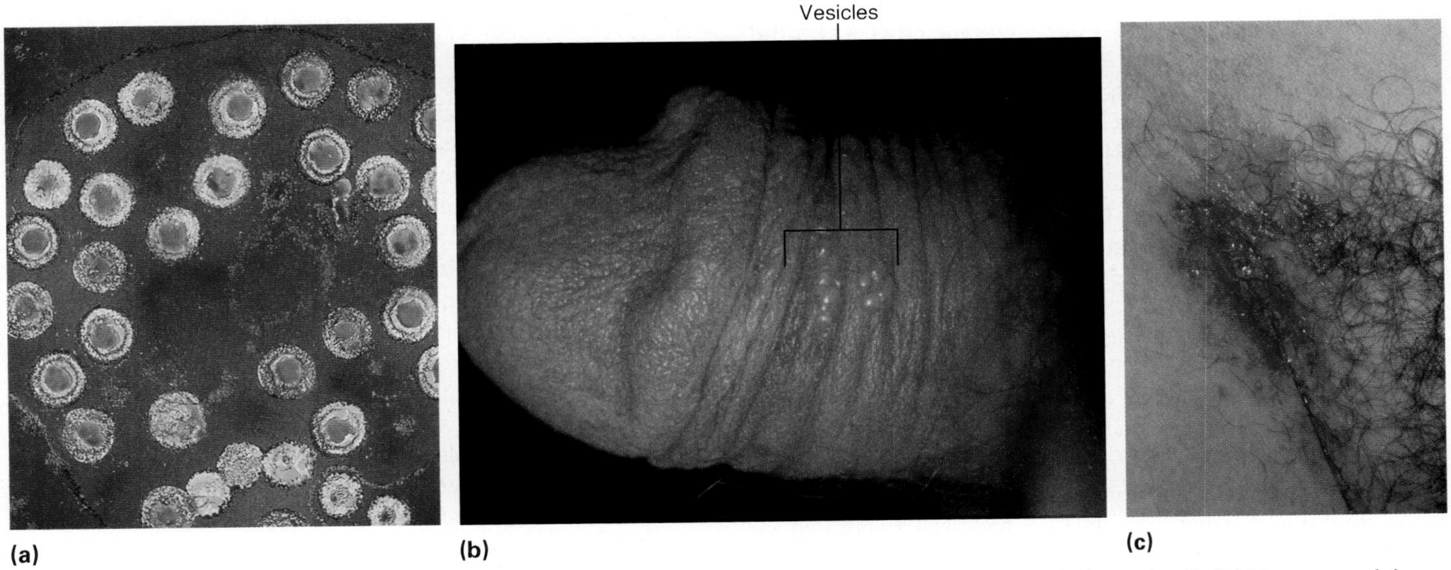

FIGURE 37.16 Genital Herpes. (a) Herpes simplex virus type 2 (yellow and green) inside an infected cell. (b) Herpes vesicles on the penis. (c) Herpes vesicles and blisters on the vaginal labia. The vesicles contain fluid that is infectious.

Figure 37.16 Micro Inquiry

Where does HSV-2 reside during latency?

as well as blindness. As a result, any pregnant female who has active or new genital herpes should have a cesarian section instead of delivering vaginally. For unknown reasons, the HSV-2 virus is also associated with a higher-than-normal rate of cervical cancer and miscarriages.

Diagnosis of HSV-2 infection is by ELISA screening of blood or serum, direct fluorescent antibody testing of tissue, and/or PCR. Although there is no cure for genital herpes, oral use of the antiviral drugs acyclovir (Zovirax or Valtrex) and famciclovir (Famvir) has proven to be effective in ameliorating the recurring blister outbreaks. Topical acyclovir is also effective in reducing virus shedding, the time until the crusting of blisters occurs, and new lesion formation.

In the United States, the incidence of genital herpes has increased so much during the past several decades that it is now a very common sexually transmitted disease. It is estimated that over 25 million Americans (20% of adults) are infected with the herpes simplex virus type 2.

Human Herpesvirus 6 Infection

Human herpesvirus 6 (HHV-6) is the etiologic agent of **exanthem subitum** [Greek *exanthema*, rash] in infants. HHV-6 is a unique member of the family *Herpesviridae* that is distinct serologically and genetically from the other herpesviruses. The virus envelope encloses an icosahedral capsid and a core containing double-stranded DNA. The disease caused by HHV-6 was originally termed **roseola infantum** and then given the ordinal designation **sixth disease** to differentiate it from other exanthems and roseolas. Exanthem subitum is a short-lived disease characterized by a high fever of 3 to 4 days' duration, after which the temperature suddenly drops to normal and a macular rash appears on the trunk and then spreads to other areas of the body. HHV-6 infects over 95% of the U.S. infant population, and most children are seropositive for HHV-6 by three years of age. It follows that nearly all individuals evaluated test positive for the evidence of HHV-6 infection. $CD4^+$ T cells are the main site of viral replication, whereas monocytes are in an infected, latent state. The tropism of HHV-6 appears to be wide, including $CD8^+$ T cells, natural killer cells, and probably epithelial cells. In adults, HHV-6 is commonly found in peripheral-blood mononuclear cells and saliva, suggesting that the infection is lifelong. Since the salivary glands are the major site of latent infection, transmission is probably by way of saliva.

HHV-6 also produces latent and chronic infections, and is occasionally reactivated in immunocompromised hosts, leading to pneumonitis. Furthermore, HHV-6 has been implicated in several other diseases (lymphadenitis and multiple sclerosis) in immunocompetent adults. Diagnosis is by immunofluorescence or enzyme immunoassay, or PCR. To date, there is no antiviral therapy or prevention.

HHV-6 provirus DNA has been found intact within human chromosomes and may be transmitted in the germline. In one study of 43 congenitally HHV-6 infected infants, 85% had inherited the virus directly in their DNA. Mapping studies have shown that HHV-6 integrates into a number of chromosomes, typically at the ends near the telomeres. It is not yet known what the impact of viral integration is.

Human Parvovirus B19 Infection

Since its discovery in 1974, ***Human parvovirus B19*** (family *Parvoviridae*, genus *Erythrovirus*) has emerged as a significant human pathogen. B19 virions are uniform, icosahedral, naked particles approximately 23 nm in diameter. Parvoviruses have a genome composed of one ssDNA molecule of about 5,000 bases. Parvoviruses are among the simplest of the DNA viruses. The genome is so small that it must resort to the use of overlapping genes to encode the very few proteins it encodes. The genome does not code for any enzymes, and the virus must use host cell enzymes for all biosynthetic processes. Thus viral DNA can only be replicated in the nucleus during the S phase of the cell cycle, when the host cell replicates its own DNA. Because the viral genome is single stranded and linear, the host DNA polymerase must be tricked into copying it. By using a self-complementary sequence at the ends of the viral DNA, the parvovirus genome folds back on itself to form a primer for replication (*see figure 25.18*). This is recognized by the host DNA polymerase and DNA replication ensues. ◀◀ *Viruses with single-stranded DNA genomes (group II) (section 25.3)*

A spectrum of disease is caused by parvovirus B19 infection, ranging from mild symptoms (fever, headache, chills, malaise) in normal persons and **erythema infectiosum** in children (**fifth disease**), to a joint disease syndrome in adults. More serious diseases include aplastic crisis in persons with sickle cell disease and autoimmune hemolytic anemia, and pure red cell aplasia due to persistent B19 virus infection in immunocompromised individuals. The B19 parvovirus can also infect the fetus, resulting in anemia, fetal hydrops (the accumulation of fluid in tissues), and spontaneous abortion. This has prompted a grass-roots awareness of B19 complications among school teachers who may be pregnant. It is assumed that the natural mode of infection is by the respiratory route. The average incubation period is 4 to 14 days. Approximately 20% of infected individuals are asymptomatic, and a smaller percentage of infected individuals have symptoms for up to 3 weeks. Infection typically results in a life-long immunity to B19.

A variety of techniques are available for the detection of the B19 virus. Antiviral antibodies appear to represent the principal means of defense against B19 parvovirus infection and disease. The treatment of individuals suffering from acute and persistent B19 infections with commercial immunoglobulins containing anti-B19 and human monoclonal antibodies to B19 is an effective therapy. As with other diseases spread by contact with respiratory secretions, frequent hand-washing is the best prevention of the disease.

Mononucleosis (Infectious)

The Epstein-Barr virus (EBV) is a member of the family *Herpesviridae*. EBV exhibits the characteristic herpesvirus morphology—all herpesviruses consist of an icosahedral capsid (approximately

125 nm in diameter) surrounded by an envelope. The capsid contains the viral double-stranded (ds) DNA. Its dsDNA exists as a linear form in the mature viron and a circular episomal form in latently infected host cells. EBV is the etiologic agent of **infectious mononucleosis (mono),** a disease whose symptoms closely resemble those of cytomegalovirus-induced mononucleosis. Because the Epstein-Barr virus occurs in oropharyngeal secretions, it can be spread by mouth-to-mouth contact (hence the terminology infectious and kissing disease) or shared drinking bottles and glasses. A person gets infected when the virus from someone else's saliva makes its way into epithelial cells lining the throat. After a brief bout of multiplication in the epithelial cells, the new viruses are shed and infect memory B cells. Infected B cells rapidly proliferate and take on an atypical appearance (Downey cells) that is useful in diagnosis (**figure 37.17**). The disease is manifested by enlargement of the lymph nodes and spleen, sore throat, headache, nausea, general weakness and tiredness, and a mild fever that usually peaks in the early evening. The disease lasts for 1 to 6 weeks and is self-limited. Like other herpesviruses, EBV becomes latent in its host.

Treatment of mononucleosis is largely supportive and includes plenty of rest. Diagnosis of mononucleosis is usually confirmed by demonstration of an increase in circulating mononuclear cells, along with a serological test for nonspecific (heterophile) antibodies, specific viral antibodies, or identification of viral nucleic acid. Several rapid tests are on the market.

The peak incidence of mononucleosis occurs in people fifteen to twenty-five years of age. Collegiate populations, particularly those in the upper socioeconomic class, have a high incidence of the disease. About 50% of college students have no immunity, and approximately 15% of these can be expected to contract mononucleosis. People in lower socioeconomic classes tend to acquire immunity to the disease because of early childhood infection. The Epstein-Barr virus may well be the most common virus in humans, as it infects 80 to 90% of all adults worldwide. EBV infections are associated with the cancers Burkitt's lymphoma in tropical Africa and nasopharyngeal carcinoma in Southeast Asia, East and North Africa, and in Inuit populations.

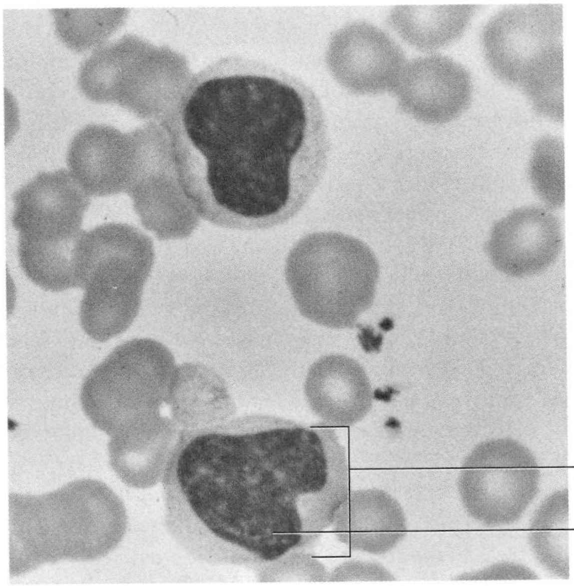

FIGURE 37.17 Evidence of Epstein-Barr Infection in the Blood Smear of a Patient with Infectious Mononucleosis. Note the abnormally large lymphocytes containing indented nuclei with light discolorations.

Viral Hepatitides

Inflammation of the liver is called **hepatitis** (pl., hepatitides; Greek *hepaticus*, liver). Currently 11 viruses are recognized as causing hepatitis. Two are herpesviruses (cytomegalovirus [CMV] and Epstein-Barr virus [EBV]) and nine are hepatotropic viruses that specifically target liver hepatocytes. EBV and CMV cause mild, self-resolving forms of hepatitis with no permanent hepatic damage. Both viruses cause the typical infectious mononucleosis syndrome of fatigue, nausea, and malaise. Of the nine human hepatotropic viruses, only five are well characterized; hepatitis G (**table 37.2**) and TTV (transfusion-transmitted virus) are more recently discovered viruses. Hepatitis A (sometimes called infectious hepatitis) and hepatitis E are transmitted by fecal-oral contamination and discussed in the section on food-borne and waterborne diseases (section 37.4). The other major types include hepatitis B (sometimes called serum hepatitis), hepatitis C (formerly non-A, non-B hepatitis), and hepatitis D (a virusoid formerly called delta hepatitis).

Hepatitis B (serum hepatitis) is caused by ***Hepatitis B virus*** (HBV), an enveloped, double-stranded circular DNA virus of complex structure. HBV is classified as an *Orthohepadnavirus* within the family *Hepadnaviridae*. Serum from individuals infected with HBV contains three distinct antigenic particles: a spherical 22 nm particle, a 42 nm spherical particle (containing DNA and DNA polymerase) called the **Dane particle,** and tubular or filamentous particles that vary in length (**figure 37.18**). The viral genome is 3.2 kb in length, consisting of four partially overlapping, open-reading frames that encode viral proteins. Viral multiplication takes place predominantly in hepatocytes. The infecting virus encases its double-shelled Dane particles within membrane envelopes coated with hepatitis B surface antigen (HBsAg). The inner nucleocapsid core antigen (HBcAg) encloses a single molecule of double-stranded HBV DNA and an active DNA polymerase. HBsAg in body fluids is (1) an indicator of hepatitis B infection, (2) used in the large-scale screening of blood for HBV, and (3) the basis for the first vaccine for human use developed by recombinant DNA technology. Diagnosis of HBV is made by detection of HBsAg in unimmunized individuals or HBcAg antibody, or detection of HBV nucleic acid by PCR.

Hepatitis B virus is normally transmitted through blood or other body fluids (saliva, sweat, semen, breast milk, urine, feces) and body-fluid–contaminated equipment (including shared intravenous needles). The virus can also pass through the placenta to the fetus of an infected mother. The number of

Table 37.2 Characteristics of Hepatitides Caused by Hepatotropic Viruses[a]

Disease	*Genome*	*Classification*	*Transmission*	*Outcome*	*Prevention*
Hepatitis A	RNA	*Picornaviridae, Hepatovirus*	Fecal-oral	Subclinical, acute infection	Killed HAV (Havrix vaccine)
Hepatitis B	DNA	*Hepadnaviridae, Orthohepadnavirus*	Blood, needles, body secretions, placenta, sexually	Subclinical, acute chronic infection; cirrhosis; primary hepatocarcinoma	Recombinant HBV vaccines
Hepatitis C	RNA	*Flaviviridae, Hepacivirus*	Blood, sexually	Subclinical, acute chronic infection; primary hepatocarcinoma	Routine screening of blood
Hepatitis D	RNA	Virusoid	Blood, sexually	Superinfection or coinfection with HBV	HBV vaccine
Hepatitis E	RNA	*Hepevirus*	Fecal-oral	Subclinical, acute infection (but high mortality in pregnant women)	Improve sanitary conditions
Hepatitis G	RNA	*Flaviviridae*	Sexually, parenterally	Chronic liver inflammation	HBV vaccine

[a] Hepatitis TTV has been discovered but not well characterized. Thus it is not included in this table.

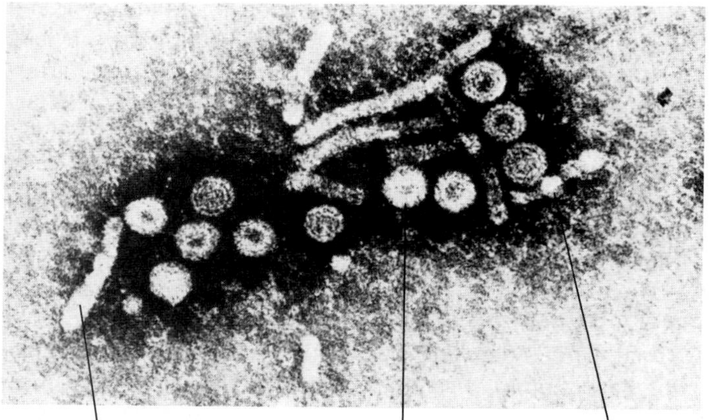

FIGURE 37.18 *Hepatitis B Virus* in Serum. Electron micrograph (×210,000) showing the three distinct types of hepatitis B antigenic particles. The spherical particles and filamentous forms are small spheres or long filaments without an internal structure, and only two of the three characteristic viral envelope proteins appear on their surface. Dane particles are the complete, infectious virion.

new HBV cases in the United States declined by 60% between 1985 and 1995, and was only 78,000 in 2001 (down from 260,000 in the 1980s). It is estimated, however, that there are currently 1.25 million chronically infected Americans. In the United States, about 5,000 persons die yearly from hepatitis-related cirrhosis and about 1,000 die from HBV-related liver cancer. (HBV is second only to tobacco as a known cause of human cancer.) Worldwide, HBV infects over 200 million people.

The clinical signs of hepatitis B vary widely. Most cases are asymptomatic. However, sometimes fever, loss of appetite, abdominal discomfort, nausea, fatigue, and other symptoms gradually appear following an incubation period of 1 to 3 months. The virus infects hepatic cells and causes liver tissue degeneration and the release of liver-associated enzymes (transaminases) into the bloodstream. This is followed by jaundice, the accumulation of bilirubin (a breakdown product of hemoglobin) in the skin and other tissues with a resulting yellow appearance. Chronic hepatitis B infection also causes the development of primary liver cancer, known as hepatocellular carcinoma.

General measures for prevention and control involve (1) excluding contact with HBV-infected blood and secretions, and minimizing accidental needle-sticks; (2) passive prophylaxis with intramuscular injection of hepatitis B immune globulin within 7 days of exposure; and (3) active prophylaxis with recombinant vaccines: Energix-B, Recombivax HB, Pediatrix, and Twinrix. These vaccines are widely used and are recommended for routine prevention of HBV in infants to eighteen-year-olds and risk groups of all ages (e.g., household contacts of HBV carriers, health-care and public safety professionals, men who have sex with other men, international travelers, hemodialysis patients). Recommended treatments for HBV include adefovir dipivoxil, alpha-interferon, and lamivudine.

Hepatitis C is caused by the enveloped ***Hepatitis C virus*** (HCV), which has an 80 nm diameter, a lipid coat, and contains a single strand of linear RNA. HCV is a member of the family *Flaviviridae*. HCV is classified into multiple genotypes. It is transmitted by contact with virus-contaminated blood, by the fecal-oral route, by in utero transmission from mother to fetus, sexually, or through organ transplantation. Diagnosis is made by ELISA, which detects serum antibody to a recombinant antigen of HCV, and nucleic acid detection by PCR. HCV is found worldwide. Prior to routine screening, HCV accounted for more than 90% of hepatitis cases developed after a blood transfusion. Worldwide, hepatitis C has reached epidemic proportions, with more than 1 million new cases reported annually. In the United States, nearly 4 million persons are infected and 25,000 new cases occur annually. Currently HCV is responsible for about 8,000 deaths annually in the United States. Furthermore, HCV is the leading reason for liver transplantation in the United States. Treatment is with Ribovirin and pegylated (coupled to polyethylene glycol) recombinant interferon-alpha (Intron A, Roferon-A). This combination therapy can rid the virus in 50% of those infected with genotype 1 and in 80% of those infected with genotype 2 or 3.

In 1977 a cytopathic hepatitis agent termed the **Delta agent** was discovered. Later it was called the hepatitis D virus (HDV) and the disease hepatitis D was designated. It is now named *Hepatitis delta virus*. HDV is a unique agent in that it is dependent on *Hepatitis B virus* to provide the envelope protein (HBsAg) for its RNA genome. Thus HDV only replicates in liver cells coinfected with HBV. Both must be actively replicating. Furthermore, the RNA of the HDV is smaller than the RNA of the smallest picornaviruses, and its circular conformation differs from the linear structure typical of animal RNA viruses. Thus its similarity to plant viroids and plant virusoids has led some to call this agent a virusoid. HDV is spread only to persons who are already infected with HBV (superinfection) or to individuals who get HBV and the virusoid at once (coinfection). The primary laboratory tools for the diagnosis of an HDV infection are serological tests for antidelta antibodies. Treatment of patients with chronic HDV remains difficult. Some positive results can be obtained with alpha interferon treatment for 3 months to 1 year. Liver transplantation is the only alternative to chemotherapy. Worldwide, there are approximately 300 million HBV carriers, and available data indicate that no fewer than 5% of these are infected with HDV. Thus because of the propensity of HDV to cause acute as well as chronic liver disease, continued incursion of HDV into areas of the world where persistent hepatitis B infection is endemic has serious implications. Prevention and control involves the widespread use of the hepatitis B vaccine.

Two other forms of hepatitis have been identified: **hepatitis F** (causing fulminant, posttransfusion hepatitis) and **hepatitis G** (a syncytial giant-cell hepatitis with viruslike particles resembling the measles virus). Hepatitis G virus (HGV) is a member of the *Flaviviridae* family. It is widely distributed in humans. HGV can be transmitted through needles or sexually. The significance of HGV infections in liver disease is not yet clear. However, infection causes chronic liver inflammation with its associated sequelae. Further virologic, epidemiological, and molecular efforts to characterize these new agents and their diseases are being undertaken.

Warts

Warts, or verrucae (Latin *verruca*, wart), are horny projections on the skin caused by human papillomaviruses. Human papillomavirus (HPV) is the name given to a group of DNA viruses that includes more than 100 different strains, some of which are oncogenic (cancer-associated) (**figure 37.19*a***). They differ in terms of the types of epithelium they infect; some infect cutaneous sites, whereas others infect mucous membranes. More than 30 of these viruses are sexually transmitted; they can cause genital warts—soft, pink cauliflower-like growths that occur on the genital area of women (figure 37.19*b*) and men. They are the most common sexually transmitted disease in the United States today.

The papillomaviruses are placed in the family *Papillomaviridae*. These viruses have nonenveloped icosahedral capsids with a double-stranded, supercoiled, circular DNA genome. At least eight distinct genotypes produce benign epithelial tumors that vary in respect to their location, clinical appearance, and histopathologic

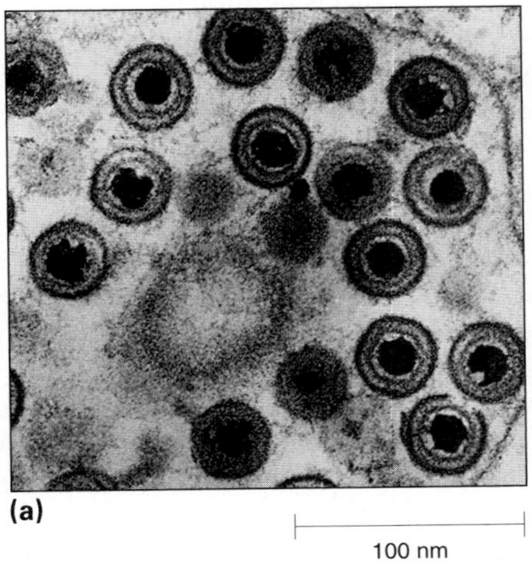

(a)

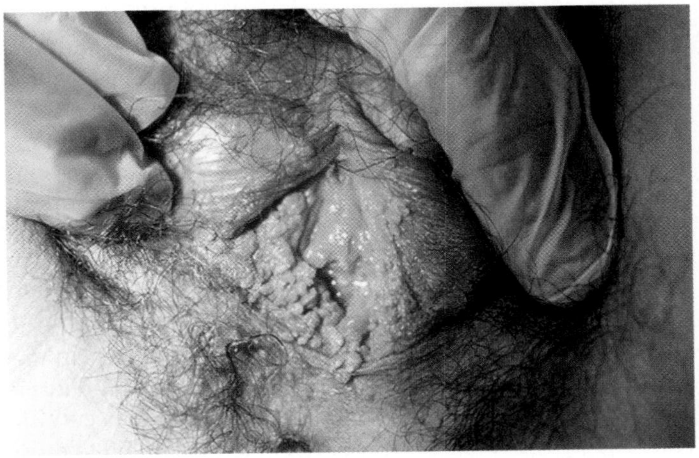

(b)

FIGURE 37.19 Human Papillomavirus (HPV). (a) Transmission electron micrograph of HPV. (b) Genital warts of the vaginal labia.

features. Warts occur principally in children and young adults, and are limited to the skin and mucous membranes. The viruses are spread between people by direct contact; autoinoculation occurs through scratching. Four major kinds of warts are **plantar warts, verrucae vulgaris, flat** or **plane warts,** and **anogenital condylomata (venereal warts).** Treatment includes physical destruction of the wart by electrosurgery, cryosurgery with liquid nitrogen or solid CO_2, laser fulguration (drying), direct application of the drug podophyllum to the wart, or injection of IFN-α (Intron A, Alferon N).

Anogenital condylomata (venereal warts) are sexually transmitted. Once HPV enters the body, the incubation period is 1 to 6 months. Genital infection with HPV is of considerable importance because specific types of genital HPV play a major role in the pathogenesis of epithelial cancers of the male and female genital tracts. HPV strains are designated by numbers and can be divided into "high-risk" (oncogenic) and "low-risk" (nononcogenic) strains. Four high-risk strains (6, 11, 16, and 18) have been reported to be responsible for 70% of cervical cancers and 90% of genital warts. Specifically they have been found in association with invasive cancers of the cervix, vulva, penis, or anus (and other sites). Other high-risk strains have been found to be associated with cancers no more than 1% of the time. These other strains cause benign or low-grade cervical cell changes and genital warts but are rarely, if ever, associated with invasive cancer.

More than 50% of sexually active men and women are infected with HPV at some point in their lives. In the United States, approximately 20 million people age fifteen to forty-nine years (mostly women, representing approximately 15% of the population) are currently infected with HPV. Approximately 6.2 million people in the United States become infected each year. Frequent sexual contact is the most consistent predictor of HPV infection; the number of sexual partners is proportionately linked to the risk of HPV infection. About half of the infected individuals are sexually active teens and young adults (fifteen to twenty-four years of age). HPV is typically transmitted through direct contact, especially during vaginal or anal sex. Other types of genital contact in the absence of penetration (oral-genital, manual-genital, and genital-genital contact) can also lead to infection with HPV, but infection by these routes is less common. In most cases, infections with HPV are not serious. Most HPV infections are asymptomatic, transient, and clear on their own without treatment. However, in some individuals, HPV infections result in genital warts weeks or months after infection.

A quadrivalent vaccine against four strains of HPV (6, 11, 16, and 18) was licensed in the United States in June 2006. The vaccine is made from noninfectious HPV-like particles and was recommended by the U.S. Advisory Committee on Immunization Practices for prophylactic use in females nine to twenty-six years of age. The vaccine has been found to be safe and to cause no serious side effects. Clinical trials in HPV-naïve women sixteen to twenty-six years of age have demonstrated 100% efficacy in preventing cervical precancers caused by the four HPV strains and nearly 100% efficacy in preventing vulvar and vaginal precancers and genital warts caused by the same four strains. While it is possible that vaccination of males with the vaccine may offer direct health benefits, currently no data support use of the HPV vaccine in males. The duration of vaccine protection is unknown, although the vaccine remains effective for at least 5 years. No evidence exists of waning immunity during that time period.

1. Describe the AIDS virus and how it cripples the immune system. What types of pathological changes can result?
2. Why do people periodically get cold sores? Describe the causative agent.
3. Why do people get the common cold so frequently? How are cold viruses spread?
4. Give two major ways in which herpes simplex virus type 2 is spread. Why do herpes infections become active periodically?
5. Describe the causative agent and some symptoms of mononucleosis and exanthem subitum.
6. What are the different causative viruses of hepatitis and how do they differ from one another? How can one avoid hepatitis? Do you know anyone who is a good candidate for infection with these viruses?
7. What kind of viruses cause the formation of warts? Describe the formation of venereal warts and a method to prevent infection.

37.4 Food-Borne and Waterborne Diseases

Food and water have been recognized as potential carriers (vehicles) of disease since the beginning of recorded history. Collectively more infectious diseases occur by these two routes than any other. A few of the many human viral diseases that are food- and waterborne are now discussed. ▶▶| *Water purification and sanitary analysis (section 42.1)*

Gastroenteritis (Viral)

Acute viral gastroenteritis (inflammation of the stomach or intestines) is caused by four major categories of viruses: rotaviruses (**figure 37.20**), adenoviruses, caliciviruses (noroviruses and sapoviruses), and astroviruses. The medical importance of these viruses is summarized in **table 37.3**. Acute viral gastroenteritis is a common illness among infants and children throughout the world, causing between 5 million and 10 million deaths per year. Viral diarrhea is an especially common cause of mortality among children less than five years of age in developing countries, with an estimated 1.5 billion episodes and 1.5 million to 2.5 million deaths estimated to occur annually. In the United States, acute diarrhea in children accounts for more than 1.5 million outpatient visits, 200,000 hospitalizations, and approximately

300 deaths per year. Current estimates also indicate that viral gastro-enteritis produces 30 to 40% of the cases of infectious diarrhea in the United States, far outnumbering documented cases of bacterial and protozoan diarrhea; the cause of approximately 40% of presumed cases of diarrhea remains unknown. The viruses responsible for gastroenteritis are transmitted by the fecal-oral route.

Noroviruses are estimated to cause 23 million cases of acute gastroenteritis globally (at least 50% of all food-borne outbreaks of gastroenteritis). At least another 20% of global cases, and the majority of severe cases of diarrhea in children, are due to rotaviruses, which are responsible for the hospitalization of approximately 55,000 children each year in the United States and the death of over 600,000 children worldwide annually. Adenoviruses (types 40 and 41), sapoviruses, and astroviruses are the other major viral agents responsible for gastroenteritis. Infection with rotaviruses and astroviruses is most common during the cooler months, whereas infection with adenovirus occurs year-round. (Bacteria-caused diarrheal diseases usually occur in the warmer months of the year.) The average incubation period for most of these viral diseases is 1 to 2 days. Viral gastroenteritis is seen most frequently in infants one to eleven months of age, where the virus attacks the epithelial cells of the upper intestinal villi, causing malabsorption, impairment of sodium transport, and diarrhea. The clinical manifestations typically range from asymptomatic to a relatively mild diarrhea with headache and fever, to a severe, watery, nonbloody diarrhea with abdominal cramps. Fatal dehydration is most common in young children. Vomiting is almost always present, especially in children. Viral gastroenteritis can be self-limiting. Treatment is designed to provide relief through the use of oral fluid replacement with isotonic liquids, analgesics, and antiperistaltic agents. Symptoms usually last for 1 to 5 days, and recovery often results in protective immunity to subsequent infection. Since 2006, a live, oral vaccine to prevent rotavirus infection (RotaTeq™) has been licensed in the United States for use in children. A second rotavirus vaccine is currently in clinical trials and has shown significant protection in children of developing countries.

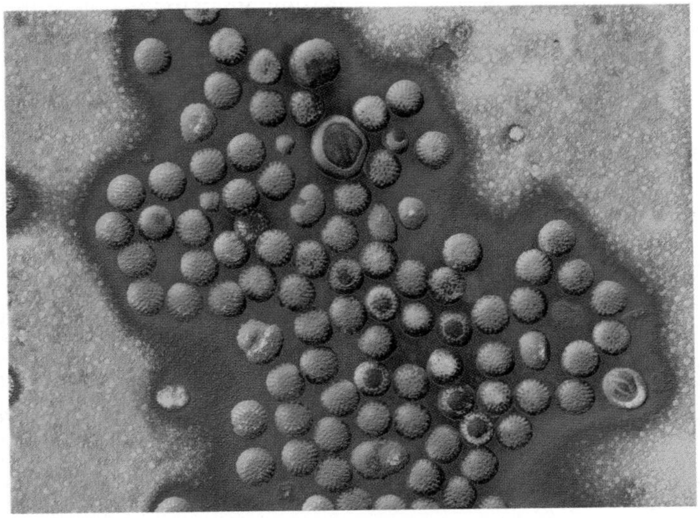

FIGURE 37.20 Rotavirus. Electron micrograph of rotaviruses (reoviruses) in a human gastroenteritis stool filtrate (×90,000). Note the spokelike appearance of the icosahedral capsids that surround double-stranded RNA within each virion.

Hepatitis A

Hepatitis A (infectious hepatitis) usually is transmitted by fecal contamination of food, drink, or shellfish that live in contaminated water and contain the virus in their digestive system. The disease is caused by ***Hepatitis A virus*** (HAV) of the genus *Hepatovirus* in the family *Picornaviridae*. We should note that while all

Table 37.3 Medically Important Gastroenteritis Viruses

Virus	*Epidemiological Characteristics*	*Clinical Characteristics*
Rotaviruses		
Group A	Endemic diarrhea in infants worldwide	Dehydrating diarrhea for 5–7 days; fever, abdominal cramps, nausea, and vomiting common
Group B	Large outbreaks in adults and children in China	Severe watery diarrhea for 3–5 days
Group C	Sporadic cases in children in Japan	Similar to group A
Norovirus	Epidemics of vomiting and diarrhea in older children and adults; occurs in families, communities, and nursing homes; often associated with shellfish, other food, or water and infected food handlers, cruise ship occurrences	Acute vomiting, fever, myalgia, and headache lasting 1–2 days, diarrhea
Sapoviruses	Pediatric diarrhea; also associated with shellfish and other foods in adults	Rotavirus-like illness in children; *Norovirus*-like illness in adults
Astroviruses	Pediatric diarrhea; reported in nursing homes	Watery diarrhea for 1–3 days
Adenoviruses	Pediatric diarrhea; also reported in military bases	Gastroenteritis, more severe in immuno-compromised adults

hepatitis viruses can cause liver disease, not all are taxonomically related. HAV is an icosahedral, linear, positive-strand RNA virus that lacks an envelope. Once in the digestive system, the viruses multiply within the intestinal epithelium. Usually only mild intestinal symptoms result. Occasionally viremia (the presence of viruses in the blood) occurs, and the viruses may spread to the liver. The viruses reproduce in the liver, enter the bile, and are released into the small intestine. This explains why feces are so infectious. After about a 4-week incubation period, symptoms develop that include anorexia, general malaise, nausea, diarrhea, fever, and chills. If the liver becomes infected, jaundice ensues. Most cases resolve in 4 to 6 weeks and yield a strong immunity, although some patients relapse, exhibiting symptoms for 6 months or more. Fortunately the mortality rate is low (less than 1%). Approximately 40 to 80% of the U.S. population has antibodies, though few are aware of having had the disease. Control of infection is by simple hygienic measures, the sanitary disposal of excreta, and the HAV vaccine. The number of new cases has been dramatically reduced since the introduction of the hepatitis A vaccine (Havrix) in the 1990s. This vaccine is recommended for travelers (*see table 36.8*) going to regions with high evidence rates of hepatitis A. However, the increasing number of sexually transmitted HAV outbreaks among men who have unprotected sex with other men is of continuing concern.

Hepatitis E

Hepatitis E is implicated in many epidemics in certain developing countries in Asia, Africa, and Central and South America. It is uncommon in the United States but is occasionally imported by infected travelers. The single chromosome, positive-strand, RNA viral genome (7,900 nucleotides) is linear. The virion is spherical, nonenveloped, and 32 to 34 nm in diameter. Based on biologic and physicochemical properties, *Hepatitis E virus* (HEV) has not been assigned to a virus family; therefore HEV may eventually be classified in a separate family.

Infection usually is associated with feces-contaminated drinking water. Presumably HEV enters the blood from the gastrointestinal tract, replicates in the liver, is released from hepatocytes into the bile, and is subsequently excreted in the feces. Like hepatitis A, an HEV infection usually runs a benign course and is self-limiting. The incubation period varies from 15 to 60 days, with an average of 40 days. The disease is most often recorded in patients who are fifteen to forty years of age. Children are typically asymptomatic or present mild signs and symptoms, similar to those of other types of viral hepatitis, including abdominal pain, anorexia, dark urine, fever, hepatomegaly, jaundice, malaise, nausea, and vomiting. Case fatality rates are low (1 to 3%), except for pregnant women (15 to 25%), who may die from fulminant hepatic failure. Diagnosis of HEV is by ELISA (IgM or IgG to recombinant HEV) or reverse transcriptase PCR. There are no specific measures for preventing HEV infections, other than those aimed at improving the level of health and sanitation in affected areas.

Poliomyelitis

Poliomyelitis (Greek *polios*, gray, and *myelos*, marrow or spinal cord), **polio,** or **infantile paralysis** is caused by *Poliovirus* and was first described in England in 1789 as a "leg-wasting" disease of children, although a pictograph from ancient Egypt clearly depicts a man with what appears to be polio (**Historical Highlights 37.1**). The virus is an enterovirus—a transient inhabitant of the gastrointestinal tract—and a member of the family *Picornaviridae* and as such is a nonenvelopled, positive-strand RNA virus. Three different *Poliovirus* subtypes have been identified. Importantly, immunity to one subtype does not confer immunity to the others; the vaccine is therefore tripartite. Like other viruses transmitted by the fecal-oral route, *Poliovirus* is very stable, especially at acidic pH, and can remain infectious for relatively long periods in food and water, its main routes of transmission. The incubation period ranges from 6 to 20 days.

Once ingested, the virus multiplies in the mucosa of the throat or small intestine. From these sites, the virus invades the tonsils and lymph nodes of the neck and terminal portion of the small intestine. Generally there are either no symptoms or a brief illness characterized by fever, headache, sore throat, vomiting, and loss of appetite. The virus sometimes enters the bloodstream, causing viremia. In most cases (more than 99%), the viremia is transient and clinical disease does not result. However, in a minority of cases, the viremia persists and the virus enters the central nervous system and causes paralytic polio. The virus has a high affinity for anterior horn motor nerve cells of the spinal cord. Once inside these cells, it multiplies and destroys the cells; this results in motor and muscle paralysis. Since the licensing of the tripartite Salk vaccine (1955) and the tripartite Sabin vaccine (1962), the incidence of polio has decreased markedly. No wild polio viruses exist in the United States. An ongoing global polio eradication effort has been very successful. However, sporadic cases are reported, mostly in areas where religious views and misinformation diminish vaccination efforts and civil wars interrupt public health efforts. Nonetheless, it is likely that polio will be the next human disease to be completely eradicated.

1. What two virus groups are associated with acute viral gastroenteritis? How do they cause the disease's symptoms?
2. Describe some symptoms of hepatitis A.
3. Why was hepatitis A called infectious hepatitis?
4. At what specific sites within the body can the poliomyelitis virus multiply? What is the usual outcome of an infection?

37.5 Zoonotic Diseases

The diseases discussed here are caused by viruses that are normally zoonotic (animal-borne). The RNA virus families *Arenaviridae, Bunyaviridae, Flaviviridae, Filoviridae,* and *Picornoviridae* represent notable examples of human viral infections found in

HISTORICAL HIGHLIGHTS

37.1 A Brief History of Polio

Like many other infectious diseases, polio is probably of ancient origin. Various Egyptian hieroglyphics dated approximately 2000 B.C. depict individuals with wasting, withered legs and arms (see **box figure**). In 1840 the German orthopedist Jacob von Heine described the clinical features of poliomyelitis and identified the spinal cord as the problem area. Little further progress was made until 1890, when Oskar Medin, a Swedish pediatrician, portrayed the natural history of the disease as epidemic in form. He also recognized that a systemic phase, characterized by minor symptoms and fever, occurred early and was complicated by paralysis only occasionally. Major progress occurred in 1908, when Karl Landsteiner and William Popper successfully transmitted the disease to monkeys. In the 1930s much public interest in polio occurred because of the polio experienced by Franklin D. Roosevelt. This led to the founding of the March of Dimes campaign in 1938; the sole purpose of the March of Dimes was to collect money for research on polio. In 1949 John Enders, Thomas Weller, and Frederick Robbins discovered that the polio virus could be propagated in vitro in cultures of human embryonic tissues of nonneural origin. This was the keystone that later led to the development of vaccines.

In 1952 David Bodian recognized that there were three distinct serotypes of the *Poliovirus*. Jonas Salk successfully immunized humans with formalin-inactivated *Poliovirus* in 1952, and this vaccine (IPV) was licensed in 1955. The live attenuated *Poliovirus* vaccine (oral polio vaccine, OPV) developed by Albert Sabin and others had been employed in Europe since 1960 and was licensed for U.S. use in 1962. Both the Salk and Sabin vaccines led to a dramatic decline of paralytic poliomyelitis in most developed countries and, as such, have been rightfully hailed as two of the great accomplishments of medical science.

Ancient Egyptian with Polio. Note the withered leg.

animal reservoirs before transmission to and between humans. Some of these viruses are exotic and rare; others are being eradicated by public health efforts. Some of the virus types are found in relatively small geographic areas; others are distributed across continents. Several of these viruses cause diseases with substantial morbidity and mortality. It is for these reasons that many are placed on the select agents list as potential bioweapons, as indicated by double asterisks (**).

**Ebola and Marburg Hemorrhagic Fevers

Viral hemorrhagic fever (VHF) is the term used to describe a severe, multisystem syndrome caused by several distinct viruses (**Disease 37.2**). Characteristically the overall host vascular system is damaged, resulting in vascular leakage (hemorrhage) and dysfunction (coagulopathy). **Ebola hemorrhagic fever** is caused by Ebola virus, first recognized near the Ebola River in the Democratic Republic of the Congo in Africa. The virus is a member of a family of negative-strand RNA viruses called the *Filoviridae*. Four Ebola subtypes are known. Ebola-Zaire, Ebola-Sudan, and Ebola-Ivory Coast cause disease in humans (**figure 37.21**). The fourth, Ebola-Reston, was first discovered in 1989 and to date has only caused disease in nonhuman primates and pigs; unlike the others, it is spread by aerosol transmission as well as by contact with body fluids.

Infection with Ebola virus is severe and approximately 80% fatal. The incubation period for Ebola hemorrhagic fever ranges from 2 to 21 days and is characterized by abrupt fever, headache, joint and muscle aches, sore throat, and weakness, followed by diarrhea, vomiting, and stomach pain. Signs of infection include

DISEASE

37.2 Viral Hemorrhagic Fevers—A Microbial History Lesson

Scientists know of several viruses lurking in the tropics that—with a little help from nature—could wreak far more loss of life than will likely result from the AIDS pandemic. Collectively these viruses produce **hemorrhagic fevers.** The viruses are passed among wild vertebrates, which serve as reservoir hosts. Arthropods transmit the viruses among vertebrates, and humans are infected when they invade the environment of the natural host. These diseases, distributed throughout the world, are known by over 70 names, usually denoting the geographic area where they were first described.

Viral hemorrhagic fevers can be fatal. Patients suffer headache, muscle pain, flushing of the skin, massive hemorrhaging either locally or throughout the body, circulatory shock, and death. The first documented cases of hemorrhagic fever occurred in the late 1960s when dozens of scientists in West Germany fell seriously ill and several died. Victims suffered from a breakdown of liver function and a bizarre combination of bleeding and blood clots. The World Health Organization traced the outbreak to a batch of fresh monkey cells the scientists had used to grow polioviruses. The cells from the imported Ugandan monkeys were infected with the lethal tropical Marburg virus (see **box figure**) and the scientists suffered from **Marburg viral hemorrhagic fever.**

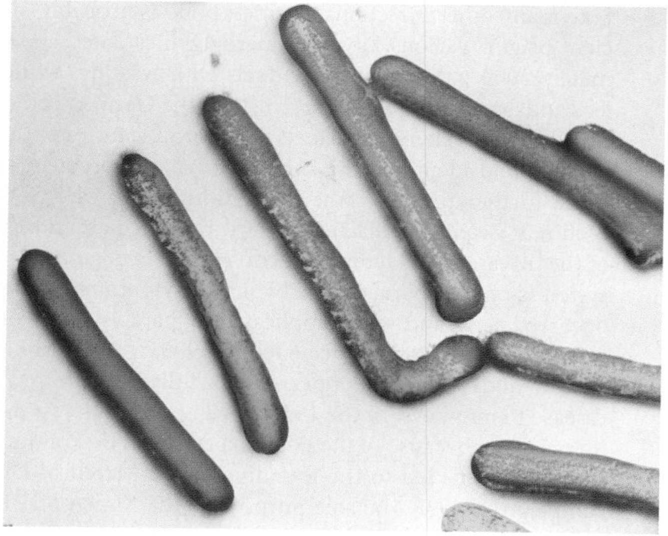

Sinister Foe. The deadly Marburg virus was first isolated in 1967 at the Institute for Hygiene and Microbiology in Marburg, Germany.

In 1977 the *Phlebovirus* causing Rift Valley fever in sheep and cattle moved from these animals into the South African population. The virus, which causes severe weakness, incapacitating headaches, damage to the retina, and hemorrhaging, then made its way to Egypt, where millions of humans became infected and thousands died.

Among the most frightening hemorrhagic outbreak was that of the Ebola virus hemorrhagic fever in Zaire and Sudan in 1976. This disease infected more than 1,000 people and left over 500 dead. It became concentrated in hospitals, where it killed many of the Belgian physicians and nurses treating infected patients. Since that time, there have been numerous additional outbreaks in which hundreds of patients and health-care workers have died. All of these outbreaks have so far occurred in Africa.

In the United States in 1989, epidemiologists provided new evidence that rats infected with a potentially deadly hemorrhagic virus are prevalent in Baltimore slums. The virus appears to be taking a previously unrecognized toll on the urban poor by causing Korean hemorrhagic fever.

In the summer of 1993, reports appeared in the news media about a mysterious illness that had caused over 30 deaths among the Navajo Nation in the Four Corners area of the southwestern United States. The CDC finally determined the causative agent to be a hantavirus, a negative-strand RNA virus that is a member of the family *Bunyaviridae*. Hantaviruses are endemic in rodents, such as deer mice, in many areas of the world. Deer mice shed the virus in their saliva, feces, and urine. Humans contract the disease when they inhale aerosolized particles containing the excreted virus. Throughout Asia and central Europe, hantaviruses cause hemorrhagic fever with renal syndrome in humans. But the type of virus found in the Southwest had not been previously recognized, and no hantavirus anywhere in the world had been associated with the clinical syndrome initially seen among the Navajo; namely, the hantavirus pulmonary syndrome in which the virus destroys the lungs. In 1993 the CDC named this virus **pulmonary syndrome hantavirus** (sometimes called *Sin Nombre virus* or no-name virus), and isolated cases have since been reported from almost every state. Prevention involves wearing gloves when handling mice and spraying the feces and urine of all mice with a disinfectant.

Although to date these epidemics have not become global, they provide a humbling vision of humankind's vulnerability. History shows that life-threatening viral hemorrhagic outbreaks often arise when humans move into unexplored terrain or when living conditions deteriorate in ways that generate new viral hosts. In each case, medical and scientific resources have been reactive, not proactive.

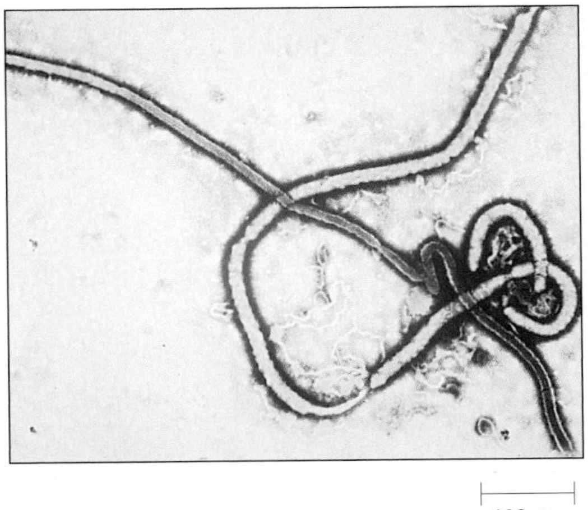

FIGURE 37.21 Ebola Virus.

rash, red eyes, bleeding, and hiccups; symptoms alerting of internal hemorrhage. The reservoir of Ebola virus appears to be at least three species of fruit bats that are native only to Africa. Exposure can also occur through contact with bodies of Ebola victims. There is no standard treatment for Ebola infection. Patients receive supportive therapy. This consists of balancing patients' fluids and electrolytes, maintaining their oxygen status and blood pressure, and treating them for any complicating infections. Experimental vaccines are currently being evaluated and show promise in nonhuman primate models.

Marburg hemorrhagic fever is caused by a genetically unique RNA virus in the *Filoviridae* family. Marburg fever is a rare, severe type of hemorrhagic fever that affects both humans and nonhuman primates. Recognition of this virus led to the creation of the family. Marburg virus was first recognized in 1967, when outbreaks of hemorrhagic fever occurred simultaneously in laboratories in Marburg and Frankfurt, Germany, and in Belgrade, Yugoslavia (now Serbia). The first people infected had been exposed to African green monkeys or their tissues. In Marburg, the monkeys had been imported for research and to prepare polio vaccine. Marburg virus is indigenous to Africa, but its specific origin and its definitive animal host are unknown. The average incubation period for Marburg hemorrhagic fever is 5 to 10 days. The disease symptoms are abrupt, marked by fever, chills, headache, and myalgia. A maculopapular rash (i.e., discolored with bumps), most prominent on the chest, back, and stomach, typically appears around the fifth day after the onset. Nausea, vomiting, chest pain, sore throat, abdominal pain, and diarrhea may also occur in infected patients. Symptoms become increasingly severe and may include jaundice, delirium, liver failure, pancreatitis, severe weight loss, shock, and multiorgan dysfunction. A specific treatment for this disease is unknown. However, supportive hospital therapy should be utilized. This includes balancing the patient's fluids and electrolytes, maintaining oxygen status and blood pressure, replacing lost blood and clotting factors, and treating for other complicating infections (Disease 37.1).

**Hantavirus Pulmonary Syndrome

Hantavirus pulmonary syndrome (HPS) is a disease caused by a negative-strand RNA virus of the *Bunyaviridae*. HPS is typically transmitted to humans by inhalation of viral particles shed in urine, feces, or saliva of infected rodents. HPS was first recognized in 1993 and has since been identified throughout the United States. Although rare, HPS is potentially deadly. Rodent control in and around the home remains the primary strategy for preventing hantavirus infection. HPS in the United States is not transmitted from person to person, nor is it known to be transmitted by rodents purchased from pet stores.

Hantaviruses have lipid envelopes that are susceptible to most disinfectants. The length of time hantaviruses can remain infectious in the environment is variable and depends on environmental conditions. Temperature, humidity, exposure to sunlight, and even the rodent's diet, which affects the chemistry of rodent urine, strongly influence viral survival. Viability of dried virus has been reported at room temperature for 2 to 3 days. Hantaviruses are shed in body fluids but do not appear to cause disease in their reservoir rodent hosts. Data indicate that viral transfer then may occur through biting, as field studies suggest that viral transmission in rodent populations occurs horizontally and more frequently between males. A specific treatment for HPS is unknown. Supportive therapy is used to treat symptoms, including balancing the patient's fluids and electrolytes, maintaining oxygen status and blood pressure, replacing lost blood and clotting factors, and treating for other complicating infections.

Rabies

Rabies (Latin *rabere*, rage or madness) is caused by a number of different strains of highly neurotropic viruses. Most belong to a single serotype in the genus *Lyssavirus* (Greek *lyssa*, rage or rabies), family *Rhabdoviridae*. The bullet-shaped virion contains a negative-strand RNA genome (**figure 37.22*a***). Rabies has been the object of human fascination, torment, and fear since the disease was first recognized. Prior to Pasteur's development of an antirabies vaccine, few words were more terrifying than the cry of "mad dog!" Improvements in prevention during the past 50 years have led to almost complete elimination of indigenously acquired rabies in the United States, where rabies is primarily a disease of feral animals and domestic cats that contact feral animals. Most wild animals can become infected with rabies, but susceptibility varies according to species. Foxes, coyotes, and wolves are the most susceptible; intermediate are skunks, raccoons, insectivorous bats, and bobcats, while opossums are quite resistant (figure 37.22*b*). Worldwide, almost all cases of human rabies are attributed to dog bites. In developing countries, where canine rabies is still endemic, rabies accounts for up to 40,000 deaths per year. Occasionally other domestic animals are responsible for transmission of rabies to humans. It should be noted, however, that not all rabid animals exhibit signs of agitation and aggression (known as furious rabies). In fact, paralysis (dumb rabies) is the more common sign exhibited by rabid animals.

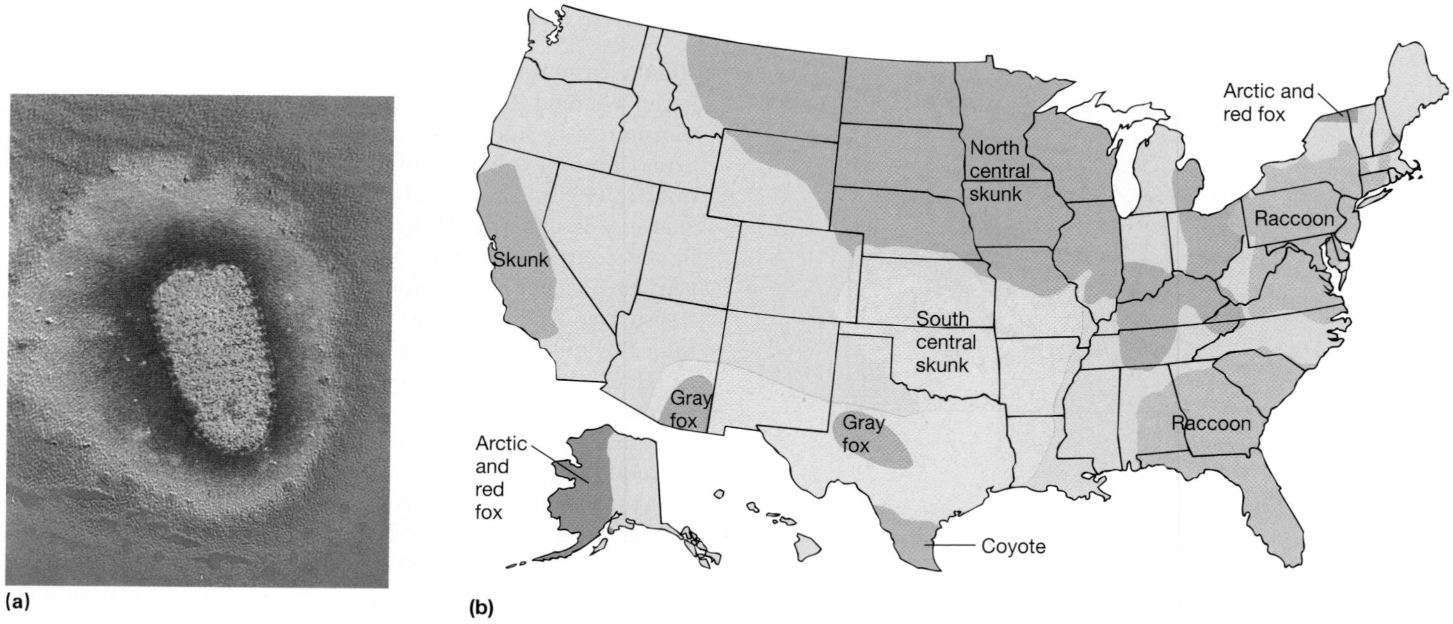

FIGURE 37.22 Rabies. (a) Electron micrograph of the rabies virus (yellow) (×36,700). Note the bullet shape. The external surface of the virus contains spikelike glycoprotein projections that bind specifically to cellular receptors. (b) In the United States, rabies is found in terrestrial animals in 10 distinct geographic areas. In each area, a particular species is the reservoir and one of five antigenic variants of the virus predominates, as illustrated by the five different colors. Although not shown, another eight viral variants are found in insectivorous bats and cause sporadic cases of rabies in terrestrial animals throughout the country. Absence of a strain does not imply absence of rabies.

Figure 37.22 Micro Inquiry

What type of host cell does the rabies virus initially infect? To where does the virus then travel?

The virus multiplies in the salivary glands of an infected host. It is transmitted to humans or other animals by the bite of an infected animal whose saliva contains the virus; by aerosols of the virus that can be spread in caves where bats dwell; or by contamination of scratches, abrasions, open wounds, and mucous membranes with saliva from an infected animal. After inoculation, a region of the virions' envelope spike attaches to the plasma membrane of nearby skeletal muscle cells, which the virus enters. Multiplication of the virus then occurs. When the concentration of the virus in the muscle is sufficient, the virus enters the nervous system through unmyelinated sensory and motor terminals; the binding site is the nicotinic acetylcholine receptor.

The virus spreads by retrograde axonal flow at 8 to 20 mm per day until it reaches the spinal cord, when the first specific symptoms of the disease—pain or paresthesia at the wound site—may occur. A rapidly progressive encephalitis develops as the virus quickly disseminates through the central nervous system. The virus then spreads throughout the body along the peripheral nerves, including those in the salivary glands, where it is shed in the saliva.

Within brain neurons, the virus produces characteristic Negri bodies, masses of viruses or unassembled viral subunits that are visible in the light microscope. In the past, diagnosis of rabies consisted solely of examining nervous tissue for the presence of these bodies. Today diagnosis is based on *d*irect immuno*f*luorescent *a*ntibody (DFA) of brain tissue, virus isolation, detection of Negri bodies, and a rapid rabies enzyme-mediated immunodiagnosis test.

Symptoms of rabies in humans usually begin 2 to 16 weeks after viral exposure and include anxiety, irritability, depression, fatigue, loss of appetite, fever, and a sensitivity to light and sound. The disease quickly progresses to a stage of paralysis. In about 50% of all cases, intense and painful spasms of the throat and chest muscles occur when the victim swallows liquids. The mere sight, thought, or smell of water can set off spasms. Consequently, rabies has been called hydrophobia (fear of water). Death results from destruction of the regions of the brain that regulate breathing. Safe and effective vaccines (*h*uman *d*iploid-*c*ell rabies *v*accine HDCV [Imovax Rabies] or *r*abies *v*accine *a*dsorbed [RVA]) against rabies are available; however, to be effective they must be given soon after the person has been infected. Veterinarians and laboratory personnel, who have a high risk of exposure to rabies, usually are immunized every 2 years and tested for the presence of suitable antibody titer. About 30,000 people annually receive this treatment. In the United States, fewer than 10 cases of rabies occur yearly in humans, although about 8,000 cases of animal rabies are reported each year from various sources (figure 37.22*b*). Prevention and control involves preexposure vaccination of dogs and cats, postexposure vaccination of humans, and preexposure vaccination of humans at special risk, including persons spending a month or more in countries where rabies is common in dogs.

Some states and countries (e.g., Hawaii and Great Britain) retain their rabies-free status by imposing quarantine periods on any entering dog or cat. If an asymptomatic, unvaccinated dog or cat bites a human, the animal is typically confined and observed by a veterinarian for at least 10 days. If the animal shows no signs of rabies in that time, it is determined to be uninfected. Animals demonstrating signs of rabies are killed and brain tissue submitted for rabies testing. Postexposure prophylaxis—rabies immune globulin for passive immunity and rabies vaccine for active immunity—is initiated to exploit the relatively long incubation period of the virus. This is usually recommended for anyone bitten by one of the common reservoir species (raccoons, skunks, foxes, and bats), unless it is proven that the animal was uninfected. Once symptoms of rabies develop in a human, death usually occurs.

1. Why are Ebola and Marburg hemorrhagic fever diseases so deadly?
2. What precautions can be taken to prevent hantavirus and *Lyssavirus* transmission to humans?
3. How does the rabies virus cause death in humans?
4. How does the rotavirus structure give rise to its name?

37.6 Prion Diseases

Prion diseases, also called **transmissible spongiform encephalopathies (TSEs),** are fatal neurodegenerative disorders that have attracted enormous attention not only for their unique biological features but also for their impact on public health. Prions (proteinaceous infectious particles) are thought to consist of abnormally folded proteins (PrP^{Sc}), which can induce normal forms of the protein (PrP^{C}) to fold abnormally (*see figure 5.24*). This group of diseases includes kuru, Creutzfeldt-Jakob disease (CJD), variant Creutzfeldt-Jakob disease (vCJD), Gerstmann-Sträussler-Scheinker disease (GSD), and fatal familial insomnia (FFI; **table 37.4**). The first of these diseases to be studied in humans was kuru, discovered in the Fore tribe of New Guinea. Carlton Gadjusek and others showed that the disease was transmitted by ritual cannibalism (especially where brains and spinal cords were eaten). The primary symptom of the human disorders is dementia, usually accompanied by manifestations of motor dysfunction such as cerebral ataxia (inability to coordinate muscle activity) and myoclonus (shocklike contractions of muscle groups). FFI is also characterized by dysautonomia (abnormal functioning of the autonomic nervous system) and sleep disturbances. These symptoms appear insidiously in middle to late adult life and last from months (CJD, FFI, and kuru) to years (GSD) prior to death. ◀◀ *Prions (section 5.7)* *Prion Diseases*

Neuropathologically these disorders produce a characteristic spongiform degeneration of the brain, as well as deposition of amyloid plaques. Prion diseases thus share important clinical, neuropathological, and cell biological features with another, more common cerebral amyloidosis, Alzheimer's disease. A familial (inherited) form of CJD has also been described, suggesting that certain genetic mutations cause the PrP^{C} protein to more easily assume the PrP^{Sc} conformation.

Classic CJD appears to arise sporadically, as the result of spontaneous mutation, with a worldwide incidence of 1 case per million people per year. It is not contagious. However, CJD can also be acquired, that is, transmitted from cattle that have bovine spongiform encephalopathy (BSE or mad cow disease). CJD that is acquired is known as variant (v) CJD, so named as it is clinically and pathologically different from classic CJD. There have been two confirmed cases of BSE in the United States (Washington, 2004, and Texas, 2005), compared to approximately 40,000 BSE cases in the United Kingdom. Cattle experimentally infected by the oral route have tested positive for the BSE agent in the brain, spinal cord, retina, dorsal root ganglia, distal ileum, and bone marrow, suggesting that (1) the BSE agent survives passage along the gastrointestinal tract, (2) the BSE agent is neurotropic, and (3) these tissues represent a source of infectious material that may be transmitted to humans and other animals. In fact, much evidence has accumulated to suggest that human vCJD can be acquired by individuals who eat meat products (especially if they contain brain and spinal cord) prepared from infected cattle. Data from the United Kingdom report 151 deaths attributed to vCJD, compared to 260 total worldwide. Estimates of the total vCJD cases (extrapolated from analysis of positive results from a tonsil and appendix tissue bank) expected in the United Kingdom by 2080 range from a few hundred to 140,000. Of additional concern is the report of four vCJD cases associated with blood transfusion in the United Kingdom. Variant CJD can also be acquired from a physician or surgeon, a medical treatment, or diagnostic procedures. Known as iatrogenic CJD, it has been transmitted by prion-contaminated human growth hormone, corneal grafts, and grafts of dura mater (tissue surrounding the brain). Donor screening and more thorough testing of grafts have decreased the frequency of prion transmission.

1. How are prions different from viruses? How are they similar?
2. In what way are spongiform encephalopathies commonly acquired?

Table 37.4 Prion Diseases of Humans

Disease	Nature of Disease
Creutzfeldt-Jakob disease (CJD) (sporadic, iatrogenic, familial, new-variant)	Spongiform encephalopathy (degenerative changes in the central nervous system)
Kuru	Spongiform encephalopathy
Gerstmann-Sträussler-Scheinker disease (GSD)	Genetic neurodegenerative disease
Fatal familial insomnia (FFI)	Genetic neurodegenerative disease with progressive, untreatable insomnia

Summary

37.1 Airborne Diseases

a. More than 400 different viruses can infect humans. These viruses can be grouped and discussed according to their mode of transmission and acquisition.

b. Most airborne viral diseases involve either directly or indirectly the respiratory system. Examples include chickenpox (varicella, **figure 37.1**), shingles (herpes zoster, **figure 37.2**), influenza (flu, **figure 37.3**), measles (rubeola, **figure 37.4**), rubella (German measles, **figure 37.5**), acute respiratory viruses such as the SARS virus (**figure 37.6**), the eradicated smallpox (variola, **figure 37.7**), and viral pneumonia.

37.2 Arthropod-Borne Diseases

a. The arthropod-borne viral diseases are transmitted by arthropod vectors from human to human or animal to human.

b. Examples arthropod-borne diseases include eastern, western, and Venezuelan equine encephalitis, and West Nile fever. All these diseases are characterized by fever, headache, nausea, vomiting, and characteristic encephalitis.

37.3 Direct Contact Diseases

a. Person-to-person contact is another way of acquiring or transmitting a viral disease.

b. Examples of direct contact diseases include AIDS (**figures 37.8–37.11**), cold sores (**figure 37.13**), the common cold (rhinovirus; **figure 37.14**), cytomegalovirus inclusion disease (**figure 37.15**), genital herpes (**figure 37.16**), *Human herpesvirus 6* infections, *human parvovirus B19* infection, certain leukemias, infectious mononucleosis, and hepatitis (**table 37.2**)—hepatitis B (serum hepatitis); hepatitis C; hepatitis D (delta hepatitis); hepatitis F; and hepatitis G.

37.4 Food-Borne and Waterborne Diseases

a. The viruses that are transmitted in food and water usually grow in the intestinal system and leave the body in the feces (**table 37.3**). Acquisition is generally by the oral route.

b. Examples of diseases caused by food-borne viruses include acute viral gastroenteritis (rotaviruses and others), infectious hepatitis A, hepatitis E, and poliomyelitis.

37.5 Zoonotic Diseases

a. Diseases transmitted from animals are zoonotic.

b. Several animal viruses can cause disease in humans. Examples of viral zoonoses include Ebola and Marburg fevers, hantavirus pulmonary syndrome, and rabies.

37.6 Prion Diseases

a. A prion disease is a pathological process caused by a transmissible agent (a prion) that remains clinically silent for a prolonged period, after which the clinical disease becomes apparent.

b. Examples of prion diseases include Creutzfeldt-Jakob disease, variant Creutzfeldt-Jakob disease, kuru, Gerstmann-Sträussler-Scheinker disease, and fatal familial insomnia (**table 37.4**). These diseases are chronic infections of the central nervous system that result in progressive degenerative changes and eventual death.

Critical Thinking Questions

1. Explain why antibiotics are not effective against viral infections. Advise a person about what can be done to relieve symptoms of a viral infection and recover most quickly. Address your advice to (a) someone who has had only a basic course in high school biology and (b) a third-grade student.
2. Several characteristics of AIDS render it particularly difficult to detect, prevent, and treat effectively. Discuss two of them. Contrast the disease with polio and smallpox.
3. From an epidemiological perspective, why are most arthropod-borne viral diseases hard to control?
4. In terms of molecular genetics, why is the common cold such a prevalent viral infection in humans?
5. Will it be possible to eradicate many viral diseases in the same way as smallpox? Why or why not?
6. Prior to the development of detection assay for HIV in the blood supply, HCV was used as a proxy. That is to say, any donated blood found to be contaminated with HCV was also assumed to be HIV positive as well. What do you think was the rationale behind the use of HCV as an indicator of HIV? Do you think this was a reasonable approach at the time?
7. In 2005 a reconstructed influenza virus containing eight genes from the 1918 Spanish influenza virus was introduced into human lung epithelial cells and mice. This virus was able to rapidly reproduce and cause sudden illness and death in mice. In 2008 these genes were replaced on an individual basis with homologues from a contemporary H1N1 influenza virus and tested for virulence. Substitution of the 1918 gene with the H1N1 hemagglutinin (HA), neuramidase (NA), or polymerase PB1 subunit severely diminished the virulence of the 1918 virus.
   a. Explain why each of these genes is so critical for infection and virulence.
   b. Do you think there would be an additive effect in loss of virulence if more than one gene were replaced? Explain your answer and discuss what this implies with regard to the emergence of another highly virulent influenza virus capable of causing a pandemic.

   Read the original paper: Pappas, C., et al. 2008. Single gene reassortments identify a critical role for PB1, HA, and NA in the high virulence of the 1918 pandemic influenza virus. *Proc. Nat. Acad. Sci., USA.* 105:3064–69.
8. Chronic wasting disease (CWD), a prion disease found in cervids (elk, deer, and moose), was first described in the 1960s and is now found in animals in at least 14 states. Despite its long history, it was unclear how the disease spread and if humans were at risk of acquiring the disease. In 2006 researchers designed and executed an experiment to determine if prions could be transmitted between cervids by contact with saliva, blood, feces, and urine. They found that CWD could be transmitted by oral contact with saliva and by blood following transfusion. This finding has raised concern among hunters and others who come in contact with cervid body fluids.

   How would you design a similar experiment to identify the transmission of scrapie between sheep and possible transmission to humans? Be sure to explain the controls you would use and how you would unequivocally document the development of disease.

   Read the original paper: Mathiason, C. K., et al. 2006. Infectious prions in the saliva and blood of deer with chronic wasting disease. *Science* 314:133.

Concept Mapping Challenge

Design a concept map that demonstrates the process by which an influenza pandemic occurs. Be sure to identify linking words to make clear the connections between the terms you identify in chapter 37.

Learn More

Learn more by visiting the text website at www.mhhe.com/willey8, where you will find a complete list of references.

38

Human Diseases Caused by Bacteria

The toll of tetanus. Sir Charles Bell's portrait (c. 1821) of a soldier wounded in the Peninsular War in Spain shows opisthotonus resulting from tetanus.

CHAPTER GLOSSARY

anthrax A highly infectious animal disease that can be transmitted to humans through skin infection, inhalation, and ingestion. Caused by *Bacillus anthracis*, a member of the order *Bacillales*.

atypical pneumonia A lung infection that can range in severity from nearly asymptomatic to serious. Often caused by *Mycoplasma pneumoniae*, a member of the class *Mollicutes*.

botulism A life-threatening intoxication resulting from a toxin produced by *Clostridium botulinum*, a member of the class *Clostridia*.

chancre The primary lesion of syphilis occurring at the site of entry of the infection.

chlamydiae Obligate intracellular bacteria that have a unique mode of reproduction.

coagulase An enzyme that induces blood clotting; it is characteristically produced by pathogenic staphylococci.

diphtheria A vaccine-preventable disease caused by *Corynebacterium diphtheriae*, a member of the suborder *Corynebacterineae*.

elementary body (EB) A small, dormant cell that serves as the agent of transmission between host cells in the chlamydial life cycle.

gas gangrene A necrotizing infection of skeletal muscle caused by *Clostridium perfringens*, a member of the class *Clostridia*.

group A streptococci (GAS) A group of human pathogens that includes members of the genus *Streptococcus* (order *Lactobacillales*) that form a capsule and produce extracellular degradative enzymes, streptokinases (enzymes that result in dissolution of blood clots), the white blood cell cytolysins streptolysin O and S, and a cell wall M protein.

impetigo A common skin infection caused by *Staphylococcus aureus* and *Streptococcus pyogenes*, both members of the class *Bacilli*.

lactic acid bacteria (LAB) Members of the order *Lactobacillales* that generate lactic acid as their major or sole fermentation product.

Lancefield grouping system A method by which streptococci can be placed in serologically distinguishable groups.

leprosy An insidious skin disease caused by *Mycobacterium leprae*, a member of the suborder *Corynebacterineae*.

***Mycobacterium avium* complex (MAC)** A disease that chiefly affects immunocompromised individuals, caused by *M. avium* and *M. intracellulare*, members of the suborder *Corynebacterineae*.

mycolic acids Complex 60 to 90 carbon fatty acids with a hydroxyl on the β-carbon and an aliphatic chain on the α-carbon; found in the cell walls of mycobacteria.

mycoplasmas Bacteria that are members of the class *Mollicutes* that lack cell walls and cannot synthesize peptidoglycan precursors; most require sterols for growth.

necrotizing fasciitis Inflammation and destruction of the sheath covering skeletal muscle caused by invasive group A streptococci.

reticulate body (RB) The cellular form in the chlamydial life cycle whose role is growth and reproduction within the host cell.

snapping division A distinctive type of binary fission resulting in an angular or a palisade arrangement of cells, which is characteristic of the genus *Corynebacterium*.

streptococcal pharyngitis (strep throat) An infection of the pharynx by group A *Streptococcus pyogenes*.

tetanus An intoxication resulting from infection with *Clostridium tetani* (class *Clostridia*); causes spastic paralysis.

toxic shock syndrome (TSS) A staphylococcal disease caused by toxic shock syndrome toxin-1, staphylococcal enterotoxin B, or enterotoxin C1 (also known as superantigens).

tuberculosis A disease that infects one-third of the world's population, caused principally by *Mycobacterium tuberculosis* but also by *M. bovis* and *M. africanum*, members of the suborder *Corynebacterineae*.

In this chapter, we continue our discussion of infectious disease by turning our attention to bacterial pathogens. These include bacteria that cause localized and systemic infections. We present examples of bacteria and the diseases that they cause by route of transmission. The microorganisms involved in dental infections are also described. Diseases caused by bacteria that are now listed as select agents (potential bioterror agents) are identified within the chapter by two asterisks (**).

Of all the known bacterial species, only a few are pathogenic to humans. Some human diseases have been only recently recognized; others have been known since antiquity. In the following sections, the more important disease-causing bacteria are discussed according to their mode of acquisition or transmission.

38.1 Airborne Diseases

Most airborne diseases caused by bacteria involve the respiratory system. Other airborne bacteria can cause skin diseases. Some of the better known of these diseases are now discussed.

Chlamydial Pneumonia

Chlamydial pneumonia is caused by *Chlamydophila pneumoniae.* The gram-negative chlamydiae are obligate intracellular parasites. They must grow and reproduce within host cells (*see figure 19.12*). Although their ability to cause human disease is widely recognized, many species grow within protists and animal cells without adverse affects. It is thought that these hosts represent a natural reservoir for the chlamydiae. Infection begins when the 0.2 to 0.6 μm diameter elementary bodies (EBs) are inhaled. The EBs have minimal metabolic activity and cannot synthesize their own proteins. They are designed exclusively for transmission and infection. Host cells phagocytose the EBs; however, they are held in inclusion bodies, where they reorganize to form reticulate bodies (RBs). The RBs are the reproductive form of the bacteria. About 8 to 10 hours after infection, the RBs undergo binary fission, and reproduction continues until the host cell dies. Interestingly, a chlamydia-filled inclusion body can become large enough to be visible by light microscopy. RBs differentiate into infectious EBs after 20 to 25 hours. The host cell lyses and releases the EBs 48 to 72 hours after infection. ◀◀ *Phylum Chlamydiae (section 19.5)*

Clinically, infections are generally mild; pharyngitis, bronchitis, and sinusitis commonly accompany some lower respiratory tract involvement. Symptoms include fever, a productive cough (respiratory secretion brought up by coughing), sore throat, hoarseness, and pain on swallowing. Infections with *C. pneumoniae* are common but sporadic; about 50% of adults have antibody to the chlamydiae. Evidence suggests that *C. pneumoniae* is primarily a human pathogen directly transmitted from human to human by droplet (respiratory) secretions. Diagnosis of chlamydial pneumonia is based on symptoms and an immunofluorescence test. Tetracycline and erythromycin are routinely used for treatment.

In seroepidemiological studies, *C. pneumoniae* infections have been linked with coronary artery disease as well as vascular disease at other sites. Following a demonstration of *C. pneumoniae*–like particles in atherosclerotic plaque tissue by electron microscopy, *C. pneumoniae* genes and antigens have been detected in artery plaque. Rarely, however, has the microorganism been recovered in cultures of atheromatous tissue (i.e., artery plaque). As a result of these findings, the possible etiologic role of *C. pneumoniae* in coronary artery disease and systemic atherosclerosis is under intense scrutiny.

Diphtheria

Diphtheria (Greek *diphthera,* membrane, and *-ia,* condition) is an acute, contagious disease caused by the gram-positive bacterium *Corynebacterium diphtheriae (see figure 22.9). C. diphtheriae* is well-adapted to airborne transmission by way of nasopharyngeal secretions because it is very resistant to drying. Diphtheria mainly affects unvaccinated, poor people living in crowded conditions. Once within the respiratory system, bacteria that carry the prophage containing the *tox* gene produce diphtheria toxin; thus, tox phage infection of *C. diphtheriae* is required for toxin production. This toxin is an exotoxin that causes an inflammatory response and the formation of a grayish pseudomembrane on the pharynx and respiratory mucosa (**figure 38.1**). The pseudomembrane consists of dead host cells and cells of *C. diphtheriae.*

Diphtheria toxin is absorbed into the circulatory system and distributed throughout the body, where it may cause destruction of cardiac, kidney, and nervous tissues by inhibiting protein synthesis. The toxin is an AB toxin, being composed of two polypeptide subunits: A and B. The A subunit consists of the catalytic domain; the B subunit is composed of the receptor and transmembrane domains (*see figure 31.11*). The receptor domain binds to the heparin-binding epidermal growth factor receptor on the surface of various eukaryotic cells. Once bound, the toxin enters the cytoplasm by endocytosis. The transmembrane domain of the toxin embeds itself into the target cell membrane, causing the catalytic domain to be cleaved and translocated into the cytoplasm. The cleaved catalytic domain becomes an active enzyme,

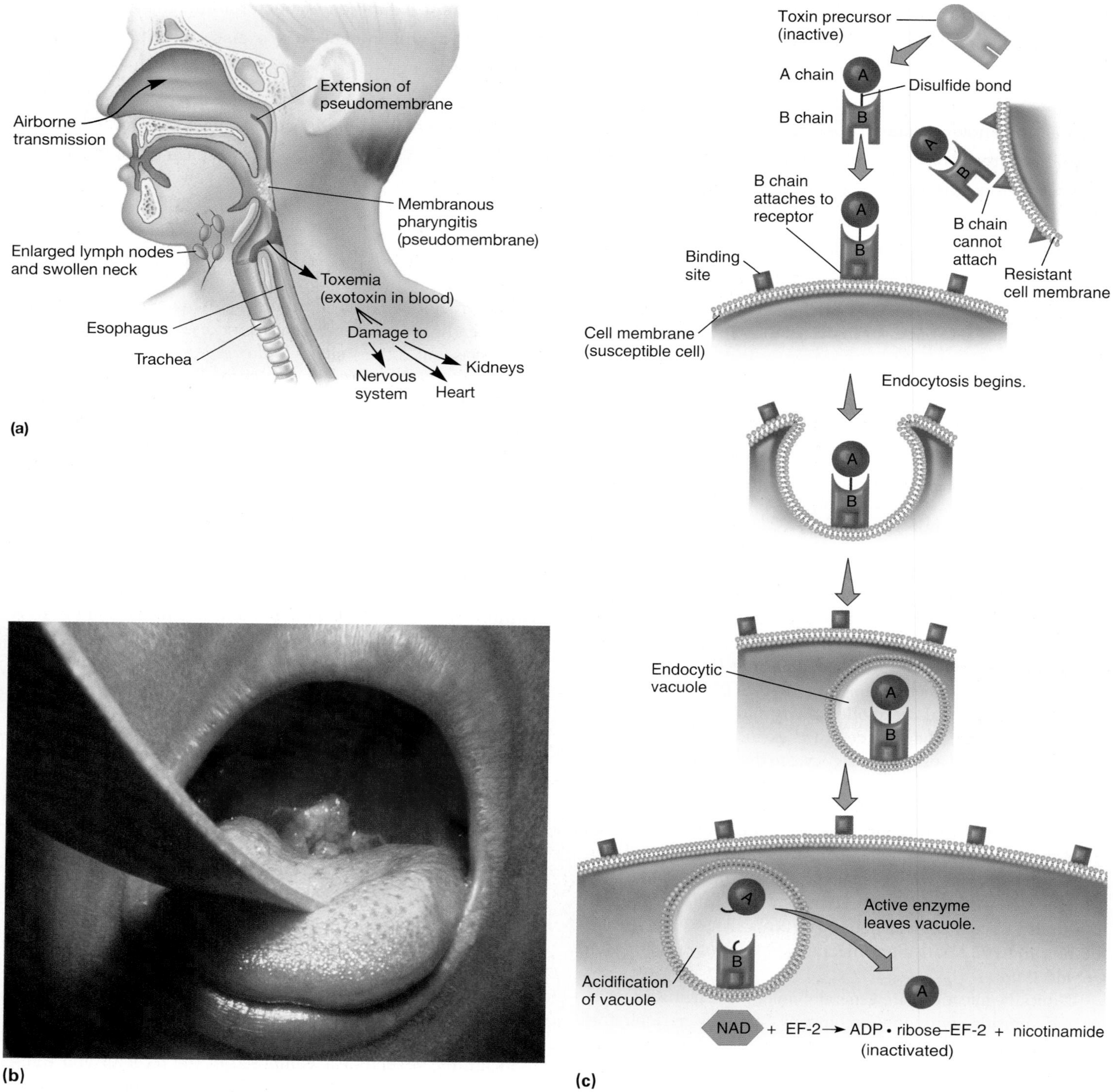

FIGURE 38.1 Diphtheria Pathogenesis. (a) Diphtheria is a well-known, exotoxin-mediated infectious disease caused by *Corynebacterium diphtheriae*. The disease is an acute, contagious, febrile illness characterized by local oropharyngeal inflammation and pseudomembrane formation. If the exotoxin gets into the blood, it is disseminated and can damage the peripheral nerves, heart, and kidneys. (b) The clinical appearance includes gross inflammation of the pharynx and tonsils marked by grayish patches (a pseudomembrane) and swelling of the entire area. (c) The B subunit of the diphtheria toxin binds to the host cell membrane facilitating entry of the A subunit, which ADP-ribosylates EF-2 to inhibit protein synthesis.

catalyzing the attachment of ADP-ribose (from NAD^+) to translation elongation factor-2 (EF-2). A single enzyme (i.e., catalytic domain) can exhaust the entire supply of cellular EF-2 within hours, resulting in protein synthesis inhibition and cell death. ◀◀ *Suborder* Corynebacterineae *(section 22.4); Exotoxins (section 31.3)*

Typical symptoms of diphtheria include a thick mucopurulent (containing both mucus and pus) nasal discharge, pharyngitis, fever, cough, paralysis, and death. (*C. diphtheriae* can also infect the skin, usually at a wound or skin lesion, causing a slow-healing ulceration termed **cutaneous diphtheria**.) Diagnosis is made by observation of the pseudomembrane in the throat and by bacterial culture. Diphtheria antitoxin is given to neutralize any unabsorbed exotoxin in the patient's tissues; penicillin and erythromycin are used to treat the infection. Prevention is by active immunization with **DPT** (diphtheria-pertussis-tetanus) vaccine and then boosted with DTap (diphtheria toxoid, tetanus toxoid, acellular *B. pertussis* vaccine) or Tdap (tetanus toxoid, reduced diphtheria toxoid, acellular pertussis vaccine, adsorbed), approved in 2005 (*see table 36.3*). Most cases involve people over thirty years of age who have a weakened immunity to the diphtheria toxin and live in tropical areas. Since 1980, fewer than six diphtheria cases have been reported annually in the United States, and most occur in nonimmunized individuals. ◀◀ *Vaccines and immunization (section 36.7)*

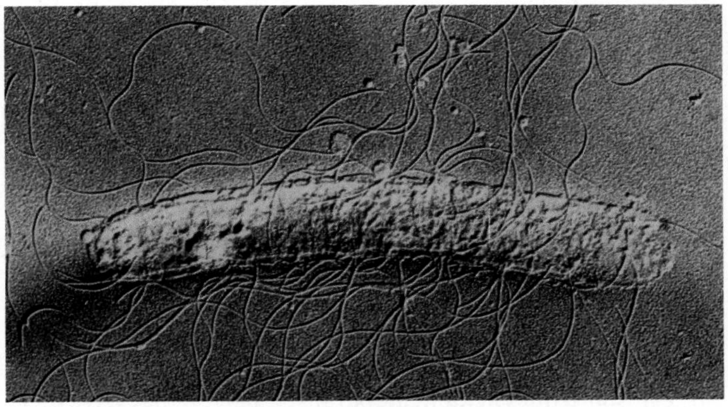

FIGURE 38.2 Legionnaires' Disease. *Legionella pneumophilia,* the causative agent of Legionnaires' disease, with many lateral flagella; SEM (× 10,000).

Figure 38.2 Micro Inquiry

What organisms serve as the environmental reservoir for *L. pneumophilia*?

Legionnaires' Disease

In 1976 the term **Legionnaires' disease,** or **legionellosis,** was coined to describe an outbreak of pneumonia that occurred at the Pennsylvania State American Legion Convention in Philadelphia. The bacterium responsible for the outbreak was *Legionella pneumophila*, a nutritionally fastidious, aerobic, gram-negative rod (**figure 38.2**). It is now known that this bacterium is part of the natural microbial community of soil and freshwater ecosystems, and it has been found in large numbers in air-conditioning systems and shower stalls. ◀◀ *Class* Gammaproteobacteria *(section 20.3)*

Infection with *L. pneumophila* and other *Legionella* spp. results from the airborne spread of bacteria from an environmental reservoir to the human respiratory system. An increasing body of evidence suggests that the environmental reservoir is a variety of free-living amoebae and ciliated protozoa, and that these intermediate hosts are the most important factor for the survival and growth of *Legionella* in nature.

Legionella spp. multiply intracellularly within the protists, just as they do within human macrophages. This might explain why there is no human-to-human spread of legionellosis, as transmission of host cells containing the intracellular bacteria is unlikely. The bacteria reside within the phagosomes of alveolar macrophages, where they multiply and produce localized tissue destruction through export of a cytotoxic exoprotease. Symptoms start 2 to 10 days after exposure and include a high fever, nonproductive cough (respiratory secretions are not brought up during coughing), headache, neurological manifestations, and severe bronchopneumonia. Diagnosis depends on isolation of the bacterium, documentation of a rise in antibody titer over time, or the presence of *Legionella* antigens in the urine as detected by a rapid test kit. Treatment begins with supportive measures and the administration of erythromycin or rifampin. Death occurs in 10 to 15% of symptomatic cases.

Prevention of Legionnaires' disease depends on the identification and elimination of the environmental source of *L. pneumophila*. Chlorination, ozonation, the heating of water, and the cleaning of water-containing devices can help control the multiplication and spread of *Legionella*. These control measures are effective because the pathogen does not appear to be spread from person to person. Since the initial outbreak of this disease in 1976, many outbreaks during summer months have been recognized in all parts of the United States. About 1,000 to 1,600 cases are diagnosed each year, and about 18,000 or more additional mild or subclinical cases are thought to occur. It is estimated that 23% of all nosocomial pneumonias are due to *L. pneumophila*, especially among immunocompromised patients.

1. Why do you think chlamydiae differentiate into specialized cell types for infection and reproduction?
2. How do humans contract chlamydial pneumonia? Pneumonia caused by Legionella?
3. What causes the typical symptoms of diphtheria? How are individuals protected against this disease?

Meningitis

Meningitis (Greek *meninx*, membrane, and *–itis*, inflammation) is an inflammation of the brain or spinal cord meninges (membranes). There are many causes of meningitis, some of which can be treated with antimicrobial agents. Several bacterial organisms that cause meningitis include: *Streptococcus pneumoniae, Neisseria meningitidis, Haemophilus influenzae* type b, Group B streptococci, *Listeria monocytogenes, Mycobacterium tuberculosis, Nocardia asteroides, Staphylococcus aureus,* and *S. epidermidis.* The accurate identification of the causative agent is essential for proper treatment of the disease. However, this can sometimes be difficult because a person may have meningitis symptoms but show no microbial agent in Gram-stained specimens and have negative cultures. In such a case, the diagnosis often is called aseptic (meaning lack of an identifiable agent) meningitis syndrome; aseptic meningitis can be caused by a virus or protozoan. Aseptic meningitis is typically more difficult to treat as therapy targeting a specific microorganism is not achieved.

Bacterial meningitis can be diagnosed by a Gram stain and culture of bacteria from cerebral spinal fluid (CSF). Rapid tests are also used. The immediate sources of the bacteria responsible for meningitis are respiratory secretions from carriers. Bacterial meningitis can be caused by various gram-positive and gram-negative bacteria. However, three organisms tend to be associated with meningitis more frequently than others: *Streptococcus pneumoniae, Neisseria meningitidis,* and *Haemophilus influenzae* (serotype b).

N. meningitidis, often referred to as the meningococcus, is a normal inhabitant of the human nasopharynx (5 to 15% of humans carry the nonpathogenic serotypes). Most disease-causing *N. meningitidis* strains belong to serotypes A, B, C, Y, and W-135. In general, serotype A strains are the cause of epidemic disease in developing countries, whereas serotype B, C, and W-135 strains are responsible for meningitis outbreaks in the United States. Infection results from airborne transmission of the bacteria, typically through close contact with a primary carrier. The disease process is initiated by pili-mediated colonization of the nasopharynx by pathogenic bacteria (**figure 38.3**). The bacteria cross the nasopharyngeal epithelium (typically through endocytosis) and invade the bloodstream (meningococcemia), where they proliferate.

Symptoms caused by *N. meningitidis* are variable, depending on the degree of bacterial dissemination. The usual symptoms of meningitis include an initial respiratory illness or sore throat interrupted by one or more of the meningeal syndromes: vomiting, headache, lethargy, confusion, and stiffness in the neck and back. Once bacterial meningitis is suspected, antibiotics are administered immediately. In fact, antibiotics (penicillin, chloramphenicol, cefotaxime, ceftriaxone, ofloxacin) are often administered prophylactically to patient contacts; if left untreated, meningitis is fatal.

Control of *N. meningitidis* infection is with vaccination and antibiotics. Two vaccines are currently available: the meningococcal polysaccharide (MPSV4) and the meningococcal conjugate vaccine (MCV4). Both vaccines are effective against serotypes A, C, Y, and W-135. The development of a vaccine effective against serotype B has required careful genomic analysis and is discussed in chapter 16 (*p. 440*). Vaccination is recommended for all college students living in residence halls. MCV4 is recommended for preteens, teens, and adults less than fifty-five years of age. MPSV4 should be used for children two to ten years of age and adults over fifty-five who are at risk.

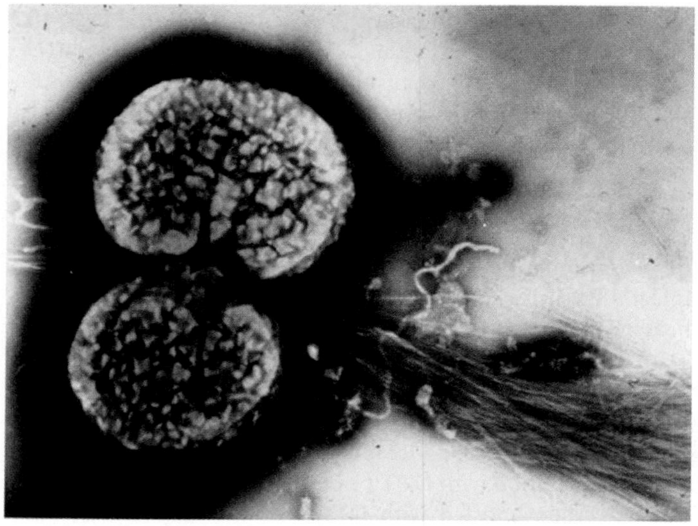

FIGURE 38.3 ***Neisseria meningitidis.*** Type IV pili are present on the cell surface as bundled filamentous appendages, facilitating transformation and attachment.

Another agent of meningitis is *H. influenzae*, a small, gram-negative bacterium. Transmission is by inhalation of droplet nuclei shed by infectious individuals or carriers. *H. influenzae* serotype b can infect mucous membranes, resulting in sinusitis, pneumonia, and bronchitis. It can disseminate to the bloodstream and cause a bacteremia. *H. influenzae* serotype b can cross into the CSF, resulting in inflammation of the meninges (meningitis). In the United States, *H. influenzae* disease (including pneumonia and meningitis) is primarily observed in children less than five years of age, with fewer than 1 case per 100,000 children occurring. Globally, it is estimated that *H. influenzae* serotype b causes at least 3 million cases of serious disease and several hundreds of thousands of deaths each year.

A sharp reduction in the incidence of *H. influenzae* serotype b infections began in the mid-1980s due to approval of the *H. influenzae* type b conjugate vaccine, rifampin prophylaxis of disease contacts, and the availability of more efficacious therapeutic agents. Since routine use of the vaccine began in the 1990s, the incidence of invasive infection among U.S. children under five years of age declined by 99%. Three to six percent of all *H. influenzae* infections are fatal. Furthermore, up to 20% of surviving patients have permanent hearing loss or other long-term sequelae. Currently, all children should be vaccinated with the *H. influenzae* type b conjugate vaccine at the age of two months.

Mycobacterium Infections

Mycobacteria cause a number of human infections that are difficult to treat. Recall that mycobacteria make cell walls that have a very high lipid content and contain waxes with 60 to 90 carbon mycolic acids. These are complex fatty acids with a hydroxyl group on the α-carbon and an aliphatic chain attached to the β-carbon (*see figure 22.11*). The presence of mycolic acids and other lipids outside the peptidoglycan layer makes mycobacteria acid-fast, which means that basic fuchsin dye cannot be removed from the cell by acid alcohol treatment. More importantly, the mycolic acids also make the mycobacteria resistant to penetration of water-soluble antibiotics. Mycobacteria are an extremely large group of bacteria that are normal inhabitants of soil, water, and house dust.

M. avium Complex

Two mycobacteria are noteworthy pathogens in the United States: *Mycobacterium avium* and *Mycobacterium intracellulare*. These mycobacteria are so closely related that they are referred to as the *M. avium* complex (MAC). Globally, *M. tuberculosis* has remained more prevalent in developing countries, whereas MAC has become the most common cause of mycobacterial infections in the United States. ◀◀ *Suborder* Corynebacterineae *(section 22.4)*

MAC are found worldwide and infect a variety of insects, birds, and animals. Both the respiratory and the gastrointestinal tracts have been proposed as entry portals for MAC; however, person-to-person transmission is not very efficient. MAC causes a pulmonary infection in humans similar to that caused by *M. tuberculosis*. Pulmonary MAC is more common in elderly persons with preexisting pulmonary disease. The gastrointestinal tract is thought to be the most common site of colonization and dissemination in AIDS patients. In the United States, disseminated infection with MAC occurs in 15 to 40% of AIDS patients with $CD4^+$ cell counts of less than 100 per mm^3; it produces disabling symptoms, including fever, malaise, weight loss, night sweats, and diarrhea. Carefully controlled epidemiological studies have shown that MAC shortens survival by 5 to 7 months among persons with AIDS. With more effective antiviral therapy for AIDS and prolonged survival, the number of cases of disseminated MAC is likely to increase substantially, and its contribution to AIDS mortality will increase. ◀◀ *Acquired immune deficiency syndrome (AIDS) (section 37.3)*

MAC can be isolated from sputum, blood, and aspirates of bone marrow. Acid-fast stains are of value in making a diagnosis. The most sensitive method for detection is the commercially available blood culture system (Wampole Laboratories). Although no drugs are currently approved by the U.S. Food and Drug Administration (FDA) for the therapy of MAC, every regimen should contain either azithromycin or clarithromycin and ethambutol as a second drug. One or more of the following can be added: clofazimine, rifabutin, rifampin, ciprofloxacin, and amikacin.

Search This: TB and Wampole Labs

Mycobacterium tuberculosis

Over a century ago, Robert Koch identified *Mycobacterium tuberculosis* as the causative agent of **tuberculosis (TB).** At the time, TB was rampant, causing one-seventh of all deaths in Europe and one-third of deaths among young adults. Today TB remains a global health problem of enormous dimension. It is estimated that one-third of the world's human population is infected (i.e., at least 2 billion people). Of the 212 countries and territories in the world, 202 (99.6% of the world population) reported TB cases in 2007; the World Health Organization (WHO) reported 9.2 million new TB cases, with approximately 7.7% being HIV positive also (**figure 38.4*a***). Worldwide, TB is caused by *M. bovis* and *M. africanum*, in addition to *M. tuberculosis*. Since AIDS has become pandemic, the number of global TB cases has increased dramatically. Because a close association exists between AIDS and TB, further spread of HIV infection among populations with a high prevalence of TB infection is resulting in dramatic increases in TB. Importantly, TB is the direct cause of death of over half of all AIDS patients worldwide.

In the United States, TB occurs most commonly among the homeless, elderly, and malnourished, or among alcoholic males, minorities, immigrants, prison populations, and Native Americans. Between 1999 and 2007, the incidence of tuberculosis in the United States steadily declined to about 13,000 cases; about 1,000 deaths are reported each year. During 2007, a total of 13,299 confirmed TB cases were reported in the United States, representing a 3.3% decline in the rate from 2006. Forty-eight percent of the national TB case total came from California, Texas, New York, and Florida. More than half (58%) of these U.S. cases were in foreign-born persons. It appears that about one-fourth to one-third of active TB cases in the United States may be due to recent transmission. The majority of active cases result from the reactivation of old, dormant infections.

Infection results when the bacteria are phagocytosed by macrophages in the lungs, where they survive the normal antimicrobial processes (figure 38.4*b*). In fact, infected macrophages often die attempting to destroy the bacteria, thus releasing viable bacteria into respiratory spaces. The incubation period is about 4 to 12 weeks, and the disease develops slowly. The symptoms of TB are fever, fatigue, and weight loss. A cough, which is characteristic of pulmonary involvement, may result in expectoration of bloody sputum.

M. tuberculosis (Mtb) does not produce classic virulence factors such as toxins, capsules, and fimbriae. Instead, Mtb has some rather unique products and properties that contribute to its virulence. The cell envelope of Mtb differs substantially from that of gram-positive and gram-negative bacteria in that it contains several unique lipids and glycolipids. In addition to mycolic acids, these include lipoarabinomannan, trehalose dimycolate, and phthiocerol dimycocerosate. These materials are directly toxic to eukaryotic cells and create a hydrophobic

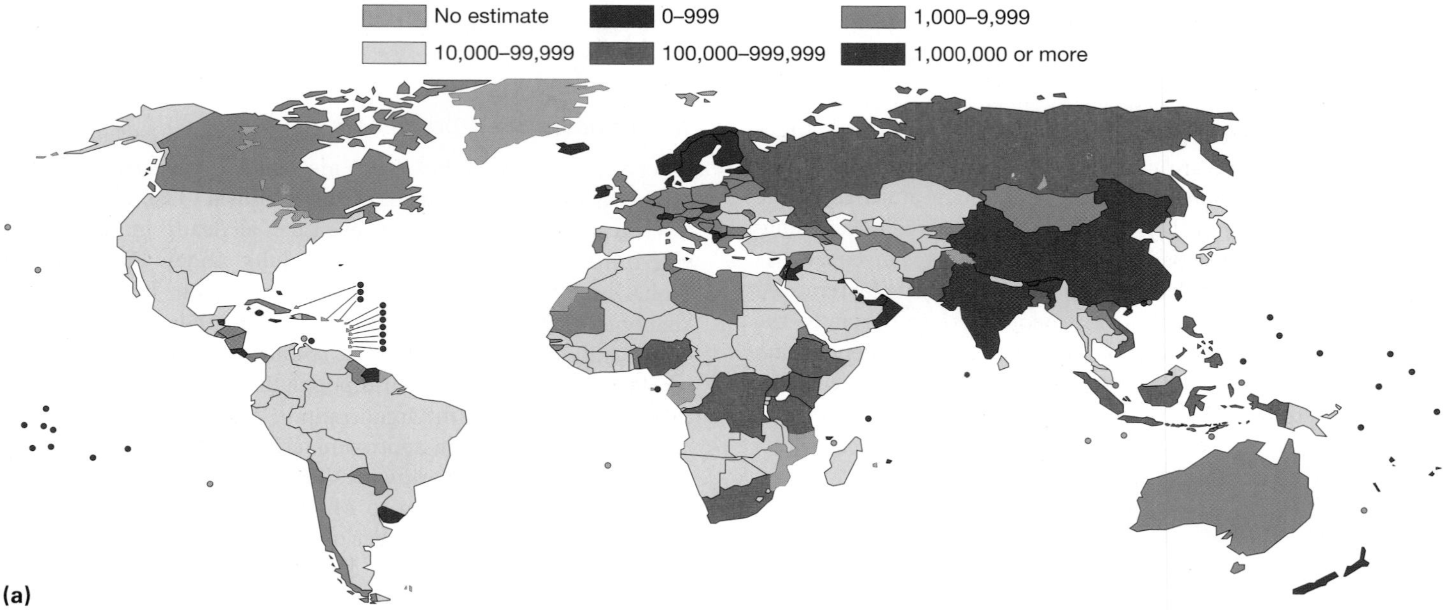

(a)

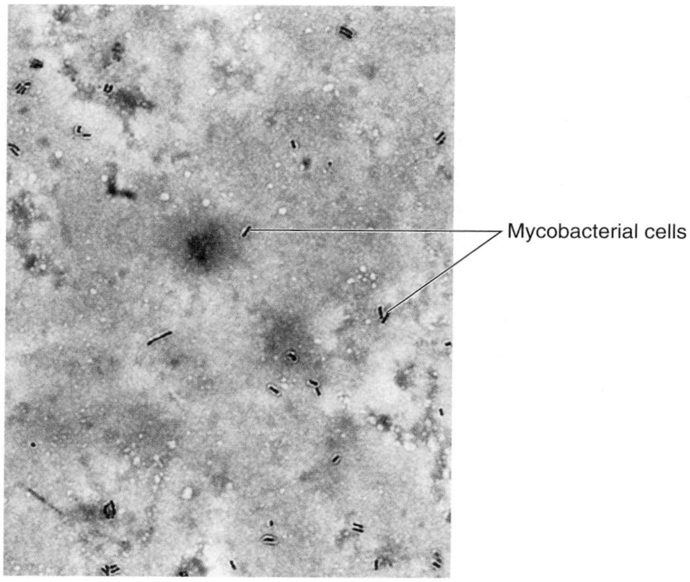

(b) *M. tuberculosis* in sputum (Ziehl-Neelson stain).

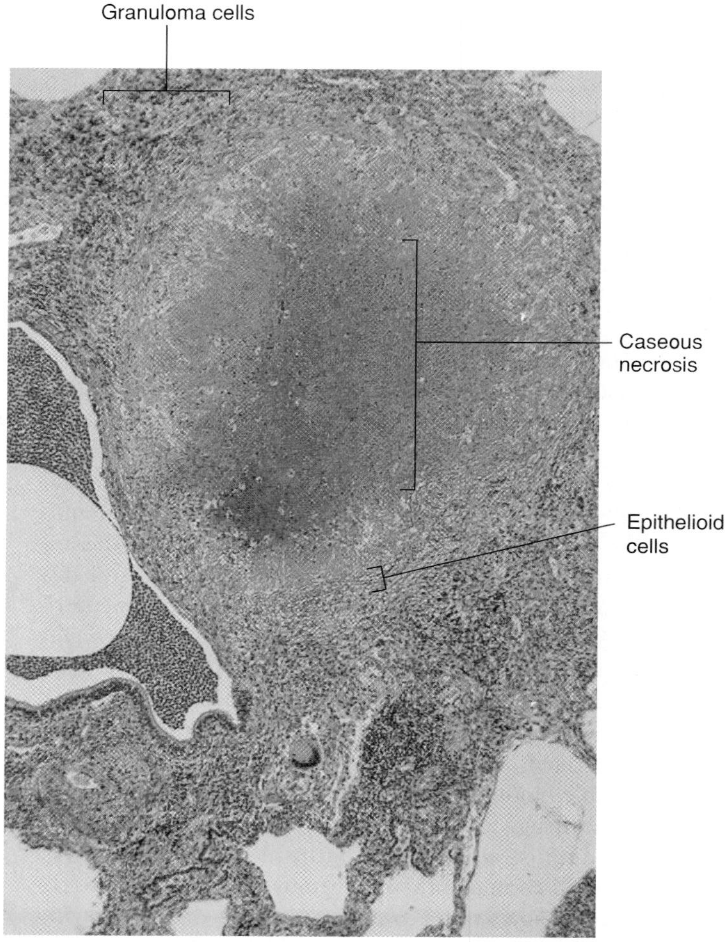

(c)

FIGURE 38.4 Tuberculosis. (a) Tuberculosis is a significant global disease. (b) Mycobacteria are recovered in the sputum of tuberculosis patients and can be identified using an acid-fast stain. (c) In the lungs, tuberculosis is identified by the tubercle, a granuloma of white blood cells, bacteria, fibroblasts, and epithelioid cells. The center of the tubercle contains caseous (cheesey) pus and bacteria.

Continued

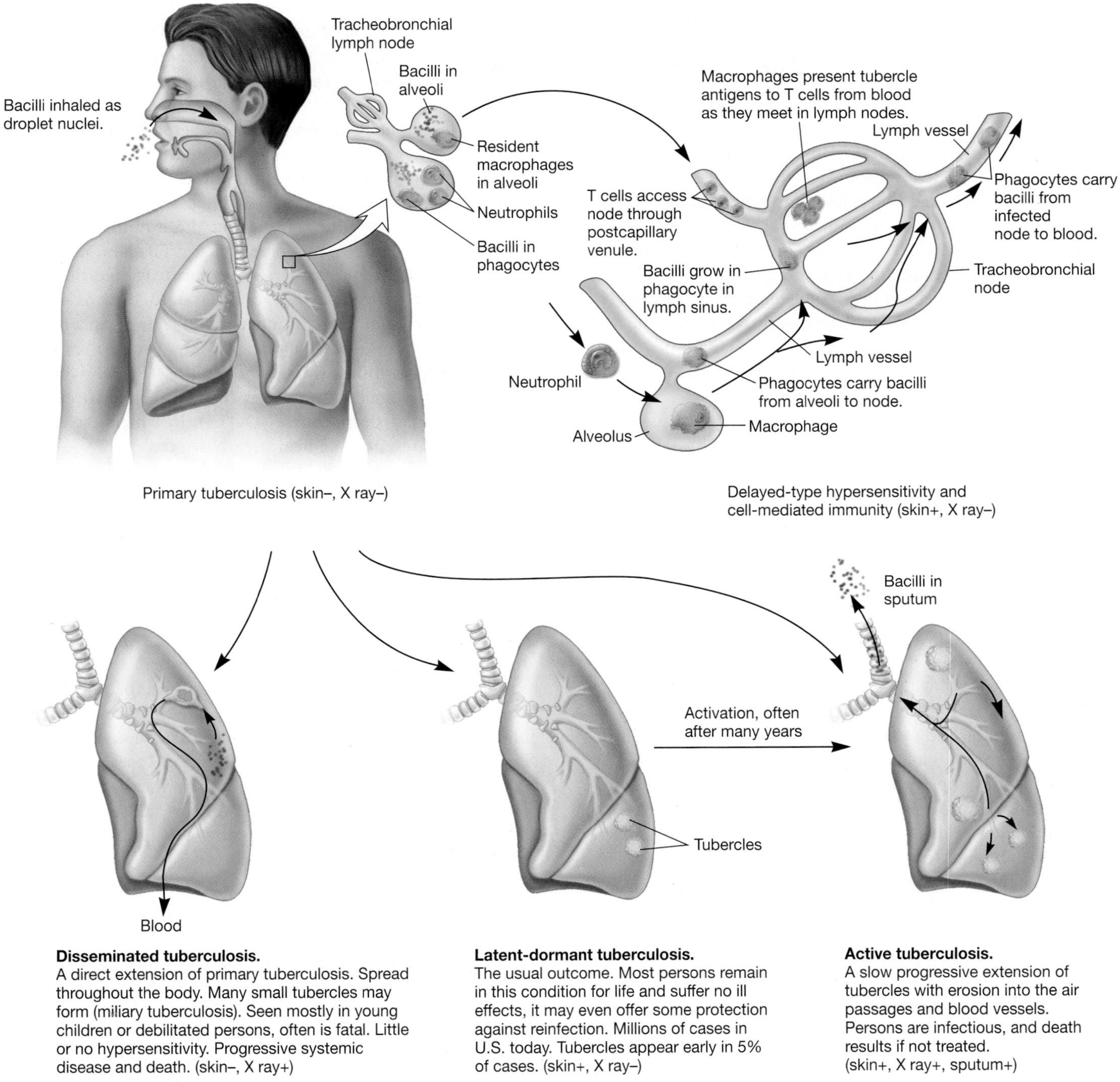
Tracheobronchial lymph node
Bacilli in alveoli
Bacilli inhaled as droplet nuclei.
Resident macrophages in alveoli
Neutrophils
Bacilli in phagocytes
Primary tuberculosis (skin–, X ray–)
Macrophages present tubercle antigens to T cells from blood as they meet in lymph nodes.
Lymph vessel
Phagocytes carry bacilli from infected node to blood.
T cells access node through postcapillary venule.
Bacilli grow in phagocyte in lymph sinus.
Tracheobronchial node
Lymph vessel
Neutrophil
Phagocytes carry bacilli from alveoli to node.
Alveolus
Macrophage
Delayed-type hypersensitivity and cell-mediated immunity (skin+, X ray–)
Bacilli in sputum
Activation, often after many years
Tubercles
Blood
Disseminated tuberculosis.
A direct extension of primary tuberculosis. Spread throughout the body. Many small tubercles may form (miliary tuberculosis). Seen mostly in young children or debilitated persons, often is fatal. Little or no hypersensitivity. Progressive systemic disease and death. (skin–, X ray+)
Latent-dormant tuberculosis.
The usual outcome. Most persons remain in this condition for life and suffer no ill effects, it may even offer some protection against reinfection. Millions of cases in U.S. today. Tubercles appear early in 5% of cases. (skin+, X ray–)
Active tuberculosis.
A slow progressive extension of tubercles with erosion into the air passages and blood vessels. Persons are infectious, and death results if not treated.
(skin+, X ray+, sputum+)

(d) *Mycobacterium tuberculosis* infection and its possible outcomes.

FIGURE 38.4 ***Continued.*** (d) The natural history of mycobacterial infection leading to tuberculosis demonstrates its public health threat.

barrier around the bacterium that facilitates impermeability and resistance to antimicrobial agents. It also protects against killing by acidic and alkaline compounds, osmotic lysis, and lysozyme. Furthermore, cell wall glycolipids associate with mannose giving Mtb control over entry into macrophages by exploiting the macrophage mannose receptors. Once inside, Mtb can inhibit phagosome-lysosome fusion by altering the phagosome membrane. Resistance to oxidative killing, inhibition of phagosome-lysosome fusion, and inhibition of diffusion of lysosomal enzymes are just some of the mechanisms that help explain the survival of Mtb inside macrophages. ◂◂ *Phagocytosis (section 32.5)*

During infection, other immune effector cells are recruited to the site of infection by cytokines released from the responding macrophages. Together, and in response to several mycobacterial products, a hypersensitivity response results in the formation of small, hard nodules called **tubercles** composed of bacteria, macrophages, T cells, and various human proteins (figure 38.4*c*). Most cases in the United States are acquired from other humans through droplet nuclei and the respiratory route (figure 38.4*d*). Tubercles are characteristic of tuberculosis and give the disease its name. The disease process usually stops at this stage, but the bacteria often remain alive within macrophage phagosomes. However, in some cases, the disease may become active, even after many years of latency.

In time, a tubercle may change to a cheeselike consistency and is then called a **caseous lesion** (figure 38.4*d*). If such lesions calcify, they are called **Ghon complexes,** which show up prominently in a chest X-ray. Sometimes the tubercle lesions liquefy and form air-filled **tuberculous cavities.** From these cavities, the bacteria can be disseminated to new foci throughout the body. This spreading is often called **miliary tuberculosis** due to the many tubercles the size of millet seeds that are formed in the infected tissue. It is also called **reactivation tuberculosis** because the bacteria have been reactivated in the initial site of infection.

TB must be treated with antimicrobial therapy. Several drugs are administered simultaneously (e.g., isoniazid [INH], plus rifampin, ethambutol, and pyrazinamide). These drugs are administered for 6 to 9 months as a way of decreasing the possibility that the bacterium develops drug resistance. However, ***m*ulti*d*rug-*r*esistant strains of *t*uberculosis (MDR-TB)** have developed and are spreading. A multidrug-resistant strain is defined as *M. tuberculosis* that is resistant to INH and rifampin, with or without resistance to other drugs. Between 2006 and 2007 in the United States, 98 new cases of MDR-TB were found in people with no previous TB history. At the same time, 19 new cases were reported in people with previous TB diagnosis. Inadequate therapy is the most common means by which resistant bacteria are acquired, and patients who have previously undergone therapy are presumed to harbor MDR-TB until proven otherwise. Recently strains of Mtb considered to cause MDR-TB have emerged that are e*x*tensively *d*rug-*r*esistant (XDR); they are resistant to INH, rifampin, and the secondary antibiotics. XDR-TB is a unique challenge to public health as first-line and second-line drug resistance severely limits treatment options. MDR-TB and XDR-TB can be fatal. Of U.S.-born persons with TB, 1.1% of them had MDR-TB in 2007, while the incidence of MDR-TB in foreign-born persons was 80%. In 2007 there were two U.S. cases of XDR-TB reported, a 50% drop from 2006. However, in countries with high rates of HIV and limited health-care resources, XDR- TB cases can be hard to track. As of April 2007, 352 cases of XDR-TB were reported in South Africa (where HIV prevalence was estimated at 10.8% in 2005), but substantially greater numbers of XDR-TB cases are expected because culture and susceptibility testing are performed on only a small fraction of patients suspected of TB infection. Thus, many drug-resistant TB cases will go undetected.

Drug-resistant TB arises because tubercle bacilli have spontaneous, predictable rates of chromosomal mutations that confer resistance to drugs. These mutations are unlinked; hence resistance to one drug is not associated with resistance to an unrelated drug. The observation that these mutations are not linked is the cardinal principle underlying TB chemotherapy. In the circumstances of monotherapy, erratic drug ingestion, omission of one or more drugs, suboptimal dosage, poor drug absorption, or an insufficient number of active drugs in a regimen, a susceptible strain of *M. tuberculosis* may become resistant to multiple drugs within a matter of months. This is why many public health organizations worldwide practice *d*irectly *o*bserved *t*reatment *s*hort course (DOTS), in which each dose is taken in the presence of a health-care worker.

Persons infected with Mtb develop cell-mediated immunity because the bacteria are phagocytosed by macrophages (i.e., it is an intracellular pathogen). This immunity involves sensitized T cells and is the basis for the tuberculin skin test. In this test, a *p*urified *p*rotein *d*erivative (PPD) of Mtb is injected intracutaneously (the Mantoux test). If the person has had TB or has been exposed to Mtb, sensitized T cells react with these proteins and a delayed hypersensitivity reaction occurs within 48 hours (*see figure 33.25*). This positive skin reaction appears as an induration (hardening) and reddening of the area around the injection site. In a young person, a positive skin test could indicate active tuberculosis. In older persons, it may result from previous disease, vaccination, or a false-positive test. In these cases, X-rays and bacterial isolation are completed to confirm the diagnosis. Other laboratory tests include microscopy of the acid-fast bacterium and commercially available DNA probes. ◂◂ *Type IV hypersensitivy (section 33.10)*

1. What are the three major types of meningitis? Why is it so important to determine which type a person has?
2. How is tuberculosis diagnosed? Describe the various types of lesions and how they are formed.
3. What is the reason for the complex antibiotic therapy used to treat tuberculosis?
4. How do multidrug-resistant strains of tuberculosis develop?

Mycoplasmal Pneumonia

Typical pneumonia has a bacterial origin (most frequently *Streptococcus pneumoniae*) with fairly consistent signs and symptoms. If the symptoms of pneumonia are different from what is typically observed, the disease is often called **atypical pneumonia.** One cause of atypical pneumonia is *Mycoplasma* *pneumoniae*, a mycoplasma with worldwide distribution. Transmission involves close contact and airborne droplets. The disease is fairly common and mild in infants and small children; serious disease is seen principally in older children and young adults. ◀◀ *Class* Mollicutes (the mycoplasmas) *(section 21.1)*

M. pneumoniae cells vary in shape, are about 0.3 to 0.8 μm in diameter, and stain gram-negative. They evolved from ancestors having gram-positive cell walls, and they now lack cell walls and cannot synthesize peptidoglycan precursors. Thus they are resistant to all β-lactam antibiotics (e.g., penicillin). Mycoplasma typically infect the upper respiratory tract and subsequently move to the lower respiratory tract, where they attach to respiratory mucosal cells. Mycoplasma then produce peroxide, which may be a toxic factor, but the exact mechanism of pathogenesis is unknown. A change in mucosal cell nucleic acid synthesis has been observed. The manifestations of this disease vary in severity from asymptomatic to a serious pneumonia. The latter is accompanied by death of the surface mucosal cells, lung infiltration, and congestion. Initial symptoms include headache, weakness, a low-grade fever, and a predominant, characteristic cough. The disease and its symptoms usually persist for weeks. The mortality rate is less than 1%.

Several rapid tests using latex-bead agglutination for *M. pneumoniae* antibodies are available for diagnosis of mycoplasmal pneumonia. When isolated from respiratory secretions, some mycoplasmas form distinct colonies with a "fried-egg" appearance on agar (*see figure 21.4*). During the acute stage of the disease, diagnosis must be made by clinical observations. Tetracyclines or erythromycin are effective in treatment. There are no preventive measures.

Pertussis

Pertussis (Latin *per*, intensive, and *tussis*, cough), sometimes called "whooping cough," is caused by the gram-negative bacterium *Bordetella pertussis*. *B. parapertussis* is a closely related species that causes a milder form of the disease. Pertussis bacteria colonize the respiratory epithelium to produce a disease characterized by fever, malaise, uncontrollable cough, and cyanosis (bluish skin color resulting from inadequate tissue oxygenation). Whooping cough gets its name from the characteristically prolonged and paroxysmal coughing that ends in an inspiratory gasp, or whoop. Pertussis is a highly contagious, vaccine-preventable disease that primarily affects children. It has been estimated that over 95% of the world's population has experienced either mild or severe symptoms of the disease. The WHO reported over 152,500 pertussis cases globally in 2007. Around 300,000 die from the disease each year. However, there are 5,000 to 7,000 cases and approximately 10 deaths annually in the United States. ◀◀ *Order* Burkholderiales *(section 20.2)*

Search This: Bordetella *pertussis and whooping cough*

Disease transmission occurs by inhalation of the bacterium in droplets released from an infectious person. The incubation period is 7 to 14 days. Once inside the upper respiratory tract, the bacteria colonize the cilia of the mammalian respiratory epithelium through fimbria-like structures called filamentous hemagglutinins that bind to complement receptors on phagocytes. Additionally, some of the components of one of the *B. pertussis* toxins (S2 and S3 subunits of the PTx toxin) assist in adherence to cilia by bridging the bacterial and host cells; S2 binds to the cilial glycolipid lactosylceramide, and S3 binds to phagocyte glycoproteins. Thus attachment is an important virulence factor in the initiation of the disease.

B. pertussis produces several toxins. The most important is the pertussis toxin (PTx). PTx is an AB exotoxin (*see figure 31.11*). The A subunit (S1) is an ADP-ribosyl transferase, similar to the diphtheria toxin. The B subunit is composed of five polypeptides (S2–S5; there are two S4 subunits) that bind to specific carbohydrates on cell surfaces. The B subunit binds to host cells, transporting the A subunit to the cell membrane, where it is inserted and released into the cytoplasm. As an enzyme, the A subunit transfers the ADP-ribosyl moiety of NAD to a membrane-bound, regulatory G protein called Gi. Gi normally inhibits eukaryotic adenylate cyclase, which catalyzes the conversion of ATP into cyclic AMP (cAMP). Thus the net effect of PTx on a cell is an increase in intracellular levels of cAMP, which accelerates mucin secretion and alters water transport affecting electrolyte balance. *B. pertussis* also produces an extracytoplasmic invasive adenylate cyclase, tracheal cytotoxin, and dermonecrotic toxin, which destroy epithelial tissue. Working together, the tracheal cytotoxin and pertussis toxin also provoke the secretory cells in the respiratory tract to produce nitric oxide, which kills nearby ciliated cells, inhibiting removal of bacteria and mucus. The secretion of a thick mucus also impedes ciliary action.

Pertussis is divided into three stages. (1) The catarrhal stage, so named because of the mucous membrane inflammation, is insidious and resembles the common cold. (2) The paroxysmal stage is characterized by prolonged coughing sieges. During this stage, the infected person tries to cough up the mucous secretions by making 5 to 15 rapidly consecutive coughs followed by the characteristic whoop—a hurried, deep inspiration. The catarrhal and paroxysmal stages last about 6 weeks. (3) The convalescent stage, when final recovery occurs, may take several months.

Laboratory diagnosis of pertussis is by culture of the bacterium, fluorescent antibody staining of smears from nasopharyngeal swabs, other antibody-based detection tests, and PCR. The development of a strong, lasting immunity takes place after an initial infection. Treatment is with erythromycin, tetracycline, or chloramphenicol. Treatment ameliorates clinical illness when begun during the catarrhal phase and may also reduce the severity of the disease when begun within 2 weeks of the onset of the paroxysmal cough. Prevention is with the DPT vaccine in children when they are two to three months old, and with the tetanus toxoid, reduced diphtheria toxoid, and acellular pertussis (Tdap) vaccine (approved in 2005; replaces the older tetanus and reduced diphtheria toxoid [Td] booster shots) for older children and adults (*see table 36.3*).

1. Describe the pneumonia caused by *M. pneumoniae.*
2. What is the mechanism by which PTx kills host cells?
3. Describe the three stages of pertussis.

Streptococcal Diseases

Pathogenic streptococci, commonly called strep, are a heterogeneous group of gram-positive bacteria. In this group, *Streptococcus pyogenes* (group A, β-hemolytic streptococci) is one of the most important bacterial pathogens. The different serotypes of **group A streptococci (GAS)** produce (1) extracellular enzymes that break down host molecules; (2) streptokinases, enzymes that activate a host-blood factor that dissolves blood clots; (3) the cytolysins streptolysin O and streptolysin S, which kill host leukocytes; and (4) capsules and M protein, which help to retard phagocytosis (**figure 38.5**). M protein, a filamentous protein anchored in the streptococcal cell membrane, facilitates attachment to host cells and prevents opsonization by complement protein C3b. It is the major virulence factor of GAS. M protein types 1, 3, 12, and 28 are commonly found in patients with streptococcal toxic shock and multiorgan failure. ◀◀ *Complement (section 32.3)*

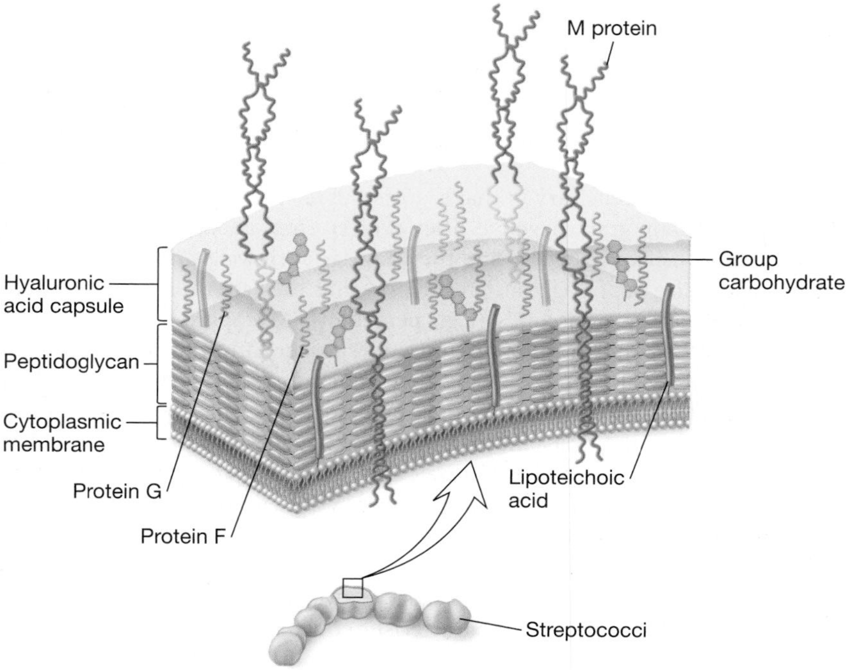

FIGURE 38.5 Streptococcal Cell Envelope. The M protein is a major virulence factor for streptococci. It facilitates bacterial attachment to host cells and has antiphagocytic activity. Protein G prevents attack by antibodies because it binds to the Fc portion, preventing the antigen-binding site from bacterial capture. Protein F is an epithelial cell attachment factor.

Figure 38.5 Micro Inquiry

How is the M protein also thought to be involved in poststreptococcal diseases?

S. pyogenes is widely distributed among humans; some people become asymptomatic carriers. Individuals with acute infections may spread the pathogen, and transmission can occur through respiratory droplets, as direct or indirect contact (**figure 38.6**). ◀◀ *Order* Lactobacillales *(section 21.3)*

Diagnosis of a streptococcal infection is based on both clinical and laboratory findings. Several rapid tests are available. Treatment is with penicillin or macrolide antibiotics. Vaccines are not available for streptococcal diseases other than streptococcal pneumonia because of the large number of serotypes. The best control measure is prevention of transmission. Individuals with a known infection should be isolated and treated. Personnel working with infected patients should follow standard aseptic procedures. Next, we discuss some of the more important human streptococcal diseases.

Streptococcal Pharyngitis

Streptococcal pharyngitis is one of the most common bacterial infections of humans and is commonly called strep throat (**figure 38.7**). The β-hemolytic, group A streptococci are spread by droplets of saliva or nasal secretions. The incubation period in humans is 2 to 4 days.

The action of the strep bacteria in the throat **(pharyngitis)** or on the tonsils **(tonsillitis)** stimulates an inflammatory response and the lysis of white and red blood cells. An inflammatory exudate consisting of cells and fluid is released from the blood vessels and deposited in the surrounding tissue, although only about 50% of patients with strep pharyngitis present with an exudate. This is accompanied by a general feeling of discomfort or malaise, fever, and headache. Prominent physical manifestations include redness, edema, and lymph node enlargement in the throat. Signs and symptoms alone are not diagnostic because viral infections have a similar presentation. Several common rapid test kits are available for diagnosing strep throat. In the absence of complications, the disease can be self-limiting and may disappear within a week. However, antibiotic treatment (penicillin G benzathine or a

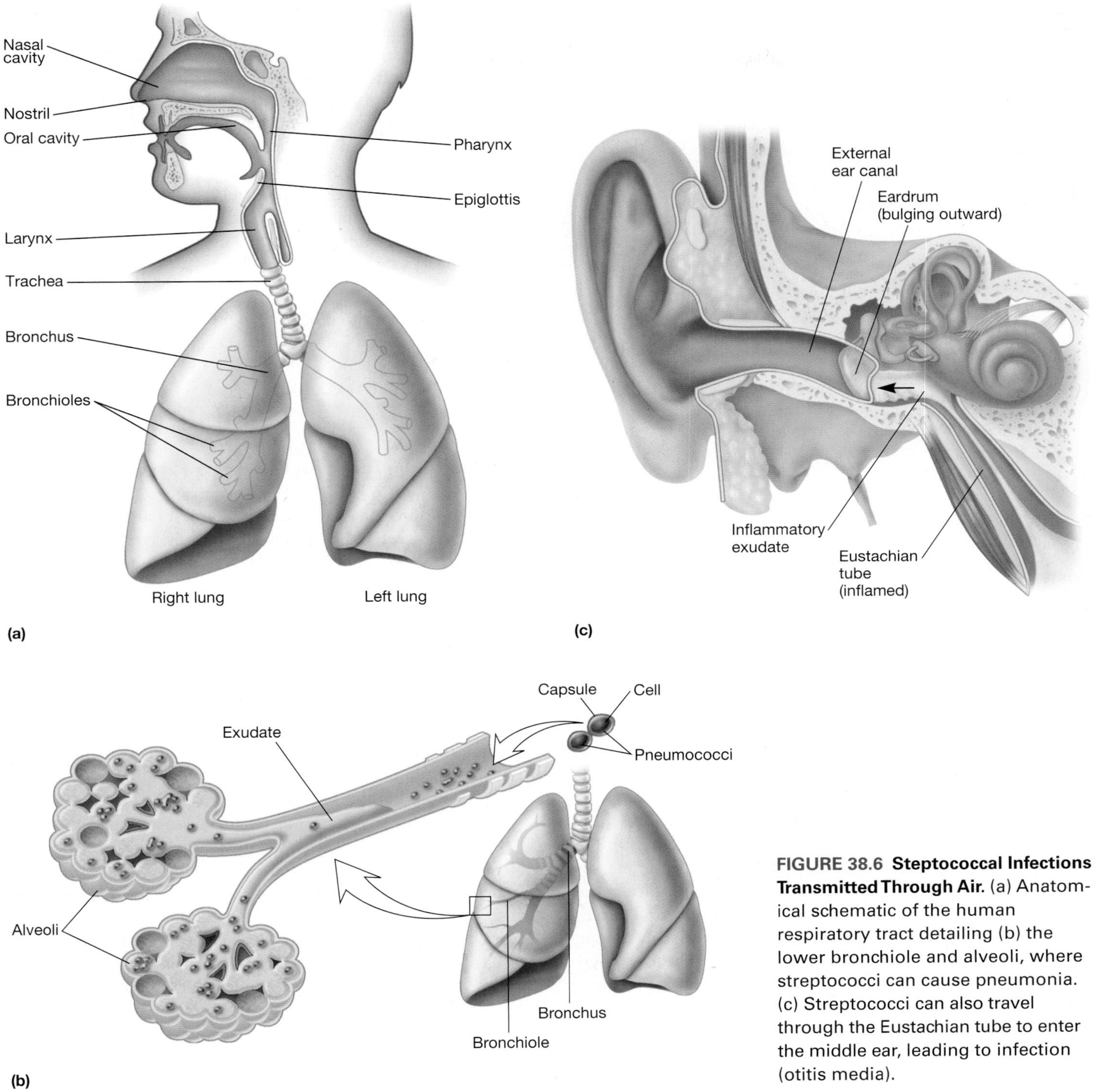

FIGURE 38.6 Steptococcal Infections Transmitted Through Air. (a) Anatomical schematic of the human respiratory tract detailing (b) the lower bronchiole and alveoli, where streptococci can cause pneumonia. (c) Streptococci can also travel through the Eustachian tube to enter the middle ear, leading to infection (otitis media).

macrolide antibiotic for penicillin-allergic people) can shorten the infection and clinical syndromes, and is especially important in children for the prevention of complications such as rheumatic fever and glomerulonephritis, as discussed in the next section. Infections in older children and adults tend to be milder and less frequent due in part to the immunity they have developed against the many serotypes encountered in early childhood.

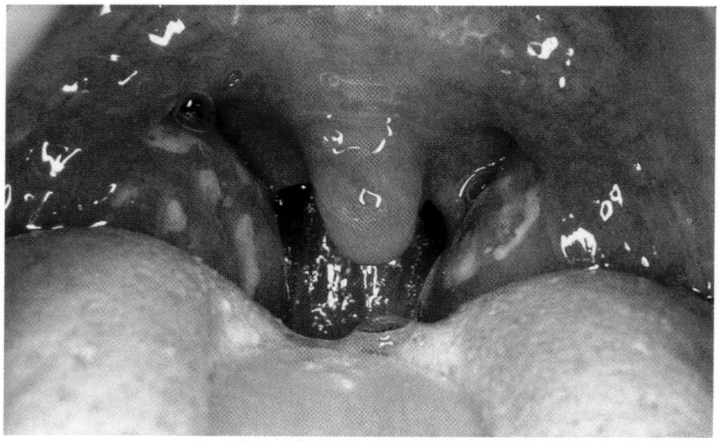

FIGURE 38.7 Streptococcal Pharyngitis (Strep Throat).

Figure 38.7 Micro Inquiry

What causes the exudate seen in strep throat? Is it seen in all cases?

Poststreptococcal Diseases

The poststreptococcal diseases are glomerulonephritis and rheumatic fever. They occur 1 to 4 weeks after an acute streptococcal infection (hence the prefix "post-"). These two nonsupporative (nonpus-producing) diseases are the most serious problems associated with streptococcal infections in the United States.

Glomerulonephritis, also called **Bright's disease,** is an inflammatory disease of the renal glomeruli—membranous structures within the kidney where blood is filtered. Damage probably results from the deposition of antigen-antibody complexes, possibly involving the streptococcal M protein, in the glomeruli. The complexes cause destruction of the glomerular membrane, allowing proteins and blood to leak into the urine. Clinically the affected person exhibits edema, fever, hypertension, and hematuria (blood in the urine). The disease occurs primarily among school-age children. Diagnosis is based on clinical history, physical findings, and confirmatory evidence of prior streptococcal infection. The incidence of glomerulonephritis in the United States is less than 0.5% of streptococcal infections. Penicillin G or erythromycin can be given for any residual streptococci. However, there is no specific therapy once kidney damage has occurred. About 80 to 90% of all cases undergo slow, spontaneous healing of the damaged glomeruli, whereas the others develop a chronic form of the disease. The latter may require a kidney transplant or lifelong renal dialysis. ◀◀ *Type III hypersensitivity (section 33.10)*

Rheumatic fever is an autoimmune disease characterized by inflammatory lesions involving the heart valves, joints, subcutaneous tissues, and central nervous system. The exact mechanism of rheumatic fever development remains unknown. However, it has been associated with prior streptococcal pharyngitis of specific M protein strains. The disease occurs most frequently among children six to fifteen years of age and manifests itself through a variety of signs and symptoms, making diagnosis difficult. In the United States, rheumatic fever has become very rare (less than 0.05% of streptococcal infections). It occurs 100 times more frequently in tropical countries. Therapy is directed at decreasing inflammation and fever, and controlling cardiac failure. Salicylates and corticosteroids are the mainstays of treatment. Although rheumatic fever is rare, it is still the most common cause of permanent heart valve damage in children.

1. Describe the streptococcal exotoxins and how they are involved in virulence.
2. Which streptococcal disease is most prevalent? Why do you think this is the case?

38.2 Arthropod-Borne Diseases

Although arthropod-borne bacterial diseases are generally rare, they are of interest either historically (plague, typhus) or because they have been newly introduced into humans (ehrlichiosis, Lyme disease). In the next sections, only diseases that occur in the United States are discussed.

Lyme Disease

Lyme disease (LD, Lyme borreliosis) was first observed and described in 1975 among people of Old Lyme, Connecticut. It has become the most common tick-borne zoonosis in the United States, with about 17,000 cases reported annually. The disease is also present in Europe and Asia.

The spirochetes responsible for this disease comprise at least three species. *Borrelia burgdorferi* appears to be the primary cause of Lyme disease in the United States, whereas *B. garinii*, and *B. afzelii* appear to cause the disease in Europe (**figure 38.8*a***). Rodents and other wild animals are the natural hosts; although adult deer ticks feed on deer, deer do not become infected. In the northeastern United States, *B. burgdorferi* is transmitted to humans by the bite of infected black-legged (or deer) ticks (*Ixodes scapularis*; figure 38.8*b*). On the Pacific Coast, especially in California, the reservoir is a dusky-footed woodrat and the tick *I. pacificus.* ◀◀ *Phylum* Spirochaetes *(section 19.6)*

Clinically, Lyme disease is a complex illness with three major stages. The initial, localized stage occurs a week to 10 days after an infectious tick bite. The illness often begins with an expanding, ring-shaped skin lesion with a red outer border and partial central clearing called erythema migrans (figure 38.8*c*). This often is accompanied by flulike symptoms (malaise and fatigue, headache, fever, and chills). Often the tick bite is unnoticed, and the skin lesion may be missed due to skin coloration or its obscure location, such as on the scalp. Thus treatment, which is usually

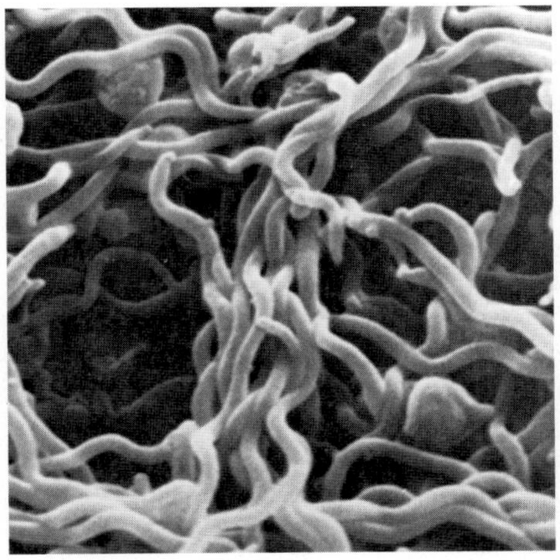

(a)

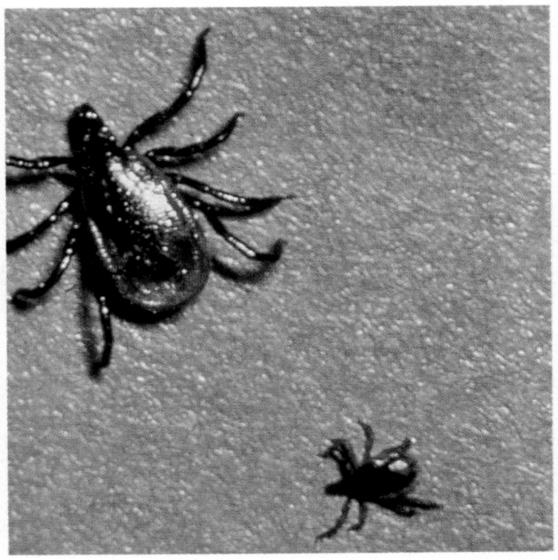

(b)

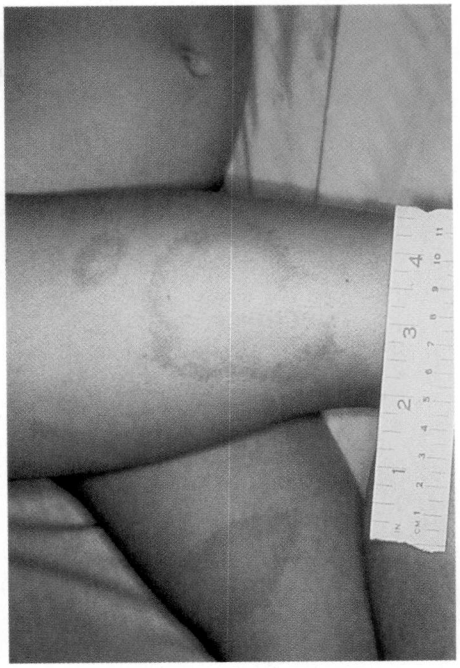

(c)

FIGURE 38.8 Lyme Disease. (a) The vector for *Borrelia burgdorferi* in the northeastern United States is the tick *Ixodes scapularis.* (b) An unengorged adult (bottom) is about the size of pinhead, and an engorged adult (top) can reach the size of a jelly bean. (c) The typical rash (erythema migrans) showing concentric rings around the initial site of the tick bite.

Figure 38.8 Micro Inquiry

What are the three stages of Lyme disease? What other spirochete disease does this resemble?

effective at this stage, may not be given because the illness is assumed to be "just the flu."

The second, disseminated stage may appear weeks or months after the initial infection. It consists of several symptoms such as neurological abnormalities, heart inflammation, and bouts of arthritis (usually in the major joints such as the elbows or knees). Current research indicates that Lyme arthritis might be an autoimmune response to major histocompatibility (MHC) molecules on cells in the synovium (joint) that are similar to the bacterial antigens. Inflammation that produces organ damage is initiated and possibly perpetuated by the immune response to one or more spirochetal proteins. Finally, like the progression of syphilis—another disease caused by a spirochete—the late stage may appear years later. Infected individuals may develop neuron demyelination with symptoms resembling Alzheimer's disease and multiple sclerosis. Behavioral changes can also occur. ◀◀ *Recognition of foreignness (section 33.4)*

Laboratory diagnosis of LD is based on (1) serological testing (Lyme ELISA or Western blot) for IgM or IgG antibodies to the pathogen, (2) detection of *Borrelia* DNA in patient specimens (especially synovial fluid) after amplification by PCR, and (3) recovery of the spirochete from patient specimens, although cultures are laborious with modest success. Treatment with amoxicillin or tetracycline early in the illness results in prompt recovery and prevents arthritis and other complications. If nervous system involvement is suspected, ceftriaxone is used because it can cross the blood-brain barrier. ◀◀ *Polymerase chain reaction (section 15.2)*

Prevention and control of LD involves environmental modifications such as clearing and burning tick habitat and the application of acaricidal compounds. The compounds destroy mites and ticks. An individual's risk of acquiring LD may be greatly reduced by education and personal protection.

**Plague

Plague (Latin *plaga*, pest) is caused by the gram-negative bacterium *Yersinia pestis*. It is transmitted from rodent to human by the bite of an infected flea, direct contact with infected animals or their products, or inhalation of contaminated airborne droplets (**figure 38.9*a***). Initially spread by contact with flea-infested animals, *Y. pestis* can spread among people by airborne transmission. Once in the human body, the bacteria multiply in the blood and lymph. Symptoms include subcutaneous hemorrhages, fever, chills, headache, extreme exhaustion, and the appearance of enlarged lymph nodes called **buboes** (hence the old name, **bubonic plague**) (figure 38.9*b*). In 50 to 70% of the untreated cases, death

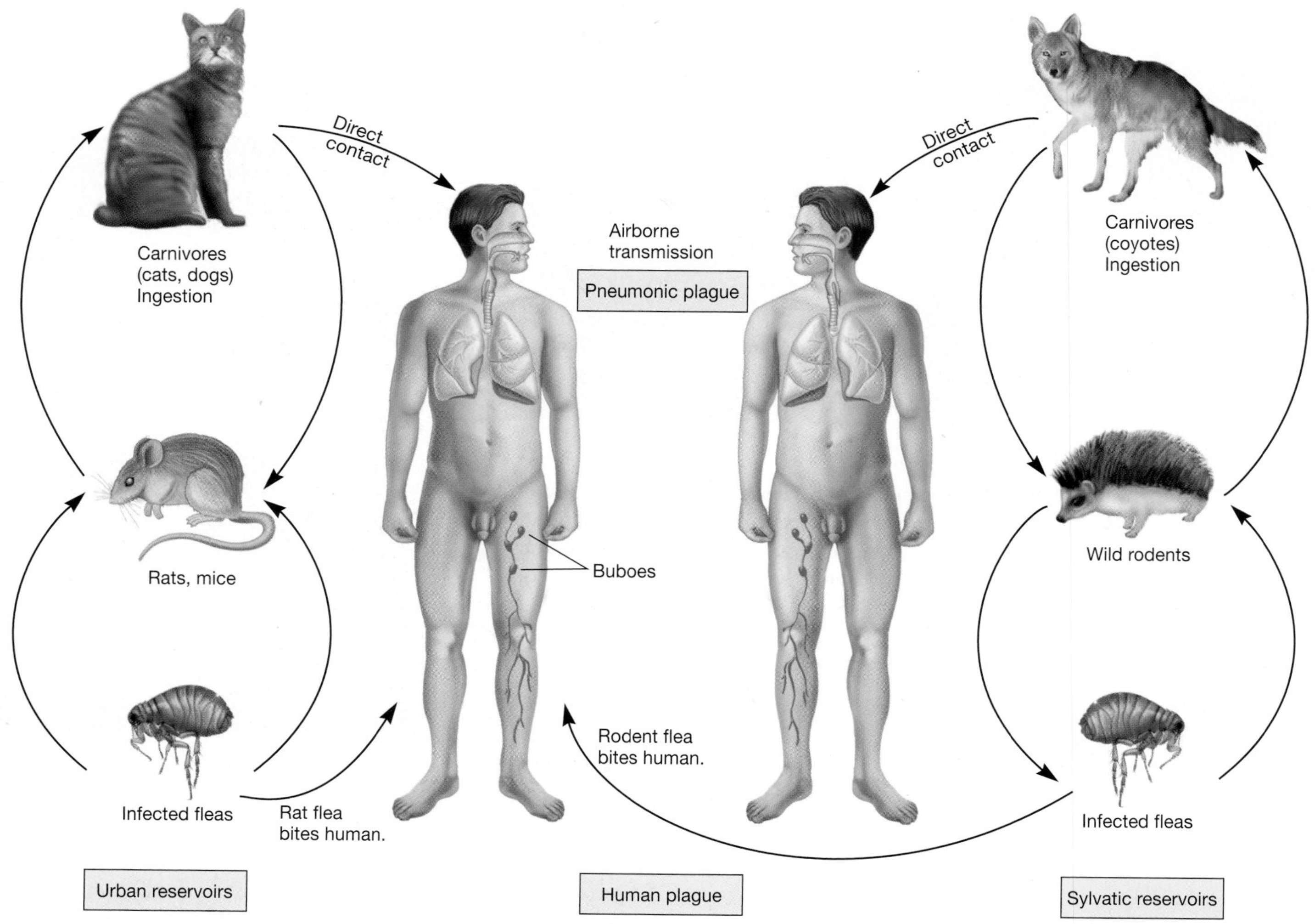

(a)

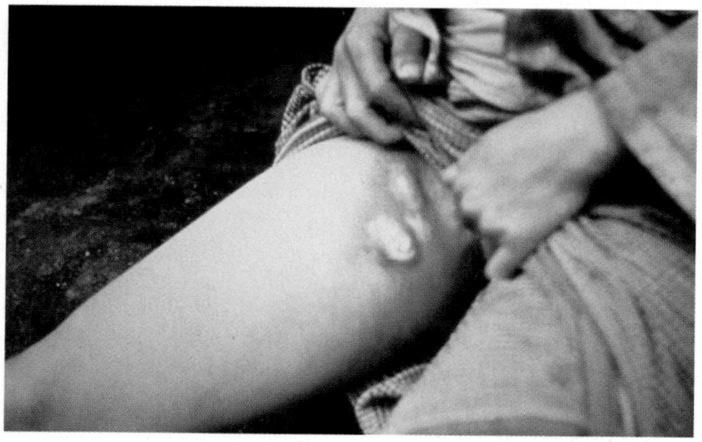

(b)

FIGURE 38.9 Plague. (a) Plague is spread to humans through (1) the urban cycle and rat fleas, (2) the sylvatic cycle, centering on wild rodents and their fleas, or (3) by airborne transmission from an infected person, leading to pneumonic plague. Dogs, cats, and coyotes can also acquire the bacterium by ingestion of infected animals. (b) Enlarged lymph nodes called buboes are characteristic of *Yersinia* infection.

follows in 3 to 5 days from toxic conditions caused by the large number of bacilli in the blood.

An important factor in the virulence of *Y. pestis* is its ability to survive and proliferate inside phagocytic cells, rather than being killed by them. One of the ways this is accomplished is by its modification of host cell behavior. Like other gram-negative pathogens, *Y. pestis* uses a type III secretion system to deliver effector proteins. *Y. pestis* secretes plasmid-encoded yersinal outer

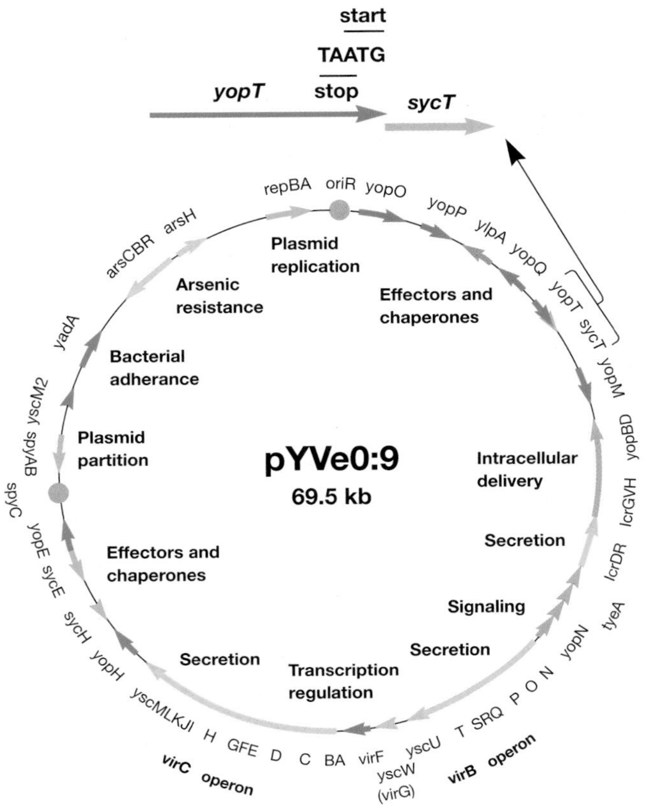

(a)

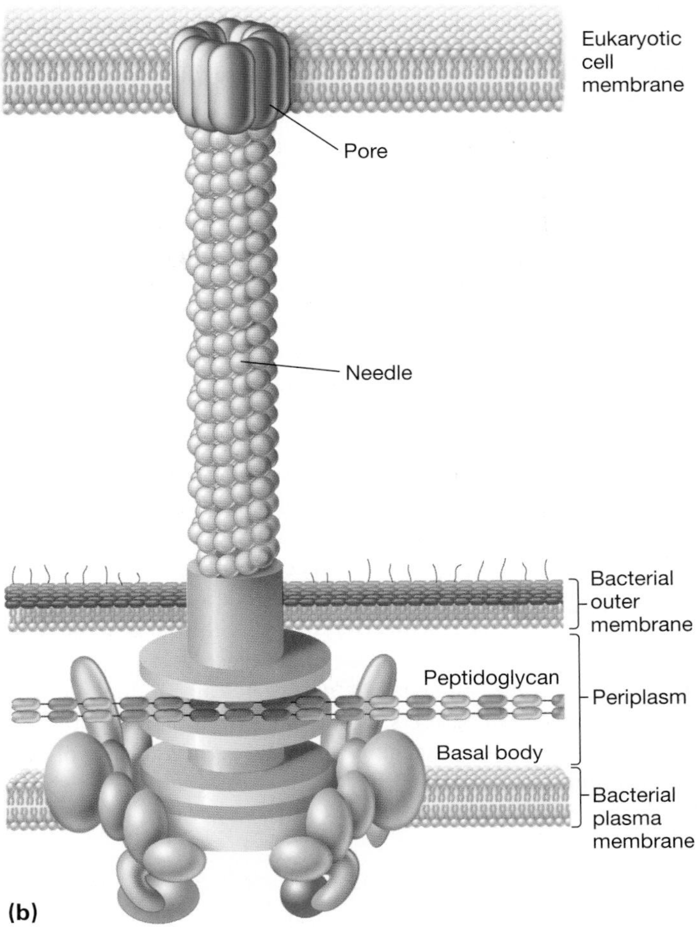

(b)

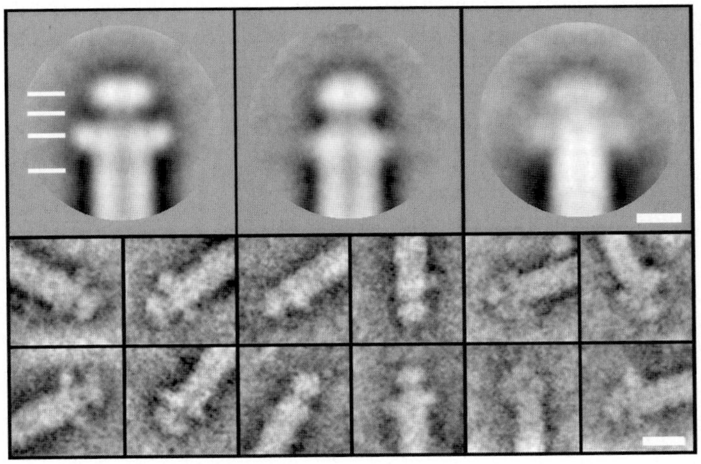
(c)

FIGURE 38.10 Type III Secretion System. (a) The type III secretion system (TTSS) and other virulence genes of *Yersinia* are encoded on the pYV plasmid. The TTSS genes encoding the *Yersinia* outer proteins (Yop) are homologous to many of the genes encoding flagellar proteins. (b) X-ray fiber diffraction studies resolve the injectisome as a helical structure. (c) Scanning tunneling electron microscopy reveals the injectisome tip, indicating how it may lock into the translocator pore on the target cell.

membrane proteins (YOPS) into phagocytic cells to counteract natural defense mechanisms and help the bacteria multiply and disseminate in the host (**figure 38.10**). ◀◀ *Order* Enterobacteriales *(section 20.3); Pathogenicity islands (section 31.3)*

In the southwestern part of the United States, plague occurs primarily in wild ground squirrels, chipmunks, mice, and prairie dogs. However, massive human epidemics occurred in Europe during the Middle Ages, where the disease was known as the

Black Death due to black-colored, subcutaneous hemorrhages. Infections now occur in humans only sporadically or in limited outbreaks. In the United States, 10 to 20 cases are reported annually; the mortality rate is about 14%.

Pneumonic plague arises (1) from primary exposure to infectious respiratory droplets from a person or animal with respiratory plague, or (2) secondary to blood-borne spread in a patient with bubonic or septicemic plague. Pneumonic plague can also arise from accidental inhalation of *Y. pestis* in the laboratory. The mortality rate for this kind of plague is almost 100% if it is not recognized within 12 to 24 hours. Obviously great care must be taken to prevent the spread of airborne infections to personnel who care for pneumonic plague patients.

Laboratory diagnosis of plague is made in reference labs, where direct microscopic examination, culture of the bacterium, and serological tests are used. Because the plague bacillus is listed as a select agent, the sentinel laboratory identification of *Y. pestis* is restricted to these few tests. Confirmation of the identity of *Y. pestis* by PCR, advanced serological studies, and other approved practices is restricted to national reference laboratories. Treatment is with streptomycin, chloramphenicol, or tetracycline, and recovery from the disease gives a good immunity. Prevention and control involve flea and rodent control, isolation of human patients, prophylaxis or abortive therapy of exposed persons, and vaccination (USP Plague vaccine) of persons at high risk.

Rocky Mountain Spotted Fever

Rocky Mountain spotted fever is caused by the rickettsia, *Rickettsia rickettsii*. Although this disease was originally detected in the Rocky Mountain area, most cases now occur east of the Mississippi River. The disease is transmitted by ticks and usually occurs in people who are or have been in tick-infested areas. There are two principal vectors: *Dermacentor andersoni*, the wood tick, which is distributed in the Rocky Mountain states and is active during the spring and early summer; and *D. variabilis*, the dog tick, which has assumed greater importance and is almost exclusively confined to the eastern half of the United States. Unlike other rickettsias, *R. rickettsii* can pass from generation to generation of ticks through their eggs in a process known as **transovarian passage.** No humans or mammals are needed as reservoirs for the continued propagation of this rickettsia in the environment.

When humans contact infected ticks, the rickettsias are deposited either on the skin (if the tick defecates after feeding) and then subsequently rubbed or scratched into the skin, or into the skin as the tick feeds. Once inside the skin, the rickettsias enter the endothelial cells of small blood vessels, where they multiply and produce a characteristic vasculitis. The bacteria induce phagocytosis by macrophages responding to the infection. The bacteria escape the phagosome and reproduce in the cytoplasm. They continue to reproduce until the host cell bursts, releasing bacteria to infect other host cells. Rickettsias are very different from most other bacteria in physiology and metabolism. Described as an energy parasite, rickettsias have a membrane carrier protein that exchanges internal ADP for external ATP. Thus bacterial growth and reproduction continues at the expense of the host cell. ◀◀ *Phagocytosis (section 32.5)*

Rocky Mountain spotted fever is characterized by the sudden onset of a headache, high fever, chills, and a skin rash (**figure 38.11**) that initially appears on the ankles and wrists, and then spreads to the trunk of the body. The symptoms reflect the accumulation of lysed host cells and toxic effects of rickettsial cell walls. If the disease is not treated, the rickettsias can destroy the blood vessels in the heart, lungs, or kidneys and cause death. Usually, however, severe pathological changes are avoided by antibiotic therapy (chloramphenicol, chlortetracycline), the development of immune resistance, and supportive therapy. Diagnosis is made through observation of symptoms and signs such as the characteristic rash and by serological tests. The best means of prevention remains the avoidance of tick-infested habitats and animals. There are roughly 400 to 800 reported cases of Rocky Mountain spotted fever annually in the United States.

1. Why are two antibiotics used to treat most rickettsial infections?
2. What is the causative agent of Lyme disease and how is it transmitted to humans? What preventative measures can an individual take to avoid infection?
3. Why was plague called the Black Death? How is it transmitted? Distinguish between bubonic and pneumonic plague.
4. Describe the symptoms of Rocky Mountain spotted fever.
5. How does transovarian passage occur?

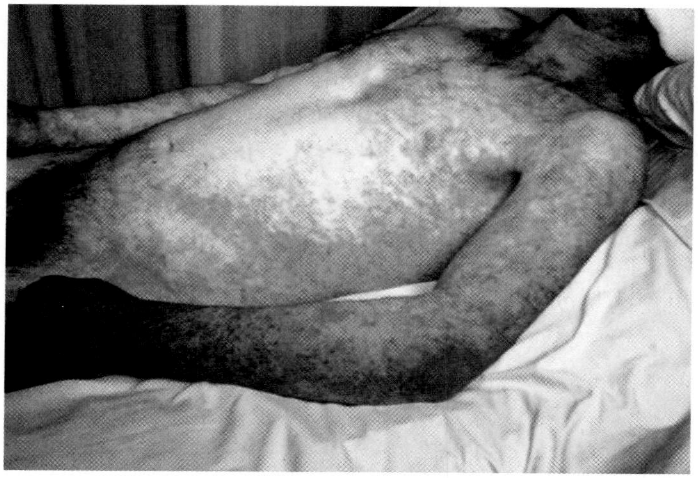

FIGURE 38.11 Rocky Mountain Spotted Fever. Typical rash occurring on the arms and chest consists of generally distributed, sharply defined macules.

38.3 Direct Contact Diseases

Most of the direct contact bacterial diseases involve the skin or underlying tissues. Others can become disseminated through specific regions of the body. Some of the better-known of these diseases are now discussed.

Gas Gangrene (Clostridial Myonecrosis)

Clostridium perfringens, C. novyi, and *C. septicum* are gram-positive, endospore-forming rods termed the histotoxic clostridia. They can produce a necrotizing infection of skeletal muscle called **gas gangrene** (Greek *gangraina*, an eating sore) or **clostridial myonecrosis** (*myo*, muscles, and *necrosis*, death); however, *C. perfringens* is the most common cause. Analysis of the *C. perfringens* genome sequence reveals that the microbe possesses the genes for fermentation with gas production but lacks genes encoding enzymes for the TCA cycle or a respiratory chain. Nonetheless, *C. perfringens* has an extraordinary doubling time of only 8 to 10 minutes when in the human host, making diagnosis and treatment of gangrene critical. ◀◀ *Class* Clostridia *(section 21.2)*

Histotoxic clostridia occur in the soil worldwide and also are part of the normal endogenous microflora of the human large intestine. Contamination of injured tissue with spores from soil containing histotoxic clostridia or bowel flora is the usual means of transmission. Infections are commonly associated with wounds resulting from abortions, automobile accidents, military combat, or frostbite. If the spores germinate in anoxic tissue, the bacteria grow and secrete α-toxin (a lecithinase), which breaks down muscle tissue by disrupting cell membranes and causing cell lysis. Growth often results in the accumulation of gas (mainly hydrogen as a result of carbohydrate fermentation) and of the toxic breakdown products of skeletal muscle tissue (**figure 38.12**). ◀◀ *Exotoxins (section 31.3)*

Clinical manifestations of gas gangrene include severe pain, edema, drainage, and muscle necrosis. The pathology arises from progressive skeletal muscle necrosis due to the effects of the toxin. Other enzymes produced by the bacteria degrade collagen and tissue, facilitating spread of the disease.

Gas gangrene is a medical emergency. Laboratory diagnosis is through recovery of the appropriate species of clostridia accompanied by the characteristic disease symptoms. Treatment is extensive surgical debridement (removal of all dead tissue), the administration of antitoxin, and antimicrobial therapy with penicillin and tetracycline. Hyperbaric oxygen therapy (the use of high concentrations of oxygen at elevated pressures) also is considered effective. The oxygen saturates the infected tissue and thereby prevents the growth of the obligately anaerobic clostridia. Prevention and control include debridement of contaminated traumatic wounds plus antimicrobial prophylaxis and prompt treatment of all wound infections. Amputation of limbs often is necessary to prevent further spread of the disease.

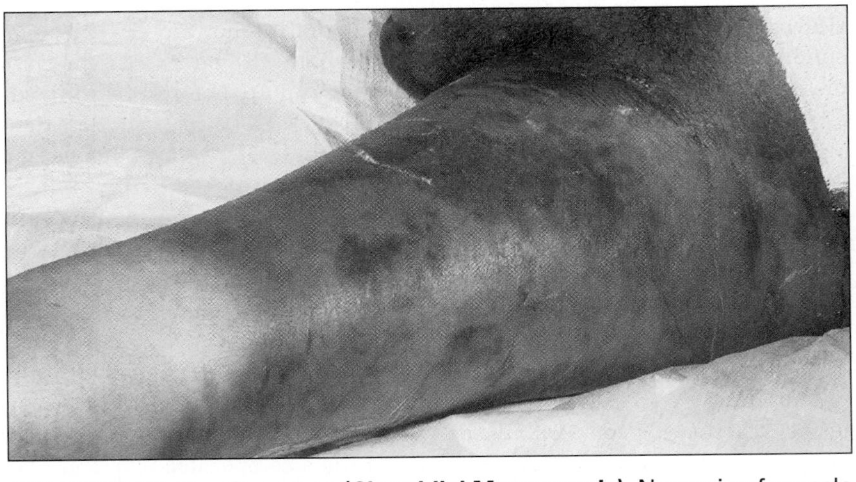

FIGURE 38.12 Gas Gangrene (Clostridial Myonecrosis). Necrosis of muscle and other tissues results from the numerous toxins produced by *Clostridium perfringens.*

Group B Streptococcal Disease

Streptococcus agalactiae, or **group B streptococcus (GBS),** is a gram-positive bacterium that causes illness in newborn babies, pregnant women, the elderly, and adults compromised by other severe illness. GBS is the most common cause of sepsis and meningitis in newborns, a frequent cause of newborn pneumonia, and a common cause of life-threatening neonatal infections. GBS in neonates is more common than rubella, congenital syphilis, and respiratory syncytial virus, for example. Nearly 75% of infected newborns present symptoms in the first week of life; premature babies are more susceptible. Most GBS cases are apparent within hours after birth. This is called "early-onset disease." GBS can also develop in infants who are one week to several months old ("late-onset disease"); however, this is very rare. Meningitis is more common with late-onset disease. Interestingly, about half of the infants with late-onset GBS disease are associated with a mother who is a GBS carrier; the source of infection for the other newborns with late-onset GBS disease is unknown. Infant mortality due to GBS disease is about 5%. Babies who survive GBS disease, particularly those with meningitis, may have permanent disabilities, such as hearing or vision loss or developmental disabilities. The Centers for Disease Control and Prevention (CDC) reports 20,000 cases of invasive GBS in 2007, with 1,475 deaths.

GBS is transmitted directly from person to person. Many people are asymptomatic GBS carriers—they are colonized by GBS but do not become ill from it. Adults can carry GBS in the bowel, vagina, bladder, or throat. Twenty to 25% of pregnant women carry GBS in the rectum or vagina. Thus a fetus may be exposed to GBS before or during birth if the mother is a carrier. GBS carriers appear to be so temporarily—that is, they do not harbor the bacteria for life.

GBS disease is diagnosed when the gram-positive, β-hemolytic, streptococcal bacterium is grown from cultures of otherwise sterile body fluids, such as blood or spinal fluid. GBS cultures may take 24 to 72 hours for strong growth to appear. Confirmation of GBS is by latex agglutination immunoassay. An FDA-approved DNA test for GBS is also available if rapid diagnosis of disease is necessary or culture results are equivocal.

GBS carriage can be detected during pregnancy by the presence of the bacterium in culture specimens of both the vagina and the rectum. The CDC recommends that pregnant women have vaginal and rectal specimens cultured for GBS in late pregnancy (35 to 37 weeks' gestation) to accurately predict whether GBS colonization needs to be treated prior to delivery. Antibiotic chemotherapy is mostly effective. GBS infections in both newborns and adults are usually treated with penicillin or ampicillin, unless resistance or hypersensitivity is indicated. A GBS vaccine based on genomic analysis has been under development. However, GBS vaccine development is problematic due to antigen-shifting serotypes; concerns over the impact of vaccination during pregnancy and potential litigation resulting from this practice have resulted in a declining interest in GBS vaccine development. ⏮ *Comparative genomics (section 16.7)*

A positive culture result means that the mother carries GBS—not that she or her baby will definitely become ill. Women who carry GBS are not given oral antibiotics before labor because antibiotic treatment at this time does not prevent GBS disease in newborns. An exception to this is when GBS is identified in urine during pregnancy. GBS in the urine indicates an active infection and is treated at the time it is diagnosed. Carriage of GBS, in either the vagina or rectum, becomes important at the time of labor and delivery or any time the placental membrane ruptures. At this time, antibiotics are effective in preventing the spread of GBS from mother to baby. GBS carriers at highest risk are those with any of the following conditions: fever during labor, rupture of membranes 18 hours or more before delivery, or labor or membrane rupture before 37 weeks.

Mycobacterial Skin Infections

Many pathogenic species of nontuberculous mycobacteria cause skin or soft-tissue infections. These diseases often are hard to diagnose as they may cause diverse and misleading signs and symptoms. Two skin-associated diseases caused by nontuberculous mycobacteria are presented here. Leprosy has a reputation from history that caused people with the disease to be shunned, rather than helped. We discuss leprosy because it is still a complicated disease that often stigmatizes its victims. In contrast, we also discuss an emerging mycobacterial infection associated with the recent trend of tattoo art.

Leprosy

Leprosy (Greek *lepros*, scaly, scabby, rough), or Hansen's disease, is a severely disfiguring skin disease caused by *Mycobacterium leprae*. The only reservoirs of proven significance are humans. The disease most often occurs in tropical countries, where there are more than 11 million cases. In 2002 there were approximately 764,000 new cases of leprosy worldwide. An estimated 4,000 cases exist in the United States, with approximately 100 new cases reported annually. Transmission of leprosy is most likely to occur when individuals are exposed for prolonged periods to infected individuals who shed large numbers of *M. leprae*. Nasal secretions probably are the infectious material for family contacts. ⏮ *Suborder* Corynebacterineae *(section 22.4)*

The incubation period for leprosy is about 3 to 5 years but may be much longer, and the disease progresses slowly. The bacterium invades peripheral nerve and skin cells, and becomes an obligately intracellular parasite. It is most frequently found in the Schwann cells that surround peripheral nerve axons and in mononuclear phagocytes. The earliest symptom of leprosy is usually a slightly pigmented skin eruption several centimeters in diameter. Approximately 75% of all individuals with this early, solitary lesion heal spontaneously because of the cell-mediated immune response to *M. leprae*. However, in some individuals, this immune response may be so weak that one of two distinct forms of the disease occurs: tuberculoid or lepromatous leprosy (**figure 38.13**).

Tuberculoid (neural) leprosy is a mild, nonprogressive form of leprosy associated with a delayed-type hypersensitivity reaction to antigens on the surface of *M. leprae*. It is characterized by damaged nerves and regions of the skin that have lost sensation and are surrounded by a border of nodules (**figure 38.14**). Afflicted individuals who do not develop hypersensitivity have a relentlessly progressive form of the disease, called **lepromatous (progressive) leprosy,** in which large numbers of *M. leprae* develop in the skin cells. The bacteria kill skin tissue, leading to a progressive loss of facial features, fingers, toes, and other structures. Moreover, disfiguring nodules form all over the body. Nerves are also infected but usually are less damaged than occurs in tuberculoid leprosy. ⏮ *Type IV hypersensitivity (section 33.10)*

Because the leprosy bacillus cannot be cultured in vitro, laboratory diagnosis is supported by the demonstration of the bacterium in biopsy specimens using direct fluorescent antibody staining. Serodiagnostic methods, such as the fluorescent leprosy antibody absorption test and ELISA, have been developed.

Treatment is long-term with the sulfone drug diacetyl dapsone, which acts by inhibiting the synthesis of dihydrofolic acid through competition with para-amino-benzoic acid (*see figure 34.12*, and rifampin, with or without clofazimine. Alternative drugs are ethionamide or protionamide. Use of *Mycobacterium* vaccine in conjunction with the drugs shortens the duration of drug therapy and speeds recovery from the disease.

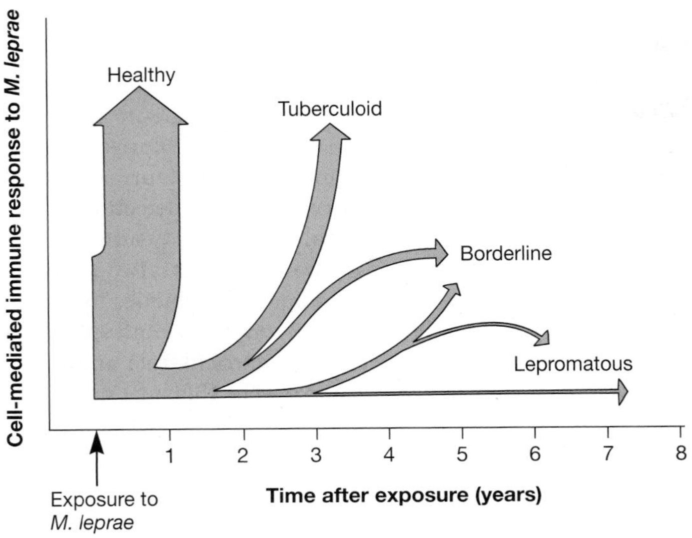

FIGURE 38.13 Development of Leprosy. A schematic representation of a hypothesis explaining the relationship between development of subclinical infection and various aspects of leprosy to the time of onset of cell-mediated immune response after initial exposure. The thickness of the lines indicates the proportion of individuals from the exposed population that is likely to be in each category.

***Figure 38.13** Micro Inquiry*

What are the chief differences between tuberculoid and lepromatous leprosy?

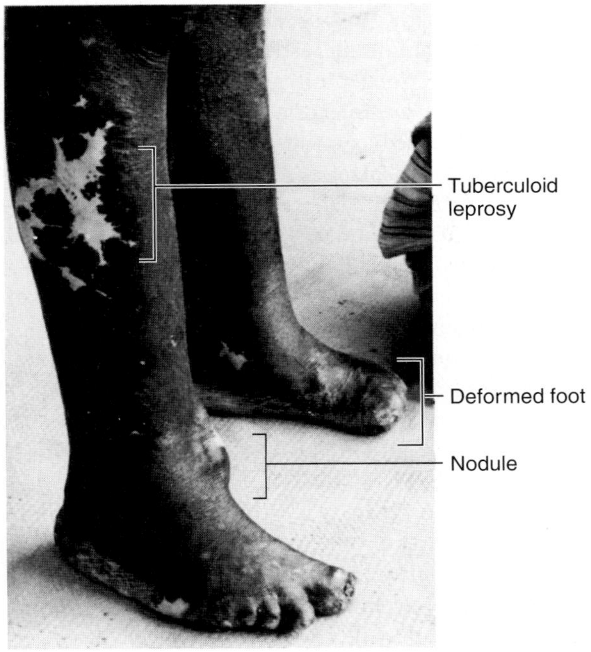

FIGURE 38.14 Leprosy. In tuberculoid leprosy, the skin manifests shallow, painless lesions.

There is good evidence that the nine-banded armadillo is an animal reservoir for the leprosy bacillus in the United States but plays no role in transmission of leprosy to humans. Identification and treatment of patients with leprosy is the key to control. Children of presumably contagious parents should be given chemoprophylactic drugs until treatment of their parents has made them noninfectious.

Tattoo-Associated Mycobacterial Infections

A recently described community outbreak of tattoo-associated mycobacterial infections reminds us that emerging diseases don't have to be exotic. *Mycobacterium chelonae* was recently identified as the elusive agent responsible for several cases of tattoo-associated infection. The index case was a forty-nine-year-old man who had over 100 small pustules and reddened papules within two-week-old tattoos. Examination of biopsy specimens were negative for bacteria, including acid-fast bacilli. Culture of the specimen revealed the rapidly growing mycobacterium. Five additional cases were also reported that had a range of skin lesions, including pink, red, and purple papules; scaly papules; granulomatous papules; lichenoid papules; pustules; and plaques. The common link to all six cases was the inking of tattoos by the same artist; all of the lesions were found to be concentrated in "gray wash" (areas of shading using grey tones) within the tattoos. The epidemiology of the outbreak was unknown at the time of this writing. However, an alert to physicians was posted to assist with any tattoo-associated skin infections that fail to respond to routine treatment.

Laboratory diagnosis for rapidly growing mycobacterial organisms is by culture and routine bacterial staining. Antibiotic sensitivity testing is also important when weighing treatment options.

Treatment of *M. chelonae* is successful with macrolide antibiotics (e.g., clarithromycin or azithromycin), orally and topically, but may be for as long as six months. Topical steroids may also be indicated to control the lesions.

1. How can humans acquire gas gangrene? Group B strep? Leprosy? Describe the major symptoms of each.
2. Why do you think the slow growth rate of *M. leprae* makes leprosy treatment more difficult?

Peptic Ulcer Disease and Gastritis

Peptic ulcer disease is caused by the gram-negative, spiral bacillus *Helicobacter pylori*. Found in gastric biopsy specimens from patients with histologic gastritis, *H. pylori* is responsible for most cases of chronic **gastritis** not associated with another known primary cause (e.g., autoimmune gastritis or eosinophilic gastritis). In addition, there are strong, positive correlations between gastric cancer rates and *H. pylori* infection rates in certain populations, and it has been classified as a Class I carcinogen by the WHO. ◀◀ *Class* Epsilonproteobacteria *(section 20.5)*

H. pylori colonizes only gastric mucus-secreting cells, beneath the gastric mucous layers, and surface fimbriae are believed to be one of the adhesins associated with this process (**figure 38.15**). *H. pylori* binds to Lewis B antigens. These antigens are part of the blood group antigens that determine blood group O. *H. pylori* also binds to the monosaccharide sialic acid, which is found in the glycoproteins on the surface of gastric epithelial cells. The bacterium moves into the mucous layer to attach to mucus-secreting cells. Movement into the mucous layer may be aided by the fact that *H. pylori* is a strong producer of urease. Urease activity is thought to create a localized alkaline environment when hydrolysis of urea produces ammonia. The increased pH may protect the bacterium from gastric acid until it is able to grow under the layer of mucus in the stomach. The potential virulence factors responsible for epithelial cell damage and inflammation probably include proteases, phospholipases, cytokines, and cytotoxins.

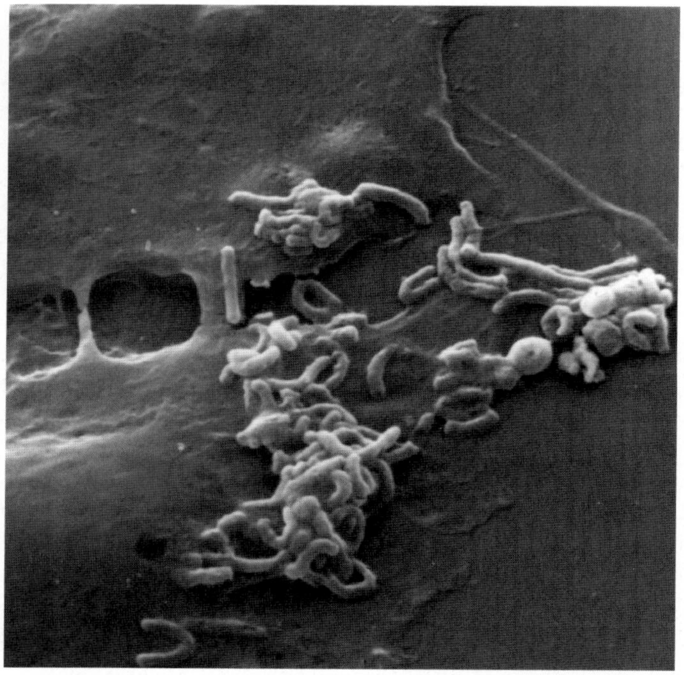

FIGURE 38.15 Peptic Ulcer Disease. Scanning electron micrograph (×3,441) of *Helicobacter pylori* adhering to gastric cells.

Figure 38.15 Micro Inquiry

How does *H. pylori* increase the local pH in its gastric microenvironment?

Laboratory identification of *H. pylori* is by culture of gastric biopsy specimens, examination of stained biopsies for the presence of bacteria, detection of serum IgG (Pyloriset EIA-G, Malakit *Helicobacter pylori*), the urea breath test, urinary excretion of (^{15}N) ammonia, stool antigen assays, or detection of urease activity in the biopsies. The goal of *H. pylori* treatment is the complete elimination of the organism. Treatment is two-pronged: use of drugs to decrease stomach acid and antibiotics to kill the bacteria. Three regimes are commonly used: bismuth subsalicylate (Pepto-Bismol) combined with metronidazole and either tetracycline or amoxicillin; clarithromycin (Biaxin), ranitidine, and bismuth citrate; or clarithromycin, amoxicillin, and lansoprazole (Prevacid).

Approximately 50% of the world's population is estimated to be infected with *H. pylori*. *H. pylori* is most likely transmitted from person to person (usually acquired in childhood), although infection from a common exogenous source cannot be completely ruled out, and some think that it is spread by food or water. Support for the person-to-person transmission comes from evidence of clustering within families and from reports of higher than expected prevalences in residents of custodial institutions and nursing homes.

Sexually Transmitted Diseases

Sexually transmitted diseases (STDs) represent a worldwide public health problem. The various viruses that cause STDs are presented in chapter 37, the responsible bacteria in this chapter, and the yeasts and protozoa in chapter 39. **Table 38.1** lists the bacteria that can be sexually transmitted and the diseases they cause.

The spread of most STDs is currently out of control. The WHO estimates that 300 million new cases of sexually transmitted diseases occur annually, with the predominant number of infections in fifteen-to-thirty-year-old individuals. In the United States, STDs remain a major public health challenge. The *2007 STD Surveillance Report* published by the CDC indicated that while substantial progress has been made in preventing, diagnosing, and treating certain STDs, 19 million new infections occur each year, nearly half of them among people aged fifteen to twenty-four. STDs also exact a tremendous economic toll, with direct medical costs estimated at $13 billion annually in the United States alone. Many cases of STDs go undiagnosed; others are not reported at all. This is particularly problematic as women with untreated chlamydia or gonorrhea infections risk infertility, and both men and women with papillomavirus infection have a greater than normal risk of cancer.

STDs were formerly called venereal diseases (from Venus, the Roman goddess of love) and may sometimes be referred to as sexually transmitted infections (STIs). Although they occur most frequently in the most sexually active age group—fifteen to thirty years of age—anyone who has sexual contact with an infected individual is at increased risk. In general, the more

Table 38.1 Summary of the Major Sexually Transmitted Diseases (STDs) Caused by Bacteria

Microorganism	*Disease*	*Comments*	*Treatment*
Calymmatobacterium granulomatis	Granuloma inguinale (donovanosis)	Rare in the U.S.; draining ulcers can persist for years	Tetracycline, erythromycin, newer quinolones
Chlamydia trachomatis	Nongonococcal urethritis (NGU); cervicitis, pelvic inflammatory disease (PID), lymphogranuloma venereum	Serovars D-K cause most of the STDs in the U.S.; lymphogranuloma venereum rare in the U.S.	Tetracyclines, erythromycin, doxycycline, ceftriaxone
Haemophilus ducreyi	Chancroid ("soft chancre")	Open sores on the genitals can lead to scarring without treatment; on the rise in the U.S.	Erythromycin or ceftriaxone
Helicobacter cinaedi, H. fennelliae	Diarrhea and rectal inflammation in men who have sex with other men	Common in immunocompromised individuals	Metronidazole, macrolides
Mycoplasma genitalium	Implicated in some cases of NGU	Only recently described as an STD	Tetracyclines or erythromycin
Mycoplasma hominis	Implicated in some cases of PID	Widespread, often asymptomatic but can cause PID in women	Tetracyclines or erythromycin
Neisseria gonorrhoeae	Gonorrhea, PID	Most commonly reported STD in the U.S.; usually symptomatic in men and asymptomatic in women; antibiotic-resistant strains	Third-generation cephalosporins and/or quinolones
Treponema pallidum subsp. *pallidum*	Syphilis, congenital syphilis	Manifests many clinical syndromes	Benzathine penicillin G
Ureaplasma urealyticum	Urethritis	Widespread, often asymptomatic but can cause PID in women and NGU in men; premature birth	Tetracyclines or erythromycin

sexual partners a person has, the more likely the person will acquire an STD.

As noted in previous chapters, some of the microorganisms that cause STDs can also be transmitted by nonsexual means. Examples include transmission by contaminated hypodermic needles and syringes shared among intravenous drug users and transmission from infected mothers to their infants. Some STDs can be cured quite easily, but others are presently difficult or impossible to cure. Because treatments are often inadequate, prevention is essential. Preventive measures are based mainly on better education of the total population and, when possible, control of the sources of infection and treatment of infected individuals with chemotherapeutic agents.

Gonorrhea

Gonorrhea (Greek *gono*, seed, and *rhein*, to flow) is an acute, infectious, sexually transmitted disease of the mucous membranes of the genitourinary tract, eye, rectum, and throat (table 38.1). It is caused by the gram-negative, diplococcus *Neisseria gonorrhoeae*. These bacteria are also referred to as **gonococci** (pl. of gonococcus; Greek *gono*, seed, and *coccus*, berry) and have a worldwide distribution. In 2007 approximately 356,000 cases were reported in the United States; the actual incidence is significantly higher, although the reported rates have remained steady for 10 years. ◀◀ *Order* Neisseriales *(section 20.2)*

Once inside the body, gonococci attach to the microvilli of mucosal cells by means of pili and protein II, which function as adhesins. This attachment prevents the bacteria from being washed away by normal cervical and vaginal discharges or by the flow of urine. They are often phagocytosed by the mucosal cells and may even be transported through the cells to the intercellular spaces and subepithelial tissue. The host tissue responds locally by the infiltration of immune cells, including mast cells, neutrophils, and antibody-secreting plasma cells. Neutrophils may contain gonococci inside vesicles as they can withstand the harsh phagosomal environment (**figure 38.16*a***). Because the gonococci are intra-cellular at this time, the host's defenses have little effect on the bacteria. Neutrophils are later replaced by fibrous tissue that may lead to urethral closing, or stricture, in males. ◀◀ *Cells, tissues, and organs of the immune system (section 32.4); Phagocytosis (section 32.5)*

In males the incubation period is 2 to 8 days. The onset consists of a urethral discharge of yellow, creamy pus, and frequent, painful urination that is accompanied by a burning sensation. In females, the cervix is the principal site infected. The disease is

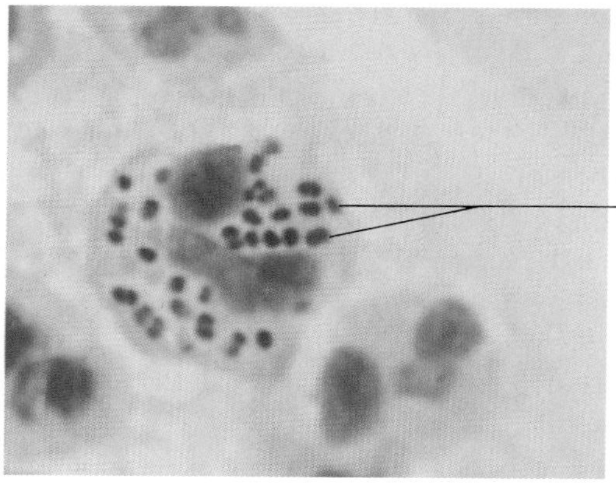

(a)

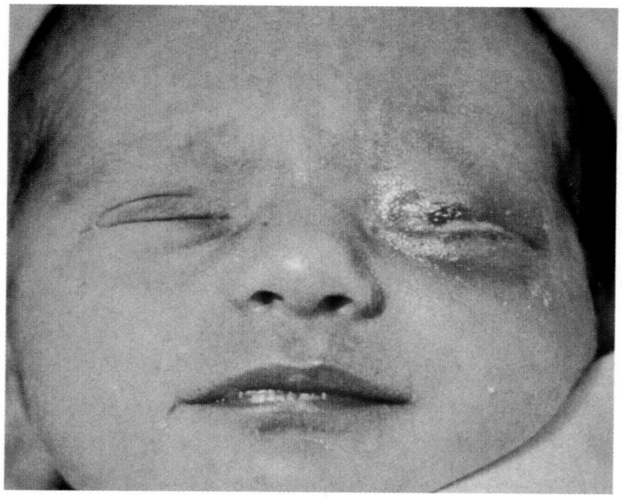

(b)

FIGURE 38.16 Gonorrhea. (a) Gram stain of male urethral exudate showing *Neisseria gonorrhoeae* (diplococci) inside a PMN; light micrograph (×500). Although the presence of gram-negative diplococci in exudates is a probable indication of gonorrhea, the bacterium should be isolated and identified. (b) Gonococcal opthalmia neonatorum in a one-week-old infant.

more insidious in females, and few individuals are aware of any symptoms. However, some symptoms may begin 7 to 21 days after infection. These are generally mild; some vaginal discharge may occur. The gonococci also can infect the Fallopian tubes and surrounding tissues, leading to **pelvic inflammatory disease (PID).** This occurs in 10 to 20% of infected females. Gonococcal PID is a major cause of sterility and ectopic pregnancies because of scar formation in the Fallopian tubes. Gonococci disseminate most often during menstruation, a time in which there is an increased concentration of free iron available to the bacteria. In both genders, disseminated gonococcal infection with bacteremia may occur. This can lead to involvement of the joints (gonorrheal arthritis), heart (gonorrheal endocarditis), or pharynx (gonorrheal pharyngitis).

Gonorrheal eye infections can occur in newborns as they pass through an infected birth canal. The resulting disease is called **ophthalmia neonatorum,** or **conjunctivitis of the newborn** (figure 38.16*b*). This was once a leading cause of blindness in many parts of the world. To prevent this disease as well as chlamydial conjunctivitis, tetracycline, erythromycin, povidone-iodine, or silver nitrate in dilute solution is placed in the eyes of newborns. This type of treatment is required by law in the United States and many other nations.

Laboratory diagnosis of gonorrhea relies on the successful growth of *N. gonorrhoeae* in culture to determine oxidase reaction, Gram-stain reaction, and colony and cell morphology. The performance of confirmation tests also is necessary. Because gonococci are very sensitive to adverse environmental conditions and survive poorly outside the body, specimens should be plated directly; when this is not possible, special transport media are necessary. A DNA probe (Gen-Probe Pace) for *N. gonorrhoeae* has been developed and is used to supplement other diagnostic techniques.

The CDC recommends single doses of ceftriaxone or cefixime to eradicate uncomplicated gonorrhea infection. Penicillin-resistant and, more recently, fluoroquinolone-resistant strains of gonococci are prevalent worldwide. Resistance of *N. gonorrhea* to penicillins, tetracyclines, spectinomycin, and fluoroquinolones is chromosomally mediated. Penicillin and tetracycline resistance is plasmid-mediated as well.

Nongonococcal Urethritis

Nongonococcal urethritis (NGU) is any inflammation of the urethra not due to the bacterium *Neisseria gonorrhoeae.* This condition is caused both by nonmicrobial factors such as catheters and drugs, and by infectious microorganisms. The most important causative bacterial agents are *Chlamydia trachomatis, Ureaplasma urealyticum,* and *Mycoplasma hominis.* Most infections are acquired sexually, and of these, approximately 50% are *Chlamydia* infections. NGU is endemic throughout the world, with an estimated 10 million Americans infected.

Symptoms of NGU vary widely. Males may have few or no manifestations of disease; however, complications can exist. These include a urethral discharge, itching, and inflammation of the male reproductive structures. Females may be asymptomatic or have a severe infection leading to pelvic inflammatory disease (**figure 38.17**). *Chlamydia* may account for as many as 200,000 to 400,000 cases of PID annually in the United States. In pregnant females, chlamydial infections are especially serious because they are directly related to miscarriage, stillbirth, inclusion conjunctivitis, and infant pneumonia.

Syphilis

Venereal syphilis (Greek *syn*, together, and *philein*, to love) is a contagious, sexually transmitted disease (table 38.1) caused by the spirochete *Treponema pallidum* subsp. pallidum (*T. pallidum*; *see figure 16.11*). **Congenital syphilis** is the disease acquired in utero from the mother. ◀◀ *Phylum Spirochaetes (section 19.6)*

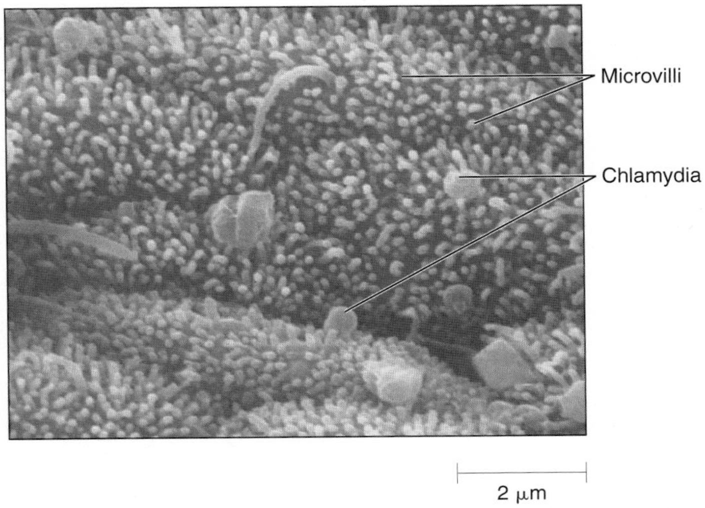

FIGURE 38.17 ***Chlamydia trachomatis*** **Attached Mucosa of a Fallopian Tube.**

T. pallidum enters the body through mucous membranes or minor breaks or abrasions of the skin. It migrates to the regional lymph nodes and rapidly spreads throughout the body. The disease is not highly contagious, and there is only about a 1 in 10 chance of acquiring it from a single exposure to an infected sex partner.

Three recognizable stages of syphilis occur in untreated individuals. In the primary stage, after an incubation period of about 10 days to 3 weeks or more, the initial symptom is a small, painless, reddened ulcer, or **chancre** (French, meaning a destructive sore), with a hard ridge that appears at the infection site and contains spirochetes (**figure 38.18*a***). Contact with the chancre during sexual contact may result in disease transmission. In about one-third of the cases, the disease does not progress further and the chancre disappears. Serological tests are positive in about 80% of the individuals during this stage (**figure 38.19**). The spirochetes typically enter the bloodstream and are distributed throughout the body.

Within 2 to 10 weeks after the primary lesion appears, the disease may enter the secondary stage, which is characterized by a highly variable skin rash (figure 38.18*b*). By this time, 100% of the individuals are serologically positive. Other symptoms during this stage include the loss of hair in patches, malaise, and fever. Both the chancre and the rash lesions are infectious.

After several weeks, the disease becomes latent. During the latent period, the disease is not normally infectious, except for

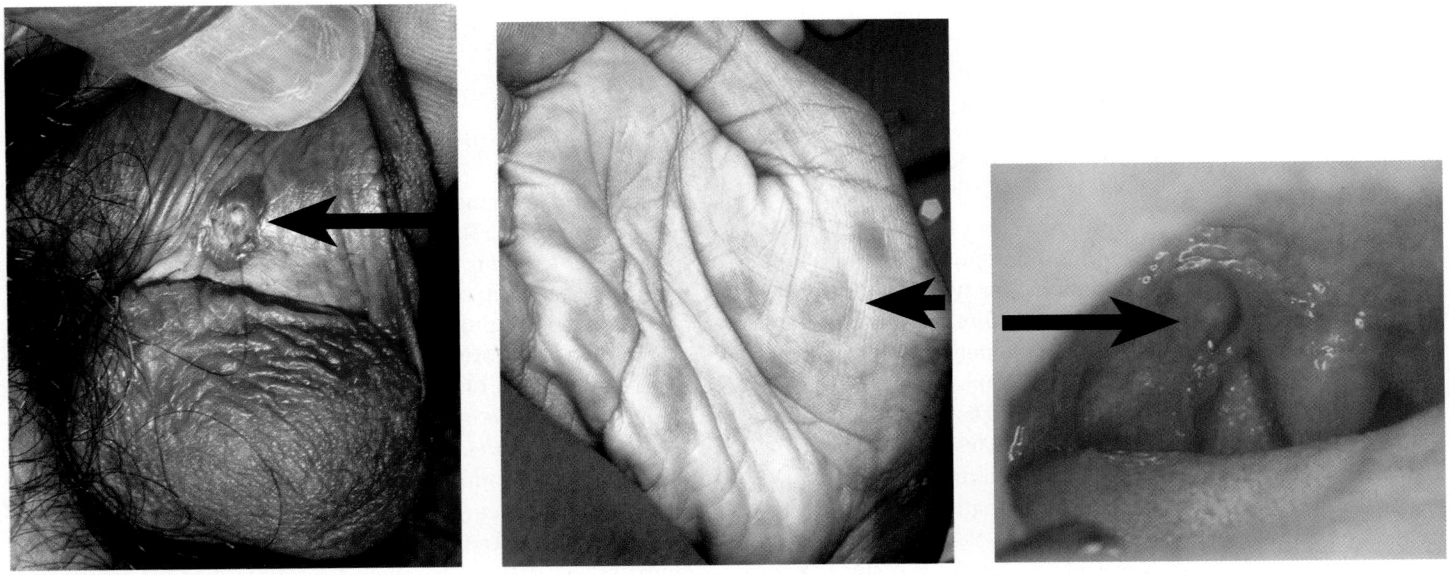

(a) (b) (c)

FIGURE 38.18 Syphilis. (a) Primary syphilitic chancre of the penis. (b) Palmar lesions of secondary syphilis. (c) Ruptured gumma and ulcer of upper hard palate of the mouth.

Figure 38.18 *Micro Inquiry*

Which of these clinical manifestations are associated with syphilis transmission? Explain.

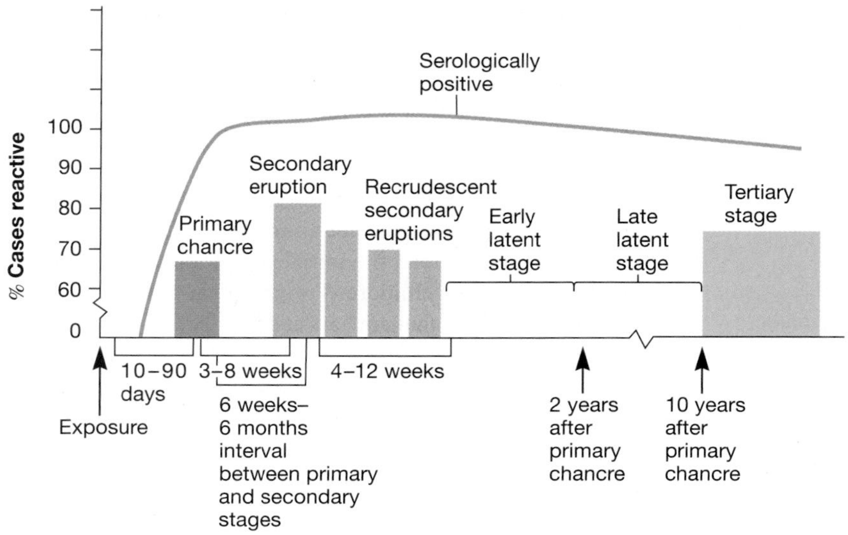

FIGURE 38.19 The Course of Untreated Syphilis.

possible transmission from mother to fetus (congenital syphilis). After many years, a tertiary stage develops in about 40% of untreated individuals with secondary syphilis. During this stage, degenerative lesions called gummas (figure 38.18c) form in the skin, bone, and nervous system as the result of hypersensitivity reactions. This stage also is characterized by a great reduction in the number of spirochetes in the body. Involvement of the central nervous system may result in tissue loss that can lead to cognitive deficits, blindness, a "shuffle" walk (tabes), or insanity. Many of these symptoms have been associated with such well-known people as Al Capone, Francisco Goya, Henry VIII, Adolf Hitler, Scott Joplin, Friedrich Nietzsche, Franz Schubert, Oscar Wilde, and Kaiser Wilhelm (**Disease 38.1**). ◄◄ *Hypersensitivities (section 33.10)*

Diagnosis of syphilis is through a clinical history, physical examination, and dark-field and immunofluorescence examination of lesion fluids (except oral lesions) for typical motile spirochetes. Because humans respond to *T. pallidum* with the formation of antitreponemal antibody and a complement-fixing reagin, serological tests are very informative. Examples include tests for nontreponemal antigens (VDRL, Venereal Disease Research Laboratories test; RPR, Rapid Plasma Reagin test) and treponemal antibodies (FTA-ABS, fluorescent treponemal antibody-absorption test; TP-PA, *T. pallidum* particle agglutination). ◄◄ *Immune complex formation (section 33.8)*

Treatment in the early stages of the disease is easily accomplished with long-acting benzathine penicillin G or aqueous procaine penicillin. Later stages of syphilis are more difficult to treat with drugs and require much larger doses over a longer period. For example, in neurosyphilis cases, treponemes occasionally survive such drug treatment. Immunity to syphilis is not complete, and subsequent infections can occur.

Prevention and control of syphilis depends on (1) public education, (2) prompt and adequate treatment of all new cases,

(3) follow-up on sources of infection and contact so they can be treated, and (4) prophylaxis (barrier protection) to prevent exposure. At present, the incidence of syphilis, as well as other sexually transmitted diseases, is rising in most parts of the world. An estimated 12 million new cases occur each year.

In the United States, the rate of primary and secondary syphilis in the civilian population declined 89.7% between 1990 and 2000. Unfortunately that rate increased yearly between 2001 and 2007, jumping 17.5% between 2006 and 2007, to 11,466 cases, primarily among men. The number of cases in women was 1,692 during the same time frame. The highest incidence is among those twenty-five to twenty-nine years of age. Importantly, syphilis incidence has increased worldwide as a secondary infection in HIV-positive individuals; conversely, untreated syphilis makes a person more susceptible to HIV infection.

1. How is *H. pylori* transmitted and what is its global prevalence?
2. What is ophthalmia neonatorum and how is it transmitted?
3. Describe the lesions of syphilis. What do the symptoms indicate about the disease progression?
4. What distinguishes gonorrhea from syphilis?
5. Infection with *N. gonorrhea* does not confer life-long protection from reinfection. Why do you think this is the case?

Staphylococcal Diseases

Staphylococci are among the most important bacteria that cause disease in humans. They are normal inhabitants of the upper respiratory tract, skin, intestine, and vagina. Staphylococci are members of a group of invasive gram-positive bacteria known as the pyogenic (pus-producing) cocci. These bacteria cause various suppurative (pus-forming) diseases in humans (e.g., boils, carbuncles, folliculitis, impetigo contagiosa, scalded-skin syndrome). The genus *Staphylococcus* consists of gram-positive cocci, 0.5 to 1.5 μm in diameter (*see figure 21.16*). The cell wall peptidoglycan and teichoic acid contribute to the pathogenicity of staphylococci. ◄◄ *Order Bacillales (section 21.3)*

Staphylococci can be divided into pathogenic and relatively nonpathogenic strains based on the synthesis of the enzyme coagulase. The coagulase-positive species *S. aureus* is the most important human pathogen in this genus. Coagulase-negative staphylococci (CoNS) such as *S. epidermidis* are nonpigmented, and are generally less invasive. However, they have increasingly been associated, as opportunistic pathogens, with serious nosocomial infections.

Staphylococci are further classified into slime producers (SP) and nonslime producers (NSP). The ability to produce

DISEASE

38.1 A Brief History of Syphilis

Syphilis was first recognized in Europe near the end of the fifteenth century. During this time, the disease reached epidemic proportions in the Mediterranean areas. According to one hypothesis, syphilis is of New World origin and Christopher Columbus (1451–1506) and his crew acquired it in the West Indies and introduced it into Spain after returning from their historic voyage. Another hypothesis is that syphilis had been endemic for centuries in Africa and may have been transported to Europe at the same time that vast migrations of the civilian population were occurring (1500). Others believe that the Vikings, who reached the New World well before Columbus, were the original carriers.

Syphilis was initially called the Italian disease, the French disease, and the great pox as distinguished from smallpox. In 1530 the Italian physician and poet Girolamo Fracastoro wrote *Syphilis sive Morbus Gallicus* (Syphilis or the French Disease). In this poem a Spanish shepherd named Syphilis is punished for being disrespectful to the gods by being cursed with the disease. Several years later, Fracastoro published a series of papers in which he described the possible mode of transmission of the "seeds" of syphilis through sexual contact.

Its venereal transmission was not definitely shown until the eighteenth century. The term venereal is derived from the name Venus, the Roman goddess of love. Recognition of the different stages of syphilis was demonstrated in 1838 by Philippe Ricord, who reported his observations on more than 2,500 human inoculations. In 1905 Fritz Schaudinn and Erich Hoffmann discovered the causative bacterium, and in 1906 August von Wassermann introduced the diagnostic test that bears his name. In 1909 Paul Ehrlich introduced an arsenic derivative, arsphenamine or salvarsan, as therapy. During this period, an anonymous limerick aptly described the course of this disease:

There was a young man from Black Bay
Who thought syphilis just went away
He believed that a chancre
Was only a canker
That healed in a week and a day.

But now he has "acne vulgaris"—
(Or whatever they call it in Paris);
On his skin it has spread
From his feet to his head,
And his friends want to know where his hair is.

There's more to his terrible plight:
His pupils won't close in the light
His heart is cavorting,
His wife is aborting,
And he squints through his gun-barrel sight.

Arthralgia cuts into his slumber;
His aorta is in need of a plumber;
But now he has tabes,
And saber-shinned babies,
While of gummas he has quite a number.

He's been treated in every known way,
But his spirochetes grow day by day;
He's developed paresis,
Has long talks with Jesus,
And thinks he's the Queen of the May.

slime has been proposed as a marker for pathogenic strains of staphylococci (**figure 38.20*a***). Slime is a viscous, extracellular glycoconjugate that allows these bacteria to adhere to smooth surfaces such as prosthetic medical devices and catheters. Scanning electron microscopy has clearly demonstrated that biofilms (figure 38.20*b*) consisting of staphylococci encased in a slimy matrix are formed in association with biomaterial-related infections (**Disease 38.2**). Slime also appears to inhibit neutrophil chemotaxis, phagocytosis, and the antimicrobial agents (*see figure 31.10*). ◀◀ *Biofilms (section 7.7)*

Staphylococci, harbored by either an asymptomatic carrier or a person with the disease (i.e., an active carrier), can be spread by the hands, expelled from the respiratory tract, or transported in or on animate and inanimate objects. Staphylococci can produce disease in almost every organ and tissue of the body (**figure 38.21**). For the most part, however, staphylococcal disease occurs in people whose defensive mechanisms have been compromised, such as patients in hospitals.

Staphylococci produce disease through their ability to multiply and spread widely in tissues and through their production of many virulence factors (**table 38.2**). Some of these factors are exotoxins, and others are enzymes thought to be involved in staphylococcal invasiveness. Many toxin genes are carried on plasmids; in some cases, genes responsible for pathogenicity reside on both a plasmid and the bacterial chromosome. The pathogenic capacity of a particular *S. aureus* strain is due to the combined effect of extracellular factors and toxins, together with the invasive properties of the strain. At one end of the disease spectrum is staphylococcal food poisoning, caused solely by the ingestion of preformed enterotoxin

DISEASE

38.2 Biofilms

Biofilms consist of microorganisms immobilized at a substratum surface and typically embedded in an organic polymer matrix of microbial origin. They develop on virtually all surfaces immersed in natural aqueous environments, including both biological (aquatic plants and animals) and abiological (concrete, metal, plastics, stones). Biofilms form particularly rapidly in flowing aqueous systems where a regular nutrient supply is provided to the microorganisms. Extensive microbial growth, accompanied by excretion of copious amounts of extracellular organic polymers, leads to the formation of visible slimy layers (biofilms) on solid surfaces. ◀◀ *Biofilms (section 7.7)*

Most of the human gastrointestinal tract is colonized by specific microbial groups (the normal indigenous microbiota) that give rise to natural biofilms. At times, these natural biofilms provide protection for pathogenic species, allowing them to colonize the host. ◀◀ *Large intestine (section 30.3)*

Insertion of a prosthetic device into the human body often leads to the formation of biofilms on the surface of the device (*see figure 7.32*). The microorganisms primarily involved are *Staphylococcus epidermidis,* other coagulase-negative staphylococci, and gram-negative bacteria. These normal skin inhabitants possess the ability to tenaciously adhere to the surfaces of inanimate prosthetic devices. Within the biofilms, they are protected from the body's normal defense mechanisms and also from antibiotics; thus the biofilm also provides a source of infection for other parts of the body as bacteria detach during biofilm sloughing.

Some examples of biofilms of medical importance include:

1. The deaths following massive infections of patients receiving Jarvik 7 artificial hearts
2. Cystic fibrosis patients harboring great numbers of *Pseudomonas aeruginosa* that produce large amounts of alginate polymers, which inhibit the diffusion of antibiotics
3. Teeth, where biofilm forms plaque that leads to tooth decay (figure 38.34)
4. Contact lenses, where bacteria may produce severe eye irritation, inflammation, and infection
5. Air-conditioning and other water retention systems where potentially pathogenic bacteria, such as *Legionella* species, may be protected by biofilms from the effects of chlorination.

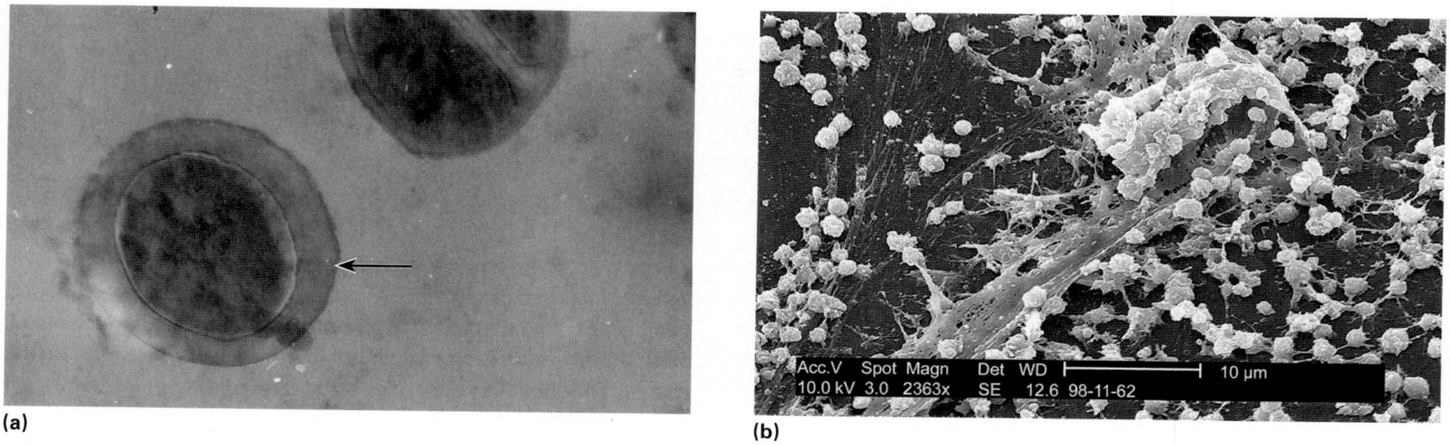

(a) (b)

FIGURE 38.20 Slime and Biofilms. (a) Cells of *S. aureus,* one of which produces a slime layer (arrowhead; transmission electron microscopy, ×10,000). (b) A biofilm on the inner surface of an intravenous catheter. Extracellular polymeric substances, mostly polysaccharides, surround and encase the staphylococci (scanning electron micrograph, ×2,363).

(table 38.2). At the other end of the spectrum are staphylococcal bacteremia and disseminated abscesses in most organs of the body.

The classic example of a staphylococcal lesion is the localized abscess of a domed pustule (**figure 38.22*a,b***). When *S. aureus* becomes established in a hair follicle, tissue necrosis results. Coagulase is produced and forms a fibrin wall around the lesion that limits the spread. Within the center of the lesion, liquefaction of necrotic tissue occurs, and the abscess spreads in the direction of least resistance. The abscess may be either a furuncle (boil) (figure 38.22*c*) or a carbuncle (figure 38.22*d*). The central necrotic tissue drains, and healing eventually occurs. However, the bacteria may spread from any focus through lymph or blood to other parts of the body.

Newborn infants and children can develop a superficial skin infection characterized by the presence of encrusted pustules (figure 38.22*e*). This disease, called impetigo contagiosa, can be caused by *S. aureus* or group A streptococci. It is contagious and can spread rapidly through a nursery or school. It usually occurs in areas where sanitation and personal hygiene are poor.

Staphylococcal scalded skin syndrome (SSSS) is another common staphylococcal disease (figure 38.22*f*). SSSS is caused by strains of *S. aureus* that produce **exfoliative toxin (exfoliatin).** This protein is usually plasmid encoded, although in some strains, the toxin gene is on the bacterial chromosome. In this disease, the epidermis peels off to reveal a red area underneath—thus the name of the disease. SSSS is seen most commonly in infants and children, and neonatal nurseries occasionally suffer large outbreaks of the disease.

Toxic shock syndrome (TSS) is a staphylococcal disease with potentially serious consequences. Most cases of this syndrome have occurred in females who used superabsorbent tampons during menstruation. These tampons can trigger a change in the normal vaginal flora, thereby enabling the growth of toxin-producing *S. aureus*. Toxic shock syndrome is characterized by low blood pressure, fever, diarrhea, an extensive skin rash, and shedding of the skin. TSS results from the massive overproduction of cytokines by T cells induced by the TSST-1 protein (or to staphylococcal enterotoxins B and C1). TSST-1 binds both class II MHC receptors and T-cell receptors, stimulating T-cell responses in the absence of specific antigen. TSST-1 and other proteins having this property are called **superantigens.** Superantigens activate 5 to 30% of the total T-cell population, whereas specific antigens activate only 0.01 to 0.1% of the T-cell population. The net effect of cytokine overproduction is circulatory collapse leading to shock and multiorgan failure. Tumor necrosis factor α (TNF-α) and interleukins (IL) 1 and 6 are strongly associated with superantigen-induced shock. Mortality rates for TSS are 30 to 70%, and morbidity due to surgical debridement and amputation is very high. Approximately 150 cases of toxic shock syndrome are reported annually in the United States. For these reasons, staphylococcal and streptococcal superantigens are categorized as select agents; their production and use are restricted, as they may be used as bioterror agents. ⏮ *Superantigens (section 33.5)*

The definitive diagnosis of staphylococcal disease can be made only by isolation and identification of the staphylococcus involved. This requires culture, catalase, coagulase, and other biochemical tests. Commercial rapid test kits also are available. There is no specific prevention for staphylococcal disease. The mainstay of treatment is the administration of specific antibiotics: penicillin, cloxacillin, methicillin, vancomycin, oxacillin, cefotaxime, ceftriaxone, a cephalosporin, or rifampin and others. Because of the prevalence of drug-resistant strains (e.g., methicillin-resistant staphylococci), all staphylococcal isolates should be tested for antimicrobial susceptibility. Cleanliness, good hygiene, and aseptic

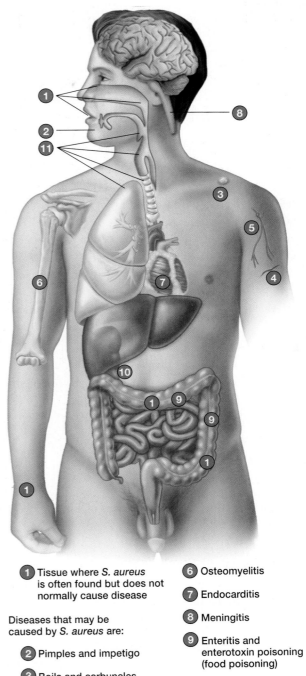

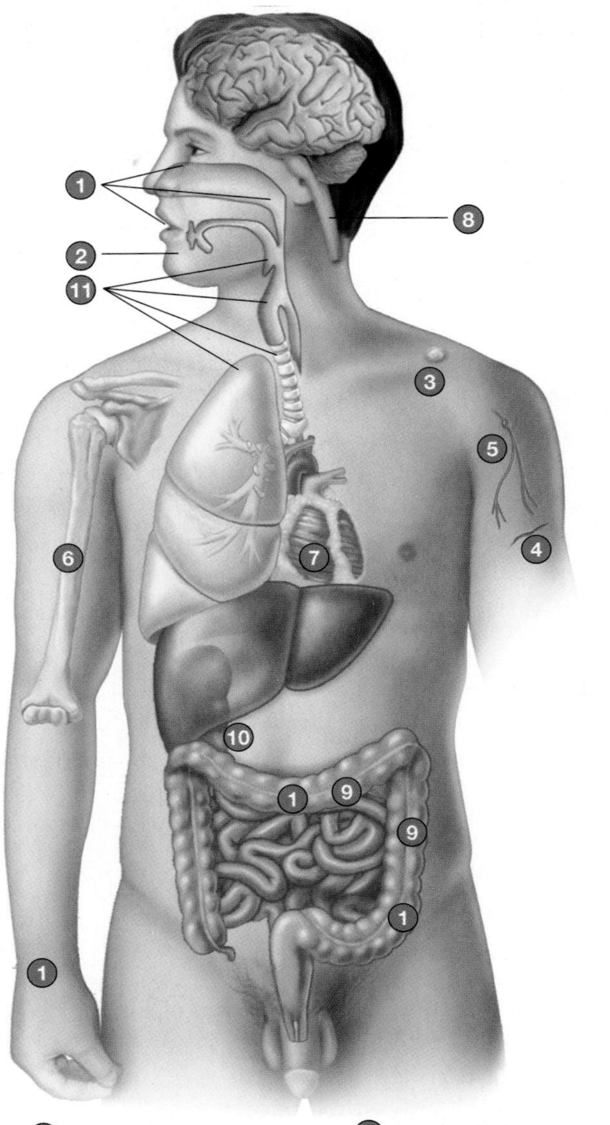

1 Tissue where *S. aureus* is often found but does not normally cause disease

Diseases that may be caused by *S. aureus* are:

2 Pimples and impetigo

3 Boils and carbuncles on any surface area

4 Wound infections and abscesses

5 Spread to lymph nodes and to blood (septicemia), resulting in widespread seeding

6 Osteomyelitis

7 Endocarditis

8 Meningitis

9 Enteritis and enterotoxin poisoning (food poisoning)

10 Nephritis

11 Respiratory infections:
Pharyngitis
Laryngitis
Bronchitis
Pneumonia

FIGURE 38.21 Staphylococcal Diseases. The anatomical sites of the major staphylococcal infections of humans are indicated by the corresponding numbers.

Table 38.2 Various Enzymes and Toxins Produced by Staphylococci

Product	*Physiological Action*
β-lactamase	Breaks down penicillin
Catalase	Converts hydrogen peroxide into water and oxygen and reduces killing by phagocytosis
Coagulase	Reacts with prothrombin to form a complex that can cleave fibrinogen and cause the formation of a fibrin clot; fibrin may also be deposited on the surface of staphylococci, which may protect them from destruction by phagocytic cells; coagulase production is synonymous with invasive pathogenic potential
DNase	Destroys DNA
Enterotoxins	Are divided into heat-stable toxins of six known types (A, B, C1, C2, D, E); responsible for the gastrointestinal upset typical of food poisoning
Exfoliative toxins A and B (superantigens)	Causes loss of the surface layers of the skin in scalded-skin syndrome
Hemolysins	Alpha hemolysin destroys erythrocytes and causes skin destruction Beta hemolysin destroys erythrocytes and sphingomyelin around nerves
Hyaluronidase	Also known as spreading factor; breaks down hyaluronic acid located between cells, allowing for penetration and spread of bacteria
Panton-Valentine leukocidin	Inhibits phagocytosis by granulocytes and can destroy these cells by forming pores in their phagosomal membranes
Lipases	Break down lipids
Nuclease	Breaks down nucleic acids
Protein A	Is antiphagocytic by competing with neutrophils for the Fc portion of specific opsonins
Proteases	Break down proteins
Toxic shock syndrome toxin-1 (a superantigen)	Is associated with the fever, shock, and multisystem involvement of toxic shock syndrome

management of lesions are the best means of control. ◀◀ *Drug resistance (section 34.9)*

Methicillin-Resistant *Staphylococcus aureus* (MRSA)

Isolates of *Staphylococcus aureus* that have become resistant to several specific antibiotics are known as **methicillin-resistant *Staphylococcus aureus* (MRSA).** MRSA is usually resistant to currently available β-lactam antibiotics, including penicillins (e.g., penicillin, amoxicillin), "antistaphylococcal" penicillins (e.g., methicillin, oxacillin), and cephalosporins (e.g., cephalexin). Otherwise healthy people can acquire MRSA infections, usually on or in the skin, such as abscesses, boils, and other pus-filled lesions. When the infection is acquired by people who have not been hospitalized (i.e., in the past year) or had a medical procedure (e.g., dialysis, surgery, catheters) recently, the infection is referred to as a *c*ommunity-*a*ssociated (CA)-MRSA infection. However, MRSA is predominantly a nosocomial infection. Screening of MRSA isolates has found that most of the strains associated with serious infections were usually associated with health care. Importantly, CA-MRSA strains are now being reported inside health-care settings. Furthermore, even though the incidence of infection varies greatly between geographically diverse areas, surveillance data report that overall MRSA rates were consistently highest among older persons (over age 65), African-Americans, and males.

MRSA can only be diagnosed by culture and sensitivity testing. However, the CDC encourages the consideration of MRSA when other skin diseases compatible with *S. aureus* infections are suspected. Similar to other infections caused by staphylococci, MRSA infections begin as a bump or reddened area on the skin that looks like a large pimple. In other words, MRSA infection would appear as fluid-filled pustules, with a yellow or white center, central point, or "head," that may be draining pus (**figure 38.23**).

Treatment is by incision and drainage, as for other purulent (pus-filled) skin infections. Antimicrobial therapy may be prescribed in addition to incision and drainage, while awaiting laboratory results. Importantly, vancomycin (an antibiotic thought to be the drug of last resort) is no longer a sure treatment for MRSA as vancomycin-resistant strains are occurring worldwide. Good aseptic technique is paramount in preventing the transmission of MRSA from the patient to health-care workers.

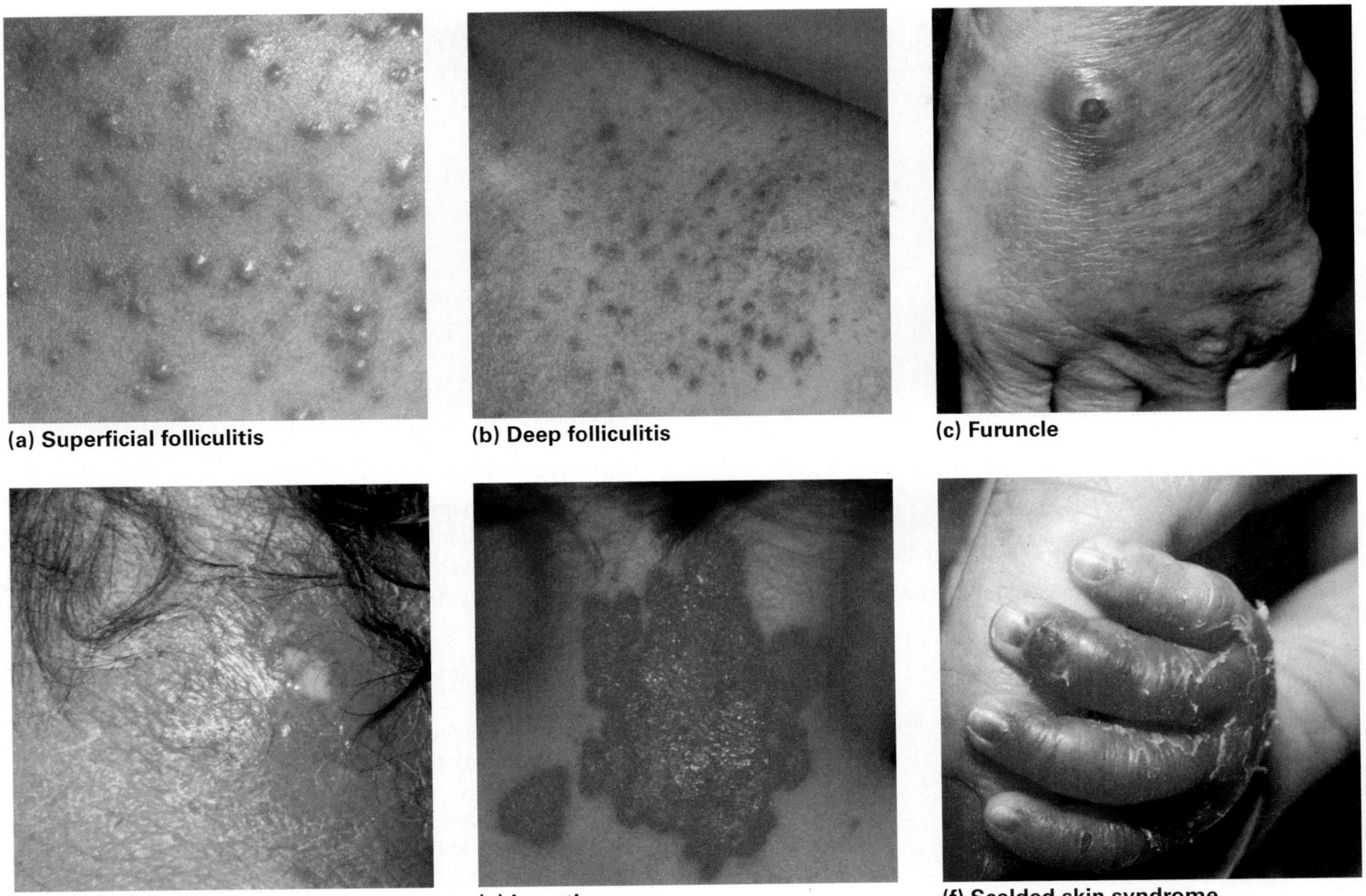

(a) Superficial folliculitis **(b) Deep folliculitis** **(c) Furuncle**
(d) Carbuncle **(e) Impetigo** **(f) Scalded skin syndrome**

FIGURE 38.22 Staphylococcal Skin Infections. (a) Superficial folliculitis in which raised, domed pustules form around hair follicles. (b) In deep folliculitis, the microorganism invades the deep portion of the follicle and dermis. (c) A furuncle arises when a large abscess forms around a hair follicle. (d) A carbuncle consists of a multilocular abscess around several hair follicles. (e) Impetigo on the neck of two-year-old male. (f) Scalded skin syndrome in a one-week-old premature male infant. Reddened areas of skin peel off, leaving scalded-looking moist areas.

Streptococcal Diseases

Streptococci that are transmitted through the air are discussed in section 38.1. Superficial cutaneous disease caused by streptococci can lead to diseases such as cellulitis, impetigo, and erysipelas. Streptococcal disease can also invasively infect the skin, spreading deeper into the tissue and often reaching the underlying muscle. This oxygenated tissue is a very rich "growth medium" for the streptococci, providing the nutrients for rapid growth and replication. We discuss both the superficial and invasive streptococcal diseases next.

Cellulitis, Impetigo, and Erysipelas

Cellulitis is a diffuse, spreading infection of subcutaneous skin tissue. The resulting inflammation is characterized by a defined area of redness (erythema) and the accumulation of fluid (edema). A number of different bacteria can cause cellulitis.

The most frequently diagnosed skin infection caused by *S. pyogenes* is **impetigo** (impetigo also can be caused by *Staphylococcus aureus*; figure 38.22*e*). Impetigo is a superficial cutaneous infection, most commonly seen in children, usually located on the face, and characterized by crusty lesions and vesicles surrounded by a red border. Impetigo is most common in late summer and early fall. The drug of choice for impetigo is penicillin; erythromycin is prescribed for those individuals who are allergic to penicillin.

Erysipelas (Greek *erythros*, red, and *pella*, skin) is an acute infection and inflammation of the dermal layer of the skin. It occurs primarily in infants and people over thirty years of age with a history of streptococcal sore throat. The skin often develops painful, reddish patches that enlarge and thicken with a sharply defined edge (**figure 38.24**). Recovery usually takes a week or longer if no treatment is given. The drugs of choice for the treatment of erysipelas are erythromycin and penicillin.

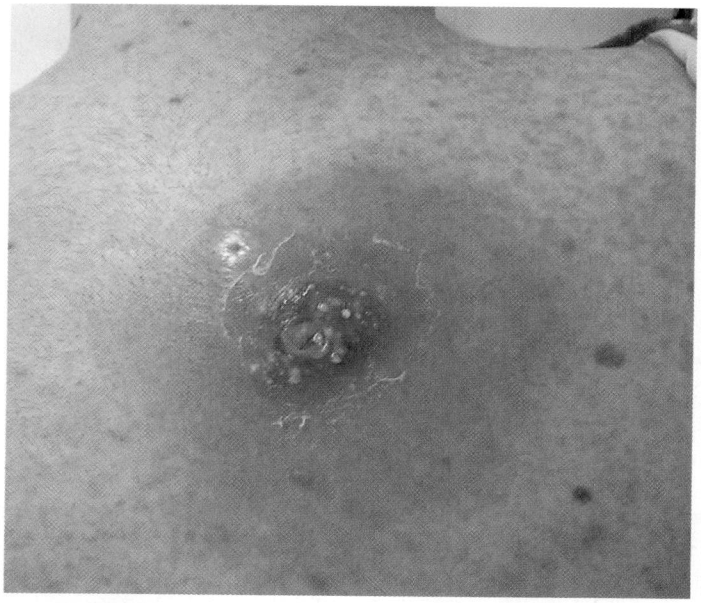

FIGURE 38.23 Methicillin-Resistant *Staphylococcus aureus* Wound.

Figure 38.23 *Micro Inquiry*

To what drugs, besides methicillin, are MRSA strains also resistant?

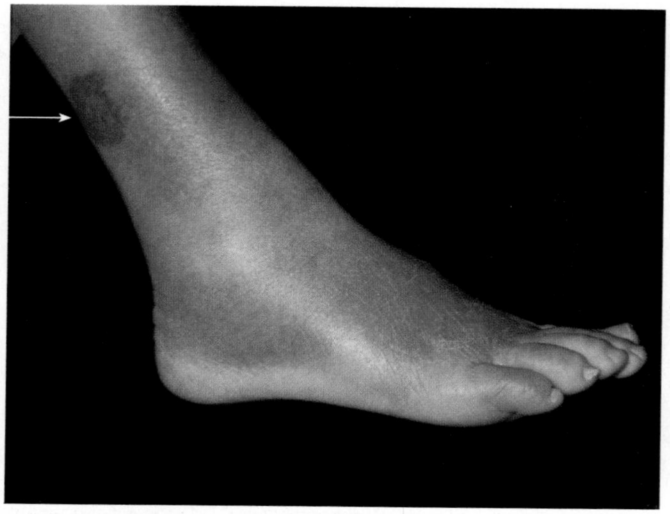

FIGURE 38.24 Erysipelas. Notice the bright, raised, rubbery lesion at the site of initial entry (white arrow) and the spread of the inflammation to the foot. The reddening is caused by toxins produced by the streptococci as they invade tissue.

Erysipelas may recur periodically at the same body site for years.

Invasive Streptococcal Infections

The development of invasive group A strep (GAS; p. 942) disease appears to depend on the presence of specific virulent strains (e.g., M-1 and M-3 serotypes) and predisposing host factors (surgical or nonsurgical wounds, diabetes, and other underlying medical problems). A life-threatening infection begins when GAS strains penetrate a mucous membrane or take up residence in a wound such as a bruise. This infection can quickly lead either to **necrotizing fasciitis** (Greek *nekrosis*, deadness, Latin *fascis*, band or bandage, and *itis*, inflammation), which destroys the sheath covering skeletal muscles, or to **myositis** (Greek *myos*, muscle, and *itis*), the inflammation and destruction of skeletal muscle and fat tissue (**figure 38.25**). Because necrotizing fasciitis and myositis arise and spread so quickly, they have been colloquially called "galloping gangrene."

Rapid treatment is necessary to reduce the risk of death due to necrotizing fasciitis, and penicillin G remains the treatment of choice. In addition, surgical removal of dead and dying tissue usually is needed in more advanced cases. It is estimated that approximately 11,000 cases of GAS infections occur annually in the United States (1,350 deaths), and between 5 and 10% of them are associated with necrotizing conditions.

One reason GAS strains are so deadly is that about 85% of them carry the genes for the production of streptococcal pyrogenic exotoxins A and B (Spe exotoxins). Exotoxin A acts as a superantigen, a nonspecific T-cell activator. This superantigen quickly stimulates T cells to begin producing abnormally large quantities of cytokines. The cytokines damage the endothelial cells that line blood vessels, causing fluid loss and rapid tissue death from a lack of oxygen. Another pathogenic mechanism involves the secretion of exotoxin B, a cysteine protease (a proteolytic enzyme that has a cysteine residue in the active site). This protease rapidly destroys tissue by breaking down proteins. ⏮ *Superantigens (section 33.5)*

Since 1986 it has been recognized that GAS infections can also trigger a toxic shocklike syndrome (TSLS), characterized by a precipitous drop in blood pressure, failure of multiple organs, and a very high fever. TSLS is caused by an GAS that produces one or more of the streptococcal pyrogenic exotoxins. TSLS has a mortality rate of over 30%.

Tetanus

Tetanus (Greek *tetanos*, to stretch) is caused by *Clostridium tetani*, an anaerobic, gram-positive, endospore-forming rod (*see figure 21.7*). The spores of *C. tetani* are commonly found in hospital environments, in soil and dust, and in the feces of many farm animals and humans. Transmission to humans is associated with skin wounds. Any break in the skin can allow *C. tetani* spores to enter. If the oxygen tension is low enough, the spores germinate

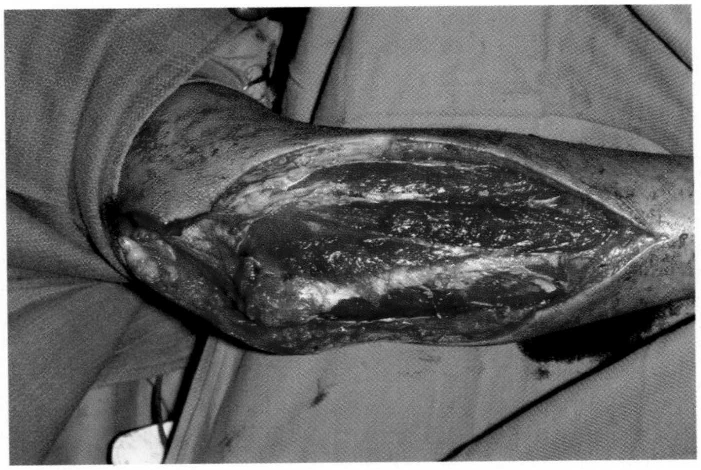

FIGURE 38.25 Necrotizing Fasciitis. Rapidly advancing streptococcal disease can lead to large, necrotic sites, sometimes with blisters that rupture and expose the dying tissue. This is often called flesh-eating disease or necrotizing fasciitis.

Figure 38.25 Micro Inquiry

What virulence factor(s) enable the rapid invasion of certain strep A strains?

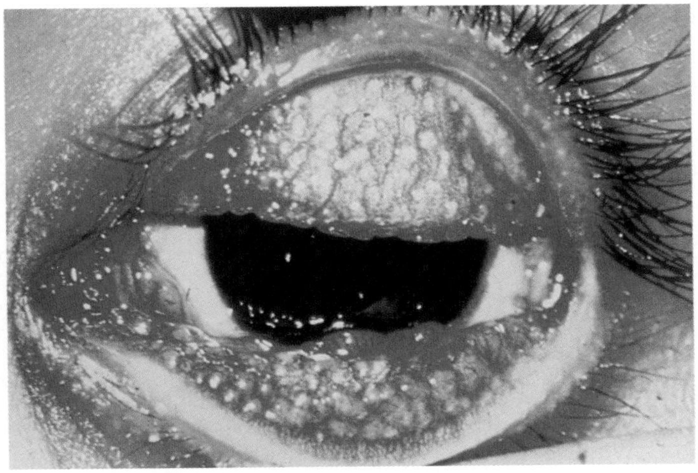

FIGURE 38.26 Trachoma. An active infection showing marked follicular hypertrophy of the eyelid. The inflammatory nodules cover the thickened conjunctiva of the eye.

and release the neurotoxin tetanospasmin. **Tetanospasmin** is an endopeptidase that selectively cleaves the synaptic vesicle membrane protein synaptobrevin. This prevents exocytosis (release from the cell) and release of inhibitory neurotransmitters (γ-aminobutyric acid and glycine) at synapses within the spinal cord motor nerves. The result is uncontrolled stimulation of skeletal muscles (spastic paralysis). A second toxin, **tetanolysin,** is a hemolysin that aids in tissue destruction. ⏮ *Class* Clostridia *(section 21.2)*

Early in the course of the disease, tetanospasmin causes tension or cramping and twisting in skeletal muscles surrounding the wound and tightness of the jaw muscles. With more advanced disease, there is trismus ("lockjaw"), an inability to open the mouth because of the spasm of the masseter muscles. Facial muscles may go into spasms, producing the characteristic expression known as risus sardonicus. Spasms or contractions of the trunk and extremity muscles may be so severe that there is boardlike rigidity, painful tonic convulsions, and backward bowing of the back so that the heels and back approach each other (opisthotonos; **chapter opener**). Death usually results from spasms of the diaphragm and intercostal respiratory muscles.

Prevention of tetanus involves the use of the tetanus toxoid. The toxoid, which incorporates an adjuvant (aluminum salts) to increase its immunizing potency, is given routinely with diphtheria toxoid and pertussis vaccine. An initial dose is normally administered a few months after birth, a second dose 4 to 6 months later, and finally a reinforcing dose 6 to 12 months after the second injection. Another booster is given between the ages of four to six years (*see table 36.3*).

Control measures for tetanus are not possible because of the wide dissemination of the bacterium in the soil and the long survival of its spores. The case fatality rate in generalized tetanus ranges from 30 to 90% because tetanus treatment is not very effective. Therefore prevention is all important and depends on (1) active immunization with toxoid, (2) proper care of wounds contaminated with soil, (3) prophylactic use of antitoxin, and (4) administration of penicillin. Around 50 cases of tetanus are reported annually in the United States—the majority of which are in intravenous drug users.

Trachoma

Trachoma (Greek *trachoma*, roughness) is a contagious disease caused by *Chlamydia trachomatis* serotypes A–C. It is one of the oldest known infectious diseases of humans and is the greatest single cause of blindness throughout the world—over 500 million people are infected and 20 million blinded each year. In endemic areas, most children are chronically infected within a few years of birth. Active disease in adults over age twenty is three times as frequent in females as in males because of mother-child contact. Although trachoma is uncommon in the United States, except among Native Americans in the Southwest, it is widespread in Asia, Africa, and South America.

Trachoma is transmitted by contact with inanimate objects such as soap and towels, by hand-to-hand contact that carries *C. trachomatis* from an infected eye to an uninfected eye, or by flies. The disease begins abruptly with an inflamed conjunctiva. This leads to an inflammatory cell exudate and necrotic eyelash follicles (**figure 38.26**). The disease usually heals spontaneously.

However, with reinfection, secondary infections, vascularization of the cornea, or pannus (tissue flap) formation, scarring of the conjunctiva can occur. If scar tissue accumulates over the cornea, blindness results.

Diagnosis of trachoma is by serologic methods. Treatment is with tetracyclines, macrolides, or doxycycline. However, prevention and control of trachoma depend more on health education and personal hygiene—such as access to clean water for washing—than on treatment.

1. Compare and contrast the diseases caused by the TSST and SSSS proteins of *S. aureus*.
2. Describe several diseases caused by staphylococci.
3. What is the difference between *S. aureus* and MRSA?
4. Why are streptococci able to cause human disease? How are superficial and invasive diseases different?
5. Explain why tetanus is potentially life-threatening. How is tetanus acquired? What are its symptoms and how do they arise?
6. How does *C. trachomatis* serotypes A–C cause trachoma? Describe how it is transmitted, and the way in which blindness may result.

38.4 Food-Borne and Waterborne Diseases

Many microorganisms that contaminate food and water can cause acute gastroenteritis—inflammation of the stomach and intestinal lining. When food is the source of the pathogen, the condition is often called **food poisoning.** Gastroenteritis can arise in two ways. The microorganisms may actually produce a **food-borne infection.** That is, they may first colonize the gastrointestinal tract and grow within it, then either invade host tissues or secrete exotoxins. Alternatively, the pathogen may secrete an exotoxin that contaminates the food and is then ingested by the host. This is sometimes referred to as a **food intoxication** because the toxin is ingested and the presence of living microorganisms is not required. Because these toxins disrupt the functioning of the intestinal mucosa, they are called **enterotoxins.** Common symptoms of enterotoxin poisoning are nausea, vomiting, and diarrhea.

Worldwide, diarrheal diseases are second only to respiratory diseases as a cause of adult death; they are the leading cause of childhood death, and in some parts of the world, they are responsible for more years of potential life lost than all other causes combined. For example, each year around 5 million children (more than 13,600 a day) die from diarrheal diseases in Asia, Africa, and South America. In the United States, estimates exceed 10,000 deaths per year from diarrhea, and an average of 500 childhood deaths are reported.

This section describes several of the more common bacterial gastrointestinal infections, food intoxications, and waterborne diseases. **Table 38.3** summarizes many of the bacterial pathogens responsible for food poisoning, and **table 38.4** lists many important water-based bacterial pathogens. The protozoa responsible for food- and waterborne diseases are covered in chapter 39. ▶▶| *Controlling food spoilage (section 40.2)*

Botulism

Food-borne **botulism** (Latin *botulus*, sausage) is a form of food poisoning. The most common source of disease is home-canned food that has not been heated sufficiently to kill contaminating *Clostridium botulinum* spores. The spores then germinate, and a toxin is produced during anaerobic vegetative growth. If the food is later eaten without adequate cooking, the active toxin causes disease.

Botulinum toxin is a neurotoxin that binds to the synapses of motor neurons (**figure 38.27**). It selectively cleaves the synaptic vesicle membrane protein synaptobrevin, thus preventing exocytosis (release from the cell) and release of the neurotransmitter acetylcholine. As a consequence, muscles do not contract in response to motor neuron activity, and flaccid paralysis results (**Techniques & Applications 38.3**). Symptoms of botulism occur within 12 to 72 hours of toxin ingestion and include blurred vision, difficulty in swallowing and speaking, muscle weakness, nausea, and vomiting. Treatment relies on supportive care and polyvalent antitoxin. If untreated, one-third of the patients die of either respiratory or cardiac failure within a few days. Fewer than 100 cases of botulism occur in the United States annually.

Infant botulism is the most common form of botulism in the United States and is confined to children under a year of age. Approximately 100 cases are reported each year. It appears that ingested spores, which may be naturally present in honey or house dust, germinate in the infant's intestine. *C. botulinum* then multiplies and produces the toxin. The infant becomes constipated, listless, generally weak, and eats poorly. Death may result from respiratory failure. It is therefore recommended that infants not be fed honey.

Campylobacter jejuni Gastroenteritis

Campylobacter infections cause more diarrhea in the United States than *Salmonella* and *Shigella* combined. Studies with chickens, turkeys, and cattle have shown that as much as 50 to 100% of a flock or herd of these birds or animals excrete *Campylobacter jejuni*. These bacteria also can be isolated in high numbers from surface waters. They are transmitted to humans by contaminated food and water, contact with infected animals, or anal-oral sexual activity. Human gastroenteritis—inflammation of the intestine—caused by *C. jejuni* (a slender, gram-negative, motile, curved rod found in the intestinal tract of animals) causes an estimated 2 million cases of ***Campylobacter* gastroenteritis (campylobacteriosis)** and subsequent diarrhea in the United States each year. |◀◀ *Class* Epsilonproteobacteria *(section 20.5)*

TECHNIQUES & APPLICATIONS

38.3 Clostridial Toxins as Therapeutic Agents—Benefits of Nature's Most Toxic Proteins

Some toxins are currently being used for the treatment of human disease. Specifically, botulinum toxin, the most poisonous biological substance known, is being used for the treatment of specific neuromuscular disorders characterized by involuntary muscle contractions. Since approval of type-A botulinum toxin (Botox) by the FDA in 1989 for three disorders (strabismus [crossing of the eyes], blepharospasm [spasmotic contractions of the eye muscles], and hemifacial spasm [contractions of one side of the face]), the number of neuromuscular problems being treated has increased to include other tremors, migraine and tension headaches, and other maladies. In 2000 dermatologists and plastic surgeons began using Botox to eradicate wrinkles caused by repeated muscle contractions as we laugh, smile, or frown. The remarkable therapeutic utility of botulinum toxin lies in its ability to specifically and potently inhibit involuntary muscle activity for an extended duration. Overall, the clostridia (currently one of the largest and most diverse genera of bacteria containing about 130 species) produce more protein toxins than any other known bacterial genus and are a rich reservoir of toxins for research and medicinal uses. For example, research is underway to use clostridial toxins or toxin domains for drug delivery, prevention of food poisoning, and the treatment of cancer and other diseases. The remarkable success of botulinum toxin as a therapeutic agent has thus created a new field of investigation in microbiology.

The incubation period for campylobacteriosis is 2 to 10 days. *C. jejuni* invades the epithelium of the small intestine, causing inflammation, and also secretes an exotoxin that is antigenically similar to the cholera toxin. Symptoms include diarrhea, high fever, severe inflammation of the intestine along with ulceration, and bloody stools. *C. jejuni* infection has also been linked to Guillain-Barre syndrome, a disorder in which the body's immune system attacks peripheral nerves, resulting in life-threatening paralysis. ◀◀ *Exotoxins (section 31.3)*

Laboratory diagnosis is by culture in an atmosphere with reduced O_2 and added CO_2. The disease is usually self-limited, and treatment is supportive; fluids, electrolyte replacement, and erythromycin may be used in severe cases. Recovery usually takes from 5 to 8 days. Prevention and control involve good personal hygiene and food-handling precautions, including pasteurization of milk and thorough cooking of poultry.

Cholera

Cholera is an acute diarrheal disease caused by infection of the intestine with the gram-negative, comma-shaped bacterium *Vibrio cholerae* (**figure 38.28**). The infection is usually mild or without symptoms in most healthy adults but sometimes can be severe. The disease is characterized by profuse watery diarrhea, vomiting, and leg cramps. These symptoms result from rapid loss of body fluids leading to dehydration and shock. Death can occur within hours when persons with rapid fluid loss are left untreated. Throughout recorded history, cholera (Greek *chole*, bile) has caused seven pandemics in various areas of the world, especially in Asia, the Middle East, and Africa. The disease is rare in the United States (0–5 cases per year) since the 1800s, but an endemic focus is believed to exist on the Gulf Coast of Louisiana and Texas.

Although there are many *V. cholerae* serogroups, only O1 and O139 have caused epidemics. *V. cholerae* O1 is divided into two serotypes, Inaba and Ogawa, and two biotypes, classic and El Tor. In 1961 the El Tor biotype emerged as an important cause of cholera pandemics, and in 1992 the newly identified strain *V. cholerae* O139 emerged in Asia. In Calcutta, India, serogroup O139 of *V. cholerae* has displaced El Tor *V. cholerae* serogroup O1, an event that has never before happened in the recorded history of cholera. Interestingly, the cholera toxin gene is carried by the lysogenic CTX filamentous bacteriophage. The receptor for the phage is the toxin coregulated pilus (TCP)—the same structure used to colonize the host's gut. Thus *V. cholerae* is an excellent example of how horizontal transfer of genes can confer pathogenicity.

Cholera is transmitted by ingesting food or water contaminated by fecal material from infected individuals. Once the bacteria enter the body, the incubation period is 12 to 72 hours. The bacteria adhere to the intestinal mucosa of the small intestine, where they are not invasive but secrete **choleragen,** a cholera toxin. Choleragen is an AB toxin (*see figure 31.11*). The A subunit enters the intestinal epithelial cells and activates the enzyme adenylate cyclase by the addition of an ADP-ribosyl group in a way similar to that employed by pertussis toxin. This results in hypersecretion of water and chloride ions while inhibiting absorption of sodium ions. The infected person loses massive quantities of fluid and electrolytes, causing abdominal muscle cramps, vomiting, fever, and watery diarrhea. The voided fluid is often referred to as "rice-water stool" because of the flecks of

Table 38.3 Bacteria That Cause Acute Bacterial Diarrhea and Food Poisoning

Organism	*Incubation Period (Hours)*	*Vomiting*	*Diarrhea*	*Fever*	*Epidemiology*
Staphylococcus aureus	1–8 (rarely, up to 18)	+++	+	–	Staphylococci grow in meats, dairy and bakery products and produce enterotoxins.
Bacillus cereus	2–16	+++	++	–	Reheated fried rice causes vomiting or diarrhea.
Clostridium perfringens	8–16	±	+++	–	Clostridia grow in rewarmed meat dishes.
Clostridium botulinum	18–24	±	Rare	–	Clostridia grow in anoxic foods and produce toxin.
Escherichia coli (enterohemorrhagic)	3–5 days	±	++	±	Generally associated with ingestion of undercooked ground beef, and unpasteurized fruit juices and cider.
Escherichia coli (enterotoxigenic strain)	24–72	±	++	–	Organisms grow in gut and are a major cause of traveler's diarrhea.
Vibrio parahaemolyticus	6–96	+	++	±	Organisms grow in seafood and in gut and produce toxin, or invade.
Vibrio cholerae	24–72	+	+++	–	Organisms grow in gut and produce toxin.
Shigella spp. (mild cases)	24–72	±	++	+	Organisms grow in superficial gut epithelium. *S. dysenteriae* produces toxin.
Salmonella spp. (gastroenteritis)	8–48	±	++	+	Organisms grow in gut.
Salmonella enterica serovar Typhi (typhoid fever)	10–14 days	±	±	++	Bacteria invade the gut epithelium and reach the lymph nodes, liver, spleen, and gallbladder.
Clostridium difficile	Days to weeks after antibiotic therapy	–	+++	+	Antibiotic-associated colitis
Campylobacter jejuni	2–10 days	–	+++	++	Infection by oral route from foods, pets. Organism grows in small intestine.
Yersinia enterocolitica	4–7 days	±	++	+	Fecal-oral transmission, food-borne, animals infected

Pathogenesis	*Clinical Features*
Enterotoxins act on gut receptors that transmit impulses to medullary centers; may also act as superantigens.	Abrupt onset, intense vomiting for up to 24 hours, recovery in 24–48 hours. Occurs in persons eating the same food. No treatment usually necessary except to restore fluids and electrolytes. With incubation period of 2–8 hours, mainly vomiting.
Enterotoxins formed in food or in gut from growth of *B. cereus*.	With incubation period of 8–16 hours, mainly diarrhea.
Enterotoxins produced during sporulation in gut; causes hypersecretion.	Abrupt onset of profuse diarrhea; vomiting occasionally. Recovery usual without treatment in 1–4 days. Many clostridia in cultures of food and feces of patients.
Toxin absorbed from gut and blocks acetylcholine release at neuromuscular junction.	Diplopia, dysphagia, dysphonia, difficulty breathing. Treatment requires clearing the airway, ventilation, and intravenous polyvalent antitoxin. Exotoxin present in food and serum. Mortality rate high.
Toxins cause epithelial necrosis in colon; mild to severe complications.	Symptoms vary from mild to severe bloody diarrhea. The toxin can be absorbed, becoming systemic and producing hemolytic uremic syndrome, most frequently in children.
Heat-labile (LT) and heat-stable (ST) enterotoxins cause hypersecretion in small intestine.	Usually abrupt onset of diarrhea; vomiting rare. A serious infection in newborns. In adults, "traveler's diarrhea" is usually self-limited in 1–3 days.
Toxin causes hypersecretion; vibrios invade epithelium; stools may be bloody.	Abrupt onset of diarrhea in groups consuming the same food, especially crabs and other seafood. Recovery is usually complete in 1–3 days. Food and stool cultures are positive.
Toxin causes hypersecretion in small intestine. Infective dose $>10^5$ vibrios.	Abrupt onset of liquid diarrhea in endemic area. Needs prompt replacement of fluids and electrolytes IV or orally. Tetracyclines shorten excretion of vibrios. Stool cultures positive.
Organisms invade epithelial cells; blood, mucus, and neutrophils in stools. Infective dose $<10^3$ organisms.	Abrupt onset of diarrhea, often with blood and pus in stools, cramps, tenesmus, and lethargy. Stool cultures are positive. Trimethoprim sulfamethoxazole, ampicillin, or chloramphenicol given in severe cases. Do not give opiates. Often mild and self-limited. Restore fluids.
Superficial infection of gut, little invasion. Infective dose $>10^5$ organisms.	Gradual or abrupt onset of diarrhea and low-grade fever. Nausea, headache, and muscle aches common. Administer no antimicrobials unless systemic dissemination is suspected. Stool cultures are positive. Prolonged carriage is frequent.
Symptoms probably due to endotoxins and tissue inflammation; infective dose $\geq 10^7$ organisms.	Initially fever, headache, malaise, anorexia, and muscle pains. Fever may reach 40°C by the end of the first week of illness and lasts for 2 or more weeks. Diarrhea often occurs, and abdominal pain, cough, and sore throat may be prominent. Antibiotic therapy shortens duration of the illness.
Toxins causes epithelial necrosis in colon; pseudomembranous colitis.	Especially after abdominal surgery, abrupt bloody diarrhea and fever. Toxins in stool. Oral vancomycin useful in therapy.
Invasion of mucous membrane; toxin production uncertain	Fever, diarrhea; PMNs and fresh blood in stool, especially in children. Usually self-limited. Special media needed for culture at 43°C. Erythromycin given in severe cases with invasion. Usual recovery in 5–8 days.
Gastroenteritis or mesenteric adenitis; occasional bacteremia; toxin produced occasionally.	Severe abdominal pain, diarrhea, fever; PMNs and blood in stool; polyarthritis, erythema nodosum, especially in children. Gentamicin used in severe cases. Keep stool specimen at 4°C before culture.

Table 38.4 Water-Borne Bacterial Pathogens

Organism	*Reservoir*	*Comments*
Aeromonas hydrophila	Free-living	Sometimes associated with gastroenteritis, cellulitis, and other diseases
Campylobacter	Bird and animal reservoirs	Major cause of diarrhea; common in processed poultry; a microaerophile
Helicobacter pylori	Free-living	Can cause gastritis, peptic ulcers, gastric adenocarcinomas
Legionella pneumophila	Free-living and associated with protozoa	Found in cooling towers, evaporators, condensers, showers, and other water sources
Leptospira	Infected animals	Hemorrhagic effects, jaundice
Mycobacterium	Infected animals and free-living	Complex recovery procedure required
Pseudomonas aeruginosa	Free-living	Swimmer's ear and related infections
Salmonella enteriditis	Animal intestinal tracts	Common in many waters
Vibrio cholerae	Free-living	Found in many waters including estuaries
Vibrio parahaemolyticus	Free-living in coastal waters	Causes diarrhea in shellfish consumers
Yersinia enterocolitica	Frequent in animals and in the environment	Waterborne gastroenteritis

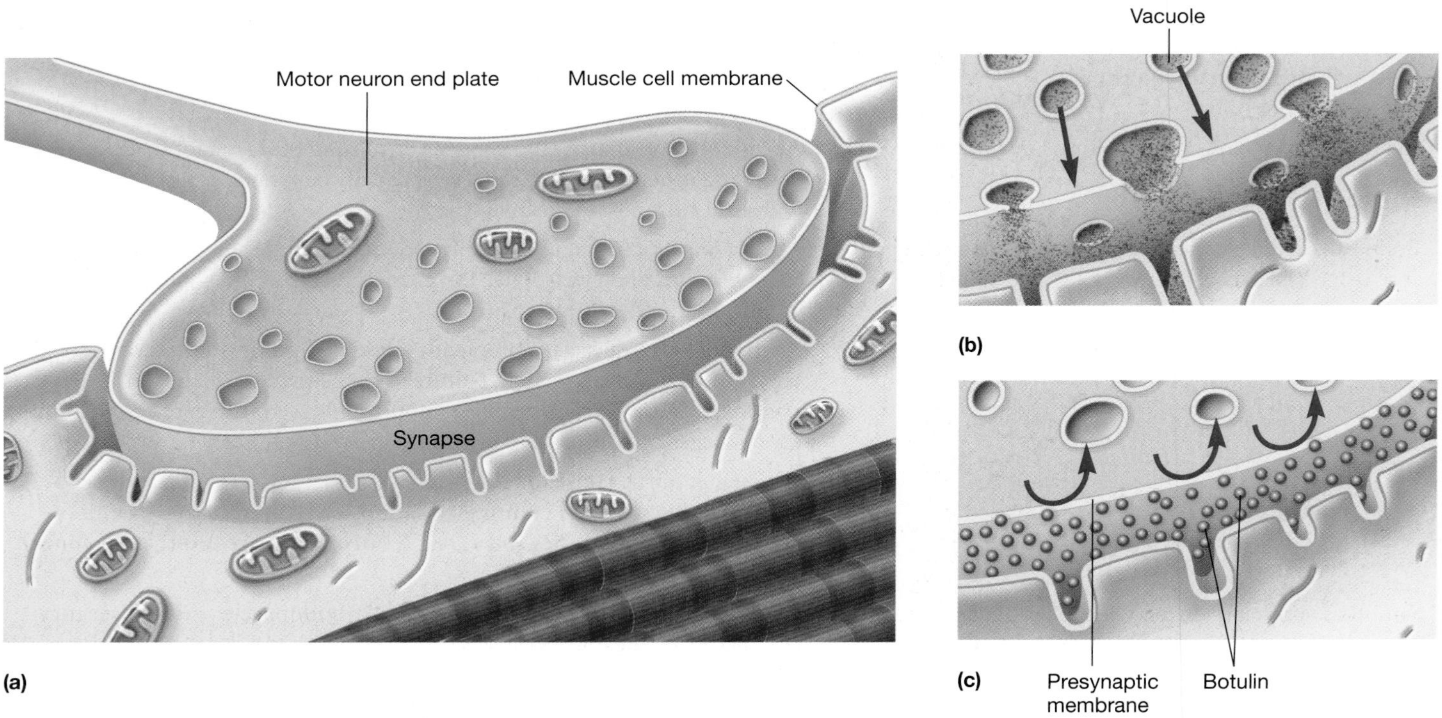

FIGURE 38.27 The Physiological Effects of Botulism Toxin. (a) The relationship between the motor neuron and the muscle at the neuromuscular junction. (b) In the normal state, acetylcholine released at the synapse crosses to the muscle and creates an impulse that stimulates muscle contraction. (c) In botulism, the toxin enters the motor end plate and attaches to the presynaptic membrane, where it blocks release of the acetylcholine. This prevents impulse transmission, and keeps the muscle from contracting.

Figure 38.27 Micro Inquiry

How do the neurological effects of botulism toxin differ from those of tetanus toxin?

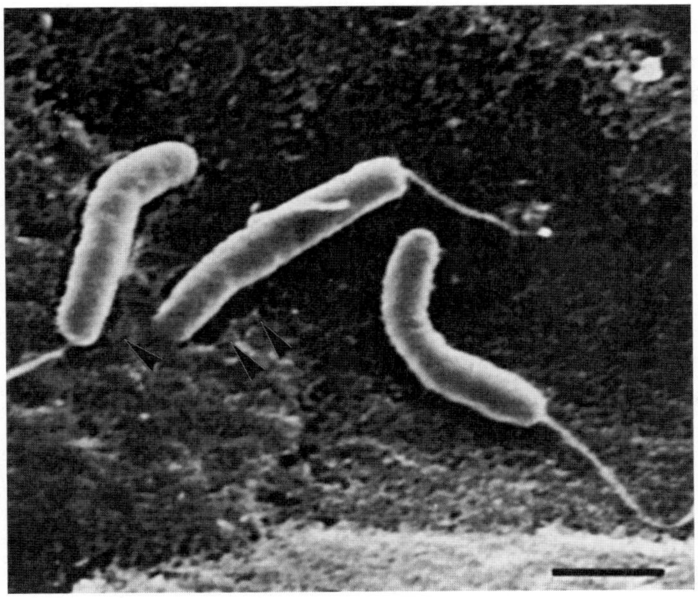

FIGURE 38.28 Cholera. *Vibrio cholerae* adhering to intestinal epithelium; scanning electron micrograph (×12,000). Notice that the bacteria are slightly curved with a single polar flagellum.

mucus floating in it. The diarrhea can be so profuse that a person can lose 10 to 15 liters of fluid during the infection. Death may result from the elevated concentrations of blood proteins, caused by reduced fluid levels, which leads to circulatory shock and collapse.

Evidence indicates that passage through the human host enhances infectivity, although the exact mechanism is unclear. Before *V. cholerae* exits the body in watery stools, some unknown aspect of the intestinal environment stimulates the activity of certain bacterial genes. These genes, in turn, seem to prepare the bacteria for ever more effective colonization of their next victims, possibly fueling epidemics. *V. cholerae* can also be free-living in warm, alkaline, and saline environments.

Treatment is by oral rehydration therapy with NaCl plus glucose to stimulate water uptake by the intestine; antibiotics may also be given. The most reliable control methods are based on proper sanitation, especially of water supplies. The mortality rate without treatment is often over 50%, but with treatment and supportive care, it is less than 1%.

1. How can *Clostridium botulinum* cause disease even when the bacteria are dead?
2. Compare the gastroenteritis caused by *Vibrio cholerae* with that of *Campylobacter jejuni*. How are they similar and different?

Escherichia coli Gastroenteritis

Millions of people travel yearly from country to country. Unfortunately, a large percentage of these travelers acquire a rapidly acting, dehydrating condition called traveler's diarrhea. This diarrhea results from an encounter with certain viruses, bacteria, or protozoa usually absent from the traveler's normal environment. One of the major causative agents is *Escherichia coli*. Interestingly, *E. coli* is undoubtedly the best-studied bacterium and the experimental organism of choice for many microbiologists. It inhabits the colon of humans and other warm-blooded animals, and it is quite useful in the identification of fecal contamination of water. *E. coli* circulates in the resident population, typically without causing symptoms due to the immunity afforded by previous exposure. Because many cells are needed to initiate infection, contaminated food and water are the major means by which they are spread. This is the basis for the popular warnings to international travelers: "Don't drink the local water" and "Boil it, peel it, cook it, or forget it."

Although the vast majority of *E. coli* strains are nonpathogenic members of the normal intestinal microbiota, some strains cause diarrheal disease by several mechanisms. Six categories or strains of diarrheagenic *E. coli* are now recognized (**figure 38.29**): enterotoxigenic *E. coli* (ETEC), enteroinvasive *E. coli* (EIEC), enterohemorrhagic *E. coli* (EHEC), enteropathogenic *E. coli* (EPEC), enteroaggregative *E. coli* (EAggEC), and diffusely adhering *E. coli* (DAEC). ◀◀ *Order* Enterbacteriales *(section 20.3)*

The **enterotoxic *E. coli* (ETEC)** strains produce one or both of two distinct enterotoxins, which are responsible for the diarrhea and distinguished by their heat stability: heat-stable enterotoxin (ST) and heat-labile enterotoxin (LT) (figure 38.29*a*). The genes for ST and LT production and for colonization factors are usually plasmid-borne and acquired by horizontal gene transfer. ST binds to a glycoprotein receptor that is coupled to guanylate cyclase on the surface of intestinal epithelial cells. Activation of guanylate cyclase stimulates the production of cyclic guanosine monophosphate (cGMP). This leads to the secretion of electrolytes and water into the lumen of the small intestine, manifested as the watery diarrhea characteristic of an ETEC infection. LT binds to specific gangliosides on epithelial cells and activates membrane-bound adenylate cyclase, which leads to increased production of cyclic adenosine monophosphate (cAMP) through the same mechanism employed by cholera toxin. Again, the result is hypersecretion of electrolytes and water into the intestinal lumen.

The **enteroinvasive *E. coli* (EIEC)** strains cause diarrhea by penetrating and multiplying within the intestinal epithelial cells (figure 38.29*b*). The ability to invade epithelial cells is associated with the presence of a large plasmid; EIEC may also produce a cytotoxin and an enterotoxin.

The **enteropathogenic *E. coli* (EPEC)** strains attach to the brush border of intestinal epithelial cells and cause a specific type of cell damage called effacing lesions (figure 38.29*c*). Effacing lesions or attaching-effacing (AE) lesions represent destruction of brush border microvilli adjacent to adhering bacteria.

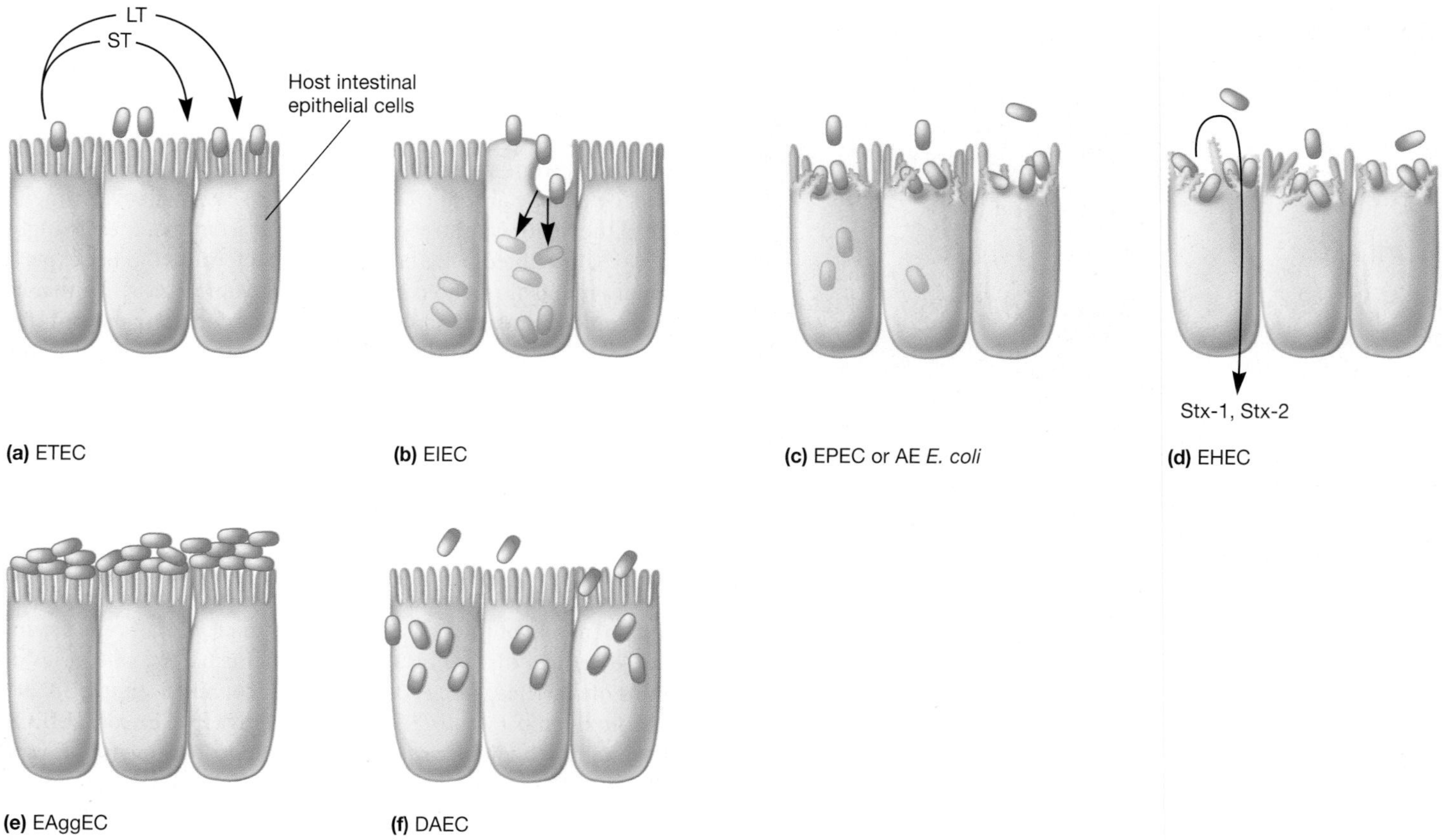

FIGURE 38.29 Six Strains of Diarrheagenic *E. coli*. Each strain of diarrhea-causing *E. coli* can be classified by the nature of its interaction with host intestinal epithelial cells.

Figure 38.29 Micro Inquiry

To which *E. coli* class does the strain O157:H7 belong?

AE lesion formation is a now known to result from the delivery of specific virulence proteins into host cells, through a type III secretion (TTS) system (figure 38.10). These virulence proteins are essential for the subversion of the intestinal epithelial host cell signal transduction pathways, in addition to directly causing AE lesions. Scanning electron microscopy has captured a filamentous TTS microinjection apparatus emanating from the EPEC, mediating EPEC adhesion to a host cell, and inserting into the host cell (**figure 38.30**). The EPEC virulence proteins are homologous to *Yersinia* TTS Yop proteins that lyse red blood cells. Cell destruction leads to the subsequent diarrhea. As a result of this pathology, the term AE *E. coli* is used to describe true EPEC strains. It is now known that AE *E. coli* is an important cause of diarrhea in children residing in developing countries.

The **enterohemorrhagic *E. coli* (EHEC)** strains carry the bacteriophage-encoded genetic determinants for Shiga-like toxin (Stx-1 and Stx-2 proteins; figure 38.29*d*). EHEC also produce AE lesions, causing hemorrhagic colitis with severe abdominal pain and cramps followed by bloody diarrhea. Stx-1 and Stx-2 (previously verotoxins 1 and 2) have also been implicated in the extraintestinal disease hemolytic uremic syndrome, a severe hemolytic anemia that leads to kidney failure. It is believed these toxins kill vascular endothelial cells. A major form of EHEC is *E. coli* O157:H7, which has caused many outbreaks of hemorrhagic colitis in the United States since it was first recognized in 1982. Currently there are an estimated 73,000 *E. coli* O157:H7 cases in the United States each year, resulting in 60 deaths. Other serotypes of *E. coli* can cause similar disease, but they do not typically carry the genes for Shiga-like toxin. However, most laboratories do not test for non-O157 strains, so the actual incidence of EHEC is underreported.

The **enteroaggregative *E. coli* (EAggEC)** strains adhere to epithelial cells in localized regions, forming clumps of bacteria

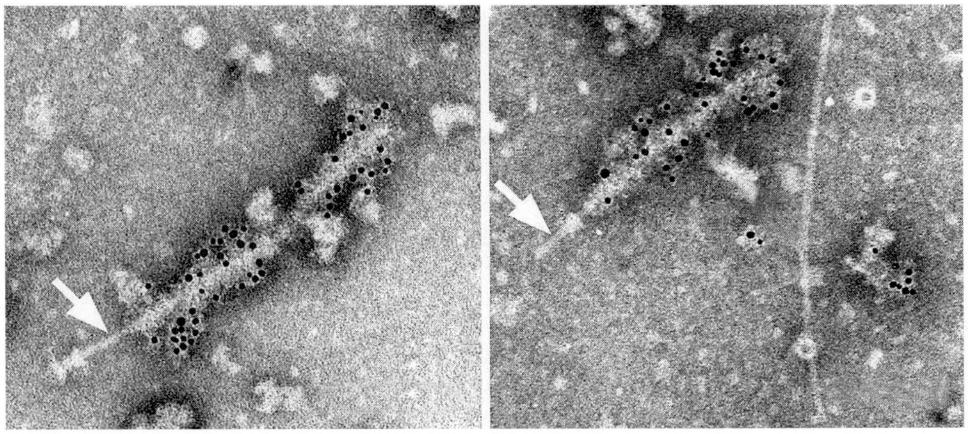

FIGURE 38.30 Type three secretion system (TTSS) of Enteropathogenic *E. coli* (EPEC). Transmission electron micrographs of TTSS structures from EPEC cells. TTSS sheath-like structures are labeled with 6 nm immunogold particles; the needle complexes, neck and stem components (white arrows) remain unlabeled.

with a "stacked brick" appearance (figure 38.29*e*). Conventional extracellular toxins have not been detected in EAggEC, but unique lesions are seen in epithelial cells, suggesting the involvement of toxins.

The **diffusely adhering *E. coli* (DAEC)** strains adhere over the entire surface of epithelial cells and usually cause disease in immunologically naïve or malnourished children (figure 38.29*f*). It has been suggested that DAEC may have an as-yet undefined virulence factor.

Diagnosis of traveler's diarrhea caused by *E. coli* is based on past travel history and symptoms. Laboratory diagnosis is by isolation of the specific type of *E. coli* from feces and identification using DNA probes, the determination of virulence factors, and the polymerase chain reaction. Treatment is with fluid and electrolytes plus doxycycline and trimethoprim-sulfamethoxazole. Recovery can be without complications except in EHEC damage to kidneys. Prevention and control involve avoiding contaminated food and water.

Sepsis and Septic Shock

Gram-negative sepsis is most commonly caused by *E. coli, Salmonella* spp., *Klebsiella* spp., *Enterobacter* spp., or *Pseudomonas aeruginosa*. Endotoxin, or more specifically, the lipid A moiety of lipopolysaccharide (LPS), an integral component of the outer membrane of gram-negative bacteria, has been implicated as a primary initiator of septic shock. Endotoxin-containing gram-negative bacteria can initiate the pathogenic cascade of sepsis leading to septic shock, although gram-positive bacteria and fungi are also known to do this. We discuss sepsis and septic shock here because (1) some microbial diseases and their effects cannot be categorized under a specific mode of transmission, and (2) we discuss many of the gram-negative pathogens in this section. ⏮ *Gram-negative cell walls (section 3.3)*

Septic shock is the most common cause of death in intensive care units and the thirteenth most common cause of death in the United States. Unfortunately, the incidence of sepsis and septic shock continues to rise: 400,000 cases of sepsis and 200,000 episodes of septic shock are estimated to occur annually in the United States, resulting in more than 100,000 deaths.

Sepsis has been redefined by physicians as the systemic response to a microbial infection. This response is manifested by two or more of the following conditions: temperature above 38°C or below 36°C; heart rate above 90 beats per minute; respiratory rate above 20 breaths per minute; leukocyte count above 12,000 cells per cubic milliliter or below 4,000 cells per cubic milliliter. **Septic shock** is sepsis associated with severe hypotension (low blood pressure) despite adequate fluid replacement.

The pathogenesis of sepsis and septic shock begins with the proliferation of the microorganism at the infection site (*see section 31.2*). The microorganism may invade the bloodstream directly or may proliferate locally and release various products into the bloodstream. These products include both structural components of the microorganisms (endotoxin, teichoic acid antigen) and exotoxins synthesized by the microorganism. All of these products can stimulate the release of the endogenous mediators of sepsis from endothelial cells, monocytes, macrophages, neutrophils, and plasma cell precursors.

The endogenous mediators have profound physiological effects on the heart, vasculature, and other body organs. Because there is no drug therapy (despite vigorous research efforts), the consequences of septic shock are either recovery or death. Death usually ensues if one or more organ systems fail completely.

1. What are the mechanisms by which *E. coli* can cause disease?
2. Compare toxigenic *E. coli* disease with invasive disease. How are they different?
3. Explain how gram-negative bacteria cause sepsis.

Salmonellosis

Salmonellosis (*Salmonella* gastroenteritis) is caused by over 2,000 *Salmonella* serovars. Based on DNA homology studies, all known *Salmonella* are thought to belong to a single species, *Salmonella enterica*, although the taxonomy of this bacterium remains controversial. The most frequently isolated serovars from humans are Typhimurium and Enteritidis. (Serovar names are not italicized and the first letter is capitalized.) An estimated 1.4 million cases of *Salmonella infections* occur annually in the

United States; of these, approximately 45,000 are culture-confirmed and reported to CDC.

The initial source of the bacterium is the intestinal tracts of birds and other animals. Humans acquire the bacteria from contaminated water or foods such as beef products, poultry, eggs, or egg products. Once the bacteria are in the body, the incubation time is only about 8 to 48 hours. The disease results when the bacteria multiply and invade the intestinal mucosa, where they produce an enterotoxin and a cytotoxin that destroys epithelial cells. Abdominal pain, cramps, diarrhea, nausea, vomiting, and fever are the most prominent symptoms, which usually persist for 2 to 5 days but can last for several weeks. During the acute phase of the disease, as many as 1 billion *Salmonella* can be found per gram of feces. Most adult patients recover, but the loss of fluids can cause problems for children and elderly people.

Other serovars of *Salmonella* are infrequently isolated from humans. We mention a few of these because of their ability to cause occasional epidemics. **Typhoid** (Greek *typhodes*, smoke) fever is caused by *Salmonella enterica* serovar Typhi and is acquired by ingestion of food or water contaminated by feces of infected humans or person-to-person contact. In earlier centuries, the disease occurred in great epidemics. The incubation period for typhoid fever is about 10 to 14 days. The bacteria colonize the small intestine, penetrate the epithelium, and spread to the lymphoid tissue, blood, liver, and gallbladder. After approximately 3 months, most individuals stop shedding bacteria in their feces. However, a few individuals continue to shed *Salmonella* for extended periods but show no symptoms. In these carriers, the bacteria continue to grow in the gallbladder and reach the intestine through the bile duct. This was almost certainly the case for "Typhoid Mary" (*see Historical Highlights 36.3*).

More recently, *S.* Agona, and *S.* Tennessee have been reported as contaminants of cereal products and peanut butter, respectively. The ongoing concern over food safety has prompted greater diligence for the detection and reporting of food contamination,so it is not so unusual to learn of foods contaminated with *Salmonella*. The CDC and U.S. Food and Drug Authority (FDA) respond quickly with efforts to identify the strain and remove contaminated foods from the supply chain. These recent foodborne contaminations involving *Salmonella* demonstrate (1) the continued potential for farmed food contamination and (2) the speed with which a public health response can occur. Of note, however, is the increasing concern regarding infection resulting from antibiotic-resistant *Salmonella*. *S.* Typhimurium is now resistant to at least five antimicrobial agents. Another serovar infrequently isolated from humans, *S.* Newport, is resistant to at least seven agents.

Shigellosis

Shigellosis, or bacillary dysentery, is a diarrheal illness resulting from an acute inflammatory reaction of the intestinal tract caused by the four species of the genus *Shigella*. About 20,000 to 25,000 cases a year are reported in the United States, and around 600,000 deaths a year worldwide are due to bacillary dysentery.

Shigella is restricted to human hosts. *S. sonnei* is the usual pathogen in the United States and Britain, but *S. flexneri* is also fairly common. The organism is transmitted by the fecal-oral route—primarily by food, fingers, feces, and flies (the four "F's")—and is most prevalent among children, especially one- to four-year-olds. The infectious dose is only around 10 to 100 bacteria. In the United States, shigellosis is a particular problem in day-care centers and crowded custodial institutions.

The shigellae are intracellular parasites that multiply within the villus cells of the colonic epithelium. The bacteria induce Peyer's patch cells to phagocytose them. After being ingested, the bacteria disrupt the phagosome membrane and are released into the cytoplasm, where they reproduce. They then invade adjacent mucosal cells. Shigellae initiate an inflammatory reaction in the mucosa. Both endotoxins and exotoxins may participate in disease progression, but the bacteria do not usually spread beyond the colonic epithelium. The watery stools often contain blood, mucus, and pus. In severe cases, the colon can become ulcerated. Virulent *Shigella* produce a heat-labile AB exotoxin known as Shiga-toxin (Sxt, formerly called verotoxin). The complete toxin molecule is composed of one A protein surrounded by five B proteins. The B proteins attach to host vascular cells, stimulating the internalization of the whole toxin. The A subunit protein is subsequently released from the B protein units and binds to host ribosomes, inhibiting protein synthesis. A specific target of the B protein seems to be the glomerular endothelium; toxin action on these cells leads to kidney failure. Additionally, like *Y. pestis* and the enteroaggregative strain of *E. coli*, shigellae also use a type III secretion system to deliver specific virulence factors to target epithelial cells. ◀◀ *Phagocytosis (section 32.5); Exotoxins (section 31.3)*

The incubation period for shigellosis usually ranges from 1 to 3 days, and the organisms are shed over a period of 1 to 2 weeks. The disease normally is self-limiting in adults and lasts an average of 4 to 7 days; in infants and young children, it may be fatal. Fluid and electrolyte replacement are usually sufficient; antibiotics may not be required in mild cases, although they can shorten the duration of symptoms and limit transmission to family members. Sometimes, particularly in malnourished infants and children, neurological complications and kidney failure occur. Antibiotic-resistant strains are becoming a problem. Prevention is a matter of good personal hygiene and maintenance of a clean water supply.

Staphylococcal Food Poisoning

Staphylococcal food poisoning is the major type of food intoxication in the United States. It is caused by ingestion of improperly stored or cooked food (particularly foods such as ham, processed meats, chicken salad, pastries, ice cream, and hollandaise sauce) in which *S. aureus* has been allowed to incubate. Because staphylococci produce heat-stable enterotoxins, the food can be extensively and properly cooked, killing the bacteria without destroying the toxin. Thirteen different enterotoxins have been identified; enterotoxins A, B, C1, C2, D, and E are the most common. Enterotoxins A and B are superantigens. Typical symptoms include severe abdominal pain, cramps, diarrhea, explosive vomiting, and nausea. The onset of symptoms is rapid (usually 1 to 8 hours) and of short duration (usually less than 24 hours). The mortality rate of staphylococcal food poisoning is negligible among healthy individuals.

Treatment of staph food poisoning primarily involves keeping hydrated while letting the toxin flush from the body. The toxemia usually resolves without medication. Death due to staph food poisoning is very rare, but it has occurred in infants, the elderly, and individuals who are immunocompromised. Prevention is through thorough hand washing during food preparation and the use of safe cooking and dining practices.

1. How is it that *Salmonella* can sometimes contaminate peanut butter?
2. Compare the intestinal diseases caused by *Salmonella* and *Shigella*. How are they the same? Different?
3. What is the origin of staphylococci that contaminate foods? How is this prevented?

38.5 Zoonotic Diseases

Diseases transmitted from animals to humans are called zoonotic diseases, or zoonoses. A number of important human pathogens begin as normal flora or parasites of animals and can often adapt to cause disease in humans. Here, we highlight a few of the more notable diseases and the agents that cause them. Recall that we identify organisms of bioterror potential by two astericks (**).

**Anthrax

Anthrax (Greek *anthrax*, coal) is a highly infectious animal disease caused by the gram-positive, endospore-forming *Bacillus anthracis*. *B. anthracis* is found worldwide and can be transmitted to humans by direct contact with infected animals (cattle, goats, sheep) or their products, especially hides. *B. anthracis* spores can remain viable in soil and animal products for decades (*see figure 3.52*). Although *B. anthracis* is one of the most molecularly monomorphic (of one shape) bacteria, it is now possible to separate all known strains into five categories (providing some clues to their geographic sites of origin) based on the number of tandem repeats in various genes.

There are three forms of anthrax. When the bacterium enters through a cut or abrasion of the skin, **cutaneous anthrax** results. Inhaling spores may result in **pulmonary anthrax,** also known as woolsorter's disease. If spores reach the gastrointestinal tract, **gastrointestinal anthrax** may result. *B. anthracis* bacteremia can develop from any form of anthrax. Discounting the 2001 bioterrorism events in the United States, more than 95% of anthrax is the cutaneous form. Spores and vegetative bacteria are detected by skin macrophages and dendritic cells. In pulmonary anthrax, the spores (1 to 2 μm in diameter) are inhaled and lodge in the alveolar spaces, where they are engulfed by alveolar macrophages. Symptoms of gastrointestinal anthrax typically occur after the ingestion of undercooked meat containing spores and include nausea, vomiting, fever, and abdominal pain.

For a successful infection, *B. anthracis* must evade the host's innate immune system by killing phagocytic macrophages. The bacteria have two principal virulence factors that allow them to evade host defenses. The virulence factors are encoded on two plasmids—one involved in the synthesis of a polyglutamyl capsule that inhibits phagocytosis and the other bearing the genes for the synthesis of its exotoxin. *B. anthracis* produces a complex exotoxin composed of three proteins: protective antigen (PA), edema factor (EF), and lethal factor (LF). Macrophages have many receptors (capillary morphogenesis protein-2) on their plasma membranes to which the PA portion of the anthrax exotoxin attaches. Attachment continues until seven PA-receptor complexes gather in a doughnut-shaped ring (**figure 38.31**). The ring acts like a syringe, boring through the plasma membrane of the macrophage. It then binds EF and LF, and the entire complex is engulfed by the macrophages' plasma membrane and shuttled to an endosome inside the cell (*see figure 4.10*). Once there, the PA molecules form a pore that pierces the endosomal membrane,

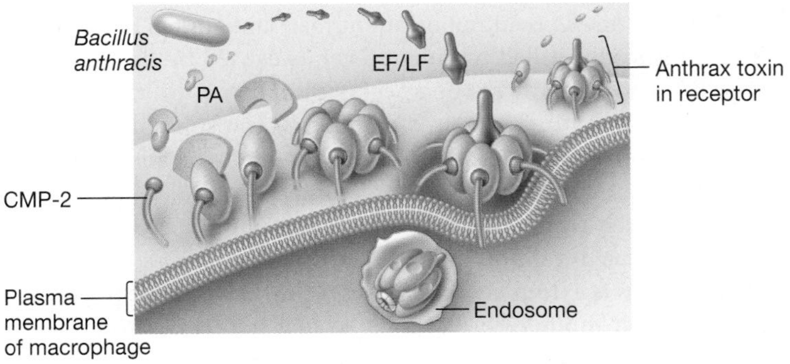

FIGURE 38.31 Anthrax. A protein called protective antigen (PA) delivers two other proteins, edema factor (EF) and lethal factor (LF), to the capillary morphogenesis protein-2 (CMP-2) receptor on the cell membrane of a target macrophage, where PA, EF, and LF are transported to an endosome. PA then delivers EF and LF from the endosome into the cytoplasm of the macrophage, where they exert their toxic effects.

Figure 38.31 Micro Inquiry

By what mechanisms do EF and LF kill macrophages?

and EF and LF enter the cytoplasm. EF has adenylate cyclase activity so intracellular cAMP increases. Toxin activity results in fluid release, or the formation of edema. Additionally LF prevents the transcription factor nuclear factor B (NFkB) from regulating numerous cytokine and other immunity genes needed to promote macrophage survival. As thousands of macrophages die, they release their lysosomal contents, leading to fever, internal bleeding, septic shock, and rapid death. Without antibiotic treatment, mortality rates approach 100% for inhalational and gastrointestinal anthrax, and between 20 and 50% for cutaneous anthrax.

Between 20,000 and 100,000 cases of anthrax are estimated to occur worldwide annually; in the United States, the annual incidence was less than 1 case per year—a rate maintained for 20 years until 2001. The 2001 occurrence of 22 cases of anthrax has spotlighted the real concern about anthrax as a weapon of bioterrorism.

Treatment of anthrax is with ciprofloxacin, penicillin, or doxycycline (the same antibiotics may be used for all three forms of anthrax) and is successful only if begun before a critical concentration of toxin has accumulated. Although antibiotics may kill the bacterium or suppress its growth, the exotoxin can still eventually kill the patient. Vaccination of animals, primarily cattle, is an important control measure. However, people with a high occupational risk, such as those who handle infected animals or their products, including hides and wool, should be immunized. U.S. military personnel also receive the vaccine.

**Brucellosis (Undulant Fever)

The genus *Brucella* is an important human and animal pathogen. **Brucellosis,** also called undulant fever, is caused by tiny, faintly staining coccobacilli of the species *B. abortus, B. melitensis, B. suis,* or *B. canis. Brucella* is commonly transmitted through consumption of contaminated animal products or abrasions of the skin from handling infected mammals (cattle, sheep, goats, pigs, and rarely from dogs). The most common route is ingestion of contaminated milk products. Direct person-to-person spread of brucellosis is extremely rare. However, infants may be infected through their mother's breast milk; sexual transmission of brucellosis has also been reported. Although uncommon, transmission of brucellosis may also occur through transplantation of contaminated blood or tissue. A controversy in the northwestern United States over the transmission of *Brucella*, endemic in the wild bison and elk populations, to otherwise *Brucella*-free cattle, has continued for many years. The loss of the brucellosis-free status has serious economic consequences. In 2008 Montana lost its brucellosis-free status. It is thought that the *Brucella*-infected cattle in that state obtained the pathogen from infected elk. ◀◀ *Class* Alphaproteobacteria *(section 20.1)*

In the United States, human *Brucella* infections (primarily *B. melitensis*) occur more frequently when individuals ingest unpasteurized milk or dairy products. Brucellosis also has occurred in laboratory workers—culturing the organisms concentrates them and increases the risk of their aerosolization. Naturally occurring cases in the United States are typically reported from California, Florida, Texas and Virginia. For the past 10 years, approximately 100 cases of brucellosis have been reported annually. Most of these cases have been in abattoir (slaughterhouse) workers, meat inspectors, animal handlers, veterinarians, and laboratorians. Important for tourists is the potential infection through unpasteurized cheeses, sometimes called "village cheeses."

Brucellosis, in the acute (8 weeks from onset) form, presents as nonspecific, flulike symptoms, including fever, sweats, malaise, anorexia, headache, myalgia (muscle pain), and back pain. In the undulant (rising and falling) form, symptoms of brucellosis include undulant fevers, arthritis, and testicular inflammation. Neurologic symptoms may occur acutely in up to 5% of the cases. In the chronic form (1 year from onset), brucellosis symptoms may include chronic fatigue, depression, and arthritis. Mortality is low, less than 2%. Treatment is usually with doxycycline and rifampin in combination for 6 weeks to prevent recurring infection. The insidious disease and significant morbidity associated with *Brucella* infection have prompted its restriction as a select agent; the public health impact, should it be used as a bioweapon, would be substantial. ◀◀ *Bioterrorism preparedness (section 36.8)*

Psittacosis (Ornithosis)

Psittacosis (ornithosis) is a worldwide infectious disease of birds that is transmissible to humans. Fewer than 50 confirmed cases are reported in the United States each year. Psittacosis was first described in association with parrots and parakeets, both of which are psittacine birds. The disease is now recognized in many other birds—among them, pigeons, chickens, ducks, and turkeys—and the general term ornithosis (Latin *ornis*, bird) is used.

Ornithosis is caused by *Chlamydophilia psittaci.* Humans contract this disease either by handling infected birds or by inhaling dried bird excreta that contains viable *C. psittaci.* Ornithosis is recognized as an occupational hazard within the poultry industry, particularly to workers in turkey-processing plants. Incubation is typically 5 to 19 days. After entering the respiratory tract, the bacteria are transported to the cells of the liver and spleen. They multiply within these cells and then invade the lungs, where they cause inflammation, hemorrhaging, and pneumonia. Endocarditis (inflammation of the heart muscle), hepatitis, and neurologic complications may occasionally occur. With antibiotic therapy, the mortality rate is about 2%.

**Q Fever

The Q in **Q fever** stands for query because the cause of the fever was not known for some time. The disease is caused by the γ-proteobacterium *Coxiella burnetii. C. burnetii* can survive outside host cells by forming a resistant, endospore-like body (**figure 38.32**). This bacterium infects both wild animals

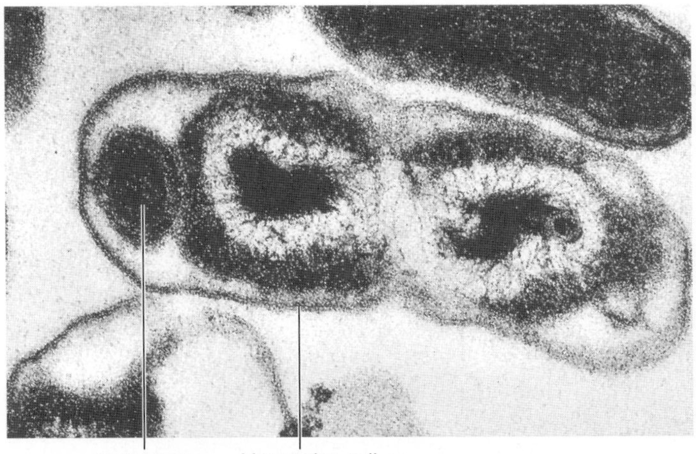

FIGURE 38.32 ***Coxiella burnetti.*** Note the unique endospore-like structure within the vegetative cell.

and livestock. Cattle, sheep, and goats are the primary reservoirs. It does not need an arthropod vector for transmission, but in animals, ticks (many species) transmit *C. burnetii*. By contrast, human transmission is primarily by inhalation of dust contaminated with bacteria from dried animal feces or urine, or consumption of unpasteurized milk. Importantly, organisms are shed in high numbers within the amniotic fluids and the placenta during birthing. The organisms are resistant to heat, drying, and many common disinfectants. These features enable the bacteria to survive for long periods in the environment. The disease can occur in epidemic form among slaughterhouse workers and sporadically among farmers and veterinarians. Each year, fewer than 30 cases of Q fever are reported in the United States. ⏮ *Class* Gammaproteobacteria *(section 20.3)*

In humans, Q fever is an acute illness characterized by the sudden onset of severe headache, malaise, confusion, sore throat, chills, sweats, nausea, chest pain, myalgia, and high fever. It is rarely fatal, but endocarditis occurs in about 10% of the cases. Five to 10 years may elapse between the initial infection and the appearance of the endocarditis. During this interval, the bacteria reside in the liver and often cause hepatitis. The fatality rate is approximately 4%.

Diagnosis of Q fever is made by national reference laboratories using PCR or culture of the bacterium and fluorescent antibody and agglutination tests. Treatment is with doxycycline, the antibiotic of choice for acute Q fever. Antibiotic treatment is most effective when initiated within the first 3 days of illness. Prevention and control involve public education, protective clothing, and vector control. An attenuated live vaccine is available from the U.S. Army for high-risk laboratory workers. Importantly, *C. burnetii* is a microorganism of concern as a biological threat agent.

**Tularemia

The gram-negative bacterium *Francisella tularensis* is widely found in animal reservoirs in the United States and causes the disease **tularemia** (from Tulare, a county in California where the disease was first described). It may be transmitted to humans by biting arthropods (ticks, deer flies, or mosquitoes), direct contact with infected tissue (rabbits), inhalation of aerosolized bacteria, or ingestion of contaminated food or water. However, tularemia is most often transmitted through contact with infected animals; it is called rabbit fever in the central United States because it is often a disease of hunters. After an incubation period of 2 to 10 days, a primary ulcerative lesion appears at the infection site, lymph nodes enlarge, and a high fever develops. ⏮ *Class* Gammaproteobacteria *(section 20.3)*

Diagnosis is made by national reference laboratories using PCR or culture of the bacterium and fluorescent antibody and agglutination tests. Treatment is with streptomycin, tetracycline, or aminoglycoside antibiotics. Prevention and control involve public education, protective clothing, and vector control. An attenuated live vaccine is available from the U.S. Army for high-risk laboratory workers. Fewer than 200 cases of tularemia are reported annually in the United States. *F. tularensis* is also a microorganism of concern as a biological threat agent. Because public health preparedness efforts in the United States have shifted toward a stronger defense against biological terrorism, public health and medical management protocols following a potential release of tularemia are now in place.

1. How can humans acquire anthrax? Brucellosis?
2. Describe the symptoms of the disease as related to the infection process for anthrax and brucellosis.
3. How is ornithosis transmitted?
4. Describe the disease of tularemia. Why is tularemia considered a potential agent of bioterrorism?

38.6 Opportunistic Diseases

Recall that the normal microbiota of a human can also become pathogenic, especially if they gain access to a tissue site that is not their regular environment or they overgrow other normal flora. Microbes that are otherwise members of the normal microbiota but become pathogens are referred to as opportunists; they cause opportunistic disease. We discuss a few examples of opportunistic disease next.

Antibiotic-Associated Colitis (Pseudomembranous Colitis)

Antibiotic-associated colitis is a spectrum of disease caused by *Clostridium difficile*. The disease spectrum (*C. difficile*–associated disease or CDAD) includes uncomplicated diarrhea, pseudomembranous colitis (a viscous collection of inflammatory cells, dead cells, necrotic tissue, and fibrin that obstructs the intestine), and toxic megacolon (inflammation that results in intestinal tissue death), which can, in some instances, lead to sepsis and even death (**figure 38.33**). *C. difficile* is an anaerobic, spore-forming bacillus that can be found in the intestines of healthy people. It is an opportunist that is usually kept in check by other intestinal microbiota. Excessive antibiotic use can eliminate some of the bacteria that suppress *C. difficile* overgrowth. The most common antibiotics associated with the development of *C. difficile* infection are amoxicillin, ampicillin, cephalosporins, and clindamycin. When antibiotics reduce the other bacteria, *C. difficile* bacteria multiply, produce toxins, and sporulate. *C. difficile* toxins include an enterotoxin (toxin A) and a cytotoxin (toxin B), both of which are responsible for the inflammation and diarrhea. The most common symptom of CDAD is watery diarrhea, consisting of three or more bowel movements per day for 2 or more days. Other common symptoms include fever, loss of appetite, nausea, and abdominal cramping or tenderness.

C. difficile is the most commonly recognized cause of diarrhea in hospitalized patients, as spores from one patient can be transmitted to others. Antibiotic use is the main modifiable risk factor for CDAD. In addition to suppressing other bacteria promoting *C. difficile* overgrowth, antibiotics increase the risk of selecting for antimicrobial-resistant strains. It is likely that alteration of the complex colonic ecology provides an environment for *C. difficile* to thrive and produce disease. Ironically, treatment of CDAD is by oral metronidazole or oral vancomycin. A newer therapy showing success is the use of a probiotic milk shake containing *Lactobacillus casei, L. bulgaricus,* and *Streptococcus thermophilus* to prevent antibiotic-associated diarrhea caused by *C. difficile.*

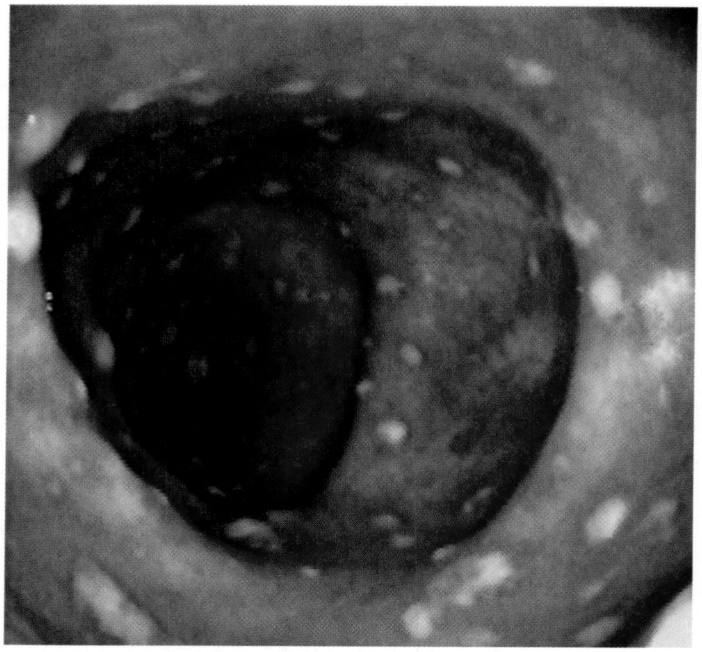

FIGURE 38.33 Pseudomembranous Colitis. Endoscopic view of *C. difficile*-associated colitis.

Figure 38.33 Micro Inquiry

What cells and tissues make up the pseudomembrane?

Bacterial Vaginosis

Bacterial vaginosis is caused by *Gardnerella vaginalis* (a gram-positive to gram-variable, pleomorphic, nonmotile rod), *Mobiluncus* spp. (gram-negative, anaerobic rods), and various other anaerobic bacteria. These microorganisms inhabit the vagina and rectum of 20 to 40% of healthy women, suggesting an opportunistic etiology; however, some consider the disease to be sexually transmitted because the organisms can be shared through intimate contact. Vaginosis of bacterial origin is typically a mild disease, although it is a risk factor for obstetric infections, various adverse outcomes of pregnancy, and pelvic inflammatory disease.

Vaginosis is characterized by a copious, frothy, fishy-smelling discharge with varying degrees of pain or itching. Diagnosis is based on this fishy odor and the microscopic observation of clue cells in the discharge. Clue cells are sloughed-off vaginal epithelial cells covered with bacteria, mostly *G. vaginalis*. Treatment for bacterial vaginosis is with metronidazole (Flagyl, MetroGel-Vaginal), a drug that kills anaerobic streptococci and the *Mobiluncus* spp. that appear to be needed for the continuation of the disease.

Dental Diseases

Some microorganisms found in the oral cavity are discussed in section 30.3 and presented in figure 30.16. Of this large number of symbiotic microbiota, only a few bacteria can be considered true opportunistic dental pathogens, or odontopathogens. These few odontopathogens are responsible for the most common bacterial diseases in humans: tooth decay and periodontal disease.

Dental Plaque

The human tooth has a natural defense mechanism against bacterial colonization that complements the protective role of saliva. The hard enamel surface selectively absorbs acidic glycoproteins (mucins) from saliva, forming a membranous layer called the acquired enamel pellicle. This pellicle, or organic covering, contains many sulfate (SO_4^{2-}) and carboxylate ($-COO^-$) groups that confer a net negative charge to the tooth surface. Because most bacteria also have a net negative charge, there is a natural repulsion between the tooth surface and bacteria in the oral cavity. Unfortunately, this natural defense mechanism breaks down when dental plaque forms.

Dental plaque is one of the most dense collections of bacteria in the body—perhaps the source of the first microorganisms from a human to be seen under a microscope, by Antony van Leeuwenhoek in the seventeenth century. Dental plaque formation begins with the initial colonization of the pellicle by *Streptococcus gordonii, S. oralis,* and *S. mitis* (**figure 38.34**). These bacteria selectively adhere to the pellicle by specific ionic, hydrophobic, and lectinlike interactions. Once the tooth surface is colonized, subsequent attachment of other bacteria results from a variety of specific coaggregation reactions (**figure 38.35**). Coaggregation is the result of cell-to-cell recognition between genetically distinct bacteria. Many of these interactions are mediated by a lectin (a carbohydrate-binding protein) on one bacterium that interacts with a complementary carbohydrate on another bacterium. The most important species at this stage are *Actinomyces viscosus, A. naeslundii,* and *S. gordonii.* After these species colonize the pellicle, a microenvironment is created that allows *S. mutans* and *S. sobrinus* to become established on the tooth surface by attaching to these initial colonizers (figure 38.35) |◀◀ *Biofilms (section 7.7)*

S. mutans and *S. sobrinus* produce extracellular enzymes (glucosyltransferases) that polymerize the glucose moiety of sucrose into a heterogeneous group of extracellular, water-soluble, and water-insoluble glucan polymers and other polysaccharides. The fructose by-product can be used in fermentation. **Glucans** are branched-chain polysaccharides composed of glucose units. They act like a cement to bind bacterial cells together, forming a plaque ecosystem, a biofilm. Once plaque becomes established, the surface of the tooth becomes anoxic. This leads to the growth of strict anaerobic bacteria (*Bacteroides melaninogenicus, B. oralis,* and *Veillonella alcalescens*), especially between opposing teeth and in the dental-gingival crevices (**figure 38.36**).

After the microbial plaque ecosystem develops, bacteria produce lactic and possibly acetic and formic acids from sucrose and other sugars. Because plaque is not permeable to saliva, the acids are not diluted or neutralized, and they demineralize the enamel to produce a lesion on the tooth. It is this chemical lesion that initiates dental decay.

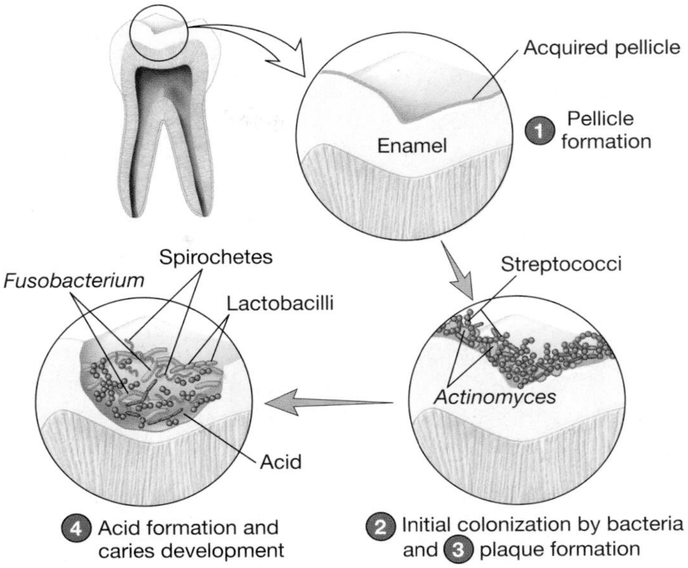

FIGURE 38.34 Stages in Plaque Development and Cariogenesis. A drawing of a microscopic view of pellicle and plaque formation, acidification, and destruction of tooth enamel.

Dental Decay (Caries)

As fermentation acids move below the enamel surface, they dissociate and react with the hydroxyapatite of the enamel to form soluble calcium and phosphate ions. As the ions diffuse outward, some reprecipitate as calcium phosphate salts in the tooth's surface layer to create a histologically sound outer layer overlying a porous subsurface area. Between meals and snacks, the pH returns to neutrality and some calcium phosphate reenters the lesion and crystallizes. The result is a demineralization-remineralization cycle.

When an individual eats fermentable foods high in sucrose for prolonged periods, acid production overwhelms the repair process, and demineralization is greater than remineralization. This leads to dental decay or **caries** (Latin, rottenness). Once the hard enamel has been breached, bacteria can invade the dentin and pulp of the tooth and cause its death.

No drugs are available to prevent dental caries. The main strategies for prevention include minimal ingestion of sucrose; daily brushing, flossing, and rinsing with mouthwashes; and professional cleaning at least twice a year to remove plaque. The use of fluorides in toothpaste, drinking water, and mouthwashes or fluoride and sealants applied professionally to the teeth protects against lactic and acetic acids and reduces tooth decay.

Periodontal Disease

Periodontal disease refers to a diverse group of inflammatory diseases that affect the periodontium and is the most common chronic infection in adults. The **periodontium** is the supporting structure of a tooth and includes the cementum, the periodontal membrane, the bones of the jaw, and the gingivae (gums). The gingiva is dense, fibrous tissue and its overlying mucous membrane that surrounds the necks of the teeth. The gingiva helps to hold the teeth in place. Disease is initiated by the formation of **subgingival plaque,** the plaque that forms at the dentogingival margin and extends down into the gingival tissue. Colonization of the subgingival region is aided by the ability of *Porphyromonas gingivalis* to adhere to substrates such as adsorbed salivary molecules, matrix proteins, epithelial cells, and bacteria in biofilms on teeth and epithelial surfaces. Binding to these substrates is mediated by *P. gingivalis* fimbrillin, the structural subunit of the major fimbriae. *P. gingivalis* does not use sugars as an energy

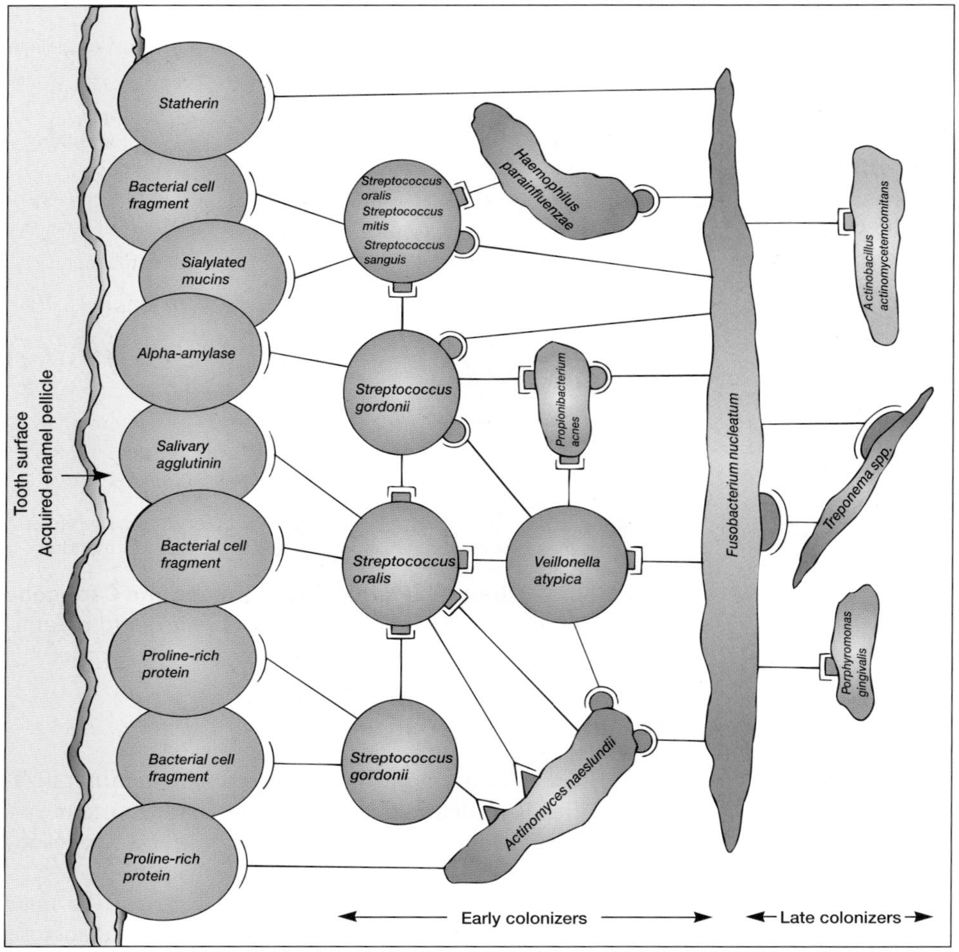

FIGURE 38.35 The Formation of Dental Plaque on a Freshly Cleaned Tooth Surface. The proposed temporal relationship of bacterial accumulation and multigeneric coaggregation during the formation of dental plaque on the acquired enamel pellicle. Early tooth surface colonizers coaggregate with each other, and late colonizers coaggregate with each other. With a few exceptions, early colonizers do not recognize late colonizers. After the tooth surface is covered with the earliest colonizers, each newly added bacterium becomes a new surface for recognition by unattached bacteria.

source but requires hemin as a source of iron and peptides for energy and growth. The bacterium produces at least three hemagglutinins and five proteases to satisfy these requirements. It is the proteases that are responsible for the breakdown of the gingival tissue. A number of other bacterial species contribute to tissue damage. The result is an initial inflammatory reaction known as **periodontitis,** which is caused by the host's immune response to both the plaque bacteria and the tissue destruction. This leads to swelling of the tissue and the formation of periodontal pockets. Bacteria colonize these pockets and cause more inflammation, which leads to the formation of a periodontal abscess; bone destruction, or periodontosis; inflammation of the gingiva, or gingivitis; and general tissue necrosis (**figure 38.37**). If the condition is not treated, the tooth may fall out of its socket. ◀◀ *Phylum Bacteroidetes (section 19.7)*

Periodontal disease can be controlled by frequent plaque removal; by brushing, flossing, and rinsing with mouthwashes; and at times, by oral surgery of the gums and antibiotics.

Streptococcal Pneumonia

Streptococcal pneumonia is now considered an opportunistic infection—in this case, it is contracted from one's own normal microbiota. It is caused by the gram-positive *Streptococcus pneumoniae*, normally found in the upper respiratory tract (figure 38.6*b*). However, disease usually occurs only in those individuals with predisposing factors such as viral infections of the respiratory tract, physical injury to the tract, alcoholism, or diabetes. About 60 to 80% of all respiratory diseases known as pneumonia are caused by *S. pneumoniae.*

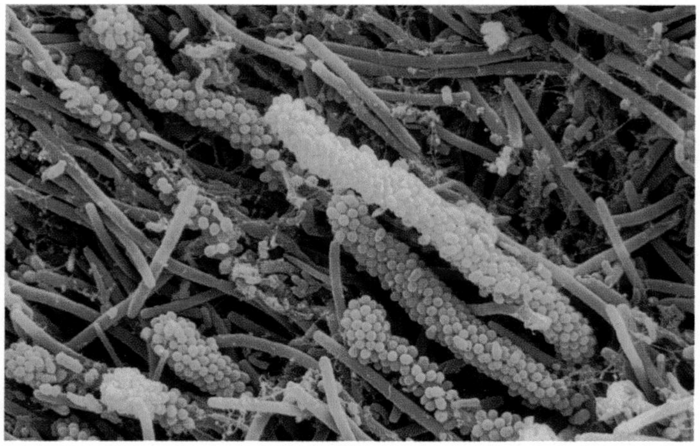

FIGURE 38.36 The Microscopic Appearance of Plaque. Scanning electron micrograph of plaque with long filamentous forms and "corn cobs" that are the mixed bacterial aggregates.

Figure 38.36 *Micro Inquiry*

How is tooth enamel demineralized by bacterial growth?

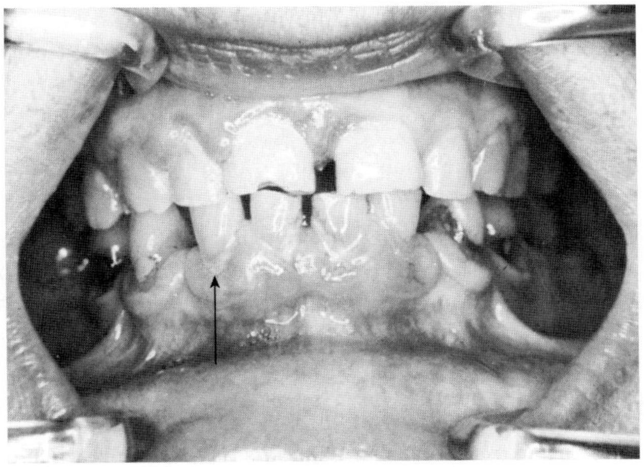

FIGURE 38.37 Periodontal Disease. Notice the plaque on the teeth (arrow), especially at the gingival (gum) margins, and the inflamed gingiva.

An estimated 150,000 to 300,000 people in the United States contract this form of pneumonia annually, and between 13,000 to 66,000 deaths result.

The primary virulence factor of *S. pneumoniae* is its capsular polysaccharide, which is composed of hyaluronic acid. The production of large amounts of hyaluronic capsular polysaccharide plays an important role in protecting the organism from ingestion and killing by phagocytes. Pathogenesis is due to the rapid multiplication of the bacteria in alveolar spaces (figure 38.6). The bacteria also produce the toxin pneumolysin, which destroys host cells. The alveoli fill with blood cells and fluid, and become inflamed. The sputum is often rust-colored because of blood coughed up from the lungs. The onset of clinical symptoms is usually abrupt, with chills; hard, labored breathing; and chest pain. Diagnosis is by chest X ray, Gram stain, culture, and tests for metabolic products. Cefotaxime, ofloxacin, and ceftriaxone have contributed to a greatly reduced mortality rate. For individuals who are sensitive to penicillin, erythromycin or tetracycline can be used.

S. pneumoniae is also associated with sinusitis, conjunctivitis, and otitis media (figure 38.6). It is an important cause of bacteremia (blood infections) and meningitis. Penicillin- and tetracycline-resistant strains of *S. pneumoniae* are now in the United States. Pneumococcal vaccines (Pneumovax 23, Pnu-Imune 23) are available for people who are at greater risk for exposure (e.g., college students and people in chronic-care facilities). The Pneumovax vaccines are pooled collections of 23 different *S. pneumoniae* capsular polysaccharides, which are effective because they generate antibodies to the capsule. When these antibodies are deposited on the surface of the capsule, they become opsonic and enhance phagocytosis (*see figure 32.5*). In 2000 a pediatric vaccine containing seven different capsular serotypes was approved for use in the United States. Preventive and control measures include immunization and adequate treatment of infected persons.

1. Describe the process by which *C. difficile*–associated disease occurs.
2. In both *C. difficile*–associated disease and bacterial vaginosis, the disease begins when the normal microbiota of the tissue site is altered. Why is this?
3. Name some common odontopathogens that are responsible for dental caries, dental plaque, and periodontal disease.

Summary

38.1 Airborne Diseases

a. A number of infectious diseases are caused by bacteria as a result of their transmission through air.

b. Examples of diseases transmitted by airborne dissemination include: chlamydial pneumonia (*Chlamydophila pneumoniae*); diphtheria (*Corynebacterium diphtheriae;* **figure 38.1**); Legionnaires' disease (*Legionella pneumophila;* **figure 38.2**); meningitis (*Neisseria meningitidis* [**figure 38.3**] and *Haemophilus influenzae* type b); mycobacterial infections (*M. avium* and *M. intracellulare* pneumonia and *M. tuberculosis;* **figure 38.4**); mycoplasmal pneumonia (*Mycoplasma pneumoniae*); pertussis (*Bordetella pertussis*); and streptococcal diseases (*Streptococcus* spp.; **figures 38.5–38.7**).

38.2 Arthropod-Borne Diseases

a. Bacteria can be transmitted to humans as a result of their interaction with arthropod vectors, such as ticks, lice, and fleas of animals.

b. Examples of arthropod-borne diseases include: Lyme disease (*Borrelia burgdorferi;* **figure 38.8**); plague (*Yersinia pestis;* **figure 38.9**); and Rocky Mountain spotted fever (*Rickettsia rickettsii;* **figure 38.11**).

38.3 Direct Contact Diseases

a. The direct contact of an uninfected human with sources of bacteria (including infected humans) can result in the transmission of bacteria and result in disease. Direct contact of skin, mucus membranes, open wounds, or body cavities can lead to bacterial colonization and disease.

b. Examples of direct contact diseases caused by bacteria include: gas gangrene or clostridial myonecrosis (*Clostridium perfringens*; **figure 38.12**); Group B streptococcal disease (*Streptococcus agalactiae*); mycobacterial infections including leprosy (*Mycobacterium leprae*; **figures 38.13** and **38.14**) and tattoo-associated infections; peptic ulcer disease (*Helicobacter pylori*; **figure 38.15**); sexually transmitted diseases such as gonorrhea (*Neisseria gonorrhoeae*; **figure 38.16**), nongonococcal urethritis (various microorganisms), and syphilis (*Treponema pallidum*; **figure 38.18**); staphylococcal diseases (*Staphylococcus aureus*; **figures 38.20–38.22**), including methicillin-resistant *Staphylococcus aureus* (**figure 38.23**); streptococcal diseases (*Streptococcus pyogenes*; **figures 38.24** and **38.25**); tetanus (*Clostridium tetani*); and trachoma (*Chlamydia trachomatis*; **figure 38.26**).

38.4 Food-Borne and Waterborne Diseases

a. Food and water can serve as vehicles that transport bacteria to humans. Ingestion of contaminated food and water often results in infections of the gastrointestinal tract (**tables 38.3 and 38.4**).

b. Some common food- and water-borne bacterial infectious diseases of the intestinal tract are botulism (*Clostridium botulinum*; **figure 38.27**), *Campylobacter* gastroenteritis (*Campylobacter jejuni*), cholera (*Vibrio cholerae*; **figure 38.28**), *Escherichia coli* gastroenteritis (**figure 38.29**), salmonellosis (various *Salmonella* serovars), typhoid fever (*Salmonella* serovar Typhi), shigellosis (*Shigella* spp.), and staphylococcal food poisoning (*Staphylococcus aureus* enterotoxins).

c. Sepsis and septic shock are most commonly caused by gram-negative bacteria, medicated by the lipid A component of the outer cell membrane.

d. Some diseases are caused by the bacterial action on the cells of the intestine. Other diseases result from bacterial toxins.

38.5 Zoonotic Diseases

a. Diseases of animals that are transmitted to humans are called zoonoses (s., zoonsis).

b. Important bacterial zoonoses include: anthrax (*Bacillus anthracis*; **figure 38.31**); brucellosis (*Brucella* spp.); psittacosis (*Chlamydophilia psittaci*); Q fever (*Coxiella burnetti*; **figure 38.32**); and tularemia (*Francisella tularensis*).

38.6 Opportunistic Diseases

a. Normal microbiota can cause disease especially when they are in tissues that are not the regular environment or they overgrow the regular environment; these are opportunists and cause opportunistic diseases.

b. Opportunistic diseases include: antibiotic-associated colitis (*Clostridium difficile*; **figure 38.33**); bacterial vaginosis (*Gardnerella vaginalis*); dental infections; and streptococcal pneumonia (*Streptococcus pneumoniae*).

c. Dental plaque formation begins on a tooth with the initial colonization of the acquired enamel pellicle by *Streptococcus gordonii*, *S. oralis*, and *S. mitis*. Other bacteria then become attached and form a plaque ecosystem (**figures 38.34–38.36**). The bacteria produce acids that cause a chemical lesion on the tooth and initiate dental decay or caries.

d. Periodontal disease is a group of diverse clinical entities that affect the periodontium. Disease is initiated by the formation of subgingival plaque, which leads to tissue inflammation known as periodontitis and to periodontal pockets. Bacteria that colonize these pockets can cause an abscess, periodontosis, gingivitis, and general tissue necrosis (**figure 38.37**).

Critical Thinking Questions

1. Describe a typhoid carrier. How does one become a carrier?
2. Account for the ease with which anaerobic clostridia can be isolated from soil and other generally aerobic niches.
3. Many consider cholera as the most severe form of gastroenteritis. Why do you think this is so?
4. Compare the three stages of syphilis and Lyme disease. Why do you think both diseases are so hard to treat once they progress beyond the primary stage?
5. While many *Vibrio cholerae* strains are found in the aquatic environment, only a small fraction of these cause human disease. When in aquatic ecosystems, *V. cholerae* is frequently found attached to the exoskeleton of zooplankton, which is made of chitin. Indeed, *V. cholerae* produces an extracellular chitinase, so zooplankton-associated growth presumably provides a good source of organic carbon and nitrogen. Growth on zooplankton and in the human gut shares the requirement for attachment proteins. Remarkably, the same protein, called GbpA, has been shown to bind to both the *N*-acetylglucosamine (GlcNAc) of zooplankton chitin and to epithelial cells. The glycoproteins and lipids on the surface of epithelial cells are commonly modified with GlcNAc.

   How would you show whether or not GbpA binds specifically to epithelial GlnNAc? How would you determine if GbpA is needed for virulence? How would you test the hypothesis that GbpA is produced by pathogenic strains but not, or to a lesser extent, by nonpathogenic strains of *V. cholerae*?

Read the original paper: Kirn, T. J., et al. 2005. A colonization factor links *Vibrio cholerae* environmental survival and human infection. *Nature* 438:863.

6. The spirochete *Borrelia burgdorferi* causes Lyme disease, which is transmitted to humans when its vector, the deer tick, takes a blood meal. When *B. burgdorferi* is in the gut of the tick, its outer surface protein A (OspA) is expressed at high levels. Once it has been transmitted to a human, OspA production diminishes. However, to complete its infection cycle and return to the tick, *B. burgdorferi* cells must increase OspA synthesis. Microbiologists wanted to determine how the spirochete senses the presence of a feeding tick and thus resumes high levels of OspA production. It was discovered that *B. burgdorferi* specifically binds the host neuroendocrine stress hormones epinephrine and norepinephrine. This enables the microbe to detect the presence of a tick, which results in upregulation of OspA.

   Discuss the role of coevolution in the development of *B. burgdorferi*'s ability to sense and respond to these host compounds. How would you show that among all the mediators of inflammation at the site of a tick bite, the spirochete responds specifically to epinephrine and norepinephrine? Your answer should include some carefully considered control experiments.

   Read the original paper: Schekelhoff, M. R., et al. 2007. *Borrelia burgdorferi* intercepts host hormonal signals to regulate expression of outer surface protein A. *Proc. Nat. Acad. Sci., USA.* 104:7247–52.

Concept Mapping Challenge

Construct a concept map using the following words and your own linking terms.

Gram-positive	Endotoxin	Invasive
Mycobacteria	Gram-negative	Exotoxin
Streptococci	Bacteria	Mycolic acid
Penicillin-resistant	Shiga toxin	Anthrax
Type III secretion virulence factor		Plague

Learn More

Learn more by visiting the text website at www.mhhe.com/willey8, where you will find a complete list of references.

39

Human Diseases Caused by Fungi and Protists

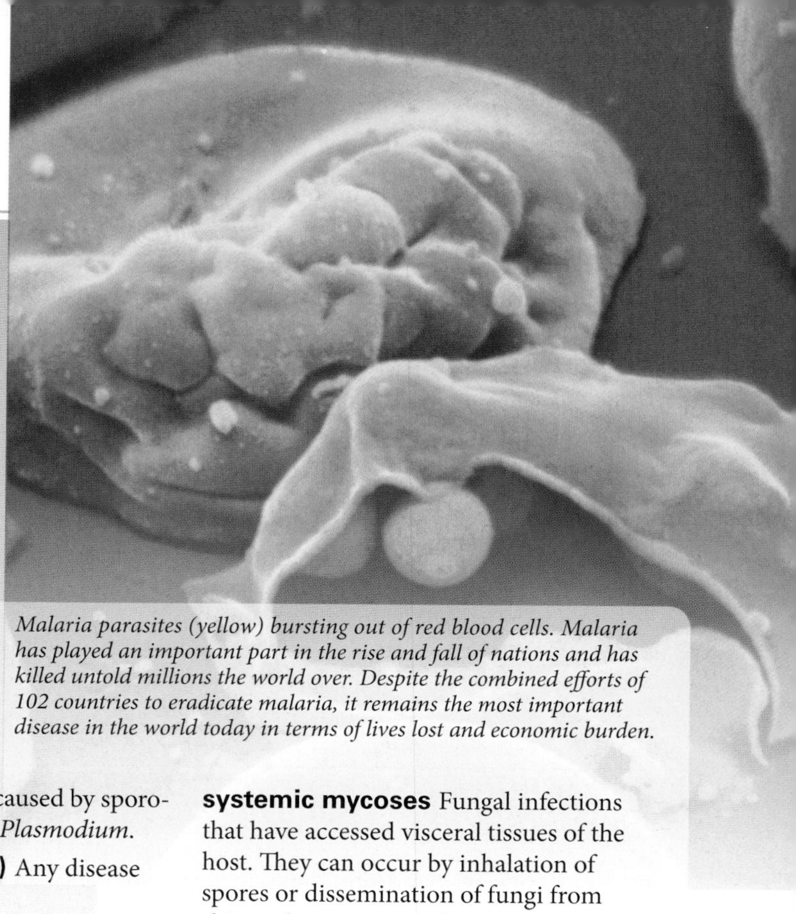

Malaria parasites (yellow) bursting out of red blood cells. Malaria has played an important part in the rise and fall of nations and has killed untold millions the world over. Despite the combined efforts of 102 countries to eradicate malaria, it remains the most important disease in the world today in terms of lives lost and economic burden.

CHAPTER GLOSSARY

amebiasis Intestinal disease caused by *Entamoeba histolytica*, an amoeboid protozoa found in water contaminated with feces.

cyst The inactive or resting stage of a protozoan.

dermatophytes Cutaneous fungi that parasitize the hair, nails, and the outer layer of the skin.

giardiasis Common intestinal disease caused by the flagellated protozoan *Giardia intestinalis.*

hemoflagellates Flagellated protozoa that are transmitted by arthropods and infect the blood and tissues of humans.

malaria Human disease caused by sporozoan protists in the genus *Plasmodium.*

mycosis (pl., mycoses) Any disease caused by a fungus.

opportunistic microbe Microorganism that is generally harmless in its normal environment but that can become pathogenic in a compromised host or when it disseminates to other host environments.

piedra Superficial fungal infection (mycosis) of the hair shaft.

sporozoite The motile, infective stage of apicomplexan protists, including *Plasmodium*, the causative agent of malaria.

systemic mycoses Fungal infections that have accessed visceral tissues of the host. They can occur by inhalation of spores or dissemination of fungi from skin and mucous membranes.

tinea (ringworm) Superficial fungal infection (mycosis) of the dead keratinocytes of skin, hair, and nails.

trophozoite The active, motile feeding stage of a protozoan.

yeast A unicellular, uninuclear fungus that reproduces either asexually by budding or fission, or sexually through spore formation.

In this chapter, we describe some of the fungi and protists that are pathogenic to humans and discuss the clinical manifestations, diagnosis, epidemiology, pathogenesis, and treatment of the diseases caused by them. The biology and diversity of these organisms are covered in chapters 4, 23, and 24. This chapter follows the format of the previous two chapters in identifying their diseases by route of transmission.

39.1 Pathogenic Fungi and Protists

Fungi are eukaryotic saprophytes that are ubiquitous in nature. Although hundreds of thousands of fungal species are found in the environment, only about 50 produce disease in humans. Medical mycology is the discipline that deals with the fungi that cause human disease. These fungal diseases, known as **mycoses** (s., mycosis; Greek *mykes*, fungus), are typically divided into five groups according to the route of infection: superficial, cutaneous, subcutaneous, systemic, and opportunistic mycoses (**table 39.1**). Superficial, cutaneous, and subcutaneous mycoses are direct contact infections of the skin, hair, and nails. Systemic mycoses are fungal infections that have disseminated to visceral tissues. Except for *Cryptococcus neoformans*, which has only a yeast form, the fungi that cause the systemic or deep mycoses are dimorphic—they exhibit a parasitic yeastlike (single-cell) phase (Y) and a saprophytic mold or mycelial (filamentous) phase (M; *see figure 4.26*). Most systemic mycoses are

Table 39.1 Examples of Some Medically Important Fungi

Group	Pathogen	Location	Disease
Superficial mycoses	*Piedraia hortae*	Scalp	Black piedra
	Trichosporon beigelii	Beard, mustache	White piedra
	Malassezia furfur	Trunk, neck, face, arms	Tinea versicolor
Cutaneous mycoses	*Trichophyton mentagrophytes, T. verrucosum, T. rubrum*	Beard hair	Tinea barbae
	Trichophyton, Microsporum canis	Scalp hair	Tinea capitis
	Trichophyton rubrum	Smooth or bare parts of the skin	Tinea corporis
	Epidermophyton floccosum, T. mentagrophytes, T. rubrum	Groin, buttocks	Tinea cruris (jock itch)
	T. rubrum, T. mentagrophytes, E. floccosum	Feet	Tinea pedis (athlete's foot)
	T. rubrum, T. mentagrophytes, E. floccosum	Nails	Tinea unguium (onychomycosis)
Subcutaneous mycoses	*Phialophora verrucosa, Fonsecaea pedrosoi*	Legs, feet	Chromoblastomycosis
	Madurella mycetomatis	Feet, other areas of body	Maduromycosis
	Sporothrix schenckii	Puncture wounds	Sporotrichosis
Systemic mycoses	*Blastomyces dermatitidis*	Lungs, skin	Blastomycosis
	Coccidioides immitis	Lungs, other parts of body	Coccidioidomycosis
	Cryptococcus neoformans	Lungs, skin, bones, viscera, central nervous system	Cryptococcosis
	Histoplasma capsulatum	Within phagocytes	Histoplasmosis
Opportunistic mycoses	*Aspergillus fumigatus, A. flavus*	Respiratory system	Aspergillosis
	Candida albicans	Skin or mucous membranes	Candidiasis
	Pneumocystis jiroveci	Lungs, sometimes brain	Pneumocystis pneumonia
	Encephalitozoon, Pleistophora, Enterocytozoon, Microsporidium	Lungs, sometimes brain	Microsporidiosis

acquired by the inhalation of spores from soil in which the mold-phase of the fungus resides. If a susceptible person inhales enough spores, an infection begins as a lung lesion, becomes chronic, and spreads through the bloodstream to other organs (the target organ varies with the species). ◀◀ *The* Fungi *(chapter 24)*

Protozoa, single-celled eukaryotic chemoorganotrophs, have become adapted to practically every type of habitat on Earth, including the human body. Many protists are transmitted to humans by arthropod vectors or by food and water vehicles. However, some protozoan diseases are transmitted by direct contact. Although fewer than 20 genera of protists cause disease in humans (**table 39.2**), their impact is formidable. For example, over 350 million cases of malaria occur in the world each year. In sub-Saharan Africa alone, malaria is responsible for the deaths of more than a million children under the age of fourteen annually. It is estimated that there are at least 8 million cases of trypanosomiasis, 12 million cases of leishmaniasis, and over 500 million cases of amebiasis yearly. There is also an increasing problem with *Cryptosporidium* and *Cyclospora* contamination of food and water supplies (**table 39.3**). More of our population is elderly, and a growing number of persons are immunosuppressed due to HIV infection, organ transplantation, or cancer chemotherapy. These populations are at increased risk for protozoan infections. ◀◀ *The protists (chapter 23)*

While seemingly different, fungi and protists share a number of phenotypic features that serve them in their ability to cause infection: microscopic size, eukaryotic physiology, cell walls or wall-like structures, alternative stages for survival outside of the host, degradative enzymes, and others. Some fungi and protists also share transmission routes. We now discuss diseases of fungi and protists based on how they are acquired by the human host.

39.2 Airborne Diseases

A number of eukaryotic pathogens are transmitted to people through air. These infectious agents are small particles that are carried by air currents and are typically inhaled by the unsuspecting host. Here, we present several examples of infectious agents that are acquired through airborne transmission.

Table 39.2 Examples of Medically Important Protozoan Infections

Initial Infection Site	*Pathogen*	*Disease*
Blood	*Plasmodium* spp. *Trypanosoma brucei* *Trypanosoma cruzi*	Malaria African sleeping sickness American trypanosomiasis
Intestinal Tract	*Cryptosporidium* spp. *Cyclospora* spp. *Entamoeba histolytica* *Giardia intestinalis* *Toxoplasma gondii*	Cryptosporidiosis Cyclosporidiosis Amebiasis Giardiasis Toxoplasmosis
Lungs	*Balamuthia mandrillaris* *Pneumocystis jiroveci* *Trichomonas tenax*	Balamuthia amebic encephalitis Pneumocystis pneumonia Pulmonary trichononiasis
Muscosal Membranes Ears, Eyes, Nose Genitals	*Acanthamoeba* spp. *Naegleria fowleri* *Trichomonas vaginalis*	Amebic meningoencephalitis Amebic meningoencephalitis Trichomoniasis
Skin	*Leishmania braziliensis* *Leishmania donovani* *Leishmania tropica*	Mucocutaneous leishmaniasis Visceral leishmaniasis Cutaneous leishmaniasis

Table 39.3 Water-Based Protozoal Reservoirs

Organism	*Reservoir*	*Comments*
Acanthamoeba	Sewage sludge disposal areas	Can cause granulomatous amebic encephalitis, keratitis, corneal ulcers
Cryptosporidium	Many species of domestic and wild animals	Causes acute enterocolitis; important with immunologically compromised individuals; cysts resistant to chemical disinfection; not antibiotic sensitive
Cyclospora cayetanensis	Waters—does not withstand drying; possibly other reservoirs	Causes long-lasting (43 days average) diarrheal illness; infection self-limiting in immunocompetent hosts; sensitive to prompt treatment with sufonamide and trimethoprim
Giardia intestinalis	Beavers, sheep, dogs, cats	Major cause of early spring diarrhea; important in cold mountain water
Naegleria fowleri	Warm water (hot tubs), swimming pools, lakes	Inhalation in nasal passages; central nervous system infection; causes primary amebic meningoencephalitis

Blastomycosis

Blastomycosis is the systemic mycosis caused by *Blastomyces dermatitidis*, a fungus that grows as a budding yeast in humans but as a mold on culture media and in the environment. It is found predominately in moist soil enriched with decomposing organic debris, as in the Mississippi and Ohio River basins. *B. dermatitidis* is endemic in parts of the south-central, southeastern, and midwestern United States. Additionally, microfoci have been reported in Central and South America and in parts of Africa. The disease occurs in three clinical forms: cutaneous, pulmonary, and disseminated. The initial infection begins when blastospores are inhaled into the lungs. The fungus can then spread rapidly, especially to the skin, where cutaneous ulcers and abscess formation occur (**figure 39.1**). *B. dermatitidis* can be isolated from pus and biopsy sections. Diagnosis requires the demonstration of thick-walled, yeastlike cells, 8 to 15 μm in diameter. Complement-fixation, immunodiffusion, and skin tests (blastomycin) are also useful. Amphotericin B (Fungizone), itraconazole (Sporanox), or ketoconazole (Nizoral) are the drugs of choice for treatment (*see figure 34.15*). Surgery may be necessary for the drainage of large abscesses. Mortality is about 5%. There are no preventive or control measures. ◀◀ *Antifungal drugs (section 34.5)*

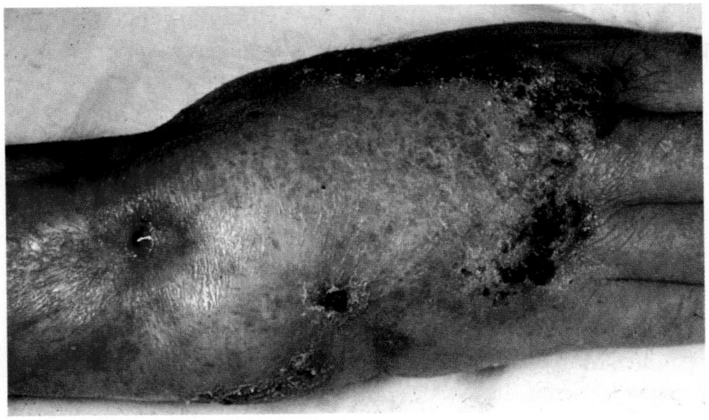

FIGURE 39.1 Systemic Mycosis. Blastomycosis of the forearm caused by *Blastomyces dermatitidis.*

Figure 39.1 Micro Inquiry

What is the morphology of *B. dermatitidis* in the environment? In the human host?

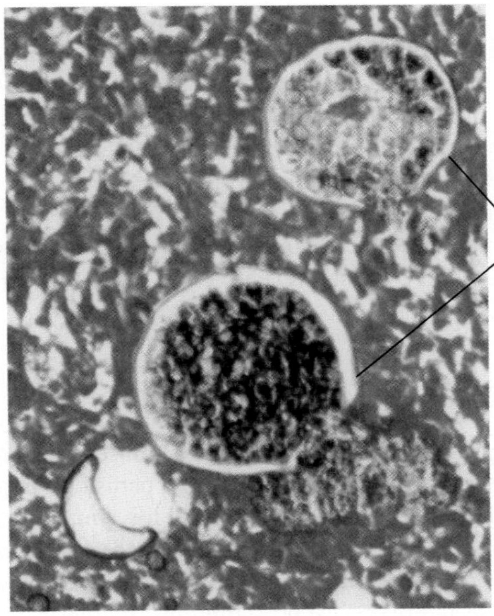

FIGURE 39.2 Systemic Mycosis: Coccidioidomycosis. *Coccidioides immitis* mature spherules filled with endospores within a tissue section; light micrograph (×400).

Coccidioidomycosis

Coccidioidomycosis, also known as valley fever, San Joaquin fever, or desert rheumatism because of the geographical distribution of the fungus, is caused by *Coccidioides immitis. C. immitis* exists in the semiarid, highly alkaline soils of the southwestern United States and parts of Mexico and South America. It has been estimated that in the United States, about 100,000 people are infected annually, resulting in 50 to 100 deaths. Endemic areas have been defined by massive skin testing with the antigen coccidioidin, where 10 to 15% positive cases are reported. In the soil and on culture media, this fungus grows as a mold that forms arthroconidia at the tips of hyphae (*see figure 4.26*). Because arthroconidia are so abundant in these endemic areas, immunocompromised individuals can acquire the disease by inhalation as they simply move through the area. Wind turbulence and even construction of outdoor structures have been associated with increased exposure and infection. In humans, the fungus grows as a yeast-forming, thick-walled spherule filled with spores (**figure 39.2**). Most cases of coccidioidomycosis are asymptomatic or indistinguishable from ordinary upper respiratory infections. Almost all cases resolve in a few weeks, and a lasting immunity results. A few infections result in a progressive chronic pulmonary disease or disseminated infections of other tissues; the fungus can spread throughout the body, involving almost any organ or site.

Diagnosis is accomplished by identification of the large spherules (approximately 80 μm in diameter) in pus, sputum, and aspirates. Culturing clinical samples in the presence of penicillin and streptomycin on Sabouraud agar (used to isolate fungi) also is diagnostic. Newer methods of rapid confirmation include the testing of supernatants of liquid media cultures for antigens, serology, and skin testing. Miconazole (Lotrimin), itraconazole, ketoconazole, and amphotericin B are the drugs of choice for treatment. Prevention involves reducing exposure to dust (soil) in endemic areas. ⏮ *Antifungal drugs (section 34.5); Identification of microorganisms from specimens (section 35.2)*

Cryptococcosis

Cryptococcosis is a systemic mycosis caused by *Cryptococcus neoformans.* This fungus always grows as a large, budding yeast in the human host but as smaller cells in the environment, where it is a saprophyte with a worldwide distribution. The closely related *C. gattii* has also been implicated in systemic mycoses. An outbreak of over 200 cases of *C. gattii* infection was reported on Vancouver Island in 1999. Nineteen cases of *C. gattii* disease have been identified in Oregon and Washington since 2004. Typically, cryptococci cause disease in the immunocompromised and often cause disease in people with AIDS, cancer, or a recent organ transplant. In fact, cryptococcosis is found in approximately 15% of AIDS patients. Unlike *C. neoformans,* however, *C. gattii* has also been found in people who do not have these risk factors.

Aged, dried pigeon droppings are an apparent source of infection. The fungus enters the body by the respiratory tract, causing a minor pulmonary infection in most that is usually transitory. In others who inhale the fungi, a pneumonia-like disease with coughing and fever occurs. Some pulmonary infections spread to the skin, bones, viscera, and the central nervous system, where they may cause headaches and mental status changes. Once the nervous system is involved, cryptococcal meningitis usually results. Diagnosis is accomplished by detection of the thick-walled, spherical yeast cells in pus, sputum, or exudate smears using India ink to define the organism (**figure 39.3**). The fungus can be easily cultured on Sabouraud dextrose agar. Identification of the fungus in body fluids is made by immunologic procedures. Treatment for *C. neoformans* includes amphotericin B with or without flucytosine. HIV/AIDS patients with cryptococcosis are treated prophylactically with fluconazole. Prevention of infection with *C. neoformans* is through avoidance of birds and areas contaminated with bird droppings. There are no preventive or control measures yet suggested for *C. gattii*.

Histoplasmosis

Histoplasmosis is primarily a disease of the lungs caused by *Histoplasma capsulatum* var. *capsulatum*, a facultative, parasitic fungus that grows intracellularly. The incubation period for symptomatic patients is approximately 10 days; symptoms include flulike responses such as fever, cough, headaches, and muscle aches. However, most of the infected may have no symptoms. It appears as a small, budding yeast in humans and on culture media at 37°C (98.6°F). At 25°C (77°F) it grows as a mold, producing small microconidia (1 to 5 μm in diameter) that are borne singly at the tips of short conidiophores (*see figure 4.26*). Large macroconidia (8 to 16 μm in diameter) are also formed on conidiophores (**figure 39.4*a***).

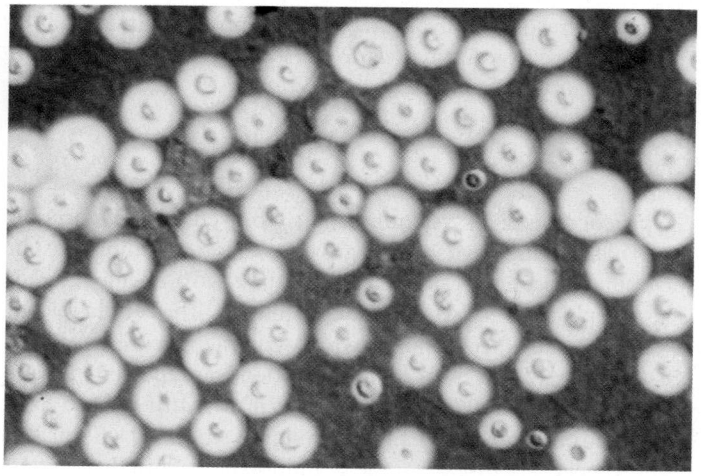

FIGURE 39.3 Systemic Mycosis: Cryptococcosis. India ink preparation showing *Cryptococcus neoformans.* Although these microorganisms are not budding, they can be differentiated from artifacts by their doubly refractile cell walls, distinctly outlined capsules surrounding all cells, and refractile inclusions in the cytoplasm; light micrograph (×150).

***Figure 39.3** Micro Inquiry*

Once this microbe is acquired by inhalation, to what other body systems can it spread?

In humans, the yeastlike form grows by budding (figure 39.4*b*). *H. capsulatum* var. *capsulatum* is found as the mycelial form in soils throughout the world and is localized in areas that have been contaminated with bird or bat excrement. The microconidia can become airborne when contaminated soil is disturbed. Microconidia are thus most prevalent where bird droppings—especially from starlings, crows, blackbirds, cowbirds, sea gulls, turkeys, and chickens—have accumulated. It is noteworthy that the birds themselves are not infected because of their high body temperature; their droppings simply provide the nutrients for this fungus. Only bats and humans demonstrate the disease and harbor the fungus.

Infection ensues when the microconidia are inhaled. Histoplasmosis is not, however, transmitted from person to person. Histoplasmosis is found worldwide, with Africa, Australia, India, and Malaysia being endemic regions. Within the United States, histoplasmosis is endemic within the Mississippi, Kentucky,

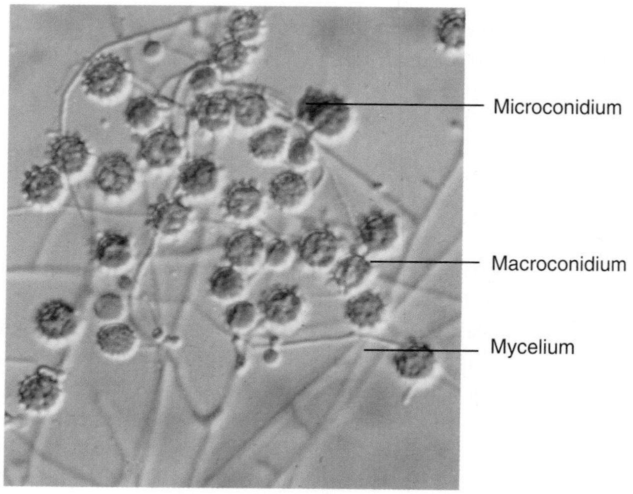

(a)

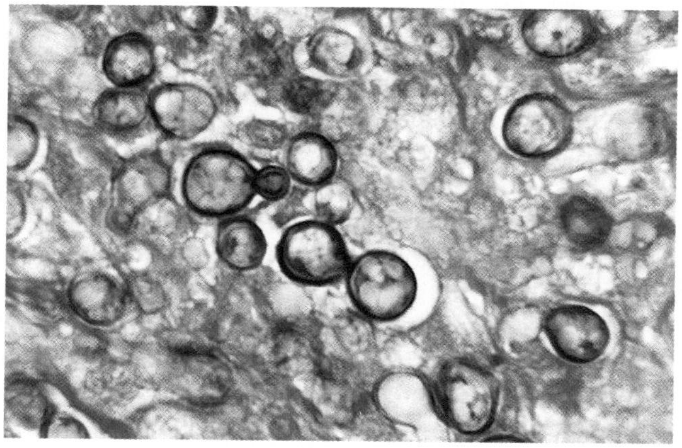

(b)

FIGURE 39.4 Morphology of *Histoplasma capsulatum* var. *capsulatum*. (a) Mycelia, microconidia, and chlamydospores as found in the soil. These are the infectious particles; light micrograph (×125). (b) Yeastlike budding of *H. capsulatum.*

Tennessee, Ohio, and Rio Grande River basins. More than 80% of the people who reside in parts of these areas have antibodies against the fungus. It has been estimated that in endemic areas of the United States, about 500,000 individuals are infected annually: 50,000 to 200,000 become ill; 3,000 require hospitalization; and about 50 die. The total number of infected individuals may be over 40 million in the United States alone. Histoplasmosis is a common disease among poultry farmers, spelunkers (people who explore caves), and bat guano miners (bat guano is used as fertilizer).

Because *H. capsulatum* grows within (pulmonary) macrophages, many organs of the body can be infected as the infected macrophages circulate throughout the body. More than 95% of "histo" cases have either no symptoms or mild symptoms such as coughing, fever, and joint pain. Lesions may appear in the lungs and show calcification; the disease may resemble tuberculosis. Most infections resolve spontaneously. Only rarely does the disease disseminate. Laboratory diagnosis is accomplished by complement-fixation tests and isolation of the fungus from tissue specimens. Most individuals with this disease exhibit a hypersensitive state that can be demonstrated by the histoplasmin skin test. Currently the most effective treatment is with amphotericin B, ketoconazole, or itraconazole. Prevention and control involve wearing protective clothing and masks before entering or working in infested habitats. Soil decontamination with 3 to 5% formalin is effective where economically and physically feasible.

39.3 Arthropod-Borne Diseases

Malaria

The most important human protozoa is *Plasmodium*, the causative agent of **malaria** (**Disease 39.1**). It has been estimated that more than 350 million to 500 million people are infected worldwide each year, and over 1 million (mostly children) die annually of malaria in Africa alone (**figure 39.5**). About 1,000 cases are reported each year in the United States, divided between returning U.S. travelers and non-U.S. citizens.

Human malaria is caused by four species of *Plasmodium: P. falciparum, P. malariae, P. vivax,* and *P. ovale*. The life cycle of plasmodial protists is shown in **figure 39.6**. The parasite first enters the bloodstream through the bite of an infected female *Anopheles* mosquito. As she feeds, the mosquito injects a small amount of saliva containing an anticoagulant along with small haploid sporozoites. The sporozoites in the bloodstream immediately enter hepatic cells of the liver. In the liver, they undergo multiple asexual fission (schizogony) and produce merozoites. After being released from the liver cells, the merozoites attach to and penetrate erythrocytes. The incubation period is as short as 7 days (frequently due to *P. falciparum*) and as long as 30 days (usually due to *P. malariae*). ◀◀ Alveolata *(section 23.5)*

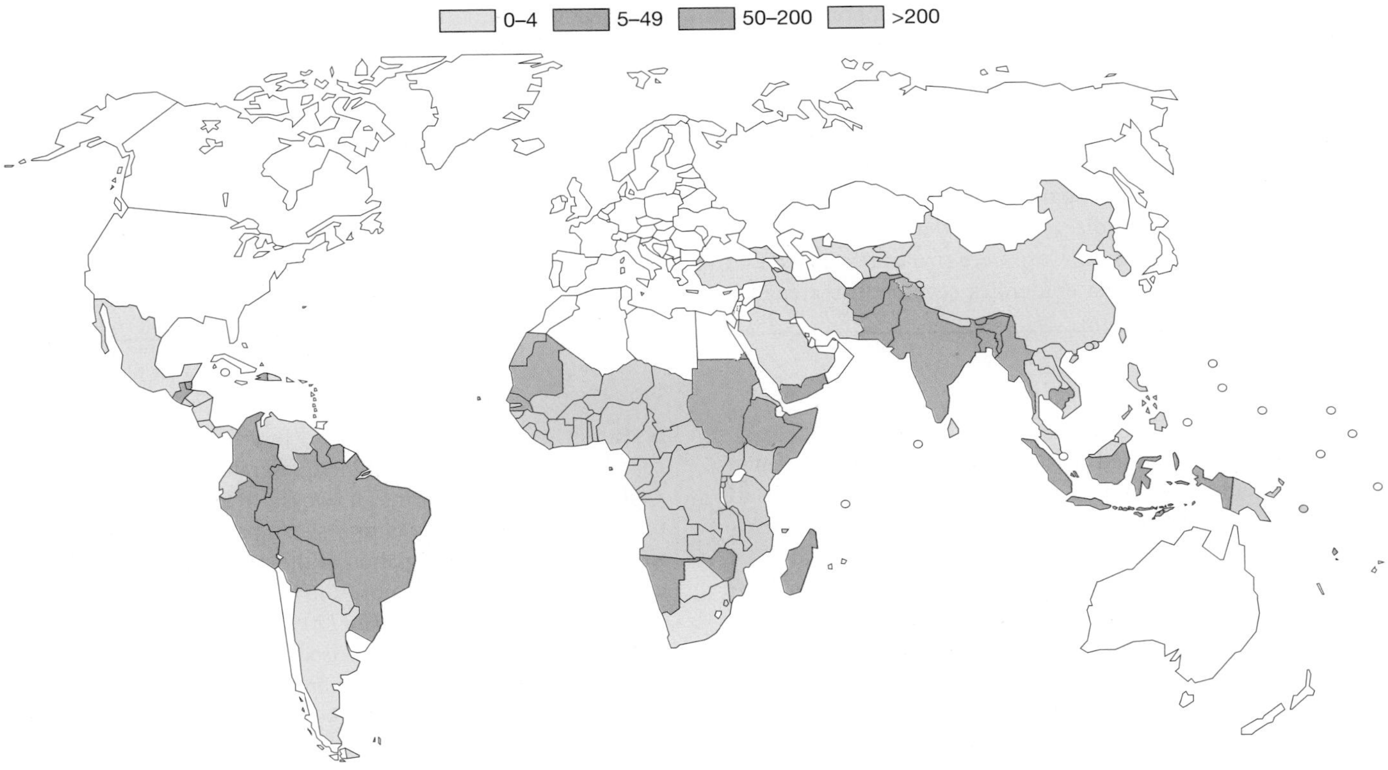

FIGURE 39.5 Geographic Distribution of Malaria by End of 2006. Estimated incidence of malaria per 1,000 population.

Source: WHO, World Malaria Report 2008.

DISEASE

39.1 A Brief History of Malaria

No other single infectious disease has had the impact on humans that malaria has had. The first references to its periodic fever and chills can be found in early Chaldean, Chinese, and Hindu writings. In the late fifth century BCE, Hippocrates described certain aspects of malaria. In the fourth century BCE, the Greeks noted an association between individuals exposed to swamp environments and the subsequent development of periodic fever and enlargement of the spleen (splenomegaly). In the seventeenth century, the Italians named the disease *mal' aria* (bad air) because of its association with the ill-smelling vapors from the swamps near Rome. At about the same time, the bark of the quinaquina (cinchona) tree of South America was used to treat the intermittent fevers, although it was not until the mid-nineteenth century that quinine was identified as the active alkaloid. The major epidemiological breakthrough came in 1880, when French army surgeon Charles Louis Alphonse Laveran observed gametocytes in fresh blood. Five years later, the Italian histologist Camillo Golgi observed the multiplication of the asexual blood forms. In the late 1890s Patrick Manson postulated that malaria was transmitted by mosquitoes. Sir Ronald Ross, a British army surgeon in the Indian Medical Service, subsequently observed developing plasmodia in the intestine of mosquitoes, supporting Manson's theory. Using birds as experimental models, Ross definitively established the major features of the life cycle of *Plasmodium* and received the Nobel Prize in 1902.

Human malaria is known to have contributed to the fall of the ancient Greek and Roman empires. Troops in both the U.S. Civil War and the Spanish-American War were severely incapacitated by the disease. More than 25% of all hospital admissions during these wars were malaria patients. During World War II, malaria epidemics severely threatened both the Japanese and Allied forces in the Pacific. The same can be said for the military conflicts in Korea and Vietnam.

In the twentieth century, efforts were directed toward understanding the biochemistry and physiology of malaria, controlling the mosquito vector, and developing antimalarial drugs. In the 1960s it was demonstrated that resistance to *P. falciparum* among West Africans was associated with the presence of hemoglobin-S (Hb-S) in their erythrocytes. Hb-S differs from normal hemoglobin-A by a single amino acid, valine, in each half of the Hb molecule. Consequently these erythrocytes—responsible for sickle cell disease—have a low binding capacity for oxygen. Because the malarial parasite has a very active aerobic metabolism, it cannot grow and reproduce within these erythrocytes.

In 1955 the World Health Organization began a worldwide malarial eradication program that finally collapsed by 1976. Among the major reasons for failure were the development of resistance to DDT by mosquitoes and the development of resistance to chloroquine by strains of *Plasmodium.* Scientists are exploring new approaches, such as the development of vaccines and more potent drugs. In 2002 the complete DNA sequences of *P. falciparum* and *Anopheles gambiae* (the mosquito that most efficiently transmits this parasite to humans in Africa) were determined. Together with the human genome sequence, researchers now have in hand the genetic blueprints for the parasite, its vector, and its victim. This has made possible a holistic approach to understanding how the parasite interacts with the human host, leading to new antimalarial strategies, including vaccine design. Overall, no greater achievement for microbiology could be imagined than the control of malaria—a disease that has caused untold misery throughout the world since antiquity and remains one of the world's most serious infectious diseases.

Once inside the erythrocyte, the *Plasmodium* begins to enlarge as a uninucleate cell termed a trophozoite. The trophozoite's nucleus then divides asexually to produce a schizont that has 6 to 24 nuclei. The schizont divides and produces mononucleated merozoites. Eventually the erythrocyte lyses, releasing the merozoites into the bloodstream to infect other erythrocytes. This erythrocytic stage is cyclic and repeats approximately every 48 to 72 hours or longer, depending on the species of *Plasmodium* involved. The sudden release of merozoites, toxins, and erythrocyte debris triggers an attack of the chills, fever, and sweats characteristic of malaria. Occasionally, merozoites differentiate into macrogametocytes and microgametocytes, which do not rupture the erythrocyte. When these are ingested by a mosquito, they develop into female and male gametes, respectively. In the mosquito's gut, the infected erythrocytes lyse and the gametes fuse to form a diploid zygote called the ookinete. The ookinete migrates to the mosquito's gut wall, penetrates, and forms an oocyst. In a process called sporogony, the oocyst undergoes meiosis and forms sporozoites, which migrate to the salivary glands of the mosquito. The cycle is now complete, and when the mosquito bites another human host, the cycle begins anew. *Malaria: Life Cycle of* Plasmodium

The pathological changes caused by malaria involve not only the erythrocytes but also the spleen and other visceral organs. Classic symptoms first develop with the synchronized release of merozoites and erythrocyte debris into the bloodstream, resulting in the malarial paroxysms—shaking chills, then burning fever followed by sweating. It may be that the fever and chills are caused partly by a malarial toxin that induces macrophages to release TNF-α and

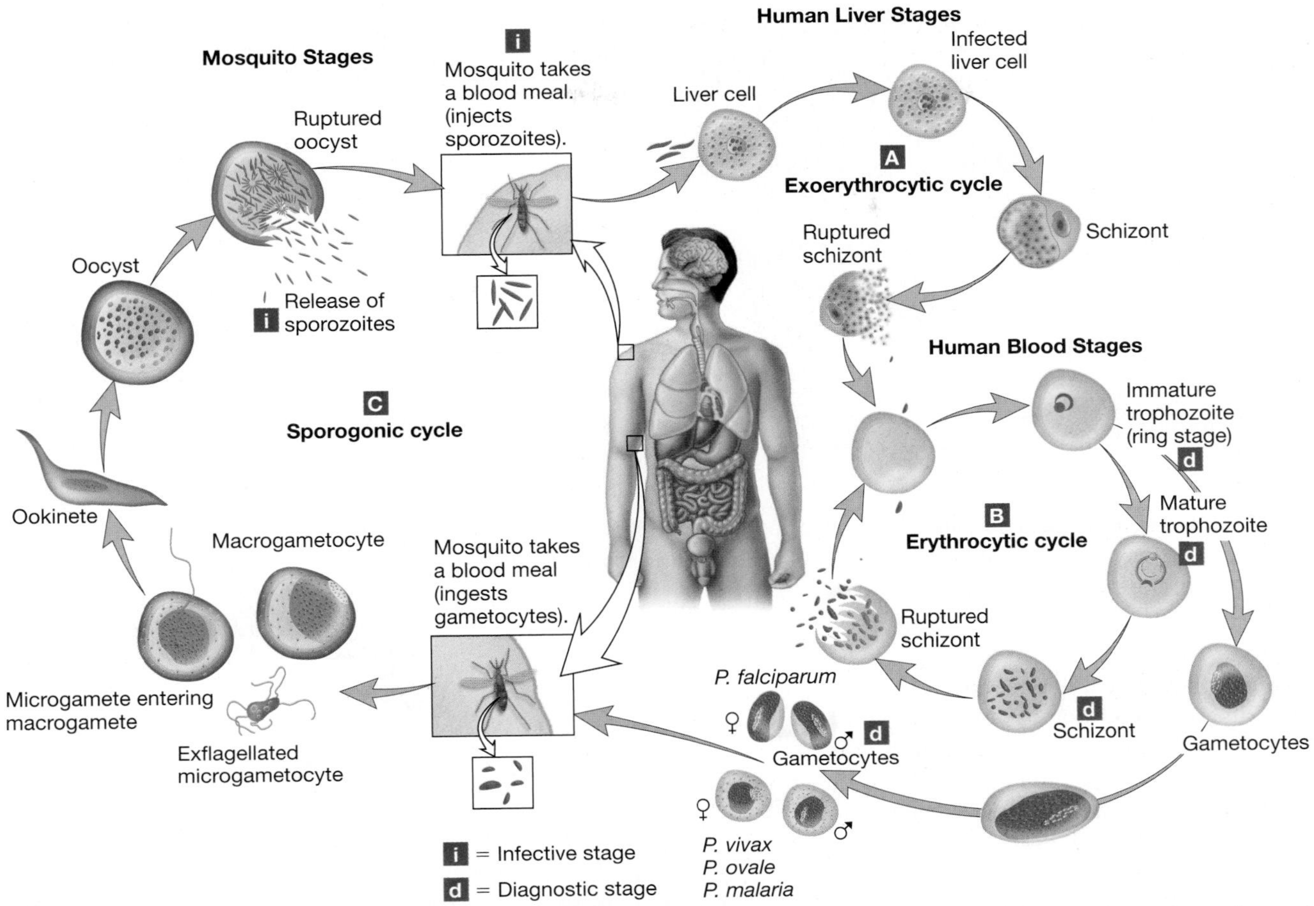

FIGURE 39.6 Malaria. Life cycle of *Plasmodium vivax*. Note the (A) exoerythrocytic cycle, (B) the erythrocytic cycle, and the (c) the sporogonic cycle.

Figure 39.6 Micro Inquiry

How is schizont formed? What is the cell type that is released from erythrocytes, and where does the protist go next?

interleukin-1. Several of these paroxysms constitute an attack. After one attack, there is a remission that lasts from a few weeks to several months, then there is a relapse. Between paroxysms, the patient feels normal. Anemia can result from the loss of erythrocytes, and the spleen and liver often hypertrophy. Children and nonimmune individuals can die of cerebral malaria. ◀◀ *Cytokines (section 32.3)*

Diagnosis of malaria is made by demonstrating the presence of parasites within Wright- or Giemsa-stained erythrocytes (**figure 39.7**). When blood smears are negative, serological testing can establish a diagnosis of malaria. Outside the United States, rapid diagnostic tests using species-specific antibodies are used for the diagnosis of malaria; these tests are not approved for use by the U.S. Food and Drug Administration (FDA). Specific recommendations for treatment are region-dependent. Treatment includes administration of chloroquine, amodiaquine, or mefloquine. These drugs suppress protozoan reproduction and are effective in eradicating erythrocytic asexual stages. Primaquine has proved satisfactory in eradicating the exoerythrocytic stages. However, because resistance to these drugs is occurring rapidly, more expensive drug combinations are now being used. One example is Fansidar, a combination of pyrimethamine and sulfadoxine. It is worth noting that individuals who are traveling to areas where malaria is endemic should receive chemoprophylactic treatment with chloroquine, atovaquone/proguanil, or primaquine (figure 39.5). ◀◀ *Identification of microorganisms from specimens (section 35.2)*

Control of malaria has been a nagging problem for some time. The best approach would be to vaccinate people in endemic areas at a very young age. However, to be effective, a vaccine must interfere with a critical malarial component or process (**figure 39.8**). A number of credible vaccines against malaria have been in clinical trials since 2005. A 2008 report indicated that one vaccine, the RTS,S vaccine, provided 65% reduction in first infections in

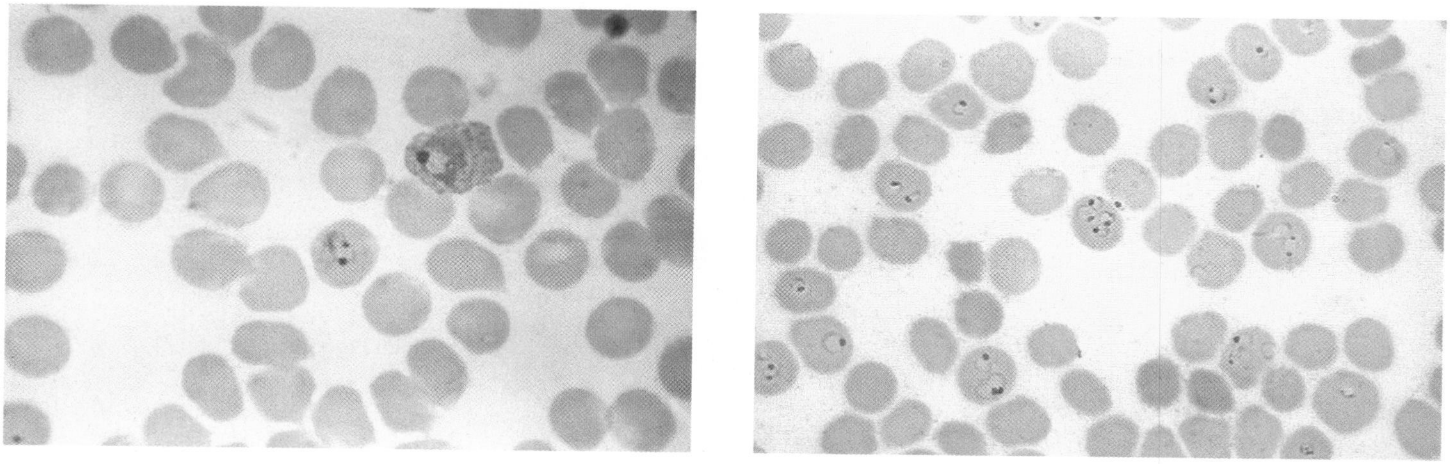

(a) (b)

FIGURE 39.7 Malaria: Erythrocytic Cycle. Trophozoites of (a) *P. vivax* and (b) *P. falciparum* in circulating erythrocytes; light micrograph (×1,100). The young trophozoites resemble small rings within the erythrocyte cytoplasm.

340 Tanzanian infants receiving three doses over six months and 53% reduction in 894 Kenyan and Tanzanian children aged five to seventeen months. Until a vaccine that can provide 90 to 95% effectiveness is approved for use, though, malaria prevention is still best attempted with the use of bed netting and insecticides. ◀◀ *Vaccines and immunization (section 36.7)*

Search This: Malarial vaccine initiative

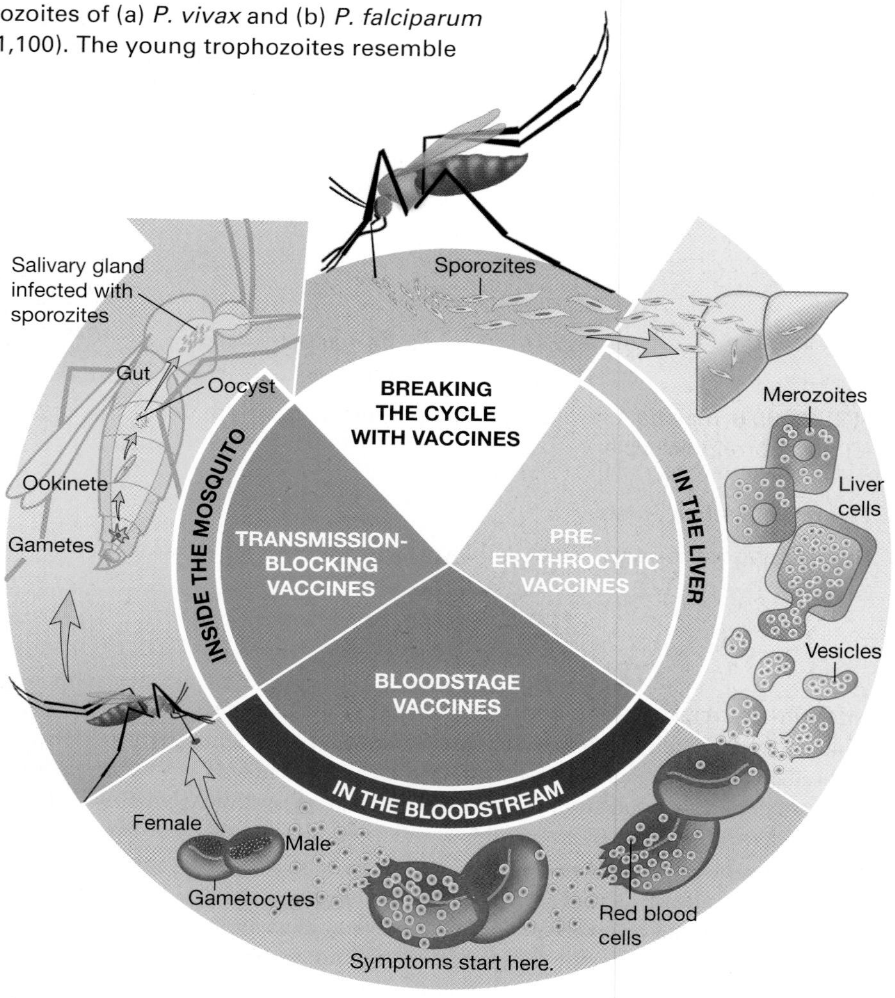

FIGURE 39.8 Malaria Vaccine Strategies.

Figure 39.8 ***Micro Inquiry***

What is currently the most effective strategy to prevent malaria?

Leishmaniasis

Leishmanias are flagellated protists that cause a group of several human diseases collectively called **leishmaniasis.** The most common forms of leishmaniasis are cutaneous (skin associated) and visceral (affecting internal organs). Worldwide, there are 1.5 million new cases of the cutaneous disease and 500,000 cases of the visceral disease each year. One-tenth of the world's population is at risk of infection; more than 90% of cutaneous disease occurs in parts of Afghanistan, Algeria, Iran, Iraq, Saudi Arabia, Syria, Brazil, and Peru, and more than 90% of visceral disease occurs in parts of India, Bangladesh, Nepal, Sudan, and Brazil. About 60,000 people die from leishmaniasis each year, and many more are permanently disfigured (**figure 39.9**).

The primary reservoirs of these parasites are canines and rodents. All species of *Leishmania* use female sand flies such as those of

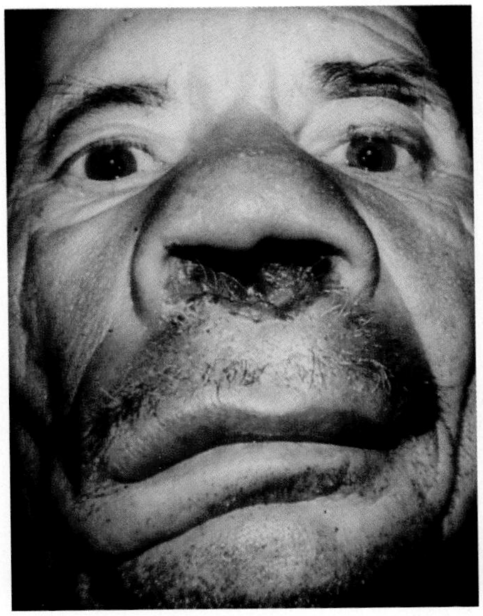

(a)

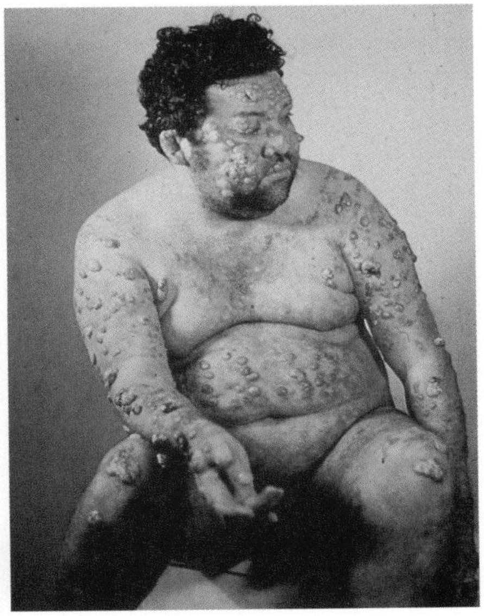

(b)

FIGURE 39.9 Leishmaniasis. (a) A person with mucocutaneous leishmaniasis, which has destroyed nasal septum and deformed the nose and lips. (b) A person with diffuse cutaneous leishmaniasis.

***Figure 39.9* Micro Inquiry**

What is the primary reservoir of *Leishmania*? What is the intermediate host?

the genera *Lutzomyia* and *Phlebotomus* as intermediate hosts. Sand flies are about one-third the size of a mosquito, so they are hard to see and hear. The leishmanias are transmitted from animals to humans or between humans by sand flies. Rare cases of transmission by blood transfusion and needle sharing have been reported. When an infected sand fly takes a human blood meal, it introduces flagellated promastigotes into the skin of the definitive (human) host (**figure 39.10**). Within the skin, the promastigotes are engulfed by macrophages, multiply by binary fission, and form small, nonmotile cells called amastigotes. These destroy the host cell and are engulfed by other macrophages in which they continue to develop and multiply. ◀◀ Euglenozoa *(section 23.2)*

Leishmania tropica and *L. mexicana* occur in the more arid regions of the Eastern Hemisphere, while *L. braziliensis* occurs in the forest regions of tropical America. They cause cutaneous leishmaniasis. *L. mexicana* is also found in the Yucatan Peninsula (Mexico) and has been reported as far north as Texas. In this disease, a relatively small, red papule forms at the site of each insect bite—the inoculation site. These papules are frequently found on the face and ears. They eventually develop into crustated ulcers (figure 39.9*b*). Healing occurs with scarring and a permanent immunity.

L. donovani is endemic in large areas within northern China, eastern India, the Sudan, the Mediterranean countries, and Latin America. Infection with *L. donovani* results in visceral disease as infected macrophages lodge in various organs where the ongoing cycle of macrophage capture of promastigotes, amastigote development, macrophage death, and amastigote release damages host organs and tissues.

Laboratory diagnosis of leishmaniasis is based on finding the protist within infected macrophages in stained smears from lesions or infected organs. Culture and serological tests are also available for diagnosis of leishmaniasis. Treatment includes pentavalent antimicrobial compounds (Pentostam, Glucantime). Sodium stibogluconate is available under an Investigational New Drug protocol from the Centers for Disease Control and Prevention (CDC) Drug Service for difficult cases. Vector and reservoir control, and aggressive epidemiological surveillance are the best options for prevention and containment of this disease. As of 2009 there were no approved vaccines against leishmaniasis, although one experimental vaccine induces a 30% skin-positive reaction that appears to be associated with decreased visceral leishmaniasis.

Trypanosomiasis

Another group of flagellated protists called **trypanosomes** cause the aggregate of diseases termed **trypanosomiasis.** *Trypanosoma brucei gambiense*, found in the rain forests of west and central Africa, is the agent of West African sleeping sickness. *T. brucei rhodesiense*, found in the upland savannas of east Africa, is the agent of East African sleeping sickness. Reservoirs for these trypanosomes are domestic cattle and wild animals, within which they cause severe malnutrition. Both species use tsetse flies (genus *Glossina*) as intermediate hosts.

Trypanosomiasis is such a problem in parts of Africa that millions of square miles are not fit for human habitation. Five hundred to 1,000 cases of East African trypanosomiasis and 12,000 new cases of West African trypanosomiasis are reported to the World Health Organization (WHO) annually. However, it is likely that the majority of cases are not reported due to the lack of public health infrastructure in endemic regions. As a result, more than 100,000 new cases per year are likely. In the United States, cases of West African trypanosomiasis are extremely rare, and only 37 cases of East African trypanosomiasis have been reported since 1967; all of these patients were travelers to Africa.

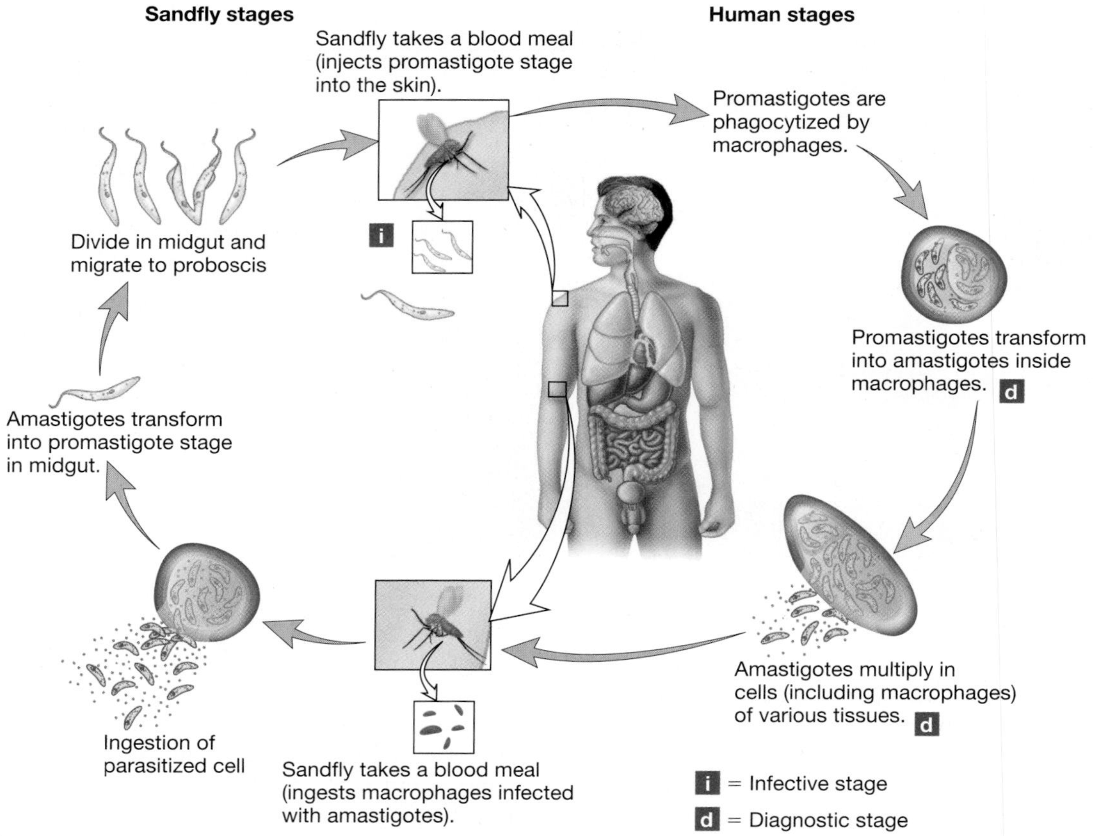
Sandfly stages
Human stages
Sandfly takes a blood meal (injects promastigote stage into the skin).
Promastigotes are phagocytized by macrophages.
Divide in midgut and migrate to proboscis
i
Promastigotes transform into amastigotes inside macrophages.
d
Amastigotes transform into promastigote stage in midgut.
Amastigotes multiply in cells (including macrophages) of various tissues.
d
Ingestion of parasitized cell
Sandfly takes a blood meal (ingests macrophages infected with amastigotes).
i = Infective stage
d = Diagnostic stage

FIGURE 39.10 Leishmaniasis. *Leishmania* life cycle. Note the promastigote-to-amastigote cycle.

The trypanomastigote parasites are transmitted through the bite of the fly to humans (**figure 39.11**; *also see figure 23.4*). The protists pass through the lymphatic system and enter the bloodstream, and replicate by binary fission as they pass to other blood fluids. The tsetse fly bite is painful and can develop into a chancre. Patients may exhibit symptoms of fever, severe headaches, extreme fatigue, muscle and joint aches, irritability, and swollen lymph nodes. Some patients may develop a skin rash. The disease gets its name from the fact that patients often exhibit lethargy—characteristically lying prostrate, drooling from the mouth, and insensitive to pain; they also exhibit progressive confusion, slurred speech, seizures, and personality changes. The protists cause interstitial inflammation and necrosis within the lymph nodes and small blood vessels of the brain and heart. *T. brucei rhodesiense* disease develops so rapidly that infected individuals often die within a year. *T. brucei gambiense* invades the central nervous system, where necrotic damage causes a variety of nervous disorders, including the characteristic sleeping sickness. Untreated victims die in 2 to 3 years.

T. cruzi causes **American trypanosomiasis (Chagas' disease),** which occurs in the tropics and subtropics of continental America (southern United States, Central America and northern South America). The parasite uses the triatomine (kissing) bug as a vector (**figure 39.12*a***). As the triatomine bug takes a blood meal, the protists are discharged in the insect's feces (figure 39.12*b*). Some trypanosomes enter the bloodstream through the wound and invade the liver, spleen, lymph nodes, and central nervous system; disease occurs immediately after infection, lasting weeks or months (acute phase). Cell invasion stimulates the trypanosome's transformation into amastigotes, resulting in clinical manifestations of infection. The bloodstream trypanomastigotes do not replicate, however, until they enter a cell (becoming amastigotes) or are ingested by the arthropod vector (becoming epimastigotes). A chronic indeterminate (asymptomatic) phase of the disease follows the acute phase. During this phase, a high percentage of heart and gastrointestinal disease occurs due to parasitized cardiac and intestinal cells, respectively. There are 16 million to 18 million new cases each year and over 50,000 deaths.

Trypanosomiasis is diagnosed by finding motile protists in fresh blood, spinal fluid, or skin biopsy and by serological testing. Treatment depends on the infecting trypanosome species and the stage of infection. Pentamidine isethionate and suramin (obtained as an Investigational New Drug protocol from the CDC) are used to treat the hemolymphatic stages of West and East African trypanosomiasis, respectively. Melarsoprol is used to treat late disease with central nervous system involvement caused by both trypanosomes. Benznidazole or nifurtimox (obtained as an Investigational New Drug protocol from the CDC) are used to treat Chagas' disease, but effectiveness is best when administered during the acute

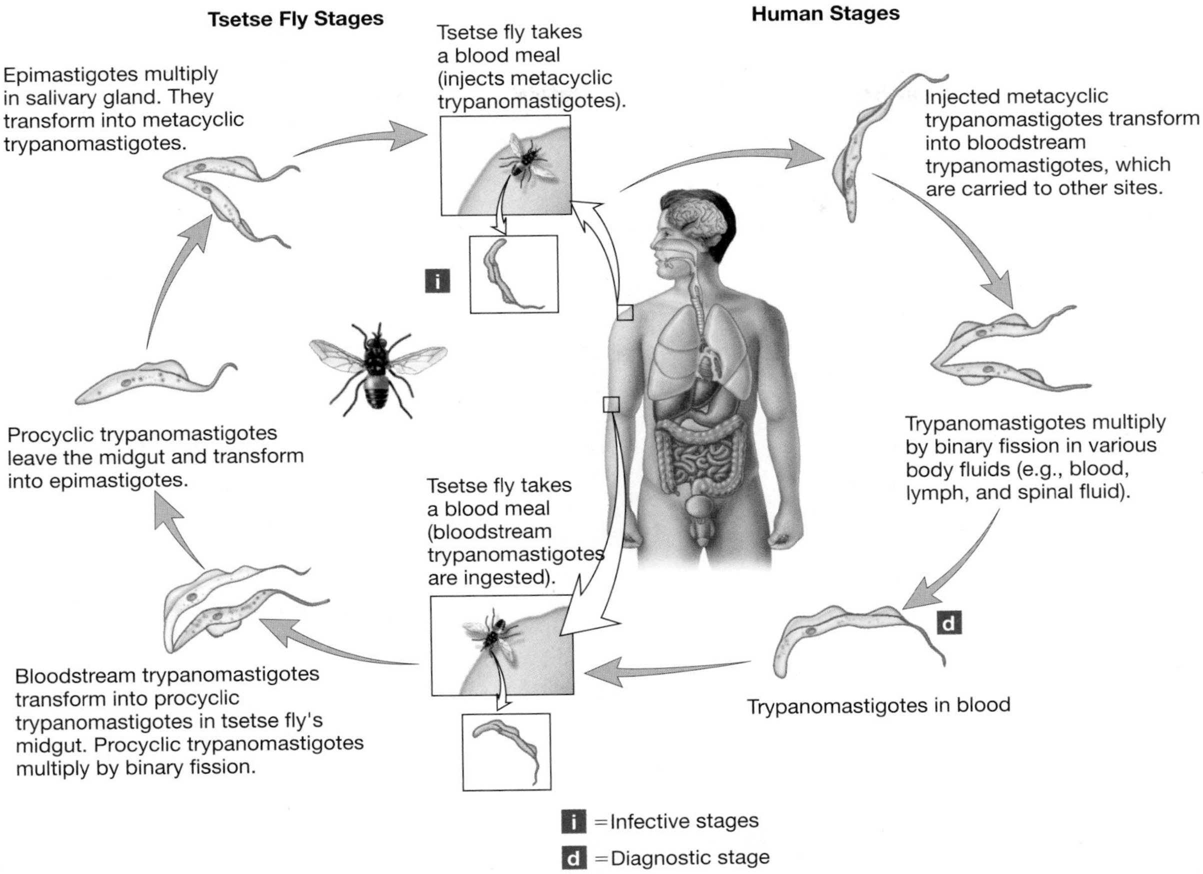
Tsetse Fly Stages
Human Stages
Tsetse fly takes a blood meal (injects metacyclic trypanomastigotes).
Epimastigotes multiply in salivary gland. They transform into metacyclic trypanomastigotes.
Injected metacyclic trypanomastigotes transform into bloodstream trypanomastigotes, which are carried to other sites.
Procyclic trypanomastigotes leave the midgut and transform into epimastigotes.
Trypanomastigotes multiply by binary fission in various body fluids (e.g., blood, lymph, and spinal fluid).
Tsetse fly takes a blood meal (bloodstream trypanomastigotes are ingested).
Bloodstream trypanomastigotes transform into procyclic trypanomastigotes in tsetse fly's midgut. Procyclic trypanomastigotes multiply by binary fission.
Trypanomastigotes in blood
i
d
i =Infective stages
d =Diagnostic stage

FIGURE 39.11 Life Cycle of *Trypanosoma brucei*. This trypanosome is transmitted to humans through the bite of the tsetse fly. The metacyclic trypanomastigotes enter the bloodstream, where they disseminate to various tissue sites. They reproduce in the human and are ingested by tsetse flies as part of the blood meal. In the fly gut, they transform into procyclic trypanomastigotes and then into epimastigotes. Epimastigotes migrate to the salivary glands, where they transform into metacyclic trypanomastigotes that can be passed on during feeding.

phase of infection. Vaccines are not useful because the parasite is able to change its protein coat (antigenic shift) and evade the immunologic response. ◀◀ Euglenozoa *(section 23.2)*

1. Besides their route of transmission, how else are human fungal diseases categorized?
2. Why are fungal infections of the lungs potentially life-threatening?
3. Why is *Histoplasma capsulatum* found in bird feces but not within the birds themselves? What are the public health implications of this?
4. What flagellated protists invade the blood? What diseases do they cause?
5. Describe the malarial life cycle. What stages of the life cycle occur in humans?
6. What malarial stages could vaccines block? What are the challenges to vaccine development?

39.4 Direct Contact Diseases

Transmission of eukaryotic pathogens by direct contact results in a number of infectious diseases of the skin and reproductive organs. The direct contact diseases caused by fungi can be classified by their penetration into the host; superficial, cutaneous, and subcutaneous fungal infections are known. Other diseases caused by eukaryotic pathogens occur through hand-to-mouth/nose exchange as well as their transfer through body fluids.

Superficial Mycoses

Superficial mycoses are extremely rare in the United States, and most occur in the tropics. The fungi responsible are limited to the outer surface of hair and skin, and hence are called superficial. Infections of the hair shaft are collectively called **piedras** (Spanish for stone because they are associated with the hard

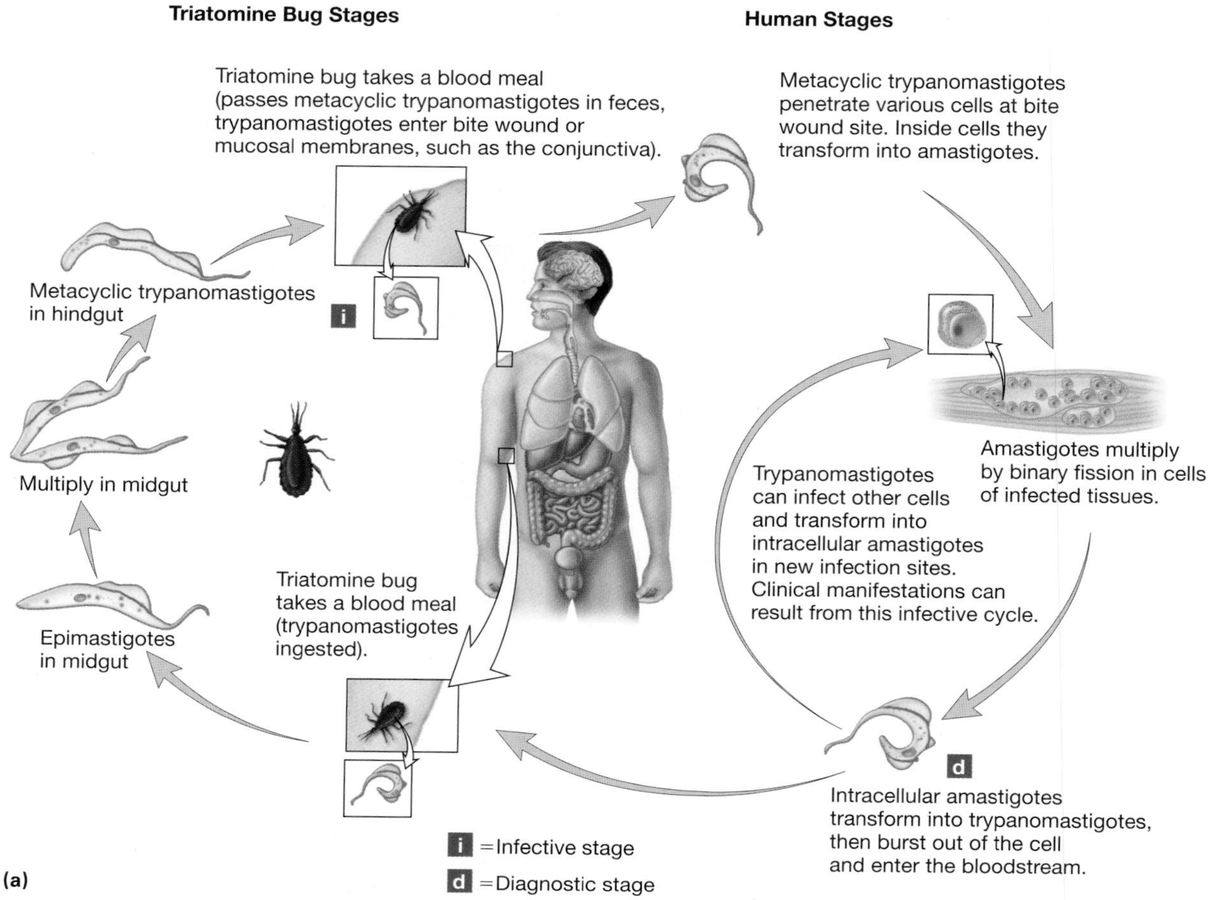
Triatomine Bug Stages
Human Stages
Triatomine bug takes a blood meal (passes metacyclic trypanomastigotes in feces, trypanomastigotes enter bite wound or mucosal membranes, such as the conjunctiva).
Metacyclic trypanomastigotes penetrate various cells at bite wound site. Inside cells they transform into amastigotes.
Metacyclic trypanomastigotes in hindgut
i
Amastigotes multiply by binary fission in cells of infected tissues.
Multiply in midgut
Trypanomastigotes can infect other cells and transform into intracellular amastigotes in new infection sites. Clinical manifestations can result from this infective cycle.
Triatomine bug takes a blood meal (trypanomastigotes ingested).
Epimastigotes in midgut
d
Intracellular amastigotes transform into trypanomastigotes, then burst out of the cell and enter the bloodstream.
i =Infective stage
d =Diagnostic stage
(a)

(b)

FIGURE 39.12 Life Cycle of *Trypanosoma cruzi*. (a) T. cruzi is transmitted to humans from the triatomine bug. (b) The metacyclic trypanomastigotes are shed in the bug feces, which enter the host through the bite wound or mucous membranes. The trypanomastigotes penetrate host cells and transform into amastigotes. After replication by binary fission, amastigotes transform into trypanomastigotes, which burst from their host cell to disseminate throughout the host. Blood-borne trypanomastigotes can be ingested by triatomine bugs as part of a blood meal, transform into epimastigotes in the bug's midgut, and transform into metacyclic trypanomastigotes in the hindgut.

nodules formed by mycelia on the hair shaft). For example, black piedra is caused by *Piedraia hortae* and forms hard, black nodules on the hairs of the scalp. White piedra is caused by the yeast *Trichosporon beigelii* and forms light-colored nodules on the beard and mustache. Some superficial mycoses are called **tineas** (Latin for grub, larva, worm); the specific type is designated by a modifying term. Tineas involve the outer layers of the skin, nails, and hair. Tinea versicolor is caused by the yeast *Malassezia furfur* and forms brownish-red scales on the skin of the trunk, neck, face, and arms. Treatment involves removal of the skin scales

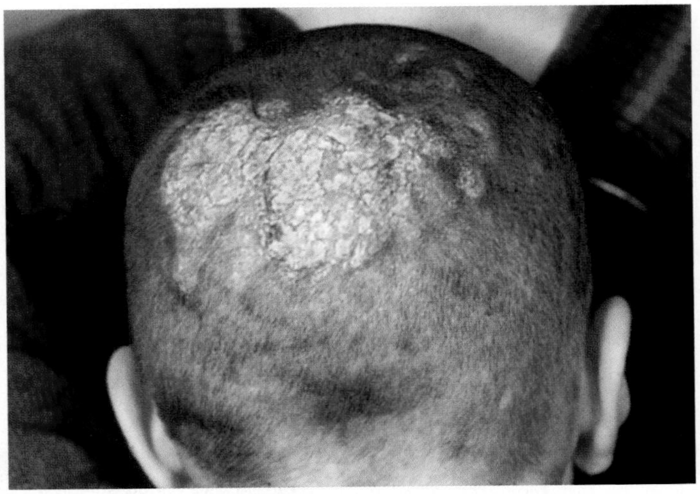

(a)

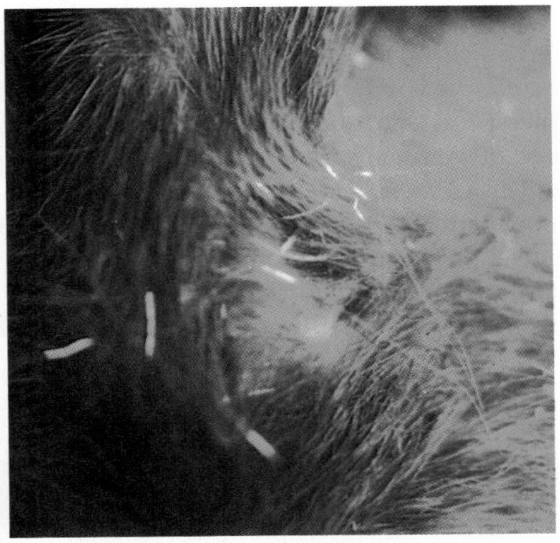

(b)

FIGURE 39.13 Cutaneous Mycosis: Tinea Capitis. (a) Ringworm of the head caused by *Microsporum audouinii.* (b) Close-up using a Wood's light (a UV lamp).

Figure 39.13 Micro Inquiry

What is the difference between a piedra and a tinea? Which is shown here?

with a cleansing agent and removal of the infected hairs. Good personal hygiene prevents these infections.

Cutaneous Mycoses

Cutaneous mycoses—also called **dermatomycoses, ringworms,** or tineas—occur worldwide and represent the most common fungal diseases in humans. Three genera of cutaneous fungi, or dermatophytes, are involved in these mycoses: *Epidermophyton, Microsporum,* and *Trichophyton.* Diagnosis is by microscopic examination of biopsied areas of the skin cleared with 10% potassium hydroxide and by culture on Sabouraud dextrose agar. Treatment is with topical ointments such as miconazole (Monistat-Derm), tolnaftate (Tinactin), or clotrimazole (Lotrimin) for 2 to 4 weeks. Griseofulvin (Grifulvin V) and itraconazole (Sporanox) are the only oral antifungal agents currently approved by the FDA for treating dermatophytoses. Infections by dermatophytes may be insidious; the desquamated (shed) skin cells can contain viable fungi for months or years. Thus any treatment plan should also include thorough cleaning of bedding, carpets, and any other substrates that could hold desquamated skin cells.

Tinea barbae (Latin *barba,* the beard) is an infection of the beard hair caused by *Trichophyton mentagrophytes* or T. *verrucosum.* It is predominantly a disease of men who live in rural areas and acquire the fungus from infected animals. Tinea capitis (Latin *capita,* the head) is an infection of the scalp hair (**figure 39.13*a***). It is characterized by loss of hair, inflammation, and scaling. Tinea capitis is primarily a childhood disease caused by *Trichophyton* or *Microsporum* species. Person-to-person transmission of the fungus occurs frequently when poor hygiene and overcrowded conditions exist. The fungus also occurs in domestic animals, which can transmit it to humans. A Wood's lamp (a UV light) can help with the diagnosis of tinea capitis because fungus-infected hair fluoresces when illuminated by UV radiation (figure 39.13*b*).

Tinea corporis (Latin *corpus,* the body) is a dermatophytic infection that can occur on any part of the skin (**figure 39.14*a***). The disease is characterized by circular, red, well-demarcated, scaly, vesiculopustular lesions accompanied by itching. Tinea corporis is caused by *Trichophyton rubrum, T. mentagrophytes,* or *Microsporum canis.* Transmission of the disease agent is by direct contact with infected animals and humans or by indirect contact through fomites (inanimate objects).

Tinea cruris (Latin *crura,* the leg) is a dermatophytic infection of the groin (figure 39.14*b*). The pathogenesis and clinical manifestations are similar to those of tinea corporis. The responsible fungi are *Epidermophyton floccosum, T. mentagrophytes,* or *T. rubrum.* Factors predisposing one to recurrent disease are moisture, occlusion, and skin trauma. Wet bathing suits, athletic supporters (jock itch), tight-fitting slacks, panty hose, and obesity are frequently contributing factors.

Tinea pedis (Latin *pes,* the foot), also known as athlete's foot, and tinea manuum (Latin *mannus,* the hand) are dermatophytic infections of the feet (figure 39.14*c*) and hands, respectively. Clinical symptoms vary from a fine scale to a vesiculopustular eruption. Itching is frequently present. Warmth, humidity, trauma, and occlusion increase susceptibility to infection. Most infections are caused by *T. rubrum, T. mentagrophytes*, or *E. floccosum*. Tinea pedis and tinea manuum occur throughout the world, are most commonly found in adults, and increase in frequency with age.

Tinea unguium (Latin *unguis*, nail) is a dermatophytic infection of the nail bed (figure 39.14*d*). In this disease, the nail becomes discolored and then thickens. The nail plate rises and separates from the nail bed. *Trichophyton rubrum* or *T. mentagrophytes* is the causative fungus.

Subcutaneous Mycoses

The dermatophytes that cause subcutaneous mycoses are normal saprophytic inhabitants of soil and decaying vegetation. Because they are unable to penetrate the skin, they must be introduced into the subcutaneous tissue by a puncture wound. Most infections involve barefooted agricultural workers. Once in the subcutaneous tissue, the disease develops slowly—often over a period of years. During this time, the fungi produce a nodule that eventually ulcerates and the organisms spread along lymphatic channels, producing more subcutaneous nodules. At times, the

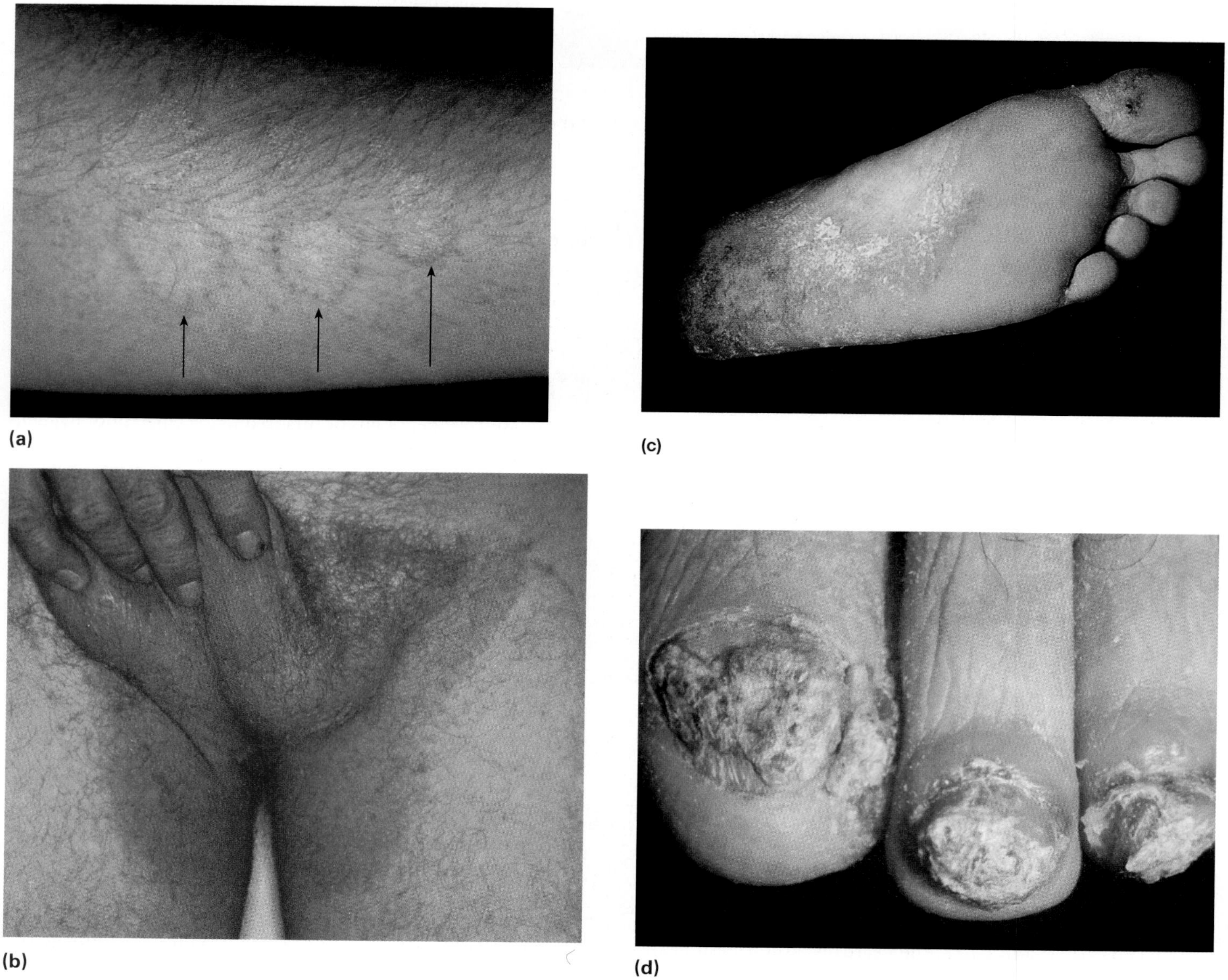

FIGURE 39.14 Cutaneous Mycosis. (a) Tinea corporis. Ringworm of the body—in this case, the forearm—caused by *Trichophyton mentagrophytes*. Notice the circular patches (arrows). (b) Tinea cruris. Ringworm of the groin caused by *Epidermophyton floccosum*. (c) Tinea pedis. Ringworm of the foot caused by *Trichophyton rubrum, T. mentagrophytes*, or *Epidermophyton floccosum*. (d) Tinea unguium. Ringworm of the nails caused by *Trichophyton rubrum*.

nodules drain to the skin surface. The administration of oral 5-fluorocytosine, iodides, or amphotericin B, and surgical excision, are the usual treatments. Diagnosis is accomplished by culture of the infected tissue.

One type of subcutaneous mycosis is **chromoblastomycosis,** in which dark brown nodules are formed. This disease is caused by the black molds *Phialophora verrucosa* and *Fonsecaea pedrosoi.* These fungi exist worldwide, especially in tropical and subtropical regions. Most infections involve the legs and feet (**figure 39.15*a***).

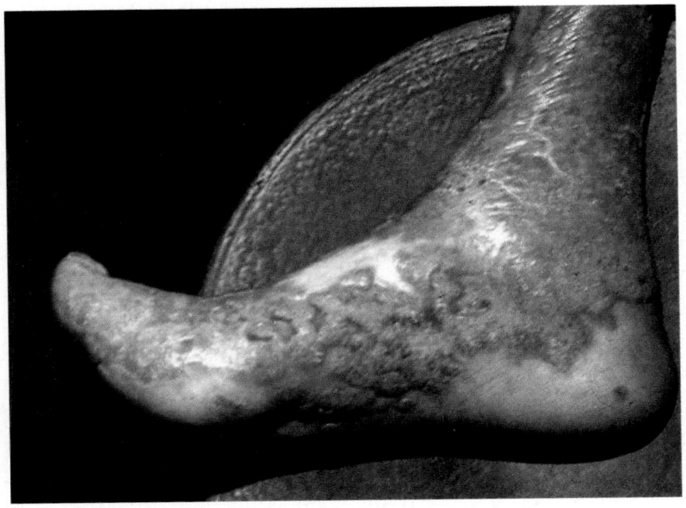

(a)

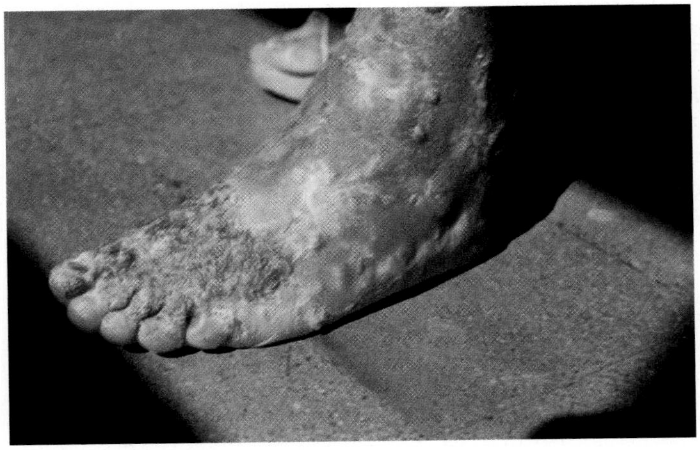

(b)

FIGURE 39.15 Subcutaneous Mycosis. (a) Chromoblastomycosis of the foot caused by *Fonsecaea pedrosoi.* (b) Eumycotic mycetoma of the foot caused by *Madurella mycetomatis.*

Figure 39.15 Micro Inquiry

How do fungi causing subcutaneous mycoses gain access to tissue below the skin surface?

Another subcutaneous mycosis is **maduromycosis,** caused by *Madurella mycetomatis*, which is distributed worldwide and is especially prevalent in the tropics. Because the fungus destroys subcutaneous tissue and produces serious deformities, the resulting infection is often called a **eumycotic mycetoma,** or fungal tumor (figure 39.15*b*). One form of mycetoma, known as Madura foot, occurs through skin abrasions acquired while walking barefoot on contaminated soil.

Sporotrichosis is the subcutaneous mycosis caused by the dimorphic fungus *Sporothrix schenckii.* The disease occurs throughout the world and is the most common subcutaneous mycotic disease in the United States. The fungus can be found in the soil; on living plants, such as barberry shrubs and roses; or in plant debris, such as sphagnum moss, baled hay, and pine-bark mulch. Infection occurs by a puncture wound from a thorn or splinter contaminated with the fungus. The disease is an occupational hazard for florists, gardeners, and forestry workers. It is not spread from person to person. After an incubation period of 1 to 12 weeks, a small, red papule arises and begins to ulcerate (**figure 39.16**). New lesions appear along lymph channels and can remain localized or spread throughout the body, producing extracutaneous sporotrichosis.

Sporotrichosis is typically treated by ingestion of potassium iodide or itraconazole (Sporanox) until the lesions are healed, usually several weeks. Preventative measures include gloves and other protective clothing, as well as avoidance of contaminated landscaping materials, especially sphagnum moss.

Toxoplasmosis

Toxoplasmosis is a worldwide disease caused by the protist *Toxoplasma gondii* that infects nearly 90% of people over the age of twelve, although most of those infected do not exhibit symptoms of the disease because the immune system usually prevents illness. Approximately 60 million U.S. citizens are infected and may have no symptoms. The natural reservoir of *T. gondii* is wild rodents,

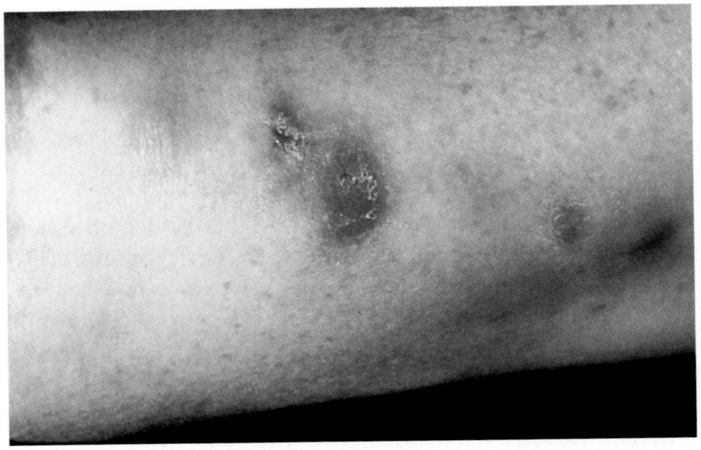

FIGURE 39.16 Subcutaneous Mycosis. Sporotrichosis of the arm caused by *Sporothrix schenckii.*

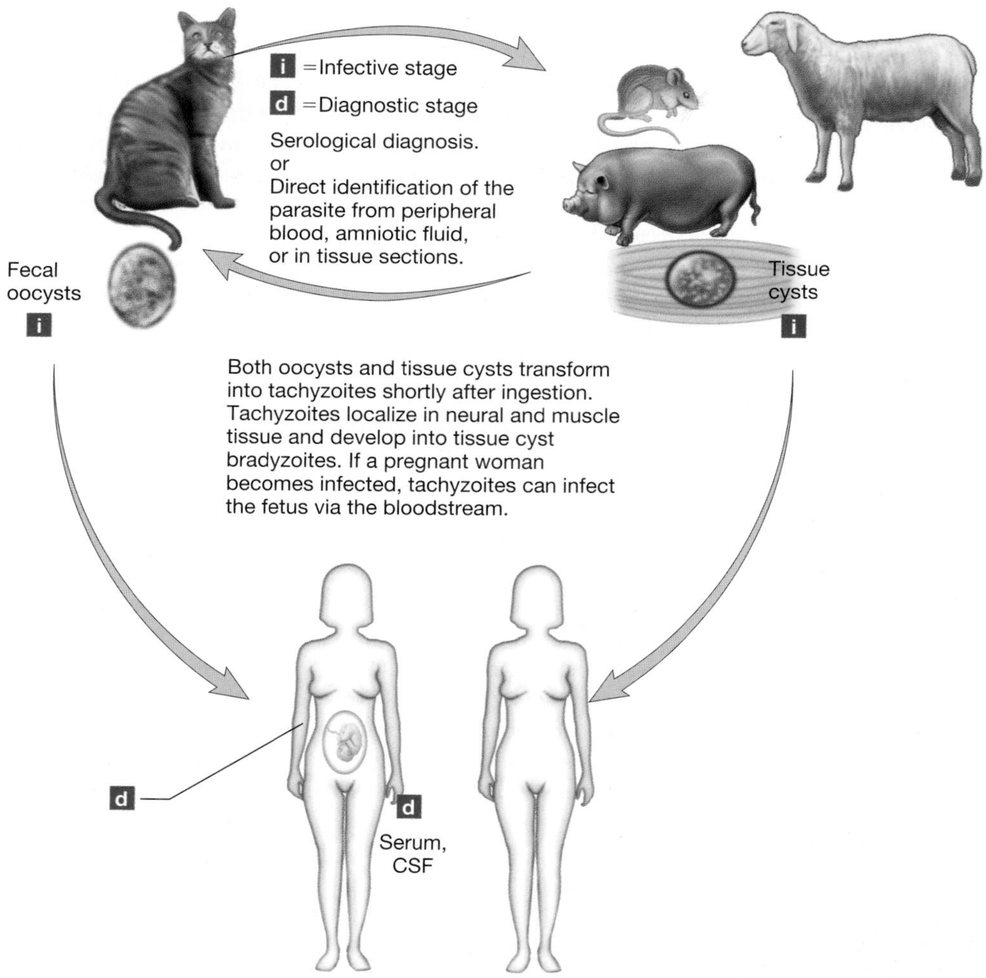

FIGURE 39.17 Life Cycle of *Toxoplasma gondii*. *Toxoplasma gondii* is a protozoan parasite of numerous mammals and birds. However, the cat, the definitive host, is required for completion of the sexual cycle. Oocysts are shed in the feces of infected animals, where they may be ingested by another host. Ingested oocysts transform into tachyzoites, which migrate to various tissue sites via the bloodstream.

Figure 39.17 Micro Inquiry

Upon entry into the human host by the nose or mouth, where do *T. gondii* colonize?

birds, and other small animals. Cats are an important factor in the transmission of *T. gondii* to humans; cats are the definitive host and are required for completion of the sexual cycle (**figure 39.17**). Kittens especially shed oocysts of the protist in their feces for several weeks after infection. The protist is ubiquitous and finds its way into the food supply; toxoplasmosis can be transmitted by the ingestion of raw or undercooked meat. In fact, the CDC lists toxoplasmosis as the third leading cause of death due to foodborne infection. Oocysts enter the human host by way of the nose or mouth, and the parasites colonize the intestine. Congenital transfer, blood transfusion, or a tissue transplant also can transmit the protist. Originally toxoplasmosis gained public notice when it was discovered that in pregnant women, the protist might also infect the fetus, causing serious congenital defects or death. Most cases of toxoplasmosis are asymptomatic. Adults usually complain of an "infectious mononucleosis-like" syndrome. In immunocompromised or immunosuppressed individuals, it frequently results in fatal disseminated disease with a heavy cerebral involvement.

Acute toxoplasmosis is usually accompanied by lymph node swelling (lymphadenopathy) with reticular cell hyperplasia (enlargement). Pulmonary necrosis, myocarditis, and hepatitis caused by tissue necrosis are common. Retinitis (inflammation of the retina of the eye) is associated with necrosis due to the proliferation of the parasite within retinal cells. *Toxoplasma* in the immunocompromised, such as AIDS or transplant patients, can produce a unique encephalitis with necrotizing lesions accompanied by inflammatory infiltrates. It continues to cause more than 3,000 congenital infections per year in the United States. ◀◀ Alveolata *(section 23.5)*

Laboratory diagnosis of toxoplasmosis is by serological tests. Epidemiologically, toxoplasmosis is ubiquitous in all higher animals. Treatment of toxoplasmosis is with a combination of pyrimethamine (Daraprim) and sulfadiazine. Prevention and control require minimizing exposure by the following: avoiding eating raw meat and eggs, washing hands after working in the soil, cleaning cat litterboxes daily, keeping household cats indoors if possible, and feeding cats commercial food.

Trichomoniasis

Trichomoniasis is a sexually transmitted disease caused by the protozoan flagellate *Trichomonas vaginalis* (*see figure 23.2b*). It is one of the most common sexually transmitted diseases, with an estimated 7.4 million cases annually in the United States and 180 million cases annually worldwide. In response to the protist, the body accumulates leukocytes at the site of the infection. In females, this usually results in a profuse, purulent vaginal discharge that is yellowish to light cream in color and characterized by a disagreeable odor. The discharge is

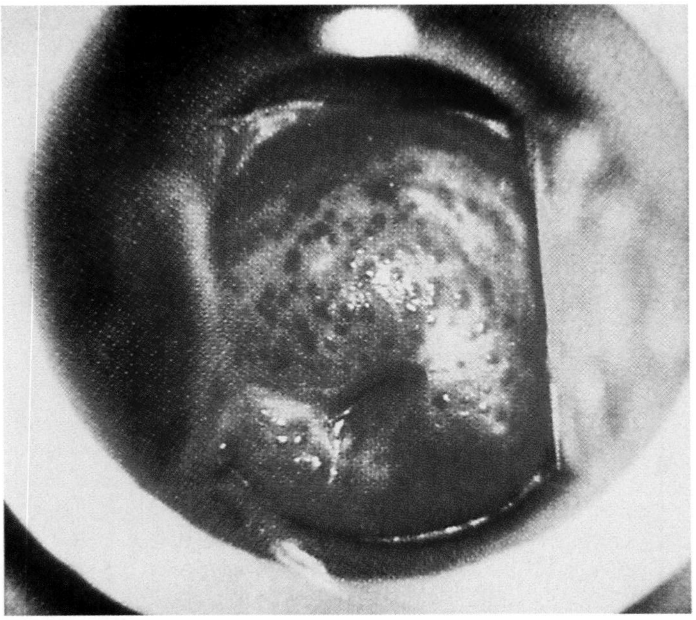

FIGURE 39.18 Trichomoniasis. *Trichomonas vaginalis* infection of the cervix causing hemorrhages, papules, and vesicles ("strawberry cervix").

accompanied by itching. The cervical mucosa is covered with punctuate hemorrhages, papules, and vesicles, referred to as "strawberry cervix" (**figure 39.18**). Males are generally asymptomatic because of the trichomonacidal action of prostatic secretions; however, at times a burning sensation occurs during urination. Diagnosis is made in females by microscopic examination of the discharge and identification of the protozoan. Infected males demonstrate protozoa in semen or urine. Treatment is by administration of metronidazole (Flagyl) to all sexual contacts. ◄◄ Parabasalia *(section 23.2)*

1. Describe two piedras that infect humans.
2. Briefly describe the major tineas that occur in humans.
3. Describe the three types of subcutaneous mycoses that affect humans.
4. In what two ways does *Toxoplasma* affect human health?
5. How would you diagnose trichomoniasis in a female? In a male?

39.5 Food-Borne and Waterborne Diseases

Food and water can serve as vehicles that transmit fungi and protozoa, as well as other infectious agents. Fungi and protists are able to survive various environmental conditions and ultimately contaminate food and water. Since the infection occurs in the intestinal tract, recovery of the infectious agent in feces is typical.

Amebiasis

It is now accepted that two species of *Entamoeba* infect humans: the nonpathogenic *E. dispar* and the pathogenic *E. histolytica*. *E. histolytica* is responsible for **amebiasis (amebic dysentery).** This very common parasite is endemic in warm climates where adequate sanitation and effective personal hygiene are lacking. Within the United States, about 3,000 to 5,000 cases are reported annually. However, it is a major cause of parasitic death worldwide; about 500 million people are infected and as many as 100,000 die of amebiasis each year. ◄◄ *Entamoebida (section 23.3)*

Infection occurs by ingestion of mature cysts from fecally contaminated water, food, or hands, or from fecal exposure during sexual contact. After excystation in the lower region of the small intestine, the metacyst divides rapidly to produce eight small trophozoites (**figure 39.19**). These trophozoites move to the large intestine, where they can invade the host tissue, live as commensals in the lumen of the intestine, or undergo encystation. In many hosts, the trophozoites remain in the intestinal lumen, resulting in an asymptomatic carrier state with cysts shed in feces.

If the infective trophozoites invade the intestinal tissues, they multiply rapidly and spread laterally, while feeding on erythrocytes, bacteria, and yeasts. The invading trophozoites destroy the epithelial lining of the large intestine by producing a cysteine protease. This protease is a virulence factor of *E. histolytica* and may play a role in intestinal invasion by degrading the extracellular matrix and circumventing the host immune response through cleavage of secretory immunoglobulin A (sIgA), IgG, and complement factors. These cysteine proteases are encoded by at least seven genes, several of which are found in *E. histolytica* but not *E. dispar*. Lesions (ulcers) are characterized by minute points of entry into the mucosa and extensive enlargement of the lesion after penetration into the submucosa. *E. histolytica* also may invade and produce lesions in other tissues, especially the liver, to cause hepatic amebiasis. However, all extraintestinal amebic lesions are secondary to those established in the large intestine. The symptoms of amebiasis are highly variable, ranging from an asymptomatic infection to fulminating dysentery (exhaustive diarrhea accompanied by blood and mucus), appendicitis, and abscesses in the liver, lungs, or brain. ◄◄ *Antibodies (section 33.7)*

Laboratory diagnosis of amebiasis can be difficult and is based on finding trophozoites in fresh, warm stools and cysts in ordinary stools. Serological testing for *E. histolytica* is available but often unreliable. The therapy for amebiasis is complex and depends on the location of the infection within the host and the host's condition. Asymptomatic carriers who are passing cysts should always be treated with iodoquinol or paromomycin because they represent the most important reservoir of the protozoan in the population. In symptomatic intestinal amebiasis, metronidazole (Flagyl) or iodoquinol (Yodoxin) are the drugs of choice. Prevention and control of amebiasis is achieved by practicing good personal hygiene and avoiding water or food that might be contaminated with human feces. Viable cysts in water can be destroyed by hyperchlorination or iodination. ◄◄ *Antiprotozoan drugs (section 34.7)*

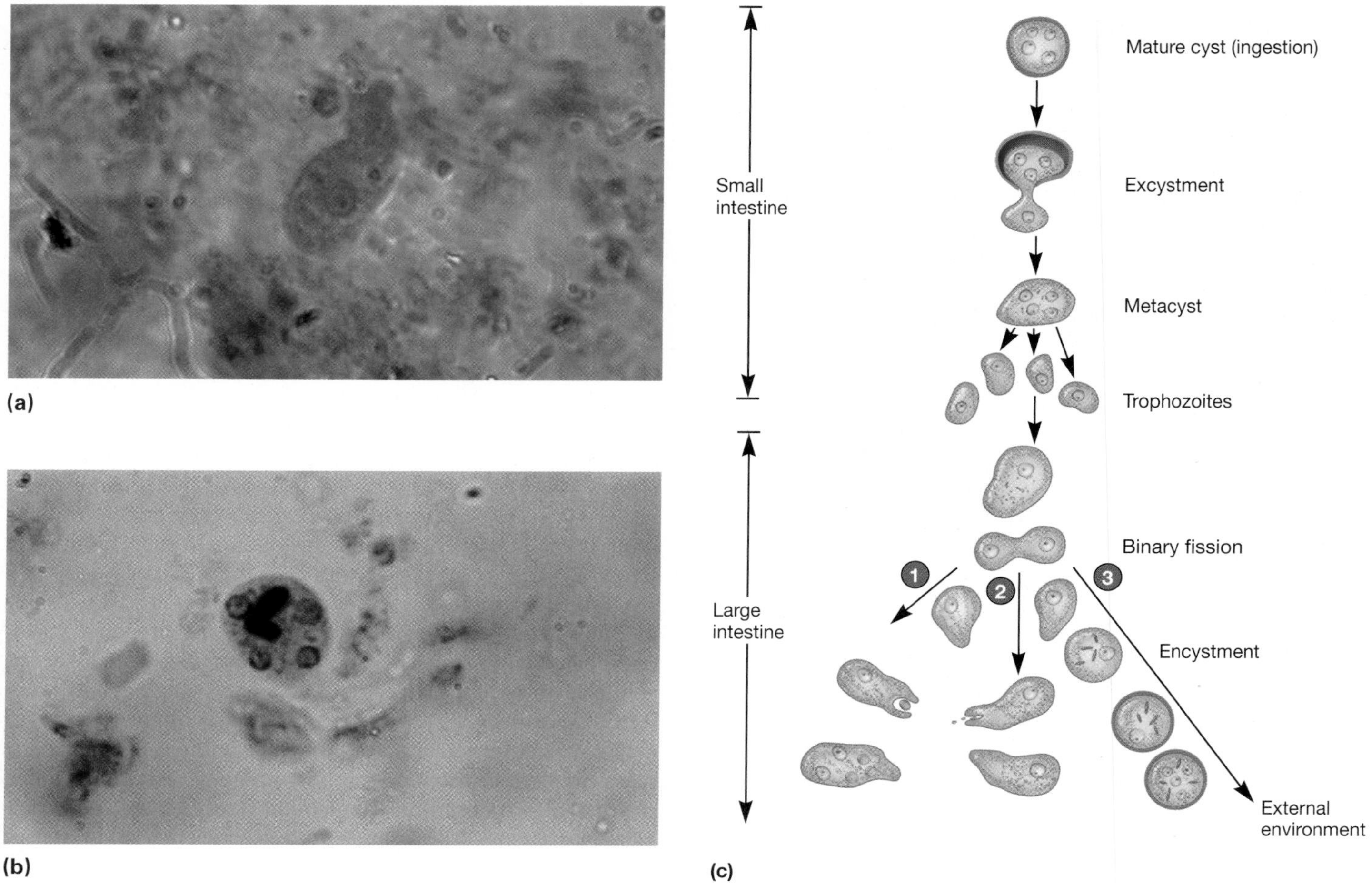

FIGURE 39.19 Amebiasis Caused by *Entamoeba histolytica*. (a) Light micrographs of a trophozoite (×1,000) and (b) a cyst (×1,000). (c) Life cycle. Infection occurs by the ingestion of a mature cyst. Excystment occurs in the lower region of the small intestine, and the metacyst rapidly divides to give rise to eight small trophozoites (only four are shown). These enter the large intestine, undergo binary fission, and may (1) invade the host tissues, (2) live in the lumen of the large intestine without invasion, or (3) undergo encystment and pass out of the host in the feces.

Amebic Meningoencephalitis and Keratitis

Free-living amoebae of the genera *Naegleria* and *Acanthamoeba* are facultative parasites responsible for causing two clinically distinct central nervous system diseases in humans: **granulomatous amoebic encephalitis** and **keratitis** (inflammation of the cornea). *Naegleria fowleri* causes primary amebic meningoencephalitis, a disease that is typically fatal in 3 to 10 days after infection. They are among the most common protists found in freshwater (including natural and treated pools and hot tubs), drinking water systems, ventilation and air conditioning systems, and moist soil. In addition, several *Acanthamoeba* spp. are known to cause encephalitis and to infect the eye, causing a chronically progressive, ulcerative *Acanthamoeba* keratitis that may result in blindness. Wearers of soft contact lenses may be predisposed to this infection and should take care to prevent contamination of their lens-cleaning and soaking solutions. Diagnosis of these infections is by demonstration of amoebae in clinical specimens. Most freshwater amoebae are resistant to commonly used antimicrobial agents. These amoebae are reported in fewer than 100 human disease cases annually in the United States, although the incidence (especially of *Acanthamoeba keratitis*) is likely higher.

Cryptosporidiosis

The first case of human **cryptosporidiosis** was reported in 1976. The protist responsible was identified as *Cryptosporidium parvum*, and classified as an emerging pathogen by the CDC. In 1993 *C. parvum* contaminated the Milwaukee, Wisconsin, water supply and caused severe diarrheal disease in about 400,000 individuals, the largest recognized outbreak of water-borne illness in U.S. history. A number of cryptosporidial species infect humans and other animals. *Cryptosporidium* ("hidden spore cysts") is found in about 90% of sewage samples, in 75% of river waters, and in 28% of drinking waters. ◀◀ Alveolata *(section 23.5)*

C. parvum is a common protist found in the intestine of many birds and mammals. When these animals defecate, oocysts are shed into the environment. If a human ingests food or water that is contaminated with the oocysts, excystment occurs within the small intestine, and sporozoites enter epithelial cells and develop into merozoites. Some of the merozoites subsequently undergo sexual reproduction to produce zygotes, and the zygotes differentiate into thick-walled oocysts. Oocyst release into the environment begins the life cycle again. A major problem for public health arises from the fact that the oocysts are only 4 to 6 μm in diameter—much too small to be easily removed by the sand filters used to purify drinking water. *Cryptosporidium* also is extremely resistant to disinfectants such as chlorine. The problem is made worse by the low infectious dose, around 10 to 100 oocysts, and the fact that the oocysts may remain viable for 2 to 6 months in a moist environment.

The incubation period for cryptosporidiosis ranges from 5 to 28 days. Diarrhea, which characteristically may be cholera-like, is the most common symptom. Other symptoms include abdominal pain, nausea, fever, and fatigue. The pathogen is routinely diagnosed by fecal concentration and acid-fast stained smears. (The thick-walled oocysts can be stained by the same method used to stain mycobacterial species.) Treatment of people with healthy immune systems is with nitazoxanide along with rehydration strategies. Although the disease usually is self-limiting in healthy individuals, patients with late-stage AIDS or who are immunocompromised in other ways may develop prolonged, severe, and life-threatening diarrhea.

Cyclosporiasis

Cyclosporiasis is caused by the unicellular protist *Cyclospora cayetanensis* (previously known as the cyanobacterium-like body). The disease is most common in tropical and subtropical environments, although it has been reported in most countries. In Canada and the United States, cyclosporiasis has been responsible for affecting over 3,600 people in at least 11 food-borne outbreaks since 1990. *Cyclospora* was first identified in 1979. A number of cyclosporiasis outbreaks have been linked to contaminated produce.

The disease presents with frequent, sometimes explosive, diarrhea. Often the patient exhibits loss of appetite, cramps, and bloating due to substantial gas production, nausea, vomiting, fever, fatigue, and substantial weight loss. Patients may report symptoms for days to weeks with decreasing frequency and then relapses. The protozoa infect the small intestine with a mean incubation period of approximately one week. Cyclosporan oocysts in freshly passed feces are not infective (**figure 39.20**). Thus direct oral-fecal transmission does not occur. Instead the oocysts must differentiate into sporozoites after days or weeks at temperatures

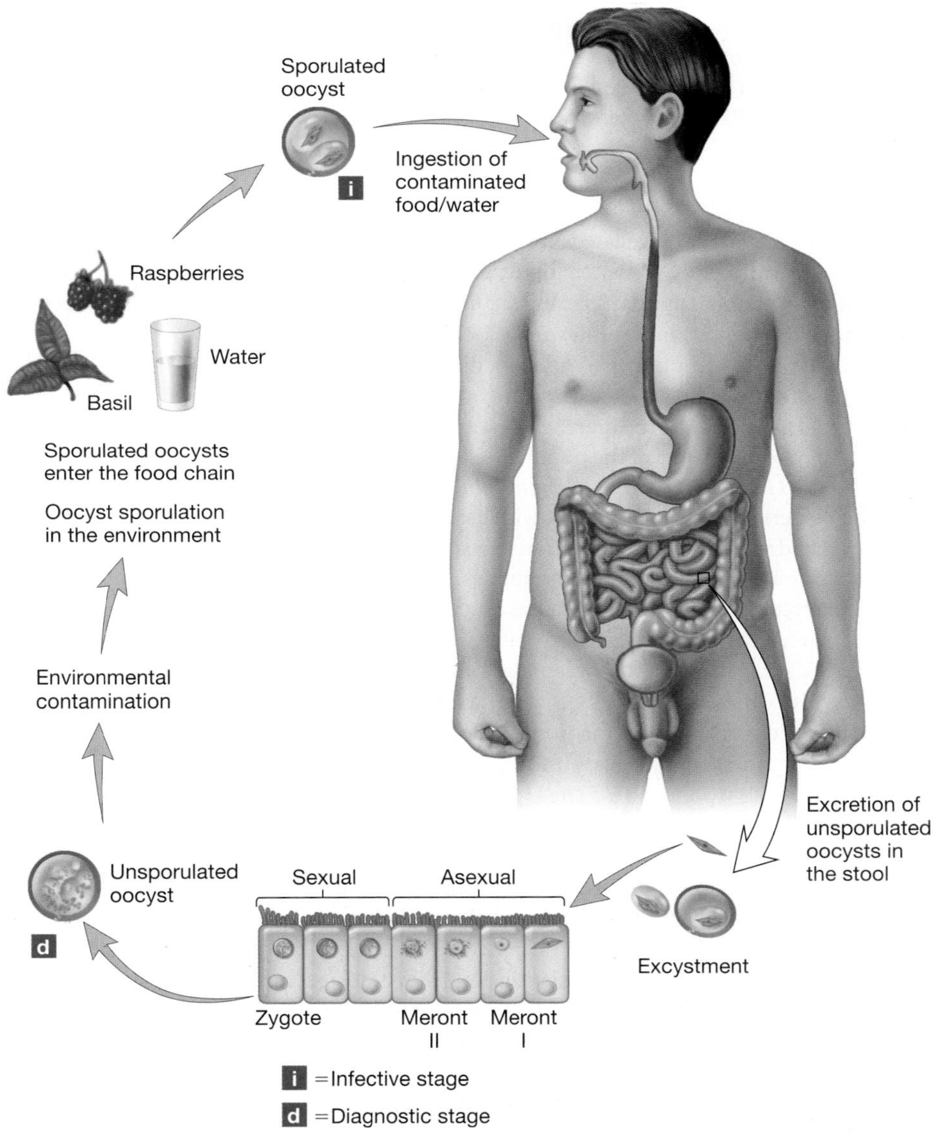

FIGURE 39.20 Life Cycle of *Cyclospora cayetanenis*. Unsporulated oocysts, shed in the feces of infected animals, sporulate and contaminate food and water. Once ingested, the sporulated oocysts excyst, penetrate host cells, and reproduce by binary fission (merogony). Sexual stages unite to form zygotes, resulting in new oocysts (unsporulated) that are shed in feces.

Some of the elements in this figure were created based on an illustration by Ortega, et al. Cyclospora cayetanensis. *In:* Advances in Parasitology: Opportunistic protozoa in humans. *San Diego: Academic Press; 1998. pp. 399–418.*

between 22 and 32°C (71.6–89.6°F). Sporozoites often enter the food chain when oocyst-contaminated water is used to wash fruits and vegetables prior to their transport to market. Once ingested, the sporozoites are freed from the oocysts and invade intestinal epithelial cells, where they replicate asexually. Sexual development is completed when sporozoites mature into new oocysts and are released into the intestinal lumen to be shed with the feces.

Laboratory diagnosis of cyclosporiasis is by the identification of oocycts in feces. Identification may require several specimens over several days. Treatment is with a combination of trimethoprim and sulfamethoxazole (Bactrim or Septra; *see figure 34.13*), and fluids to restore water lost through diarrhea. Prevention of cyclosporiasis is by avoidance of contaminated food and water. No vaccine is available.

Giardiasis

Giardia intestinalis is a flagellated protist (discovered by Antony van Leeuwenhoek in 1681, when he examined his own stools) found in more than 40 animal species (*see figure 23.2a*). *G. intestinalis* causes the very common intestinal disease **giardiasis,** which is worldwide in distribution and affects children more seriously than adults. In the United States, this protist is the most common cause of epidemic waterborne diarrheal disease (about 30,000 cases yearly). Approximately 7% of the population are asymptomatic carriers who shed cysts in their feces. *G. intestinalis* is endemic in child day-care centers in the United States, with estimates of 5 to 15% of diapered children being infected. Transmission occurs most frequently by cyst-contaminated water supplies. Epidemic outbreaks have been recorded in wilderness areas, suggesting that humans may be infected from pristine stream water with *Giardia* harbored by rodents, deer, cattle, or household pets. This implies that human infections also can be a zoonosis. As many as 200 million humans may be infected worldwide by ingesting as few as 10 cysts.

Following ingestion, the cysts undergo excystment in the duodenum, forming trophozoites. The trophozoites inhabit the upper portions of the small intestine, where they attach to the intestinal mucosa by means of their sucking disks (**figure 39.21**). The ability of the trophozoites to adhere to the intestinal epithelium accounts for the fact that they are rarely found in stools. It is thought that the trophozoites feed on mucous secretions and reproduce to form such a large population that they interfere with nutrient absorption by the intestinal epithelium. Giardiasis varies in severity, and asymptomatic carriers are common. Symptoms of giardiasis typically begin 1 week after infection and can last 6 to 8 weeks or longer if untreated. The disease can be acute or chronic. Acute giardiasis is characterized by severe diarrhea, epigastric pain, cramps, voluminous flatulence ("passing gas"), and anorexia. Chronic giardiasis is characterized by intermittent diarrhea, with periodic appearance and remission of symptoms.

Laboratory diagnosis is based on the identification of trophozoites—only in the severest of diarrhea—or cysts in stools. A commercial ELISA test is also available for the detection of *G. intestinalis* antigen in stool specimens. Quinacrine hydrochloride (Atabrine) and metronidazole (Flagyl) are the drugs of choice

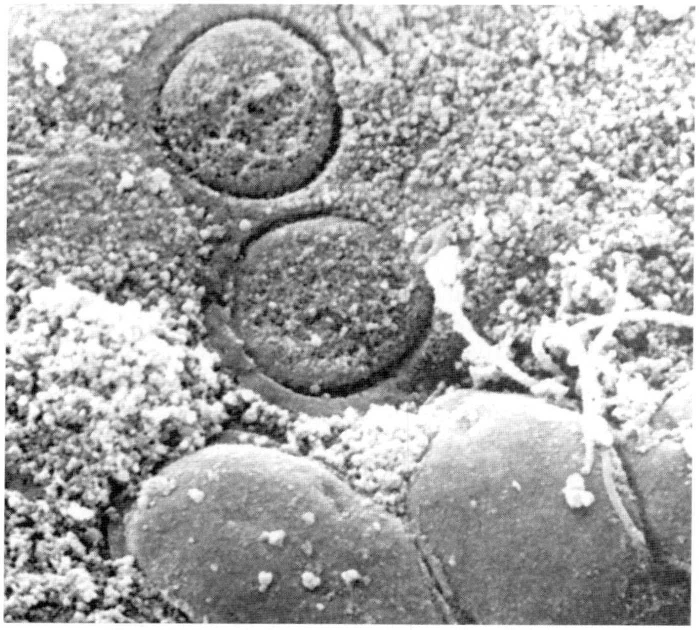

FIGURE 39.21 Giardiasis. Upon detachment from the epithelium, the protozoa often leave clear impressions on the microvillus surface (upper circles); scanning electron micrograph.

for adults, and furazolidone is used for children because it is available in a liquid suspension. Prevention and control involve proper treatment of community water supplies, especially the use of slow sand filtration because the cysts are highly resistant to chlorine treatment. ▶▶| *Water purification and sanitary analysis (section 42.1); Wastewater treatment (section 42.2)*

Search This: Dogs and Giardia

39.6 Opportunistic Diseases

Recall that an opportunistic microorganism can be a member of the resident microbiota that is generally harmless in its normal environment; it becomes pathogenic when (1) introduced to a new environment, (2) outcompetes the other resident microbes, or (3) the host becomes (immuno-) compromised. A compromised host is seriously debilitated and has a lowered resistance to infection. There are many causes of this condition: malnutrition, alcoholism, cancer, diabetes, another infectious disease (e.g., HIV/AIDS), trauma from surgery or injury, an altered microbiota from the prolonged use of antibiotics (e.g., in vaginal candidiasis), and immunosuppression (e.g., by drugs, hormones, genetic deficiencies, cancer chemotherapy, and old age). Opportunistic mycoses may start as normal microbiota or ubiquitous environmental contaminants. The importance of opportunistic fungal pathogens is increasing because of the expansion of the immunocompromised patient population.

Aspergillosis

Of all the fungi that cause disease in human hosts, none is as widely distributed in nature as *Aspergillus*. A filamentous fungus, *Aspergillus* is omnipresent—it is found wherever organic debris occurs, especially in soil, decomposing plant matter, household dust, building materials, some foods, and water. In fact, it is almost impossible to avoid daily inhalation of *Aspergillus* spores (**figure 39.22**). In people with healthy immune systems, *Aspergillus* does not cause disease. However, infection results when healthy immune functions deteriorate.

Aspergillus fumigatus is the usual cause of pulmonary **aspergillosis.** *A. flavus* is more common in some locations such as hospitals than *A. fumigatus* and is apparently more common in the air. *A. flavus* is the leading cause of invasive disease of immunosuppressed patients and the most common cause of superficial infection. Invasive disease typically results in pulmonary infection (with fever, chest pain, and cough) that disseminates to the brain, kidney, liver, bone, or skin. In immunocompetent patients, *Aspergillus* typically causes allergic sinusitis, allergic bronchitis, or a milder, localized bronchopulmonary infection. Invasive aspergillosis has a mortality rate of 50 to 100%. ◀◀ *Ascomycota (section 24.5)*

The major portal of entry for *Aspergillus* is the respiratory tract. Inhalation of conidiospores can lead to several types of pulmonary aspergillosis. One type is allergic aspergillosis. Infected individuals may develop an immediate allergic response and suffer asthma attacks when exposed to fungal antigens on the conidiospores. In bronchopulmonary aspergillosis, the major clinical manifestation of the allergic response is a bronchitis resulting from both type I and type III hypersensitivities. Although tissue invasion seldom occurs in bronchopulmonary aspergillosis, *Aspergillus* often can be cultured from the sputum. A most common manifestation of pulmonary involvement is the occurrence of colonizing aspergillosis, in which *Aspergillus* forms colonies within the lungs that develop into "fungus balls" called aspergillomas. These consist of a tangled mass of hyphae growing in a circumscribed area. From the pulmonary focus, the fungus may spread, producing disseminated aspergillosis in a variety of tissues and organs (figure 39.22*b*). In patients whose resistance is severely compromised, invasive aspergillosis may occur and fill the lung with fungal hyphae. ◀◀ *Hypersensitivities (section 33.10)*

Laboratory diagnosis of aspergillosis depends on identification, either by direct examination of pathological specimens or by isolation and characterization of the fungus. An enzyme immunoassay that detects galactomannan (an exoantigen of *Aspergillus*) can be used to screen suspected cases of aspergillosis. Successful therapy depends on treatment of the underlying disease so that host resistance increases. Treatment is with voriconazole and itraconazole. ◀◀ *Antifungal drugs (section 34.5)*

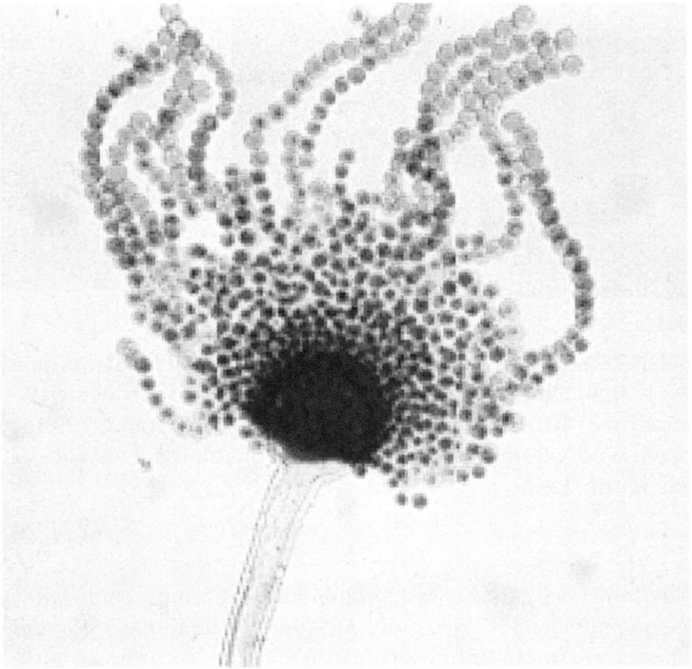

(a)

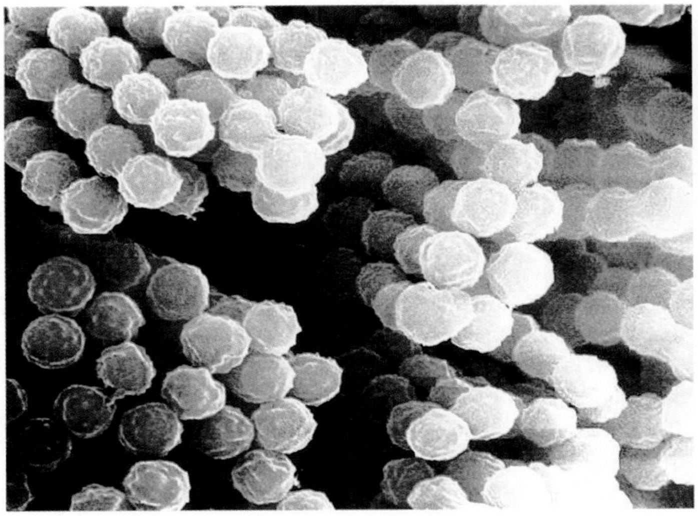

(b)

FIGURE 39.22 Cause of an Opportunistic Mycosis. (a) Conidiophores of *Aspergillus flavus.* (b) SEM of *Aspergillus* spores.

Candidiasis

Candidiasis is the mycosis caused by the dimorphic fungi *Candida albicans* (**figure 39.23*a***) and *C. glabrata*. In contrast to the other pathogenic fungi, *C. albicans* and *C. glabrata* are members of the normal microbiota within the gastrointestinal tract, respiratory tract, vaginal area, and mouth. In healthy individuals, they do not produce disease because growth is suppressed by other microbiota and other host resistance mechanisms. However, if

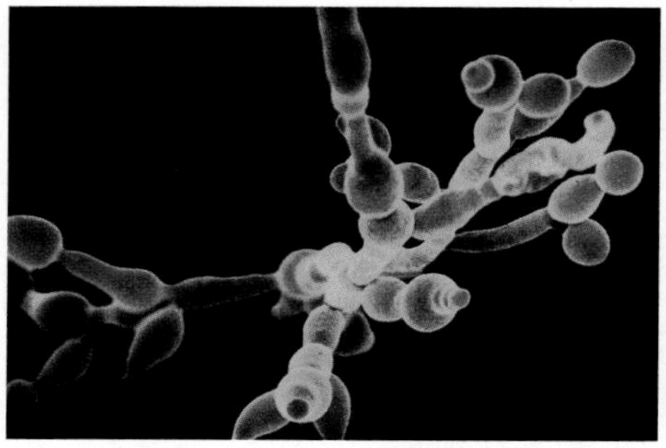

(a)

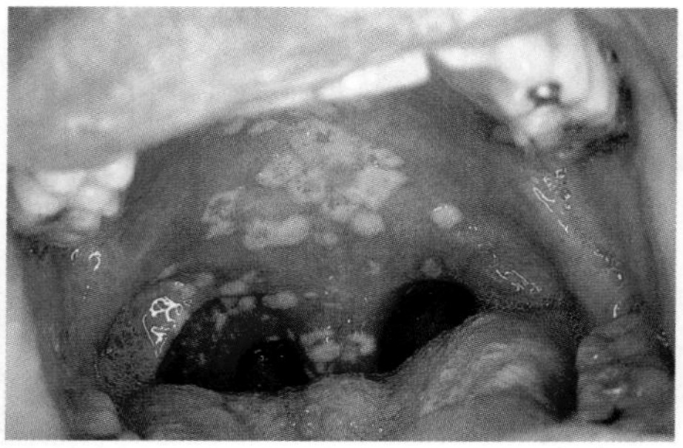

(b)

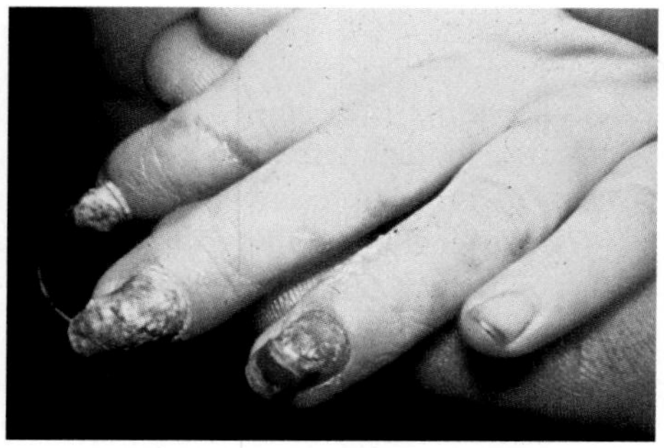

(c)

FIGURE 39.23 Opportunistic Mycoses Caused by *Candida albicans*. (a) Scanning electron micrograph of the yeast form (×10,000). Notice that some of the cells are reproducing by budding. (b) Thrush, or oral candidiasis, is characterized by the formation of white patches on the mucous membranes of the tongue and elsewhere in the oropharyngeal area. These patches form a pseudomembrane composed of spherical yeast cells, leukocytes, and cellular debris. (c) Paronychia and onychomycosis of the hands.

anything upsets the normal microbiota and immune competency, *Candida* may multiply rapidly and produce candidiasis. *Candida* species are important nosocomial pathogens. In some hospitals, they may represent almost 10% of nosocomial bloodstream infections. Because *Candida* can be transmitted sexually, it is also listed by the CDC as a sexually transmitted disease.

No other mycotic pathogen produces as diverse a spectrum of disease in humans as does *Candida*. Most infections involve the skin or mucous membranes. This occurs because *Candida* is a strict aerobe and finds such surfaces very suitable for growth. Cutaneous involvement usually occurs when the skin becomes overtly moist or damaged.

Oropharyngeal candidiasis, or **thrush** (figure 39.23*b*), is a fairly common disease in newborns, persons with dentures, and people who use inhaled steroids. A weakened immune system predisposes people to infections with *Candida*. *Candida* appears as many small, white flecks that cover the tongue and mouth. At birth, newborns do not have a normal microbiota in the oropharyngeal area. If the mother's vaginal area is heavily colonized with *Candida*, the upper respiratory tract of the newborn becomes colonized during passage through the birth canal. Thrush occurs because growth of *Candida* cannot be inhibited by the other microbiota. Once the newborn has developed his or her own normal oropharyngeal microbiota, thrush becomes uncommon. Paronychia and onychomycosis are associated with *Candida* infections of the subcutaneous tissues of the digits and nails, respectively (figure 39.24*c*). These infections usually result from continued immersion of the appendages in water.

Intertriginous candidiasis involves those areas of the body, usually opposed skin surfaces, that are warm and moist: axillae, groin, skin folds. Napkin (diaper) candidiasis is typically found in infants whose diapers are not changed frequently and therefore are not kept dry. **Candidal vaginitis** can result as a complication of diabetes, antibiotic therapy, oral contraceptives, pregnancy, or any other factor that compromises the female host. Normally the omnipresent lactobacilli can control *Candida* by the low pH they create. However, if their numbers are decreased by any of the aforementioned factors, *Candida* may proliferate, causing a curdlike, yellow-white discharge from the vaginal area. *Candida* can be transmitted to males during intercourse and lead to **balanitis;** thus it also can be considered a sexually transmitted disease. Balanitis is a *Candida* infection of the male glans penis and occurs primarily in uncircumcised males. The disease begins as vesicles on the penis that develop into patches and are accompanied by severe itching and burning.

Diagnosis of candidiasis is sometimes difficult because (1) this fungus is a frequent secondary invader in diseased hosts, (2) a mixed microbiota is most often found in the diseased tissue, and (3) no completely specific immunologic procedures for the identification of *Candida* currently exist. Mortality is almost 50% when *Candida* invade the blood or disseminate to visceral organs, as occasionally seen in immunocompromised patients.

There is no satisfactory treatment for candidiasis. Cutaneous lesions can be treated with topical agents such as sodium caprylate, sodium propionate, gentian violet, nystatin, miconazole, and trichomycin. Clotrimazole lozenges and nystatin swishing are used to treat oropharyngeal candidiasis. Ketoconazole, amphotericin B, fluconazole, itraconazole, and flucytosine also can be used for systemic candidiasis.

Microsporidiosis

Microsporidia is a term used to describe obligate, intracellular fungi that belong to the phylum *Microspora*. **Microsporidiosis** is an emerging infectious disease, found mostly in HIV patients. The taxonomy of these microorganisms is unsettled and many still consider them protists. More than 1,200 species have been catalogued, in 143 genera; at least 14 species are known to be human pathogens. Several domestic and feral animals appear to be reservoirs for several species that infect humans. Increasingly recognized as opportunistic infectious agents, microsporidia infect a wide range of vertebrate and invertebrate hosts. One unifying characteristic of the microsporidia is their production of a highly resistant spore, capable of surviving long periods of time in the environment. Spore morphology varies with species, but most superficially resemble enteric bacteria. Microsporidial spores recovered from human infections are oval to rodlike, measuring 1 to 4 μm. Microsporidia also possess a unique organelle known as the polar tubule, which is coiled within the spore (**figure 39.24**; *also see figure 24.15*). ◀◀ Microsporidia *(section 24.7)*

Infection of a host cell results when the microsporidia extrudes its polar tubule from within the spore. Contact with a eukaryotic cell membrane allows the polar tubule to bore through the membrane. A sudden increase in spore calcium results in the injection of the sporoplasm (cytoplasm-like contents) through the polar tubule into the host cell. Inside the cell, the sporoplasm condenses and undergoes asexual multiplication. Multiplication of the sporoplasm is species-specific, occurring within the host cytoplasm or within a vacuole, and is usually completed through binary (merogony) or multiple (shizogony) fission. Once the sporoplasm has multiplied, it directs the development of new spores (sporogony) by first encapsulating meronts or shizonts (products of asexual multiplication) within a thick spore coating. New spores increase until the host cell membrane ruptures, releasing the progeny.

Microsporidia infections can result in a wide variety of patient symptoms. These include hepatitis, pneumonia, skin lesions, diarrhea, weight loss, and wasting syndrome. Diagnosis is based on clinical manifestations and identification of microsporidia in fecal smears stained by the "Quick-Hot Gram Chromotrope technique," the Chromotrope 2R method (as Gram-stained or Giemsa-stained specimens are somewhat difficult). Where possible by electron microscopy, identification can be made based on the characteristic polar tubule coiled within a spore. Molecular identification is also possible using PCR and primers to the small subunit ribosomal RNA (*see figure 17.3*). Treatment of microsporidiosis is not well defined. However, some successes have been reported with the use of albendazole (which inhibits tubulin and ATP synthesis in helminths), metronidazole, or thalidomide (a toxic, immunosuppressant drug).

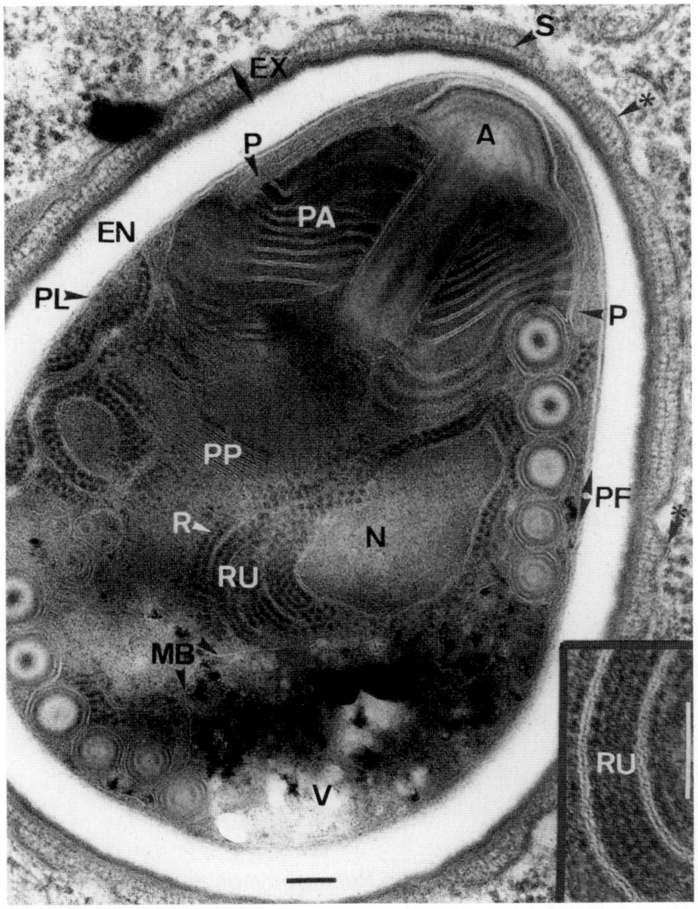

FIGURE 39.24 Microsporidia. Transmission electron micrograph of an *Episeptum inversum* spore. The infection apparatus has three components: PF, the polar filament, which is attached to A, the anchoring disc; and PA and PP, the anterior and posterior parts of the polaroplast, a system of membrane-bound sacs. Two wall layers are seen: EX, an external electron-dense exospore layer; and EN, a wide endospore layer containing chitin, as well as MB, an internal plasma membrane. N, nucleus; R, ribosomes; RU, rough endoplasmic reticulum (see inset); S, septum of the exospore coat; V, posterior vacuole; P, polar sac; and * outermost layer of exospore.

Pneumocystis Pneumonia

Pneumocystis is a fungus found in the lungs of a wide variety of mammals. Although it was previously classified as a protist, its rRNA, DNA sequences (from several genes), and biochemical analyses have shown that *Pneumocystis* is more closely related to fungi than to protists. The life cycle of *Pneumocystis* is presented in **figure 39.25**, although some aspects of its development are not well known. The disease that this fungus causes has been called *Pneumocystis* pneumonia or *Pneumocystis carinii* pneumonia (PCP). Its name was changed to *Pneumocystis jiroveci* in honor of the Czech parasitologist, Otto Jirovec, who first described this pathogen in humans. In recent medical literature, the acronym PCP for the disease has been retained despite the loss of the old species epithet (*carinii*). PCP now stands for *Pneumocystis* pneumonia.

Serological data indicate that most humans are exposed to *Pneumocystis* by age three or four. However, PCP occurs almost exclusively in immunocompromised hosts. Extensive use of immunosuppressive drugs and irradiation for the treatment of cancers and following organ transplants accounts for the formidable prevalence rates for PCP. This pneumonia also occurs in premature, malnourished infants and in more than 80% of AIDS patients. Both the organism and the disease remain localized in the lungs—even in fatal cases. Within the lungs, *Pneumocystis* causes the alveoli to fill with a frothy exudate.

Laboratory diagnosis of *Pneumocystis* pneumonia can be made definitively only by microscopically demonstrating the presence of the microorganisms in infected lung material or by a PCR analysis. Treatment is by means of oxygen therapy and either a combination

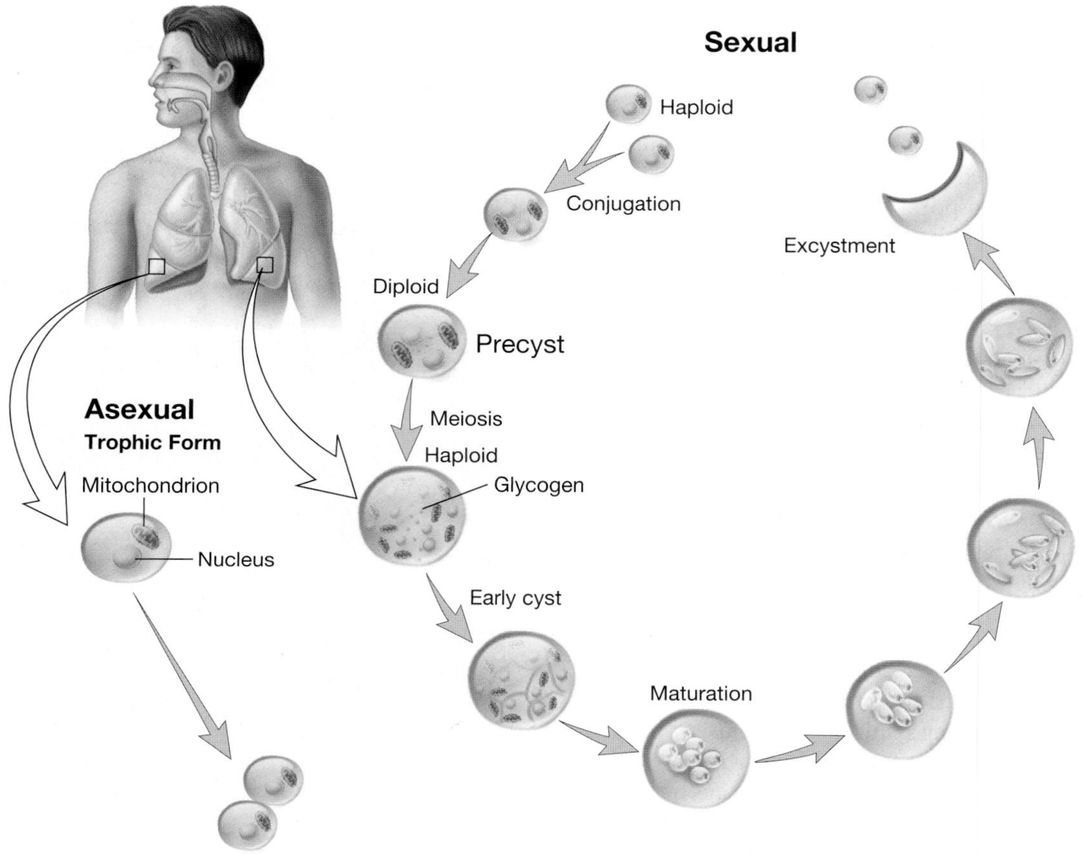

FIGURE 39.25 Life Cycle of *Pneumocystis jiroveci*. The *Pneumocystis* fungus exhibits both sexual and asexual reproductive stages. Sexual reproduction occurs when haploid cells undergo conjugation followed by meiosis. Asexual reproduction occurs by binary fission. Both forms of reproduction can occur within the human host.

of trimethoprim and sulfamethoxazole (Bactrim, Septra), atovaquone (Mepron), or trimetrexate (Neutrexin). Prevention and control are through prophylaxis with drugs in susceptible persons.

1. How do infections caused by *Entamoeba histolytica* occur?
2. What is the most common cause of epidemic waterborne diarrheal disease?
3. Describe in detail the life cycle of the malarial parasite.
4. Why are some mycotic diseases of humans called opportunistic mycoses?
5. What parts of the human body can be affected by *Candida* infections?
6. Describe the infection process used by microsporidia.
7. When is *Pneumocystis* pneumonia likely to occur in humans?

Summary

39.1 Pathogenic Fungi and Protists

a. Human fungal diseases, or mycoses, can be divided into five groups according to the level and mode of entry into the host. These are the superficial, cutaneous, subcutaneous, systemic, and opportunistic mycoses (**table 39.1**).

b. Protozoa are responsible for some of the most serious human diseases that affect hundreds of millions of people worldwide (**table 39.2**) and can also be grouped by route of disease transmission.

39.2 Airborne Diseases

a. Most systemic mycoses that occur in humans are acquired by inhaling the spores from the soil where the free-living fungi are found.
b. Four types can occur in humans: blastomycosis (**figure 39.1**), coccidioidomycosis (**figure 39.2**), cryptococcosis, (**figure 39.3**), and histoplasmosis (**figure 39.4**).

39.3 Arthropod-Borne Diseases

a. The most important human pathogen among the sporozoa is *Plasmodium*, the causative agent of malaria (**figures 39.5–39.8**). Human malaria is caused by four species of *Plasmodium: P. falciparum, P. vivax, P. malariae,* and *P. ovale.*
b. The flagellated protozoa that are transmitted by arthropods and infect the blood and tissues of humans are called hemoflagellates. Two major groups occur: the leishmanias, which cause the diseases collectively termed leishmaniasis (**figures 39.9** and **39.10**), and the trypanosomes (**figures 39.11** and **39.12**), which cause trypanosomiasis.

39.4 Direct Contact Diseases

a. Superficial mycoses of the hair shaft are collectively called piedras. Two major types are black piedra and white piedra. Tinea versicolor is a third common superficial mycosis.
b. The cutaneous fungi that parasitize the hair, nails, and outer layer of the skin are called dermatophytes, and their infections are termed dermatophytoses, ringworms, or tineas. At least seven types can occur in humans: tinea barbae (ringworm of the beard), tinea capitis (ringworm of the scalp; **figure 39.13**), tinea corporis (ringworm of the body; **figure 39.14*a***), tinea cruris (ringworm of the groin; **figure 39.14*b***), tinea pedis (ringworm of the feet; **figure 39.14*c***), tinea manuum (ringworm of the hands), and tinea unguium (ringworm of the nails; figure **39.14*d***).
c. The dermatophytes that cause the subcutaneous mycoses are normal saprophytic inhabitants of soil and decaying vegetation. Three types of subcutaneous mycoses can occur in humans: chromoblastomycosis (**figure 39.15*a***), maduromycosis (**figure 39.15*b***), and sporotrichosis (**figure 39.16**).
d. Toxoplasmosis is a disease caused by the protozoan *Toxoplasma gondii.* It is one of the major causes of death in AIDS patients (**figure 39.17**).
e. Trichomoniasis is a sexually transmitted disease caused by the protozoan flagellate *Trichomonas vaginalis. T. vaginalis* causes bleeding and pustules (**figure 39.18**).

39.5 Food-Borne and Waterborne Diseases

a. *Entamoeba histolytica* is the amoeboid protozoan responsible for amebiasis. This is a very common disease in warm climates throughout the world. It is acquired when cysts are ingested with contaminated food or water (**figure 39.19**).
b. Freshwater parasites such as *Naegleria* and *Acanthamoeba* can cause primary amebic meningoencephalitis.
c. *Cryptosporidium parvum* is a common coccidial apicomplexan parasite that causes severe diarrheal disease. It is acquired from contaminated food or water.
d. *Cyclospora cayetanensis* is shed in the feces of a current host and can only infect the gastrointestinal tract of a new host after it has developed at 22 to 32°C for several days or weeks. It can be acquired from contaminated water used to rinse fruit or vegetables (**figure 39.20**).
e. *Giardia intestinalis* is a flagellated protozoan that causes the common intestinal disease giardiasis (**figure 39.21**). This disease is distributed throughout the world, and in the United States, it is the most common cause of waterborne diarrheal disease.

39.6 Opportunistic Diseases

a. An opportunistic organism is one that is generally harmless in its normal environment but can become pathogenic in a compromised host.
b. The most important opportunistic mycoses affecting humans include systemic aspergillosis (**figure 39.22**), candidiasis (**figure 39.23*a***), and *Pneumocystis* pneumonia (**figure 39.25**).
c. An emerging opportunistic fungal disease, especially of the immunocompromised, is caused by the group of unique microbes known as the microsporidia. They live as a spore form and have a unique organelle used for infecting new host cells (**figure 39.24**).

Critical Thinking Questions

1. Compare and contrast treatment of diseases caused by fungi with those caused by viruses or bacteria.
2. What is one distinct feature of fungi that could be exploited for antibiotic therapy?
3. Why do you think most fungal diseases in humans are not contagious?
4. Trypanosomes are notorious for their ability to change their surface antigens frequently. Given the kinetics of a primary immune response (primary antibody production), how often would the surface antigen need to be changed to stay "ahead" of the antibody specificity? Why shouldn't it change the expression every time transcription occurs?

5. The initial, nonsymptomatic liver stage (LS) of *Plasmodium* is considered very promising for prophylactic drug and vaccine development (figure 39.8). However, technical obstacles in obtaining LS protists has hindered analysis of gene and protein expression. Tarun and her colleagues developed a rodent model whereby LS protists could be studied. They identified a set of proteins expressed only in the LS, including enzymes involved in metabolic pathways active only in LS, as well as exported proteins.

   Examine the *Plasmodium* life cycle in figure 39.6. Why do you think the metabolic requirements of LS protists are so unique? Having identified these proteins, what would be the next step in identifying drugs that might be effective against LS protozoa? How might this information assist in vaccine development?

   Read the original paper: Tarun, A. S., et al. 2008. A combined transcriptome and proteome survey of malaria parasite liver stages. *Proc. Nat. Acad. Sci., USA* 105:305.

6. Infection with the fungal pathogen *Cryptococcus* can be devastating to immunocompromised individuals, yet not much is known about its natural habitat. To explore its association with plants, model systems were developed between *C. gatti* and the mustard plant *Arabidopsis* and *Eucalyptus*. It was discovered that *C. gatti* grows well on these plants and is a plant pathogen only when plants are infected with strains of opposite mating types. In addition, two compounds produced by the plants, myo-inositol and indole acetic acid, were found to promote fungal mating.

   Why is it important to understand the fungal niche in nature? List two questions these results raise concerning the transmission of *Cryptococcus* to human hosts. How would you go about addressing these questions?

   Read the original paper: Xue, C., et al. 2007. The human fungal pathogen *Cryptococcus* can complete its sexual life cycle during a pathogenic association with plants. *Cell Host Microbe.* 1:263.

Concept Mapping Challenge

Construct a concept map using the following words and your own linking terms.

Fungus	Protozoan		Parasite	Water	Skin
Eukaryote	Opportunist		Spores	Intestine	
Mycosis	Cyst	Tinea	Flagella	Malaria	Amoeba

Learn More

Learn more by visiting the text website at www.mhhe.com/willey8, where you will find a complete list of references.

40

Microbiology of Food

Large tanks used for wine production. Fermentations can be carried out in such open-air units in temperate regions. After completion of the fermentation, the fresh wine will be transferred to barrels for storage and aging.

CHAPTER GLOSSARY

aflatoxin A polyketide secondary fungal metabolite that can cause cancer.

bacteriocin A peptide produced by a bacterium that kills other closely related bacteria. Nisin is a bacteriocin used to prevent spoilage of certain foods.

enology The science of wine making.

food-borne infection Gastrointestinal illness caused by ingestion of microorganisms, followed by their growth in the host. Symptoms arise from tissue invasion and toxin production.

food intoxication Food poisoning caused by microbial toxins produced in a food prior to consumption. The presence of living bacteria is not required.

fumonisin A family of toxins produced by mold belonging to the genus *Fusarium;* primarily affects corn and is known to be hepato- and nephrotoxic in animals.

GRAS The designation "generally *re*garded *as* safe" is given to food additives and some preservatives.

lactic acid bacteria (LAB) Strictly fermentative, gram-positive bacteria that produce lactic acid as the primary fermentation end product.

pasteurization The process of heating milk and other liquids to destroy microorganisms that can cause spoilage or disease.

probiotic Living microorganisms that confer a health benefit to the host when consumed or administered in sufficient quantities.

putrefaction The microbial decomposition of organic matter, especially the anaerobic breakdown of proteins, with the production of foul-smelling compounds such as hydrogen sulfide and amines.

starter culture An inoculum consisting of a mixture of carefully selected microorganisms used to start a commercial fermentation.

Foods, microorganisms, and humans have had an interesting association that developed long before recorded history. On the one hand, microorganisms can be used to transform raw foods into gastronomic delights including chocolate, cheeses, pickles, sausages, soy sauce, wines, and beers. On the other hand, microorganisms can degrade food quality and lead to spoilage. Sometimes a fine line exists between the microbial enhancement and degradation of foods. The detection and control of pathogens and food spoilage microorganisms are important parts of food microbiology. In this chapter, we consider the two opposing roles of microorganisms in food production and preservation.

40.1 Microbial Growth and Food Spoilage

Foods, because they are nutrient-rich, are excellent environments for the growth of microorganisms. Microbial growth is controlled by factors related to the food itself, called intrinsic factors, by extrinsic factors, which include the environment where the food is stored and any preservatives that may have been added (**figure 40.1**). Growth of microbes in foods can alter the food in numerous ways. In some cases, these changes increase the shelf life of the food by making a new food product (e.g., the fermentation of milk to make yogurt). Such changes are described in more detail in section 40.5. In this section, we focus on those changes that spoil food.

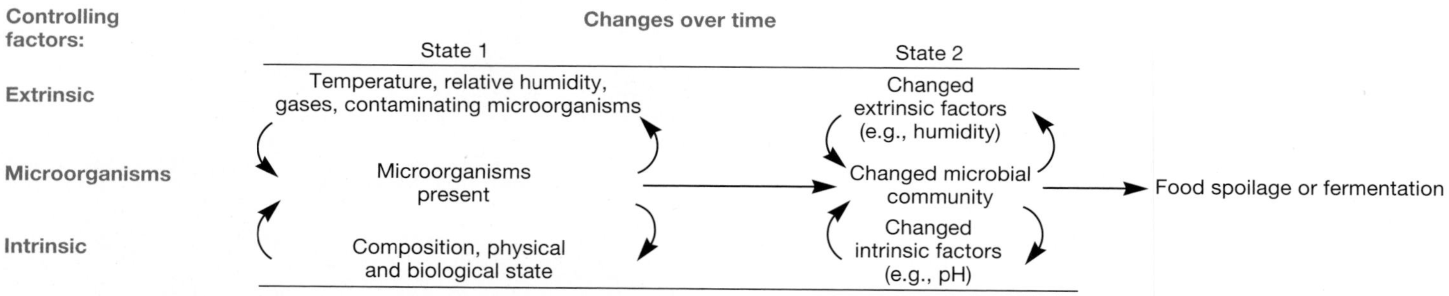

FIGURE 40.1 Intrinsic and Extrinsic Factors. A variety of intrinsic and extrinsic factors can influence microbial growth in foods. Time-related successional changes occur in the microbial community and the food.

Table 40.1 Differences in Spoilage Processes in Relation to Food Characteristics

Substrate	*Food Example*	*Chemical Reactions or Processes*[a]	*Typical Products (and Effects)*
Pectin	Fruits	Pectinolysis	Methanol, uronic acids (loss of fruit structure, soft rots)
Proteins	Meat	Proteolysis, deamination	Amino acids, peptides, amines, H_2S, ammonia, indole (bitterness, souring, bad odor, sliminess)
Carbohydrates	Starchy foods	Hydrolysis, fermentations	Organic acids, CO_2, mixed alcohols (souring, acidification)
Lipids	Butter	Hydrolysis, fatty acid degradation	Glycerol and mixed fatty acids (rancidity, bitterness)

[a]Other reactions also occur during the spoilage of these substrates.

Intrinsic Factors

Food composition is a critical intrinsic factor that influences microbial growth (**table 40.1**). If a food consists primarily of carbohydrates, fungal growth predominates and spoilage does not result in major odors. Thus foods such as breads, jams, and some fruits and vegetables first show spoilage by fungal growth. Molds have enzymes that weaken and penetrate the protective outer skin of fruit and vegetables. Once molds degrade carbohydrates within the skin, bacteria can colonize the food. The colonizing bacteria include those that cause soft rots, such as *Erwinia carotovora,* which produce hydrolytic enzymes such as pectinase. The oxic conditions enables aerobes and facultative anaerobes to contribute to decomposition.

Molds can rapidly grow on grains and corn when these products are stored in moist conditions. The moldy bread pictured in **figure 40.2*a*** shows extensive fungal hyphal development and sporulation. The green growth most likely is *Penicillium;* the black growth is characteristic of *Rhizopus stolonifer* (*see figure 24.4*). Contamination of grains by the ascomycete *Claviceps purpurea* causes **ergotism,** a toxic condition (*see figure 24.10*). Hallucinogenic alkaloids produced by this fungus can lead to altered behavior, abortion, and death if infected grains are eaten. Molds are also a problem for tomatoes. Even the slightest bruising of the tomato skin, exposing the interior, will result in rapid fungal growth. This affects the quality of tomato products, including tomato juices and ketchup.

In contrast, when foods contain large amounts of proteins or fats (e.g., meat and dairy products), bacterial growth can produce a variety of foul odors. The production of short-chained fatty acids from fats renders butter rancid. The anaerobic breakdown of proteins is called **putrefaction.** It yields foul-smelling amine compounds such as cadaverine (imagine the origin of that name) and putrescine (**figure 40.3**). Unpasteurized milk undergoes a predictable, four-step microbial succession during spoilage: acid production by *Lactococcus lactis* subspecies *lactis* is followed by additional acid production associated with the growth of more acid-tolerant organisms such as *Lactobacillus.* At this point, yeasts and molds become dominant and degrade the accumulated lactic acid, and the acidity gradually decreases. Eventually protein-digesting bacteria become active, resulting in a putrid odor and bitter flavor. The milk, originally opaque, eventually becomes clear because the proteins and fats have become coagulated (**figure 40.4**).

The pH and oxidation-reduction (redox) potential of a food also are critical. A low pH favors the growth of yeasts and molds. In neutral or alkaline pH foods, such as meats, bacteria are more dominant in spoilage and putrefaction. Furthermore, when meat products, especially broths, are cooked, they often have lower oxidation-reduction potentials—that is, they present a reducing environment for microbial growth. These products, with their readily available amino acids, peptides, and growth factors, are ideal media for the growth of anaerobes, including *Clostridium.*

(a)

(b)

FIGURE 40.2 Food Spoilage. When foods are not stored properly, microorganisms can cause spoilage. Typical examples are fungal spoilage of (a) bread and (b) corn. Such spoilage of corn is called ear rot and can result in major economic losses.

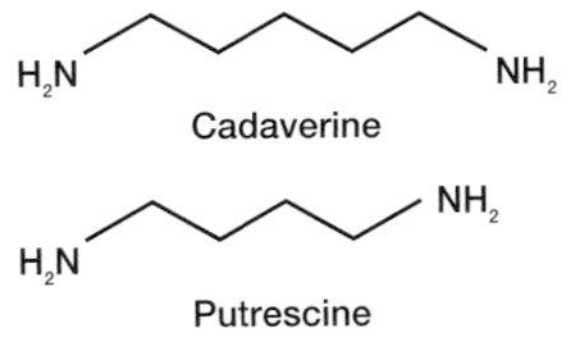

FIGURE 40.3 Cadaverine and Putrescine. These foul-smelling amines are the products of protein degradation.

The physical structure of a food also can affect the course and extent of spoilage. Grinding and mixing of foods such as sausage and hamburger increase the food surface area and distribute contaminating microorganisms throughout the food. This can result in rapid spoilage if such foods are stored improperly. In addition, some spoilage microorganisms have specialized enzymes that help them penetrate protective peels and rinds, especially after the fruits and vegetables have been bruised.

FIGURE 40.4 Spoilage of a Dairy Product. Fresh (left) and curdled (right) milk are shown. The spoilage process has produced acidic conditions that have denatured and precipitated the milk casein to yield separated curds and whey.

Many foods contain natural antimicrobial substances, including complex chemical inhibitors and enzymes. Coumarins found in fruits and vegetables have antimicrobial activity. Cow's milk and eggs also contain antimicrobial substances. Eggs are rich in the enzyme lysozyme that can lyse the cell walls of contaminating gram-positive bacteria. Herbs and spices often possess significant antimicrobial substances; generally fungi are more sensitive to these compounds than most bacteria. Sage and rosemary are two of the most antimicrobial spices. Aldehydic and phenolic compounds that inhibit microbial growth are found in cinnamon, mustard, and oregano. Other important inhibitors are garlic, which contains allicin; cloves, which have eugenol; and basil, which contains rosmarinic acid. Nonetheless, spices can sometimes contain pathogenic and spoilage organisms. Enteric bacteria, *Bacillus cereus, Clostridium perfringens,* and *Salmonella* spp. have been detected in spices. Microorganisms can be eliminated or reduced by ethylene oxide sterilization. This treatment can result in *Salmonella*-free spices and herbs and a 90% reduction in the levels of general spoilage organisms. ◀◀ *Chemical control agents (section 8.5)*

Extrinsic Factors

Extrinsic factors are more easily controlled than intrinsic factors and are therefore frequently the basis for food preservation. Temperature is an important extrinsic factor in determining whether a food will spoil. For instance, spoilage problems can occur with minimally processed, concentrated frozen citrus products. These are prepared with little or no heat treatment, and the gram-positive bacteria *Lactobacillus* and *Leuconostoc* spp., which convert citrate to diacetyl, may be present along with the yeasts *Saccharomyces* and *Candida*. When frozen, concentrated juices are thawed and diluted with water, these microorganisms can begin the spoilage process. Ready-to-serve juices are routinely pasteurized, even though this results in some flavor loss (section 40.2).

Humidity is also a key factor in controlling food spoilage. At higher relative humidities, microbial growth is initiated more rapidly, even at lower temperatures (especially when refrigerators are not maintained in a defrosted state). When drier foods are placed in moist environments, moisture absorption can occur on the food surface, promoting microbial growth.

The atmosphere in which food is stored also is important. This is especially true with shrink-wrapped foods because many plastic films allow oxygen diffusion, which results in increased growth of surface-associated microorganisms. Excess CO_2 can decrease the solution pH, inhibiting microbial growth. Storing meat in a high CO_2 atmosphere inhibits gram-negative bacteria, resulting in a population dominated by the lactobacilli.

1. What are some intrinsic factors that influence food spoilage and how do they exert their effects?
2. What are the effects of food composition on spoilage processes?
3. What fungal genus produces ergot alkaloids? What conditions are required for the synthesis of these substances?
4. Why might sausage and other ground meat products provide a better environment for the growth of food spoilage organisms than solid cuts of meats?
5. What extrinsic factors can determine whether food spoilage will occur?

40.2 Controlling Food Spoilage

The development of agricultural societies brought a decreasing dependence on hunting and gathering, and the need to preserve surplus foods became essential to human survival. The use of salt as a meat preservative, the production of wines, and the preservation of fish and meat by smoking were introduced in Near Eastern civilization as early as 3000 BCE. But it was not until the nineteenth century that microbial spoilage of food was studied systematically. Louis Pasteur established the modern era of food microbiology in 1857, when he showed that microorganisms cause milk spoilage. Pasteur's work in the 1860s proved that heat could be used to control spoilage organisms in wines and beers. ◀◀ *Microbiology and its origins (section 1.3)*

Foods can be preserved by a variety of methods. The goal of each method is to eliminate or reduce the populations of spoilage and disease-causing microorganisms while maintaining food quality. We now briefly discuss some of these techniques. *Food Spoilage*

Removal of Microorganisms

Microorganisms can be removed from water, wine, beer, juices, soft drinks, and other liquids by filtration. This can keep bacterial populations low or eliminate them entirely. Removal of large particulates by prefiltration and centrifugation maximizes filter life and effectiveness. Several major brands of beer are filtered rather than pasteurized to better preserve flavor and aroma.

Low Temperature

Refrigeration at 5°C retards microbial growth, although with extended storage, microorganisms eventually grow and produce spoilage. Slow microbial growth at temperatures below 10°C has been described, particularly with fruit juice concentrates, ice cream, and some fruits. Of particular concern is the growth of *Listeria,* which can grow at temperatures used for refrigeration. It should be kept in mind that refrigeration slows the metabolic activity of most microbes, but it does not lead to significant decreases in overall microbial populations.

High Temperature

Controlling microbial populations in foods by means of high temperatures can significantly limit disease transmission and spoilage. Heating processes, first used by Nicholas Appert in 1809, provide a safe means of preserving foods, particularly when carried out in commercial canning operations (**figure 40.5**). Canned food is heated in special containers called retorts at about 115°C for intervals ranging from 25 to over 100 minutes. The precise time and temperature depend on the nature of the food. Sometimes canning does not kill all microorganisms but only those that will spoil the food (e.g., remaining bacteria are unable to grow due to acidity of the food). After heat treatment, the cans are cooled as rapidly as possible, usually with cold water. Quality control and processing effectiveness are sometimes compromised, however, in home processing of foods, especially with less acidic (i.e., those with pH values greater than 4.6) products such as green beans or meats. ◀◀ *Physical control methods (section 8.4)*

Despite efforts to eliminate spoilage microorganisms during canning, canned foods may become spoiled. This may be due to spoilage before canning, underprocessing during canning, and leakage of contaminated water through seams during cooling. Spoiled food can be altered in such characteristics as color, texture, odor, and taste. Organic acids, sulfides, and gases (particularly CO_2 and H_2S) may be produced. If spoilage microorganisms produce gas, both ends of the can will bulge outward. Sometimes the swollen ends can be moved by thumb pressure (soft swells); in other cases, the gas pressure is so great that the ends cannot be dented by hand (hard swells). However, swelling is not always due to microbial spoilage; acid in low pH foods may react with iron in the can to release hydrogen and generate a hydrogen swell. ◀◀ *Botulism (section 38.4)*

FIGURE 40.5 Food Preparation for Canning. Microbial control is important in processing and preserving many foods. Here, a worker is pouring peas into a large, clean vat during the preparation of vegetable soup. After preparation, the soup is transferred to cans. Each can is heated for a short period, sealed, processed at temperatures from 110 to 121°C in a canning retort to destroy spoilage microorganisms, and finally cooled.

Figure 40.5* *Micro Inquiry

Canned products such as vegetable soup often contain salt. Other than serving as a flavoring, what is the purpose of salt in the canned product?

Pasteurization involves heating food to a temperature that kills disease-causing microorganisms and substantially reduces the levels of spoilage organisms. When processing milk, beers, and fruit juices by conventional low-temperature holding (LTH) pasteurization, the liquid is maintained at 62.8°C for 30 minutes. Products can also be held at 72°C for 15 seconds, a high-temperature, short-time (HTST) process; milk can be treated at 138°C for 2 seconds for ultra-high-temperature (UHT) processing. Shorter-term processing results in improved flavor and extended product shelf life. Both HTST and UHT require stainless steel heat exchangers specifically designed to rapidly heat and cool the liquid being pasteurized. The duration of pasteurization is based on the statistical probability that the number of remaining viable microorganisms will be below a certain level after a particular heating time at a specific temperature. These calculations are discussed in detail in section 8.4. *Food Pathogens and Temperature*

Water Availability

The presence and availability of water also affect the ability of microorganisms to colonize foods. Simply by drying a food, one can slow or prevent spoilage processes. Water can be made less available by adding solutes such as sugar and salt. Water availability is measured in terms of **water activity** (a_w). This represents the ratio of relative humidity of the air over a test solution compared with that of distilled water, which has an a_w of 1. When large quantities of salt or sugar are added to food, most microorganisms are dehydrated by the hypertonic conditions and cannot grow (**table 40.2**). Even under these adverse conditions, osmophilic and xerophilic microorganisms may spoil food. Osmophilic (Greek *osmus*, impulse, and *philein*, to love) microbes grow best in or on media with a high osmotic concentration (e.g., jams and jellies), whereas xerophilic (Greek *xerosis*, dry, and *philein*, to love) microorganisms prefer a low a_w environment (e.g., dried fruits, cereals) and may not grow under high a_w conditions. Dehydration, such as lyophilization to produce freeze-dried foods, is a common means of preventing microbial growth. Although freeze-drying increases the osmolarity of the product while at the same time reducing the a_w, microbial growth is retarded because most xerophiles are not osmophilic and vice versa.

Table 40.2 Approximate Minimum Water Activity Tolerated by Microbes

Organisms	a_w
Groups	
Most spoilage bacteria	0.9
Most spoilage yeasts	0.88
Most spoilage molds	0.80
Halophilic bacteria	0.75
Xerophilic molds	0.61
Osmophilic yeasts	0.61

Adapted from James M. Jay. 2000. *Modern Food Microbiology*, 6th edition. Reprinted by permission of Aspen Publishers, Inc. Gaithersburg, MD. Tables 3–5, p. 42.

Chemical-Based Preservation

Various chemical agents can be used to preserve foods; these substances are closely regulated by the U.S. Food and Drug Administration (FDA) and are listed as being "generally recognized as safe" or **GRAS** (**table 40.3**). They include simple organic acids, sulfite, ethylene oxide as a gas sterilant, sodium nitrite, and ethyl formate. These chemical agents may damage the microbial plasma membrane or denature various cell proteins. Other compounds interfere with the functioning of nucleic acids, thus inhibiting cell reproduction.

Search This: FDA magazine/GRAS

Sodium nitrite is an important chemical used to help preserve ham, sausage, bacon, and other cured meats by inhibiting the growth of *Clostridium botulinum* and the germination of its

Table 40.3 Major Groups of Chemicals Used in Food Preservation

Preservatives	*Approximate Maximum Use*	*Organisms Affected*	*Foods*
Propionic acid/propionates	0.32%	Molds	Bread, cakes, some cheeses; inhibitor of ropy bread dough
Sorbic acid/sorbates	0.2%	Molds	Hard cheeses, figs, syrups, salad dressings, jellies, cakes
Benzoic acid/benzoates	0.1%	Yeasts and molds	Margarine, pickle relishes, apple cider, soft drinks, tomato ketchup, salad dressings
Parabens[a]	0.1%	Yeasts and molds	Bakery products, soft drinks, pickles, salad dressings
SO_2/sulfites	200–300 ppm	Insects and microorganisms	Molasses, dried fruits, wine, lemon juice (not used in meats or other foods recognized as sources of thiamine)
Ethylene/propylene oxides	700 ppm	Yeasts, molds, vermin	Fumigant for spices, nuts
Sodium diacetate	0.32%	Molds	Bread
Dehydroacetic acid	65 ppm	Insects	Pesticide on strawberries, squash
Sodium nitrite	120 ppm	Clostridia	Meat-curing preparations
Caprylic acid	—	Molds	Cheese wraps
Ethyl formate	15–200 ppm	Yeasts and molds	Dried fruits, nuts

From James M. Jay. 2000. Modern Food Microbiology, *6th edition. Reprinted by permission of Aspen Publishing, Frederick, MD.*

[a] *Methyl, propyl, and heptyl esters of p-hydroxybenzoic acid.*

spores. This protects against botulism and reduces the rate of spoilage. Besides increasing meat safety, nitrite decomposes to nitric acid, which reacts with heme pigments to keep the meat red in color. Concern about nitrite arises from the observation that it can react with amines to form carcinogenic nitrosamines.

Low pH can also be used to hinder microbial spoilage. For example, acetic and lactic acids inhibit growth of *Listeria* spp. Organic acids (1–3%) can be used to treat meat carcasses, and poultry can be cleansed with 10% lactic acid/sodium lactate buffer (pH 3) prior to packaging. In addition, low pH can increase the activity of other chemical preservatives. Sodium propionate is most effective at lower pH values, where it is primarily undissociated. Breads, with their low pH values, often contain sodium propionate as a preservative. The use of citric acid is also common. ▶▶| *Organic acids (section 41.3)*

Radiation

The major method used for radiation sterilization of food is gamma irradiation from a cobalt-60 source; however, cesium-137 is used in some facilities. Gamma radiation has excellent penetrating power but must be used with moist foods because radiation is effective only if it can generate peroxides from water in the microbial cells, resulting in oxidation of sensitive cellular constituents. This process of **radappertization,** named after Nicholas Appert, can extend the shelf life of seafoods, fruits, and vegetables. Unlike canning, radappertization does not heat the food product. To sterilize meat products, commonly 4.5 to 5.6 megarads are used.

Electron beams can also be used to irradiate foods. The electrons are generated electrically, so unlike gamma radiation, they can be turned on only when needed. On the other hand, electron beams do not penetrate food items as deeply as does gamma radiation. It is important to note that regardless of the radiation source (gamma rays or electron beams), the food itself does not become radioactive.

Microbial Product–Based Inhibition

Interest is increasing in the use of bacteriocins for the preservation of foods. **Bacteriocins** are bactericidal proteins that kill closely related microbes by binding to specific sites on the target cell. They often affect cell membrane integrity and function. The only bacteriocin currently approved for food preservation is nisin, a small amphiphilic peptide produced by some strains of *Lactococcus lactis.* It is nontoxic to humans and affects other gram-positive bacteria by binding to lipid II during peptidoglycan synthesis and forming pores in the plasma membrane (*see figure 11.13*). Nisin was added to the GRAS list of food additives by the FDA about 40 years ago. This means that it can be added directly to foods such as packaged meats, cheese, eggs, and canned vegetables. Nisin is effective in low-acid foods, where it improves inactivation of *Clostridium botulinum* during the canning process and inhibits germination of any surviving spores. |◀◀ *Bacteriocins (section 32.3)*

In 2006 the FDA approved the use of a preparation of six strains of **bacteriophages** that specifically infect and kill *Listeria* *monocytogenes*. The phages are present in a spray designed for use on ready-to-eat meats such as hot dogs and lunch meats. The spray is applied to the surface of the meats prior to packaging. The phages are present in equal concentration; using multiple phage types significantly reduces the development of phage-resistant strains.

Packaging

Gases present in the atmosphere within a food storage package can have a significant affect on microbial growth. This has led to the development of **modified atmosphere packaging (MAP).** Modern shrink-wrap materials and vacuum technology make it possible to package foods with controlled atmospheres. These materials are largely impermeable to oxygen. This prolongs shelf life by a factor of two to five times compared to the same product packaged in air. With a CO_2 content of 60% or greater in the atmosphere surrounding a food, spoilage fungi will not grow, even if low levels of oxygen are present. Recently it has been found that high-oxygen MAP also may be effective. This is due to the formation of superoxide (O_2^-) anions inside cells under these conditions. The superoxide anions are then transformed to highly toxic peroxide and hydroxyl radicals, resulting in antimicrobial effects. Some products currently packaged using MAP technology include delicatessen meats and cheeses, pizza, grated cheese, some bakery items, and dried products such as coffee.

Scientists at the U.S. Department of Agriculture (USDA) have recently tested a packaging system embedded with nisin. The bacteriocin is uniformly distributed in the packaging from which it is slowly released during the shelf-life of the product. The packaging is biodegradable, consisting of polylactic acid (PLA) and pectin. PLA is already used to package water, juice, and yogurt. It is derived from wood and corn, so it is a "green" alternative to the petrochemical-based plastics more commonly used. Interestingly, the addition of pectin to the packaging material increases the effectiveness of nisin. Thus this innovative packaging addresses two pressing societal concerns: food safety and decreasing dependence on petroleum.

1. Describe the major approaches used in food preservation.
2. What types of chemicals can be used to preserve foods?
3. Under what conditions can gamma radiation be used to control microbial populations in foods and food preparation? What is radappertization?
4. How does nisin function? What bacterial genus produces this important peptide?
5. Consider the fact that *Listeria* phages must be propagated in the pathogen prior to producing an anti-*Listeria* spray. What safeguards do you think must be instituted to ensure the safety of the spray?
6. What are the major gases involved in MAP? How are their concentrations varied to inhibit microbial growth?

40.3 Types of Food-Borne Disease

Food-borne illnesses impact the entire world. According to the Centers for Disease Control and Prevention (CDC), there are about 76 million cases of food-related diseases annually in the United States. Of these, only 14 million (about 18%) can be attributed to known pathogens. Food-borne diseases result in 325,000 hospitalizations and at least 5,000 deaths per year. Since 1942 the number of recognized food-borne pathogens has increased over fivefold. In most cases, these "new" pathogens are simply agents that now can be described, based on an improved understanding of microbial diversity. Recent estimates indicate that noroviruses, *Campylobacter jejuni,* and *Salmonella* are the major causes of food-borne diseases. In addition, *Escherichia coli* O157:H7 and *Listeria* are important food-related pathogens. ◀◀ *Food-borne and waterborne diseases (section 38.4)*

There are two primary types of food-related diseases: food-borne infections and food intoxications. All of these food-borne diseases are associated with poor hygienic practices. Whether by water or food transmission, the fecal-oral route is key, with the food providing the vital link between hosts. Fomites, such as faucets, drinking cups, and cutting boards, also play a role in the maintenance of the fecal-oral route of contamination.

Search This: CDC/food-borne illness

Food-Borne Infection

A **food-borne infection** involves the ingestion of the pathogen, followed by growth in the host, including tissue invasion or the release of toxins. The most common microbes involved are summarized in **table 40.4** (*also see tables 37.5 and 38.6*). These pathogens and others are discussed in detail in chapters 37 to 39. Food-borne infections can cause serious outbreaks with wide-ranging consequences. Here, we describe one particular outbreak of food-borne infection that illustrates this.

Listeriosis, caused by *Listeria monocytogenes,* was responsible for the largest meat recall in U.S. history. In 2002 a seven-state listeriosis outbreak was linked to deli meats and hot dogs produced at a single meat-processing plant in Pennsylvania. Pregnant women, the young and the old, and immunocompromised individuals are especially vulnerable to *L. monocytogenes* infections. In this outbreak, 7 deaths, 3 stillbirths, and 46 illnesses were caused by consumption of contaminated meats. Microbiologists matched the strain of *L. monocytogenes* found in the contaminated food products with samples obtained from floor drains in a Wampler, Inc., packaging plant. This prompted the recall of 27.4 million pounds of meats that had been distributed over a 5-month period to stores, restaurants, and school lunch programs. Following the outbreak, the plant closed for a month and the Wampler brand name was phased out. This episode prompted the USDA to step up its environmental testing program for *L. monocytogenes* so that it now tests plants that do not regularly submit data to the

Table 40.4 Some Food-Borne Bacteria That Cause Acute Bacterial Vomiting and Diarrhea

Organism	*Incubation Period (Hours)*	*Vomiting*	*Diarrhea*	*Fever*	*Food Source*
Staphylococcus aureus	1–8 (rarely, up to 18)	+++[a]	+	–[b]	Meats, dairy, and bakery products, sprouts, carrots, lettuce, parsley, and radishes
Bacillus cereus	2–16	+++	++	–	Reheated fried rice, sprouts, cucumber
Clostridium perfringens	8–16	±[c]	+++	–	Rewarmed meat dishes
Clostridium botulinum	18–24	±	Rare	–	Canned goods contaminated during processing or packaging
Escherichia coli (enterohemorrhagic)	3–5 days	±	++	±	Undercooked ground beef, unpasteurized fruit juices and cider, and raw vegetables such as sprouts, lettuce, and celery
Escherichia coli (enterotoxigenic strain)	24–72	±	++	–	Contaminated drinking water; major cause of traveler's diarrhea.
Vibrio parahaemolyticus	6–96	+	++	±	Shellfish, particularly clams and oysters
Vibrio cholerae	24–72	+	+++	–	Contaminated drinking water, as well as cabbage, coconut milk, and lettuce
Shigella spp.	24–72	±	++	+	Celery, melon, lettuce and other greens, parsley, sprouts
Salmonella spp. (gastroenteritis)	8–48	±	++	+	Many fruits and vegetables (including celery, green onions, lettuce and other greens, strawberries, tomatoes, melon), eggs and egg products, poultry
Salmonella enterica serovar Typhi (typhoid fever)	10–14 days	±	±	++	Usually spread from a heathy carrier to food via fecal-oral transmission.
Campylobacter jejuni	2–10 days	–	+++	++	Poultry, shellfish, green onions, lettuce, mushrooms, potatoes, peppers, spinach
Yersinia enterocolitica	4–7 days	±	++	+	Fecal-oral transmission from carrier to non-carrier

Adapted from Geo. F. Brooks, et al., *Medical Microbiology,* 21st edition. Copyright 1998 Appleton & Lange, Norwalk, CT. Reprinted by permission.
[a] + indicates condition is present, number of symbols indicates severity
[b] – indicates condition is absent
[c] ± indicates condition sometimes occurs

USDA. It also performs surprise inspections of those that do. The USDA advises people at risk of contracting listeriosis to avoid eating soft cheeses (e.g., feta, Brie, Camembert), refrigerated smoked meats such as lox, as well as deli meats and undercooked hot dogs. In 2006 the Wampler plant in Pennsylvania was closed, having never recovered from the $100 million cost and the damage to its reputation caused by the 2002 outbreak. Thus a single outbreak resulted in death, illness, bankruptcy, and changes to the nature of food safety inspections.

Food Intoxications

Microbial growth in food products also can result in **food intoxication** (table 40.4). Intoxication produces symptoms shortly after the food is consumed because growth of the disease-causing microorganism is not required. Toxins produced in the food can be associated with microbial cells or can be released from the cells. For instance, most *Staphylococcus aureus* strains cause a staphylococcal enteritis related to the synthesis of extracellular toxins. These are heat-resistant proteins, so heating does not usually render the food safe. The effects of the toxins are quickly felt, with disease symptoms occurring within 2 to 6 hours. The main reservoir of *S. aureus* is the human nasal cavity. Frequently *S. aureus* is transmitted to a person's hands and then is introduced into food during preparation. Growth and enterotoxin production usually occur when contaminated foods are held at room temperature for several hours.

Three gram-positive rods are known to cause food intoxications: *Clostridium botulinum, C. perfringens,* and *Bacillus cereus. C. botulinum* poisoning is discussed in chapter 38. However, here we note that baked potatoes served in aluminum foil can, even after washing, be contaminated with *C. botulinum*, which naturally occurs in the soil. If the foil-covered potatoes are not heated sufficiently in the baking process, surviving clostridia can proliferate after removal of the potatoes from the oven and rapidly produce toxins.

The enterohemorrhagic microbe *E. coli* O157:H7 produces a toxin known as Shiga-like toxin, which was acquired by horizontal gene transfer from another enteric, *Shigella*. Shiga-like toxin can lead to a potentially life-threatening complication known as hemolytic uremic syndrome, most commonly seen in children and the elderly. The first U.S. outbreaks of *E. coli* O157:H7 (in the 1980s) occurred in individuals who had consumed undercooked hamburgers. This microbe inhabits the intestines of a variety of mammals, and ground beef contaminated with cattle feces during slaughter and processing continues to be a common cause of these outbreaks.

More recently other types of foods have been found to be contaminated with *E. coli* O157:H7. One example is minimally processed, ready-to-eat fresh salad greens, which have become a popular convenience and represent a multibillion-dollar business in the United States and Europe. Prior to packaging, the produce is repeatedly washed, then dried and packed into plastic bags. When transported, the produce must be kept cool. Unfortunately, this healthy approach to consuming vegetables is not always without risk. Between 2002 and 2006, there were 10 outbreaks of enterohemorrhagic *E. coli* O157:H7 traced to minimally processed produce. The largest and most well publicized was the spinach-associated outbreak in the summer and fall of 2006, which sickened more than 200 people. This outbreak prompted intense investigation, and it was determined that the contamination occurred while the spinach was still in the field, or preharvest. Surveillance studies determined that the most likely source of the 2006 outbreak was feral swine that burrowed under fences and traveled across the spinach fields. This finding raises more questions than it answers, including: how can thousands of agricultural fields be protected from small, wild animals; why didn't the postharvest washing procedure cleanse the spinach of the *E. coli*; and what factors were in play that allowed the initial bacterial inoculum on the plants to grow by at least a factor of 10? While there are currently no answers to these questions, both government- and industry-funded studies are under way to ensure that minimally processed produce remains a convenience, not a sporadic source of disease. |◀◀ Escherichia coli *gastroenteritis (section 38.4)*

Some fungus-derived toxins have a more insidious health effect. Fungal carcinogens include the aflatoxins and fumonisins. **Aflatoxins** are produced most commonly in moist grains and nut products. Aflatoxins were discovered in 1960, when 100,000 turkey poults (young turkeys) died after eating fungus-infested peanut meal. *Aspergillus flavus* was found in the contaminated peanut meal, together with alcohol-extractable toxins termed aflatoxins. These planar, ringed compounds intercalate with nucleic acids and act as potent frameshift mutagens and carcinogens. This occurs primarily in the liver, where they are converted to unstable derivatives. Currently, a total of 18 aflatoxins are known. The most important are shown in **figure 40.6**. Of these, aflatoxin B1 is the most common and the most potent carcinogen. Aflatoxins B1 and B2, after ingestion by lactating animals (e.g., dairy cows), are modified in the animal body to yield the aflatoxins M1 and M2. If cattle consume aflatoxin-contaminated feeds, aflatoxins also can thus appear in milk and dairy products. Besides their importance in grains, aflatoxins have also been found in beer, cocoa, raisins, peanut butter, and soybean meal.

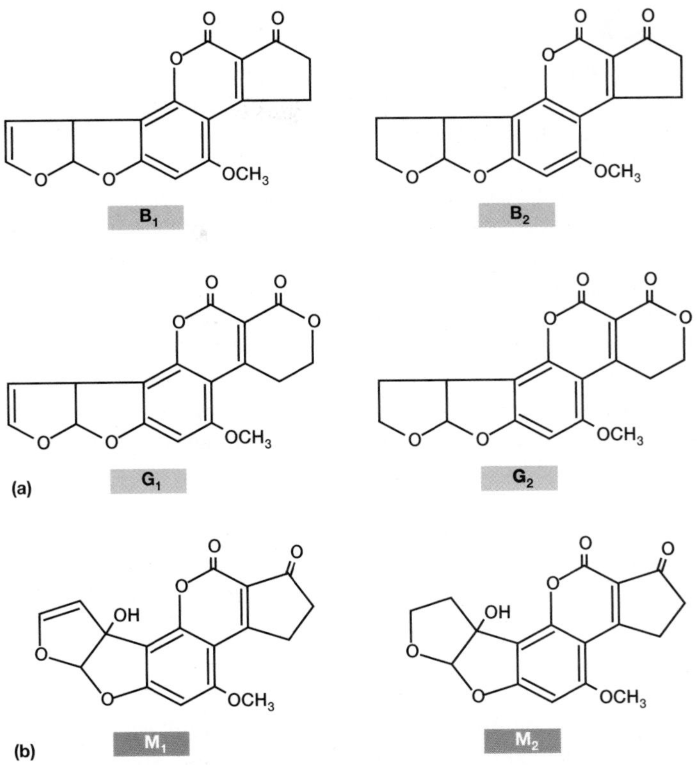

FIGURE 40.6 Aflatoxins. When *Aspergillus flavus* and related fungi grow on foods, carcinogenic aflatoxins can be formed. These have four basic structures. (a) The letter designations refer to the color of the compounds under ultraviolet light after extraction from the grain and separation by chromatography. The B_1 and B_2 compounds fluoresce with a blue color, and the G_1 and G_2 appear green. (b) The two type M aflatoxins are found in the milk of lactating animals that have ingested type B aflatoxins.

Figure 40.6 Micro Inquiry

What foods are most prone to aflatoxin contamination?

Diet appears to be related to aflatoxin exposure: the average aflatoxin intake in the typical European-style diet is 19 ng/day, whereas for some Asian diets it is estimated to be 103 ng/day. Aflatoxin sensitivity also can be influenced by prior disease exposure. Individuals who have had hepatitis B have a 30-times higher risk of liver cancer upon exposure to aflatoxins than individuals who have not had this disease. This association illustrates an emerging link between inflammation and cancer. It has been observed that prevention of hepatitis B infections by vaccination and reduction of carrier populations help control the potential effects of aflatoxins in foodstuffs.

Search This: Aflatoxins/ehso

The **fumonisins** are fungal contaminants of corn that were first isolated in 1988. These are produced by *Fusarium moniliforme* and cause leukoencephalomalacia in horses (also called "blind staggers"—it is fatal within 2 to 3 days), pulmonary edema in pigs, and esophageal cancer in humans. The fumonisins inhibit ceramide synthase, a key enzyme for the proper use of fatty substances in the cell. This disrupts the synthesis and metabolism of sphingolipids, important compounds that influence a wide variety of cell functions. At least 10 different fumonisins exist; the basic structures of fumonisins FB1 and FB2 are shown in **figure 40.7**. Corn and corn-based feeds and foods, including cornmeal and corn grits, can be contaminated. Thus it is extremely important to store corn and corn products under dry conditions, where these fungi cannot develop.

Other eukaryotic microorganisms also synthesize potent toxins. For example, algal toxins contaminate fish and thereby affect the health of marine animals higher in the food chain; they also contaminate shellfish and finfish, which are later consumed by humans. These toxins are produced during harmful algal blooms, which are discussed in section 28.2.

1. What is the difference between a food-borne infection and a food intoxication? Which do you think is harder to prevent? Explain your answer.
2. What practical recommendations regarding food preparation could you make to someone concerned about *E. coli* O157:H7 contamination?
3. Aflatoxins are produced by which fungal genus? How do they damage animals that eat the contaminated food?
4. What microbial genus produces fumonisins and why are these compounds of concern? If improperly stored, what are the major foods and feeds in which these chemicals might be found?

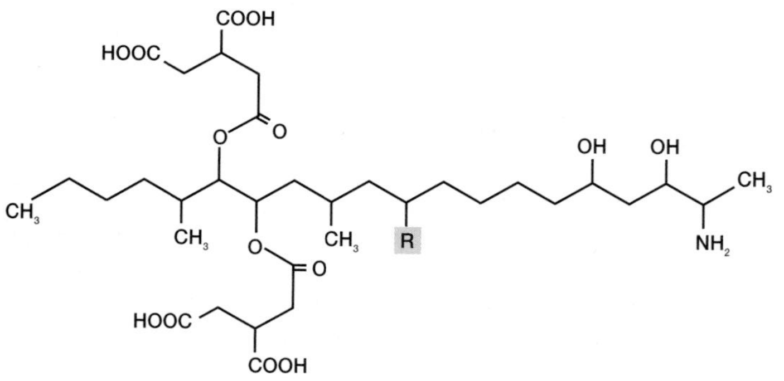

FIGURE 40.7 Fumonisin Structure. The basic structure of fumonisins FB1 and FB2 produced by *Fusarium moniliforme*, a fungal contaminant that can grow in improperly stored corn. A total of at least 10 different fumonisins have been isolated. These are strongly polar compounds that cause diseases in domestic animals and humans. FB1, R = OH; FB2, R = H.

40.4 Detection of Food-Borne Pathogens

As recent outbreaks of food-borne illnesses emphasize, the need to protect the public from microbial contamination in the food supply is paramount. Several guiding principles must be considered in developing the technologies, protocols, and policies used to keep foods safe. These include (1) specificity and sensitivity for the given pathogen, (2) speed, and (3) simplicity (e.g., foods should be tested without the need for a lot of sample preparation). Furthermore, the goal of each detection strategy should be clearly defined. Ideally "testing to prevent," that is, to confirm food is safe before it leaves the farm, is the goal. When this is unreasonable due to logistics and cost, then the next level is "testing to protect," which involves analysis before the food is accessible for consumption. These strategies are designed to avoid "testing to recover," when an outbreak has occurred and the origin of the of the contaminated food must be identified.

Unfortunately it is impossible to test every food item for every potential pathogen. Apart from the logistical implications, such an approach is economically untenable. For example, if the U.S. dairy industry were to test all dairy products daily at each farm with a \$5-per-sample assay, it would cost about \$150 million annually. Checking each milk tanker truck as it enters a processing plant reduces the cost to about \$21 million—a savings of \$129 million each year. So the challenge in developing a testing strategy involves identifying which pathogens to detect, as well as when and where they will be detected.

Despite the use of molecular techniques in other branches of microbiology, food-borne pathogens are still most commonly identified by standard culture techniques (e.g., most probable number; *see figure 27.3*), followed by immunological verification using fluorescent antibody or *e*nzyme-*l*inked *i*mmuno*s*orbant *a*ssays (ELISAs; *see figure 35.12*). Unfortunately, these methods require days and identification is often complicated by the low numbers of pathogens compared with the background microflora. Furthermore, the varied chemical and physical composition of foods can make isolation difficult. For instance, *Vibrio vulnificus*, which can contaminate shellfish, must be detected by plating on one of several selective media followed by biochemical or molecular confirmation. Not only is this approach slow, it also does not take into account potential variations in strain virulence. Thus the development of accurate, reproducible, and field-portable approaches is a paramount goal for food microbiologists. The introduction of molecular techniques has been stymied by the need to standardize all protocols. ◀◀ *Measurement of microbial growth (section 7.4); Culturing techniques (section 27.1); Clinical microbiology and immunology (chapter 35)*

Despite the difficulty of introducing molecular techniques, they are gradually replacing or are being used in conjunction with culture-based approaches. Molecular methods are valuable for a number of reasons. These include the ability to detect (1) the presence of a specific pathogen; (2) viruses that cannot be grown conveniently; (3) microbes that are present in very small numbers; and (4) slow-growing or nonculturable pathogens. In addition, results are often obtained much faster using molecular approaches than with culture-based methods. For instance, although the important pathogen *E. coli* O157:H7 can be isolated and identified using selective culture media, the application of multilocus sequence typing, serotype-specific probes, and polymerase chain reaction (PCR) enables the detection of a few target cells in large populations of background microorganisms. Indeed, by using PCR, as few as 10 toxin-producing *E. coli* cells can be detected in a population of 100,000 cells isolated from soft-cheese samples. Furthermore, it is possible to confirm bacterial contamination within 24 hours, whereas at least 3 to 4 days are needed for presumptive identification with culture procedures. Multiple pathogens can be detected in the same food sample through the recovery of PCR products of differing sizes, which can be separated electrophoretically. For instance, it is possible to detect *Campylobacter jejuni* and *Arcobacter butzleri* in the same sample within 8 hours. ◀◀ *Polymerase chain reaction (section 15.2); Techniques for determining microbial taxonomy and phylogeny (section 17.3)*

A major advance in the detection of food-borne pathogens is the use of standardized patterns of pathogen DNA following endonuclease digestion, or "food-borne pathogen fingerprinting." The CDC has established a program called **PulseNet** in which pulsed-field gel electrophoresis (PFGE) is used under carefully controlled and duplicated conditions to determine the distinctive DNA pattern of each bacterial pathogen (**figure 40.8**). PFGE is very similar to standard DNA gel electrophoresis but is capable of separating very large fragments of DNA. This requires periodic switching of the voltage supply in three different directions: the central axis of the gel, and two that run at a 120° angle on either side. The duration of each pulse is the same but can be increased over the duration of the run. Because PFGE can resolve large (Mb) pieces of DNA, whole genomes can be cut with restriction enzymes that cleave in only a few places to generate a strain-specific fingerprint. The patterns obtained for each strain of food-borne pathogen can be found in a national database maintained by the CDC. With this uniform procedure, it is possible to link pathogens associated with disease outbreaks in different parts of the world to a specific food source. For example, a *Shigella* outbreak in three different areas of North America was traced to Mexican parsley that had been tainted with polluted irrigation water. This program has resulted in more rapid establishment of epidemiological linkages and a decreased occurrence of many of these important food-borne diseases. ◀◀ *Restriction enzymes (section 15.1); Gel electrophoresis (section 15.3)*

Search This: PulseNet

Other molecular techniques are constantly being developed. Here, we describe just a few. (1) PCR can be combined with restriction fragment length polymorphism (RFLP) analysis so that

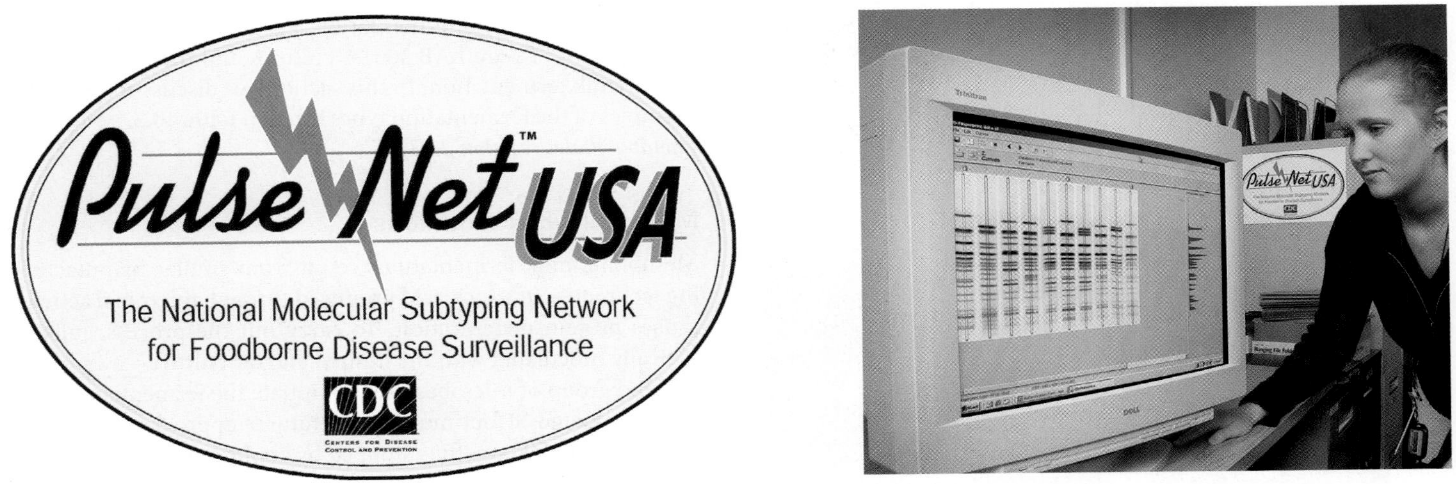

(a) (b)

FIGURE 40.8 **PulseNet.** This network operates throughout the United States to identify microbes causing food-borne disease using pulsed-field gel electrophoresis of genomes extracted from contaminating microbes. It is overseen by the Centers for Disease Control and Prevention, which coordinates local public health and food regulatory agency laboratories to facilitate early identification of common-source food-borne disease outbreaks.

Figure 40.8 Micro Inquiry

Why is it important that PulseNet coordinate the identities of food-borne pathogens on a national level?

rather than visualizing genomic fragments that have been cleaved, PCR amplifies the variable regions to generate a cleaner genomic fingerprint in a relatively short period of time. (2) Species-specific PCR primers have been developed for a number of pathogens. For instance, different species of *Campylobacter* can be detected by amplifying a region of gene that encodes gyrase B (**figure 40.9**). (3) Microarray technology has also been applied to food safety. Oligonucleotides representing housekeeping and virulence genes are used as probes to which DNA extracted directly from the food item can be hybridized. Note that unlike most microarray procedures, it is not necessary to make cDNA from mRNA because one is looking only for the presence or absence of a given pathogen's genome, not relative levels of gene expression. Microarrays are especially attractive because, unlike some other molecular techniques (e.g., PFGE), there is no need to first enrich for the pathogen by culture-based techniques. The pressing need for food safety combined with the rapid pace of innovation ensures the introduction of newer, faster, and highly accurate techniques will continue. |◀◀ *Microarray analysis (section 16.4); Techniques for determining microbial taxonomy and phylogeny (section 17.3)*

1. What advantages do molecular techniques have over culture-based approaches in the detection of food-borne pathogens?
2. How is PulseNet used in the surveillance of food-borne diseases?
3. Why is it advantageous to omit the pre-enrichment step in microarray and PCR-based approaches?
4. Compare how you might implement the testing of a meat product (e.g., hot dogs) with that of a vegetable (e.g., spinach). Consider the pathogen(s) to be detected, in addition to when and where testing should occur.

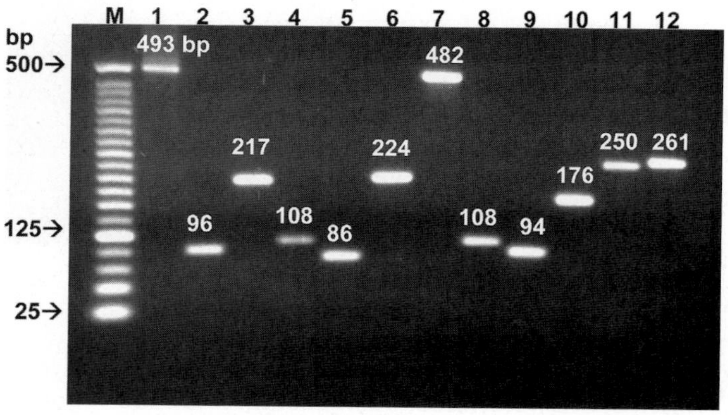

FIGURE 40.9 The Use of PCR to Identify Food-Borne Pathogens. This gel shows PCR products from different *Campylobacter* spp. that were generated using primers designed to amplify the *gyrB* (gyrase) gene. Each primer pair is specific for only one species. Fragment size in base pairs is listed above each band. M, molecular weight marker.

40.5 Microbiology of Fermented Foods

Fermentation has been used to process food for thousands of years; the earliest records date as far back as 6000 BCE. Microbial growth, either of natural or inoculated populations, causes chemical or textural changes to form a product that can be stored for extended periods. The fermentation process also is used to create new, pleasing food flavors and odors—such as chocolate (**Techniques & Applications 40.1**).

The major fermentations used in food microbiology are the lactic, propionic, and alcoholic fermentations. These fermentations are carried out with a wide range of microbes, many of which have not been characterized. |◀◀ *Fermentation (section 10.7)*

Fermented Milks

Throughout the world, at least 400 different fermented milks are produced, representing a total annual economic value of more than 55 billion U.S. dollars. The majority of fermented milk products rely on **lactic acid bacteria (LAB),** which include species belonging to the genera *Lactobacillus, Lactococcus, Leuconostoc,* and *Streptococcus* (**figure 40.10**). These are low G + C gram-positive bacteria that tolerate acidic conditions, are nonsporing, and are aerotolerant with a strictly fermentative metabolism. The art of fermentation developed long before the science; it was not until the mid-nineteenth century that Louis Pasteur discovered lactic acid fermentation. Pasteur's work enabled the development of pure LAB starter cultures and the industrialization of milk fermentation. In this section, we discuss some major examples of the fermentation types listed in **table 40.5.** |◀◀ *Order Lactobacillales (section 21.3)*

Mesophilic Fermentations

Mesophilic milk fermentations result from similar manufacturing techniques in which acid produced through microbial activity causes protein denaturation. To carry out the process, milk is typically inoculated with the desired **starter culture**—a carefully selected group of microbes used to initiate the fermentation. It is then incubated at optimum temperature (approximately 20 to 30°C). Microbial growth is stopped by cooling, and *Lactobacillus* spp. and *Lactococcus lactis* cultures are used for aroma and acid production. *Lactococcus lactis* subspecies *diacetilactis* converts milk citrate to diacetyl, which gives a richer, buttery flavor to the finished product. The use of these microorganisms with skim milk produces cultured buttermilk, and when cream is used, sour cream is the result.

Thermophilic Fermentations

Thermophilic fermentations are carried out at temperatures around 45°C. An important example is yogurt production. Yogurt is one of the most popular fermented milk products in the

TECHNIQUES & APPLICATIONS

40.1 Chocolate: The Sweet Side of Fermentation

Chocolate could be characterized as the "world's favorite food," but few people realize that fermentation is an essential part of chocolate production. The Aztecs were the first to develop chocolate fermentation, serving a chocolate drink made from the seeds of the chocolate tree, *Theobroma cocao* (Greek *theos,* god, and *broma,* food, or "food of the gods"). Chocolate trees now grow in West Africa as well as South America.

The process of chocolate fermentation has changed very little over the past 500 years. Each tree produces large pods that each hold 30 to 40 seeds in a sticky pulp (**box figure**). Ripe pods are harvested and slashed open to release the pulp and seeds. The sooner the fermentation begins, the better the product, so fermentation occurs on the farm where the trees are grown. The seeds and pulp are placed in "sweat boxes" or in heaps in the ground and covered, usually with banana leaves.

(a)

(b)

Cocoa Fermentation. (a) Cocoa pods growing on the cocoa tree. Each pod is 13 to 15 cm in length and contains 30 to 40 seeds in a sticky white pulp. (b) Seeds and pulp are fermented in boxes covered with banana leaves for 5 to 7 days and then dried in the sun, as shown here. Chocolate cannot be produced without fermentation.

Like most fermentations, this process involves a succession of microbes. First, a community of yeasts, including *Candida rugosa* and *Kluyveromyces marxianus,* hydrolyzes the pectin that covers the seeds and ferments sugars to release ethyl alcohol and CO_2. As the temperature and the alcohol concentration increase, the yeasts are inhibited and lactic acid bacteria increase in number. The mixture is stirred to ensure an even temperature distribution. Lactic acid production drives the pH down; this encourages the growth of bacteria that produce acetic acid as a fermentation end product. Acetic acid is critical to the production of fine chocolate because it kills the sprout inside the seed and releases enzymes that cause further degradation of proteins and carbohydrates, contributing to the overall taste of the chocolate. In addition, acetate esters, derived from acetic acid, are important for the development of good flavor. Fermentation takes 5 to 7 days. An experienced chocolate grower knows when the fermentation is complete: if it is stopped too soon, the chocolate will be bitter and astringent; if it lasts too long, microbes start growing on the seeds instead of in the pulp. "Off-tastes" arise when the gram-positive bacterium *Bacillus* and the filamentous fungi *Aspergillus, Penicillium,* and *Mucor* hydrolyze lipids in the seeds to release short-chain fatty acids. As the pH begins to rise, bacteria of the genera *Pseudomonas, Enterobacter,* and *Escherichia* also contribute to bad tastes and odor.

After fermentation, the seeds, now called beans, are spread out to dry. Ideally this is done in the sun, although drying ovens are also used. The dried beans are brown and lack pulp. They are bagged and sold to chocolate manufacturers, who first roast the beans to further reduce the bitter taste and kill most of the microbes (some *Bacillus* spores may remain). The beans are then ground and the nibs—the inner part of each bean—are removed. The nibs are crushed into a thick paste called a chocolate liquor, which contains cocoa solids and cocoa butter but no alcohol. Cocoa solids are brown and have a rich flavor, and cocoa butter has a high fat content and is off-white in color. The two components are separated, and the cocoa solids can be sold as cocoa for baking and hot chocolate, while the cocoa butter is used to make white chocolate or sold to cosmetics companies for use in lipsticks and lotions. However, the bulk of these two components will be used to make chocolate. The cocoa solids and butter are reunited in controlled ratios, and sugar, vanilla,

Continued

TECHNIQUES & APPLICATIONS

40.1 Chocolate: The Sweet Side of Fermentation *(Continued)*

and other flavors are added. The better the fermentation, the less sugar needs to be added (and the more expensive the chocolate will be).

The final product, delicious chocolate, is a combination of over 300 different chemical compounds. This mixture is so complex that no one has yet been able to make synthetic chocolate that can compete with the natural fermented plant (note that artificial vanilla is readily available). Microbiologists and food scientists are studying the fermentation process to determine the role of each microbe. But like the chemists, they have had little luck in replicating the complex, imprecise fermentation that occurs on cocoa farms. In fact, the finest, most expensive chocolate starts as cocoa on farms where the details of fermentation have been handed down through generations. Chocolate production is truly an art as well as a science, while eating it is simply divine.

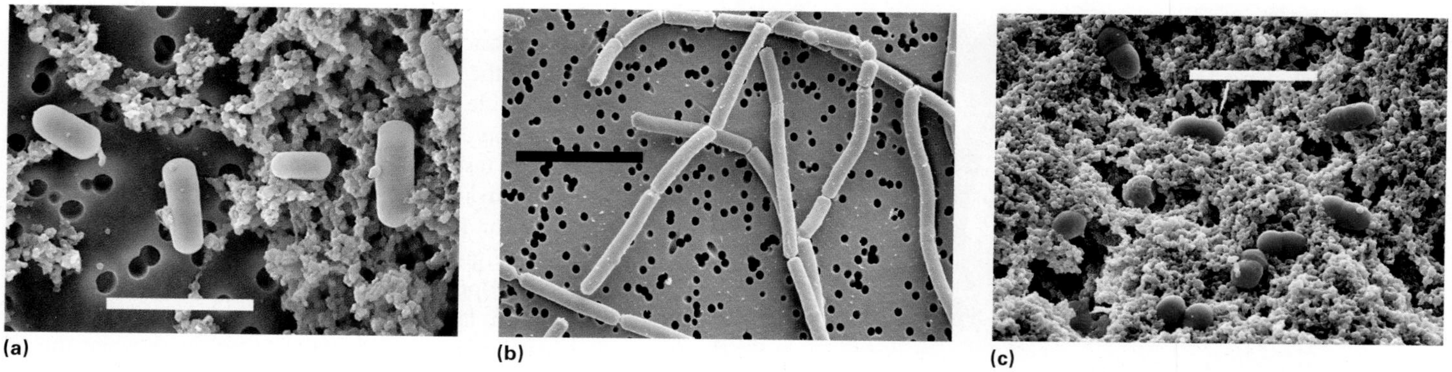

FIGURE 40.10 Lactic Acid Bacteria (LAB). Colorized scanning electron micrographs of LAB used as starter cultures. (a) *Lactobacillus helveticus.* (b) *Lactobacillus delbrueckii* subspecies *bulgaricus.* (c) *Lactococcus lactis.* The bacteria are supported by filters, seen as holes in the background. Scale bar = 5 μm.

Figure 40.10 Micro Inquiry

Can you name at least two features that make these bacteria well suited for milk fermentations?

United States and Europe. In commercial production, nonfat or low-fat milk is pasteurized, cooled to 43°C or lower, and inoculated with a 1:1 ratio of *Streptococcus salivarius* subspecies *thermophilus* (*S. thermophilus*) and *Lactobacillus delbrueckii* subspecies *bulgaricus* (*L. bulgaricus*). *S. thermophilus* grows more rapidly at first and renders the milk anoxic and weakly acidic. *L. bulgaricus* then acidifies the milk even more. Acting together, the two species ferment almost all of the lactose to lactic acid and flavor the yogurt with diacetyl (*S. thermophilus*) and acetaldehyde (*L. bulgaricus*). Many yogurts now contain probiotic bacterial strains, but these do not necessarily contribute to the fermentation process. Fruits or fruit flavors are pasteurized separately and then combined with the yogurt. Freshly prepared yogurt contains about 10^9 bacteria per gram.

Probiotics

The health benefits of fermented foods such as yogurt have been touted for a great number of years. However, only recently have rigorous studies explored the effects of certain bacteria that are either commensals or mutualists in the human intestine. In an effort to standardize the use of the term **probiotic**, the Food and Agricultural Organization of the United Nations–World Health Organization (FAO-WHO) defines probiotics as "live microorganisms, which when administered in adequate amounts, confer a health benefit to the host." FAO-WHO guidelines published in 2002 provide specific requirements that should be met for products to meet this standard (**figure 40.11**). Because these are voluntary guidelines, the consumer should

Table 40.5 Major Categories and Examples of Fermented Milk Products

Category	Typical Examples
I. Lactic fermentations	
Mesophilic	Buttermilk, cultured buttermilk, *långofil*, *tëtmjolk*, *ymer*
Thermophilic	Yogurt, *laban*, *zabadi*, *labneh*, *skyr*, Bulgarian buttermilk
Probiotic	Biogarde, Bifighurt Acidophilus milk, *yakult*, Cultura-AB
II. Yeast-lactic fermentations	Kefir, *koumiss*, acidophilus-yeast milk
III. Mold-lactic fermentations	*Viili*

Source: Table 3.1, p. 58. In B. A. Law, editor, 1997. Microbiology and Biochemistry of Cheese and Fermented Milk, *2nd ed. New York: Chapman and Hall.*

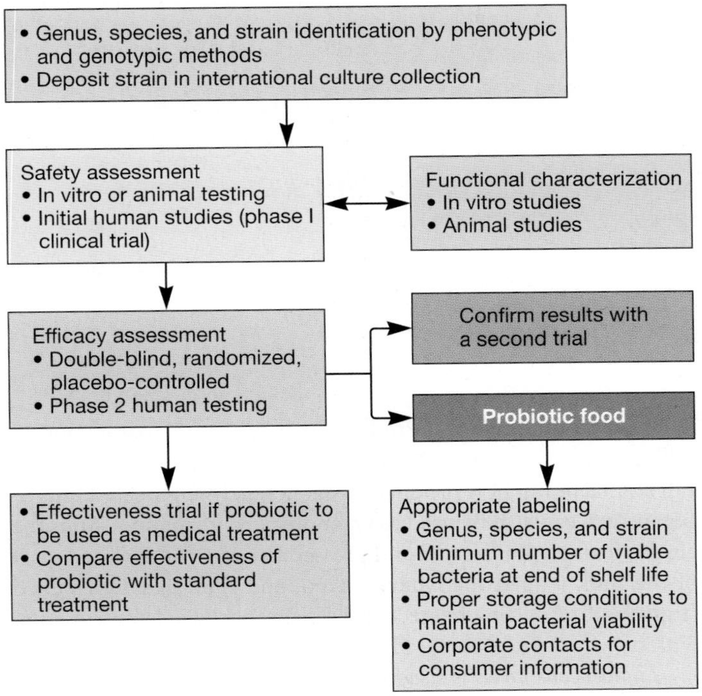

FIGURE 40.11 The Development of a Probiotic Food.

be aware that some claims of probiotic activity do not meet these criteria or have not been supported by scientific, clinical studies.

Microorganisms such as *Lactobacillus* and *Bifidobacterium* are being used to develop new probiotic foods. The possible health benefits of the use of such microbial dietary adjuvants include immunomodulation, control of diarrhea, anticancer effects, and possible improvement of Crohn's disease (inflammatory bowel disease). Probiotics have become a more attractive treatment option in cases of enteric disease because the rate of antibiotic resistance among pathogens continues to climb. In addition, disease ecologists have come to recognize that some intestinal microflora can be a contributing factor for certain conditions (e.g., Crohn's disease). ◀◀ *Normal microbiota of the human body (section 30.3)*

Acidophilus milk is produced by using *Lactobacillus acidophilus*. *L. acidophilus* may modify the microbial flora in the lower intestine, thus improving general health, and it often is used as a dietary adjunct, especially for lactose-intolerant persons. Many microorganisms in fermented dairy products stabilize the bowel microflora, and some appear to have antimicrobial properties. The exact nature and extent of health benefits of consuming fermented milks may involve minimizing lactose intolerance, lowering serum cholesterol, and possibly exhibiting anticancer activity. Several lactobacilli have antitumor compounds in their cell walls. Such findings suggest that diets including lactic acid bacteria, especially *L. acidophilus*, may contribute to the prevention of colon cancer.

Another interesting group used in milk fermentations are the bifidobacteria. The genus *Bifidobacterium* contains irregular, nonsporing, gram-positive rods that may be club-shaped or forked at the end. Bifidobacteria are nonmotile, anaerobic, and ferment lactose and other sugars to acetic and lactic acids. They are typical residents of the human intestinal tract, and many beneficial properties are attributed to them. Bifidobacteria are thought to help maintain the normal intestinal balance, while improving lactose tolerance; to possess antitumorigenic activity; and to reduce serum cholesterol levels. In addition, some believe that they promote calcium absorption and the synthesis of B-complex vitamins. It has also been suggested that bifidobacteria reduce or prevent the excretion of rotaviruses, a cause of diarrhea among children. *Bifidobacterium*-amended fermented milk products, including yogurt, are commercially available.

Search This: Probiotics/NIH

Yeast-Lactic Fermentation

Yeast-lactic fermentations include kefir, a product with an ethanol concentration of up to 2%. This unique fermented milk originated in the Caucasus Mountains and is produced east into Mongolia. Kefir products tend to be foamy due to active carbon dioxide production. This process is based on the use of kefir "grains" as an inoculum. These are coagulated lumps of casein (milk protein) that contain yeasts, lactic acid bacteria, and acetic acid bacteria, which are recovered at the end of the fermentation. Originally, kefir was produced in leather sacks hung by the front door during the day, and passersby were expected to push and knead the sack to mix and stimulate the fermentation. Fresh milk could be added occasionally to replenish nutrients and maintain activity.

Table 40.6 Major Types of Cheese and Microorganisms Used in Their Production

	Contributing Microorganisms[a]	
Cheese (Country of Origin)	*Earlier Stages of Production*	*Later Stages of Production*
Soft, unripened		
Cottage	*Lactococcus lactis*	*Leuconostoc cremoris*
Cream	*Lactococcus cremoris, L. diacetylactis, Streptococcus salivarius* subspecies *thermophilus, L. delbrueckii* subspecies *bulgaricus*	
Mozzarella (Italy)	*S. thermophilus, L. bulgaricus*	
Soft, ripened		
Brie (France)	*Lactococcus lactis, Lactococcus cremoris*	*Penicillium camemberti, P. candidum, Brevibacterium linens*
Camembert (France)	*L. lactis, Lactococcus cremoris*	*Penicillium camemberti, B. linens*
Semisoft		
Blue, Roquefort (France)	*L. lactis, Lactococcus cremoris*	*P. roqueforti*
Brick, Muenster (United States)	*L. lactis, Lactococcus cremoris*	*B. linens*
Limburger (Belgium)	*L. lactis, Lactococcus cremoris*	*B. linens*
Hard, ripened		
Cheddar, Colby (Britain)	*L. lactis, Lactococcus cremoris*	*Lactobacillus casei, L. plantarum*
Swiss (Switzerland)	*L. lactis, L. helveticus, S. salivarius* subspecies *thermophilus*	*Propionibacterium shermanii, P. freudenreichii*
Very hard, ripened		
Parmesan (Italy)	*L. lactis, Lactococcus cremoris, S. salivarius* subspecies *thermophilus*	*L. delbrueckii* subspecies *bulgaricus*

[a] *Lactococcus lactis* stands for *L. lactis* subspecies *lactis. Lactococcus cremoris* is *L. lactis* subspecies *cremoris,* and *Lactococcus diacetilactis* is *L. lactis* subspecies *diacetilactis.*

Mold-Lactic Fermentation

Mold-lactic fermentation results in a unique Finnish fermented milk called *viili*. The milk is placed in a cup and inoculated with a mixture of the fungus *Geotrichium candidum* and lactic acid bacteria. The cream rises to the surface, and after incubation at 18 to 20°C for 24 hours, lactic acid reaches a concentration of 0.9%. The fungus forms a velvety layer across the top of the final product, which also can be made with a bottom fruit layer.

Cheese Production

Cheese is one of the oldest foods, probably developed roughly 8,000 years ago. About 2,000 distinct varieties of cheese are produced throughout the world, representing approximately 20 general types and an annual economic value of about 75 billion U.S. dollars (**table 40.6**). Often cheeses are classified based on texture or hardness as soft cheeses (cottage, cream, Brie), semisoft cheeses (Muenster, Limburger, blue), hard cheeses (cheddar, Colby, Swiss), or very hard cheeses (Parmesan). All cheese results from a lactic acid fermentation of milk, which results in coagulation of milk proteins and formation of a curd. Rennin, an enzyme from calf stomachs but now produced by genetically engineered microorganisms, can also be used to promote curd formation. After the curd is formed, it is heated and pressed to remove the watery part of the milk (called the whey), salted, and then usually ripened (**figure 40.12**). The cheese curd can be packaged for ripening with or without additional microorganisms.

Lactococcus lactis is used as a starter culture for a number of cheeses. Starter culture density is often over 10^9 colony-forming units (CFUs) per gram of cheese curd before ripening. However, the high salt, low pH, and the temperatures that characterize the cheese microenvironment reduce these numbers rather quickly. This enables other bacteria, sometimes called nonstarter lactic acid bacteria (NSLAB), to grow; their numbers can reach 10^7 to 10^9 CFUs/g after several months of aging. Thus both starter and nonstarter LAB contribute to the final taste, texture, aroma, and appearance of the cheese.

In some cases, molds are used to further enhance the cheese. Obvious examples are Roquefort and blue cheese. For these

FIGURE 40.12 Cheddar Cheese Production. Cheddar, a village in England, has given its name to a cheese made in many parts of the world. "Cheddaring" is the process of turning and piling the curd to expel whey and develop the desired cheese texture.

cheeses, *Penicillium roqueforti* spores are added to the curds just before the final cheese processing. Sometimes the surface of an already formed cheese is inoculated at the start of ripening; for example, Camembert cheese is inoculated with spores of *Penicillium camemberti.*

The final hardness of the cheese is partially a function of the length of ripening. Soft cheeses are ripened for only about 1 to 5 months, whereas hard cheeses need 3 to 12 months, and very hard cheeses such as Parmesan require 12 to 16 months of ripening. The ripening process also is critical for Swiss cheese; gas production by *Propionibacterium* contributes to final flavor development and hole or eye formation in this cheese. Some cheeses, such as Limburger, are soaked in brine to stimulate the development of specific fungi and bacteria.

Search This: National Dairy Council

Meat and Fish

A variety of meat products can be fermented: sausage, country-cured hams, salami, cervelat, Lebanon bologna, fish sauces (processed by halophilic *Bacillus* species), *izushi,* and *katsuobushi. Pediococcus acidilactici* and *Lactobacillus plantarum* are most often involved in sausage fermentations. *Izushi* is based on the fermentation of fresh fish, rice, and vegetables by *Lactobacillus* spp.; *katsuobushi* results from the fermentation of tuna by *Aspergillus glaucus.* These latter two fermentations originated in Japan.

1. What are the major types of milk fermentations?
2. Briefly describe how buttermilk, sour cream, and yogurt are made.
3. What is unique about the morphology of *Bifidobacterium*? Why is it used in milk fermentation?
4. What major steps are used to produce cheese? How is the cheese curd formed in this process? What is whey? How does Swiss cheese get its holes?
5. Which fungal genus is often used in cheese making?

Wines and Champagnes

Wine production, the focus of **enology** (Greek *oinos,* wine, and *ology,* the science of), starts with the collection of grapes, continues with their crushing and the separation of the liquid, called **must,** before fermentation, and concludes with a variety of storage and aging steps (**figure 40.13**). All grapes have white juices. To make a red wine from a red grape, the grape skins are allowed to remain in contact with the must before fermentation to release their skin-coloring components. Wines can be produced by using the natural grape skin microorganisms, but this natural mixture of bacteria and yeasts gives unpredictable fermentation results. To avoid this, fresh must is treated with a sulfur dioxide fumigant and a desired strain of the yeast *Saccharomyces cerevisiae* or *S. ellipsoideus* is added. After inoculation, the juice is fermented for 3 to 5 days at temperatures between 20 and 28°C. Depending on the alcohol tolerance of the yeast strain (the alcohol eventually kills the yeast that produced it), the final product may contain 10 to 14% alcohol. Clearing and development of flavor occur during the aging process. The malolactic fermentation is an important part of wine production. Grape juice contains high levels of organic acids, including malic and tartaric acids. If the levels of these acids are not decreased during the fermentation process, the wine will be too acidic and have poor stability and "mouth feel." This essential fermentation is carried out by the bacteria *Leuconostoc oenos, L. plantarum, L. hilgardii, L. brevis,* and *L. casei.* The activities of these microbes transform malic acid (a four-carbon tricarboxylic acid) to lactic acid (a three-carbon monocarboxylic acid) and carbon dioxide. This results in deacidification, improvement of flavor stability, and, in some cases, the possible accumulation of bacteriocins in the wines. ◀◀ Ascomycota *(section 24.5); Fermentation (section 10.7)*

A critical part of wine making involves the choice of whether to produce a dry (no remaining free sugar) or a sweeter (varying amounts of free sugar) wine. This can be controlled by regulating the initial must sugar concentration. With higher levels of sugar, alcohol will accumulate and inhibit the fermentation before the sugar can be completely used, thus producing a sweeter wine. During final fermentation in the aging process, flavoring compounds accumulate and influence the bouquet of the wine.

Microbial growth during the fermentation process produces sediments, which are removed during **racking.** Racking can be

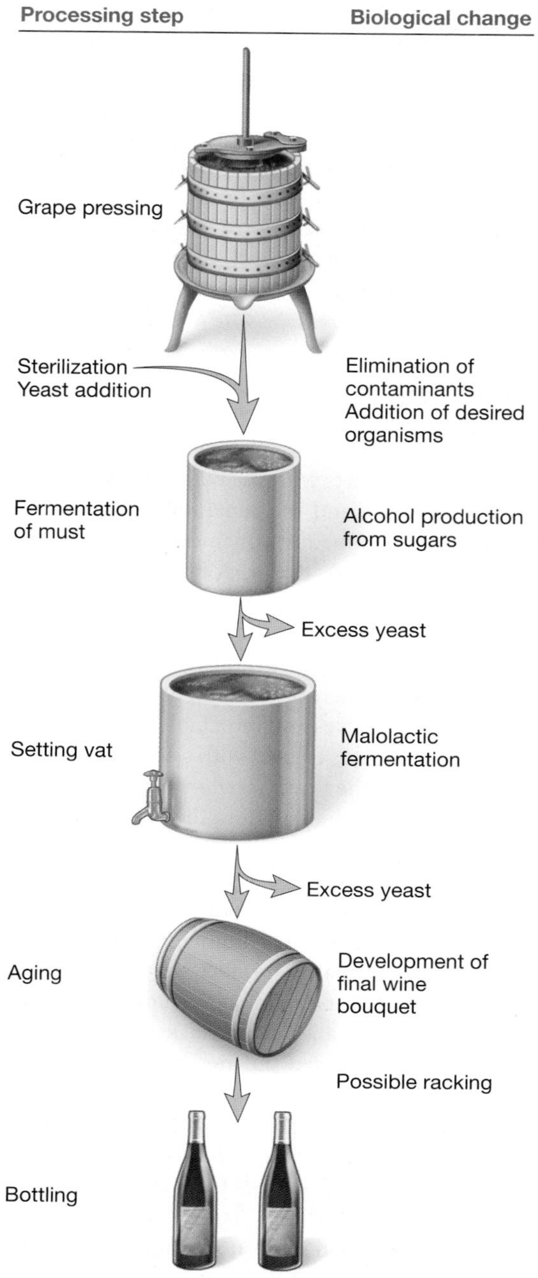

FIGURE 40.13 Wine Making. Once grapes are pressed, the sugars in the juice (the must) can be immediately fermented to produce wine. Must preparation, fermentation, and aging are critical steps.

carried out at the time the fermented wine is transferred to bottles or casks for aging or after the wine is placed in bottles.

Natural champagnes are produced when the fermentation is continued in bottles to produce carbon dioxide and a naturally sparkling wine. Sediments that remain are collected in the necks of inverted champagne bottles after the bottles have been carefully turned. The necks of the bottles are then frozen and the corks removed to disgorge the accumulated sediments. The bottles are refilled with clear champagne from another disgorged bottle, and the product is ready for final packaging and labeling.

Beers and Ales

Beer and ale production uses cereal grains such as barley, wheat, and rice. The complex starches and proteins in these grains must be hydrolyzed to a more readily usable mixture of simpler carbohydrates and amino acids. This process, known as **mashing,** involves germination of the barley grains and activation of their enzymes to produce a **malt** (**figure 40.14**). The malt is then mixed with water and the desired grains, and the mixture is transferred to the mash tun or cask in order to hydrolyze the starch to usable carbohydrates. Once this process is completed, the **mash** is heated with **hops** (dried flowers of the female vine *Humulus lupulus*), which were originally added to the mash to inhibit spoilage microorganisms (**figure 40.15**). The hops also provide flavor and assist in clarification of the liquid portion called the wort. In this heating step, the hydrolytic enzymes are inactivated and the wort can be **pitched**—inoculated—with the desired yeast.

Most beers are fermented with bottom yeasts, related to *Saccharomyces pastorianus,* which settle at the bottom of the fermentation vat. The beer flavor also is influenced by the production of small amounts of glycerol and acetic acid. Bottom yeasts require 7 to 12 days of fermentation to produce beer with a pH of 4.1 to 4.2. With a top yeast, such as *Saccharomyces cerevisiae,* the pH is lowered to 3.8 to produce ales. Freshly fermented (green) beers are aged, and when they are bottled, CO_2 is usually added. Beer can be pasteurized at 40°C or higher or sterilized by passage through membrane filters to minimize flavor changes.

Distilled Spirits

Distilled spirits are produced by an extension of beer production processes. Rather than starting with fresh grain, a **sour mash** is used. This is the liquid fermented grain collected from a previously prepared batch of alcohol. The mash is inoculated with a homolactic (lactic acid is the major fermentation product) bacterium such as *Lactobacillus delbrueckii* subspecies *bulgaricus* (figure 40.10*b*), which can lower the mash pH to around 3.8 in 6 to 10 hours. This limits the development of undesirable organisms. It is boiled, and the volatile components are condensed to yield a product with a higher alcohol content than beer. Rye and bourbon are examples of whiskeys. Rye whiskey must contain at least 51% rye grain, and bourbon must contain at least 51% corn. Scotch whiskey is made primarily of barley. Vodka and grain alcohols are also produced by distillation. Gin is vodka to which resinous flavoring agents—often juniper berries—have been added to provide a unique aroma and flavor.

Production of Breads

Bread is one of the most ancient of human foods. The use of yeasts to leaven bread is carefully depicted in paintings from ancient Egypt, and a bakery at the Giza Pyramid area, from the year 2575 BCE, has been excavated. In bread making, yeast growth is carried out under oxic conditions. This results in increased CO_2 production and minimum alcohol accumulation. The fermentation of bread involves several steps: α- and β-amylases present in the moistened dough release maltose and glucose from starch. Then a baker's strain of the yeast *Saccharomyces cerevisiae,* which produces maltase, invertase, and zymase enzymes, is added. The CO_2 produced by the yeast results in the light texture of many breads, and traces of fermentation products contribute to the final flavor. Usually bakers add sufficient yeast to allow the bread to rise within 2 hours—the longer the rising time, the more additional growth by contaminating bacteria and fungi can occur, making the product less desirable. For instance, bread products can be spoiled by *Bacillus* species that produce ropiness. If the dough is baked after these organisms have grown, stringy and ropy bread will result, leading to decreased consumer acceptance.

Other Fermented Foods

Many other plant products can be fermented, as summarized in **table 40.7**. These include *sufu,* which is produced by the fermentation of tofu, a chemically coagulated soybean milk product. To carry out the fermentation, the tofu curd is cut into small chunks and dipped into a solution of salt and citric acid. After the cubes are heated to pasteurize their surfaces, the fungi *Actinimucor elegans* and some *Mucor* species are added. When a white mycelium develops, the cubes, now called *pehtze,* are aged in salted rice wine. This product has achieved the status of a delicacy in many parts of Asia. Another popular product is tempeh, a soybean mash fermented by *Rhizopus.*

Sauerkraut or sour cabbage is produced from shredded cabbage. Usually the mixed microbial community of the cabbage is

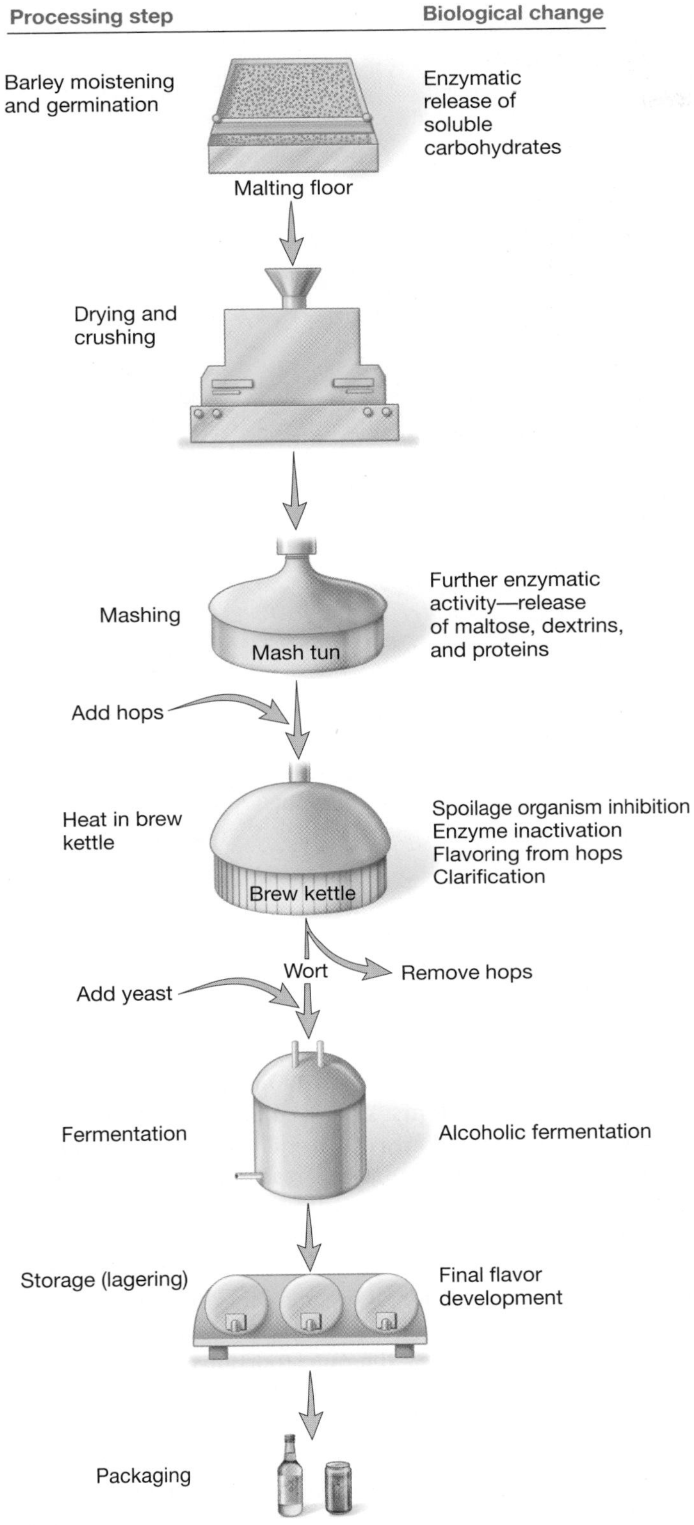

FIGURE 40.14 Producing Beer. To make beer, the complex carbohydrates in the grain must first be transformed into a fermentable substrate. Beer production thus requires the important steps of malting and the use of hops and boiling for clarification, flavor development, and inactivation of malting enzymes to produce the wort.

FIGURE 40.15 Brew Kettles Used for Preparation of Wort. In large-scale processes, copper brew kettles can be used for wort preparation, as shown here at the Carlsberg Brewery, Copenhagen, Denmark.

Table 40.7 Fermented Foods Produced from Fruits, Vegetables, Beans, and Related Substrates

Foods	*Raw Ingredients*	*Fermenting Microorganisms*	*Area*
Coffee	Coffee beans	*Erwinia dissolvens, Saccharomyces* spp.	Brazil, Congo, Hawaii, India
Gari	Cassava	*Corynebacterium manihot, Geotrichum* spp.	West Africa
Kenkey	Corn	*Aspergillus* spp., *Penicillium* spp., lactobacilli, yeasts	Ghana, Nigeria
Kimchi	Cabbage and other vegetables	Lactic acid bacteria	Korea
Miso	Soybeans	*Aspergillus oryzae, Zygosaccharomyces rouxii*	Japan
Ogi	Corn	*Lactobacillus plantarum, Lactococcus lactis, Zygosaccharomyces rouxii*	Nigeria
Olives	Green olives	*Leuconostoc mesenteroides, Lactobacillus plantarum*	Worldwide
Ontjom	Peanut presscake	*Neurospora sitophila*	Indonesia
Peujeum	Cassava	Molds	Indonesia
Pickles	Cucumbers	*Pediococcus cerevisiae, L. plantarum, L. brevis*	Worldwide
Poi	Taro roots	Lactic acid bacteria	Hawaii
Sauerkraut	Cabbage	*L. mesenteroides, L. plantarum, L. brevis*	Worldwide
Soy sauce	Soybeans	*Aspergillus oryzae* or *A. soyae, Z. rouxii, Lactobacillus delbrueckii*	Japan
Sufu	Soybeans	*Actinimucor elegans, Mucor* spp.	China
Tao-si	Soybeans	*A. oryzae*	Philippines
Tempeh	Soybeans	*Rhizopus oligosporus, R. oryzae*	Indonesia, New Guinea, Surinam

Adapted from Jay, James M. 2000. Modern Food Microbiology, *6th ed. Reprinted by permission of Aspen Publishing, Frederick, MD.*

used. A concentration of 2.2 to 2.8% sodium chloride restricts the growth of gram-negative bacteria while favoring the development of the lactic acid bacteria. The primary microorganisms contributing to this product are *Leuconostoc mesenteroides* and *Lactobacillus plantarum*. A predictable microbial succession occurs in sauerkraut's development. The activities of the lactic acid-producing cocci usually cease when the acid content reaches 0.7 to 1.0%. At this point, *Lactobacillus plantarum* and *Lactobacillus brevis* continue to function. The final acidity is generally about pH 1.7, with lactic acid comprising 1.0 to 1.3% of the total acid in a satisfactory product.

Pickles are produced by placing cucumbers and components such as dill seeds in casks filled with a brine. The sodium chloride concentration begins at 5% and rises to about 16% in 6 to 9 weeks. The salt not only inhibits the growth of undesirable bacteria but also extracts water and water-soluble constituents from the cucumbers. These soluble carbohydrates are converted to lactic acid. The fermentation, which can require 10 to 12 days, involves the growth of the gram-positive bacteria *L. mesenteroides, Enterococcus faecalis, Pediococcus acidilactici, L. brevis,* and *L. plantarum. L. plantarum* plays the dominant role in this fermentation process. Sometimes, to achieve more uniform pickle quality, natural microorganisms are first destroyed and the cucumbers are fermented using pure cultures of *P. acidilactici* and *L. plantarum.*

Grass, chopped corn, and other fresh animal feeds, if stored under moist anoxic conditions, will undergo a mixed acid fermentation that produces pleasant-smelling **silage.** Trenches or more traditional vertical steel or concrete silos are used to store the silage. The accumulation of organic acids in silage can cause rapid deterioration of silos. Older wooden stave silos, if not properly maintained, allow the outer portions of the silage to become oxic, resulting in spoilage of a large portion of the plant material.

1. Describe and contrast the processes of wine and beer production.
2. How do champagnes differ from wines?
3. Describe how distilled spirits such as whiskey are produced.
4. What microorganisms are most important in sauerkraut and pickle fermentations?

40.6 Microorganisms as Foods and Food Amendments

A variety of bacteria, yeasts, and other fungi have been used as animal and human food sources. Mushrooms (e.g., *Agaricus bisporus*) are one of the most important fungi used directly as a food source. Large caves provide optimal conditions for their production (**figure 40.16**). Another popular microbial food supplement is the cyanobacterium *Spirulina.* It is used as a food source in Africa and is sold in North American health food stores as a dried cake or powdered product.

Probiotic microbes can also be used as food amendments. Such microbes, primarily *Lactobacillus acidophilus,* are used in beef cattle feed. When the bacteria are sprayed on feed, the cattle that eat it appear to have markedly lower (60% in some experiments) carriage of the pathogenic *E. coli* strain O157:H7. This can make it easier to produce beef that will meet current standards for microbiological quality at the time of slaughter.

Probiotics are also used successfully with poultry. For instance, the USDA has designated a probiotic *Bacillus* strain for use with chickens as GRAS (p. 1013). Feeding chickens a strain of *Bacillus subtilis* leads to increased body weight and feed conversion. There is also a reduction in coliforms and *Campylobacter* in the processed carcasses. It has been suggested that this probiotic decreases the need for antibiotics in poultry production and pathogen levels on farms. *Salmonella* can be controlled by spraying a patented blend of 29 bacteria, isolated from the chicken cecum, on day-old chickens. As they preen themselves, the chicks ingest the bacterial mixture, establishing a functional microbial community in the cecum and limiting *Salmonella* colonization of the gut in a process called competitive exclusion. In 1998 this product, called PREEMPT, was approved for use in the United States by the FDA.

FIGURE 40.16 Mushroom Farming. Growing mushrooms requires careful preparation of the growth medium and control of environmental conditions. The mushroom bed is a carefully developed compost, which can be steam-sterilized to improve mushroom growth.

1. What conditions are needed to have most efficient production of edible mushrooms?
2. How are probiotics used in agriculture?

Summary

40.1 Microbial Growth and Food Spoilage

a. Most foods, especially when raw, provide an excellent environment for microbial growth. This growth can lead to spoilage or preservation, depending on the microorganisms present and environmental conditions.

b. The course of microbial development in a food is influenced by the intrinsic characteristics of the food itself—pH, salt content, substrates present, water presence, and availability—and extrinsic factors, including temperature, relative humidity, and atmospheric composition.

c. Microorganisms can spoil meat, dairy products, fruits, vegetables, and canned goods in several ways. Spices, with their antimicrobial compounds, sometimes protect foods.

40.2 Controlling Food Spoilage

a. Foods can be preserved in a variety of physical and chemical ways, including filtration, alteration of temperature (cooling, pasteurization, sterilization), drying, the addition of chemicals, radiation, and fermentation.

b. Modified atmosphere packaging (MAP) is used to control microbial growth in foods and to extend product shelf life. This process involves decreased oxygen and increased carbon dioxide levels in the space between the food surface and the wrapping material.

c. Interest is increasing in using bacteriocins for food preservation. Nisin, a product of *Lactococcus lactis,* is a major substance approved for use in foods.

40.3 Types of Food-Borne Disease

a. Foods can be contaminated by pathogens at any point in the food production, storage, or preparation processes. Pathogens such as *Salmonella, Campylobacter, Listeria,* and *E. coli* can be transmitted by the food to the susceptible consumer, where they grow and cause a food-borne infection (**table 40.4**).

b. If the pathogen grows in the food before consumption and forms toxins that affect the food consumer without further microbial growth, the disease is a food intoxication. Examples are intoxications caused by *Staphylococcus, Clostridium,* and *Bacillus.* Fungi that grow in foods, especially cereals and grains, can produce important disease-causing chemicals, including the carcinogens aflatoxins (**figure 40.6**) and fumonisins (**figure 40.7**).

40.4 Detection of Food-Borne Pathogens

a. Detection of food-borne pathogens is a major part of food microbiology. Many assays rely on culture-based approaches.

b. The use of immunological and molecular techniques such as DNA and RNA hybridization, PCR, and pulsed-field gel electrophoresis often makes it possible to link disease occurrences to a common infection source. PulseNet programs are used to coordinate these control efforts (**figures 40.8** and **40.9**).

40.5 Microbiology of Fermented Foods

a. Dairy products can be fermented to yield a wide variety of cultured milk products. These include mesophilic, therapeutic probiotic, thermophilic, yeast-lactic, and mold-lactic products.

b. Growth of lactic acid–forming bacteria, often with the additional use of rennin, can coagulate milk solids. These solids can be processed to yield a wide variety of cheeses, including soft unripened, soft ripened, semisoft, hard, and very hard types (**table 40.6**). Both bacteria and fungi are used in these cheese production processes.

c. Wines are produced from pressed grapes and can be dry or sweet, depending on the level of free sugar that remains at the end of the alcoholic fermentation (**figure 40.13**). Champagne is produced when the fermentation, resulting in CO_2 formation, is continued in the bottle.

d. Beer and ale are produced from cereals and grains. The starches in these substrates are hydrolyzed, in the processes of malting and mashing, to produce a fermentable wort. *Saccharomyces cerevisiae* is a major yeast used in the production of beer and ale (**figure 40.14**).

e. Many plant products can be fermented with bacteria, yeasts, and molds. Important products are breads, soy sauce, *sufu,* and tempeh (**table 40.7**). Sauerkraut and pickles are produced in a fermentation process in which natural populations of lactobacilli play a major role.

40.6 Microorganisms as Foods and Food Amendments

a. Microorganisms themselves can serve as an important food source. Mushrooms are one of the most important fungi used as a food source. *Spirulina,* a cyanobacterium, also is a popular food source sold in specialty stores.

b. Many microorganisms, including some of those used to ferment milks, can be used as food amendments or microbial dietary adjuvants. Several types of probiotic microorganisms are used successfully in poultry production.

Critical Thinking Questions

1. Fresh lemon slices are often served with raw or steamed seafood (oysters, crab, shrimp). From a food microbiology perspective, provide an explanation for their being served. Are there other examples of ingredients or additions used in either cooking or serving foods that not only enhance flavor but also might have an antimicrobial strategy? Consider the example of marinades.

2. You are going through a salad line in a cafeteria at the end of the day. Which types of foods would you tend to avoid and why?

3. Why were aflatoxins not discovered before the 1960s? Do you think this was the first time they had grown in a food product to cause disease?

4. What advantage might the Shiga-like toxin give *E. coli* O157:H7? Can we expect to see other "new" pathogens appearing, and what should we do, if anything, to monitor their development?

5. Keep a record of what you eat for a day or two. Determine if the food, beverages, and snacks you ate were produced (at any level) with the aid of microorganisms. Indicate at what levels microorganisms were deliberately used. Be sure to consider ingredients such as citric acid, which is produced at the industrial level by several species of fungi.

6. Following recent outbreaks of *E. coli* O157:H7 associated with leafy vegetables, there has been a need to better understand the growth of this pathogen on vegetables. To that end, a study was performed to compare the colonization and growth of *E. coli* O157:H7 on lettuce. In this study, the success of the microbe on intact leaves was compared with regions that had been mechanically damaged by shredding or

bruising. Indeed, even within 4 hours after inoculation, there was a marked increase in microbial growth on the injured surfaces of the lettuce.

Why do you think this was the case? If you were to perform this study (or one similar to it), how would you quantify *E. coli* colonization? If you were able to repeat and extend these findings, what kinds of recommendations might you make to the vegetable producers? To consumers?

Read the original paper: Brandl, M. T. 2008. Plant lesions promote the rapid multiplication of *Escherichia coli* O157:H7. *Appl. Environ. Microbiol.* 74:5285.

7. During cheese production, LAB convert lactose to lactate and casein (milk protein) to amino acids. Lactate and amino acids then become the substrates for further microbial growth, which results in aroma production, and deacidification of the cheese. The yeast *Yarrowia lipolytica* grows on the surface of many cheeses; it is capable of both lactate and amino acid catabolism. When grown on a lactate plus amino acid medium, *Y. lipolytica* preferentially consumes amino acids. Amino acid degradation results in the release of ammonia, which increases the pH.

Draw a flow chart that shows the LAB fermentation of milk, followed by the growth of *Y. lipolytica*. Indicate which substrates are consumed first and what happens to the pH. Based on this simplified scenario, why do you think most cheeses involve the activity of more than one yeast species?

Read the original paper: Mansour, S., et al. 2008. Lactate and amino acid catabolism in the cheese ripening yeast *Yarrowia lipolytica*. *Appl. Environ. Microbiol.* 74:6505.

Concept Mapping Challenge

Provide your own linking words to develop a concept map using the terms listed below. Your map should display the dichotomy between the prevention of food spoilage and the use of microbes to manufacture foods.

Food spoilage	Wine	Fermentation	PCR	Yogurt	Beer
Prevention	ELISA	PulseNet	Pasteurization	Yogurt	Food production
			Probiotics	LAB	Low temperature
			Detection	Cheese	Culture-based techniques
			Chemical/GRAS		

Learn More

Learn more by visiting the text website at www.mhhe.com/willey8, where you will find a complete list of references.

41

Industrial Microbiology

A fermenter, or cell culture vessel, used to mass-produce industrial products such as pharmaceuticals.

CHAPTER GLOSSARY

anaerobic digester A system in which biomass is degraded almost entirely to methane and CO_2 by an anaerobic microbial consortium.

biofuels Compounds made by microbes that can be used to provide energy to vehicles and other machines. Common biofuels include ethanol and hydrogen.

bioinsecticides Biological agents such as microbes or their components that can be used to kill insects; also called **biopesticides.**

combinatorial biosynthesis The manipulation of genes encoding enzymes with the goal of changing an industrially important product.

directed evolution A group of techniques that use the in vitro manipulation of genes and chromosomes to alter or create an industrially important product.

fermenter A large vessel used to grow microbes in industrial settings. Physical and biological parameters are carefully controlled to maximize output of the microbial product.

heterologous gene expression The transcription and translation of a gene that has been introduced (cloned) into an organism that does not otherwise possess the gene.

high-throughput screening A system that combines liquid handling devices, robotics, computers, data processing, and a sensitive detection system to rapidly screen thousands of molecules for a single capability.

microbial fuel cell A means of generating power in which biomass is oxidized by microbes and the electrons are transferred directly to an anode so that electricity can be generated.

natural product A compound produced by an organism that has an industrial or medical application.

scale-up The process of increasing the volume of a culture from a small, experimental system to a large-volume system for the production of an industrially important metabolite.

secondary metabolites Cellular products that are synthesized during late logarithmic or stationary phase of growth that are not essential for normal metabolism and growth. Antibiotics, mycotoxins, and many secreted proteins of industrial importance are considered secondary metabolites.

site-directed mutagenesis A molecular genetic technique whereby specific nucleotides can be changed to nucleotides designated by the experimentalist.

The genetic diversity that abounds in the microbial world represents a vast, untapped resource of medically and industrially important molecules. Industrial microbiology harnesses the capacity of microbes to synthesize compounds that have important applications in medicine, agriculture, food preparation, and other industrial processes. These compounds are commonly called **natural products**. As discussed in chapter 34, the majority of antimicrobial compounds are natural products. However, it is not as well appreciated that about 60% of synthetic chemicals rely on microbial catalyzed reactions during their production. In this chapter, we discuss how microorganisms (or their genes) are optimized for industrial manufacturing. We then review some of the more important products of industrial microbiology to illustrate the diversity of products and means by which they are produced.

41.1 Microorganisms Used in Industrial Microbiology

Until relatively recently, microbes used in industrial microbiology were cultured from natural materials such as soil samples, waters, and spoiled bread and fruit. Cultures from all areas of the world continue to be examined to identify new strains with desirable characteristics. However, because most of these microbes will resist growth under standard laboratory conditions, techniques for enrichment of natural samples and further purification are constantly evolving. For instance, the microdroplet culture technique described in chapter 27 (*p. 661*) has been roboticized so that many hundreds of nutrient combinations can be efficiently tested for their capacity to support growth. Here, we describe some of the approaches used to obtain microbes for industrial purposes.

Genetic Manipulation of Microorganisms

Genetic manipulations are used to produce microorganisms with new and desirable characteristics. The classical methods of genetic exchange coupled with recombinant DNA technology, genomics, and systems biology play a vital role in the development of industrially important microbial products.

Mutagenesis

Once a microbe is found to produce a promising compound, a variety of techniques can be used to increase product yield. These include chemical, ultraviolet light, and transposon mutagenesis. For example, the first cultures of *Penicillium notatum,* which had to be grown in stationary vessels, produced low concentrations of penicillin. In 1943 strain NRRL 1951 of *Penicillium chrysogenum* was isolated and further improved through chemical and UV mutagenesis (**figure 41.1**). Today most penicillin is produced with *Penicillium chrysogenum* grown in aerobic, stirred fermenters, yielding 55 times more penicillin than the original static cultures. Microbial strains used to generate the product of interest are called **production strains.** ⏮ *Mutations: Their chemical basis and effects (section 14.1)*

Protoplast Fusion

Protoplast fusion is another technique that can be used to promote genetic variability in microorganisms and plant cells. In this method, protoplasts—cells lacking a cell wall—are prepared by growing microbes in an isotonic solution in the presence of enzymes that degrade the cell wall (e.g., lysozyme for bacteria, chitinase for yeasts). The protoplasts of cells of the same or even different species can be fused (joined) during coincubation. The cell wall is then regenerated using osmotic stabilizers such as sucrose. Protoplast fusion is inherently mutagenic and is especially useful for prompting recombination. For example, when protoplasts of *Penicillium roquefortii* were fused with those of *P. chrysogenum,* new industrially useful *Penicillium* strains were created.

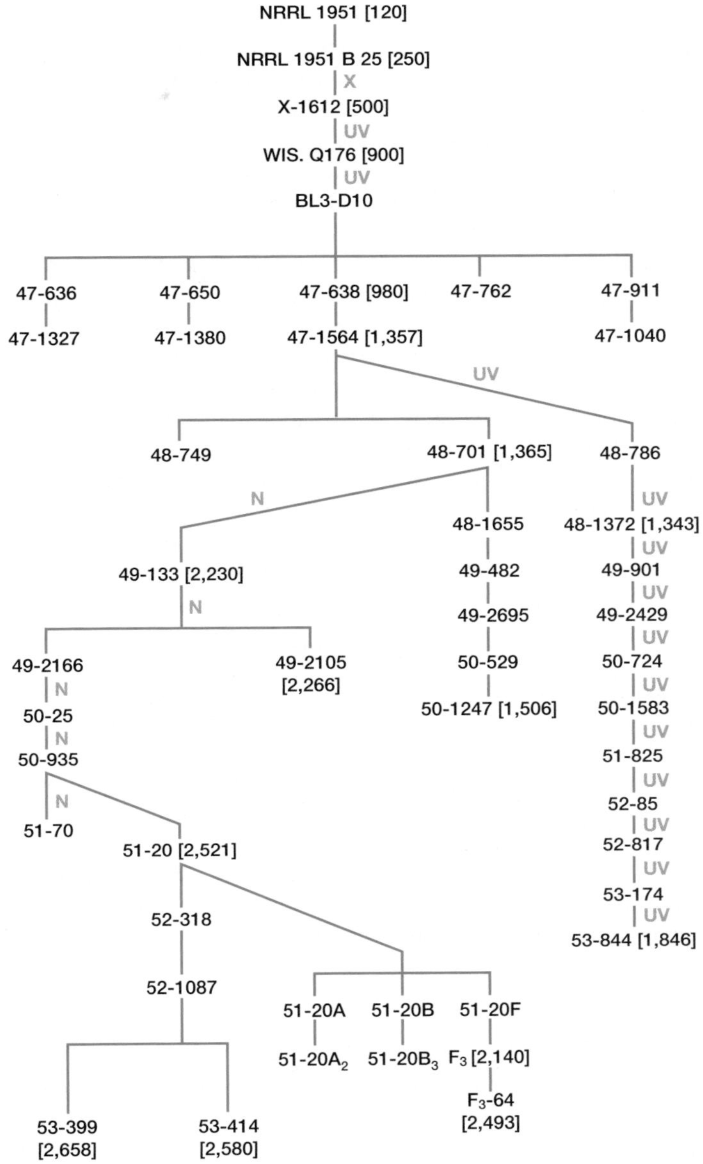

FIGURE 41.1 Mutation Makes It Possible to Increase Fermentation Yields. A "genealogy" of the mutation processes used to increase penicillin yields with *Penicillium chrysogenum* using X-ray treatment (X), UV treatment (UV), and mustard gas (N). By using these mutational processes, the yield was increased from 120 International Units (IU) to 2,580 IU, a 20-fold increase. Unmarked transfers were used for mutant growth and isolation. Yields in International Units/mL in brackets.

Genetic Transfer Between Different Organisms

The transfer and expression of genes between different organisms can give rise to novel products, may change the way in which genes of interest are regulated, or may even change the posttranslational modification of the protein product. Whereas protoplast fusion results in the in vivo recombination between microbial genomes, in vitro genetic transfer of specific genes enables a more targeted approach to strain construction. When functional genes from one organism are cloned into another in which they are transcribed and translated, it is called **heterologous gene expression.** Heterologous gene expression enables the production of specific proteins and peptides without contamination by other products that might be synthesized in the original organism. This approach can decrease the time and cost of recovering and purifying a product. For example, pediocin is a bacteriocin (antibacterial peptide) normally produced by the bacterium *Pediococcus pentosaceus.* However, the gene for this peptide is expressed in a yeast during wine fermentation to control bacterial contaminants. Conversely, yeast proteins are sometimes cloned into and expressed in bacteria so that the protein is not glycosylated, that is decorated with sugar moieties. Another major advantage of engineered protein production is that only biologically active stereoisomers are produced. This specificity is required to avoid the possible harmful side effects of inactive stereoisomers. Examples of a few heterologous systems are shown in **table 41.1**.

Modification of Gene Expression

In addition to inserting new genes in organisms, it also is possible to increase product yield by modifying regulatory molecules or the DNA sites to which they bind. For instance, the tight regulation of secondary metabolites such as antibiotics can be diminished by promoter mutations. The goal is to prompt microbes to begin antibiotic synthesis earlier, or at a higher level than the parental strain, or both. These approaches make it possible to overproduce a wide variety of products, such as antibiotics, amino acids, and enzymes of industrial importance.

Directed Evolution

As demonstrated in figure 41.1, mutants that overproduce a product or had some other desirable trait were traditionally obtained by "brute force" mutation. That is, mutations that conferred the trait were allowed to accumulate in a microbial strain after repeated rounds of mutagenesis and transfer from culture to culture. Production strains created in this way have an ill-defined genotype quite different from the parental strain. **Directed evolution** takes a more rational approach to the construction of production strains because only genes of interest are targeted for mutagenesis. In most cases, this requires that the gene sequences be altered in vitro, then replaced in the original strain or a heterologous host that may be better suited to produce the industrial compound at high levels.

Search This: Directed evolution/caltech

Several technical advancements were key to the development of direct evolution methodologies. The first is **site-directed mutagenesis.** In this technique, the nucleotide sequence of a specific gene is altered. The gene of interest is first cloned into a plasmid and introduced into a host microbial strain. In the meantime, an oligonucleotide of about 20 nucleotides that contains the desired nucleotide sequence change is synthesized. The recombinant plasmid is purified, and PCR is performed in which the altered oligonucleotide binds to the single-stranded copy of the wild type gene (**figure 41.2**). The polymerase then extends the oligonucleotide and replicates the remainder of the target gene and plasmid. This produces a double-stranded plasmid, but only one strand bears the mutation. Following PCR amplification, the parental copy of the plasmid is eliminated by digestion with a nuclease that degrades only DNA that has been methylated. Because the parent plasmid was obtained from a live cell that methylates its DNA and PCR synthesizes unmethylated DNA, only the original, wild-type plasmid is digested. The mutated version of the plasmid can now be introduced into a host microbe. In this way, minor amino acid substitutions have been found to lead to unexpected changes in protein characteristics, resulting in new products such as more environmentally resistant enzymes.

Site-directed mutagenesis changes only a few nucleotides at a time. A variety of other directed evolution techniques have been

Table 41.1 Heterologous Gene Expression to Improve Processes and Products

Property or Product Transferred	*Microorganism Used*	*Combinatorial Process*
Ethanol production	*Escherichia coli*	Integration of pyruvate decarboxylase and alcohol dehydrogenase II from *Zymomonas mobilis*
1,3-Propanediol production	*E. coli*	Introduction of genes from the *Klebsiella pneumoniae dha* region into *E. coli* makes possible anaerobic 1,3-propanediol production
Cephalosporin precursor synthesis	*Penicillium chrysogenum*	Production of 7-ADC and 7-ADCA[a] precursors by incorporation of the expandase gene of *Cephalosoporin acremonium* into *Penicillium* by transformation
Lactic acid production	*Saccharomyces cerevisiae*	A muscle bovine lactate dehydrogenase gene (LDH-A) expressed in *S. cerevisiae*

[a] 7-ADC = 7-aminocephalosporanic acid; 7-ADCA = 7-aminodecacetoxycephalosporonic acid.

Adapted from Ostergaard, S., et al. 2000. Metabolic engineering of Saccharomyces cerevisiae. Microbiol. Mol. Biol. Rev. *64(1):34–50.*

developed that can make larger changes to the gene of interest; some of these are listed in **table 41.2.** Careful inspection of table 41.2 reveals that many natural products are not proteins and therefore cannot be modified by simply mutating a single, structural gene. Instead, altering these nonprotein products requires an understanding of the biosynthetic pathway by which they are catalytically assembled. The genetic manipulation of enzymes with the goal of changing an industrially important product is called **combinatorial biosynthesis.** Currently combinatorial biosynthesis is used to make chiral intermediates for synthetic chemicals and analogs of natural products that are difficult to chemically synthesize. Work is in progress to harness this technology to develop libraries of novel microbial products that could be screened for a variety of medicinal and industrial purposes.

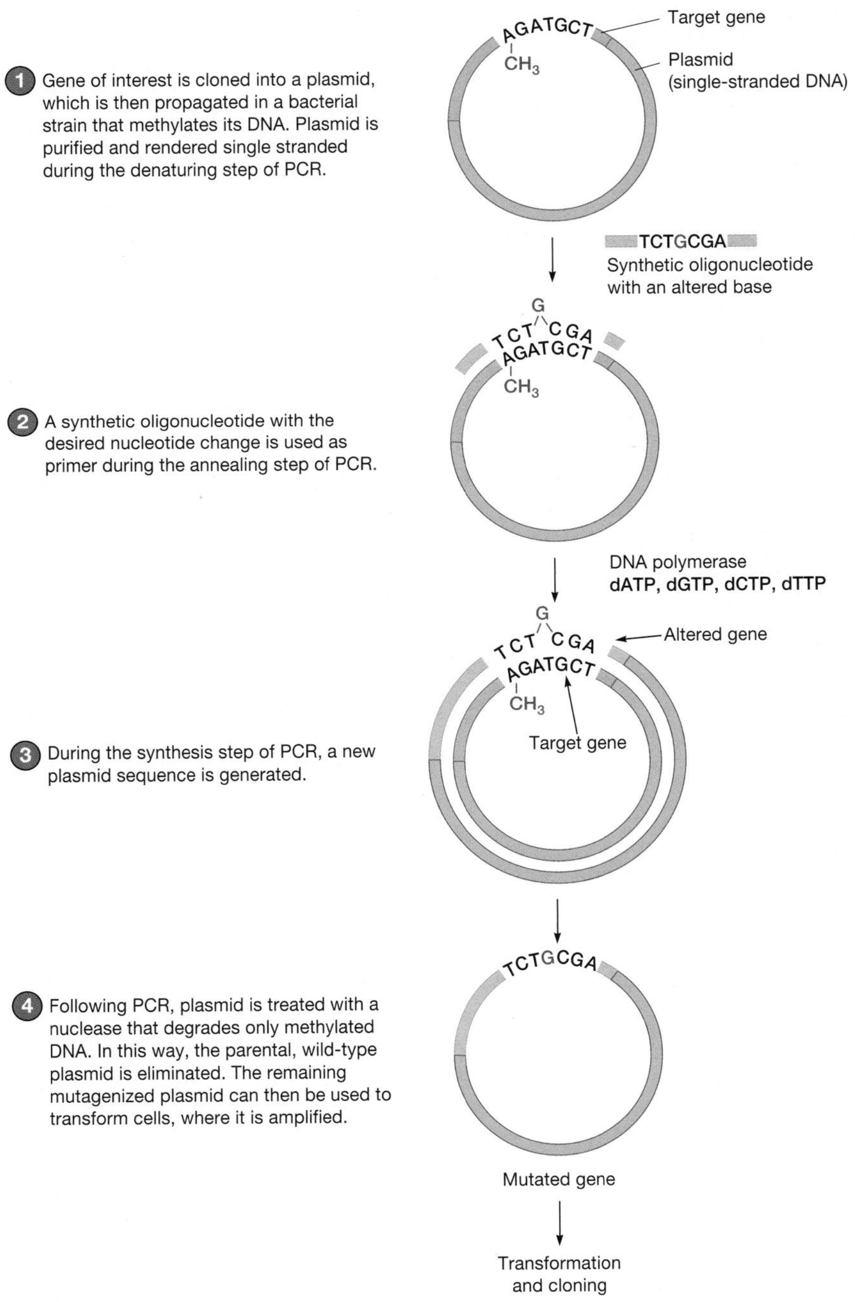

1 Gene of interest is cloned into a plasmid, which is then propagated in a bacterial strain that methylates its DNA. Plasmid is purified and rendered single stranded during the denaturing step of PCR.
AGATGCT
CH_3
Target gene
Plasmid (single-stranded DNA)
TCTGCGA
Synthetic oligonucleotide with an altered base
2 A synthetic oligonucleotide with the desired nucleotide change is used as primer during the annealing step of PCR.
DNA polymerase
dATP, dGTP, dCTP, dTTP
Altered gene
Target gene
3 During the synthesis step of PCR, a new plasmid sequence is generated.
4 Following PCR, plasmid is treated with a nuclease that degrades only methylated DNA. In this way, the parental, wild-type plasmid is eliminated. The remaining mutagenized plasmid can then be used to transform cells, where it is amplified.
Mutated gene
Transformation and cloning

FIGURE 41.2 Site-Directed Mutagenesis. A synthetic oligonucleotide is used to add a specific mutation to a gene.

Certain types of natural products are particularly amenable to combinatorial biosynthesis. For instance, polyketide and certain peptide antibiotics have become model systems for developing this technique because the biosynthetic enzymes are grouped into large modular complexes, which function in an assembly-line fashion (**figure 41.3**). Each module within the complex catalyzes a specific type of reaction. For instance, the anticancer agents epothilones C and D are synthesized by a combination of polyketide synthase (PKS) and nonribosomal peptide synthetase (NRPS) enzymes. The first module is responsible for binding a starter unit and transferring it to the first extension module. Each extension module modifies the intermediate and hands it off to the next module in line. The intermediate is modified by each module until it reaches the last module and is released. Combinatorial biosynthesis involves the mixing and matching of domains from hundreds of such modular multienzyme complexes. Many soil bacteria, such as members of the genera *Streptomyces* and *Bacillus*, are known to possess PKS and NRPS genes. Biotechnologists have only recently begun mining other microbes such as the myxobacteria for these genes.

The recognition that RNA is a versatile molecule that can have catalytic activity and control gene expression has given rise to a new class of therapeutic agents. Noting that RNA can fold into complex three-dimensional structures, molecular biologists tested the hypothesis that RNA molecules could be developed that would bind to and thereby alter the activity of specific cellular targets. To accomplish this, a protocol has been designed that uses in vitro evolution to generate a library of up to 10^{15} different RNA molecules of interest. This procedure is called "systematic evolution of ligands by exponential enrichment" (SELEX) and is outlined in **figure 41.4.** Recall that a molecule that binds to a specific target or receptor is called a ligand, so in this case, the goal is to design an RNA ligand that has biological activity because it fits very tightly into a cleft or groove in a specific target molecule. This class of engineered RNAs have been given the name **aptamers** (Latin *aptus*, means "fitting"). In 2005 the first aptamer-based therapy, for the progressive vision disease macular degeneration, began clinical use. ◀◀ *Ribozymes (section 9.8)*

Table 41.2 Some Directed Evolution Technologies

Technology	*Overview of Approach*	*Examples of Products Generated*
Genome-based strain reconstruction	A new strain is constructed based on the genotype of a production strain previously generated by "brute force." The new strain possesses only the mutations that are required for overproduction of the product.	Lysine
Metabolic pathway engineering	Rational approach to product improvement by either modification of specific biochemical reactions in a pathway, e.g., by site directed mutagenesis of biosynthetic enzyme genes; or the use of recombinant DNA technology to introduce new genes into the production strain. The goal is to optimize the flux of metabolites so the pathway operates most efficiently and the highest product yields possible are obtained.	Antibiotics: cephamycin C, neomycin, spiramycin, erythromycin L-lysine, aromatic amino acids, ethanol, vitamins
Systems biology	Genomewide and whole-cell measurements are used to identify multiple factors that influence product biosynthesis (and secretion, if relevant). The use of mathematical models is required to predict potential targets, that when modified, may increase product yield.	Levostatin
Assembly of designed oligonucleotides (ADO)	If the gene of interest contains a conserved sequence flanking the region to be mutated, oligonucleotides can be designed that include the conserved region and a mixture of point mutations in the region to be mutated. This combination of different (called degenerate) oligonucleotides is used to create variants by in vitro homologous recombination.	Lipases
Error-prone PCR	Variants of a gene are produced by using a DNA polymerase that is particularly error-prone during PCR.	Lycopene
DNA shuffling	Similar genes from different species are randomly fragmented and pooled, and new DNA fragments are generated by in vitro homologous recombination. Progeny sequences encoding desirable traits are identified; these new genes are then shuffled (bred) over and over again, creating new progeny that contain multiple desirable mutations.	Cephalosporinase
Whole genome shuffling	Similar to DNA shuffling described above, except entire genomes are recombined. Protoplast fusion may be used to obtain a single cell with the genomes of two microbial species so that recombination can occur.	Lactic acid

Modified from Adrio, J. L., and Demain, A. L. 2006; Genetic improvement of processes yielding microbial products. FEMS Microbiol. Rev. *30:187–214.*

All directed evolution methodologies are designed to produce new genes, proteins, or RNAs that will create hundreds, if not thousands, of variants. Sorting through so many new candidate molecules requires a specific assay that assesses the desired trait, such as stability under reducing conditions or increased ability to bind a specific molecule. Selecting the best variant could represent a tremendous workflow bottleneck; however, the development of **high-throughput screening (HTS)** enables the rapid selection of a subset of desirable molecules from tens of thousands of candidates. HTS employs a combination of robotics and computer analysis to screen samples (proteins, natural compounds, and whole cells) for a specific trait, usually in 96-well microtiter plates. The combination of directed evolution approaches and HTS has propelled industrial microbiology to a new level of exploration and efficiency not previously envisioned.

Metagenomics

As we have learned, techniques for growing most microbes found in nature have not yet been developed. While having a pure culture of the microorganism that produces an interesting compound is always the best approach, industrial microbiologists would be at a great disadvantage if they limited their search for products to only those encoded by genes found in microbes that can be grown in the laboratory. Indeed, this would exclude over 90% of microbial gene products. The example of antibiotics made by actinomycetes is instructive. Worldwide, the top 10 cm of many soils are estimated to contain as many as 10^{25} actinomycetes, but over the past 50 years, only about 10^7 have been screened for antibiotic production. At this rate, it would take another 128 years to screen the remaining actinomycetes, assuming they can be as easily cultured as the first 10^7. Industrial microbiologists have therefore turned to metagenomics to

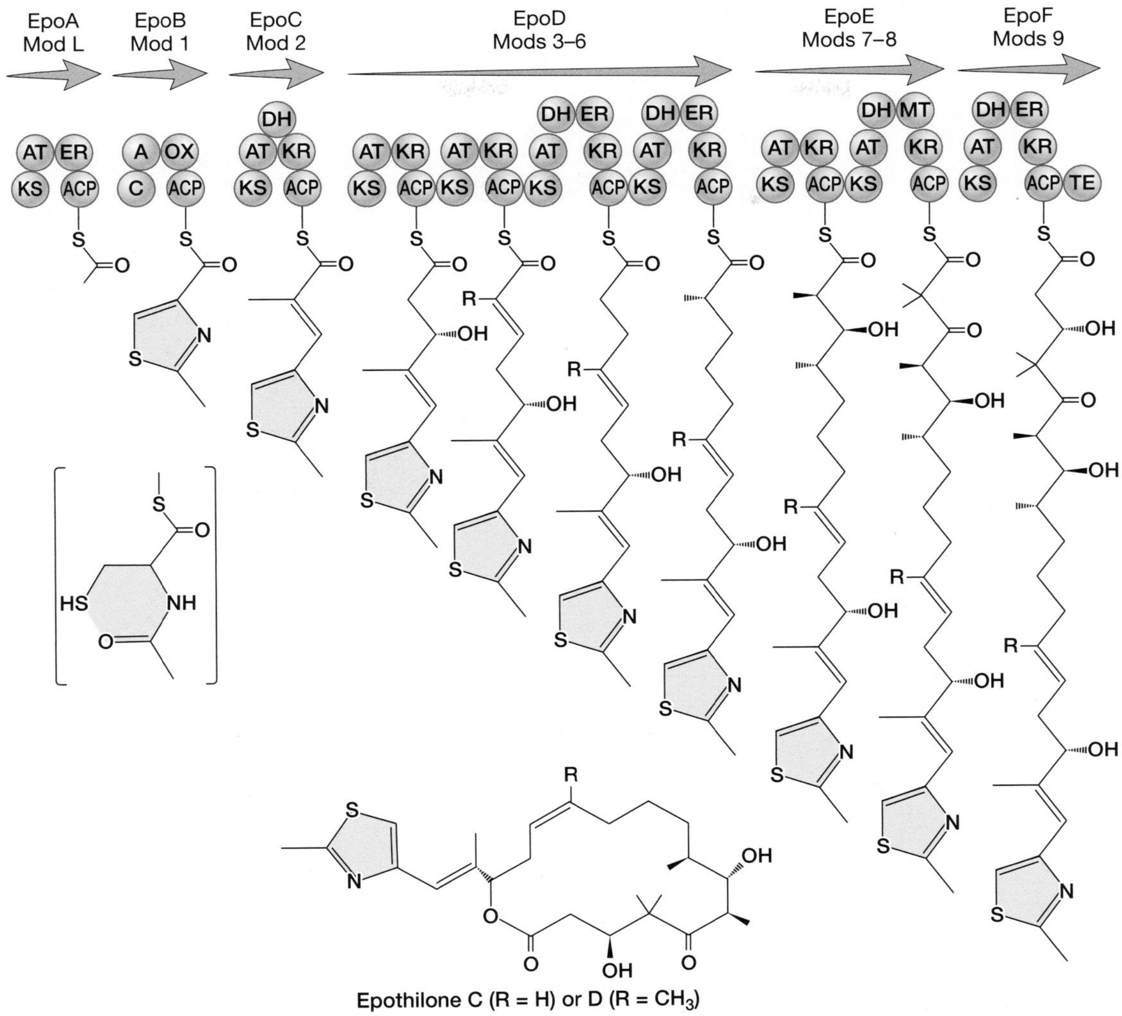

FIGURE 41.3 Modular Assembly of the Anticancer Agents Epothilone C and D. The genes encoding the enzymes that assemble this polyketide are found on the genome of the myxobacterium *Sorangium cellulosum.* They are arranged in the order of synthesis. The catalytic domains in the corresponding enzyme are shown as circles. These domains include keto-acyl synthase (KS), acyltransferase (AT), enoyl reductase (ER), acyl carrier protein (ACP), adenylation (A), condensation (C), oxidation (O), dehydratase (DH), ketoreductase (KR), and thioesterase (TE). The progression of synthesis of the molecule is shown, as is the mature structure.

Figure 41.3 Micro Inquiry

How are modular genes used in combinatorial biosynthesis?

sample the genetic diversity from a number of environments. Certain areas of the globe have been dubbed "biodiversity hot spots," and these are of particular interest when exploring genetic diversity. The process of exploring nature for new and potentially useful organisms and their products is called **bioprospecting.** ◀◀ *Metagenomics (section 16.8)*

Search This: Biodiversity hot spots/interactive map

Metagenomic libraries from natural samples are generally screened in two ways (*see figure 16.21*). First, all the cloned DNA from the environment can be sequenced and compared to genes that have already been characterized. In this way, new versions of known genes can be identified. Using this approach, the identification of a new gene depends upon its similarity to genes that are already known; so, many genes for which no function can be deduced will not be further pursued. But what if one of these

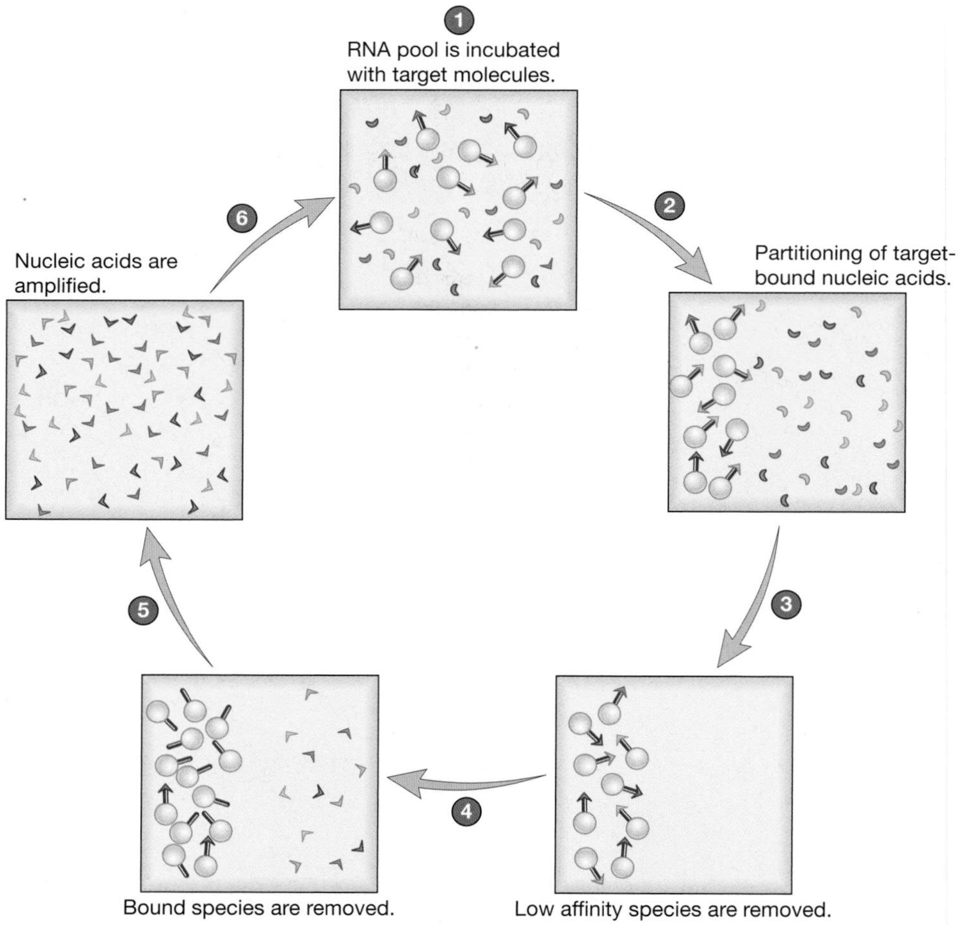
1
RNA pool is incubated with target molecules.
2
Partitioning of target-bound nucleic acids.
3
Low affinity species are removed.
4
Bound species are removed.
5
Nucleic acids are amplified.
6

FIGURE 41.4 Directed Evolution of RNA Molecules. Systematic evolution of ligands by exponential enrichment (SELEX) starts with (1) an assortment of RNA molecules incubated with target molecule. (2) RNAs that have bound the target are separated from those that have not (3) removal of RNAs that have not bound tightly to the target molecule (4) RNAs that have bound tightly to the target are removed from the target so that (5) they can be amplified. (6) This pool of tight-binding RNAs is now subject to repeated rounds of SELEX so that a molecule with the highest binding affinity can evolve.

***Figure 41.4** Micro Inquiry*

Why is SELEX considered an "enrichment" technique?

novel genes encodes a previously undiscovered protein with a highly desirable activity? A functional screen of the metagenomic library can be used to explore this possibility. In a functional screen, the genes are expressed in microbes such as *Escherichia coli* or *Bacillus subtilis*, which are then screened for the acquisition of a new, specific function. Of course, these two lines of inquiry are not mutually exclusive. This is demonstrated by the use of both approaches to analyze a termite's gut metagenome for new enzymes that might be useful in converting wood into biofuel. Keep in mind that once interesting genes are identified, they can undergo directed evolution techniques to optimize products for industrial use. ◀◀ *Microorganism-insect mutualisms (section 30.1)*

1. Why is the recovery of previously uncultured microorganisms from the environment an important goal?
2. Define protoplast fusion and list types of microorganisms used in this process.
3. What is combinatorial biology and what is the basic approach used in this technique? Besides antibiotics, what other types of biologically active molecules do you think might be created using combinatorial biology?
4. What types of recombinant DNA techniques are being used to modify gene expression in microorganisms?
5. What is high-throughput screening and why has it become so important?

41.2 Growing Microbes in Industrial Settings

Once a microbe has been identified and optimized for the biosynthesis of an important product, it must be grown using specifically designed media under carefully controlled conditions. The development of appropriate culture media and the growth of microorganisms under industrial conditions are the subjects of this section.

It is first necessary to clarify terminology. The term **fermentation,** used in a physiological sense in earlier sections of the book, is employed differently in the context of industrial microbiology and biotechnology. To industrial microbiologists, fermentation means the mass culture of microorganisms (or plant and animal cells). Industrial fermentations require the development of appropriate culture media and the transfer of small-scale technologies to a much larger scale. In fact, the success of an industrially important microbial product rests on the process of **scale-up.** This is when a procedure developed in a small flask is modified for use in a large fermenter. The microenvironment of the small culture must be maintained despite increases in culture volume. If a successful transition is made from a process originally developed in a 250-milliliter Erlenmeyer flask to a 100,000-liter reactor, then the process of scale-up has been carried out successfully.

Microorganisms are often grown in stirred fermenters or other mass culture systems. Stirred **fermenters** can range in size from 3 liters to 100,000 liters or larger, depending on production requirements. A typical industrial stirred fermentation unit is illustrated in **figure 41.5** (also see **chapter opener**). Not only must the medium be sterilized, but aeration, pH adjustment, sampling, and process monitoring must be carried out under rigorously controlled conditions. When required, foam control agents must be added, especially with high-protein media. Computers are used to monitor outputs from probes that determine microbial biomass, levels of critical metabolic products, pH, input and exhaust gas composition, and other parameters. Environmental conditions can be changed or held constant over time, depending on the requirements of the particular process.

Frequently a critical component in the medium, often the carbon source, is added continuously—a process called **continuous feed**—so that the microorganisms will not have excess substrate available at any given time. This is particularly important with glucose and other carbohydrates. If excess glucose is present at the beginning of a fermentation, it can be catabolized to yield ethanol, which is lost as a volatile product and reduces the final yield. This can occur even under oxic conditions.

Other culturing approaches exist, among them continuous culture techniques using chemostats. These can sometimes markedly improve cell outputs and rates of substrate use because microorganisms can be maintained in logarithmic phase. However, continuous active growth is undesirable in many industrial processes because the product of interest is often not made during exponential growth. Microbial products can be classified as primary and secondary metabolites. **Primary metabolites** consist of compounds related to the synthesis of microbial cells during balanced growth. They include amino acids, nucleotides, and fermentation end products such as ethanol and organic acids. In addition, industrially useful enzymes, either associated with the microbial cells or exoenzymes, often are synthesized by microorganisms during log phase growth. **Secondary metabolites** usually accumulate during the period of nutrient limitation or waste product accumulation that follows the active growth phase. These compounds have only a limited relationship to the synthesis of cell materials and normal growth. Most antibiotics and the mycotoxins fall into this category. ◀◀ *Growth curve (section 7.3); Continuous culture of microorganisms (section 7.5)*

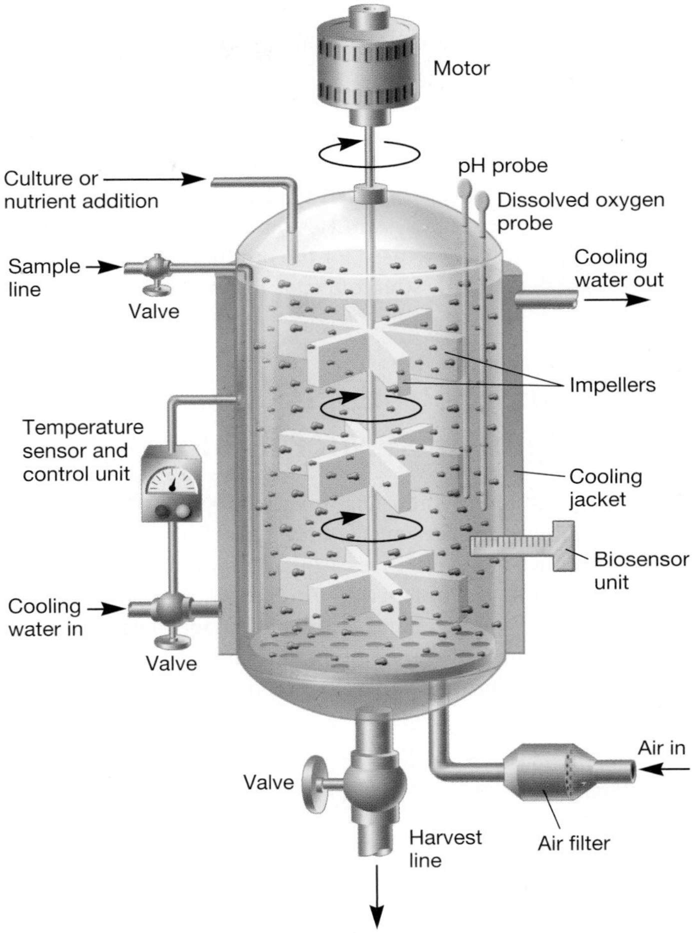

FIGURE 41.5 An Industrial Fermenter. This unit can be run under oxic or anoxic conditions, and nutrient additions, sampling, and fermentation monitoring can be carried out under aseptic conditions. Biosensors and infrared monitoring can provide real-time information on the course of the fermentation. Specific substrates, metabolic intermediates, and final products can be detected.

1. What is the objective of the scale-up process and why is it so critical?
2. What parameters can be monitored in a modern, large-scale industrial fermentation?
3. Which of the following do you think are secondary metabolites: pyruvate, a secreted amylase, erythromycin, leucine? What kind of system (fermenter versus chemostat) would most likely be used for the production of the streptomycete antibiotic streptomycin? Explain your answer.

41.3 Major Products of Industrial Microbiology

Products generated by industrial microbiology have profoundly changed our lives and lifespans. They include industrial and agricultural products, food additives, products for human and animal health, and biofuels (**table 41.3**). Antibiotics and other natural products used in medicine and health have made major contributions to the improved well-being of animal and human populations. Only major products in each category are discussed here.

Table 41.3 Major Microbial Products and Processes of Interest in Industrial Microbiology

Substances	***Microorganisms***
Industrial Products	
Ethanol (from glucose)	*Saccharomyces cerevisiae*
Ethanol (from lactose)	*Kluyveromyces fragilis*
Acetone and butanol	*Clostridium acetobutylicum*
2,3-butanediol	*Enterobacter, Serratia*
Enzymes	*Aspergillus, Bacillus, Mucor, Trichoderma*
Agricultural Products	
Gibberellins	*Gibberella fujikuroi*
Food Additives	
Amino acids (e.g., lysine)	*Corynebacterium glutamicum*
Organic acids (citric acid)	*Aspergillus niger*
Nucleotides	*Corynebacterium glutamicum*
Vitamins	*Ashbya, Eremothecium, Blakeslea*
Polysaccharides	*Xanthomonas*
Medical Products	
Antibiotics	*Penicillium, Streptomyces, Bacillus*
Alkaloids	*Claviceps purpurea*
Steroid transformations	*Rhizopus, Arthrobacter*
Insulin, human growth hormone, somatostatin, interferons	*Escherichia coli, Saccharomyces cerevisiae,* and others (recombinant DNA technology)
Biofuels	
Hydrogen	Photosynthetic microorganisms
Methane	*Methanobacterium*
Ethanol	*Zymomonas, Thermoanaerobacter*

Antibiotics

Many antibiotics are produced by microorganisms, predominantly by actinomycetes in the genus *Streptomyces* and by filamentous fungi. The synthesis of penicillin and its derivatives illustrates the critical role of medium formulation and environmental control in the production of these important compounds. Although penicillin is less widely used because of antibiotic resistance, the growth of *Penicillium chrysogenum* is an excellent example of a fermentation for which careful adjustment of the medium composition is used to achieve maximum yields. Provision of the slowly hydrolyzed disaccharide lactose, in combination with limited nitrogen availability, stimulates a greater accumulation of penicillin after growth has stopped (**figure 41.6**). The same result can be achieved by using a slow, continuous feed of glucose. If a particular penicillin is needed, the specific precursor is added to the medium. For example, phenylacetic acid is added to maximize production of penicillin G, which has a benzyl side chain *(see figure 34.5)*. The fermentation pH is maintained around neutrality by the addition of sterile alkali, which ensures maximum stability of the newly synthesized penicillin. Once the fermentation is completed, normally in 6 to 7 days, the broth is separated from the fungal mycelium and processed by absorption, precipitation, and crystallization to yield the final product. This basic product can then be modified by chemical procedures to yield a variety of semisynthetic penicillins. Manufacture of semisynthetic β-lactam antibiotics, such as ampicillin and methicillin, begins with natural penicillin. The β-lactam ring is preserved, but the side chain is modified to provide a broader spectrum of activity. ◀◀ *Inhibitors of cell wall synthesis (section 34.4)*

Amino Acids

Amino acids such as lysine and glutamic acid are used in the food industry as nutritional supplements in bread products and as flavor-enhancing compounds such as monosodium glutamate (MSG). Amino acid production is typically carried out by regulatory mutants, which have a reduced ability to limit synthesis of a specific amino acid or a key intermediate. Wild-type microorganisms avoid overproduction of biochemical intermediates by the careful regulation of cellular metabolism. Production of glutamic acid and several other amino acids is carried out using mutants of *Corynebacterium glutamicum* that

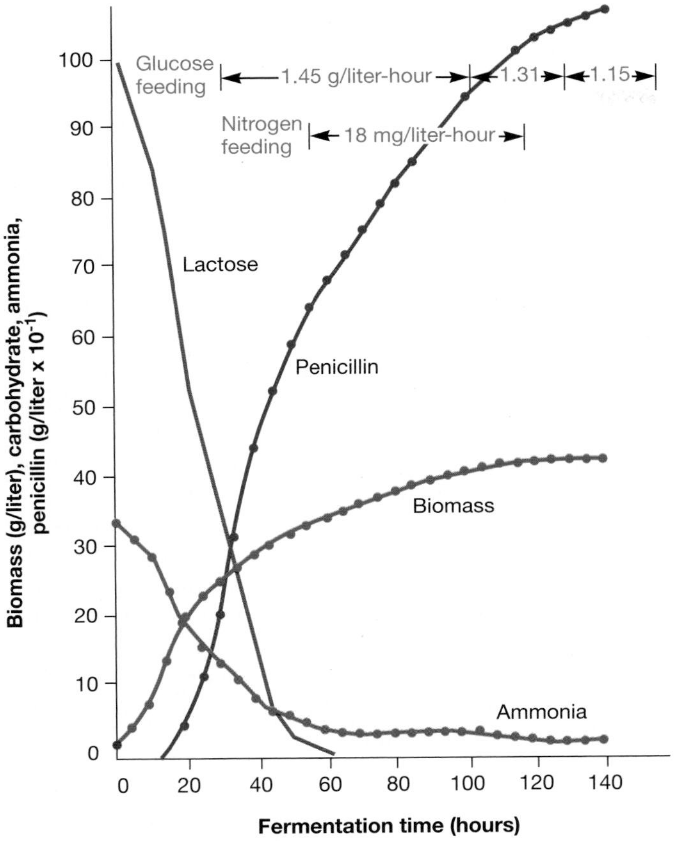

FIGURE 41.6 Penicillin Fermentation Involves Precise Control of Nutrients. The synthesis of penicillin begins when nitrogen from ammonia becomes limiting. After most of the lactose (a slowly catabolized disaccharide) has been degraded, glucose (a rapidly used monosaccharide) is added along with a low level of nitrogen. This stimulates maximum transformation of the carbon sources to penicillin. The scale factor is presented using the convention recommended by the American Society for Microbiology. That is, a number on the axis should be multiplied by 0.10 to obtain the true value.

Figure 41.6 Micro Inquiry

Why are the cells first fed lactose and then glucose?

lack or have only a limited ability to process the TCA cycle intermediate α-ketoglutarate to succinyl-CoA (*see appendix II*). A controlled low biotin level and the addition of fatty acid derivatives results in increased membrane permeability and excretion of high concentrations of glutamic acid. The impaired bacteria use the glyoxylate pathway to meet their needs for essential biochemical intermediates, especially during the growth phase (*see figure 11.28*). After growth becomes limited because of changed nutrient availability, an almost complete molar conversion (or 81.7% weight conversion) of isocitrate to glutamate occurs. ◀◀ *Synthesis of amino acids (section 11.5)*

Organic Acids

Citric, acetic, lactic, fumaric, and gluconic acids are major products used in food preparation as well as other industries. Until microbial processes were developed, the major source of citric acid was citrus fruit. Organic acid production by microorganisms is important in industrial microbiology and illustrates the effects of trace metal levels on organic acid synthesis and excretion.

Citric acid fermentation involves limiting the amounts of trace metals such as manganese and iron to stop *Aspergillus niger* growth at a specific point in the fermentation. The medium often is treated with ion exchange resins to ensure low and controlled concentrations of available metals and is carried out in oxygenated, stirred fermenters. Generally, high sugar concentrations (15 to 18%) are used, and copper has been found to counteract the inhibition of citric acid production by iron above 0.2 parts per million. The success of this fermentation depends on the regulation and functioning of the glycolytic pathway and the tricarboxylic acid cycle. After the active growth phase, when the substrate level is high, citrate synthase activity increases and the activities of aconitase and isocitrate dehydrogenase decrease. This results in citric acid accumulation and excretion by the stressed microorganism.

Biopolymers

Biopolymers are microbially produced polymers, primarily polysaccharides, used to modify the flow characteristics of liquids and to serve as gelling agents. These are employed in many areas of the pharmaceutical and food industries.

Biopolymers include (1) dextrans, which are used as blood expanders and absorbents; (2) *Erwinia* polysaccharides used in paints; (3) polyesters, derived from *Pseudomonas oleovorans*, which are used for specialty plastics; (4) cellulose microfibrils, produced by an *Acetobacter* strain, that serve as a food thickener; (5) polysaccharides such as scleroglucan used by the oil industry as drilling mud additives; and (6) xanthan polymers, which have a variety of applications as a food additive as well to enhance oil by thickening drilling mud. This use of xanthan gum, produced by *Xanthomonas campestris*, represents a large market for this microbial product.

Biosurfactants

Biosurfactants are amphiphilic molecules that possess both hydrophobic and hydrophilic regions; thus they partition at the interface between fluids that differ in polarity, such as oil and water. For this reason, they are used for emulsification, increasing detergency, wetting and phase dispersion, as well as solubilization. These properties are especially important in bioremediation, oil spill dispersion, and enhanced oil recovery. The most widely used microbially produced biosurfactants are glycolipids. These are carbohydrates that bear long-chain aliphatic acids or hydroxy aliphatic acids. They can be isolated as extracellular products from a variety of microorganisms, including pseudomonads and yeasts.

Many biosurfactants also have antibacterial and antifungal activity. The amphiphilic nature of these molecules promotes their ability to disrupt plasma membranes. Some biosurfactants even inactivate enveloped viruses. In addition, their capacity to prevent microbial adhesion to sites of invasion has generated interest in their use as potential protective agents. Although these and other medical applications of biosurfactants are still being investigated, these molecules hold promise for the future. They can be genetically engineered with relative ease, and the continued increase in resistance to antimicrobial agents has given rise to a tremendous need for new products.

Bioconversion Processes

Bioconversions, also known as **microbial transformations** or **biotransformations,** are minor changes in molecules, such as the insertion of a hydroxyl or keto function, or the saturation or desaturation of a complex cyclic structure, that are carried out by nongrowing microorganisms. The microorganisms thus act as biocatalysts. Bioconversions have many advantages over chemical procedures. A major advantage is stereochemical; the biologically active form of a product is made. Enzymes also carry out very specific reactions under mild conditions, and larger water-insoluble molecules can be transformed. Unicellular bacteria, actinomycetes, yeasts, and molds have been used in various bioconversions. The enzymes responsible for these conversions can be intracellular or extracellular. Cells can be produced in batch or continuous culture and then dried for direct use, or they can be prepared in more specific ways to carry out desired bioconversions.

A typical bioconversion is the hydroxylation of a steroid (**figure 41.7**). In this example, the water-insoluble steroid is dissolved in acetone and then added to the reaction system that contains the pregrown microbial cells. The course of the modification is monitored, and the final, soluble product is extracted from the medium and purified.

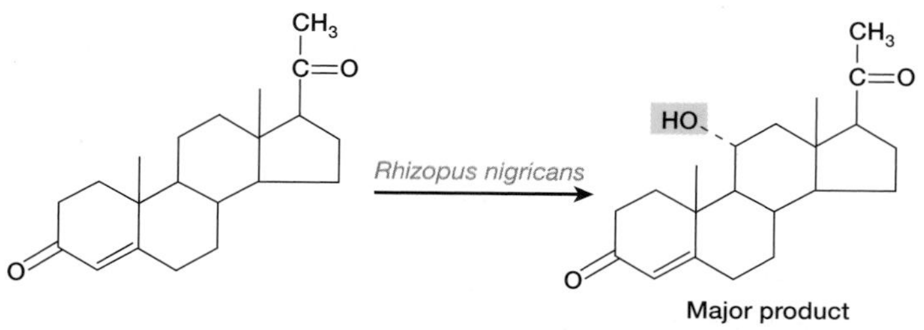

FIGURE 41.7 Biotransformation to Modify a Steroid. Hydroxylation of progesterone in the 11α position by *Rhizopus nigricans.* The steroid is dissolved in acetone before addition to the pregrown fungal culture.

Microbial Energy Conversion

Global energy consumption is predicted to increase by 71% between 2003 and 2030. Currently the world relies on petroleum to fuel almost all its transportation, and coal and natural gas (methane) are used in abundance to generate power and heat buildings. These fossil fuels need to be phased out of use, not only because they are finite resources but also because together they are the major contributors of greenhouse gases and global climate change. Many agree that renewable energy sources based on microbial transformation of organic material are the most viable approach to solve our global energy and climate crises. **Microbial energy conversion** includes the microbial transformation of organic materials into **biofuels**, such as ethanol and hydrogen, that can be burned to fuel cars or other machines. Microbial fuel cells are another form of microbial energy conversion. These are bioreactors that use microbes to change the chemical energy stored in organic matter directly into electricity. In this section, we discuss the microbial production of the biofuels ethanol, hydrogen, and methane, and the emerging field of microbial fuel cells. ◀◀ *Global climate change (section 26.2)*

Search This: Michaelbluejay/electricity

The U.S. Department of Energy has set a goal of substituting 30% of gasoline with biofuels by 2030. Currently ethanol is used as an additive to gasoline because it is easy to store and can be burned in cars designed to use pure gasoline. This is why gasoline sold in the United States contains at least 10% ethanol. Corn is currently the substrate most commonly used for generating ethanol. Its fermentation into ethanol involves the microbial degradation of plant starch using amylases and amyloglucosidases. The resulting sugars are then fermented to ethanol. However, the use of corn to make bioethanol has had unintended consequences, chiefly in the form of higher worldwide food prices. The use of crop residues could significantly boost biofuel yields and ease the cost of corn and corn-based foods. Crop residues are the plant material left in the field after harvest. This material consists of cellulose and hemicellulose—polymers of five different hexoses and pentoses: glucose, xylose, mannose, galactose, and arabinose. While no microorganism naturally ferments all five sugars, a *Saccharomyces cerevisiae* strain has been engineered to ferment xylose and an *E. coli* strain that expresses *Zymomonas mobilis* genes is able to ferment all these sugars. Another area of research focuses on degrading the cellulose and hemicellulose to release these monomers. This is commonly done by heating the plant material and treating it with acid, which is energy-intensive and corrosive. Genome analysis of the yeast *Pichia stipitis* strain Pignal uncovered the genes necessary for lignocellulose digestion and the ethanolic fermentation of xylose. Bioprospecting for enzymes from thermoacidophiles is also ongoing in an effort to replace the harsh thermochemical approach with a biological treatment.

Even if cheaper plant material were found for ethanol production, ethanol as a biofuel has several disadvantages. One is that it absorbs water and cannot be shipped through existing pipelines because they almost always contain some water. So once ethanol has been generated and then again after its shipping, it must be distilled to remove the water. Furthermore, ethanol contains far less energy than other fuels, including gasoline and higher molecular weight alcohols such as butanol, which are easier to generate. In fact, some analysts suggest that the entire ethanol manufacturing process may consume more energy than the product can yield; that is, ethanol is not a carbon-neutral fuel. This has motivated a number of small biotechnology companies to engineer microbes to produce other alcohols and even hydrocarbons. This is certainly a microbial industry with tremendous potential for growth.

In many ways, hydrogen (H_2) as a biofuel compares very favorably to ethanol and other fuels. In fact, H_2 has about three times more potential energy per unit weight than gasoline; it is thus the highest energy-content fuel available. However, until a liquid fuel with a high hydrogen content is developed, the storage and distribution of hydrogen will remain a problem. Also, unlike ethanol, hydrogen cannot simply be mixed with gasoline and used in today's vehicles. Nonetheless, because of its high energy yield, among other factors, there is great interest in developing H_2 as a fuel source, and the concept of a "hydrogen economy" has been promoted.

A diverse group of microbes produce H_2, thus making its manufacture quite flexible. Like ethanol, H_2 can be produced as a product of fermentation; it is also produced by oxygenic photosynthetic microbes (cyanobacteria and algae) and anoxygenic photoheterotrophs. Two different enzymes (thus processes) generate H_2: hydrogenase and nitrogenase. Both are keenly sensitive to oxygen, so H_2 production is an anaerobic process. The use of fermentative microbes as an industrial H_2 source has not been pursued, probably because only one-third of the electrons involved in fermentation wind up in H_2 (*see figure 20.29*). In addition, the oxidation of organic substrates yields a lot of CO_2, making fermentation contrary to the goals of clean energy production. Oxygenic photosynthetic microbes produce H_2 at night, when they oxidize cellular storage products such as poly-β-hydroxybutyrate. The use of aquatic phototrophs to generate H_2 holds promise because there is plenty of substrate (light and CO_2). However, hydrogenase is strongly inhibited by its product, so H_2 must be removed from the system as soon as it is created. By contrast, H_2 production by anoxygenic photoheterotrophic bacteria can occur via cellular hydrogenases or nitrogenases. It is the latter that has sparked interest among industrial microbiologists. This is because in an atmosphere that has been artificially purged of all N_2, nitrogenase will form only H_2. This process requires a tremendous amount of ATP and reductant, but unlike hydrogenase, nitrogenase is not inhibited by the accumulation of H_2, and electron conversion to H_2 reaches efficiencies near 100 percent. Mutants of the purple α-proteobacterium *Rhodopseudomonas palustris* have been isolated that are not subject to normal regulation, so that high levels of H_2 are produced constitutively, even in the presence of N_2 gas. Efforts are now directed at addressing the availability of the two substrates that might prove rate limiting. Specifically, ATP levels need to be increased by boosting photophosphorylation, and a more efficient means of providing electrons, perhaps as inorganic compounds, needs to be developed.

Methane (CH_4) is the major component of natural gas, which currently meets much of the global energy demand. Because natural gas is a fossil fuel, it has the same problems as petroleum: it is a finite resource and burning methane releases CO_2. Methane can be produced by archaea in a controlled bioreactor, called an **anaerobic digester.** When this is the case, the CH_4 is derived from recently consumed CO_2, so when burned, there is no net increase in atmospheric CO_2 (**figure 41.8;** *also see table 42.6*). Anaerobic digesters have long been a low-tech way of producing "biogas"—a mixture of CO_2 and CH_4 that can be burned much like natural gas to provide heat and power. A groundswell of interest in biogas is illustrated by its growth in United States (figure 41.8*b*). In 2000 only a negligible amount was generated; by 2007 farms using small bioreactors produced 2.2×10^8 kilowatt hours (kWh) of electricity (an average central air conditioner uses about 84 kWh per day). It should be no surprise that commercial microbial production of CH_4 now exists, with efficiencies of 50 to 100% of organic substrate transformed to CH_4. Regardless of whether the methane is produced in a farmyard or municipal bioreactor, it depends on a consortium of fermentative bacteria to produce the acetate, CO_2, and H_2 that methanogenic archaea convert to CH_4. ⏮ *Methanogens (section 18.3); Global climate change (section 26.2); Methane-based mutualisms (section 30.1);* ⏭ *Wastewater treatment processes (section 42.2)*

Search This: Anaerobic digester/mrec

Another area of exploration is the use of methanogenic consortia to convert crude oil to methane. Current oil recovery methods extract only about one-third of the oil from geological deposits, with the remainder left behind. It is now understood that small numbers of anaerobic microbes have been degrading hydrocarbons in oil fields for millennia. Research is underway to better understand this community of microbes in an effort to accelerate the transformation of inaccessible crude oil to methane. Along with methanogenic archaea, the γ-proteobacterium *Syntrophus aciditrophicus* has been isolated from oil fields and appears to be key for archaeal methane production.

The last form of energy production we discuss bypasses the need for biofuel production. Some microbes can oxidize organic material and pass the electrons directly to an electrode. A **microbial fuel cell** captures these electrons to generate electricity. To accomplish this, the microbes must be continuously fed a rich diet of organic substrates so that very little biosynthesis is required (recall that anabolic reactions require reducing power as well as ATP). The basic design of a microbial fuel cell

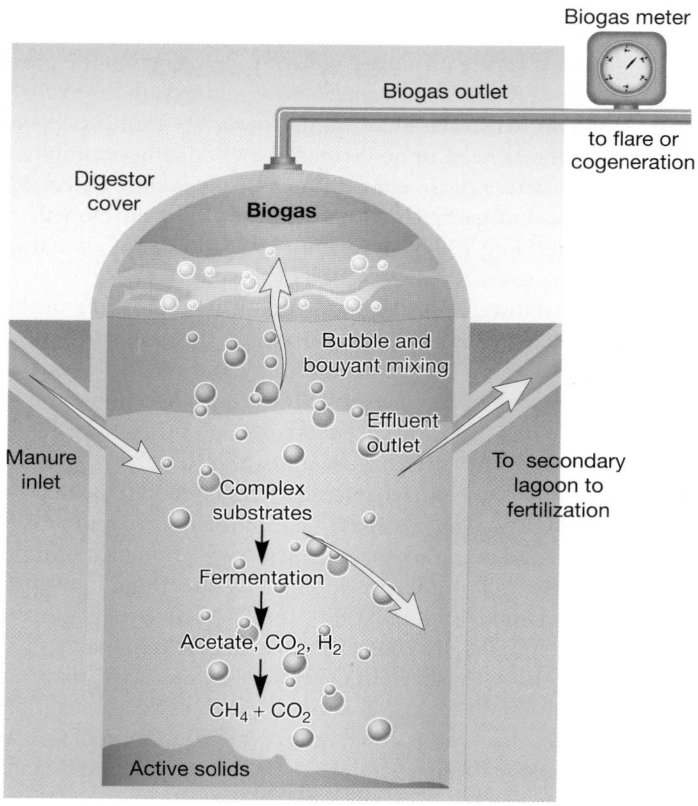

(a)

(b)

FIGURE 41.8 Anaerobic Digesters Generate Methane for Local or Municipal Use. (a) The basic design of a farm's anaerobic digester includes an inlet for the delivery of animal waste. Fermentation of complex substrates yields CO_2 and H_2, which archaea convert to CH_4. The effluent outlet functions to optimize the retention time (usually from 22 to 28 days) to maximize CH_4 capture. (b) Interest in the use of anaerobic digesters on farms has increased in recent years, as demonstrated by this advertisement for an anaerobic digester workshop sponsored by the U.S. Department of Agriculture.

is relatively simple (**figure 41.9**). It consists two chambers, one oxic and the other anoxic, divided by a membrane through which protons can pass. Biomass is delivered to the bacteria in the anoxic side. The oxidation of this organic matter yields protons and electrons. The protons diffuse across the membrane to the oxic chamber. The electrons are deposited on an anode that connects the two chambers. Electrons flow toward the cathode but can be used to generate electricity while in transit. A catalyst in the oxic chamber combines the protons, electrons, and oxygen to generate water.

Many different types of heterotrophic microbes can produce power in fuel cells. The only requirement is that they oxidize organic matter anaerobically. In fact, communities of microbes can be employed. The largest constraint to obtaining maximum efficiency appears to be delivery of electrons to the electrode. There are several ways this can be accomplished. Fermentation yields H_2, which is transferred to the anode, although somewhat inefficiently. Often a chemical mediator is used to shuttle electrons from inside the cell, across the cellular membrane, to the anode (figure 41.9). Still other microbes, such as the γ-proteobacterium *Shewanella oneidensis,* produce "nanowires" to transmit electrons to the anode (in nature, these nanowires shuttle electrons to external electron acceptors such as Fe^{3+}; *see figure 20.27*). Microbial fuel cell development is still ongoing, and great strides have been made to increase power output, which has increased orders of magnitude in just a decade. It is envisioned that the first application will most likely be in circumstances in which energy requirements are small but batteries cannot be used. Environmental instruments that need to be left in place for years at a time are good candidates. In addition, because microbial fuel cells are self-contained, they are attractive for developing countries that lack well-developed power grids. ⏮ *Order* Alteromonadales *(section 20.3); Order* Desulfuromonales *(section 20.4);* ⏭ *Biodegradation and bioremediation processes (section 42.3)*

Search This: Microbial fuel cells

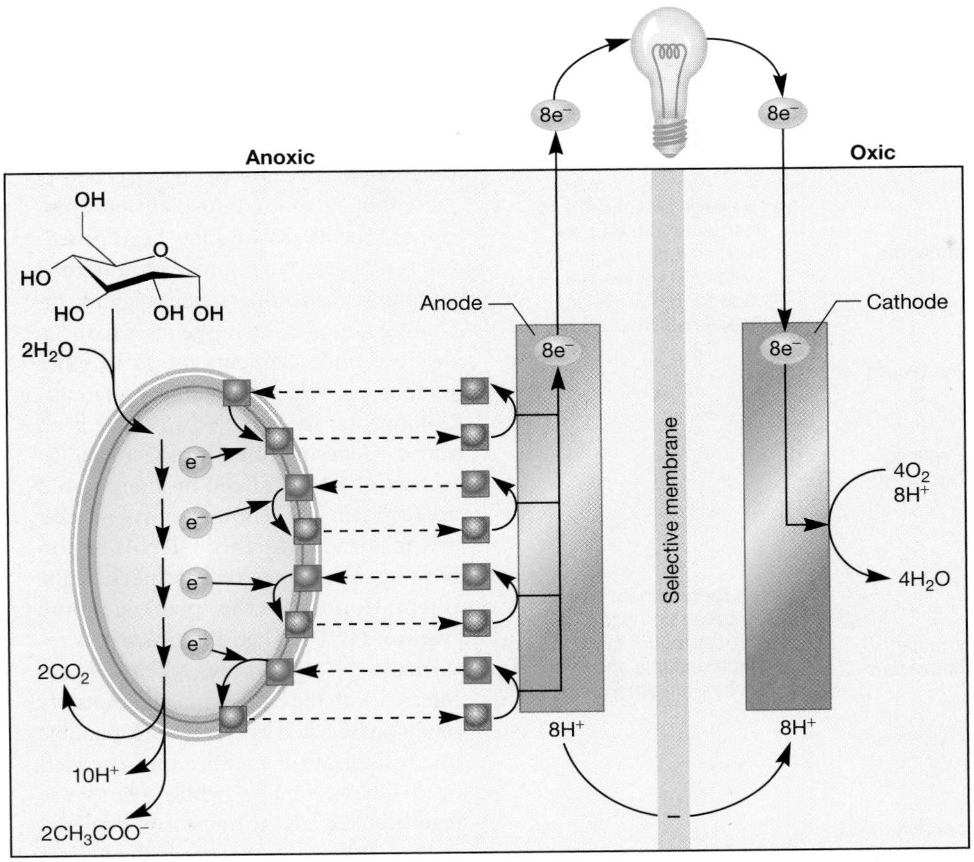

(a)

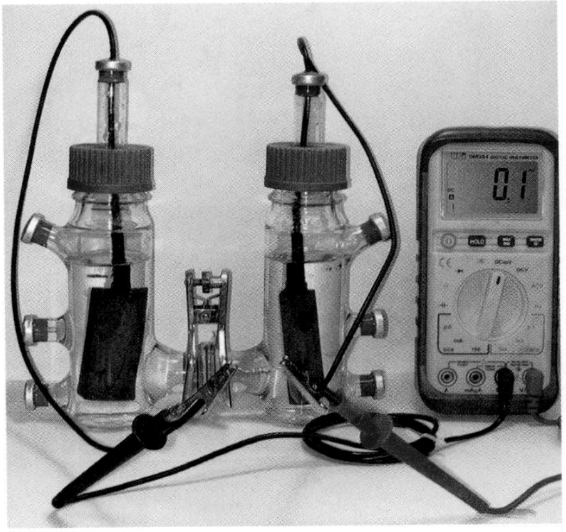

(b)

FIGURE 41.9 Microbial Fuel Cells. (a) Sugar is catabolized anaerobically by microbes in the anoxic chamber. The resulting protons flow to the oxic chamber, and the electrons are shuttled from the cell membrane to an anode, where they travel to the oxic chamber. During electron transit, they can be used to generate electricity. Green boxes represent oxidized electron shuttles and red boxes represent reduced shuttles. (b) A prototypical two-chambered microbial fuel cell.

Figure 41.9 Micro Inquiry

Why must the membrane between the two chambers be selectively permeable?

1. What critical limiting factors are used in the fermentation of penicillin? Why do you think the microbe must be stressed to produce high levels of drug?
2. What are regulatory mutants and how were they used to increase the production of glutamic acid by *Corynebacterium*?
3. What is the principal limitation created to stimulate citric acid accumulation by *Aspergillus niger*? In this case, is citric acid a primary or secondary metabolite?
4. Discuss the major uses for biopolymers and biosurfactants.
5. What are bioconversions or biotransformations? What kinds of molecular changes result from these processes?
6. What is an anaerobic digester? Where can anaerobic digesters most easily be used to meet energy needs?
7. How does a microbial fuel cell work? Why is it important to supply a high level of organic nutrients to the bacteria in the fuel cell?

41.4 Agricultural Biotechnology

A plasmid from the plant pathogenic bacterium *Agrobacterium tumefaciens* is chiefly responsible for the successful genetic engineering of plants. In natural situations, infection of dicotyledonous (dicot) plant cells by the bacterium transforms the cells into tumor cells, and crown gall disease develops. The tumor forms because of the insertion of bacterial genes into the plant cell genome, and only strains of *A. tumefaciens* possessing a large conjugative plasmid called the Ti (*t*umor *i*nducing) plasmid are pathogenic *(see figures 29.12 and 29.13)*. This is because the Ti plasmid transfers a region known as T-DNA to the plant cell. T-DNA is very similar to a transposon and is inserted into the chromosomes at various sites and stably maintained in the cell nucleus. ◀◀ *Transposable elements (section 14.5)*

When the molecular nature of crown gall disease was recognized, it became clear that the Ti plasmid and its T-DNA had great potential as a vector for the insertion of recombinant DNA into plant chromosomes. The Ti plasmid has been genetically engineered to improve its utility as a cloning vector. Nonessential regions, including tumor-inducing genes, have been deleted, while elements found in other vectors (e.g., selectable marker and multicloning site; *see figure 15.10*) have been added. Genes required for the actual infection of the plant cell by the plasmid are retained. The gene or genes of interest are inserted into the T-DNA region between the direct repeats needed for integration into the plant genome (**figure 41.10**). Then the plasmid is returned to *A. tumefaciens,* plant cells are infected with the bacterium, and transformants are selected by screening for antibiotic resistance (or another trait encoded by T-DNA). Finally, whole plants are regenerated from the transformed cells.

While the *A. tumefaciens* Ti plasmid can be used to modify dicots such as potato, tomato, celery, lettuce, and alfalfa, it cannot be transferred naturally to monocots such as corn, wheat, and other grains. However, genetic modification of the Ti plasmid so that it can be productively transferred to monocots is an active area of research. The creation of new procedures for inserting DNA into plant cells may well lead to the use of recombinant DNA techniques with many important crop plants.

Another microbial application to agriculture involves the use of bacteria, fungi, and viruses as **bioinsecticides** and **biopesticides.** These are defined as biological agents, i.e., microbes or their components, that can be used to kill susceptible insects. Bacterial agents include a variety of *Bacillus* species; however, *B. thuringiensis* is most widely used. This bacterium is only weakly toxic to insects as a vegetative cell, but during sporulation, it produces an intracellular protein toxin crystal, the parasporal body, that acts as a microbial insecticide for specific insect groups (*see figure 21.14*). The parasporal crystal, after exposure to

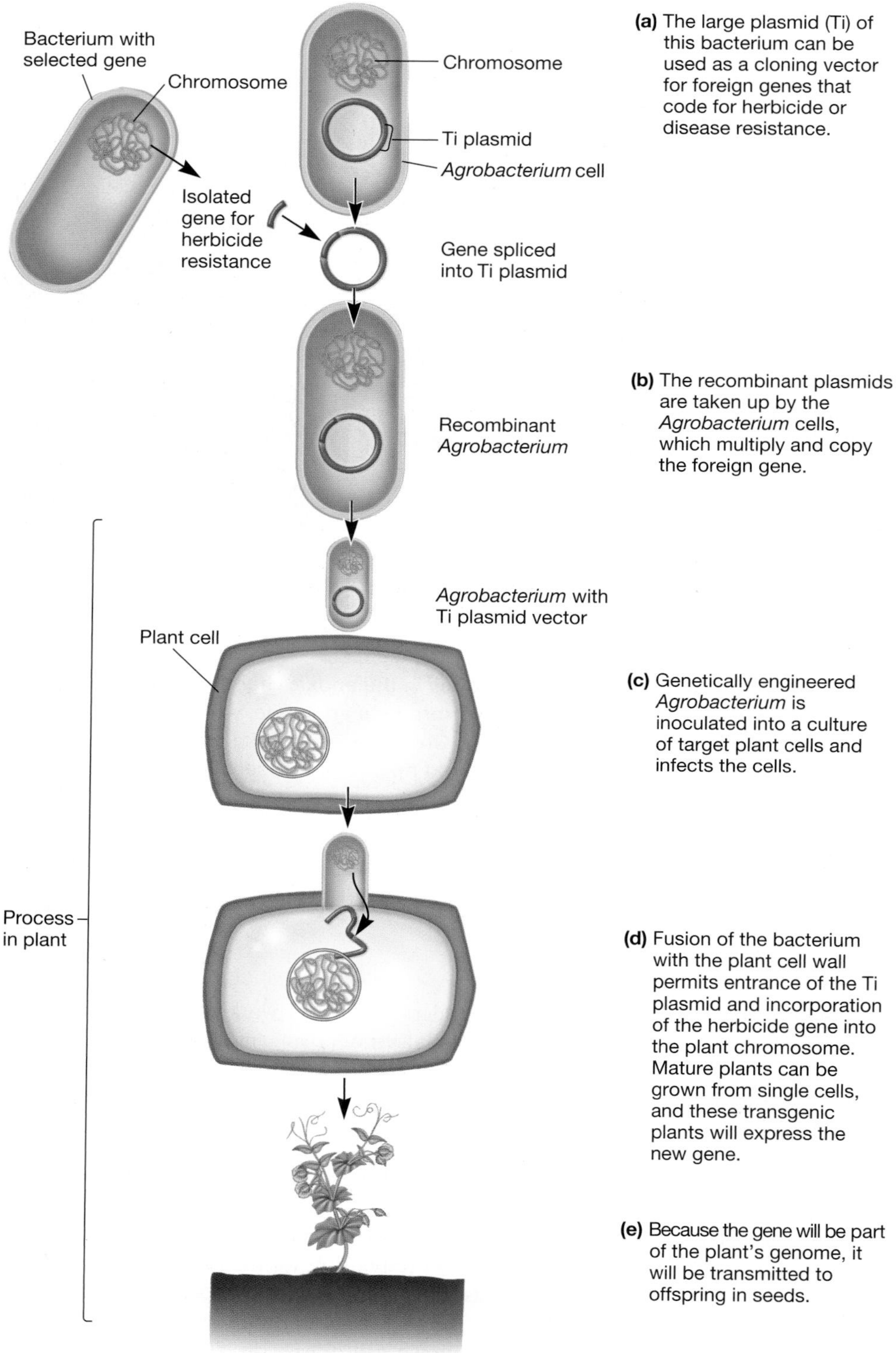

Bacterium with selected gene
Chromosome
Chromosome
Ti plasmid
Agrobacterium cell
Isolated gene for herbicide resistance
Gene spliced into Ti plasmid
Recombinant *Agrobacterium*
Agrobacterium with Ti plasmid vector
Plant cell
Process in plant
(a) The large plasmid (Ti) of this bacterium can be used as a cloning vector for foreign genes that code for herbicide or disease resistance.
(b) The recombinant plasmids are taken up by the *Agrobacterium* cells, which multiply and copy the foreign gene.
(c) Genetically engineered *Agrobacterium* is inoculated into a culture of target plant cells and infects the cells.
(d) Fusion of the bacterium with the plant cell wall permits entrance of the Ti plasmid and incorporation of the herbicide gene into the plant chromosome. Mature plants can be grown from single cells, and these transgenic plants will express the new gene.
(e) Because the gene will be part of the plant's genome, it will be transmitted to offspring in seeds.

FIGURE 41.10 Bioengineering of Plants. Most techniques employ a genetically modified strain of the natural tumor-producing bacterium called *Agrobacterium tumefaciens.*

alkaline conditions in the insect hindgut, fragments to release protoxin. After this reacts with a protease enzyme, the active toxin is produced. Six of the active toxin units integrate into the plasma membrane to form a hexagonal-shaped pore through the midgut cell, as shown in **figure 41.11**. This leads to the loss of osmotic balance and ATP, and finally to cell lysis. ◀◀ *Order* Bacillales *(section 21.3)*

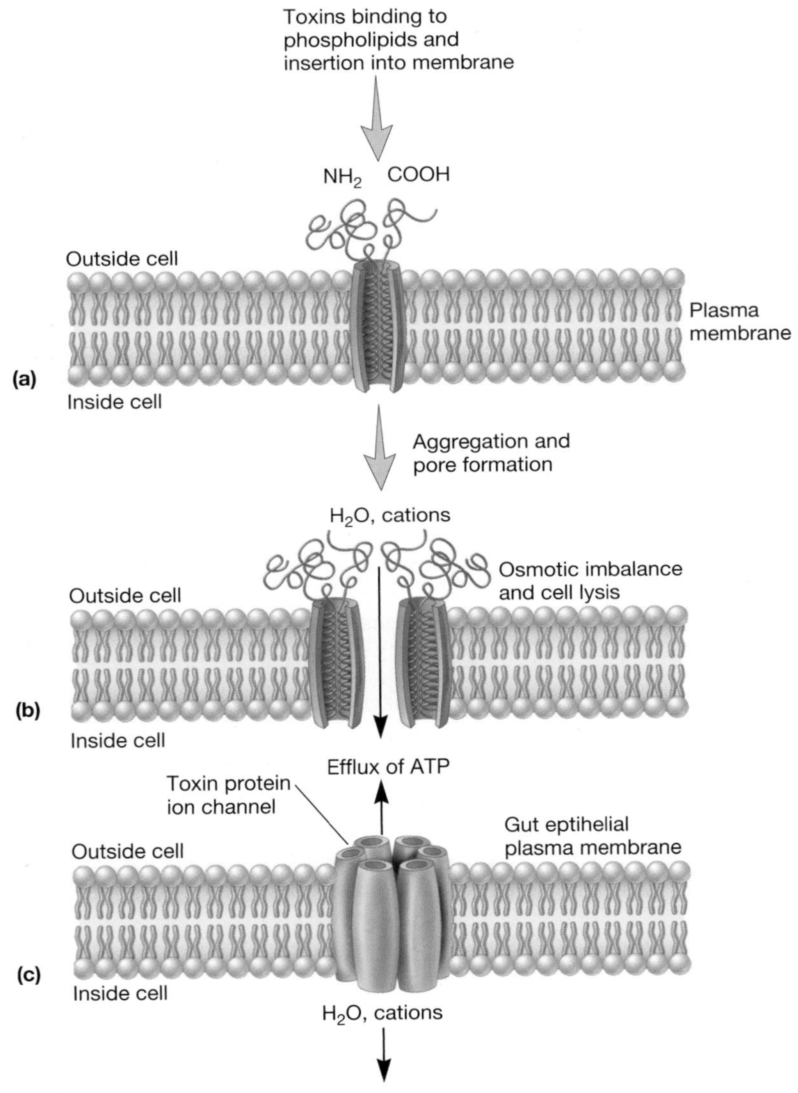

FIGURE 41.11 The Mode of Action of the *Bacillus thuringiensis* Toxin. (a) Insertion of the 68 kDa active toxin molecules into the membrane. (b) Aggregation and pore formation, showing a cross section of the pore. (c) Final hexagonal pore, which causes an influx of water and cations as well as a loss of ATP, resulting in cell imbalance and lysis.

Figure 41.11 Micro Inquiry

Why is it advantageous to have the Bt toxin inactive until it reaches the insect gut?

B. thuringiensis has become an important industrial microbe. The toxin is widely marketed as the insecticide **Bt,** and the toxin genes can be used to genetically modify plants. The manufacture of Bt begins by the growth of *B. thuringiensis* in fermenters. The spores and crystals are released into the medium when cells lyse. The medium is then centrifuged and converted to a dust or wettable powder for application to plants. This insecticide has been used on a worldwide basis for over 40 years. Unlike chemical insecticides, Bt does not accumulate in the soil or in nontarget animals. Rather, it is readily lost from the environment by microbial and abiotic degradation.

Genetically modified (GM) crops have become big business, having been sown in over 100 million hectares in 22 countries. Genes that confer resistance to herbicides and insects have been introduced into a variety of crops, including soybeans, cotton, and corn. Soybeans have also been engineered to have a lower saturated fat content. Other examples of genetically engineered commercial crops are canola, potato, squash, and tomato. Unlike other *g*enetically *m*odified *o*rganisms (GMOs), transgenic plants expressing the *B. thuringiensis* toxin gene, *cry* (for *cry*stal), have generally been well accepted. In 1996 commercialized Bt-corn, Bt-potato, and Bt-cotton were introduced into the United States, and soon farmers in other countries such as Australia, Canada, China, France, Indonesia, Mexico, Spain, and Ukraine followed. The widespread acceptance of these plants reflects the history of safe application of Bt as an insecticide without adverse environmental or health impacts. In addition, it is well understood that the Cry protein can only be activated in the target insect. Long-term studies have shown that Bt is nontoxic to mammals and is not an allergen in humans. One potential problem, the horizontal gene flow of the *cry* gene to weeds and other plants, has not been reliably demonstrated.

1. What is the Ti plasmid and how is it modified for the genetic modification of plants?
2. How is the Bt toxin produced and why is it so widely accepted?
3. Can you think of other traits that might be useful, either to the farmer or the consumer, that could be introduced into plants?

41.5 Microbes as Products

So far, we have discussed the use of microbial products to meet defined goals. However, microbial cells can be marketed as valuable products. Perhaps the most common example is the inoculation of legume seeds with rhizobia to ensure efficient nodulation and nitrogen fixation, as discussed in chapter 29. Here, we introduce several other microbes and microbial structures that are of industrial or agricultural relevance.

Diatoms have aroused the interest of nanotechnologists. These photosynthetic protists produce intricate silica shells that differ according to species (**figure 41.12**). Nanotechnologists are interested in diatoms because they create precise structures at the micrometer scale. Three-dimensional structures in nanotechnology are currently built plane by plane, and meticulous care must be taken to etch each individual structure to its final, exact shape. Diatoms, on the other hand, build directly in three dimensions and do so while growing exponentially. There have been a number of ideas and approaches to harness these microbial "factories," but one technique is especially fascinating. Diatoms shells are incubated at 900°C in an atmosphere of magnesium for several hours. Amazingly, this results in an atom-for-atom substitution of silicon with magnesium without loss of three-dimensional structure. Thus silicon oxide, which is of little use in nanotechnology, is converted to highly useful magnesium oxide. The diatom shells thus become attractive for the production of products that require miniaturized scales with a large surface area, such as gas and other sensing devices. ◀◀ Stramenopila *(section 23.5)*

Magnetotactic bacteria are also of interest to nanotechnologists. Magnetosomes are formed by certain bacteria that convert iron to magnetite. The sizes and shapes of the magnetosomes differ among species, but like diatom shells, they are perfectly formed despite the fact that they are only tens of nanometers in diameter. Although tiny, magnetic beads can be chemically synthesized, they are not as precisely formed and lack the membrane that surrounds magnetosomes. This membrane enables the attachment of useful biological molecules such as enzymes and antibodies. Potential applications for magnetosomes currently under investigation include their use as a contrast medium to improve magnetic resonance tomography (MRI) and as biological probes to detect cancer at early stages. ◀◀ *Inclusions (section 3.5)*

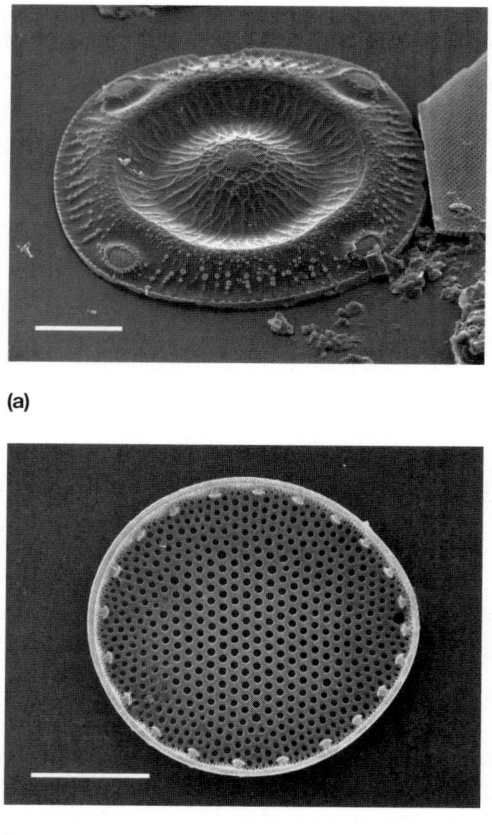

FIGURE 41.12 Marine Diatom Surface Features. (a) *Glyphodiscus stellatus;* scale bar is 20 μm. (b) *Roperia tesselata;* scale bar is 10 μm.

Another rapidly developing area of biotechnology is that of **biosensor** production. In this field of bioelectronics, living microorganisms (or their enzymes or organelles) are linked with electrodes, and biological reactions are converted into electrical currents (**figure 41.13**). Biosensors are used to measure specific components in beer, monitor pollutants, detect flavor compounds in food, and study environmental processes such as changes in biofilm concentration gradients. Using biosensors, it is possible to measure the concentration of substances from many different environments. Recently biosensors have been developed using immunochemical-based detection systems. These biosensors detect pathogens, herbicides, toxins, proteins, and DNA. Since the bioterrorism attacks of 2001, the U.S. government has stepped up funding for research and development of biosensors capable of detecting minute levels of potential airborne pathogens. Many of these biosensors are based on the use of a streptavidin-biotin recognition system (*see Techniques & Applications 15.1*).

Search This: Biosensors/ornl

1. Why are diatoms and magnetotactic bacteria of interest to nanotechnologists? What do you think might be special challenges when growing these microbes for industrial applications?
2. What are biosensors and how do they detect substances?
3. In what areas are biosensors being used to assist in chemical and biological monitoring efforts?

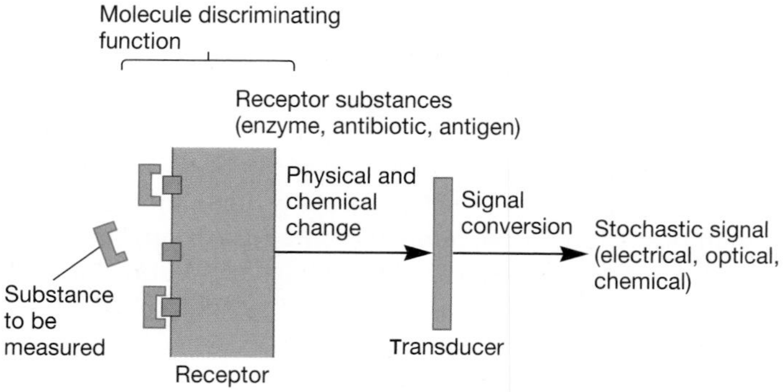

FIGURE 41.13 Biosensor Design. Biosensors are finding increasing applications in medicine, industrial microbiology, and environmental monitoring. In a biosensor, a biomolecule or whole microorganism carries out a biological reaction, and the reaction products are used to produce an electrical signal.

Summary

41.1 Microorganisms Used in Industrial Microbiology

a. Historically, microorganisms used in industry were isolated from nature and modified by the use of classic mutation techniques. Biotechnology involves the use of molecular techniques to modify and improve microorganisms.
b. Finding new microorganisms in nature for use in biotechnology is a continuing challenge. Only about 1% of the observable microbial community has been grown, but major advances in growing uncultured microbes are being made.
c. Selection and mutation continue to be important approaches for identifying new microorganisms (**figure 41.1**).
d. Site-directed mutagenesis and protein engineering are used to modify genes and their expression. These approaches are leading to new and often different products with new properties (**figures 41.2** and **41.3**).
e. Directed evolution includes a suite of techniques designed to target specific molecules for change with the goal of optimizing their function in a particularly useful way (**table 41.2** and **figure 41.4**).

41.2 Growing Microbes in Industrial Settings

a. Microorganisms can be grown in controlled environments of various types using fermenters and other culture systems (**figure 41.5**).
b. If defined constituents are used, growth parameters can be chosen and varied in the course of growing a microorganism. This approach is used particularly for the production of amino acids, organic acids, and antibiotics.

41.3 Major Products of Industrial Microbiology

a. A wide variety of compounds are produced in industrial microbiology that impact our lives in many ways. These include antibiotics, amino acids, organic acids, biopolymers, and biosurfactants.
b. Microorganisms also can be used as biocatalysts to carry out specific chemical reactions.
c. The field of microbial energy conversion has become extremely relevant. This includes the production of biofuels and the development of microbial fuel cells (**figures 41.8** and **41.9**).

41.4 Agricultural Biotechnology

a. Many plants are genetically modified through the use of the Ti plasmid of *Agrobacterium tumefaciens.* The plasmid can be engineered to introduce genes whose products confer new traits to plants (**figure 41.10**).
b. *Bacillus thuringiensis* is an important biopesticide, and the Bt gene has been incorporated into several important crop plants (**figure 41.11**).

41.5 Microbes as Products

a. Microorganisms are being used in a wide range of biotechnological applications such as nanotechnology and biosensors (**figures 41.12** and **41.13**).

Critical Thinking Questions

1. You have discovered a secondary metabolite made by an unusual fungus that has great industrial potential. Unfortunately, the fungus makes very little of this product, so you must develop a scheme to increase the yield. Discuss the pros and cons of each of the following strategies: (a) mutagenesis of the fungus followed by repeated passaging until a high-producing strain evolves, (b) heterologous expression of the genes and regulators responsible for producing this product in a bacterial host, (c) heterologous expression of the genes responsible for producing this product in a bacterial host but using a bacterial promoter that will drive constitutive expression of the product.

2. Brewer's yeast, *Saccharomyces cerevisiae,* does not metabolize lactose, but an engineered strain of *S. cerevisiae* able to consume lactose is of industrial interest. This is because such a strain can convert whey (an unusable by-product of cheese production) to ethanol. Such a strain has been engineered by cloning the genes that encode β-galactosidase and lactose permease from another yeast, *Kluyveromyces lactis,* into *S. cerevisiae.* However, the recombinant *S. cerevisiae* expresses the *K. lactis* genes only at a low level, so very little lactose is consumed. In an effort to increase the expression of these heterologous genes, a directed evolution approach was applied.

   Suggest one directed evolution technique that could be applied in vivo (on the entire organism) and one that could be used in vitro, that is, on the two genes from *K. lactis.* Which do you think would be easier to carry out? Which do you think would be more effective?

   Read the original paper: Guimaraes, P. M. R., et al. 2008. Adaptive evolution of a lactose-consuming *Saccharomyces cerevisiae* recombinant. *Appl. Environ. Microbiol.* 74:1748.

3. Because hydrogen holds great promise as a fuel, microbiologists are exploring a number of potential sources. *E. coli* has three hydrogenases that are involved in the fermentation of pyruvate to formate and ultimately to H_2 and CO_2 (*see figure 20.29*). By mutating the regulators of these hydrogenases, an *E. coli* strain was engineered that has a 141-fold increase in H_2 production from formate.

   Now that the regulators have been mutated, how might the structural genes be altered to increase H_2 production even further? If you were to apply a directed evolution technique, how would you select for the desired phenotype?

   Read the original paper: Maeda, T., et al. 2008. Metabolic engineering to enhance hydrogen production. *Microbial Biotechnol.* 1:30.

4. The detection of unexploded land mines is a horrific problem. Unexploded ordinances emit a small amount of nitroaromatic compounds, affording the opportunity to develop a biosensor. Recently a *Pseudomonas putida* strain was constructed that had a mutated toluene responsive regulator. When activated by bi- and trinitro-substituted toluenes, the regulator triggers the expression of genes transcriptionally fused to GFP (*see figure 15.15*). When spread on soil spotted with nitrotoluenes, there is sufficient activation of GFP-generated light that the location of the chemicals is evident.

   Draw a flow chart of how the signal is emitted, received, and transduced into a light signal that would enable you to explain this biosensor to military personnel.

   Read the original paper: Garmendia, J., et al. 2008. Tracing explosives in soil with transcriptional regulators for *Pseudomonas putida* evolved for responding to nitrotoluenes. *Microbial Biotechnol.* 1:236.

Concept Mapping Challenge

Provide your own linking words to construct a concept map using the following words:

Ethanol	Microbial fuel cell	Directed evolution
Methane	Aptamers	Fermentation
Antibiotics	Bioprospecting	Amino acids
Hydrogen	Primary metabolite	Production strain
Biofuels	Anaerobic digester	Secondary metabolite
Organic acids	Site direction mutagenesis	
Combinatorial biosynthesis		

Learn More

Learn more by visiting the text website at www.mhhe.com/willey8, where you will find a complete list of references.

42

Applied Environmental Microbiology

Biodegradation often can be facilitated by changing environmental conditions. Polychlorinated biphenyls (PCBs) are widespread industrial contaminants that accumulate in anoxic river muds. Although reductive dechlorination occurs under these conditions, oxygen is required to complete the degradation process. In this experiment, muds are being aerated to allow the final biodegradation steps to occur.

CHAPTER GLOSSARY

activated sludge Solid matter or sediment composed of actively growing microbes that participate in the aerobic portion of a biological sewage treatment process.

bioaugmentation Addition of pregrown microbial cultures to an environment to perform a specific task, for example, nitrogen fixation.

biochemical oxygen demand (BOD) The amount of oxygen used by organisms under certain standard conditions; it provides an index of the amount of microbially oxidizable organic matter present.

biodegradation The breakdown of a complex chemical through biological processes.

bioremediation The use of biologically mediated processes to remove pollutants from contaminated environments.

bulking sludges Sludges produced in sewage treatment that do not settle properly, usually due to the development of filamentous microorganisms.

chemical oxygen demand (COD) The amount of chemical oxidation required to convert organic matter in water and wastewater to CO_2.

cometabolism The modification of a compound not used for growth by a microorganism, which occurs in the presence of another organic material that serves as a carbon and energy source.

disinfection by-products (DBPs) Chlorinated organic compounds such as trihalomethanes formed during chlorine use for water disinfection. Many are carcinogens.

indicator organism An organism whose presence suggests the quality of a substance or environment, for example, the potential presence of pathogens.

maximum containment level goal (MCLG) The number of pathogens that can be present in a water supply that will prevent adverse health effects and provide a margin of safety.

membrane filter technique The use of a thin, porous filter to collect microorganisms from water, air, and food.

most probable number (MPN) The statistical estimation of the probable population in a liquid by diluting and determining end points for microbial growth.

primary treatment The first step of sewage treatment, in which physical settling and screening are used to remove particulate materials.

reductive dehalogenation The cleavage of carbon-halogen bonds by anaerobic bacteria.

rhizoremediation The detoxification of soil contaminants by microbes that are stimulated by the release of nutrients, oxygen, and other factors by plant roots.

secondary treatment The biological degradation of dissolved organic matter in the process of sewage treatment.

septic tank A tank used to process home sewage. Solid material settles out and is partially degraded by anaerobic bacteria as sewage slowly flows through the tank. The outflow is further treated or dispersed in oxic soil.

settling basin A basin used during water purification to chemically precipitate fine particles, microorganisms, and organic material by coagulation or flocculation.

slow sand filter A bed of sand through which water slowly flows; the gelatinous microbial layer on the sand grain surfaces removes waterborne microorganisms.

sludge A general term for the precipitated solid matter produced during water and sewage treatment.

tertiary treatment The removal from sewage of inorganic nutrients, heavy metals, viruses, and so forth by chemical and biological means after microbes have degraded dissolved organic material during secondary sewage treatment.

wastewater treatment The use of physical and biological processes to remove particulate and dissolved material from sewage and to control pathogens.

In this final chapter, we use the term "applied environmental microbiology" to refer to the use of microbes in their natural environment to perform processes useful to humankind. Such processes include wastewater treatment and bioremediation. In developed countries, the processes and products of applied microbiology are taken for granted, but this is not true globally. For instance, clean drinking water and sanitary treatment of contaminated water are beyond reach for an alarmingly high number of people. According to the World Health Organization, over 1 billion people worldwide do not have access to safe, drinkable water, and about 40% of the world's population lacks basic sanitation.

We begin this chapter by presenting ways in which water can be purified so that it can be consumed without fear of disease transmission. We then describe several approaches to treat wastewater to keep our rivers, streams, lakes, and groundwater—often the source of drinking water—clean. The remainder of the chapter focuses on the use of microbes as tools to clean up toxic chemicals. Many concepts presented throughout the text are integrated in these important subjects.

42.1 Water Purification and Sanitary Analysis

Many important human pathogens are maintained in association with living organisms other than humans, including many wild animals and birds. The U.S. Environmental Protection Agency (EPA) has designated certain bacterial, viral, and protozoan pathogens that can survive (or thrive) in water as "microbial containment candidates" because they represent particularly severe health risks (**table 42.1**). When waters are used for recreation or are a source of food, the possibility for disease transmission exists. In many cases, such waters are a source of drinking water.

Water purification is a critical link in controlling disease transmission by water. The purification method depends on the volume, the source, and the initial quality of water. Many municipalities draw their water from surface sources, such as reservoirs. Surface water is usually purified by a process that consists of at least three or four steps (**figure 42.1**). It is often helpful to hold the water for a specified duration to allow larger particles to settle out in a **sedimentation basin.** The partially clarified water is then moved to a **settling basin** and mixed with chemicals such as alum (aluminium sulfate) and lime to facilitate further precipitation. Sometimes coagulants are also added in the initial sedimentation phase. This procedure is called **coagulation** or flocculation and removes some microorganisms, organic matter, toxic contaminants, and suspended fine particles. The coagulated particles are called **flocs.** The water is further purified by passing it through **rapid sand filters.** Here, sand of roughly 1 mm in diameter physically traps fine particles and flocs and removes up to

Table 42.1 Microbial Water Contaminant Candidates Listed by the EPA

Organism	*Concern*	*Cross-referenced Pages*
Caliciviruses	Viruses that cause relatively mild self-limiting nausea, vomiting, diarrhea. Includes *Norovirus*.	922
Campylobacter jejuni	Causes nausea, vomiting, and diarrhea; usually self-limiting	964
Entamoeba histolytica	Protozoan causes short-term as well as long-lasting gastrointestinal (GI) illness.	999
E. coli 0157:H7	Enterohemorrhagic *E. coli* that causes severe GI illness; can also cause kidney failure	969
Helicobacter pylori	Bacterium that causes gastric ulcers and cancer	951
Hepatitis A virus	Causes liver disease leading to jaundice	923
Legionella pneumophila	Found in hot water systems; causes bacterial pneumonia when inhaled. Usually limited to elderly patients.	935
Naegleria fowleri	Protozoan found in warm, surface and groundwater; causes primary ameobic meningoencephalitis	1000
Salmonella enterica	Causes GI illness; severity depends on subspecies.	971
Shigella sonnei	Causes GI illness, including bloody diarrhea	972
Vibrio cholerae	Causes cholera	965

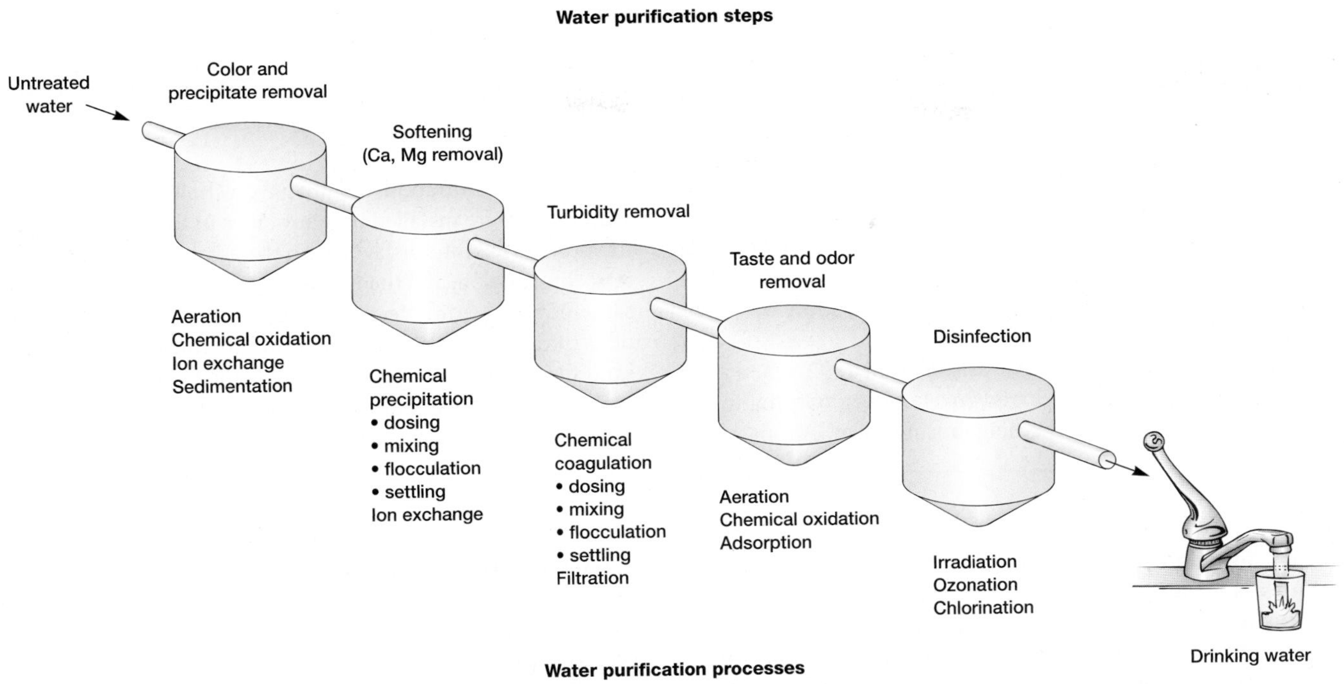

FIGURE 42.1 Water Purification. Several alternatives can be used for drinking water treatment, depending on the initial water quality. A major concern is disinfection: chlorination can lead to the formation of *d*isinfection *b*y-*p*roducts (DBPs), including potentially carcinogenic *tri*h*alom*ethanes (THMs).

Figure 42.1 Micro Inquiry

What is the difference between chemical precipitation and chemical coagulation?

99% of the bacteria. After filtration, the water is disinfected. This step usually involves chlorination, but ozonation is becoming increasingly popular. When chlorination is employed, the chlorine dose must be large enough to leave residual free chlorine at a concentration of 0.2 to 2.0 milligrams per liter. A concern is the creation of **disinfection by-products (DBPs)** such as trihalomethanes (THMs), formed when chlorine reacts with organic matter. Because some DBPs are carcinogenic, the EPA regulates the concentrations of specific DBPs permissible in municipal water systems.

These purification processes remove or inactivate disease-causing bacteria and indicator organisms (coliforms). Unfortunately the use of coagulants, rapid filtration, and chemical disinfection often does not remove *Giardia* cysts, *Cryptosporidium* oocysts, *Cyclospora*, and viruses. *Giardia*, a cause of human diarrhea, is now recognized as the most commonly identified waterborne pathogen in the United States (*see figure 23.2*). *Cryptosporidium* is also a significant problem. This protist is smaller than *Giardia* and is even more difficult to remove from water, in part because it is resistant to chlorine and other disinfectants. A major source of *Giardia* and *Cryptosporidium* contamination is the Canada goose, a migratory bird with an expanding population. Water supplies that receive terrestrial runoff are therefore at particular risk of contamination by these protists.

In the United States, the EPA has developed regulations called the Long Term 2 Enhanced Surface Water Treatment Rule (LT2 rule) as a means to ensure protection against these pathogens. The LT2 rule applies to all public water systems that draw from surface waters or groundwater that is fed by surface waters. It sets a **maximum containment level goal (MCLG)** for specific pathogens. The EPA defines MCLGs as "health goals set at a level at which no known or anticipated adverse effects on health of persons occur and which allows an adequate margin on safety." MCLGs are set at zero for *Giardia, Legionella,* viruses, and *Cryptosporidium.* In addition, the Total Coliform Rule, which applies to all public water systems regardless of supply source, sets an MCLG of zero for total and fecal coliform bacteria. To meet these MCLGs, water systems found to be contaminated must develop additional water treatment to provide 99.9% inactivation of the microbes. The precise treatment depends on the level of contamination but generally involves filtration techniques such as **slow sand filters.** This treatment involves the slow passage of water through a bed of sand in which a microbial layer (biofilm) covers the surface of each sand grain. Waterborne microorganisms are

TECHNIQUES & APPLICATIONS

42.1 Waterborne Diseases, Water Supplies, and Slow Sand Filtration

Slow sand filtration, in which drinking water is passed through a sand filter that develops a layer of microorganisms on its surface, has had a long and interesting history. After London's severe cholera epidemic of 1849, Parliament required that the entire water supply of London be passed through slow sand filters before use.

The value of this process was shown in 1892, when a major cholera epidemic occurred in Hamburg, Germany, and 10,000 lives were lost. The neighboring town, Altona, which used slow sand filtration, did not have a cholera epidemic. Slow sand filters were installed in many cities in the early 1900s, but the process fell into disfavor with the advent of rapid sand filters, chlorination, and the use of coagulants such as alum. Slow sand filtration, a time-tested process, is regaining favor because of its filtration effectiveness and lower maintenance costs. Slow sand filtration is particularly effective for the removal of *Giardia* cysts. For this reason, slow sand filtration is used in many mountain communities where *Giardia* is a problem.

removed by adhesion to the gelatinous surface microbial layer (**Techniques & Applications 42.1**). Coagulation and filtration reduce virus levels about 90 to 99%. Further inactivation of viruses can be achieved by chemical oxidants, high pH, and photo-oxidation; the EPA calls for a 99.99% reduction of viruses. ◀◀ *Biofilms (section 7.7)*

In most rural communities, as well as some larger regions, such as Cape Cod, Massachusetts, groundwater from wells and springs is the source of drinking (potable) water. This water is typically of high quality and requires aeration, rapid sand filtration, and disinfection. In areas such as these, it is especially important that groundwater quality is protected (p. 1069).

Search This: EPA safewater

1. What steps are usually taken to purify drinking water?
2. Why is chlorination, although beneficial in terms of bacterial pathogen control, of environmental concern?
3. How can *Giardia* and *Cryptosporidium* be removed from a water supply?

Sanitary Analysis of Waters

Once water has been purified, it must be deemed potable. Monitoring and detecting indicator and disease-causing microorganisms are major parts of sanitation microbiology. Bacteria from the intestinal tract generally do not survive in aquatic environments or are under physiological stress and gradually lose their ability to form colonies on differential and selective media. Procedures have been developed to attempt to "resuscitate" these stressed coliforms using selective and differential media.

Although a wide range of viral, bacterial, and protozoan diseases result from the contamination of water with human and other animal fecal wastes (table 42.1), the detection of **indicator organisms** as an index of possible water contamination by human pathogens has long been the standard approach to monitoring drinking water safety. Researchers are still searching for the "ideal" indicator organism to use in sanitary microbiology. Among the suggested criteria for such an indicator microbe are:

1. The indicator bacterium should be suitable for the analysis of all types of water: tap, river, ground, impounded, recreational, estuary, sea, and waste.
2. It should be present whenever enteric pathogens are present.
3. It should survive longer than the hardiest enteric pathogen.
4. It should not reproduce in the contaminated water, thereby producing an inflated value.
5. It should be harmless to humans.
6. Its level in contaminated water should have some direct relationship to the degree of fecal pollution.
7. The assay procedure for the indicator should have great specificity; in other words, other bacteria should not give positive results. In addition, the procedure should have high sensitivity and detect low levels of the indicator.
8. The testing method should be easy to perform.

Coliforms, including *Escherichia coli,* are members of the family *Enterobacteriaceae.* These bacteria make up about 10% of the intestinal microorganisms of humans and other animals, and have found widespread use as indicator organisms. They lose viability in freshwater at slower rates than most of the major intestinal bacterial pathogens. When such "foreign" enteric indicator bacteria are not detectable in a specific volume (generally 100 milliliters) of water, the water is considered potable (Latin *potabilis,* fit to drink). ◀◀ *Order* Enterobacteriales *(section 20.3)*

The coliform group includes *E. coli, Enterobacter aerogenes,* and *Klebsiella pneumoniae.* Coliforms are defined as facultatively anaerobic, gram-negative, nonsporing, rod-shaped bacteria that ferment lactose with gas formation within 48 hours at 35°C. The original test for coliforms involved the presumptive, confirmed, and completed tests, as shown in **figure 42.2**. The presumptive

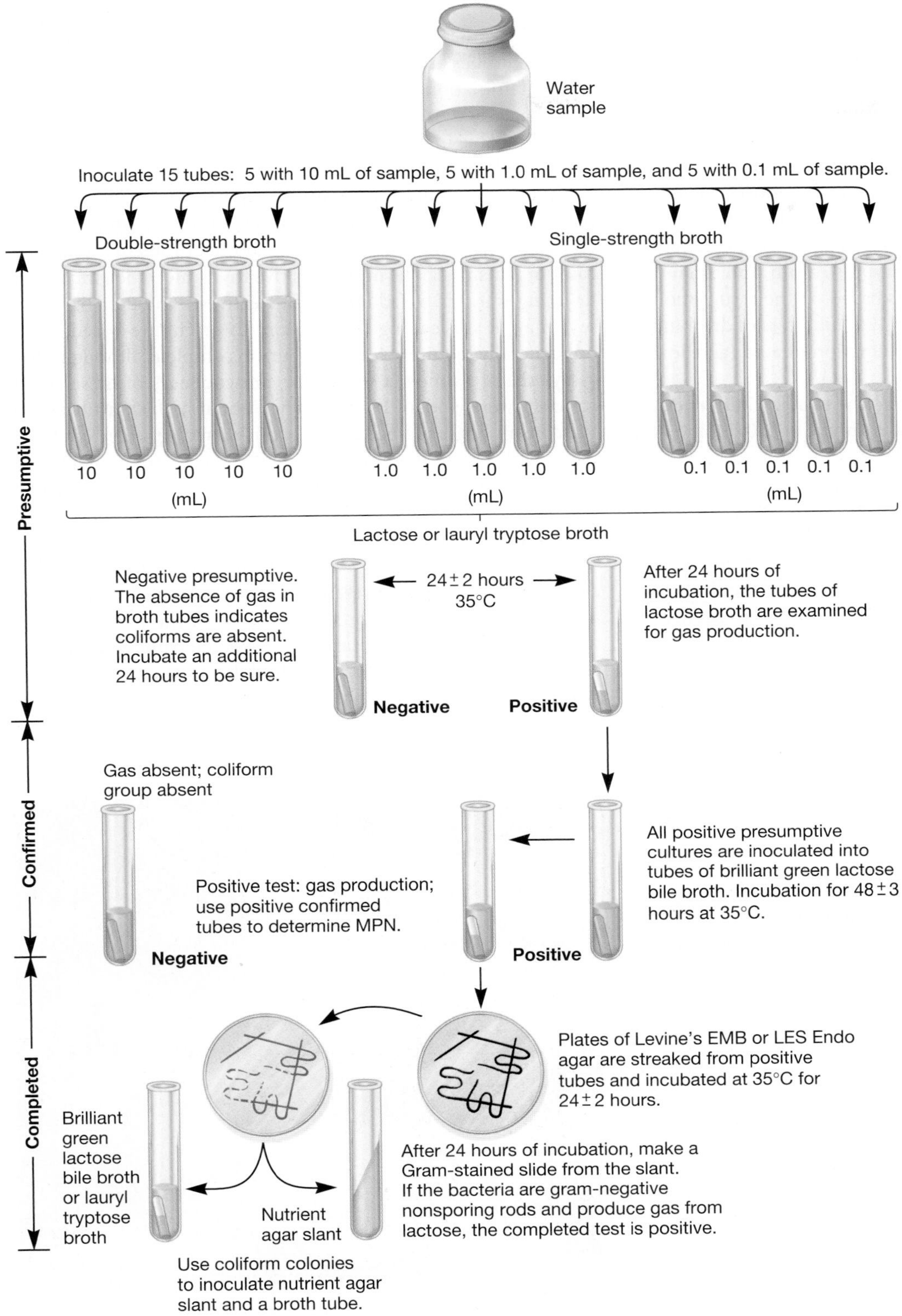
Water sample
Inoculate 15 tubes: 5 with 10 mL of sample, 5 with 1.0 mL of sample, and 5 with 0.1 mL of sample.
Double-strength broth
Single-strength broth
10 10 10 10 10
(mL)
1.0 1.0 1.0 1.0 1.0
(mL)
0.1 0.1 0.1 0.1 0.1
(mL)
Lactose or lauryl tryptose broth
Presumptive
Negative presumptive. The absence of gas in broth tubes indicates coliforms are absent. Incubate an additional 24 hours to be sure.
24±2 hours 35°C
After 24 hours of incubation, the tubes of lactose broth are examined for gas production.
Negative
Positive
Confirmed
Gas absent; coliform group absent
Positive test: gas production; use positive confirmed tubes to determine MPN.
All positive presumptive cultures are inoculated into tubes of brilliant green lactose bile broth. Incubation for 48±3 hours at 35°C.
Negative
Positive
Completed
Plates of Levine's EMB or LES Endo agar are streaked from positive tubes and incubated at 35°C for 24±2 hours.
Brilliant green lactose bile broth or lauryl tryptose broth
Nutrient agar slant
After 24 hours of incubation, make a Gram-stained slide from the slant. If the bacteria are gram-negative nonsporing rods and produce gas from lactose, the completed test is positive.
Use coliform colonies to inoculate nutrient agar slant and a broth tube.

FIGURE 42.2 The Multiple-Tube Fermentation Test. The multiple-tube fermentation technique has been used for many years for the sanitary analysis of water. Lactose broth tubes are inoculated with different water volumes in the presumptive test. Tubes that are positive for gas production are inoculated into brilliant green lactose bile broth in the confirmed test, and positive tubes are used to calculate the most probable number (MPN) value. The completed test is used to establish that coliform bacteria are present.

step is carried out by means of tubes inoculated with three different sample volumes to give an estimate of the ***most probable number*** **(MPN)** of coliforms in the water (*see figure 27.3*). The complete process, including the confirmed and completed tests, requires at least 4 days of incubations and transfers.

The coliforms include a wide range of bacteria whose primary source may not be the intestinal tract. To address this, tests have been developed that assay for the presence of **fecal coliforms.** These are coliforms derived from the intestine of warm-blooded animals, which grow at the more restrictive temperature of 44.5°C. To test for coliforms and fecal coliforms, and more effectively recover stressed coliforms, a variety of simpler and more specific tests has been developed.

The **membrane filter technique** has become a common and preferred method of evaluating the microbiological characteristics of water. The water sample is passed through a membrane filter. The filter with its trapped bacteria is transferred to the surface of a solid medium or to an absorptive pad containing the desired liquid medium. Use of the proper medium enables the rapid detection of total coliforms, fecal coliforms, or fecal enterococci (p. 1056) by the presence of their characteristic colonies. Samples can be placed on a less selective resuscitation medium or incubated at a less stressful temperature prior to growth under the final set of selective conditions. An example of a resuscitation step is the use of a 2-hour incubation on a pad soaked with lauryl sulfate broth, as is carried out in the LES Endo procedure. A resuscitation step often is needed with chlorinated samples, where the microorganisms are especially stressed. The advantages and disadvantages of the membrane filter technique are summarized in **table 42.2**. Membrane filters have been widely used with water that does not contain high levels of background organisms, sediment, or heavy metals. ◀◀ *Filtration (section 8.4)*

More simplified tests for detecting coliforms and fecal coliforms are now available. The ***presence-absence* (P-A) test** can be used for coliforms. This is a modification of the most probable number procedure in which a larger water sample (100 milliliters) is incubated in a single culture bottle with a triple-strength broth containing lactose broth, lauryl tryptose broth, and bromocresol purple indicator. The P-A test is based on the assumption that no coliforms should be present in 100 milliliters of drinking water. A positive test results in the production of acid (a yellow color), which requires confirmation.

To test for both coliforms and *E. coli*, the related **Colilert defined substrate test** can be used. A water sample of 100 milliliters is added to a specialized medium containing *o*-nitrophenyl-β-D-galactopyranoside (ONPG) and 4-methylumbelliferyl-β-D-glucuronide (MUG) as the only nutrients. If coliforms are present, the medium will turn yellow within 24 hours at 35°C due to the hydrolysis of ONPG, which releases *o*-nitrophenol, as shown in **figure 42.3**. To check for *E. coli*, the medium is observed under long-wavelength UV light for fluorescence. When *E. coli* is present, MUG is modified to yield a fluorescent product. If the test is negative for the presence of coliforms, the water is considered acceptable for human consumption. Current standards dictate that water be free of coliforms and fecal coliforms. If coliforms are present, fecal coliforms or *E. coli* must be tested for.

Table 42.2 The Membrane Filter Technique for Evaluation of the Microbial Quality of Water
Advantages
Reproducible
Single-step results often possible
Filters can be transferred between different media.
Large volumes can be processed to increase assay sensitivity.
Time savings are considerable.
Ability to complete filtrations on site
Lower total cost in comparison with MPN procedure
Disadvantages
High-turbidity waters limit volumes sampled.
High populations of background bacteria cause overgrowth.
Metals and phenols can adsorb to filters and inhibit growth.

Source: Data from Clesceri, L. S., et al., 1998. Standard Methods for the Examination of Water and Wastewater, *20th ed. pp. 9–56. American Public Health Association, Washington, D.C.*

Other indicator microorganisms include **fecal enterococci.** These microbes are increasingly being used as an indicator of fecal contamination in brackish and marine water. In saltwater, these bacteria die at a slower rate than fecal coliforms, providing a more reliable indicator of possible recent pollution. ◀◀ *Order* Lactobacillales *(section 21.3)*

The use of indicator organisms as a means of detecting all pathogenic microbes is limited. Some pathogens simply do not co-occur with indicator organisms. Another problem is that indicator organisms can be more sensitive to disinfectants; thus when drinking water is tested after such treatment, indictor-based assays may be misleading. In addition, these culture-based methods generally require at least 18 hours to complete, although some assays require almost a week. The consequences of these shortcomings were evident in 1993, when about 400,000 Milwaukee, Wisconsin, residents became ill from drinking water contaminated with *Cryptosporidium.* Currently the Centers for Disease Control and Prevention estimates that in the United States, a million people per year are sickened from drinking impure water and another 1,000 die annually.

It is clear that effective, reliable, rapid, and quantitative methods for determining the presence or absence of specific pathogens are needed. Molecular approaches can be all these things—with the ability to quickly detect multiple, specific pathogens in a single water sample. Some of the methodologies that have been developed (and where they are described in this text) include the

FIGURE 42.3 The Defined Substrate Test. This much simpler test is used to detect coliforms and fecal coliforms in single 100-milliliter water samples. The medium uses ONPG and MUG as defined substrates. (a) Uninoculated control. (b) Yellow color due to the presence of coliforms. (c) Fluorescent reaction due to the presence of fecal coliforms.

flow cytometry (section 35.3), fluorescent in situ hybridization (FISH; section 27.2), quantitative PCR (section 15.2), and microarrays (sections 16.4 and 27.3). One example is *Legionella pneumophila* contamination, which is routinely detected by PCR amplification. Results are reported in genome units per liter, but the equivalence of these units to colony forming units is unclear. The EPA is evaluating a number of assays that employ molecular techniques so that standard protocols can be established for nationwide drinking water analysis. However, in many communities, molecular techniques are already routinely used to detect coliforms in waters and other environments, including foods. Following a short enrichment step, PCR amplification using primers for the 16S rRNA gene found in coliforms can detect one colony-forming unit (CFU) of *E. coli* per 100 milliliters of water. This technique can also be used to differentiate between nonpathogenic and enterotoxigenic strains, including the Shiga-like toxin–producing *E. coli* O157:H7.

1. What is an indicator organism, and what properties should it have?
2. How is a coliform defined? How does this definition relate to presumptive, confirmed, and completed tests?
3. How does one differentiate between coliforms and fecal coliforms in the laboratory?
4. Why do you think workers who perform these tests must practice excellent personal hygiene and be specifically trained in the implementation of these techniques?

42.2 Wastewater Treatment

The flip side of drinking water sanitation and analysis is wastewater management. Wastewater includes domestic and industrial sewage, industrial and agricultural effluent, and street runoff that is collected in storm sewers. These waters often contain high levels of organic matter, heavy metals, nutrients, and particulates (**table 42.3**). Depending on the effort given to this task, it may still produce waters containing unacceptable levels of nutrients and microorganisms. Thus the process of wastewater treatment, when performed at a municipal level, must be monitored to ensure that waters released into the environment do not pose environmental and health risks. Our discussion of wastewater treatment must therefore begin with the means by which water quality is monitored. We then discuss large-scale wastewater treatment processes, followed by home treatment systems.

Measuring Water Quality

Carbon removal during wastewater treatment is measured several ways, including (1) as **total organic carbon (TOC),** (2) as chemically oxidizable carbon by the **chemical oxygen demand (COD)** test, or (3) as biologically usable carbon by the **biochemical oxygen demand (BOD)** test. The TOC includes all carbon, whether or not it can be used by microorganisms. It is measured by oxidizing the organic matter in a sample to CO_2 at high temperature in an oxygen stream. The CO_2 produced is measured by infrared or potentiometric techniques. The COD gives a similar measurement, except that lignin (*see figure 29.2*) often will not react with the oxidizing chemical, such as permanganate, that is used in this procedure.

The BOD test measures only that portion of total carbon oxidized by microorganisms in a 5-day period at 20°C. The BOD test measures the amount of dissolved O_2 needed for microbial oxidation of biodegradable organic material. When O_2 consumption is determined, O_2 must be present in excess so as not to limit oxidation of other nutrients such as nitrogen and phosphorus. To achieve this, the wastewater sample is diluted to ensure that at least 2 milligrams per liter of O_2 are used while at least 1 milligram per liter of O_2 remains in the test bottle. Ammonia released during organic matter oxidation can also exert an O_2 demand in the BOD test. To prevent this "bottle effect", nitrification is often

Table 42.3 Common Contaminants of Wastewater

Contaminant	*Problems Caused by Contaminant*
Suspended solids	Can cause deposition of sludge leading to anoxic conditions when discharged into aquatic environments
Biodegradable organic compounds	Includes proteins, fats, and carbohydrates that when discharged into streams and rivers cause increased heterotrophic growth leading to anoxic conditions
Pathogenic microorganisms	Cause infectious disease
Priority pollutants	Include organic and inorganic compounds that may be toxic, carcinogenic, mutagenic, or teratogenic
Refractory organic compounds	Organic compounds such as phenols, surfactants, and agricultural pesticides that resist conventional wastewater treatment
Heavy metals	Usually discharged by industry; discharge into rivers and streams is toxic to all trophic levels.
Dissolved inorganic constituents	Include calcium, sodium, and sulfate that may be added to domestic water supplies and may have to be removed prior to wastewater reuse

Source: Modified from Metcalf and Eddy, Inc. Wastewater Engineering, *4th ed. McGraw-Hill, 2002.*

inhibited by the addition of 2-chloro-6-(trichloromethyl) pyridine (nitrapyrin). Except when treated effluents are analyzed, nitrification is not a major concern.

In terms of speed, the TOC is fastest, but it is less informative with regard to biological processes. The COD is slower and involves the use of wet chemicals with higher waste chemical disposal costs. TOC, COD, and BOD provide different but complementary information on the carbon in a water sample. It is critical to note that these measurements are concerned only with carbon removal. They do not directly address the removal of minerals such as nitrate, phosphate, and sulfate. These minerals impact cyanobacterial and algal growth in lakes, rivers, and oceans by contributing to the process of eutrophication. ◀◀ *Nutrient cycling in marine and freshwater environments (section 28.1)*

1. Compare TOC, COD, and BOD: how are these similar and different?
2. What factors can lead to a nitrogen oxygen demand (NOD) in water?
3. What minerals can contribute to eutrophication?

Wastewater Treatment Processes

An aerial photograph of a modern sewage treatment plant is shown in **figure 42.4*a***. Wastewater treatment involves a number of steps that are spatially segregated (figure 42.4*b*). The first three steps are called primary, secondary, and tertiary treatment (**table 42.4**). At the end of the process, the water is usually chlorinated (itself an emerging environmental and human health problem) before it is released.

Primary treatment prepares the wastewater for treatment by physically removing much of the solid material. This can be accomplished by several approaches, including screening, precipitation of small particulates, and settling in basins or tanks (**table 42.5**). The resulting solid material is usually called **sludge.**

Table 42.4 Major Steps in Primary, Secondary, and Tertiary Treatment of Wastes

Treatment Step	*Processes*
Primary	Removal of insoluble particulate materials by settling, screening, addition of alum and other coagulation agents, and other physical procedures
Secondary	Biological removal of dissolved organic matter
	Trickling filters
	Activated sludge
	Lagoons
	Extended aeration systems
	Anaerobic digesters
Tertiary	Biological removal of inorganic nutrients
	Chemical removal of inorganic nutrients
	Virus removal/inactivation
	Trace chemical removal

(a)

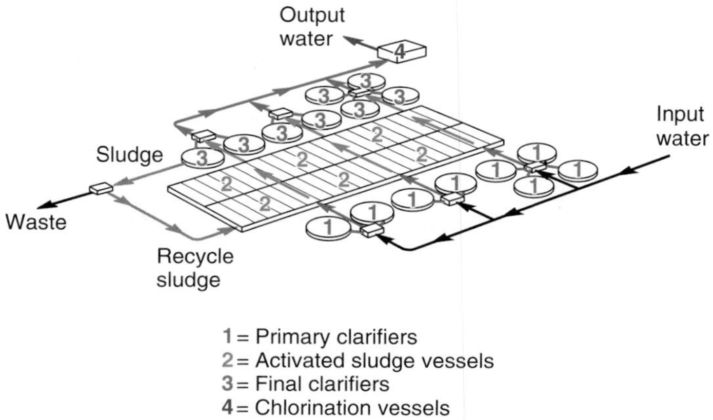

(b)

FIGURE 42.4 Aerial View of a Conventional Sewage Treatment Plant. Sewage treatment plants allow natural processes of self-purification that occur in rivers and lakes to be carried out under more intense, managed conditions in large concrete vessels. (a) A plant in New Jersey. (b) A diagram of flows in the plant.

Figure 42.4 Micro Inquiry

This plant does not use an anaerobic digester. What would be the advantage of this additional step?

Table 42.5 Wastewater Treatment Methods

Nature of Operation	*Process*	*Brief Explanation*
Physical Operations	Screening	Mechanical removal of large suspended materials with parallel bars, rods, wires, grating, wire mesh, or perforated plates
	Comminution	Comminutors are cutting instruments used to pulverize large floating material before wastewater enters primary settling tank.
	Flow equalization	A basin or basins used to adjust the rate of wastewater entry into secondary and tertiary treatment facilities
	Sedimentation	Also called clarification, this widely used procedure is based on the gravitational separation of large particles. Used in primary settling basins and to separate biological floc in activated sludge settling basins. Also used to precipitate chemically treated particulates when chemical coagulation is employed.
	Flotation	The removal of particles by bubbling air or other gas into the water. Particles float to the surface and are collected by skimming; removes fine particles more efficiently than sedimentation. Chemical additives can be added to improve particle removal.
	Granular-medium filtration	Wastewater is passed through a filter bed composed of granular material such as sand; used to remove suspended solids and chemically precipitated phosphours.
Chemical Operations	Chemical precipitation	Addition of coagulants such as alum, ferric chloride, ferric sulfate, or lime to wastewater prior to sedimentation; facilitates the flocculation of finely divided solids into flocs that will settle
	Adsorption	Soluble material is collected by adsorption of activated carbon; usually performed after biological treatment to remove remaining dissolved organic carbon.
	Disinfection	Elimination of pathogens (and other microbes) by a variety of methods including UV light, heat, gamma radiation, and chemical agents such as chlorine, bromine, ozone, phenolic compounds, and others
	Dechlorination	Removal of all chlorine residues before discharge into receiving waters. These compounds are reactive and potentially dangerous.
Biological Operations	Activated sludge process	An aerobic, continuous flow system that uses microbes to aerobically degrade organic matter; occurs after primary settling. Material is mechanically mixed to maintain aeration and prevent settling (figure 42.6*a*).
	Aerated lagoon	A basin 1 to 4 meters in depth where wastewater is treated in a process similar to activated sludge treatment. Turbulence created by aeration keeps the contents of the lagoon mixed.
	Trickling filters	Most commonly used aerobic attached-growth biological treatment. A large basin filled with rock or plastic packing material provides a substrate for microbial biofilm growth and increased contact area for wastewater (figure 42.6*b*)
	Rotating biological contractors	Attached growth process that features large cylinders partially submerged in wastewater. Biofilms form on the cylinders, which are continuously rotated.
	Stabilization ponds	Shallow basin that is mixed naturally or mechanically. Aerobic ponds treat soluble organic wastes and effluents from wastewater treatment facilities; anaerobic ponds stabilize organic wastes. Wastewater retention times from 1 to 3 months.
	Anaerobic digestion	Used for treatment of sludge and wastewater with high organic content *(see figure 41.8)*
	Biological nutrient removal	Removal of nitrogen by nitrification-denitrification and anammox reaction (*see figure 19.11*) and phosphorus, often by proprietary processes (e.g., A/O, PhoStrip, and SBR process)

Modified from Water-Water Treatment Technologies: A General Review. New York: United Nations, 2003

In general, the goal of primary treatment is to generate an effluent that can be further purified by biological treatment and produce sludge that can be treated before eventual disposal.

Secondary treatment promotes the biological transformation of dissolved organic matter into microbial biomass and carbon dioxide. The incoming effluent has a relatively high BOD, but secondary treatment reduces this by about 90 to 95%. In addition, many bacterial pathogens are removed by this process. Several approaches can be used in secondary treatment to remove dissolved organic matter. When microbial growth is completed, under ideal conditions the microorganisms will aggregate and form stable flocs that settle. Minerals in the water also may be tied up in microbial biomass. As shown in **figure 42.5*a***, a healthy settleable floc is compact. In contrast, poorly formed flocs have a network of filamentous microbes that retard settling (figure 42.5*b*).

When these processes occur with lower O_2 levels or with a microbial community that is too young or too old, unsatisfactory floc formation and settling can occur. The result is a **bulking sludge,** caused by the massive development of filamentous bacteria such as *Sphaerotilus* and *Thiothrix,* together with many poorly characterized filamentous organisms. These important filamentous bacteria form flocs that do not settle well and thus produce effluent quality problems. ◀◀ *Order* Burkholderiales *(section 20.2); Order* Thiotrichales *(section 20.3)*

An aerobic **activated sludge** system (**figure 42.6*a***) involves the horizontal flow of materials with recycling of sludge—the active biomass that is formed when organic matter is oxidized and degraded by microorganisms. Activated sludge systems can be designed with variations in mixing. In addition, the ratio of organic matter added to the active microbial biomass can be varied. A low-rate system (low nutrient input per unit of microbial biomass), with slower growing microorganisms, will produce an effluent with low residual levels of dissolved organic matter. A high-rate system (high nutrient input per unit of microbial biomass), with faster growing microorganisms, will remove more dissolved organic carbon per unit time but produce a poorer quality effluent.

Aerobic secondary treatment is often carried out with a **trickling filter** (figure 42.6*b*). The waste effluent is passed over rocks or other solid materials upon which microbial biofilms have developed, and the microbial community degrades the organic waste. A sewage treatment plant can be operated to produce less sludge by employing the **extended aeration** process (figure 42.6*c*). Microorganisms grow on the dissolved organic matter, and the newly formed microbial biomass is eventually consumed to meet maintenance energy requirements. This requires extremely large aeration basins and long aeration times. In addition, with the biological self-utilization of the biomass, minerals originally present in the microorganisms are again released to the water.

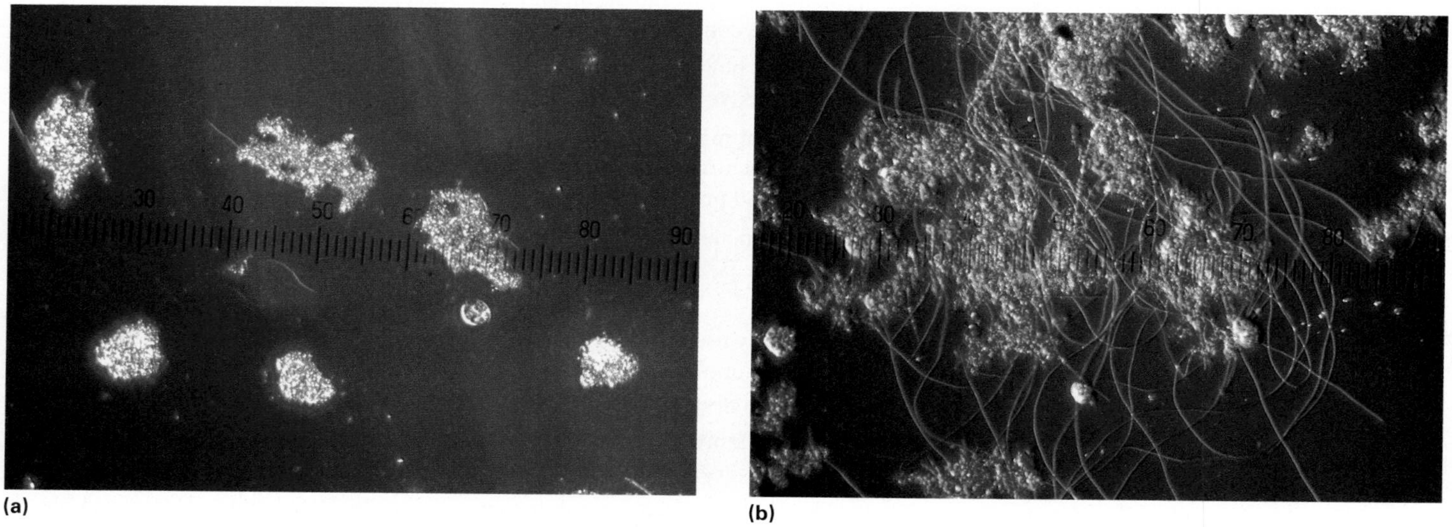

(a) (b)

FIGURE 42.5 Proper Floc Formation in Activated Sludge. Microorganisms play a critical role in the functioning of activated sludge systems. (a) The operation is dependent on the formation of settleable flocs. (b) If the plant does not run properly, flocs form that settle poorly because of their open, porous structure. They form as a result of low aeration, sulfide, and acidic organic substrates. As a consequence, they are released as organic material with the treated water, which lowers the quality of the final effluent.

Figure 42.5 Micro Inquiry

Why is proper floc formation critical to the function of a treatment facility?

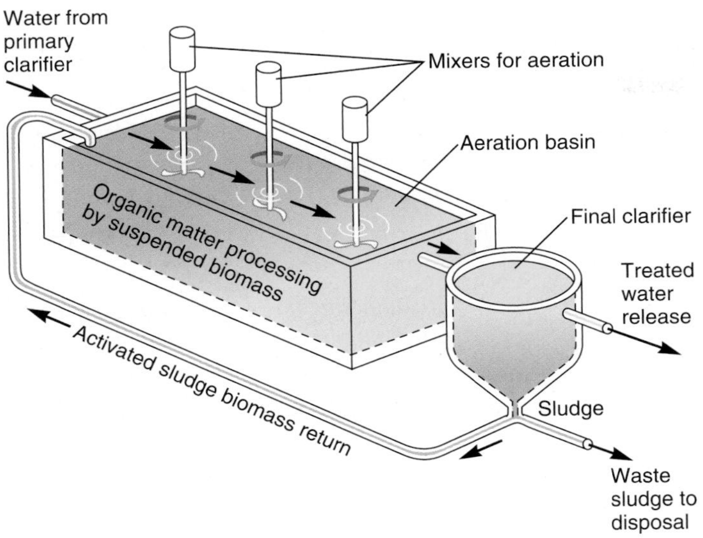

(a) Aerobic activated sludge system

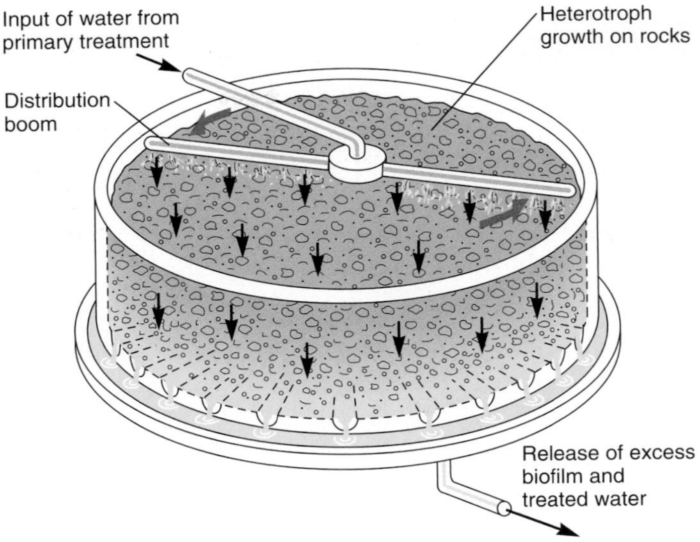

(b) Trickling filter

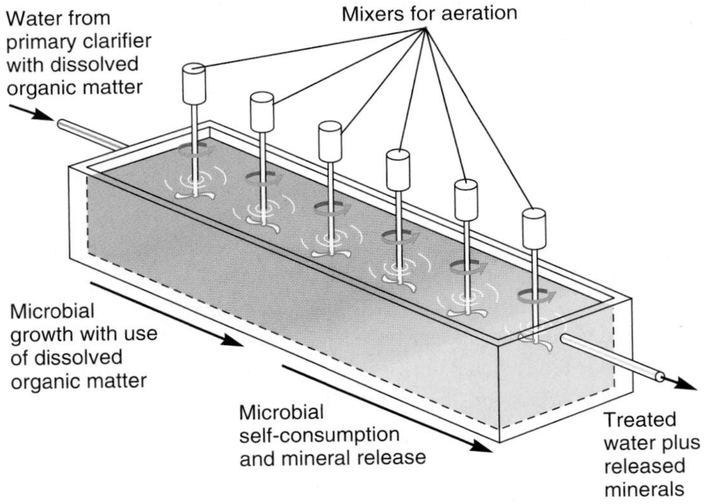

(c) Extended aeration system

FIGURE 42.6 Aerobic Secondary Sewage Treatment. (a) Activated sludge with microbial biomass recycling. The biomass is maintained in a suspended state to maximize oxygen, nutrient, and waste transfer processes. (b) Trickling filter, where waste water flows over biofilms attached to rocks or other solid supports, resulting in transformation of dissolved organic matter to new biofilm biomass and carbon dioxide. Excess biomass and treated water flow to a final clarifier. (c) An extended aeration process, where aeration is continued beyond the point of microbial growth, allows the microbial biomass to self-consume due to microbial maintenance energy requirements. (The extended length of the reactor allows this process of biomass self-consumption to occur.) Minerals originally incorporated in the microbial biomass are released to the water as the process occurs.

All aerobic processes produce excess microbial biomass, or sewage sludge, which contains many recalcitrant organics. Often the sludges from aerobic sewage treatment, together with the materials that settled in primary treatment, are further treated by anaerobic digestion. **Anaerobic digesters** are large tanks designed to operate with continuous input of untreated sludge and removal of the final, stabilized sludge product (*see figure 41.8*). Methane is vented and often burned for heat and electricity production. This digestion process involves three steps: (1) the fermentation of the sludge components to form organic acids, including acetate; (2) production of the methanogenic substrates: acetate, CO_2, and hydrogen; and finally, (3) methanogenesis by methane producers. These methanogenic processes involve critical balances between electron acceptors and donors (**table 42.6**). To function most efficiently, the hydrogen concentration must be maintained at a low level. If hydrogen and organic acids accumulate, the drop in pH will inhibit methane production, resulting in a stuck digester. Often lime is added to neutralize the pH and restore methanogen function. ◀◀ *Methanogens (section 18.3)*

Anaerobic digestion has many advantages. Most of the microbial biomass produced in aerobic growth is used for methane production in the anaerobic digester. Also, because the process of methanogenesis is energetically very inefficient, the microbes must consume about twice the nutrients to produce an equivalent biomass as that of aerobic systems. Consequently less sludge is produced and it can be easily dried. Dried sludge removed from well-operated anaerobic systems can even be sold as organic garden fertilizer. However, sludge from poorly managed systems may

Table 42.6 Sequential Reactions in the Anaerobic Digester

Process Step	*Substrates*	*Products*	*Major Microorganisms*	
Fermentation	Organic polymers	Butyrate, propionate, lactate, succinate, ethanol, acetate,[a] H_2,[a] CO_2[a]	*Clostridium* *Bacteroides* *Peptostreptococcus*	*Peptococcus* *Eubacterium* *Lactobacillus*
Acetogenic reactions	Butyrate, propionate, lactate, succinate, ethanol	Acetate, H_2, CO_2	*Syntrophomonas* *Syntrophobacter*	
Methanogenic reactions	Acetate H_2 and HCO_3^-	$CH_4 + CO_2$ CH_4	*Methanosarcina* *Methanothrix* *Methanobrevibacter* *Methanomicrobium*	*Methanogenium* *Methanobacterium* *Methanococcus* *Methanospirillum*

[a]Methanogenic substrates produced in the initial fermentation step.

contain high levels of heavy metals and other contaminants, which require removal prior to release into the environment.

Tertiary treatment further purifies wastewaters. It is particularly important to remove nitrogen and phosphorus compounds that can promote eutrophication. It also removes heavy metals, biodegradable organics, and many remaining microbes, including viruses. Organic pollutants can be removed with activated carbon filters. Phosphate usually is precipitated as calcium or iron phosphate (e.g., by the addition of lime). To remove phosphorus, oxic and anoxic conditions are used alternately in a series of treatments, and phosphorus accumulates in microbial biomass as polyphosphate. Excess nitrogen may be removed by "stripping," volatilization of NH_3 at high pH. Ammonia itself can be chlorinated to form dichloramine, which is then converted to molecular nitrogen. In some cases, microbial processes can be used to remove nitrogen and phosphorus. A widely used process for nitrogen removal is denitrification. During disinfection, nitrate, produced by microbes under aerobic conditions, is used as an electron acceptor under conditions of low oxygen with organic matter added as an energy source. Nitrate reduction yields nitrogen gas (N_2) and nitrous oxide (N_2O) as the major products. In addition to denitrification, the anammox process is also important. In this reaction, ammonium ion (used as the electron donor) reacts with nitrite (the electron acceptor) to yield N_2. The anammox process can convert up to 80% of the initial ammonium ion to N_2 gas. Tertiary treatment is expensive and is usually not employed except where necessary to prevent obvious ecological disruption. ◀◀ *Phylum* Planctomycetes *(section 19.4); Nitrogen cycle (section 26.1)*

An innovative means of wastewater treatment is the use of constructed wetlands. In these systems, wastewater that has undergone primary treatment to remove large particulates is introduced into a human-made wetland that features floating, emergent, or submerged plants, as shown in **figure 42.7**. The aquatic plants provide nutrients in the root zone, which support microbial growth. Especially with emergent plants, the root zone can be maintained in an anoxic state so that bacteria such as *Desulfovibrio* can grow and produce sulfide, which can trap metals. Constructed wetlands are increasingly employed in the treatment of liquid wastes in small communities and rural regions. It is also used for bioremediation, which is discussed in section 42.3. Constructed wetlands also are being used to treat acid mine drainage and industrial wastes in many parts of the world.

Search This: Wastewater management/EPA

1. Explain how primary, secondary, and tertiary treatments are accomplished.
2. What is bulking sludge? Name several important microbial groups that contribute to this problem.
3. What are the steps of organic matter processing that occur in anaerobic digestion? Why is acetogenesis such an important step?
4. After anaerobic digestion is completed, why is sludge disposal of concern?
5. Why might different aquatic plant types be used in constructed wetlands?

Home Treatment Systems

Groundwater—the water in gravel beds and fractured rocks below the surface soil—is a widely used but often unappreciated resource. In the United States, groundwater supplies at least 100 million people with drinking water, and in rural and suburban areas beyond municipal water distribution systems, 90 to 95% of all drinking water comes from this source.

Unfortunately our dependence on groundwater has not resulted in a corresponding understanding of microorganisms and microbiological processes that occur in this environment. Pathogenic microorganisms and dissolved organic matter are removed from water during subsurface passage through adsorption and trapping by fine sandy materials, clays, and organic matter.

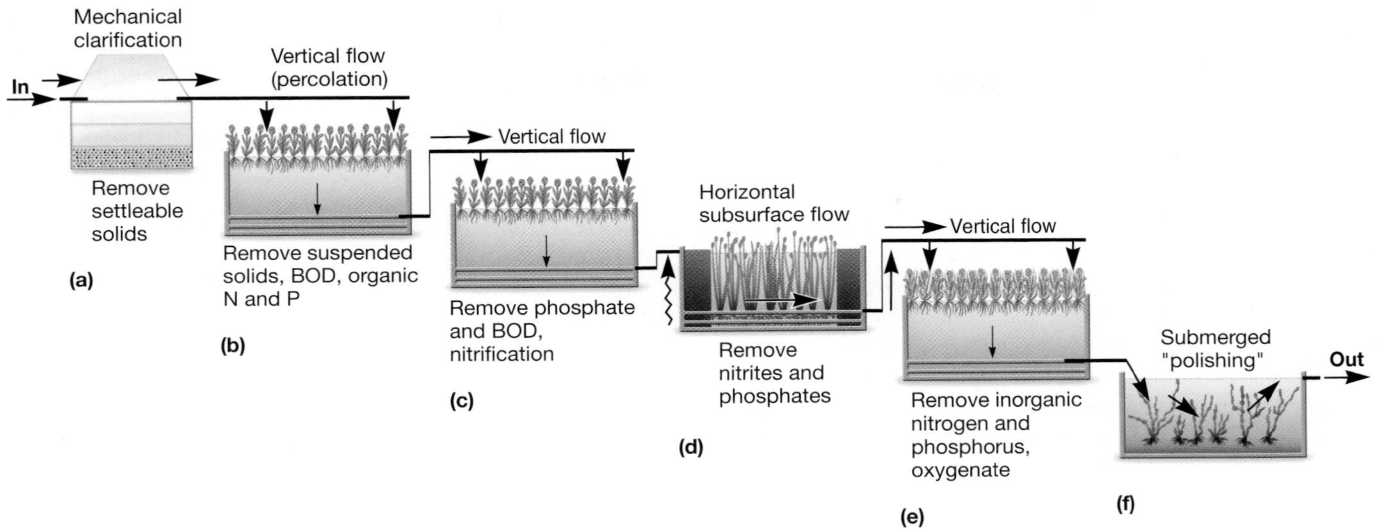

FIGURE 42.7 Constructed Wetland for Wastewater Treatment. Multistage constructed wetland systems can be used for organic matter and phosphate removal. (a) Free-floating macrophytes (b,c,e), such as duckweed and water hyacinth, can be used for a variety of purposes. Emergent macrophytes (d), such as bulrush, allow surface flow as well as vertical and horizontal subsurface flow. Submerged vegetation (f), such as waterweed, allows final "polishing" of the water. These wetlands also can be designed for nitrification and metal removal from waters.

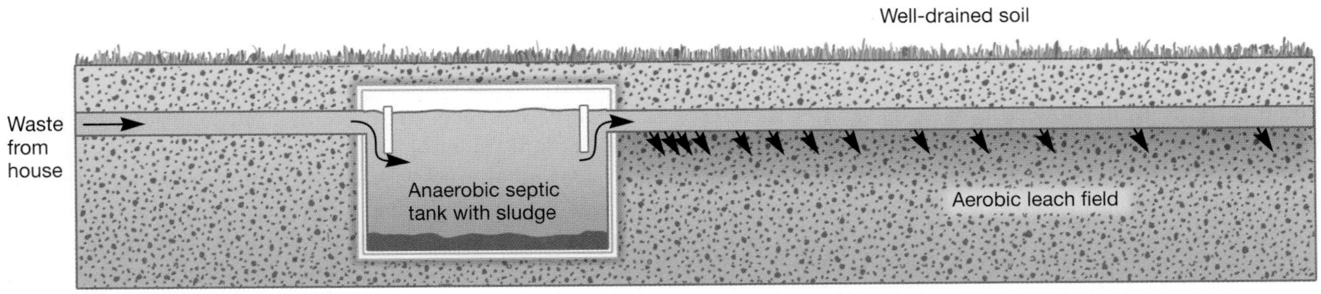

FIGURE 42.8 The Conventional Septic Tank Home Treatment System. This system combines an anaerobic waste liquefaction unit (the septic tank) with an aerobic leach field. Biological oxidation of the liquefied waste takes place in the leach field, unless the soil becomes flooded.

***Figure 42.8** Micro Inquiry*

Why would flooding halt biological oxidation of the waste in the leach field?

Microorganisms associated with these materials—including predators such as protozoa—can use the trapped pathogens as food. This results in purified water with a lower microbial population.

This combination of adsorption and biological predation is also used in home treatment systems (**figure 42.8**). In the absence of municipal sewage systems, a **septic tank** is most commonly used. Conventional septic tank systems include an anaerobic liquefaction and digestion step that occurs in the septic tank, which functions as a simple anaerobic digester. This is followed by organic matter adsorption and entrapment of microorganisms in an aerobic leach field, where biological oxidation occurs. Not all sites are suitable for septic systems. Prior to septic tank installation, a percolation test (or "perc test") must be performed. Perc tests determine if the surrounding soil will be able to absorb the liquid produced by the septic tank. Local governments develop specific protocols for perc tests, which are often witnessed when performed since the outcome of a perc test often determines if a piece of real estate can be developed.

A septic tank may not operate correctly for several reasons. If the retention time of the waste in the septic tank is too short, undigested solids move into the leach field, plugging the system. If the leach field floods and becomes anoxic, biological oxidation does not occur, and effective treatment ceases. Other problems can occur, especially when a suitable soil is not present and the septic tank outflow from a conventional system drains too rapidly to the deeper subsurface. Fractured rocks and coarse gravel materials provide little effective adsorption or filtration. This may result in the contamination of well water with pathogens and the transmission of disease. In addition, nitrogen and phosphorus from the waste can pollute the groundwater. This leads to nutrient enrichment of ponds, lakes, rivers, and estuaries as the subsurface water enters these environmentally sensitive ecosystems.

Domestic and commercial on-site septic systems are now being designed with nitrogen and phosphorus removal steps. Nitrogen is usually removed by nitrification and denitrification, with organic matter provided by sawdust or a similar material. A reductive iron dissolution process can be used for phosphorus removal. Nitrogen and phosphorus releases from septic systems must be controlled to prevent eutrophication of local surface waters. This has resulted in an increased emphasis on the use of these and similar technologies. ◀◀ *Nutrient cycling in marine and freshwater environments (section 28.1)*

Subsurface zones also can become contaminated with pollutants from other sources. Land disposal of sewage sludges, illegal dumping of septic tank sludge, improper toxic waste disposal, and runoff from agricultural operations all contribute to groundwater contamination with chemicals and microorganisms. Many pollutants that reach the subsurface will persist and may affect the quality of groundwater for extended periods. Much research is being conducted to find ways to treat groundwater in place—in situ treatment. As will be discussed in section 42.3, microorganisms and microbial processes are critical in many of these remediation efforts.

1. What factors can limit microbial activity in subsurface environments? Consider the energetic and nutritional requirements of microorganisms in your answer.
2. How, in principle, are a conventional septic tank and leach field system supposed to work? What factors can reduce the effectiveness of this system?
3. What type of wastewater treatment system does your hometown use?

42.3 Biodegradation and Bioremediation by Natural Communities

The metabolic activities of microbes can also be exploited in complex natural environments such as waters, soils, or high organic matter–containing composts where the physical and nutritional conditions for microbial growth cannot be completely controlled and a largely unknown microbial community is present. Examples are (1) the use of microbial communities to carry out biodegradation, bioremediation, and environmental maintenance processes; and (2) the addition of microorganisms to soils or plants for the improvement of crop production. We discuss both of these applications, with an emphasis on bioremediation in this section.

Biodegradation and Bioremediation Processes

Before discussing biodegradation processes carried out by natural microbial communities, it is important to consider definitions. **Biodegradation** has at least three outcomes (**figure 42.9**): (1) a minor change in an organic molecule, leaving the main structure still intact; (2) fragmentation of a complex organic molecule in such a way that the fragments could be reassembled to yield the original structure; and (3) complete mineralization, which is the transformation of organic molecules to inorganic forms. ◀◀ *Biogeochemical cycling (section 26.1)*

The removal of toxic industrial products in soils and aquatic environments is a daunting and necessary task. Compounds such as perchloroethylene (PCE), trichloroethylene (TCE), and polychlorinated biphenyls (PCBs) are common contaminants. These compounds adsorb onto organic matter in the environment, making decontamination using traditional approaches difficult or ineffective. The use of microbes to transform these contaminants to nontoxic degradation products is called **bioremediation.** To understand how bioremediation takes place at the level of an ecosystem, we first must consider the biochemistry of biodegradation.

Degradation of complex compounds requires several discrete stages, usually performed by different microbes. Initially contaminants are converted to less toxic compounds that are more readily degraded. The first step for many contaminants, including organochloride pesticides, alkyl solvents, and aryl halides, is **reductive dehalogenation.** This is the removal of a halogen substituent (e.g., chlorine, bromine, fluorine) while at the same time adding electrons to the molecule. This can occur in two ways. In hydrogenolysis, the halogen substituent is replaced by a hydrogen atom (**figure 42.10*a***). Alternatively dihaloelimination removes two halogen substituents from adjacent carbons while inserting an additional bond between the carbons (figure 41.10*b*). Both processes require an electron donor. The dehalogenation of PCBs uses electrons derived from water; alternatively hydrogen can be the electron donor for the dehalogenation of different chlorinated compounds. Major genera that carry out this process include *Desulfitobacterium, Dehalospirillum,* and *Desulfomonile.*

Reductive dehalogenation usually occurs under anoxic conditions. In fact, humic acids (polymeric residues of lignin decomposition that accumulate in soils and waters) have been found to play a role in anaerobic biodegradation processes. They can serve as terminal electron acceptors under what are called "humic acid-reducing conditions." The use of humic acids as electron acceptors has been observed with the anaerobic dechlorination of vinyl chloride and dichloroethylene. Once the anaerobic dehalogenation steps are completed, degradation of the main structure of many pesticides and other xenobiotics (foreign compounds) often proceeds more rapidly in the presence of O_2. Thus the degradation of halogenated toxic compounds generally requires the action of several microbial genera, sometimes referred to as a consortium.

Structure and stereochemistry are critical in predicting the fate of a specific chemical in nature. When a constituent is in the *meta* as opposed to the *ortho* position, the compound will be degraded at a much slower rate. This stereochemical difference is the reason that the common lawn herbicide 2,4-dichlorophenoxyacetic acid (2,4-D), with a chlorine in the *ortho* position, will be largely degraded in a single summer. In contrast, 2,4,5-trichlorophenoxyacetic acid, with a constituent in the *meta* position, will persist in the soils for several years and thus is used for long-term brush control.

An important factor that influences biodegradation is a compound's chirality or handedness. Microorganisms often can degrade only one isomer of a substance; the other isomer will remain in the environment. At least 25% of herbicides are chiral. Thus it is critical to add the herbicide isomer that is effective and also degradable. Studies have shown that microbial communities in different environments will degrade different isomers. Changes in environmental conditions and nutrient supplies can alter the patterns of chiral form degradation.

The populations that make up a microbial community can change in response to the addition of inorganic or organic substrates. If a particular compound, such as a herbicide, is added repeatedly to a microbial community, the community changes and faster rates of degradation can occur—a process of acclimation (**figure 42.11**). The adaptive process often is so effective that enrichment culture–based approaches can be used to isolate organisms with a desired set of capabilities. For example, a microbial community can become so efficient at herbicide degradation that herbicide effectiveness is diminished. To counteract this process, farmers may have to switch herbicides to alter the microbial community.

Degradation processes that occur in soils also can be used in large-scale degradation of hydrocarbons or wastes from agricultural operations. The waste material is incorporated into the soil or allowed to flow across the soil surface, where degradation occurs. While this is often very effective, sometimes such degradation processes do not reduce environmental problems because the partial degradation or modification of an organic compound may not reduce toxicity. An example of this problem is the microbial metabolism of 1,1,1-trichloro-2,2-bis-(p-chlorophenyl)

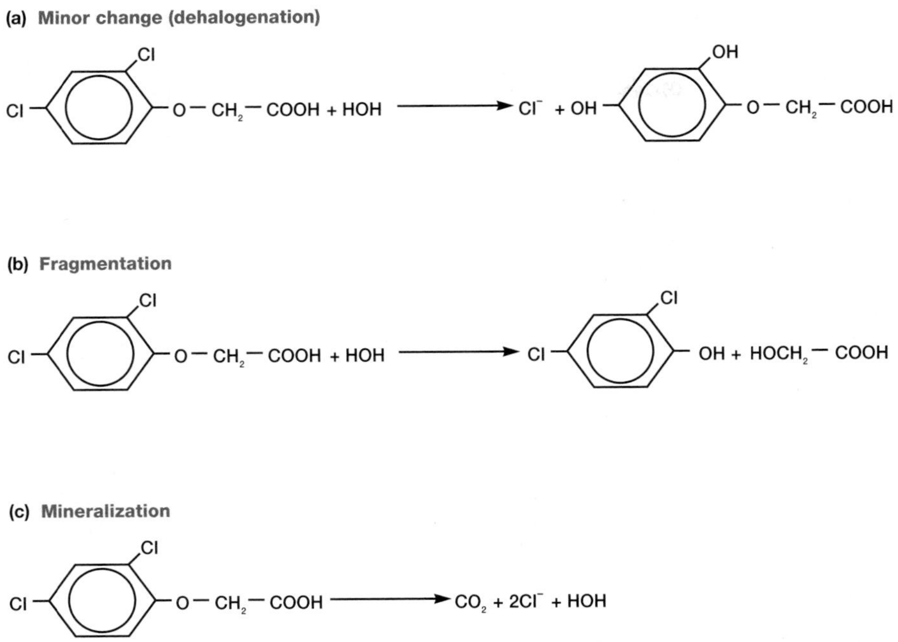

FIGURE 42.9 Biodegradation Has Several Meanings. Biodegradation is a term that can be used to describe three major types of changes in a molecule: (a) a minor change in the functional groups attached to an organic compound, such as the substitution of a hydroxyl group for a chlorine group; (b) an actual breaking of the organic compound into organic fragments in such a way that the original molecule could be reconstructed; (c) the complete degradation of an organic compound to minerals.

Figure 42.9 Micro Inquiry

Why does dehalogenation increase the likelihood that the carbon compound will be degraded?

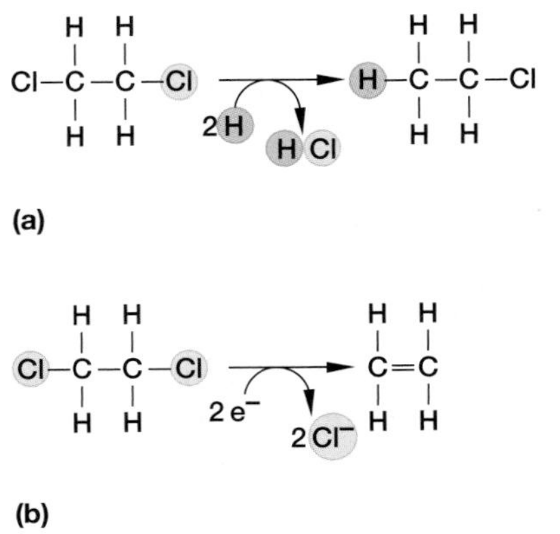

FIGURE 42.10 Reductive Dehalogenation. In these two examples 1,2 dichlorobenzoate is dehalogenated by (a) alkyl hydrogenolysis and (b) dihaloelimination. Note that although the substrate is the same, these strategies result in different products.

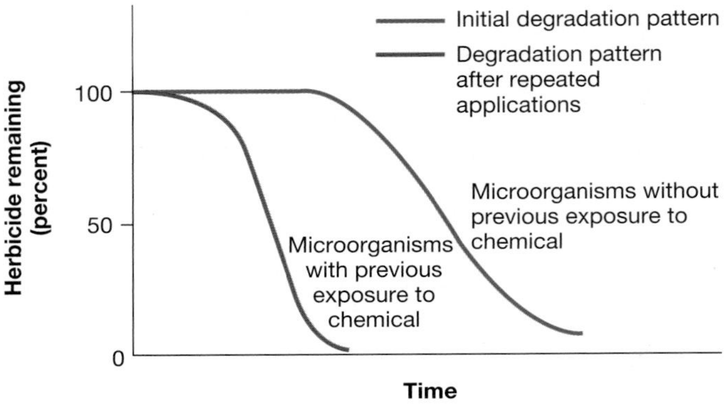

FIGURE 42.11 Repeated Exposure and Degradation Rate. Addition of an herbicide to a soil can result in changes in the degradative ability of the microbial community. Relative degradation rates for an herbicide after initial addition to a soil and after repeated exposure to the same chemical.

Figure 42.11 Micro Inquiry

Why does the rate of herbicide degradation increase with continued application?

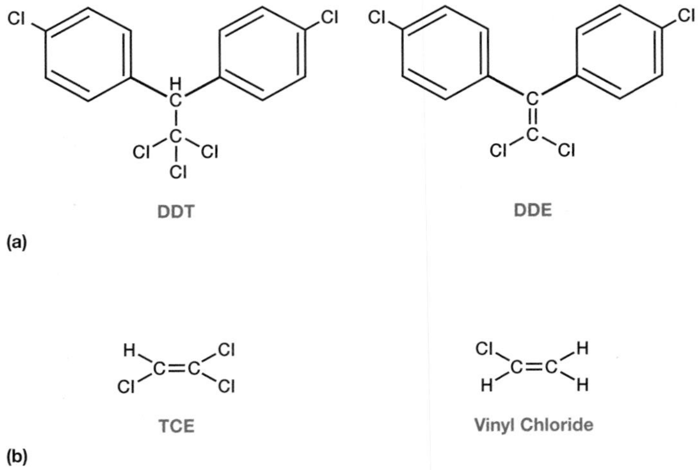

FIGURE 42.12 Toxic Degradation Products. The parent compounds DDT and TCE are metabolized to the toxins DDE and vinyl chloride, respectively.

ethane (DDT), a pesticide once commonly used in the United States. Degradation removes a chlorine function to give 1,1-dichloro-2,2-bis-(p-chlorophenyl) ethylene (DDE), which is still of environmental concern (**figure 42.12*a***). Another important example is the degradation of trichloroethylene (TCE), a widely used solvent. If this is degraded under anoxic conditions, the dangerous carcinogen vinyl chloride can be formed (figure 42.12*b*).

An emerging approach to the bioremediation of contaminants in soils is the use of microbial fuel cells (*see figure 41.9*). One of the major challenges to the microbial degradation of subsurface organic compounds and heavy metals is the provision of an electron acceptor that will participate in the oxidation of the contaminant. For instance, the biodegradation of petroleum in soils is frequently limited because it is very difficult and expensive to introduce oxygen into sediments. As reviewed in chapter 41, a microbial fuel cell converts the chemical energy released during substrate oxidation to electricity. The basic design of a microbial fuel cell includes two chambers divided by a membrane through which protons can pass. Electrons generated by microbial oxidation of a substrate in one chamber are transferred to an anode that connects the two chambers. Protons that are released during oxidation diffuse across the membrane to the other chamber. A catalyst then combines the protons, electrons, and oxygen to generate water. Microbial fuel cells can be inserted in sediments such that the anode can serve as the terminal electron acceptor by bacteria in the community during biodegradation. Several innovative applications of microbial fuel cells have been demonstrated, including the oxidation of toluene by *Geobacter metallireducens*, which uses the fuel cell electrode as an electron acceptor. ◀◀ *Order* Desulfuromonales *(section 20.4); Microbial energy conversion (section 41.3)*

1. Give alternative definitions for the term biodegradation.
2. What is reductive dehalogenation? Describe humic acids and the role they can play in anaerobic degradation processes.
3. Why is the *meta* effect important for understanding biodegradation?
4. Discuss chirality and its importance for understanding biodegradation.

Stimulating Biodegradation

Bioremediation usually involves stimulating the degradative activities of microorganisms already present in contaminated waters or soils because existing microbial communities often cannot carry out biodegradation processes at a desired rate due to limiting physical or nutritional factors. For example, biodegradation may be limited by low levels of oxygen, nitrogen, phosphorus, and other needed nutrients. In these cases, it is necessary to determine the limiting factors and supply the needed materials or modify the environment. Often the addition of easily metabolized organic matter such as glucose increases biodegradation of recalcitrant compounds that are usually not used as carbon and energy sources by microorganisms. This process, termed **cometabolism,** can be carried out by simply adding easily catabolized organic matter and the compound to be degraded to a complex microbial community.

Stimulating Hydrocarbon Degradation in Waters and Soils

Experience with oil spills in marine environments illustrates how effective stimulating natural microbial communities can be. Hydrocarbons lack phosphorus and nitrogen, so these elements will limit degradation. One of the first demonstrations that simple fertilization will accelerate hydrocarbon degradation by natural populations occurred in 1989 following the massive *Exxon-Valdez* oil spill in Prince Williams Sound, Alaska. As seen in **figure 42.13**, the addition of nitrogen and phosphorus-containing fertilizer significantly diminished oil in the coastal sediments. Since that time, it has been recognized that another key variable in promoting bioremediation is the degree of contact between microorganisms and the hydrocarbon substrate. To increase hydrocarbon accessibility by microbes, pellets containing nutrients and an oleophilic (hydrocarbon soluble) preparation are often used. This technique accelerates the degradation of different crude oil slicks by 30 to 40%, in comparison with control oil slicks where the additional nutrients are not available. More recent research also focuses on the entire microbial community involved in the bioremediation process. Oil, which is a complex mixture of organic molecules, is degraded by a variety of bacteria (so far, archaea do not appear to degrade oil), but their capacity to do so is influenced by rates of predation by viruses and protists, the rate of nutrient recycling, and even the release of biosurfactants by other microbes (**figure 42.14**).

Polychlorinated biphenols (PCBs) are also of great concern but often can be degraded by promoting the activities of natural microbial communities. This requires a two-stage process. First, partial dehalogenation of the PCBs occurs naturally under anoxic conditions. Then the muds are aerated to promote the complete degradation of the less chlorinated residues (**chapter opener**).

FIGURE 42.13 Bioremediation of the *Exxon Valdez* Oil Spill. Fertilization of coastline with nitrogen and phosphorus-containing fertilizer increased the rate at which oil was degraded. The quadrant on the left was not treated; the quadrant on the right was fertilized.

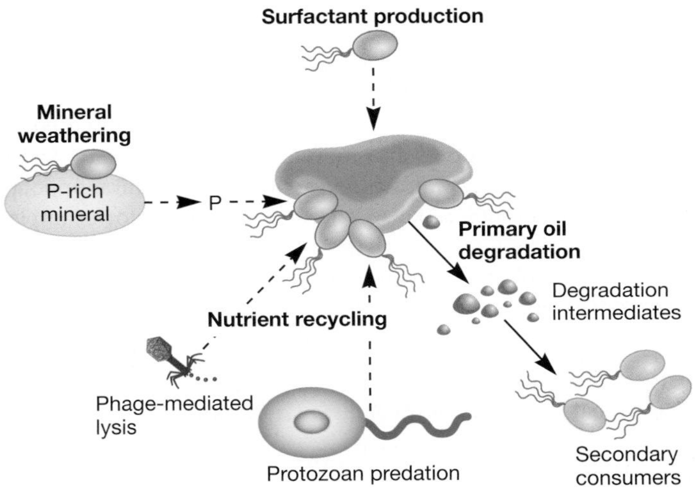

FIGURE 42.14 Oil Degradation Requires a Microbial Community. Primary oil degradation involves bacteria that begin the oil degradation process (blue microbes). These bacteria release a variety of degradation products that are further broken down by secondary consumers (yellow microbes). Dashed lines represent indirect interactions that influence the rate of bioremediation, such as phage lysis and protozoan grazing.

Search This: Bioremediation/USGS

Phytoremediation

Phytoremediation—the use of plants to stimulate the extraction, degradation, adsorption, stabilization, or volatilization of soil contaminants—is becoming an important part of bioremediation technology. The introduction of bacterial genes into plants can enhance phytoremediation. For example, using cloning techniques with *Agrobacterium,* the *merA* and *merB* genes have been integrated into the mustard plant *Arabidopsis thaliana,* making it possible to transform extremely toxic organic mercury (e.g., methylmercury) to elemental mercury, which is less of an environmental hazard. Transgenic tobacco plants have been constructed that express tetranitrate reductase, an enzyme from an explosive-degrading bacterium, thereby enabling the transgenic plants to degrade nitrate ester and nitro aromatic explosives. ◀◀ *Agricultural biotechnology (section 41.4)*

Microbes that associate with plants in the rhizosphere can also be employed in bioremediation strategies. This approach, sometimes called **rhizoremediation,** involves stimulation of microbial growth by the exudates released by plant roots. Rhizoremediation can be optimized through enrichment and molecular genetic techniques. A recent example involves the pesticide lindane, which exists in nature as γ-hexachlorocyclohexane (HCH) and its toxic isomers α-, β-, and δ-HCH. Microbes found in the rhizosphere of lindane-contaminated soils were enriched by growth in the presence of all four HCH isomers. Repeated rounds of coculture with plants in soils to which HCH isomers were added resulted in the evolution of strains of *Sphingomonas* that efficiently colonized roots and degraded these toxins.

Finally, endophytic bacteria (i.e., bacteria growing within the plant roots) can be engineered for bioremediation. For example, this was used to degrade the solvent toluene in soil. Conjugation was used to introduce a plasmid bearing the genes for toluene degradation into a strain of *Burkholderia cepacia.* When the new toluene-degrading *B. cepacia* bacteria colonize yellow lupine roots, the host plant is protected from the toxic effects of toluene because its endophytic bacterial population locally degrades the toxin. Experiments such as these suggest that the field of phytoremediation will continue to provide important new approaches to decontaminating soils.

Metal Bioleaching

Bioleaching is the use of microorganisms that produce acids from reduced sulfur compounds to create acidic environments that solubilize desired metals for recovery. This approach is used to recover metals from ores and mining tailings with metal levels too low for smelting. Bioleaching carried out by natural populations of *Leptospirillum*-like species and thiobacilli, for example, allow recovery of up to 70% of the copper in low-grade ores. This involves the biological oxidation of copper present in these ores to produce soluble copper sulfate. In fact, this process is now used for copper mining.

1. What is cometabolism and why is this important for degradation processes?
2. What factors must one consider when attempting to stimulate the microbial degradation of an oil spill in a marine environment?
3. What role can microorganisms play in phytoremediation?
4. How is bioleaching carried out, and what microbial genera are involved?

42.4 Bioaugmentation

The acceleration of microbiological processes by the addition of known active microorganisms to soils, waters, or other complex systems is called **bioaugmentation.** For example, commercial culture preparations are available to facilitate silage formation and to improve septic tank performance.

With the development of the "superbug" by A. M. Chakrabarty in 1974, there was initial excitement as it was hoped that such an improved microorganism might be able to degrade hydrocarbon pollutants very effectively. Chakrabarty's superbug was a laboratory pseudomonad that had been transformed with plasmids encoding enzymes needed for efficient degradation of several hydrocarbon compounds. However, a critical point not considered was the actual location, or microhabitat, in which the microbe had to survive and function. Engineered microorganisms were added to soils and waters with the expectation that rates of degradation would be stimulated as these microorganisms established themselves. While such additions usually led to short-term increases in rates of the desired activity, after a few days, the microbial community responses were similar in treated and control systems. The lack of effectiveness of such added cultures was due to at least three factors: (1) the attractiveness of laboratory-grown microorganisms as a food source for predators such as soil protozoa, (2) the inability of these added microorganisms to contact the compounds to be degraded, and (3) the failure of the added microorganisms to survive and compete with indigenous microorganisms. Such a modified microorganism may be less fit to compete and survive because of the additional energetic burden required to maintain the extra DNA.

Additions of a microbe to natural environments can be more successful if the microorganism is added together with a microhabitat that gives the organism physical protection, as well as possibly supplying nutrients. This makes it possible for the microorganism to survive in spite of the intense competitive pressures that exist in the natural environment, including pressure from protozoan predators. Microhabitats may be either living or inert. Specialized living microhabitats include the surface of a seed, a root, or a leaf. Here, higher nutrient fluxes and rates of initial colonization by the added microorganisms protect against the fierce competitive conditions in the natural environment. For example, to ensure that the nitrogen-fixing microbe *Rhizobium* is in close association with the legume, seeds are coated with the microbe using an oil-organism mixture or the bacteria are placed in a band under the seed, where the newly developing primary root will penetrate. ◀◀ *The Rhizobia (section 29.3)*

Microorganisms can be added to natural communities together with protective inert microhabitats. As an example, if microbes are added to a soil with microporous glass, the survival of added microorganisms can be markedly enhanced. Other microbes have been observed to create their own microhabitats. Microorganisms in the water column overlying PCB-contaminated sand-clay soils have been observed to create their own "clay hutches" by binding clays to their outer surfaces with exopolysaccharides. Thus the application of principles of microbial ecology can facilitate the successful management of microbial communities in nature.

1. What factors might limit the ability of microorganisms, after addition to a soil or water, to persist and carry out desired functions?
2. What types of microhabitats can be used with microorganisms when they are added to a complex natural environment?

Summary

42.1 Water Purification and Sanitary Analysis

a. Water purification can involve the use of sedimentation, coagulation, chlorination, and rapid and slow sand filtration. Chlorination may lead to the formation of organic disinfection by-products, including trihalomethane compounds, which are potential carcinogens (**figure 42.1**).

b. *Cryptosporidium, Cyclospora,* viruses, and *Giardia* are of concern, as conventional water purification and chlorination will not always ensure their removal and inactivation to acceptable limits.

c. Indicator organisms are used to assess the presence of pathogenic microorganisms. Most probable number (MPN) and membrane filtration procedures are employed to estimate the number of indicator organisms present. The presence-absence (P-A) test is used for coliforms and defined substrate tests, for coliforms and *E. coli* (**table 42.2; figures 42.2** and **42.3**).

d. Molecular techniques based on the polymerase chain reaction (PCR) can be used to detect waterborne pathogens such as Shiga-like toxin–producing *E. coli* O157:H7 when a preenrichment step is used.

42.2 Wastewater Treatment

a. The biochemical oxygen demand (BOD) test is an indirect measure of the organic matter that can be oxidized by the aerobic microbial community. The chemical oxygen demand (COD) and total organic carbon (TOC) tests provide information on carbon that is not biodegraded in the 5-day BOD test.

b. Conventional sewage treatment is a controlled intensification of natural self-purification processes, and it can involve primary, secondary, and tertiary treatment steps (**figure 42.4; tables 42.4** and **42.5**).

c. Constructed wetlands involve the use of aquatic plants (floating, emergent, submerged) and their associated microorganisms for the treatment of liquid wastes (**figure 42.7**).

d. Home treatment systems operate on general self-purification principles. The conventional septic tank (**figure 42.8**) provides anaerobic liquefaction and digestion, whereas the aerobic leach field allows oxidation of the soluble effluent. These systems are now designed to provide nitrogen and phosphorus removal to lessen impacts of on-site sewage treatment systems on vulnerable marine and freshwaters.

42.3 Biodegradation and Bioremediation by Natural Communities

a. Microbial growth in complex natural environments such as soils and waters is used to carry out environmental management processes, including bioremediation, plant inoculation, and other related activities.

b. Biodegradation is a critical part of natural systems mediated largely by microorganisms. This can involve minor changes in a molecule, fragmentation, or mineralization (**figure 42.9**).

c. Biodegradation can be influenced by many factors, including oxygen levels, humic acids, and the presence of readily usable organic matter. Reductive dehalogenation occurs best under anoxic conditions, and the presence of organic matter can facilitate modification of recalcitrant compounds in the process of cometabolism (**figure 42.10**).

d. The structure of organic compounds influences degradation. If constituents are in specific locations on a molecule, as in the *meta* position, or if varied structural isomers are present, degradation can be affected.

e. Degradation management can be carried out in place, whether this be large marine oil spills, soils, or the subsurface. Such large-scale efforts usually involve the use of natural microbial communities (**figures 42.13** and **42.14**).

f. Plants can be used to stimulate biodegradation processes during phytoremediation. This can involve extraction, filtering, stabilization, and volatilization of pollutants.

42.4 Bioaugmentation

a. Microorganisms can be added to environments that contain complex microbial communities with greater success if living or inert microhabitats are used.

b. *Rhizobium* is an important example of a microorganism added to a complex environment using a living microhabitat (the plant root).

Critical Thinking

1. You wish to develop a constructed wetland for removing metals from a stream at one site, and at another site, you wish to treat acid mine drainage. How might you approach each of these problems?

2. What alternatives, if any, can one use for protection against microbiological infection when swimming in polluted recreational water? Assume that you are part of a water rescue team.

3. What possible alternatives could be used to eliminate N and P releases from sewage treatment systems? What suggestions could you make that might lead to new technologies?
4. Why, when a microorganism is removed from a natural environment and grown in the laboratory, will it usually not be able to effectively colonize its original environment if it is added back? Consider the nature of growth media used in the laboratory in comparison to growth conditions in a soil or water when attempting to understand this fundamental problem in microbial ecology.
5. Biofilms are known to form on surfaces in drinking water distribution networks. It is therefore important to understand how such biofilms develop and how long they last. A recent study followed the development of biofilms populated by the protists *Giardia* and *Cryptosporidium,* as well as a strain of *Poliovirus* and several bacteriophages. Attachment to, growth within, and detachment from model biofilms were measured in tap water flowing through a special reactor designed to simulate the type of turbulence found in a water distribution network.

   Why were the investigators especially interested in biofilm stability? Also, why were they interested in bacteriophages? What special considerations do you think need to be addressed when studying protists with relatively complex life cycles? What method do you think would be most appropriate to detect the protists? What about detecting the viruses?

   Read the original paper: Helmi, K., et al. 2008. Interactions of *Cryptosporidium parvum, Giardia lamblia,* vaccinal poliovirus type 1, and bacteriophages fX174 and MS2 with a drinking water biofilm and a wastewater biofilm. *Ann. Rev. Microbiol.* 74:2079.
6. Lindane (γ-hexachlorocyclohexane) is an organochlorine pesticide that was once widely used but is now tightly regulated and rarely used in most countries due to its extended persistence, toxicity, and capacity to bioaccumulate. Isolates of the α-proteobacterium *Sphingomonas* can degrade lindane, but bioremediation efforts using fertilization or bioaugmentation have not been very successful. It is argued that bioaugmentation might be successful if the bacteria were protected from predation, intermicrobial competition, and starvation. One way to accomplish this is to encase the bacteria in a nutrient-rich matrix that enables the diffusion of gases and liquids (and lindane) in and waste products out. When this approach was tested using a *Sphingomonas* strain isolated from lindane-contaminated soils, biodegradation continued twice as long as control degradation in which bacterial cells were added without any protection.

   Why do you think the investigators used a *Sphingomonas* strain recently isolated from a lindane-contaminated site, rather than a genetically engineered strain? How would you determine which nutrient to add to the protective matrix that would result in maximal lindane degradation? In addition to the use of a protective matrix, the *Sphingomonas* cells were added slowly at a relatively constant rate over the course of the test period (about 20 days). What hypothesis do you think the investigators were testing when they designed experiments to compare slow-release inoculation with single high-density inoculation? What do you think was the outcome?

   Read the original paper: Mertens, B., et al. 2006. Slow-release inoculation allows sustained biodegradation of γ-hexachlorocyclohexane. *Appl. Environ. Microbiol.* 72:622.

Concept Mapping Challenge

Use your own linking words to construct a concept map that illustrates the principles of water analysis and purification. Use the following terms:

Sanitary analysis	P-A test	Trickling filter	Secondary treatment	Activated sludge	Bioremediation
Fecal enterococci	Sludge	Primary treatment	Tertiary treatment	Perc test	Indicator organisms
			Septic systems	MPN	Fecal coliforms
			Reductive halogenation		Rhizoremediation
			Wastewater treatment		

Learn More

Learn more by visiting the text website at www.mhhe.com/willey8, where you will find a complete list of references.

Appendix I

A Review of the Chemistry of Biological Molecules

Appendix I provides a brief summary of the chemistry of organic molecules with particular emphasis on the molecules present in microbial cells. Only basic concepts and terminology are presented; introductory textbooks in biology and chemistry should be consulted for a more extensive treatment of these topics.

ATOMS AND MOLECULES

Matter is made of elements that are composed of atoms. An element contains only one kind of atom and cannot be broken down to simpler components by chemical reactions. An atom is the smallest unit characteristic of an element and can exist alone or in combination with other atoms. When atoms combine, they form molecules. Molecules are the smallest particles of a substance. They have all the properties of the substance and are composed of two or more atoms.

Although atoms contain many subatomic particles, three directly influence their chemical behavior—protons, neutrons, and electrons. The atom's nucleus is located at its center and contains varying numbers of protons and neutrons (**figure AI.1**). Protons have a positive charge, and neutrons are uncharged. The mass of these particles and the atoms that they compose is given in terms of the atomic mass unit (AMU), which is 1/12 the mass of the most abundant carbon isotope. Often the term *dalton* (Da) is used to express the mass of molecules. It also is 1/12 the mass of an atom of ^{12}C or 1.661×10^{-24} grams. Both protons and neutrons have a mass of about 1 dalton. The atomic weight is the actual measured weight of an element and is almost identical to the mass number for the element, the total number of protons and neutrons in its nucleus. The mass number is indicated by a superscripted number preceding the element's symbol (e.g., ^{12}C, ^{16}O, and ^{14}N).

Negatively charged particles called electrons circle the atomic nucleus (figure AI.1). The number of electrons in a neutral atom equals the number of its protons and is given by the atomic number, the number of protons in an atomic nucleus. The atomic number is characteristic of a particular type of atom. For example, carbon has an atomic number of six, hydrogen's number is one, and oxygen's is eight (**table AI.1**).

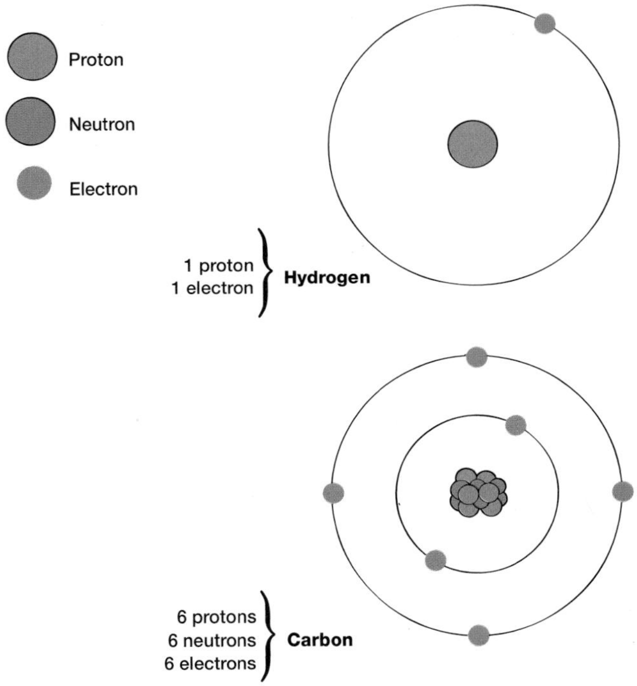

FIGURE AI.1 Diagrams of Hydrogen and Carbon Atoms. The electron orbitals are represented as concentric circles.

Table AI.1 Atoms Commonly Present in Organic Molecules

Atom	Symbol	Atomic Number	Atomic Weight	Number of Chemical Bonds
Hydrogen	H	1	1.01	1
Carbon	C	6	12.01	4
Nitrogen	N	7	14.01	3
Oxygen	O	8	16.00	2
Phosphorus	P	15	30.97	5
Sulfur	S	16	32.06	2

The electrons move constantly within a volume of space surrounding the nucleus, even though their precise location in this volume cannot be determined accurately. This volume of space in which an electron is located is called its orbital. Each orbital can contain two electrons. Orbitals are grouped into shells of different energy that surround the nucleus. The first shell is closest to the nucleus and has the lowest energy; it contains only one orbital. The second shell contains four orbitals, one circular and three shaped like dumbbells (**figure AI.2*a***). It can contain up to eight electrons. The third shell has even higher energy and holds more than eight electrons. Shells are filled beginning with the innermost and moving outward. For example, carbon has six electrons, two in its first shell and four in the second (figures AI.1 and AI.2*b*). The electrons in the outermost shell are the ones that participate in chemical reactions. The most stable condition is achieved when the outer shell is filled with electrons. Thus the number of bonds an element can form depends on the number of electrons required to fill the outer shell. Since carbon has four electrons in its outer shell and the shell is filled when it contains eight electrons, it can form four covalent bonds (table AI.1).

CHEMICAL BONDS

Molecules are formed when two or more atoms associate through chemical bonding. Chemical bonds are attractive forces that hold together atoms, ions, or groups of atoms in a molecule or other substance. Many types of chemical bonds are present in organic molecules; three of the most important are covalent bonds, ionic bonds, and hydrogen bonds.

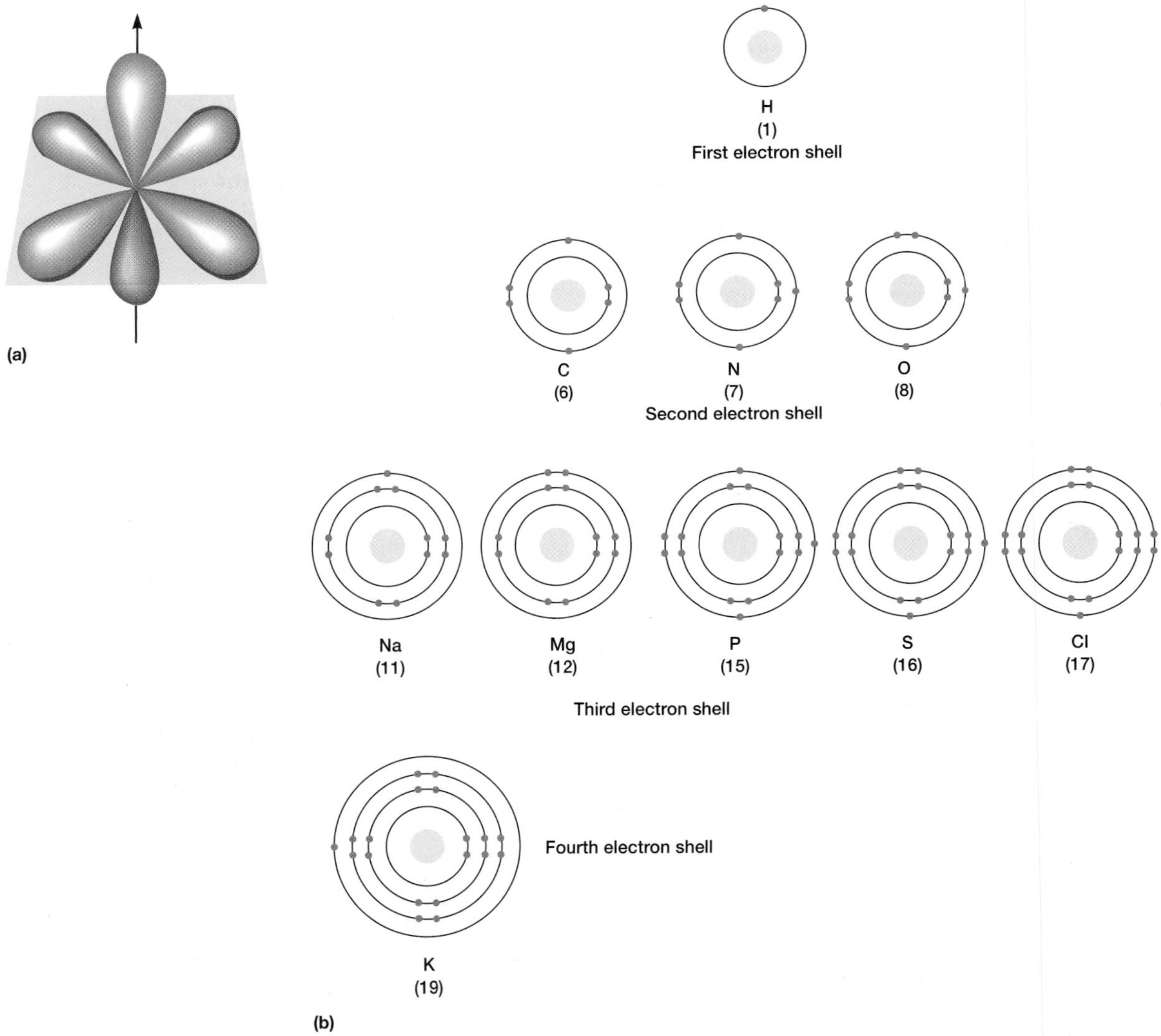

FIGURE AI.2 Electron Orbitals. (a) The three dumbbell-shaped orbitals of the second shell. The orbitals lie at right angles to each other. (b) The distribution of electrons in some common elements. Atomic numbers are given in parentheses.

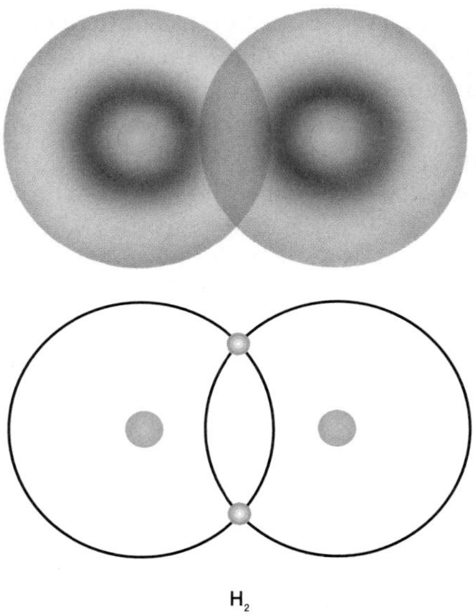

FIGURE AI.3 **The Covalent Bond.** A hydrogen molecule is formed when two hydrogen atoms share electrons.

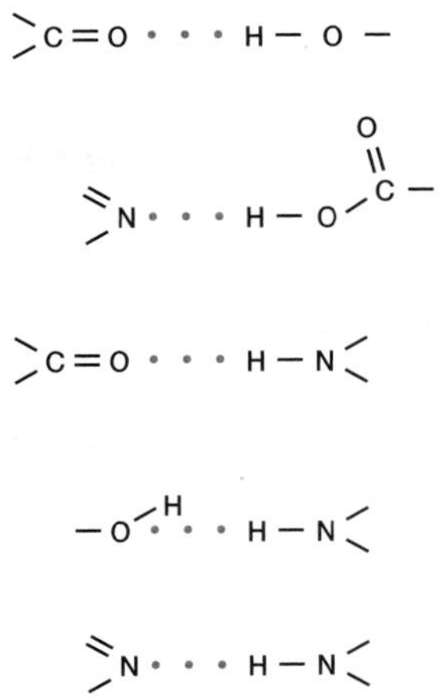

FIGURE AI.4 **Hydrogen Bonds.** Representative examples of hydrogen bonds present in biological molecules.

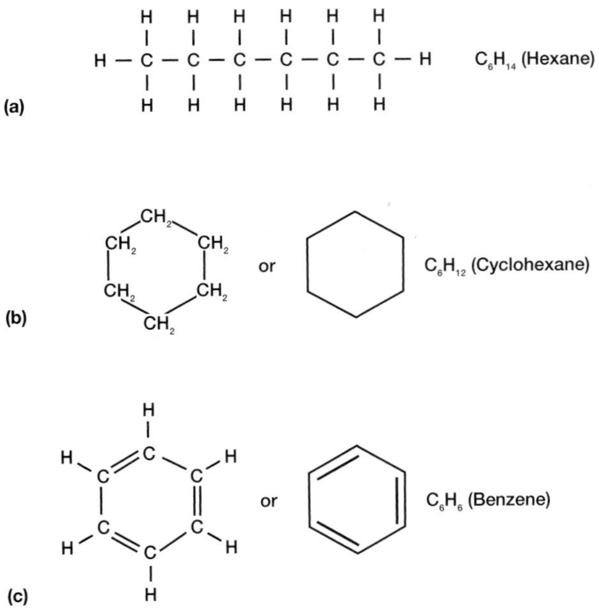

FIGURE AI.5 **Hydrocarbons.** Examples of hydrocarbons that are (a) linear, (b) cyclic, and (c) aromatic.

In covalent bonds, atoms are joined together by sharing pairs of electrons (**figure AI.3**). If the electrons are equally shared between identical atoms (e.g., in a carbon-carbon bond), the covalent bond is strong and nonpolar. When two different atoms such as carbon and oxygen share electrons, the covalent bond formed is polar because the electrons are pulled toward the more electronegative atom, the atom that more strongly attracts electrons (the oxygen atom). A single pair of electrons is shared in a single bond; a double bond is formed when two pairs of electrons are shared.

Atoms often contain either more or fewer electrons than the number of protons in their nuclei. When this is the case, they carry a net negative or positive charge and are called ions. Cations carry positive charges and anions have a net negative charge. When a cation and an anion approach each other, they are attracted by their opposite charges. This ionic attraction that holds two groups together is called an ionic bond. Ionic bonds are much weaker than covalent bonds and are easily disrupted by a polar solvent such as water. For example, the Na^+ cation is strongly attracted to the Cl^- anion in a sodium chloride crystal, but sodium chloride dissociates into separate ions (ionizes) when dissolved in water. Ionic bonds are important in the structure and function of proteins and other biological molecules.

When a hydrogen atom is covalently bonded to a more electronegative atom such as oxygen or nitrogen, the electrons are unequally shared and the hydrogen atom carries a partial positive charge. It will be attracted to an electronegative atom such as oxygen or nitrogen, which carries an unshared pair of electrons; this attraction is called a hydrogen bond (**figure AI.4**). Although an individual hydrogen bond is weak, there are so many hydrogen bonds in proteins and nucleic acids that they play a major role in determining protein and nucleic acid structure.

ORGANIC MOLECULES

Most molecules in cells are organic molecules, molecules that contain carbon. Since carbon has four electrons in its outer shell, it tends to form four covalent bonds in order to fill its outer shell with eight electrons. This property makes it possible to form chains and rings of carbon atoms that also can bond with hydrogen and other atoms (**figure AI.5**). Although adjacent carbons usually are connected by single bonds, they may be joined by double or triple bonds. Rings that have alternating single and double bonds, such as the benzene ring, are called aromatic rings. The hydrocarbon chain or ring provides a chemically inactive skeleton to which more reactive groups of atoms may be attached. These reactive groups with specific properties are known as functional groups. They usually contain atoms of oxygen, nitrogen, phosphorus, or sulfur (**figure AI.6**) and are largely responsible for most characteristic chemical properties of organic molecules.

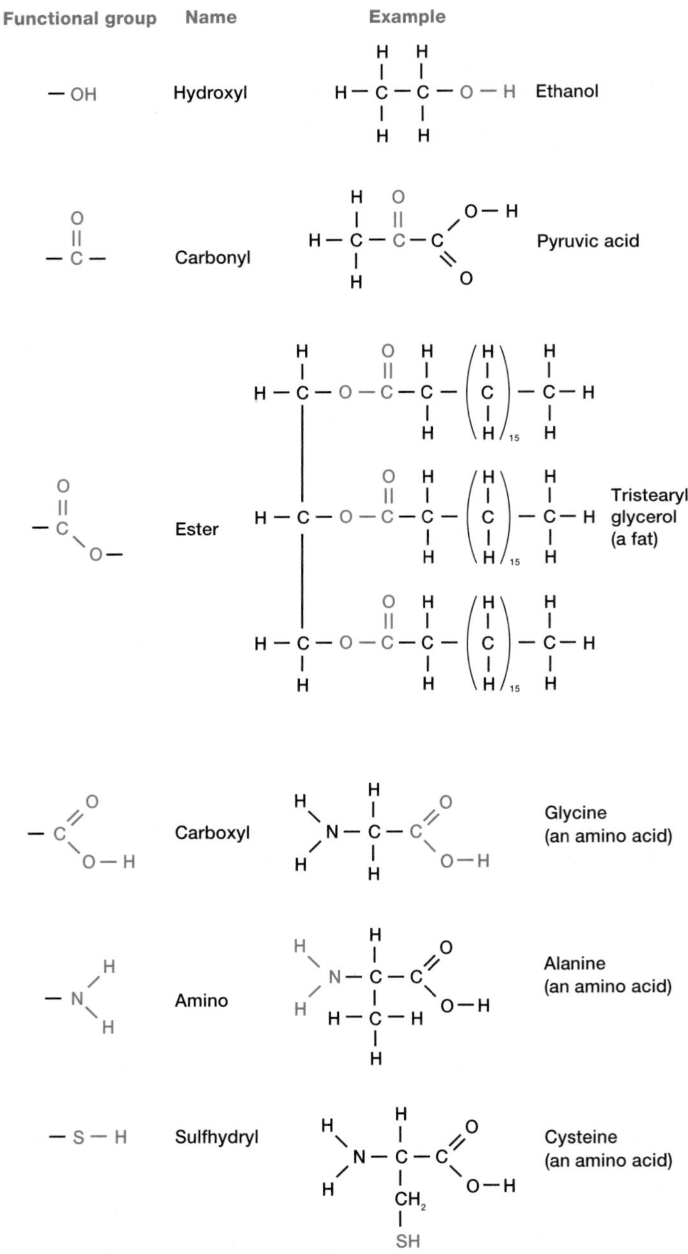

FIGURE AI.6 Functional Groups. Some common functional groups in organic molecules. The groups are shown in red.

Organic molecules are often divided into classes based on the nature of their functional groups. Ketones have a carbonyl group within the carbon chain, whereas alcohols have a hydroxyl on the chain. Organic acids have a carboxyl group, and amines have an amino group (**figure AI.7**).

Organic molecules may have the same chemical composition and yet differ in their molecular structure and properties. Such molecules are called isomers. One important class of isomers is the stereoisomers. Stereoisomers have the same atoms arranged in the same nucleus-to-nucleus sequence but differ in the spatial arrangement of their atoms. For example, an amino acid such as alanine can form stereoisomers (**figure AI.8**). L-Alanine and other L-amino acids are the stereoisomer forms normally present in proteins.

CARBOHYDRATES

Carbohydrates are aldehyde or ketone derivatives of polyhydroxy alcohols. The smallest and least complex carbohydrates are the simple sugars or monosaccharides. The most common sugars have five or six carbons. A sugar in its ring form has two isomeric structures, the α and β forms, that differ in the orientation of the

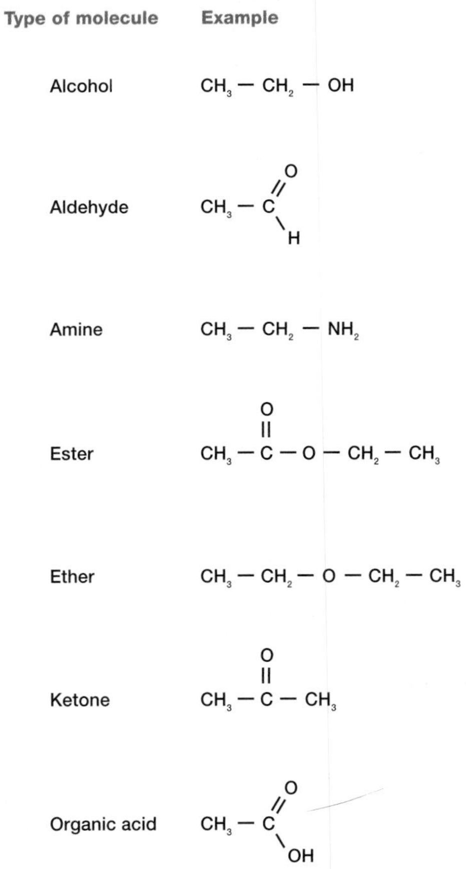

FIGURE AI.7 Types of Organic Molecules. These are classified on the basis of their functional groups.

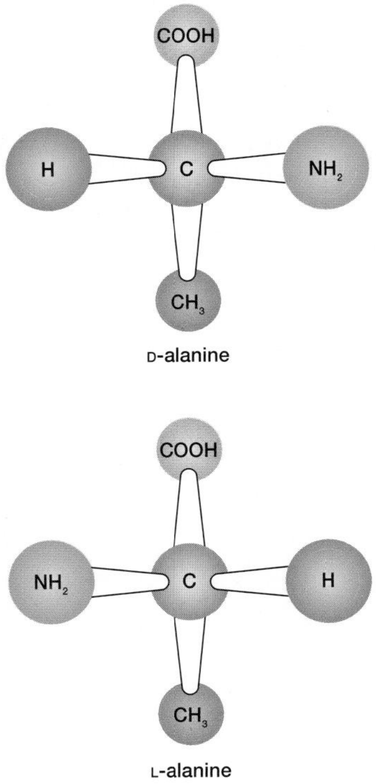

FIGURE AI.8 The Stereoisomers of Alanine. The α-carbon is in gray. L-alanine is the form usually present in proteins.

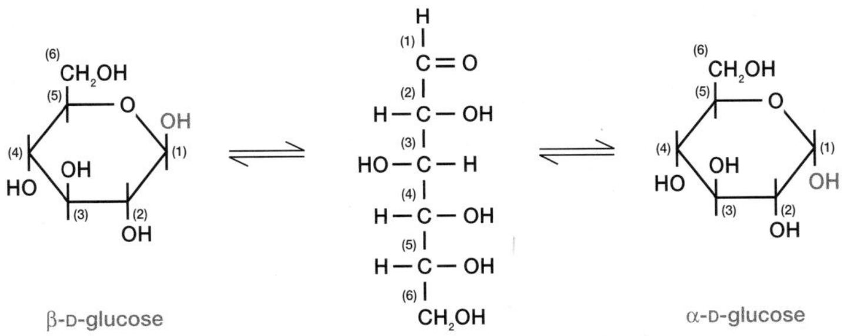

FIGURE AI.9 The Interconversion of Monosaccharide Structures. The open chain form of glucose and other sugars is in equilibrium with closed ring structures (depicted here with Haworth projections). Aldehyde sugars form cyclic hemiacetals, and keto sugars produce cyclic hemiketals. When the hydroxyl on carbon one of cyclic hemiacetals projects above the ring, the form is known as a β form. The α form has a hydroxyl that lies below the plane of the ring. The same convention is used in showing the α and β forms of hemiketals such as those formed by fructose.

hydroxyl on the aldehyde or ketone carbon, which is called the anomeric or glycosidic carbon (**figure AI.9**). Microorganisms have many sugar derivatives in which a hydroxyl is replaced by an amino group or some other functional group (e.g., glucosamine).

Two monosaccharides can be joined by a bond between the anomeric carbon of one sugar and a hydroxyl or the anomeric carbon of the second (**figure AI.10**). The bond joining sugars is a glycosidic bond and may be either α or β depending on the orientation of the anomeric carbon. Two sugars linked in this way constitute a disaccharide. Some common disaccharides are maltose (two glucose molecules), lactose (glucose and galactose), and sucrose (glucose and fructose). If 10 or more sugars are linked together by glycosidic bonds, a polysaccharide is formed. For example, starch and glycogen are common polymers of glucose that are used as sources of carbon and energy (**figure AI.11**).

LIPIDS

All cells contain a heterogeneous mixture of organic molecules that are relatively insoluble in water but very soluble in nonpolar solvents such as chloroform, ether, and benzene. These molecules are called lipids. Lipids vary greatly in structure and include triacylglycerols, phospholipids, steroids, carotenoids, and many other types. Among other functions, they serve as membrane components, storage forms for carbon and energy, precursors of other cell constituents, and protective barriers against water loss.

Most lipids contain fatty acids, which are monocarboxylic acids that often are straight chained but may be branched. Saturated fatty acids lack double bonds in their carbon chains, whereas unsaturated fatty acids have double bonds. The most common fatty acids are 16 or 18 carbons long.

Two good examples of common lipids are triacylglycerols and phospholipids. Triacylglycerols are composed of glycerol esterified to three fatty acids (**figure AI.12*a***). They are used to store carbon and energy. Phospholipids are lipids that contain at least one phosphate group and often have a nitrogenous constituent as well. Phosphatidylethanolamine is an important phospholipid frequently present in bacterial membranes (figure AI.13*b*). It is composed of two fatty acids esterified to glycerol. The third glycerol hydroxyl is joined with a phosphate group, and ethanolamine is attached to the phosphate. The resulting lipid is very asymmetric with a hydrophobic nonpolar end contributed by the fatty acids and a polar, hydrophilic end. In cell membranes, the hydrophobic end is buried in the interior of the membrane, while the polar-charged end is at the membrane surface and exposed to water.

PROTEINS

The basic building blocks of proteins are amino acids. An amino acid contains a carboxyl group and an amino group on its alpha carbon (**figure AI.13**). About 20 amino acids are normally found in proteins; they differ from each other with respect to their side chains. In proteins, amino acids are linked together by peptide bonds between their carboxyls and α-amino groups to form linear polymers called peptides. If a peptide contains more than 50 amino acids, it usually is called a polypeptide. Each protein is composed of one or more polypeptide chains and has a molecular weight greater than about 6,000 to 7,000.

Proteins have three or four levels of structural organization and complexity. The primary structure of a protein is the sequence of the amino acids in its polypeptide chain or chains. The structure of the polypeptide chain backbone is also

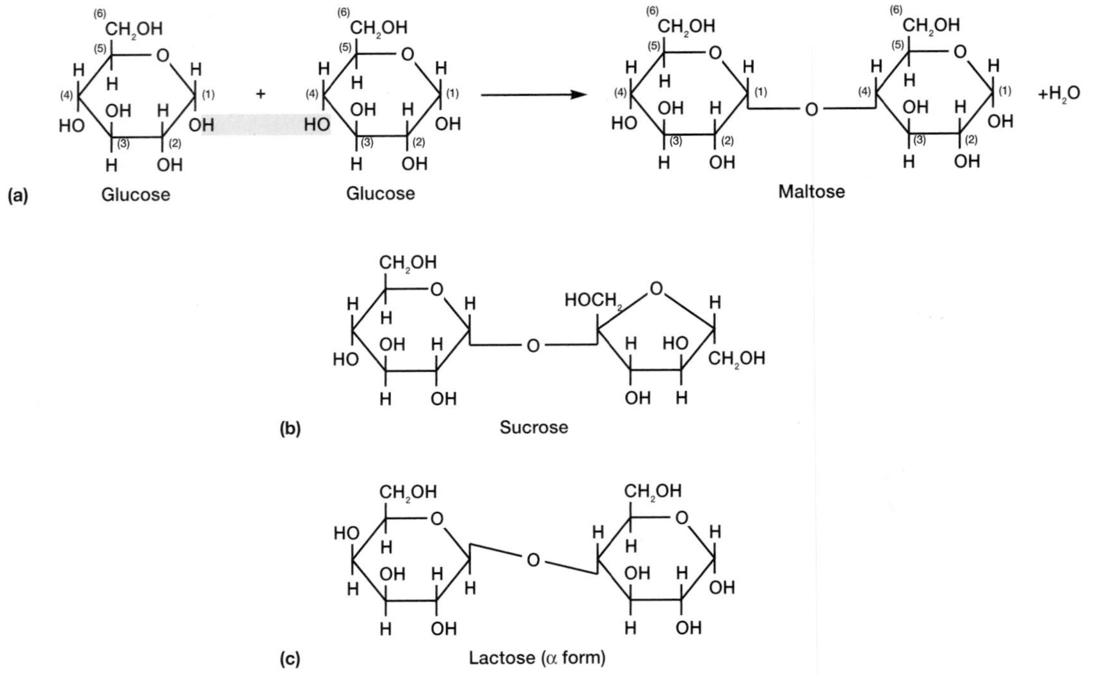

FIGURE AI.10 Common Disaccharides. (a) The formation of maltose from two molecules of an α-glucose. The bond connecting the glucose extends between carbons one and four, and involves the α form of the anomeric carbon. Therefore, it is called an α (1→4) glycosidic bond. (b) Sucrose is composed of a glucose and a fructose joined to each other through their anomeric carbons, and αβ (1→2) bond. (c) The milk sugar lactose contains galactose and glucose joined by a β (1→4) glycosidic bond.

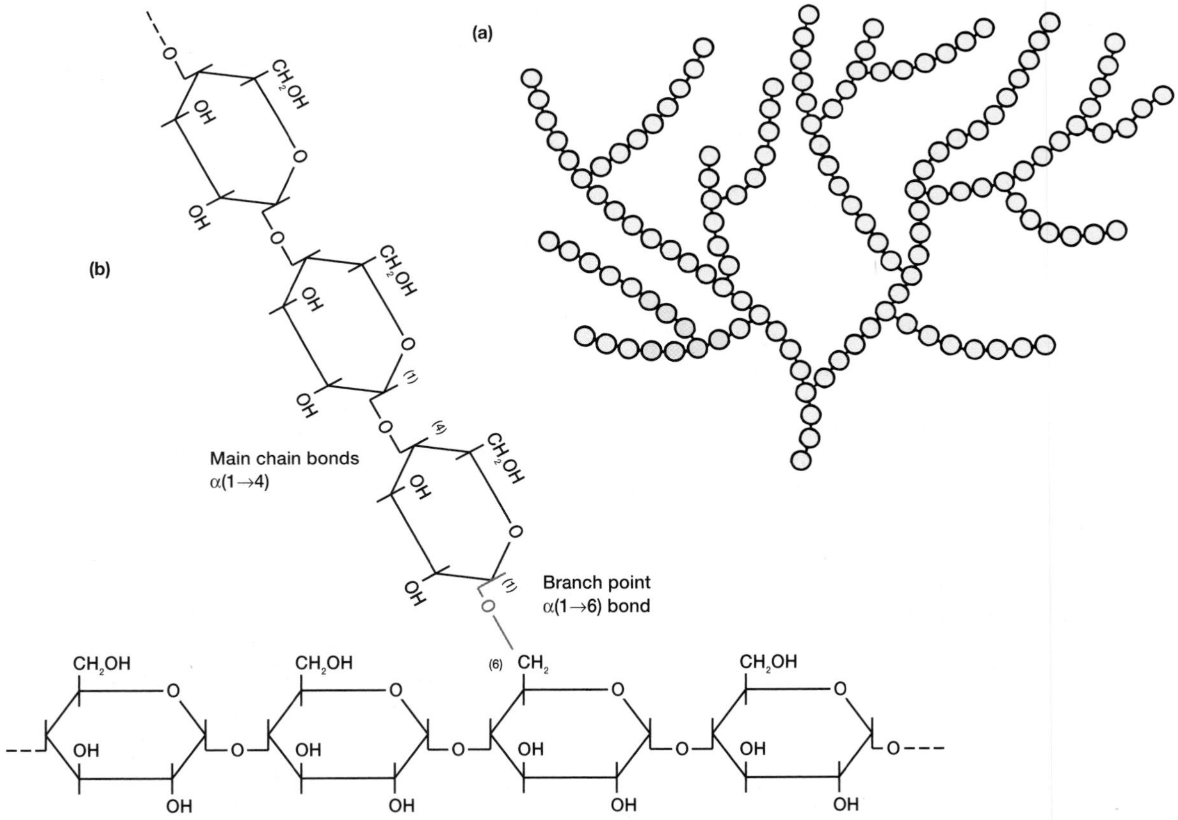

FIGURE AI.11 Glycogen and Starch Structure. (a) An overall view of the highly branched structure characteristic of glycogen and most starch. The circles represent glucose residues. (b) A close-up of a small part of the chain (shown in blue in part *a*) revealing a branch point with its α (1→6) glycosidic bond, which is colored blue.

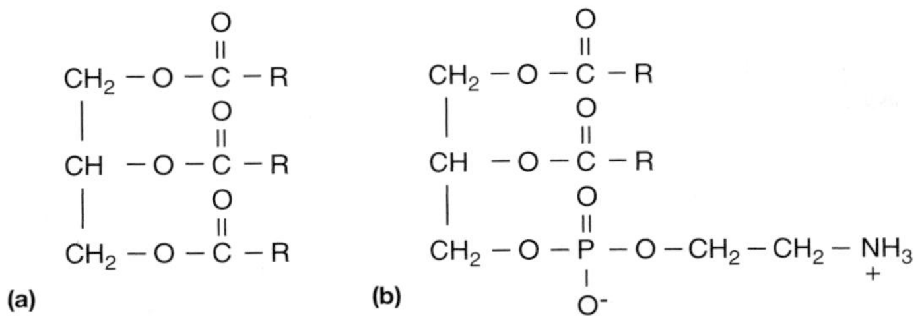

FIGURE AI.12 Examples of Common Lipids. (a) A triacylglycerol or neutral fat. (b) The phospholipid phosphatidylethanolamine. The R groups represent fatty acid side chains.

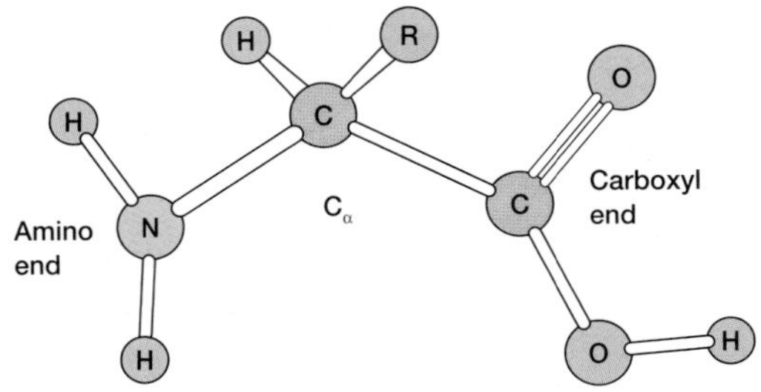

FIGURE AI.13 L-Amino Acid Structure. The uncharged form is shown.

considered part of the primary structure. Each different polypeptide has its own amino acid sequence that is a reflection of the nucleotide sequence in the gene that codes for its synthesis. The polypeptide chain can coil along one axis in space into various shapes such as the α-helix (**figure AI.14**). This arrangement of the polypeptide in space around a single axis is called the secondary structure. Secondary structure is formed and stabilized by the interactions of amino acids that are fairly close to one another on the polypeptide chain. The polypeptide with its primary and secondary structure can be coiled or organized in space along three axes to form a more complex, three-dimensional shape (**figure AI.15**). This level of organization is the tertiary structure (**figure AI.16**). Amino acids more distant from one another on the polypeptide chain contribute to tertiary structure. Secondary and tertiary structures are examples of conformation, molecular shape that can be changed by bond rotation and without breaking covalent bonds. When a protein contains more than one polypeptide chain, each chain with its own primary, secondary, and tertiary structure associates with the other chains to form the final molecule. The way in which polypeptides associate with each other in space to form the final protein is called the protein's quaternary structure (**figure AI.17**). The final conformation of a protein is ultimately determined by the amino acid sequence of its polypeptide chains.

Protein secondary, tertiary, and quaternary structure is largely determined and stabilized by many weak noncovalent forces such as hydrogen bonds and ionic bonds. Because of this, protein shape often is very flexible and easily changed. This flexibility is very important in protein function and in the regulation of enzyme

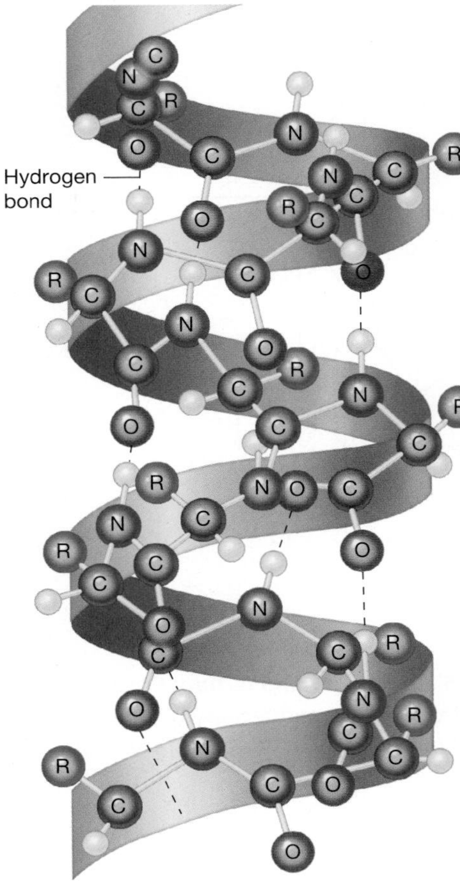

FIGURE AI.14 The α-Helix. A polypeptide twisted into one type of secondary structure, the α-helix. The helix is stabilized by hydrogen bonds joining peptide bonds that are separated by three amino acids.

activity. Because of their flexibility, however, proteins readily lose their proper shape and activity when exposed to harsh conditions. The only covalent bond commonly involved in the secondary and tertiary structure of proteins is the disulfide bond. The disulfide bond is formed when two cysteines are linked through their sulfhydryl groups. Disulfide bonds generally strengthen or stabilize protein structure.

NUCLEIC ACIDS

The nucleic acids, deoxyribonucleic acid (DNA) and ribonucleic acid (RNA), are polymers of deoxyribonucleosides and ribonucleosides joined by phosphate groups. The nucleosides in DNA contain the purines adenine and guanine, and the pyrimidine bases thymine and cytosine. In RNA the pyrimidine uracil is substituted for thymine. Because of their importance for genetics and molecular biology, the chemistry of nucleic acids is introduced earlier in the text. The structure and synthesis of purines and pyrimidines are discussed in chapter 11 (section 11.6). The structures of DNA and RNA are described in chapter 12 (section 12.3).

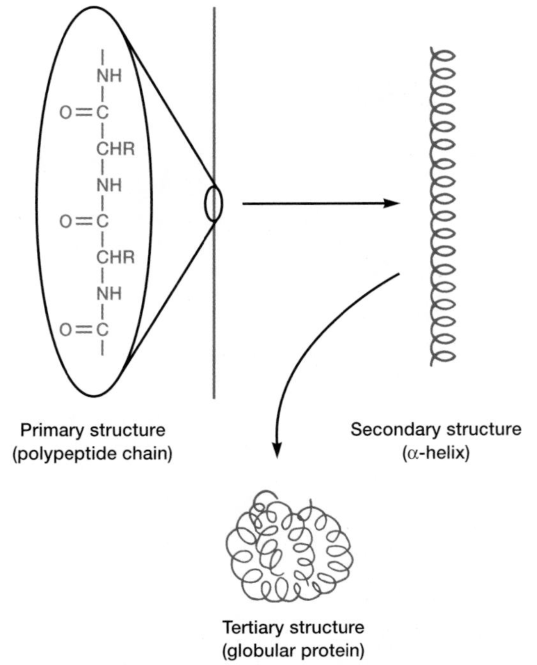

FIGURE AI.15 Secondary and Tertiary Protein Structures. The formation of secondary and tertiary protein structures by folding a polypeptide chain with its primary structure.

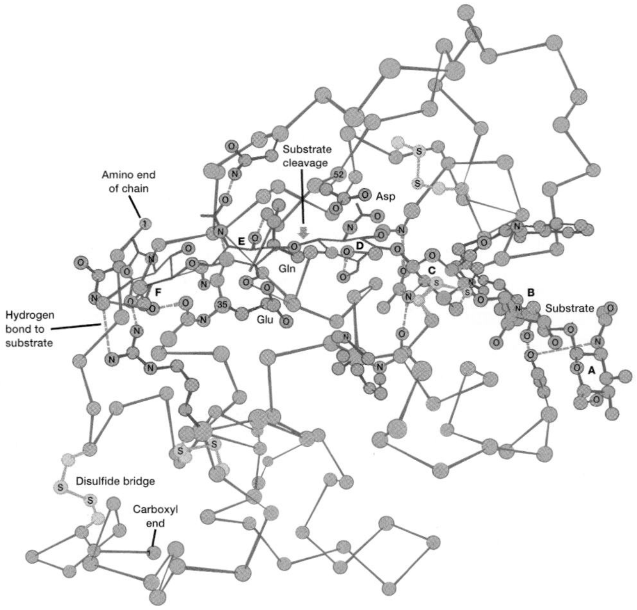

FIGURE AI.16 Lysozyme. The tertiary structure of the enzyme lysozyme. (a) A diagram of the protein's polypeptide backbone with the substrate hexasaccharide shown in blue. The point of substrate cleavage is indicated.

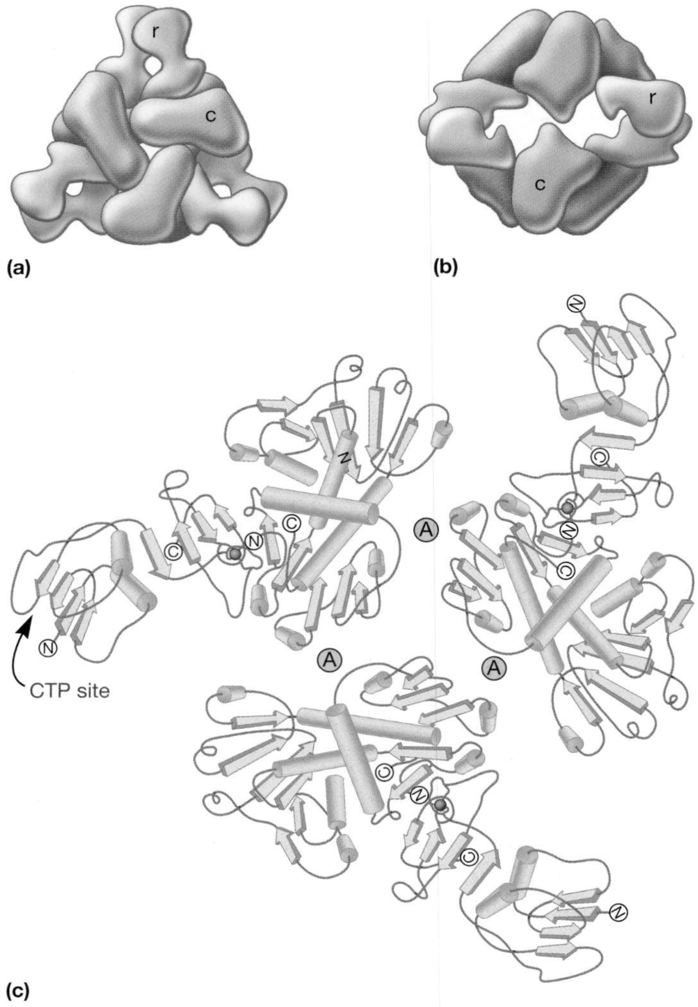

FIGURE AI.17 An Example of Quaternary Structure. The enzyme aspartate carbamoyltransferase from *Escherichia coli* has two types of subunits, catalytic and regulatory. The association between the two types of subunits is shown: (a) a top view, and (b) a side view of the enzyme. The catalytic (C) and regulator (r) subunits are shown in different colors. (c) The peptide chains shown when viewed from the top as in (a). The active sites of the enzyme are located at the positions indicated by A.

(a and b) Adapted from Krause, et al., in Proceedings of the National Academy of Sciences, *V. 82, 1985, as appeared in* Biochemistry, *3d edition, by Lubert Stryer. Copyright © 1975, 1981, 1988. Reprinted with permission of W. H. Freeman and Company. (c) Adapted from Kantrowitz, et al., in* Trends in Biochemical Science, *V. 5, 1980, as appeared in* Biochemistry, *3d edition, by Lubert Stryer. Copyright © 1975, 1981, 1988. Reprinted with permission of W. H. Freeman and Company.*

Appendix II

Common Metabolic Pathways

This appendix contains a few of the more important pathways discussed in the text, particularly those involved in carbohydrate catabolism. Enzyme names and final end products are given in color. Consult the text for a description of each pathway and its physiologic role.

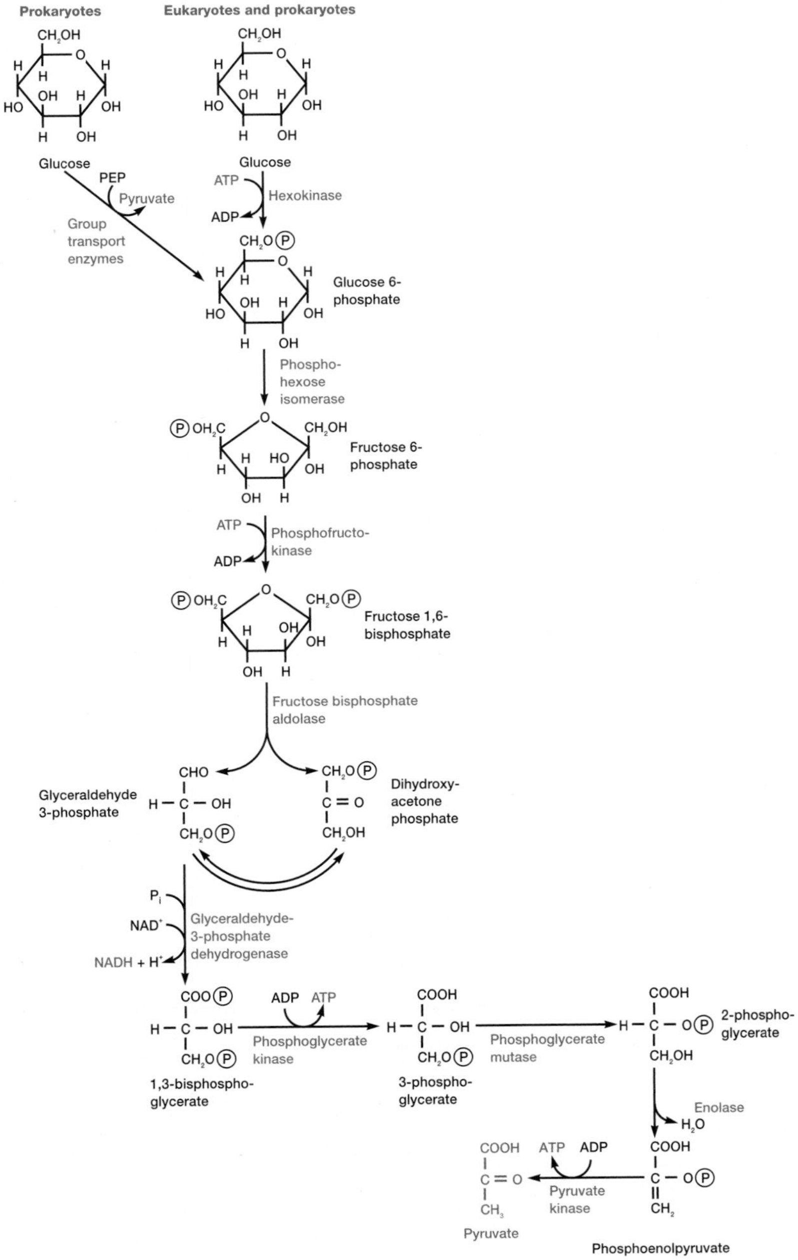

FIGURE AII.1 The Embden-Meyerhof Pathway. This pathway converts glucose and other sugars to pyruvate and generates NADH and ATP. In some prokaryotes, glucose is phosphorylated to glucose 6-phosphate during group translocation transport across the plasma membrane.

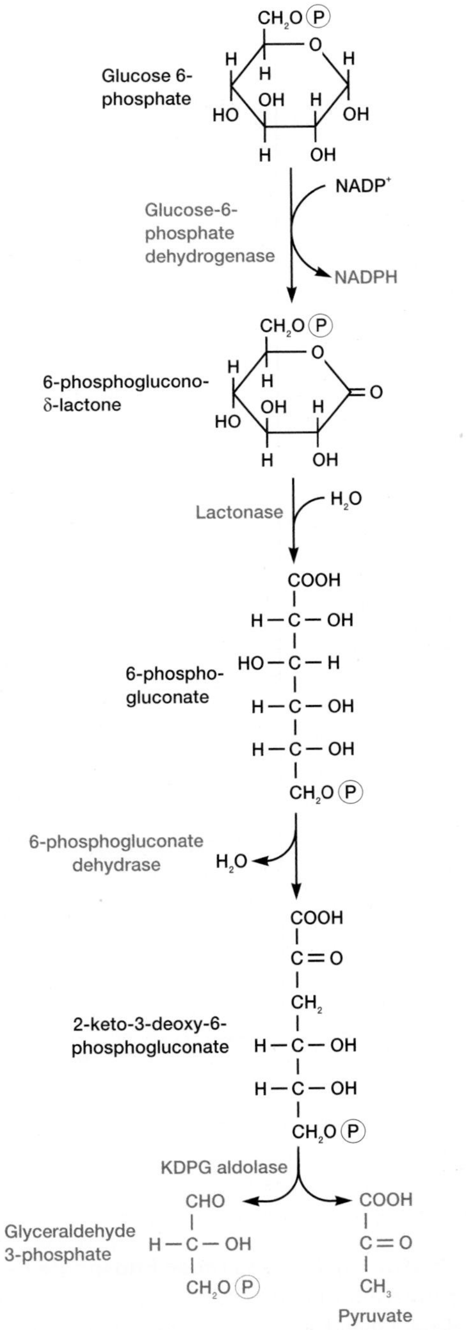

FIGURE AII.2 The Entner-Doudoroff Pathway.

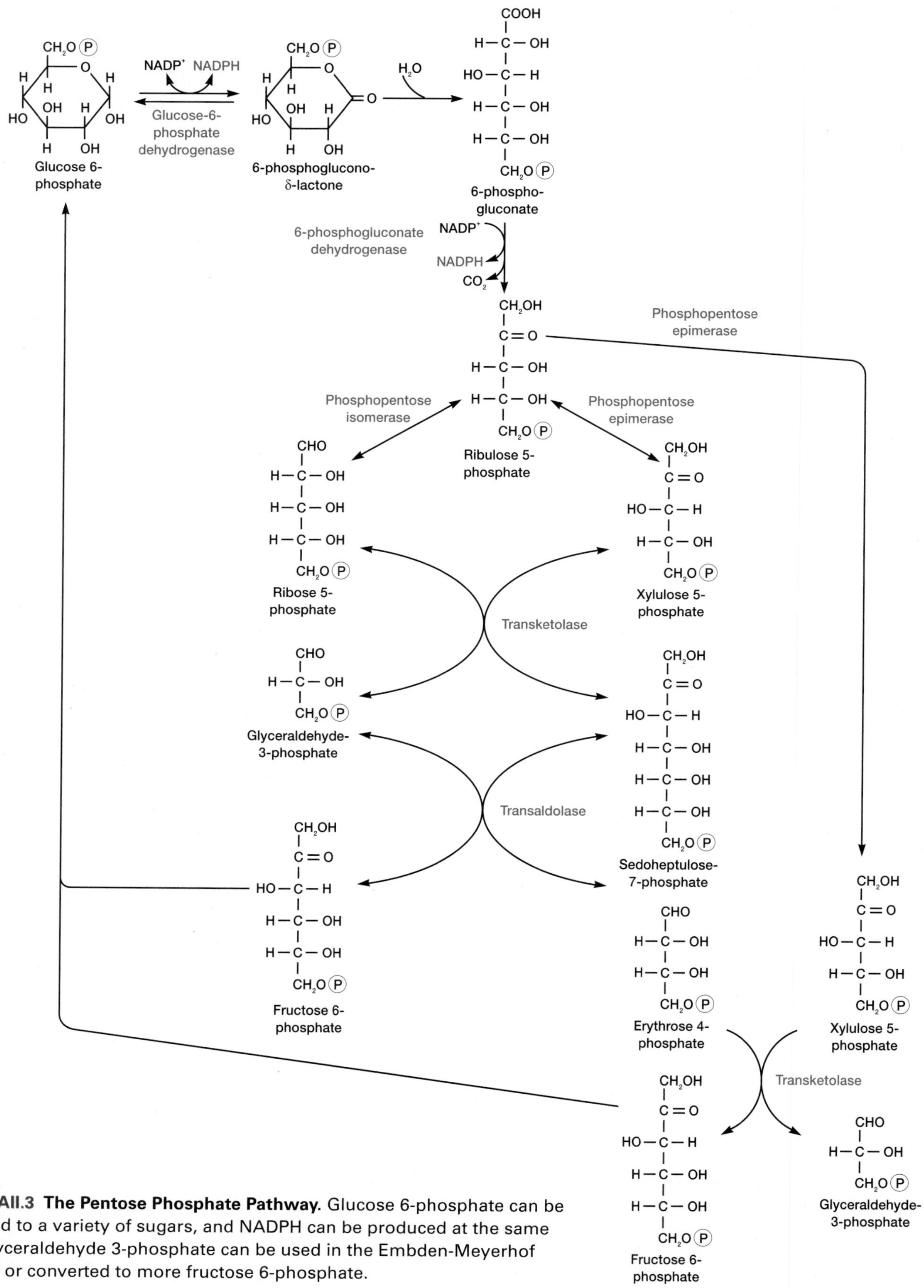

FIGURE AII.3 The Pentose Phosphate Pathway. Glucose 6-phosphate can be converted to a variety of sugars, and NADPH can be produced at the same time. Glyceraldehyde 3-phosphate can be used in the Embden-Meyerhof pathway or converted to more fructose 6-phosphate.

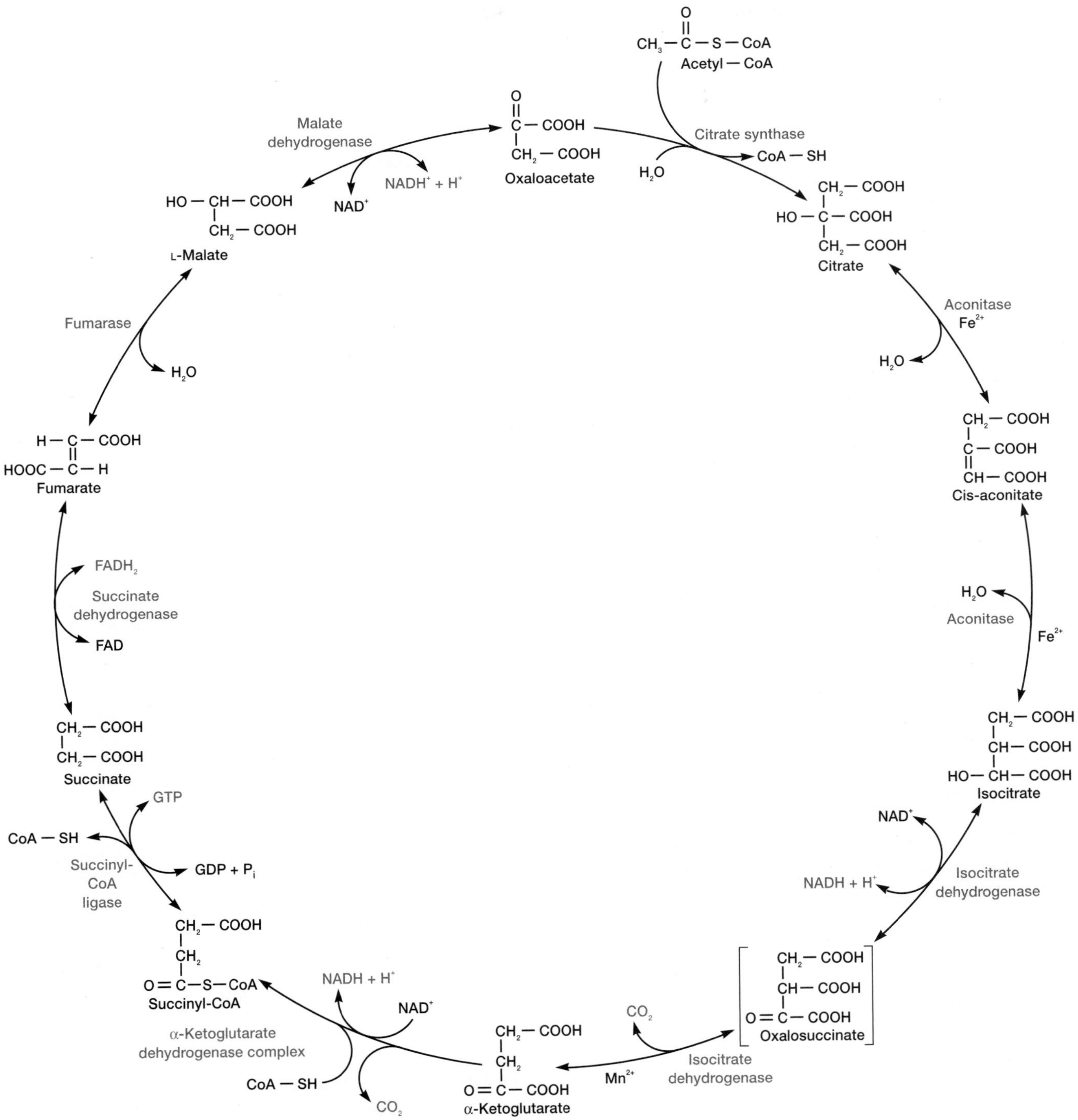

FIGURE AII.4 The Tricarboxylic Acid Cycle. Cis-aconitate and oxalosuccinate remain bound to aconitase and isocitrate dehydrogenase. Oxalosuccinate has been placed in brackets because it is so unstable.

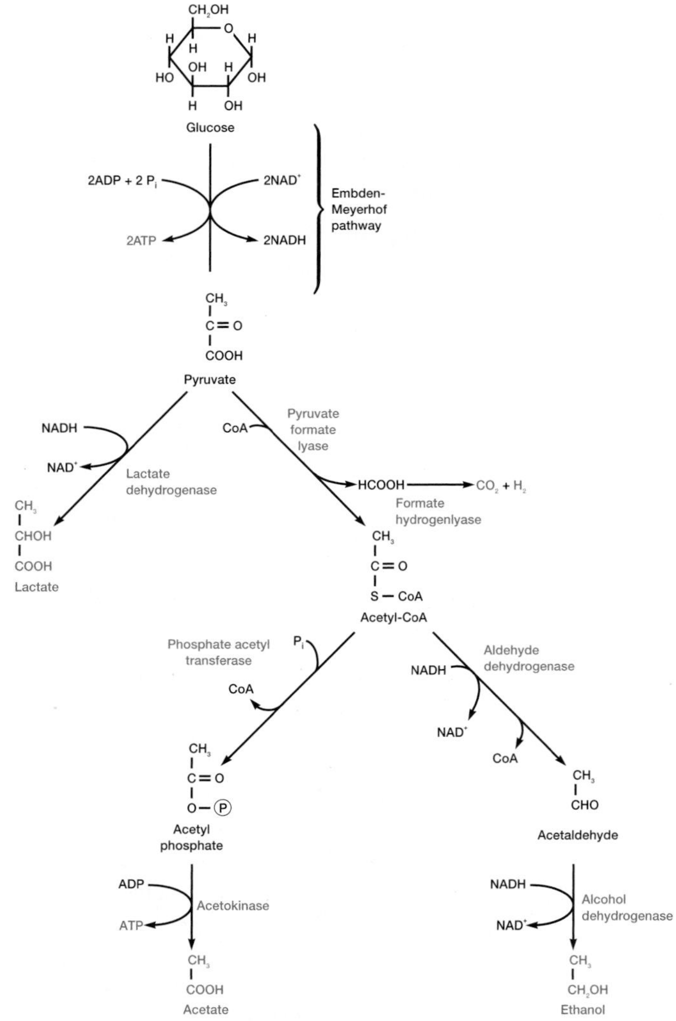

FIGURE AII.5 The Mixed Acid Fermentation Pathway. This pathway is characteristic of many members of the *Enterobacteriaceae* such as *Escherichia coli.*

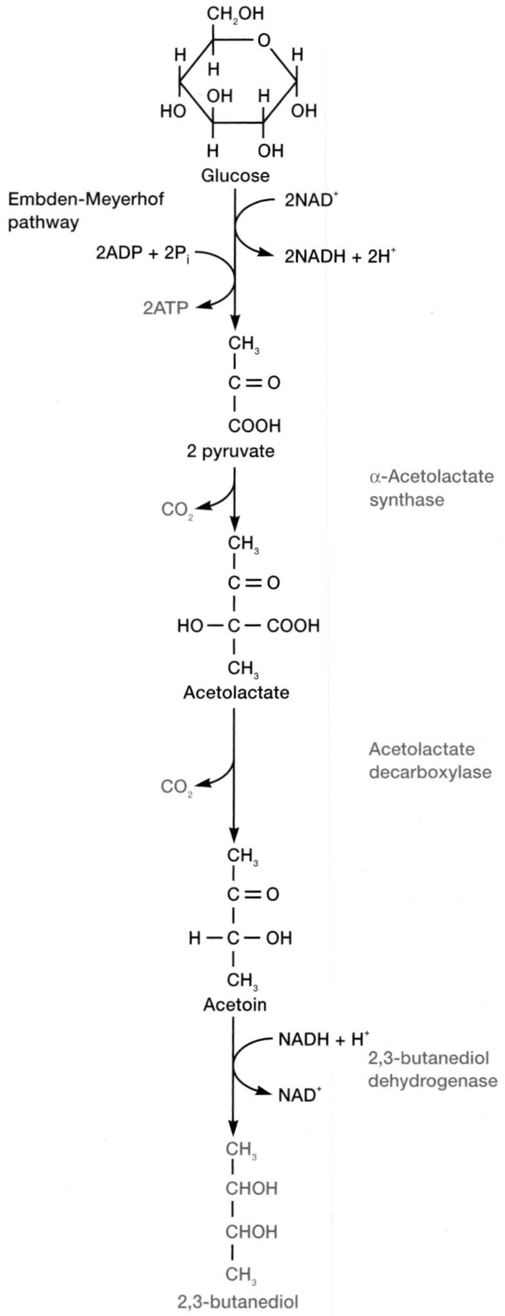

FIGURE AII.6 The Butanediol Fermentation Pathway. This pathway is characteristic of members of the *Enterobacteriaceae* such as *Enterobacter.* Other products may also be formed during butanediol fermentation.

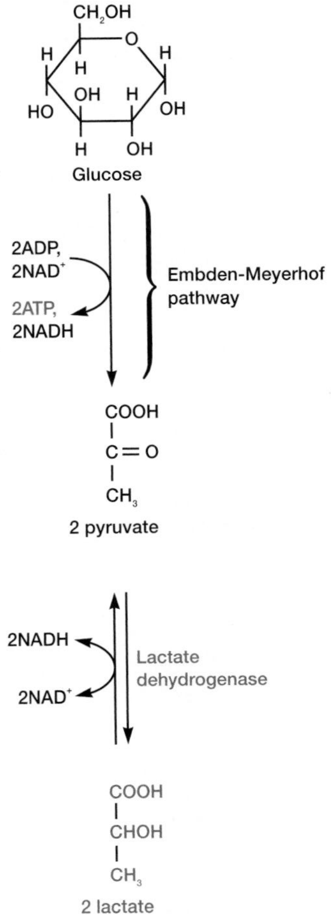

(a)

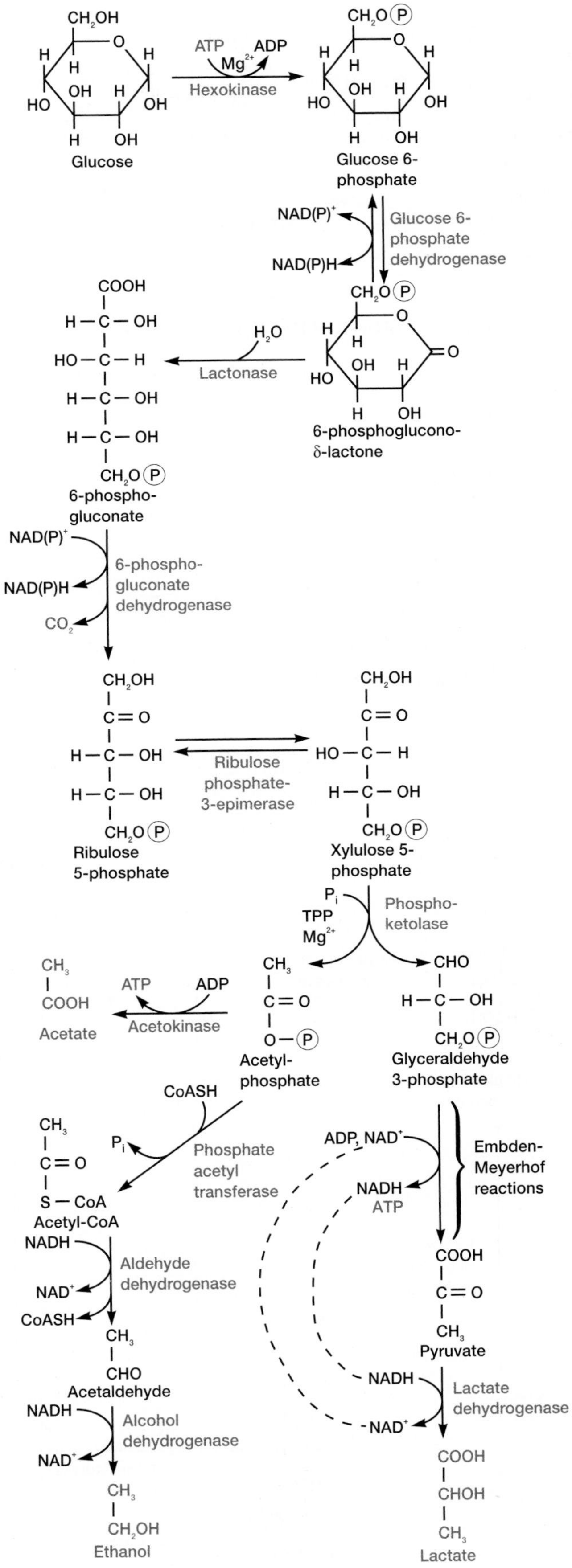

(b)

FIGURE AII.7 Lactic Acid Fermentations. (a) Homolactic fermentation pathway. (b) Heterolactic fermentation pathways.

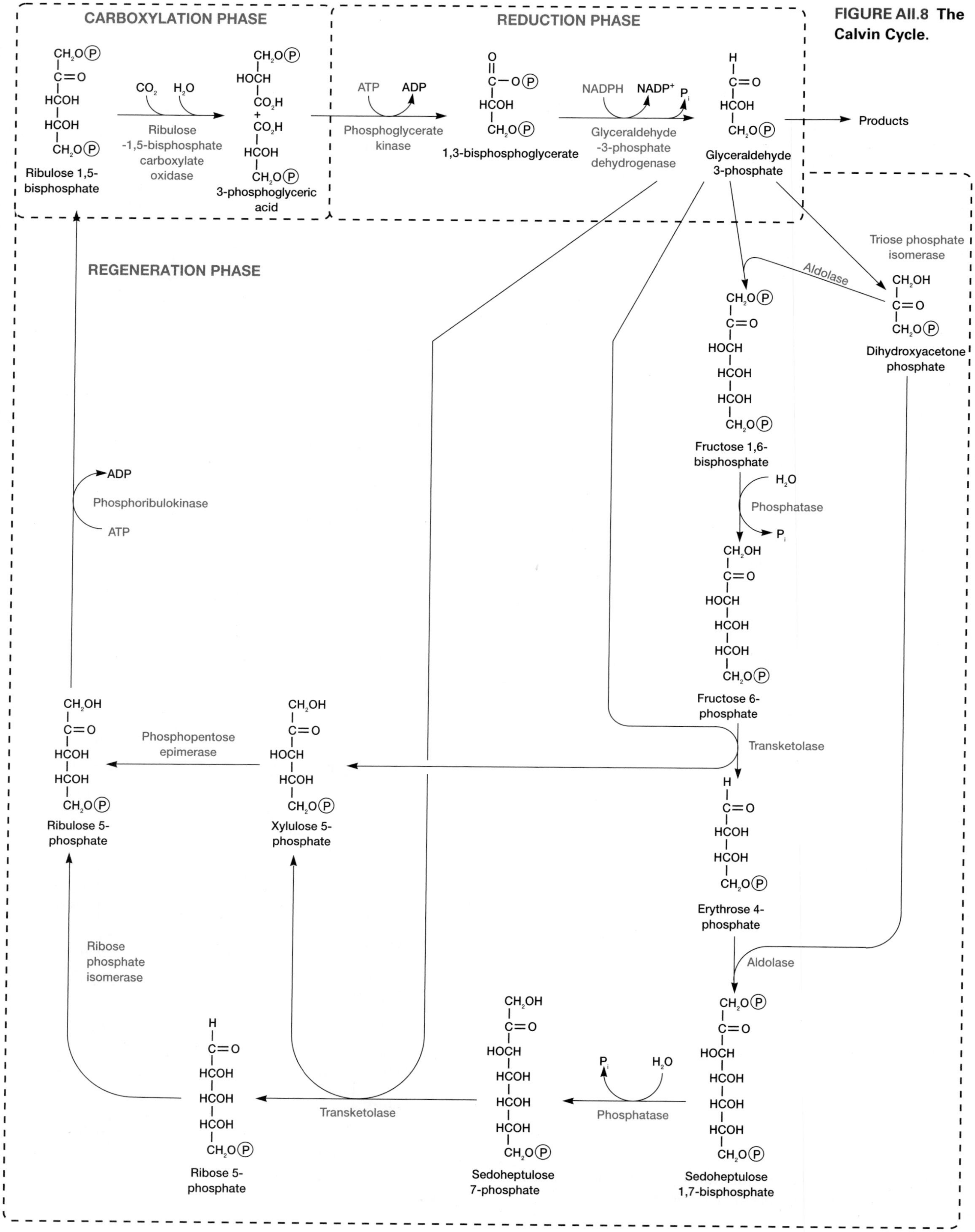

FIGURE AII.8 The Calvin Cycle.

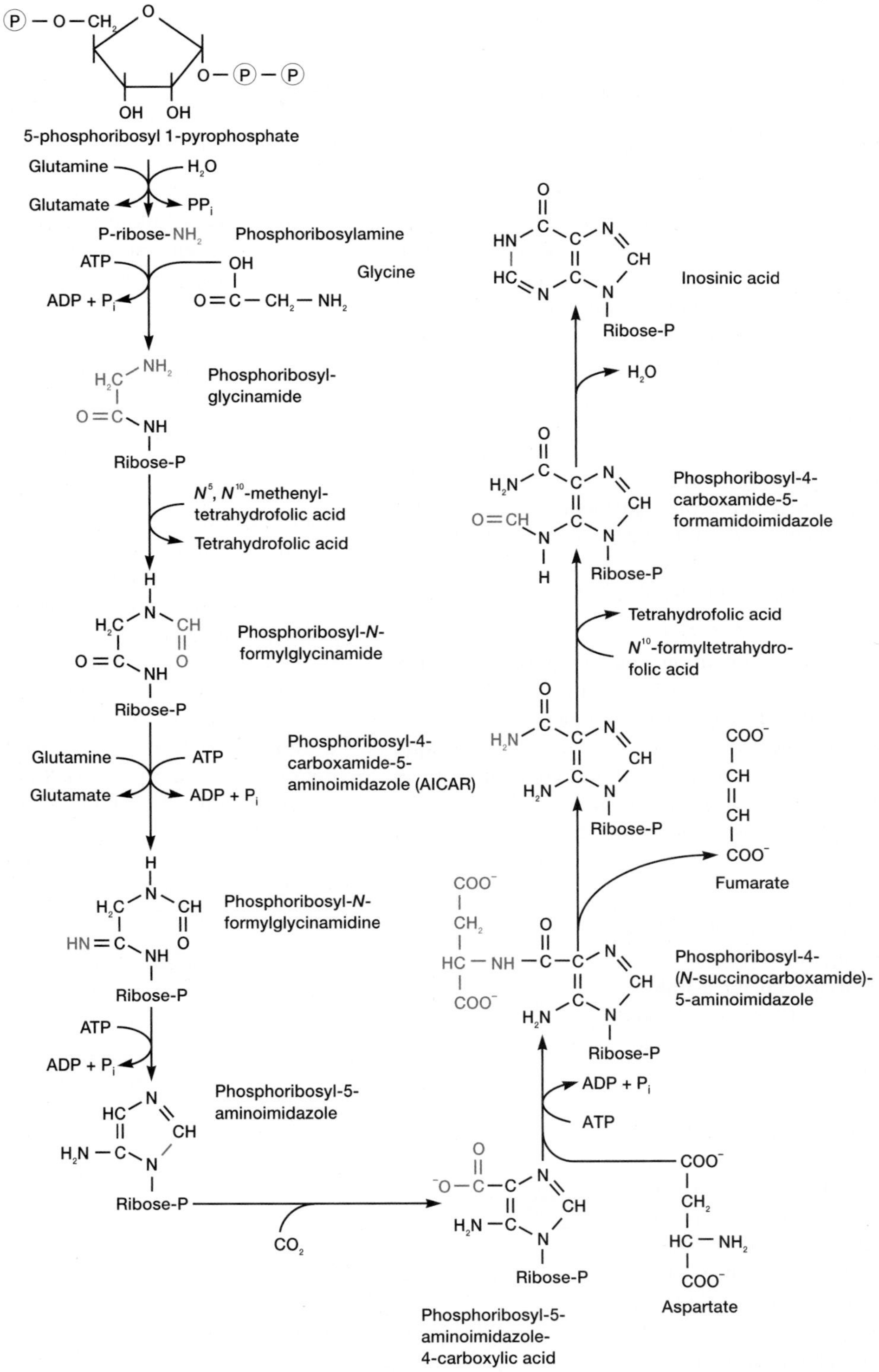

FIGURE AII.9 The Pathway for Purine Biosynthesis. Inosinic acid is the first purine end product. The purine skeleton is constructed while attached to a ribose phosphate.

Appendix III

Concept Mapping

Concept maps are visual tools for presenting and organizing information. They can take the place of an outline, though they often contain much more meaning because they can show connections and interconnections between information; this is not as easily done in an outline. Concept maps are also very flexible; there is a nearly infinite number of ways a concept map can be constructed. Concept maps also can be used by the individual making the map to express his or her understanding of concepts and their connections with each other. Because they are a visual representation of information, concept maps are also valuable study aids for students. In this role, they help the student understand the information as the student creates the map, and then they can be used to review information before a test.

As can be seen in the concept map shown here, concept maps are made of two basic components:

1. Boxes or circles, each containing a single concept, which is usually a noun. The boxes can be arranged in numerous ways, varying from hierarchical forms to more free-form arrangements. In the hierarchical forms, the most general concept, and usually the main focus of the concept map, is located at the top of the page. The concepts below this become increasingly more specific as the viewer moves down the page.
2. Connecting lines that join each concept box to at least one other box. Each connecting line has a word or a phrase associated with it—a linking word or phrase. These words are almost never nouns but are verbs, adjectives, or adverbs. The connecting line often is drawn as an arrow, illustrating the direction of the relationship. Thus it is possible to "read" the two concepts and the linking words as a sentence. For instance in the concept map here, the connection between "cell wall" and "gram-positive" can be used to form the sentence: "The cell wall, if present, may be gram-positive." Likewise, the connection between "cell wall" and "gram-negative" forms the sentence: "The cell wall, if present, may be gram-negative." However, because the concepts of gram-positive and gram-negative (as well as acid-fast) are placed close together and connected to "cell wall" by the same linking phrase, the viewer of the map can quickly see that cell walls, if present, may be either gram-positive, gram-negative, or acid-fast.

The power of concept maps lies in their ability to illustrate which concepts are bigger (and more important) and which are details. They also illustrate that multiple concepts may be connected. This is especially useful for complex subjects such as microbiology. The main task for the creator of the map is to find the right linking phrase to show connections between concepts. This takes practice and a deeper understanding of the information than simply memorizing definitions of terms.

In this text, you are asked at the end of most chapters to create a concept map using terms provided. You will begin by drawing the boxes and placing the terms in them in a way that makes sense to you. Then, you will need to add linking words or phrases to your map. You will have generated ideas for this as you arranged the terms when you began the map. You may find your first experiences with concept mapping to be frustrating. But most students report that when they have invested some time in their first few concept maps, they can never go back to organizing information in other ways. Maps can make the time you invest in studying more effective. And creating concept maps with a partner or a group is a great way to review material in a meaningful way. Give concept maps a try, and let your creative side show!

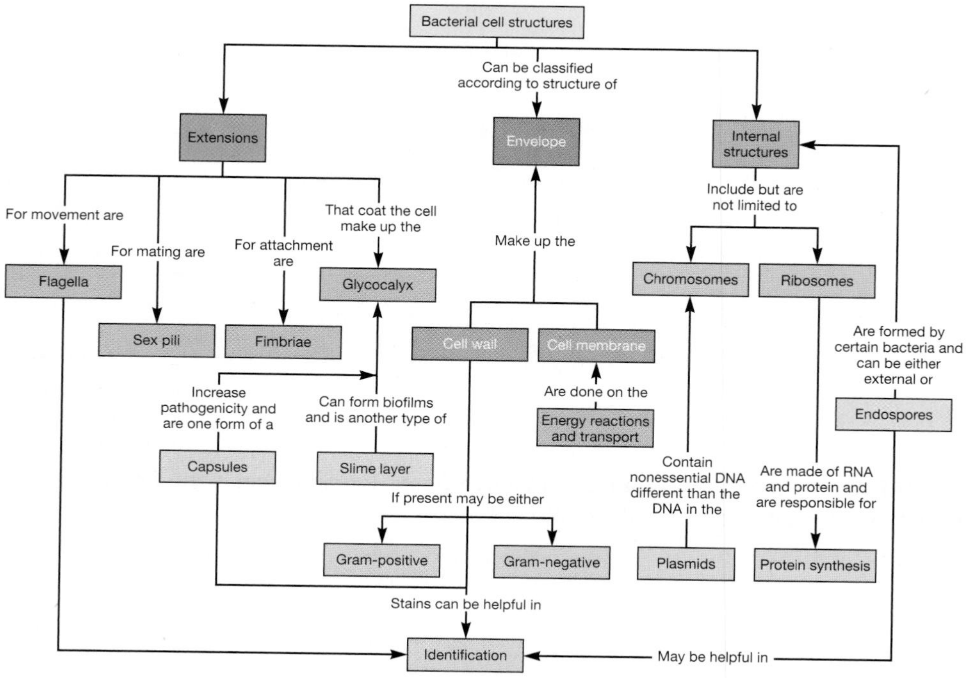

Glossary

Pronunciation Guide

Some of the boldface terms in this glossary are followed by a phonetic spelling in parentheses. These pronunciation aides usually come from *Dorland's Illustrated Medical Dictionary.* The following rules are taken from this dictionary and will help in using its phonetic spelling system.

1. An unmarked vowel ending a syllable (an open syllable) is long; thus, *ma* represents the pronunciation of *may; ne,* that of *knee; ri,* of *wry; so,* of *sew; too,* of *two;* and *vu,* of *view.*
2. An unmarked vowel in a syllable ending with a consonant (a closed syllable) is short; thus, *kat* represents *cat; bed, bed; hit, hit; not, knot; foot, foot;* and *kusp, cusp.*
3. A long vowel in a closed syllable is indicated by a macron; thus, *māt* stands for *mate; sēd,* for *seed; bīl,* for *bile; mōl,* for *mole; fūm,* for *fume;* and *fo͞ol,* for *fool.*
4. A short vowel that ends or itself constitutes a syllable is indicated by a breve; thus, *ĕ-fekt′* for *effect, ĭ-mūn′* for *immune,* and *ŭ-klo͞od′* for *occlude.*

Primary (′) and secondary (″) accents are shown in polysyllabic words. Unstressed syllables are followed by hyphens.

Some common vowels are pronounced as indicated here.

ə	sofa	ĕ	met	ŏ	got
ā	mate	ī	bite	ū	fuel
ă	bat	ĭ	bit	ŭ	but
ē	beam	ō	home		

A

AB toxins Exotoxins composed of two parts (A and B). The B portion is responsible for toxin binding to a cell but does not directly harm it; the A portion enters the cell and disrupts its function.

ABC protein secretion pathway *See* ATP-binding cassette transporters.

accessory pigments Photosynthetic pigments such as carotenoids and phycobiliproteins that aid chlorophyll in trapping light energy.

acellular slime mold *See* plasmodial slime mold

acetyl-CoA pathway A biochemical pathway used by methanogens to fix CO_2 and by acetogens to generate acetic acid.

acetyl-coenzyme A (acetyl-CoA) A combination of acetic acid and coenzyme A that is energy rich; it is produced by many catabolic pathways and is the substrate for the tricarboxylic acid cycle, fatty acid biosynthesis, and other pathways.

acid-fast Refers to bacteria (e.g., mycobacteria) that cannot be easily decolorized with acid alcohol after being stained with dyes such as basic fuchsin.

acid-fast staining A staining procedure that differentiates between bacteria based on their ability to retain a dye when washed with an acid alcohol solution.

acidic dyes Dyes that are anionic or have negatively charged groups such as carboxyls.

acidophile (as′id-o-fīl″) A microorganism that has its growth optimum between about pH 0 and 5.5.

acquired immune tolerance The ability to produce antibodies against nonself antigens while "tolerating" (not producing antibodies against) self antigens.

acquired immunity Refers to the type of specific (adaptive) immunity that develops after exposure to a suitable antigen or is produced after antibodies are transferred from one individual to another.

actinobacteria (ak″tĭ-no-bak-tē r-e-ah) A group of gram-positive bacteria containing the actinomycetes and their high G + C relatives.

actinomycete (ak″tĭ-no-mi′sēt) An aerobic, gram-positive bacterium that forms branching filaments (hyphae) and asexual spores.

actinorhizae Associations between actinomycetes and plant roots.

activated sludge Solid matter or sediment composed of actively growing microorganisms that participate in the aerobic portion of a biological sewage treatment process.

activation energy The energy required to bring reacting molecules together to reach the transition state in a chemical reaction.

activator-binding site *See* activator protein.

activator protein A transcriptional regulatory protein that binds to a specific site on DNA (activator-binding site) and enhances transcription initiation.

active carrier An individual who has an overt clinical case of a disease and who can transmit the infection to others.

active immunization The induction of immunity by natural exposure to a pathogen or by vaccination.

active site The part of an enzyme that binds the substrate to form an enzyme-substrate complex and catalyze the reaction. Also called the catalytic site.

active transport The transport of solute molecules across a membrane against a gradient; it requires a carrier protein and the input of energy.

acute carrier *See* casual carrier.

acute infections Infections with a rapid onset that last for a relatively short time.

acute viral gastroenteritis An inflammation of the stomach and intestines caused by viruses.

acyclovir (a-si′klo-vir) A synthetic purine nucleoside derivative with antiviral activity against herpes simplex virus.

***N*-acylhomoserine lactone** *See* quorum sensing.

adaptive mutation Mutations that arose during the evolution of a species and help the species survive in its current environment. The selective value of these mutations causes them to become prevalent in the population.

adenine (ad′e-nēn) A purine derivative, 6-aminopurine, found in nucleosides, nucleotides, coenzymes, and nucleic acids.

adenine arabinoside (vidarabine) An antiviral agent used especially to treat keratitis and encephalitis caused by the herpes simplex virus.

adenosine diphosphate (ADP) (ah-den′o-sēn) The nucleoside diphosphate usually formed upon the breakdown of ATP when it provides energy for work.

adenosine 5′-triphosphate (ATP) A high energy molecule that serves as the cell's major form of energy currency.

adhesin (ad-he′zin) A molecular component on the surface of a microorganism that is involved in adhesion to a substratum (e.g., host tissue) or cell.

adjuvant Material added to an antigen in a vaccine preparation that increases its immunogenicity.

aerial mycelium The mat of hyphae formed by actinomycetes or fungi that grows above the substrate, imparting a fuzzy appearance to colonies.

aerobe An organism that grows in the presence of atmospheric oxygen.

aerobic anoxygenic photosynthesis (AAnP) Photosynthetic process in which electron donors such as organic matter or sulfide are used under aerobic conditions.

aerobic respiration A metabolic process in which molecules, often organic, are oxidized with oxygen as the final electron acceptor.

aerotolerant anaerobes Microbes that grow equally well whether or not oxygen is present.

aflatoxin (af″lah-tok′sin) A polyketide secondary fungal metabolite that can cause cancer.

agar (ahg′ar) A complex sulfated polysaccharide, usually from red algae, that is used as a solidifying agent in the preparation of culture media.

agglutinates The visible aggregates or clumps formed by an agglutination reaction.

agglutination reaction The formation of an insoluble immune complex by the cross-linking of cells or particles.

agglutinin The antibody responsible for an agglutination reaction.

airborne transmission The type of transmission in which the pathogen is suspended in the air and travels over a meter or more from the source to the host.

akinetes (ā-kin′ēts) Specialized, nonmotile, dormant, thick-walled resting cells formed by some cyanobacteria.

alcoholic fermentation A fermentation process that produces ethanol and CO_2 from sugars.

alga (al′gah) A common term for several unrelated groups of photosynthetic eukaryotic microorganisms lacking multicellular sex organs (except for the charophytes) and conducting vessels. Most are now considered protists.

algicide (al′jĭ-sīd) An agent that kills algae.

alignment In the comparison of gene, RNA, and protein sequences, the arrangement of sequences such that a base-by-base (or amino acid by amino acid) comparison can be made and used to calculate the degree of similarity between sequences.

alkaliphile (alkalophile) A microorganism that grows best at pHs from about 8.5 to 11.5.

allele An alternative form of a gene.

allergen An antigen that induces an allergic response.

allergic contact dermatitis An allergic reaction caused by haptens that combine with proteins in the skin to form the allergen that stimulates the immune response.

allergy *See* Type I hypersensitivity.

allochthonous (ăl′ək-thə-nəs) Substances not native to a given environment (e.g., nutrient influx into freshwater ecosystems).

allograft A transplant between genetically different individuals of the same species.

allosteric effector *See* allosteric enzyme.

allosteric enzyme An enzyme whose activity is altered by the noncovalent binding of a small effector (allosteric effector) at a regulatory site separate from the catalytic site; effector binding causes a conformational change in the enzyme's catalytic site, which leads to enzyme activation or inhibition.

alpha hemolysis *See* hemolysis.

alternative complement pathway An antibody-independent pathway of complement activation.

alternative splicing The use of different exons during RNA splicing to generate different polypeptides from the same gene.

alveolar macrophage A vigorously phagocytic macrophage located on the epithelial surface of the lung alveoli, where it ingests inhaled particulate matter and microorganisms.

amantadine (ah-man′tah-den) An antiviral agent used to prevent type A influenza infections.

amensalism A relationship in which the product of one organism has a negative effect on another organism.

amino acid activation The preparatory step of protein synthesis in which amino acids are attached to transfer RNA molecules. The reaction is catalyzed by aminoacyl-tRNA synthetases.

aminoacyl or **acceptor site (A site)** The ribosomal site that contains an aminoacyl-tRNA at the beginning of the elongation cycle during protein synthesis.

aminoacyl-tRNA synthetases *See* amino acid activation.

aminoglycoside antibiotics A group of antibiotics synthesized by *Streptomyces* and *Micromonospora* that contain a cyclohexane ring and amino sugars; all aminoglycoside antibiotics bind to the small ribosomal subunit and inhibit protein synthesis.

amoeboid movement Moving by means of cytoplasmic flow and the formation of pseudopodia.

amphibolic pathways Metabolic pathways that function both catabolically and anabolically.

amphipathic Term describing a molecule that has both hydrophilic and hydrophobic regions (e.g., phospholipids).

amphitrichous (am-fit′rĕ-kus) A cell with a single flagellum at each end.

amphotericin B (am″fo-ter′i-sin) An antifungal agent produced by *Streptomyces nodosus.*

amplification In reference to the polymerase chain reaction, the process of making large quantities of a particular DNA sequence.

anabolism The synthesis of complex molecules from simpler molecules with the input of energy and reducing power.

anaerobe (an-a′er-ōb) An organism that grows in the absence of free oxygen.

anaerobic digestion (an″a-er-o′bik) The microbiological treatment of sewage wastes under anaerobic conditions to produce methane.

anaerobic respiration An energy-conserving process in which the electron transport chain acceptor is a molecule other than oxygen.

anagenesis Changes in gene frequencies and distribution among species; the accumulation of small genetic changes within a population that introduces genetic variability but is not enough to result in either speciation or extinction.

anammoxosome The compartment within a planctomycete cell in which the anammox reaction occurs.

anammox reaction The coupled use of nitrite as an electron acceptor and ammonium ion as an electron donor under anoxic conditions to yield nitrogen gas.

anamnestic response (an″am-nes′tik) The recall, or the remembering, by the immune system of a prior response to a given antigen.

anaphylaxis (an″ah-fĭ-lak′sis) An immediate (type I) hypersensitivity reaction following exposure of a sensitized individual to the appropriate antigen.

anaplasia The reversion of an animal cell to a more primitive, undifferentiated state.

anaplerotic reactions (an′ah-plĕ-rot′ik) Reactions that replenish depleted tricarboxylic acid cycle intermediates.

anergy (an′ər-je) A state of unresponsiveness to antigens. Absence of the ability to generate a sensitivity reaction to substances that are expected to be antigenic.

anoxic Without oxygen present.

anoxygenic photosynthesis Photosynthesis that does not oxidize water to produce oxygen.

antibiotic A microbial product or its derivative that kills susceptible microorganisms or inhibits their growth.

antibody (immunoglobulin) A glycoprotein made by plasma cells in response to the introduction of an antigen.

antibody affinity The strength of binding between an antigen and an antibody.

antibody-dependent cell-mediated cytotoxicity (ADCC) The killing of antibody-coated target cells by cells with Fc receptors that recognize the Fc region of the bound antibody.

antibody-mediated immunity *See* humoral immunity.

antibody class switching The process by which plasma cells change their expression from one isotype of immunoglobulin to a second isotype.

antibody titer An approximation of the antibody concentration.

anticodon The three bases on a tRNA that are complementary to the codon on mRNA.

antigen A substance (such as a protein, nucleoprotein, polysaccharide, or sometimes a glycolipid) to which lymphocytes respond.

antigen-binding fragment (Fab) "Fragment antigen binding." A monovalent antigen-binding fragment of an immunoglobulin molecule that consists of one light chain and part of one heavy chain, linked by interchain disulfide bonds.

antigen-presenting cells (APCs) Cells that take in protein antigens, process them, and present antigen fragments to T cells in conjunction with MHC molecules so that the cells are activated; includes macrophages, B cells, and dendritic cells.

antigen processing The hydrolytic digestion of antigens to produce antigen fragments, which are collected by class I or class II MHC molecules and presented on the surface of a cell.

antigenic determinant site *See* epitope.

antigenic drift A small change due to mutation in the antigenic character of an organism that allows it to avoid attack by the immune system.

antigenic shift A major change in the antigenic character of an organism that makes it unrecognized by host immune mechanisms.

antigenic variation The capacity of some microbes to change surface proteins when in a host organism, thereby evading the host immune system.

antimetabolite A compound that blocks metabolic pathway function by competitively inhibiting a key enzyme because it closely resembles the normal enzyme substrate.

antimicrobial agent An agent that kills microorganisms or inhibits their growth.

antiport Coupled transport of two molecules in which one molecule enters the cell as the other leaves the cell.

antisense RNA A single-stranded RNA with a base sequence complementary to a segment of a target RNA molecule; when bound to the target RNA, the target's activity is altered.

antisepsis The prevention of infection or sepsis.

antiseptic Chemical agent applied to tissue to prevent infection by killing or inhibiting pathogens.

antiserum A serum that contains antibodies to a particular antigen.

antitoxin An antibody to a microbial toxin, usually a bacterial exotoxin, that combines specifically with the toxin to neutralize the toxin.

apical complex A set of organelles characteristic of members of the protist subdivision *Apicomplexa*: includes polar rings, subpellicular microtubules, conoid, rhoptries, and micronemes.

apicomplexan (a′pĭ-kom-plek′san) A protist that lacks special locomotor organelles but has an apical complex and a spore-forming stage. It is either an intra- or extracellular parasite of animals; a member of the taxon *Apicomplexa*.

apicoplast (a′pĭ-ko-plast) Organelles in apicomplexan protists that derive from cyanobacterial endosymbionts. Apicoplasts, however, are not involved in phototrophy; rather, they are thought to be the site of fatty acid, isoprenoid, and heme biosythesis.

apoenzyme (ap″o-en′zīm) The protein part of an enzyme that also has a nonprotein component.

apoptosis (ap″o-to′sis) Programmed cell death. A physiological suicide mechanism that results in fragmentation of a cell into membrane-bound particles that are eliminated by phagocytosis.

aporepressor The inactive form of a repressor protein; the repressor becomes active when the corepressor binds to it.

appressorium A flattened region of hypha found in some plant-infecting fungi that aids in penetrating the host plant cell wall. Occurs in both pathogenic fungi and nonpathogenic mycorrhizal fungi.

aptamer (ap′tŭm-er) An RNA molecule that has been engineered to bind to a target protein, thereby modifying the activity of the protein.

arbuscular mycorrhizal (AM) fungi The mycorrhizal fungi in a fungus-root association that penetrate the outer layer of the root, grow intracellularly, and form characteristic much-branched hyphal structures called arbuscules.

arbuscules *See* arbuscular mycorrhizal fungi.

arthroconidia (arthrospores) Fungal spores formed by fragmentation.

artificially acquired active immunity The type of immunity that results from immunizing an animal with a vaccine. The immunized animal produces its own antibodies and activated lymphocytes.

artificially acquired passive immunity The temporary immunity that results from introducing into an animal antibodies that have been produced either in another animal or by in vitro methods.

artemisinin (ar′te-mĭ-sin-in) Drug used in malaria-endemic areas to reduce parasitemia.

ascocarp A multicellular structure in ascomycetes lined with specialized cells called asci.

ascogenous hypha A specialized hypha that gives rise to one or more asci.

ascomycetes (as″ko-mi-se′tēz) A taxon of fungi that form ascospores.

ascospore (as′ko-spor) A spore contained or produced in an ascus.

ascus (pl., **asci**) A specialized cell, characteristic of the ascomycetes, in which two haploid nuclei fuse to produce a zygote, which often immediately divides by meiosis; at maturity, an ascus contains ascospores.

aseptate (coenocytic) hypha A multinucleate hypha (filament) that lacks cross walls.

assimilatory reduction (e.g., **assimilatory nitrate reduction** and **assimilatory sulfate reduction**) The reduction of an inorganic molecule to incorporate it into organic material. No energy is conserved during this process.

associative nitrogen fixation Nitrogen fixation by bacteria in the plant root zone (rhizosphere).

atomic force microscope *See* scanning probe microscope.

atopic reaction A type I hypersensitivity response caused by environmental allergens.

ATP-binding cassette transporters (ABC transporters) Transport systems that use ATP hydrolysis to drive translocation across the plasma membrane, can be used for nutrient uptake (ABC importer) or export of substances (ABC exporter), including protein secretion (ABC protein secretion pathway).

ATP synthase An enzyme that catalyzes synthesis of ATP from ADP and P_i, using energy derived from the proton motive force.

attenuated vaccine Live, nonpathogenic organisms used to activate adaptive immunity. The nonpathogenic organisms grow in the vaccinated individual without producing serious clinical disease while stimulating lymphocytes to produce antibody and activated T cells.

attenuation 1. A mechanism for the regulation of transcription termination of some bacterial operons by aminoacyl-tRNAs. 2. A procedure that reduces or abolishes the virulence of a pathogen without altering its immunogenicity.

attenuator A factor-independent transcription termination site in the leader sequence that is involved in attenuation.

autochthonous (ô-tŏk-thə-nəs) Substances (nutrients) that originate in a given environment. *See also* allochthonous.

autoclave An apparatus for sterilizing objects by the use of steam under pressure.

autogenous infection (aw-toj′e-nus) An infection that results from a patient's own microbiota, regardless of whether the infecting organism became part of the patient's microbiota subsequent to admission to a clinical care facility.

autoimmune disease A disease produced by the immune system attacking self antigens.

autoimmunity A condition characterized by the presence of serum autoantibodies and self-reactive lymphocytes that may be benign or pathogenic.

autoinducer With respect to quorum sensing, a small molecule that induces synthesis of the enzyme responsible for the autoinducer's synthesis. *See* quorum sensing.

autoinduction In general terms, any molecule that increases its own level of production. *See* autoinducer and quorum sensing.

autolysins (aw-tol′ĭ-sins or aw″to-lye′sins) Enzymes that partially digest peptidoglycan in growing bacteria so that the cell wall can be enlarged.

autophagosome *See* macroautophagy.

autoradiography A procedure that detects radioactively labeled materials using a photographic process.

autotroph (aw′to-trōf) An organism that uses CO_2 as its sole or principal source of carbon.

auxotroph (awk′so-trōf) An organism with a mutation that causes it to lose the ability to synthesize an essential nutrient; because of the mutation, the organism must obtain the nutrient or a precursor from its surroundings.

avidity The combined strength of binding between an antigen and all the antibody-binding sites.

axenic (a-zen′ik) Not contaminated by any foreign organisms; the term is used in reference to pure microbial cultures or to germfree animals.

axial fibrils *See* periplasmic flagella.

axial filament *See* periplasmic flagella.

axopodium A thin, needlelike type of pseudopodium with a central core of microtubules.

azidothymidine (AZT) Thymidine analogue used to treat HIV infection.

B

bacille Calmette-Guerin (BCG) An attenuated form of *Mycobacterium tuberculosis* used in some countries as a vaccine for tuberculosis.

bacillus (bah-sil′lus) A rod-shaped bacterium or archaeon.

bacteremia The presence of viable bacteria in the blood.

bacterial artificial chromosome (BAC) A cloning vector constructed from the *Escherichia coli* F-factor plasmid.

bactericide An agent that kills bacteria.

bacteriochlorophyll (bak-te″re-o-klo′ro-fil) A modified chlorophyll that serves as the primary light-trapping pigment in purple and green photosynthetic bacteria and heliobacteria.

bacteriocin (bak-te′re-o-sin) A protein produced by a bacterial strain that kills other closely related bacteria.

bacteriophage (bak-te′re-o-fāj″) A virus that uses bacteria as its host; often called a phage.

bacteriophage (phage) typing A technique in which strains of bacteria are identified based on their susceptibility to specific bacteriophages.

bacteriorhodopsin A transmembranous protein to which retinal is bound; it functions as a light-driven proton pump resulting in photophosphorylation without chlorophyll or bacteriochlorophyll. Found in the purple membrane of halophilic archaea.

bacteriostatic Inhibiting the growth and reproduction of bacteria.

bacteroid A modified, often pleomorphic, bacterial cell within the root nodule cells of legumes; it carries out nitrogen fixation.

baculoviruses Common name for viruses belonging to the family *Baculoviridae*.

baeocytes Small, spherical, reproductive cells produced by pleurocapsalean cyanobacteria through multiple fission.

balanced growth Microbial growth in which all cellular constituents are synthesized at constant rates relative to each other.

barophilic (bar″o-fil′ik) or **barophile** Organisms that prefer or require high pressures for growth and reproduction.

barotolerant Organisms that can grow and reproduce at high pressures but do not require them.

basal body The structure at the base of flagella that attaches them to the cell.

base analogs Molecules that resemble normal DNA nucleotides and can substitute for them during DNA replication, leading to mutations.

base excision repair *See* excision repair.

basic dyes Dyes that are cationic, or have positively charged groups, and bind to negatively charged cell structures.

basidiocarp (bah-sid′e-o-karp″) The fruiting body of a basidiomycete that contains the basidia.

basidiomycetes (bah-sid″e-o-mi-se′tēz) A division of fungi in which the spores are borne on club-shaped organs called basidia.

basidiospore (bah-sid′e-ō-spōr) A spore borne on the outside of a basidium following karyogamy and meiosis.

basidium (bah-sid′e-um; pl., **basidia**) A structure formed by basidiomycetes; it bears on its surface basidiospores (typically four) that are formed following karyogamy and meiosis.

basophil A weakly phagocytic white blood cell in the granulocyte lineage. It synthesizes and stores vasoactive molecules (e.g., histamine) that are released in response to external triggers (*see* mast cell).

batch culture A culture of microorganisms produced by inoculating a closed culture vessel containing a single batch of medium.

B cell (B lymphocyte) A type of lymphocyte that, following interaction with antigen, becomes a plasma cell, which synthesizes and secretes antibody molecules involved in humoral immunity.

B-cell receptor (BCR) A transmembrane immunoglobulin complex on the surface of a B cell that binds an antigen and stimulates the B cell. It is composed of a membrane-bound immunoglobulin, usually IgD or a modified IgM, complexed with another membrane protein (the Ig-α /Ig-β heterodimer).

benthic Pertaining to the bottom of the sea or other body of water.

beta hemolysis *See* hemolysis.

β-lactam Referring to antibiotics containing a β-lactam ring (e.g., penicillins and cephalosporins).

β-lactam ring The cyclic chemical structure composed of one nitrogen and three carbon atoms. It has antibacterial activity, interfering with bacterial cell wall synthesis.

β lactamase An enzyme that hydrolyzes the β-lactam ring, rendering the antibiotic inactive. Sometimes called penicillinase.

β-oxidation pathway The major pathway of fatty acid oxidation to produce NADH, $FADH_2$, and acetyl coenzyme A.

binal symmetry The symmetry of some virus capsids (e.g., those of complex phages) that is a combination of icosahedral and helical symmetry.

binary fission Asexual reproduction in which a cell or an organism separates into two identical daughter cells.

binomial system The nomenclature system in which an organism is given two names; the first is the capitalized generic name, and the second is the uncapitalized specific epithet.

bioaugmentation Addition of pregrown microbial cultures to an environment to perform a specific task.

biocatalysis The use of enzymes or whole microbes to perform chemical transformations on natural products.

biochemical oxygen demand (BOD) The amount of oxygen used by organisms in water under certain standard conditions; it provides an index of the amount of microbially oxidizable organic matter present.

biocide Chemical or physical agent, usually with a broad spectrum of biological activity, that inactivates microorganisms.

biocrime The use of biological materials (organisms or their toxins) to subvert societal goals or laws.

biodegradation The breakdown of a complex chemical through biological processes that can result in minor loss of functional groups, fragmentation into smaller constituents, or complete breakdown to carbon dioxide and minerals.

biofilms Organized microbial communities encased in extracellular polymeric substances (EPS) and associated with surfaces, often with complex structural and functional characteristics.

biofuels Energy-dense compounds (i.e., fuels) produced by microbes, usually by the degradation of plant substrates.

biogeochemical cycling The oxidation and reduction of substances carried out by living organisms and abiotic processes that results in the cycling of elements within and between different parts of the ecosystem.

bioinformatics The interdisciplinary field that manages and analyzes large biological data sets, including genome and protein sequences.

bioinsecticide A microbe or microbial product used to kill or disable unwanted insect pests.

biologic transmission A type of vector-borne transmission in which a pathogen goes through some morphological or physiological change within the vector.

biological safety cabinets Cabinets that use HEPA filters to project a curtain of sterile air across its opening, preventing microbes from entering or exiting.

bioluminescence The production of light by living cells, often through the oxidation of molecules by the enzyme luciferase.

biomagnification The increase in concentration of a substance in higher-level consumer organisms.

biopesticide A microorganism or another biological agent used to control a specific pest.

bioprospecting The collection, cataloging, and analysis of organisms, including microorganisms and plants, with the intent of finding a useful application and/or to document biodiversity.

bioremediation The use of biologically mediated processes to remove or degrade pollutants from specific environments.

biosensor A device for the detection of a particular substance (an analyte) that combines a biological receptor with a physicochemical detector.

biosynthesis *See* anabolism.

biosynthetic reactions *See* anabolism.

biotechnology The process in which living organisms are manipulated, particularly at the molecular genetic level, to form useful products.

bioterrorism The intentional or threatened use of microorganisms or toxins from living organisms to produce death or disease in humans, animals, and plants.

biotransformation (microbial transformation) The use of living organisms to modify substances that are not normally used for growth.

biovar Variant strains of microbes characterized by biological or physiological differences.

BLAST (basic local alignment search tool) A publicly available computer program that identifies regions of similarity between nucleotide or amino acid sequences. It compares input or query sequences with sequence databases and determines the statistical significance of matches. Homologies can be used to infer gene or protein function and evolutionary relationships.

blastospores Fungal spores produced from a vegetative mother cell by budding.

bloom The growth of a single microbial species or a limited number of species in an aquatic environment, usually in response to a sudden pulse of nutrients. *See* harmful algal bloom.

B lymphocyte *See* B cell.

bright-field microscope A microscope that illuminates the specimen directly with bright light and forms a dark image on a brighter background.

broad-spectrum drugs Chemotherapeutic agents that are effective against many different kinds of pathogens.

bronchial-associated lymphoid tissue (BALT) *See* mucosal-associated lymphoid tissue

B7 (CD 80) protein Single-chain glycoprotein expressed on activated T cells, B cells, dendritic cells, and macrophages (peripheral blood monocytes express low levels) that binds CD28 on T cells, resulting in second signal activation.

Bt An insecticide that consists of a toxin produced by *Bacillus thuringiensis.*

bubo (bu′bo) A tender, inflamed, enlarged lymph node that results from a variety of infections.

budding A vegetative outgrowth of yeast and some bacteria as a means of asexual reproduction; the daughter cell is smaller than the parent.

bulking sludge Sludge produced in sewage treatment that does not settle properly, usually due to the development of filamentous microorganisms.

butanediol fermentation A type of fermentation most often found in the family *Enterobacteriaceae* in which 2,3-butanediol is a major product; acetoin is an intermediate in the pathway and may be detected by the Voges-Proskauer test.

C

calorie The amount of heat needed to raise one gram of water from 14.5 to 15.5°C.

Calvin cycle The main pathway for the fixation (reduction and incorporation) of CO_2 into organic material by photoautotrophs and chemolithoautotrophs.

cancer A malignant tumor that expands locally by invasion of surrounding tissues and systemically by metastasis.

capsid The protein coat or shell that surrounds a virion's nucleic acid.

capsomer The ring-shaped morphological unit of which icosahedral capsids are constructed.

capsule A layer of well-organized material, not easily washed off, lying outside the cell wall.

carbonate equilibrium system The interchange among CO_2, HCO_3^- and CO_3^{2-} that keeps oceans buffered between pH 7.6 to 8.2.

carbon to nitrogen ratio The ratio of carbon to nitrogen in a soil can be used to predict rates of organic decomposition and thus the cycling of nutrients in the system.

carboxysomes Polyhedral inclusions that contain the CO_2-fixation enzyme ribulose 1,5-bisphosphate carboxylase/oxygenase and are a site where CO_2 fixation occurs. They are considered to be a type of microcompartment.

cardinal temperatures The minimal, maximal, and optimum temperatures for growth.

carotenoids (kah-rot′e-noids) Pigment molecules, usually yellowish in color, that are often used to aid chlorophyll in trapping light energy during photosynthesis.

carrier An infected individual who is a potential source of infection for others and plays an important role in the epidemiology of a disease.

caseous lesion (ka′se-us) A lesion resembling cheese or curd; cheesy. Most caseous lesions are caused by *Mycobacterium tuberculosis.*

casual carrier An individual who harbors an infectious organism for only a short period.

catabolism That part of metabolism in which larger, more complex molecules are broken down into smaller, simpler molecules with the release of energy.

catabolite activator protein (CAP) A protein that regulates the expression of genes encoding catabolite repressible enzymes; also called cyclic AMP receptor protein (CRP).

catabolite repression Inhibition of the synthesis of several catabolic enzymes by a preferred carbon and energy source (e.g., glucose).

catalase An enzyme that catalyzes the destruction of hydrogen peroxide.

catalyst A substance that accelerates a reaction without being permanently changed itself.

catalytic site *See* active site.

catalyzed reporter deposition-fluorescent in situ hybridization (CARD-FISH) A FISH technique in which the signal is amplified so that low levels of fluorescence can be detected. It is used in microbial ecology to detect single cells in a mixed population. *See* fluorescent in situ hybridization (FISH).

catenanes (kăt′ə-nāns′) Circular, covalently closed nucleic acid molecules that are locked together like the links of a chain.

cathelicidins Antimicrobial cationic peptides that are produced by a variety of cells (e.g., neutrophils, respiratory epithelial cells, and alveolar macrophages).

caveolin-dependent (kav-e-o′lē) **endocytosis** *See* endocytosis.

CD4$^+$ cell *See* T-helper cell.

CD8$^+$ cell *See* cytotoxic T lymphyocyte.

CD95 pathway *See* Fas-FasL pathway.

cell cycle The sequence of events in a cell's growth-division cycle between the end of one division and the end of the next.

cell envelope The plasma membrane plus all other external layers.

cell-mediated immunity The type of immunity that results from T cells coming into close contact with foreign cells or infected cells to destroy them.

cellular slime molds Protists with a vegetative phase consisting of amoeboid cells that aggregate to form a multicellular pseudoplasmodium. They belong to the taxon *Dictyostelia;* formerly considered fungi.

cellulose The major structural carbohydrate of plant cell walls; a linear $\beta(1 \rightarrow 4)$ glucan.

cell wall The strong structure that lies outside the plasma membrane; it supports and protects the membrane and gives the cell shape.

central metabolic pathways Those pathways central to the metabolism of an organism because they function catabolically and anabolically (e.g., glycolytic pathways and tricarboxylic acid cycle).

central tolerance The process by which immune cells are rendered inactive.

cephalosporin (sef″ah-lo-spōr′in) A group of β-lactam antibiotics derived from the fungus *Cephalosporium* that share the 7-aminocephalosporanic acid nucleus.

chain termination DNA sequencing method A DNA-sequencing method that uses dideoxynucleotides to cause termination of DNA replication at random sites.

chancre (shang′ker) The primary lesion of syphilis occurring at the site of entry of the pathogen.

chaperone proteins Proteins that assist in the folding and stabilization of other proteins. Some are also involved in directing newly synthesized proteins to protein secretion systems or to other locations in the cell.

chemical fixation *See* fixation.

chemical oxygen demand (COD) The amount of chemical oxidation required to convert organic matter in water and wastewater to CO_2.

chemiosmotic hypothesis (kem″e-o-os-mot′ik) The hypothesis that proton and electrochemical gradients are generated by electron transport and then used to drive ATP synthesis by oxidative phosphorylation or photophosphorylation.

chemoheterotroph (ke″mo-het′er-o-trōf″) *See* chemoorganoheterotroph.

chemokine A type of cytokine that stimulates chemotaxis and chemokinesis.

chemolithoautotroph A microorganism that oxidizes reduced inorganic compounds to derive both energy and electrons; CO_2 is the carbon source. Also called chemolithotrophic autotroph.

chemolithoheterotroph A microorganism that uses reduced inorganic compounds to derive both energy and electrons; organic molecules are used as the carbon source.

chemolithotroph A microorganism that uses reduced inorganic compounds as a source of energy and electrons.

chemoorganoheterotroph An organism that uses organic compounds as sources of energy, electrons, and carbon for biosynthesis. Also called chemoheterotroph and chemoorganotrophic heterotroph.

chemoreceptors Proteins in the plasma membrane or periplasmic space that bind chemicals and trigger the appropriate chemotactic response.

chemostat A continuous culture apparatus that feeds medium into the culture vessel at the same rate as medium containing microorganisms is removed; the medium in a chemostat contains one essential nutrient in a limiting quantity.

chemotaxins Agents that cause white blood cells to migrate up its concentration gradient.

chemotaxis The pattern of microbial behavior in which the microorganism moves toward chemical attractants and away from repellents.

chemotherapeutic agents Compounds used in the treatment of disease.

chemotherapy The use of chemical agents to treat disease.

chemotrophs Organisms that obtain energy from the oxidation of chemical compounds.

chiral (ki'rəl) Having handedness: consisting of one or another stereochemical form.

chitin (ki'tin) A tough, resistant, nitrogen-containing polysaccharide found in the walls of certain fungi, the exoskeleton of arthropods, and the epidermal cuticle of other surface structures of certain protists and animals.

chlamydiae (klə-mid'e-e) Members of the genera *Chlamydia* and *Chlamydiophila:* gram-negative, coccoid cells that reproduce only within the cytoplasmic vesicles of host cells.

chloramphenicol (klo"ram-fen'ĭ-kol) A broad-spectrum antibiotic that is produced by *Streptomyces venezuelae* or synthetically; it binds to the large ribosomal subunit and inhibits the peptidyl transferase reaction.

chlorophyll The green photosynthetic pigment that consists of a large tetrapyrrole ring with a magnesium atom in the center.

chloroplast A eukaryotic plastid that contains chlorophyll and is the site of photosynthesis.

chlorosomes Elongated, intramembranous vesicles found in the green sulfur and nonsulfur bacteria; they contain light-harvesting pigments. Sometimes called chlorobium vesicles.

choleragen (kol'er-ah-gen) The cholera toxin.

chromatic adaptation The ability of cyanobacteria to change the relative concentrations of light-harvesting pigments in response to changes in wavelengths of incident light.

chromatin The DNA-containing portion of the eukaryotic nucleus; the DNA is almost always complexed with histones. It can be very condensed (heterochromatin) or more loosely organized and genetically active (euchromatin).

chromatin immunoprecipitation (ChIP) A technique that enables the identification of protein-DNA interactions.

chromogen (kro'me-jen) A colorless substrate that is acted on by an enzyme to produce a colored end product.

chromophore group A chemical group with double bonds that absorbs visible light and gives a dye its color.

chromosomes The bodies that have most or all of the cell's DNA and contain most of its genetic information (mitochondria and chloroplasts also contain DNA).

chronic carrier An individual who harbors a pathogen for a long time.

chytrids A term used to describe the *Chytridiomycota,* fungi that produce motile zoospores with single, posterior, whiplash flagella.

cidal Causing death.

cilia Threadlike appendages extending from the surface of some protists that beat rhythmically to propel them; cilia are membrane-bound cylinders with a complex internal array of microtubules, usually in a 9 + 2 pattern.

ciliates (*Ciliophora*) Protists that move by rapidly beating cilia and belong to the *Alveolata* (supergroup *Chromalveolata*).

citric acid cycle *See* tricarboxylic acid cycle.

class I MHC molecule *See* major histocompatibility complex.

class II MHC molecule *See* major histocompatibility complex.

class switching *See* antibody class switching.

classical complement pathway The antibody-dependent pathway of complement activation; it leads to the lysis of pathogens and stimulates phagocytosis and other host defenses.

classification The arrangement of organisms into groups based on mutual similarity or evolutionary relatedness.

clathrin-dependent endocytosis *See* endocytosis.

clonal selection The process by which an antigen binds to the best-fitting B-cell receptor, activating that B cell, resulting in the synthesis of antibody and clonal expansion.

clone (1) A group of genetically identical cells or organisms derived by a sexual reproduction from a single parent. (2) A DNA sequence that has been isolated and replicated using a cloning vector.

cloning vector A DNA molecule that can replicate independently of the host chromosome and maintain a piece of inserted foreign DNA, such as a gene, into a recipient cell. It may be a plasmid, phage, cosmid, or artificial chromosome.

club fungi Common name for the *Basidiomycota.*

clue cells Vaginal epithelial cells covered with bacteria.

cluster of differentiation molecules (CDs) Functional cell surface proteins that can be used to identify leukocyte subpopulations (e.g., interleukin-2 receptor, CD4, CD8, CD25, and intercellular adhesion molecule-1).

coagulase (ko-ag'u-las) An enzyme that induces blood clotting; it is characteristically produced by pathogenic staphylococci.

coagulation The process of adding chemicals to precipitate impurities from water during purification; also called flocculation.

coated vesicles The clathrin-coated vesicles formed by clathrin-dependent endocytosis.

coccolithophore Photosynthetic protists belonging to the group *Stramenopila,* supergroup *Chromalveolata.* They are characterized by coccoliths—intricate cell walls made of calcite.

coccus (kok'us, pl., **cocci,** kok'si) A roughly spherical bacterial or archaeal cell.

code degeneracy The presence of more than one codon for each amino acid.

coding sequences (CDS) In genomic analysis, open reading frames presumed to encode proteins but not tRNA or rRNA.

codon (ko'don) A sequence of three nucleotides in mRNA that directs the incorporation of an amino acid during protein synthesis or signals the stop of translation.

coenocytic (se"no-sit'ik) Refers to a multinucleate cell or hypha formed by repeated nuclear divisions not accompanied by cell divisions.

coenzyme A loosely bound cofactor that often dissociates from the enzyme active site after product has been formed.

coenzyme Q (CoQ, ubiquinone) *See* electron transport chain.

coevolution The evolution of two interacting organisms such that they optimize their survival with one another; that is, they develop reciprocal adaptations.

cofactor The nonprotein component of an enzyme; it is required for catalytic activity.

colicin (kol'ĭ-sin) A plasmid-encoded protein that is produced by enteric bacteria and binds to receptors on the cell envelope of sensitive target bacteria, where it may cause lysis or attack specific intracellular sites such as ribosomes.

coliform A gram-negative, nonsporing, facultative rod that ferments lactose with gas formation within 48 hours at 35°C.

Colilert defined substrate test An assay designed to detect the presence of coliforms in water.

colonization The establishment of a site of microbial reproduction on an inanimate surface or organism without necessarily resulting in tissue invasion or damage.

colony An assemblage of microorganisms growing on a solid surface; the assemblage often is directly visible.

colony forming units (CFU) The number of microorganisms that form colonies when cultured using spread plates or pour plates, an indication of the number of viable microorganisms in a sample.

colony stimulating factor (CSF) A protein that stimulates the growth and development of specific cell populations (e.g., granulocyte-CSF stimulates granulocytes to be made from their precursor stem cells).

colorless sulfur bacteria A diverse group of nonphotosynthetic proteobacteria that can oxidize reduced sulfur compounds such as hydrogen sulfide. Many are chemolithotrophs and derive energy from sulfur oxidation.

combinatorial biosynthesis The creation of novel metabolic processes by transferring and expressing genes between organisms.

cometabolism The modification of a compound not used for growth by a microorganism, which occurs in the presence of another organic material that serves as a carbon and energy source.

commensal Living on or within another organism without injuring or benefiting the other organism.

commensalism A type of symbiosis in which one individual gains from the association (the commensal) and the other is neither harmed nor benefited.

common-source epidemic An epidemic characterized by a sharp rise to a peak and then a rapid but not as pronounced decline in the number of individuals infected; it usually involves a single contaminated source from which individuals are infected.

common vehicle transmission The transmission of a pathogen to a host by means of an inanimate medium or vehicle.

communicable disease A disease associated with a pathogen that can be transmitted from one host to another.

community An assemblage of different types of organisms or a mixture of different microbial populations.

community genome arrays Microarrays in which probes are prepared from genes that represent a selection of cultured microbes; used for determining the presence or absence of these microbes in a sampled natural environment.

comparative genomics Comparison of genomes from different organisms to discern differences and similarities.

compartmentation The differential distribution of enzymes and metabolites to separate cell structures or organelles.

compatible solute A low molecular weight molecule used to protect cells against changes in solute concentrations (osmolarity) in their habitat; it can exist at high concentrations within the cell and still be compatible with metabolism and growth.

competent cell A cell that can take up free DNA fragments and incorporate them into its genome during transformation.

competition An interaction between two organisms attempting to use the same resource (nutrients, space, etc.).

competitive exclusion principle Two competing organisms overlap in resource use, which leads to the exclusion of one of the organisms.

competitive inhibitor A molecule that inhibits enzyme activity by binding the enzyme's active site.

complementarity determining regions (CDRs) Hypervariable regions in an immunoglobulin protein that form the three-dimensional epitope binding sites.

complementary DNA (cDNA) A DNA copy of an RNA molecule (e.g., a DNA copy of an mRNA).

complement system A group of plasma proteins that plays a major role in an animal's immune response.

complex medium Culture medium that contains some ingredients of unknown chemical composition.

complex viruses Viruses with capsids having a complex morphology that is neither icosahedral nor helical.

compromised host A host with lowered resistance to infection and disease.

concatemer A long DNA molecule consisting of several genomes linked together in a row.

conditional mutations Mutations with phenotypes that are expressed only under certain environmental conditions.

confocal scanning laser microscope (CSLM) A light microscope in which monochromatic laser derived light scans across the specimen at a specific level. Stray light from above and below the plane of focus is blocked out to give an image with excellent contrast and resolution.

conidiospore (ko-nid′e-o-spōr) An asexual, thin-walled spore borne on hyphae and not contained within a sporangium; it may be produced singly or in chains.

conidium (ko-nid′e-um; pl., **conidia**) *See* conidiospore.

conjugants Complementary mating types among protists that participate in a form of sexual reproduction called conjugation.

conjugate redox pair The electron acceptor and electron donor of a redox half reaction.

conjugation (1) The form of gene transfer and recombination in *Bacteria* and *Archaea* that requires direct cell-to-cell contact. (2) A complex form of sexual reproduction commonly employed by protists.

conjugative plasmid A plasmid that carries the genes that enable its transfer to other bacteria during conjugation (e.g., F plasmid).

conjugative transposon A mobile genetic element that is able to transfer itself from one bacterium to another by conjugation.

consensus sequence A commonly occurring sequence of nucleotides within a genetic element.

conserved hypothetical proteins In genomic analysis, proteins of unknown function for which at least one match exists in a database.

consortium A physical association of two different organisms, usually beneficial to both organisms.

constant region (C_L and C_H) The part of an antibody molecule that does not vary greatly in amino acid sequence among molecules of the same class, subclass, or type.

constitutive gene A gene that is expressed at nearly the same level at all times.

constitutive mutant A strain that produces an inducible enzyme continually, regardless of need, because of a mutation in either the operator or regulator gene.

constructed wetlands Intentional creation of marshland plant communities and their associated microorganisms for environmental restoration or to purify water by the removal of bacteria, organic matter, and chemicals as the water passes through the aquatic plant communities.

consumer An organism that feeds directly on living or dead animals by ingestion or by phagocytosis.

contact transmission Transmission of the pathogen by contact of the source or reservoir of the pathogen with the host.

contigs In genomic sequencing, fragments of DNA with identical sequences at the ends; that is, the ends of the fragments overlap in sequence.

continuous culture system A culture system with constant environmental conditions maintained through continual provision of nutrients and removal of wastes. *See* chemostat and turbidostat.

continuous feed The constant addition of a carbon source during industrial fermentation.

contractile vacuole (vak′u-ōl) In protists and some animals, a clear fluid-filled vacuole that takes up water from within the cell and then contracts, releasing it to the outside through a pore in a cyclical manner. It functions primarily in osmoregulation and excretion.

convalescent carrier An individual who has recovered from an infectious disease but continues to harbor large numbers of the pathogen.

cooperation A positive but not obligatory interaction between two different organisms.

coral bleaching The loss of photosynthetic pigments by either physiological inhibition or expulsion of the coral photosynthetic endosymbiont, zooxanthellae, a dinoflagellate.

core genome The common set of genes found in all genomes in a species or other taxon.

core polysaccharide *See* lipopolysaccharide.

corepressor A small molecule that binds a transcription repressor protein, thereby activating the repressor and inhibiting the synthesis of a repressible enzyme.

cosmid A plasmid vector with lambda phage *cos* sites that can be packaged in a phage capsid; it is useful for cloning large DNA fragments.

crenarchaeol Cyclopentane ring–containing lipids unique to nonthermophilic *Crenarchaeota*.

cristae (kris′te) Infoldings of the inner mitochondrial membrane.

cryptins Antimicrobial peptides produced by Paneth cells in the intestines.

crystallizable fragment (Fc) The stem of the Y portion of an antibody molecule. Cells such as macrophages bind to the Fc region; it also is involved in complement activation.

curing The loss of a plasmid from a cell.

cyanobacteria A large group of gram-negative, morphologically diverse bacteria that carry out oxygenic photosynthesis using a system like that present in photosynthetic eukaryotes.

cyanophycin Inclusions that store nitrogen found in cyanobacteria.

cyclic photophosphorylation (fo″to-fos″for-ĭ-la′shun) The formation of ATP when light energy is used to move electrons cyclically through an electron transport chain during photosynthesis.

cyst A general term used for a specialized microbial cell enclosed in a wall. Cysts are formed by protists and a few bacteria. They may be dormant, resistant structures formed in response to adverse conditions or reproductive cysts that are a normal stage in the life cycle.

cytochromes (si′to-krōms) *See* electron transport chain.

cytokine (si′to-kīn) A general term for proteins released by a cell in response to inducing stimuli, which are mediators that influence other cells. Produced by lymphocytes, monocytes, macrophages, and other cells.

cytokinesis Processes that apportion the cytoplasm and organelles, synthesize a septum, and divide a cell into two daughter cells during cell division.

cytopathic effect The observable change that occurs in cells as a result of viral replication.

cytoplasm All material in the cell enclosed by the plasma membrane, with the exception of the nucleus in eukaryotic cells.

cytoproct A specific site in certain protists (e.g., ciliates) where digested material is expelled.

cytosine (si′to-sēn) A pyrimidine 2-oxy-4-aminopyrimidine found in nucleosides, nucleotides, and nucleic acids.

cytoskeleton A network of structures made from filamentous proteins (e.g., actin and tubulin) and other components in the cytoplasm of cells.

cytostome A permanent site in a ciliate protist in which food is ingested.

cytotoxic T lymphocyte (CTL) Effector T cell that can lyse target cells bearing class I MHC molecules complexed with antigenic peptides. CTLs arise from antigen-activated Tc cells.

cytotoxin A toxin that has a specific toxic action upon cells; cytotoxins are named according to the cell for which they are specific (e.g., nephrotoxin).

D

Dane particle A 42-nm spherical particle that is one of three seen in *Hepatitis B virus* infections. The Dane particle is the complete virion.

dark-field microscopy Microscopy in which the specimen is brightly illuminated while the background is dark.

dark reaction *See* photosynthesis.

dark reactivation The excision and replacement of thymine dimers in DNA that occurs in the absence of light.

deamination The removal of amino groups from amino acids.

decimal reduction time (*D* or *D* value) The time required to kill 90% of the microorganisms or spores in a sample at a specified temperature.

decomposer An organism that breaks down complex materials into simpler ones, including the release of simple inorganic products.

defensins Specific peptides produced by neutrophils that permeabilize the outer and plasma membranes of certain microorganisms, thus killing them.

defined (synthetic) medium Culture medium made with components of known composition.

denaturation A change in either protein shape or nucleic-acid structure.

denaturing gradient gel electrophoresis (DGGE) A technique by which DNA is rendered single stranded (denatured) while undergoing electrophoresis so that DNA fragments of the same size can be separated according to nucleotide sequence.

dendritic cell An antigen-presenting cell that has long membrane extensions resembling the dendrites of neurons.

dendrogram A treelike diagram that is used to graphically summarize mutual similarities and relationships between organisms.

denitrification The reduction of nitrate to gaseous products, primarily nitrogen gas, during anaerobic respiration.

deoxyribonucleic acid (DNA) (de-ok″se-ri″bo-nu-kle′ik) A polynucleotide that constitutes the genetic material of all cellular organisms. It is composed of deoxyribonucleotides connected by phosphodiester bonds.

depth filters Filters composed of fibrous or granular materials that are used to decrease microbial load and sometimes to sterilize solutions.

dermatophyte (der′mah-to-fit″) A fungus parasitic on the skin.

desensitization The process by which a sensitized or hypersensitive individual is made insensitive or nonreactive to a sensitizing agent (e.g., an allergen).

detergent An organic molecule, other than a soap, that serves as a wetting agent and emulsifier; it is normally used as cleanser, but some may be used as antimicrobial agents.

diatoms (*Bacillariophyta*) Photosynthetic protists with siliceous cell walls called frustules belonging to the group *Stramenopila*, supergroup *Chromalveolata.*

diauxic growth (di-awk′sik) A biphasic growth response in which a microorganism, when exposed to two nutrients, initially uses one of them for growth and then alters its metabolism to make use of the second.

differential interference contrast (DIC) microscope A light microscope that employs two beams of plane polarized light. The beams are combined after passing through the specimen and their interference is used to create the image.

differential media Culture media that distinguish between groups of microorganisms based on differences in their growth and metabolic products.

differential staining Staining procedures that divide bacteria into separate groups based on staining properties.

diffusely adhering *E. coli* (DAEC) Strains of *Escherichia coli* that adhere over the entire surface of epithelial cells and usually cause diarrheal disease in immunologically naive and malnourished children.

dikaryotic stage (di-kar-e-ot′ik) In fungi, having pairs of nuclei within cells or compartments. Each cell contains two separate haploid nuclei, one from each parent.

dilution susceptibility tests A method by which antibiotics are evaluated for their ability to inhibit bacterial growth in vitro. A standardized concentration of bacteria is added to serially diluted antibiotics and incubated. Tubes lacking additional bacterial growth suggest antibiotic concentrations that are bacteriocidal or bacteriostatic.

dimorphic fungi Fungi able to switch back and forth between a yeast form and a filamentous form.

dinoflagellate (di″no-flaj′e-lāt) A photosynthetic protist characterized by two flagella used in swimming in a spinning pattern; belong to the group *Alveolata*, supergroup *Chromalveolata.*

diplococcus (dip″lo-kok′us) A pair of cocci.

direct immunofluorescence Technique for imaging of microscopic agents using fluorescently labeled antibodies to bind to the agent.

direct repair A type of DNA repair mechanism in which a damaged nitrogenous base is returned to its normal form (e.g., conversion of a thymine dimer back to two normal thymine bases).

directed evolution A variety of approaches used to mutate a specific gene or genes with the aim of increasing product yield or generating new molecules.

disease A deviation or interruption of the normal structure or function of any part of the body that is manifested by a characteristic set of symptoms and signs.

disease syndrome A set of signs and symptoms that are characteristic of a disease.

disinfectant An agent, usually chemical, that disinfects; normally, it is employed only with inanimate objects.

disinfection The killing, inhibition, or removal of microorganisms that may cause disease. It usually refers to the treatment of inanimate objects with chemicals.

disinfection by-products (DBPs) Chlorinated organic compounds such as trihalomethanes formed during chlorine use for water disinfection. Many are carcinogens.

dissimilatory reduction (e.g., **dissimilatory nitrate reduction** and **dissimilatory sulfate reduction**) The use of a substance as an electron acceptor for an electron transport chain. The acceptor (e.g., sulfate or nitrate) is reduced but not incorporated into organic matter.

dissolved organic matter (DOM) Nutrients that are available in the soluble, or dissolved, state.

divisome A collection of proteins that aggregate at the region in a dividing microbial cell where a septum will form.

DNA amplification The process of increasing the number of DNA molecules, usually by using the polymerase chain reaction.

DNA-dependent DNA polymerase An enzyme that catalyzes the synthesis of DNA using a DNA molecule as a template.

DNA-dependent RNA polymerase An enzyme that catalyzes the synthesis of RNA using a DNA molecule as a template.

DNA-DNA hybridization (DDH) An assay used to determine genetic similarity between two organisms. The single-stranded DNA of one organism is hybridized to that of another, and the percent that hybridizes is calculated.

DNA gyrase A topoisomerase enzyme that relieves tension generated by the rapid unwinding of DNA during DNA replication.

DNA ligase An enzyme that joins two DNA fragments together through the formation of a new phosphodiester bond.

DNA microarrays Solid supports that have DNA attached in organized arrays and are used to evaluate gene expression.

DNA polymerase (pol-im′er-ās) An enzyme that synthesizes new DNA using a parental nucleic acid strand (usually DNA) as a template.

DNA reassociation A procedure in which the number of microbial genomes in a natural environment is estimated by determining the length of time required for all the DNA, rendered single stranded by heating, to reanneal, or become double stranded upon cooling.

DNA vaccine A vaccine that contains DNA, which encodes antigenic proteins.

double diffusion agar assay (Öuchterlony technique) An immunodiffusion reaction in which both antibody and antigen diffuse through agar to form stable immune complexes, which can be observed visually.

double-stranded break model A model for reciprocal homologous recombination that involves the formation of breaks in both strands of DNA and the activity of the RecA protein.

doubling time *See* generation time.

DPT (diphtheria-pertussis-tetanus) vaccine A vaccine containing three antigens that is used to immunize people against diphtheria, pertussis (whooping cough), and tetanus.

droplet nuclei Small particles (0 to 4 μm in diameter) that represent what is left from the evaporation of larger particles (10 μm or more in diameter) called droplets.

drug inactivation Physical or chemical alteration of chemotherapeutic agent resulting in its loss of biological activity.

***D* value** *See* decimal reduction time.

E

early mRNA Messenger RNA produced early in a viral infection that codes for proteins needed to take over the host cell and manufacture viral nucleic acids.

ecosystem A self-regulating biological community and its associated physical and chemical environment.

ecotype A population of an organism that is genetically very similar but ecologically distinct from others of the same species.

ectomycorrhizae A mutualistic association between fungi and plant roots in which the fungus surrounds the root tip with a sheath.

ectoplasm In some protists, the cytoplasm directly under the cell membrane (plasmalemma) is divided into an outer gelatinous region, the ectoplasm, and an inner fluid region, the endoplasm.

ectosymbiosis A type of symbiosis in which one organism remains outside of the other organism.

effacing lesion The type of lesion caused by enteropathogenic strains of *Escherichia coli* (EPEC) when the bacteria destroy the brush border of intestinal epithelial cells.

efflux pumps Membrane-associated protein assemblies that transport amphiphilic molecules from cytoplasm to the extracellular environment.

electron acceptor A compound that accepts electrons in an oxidation-reduction reaction. Often called an oxidizing agent or oxidant.

electron cryotomography A specialized electron microscopy procedure that involves rapid freezing of intact specimens, maintenance of the specimen in a frozen state while being examined, and imaging from different angles. The information from each angle is used to create a three-dimensional reconstruction.

electron donor An electron donor in an oxidation-reduction reaction. Often called a reducing agent or reductant.

electron shuttle A substance or structure along which electrons are passed from the electron transport chain to an external terminal electron acceptor.

electron transport chain (ETC) A series of electron carriers that operate together to transfer electrons from donors to acceptors such as oxygen. Molecules involved in electron transport include nicotinamide adenine dinucleotide (NAD^+), NAD phosphate ($NADP^+$), cytochromes, heme proteins, nonheme proteins (e.g., iron-sulfur proteins and ferredoxin), coenzyme Q, flavin adenine dinucleotide (FAD), and flavin mononucleotide (FMN). Also called an electron transport system.

electrophoresis (e-lek″tro-fo-re′sis) A technique that separates substances through differences in their migration rate in an electrical field.

electroporation The application of an electric field to create temporary pores in the plasma membrane in order to render a cell temporarily transformation competent.

elementary body (EB) A small, dormant body that serves as the agent of transmission between host cells in the chlamydial life cycle.

elongation cycle The cycle in protein synthesis that results in the addition of an amino acid to the growing end of a peptide chain.

elongation factors Proteins that function in the elongation cycle of protein synthesis.

Embden-Meyerhof pathway (em′den mi′er-hof) A pathway that degrades glucose to pyruvate (i.e., a glycolytic pathway).

encystment The formation of a cyst.

endemic disease A disease that is commonly or constantly present in a population, usually at a relatively steady low frequency.

endergonic reaction (end″er-gon′ik) A reaction that does not spontaneously go to completion as written; the standard free energy change is positive, and the equilibrium constant is less than one.

endocytic pathways *See* endocytosis.

endocytosis The process in which a cell takes up solutes or particles by enclosing them in vesicles pinched off from its plasma membrane. It often occurs at regions of the plasma membrane coated by proteins such as clathrin and caveolin. Endocytosis involving these proteins is called clathrin-dependent endocytosis and caveolin-dependent endocytosis, respectively. When the substance endocytosed first binds to a receptor, the process is called receptor-mediated endocytosis.

endogenous antigen processing Antigen degradation by cytoplasmic proteosomes and subsequent complexing with class I MHC molecules for presentation on the processing cell's surface.

endogenous infection An infection by a member of an individual's own normal body microbiota.

endogenous pyrogen A host-derived chemical mediator (e.g., interleukin-1) that acts on the hypothalamus, stimulating a rise in core body temperature (i.e., it stimulates the fever response).

endomycorrhizae Referring to a mutualistic association of fungi and plant roots in which the fungus penetrates into the root cells.

endophyte A microorganism living within a plant but not necessarily parasitic on it.

endoplasm *See* ectoplasm.

endoplasmic reticulum (ER) A system of membranous tubules and flattened sacs (cisternae) in the cytoplasm of eukaryotic cells. Rough endoplasmic reticulum (RER) bears ribosomes on its surface; smooth endoplasmic reticulum (SER) lacks them.

endosome A membranous vesicle formed by endocytosis. It undergoes a maturational process that starts with early endosomes and proceeds to late endosomes and finally lysosomes.

endospore An extremely heat- and chemical-resistant, dormant, thick-walled spore that develops within some gram-positive bacteria. It has a complex structure that includes (from outermost to innermost) exosporium, spore coat, cortex, spore cell wall, and spore core.

endosymbiosis A type of symbiosis in which one organism is found within another organism.

endosymbiotic hypothesis The hypothesis that mitochondria, hydrogenosomes, and chloroplasts arose from bacterial endosymbionts of ancestral eukaryotic cells.

endotoxin The lipid A component of lipopolysaccharide (LPS) that is released from gram-negative bacterial cell walls when the bacteria die. Nanogram quantities can induce fever, activate complement and coagulation cascades, act as a mitogen to B cells, and stimulate cytokine release from a variety of cells. Systemic effects of endotoxin are referred to as endotoxic shock.

end product inhibition *See* feedback inhibition.

energy The capacity to do work or cause particular changes.

enhancer A site in the DNA to which a eukaryotic activator protein binds.

enology The science of wine making.

enrichment culture The growth of specific microbes from natural samples by including selective features to promote growth of the desired microbes while selecting against other microbes that might be in the same sample.

enteric bacteria (enterobacteria) Members of the family *Enterobacteriaceae*; also can refer to bacteria that live in the intestinal tract.

enteroaggregative *E. coli* (EAggEC) A toxin-producing strain of *Escherichia coli* associated with persistent watery, bloody diarrhea, and cramping in young children. EAggEC adhere to the intestinal mucosa and cause nonbloody diarrhea without invading or causing inflammation.

enterohemorrhagic *E. coli* (EHEC) (en′tər-o-hem″ə-raj′ik) EHEC strains of *Escherichia coli* (O157:H7) produce several cytotoxins that provoke fluid secretion and diarrhea.

enteroinvasive *E. coli* (EIEC) EIEC strains of *Escherichia coli* cause diarrhea by penetrating and binding to the intestinal epithelial cells. EIEC may also produce a cytotoxin and enterotoxin.

enteropathogenic *E. coli* (EPEC) EPEC strains of *Escherichia coli* attach to the brush border of intestinal epithelial cells and cause effacing lesions that lead to diarrhea.

enterotoxigenic *E. coli* (ETEC) ETEC strains of *Escherichia coli* that produce two plasmid-encoded enterotoxins: heat-stable enterotoxin (ST) and heat-labile enterotoxin (LT).

enterotoxin A toxin specifically affecting the cells of the intestinal mucosa, causing vomiting and diarrhea.

enthalpy Heat content of a system.

Entner-Doudoroff pathway A pathway that converts glucose to pyruvate and glyceraldehyde 3-phosphate by producing 6-phosphogluconate and then dehydrating it (i.e., a glycolytic pathway).

entropy A measure of the randomness or disorder of a system; a measure of that part of the total energy in a system that is unavailable for useful work.

envelope In virology, an outer membranous layer that surrounds the nucleocapsid in some viruses.

enveloped virus A virus that consists of a nucleocapsid enclosed within an envelope.

environmental genomics *See* metagenomics.

environmental microbiology The study of microbial processes that occur in soil, water, and other natural habitats. Specific microorganisms are not necessarily investigated; rather, the cumulative impact of the microbial community on the specific biome, as well as global implications, are assessed.

enzyme A protein catalyst with specificity for the reaction catalyzed and its substrates.

enzyme-linked immunosorbent assay (ELISA) A serological assay in which bound antigen or antibody is detected by another antibody that is conjugated to an enzyme. The enzyme converts a colorless substrate to a colored product reporting the antibody capture of the antigen.

eosinophil (e″o-sin′o-fil) A phagocytic, polymorphonuclear leukocyte that has a two-lobed nucleus and cytoplasmic granules that stain yellow-red.

epidemic A disease that suddenly increases in occurrence above the normal level in a given population.

epidemiologist A person who specializes in epidemiology.

epidemiology The study of the factors determining and influencing the frequency and distribution of disease, injury, and other health-related events and their causes in defined human populations.

epifluorescence microscopy The illumination of a microscope sample with a high-intensity light from above. The light is passed through an exciter filter so that when light of a specific wavelength is focused on the specimen, it emits light, thereby making the sample visible as a specific color on a black background.

epilimnion The upper, warmer layer of water in a stratified lake.

epiphyte An organism that grows on the surface of plants.

episome A plasmid that can either exist independently of the host cell's chromosome or be integrated into it.

epitheca (ep″ĭ-the′kah) The larger of two halves of a diatom frustule (shell).

epitope (ep′i-tōp) An area of an antigen that stimulates the production of, and combines with, specific antibodies; also known as the antigenic determinant site.

equilibrium The state of a system in which no net change is occurring and free energy is at a minimum; in a chemical reaction at equilibrium, the rates in the forward and reverse directions exactly balance each other out.

equilibrium constant (Keq) A value that relates the concentration of reactants and products to each other when a reaction is at equilibrium.

ergot The dried sclerotium of *Claviceps purpurea.* Also, an ascomycete that parasitizes rye and other higher plants, causing an animal disease called ergotism.

erythromycin (ĕ-rith″ro-mi′sin) An intermediate spectrum macrolide antibiotic produced by *Saccharopolyspora erythraea.*

eukaryotic cells Cells that have a membrane-delimited nucleus and differ in many other ways from bacterial and archaeal cells; protists, fungi, plants, and animals are all eukaryotic.

euglenids (u-gle′nids) A group of protists (super group *Excavata*) that includes chemoorganotrophs and photoautotrophs with chloroplasts containing chlorophyll *a* and *b.* They usually have a stigma and one or two flagella emerging from an anterior reservoir.

eutrophic Nutrient enriched.

eutrophication The enrichment of an aquatic environment with nutrients.

evolutionary distance A quantitative indication of the number of positions that differ between two aligned macromolecules, and presumably a measure of evolutionary similarity between molecules and organisms.

excision repair A type of DNA repair mechanism in which a section of a strand of damaged DNA is excised and replaced, using the complementary strand as a template. Two types are recognized: base excision repair and nucleotide excision repair.

excystment The escape of one or more cells or organisms from a cyst.

exergonic reaction (ek″ser-gon′ik) A reaction that spontaneously goes to completion as written; the standard free energy change is negative, and the equilibrium constant is greater than one.

exfoliative toxin (exfoliatin) An exotoxin produced by *Staphylococcus aureus* that causes the separation of epidermal layers and the loss of skin surface layers. It produces the symptoms of the scalded skin syndrome.

exit site (E site) The location on a ribosome to which an empty (uncharged) tRNA moves from the P site before it finally leaves during protein synthesis.

exoenzymes Enzymes that are secreted by cells.

exogenous antigen processing Antigen degradation by phagolysosome enzymes and subsequent complexing with class II MHC molecules for presentation on the processing cell's surface.

exon The region in a split (interrupted) gene that codes for pre-mRNA and is retained in mature mRNA.

exospore A spore formed outside the mother cell; often observed in actinobacteria.

exotoxin A heat-labile, toxic protein produced by a bacterium and usually released into the bacterium's surroundings.

exponential (log) phase The phase of the growth curve during which the microbial population is growing at a constant and maximum rate, dividing and doubling at regular intervals.

expressed sequence tag (EST) A partial gene sequence unique to a gene that can be used in microarray analysis to determine when a gene is expressed.

expression vector A cloning vector used to express a recombinant gene in host cells; the gene is transcribed and its protein synthesized.

extant organism An organism that exists on Earth today.

exteins Polypeptide sequences of precursor self-splicing proteins that are joined together during formation of the final, functional protein. They are separated from one another by intein sequences.

extended aeration process An approach that can be used during secondary sewage treatment to reduce the amount of sludge produced.

extinction culture technique A method for obtaining an axenic (pure) culture in which natural samples are serial diluted so that the final dilution has between 1 and 10 cells (i.e., diluted to extinction).

extreme environment An environment in which physical factors such as temperature, pH, salinity, and pressure are outside of the normal range for growth of most microorganisms; these conditions allow unique organisms to survive and function.

extreme halophiles *See* halobacteria.

extremophiles Microorganisms that grow in extreme environments.

extrinsic factor An environmental factor such as temperature that influences microbial growth in food.

F

facilitated diffusion Diffusion across the plasma membrane that is aided by a carrier protein.

facultative anaerobes Microorganisms that do not require oxygen for growth, but grow better in its presence.

facultative psychrophile (fak′ul-ta″tiv si′kro-fīl) *See* psychrotroph.

Fas-FasL pathway One of two pathways used by cytotoxic T cells to kill target cells (e.g., those infected with a virus). Fas is another name for a protein called CD95 receptor, which is located on the target cell. FasL is the ligand to the CD95 receptor and is found on the cytotoxic T cell.

fatty acid synthase The multienzyme complex that makes fatty acids.

fecal coliform Coliforms whose normal habitat is the intestinal tract and that can grow at 44.5°C. They are used as indicators of fecal pollution of water.

fecal enterococci (en″ter-o-kok′si) Enterococci found in the intestine of humans and other warm-blooded animals.

feedback inhibition A negative feedback mechanism in which an end product inhibits the activity of an enzyme in the pathway leading to its formation.

fermentation (1) An energy-yielding process in which an organic molecule is oxidized without an exogenous electron acceptor. Usually pyruvate or a pyruvate derivative serves as the electron acceptor. (2) The growth of microbes in very large volumes for the production of industrially important products.

fermenter In industrial microbiology, the large vessel used to culture microorganisms.

ferredoxin *See* electron transport chain.

fever A complex physiological response to disease mediated by pyrogenic cytokines and characterized by a rise in core body temperature and activation of the immune system.

F factor The fertility factor, a plasmid that carries genes for bacterial conjugation and makes its *Escherichia coli* host the gene donor during conjugation.

filopodia Long, narrow pseudopodia found in certain amoeboid protists.

fimbria (fim′bre-ah; pl., **fimbriae**) A fine, hairlike protein appendage on many gram-negative bacteria, some archaea, and some fungi; also called pili. They attach cells to surfaces, and some are involved in twitching motility.

final host The host on or in which a parasite either attains sexual maturity or reproduces.

first law of thermodynamics Energy can be neither created nor destroyed (although it can be changed in form or redistributed).

fixation (1) The process in which the internal and external structures of cells and organisms are preserved and fixed in position. Two methods are commonly used: heat fixation and chemical fixation. (2) The conversion of inorganic, gaseous elements such as carbon (CO_2) and nitrogen (N_2) to organic forms.

flagellin (flaj′ĕ-lin) A family of related proteins in motile bacteria used to construct the filament of a flagellum. An unrelated group of proteins, also called flagellins, are used to construct archaeal flagellar filaments.

flagellum (flah-jel′um; pl., **flagella**) A threadlike appendage on many cells that is responsible for their motility. Bacterial flagella are composed of a flagellar filament, flagellar hook, and basal body.

flavin adenine dinucleotide (FAD) (fla′vin ad′ĕ-nēn) An electron carrying cofactor often involved in energy production (for example, in the tricarboxylic acid cycle and the β-oxidation pathway). *See also* electron transport chain.

flavin mononucleotide *See* electron transport chain.

flow cytometry A tool for defining and enumerating cells using a capillary tube to control cell movement and a laser to detect cell size and morphology.

fluid mosaic model The model of cell membranes in which the membrane is a lipid bilayer with integral proteins buried in the lipid and peripheral proteins more loosely attached to the membrane surface.

fluorescence microscope A microscope that exposes a specimen to light of a specific wavelength and then forms an image from the fluorescent light produced.

fluorescent in situ hybridization (FISH) A technique for identifying certain genes or organisms in which specific DNA fragments are labeled with fluorescent dye and hybridized to the chromosomes of interest.

fluorescent light The light emitted by a substance when it is irradiated with light of a shorter wavelength.

fluorochrome A fluorescent dye.

fomite (pl., **fomites**) An object that is not in itself harmful but is able to harbor and transmit pathogenic organisms. Also called fomes.

food-borne infection Gastrointestinal illness caused by ingestion of microorganisms, followed by their growth within the host.

food intoxication Food poisoning caused by microbial toxins produced in a food prior to consumption. The presence of living bacteria is not required.

food poisoning A general term usually referring to a gastrointestinal disease caused by the ingestion of food contaminated by pathogens or their toxins.

forward mutation A mutation from the wild type to a mutant form.

F′ plasmid An F plasmid that carries some bacterial genes and transmits them to recipient cells when the F′ cell carries out conjugation.

fragmentation A type of asexual reproduction among filamentous microbes in which hyphae break into two or more parts, each of which forms new hyphae.

frameshift mutations Mutations arising from the loss or gain of a base or DNA segment, leading to a change in the codon reading frame and thus a change in the amino acids incorporated into protein.

free energy change The total energy change in a system that is available to do useful work as the system goes from its initial state to its final state at constant temperature and pressure.

fruiting body A specialized structure that holds sexually or asexually produced spores; found in fungi and in some bacteria (e.g., the myxobacteria).

frustule (frus′tūl) A silicified cell wall in the diatoms.

fueling reactions Chemical reactions that supply the ATP, precursor metabolites, and reducing power needed for biosynthesis.

fumonisins A family of toxins produced by molds belonging to the genus *Fusarium.* It primarily affects corn and it is known to be hepato- and nephrotoxic in animals.

functional gene array Microarrays designed to determine potential microbial activities in natural environments. The probes on a functional gene array represent genes whose products are involved in specific processes, for example, nitrogen fixation, methane oxidation, etc.

functional genomics Genomic analysis concerned with determining the way a genome functions.

functional proteomics Analysis concerned with determining the function of proteins produced by a cell.

fungicide An agent that kills fungi.

fungistatic Inhibiting the growth and reproduction of fungi.

fungus (pl., **fungi**) Achlorophyllous, heterotrophic, spore-bearing eukaryotes with absorptive nutrition and a walled thallus; sometimes called the "true fungi" or *Eumycota.*

G

G + C content The percent of guanines and cytosines present in a genome; this percent is used for taxonomic purposes.

gametangium (gam-ĕ-tan′je-um; pl., **gametangia**) A structure that contains gametes or in which gametes are formed.

gamonts Gametic cells formed by protists when they undergo sexual reproduction.

gas vacuole A gas-filled vacuole found in cyanobacteria and some other aquatic bacteria and archaea that provides flotation. It is composed of gas vesicles, which are made of protein.

gas vesicle *See* gas vacuole.

gel electrophoresis The separation of molecules according to charge and size through a gel matrix.

Gell-Coombs classification System used to codify hypersensitivity reactions.

gene A DNA segment or sequence that codes for a polypeptide, rRNA, or tRNA.

gene cassette DNA sequence that encodes one to a few genes for a complete biochemical function.

generalized transduction The transfer of any part of a bacterial or archaeal genome when the DNA fragment is packaged within a virus's capsid by mistake.

general secretion pathway (GSP) *See* Sec system.

generation (doubling) time The time required for a microbial population to double in number.

genetic drift Changes in the genotype of an organism that collect over time. These changes are generally neutral mutations that occur and are maintained without selection.

genetic engineering The deliberate modification of an organism's genetic information by changing its nucleic acid genome.

genome The full set of genes present in a cell or virus; all the genetic material in an organism.

genome annotation The process of determining the location and potential function of specific genes and genetic elements in a genome sequence.

genomic analysis *See* genomics.

genomic fingerprinting A series of techniques based on restriction enzyme digestion patterns that enable the comparison of microbial species and strains, and is thus useful in taxonomic identification.

genomic island A region in a genome that was introduced in an ancestral microbe by horizontal gene transfer. Genomic islands often bring new phenotypic traits to the microbe; *see* pathogenicity island.

genomic library A collection of clones that contains fragments that represent the complete genome of an organism.

genomic reduction The decrease in genomic information that occurs over evolutionary time as an organism or organelle becomes increasingly dependent on another cell or a host organism.

genomics The study of the molecular organization of genomes, their information content, and the gene products they encode.

genotype The specific set of alleles carried in the genome of an organism.

genotypic classification The use of genetic data to construct a classification scheme for the identification of an unknown species or the phylogeny of a group of microbes.

genus A well-defined group of one or more species that is clearly separate from other organisms.

geographic information system (GIS) Use of hardware, software, and data for the capture, management, and display of geographically referenced information.

germination The stage following spore activation in which the spore breaks its dormant state. Germination is followed by outgrowth.

Ghon complex (gon) The initial focus of infection in primary pulmonary tuberculosis.

gliding motility A type of motility in which a microbial cell glides along a solid surface.

global regulatory systems Regulatory systems that simultaneously affect many genes.

glomeromycetes Common name for members of the fungal taxon *Glomeromycota.*

glucans Polymers of D-glucose.

gluconeogenesis (gloo″ko-ne″o-jen′e-sis) The synthesis of glucose from noncarbohydrate precursors such as lactate and amino acids.

glutamine synthetase–glutamate synthase (GS-GOGAT) system A mechanism used by many microbes to incorporate ammonia.

glycocalyx (gli″ko-kal′iks) A network of polysaccharides extending from the surface of bacteria and other cells.

glycogen (gli′ko-jen) A highly branched polysaccharide containing glucose, which is used to store carbon and energy.

glycolysis (gli-kol′ĭ-sis) The conversion of glucose to pyruvic acid by use of the Embden-Meyerhof pathway, pentose phosphate pathway, or Entner-Doudoroff pathway.

glycolytic pathway A pathway that converts glucose to pyruvic acid (e.g., Embden-Meyerhof pathway).

glycomics Analysis of all the carbohydrates produced by a cell under specific circumstances.

glyoxylate cycle A modified tricarboxylic acid cycle in which the decarboxylation reactions are bypassed by the enzymes isocitrate lyase and malate synthase; it is used to convert acetyl-CoA to succinate and other metabolites.

gnotobiotic (no″to-bi-ot′ik) Animals that are germfree (microorganism free) or live in association with one or more known microorganisms.

Golgi apparatus (gol′je) A membranous eukaryotic organelle composed of stacks (dictyosomes) of flattened sacs (cisternae) that is involved in packaging and modifying materials for secretion and many other processes.

gonococci (gon′o-kok′si) Bacteria of the species *Neisseria gonorrhoeae*—the organism causing gonorrhea.

Gram stain A differential staining procedure that divides bacteria into gram-positive and gram-negative groups based on their ability to retain crystal violet when decolorized with an organic solvent such as ethanol.

granulocyte A type of white blood cell that stores preformed molecules (enzymes and antimicrobial proteins) in vacuoles near the cell membrane.

granuloma Term applied to nodular inflammatory lesions containing phagocytic cells.

granzyme Enzyme found in the granules of cytotoxic T cells (CTLs) that causes targets of CTLs cells to undergo apoptosis.

GRAS An acronym for generally regarded as safe: a category of food additives recognized in the United States and by the United Nations.

great plate count anomaly (GPCA) The discrepancy between the number of viable microbial cells and the number of colonies that can be cultivated from the same natural sample.

green fluorescent protein (GFP) A protein from the jellyfish *Aequorea victoria* that fluoresces when illuminated at a specific wavelength. The gene for GFP can be fused to gene promoters to determine when a gene is expressed (transcriptional fusion). When fused to the coding sequence of a gene so that a chimeric protein is produced, GFP fluorescence reveals the localization of the protein (translational fusion).

greenhouse gases Gases (e.g., CO_2, CH_4) released from Earth's surface through chemical and biological processes that interact with the chemicals in the stratosphere to decrease radiational cooling of Earth. This leads to global warming.

green nonsulfur bacteria Anoxygenic photosynthetic bacteria that contain bacteriochlorophylls *a* and *c;* usually photoheterotrophic and display gliding motility. Include members of the phylum *Chloroflexi.*

green sulfur bacteria Anoxygenic photosynthetic bacteria that contain bacteriochlorophylls *a,* plus *c, d,* or *e;* photolithoautotrophic; use H_2, H_2S, or S as electron donor. Include members of the phylum *Chlorobi.*

griseofulvin (gris″e-o-ful′vin) An antibiotic from *Penicillium griseofulvum* given orally to treat chronic dermatophytic infections of skin and nails.

group A streptococcus (GAS) A gram-positive, coccus-shaped bacterium often found in the throat and on the skin of humans, having the A group of surface carbohydrate.

group B streptococcus (GBS) A gram-positive, coccus-shaped bacterium found occasionally on mucous membranes of humans, having the B group of surface carbohydrate.

group translocation A transport process in which a molecule is moved across a membrane by carrier proteins while being chemically altered at the same time (e.g., phosphoenolpyruvate: sugar phosphotransferase system).

growth factors Organic compounds that must be supplied in the diet for growth because they are essential cell components or precursors of such components and cannot be synthesized by the organism.

guanine (gwan′in) A purine derivative, 2-amino-6-oxypurine found in nucleosides, nucleotides, and nucleic acids.

gumma (gum′ah) A soft, gummy tumor occurring in tertiary syphilis.

gut-associated lymphoid tissue (GALT) *See* mucosal-associated lymphoid tissue.

H

halobacteria (extreme halophiles) A group of archaea that depend on high NaCl concentrations for growth and do not survive at a concentration below about 1.5 M NaCl.

halophile A microorganism that requires high levels of sodium chloride for growth.

halotolerant The ability to withstand large changes in salt concentration.

hapten A molecule not immunogenic by itself that, when coupled to a macromolecular carrier, can elicit antibodies directed against itself.

harborage transmission The mode of transmission in which an infectious organism does not undergo morphological or physiological changes within the vector.

harmful algal bloom (HAB) In aquatic ecosystems, the growth of a single population of phototroph, either a protist (e.g., diatom, dinoflagellate) or a cyanobacterium, that produces a toxin that is poisonous to other organisms, sometimes including humans. Alternatively, in the absence of toxin production, the concentration of bloom microbe reaches levels that are inherently harmful to other organisms, such as filter-feeding bivalves.

Hartig net The area of nutrient exchange between ectomycorrhizal fungal hyphae and plant host cells.

healthy carrier An individual who harbors a pathogen but is not ill.

heat fixation *See* fixation.

heat-shock proteins Proteins produced when cells are exposed to high temperatures or other stressful conditions. They protect the cells from damage and often aid in the proper folding of proteins.

helical capsid A viral capsid in the form of a helix.

helicases Enzymes that use ATP energy to unwind DNA ahead of the replication fork.

hemadsorption The adherence of red blood cells to the surface of something, such as another cell or a virus.

hemagglutination (hem″ah-gloo″tĭ-na′shun) The agglutination of red blood cells by hemagglutinins (e.g., antibodies or components of virus capsids).

hemagglutination assay A testing procedure based on a hemagglutination reaction.

hemagglutinin (HA) (1) Proteins that cause red blood cells to clump together (hemagglutination). (2) One of the envelope spikes of influenza viruses. They are the basis for identifying different influenza virus strains.

hematopoesis The process by which blood cells develop into specific lineages from stem cells. Red and white blood cells and platelets develop from this process.

hemolysin (he-mol′ĭ-sin) A substance that causes hemolysis.

hemolysis The disruption of red blood cells and release of their hemoglobin. There are several types of hemolytic reactions when bacteria grow on blood agar. In α-hemolysis, a greenish zone of incomplete hemolysis forms around the colony. A clear zone of complete hemolysis without any obvious color change is formed during β-hemolysis.

hepadnaviruses Common name for viruses belonging to the family *Hepadnaviridae.*

herd immunity The resistance of a population to infection and spread of an infectious agent due to the immunity of a high percentage of the population.

heterocysts Specialized cells of cyanobacteria that are the sites of nitrogen fixation.

heteroduplex DNA A double-stranded stretch of DNA formed by two slightly different strands that are not completely complementary.

heterokont flagella A pattern of flagellation found in the protist subdivision *Stramenopila,* featuring two flagella, one extending anteriorly and the other posteriorly.

heterolactic fermenters Microorganisms that ferment sugars to form lactate and other products such as ethanol and CO_2.

heterologous gene expression The cloning, transcription, and translation of a gene that has been introduced (cloned) into an organism that normally does not possess the gene.

heterotroph An organism that uses reduced, preformed organic molecules as its principal carbon source.

heterotrophic nitrification Nitrification carried out by chemoheterotrophic microorganisms.

hexose monophosphate pathway *See* pentose phosphate pathway.

Hfr conjugation Conjugation involving an Hfr strain and an F^- strain.

Hfr strain A strain of *Escherichia coli* that donates its genes with high frequency to a recipient cell during conjugation because the F factor is integrated into the donor's chromosome.

hierarchical cluster analysis The organization of microarray data such that induced and repressed genes, or genes of similar function, are grouped separately.

hierarchical oligonucleotide primer extension A technique that enables the detection of multiple rRNA genes in a single sample.

high-efficiency particulate (HEPA) filter Thick, fibrous filter constructed to remove 99.97% of particles that are 0.3 μm or larger.

high-nitrate, low-chlorophyll (HNLC) Certain areas of the open ocean that paradoxically have a high concentration of nutrients but a limited population of photosynthetic microbes as measured by chlorophyll. Many of these ecosystems are limited by micronutrients such as iron and silica.

high-throughput screening (HTS) A system that combines liquid handling devices, robotics, computers, data processing, and a sensitive detection system to screen thousands of compounds for a single capability. It is often used by pharmaceutical companies to identify natural products that have potentially useful applications.

histatin An antimicrobial peptide composed of 24 to 38 amino acids, heavily enriched with histidine, that targets fungal mitochondria.

histone A small basic protein with large amounts of lysine and arginine that is associated with eukaryotic DNA in chromatin. Related proteins are observed in many archaeal species, where they form archaeal nucleosomes.

HIV protease inhibitor A drug that prevents HIV protease enzymes from cleaving HIV polyproteins into mature proteins required for HIV virion assembly.

holdfast A structure produced by some bacteria (e.g., *Caulobacter*) that attaches them to a solid object.

holoenzyme A complete enzyme consisting of the apoenzyme plus a cofactor. Also refers to a complete enzyme consisting of all protein subunits (e.g., RNA polymerase holoenzyme).

holozoic nutrition Acquisition of nutrients (e.g., bacteria) by endocytosis and the subsequent formation of a food vacuole or phagosome.

homolactic fermenters Organisms that ferment sugars almost completely to lactic acid.

homologous recombination Recombination involving two DNA molecules that are very similar in nucleotide sequence; it can be reciprocal or nonreciprocal.

hopanoids Lipids found in bacterial membranes that are similar in structure and function to the sterols found in eukaryotic membranes.

hops Dried flowers of the plant *Humulus lupulus* used in the preparation of beer and lager.

horizontal (lateral) gene transfer (HGT/LGT) The process by which genes are transferred from one mature, independent organism to another. In *Bacteria* and *Archaea,* transformation, conjugation, and transduction are the mechanisms by which HGT can occur.

hormogonia Small motile fragments produced by fragmentation of filamentous cyanobacteria; used for asexual reproduction and dispersal.

host An organism that harbors another organism.

host-parasite relationship The symbiosis between a pathogen and its host. The term parasite is used to imply pathogenicity in this context.

host restriction The degradation of foreign genetic material by nucleases after the genetic material enters a host cell.

housekeeping genes Genes that encode proteins that function throughout most of the life cycle of an organism (e.g., enzymes of the Embden-Meyerhof pathway).

human immunodeficiency virus (HIV) A lentivirus of the family *Retroviridae* that is the cause of AIDS.

human leukocyte antigen complex (HLA) A collection of genes on chromosome 6 that encodes proteins involved in host defenses; also called major histocompatibility complex.

humoral (antibody-mediated) immunity The type of immunity that results from the presence of soluble antibodies in blood and lymph.

hybridoma (hi″brĭ-do′mah) A fast-growing cell line produced by fusing a cancer cell (myeloma) to another cell, such as an antibody-producing cell.

hydrogen hypothesis A hypothesis that considers the origin of the eukaryotes through the development of the hydrogenosome. It suggests the organelle arose as the result of an endosymbiotic anaerobic bacterium that produced CO_2 and H_2 as the products of fermentation.

hydrogenosome An organelle found in some anaerobic protists that produce ATP by fermentation.

hydrophilic A polar substance that has a strong affinity for water (or is readily soluble in water).

hydrophobic A nonpolar substance lacking affinity for water (or which is not readily soluble in water).

3-hydroxypropionate cycle A pathway used by green nonsulfur bacteria and some archaea to fix CO_2.

hyperendemic disease A disease that has a gradual increase in occurrence beyond the endemic level but not at the epidemic level in a given population; also may refer to a disease that is equally endemic in all age groups.

hyperferremia Excessive iron in the blood.

hypersensitivity A condition in which the body reacts to an antigen with an exaggerated immune response that usually harms the individual. Also termed an allergy.

hyperthermophile (hi″per-ther′mo-fīl) A microbe that has its growth optimum between 85°C and about 120°C. Hyperthermophiles usually do not grow well below 55°C.

hypha (hi′fah; pl., **hyphae**) The unit of structure of most fungi and some bacteria; a tubular filament.

hypoferremia Deficiency of iron in the blood.

hypolimnion The colder, bottom layer of water in a stratified lake.

hypotheca The smaller half of a diatom frustule.

hypoxic (hi pok′sik) Having a low oxygen level.

I

icosahedral capsid A viral capsid that has the shape of a regular polyhedron having 20 equilateral triangular faces and 12 corners.

identification The process of determining that a particular isolate or organism belongs to a recognized taxon.

IgA Immunoglobulin A; the class of immunoglobulins that is present in dimeric form in many body secretions (e.g., saliva, tears, and bronchial and intestinal secretions) and protects body surfaces. IgA also is present in serum.

IgD Immunoglobulin D; the class of immunoglobulins found on the surface of many B lymphocytes; thought to serve as an antigen receptor in the stimulation of antibody synthesis.

IgE Immunoglobulin E; the immunoglobulin class that binds to mast cells and basophils, and is responsible for type I or anaphylactic hypersensitivity reactions such as hay fever and asthma. IgE is also involved in resistance to helminth parasites.

IgG Immunoglobulin G; the predominant immunoglobulin class in serum. Has functions such as neutralizing toxins, opsonizing bacteria, activating complement, and crossing the placenta to protect the fetus and neonate.

IgM Immunoglobulin M; the class of serum antibody first produced during an infection. It is a large, pentameric molecule that is active in agglutinating pathogens and activating complement. The monomeric form is present on the surface of some B lymphocytes.

immobilization The incorporation of a simple, soluble substance into the body of an organism, making it unavailable for use by other organisms.

immune complex The product of an antigen-antibody reaction, which may also contain components of the complement system.

immune surveillance The process by which cells of the immune system police the host for nonself antigens.

immune system The defensive system in an animal consisting of the nonspecific (innate) and specific (adaptive) immune responses. It is composed of widely distributed cells, tissues, and organs that recognize foreign substances and microorganisms and acts to neutralize or destroy them.

immunity Refers to the overall general ability of a host to resist a particular disease; the condition of being immune.

immunization The deliberate introduction of foreign materials into a host to stimulate an adaptive immune response. *See* vaccine.

immunoblotting The electrophoretic transfer of proteins from polyacrylamide gels to filters to demonstrate the presence of specific proteins through reaction with labeled antibodies.

immunodeficiency The inability to produce a normal complement of antibodies or immunologically sensitized T cells in response to specific antigens.

immunodiffusion A technique involving the diffusion of antigen or antibody within a semisolid gel to produce a precipitin reaction. Often both the antibody and antigen diffuse through the gel; sometimes an antigen diffuses through a gel containing antibody.

immunoelectrophoresis (ĭ-mu″no-e-lek″tro-fo-re′sis; pl., **immunoelectrophoreses**) The electrophoretic separation of protein antigens followed by diffusion and precipitation in gels using antibodies against the separated proteins.

immunofluorescence A technique used to identify particular antigens microscopically in cells or tissues by the binding of a fluorescent antibody conjugate.

immunoglobulin (Ig) (im″u-no-glob′u-lin) *See* antibody.

immunology The study of host defenses against invading foreign materials, including pathogenic microorganisms, transformed or cancerous cells, and tissue transplants from other sources.

immunopathology The study of diseases or conditions resulting from immune reactions.

immunoprecipitation A reaction involving soluble antigens reacting with antibodies to form a large aggregate that precipitates out of solution.

immunotoxin A monoclonal antibody that has been attached to a specific toxin or toxic agent (antibody + toxin = immunotoxin) and can kill specific target cells.

inactivated vaccines Immunogenic materials, once viable infectious agents, that are prepared so as to not cause infectious disease; also known as killed vaccines.

inclusions (1) Granules, crystals, or globules of organic or inorganic material in the cytoplasm of bacteria. (2) Clusters of viral proteins or virions within the nucleus or cytoplasm of virus-infected cells.

incubation period The period after pathogen entry into a host and before signs and symptoms appear.

incubatory carrier An individual who is incubating a pathogen but is not yet ill.

indel A blending of the words *insertion* and *deletion*. These types of genetic mutations are often considered together when examining genomes or other nucleotide sequences for mutations that are taxonomically useful.

index case The first disease case in an outbreak or epidemic.

indicator organism An organism whose presence indicates the condition of a substance or environment, for example, the potential presence of pathogens. Coliforms are used as indicators of fecal pollution.

indirect immunofluorescence Technique for imaging microscopic agents using antibodies to target the agent and a second, fluorochrome-labeled reagent to bind to the antibodies.

induced mutations Mutations caused by exposure to a mutagen.

inducer A small molecule that stimulates the synthesis of an inducible enzyme.

inducible enzyme An enzyme whose level rises in the presence of a small molecule that stimulates its synthesis or activity.

inducible gene A gene that encodes a product whose level can be increased by a molecule, often the substrate of the metabolic pathway in which the enzyme functions.

induction (1) In virology, the events that trigger a virus to switch from a lysogenic mode to a lytic pathway. (2) In genetics, an increase in gene expression.

infection The invasion of a host by a microorganism with subsequent establishment and multiplication of the agent. An infection may or may not lead to overt disease.

infection thread A tubular structure formed during the infection of a root by nitrogen-fixing bacteria. The bacteria enter the root by way of the infection thread and stimulate the formation of the root nodule.

infectious disease Any change from a state of health in which part or all of the host's body cannot carry on its normal functions because of the presence of an infectious agent or its products.

infectious disease cycle (chain of infection) The chain or cycle of events that describes how an infectious organism grows, reproduces, and is disseminated.

infectious dose 50 (ID_{50}) Refers to the dose or number of organisms that will infect 50% of an experimental group of hosts within a specified time period.

infectivity Infectiousness; the state or quality of being infectious or communicable.

inflammation A localized protective response to tissue injury or destruction. Acute inflammation is characterized by pain, heat, swelling, and redness in the injured area.

initiator codon The first codon, usually AUG, in the coding region of mRNA.

innate or **natural immunity** *See* nonspecific resistance.

insertion sequence A transposable element that contains genes only for those enzymes, such as transposase, that are required for transposition.

in silico analysis The study of physiology or genetics through the examination of nucleic acid and amino acid sequence. *See* bioinformatics.

in situ reverse transcriptase (ISRT)-FISH A method that uses fluorescently labeled probes to identify mRNA transcripts in natural samples. *See* fluorescent in situ hybridization (FISH).

integral proteins *See* plasma membrane.

integrase An enzyme observed in some viruses; it catalyzes the integration of provirus DNA into the host chromosome.

integration The incorporation of one DNA segment into a second DNA molecule to form a hybrid DNA; occurs during such processes as genetic recombination, episome incorporation into host DNA, and provirus insertion into the host chromosome.

integrins (in′tə-grinz) Cellular adhesion receptors that mediate cell-cell and cell-substratum interactions, usually by recognizing linear amino acid sequences on protein ligands.

integron A genetic element with an attachment site for site-specific recombination and an integrase gene. It can capture genes and gene cassettes.

inteins Internal intervening sequences of precursor self-splicing proteins that are removed during formation of the final protein.

intercalating agents Molecules that can be inserted between the stacked bases of a DNA double helix, thereby distorting the DNA and inducing insertion and deletion (i.e., frameshift) mutations.

interdigitating dendritic cell Special dendritic cells in the lymph nodes that function as potent antigen-presenting cells and develop from Langerhans cells.

interferon (IFN) (in″tər-fēr′on) A set of cytokines that stimulate cells to produce antiviral proteins.

Others regulate growth, differentiation, or function of a variety of immune system cells.

interleukin (in″tər-loo′kin) A cytokine produced by macrophages and T cells that regulates growth and differentiation, particularly of lymphocytes. Interleukins promote cellular and humoral immune responses.

intermediate filaments Small protein filaments, about 8 to 10 nm in diameter, in the cytoplasm of eukaryotic cells that are important in cell structure.

intermediate host The host that serves as a temporary but essential environment for development of a parasite and completion of its life cycle.

internal transcribed spacer region (ITSR) The region between SSU rRNA genes found in many microbial genomes that is conserved and transcribed. It can be used in taxonomic surveys of natural habitats.

interspecies hydrogen transfer The linkage of hydrogen production from organic matter by anaerobic heterotrophic microorganisms to the use of hydrogen by archaea in the reduction of carbon dioxide to methane.

intoxication A disease that results from the entrance of a specific toxin into the body of a host. The toxin can induce the disease in the absence of the toxin-producing organism.

intraepidermal lymphocytes T cells found in the epidermis of the skin that express the $\gamma\delta$ T-cell receptor.

intranuclear inclusion body A structure found within cells infected with cytomegalovirus.

intrinsic factors Food-related factors such as moisture, pH, and available nutrients that influence microbial growth.

intron A noncoding intervening sequence in a split (interrupted) gene that codes for pre-mRNA and is missing from the final RNA product.

invasiveness The ability of a microorganism to enter a host, grow and reproduce within the host, and spread throughout its body.

iodophor An antimicrobial agent consisting of an organic compound complexed with iodine.

ionizing radiation Radiation of very short wavelength and high energy that causes atoms to lose electrons (i.e., ionize).

iron-sulfur (Fe-S) protein *See* electron transport chain.

isoelectric focusing An electrophoretic technique in which proteins are separated based on their isoelectric point.

isoelectric point The pH value at which a protein or other molecule no longer carries a net charge.

isoenzyme An enzyme that carries out the same catalytic function but differs in terms of its amino acid sequence, regulatory properties, or other characteristics.

isotope fractionation The preferential use of one stable isotope over another by microbes carrying out metabolic processes.

isotype A variant form of an immunoglobulin that differs in the constant region of the heavy and light chains and occurs in every normal individual of a particular species.

J

J chain A polypeptide present in polymeric IgM and IgA that links the subunits together.

joule The SI (International System of Units) unit of measure for energy or work. One calorie is equivalent to 4.1840 joules.

K

kallikrein An enzyme that acts on kininogen, releasing the active bradykinin protein.

keratinocytes (kĕ-rat′ĭ-no-sīt) Cells of the skin that express keratin.

kinetoplast (ki-ne′to-plast) A special structure in the mitochondrion of certain protists. It contains the mitochondrial DNA.

kinetosome Intracellular microtubular structure that serves as the base of cilia in ciliated protists. Similar in structure to a centriole. Also referred to as a basal body.

Kirby-Bauer method A disk diffusion test to determine the susceptibility of a microorganism to chemotherapeutic agents.

Koch's postulates A set of rules for proving that a microorganism causes a particular disease.

Krebs cycle *See* tricarboxylic acid (TCA) cycle.

L

labyrinthulids A subgroup of the stramenopiles characterized by having heterokont flagellated zoospores.

lactic acid bacteria (LAB) Strictly fermentative, gram-positive bacteria that produce lactic acid as the primary fermentation end product; used in the fermentation of dairy products.

lactic acid fermentation A fermentation that produces lactic acid as the sole or primary product.

lactoferrin An iron-sequestering protein released from macrophages and neutrophils into plasma.

lager Pertaining to the process of aging beers to allow flavor development.

lagging strand The strand of DNA synthesized discontinuously during DNA replication.

lag phase A period following the introduction of microorganisms into fresh culture medium when there is no increase in cell numbers or mass during batch culture.

Lancefield system (group) One of the serologically distinguishable groups (e.g., group A and group B) into which streptococci can be divided.

Langerhans cell Cell found in the skin that internalizes antigen and moves in the lymph to lymph nodes, where it differentiates into a dendritic cell.

late mRNA Messenger RNA produced later in a viral infection, which codes for proteins needed in capsid construction and virus release.

latent viral infections Viral infections in which the virus stops reproducing and remains dormant for a period before becoming active again.

lateral gene transfer *See* horizontal gene transfer.

leader sequence A sequence in a gene that lies between the promoter and the start codon. It is transcribed, becoming a nontranslated sequence at the 5′ end of mRNA. It aids in initiation and regulation of transcription.

leading strand The strand of DNA that is synthesized continuously during DNA replication.

lectin complement pathway An antibody independent pathway of complement activation that is initiated by microbial lectins (proteins that bind carbohydrates) and includes the C3–C9 components of the classical pathway.

leghemoglobin A heme-containing pigment produced in leguminous plants. It is similar in structure to vertebrate hemoglobin and functions to protect nodule-forming, nitrogen-fixing bacteria from oxygen, which would poison their nitrogenase.

leishmanias (lēsh″ma′ne-ăs) Trypanosomal protists of the genus *Leishmania.*

lentic An aquatic system that features slow-moving or still waters.

lethal dose 50 (LD_{50}) Refers to the dose or number of organisms that will kill 50% of an experimental group of hosts within a specified time period.

leukocidin (loo″ko-si′din) A microbial toxin that can damage or kill leukocytes.

leukocyte (loo′ko-sĭt) Any white blood cell.

lichen (li′ken) A symbiotic association of a fungus and either photosynthetic protists or cyanobacteria.

light reactions *See* photosynthesis.

lignin A complex organic molecule that is an important structural component of woody plants.

limnology The study of the biological, chemical, and physical aspects of freshwater systems.

lipid A The lipid component of a lipopolysaccharide; also called endotoxin.

lipidomics Determination of an organism's lipid content under specific circumstances.

lipopolysaccharide (LPS) (lip″o-pol″e-sak′ah-rīd) A molecule containing both lipid and polysaccharide, which is important in the outer membrane of the gram-negative cell wall.

lithoheterotrophy The metabolic strategy characterized by the use of inorganic chemicals as an electron and energy source and organic carbon as a source of carbon.

lithotroph An organism that uses reduced inorganic compounds as its electron source.

littoral zone The coastal region of a lake, stream, or ocean.

lobopodia Rounded pseudopodia found in some amoeboid protists.

localized anaphylaxis (atopic reaction) Type I (immediate) hypersensitivity that remains contained in one tissue space (e.g., hay fever, food allergies, etc.).

log phase *See* exponential phase.

lophotrichous (lo-fot′rĭ-kus) A cell with a cluster of flagella at one or both ends.

lotic An aquatic system that features fast-moving, free-running waters.

lymph node A small secondary lymphoid organ that contains lymphocytes, macrophages, and dendritic cells. It serves as a site for (1) filtration and removal of foreign antigens and (2) activation and proliferation of lymphocytes.

lymphocyte A nonphagocytic, mononuclear leukocyte that is an immunologically competent cell, or its precursor. Lymphocytes are present in the blood, lymph, and lymphoid tissues. *See* B cell and T cell.

lymphokine A glycoprotein cytokine (e.g., IL-1) secreted by activated lymphocytes, especially sensitized T cells.

lyophilization Freezing and dehydrating samples as a means of preservation. Commonly referred to as freeze-drying.

lysis (li′sis) The rupture or physical disintegration of a cell.

lysogenic (li-so-jen′ik) *See* lysogens.

lysogenic conversion A change in the phenotype of a bacterium due to the presence of a prophage.

lysogens (li′so-jens) Bacterial and archaeal cells that carry a provirus and can produce viruses under the proper conditions.

lysogeny (li-soj′e-ne) The state in which a viral genome remains within a bacterial or archaeal cell after infection and reproduces along with it, rather than taking control of the host cell and destroying it.

lysosome A spherical membranous eukaryotic organelle that contains hydrolytic enzymes and is responsible for the intracellular digestion of substances.

lysozyme An enzyme that degrades peptidoglycan by hydrolyzing the $\beta(1 \rightarrow 4)$ bond that joins *N*-acetylmuramic acid and *N*-acetylglucosamine.

lytic cycle (lit′ik) A viral life cycle that results in the lysis of the host cell.

M

macroautophagy Digestion of cytoplasmic components that involves enclosing the material (e.g., an organelle) in a double-membrane structure called an autophagosome. The autophagosome delivers the material to a lysosome for digestion.

macroelement A nutrient that is required in relatively large amounts (e.g., carbon and nitrogen).

macroevolution Major evolutionary change leading to either speciation or extinction.

macrolide antibiotic An antibiotic containing a macrolide ring—a large lactone ring with multiple keto and hydroxyl groups—linked to one or more sugars.

macromolecule A large molecule that is a polymer of smaller units joined together.

macromolecule vaccine A vaccine made of specific, purified macromolecules derived from pathogenic microorganisms.

macronucleus The larger of the two nuclei in ciliate protists. It is normally polyploid and directs the routine activities of the cell.

macrophage The name for a large, mononuclear phagocytic antigen-presenting cell, present in blood, lymph, and other tissues.

madurose The sugar derivative 3-*O*-methyl-D-galactose, which is characteristic of several actinomycete genera that are collectively called maduromycetes.

magnetosomes Magnetite particles in magnetotactic bacteria that are tiny magnets and allow the bacteria to orient themselves in magnetic fields.

maintenance energy The energy a cell requires simply to maintain itself or remain alive and function properly. It does not include the energy needed for either growth or reproduction.

major histocompatibility complex (MHC) A chromosome locus encoding the histocompatibility antigens and other components of the immune system. Class I MHC molecules are cell surface glycoproteins present on all nucleated cells; class II MHC glycoproteins are on antigen-presenting cells.

malt Grain soaked in water to soften it, induce germination, and activate its enzymes. The malt is then used in brewing and distilling.

mannose-binding protein (MBP) Also known as mannose-binding lectin, a serum protein that binds to mannose residues in microbial cell walls so as to initiate complement activation.

marine snow Organic matter that sinks out of the photic zone of the ocean. These flocculent particles are composed of fecal pellets, diatom frustules, and other materials that are not rapidly degraded.

mash The soluble materials released from germinated grains and prepared as a microbial growth medium.

mashing The process in which cereals are mixed with water and incubated to degrade their complex carbohydrates (e.g., starch) to more readily usable forms such as simple sugars.

mass spectrometry A type of spectrometry that determines mass-to-charge ratio of ions formed from the molecule being analyzed. The ratio can be used to identify structures and determine sequences of proteins.

mast cell A white blood cell that produces vasoactive molecules (e.g., histamine) and stores them in vacuoles near the cell membrane where they are released upon cell stimulation by external triggers.

mating type A strain of a eukaryotic organism that can mate sexually with another strain of the same species. Commonly refers to strains of fungal mating types (MAT) (e.g., MAT α and MAT **a** of *Saccharomyces cerevisiae).*

maximum containment level goal The number of pathogens that can be present in a water supply that will prevent adverse health effects and provide a margin of safety.

M cell Specialized cell of the intestinal mucosa and other sites (e.g., urogenital tract) that delivers antigen from its apical face to lymphocytes clustered within the pocket in its basolateral face.

mean generation (doubling) time The time it takes a population to double in size.

mean growth rate (μ) The rate of microbial population growth expressed in terms of the number of generations per unit time.

medical mycology The discipline that deals with the fungi that cause human disease.

meiosis (mi-o′sis) The type of cell division by which a diploid cell divides and forms four haploid cells.

melting temperature (T_m) The temperature at which double-stranded DNA separates into individual strands; it is dependent on the G + C content of the DNA and is used to compare genetic material in microbial taxonomy.

membrane attack complex (MAC) The complex of complement proteins (C5b–C9) that creates a pore in the plasma membrane of a target cell and leads to cell lysis.

membrane-disrupting exotoxin A type of exotoxin that lyses host cells by disrupting the integrity of the plasma membrane.

membrane filter Porous material that retains microorganisms as the suspending liquid passes through the pores.

membrane filter technique The use of a thin porous filter made from cellulose acetate or some other polymer to collect microorganisms from water, air, and food.

memory cell An inactive lymphocyte derived from a sensitized B or T cell capable of an accentuated response to a subsequent antigen exposure.

meningococcus Common name for *Neisseria meningitidis,* one of the causative agents of bacterial meningitis.

mesophile A microorganism with a growth optimum around 20 to 45°C, a minimum of 15 to 20°C, and a maximum about 45°C or lower.

messenger RNA (mRNA) Single-stranded RNA synthesized from a nucleic acid template (DNA in cellular organisms, RNA in some viruses) during transcription; mRNA binds to ribosomes and directs the synthesis of protein.

metabolic channeling The localization of metabolites and enzymes in different parts of a cell.

metabolic control engineering Modification of the controls for biosynthetic pathways without altering the pathways themselves in order to improve process efficiency.

metabolic pathway engineering (MPE) The use of molecular techniques to improve the efficiency of pathways that synthesize industrially important products.

metabolism The total of all chemical reactions in the cell; almost all are enzyme catalyzed.

metabolomics Determination of the small metabolites present in a cell under specific circumstances.

metachromatic granules Granules of polyphosphate in the cytoplasm of some bacteria that appear a different color when stained with a blue basic dye. They are storage reservoirs for phosphate. Sometimes called volutin granules.

metagenomics Also called environmental or community genomics, metagenomics is the study of genomes recovered from environmental samples (including the human body) without first isolating members of the microbial community and growing them in pure cultures.

metalimnion In lakes, the region of the water column between the hypolimnion and the epilimnion.

metaproteomics The emerging field of examining all the proteins produced in a given microbial habitat at a specific time.

metastasis (mĕ-tas′tah-sis) The transfer of a disease such as cancer from one organ to another not directly connected with it.

methane hydrate Pools of trapped methane that accumulates in latticelike cages of crystalline water 500 m or more below the sediment surface in many regions of the world's oceans.

methanogenesis The production of methane by certain *Euryarchaeota*.

methanogens (meth′ə-no-jens″) Strictly anaerobic archaea that derive energy by converting CO_2, H_2, formate, acetate, and other compounds to either methane or methane and CO_2.

methanotroph A microbe that has the ability to grow on methane as the sole carbon source.

methanotrophic bacteria The ability to grow on methane as the sole carbon source.

methicillin-resistant *Staphylococcus aureus* (MRSA) Strains of the gram-positive bacterium *S. aureus* that are resistant to all β-lactam antibiotics, including methicillin.

methylotroph A bacterium that uses reduced one-carbon compounds such as methane and methanol as its sole source of carbon and energy.

metronidazole Antimicrobial chemotherapeutic agent of the nitromidazole family having biological activity on anaerobic bacteria and protozoa.

Michaelis constant (K_m) (mĭ-ka′lis) A kinetic constant for an enzyme reaction that equals the substrate concentration required for the enzyme to operate at half maximal velocity.

microaerophile (mi″kro-a′er-o-fīl) A microorganism that requires low levels of oxygen for growth, around 2 to 10%, but is damaged by normal atmospheric oxygen levels.

microautoradiography A technique whereby the uptake of a radioactive substrate by single cells is determined.

microbial ecology The study of microbial populations and communities in natural habitats, such as soils, water, food, etc.

microbial energy conversion The processes by which microbes are used to generate useful forms of energy, such a biofuel (ethanol, H_2, methanol) production, or electricity through microbial fuel cells.

microbial flora (microbiota) The microbes found in a habitat.

microbial fuel cell An apparatus designed to capture electrons produced during microbial respiration and convert these electrons to electricity.

microbial loop The cycling of organic matter synthesized by photosynthetic microorganisms among other microbes, such as bacteria and protozoa. This process "loops" organic nutrients, minerals, and carbon dioxide back for reuse by the primary producers and makes the organic matter unavailable to higher consumers.

microbial mat A firm structure of layered microorganisms with complementary physiological activities that can develop on surfaces in aquatic environments.

microbial transformation *See* biotransformation.

microbiology The study of organisms that are usually too small to be seen with the naked eye. Special techniques are required to isolate and grow them.

microbiome All the microbes that live on and in an organism; *see* metagenomics.

microbivory The use of microorganisms as a food source by organisms (e.g., protists) that can ingest or phagocytose them.

microcosms Small incubation chambers with conditions that mimic those in a natural setting.

microdroplet culture A technique designed to obtain an axenic culture whereby cells from a microbial community are encapsulated in a gel matrix, which when emulsified generates a porous microdroplet that houses a single cell. The microdroplets can be incubated together, thereby exchanging small molecules. Following incubation, microdroplets containing colonial growth can be separated by flow cytometry.

microelectrode An electrode with a tip that is sufficiently small to perform nondestructive measurements at intervals of less than 1 mm.

microenvironment The immediate environment surrounding a microbial cell or other structure, such as a root.

microevolution *See* anagenesis.

microfilaments Protein filaments, about 4 to 7 nm in diameter, that are present in the cytoplasm of eukaryotic cells and play a role in cell structure and motion.

micronucleus The smaller of the two nuclei in ciliate protists. Micronuclei are diploid and involved only in genetic recombination and the regeneration of macronuclei.

micronutrients Nutrients such as zinc, manganese, and copper that are required in very small quantities for growth and reproduction. Also called trace elements.

microorganism An organism that is too small to be seen clearly with the naked eye and is often unicellular, or if multicellular, does not exhibit a high degree of differentiation.

microsporidia Primitive, fungal, obligate intracellular parasites of animals, primarily vertebrates. The infectious spore (0.5 to 2.0 μm) contains a coiled polar tubule used for injecting the spore contents into host cells where the sporoplasm undergoes mitotic division, producing more spores.

microtubules (mi″kro-tu′buls) Small cylinders, about 25 nm in diameter, made of tubulin proteins and present in the cytoplasm and flagella of eukaryotic cells; they are involved in cell structure and movement.

mineralization The conversion of organic nutrients into inorganic material during microbial growth and metabolism.

mineral soil Soil that contains less than 20% organic carbon.

minimal inhibitory concentration (MIC) The lowest concentration of a drug that will prevent the growth of a particular microorganism.

minimal lethal concentration (MLC) The lowest concentration of a drug that will kill a particular microorganism.

minus (negative) strand A viral single-stranded nucleic acid that is complementary to the viral mRNA.

mismatch repair A type of DNA repair in which a portion of a newly synthesized strand of DNA containing mismatched base pairs is removed and replaced, using the parental strand as a template.

missense mutation A single base substitution in DNA that changes a codon for one amino acid into a codon for another.

mitochondrion (mi″to-kon′dre-on) The eukaryotic organelle that is the site of electron transport, oxidative phosphorylation, and pathways such as the Krebs cycle; it provides most of a nonphotosynthetic cell's energy under aerobic conditions.

mitosis A process that takes place in the nucleus of a eukaryotic cell and results in the formation of two new nuclei, each with the same number of chromosomes as the parent.

mitosome An organelle found in certain protists that lack mitochondria or hydrogenosomes. The presence of proteins with iron-sulfur clusters that resemble those found in mitochondria suggests that these protists may have once had mitochondria.

mixed acid fermentation A type of fermentation carried out by members of the family *Enterobacteriaceae* in which ethanol and a complex mixture of organic acids are produced.

mixotrophy A nutritional type that is a mixture of different types of metabolism (e.g., phototrophy and chemoorganotrophy).

modified atmosphere packaging (MAP) Addition of gases such as nitrogen and carbon dioxide to packaged foods to inhibit the growth of spoilage organisms.

modulon A set of operons that are controlled by a common regulatory protein and by a unique regulatory protein, specific for each operon.

mold Any of a large group of fungi that exist as multicellular filamentous colonies; also the deposit or growth caused by such fungi. Molds typically do not produce macroscopic fruiting bodies.

molecular chaperones *See* chaperone proteins.

monoclonal antibody (mAb) An antibody of a single type that is produced by a population of genetically identical plasma cells (a clone); a monoclonal antibody is typically produced from a cell culture derived from the fusion of a cancer cell and an antibody-producing cell (i.e., hybridoma).

monocyte A mononuclear phagocytic leukocyte; precursor of macrophages and dendritic cells.

monocyte-macrophage system The collection of fixed phagocytic cells (including macrophages, monocytes, and specialized endothelial cells) located in the liver, spleen, lymph nodes, and bone marrow; an important component of the host's general nonspecific (innate) defense against pathogens.

monokine A generic term for a cytokine produced by macrophages or monocytes.

monotrichous (mon-ot′rĭ-kus) Having a single flagellum.

morbidity rate Measures the number of individuals who become ill as a result of a particular disease within a susceptible population during a specific time period.

mordant A substance that helps fix dye on or in a cell.

morphovar A variant strain of a microbe characterized by morphological differences.

mortality rate The ratio of the number of deaths from a given disease to the total number of cases of the disease.

most probable number (MPN) The statistical estimation of the probable population in a liquid by diluting and determining end points for microbial growth.

mucociliary blanket The layer of cilia and mucus that lines certain portions of the respiratory system; it traps microorganisms and then transports them by ciliary action away from the lungs.

mucosal-associated lymphoid tissue (MALT) Organized and diffuse immune tissues found as part of the mucosal epithelium. It can be specialized to the gut (GALT) or the bronchial system or (BALT).

multicloning site (MCS) A region of DNA on a cloning vector that has a number of restriction enzyme recognition sequences to facilitate the introduction, or cloning, of a gene.

multilocus sequence analysis (MLSA) The application of multilocus sequence typing to a wider taxonomic group of microbes; that is, to more than one species.

multilocus sequence typing (MLST) A method for genotypic classification of bacteria or archaea within a single genus using nucleotide differences among five to seven housekeeping genes.

multiple displacement amplification (MDA) A technique used to amplify very low concentrations of DNA; it can yield enough DNA from a single cell to perform DNA sequencing.

multiplex PCR The use of multiple primers in a single polymerase chain reaction so that multiple products, each of a unique size, are generated.

murein *See* peptidoglycan.

must The juices of fruits, including grapes, that can be fermented for the production of alcohol.

mutagen (mu'tah-jen) A chemical or physical agent that causes mutations.

mutation A permanent, heritable change in the genetic material.

mutualism A type of symbiosis in which both partners gain from the association and are metabolically dependent on each other.

mutualist An organism associated with another in an obligatory relationship that is beneficial to both.

mycelium (mi-se'le-um) A mass of branching hyphae found in fungi and some bacteria.

mycobiont The fungal partner in a lichen.

mycolic acids Complex 60 to 90 carbon fatty acids with a hydroxyl on the β-carbon and an aliphatic chain on the α-carbon; found in the cell walls of mycobacteria.

mycologist A person specializing in mycology; a student of mycology.

mycology The science and study of fungi.

mycoplasma Bacteria that are members of the phylum *Firmicutes,* class *Mollicutes,* and order *Mycoplasmatales;* they lack cell walls and cannot synthesize peptidoglycan precursors; most require sterols for growth.

mycorrhizal fungi Fungi that form stable, mutualistic relationships on (ectomycorrhizal) or in (endomycorrhizal) the roots of vascular plants.

mycosis (mi-ko'sis; pl., **mycoses**) Any disease caused by a fungus.

mycotoxicology (mi-ko'tok"si-kol'o-je) The study of fungal toxins and their effects on various organisms.

myeloma cell (mi"e-lo'mah) A tumor cell that is similar to the cell type found in bone marrow. Also, a malignant, neoplastic plasma cell that produces large quantities of antibodies and can be readily cultivated.

myositis (mi"o-si'tis) Inflammation of a striated or voluntary muscle.

myxobacteria A group of gram-negative, aerobic soil bacteria characterized by gliding motility, a complex life cycle with the production of fruiting bodies, and the formation of myxospores.

myxospores (mik'so-spōrs) Special dormant spores formed by the myxobacteria.

N

***N*-acylhomoserine lactone** *See* quorum sensing.

naked amoeba An amoeboid protist that lacks a cell wall or other supporting structures.

nanowires Pililike appendages that transfer electrons from the terminal point in the electron transport chain to an external metal surface.

narrow-spectrum drugs Chemotherapeutic agents that are effective only against a limited variety of microorganisms.

natural attenuation The decrease in the level of an environmental contaminant that results from natural chemical, physical, and biological processes.

natural classification A classification system that arranges organisms into groups whose members share many characteristics and reflect as much as possible the biological nature of organisms.

natural killer (NK) cell A type of white blood cell that has a lineage independent of the granulocyte, B-cell, and T-cell lineages; part of the innate immune system.

naturally acquired active immunity The type of active immunity that develops when an individual's immunologic system comes into contact with an appropriate antigenic stimulus during the course of normal activities; it often arises as the result of recovering from an infection and lasts a long time.

naturally acquired passive immunity The type of temporary immunity that involves the transfer of antibodies from one individual to another.

natural products Industrially important compounds, such as antibiotics, produced by microbes.

negative selection In immunology, the process by which lymphocytes that recognize host (self) antigens undergo apoptosis or become anergic (inactive).

negative staining A staining procedure in which a dye is used to make the background dark while the specimen is unstained.

negative transcriptional control Regulation of transcription by a repressor protein. When bound to the repressor-binding site, transcription is inhibited.

Negri bodies (na'gre) Masses of viruses or unassembled viral subunits found within the neurons of rabies-infected animals.

neoplasia Abnormal cell growth and reproduction due to a loss of regulation of the cell cycle; produces a tumor in solid tissues.

neuraminidase An enzyme that cleaves the chemical bond linking neuraminic acids to the sugars present on the surface of animal cells; in virology, one type of envelope spike on influenza viruses has neuraminidase activity and is used to identify different strains.

neurotoxin (nu"ro-tok'sin) A toxin that is poisonous to or destroys nerve tissue.

neutralization Binding of specific immunoglobulins (antitoxins) to toxins, effectively inhibiting their biological activity.

neutrophil A mature white blood cell in the granulocyte lineage. It has a nucleus with three to five lobes and is very phagocytic.

neutrophile A microorganism that grows best at a neutral pH range between pH 5.5 and 8.0.

niche The function of an organism in a complex system, including place of the organism, the resources used in a given location, and the time of use.

nicotinamide adenine dinucleotide (NAD^+) (nik"o-tin'ah-mīd) An electron-carrying coenzyme; it is particularly important in catabolic processes and usually transfers electrons from an electron source to an electron transport chain.

nicotinamide adenine dinucleotide phosphate ($NADP^+$) An electron-carrying coenzyme that most often participates as an electron carrier in biosynthetic metabolism.

nitrification The oxidation of ammonia to nitrate.

nitrifying archaea Mesophilic crenarchaea that are capable of oxidizing ammonium to nitrite.

nitrifying bacteria Chemolithotrophic, gram-negative bacteria that are members of several families within the phylum *Proteobacteria* that either oxidize ammonia to nitrite or nitrite to nitrate.

nitrogenase (ni'tro-jen-ās) The enzyme that catalyzes biological nitrogen fixation.

nitrogen fixation The metabolic process in which atmospheric molecular nitrogen (N_2) is reduced to ammonia; carried out by cyanobacteria, *Rhizobium,* methanogens, and other nitrogen-fixing bacteria and archaea.

nitrogen oxygen demand (NOD) The demand for oxygen in sewage treatment, caused by nitrifying microorganisms.

nocardioforms Bacteria that resemble members of the genus *Nocardia;* they develop a substrate mycelium that readily breaks up into rods and coccoid elements (a quality sometimes called fugacity).

Nod factors Signaling compounds that alter gene expression in the plant host of a rhizobium.

nomenclature The branch of taxonomy concerned with the assignment of names to taxonomic groups in agreement with published rules.

noncompetitive inhibitor A chemical that inhibits enzyme activity by a mechanism that does not involve binding the active site of the enzyme.

noncyclic photophosphorylation (fo"to-fos"for-i-la'shun) The process in which light energy is used to make ATP when electrons are moved from water to $NADP^+$ during oxygenic photosynthesis; both photosystem I and photosystem II are involved.

noncytopathic virus A virus that does not kill its host cell by viral release-induced lysis.

nonenveloped (naked) virus A virus composed only of a nucleocapsid (i.e., lacking an envelope).

nonheme iron protein *See* electron transport chain.

nonsense (stop) codon A codon that does not code for an amino acid but is a signal to terminate protein synthesis.

nonsense mutation A mutation that converts a sense codon to a nonsense (stop) codon.

nonspecific immune response (innate or **natural immunity)** *See* nonspecific resistance.

nonspecific resistance Refers to those general defense mechanisms that are inherited as part of the innate structure and function of each animal; also known as nonspecific, innate, or natural immunity.

normal microbiota (indigenous microbial population, microflora, microbial flora) The microorganisms normally associated with a particular tissue or structure.

nosocomial infection (nos″o-ko′me-al) An infection that is acquired during the stay of a patient in a hospital or other type of clinical care facility.

nuclear envelope The complex double-membrane structure forming the outer boundary of the eukaryotic nucleus. It is covered by nuclear pores through which substances enter and leave the nucleus.

nuclear pore complex The nuclear pore plus about 30 proteins that form the pore and are involved in moving materials across the nuclear envelope.

nucleic acid hybridization The process of forming a hybrid double-stranded DNA molecule using a heated mixture of single-stranded DNAs from two different sources; if the sequences are fairly complementary, stable hybrids will form.

nucleocapsid (nu″kle-o-kap′sid) The viral nucleic acid and its surrounding capsid; the basic unit of virion structure.

nucleoid An irregularly shaped region in a bacterial or archaeal cell that contains its genetic material.

nucleolus (nu-kle′o-lus) The organelle, located within the nucleus and not bounded by a membrane, that is the location of ribosomal RNA synthesis and the assembly of ribosomal subunits.

nucleoside (nu′kle-o-sīd″) A combination of ribose or deoxyribose with a purine or pyrimidine base.

nucleosome (nu′kle-o-sōm″) A complex of histones and DNA found in eukaryotic chromatin and some archaea; the DNA is wrapped around the surface of the beadlike histone complex.

nucleotide (nu′kle-o-tīd) A combination of ribose or deoxyribose with phosphate and a purine or pyrimidine base; a nucleoside plus one or more phosphates.

nucleotide excision repair *See* excision repair.

nucleus The eukaryotic organelle enclosed by a double-membrane envelope that contains the cell's chromosomes.

numerical aperture The property of a microscope lens that determines how much light can enter and how great a resolution the lens can provide.

nutrient A substance that supports growth and reproduction.

nystatin (nis′tah-tin) A polyene antibiotic from *Streptomyces noursei* that is used in the treatment of *Candida* infections of the skin, vagina, and alimentary tract.

O

O antigen A polysaccharide antigen extending from the outer membrane of some gram-negative bacterial cell walls; it is part of the lipopolysaccharide.

obligate aerobes Organisms that grow only in the presence of oxygen.

obligate anaerobes Microorganisms that cannot tolerate the presence of oxygen and die when exposed to it.

oceanography The study of the biological, chemical, and physical aspects of the oceans.

Okazaki fragments Short stretches of polynucleotides produced during discontinuous DNA replication.

oligonucleotide A short fragment of DNA or RNA, usually artificially synthesized, used in a number of molecular genetic techniques such as DNA sequencing, polymerase chain reaction, and Southern blotting.

oligonucleotide signature sequence Short, conserved nucleotide sequences that are specific for a phylogenetically defined group of organisms. The signature sequences found in small subunit rRNA molecules are most commonly used.

oligotrophic environment (ol″ĭ-go-trof′ik) An environment containing low levels of nutrients, particularly nutrients that support microbial growth.

oncogene (ong″ko-jēn) A gene whose activity is associated with the conversion of normal cells to cancer cells.

oncovirus A virus known to be associated with the development of cancer.

oocyst (o′o-sist) Cyst formed around a zygote of malaria and related protozoa.

öomycetes (o″o-mi-se′tēz) A collective name for protists also known as water molds. Formerly thought to be fungi.

open reading frame (ORF) A sequence of DNA thought to encode a protein because it is not interrupted by a stop codon and has an apparent promoter and ribosome binding site at the 5′ end and a terminator at the 3′ end. It is usually determined by nucleic acid sequencing studies.

operational taxonomic unit (OTU) A term that includes all sources of data used to construct phylogenetic trees; that is, it includes species, strains, and genomic sequences from organisms not yet grown in culture.

operator The segment of DNA in a bacterial operon to which the repressor protein binds; it controls the expression of the genes adjacent to it.

operon The sequence of bases in DNA that contains one or more structural genes together with the operator or activator-binding site controlling their expression.

opportunistic microorganism or **pathogen** A microorganism that is usually free living or a part of the host's normal microbiota but may become pathogenic under certain circumstances, such as when the immune system is compromised.

opsonization (op″so-ni-za′shun) The coating of foreign substances by antibody, complement proteins, or fibronectin to make the substances more readily recognized by phagocytic cells.

optical tweezer The use of a focused laser beam to drag and isolate a specific microorganism from a complex microbial mixture.

organelle A structure within or on a cell that performs specific functions and is related to the cell in a way similar to that of an organ to the body of a multicellular organism.

organic soil A soil that has at least 20% organic carbon.

organotrophs Organisms that use reduced organic compounds as their electron source.

origin of replication A site on a chromosome or plasmid where DNA replication is initiated.

orthologue A gene found in the genomes of two or more different organisms that share a common ancestry. The products of orthologous genes are presumed to have similar functions.

O side chain *See* O antigen.

osmophilic microorganisms Microorganisms that grow best in or on media of high solute concentration.

osmotolerant Organisms that grow over a fairly wide range of water activity or solute concentration.

osmotrophy A form of nutrition in which soluble nutrients are absorbed through the cytoplasmic membrane; found in bacterial and archaeal cells, fungi, and some protists.

Öuchterlony technique *See* double diffusion agar assay.

outbreak The sudden, unexpected occurrence of a disease in a given population.

outer membrane A membrane located outside the peptidoglycan layer in the cell walls of gram-negative bacteria.

oxidation-reduction (redox) reactions Reactions involving electron transfers; the electron donor (reductant) gives electrons to an electron acceptor (oxidant).

oxidative burst The generation of reactive oxygen species, primarily superoxide anion (O_2^-) and hydrogen peroxide (H_2O_2) by a plant or an animal, in response to challenge by a potential bacterial, fungal, or viral pathogen. *See also* respiratory burst.

oxidative phosphorylation (fos″for-ĭ-la′-shun) The synthesis of ATP from ADP using energy made available during electron transport initiated by the oxidation of a chemical energy source.

oxygenic photosynthesis Photosynthesis that oxidizes water to form oxygen; the form of photosynthesis characteristic of plants, protists, and cyanobacteria.

P

palisade arrangement Angular arrangements of bacteria whereby cells remain partially attached after snapping division.

pandemic An increase in the occurrence of a disease within a large and geographically widespread population (often refers to a worldwide epidemic).

Paneth cell (pah′net) A specialized epithelial cell of the intestine that secretes hydrolytic enzymes and antimicrobial proteins and peptides.

pan-genome The collection of genes found in all strains that belong to a single species or other taxonomic group.

paralogue Two or more genes in the genome of a single organism that arose through duplication of a common ancestral gene. The products of paralogous genes may have slightly different functions.

paraphyletic A taxonomic group that includes some, but not all, descendants of a single common ancestor.

parasite An organism that lives on or within another organism (the host) and benefits from the association while harming its host. Often the parasite obtains nutrients from the host.

parasitism A type of symbiosis in which one organism benefits from the other and the host is usually harmed.

parasporal body An intracellular, solid protein crystal made by the bacterium *Bacillus thuringiensis.* It is the basis of the bacterial insecticide Bt.

parenteral route (pah-ren′ter-al) A route of drug administration that is nonoral (e.g., by injection).

parfocal A microscope that retains proper focus when the objectives are changed.

parsimony analysis A method for developing phylogenetic trees based on the estimation of the minimum number of nucleotide or amino acid sequence changes needed to give the sequences being compared.

particulate organic matter (POM) Nutrients that are not dissolved or soluble, generally referring to aquatic ecosystems. This includes microorganisms or their cellular debris after senescence or viral lysis.

passive diffusion The process in which molecules move from a region of higher concentration to one of lower concentration as a result of random thermal agitation.

passive immunization Temporary immunity brought about by the transfer of immune products, such as antibodies or sensitized T cells, from an immune vertebrate to a nonimmune one.

pasteurization The process of heating liquids to destroy microorganisms that can cause spoilage or disease.

pathogen (path′o-jən) Any virus, bacterium, or other agent that causes disease.

pathogen-associated molecular pattern (PAMP) Conserved molecular structures that occur in patterns on microbial surfaces. The structures and their patterns are unique to particular types of microorganisms and invariant among members of a given microbial group.

pathogenicity (path″o-je-nis′ĭ-te) The condition or quality of being pathogenic, or the ability to cause disease.

pathogenicity island A 10 to 200 Kb segment of DNA in some pathogens that contains the genes responsible for virulence; often it codes for a type III secretion system that allows the pathogen to secrete virulence proteins and damage host cells.

pathogenic potential The degree that a pathogen causes morbid signs and symptoms.

pattern recognition receptor (PRR) A receptor found on macrophages and other phagocytic cells that binds to pathogen-associated molecular patterns on microbial surfaces.

PCR bias Artifactual or misleading results obtained when microbial populations are assessed by first amplifying genes or genomes by the polymerase chain reaction (PCR). This occurs because not all DNA or RNA is amplified at the same rate.

pelagic zone The central region of a lake or ocean not influenced by events or nutrients found in the coastal regions.

pellicle (pel′ĭ-k′l) A relatively rigid layer of proteinaceous elements just beneath the plasma membrane in many protists. The plasma membrane is sometimes considered part of the pellicle.

penicillins A group of antibiotics containing a β-lactam ring.

pentose phosphate pathway A glycolytic pathway that oxidizes glucose 6-phosphate to ribulose 5-phosphate and then converts it to a variety of three to seven carbon sugars.

peplomer (spike) (pep′lo-mer) A protein or protein complex that extends from the viral envelope and often is important in virion attachment to the host cell surface.

peptide interbridge A short peptide chain that connects the tetrapeptide chains in the peptidoglycan of some bacteria.

peptidoglycan (pep″tĭ-do-gli′kan) A large polymer composed of long chains of alternating *N*-acetylglucosamine and *N*-acetylmuramic acid residues. The polysaccharide chains are linked to each other through connections between tetrapeptide chains attached to the *N*-acetylmuramic acids. It provides much of the strength and rigidity possessed by bacterial cell walls; also called murein.

peptidyl or **donor site (P site)** The site on the ribosome that contains the peptidyl-tRNA at the beginning of the elongation cycle during protein synthesis.

peptidyl transferase The ribozyme that catalyzes the transpeptidation reaction in protein synthesis; in this reaction, an amino acid is added to the growing peptide chain.

peptones Water-soluble digests or hydrolysates of proteins that are used in the preparation of culture media.

perforin pathway The cytotoxic pathway that uses perforin protein, which polymerizes to form membrane pores that help destroy cells during cell-mediated cytotoxicity.

period of infectivity Refers to the time during which the source of an infectious disease is infectious or is disseminating the pathogen.

peripheral protein *See* plasma membrane.

peripheral tolerance The inhibition of effector lymphocyte activity resulting in the lack of a specific response.

periplasm (per′ĭ-plaz-əm) The substance that fills the periplasmic space.

periplasmic flagella The flagella that lie under the outer sheath and extend from both ends of the spirochete cell to overlap in the middle and form the axial filament. Also called axial fibrils, axial filaments, and endoflagella.

periplasmic space The space between the plasma membrane and the outer membrane in gram-negative bacteria, and between the plasma membrane and the cell wall in gram-positive bacteria. A similar space is sometimes observed between the plasma membrane and the cell wall of some archaea.

peristalsis The muscular contractions of the gut that propel digested foods and waste through the intestinal tract.

peritrichous (pĕ-rit′rĭ-kus) A cell with flagella distributed over its surface.

permease (per′me-ās) A membrane-bound carrier protein or a system of two or more proteins that transports a substance across the membrane.

peronosporomycetes New name for the öomycetes, a subgroup of the *Stramenopila,* distinguished by the formation of large egg cells when they reproduce.

Petri dish (pe′tre) A shallow dish consisting of two round, overlapping halves that is used to grow microorganisms on solid culture medium.

phage (fāj) *See* bacteriophage.

phagocytic vacuole (fag″o-sit′ik vak′u-ol) A membrane-delimited vacuole produced by cells carrying out phagocytosis. It is formed by the invagination of the plasma membrane and contains solid material.

phagocytosis (fag″o-si-to′sis) The endocytotic process in which a cell encloses large particles in a phagocytic vacuole and engulfs them.

phagolysosome (fag″o-li′so-sōm) The vacuole that results from the fusion of a phagosome with a lysosome.

phagosome A membrane-enclosed vacuole formed by the invagination of the cell membrane during endocytosis.

phagovar (fag′o-var) A specific phage type.

phase-contrast microscope A microscope that converts slight differences in refractive index and cell density into easily observed differences in light intensity.

phenetic system A classification system that groups organisms together based on the similarity of their observable characteristics.

phenol coefficient test A test to measure the effectiveness of disinfectants by comparing their activity against test bacteria with that of phenol.

phenotype Observable characteristics of an organism.

phenotypic rescue The identification of a cloned gene based on its ability to complement a genetic deficiency in the host cell.

phosphatase (fos′fah-tās″) An enzyme that catalyzes the hydrolytic removal of phosphate from molecules.

phosphoenolpyruvate: sugar phosphotransferase system (PTS) An important group translocation system used by many bacteria. As a sugar is transported into the cell, the hydrolysis of a high-energy phosphate bond fuels its import, and the sugar is modified by the covalent attachment of the phosphoryl group.

phospholipase An enzyme that hydrolyzes a specific ester bond in phospholipids.

phosphorelay system A mechanism for regulating either protein activity or transcription. It involves the transfer of a phosphate from one protein to another.

photic zone The illuminated area of an aquatic habitat from the surface to the depth to which the rate of photosynthesis equals that of respiration.

photolithoautotroph An organism that uses light energy, an inorganic electron source (e.g., H_2O, H_2, H_2S), and CO_2 as its carbon source. Also called photolithotrophic autotroph or a photoautotroph.

photoorganoheterotroph A microorganism that uses light energy, organic electron sources, and organic molecules as a carbon source. Also called photoorganotrophic heterotroph.

photophosphorylation The synthesis of ATP from ADP using energy made available by the absorption of light.

photoreactivation The process in which blue light is used by a photoreactivating enzyme to repair thymine dimers in DNA.

photosynthate Nutrient material that leaks from phototrophic organisms and contributes to the pool of dissolved organic matter that is available to microorganisms.

photosynthesis The trapping of light energy and its conversion to chemical energy (light reactions), which is then used to reduce CO_2 and incorporate it into organic forms (dark reactions).

photosystem I The photosystem in eukaryotic cells and cyanobacteria that absorbs longer wavelength light, usually greater than about 680 nm, and transfers the energy to chlorophyll P700 during photosynthesis; it is involved in both cyclic and noncyclic photophosphorylation.

photosystem II The photosystem in eukaryotic cells and cyanobacteria that absorbs shorter wavelength light, usually less than 680 nm, and transfers the energy to chlorophyll P680 during photosynthesis; it participates in noncyclic photophosphorylation.

phototaxis The ability of certain phototrophic organisms to move in response to a light source. In positive phototaxis, the microbe moves toward the light; in negative, it moves away.

phototrophs Organisms that use light as their energy source.

phycobiliproteins Photosynthetic pigments found in cyanobacteria that are composed of proteins with attached tetrapyrroles.

phycobilisomes Special particles on the membranes of cyanobacteria that contain photosynthetic pigments.

phycobiont (fi″ko-bi′ont) The photosynthetic protist or cyanobacterial partner in a lichen.

phycocyanin (fi″ko-si′an-in) A blue phycobiliprotein pigment used to trap light energy during photosynthesis.

phycoerythrin (fi″ko-er′i-thrin) A red photosynthetic phycobiliprotein pigment used to trap light energy.

phycology (fi-kol′o-je) The study of algae.

phyllosphere The surface of plant leaves.

phylochip *See* phylogenetic oligonucleotide arrays.

phylogenetic or **phyletic classification system** (fi″lo-jĕ-net′ik, fi-let′ik) A classification system based on evolutionary relationships, rather than the general similarity of characteristics.

phylogenetic oligonucleotide arrays Microarrays that contain rRNA sequences as probes and are used as a means to detect complementary rRNA sequences from the environment.

phylogenetic tree A graph made of nodes and branches, much like a tree in shape, that shows phylogenetic relationships between groups of organisms and sometimes also indicates the evolutionary development of groups.

phylogeny The evolutionary development of a species.

phylotype A taxon that is characterized only by its nucleic acid sequence; generally discovered during metagenomic analysis.

physical map In genomics, this refers to a diagram of a chromosome, gene, or other genetic element that shows restriction endonuclease recognition sites. Specific genes and other genetic elements (e.g., origin of replication) may also be noted.

phytoplankton A community of floating photosynthetic organisms, composed of photosynthetic protists and cyanobacteria.

phytoremediation The use of plants to stimulate the extraction, degradation, adsorption, stabilization, or volatilization of soil contaminants.

picoplankton Planktonic microbes between 0.2 and 2.0 μm in size, including the cyanobacterial genera *Prochlorococcus* and *Synechococcus*.

piezophilic (pi-e-zo-fil′ic) Describing organisms that prefer or require high pressures for growth and reproduction.

pili (s., **pilus**) *See* fimbria.

pitching Pertaining to inoculation of a nutrient medium with yeast, for example, in beer brewing.

plankton Free-floating, mostly microscopic microorganisms that can be found in almost all waters.

plaque A clear area in a lawn of host cells that results from their lysis by viruses.

plaque assay A method used to determine the number of infectious virions in a population of viruses.

plaque-forming unit The unit of measure of a plaque assay. It usually represents a single infectious virion.

plasma cell A mature, differentiated B lymphocyte that synthesizes and secretes antibody.

plasmalemma The plasma membrane in protists.

plasma membrane The selectively permeable membrane surrounding the cell's cytoplasm; also called the cell membrane, plasmalemma, or cytoplasmic membrane. For most cells it is a lipid bilayer (some archaea have a lipid monolayer) with proteins embedded in it (integral proteins) and associated with the surface (peripheral proteins).

plasmid A double-stranded DNA molecule that can exist and replicate independently of the chromosome. A plasmid is stably inherited but is not required for the host cell's growth and reproduction.

plasmid fingerprinting A technique used to identify microbial isolates as belonging to the same strain because they contain the same number of plasmids with the identical molecular weights and similar phenotypes.

plasmodial (acellular) slime mold A member of the protist division *Amoebozoa (Myxogastria)* that exists as a thin, streaming, multinucleate mass of protoplasm (plasmodium), which creeps along in an amoeboid fashion.

plasmodium (pl., **plasmodia**) (1) A stage in the life cycle of myxogastria protists; a multinucleate mass of protoplasm surrounded by a membrane. (2) A parasite of the genus *Plasmodium*.

plasmolysis (plaz-mol′ĭ-sis) The process in which water osmotically leaves a cell, which causes the cytoplasm to shrivel up and pull the plasma membrane away from the cell wall.

plastid A cytoplasmic organelle of algae and higher plants that contains pigments such as chlorophyll, stores food reserves, and often carries out processes such as photosynthesis.

plate wash polymerase chain reaction A technique for identifying the culture conditions that support growth of a specific microbe or type of microbe. Following growth of natural isolates under a variety of conditions, polymerase chain reaction (PCR) is used to determine which condition supported the growth of the desired microbe.

pleomorphic (ple″o-mor′fik) Refers to cells or viruses that are variable in shape and lack a single, characteristic form.

plus (positive) strand A viral nucleic acid strand that is equivalent in base sequence to the viral mRNA.

point mutation A mutation that affects only a single base pair.

polar flagellum A flagellum located at one end of an elongated cell.

poly-β-hydroxybutyrate (PHB) (hi-drok″se-bu′tĭ-rāt) A linear polymer of β-hydroxybutyrate used as a reserve of carbon and energy by many bacteria.

polyhistidine tagging (His-tagging) One of several methods by which a protein is fused to a series of amino acids, in this case, six histidine residues, thereby facilitating the protein's purification through affinity chromatography.

polymerase chain reaction (PCR) An in vitro technique used to synthesize large quantities of specific nucleotide sequences from small amounts of DNA. It employs oligonucleotide primers complementary to specific sequences in the target gene and special heat-stable DNA polymerases (e.g., Taq polymerase).

polymorphonuclear leukocyte (pol″e-mor″fo-noo′kle-ər) A leukocyte that has a variety of nuclear forms.

polyphasic taxonomy An approach in which taxonomic schemes are developed using a wide range of phenotypic and genotypic information.

polyprotein A large polypeptide that contains several proteins, which are released when the polyprotein is cut by proteases. Polyproteins are often produced by viruses having positive-strand RNA genomes.

polyribosome (pol″e-ri′bo-sōm) A complex of several ribosomes translating a single messenger RNA.

population An assemblage of organisms of the same type.

porin proteins Proteins that form channels across the outer membrane of gram-negative bacterial cell walls through which small molecules enter the periplasm.

positive transcriptional control Control of transcription by an activator protein. When the activator is bound to the activator-binding site, the level of transcription increases.

postherpetic neuralgia The severe pain after a herpes infection.

posttranscriptional modification The processing of the initial RNA transcript (pre-mRNA) to form mRNA. *See* pre-mRNA.

posttranslational regulation Alteration of the activity of a protein after it has been synthesized by a structural modification, such as phosphorylation.

potable Refers to water suitable for drinking.

pour plate A Petri dish of solid culture medium with isolated microbial colonies growing both on its surface and within the medium that has been prepared by mixing microorganisms with cooled, still-liquid medium and then allowing the medium to harden.

precipitation (or **precipitin**) **reaction** The reaction of an antibody with a soluble antigen to form an insoluble precipitate.

precipitin The antibody responsible for a precipitation reaction.

precursor metabolites Intermediates of glycolytic pathways, TCA cycle, and other pathways that serve as starting molecules for biosynthetic pathways that generate monomers and other building blocks needed for synthesis of macromolecules.

pre-mRNA In eukaryotes, the RNA transcript of DNA made by RNA polymerase II; it is processed to form mRNA.

presence-absence (P-A) test An assay designed to detect the presence of coliforms in water; a positive result must be confirmed by a different assay.

prevalence rate Refers to the total number of individuals infected at any one time in a given population regardless of when the disease began.

Pribnow box A base sequence in bacterial promoters that is recognized by RNA polymerase and is the site of initial polymerase binding.

primary (frank) pathogen Any organism that causes a disease in the host by direct interaction with or infection of the host.

primary metabolites Microbial metabolites produced during active growth of an organism.

primary producer Photoautotrophic and chemoautotrophic organisms that incorporate carbon dioxide into organic carbon and thus provide new biomass for the ecosystem.

primary production The incorporation of carbon dioxide into organic matter by photosynthetic organisms and chemoautotrophic organisms.

primary treatment The first step of sewage treatment, in which physical settling and screening are used to remove particulate materials.

primase *See* primosome.

primers *See* polymerase chain reaction and primosome.

primosome A complex of proteins that includes the enzyme primase, which is responsible for synthesizing the RNA primers needed for DNA replication.

prion (pre′on) An infectious agent consisting only of protein; prions cause a variety of spongiform encephalopathies such as scrapie in sheep and goats.

probe A short, labeled nucleic acid segment complementary in base sequence to part of another nucleic acid that is used to identify or isolate the particular nucleic acid from a mixture through its ability to bind specifically with the target nucleic acid.

probiont The hypothesized early RNA-based biological entity that gave rise to DNA-based cellular life-forms.

probiotic A living microorganism that may provide health benefits beyond its nutritional value when ingested.

prochlorophytes A group of cyanobacteria having both chlorophyll *a* and *b* but lacking phycobilins.

prodromal stage (pro-dro′məl) The period during the course of a disease when there is the appearance of signs and symptoms, but they are not yet distinctive enough to make an accurate diagnosis.

production strains Microbial strains used in industrial processes that have been engineered to generate a high yield of the desired product.

programmed cell death (1) In eukaryotes, apoptosis. (2) In some bacteria, a mechanism proposed to account for the decline in cell numbers during the death phase of a growth curve. *See* apoptosis.

prokaryotic cells (pro″kar-e-ot′ik) Cells having a type of structure characterized by the lack of a true, membrane-enclosed nucleus. All known members of *Archaea* and most members of *Bacteria* exhibit this type of cell structure; some members of the bacterial phylum *Planctomycetes* have a membrane surrounding their genetic material.

prokaryotic species A collection of bacterial or archaeal strains that share many stable properties and differ significantly from other groups of strains.

promoter The region on DNA at the start of a gene that RNA polymerase binds to before beginning transcription.

proofreading The ability of enzymes such as DNA polymerase to check their products to ensure that the correct product is made. For instance, DNA polymerase checks newly synthesized DNA to ensure that the correct nucleotides have been added.

propagated epidemic An epidemic that is characterized by a relatively slow and prolonged rise and then a gradual decline in the number of individuals infected. It usually results from the introduction of an infected individual into a susceptible population and transmission of the pathogen from person to person.

prophage (pro′fāj) **(provirus)** (1) The latent form of a temperate bacterial or archaeal virus that remains within the lysogen, usually integrated into the host chromosome. (2) The form of a eukaryotic virus that remains within the host cell during a latent infection. Also refers to a retroviral genome after it has been integrated into the host's chromosome.

prostheca (pros-the′kah) An extension of a bacterial cell, including the plasma membrane and cell wall, that is narrower than the mature cell.

prosthetic group A tightly bound cofactor that remains at the active site of an enzyme during its catalytic activity.

protease (pro′te-ās) An enzyme that hydrolyzes proteins to their constituent amino acids. Also called a proteinase.

proteasome A large, cylindrical protein complex that degrades ubiquitin-labeled proteins to peptides in an ATP-dependent process. Also called 26S proteasome.

protein engineering The rational design of proteins by constructing specific amino acid sequences through molecular techniques, with the objective of modifying protein characteristics.

protein modeling The process by which the amino acid sequence of a protein is analyzed using software designed to predict the protein's three-dimensional structure.

protein splicing The posttranslational process in which part of a precursor polypeptide is removed before the mature polypeptide folds into its final shape; it is carried out by self-splicing proteins that remove inteins and join the remaining exteins.

proteome The complete collection of proteins that an organism produces.

proteomics The study of the structure and function of cellular proteins.

proteorhodopsin A rhodopsin molecule first identified in marine proteobacteria but since found in a variety of microbes. It enables rhodopsin-based phototrophy. *See* bacteriorhodpsin.

protist Unicellular and sometimes colonial eukaryotic organisms that lack cellular differentiation into tissues. Many chemoorganotrophic protists are referred to as protozoa; many phototrophic protists are referred to as algae.

protistology The study of protists.

protomer An individual subunit of a viral capsid; a capsomer is made of protomers.

proton motive force (PMF) The force arising from a gradient of protons and a membrane potential that powers ATP synthesis and other processes.

proto-oncogene The normal cellular form of a gene that when mutated or overexpressed results in or contributes to malignant transformation of a cell.

protoplast (1) The plasma membrane and everything within it. (2) A bacterial, archaeal, or fungal cell with its cell wall completely removed. It is spherical in shape and osmotically sensitive.

protoplast fusion The joining of cells that have had their walls weakened or completely removed.

prototroph A microorganism that requires the same nutrients as the majority of naturally occurring members of its species.

protozoa (pro″to-zo′a) A common term for a group of unrelated unicellular, chemoorganotrophic protists.

protozoology The study of protozoa.

proviral DNA Viral DNA that has been integrated into host cell DNA. In retroviruses, it is the double-stranded DNA copy of the RNA genome.

pseudogene A degraded, nonfunctional gene.

pseudomurein A complex polysaccharide observed in the cell walls of some archaea; it resembles peptidoglycan (murein) in structure and chemical make-up; also called pseudopeptidoglycan.

pseudopodium or **pseudopod** (soo″do-po′de-um) A nonpermanent cytoplasmic extension of the cell by which amoeboid protists move and feed.

psychrophile (si′kro-fīl) A microorganism that grows well at 0°C and has an optimum growth temperature of 15°C or lower and a temperature maximum around 20°C.

psychrotroph (psychrotolerant) A microorganism that grows at 0°C but has a growth optimum between 20 and 30°C and a maximum of about 35°C.

PulseNet A program developed and overseen by the Centers for Disease Control and Prevention in which pulse-field gel electrophoresis is used to characterize pathogens found to contaminate food in an effort to detect and track these microbes.

punctuated equilibria The observation based on the fossil record that evolution does not proceed at a slow and linear pace but rather is periodically interrupted by rapid bursts of speciation and extinction driven by abrupt changes in environmental conditions.

pure culture A population of cells that are identical because they arise from a single cell.

purine (pu′rēn) A heterocyclic, nitrogen-containing molecule with two joined rings that occurs in nucleic acids and other cell constituents (e.g., adenine and guanine).

purple membrane An area of the plasma membrane of *Halobacterium* that contains bacteriorhodopsin and is active in trapping light energy.

purple nonsulfur bacteria Anoxygenic photosynthetic bacteria belonging to the α- and β-proteobacteria.

purple sulfur bacteria Anoxygenic photosynthetic bacteria belonging to the γ-proteobacteria.

putrefaction The microbial decomposition of organic matter, especially the anaerobic breakdown of proteins, with the production of foul-smelling compounds such as hydrogen sulfide and amines.

pyogenic Pus-producing.

pyrenoid (pi′rĕ-noid) The differentiated region of the chloroplast that is a center of starch formation in some photosynthetic protists.

pyrimidine (pi-rim′i-dēn) A nitrogen-containing molecule with one ring that occurs in nucleic acids and other cell constituents; the most important pyrimidines are cytosine, thymine, and uracil.

pyrosequencing A twenty-first–century DNA sequencing technique in which the synthesis of a new strand of DNA based on a template strand is followed so that the identity of each nucleotide added is detected by the release of light when the nucleotide triphosphate is cleaved to release pyrophosphate.

Q

Q cycle The process associated with proton movement across the membrane at complex III of an electron transport chain.

Quellung reaction The increase in visibility or the swelling of the capsule of a microorganism in the presence of antibodies against capsular antigens.

quinine drugs Antimicrobial chemotherapeutic agents derived from the bark of the cinchona tree and used to treat malaria.

quinolones A class of broad-spectrum, bacteriocidal antibiotics, derived from nalidixic acid, that bind to DNA gyrase, inhibiting DNA replication.

quorum sensing The process in which bacteria monitor their own population density by sensing the levels of signal molecules (e.g., *N*-acyl-homoserine lactone) that are released by the microorganisms. When these signal molecules reach a threshold concentration, quorum-dependent genes are expressed.

R

racking The removal of sediments from wine bottles.

radappertization The use of gamma rays from a cobalt source for control of microorganisms in foods.

radioimmunoassay (RIA) (ra″de-o-im″u-no-as′a) A very sensitive assay technique that uses a purified radioisotope-labeled antigen or antibody to compete for antibody or antigen with unlabeled standard antigen or test antigen in experimental samples to determine the concentration of a substance in the samples.

rapid sand filters A step in drinking water purification in which particulates are removed by passage over sand.

rational drug design An approach to new drug development based on a specific cellular macromolecule or target.

reaction-center chlorophyll pair The two chlorophyll molecules in a photosystem that are energized by the absorption of light and release electrons to an associated electron transport chain, thus initiating energy conservation by photophosphorylation.

reactive nitrogen intermediate (RNI) Charged nitrogen radicals.

reactive oxygen intermediate (ROI) Charged oxygen radicals; also called reactive oxygen species (ROS).

reading frame The way in which nucleotides in DNA and mRNA are grouped into codons for reading the message contained in the nucleotide sequence.

readthrough A phenomenon in which a ribosome ignores a stop codon and continues protein synthesis. This generates a different protein than that produced if the ribosome had stopped at the stop codon.

reagin (re′ah-jin) Antibody that mediates immediate hypersensitivity reactions. IgE is the major reagin in humans.

real-time PCR A type of polymerase chain reaction (PCR) that quantitatively measures the amount of template in a sample as the amount of fluorescently labeled amplified product.

RecA protein A protein found in all domains of life; it functions in DNA repair and recombination processes.

receptor-mediated endocytosis *See* endocytosis.

receptors Proteins that bind signaling molecules (ligands), thereby initiating cellular responses. Many viruses use host receptors to gain access to the cell.

recognition sites Specific nucleotide sequences to which DNA-binding proteins (e.g., regulatory proteins, sigma factors, restriction enzymes) are able to bind.

recombinant DNA technology The techniques used in carrying out genetic engineering; they involve the identification and isolation of a specific gene, the insertion of the gene into a vector such as a plasmid to form a recombinant molecule, and the production of large quantities of the gene and its product.

recombinants Organisms produced following a recombination event.

recombinant-vector vaccine A vaccine produced by the introduction of one or more of a pathogen's genes into attenuated viruses or bacteria. The attenuated virus or bacterium serves as a vector, replicating within the vertebrate host and expressing the gene(s) of the pathogen. The pathogen's antigens induce an immune response.

recombination The process in which a new recombinant chromosome is formed by combining genetic material from two different nucleic acids. In some cases, the nucleic acids are from two different organisms.

recombinational repair A DNA repair process that repairs damaged DNA when there is no remaining template; a piece of DNA from a sister molecule is used.

Redfield ratio The carbon-nitrogen-phosphorus ratio of marine phytoplankton. This ratio is important for predicting limiting factors for microbial growth.

red tides Population blooms of dinoflagellates that release red pigments and toxins, which can lead to paralytic shellfish poisoning. *See* harmful algal blooms.

redox potential The tendency of a chemical compound to accept electrons; more formally called reduction/oxidation potential.

reducing power Molecules such as NADH and NADPH that temporarily store electrons. The stored electrons are used in anabolic reactions such as CO_2 fixation and the synthesis of monomers (e.g., amino acids).

reductive amination A mechanism used by many microbes to incorporate ammonia.

reductive dehalogenation The cleavage of carbon-halogen bonds by anaerobic bacteria that creates a strong electron-donating environment.

reductive TCA cycle A pathway used by some chemolithotrophs and some anoxygenic phototrophs to fix CO_2. It is essentially the TCA cycle running in the reverse direction; also called reverse citric acid cycle.

refractive index A measure of how much a substance deflects a light ray from a straight path as it passes from one medium (e.g., glass) to another (e.g., air). It is calculated as the ratio of the velocity of light passing through the first medium to that of light passing through the second medium.

regulator T cell T cells that control the development of effector T cells and B cells. Two types exist: T-helper cells ($CD4^+$ cells) and T-suppressor cells.

regulatory mutants Mutant organisms that have lost the ability to limit synthesis of a product, which normally occurs by regulation of activity of an earlier step in the biosynthetic pathway.

regulatory transcription factors Eukaryotic proteins that specifically regulate transcription of one or more genes.

regulon A collection of genes or operons that is controlled by a common regulatory protein.

replica plating A technique for isolating mutants from a population by plating cells from each colony growing on a nonselective agar medium onto plates with selective media or environmental conditions, such as the lack of a nutrient or the presence of an antibiotic or a phage.

replicase An RNA-dependent RNA polymerase used to replicate the genome of an RNA virus.

replication The process in which an exact copy of parental DNA (or viral RNA) is made with the parental molecule serving as a template.

replication fork The Y-shaped structure where DNA is replicated. The arms of the Y contain template strand and a newly synthesized DNA copy.

replicative form (RF) A double-stranded nucleic acid that is formed from a single-stranded viral genome and used to synthesize new copies of the genome.

replicative transposition A transposition event in which a replicate of the transposon remains in its original site.

replicon A unit of the genome that contains an origin of replication and in which DNA is replicated.

replisome A large protein complex that copies the DNA double helix to form two daughter chromosomes.

reporter microbes Microbes that have been engineered to have certain genetic characteristics; they can be used to characterize the physical microenvironment.

repressible gene A gene that encodes a protein whose level drops in the presence of a small molecule. If the gene product is a biosynthetic enzyme, the small molecule is often an end product of the metabolic pathway in which it functions.

repressor protein A protein coded for by a regulator gene that can bind to a repressor-binding site and inhibit transcription.

reservoir A site, alternate host, or carrier that normally harbors pathogenic organisms and serves as a source from which other individuals can be infected.

reservoir host An organism other than a human that is infected with a pathogen that can also infect humans.

residual body A lysosome after digestion of its contents has occurred. It contains undigested material.

resistance factor *See* R plasmid.

resolution The ability of a microscope to separate or distinguish between small objects that are close together.

respiration An energy-yielding process in which the energy substrate is oxidized using an exogenous or externally derived electron acceptor.

respiratory burst An increase in respiratory activity that occurs when an activated phagocytic cell increases its oxygen consumption to generate highly toxic oxygen products such as singlet oxygen, superoxide radical, hydrogen peroxide, hydroxyl radical, and hypochlorite within the phagolysosome.

response-regulator protein *See* two-component signal transduction system.

restriction A process used by bacteria to defend the cell against viral infection. It is accomplished by enzymes that cleave the viral DNA.

restriction enzymes Enzymes produced by bacterial cells that cleave DNA at specific points. They evolved to protect bacteria from viral infection and are used in carrying out genetic engineering.

restriction fragment length polymorphism (RFLP) Fragments of DNA from a gene or genetic element that are of different sizes following endonuclease cleavage. The differences can be used to determine genetic similarity among organisms.

reticulate body (RB) The cellular form in the chlamydial life cycle that grows and reproduces within the host cell.

reticulopodia Netlike pseudopodia found in certain amoeboid protists.

retroviruses A group of viruses with RNA genomes that carry the enzyme reverse transcriptase and form a DNA copy of their genome during their life cycle.

reverse electron flow An energy-consuming process used by some chemolithotrophs and phototrophs to generate reducing power.

reverse transcriptase (RT) A multifunctional enzyme used by retroviruses and hepadnaviruses during their life cycles. It has RNA-dependent DNA polymerase, DNA-dependent DNA polymerase, and RNAase activity. Its function in these viruses is to synthesize double-stranded DNA from single-stranded RNA; this is a reverse of the flow of genetic information in cells, which proceeds from DNA to RNA.

reverse vaccinology The use of genomics to identify target molecules on a virus or bacterium that would be good candidates for vaccine development.

reversible covalent modification A mechanism of enzyme regulation in which the enzyme's activity is either increased or decreased by the reversible covalent addition of a group such as phosphate or AMP to the protein.

reversion mutation A mutation in a mutant organism that changes its phenotype back to the wild type.

rhizobia Any one of a number of α- and β-proteobacteria that form symbiotic nitrogen-fixing nodules on the roots of leguminous plants.

rhizomorph A macroscopic, densely packed thread consisting of individual cells formed by some fungi. A rhizomorph can remain dormant or serve as a means of fungal dissemination.

rhizoplane The surface of a plant root.

rhizoremediation The detoxification of soil contaminants by microbes that are stimulated by the release of nutrients, oxygen, and other factors by plant roots.

rhizosphere A region around the plant root where materials released from the root increase the microbial population and its activities.

ribonucleic acid (RNA) (ri″bo-nu-kle′ik) A polynucleotide composed of ribonucleotides joined by phosphodiester bonds.

ribosomal frame-shifting A phenomenon in which a ribosome reaches a stop codon and shifts the reading frame of the message, thus generating a different protein than it would have if it had stopped at the stop codon.

ribosomal RNA (rRNA) The RNA present in ribosomes. They contribute to ribosome structure and are also directly involved in the mechanism of protein synthesis.

ribosome (ri′bo-sōm) The organelle where protein synthesis occurs; the message encoded in mRNA is translated here.

ribosome-binding site (RBS) Nucleotide sequences on mRNA recognized by a ribosome, enabling the ribosome to orient properly on the mRNA and begin translation at the appropriate codon.

riboswitch A site in the leader of an mRNA molecule that interacts with a metabolite or other small molecule, causing the leader to change its folding pattern. In some riboswitches, this change can alter transcription; in others, it affects translation.

ribotyping The use of conserved rRNA sequences to probe chromosomal DNA for typing bacterial strains.

ribozyme An RNA molecule with catalytic activity.

ribulose 1,5-bisphosphate carboxylase/oxygenase (RuBisCO) The enzyme that catalyzes incorporation of CO_2 in the Calvin cycle.

rimantadine An antiviral chemotherapeutic agent that is used specifically to inhibit replication of influenza A.

RNA-dependent DNA polymerase An enzyme that synthesizes DNA using an RNA template; also called reverse transcriptase.

RNA-dependent RNA polymerase An enzyme that synthesizes RNA using an RNA template.

RNA polymerase An enzyme that catalyzes the synthesis of RNA.

RNA polymerase core enzyme The portion of bacterial RNA polymerase that synthesizes RNA. It consists of all subunits except sigma factor.

RNA polymerase holoenzyme Bacterial RNA polymerase core enzyme plus sigma factor.

RNA silencing A response made by eukaryotic cells that protects them from infections caused by double-stranded RNA viruses.

RNA world The theory that posits that the first self-replicating molecule was RNA and this led to the evolution of the first primitive cell.

rolling-circle replication A mode of DNA replication in which the replication fork moves around a circular DNA molecule, displacing a strand to give a 5′ tail that is also copied to produce a new double-stranded DNA.

root nodule Gall-like structures on roots that contain endosymbiotic nitrogen-fixing bacteria (e.g., *Rhizobium* or *Bradyrhizobium*).

rough endoplasmic reticulum (RER) *See* endoplasmic reticulum.

R plasmids (R factors) Plasmids bearing one or more drug-resistance genes.

rumen The expanded upper portion or first compartment of the stomach of ruminants.

ruminant An herbivorous animal that has a stomach divided into four compartments and chews a cud consisting of regurgitated, partially digested food.

S

sac fungi Common name for the *Ascomycota.*

salt wedge A characteristic salinity profile observed in coastal regions where freshwater and salt water mix.

sanitization Reduction of the microbial population on an inanimate object to levels judged safe by public health standards.

saprophyte (sap′ro-fīt) An organism that takes up nonliving organic nutrients in dissolved form and usually grows on decomposing organic matter.

SAR11 The most abundant group of microbes on Earth, this lineage of α-proteobacteria has been found in almost all marine ecosystems studied.

scaffolding proteins Proteins that are used to aid capsid construction that are not present in the mature virion.

scale-up The process of increasing the volume of a culture from a small, experimental system to a large-volume system for the production of an industrially important metabolite.

scanning electron microscope (SEM) An electron microscope that scans a beam of electrons over the surface of a specimen and forms an image of the surface from the electrons that are emitted by it.

scanning probe microscope A microscope used to study surface features by moving a sharp probe over the object's surface (e.g., atomic force and scanning tunneling microscopes).

scanning tunneling microscope *See* scanning probe microscope.

sclerotia (sklĕ-ro′she-ə) Compact masses of hyphae produced by some filamentous fungi that permit overwintering. In the spring, they germinate to produce either additional hyphae or conidia.

Sec system A system found in all domains of life that can transport proteins through the plasma membrane or insert them into the membrane.

secondary metabolites Products of metabolism that are synthesized after growth has been completed. Antibiotics are considered secondary metabolites.

secondary treatment The biological degradation of dissolved organic matter in the process of sewage treatment; the organic material is either mineralized or changed to settleable solids.

second law of thermodynamics Physical and chemical processes proceed in such a way that the entropy of the universe (the system and its surroundings) increases to the maximum possible.

secretory IgA (sIgA) The primary immunoglobulin of the secretory immune system. *See* IgA.

secretory pathway The process used by eukaryotic cells to synthesize proteins and lipids, followed by secretion or delivery to organelles or the plasma membrane.

secretory vacuole In protists and some animals, these organelles usually contain specific enzymes that perform various functions such as excystment. Their contents are released to the cell exterior during exocytosis.

sedimentation basin A containment vessel used during surface water purification. Following the addition of material to facilitate the formation of coagulated impurities, the water is left until such material has precipitated, forming sediment at the bottom of the basin.

segmented genome A viral genome that is divided into several parts or fragments, each usually coding for a single polypeptide.

selectable marker A gene whose wild-type or mutant phenotype can be determined by growth on specific media.

selectins A family of cell adhesion molecules that are displayed on activated endothelial cells; they mediate leukocyte binding to the vascular endothelium.

selective media Culture media that favor the growth of specific microorganisms; this may be accomplished by inhibiting the growth of undesired microorganisms.

selective toxicity The ability of a chemotherapeutic agent to kill or inhibit a microbial pathogen while damaging the host as little as possible.

self-assembly The spontaneous formation of a complex structure from its component molecules without the aid of special enzymes or factors.

sensor kinase protein *See* two-component signal transduction system.

sensory rhodopsin Microbial rhodopsin that senses the spectral quality of light. Found in halophilic archaea (SRI and SRII) and cyanobacteria. *See* bacteriorhodopsin.

sepsis Systemic response to infection manifested by two or more of the following conditions: temperature $>38°C$ or $<36°C$; heart rate >90 beats per min; respiratory rate >20 breaths per min, or pCO_2 <32 mm Hg; leukocyte count $>12{,}000$ cells per ml^3 or $>10\%$ immature (band) forms. Sepsis also has been defined as the presence of pathogens or their toxins in blood and other tissues.

septate Divided by a septum or cross wall; also with more or less regular occurring cross walls.

septate hyphae Fungal hyphae (filaments) having cross walls.

septation The process of forming a cross wall between two daughter cells during cell division.

septicemia (sep″tĭ-se′me-ah) A disease associated with the presence in the blood of pathogens or bacterial toxins.

septic shock Sepsis associated with severe hypotension despite adequate fluid resuscitation, along with the presence of perfusion abnormalities that may include but are not limited to lactic acidosis, oliguria, or an acute alteration in mental status.

septic tank A tank used to process domestic sewage. Solid material settles out and is partially degraded by anaerobic bacteria as sewage slowly flows through the tank. The outflow is further treated or dispersed in oxic soil.

septum (pl., **septa**) A partition that is used to separate cellular material. Septa are formed during cell division to divide the two daughter cells. They also separate cells in filamentous organisms (e.g., some molds and algae) and a developing endospore from its vegetative mother cell.

sequencing by ligation A twenty-first–century DNA sequencing technique that relies on repeated rounds of ligation and cleavage of short, fluorescently labeled oligonucleotides to the fragments. Often referred to as SOLiD (supported oligo ligation detection) technology.

serology The branch of immunology that is concerned with in vitro reactions involving one or more serum constituents (e.g., antibodies and complement).

serotyping A technique or serological procedure used to differentiate between strains (serovars or serotypes) of microorganisms that have differences in the antigenic composition of a structure or product.

serovar A variant strain of a microbe that has distinctive antigenic properties.

serum (pl., **serums** or **sera**) The clear, fluid portion of blood lacking both blood cells and fibrinogen. It is the fluid remaining after coagulation of plasma, the noncellular liquid fraction of blood.

serum resistance The type of resistance that occurs with bacteria such as *Neisseria gonorrhoeae* because the pathogen interferes with membrane attack complex formation during the complement cascade.

settling basin A basin used during water purification to chemically precipitate out fine particles, microorganisms, and organic material by coagulation or flocculation.

sex pilus (pĭ′lus) A thin protein appendage required for bacterial mating or conjugation. The cell with sex pili donates DNA to recipient cells.

sheath A hollow, tubelike structure surrounding a chain of cells and present in several genera of bacteria.

Shine-Dalgarno sequence A segment in the leader of bacterial mRNA and some archaeal mRNA that binds to a sequence on the 16S rRNA of the small ribosomal subunit. This helps properly orient the mRNA on the ribosome.

shuttle vector A DNA vector (e.g., plasmid, cosmid) that has two origins of replication, each recognized by a different microorganism. Thus the vector can replicate in both microbes.

siderophore (sid′er-o-for″) A small molecule that complexes with ferric iron and supplies it to a cell by aiding in its transport across the plasma membrane.

sigma factor A protein that helps bacterial RNA polymerase core enzyme recognize the promoter at the start of a gene.

sign An objective change in a diseased body that can be directly observed (e.g., a fever or rash).

signal 1 Primary lymphocyte activation that occurs when antigen is presented to the lymphocyte by its appropriate MHC molecule on the presenting cell.

signal 2 Confirmatory event causing lymphocytes stimulated by signal 1 to complete activation and thus become effector cells.

signal peptide The amino-terminal sequence on a peptide destined for transport across the plasmsa membrane. It delays protein folding and is recognized in cells by the Sec system.

silage Fermented plant material with increased palatability and nutritional value for animals, which can be stored for extended periods.

silencer A site in the DNA to which a eukaryotic repressor protein binds.

silent mutation A mutation that does not result in a change in the organism's proteins or phenotype even though the DNA base sequence has been changed.

simple (cut-and-paste) transposition A transposition event in which the transposable element is cut out of one site and ligated into a new site.

single-cell genomic sequencing A method whereby the genome of a single microbial cell is copied many times by the multiple displacement amplification (MDA) technique. The genome is then sequenced by a post-Sanger technique such as pyrosequencing.

single nucleotide polymorphism (SNP) Difference of one base in genes or genetic elements that are otherwise highly conserved among members of a given taxon.

single radial immunodiffusion (RID) assay An immunodiffusion technique that quantitates antigens by following their diffusion through a gel containing antibodies directed against the test antigens.

site-directed mutagenesis The in vitro process by which a specific nucleotide change is made to a particular gene.

site-specific recombination Recombination of nonhomologous genetic material with a chromosome at a specific site.

skin-associated lymphoid tissue (SALT) The lymphoid tissue in the skin that forms a first-line defense as a part of nonspecific (innate) immunity.

S-layer A regularly structured layer composed of protein or glycoprotein that lies on the surface of many bacteria and archaea.

slime The viscous extracellular glycoproteins or glycolipids produced by some bacteria. *See* slime layer.

slime layer A layer of diffuse, unorganized, easily removed material lying outside an archaeal or bacterial cell wall.

slime mold A common term for members of the protist divisions *Myxogastria* and *Dictyostelia.*

slow sand filter A bed of sand through which water slowly flows; the gelatinous microbial layer on the sand grain surface removes waterborne microorganisms, particularly *Giardia.*

sludge A general term for the precipitated solid matter produced during water and sewage treatment; solid particles composed of organic matter and microorganisms that are involved in aerobic sewage treatment (activated sludge).

small RNAs (sRNAs) Small regulatory RNA molecules that do not function as messenger, ribosomal, or transfer RNAs.

small subunit rRNA (SSU rRNA) The rRNA associated with the small ribosomal subunit: 16S rRNA in bacterial and archaeal cells and 18S rRNA in eukaryotes.

smooth endoplasmic reticulum *See* endoplasmic reticulum.

snapping division A distinctive type of binary fission resulting in an angular or a palisade arrangement of cells, which is characteristic of the genera *Arthrobacter* and *Corynebacterium.*

soil organic matter (SOM) The organic material within a soil; organic material is important because it helps to retain nutrients, maintain soil structure, and hold water for plant use.

SOLEXA sequencing A twenty-first–century DNA sequencing technique in which DNA fragments immobilized in a hairpin-like fashion are used as a template for the synthesis of a new strand that incorporates fluorescently labeled nucleotides. Synthesis takes place in a flow cell so that reagents can be removed after the addition of each new nucleotide is detected.

SOS response A complex, inducible process that allows bacterial cells with extensive DNA damage to survive, although often in a mutated form; it involves cessation of cell division, upregulation of several DNA repair systems, and induction of translesion DNA synthesis.

source The location or object from which a pathogen is immediately transmitted to the host, either directly or through an intermediate agent.

sour mash Liquid mixture of fermented grain that is reserved from one batch of distilled spirits to start the fermentation of a new batch; used in the production of whiskey and bourbon.

Southern blotting technique The procedure used to isolate and identify DNA fragments from a complex mixture. The isolated, denatured fragments are transferred from an agarose gel to a nylon filter and identified by hybridization with probes.

specialized transduction A transduction process in which only a specific set of bacterial or archaeal genes is carried to a recipient cell by a temperate virus; the cell's genes are acquired because of a mistake in the excision of a provirus during the lysogenic life cycle.

species Species of higher organisms are groups of interbreeding or potentially interbreeding natural populations that are reproductively isolated. Archaeal and bacterial species are often defined as collections of strains that have many stable properties in common and differ significantly from other strains. However, there is currently considerable debate about the best definition of archaeal and bacterial species.

specific immune response (acquired, adaptive, or specific immunity) *See* acquired immunity.

spheroplast (sfēr′o-plast) A spherical cell formed by the weakening or partial removal of the rigid cell wall component (e.g., by penicillin treatment of gram-negative bacteria). Spheroplasts are usually osmotically sensitive.

spike *See* peplomer.

spirillum (spi-ril′um) A rigid, spiral-shaped bacterium.

spirochete (spi′ro-kēt) A flexible, spiral-shaped bacterium with periplasmic flagella.

spleen A secondary lymphoid organ where old erythrocytes are destroyed and blood-borne antigens are trapped and presented to lymphocytes.

spliceosome A complex of proteins that carries out RNA splicing.

split (interrupted) gene A structural gene with DNA sequences that code for the final RNA product (exons) separated by regions coding for RNA absent from the mature RNA (intervening sequences or introns).

spongiform encephalopathies Degenerative central nervous system diseases in which the brain has a spongy appearance; they are due to prions.

spontaneous generation An early belief that living organisms could develop from nonliving matter.

spontaneous mutations Mutations that occur due to errors in replication or the occurrence of lesions in the DNA (e.g., apurinic sites).

sporadic disease A disease that occurs occasionally and at random intervals in a population.

sporangiospore (spo-ran′je-o-spōr) A spore borne within a sporangium.

sporangium (spo-ran′je-um; pl., **sporangia**) A saclike structure or cell, the contents of which are converted into one or more spores.

spore A differentiated, specialized cell that can be used for dissemination, for survival of adverse conditions because of its heat and dessication resistance, and/or for reproduction. Spores are usually unicellular and may develop into vegetative organisms or gametes. They may be produced asexually or sexually.

sporogenesis (spor′o-jen′ĕ-sis) *See* sporulation.

sporozoite The motile, infective stage of apicomplexan protists, including *Plasmodium,* the causative agent of malaria.

sporulation The process of spore formation.

spotted arrays *See* DNA microarrays.

spread plate A Petri dish of solid culture medium with isolated microbial colonies growing on its surface that has been prepared by spreading a dilute microbial suspension evenly over the agar surface.

sputum The mucus secretion from the lungs, bronchi, and trachea that is ejected (expectorated) through the mouth.

stable isotope analysis The use of stable (nonradioactive) isotopes to determine the fate of a given nutrient or compound in the environment; commonly used in nutrient cycling analyses.

stable isotope probing A technique that uses stable isotopes to examine nutrient cycling and identify the microbes involved.

stalk A bacterial appendage produced by the cell and extending from it.

standard free energy change The free energy change of a reaction at 1 atmosphere pressure when all reactants and products are present in their standard states; usually the temperature is 25°C.

standard reduction potential A measure of the tendency of an electron donor to lose electrons in an oxidation-reduction (redox) reaction. The more negative the reduction potential of a compound, the better electron donor it is.

starter culture An inoculum, consisting of a mixture of microorganisms, used to start a commercial fermentation.

static Inhibiting or retarding growth.

stationary phase The phase of microbial growth in a batch culture when population growth ceases and the growth curve levels off.

stem-nodulating rhizobia Rhizobia that produce nitrogen-fixing structures above the soil surface on plant stems.

sterilization The process by which all living cells, spores, viruses, and viroids are either destroyed or removed from an object or habitat.

Stickland reaction A type of fermentation in which an amino acid serves as the energy source and another amino acid serves as the electron acceptor.

sticky ends The complementary single-stranded ends of double-stranded DNA that result from cleavage with certain restriction endonuclease enzymes. These single-stranded ends can be used to introduce new fragments of DNA to generate a recombinant molecule.

stigma A photosensitive region on the surface of certain protists that is used in phototaxis.

strain A population of organisms that descends from a single organism or pure culture isolate.

streak plate A Petri dish of solid culture medium with isolated microbial colonies growing on its surface that has been prepared by spreading a microbial mixture over the agar surface, using an inoculating loop.

streptolysin-O (SLO) (strep-tol′ĭ-sin) A specific hemolysin produced by *Streptococcus pyogenes* that is inactivated by oxygen (hence, the "O" in its name). SLO causes beta-hemolysis of blood cells on agar plates incubated anaerobically.

streptolysin-S (SLS) A product of *Streptococcus pyogenes* that is bound to the bacterial cell but may sometimes be released. SLS causes beta hemolysis on aerobically incubated blood-agar plates and can act as a leukocidin by killing white blood cells that phagocytose the bacterial cell to which it is bound.

streptomycete Any high G + C gram-positive bacterium of the genera *Kitasatospora, Streptomyces,* and *Streptoverticillium.* The term is also often used to refer to other closely related families including *Streptosporangiaceae* and *Nocardiopsaceae.*

streptomycin (strep′to-mi″sin) A bactericidal aminoglycoside antibiotic produced by *Streptomyces griseus.*

strict anaerobes *See* obligate anaerobes.

stringent response A change made in tRNA, rRNA, and amino acid biosynthetic enzyme production in response to amino acid starvation. The levels of tRNA and rRNA produced are decreased while the level of specific amino acid biosynthetic enzymes is increased.

stroma (stro′mah) The chloroplast matrix that is the location of the photosynthetic carbon dioxide fixation reactions.

stromatolite (stro″mah-to′līt) Domelike microbial mat communities consisting of filamentous photosynthetic bacteria and occluded sediments (often calcareous or siliceous). They usually have a laminar structure.

structural gene A gene that codes for the synthesis of a polypeptide or polynucleotide (i.e., rRNA, tRNA) with a nonregulatory function.

structural genomics Genomic analysis that involves sequencing the genome, identifying genes and potential genes, and predicting three-dimensional protein structure.

structural proteomics Proteomic analysis that focuses on determining the three-dimensional structure of many proteins and using these to predict the structure of other proteins and protein complexes.

subgenomic mRNA Viral mRNA that is smaller than the genomic RNA of an RNA virus.

substrate (1) A reacting molecule in a chemical reaction that is bound by an enzyme. (2) A surface upon which microbes grow.

substrate-level phosphorylation The synthesis of ATP from ADP by phosphorylation coupled with the exergonic breakdown of a high-energy organic substrate molecule.

substrate mycelium In the actinomycetes and fungi, hyphae that are on the surface and may penetrate into the solid medium on which the microbes are growing.

subsurface biosphere The region below the plant root zone where microbial populations can grow and function.

subunit vaccines Immunogenic materials derived from whole infectious agents; examples include bacterial capsules, DNA, etc.

sulfate- or sulfur-reducing bacteria (SRB) Anaerobic bacteria that use sulfate or sulfur or both as a terminal electron acceptor during respiration, thereby performing dissimilatory sulfate reduction.

sulfate reduction The use of sulfate as an electron acceptor, which results in the accumulation of reduced forms of sulfur such as sulfide or incorporation of sulfur into organic molecules, usually as sulfhydryl groups.

sulfonamide (sul-fon′ah-mīd) A chemotherapeutic agent that has the SO_2-NH_2 group and is a derivative of sulfanilamide.

superantigens Toxic bacterial proteins that stimulate the immune system much more extensively than do normal antigens.

superinfection A new bacterial or fungal infection of a patient that is resistant to the drug(s) being used for treatment.

superoxide dismutase (SOD) (dis-mu′tās) An enzyme found in all aerobes, facultative anaerobes, and microaerophiles that catalyzes the conversion of the radical oxygen species superoxide to hydrogen peroxide and molecular oxygen (O_2).

supportive media Culture media that are able to sustain the growth of many different kinds of microorganisms.

suppressor mutation A mutation that overcomes the effect of another mutation and produces the normal phenotype.

Svedberg unit (sfed′berg) The unit used in expressing the sedimentation coefficient; the greater a particle's Svedberg value, the faster it travels in a centrifuge.

symbiont Any organism that has a specific relationship with another that can be characterized as mutualism, cooperation, or commensalism.

symbiosis The living together or close association of two dissimilar organisms, each of these organisms being known as a symbiont.

symbiosome The organelle-like compartment that houses bacteroids that fix nitrogen during rhizobium nodulation.

symport Linked transport of two substances in the same direction.

symptom A change during a disease that a person subjectively experiences (e.g., pain, fatigue, or loss of appetite). Sometimes the term symptom is used more broadly to include any observed signs.

syncytium A large, multinucleate cell formed by the fusion of numerous cells.

syndrome *See* disease syndrome.

syngamy The fusion of haploid gametes.

synteny Partial or complete conservation of gene order observed when the genomes of two closely related organisms are compared.

synthetic medium *See* defined medium.

syntrophism (sin′trōf-izəm) The association in which the growth of one organism either depends on, or is improved by, the provision of one or more growth factors or nutrients by a neighboring organism. Sometimes both organisms benefit.

systematic epidemiology The field of epidemiology that focuses on the ecological and social factors that influence the development of emerging and reemerging infectious diseases.

systematics The scientific study of organisms with the ultimate objective of characterizing and arranging them in an orderly manner; often considered synonymous with taxonomy.

systemic anaphylaxis Type 1 (immediate) hypersensitivity that involves multiple tissue sites and results in shock. It is often fatal if not treated.

T

tandem mass spectrometry The analysis of molecules using two mass spectrometers sequentially.

Taq polymerase A thermostable DNA polymerase used in the polymerase chain reaction.

target modification Alteration of drug binding sites so as to confer resistance to that drug.

Tat system A system used to transport folded proteins across the plasma membrane; in gram-negative bacteria, the proteins are then delivered to a type II secretion system for translocation across the outer membrane.

taxon A group into which related organisms are classified.

taxonomy The science of biological classification; it consists of three parts: classification, nomenclature, and identification.

TB skin test Tuberculin hypersensitivity test for a previous or current infection with *Mycobacterium tuberculosis.*

T cell (T lymphocyte) A type of lymphocyte that matures into an immunologically competent cell under the influence of the thymus.

T-cell antigen receptor (TCR) The receptor on the T cell surface consisting of two antigen-binding peptide chains; it is associated with a large number of other glycoproteins. Binding of antigen to the TCR, usually in association with MHC, activates the T cell.

T-cell receptor complexes *See* T-cell antigen receptor.

T-dependent antigen An antigen that effectively stimulates B-cell response only with the aid of T-helper cells.

T DNA A portion of the *Agrobacterium tumefaciens* Ti plasmid that is integrated into the host plant's chromosome.

teichoic acids (ti-ko′ik) Polymers of glycerol or ribitol joined by phosphates; they are found in the cell walls of gram-positive bacteria.

teichoplanin Antibacterial agent that interferes with cell wall synthesis.

teliospores Diploid spores formed by fungi of the *Ustilaginomycota.*

telomerase An enzyme in eukaryotes that is responsible for replicating the ends of chromosomes.

temperate bacteriophages (viruses) Bacterial and archaeal viruses that can establish a lysogenic relationship, rather than immediately lysing their hosts.

template strand A DNA or RNA strand that specifies the base sequence of a new complementary strand of DNA or RNA.

terminator A sequence that marks the end of a gene and stops transcription.

terminus (1) The end of a molecule (e.g., amino terminus of a protein). (2) In reference to DNA replication, the site on a bacterial or archaeal chromosome where DNA replication stops.

tertiary treatment The removal from sewage of inorganic nutrients, heavy metals, viruses, and so forth by chemical and biological means after microbes have degraded dissolved organic material during secondary sewage treatment.

testate amoeba An amoeba having a covering (test) made either by the protist or from materials collected from the environment.

tetanolysin (tet″ah-nol′ĭ-sin) A hemolysin that aids in tissue destruction and is produced by *Clostridium tetani.*

tetanospasmin (tet″ah-no-spaz′min) The neurotoxic component of the tetanus toxin, which causes the muscle spasms of tetanus.

tetracyclines (tet″rah-si′klēns) A family of antibiotics with a common four-ring structure that are isolated from the genus *Streptomyces* or produced semisynthetically; all are related to chlortetracycline or oxytetracycline.

T_H0 cell Precursor to T_H1, T_H17, and T_H2 cells.

T_H 1 cell A $CD4^+$ T-helper cell that secretes interferon-gamma, interleukin-2, and tumor necrosis factor, influencing phagocytic cells and cytotoxic T cells.

T_H2 cell A $CD4^+$ T-helper cell that secretes interleukin (IL)-4, IL-5, IL-6, IL-9, IL-10, and IL-13, influencing growth and differentiation of B cells.

thallus (thal′us) A type of body that is devoid of root, stem, or leaf; characteristic of fungi.

T-helper (T_H) cell A cell that is needed for T-cell–dependent antigens to be effectively presented to B cells. It also promotes cell-mediated immune responses.

therapeutic index The ratio between the toxic dose and the therapeutic dose of a drug, used as a measure of the drug's relative safety.

thermal death time (TDT) The shortest period of time needed to kill all the organisms in a microbial population at a specified temperature and under defined conditions.

thermoacidophiles Microbes that grow best at acidic pHs and high temperatures.

thermocline A layer of water characterized by a steep temperature gradient.

thermocycler The instrument in which the polymerase chain reaction is performed.

thermodynamics A science that analyzes energy changes in a system.

thermophile (ther′mo-fīl) A microorganism that can grow at temperatures of 55°C or higher; the minimum is usually around 45°C.

3′, 5′-cyclic adenosine monophosphate (cAMP) A small nucleotide that is involved in regulating a number of cellular processes in bacteria and other organisms. In some bacteria, it functions in catabolite repression.

thylakoid (thi′lah-koid) A flattened sac in the chloroplast stroma that contains photosynthetic pigments and the photosynthetic electron transport chain.

thymic selection A process occurring in the thymus in which T-cell precursors are evaluated for functionality.

thymine (thi′min) The pyrimidine 5-methyluracil that is found in nucleosides, nucleotides, and DNA.

thymus (thi′məs) A primary lymphoid organ in the chest that is necessary in early life for the development of immunological functions, including T-cell maturation.

T-independent antigen An antigen that triggers a B cell into immunoglobulin production without T-cell cooperation.

tinea (tin′e-ah) A name applied to many different kinds of superficial fungal infections of the skin, nails, and hair, the specific type (depending on characteristic appearance, etiologic agent, and site) usually designated by a modifying term.

Ti plasmid Tumor-inducing plasmid found in plant pathogenic species of the bacterium *Agrobacterium.*

titer (ti′ter) Reciprocal of the highest dilution of an antiserum that gives a positive reaction in the test being used.

T lymphocyte *See* T cell.

toll-like receptor (TLR) A type of pattern recognition receptor on phagocytes such as macrophages that triggers the proper response to different classes of pathogens. It signals the production of transcription factor NFκB, which stimulates formation of cytokines, chemokines, and other defense molecules.

topoisomerases Enzymes that change the topology of DNA molecules by transiently breaking one or both strands. They play an important role in DNA replication.

toxemia (tok-se′me-ah) The condition caused by toxins in the blood of the host.

toxigenicity (tok″sĭ-jĕ-nis′i-tē) The capacity of an organism to produce a toxin.

toxin A microbial product or component that injures another cell or organism. Often the term refers to a protein, but toxins may be lipids and other substances.

toxin neutralization *See* neutralization.

toxoid A bacterial exotoxin that has been modified so that it is no longer toxic but will still stimulate antitoxin formation when injected into a person or animal.

trailer sequence That portion of a gene downstream of the coding region. It is transcribed but not translated.

transaminase An enzyme that catalyzes a transamination reaction.

transamination The transfer of an amino group from an amino acid to an α-keto acid acceptor. Transamination is used both for degradation of amino acids and in their synthesis.

transcriptase (trans-krip′tās) An enzyme that catalyzes transcription; in viruses with RNA genomes, this enzyme is an RNA-dependent RNA polymerase.

transcription The process in which single-stranded RNA with a base sequence complementary to the template strand of DNA or RNA is synthesized.

transcriptome All the messenger RNA that is transcribed from the genome of an organism under a given set of circumstances.

transduction The transfer of genes between bacterial or archaeal cells by viruses.

transfer host A host that is not necessary for the completion of a parasite's life cycle but is used as a vehicle for reaching a final host.

transfer RNA (tRNA) A small RNA that binds an amino acid and delivers it to the ribosome for incorporation into a polypeptide chain during protein synthesis.

transformation (1) A mode of gene transfer in *Bacteria* and *Archaea* in which a piece of free DNA is taken up by a cell and integrated into its genome.

(2) In recombinant DNA technology, the uptake of free DNA by a cell. (3) The conversion of a cell into a malignant, cancerous cell.

transgenic animal or **plant** An animal or plant that has gained new genetic information by the insertion of foreign DNA.

transient carrier *See* casual carrier.

transition mutations Mutations that involve the substitution of a different purine base for the purine present at the site of the mutation or the substitution of a different pyrimidine for the normal pyrimidine.

transition-state complex An intermediate in a chemical reaction that is a complex of the reactants and resembles both the reactants and the products.

translation Protein synthesis; the process by which the genetic message carried by mRNA directs the synthesis of polypeptides with the aid of ribosomes and other cell constituents.

translesion DNA synthesis A type of DNA synthesis that occurs during the SOS response. DNA synthesis occurs without an intact template and generates mutations as a result.

translocation During the elongation cycle of protein synthesis, the movement of the ribosome relative to the mRNA that repositions the A site of the ribosome over the next codon to be translated.

transmission electron microscope (TEM) A microscope in which an image is formed by passing an electron beam through a specimen and focusing the scattered electrons with magnetic lenses.

transovarian passage (trans″o-va′re-an) The passage of a microorganism such as a rickettsia from generation to generation of hosts through tick eggs. No humans or other mammals are needed as reservoirs for continued propagation.

transpeptidation (1) The reaction that forms the peptide cross-links during peptidoglycan synthesis. (2) The reaction that forms a peptide bond during the elongation cycle of protein synthesis.

transposable element A small DNA molecule that carries the genes for transposition and thus can move around the genome.

transposase An enzyme that catalyzes the movement of a mobile genetic element.

transposition The movement of a piece of DNA around the chromosome.

transposon (tranz-po′zon) A mobile genetic element that encodes transposase, which is needed for transposition, and contains other genes that are not required for transposition.

transversion mutations Mutations that result from the substitution of a purine base for the normal pyrimidine or a pyrimidine for the normal purine.

triacylglycerol A lipid consisting of glycerol to which three fatty acids are attached.

tricarboxylic acid (TCA) cycle The cycle that oxidizes acetyl coenzyme A to CO_2 and generates NADH and $FADH_2$ for oxidation in an electron transport chain; the cycle also supplies precursor metabolites for biosynthesis.

trichome (tri′kōm) A row or filament of microbial cells that are in close contact with one another over a large area.

trickling filter A bed of rocks covered with a microbial film that aerobically degrades organic waste during secondary sewage treatment.

trimethoprim A synthetic antibiotic that inhibits production of folic acid by binding to dihydrofolate reductase. Trimethoprim has a wide spectrum of activity and is bacteriostatic.

trophozoite (trof″o-zo′īt) The active, motile feeding stage of a protozoan organism; in the malarial parasite, the stage of schizogony between the ring stage and the schizont.

tropism (1) The movement of living organisms toward or away from a focus of heat, light, or other stimulus. (2) The selective infection of certain organisms or host tissues by a virus; results from the distribution of the specific receptor for a virus in different organisms or certain tissues of the host.

trichocyst A small organelle beneath the pellicle of some protists. It can be extruded to function as either an anchoring device or as a form of defense.

trypanosome (tri-pan′o-sōm) A protozoan of the genus *Trypanosoma.* They are parasitic flagellate protozoa that often live in the blood of humans and other vertebrates and are transmitted by insect bites.

tubercle (too′ber-k′l) A small, rounded nodular lesion produced by *Mycobacterium tuberculosis.*

tuberculous cavity An air-filled cavity that results from a tubercle lesion caused by *M. tuberculosis.*

tumble Random turning or tumbling movements made by bacteria when they stop moving in a straight line.

tumor A growth of tissue resulting from abnormal new cell growth and reproduction (neoplasia).

tumor necrosis factor (TNF) A cytokine produced by activated macrophages, originally named for its cytotoxic effect on tumor cells.

tumor-suppressor proteins Proteins that regulate cell cycling or repair DNA. Their inactivation can contribute to the development of cancer.

turbidostat A continuous culture system equipped with a photocell that adjusts the flow of medium through the culture vessel to maintain a constant cell density or turbidity.

twitching motility A type of bacterial motility characterized by short, intermittent jerky motions due to extension and retraction of type IV pili.

two-component signal transduction system A regulatory system that uses the transfer of phosphoryl groups to control gene transcription and protein activity. It has two major components: a sensor kinase and a response regulator protein.

two-dimensional gel electrophoresis An electrophoretic procedure that separates proteins using two procedures: isoelectric focusing and SDS-PAGE.

tyndallization The process of repeated heating and incubation to destroy bacterial spores.

type I hypersensitivity A form of immediate hypersensitivity arising from the binding of antigen to IgE attached to mast cells, which then release anaphylaxis mediators such as histamine. Examples: hay fever, asthma, and food allergies.

type II hypersensitivity A form of immediate hypersensitivity involving the binding of antibodies to antigens on cell surfaces followed by destruction of the target cells (e.g., through complement attack, phagocytosis, or agglutination).

type III hypersensitivity A form of immediate hypersensitivity resulting from the exposure to excessive amounts of antigens to which antibodies bind. These antibody-antigen complexes activate complement and trigger an acute inflammatory response with subsequent tissue damage.

type IV hypersensitivity A delayed hypersensitivity response (it appears 24 to 48 hours after antigen exposure). It results from the binding of antigen to activated T lymphocytes, which then release cytokines and trigger inflammation and macrophage attacks that damage tissue. Type IV hypersensitivity is seen in contact dermatitis from poison ivy, leprosy, and tertiary syphilis.

type I secretion system *See* ABC protein secretion pathway.

type III secretion system A system in gram-negative bacteria that secretes virulence factors and injects them into host cells.

type IV secretion system A system in gram-negative bacteria that secretes proteins; some systems transfer DNA during bacterial conjugation.

type V secretion system A pathway in gram-negative bacteria that translocates proteins across the outer membrane after they have been translocated across the plasma membrane by the Sec system.

type strain The microbial strain that is the nomenclatural type or holder of the species name. A type strain will remain within that species should nomenclature changes occur.

U

ultraviolet (UV) radiation Radiation of fairly short wavelength, about 10 to 400 nm, and high energy.

unbalanced growth A type of growth observed in response to changes in a microbe's environment. The rates of synthesis of cell components vary relative to each other. *See* balanced growth.

universal phylogenetic tree A phylogenetic tree that considers the evolutionary relationships among organism from all three domains of life: *Bacteria, Archaea,* and *Eukarya.*

uracil (u′rah-sil) The pyrimidine 2,4-dioxypyrimidine, which is found in nucleosides, nucleotides, and RNA.

use dilution test Method of evaluating disinfectant effectiveness at a recommended dilution.

V

vaccine A preparation of either killed microorganisms; living, weakened (attenuated) microorganisms; or inactivated bacterial toxins (toxoids) administered to induce development of the immune response and protect the individual against a pathogen or a toxin.

vaccinomics The application of genomics and bioinformatics to vaccine development.

valence The number of antigenic determinant sites on the surface of an antigen or the number of antigen-binding sites possessed by an antibody molecule.

vancomycin A glycopeptide antibiotic obtained from *Nocardia orientalis* that is effective only against gram-positive bacteria and inhibits peptidoglycan synthesis.

variable region (V_L and V_H) The region at the N-terminal end of immunoglobulin heavy and light chains whose amino acid sequence varies between antibodies of different specificity. Variable regions form the antigen-binding site.

vasculitis (vas″ku-li′tis) Inflammation of a blood vessel.

vector (1) In genetic engineering, another name for a cloning vector. A DNA molecule that can replicate (a replicon) and transport a piece of inserted foreign DNA, such as a gene, into a recipient cell. It may be a plasmid, phage, cosmid or artificial chromosome. (2) In epidemiology, it is a living organism, usually an arthropod or other animal, that transfers an infective agent between hosts.

vector-borne transmission The transmission of an infectious pathogen between hosts by means of a vector.

vehicle An inanimate substance or medium that transmits a pathogen.

viable but nonculturable (VBNC) microorganisms Microbes that have been determined to be living in a specific environment but are not actively growing and cannot be cultured under standard laboratory conditions.

vibrio (vib′re-o) A rod-shaped bacterial cell that is curved to form a commalike shape or an incomplete spiral.

viral hemagglutination The clumping or agglutination of red blood cells caused by some viruses.

viral neutralization An antibody-mediated process in which IgG, IgM, and IgA antibodies bind to some viruses during their extracellular phase and inactivate or neutralize them.

viremia (vi-re′me-ə) The presence of viruses in the blood stream.

viricide (vir′i-sīd) An agent that inactivates viruses so that they cannot reproduce within host cells.

virion A complete virus particle that represents the extracellular phase of the virus life cycle; at the simplest, it consists of a protein capsid surrounding a single nucleic acid molecule.

virioplankton Viruses that occur in waters; high levels are found in marine and freshwater environments.

viroid An infectious agent that is a single-stranded RNA not associated with any protein; the RNA does not code for any proteins and is not translated.

virology The branch of microbiology that is concerned with viruses and viral diseases.

virulence The degree or intensity of pathogenicity of an organism as indicated by case fatality rates and/or ability to invade host tissues and cause disease.

virulence factor A product, usually a protein or carbohydrate, that contributes to virulence or pathogenicity.

virulent (bacteriophages) viruses Viruses that lyse their host cells at the end of the viral life cycle.

virus An infectious agent having a simple acellular organization with a protein coat and a nucleic acid genome, lacking independent metabolism, and multiplying only within living host cells.

viruslike particles (VLP) Presumptive viruses detected by electron microscopy or fluorescence microscopy; because these approaches do not demonstrate that these entities can infect host cells, they cannot be considered viruses.

virusoid An infectious agent that is composed only of single-stranded RNA that encodes some but not all proteins required for its replication; can be replicated and transmitted to a new host only if it infects a cell also infected by a virus that serves as a helper virus.

vitamin An organic compound required by organisms in minute quantities for growth and reproduction because it cannot be synthesized by the organism; vitamins often serve as enzyme cofactors or parts of cofactors.

volutin granules (vo-lu′tin) *See* metachromatic granules.

W

wart An epidermal tumor of viral origin.

wastewater treatment The use of physical and biological processes to remove particulate and dissolved material from sewage and to control pathogens.

water activity (a_w) A quantitative measure of water availability in the habitat; the water activity of a solution is one-hundredth its relative humidity.

white blood cell (WBC) Blood cell having innate or acquired immune function. These cells are named for the white or buffy layer in which they are found when blood is centrifuged.

whole-cell vaccine A vaccine made from complete pathogens, which can be of four types: inactivated viruses; attenuated viruses; killed microorganisms; and live, attenuated microbes.

whole-genome shotgun sequencing An approach to genome sequencing in which the complete genome is broken into random fragments, which are then individually sequenced. Finally, the fragments are placed in the proper order using computer programs.

Widal test (ve-dahl′) A test involving agglutination of typhoid bacilli when they are mixed with serum containing typhoid antibodies from an individual having typhoid fever; used to detect the presence of *Salmonella typhi* and *S. paratyphi.*

wild type Prevalent form of a gene or phenotype.

Winogradsky column A glass column with an anoxic lower zone and an oxic upper zone, which allows growth of microorganisms under conditions similar to those found in a nutrient-rich lake.

wobble The loose base pairing between an anticodon and a codon at the third position of the codon.

wort The filtrate of malted grains used as the substrate for the production of beer and ale by fermentation.

X

xenograft (zen″o-graft) A tissue graft between animals of different species.

xerophilic microorganisms (ze″ro-fil′ik) Microorganisms that grow best under low a_w conditions, and may not be able to grow at high a_w values.

Y

yeast A unicellular, uninuclear fungus that reproduces either asexually by budding or fission, or sexually through spore formation.

yeast artificial chromosome (YAC) Engineered DNA that contains all the elements required to propagate a chromosome in yeast and which is used to clone foreign DNA fragments in yeast cells.

YM shift The change in shape by dimorphic fungi when they shift from the yeast (Y) form in the animal body to the mold or mycelial form (M) in the environment.

Z

zidovudine A drug that inhibits nucleoside reverse transcriptase (also known as AZT, ZDV, or retrovir) and is used as an anti-HIV treatment.

zoonosis (zo″o-no′sis; pl., **zoonoses**) A disease of animals that can be transmitted to humans.

zooxanthella (zo″o-zan-thel′ah) A dinoflagellate found living symbiotically within cnidarians (corals) and other invertebrates.

***z* value** The increase in temperature required to reduce the decimal reduction time to one-tenth of its initial value.

zygomycetes (zi″go-mi-se′tez) A division of fungi that usually has a coenocytic mycelium with chitinous cell walls. Sexual reproduction normally involves the formation of zygospores.

zygospore A thick-walled, sexual, resting spore characteristic of the zygomycetous fungi.

zygote The diploid (2n) cell resulting from the fusion of male and female gametes.

Photo Credits

Chapter 1

Opener: © Bettmann/Corbis; **1.3a:** Janice Haney Carr/CDC; **1.3b:** © The McGraw-Hill Companies, Inc./Don Rubbelke photographer; **1.3c:** © CNRI/Photo Researchers, Inc.; **1.3d:** © Wayne P. Armstrong, Palomar College; **1.3e:** Dr. Amy Gehring; **1.3f:** Dr. Peter Siver/Getty Images; **1.3g:** © Lauritz Jensen/Visuals Unlimited; **1.4:** J. William Schopf. Reprinted with permission from *Science* 260 © 1993 Sept. 30, fig. 4 a, f, g p. 643 AAAS; **1.8:** Dirk Wiersma/SPL/Photo Researchers, Inc; **1.11a:** © Bettmann/Corbis; **1.11b(both):** © Kathy Park Talaro/Visuals Unlimited; **1.11c:** © Science VU/Visuals Unlimited; **1.12:** © Pixtal/age Fotostock RF; **1.14:** © Bettmann/Corbis.

Chapter 2

Opener: © Alfred Pasieka/Peter Arnold, Inc.; **2.3:** Courtesy of Leica, Inc.; **2.7a:** © Charles Stratton/Visuals Unlimited; **2.7b:** © Stephen Durr; **2.8a:** ASM Microbelibrary.org. Photo micrograph by William Ghiorse; **2.8b:** © Stephen Durr; **2.8c:** © M. Abbey/Visuals Unlimited; **2.11:** © Stephen Durr; **2.13a:** Courtesy of Molecular Probes, Eugene, OR; **2.13b:** © Evans Roberts; **2.14:** Image courtesy of Rut Carballido-Lo'pez and Jeff Errington; **2.17a:** © Kathy Park Talaro; **2.17b:** Harold J. Benson; **2.19a:** © Harold Fisher; **2.19b:** © Fred Hossler/Visuals Unlimited; **2.19c:** © Manfred Kage/Peter Arnold, Inc.; **2.19d:** © Steven P. Lynch; **2.19e:** © David B. Fankhauser, University of Cincinnati Clermont College *http://biology.clc.uc.edu/Fankhauser/*; **2.21a:** © George J. Wilder/Visuals Unlimited; **2.21b:** © Biology Media/Photo Researchers, Inc.; **2.22:** © William Ormerod/Visuals Unlimited; **2.24a:** © Harold Fisher; **2.24b:** © Fred Hossler/Visuals Unlimited; **2.25:** Courtesy of E.J. Laishley University of Calgary; **2.27:** CDC/Janice Carr; **2.28a:** J.S. Poindexter, J.G. Hagenzieker, *Journal of Bacteriology*, 1982, Volume 150, No. 1 Novel peptidoglycans in Caulobacter and Asticcacaulis spp. page 344, fig. 8A; **2.28b:** Grant J. Jensen and Ariane Briegel, *Current Opinion in Structural Biology*, April, 2007, How electron cryotomography is opening a new window onto prokaryotic ultrastructure Elsevier; **2.29:** © Driscoll, Yougquist & Baldeschwieler, Cal-tech/SPL/Photo Researchers, Inc.; **2.31a–b:** From Simon Scheuring (Scheuring S., Ringler P., Borgnia M., Stahlberg H., Muller D.J., Agre P., Engel A., High resolution AFM topographs of the Escherichia coli water channel aquaporin, Z. *Embo J.* 1999, 18:4981–4987.

Chapter 3

Opener: © M. Dworkin-H. Reichenbach/Phototake; **3.1a:** Photo Researchers, Inc.; **3.1b:** © Bruce Iverson; **3.1c:** © Arthur M. Siegelman/Visuals Unlimited; **3.2a:** Centers for Disease Control; **3.2b:** © Thomas Tottleben/Tottleben Scientific Company; **3.2c:** CDC/NCID/HIP/Janice Carr; **3.2d:** Dr. Amy Gehring; **3.2e:** Reprinted from *The Shorter Bergey's Manual of Determinative Bacteriology*, 8e, John G. Holt, Editor, 1977 © Bergey's Manual Trust. Published by Williams & Wilkins Baltimore, MD; **3.2f:** From J.T. Staley, M.P. Bryant, N. Pfenning and J.G. Holt (Eds.), *Bergey's Manual of Systematic Bacteriology*, Vol. 3. © 1989 Williams and Wilkins Co., Baltimore, Robinson, Dept. of Micro, U. of Cal., LA; **3.2g:** From Walther Stoeckenius: *Walsby's Square Bacterium Fine Structures of an Orthogonal Procaryote*; **3.4:** Dr. Leon J. LeBeau; **3.10(both):** © T.J. Beveridge/Biological Photo Service; **3.14:** Courtesy of M.R. J. Salton, NYU Medical Center; **3.20b:** From M. Kastowsky, T. Gutberlet, and H. Bradaczek, *Journal of Bacteriology*, 774:4798–4806, 1992; **3.23a–b:** © John D. Cunningham/Visuals Unlimited; **3.24:** © George Musil/Visuals Unlimited; **3.25:** Dr. Robert G.E. Murray; **3.30a:** Dr. Joseph Pogliano; **3.30b–c:** Image courtesy of Rut Carballido-Lo'pez and Jeff Errington; **3.30d:** Dr. Chistine Jacobs-Wagner; **3.31a:** American Society for Microbiology; **3.31b:** Reprinted from *The Shorter Bergey's Manual of Determinative Bacteriology*, 8e, John G. Holt, Editor, 1977 © Bergey's Manual Trust. Published by Williams & Wilkins Baltimore, MD; **3.33a:** © Ralph A. Slepecky/Visuals Unlimited; **3.34:** Reprinted from *The Shorter Bergey's Manual of Determinative Bacteriology*, 8e, John G. Holt, Editor, 1977 © Bergey's Manual Trust. Published by Williams & Wilkins Baltimore, MD; **3.35:** Michael Schmid; **3.36a:** Courtesy of Daniel Branton, Harvard University; **3.36b:** Reprinted from *Biophysical Journal* Vol. 86, Marina Belenky, Rebecca Meyers and Judith Herzfeld, Subunit Structure of Gas Vesicles: A MALDI-TOF Mass Spectrometry Study, January 2004, with permission from Elsevier; **3.37a:** Y. Gorby; **3.37b:** Zhuo Li and Grant Jensen; **3.38d:** Harry Noller, University of California, Santa Cruz; **3.39a:** © CNRI/SPL. Photo Researchers, Inc.; **3.39b:** © Dr. Gopal Murti/SPL/Photo Researchers, Inc.; **3.39c:** Ohniwa R. Morikawa K, Kim J, Kobori T, Hizume K, et al. 2007. *Microsec. Microanal.* 13:3–12; **3.40:** © Fred Hossler/Visuals Unlimited; **3.41a–b:** © E.C.S. Chan/Visuals Unlimited; **3.41c:** © George J. Wilder/Visuals Unlimited; **3.42a:** Courtesy of Dr. Julius Adler; **3.46:** D.R. Thomas, N.R. Francis, C. Xu, and De.J. DeRosier; **3.47b:** Courtesy of Dr. Jacques Izard, Forsyth Institute, Boston; **3.49a–b:** Courtesy of Dr. Julius Adler; **3.52a:** © CNRI/Photo Researchers, Inc.; **3.52b:** © Dr. Tony Brain/Photo Researchers, Inc.; **3.53(all):** From A.N. Barker et al. (Eds.), *Spore Research*, 1974, pages 161–174, 1971 Academic Press; **3.54:** American Society for Microbiology.

Chapter 4

Opener: © Natural History Museum, London; **4.1a:** © Eric Grave/Photo Researchers, Inc.; **4.1b:** © PhotoLink/Getty Images RF; **4.1c:** © Arthur M. Siegelman/Visuals Unlimited; **4.1d:** © John D. Cunningham/Visuals Unlimited; **4.1e:** © Tom E. Adams/Visuals Unlimited; **4.2:** European Molecular Biology Laboratory (EMBL); **4.6:** Amberg DC Three-dimensional imaging of the yeast actin cytoskeleton through the budding cell cycle. (1998) *Mol. Biol. Cell*; 9:3259–3262. American Society for Cell Biology; **4.7:** © Manfred Schliwa/Visuals Unlimited; **4.8a:** © Henry C. Aldrich/Visuals Unlimited; **4.12a:** Courtesy of Dr. Garry T. Cole, Univ. of Texas Austin; **4.12b:** Beck et al. *Nature* (2007) 449: 611–615 fig. 3b p. 613; **4.13:** © Don Fawcett/Visuals Unlimited; **4.14:** H. Gao et al., *PNAS* July 19, 2005 Vol. 102 no. 29, fig. 1, p. 10207; **4.16b:** *International Review of Cytology*, Vol. 68, p. 111. Academic Press, 1980; **4.18a:** FRAENKEL-CONRAT, HEINZ; KIMBALL, PAUL C.; LEVY, JAY A., *VIROLOGY*, 2nd, © 1988. Electronically reproduced by permission of Pearson Education, Inc., Upper Saddle River, New Jersey.; **4.20:** National Research Council of Canada; **4.21:** © Karl Aufderheide/Visuals Unlimited; **4.22a:** © K.G. Murti/Visuals Unlimited; **4.23a:** Vincent A. Fischetti, Ph.D, Rockefeller University, *www.rockefeller.edu/vaf*; **4.23b:** © W.L. Dentler/Biological Photo Service; **4.26c:** Courtesy of Dr. Garry T. Cole, Univ. of Texas at Austin.

Chapter 5

Opener: © Oliver Meckes/Photo Researchers, Inc.; Page 116 (a–d): Haring, et al. Independent virus development outside a host. *Nature* Vol. 436 Nature Publishing 2005 Figure 1, p. 1101; **5.3a:** Robert G. Milne, Plant Virus Institute, National Research Council, Italy; **5.3c:** Courtesy of Gerald Stubbs and Keiichi Namba, Vanderbilt University; and Donald Caspar, Brandeis University; **5.4b:** Centers for Disease Control/Dr. F.A. Murphy; **5.5a:** © R. Feldman-Dan McCoy/Rainbow; **5.5b:** Courtesy of Harold Fisher, University of Rhode Island and Robley Williams, University of California at Berkeley; **5.5c:** © Science VU-NIH, R Feldman/Visuals Unlimited; **5.6b:** Marek Cyrklaff; **5.7b:** © Harold Fisher; **5.8a:** © Chris Bjornberg/Photo Researchers, Inc.; **5.8b:** © Dr. Linda Stannard, UCT/Photo Researchers, Inc.; **5.12:** © K.G. Murti/Visuals Unlimited; **5.13:** © Lee D. Simon/Photo Researchers, Inc.; **5.17:** © Terry Hazen/Visuals Unlimited; **5.18:** © Charles Marden Fitch.

Chapter 6

Opener: Dr. Eshel Ben-Jacob; **6.1a:** © John D. Cunningham/Visuals Unlimited; **6.1b:** From *ASM News* 53(2): cover, 187, American Society for Microbiology. Photo by H. Kaltwasser; **6.2a:** © Woods Hole Oceanographic Institution; **6.2b:** Image courtesy Mark Schneegurt; **6.9a–b, 6.10b, 6.11b:** © Kathy Park Talaro; **6.13b:** Image courtesy Mark Schneegurt; **6.13c:** Dr. Eshel Ben-Jacob.

Chapter 7

Opener: © Science VU-D Foster, WHOI/Visuals Unlimited; **7.2a:** Reprinted *from The Shorter Bergey's Manual of Determinative Bacteriology*, 8e, John G. Holt, Editor, 1977 © Bergey's Manual Trust. Published by Williams & Wilkins Baltimore, MD; **7.2b:** Dr. John B. Waterbury, Wood Hole Oceanographic Inst.; **7.2c:** Nora Ausmees, University of Uppsala; **7.5a:** Kenn Gerdes; **7.6:** Reprinted with permission from the American Society for Microbiology (*Microbe* 2008, 3(7): 329–36); **7.20a:** Courtesy of Nalge Company; **7.20b:** © B. Otero/Visuals Unlimited; **7.20c:** Courtesy of Nalge Company; **7.29:** Coy Laboratory Products, Inc.; **7.32a:** Dr. Joachim Reitner; **7.32b:** Courtesy Nicholas G. Sotereanos, M.D.; **7.36a:** Chris Frazee, UW–Madison; **7.36b:** Courtesy Dr. Margaret Jean McFall-Ngai.

Chapter 8

Opener: James Gathany/CDC; **8.4a:** Courtesy of AMSCO Scientific, Apex, NC; **8.5:** © Raymond B. Otero/Visuals Unlimited; **8.7:** © Fred Hossler/Visuals Unlimited; **8.8a:** Photo provided by ThermoForma of Marietta, Ohio; **8.12a:** Andersen Products, *www.anpro.com.*

Chapter 9

Opener: Courtesy of David Eisenberg, UCLA; **9.13:** MC. Peitsch, DR. Stampf, TNC. Wells, JL. Sussman The Swiss-3D Image collection and PDB-Browser on the World-Wide Web. *Trends in Biochemical Sciences* 20: 82–84 (1995).; **9.17a–b:** John Wiley & Sons. Courtesy of Donald Voet.

Chapter 10

Opener: © T.E. Adams/Visuals Unlimited; **10.15(inset):** *Journal of Molecular Biology*, Doreen Matthies, Laura Preiss, Adriana L. Klyszejko, Daniel J. Muller, Gregory M. Cook, Janet Vonck and Thomas Meier, The c13 Ring from a Thermoalkaliphilic ATP Synthase Reveals an Extended Diameter Due to a Special Structural Region, Vol. 388, p. 611–618. Copyright (2009), with permission from Elsevier.

Chapter 11

Opener: © John D. Cunningham/Visuals Unlimited.

Chapter 12

Opener: Kriegstein, J.J. & Hogness, D.S. Mechanism of DNA replication in Drosophila chromosomes: structure of replication forks and evidence for bidirectionality, *PNAS* 1974 Vol. 71. f.2 p. 137; **12.10b:** From Cold Spring Harbor Symposia of Quantitative Biology, 28, p. 43 (1963); **12.24:** Reprinted with permission from Murakami, K.S.; Masuda, S.; and Darst, S.A. *Science* 296, May 17, 2002. Copyright 2002 by the AAAS. Image courtesy Seth Darst; **12.34a:** Steven McKnight and Oscar L. Miller, Department of Biology, University of Virginia; **12.37a–c:** Reprinted figures 2b,f–g from Yusupov, M.M.; Yusupov, G; Baucom, A; Lieberman, K; Earnest, T; Cate, J; and Noller, H. *Science* 292:883–896. May 4, 2001. Copyright 2001, AAAS.

Chapter 13

Opener & 13.6b: Lewis et al. Crystal Structure of the Lactose Operon Repressor and Its Complexes with DNA and Inducer, *Science*, 1996 Mar 1, Vol. 271 cover; **13.15b:** Courtesy Prof. Janine Maddock; **13.19b:** From S.C. Schultz, G.C. Shields and T.A. Steitz, Crystal structure of a CAP-DNA Complex. The DNA is Bent by 90 degrees, *Science* 253: 1001–1007, Aug 30, 1991. Copyright 1991, AAAS.

Chapter 14

Opener: © Oliver Meckes/Photo Researchers, Inc.; **14.19:** Courtesy of Charles C. Brinton, Jr. and Judith Carnahan.

Chapter 15

Opener: © Argus Fotoarchiv/Peter Arnold, Inc.; **15.2:** Provided by Dr. A.K. Aggarwal (Mount Sinai School of Medicine) from Newman et al. *Science* 269, 656–663 (1995). © 1995 by the AAAS; **15.9b:** © Kathy Park Talaro; **15.11a–b:** © Huntington Potter and David Dressler/Time Life Pictures/Getty Images; **15.11c:** Reprinted with permission of Edvotek, Inc. *www.edvotek.com*; **15.15a–b:** Ausmees, N., Wahlstedt, H., Bagchi, S., Elliot, M.A., Buttner, M.J., and Flärdh, K. (2007) SmeA, a small membrane protein with multiple functions in Streptomyces sporulation including targeting of a SpoIIIE/FtsK-like protein to cell division septa. *Mol. Microbiol.* 65: 1458–1473.

Chapter 16

Opener: © Alfred Pasieka/Photo Researchers, Inc.; **16.10:** Reprinted with permission from Fleishchman, R.E. et al. 1995. Whole-Genome Random Sequencing and Assembly of Haemophilus Influenzae Rd., *Science* 269:496–512, fig. 1, page 507. © 1995 by the AAAS. Photo by the Institute for Genomic Research; **16.13:** Liu, Zhou, Omelchenko, et al. Transcritomedyamics, *PNAS*, April 2003, vol. 100: 4191–96, fig. 4193; **16.14b:** © Tyne/Simon Fraser/Photo Researchers, Inc.; **16.21a–d:** Jo Handelsman, *Metagenomics* MMBR, ASM Press Dec 2004, vol., 68, fig. 3 page 674.

Chapter 17

Opener: Reprinted with permission from Schulz; H.N., Brinkhoff, T., Ferdelman, T.G., Hernandez Marine, M., Teske, A., and Jorgensen, B.B. 1999. Dense Populations of a Giant Sulfur Bacterium in Namibian Shelf Sediments, *Science* 284, 493–495, fig 1. © 1999 American Association for the Advancement of Science. Image courtesy of Heide Schulz; **17.14(Euryarchaeota):** © SPL/Photo Researchers; (Aquificae): © Dr. R. Rachel and Prof. Dr. K. O. Stetten, University of Regensburg, Lehrstuhl fuer Mikrobiogie, Regensburg, Germany; (Bacilli): © Andrew Syred/SPL/Photo Researchers; (Spirochaetes): © Alfred Paseika/SPL/Photo Researchers; (Cyanobacteria): © The McGraw-Hill Companies, Inc./Don Rubbelke photographer; (Beta): © Science VU/S. Watson/Visuals Unlimited; (Gamma): © Dennis Kunkel Microscopy Inc.; (Delta): Prof. Dr. Hans Reichenbach, Helmholtz Centre for Infection Research, Braunschweig.

Chapter 18

Opener: Lansing Prescott; **18.3:** Karl O. Stetter, M. Hohn; **18.5:** Prepared by Stephen D. Bell, MRC Cancer Cell Unit, UK; **18.9:** Lansing Prescott; **18.10a–b:** Fig. 1A & 1C from Kashefi

& Lovley, *Science* 301:934 (Aug. 2004). Copyright by the AAAS; **18.13a:** © Friederich Widdell/Visuals Unlimited; **18.13b:** From J.T. Staley, M.P. Bryant, N. Pfenning and J.G. Holt (Eds), *Bergey's Manual of Systematic Bacteriology*, Vol. 3. © 1989 Williams and Wilkins Co., Baltimore; **Box 18.1:** From *PNAS*, Vol. 99, pages 7663–7668, fig. 1, Orphan et al., May 2002. Image courtesy of Christopher H. House; **18.16a:** Dr. Mike Dyall-Smith; **18.16b:** © Wayne P. Armstrong, Palomar College.

Chapter 19

Opener: Wadsworth Center, New York State Department of Health; **19.2:** R. Huber, H. Konig, K.O. Stetter; **19.3a:** Gross L (2007) Paradox Resolved? The Strange Case of the Radiation-Resistant Bacteria. PLoS Biol 5(4): e108 doi:10.1371/journal.pbio.0050108); **19.3b:** *From Bergey's Manual of Systematic Bacteriology*, 2/e, vol. 1, Figure B4.4, part A page 400, Family I. Deinococcaceae, Battista, J.R., and Rainey, F.A. © Springer 2001. Image courtesy of John R. Battista; **19.5a:** © Dr. T.J. Beveridge/Visuals Unlimited; **19.6a:** Joanne M. Willey, Ph.D.; **19.7a:** © M.I. Walker/Photo Researchers, Inc.; **19.7b:** © T.E. Adams/Visuals Unlimited; **19.7c:** © Ron Dengler/Visuals Unlimited; **19.7d:** © T.E. Adams/Visuals Unlimited; **19.8a:** Jason K. Oyadomari, *www.keweenawalgae.mtu.edu;* **19.8b:** Micrographia; **19.9:** © John D. Cunningham/Visuals Unlimited; **19.10a–b:** Image courtesy John A. Fuerst and Richard I. Webb, from Novel Compartmentalisation in Planctomycete Bacteria, *Microsc & Microanal*, 2004 10 (Suppl 2); **19.12a–b:** Dr. Peter Braun, Max Plank Institute for Infection Biology, Dept. of Molecular Biology, Berlin, Germany; **19.13a–b:** Reprinted from *The Shorter Bergey's Manual of Determinative Bacteriology*, 8e, John G. Holt, Editor, 1977 © Bergey's Manual Trust. Published by Williams & Wilkins Baltimore, MD; **19.14a2, c–d:** From S.C. Holt, Anatomy and chemistry of spirochetes. *Microbiological Reviews* 42(1):122, 1978 American Society for Microbiology; **19.16a–c:** From M.P. Starr, et al. (Eds), *The Prokaryotes*, Springer-Verlag.

Chapter 20

Opener: © Dennis Kunkel Microscopy, Inc.; **20.3b:** Reprinted from *The Shorter Bergey's Manual of Determinative Bacteriology*, 8e, John G. Holt, Editor, 1977 © Bergey's Manual Trust. Published by Williams & Wilkins Baltimore, MD; **20.3c:** Xiche Hu, Thorsten Ritz, Ana Damjanovic, Felix Autenrieth, and Klaus, Schulten. Photosynthetic apparatus of purple bacteria. *Quarterly Reviews of Biophysics*, 35:1–62, 2002. Cambridge University Press; **20.5a–b:** Berleman et al., Hypercyst Mutants in Rhodospirillum centenum Identify Regulatory Loci Involved in Cyst Cell Differentiation *J. Bacteriol.* 2004 186, fig. 1, pg. 5836; **20.6a:** David J. Silverman, Anna Waddell In Vitro Studies of Rickettsia-Host Cell Interactions: Ultrastructural Study of Rickettsia prowazekii-Infected Chicken Embryo Fibroblasts, *Infect Immun.* 1980 August; 29(2): 778–790; **20.6b:** ASM MicrobeLibrary.org © Walker; **20.7:** From J.T. Staley, M.P. Bryant, N. Pfenning and J.G. Holt (Eds), *Bergey's Manual of Systematic Bacteriology*, Vol. 3. © 1989 Williams and Wilkins Co., Baltimore; **20.9a–b:** Courtesy of Jeanne S. Poindexter, Long Island Univ.; **20.9c:** From J.T. Staley, M.P. Bryant, N. Pfenning and J.G. Holt (Eds), *Bergey's Manual of Systematic Bacteriology*, Vol. 3. © 1989 Williams and Wilkins Co., Baltimore; **20.11a:** From N.R. Krieg and J.G. Holt (Eds), *Bergeys Manual of Systematic Bacteriology*, Vol. 1, 1984. Williams and Wilkins Co., Baltimore; **20.11c:** © John Kaprielian, The National Audubon Society Collection/Photo Researchers; **20.12:** © John D. Cunningham/Visuals Unlimited; **20.13 (Ammonia-Oxidizing Bacteria):** Dr. John B. Waterbury, Wood Hole Oceanographic Inst.; (Ammonia-Oxidizing Archaea): Dr. Martin Könneke; (Nitrite-Oxidizing Bacteria): © Woods Hole Oceanographic Institution; **20.15a:** W L van Veen, E G Mulder, and M H Deinema The Sphaerotilus-Leptothrix group of bacteria, *Microbiol. Mol. Biol. Rev.* 1978 42: 329–356, fig. 6, p. 334. American Society of Microbiology; **20.15b:** Mulder, E.G. & van Veen, W.L., Investigations on the Sphaeerotilus-Leptothrix group, Antonie von Leeuwenhoek *Journal of Microbiology and Serology* 29: 121–153. Kluwer Publishers; **20.19:** From M.P. Starr, et al. (Eds), *The Prokaryotes*, Springer-Verlag; **20.20a:** From M.P. Starr, et al. (Eds*), The Prokaryotes*, Springer-Verlag; **20.20b:** From J.T. Staley, M.P. Bryant, N. Pfenning and J.G. Holt (Eds), *Bergey's Manual of Systematic Bacteriology*, Vol. 3. © 1989 Williams and Wilkins Co., Baltimore; **20.21:** From ASM News 53(2): cover, 187, American Society for Microbiology. Photo by H. Kaltwasser; **20.22:** Image courtesy Mark Schneegurt; **20.23:** Reprinted with permission from Schulz; H.N., Brinkhoff, T., Ferdelman, T.G., Hernandez Marine, M., Teske, A., and Jorgensen, B.B. 1999. Dense Populations of a Giant Sulfur Bacterium in Namibian Shelf Sediments, *Science* 284, 493–495, fig 1. © 1999 American Association for the Advancement of Science. Image courtesy of Heide Schulz; **20.27b:** Bruce Arey and Yuri Gorby; **20.27c:** Photo courtesy of PNNL; **20.28a:** © Kenneth Lucas, Steinhart Aquarium/Biological Photo Service; **20.28b:** Courtesy of James G. Morin, University of California Los Angeles; **20.33b–c:** Courtesy Dr. Jeffrey C. Burnham; **20.34b:** Jerry M. Kuner and Dale Kaiser, Fruiting body morphogenesis in submerged cultures of Myxococeus xauthus, *J. Bacteriol.* 151, 458–461, 1982; **20.35a–b:** © M. Dworkin-H. Reichenbach/Phototake; **20.35c:** © Patricia L. Grillione/Phototake; **20.37:** Annette Summers Engel, Ph.D.

Chapter 21

Opener: © Dennis Kunkel Microscopy, Inc.; **21.3:** © David M. Phillips/Visuals Unlimited; **21.4:** © Michael G. Gabridge/Visuals Unlimited; **21.7:** © Alfred Pasieka/Peter Arnold, Inc.; **21.10:** © F. Widdel/Visuals Unlimited; **21.12a:** © Arthur M. Siegelman/Visuals Unlimited; **21.12b:** Courtesy of Molecular Probes, Inc.; **21.13a–b:** Branda et al., Proc. Nat. Acad. *Science* 25 Dept. 2001, vol. 98, fig 1A, p. 11623 Copyright National Academy of Sciences, U.S.A.; **21.14a:** Courtesy of Dr. A.A. Yousten; **21.14b:** From H. de Barjac & J.F. Charles, Une nouvelle toxine active sur les moustiques, presente dans des inclusions cristallines produites par Bacillus sphaericus. C.R. Acad. Sci. Paris ser. II: 296:905–910, 1983; **21.15a:** From M.P. Starr, et al. (Eds.*)* *The Prokaryotes*, Springer Verlag; **21.15b:** From S.T. Williams, M.E. Sharpe and J.G. Holt (Eds.), *Bergey's Manual of Systematic Bacteriology*, Vol. 4, © 1989 Williams and Wilkins Co., Baltimore; **21.16a:** Joanne M. Willey, Ph.D.; **21.16b:** © Photo Researchers, Inc.; **21.18:** From M.P. Starr, et al. (Eds.), *The Prokaryotes*, Springer-Verlag; **21.19a:** © Thomas Tottleben/Tottleban Scientific Company; **21.19b:** © Photo Researchers, Inc.; **21.20a:** © Kathy Park Talaro/Visuals Unlimited; **21.20b:** © Evans Roberts; **21.20c:** © Fred E. Hossler/Visuals Unlimited.

Chapter 22

Opener: Joanne M. Willey, Ph.D; **22.2a:** Photo © Dr. Alisa A. Gaskell; **22.3a–d:** From S.T. Williams, M.E. Sharpe and J.G. Holt (Eds.), *Bergey's Manual of Systematic Bacteriology*, Vol. 4, © 1989 Williams and Wilkins Co., Baltimore; **22.6a:** © E.C.S. Chan/Visuals Unlimited; **22.6b:** © David M. Phillips/Visuals Unlimited; **22.7:** © Thomas Tottleben/Tottleben Scientific Company; **22.8a–d:** From J.G. Holt et al. (Eds.) *Bergey's Manual of Systematic Bacteriology*, Vol. 2, © 1986. Williams and Wilkins Co., Baltimore; **22.9:** © Grant Heilman Photography; **22.10:** © John D. Cunningham/Visuals Unlimited; **22.12:** From J.G. Holt et al. (Eds), Bergey's Manual of Systematic Bacteriology, 1986 Williams and Wilkins Co., Baltimore; **22.13b:** From Dr. Akio Seino, Hakko to Kogyo (Fermentation and industry) 41 (3):3–4, 1983. Japan Bioindustry Association; **22.13d:** From S.T. Williams, M.E. Sharpe and J.G. Holt (Eds.), *Bergey's Manual of Systematic Bacteriology*,

Vol. 4, © 1989 Williams and Wilkins Co., Baltimore; **22.14a–b:** Joanne M. Willey, Ph.D.; **22.15a–c:** From S.T. Williams, M.E. Sharpe and J.G. Holt (Eds.), *Bergey's Manual of Systematic Bacteriology*, Vol. 4, © 1989 Williams and Wilkins Co., Baltimore; **22.16a:** © Christine L. Case/Visuals Unlimited; **22.16b:** © Sherman Thompson/Visuals Unlimited; **22.17a–b:** From S.T. Williams, M.E. Sharpe and J.G. Holt (Eds.), *Bergey's Manual of Systematic Bacteriology*, Vol. 4, © 1989 Williams and Wilkins Co., Baltimore; **22.18a–b:** David Benson at the University of Connecticut; **22.19:** Staley, *Bergey's Manual Systematic Bacteriology*, Vol. 2, page 1418, figure 15.96a. Courtesy Prof. Bruno Biavati, Instituto Di Microbiologia.

Chapter 23

Opener: © Claude Nuridsany & Marie Perennou/Science Photo Library/Photo Researchers, Inc.; **23.2a:** © Jerome Paulin/Visuals Unlimited; **23.2b:** © David Phillips/Mediscan Visuals Unlimited; **23.4a:** © Manfred Kage/Peter Arnold, Inc.; **23.4b:** © Edward S. Ross; **23.6b:** © B. Beatty/Visuals Unlimited; **23.6c:** Edward Degginger/Bruce Coleman/Photoshot; **23.6d:** © Sherman Thompson/Visuals Unlimited; **23.7b–d:** © Carolina Biological Supply/Phototake; **23.7e:** © David Scharf/Peter Arnold, Inc.; **23.8b:** © Visuals Unlimited; **23.9:** © Manfred Kage/Peter Arnold, Inc.; **23.10:** © Richard Rowan/Photo Researchers, Inc.; **23.11a:** Lee W. Wilcox; **23.11b:** © Claude Taylor, III; **23.11c:** Lee W. Wilcox; **23.12a–b:** © Eric Grave/Photo Researchers, Inc.; **23.12c:** © Cabisco/Phototake; **23.16a:** © Dr. Anne Smith/SPL/Photo Researchers, Inc.; **23.16b:** © Jim Hinsch/Photo Researchers, Inc.; **23.18:** © Natural History Museum, London; **23.19a:** © M.I. Walker/Photo Researchers, Inc.; **23.19b:** © John D. Cunningham/Visuals Unlimited; **23.19c:** © Manfred Kage/Peter Arnold, Inc.; **23.19d:** © John D. Cunningham/Visuals Unlimited.

Chapter 24

Opener: Reprinted with permission from Schulz; H.N., Brinkhoff, T., Ferdelman, T.G., Hernandez Marine, M., Teske, A., and Jorgensen, B.B. 1999. Dense Populations of a Giant Sulfur Bacterium in Namibian Shelf Sediments, *Science* 284, 493–495, fig 1. © 1999 American Association for the Advancement of Science. Image courtesy of Heide Schulz; **24.2a:** © Dean A. Glawe/Biological Photo Service; **24.2b:** © Carolina Biological Supply Company/Phototake; **24.2c:** © Agriculture and Agri-Food Canada; **24.2d:** © Michael & Patricia Fogden; **24.2e:** Robert W. Seagull; **24.2f:** Dr. James Becnel, USDA ARS Gainesville, FL and Society of Invertebrate Pathology; **24.3a–b:** Martha Powell; **24.4a:** © Carolina Biological Supply Company/Phototake; **24.5a:** © Eye of Science/Photo Researchers; **24.5b:** © Dr. Gerald Van Dyke/Visuals Unlimited; **24.6a–b:** © John D. Cunningham Visuals Unlimited; **24.7a:** © J. Forsdyke, Gene Cox/SPL/Photo Researchers, Inc.; **24.8:** © David M. Phillips/Visuals Unlimited; **24.10:** © Hans Pfletschinger/Peter Arnold, Inc.; **24.11a–b:** From Andersson, M.A. et al., Bacteria, Molds and Toxins in Water-damaged Building Materials, 1997 63(2)387–393, fig. 1, p. 388, *Applied and Environmental Microbiology*, American Society for Microbiology. Image courtesy of Maria Anderson and Mirja; **24.12a:** © Alexandra Lowry/The National Audubon Society Collection/Photo Researchers; **24.13:** JK Pataky, University of Illinois.

Chapter 25

Opener: © Science Photo Library/Photo Researchers, Inc.; **25.5f:** © Lee D. Simon/Photo Researchers, Inc.; **25.6b:** George Chapman, Georgetown University; **25.10:** © M. Wurtz/Photo Researchers, Inc.; **25.14:** © CDC/Science Source/Photo Researchers; **25.20:** © Eye of Science/Photo Researchers.

Chapter 26

Opener: Reprinted with permission from Edwards, K.J., Bond, P.L., Gihring, T.M., and Banfield, J.F. An Archael Iron-oxidizing Extreme Acidophile Important in: Acid Mine Drainage, *Science* 287: 1796–2799. (10 March, 2000) Figure 3A, page 1798. © 2000 American Association for the Advancement of Science; Image courtesy of K.J. Edwards; **26.13:** © Michael & Patricia Fogden/Minden Pictures.

Chapter 27

Opener: © Dan McCoy/Rainbow; **27.2:** Courtesy of Molecular Probes, Inc.; **27.5:** Frohlich, J., and H. Koening, 1999. Rapid Isolation of Single Microbial Cells from Mixed Natural and Laboratory Populations with Aid of a Micromanipulator System, *Applied Microbiology* 2:249–257, fig. 4, p. 253. Urban and Fisher Verlag. Photo courtesy Dr. Helmut Koenig; **27.6:** Joanne M. Willey, Ph.D.; **27.8:** Reprinted with permission from *Nature* 417:63 H. Huber et al. © 2002 *Nature*; Prof. Dr. K.O. Stetter, Dr.R. Rachel, Dr. H. Huber, University of Regensburg, Germany; **27.14:** from Ecophysiological Interaction between Nitrifying Bacteria and Heterotrophic Bacteria in Autotrophic Nitrifying Biofilms as Determined by Microautoradiography-Fluorescence In Situ Hybridization Tomonori Kindaichi, Tsukasa Ito, and Satoshi Okabe Applied and Environmental.

Chapter 28

Opener: Photo courtesy of Stephane Blain of CNRS/Université Pierre et Marie Curie, Paris 6; **28.2b:** Jackie Parry; **28.2c:** © Eye of Science/Photo Researchers, Inc.; **28.4:** Burkholder Laboratory, North Carolina State University & Sea Grant National Media Relations; **28.6:** NOAA; **28.8a:** Photo by Susumu Honjo, Woods Hole Oceanographic Institution; **28.11:** ESA; (inset): © Natural History Museum, London; **38.15b–c:** Reprinted with permission from Vincent et al., *Science* 286: 2094 (1999). Copyright 2006 AAAS.

Chapter 29

Opener: Nigel Cattlin/Photo Researchers, Inc.; **29.6:** N.C. Schenck, *Methods & Principles of Micorrhizal Research*, © 1992 American Phytopathological Society. Photo courtesy of Dr. Hugh Wilcox; **29.9d:** Courtesy of Ray Tully, U.S. Department of Agriculture; **29.9f:** Alfalfa Root Nodule Invasion Efficiency Is Dependent on Sinorhizobium meliloti Polysaccharides. Pellock BJ, Cheng HP, Walker GC. *J Bacteriol.* 2000 Aug; 182(15):4310–8.; **29.9i:** © John D. Cunningham/Visuals Unlimited; **29.9j:** © Kingsley R. Stern; **29.10a:** Bernard Dreyfus; **29.11:** © John D. Cunningham/Visuals Unlimited; **29.12:** Courtesy of Dr. Sandor Sule, Plant Protection Institute, Hungary Academy of Sciences.

Chapter 30

Opener: © Science VU/WHOI/Visuals Unlimited; **30.2a:** © M. Abbey/Visuals Unlimited; p. 716: © T. Wenseleers; p. 717: World Health Organization; **30.3a:** © Bob DeGoursey/Visuals Unlimited; **30.5a:** © WHOI/Visuals Unlimited; **30.9:** Craig Cary, University of Delaware; **30.11a–c:** James E. Berleman and John R. Kirby Multicellular Development in Myxococcus xanthus Is Stimulated by Predator-Prey Interactions *J. Bacteriol.* 2007 189: 5675–5682; **30.12a–b:** Lee W. Wilcox; **30.13a:** © John Durham/SPL/Photo Researchers; **30.13c:** Currie et al., *Science* Vol. 311: 81–83, Jan 6 2006. Image courtesy Cameron Currie. Also adapted from Schultz T.R, Meier R. A phylogenetic analysis of the fungus-growing ants (Hymenoptera: Formicidae: Attini) based on morphological characters of the larvae. *Syst. Entomol.* 1995; 20:337–370 & Phylogeny of Fungus-Growing Ants (Tribe Attini) Based on mtDNA Sequence and Morphology James K. Wetterer, Ted R. Schultz, and Rudolf Meier *Molecular Phylogenetics and Evolution* Vol. 9, No. 1, February, pp. 42–47, 1998; **30.14b:** © Science Source/Photo Researchers.

Chapter 31

Opener: © Alfred Pasieka/Peter Arnold; **31.5:** © Runk/Schoenberger/Grant Heilman Photography; **31.8a:** © Stan Elms/Visuals Unlimited; **31.8b:** © John C. Cunningham/Visuals Unlimited; **31.8c:** © Science Photo Library/Photo Researchers, Inc.; **31.9a:** Daniel A. Portnoy, from 1989 *Journal of Cell Biology*, Rockefeller Press; **31.9b:** Stevens, M.P. et al., *Molecular Microbiology* 56: 40–53, 2005. John Wiley and Sons.

Chapter 32

Opener: © Jim Dowdalls/Photo Researchers, Inc.; **32.4b:** © Ellen R. Dirksen/Visuals Unlimited; **32.7b:** © Rosse et al., 1966, Originally published in *The Journal of Experimental Medicine*, 123:969–984; **32.12:** Lennart Nilsson/Albert Bonniers Forlag AB; **32.13:** © David Scharf/Peter Arnold.

Chapter 33

Opener: © Science Source/Photo Researchers, Inc.; **33.3(Infection, Maternal antibody, Vaccination):** © PhotoDisc RF/Getty; (Immune globulin therapy): © Creatas/PictureQuest; **33.4c–d:** Courtesy of Dr. Paul Travers; **33.7c–d:** Courtesy of Dr. Gilla Kaplan, PHRI, UMDNJ; **33.11b:** © R. Feldman-Dan McCoy/Rainbow; **33.20:** © Stan Elms/Visuals Unlimited; **33.25:** © Kathy Park Talaro.

Chapter 34

Opener: © SPL/Photo Researchers; **34.1:** Christine L. Case, Skyline College; **34.2a:** Courtesy and © Becton, Dickinson and Company; **34.2b:** © Lauritz Jensen/Visuals Unlimited; **34.4:** Etest® is a registered trademark of bioMérieux S.A. or one of its subsidiaries; **34.12b:** Alex MacKerell, Ph.D.

Chapter 35

Opener & 35.2a–c: © Raymond B. Otero/Visuals Unlimited; **35.3c:** Millipore Corporation; **35.5a–b:** Analytab Products, A Division of Bio-Merieux, Inc.; **35.7:** Centers for Disease Control and Prevention; **35.8:** © Raymond B. Otero/Visuals Unlimited; **35.9c:** MicrobeLibrary Resources/American Society for Microbiology; **35.12b:** © Hank Morgan/Science Source/Photo Researchers; **35.13:** Genelabs Diagnostics Pte Ltd; **35.15c:** Immuno-Mycologics, Inc.; **35.16d:** N.R. Rose, E.D. de Macario, J.L. Fahey, H. Friedman, and G.M. Penn (eds), *Manual of Clinical Laboratory Immunology*, 4/e, 1992, fig. 8b, p. 92, *Immunoelectrophoresis* American Society of Microbiology.

Chapter 36

Opener: CDC; p. 888: © Historic VU-NHI/Visuals Unlimited.

Chapter 37

Opener(left): Jim Gathany, Centers for Disease Control and Prevention; **(right):** Cynthia Goldsmith, Centers for Disease Control and Prevention; **37.1b:** © John D. Cunningham/Visuals Unlimited; **37.2c:** © Carroll H. Weiss/Camera M.D. Studios; **37.4e:** Armed Forces Institute of Pathology; **37.5:** © Carroll H. Weiss/Camera M.D. Studios; **37.6a:** Dr. Fred Murphy, 1975, Centers for Disease Control and Prevention; **37.7b:** Armed Forces Institute of Pathology; **37.12:** © Science VU/Visuals Unlimited; **37.13:** © Carroll H. Weiss/Camera M.D. Studios; **37.14:** Custom Medical Stock Photo; **37.15:** Courtesy of Dan Wiedbrauk, Ph.D, Warde Medical Laboratory, Ann Arbor, Michigan; **37.16a:** © CDC/Science Source/Photo Researchers; **37.16b:** © Carroll H. Weiss/Camera M.D. Studios; **37.16c:** © Dr. P. Marazzi/Photo Researchers, Inc.; **37.17:** Barbara O'Connor; **37.18:** Courtesy of The National Institute of Health; **37.19a:** © CDC/Science Source/Photo Researchers, Inc.; **37.19b:** © Kenneth Greer/Visuals Unlimited; **37.20:** © Eye of Science/Photo Researchers; p. 925: © Bettmann/Corbis; p. 926: © Photo Researchers, Inc.; **37.21:** Courtesy of Dr. Frederick A. Murphy, Centers for Disease Control and Prevention; **37.22a:** Science Photo/Custom Medical Stock.

Chapter 38

Opener: Courtesy of The Royal College of Surgeons Museum, Edinburg, Scotland; **38.1b & 38.2:** Centers for Disease Control and Prevention; **38.3:** Tone Tønjum, with permission; **38.4b:** © Dr. Leonid Heifets, National Jewish Medical Research Center; **38.4c:** © John Cunningham/Visuals Unlimited; **38.7:** © Pulse Picture Library/CMP Images/Phototake; **38.8a:** From ASM News 55(2) cover, 1986. American Society for Microbiology; **38.8b:** © CDC/Peter Arnold, Inc.; **38.8c & 38.9b:** Centers for Disease Control and Prevention; **38.10c:** Prof. Guy R. Cornelis; **38.11:** Department of Health & Human Services, courtesy of Dr. W. Burgdorfer; **38.12:** Wellcome Images; **38.14:** © Science VU-WHO/Visuals Unlimited; **38.15:** From V. Neman-Simha and F. Megrud, In Vitro Model for Capylobacter pylori Adherence Properties, *Infection and Immunity*, 56(12):3329–3333, Dec. 1988. American Society for Microbiology; **38.16a:** © Arthur M. Siegelman/Visuals Unlimited; **38.16b:** James Bingham, *Pocket Guide for Clinical Medicine.* © 1984 Williams and Wilkins Co., Baltimore, MD; **38.17:** Courtesy Morris D. Cooper, Ph.D., Professor of Medical Microbiology, Southern Illinois University School of Medicine, Springfield, Il.; **38.18a–c:** © Carroll H. Weiss/Camera M.D. Studios; **38.20a:** From R. Baselga et al., Staphylococcus Aureus: Implications in Colonization and Virulence, *Infection and Immunity*, 61(11) L4857–4862, 1993. © American Society for Microbiology; **38.20b:** Janice Carr, Centers for Disease Control and Prevention; **38.22a–f:** © Carroll H. Weiss/Camera M.D. Studios; **38.23:** Gregory Moran, M.D.; **38.24:** © Carroll H. Weiss/Camera M.D. Studios; **38.25:** © Biomedical Communications/Custom Medical Stock Photo; **38.26:** Armed Forces Institute of Pathology; **38.28:** From Jacob S. Teppema, In Vivo Adherence and Colonization of Vibrio cholerae Strains That Differ in Hemagglutinating Activity and Motility. *Journal of Infection and Immunity*, 55(9)2093–2102, Sept. 1987. Reprinted by permission of American Society for Microbiology; **38.30:** from Supermolecular structure of the enteropathogenic Escherichia coli type III secretion system and its direct interaction with the EspA-sheath-like structure PNAS September 25, 2001, vol. 98, no. 20 11638–11643; **38.32:** McCaul and Williams Development Cycle of C. Burnetii, *Journal of Bacteriology*, 147: 1063, 1981. Reprinted with permission of American Society for Microbiology; **38.33:** Dr. Paul L Beck, University of Calgary; **38.36:** © Stanley Flegler/Visuals Unlimited; **38.37:** © E.C.S. Chan/Visuals Unlimited.

Chapter 39

Opener: © Lennart Nilsson/Albert Bonniers Forlag AB; **39.1:** Reprinted by permission of Upjohn Co. from E.S. Beneke et al., 1984 Human Mycosis in Microbiology; **39.2:** © E.C.S. Chan/Visuals Unlimited; **39.3 & 39.4a:** © Arthur M. Siegelman/Visuals Unlimited; **39.4b–39.7b:** CDC; **39.9a–b:** Armed Forces Institute of Pathology; **39.12b:** CDC; **39.13a–b:** © Everett S. Beneke/Visuals Unlimited; **39.14a–b:** © Carroll H. Weiss/Camera M.D. Studios; **39.14c:** Image courtesy of the Centers of Disease Control and Prevention; **39.14d:** © Carroll H. Weiss/Camera M.D. Studios; **39.15a:** © Everett S. Beneke/Visuals Unlimited; **39.15b:** Reprinted by permission of Upjohn Co. from E.S. Beneke et al., 1984 Human Mycosis in Microbiology; **39.16:** © Everett S. Beneke/Visuals Unlimited; **39.18:** Image courtesy of the Centers of Disease Control and Prevention; **39.19a:** © Lauritz Jensen/Visuals Unlimited; **39.19b:** © Robert Calentine/Visuals Unlimited; **39.21:** From M. Schaechter, G. Medoff, & D. Schlessinger (Eds.) *Mechanisms of Microbial Disease*, 1989. Williams and Wilkins; **39.22a:** Aspergillus flavus: human pathogen, allergen and mycotoxin producer M. T. Hedayati1, A. C.

Pasqualotto, P. A. Warn, P. Bowyer and D. W. Denning Microbiology 153 (2007), 1677–1692; **39.22b:** Centers for Disease Control and Prevention; **39.23a:** © Everett S. Beneke/Visuals Unlimited; **39.23b:** Centers for Disease Control and Prevention; **39.23c:** © Everett S. Beneke/Visuals Unlimited; **39.24:** Archiv fur Protistenkunde (1986, Vol. 131, 257–279), image courtesy J.I. Ronny Larsson.

Chapter 40

Opener: © Christiana Dittmann/Rainbow; **40.2a:** © Tom E. Adams/Peter Arnold, Inc.; **40.2b:** © Martha Powell/Visuals Unlimited; **40.4:** Donald Klein; **40.5:** Photo by Mark Seliger, Courtesy of Campbell Soup Company; **40.8a–b:** Centers for Disease Control and Prevention; **40.9:** Kawasaki et al. Species-Specific Identification of Campylobacters by PCR-Restriction Fragment Length Polymorphism and PCR Targeting of the Gyrase B Gene *Applied and Environmental Microbiology* Vol. 74, April 2008, fig. 1 (p. 2531) & fig. 2 (p. 2532); **Box 40.1a–b:** The Fairtrade Foundation; **40.10a–c:** Jeffery Broadbent, Utah State University; **40.12:** © Joe Munroe/Photo Researchers, Inc.; **40.15:** © Vance Henry/Nelson Henry; **40.16:** © Stanley Flegler/Visuals Unlimited.

Chapter 41

Opener: © J.T. MacMillan; **41.8b:** Ron Nichols, USDA-NRCS; **41.9b:** Derek R. Lovley and Kelly P. Nevin; **41.12a:** Mary Ann Tiffany; **41.12b:** From Drum & Gordon, *Trends Biotechnology*, Vol. 21, No 8, 2003 pp. 235–328 Figure 2, p. 326, Elsevier.

Chapter 42

Opener: Courtesy of General Electric Research and Development Center; **42.3a–b:** Donald A. Klein; **42.3c:** Ethan Shelkey and Joanne M. Willey, Ph.D.; **42.4a:** From D. Jenkins, et al. *Manual of the Causes & Control of Activated Sludge Bulking & Forming*, 1986, US Environmental Protection Agency; **42.5a–b:** Cindy Wright-Jones, City of Ft. Collins, CO; **42.13:** © EPA/Visuals Unlimited.

Line Art/Tables/Illustrations/Photo Illustrations

Chapter 1

1.2: Figure from "A molecular view of microbial…" by Norman Pace from SCIENCE, Vol. 276, 1997, p. 735, figure 1. Coypright © 1997 AAAS. Reprinted by permission of AAAS.

Chapter 6

Table 6.4: Adapted from A.D. Brown, "Microbial Water Stress" in BACTERIOLOGICAL REVIEWS, 40(4):803–846. Copyright © 1976 by the American Society for Microbiology. Reprinted by permission.

Chapter 8

Table 8.3: Source: From Seymour S. Block, DISINFECTION, STERILIZATION AND PRESERVATION. Copyright © 1983 Lea & Febiger, Malvern, PA, 1983. Philadelphia, PA: Lippincott Williams Wilkins. **Table 8.4:** Source: From Seymour S. Block, DISINFECTION, STERILIZATION AND PRESERVATION. Copyright © 1983 Lea & Febiger, Malvern, PA, 1983. Philadelphia, PA: Lippincott Williams Wilkins.

Chapter 16

16.5: Ronaghi, M., *Genome Research*, 2001, Cold Spring Harbor Press, figure 4, p. 6. Reprinted by permission of Cold Spring Harbor Laboratory Press. **16.19:** Guerrero et al., BMC Evolutionary Biol. 2005, 5:55, figure a+d.

Chapter 18

18.11: She et al., *Proc. Nat. Acad. Sci*, July 3, 2001, Vol. 98, figure 1, p. 7836. Copyright © 2001 National Academy of Sciences, U.S.A. Reprinted by permission. **18.18:** "Genome Sequence of Halobacterium species NRC-1" by Wailap Victor Ng et al. from PNAS, October 3, 2000, figure 1, page 12178. Copyright © 2000 National Academy of Sciences, U.S.A. Reprinted by permission.

Chapter 20

20.17: "Complete Genome Sequence of the Ammonia-Oxidizing Bacterium and Obligate Chemolithoautotroph Nitrosomonas europaea" by Patrick Chain and Daniel Arp from *J. Bacteriol.*, Vol. 185, May 2003, p. 276, figure 3. Reprinted by permission of American Society for Microbiology. **20.26:** Scott et al., *PloS Biology*, December 2006, Vol. 4, figure 2, p. 2199. **Figure 20.37:** From Dan Kearns, *Science Magazine*, 2007, February 9, figure 1, p. 774. Copyright © 2007 AAAS. Reprinted by permission of AAAS.

Chapter 21

21.9: Bruggemann et al., *Proc. National Acad. Sci*, February 2003, Vol. 100:1316–21, figure 4, p. 1320. Copyright © 2003 National Academy of Sciences, U.S.A. Reprinted by permission.

Chapter 24

24.1: "Evolutionary biology: A kingdom revisited" by Bruns, Tom, NATURE, Vol. 444, 2006, p. 759, figure 1. Reprinted by permission of Nature Publishing Group.

Chapter 27

27.15: "Metaproteomics: studying functional gene expression in microbial ecosystems" by Paul Wilmes and Philip L. Bond from TRENDS IN MICROBIOLOGY, February 2006, Vol. 14, Issue 2, figure 1, page 95. Copyright © 2006. Reprinted by permission of Elsevier.

Chapter 28

Chapter 28 Box: Hugh Powell, *Oceanus*, Vol. 46, Jan 2008, Woods Hole, p. 11. Reprinted by permission of Woods Hole Oceanographic Institution. **Figure 28.1:** Table 2.1: "Physical and Biological Factors Affecting the Microbial Communities in Different Freshwater Habitats" from page 52 of FRESHWATER MICROBIOLOGY by D.C. Sigee. Copyright © 2005. Reprinted by permission of Wiley-Blackwell.

Chapter 29

29.4: "Metagenomic and Small-Subunit rRNA Analyses Reveal the Genetic Diversity of Bacterial, Archaea, Fungi, and Viruses in Soil" by Noah Fierer and Robert B. Jackson from *Appl. Env. Microbiol.* November 2007, Vol. 73, figure 1, p. 7060. Reprinted by permission of American Society for Microbiology.

Chapter 36

Figure **36.9:** Global trends in emerging infectious diseases by Jones et al., *Nature*, Vol. 451, 2008, figure 2, p. 992.

Chapter 37

37.6: Li et al., SCIENCE, Vol. 309, 2005, figure 2, p. 1866. Copyright © 2005 AAAS. Reprinted by permission of AAAS. **37.7a:** Esposito et al., SCIENCE, Vol. 313, 2006, figure 1, p. 808. Copyright © 2006 AAAS. Reprinted by permission of AAAS. **37.9:** UNAIDS.org/pub/globalreport/2008/GR08_2007_HIVPrevWallMap_GR08_en.jpg. Reproduced by kind permission of UNAIDS.

Chapter 38

38.4: Tuberculosis from WHO, 2006. Reprinted by permission of World Health Organization.

Chapter 39

39.5: World Malaria Report, figure 3.3. Reprinted by permission of World Health Organization.

Chapter 41

41.3: From Combinatorial Biosynthesis Yields Novel Analogs of Natural Products by Menzella and Reeves from MICROBE, Vol. 2, Number 9, 2007, p. 431.

Index

Page numbers followed by t and f designate tables and figures, respectively; page numbers in *italics* refer to definitions; page numbers in **boldface** refer to major discussions; pages numbers with A refer to appendix.

A

D

F

G

H

J

K

L

M

O

Q

R

S

T

U

V

Microorganism Pronunciation Guide

The pronunciation of each name is given in parentheses. The phonetic spelling system is explained at the beginning of the glossary (p. G-1)

Bacteria

Acetobacter (ah-se"to-bak'ter)
Acinetobacter (as"ĭ-net"o-bak'ter)
Actinomyces (ak"tĭ-no-mi'sēz)
Agrobacterium (ag"ro-bak-te're-um)
Alcaligenes (al"kah-lij'ĕ-nēz)
Anabaena (ah-nab'ē-nah)
Arthrobacter (ar"thro-bak'ter)
Bacillus (bah-sil'lus)
Bacteroides (bak"tĕ-roi'dēz)
Bdellovibrio (del"o-vib're-o)
Beggiatoa (bej"je-ah-to'ah)
Beijerinckia (bi"jer-ink'e-ah)
Bifidobacterium (bi"fid-o-bak-te're-um)
Bordetella (bor"dĕ-tel'lah)
Borrelia (bō-rel'e ah)
Brucella (broo-sel'lah)
Campylobacter (kam"pĭ-lo-bak'ter)
Caulobacter (kaw"lo-bak'ter)
Chlamydia (klah-mid'e-ah)
Chlorobium (klo-ro'be-um)
Chromatium (kro-ma'te-um)
Citrobacter (sit"ro-bak'ter)
Clostridium (klo-strid'e-um)
Corynebacterium (ko-ri"ne-bak-te're-um)
Coxiella (kok"se-el'lah)
Cytophaga (si-tof'ah-gah)
Desulfovibrio (de-sul"fo-vib're-o)
Enterobacter (en"ter-o-bak'ter)
Erwinia (er-win'e-ah)
Escherichia (esh"er-i'ke-ah)
Flexibacter (flek"sĭ-bak'ter)
Francisella (fran-sĭ-sel'ah)
Frankia (frank'e-ah)
Gallionella (gal"le-o-nel'ah)
Haemophilus (he-mof'ĭ-lus)
Halobacterium (hal"o-bak-te're-um)
Hydrogenomonas (hi-dro"jĕ-no-mo'nas)
Hyphomicrobium (hi"fo-mi-kro'be-um)
Klebsiella (kleb"se-el'lah)
Lactobacillus (lak"to-bah-sil'lus)
Legionella (le"jun-el'ah)
Leptospira (lep"to-spi'rah)
Leptothrix (lep'to-thriks)
Leuconostoc (loo"ko-nos'tok)
Listeria (lis-te're-ah)
Methanobacterium (meth"ah-no-bak-te're-um)
Methylococcus (meth"il-o-kok'-us)
Methylomonas (meth"il-o-mo'nas)
Micrococcus (mi"kro-kok'us)
Mycobacterium (mi"ko-bak-te're-um)
Mycoplasma (mi"ko-plaz'mah)
Neisseria (nīs-se're-ah)
Nitrobacter (ni"tro-bak'ter)
Nitrosomonas (ni-tro"so-mo'nas)
Nocardia (no-kar'de-ah)
Pasteurella (pas"tĕ-rel'ah)
Photobacterium (fo"to-bak-te're-um)
Propionibacterium (pro"pe-on"e-bak-te're-um)
Proteus (pro'te-us)
Pseudomonas (soo"do-mo'nas)
Rhizobium (ri-zo'be-um)
Rhodopseudomonas (ro"do-soo"do-mo'nas)
Rhodospirillum (ro"do-spi-ril'um)
Rickettsia (rĭ-ket'se-ah)
Salmonella (sal"mo-nel'ah)
Sarcina (sar'sĭ-nah)
Serratia (sĕ-ra'she-ah)
Shigella (shĭ-gel'ah)
Sphaerotilus (sfe-ro'tĭ-lus)
Spirillum (spi-ril'um)
Spirochaeta (spi"ro-ke'tah)
Spiroplasma (spi"ro-plaz'mah)
Staphylococcus (staf"ĭ-lo-kok'us)
Streptococcus (strep"to-kok'us)
Streptomyces (strep"to-mi'sēz)
Sulfolobus (sul"fo-lo'bus)
Thermoactinomyces (ther"mo-ak"tĭ-no-mi'sēz)
Thermoplasma (ther"mo-plaz'mah)
Thiobacillus (thi"o-bah-sil'us)
Thiothrix (thi'o-thriks)
Treponema (trep"o-ne'mah)
Ureaplasma (u-re'ah-plaz"ma)
Veillonella (va"yon-el'ah)
Vibrio (vib're-o)
Xanthomonas (zan"tho-mo'nas)
Yersinia (yer-sin'e-ah)
Zoogloea (zo"o-gle'ah)